THE BIOGRAPHICAL DICTIONARY OF SCIENTISTS

First published in Great Britain in six volumes as
The Biographical Dictionary of Scientists:
Biologists (1983), *Chemists* (1983),
Astronomers (1984), *Physicists* (1984),
Engineers and Inventors (1985),
Mathematicians (1985)

Second edition published 1994 in Great Britain
by Helicon Publishing Limited
Second edition published 1994 in the United States of America
by Oxford University Press, Inc.,
200 Madison Avenue, New York, NY 10016

Oxford is a registered trademark of Oxford University Press

Library of Congress Cataloging-in-Publication Data

The Biographical dictionary of scientists / consultant editor,
Roy Porter. -- 2nd ed.
p. cm.
Includes index.
ISBN 0-19-521083-2
1. Scientists--Biography--Dictionaries. I. Porter, Roy, 1946–

Q141.B528 1994
509.2'2--dc20 94–10982
[B]   CIP

Printing (last digit): 9 8 7 6 5 4 3

Printed and bound in Great Britain by
The Bath Press Ltd, Bath

THE BIOGRAPHICAL DICTIONARY OF SCIENTISTS

SECOND EDITION

CONSULTANT EDITOR
Roy Porter

OXFORD UNIVERSITY PRESS
New York   1994

Contents

CONTRIBUTORS

Neil Ardley, Gareth Ashurst, Jim Bailey, Mary Basham, Alan Bishop, William Cooksey, David Cowey, Michael Darton, Keld Fenwick, Lorraine Ferguson, Judy Garlick, Richard Gulliver, Ian Harvey, Nicolas Law, Lloyd Lindo, Robert Matthews, Nigel Morrison, Robert Mortimer, Patricia Nash, Valerie Neal, Jon and Lucia Osborne, Adam Ostaszewski, Caroline Overy, Roy Porter, Helen Rapson, Peter Rodgers, Mary Sanders, Martin Sherwood, Robert Smith, Robert Stewart, E. M. Tansey, Christopher Tunney, Zusa Vrbova, David Ward

EDITORS

Consultant Editor
Roy Porter

Project Editors
Peter Lafferty
Sara Jenkins-Jones

Text Editors
Jane Anson
Catherine Thompson
Louise Jones

Proofreaders
Jonathan Wainwright
Catherine Thompson
Avril Cridlan

Researcher
Anna Farkas

Production
Tony Ballsdon

Page Make-up
Helen Bird

Art Editor
Terence Caven

Illustrations
Taurus Graphics

Introduction

Science is sometimes viewed as impersonal: a method, a system, a technique for generating knowledge. But it is highly personal. The story of science is the story of the individuals who have discovered its truths.

This volume details the lives of the men and women responsible for the world's greatest scientific achievements. Its form – a biographical dictionary, alphabetically arranged – may seem straightforward and commonsensical. But not so. For it inevitably raises the question of the role of the individual in history, and that has been fiercely disputed.

The idea that a single mind can produce a superlative novel, opera, poem or painting is uncontroversial – all apart from the most radical literary critics are happy to talk of Jane Austen, Richard Wagner, John Keats or Pablo Picasso and their output. Yet debate rages about the comparable notion in science. The Great Man (or Woman) theory of scientific creativity is widely contested, and has been dismissed as outmoded. There are many historians and sociologists who would say that the discovery of the law of gravity (to take one signal instance) should not particularly be credited to Sir Isaac Newton. Many would say that the work of precursors should be credited – it was, after all, Newton himself who likened himself to a dwarf standing on a giant's shoulders. Marxists might insist that the theory of gravity should be seen less as an idea created by some lightning-flash in Newton's head than a necessary response to the industrial and technological needs of early capitalism – if Newton hadn't formulated the law, someone else would have done. Others would argue that it was an expression of the *Zeitgeist* or spirit of the age. Social and institutional historians would see that scientific breakthrough as the collective product of the shared labours of the late seventeenth-century scientific community or of the Fellows of the Royal Society. Many historians nowadays are of the opinion that to attribute creativity in science to individual genius is the fallacious and obsolete legacy of nineteenth century individualism and Romanticism.

The biographical arrangement of this dictionary is not intended as a reassertion of the individualistic theory of creativity. Rather it is designed to spell out and make accessible the multi-dimensional nature of science as thought and activity: meditation, observation, experimentation, exploration, debate, synthesis. Some of these are mainly solitary activities; others are deeply social, cooperative and competitive. Abundant attention is given, of course, to the experimental and theoretical breakthroughs made by the men and women whose careers are charted in the twelve hundred concise, factually well-documented biographies that follow in this work, from Abbe to Zworykin. Entries explore the logic of their trains of thought, their flights of imagination, their demolition of traditional patterns of argument. But coverage is also given to the collective social lives of these scientists and doctors, thinkers and tinkers. The accounts below explore their positions in nation and religious cultures, their education, apprenticeships and training, the societies and institutions which they founded or to which they belonged, their colleagues, friends – and foes! Distinguished scientists are seen to be children of their times, but also the progenitors of new schools of thought, in a continuum stretching from past to present. Some achieved greatness in their own lifetimes. Others, like Semmelweis for instance, led lives of comparative obscurity or were ridiculed in their own day, and have won fame only posthumously.

The first function of a dictionary like this is, of course, to provide basic information, clearly, accurately, helpfully. But it is also to point to relationships, to indicate new connexions. It must contain the information the reader was seeking. But it must also whet the appetite by revealing new and unexpected knowledge. This Dictionary eschews narrowness and avoids the mistake of talking about individuals merely as *scientists* – after all, that technical professional term dates only from the 1830s. Entries show how such-and-such a person was not merely (say) a pre-eminent physicist or a celebrated chemist, but also politically active, or a major poet, a deep humanitarian, a man or woman of profound religious convictions – or, in some cases, a dabbler in magic or an industrial magnate.

The individuals biographized in the following pages all have something in common: – all made major contributions to the various scientific disciplines, astronomy, botany, biology, chemistry, cosmology, engineering, exploration, geology, mathematics, physics, physiology, and so forth. But a browse through the pages will equally highlight the differences, the wide range of skills and outlooks former scientists have possessed. Science has been the work both of men and women – in former centuries mainly a male preserve, but more latterly women have been more fully represented in its ranks (and hence in this Dictionary). Scientific leaders have come from the Old and the New Worlds; they have been professionals and amateurs, full-timers and part-timers. In origins they have been princes and paupers, professors, obscure surveyors and lens-grinders. Science has been pursued for the love of Truth, the love of God, the solution of technical problems, a passion for beauty, the desire for money or fame. Many have been driven by some inner demon. Many of the scientists represented in the pages below have attempted, not just to expand the frontiers of science but also to reflect – philosophically, autobiographically, or even poetically – upon scientific thinking, the scientific process and the scientific personality.

Nowadays some fear that science has become rigid, formal, bureaucratized: nothing but a mechanized process for churning out findings and results and solving technical problems. The lives of the scientists documented below would prove an antidote to such pessimism. They afford abundant evidence of the visions and goals that have driven the pursuit of science in history.

Roy Porter
Professor of the History of Medicine
Wellcome Institute for the History of Medicine

Historical Reviews
of the Sciences

Astronomy

Astronomy is the oldest of all sciences, its origins dating back at least several thousands of years. The word astronomy comes from the two Greek words *astron* 'star' and *nomos* 'law' and perhaps the first attempt at understanding the laws that governed the stars and their apparent movement across the night sky was prompted by the need to produce an accurate calendar.

In order to predict the flooding of the river Nile, and hence the time when the surrounding lands would be fertile enough for crops to be planted, the Egyptians made observations of the brightest star in the night sky, Sirius. It was discovered that the date when this star (called Sothis by the Egyptians) could first be seen in the dawn sky (the heliacal rising) enabled the date of the flooding to be calculated. This also enabled the length of the year to be calculated quite accurately; by 2780 BC the Egyptians knew that the time between successive heliacal risings was about 365 days. More accurate observations enabled them to show that the year was about $365\frac{1}{4}$ days long, with a slight difference of 20 minutes between the sidereal year (the time between successive appearances of a star in the same position in the sky) and the tropical year (the time between successive appearances of the Sun on the vernal equinox).

Although evidence suggests that the grouping of stars into constellations was done before a fairly accurate calendar was drawn up, this latter achievement was probably the first scientific act carried out in the field of astronomy.

The prediction of phenomena, which is the fundamental activity of any science, was also being carried out by other ancient civilizations, such as the Chinese. Some historians suggest that the existence in Europe and elsewhere of megalithic sites such as Stonehenge in England and Carnac in France (some of which date back to almost 3000 BC) shows that even minor early civilizations could calculate their calendars and, possibly, predict basic astronomical events such as eclipses.

Certainly by 500 BC the prediction of eclipses had become quite accurate. Thales of Miletus (*c*.624–*c*.547 BC), a Greek philosopher, ended a war between the Medes and the Lydians by accurately predicting the occurrence of an eclipse of the Sun on 28 May 585 BC. His prediction was probably based on the Saros, a period of about 18 years, after which a particular sequence of solar (and lunar) eclipses recurs. This interval was known to Babylonian and Chaldean astronomers long before Thales' time.

The Greeks and early astronomy

The next 800 years of astronomy were dominated by the Greeks. Anaximander (*c*.611–547 BC), a pupil of Thales, helped bring the knowledge of the ancient Egyptians and others to Greece and introduced the sundial as a timekeeping device. He also pictured the sky as a sphere with the Earth floating in space at the centre.

Anaxagoras (*c*.500–428 BC) made great advances in astronomical thought, with his correct explanation of the cause of lunar and solar eclipses. In addition, he considered the Moon to be illuminated by reflected light and all material in the heavens to be composed of the same material; a rocky meteorite falling on the Aegean coast in 468 BC may have brought him to this conclusion. All these thoughts are now known to be broadly correct.

The Greek philosopher Plato (*c*.427–*c*.347 BC) effectively cancelled out these advances with his insistence on the perfection of the heavens, which according to him implied that all heavenly bodies must follow the perfect curve (the circle) across the sky. This dogma cast its dark shadow over astronomical thought for the next 2,000 years. Both Eudoxus (*c*.406–*c*.355 BC) and Callippus (370–300 BC) tried to convert observations into proof for this idea: each planet was set into a sphere, so that the universe took on an onion-ring appearance, with the Earth at the centre. More spheres had to be added to make the theory even approximate to the observations, and by the time of Callippus there were 34 such spheres. Even so, the theory did not match observations.

It was Heraklides of Pontus (388–315 BC) who noted that the apparent motion of the stars during one night might be the result of the daily rotation of the Earth, not the stars, on its axis. He also maintained that Mercury and Venus revolved around the Sun, but he held to the geocentric belief that the Earth was the centre of the universe.

Aristarchos of Samos (c. 320–250 BC) took these ideas one step further. In c. 260 BC he put forward the heliocentric theory of the Solar System, which puts the Sun at the centre of the system of the six planets then known, with the stars infinitely distant. This latter conclusion was based on the belief that the stars were motionless, their apparent motion resulting simply from the Earth's daily rotation. Aristarchos still maintained that the planets moved in Platonically perfect circles, however.

By noting that when the Moon was exactly half-illuminated it must lie at the right angle of a triangle formed by lines joining the Earth, Sun and Moon, Aristarchos was able to make estimates of the relative sizes and distances of the Sun and Moon. Unfortunately, although his theory was correct, the instruments he needed were not available, and his results were highly inaccurate. They were good enough, however, to show that the Sun was more distant and much larger than the Moon. Although this provided indirect support for the heliocentric theory, since it seemed logical for the small Earth to orbit the vast Sun, the geocentric dogma prevailed for another 1,800 years.

Greek astronomy had its last success between 240 BC, when Eratosthenes (c. 276–c. 194 BC) made his calculation of the Earth's size, and c. AD 180, when Ptolemy died. In between those two events, Hipparchus replaced the Eudoxian theory of concentric spheres with an even more contrived arrangement based on the ideas of Apollonius. The Sun and planets were all considered to revolve upon a small wheel, or epicycle, the centre of which revolved around the Earth (the centre of the universe) on a larger circle (the deferent). Using this theory, Ptolemy (or Claudius Ptolemaeus) was able to predict the position of the planets to within 1°, i.e. about two lunar diameters.

Having begun so well, Greek astronomy ended somewhat dismally, preferring dogma to observable evidence and so failing to come to terms with the heliocentric theory of Aristarchos.

It was not until some 600 years after the death of Ptolemy that astronomy started once again to move forward. The lead was taken by the Arabs. Their great mathematical skill and ingenuity with instruments enabled them to refine the observations and theories of the Greeks and produce better star maps, which were becoming increasingly useful for navigation, one of the spurs to astronomical research for many centuries to come. During the Middle Ages, nevertheless, European astronomers did little more than tinker with Ptolemy's epicycles. The Roman Catholic Church decided that the geocentric theory was the only one compatible with the Scriptures, thus making anyone who attempted to put the Sun at the centre of the Solar System guilty of heresy, which was punishable by death.

It was, however, the publication of one book that made Europe the centre of astronomical development, after the science had effectively stood still for many centuries.

European astronomical thought

In 1543 the Polish astronomer Nicolaus Copernicus (1473–1543) published a book entitled *De Revolutionibus Orbium Coelestium*, in which he demonstrated that by placing the Sun at the centre of the Solar System, with the planets orbiting about it, it was possible to account for the apparent motion of the planets in the sky much more neatly than by geocentric theory. To explain the phenomenon of retrogression, for example, cosmologists had been compelled to add complication upon complication to Ptolemy's theory. With the heliocentric theory it could be explained simply as the result of a planet's distances from the Sun.

The 'Copernican revolution' was far from a complete break with the past, nevertheless. Copernicus continued to believe that all celestial motion must be circular, so that his model retained the epicycles and deferents of Ptolemy. But whereas Aristarchos' heliocentric belief had been left to moulder for centuries, Copernicus' ideas were picked up by other cosmologists who, by the end of the sixteenth century, had far more accurate observations at their disposal.

The supplier of these more accurate observations was the Danish astronomer Tycho Brahe (1546–1601), who, using only the naked eye, was able to make observations accurate to two minutes of arc, five times more precise than those of Ptolemy. The effect was enormous. Calendar reform took place, with the Gregorian calendar, now used throughout the Western world, being instituted in 1582. More important for astronomy, Brahe's observations were used by his German assistant, Johannes Kepler (1571–1630), to establish the heliocentric theory of the Solar System once and for all. Out went the complicated systems of epicycles and the dogma of the perfect circle; for Kepler's idea that the planets followed elliptical paths around the Sun, which itself sat at one focus of the ellipses, accorded nicely with the observations of Brahe.

In *Astronomia Nova*, published in 1609, Kepler enunciated the first two laws of planetary motion: the first stated that each planet moved in an elliptical orbit; the second that it did so in such a way that the line joining it to the Sun sweeps out equal areas in equal times. Kepler's third law, that the cube of the distances of a planet from the Sun is proportional to the time required for it to complete one orbit, was announced in 1618.

At the same time as Kepler was making these theoretical breakthroughs, the invention of the telescope by the Dutch optician Hans Lippershey (1587–1619) in 1608 was effecting a revolution in observation. Galileo Galilei (1564–1642) quickly put the instrument to use. In the years 1609 and 1610 he discovered the phases of Venus (showing that planet to be orbiting inside the path of the Earth) and identified four satellites of Jupiter; he also established the stellar nature of the Milky Way. As if to underline the imperfect nature of the heavens, contrary to the Platonic dogma that had crippled astronomy for so long, Galileo also discovered spots on the Sun and mountains on the 'perfect' sphere of the Moon.

The next major breakthrough was again a theoretical one, the discovery by Isaac Newton (1642–1727) of the law of universal gravitation in 1665. Gravity is the single most important force in astronomy and Newton's discovery enabled him to deduce all three of Kepler's laws.

By nature a somewhat reticent man, Newton did not publish his discoveries, in the *Philosophiae Naturalis Principia Mathematica*, until 1687, after the prompting of his friend Edmond Halley (1656–1742). Newton showed that his law could account even for small effects such as the precession of the equinoxes discovered by Hipparchus and that the slight deviations of the planets from their Keplerian orbits were the result of their mutual gravitational attraction. By applying this perturbations theory to the Earth–Sun–Moon system, Newton was able to solve problems about the various motions that had baffled Kepler and his predecessors.

Newton also made a significant contribution to observational astronomy in 1668, when he built the first reflecting telescope, with optics that were free of some of the defects of the refractors then in use.

Seven years later, in 1675, Charles II founded the Royal Observatory at Greenwich, essentially to find an accurate way of determining longitude for British ships involved in overseas exploration and colonization. The line of zero longitude was set at Greenwich, and the First Astronomer Royal, John Flamsteed (1646–1719), drew up a new star catalogue with positions accurate to 20 seconds of arc. Published in 1725, it was the first map of the telescopic age.

Careful observer as Flamsteed was, he failed to notice anything strange about the star which he noted in his catalogue as 34 Tauri. Its true nature was discovered by William Herschel (1738–1822), using the best reflector then in existence, in 1781. The 'star' was in fact a previously undiscovered planet, and was named Uranus. Its discovery doubled the dimensions of the known Solar System.

Beginnings of astrophysics

The opening of the nineteenth century marked the beginning of one of the most important branches of astronomical observation: spectroscopy. In 1802 William Wollaston (1766–1828) discovered that the Sun's spectrum was crossed by a number of dark lines, and in 1815 Joseph von Fraunhofer (1787–1826) made a detailed map of these lines. Fraunhofer noticed that the spectra of stars were slightly different from those of the Sun, but the enormous significance of this observation was not grasped until half a century later by Gustav Kirchhoff (1824–1887). In the meantime, Fraunhofer's skill as a telescope-maker enabled Friedrich Bessel (1784–1846) to determine the distance to the star 61 Cygni; he found it to lie at a distance of about six light years, a term that had then became commonly used as a measure of stellar distance.

Although Kepler's laws enabled the calculation of the relative sizes of the planetary orbits to be obtained, a figure in absolute terms for the mean Earth–Sun separation was still needed. In the nineteenth century much effort was expended on this task, using the method suggested by Kepler of observing transits of Venus from widely separated places in order to take parallax observations. By the middle of the century a figure within 2% of the correct value had been obtained.

The year 1846 saw another triumph for Newton's theory of gravitation, when John Adams (1819–1892) and Urbain Leverrier (1811–1877) predicted the position of an as-yet-unseen planet. Their prediction was based on the observed discrepancies in the motion of Uranus – discrepancies which the two astronomers took to be caused by another, massive planet orbiting beyond the path of Uranus. By calculating the new planet's orbit and its position at certain times, Johann Galle (1812–1910) spotted it on 23 September 1846, less than two lunar diameters from the predicted spot. The new planet was later named Neptune.

Fraunhofer's study of the dark lines in the solar spectrum bore fruit in 1859 with Kirchoff's explanation of them. The lines were absorption lines and were the result of the presence in the Sun of certain chemical elements. Kirchoff's discovery made it possible to determine the chemical composition of the Sun from Earth-bound observations. His work was extended by William Huggins (1824–1910), who was able to show that the stars were made of similar elements to those found on the Earth, thus supporting the 2,300-year-old contention of Anaxagoras. Huggins used the new invention of photography to record the spectra. He also made the first measurement of stellar red-shift, to determine the relative motion of stars towards or away from the Earth. These developments were crucial to the future development of astronomy and astrophysics.

By the end of the nineteenth century, photography had begun to take over from naked-eye observations of the universe. Stars, comets, nebulae and the Andromeda galaxy had all been photographed by 1900. Spectroscopy had also been used on all these objects, and the composition, motion and distance of many of them determined. This latter achievement was possible because of the discovery by Ejnar Hertzsprung (1873–1967) of a relationship between the spectrum of a star and its intrinsic luminosity. The relation was also found by Henry Russell (1877–1957) in the USA and published nine years after Hertzsprung in 1914. As a result, the diagram plotting the luminosities of stars against their spectra is called the Hertzsprung–Russell diagram. Its importance lies in its ability to show how stars evolve and how distant a star of a given spectrum and apparent brightness is.

Hertzsprung was able, by means of his diagram, to calculate the distance of a Cepheid variable star. As Henrietta Leavitt (1868–1921) discovered in 1904, variable stars exhibit a relationship between their intrinsic luminosity and their period of variability, so that a measurement of the period above can give the Cepheid's distance. As such a measurement could be carried out even on Cepheids in other galaxies, the distances of these galaxies could now be calculated by means of Leavitt's period–luminosity law, published in 1912.

The origins of the universe

In 1916 Albert Einstein published his 'Foundation of the General Theory of Relativity' in *Annalen der Physik*. Essentially a theory of gravitation, it marked the greatest theoretical advance in our understanding of the universe since Newton's *Principia* and like Newton's theory it had far-reaching implications for astronomy.

Einstein's theory immediately cleared up a long-standing problem concerning the orbit of Mercury, which was slowly rotating at the rate of about 43 seconds of arc per century. Einstein's theory showed this to be a result of effects arising from the high orbital velocity of Mercury and the intense gravitational field of the Sun. The theory also made two predictions. First, that in the presence of an intense gravitational field, light should be red-shifted to longer wavelengths as it struggled to escape the field. In 1925 this was found by Walter Adams (1876–1956) to be true of the spectrum of the white-dwarf companion of Sirius. Second, that according to general relativity, light should be bent by the space-time curvature pictured in the theory as being the cause of gravitation. Observations of a solar eclipse in 1919 showed the light of stars close to the position of the Sun was indeed bent by the amount predicted by Einstein.

When he applied his theory to the entire universe, in an attempt to reach a universal understanding of dynamics, Einstein was dismayed to discover that in its pure form it would only apply to a universe that is in overall motion. Such a prediction was contrary to the contemporary belief that, apart from the individual motions of the stars within galaxies, the universe was static.

However, by applying spectroscopy to the light of distant galaxies, Vesto Slipher (1875–1969) was able to show that the galaxies were in a state of recession. Surprisingly, it was not until 1924 that proof that the galaxies were star-systems separate from our own galaxy was given by Edwin Hubble (1889–1953). Armed with this knowledge, Hubble was then able to show that the universe as a whole was expanding, and that the rate at which a galaxy appeared to be moving away from Earth was proportional to its distance from us (as determined by the Cepheid law). Thus, Einstein's theory was correct in predicting an expanding universe.

By the end of the 1920s the idea that the universe was born in a 'Big Bang', as proposed by Abbé George Lemaître (1894–1966) in 1927, had become, as it still remains, the established dogma of cosmology. Estimates based on Hubble's law currently set the date of the creation at about 15,000 million years ago.

Will the universe expand forever? This depends upon the mean density of matter in the universe. If the density exceeds a certain critical value (roughly equal to three protons per 1 cu m/35 cu ft) the expansion will halt sometime in the future and the universe will collapse to a 'big crunch'. With this in mind, astronomers have tried to estimate the mean density of matter, but have been hampered by the fact that most of the material in the universe is invisible; this conclusion was reached in the 1980s by considering the gravitational pulls on galaxies. The hunt for the invisible 'dark matter' continues in the 1990s. Modern theories of particle physics have thrown up a wealth of possible dark matter candidates. These theories also allow cosmologists to discuss events in the early Big Bang, possibly as far back as 10^{-43} sec after the initial event. One theory dealing with the very early universe was developed by US cosmologist Alan Guth (1947–), in which the universe is supposed to have undergone a brief phase of accelerated expansion (called inflation) around 10^{-35} sec after its birth.

The year 1930 saw the discovery by photographic means of a ninth planet in the Solar System, Pluto. After a painstaking search involving millions of star images, Clyde Tombaugh (1906–) detected the tiny speck of light on plates taken at the Lowell observatory in Arizona. In 1938 the long-standing problem of the power-source of the Sun and stars was finally solved by Hans Albrecht Bethe (1906–) and Carl Friedrich von Weizsäcker (1912–). They found that the vast outpourings of energy were attributable to the fusing of hydrogen nuclei deep within stars, the process being so efficient that luminosity could be sustained for thousands of millions of years.

The universe in a new light

Experiments in America by Karl Jansky (1905–1950) and Grote Reber (1911–) in the 1930s heralded the start of a new era in astronomical observation, marked by the use of wavelengths other than light, in particular radio waves. Solar radio emission was detected in 1942 and following a theoretical prediction by Hendrik van de Hulst (1918–) in the Netherlands in 1944, interstellar hydrogen radio emission was detected and this was used to produce a map of our Galaxy which was not limited to those regions not obscured by light-absorbing dust.

A radically different theory of the universe, which was free of the Big Bang and its attendant difficulties was proposed by Thomas Gold (1920–), Hermann Bondi (1919–) and Fred Hoyle (1915–) in 1948. Called the steady-state theory, it pictured the universe as being in a constant state of expansion, with new matter being created to compensate for the dilution caused by the Hubble recession. The theory aroused much criticism, but it was not until 1965 that the theory was considered by many to have been finally disproved.

In that year, experiments by Arno Penzias (1933–) and Robert Wilson (1936–) in America resulted in the discovery that the universe appears to contain an isotropic background of microwave radiation. On the Big-Bang picture, that can be interpreted as the red-shifted remnant of the radiation generated in the original Big Bang. Since it is difficult to reproduce the properties of this background on the steady-state theory, the theory was abandoned by most astronomers.

Radioastronomy, which had progressed far in attempts to discover the true nature of the universe by observations of distant galaxies, had two major successes in the 1960s. The first was the discovery of quasi-stellar objects (quasars), and their identification as extremely remote yet very powerful sources by Maarten Schmidt (1929–) in 1963. Their power-source remains enigmatic, but the current belief is that they are galactic centres which contain massive black holes, sucking material into themselves. The second discovery was made by Jocelyn Bell Burnell (1943–) at Cambridge in 1967. Rapid bursts of radio energy at extremely regular intervals were picked up and interpreted as being generated by a rapidly rotating magnetized neutron star, or pulsar. Such an object may be the result of a supernova explosion. In February 1987, a supernova SN1987A exploded in the Small Magellanic Cloud, the first naked-eye supernova since 1604. Observations of SN1987A allowed refinement of supernova theory. In particular, observation of neutrinos from the explosion confirmed that neutrinos carry off much of the energy released in the explosion.

Techniques have been developed in which the outputs from two or more radio telescopes are combined to allow better resolution than is possible with a single dish. In aperture synthesis, several dishes are linked together to simulate the performance of one very large single dish. This technique was pioneered by Martin Ryle (1918–) at Cambridge, England. Very long baseline interferometry uses radio telescopes spread across the world to resolve minute details in radio sources.

Satellite and space-probe astronomy

The late 1960s and early 1970s were marked by landings on the Moon by the American Apollo mission, beginning in 1969 with the expedition of *Apollo 11*. Throughout the 1960s and 1970s, uncrewed probes to the planets revealed more about them than had been discovered in all the previous centuries of study combined. The missions were either fly-bys, beginning with the American *Mariner 2* probe of Venus in 1962, or landings, such as the descent of the Russian *Venera 4* on to the surface of Venus in 1967.

Detailed maps of the four terrestrial planets, Mercury, Venus, Earth and Mars, have now been made, and American Pioneer and Voyager probes passed by Jupiter and Saturn, taking in much of their satellite systems, in the 1970s and early 1980s. *Voyager 2* flew past Uranus in 1986, and Neptune in 1989. Earth-bound observations, which have always been hampered by the interference of a turbulent and polluted atmosphere, are now being supplemented by orbiting observatories operating at a wide range of wavelengths. The first X-ray observatory, UHURU, was launched in 1971, and it detected many sources of X-rays both within and beyond our Galaxy.

Other highlights of probe and satellite astronomy include the 1986 rendezvous of the *Giotto* probe with Halley's comet, the 1990 launch of the Hubble Space Telescope, and the 1990 arrival of *Magellan* at Venus. In 1991 the *Galileo* probe flew to within 1,600 km/994 mi of the asteroid Gaspra. In 1992 *Giotto* flew at a speed of 14 km/8.7 mi per second to within 200 km/124 mi of comet Grigg-Skjellerup, at a distance of 240 million km/150 million mi from Earth (13 light minutes away). On 25 September 1992 the *Mars Observer* was launched from Cape Canaveral, the first US mission to Mars tefor 17 years. Unfortunately, the craft was lost in 1993 as it was about to go into orbit around the planet. On 28 August 1993 *Galileo* flew past the asteroid Ida.

New telescopes

The 1990s saw the installation of an array of outstanding telescopes, incorporating radical new technology. The Keck telescope on Mauna Kea, Hawaii, with a mirror consisting of 36 hexagonal sections, came into operation. The New Technology Telescope, La Silla, Chile, has a 3.5-m/11-ft, computer-controlled flexible mirror system which adjusts its shape 100 times per second to achieve better viewing. The UKIT (United Kingdom Infrared Telescope), Mauna Kea, Hawaii, has a single mirror 374 cm/147 in across. Its performance is so good that it is used for observing with visible light as well as with infrared.

Biology

Like so much else, the systematic study of living things began with the Greeks. Earlier cultures such as those in Egypt and Babylon in the Near East, and the early Indian and Chinese civilizations in Asia, had their own approaches to the study of Nature and its products. But it was the Greeks who fostered the attitudes of mind and identified the basic biological problems and methods from which modern biology has grown. Inquiry begun by Greek philosophers such as Alcmaeon (born c. 535 BC) and Empedocles (c. 490–430 BC) culminated in the biological work of Aristotle (384–320 BC) and the medical writings of Hippocrates (c. 460–c. 377 BC) and his followers.

Natural philosophy

Aristotle was an original thinker of enormous power and energy who wrote on physics, cosmology, logic, ethics, politics and many other branches of knowledge. He also wrote several biological works which laid the foundations for comparative anatomy, taxonomy (classification) and embryology. He was particularly fascinated by sea creatures; he dissected many of these as well as studying them in their natural habitats. Aristotle's approach to anatomy was functional: he believed that questions about structure and function always go together and that each biological part has its own special uses. Nature, he insisted, does nothing in vain. He therefore thought it legitimate to enquire about the ultimate purposes of things. This teleological approach has persisted in biological work until the twentieth century. In addition, Aristotle studied reproduction and embryological development, and he established many criteria by which animals could be classified. He believed that animals could be placed on a vertical, hierarchical scale ('scale of being'), extending from man down through quadrupeds (four-legged animals), birds, snakes and fishes to insects, molluscs and sponges. His investigations suggested to him that animals had increased in complexity throughout the ages, an idea of evolution such as that put forward by Darwin 2,200 years later. One of Aristotle's pupils, Theophrastus (c. 372–c. 287 BC), carried out detailed studies of plants and founded for botany many of the fundamentals that Aristotle had established for zoology.

What Aristotle achieved for biology, Hippocrates and his followers contributed for medicine: they established a naturalistic framework for thinking about health and disease. Unlike earlier priests and doctors, they did not regard illness as the result of sin, or as a divine punishment for misdeeds. They were keen observers whose most influential explanatory framework saw disease as the result of an imbalance in one of the four physiologically active humours (blood, phlegm, black bile and yellow bile). The humours were schematically related to the four elements (earth, air, fire and water) of Greek natural philosophy. Each person was supposed to have his or her own dominant humour, although different humours tended to predominate at different times of life (such as youth and old age) or seasons of the year. The therapy of Hippocrates aimed at restoring the ideal balance through diet, drugs, exercise, change of life style, and so on.

Galen's influence on medicine

After the Classical period of Greek thought, the most important biomedical thinker was Galen (c. AD 130–c. 200), who combined Hippocratic humoralism with Aristotle's tendency to think about the ultimate purposes of the parts of the body. Galen should not be blamed for the fact that later doctors thought that he had discovered everything, so that they had no need to look at biology and medicine for themselves. In fact, Galen was a shrewd anatomist and the most brilliant experimental physiologist of antiquity. For most of that time, human dissection was prohibited and Galen learnt his anatomy from other animals such as pigs, elephants and apes. It took more than a thousand years before it was discovered that some of the structures which he had accurately described in animals (such as the five-lobed liver and the rete mirabile network of veins in the brain) were not present in the human body. Like that of Aristotle, Galen's anatomy was functional, and often his tendency to speculate went further than sound observation would have permitted, as when he postulated invisible pores in the septum of the heart which were supposed to allow blood to seep from the right ventricle to the left.

Following Galen's death at the end of the second century and the collapse of the Roman Empire, biology and medicine remained stagnant for a thousand years. Most Classical writings were lost to the European West, to be preserved and extended in Constantinople and other parts of the Islamic Empire. From the twelfth century these texts began to be rediscovered in southern Europe – particularly in Italy, where universities were also established. For a while scholars were content merely to translate and comment on the works of men such as Aristotle and Galen, but eventually an independent spirit of enquiry arose in European biology and medicine. Human dissections were routinely performed from the fourteenth century and anatomy emerged as a mature science from the fervent activity of Andreas Vesalius (1514–1564), whose *De Humani Corporis Fabrica* (1543; *On the Fabric of the Human Body*) is one of the masterpieces of the scientific revolution. His achievement was to examine the body itself rather than relying simply on Galen; the illustrations in his work are simultaneously objects of scientific originality and of artistic beauty. The rediscovery of the beauty of the human body by Renaissance artists encouraged the study of anatomy by geniuses such as Leonardo da Vinci (1452–1519). Shortly afterwards, William Harvey (1578–1657) discovered the circulation of the blood and established physiology on a scientific footing. His little book *De Motu Cordis* (1628; *On the Motion of the Heart*) was the first great work on experimental physiology since the time of Galen. The eccentric wandering doctor Paracelsus (1493–1541) had also deliberately set aside the teachings of Galen and other Ancients in favour of a fresh approach to Nature and medicine and to the search for new remedies for disease.

The age of discovery

While these achievements were happening in medicine, anatomy and physiology, other areas of biology were not stagnating. Voyages of exploration alerted naturalists to the existence of many previously unknown plants and animals and encouraged them to establish sound principles of classification, to create order out of the apparently haphazard profusion of Nature. Zoological and botanical gardens began to be established so that the curious could view wonderful creatures like the rhinoceros and the giraffe. And just as the Great World (Macrocosm) was revealing plants and animals unknown in Europe only a short time before, so the invention of the microscope in the seventeenth century gave scientists the opportunity of exploring the secrets of the Little World (Microcosm). The microscope permitted Anton van Leeuwenhoek (1632–1723) to see bacteria, protozoa and other tiny organisms; it enabled Robert Hooke (1635–1703) to observe in a thin slice of cork regular structures which he called 'cells'. And it aided Marcello Malpighi (1628–1664) to complete the circle of Harvey's concept of the circulation of the blood by first seeing it flowing through capillaries, the tiny vessels that connect the arterial and venous systems. Many of these microscopical discoveries were communicated to the Royal Society of London, one of several scientific societies established during the mid-seventeenth century.

The full potential of the microscope as a biological tool had to wait for technical improvements effected in the early nineteenth century. But it also led scientists along some blind alleys of theory. Observations of sperm 'swimming' in seminal fluid provided some presumed evidence for a theory that was much debated during the eighteenth century, concerning the nature of embryological development. Aristotle had thought that the body's organs (heart, liver, stomach and so on) only gradually appear once conception has initiated the growth of the embryo. Later scientists, including William Harvey, extended Aristotle's theory with new observations. But now the visualization of moving sperm suggested that some miniature, but fully formed organism was already present in the reproductive fluids of the male or female. The tiny 'homunculi' were thought to be stimulated to growth by fertilization. If the homunculus was always there, it followed that its own reproductive parts contained all its future offspring, which in turn contained all their future offspring and so on, back to Adam and Eve (depending on whether the male or the female was postulated as the carrier). This doctrine, called preformationalism, was held by most eighteenth-century biologists, including Albrecht von Haller (1708–1777) and Lazzaro Spallanzani (1729–1799), two of the century's greatest scientists. Both men, like virtually all scientists of the period, were devout Christians, and preformationalism did not conflict with their belief that God established regular, uniform laws which governed the development and functions of living things. They did not believe that inert matter could join together by accident to make a living organism. They rejected, for instance, the possibility of spontaneous generation, and Spallanzani devised some ingenious experiments designed to show that maggots found in rotting meat or the teeming life discoverable after jars of water are left to stand did not spontaneously generate. Haller, Spallanzani, John Hunter (1728–1793) and most other great eighteenth-century experimentalists held that the actions of living things could not be understood simply in terms of the laws of physics and chemistry. They were Vitalists, who believed that unique characteristics separated living from non-living matter. Man's special attributes were often ascribed to the soul, and lower animals and plants were thought to possess more primitive animal and vegetable souls, respectively, which gave them basic biological capacities such as reproduction, digestion, movement and so on.

Taxonomy and classification

Experimental biology was well established in the eighteenth century; another great area of eighteenth-century activity was classification. Again inspired by Aristotle, and drawing on the work of previous biologists such as John Ray (1627–1705), the Swedish botanist Carolus Linnaeus (1707–1778) spent a lifetime trying to bring order to the ever increasing number of plants and animals uncovered by continued exploration of the earth and its oceans. His *Systema Naturae* (1735; *System of Nature*) was the first of many books in which he elaborated a philosophy of taxonomy and established the convention of binomial nomenclature still followed today. In this convention, all organisms are identified by their genus and species; thus human beings in the Linnean system are *Homo sapiens*. Depending on the nature of the characteristics examined, however, plants and animals could be placed in a variety of groups, ranging from the kingdom at the highest level through phyla, classes, orders and families and so on beyond the species to the variety and, finally, the individual. Naturalists had traditionally accepted that the species was the most significant taxonomic category, Christian doctrine generally holding that God had specially created each individual species. It was also assumed that the number of species existing was fixed during the Creation, as described in the Book of Genesis – no new species having been created and none becoming extinct. Linnaeus, however, believed that God had created genera and that it was possible that new species had emerged during the time since the original Creation.

Palaeontology and evolution

Some eighteenth-century naturalists such as Georges Buffon (1707–1788) began to suggest that the Earth and its inhabitants were far older than the 6,000 or so years inferred from the Bible. General acceptance of a vastly increased age of the Earth, and of the reality of biological extinction, awaited the work of early nineteenth-century scientists such as Georges Cuvier (1769–1832), whose reconstructions of the fossil remains of large vertebrates like the mastodon and dinosaurs found in the Paris basin and elsewhere so stirred both the popular and scientific imaginations of his day. Despite Cuvier's work on the existence of life on Earth for perhaps millions of years, he firmly opposed the notion that these extinct creatures might be the ancestors of animals alive today. Rather he believed that the extent to which any species might change (variability) was fixed and that species themselves could not change much over time. His contemporary and scientific opponent, Jean Baptiste Lamarck (1744–1829), argued, however, that species do change over time. He insisted that species never become extinct; instead they are capable of change as new environmental conditions and new needs arise. According to this argument, the ancestors of the giraffe need not have had such a long neck, which instead might have slowly developed as earlier giraffe-like creatures stretched their necks to feed on higher leaves. Lamarck believed that physical characteristics and habits acquired after birth could – particularly if repeated from generation to generation – become inherited and thus inborn in the organism's offspring. We still call the doctrine of the inheritance of acquired characteristics 'Lamarckism', although most naturalists before Lamarck had already believed it. It continued to be generally accepted (for instance, by Charles Darwin) until late in the nineteenth century.

The debates between Cuvier and Lamarck were part of the new possibilities opened up by the revolution in thinking about the age of the Earth and of life on it. During the same fertile period, the systematic use of improved microscopes revolutionized the way in which biologists conceived organisms. In the closing years of the eighteenth century a French pathologist named Xavier Bichat (1771–1802), aided only with a hand lens, developed the idea that organs such as the heart and liver are not the ultimate functional units of animals. He postulated that the body can be divided into different kinds of tissues (such as nervous, fibrous, serous and muscular tissue) which make up the organs. Increasingly, biologists and doctors began thinking in terms of smaller functional units, and microscopists such as Robert Brown (1773–1858) began noticing regular structures within these units, which we now recognize as cells. Brown called attention to the nucleus in the cells of plants in 1831 and by the end of the decade two German scientists, Matthias Schleiden (1804–1881) and Theodor Schwann (1810–1882) systematically developed the idea that all plants and animals are composed of cells. The cell theory was quickly established for adult organisms, but in certain situations – such as the earliest stages of embryological development or in the appearance of 'pus' cells in tissues after inflammation or injury – it appeared that new cells were actually crystallized out of an amorphous fluid which Schwann called the 'blastema'. The notion of continuity of cells was enlarged upon by the pathologist Rudolf Virchow (1821–1902), who summarized it in his famous slogan 'All cells from cells'. The cell theory gave biologists and physicians a new insight into the architecture and functions of the body in health and disease.

Microorganisms and disease

Concern with one-celled organisms also lay behind the work of Louis Pasteur (1822–1895), which helped to establish the germ theory of disease. Pasteur trained as a chemist, but his researches into everyday processes such as the souring of milk and the fermentation of beer and wine opened for him a new understanding of the importance of yeast, bacteria and other microorganisms in our daily lives. It was Pasteur who finally convinced scientists that these tiny organisms do not spontaneously generate on foods and other dead matter; our skin, the air and everything we come into contact with can be a source of their spores. After reading about Pasteur's work Joseph Lister (1827–1912) first conceived the idea that by keeping away these germs (as they were eventually called) from the wounds made during surgical operations, healing would be much faster and post-operative infection would be much less common. When in 1867 Lister published the first results using his new technique, antiseptic surgery was born. He spent much time developing the methods, which were taken up by other surgeons who soon realized that it was better to prevent infection altogether (asepsis) by carefully sterilizing their hands, instruments and dressings. By the time Lister died, he was world famous and surgeons were performing operations that would have been impossible without his work.

After Lister drew attention to the importance of Pasteur's discoveries for medicine and surgery, Pasteur himself showed the way in which germs cause not only wound infections, but also many diseases. He first studied a disease of silkworms which was threatening the French silk industry; he then turned his attention to other diseases of farm animals and human beings. In the course of this research, he discovered that under certain conditions an organism could be grown which, instead of causing a disease, actually prevented it. He publicly demonstrated these discoveries for anthrax, then a common disease of sheep, goats and cattle which sometimes also affected human beings. He proposed to call this process of protection vaccination, in honour of Edward Jenner (1749–1823), who in 1796 had shown how inoculating a person with cowpox (vaccinia) can protect against the deadly smallpox. Pasteur's most dramatic success came with a vaccine against rabies, a much-dreaded disease occasionally contracted after a bite from an animal infected with rabies.

By the 1870s, other scientists were investigating the role played by germs in causing disease. Perhaps the most important of them was Robert Koch (1843–1910), who devised many key techniques for growing and studying bacteria, and who showed that tuberculosis and cholera – prevalent diseases of the time – were caused by bacteria. Immunology, the study of the body's natural defence mechanisms against invasion by foreign cells, was pioneered by another German, Paul Ehrlich (1854–1915), who also began looking for drugs that would kill disease-causing organisms without being too dangerous for the patient. His first success, a drug named Salvarsan, was effective in the treatment of syphilis. Ehrlich's hopes in this area were not fully realized immediately, and it was not until the 1930s that the synthetic sulpha drugs, also effective against some bacterial diseases, were developed by Gerhard Domagk (1895–1964) and others. Slightly earlier Alexander Fleming (1881–1955) had noticed that a mould called *Penicillium* inhibited the growth of bacteria on cultures.

Fleming's observation was investigated during World War II by Howard Florey (1898–1968) and Ernst Chain (1906–1979), and since then many other antibiotics have been discovered or synthesized. But antibiotics are not effective against diseases caused by viruses; such infections can, however, often be prevented using vaccines. An example is poliomyelitis, vaccines against which were developed in the 1950s by Albert Sabin (1906–1993) and Jonas Salk (1914–   ).

Evolution and genetics

Many of these advances in modern medical science are a direct continuation of discoveries made in the nineteenth century, although of course we now know much more about bacteria and other pathogenic microorganisms than did Pasteur and Koch. Another nineteenth-century discovery which is still being intensively investigated is evolution. Charles Darwin (1809–1882) was not the first to suggest that biological species can change over time, but his book *On the Origin of Species* (1859) first presented the idea in a scientifically plausible form. As a young man, Darwin spent five years (1831–1836) on HMS *Beagle*, during which he studied fossils, animals and geology in many parts of the world, particularly in South America. By 1837 he had come to believe the fact of evolution; in 1838 he hit upon its mechanism: natural selection. This principle makes use of the fact that organisms produce more offspring than can survive to maturity. In this struggle for existence, those offspring with characteristics best suited to their particular environment will tend to survive. In this way, Nature can work on the normal variation which plants and animals show and, under changing environmental conditions, significant change can occur through selective survival.

Darwin knew that his ideas would be controversial so he initially imparted them to only a few close friends, such as the geologist Charles Lyell (1797–1875) and the botanist Joseph Hooker (1817–1911). For 20 years he continued quietly to collect evidence favouring the notion of evolution by natural selection, until in 1858 he was surprised to receive a short essay from Alfred Russel Wallace (1823–1913), then in Malaya, perfectly describing natural selection. Friends arranged a joint Darwin–Wallace publication, and then Darwin abandoned a larger book he was writing on the subject to prepare instead *On the Origin of Species*. In it he marshalled evidence from many sources, including palaeontology, embryology, geographical distribution, ecology (a word coined only later) and hereditary variation. Darwin did not have a very clear idea of how variations occur, but his work convinced a number of scientists, including Thomas Huxley (1825–1895), Ernst Haeckel (1834–1919) and Francis Galton (1822–1911), Darwin's cousin. Huxley became Darwin's chief publicist in Britain, Haeckel championed Darwin's ideas in Germany, and Galton quietly absorbed the evolutionary perspective into his own work in psychology, physical anthropology and the use of statistics and other forms of mathematics in the life sciences.

Meanwhile, unknown to Darwin (and largely unrecognized during his lifetime), a monk named Gregor Mendel (1822–1884) was elucidating the laws of modern genetics through his studies of inheritance patterns in pea plants and other common organisms. Mendel studied characteristics that were inherited as a unit; this enabled scientists to understand such phenomena as dominance and recessiveness in these units, called genes in 1909 by the Danish biologist W. L. Johannsen (1859–1927). By 1900, when Hugo de Vries (1848–1935), William Bateson (1861–1926) and others were recognizing the importance of Mendel's pioneering work, much more was known about the microscopic appearances of cells both during adult division (mitosis) and reduction division (meiosis). In addition, August Weismann (1834–1914) had developed notions of the continuity of the inherited material (which he called the 'germ plasm') from generation to generation, thus suggesting that acquired characteristics are not inherited. Only a few scientists in the twentieth century, such as the Soviet botanist Trofim Lysenko (1898–1976), have continued to believe in Lamarckism, for modern genetics has accumulated overwhelming evidence that characteristics such as the loss of an arm or internal muscular development do not change the make-up of reproductive cells.

It is now believed that new inheritable variations occur when genes mutate. The study of this process and of the factors (such as X-rays and certain chemicals) that can make the occurrence of mutations likely was pioneered by geneticists such as Thomas Hunt Morgan (1866–1945) and Herman Muller (1890–1967). They did much of their work with fruit flies (*Drosophila*). While these and other scientists were showing that genes are located on chromosomes – strands of darkstaining material in the nuclei of cells – other researchers were trying to determine the exact nature of the hereditary substance itself. Originally it was thought to be a protein, but in 1953 Francis Crick (1916–) and James Watson (1928–) were able to show that it is dioxyribonucleic acid (DNA). Their work was an early triumph of molecular biology, a branch of the science that has grown enormously since the 1950s. Scientists now know a great deal about how DNA works. Among those who have contributed are George Beadle (1903–1989), Edward Tatum (1909–1975), Jacques Monod (1910–1976), Joshua Lederberg (1925–) and Maurice Wilkins (1916–). Molecular biologists and chemists have also been concerned with determining the structures of many other large biological molecules, such as the muscle protein myoglobin by John Kendrew (1917–) and Max Perutz (1914–). In their researches they often have to interpret the diffraction patterns produced when X-rays pass through these complex molecules, a technique pioneered by Dorothy Hodgkin (1910–).

Modern biological sciences

Molecular biology is only one of several new biological disciplines to be developed during the past century. The oldest of these, biochemistry, was established in Britain by Frederick Gowland Hopkins (1861–1947), who, along with Casimir Funk (1884–1967) and Elmer McCollum (1879–1967), is remembered for his fundamental work in the discovery of vitamins, substances that help to regulate many complex bodily processes. Other biochemists such as Carl Cori (1896–1984) and his wife Gerty Cori (1896–1957) have studied the ways in which organisms make use of the energy gained when food is broken down. Many of these internal processes are also moderated by the action of hormones, one important example of which is insulin, discovered in the 1920s by Frederick Banting (1891–1941) and others.

Modern biologists also often use physics in their work, and biophysics is now an important discipline in its own right. Archibald Hill (1886–1977) and Otto Meyerhof (1884–1951) pioneered in this area with their work on the release of heat when muscles contract. More recently, Bernhard Katz (1911–) has used biophysical techniques in studying the events at the junctions between muscles and nerves, and at the junctions between pairs of nerves (synapses). The events at synapses are initiated by the release of chemical substances such as adrenaline and acetylcholine, as was demonstrated by Henry Dale (1875–1968) and Otto Loewi (1873–1961). The way in which nerve impulses move along the nerve axon has been investigated by Alan Hodgkin (1914–) and Andrew Huxley (1917–). For this work, they made use of the giant axon of the squid, an experimental preparation whose importance for biology was first shown by John Young (1907–). The complicated way in which the nervous system operates as a whole was first rigorously investigated by Charles Sherrington (1857–1952).

Another area of fundamental importance in modern biology and medicine is immunology. For instance, the discovery by Karl Landsteiner (1868–1943) of the major human blood group system (A, B and O) permitted safe blood transfusions. The development of the immune system – and the way in which the body recognizes foreign substances ('self' and 'not-self') – has been investigated by such scientists as Peter Medawar (1915–1987) and Frank Burnet (1899–1985). Much of this knowledge has been important to transplant surgery, pioneered for kidneys by Roy Calne (1930–) and for hearts by Christiaan Barnard (1922–).

Genetic engineering

Genetic engineering consists of a collection of methods used to manipulate genes. The first of these techniques dates back to 1952 when Joshua Lederberg discovered that bacteria exchange genetic material contained in a body he called a plasmid. The next year William Hayes established that plasmids were rings of DNA free from the main DNA in the chromosome of the bacteria. The next step was taken by Werner Arber, who studied viruses (called bacteriophages) which infect bacteria. He found that bacteria resist phages by splitting the phage DNA using enzymes. By 1968 Arber had discovered the enzymes produced by bacteria that split DNA at specific locations. In addition, he found that different genes that have been split at the same location by one of the restriction enzymes, as they are called, will recombine when placed together in the absence of the enzyme. The resulting product is called recombinant DNA.

In 1973 Stanley H. Cohen and Herbert W. Brown combined restriction enzymes with plasmids in the first genetic engineering experiment. They cut a chunk out of a plasmid found in the bacterium

Escherichia coli and inserted in the gap a gene created from a different bacterium. In 1976 Har Gobind Khorana (1922–   ) and co-workers constructed the first artificial gene to function naturally when inserted into a bacterial cell. In 1985 the first human cancer gene was isolated by researchers in Massachusetts, USA. The Human Genome Organization was established in Washington, DC, in 1988 with the aim of mapping the complete sequence of human DNA.

Biological research in the twentieth century

There are far more biologists at work today than ever before and, like other branches of science, biology is constantly expanding and changing. Increasingly, new developments result from a team effort, rather than from the work of an individual scientist. Some areas of research – such as biological warfare and genetic engineering – are controversial. Some biomedical achievements, such as pesticides, antibiotics, vaccines and a better knowledge of disease prevention, have contributed to the potential problems of over-population. Science is a human creation, and as such can be used for good or evil. As the philosopher Francis Bacon said: 'Knowledge is power'. A knowledge of biology is necessary for everyone, not only because it has much to teach us about ourselves, but also because biology and the other sciences are increasingly significant aspects of modern life.

Chemistry

Chemistry seems to have originated in Egypt and Mesopotamia several thousand years before Christ. Certainly by about 3000 BC the Egyptians had produced the copper–tin alloy known as bronze, by heating the ores of copper and tin together, and this new material was soon common enough to be made into tools, ornaments, armour and weapons. The Ancient Egyptians were also skilled at extracting juices and infusions from plants, and pigments from minerals, which they used in the embalming and preserving of their dead. By 600 BC the Greeks were also becoming a settled and prosperous people with leisure time in which to think. They began to turn their attention to the nature of the universe and to the structure of its materials. They were thus the first to study the subject we now call chemical theory. Aristotle (384–322 BC) proposed that there were four elements – earth, air, fire and water – and that everything was a combination of these four. They were thought to possess the following properties: earth was cold and dry, air was hot and moist, fire was hot and dry, and water was cold and moist. The idea of the four elements persisted for 2,000 years. The Greeks also worked out, at least hypothetically, that matter ultimately consisted of small indivisible particles, *atomos* – the origin of our word 'atom'.

From the Egyptians and the Greeks comes *khemeia*, alchemy and eventually chemistry as we know it today. The source of the word *khemeia* is debatable, but it is certainly the origin of the word chemistry. It may derive from the Egyptians' word for their country *Khem*, 'the black land'. It may come from the Greek word *khumos* (the juice of a plant), so that *khemeia* is 'the art of extracting juices'; or from the Greek *cheo* 'I pour or cast', which refers to the activities of the metal workers. Whatever its origin, the art of *khemeia* soon became akin to magic and was feared by the ordinary people. One of the greatest aims of the subject involved the attempts to transform base metals such as lead and copper into silver or gold. From the four-element theory, it seemed that it should be possible to perform any such change, if only the proper technique could be found.

The Arabs and alchemy

With the decline of the Greek empire *khemeia* was not pursued and little new was added to the subject until it was embraced by the increasingly powerful Arabs in the seventh century AD. Then for five centuries *al-kimiya*, or alchemy, was in their hands. The Arabs drew many ideas from the *khemeia* of the Greeks, but they were also in contact with the Chinese – for example, the idea that gold possessed healing powers came from China. They believed that 'medicine' had to be added to base metals to produce gold, and it was this medicine that was to become the philosopher's stone of the later European alchemists. The idea that not only could the philosopher's stone heal 'sick' or base metals, but that it could also act as the elixir of life, was also originally Chinese. The Arab alchemists discovered new classes of chemicals such as the caustic alkalis (from the Arabic *al-qalíy*) and they improved technical procedures such as distillation.

Western Europe had its first contact with the Islamic world as a result of the Crusades. Gradually the works of the Arabs – handed down from the Greeks – were translated into Latin and made available to European scholars in the twelfth and thirteenth centuries. Many men spent their lives trying in vain to change base metals into gold; and many alchemists lost their heads for failing to supply the promised gold. Throughout the fifteenth to seventeenth centuries their symbolism became more and more complex. In 1689 the preparation of silver chloride, by dissolving metallic silver in nitric acid and then adding hydrochloric acid, was described like this:

Recipe ☽, in Ω ⊙ solve, cum Ω ⊖ precipitata, filtra.

☽ = Moon = silver (the same symbols were used for the metals and their associated planets),
⊙ = nitre, ⊖ = mineral acid, and Ω = spirit.

A new era in chemistry began with the researches of Robert Boyle (1627–1691), who carried out many experiments on air. These experiments were the beginning of a long struggle to find out what air had to do with burning and breathing. From Boyle's time onwards, alchemy became chemistry and it was realized that there was more to the subject than the search for the philosopher's stone.

Chemistry as an experimental science

During the 1700s the phlogiston theory gained popularity. It went back to the alchemists' idea that combustible bodies lost something when they burned. Metals were thought to be composed of a calx (different for each) combined with phlogiston, which was the same in all metals. When a candle burned in air, phlogiston was given off. It was believed that combustible objects were rich in phlogiston and what was left after combustion possessed no phlogiston and would therefore not burn. Thus wood possessed phlogiston but ash did not; when metals rusted, it was considered that the metals contained phlogiston but that its rust or calx did not. By 1780 this theory was almost universally accepted by chemists. Joseph Priestley (1733–1804) was a supporter of the theory and in 1774 he had succeeded in obtaining from mercuric oxide a new gas which was five or six times purer than ordinary air. It was, of course, oxygen but Priestley called it 'dephlogisticated air' because a smouldering splint of wood thrust into an atmosphere of this new gas burst into flames much more readily than it did in an ordinary atmosphere. He took this to mean that the gas must be without the usual content of phlogiston, and was therefore eager to accept a new supply.

It was Antoine Lavoisier (1743–1794) who put an end to the phlogiston theory by working out what was really happening in combustion. He repeated Priestley's experiments in 1775 and named the dephlogisticated air oxygen. He realized that air was not a single substance but a mixture of gases, made up of two different gases in the proportion of 1 to 4. He deduced that one-fifth of the air was Priestley's dephlogisticated air (oxygen), and that it was this part only that combined with rusting or burning materials and was essential to life. Oxygen means 'acid-producer' and Lavoisier thought, erroneously, that oxygen was an essential part of all acids. He was a careful experimenter and user of the balance, and from his time onwards experimental chemistry was concerned only with materials that could be weighed or otherwise measured. All the 'mystery' disappeared and Lavoisier went on to work out a logical system of chemical nomenclature, much of which has survived to the present day.

Early in the nineteenth century many well-known chemists were active. Claude Berthollet (1748–1822) worked on chemical change and composition, and Joseph Gay-Lussac (1778–1850) studied the volumes of gases that take part in chemical reactions. Others included Berzelius, Cannizzaro, Avogadro, Davy, Dumas, Kolbe, Wöhler and Kekulé. The era of modern chemistry was beginning.

Atomic theory and new elements

An English chemist, John Dalton (1766–1844), founded the atomic theory in 1803 and in so doing finally crushed the belief that the transmutation to gold was possible. He realized that the same two elements can combine with each other in more than one set of proportions, and that the variation in combining proportions gives rise to different compounds with different properties. For example, he determined that one part (by weight) of hydrogen combined with eight parts of oxygen to form water, and if it was assumed (incorrectly) that a molecule of water consisted of one atom of hydrogen and one atom of oxygen, then it was possible to set the mass of the hydrogen atom arbitrarily at 1 and call the mass of oxygen 8 (on the same scale). In this way Dalton set up the first table of atomic weights (now called relative atomic masses), and although this was probably his most important achievement, it contained many incorrect assumptions. These errors and anomalies were researched by Jöns Berzelius (1779–1848), who found that for many elements the atomic weights were not simple multiples of that of hydrogen. For many years, oxygen was made the standard and set at 16.000 until the mid-twentieth century, when carbon (= 12.000) was adopted. Berzelius suggested representing each element by a symbol consisting of the first one or two letters of the name of the element (sometimes in Latin) and these became the chemical symbols of the elements as still used today.

At about the same time, in 1808, Humphry Davy (1778–1829) was using an electric current to obtain from their oxides elements that had proved to be unisolatable by chemical means: potassium, sodium, magnesium, barium and calcium. His assistant, Michael Faraday (1791–1867), was to become even better known in connection with this technique, electrolysis. By 1830, more than 50 elements had been isolated; chemistry had moved a long way from the four elements of the Ancient Greeks, but their properties seemed to be random. In 1829 the German chemist Johann Döbereiner (1780–1849) thought that he had observed some slight degree of order. He wondered if it was just coincidence that the properties of the element bromine seemed to lie between those of chlorine and iodine, but he went on to notice a similar gradation of properties in the triplets calcium, strontium and barium and with sulphur, selenium and tellurium. In all of these examples, the atomic weight of the element in the middle of the set was about half-way between the atomic weights of the other two elements. He called these groups 'triads', but because he was unable to find any other such groups, most chemists remained unimpressed by his discovery. Then in 1864 John Newlands (1837–1898) arranged the elements in order of their increasing atomic weights and found that if he wrote them in horizontal rows, and started a new row with every eighth element, similar elements tended to fall in the same vertical columns. Döbereiner's three sets of triads were among them. Newlands called this his 'Law of Octaves' by analogy with the repeating octaves in music. Unfortunately there were many places in his chart where obviously dissimilar elements fell together and so it was generally felt that Newland's similarities were not significant but probably only coincidental. He did not have his work published.

In 1862 a German chemist, Julius Lothar Meyer (1830–1895), looked at the volumes of certain fixed weights of elements, and talked of atomic volumes. He plotted the values of these for each element against its atomic weight, and found that there were sharp peaks in the graph at the alkali metals – sodium, potassium, rubidium and caesium. Each part of the graph between the peaks corresponded to a 'period' or horizontal row in the table of the elements, and it became obvious where Newlands had gone wrong. He had assumed that each period contained only seven elements; in fact the later periods had to be longer than the earlier ones. By the time Meyer published his findings, he had been anticipated by the Russian chemist Dmitri Mendeleyev (1834–1907), who in 1869 published his version of the periodic table, which remains largely unchanged today. He had the insight to leave gaps in his table for three elements which he postulated had not yet been discovered, and was even able to predict what their properties would be. Chemists were sceptical, but within 15 years all three of the 'missing' elements had been discovered and their properties were found to agree with Mendeleyev's predictions.

The beginnings of physical chemistry

Until the beginning of the nineteenth century, the areas covered by the subjects of chemistry and physics seemed well-defined and quite distinct. Chemistry studied changes where the molecular bonding structure of a substance was altered, and physics studied phenomena in which no such change occurred. Then in 1840 physics and chemistry merged in the work of Germain Hess (1802–1850). It had been realized that heat – a physical phenomenon – was produced by chemical reactions such as the burning of wood, coal and oil, and it was gradually becoming clear that all chemical reactions involved some sort of heat transfer. Hess showed that the quantity of heat produced or absorbed when one substance was changed into another was the same no matter by which chemical route the change occurred, and it seemed likely that the law of conservation of energy was equally applicable to chemistry and physics. Thermochemistry had been founded and work was able to begin on thermodynamics. Most of this research was done in Germany and it was Wilhelm Ostwald (1853–1932), towards the end of the nineteenth century, who was responsible for physical chemistry developing into a discipline in its own right. He worked on chemical kinetics and catalysis in particular, but was the last important scientist to refuse to accept that atoms were real – there was at that time still no direct evidence to prove that they existed. Other contemporary chemists working in the new field of physical chemistry included Jacobus van't Hoff (1852–1911) and Svante Arrhenius (1859–1927). Van't Hoff studied solutions and showed that molecules of dissolved substances behaved according to rules analogous to those that describe the behaviour of gases. Arrhenius carried on the work which had been begun by Davy and Faraday on solutions that could carry an electric current. Faraday had called the current-carrying particles 'ions', but nobody had worked out what they were. Arrhenius suggested that they were atoms or groups of atoms which bore either a positive or a negative electric charge. His theory of ionic dissociation was used to explain many of the phenomena in electrochemistry.

Towards the end of the nineteenth century, mainly as a result of the increasing interest in the physical side of chemistry, gases came under fresh scrutiny and some errors were found in the law that had been proposed three centuries earlier by Robert Boyle. Henri Regnault (1810–1878), James Clerk Maxwell (1831–1879) and Ludwig Boltzmann (1844–1906) had all worked on the behaviour of gases, and the kinetic theory of gases had been derived. Taking all their findings into account, Johannes van der Waals (1837–1923) arrived at an equation that related pressure, volume and temperature of gases and made due allowance for the sizes of the different gas molecules and the attractions between them. By the end of the century William Ramsay (1852–1916) had begun to discover a special group of gases – the inert or rare gases – which have a valency (oxidation state) of zero and which fit neatly into the periodic table between the halogens and the alkali metals.

Organic chemistry becomes a separate discipline

Meanwhile the separate branches of chemistry were emerging and organic substances were being distinguished from inorganic ones. In 1807 Berzelius had proposed that substances such as olive oil and sugar, which were products of living organisms, should be called organic, whereas sulphuric acid and salt should be termed inorganic. Chemists at that time had realized that organic substances were easily converted into inorganic substances by heating or in other ways, but it was thought to be impossible to reverse the process and convert inorganic substances into organic ones. They believed in Vitalism – that somehow life did not obey the same laws as did inanimate objects and that some special influence, a 'vital force', was needed to convert inorganic substances into organic ones. Then in 1828 Friedrich Wöhler (1800–1882) succeeded in converting ammonium cyanate (an inorganic compound) into urea. In 1845 Adolf Kolbe (1818–1884) synthesized acetic acid, squashing the Vitalism theory for ever. By the middle of the nineteenth century organic compounds were being synthesized in profusion; a new definition of organic compounds was clearly needed, and most organic chemists were working by trial and error. Nevertheless there was a teenage assistant of August von Hofmann (1818–1892), called William Perkin (1838–1907), who was able to retire at the age of only 35 because of a brilliant chance discovery. In 1856 he treated aniline with potassium chromate, added alcohol, and obtained a beautiful purple colour, which he suspected might be a dye (later called aniline purple or mauve). He left school and founded what became the synthetic dyestuffs industry.

Then in 1861 the German chemist Friedrich Kekulé (1829–1886) defined organic chemistry as the chemistry of carbon compounds and this definition has remained, although there are a few carbon compounds (such as carbonates) which are considered to be part of inorganic chemistry. Kekulé suggested that carbon had a valency of four, and proceeded to work out the structures of simple organic compounds on this basis. These representations of the structural formulae showed how organic molecules were generally larger and more complex than inorganic molecules. There was still the problem of the structure of the simple hydrocarbon benzene, C_6H_6, until 1865 when Kekulé suggested that rings of carbon atoms might be just as possible as straight chains. The idea that molecules might be three-dimensional came in 1874 when Van't Hoff suggested that the four bonds of the carbon atom were arranged tetrahedrally. If these four bonds are connected to four different types of groups, the carbon atom is said to be asymmetric and the compound shows optical activity – its crystals or solutions rotate the plane of polarized light. Viktor Meyer (1848–1897) proposed that certain types of optical isomerism could be explained by bonds of nitrogen atoms. Alfred Werner (1866–1919) went on to demonstrate that this principle also applied to metals such as cobalt, chromium and rhodium, and succeeded in working out the necessary theory of molecular structure, known as coordination theory. This new approach allowed there to be structural relationships within certain fairly complex inorganic molecules, which were not restricted to bonds involving ordinary valencies. It was to be another 50 years before enough was known about valency for both Kekulé's theory and Werner's to be fully understood, but by 1900 the idea was universally accepted that molecular structure could be represented satisfactorily in three dimensions.

Kekulé's work gave the organic chemist scope to alter a structural formula stage by stage, to convert one molecule into another, and modern synthetic organic chemistry began. Richard Willstätter (1872–1942) was able to work out the structure of chlorophyll, and Heinrich Wieland (1877–1957) determined the structures of steroids. Paul Karrer (1889–1971) elucidated the structures of the carotenoids and other vitamins and Robert Robinson (1886–1975) tackled the alkaloids – he worked out the structures of morphine and strychnine. The alkaloids have found medical use as drugs, as have many other organic compounds. The treatment of disease by the use of specific chemicals is known as chemotherapy and was founded by the bacteriologist Paul Ehrlich (1854–1915). The need for drugs to combat disease and infection during World War II spurred on research, and by 1945 the antibiotic penicillin, first isolated by Howard Florey (1898–1968) and Ernst Chain (1906–1979), was being produced in quantity. Other antibiotics such as streptomycin and the tetracyclines soon followed.

Some organic molecules contain thousands of atoms; some, such as rubber, are polymers and others, such as haemoglobin, are proteins. Synthetic polymers have been made which closely resemble natural rubber; the leader in this field was Wallace Carothers (1896–1937), who also invented nylon. Karl Ziegler (1898–1973) and Giulio Natta (1903–1979) worked out how to prevent branching during polymerization, so that plastics, films and fibres can now be made more or less to order. However, work on the make-up of proteins had to wait for the development of chemical techniques such as chromatography (by Mikhail Tswett (1872–1919) and by Archer Martin (1910–) and Richard Synge (1914–)) and electrophoresis (Arne Tiselius (1902–1971)). In the forefront of molecular biological research are Frederick Sanger (1918–), John Kendrew (1917–) and Max Perutz (1914–). One technique that has been essential for their work is X-ray diffraction, and for the background to this development we have to return to the area of research between chemistry and physics at the beginning of the present century.

Modern atomic theory

Ever since Faraday had proposed his laws of electrolysis, it had seemed likely that electricity might be carried by particles. The physicist Ernest Rutherford (1871–1937) decided that the unit of positive charge was a particle quite different from the electron, which was the unit of negative charge, and in 1920 he suggested that this fundamental positive particle be called the proton. In 1895 Wilhelm Röntgen (1845–1923) discovered X-rays, but other radiation components – alpha and beta rays – were found to be made up of helium nuclei and electrons, respectively. In about 1902 it was proved, contrary to all previous ideas, that radioactive elements changed into other elements, and by 1912 the complicated series of changes of these elements had been worked out. In the course of this research, Frederick Soddy (1877–1956) realized that there could be several atoms differing in mass but having the same properties. They were called isotopes and we now know that they differ in the number of neutrons which they possess, although the neutron was not to be discovered until 1932, by the physicist James Chadwick (1891–1974).

Rutherford evolved the theory of the nuclear atom, which suggested that sub-atomic particles made up the atom, which had until that time been considered to be indivisible. The question now was, how did the nuclear atom of one element differ from that of another? In 1909 Max von Laue (1879–1960) began a series of brilliant experiments. He established that crystals consist of atoms arranged in a geometric structure of regularly repeating layers, and that these layers scatter X-rays in a set pattern. In so doing, he had set the scene for X-ray crystallography to be used to help to work out the structures of large molecules for which chemists had not been able to determine formulae.

In 1913 the young scientist Henry Moseley (1887–1915) found that there were characteristic X-rays for each element and that there was an inverse relationship between the wavelength of the X-ray and the atomic weight of the element. This relationship depended on the size of the positive charge on the nucleus of the atom, and the size of this nuclear charge is called the atomic number. Mendeleyev had arranged his periodic table by considering the valencies of the elements, in sequence of their atomic weights, but the proper periodic classification is by atomic number. It was now possible to predict exactly how many elements were still to be discovered. Since the proton is the only positively charged particle in the nucleus, the atomic number is equal to the number of protons; the neutrons contribute to the mass but not to the charge. For example, a sodium atom, with an atomic number of 11 and an atomic weight (relative atomic mass) of 23, has 11 protons and 12 neutrons in its nucleus.

Isotopes and biochemistry

The new electronic atom was also of great interest to organic chemists. It enabled theoreticians such as Christopher Ingold (1893–1970) to try to interpret organic reactions in terms of the movements of electrons from one point to another within a molecule. Physical chemical methods were being used in organic chemistry, founding physical organic chemistry as a separate discipline. Linus Pauling (1911–), a chemist who was to suggest in the 1950s that proteins and nucleic acids possessed a helical shape, worked on the wave properties of electrons, and established the theory of resonance. This idea was very useful in establishing that the structure of the benzene molecule possessed 'smeared out' electrons and was a resonance hybrid of the two alternating double bond/single bond structures. The concept of atomic number was clarified by Francis Aston (1877–1945) with the mass spectrograph. This instrument used electric and magnetic fields to deflect ions of identical charge by an extent that depended on their mass – the greater the mass of the ion, the less it was deflected. He found for instance that there were two kinds of neon atoms, one of mass 20 and one of mass 22. The neon-20 was ten times as common as the neon-22, and so it seemed reasonable that the atomic weight of the element was 20.2 – a weighted average of the individual atoms and not necessarily a whole number. In some cases, the weighted average (atomic weight) of a particular atom may be larger than that for an atom of higher atomic number. This explains the relative positions of iodine and tellurium in the periodic table, which Mendeleyev had placed correctly without knowing why.

In 1931 Harold Urey (1893–1981) discovered that hydrogen was made up of a pair of isotopes, and he named hydrogen-2 deuterium. In 1934 it occurred to the physicist Enrico Fermi (1901–1954) to bombard uranium (element number 92, the highest atomic number known at that time) in order to see whether he could produce any elements of higher atomic numbers. This approach was pursued by Glenn T. Seaborg (1912–) and the transuranic elements were discovered, going up from element 94 to 104 but becoming increasingly difficult to form and decomposing again more rapidly with increasing atomic number.

The area between physics and chemistry has been replaced by a common ground where atoms and molecules are studied together with the forces that influence them. A good example is the discovery in the early 1990s of a new form of carbon, with molecules called buckyballs, consisting of 60 carbon atoms arranged in 12 pentagons and 20 hexagons to form a perfect sphere.

The boundary between chemistry and biology has also become less well defined and is now a scene of intense activity, with the techniques of chemistry being applied successfully to biological problems. Electron diffraction, chromatography and radioactive tracers have all been used to help discover what living matter is composed of, although it is possible that these investigations in biology are only now at the stage that atomic physics was at the beginning of this century. It was Lavoisier who said that life is a chemical function, and perhaps the most important advance of all is towards understanding the chemistry of the cell. Biochemical successes of recent years include the synthesis of human hormones, the development of genetic fingerprinting, and the use of enzymes in synthesis. The entire field of genetic engineering is essentially biochemistry.

In the 1960s, Elias J. Corey (1928–) made a breakthrough in organic chemistry when he developed retrosynthesis, a powerful tool for building complex molecules from smaller, cheaper and more readily available ones. Retrosynthesis can be used to picture a molecule like a jigsaw, working backwards to find reactive components to complete the puzzle. Modern chemists use retrosynthesis to design everything from insect repellents to better drugs.

Engineering and Technology

Our existence today is powerful evidence of our ability to invent. Were it not for the invention of simple tools made from sharp-edged stones some two million years ago, it is doubtful whether our relatively weak and slow ancestors would have survived for long. Such tools enabled early humans to fight off predators and to hunt for food.

As well as being surrounded by potentially hostile animals, early humans were at the mercy of the climate. It was the second of the major inventions of prehistory, a means of creating fire, that enabled them to survive the ice ages. *Homo erectus* was using this to live through the second ice age some 400,000 years ago.

These two inventions served our ancestors well for an extremely long time. Not until the formation of Jericho, the world's first walled town, c. 7000 BC does another major invention reveal itself, in the development of pottery. Copper, and the alloy made from copper and tin called bronze, also appeared at around that time.

Behind the foundation of a settled community in Jericho were the beginnings of organized agriculture, and it was this that provided the stimulus for the development on increasingly sophisticated tools, such as the plough and the sickle.

That most famous, and significant, of inventions of the ancient world, the wheel, first appeared around 3000 BC, in what is now southern Russia. These early wheels were solid, wooden and fixed to sleds which had previously been dragged across the ground. Although its first use was in transporting heavy loads, the wheel and axle combination later became a feature of milling devices, and irrigation systems. Sumerian and Assyrian engineers used wheel-driven water-drawing devices in irrigation networks which are still in use today.

For several thousand years, these early inventions were enlarged and improved upon, without any major advances being made. By the time Archimedes (c. 287–212 BC) was investigating the principle of the lever and producing his famous helical screw, the scene of the most significant developments had shifted from the Middle East to Greece. The measurement of time in particular remained a continuing challenge, and by the second century BC, Ctesibius of Alexandria had developed the Egyptian clepsydra (water-clock) to give accuracy not surpassed until well into the Middle Ages.

Hero of Alexandria (c. AD 60) was the last of the Greek technologists, his most famous invention being of the aeolipile, a primitive steam turbine that gave a hint, as early as the first century BC, of the potency of steam as a powersource.

Roman and Chinese influences

Despite their astonishing ability in geometry, physics and mathematics, the Greeks were unable to make the advance which transformed architecture and civil engineering: the arch. The pre-Roman Etruscans used the semicircular arch as an architectural feature, but it was the Romans who put the arch to full use. Because of its ability to spread stresses more evenly, the arch allowed greater spans in buildings than the Greeks' simple pillar-and-beam arrangements. Aqueducts comprising 6-m/20-ft-wide arches were possible as early as 142 BC, as evidenced by the Pons Aemilius in Italy.

Combining the structural economy of the arch with the availability of good cement the Romans set up an infrastructure and communications network that gave them a standard of living hitherto unprecedented in history. It also enabled them to extend that standard, and maintain it by its armies, over a similarly unprecedented expanse of the world.

One major difficulty facing the Romans was the shortage of labour. As a result of the lack of an efficient means of harnessing animals to ploughs and other implements, the Romans had to use the weaker human to provide the power. It was the Chinese who first produced an efficient animal harness, freeing humans from such drudgery, and enabling animals to be used in tandem to haul great loads. The harness did not reach the West, however, until the ninth century, well after the fall of the

Roman Empire. China was the birthplace of a number of other major inventions: of paper from pulp (by Ts'ai Lun *c.* AD 100), of the magnetic compass and of gunpowder, about 500 and 850 respectively.

The development of Western technology

It was also around this time that northwestern Europe began its climb to ascendency in technology that it has held on to for centuries since. The poorer climate of this region, combined with the need to develop a new form of agriculture, was responsible for the emergence around the eighth century of the crop rotation methods still used today. The wind was put to use in both sea-going vessels and land-based mills. The region grew more populous and, by the eleventh century, northern Europeans were moving their influence into the Mediterranean and Middle East.

As the societies grew, becoming more complex, the need for metals for housing, tools, equipment and coinage increased concomitantly. This caused a renewed interest in the extraction and treatment of ores, which were also in increasing demand by rulers anxious to develop weapons capable of keeping their rivals at bay. Many of these rulers, notably in fifteenth-century Italy, employed engineers to come up with new systems for both defence and attack. Undoubtedly the most famous such engineer was Leonardo da Vinci (1452–1519), military engineer to the Duke of Milan. Among the thousands of pages of da Vinci's notes are to be found an astonishing number of prescient plans for modern-day inventions: tanks, submarines, helicopters and a whole range of firearms.

The Renaissance was ushered in by one of the most influential inventions in history: the development of the movable type printing press by Johannes Gutenberg (*c.* 1398–1468) of Mainz, Germany. The Gutenberg Bible, the first book to be printed using this process, appeared in 1454. The Englishman William Caxton set up a press in England in 1476. As well as disseminating religious knowledge on a far larger scale, the invention enabled the speeding up of the transfer of technological advances from one country to another.

Steam power and the Industrial Revolution

The ever increasing use of metals made mining the focus of much effort in the fifteenth century, one of the biggest problems being that of adequately draining the mines. Pumping water fast enough and in sufficient volume was also a problem facing those wanting to create more agricultural land from poorly drained areas, and to supply towns with their needs. By the mid-seventeenth century, a number of patents had been granted to water pumps which used a new, remarkably versatile source of power: steam.

Thomas Savery (*c.* 1650–1715) demonstrated an engine for 'raising of water and occasioning motion to all sorts of mill works, by the impellant force of fire', to the Royal Society of London in 1699. He formed a partnership with Thomas Newcomen (1663–1729) and together they used the power of steam, in the form of a beam piston engine, to maintain the mines of Staffordshire, Cornwall and Newcastle in workable condition.

Newcomen steam engines grew in popularity during the eighteenth century, and it was while repairing a model one used on the physics course at Glasgow University in 1763 that the young James Watt (1736–1819) saw how major improvements could be made, improvements that led to Watt's condenser engine completely replacing the earlier model by 1800.

Through its predominance as a manufacturing market, Britain was able to reap rich rewards but competition from overseas urged on the hunt for greater efficiency of production. Technologists were at the forefront of this effort, with John Kay's (1704–1780) invention of the flying shuttle for the production of textiles leading to James Hargreaves' (1720–1778) spinning jenny. Arkwright's (1732–1792) spinning machine was the centrepiece of the first cotton factory of 1771, and semi-automated mass production as a technique was born. The Industrial Revolution, which is arguably still under way, had begun.

The end of the eighteenth century saw Richard Trevithick (1771–1833) experimenting with the use of steam to provide motive power for boats, road vehicles and locomotives on steel rails. Dogged by bad luck, his ideas did not achieve the success or acclaim they deserved, and the world had to wait longer than it should have for the full exploitation of steam in such applications. French engineers had more success initially, with Nicholas Joseph Cugnot (1725–1804) developing a three-wheeled steam-powered tractor, around 1770, capable of 6 kph/3.5 mph. The Montgolfier brothers Joseph-Michael

(1740–1810) and Jacques Etiènne (1745–1799) were responsible for the first sustained human flight, aboard a hot-air balloon, in 1763.

By the time Trevithick left England for Peru, his high-pressure engine had shown that steam was amply capable of providing a mobile power source. He returned to find his place as the greatest engineer in the new technology usurped by others. Most famous of these was George Stephenson (1781–1848), whose steam locomotives were responsible for the setting up of the first practical passenger railway ever built, in 1825, between Stockton and Darlington.

Rapid communication, which remains a hallmark of an advanced technological society, became of increasing importance as the nineteenth century wore on, both in peace-time and in war. Thomas Telford (1757–1834) became famous for his canals and aqueduct building, enabling very large loads to be transported using little motive power. In France, the quality of the roads built by Pierre Trésaguet (1716–1796) made the rapid strikes of Napoleon's armies possible. By contrast, Britain's roads were in an appalling state through years of neglect and the operation of the Turnpike Trusts. The scientific approach to road construction devised by John McAdam (1756–1836) radically improved this aspect of Britain's infrastructure.

Steam power also found its way into ocean-going vessels. Robert Fulton (1765–1815) returned from Europe, where he had seen what steam engines were capable of, to set up the first regular steamship service, between New York and Albany, in 1807. Whether steam-powered ships were capable of a longer journey, in particular across the Atlantic Ocean, was a major debating point when Isambard Kingdom Brunel's (1806–1859) *Great Western* succeeded in sailing to New York without refuelling, in 1838. Brunel went on to design and launch the first iron ship (and the first to use a screw propellor, rather than paddle wheels) the *Great Britain*, and the colossal *Great Eastern*, whose steadiness and manoeuvrability were put to use in a key event in another field of technology altogether: the laying of the Atlantic telegraph cable in 1865.

Such very long distance communication was made possible by advances in understanding of electricity, and the translation of this into devices that could transmit and receive messages at the speed of light. The first successful system was brought out by Charles Wheatstone (1802–1875), in which electrical signals deflected magnetized needles indicating letters of the alphabet. The development of a communication system using a code of dots and dashes to represent the letters by Samuel Morse (1791–1872) proved so successful it is still in use today.

The outbreak of the Crimean War involving the British, French and Turkish against the Russians led Henry Bessemer (1813–1898) to develop the method for removing impurities, in particular carbon, from molten iron, thus enabling steel to be produced cheaply in the quantities required by both military and civilian engineers. The war also resulted in the development of the first wrought-iron breech-loading gun by William Armstrong (1847–1908).

The wars affecting the United States also assisted the development of weapons on that side of the Atlantic. The war against Mexico, which broke out in 1846, accelerated the revolution in small arms manufacture initiated by Samuel Colt (1814–1862), who had produced the first revolver in 1836. The Civil War of 1861–1865 prompted Richard Gatling (1818–1903) to develop the rapid-fire gun that bears his name.

The internal combustion engine

It was around the middle of the nineteenth century that attention began to shift from steam to gas and other combustible materials as a means of providing motive power. As early as 1833, an engine that ran on an inflammable mixture of gas and air had been described, and a number of the fundamental principles of the fuel-powered engine had been described by the time Jean Lenoir (1822–1900) began building engines using the system which operated smoothly in 1860. Together, Nikolaus Otto (1832–1891) and Eugen Langen managed by 1877 to solve the basic problems facing the development of the four-stroke internal combustion engine. This work led to the development of the modern motor car, and of powered flight. Gottlieb Daimler (1834–1900) was to join the pair as an engineer, leaving in 1883 to develop lighter, more efficient high-speed engines capable of driving cycles and boats, as well as automobiles.

Rudolph Diesel (1858–1913) experimented with internal combustion engines during the 1890s; by using the heat developed by compression of the fuel–air mixture, rather than a spark from an ignition system, to ignite the mixture, Diesel succeeded in producing an engine that could use cheaper fuels than the Otto cycle engines. However, the high pressures produced in the engines required the use of very heavy gauge metal, with consequent weight–power ratio problems. Later advances in metallurgy enabled this disadvantage to be significantly reduced, with the result that the Diesel engine is still used in a wide range of vehicles today.

Work in the USA, as well as in Germany, succeeded in bringing the power of the internal combustion engine to bear on the problem of powered flight. Early work on the flow of air over gliders by Otto Lilienthal (1848–1896) and others established a body of knowledge needed to supplement the work of earlier enthusiasts such as George Cayley (1773–1857), who had defined the basic aerodynamic forces acting on a wing as early as 1799.

Internal combustion engines, coupled to balloons to form airships, were in use by the turn of the century. The first flight of a heavier-than-air machine powered by a light, efficient internal combustion engine was constructed by the Wright brothers Orville (1871–1948) and Wilbur (1867–1912) on 17 December 1903. As well as finding a suitable engine for such a machine, the Wrights had succeeded in solving the problem of controlling the aeroplane in all three axes.

Electricity as a source of power

Despite the success of the internal combustion engine in powering a wide range of machines, the use of water and steam as power sources remained important in another major field of technology: the generation of electricity. Water had been used to drive turbines of increasing efficiency for many years when Charles Parsons (1854–1931) adapted the basic design of a water turbine to enable a jet of steam to impart its kinetic energy to a series of turbine blades which then rotate. By combining this rotation with the ability of a dynamo to convert rotary motion into electric power, the electric generator was born. The first ever turbine-powered generating station was set up in 1888, using four Parsons turbines each developing 75 kW/100 hp. Direct use of the mechanical power developed by steam was made in Parson's 44 tonne/44.7 ton *Turbinia*, whose turbine engine developed 1.5 MW/2,000 hp, enabling it to travel at 60 kph/37 mph in 1897.

Work by Joseph Swan (1828–1914) in England and Thomas Edison (1847–1931) in the United States finally resulted in the creation of a long-lasting light source powered by electricity – the filament lamp – around 1880.

While Germany and the United States were quick to use electrical power to bring about a revolution in their industrial processes, the availability of cheap labour and concentration on waning industries based on traditional raw material inhibited the adoption of electrical power in Britain. Concentration on telegraphic technology and the generation of electric illumination did have an indirect advantage, however. The invention of the two-electrode electric valve, the diode, by John Fleming (1849–1945) provided a new outlet for the vacuum bulb technology. Such developments as radio communication, radar, television and the computer all benefited from this.

Communication over long distances without the use of cables – 'wireless' communication – had been a practical possibility from the day when the electromagnetic wave physicist James Clerk Maxwell's (1831–1879) theory combining electrical and magnetic phenomena had been investigated by Heinrich Hertz (1857–1894) in 1888. Both transmitters and detectors of these radio waves were developed until Guglielmo Marconi (1874–1937) succeeded in transmitting messages over a few yards using electromagnetic waves in 1895. By 1901, he had succeeded in sending signals right across the Atlantic.

More sophisticated communication was made possible by Lee De Forest (1873–1961) and Reginald Fessenden (1866–1932), and their invention of the triode amplifier and amplitude modulation respectively. These advances enabled speech and sound to be transmitted over very long distances, and gave birth to modern communications.

Developments of the war years

World War I (1914–1918) saw the use of technology on an unprecedented scale. Although many of the advances then simply led to the deaths of hundreds of thousands of troops, many later found major applications in peacetime. An excellent example of this is provided by the development of the nitrogen fixation process to an industrial scale by Karl Bosch (1874–1940). This enabled the Germans to manufacture explosives such as TNT without relying on foreign imports of nitrogen-bearing materials, capable of being blockaded by the Allies. In peacetime, the process allowed the cheap manufacture of fertilizers, equally vital to the survival of a country.

The war also had a profound effect on the aircraft industry. Starting the war as chiefly reconnaissance vehicles, the aircraft became directly involved in the fighting by the end, and mass production of tens of thousands became necessary. Governments spent money on research, accelerating advances in aerodynamics and power systems . Civil aviation, begun by the Germans before the war, benefited, initially using modified military aircraft: the famous Imperial Airways started up in 1924.

As with the automobile, engineers started to look at new power sources for the aircraft. The use of gas turbines was put forward in 1926, a suggestion turned into reality by Frank Whittle (1907–) in 1930. By combining a gas turbine with a centrifugal compressor, he created the jet engine.

The inter-war years saw considerable advances in rocket technology: an area of engineering that was to enable humans to leave the planet of their birth. The Chinese had used solid-fuelled rockets in battles as early as 1232; their direct ancestors are still to be seen strapped to the central booster of the Space Shuttle. It was Konstantin Tsiolkovskii (1857–1935) who pointed out that liquid propellants had distinct advantages of power and controllability over solid fuels. The American astronautics pioneer Robert Goddard (1882–1949) succeeded in launching the first liquid-fuelled rocket in 1926. Just 35 years later, Soviet engineers used a liquid-filled booster to send the first human being into earth orbit. Eight years after that a human being set foot on another celestial body for the first time.

By the 1920s, a number of devices born in research laboratories had become established as massively popular forms of entertainment. The work of George Eastman (1854–1932), Thomas Edison (1847–1931) and others brought photography and sound-and-motion 'movie' pictures to millions. A working system of television was devised by John Baird and shown in 1925, while the modern electronic system later adopted as standard for television was demonstrated by Vladimir Zworykin (1889–1982) in 1929. The British Broadcasting Corporation's forebear, the British Broadcasting Company, was formed in 1922, transmitting radio programmes to the public on a national scale. In 1936, experimental television broadcasts were made by the BBC from Alexandra Palace in what is now north London.

The growth in the use of electrical power put greater emphasis on ways of generating it cheaply. The United States in particular built many large storage dams, producing electricity by hydroelectric turbine technology. By 1920 some 40% of electricity in the USA was generated by this means. But developments in particle physics during these years were beginning to show that the fundamental constituents of matter would be capable of providing another, far more concentrated, form of energy: atomic power. By 1939 and the outbreak of World War II, a number of physicists had begun to appreciate the possibilities offered by 'chain reactions' involving the fission of unstable chemical elements such as uranium.

The war itself again proved to be a sharp stimulus for the refinement of old ideas and the development of new ones. Radar, devised by Robert Watson-Watt (1892–1973) in 1935, was developed into a national defence system against aircraft that were growing ever faster and more deadly. More sophisticated weapons and ever more complex message-encoding systems resulted in the development of early electronic computers. These were needed to rapidly sift through data and perform arithmetical operations upon it, and also to carry out numerical integrations which were particularly useful in the precise calculation of the trajectories for artillery shells. The pioneering work on mechanical computing machines by Charles Babbage (1792–1871) in the early 1800s was transformed by the introduction of electronic devices by Vannevar Bush (1890–1974) and others during the war years.

The inter-war ideas of jet and rocket propulsion were used in the development of fighter aircraft and missiles such as the 'V' (*Vergeltungswaffe*) 1 and 2, powered by ram-jet and liquid fuel respectively. Most devastating of all was the use of the chain reaction of atomic energy in an uncontrolled explosive device against the Japanese in 1945. Although the use of the atomic bomb finally ended World War II, the world still lives under the threat of their use, in still more deadly form, to this day. The use of the first atomic bombs has tended to overshadow another event in the development of atomic power that took place during the war. This was the setting up of the first controlled atomic chain reaction in 1942 by a team of scientists at the University of Chicago, which paved the way for the peaceful use of atomic power to generate electricity.

The age of the microchip

While the demand for yet more electric power grew among industrial nations mass-producing cars, ships and aircraft, ways of reducing the complexity and power consumption of electronic devices such as computers, radios and televisions were being sought. Most crucial of these was the invention of the transistor, by John Bardeen (1908–1991), William Shockley (1910–1989) and Walter Brattain (1902–1987) at Bell Laboratories in America, in 1948. These tiny semiconductor-based devices could achieve the rectification and amplification of the thermionic valves of the pre-war years at a fraction of the power consumption. The use of such devices in still smaller form allowed the miniaturization of electronic devices to continue to the stage where the computing power of a whole room of electric devices and thousands of valves can now be contained on a 5 sq mm/0.2 sq in slice of silicon.

Computing power packed into smaller volumes has been the driving force of many areas of technology over the past 40 years. It made possible the era of human space flight, where keeping the mass of all components to a minimum is vital. Telecommunication satellites, such as *Telstar* (which in 1962 transmitted the first live television pictures across the Atlantic), weather satellites and planetary probes were all made possible by the semiconductor breakthrough. New materials capable of withstanding the rigours of space were also produced, many of which found uses back on Earth.

The ancient human desire to be rid of drudgery and physical labour is also becoming increasingly close to realization as a result of such devices. Robots are now used in mass-production industries, such as car manufacturing. The use of robots today is the result of advances in the two fields of mechanization and control over the past 200 years. Oliver Evans (1755–1819) and Joseph Jacquard (1752–1834) had devised automatic textile production systems by the turn of the eighteenth century. Jacquard's use of punched cards by which the looms for weaving could be programmed has proved particularly prescient. Control methods such as feedback (which had been employed by Watt in his centrifugal steam governor), and servo mechanisms were also crucial to the development of the modern robots. By combining these ideas with the new electronic technology, the first industrial robots made their appearance in the United States in 1961, where they tended die-casting equipment. Another development of the 1960s was the laser, invented by Theodore Maiman (1927–), and often used in robotic welders. Holograms, or three-dimensional pictures, became practicable after the development of laser technology.

Despite the proven ability of robots to carry out welding, spraying and other manual tasks, there are still areas where humans remain the best solution; visual inspection of products is an example. The reason for this is that despite the apparent simplicity of such tasks, the computing power necessary to enable a robot to inspect and work on products is enormous. Not until still more powerful computers have been successfully coupled to the robot will it reach its full potential.

What computing power is available today has proved invaluable, nonetheless. Computer-aided design and manufacture have enabled new ideas in fields from architecture to aircraft manufacture to be tried out, tested and produced far more quickly and cheaply. The influence of the computer is felt in everyday life, from the diagnosis of disease by the CAT scanner invented by Godfrey Hounsfield (1919–) to the production of bank statements.

Power sources for the future

Even if, between them, the robot and the computer free us from manual labour, power will still be needed in vast quantities to process the raw material from which goods are produced. There is still, therefore, considerable interest in finding ways of generating cheap power. Using nuclear fission has proved only a partial answer to the question of what will replace the burning of hydrocarbons such as coal to generate electricity. Public concern about both its inherent safety and the toxic waste produced has cast a shadow over the long-term future of fission-generated electricity.

Engineers and physicists in Europe, the United States, the former Soviet Union and Japan are currently studying the generation of power by nuclear fusion. Using hydrogen and its isotopes derived from sea water, they hope to be able to mimic the reactions that have kept the Sun burning for thousands of millions of years. The engineering difficulties presented by trying to keep a plasma stable at a temperature of 100 million degrees are immense, but in 1991 fusion power came a step nearer when the Joint European Torus (JET) at Culham, England, produced a substantial amount of fusion power for the first time.

Others are turning to less exotic sources of energy, such as the wind, solar energy and tides, to find better ways of exploiting them and generating cheap, clean power. The world's largest photovoltaic power station was plugged into the power grid at Davis, California, USA, in 1993. A wind farm at Altmont Paso, California, uses 300 wind turbines, the largest generating 750 kW/1,000 hp, to supply power to nearby Los Angeles.

Geology

Throughout history humans have sought to control and understand their environment. Practical activities like agriculture and quarrying naturally lead to enhanced knowledge, and science suggests further ways of utilizing the Earth.

Growing interaction with the Earth has been important in the development of numerous sciences – not just geology but cosmogony and geophysics; alchemy and chemistry; mineralogy and crystallography; meteorology, physical geography, topography and oceanography; natural history, biology and ecology. Distinct investigation of the Earth itself – geology – has been a recent development. Geology (literally 'Earth-knowledge') does not date back more than two hundred years.

Antiquity

Scientific thinking about the Earth grew out of traditions of thought which took shape in the Middle East and the Eastern Mediterranean. Early civilization needed to adapt to the seasons, to deserts and mountains, volcanoes and earthquakes. Yet inhabitants of Mesopotamia, the Nile Valley, and the Mediterranean littoral had experience of only a fraction of the Earth. Beyond lay *terra incognita*. Hence legendary alternative worlds were conjured up in myths of burning tropics, lost continents and unknown realms where the gods lived.

The first Greek philosopher about whom much is known was Thales of Miletus (*c.*624–547 BC). He postulated water as the primary ingredient of material nature. Thales' follower, Anaximander, believed the universe began as a seed which grew; and living things were generated by the interaction of moisture and the Sun. Xenophanes (*c.*570–*c.*480 BC) is credited with a cyclic worldview: eventually the Earth would disintegrate, returning to a watery state.

Like many other Greek philosophers, Empedocles (*c.*490–430 BC) was concerned with change and stability, order and disorder, unity and plurality. The terrestrial order was dominated by strife. In the beginning, the Earth had brought forth living structures more or less at random. Some had died out. The survivors became the progenitors of modern species.

The greatest Greek thinker was Aristotle. He considered the world was eternal. Aristotle drew attention to natural processes continually changing its surface features. Earthquakes and volcanoes were due to the wind coursing about in underground caves. Rivers took their origin from rain. Fossils indicated that parts of the Earth had once been covered by water.

In the second century AD, Ptolemy composed a geography that summed up the Ancients' learning. Ptolemy accepted that the equatorial zone was too torrid to support life, but he postulated an unknown land mass to the South, the *terra australis incognita*. Antiquity advanced a *geocentric* and *anthropocentric* view. The planet had been designed as a habitat for humans. A parallel may be seen in the Judaeo-Christian cosmogony.

The centuries from Antiquity to the Renaissance accumulated knowledge on minerals, gems, fossils, metals, crystals, useful chemicals and medicaments, expounded in encyclopedic natural histories by Pliny (23–79) and Isidore of Seville (560–636). The great Renaissance naturalists were still working within this 'encyclopedic' tradition. The most eminent was Conrad Gesner, whose *On Fossil Objects* was published in 1565, with superb illustrations. Gesner saw resemblances between 'fossil objects' and living sea creatures.

At the same time, comprehensive philosophies of the Earth were being elaborated, influenced by the Christian revelation of Creation as set out in Genesis. This saw the Earth as recently created. Bishop Ussher (1581–1656) in his *Sacred Chronology* (1660), arrived at a creation date for the Earth of 4004 BC. In Christian eyes, time was directional, not cyclical. God had made the Earth perfect but, in response to Original Sin, he had been forced to send Noah's Flood to punish people by depositing them in a harsh environment, characterized by the niggardliness of Nature. This physical decline would continue until God had completed his purposes with humans.

The scientific revolution

The sixteenth and seventeenth centuries brought the discovery of the New World, massive European expansion and technological development. Scientific study of the Earth underwent significant change. Copernican astronomy sabotaged the old notion that the Earth was the centre of the system. The new mechanical philosophy (Descartes, Gassendi, Hobbes, Boyle and Hooke) rejected traditional macrocosm–microcosm analogies and the idea that the Earth was alive. Christian scholars adopted a more rationalist stance on the relations between Scripture and scientific truth. The possibility that the Earth was extremely old arose in the work of savants like Robert Hooke. For Enlightenment naturalists, the Earth came to be viewed as a machine, operating according to fundamental laws.

The old quarrel as to the nature of fossils was settled. Renaissance philosophies had stressed the living aspects of Nature. Similarities between fossils and living beings seemed to prove that the Earth was capable of growth. Exponents of the mechanical philosophy denied these generative powers. Fossils were petrified remains, rather like Roman coins, relics of the past, argued Hooke. Such views chimed with Hooke's concept of major terrestrial transformations and of a succession of fauna and flora now perished. Some species had been made extinct in great catastrophes.

This integrating of evidence from fossils and strata is evident in the work of Nicolaus Steno (1638–1686). He was struck by the similarity between shark's teeth and fossil *glossopetrae*. He concluded that the stones were petrified teeth. On this basis, he posited six successive periods of Earth's history. Steno's work is one of the earliest 'directional' accounts of the Earth's development that integrated the history of the globe and of life. Steno treated fossils as evidence for the origin of rocks.

The Enlightenment

Mining schools developed in Germany. German mineralogists sought an understanding of the order of rock formations which would be serviceable for prospecting purposes. Johann Gottlob Lehmann (1719–1776) set out his view that there were fundamental distinctions between the various *Ganggebürgen* (masses formed of stratified rock). These distinctions represented different modes of origin, strata being found in historical sequence. Older strata had been chemically precipitated out of water, whereas more recent strata had been mechanically deposited.

Abraham Gottlob Werner (1749–1817) was appointed in 1775 to the Freiberg Akademie. He was the most influential teacher in the history of geology. Werner established a well-ordered, clear, practical, physically based stratigraphy. He proposed a succession of the laying down of rocks, beginning with 'Primary Rocks' (precipitated from the water of a universal ocean), then passing through 'Transition', 'Flötz' (sedimentary), and finally 'Recent' and 'Volcanic'. The oldest rocks had been chemically deposited; they were therefore crystalline and without fossils. Later rocks had been mechanically deposited. Werner's approach linked strata to Earth history.

Thanks to the German school, but also to French observers like Guettard, Lavoisier and Dolomieu, to Italians such as Arduino, to Swedes like Bergman, stratigraphy was beginning to emerge in the eighteenth century.

Of course, there were many rival classifications and all were controversial. In particular, battle raged over the nature of basalt: was it of aqueous or igneous origin? The Wernerian or Neptunist school saw the Earth's crust precipitated out of aqueous solution. The other, culminating in Hutton, asserted the formation of rock types from the Earth's central heat.

A pioneer of this school was Buffon (1707–1778). He stressed ceaseless transfigurations of the Earth's crust produced by exclusively natural causes. In his *Epochs of Nature* (1779) he emphasized that the Earth had begun as a fragment thrown off the Sun by a collision with a comet. Buffon believed the Earth had taken at least 70,000 years to reach its present state. Extinction was a fact, caused by gradual cooling. The seven stages of the Earth explained successive forms of life, beginning with gigantic forms, now extinct, and ending with humans.

Though a critic of Buffon, James Hutton shared his ambitions. Hutton (1726–1797) was a scion of the Scottish Enlightenment, being friendly with Adam Smith and James Watt. In his *Theory of the Earth* (1795), Hutton demonstrated a steady-state Earth, in which natural causes had always been of the same kind as a present, acting with precisely the same intensity ('uniformitarianism'). There was 'no vestige of a beginning, no prospect of an end'. All continents were gradually eroded by rivers and weather. Debris accumulated on the sea bed, to be consolidated into strata and thrust upwards by the central heat to form new continents. Hutton thus postulated an eternal balance between uplift and erosion. All the Earth's processes were gradual. The Earth was incalculably old. His maxim was that 'the past is the key to the present'.

Hutton's theory was much attacked in its own day. Following the outbreak of the French Revolution in 1789, conservatives saw all challenges to the authority of the Bible as socially subversive. Their writings led to ferocious 'Genesis versus Geology' controversies in England.

The nineteenth century

New ideas about the Earth brought momentous social, cultural and economic reverberations. Geology clashed with traditional religious dogma over Creation. Modern state-funded scientific education and research organizations emerged. German universities pioneered scientific education. The Geological Survey of Great Britain was founded, after Henry De la Beche (1796–1855) obtained state finance for a geological map of southwest England. De la Beche's career culminated in the establishment of a Mines Record Office and the opening in 1851 of the Museum of Practical Geology and the School of Mines in London.

Specialized societies were founded. The Geological Society of London dates from 1807. In the United States, government promoted science. Various states established geological surveys, New York's being particularly productive. The US Geological Survey was founded in 1879, under Clarence King and later John Wesley Powell. In 1870 Congress appointed Powell to lead a survey of the natural resources of the Utah, Colorado and Arizona area.

Building on Werner, the great achievement of early nineteenth-century geology lay in the stratigraphical column. After 1800, it was perceived that mineralogy was not the master key. Fossils became regarded as the indices enabling rocks of comparable age of origin to be identified. Correlation of information from different areas would permit tabulation of sequences of rock formations, thereby displaying a comprehensive picture of previous geological epochs.

In Britain the pioneer was William 'Strata' Smith (1769–1839). Smith received little formal education and became a canal surveyor and mining prospector. By 1799 he set out a list of the secondary strata of England. This led him to the construction of geological maps. In 1815 he brought out *A Delineation of the Strata of England and Wales*, using a scale of five miles to the inch. Between 1816 and 1824 he published *Strata Identified by Organized Fossils*, which displayed the fossils characteristic of each formation.

Far more sophisticated were the French naturalists Georges Cuvier (1769–1832) and Alexandre Brongniart (1770–1837), who worked on the Paris basin. Cuvier's contribution lay in systematizing the laws of comparative anatomy and applying them to fossil vertebrates. He divided invertebrates into three phyla, and conducted notable investigations into fish and molluscs. In *Researches on the Fossil Bones of Quadrupeds* (1812), he reconstructed such extinct fossil quadrupeds as the mastodon, applying the principles of comparative anatomy. Cuvier was the most influential palaeontologist of the nineteenth century.

Fossils, in Cuvier's and Brongniart's eyes, were the key to the identification of strata and Earth history. Cuvier argued for occasional wholesale extinctions caused by geological catastrophes, after which new flora and fauna appeared by migration or creation. Cuvier's *Discours sur les révolutions de la surface du globe* (1812) became the foundation text for catastrophist views.

Clarification of older rock types was achieved by Adam Sedgwick (1785–1873) and Roderick Murchison (1792–1871). Sedgwick unravelled the stratigraphic sequence of fossil-bearing rocks in North Wales, naming the oldest of them the Cambrian period (now dated at 500–570 million years ago). Further south, Murchison delineated the Silurian system amongst the *grauwacke*. Above the Silurian, the Devonian was framed by Sedgwick, Murchison and De la Beche. Shortly afterwards, Charles Lapworth developed the Ordovician.

Werner's retreating-ocean theory was quickly abandoned, as evidence accumulated that mountains had arisen not by evaporation of the ocean, but through processes causing elevation and depression of the surface. This posed the question of the rise and fall of continents. Supporters of 'catastrophes', argued that terrestrial upheavals had been sudden and violent. Opposing these views, Charles Lyell advocated a revised version of Hutton's gradualism. Lyellian uniformitarianism argued that both uplift and erosion occurred by natural forces.

Expansion of fieldwork undermined traditional theories based upon restricted local knowledge. The retreating-ocean theory collapsed as Werner's students travelled to terrains where the proof of uplift was self-evident.

Geologists had to determine the earth movements that had uplifted mountain chains. Chemical theories of uplift yielded to the notion that the Earth's core was intensely hot, by consequence of the planet commencing as a molten ball. Many hypotheses were advanced. In 1829, Elie de Beaumont published *Researches on Some of the Revolutions of the Globe*, which linked a cooling Earth to sudden uplift: each major mountain chain represented a unique episode in the systematic crumpling of the crust. The Earth was like an apple whose skin wrinkled as the interior shrank through moisture loss. The idea of horizontal (lateral) folding was applied in America by James Dwight Dana to explain the complicated structure of the Appalachians. Such views were challenged by Charles Lyell in his bid to prove a steady-state theory. His classic *Principles of Geology* (1830–33) revived Hutton's vision of a uniform Earth that precluded cumulative, directional change in overall environment; Earth history proceeded like a cycle, not like an arrow. In *Principles of Geology*, Lyell thus attacked diluvialism and catastrophism by resuscitating Hutton's vision of an Earth subject only to changes currently discernible. Time replaced violence as the key to geomorphology.

Lyell discounted Cuvier's apparent evidence for the catastrophic destruction of fauna and flora populations. For over 30 years he opposed the transmutation of species, reluctantly conceding the point at last only in deference to his friend, Charles Darwin, and the cogency of Darwin's *On the Origin of Species* (1859).

Ice ages

Landforms presented a further critical difficulty. Geologists had long been baffled by beds of gravel and 'erratic boulders' strewn over much of Northern Europe and North America. Bold new theories in the 1830s attributed these phenomena to extended glaciation. Jean de Charpentier and Louis Agassiz contended that the 'diluvium' had been moved by vast ice sheets covering Europe during an 'ice age'. Agassiz's *Studies on Glaciers* (1840) postulated a catastrophic temperature drop, covering much of Europe with a thick covering of ice that had annihilated all terrestrial life.

The ice-age hypothesis met opposition but eventually found acceptance through James Geikie, James Croll and Albrecht Penck. Syntheses were required. The most impressive unifying attempt came from Eduard Suess. His *The Face of the Earth* (1885–1909) was a massive work devoted to analysing the physical agencies contributing to the Earth's geographical evolution. Suess offered an encyclopedic view of crustal movement, the structure and grouping of mountain chains, of sunken continents, and the history of the oceans. He made significant contributions to structural geology.

Suess disputed whether the division of the Earth's relief into continents and oceans was permanent, thus clearing the path for the theory of continental drift. Around 1900, the US geologist and cosmologist Thomas C. Chamberlain proposed a different synthesis: the Earth did not contract; its continents were permanent. Continents, Chamberlin argued, were gradually filling the oceans and thereby permitting the sea to overrun the land.

The twentieth century

By 1900, study of the Earth had become fragmented into specialisms like stratigraphy, mineralogy, crystallography, sedimentology, petrography and palaeobotany, and there was no univerally accepted unifying research programme. Geophysics increasingly provided intellectual coherence. Geophysics emerged as a distinct discipline in the late nineteenth century. Study of the Earth's magnetic field came to early prominence. In 1919 the American Geophysical Union was formed, and 1957 was designated the International Geophysical Year. The modern term 'earth sciences', to some degree replacing geology, marks the triumph of geophysics.

Nineteenth-century fieldwork had set the agenda for an enduring tradition of stratigraphic surveying and investigation of landforms. These traditions continued to yield valuable harvests. Immensely influential was the US geomorphologist William Morris Davis. Davis developed the organizing concept of the cycle of erosion. He proposed a stage-by-stage life-cycle for a river valley, marked by youth (steep-sided V-shaped valleys), maturity (flood-plain floors), and old age, as the river valley was imperceptibly worn down into the rolling landscape he termed a 'peneplain'.

Nineteenth-century geology built on the idea of the cooling Earth. Kelvin's estimates of the Earth's age suggested a relatively low antiquity, but this was soon challenged from within physics itself, for in 1896 the discovery of radioactivity revealed a new energy source unknown to Kelvin. In *The Age of the Earth* (1913), Arthur Holmes pioneered the use of radioactive decay methods for rock-dating. By showing the Earth had cooled far more slowly than Kelvin asserted, the new physicists undermined the 'wrinkled apple' analogy.

Amidst such challenges Alfred Wegener went further and declared that continental rafts might actually slither across the Earth's face. From 1910 Wegener developed a theory of continental drift. Empirical evidence for such displacement lay, he thought, in the close jigsaw-fit between coastlines on either side of the Atlantic, and notably in palaeontological similarities between Brazil and Africa. He was also convinced that geophysical factors would corroborate wandering continents. Wegener supposed that a united supercontinent, Pangaea, had existed in the Mesozoic. This had developed numerous fractures and had drifted apart some 200 million years ago. During the Cretaceous, South America and Africa had largely been split, but not until the end of the Quaternary had North America and Europe finally separated. Australia had been severed from Antarctica during the Eocene.

What had caused continental drift? Wegener offered a choice of possibilities. One was a westwards tidal force caused by the Moon. The other involved a centrifugal effect propelling continents away from the poles towards the Equator (the 'flight from the pole'). In its early years, drift theory won few champions, and in the English-speaking world reactions were especially hostile. A few geologists were intrigued by drift, especially the South African, Alexander Du Toit, who adumbrated the similarities in the geologies of South America and South Africa, suggesting they had once been contiguous. In *Our Wandering Continents* (1937), Du Toit maintained that the southern continents had formed the supercontinent of Gondwanaland.

The most ingenious support for drift came, however, from the British geophysicist Arthur Holmes. Assuming radioactivity produced vast quantities of heat, Holmes argued for convection currents within the crust. Radioactive heating caused molten magma to rise to the surface, which then spread out in a horizontal current before descending back into the depths when chilled. Such currents provided a new mechanism for drift.

The real breakthrough required diverse kinds of evidence accumulating from the 1940s, especially through oceanography and palaeomagnetism. Advances in palaeomagnetism arose from controversies over origins of the Earth's magnetic field. The evidence for changing directions of the magnetic field recorded by the rocks was linked to a baffling anomaly: in many cases the direction of the field seemed to be reversed. This led geophysicists to suspect that the terrestrial magnetic field occasionally switched. Over millions of years, there would be intermittent reversal events in which the North and South magnetic poles would alternate. Remanent magnetization would record these events, and, if the rocks could be dated sufficiently precisely, a complete register of reversals could be traced against the geological record.

By 1960, US scientists had refined the radiometric technique for dating rocks, deploying especially the potassium–argon method. A group at Berkeley developed a timescale of reversals for the Pleistocene era; Australian scientists produced their own scale, based on the dating of Hawaiian lava flows. Oceanography was developing too. In the UK the work of William Maurice Ewing (1906–1974) was especially significant. Ewing ascertained that the crust under the ocean is much thinner than the continental shelf. Ewing also demonstrated that mid-ocean ridges were common to all oceans. Ewing's work demonstrated that far from being ancient, ocean rocks were recent.

The US geophysicist Harry Hess (1906–1969) played a key role in promoting the new theories, viewing the oceans as the major centre of activity. The new crust was produced in the ridges, whereas trenches marked the sites where old crust was subtended into the depths, completing the convection current's cycle. Carried by the horizontal motion of the convection current, continents would glide across the surface. Constantly being formed and destroyed, the ocean floors were young; only continents – too light to be drawn down by the current – would preserve testimony of the remote geological past. Support came from John Tuzo Wilson (1908–1993), a Canadian geologist, who provided backing for the sea-floor spreading hypothesis. A dramatic new line of evidence, developed by Drummond Hoyle Matthews (1931–   ) and Fred Vine (1939–1988) of Cambridge University, confirmed sea-floor spreading.

The majority of earth scientists accepted the new plate tectonics model with remarkable rapidity. In the mid-1960s, a full account of plate tectonics was expounded. The Earth's surface was divided into six major plates, the borders of which could be explained by way of the convection-current theory. Deep earthquakes were produced where one section of crust was driven beneath another, the same process also causing volcanic activity in zones like the Andes. Mountains on the western edge of the North and South American continents arose from the fact that the continental 'raft' is the leading edge of a plate, having to face the oncoming material from other plates being forced beneath them. The Alps and Himalayas are the outcome of collisions of continental areas, each driven by a different plate system. Geologists of the late 1960s and 1970s undertook immense reinterpretation of their traditional doctrines. Well-established stratigraphical and geomorphological data had to be redefined in terms of a new the forces operating in the crust. Tuzo Wilson's *A Revolution in Earth Science* (1967) was a persuasive account of the plate tectonics revolution.

Geology is remarkable for having undergone such a dramatic and comprehensive conceptual revolution within recent decades. The fact that the most compelling evidence for the new theory originated from the new discipline of ocean-based geophysics has involved considerable revaluing of skills and priorities within the profession. Above all, the ocean floor now appears to be the key to understanding of the Earth's crust, in a way that Wegener never appreciated.

Mathematics

Most ancient civilizations had the means to make accurate measurements, to recode them in writing, and to use them in calculations involving elementary addition and subtraction. And for many of them that was apparently sufficient. It seems, for example, that the ancient Egyptians relied on simple addition and subtraction even for calculations of area and volume – although their number system was founded upon base 10, as ours is today. They certainly never thought of mathematics as a subject of potential interest or study for its own sake.

Not so the Babylonians. Contemporaries of the Egyptians, they nevertheless had a more practical form of numerical notation and were genuinely interested in improving their mathematical knowledge. (Perversely, however, their system used base 10 up to 59, after which 60 became a new base; one result of this is the way we now measure time and angles.) By about 1700 BC the Babylonians not only had the four elementary algorithms – the rules for addition, subtraction, multiplication and division – but also had made some progress in geometry. They knew what we now call Pythagoras' theorem, and had formulated further theorems concerning chords in circles. This even led to a rudimentary understanding of algebraic functions.

The ancient Greeks
Until the very end of their own civilization, the ancient Greeks had little use for algebra other than within a study of logic. After all, to them learning was as interdisciplinary as possible. Even Thales of Miletus (c.624–c.547 BC), regarded as the first named mathematician, considered himself a philosopher in a school of philosophers; mathematics was peripheral. The Greeks' attitude of scientific curiosity was, however, to result in some notable advances in mathematics, especially in the endeavour to understand why and how algorithms worked, theorems were consistent, and calculations could be relied on. It led in particular to the notion of mathematical proof, in an elementary but no less factual way. Pythagoras (c.580–c.500 BC), having proved the theorem now called after him, imbued mathematics with a kind of religious mystique on the basis of which he became a rather unsuccessful social reformer. Others became fascinated by solving problems using a ruler and compass, in which an outline of the concept of an irrational number (such as π) inevitably appeared. Further investigations of curves followed, and resulted in the first suggestions of what we now call integration. Such geometrical studies were often applied to astronomy. A corpus of various kinds of mathematical knowledge was beginning to accumulate.

The man who recorded much of it was Euclid (lived c.300 BC). His work *The Elements* is intended as much as a history of mathematics as a compendium of knowledge, and was massive therefore in both scope and production. It contained many philosophical elements (as we would now define them) and astronomical hypotheses, but the exposition of the mathematical work was masterly, and became the style of presentation emulated virtually to this day. Euclid's geometry, especially, became the standard for millennia: mathematicians still distinguish between Euclidean and non-Euclidean geometry. He included even discussion and ideas on spherical geometry. Unfortunately, some of *The Elements* was lost, including the work on conic sections.

Conic sections seem to have been a source of fascination to many ancient Greek mathematicians. Archimedes (c.287–212 BC), one of the most practical men of all time, used the principle of conic sections in an investigation into how to solve problems of an algebraic nature. A little later, Apollonius of Perga (c.245–c.190 BC), wrote definitely of the subject, in great detail, adducing a considerable number of associated theorems and including relevant proofs. The significance of part of this extra material was established only at the end of the nineteenth century.

The Romans and their successors
After about 150 BC, the study of astronomy dominated the scientific world. Consequently, for a while,

little mathematical progress was made except in the context of the cosmological theories of the time. (There was accordingly some significant research into spherical geometry and spherical trigonometry.) It was then too that Roman civilization briefly flourished and began to recede – again with little effect on the status of mathematics. Surprisingly, however, after about 400 years, the Alexandrian Diophantus (lived third century AD) devised something of extreme originality: the algebraic variable, in which a symbol stands for an unknown quantity. Equations involving such indeterminates – Diophantus included one indeterminate per equation, needing thus only one symbol – are now commonly called Diophantine equations.

Alexandria thus became the centre for mathematical thought at the time. Very shortly afterwards, Pappus (lived *c.*AD 300) deemed it time again for a compilation of all known mathematical knowledge. In *The Collection* he revised, edited and expanded the works of all the classic writers and added many of his own proofs and theorems, including some well-known problems he left unsolved. It is this work more than any other that ensured the survival of the mathematics of the Greeks until the Renaissance about a thousand years later.

In the meantime the initiative was taken by the Arabs, whose main sphere of influence was, significantly, farther east. They were thus in contact with Persian and Indian scientific schools, and accustomed to translating learned texts. Both Greek and Babylonian precepts were assimilated and practised – the best known proponent was Al-Khwarizmi (*c.*780–*c.*850), whose work was historically important to later mathematicians in Europe. The Arabs devised accurate trigonometrical tables (primarily for astronomical research) and continued the development of spherical trigonometry; they also made advances in descriptive geometry.

It was through his learning in the Arab markets of Algeria that the medieval merchant from Pisa Leonardo Fibonacci (or Leonardo of Pisa; *c.*1180–*c.*1250), brought much of contemporary mathematics back to Europe. It included – only then – the use of the 'Arabic' numerals 1 to 9 and the 'zephirum' (0), the innovation of partial numbers or fractions, and many other features of both geometry and algebra. From that time, hundreds of translators throughout Europe (especially in Spain) worked on Latin versions of Arab works and transcriptions. Only when Europe had regained all the knowledge and, so to speak, updated itself could genuine development take place. The effort took nearly 400 years before any truly outstanding advances were made – but may be said to be directly responsible for the overall updating and advance in science that then came about, known as the Renaissance.

The sixteenth century
One of the first instances of genuine progress in mathematics was the means of solving cubic equations, although acrimonious recriminations over priority surrounded its initial publication. One particularly charismatic contender – Niccolo Fontana (*c.*1499–1557), usually known as Tartaglia – besides being a military physicist, remained an inspirational figure in the propagation of mathematics. The means of solving quartic equations was discovered soon afterwards.

Within another 20 years, the French mathematician François Viète (1540–1603) was improving on the systematization of algebra in symbolic terms and expounding on mathematical (as opposed to astronomical) applications of trigonometry. It was he, if anyone, who initiated the study of number theory as an independent branch of mathematics. At the time of Viète's death, Henry Briggs (1561–1630) in England was already Professor of Geometry; a decade later he combined with John Napier (1550–1617), the deviser of 'Napier's Bones', to produce the first logarithm tables using the number 10 as its base, a means of calculation commonly used until the late 1960s but now outmoded by the computer and pocket calculator. Simultaneously, the astronomer Johannes Kepler was publishing one of the first works to consider infinitesimals, a concept that would lead later to the formulation of the differential calculus.

The seventeenth century
It was in France that the scope of mathematics was then widened by a group of great mathematicians. Most of them met at the scientific discussions run by the director of the convent of Place Royale in Paris, Father Marin Mersenne (1588–1648). To these discussions sometimes came the philosopher-mathematician René Descartes (1596–1675), the lawyer and magistrate Pierre de Fermat

(1601–1675), the physicist and mathematician Blaise Pascal (1623–1662), and the architect and mathematician Gérard Desargues (1591–1661). Descartes was probably the foremost of these in terms of mathematical innovation, although it is thought that Fermat – for whom mathematics was an absorbing but part-time hobby – had a profound influence upon him. His greatest contribution to science was in virtually founding the discipline of analytical (coordinate) geometry, in which geometrical figures can be described by algebraic expressions. He applied the tenets of geometry to algebra, and was the first to do so, although the converse was not uncommon.

Unfortunately, Descartes so much enjoyed the reputation his mathematical discoveries afforded him that he began to envy anyone who then also achieved any kind of mathematical distinction. He therefore regarded Desargues – who published a well-received work on conics – not only as competition but actually as retrogressive. And when Pascal then publicly championed Desargues (whom Descartes had openly ridiculed), putting forward an equally accepted form of geometry now known as projective geometry, matters became more than merely unfriendly.

In the meantime, Fermat took no sides, studied both types of geometry, and was in contact with several other European mathematicians. In particular, he used Descartes' geometry to derive an evaluation of the slope of a tangent, finding a method by which to compute the derivative and thus being considered by many the actual formulator of the differential calculus. Part of his study was of tangents as limits of secants. With Pascal he investigated probability theory, and in number theory he independently devised many theorems, one of them now famous as Fermat's last theorem.

It is now known that at about this time in Japan, a mathematician called Seki Kowa (c. 1642–1708) was independently discovering many of the mathematical innovations also being formulated in the West. Even more remarkably, he managed to change the social order of his time in order to popularize the subject.

Three years after Pascal died a religious recluse haunted by self-doubt, Isaac Newton (1642–1727), was obliged by the spread of the plague to his university college in Cambridge to return home to Woolsthorpe in Lincolnshire and there spend the next year and a half in scientific contemplation. One of his first discoveries was what is now called the binomial theorem, which led Newton to an investigation of infinite series, which in turn led to a study of integration and the notion that it might be achieved as the opposite of differentiation. He arrived at this conclusion in 1666 – but did not publish it. More than seven years later, in Germany, Gottfried Leibniz (1646–1716) – who had possibly read the works of Pascal – arrived at exactly the same conclusion, and did publish it. He received considerable acclaim in Europe, much to Newton's annoyance – and a priority argument was very quickly in process. Naively, Leibniz submitted his claim for priority to a committee on which Newton was sitting, so the outcome was a surprise to no one else – but it was in fact Leibniz's notation system that was eventually universally adopted. And it was not until 1687 that Newton's studies on calculus were published within his massive *Principia Mathematica*, which also included much of his investigations into physics and optics. Leibniz went on to try to develop a mathematical notation symbolizing logic, but although he made good initial progress it met with little general interest, and despite his energy and status he died a somewhat lonely and forlorn figure.

Another who died in even worse straits was an acquaintance of both Newton and Leibniz: Abraham de Moivre (1667–1754), a Huguenot persecuted for his religious background to the extent that he could find no professional position despite being a first-class and innovative mathematician. He met his end broken by poverty and drink – but not before he had formulated game theory, reconstituted probability theory, and set the business of life insurance on a firm statistical basis.

Leibniz's work on calculus was greatly admired in Europe, and particularly by the great Swiss mathematician family domiciled in Basle: the Bernoullis. The eldest of three brothers, Jacques (or Jakob; 1654–1705), actually corresponded with Leibniz; the youngest, Jean (or Johann; 1667–1748), was recommended by the physicist Christiaan Huygens to a professorship at Groningen. Both brothers were fascinated by investigating possible applications of the new calculus. Unhappily, their study of special curves (particularly cycloids) using polar coordinates proceeded independently along identical lines and resulted in considerable animosity between them. When Jacques died, however, Jean succeeded him at Basle, where he educated his son Daniel (1700–1782) – also a brilliant mathematician – whose great friends were Leonhard Euler (1707–1783) and Gabriel Cramer (1704–1752).

The eighteenth century

Euler may have been the most prolific mathematical author ever. He had amazing energy, a virtually photographic memory and a gift for mental calculation that stood him in good stead late in life when he became totally blind. Not since Descartes had anyone contributed so innovatively to mathematical analysis – Euler's *Introduction* (1748) is considered practically to define in textbook fashion the modern understanding of analytical methodology, including especially the concept of a function. Other works introduced the calculus of variations and the now familiar symbols π, e and i, and systematized differential geometry. He also popularized the use of polar coordinates, and explained the use of graphs to represent elementary functions.

It was his friend Daniel Bernoulli (1700–1782) who had originally managed to secure a position for him in St Petersburg. When, in 1766, Euler returned there from a post at the Prussian Academy, his place in Berlin was taken by the Frenchman Joseph Lagrange (1736–1813), whose ideas ran almost parallel with Euler's. In many ways Lagrange was equally as formative in the popularizing of mathematical analysis, for although he might not have been as energetic or outrightly creative as Euler, he was far more concerned with exactitude and axiomatic rigour, and combined with this a strong desire to generalize. The publication of his studies of number theory and algebra were thus models of precise presentation, and his mathematical research into mechanics began a process of creative thought that has not ceased since. One immediate result of the latter was to inspire his friend and fellow-Frenchman Jean le Rond d'Alembert (1717–1783) to great achievements in dynamics and celestial mechanics. It was d'Alembert who first devised the theory of partial differential equations.

Towards the end of Lagrange's life, when he was already ailing, he became Professor of Mathematics at the institution which for the next 50 years at least was to exercise considerable influence over the progress of mathematics; the newly established Ecole Polytechnique in Paris. Two of his contemporaries there were Pierre Laplace (1749–1827) and Gaspard Monge (1746–1818). Laplace became famous for his astronomical calculations, Monge for his textbook on geometry; both were acquaintances of Napoleon Bonaparte – as was Joseph Fourier (1768–1830), the physicist who demonstrated that a function could be expanded in sines and cosines through a series now known as the Fourier series.

It was one of Gaspard Monge's pupils – Jean-Victor Poncelet (1788–1867) – who first popularized the notion of continuity and outlined contemporary thinking on the principle of duality. And it was one of Laplace's colleagues (whom he disliked), Adrien Legendre (1752–1833), who took over where Lagrange left off, and researched into elliptic functions for more than 40 years, eventually deriving the law of quadratic reciprocity and, in number theory, proving that π is irrational.

The nineteenth century

Legendre's investigations into elliptic integrals were outdated almost as soon as they were published by the work of the Norwegian Niels Abel (1802–1829) and the German Karl Jacobi (1804–1851). Jacobi went on to make important discoveries in the theory of determinants: he was a great interdisciplinarian. The tragically shortlived Abel has probably had the longer-lasting influence, in that he devised the functions now named after him. He was unlucky, too, in that his proof, that in general roots cannot be expressed in radicals was discovered simultaneously and independently by the equally tragic Evariste Galois (1811–1832), who only just had time before his violent death to initiate the theory of groups. Further progress in function theory was made by Augustin Cauchy (1789–1857), a prolific mathematical writer who in his works pioneered many modern mathematical methods, developing in particular the use of limits and continuity. He also originated the theory of complex variables, based at least partly on the work of Jean Argand (1767–1822), who had succeeded in representing complex numbers by means of a graph.

By this time, however, the centre of mathematics in Europe was undoubtedly Göttingen, where the great Karl Gauss (1777–1855) had long presided. Sometimes compared with Archimedes and Newton, Gauss was indisputably not only a mathematical genius who made a multitude of far-reaching discoveries – particularly in geometry and statistical probability – but was also an exceptionally inspirational teacher who inculcated in his pupils the need for meticulous attention to proofs. Late in his tenure at Göttingen, three of his pupils/colleagues were Lejeune Dirichlet (1805–1859), Bernhard Riemann (1826–1866) and Julius Dedekind (1831–1916). There could not have been a more influential quartet in the history of mathematics: the work of all four provides the basis for a major part of modern mathematical knowledge.

Gauss himself was most interested in geometry. Jakob Steiner (1796–1863) in Germany was trying to remove geometry from the 'taint' of analysis as propounded by the French, but Gauss went further and decided to investigate geometry outside the scope of that described by Euclid. It was a momentous decision – made almost simultaneously and quite independently by Nikolai Lobachevsky (1792–1856) and János Bolyai (1802–1860). Between them they thus derived non-Euclidean geometry. The ramifications of this were widespread and fast-moving. In Ireland William Hamilton (1805–1865) suggested the concept of n-dimensional space; in Germany Hermann Grassmann (1809–1877) not only defined it but went on to use a form of calculus based on it. But it was Gauss's own pupil, Riemann, who really became the archapostle of the subject. He invented elliptical hyperbolic geometries, introduced 'Riemann surfaces' and redefined conformal mapping (transformations) explaining his innovations with such enthusiasm and accuracy that the modern understanding of time and space now owes much to his work.

Meanwhile Dirichlet – who succeeded Gauss when the great man died and himself became an influential teacher – and Dedekind concentrated more on number theory. Dirichlet slanted his teaching of mathematics towards applications in physics, whereas Dedekind was determined to arrive at a philosophical interpretation of the concept of numbers. Such an interpretation was thought likely to be of use in the contemporary search for a mathematical basis for logic. George Boole (1815–1864) had already attempted to create a form of algebra intended to represent logic that, although not entirely successful, was stimulating to others.

As the study of geometry expanded rapidly, the importance of algebra also increased accordingly. Riemann was influential; Karl Weierstrass (1815–1897) provided important redefinitions in function theory; but in algebraic terms development was next most instigated by the Englishman Arthur Cayley (1821–1895) who discovered the theory of algebraic invariants even as he carried out research into n-dimensional geometry. The principles of topology were being established one by one even though the branch itself was not yet complete. Sophus Lie (1842–1899) made important contributions to geometry and to algebra – and indirectly to topology – with the concept of continuous groups and contact transformations, and Cayley went on to invent the theory of matrices. Gaston Darboux (1842–1917) revised popular thinking about surfaces. Felix Klein (1849–1925) – an influential figure in his time – unified all the geometries within his Erlangen Programme (1872). But it was Felix Hausdorff (1868–1942) who is actually credited with the formulation of topology.

Dedekind finally achieved his goal and axiomatized the concept of numbers – only for his axioms to be (albeit apologetically and acknowledgedly) 'stolen' from him by Giuseppe Peano (1858–1932). The axioms, however, may have inspired – among others – Hausdorff to conceive the idea of point sets in topology, and Georg Cantor (1843–1918) to define set theory (the basis on which most mathematics is taught in schools today) and transfinite numbers, and certainly caused a revival of interest in number theory generally. Immanuel Fuchs (1833–1902) reformulated much of function theory while attempting to refine Riemann's method for solving differential equations. His pupil, Henri Poincaré (1854–1912) – similarly fascinated by Riemann's work – made many conjectures that were later useful in the investigation of topology and of space and time, but less successfully spent years researching into what are now called integral equations, only to discover after they were finally axiomatized by Ivar Fredholm (1866–1927) that he had done all the work without perceiving the answer.

The twentieth century

A different result of the Peano axioms was a renewal of the quest to find a relationship between mathematics and logic. Another system of symbolic logic had been devised by Gottlob Frege (1848–1925), whose pride was turned to ashes when Bertrand Russell (1872–1970) pointed out to him an internal, and fundamental, inconsistency. Russell, with his pupil and friend Alfred North Whitehead (1861–1947), attended lectures given by Peano; together they then published a large work on the foundations of mathematics, entitled *Principia Mathematica*. It had an immediate impact, and remained influential. Other prominent figures in the philosophy of mathematics at the time included Hermann Weyl (1885–1935) and Jacques Herbrand (1908–1931).

The search for meaning in mathematics was not solely philosophical, however. One of Fredholm's pupils was David Hilbert (1862–1943), possibly the latest of the truly great mathematicians. A genuine polymath and an enthusiastic teacher, he expanded virtually all branches of mathematics, especially in the interpretation of geometric structures implied by infinite-dimensional space. He too was involved in the debate over the primary nature of mathematics, formal or intuitional. But all philosophical theories were dealt a heavy blow by the theorem formulated in 1930 by Kurt Gödel (1906–1978). This stated that the overall consistency (completeness) of mathematics cannot itself be proved mathematically – which means that the foundations of mathematics must forever remain impenetrable.

The days of debate were over; Hilbert went on with his work. Mathematics became gradually either more theoretical or more practical. Theoretically interest swung towards finding features in common between disparate mathematical structures. Henri Lebesgue (1875–1941) devised a concept of measure that contributed greatly to the theory of abstract spaces. Andrei Kolmogorov (1903–) and others not only related this to probability theory but thereby to problems of statistical mechanics and the clarification of the ergodic theorems provided by George Birkhoff (1884–1944) in 1932. In algebraic topology, René Thom (1923–) categorized surfaces. It is worthy of note that thereafter most modern mathematics has concerned itself with such abstract mathematical structures or concepts as fields, rings or ideals.

The study of statistics and probability was also taken up with new enthusiasm for more practical applications. Karl Pearson (1857–1936) refined Gauss's ideas to derive the notion of standard deviation. Agner Erlang (1878–1929) used probability theory in a highly practical way to aid the efficiency of the circuitry of Copenhagen's telephone system. Alonso Church (1903–) defined a 'calculable function' and by so doing clarified the nature of algorithms. Following this, George Dantzig (1914–) was able to set up complex linear programs for computers. Such progress is being maintained, sometimes now as a result of using the machines themselves to devise further advances.

Computer scientists have devised symbolic computation systems which manipulate algebraic expressions in the same way that a human mathematician would do, only faster and more accurately. The result might be called 'computer-assisted mathematics'. A good example is the proof in 1972 of the four-colour theorem by Kenneth Appel and Wolfgang Haken. In 1850 Francis Guthrie conjectured that no more than four colours need be used in order to ensure that no two adjacent colours on a map share the same colour. Mathematicians quickly proved that five colours would suffice, but had no success in reducing the number to four. A direct attack by computer would not be possible, for how could a computer consider all possible maps? But Appel and Haken came up with a list of 1936 particular maps, and showed that if each had a rather complicated property, then the conjecture must be true. They then checked this property, case by case, on a computer, taking about 1,200 hours.

Another modern computer-based development is chaos theory, a theory of nonlinear dynamic systems. The central discovery, made in 1961 by US meteorologist Edward Lorenz, is that random behaviour can arise in systems whose mathematical description contains no hint whatever of randomness. The geometry of chaos can be explored using theoretical techniques such as topology but the most vivid pictures are obtained using computer graphics. The geometric structures of chaos are called fractals; they have the same detailed form on all scales of magnification. Frenchman Benoit Mandelbrot produced the first fractal images in 1962, using a computer that repeated the same mathematical pattern over and over again. In 1975 US mathematician Mitchell Feigenbaum discovered a new universal constant (approximately 4.669201609103) which is important in chaos theory. Order and chaos, traditionally seen as opposites, are now viewed as two aspects of the same basic process, the evolution of a system in time.

In 1980 mathematicians completed the classification of all finite and simple groups. The classification has taken over a hundred mathematicians more than 35 years to complete, and covers over 14,000 pages in mathematical journals. In 1989 a team of US computer mathematicians discovered the highest known prime number (the number contains 65,087 digits).

Physics

Physics is a branch of science in which the theoretical and the practical are firmly intertwined. It has been so since ancient times, as physicists have striven to interpret observation or experiment in order to arrive at the fundamental laws that govern the behaviour of the universe. Physicists aim to explain the manifestations of matter and energy that characterize all things and processes, both living and inanimate, extending from the grandest of galaxies down to the most intimate recesses of the atom.

The history of physics has not been a straight and easy road to enlightenment. The exploration of new directions sometimes leads to dead ends. New ways of looking at things may result in the over-throw of a previously accepted system. Not Aristotle's system, nor Newton's, nor even Einstein's was 'true'; rather statements, or 'laws', in physics satisfy contemporary requirements or – in the existing state of knowledge – contemporary possibilities. The question that physicists ask is not so much 'Is it true?' as 'Does it work?'

Physics has many strands – such as mechanics, heat, light, sound, electricity and magnetism – and, although they are often pursued separately, they are also all ultimately interdependent. To pursue the history of physics, therefore, it is necessary to follow several separate chains of discovery and then to find the links between them. The story is of frustration and missed opportunities as well as of genius and perseverence. But however complex it may appear, all physicists seek or have sought to play a part in the evolution of an ultimate explanation of all the effects that occur throughout the universe. That goal may be unattainable but the thrust towards it has kept physics as alive and vital today as it was when it originated in ancient times.

Force and motion

The development of an understanding of the nature of force and motion was a triumph for physics, one which marked the evolution of the scientific method. As in most other branches of physics, this development began in ancient Greece.

The earliest discovery in physics, apart from observations of effects like magnetism, was the relation between musical notes and the lengths of vibrating strings. Pythagoras (c.582–c.500 BC) found that harmonious sounds were given by strings whose lengths were in simple numerical ratios, such as 2:1, 3:2 and 4:3. From this discovery the belief grew that all explanations could be found in terms of numbers. This was developed by Plato (c.427–c.347 BC) into a conviction that the cause underlying any effect could be expressed in mathematical form. The motion of the heavenly bodies, Plato reasoned, must consist of circles, since these were the most perfect geometric forms.

Reason also led Democritus (c.460–c.370 BC) to propose that everything consisted of minute indivisible particles called atoms. The properties of matter depend on the characteristics of the atoms of which it is composed, and the atoms combine in ways that are determined by unchanging fundamental laws of nature.

A third view of the nature of matter was given by Aristotle (384–322 BC), who endeavoured to interpret the world as he observed it, without recourse to abstractions such as atoms and mathematics. Aristotle reasoned that matter consisted of four elements – earth, water, air and fire – with a fifth element, the ether, making up the heavens. Motion occurred when an object sought its rightful place in the order of elements, rocks falling through air and water to the earth, air rising through water as bubbles and fire through air as smoke.

There was value in all these approaches and physics has absorbed them all to some degree. Plato was essentially correct; only his geometry was wrong, the planets following elliptical, not circular, orbits. Atoms do exist as Democritus foretold and they do explain the properties of matter. Aristotle's emphasis on observation (though not his reasoning) was to be a feature of physics and many other sciences, notably biology, of which he may be considered the founder.

These ideas were, however, mainly deductions based solely on reason. Few of them were given the test of experiment to prove that they were right. Then came the achievements of Archimedes (c. 287–212 BC), who discovered the law of the lever and the principle of flotation by measuring the effects that occur and deduced general laws from his results. He was then able to apply his laws, building pulley systems and testing the purity of the gold in King Hieron's crown by a method involving immersion.

Archimedes thus gave physics the scientific method. All subsequent principal advances made by physicists were to take the form of mathematical interpretations of observations and experiments. Archimedes developed the method in founding the science of statics – how forces interact to produce equilibrium. But an understanding of motion lay a long way off. In the centuries following the collapse of Greek civilization in around AD 100, physics marked time. The Arabs kept the Greek achievements alive, but they made few advances in physics, while in Europe the scientific spirit was overshadowed by the 'Dark Ages'. Then in about 1200, the spirit of enquiry was rekindled in Europe by the import of Greek knowledge from the Arabs. Unfortunately, progress was hindered somewhat by the fact that Aristotle's ideas, particularly his views on motion, prevailed. Aristotle had assumed that a heavy object falls faster than a light object simply because it is heavier. He also argued that a stone continues to move when thrown because the air displaced by the stone closes behind it and pushes the stone. This explanation derived from Aristotle's conviction that nature abhors a vacuum (which is why he placed a fifth element in the heavens).

Aristotle's ideas on falling bodies were probably first disproved by Simon Stevinus (1548–1620), who is believed to have dropped unequal weights from a height and found that they reached the ground together. At about the same time Galileo (1564–1642) measured the speeds of 'falling' bodies by rolling spheres down an inclined plane and discovered the laws that govern the motion of bodies under gravity. This work was brought to a brilliant climax by Isaac Newton (1642–1727), who in his three laws of motion achieved an understanding of force and motion, relating them to mass and recognizing the existence of inertia and momentum. Newton thus explained why a stone continues to move when thrown; and he showed the law of falling bodies to be a special case of his more general laws. Newton went on to derive from existing knowledge of the motion and dimensions of the Earth–Moon system a universal law of gravitation, one which provided a mathematical statement for the laws of planetary motion discovered empirically by Johannes Kepler (1571–1630).

Newton's laws of motion and gravitation, which were published in 1687, were fundamental laws which sought to explain all observed effects of force and motion. This triumph of the scientific method heralded the Age of Reason – not the Greek kind of reasoning, but a belief that all could be explained by the deduction of fundamental laws upheld by observation or experiment. It was to result in an explosion of scientific discovery in physics that has continued to the present day. In the field of force and motion, important advances were made with the discovery of the law governing the pendulum and the principle of conservation of momentum by Christiaan Huygens (1629–1695) and the determination of the gravitational constant by Henry Cavendish (1731–1810).

The behaviour of matter
Physics is basically concerned with matter and energy, and investigation into the behaviour of matter also originated in ancient Greece with Archimedes' work concerning flotation. As with force and motion, Simon Stevinus made the first post-Greek advance with the discovery that the pressure of a liquid depends on its depth and area. This achievement was developed by Blaise Pascal (1623–1662), who found that pressure is transmitted throughout a liquid in a closed vessel, acting perpendicularly to the surface at any point. Pascal's principle is the basis of hydraulics. Pascal also investigated the mercury barometer invented in 1643 by Evangelista Torricelli (1608–1647) and showed that air pressure supports the mercury column and that there is a vacuum above it, thus disproving Aristotle's contention that a vacuum cannot exist. The immense pressure that the atmosphere can exert was subsequently demonstrated in several sensational experiments by Otto von Guericke (1602–1686).

Solid materials were also investigated. The fundamental law of elasticity was discovered by Robert Hooke (1635–1703) in 1678 when he found that the stress (force) exerted is proportional to the strain (elongation) produced. Thomas Young (1773–1829) later showed that a given material has a constant, known as Young's modulus, that defines the strain produced by a particular stress.

The effects that occur with fluids (liquids or gases) in motion were then explored. Daniel Bernoulli

(1700–1782) established hydrodynamics with his discovery that the pressure of a fluid depends on its velocity. Bernoulli's principle explains how lift occurs and led eventually to the invention of heavier-than-air flying machines. It also looked forward to ideas of the conservation of energy and the kinetic theory of gases. Other important advances in our understanding of fluid flow were later made by George Stokes (1819–1903), who discovered the law that relates motion to viscosity, and Ernst Mach (1838–1916) and Ludwig Prandtl (1875–1953), who investigated the flow of fluids over surfaces and made discoveries vital to aerodynamics.

The effects of light
The Greeks were aware that light rays travel in straight lines, but they believed that the rays originate in the eyes and travel to the object that is seen. Euclid (lived c. 300 BC), Hero (lived AD 60) and Ptolemy (lived 2nd century AD) were of this opinion although, recognizing that optics is essentially a matter of geometry, they discovered the law of reflection and investigated refraction.

Optics made an immense stride forward with the work of Alhazen (c. 965–1038), who was probably the greatest scientist of the Middle Ages. Alhazen recognized that light rays are emitted by a luminous source and are then reflected by objects into the eyes. He studied images formed by curved mirrors and lenses and formulated the geometrical optics involved. Alhazen's discoveries took centuries to filter into Europe, where they were not surpassed until the seventeenth century. The refracting telescope was then invented in Holland in 1608 and quickly improved by Galileo and Kepler, and in 1621 Willebrord Snell (1580–1626) discovered the laws that govern refraction.

The next major steps forward were taken by Newton, who not only invented the reflecting telescope in 1668, but a couple of years earlier found that white light is split into a spectrum of colours by a prism. Newton published his work in optics in 1704, provoking great controversy with his statement that light consists of a stream of particles. Huygens had put forward the view that light consists of a wave motion, an opinion reinforced by the discovery of diffraction by Francesco Grimaldi (1618–1663). Such was Newton's reputation, however, that the particulate theory held sway for the following century. In 1801 Young discovered the principle of interference, which could be explained only by assuming that light consisted of waves. This was confirmed in 1821, when Augustin Fresnel (1788–1827) showed from studies of polarized light, which had been discovered by Etienne Malus (1775–1812) in 1808, that light is made up of a transverse wave motion, not longitudinal as had previously been thought.

Newton's discovery of the spectrum remained little more than a curiosity until 1814, when Joseph von Fraunhofer (1787–1826) discovered that the Sun's spectrum is crossed by the dark lines now known as Fraunhofer lines. Fraunhofer was unable to explain the lines, but he did go on to invent the diffraction grating for the production of high-quality spectra and the spectroscope to study them. An explanation of the lines was provided by Gustav Kirchhoff (1824–1887), who in 1859 showed that they are caused by elements present in the Sun's atmosphere. With Robert Bunsen (1811–1899), Kirchhoff discovered that elements have unique spectra by which they can be identified, and several new elements were found in this way. In 1885, Johann Balmer (1825–1898) derived a mathematical relationship governing the frequencies of the lines in the spectrum of hydrogen. This later proved to be a crucial piece of evidence for revolutionary theories of the structure of the atom.

Meanwhile, several scientists investigated the phenomenon of colour, notably Young, Hermann von Helmholtz (1821–1894) and James Clerk Maxwell (1831–1879). Their research led to the establishment of the three-colour theory of light, which showed that the eye responds to varying amounts of red, green and blue in light and mixes them to give particular colours. This led directly to colour photography and other methods of colour reproduction used today.

The velocity of light was first measured accurately in 1862 by Jean Foucault (1819–1868), who obtained a value within 1 per cent of the correct value. This led to a famous experiment performed by Albert Michelson (1852–1931) and Edward Morley (1838–1923) in which the velocity of light was measured in two directions at right angles. Their purpose was to test the theory that a medium called the ether existed to carry light waves. If it did exist, then the two values obtained would be different. The Michelson–Morley experiment, performed in 1881 and then again in 1887, yielded a negative result both times (and on every occasion since), thus proving that the ether does not exist.

More important, the Michelson–Morley experiment showed that the velocity of light is constant regardless of the motion of the observer. From this result, and from the postulate that all motion is relative, Albert Einstein (1879–1955) derived the special theory of relativity in 1905. The principal conclusion of special relativity is that in a system moving relative to the observer, length, mass and time vary with the velocity. The effects become noticeable only at velocities approaching light; at slower velocities, Newton's laws hold good. Special relativity was crucial to the formulation of new ideas of atomic structure and it also led to the idea that mass and energy are equivalent, an idea used later to explain the great power of nuclear reactions. In 1915 Einstein published his general theory of relativity, in which he showed that gravity distorts space. This explained an anomaly in the motion of Mercury, which does not quite obey Newton's laws, and it was dramatically confirmed in 1919 when a solar eclipse revealed that the Sun's gravity was bending light rays coming from stars.

Electricity and magnetism

The phenomena of electricity and magnetism are believed to have been first studied by the ancient Greek philosopher Thales (*c.*624–*c.*547 BC), who was considered by the Greeks to be the founder of their science. Thales found that a piece of amber picks up light objects when rubbed, the action of rubbing thus producing a charge of static electricity. The words 'electron' and 'electricity' came from this discovery, *elektron* being the Greek word for amber. Thales also studied the similar effect on each other of pieces of lodestone, a magnetic mineral found in the region of Magnesia. It is fitting that the study of electricity and magnetism originated together, for the later discovery that they are linked was one of the most important ever made in physics.

No further progress was made, however, for nearly 2,000 years. The strange behaviour of amber remained no more than a curiosity, though magnets were used to make compasses. From this, Petrus Peregrinus (lived 13th century) discovered the existence of north and south poles in magnets and realized that they attract or repel each other. William Gilbert (1544–1603) first explained the Earth's magnetism and also investigated electricity, finding other substances besides amber that produce attraction when rubbed. Then Charles Du Fay (1698–1739) discovered that substances charged by rubbing may repel as well as attract in a similar way to magnetic poles and Benjamin Franklin (1706–1790) proposed that positive and negative charges are produced by the excess or deficiency of electricity. Charles Coulomb (1763–1806) measured the forces produced between magnetic poles and between electric charges and found that they both obey the same inverse square law.

A major step forward was taken in 1800, when Alessandro Volta (1745–1827) invented the battery. A source of current electricity was now available and in 1820 Hans Oersted (1777–1851) found that an electric current produces a magnetic field. This discovery of electromagnetism was immediately taken up by Michael Faraday (1791–1867), who realized that magnetic lines of force must surround a current. This concept led him to discover the principle of the electric motor in 1821 and electromagnetic induction in 1831, the phenomenon in which a changing magnetic field produces a current. This was independently discovered by Joseph Henry (1797–1878) at the same time.

Meanwhile, important theoretical developments were taking place in the study of electricity. In 1827, André Ampère (1775–1836) discovered the laws relating magnetic force to electric current and also properly distinguished current from tension, or electromotive force (emf). In the same year, Georg Ohm (1789–1854) published his famous law relating current, emf and resistance. Kirchhoff later extended Ohm's law to networks, and he also unified static and current electricity by showing that electrostatic potential is identical to emf.

In the 1830s, Karl Gauss (1777–1855) and Wilhelm Weber (1804–1891) defined a proper system of units for magnetism; later they did the same for electricity. In 1845 Faraday found that materials are paramagnetic or diamagnetic, and Lord Kelvin (1824–1907) developed Faraday's work into a full theory of magnetism. An explanation of the cause of magnetism was finally achieved in 1905 by Paul Langevin (1872–1946), who ascribed it to electron motion.

Electricity and magnetism were finally brought together in a brilliant theoretical synthesis by James Clerk Maxwell. From 1855 to 1873 Maxwell developed the theory of electromagnetism to show that electric and magnetic fields are propagated in a wave motion and that light consists of such an electromagnetic radiation. Maxwell predicted that other similar electromagnetic radiations must exist and, as a result, Heinrich Hertz (1857–1894) produced radio waves in 1888. X-rays and gamma rays were discovered accidentally soon after.

The nature of heat and energy

The first step towards measurement – and therefore an understanding – of heat was taken by Galileo, who constructed the first crude thermometer in 1593. Gradually these instruments improved and in 1714 Daniel Fahrenheit (1686–1736) invented the mercury thermometer and devised the Fahrenheit scale of temperature. This was replaced in physics by the Celsius or centigrade scale proposed by Anders Celsius (1701–1744) in 1742.

At this time, heat was considered to be a fluid called caloric that flowed into or out of objects as they got hotter or colder, and even after 1798 when Count Rumford (1753–1814) showed the idea to be false by his observation of the boring of cannon, it persisted. Earlier, Joseph Black (1728–1799) had correctly defined the quantity of heat in a body and the latent heat and specific heat of materials, and his values had been successfully applied to the improvement of steam engines. In 1824, Sadi Carnot (1796–1832), also a believer in the caloric theory, found that the amount of work that can be produced by an engine is related only to the temperature at which it operates.

Carnot's theorem, though not invalidated by the caloric theory, suggested that, since heat gives rise to work, it was likely that heat was a form of motion, not a fluid. The idea also grew that energy may be changed from one form to another (i.e. from heat to motion) without a change in the total amount of energy involved. The interconvertibility of energy and the principle of the conservation of energy were established in the 1840s by several physicists. Julius Mayer (1814–1878) first formulated the principle in general terms and obtained a theoretical value for the amount of work that may be obtained by the conversion of heat (the mechanical equivalent of heat). Helmholtz gave the principle a firmer scientific basis and James Joule (1818–1889) made an accurate experimental determination of the mechanical equivalent. Rudolf Clausius (1822–1888) and Kelvin developed the theory governing heat and work, thus founding the science of thermodynamics. This enabled Kelvin to propose the absolute scale of temperature that now bears his name.

The equivalence of heat and motion led to the kinetic theory of gases, which was developed by John Waterston (1811–1883), Clausius, Maxwell and Ludwig Boltzmann (1844–1906) between 1845 and 1868. It gave a theoretical description of all effects of heat in terms of the motion of molecules.

During the nineteenth century it also came to be understood that heat may be transmitted by a form of radiation. Pioneering theoretical work on how bodies exchange heat had been carried out by Pierre Prévost (1751–1839) in 1791, and the Sun's heat radiation had been discovered to consist of infrared rays by William Herschel (1738–1822) in 1800. In 1862 Kirchhoff derived the concept of the perfect black body – one that absorbs and emits radiation at all frequencies. In 1879 Josef Stefan (1835–1893) discovered the law relating the amount of energy radiated by a black body to its temperature, but physicists were unable to relate the frequency distribution of the radiation to the temperature. This increases as the temperature is raised, causing an object to glow red, yellow and then white as it gets hotter. Lord Rayleigh (1842–1919) and Wilhelm Wien (1864–1928) derived incomplete theories of this effect, and then in 1900 Max Planck (1858–1947) showed that it could be explained only if radiation consisted of indivisible units, called quanta, whose energy was proportional to their frequency.

Planck's theory revolutionized physics. It showed that heat radiation and other electromagnetic radiations including light must consist of indivisible particles of energy and not of waves as had previously been thought. In 1905 Einstein explained the photoelectric effect using quantum theory, and the theory was experimentally confirmed by James Franck (1882–1964) in the 1920s.

Another advance in the study of heat that took place in the same period was the production of low temperatures. In 1852 Joule and Kelvin found the effect – named after them – that is used to produce refrigeration by adiabatic expansion of a gas, and James Dewar (1842–1923) developed this effect into a practical method of liquefying gases from 1877 onwards. Heike Kamerlingh-Onnes (1853–1926) first produced temperatures within a degree of absolute zero and in 1911 he discovered superconductivity. A theoretical explanation of superconductivity had to await the work of John Bardeen (1908–1991), Leon Cooper (1930–) and John Schrieffer (1931–). Their ideas, the 'BCS theory', explained superconductivity as the result of electrons coupling in pairs, called Cooper pairs, that do not undergo scattering by collision with atoms in a conductor. In 1986 IBM researchers in Zurich, Georg Bednorz (1950–) and Alex Muller (1927–), produced superconductivity in metallic ceramics at relatively high temperatures, around 35K. The theoretical explanation of high-temperature superconductivity was still being developed in the early 1990s.

Sound

Sound is the one branch of physics that was well established by the Greeks, especially by Pythagoras. They surmised, correctly, that sound does not travel through a vacuum, a contention proved experimentally by Guericke in 1650. Measurements of the velocity of sound in air were made by Pierre Gassendi (1592–1655) and in other materials by August Kundt (1839–1894). Ernst Chladni (1756–1827) studied how the vibration of surfaces produces sound waves, and in 1845 Christian Doppler (1803–1853) discovered the effect relating the frequency (pitch) of sound to the relative motion of the source and observer. The Doppler effect is also produced by light and other wave motions and has proved to be particularly valuable in astronomy.

The structure of the atom

The existence of atoms was proved theoretically by chemists during the nineteenth century, but the first experimental demonstration of their existence and the first estimate of their dimensions was made by Jean Perrin (1870–1942) in 1909.

The principal direction taken in physics in this century has been to determine the inner structure of the atom. It began with the discovery of the electron in 1897 by J. J. Thomson (1856–1940), who showed that cathode rays consist of minute indivisible electric particles. The charge and mass of the electron were then found by John Townsend (1868–1937) and Robert Millikan (1868–1953).

Meanwhile, another important discovery had been made with the detection of radioactivity by Antoine Becquerel (1852–1908) in 1896. Three kinds of radioactivity were found; these were named alpha, beta and gamma by Ernest Rutherford (1871–1937). Becquerel recognized in 1900 that beta particles are electrons. In 1903 Rutherford explained that radioactivity is caused by the breakdown of atoms. In 1909 he identified alpha particles as helium nuclei, and in association with Hans Geiger (1882–1945) produced the nuclear model of the atom in 1911, proposing that it consists of electrons orbiting a nucleus. Then in 1914 Rutherford identified the proton and in 1919 he produced the first artificial atomic disintegration by bombarding nitrogen with alpha particles.

Rutherford's pioneering elucidation of the basic structure of the atom was aided by developments in the use of X-rays, which had been discovered in 1895 by Wilhelm Röntgen (1845–1923). In 1912 Max von Laue (1879–1960) produced diffraction in X-rays by passing them through crystals, showing X-rays to be electromagnetic waves, and Lawrence Bragg (1890–1971) developed this method to determine the arrangement of atoms in crystals. His work influenced Henry Moseley (1887–1915), who in 1914 found by studying X-ray spectra that each element has a particular atomic number, equal to the number of protons in the nucleus and to the number of electrons orbiting it.

In 1913, Niels Bohr (1885–1962) achieved a brilliant synthesis of Rutherford's nuclear model of the atom and Planck's quantum theory. He showed that the electrons must move in orbits at particular energy levels around the nucleus. As an atom emits or absorbs radiation, it moves from one orbit to another and produces or gains a certain number of quanta of energy. In so doing. the quanta give rise to particular frequencies of radiation, producing certain lines in the spectrum of the radiation. Bohr's theory was able to explain the spectral lines of hydrogen, found earlier by Balmer.

These discoveries, made so quickly, seemed to achieve an astonishingly complete picture of the atom, but more was to come. In 1923, Louis de Broglie (1892–1987) described how electrons could behave as if they made up waves around the nucleus. This discovery was developed into a theoretical system of wave mechanics by Erwin Schrödinger (1887–1961) in 1926 and experimentally confirmed in the following year. It showed that electrons exist both as particles and waves. Furthermore it reconciled Planck's quantum theory with classical physics by indicating that electromagnetic quanta or photons, which were named and detected experimentally in X-rays by Arthur Compton (1892–1962) in 1923, could behave as waves as well as particles. A prominent figure in the study of atomic structure was Werner Heisenberg (1901–1976), who showed in 1927 that the position and momentum of the electron in the atom cannot be known precisely, but only found with a degree of probability or uncertainty. His uncertainty principle follows from wave–particle duality and it negates cause and effect, an uncomfortable idea in a science that strives to reach laws of universal application.

The next step was to investigate the nucleus. A series of discoveries of nuclear particles accompanying the proton were made, starting in 1932 with the discovery of the positron by Carl Anderson (1905–1991) and the neutron by James Chadwick (1891–1974).

This work was aided by the development of particle accelerators, beginning with the voltage multiplier built by John Cockcroft (1897–1967) and Ernest Walton (1903–), which achieved the first artifical nuclear transformation in 1932. It led to the discovery of nuclear fission by Otto Hahn (1879–1968) in 1939 and the production of nuclear power by Enrico Fermi (1901–1954) in 1942.

Modern physics

Much of modern physics has been concerned with the behaviour of elementry particles. The first major theory in this area was quantum electrodynamics (QED), developed by US physicists Richard Feynman (1918–1988) and Julian Schwinger (1918–), and by Japanese physicist Sin-Itiro Tomonaga (1906–1979). This theory describes the interaction of charged subatomic particles in electric and magnetic fields. It combines quantum theory and relativity and considers charged particles to interact by the exchange of photons. QED is remarkable for the accuracy of its predictions – for example, it has been used to calculate the value of some physical quantities to an accuracy of ten decimal places, a feat equivalent to calculating the distance between New York and Los Angeles to within the thickness of a hair.

By 1960 the existence of around 200 elementary particles had been established, some of which did not behave as theory predicted. They did not decay into other particles as quickly as theory predicted, for example. To explain these anomalies, US theoretical physicist Murray Gell-Mann (1929–) developed a classification for elementary particles, called the eightfold way. This scheme predicted the existence of previously undetected particles. The omega-minus (Ω^-) particle found in 1964 confirmed the theory. In the same year Gell-Mann suggested that some elementary particles were made up of smaller particles called quarks which could have fractional electric charges. This idea explained the eightfold classification and now forms the basis of the standard model of elementary particles and their interactions. Quantum chromodynamics is the mathematical theory, similar in many ways to quantun electrodynamics, which describes the interactions of quarks by the exchange of particles called gluons. The mathematics involved is very complex and although a number of successful predictions have been made, as yet the theory does not compare in accuracy with QED.

The success of the mathematical methods of quantum electrodynamics and quantum chromodynamics encouraged others to use these methods to explain the behaviour of elementary particles left out of the QCD and QED schemes. Abdus Salam (1926–), Steven Weinberg (1933–) and Sheldon Glashow (1932–) demonstrated that at high energies the electromagnetic and weak nuclear force could be regarded as aspects of a single combined force, the electroweak force. This was confirmed in 1983 by the discovery of new particles predicted by the theory. Efforts are now being made to include the strong nuclear force in the scheme, producing a grand unified theory or GUT.

A theory that relates the two classes of elementary particles, the fermions and the bosons, is supersymmetry. According to supersymmetry, each fermion particle has a boson partner particle, and vice versa. It has not been possible to marry up all the known fermions with the known bosons and so the theory postulates the existence of as-yet-undiscovered fermions, such as the photino (partners of the photon), gluino (partners of the gluon), and so on. Using these ideas, it has been possible to develop a theory of gravity – called supergravity – that extends Einstein's work and considers gravitational, nuclear and electromagnetic forces to be manifestations of an underlying superforce.

In the 1980s a mathematical theory called superstring theory was developed to take these ideas further. In string theory, the fundamental objects of the universe were not point-like particles but extremely small string-like objects. These objects exist in a universe of ten dimensions, although for reasons which are not yet understood, only three space dimensions and one time dimension are discernible. There are many unresolved difficulties, but some physicists think that string theory and supersymmetry are crucial ingredients in the 'theory of everything'.

Biographies

A

Abbe Ernst 1840–1905 was a German physicist who, working with Carl Zeiss, greatly improved the design and quality of optical instruments, particularly the compound microscope. He indirectly had a great influence in various physical sciences, particularly in biology where the improved resolving power of his instruments permitted researchers to observe microorganisms and internal cellular structures for the first time.

Abbe was born in Eisenach, Thuringia, Germany, on 23 January 1840, the son of a spinning-mill worker. On a scholarship provided by his father's employers he attended high school (graduating in 1857) and then went to study physics at the University of Jena. He gained his doctorate from Göttingen University in 1861 and three years later became a lecturer in mathematics, physics and astronomy at Jena, being appointed Professor in 1870. In 1866 he began his association with Carl Zeiss, an instrument manufacturer who supplied optical instruments to the university and repaired them. Abbe was appointed Director of the Astronomical and Meteorological Observatory at Jena in 1878. Two years earlier he had become a partner in Zeiss's firm, and in 1881 Abbe invited Otto Schott (1851–1935), who had studied the chemistry of glasses and manufactured them, to go to Jena, and the famous company of Schott and Sons was founded in 1884. On the death of Zeiss in 1888 Abbe became the sole owner of the Zeiss works. He established the Carl Zeiss Foundation in 1891, and in 1896 he formalized the association between the Zeiss works and Jena University by making the company a cooperative, with the profits shared between the workers and the university. He died in Jena on 14 January 1905.

The success of the Jena enterprise arose largely from the right combination of talents: Zeiss as the manufacturer; Abbe as the physicist/theoretician, who performed the mathematical calculations for designing new lenses; and Schott the chemist, who formulated and made the special glasses needed by Abbe's designs. Abbe worked out why, contrary to expectation, the definition of a microscope decreases with a reduction in the aperture of the objective; he found that the loss in resolving power is a diffraction effect. He calculated how to overcome spherical aberration in lenses – by a combination of geometry and the correct types of glass.

He also explained the phenomenon of coma (first recognized in 1830 by Joseph Jackson Lister (1786–1869), the father of the famous surgeon), in which even a corrected lens displays aberration when the object is slightly off the instrument's axis. It was overcome by applying Abbe's 'sine condition', producing the so-called aplanatic lens. Finally Abbe calculated how to correct chromatic aberration, using Schott's special glasses and, later, fluorite to make microscope objective lenses, culminating in the apochromatic lens system of 1886.

In 1872 he developed the Abbe substage condenser for illuminating objects under high-power magnification. Among his other inventions were a crystal refractometer and, developed with Armand Fizeau, an optical dilatometer for measuring the thermal expansion of solids.

Abel Niels Henrik 1802–1829 was a Norwegian mathematician who, in a very brief career, became the first to demonstrate that an algebraic solution of the general equation of the fifth degree is impossible.

Abel was born at Finnöy, a small island near Stavanger, on 5 August 1802. He was educated by his father, a Lutheran minister, until the age of twelve, when he was enrolled in the Cathedral school at Christiania (Oslo). There his flair for mathematics received little encouragement until he came under the tutelage of Bernt Holmboe in 1817. Holmboe put Abel in touch with Euler's calculus texts and introduced him to the work of Lagrange and Laplace. Abel's imagination was fired by algebraic equations theory and by the time that he left school in 1821 to enter the University of Oslo he had become familiar with most of the body of mathematical literature then known. In particular he had, during his last year at school, began to work on the baffling problem of the quintic equation, or general equation of the fifth degree, unsolved since it had been taken up by Italian mathematicians early in the sixteenth century.

Because of his father's death in 1820, Abel arrived at the university virtually penniless. Fortunately, his talent was apparent, and he was given free rooms and financial support by the university. Since the university offered no courses in advanced mathematics, most of Abel's research was done on his own initiative. In 1823 he published his first paper. It was an unimportant discussion of functional equations, but another paper published in that year heralded the arrival of a highly original new mind in the world of mathematics, although it went unregarded at the time. In it Abel provided the first solution in the history of mathematics of an integral equation. All the while he remained obsessed by the problem of the quintic equation. During his last year at school he had sent to the Danish mathematician, Ferdinand Deger, his 'solution' to the problem, only to receive from Deger the advice to abandon that 'sterile' question and turn his mind to elliptic transcendentals (elliptic integrals). Deger was kind enough, even so, to ask Abel for examples of his solution and this request proved to be fruitful. For when Abel began to construct examples he discovered that it was no solution at all. He therefore wrote a paper demonstrating that a radical expression to represent a solution to fifth- or higher-degree equations was impossible. After three centuries a niggling question had been resolved. Yet when Abel sent his demonstration to Gauss he received no reply. Nor was anyone else much interested and Abel was forced to publish the paper himself.

In 1825, taking advantage of a government grant to enable scholars to study foreign languages abroad, Abel went to Berlin. There he met Leopold Crelle, the privy councillor and engineer much taken with problems in mathematics. Together they brought out the first issue of *Crelle's Journal*, which was to become the leading nineteenth-century German organ of mathematics. (The first issue consisted almost entirely of seven papers by Abel.) A year later Abel moved on to Paris, where he wrote his famous paper, 'Mémoire sur une propriété générale d'une classe très-étendue de fonctions transcendantes'. It dealt with the sum of the integrals of a given algebraic function and presented the theorem that any such sum can be expressed as a fixed number of these integrals with integration arguments that are algebraic functions of the original arguments. Abel sent the manuscript to the French Academy of Sciences and was deeply disappointed when the referees – who alleged that the manuscript was illegible! – did not publish it. He returned to Berlin a disheartened man. Low in funds and unable to get a post at the university there, he accepted Crelle's offer to edit the journal. In 1827 he published the longest paper of his career, the 'Recherche sur les fonctions elliptiques'. He also suffered his first attack of the tuberculosis that was to kill him. At the end of the year he returned to Norway, where he lived in gradually deteriorating health until his death on 6 April 1829.

Abel, in addition to this work on quintic equations, transformed the theory of elliptic integrals by introducing elliptic functions, and this generalization of trigonometric functions became one of the favourite topics of 19th-century mathematics. It led eventually to the theory of complex multiplication, with its important implications for algebraic number theory. He also provided the first stringent proof of the binomial theorem. A number of useful concepts in modern mathematics, notably the Abelian group and the Abelian function, bear his name. Yet it was only after his death that his achievement was publicly acknowledged. In 1830 the French Academy awarded him the Grand Prix, which he shared with Karl Jacobi (1804–1851), the German mathematician who had (independently) made important discoveries about elliptic functions. And eleven years later the Academy finally came round to publishing the 'illegible' manuscript of 1826.

Abetti Giorgio 1882–1982 was an Italian astrophysicist best known for his studies of the Sun.

Abetti was born in Padua on 5 October 1882. He studied at the universities of Padua and Rome and earned a PhD in the physical sciences. In 1921 he was appointed Professor at the University of Florence, where he remained until his retirement in 1957. From 1921 until 1952 he was Director of the Arcetri Observatory in Florence. During his tenure at Florence, Abetti travelled to Cairo in 1948 to serve as Visiting Professor at the university there, and he toured the United States in 1950.

Abetti's research contribution quickly earned him a prominent and respected position among Italian scientists. He was awarded the Silver Medal of the Italian Geographical Society in 1915, the Reale Prize of the Academy of Lincei in 1926, and the Janssen Gold Medal of the Ministry of Public Instruction in 1937. He was a member of the Socio Nazionale, the Academy of Lincei in Rome, and the Royal Society of Edinburgh and the Royal Astronomical Society in Britain.

Abetti's research was in the field of astrophysics, with particular emphasis on the Sun. He participated in numerous expeditions to observe eclipses of the Sun, and led one such expedition to Siberia to observe the total solar eclipse of 19 June 1936. He was well known for his influential popular text on the Sun, and he wrote a handbook of astrophysics, published in 1936, and a popular history of astronomy, which appeared in 1963.

Adams John Couch 1819–1892 was an English astronomer who was particularly skilled mathematically. His ability to deal adeptly with complex calculations helped him to discover, independently of Urbain Leverrier, the planet Neptune.

Adams was born in Landeast, Cornwall, on 5 June 1819. His mathematical talents and interest in astronomy were apparent from an early age, and in 1839 he won a scholarship to Cambridge University. He graduated with top honours in 1843 and took up a Fellowship at St John's College. When this lapsed in 1853 he was given a life Fellowship at Pembroke College. St Andrew's University in Aberdeen appointed Adams to the Chair of Mathematics in 1858, but he returned to Cambridge a year later to become Lowdean Professor of Astronomy and Geometry, a post he held until his death.

The initial public acclaim for the discovery of Neptune went to Leverrier, but Adams nevertheless received many honours. He was awarded the Royal Society's Copley Medal (its highest honour) in 1848 and was elected Fellow of the Royal Society a year later. He was made a member of the Royal Astronomical Society and served it twice as President. He was awarded its Gold Medal for his later research into lunar theory. He succeeded James Challis as Director of the Cambridge Observatory, but he was always a modest man. He declined the offer of a knighthood from Queen Victoria, and also turned down the position of Astronomer Royal, pleading old age. He died in Cambridge on 21 January 1892.

The planet Uranus was discovered by William Herschel in 1781. Its path was carefully studied during the first orbit after discovery, and it soon became clear that early predictions of the motions of Uranus were incorrect. On the basis of Isaac Newton's gravitational theory, certain aberrations in the orbit were accounted for as the result of perturbations caused by Jupiter and Saturn, but these were insufficient to explain the magnitude of Uranus's deviation from its predicted orbital path. This suggested that either the gravitational theory was incorrect or that an as yet undetected planet lay beyond the orbit of Uranus. The mathematical calculations necessary to solve the mystery were taken up independently by Adams and Leverrier. Adams had become interested in this problem while still an undergraduate, but it was not until he had completed his studies in 1843 that he had the time to focus his full attention on it. By 1845 he had determined the position and certain characteristics of this hypothetical planet. He attempted to convey the information to the new Astronomer Royal, George Airy, but the significance of Adams's findings was not fully appreciated. A search for the new planet was not instigated for nearly a year and was carried out by Challis at Cambridge.

Meanwhile, in France, Leverrier had followed the same lines of thought as Adams. The Frenchman sent his figure to Johann Galle at the Berlin Observatory. It was Leverrier's good fortune that Galle had just received a new and improved map of the sector of the sky in which the planet could be located. As a result, Galle was able to find the planet, which was later named Neptune, within a few hours of beginning his search on 25 September 1846. It later transpired that Challis had observed the new planet on a number of occasions, but had failed to recognize that it was new because of his inferior maps. The discovery of Neptune was credited to Leverrier, although not without much nationalistic acrimony on both sides of the Channel.

Adams's later work included research into lunar theory and terrestrial magnetism, as well as observations of the Leonid meteor shower. Later, he improved the findings of Pierre Laplace. This resulted in a reduction of 50% in the then current value for the secular acceleration of the Moon's mean motion.

Adams's contributions to observational astronomy, as well as his improvements to the accuracy of many mathematical constants, made him deeply respected – not only for the value of his work, but also for his modest attitude towards his achievements.

Adams Walter Sydney 1876–1956 was a US astronomer who was particularly interested in stellar motion and luminosity. He developed spectroscopy as a valuable tool in the study of stars and planets.

Adams was born on 20 December 1876 at Antioch, Syria, where his parents were serving as missionaries. His early education was provided by his parents, who taught him much about ancient history and classical languages. In 1885 his parents returned to the United States for the sake of their children's education. When Adams entered Dartmouth College in Massachusetts he had to choose between his love of classics and mathematical sciences. He graduated in 1898 and went to the University of Chicago for his postgraduate studies.

He studied celestial mechanics, publishing a paper during his first year on the polar compression of Jupiter. The next year was spent under George Hale at the Yerkes Observatory, where Adams made a number of studies including one on the measurement of radial velocity. He went to Munich the following year. Hale then invited him to return to Yerkes, which he did and Adams spent the next three years working on stellar spectroscopy.

In 1904 he assisted Hale in the establishment of the Mount Wilson Observatory above Pasadena in California. Mount Wilson gradually became a renowned research centre. Adams served as Deputy Director, under Hale, from 1913 to 1923, when he took over as Director. Adams was a member of many scientific organizations in both the United States and Europe. He was honoured for his achievements by being elected President of the American Astronomical Society in 1931. He died on 11 May 1956.

Adams's early work on radial velocities had used the 100-cm/40-in refractor at Yerkes, at the time the largest in the world. At Mount Wilson he was able to use larger and more sophisticated equipment. The first area that Adams investigated at Mount Wilson was the spectra obtained from sunspots as compared to those obtained from the rest of the solar disc and from laboratory sources. He found that the temperature, pressure and density of a source affects the relative intensities of its spectral lines. This and other information enabled him to demonstrate that sunspots have a lower temperature than the rest of the solar disc. Adams also used Doppler displacements to study the rotation of the Sun.

In 1914 Adams turned to the spectroscopy of other stars. He found that luminosity and the relative intensities of particular spectral lines could distinguish giant stars from dwarf stars. Spectra could also be used to study the physical properties, motions and distances of stars. This use of the intensity of spectral lines to determine the distance of stars has been termed spectroscopic parallax.

Adams was involved in a long-term project with other astronomers to determine the absolute magnitudes of stars; together, they found the value for 6,000 stars. A second long-term collaborative project was the determination of the radial velocities of more than 7,000 stars. This work led to an improved understanding of the behaviour and evolution of stars.

In 1915 Adams made a spectroscopic study of the small companion star of Sirius, Sirius B. He identified it as a white dwarf containing about 80% of the mass of the Sun in a volume approximately the same as that of the Earth and thus having a density more than 40,000 times that of water. Adams demonstrated that the companion star was hotter than our Sun and not cold, as everyone had assumed. Arthur Eddington suggested in 1920 that if Sirius B was indeed so dense it would produce a powerful gravitational field and show a red shift (as predicted by Albert Einstein's general theory of relativity). In 1925 Adams reported a displacement of 21 km sec^{-1}/13 mi sec^{-1}, thus confirming Einstein's theory.

During the 1920s and 1930s Adams studied the atmosphere of Mars and Venus, reporting in 1932 the presence of carbon dioxide in the atmosphere of Venus and, in 1934, the occurrence of oxygen in concentrations of less than 0.1% on Mars. He was involved in many other research projects, and he also made an important contribution in his capacity as Director of the Observatory. He was responsible for the design and installation of the 254-cm/100-in and 508-cm/200-in telescopes at Mount Wilson and Palomar. Adams was a fine scholar and administrator. Astronomy matured as a science during his active research years, a development to which he was an important contributor.

Addison Thomas 1793–1860 was a British physician and endocrinologist who was the first to correlate a collection of symptoms with pathological changes in an endocrine gland. He described a metabolic disorder caused by a deficiency in the secretion of hormones from the adrenal glands (caused, in turn, by atrophy of the adrenal cortex), a condition now called Addison's disease. He is also known for his discovery of what is now called pernicious (or Addison's) anaemia.

Addison was born in April 1793 in Longbenton, Northumberland, and studied medicine at Edinburgh University, graduating in 1815. He then moved to London, where he was appointed a surgeon at the Lock Hospital. He also studied dermatology under Thomas Bateman (1778–1821) during his first years in London. In about 1820 Addison entered Guy's Hospital as a student, despite being a fully qualified physician, and remained there in various positions for the rest of his life, becoming Assistant Physician in 1824, Lecturer in Materia Medica in 1827, and a full Physician in 1837. While at Guy's Hospital, Addison collaborated with Richard Bright, who also made important contributions to medicine. Addison's mental health deteriorated and he committed suicide on 29 June 1860 in Brighton, Sussex.

Addison gave a preliminary account of the condition now known as Addison's disease in 1849 in a paper entitled 'On anaemia: disease of the suprarenal capsules', which he read to the South London Medical Society. The paper went unnoticed, despite which Addison extended his original account in *On the Constitutional and Local Effects of Disease of the Suprarenal Capsules* (1855), in which he gave a full description of Addison's disease (characterized by

abnormal darkening of the skin, progressive anaemia, weakness, intestinal disturbances and weight loss) and differentiated it from pernicious anaemia (characterized by anaemia, intestinal disturbances, weakness, and tingling and numbness in the extremities). He also pointed out that Addison's disease is caused by atrophy of the suprarenal capsules (later called the adrenal glands). (Pernicious anaemia is caused by failure of the stomach's secretion of intrinsic factor which, in turn, causes inadequate absorption of vitamin B_{12}.)

Addison also described xanthoma (flat, soft spots that appear on the skin, usually on the eyelids) and wrote about other skin diseases, tuberculosis, pneumonia and the anatomy of the lung. In collaboration with John Morgan (1797–1847), he wrote *An Essay on the Operation of Poisonous Agents Upon the Living Body* (1829), the first book on this subject to be published in English. And in 1839 appeared the first volume of *Elements of the Practice of Medicine*, written by Addison and Richard Bright. In this volume – which was, in fact, written almost entirely by Addison, Bright was to have been the principal contributor to the second volume, which was never published – Addison gave the first full description of appendicitis.

Adler Alfred 1870–1937 was an Austrian psychiatrist who broke away from the theories of Sigmund Freud, setting up the Individual Psychology Movement. He placed 'inferiority feeling' at the centre of his theory of neuroses.

Adler was born in Vienna on 7 February 1870, the son of a corn merchant. He obtained his MD from the University of Vienna in 1895 and worked for two years as a physician at Vienna General Hospital. His interests soon turned towards mental disorders, and by 1902 he had made contact with Freud. He played a major part in the development of the psychoanalytical movement, and was President of the Vienna Psychoanalytical Society. But by 1907 he had shifted his theory away from Freud's emphasis on infantile sexuality towards power as the origin of neuroses; in 1911 Adler, and a number of others, left the Freudian circle and founded the Individual Psychology Movement. By the late 1920s Adler was making many trips to the United States, where he proved to be a popular lecturer; in 1927 he became a visiting professor at Columbia University. In 1935 he decided to make the United States his permanent home and he became Professor of Psychiatry at Long Island College of Medicine, New York. He died from a heart attack on 28 May 1937 in Aberdeen, Scotland, during a lecture tour.

The essence of Adler's theories differed from Freud's in that he thought that power not sex was the important factor in neurotic disorders. He popularized the term 'inferiority feeling' – later changed to 'complex' – and felt that much neurotic behaviour is a result of feelings of inadequacy or inferiority caused by, for instance, being the youngest in a family or being a child who is trying to compete in an adult world. In an attempt to overcome these inferiority feelings the patient overcompensates, often at the expense of normal social behaviour or, as Adler put it, 'social interest'.

Adler's belief led on to his idea that a person can realize this ambition alone, which has consequences on the way in which a psychiatrist helps a patient if help is needed. His impact was less forceful than that of Carl Jung or Freud, and even though his psychology made good sense it lacked adequate definition and rigour of method. Adler summarized his theories in *Practice and Theory of Individual Psychology* (1927).

Adler's more practical work included the setting up of a system of child guidance services in the schools in Vienna, which lasted until 1934, when they were closed by the Austrian Fascist government.

Adrian Edgar Douglas, 1st Baron Adrian of Cambridge 1889–1977 was a British physiologist known for his experimental research in electrophysiology and, in particular, nerve impulses. He was one of the first scientists to study the variations in electrical potential of nerve impulses amplified by thermionic valve amplifiers, and was also one of the first to study the electrical activity of the brain. He shared the 1932 Nobel Prize in Physiology or Medicine with Charles Sherrington (1857–1952) for his work on neurons and their processes.

Adrian was born in London on 30 November 1889, the son of a lawyer. After attending Westminster School, he won a scholarship to Trinity College, Cambridge, where he studied natural sciences. During World War I he went to St Bartholomew's Hospital in London to study medicine, graduating in 1915 and then working on nerve injuries and shell shock (battle fatigue) at Queen's Square and later at the Connaught Military Hospital in Aldershot. Adrian was much in demand as a lecturer from 1919 – when he became a Cambridge University lecturer – especially on such subjects as sleep, dreams, hysteria and multiple sclerosis. From 1937 until 1951 he was Professor of Physiology at Cambridge. He was awarded the Order of Merit in 1942 and was President of the Royal Society from 1950 to 1955. In that same year he was raised to the peerage and took the barony of Cambridge. He was Master of Trinity College, Cambridge, from 1951 until he retired in 1965. He died in London on 4 August 1977.

In 1912 nothing was known about electrochemical transmission within the nervous system. Adrian's most ambitious work in the pre-war years was to attempt to prove that the intensity of a nerve impulse at any point in a normal nerve is independent of the stimulus or of any change in intensity that may have occurred elsewhere.

The mechanism of muscular control was a subject of clinical interest in wartime, and Adrian studied and wrote on the electrical excitation of normal and denervated muscle. He showed that with normal muscle the time factor in excitation, known as the chronaxie, is very short and that it increases by a factor of 100 in denervated muscle after the nerve endings have degenerated. With L. R. Yealland of Queen's Square he worked on the application of a method of treatment based on suggestion, re-education and discipline in cases of hysterical disorders such as mutism, deafness and paralysis of limbs.

Between 1925 and 1933 Adrian successfully recorded trains of nerve impulses travelling in single sensory or motor nerve fibres. This work was a turning point in the history of physiology. He began to use valve amplifiers and found that in a single nerve fibre the electrical impulse does not change with the nature or strength of the stimulus. He also discovered that some sense organs, such as those concerned with touch, rapidly adapt to a steady stimulus whereas others, such as muscle spindles, adapt slowly or not at all. His work at this time included the recording of optic nerve impulses in the conger eel, investigations of the action of light on frogs' eyes, and researching the problem of pain and the responses of animals to speech.

Between 1933 and 1946 he worked on the ways in which the nervous system generates rhythmic electrical activity. He was one of the first scientists to use extensively the recently devised electroencephalograph (EEG) – a system of recording brain waves – to study the electrical activity of the brain. This system has since proved an invaluable diagnostic aid – for example, in the diagnosis of epilepsy and the location of cerebral lesions. The last 20 years of his research life, from 1937 to 1959, were spent studying the sense of smell.

In 1932 Adrian published *The Mechanism of Nervous Action* and in 1947 *The Physical Background of Perception*, based on the Waynflete Lectures he gave in Oxford the preceding year.

Agassiz Jean Louis Rodolphe 1807–1873 was a Swiss palaeontologist who developed the idea of the ice age.

The son of a Protestant pastor, Agassiz was born at Motier, Switzerland. He received his medical and scientific training at Zurich, Heidelberg, Munich and Paris, where he fell under the spell of Cuvier and embraced his pioneering application of the techniques of comparative anatomy to palaeontology. His momentous *Researches on Fossil Fish* 1833–44 used Cuvierian comparative anatomy to describe and classify over 1,700 species; ichthyology was his preferred specialty. Travelling in 1836 in his native Alps, he developed the novel idea that glaciers, far from being static, were in a constant state of almost imperceptible motion. Finding rocks that had been shifted or abraded, presumably by glaciers, he inferred that in earlier times much of northern Europe had been covered with ice sheets. Agassiz's *Studies on Glaciers* 1840 developed the original concept of the ice age, which he viewed as a cause of extinction, demarcating past flora and fauna from those of the present. Agassiz's geological principles thus entailed a mode of catastrophism, fundamentally opposed to the extreme uniformitarianism promoted by Charles Lyell.

Agassiz rose to became professor at Harvard in 1847, where he worked for the rest of his career, founding the Museum of Comparative Zoology and building a huge collection (a quite obsessional collector, his accumulating habits ruined his purse and wrecked his marriage). His last major project was the *Contributions to the Natural History of the United States* 1857–62, an exhaustive study of the American natural environment. Agassiz was convinced that North America too had been the setting of glaciation. Like his mentor Cuvier, Agassiz was always anti-evolutionist. Adducing religious, philosophical and palae-ontological arguments, he proved one of the staunchest adversaries of Darwin's theory of descent by natural selection.

Agricola Georgius 1494–1555 was a German mineralogist who pioneered mining technology.

Born at Glauchau in Saxony, Georg Bauer – 'Agricola' is the Latinized form, meaning peasant farmer – trained in medicine first in Leipzig and later in Italy. He served for many years as town physician in Joachimstal. Involvement with the medicinal use of minerals sparked curiosity for the products of the Earth, and he soon developed an interest in the local mining ventures. Agricola quickly made himself an authority on mining, metal extraction, smelting, assaying and related chemical processes. His *The Nature of Fossils* 1546 advances one of the first comprehensive classifications of minerals. Familiar with previous writers on mining like Bermannus, Agricola went on to explore the origins of rocks, mountains and volcanoes. His most renowned work, *On Metals/De Re Metallica* 1556 is an indispensable survey, lavishly illustrated with woodcuts, of the smelting and chemical technology of the time and of the state of the mining industry. Drawing intelligently upon the *Pirotechnia* of the Italian Vannucio Biringuccio, Agricola's work became a standard text, being translated into Italian and German.

Aiken Howard Hathaway 1900–1973 was a US computer and data-processing pioneer who invented the Harvard Mark I and Harvard Mark II computers, the prototypes of modern digital computers.

Aiken was born at Hoboken, New Jersey, on 9 March 1900. He studied engineering at the University of Wisconsin and graduated in 1923. He then took a job with the Madison Gas and Electric Company, where he remained until 1927, when he went to Chicago to work for the Westinghouse Electric Manufacturing Company. In 1931 he left Westinghouse to take up a research post in the department of phyics at the University of Chicago.

The rest of the decade he spent in research, both at Chicago and at Harvard. He received his PhD from Harvard in 1939 and was appointed an instructor in physics and communication engineering. He quickly rose to become a full professor in applied mathematics and remained at Harvard until 1961, when he was appointed Professor of Information Technology at the University of Miami. He died at St Louis, Missouri, on 14 March 1973.

When Aiken began his research into computer technology in the 1930s the subject was still in its infancy. Simple, manual calculating machines had been in use since the mid-seventeenth century, but they were too elementary and too slow to meet the military and industrial requirements of the twentieth century.

It was, indeed, the US Navy which started Aiken on the career for which he became world famous. His early research at Harvard was sponsored by the Navy Board of Ordnance and in 1939 he and three other engineers from the International Business Machines Corporation (IBM) were placed under contract by the Navy to develop a machine capable of performing both the four basic operations of addition, subtraction, multiplication and division and also referring to stored, tabulated results.

Aiken played the central role in the development for the US Navy of the first Automatic Sequence Controlled Calculator in the world, the Harvard Mark I, which was completed in 1944. It was principally a mechanical device, although it had a few electronic features; it was 15 m/49 ft long, 2.5 m/8 ft high and weighed more than 30 tonnes. Addition took 0.3 sec, multiplication 4 sec. It was able to manipulate numbers of up to 23 decimal places and to store 72 of them. Information was fed into the machine by tape or punched cards and produced output in a similar form. Its chief functions were to produce mathematical tables and to assist the ballistics and gunnery divisions of the military.

On the completion of the Mark I, Aiken was posted to the Naval Proving Ground at Dahlgren, Virginia, to begin working on improving his invention. There the Mark II was completed in 1947. It was a fully electronic machine containing 13,000 electronic relays. It was also much faster than its predecessor, requiring only 0.2 sec for addition and 0.7 sec for multiplication. Moreover it could store 100 ten-digit figures and their signs.

For his great achievements, sufficient to earn him the name of the pioneer of modern computers, Aiken was given the rank of Commander in the United States Navy Research Department.

Airy George Biddell 1801–1892 was a British astronomer who, as Astronomer Royal for 46 years, was responsible for greatly simplifying the systematization of astronomical observations and for expanding and improving the Royal Observatory at Greenwich.

Airy was born in Alnwick, Northumberland, on 27 July 1801. His father, a collector of taxes and excise duties, was periodically transferred from one part of the country to another, with the result that his son was educated in a number of places. From 1814 to 1819 he attended Colchester Grammar School, where he was noted for his incredible memory (on one occasion he recited from memory 2,394 lines of Latin verse). In 1819 he became a student at Trinity College, Cambridge, and three years later took a scholarship there. He graduated in 1823 at the top of his class in mathematics. The following year he was elected a Fellow of Trinity College and became an assistant tutor in mathematics. The physics of light and optics began to interest him and he was the first to describe the defect of vision – later termed astigmatism – from which he also suffered. In 1826 he became Professor of Mathematics at Cambridge and in the same year, having become interested in astronomy, published *Mathematical Tracts on Physical Astronomy*, which became a standard work. He was elected Professor of Astronomy and Director of the Cambridge Observatory in 1828, and was then appointed Astronomer

Royal in 1835, a post which he held until 1881. During this period he sat on many commissions and supervised the cataloguing of geographical boundaries. He was awarded the Copley and Royal medals by the Royal Society and was its President from 1827 to 1873. He was five times President of the Royal Astronomical Society, twice receiving its Gold Medal, and received various honorary degrees. He died in Greenwich, London, on 2 January 1892.

While Airy was Director of the Cambridge Observatory, it flourished under his control; he introduced a much improved system of meridian observations and set the example of reducing them in scale before publishing them. As Astronomer Royal, Airy had the Royal Observatory at Greenwich re-equipped and many innovations were made. He supervised the gigantic task of reducing in scale all the planetary and lunar observations made at Greenwich between 1750 and 1830. In 1847 he had erected the altazimuth (an instrument he devised to calculate altitude and azimuth) for observing the Moon in every part of the sky. Airy also introduced new departments to the Observatory; in 1838 he created one for magnetic and meteorology data and, in 1840, a system of regular two-hourly observations was begun. Other innovations included photographic registration in 1848, transits timed by electricity in 1854, spectroscopic observations from 1868 and a daily round of sunspots using the Kew heliograph in 1873.

As an expert mathematician Airy's skills were required in the exact mapping of geographical boundaries: he was responsible for establishing the border between Canada and the United States and later of the Oregon and Maine boundaries. He also established exact determinations of the longitudes of Valencia, Cambridge, Edinburgh, Brussels and Paris. Airy's scientific expertise was also called on during the launch of the steamship *Great Eastern*, the laying of the transatlantic telegraph cable, and the construction of the chimes of the clock in the tower of the Houses of Parliament ('Big Ben'). During 1854 he supervised several experiments in Harton Colliery, South Shields, to measure the change in the force of gravity with distance below the Earth's surface.

In spite of all his additional duties Airy never allowed his work with the Royal Observatory to suffer and it was arguably due to his enthusiasm and hard work that the Greenwich Observatory grew in importance both nationally and internationally.

Aitken Robert Grant 1864–1951 was a US astronomer whose primary contribution to astronomy was the discovery and observation of thousands of double stars.

Aitken was born in Jackson, California, on 31 December 1864. He took his degree at Williams College, where he forsook his earlier plans to enter the ministry for his interest in astronomy. He taught at Livermore College from 1888 until 1891, when he was made Professor of Mathematics at the University of the Pacific. From 1895 onwards he worked at the Lick Observatory on Mount Hamilton, first as Assistant Astronomer and ultimately as Director of the Observatory from 1930 until his retirement in 1935.

Aitken's work on binary systems brought him him widespread recognition and many honours. He was a member of numerous professional bodies, often holding positions of responsibility within them. These included the chairmanship (from 1929 to 1932) of the Astronomy section of the National Academy of Sciences. He died in Berkeley, California, on 29 October 1951.

At first Aitken's research at Lick was in many fields, but his interest soon focused on double stars. He began a mammoth survey of double stars in 1899 and this was not finished until 1915. During the early years of the project he was assisted by W. J. Hussey, and between them they discovered nearly 4,500 new binary systems. Their primary tool was the 91-cm/36-in refractor. Aitken then began a thorough statistical examination of this vast amount of information, which he published first in 1918 and then revised in 1935. His work lay not merely in the discovery of new binary stars, but also in determining their motions and orbits.

Aitken's other famous contribution was his revision of S. W. Burnham's catalogue of double stars, first published in 1906. This was completed in 1927. Aitken was also interested in, and contributed to, the popularization of astronomy, especially after his retirement.

Alder Kurt 1902–1958 was a German organic chemist who with Otto Diels (1876–1954) developed the diene synthesis, a fundamental process that has become known as the Diels–Alder reaction. It is used on organic chemistry to synthesize cyclic (ring) compounds, including many that can be made into plastics and others – which normally occur only in small quantities in plants and other natural sources – that are the starting materials for various drugs and dyes. This outstanding achievement was recognized by the award of the 1950 Nobel Prize for Chemistry jointly to Alder and Diels.

Alder was born on 10 July 1902 in the industrial town of Königshütte (Krolewska Huta) in Upper Silesia, which was then part of Germany (it is now in Poland). He was the son of a schoolteacher and began his education in his home town. When the region became part of Poland at the end of World War I, the Alder family moved to Berlin. There Kurt Alder finished his schooling and went on to study chemistry, first at the University of Berlin and later at Kiel, where he worked under Otto Diels. Alder became a chemistry reader at Kiel in 1930 and a professor in 1934, but two years later he began a four-year period in industry as Research Director of I. G. Farben at Leverkusen on the northern outskirts of Cologne. He returned to academic life in 1940 as Professor and Director of the Chemical Institute at the University of Cologne, where he remained for the rest of his life. He died in Cologne on 20 June 1958.

The first report of the diene synthesis, stemming from work in Diels's laboratory at Kiel, was made in 1928. The Diels–Alder reaction involves the adding of an organic compound that has two double bonds separated by a single bond (called a conjugated diene) to a compound with only one, activated double bond (termed a dienophile). A common example of a conjugated diene is butadiene (but-1,2:3,4-diene) and of a dienophile is maleic anhydride (*cis*-butenedioic anhydride). These two substances react readily to form the bicyclic compound tetrahydrophthalic anhydride (cyclohexene-1:2-dicarboxylic anhydride) – this was one of the reactions originally reported by Diels and Alder in 1928.

Azo-diesters, general formula $RCO_2.N:N.CO_2R$, also act as dienophiles in the reaction, as can other unsaturated acids and their esters. With a cyclodiene, the synthesis yields a bridged cyclic compound. One or two reactions of this type had been reported in the early 1900s, but Diels and Alder were the first to recognize its widespread and general nature. They also demonstrated the ease with which it takes place and the high yield of the product – two vitally important factors for successful organic synthesis. In association with Diels, and later with his own students, Alder continued to study the general conditions of the diene synthesis and the overall scope of the method for synthetic purposes. In his Nobel Prize address, Alder listed more than a dozen different dienes of widely differing structure that participate in the reaction. He also showed that the reaction is equally general with respect to dienophiles, provided that their double bonds are activated by a nearby group such as carboxyl, carbonyl, cyano, nitro or ester. Many of the compounds studied were prepared for the first time in Alder's laboratory.

Alder was a particularly able stereochemist, and he showed that the diene addition takes place at double bonds with a *cis* configuration – that is, where the two groups substituting the double bond are both on the same side of that bond, as opposed to the *trans* isomer, with the groups on opposite sides:

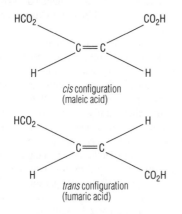

cis configuration
(maleic acid)

trans configuration
(fumaric acid)

That is why maleic acid (*cis*-butenedicarboxylic acid) reacts whereas its isomer fumaric acid (*trans*-butenedicarboxylic acid) does not. The stereospecific nature of the Diels–Alder reaction has thus become useful in structural studies for the detection of conjugated double bonds.

The bridged ring, or bicyclic, compounds formed by using cyclic dienes (such as cyclopentadiene, above) are closely related to many naturally occuring organic compounds such as camphor and pinene, which belong to the

group known as terpenes. The diene synthesis stimulated and made easier the understanding of this important group of natural products by providing a means of synthesizing them. Indeed, the ease with which the reaction takes place suggests that it may be the natural biosynthetic pathway. It has been found to be relevant in connection with quinone (vitamin K) – whose synthetic analogues are used to stimulate blood clotting – and anthraquinone type dyes now used universally. Many other commercial products have been made possible by Alder's work, including drugs, insecticides, lubricating oils, synthetic rubber and plastics. He made a great contribution to synthetic organic chemistry at a time when it was effecting a great transition in industry and science.

Aleksandrov Pavel Sergeevich 1896–1982 was a Soviet mathematician who was a leading expert in the field of topology and one of the founders of the theory of compact and bicompact spaces.

He was born in Bogorodsk (now Noginok), near Moscow, on 7 May 1896. He studied mathematics at Moscow University, graduating in 1917, and was appointed a lecturer there in 1921. In 1929 he was appointed Professor of Mathematics. From 1932 to 1964 he was President of the Moscow Mathematical Society. He received five Orders of Lenin and was awarded the State Prize in 1942.

Although he began his career by studying set theory and the theory of functions, Aleksandrov worked principally in the development of topology. He introduced many of the basic concepts of this relatively new branch of mathematics, notably the notion that an arbitrarily general topological space can be approximated to an arbitrary degree of accuracy by simple geometric figures such as polyhedrons. Of great importance, too, were his investigations into that branch of topology known as homology, which examines the relationships between the ways in which spatial structures are dissected. He formulated the theory of essential mappings and the homological theory of dimensionality, which led to a number of basic laws of duality relating to the topological properties of an additional part of space.

Aleksandrov was always greatly interested in the dissemination of mathematical knowledge and in broad collaboration in seeking it. Much of his topological work was done within a group of colleagues and students whom he gathered round him. Being one of the few Soviet scientists given great freedom to travel abroad, he, by his numerous visits to European universities, did much to carry new ideas back and forth between the East and the West. His passion for international cooperation led him to supervise the publication of an English–Russian dictionary of mathematical terminology in 1962.

Alfvén Hannes Olof Gösta 1908–   is a Swedish astrophysicist who has made fundamental contributions to plasma physics, particularly in the field of magnetohydrodynamics (MHD) – the study of plasmas in magnetic fields. For his pioneering work in this area he shared the 1970

Nobel Prize for Physics with the French physicist Louis Néel (1904–).

Alfvén was born in Norrköping, Sweden, on 30 May 1908 and was educated at the University of Uppsala, from which he gained his PhD in 1934. In 1940 he joined the Royal Institute of Technology, Stockholm, becoming Professor of Electronics in 1945 then Professor of Plasma Physics in 1963, this latter chair having been specially created for him. In 1967, however, after disagreements with the Swedish government, he obtained a professorship at the University of California, San Diego. Later he divided his time between this university and the Royal Institute.

Alfvén made his most important contributions in the late 1930s and early 1940s. Investigating the interactions of electrical and magnetic fields with plasmas (highly ionized gases containing both free positive ions and free electrons) in an attempt to explain sunspots, he formulated the frozen-in-flux theorem, according to which a plasma is – under certain conditions – bound to the magnetic lines of flux passing through it; later he used this theorem to explain the origin of cosmic rays. In 1939 he went on to propose a theory to explain aurorae and magnetic storms, a theory that greatly influenced later ideas about the Earth's magnetosphere. He also devised the guiding centre approximation, a widely used technique that enables the complex spiral movements of a charged particle in a magnetic field to be calculated relatively easily.

Three years later, in 1942, he postulated that a form of electromagnetic wave would propagate through plasma; other scientists later observed this phenomenon in plasmas and in liquid metals. Also in 1942 Alfvén developed a theory of the origin of the planets in the Solar System. In this theory (sometimes called the Alfvén theory) he hypothesized that planets were formed from the material captured by the Sun from an interstellar cloud of gas and dust. As the atoms were drawn towards the Sun they became ionized and influenced by the Sun's magnetic field. The ions then condensed into small particles which, in turn, coalesced to form the planets, this process having occurred in the plane of the solar equator. This theory did not adequately explain the formation of the inner planets but it was important in suggesting the role of MHD in the genesis of the Solar System.

Although Alfvén has studied MHD mainly in the context of astrophysics, his work has been fundamental to plasma physics and is applicable to the use of plasmas in experimental nuclear fusion reactors.

Alhazen c. 965–1038 was an Arabian scientist who made significant advances in the theory and practice of optics. He was probably the greatest scientist of the Middle Ages and his work remained unsurpassed for nearly 600 years until the time of Johannes Kepler. His Arabic name was Abu Alī al-Hassan ibn al-Haytham.

Alhazen was born in Basra (Al Basra, now in Iraq) in about 965. He made many contributions to optics, contesting the Greek view of Hero and Ptolemy that vision involves rays that emerge from the eye and are reflected by objects viewed. Alhazen postulated that light rays originate in a flame or in the Sun, strike objects, and are reflected by them into the eye. He studied lenses and mirrors, working out that the curvature of a lens accounts for its ability to focus light. He measured the refraction of light by lenses and its reflection by mirrors, and formulated the geometric optics of image formation by spherical and parabolic mirrors. He used a pin-hole as a 'lens' to construct a primitive camera obscura. He also tried to account for the occurrence of rainbows, appreciating that they are formed in the atmosphere, which he estimated extended for about 15 km/ 9 mi above the ground. He wrote many scientific works, the chief of them being *Opticae Thesaurus* which was published in 1572 from a thirteenth-century Latin translation.

Alhazen spent part of his life in Egypt, where he fell foul of the tyrannical (and reputedly mad) Caliph al-Hakim. In a foolhardy attempt to impress the caliph, Alhazen claimed he could devise a method of controlling the flooding of the River Nile. To escape the inevitable wrath of the caliph for non-fulfilment of the promise, Alhazen pretended to be mad himself and had to maintain the charade for many years until 1021, when al-Hakim died. Alhazen died in Cairo in 1038.

Alpher Ralph Asher 1921– is a US scientist who carried out the first quantitative work on nucleosynthesis and was the first to predict the existence of primordial background radiation.

Alpher, the youngest of four children and the son of a building contractor, was born in Washington, DC, in 1921. His initial interest in science was stimulated by his English teacher, Matilde Eiker, who was also an amateur astronomer, and by his chemistry teacher, Sarah Branch. Due to economic circumstances and the advent of World War II, Alpher was forced to continue his education as a night school student, receiving his BSc from George Washington University in 1943 and his PhD in 1948. His PhD research topic was nucleosynthesis in a Big-Bang universe, which was carried out under the supervision of George Gamow. During World War II Alpher worked at the Naval Ordnance Laboratory and after the war he joined the Applied Physics Laboratory of Johns Hopkins University. Here he took part in a varied research programme that, besides cosmology, included cosmic ray physics and guided missile aerodynamics. In 1955 Alpher took up a post at the Central Electric Research Laboratory in the USA where besides his professional duties he continued his vocational involvement in cosmological research.

Having graduated from George Washington University, Alpher worked with George Gamow and Robert Herman on a series of papers that sought to explain physical aspects of the Big-Bang theory of the universe. In 1948 Alpher and Gamow published the results of their work on nucleosynthesis in the early universe. They included the name of Hans Bethe as a co-author of this paper, so that their new theory became popularly known as the alpha-beta-gamma theory, appropriate for a theory on the beginning of the universe. Also in 1948, Alpher, together with Robert Herman, predicted the existence of the pervasive relic cosmic black-body radiation. They postulated that this radiation must exist, having originated in the early stages of the Big Bang with which the universe is thought to have begun. This primordial radiation was detected by Arno A. Penzias and Robert W. Wilson in 1965 and was found to have a temperature of 3K (–270°C/–454°F). Alpher and Herman had originally theorized that the radiation would have a temperature of approximately 5K (–268°C/–450°F), which was remarkably close to the actual value observed.

The existence of this low-temperature radiation that permeates the entire universe is now regarded as one of the major pieces of evidence for the validity of the Big-Bang model of the universe; thus Alpher's early cosmological work has had a profound impact towards our understanding of the nature of the universe.

al-Sufi 903–986 was a Persian astronomer whose importance lies in his compilation of a valuable catalogue of 1,018 stars with their approximate positions, magnitudes, and colours.

Little is known about the life of al-Sufi, but it has been established that he was a nobleman whose love of his country's folklore and mythology and interest in mathematics led him to the study of astronomy.

Alter David 1807–1881 was a US inventor and physicist whose most important contribution to science was in the field of spectroscopy.

Alter was born in Westmoreland County, Pennsylvania, on 3 December 1807. He had little early schooling but in 1828 he entered the Reformed Medical College, New York City, to study medicine, in which he graduated in 1831. Thereafter he spent the rest of his life experimenting and making inventions, working alone and using home-made apparatus. He died in Freeport, Pennsylvania, on 18 September 1881.

Alter made his most important contribution to physics in 1854, when he put forward the idea that each element has a characteristic spectrum, and that spectroscopic analysis of a substance can therefore be used to identify the elements present. He also investigated the Fraunhofer lines in the solar spectrum. Although the significance of this work was not recognized in the United States at that time, his idea was experimentally verified a few years later (about 1860) by Robert Bunsen (1811–1899) and Gustav Kirchhoff (1824–1887) and today spectroscopic analysis is extensively used in chemistry for identifying the component elements of substances and in astronomy for determining the compositions of stars.

Alter devoted most of his life, however, to making inventions, which included a successful electric clock, a model for an electric locomotive (which was not put into production), a new process for purifying bromine, an electric telegraph that spelled out words with a pointer, and a method of extracting oil from coal (which was not put into commercial practice because of the discovery of oil in Pennsylvania).

Alvarez Luis Walter 1911–1988 was a US physicist who won the 1968 Nobel prize for developing the liquid hydrogen bubble chamber and detecting new resonant states in particle physics. Discoveries made with the hydrogen bubble chamber were instrumental in the prediction of quarks. Alvarez also made many other breakthroughs in fundamental physics, accelerators and radar, and was well known for his studies of the pyramids and suggestion that a meteor impact led to the extinction of the dinosaurs.

Alvarez was born on 13 June 1911 in San Francisco. He went to the University of Chicago to study chemistry but changed to physics and stayed there to complete a PhD. His first major discovery, with Arthur Compton, was the discovery of the 'east–west' effect in cosmic rays, proving them to be positive.

He then moved to Ernest Lawrence's Radiation Laboratory at the University of California, Berkeley, where he spent the rest of his career. There he discovered that the capture of electrons by the nucleus of an atom is a beta-decay process, and that helium-3 was stable but hydrogen-3 (tritium) was not. Alvarez also made important contributions to the study of the spin dependence of nuclear forces and, working with Felix Bloch, measured the magnetic moment of the neutron.

During the war he moved to the Massachusetts Institute of Technology where he developed the VIXEN radar for the airborne detection of submarines, phased-array radars, and the ground-controlled approach (GCA) radar that enabled aircraft to land in conditions of poor visibility. Alvarez received the US government's most prestigious aviation award, the Collier Trophy, for these achievements. Alvarez later worked on the atomic bomb project – with Enrico Fermi at Chicago and in the explosives division at Los Alamos – and participated in the Hiroshima mission.

After the war he returned to Berkeley and built the first practical linear accelerator (a 32-MeV proton linac) and invented the tandem electrostatic accelerator. He also devised, but never built, the microtron for accelerating electrons. During the Korean War, Alvarez and Lawrence became convinced that the US needed to produce its own plutonium and built another accelerator to 'breed' plutonium. This machine was later used for nuclear physics.

In 1953 Alvarez changed direction once again when he met Donald Glaser, inventor of the bubble chamber detector for particle physics (and winner of the 1960 Nobel prize). Glaser had been using a small (about one inch) glass bulb full of diethyl ether. Alvarez decided to build a massive (72-in) chamber containing liquid hydrogen. His next idea was to automate the analysis of the particle tracks captured in the chamber. He also developed automatic scanning and measuring equipment whose output could be stored on punched cards and then analysed using computers. Alvarez and co-workers used the bubble chamber to discover a large number of new short-lived particles ('resonances') including the K (the first meson resonance) and the ω (omega) meson. These experimental findings were crucial in the development of the 'eightfold way' model of elementary particles, and subsequently the theory of quarks, by Murray Gell-Mann. The techniques developed by Alvarez became standard in high-energy laboratories all over the world.

In later life Alvarez moved away from conventional physics, using cosmic rays to search for hidden chambers in the Egyptian pyramids and his knowledge of shock waves to study the Kennedy assassination. Best-known of these researches was the discovery he made with his son Walter, a geologist, of unexpectedly high concentrations of an isotope of iridium in the thin layer of clay separating Cretaceous and Tertiary rocks. Alvarez postulated that the iridium must have come from a giant meteorite impact some 65 million years ago, and that the resulting dust in the atmosphere must have so changed the climate that the dinosaurs, who lived at that time, must have become extinct. The first half of the hypothesis is now widely accepted. Alvarez died on 1 September 1988.

Ambatzumian Victor Amazaspovich 1908– is an astronomer from the former Soviet Union whose chief contribution has been to the theory of stellar origins.

Few biographical details are known, but Ambatzumian was appointed Head of the Byurakan Observatory in 1944, having taught at the University of Leningrad. He proposed the manner in which enormous catastrophes might take place within stars and galaxies during their evolution.

The radio source in Cygnus had been associated with what appeared to be a closely connected pair of galaxies, and it was generally supposed that a galactic collision was taking place. If this were the case, such phenomena might account for many extra-galactic radio sources. Ambatzumian, however, presented convincing evidence in 1955 of the errors of this theory. He suggested instead that vast explosions occur within the cores of galaxies, analogous to supernovae, but on a galactic scale.

Ampère André Marie 1775–1836 was a French physicist, mathematician, chemist and philosopher, famous for founding the science of electromagnetics (which he named electrodynamics) and who gave his name to the unit of electric current.

Ampère was born in Polémieux, near Lyons, on 22 January 1775. The son of a wealthy merchant, he was tutored privately and was, to a great extent, self-taught. His genius was evident from an early age, particularly in mathematics, which he taught himself and had mastered to an extremely high level by the age of about 12. The later part of his youth, however, was severely disrupted by the French Revolution. In 1793 Lyons was captured by the Republican army and his father – who was both wealthy and a city official – was guillotined. Ampère taught mathematics at a school in Lyons from 1796 to 1801, during which period he married (in 1799); in the following year his wife gave birth to a son, Jean-Jacques-Antoine, who later became an eminent historian and philologist. In 1802 Ampère was appointed Professor of Physics and Chemistry at the Ecole Centrale in Bourg then, later in the same year, Professor of Mathematics at the Lycée in Lyons. Two years later his

wife died, a blow from which Ampère never really recovered – indeed, the epitaph he chose for his gravestone was *Tandem felix* ('Happy at last'). In 1805 he was appointed an assistant lecturer in mathematical analysis at the Ecole Polytechnique in Paris where, four years later, he was promoted to Professor of Mathematics. Meanwhile his talent had been recognized by Napoleon, who in 1808 appointed him Inspector-General of the newly formed university system, a post he retained until his death. In addition to his professorship and inspector-generalship, Ampère taught philosophy at the University of Paris in 1819, became Assistant Professor of Astronomy in 1820 and was appointed to the Chair in Experimental Physics at the Collège de France in 1824 – an indication of the breadth of his talents. He died of pneumonia on 10 June 1836 while on an inspection tour of Marseille.

Ampère's first publication was an early contribution to probability theory – *Considérations sur la théorie mathématique de jeu* (1802; *Considerations on the Mathematical Theory of Games*) in which he discussed the inevitability of a player losing a gambling game of chance against an opponent with vastly greater financial resources. It was on the strength of this paper that he was appointed to the professorship at Lyons and later to a post at the Ecole Polytechnique in Paris.

In the period between his arrival in Paris in 1805 and his famous work on electromagnetism in the 1820s, Ampère studied a wide range of subjects, including psychology, philosophy, physics and, more important, chemistry. His work in chemistry was both original and topical but in almost every case public recognition went to another scientist; for example, his studies on the elemental nature of chlorine and iodine were credited to Humphry Davy (1778–1829). Ampère also suggested a method of classifying elements based on a comprehensive assessment of their chemical properties, anticipating to some extent the development of the periodic table later in that century. And in 1814 he independently arrived at what is now known as Avogadro's hypothesis of the molecular constitution of gases. He also analysed Boyle's law in terms of the isothermal volume and pressure of gases.

Despite these considerable and varied achievements, Ampère's fame today rests almost entirely on his even greater work on electromagnetism, a discipline that he, more than any other single scientist, was responsible for establishing. His work in this field was stimulated by the finding of the Danish physicist Hans Christian Oersted that an electric current can deflect a compass needle – i.e. that a wire carrying a current has a magnetic field associated with it. On 11 September 1820 Ampère witnessed a demonstration of this phenomenon given by Dominique Arago at the Academy of Sciences and, like many other scientists, was prompted to hectic activity. Within a week of the demonstration he had presented the first of a series of papers in which he expounded the theory and basic laws of electromagnetism (which he called electrodynamics to differentiate it from the study of stationary electric forces, which he called electrostatics). He showed that two parallel wires carrying current in the same direction attract each other, whereas when the currents are in opposite directions, mutual repulsion results. He also predicted and demonstrated that a helical 'coil' of wire (which he called a solenoid) behaves like a bar magnet while it is carrying an electric current.

In addition, Ampère reasoned that the deflection of a compass needle caused by an electric current could be used to construct a device to measure the strength of the current, an idea that eventually led to the development of the galvanometer. He also realized the difference between the rate of passage of an electric current and the driving force behind it; this has been commemorated in naming the unit of electric current the ampere or amp (a usage introduced by Lord Kelvin in 1883). Furthermore, he tried to develop a theory to explain electromagnetism, proposing that magnetism is merely electricity in motion. Prompted by Augustus Fresnel (one of the originators of the wave theory of light), Ampère suggested that molecules are surrounded by a perpetual electric current – a concept that may be regarded as a precursor of the electron shell model.

The culmination of Ampère's studies came in 1827, when he published his famous *Mémoire sur la théorie mathématique des phénomènes électrodynamiques uniquement déduite de l'expérience (Notes on the Mathematical Theory of Electrodynamic Phenomena Deduced Solely from Experiment)*, in which he enunciated precise mathematical formulations of electromagnetism, notably Ampère's law – an equation that relates the magnetic force produced by two parallel current-carrying conductors to the product of their currents and the distance between the conductors. Today Ampère's law is usually stated in the form of calculus: the line integral of the magnetic field around an arbitrarily chosen path is proportional to the net electric current enclosed by the path.

Ampère produced little worthy of note after the publication of his *Mémoire* but his work had a great impact and stimulated much further research into electromagnetism.

Amsler-Laffon Jakob 1823–1912 was a Swiss mathematical physicist who designed and manufactured precision instruments for use in engineering.

Amsler-Laffon was born Jakob Amsler at Stalden bei Brugg on 16 November 1823 and educated locally until the age of 19, when he went to Jena to study theology. Theology absorbed his interest for only one year, however, and in 1844 he went to the university at Königsberg to study mathematics and physics. He received his doctorate in 1848, after which he worked briefly at the observatory in Geneva. The next eight years were spent in teaching, first at the University of Zurich (1849–1851) and then at the Gymnasium in Schaffhausen. But in the middle of this period Amsler abandoned his interest in pure science and became involved in the design of scientific instruments. He established a factory for the manufacture of his designs at Schaffhausen in 1854. In that year he also married Elise Laffon, the daughter of a drugs manufacturer, and added her surname to his own. He devoted himself to his factory

for the rest of his life, until his death at Schaffhausen on 3 January 1912.

Amsler-Laffon's best idea was his first, the design for an improved tool to measure areas inside curves. This was his polar planimeter. Earlier models of tools to measure the surface of spheres had been based on Cartesian coordinates, but they were bulky and expensive. Amsler-Laffon's design, based on a polar coordinate system, was not only more delicate and more flexible than its predecessors, but was also much cheaper to manufacture. It could be used in the determination of Fourier coefficients and was thus particularly valuable to shipbuilders and railway engineers. By the time he died, his factory had produced more than 50,000 polar planimeters.

Anaximander the Elder *c.*611–547 BC was an Ionian natural philosopher who seems to have formulated some basic natural philosophical views, often in opposition to his teacher, Thales. Though his writings are lost, he is credited by later writers with having developed major new ideas at the dawn of natural philosophy. He was the first Greek to handle a gnomon – a sundial with a vertical needle (it had been evolved earlier in the Middle East). Use of the gnomon enabled him, it seems, to ascertain the lengths and angles of shadows, and thereby to measure the length of the years (and thereby) of the seasons, by settling the times of the equinoxes and the solstices.

Anaximander viewed the universe as boundless; it was composed of an elementary substance of unspecified attributes. Recognizing that the Earth's surface was curved, he maintained it to be cylindrical, with its axis running east to west. With this cylindrical shape, the Earth's height was one third of its breadth, and it floated in space, poised motionless. Anaximander explained the starry heavens by presuming the Earth was encircled by bands of condensed air with vents in it; at these points the fire that was assumed to be trapped within them became visible. He regarded the relative distances of those stars to be an important scientific question. He taught that perpetual rotation in the universe created cosmic order by sorting heavier from lighter materials and packing them in concentric layers. He may have been the first Greek to map the whole known world.

Anaximander has been credited with naturalistic views of the origins of life and of humankind. It is said that he taught that life originated with primitive forms in the oceans , which in time adjusted themselves to living on dry land. Human life itself had arisen from the lower animals. Overall, he seems to have shared the early Greek philosophical urge to explain all Creation within a tiny number of general laws.

Anderson Carl David 1905–1991 was a US physicist who did pioneering work in particle physics, notably discovering the positron – the first antimatter particle to be found – and the muon (or mu-meson). He received many honours for his work, including the 1936 Nobel Prize for Physics, which he shared with Victor Hess.

Anderson was born in New York City on 3 September

1905, the son of Swedish immigrants. He was educated at the California Institute of Technology, from which he gained a BSc in physics and engineering in 1927 and a PhD in 1930. Thereafter he remained at the Institute for the rest of his career, as a Research Fellow from 1930 to 1933, Assistant Professor of Physics from 1933 to 1939, and Professor of Physics from 1939 until his retirement in 1976. After retiring he was made an Emeritus Professor of the California Institute of Technology.

Anderson's first research – performed for his doctoral thesis – was a study of the distribution of photoelectrons emitted from various gases as a result of irradiation with X-rays. Then, as a member of Robert Millikan's research team, he began in 1930 to study gamma rays and cosmic rays, extending the work originally published by Skobelzyn, who had photographed tracks of cosmic rays made in a cloud chamber. Anderson devised a special type of cloud chamber that was divided by a lead plate in order to slow down the particles sufficiently for their paths to be accurately determined. Using this modified chamber he measured the energies of cosmic and gamma rays (by measuring the curvature of their paths) in strong magnetic fields (up to about 24,000 gauss, or 2.4 teslas). In 1932, in the course of this investigation, Anderson reported that he had found positively charged particles that occurred as abundantly as did negatively charged particles, and that in many cases several negative and positive particles were simultaneously projected from the same centre. Anderson initially thought that the positive particle was a proton but, after determining that its mass was similar to that of an electron, concluded that it was a positive electron; he then suggested the name positron for this antimatter particle. Working with Neddermeyer, Anderson also showed that positrons can be produced by irradiation of various materials with gamma rays. In 1932 and 1933 other established scientists – notably Patrick Blackett, James Chadwick and the Joliot-Curies – independently confirmed the existence of the positron and, later, elucidated some of its properties.

In 1936 Anderson contributed to the discovery of another fundamental particle, the muon. While studying tracks in a cloud chamber, he noticed an unusual track that seemed to have been made by a particle intermediate in mass between an electron and a proton. Initially it was thought that this new particle was the one whose existence had previously been predicted by Hideki Yukawa (his hypothetical particle was postulated to hold the nucleus together and to carry the strong nuclear force). Anderson named the particle he had discovered the mesotron, which later became shortened to meson. Further studies of the meson, however, showed that it did not readily interact with the nucleus and therefore could not be the particle predicted by Yukawa.

In 1947 Cecil Powell discovered another, more active type of meson that proved to be Yukawa's predicted particle. Anderson's particle – the role of which is still unclear – is now called the muon (or mu-meson) to distinguish it from Powell's particle, which is called the pion (or pi-meson).

Anderson Philip Warren 1923–   is a US physicist who shared the 1977 Nobel Prize for Physics with Nevill Mott and John Van Vleck for his theoretical work on the behaviour of electrons in magnetic, noncrystalline solids.

Anderson was born in Indianapolis on 13 December 1923 and was educated at Harvard University, from which he gained his BS in 1943, MS in 1947 and PhD in 1949; he did his doctoral thesis under Van Vleck. Anderson's studies were interrupted by military service during part of World War II: from 1943 to 1945 he worked at the Harvard Naval Research Laboratories in Washington, DC, becoming a Chief Petty Officer in the United States Navy. After obtaining his doctorate, in 1949, he joined the Bell Laboratories in New Jersey, becoming Consulting Director of Physics Research in 1976. In addition to this appointment he was appointed Joseph Henry Professor of Physics at Princeton University in 1975. Anderson has also visited several foreign universities: he was a Fulbright Lecturer at Tokyo University from 1952 to 1953; an Overseas Fellow at Churchill College, Cambridge, from 1961 to 1962; and Visiting Professor of Theoretical Physics at Cambridge from 1967 to 1975. He was a Fellow of Jesus College, Cambridge, from 1969 to 1975, and was made an Honorary Fellow in 1978.

Anderson has made many varied contributions, although most of his work has been in solid state physics. While studying under Van Vleck at Harvard, he investigated the pressure-broadening of spectral lines in spectroscopy and developed a method of deducing details of molecular interactions from the shapes of spectral peaks. In the late 1950s he devised a theory to explain superexchange – the coupling of the spins of two magnetic ions in an antiferromagnetic material through their interaction with a nonmagnetic anion situated between them – and then went on to apply the Bardeen–Cooper–Schrieffer (BCS) theory to explain the effects of impurities on the properties of superconductors. In the early 1960s he investigated the interatomic effects that influence the magnetic properties of metals and alloys, devising a theoretical model (now called the Anderson model) to describe the effect of the presence of an impurity atom in a metal. He also developed a method of describing the movements of impurities within crystalline substances; this method is now known as Anderson localization.

In addition, Anderson has studied the relationship between superconductivity, superfluidity and laser action – all of which involve coherent waves of matter or energy – and predicted the existence of resistance in superconductors. Of more immediate widespread practical application, however, is his work on the semiconducting properties of inexpensive, disordered glassy solids; his studies of these materials indicate that they could be used instead of the expensive crystalline semiconductors now used in many electronic devices, such as computer memories, electronic switches and solar energy converters.

Andrews Thomas 1813–1885 was an Irish physical chemist, best known for postulating the idea of critical temperature and pressure from his experimental work on the liquefaction of gases, which demonstrated the continuity of the liquid and gaseous states. He also studied heats of chemical combination and was the first to establish the composition of ozone, proving it to be an allotrope of oxygen.

Andrews was born in Belfast on 19 December 1813, the son of a linen merchant. He attended five universities (acting on the advice of a physician friend of his father), beginning at the age of 15 at the University of Glasgow; after only a year there he published two scientific papers. Then in 1830 he went to Paris where, like his contemporary Louis Pasteur (1822–1895), he studied under the French organic chemist Jean Baptiste Dumas. He also became acquainted with several famous scientists of that time, including Joseph Gay-Lussac and Henri Becquerel. He returned to Ireland, but to Trinity College Dublin, and went from there via Belfast to Edinburgh. In 1835 he graduated from Edinburgh as a qualified doctor and surgeon, with a thesis on the circulation and properties of blood.

Even while following his medical studies, Andrews continued to experiment in chemistry, although he declined professorships in that subject at both the Richmond School of Medicine and at the Park Street School of Medicine, Dublin, preferring to devote his time to the private medical practice he had established in Belfast. He did, however, lecture on chemistry for a few hours each week at the Royal Belfast Academical Institution. It was during this time that he began work on a study of the heats of chemical combination, and in 1844 his paper on the thermal changes that accompany the neutralization of acids by bases won him the Royal Medal of the Royal Society (of which he became a Fellow five years later). By this time he was one of the leading scientific figures in the British Isles. In 1845 he was appointed Vice-President designate of the projected Queen's College, Belfast, in order that he might contribute to its foundation and philosophy, and in 1849 he became its Professor of Chemistry. He held both posts until 1879, when ill-health forced his retirement. He died in Belfast six years later, on 26 November 1885.

Andrews's research was concentrated into three main channels: the heat of chemical combination, ozone, and changes in physical state. He was only one of many mid-nineteenth century investigators of thermochemistry: his Russian contemporary Germain Hess was carrying out similar experiments at St Petersburg. Andrews's chief contribution was the direct determination of heats of neutralization and of formation of halides (chlorides, bromides and iodides), but the law of constant heat summation was finally worked out by and is now named after Hess.

Before Andrews began to study ozone, it was postulated that the gas was either a 'compound' of oxygen or that it was an oxide of hydrogen that contained a larger proportion of oxygen than does water. Andrews proved conclusively that ozone is an allotrope of oxygen, that from whatever source it is 'one and the same body, having identical properties and the same constitution, and is not a compound body but oxygen in an altered or allotropic

condition'. Ozone is triatomic, with molecules represented by the formula O_3.

Many other scientists had tried to explain the relationship between gases and liquids, but none had really come to grips with the fundamentals. It is for his meticulous experimental work in this area that Andrews is best remembered. He constructed elaborate equipment in which he initially investigated the liquefaction of carbon dioxide, exploring the state of the substance (gas or liquid) over a wide range of temperatures and pressures. By 1869 he had concluded that if carbon dioxide is maintained at any temperature above 30.9°C/87.6°F, it cannot be condensed into a liquid by any pressure no matter how great. This discovery of a critical temperature (or critical point) soon enabled other workers – such as Raoul Pictet (1846–1929) in Geneva and Louis Cailletet (1832–1913) in France, both of whom independently in 1877 liquefied oxygen – to liquefy gases that had previously been thought to be 'non-condensible', the so-called permanent gases. Hydrogen, nitrogen and air were also liquefied by applying pressure to the gases once they had been cooled to below their critical temperatures. Andrews also worked out sets of pressure-volume isotherms at temperatures above and below the critical temperature, and brought a sense of order to what had previously been a chaotic branch of physical chemistry.

Ångström Anders Jonas 1814–1874 was a Swedish physicist and astronomer, one of the early pioneers in the development of spectroscopy.

Ångström was born at Lögdö, Sweden, on 13 August 1814, the son of a chaplain. He was educated at the University of Uppsala, which awarded him his doctorate in physics in 1839; he began to lecture there in the same year. In 1843 he was appointed an observer at the Uppsala Observatory. In 1858 he was elected to the Chair of Physics at the University, a post which he held until his death, at Uppsala, on 21 June 1874.

Ångström's first important work was an investigation into the conduction of heat and his first important result was to devise a method of measuring thermal conductivity, which demonstrated it to be proportional to electrical conductivity.

Then, in 1853, he published his most substantial and influential work, *Optical Investigations*, which contains his principle of spectrum analysis. Ångström had studied electric arcs and discovered that they yield two spectra, one superimposed on the other. The first was emitted from the metal of the electrode itself, the second from the gas through which the spark passed. By applying Euler's theory of resonances Ångström was then able to demonstrate that a hot gas emits light at the same frequency as it absorbs it when it is cooled.

Ångström's early work provided the foundation for the spectrum analysis to which he devoted the rest of his career. He was chiefly interested in the Sun's spectrum, although in 1867 he investigated the spectrum of the aurora borealis, the first person to do so. In 1862 he announced his

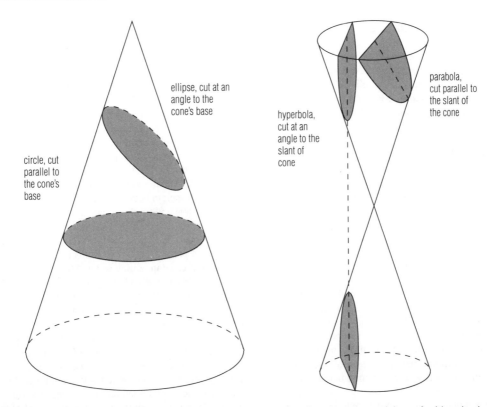

circle, cut
parallel to
the cone's
base

ellipse, cut at an
angle to the
cone's base

hyperbola,
cut at an
angle to the
slant of
cone

parabola,
cut parallel to
the slant of
the cone

Apollonius of Perga *Conic sections were studied avidly by ancient Greek mathematicians. Archimedes in particular wrote several contemplative articles on the subject – but it was Apollonius whose work became the definitive description.*

inference (in fact, it amounted to a discovery) that hydrogen was present in the Sun. In 1868 he published *Researches on the Solar System*, a famous work in which he presented measurements of the wavelengths of more than 100 Fraunhofer lines. The lines were measured to six significant figures in units of 10^{-8} cm/3.9×10^{-9} in. The unit of measure for wavelength of light, called the angstrom (Å, equal to 10^{-10} m/3.3×10^{-10} ft), was officially adopted by 1907.

Another of Ångström's important contributions was his map of the normal solar spectrum, published in 1869, which remained a standard reference tool for 20 years.

Antoniadi Eugène Marie 1870–1944 was a Turkish-born French astronomer who had a particular interest in Mars and later became an expert also on the scientific achievements of ancient civilizations.

Antoniadi was born in Istanbul (then Constantinople) in 1870. He became interested in astronomy as a young man and in 1893 he went to Juvisy-sur-Orge in France, where he worked at the Observatory with Nicolas Flammarion. He later moved to Meudon where he continued his research at the observatory there. He became a French citizen in 1928, and was appointed Director of the Mars Section of the British Astronomical Association. He died in Meudon on 10 February 1944.

Antoniadi began to make astronomical observations in 1888 at home and while visiting the Greek Islands. He was interested in the nearby celestial system – the planets of our Solar System – but was frustrated by the primitive instruments available to him. When he moved to Juvisy he was able to use the 42-cm/16.5-in telescope there with which, in 1893, he and Flammarion observed faint spots on the surface of Saturn. This observation stimulated a vigorous debate with the American astronomer, Edward Barnard, who claimed the spots to be illusory. Antoniadi and Flammarion were vindicated when in 1902 Barnard discovered one of these spots himself.

Antoniadi's chief interest was, however, the planet Mars. When he was at Meudon Observatory, he took advantage of a favourable opposition of Mars to observe it using the 84-cm/33-in telescope. He detected an apparent spot on the planet's surface, but soon realized that it was due merely to an optical effect caused by the diffraction of light by the Earth's atmosphere. His scepticism was not easy to announce, because there was at that time great interest in Giovanni Schiaparelli's suggestion, seized upon by astronomers such as Percival Lowell, that there was an intricate pattern of canals on the surface of Mars suggestive of advanced technology. Antoniadi eventually proposed that these 'canals' were also an optical illusion, produced by the eye's linking of many tiny surface details into an apparently meaningful pattern. In 1924 he was able, however, to confirm Schiaparelli's value for the rotational period of Mars.

Antoniadi's later work included research into the angle of the axis of rotation of Venus and the behaviour and properties of the planet Mercury. He published a book on the planet (*La planète Mercure* 1934) and then turned to a study of the history of astronomy and, in particular, to the work of the ancient Greek and Egyptian astronomers.

Apollonius of Perga *c*.245–*c*.190 BC was the last of the great Greek mathematicians, whose treatise on conic sections represents the final flowering of Greek mathematics.

Apollonius was born in the reign of Ptolemy Euergetes, King of Egypt (247–222 BC), in the Greek town of Perga in southern Asia Minor (now part of Turkey). Little is known of his life. It is thought that he may have studied at the school established by Euclid at Alexandria, especially since much of his work was built on Euclidean foundations.

Apollonius' fame rests on his eight-volume treatise, *The Conics*, seven volumes of which are extant. The first four books consisted of an introduction and a statement of the state of mathematics provided by his predecessors. In the last four volumes Apollonius put forth his own important work on conic sections, the foundation of much of the geometry still used today in astronomy, ballistic science and rocketry.

Apollonius described how a cone could be cut so as to produce circles, ellipses, parabolas and hyperbolas; the last three terms were coined by him. He investigated the properties of each and showed that they were all interrelated because, as he stated, 'any conic section is the locus of a point which moves so that the ratio of its distance, f, from a fixed point (the focus) to its distance, d, from a straight line (the directrix) is constant'. Whether this constant, e, is greater than, equal to, or less than 1 determines which of the three types of curve the function represents. For a hyperbola $e > 1$; for a parabola $e = 1$; and for an ellipse $e < 1$. At the time, Apollonius' discoveries lay in the realm of pure mathematics; it was only later that their immensely valuable application became apparent, when it was discovered that conic sections form the paths, or loci, followed by planets and projectiles in space.

Other than *The Conics*, only one treatise of Apollonius survives; it is entitled *Cutting off a Ratio*. It was found written in Arabic and was translated into Latin in 1706, but is of little mathematical significance.

Apollonius' brilliant concept of geometry was a milestone in the understanding of mechanics, navigation and astronomy. Above all, his work on epicircles and ellipses played a major part in Ptolemy's working out of the cosmology which dominated Western astronomy from the second century to the sixteenth century.

Appert Nicolas *c*.1750–1841 was a French confectioner and inventor who originated the modern process of thermal sterilization of food in sealed containers; he has become known as the pioneer of canning.

Appert was born *c*.1750 in France, at Chalons-sur-Marne, just east of Paris. He was self-educated and, as the son of an innkeeper, at an early age learned methods of brewing and pickling. He served his apprenticeship as a chef and confectioner at the Palais Royal Hotel in Chalons and was later employed by the Duke and Duchess of Deux-Ponts. By 1780 he had settled in Paris where he became widely known as a confectioner. By the 1790s the feeding of Napoleon's armies and the Navy was becoming a problem and new methods of preventing food decay were needed urgently. In 1795, the French Directory offered a prize for a practical method of preserving food, which encouraged Appert to begin a 14-year period of experimentation. In 1804, with the financial backing of de la Reyniere, he opened the world's first canning factory – the House of Appert – in Massy, south of Paris. By 1809 he had succeeded in preserving certain foods in glass bottles that had been immersed in boiling water. After his methods had been favourably tested and approved by the French Navy and the Consulting Bureau of Arts and Manufacturers, he was awarded the prize of 12,000 francs on 30 January 1810. One of the conditions of the award was that he should write and publish a detailed description of the processes he used. In 1811 he published *The Art of Preserving all Kinds of Animal and Vegetable Substances for Several Years*. The following year he was presented with a gold medal by the Society for the Encouragement of National Industry, and ten years later was given the title 'Benefactor of Humanity'. The fall of Napoleon, however, marked the end of his financial success, and his factories were destroyed by enemy action. He died in poverty in Massy on 3 June 1841. The cannery he founded continued to operate until 1933.

At the time Appert began his investigations into the preservation of perishable foods, chemistry was in its infancy and bacteriology was unknown. The only reference Appert had to similar work was to that of Lazzaro Spallanzani (1729–1799) in 1765, on the preservation of food by heat sterilization. Appert's final successful results were produced, therefore, by trial and error – with a little insight. He based his methods on the heating of food to temperatures above 100°C/212°F, using an autoclave (which he perfected) and then sealing the food container to prevent putrefaction. Initially, he used glass jars and bottles stoppered with corks and reinforced with wire and sealing wax, but in 1822 he changed to cylindrical tin-plated steel cans. He experimented with about 70 foods until he achieved his objective.

In addition to his work on food preservation, Appert was also responsible for the invention of the bouillon cube and he devised a method of extracting gelatin from bones without using acid; he also popularized the use of cylindrical containers for preserved foods. Appert's work was the foundation for the development of the modern canning industry although he himself could give no scientific explanation for the effectiveness of his methods. It was not until about 1860 that the biological causes of food decay became known as a result of the research begun by Louis Pasteur.

Appleton Edward Victor 1892–1965 was a British physicist famous for his discovery of the Appleton layer of the ionosphere which reflects radio waves and is therefore important in communications. He received many honours for his work, including a knighthood in 1941 and the Nobel Prize for Physics in 1947.

Appleton was born on 6 September 1892 in Bradford, Yorkshire. He attended Barkerend Elementary School from 1899 to 1903, then won a scholarship to Hanson Secondary School. A gifted student, he also won a scholarship to St John's College, Cambridge, in 1910 and graduated with first class honours in 1913. After a short period of postgraduate research with William Henry Bragg, Appleton became a signals officer in the Royal Engineers with the outbreak of World War I in 1914. This aroused his interest in radio and he began to investigate radio propagation when he returned to Cambridge after the war. In 1919 he was elected a Fellow of St John's College, and in the following year was appointed an Assistant Demonstrator at the Cavendish Laboratory. In 1924, when only 32 years old, he was appointed Wheatstone Professor of Physics at King's College, London, a post he held until 1936, when he was made Jacksonian Professor of Natural Philosophy at Cambridge. In 1939 he became Secretary of the Department of Scientific and Industrial Research, in which position he gained a reputation as an adviser on government scientific policy. During World War II, he was involved in the development of radar and of the atomic bomb. In 1949 he was made Principal and Vice-Chancellor of Edinburgh University, a position he held until his death (in Edinburgh) on 21 April 1965. While at Edinburgh, he founded the *Journal of Atmospheric Research*, which became known as 'Appleton's journal'; he remained its Editor-in-Chief for the rest of his life. Appleton began his research on radio when he returned to Cambridge after World War I. Initially he investigated (with Balthazar van der Pol Jr) thermionic vacuum tubes – on which he wrote a monograph in 1932 – then in the early 1920s he turned his attention to studying the fading of radio signals, a phenomenon he had encountered while a signals officer during World War I.

The first transatlantic radio transmission had been made by Guglielmo Marconi in 1901, and to explain why this was possible (that is, why the radio waves 'bent' around the Earth and did not merely go straight out into space) Oliver Heaviside and Arthur Kennelly postulated the existence of an atmospheric layer of charged particles (now called the Kennelly–Heaviside layer or E layer) that reflected the radio waves. Working with the New Zealand graduate student Miles Barnell – and using the recently set up BBC radio transmitters – Appleton proved the existence of the Kennelly–Heaviside layer. By periodically varying the frequency of the BBC transmitter at Bournemouth and measuring the intensity of the received transmission 100 km/62 mi away, Appleton and Barnell found that there was a regular 'fading in' and 'fading out' of the signals at night but that this effect diminished considerably at dawn as the Kennelly–Heaviside layer broke up. They also noticed, however, that radio waves continued to be reflected by the atmosphere during the day but by a higher level ionized layer. By 1926 this layer, which Appleton measured at about 250 km/155 mi above the Earth's surface (the first distance measurement made by means of radio), became generally known as the Appleton layer (it is now also known as the F layer).

Appleton continued his studies of the ionosphere (as the charged layers of the atmosphere above the stratosphere are called), showing how they are affected by the position of the Sun and by changes in the sunspot cycle. He also calculated their reflection coefficients, electron densities and their diurnal and seasonal variations. Further, he showed that the Appleton layer is strongly affected by the Earth's magnetic field and that although further above the Earth, it has a greater density and temperature than does the Kennelly–Heaviside layer.

Appleton's research into the atmosphere was of fundamental importance to the development of radio communications, and his experimental methods were later used by the British physicist Robert Watson-Watt – with whom Appleton had collaborated on several projects – in his development of radar.

Arago Dominique François 1786–1853 was a French scientist who made contributions to the development of many areas of physics and astronomy, the breadth of his work compensating for the absence of a single product of truly outstanding quality. He was closely involved with André Ampère (1775–1836) in the development of electromagnetism and with Augustin Fresnel (1788–1827) in the establishment of the wave theory of light. Arago's political commitment demanded much time during his latter years, but he maintained a continuous flow of scientific investigations until almost the end of his life.

Arago was born in Estagel, France on 26 February 1786. He studied at the Ecole Polytechnique in Paris and was then appointed to the Bureau of Longitudes. He travelled to the south of France and Spain with Jean Biot (1774–1862) in 1806, where they intended to measure an arc of the terrestrial meridian. Biot returned to France in 1807, but Arago continued his work amidst a deteriorating political situation. His return to France in 1809 was somewhat enlivened by a shipwreck and his subsequent near-escape from being sold into slavery in Algiers!

In the same year, Arago was elected to the membership of the French Academy of Sciences and became professor of analytical geometry at the Ecole Polytechnique, a post he held until 1830. He became a Foreign Member of the Royal Society of London in 1818, which awarded him the Copley Medal in 1825 for his work on electromagnetism.

The year 1830 was one of several changes for Arago. He resigned his post at the Ecole Polytechnique and succeeded Jean Fourier (1768–1830) as permanent secretary to the Academy of Sciences. He also became director of the Paris Observatory and deputy for Pyrénées Orientales, a commitment he retained until 1852.

Arago's political affiliation was with the extreme left,

and the political turbulence of 1848 saw him elected to a ministerial position in the Provisional Government. It was under his administration that slavery was abolished in the French colonies. He resigned his post of astronomer in 1852 upon the coronation of Emperor Napoleon III, refusing to take an oath of allegiance to the Emperor. Arago's reputation protected him but he died soon afterwards in Paris on 2 October 1853.

Arago's one area of sustained scientific effort was the study of the nature of light. The controversy over the question of whether light behaves as a stream of particles or as a wave motion was one of the most hotly debated of the time. Arago initially sided with Biot in the particulate camp, but later took the other view with Baron Humboldt (1769–1859), Fresnel and others. In 1811, he invented the polariscope, with which he was able to measure the degree of polarization of light rays. From 1815, Arago worked with Fresnel on polarization and was able to elucidate the fundamental laws governing it. Fresnel's mathematical expertise complemented Arago's experimental ability in establishing the wave theory of light though, because of difficulties in explaining its transmission through the ether, Arago could not accept Fresnel's assertion that light moves in transverse waves.

In 1838, Arago published a method for determining the speed of light using a rotating mirror. He was interested to find out how the speed of light is affected by travelling through a medium such as water that is dense in comparison with air. The experiment was a crucial one in determining whether light is a wave motion or not, but sadly difficulties with his laboratory equipment, the 1848 Revolution and finally the loss of his eyesight in 1850 prevented Arago from completing the experiment. Leon Foucault (1819–1869) was able to do this for him, and he obtained results which confirmed the wave theory before Arago's death.

In 1820, Arago announced to the Academy of Sciences some observations on the effect of an electric current on a magnet which had been obtained by Hans Oersted (1777–1851). Arago himself turned to the study of electromagnetism, inspiring Ampère to do the same and producing many interesting results himself. For example, he found in 1820 that an electric current produces temporary magnetization in iron, a discovery crucial to the later development of electromagnets, electric relays, and loudspeakers. Then, in 1824, he discovered that a rotating nonmagnetic metal disc, for example of copper, deflects a magnetic needle placed above it. This was a demonstration of electromagnetic induction, which would be explained by Michael Faraday (1791–1867) in 1831.

Arago also investigated the compressibility, density, diffraction and dispersion of gases; the speed of sound, which he found to be 331.2 m/1,087 ft per second; lightning, of which he found four different types; and heat. His studies in astronomy included investigations of the solar corona and chromosphere, measurements of the diameters of the planets, and a theory that light interference is responsible for the 'twinkling' of stars.

Arago's energy and enthusiasm, while directing him into a multiplicity of endeavours, acted as a catalyst in the achievement of several fundamental advances in the study of light and electromagnetism.

Archimedes *c*. 287–212 BC is generally considered to be the greatest mathematician and physicist of the ancient world. The details of his personal life, and many of the stories surrounding his achievements are of dubious authenticity. A biography of his life written by Heracleides (a friend of Archimedes) has been lost, so modern historians of science have to rely on the mathematical treatises that Archimedes is known to have published and on accounts of his life by Greeks who lived after his time. The reliability of at least parts of these accounts must be questioned, particularly those dealing with Archimedes' military exploits against the Romans during their siege of his home town Syracuse shortly before his death.

Archimedes is believed to have been born in 287 BC in Syracuse, Sicily, then a Greek colony. This date is based on the claim of a twelfth-century historian that Archimedes survived to the age of 75, for the date of his death in 212 BC is not questioned. His father was Phidias, an astronomer, and the family was a noble one possibly related to that of King Hieron II of Syracuse.

Alexandria was a great centre of learning for mathematicians, and Archimedes travelled there to study under Conon (lived *c*.250 BC) and other mathematicians who had in turn studied under Euclid (lived *c*.300 BC). Unlike most other disciples, Archimedes did not remain in Alexandria but returned to his home where he devoted the rest of his life to the serious study of mathematics and physics and, by way of recreation, to the design of a variety of mechanical devices which brought him great fame.

He chose only to publish the results of his scientific studies since only these, in his eyes, were worthy of serious consideration. His ability to invent and construct machinery was exploited during the Roman's siege of Syracuse from 215 to 212 BC. The siege is reported to have lasted such a long time because Archimedes was able to hold the Roman fleet at bay with a series of weapons which allegedly set fire to the ships or even caused them to capsize.

When the Roman sack of the city eventually came, Archimedes comported himself as the truly disinterested philosopher that he was. A Roman soldier came across him kneeling over a mathematical problem which he was examining, scratched out in the sand at the marketplace. Despite orders that Archimedes be taken alive and treated well, the impatient soldier – unable to remove him from his study – killed him on the spot.

Archimedes had decreed that his gravestone be inscribed with a cylinder enclosing a sphere together with the formula for the ratio of their volumes, this being the discovery that he regarded as his greatest achievement. His wish was evidently granted for Cicero found this gravestone in 75 BC, possibly confirming Plutarch's view that Archimedes himself thought highly only of his theoretical endeavours and disdained his practical inventions.

In physics, Archimedes is best known for his establishment of the sciences of statics and hydrostatics. In the field of statics, he is credited with working out the rigorous mathematical proofs behind the law of the lever. The lever had been used by other scientists, but it was Archimedes who demonstrated mathematically that the ratio of the effort applied to the load raised is equal to the inverse ratio of the distances of the effort and load from the pivot, or fulcrum, of the lever. Archimedes is credited with having claimed that if he had a sufficiently distant place to stand, he could use a lever to move the world.

This claim is said to have given rise to a challenge from King Hieron to Archimedes to show how he could move a truly heavy object with ease, even if he couldn't move the world. In answer to this, Archimedes developed a system of compound pulleys. According to Plutarch's Life of Marcellus (who sacked Syracuse), Archimedes used this to move with ease a ship which had been lifted with great effort by many men out of the harbour onto dry land. The ship was laden with passengers, crew and freight, but Archimedes – sitting at a distance from the ship – was reportedly able to pull it over the land as though it were gliding through water.

The best known result of Archimedes' work on hydrostatics is the so-called Archimedes principle, which states that a body immersed in water will displace a volume of fluid which weighs as much as the body would weigh in air. Archimedes is said to have discovered this famous principle when causing water to overflow from a bath. He was so overjoyed at the idea that he ran naked through the town crying 'Eureka!' which, roughly translated, means 'I've got it!'. His interest in the problem is said to have been stimulated by the problem of determining whether King Hieron's new crown was pure gold or not. The king had ordered that this be checked without damaging the crown

in any way. Archimedes realized that if the gold had been mixed with silver (which is less dense than gold), the crown would have a greater volume and therefore displace more water than an equal weight of pure gold. The story goes that the crown was in fact found to be impure, and that the unfortunate goldsmith was executed.

Among Archimedes' inventions was a design for a model planetarium able to show the movement of the Sun, Moon, planets and possibly constellations across the sky. According to Cicero, this was captured as booty by Marcellus after the sack of Syracuse. The Archimedes screw, which is an auger used to raise water for irrigation, is also credited to Archimedes, who is supposed to have invented it during his days in Egypt though he may simply have borrowed the idea from others in Egypt.

Archimedes wrote many mathematical treatises, some of which still exist in altered forms in Arabic. Among the areas he investigated were the value for π. Archimedes' approximation was more accurate than any previous estimate – he stated that the value for π lay between 223/71 and 220/70, the average of these two values being less than three parts in ten thousand different from the modern approximation for π. He also examined the expression of very large numbers, using a special notation to estimate the number of grains of sand in the universe. The result – 10^{63} – may be far from accurate, but it showed that large numbers could be considered and handled effectively. Archimedes also evolved methods to solve cubic equations and to determine square roots by approximation, as well as formulas for the determination of the surface areas and volumes of curved surfaces and solids. In the latter area, Archimedes' work anticipated the development of integral calculus, which did not come for another 2,000 years. In his mathematical proofs Archimedes employed the methods of exhaustion and *reductio ad absurdum*, which were perhaps originally developed by Eudoxus and were also used by Euclid.

Surprisingly, in view of his reputation, Archimedes' work was not widely known in antiquity and little advance was made on it. It was preserved by Byzantium and Islam, from where it spread to Europe from the twelfth century onwards. Archimedes then had a profound effect on the history of science, for his method of finding mathematical proof to substantiate experiment and observation became the method of modern science introduced by Simon Stevinus, Johannes Kepler, Galileo and Evangelista Torricelli among others in the late sixteenth and early seventeenth centuries.

Argand Jean Robert 1768–1822 was a Swiss mathematician who invented a method of geometrically representing complex numbers and their operations.

Argand was born in Geneva on 18 July 1768. Almost nothing is known of his life except that he was working in Paris as a bookseller and living there with his wife, son and daughter in 1806, the year in which he published his method. He appears to have been entirely self-taught as a mathematician and to have had no contact with any other

Archimedes *The Archimedes screw, a spiral screw turned inside a cylinder, was once commonly used to lift water from canals. The screw is still used to lift water in the Nile delta in Egypt, and is often used to shift grain in mills and powders in factories.*

mathematicians of any standing. By dint of his diligent pursuit of what was for him simply a hobby he hit upon a happy idea at just the right time in the history of mathematics. He devised his method in 1806 and thereafter did nothing again mathematically important. He died in obscurity in Paris on 13 August 1822.

The idea of giving geometric representation to complex numbers had already been worked out by the Norwegian mathematician Caspar Wessel (1745–1818) and by Gauss when, in 1806, an anonymous book, *Essai sur une manière de représenter les quantités imaginaires dans les constructions géométriques*, was published in a small, privately printed edition. But neither Wessel nor Gauss had published his idea, so that Argand, the anonymous author, deserves the credit for the system which the book propounded, sometimes wrongly attributed to Gauss, but properly known as the 'Argand diagram'. Yet Argand's name might never have come to light but for a curious set of circumstances. Before publishing his book he had outlined his ideas to Adrien Legendre. Some years later Legendre mentioned the system in a letter to the brother of J. R. Français, a lecturer at the Imperial College of Artillery at Paris. Français discussed the new system in a paper published in the journal *Annales de mathématiques* in 1813 and, acting from curiosity and kindness, asked the anonymous inventor of it to come forward and make himself known. Argand did so and a paper of his was published in a later issue of the *Annales* for that year.

In his book Argand adopted Descartes's practice of calling all multiples of $\sqrt{-1}$ 'imaginary'. That pure imaginary numbers might be represented by a line perpendicular to the axis of real numbers had been suggested in the seventeenth century by John Wallis. Argand went further, to demonstrate that real and imaginary parts of a complex number could be represented as rectangular coordinates. It had for some time been usual to picture real numbers – negative and positive – as corresponding to points on a straight line. Of the models studied by Argand, and discussed in his book, one made use of weights, progressively removed from a beam balance, to represent the generation of negative numbers by repeated subtractions. Another simply subtracted portions from a sum of money. Argand argued that these models showed that distance could be considered separately from direction in constructing a geometrical representation, such distance being 'absolute'. Furthermore, whether a negative quantity were considered to be real or imaginary depended on the sort of quantity that is being measured. Argand was thus able to use these distinctions – between direction and absolute distance, and between real and imaginary negative quantities – to construct his diagram.

The diagram is a graphic representation of complex numbers of the form $a + bi$, in which a and b are real numbers and i is $\sqrt{-1}$. One axis represents the pure imaginary numbers (those belonging to the bi category) and the other the real numbers (those belonging to the a category); it is thus possible to plot a complex number as a set of coordinates in the field defined by the two axes.

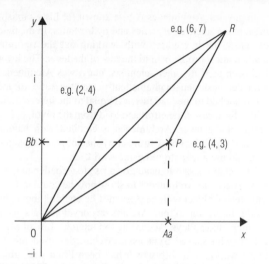

Argand *The Argand diagram is a graphic representation of complex numbers and their additon. A complex number of the form a + bi, in which a and b are real numbers and i is the square root of –1, is represented by the point (a, bi). Complex numbers are added by means of a parallelogram construction.*

Argand was an amateur and had a somewhat patchy knowledge of mathematics. It is clear, for instance, from his discussion of the central problem, whether all rational functions, $f(a + bi)$, could be reduced to the form $A + Bi$, where i, b, A and B are real, that he understood the work of Lagrange, Euler and d'Alembert; but he also revealed that he did not know that Euler had so reduced $\sqrt{-1}^{\sqrt{-1}}$, since he cited that expression as one that could not be reduced. It is also true that, had Argand never lived, his idea would have come to light through the work of Gauss at about the same time. It would nevertheless be a grave injustice to deny him his honourable, if minor, place in the history of mathematics.

Argelander Friedrich Wilhelm August 1799–1875 was a Prussian astronomer whose approach to the subject was one of great resourcefulness and thoroughness. His most enduring contribution was the publication of the *Bonner Durchmusterung* (Bonn Survey, 1859–1862) of more than 300,000 stars in the northern hemisphere. Its value is such that it was reprinted as recently as 1950.

Argelander was born on 22 March 1799 in Memel, East Prussia (now Klaipeda in Lithuania). He studied in Elbing and Königsberg. His initial plan had been to study economics and politics, but lectures by Friedrich Bessel soon fired his interest in astronomy. He worked under Bessel and in 1822 was awarded a PhD for a thesis in which he described work he had done as part of Bessel's systematic evaluation of bright stars in part of the northern hemisphere. In the same year Argelander earned the title of Lecturer.

In 1823 he went to Åbo (Turku) in Finland. He worked as an astronomical observer there, under difficult conditions, for four years until the observatory was destroyed by a fire that swept the town in 1827. He became Professor of Astronomy at the University of Helsinki in 1828, and from 1832 until 1836 he was Director of the observatory there.

During the upheavals that followed the Napoleonic Wars, the young princes of the Prussian Kingdom had lived for a few years with the Argelander family. In 1836, when Argelander went to the University of Bonn as Professor of Astronomy, the grateful crown prince (who later became King Friedrich Wilhelm IV), promised Argelander a magnificent new observatory, which was eventually constructed under Argelander's supervision. Meanwhile, he continued his own studies under primitive conditions.

Argelander's work was of such impeccable standard that he came to earn an impressive international reputation. He was elected to virtually every prominent European scientific academy and was a member of scientific organizations in the United States. He was an active member of the Astronomische Gesellschaft, serving as chairman of its governing body from 1864 until 1867. He died in Bonn on 17 February 1875.

Argelander's early astronomical studies were a continuation of Bessel's work on the mapping of stellar positions, so that his systematic approach was established from the beginning. After his move to Åbo, Argelander began to concentrate on the proper motion of stars, that is the movement of stars measured in seconds of arc per year, relative to one another. Argelander considered this movement by analogy to a ship moving among a fleet: the farther vessels appear to be almost stationary relative to those nearby, although they are all moving. He studied the proper motion of more than 500 stars and was able to publish the most accurate catalogue of the day on the subject.

His next major area of study was a continuation of a preliminary investigation done by William Herschel in 1783 on the movement of the Sun through the cosmos. Herschel's study had involved the proper motion of only seven stars. Argelander realized that observations of many more stars would be necessary before a firm conclusion about the direction of movement of the Sun, if indeed there were any, could be made. Only a large quantity of data would enable the dual effects of the movement of 'fixed' stars and the movement of the Sun to be distinguished. Argelander's conclusions, based on nearly 400 stars, confirmed Herschel's results. He found that the Sun is indeed moving towards the constellation of Hercules.

In 1843 Argelander published his *Uranometrica Nova*, based on studies made exclusively with the naked eye, because the Bonn Observatory was still under construction. The most important innovation in this study was the introduction of the 'estimation by steps' method for determining stellar magnitudes. It relied exclusively on the sensitivity of the trained eye in comparing the brightness of neighbouring stars.

In 1850 the system was elaborated by N. G. Pogson, who found that each step along the scale meant a change in brightness 2.5 fold. Bright stars have low numbers, for example 1 or 2, and dim stars have high numbers, such as 9. Stars of magnitude 7 and above are not visible to the naked eye. Extremely bright bodies have magnitudes with negative values, such as the full Moon with a value of −11 or the Sun with a value of −26.7.

Argelander's next project was an extension of Bessel's study of stars in the northern sky. At first he neglected stars up to a certain magnitude, which limited the usefulness of his data since it made them inadequate for statistical analysis. This was remedied during the late 1850s, with the help of E. Schönfeld and A. Kruger, and resulted in the publication of the *Bonner Durchmusterung*. This catalogued the position and brightness of nearly 324,000 stars, and although it was the last major catalogue to be produced without the aid of photography, it represents the cornerstone of later astronomical work. Argelander also initiated a mammoth project that required the cooperation of many observatories and was aimed at improving the accuracy of positional data recorded in the survey.

Argelander's work was characterized by its grand scope and admirable thoroughness. His contributions were fundamental to many later astronomical studies.

Aristarchos *c.* 320–*c.* 250 BC was a mathematician and astronomer of great renown in ancient Greece.

Aristarchos was born on the island of Samos in about 320 BC. He was born before Archimedes (*c.* 287–*c.* 212 BC), although Aristarchos and Archimedes certainly knew of each other. Little is known of Aristarchos' life, but it is thought most likely that he studied in Alexandria under Strato of Lampsacos (*c.* 340–270 BC), before the latter succeeded Theophrastus (372–287 BC) as head of the Athenian Lyceum (originally founded by Aristotle) in 287 BC. Aristarchos died in Alexandria, *c.* 250 BC.

The only work by Aristarchos that still exists is *On the Magnitude and Distances of the Sun and Moon*. This document describes the first attempt, by means of simple trigonometry, to measure these sizes and distances. The measurements were not very accurate and they were improved upon a century later by Hipparchus. It was, nevertheless, probably as a consequence of the figures Aristarchos obtained – however inaccurate – that he first began to conceive of his revolutionary cosmological model. In the text of his essay Aristarchos described a right-angled triangle with the Earth, the half-illuminated Moon and the Sun at the corners; the Moon was positioned at the right angle, and the hypotenuse ran between the Sun and the Earth. Aristarchos reasoned that since the angle at the Earth corner could be measured, the angle at the Sun could be deduced.

Based on six astronomical hypotheses Aristarchos obtained eighteen propositions which described, among other things, the Sun–Earth and Moon–Earth distances in terms of Earth-diameters. The circumference (and therefore the diameter) of the Earth had already been calculated with considerable accuracy; in addition, his basic mathematics was sound, but because Aristarchos was unable to make accurate measurements he arrived at incorrect results. Not only did he have difficulty in knowing exactly when the Moon was half-illuminated (when it forms a right

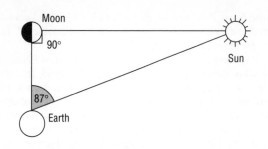

Aristarchos *Aristarchos calculated the distance between the Earth and the Sun (in terms of the Earth–Moon distance) by measuring the angle to the Sun when it was exactly half full (and its angle to the Sun was 90˚).*

angle with the Sun and Earth), he also underestimated the angle formed by the Sun, the Earth and the Moon at the Earth corner.

This miscalculation led to the erroneous conclusion that the distance between the Sun and the Earth was only 18 to 20 times the distance between the Moon and the Earth; in fact, the correct multiple is 397 times. Aristarchos also grossly miscalculated the diameter of the Sun as being only seven times that of the Earth. Even so, it still struck him as strange that a larger body should orbit a smaller one, as was assumed by the geocentric cosmological model.

Heraklides of Pontus (*c*. 388–315 BC) had earlier proposed the idea, which had been accepted by many, that Mercury and Venus orbit the Sun. Aristarchos carried the argument further and suggested that the Earth also orbits the Sun. Aristarchos' model, the first heliocentric one to be proposed, described the Sun and the fixed stars as stationary in the cosmos, and the planets – including the Earth – as travelling in circular orbits around the Sun. He further stated that the apparent daily rotation of the sphere of stars is due to the Earth's rotation on its axis as it travels along its orbit. He anticipated the most powerful argument against his theory by stating that the reason no stellar parallax (change in position of the stars) was observed from one extreme of the orbit to the other is that even the diameter of the Earth's orbit is insignificant in relation to the vast dimensions of the universe.

Aristarchos' model is recorded in letters, in a book by Archimedes, and in the writings of Plutarch. Copernicus certainly knew of Aristarchos' heliocentric model, but he deliberately suppressed reference to it, perhaps so as not to compromise his claims to originality.

Aristarchos is said to have carried out other astronomical research, including an observation of the summer solstice in 281 BC (according to Ptolemy). He is also thought to have designed the skaphe – an improved sundial which consisted of a hollow hemisphere with a vertical needle, protruding from the base, which cast shadows to indicate the time.

We know today that Aristarchos was substantially correct in his views. However, his theory made little impact on his contemporaries, since the powerful philosophical, religious and astronomical ideas of the time were all based on a geocentric view of the universe. Even Aristarchos' initial measurements of the distances and sizes of the Earth, Moon and Sun were probably based on a geocentric model. The new model, however, demanded that the universe be considered to have dimensions which exceeded the imagination of and were unacceptable to Aristarchos' fellow cosmologists. He was accused of impiety, but perhaps even more damning was the inability of his model to account for a number of astronomical anomalies and the unequal duration of the seasons. It was never suspected that the introduction of elliptical rather than circular orbits would have gone a long way to resolving these difficulties.

Virtually alone among the Greek astronomers, Aristarchos proposed a cosmological model based not on mathematical harmony, but on observed physical 'reality'. This achievement is something of a paradox since he is generally credited with being a mathematician, rather than a descriptive astronomer.

Aristotle 384 BC–322 BC was a Greek polymath, one of the most imaginative and systematic thinkers in history, whose writings embraced virtually every aspect of contemporary thought, including cosmology.

Aristotle was born at Stagirus, a port on the Chalcidic peninsula of Macedonia, in 384 BC. His father, Nichomachus, was court physician to Amyntas III (sometimes called Amyntas II), King of Macedonia, and it seems probable that he introduced Aristotle to the body of medical and biological knowledge at an early age. Nichomachus died in Aristotle's youth and Aristotle was placed in the care of a ward, who sent him to Athens in 367 BC to study at Plato's Academy. Plato's death in 348/347 BC coincided with a wave of anti-Macedonian fervour in Athens, a combination of events which induced Aristotle to leave the city and go on an extensive tour of Asia Minor, where for the first time he engaged in a serious study of natural history.

In 342 BC King Philip II invited Aristotle to the Macedonian court to become tutor to the crown prince, the future Alexander the Great. Shortly after Alexander came to the throne in 336 BC, Aristotle returned to Athens, where he established his own school, the Lyceum (known also as the Peripatetic School from Aristotle's habit of lecturing while walking in the garden) in 335 BC. At the Lyceum Aristotle established a zoo (stocked with animals captured during Alexander's Asian campaigns) and a library. The latter formed the basis of the great library established in Alexandria by the Ptolemies. The death of Alexander in 323 BC and another upsurge of anti-Macedonian sentiment in Athens suddenly made Aristotle's position uncomfortable. Largely because of his association with Antipater, the Macedonian regent and general, Aristotle was politically suspect. He was charged with impiety and, rather than suffer the fate of Socrates, he withdrew to Chalcis (now Khalkis), north of Athens, where he died in 322 BC.

Aristotle's writings, which have come down to us only in later, edited versions of his notes, lectures and publications, cover philosophy, logic, politics, physics, biology and cosmology. Among his many scientific and philosophical treatises are the *Organon*, a collection of treatises on logic; the *Physica*, on natural science; the *Historia Animalium*, a classification of animals; and *De Incessu Animalium*, on the progression of animals. His major writings on cosmology, or astronomy, are brought together in the four-volume text, *De caelo* (*Of the Heavens*). Aristotle rejected the notion of infinity and the notion of a vacuum. A vacuum he held to be impossible because an object moving in it would meet no resistance and would therefore attain infinite velocity. Space could not be infinite, because in Aristotle's view, adopted from the work of Eudoxus (*c*.406 BC–*c*.305 BC) and Callippus (*c*.370 BC–*c*.300 BC), the universe consisted of a series of concentric spheres which rotated around the centrally placed, stationary Earth. If the outermost sphere were an infinite distance from the Earth, it would be unable to complete its rotation within a finite period of time, in particular within the 24-hour period in which the stars, fixed, as Aristotle believed, to the sphere, rotated around the Earth.

Aristotle's cosmos – geocentric and broadly speaking mechanical, not dynamic – differed only in details from the model proposed by Eudoxus and Callippus. Callippus posited 33 spheres; Aristotle added 22 new spheres, then amalgamated some of them, to reach a total of 49. This clumsy model, which was unable to account even for eclipses, was partly replaced by the Ptolemaic system based on epicycles. In Aristotle's system the outermost sphere contained the fixed stars. Then followed the spheres of Saturn, Jupiter, Mars, the Sun, Venus, Mercury, and, closest to the Earth, the Moon. Each of these had several spheres in order to account for all their movements. The outermost sphere was controlled by divine influence and indirectly it determined the movement of all the inner spheres. The original motive power of the universe was thus removed from the centre, where the Pythagoreans had placed it.

According to Aristotle's laws of motion, bodies moved upwards or downwards in straight lines. Of Empedocles' four natural elements in the universe, earth and water fell, air and fire rose. To explain the motion of the heavenly spheres, therefore, Aristotle introduced a fifth element, ether, whose natural movement was circular. Aristotle thus posited that the laws of motions governing the celestial bodies above the Moon were different from the laws which governed bodies beneath the Moon.

Aristotle's work in astronomy also included proving that the Earth was spherical. He observed that the Earth cast a circular shadow on the Moon during an eclipse and he pointed out that as one travelled north or south, the stars changed their positions. Since it was not necessary to travel very far to observe this effect, it was clear that the Earth was a sphere, and a rather small one at that. As a result Aristotle was able to make a tolerably fair estimate of the Earth's diameter, overestimating it by only 50%.

Arkwright Richard 1732–1792 was a prominent English pioneer in Britain's Industrial Revolution (1733–*c*.1840).

His spinning machinery, which replaced and increased the speed and efficiency of handspinning, transformed the textile industry.

Born in Preston, Lancashire, England, on 23 December 1732, Arkwright was the youngest of seven surviving children of poor parents. He had little formal education but was taught to read by an uncle and educated himself until quite late in life.

Arkwright first worked as a barber–wigmaker using his own process for preparing and dyeing wigs. When the wig trade declined he turned his attention to engineering, and especially to the textile industry which was then being industrialized following John Kay's invention of the flying shuttle (1733), the introduction of a carding machine (1760) and of James Hargreaves's spinning jenny (1767, patented 1770). The jenny partly did away with old-fashioned handwheel spinning, but only for spinning the weft. The roving process was still performed by hand, since the Jenny did not have sufficient strength for the warp, the longitudinal threads.

In 1767, Arkwright, probably unaware of Hargreaves's work, began to develop a machine which could spin by rollers. With this, Arkwright applied a new principle rather than simply mechanizing handwheel methods. His spinning frame consisted of four pairs of rollers acting by tooth and pinion. The top roller was covered with leather, to enable it to take hold of the cotton material. The lower was fluted longitudinally to let the cotton pass through it. One pair of rollers revolved more quickly than its corresponding pair, and so the rove was drawn to the requisite fineness for twisting. This was accomplished by spindles, or flyers, placed in front of each set of rollers.

Arkwright's spinning frame was patented in 1769. He set up a small factory at Nottingham, followed two years later by another at Cromford, Derbyshire. His partners there included Jedediah Strutt and Samuel Need, owners of a patent for the manufacture of ribbed stockings. At Cromford, the machines were run by water power, and the spinning frame therefore became known as the 'water' frame: it was later renamed 'throstle' with the advent of steam power in 1785.

The stockings woven on the spinning frame, from yarn of hard, firm texture and smooth consistency, were superior to those woven from handspun cotton. Other manufacturers who had declined to use the yarn were later to regret their decision. In 1773 Arkwright produced the first cloth made entirely from cotton, when he used thread as warp for the manufacture of calico: previously, the warp was of linen, and only the weft was of cotton.

The new material attracted enthusiastic demand, especially after 1774, when a special act of Parliament was passed exempting Arkwright's goods from the double duty imposed on cottons by an act of 1736. (The 1736 Act had been designed to protect woollen manufacturers against the calico cottons of India, where the British East India Company was active in trade.)

By 1775, Arkwright's factories, sited countrywide, were using a sequence of machines to carry out all cotton manufacturing operations, from carding, drawing and roving to spinning. The prosperity of the cotton industry was thus assured, and Arkwright's inventions were later adapted, with equal success, for the woollen and worsted trades. With his comprehensive machinery, Arkwright was an improver and adapter rather than an inventor. To a great extent he incorporated the ideas of others, including fundamental and original ideas, which was why the comprehensive patent he took out in 1775 was later rescinded. This did not detract from Arkwright's achievement, however, and in 1786 his work was acknowledged by King George III with a knighthood. Further recognition followed in 1787, when Arkwright was made High Sheriff of Derbyshire.

Arkwright was the archetypal product of a new industrial age which enabled men of ingenuity and determination to rise to eminence from a state of poverty. Apart from his technical achievements, he was also notable for his organization of large-scale factory production and the division of labour between machines and work force. By 1782 Arkwright employed 5,000 workers. He died, rich and respected, at Cromford on 3 August 1792.

Armstrong Edwin Howard 1890–1954 was a US electronics engineer who worked almost exclusively in radio, developing the superheterodyne receiver and frequency modulation (FM) radio transmission.

Armstrong was born in New York City on 18 December 1890, the son of a publisher. As a child he had an intense interest in machines and mechanisms and by the age of 14, stimulated by reports of the work of Guglielmo Marconi (1874–1937), he was constructing wireless (radio) circuits for transmitters and receivers. After graduating from High School he attended the engineering department of Columbia University, New York, gaining a degree in electrical engineering in 1913. He became an instructor at Columbia, and during World War I he worked in the Laboratories of the US Signal Corps in Paris. In 1918, after the end of the war, he returned to Columbia University as an assistant to the physicist Michael Pupin (1858–1935), whom he eventually succeeded as Professor of Electrical Engineering in 1934. During World War II he again carried out research for the military. His inventions earned him much money – he became a millionaire – but also involved him in expensive, protracted litigation and promotional costs. His health declined and he became depressed; at the end of January 1954 he committed suicide by jumping from the window of his New York apartment.

In 1912, while he was still a student at Columbia University, Armstrong designed and built a regenerative, or feedback, radio receiving circuit that made use of the amplifying ability of the triode thermionic valve, called the Audion by its inventor Lee De Forest in 1906. The same circuit acted as an oscillator (transmitter) at high amplifications. He patented it, and for 14 years fought a series of legal battles about priority with De Forest. Armstrong finally lost the struggle in the Supreme Court, although the legal interpretation was not accepted by most of the scientists and engineers of the time who continued to uphold Armstrong's claim and went on to award him the Franklin Medal for his invention.

The superheterodyne receiving circuit was developed by Armstrong during World War I in an attempt to make a receiver that could detect the presence of enemy aircraft by means of the electromagnetic (radio) waves given off by the sparking of the ignition systems of their engines. Again he patented the idea, and sold patent rights and licences to major manufacturers, including RCA in the United States. During the 1920s, earnings from these sources made him a rich man.

At that time, radio broadcasting used amplitude modulation (AM), in which the audio signal varies the amplitude of a transmitted carrier wave. Electrical disturbances in the atmosphere caused by storms or electrical machinery also modulate the carrier wave, resulting in static interference at the receiver. In 1933 Armstrong, working with Pupin, developed and patented a method of radio broadcasting in which the transmitted signal is made to modulate the frequency of the carrier wave over a wide waveband. This method, called frequency modulation (FM), is unaffected by static and is capable of high-fidelity sound reproduction; it remains the basis of quality radio, television, microwave and satellite transmissions, although the high frequencies used are generally limited to line-of-sight distances. FM broadcasting did not gain ground until after World War II, and then only after Armstrong had used much of his fortune in promoting it and in another round of legal actions about patent rights. Frustration and disenchantment with the attendant commercial, legal and financial problems probably led to his suicide.

Armstrong William George 1810–1900 was the British engineer and inventor who revolutionized the manufacture of big guns in the mid-nineteenth century. He was also responsible for pioneering developments in hydraulic equipment.

Armstrong was born in Newcastle-upon-Tyne on 26 November 1810. He attended private schools in Newcastle and Whickham, and later completed his formal education at a grammar school in Bishop Auckland. Following this, he studied law in London. In 1833 he returned to Newcastle and was engaged in private practice as a solicitor. In his spare time he carried out numerous scientific experiments.

In 1839 he constructed an overshot water wheel and soon afterwards he designed a hydraulic crane, which contained the germ of the ideas underlying all the hydraulic machinery for which he subsequently became famous. Abandoning his law practice in 1847, he founded an engineering works at Elswick to specialize in building hydraulic cranes. In 1850 he invented the hydraulic pressure accumulator and in 1854 he designed submarine mines for use in the Crimean War. A year later Armstrong designed a three-pounder gun with a barrel made of wrought iron wrapped round an inner steel tube. In 1859 he established the Elswick Ordnance Company for making the Armstrong gun for the British Army. In the same year, he was appointed Chief Engineer of Rifled Ordnance at Woolwich, and received a knighthood. However, a government decision to revert to the production and use of muzzle-loading guns led to his resignation from Woolwich. He improved the design of his original gun and, by 1880, had completed the design of a 150-mm/6-in breech-loading gun with a wire-wound cylinder. This design was adopted by the British government and was the prototype of all subsequent artillery.

In 1882 Armstrong established a new shipbuilding yard at Elswick for the construction of warships. Five years later he was raised to the peerage and created Baron Armstrong of Cragside. He also entered into partnership with Joseph Whitworth (1803–1887) to form the famous Armstrong–Whitworth company at Openshaw, Manchester.

Although Armstrong initially chose law as his career, he soon switched to engineering. Even before he had abandoned his law practice, he had constructed an overshot water wheel, having become engrossed with water wheels as a source of power when, on a fishing holiday in Dentdale, he saw an overshot water wheel and observed that only about a twentieth of the energy available was being utilized. His original invention of 1839 was improved upon by his production of a rotary water motor.

His invention of a hydroelectric machine, which generated electricity by steam escaping through nozzles from an insulated boiler, originated in a report he read of a colliery engineman who noticed that he had received a sharp electric shock on exposing one hand to a jet of steam issuing from a boiler with which his other hand was in direct contact.

Armstrong's first hydraulic crane depended simply on the pressure of water acting directly on a piston in a cylinder. The resulting movement of the piston produced a corresponding movement through suitable gears. The first example was erected on the quay at Newcastle in 1846 – the pressure being obtained from the ordinary water mains of the town. The merits and advantages of this device soon became widely appreciated.

In 1850 a hydraulic installation was required for a new ferry station at New Holland on the Humber estuary. The absence of a water mains of any kind, coupled with the prohibitive cost of a special reservoir because of the character of the soil, made it necessary for Armstrong to invent a fresh piece of apparatus. This became known as the hydraulic accumulator. It consisted of a large cylinder containing a piston that could be loaded to any desired pressure – the water being pumped in below it by a steam engine or other prime mover. With various modifications, this device made possible the installation of hydraulic power in almost any situation. In particular it could be used on board ship, and its application to the manipulation of heavy naval guns made it among the most important of Armstrong's inventions.

The Elswick works had originally been founded for the manufacture of this hydraulic machinery, but it was not long before it became the birthplace of a revolution in gun-making. Modern artillery dates from 1855 when Armstrong's first gun made its appearance. This weapon embodied all the essential features which distinguish the ordnance of today from the cannon of the Middle Ages.

There had been little change in heavy guns for 500 years. Their barrels were cast in bronze and they were loaded from the muzzle. Armstrong used the advances of the nineteenth century in metallurgy and chemistry, together with his own inventive genius. His gun was built up of rings of metal shrunk upon an inner steel barrel. It was loaded at the breech and it was rifled, and it threw not a round ball but an elongated shell. The British Army adopted the Armstrong gun in 1859.

Britain had thus originated an armament superior to that possessed by any other nation at that time. But in 1863, when defects in the breech mechanism caused a disagreement with the government over this decision, Armstrong resigned his Woolwich appointment and devoted his time to improving the original design. He succeeded and, after 17 years (in 1880), the government acknowledged his improvements, which included the use of steel wire. He perceived that to coil many turns of wire round an inner barrel was a logical extension of the large hooped method that had been used in the gun designed by him in 1855.

Arp Halton Christian 1927– is a US astronomer known for his work on the identification of galaxies.

Arp was born on 21 March 1927 in New York City. He was educated at Harvard Universtiy, where he obtained his BA in 1949. Four years later he gained a PhD at the California Institute of Technology and became a Carnegie Fellow at the Mount Wilson and Palomar Observatory in 1953. He was a Research Associate at the University of Indiana from 1955 to 1957, and for the next eight years was Assistant Astronomer at the Mount Wilson and Palomar Observatory, of the Carnegie Institute in Washington and at the California Institute of Technology. From 1965 to 1969 he was Astronomer at those institutions. In 1969 he became Astronomer at the Hale Observatory in California and in 1960 a Visiting Professor of the National Science Foundation. He is a member of several astronomical associations and was Chairman of the Los Angeles Chapter of the Federation of American Scientists and of Sigma XI, in 1965. He also received several awards for his achievements in the field of astronomy.

In 1956, while at Indiana University, Arp established the ratio between the absolute magnitude of novae at maximum brightness and the speed of decline of magnitude. Since then he has published several papers and in 1965 wrote the *Atlas of Peculiar Galaxies*.

During his research on globular clusters, globular cluster variable stars, novae, Cepheid variables, extragalactic nebulae and so on, Arp has attempted to relate the listings of galaxies to radio sources and has compiled a catalogue of radio sources from the galaxies shown in the Palomar sky atlas; the optical identification of these sources can now be done fairly accurately.

The most interesting of the radio sources found so far are quasars (quasi-stellar objects), which are characterized by a strong emission in the ultraviolet part of the spectrum. Arp is working with other astronomers on the question whether the red shifts in the spectrum of quasars are due to the general expansion of the universe. If this is proved to be true, then quasars are among the oldest and most remote objects in the universe.

Arp has also carried out the first photometric work on the Magellanic Clouds – the nearest extragalactic system. In his research on pulsating novae, particularly those in the Andromeda Nebula, Arp has demonstrated that there is a close relationship between the maximum magnitude and the luminosity of novae so that it is now possible to obtain absolute luminosities for novae fairly easily using light curves (graphs relating apparent magnitude to time).

Arp has attempted also to obtain better data on RV Tauri stars – variables that are much brighter than other cluster stars, many of which lie well outside the clusters. In doing this research he investigated the whole series of variable stars with periods of more than one day, in a number of globular clusters, and related their magnitudes with those of cluster-type variables in the same cluster. Arp's results are referred to as the zero-point of the cluster-type variables.

In the continuing search for explanations for and identification of phenomena in the universe, Arp has done a great deal to aid the classification of information as well as to provide a basis from which other astronomers may work to increase knowledge of these and other, as yet unexplained phenomena.

Arrhenius Svante August 1859–1927 was a Swedish physical chemist who first explained that in an electrolyte (a solution of a chemical dissolved in water) the dissolved substance is dissociated into electrically charged ions. The electrolyte conducts electricity because the ions migrate through the solution. Although later modified, Arrhenius's theory of conductivity in solutions has stood the test of time. It was a major contribution to physical chemistry, ultimately acknowledged by the award to Arrhenius of the 1903 Nobel Prize for Chemistry.

Arrhenius was born in Uppsala on 19 February 1859, the son of a surveyor and estate manager who was also a supervisor at the local university. He was a brilliant student, entering Uppsala University at the age of 17 to study chemistry, physics and mathematics. After graduating in 1878, he stayed on to write his doctoral thesis, but became dissatisfied with the teaching at Uppsala and went to Stockholm to study solutions and electrolytes under Erik Edlund (1819–1888).

In 1884 he submitted his thesis, which contained the basis of the dissociation theory of electrolytes (although he did not at that time use the term dissociation), together with many other novel theories which aroused only suspicion and doubt in his superiors. The largely theoretical document was not welcomed by the academics, who were devoted experimentalists; Arrhenius later boasted that he had never performed an exact experiment in his life and preferred to take a general view of relationships from the

results of many approximate experiments. The thesis (written in French) was awarded only a fourth class, the lowest possible pass, but Arrhenius sent copies of it to several eminent chemists, including Friedrich Wilhelm Ostwald (at Riga), Rudolf Clausius (Bonn), Lothar Meyer (Tübingen) and Jacobus van't Hoff (Amsterdam). Ostwald's offer to Arrhenius of an academic appointment moderated the scepticism of the Uppsala authorities, who finally offered him a position and, later, a travelling fellowship. This enabled him to spend some time with other scientists working in the same field, such as Friedrich Kohlrausch and Hermann Nernstat Würzburg, Ludwig Boltzmann at Graz and Jacobus van't Hoff, whose solution theories paralleled that of Arrhenius.

In 1891 Arrhenius was offered a professorship at Giessen in Germany as successor to Justus von Liebig, but he declined in preference for an appointment at the Royal Institute of Technology in Stockholm. Four years later he became professor of physics at a time when his work was attracting the attention of scientists throughout Europe, if not in his native Sweden – his election to the Swedish Academy of Sciences had to wait another six years until 1901 (the same year in which van't Hoff was awarded the first Nobel Prize for Chemistry). He again declined a German professorship, in Berlin, in 1905 and took instead the specially created post of Director of the Nobel Institute of Physical Chemistry (Stockholm), where he remained until shortly before his death on 2 October 1927.

The Arrhenius theory of electrolytes (1887) is concerned with the formation, number, and speed of ions in solution. The key to the theory is the behaviour of the dissolved substance (solute) and the liquid (solvent), both of which are capable of dissociating into ions. It postulates that there is an equilibrium between undissociated solute molecules and its ions, whose movement or migration can conduct an electric current through the solution. Its chief points may be summarized as follows:

(a) An electrolytic solution contains free ions (i.e. dissociation takes place even if no current is passed through the solution).

(b) Conduction of an electric current through such a solution depends on the number and speed of migration of the ions present.

(c) In a weak electrolyte, the degree of ionization (dissociation) increases with increasing dilution.

(d) In a weak electrolyte at infinite dilution, ionization is complete.

(e) In a strong electrolyte, ionization is always incomplete because the ions impede each other's migration; this interference is less in dilute solutions of strong electrolytes.

Apart from (e), regarding strong electrolytes (a difficulty that was not resolved until the work of Peter Debye and Erich Hückel in the 1920s), Arrhenius's theory is still largely accepted.

In another notable achievement, Arrhenius adapted van't Hoff's work on the colligative properties of non-electrolyte solutions. He found that solutions of salts, acids and bases – electrolytes – possess greater osmotic pressures, higher vapour pressures and lower freezing points than van't Hoff's calculated values but explained the discrepancies in terms of ionic dissociation by taking into account the number of solute ions (as opposed to molecules) present. In 1889 Arrhenius suggested that a molecule will take part in a chemical reaction on collision only if it has a higher than average energy – that is, if it is activated. As a result, the rate of a chemical reaction is proportional to the number of activated molecules (not to the total number of molecules, or concentration) and can be related to the activation energy.

After 1905 Arrhenius widened his research activities. For example, he applied the laws of theoretical chemistry to physiological problems (particularly immunology); once again initial criticism was replaced by universal acceptance. With N. Ekholm he published papers on cosmic physics concerning the Northern Lights, the transport of living matter ('spores') through space from one planet to another, and the climatic changes of the Earth over geological time – pointing out the 'greenhouse effect' brought about by carbon dioxide in the atmosphere. Arrhenius became more and more respected by the world of science and was much sought after for meetings, lectures and discussions throughout the world. In his latter years he had to rise at 4 a.m. in order to maintain his scientific activities, and this consistent hard work probably contributed to his death at the age of 68.

Artin Emil 1898–1962 was an Austrian mathematician who made important contributions to the development of class field theory and the theory of hyper-complex numbers.

He was born in Vienna on 3 March 1898, but his secondary education was at Reichenburg in Bohemia (now part of Czechoslovakia). His undergraduate study at the University of Vienna was interrupted when he was called up for military service in World War I; in 1919 he went to the University of Leipzig, where he received his PhD in 1921. From 1923 to 1937 he lectured at the University of Hamburg in mathematics, mechanics and the theory of relativity. He and his family emigrated to the United States in 1937. There Artin lectured at the University of Indiana (1938–1946) and Princeton (1946–1958). In 1958 he returned to Hamburg, where he died on 20 December 1962.

Artin's early work was concentrated on the analytical and arithmetical theory of quadratic number fields and in the 1920s he made a number of major advances in this field. In his doctoral thesis of 1921 he formulated the analogue of the Riemann hypothesis about the zeros of the classical zeta function, studying the quadratic extension of the field of rational functions of one variable over finite constant fields, by applying the arithmetical and analytical theory of quadratic numbers over the field of natural numbers. Then in 1923, in the most important discovery of his

career, he derived a functional equation for his new-type L-series. The proof of this he published in 1927, thereby providing, by the use of the theory of formal real fields, the solution to Hilbert's problem of definite functions. The proof produced the general law of reciprocity – Artin's phrase – which included all previously known laws of reciprocity going back to Gauss and which became the fundamental theorem in class field theory. Between the statement of his theory in 1923 and its publication in 1927, Artin made two other important theoretical advances. His theory of braids, given in 1925, was a major contribution to the study of nodes in three-dimensional space. A year later, in collaboration with Schrier, he succeeded in treating real algebra in an abstract manner, defining a field as real-closed if it itself was real but none of its algebraic extensions were. He was then able to demonstrate that a real-closed field could be ordered in an exact manner and that, in it, typical laws of algebra were valid.

Although the fires of his genius burned less brightly after 1930, Artin continued to work at a high level. In 1944 his discovery of rings with minimum conditions for right ideals – now known as Artin rings – was a fertile addition to the theory of associative ring algebras, and in 1961 he published his *Class Field Theory*, a rounded summation of his life's work as one of the leading creators of modern algebra.

Aston Francis William 1877–1945 was a British chemist and physicist who developed the mass spectrograph, which he used to study atomic masses and to establish the existence of isotopes. For his unique contribution to analytic chemistry and the study of atomic theory he was awarded the 1922 Nobel Prize for Chemistry.

Aston was born on 1 September 1877 at Harbourne, Birmingham, the son of a merchant. He went to school at Malvern College, where he excelled at mathematics and science, and then to Mason College (which later became the University of Birmingham) to study chemistry. There from 1898 to 1900 he studied optical rotation with P. F. Frankland (1858–1946). Aston then left academic life for three years to work with a firm of brewers, although during his spare time he continued to experiment with discharge tubes. He returned to Mason College in 1903 to study gaseous discharges, before moving in 1909 to the Cavendish Laboratory, Cambridge, where J. J. Thomson (1856–1940) was also investigating positive rays from discharge tubes. Thomson and Aston examined the effects of electric and magnetic fields on positive rays, showing that the rays were deflected – one of the basic principles of the mass spectrograph.

Aston's researches were interrupted by World War I, during which he worked at the Royal Aircraft Establishment, Farnborough, on the treatment of aeroplane fabrics using dopes (lacquers). He escaped injury in 1914 after crashing in an experimental aircraft. After the war he returned to Cambridge and improved his earlier equipment, and the mass spectrograph was born. He went on to refine the instrument and apply it to a study of atomic masses and isotopes. Aston continued to live and work in Cambridge, where he died on 20 November 1945.

Between 1910 and 1913, Thomson and Aston showed that the amount of deflection of positive rays in electric and magnetic fields depends on their mass. The deflected rays were made to reveal their positions by aiming them at a photographic plate. Thomson was seeking evidence for an earlier theory of William Crookes that the nonintegral atomic mass of neon (20.2) was caused by the presence of two very similar but different atoms. Each was expected to have a whole number atomic mass, but their mixture would result in an 'average' nonintegral value. The Cambridge scientists found that the two paths of deflected positive rays from a neon discharge tube were consistent with atomic masses of 20 and 22. Aston attempted to fractionate neon and in 1913 made a partial separation by repeatedly diffusing the gas through porous pipeclay.

Aston's first improvement to the apparatus, made in 1919, caused the positive-ray deflections by both electric and magnetic fields to be in the same plane. The image produced on the photographic plate became known as a mass spectrum, and the instrument itself as a mass spectrograph.

Its principle is relatively simple. A beam of positive ions is produced by an electric discharge tube (in which a high voltage is passed between electrodes in a glass tube containing rarefied gas), which has holes in its cathode to let the accelerated ions pass through. The beam passes between a pair of electrically charged plates, whose electric field deflects the moving ions according to their charge-to-mass ratio, e/m (where e is the charge on the ion – usually 1 or 2 – and m is its mass). Lighter ions are deflected most, whereas those of largest mass are deflected least. The now separate ion streams then pass through a magnetic field arranged at right angles to the electric field, which deflects them still further in the same plane. The streams strike a photographic plate, where they expose a series of lines that constitute the mass spectrum. The position of a line depends on the ion's mass, and its intensity depends on the relative abundance of that ion in the original beam.

The work with neon established that two spectral lines are produced on the plate, one about nine times darker (on a positive print) than the other. Calculation showed that these correspond to two types of ions of atomic masses 20 and 22. There are nine times as many of the former as of the latter, giving a weighted average atomic mass of about 20.2 (the value originally reported in 1898 by William Ramsay and Morris Travers, the discoverers of neon). Aston stated that there must be two kinds of neon atoms which differ in mass but not in chemical properties, i.e. that naturally occurring neon gas consists of two isotopes.

Over the next few years Aston examined the isotopic composition of more than 50 elements. Most were found to have isotopes – tin has ten with atomic masses that are whole numbers (integers). In 1920, using the first mass spectrograph, he determined the mass of a hydrogen atom and found it to be 1% greater than a whole number (1.01). (Twelve years later in the United States Harold Urey discovered deuterium, an isotope of hydrogen with mass 2.)

With an improved spectrograph, accurate to 1 part in 10,000, Aston confirmed that some other isotopes also show small deviations from the whole-number rule. The slight discrepancy is the packing fraction.

Aston's interests also included astronomy, particularly observations of the Sun and its eclipses. His knowledge of photography made him a valuable member of the expeditions that studied eclipses in Sumatra (1925), Canada (1932) and Japan (1936). But Aston will be remembered for his development of the mass spectrograph, which became an essential tool in the study of nuclear physics and later found application in the determination of the structures of organic compounds.

Atkinson Robert D'escourt 1898–1982 was an English astronomer and inventor.

Atkinson was born on 11 April 1898 at Rhayader in Wales. He was educated at Oxford University, where he obtained a BA in 1922, and then at the University of Göttingen in Germany, where he gained a PhD in 1928. He was a demonstrator in physics at the Clarendon Laboratory at Oxford from 1922 to 1926 and an assistant at the Technical University in Berlin from 1928 to 1929. He became Assistant Professor at Rutgers University in New Jersey (1922–34) and then Associate Professor (1964–73) and later Adjunct Professor of Astronomy there.

Atkinson was a member of the Harvard University/Massachusetts Institute of Technology eclipse expedition to the Soviet Union in 1936, and during World War II he was with the mine design department of the British Admiralty. From 1944 to 1946 he served with Ballistic Research Laboratory in Maryland.

Between 1952 and 1955 Atkinson designed the astronomical clock at York Minster in England, and he designed a standard time sundial at Indiana University in 1977. He was a member of the British National Committee for Astronomy (1960–62), the American Physical Society, the American Astronomical Society, the Royal Astronomical Society, the British Astronomical Association and the Royal Institute of Navigation. He was awarded the Royal Commission Award to Inventors in 1948 and the Eddington Medal of the Royal Astronomical Society in 1960. In 1977 the International Astronomical Union named a minor planet (1,827 Atkinson) in his honour.

Atkinson's research was in the field of atomic synthesis, stellar energy and positional astronomy. He was also deeply involved in instrument design.

Many scientists had been concerned with the problem of discovering how the Sun has maintained a reasonably steady yet high rate of radiation for at least three billion years. The problem essentially was that there was no known physical or chemical process that could generate radiation from the materials that make up the Sun at so great a rate over so long a period of time, nor was there enough energy released by the contraction of the Sun under its own gravitation. Astronomers therefore began to look elsewhere for a process that could explain the mystery. In 1924 Arthur Eddington, whose field of study was the internal make-up of stars, was computing what conditions must be like beneath the surface of stars in order that the basic laws of physics be obeyed. Eddington suggested that the only possibility was a process whereby atoms were broken down inside the central core of a star, converting matter into energy. He was supported in this view by Atkinson, who in 1932 was working at the Royal Greenwich Observatory. In that year there were new results from the physicists Ernest Rutherford, John Cockcroft and Ernest Walton, who had just succeeded in splitting the central core, or nucleus, of an atom. Atkinson was able to work out a theoretical model of the way in which matter could be annihilated. Not only did he determine the amount of energy released from atomic reactions within stars, but he was also able to suggest the kinds of reactions necessary to produce the vast quantities of radiation required.

With an enormous amount of research into nuclear physics carried out since World War II, astronomers now have a much greater insight into how stars evolve and how long their evolution takes. There are still many questions to be answered, but it is now thought that energy is generated in the Sun by a process of nuclear fusion. The nucleus of an atom fuses with the nuclei of other atoms, forming a new and heavier atom and at the same time releasing a vast amount of energy. Calculations have shown that this energy is more than enough to keep a star like the Sun radiating for billions of years.

Robert Atkinson's contributions were fundamental to our basic understanding of how stars like the Sun work and how they evolve.

Attenborough David Frederick 1926– is a British naturalist, film-maker and author who is best known for his wildlife films, which have brought natural history to a wide general audience.

Attenborough was born on 8 May 1926. He was educated at Wyggeston Grammar School in Leicester and Clare College, Cambridge, where he read zoology. From 1947 to 1949 he did two years' military service in the Royal Navy, after which he became an editorial assistant in an educational publishers. In 1952 he joined the BBC Television Service as a trainee producer, then in 1954 he went on his first expedition, to West Africa. During the next ten years he made annual trips to film and study wildlife and human cultures in remote parts of the world; these expeditions were recorded in the *Zoo Quest* series of television programmes and books. From 1965 to 1968 Attenborough was Controller of BBC2 and of the BBC Television Service, then from 1969 to 1972 he was Director of Television Programmes for the BBC and a member of its board of management. Despite the large amount of administrative work involved in these posts, he still managed to undertake several filming expeditions. His next major achievement was the television series *Life on Earth*, which was first shown in 1979. In this huge project, which took three years to complete, Attenborough attempted to outline the development of life on Earth – from its very beginnings to the present day – using plants and animals found today to illustrate this evolution. The series and its associated book (which also first appeared in 1979) met with great popular and critical acclaim and set new standards for the presentation of natural history to nonspecialists. He later made two further, equally successful television series: *The Living Planet* (1983), which dealt with ecology and the environment, and *The Trials of Life* (1990), which described life cycles. He was knighted in 1985.

Audubon John James Laforest 1785–1851 was a French-born American ornithologist who painted intricately detailed studies of birds and animals. He was also an ardent conservationist.

Audubon was born at Les Cayes, Santo Domingo (now Haiti), on 26 April 1785. He was the illegitimate son of a French sea captain and a Creole woman, who died soon after his birth. His father, who was also a planter, sent him back to France to his home near Nantes, where Audubon and his half-sister were taken in by the captain's wife, who had no children of her own. The couple legally adopted him in 1794. The young Audubon acquired a deep interest for natural history, painting and music. He was educated locally and in Paris, where he had six months' tuition at the studio of the well-known painter Jacques Louis David. By 1803 young men in France were being conscripted for Napoleon's army, but the 18-year-old Audubon avoided the draft by making a timely emigration to the United States to take up the running of his father's properties near Philadelphia. In 1808 he married and opened a store in Louisville, Kentucky, but he was a casual and poor businessman, his time being so passionately absorbed by nature; he was even imprisoned for debt in 1819 and declared a bankrupt.

Despite his financial problems, Audubon maintained a successful marriage and travelled throughout the United States collecting and painting the wildlife around him, while his wife worked as a teacher and governess to help to support them. He also painted portraits and even street signs, and gave lessons in drawing and French. By 1825 he had compiled his beautiful set of bird paintings, but American publishers were not interested. The following year Audubon set sail for England, where the Havells engraved his plates, which he then published by subscription. His talent was lauded and he attracted much publicity by appearing in English society wearing the outlandish clothes so suitable for his travels in the wilds of the United States. Even so, he was elected a Fellow of the Royal Society in 1830. After 13 years in Britain he went back to the United States. He died in New York on 27 January 1851.

Before Audubon most painters of birds used stylized techniques; stuffed birds were often used as subjects. Audubon painted from life and his compositions were startling, his detail minute. *The Birds of America* was published in Britain in 87 parts between 1827 and 1838. On his return to the United States in 1839 he published a bound edition of the plates with additions. He illustrated *Viviparous Quadrupeds of North America* (1845–1848), compiling the text (1846–1854) with his sons and John Bachman. Audubon was one of the earliest naturalists to pioneer conservation, and the various Audubon societies of today are named in his honour.

Auer Carl 1858–1929. Austrian chemist and engineer; *see* von Welsbach, Freiherr

Avery Oswald Theodore 1877–1955 was a Canadian-born US bacteriologist whose work on transformation in bacteria established that DNA (deoxyribonucleic acid) is responsible for the transmission of heritable characteristics. He also did pioneering research in immunology – again working with bacteria – proving that carbohydrates play an important part in immunity. Avery's achievements gained him many honours, including election to the National Academy of Science in the United States and to the Royal Society of London.

Avery was born on 21 October 1877 in Halifax, Nova Scotia, the son of a clergyman, but spent most of his life in New York City, where he was taken by his father in 1887. After qualifying in medicine in 1904 at Columbia University, Avery spent a brief period as a clinical physician but soon moved to the Hoagland Laboratory in Brooklyn in order to research and lecture in bacteriology and immunology. In 1913 he transferred to the Rockefeller Institute Hospital in New York, where he remained until he retired in 1948. Avery died in Nashville, Tennessee, having moved there on his retirement.

Avery's work on transformation – a process by which heritable characteristics of one species are incorporated into another species – is generally considered to be his most important contribution and was stimulated by the research of F. Griffith, who in 1928 published the results of his studies on *Diplococcus pneumoniae*, a species of bacterium that causes pneumonia in mice. Griffith found that mice contracted pneumonia and died when they were injected with a mixture of an encapsulated strain of dead *Diplococcus pneumoniae* (living encapsulated bacteria are lethal to mice) and living, unencapsulated bacteria (which have no protective outer capsule to resist antibodies and therefore do not cause pneumonia), despite the fact that, separately, each of the mixture's components is harmless. From the corpses he then isolated virulent, living, encapsulated bacteria. These findings led Griffith to postulate that a transforming principle from the dead, encapsulated bacteria had caused capsule development in the living, unencapsulated bacteria, thereby making them virulent. Moreover, when these living bacteria reproduced, the offspring were encapsulated, suggesting that the transforming principle had become incorporated into their genetic constitution.

Initially, Avery dismissed Griffith's findings but when they were supported by later studies, Avery and his colleagues Colin MacLeod and Maclyn McCarthy began investigating the nature of the transforming principle. They started experimenting in the early 1940s, working on *Diplococcus pneumoniae*. They obtained a pure sample of the virulent, living, encapsulated bacteria which were killed by heat treatment. The bacteria's protein and polysaccharide (which makes up the capsule and is also found within the cells) were then removed and the remaining portion was added to living, unencapsulated pneumococcus. It was found that the progeny of these bacteria had capsules, so the active transforming principle still remained and was neither a protein nor a polysaccharide. Finally Avery extracted and purified the transforming principle and used various chemical, physical and biological techniques to identify it. His analysis proved conclusively that DNA was the transforming principle responsible for the development of polysaccharide capsules in the unencapsulated bacteria.

Avery's discovery (which he published in 1944) was extremely important because for the first time it had been proved that DNA controls the development of a cellular feature – in this case the polysaccharide capsule – and implicated DNA as the basic genetic material of the cell. Other researchers later confirmed that DNA controls the development of cellular features in different organisms, and also established that it is the fundamental molecule involved in heredity. Moreover, Avery's work stimulated interest in DNA, eventually leading to the determination of its structure and method of replication by Francis Crick and James Watson in the early 1950s.

Avery's early work also involved pneumococci, but was in the field of immunology. He demonstrated that pneumococci bacteria could be classified according to their immunological response to specific antibodies and that this immunological specificity is due to the particular polysaccharides that constitute the capsule of each bacterial type. This research established that polysaccharides play an important part in immunity and led to the development of sensitive diagnostic tests to identify the various types of pneumococcus bacteria.

Avogadro Amedeo, Conte de Quaregna 1776–1856 was an Italian scientist who shares with his contemporary Claud Berthollet (1748–1822) the honour of being one of the founders of physical chemistry. Although he was a professor of physics, he acknowledged no boundary between physics and chemistry and based most of his findings on a mathematical approach. Principally remembered for the hypothesis subsequently known as Avogadro's law (which states that, at a given temperature, equal volumes of all gases contain the same number of molecules), he gained no recognition for his achievement during his lifetime. He lived in what was a scientific backwater, with the result that his writings received scant examination or regard from the leading authorities of his day.

Avogadro was born in Turin on 9 June 1776. He began his career in 1796 by obtaining a doctorate in law and for the next three years practised as a lawyer. In 1800 he began to take private lessons in mathematics and physics, made impressive progress and decided to make the natural sciences his vocation. He was appointed as a demonstrator at the Academy of Turin in 1806 and Professor of Natural Philosophy at the College of Vercelli in 1809, and when in 1820 the first professorship in mathematical physics in

Italy was established at Turin, Avogadro was chosen for the post. Because of the political turmoil at that time the position was subsequently abolished, but calmer times permitted its re-establishment in 1832 and two years later Avogadro again held the appointment. He remained at Turin until his retirement in 1850. When he died there on 9 July 1856 his European contemporaries still regarded him as an incorrigibly self-deluding provincial.

In 1809 Joseph Gay-Lussac had discovered that all gases, when subjected to an equal rise in temperature, expand by the same amount. Avogadro therefore deduced (and announced in 1811) that at a given temperature all gases must contain the same number of particles per unit volume. He also made it clear that the gas particles need not be individual atoms but might consist of molecules, the term he introduced to describe combinations of atoms. No previous scientists had made this fundamental distinction between the atoms of a substance and its molecules.

Using his hypothesis Avogadro provided the theoretical explanation of Gay-Lussac's law of combining volumes. It had already been observed that the electrolysis of water (to form hydrogen and oxygen) produces twice as much hydrogen (by volume) as oxygen. He reasoned that each molecule of water must contain hydrogen and oxygen atoms in the proportion of 2 to 1. Also, because the oxygen gas collected weighs eight times as much as the hydrogen, oxygen atoms must be 16 times as heavy as hydrogen atoms. It also follows from Avogadro's hypothesis that a molar volume of any substance (i.e. the volume whose mass is one gram molecular weight) contains the same number of molecules. This quantity, now known as Avogadro's number or constant, is equal to 6.02252×10^{23}.

Leading chemists of the day paid little attention to Avogadro's hypothesis, with the result that the confusion between atoms and molecules and between atomic weights and molecular weights continued for nearly 50 years. In 1858, only two years after Avogadro's death, his fellow Italian Stanislao Cannizzaro showed how the application of Avogadro's hypothesis could solve many of the major problems in chemistry. At the Karlsruhe Chemical Congress of 1860 Avogadro's 1811 paper was read again to a much wider and more receptive audience of distinguished scientists. One of the most impressed was the young German chemist Julius Lothar Meyer. He found this final establishment of order in place of conflicting theories one of the great stimuli that eventually led him in 1870 to produce his most detailed exposition of the periodic law. A year later his namesake Viktor Meyer used Avogadro's law as his principal yardstick in theoretically explaining the nature of vapour density.

It is interesting to analyse why such a fundamental and potentially useful work as Avogadro's lay fallow for nearly half a century. Various factors contributed to the delay. To begin with, Avogadro did not support his hypothesis with an impressive display of experimental results. He never acquired, nor did he deserve, a reputation for accurate experimental work; his contemporaries did not therefore regard him as a brilliant theoretician, merely as a careless experimenter. Also Avogadro extended his hypothesis to solid elements – and lacking experimental evidence he relied on analogy. So that whereas he was correct in considering molecules of oxygen and hydrogen to be diatomic, he had little justification for making a similar assumption about carbon and sulphur. His speculative treatment of metals in the vapour state (in his second paper of 1814) did little to advance his cause, revealing an excess of theorizing at the cost of attention to detail.

Furthermore, Avogadro's idea of a diatomic molecule was at odds with the dominant dualistic outlook of Jöns Berzelius. According to the principles of electrochemistry, two atoms of the same element would have similar electric charges and therefore repel rather than attract each other (to form a molecule). During the 50 years after Avogadro's original hypothesis most activity was being devoted to organic chemistry, whose analysis and classification was based chiefly on weights, not volumes. And even when Avogadro's work was translated and published, it tended to appear in obscure journals, perhaps as a result of his modesty and his geographical isolation from the mainstream of the chemistry of his time.

Ayrton William Edward 1847–1908 was a British physicist and electrical engineer who invented many of the prototypes of modern electrical measuring instruments. He also created in 1873 the world's first laboratory for teaching applied electricity in Tokyo.

Ayrton was born in London on 14 September 1847, the son of a barrister. He was educated at University College School from 1859 to 1864, when he entered University College, London. In 1867 he obtained an honours degree in mathematics and joined the Indian Telegraph Service. He was sent to Glasgow to study electricity under William Thomson (later Lord Kelvin) and, after practical study at the works of the Telegraph and Maintenance Company, he went to Bombay in 1868 as an assistant superintendent.

In 1872 he returned to England and was placed in charge of the Great Western telegraph factory under Thomson and Fleeming Jenkin. A year later he accepted the chair of Physics and Telegraphy at the new Imperial Engineering College in Tokyo – founded by the Japanese government and at that time the world's largest technical university. There he created the first laboratory in the world for teaching applied electricity.

In 1878 Ayrton returned to England and in 1879 he became Professor at the City and Guilds of the London Institute for the advancement of technical education. His first class consisted of one man and a boy. From 1881 to 1884 he acted as Professor of Applied Physics at the new Finsbury Technical College and became the first Professor of Physics and Electrical Engineering at the new Central Technical College – now the City and Guilds College, South Kensington. He held this post until he died in London on 8 November 1908.

In 1881 Ayrton and John Perry invented the surface-contact system for electric railways, which, together with Fleeming Jenkin, they applied to 'telpherage' (a system of overhead transport) and a line based on this system was installed at Glynde in Sussex. In that year (1882) they also brought out the first electric tricycle.

There followed a whole series of new, portable electrical measuring instruments including the ammeter (so-named by its inventors), an electric power meter, various forms of improved voltmeters, and an instrument used for measuring self and mutual induction. In this, great use was made of an ingeniously devised flat spiral spring which yields a relatively large rotation for a small axial elongation. These instruments served as the prototypes for the many electrical measuring instruments which came into use in countries all over the world as electrical power became generally employed for domestic and industrial purposes. Ayrton's instruments gave the electrical engineer the means of measuring almost every electrical quantity he had to deal with, and his electric meter was the only one to be awarded prizes at the Paris Exhibition in 1899.

Besides his contribution to the advancement of the practical aspect of electrical engineering, Ayrton was also a great teacher of the subject. His system of teaching was adopted and extended throughout the profession. He published many scientific papers and a book entitled *Practical Electricity*.

His wife Bertha also achieved renown for her researches into the electric arc and she had the distinction of becoming the first woman member of the Institute of Electrical Engineers.

B

Baade Walter 1893–1960 was a German-born US astronomer who is known for his discovery of stellar populations and whose research proved that the observable universe is larger than originally believed.

Baade was born in Shröttinghausen on 24 March 1893, the son of a schoolteacher. He studied at Münster and at Göttingen Universities, obtaining a PhD from the latter in 1919. For the next 11 years he worked at Hamburg University in the Bergedorf Observatory. In 1931 he emigrated to America and joined the staff of the Mount Wilson Observatory, Pasadena. He left in 1948 and went to the nearby Mount Palomar Observatory where he worked until 1958, when he returned to Germany. The following year he became Gauss Professor at Göttingen University. He died there on 25 June 1960.

The important contributions to astronomy that Baade made were numerous. In 1920 he discovered the most distant known asteroid (minor planet), Hidalgo, whose orbit goes out as far as that of Saturn. In 1948 he found the innermost asteroid, Icarus, whose orbit comes within 18 million miles of the Sun, even closer than Mercury. At Mount Wilson Baade worked with Fritz Zwicky and Edwin Hubble on supernovae and galactic distances.

During the wartime blackout of Los Angeles, in 1943, Baade made his most important discovery. He made use of the enforced darkness to study the Andromeda galaxy with a 2.5-m/8.2-ft reflecting telescope. Until that time, Hubble had managed to view only the bright blue giant stars in the spiral arms of the galaxy, and a bright haze in its centre. Baade was able to observe, for the first time, some of the stars in the inner regions of the galaxy and he found that the most luminous stars towards the centre are not blue-white but reddish. He proposed that there exist two groups of stars with differing structures and origins. The bluish stars on the edge of the galaxy, called Population I stars, were distinguished from the reddish ones in the inner regions, Population II. Population I stars are young and formed from the dusty material of the spiral arms – hydrogen, helium and heavier elements; Population II stars are old and were created near the nucleus and contain fewer heavy elements.

After the War, the 5-m/200-in reflecting telescope was built at Mount Wilson and Baade continued his research using this instrument. He found that both stellar populations contain Cepheid variable stars (of which there are more than 300 in Andromeda). He also discovered that the period–luminosity curve established for Cepheid variables by Harlow Shapley and Henrietta Leavitt applied only to Population II Cepheids. This discovery meant that two types of Cepheids existed.

Baade's findings had even greater implications. In the 1920s the distances of the outer galaxies had been calculated by Hubble using light curves for Population I Cepheids. Baade decided that these calculations were incorrect and redrew a period–luminosity curve for Population I Cepheids, revealing that they are much brighter than had been previously thought. His discovery showed that the dim bluish-white Cepheids seen in the spiral arms of Andromeda were much farther away than believed and that Andromeda was not 800,000, but more than 2 million, light years distant. Also, since Hubble had used his incorrect distance of Andromeda to gauge the size of the universe, the new findings meant that the universe and all the extragalactic distances in it were at least double the size previously calculated. The increased distance of the Andromeda galaxy and others meant that to appear so bright they must be larger than had been thought. Astronomers realized that our Galaxy is smaller than Andromeda and not the pre-eminent galaxy they imagined.

The enlarged scale of the universe stimulated the construction of far-reaching radio telescopes. Astronomers had known that a strong radio source existed in the sky but it could not be located with the 5-m/16.4-ft telescope. With a radio telescope, however, Baade discovered the source to be a distorted galaxy colliding with another galaxy in the constellation Cygnus. The interstellar dust created by the collision and the resulting radio waves could be detected clearly even though they were 260 million light years away.

Baade contributed a great deal to our knowledge of the universe. His main interests were extragalactic nebulae as stellar systems but he also studied variable stars in our own

Galaxy, in globular clusters and in the Andromeda nebula, and by doing so stimulated interest in the theory of stellar interiors as the basis of theoretical interpretations of stellar evolution.

Babbage Charles 1792–1871 was one of the greatest pioneers of mechanical computation.

Babbage was born at Totnes, in Devonshire, on 26 December 1792. His father was a banker who left him a large inheritance, and throughout his life Babbage was financially secure. He developed an interest in mathematics as a young boy and in 1810 went to Cambridge University to study mathematics. There he became a close friend of the future astronomer, John Herschel (1792-1871), and, convincing himself that Herschel would be placed Senior Wrangler in the honours examinations, chose not to be placed second and took only a pass degree, in 1814. While at Cambridge Babbage, Herschel and other undergraduates founded the Analytical Society, and it was in the society's rooms one evening, that Babbage is recorded as saying (while looking over some error-filled logarithm tables) 'I am thinking that all these mathematical tables might be calculated by machinery'.

The year after he left Cambridge Babbage wrote three papers on 'The calculus of functions' for publication by the Royal Society and in the following year, 1816, he was elected a Fellow of the Society. Some time later, while making a tour of France, he examined the famous French logarithms which had been recently calculated and which were the most accurate tables then known. They had, however, required the combined efforts of nearly 100 clerks and mathematicians. Mathematical tables of all kinds, including logarithmic tables, were of great use for astronomical, commercial and, especially, navigational purposes. On his return to London, therefore, Babbage set about developing his ideas for a cheaper and more accurate method of producing tables by mechanical computation and automatic printing.

By 1822 he had ready a small calculating machine able to compute squares and the values of quadratic functions. It worked on the method of differences, an example of which involves subtracting one square from the preceding one to obtain a first difference, and then the first difference from the next above it to obtain a second difference, which is always 2.

By working backwards, adding the second difference to the first difference, and then adding the higher of the squares used in the subtraction which produced the first difference, the next square is always obtained. Using second-order differences of this kind Babbage's first machine could produce figures to six places of decimals.

With the backing of the Royal Society, Babbage was able to persuade the government to support his work to devise a much larger difference engine to calculate navigational and other tables. He was elected to the Lucasian Chair of Mathematics at Cambridge in 1826, a post which he held until 1835, but he continued to live in London and did not perform the usual professorial duties of lecturing

and teaching. He devoted himself entirely to his machine, which he proposed to make work to sixth-order differences and 20 places of decimals. The construction of the large difference engine was a laborious and lengthy operation. New tools had to be designed to previously unknown tolerances, many of them being made for Babbage by the great pioneer of precision engineering, Joseph Whitworth (1803-1887). The project cost the government £20,000, but it was eventually abandoned, partly because of problems of friction (the arch-enemy of the engineer), partly because of personality clashes, but chiefly because before it was completed Babbage had hit upon a better idea.

That idea was the analytical engine. The difference engine could perform only one function, once it was set up. The analytical engine was intended to perform many functions; it was to store numbers and be capable of working to a programme. In order to achieve this, Babbage borrowed the idea of using punched cards from Joseph Jacquard, who had invented such cards in 1801 to program carpet-making looms to weave a pattern. Babbage's machine, begun in 1833, was to have a mill to carry out arithmetical operations, a memory unit to store 1,000 numbers of 50 digits, and the program cards, linked together to direct the machine. The cards were of three kinds: those to supply the store with numbers, those to transfer numbers from mill to store or store to mill, and those to direct the four basic arithmetical operations. In order to unite the programs for his cards Babbage devised a new mathematical notation.

Babbage's machine had not been completed by the time of his death, on 18 October 1871. And although his son H. P. Babbage carried on the enterprise between 1880 and 1910, the fact is that its complexity was beyond the engineering expertise of the day and its very conception beyond the grasp of society to spend the amount of time and money required to build it. Nevertheless, it was Babbage's machine – although, being decimal, not binary, and requiring the use of wheels, not strictly digital in the modern sense – which Howard Aiken used as the basis for his development of the Harvard Mark I Calculator.

(In 1991, the Science Museum, London, completed Babbage's second difference engine and found that it could evaluate polynomials up to the seventh power with a 30-digit accuracy.)

Babbage was a true representative of the Victorian age, the age of the steam-engine and the Industrial Revolution. He wished to harness science to the practical improvement of society. Hence, dissatisfied by the loftiness of the Royal Society, he had a hand in founding the Astronomical Society (1820), the British Association for the Advancement of Science (1831) and the Statistical Society of London (1834). The same passion for improvement led him to investigate the operation of the Post Office and, on finding that most of its costs derived from the handling of letters, not their transport, to recommend (what Rowland Hill introduced as the penny post) that it should simplify its procedure by introducing a single rate. But his greatest idea, the mechanical computation of tables, remained ahead of the practical possibilities of his time.

Babcock George Herman 1832–1893 was the US co-inventor of the first polychromatic printing press, but is chiefly remembered for the Babcock–Wilcox steam boiler, devised with his partner, Stephen Wilcox.

Born at Unadilla Forks, near Otego, New York, on 1 June 1832 Babcock inherited his engineering and mechanical expertise from both sides of his family. When he was 12, the Babcock family moved to Westerly, Rhode Island.

Babcock first went to work with his father, Asher, in daguerrotype and job printing for newpapers. It was during this period up to 1854, that Babcock and his father invented the polychromatic printing press. The father-and-son team also invented a job printing press which is still manufactured today.

Moving to Brooklyn, New York, in 1860, Babcock held many posts – first in the offices of a patents solicitor then with the Mystic Iron Works and afterwards as chief draughtsman at the Hope Iron Works, Providence, Rhode Island.

At Providence, Babcock and Stephen Wilcox met again, and became involved in steam engineering. Together they tackled the contemporary problems of steam boilers. Since their introduction in the 18th century fire-tube boilers had been bedevilled by structural problems. Explosions – often fatal – were a constant danger when large quantities of water and steam were contained in boilers not built to withstand the strains imposed upon them. In the 19th century the growing use of engines operating at ever higher pressures increased the incidence of such disasters.

The water-tube boiler was developed as a safer replacement and it was to the improvement of this early model that Babcock and Wilcox applied themselves. Their design for a sectionally headed boiler was one of the earliest with automatic cut-off being based on a safety water tube patented by Wilcox in 1856. This was the first engine to have front and rear water spaces connected by slanting tubes, the steam space being situated above. The Babcock–Wilcox boiler patented in 1867 used cast-iron steam generating tubes placed in vertical rows over a grate. Steel or wrought-iron tubes connected them with headers leading to the separating drum where the water was separated from the steam and recirculated.

The new boiler, considerably more powerful than its predecessors, was able to withstand very high pressures and also ensured a high standard of protection against explosions. It was first manufactured at Providence and then in New York, where the firm of Babcock and Wilcox was incorporated in 1881. The boilers were also built in New Jersey, at Elizabethport, and at a specially designed plant at Bayonne.

Patent restrictions on the design of steam boilers made the Babcock–Wilcox model expensive, and it fell behind the competition when the patents expired and the cheap engines invaded the market. The high quality of the boiler, however, ensured its continuation and the design was still being used 70 years after its inception. The firm of Babcock and Wilcox still exists, and today manufactures modern high quality steam boilers.

Babcock actively promoted his company's product by lecturing at technical institutes and colleges, an activity for which his impressive presence and personality made him especially suitable. In 1887, Babcock was president of the American Society of Mechanical Engineers, and was also seen frequently at meetings of technical societies. A sincerely religious man, a member of the Seventh Day Baptists, Babcock took great personal interest in the welfare of his employees and in his public duties: these included the presidencies of the Board of Trustees of Alfred University and, at his home of Plainfield, New Jersey, of the Board Education and the Public Library.

Babcock was alert and energetic to the end of his life. He died on 16 December 1893.

Babcock Harold Delos 1882–1968 was a US astronomer and physicist whose most important contributions were to spectroscopy and the study of solar magnetism.

Babcock was born on 24 January 1882 in Edgerton, Wisconsin. Much of his schooling was done at home, but in 1901 he enrolled at the College of Electrical Engineering of the University of California in Berkeley, and he was awarded a BA in 1907. By that time he had already taken up employment at the National Bureau of Standards. In 1908 Babcock taught a course in physics at the University of California. A year later George Hale invited him to work at the Mount Wilson Observatory.

The only breaks in Babcock's service at Mount Wilson between 1909 and 1948 came during the two World Wars. During World War I Babcock worked for the Research Information Service of the National Research Council; during World War II he served as a consultant on several programmes, including the Manhattan Project. In 1933 he was elected to the National Academy of Sciences. Babcock retired formally in 1948, but he remained active in the supervision of the ruling engine for the 508-cm/200-in Hale telescope. He died on 8 April 1968.

As a child, Babcock had been interested in electricity, radio and photography. In his first job, at the National Bureau of Standards, he investigated the problems concerning electrical resistance. His early astronomical work was in stellar photography, as part of an international research programme on the structure of the Galaxy being coordinated by Jacobus Kapteyn. Babcock also collaborated with Walter Adams in his spectroscopic studies.

The Sun was always a subject of Babcock's particular interest. He made an investigation of the Zeeman effect (whereby a magnetic field causes a substance's spectral lines to be split) in chromium and vanadium – important elements in the solar spectrum. Babcock's next major concern was the establishment of a standard spectrum for iron. This study was part of a large programme to determine standards for astronomical spectra.

During the 1920s much of Babcock's research dealt with the production of a revised table of wavelengths for the solar spectrum. In 1928 he published a list which included 22,000 spectral lines, and this list was extended in 1947 and again in 1948.

Babcock was also the director of a project aimed at devising an engine that could reliably and accurately rule gratings for the new telescope at Mount Wilson. The superior Babcock gratings that were eventually produced were installed in all of the observatory's spectrographs, including that of the 508-cm/200-in telescope.

There had been considerable interest among astronomers in the measurement of the Sun's magnetic field, although no reliable information had been obtained by 1938, when Babcock turned to the problem. He had little success until 1948 when, in collaboration with his son H. W. Babcock, the solar magnetic field was measured. They used an instrument of their own design which was dubbed the 'solar magnometer' and which exploited the Zeeman effect to produce a continuously changing record of the Sun's local magnetic fields. These surface fields are only weak, but they could be observed satisfactorily with the new instrument. They also studied the Sun's general magnetic field and the relationship between sunspots and local magnetic fields.

Babcock was a skilled observational astronomer with a flair for the more practical side of his subject. His contributions to solar spectroscopy and magnetism were original and thorough.

Bacon Francis, Baron Verulam, Viscount St Albans 1561–1626 was a distinguished English statesman and philosopher of science.

Son of Sir Nicholas Bacon, courtier and Keeper of the Great Seal, Francis Bacon first attended Trinity College, Cambridge, and was trained in the law with a view to following the same path as his father. Politically unsuccessful under Elizabeth I, he finally gained a succession of offices under James I, eventually becoming Lord High Chancellor in 1618. That same year, he confessed to taking bribes, was fined £40,000 (which was later remitted by the king), and spent four days in the Tower of London. Thereafter he was able to devote himself wholeheartedly to his scientific interests. (The Baconian theory, originated by James Willmot in 1785, suggesting that the works of Shakespeare were written by Bacon, is not taken seriously by scholars.)

In works like the *Advancement of Learning* 1605 and the *Novum Organum* 1620, Bacon advanced challenging views of scientific method. He fiercely criticized Aristotle and the deductive mode supposedly followed ever since under the mental tyranny of scholasticism. He instead promoted 'induction', laying emphasis on the exhaustive collection of empirical data and its rigorous processing via a logical mill until general causes and conclusions were almost mechanically ground out. Bacon deprecated dogmatic *a priori* reason and the wayward play of fancy: true science must be built on the solid foundation of fact. Though Bacon is barely remembered for his own scientific investigations, his ideas proved extraordinarily influential. His insistence upon the primacy of sense experience underpinned the commitment of a later generation of English scientists like Boyle and Newton to rejection of Continental rationalism, and suggested the Royal Society's 'experience first' motto

of *nullius in verba* (on the authority of no person). His opinion that scientific knowledge should be applied to utilitarian purposes ('for the glory of God and the relief of man's estate') appealed to the reforming minds of the Enlightenment. Until late in the nineteenth century, Lord Verulam and his faith in facts were regularly held aloft as the banner of English science.

Bacon Roger *c*.1220–*c*.1292 was an English philosopher and scientist who was among the first medieval scholars to realize and promote the value of experiment in reaching valid conclusions.

It is not certain when or where Bacon was born. One tradition says that he was born at Bisley in Gloucestershire, while another says he came from Ilchester in Somerset. His date of birth can only be guessed from his writing, in 1267, that he had learnt the alphabet 40 years before. Bacon was educated in philosophy and mathematics at Oxford, and then lectured at the Faculty of Arts in Paris from about 1241. After about 1247, he returned to Oxford where he was inspired by the English scholar Robert Grosseteste (*c*.1168–1253) to cultivate the 'new' branches of learning – languages, mathematics, optics, alchemy and astronomy. In about 1257, he entered the Franciscan Order and, while a friar, appealed to Pope Clement IV to allow sciences a higher status in the curriculum of university studies. At the pope's request, Bacon in 1266–67 wrote three volumes of an encyclopedia of all the known sciences: *Opus Majus, Opus Minus* and *Opus Tertium*. In about 1268, he published fragments of another encyclopedia – *Communia Naturalium* and *Communia Mathematica*, and in 1272, *Compendium Philosophiae*. Following the death of Pope Clement IV in 1268, Bacon lost papal protection and was imprisoned by the Franciscans at some time from 1277 to 1279, possibly for criticisms of the order's educational practices. Bacon languished in prison for as long as 15 years. His last work, on theology, was published in 1292. He is said to have died in the same year and was buried in the Franciscan church in Oxford.

Bacon in his writings states that he spent huge sums of money in acquiring 'secret' books, in constructing instruments and tables, training assistants and in getting to know knowledgeable people. The experiments he performed and the knowledge he acquired earned him the title of *doctor mirabilis*. However, little, if any, of his work was original. His main contribution was in optics, in which Robert Grosseteste acquainted Bacon with the work of Alhazen (*c*.965–1038). Bacon advocated the use of lenses as magnifying glasses to aid weak sight, and speculated that lenses might be used to make an instrument of great magnifying power – an anticipation of the telescope.

Bacon also described some of the properties of gunpowder in a volume written sometime before 1249, but there is no indication that he had any idea of its possible use as a propellant, only that it would explode and so could be used in warfare. He described diving apparatus, in a passage where he referred to 'instruments whereby men can walk on sea or river beds without danger to themselves' in 1240.

Bacon was an ardent supporter of calendar reform, suggesting the changes necessary to improve the calendar that were carried out by Pope Gregory XIII in 1582. He also promoted the use of latitude and longitude in map-making and, by estimating the distance of India from Spain, may have inspired Columbus's voyage to America, which was a search for a shorter westward route to India.

Bacon is sometimes referred to as one of the 'forerunners' of science, but investigation does not show his scientific contribution to be very great. His importance in the history of science is due to his insistence on the use of experiment to prove an argument, a method that was not to flower until some three centuries later. Bacon also referred to the laws of reflection and refraction as being 'natural laws' – that is, of universal application – a concept vital to the development of science.

Baekeland Leo Hendrik 1863–1944 was a Belgian-born US industrial chemist famous for the invention of Bakelite, the first commercially successful thermosetting plastic resin.

Baekeland was born in Ghent on 14 November 1863. He was a brilliant pupil at school and at the age of only 16 won a scholarship to the University of Ghent, from which he graduated in 1882. Two years later, aged only 21, he was awarded his doctorate after studying electrochemistry at Charlottenburg Polytechnic. In 1887 he became Professor of Physics and Chemistry at the University of Bruges, and in 1888 returned to Ghent as Assistant Professor of Chemistry. The next year he got married and then went on a tour of the United States for his honeymoon, with financial help from a travelling scholarship. He decided to settle in the United States and took a job as a photographic chemist, setting up as a consultant in his own laboratory in New York City in 1891.

He returned briefly to Europe in 1900 to study at the Technische Hochschule at Charlottenburg, near Berlin. His development of Bakelite came after he had gone back to the United States, and was announced in 1909 – the year in which he founded the General Bakelite Corporation, later to become part of the Union Carbide and Carbon Company. Baekeland continued his chemical researches and in 1924 was elected President of the American Chemical Society. He died in Beacon, New York, on 23 February 1944.

Baekeland's first chemical invention was a type of photographic printing paper, which he called Velox, which could be developed under artificial light. He began manufacturing it in 1893 at Yonkers, New York, and in 1899 George Eastman's Kodak Corporation bought the invention and the manufacturing company (the Nepera Chemical Co.) from Baekeland for 1 million dollars.

Johann von Baeyer had discovered the powdery resin formed by the condensation reaction between formaldehyde (methanal) and phenol in 1871, and in the early 1900s Baekeland began to investigate it as a possible substitute for shellac (a resin derived from secretions of the lac insect of southern Asia, used in varnishes, polishes, and leather dressings). He could find no solvent for the resin, but discovered that it can be produced in a hard, machinable form that can also be moulded by casting under heat and pressure (compression moulding). On initial heating the material melts (becomes plastic) and then sets extremely hard and will not melt on further heating. Bakelite is a good insulator, and soon found use in the manufacture of electrical fittings such as plugs and switches. The General Bakelite Corporation merged with two other companies in 1922 and seven years later was incorporated into the Union Carbide group.

Baeyer Johann Friedrich Wilhelm Adolf von 1835–1917 was a German organic chemist famous for developing methods of synthesis, the best known of which is his synthesis of the dye indigo. His major contribution to the science was acknowledged by the award of the 1905 Nobel Prize for Chemistry.

Baeyer was born in Berlin on 31 October 1835, the son of the Prussian general Johann Jacob von Baeyer, who later became head of the Berlin Geodetic Institute. He began his university career at Heidelberg in 1853 where he studied chemistry under Robert Bunsen and Friedrich Kekulé. He gained his PhD at Berlin in 1858 after working in the laboratory of August Hofmann, and two years later took an appointment as a teacher at a technical school in Berlin. He became Professor of Chemistry at the University of Strasbourg in 1872 and three years later was appointed to succeed Justus von Liebig as Professor of Chemistry at Munich, where he stayed for the rest of his career. He died at Starnberg, near Munich, on 20 August 1917.

Baeyer began his researches in the early 1860s with studies of uric acid, which led in 1863 to the discovery of barbituric acid, later to become the parent substance of a major class of hypnotic drugs.

In 1865 he turned his attention to dyes. His student Karl Graebe (1841–1927) synthesized alizarin in 1868 (at the same time as William Perkin), and in 1871 Baeyer discovered phenolphthalein and fluorescein. He also found the resinous condensation product of phenol and formaldehyde (methanal), which Leo Baekeland later developed into the first thermosetting plastic, Bakelite.

In 1883 Baeyer determined the structure of indigo by reducing it to indole using powdered zinc. He had already (1880) devised a method for its synthesis, which was more lengthy than the commercial method later used when synthetic indigo began to be manufactured in 1890. In 1888 he carried out the first synthesis of a terpene.

His work with ring compounds and the highly unstable polyacetylenes led him to consider the effects of carbon–carbon bond angles on the stability of organic compounds. He concluded that the more a bond is deformed away from the ideal tetrahedral angle, the more unstable it is; this is known as Baeyer's strain theory. It explains why rings with five or six atoms are much more common, and stable, than those with fewer or more atoms in the ring. He also noticed that the aromatic character of the six-carbon benzene and its analogues is lost on reduction and saturation of the carbon atoms.

Baily Francis 1774–1844 was a British astronomer who is best known for his discovery of the phenomenon called 'Baily's beads'.

Baily was born in Newbury, Berkshire, on 28 April 1774. He began a seven-year apprenticeship in 1788 with a firm of merchant bankers in London, but as soon as his apprenticeship ended he set out to explore unsettled parts of North America. On his return to England in 1798 he became a stockbroker, and he was very successful. Astronomy, however, took up an increasingly important part of his life. He was a founder (in 1820) and first vice-president of the Astronomical Society of London (later the Royal Astronomical Society); he was elected a Fellow of the Royal Society in 1821. He finally gave up his job as a stockbroker in 1825 and became a full-time astronomer. He was a member of numerous scientific bodies and received several distinguished awards, among them two gold medals from the Royal Astronomical Society. He died in London on 30 August 1844.

Baily began to publish his astronomical observations in 1811. He was the author of an accurate revised star catalogue in which he plotted the positions of nearly 3,000 stars. These positions were used for the determination of latitude, and for this work Baily was awarded the Astronomical Society's Gold Medal in 1827.

In 1836, on 15 May, Baily observed a total eclipse of the Sun from Scotland. He noticed that immediately before the Sun completely disappeared behind the Moon (and also just as it began to emerge from behind the Moon) light from the Sun appeared as a discontinuous line of brilliant spots. These 'spots' have been named 'Baily's beads' and are caused by sunlight showing through between the mountains on the Moon's horizon as it moves across the Sun's disc. Baily travelled to Italy in 1842 and was again able to see his 'beads' during a solar eclipse.

Baily did other research, including a redetermination of the mean density of the Earth using the methods of Henry Cavendish. He also measured the Earth's elliptical shape. He earned his second gold medal from the Astronomical Society for these studies.

Baily's sighting was not the first of the 'bead' phenomenon, but his description of it and of the rest of the 1836 eclipse was so exciting that it sparked greatly renewed interest in eclipses, which persists to this day.

Bainbridge Kenneth Tompkins 1904– is a US physicist best known for his work on the development of the mass spectrometer.

Bainbridge was born in Cooperstown, New York on 27 July 1904. He was educated at the Massachusetts Institute of Technology and at Princeton University, where he gained his MA in 1927 and a PhD in 1929. After working at the Bartol Research Foundation and at the Cavendish Laboratory in Cambridge, England, he became Assistant Professor of Physics at Harvard University in 1934, Associate Professor in 1938 and Professor from 1946 onwards. He has been Professor Emeritus since 1975. Bainbridge has held a number of important concurrent appointments. He

worked in the radiation laboratory at MIT from 1940 to 1943 and then in the Los Alamos laboratory until 1945, when he directed the first atomic bomb test. He was awarded the Presidential Certificate of Merit for his work on radar in 1948.

The mass spectrometer or mass spectrograph was invented by Francis Aston (1877–1945) in 1919. In it, a beam of ions is deflected by electric and magnetic fields so that ions of the same mass are brought to a focus at the same point, enabling the masses of the ions in the beam to be determined. It proved the existence of isotopes. Aston's mass spectrometer was a velocity-focusing machine, meaning that it focused beams of varying velocity but not varying direction. A direction-focusing machine, which focused beams of uniform velocity but varying direction, was developed by Arthur Dempster (1886–1950). Bainbridge's achievement in mass spectroscopy was to develop a double-focusing machine in 1936. It used successive electric and magnetic fields arranged in such a way that ion beams which are non-uniform in both direction and velocity can be brought to a focus. By the mid-1950s, instruments based on this principle were able to separate ions which differ in mass by only one part in 60,000.

Bainbridge is responsible for many of the innovations in the modern mass spectrometer, one of the most useful of analytical tools in physics, chemistry, geology, meteorology, biology and medicine.

Baird John Logie 1888–1946 was a Scottish inventor who was the first person to televise an image, using mechanical (nonelectronic) scanning. He also gave the first demonstration of colour television.

Baird was born in Helensburgh, Dunbartonshire, on 13 August 1888. He was educated at Larchfield Academy and later took an engineering course at the Royal Technical College, Glasgow. He then studied at Glasgow University, but World War I interrupted his final year there. Rejected as physically unfit for military service, Baird became a superintendent engineer with the Clyde Valley Electrical Power Company. In 1918 he gave up engineering because of ill health and set himself up in business, marketing successfully such diverse products as patent socks, confections and soap in Glasgow, London and the West Indies. However, persistent ill health led to a complete physical and nervous breakdown in 1923, and forced him to retire to Horsham in Sussex.

Many inventors had patented their ideas about television – the electrical transmission of images in motion simultaneously with accompanying sound – but only a few, including Baird, pursued a practical study of the problem based on the use of mechanical scanners. In 1907 Boris Rosing had proposed that, in a television system which used mirror scanning in the camera, a cathode-ray tube with a fluorescent screen should be fitted into the receiver. In 1911 A. A. Campbell Swinton had suggested that magnetically deflected cathode-ray tubes should be used both in the camera and the receiver.

On his retirement, Baird concentrated on solving the

problems of television. Having little money, his first apparatus was crude and makeshift, set up on a washstand in his attic room. A tea-chest formed the base of his motor, a biscuit tin housed the projection lamp, and cheap cycle-lamp lenses were incorporated into the design. The whole contraption was held together by darning needles, pieces of string and scrap wood. Yet within a year he had succeeded in transmitting a flickering image of the outline of a Maltese cross over a distance of a few metres.

Baird took his makeshift apparatus to London where, in one of two attic rooms in Soho, he proceeded to improve it. In 1925 he achieved the transmission of an image of a recognizable human face and the following year, on 26 January, he gave the world's first demonstration of true television before an audience of about 50 scientists at the Royal Institution, London. Baird used a mechanical scanner which temporarily changed an image into a sequence of electronic signals that could then be reconstructed on a screen as a pattern of half-tones. The neon discharge lamp Baird used offered a simple means for the electrical modulation of light at the receiver. His first pictures were formed of only 30 lines repeated approximately 10 times a second. The results were crude but it was the start of television as a practical technology.

By 1927, Baird had transmitted television over 700 km/435 mi of telephone line between London and Glasgow and soon after made the first television broadcast using radio, between London and the SS *Berengaria* , halfway across the Atlantic Ocean. He also made the first transatlantic television broadcast between Britain and the United States when signals transmitted from the Baird station in Coulson, Kent, were picked up by a receiver in Hartsdale, New York.

By 1928 Baird had succeeded in demonstrating colour television. The simplest way to reproduce a colour image is to produce the red, blue, and green primary images separately and then to superimpose them so that the eye merges the three images into one full-colour picture. Baird used three projection tubes arranged so that each threw a picture on to the same screen. By using only one amplifier chain and one cathode-ray tube which sequentially amplified the red, blue and green signals, he overcame the problem of overregistration of the three images and matched the three channels. He used two rotating discs, each with segments of red, green and blue light filters, rotating in synchronism before the camera tube and the receiver tube. Each primary-coloured filter remained over the tube face for the period of one field. Although partly successful, Baird's method had two major drawbacks. One was that the picture being transmitted consisted mainly of tones of one hue, say green, then the other two fields (red and blue) showed as black and each green field was succeeded by black ones, resulting in excessive flickering. The other was that the system required three times the bandwidth available.

Baird's black-and-white system was used by the BBC in an experimental television service in 1929. At first, the sound and vision were transmitted alternately, but by 1930 it was possible to broadcast them simultaneously. In 1936,

when the public television service was started, his system was threatened by one promoted by Marconi-EMI. The following year the Baird system was dropped in favour of the Marconi electronic system, which gave a better definition.

Despite his bitter disappointment, Baird continued his experimental work in colour television. By 1939 he had demonstrated colour television using a cathode-ray tube which he had adapted as the most successful method for producing a well-defined and brilliant image. Baird's inventive and engineering abilities were widely recognized. In 1937, he became the first British subject to receive the Gold Medal of the International Faculty of Science. The same year, he was elected Fellow of the Royal Institute of Edinburgh, where a plaque was erected to commemorate his demonstration of true television in 1926. Baird also became an Honorary Fellow of the Royal Society of Edinburgh, Fellow of the Physical Society and Associate of the Royal Technical College.

He continued his research on stereoscopic and large screen television until his death, at Bexhill-on-Sea, Sussex, on 14 June 1946.

Baker Alan 1939–   is a British mathematician whose chief work has been devoted to the study of transcendental numbers.

He was born in London on 19 August 1939 and studied mathematics at the University of London, where he received his BSc in 1961. He then did graduate work at Cambridge and was awarded a PhD in 1964. He remained at Trinity College, Cambridge, for the next ten years, as a research fellow (1964–1968) and as Director of Studies in Mathematics (1968–1974). In 1974 he became Professor of Pure Mathematics in the university. He was elected a Fellow of the Royal Society in 1973. Visiting professorships have taken him to many parts of the United States and Europe. In 1978 he was honoured by the appointment as the first Turán lecturer of the János Bolyai Mathematical Society in Hungary and in 1980 he was elected a Foreign Fellow of the Indian National Science Academy.

Since his research days as a young man, Baker has been chiefly interested in transcendental numbers (numbers which cannot be expressed as roots or as the solution of an algebraic equation with rational coefficients). In 1966 he extended Joseph Liouville's original proof of the existence of transcendental numbers by means of continued fractions, by obtaining a result on linear forms in the logarithms of algebraic numbers. This solution opened the way to the resolution of a wide range of diophantine problems and in 1967 Baker used his results to provide the first useful theorems concerning the theory of these problems. He obtained explicit upper bounds to Thue's equation, $F(x,y) = m$, where F denotes a binary irreducible form, and also to Mordell's equation, $y^2 = x^3 + k$. In 1969 he achieved the same result for the hyperelliptic equation, $y^2 = f(x)$. For this work he was awarded the Fields Medal at the International Congress of Mathematicians at Nice in 1970.

Apart from individual papers, Baker's most important

publication is *Transcendental Number Theory* (1975). Baker's work has greatly enriched the many branches of mathematics influenced by the development of transcendental number theory. It has led to an important new series of results on exponential diophantine equations: his theory provides bounds of 10^{500} or more and it has been shown that these are sufficient in simple cases to calculate the complete list of solutions. The theory has also been used to solve some classical problems of Gauss, to assist in the approximation of algebraic numbers by rationals (an investigation begun by Liouville in 1844), and to inspire new lines of research in elliptic and Abelian functions. Baker is, therefore, at the very forefront of contemporary work on number theory.

Baker Benjamin 1840–1907 was an English engineer most famous for his design of the Forth Rail Bridge in Scotland.

Baker was born on 31 March 1840 at Keyford, Frome, Somerset. He attended the grammar school at Cheltenham until he was 16 when he was apprenticed at Neath Abbey ironworks. In 1860 he left this firm and became assistant to William Wilson, who designed Victoria Station in London. Two years later he joined the staff of Sir John Fowler (becoming his partner in 1875) and became particularly involved with the construction of the Metropolitan and District lines of the London Underground. This project demanded considerable ingenuity to overcome the many hazards caused by difficult soils, underground water and the ruins of Roman and other civilizations. Baker incorporated an ingenious energy conservation measure in the construction of the Central Line: he dipped the line between stations to reduce the need both for braking to a halt and for the increase in power required to accelerate away.

During the 1870s there was much interest in extending the Scottish East Coast Railway from Edinburgh to Dundee. This required the building of bridges across the Forth and Tay. The original Tay Bridge, which had been built by Sir Thomas Bouch of wrought iron lattice girders, collapsed under an express train during a storm on the night of 28 December 1879. This structure had been a considerable accomplishment, and indeed Bouch had built many successful bridges to similar designs. However, it had been built without sufficient allowance for the force of the wind when the train was on the centre span. In addition, certain elements in its design were unsuitable, and inadequate supervision had blighted its construction. When the Tay disaster struck Bouch had already begun work on a bridge across the Forth.

It was only with great reluctance that the government authorities allowed an attempt to build a bridge across the Forth, but this time Baker was in charge. Baker's design had several features that made a repetition of the disaster unlikely. Mild steel, which was considerably stronger than the same weight of wrought iron, had become available through the new Siemens open-hearth process. This meant that the structure would be relatively light. Better estimates

of the wind force that structures were required to withstand in storms were also available. Nevertheless, a rail bridge was a very difficult undertaking since the Forth was 60 m/200 ft deep. A site allowing the use of the little island of Inchgarvie as a foundation for the central pier was chosen; the two main spans were then each of 521 m/1,710 ft. A cantilever structure, which supports the bridge platform by projecting girders, was used since a suspension bridge, in which the platform hangs from catenaries, was not considered sufficiently stable in high winds for a railway. Indeed Bouch's original design was for a stiffened suspension bridge, but this was abandoned after the Tay Bridge disaster.

The bridge was opened on 4 March 1890 by Edward VII when he was Prince of Wales. It has been in service ever since. For this achievement Baker was knighted by Queen Victoria.

Although the Forth Bridge made Baker famous, he has many other projects, both in Britain and abroad, to his credit. He worked on the Hudson River tunnel and at the docks at Hull, and played a prominent part in engineering development work in Egypt – he was involved with the Aswan Dam which realized the dream of desert irrigation. A little while before he began work on the Forth Bridge, he designed the large wrought iron vessel in which Cleopatra's Needle, the obelisk on the Thames Embankment, was brought to England. The Needle was lost at sea, but when found later it was safely preserved within the hull of Baker's ship.

In his later years. Baker built up a large practice and was held in great esteem as a successful engineer who respected the theory of engineering. Although he had only a little formal education in this, he expressed it with great practical ability and artistry. He was made a Fellow of the Royal Society in 1880. He died, a bachelor, on 19 May 1907.

Balmer Johann Jakob 1825–1898 was a Swiss mathematics teacher who devised mathematical formulae that give the frequencies of atomic spectral lines.

Balmer was born in Lausen, near Basle, on 1 May 1825, the eldest son of a farmer who was also a member of the local administration of Basle. He went to school in Liestal, Basle, then in 1844 entered the Technische Hochschule in Karlsruhe, where he studied mathematics, geology and architecture. He subsequently spent a short time at the University of Berlin, then in 1846 returned to his old school in Basle to teach technical drawing while working towards his PhD, which he was awarded in 1849 by the University of Basle. In the following year he began teaching mathematics and Latin at a girls' school in Basle, a post he held for 40 years, until his retirement in 1890. From 1865 to 1890 he was also a part-time lecturer at Basle University, teaching projective geometry, which was the subject of a textbook he published in 1887. Balmer died in Basle on 12 March 1898.

Balmer was a mathematician by education and vocation and was not trained in physics. Nevertheless, he became interested in spectroscopy and – encouraged by J. E.

Hagenbach-Bischoff, a professor at Basle University – began investigating the apparently random distribution of spectral lines. Spectroscopy was a relatively new discipline in the last quarter of the nineteenth century and, although it was known that excited molecules of a substance emit discrete, characteristic spectral lines, many scientists had failed to find a mathematical relationship between these lines. But in 1885 Balmer – then aged 60 – published an equation that described the four visible spectral lines of hydrogen (all that were then known) and also predicted the existence of a fifth line at the limit of the visible spectrum, which was soon detected and measured. He further predicted the existence of other hydrogen spectral lines beyond the visible spectrum. These lines were later found and named after their discoverers: the Lyman series (in the ultraviolet part of the spectrum), the Paschen series (in the infrared part of the spectrum), the Brackett series (infrared) and the Pfund series (infrared). The five lines in the visible part of the hydrogen spectrum are now known as the Balmer series, and are described by the formula:

$$v = R \left(1/2^2 - 1/n^2 \right)$$

where v is the wave number, R is the Rydberg constant, and $n = 3, 4, 5 \ldots$

Balmer arrived at his formula (which has been modified only slightly in the light of later findings) solely from empirical evidence and offered no explanation as to why it gave the correct results. But it was later of crucial importance to Niels Bohr, who used it to support his model of atomic structure. Moreover, not until atomic theory had developed still further was a full explanation of Balmer's equation possible.

Balmer's last paper, which was published in 1897, contained equations to describe the spectral lines of helium and lithium.

Banks Joseph 1743–1820 was a British naturalist who, although making relatively few direct contributions to scientific knowledge himself, did much to promote science, both in Britain and internationally.

Banks was born on 13 February 1743 in London, the son of William Banks of Revesby Abbey in Lincolnshire. Born into a wealthy family, Banks was educated at Harrow and Eton public schools and then at Oxford University. At that time the university curriculum was biased towards the classics, but Banks was more interested in botany so he employed Israel Lyons (1739–1775), a botanist from Cambridge University, as a personal tutor in the subject. After graduating in 1763, Banks moved to London in order to meet other scientists. Meanwhile his father had died in 1761, leaving Banks a large fortune, which he inherited when he came of age in 1764. In 1776 he made his first voyage, to Labrador and Newfoundland, as naturalist on a fishery protection ship. He collected many plant specimens during the trip and, on his return to England, was elected to the Royal Society of London.

In 1768 preparations were being made for an expedition to the southern hemisphere to observe the transit of Venus in 1769. Banks obtained the position of naturalist on the voyage and accompanied by several artists and an assistant botanist, Daniel Solander (1736–1782), set sail in the *Endeavour* – commanded by Captain James Cook – in 1768; Banks paid for his assistants and all the equipment he needed out of his own pocket, at a cost of about £10,000. After the astronomical observations had been completed (the transit was observed from Tahiti), the expedition proceeded on its second objective, to search for the large southern continent that was then thought to exist. During this part of the voyage the expedition explored the coasts of New Zealand and Australia. Banks's plant-collecting activities at the first landing place in Australia (near present-day Sydney) gave rise to the name of the area – Botany Bay. He also studied the Australian fauna, discovering that almost all of the mammals are marsupials, which are less highly developed (in evolutionary terms) than are the placental mammals found on other continents. The expedition returned to England in 1771 and Banks brought back a vast number of plant specimens, more than 800 of which were previously unknown. (Banks kept a journal of the expedition, part of which was published, although not until long after his death, but he did not write an account of his scientific findings on the voyage.) On his return, Banks found himself a celebrity and was summoned to Windsor Castle to give a personal account of his travels to King George III; this visit was the start of a lifelong friendship with the king, which helped Banks to establish many influential contacts.

In 1772 Banks went on his last expedition, to Iceland, where he studied geysers. In 1778 he was elected President of the Royal Society (perhaps because of his influence in high places), an office he held until his death 42 years later. As President, Banks re-established good relations between the Royal Society and the King, who had previously quarrelled with the Society over the issue of the best shape for the ends of lightning conductors. He also brought several wealthy patrons into the Society and helped to develop its international reputation.

As a result of the friendship between Banks and George III, the Royal Botanic Gardens at Kew – of which Banks was the honorary director – became a focus of botanical research. Banks sent plant collectors to many countries in an attempt to establish at Kew as many different species as possible. He also conceived of Kew as a major centre for the practical use of plants, to which end he initiated several important projects, including the introduction of the tea plant into India from its native China, and the transport of the breadfruit tree from Tahiti to the West Indies. This latter project, however, was initially unsuccessful because of the famous mutiny on the *Bounty*, which was carrying the breadfruit trees. At George III's request, Banks also played an active part in importing merino sheep into Britain from Spain; after initial difficulties, the breed was later successfully introduced into Australia.

Banks's voyage to Australia on the *Endeavour* stimulated a lifelong interest in the country's affairs, and he was instrumental in establishing the first colony at Botany Bay in 1788. Thereafter he greatly assisted the growth of the colony and was in regular correspondence with its various governors.

Banks was a generous patron who gave financial assistance to several talented young scientists, notably Robert Brown (1773–1858), who later became an eminent botanist although he is better known today as the discoverer of Brownian motion. Banks also made his large home in Soho Square, London, a renowned meeting place for scientists and prominent figures from other fields. In addition, his international prestige did much to promote the exchange of ideas among scientists in many countries. He also obtained safe passages for many scientists during the American War of Independence and during the Napoleonic Wars, and petitioned on behalf of scientists who had been captured.

Banks received many honours during his life, including a baronetcy in 1781 and membership of the Privy Council in 1797. When he died on 19 June 1820 in Isleworth, near London, he left an extensive natural history library and a collection of plants regarded as one of the most important in existence; both are now housed in the British Museum.

Banneker Benjamin 1731–1806 was an American mathematician, astronomer and surveyor who is chiefly known for his almanacs published in the 1790s.

Son of a freed slave, Benjamin Banneker was born on 9 November 1731 near the Patapsco River in Baltimore County. He received no formal education apart from attending a nearby Quaker school for several seasons, where he showed a great interest and ability in mathematics. Throughout most of his life he worked on his parents' tobacco farm, managing it after the death of his father in 1759, and taught himself mathematics and astronomy. He retired from farming in 1790 to devote all his time to his studies and remained at the farm until his death on 9 October 1806. Banneker became known in 1753, at the age of 21 when, having studied only a pocket watch, he constructed a striking clock. This was the first clock of its kind in America and operated for more than 40 years. He was better known, however, for his almanacs which were published from 1792 to 1797. Having borrowed instruments and texts on astronomy from his neighbour George Ellicott in 1789, he taught himself the subject and learned to calculate an ephemeris and to make projections for lunar and solar eclipses. He compiled an ephemeris for 1791 which he incorporated into an almanac but which was not published; the following year, however, his ephemeris was published as *Benjamin Banneker's Pennsylvania, Delaware, Maryland and Virginia Almanack and Ephemeris, for the year of our Lord, 1792*. Banneker sent a manuscript copy to the Secretary of State, Thomas Jefferson, accompanied by a 12-page letter defending the mental capacities of black people and urging the abolition of slavery; this, with its acknowledgement by Jefferson, was reprinted in the almanac for 1793 and formed part of a long correspondence between the two men. He continued to complete the ephemerides for almanacs each year until 1804, although the later ones were never published. He also wrote a dissertation on bees and did a study of locust plague cycles. As

well as working on his almanacs, in 1791 Banneker was appointed as scientific assistant to Major Andrew Ellicott, George Ellicott's cousin, to survey the Federal Territory for the establishment of the new capital, Washington. Working with Major Pierre Charles L'Enfant they defined the boundaries, streets and the major buildings of the new city, and it is said that when L'Enfant left America, taking his plans with him, Banneker reproduced them from memory in two days. In 1970, Banneker Circle, adjoining L'Enfant Plaza, in Washington, DC, was named after him.

Banting Frederick Grant 1891–1941 was a Canadian physiologist who discovered insulin, the hormone responsible for the regulation of the sugar content of the blood and an insufficiency of which results in the disease diabetes mellitus. For this achievement he was awarded the Nobel Prize for Physiology or Medicine in 1923, which he shared with the Scottish physiologist John Macleod (1876–1935).

Banting was born in Alliston, Ontario, on 14 November 1891, the son of a farmer. He went to the University of Toronto in 1910 to study for the ministry, but changed to medicine and obtained his medical degree in 1916. He served overseas as an officer in the Canadian Medical Corps during World War I, and was awarded the Military Cross for gallantry in 1918. After the war he held an appointment at the University of Western Ontario, but in 1921 returned to the University of Toronto to carry out research into diabetes. In 1930 the Banting and Best Department of Medical Research was opened at the University of Toronto, of which Banting became Director. He was knighted in 1934. While serving as a major in the Canadian Army Medical Corps in 1941 he was killed in an air crash at Gander, Newfoundland.

At the University of Western Ontario, Banting became interested in diabetes, a disease (often fatal at that time) characterized by a high level of blood sugar (glucose) and the appearance of glucose in the urine. In 1889 Mehring and Minkowski had shown that the pancreas was somehow involved in diabetes because the removal of the pancreas from a dog resulted in its death from the disease within a few weeks. Other workers had investigated the effect of tying off the pancreatic duct in rabbits, which resulted in atrophy of the pancreas apart from small patches of cells – the islets of Langerhans. The rabbits did not, however, develop the diuretic condition of sugar in the urine.

It had therefore been suggested that a hormone, called insulin (from the Latin for 'island'), might be concerned in glucose metabolism and that its source might be the islets of Langerhans. But efforts to isolate the hormone from the pancreas failed because the digestive enzymes produced by the pancreas broke down the insulin when the gland was processed.

In 1921 Banting went to discuss his ideas on the matter with John Macleod, the Chairman of the Physiology Department (and an expert on the metabolism of carbohydrates) at the University of Toronto. Macleod was unenthusiastic but agreed to find Banting a place for research in his laboratories.

Banting reasoned that if the pancreas were destroyed but the islets of Langerhans were retained, the absence of digestive enzymes would allow them to isolate insulin. With Charles Best (1899–1978), one of his undergraduate students, he experimented on dogs. They put several of the animals into two groups; each dog in one group had the pancreatic duct tied, and those in the other group were depancreatized. After several weeks, they removed the degenerated pancreases from the dogs of the first group, extracted the glands with saline and injected the extract into the dogs of the second group, which by then had diabetes and were in poor condition. They took regular blood samples from the diabetic dogs and found that the sugar content dropped steadily as the condition of the dogs improved.

These results encouraged Banting. He obtained foetal pancreatic material from an abattoir, thinking that it might contain more islet tissue. With his assistants he set about extracting an active product, but purification proved to be very difficult. Eventually reasonably pure insulin was produced and commercial production of the hormone started. By January 1922 it was ready for use in the Toronto General Hospital. The first patient was a 14-year-old diabetic boy who showed rapid improvement after treatment. They also discovered that the dose of insulin could be reduced by regulating the amount of carbohydrate in the patient's diet.

When Banting and Macleod were awarded the Nobel prize for this work Banting, feeling strongly that Best had made a valuable contribution, divided his share of the money with him. Banting's discovery of insulin and his attempts to purify the crude material led eventually to the commercial production of insulin, which has saved the lives of many diabetics.

Bardeen John 1908–1991 was a US physicist whose work on semiconductors, together with William Shockley (1910–1989) and Walter Brattain (1902–1987), led to the first transistor, an achievement for which all three men shared the 1956 Nobel Prize for Physics. Bardeen also gained a second Nobel Prize for Physics in 1972, which he then shared with John Robert Schrieffer (1931–) and Leon Cooper (1930–) for a complete theoretical explanation of superconductivity.

Bardeen, who is the only person thus far to be awarded two Nobel Prizes for Physics, was born in Madison, Wisconsin, on 23 May 1908. He began his career with a BS in electrical engineering at the University of Wisconsin awarded in 1928. For the next two years, he was a graduate assistant working on mathematical problems of antennas and on applied geophysics, but it was at this stage that he first became acquainted with quantum mechanics. He moved to Gulf Research in Pittsburgh, where he worked on the mathematical modelling of magnetic and gravitational oil-prospecting surveys, but the lure of pure scientific research became increasingly strong, and in 1933 he gave up his industrial career to enrol for graduate work with Eugene Wigner (1902–) at Princeton, where he was introduced to the fast-developing field of solid state

physics. His early studies of work functions, cohesive energy and electrical conductivity in metals were carried out at Princeton, Harvard and the University of Minnesota. From 1941 to 1945, he returned to applied physics at the Naval Ordnance Laboratory in Washington, DC, where he studied ship demagnetization and the magnetic detection of submarines.

In 1945 Bardeen joined the Bell Telephone Laboratory, and his work on semiconductors there led to the first transistor, an achievement rewarded with his first Nobel prize in conjunction with Walter Brattain and William Shockley. He moved to the University of Illinois in 1951, where in 1957, along with Bob Schrieffer and Leon Cooper, he developed the microscopic theory of superconductivity (the BCS theory) that was to gain him a second Nobel prize in 1972. After 1975, he was an emeritus professor at Illinois, concentrating on theories for liquid helium-3 which have analogies with the BCS theory.

The electrical properties of semiconductors gradually became understood in the late 1930s with the realization of the role of low concentrations of impurities in controlling the number of mobile charge carriers. Current rectification at metal–semiconductor junctions had long been known, but the natural next step was to produce amplification analagous to that achieved in triode and pentode valves. A group led by William Shockley began a programme to control the number of charge carriers at semiconductor surfaces by varying the electric field. John Bardeen interpreted the rather small observed effects in terms of surface trapping of carriers, but he and Walter Brattain successfully demonstrated amplification by putting two metal contacts 0.05 mm/0.002 in apart on a germanium surface. Large variations of the power output through one contact were observed in response to tiny changes in the current through the other. This so-called point contact transistor was the forerunner of the many complex devices now available through silicon chip technology.

Ever since 1911, when Heike Kamerlingh-Onnes (1853–1926) first observed zero electrical resistance in some metals below a critical temperature, physicists sought a microscopic interpretation of this phenomenon of superconductivity. The methods that proved successful in explaining the electrical properties of normal metals were unable to predict the effect. At very low temperatures, metals were still expected to have a finite resistance due to scattering of mobile electrons by impurities. Bardeen, Cooper and Schrieffer overcame this problem by showing that electrons pair up through an *attractive* interaction, and that zero resistivity occurs when there is not enough thermal energy to break the pair apart.

Normally electrons repel one another through the Coulomb interaction, but a net attraction may be possible when the electrons are imbedded in a crystal. The ion cores in the lattice respond to the presence of a nearby electron, and the motion may result in another electron being attracted to the ion. The net effect is an attraction between two electrons through the response of the ions in the solid. The pairs condense out in a kind of phase transition below a temperature

T_c (typically of the order of a few kelvin), and the involvement of the ions of the lattice in the interaction is confirmed by the dependence of T_c on the isotopic mass in the metal, i.e. T_c varies with variation of atomic vibration frequencies in the solid. The requirement of a finite energy to break up a so-called Cooper pair leads to a small energy gap below which excitations are not possible.

The BCS theory is amazingly complete, and explains all known properties associated with superconductivity. Although applications of superconductivity to magnets and motors were possible without the BCS theory, the theory is important for strategies to increase T_c as high as possible – if T_c could be raised above liquid nitrogen temperature, the economics of superconductivity would be transformed. In addition, the theory was an essential prerequisite for the prediction of Josephson tunnelling with its important applications in magnetometers, computers and determination of the fundamental constants of physics.

Both of Bardeen's achievements have important consequences in the field of computers. The invention of the transistor led directly to the development of the integrated circuit and then the microchip, which is making computers both more powerful and more practical. Superconductivity enables the basic arithmetic and logic operations of computers to be carried out at much greater speeds and may lead to the development of artificial intelligence.

Barkla Charles Glover 1877–1944 was a British physicist who made important contributions to our knowledge of X-rays, particularly the phenomenon of X-ray scattering. He received many honours for his work on ionizing radiation, including the 1917 Nobel Prize for Physics.

Barkla was born on 7 June 1877 in Widnes, Lancashire. After studying at the Liverpool Institute, he went to University College, Liverpool, in 1895, where he studied under Oliver Lodge. He obtained his BSc in 1898 and his MSc in the following year. In the autumn of 1899 he went to Trinity College, Cambridge, where he researched under J. J. Thomson at the Cavendish Laboratory, but transferred to King's College, Cambridge, in 1901 in order to sing in the choir. Nevertheless, he refused the offer of a choral scholarship, which would have enabled him to remain at Cambridge, and in 1902 returned to University College, Liverpool, as Oliver Lodge Fellow. In 1904 he was awarded a DSc and was regularly promoted, being appointed a Special Lecturer in 1907. He was made Professor of Physics at King's College, London, in 1909 and remained there until 1913, when he became Professor of Natural Philosophy at Edinburgh University, a position he held until his death on 23 October 1944.

Barkla's first major piece of research, which was carried out while he was a student at the Cavendish Laboratory, involved measuring the speed at which electromagnetic waves travel along wires of different thickness and composition. During his third year at Cambridge, he began investigating secondary radiation – that is, the effect whereby a substance subjected to X-rays re-emits secondary X-radiation – a subject that he spent most of his

subsequent career studying. He published his first paper on this phenomenon in 1903. In this paper he announced his finding that for gases of elements with a low atomic mass the secondary scattered radiation is of the same average wavelength as that of the primary X-ray beam to which the gas is subjected. He also found that the extent of such scattering is proportional to the atomic mass of the gas concerned. The more massive an atom, the more charged particles it contains and it is these charged particles that are responsible for the X-ray scattering. Thus Barkla's work was one of the first indications of the importance of the amount of charge in an atom (rather than merely its atomic mass) in determining an element's position in the periodic table – a significant early step in the evolution of the concept of atomic number.

In 1904 Barkla found that, unlike the low atomic mass elements, the heavy elements produced secondary radiation of a longer wavelength than that of the primary X-ray beam. He also showed that X-rays can be partially polarized, thus proving that they are a form of transverse electromagnetic radiation, like visible light.

In 1907 Barkla began his most important research into X-rays, working at Liverpool with C. A. Sadler. They found that secondary radiation is homogeneous and that the radiation from the heavier elements is of two characteristic types. They also showed that these characteristics (that is, of a specific wavelength) radiations are emitted only after a heavy element is exposed to X-radiation 'harder' (that is, of shorter wavelength, and therefore more penetrating) than its own characteristic emissions. This finding was the first indication that X-ray emissions are monochromatic.

Barkla named the two types of characteristic emissions the K-series (for the more penetrating emissions) and the L-series (for the less penetrating emissions). He later predicted that other series of emissions with different penetrances might exist, and an M-series – radiation with even lower penetrance than that of the K-series – was subsequently discovered. After about 1916, however, Barkla devoted his research to investigating a J-series of extremely penetrating radiations, the existence of which he had suggested. But the results of these studies could not be confirmed by other workers and the existence of the J-series was not – and still is not – part of accepted theory. Nevertheless Barkla continued to adhere to his theory of the J-phenomenon, as a result of which he became increasingly isolated from the rest of the scientific community during his later years.

Barnard Christiaan Neethling 1922– is the South African cardiothoracic surgeon who performed the first human heart transplant, on 3 December 1967 at Groote Schuur Hospital in Cape Town.

Barnard was born on 8 November 1922 in Beaufort West, South Africa. He attended the local high school and then in 1940 went to the University of Cape Town, where he received his medical degree in 1946. After working in private practice from 1948 to 1951, he became Senior Resident Medical Officer at the City Hospital in Cape Town for two years. In 1953 Barnard was elected Registrar at Groote Schuur Hospital, later to be the place of his most important work. In 1956 he was awarded a scholarship to the University of Minnesota, Minneapolis, USA; he returned to South Africa two years later, taking with him a heart–lung machine. He became Director of Surgical Research at Groote Schuur and the University of Cape Town, and from 1961 to 1983 was Head of Cardiothoracic Surgery. In 1985, he was appointed Senior Consultant and Scientist in Residence at the Oklahoma Heart Centre, Oklahoma City, USA.

Barnard's early research involved experiments with heart transplants in dogs. His success convinced him that similar operations could be performed on human patients. In December 1967 Denise Duvall, a 25-year-old woman, was critically injured in a road accident in Cape Town, and after it was established that her brain was irreparably damaged, permission was obtained for her heart to be donated for transplant purposes. The recipient was a man in his fifties, Louis Washkansky, whose heart – in Barnard's words – was 'shattered and ruined'.

X-ray motion pictures (angiograms) were taken of Washkansky's heart by injecting radio-opaque dye into each side of it using catheters inserted into the veins and arteries. These films were taken to prepare the surgical team for the operation on Washkansky. Once the donor heart had been removed, Barnard cut away part of it so that it would fit what remained of the recipient's heart. Two holes were made in the donor heart, one through which the venae cavae could enter and one for the pulmonary veins. The edges of these holes were stitched onto the waiting part of Washkansky's heart. The difference in size of the hearts did not matter because the openings in the donor heart could be enlarged to match the recipient's heart.

Surgically this first transplant was a success, but Washkansky died 18 days after the operation from double pneumonia – probably contracted as a result of the immunosuppressive drugs administered to him to prevent his body rejecting the new heart. Barnard continued to perform heart transplants, improving his methods all the time. Unfortunately the number of operations performed by him decreased because of the worsening arthritis in his hands.

Open-heart surgery was first introduced in South Africa by Barnard, and he further developed cardiothoracic surgery by new designs for artificial heart valves. His other achievements have included the discovery that intestinal artresia – a congenital deformity in the form of a hole in the small intestine – is the result of an insufficient supply of blood to the foetus during pregnancy. It was a fatal defect before Barnard developed the corrective surgery.

Barnard's techniques for heart transplant surgery have been adopted and developed by many surgeons, and as the methods improve they can give a new lease of life to those suffering from fatal heart conditions.

Barnard Edward Emerson 1857–1923 was a US observational astronomer whose keen vision and painstaking thoroughness made him an almost legendary figure.

Barnard was born in Nashville, Tennessee, USA, on 16 December 1857. His family was poor and by the time he was nine years old Barnard had begun to work as an assistant in a photographic studio. The techniques he learned were to be invaluable in his later career. Barnard's fascination with astronomy led him to take a job in the observatory at Vanderbilt University. He took some courses but spent most of his time using the telescopes.

In 1877 Barnard went to California in order to work at the Lick Observatory when it opened in 1888. He was awarded a DSc from Vanderbilt in 1893, although he had never formally graduated. In 1895 he took up the chair of Practical Astronomy at the University of Chicago and became Astronomer at the Yerkes Observatory. He participated in the expedition to Sumatra to observe the solar eclipse of 1901.

Barnard's many discoveries brought him worldwide respect and many honours. He received awards from the most prestigious scientific organizations, and was elected to their membership. He died in Williams Bay, Wisconsin, on 6 February 1923.

Barnard's early astronomical studies were made with a 12.7- cm/5-in telescope which he purchased in 1878. It was for the discovery of comets that Barnard first began to establish a reputation. He discovered his first comet on 5 May 1881 and by 1892 he had found 16. He also investigated the surface features of Jupiter, the *Gegenschein* (a faint patch of light visible only at certain times of the year and whose nature is still not certain), nebulae and other celestial bodies.

His most dramatic discovery came on 9 September 1892 when, by blocking out the glow of the parent planet, Barnard discovered the fifth satellite of Jupiter and the first to be found since the four Galilean satellites. This was the last satellite to be discovered without the aid of photography. The fifth moon orbits inside all the others, which now number more than 20.

Barnard's later discoveries included the realization that the apparent voids in the Milky Way are in fact dark nebulae of dust and gas and the sighting in 1916 of the so-called 'Barnard's run-away star', which has a proper motion of 10 seconds of arc per year (faster than any star known until 1968).

Although Barnard's lack of mathematical flair prevented him from making profound contributions to theoretical advances in astronomy, he was one of the most eminent observational astronomers of his time.

Barr Murray Llewellyn 1908– is a Canadian anatomist and geneticist known for his research into defects of the human reproductive system, and particularly chromosomal defects.

Barr was born in Belmont, Ontario, on 20 June 1908, the son of a farmer (who was originally from Ireland). He attended the University of Western Ontario, where he gained his BA in 1930, his MD in 1933 and his MSc in 1938. Apart from serving as a medical officer with the Royal Canadian Air Force during World War II, he spent

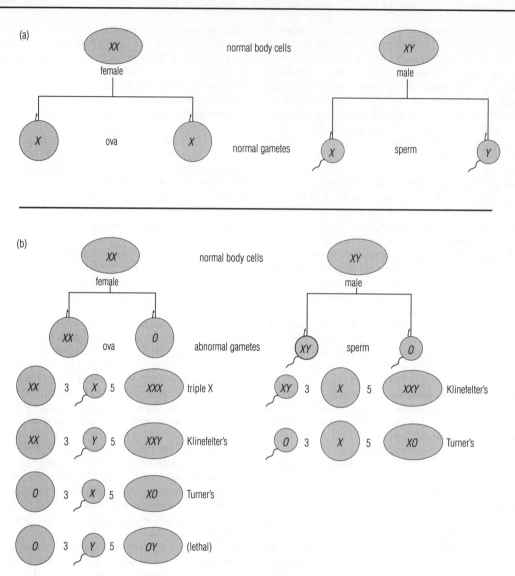

Barr *(a) Meiosis with the formation of normal gametes; (b) chromosomal defects arise from fertilization of abnormal gametes by normal gametes.*

his entire career at the University of Western Ontario, where he became Head of the Department of Anatomy in the Health Sciences Centre.

In 1949, working with Ewart Bertram, Barr noticed that the nuclei of nerve cells in females have a mass of chromatin (the nucleoprotein of chromosomes, which stains strongly with basic dyes) whereas those in males do not. He also found that this sex difference occurs in the cells of most mammals.

Improvements in cell culture methods made the closer examination of human chromosomes possible; for example, in 1954 it was discovered that the chromosome number in human beings is 46 and not 48, as was previously thought. From Barr's investigations, the sex chromatin (called the Barr body) is now known to be one of the two

X-chromosomes in the cells of females; it is more condensed than the other chromosomes and is genetically inactive. The other X-chromosome in females is attenuated and genetically active in resting cells.

Before the discovery of sex chromatin, the nature of the sex chromosome complex (XX female, or XY male) could be detected at cell level only by direct examination of chromosomes in dividing cells. The new comparative method using stained sex chromatin offered a much needed investigative and diagnostic procedure for patients with developmental anomalies of the reproductive system. Abnormalities of the sex chromosome complex often result in disorders such as Turner's syndrome (usually in females) and Klinefelter's syndrome. The former occurs when the gamete lacks a second sex chromosome so that the complex is XO. Klinefelter's syndrome, occurring in males only, results usually from a chromosomal complex of XXY.

Barr and his colleagues also devised a buccal smear test by rubbing the lining of the patient's mouth (the buccal cavity) and examining the cells obtained for chromosomal defects. This test is now used extensively to screen patients, including new-born babies, and has proved useful in the differential diagnosis of several kinds of hermaphroditism. Barr's research has thus been invaluable in simplifying diagnostic tests for chromosomal defects.

Barrow Isaac 1630–1677 was an English mathematician, physicist, classicist and Anglican divine, one of the intellectual luminaries of the Caroline period.

Barrow was born in London in October, 1630. His father, a linendraper to Charles I, sent him to Charterhouse as a day boy, but there he achieved little beyond gaining a reputation as a bully, and he was removed to Felstead School. In 1643 he was entered for Peterhouse, Cambridge, where an uncle was a Fellow, but by the time that he went up to university in 1645 his uncle had moved to Trinity College and it was there that Barrow entered as a pensioner. He received his BA in 1648 and a year later was elected a Fellow of Cambridge. In 1655 his former tutor, Dr Dupont, retired from the Regius Professorship of Greek; he wished Barrow, his former pupil, to succeed him. But the appointment was not offered to Barrow. His reputation then was more for mathematics than classics, and he was very young. Even so, it may be true that he was barred from the chair by Cromwell's intervention. Certainly Barrow had never concealed his royalist opinions and he was out of sympathy with the prevailing republican air of Cambridge. He decided to leave England for a tour of the continent and to help finance his trip sold his library.

He remained abroad for five years, returning to England only with the restoration of Charles II in 1660. Immediately he took Anglican orders and was elected to the Regius Professorship previously denied him. He was also appointed Professor of Geometry at Gresham College, London, and in 1663 became first Lucasian Professor of Mathematics at Cambridge. In 1669 he resigned the Cambridge Chair in favour of Isaac Newton and a year later was made a DD by royal mandate. During the 1660s his lectures on mathematics at Gresham College formed the basis of his mathematical reputation; thereafter his energies were devoted more to theology and preaching. He was made Master of Trinity in 1675 and died two years later, on 4 May 1677.

In his time Barrow was considered second only to Newton as a mathematician. Now he is best remembered as one of the greatest Caroline divines, whose sermons and treatises, especially the splendid *Treatise on the Pope's Supremacy* (1680), have gained a permanent place in ecclesiastical literature. Certainly he was an admirable teacher of mathematics, and although it is not true that Newton was his pupil, Newton attended his lectures and later formed a fruitful friendship with him. His mathematical importance

is slight, the *Lectiones Mathematicae*, delivered at Gresham between 1663 and 1666 and published in 1669, being marred by his insistence that algebra be separated from geometry and his desire to relegate algebra to a subsidiary branch of logic. His geometry lectures were read by few and had little influence.

More important were his lectures on optics. Most of his work in this field was immediately eclipsed by Newton's, but there is no doubt that Newton was greatly inspired by Barrow's work in the field and to Barrow is due the credit for two original contributions: the method of finding the point of refraction at a plane interface, and his point construction of the diacaustic of a spherical interface. Barrow was a man of great powers of concentration and original thought. If he failed to reach the highest class of mathematics, the reason may well be that he spread his intellectual interests so broadly.

Bartlett Neil 1932–   is a British-born chemist who achieved fame by preparing the first compound of one of the rare gases, previously thought to be totally inert and incapable of reacting with anything.

Bartlett was born in Newcastle-upon-Tyne on 15 September 1932. He attended the University of Durham, gaining his PhD in 1957. A year later he took an appointment at the University of British Columbia, Canada, and in 1966 became Professor of Chemistry at Princeton in the United States. In 1969 he moved to a similar position at the University of California, Berkeley.

In Canada in the early 1960s Bartlett was working with the fluorides of the platinum metals. He prepared platinum hexafluoride, PtF_6, and found that it is extremely reactive. It reacts with oxygen, for example, to form the ionic compound $O_2^+ PtF_6^-$. In 1962 he reacted platinum hexafluoride with xenon, the heaviest of the stable rare gases, and obtained xenon platinofluoride (xenon fluoroplatinate, $XePtF_6$), the first chemical compound of a rare gas. Other compounds of xenon followed, including xenon fluoride (XeF_4) and xenon oxyfluoride ($XeOF_4$). Other chemists soon made compounds of krypton and radon. It is for this reason that this and other modern books use the term 'rare gases' to describe the helium group of elements, not the former terms 'inert gases' or 'noble gases', for inert or noble they no longer are, due to the pioneer work of Bartlett.

Barton Derek Harold Richard 1918–   is a British organic chemist whose chief work concerns the stereochemistry of natural compounds. He showed that their biological activity often depends on the shapes of their molecules and the positions and orientations of key functional groups. For this achievement he shared the 1969 Nobel Prize for Chemistry with the Norwegian Odd Hassel (1897–   ).

Barton was born in Gravesend on 8 September 1918. He was educated at Tonbridge School and graduated from Imperial College, London, in 1940, gaining a PhD in organic chemistry two years later. He has held various professorships: at Birkbeck College, London (1953–1955), Glasgow (1955–1970) and Imperial College, London (1970–1978). In 1978 he became Emeritus Professor of Organic Chemistry at the University of London, the same year that he was appointed Director of the Institute for the Chemistry of Natural Substances at Gif-sur-Yvette in France. He was knighted in 1972.

While lecturing in the United States at Harvard between 1949 and 1950, Barton studied the different rates of reaction of certain steroids and their triterpenoid isomers (substances with the same composition but differing in the way their atoms are joined and arranged in space). He deduced that the difference in the spatial orientation of their functional groups accounts for their behaviour, and so developed a new field in organic chemistry which became known as conformational analysis. Barton realized that in a complex system where the conformation is fixed, the reactivity of a given group depends on whether it is attached to the main molecule in an axial or an equatorial position. He discovered important correlations between the chemical reactivity and conformation of various groups in steroids and terpenes (which are structurally very similar).

Barton went on to examine many natural products, including phenols. For example, in 1956 he challenged the generally accepted structure of the substance known as Plummerer's ketone, showing how it could be formed by the oxidative coupling of two phenolic residues.

He realized the biosynthetic importance of this reaction, concluding that the structures of many phenols and alkaloids could be explained and predicted. He devised new ways of preparing oxyradicals and studied various natural products that contain the dienone group, predicting that if a hydrogen atom of the same molecule is spatially orientated near to a generated oxyradical, intermolecular elimination of the hydrogen atom is preferred.

Barton also studied photochemical routes and unravelled the complex transformations that take place during photolysis. In 1959 (at Cambridge, Massachusetts) he devised a simple synthesis of the naturally occurring hormone aldosterone. He also worked on the antibiotics tetracycline and penicillin. He has thus contributed greatly to the study of natural products and their formation, and has enabled a rational interpretation to be made of much stereochemical information.

Bates Henry Walter 1825–1892 was a British naturalist and explorer whose discovery of a type of mimicry (called Batesian mimicry) lent substantial support to Charles Darwin's theory of natural selection.

Bates was born on 8 February 1825 in Leicester, the son of a clothing manufacturer. He received little formal education, leaving school when he was 13 years old to work in his father's stocking factory, but he was interested in natural history and devoted much of his spare time to private study. In 1844 he met Alfred Wallace and aroused in him an interest in entomology, and they planned a joint venture to the Amazon region of South America to study and collect its flora and fauna, aiming to pay their expenses by selling the specimens they collected. In 1848 they arrived in Brazil at Para (also called Belem) near the mouth of the River Amazon, and for the next two years they worked together, thereafter separately. Wallace returned to England in 1852 but Bates remained in South America until 1859, during which time he explored much of the River Amazon. After returning to England, he spent several years organizing the specimens he had collected and writing about his observations, discoveries and explorations. In 1864 he was appointed Assistant Secretary to the Royal Geographical Society in London, a post he held until his death in London, on 16 February 1892.

During his Amazon exploration, Bates collected a vast number of specimens, including more than 14,000 species of insects, more than half of which were previously unknown. He travelled continually up and down the Amazon waterways, usually spending only a few days at each stopping-place to collect specimens. These had to be prepared and preserved – a difficult task in the hot, humid conditions of the Amazon rainforest – before being sent to his agent in England for sale. He also managed to find time to write to Charles Darwin, who used his findings in the development of his theory of natural selection.

After returning to England, Bates presented a paper to the Linnean Society in 1861 entitled 'Contributions to an insect fauna of the Amazon Valley', in which he outlined his observations of mimicry in insects. He had discovered that several different species of butterflies have almost identical patterns of colours on their wings, and that some are distasteful to bird predators whereas others are not. Further, he suggested that the latter types, influenced by natural selection, mimic the distasteful species and thus increase their chances of survival. This form of mimicry is now called Batesian mimicry. (Subsequently Fritz Müller (1821–1897), a German-born Brazilian zoologist, discovered that, in some cases, two or more equally poisonous or distasteful species share a similar colour pattern, thereby reinforcing the warning that each gives to predators, a phenomenon known as Müllerian mimicry.)

Bates's paper was well received and he was asked to write an account of his experiences in South America. The result was *The Naturalist on the River Amazon* (1863), a two-volume work (with an introduction by Charles Darwin) in which Bates described both his explorations and his scientific findings. The book quickly sold out and a second edition was published. But in this second edition, which has been reprinted many times, much of the scientific material was omitted, which has tended to diminish Bates's reputation as a scientist.

Bateson William 1861–1926 was a British geneticist who was one of the founders of the science of genetics (a term he introduced), and a leading proponent of Mendelian views after the rediscovery in 1900 of Gregor Mendel's work on heredity. Bateson also made important contributions to embryology.

Bateson was born on 8 August 1861 in Whitby, Yorkshire. He was educated at Rugby School and St John's

College, Cambridge, from which he graduated in natural sciences in 1883. He then travelled to the United States, remaining there for two years doing embryological research. During this period he met W. K. Brooks of Johns Hopkins University, who interested Bateson in evolution, which he spent the rest of his life studying. On his return to Britain, Bateson spent several years investigating the fauna of salt lakes and undertaking other research into evolution and heredity. In 1908 he became the first Professor of Genetics at Cambridge University, but left this post in 1910 to be the Director of the newly established John Innes Horticultural Institution at Merton, Surrey, where he remained until his death on 8 February 1926 in Merton. In addition to his directorship, Bateson was Fullerian Professor of Physiology at the Royal Institution from 1912 to 1914 and a trustee of the British Museum from 1922.

Bateson's first important research was his embryological work in the United States. Studying the small, worm-like marine creature *Balanoglossus*, he discovered that although its larval stage is similar to that of the echinoderms, it also possesses a dorsal nerve cord and the beginnings of a notochord. Thus he demonstrated that *Balanoglossus* is a primitive chordate, which was the first indication that chordates had evolved from echinoderms – a theory now widely accepted.

His interest having turned to evolution while in the United States, Bateson spent the years immediately following his return to Britain investigating the fauna of the salt lakes of Europe, central Asia, and northern Egypt. The result of these studies was his book, *Material for the Study for Variation* (1894), in which he put forward his theory of discontinuity to explain the long process of evolution. According to this theory, species do not develop in a predictable sequence of very gradual changes but instead evolve in a series of discontinuous 'jumps'. This theory was unacceptable to the traditional, biometrical evolutionists who maintained that there were no breaks in nature's pattern, and so Bateson began a series of breeding experiments to find corroborative evidence for his theory. When Mendel's work on heredity was rediscovered in 1900, Bateson translated Mendel's paper into English (as *Experiments on Hybrid Plants*), and found that Mendel's work provided him with the supportive evidence he was seeking for his discontinuity theory. Bateson also assumed the task of publicizing and defending the highly controversial discoveries of Mendel. The long debate finally culminated in 1904 when, as President of the Zoological Section of the British Association, Bateson succeeded in vindicating Mendel's findings at a meeting in Cambridge.

Thereafter Bateson continued with his breeding experiments, the results of which he described in his *Mendel's Principles of Heredity* (1908). In this book he showed that certain traits are consistently inherited together, an apparent contradiction to Mendel's findings; this phenomenon is now known to result from genes being situated close together on the same chromosome – a phenomenon called linkage. Towards the end of his life Bateson proposed his own vibratory theory of inheritance, based on the physical laws of force and motion, but this theory has met with little acceptance from other scientists.

Bayliss William Maddock 1860–1924 was a British physiologist who discovered the digestive hormone secretin, the first hormone to be found, and investigated the peristaltic movements of the intestine. He received many honours in recognition of his work, including the Royal Medal (1911) and Copley Medal of the Royal Society (1919), and in 1922 he was knighted.

Bayliss was born in Wolverhampton, Staffordshire, on 2 May 1860. He began studying medicine at University College, London, from 1881, but turned to physiological research and entered Wadham College, Oxford, in 1885. After graduation he returned to University College where he began his main research. In 1893 he married the sister of Ernest Starling, the man with whom he worked during his major discoveries. In 1903 he became a Fellow of the Royal Society, and in 1912 a professorship of general physiology was created for him at University College. He was a longstanding member of the Physiological Society, first as its Secretary and then as its Treasurer. He died in Hampstead, London, on 27 August 1924.

Bayliss discovered the hormone secretin in 1902 when working with his brother-in-law Ernest Starling. He made an extract of a piece of the inner lining (mucosa) of the duodenum, which had already had hydrochloric acid introduced to it. When the extract was injected into the bloodstream the pancreas was stimulated to secrete digestive juices. Bayliss tried injecting hydrochloric acid intravenously, but no pancreatic secretion occurred. He then severed the nerves serving a loop of duodenum so that it was isolated from the pancreas except via the blood supply. Acid was introduced into the duodenum, and the pancreas produced secretions. Bayliss thus concluded that as hydrochloric acid (from the stomach's digestive juices) passes into the duodenum during the normal digestive process, the duodenal mucosa release a chemical (the hormone secretin) into the bloodstream which, in turn, makes the pancreas secrete its juices. The role of hormones in physiology is now commonly accepted, but at the time Bayliss's major discovery was a breakthrough. He then went on to study the activation of enzymes, particularly the pancreatic enzyme trypsin.

Bayliss and Starling also worked on the nerve supply to the intestines, and on pressures within the venous and arterial systems. Bayliss did independent research into vasomotor reflexes. His method of treating patients suffering from surgical shock with saline to replace blood loss was widely used during World War I on injured troops. In 1915 he published *Principles of General Physiology*, which rapidly became a standard work.

Beadle, Tatum, and Lederberg George Wells Beadle 1903–1989, Edward Lawrie Tatum 1909–1975 and Joshua Lederberg 1925–   shared the 1958 Nobel Prize for Physiology or Medicine for their pioneering work in the field of biochemical genetics.

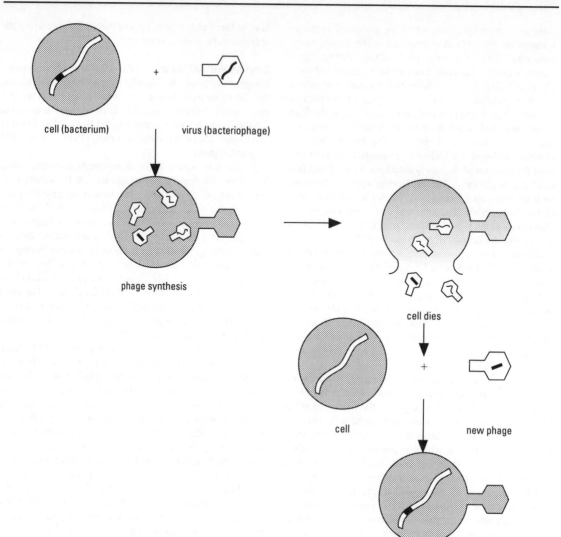

cell (bacterium)   virus (bacteriophage)

phage synthesis

cell dies

cell   new phage

DNA transferred by transduction

Beadle, Tatum, and Lederberg *Beadle showed how a bacteriophage can bring about the transfer of DNA by transduction.*

Beadle was born on 22 October 1903 in Wahoo, Nebraska, and was educated at the University of Nebraska, graduating in 1926. After obtaining his doctorate in genetics at Cornell University, New York, in 1931, he went to the laboratory of Thomas Hunt Morgan at the California Institute of Technology, where he researched into the genetics of the fruit fly (*Drosophila melanogaster*). In 1935 he went to Paris, where he continued his work on *Drosophila* at the Institut de Biologie Physico-Chimique, collaborating with Boris Ephrussi. Beadle returned to the United States in 1936 and taught genetics for a year at Harvard University. From 1937 to 1946 he was Professor of Biology at Stanford University, California, and it was during this period that he collaborated with Tatum on the work that was to gain them the Nobel prize. Beadle was appointed Professor and

Chairman of the Division of Biology at the California Institute of Technology in 1946 and remained there until 1961, when he became Chancellor of the University of Chicago. In 1968 he retired from the university in order to direct the American Medical Association's Institute for Biomedical Research.

Tatum was born on 14 December 1909 in Boulder, Colorado, and was educated at the University of Wisconsin (where his father was head of the pharmacology department), from which he graduated in 1931 and gained his doctorate in 1934. From 1937 to 1941, after a period of postdoctoral study in the Netherlands, he was a research assistant at Stanford University, where he worked with Beadle. Tatum joined the faculty of Yale University in 1945 – as Associate Professor of Botany until 1946 then as

Professor of Microbiology from 1946 to 1948. While at Yale he worked with Lederberg. Tatum returned to Stanford University in 1948 as Professor of Biology, becoming Professor of Biochemistry there in 1956. In 1957 he went to the Rockefeller Institute for Medical Research (now Rockefeller University), New York City. Tatum died on 5 November 1975 in New York City.

Lederberg was born on 23 May 1925 in Montclair, New Jersey. He studied at Columbia University, graduating in 1944, then did postgraduate work at Yale University, gaining his doctorate in 1947. It was while he was at Yale that Lederberg worked with Tatum. In 1947 Lederberg joined the University of Wisconsin, initially as Assistant Professor of Genetics then as Professor of Medicine and Genetics and finally as Chairman of the Genetics Department. In 1959 he moved to Stanford University, where he was Professor of Genetics and Biology and Chairman of the Genetics Department and, from 1962, Director of the Kennedy Laboratories of Molecular Medicine.

The research that eventually led to the award of a Nobel prize had its origins in Beadle's early work on *Drosophila*. While at Morgan's laboratory, Beadle realized that genes influence heredity by chemical means. Later, working with Ephrussi in Paris, he showed that the eye colour of *Drosophila* is a result of a series of chemical reactions under genetic control. Because of the complexity of the relationship between genes and metabolic processes in *Drosophila* – itself a relatively simple organism – when Beadle returned to the United States and continued his research at Stanford, he used the red bread mould *Neurospora crassa*, which is even simpler than *Drosophila*. *Neurospora* can be cultured in a medium containing sugar and biotin as the only organic components, plus a few inorganic salts. Working with Tatum, Beadle subjected colonies of *Neurospora* to X-rays and studied the changes in the nutritional requirements of, and therefore enzymes formed by, the mutant *Neurospora* produced by the irradiation. By repeating the experiment with various mutant strains and culture mediums, Beadle and Tatum deduced that the formation of each individual enzyme is controlled by a single, specific gene. This one-gene-one-enzyme concept found wide applications in biology and virtually created the science of biochemical genetics.

After Tatum moved to Yale in 1945, he applied the mutation-inducing technique developed by Beadle and himself to bacteria. Using *Escherichia coli*, Tatum, working with Lederberg, showed that genetic information can be passed from one bacterium to another. The discovery that a form of sexual reproduction can occur in bacteria meant that these organisms could be used for research much more extensively than previously thought possible and, in fact, bacteria are now as important as *Drosophila* and *Neurospora* in genetic research.

Tatum's collaboration with Lederberg ended in 1947, when Lederberg went to the University of Wisconsin and continued to research into bacterial genetics. In 1952 he published a paper in which he revealed that bacteriophages can transfer genetic material from one bacterium to another, a phenomenon Lederberg termed transduction. This discovery further increased the usefulness of bacteria in genetic research. Later Lederberg diversified into other fields of investigation, devising means of identifying and classifying organic compounds by mathematical graph theory and developing statistical methods of studying human biology.

Beaufort Francis 1774–1857 was a British hydrographer famous for his meteorological developments.

Born in Ireland, the son of a clergyman of Huguenot origin, Beaufort followed his father in developing an early love of geography and topography. He joined the East India Company in 1789; a year later he enlisted in the Royal Navy, and spent the next 20 years in active service.

In 1806 Beaufort drew up the wind scale named after him. This ranged from 0 for dead calm up to storm force 13. He specified the amount of sail that a full-rigged ship should carry under the various wind conditions. It was taken up in the early 1830s by Robert FitzRoy (the captain of the *Beagle* on which Darwin sailed round the world) and officially adopted by the Admiralty in 1838. Modifications were made to the scale when sail gave way to steam.

Beaufort did major surveying work especially around the Turkish coast in 1812. In 1829 he became hydrographer to the Royal Navy, promoting voyages of discovery such as that of Joseph Hooker with the *Erebus*.

Beaumont William 1785–1853 was a US surgeon who did important early work on the physiology of the human stomach by taking advantage of a bizarre surgical case that he treated early in his career. He established that digestion is a chemical process, and his work encouraged other researchers to study the physiology of digestion.

Beaumont was born in Lebanon, Connecticut, on 21 November 1785. He was the son of a farmer, and worked as a schoolteacher in Champlain, New York, before going to study medicine at St Albans, Vermont. He was granted a licence to practise, becoming an assistant surgeon in the army in 1812. He resigned his commission after three years and practised in Plattsburgh, resuming his army career in 1820 when he was sent to the frontier post of Fort Mackinac, Michigan. In 1834 he was transferred to St Louis, Missouri, and he worked there for the rest of his life, although he left the army in 1839. During this time (1837) he became Professor of Surgery at St Louis University. He died in St Louis on 25 April 1853.

On 6 June 1822 a young French Canadian trapper named Alexis St Martin was accidentally shot in the left side from the back at close range, causing severe injury where the shot passed right through his abdomen. Beaumont was at Fort Mackinac, and treated the trapper swiftly and skilfully. The young man survived although he retained a permanent traumatic fistula, or hole, between his stomach and the outside of his abdomen. Beaumont looked after St Martin for two years, and in 1825 began a series of experiments and observations on the behaviour of the human stomach under various circumstances. Through the fistula he was able to extract and analyse gastric juice and stomach contents at various stages of digestion, observe changes in secretions, and note the muscular movements of the stomach. But he so hounded the unfortunate St Martin that he left (and eventually outlived Beaumont by nearly 30 years). Beaumont published his findings in *Experiments and Observations on the Gastric Juice* (1833).

Beaumont's work predated by many years the use of endoscopic examinations of the stomach and was the first well-documented and accurate observation of the digestive processes of a living human being.

Becquerel (Antoine) Henri 1852–1908 was a French physicist who discovered radioactivity in 1896, an achievement for which he shared the 1903 Nobel Prize for Physics with Marie Curie (1867–1934) and Pierre Curie (1859–1906). The Curies did not participate in Becquerel's discovery but investigated radioactivity and gave the phenomenon its name.

Becquerel was born in Paris on 15 December 1852 and educated at the Ecole Polytechnique and Ecole des Ponts et Chaussées, where he received a training in engineering. In 1875, he began private scientific research, investigating the behaviour of polarized light in magnetic fields and in crystals, linking the degree of rotation to refractive index. Both Becquerel's grandfather and father were respected physicists with positions at the Museum of Natural History and other institutions. On their deaths, in 1878 and 1891 respectively, Becquerel succeeded to their posts. He became a member of the Academy of Sciences in 1889, a professor at the Museum in 1892 and at the Ecole Polytechnique in 1895.

Becquerel then began the work for which he is remembered, not necessarily because of his position but because of the discovery of X-rays made by Wilhelm Röntgen (1845–1923) early in 1896. This prompted Becquerel to investigate fluorescent crystals for the emission of X-rays, and in so doing he accidentally discovered radioactivity in uranium salts in the same year. Marie and Pierre Curie then searched for other radioactive materials, which led them to the discovery of polonium and radium in 1898.

Becquerel subsequently investigated the radioactivity of radium, and showed in 1900 that it consists of a stream of electrons. In the same year, Becquerel also obtained evidence that radioactivity causes the transformation of one element into another. Following his award of the 1903 Nobel Prize for Physics jointly with the Curies, Becquerel became Vice-President (1906) and President (1908) of the Academy of Sciences. He died soon after on 25 August 1908 in Brittany.

Becquerel's discovery of radioactivity was prompted by the mathematician Henri Poincaré (1854–1912), who told Becquerel that X-rays were emitted from a fluorescent spot on the glass cathode-ray tube used by Röntgen. This immediately suggested to Becquerel that X-rays might be produced naturally by fluorescent crystals, with which he was familiar through his father's interest in fluorescence. He therefore placed some crystals of potassium uranyl sulphate on a photographic plate wrapped in paper, and put it in sunlight to make the crystals fluoresce. When he developed the plate, Becquerel found it to be fogged, showing that a radiation resembling X-rays had penetrated the paper and exposed the plate. Becquerel then tried to repeat the experiment to make further investigations, but the weather was cloudy and the uranium crystals would not fluoresce as there was no sunlight. He put a wrapped plate and the crystals into a drawer and waited. The weather did not improve and Becquerel impatiently decided to develop the plate. To his astonishment, the plate had been strongly exposed to radiation. Clearly it was not connected with fluorescence, but was emitted naturally by the crystals all the time.

Becquerel studied the radiation and found that it behaved like X-rays in penetrating matter and ionizing air. He showed that it was due to the presence of uranium in the crystals, and subsequently found that a disc of pure uranium metal is highly radioactive. This led Marie and Pierre Curie to isolate the elements polonium and radium, which is even more radioactive. Becquerel later subjected the radiation from radium to magnetic fields and was able to prove by the amount of deflection that it must consist of the electrons that had been discovered by J. J. Thomson (1856–1940) in 1897. Becquerel also discovered that radioactivity could be removed from a radioactive material by chemical action, but that the material subsequently regained its radioactivity.

Becquerel's discovery of radioactivity and its investigation by himself and the Curies caused a revolution in physics. It marked the beginning of nuclear physics by showing that atoms, and then nuclei within atoms, are made up of smaller particles. Furthermore, the spontaneous regeneration of radioactivity observed by Becquerel was evidence that one element can be transformed into another with the production of energy. A full explanation of radioactivity was achieved by Ernest Rutherford (1871–1937), leading eventually to nuclear fission and the production of nuclear energy.

Beg Ulugh 1394–1449 was a title meaning 'great prince' and the name by which Muhammad Taragay, mathematician and astronomer, came to be known in later life.

Beg was born at Sulaniyya in Central Asia (Persia) on 22 March 1394 and was brought up at the court of his grandfather Timur (Tamerlane). At the age of 15 Ulugh Beg became ruler of the city of Samarkand and the province of Maverannakhr. Although his grandfather was interested in conquest, Ulugh Beg's leanings were towards science and, in particular, astronomy. In 1420 he founded an institution of higher learning, or 'madrasa', in Samarkand. It specialized in astronomy and higher mathematics. Four years later he built a three-storey observatory and a 'Fakhrī' sextant of sufficiently large dimensions to enable very accurate observations to be made. The institution and observatory were advanced for the time and consequently the work of Ulugh Beg and his hand-picked team of scientists held good for many centuries. In 1447 he succeeded his father, Shahrukh, to the Timurid throne, but he met a tragic and violent death when he was murdered at the instigation of his own son on 27 October 1449.

The observatory was reduced to ruins by the beginning of the sixteenth century and its precise location remained unknown until 1908, when the archaeologist V. L. Vyatkin found its remains. The main instrument proved to be the Fakhrī sextant, the arc of which was placed in a trench about 2 m/16.5 ft wide. The trench itself was dug into a hillside along the line of the meridian. One of the preserved artefacts is a piece of the arc consisting of two walls faced with marble and 51 cm/20 in apart. Other instruments used at the observatory included an armillary sphere, a triquetram and a 'shamila', an instrument serving as astrolabe and quadrant.

The Fakhrī sextant was used mainly for determining the basic constants of astronomy by observing the Sun and, in particular, the Moon and planets. Since the radius of the arc was 40.4 m/132.6 ft, the divisions of the arc were correspondingly large, allowing for very accurate measurements to be made. By observing the altitude of the Sun at noon every day, Ulugh Beg was able to deduce the Sun's meridianal height, its distance from the zenith and the inclination of the ecliptic. The value that he obtained for the inclination of the ecliptic differs by only 32 sec from the true value for his time.

The *Zij* of Ulugh Beg and his school is a large work that was originally written in the Tadzhik language. It consists of a theoretical section and the results of observations made at the Samarkand Observatory. Included in the work are tables of calendar calculations, of trigonometry, and of the positions of planets, as well as a star catalogue.

Ulugh Beg and his collaborator Alkashi took great pains to determine accurately the sine of 1° by two independent methods. The tables give the values of sines and tangents for every minute to 45°, and for every 5 minutes between 45° and 90°. Cotangents are given for every degree. The values in the tables differ from the true values by a maximum of only one digit in the ninth decimal place.

The great accuracy to which the school worked is also evident in the values obtained for the movements of the planets Saturn, Jupiter, Mars, Venus, and Mercury. The differences between Beg's data and that of modern times are amazingly small, the discrepancies being within the limits of 2 to 5 sec for the first four and 10 sec at the most for Mercury. The somewhat larger discrepancy for the latter is attributable to Mercury's being smaller and having a higher orbital velocity and a greater eccentricity of orbit, which makes it more difficult to observe with the naked eye.

The catalogue of stars in the *Zij* contains 1,012 stars and includes 992 fixed stars whose positions Beg re-determined with unusual precision. This was the first star catalogue to be produced since that of al-Sufi, nearly five centuries earlier. Its great value lies in the fact that it was original, even though Beg was influenced by Ptolemy in the coordinates he used.

An expedition headed by T. N. Kari-Niazov discovered the tomb of Ulugh Beg in Samarkand in 1941. It was found that Ulugh Beg had been laid to rest fully clothed – a sign that, according to the Islamic religion, Ulugh Beg had been deemed a martyr, therefore a testament to his great contribution to the advancement of science, particularly astronomy.

Behring Emil Adolph von 1854–1917 was a German physician and immunologist who won the first Nobel Prize for Physiology or Medicine for his work on serum therapy against diphtheria and tetanus.

He was born on 15 March 1854 in Hansdorf, Prussia (now in Poland), where his father was a schoolteacher. Behring was educated at the Friedrich Wilhelm Institute of Military Medicine in Berlin, where he received a free medical training in return for subsequent service in the army medical corps. While still in training, Behring became interested in the possibility of preventing disease by hygienic and disinfectant methods, these concerns received further promotion whilst serving with a cavalry regiment in Pozen, now in Poland. During the next few years he combined clinical military duties with further research and training, especially in the use of disinfectants. In 1889 after completing his army service he became the assistant of the bacteriologist Robert Koch (1843–1910) at his Institute for Hygiene in Berlin. There Behring was joined by a young Japanese worker Shibasaburo Kitasato (1842–1931) and they discovered that healthy guinea pigs injected with serum prepared from the blood of animals infected with diphtheria remained healthy even when subsequently injected directly with diphtheria bacillus. Behring recognized that there was a substance in the transferred serum that rendered the poisonous toxins harmless, and this substance he termed antitoxin. From this Behring and Kitasato developed safe clinical preparations of their antitoxin serum and defined effective dosages for this new serum therapy. In its first year of use in Berlin children's hospitals in 1891, mortality from diphtheria dropped by more than two-thirds. A modification of the technique, by the French scientist Emil Roux (1853–1933) to produce antitoxin serum from horses rather than from guinea pigs improved large-scale production, and serum therapy for diphtheria and also tetanus became available worldwide. Behring adopted Roux's procedure and was supported in the production of serum by a German chemical firm. In 1893 he was appointed Professor of Hygiene in Berlin but the follwing year, in the wake of his deteriorating relationship with Koch, he moved to the University of Halle and in the year after that to the University of Marburg, in an attempt to establish his own research institute. In 1901 he was awarded the first-ever Nobel Prize for Physiology or Medicine, and was elevated to the Prussian nobility. Throughout this period his research continued into developing methods of standardizing serum antitoxins and he began to investigate the problem of tuberculosis. He recognized that the bovine and human forms of the disease were caused by the same microorganism, thus identifying the danger to humans of drinking contaminated milk. He also introduced early vaccination techniques against diphtheria and tuberculosis, and during World War I his tetanus vaccines saved the lives of millions of German soldiers, for which Behring was, most unusually for a civilian, awarded an Iron Cross. He died in Marburg on 31 March 1917, honoured and remembered as a great benefactor to children whose lives were saved by his antitoxins and vaccines against diphtheria and tetanus.

Beijerinck Martinus Willem 1851–1931 was a Dutch botanist who in 1898 published his finding that an agent smaller than bacteria could cause diseases, an agent that he called a virus (the Latin word for poison).

Beijerinck was born in Amsterdam on 16 March 1851. His earliest scientific interest was botany, but he graduated in 1872 with a diploma in chemical engineering from the Delft Polytechnic School, where one of his friends was Jacobus van't Hoff (who later won the first Nobel Prize for Chemistry in 1901). After graduating, Beijerinck taught botany to provide himself with a living while he studied for his doctorate, which he gained in 1877. He then became interested in bacteriology and took a job as a bacteriologist with an industrial company. In order to learn more about the subject, he travelled extensively throughout Europe. In 1895 he returned to the Delft Polytechnic School, where he taught and carried out research for the rest of his career. He died on 1 January 1931 in Gorssel in the Netherlands.

In the early 1880s Beijerinck began studying the disease that stunts the growth of tobacco plants and mottles their leaves in a mosaic pattern (now called the tobacco mosaic

virus disease). He tried to find a causative bacterium but was unsuccessful. This research stimulated his interest in bacteriology, however, and led to his taking a job as an industrial bacteriologist. While working in this capacity he discovered one of the types of nitrogen-fixing bacteria that live in the nodules on the roots of leguminous plants.

After he returned to academic life in 1895, Beijerinck resumed his study of the tobacco mosaic disease and again tried to isolate a causative agent. He pressed out the juice of infected tobacco leaves and found that the juice alone was able to infect healthy plants, but he could not detect a bacterial pathogen in the juice nor could he culture a microorganism from it. Furthermore, he found that the juice remained infective after he had passed it through a filter that removed even the smallest bacteria. Beijerinck was also certain that the causative agent was not a toxin because he could infect a healthy plant and from that plant infect another healthy plant, continuing this process indefinitely – therefore the infective agent had to be capable of reproduction.

Louis Pasteur had earlier postulated the existence of pathogens too small to be visible under the microscope; and Dimitri Ivanovski, a Russian bacteriologist, had in 1892 observed that tobacco mosaic disease could be transmitted by a filtered juice but he thought that there was a flaw in his filter and still believed the disease to be bacterial. It was Beijerinck who first published his findings and stated that the tobacco mosaic disease is caused by

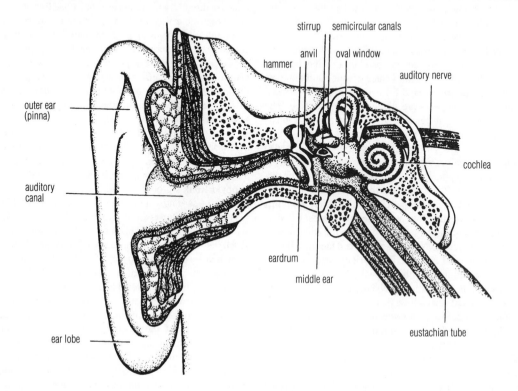

Békésy *Békésy's study of human hearing resulted in an analysis of the mechanism of the cochlea and the role of the basilar membrane within it.*

a non-bacterial pathogen. He believed that the filtered juice of the infected plants was itself alive, and he called the causative agent a filterable virus. Thus Beijerinck was the first to recognize the existence of a class of pathogens now known to cause a wide range of diseases in animals and plants, as well as in human beings. He was, however, mistaken in his belief that the virus was a liquid; in 1935 the American biochemist Wendell Stanley (1904–1971) demonstrated that viruses are particulate.

Békésy Georgvon 1899–1972 was a Hungarian-born US scientist who resolved the longstanding controversy on how the inner ear functions. For his discovery concerning the mechanism of stimulation within the cochlea, he received the 1961 Nobel Prize for Physiology or Medicine (the first physicist to do so).

Békésy was born in Budapest on 3 June 1899, where his father was a member of the diplomatic service. He went to the University of Bern in 1916, graduated in 1920, and then enrolled at the University of Budapest where he took his PhD in physics in 1923. For the next 23 years he worked in the laboratories of the Hungarian Telephone System. During this time he was also employed at the central laboratories of Siemens and Halske AC in Berlin, and from 1932 to 1939 he was a lecturer at the University of Budapest. He was appointed Special Professor there from 1939 to 1940 and a full Professor from 1940 to 1946. In that year he emigrated to Sweden, disturbed by the Soviet occupation of Hungary, and worked at the Karolinska Institute in Stockholm. He held the title of Research Professor there, although he did not actually take up the post because, in 1947, he emigrated to the United States to become a research lecturer at the Psycho-Acoustic Laboratory at Harvard. From 1949 to 1966 he served as a Senior Research Fellow in Psychophysics. He then went to the University of Hawaii where he took up the appointment of Professor of Sensory Sciences, and remained there until his death in Honolulu on 13 June 1972.

While working as a telecommunications engineer for the Hungarian Telephone System, Békésy decided that to determine what frequency range a new cable should be able to carry, he would investigate how the human ear actually receives sound. He researched the functioning of the eardrum by gluing two mirrors to it and beaming light and sound into the ear. In this way he was able to observe reflections of the movements of the membrane when it was activated by sound waves.

In another series of experiments, Békésy observed how the auditory ossicles in the middle ear – the hammer, anvil and stirrup – pick up the vibrations transmitted to them by the eardrum, and how they relay these messages to the cochlea in the inner ear. It had long been known that nerves in the cochlea pick up sound signals and transmit them along the auditory nerve to the brain for interpretation. It was also known that the cochlea consists of a spiral-shaped channel through which runs a thin partition known as the basilar membrane, containing groups of fine fibres. Hermann von Helmholtz had postulated a theory, which was

widely accepted at that time, that each fibre had a natural period of vibration and responded only to sounds that vibrated at that period. He claimed that each group of fibres stimulated different nerve endings, which thus enabled the brain to distinguish specific frequencies.

Békésy's painstaking and original work disproved the Helmholtz theory. He constructed models of the cochlea and also worked with cadavers whose auditory mechanisms he stimulated electrically. Békésy had to design new instruments and develop new techniques in order to experiment on the delicate cochlea. He devised extremely fine drills and probes, and the scissors he used had blades of only a few hundredths of a millimetre long. He reached the cochlea by grinding a small opening in the skull, and revealing part of the basilar membrane. By substituting a saline solution containing fine aluminium particles for the fluid in the cochlea and by using stroboscopic illumination, he was able to observe and measure for the first time a phenomenon he called the 'travelling wave'.

Békésy found that the innermost ossicle, the stirrup, acts as a lid on an opening in the cochlea, called the oval window. As sound vibrations cause the stirrup to move it exerts pressure on the fluid within the cochlea, and these vibrations are transmitted to the basilar membrane in the form of travelling waves. He established that as the waves pass along the basilar membrane the entire membrane vibrates. Each wave causes maximum vibration at different sections of the membrane according to its frequency. High-frequency waves produced by high-pitched sounds reach their peak on that part of the basilar membrane nearest the stirrup. Low-frequency waves, from low sounds, attain a maximum amplitude farther along the membrane.

Later, Békésy extended his interests to visual and tactile sensations, which he measured and recorded. He developed an audiometer – an instrument that determines whether deafness is caused by damage to the brain or to the ear – so that the appropriate treatment could be determined at an earlier stage.

Békésy also devised a means by which he enabled the skin to 'hear'. His apparatus consisted of a greatly enlarged version of a cochlea – a long tube filled with fluid and with a membrane running along its length. If the forearm of a person were pressed against the membrane, the skin of the arm felt the high and low sounds sent through the tube at distinctly different positions along the arm. Békésy's experiments pointed the way to methods which could enable the totally deaf to hear through tactile sensations, and improved modern interpretations of deafness.

Bell Alexander Graham 1847–1922 was the Scottish-born US scientist who invented the telephone. He became the first man to transmit the human voice by electrical means and the telephone system he initiated is now known and used worldwide.

Bell was born in Edinburgh on 3 March 1847. Until the age of 11 he was taught at home by his mother. After graduating from the Edinburgh Royal High School at the age of 15, he went to London and lived with his grandfather, who was widely known as a speech tutor. From him Bell gained a knowledge of the mechanisms of speech and sound. In 1867, after a year of teaching and studying at Bath, he became an assistant to his father, who had originated the phonetic 'visible speech' system for teaching the deaf.

After the death of his second brother in 1870 the family moved to Brantford, Ontario, for health reasons. In April 1871, Bell started giving instruction in 'visible speech' to teachers of the deaf in and around Boston, Massachusetts, and by 1873 he was Professor of Vocal Physiology at Boston University, a post he held for four years.

Bell's interest in speech and sound began at a very early age. As a boy he had constructed an automaton – using rubber, cotton and a bellows – simulating the human organs of speech. He had even experimented with the throat of his pet Skye terrier, attempting to turn cooperative growls into words.

In 1874 he was granted patents on a multiple or harmonic telegraph for sending two or more messages simultaneously over the same wire. Elisha Grey had already developed the telegraph and, at this time, several men had become aware of the possibility of transmitting the human voice by electtrical means. Bell, with Thomas A. Watson as his assistant, had devised a transmitter for all the various voice frequencies by placing a magnetized reed in the centre of a circular diaphragm that was set to vibrate at the human voice. The vibrating reed induced varying electrical oscillations in an associated electromagnet. The same device, at the distant end of the circuit, functioned as a receiver.

The following year, Bell and Watson improved on this, and on 10 March 1876 Bell became the first person ever to transmit speech from one point to another by electrical means. The message was 'Mr Watson, come here; I want you'. The first message travelled only the distance from one room to the next, but two months later the first 'long distance' voice message travelled 13 km/8 mi from Paris to Brantford, Ontario; before the end of the year, this distance had been increased to 229 km/143 mi. The patent for the telephone invented by Bell was granted in 1876. From then on, the telephone system spread rapidly across the country.

With the financial independence he had gained through the success of his telephone system Bell was able to set up a laboratory at Baddeck Day, on the Bras d'Or lakes of Cape Breton Island, Nova Scotia. Among the first fruits of his work here was the photophone, which used the photoresistive properties of selenium crystals to apply the telephone principle to transmitting words in a beam of light. Bell thus achieved the first wireless transmission of speech and from these principles evolved the photoelectric cell and other developments. Had coherent light (lasers) or optical fibres been available to him, Bell might have been able to develop the photophone into something useful. As it was, he seemed to recognize its potential instinctively for in 1921, in an age of intercontinental radio, he declared that the photophone was his greatest invention. The spectrophone, for carrying out spectrum analysis by means of sound, followed.

Bell established the Volta Laboratory in Washington and patented the gramophone and wax recording cylinder which were commercially successful improvements on Thomas Edison's first phonograph and cylinders of metal foil. The laboratory also experimented with flat disc records, electroplating records and impressing permanent magnetic fields on records – the embryonic tape recorder. Bell's entire share of $200,000 from the sale of some of his patents served to guarantee the perpetuation of the Volta Bureau which he had formed for all conceivable kinds of research into deafness.

In 1881, Bell developed two telephonic devices for locating metallic masses (usually bullets) in the human body. One, an induction balance method, was first tried out on President Garfield, who was assassinated in 1881, while the other, a probe, was widely used until the advent of X-rays. At the Baddeck Bay laboratory, he built hydrofoil speed boats and sea-water converting units. He also made tetrahedral kites capable of carrying a person.

Some idea of the true breadth of Bell's inventiveness can be gained by his lesser-known developments. These included an air-cooling system, a special strain of sheep (which he claimed produced twin or triplet lambs in more than half the births), the forerunner of the iron lung, and a sorting machine for punch-coded census cards.

In 1882 Bell became a citizen of the United States and moved to Washington, DC. In 1888, he became a founder member of the National Geographic Society.

The commercial success of his many inventions made Bell a wealthy man. He was also the recipient of many awards. In 1876, he received the Gold Medal awarded by the judges at the Centennial Exhibition in Philadephia, where his prototype telephone was first shown. In 1880 he received the Volta Prize for France and was made an officer of the French Legion of Honour. From 1898 to 1903 he was President of the National Geographic Society, and in 1898 he was appointed Regent of the Smithsonian Institution.

In 1917 a massive granite and bronze memorial was erected in Bell's honour at Brantford, Ontario. After his death on 2 August 1922 at his winter home at Baddeck Bay, a museum was built by the Canadian government at Baddeck as a permanent reminder of his achievements.

Bell Charles 1774–1842 was a British anatomist and surgeon who carried out pioneering research on the human nervous system. He gave his name to Bell's palsy, an extracranial paralysis of the facial nerve (the VIIth cranial nerve – not the same as the long thoracic nerve of Bell which he also named, and which supplies a muscle in the chest wall).

Bell was born in Edinburgh in November 1774. His brother, John Bell, was a renowned surgeon who taught him anatomy. After qualifying in 1799 Bell became a surgeon at the Edinburgh Royal Infirmary. He went to London in 1804 as a lecturer, and in 1812 was appointed surgeon at the Middlesex Hospital, London. He became Professor of Anatomy and Surgery at the Royal College of Surgeons in 1824 and four years later was invited to become the first

Principal of the Medical School at University College, London. He was knighted in 1831. In 1836 he became Professor of Surgery at the University of Edinburgh. He died at Hallow, Worcestershire, on 28 April 1842.

Bell carried out meticulous dissections and made the important discovery that nerves are composite structures, each with separate fibres for sensory and motor functions. His findings first appeared in a short essay *Idea of a New Anatomy of the Brain* (1811); his main written work was *The Nervous System of the Human Body* (1830). The chief significance of Bell's discovery was the impetus it gave to other researchers in neurology.

Bell Patrick 1799–1869 was the Scottish clergyman who invented one of the first successful reaping machines. It is probable that later commercially successful machines owed much to his pioneering work.

Bell was born in April, 1799, on a farm of which his father was a tenant in the parish of Auchterhouse, a few miles northwest of Dundee, Scotland. He studied for the ministry at St Andrew's University, and it was there in 1827 that he turned his attention to the construction of a machine which he (like many other inventors) thought would considerably reduce the labour of the grain harvest.

Harvesting equipment has progressed most since the beginning of the nineteenth century when Bell made his important contribution. Since then, methods have advanced from the use of the primitive sickle to the self-propelled combine-harvester. It is said that the grain-harvesters of Egypt in 3000 BC could have been used by most farm workers 4,800 years later without any additional training being needed.

It was the invention of the reaper that opened the way to the complete mechanization of the grain harvest, and many attempts were made from the end of the eighteenth century to cut corn by machine.

The way most of these machines performed has been lost to history and it was not until Bell's machine appeared that the record becomes clearer. He started trials in deep secrecy inside a barn on a crop which had been planted by hand, stalk by stalk. In 1828, he and his brother carried out night-time trials which were a success, leading them to exhibit the machine the following year. In the years to 1832 at least 20 machines were produced, 10 of them cutting 130 hectares (320 acres) in Britain and 10 going for export. Six reapers were exhibited at the Great Fair in New York in 1851. That year also saw the Great Exhibition in London, at which reapers by Hussey and McCormick were exhibited, both of them showing similarities to the Bell machines. This was the turning point for mechanical harvesting; mechanization gradually invaded farms and the sickle and scythe were virtually ousted by the twentieth century.

Bell's reaper was pushed from behind by a pair of horses and the standing cereals were brought on to the reciprocating cutter bar by horizontally revolving rods similar to the reels seen on modern combine harvesters. The cut cereal fell on to an inclined rotating canvas cylinder and was sheaved and stooked by hand. One of Bell's machines was used on his brother's farm for many years until, in 1868, it was bought by the museum of the Patents Office where it was afterwards kept. He did not take out a patent and the design was improved upon and reintroduced as the 'Beverly Reaper' in 1857.

In recognition of his services to agriculture, Bell was presented with £1,000 and a commemorative plate by the Highland Society, the money being raised mainly from the farmers of Scotland. He also received the honorary degree of LLD from the University of St Andrews.

Although Bell did not achieve the fame of others such as McCormick, his work was of fundamental value and importance in the successful advancement of agricultural methods to the now well-established mechanized farming used today.

Bell died in 1869 in the parish of Carmylie, Arbroath, of which he had been ordained Minister in 1843.

Bell Burnell (Susan) Jocelyn 1943–   is a British astronomer who discovered pulsating radio stars – pulsars – an important astronomical discovery of the 1960s.

Jocelyn Bell was born in Belfast, Northern Ireland, on 15 July 1943. The Armagh Observatory, of which her father was architect, was sited near her home and the staff there were particularly helpful and offered encouragement when they learned of her early interest in astronomy. From 1956 to 1961 she attended the Mount School in York. She then went to the University of Glasgow, receiving her BSc degree in 1965.

In the summer of 1965 she began to work for her PhD under the supervision of Antony Hewish at the University of Cambridge. It was during the course of this work that the discovery of pulsars was made. In 1968 she married Martin Burnell, and, in the same year, having completed her doctorate at Cambridge, went on to work in gamma-ray astronomy at the University of Southampton. From 1974 to 1982 she worked at the Mullard Space Science Laboratory in X-ray astronomy. In 1982 she was appointed a Senior Research Fellow at the Royal Observatory, Edinburgh, where she worked on infrared and optical astronomy. She was head of the James Clerk Maxwell Telescope section, responsible for the British end of the telescope project based in Hawaii. In 1991 she was appointed Professor of Physics at the Open University, Milton Keynes. A winner of the Royal Astronomical Society's prestigious Herschel medal, she has made significant contributions in the fields of X-ray and gamma-ray astronomy.

Jocelyn Bell spent her first two years in Cambridge building a radio telescope that was specially designed to track quasars – her PhD research topic. The telescope that she and her team built had the ability to record rapid variations in signals. It was also nearly 2 hectares/4.5 acres in area, equivalent to a dish of 150 m/500 ft in diameter, making it an extremely sensitive instrument. The sky survey began when the telescope was finally completed in 1967 and Bell was given the task of analysing the signals received. One day, while scanning the charts of recorded signals, she noticed a rather unusual radio source that had occurred during the night and been picked up in a part of the sky that was opposite in direction to the Sun. This was curious because strong variations in the signals from quasars are caused by solar wind and are usually weak during the night. At first she thought that the signal might be due to a local interference, but after a month of further observations it became clear that the position of the peculiar signals remained fixed with respect to the stars, indicating that it was neither terrestrial nor solar in origin. A more detailed examination of the signal showed that it was in fact composed of a rapid set of pulses that occurred precisely every 1.337 sec. The pulsed signal was as regular as the most regular clock on Earth.

One attempted explanation of this curious phenomenon was that it represented an interstellar beacon sent out by extraterrestrial life on another star and so initially it was nicknamed, LGM, for Little Green Men. Within a few months of noticing this signal, however, Bell located three other similar sources. They too pulsed at an extremely regular rate but their periods varied over a few fractions of a second and they all originated from widely spaced locations in our Galaxy. Thus it seemed that a more likely explanation of the signals was that they were being emitted by a special kind of star – a pulsar.

Since the astonishing discovery was announced, other observatories have searched the heavens for new pulsars and some 300 are now known to exist, their periods ranging from hundredths of a second to four seconds. It is thought that neutron stars are responsible for the signal. These are tiny stars, only about 7 km/10 mi in diameter, but they are incredibly massive. The whole star and its associated magnetic field are spinning at a rapid rate and the rotation produces the pulsed signal.

Beltrami Eugenio 1835–1899 was an Italian mathematician whose work ranged over almost the whose field of pure and applied mathematics, but whose fame derives chiefly from his investigations into theories of surfaces and space of constant curvature, and his position as the modern pioneer of non-Euclidean geometry.

Beltrami was born at Cremona on 16 November 1835. From 1853 to 1856 he studied mathematics at the University of Pavia. After graduating he was engaged as secretary to a railway engineer, but in his spare time he continued his research in mathematics and in 1862 he published his first paper, an analysis of the differential geometry of curves. In the same year he was appointed to the Chair of Complementary Algebra and Analytical Geometry at the University of Bologna. The rest of his life was spent in the academic world: until 1864 at Bologna, from 1864 to 1866 as Professor of Geodesy at Pisa, at Bologna again from 1866 to 1873, as Professor of Rational Mechanics, at the new University of Rome from 1873 to 1876 (in the same Chair), and at Pavia from 1876 to 1891 as Professor of Mathematical Physics. In 1891 he returned to Rome, where he continued to teach until his death on 4 June 1899. A year before he died he was appointed President of the Accademia dei Lincei and was made a member of the Italian senate.

Beltrami's career may be divided into two parts. After 1872 he devoted himself to topics in applied mathematics, but in the earlier period he worked chiefly in pure mathematics, on problems in the differential geometry of curves and surfaces. Work had been done in this field earlier, notably by Saccheria a century before and by Nikolai Lobachevsky in the first half of the nineteenth century. But their work had had little influence, since their contemporaries were unimpressed by the possibilities of non-Euclidean geometry. The publication of Beltrami's paper, 'Saggio di interpretazione della geometria non-euclidia', in 1868 is therefore a landmark in the history of mathematics. It advanced a theory of hyperbolic space that laid the analytical base for the development of non-Euclidean geometry.

In an earlier paper of 1865, Beltrami had shown that on surfaces of constant curvature (and only on them) the formula $ds^2 = E du^2 + 2F du dv + G dv^2$ can be written such that the geodesics are represented by linear expressions in u and v. For positive curvature R^{-2} the formula would be $ds^2 = R^2 (v^2 + a^2) du^2 - 2uv du dv + (u^2 + a^2) dv^2 \times (u^2 + v^2 + a^2)^{-2}$, the geodesics behaving like the great circles of a sphere. In 1868 he went further and showed that by changing R to iR and a to ia, a new formula for ds^2 was obtained, one which defined the surfaces of constant curvature $-R^{-2}$ and which presented a new type of geometry for the geodisics of constant curvature inside the region $u^2 + v^2 < a^2$.

These demonstrations were not greatly different from what Lobachevsky had shown 40 years earlier, but what Beltrami did was to present the theories in terms which were acceptable within the existing Euclidean framework of the subject. He demonstrated that the concepts and formulae of Lobachevsky's geometry are realized for geodesics on surfaces of constant negative curvature. He showed also that there are rotation surfaces of this kind – and to these he gave the name 'pseudospherical surfaces'. He also demonstrated the usefulness of employing differential parameters in surface theory, thereby beginning the use of invariant methods in differential geometry.

After 1872 Beltrami switched his attention to questions of applied mathematics, especially problems in elasticity and electromagnetism. His paper 'Richerche sulle cinematice dei fluidi' (1872) was an important development in the field of elasticity. But his lasting fame rests on his signal achievement in overcoming the prevailing mid-nineteenth-century suspicions of non-Euclidean geometry and, by bringing it into the mainstream of mathematical thought, opening wide-ranging fields of new inquiry.

Benz Karl Friedrich 1844–1929 was the German engineer who designed and built the first commercially successful motor car.

Benz was born in Karlsruhe, Germany, on 26 November 1844, the son of an engine driver, Johann Georg Benz, who died when Karl was two. He was educated at the gymnasium and polytechnic in his home town and began his career as an ordinary worker in a local machine shop when he was 21. Benz appears to have consciously shaped his career, moving from mechanical work on steam engines to design work with a more general engineering company in Mannheim, and then to a larger firm of engineers and iron-founders. In 1871 he returned to Mannheim, married Berta Ringer the following year, and opened a small engineering works in partnership with August Ritter.

After severe financial difficulties, Benz, now on his own, produced a two-stroke engine of his own design in 1878 and five years later, after more financial difficulties, attracted enough support to found a new firm, Benz and Co. In 1885, he produced what is generally recognized to have been the first vehicle successfully propelled by an internal-combustion engine.

The motor car he produced stemmed from over 150 years of experimental work by many engineers working in many different fields. Benz was the first person to bring together the many threads to produce and exploit a commercially viable road vehicle. The evolution of the modern motor car really began when Joseph Cugnot built his steam-driven gun carriage in 1769. Many other steam-driven vehicles followed, including that of Trevithick (1802). Whether the Belgian, Lenoir, or the Austrian, Marcus, was the first to produce a carriage driven by an internal combustion engine is a matter for dispute. Lenoir's machine of 1862 was driven by one of his gas engines and is said to have moved at 5 kph/3 mph. Marcus's machine is said to have run in 1868 and was certainly exhibited at the Vienna Exhibition in 1873. Neither, however, came to anything: Lenoir's had to carry about its own supply of town gas and was immensely heavy, while the Marcus engine had no clutch and was difficult to start.

In 1878 Benz produced his first two-stroke, 0.75 kW/1 hp engine in the factory he had founded at Mannheim in 1872. The commercial success of this enabled him to found a new company, Benz and Co., and to experiment with the construction of motor vehicles as well as engines. As it happened, the timing was critical because Otto and Alphonse Beau de Rochas had each independently developed four-stroke engines. As a result of litigation between the two inventors, the Otto cycle became available to Benz and he was thus able to build a suitable power unit for his three-wheeled Tri-car.

In the spring of 1885, he produced what is generally regarded as the world's first vehicle successfully propelled by an internal combustion engine. It used an Otto cycle four-stroke engine which gave about 0.56 kW/0.75 hp at 250 rpm and achieved a speed of up to 5 kph/3 mph during a journey of 91 m/299 ft on private ground adjoining the Mannheim workshop. Benz firmly believed that this vehicle would be a completely new system and not simply a carriage with a motor replacing the horse. The engine had a massive fly-wheel and was mounted horizontally in the rear, using electric ignition by coil and battery. The cooling system consisted simply of a cylinder jacket in which the water boiled away, being topped up as necessary. It had a carburettor of Benz's own design which vaporized the fuel over a hot spot.

By the autumn, the prototype Tri-car was covering 1 km/0.6 mi at 12 kph/7.5 mph and had become a familiar sight on the streets of the town. The production model Tri-car appeared in 1886–87 and had a 1 kW/1.5 hp single-cylinder engine. The following year a Tri-car with an occasional extra seat and a 1.5 kW/2 hp twin-cylinder engine appeared. Although the three-seat version was available for the Munich exhibition of 1888, Benz decided to exhibit a two-seater and won a gold medal for it.

Like most innovators, Benz had to contend with apathy and official hostility. There was also little demand for his Tri-cars at that time because of the public's general rejection of such 'monsters' and the severe restrictions placed on their use on public roads. British law, for instance, framed mainly with road-running steam engines in mind, required that all such vehicles be preceded by a man carrying a red flag and move not faster than 6.4 kph/4 mph. This kind of control did nothing to promote sales. However, there was sufficient financial interest shown in France to enable Benz to improve his vehicles still further.

Benz laid down his first four-wheeled prototype in 1891 and by 1895, he was building a range of four-wheeled vehicles that were light, strong, inexpensive and simple to operate. These automobiles had engines of 1–5.5 kW/1.5–6 hp and ran at speeds of about 24 kph/15 mph. Between 1897 and 1900 improved models appeared in increasing numbers – in particular, the Benz Velo 'Comfortable' of which over 4,000 were sold. At £135 each, they found a steady sale.

Benz and Co. was now a thriving concern and, in 1899, it was turned into a limited company. Although Benz retired from the board at the time of the transformation, the company he founded grew to become world famous for its production of high performance cars. Under the gifted Hans Nibel as chief designer, the firm produced very successful racing cars and luxury limousines. In 1926, the company merged with the other famous German firm of Daimler – a firm which had developed at the same time as Benz and Co., along very similar lines and from similar beginnings – to form the world-famous Daimler-Benz company.

Benz died on 4 April 1929, at Ladenburg.

Berg Paul 1926– is a US molecular biologist who shared (with Walter Gilbert and Frederick Sanger) the 1980 Nobel Prize for Chemistry for his work in genetic engineering, particularly for developing DNA recombinant techniques that enable genes from simple organisms to be inserted into the genetic material of other simple organisms. He is also well known for advocating restrictions on genetic engineering research because of the unpredictable, even dangerous, consequences that might result from uncontrolled DNA recombinant experiments.

Berg was born on 30 June 1926 in New York City. He was educated at Pennsylvania State University, from which he graduated in 1948, and at the Western Reserve University, from which he obtained his doctorate in 1952. From 1952 to 1954 he was an American Cancer Society Research

Fellow at the Institute of Cytophysiology in Copenhagen and at the School of Medicine at Washington University, St Louis. Between 1955 and 1974 he held several positions at Washington University: from 1955 to 1959 he was an assistant then associate professor in the Microbiology Department of the School of Medicine; from 1959 to 1969 he was Professor of Microbiology; and from 1969 to 1974 he was Chairman of the Microbiology Department. In 1970 he was also appointed Willson Professor of Biochemistry at the Medical Center of Stanford University, California.

Berg's early work concerned the mechanisms involved in intracellular protein synthesis. In 1956 he identified an RNA molecule (later known as a transfer RNA) that is specific to the amino acid methionine. He then began his Nobel-prize-winning work in which he perfected a method for making bacteria accept genes from other bacteria. This genetic engineering technique for DNA recombination can be extremely useful for creating strains of bacteria to manufacture specific substances, such as interferon. But there are also considerable dangers in the controlled use of these methods – a new, highly virulent pathogenic microorganism might accidentally be created, for example. Berg became aware of this danger and campaigned for strict controls on certain types of genetic engineering experiments. As a result, an international conference was held in California, followed by the publication in 1976 of guidelines to restrict genetic engineering research.

Berg has also studied how viral and cellular genes interact to regulate growth and reproduction, and has investigated the mechanisms of gene expression in higher organisms.

Bergius Friedrich Karl Rudolf 1884–1949 was a German industrial chemist famous for developing a process for the catalytic hydrogenation of coal to convert it into useful hydrocarbons such as petrol and lubricating oil. For this achievement he shared the 1931 Nobel Prize for Chemistry with Carl Bosch (1874–1940).

Bergius was born in Goldschmieden, near Breslau, Silesia (now in Poland), on 11 October 1884, the son of the owner of a chemical factory. He studied chemistry at the universities of Breslau and Leipzig, gaining his doctorate in 1907. He did postdoctoral research with Herman Walther Nernst at Berlin and then at Karlsruhe Technische Hochschule with Fritz Haber, who introduced him to high-pressure reactions. From 1909 he was Professor of Chemistry at the Technische Hochschule in Hannover. He then founded a private research laboratory in Hannover and in 1914 went to work for Goldschmidt AG in Essen, where he remained until the end of World War II in 1945. He lived for a while in Austria, then went to Spain, before finally settling in Argentina in 1948, where he held an appointment as a technical adviser to the government. He died in Buenos Aires on 30 March 1949.

In 1912 Bergius worked out a pilot scheme for using high pressure, high temperature and a catalyst to hydrogenate coal dust or heavy oil to produce paraffins (alkanes) such as petrol and kerosene. The commercial process went into production in the mid-1920s, and became important to Germany during World War II as an alternative source of supply of petrol and aviation fuel. The process yielded nearly 1 tonne of petrol from 4.5 tonnes of coal. He also discovered a method of producing sugar and alcohol from simple substances made by breaking down the complex molecules in wood; the rights to the process were purchased by the German government in 1936. He continued this work in Argentina, and found a way of making fermentable sugars and thus cattle food from wood.

Bernard Claude 1813–1878 was a French physiologist whose research and teaching were vitally important in founding experimental physiology as a separate discipline, distinct from anatomy, in the mid nineteenth century.

Bernard was born on 12 July 1813 in St Julien, in the Beaujolais region of France. The son of a wine grower, Bernard originally wanted to be a playwright. On the advice of a theatre critic he started to study medicine, qualified in 1839 and became a research assistant to the physiologist François Magendie at the Collège de France in Paris. He graduated MD in 1843 but never practised medicine, preferring to develop his career in experimental physiology. He experienced great difficulties in obtaining suitable positions at the beginning of his career but in 1854 a chair of general physiology was created for him at the Faculty of Sciences in Paris, and the following year he succeeded Magendie to become Professor of Medicine at the Collège de France. He received numerous honours during his lifetime including the Légion d'Honneur, and after his death on 10 February 1878 was given a state funeral. Bernard performed a series of important experiments on the physiology of digestion, showing that pancreatic secretions were important in fat metabolism; revealing the importance of the digestive activities of the small intestine, and investigating the mechanisms of nervous control of gastric secretion. He also discovered glycogen from experiments on the perfused liver; revealed the function of nerves which control the dilation or contraction of blood vessels, and investigated the physiology of fetal tissues and the nutritive role of the placenta. He made major investigations into the role of drugs such as curare and opium alkaloids and their effects on the nervous system. His most important contribution to physiological theory was the concept of the 'milieu intérieur', that life requires a consistent internal environment that is maintained by physiological mechanisms. Bernard was an ardent teacher of the new experimental physiology and young physiologists from around the world went to Paris to train in his laboratory. His major didactic work was *The Introduction to the Study of Experimental Medicine* (1865), which provided a comprehensive treatise on the role of experimental research as the basis for medicine.

Bernays Paul 1888–1977 was a British-born Swiss mathematician who was chiefly interested in the connections between logic and mathematics, especially in the field of set theory.

Bernays was born in London on 17 October 1888, but grew up in Berlin, where he attended the Köllnisches Gymnasium from 1895 to 1907. As an undergraduate he studied mathematics, philosophy and theoretical physics at the universities of Berlin and Göttingen. In 1912 he presented his post-doctoral thesis, called an *Habilitationschrift*, at Zurich, where he continued to do research until 1917. In that year he was invited to become David Hilbert's assistant at Göttingen. In 1919 he received his *venia legendi*, or right to lecture in the university, at Göttingen and he remained there as a lecturer without tenure until 1933. In that year he became one of the early Jewish victims of the Nazi regime in Germany, when his *venia legendi* was withdrawn. Hilbert employed him privately for six months, but Bernays then decided to take advantage of his father's adopted Swiss nationality and move to Switzerland. For several years he had to make do with short-term teaching appointments at the Technical High School in Zurich, before that institute granted him a *venia legendi* in 1939. He became an extraordinary professor there in 1945 and joined the editorial board of the philosophical journal *Dialectica*. In the 1950s and 1960s he was several times a visiting professor at the University of Pennsylvania and the Princeton Institute for Advanced Study.

Bernays's early interests ranged over a wide area of problems in mathematics. His doctoral thesis of 1912 was on the analytic number theory of binary quadratic forms and his *Habilitationschrift*, later in the same year, dealt with function theory, in particular Picard's theorem. He then became interested in 'axiomatic thoughts', and it was after hearing Bernays lecture on this subject at Zurich in the autumn of 1917 that Hilbert invited him to Göttingen as his assistant to work on the foundations of arithmetic. There he wrote his *Habilitationschrift* on the axiomatics of the propositional calculus in Bertrand Russell's *Principia Mathematica*; this was published in abridged form in 1926.

Bernays's most enduring work was in the field of mathematical logic and set theory. He first presented his principles of axiomatization in a talk to the Mathematical Society at Göttingen in 1931, but hesitated to publish his opinions because he was troubled by the thought that axiomatization was an artificial activity. His fullest treatment of the subject was given in lectures at the Princeton Institute for Advanced Study in 1935–36.

Bernays made a significant contribution to the theory of sets and classes. In his treatment of the subject, classes are not given the status of real mathematical objects. This represents a fundamental divergence from the theory of John von Neumann. Bernays modified the Neumann system of axioms to remain closer to the original Zermelo structure. He also used some of the set-theoretic concepts of Friedrich Schröder's (1841–1902) logic and some of the concepts of the *Principia Mathematica* which have become familiar to logicians. In Bernays's theory there are two kinds of individuals, 'sets' and 'classes': a 'set' is a multitude forming a real mathematical object, whereas a 'class' is a predicate to be regarded only with respect to its extension.

For all his own doubts about the validity of axiomatization, Bernays's arrangement of sets and classes is now widely believed to be the most useful, and by his study of the work of such men as Neumann, Hilbert, and Abraham Fraenkel, he made a major contribution to the modern development of logic.

Bernoulli Daniel 1700–1782 was a Swiss natural philosopher and mathematician, whose most important work was in the field of hydrodynamics and whose chief contribution to mathematics was in the field of differential equations.

Bernoulli was born at Groningen in the Netherlands on 9 February 1700, the son of the mathematician Jean Bernoulli and nephew of the mathematician Jacques Bernoulli. He was educated in Basle, Switzerland, where in 1705 his father had assumed the professorship of mathematics that became vacant on the death of Daniel's uncle. He first studied philosophy and logic, obtaining his baccalaureate by the age of 15 and his master's degree by the age of 16. He had also by then been given some mathematical training by his father and uncle, but in 1717 he started to study medicine. He received his doctorate for a thesis on the action of the lungs in 1721. His interest in mathematics seems then to have quickened. In 1724 he published his *Exercitationes Mathematicae* and in 1725 he was appointed to the Chair of Mathematics at the St Petersburg Academy. Bernoulli left Russia in 1732 and in the following year became Professor of Anatomy and Botany at the University of Basle. He remained there, although after 1750 as Professor of Natural Philosophy, until his retirement in 1777. He died in Basle on 17 March 1782.

Bernoulli was a scientific polymath. During his career he won ten prizes from the French Academy, for papers on subjects which included marine technology, oceanology, astronomy and magnetism. In physics, Bernoulli made an outstanding contribution in *Hydrodynamica*, which is both a theoretical and practical study of equilibrium, pressure and velocity in fluids. Bernoulli showed, in the principle given his name, that the pressure of a fluid depends on its velocity, the pressure decreasing as the velocity increases. This effect is a consequence of the conservation of energy and Bernoulli's principle is an early formulation of the idea of conservation of energy. *Hydrodynamica* also contains the first attempt at a thorough mathematical explanation of the behaviour of gases by assuming they are composed of tiny particles, producing an equation of state that enabled Bernoulli to relate atmospheric pressure to altitude, for example. This was the first step towards the kinetic theory of gases achieved a century later.

In mathematics, his interests probably stemmed, in part at least, from his close friendship with Euler and d'Alembert, who introduced him to problems associated with vibrating strings, in which connection Bernoulli did much of his work on partial differential equations. Perhaps his single most striking achievement in pure mathematics was to solve the differential equation of J. F. Riccati (1676–1754). Other achievements were to demonstrate how the differential calculus could be used in problems of probability and to do some pioneering work in trigonometrical series and the computation of trigonometrical functions. He also showed the shape of the curve known as the lemniscate $(x^2 + y^2)^2 = a^2(x^2 - y^2)$, where a is constant and x and y are variables.

Bernoulli was a competent mathematician and a first-rate physicist, and his use of mathematics in the investigation of problems in physics makes him one of the founders of mathematical physics.

Bernoulli Jacques 1654–1705 and Jean 1667–1748 were Swiss mathematicians, each of whom did important work in the early development of calculus.

The brothers were born at Basle – Jacques on 27 December 1654 and Jean on 7 August 1667. Although Jacques was originally trained in theology and expected to pursue a career in the Church, he made himself familiar with higher mathematics – especially the work of Descartes, John Wallis, and Isaac Barrow – and on a trip to England in·1676 met Robert Boyle (1627–1691) and other leading scientists. He then decided to devote himself to science, became particularly interested in comets (which he explained by an erroneous theory in 1681) and in 1682 began to lecture in mechanics and natural philosophy at the University of Basle. During the next few years he came to know the work of Leibniz and to begin a correspondence with him. In 1687 he was made Professor of Mathematics at Basle and he held the Chair until his death, at Basle, on 10 August 1705.

Jean Bernoulli originally studied medicine, but he was instructed in mathematics by his elder brother, and before he received his doctorate for a thesis on muscular movement he had already spent some time in Paris (1691) giving private tuition in mathematics. In 1694, the year in which his doctorate was awarded, he was appointed Professor of Mathematics at the University of Groningen, where he remained until 1705, when he succeeded his brother in the Chair at Basle. In 1730 he was awarded a prize by the French Academy of Sciences for a paper which sought (unsuccessfully, as d'Alembert was able to show) to reconcile Descartes's vortices with Kepler's third law. His son Daniel Bernoulli became one of the first mathematical physicists. Jean Bernoulli died on 1 January 1748.

Both Jacques and Jean wrote papers on a wide variety of mathematical and physical subjects, but their chief importance in mathematical history rests on their work on calculus and on probability theory. It is often difficult to separate their work, even though they never published together. For example, the brothers published similar solutions to the problem of the catenary at around the same time.

Jacques's most important papers were those on transcendental curves (1696) and isoperimetry (1700, 1701) – it is here that the first principles of the calculus of variations are to be found. It is probable that these papers owed something to collaboration with Jean. His other great achievement was his treatise on probability, *Ars Conjectandi*, which contained both the Bernoulli numbers and the Bernoulli theorem and which was not published until 1713, eight years after his death.

Together the two brothers advanced knowledge of the calculus not simply by their own work but also by giving spirited public support to Leibniz in his famous quarrel with Newton and thereby helping to establish the ascendancy of the Leibnizian calculus on the continent.

Bernstein Jeremy 1929–   is a US mathematical physicist who is well known for his 'popularizing' books on various topics of pure and applied science for the lay reader.

Bernstein was born at Rochester, New York, on 31 December 1929. He received his BA from Harvard in 1951 and his PhD from the same university in 1955. From 1957 to 1960 he was attached to the Institute for Advanced Study at Harvard. In 1962 he was appointed an associate professor in physics at New York University. In 1967 he became a Professor of Physics at the Stevens Institute of Technology at Hoboken, New Jersey. He has also for a number of years been a consultant to the Rand Corporation and to the General Atomic Company.

Bernstein's most important advanced work has been in the field of elementary particles and their currents. In particular he has sought to give a mathematical analysis and description to the behaviour of elementary particles. But although he is a competent mathematician who has worked at the very frontier of elementary particle theory, his greatest gift is the ability to write lucidly about difficult subjects for the non-specialist. Since 1962 he has been on the staff of the urbane magazine *The New Yorker*. His best known publication for that magazine was 'The analytical engine: computers, past, present and future', a witty guide to the history and theory of computers. He has also published a general survey of the historical progress of scientific knowledge, *Ascent* (1965), and a biography of Albert Einstein (1973).

It is a rare day when a mathematician both conducts research at the highest level of specialist theory and writes clearly and entertainingly for the public, and Bernstein's singular achievement in this regard was suitably recognized when he was awarded the Westinghouse prize for scientific writing in 1964.

Berthelot Pierre Eugène Marcelin 1827–1907 was a French chemist best known for his work on organic synthesis and in thermochemistry.

Berthelot was born in Paris on 27 October 1827, the son of a doctor. At first he studied medicine at the Collège de France, graduating in 1851; he then took up the study of chemistry. He worked at the Collège as assistant to his former tutor, Antoine Balard, under Jean Baptiste Dumas and Henri Regnault, gaining his doctorate in 1854 for a thesis on the synthesis of natural fats, which extended the work of Michel Chevreul. From 1859 to 1865 he was Professor of Organic Chemistry at the Ecole Supérieur de Pharmacie and in 1865 he returned to the Collège de France to take up a similar appointment, which he retained until his death. In 1870–1871, during the siege of Paris in the Franco-Prussian War, he was consulted about the defence of the capital and became President of the Scientific Defence Committee, and supervised the manufacture of guns and explosives. Thereafter he took an increasing part in politics. He became Inspector of Higher Education in 1876, President of the Committee on Explosives in 1878, a Senator in 1881, and Minister for Public Instruction in 1886; he was Foreign Minister from 1895 to 1896. In 1889 he succeeded Louis Pasteur as Secretary of the French Academy of Sciences. He died in Paris on 18 March 1907, on the same day as his wife.

All of Berthelot's early research concerned organic synthesis. He first studied alcohols, showing in 1854 that glycerol is a triatomic alcohol; he combined it with fatty (aliphatic) acids to make fats, including fats that do not occur naturally. This work provided increasing justification for the view that organic chemistry deals with all the compounds of carbon (including Berthelot's synthetic fats) and not just compounds formed and found in nature. He continued his research by investigating sugars, which he identified as being both alcohols and aldehydes. Using crude but effective methods he also synthesized many simple organic compounds, including methane, methyl alcohol (methanol), formic acid (methanoic acid), ethyl alcohol (ethanol), acetylene (ethyne) and benzene; he also made naphthalene and anthracene. His work during the 1850s was summed up in his book *Chimie organique fondée sur la synthèse* (1860).

Berthelot began his studies of thermochemistry in 1864. Paralleling the work of Germain Hess he measured the heat changes during chemical reactions, inventing the bomb calorimeter to do so and to study the speeds of explosive reactions. He introduced the term exothermic to describe a reaction that evolves heat, and endothermic for a reaction that absorbs heat. In 1878 he published *Mécanique chimique* followed by *Thermochimie* (1897), which put the science of thermochemistry on a firm footing.

In 1883 Berthelot established an experimental farm at Meudon, southwest of Paris. He discovered that some plants can absorb atmospheric nitrogen, investigated the action of nitrifying bacteria, and began to determine the details of the nitrogen cycle. But Berthelot was not a theorist; he was at his best carrying out practical work in the laboratory, and even led the opposition to the theory of atoms and molecules championed by Stanislao Cannizzaro in the early 1860s.

Berthollet Claude Louis 1748–1822 was a French chemist with a wide range of interests, the most significant of which concerned chemical reactions and the composition of the products of such reactions. He proposed that reactivity depends on the masses of the reactants (similar to the modern law of mass action) but that the composition of the product or products can vary, depending on the proportions of the reacting substances (contrary to the law of definite proportions). He was a champion of his contemporary Antoine Lavoisier – although not of his political views – and had a desire to put science at the service of humanity's practical needs.

Berthollet was born of French parents on 9 December 1748 (six years to the day after Karl Scheele) in the then Italian region of Savoy. In 1768 he qualified as a physician at the University of Turin, moving to Paris four years later to study chemistry under Pierre Macquer (1718–1784) while continuing his medical studies, receiving his French qualification in 1778. While private physician to Mme de Montesson in the household of the duc d'Orléans he carried out research in the laboratory at the Palais Royale.

After the death of Macquer in 1784 Berthollet was appointed inspector of dyeworks and director of the Gobelins tapestry factory. In 1787 he collaborated with Lavoisier on the publication of *Méthode de nomenclature chimique*, which incorporated the principles of the 'new chemistry' of Lavoisier. He taught chemistry to Napoleon and went with him to Egypt in 1798. There he observed the high concentration of sodium carbonate (soda) by Lake Natron on the edge of the desert. He reasoned that, under the prevailing physical conditions, sodium chloride in the upper layer of soil had reacted with calcium carbonate from nearby limestone hills – the beginning of his theory that chemical affinities are affected by physical conditions, in this case the heat and high concentration of calcium carbonate. He became a senator in 1804 but ten years later voted against Napoleon, and after the reformation he became a count. Berthollet died on 6 November 1822 at Arcueil, near Paris.

Berthollet's proposal that chemical compounds do not have a constant composition brought him into conflict with Louis Proust, who in 1799 put forward his law of definite proportions (which states that 'all pure samples of the same chemical compound contain the same elements combined together in the same proportions by weight'). It turned out that Berthollet's severe (but nonacrimonious) criticisms of Proust were based on imprecise distinctions between compounds, solutions and mixtures, as well as on the inaccurate analyses of impure compounds. For example, he suggested that lead and oxygen could combine in almost any proportion, but it is now known that he was making and analysing mixtures of various lead oxides. Although the controversy between Berthollet and Proust ended mainly in Proust's favour, Berthollet's views were not entirely wrong, although at that time they were based on false evidence. Nonstoichiometric compounds (also called Berthollide compounds), with a variable composition, have been studied since 1930. For example, it is now believed that lattice deficiencies in iron can account for ferrous sulphide – iron (II) sulphide, FeS – with compositions that vary between $FeS_{1.00}$ and $FeS_{1.14}$ (or $Fe_{1.00}S$ and $Fe_{0.88}S$).

Berthollet also found himself disagreeing with Lavoisier, and would not concur with the theory that all acids contain oxygen. In this he was correct, but shared with Karl Scheele the false assumption that chlorine was not an element but consists of oxygenated hydrochloric acid. He did, however, introduce the use of chlorine as a bleaching agent (which led him to devise a volumetric analytical method for estimating the chlorine content of a bleaching solution by titration against a standard solution of indigo). He also investigated chlorates, suggesting that the oxidizing properties of potassium chlorate could make it a replacement for potassium nitrate in gunpowder and so produce a more powerful explosive. A public demonstration of this idea in 1778 resulted in disaster and the deaths of some of the onlookers.

In other wide-ranging studies Berthollet devised a method of smelting iron and making steel; he correctly determined the compositions of ammonia, prussic (hydrocyanic) acid and sulphuretted hydrogen (hydrogen sulphide); and he made various discoveries in organic chemistry. On balance, Berthollet was right more often than he was wrong – and even when he was wrong, his arguments with Proust and Lavoisier stimulated chemical thought, ultimately to the benefit of the science.

Berthoud Ferdinand 1727–1807 was a Swiss clockmaker and a maker of scientific instruments. He improved the work of John Harrison, devoting 30 years' work to the perfection of the marine chronometer, giving it practically its modern form.

Son of Jean Berthoud, the architect and judiciary, Berthoud was born near Couvet. He was apprenticed to his brother, Jean-Henri, at the age of 14. In 1745 he went to Paris and in 1764 was appointed Horologer de la Marine. He made over 70 chronometers using a wide variety of mechanisms, and wrote ten volumes on the subject, many of them of considerable importance in his field.

In the early eighteenth century clockmakers devoted much effort to the construction of chronometers that could be used at sea. This was because if navigators did not know the time at the zero meridian it was impossible for them to plot their precise position. Since the different astronomical methods of measuring longitude gave inadequate results, the solution seemed to lie in finding ways of making very accurate clocks whose workings would not be disturbed by the motion of the ship.

The English clockmaker John Harrison (1693–1776) was undoubtedly the first maker of a timepiece that could be used satisfactorily at sea. In 1735 he completed his first marine chronometer; this was tested at sea and gave results which were good enough for further financial help to be given to its maker but not good enough for the reward offered under an act passed by the British Parliament in 1714. Harrison's chronometer number 4 was completed in 1759 and was tested at sea. While this met the requirements of the act, the responsible authorities in England, the Board of Longitude, still demanded numerous trials. Tested over long distances, his chronometer consistently showed variations, involving errors in the calculation of longitudes, that were less than the maximum of half a degree stipulated by the Board of Longitude. It was not until 1772, when Harrison was 80, and after long years of argument that he received the whole award.

Harrison had already obtained remarkable results when Berthoud began his investigations, which were conducted mainly from 1760 to 1768. He was a skilled clockmaker by the time he began to devote himself to the problem of marine chronometers and his work was guided by wide experience as well as his outstanding practical ability.

The range of parts constructed by Berthoud is large but the complete clocks made for his experiments numbered only seven or eight. His clocks numbers 1 and 2 were made in 1760 and 1763 respectively. Though number 2 was noticeably smaller than the first, these two instruments were still rather bulky. Berthoud put two circular balance wheels oscillating in opposite directions into these designs, which connected to one another by means of a toothed wheel. Their axes rested on roller bearings. A bimetallic grid iron, already widely used by clockmakers since its invention by Harrison, compensated for variations in the length of the spiral.

The escapement was made more complicated in clock number 2, in which Berthoud introduced an equalizing remontoir to compensate for errors caused by variations in the driving force of the escapements. A spring barrel fitted with a fuse was initially chosen as the source of motive power, a system still used in modern marine chronometers. However, Berthoud temporarily abandoned this system soon afterwards. In his third watch, also made in 1763, Berthoud used a bimetallic strip for thermal compensation and a single balance wheel. Retaining the single balance wheel suspended by a wire and guided by rollers, he then returned to the gridiron method of compensation for the construction of his subsequent clocks. Numbers 6 and 8 (1767) of these ensured his success.

In this series the motive power was produced by weights placed on a vertical metal plate which descended the length of three brass columns. An escapement with ruby cylinders was used, all parts being relatively compact. The mechanism was above the gridiron, the two occupying the top part of the clock, which was enclosed in a long glass cylinder. The three brass columns guiding the descent of the weight took up almost the whole height of this cylinder. A horizontal dial covered the glass box.

Clocks 6 and 8 were tested at sea in 1768 and 1769 and showed variations in their working of the order of from 5–20 sec a day.

We known of only five chronometers made by John Harrison, who did not reach his goal until his last years. Berthoud, on the other hand, was able to continue working for another 30 years after his initial success. Abandoning the roller suspension of the balance wheel, he adopted a pivot suspension. He came back to compensation by bimetallic strips, eliminating the cumbersome gridiron and used a balance wheel compensated by four small weights. Two of these weights were fixed and the two others were carried by a bimetallic strip which moved towards or away from the arbor according to variations in temperature. Berthoud also eventually returned to the spring drive. His nephew, Pierre-Louis Berthoud (1754–1813) succeeded him in his work as a marine clock maker and began to industrialize the construction of chronometers as an industry in France.

Berzelius Jöns Jakob 1779–1848 was a Swedish chemist, one of the founders of the science in its modern form. He contributed to atomic theory, devised chemical symbols, determined atomic weights, and discovered or had a hand in the discovery of several new elements. He became renowned as a teacher and gained a reputation as a world authority; such was his influence that other scientists were wary of contradicting him. Indeed the obstinacy with which he clung to his own theories, especially in later life, may well have retarded progress in some areas despite his many magnificent achievements, particularly in theoretical chemistry.

Berzelius was born on 20 August 1779 at Vaversunda, Ostergotland, of an ancient Swedish family that had long associations with the church. His father, a teacher at the gymnasium in nearby Linköping, died when Berzelius was only four years old and his mother married a pastor, Anders Ekmarck. Berzelius and his sister were brought up with the five Ekmarck children and educated by their step-father and by private tutors. When in 1788 his mother also died, the nine-year-old Berzelius moved again to live with a maternal uncle. He attended his father's old school but quarrelled with his cousins and six years later left the family to become a tutor on a nearby farm, where he developed a strong interest in collecting and classifying flowers and insects. He had been destined for the priesthood, but decided in 1796 to study natural sciences and medicine at Uppsala University.

Berzelius had to interrupt his studies to earn some money. His step-brother aroused in him an interest in chemistry, but he received little encouragement at the university and began to experiment on his own. In the summer of 1800 he was introduced to Hedin, the chief physician at the Medivi mineral springs, and began his first scientific work – an analysis of the mineral content of the water. Hedin also secured for Berzelius the unpaid post of assistant to the Professor of Medicine and Pharmacy at the College of Medicine in Stockholm, and when the professor died in 1807 Berzelius was given the post. The College became an independent medical school, the Karolinska Institute, in 1810, the same year that Berzelius was appointed President of the Swedish Academy of Sciences. On his wedding day in 1835 he was made a baron by the King of Sweden, in recognition of his status as the most influential chemist of the era. He died on 7 August 1848.

As an experimenter Berzelius was meticulous. Papers he published between 1810 and 1816 describe the preparation, purification and analysis of about 2,000 chemical compounds. In the course of this work he improved many existing methods and developed new techniques. Quantitative analysis on such a broad scale established beyond doubt Dalton's atomic theory and Proust's law of definite proportions. It also laid the foundation of Berzelius's determination of the atomic weights of the 40 elements known at that time – a prodigious task in which he was aided by the work of several contemporaries, such as Eilhard Mitscherlich (isomorphism), Pierre Dulong and Alexis Petit (specific heats), and Joseph Gay-Lussac (combining volumes); Mitscherlich and Dulong were his former students. In common with other scientists of the time,

Berzelius rejected Avogadro's hypothesis, which led him to confuse some atomic weights and molecular weights. Nevertheless most of the atomic weights in the table he published in 1828 closely correspond to the modern accepted values.

His dealing with such a variety of elements and compounds led Berzelius to simplify chemical symbols. Alchemists had used an elaborate pictorial presentation, and John Dalton had devised a system based on circular symbols that could be combined to represent compounds. Berzelius discarded both of these and introduced the notation still used today, in which letters (sometimes derived from Latin names) represent the elements and, combined with numbers if necessary, constitute the chemical formulae of compounds.

The invention of the voltaic cell at the beginning of the nineteenth century opened up a new field of research – electrochemistry. Berzelius's work in this area, although by no means as comprehensive as that of Humphry Davy, lent support to his dualistic theory, which stated that compounds consist of electrically and chemically opposed parts. He considered the parts to be stable groups of atoms, which he called 'radicals'. Although Berzelius's extension of the theory was later proved to be incorrect, it did contain the basis of the modern theory of ionic compounds.

Berzelius also made his mark in the discovery of new elements. With the German chemist Wilhelm Hisinger he found cerium in 1803. In 1815 he believed that he had isolated a second new element from a mineral specimen and named it thorium. Subsequent experiments revealed that the 'element' could be broken down into yttrium and phosphorus. His disappointment was softened by his discovery of selenium in 1818, silicon in 1824, and in 1829 a fourth new element, extracted from its ore by reduction with potassium, which he called thorium.

Several chemical terms in use today were coined by Berzelius. He noted that some reactions appeared to work faster in the presence of another substance which itself did not appear to change, and postulated that such a substance contained a 'catalytic force'. Platinum, for example, was well-endowed with such a force because it was capable of speeding up reactions between gases. Although he appreciated the nature of catalysis, Berzelius was unable to give any real explanation of the mechanism.

In the early nineteenth century it became apparent that elements could be grouped by similar chemical properties. Chlorine, bromine and iodine formed such a grouping. Each of these elements could be found as salts in sea water, consequently Berzelius invented the name 'halogens' (salt formers) to collectively describe the family.

A different branch of chemistry, concerned with substances derived from living things, was capturing the interest of scientists of the day. Berzelius referred to this new sphere as 'organic chemistry' and expounded the belief that organic compounds arose from the operation of a 'vital force' in the living cell; synthesis was therefore impossible. Then in 1828 Friedrich Wöhler (previously a student of Berzelius) prepared the organic compound urea

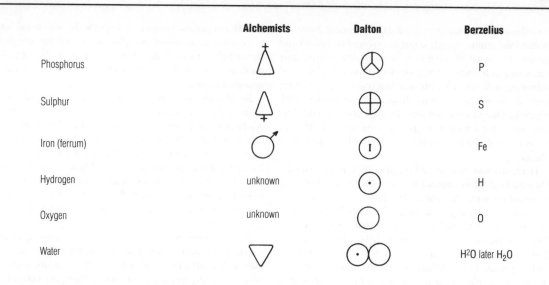

	Alchemists	Dalton	Berzelius
Phosphorus			P
Sulphur			S
Iron (ferrum)			Fe
Hydrogen	unknown		H
Oxygen	unknown		O
Water			H^2O later H_2O

Berzelius *Berzelius's system of chemical notation replaced the unwieldy pictorial symbols devised by the alchemists and by John Dalton.*

from the inorganic salt ammonium cyanate. With great reluctance Berzelius eventually abandoned his vital force theory.

The word 'isomerism' was also introduced by Berzelius, to describe substances that have the same chemical composition but different physical properties. He encountered the phenomenon when working with the salts of racemic and tartaric acids. Later Mitscherlich showed that, of the two, only tartarates rotate the plane of polarized light (are optically active); in 1848 Louis Pasteur resolved the racemates, which are an equimolecular mixture of two optically active tartarates (and have given their name to all such combinations, which are now known as racemic mixtures).

Throughout his career Berzelius recognized the importance to chemical progress of disseminating information. His *Textbook of Chemistry*, first published in 1803, was received with acclaim and was soon accepted as the definitive work for students of the time. In addition to publishing numerous research papers of his own, he collated the work of other chemists and acted as editor of an annual review of chemistry which was published between 1821 and 1849.

Bessel Friedrich Wilhelm 1784–1846 was a German astronomer who in 1838 first observed stellar parallax, and who set new standards of accuracy for positional astronomy. His measurement of the positions of about 50,000 stars enabled the first accurate calculation of interstellar distances to be made. Bessel was the first person to measure the distance of a star other than the Sun. From the parallax observation of 61 Cygni he calculated the star to be about six light years distant, thus setting a new lower limit for the scale of the universe.

The son of a government employee, Bessel was born on 22 July 1784 in Minden. He began work at the age of 15 as an apprentice in an exporting company. During this period in his life, an unhappy one, he dreamed of escape and

decided to travel. With this end in view, he studied languages, geography and the principles of navigation: this led to an interest in mathematics and, eventually, astronomy. In 1804 he wrote a paper on Halley's Comet in which he calculated the comet's orbit from observations made over a period of about a year. He sent the paper to Heinrich Olbers, who was so impressed that he arranged for its publication and obtained a post for Bessel as an assistant at Lilienthal Observatory. There Bessel worked under the early lunar observer Johann Schröter (1745–1816). After only four years the Prussian government commissioned Bessel to construct the first large German observatory at Königsberg, where in 1810 he was appointed Professor of Astronomy. Bessel's whole life was devoted first to the completion of the observatory (1813) and then its direction until his death in 1846.

Bessel's work laid the foundations of a more accurate calculation of the scale of the universe and the sizes of stars, galaxies and clusters of galaxies than any previous method had done. In addition, he made a fundamental contribution to positional astronomy (the exact measurement of the position of celestial bodies), to celestial mechanics (the movements of stars) and to geodesy (the study of the Earth's size and shape). Bessel enlarged the resources of pure mathematics by his introduction and investigation of what are now known as Bessel functions, which he first used in 1817 to determine the motions of three bodies moving under mutual gravitation. Seven years later he developed Bessel functions more fully for the study of planetary perturbations. Bessel played a great part in the final establishment of a scale for the universe in terms of the Solar System and terrestrial distances. These naturally depended upon accurate measurement of the distances of the nearest stars from the Earth.

Bessel's contributions to geodesy included a correction in 1826 to the seconds pendulum, the length of which is

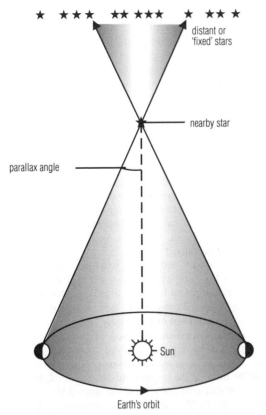

distant or 'fixed' stars

nearby star

parallax angle

Sun

Earth's orbit

Bessel *Bessel developed the technique of stellar parallax, by means of which distances to nearby stars can be calculated by observing their apparent change in position when viewed from opposite ends of a long baseline, such as a diameter of the Earth's orbit.*

precisely calculated so that it requires exactly one second for a swing. Between 1831 and 1832 he directed geodetical measurements of meridian arcs in East Prussia, and in 1841 he deduced a value of 1/299 for the ellipticity of the Earth, or the amount of elliptical distortion by which the Earth's shape departs from a perfect sphere. He was the first to make effective use of the heliometer, an instrument designed for measuring the apparent diameter of the Sun.

Bessel also introduced corrected observations for the so-called personal equation, a statistical bias in measurement that is characteristic of the observer himself and must be eliminated before results can be considered reliable. He concerned himself greatly with accuracy, to the point of making a systematic study of the causes of instrumental errors. His own corrected observations were far more accurate than previous ones and his methods pointed the way to a great advancement in the study of the stars.

Bessel's later achievements were possible only because he first established the real framework of the scale of the universe through his accurate measurement of the positions and motions of the nearest stars, making corrections for errors caused by imperfections in his telescope and by disturbances in the atmosphere. Having established exact

positions for about 50,000 stars, he was ready to observe exceedingly small but highly significant motions among them. Choosing 61 Cygni, a star barely visible to the naked eye, Bessel showed that the star apparently moved in an ellipse every year. He explained that this back-and-forth motion, called parallax, could only be caused by the motion of the Earth around the Sun. His calculation indicated a distance from Earth to 61 Cygni of 10.3 light years. When Olbers received these conclusions, on his 80th birthday, he thanked Bessel, who, he said, 'put our ideas about the universe on a sound basis'.

One of Bessel's major discoveries was that two bright stars, Sirius and Procyon, execute minute motions that could be explained only by the assumption that they had invisible companions to disturb their motions. The existence of such bodies, now called Sirius B and Procyon B, was confirmed with more powerful telescopes after Bessel's death. He also contributed to the discovery of the planet Neptune. He published a paper in 1840 in which he called attention to small irregularities in the orbit of Uranus which he had observed and which were caused, he suggested, by an unknown planet beyond. Bessel's minor publications numbered more than 350 and his major ones included a multivolume series *Astronomische Beobachtungen auf der K. Sternwarte zu Königsberg (1815–1844)*.

Realizing early in his life where his potential lay, Bessel succeeded in achieving it in a profession entirely different from the one in which he had started out. Olbers said that the greatest service he himself had rendered astronomy was that he recognized and furthered Bessel's genius.

Bessemer Henry 1813–1898 was a British engineer and inventor of the Bessemer process for the manufacture of steel, first publicly announced at a meeting of the British Association in 1856. His process was the first cheap, large-scale method of making steel from pig-iron. Bessemer had many other inventions to his credit but none is comparable to that with which his name is linked and which earned him a fortune.

With the French Revolution, the French Huguenot Bessemer family moved to England and their son Henry was born at Charlton, near Hitchin, in Hertfordshire, on 19 January 1813. He inherited much skill and enterprise from his father who was associated with the firm founded by William Caslon, one of the pioneers in the development of movable type for printing. His earliest years were spent in his father's workshop where he found every chance to develop his inclinations as an inventor. At the age of 17 he went to London where he put to use his knowledge of easily fusible metals and casting in the production of artwork. His work was noticed and he was invited to exhibit at the Royal Academy. In about 1838 Bessemer invented a typesetting machine and a little later he perfected a process for making imitation Utrecht velvet. In this his combined mechanical skill and artistic capacity proved useful, for he not only had to design all the machinery but he also had to engrave the embossing rolls.

In about 1840, encouraged by his great friend the printer

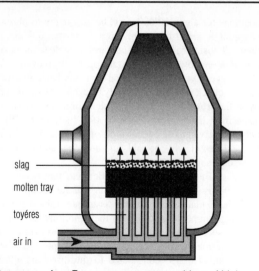

slag

molten tray

toyéres

air in

Bessemer *In a Bessemer converter, a blast of high-pressure air oxidizes impurities in molten iron and converts it to steel.*

Thomas de la Rue, Bessemer turned his attention to the manufacture of bronze powder and gold paint. At that time Germany had the monopoly in this industry, having learned its secrets from China and Japan. Bessemer's product was at least equal to that from Germany and one-eighth of the price. Between 1849 and 1853 he became interested in the process of sugar refining and obtained no less than 13 patents for machinery for this. He also invented a new method of making lead pencils. But it was the bronze powder and gold paint process, the secrets of which were kept in the Bessemer family for 40 years, that provided the capital needed to set up the small ironworks in St Pancras, London, where his experiments led to the invention of the Bessemer steel-making process. This was his greatest achievement and at the time was of enormous industrial importance.

With the Crimean War of the early 1850s Bessemer turned his energies to the problems of high gas pressures in guns which were probably caused by early attempts at rifling. When he offered his services to the British military commanders they showed no interest, so Bessemer turned to Britain's ally, France. The French military, which had used rifled weapons intermittently for years, expressed interest, and Napoleon III encouraged Bessemer to experiment further. With weapons that fire from rifled barrels, the projectile has to fit tightly or it is not set spinning by the helical grooves in the barrel. The tight fit leads to high pressures, and many early weapons exploded, killing the gun crew. Bessemer set out to find a form of iron strong enough to resist these pressures.

At that time steel was expensive and cast iron, although very hard, was extremely brittle. The carbon in cast iron could be laboriously removed to form practically pure wrought iron which was not brittle but ductile. Steel, with a carbon content between that of cast iron and wrought iron, was both hard and tough. But, to make steel, cast iron

had to be converted into wrought iron and carbon had then to be added – a laborious and time-consuming process. Bessemer considered the contemporary method of converting cast iron to wrought iron. A carefully measured quantity of iron ore was added to cast iron and then heated to the molten stage, when the oxygen in the iron ore combined with the carbon in the cast iron to form carbon monoxide (which was burned off), and leaving pure iron. Bessemer considered adding the oxygen directly as a blast of air, although it seemed likely that the cold air would cool and therefore solidify the molten iron. When he tried the process, however, he found that the reverse was true: the blast of air burned off the carbon and other impurities and the heat generated served to keep the iron molten – indeed the temperature was raised. By stopping the process at the right time he found that he could produce steel without the intermediate wrought iron stage, and that its cost was reduced dramatically.

Ironmakers were enthusiastic when in 1856 Bessemer announced his discovery, and vast sums of money were invested in equipment for the new process. But instead of becoming a hero, Bessemer was derided, because the steel that was produced was of a very poor grade. In his original experiments Bessemer had used phosphorus-free ore, while the ironmakers had used ore containing phosphorous. Bessemer assured the ironmakers that good steel could be made if the appropriate ore was used, but they would not listen. In 1860 Bessemer erected his own steel works in Sheffield, importing phosphorous-free iron ore from Sweden. The high-grade steel he made was sold for a fraction of the current price. Ironmasters, feeling the sting of competition, applied for licences and the royalties made Bessemer a very rich man. He retired at the comparatively early age of 56 but continued inventing. One of his later schemes was a stabilized saloon for an ocean-going ship, the theory being that passengers prone to seasickness would not feel the motion of the vessel. The scheme was not a success and Bessemer had his fingers severely burnt, losing around £34,000 in the venture.

The recognition Bessemer received for his steel-making process was, however, richly deserved for his method of making cheap steel benefited the whole world. He was not allowed to accept the Legion of Honour offered to him by the French Emperor, but he received many other honours and became a Fellow of the Royal Society in 1879. He was knighted in the same year for services to the Inland Revenue 40 years before when, before he was 20, he played an important part in the construction of the first method for stamping deeds, a method which was adopted by the British government. He died at Denmark Hill, London, on 15 March 1898. No fewer than six towns in the United States were named after him.

Bethe Hans Albrecht 1906– is a German-born US physicist and astronomer, famous for his work on the production of energy within stars. He was awarded the 1976 Nobel Prize for Physics for his work on energy production on stars.

Bethe was born in Strasbourg (now in France) on 2 July 1906, the son of a university professor. He was educated at the universities of Frankfurt and Munich, and gained a PhD from the latter in 1928. From 1928 to 1929 he was an instructor in physics at the University of Frankfurt, and then at Stuttgart. He went on to lecture at the universities of Munich and Tübingen (1930–33).

With the rise to power of Adolf Hitler in Germany, Bethe moved to Britain in 1933 and spent a year at the University of Manchester. He was a Fellow of Bristol University from 1934 to 1935, becoming Assistant Professor in 1935 and Professor in 1937. He held this position until 1975, when he went to the United States to become John Wendell Anderson Professor of Physics at Cornell University. He later became a naturalized US citizen.

From 1943 to 1946 Bethe was Chief of the Theoretical Physics Division of the Los Alamos Science Laboratory in New Mexico and he has been a consultant to Los Alamos since 1947. He has been a leading voice in emphasizing the social responsibility of the scientist and after World War II he served as part of the American delegation in Geneva during long negotiations with the Soviet Union on the control of nuclear weapons. He holds honorary doctorates from a large number of universities all over the world. He was awarded the Morrison Prize of the New York Academy of Sciences in 1938 and 1940, the American Medal of Merit in 1946, the Draper Medal of the National Academy of Sciences in 1948, the Planck Medal in 1955 and 1961, and the Fermi Award for his part in the development and use of atomic energy. He was made a Foreign Member of the Royal Society in 1957. Recently he won the National Medal of Science (1976). He is a Fellow of the American Physical Society and served as its president in 1954.

When Bethe went to the UK in the 1930s, he worked out how high-energy particles emit radiation when they are deflected by an electromagnetic field. This work was important to cosmic ray studies. In 1938 he made his most important contribution to science when he worked out the details of how nuclear mechanisms power stars. Carl von Weizsäcker was independently reaching the same conclusions in Germany. These nuclear mechanisms were to answer the questions that had concerned Helmholtz and Kelvin 75 years earlier.

Bethe's mechanism began with the combining of a hydrogen nucleus (a proton) with a carbon nucleus. This initiates a series of reactions, at the end of which the carbon nucleus is regenerated and four hydrogen nuclei are converted into a helium nucleus. Hydrogen acts as the fuel of the star and helium is the 'ash'; carbon serves as a catalyst. As stars like the Sun are mostly made up of hydrogen, there is ample fuel to last for thousands of millions of years. The amount of helium present indicates that the Sun has already existed for billions of years.

Bethe later proposed a second scheme that involves the direct combination of hydrogen nuclei to form helium in a series of steps that proceed at lower temperatures. When hydrogen is converted into helium, either directly or by means of the carbon, nearly 1% of the mass of the hydrogen is converted into energy. Even a small amount of mass produces a great deal of energy, and the loss of mass in the Sun is enough to account for its vast and seemingly eternal radiation of energy.

With the discovery of neutron stars, Bethe turned again to astrophysical research in 1970. These stars are held by gravity at such a high density that protons fuse with electrons to produce neutrons which constitute nearly all the matter.

Although Bethe was primarily concerned with the rapidly developing subject of atomic and nuclear processes, he has also investigated the calculation of electron densities in crystals, using classical mathematical methods, and the order–disorder states in alloys. He has also concerned himself with the operational conditions in nuclear reactors and the detection of underground explosions by means of seismographic records. Apart from his contributions to atomic theory, therefore, his concern with the implications of its practical application have earned him worldwide respect within the field.

Betti Enrico 1823–1892 was an Italian mathematician who was the first to provide a thorough exposition and development of the theory of equations formulated by Evariste Galois.

Betti was born near the Tuscan town of Pistoia on 21 October 1823. His father died when he was very young and he received his early education from his mother. After obtaining his BA in the physical and mathematical sciences at the University of Pisa, he taught for a time at a secondary school at Pistoia before being appointed to a professorship at Pisa in 1856; he held the professorship for the rest of his life. Betti was also much interested in politics. He had fought against Austria at the battles of Curtatone and Montanara during the first wars of Italian independence and in 1862 he became a member of the new independent Italian parliament. He entered the government as Under-Secretary of State for Education in 1874 and served in the senate after 1884. He died at Pisa on 11 August 1892.

Betti's early, and most important, work was on algebra and the theory of equations. In papers published in 1852 and 1855 he gave proofs of most of Galois's major theorems. In so doing he became the first mathematician to resolve integral functions of a complex variable into their primary factors. He also developed the theory of elliptical functions, demonstrating (in a paper of 1861) the theory of elliptical functions which is derived from constructing transcendental entire functions in relation to their zeros by means of infinite products. In thus providing formal demonstration to Galois's statements and in drawing out some of their implications, Betti greatly advanced the transition from classical to abstract algebra.

In 1863 a change occurred in Betti's mathematical interests. In that year the German mathematician Bernhard Riemann went to Pisa. He had only three years to live, but he became a close friend of Betti and directed Betti's mind to mathematical physics, especially to problems of potential theory and elasticity. Betti studied George Green's methods of attempting to integrate Laplace's equation (the foundation of the theory of potentials) and applied them to the study of elasticity and heat. The result was the paper of 1878 in which he gave the law of reciprocity in elasticity theory which became known as Betti's theorem. Along the way, conducting research into 'analysis situs' in hyperspace in 1871, he also did valuable work on numbers characterizing the connection of a variety, these later becoming known as 'Betti numbers'.

Betti also played a principal part in the expansion of mathematics teaching in Italian schools, in particular lending his enthusiastic advocacy to the restoration of Euclid to a central place in the secondary-school curriculum. His lectures at the University of Pisa also inspired a generation of Italian mathematicians, the most famous of them being Vito Volterra (1860–1940).

Bhabha Homi Jehangir 1909–1966 was an Indian theoretical physicist who made several important explanations of the behaviour of subatomic particles. He was also responsible for the development of research and teaching of advanced physics in India, and for the establishment and direction of the nuclear power programme in India. He commanded wide respect in the international scientific community both for his contributions in the scientific sphere and also for his formidable skills as an administrator. He held many positions of responsibility at home and abroad, particularly in organizations concerned with the development of peaceful uses of atomic energy.

Bhabha was born in Bombay on 30 October 1909. He attended a number of schools in Bombay before entering Gonville and Caius College, Cambridge, in 1927 to study mechanical engineering. There his tutor in mathematics was Paul Dirac (1902–1984), who originated the relativistic electron theory that led to the prediction of antiparticles. Bhabha became fascinated by mathematics and theoretical physics and so, after earning his first class honours degree in 1930, he began to do research at the Cavendish Laboratory in Cambridge. He was also then able to tour Europe, meeting scientists of such importance as Wolfgang Pauli (1900–1958) and Enrico Fermi (1901–1954) and also visiting Niels Bohr's laboratory in Denmark. He was awarded his PhD in 1935, and remained at Cambridge until 1939, when he returned to India for a holiday.

He was still in India when World War II broke out. As he was unable to return to Cambridge, a readership at the Bangalore Institute of Science was created for him and he was put in charge of a department investigating cosmic rays. The renowned physicist C. V. Raman (1888–1970) was the director of the Institute, and had a profound influence on him. Bhabha determined to remain in India and advance the development of science and technology there.

In 1944 he proposed the establishment of a centre for the training of scientists in advanced physics and for the cultivation of research in that field as a stimulus for industrial development. The Tata Institute of Fundamental Research was established at Bombay in 1945 with Bhabha as director, a position he held until his death.

Bhabha later took on many responsibilities in the establishment of the Indian atomic energy programme and also became a figure of great importance in the international scientific community, serving as president of the United Nations Conference on the Peaceful Uses of Atomic Energy, first held in Geneva in 1955, and from 1960 to 1963 as president of the International Union of Pure and Applied Physics. He died tragically on 24 January 1966 in a plane crash on Mont Blanc.

Bhabha studied at the Cavendish Laboratory at a time of great advances in the understanding of the structure and properties of matter. He made major contributions to the early development of quantum electrodynamics, a part of high-energy physics. His first paper concerned the absorption of high-energy gamma rays in matter. A primary gamma ray dissipates its energy in the formation of electron showers. In 1935, Bhabha became the first person to determine the cross-section (and thus the probability) of electrons scattering positrons. This phenomenon is now known as Bhabha scattering.

Bhabha also studied cosmic rays and in 1937 suggested that the highly penetrating particles detected at and below ground level could not be electrons. In 1946, they were in fact found to be mu-mesons. Bhabha also put forward a theory proposing the existence of vector mesons, which were later identified in nuclear interactions. In 1938, he suggested a classic method of confirming the time dilation effect of the special theory of relativity by measuring the lifetimes of cosmic ray particles striking the atmosphere at very high speeds. Their lifetimes were found to be prolonged by exactly the amount predicted by relativity.

Bhabha was a skilled theoretician and a master administrator. His love for his native land stimulated him to the cultivation of educational and research facilities of the highest standard, and furthermore to the awakening of his government's awareness of the potential importance of atomic energy. He did everything in his power to ensure that India would become self-sufficient in all stages of the nuclear cycle so as not to be dependent on other nations. He spoke for all developing nations when he said in 1964 at the Third United Nations Conference on the Peaceful Uses of Atomic Energy that 'no power is as expensive as no power'.

Bichat Marie Francois Xavier 1771–1802 was one of the leading French physicians of the Revolutionary era.

Born in Thoirette, Jura, a doctor's son, Bichat also studied medicine, first in Lyon, but his education was soon interrupted by military service in the Revolution. In 1793, at the height of the Terror, Bichat settled in Paris. From 1797 he taught medicine, working from 1801 at the Hotel Dieu, Paris's huge general hospital for the poor.

His greatest contribution to medicine and physiology stemmed from his insight that the diverse organs of the body contained particular tissues or (his favourite word) 'membranes'. He described 21 such membranes, including muscle, connective and nerve tissue. Bichat maintained that in the case of a diseased organ, it was generally not the whole organ but only certain tissues that were affected. While establishing the centrality of the study of tissues (histology), Bichat distrusted the microscope and made little use of it; his analysis of tissues consequently did not include any understanding of their cellular structure.

Executing his researches with enormous fervour during the last years of his short life – he performed over 600 postmortems – Bichat may be seen as a link between the morbid anatomy of Morgagni and the later cell pathology of Virchow. His lasting importance lay in the simplification he brought to anatomy and physiology, by showing that the complex structures of organs were to be understood in terms of their elementary tissues.

Bickford William 1774–1834 was an English leather merchant who invented the miner's safety fuse. He made a major contribution to safety and productivity in the mines and quarries, and for many years after electric ignition was introduced in 1952 the majority of charges were set off using fuses not very different from the one patented by him in 1831.

Bickford was born in Devonshire, and having tried unsuccessfully to carry on a currier's business (dressing and colouring tanned leather) in Truro, he set up as a leather merchant in Tuckingmill near Camborne in Cornwall. He was deeply distressed by the high casualty rate and terrible injuries suffered by local tin miners, and set out to discover a safe means of igniting charges. His first attempt failed, but his second resulted in a reliable fuse.

Gunpowder had been used for blasting since the early 1600s. The powder was put into a brass ball or 'pulta', the outside being covered with cotton soaked in saltpetre and dipped in molten pitch and sulphur. The powder was fired from outside through a small hole drilled through the brass case. Once the pitch was lit, the pulta was pushed into a crack in the rock.

The first borehole blasts were carried out by Casper Weindl in a mine near Ober-Biberstollen, about 75 miles north of Budapest. In England, gunpowder was probably first used for blasting at the Ecton copper mine near the Derbyshire–Staffordshire border. The shot holes were filled with powder and then 'stemmed', or, as a contemporary account puts it: blocked 'by stones and rubbish ramm'd in (except a little place that is left for a Train) the powder by the help of that train being fir'd'. The method had spread to quarries by the early 1670s.

By the end of the eighteenth century an extremely popular type of fuse made of goose quills was being used. The quills were cut so that they could be inserted one into the other and then filled with powder. Such fuses could be ignited directly, that is without any delaying element such as the sulphur mannikin. However, the quills were often broken or pushed apart during stemming so that an irregular or even a damp column of powder was left to act as a very uncertain fuse. Quill fuses frequently apparently failed and then rekindled so that the miner, who went to inspect the apparently extinct fuse, was injured in the blast.

Bickford's safety fuse provided a dependable means for conveying flame to the charge so that the danger of such hang fires was virtually eliminated. Its timing (the time required for a given length to burn) was more accurate and consistent than that of its predecessors, and it had much better resistance to water and to general abuse. The burning section was protected by the stemming in each shot hole so that several holes could be fired at a time without the fusing of the last being destroyed by the blast from the first.

This and other techniques which the new fuse allowed increased not only safety, but also productivity, quickly making it popular among both miners and management. Many different types were subsequently made for volley firing, for use in flammable atmospheres and for other applications. Bickford went into partnership with Thomas Davey, a working miner and a Methodist class leader, to construct the machinery for fuse production.

The first method used a funnel which trickled black powder into the centre of 12 yarns as they were spun. The process was discontinuous, producing 20 m/65 ft lengths of semifuse which were then 'countered' by twisting on a second set of yarns in the opposite direction to make a second 'rod' which would not unwind. The fuse was then covered with a layer of tar and resin. Later the process became continuous so that the length produced was limited only by interruptions in the process. This method, with refinements, is still in use today.

Bickford fell seriously ill in 1832 and took no further part in the exploitation of his invention. After his death in 1834 a fuse-making factory was built at Tuckingmill and continued in production until July 1961. Factories were later set up in Lancashire and also in America, France, Saxony, Austrio-Hungary, and Australia, the organization becoming one of the major units of Explosive Trades Limited, later to become Imperial Chemical Industries.

Bigelow Erastus Brigham 1814–1879 was the US inventor-industrialist who devised, among many such machines, the first power loom for weaving ingrain carpets and a loom for manufacturing Wilton and Brussels carpets.

Bigelow's beginnings at West Boylston, Massachusetts, where he was born on 2 April 1814, were so impoverished that he was forced to go out to work at the age of ten. Bigelow's father, a small farmer, also worked as a chairmaker and wheelwright. Until 1834, Bigelow did any work he could find – on neighbouring farms, playing the violin in the church orchestra or at local country dances, as a store clerk and as a stenography teacher; nevertheless he hankered after a formal education. He had hopes of entering Harvard University as a medical student and afterwards, of a professorial or literary career.

Though these hopes never materialized, Bigelow's natural talent for mechanics and mathematics could not be suppressed. At the age of eight he mastered arithmetic without the aid of a teacher. Bigelow produced his first loom for weaving coach lace to trim stagecoach upholstery in 1837, at the age of 23. This was followed by more looms for the manufacture of ginghams, silk brocatelle, counterpanes and other figured fabrics.

In about 1839 Bigelow met Alexander Wright, a Scots mechanic settled in the United States who had recently set up a small carpet-weaving business with three looms and twenty workers. Wright had heard of Bigelow's mechanized coach-lace loom and encouraged Bigelow to look into the possibility of inventing a power loom for ingrain-type carpets. Wright discovered that, despite 40 years of experiment in Europe, no satisfactory machine with this capability had yet been devised.

Like his predecessors, Bigelow faced many complex problems: how to set needles for the mechanical interweaving of up to three piles at a time, how to provide accurate timing for the take-up beam of the fabric, how to create a firm and even selvage, how to control the smooth surface with repeating patterns of uniform length so that they matched when seamed, and how to control the timing of several shuttle boxes.

The resultant loom had a long gestation and most of the cost was borne by the Clinton Company which Bigelow and his brother Horatio set up in 1843 to manufacture ginghams. In exchange, the company required exclusive rights to the machine which, after several modifications was at last ready for use, late in 1846.

Bigelow's power loom proved capable of producing up to 16.5 sq m/20 sq yd a day of two-ply goods and 11.7 sq m/14 sq yd of three-ply. It was a great improvement on his earlier machines, which could produce only the simplest of patterns, allowing patterns with large flowers and sweeping foliage, asymmet-rically arranged, to be produced. There was some popular criticism about the 'unnaturalness of walking on flowers' and geometric patterns were suggested instead. However, patterns with large floriated scrolls of the type woven by Bigelow's machines were ideal for incorporating the serrations around the leaf and flow-eredges which 'hid' the uneven joins of the interchange of coloured threads.

Bigelow's power loom for weaving Brussels and Wilton carpets was developed between 1845 and 1851 and together with his other machines brought the Clinton Company great riches. The town of Clinton, Massachusetts, which had started humbly as a factory village, grew up around the Bigelow plant. Later, other Bigelow mills were established at Lowell, Massachusetts, and Humphreysville, Connecticut.

Bigelow's looms transformed the carpet industry in the United States which until then had been outpaced by Britain, where weaving skills were greater and labour less costly. In fact, in the mid nineteenth century, the United States imported more British carpeting – an average of 550,000 sq m/660,000 sq yd a year – than was produced by all US factories put together. Ironically, when imported into Britain, Bigelow's looms gave the British an advantage over their French rivals because of their cost-cutting and because carpets could now be made of virtually any colour, to virtually any pattern.

Bigelow, who died in Boston on 6 December 1879, helped found the Massachusetts Institute of Technology in 1861. Six volumes of his English patents, with the original drawings, are preserved by the Massachusetts Historical Society.

Binet Alfred 1857–1911 was a French psychologist who is best known for his pioneering work on the development of mental testing, particularly the testing of intelligence.

Binet was born on 8 July 1857 in Nice. He went to Paris in 1871 to study law, but became interested in the work of the neurologist Jean-Martin Charcot on hypnosis and abandoned law in 1878 to study first neurology and later psychology at the Salpêtrière Hospital in Paris. He remained there until 1891, when he went to work in the physiological psychology laboratory at the Sorbonne in Paris. He became Director of the laboratory in 1895 and held this post until his death on 18 October 1911.

Binet was principally interested in applying experimental techniques to the measurement of intellectual abilities and, after joining the Sorbonne, began to devise various tests in an attempt to gain an objective measure of mental ability. After experimenting with various combinations of different tests Binet, with Théodore Simon, published in 1905 the Binet–Simon intelligence test. This test, which was designed to measure intellectual ability in children, required the subject to perform such tasks as naming objects, copying designs and rearranging disordered patterns. The subject was then given a mental-age score according to how well he or she had performed in the tests compared with previously established norms for various age groups. This was one of the first attempts at objectively measuring intelligence and, because of its usefulness, quickly became adopted in France and other countries. The original Binet–Simon test subsequently underwent many revisions, probably the most notable of which was the scoring of intelligence as an intelligence quotient (IQ) – calculated as the ratio of mental age to chronological age, multiplied by 100 – introduced in 1916 by Lewis Terman (1877–1956), an American psychologist working at Stanford University, in the adaptation known as the Stanford–Binet test. (The Stanford–Binet test was later revised several times and, in its latest form, is still used today.)

Binet wrote several books on mental processes and reasoning ability, notably *L'Etude expérimentale de l'intelligence* (1903; *Experimental Study of Intelligence*), a study of the mental abilities of his two daughters and, with Simon, *Les enfants anormaux* (1907; translated in 1914 as *Mentally Defective Children*). In addition he devised several tests that involved interpreting a subject's response to various visual stimuli (such as inkblots and pictures), the forerunners of some types of modern personality tests. Binet also studied and wrote about hypnosis and hysteria.

Birkhoff George David 1884–1944 was one of the most distinguished of early twentieth-century mathematicians, who made fundamental contributions to the study of dynamics and formulated the 'weak form' of the ergodic theorem.

Birkhoff was born at Overisel, Michigan, USA, on 21 March 1884. He studied at the Lewis Institute (now the

Illinois Institute of Technology) from 1896 to 1902, when he entered the University of Chicago. He then went to Harvard, where he received his BA in mathematics in 1905. In 1907 he received a PhD for his thesis on boundary problems from the University of Chicago. He taught at the University of Michigan and Princeton before being appointed an assistant professor at Harvard in 1912. He was a full professor at Harvard from 1919 until his death. Birkhoff was president of the American Mathematical Society in 1925 and of the American Association for the Advancement of Science in 1937. He was awarded the Bocher Prize in 1923 for his work on dynamics and the AAAS Prize in 1926 for his investigation of differential equations. He died at Cambridge, Massachusetts, on 12 November 1944.

Birkhoff's early work was on integral equations and boundary problems, and his investigations led him into the field of differential and difference equations. He developed a system of differential equations which is still inspiring research, which his work on difference equations was notable for the prominence which he gave to the use of matrix algebra.

Birkhoff's high reputation derives chiefly, however, from his investigation of the theory of dynamical systems such as the Solar System. After grounding himself thoroughly in Jules Poincaré's celestial mechanics, he began to examine the motion of bodies in the light of his work on asymptotic expansions and boundary value problems of linear differential equations. In 1913 – one of those exhilarating moments in mathematics – he proved Poincaré's last geometric theorem on the three-body problem, a problem with which Poincaré had grappled unsuccessfully. Birkhoff's formulation ran as follows: 'Let us suppose that a continuous one-to-one transformation T takes the ring R, formed by concentric circles C_a and C_b of radii a and b ($a >> 0$), into itself in such a way as to advance the point of C_a in a positive sense, and the point of C_b in a negative sense, and at the same time preserve areas. Then there are at least two invariant points.'

With John Von Neumann (1903–1957), Birkhoff was chiefly responsible for establishing, in the 1930s, the modern science of ergodics. He arrived, indeed, at the statement of his 'positive ergodic theorem', or what is known as the 'weak form' of ergodic theory, just before Neumann published his 'strong form' of it. By using the Lebesque measure theory Birkhoff transformed the Maxwell–Boltzmann hypothesis of the kinetic theory of gases, which was undermined by the number of exceptions found to it, into a vigorous principle.

Birkhoff also made a number of valuable contributions to related problems in other fields to which he was led by his ergodic investigations. One such was his paper of 1938, 'Electricity as a fluid' which, although consistent with Einstein's special relativity, found no need of the general curvilinear coordinates of the general theory of relativity. Throughout his life Birkhoff continued to argue that Einstein's general relativity was an unhelpful theory and his 1938 paper did much to provoke thought and research on the subject.

Few twentieth-century mathematicians achieved more than Birkhoff and he was, in addition, the most important teacher of his generation. Many of the United States's leading mathematicians did their doctoral or postdoctoral research under his direction. His standing, generally acknowledged, as the most illustrious American mathematician of the early twentieth century is deserved.

Bjerknes Vilhelm Firman Koren 1862–1951 was the Norwegian scientist who created modern meteorology.

Bjerknes came from a talented family. His father was professor of mathematics at the Christiania University (now Oslo) and a highly influential geophysicist who clearly shaped his son's studies. Bjerknes held chairs at Stockholm and Leipzig before founding the Bergen Geophysical Institute in 1917.

By developing hydrodynamic models of the oceans and the atmosphere, Bjerknes made momentous contributions that transformed meteorology into an accepted science. Not least, he showed how weather prediction could be put on a statistical basis, dependent on the use of mathematical models.

During World War I, Bjerknes instituted a network of weather stations throughout Norway; coordination of the findings from such stations led him and his co-workers to develop the highly influential theory of polar fronts, on the basis of the discovery that the atmosphere is made up of discrete air masses displaying dissimilar features. Bjerknes coined the word 'front' to delineate the boundaries between such air masses. Among much else, the 'Bergen frontal theory' explained the generation of cyclones over the Atlantic, at the junction of warm and cold air wedges. Bjerknes's work gave modern meteorology its theoretical tools and methods of investigation.

Black Joseph 1728–1799 was a Scottish physicist and chemist whose most important contribution to physics was his work on thermodynamics, notably on latent heats and specific heats. He is classed with Henry Cavendish and Antoine Lavoisier as one of the pioneers of modern chemistry. He is remembered for his discovery of carbon dioxide, which he called 'fixed air'.

Black was born in Bordeaux, France, on 16 April 1728. His father was from Belfast, but of Scottish descent, and was working in Bordeaux in the wine trade. Black went to Belfast to be educated, and then on to Glasgow University to study natural sciences and medicine. He moved to Edinburgh in 1751 to finish his studies, gaining his doctor's degree in 1754. Two years later he took over from William Cullen, his chemistry teacher at Glasgow, and was also offered the chair in anatomy. He soon changed this position for that of professor of medicine, and also practised as a physician. In 1766 he again followed Cullen as the Professor of Chemistry at Edinburgh University. Black died on 10 November 1799.

In Black's doctorate he described investigations in 'causticization' and indicated the existence of a gas distinct from common air, which he detected using a balance. He was therefore the founder of quantitative pneumatic chemistry and preceded Antoine Lavoisier in his experiments. In a more detailed account of his work, published in 1756, Black described how carbonates (which he called mild alkalis) become more alkaline (are causticized) when they lose carbon dioxide, whereas the taking up of carbon dioxide reconverts caustic alkalis into mild alkalis. Black identified 'fixed air' (carbon dioxide) but did not pursue this work. He also discovered that it behaves like an acid, is produced by fermentation, respiration and the combustion of carbon, and had guessed that it is present in the atmosphere. He also discovered the bicarbonates (hydrogen carbonates).

Until about 1760 Black devoted his research to chemistry but thereafter most of his work was in physics. He noticed that when ice melts it absorbs heat from its surroundings without itself undergoing a change in temperature, from which he argued that the heat must have combined with the ice particles and become latent. In 1761 he experimentally verified this hypothesis, thereby establishing the concept of latent heat (in the example of melting ice, the latent heat of fusion). In the following year he determined the latent heat of formation of steam (the latent heat of vaporization). Also in 1762 he described his work to a literary society in Glasgow, but did not publish his findings. In addition, he observed that equal masses of diferent substances require different quantities of heat to change their temperatures by the same amount, an observation that established the concept of specific heats (relative heat capacities).

Black Max 1909–   is a Russian-born US philosopher and mathematician, one of whose concerns has been to investigate the question, 'What is mathematics?'.

Black was born at Baku, Azerbaijan, on 24 February 1909, and he received his higher education in England, where he studied philosophy, gaining his BA at Cambridge University in 1930 and his PhD from London University in 1939. From 1936 to 1940 he was a lecturer at the University of London Institute of Education. He went to the United States in 1940 to take up a post in the department of philosophy at the University of Illinois. He became a naturalized US citizen in 1948. He moved from the University of Illinois to Cornell in 1946 and was Susan Lin-Sage Professor of Philosophy and Humane Letters there from 1954 to 1977, when he retired. In 1970 he was Vice-President of the International Institute of Philosophy.

Black's analysis led him to describe mathematics as the study of all structures whose form may be expressed in symbols. Within that broad spectrum there are three main schools of mathematics: the logical, the formalist, and the intuitional. The logical considers that all mathematical concepts, such as numbers or differential coefficients, are capable of purely logical definition, so that mathematics becomes a branch of logic. The formalist, rejecting the notion that all mathematics can be expressed as logical concepts, looks upon mathematics as the science of the structure of objects and concerns itself with the structural properties of symbols, independent of their meaning. The formalist approach has been especially fruitful in its application to geometry. The third school, the intuitional, by laying less emphasis on symbols and more on thought, considers mathematics to be grounded on the basic intuition of the possibility of constructing an infinite series of numbers. This approach has had most influence in the theory of sets of points.

Black has thus done little work in mathematics itself, but his writings, such as *The Nature of Mathematics* (1950) and *Problems of Analysis* (1954), have been a major contribution to the philosophy of mathematics.

Blackett Lord Patrick Maynard Stuart 1897–1974 was a British physicist who made the first photograph of an atomic transmutation and developed the cloud chamber into a practical instrument for studying nuclear reactions. This achievement gained him the 1948 Nobel Prize for Physics. Blackett also discovered the phenomenon of pair production of positrons and electrons in cosmic rays.

Blackett was born in Croydon, Surrey, England, on 18 November 1897. He did not initially go to university, but joined the Royal Navy in 1912 as a naval cadet. During World War I, he took part in the battles of the Falkland Islands and Jutland, and designed a revolutionary new gunsight. Attracted to science, Blackett resigned from the Navy after the war ended and embarked on a science course at Cambridge University, obtaining a BA degree in 1921. Continuing at the university, Blackett started research with cloud chambers under Ernest Rutherford (1871–1937) in the Cavendish Laboratory. In 1924, Blackett succeeded in obtaining the first photographs of an atomic transmutation, which was of nitrogen into an oxygen isotope. He continued to develop the cloud chamber and in 1932, with assistance from Guiseppe Occhialini, he designed a cloud chamber in which photographs of cosmic rays were taken automatically. Early in 1933, the device confirmed the existence of the positron (positive electron) proposed by Carl Anderson (1905–1991).

Blackett became Professor of Physics at Birkbeck College, London, in 1933 and continued his cosmic ray studies, demonstrating in 1935 the formation of showers of positive and negative electrons from gamma rays in approximately equal numbers. In 1937, he succeeded Lawrence Bragg (1890–1971) at the University of Manchester, continuing with his cosmic ray studies. During World War II, Blackett initiated the main principles of operational research, and afterwards returned to university life. At Manchester, his research team produced many important discoveries in studies involving cosmic radiation. Particles with a lifespan of 10^{-10} sec were discovered. They became known as strange particles and included the negative cascade hyperon.

As a consequence of his research into cosmic rays, Blackett became interested in the history of the Earth's magnetic field and turned to the study of rock magnetism. He was appointed Head of the Physics Department at

Imperial College, London, in 1953, where he built up a research team specializing in rock magnetism. On Blackett's advice, the government formed a Ministry of Technology in 1974, and he became President of the Royal Society in 1965 and a life peer in 1969. Blackett died on 13 July 1974.

In 1919, Rutherford explained the transmutation of elements from experiments in which he bombarded nitrogen with alpha particles. The oxygen atoms and protons produced were detected by scintillations in a screen of zinc sulphide. Blackett used a cloud chamber to photograph the tracks formed by these particles, taking more than 20,000 pictures and recording some 400,000 tracks. Of these, eight showed that a nuclear reaction had taken place, confirming Rutherford's explanation of transmutation.

To reduce the huge number of observations, Blackett and Occhialini invented a cloud chamber that was automatically triggered by the arrival of a cosmic ray likely to cause a nuclear reaction in the chamber. Two Geiger Counters were attached to the chamber, such that the passage of a cosmic ray would create a current in the counters and trigger the operation of the chamber.

Pair production (the formation of showers of positrons and electrons from gamma rays), which Blackett discovered, was the first evidence that matter may be created from energy as Albert Einstein (1879–1955) had predicted in his special theory of relativity. The two particles subsequently rejoin and annihilate each other to form a gamma ray, thus converting mass back into energy.

Blakemore Colin 1944– is an English physiologist who has made advanced studies of how the brain works, especially in connection with memory and the senses.

Blakemore was born in Stratford-on-Avon on 1 June 1944 and educated at King Henry VIII School in Coventry. In 1962 he won a scholarship to Corpus Christi College, Cambridge, to study natural sciences and in particular, medicine; he graduated in 1965. Blakemore then declined a scholarship to St Thomas's Hospital, London, and went instead to study physiological optics at the Neurosensory Laboratory at the University of California in Berkeley; he obtained his PhD in 1968. He was a demonstrator at Cambridge University from 1967 until 1972, when he was appointed Lecturer in Physiology, a position he held until 1979. He then went to work at Oxford University as Waynflete Professor of Physiology.

In his experimental work, Blakemore has shown that cells in the visual cortex of the brain of a new-born kitten are able to detect visual outlines. But if the kitten is kept at a critical period in an environment with only, say, vertical lines, it will later prove to have in the cortex only cells that can 'recognize' these patterns and not others. Blakemore suggests that it is possible that the inherited DNA of genes already contains the capacity to synthesize RNA, the protein that is involved in the storage of any new remembrance.

Blakemore is known for his ability to explain complex science to the layman. This talent was demonstrated, for example, by his award-winning educational film *The Visual Cortex of the Cat*, in his Reith lectures of 1976, and in his publication *Mechanics of the Mind* (1977). In this book he explains the mechanics of sensation, sleep, memory and thought, and discusses the philosophical questions of human consciousness, the evolution of thinking about body and mind, the relationship between art and perception, and the origin and function of language. He argues that an individual's system of knowledge, expertise and ethical standards has evolved gradually and that the resulting 'collective mind' is a functional extension of all the human brains that have contributed to it.

Bloch Konrad Emil 1912– is a German-born US biochemist whose best-known work has been concerned with the biochemistry and metabolism of fats (lipids), particularly reactions involving cholesterol.

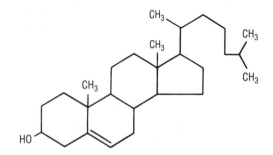

Bloch *Structural formula of cholesterol.*

His research into the biosynthesis of cholesterol has produced a better understanding of this complex substance, whose presence in the human body is of supreme importance but whose excess is thought to be dangerous. For his work on the mechanism of cholesterol and fatty acid metabolism he shared the 1964 Nobel Prize for Medicine with Feodor Lynen.

Bloch was born on 21 January 1912 at Niesse, Germany, and educated at the Munich Technische Hochschule, from which he graduated as a chemical engineer in 1934. In 1936 he emigrated to the United States where two years later he received his doctorate in biochemistry from Columbia University. Bloch became a US citizen in 1944 while serving on the faculty of the Columbian University College of Physicians and Surgeons. He joined the staff of the University of Chicago in 1946 and became Professor of Biochemistry there in 1952. He moved to a similar position at Harvard two years later and in 1956 became a member of the National Academy of Sciences.

Cholesterol, the most abundant sterol in animal tissue, was discovered in 1812 and is now one of the best known (and commonest) steroids in the human body. It occurs either free or as esters of fatty acids, being found in practically all tissues but most abundant in the brain, nervous tissue and adrenal glands and, to a lesser extent, in the liver, kidneys and skin. Its name arose because of its occurrence in gallstones (from the Greek for 'bile solid'). Cholesterol has the molecular formula $C_{27}H_{46}O$ and, like most steroids, a structure based on phenanthrene. It is important to the body because it is a component of all cell membranes and is the metabolic precursor of many compounds with various physiological functions, including vitamin D, cortisone and the male and female sex hormones.

In the early 1940s Bloch took the first steps that were to lead to today's extensive knowledge of cholesterol. Working first with David Rittenberger and later with Henry Little, he demonstrated that carbon atoms of carbon-labelled acetate (ethanoate) fed to rats was incorporated into cholesterol in the animals' livers. Using acetic acid (ethanoic acid) labelled with deuterium he showed for the first time that it is this acid, a compound having only two carbon atoms, which is the major precursor of cholesterol. This discovery was the first of a long series that elucidated the biological synthesis of the steroid. Later, using acetic acid labelled at one carbon atom with radioactive carbon-14, Bloch demonstrated which of the 27 carbon atoms in cholesterol is derived from each of the two carbon atoms of acetic acid.

The overall conversion of acetic acid to cholesterol requires 36 distinct chemical transformations, which occur in various tissues but principally in the liver. The route begins with the conversion of three molecules of acetic acid to form a five-carbon compound and carbon dioxide. Then six of the five-carbon compounds combine to form the long-chain unsaturated 30-carbon compound squalene which, after cyclization, forms langesterol; the langesterol is finally converted to cholesterol.

Bloch's work undoubtedly paved the way to the successful tracing of the numerous metabolic changes that take place in the biosynthesis of cholesterol. It has important applications to medicine, because it is now thought that high levels of cholesterol in the bloodstream can cause it to be deposited on the inner walls of arteries (arteriosclerosis), where it narrows the vessels and increases the chances of blood clotting.

Bode Johann Elert 1749–1826 was a German mathematician and astronomer who contributed greatly to the popularization of astronomy.

Bode was born in Hamburg on 19 January 1747 into a well-educated family. He taught himself astronomy and was skilled mathematically. He was publishing astronomical treatises while still in his teens, one of which remained in print for nearly a century.

In 1772 Bode joined the Berlin Academy as a mathematician, overseeing the publication of the Academy's yearbook and ensuring the accuracy of its mathematical content; he worked on all the yearbooks from 1776 to 1829. He was appointed Director of the Astronomical Observatory in 1786. He supervised the renovation of the Observatory, but he was unable to bring the standard of work there up to that of many other observatories because of the relatively simple equipment at his disposal. In 1784 he was appointed Royal Astronomer and elected to the Berlin Academy. He retired as Director of the Observatory in

1825 and died in Berlin on 23 November 1826.

Bode's early work at the Academy concerned the improvement of the accuracy of the mathematical content of the yearbook: the low standard had been depressing sales, upon which the Academy's finances largely relied. The yearbook's popularity soon increased. In addition to astronomical tables it included information about observations and scientific developments elsewhere in the world.

Bode also worked on the compilation of two atlases, the *Vorstellung der Gestirne* and the *Uranographia*, which was a massive work describing the positions of more than 17,000 stars and including for the first time some of the celestial bodies discovered by William Herschel. It was Bode who named Herschel's new planet 'Uranus'.

Bode is best known for the law named after him, even though he did not first state it, but merely popularized work already done by Johann Titius (1729–1796). The law, also known as the Titius–Bode rule, is a mathematical formula which approximately described the distances of all then known planets from the Sun. The series had no basis in theory, but it was accepted as an important finding at the time because of the near-mystical reverence attached to numbers and geometric progressions in descriptions of the universe. The discovery of the planet Neptune by Urbain Leverrier and John Couch Adams in 1846 disrupted the series and it lost its value.

Bode's main contribution to astronomy lay in the spreading of information about the subject to people from a wide range of backgrounds.

Boerhaave Hermann 1668–1738 was a Dutch physician and chemist of tremendous learning who dominated and greatly influenced various branches of science in Europe.

Boerhaave was born in Verhout, near Leiden, on 31 December 1668, the son of a minister. He intended entering the Church, and in 1684 went to the University of Leiden to study theology. While there he also studied philosophy, botany, languages, chemistry and medicine. He qualified in natural philosophy in 1687 and gained his PhD two years later. Medicine and chemistry became his predominant interests and he entered the University of Haderwijk, from which he graduated in medicine in 1693. He went back to Leiden in 1701 as a physician, and also began teaching. In 1709 he took the Chair of Medicine and Botany at Leiden, and was also made a Professor of Physic in 1714, as well as Professor of Chemistry in 1718. He was elected to the French Academy in 1728, and became a Fellow of the Royal Society in 1730. He died in Leiden on 23 September 1738.

For a man of such immense academic distinction and knowledge, Boerhaave made few original discoveries, although he did describe the structure and function of the sweat glands, and was the first to realize that smallpox is spread by contact. He was, however, an excellent tutor and re-established the technique of clinical teaching, taking his students to the bedsides of his patients. During his time at Leiden it became a famous centre of medical knowledge, attracting students from throughout Europe.

Boerhaave's writings remained authoritative works for nearly a century. In 1708 he published a physiology textbook *Institutiones Medicae* (a classification of diseases with their causes and treatment), followed by the *Book of Aphorisms* in the next year. In 1710 an *Index Plantarum* was published, followed by *Historia Plantarum* – a collection of his botanical lectures compiled by his ex-students. In 1724 his students published *Institutiones et Experimenta Chemiae*, a breakdown of Boerhaave's lectures on chemistry. He produced the official version of the lectures in 1732 called *Elementia Chemiae*, which presented a clear and precise approach to the chemistry of the day and remains his most famous work.

Bohr, Mottelson, and Rainwater Aage Niels Bohr 1922– , Ben Roy Mottelson, 1926– , and (Leo) James Rainwater, 1917– are a group of Danish and US physicists who shared the 1975 Nobel Prize for Physics for their work on the structure of the atomic nucleus.

Aage Bohr is the son of Niels Bohr (1885–1962) and was born in Copenhagen on 19 June 1922 (the year that his father won the Nobel Prize for Physics). He was educated at the University of Copenhagen, and then worked for the Department of Science and Industrial Research in London from 1943 to 1945. From 1946 he worked at his father's Institute of Theoretical Physics in Copenhagen, and since 1956 he has been Professor of Physics at the University of Copenhagen. He was Director of the Niels Bohr Institute (formerly the Institute of Theoretical Physics) from 1963 to 1970 and director of Nordita (Nordic Institute for Theoretical Atomic Physics) from 1975 to 1981. In addition to the Nobel Prize for Physics awarded in 1975, Bohr won the Atoms for Peace Award in 1969 and the Rutherford Medal in 1972. He has published *Rotational States of Atomic Nuclei* (1954) and *Nuclear Structure*, vol. I (1969) and vol. II (1975) jointly with Ben Mottelson.

Ben Mottelson was born in Chicago, Illinois, on 9 July 1926. He was educated at Purdue University, where he gained a PhD in 1950. From 1950 to 1953, he was at the Institute of Theoretical Physics in Copenhagen, and he then held a position at CERN (European Centre for Nuclear Research) in a theoretical study group formed in Copenhagen. He has been professor at Nordita in Copenhagen since 1957. Mottelson became a Danish citizen in 1971, and has published several books and many scientific papers jointly with Bohr.

James Rainwater was born on 9 December 1917 in Council, Idaho. He gained a physics degree at the California Institute of Technology in 1939 and then read for his advanced degree at Columbia University. He remained there, rising to become Professor of Physics in 1952, a position he still holds. During the period 1942 to 1946, Rainwater worked for the Office of Scientific Research and Development (OSRD) and then the Manhattan Project, and he was Director of the Nevis Cyclotron Laboratory from 1951 to 1953 and again from 1956 to 1961.

Aage Bohr, Ben Mottelson and James Rainwater shared the 1975 Nobel Prize for Physics for 'the discovery of the connection between the collective motion and the particle motion in atomic nuclei and the development of the theory of the structure of the atomic nucleus, based on this connection'. The three men had been working in loose collaboration for nearly 25 years. Niels Bohr had proposed that the particles in the nucleus of an atom are arranged like molecules in a drop of liquid. Exceptions to this rule were found and this led to the belief that the nuclear particles are arranged in concentric shells. Further studies showed that perhaps these shells were not spherically symmetrical. This suggestion was both unexpected and unattractive. In 1950, Rainwater wrote a paper in which he observed that most of the nuclear particles form an inner nucleus, while the other particles form an outer nucleus. Each set of particles is in constant motion at very high velocity and the shape of each set affects the other set. He postulated that if some of the outer particles moved in similar orbits, this would create unequal centrifugal forces of enormous power, which could be strong enough to permanently deform an ideally symmetrical nucleus.

At that time Aage Bohr was working with Rainwater as a visiting professor at the University of Columbia and was impressed by this theory. He began to work out a more detailed explanation of Rainwater's work after he returned to Denmark, and during the next three years he published results that he and his associate Ben Mottelson had obtained experimentally. These results proved the theoretical work.

The work of these three men achieved a deep understanding of the atomic nucleus and paved the way for nuclear fusion.

Bohr Niels Henrik David 1885–1962 was a Danish physicist who established the structure of the atom. For this achievement he was awarded the 1922 Nobel Prize for Physics. Bohr made another very important contribution to atomic physics by explaining the process of nuclear fission.

Bohr was born in Copenhagen on 7 October 1885. His father, Christian Bohr, was Professor of Physiology at the University of Copenhagen and his younger brother Harald became an eminent mathematician. Niels Bohr was a less brilliant student than his brother but a careful and thorough investigator. His first research project, completed in 1906, resulted in a precise determination of the surface tension of water and gained him the gold medal of the Academy of Sciences. In 1911, he was awarded his doctorate for a theory accounting for the behaviour of electrons in metals.

In the same year, Bohr went to Cambridge, England, to study under J. J. Thomson (1856–1940), who showed little interest in Bohr's electron theory so, in 1912, Bohr moved to Manchester to work with Ernest Rutherford (1871–1937), who was making important investigations into the structure of the atom. Bohr developed models of the atom in which electrons are disposed in rings around the nucleus, a first step towards an explanation of atomic structure.

Bohr returned to Copenhagen as a lecturer at the University in 1912, and in 1913 developed his theory of atomic structure by applying quantum theory to the observations of radiation emitted by atoms. He then went back to Manchester to take up a lectureship offered by Rutherford, enabling him to continue his investigations in ideal conditions. However, the authorities in Denmark enticed him back with a professorship and then built the Institute of Theoretical Physics in Copenhagen for Bohr. He became Director of the Institute in 1920, holding this position until his death. The Institute rapidly became a centre for theoretical physicists from throughout the world, and such figures as Wolfgang Pauli (1900–1958) and Werner Heisenberg (1901–1976) developed Bohr's work there, resulting in the theories of quantum and wave mechanics that more fully explain the behaviour of electrons within atoms.

The year 1922 marked not only the award of the Nobel Prize for Physics but also a triumphant vindication of Bohr's atomic theory, which he used to predict the existence of a hitherto unknown element. The element was discovered at the Institute and given the name hafnium.

In the 1930s, interest in physics turned towards nuclear reactions and in 1939 Bohr proposed his liquid-droplet model for the nucleus that was able to explain why a heavy nucleus could undergo fission following the capture of a neutron. Working from experimental results, Bohr was able to show that only the isotope uranium-235 would undergo fission with slow neutrons.

When Denmark was occupied by the Germans in 1940 early in World War II, Bohr took an active part in the resistance movement. In 1943, he escaped to Sweden with his family in a fishing boat – not without danger – and then went to England and on to the United States. He became involved in the development of the atomic bomb, helping to solve the physical problems involved, but later becoming a passionate advocate for the control of nuclear weapons. Among his efforts to persuade statesmen to adopt rational and peaceful solutions was a famous open letter addressed to the United Nations in 1950 pleading for an 'open world' of free exchange of people and ideas.

In 1952 Bohr was instrumental in creating the European Centre for Nuclear Research (CERN), now at Geneva, Switzerland. He died in Copenhagen on 18 November 1962. In addition to his scientific papers, Bohr published three volumes of essays: *Atomic Theory and the Description of Nature* (1934), *Atomic Physics and Human Knowledge* (1958) and *Essays 1958–1962 on Atomic Physics and Human Knowledge* (1963).

Bohr's first great inspiration came from working with Rutherford, who had proposed a nuclear theory of atomic structure from his work on the scattering of alpha rays in 1911. It was not, however, understood how electrons could continually orbit the nucleus without radiating energy, as classical physics demanded. Ten years earlier Max Planck (1858–1947) had proposed that radiation is emitted or absorbed by atoms in discrete units or quanta of energy. Bohr applied this quantum theory to the nuclear atom to explain why elements emit radiation at precise frequencies that give set patterns of spectral lines. He postulated that an atom may exist in only a certain number of stable states, each with a certain amount of energy; the emission or absorption of energy may occur only with a transition from one stable state to another. Electrons normally orbit the nucleus without emitting or absorbing energy. When a transition occurs, an electron moves to a lower or higher orbit depending on whether it emits or absorbs energy. In so doing, a set number of quanta of energy are emitted or absorbed at a particular frequency. Bohr developed these ideas to show that the nuclei of atoms are surrounded by shells of electrons, each assigned particular sets of quantum numbers according to their orbits. Bohr's theory was used to determine the frequencies of spectral lines produced by elements and succeeded brilliantly. It also enabled him to explain the groups of the periodic table in terms of elements with similar electron structures, which led to the prediction and discovery of hafnium.

In developing a model for the nucleus, Bohr conceived of the nuclear particles being pulled together by short-range forces rather as the molecules in a drop of liquid are attracted to one another. The extra energy produced by the absorption of a neutron may cause the nuclear particles to separate into two groups of approximately the same size, thus breaking the nucleus into two smaller nuclei – as happens in nuclear fission. The model was vindicated when Bohr correctly predicted the differing behaviour of nuclei of uranium-235 and uranium-238 from the fact that the number of neutrons in each nucleus is odd and even respectively.

Niels Bohr gained not only a love of science from his father but also a philosophical insight into the nature of knowledge that enabled him to question accepted theories and seek new explanations. By reconciling Rutherford's nuclear model of the atom with Planck's quantum theory, he was able to produce a valid model for the atom completely at odds with classical physics. However, this did not prevent him from using a classical model to explain the structure and behaviour of the nucleus. Our present knowledge of the atom and the nucleus thus rests on the fundamental discoveries made by Bohr's restless and ingenious mind.

Bok Bart 1906–1992 was a Dutch astrophysicist best known for his discovery of the small, circular dark spots in nebulae (now Bok's globules).

Bok was born in the Netherlands in 1906 and educated at the University of Leiden (1924–1926) and the University of Groningen (1927–1929). He went to the United States in 1929, having gained a Robert Wheeler Wilson Fellowship in Astronomy at Harvard. He gained his PhD in 1932 and remained at Harvard as Assistant Professor (1933–1939), Associate Professor (1939–1946) and Robert Wheeler Wilson Professor of Astronomy (1947–1957). In 1957 he went to Australia as Professor and Head of the Department of Astronomy at the Australian National University until 1966, being also Director of the Mount Stromlo Observatory near Canberra at the same time. Bok's next appointment was as Professor of Astronomy at the University of Arizona (1966–1974). He became Professor Emeritus on his retirement.

Bok served as President of a Commission in the International Astronomical Union and President of the American Astronomical Society (1972–1974). He was a member of a number of learned societies and received several medals. He published *The Distribution of the Stars in Space* with Priscilla Bok in 1937 and *The Milky Way* with F. W. Wright in 1944.

Photographs had shown that the Milky Way was dotted with dark patches or nebulae. Bok also discovered small, circular dark spots, which were best observed against a bright background. Measurements of their dimensions and opacity suggested that their masses were similar to that of the Sun. Bok suggested that the globules were clouds of gas in the process of condensation and that stars might be in the early stages of formation there. His work thus broadened our understanding of the nature of stellar birthplaces.

Boksenberg Alexander 1936– is an astronomer who devised a new kind of light-detecting system that can be attached to telescopes and so vastly improve their optical powers. His image photon counting system (IPCS) has revolutionized observational astronomy, enabling Boksenberg and others to study distant quasars, which may help towards a deeper understanding of the early phases and nature of the universe.

Boksenberg was born on 18 March 1936, the elder of two sons. He attended the Stationers' Company's School in London and was encouraged by his parents, who owned a shop, to study for a place at university. He gained a BSc in physics from London University and in 1957 began research at University College, London, into the physics of atomic collisions, for which he was awarded his PhD in 1961. Boksenberg then joined a research group at University College which was studying the ultraviolet spectra of stars using rocket- and satellite-borne instruments. It was during this period that he became interested in applying his knowledge of physics to astronomy. He also saw the need to improve the instrumentation being carried aboard space vehicles and began to specialize in image-detecting systems. He became a lecturer in physics in 1965, and in 1968 his innovative work on detectors led to his involvement in the British design team producing the instrumentation for the Anglo-Australian 4-m/13-ft telescope then being constructed at Sydney Springs in New South Wales, Australia. He devised a fundamentally new approach to optical detection in astronomy with his image photon counting system.

In 1969 Boksenberg set up his own research group at University College to work on two main topics: optical astronomy, mainly using IPCSs he built for the Anglo-Australian telescope and for the 5-m/16.4-ft Hale telescope at Mount Palomar, California; and ultraviolet astronomy, using instrumentation he designed for use on high-altitude balloon-borne platforms and on satellites, particularly the International Ultraviolet Explorer (IUE) satellite observatory. In 1975 he was promoted to reader in physics, a year later he was awarded the first Senior Fellowship of the United Kingdom Science Research Council, and in 1978 became professor of physics and a Fellow of the Royal

Society. In 1981 he was appointed Director of the Royal Greenwich Observatory. In 1982 he was awarded an Honorary Doctorate by the Paris Observatory.

During his time at University College, Boksenberg's work in ultraviolet astronomy using balloons yielded the first results for the important parameters: electron density and metal abundance in the local interstellar medium. For the European observatory satellite TD-1A, launched in 1972, he conceived a simple means of using the sky-scanning action of the satellite to produce a passive spectrum-scanning of the main ultraviolet sky-survey instrument, which greatly increased the scientific value of the project.

Boksenberg also designed and worked on a new ultraviolet television detector system for the International Ultraviolet Explorer satellite launched in 1978, and led the pioneering work on the complex Sun-baffle systems for TD-1A and IUE to enable the telescopes to observe faint astronomical objects in full orbital sunlight. Both satellites were widely used internationally and contributed greatly to the advancement of astronomy. Boksenberg's own main contribution was to a study of galactic haloes and the nuclei of active galaxies and quasars using the IUE.

Boksenberg's development of the IPCS sprang from his considerations of the workings of the human eye, which operates not only as an optical device but also relies on the correcting effect of the processing and memory functions of the retina and the brain. Rather than recording light with a photographic emulsion, he saw the potential of literally detecting the locations of the individual photons of light collected by a telescope from the faint astronomical object being studied, and building up the required image in a computer memory. By using an image intensifier coupled to a television camera he detected and amplified photons by a factor of 10^7. He then treated the signals in a special electronic processor, which passed the photon locations to a computer with a large digital memory, both to store the accumulating image and to present the incoming results as an instantaneous picture – a great advantage to an astronomer who would otherwise have to wait for a photographic image to form and be processed. The picture on the screen appears as an accumulating series of dots, each dot representing a photon which is counted, analysed and stored by the computer.

By 1973 Boksenberg's IPCS was ready to be tested and in the autumn of that year he went to Mount Palomar and attached it to the spectrograph at the Coudé focus of the largest telescope in the world at that time – the 5-m/16.4-ft Hale telescope. He used it in collaboration with Wallace Sargent of the California Institute of Technology to observe absorption lines in the spectra of quasars. Within 30 minutes they saw spectra that would normally take three nights or more of exposure time to reveal. Sargent, an established scientist and astronomer, immediately recognized the enormous validity and potential of Boksenberg's photon detector, and on this basis his technique was generally accepted and applied to all modern telescopes, including the 2.4-m/7.8-ft Space Telescope.

Boksenberg, being interested in the overall nature of the universe, continued to collaborate with Sargent, mainly using the 5-m/16.4-ft Hale telescope and the IPCS in the study of the most distant quasars. The radiation from these has taken billions of years to reach us and so by studying their light, emitted way back in time, Boksenberg and Sargent use quasars to elucidate the early nature of the universe. Quasars also provide clues to the story of galactic evolution because they seem to be intimately connected with the central cores of galaxies.

Since 1973 Boksenberg and Sargent have been particularly interested in studying the absorption lines in the spectra of quasars, which they discovered are not a manifestation of the quasar itself but a reflection of the state of the universe – galaxies and intergalactic gas – that exists between the quasar and the Earth. They can thus provide direct information on the nature and evolution of the universe. As Director of the Royal Greenwich Observatory, Boksenberg is responsible for building a major new British observatory on Las Palmas in the Canary Islands. He is also taking part in plans to build the next generation of telescope. With the development of idealized light-detecting systems, such as the IPCS, using electronic devices to receive and record photons, optical astronomy has experienced a major resurgence and there is a recognized need to increase telescope aperture beyond the few metres currently available at each of the world's major observatories. The most favourable design of future ground-based optical telescopes, in terms of minimal cost and most efficient observational powers, seems to be the 'multimirror' candidate – a telescope composed of separate mirrors that combine to produce a light-collecting area that can be equivalent to a single mirror of about 20 m/66 ft in diameter.

Boltzmann Ludwig 1844–1906 was an Austrian theoretical physicist who contributed to the development of the kinetic theory of gases, electromagnetism and thermodynamics. His work in these fields led him to consider phenomena in terms of probability theory and atomic events, which led to the establishment of the branch of physics now known as statistical mechanics.

Boltzmann was born in Vienna on 20 February 1844 and educated at Linz and Vienna. He studied at the University of Vienna under Josef Stefan (1835–1893) and Josef Loschmidt (1821–1895), and received his PhD in 1866 from that institution. In 1867 Boltzmann became an assistant at the Physikalisches Institut in Vienna, after which he held a series of professorial posts. He was Professor of Theoretical Physics at the University of Graz from 1869 to 1873, of Mathematics at the University of Vienna from 1873 to 1876, and of Experimental Physics at Graz from 1876 to 1879. He then became director of the Physikalisches Institut, moving in 1899 to Munich to become Professor of Theoretical Physics at the University there. In 1894 went back to Vienna to succeed Stefan as Professor of Theoretical Physics. During these years there was a heated and sometimes unpleasant scientific debate between the 'atomists', championed by Boltzmann, and the 'energeticists', one of whose spokesmen was Wilhelm Ostwald (1853–1932). Ernst Mach (1838–1916), one of the 'energeticists', tried to reconcile the two schools of thought in an attempt to tone down the debate.

Boltzmann moved to the University of Leipzig in 1900, where he served as Professor of Theoretical Physics. He suffered from depression and made an unsuccessful suicide attempt. He returned to his previous post at Vienna in 1902, with the added honour of the chair of natural philosophy from which Mach had just retired because of ill health. In 1904 Boltzmann travelled to the United States, lecturing at the World's Fair in St Louis and visiting Stanford and Berkeley, two prestigious campuses of the University of California. He still felt depressed at the hostility he sensed from some of his scientific colleagues, despite the fact that the revolutionary discoveries being made at the time about the structure of atoms and the properties of matter would ultimately support his theories. He committed suicide at Duino, near Trieste, on 5 September 1906, shortly before the discovery of Brownian motion, which led to a near-universal acceptance of his kinetic and statistical theories.

Mechanics, dynamics and electromagnetism were all being developed when Boltzmann was a student. He made contributions to all of these areas during the course of his career. One of his most important sources of stimulation was the work of James Clerk Maxwell (1831–1879), whose investigations in the field of electromagnetism were not well known by European scientists. Boltzmann contributed greatly to the dissemination of Maxwell's work, particularly through a book on the subject published in 1891.

Boltzmann wrote his first paper on the kinetic theory of gases in 1859 while he was only a student. In 1868 he published a paper on thermal equilibrium in gases, citing and extending Maxwell's work on this subject. He was concerned with the distribution of energy among colliding gas molecules. He derived an exponential formula to describe the distribution of molecules which relates the mean total energy of a molecule to its temperature. It includes the constant k, now known as the Boltzmann constant, which is equal to 1.38×10^{-23} J K^{-1}. This constant has become a fundamental part of virtually every mathematical formulation of a statistical nature in both classical and quantum physics.

Another important area of Boltzmann's investigation was thermodynamics. The second law of thermodynamics had been formulated in 1850 by Rudolph Clausius (1822–1888) and William Thomson (Lord Kelvin; 1824–1907). Boltzmann sought to find a mathematical description for the demonstration of the tendency of a gas to reach a state of equilibrium as the most probable state. In 1877 he published his famous equation, $S = k\log W$ (which was later engraved on his tombstone) describing the relation between entropy and probability.

Boltzmann also determined a theoretical derivation for Stefan's experimentally derived law of black-body radiation. Stefan had shown that the energy radiated by a black body is proportional to the fourth power of its absolute temperature, and this is now often known as the Stefan–Boltzmann law. Boltzmann's investigation of the phenomenon was based on the second law of thermodynamics and Maxwell's electromagnetic theory.

Boltzmann's most enduring contribution arose from his pioneering work in the field of statistical mechanics. He had begun his study of the equipartition of energy, which resulted in the Maxwell–Boltzmann distribution law, early in his career. He demonstrated that the average amount of energy required for atomic motion in all directions is equal, and so formulated an equation for the distribution of atoms due to collision. This led to the foundation of statistical mechanics. This discipline holds that macroscopic properties of matter such as conductivity and viscosity can be understood and are determined by the cumulative properties of the constituent atoms. Boltzmann held that the second law of thermodynamics should be considered from this viewpoint. Boltzmann's work on the relationship between probability and entropy and on the inevitable and irreversible processes that must therefore occur, influenced Willard Gibbs (1839–1903), Max Planck (1858–1947) and others.

Boltzmann was a theoretician with great intuitive powers and vision. It is a tragedy that he did not live to see his work vindicated and the remarkable advances achieved by the 'atomists' of the twentieth century.

Bolyai (Farkas) Wolfgang 1775–1856 and János 1802–1860 were Hungarian mathematicians, father and son, the younger of whom was one of the founders of non-Euclidean geometry.

Wolfgang Bolyai was born at Nagyszeben, in Hungary (now Sibiu, Romania), on 9 February 1775. He studied mathematics at the University of Göttingen, where he fell into friendship with Karl Gauss. For the rest of his life the elder Bolyai was a professional mathematician, teaching at the Evangelical Reformed College at Nagyszeben and then at the college at Marosvásárhely (now Tirgu-Mures, Romania) until his retirement in 1853. He died at Marosvásárhely on 20 November 1856.

János Bolyai was born at Koloszvár, in Hungary (now Cluj, Romania), on 15 December 1802. He was first taught by his father and at an early age showed a marked talent both for mathematics and for playing the violin. At the age of 13, by which time he had mastered the calculus and analytical mechanics, he entered the college at Marosvásárhely where his father was teaching. He remained there for five years, concentrating on mathematics, but also becoming an adept swordsman. Then, in 1818, against the wishes of his father (who wanted him to study under Gauss at Göttingen), he entered the Royal College of Engineers at Vienna. He graduated in 1822 and joined the army engineering corps, rising eventually to the rank of lieutenant, second class. Increasingly he fell victim to attacks of fever until, in 1833, he was retired from the army with a small pension. He returned to his father's house at Marosvásárhely and lived there, a semi-invalid, until his death on 27 January 1860.

The mathematical lives of the father and son were closely entwined. It was because of his son's growing interest in higher mathematics that Wolfgang was inspired to write the book on which his posthumous fame rests, the *Tentatum Juventem*, or *Attempt to Introduce Studious Youth into the Elements of Pure Mathematics*. Completed in 1829, but not published until 1832, it was a brilliantly suggestive survey of mathematics, although (and to Wolfgang's chagrin) overlooked by his contemporaries. Of greater importance for his son's future was Wolfgang's obsession with the hoary problem of finding a proof for Euclid's fifth postulate, that there is only one line through a point outside another line which is parallel to it, or, in layman's language, that parallel lines do not meet.

In 1804 Wolfgang thought that he had found a proof of the axiom. He sent it to Gauss, who pointed out a flaw in the argument. Undaunted Wolfgang continued his quest, and János caught his enthusiasm. By about 1820, however, János had become convinced that a proof was impossible; he began instead to construct a geometry which did not depend upon Euclid's axiom. Over the next three years he developed a theory of absolute space in which several lines pass through the point P without intersecting the line L. He developed his formula relating the angle of parallelism of two lines with a term characterizing the line. In his new theory Euclidean space was simply a limiting case of the new space, and János introduced his formula to express what later became known as the space constant.

János described his new geometry in a paper of 1823 called 'The absolute true science of space'. His father, unable to grasp its revolutionary meaning, rejected it, but sent it to Gauss. To the surprise of both father and son Gauss replied that he had been thinking along the same lines for more than 25 years. Gauss had published nothing on the matter, however, and János's paper was printed as an appendix to his father's *Tentatum* in 1832.

János's paper was a thorough and consistent exposition of the foundations of non-Euclidean geometry. He was therefore cast down when its publication received little attention and when he discovered that Nikolai Lobachevsky had published his account of a very similar geometry (also ignored) in 1829.

Deeply disappointed at their lack of recognition, the Bolyais retired into semi-seclusion. In 1837 the failure of their joint paper to win the Jablonov Society's prize plunged them deeper into dejection. Thereafter Wolfgang did no serious mathematical research, and although János dabbled in problems connected with the relationship between pure and spherical trigonometry, both died in relative obscurity. It was not until 30 years later that the work of Beltrami and Klein at last put into proper perspective János Bolyai's place as one of the pioneers of modern mathematics.

Bolzano Bernardus Placidus Johann Nepomuk 1781–1848 was a Czech philosopher and mathematician who made a number of contributions to the development of several branches of mathematics.

Bolzano was born at Prague on 5 October 1781. He went to the University of Prague, where he studied philosophy, mathematics and physics until 1800, when he entered the theology department. He was ordained in 1804. He did not abandon his mathematical interests, however, and in 1804 he was recommended for the chair of mathematics at the university. In 1805 he was appointed as the first professor to the new chair of philosophy at Prague. For the next 14 years he lectured mainly on ethical and social questions, although also on the links between mathematics and philosophy.

He was much admired by the students not only for his intellectual abilities but also for the forthright expression of liberal and Czech nationalist views. He became the dean of the philosophy faculty in 1818. But by then his opinions were bringing him into disfavour with the Austro-Hungarian authorities and in 1819, despite the backing of the Catholic hierarchy, he was suspended from his professorship and forbidden to publish. A five-year struggle ensued, ending only in 1824 when Bolzano, resolute in his refusal to sign an imperial order of 'recantation', resigned his chair. He retired to a small village in southern Bohemia. He returned to Prague in 1842 and died there on 18 December 1848.

Owing to the opposition of the imperial authorities, most of Bolzano's work remained in manuscript during his lifetime. It was not until the publication of the manuscripts in 1962 that the range and importance of his research was fully appreciated. Early in his career he worked on the theory of parallels, based on Euclid's fifth postulate. He found several faults in Euclid's method, but until the development of topology nearly a century later, these difficulties could not be resolved. Bolzano also formulated a proof of the binomial theorem and, in one of his few works published in his lifetime (1817), attempted to lay down a rigorous foundation of analysis. One of the most interesting parts of the book was his definition of continuous functions.

During the 1830s Bolzano concentrated on the study of real numbers. He also formulated a theory of real functions and introduced the nondifferentiable 'Bolzano function'. He was also able to prove the existence and define the properties of infinite sets, work later of much use to Julius Dedekind when he came to produce his definition of infinity in the 1880s.

Bolzano fell short of making any really fundamental breakthroughs in mathematics, but he was one of the most accomplished and wide ranging of nineteenth-century mathematicians.

Bond George Phillips 1825–1865 was a US astronomer whose best work was on the development of astronomical photography as an important research tool. His research was carried out exclusively at the observatory founded by his father, William Cranch Bond.

Bond was born in Dorchester, Massachusetts, on 20 May 1825. He obtained a BA at Harvard and immediately began to work at the Harvard College Observatory. He served first as Assistant Astronomer to his father and then, upon the latter's death, as Director of the Observatory. In 1865, Bond became the first citizen of the United States to be awarded the Gold Medal of the Royal Astronomical Society of London, for his beautifully produced text on Donati's Comet. He died in Cambridge, Massachusetts, on 2 February 1865.

The discovery of Hyperion (Saturn's eighth satellite) and the Crêpe Ring around Saturn were Bond's first major findings with his father. Since it was possible to see stars through the Crêpe Ring (a dim ring inside the two bright rings), Bond concluded that the rings were liquid, not solid, as most astronomers believed at the time.

During the late 1840s the Bonds worked on developing photographic techniques for astronomy. Since taking pictures even in full daylight often required long exposures in those days, photography at night was an arduous process indeed. Poor-quality daguerrotypes of the Moon had been taken in the early 1840s, but by 1850 the Bonds were able to take pictures of impressive quality. Improved techniques enabled Bond and Fred Whipple to take a picture of Vega, the first star to be photographed, in 1850. In 1857 Bond also became the first man to photograph a double star, Mizar, with the aid of wet collodion plates. Bond suggested that a star's magnitude could be quantitatively determined by measuring the size of the image it made. A bright star would affect a greater area of silver grains.

Bond also made numerous studies of comets. He discovered 11 new comets and made calculations on the factors affecting their orbits. He is, however, best remembered for his work on photography and, indeed, is often credited as the pioneer of techniques of astronomical photography.

Bond William Cranch 1789–1858 was a US astronomer who, with his son, George Phillips Bond, established the Harvard College Observatory as a centre of astronomical research.

Bond was born into a poor family in Falmouth, Maine, on 9 September 1789. He was needed in the family business and he received little formal education. He worked in the shop as a watchmaker and displayed remarkable manual dexterity and mechanical ingenuity. The solar eclipse of 1806 was the stimulus that introduced him to the study of astronomy, which became an ever more absorbing hobby. Bond was one of the independent observers who discovered the comet of 1811. He was commissioned by Harvard College to investigate the equipment at observatories in England during a trip he made there in 1815.

In the absence of an observatory in the vicinity of his home, Bond had converted one of the rooms in his house. It became the best private observatory of his day and in 1839 Harvard invited him to move it into their premises (although he was not offered any stipend for doing so). Bond thus became the first Director of the Harvard College Observatory, a post he held until his death. He was awarded the formal title of Observer, and given an honorary MA in 1842.

Public interest in astronomy was aroused by the comet of 1843 and the Observatory received sufficient funding to equip itself with a 38-cm/15-in refracting telescope. Bond continued to make observations and to design equipment until his death in Cambridge, Massachusetts, on 29 January 1859.

In addition to his early observations of comets and other celestial bodies, Bond established an international reputation for his work on chronometers. He worked not only on their design, but also on fixing the rate of the mechanisms of chronometers used, for example, in navigation.

From the Observatory he also studied the Solar System, sunspots and the nebulae in the constellations of Orion and Andromeda. It is difficult to distinguish the contributions made by Bond from those of his son during the later years of the career of Bond senior. They collaborated on the development of photographic techniques for the purposes of astronomy, succeeding in obtaining superior photographs of the Moon (exhibited in London in 1851 to enthusiastic audiences) and took the first photographs of stars.

The two Bonds discovered Hyperion (the eighth satellite of Saturn) in 1848 and the Crêpe Ring around Saturn in 1850. This is the faint ring inside the two bright rings of Saturn. Their observation that stars could be seen through the Crêpe Ring led to their conclusion that the rings of Saturn are not solid. The Crêpe Ring had probably been observed already by other astronomers, including H. Kater in 1825 and Frederich Struve in 1826. William Lassell, who missed being credited for the discovery of Hyperion by finding it a few days after the Bonds, also observed the Crêpe Ring shortly after it was described by them.

William Cranch Bond's chief contributions to astronomy were his careful observations and innovative designs, and his founding of an eminent research centre.

Bondi Hermann 1919– is an Austrian-born scientist who was trained as a mathematician, but who went on to make important contributions to many disciplines both as a research scientist and as an enthusiastic administrator. He is best known in astronomy for his development, with Thomas Gold (1920–) and Fred Hoyle (1915–), of the steady-state theory concerning the origin of the universe.

Bondi was born on 1 November 1919 in Vienna. At home he developed an early interest in mathematics. He taught himself the rudiments of calculus and theoretical physics and, after briefly meeting Arthur Eddington, who was visiting Vienna, he decided to go to Cambridge University to study mathematics. He was recognized as a mathematician of considerable talent and was awarded an exhibition in his first year. He earned a BA in 1940, despite being caught up in the General Order of May 1940 that required 'enemy aliens' resident in the United Kingdom to be interned for security reasons. While in internment Bondi met Thomas Gold; they were to become close friends and scientific associates.

Bondi returned to Cambridge in 1941 to become a research student. He began to do naval radar work for the British Admiralty in 1942 and through this work he met Fred Hoyle. Gold soon joined them and, inspired by Hoyle,

who was already an established astrophysicist, the three discussed cosmology and related subjects in their spare time. This collaboration continued after the war.

Trinity College elected Bondi to a Fellowship in 1943 on the basis of his first astrophysical research. He acquired British citizenship in 1947 and became an assistant lecturer (1945–48) and University Lecturer (1943–54). He was Visiting Professor to Cornell University in 1951 and he went to Harvard in 1953. After a tour of American observatories, he returned to England to take up the Chair of Applied Mathematics at King's College, London. He was elected Fellow of the Royal Society in 1959.

He has held advisory posts in the Ministry of Defence, in particular on the National Space Committee, the European Space Research Organization, the Department of Energy, and the Natural Environment Research Council. Bondi is the author of a number of books on cosmology and allied subjects. His scientific contributions have been recognized by his election to prominent scientific organizations and his public service by honours such as a knighthood (1973).

Bondi is perhaps best known for his proposal, together with Gold and Hoyle, of the steady-state theory. This is a cosmological model that explains the expansion of the universe not as a consequence of a singularity (as proposed by the Big-Bang model), but as a feature of the universe as it has always been and always will be. The model requires that matter be continually created – albeit at a rate of only 1 gram per cubic decimetre per 10^{36} years – in order to keep the density of matter in the universe constant. The steady-state model was felt not only to resolve an apparent discrepancy between the age of the universe and of our Earth, but also to be a simpler theory than the Big-Bang model. The steady-state theory was not in any way contradicted by facts then available.

The model created something of a sensation and stimulated much debate for, while its ideas were revolutionary, it was fully compatible with existing knowledge. The orthodox model of the day placed the origin of all elements in an early superhot stage of the universe. Hoyle stated the now universally accepted theory of the origin of the elements in observed types of stars. This was the greatest triumph of the steady-state theory, but evidence against the theory in general soon began to accumulate. In 1955 came evidence that the universe had once been denser than it is today. In 1965 came the identification of a universal 'background' radiation that was readily accounted for as a remainder of an early hot state of the universe. A further severe difficulty for the steady-state theory is that the universe seems to contain more helium than the theory predicts. Thus, today few scientists regard the steady-state theory as a serious competitor to the Big-Bang theory of the origin of the universe.

Bondi's other contributions have been to the study of stellar structure, relativity and gravitational waves. He demonstrated that gravitational waves are compatible with and are indeed a necessary consequence of the general theory of relativity. He was also able to describe the likely characteristics and physical properties of gravitational waves.

Since the 1960s, Bondi has been primarily concerned with more administrative duties. He has organized the rebuilding of King's College and the establishment of the Anglo-Australian telescope. He was also an adviser on the Thames Barrage project. Bondi is a versatile and talented scientist, whose contributions are notable for their originality, insight and scope.

Boole George 1815–1864 was a British mathematician who, by being the first to use symbolic language and notation for purely logical processes, founded the modern science of mathematical logic.

He was born in Lincoln on 2 November 1815. He received little formal education, although for a time he attended a national school in Lincoln and also a small commercial school. His interest in mathematics appears to have been kindled by his father, a cobbler with a keen amateur interest in mathematics and the making of optical instruments. Boole also taught himself Greek, Latin, French, German, and Italian. At the age of 16 he became a teacher at a school in Lincoln; he subsequently taught at Waddington; then, at the age of 20, he opened his own school. All the while his spare time was devoted to studying mathematics, especially Newton's *Principia* and Lagrange's *Mécanique analytique*. He was soon contributing papers to scientific journals and in 1844 he was awarded a Royal Society medal. In 1849, despite his lack of a university education, he was appointed Professor of Mathematics at the newly-founded Queen's College at Cork, in Ireland. He held the chair until his death, at Cork, on 8 December 1864.

Boole's first essay into the field of mathematical logic began in 1844 in a paper for the *Philosophical Transactions* of the Royal Society. In it Boole discussed ways in which algebra and calculus could be combined, and the discussion led him to the discovery that the algebra he had devised could be applied to logic. In a pamphlet of 1847 he announced, against all previous accepted divisions of human knowledge, that logic was more closely allied to mathematics than to philosophy. He argued not only that there was a close analogy between algebraic symbols and those that represented logical forms but also that symbols of quantity could be separated from symbols of operation. These were the leading ideas which received their fuller treatment in Boole's greatest work, *An Investigation of the Laws of Thought on which are founded the Mathematical Theories of Logic and Probabilities*, published in 1854.

It is not quite true to say that Boole's book reduced logic to a branch of mathematics; but it did mark the birth of the algebra of logic, later known as Boolean algebra. The basic process of Boole's system is continuous dichotomy. His algebra is essentially two valued. By subdividing objects into separate classes, each with a given property, it enables different classes to be treated according to the presence or absence of the same property. Hence it involves just two numbers, 0 and 1. This simple framework has had far-reaching practical effects. Applying it to the concept of 'on' and 'off' eventually produced the modern system of telephone switching, and it was only a step beyond this to the application of the binary system of addition and subtraction in producing the modern computer.

Later mathematicians modified Boole's algebra. Friedrich Frege (1848–1945), in particular, improved the scope of mathematical logic by introducing new symbols, whereas Boole had restricted himself to those symbols already in use. But Boole was the true founder of mathematical logic; it was on the foundations that he laid that Bertrand Russell and Alfred Whitehead attempted to build a rigidly logical structure of mathematics.

Booth Herbert Cecil 1871–1955 was an English mechanical and civil engineer best known for his invention of the vacuum cleaner in 1901.

Booth was born on 4 July 1871 in Gloucester, England, where he was educated at the college school and, later, the county school. He studied engineering at the City and Guilds Institute between 1889 and 1892.

In 1901, the same year in which he formed his own engineering consultancy, Booth conceived the principle of his vacuum cleaner after witnessing a somewhat self-defeating operation at St Pancras Station, London – the cleaning of a Midland Railway train carriage by means of compressed air which simply blew a great cloud of dust around, allowing it to settle elsewhere. Booth realized that a machine to suck in the dust and trap it so that it could be disposed of afterwards would circumvent this unhygienic problem. Booth demonstrated his idea to his companions at St Pancras by placing his handkerchief over his mouth and sucking in his breath. A ring of dust particles appeared where his mouth had been. There was also, of course, a use for Booth's invention in the cleaning of houses, which was then done with handbrushes or with simple sweepers which pushed dust into a box, or with feather dusters which just removed dirt from one place to another. Booth's cleaner replaced these methods of cleaning carpets, upholstery and surfaces with the air suction pump principle. In his machine, one end of the tube was connected to the pump, while the other, with nozzle attached, was pushed over the surface being cleaned. The cleaner incorporated an air filter to cleanse the air passing through and also served to collect the dust.

Because of the large size and high price of early vacuum cleaners and the fact that few houses had mains electricity, Booth initially offered cleaning services rather than machine sales. The large vacuum cleaner, powered by petrol or electric engine and mounted on a four-wheeled horse carriage was parked in the street outside a house while large cleaning tubes were passed in through the windows. The machine was such a novelty that society hostesses held special parties at which guests watched operatives cleaning carpets or furniture. Transparent tubes were provided so that the dust could be seen departing down them. Booth's machines received a great popular boost when they were used to clean the blue pile carpets laid in Westminster Abbey for the coronation of King Edward VII in 1902.

Smaller, more compact indoors vacuum cleaners

followed, but until the first electrically powered model appeared in 1905, two people were required to operate them – one to work the pump by bellows or a plunger, the other to handle the cleaning tube. Booth formed his British Vacuum Cleaner Company in 1903, running it in parallel with his equally successful engineering consultancy. His work in the latter area had begun with a post as a draughtsman with Britain's premier marine engine company, Maudslay, Sons and Field. There, Booth worked on the design of engines for two Royal Navy battleships. At this time Booth's keen insight into technological problems and his original flair were noticed by W. B. Bassett, one of Maudslay's directors. Bassett was interested in great wheels, and in 1894 he chose Booth to work first on the wheel at Earl's Court, London, and afterwards on similar structures in Blackpool and Vienna, and on the 61-m /200-ft diameter great wheel in Paris.

Booth's design principles were, in essence the same as those governing the design of modern long-span suspension bridges. Later in 1902, Booth directed the erection of the Connel Ferry Bridge over Loch Etive, Scotland.

From 1903 until his retirement in 1952, three years before his death at Croydon, Surrey, on 14 January 1955, Booth remained chairman of the British Vacuum Cleaner Company. He took special interest in the industrial potential of his famous invention and personally pioneered the development of cleaning installations in large industrial establishments. Both here and in the domestic context – even though he regarded the home vacuum cleaner as something of a toy – Booth was most gratified by the fact that his machines allowed higher standards of hygiene and counteracted dust-related diseases.

Borel Emile Félix-Edouard-Justin 1871–1956 was a French mathematician whose lasting reputation derives from his rationalization of the theory of functions of real variables.

Borel was born in Saint-Affrique on 7 January 1871. At an early age he showed such a strong aptitude for mathematics that he was sent away from his native village to a lycée at Montauban. In 1890 he entered the Ecole Polytechnique in Paris, where he so distinguished himself that on his graduation in 1893 he was appointed to the faculty of mathematics at the University of Lille. In 1894 he received his DSc from the Ecole Normale Supérieure. For the next few years Borel proved himself to be a prolific writer of highly valuable papers and in 1909 the Sorbonne created a chair in function theory especially for him. A year later he took on the additional duty of becoming deputy director, in charge of science, at the Ecole Normale Supérieure.

Borel's professional career was interrupted by World War I, during which he took part in scientific and technical missions on the front. The war also marked a turning-point in his life. When it was over he took less interest in pure mathematics and more in applied science. He also became involved in politics. From 1924 to 1936 he was a Radical-Socialist member of the national Chamber of Deputies,

serving as Minister of the Navy in 1925. He was also in these years one of the moving spirits behind the establishment of the National Centre for Scientific Research (he received the institute's first gold medal in 1955). He was, too, one of the founding members of the Henri Poincaré Institute, serving as its first director from 1928 until his death. In 1936 Borel left active politics and four years later he retired from his chair at the Sorbonne. World War II drew him back into public life: in 1940 he was taken briefly into custody by the occupying German forces and on his release he joined the Resistance movement. In 1945 he was awarded the Resistance Medal. Thereafter Borel lived in retirement until his death, in Paris, on 3 February 1956.

Borel's first papers appeared in 1890 and it was in the 1890s that he did his most important work – on probability, the infinitesimal calculus, divergent series and, most influential of all, the theory of measure. In 1896 he created a minor sensation by providing a proof of Picard's theorem, an achievement which had eluded a host of mathematicians for nearly 20 years. In the 1920s he wrote on the subject of game theory, before John Von Neumann (generally credited with being the founder of the subject) first wrote on it in 1928. But he will be remembered, above all, for his theory of integral functions and his analysis of measure theory and divergent series. It is this work which established him, alongside Henri Lebesgue (1875–1941), as one of the founders of the theory of functions of real variables.

Born Max 1882–1970 was a German-born British physicist who pioneered quantum mechanics, the mathematical explanation of the behaviour of an electron in an atom. His version of quantum mechanics had a short reign as the most popular explanation and was soon displaced by the equivalent and more convenient wave mechanics of Erwin Schrödinger (1887–1961), which is in use today. The importance of Born's work was eventually recognized, however, with the award of the 1954 Nobel Prize for Physics, which he shared with Walther Bothe (1891–1957). (Bothe's share of the award was for his work on the detection of cosmic rays and was not connected with Born's achievements in quantum mechanics.) Born also made important advances in the understanding of the dynamics of crystal lattices.

Born was born in Breslau, Germany (now Wroclaw, Poland), on 11 December 1882 into a wealthy Jewish family. His father was Professor of Anatomy at the University of Breslau, which Born later attended as well as the University of Göttingen, gaining his doctorate in physics and astronomy at Göttingen in 1907.

After periods of military service and various academic positions in Göttingen and Berlin, Born was appointed Professor of Physics at Frankfurt-am-Main in 1919. During this period, he became an expert on the physics of crystals. Following a meeting with Fritz Haber (1868–1934), who developed the Haber process for the synthesis of ammonia, Born became aware that little was known about the calculation of chemical energies. Haber encouraged Born to apply his work on the lattice energies of crystals (the energy given out when gaseous ions are brought together to form a solid crystal lattice) to the formation of alkali metal chlorides. Born was able to determine the energies involved in lattice formation, from which the properties of crystals may be derived, and thus laid one of the foundations of solid-state physics.

In 1921, Born moved from Frankfurt to the more prestigious university at Göttingen, again as professor. Most of his research was then concerned with quantum mechanics, in particular with the electronic structure of atoms. He made Göttingen a leading centre for theoretical physics and together with his students and collaborators – notably Werner Heisenberg (1901–1976) – he devised a system called matrix mechanics that accounted mathematically for the position of the electron in the atom. He also devised a technique, called the Born approximation method, for computing the behaviour of subatomic particles that is of great use in high-energy physics.

In 1933, with the rise to power of Adolf Hitler, Born left Germany for Cambridge, and in 1936 he became Professor of Natural Philosophy at Edinburgh University. He became a British citizen in 1939, and remained at Edinburgh until 1953, when he retired to Germany. The following year, Born received his Nobel award which, somewhat ironically, was for his discovery published in 1926 that the wave function of an electron is linked to the probability that the electron is to be found at any point. Born died at Göttingen on 5 January 1970.

Born was inspired by Niels Bohr (1885–1962) to seek a mathematical explanation for Bohr's discovery that the quantum theory applies to the behaviour of electrons in atoms. In 1924 Born coined the term 'quantum mechanics' and the following year built on conceptual work by Werner Heisenberg to relate the position and momentum of the electron in a system called matrix mechanics. Wolfgang Pauli (1900–1958) immediately used the system to calculate the hydrogen spectrum and the results were correct. But about a year after this discovery, in 1926, Erwin Schrödinger expressed the same theory in terms of wave mechanics, which was not only more acceptable mathematically but enabled physicists to visualize the position of the electron as a wave motion around the nucleus, unlike Born's purely mathematical treatment. Born, however, was able to use wave mechanics to make a statistical interpretation of the quantum theory and expressed the probability that an electron will be found at a particular point as the square of the value of the amplitude of the wave at that point. This was the discovery for which he was belatedly awarded a Nobel prize.

Bosch Karl 1874–1940 was a German chemist who developed the industrial synthesis of ammonia, leading to the cheap production of agricultural fertilizers and of explosives.

Bosch was born on 27 August 1874 in Cologne, the eldest son of an engineer. He showed an early aptitude for the sciences and, following a year gaining practical work experience as a metal-worker, he went to the University of Leipzig, where he studied chemistry under the noted organic chemist Johannes Wislicenus; he took his doctorate in 1898.

The year after graduating Bosch took a job with the major German chemicals company, Badische Anilin & Sodafabrik (BASF), and by 1902 was working on methods of 'fixing' the nitrogen present in the Earth's atmosphere, a subject for which he was to become famous. At that time, the only major sources of nitrogen compounds essential for the production of fertilizers and explosives were in the natural deposits of nitrates in Chile, thousands of miles from industrial Europe. Industrial production of ammonia using the nitrogen in the air would end this dependency on foreign sources.

In 1908 Bosch learned of the work of Fritz Haber (1868–1934), who had also been considering the problem. Haber had studied ways of combining nitrogen with hydrogen to form ammonia under the influence of high pressure and metal catalysts. Bosch's employers seized on the work, with its promise of endless supplies of nitrogen compounds, and made him responsible for making Haber's process commercially viable.

He set up a team of chemists and engineers to study the processes Haber had succeeded in producing under laboratory conditions, and to scale them up. For example, the carbon-steel vessels used by Haber to combine to nitrogen and hydrogen were attacked by the hydrogen, causing failure under the high temperature and pressure conditions needed for the reaction to take place. Alloy steel replacements were brought in by Bosch and his team. To produce the required volumes of hydrogen, Bosch also introduced the water-gas shift reaction, where carbon monoxide is combined with steam to produce carbon dioxide and hydrogen. Different catalysts were also investigated.

By the time World War I broke out in 1914, Bosch had completed what was then the largest ever feat of chemical engineering, and the BASF ammonia plant at Oppau was producing 36,000 tonnes of the material in sulphate form using the Haber–Bosch process. Following the heavy demands for its product from the military, the plant was expanded, and another much larger factory was set up in Leuna. At the end of the war, Bosch acted as technical advisor to the German delegation at the armistice and peace conferences.

Following World War I, BASF continued to profit from the work of Bosch, remaining leaders in the technology during the 1920s. Despite further involvement in scientific work on the synthesis of methyl alcohol and of petrol from coal tar, Bosch himself became increasingly involved in adminstration. He was chairman of the vast industrial conglomerate IG Farbeninindustrie AG after its formation from the merger of BASF with other major German industrial concerns in 1925. By 1935, he was chairman of its supervisory board. Illness prevented his close involvement in the group in later years (which were darkened by the rise of the Nazis).

Bosch was also recognized for his work by the scientific community, which bestowed numerous honorary degrees on him, and he received, in 1931, the Nobel Prize for Chemistry, jointly with his compatriot Friedrich Bergius for their work on high-pressure synthesis reactions.

An essentially withdrawn character, Bosch shunned public appearances, and published little. He died in Heidelberg on 26 April 1940.

Boulton Matthew 1728–1809 was a British manufacturer whose financial support and enthusiasm were of importance in the promotion of James Watt's steam-engine.

Boulton was born on 3 September in Birmingham, and it was near there in 1762 that he built his Soho factory. He produced small metal articles such as buckles, buttons, gilt and silver wares, Sheffield plate and the like, having succeeded to his father's business of silver stamping three years earlier. The Soho factory was original in combining workshops of different trades with a warehouse and merchanting lines. Boulton wished to obtain not only the best workmen but the finest artistry for his products, and to this end sent out agents to procure him the best examples of art work not only in metal but also in pottery and other materials.

The growth of the factory led to an increased need for a motive power other than water, which was poorly supplied at Soho. This resulted in a meeting with James Watt, who had just developed a steam-engine which Boulton was convinced would prove the answer to his wants. In 1769, when Watt's partner became bankrupt, Boulton took his place. Six years later the two men became partners in a steam-engine business, obtaining a 25-year extension of the patent. The inventiveness of Watt and the commercial enterprise of Boulton secured the steam-engine's future, but not before Boulton had brought himself to the verge of bankruptcy through his support. Helped by the engineer William Murdoch, they established the steam-engine by erecting pumps in machines to drain the Cornish tin mines. Boulton foresaw a great industrial demand for steam power and urged Watt to develop the double-action rotative engine patented in 1782 and the Watt engine (1788) to drive the lapping machines in his factory. The testing period of the steam-engine was a long one and Boulton was more than 60 years old before it began to make a profitable return.

In 1786 Boulton applied steam power to coining machines, obtaining a patent in 1790. So successful was the process that as well as his home market Boulton supplied coins to foreign governments and to the East India Company. In 1797 he was commissioned to reform the copper currency of the realm. One result of this highly successful venture was to make the counterfeiting of coins much more difficult. Boulton also supplied machines for the Royal Mint near Tower Hill, London, and these continued in efficient operation until 1882.

In 1785 Boulton was made a Fellow of the Royal Society, but submitted no papers to it subsequently. A friendly and generous man, he was acquainted with all the leading scientists of his day. He died in Birmingham on 17 August 1809.

Bourbaki Nicolas is the pseudonym taken by a group of mathematicians, most of them French, who began to publish collectively and anonymously in the late 1930s. The group, which at any one time contained about 20 members, was centred at the Ecole Normale Supérieure in Paris. A few Americans have at times been members. The group's effort to persuade people that Bourbaki was a real person failed, most notably when his application for membership of the American Mathematical Society was rejected.

The origin of the pseudonym is not known for certain, but it is believed to have been taken from the French general, Charles Bourbaki (1816–1897). During the siege of Metz in 1870 Bourbaki was duped by the Germans into going to England to be present at the signing of peace terms between Germany and France, and the story of this trick was told in a pamphlet entitled 'Quel est votre nom?' It has been suggested that the name 'Nicolas' was intended to signify that the group was bringing mathematical gifts to the world.

The group's object was to provide a definitive survey of mathematics, or, at least, of those subjects worthy of the Bourbaki's transcendant talents. Their work appeared in instalments, the first in 1939 and the thirty-third and last in 1967. All their work was subjected to continual revision and updating. The work as a whole is called *Elements of Mathematics*. It can be understood only by persons possessing expert knowledge of higher mathematics and follows a pattern markedly different from traditional introductions to mathematics. The order of topics dealt with is set theory, then (abstract) algebra, followed by general topology, functions of a real variable (including ordinary calculus), topological vector spaces and general theory of integration.

For a number of reasons the group's work has been very influential. It gave the first systematic account of a number of topics that had previously lain scattered in learned journals. The precision of style and the order in which topics were discussed recommended the work to professional mathematicians. And there was great interest in the group's axiomatic method – a method that was at any rate gaining currency in the 1930s. Owing to the popularity of the work, mathematicians had to learn 'Bourbaki's' terminology and it is the mark of the group's impact that many of the terms have passed into the language of mathematical research.

Bourdon Eugène 1808–1884 was a French instrument maker who developed the first compact and reliable high-pressure gauge.

Born in Paris on 8 April 1808, Bourdon was set to follow his father and become a merchant, but more practical inclinations led him to set up, at the age of 24, his own instrument and machine shop. By 1835 he had established himself at Faubourg du Temple in Paris, where he was to work for the next 37 years.

Bourdon was born into the age of steam, with the opening of George Stephenson's Liverpool to Manchester Railway taking place in 1830. Bourdon's work reflected this: in the same year that his new instrument works opened

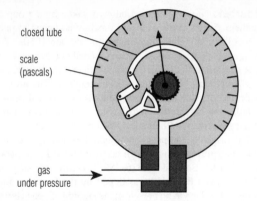

closed tube

scale
(pascals)

gas
under pressure

Bourdon *In a Bourdon gauge, liquid (or gas) under
pressure flows into a tapering, curved tube closed at one
end. The tube tends to straighten, causing movement of
levers and gears which turn a pointer round a scale
calibrated in pressure.*

in 1832, he presented a model steam-engine (complete with
glass cylinders) to the Société d'Encouragement pour l'In-
dustrie National. He was to make more than 200 small
steam-engines at his works, chiefly for demonstration
purposes.

Following Richard Trevethick's work with high-pressure steam-engines around the turn of the previous century
(Watt's earlier engines worked at atmospheric pressure),
the problem of measuring high pressures in such engines
became apparent. Watt had used a fairly imprecise device
for showing the variation of pressure in his relatively low-
pressure systems, but the major difficulty lay at the higher
pressures. Because one atmosphere pressure is equivalent
to 760 mm/30 in of mercury in a U-tube manometer, there
was clearly a problem of providing a compact, yet accurate,
pressure gauge.

Bourdon's solution to the problem was simple and
ingenious, and is described in the patent for this 'metallic
manometer' of 18 June 1849. An ordinary bicycle pump
can be used to demonstrate the principle. By pumping air
into the rubber connection dangling at the end of the pump,
the connector becomes horizontal. This is because the pressure in the tube exceeds that outside (which is usually
atmospheric), and the initially oval cross section of the tube
becomes circular, causing the tube to straighten.

In the Bourdon gauge, a metal tube, sealed at one end, is
subjected at the other to the pressure through a link to a
quadrant rack, geared by a small pinion fixed to the back of
a pointer. This moves across the calibrated scale. By using
tubes of suitable thickness, a wide range of pressures can
be measured using a variety of guages, from fractions of an
atmosphere to as much as $8,000$ atm$/8 \times 10^8$ Pa, in both liquids and gases. It remains the most widely used gauge for
measuring so wide a range of fluid pressures.

Bourdon was, apparently, led to his invention by noticing the change that occurred in a lead cooling coil under
internal pressure. Just two years after its invention, the
Bourdon gauge had become so renowned for its practical-

ity and versatility that it won for its inventor the Legion of
Honour.

By 1872, at the age of 64, Bourdon decided to pass over
the management of his business to his sons. Nevertheless,
he continued to experiment with instruments during the
following 14 years, until his death on 29 September 1884
from a fall while testing a new form of anemometer.

Boveri Theodor Heinrich 1862–1915 was a German biologist who performed valuable early work on chromosomes. Most of his life's work was devoted to investigating
those processes by which an individual develops from the
reproductive material of its parents.

Boveri was born in Bamberg on 12 October 1862, the
second of four sons of a physician. He was educated at
Bamberg from 1868 to 1875 and then for the next six years
at the Realgymnasium in Nuremberg. He went to Munich
University to study history and philosophy, but soon
decided to change to natural sciences, and gained his doctorate in 1885. He then began a five-year fellowship at the
Zoological Institute in Munich doing research into cytology under Richard Hertwig (1850–1937), and became a
lecturer in zoology and comparative anatomy in 1887. He
was appointed Professor of Zoology and Comparative
Anatomy as well as Director of the Zoological-Zootomical
Institute in Würzburg in 1893. Four years later he married
Marcella O'Grady, an American biologist, and they had
one child, Margret, who became a writer and journalist.
Apart from visits to the zoological station at Naples, Boveri
spent the rest of his working life at Würzburg. He declined
the position offered to him in 1913 of Director of the Kaiser
Wilhelm Institute for Biology in Berlin because of his failing health. He died in Würzburg on 15 October 1915, after
successive physical breakdowns and recurrent depression.

Boveri's first piece of research was for his thesis in 1885,
which had been on the structure of nerve fibres, but soon
afterwards his interest was turned towards cell biology by
Hertwig, when he worked as his assistant in Munich.

In the early 1880s the Belgian cytologist Edouard van
Beneden was investigating eggs of the roundworm *Ascaris
megalocephala*, which inhabits the intestine of horses. He
discovered that the chromosomes of the offspring are
derived in equal numbers from the nuclei of the ovum and
spermatozoon, that is equally from the two parents. It was
known that the fusion of the nuclei of the egg and the spermatozoon is the essential feature of fertilization, and that
this fusion eventually leads to the creation of all the nuclei
of the body. In 1884 it was concluded that the cell nucleus
holds the fundamental elements of heredity. It was also
shown that the chromosomes split into daughter cells and
that the chromosome number is constant for each species.

Inspired by the experiments of Beneden, Boveri did
some of his own and described some aspects of the development of eggs and the formation of polar bodies (minute
bodies which are produced during the division of an unfertilized ovum). He then demonstrated that the nuclei of
Ascaris eggs contain finger-shaped lobes which are, in fact,
chromosomes (each egg contains only two to four

chromosomes), and also that chromosomes are separated into the daughter cells during division following fertilization by a central piece in the spermatozoon, which he termed the centrosome.

In 1889 Boveri experimented on sea-urchins' eggs, fertilizing nucleated and nonnucleated fragments, and found that both types could develop normally. He also discovered that those occasional nonfertilized fragments containing only a nucleus were also able to develop normally. He showed that at fertilization the egg and the spermatozoon incorporate the same number of chromosomes each in the creation of the new individual.

He was eventually able to demonstrate that cytoplasm plays an important part in development and went on to show that it is not a specific number, but a specific assortment of chromosomes that is responsible for normal development, indicating that individual chromosomes possess different qualities.

Boveri's powers of observation as a microscopist were remarkable, but so were his theories. In 1914, he theorized that tumours may become malignant as the result of abnormal chromosome numbers, and was the first to view the tumour as a cell problem. He also tried to explain, on the basis of an irregular chromosome distribution, a condition in bees in which male and female characteristics are mosaically distributed in different parts of the body. He discovered segmental excretory organs in *Amphioxus* (an organism believed to be close to the type from which the vertebrates evolved).

Boveri enriched biological science with some fundamental discoveries and fruitful new conceptions. His theory of chromosomal individuality provided the working basis of nearly all cytological interpretations of genetic phenomena and is still true today.

Bowden Frank Philip 1903–1968 was an Australian physicist and chemist who worked mainly in Britain. He began his research career in electrochemistry, but early on became interested in surface studies and went on to make major contributions to the study of friction, lubricants and surface erosion as well as other related subjects.

Bowden was born in Hobart, Tasmania, on 2 May 1903. He received his BSc at the University of Tasmania in 1924. He continued his studies there and then in 1927 went to Cambridge University, England, where he began research in the department of colloid science. In 1935 Bowden became director of studies in natural sciences at Gonville and Caius College. He was then made a lecturer at Cambridge in 1937, a post he held until 1946.

In 1939 Bowden embarked on a lecture tour of the United States. He decided to return to Britain via Australia, and he was still there when World War II broke out in 1945. He was appointed head of the lubricants and bearings section for the Council for Scientific and Industrial Research at Melbourne University (which conducted research of great value to the war effort). When the war ended Bowden resigned his post in order to return to Cambridge to set up a similar research group there, and from 1946 to 1965 he

was director of the laboratory for physics and chemistry of solids. A chair in surface physics was created for him in 1966. Bowden died in Cambridge on 3 September 1968.

Bowden's work in Tasmania and his early work at Cambridge was in the field of electrochemistry; specifically he was interested in the establishment of the hydrogen overpotential and in the effects of impurities on this process. He also worked on electrode kinetics. Although his electrochemical investigations continued until 1939, he began to publish papers on friction as early as 1931.

His early work in this topic demonstrated that the area of contact between two solid surfaces depends not only on details of their surface features but also on the magnitude of the load applied to them and on the hardness of the two surfaces. Since even a clean smooth metallic surface is irregular on an atomic scale, the area of contact where strong intermetallic forces might form is small. Therefore, sliding produces friction over only a small fraction of the total area of the surface, but can produce exceedingly high temperatures and even induce melting in hot spots. Another implication of this work was that a lubricant might be decomposed by heat at exactly the place where it is most needed.

Bowden's interest in skiing led him to realize that the thin layer of water between a ski or an ice-skate and the snow or ice is produced not by pressure due to the weight on them but to friction-induced heat caused by irregularities in the sliding surfaces. A variety of industries saw applications for Bowden's research in more serious areas, and his war research covered a broad range of areas of military significance. Machine and tool lubricants, flame-throwing fuels, the accurate measurement of shell velocities for gun calibration and the casting of aircraft bearings all came within their scope. Several ways of creating hot spots in a variety of materials were investigated, which had applications in the prevention of explosions due to frictional heating in munitions factories.

These research topics continued to be investigated by the research group which Bowden built up in Cambridge after the war. They took advantage of new techniques such as scanning electron microscopy, electron diffraction, and high-speed photography in their investigations. The development of PTFE-coated metal appliances and devices was greatly accelerated by results obtained by Bowden and his team.

Bowen Ira Sprague 1898–1973 was a US astrophysicist who is best known for his study of the spectra of planetary nebulae. He showed that strong green lines in such spectra are due to ionized oxygen and nitrogen under extreme conditions not found on Earth.

Bowen was born in New York State in 1898 and graduated from Oberlin College, Ohio, in 1919. He was an assistant in the physics department at the University of Chicago from 1919 to 1921, when he joined the California Institute of Technology. He gained his doctorate in 1926 and subsequently held the posts of Instructor (1921–26), Assistant Professor (1926–28), Associate Professor

(1928–30) and Professor (1931–45). From 1946 to 1964 he was Director of the Mount Wilson and Palomar observatories. Bowen was a member of a number of learned societies and was awarded several medals. He was elected to the National Academy of Sciences in 1936.

In the 1860s William Huggins noticed strong green lines in the spectra of planetary nebulae. These were attributed either to complex atomic spectra or to an element previously unknown and given the name 'nebulium'.

With a greater understanding of the way in which spectral lines occur, astronomers began to doubt if nebulium really existed. They suspected that the spectral lines might be produced by a gas of extremely low density.

A spectral line is produced when an electron in an atom transfers itself from one energy level to another. Spectral analysis can determine the energy levels between which the electrons are moving, since strong lines are produced where it takes place easily ('permitted' transitions) and weak lines where it takes place with difficulty ('forbidden' transitions).

Bowen suggested that the strong green lines in the spectra of planetary nebulae might be caused not by permitted transitions in the hypothetical nebulium, but by forbidden transitions in known elements under conditions not produced in the laboratory. He calculated that the wavelengths of three of the spectral lines of nebulium were the same as those that would be produced from transitions within the lowest energy levels of doubly ionized oxygen, O(III). He compared his calculated wavelengths of forbidden transitions with those observed in the spectra of the nebulae and found that the strongest lines were produced by forbidden transitions of singly and doubly ionized oxygen, O(II) and O(III), and singly ionized nitrogen, N(II).

In 1938 Bowen constructed an ingenious piece of apparatus known as the image slicer for use with the slit spectrograph. His unmasking of nebulium led to the identification of other puzzling spectral lines, particularly those associated with the corona of the Sun, previously attributed to another hypothetical element, 'coronium'. Further research into the chemical composition and physical properties of the Sun and other celestial bodies was stimulated by Bowen's work.

Bowen Norman Levi 1887–1956 was a Canadian geologist whose work helped found modern petrology.

Born in Kingston, Ontario, Bowen was educated at the local Queen's University. Developing a love for laboratory geology, for his graduate studies he moved to the recently founded Geophysical Laboratory in Washington, DC, from 1912 publishing a steady stream of findings on the experimental melting and crystallization behaviour of silicates and similar mineral substances. In the course of more than 40 years of high-quality research centred on crystallization experiments, Bowen made distinguished contributions to modern understanding of the chemical and petrographical properties of igneous rocks.

Working alongside O. F. Tuttle and J. F. Schairer, Bowen in particular demonstrated the physicochemical principles

governing the formation of magmas by partial melting and the fractional crystallization of magmas. Thanks to such research, igneous petrology ceased to be the essentially descriptive science created in the late nineteenth century by Rosenbusch and Zirkel and assumed a new experimental vigour and a sounder basis in physics and chemistry. Bowen particularly highlighted the importance of study of the evolution of magma, setting out his views in *The Evolution of Igneous Rocks* 1928, and being known as the head of the 'magmatist school' of Canadian geology.

Boyle Robert 1627–1691 was an English natural philosopher and one of the founders of modern chemistry. He is best remembered for the law named after him which states that, at a constant temperature, the volume of a given mass of gas varies inversely as the pressure upon it. He was instrumental in the founding of the Royal Society and a pioneer in the use of experiment and the scientific method.

Boyle was born on 25 January 1627 in Lismore Castle, Ireland, the fourteenth child and seventh son of the Earl of Cork. He learned to speak French and Latin as a child and was sent to Eton College at the early age of eight. In 1641 he visited Italy, returning to England in 1644. He joined a group, which he called the 'Invisible College', whose aim was to cultivate the 'new philosophy' and which met at Gresham College, London, and in Oxford, where Boyle went to live in 1654. The Invisible College became, under a charter granted by Charles II in 1662, the Royal Society of London for Improving Natural Knowledge, and Boyle was a member of its first council. (He was elected President of the Royal Society in 1680, but declined the office.) He moved to London in 1668 where he lived with his sister for the rest of his life. He died there on 30 December 1691.

Boyle's most active research was carried out while he lived in Oxford. By careful experiments he established Boyle's law. He determined the density of air and pointed out that bodies alter in weight according to the varying buoyancy of the atmosphere. He compared the lower strata of the air to a number of sponges or small springs which are compressed by the weight of the layers of air above them. In 1660 these findings were published in a book, shortened title *The Spring of Air*, and gave us the word elastic in its present meaning.

A year later Boyle published *The Sceptical Chymist* in which he criticized previous researchers for thinking that salt, sulphur and mercury were the 'true principles of things'. He advanced towards the view that matter is ultimately composed of 'corpuscles' of various sorts and sizes, capable of arranging themselves into groups, and that each group constitutes a chemical substance. He successfully distinguished between mixtures and compounds and showed that a compound can have very different qualities from those of its constituents.

Also in about 1660 Boyle studied the chemistry of combustion, with the assistance of his pupil Robert Hooke (1635–1703). They proved using an air pump that neither charcoal nor sulphur burns when strongly heated in a vessel exhausted of air, although each inflames as soon as air is re-admitted. Boyle then found that a mixture of either substance with saltpetre (potassium nitrate) catches fire even when heated in a vacuum and concluded that combustion must depend on something common to both air and saltpetre. Further experiments involved burning a range of combustible substances in a bell jar of air enclosed over water. But it was left to Joseph Priestley in 1774 to discover the component of air that vigorously supports combustion, which three years later Antoine Lavoisier named oxygen.

The term 'analysis' was coined by Boyle and many of the reactions still used in qualitative work were known to him. He also introduced certain plant extracts, notably litmus, for the indication of acids and bases. In 1667 he was the first to study the phenomenon of bioluminescence, when he showed that fungi and bacteria require air (oxygen) for luminescence, becoming dark in a vacuum and luminescing again when air is re-admitted. In this he drew a comparison between a glowing coal and phosphorescent wood, although oxygen was still not known and combustion not properly understood. Boyle also seems to have been the first to construct a small portable box-type camera obscura in about 1665. It could be extended or shortened like a telescope to focus an image on a piece of paper stretched across the back of the box opposite the lens.

In 1665 Boyle published the first account in England of the use of a hydrometer for measuring the density of liquids. The instrument he described is essentially the same as those in use today. He can also be credited with the invention of the first match. In 1680 he found that by coating coarse paper with phosphorus, fire was produced when a sulphur-tipped splint was drawn through a fold in the treated paper. Boyle experimented in physiology, although 'the tenderness of his nature' prevented him from performing actual dissections! He also carried out experiments in the hope of changing one metal into another, and was instrumental in obtaining in 1689 the repeal of the statute of Henry IV against multiplying gold and silver.

Besides being a busy natural philosopher, Boyle was interested in theology and in 1665 would have received the provostship of Eton had he taken orders. He learned Hebrew, Greek and Syriac in order to further his studies of the scriptures, and spent large sums on biblical translations. He founded the Boyle Lectures for proving the Christian religion against 'notorious infidels such as atheists, theists, pagans, Jews and Mahommedans' (but made the proviso that controversies between Christians were not to be mentioned).

Boyle accomplished much important work in physics, with Boyle's law, the role of air in propagating sound, the expansive force of freezing water, the refractive powers of crystals, the density of liquids, electricity, colour, hydrostatics and so on. But his greatest fondness was researching in chemistry, and he was the main agent in changing the outlook from an alchemical to a chemical one. He was the first to work towards removing the mystique and making chemistry into a pure science. He questioned the basis of the chemical theory of his day and taught that the proper object of chemistry was to determine the compositions of substances. His great merit as a scientific investigator was that he carried out the principles of Bacon, although he did not consider himself to be a follower of him or any other teacher. After his death, his natural history collections passed as a bequest to the Royal Society.

Boys Charles Vernon 1855–1944 was a British inventor and physicist who is probably best known for designing a very sensitive apparatus (based on Henry Cavendish's earlier experiment) to determine Newton's gravitational constant and the mean density of the Earth. He received many honours for his work, including a knighthood in 1935.

Boys, the son of a clergyman, was born in Wing, Rutland, on 15 March 1855. He was educated at Marlborough College then in 1873 went to the Royal School of Mines, London, where he was taught physics by R. Guthrie (1833–1886) and chemistry by Edward Frankland (1825–1899). Although he received little formal instruction in mathematics, Boys taught himself calculus and designed an integrating machine that mechanically drew graphs of antiderivatives. He graduated in mining and metallurgy in 1876 and then, after a brief period working in a colliery, became an assistant to Guthrie, his former teacher, at the Royal College of Science in London. He was appointed Assistant Professor of Physics at the Royal College in 1889 but, opportunities for further promotion being rare, he resigned in 1897 to work for the Metropolitan Gas Board as a referee. In 1920 he became one of the three gas referees that served all of Britain. He held this post until his formal retirement in 1939, although he continued to serve as an adviser to the Gas Board until 1943. In addition to his official appointments, from 1893 Boys pursued a lucrative business. However, he had sufficient spare time to develop his scientific interests and to write several popular science books on subjects as diverse as soap bubbles, weeds, and natural logarithms. He was also President of the Royal Society from 1916 to 1917. He died in Andover, Hampshire, on 30 March 1944.

Boys is known mainly as an inventor and designer of scientific instruments, in which area his most important contribution was probably his improved, extremely sensitive torsion balance, which he used to elaborate on Henry Cavendish's experiment to determine the gravitational constant and the mean density of the Earth. The novel feature of Boys's apparatus was his use of an extremely fine quartz fibre – ingeniously made by firing an arrow to the end of which was attached molten quartz – in the torsion balance. The properties of these fibres are such that Boys's balance was more sensitive, smaller and quicker to react than was Cavendish's original apparatus (which used copper wire). Testing his improved design in the basement of the Clarendon Laboratories in Oxford in 1895, Boys determined Newton's gravitational constant and calculated the mean density of the Earth as 5.527 times the density of water.

Boys's other work included the invention of a 'radiomicrometer' (1890) to detect infrared radiation. This extremely sensitive instrument, which consists of a thermo-couple and a galvanometer, was intended to measure the heat radiated by the planets in the Solar System. Using this device he calculated that Jupiter's surface temperature did not exceed 100°C/212°F; it is now thought to be less than −130°C/−202°F. He also developed a special camera to record fast-moving objects such as bullets and lightning flashes; designed a scientific toy using soap bubbles (called the rainbow cup, this device was later manufactured); and devised a calorimeter for determining the calorific value of coal gas (this subsequently became the standard instrument used to measure the calorific value of fuel gas in Britain). In addition, he performed a series of experiments on soap bubbles, which increased knowledge of surface tension and of the properties of thin films.

Bradley James 1693–1762 was an English astronomer of great perception and practical skill. He was the third Astronomer Royal and the discoverer of nutation and the aberration of light, both essential steps towards modern research into positional astronomy.

Bradley was born in Sherborne, Dorset, in March 1693. He entered Balliol College, Oxford, in 1711 and studied theology. He gained a BA in 1714, but he had by this time developed a fascination for astronomy through contact with his uncle, J. Pound, who was an amateur astronomer and a friend of Edmond Halley. Bradley pursued his interest in astronomy after graduating and was made a Fellow of the Royal Society in 1718. In 1719 he became a vicar in Bridstow, and soon after was appointed chaplain to the Bishop of Hertford. He resigned his position in 1721 to become Savilian Professor of Astronomy at Oxford.

From then on Bradley devoted his whole career to astronomy. He lectured at the University until shortly before his death and pursued an active research programme. His most brilliant work was done during the 1720s, but he published material of exceptionally high standard throughout his life. In 1742, upon Halley's death, Bradley was appointed Astronomer Royal. In that position, he sought to modernize and re-equip the observatory at Greenwich, and he embarked upon an extensive programme of stellar observation. He was awarded the Copley Medal of the Royal Society in 1748 and served on the Society's Council from 1752 until 1762. He was a member of scientific academies in several European countries.

Bradley was unusually reluctant to publish his results until he had confirmed his ideas over periods of observation that sometimes exceeded 20 years. His catalogue of more than 60,000 observations made during the last years of his career was eventually published in two volumes in 1798 and 1805. He died in Chalford, Gloucestershire, on 13 July 1762.

Bradley's earliest astronomical observations were concerned with the determination of stellar parallax. Such measurement was the goal of many astronomers of his day because it would confirm Copernicus' hypothesis that the Earth moved around the Sun. Copernicus himself (echoing Aristarchus 1,800 years before him) had stated that this parallax could not be detected because even the distance from one end of the Earth's orbit to the other was negligible compared with the enormous distance of the stars themselves. Nevertheless, Bradley and his contemporaries sought to observe parallactic displacement of the nearer stars compared to those at greater distances.

Bradley worked in 1725–26 with S. Molyneux at the latter's private observatory in Kew. They chose to observe Gamma Draconis, and found that within a few days there did seem to be a displacement of the star. However, the displacement was not only too large, but was in a different direction from that which would have been expected from parallactic displacement. Bradley studied this displacement of Gamma Draconis and other stars for more than a year and observed that this was the general effect. It took him some time to realize that the displacement was simply a consequence of observing a stationary object from a moving one, namely the Earth. The telescope needed to be tilted slightly in order to compensate for the movement of the Earth on its orbit around the Sun.

Bradley called this effect the 'aberration' of light, and he measured its angle to be between 20 and 20.5 sec (the modern value being 20.47 sec). From the size of this angle Bradley was able to obtain an independent determination of the velocity of light (308,300 km/191,578 mi per second compared with the modern value of 299,792 km/186,291 mi per second), confirming Ole Römer's work of 1769. A conclusion of more immediate significance, however, was that the Copernican concept of a moving Earth had been confirmed. Bradley had failed to find the proof through measuring parallactic displacement, but he had proved it by means of aberration.

This discovery allowed Bradley to produce more accurate tables of stellar positions, but he found that even when he considered the effect of aberration his observations on the distances of stars were still variable. He studied the distribution of these variations and deduced that they were caused by the oscillation of the Earth's axis, which in turn was caused by the gravitational interaction between the Moon and the Earth's equatorial bulge, so that the orbit of the Moon was sometimes above the ecliptic and sometimes below it. Bradley named this oscillation 'nutation', and he studied it during the entire period of the revolution of the nodes of the lunar orbit (18.6 years) from 1727 to 1748. At the end of this period the positions of the stars were the same as when he started.

Perhaps as a result of his knowledge of Römer's work on the determination of the speed of light using the Jovian satellites, Bradley then turned his attention to a study of Jupiter. He measured its diameter and studied eclipses of its satellites.

The fruits of Bradley's observations were more accurate than those of his predecessors because of his discovery of the effects of aberration and nutation. Bradley was a skilful astronomer with unusual talents in both the practical and theoretical aspects of the subject.

Bragg William Henry 1862–1942 and (William) Lawrence 1890–1971, father and son, were British physicists who pioneered and perfected the technique of X-ray diffraction in the study of the structure of crystals. For this work they were jointly awarded the 1915 Nobel Prize for Physics.

William Henry Bragg was born at Westward, Cumberland, on 2 July 1862. In 1881 he went to Cambridge University, obtaining a first-class degree in mathematics in 1885. He was immediately appointed Professor of Mathematics and Physics at the University of Adelaide, South Australia. It is said that he wondered why he was offered this appointment since he knew little physics, although he put this to rights by reading important texts on the long sea journey to Australia. For many years William Henry Bragg did no research but concentrated on lecturing, although he did apprentice himself to a firm of instrument makers and subsequently made all the apparatus he required for practical laboratory teaching. Then in 1904 Bragg became president of the physics section of the Australian Association for the Advancement of Science and his opening address – on radioactivity – acted as the stimulus to begin original research. He found that radium produces alpha particles with a variety of energy ranges, confirming the theory of Ernest Rutherford (1871–1937) that the various disintegration products of the radioactive series should produce radioactivity of differing energy. Rutherford proposed Bragg for a fellowship of the Royal Society and in 1908 Bragg was appointed Professor of Physics at Leeds University, returning to Britain in 1909.

Bragg began to work on X-rays and, together with his son Lawrence, became convinced that X-rays behave as an electromagnetic wave motion. Using the skills at instrument making he had gained in Australia, Bragg constructed the first X-ray spectrometer in 1913. Both men used it to determine the structures of various crystals on the basis that X-rays passing through the crystals are diffracted by the regular array of atoms within the crystal.

During World War I, the Braggs' work on X-rays virtually ceased and William Henry Bragg devoted himself mainly to submarine detection. Both father and son were, however, awarded the 1915 Nobel Prize for Physics for their discovery, and in the same year William Henry Bragg became Professor of Physics at University College, London. After the war, he turned to the X-ray analysis of organic crystals. William Henry Bragg was knighted in 1920, and became Director of the Royal Institution in 1923 (instituting the famous series of Christmas lectures there) and then President of the Royal Society in 1935. He died in London on 10 March 1942.

William Lawrence Bragg was born at Adelaide on 31 March 1890. He studied mathematics at Adelaide University and continued in this subject at Trinity College, Cambridge. Then in 1910 he switched to physics at his father's suggestion, becoming interested in the X-ray work of Max von Laue (1879–1960), who claimed to have observed X-ray diffraction in crystals, and repeated many of von Laue's experiments. Lawrence Bragg was able to determine an equation now known as Bragg's law that enabled both him and his father to deduce the structure of

crystals such as diamond, using the X-ray spectrometer built by his father. In recognition of this achievement, Lawrence Bragg shared the 1915 Nobel Prize for Physics with his father, becoming the youngest person (at age 25) ever to receive the prize.

Lawrence Bragg then went on to determine the structures of such inorganic substances as silicates. In 1919, he became Professor of Physics at the University of Manchester and from 1938 to 1954 was Professor of Physics at Cambridge. He was knighted in 1941. Like his father before him, Bragg became Director of the Royal Institution in 1954 and he also devoted much energy to the popularization of science. Lawrence Bragg retired in 1966 and died in Ipswich on 1 July 1971.

Until the Braggs applied X-rays to the study of crystal structure, crystallography had not been concerned with the internal arrangement of atoms but only with the shape and number of crystal surfaces. The Braggs' work immediately gave a method of determining the positions of atoms in the lattices making up the crystals because they were able to relate the distance between the crystal planes or layers of atoms d to the angle of incidence of the X-rays θ and their wavelength λ, obtaining the simple equation $n\lambda = 2d \sin\theta$, where n is an integer. From the diffraction patterns obtained by passing X-rays through crystals, the Braggs were able to determine the dimensions of the crystal planes and thus the structure of a crystal. Furthermore, it gave a method for accurate determination of X-ray wavelengths.

The Braggs' pioneering discovery led to an understanding of the ways in which atoms combine with each other and also revolutionized mineralogy and later molecular biology, in which X-ray diffraction was crucial to the elucidation of the structure of DNA.

Brahe Tycho 1546–1601 sometimes known by his first name only, was a Danish astronomer who is most noted for his remarkably accurate measurements of the positions of stars and the movements of the planets.

Tycho was born of aristocratic parents in Knudstrup in 1546. He was brought up by his paternal uncle, from whom he learnt Latin, and in early life he studied law and philosophy. A political career was planned for him, but in 1560 Tycho observed a solar eclipse and was so fascinated by what he saw that he spent the rest of his life studying mathematics and astronomy.

Being of a noble family, Tycho did not need a university degree to establish himself in a profession, but he attended the University of Copenhagen and studied ethics, music, natural sciences, philosophy and mathematics. From the beginning of his astronomical career he made a series of significant observations. Having seen the eclipse, he obtained a copy of Stadius' *Ephemerides*, which was based on the Copernican system. Observing a close approach of Jupiter and Saturn in 1563, Tycho noticed that it occurred a month earlier than predicted. He set about the preparation of his own tables. In 1564 he began observing with a radius, or cross-staff consisting of an arm along which could slide the centre of a crosspiece of half its length. Both arms were graduated and there was a fixed sight at the end of the larger arm which was held near the eye. To measure the angular distance between two objects, Tycho set the shorter arm at any gradation of the longer arm and moved a sight along the shorter arm until he saw the two objects through it and a sight at the centre of the transversal arm. The required angle was then obtained from the gradations and a table of tangents.

When his uncle died in 1565, Tycho travelled and studied at Wittenburg and Rostock, where he graduated from the university in 1566. While he was at Rostock, it is said that he lost the greater part of his nose in a duel with another nobleman over a point of mathematics, and thereafter wore a false nose made of silver. After making a number of observations in Rostock, he moved to Basle before entering the intellectual life of Augsburg in 1569. Having returned home because of his father's ill health, Tycho noticed one night in November 1572 a star in the constellation of Cassiopeia that was shining more brightly than all the others and which had not been there before. With a special sextant of his own making, Tycho observed the star until March 1574, when it ceased to be visible. His records of its variations in colour and magnitude identify it as a supernova.

In 1576 King Frederick II offered Tycho the island of Hven for the construction of an observatory. This was the first of its kind in history. Tycho's reputation grew and scholars from throughout Europe visited him.

Having observed a great comet in 1577, Tycho refuted Aristotle's theory of comets. He concluded that certain celestial bodies were supralunar, having no parallax and remaining stationary like fixed stars. Many other scientists had abandoned the Aristotelian theory in favour of the belief that something new could be created in the heavens and not necessarily out of the substances of the Earth. Tycho claimed that Aristotle's 'proof' had been based on meditation, not mathematical observation or demonstration. Tycho's main objective became to determine the comet's distance from the Earth. He was also concerned with its physical appearance – colour, magnitude and the direction of the tail.

He came to the conclusion that the comet's orbit must be elongated, a controversial suggestion indeed since it meant that the comet must have passed through the various planetary spheres, and it could not do that unless the planetary spheres did not exist. This possibility went against Tycho's most cherished beliefs. He could not abandon the ideas of his Greek predecessors, although he was the last great astronomer to reject the heliocentric theory of Copernicus. He tried to compromise, suggesting that, with the exception of the Earth, all the planets revolved around the Sun.

He prepared tables of the motion of the Sun and determined the length of a year to within less than a second, making calendar reform inevitable. In 1582 ten days were dropped, the Julian year being longer than the true year. To prevent further accumulations, the Gregorian calendar was adopted thereafter.

Tycho lost his patronage on the death of the King and he left for Germany in 1597. He settled in Prague at the invitation of the Emperor and found a new assistant, Johannes Kepler. Kepler loyally accepted and propounded Tycho's tables and data and continued his work with what were to be results of great importance. Many of Tycho's great contributions to science live on and he is remembered in particular for the improvements he made to almost every important astronomical measurement.

Bramah Joseph 1748–1814 was one of the most outstanding British engineers of his day. He took out patents for 18 inventions, of which the hydraulic press was probably to be the most significant. Nevertheless, his training of a whole generation of engineers in the craft of precision engineering at the dawn of the Industrial Revolution was probably an even greater legacy to his country than any of his individual inventions.

Bramah was born Joe Brammer on 13 April 1748, in Stainborough, near Barnsley, Yorkshire. He began his employment by working on his father's farm but the age of 16 was made lame in an accident. He became apprenticed to a carpenter and cabinet maker, and on completing his apprenticeship made his way to London to set up his own business. A stream of inventions followed, one of the most useful being the flushing water closet he produced in 1778. When the patent was taken out he changed the spelling of his name to the more fashionable sounding Bramah. In 1784 he patented his most celebrated invention, the Bramah lock. Designed to foil thieves, a specimen of the lock was exhibited in a shop window in 1784 with a 200-guinea reward for anyone who could succeed in picking it. Bramah kept the money for the rest of his lifetime, 67 years passing before the lock was finally opened by a mechanic after 51 hours' work. Such an effective lock could be produced only by high precision engineering and machine tools of the finest quality.

To assist him with his work Bramah took into his employ a young blacksmith named Henry Maudslay (the later friend and partner of Sir Mark Isambard Brunel), whose mechanical skill was at least equal to Bramah's. In 1795 Bramah produced the hydraulic press. This device makes use of Pascal's law: pressure exerted upon the smaller of two pistons results in a greater force on the larger one, both cylinders being connected and filled with liquid. For an important part of his press, the seal which ensured watertightness between the plunger and the cylinder in which it worked, he was particularly indebted to Maudslay, who, then only 19, produced the leather U-seal which expands under the fluid pressure.

The possibilities water offered as a means of propulsion were always in the forefront of Bramah's mind. In 1785 he suggested the locomotion of ships by means of screws; in 1790 and 1793 he constructed the hydraulic transmission of power. Among Bramah's other inventions were a machine for numbering bank notes, a beer pump, and machines for making paper and for the manufacture of aerated waters. He also produced a machine which made nibs for pens.

Bramah died in London on 9 December 1814 and was buried in Paddington churchyard. His press laid the foundation for a whole technology, applications of which include the car-jack, presses for baling waste paper and metal, and the hydraulic braking system for cars and other vehicles which ensures that the brakes on all the wheels operate simultaneously and evenly. Massive girders could be jacked into place, providing a powerful tool for such bridge-builders as Robert Stephenson, who used hydraulics to position the massive tubular spans of the bridges over the river Conway and the Menai Straits. Extrusion and forging presses still use the principles for which Bramah laid the working foundations.

Branly Edouard Eugène Désiré 1844–1940 was a French physicist and inventor who preceded Guglielmo Marconi in performing experiments resulting in the invention of wireless telegraphy and radio.

Branly was born in Amiens on 23 October 1844, the son of a teacher. He entered the Ecole Normale Supérieure in Paris in 1865, from which he gained his licence in physical and mathematical sciences in 1867 and his Agrégé in physics and natural science in 1868. He then joined the staff of the Lycée de Bourges but almost immediately afterwards returned to Paris to take charge of the physics laboratory at the Sorbonne, becoming its adjunct director in 1870. He then submitted his doctoral thesis to the Sorbonne, gaining his Docteur (medical) qualification in 1872 and his Docteur des Sciences in 1873. In 1876 he was appointed Professor of Physics at the Ecole Supérieure de Sciences of the Catholic Institute in Paris through the influence of Abbot Hulst, who was then working at the Catholic Institute. From 1897 to 1916 Branly then worked as a medical professor of electrotherapy. It was not until 1898 that his work was recognized, with the award of the Houllevignes Prize by the French Academy of Sciences. Thereafter he received many other honours and, after his death in Paris on 24 March 1940, a national funeral was held for him.

Branly researched in various subjects, notably electricity, electrostatics (on which he wrote his doctoral thesis), magnetism and electrical dynamics. His first task at the Catholic Institute, however, was to set up (with limited financial resources) a physics laboratory. (It was not until Frère Coty made a large donation in 1932 that the physics laboratory was adequately equipped, and all of Branly's work was carried out in very poor laboratory conditions.) His most important work – on wireless telegraphy – was performed in 1899, when he demonstrated the coherer, an invention of his that enabled radio waves from a distant transmitter to be detected. Once he had established the principle, however, he did not develop it further and the practical points were later taken up by Marconi, who was to share the 1909 Nobel Prize for Physics with Karl Braun for the invention of wireless telegraphy.

Braun Emma Lucy 1889–1971 was a US botanist, an early pioneer in recognizing the importance of plant ecology and conservation.

She was born in Cincinnati on 19 April 1889 and with her elder sister Annette (1884–1978) was encouraged by her parents in both informal nature study and formal academic work. She graduated from the University of Cincinnati, gaining a master's degree in geology in 1912, and was awarded a PhD in botany in 1914, three years after her sister had also achieved the degree. She remained in academic positions at the university, becoming Professor of Plant Ecology in 1946, until early retirement in 1948. She lived with her sister, an entomologist, and continued research work until the end of her life, the two setting up a home laboratory and an experimental garden. She died in her Cincinatti home on 5 March 1971.

Although Braun did produce laboratory-based work, her major advances were made from field studies. Her work in ecology concentrated on the vegetation of a selected variety of habitats in Ohio and Kentucky. An early taxonomic study provided a detailed catalogue of the flora of the Cincinnati region, which she then compared with that of the same region a century earlier. This approach became very influential for analysing regional changes in flora over a period of time.

Her field studies of plant distribution combined with her interests in geology also led her to consider several innovative theories in the evolution of forest communities and their survival during periods of glaciation. Her laborious studies of the forests of the area over almost 30 years were distilled in a classic work *Deciduous Forests of Eastern North America* (1950).

Braun also contributed to the growing conservation movement, stressing the importance of preserving natural habitats. She established a local Wild Flower Preservation Society and wrote and campaigned to save natural areas and to create nature reserves. She became the President of the Ohio Academy of Science and the Ecological Society of America, the first woman to achieve both positions, and was honoured by the Botanical Society of America.

Braun Karl Ferdinand 1850–1918 was a German physicist who is best known for his improvements to Guglielmo Marconi's system of wireless telegraphy, for which he shared (with Marconi) the 1909 Nobel Prize for Physics. He also made other important contributions to science, including the discovery of crystal rectifiers and the invention of the oscilloscope.

Braun was born in Fulda, Germany, on 6 June 1850. He was educated at the universities of Marburg and Berlin, gaining his doctorate from the latter in 1872. His first job was assistant to Quinke at Würzburg University, after which he successively held positions at the universities of Leipzig, Marburg, Karlsruhe, Tübingen, and Strasbourg; while at Strasbourg (from 1880 to 1883) he founded the Institute of Physics. In 1883 he was elected to the Chair in Physics at the Karlsruhe Technische Hochschule, a post he held until 1885, when he became Professor of Physics at Tübingen University. In 1895 he returned to Strasbourg University as Professor of Physics and Director of the Institute of Physics. In 1917 he went to the United States to testify in litigation about radio patents, but when the United States entered World War I he was detained as an alien in New York City and died there shortly afterwards, on 20 April 1918.

Braun began to study radio transmission in the late 1890s in an attempt to increase the transmitter range to more than 15 km/9 mi (the maximum range then possible). He thought that the range could be increased by increasing the transmitter's power but, after working on Hertz oscillators, found that lengthening the spark gap to increase the power output was effective only up to a certain limit, beyond which power output actually decreased. He therefore devised a sparkless antenna (aerial) circuit in which the power from the transmitter was magnetically coupled (using electromagnetic induction) to the antenna circuit, instead of the antenna being connected directly in the power circuit. This invention, which Braun patented in 1899, greatly improved radio transmission, and the principle of magnetic coupling has since been applied to all similar transmission systems, including radar and television. Later Braun developed directional antennas.

Braun also discovered crystal rectifiers. In 1874 he published a paper describing his research on mineral metal sulphides, some of which, he found, conduct electricity in one direction only. His discovery did not find an immediate application, but crystal rectifiers ('cat's whisker') were used in the crystal radio receivers of the early twentieth century, until they were superseded by more efficient valve circuits.

Braun's other principal contribution to science was his invention of the oscilloscope (1895), which he used to study high-frequency alternating currents. In his oscilloscope – which was basically an adaptation of the cathode-ray tube – Braun used an alternating voltage applied to deflection plates to deflect an electron beam within a cathode-ray tube. The Braun tube (as his invention was initially called) became a valuable laboratory instrument and was the forerunner of more sophisticated oscilloscopes and of the modern television and radar display tubes.

Bredig Georg 1868–1944 was a German physical chemist who contributed to a wide range of subjects within his discipline but is probably best known for his work on colloids and catalysts.

Bredig was born on 1 October 1868 in Glogau, Lower Silesia (now Glogow, Poland). After qualifying he went to work as an assistant in the laboratory of the great German chemist Friedrich Ostwald in Leipzig, and it was there that he did much of his significant work. He held a series of academic appointments in physical chemistry: at Heidelberg, Germany (1901–10); Zurich, Switzerland (1910); and from 1911 at the Karlsruhe Hochschule, Germany. He went to the United States in 1940, and died in New York City on 24 April 1944.

While working with Ostwald, Bredig collaborated with him on the accumulation of experimental data with which to validate Ostwald's dilution law (which states that, for a binary electrolyte, the equilibrium constant K_c of a chemical reaction has the same value at all dilutions). The equilibrium constant depends not on the dilution, but on the chemical nature of the particular acid or base. Ostwald confirmed the law for 250 acids, and Bredig provided the comparable data for 50 bases.

The variation in the relative atomic mass (atomic weight) of lead from various sources, the transition metals (on which he worked with Jacobus van't Hoff), and catalytic action formed other areas of his research. Bredig also supervised overseas chemistry students, such as the British chemist Nevil Sidgewick.

In the field of catalysis Bredig's particular study was the catalytic action of colloidal platinum and the 'poisoning' of catalysts by impurities. His most important contribution was a method of preparing colloidal solutions (lyophobic sols) using an electric arc, which he devised in 1898.

There are two ways of producing particles of colloidal size: larger particles can be broken down (dispersed) or smaller particles can be made to aggregate. Bredig's arc method is a dispersion technique. An electric arc is struck between metal electrodes immersed in a suitable electrolyte – for example, platinum, gold, or silver electrodes in distilled water containing an alkali. The colloidal particles are thought to be produced mainly by rapid condensation of the vapour of the arc, and they may be in the form of the metal or its oxide.

A later extension of the method developed by Theodor Svedburg in the early 1900s uses an alternating current and produces sols of greater purity.

Brenner Sydney 1927– is a South African-born British molecular biologist noted for his work in the field of genetics.

Brenner was born on 13 January 1927 at Germiston, South Africa, the son of an emigrant from Lithuania. He was educated there and studied at the University of the Witwatersrand, where he gained his MSc in 1947 and his MB and BCh in 1951. He then went to Britain and studied for a PhD at Oxford, which he received in 1954. In that year he also worked in the Virus Laboratory of the University of California in Berkeley, and from 1955 to 1957 was a lecturer in physiology at the Witwatersrand University. From 1957 he researched in the Molecular Biology Laboratory of the Medical Research Council, Cambridge, and in 1980 was appointed director of that establishment.

Brenner's first research was on the molecular genetics of very simple organisms. Since then he has spent seven years on one of the most elaborate efforts in anatomy ever attempted, investigating the nervous system of nematode worms and comparing the nervous systems of different mutant forms of the animal. The nematode that lives in the soil and which feeds on or in roots can be as little as 0.5 mm/0.02 in in length. Brenner's experiments have included cutting a soil nematode into 20,000 extremely thin slices and, one at a time, projecting a long succession of electron micrographs onto a screen. The animal's nerves are traced in each picture by an electronic pen which automatically feeds information to a computer for storage.

Brenner's reason for gathering and processing all this information is to compare the 'wiring' of the nervous system of normal nematodes with that of mutant ones which show peculiarities in behaviour. About 100 genes are involved in constructing the nervous system of a nematode and most of the mutations that occur affect the overall design of a section of the nervous system. These genes are therefore an organizational type and regulate the routing of the nervous system during the growth of the animal. The nematode is a simple animal although its make-up is extremely complicated. The amount of effort that has been put into this study indicates how much biologists still have to find out about the exact organization of living tissues.

Brenner is also interested in tumour biology and in the use of genetic engineering for purifying proteins, cloning genes and synthesizing amino acids. His experiments have given and still continue to give a great impetus to molecular biology.

Brewster David 1781–1868 was a British physicist who investigated the polarization of light, discovering the law named after him for which he was awarded the Rumford Medal by the Royal Society in 1819. He also helped to popularize science in his writings, and is perhaps best known as the inventor of the kaleidoscope.

Brewster was born in Jedburgh, Scotland, on 11 December 1781. His education was extended by reading his father's university notes on physics, his services as secretary to the local minister Thomas Somerville, and his friendship with James Veitch, an amateur astronomer. Brewster entered the University of Edinburgh in 1794 but never took his degree. He continued his studies, this time in divinity, and was awarded an honorary MA in 1800. He was later licensed to preach but was not ordained, and turned to publishing and teaching as the means of making a living.

One of Brewster's major concerns was increasing the public awareness of the importance of science. He edited a number of scientific periodicals and wrote many books and articles on science, including entries in the *Encyclopaedia Britannica*. He was also instrumental in the foundation of several academic organizations including the Edinburgh School of Arts in 1821, the Royal Scottish Society for Arts in 1821 and the British Association for the Advancement of Science in 1831.

His achievements in optics were recognized by his election to many prestigious international scientific societies, including the Royal Society in 1815, and the French Institute. Brewster was knighted in 1832 and in 1859 made principal and later vice-chancellor of Edinburgh University. He died on 10 February 1868 in Allerby in Scotland.

With James Veitch, Brewster built many optical devices such as microscopes and sundials, developing an expertise that resulted in the invention of the kaleidoscope in 1816. In trying to improve lenses for telescopes, he became interested in optics and particularly in the polarization of reflected and refracted light. In 1813 he was able to demonstrate, by studying the polarization of light passing through a succession of glass plates, that the index of refraction of a particular medium determines the tangent of the angle of polarization for light that transverses it. Brewster then sought an expression for the polarization of light by reflection and found, in 1815, that the polarization of a beam of reflected light is greatest when the reflected and refracted rays are at right angles to each other. This is known as Brewster's law, and it may be stated in the form that the tangent of the angle of polarization is numerically equal to the refractive index of the reflecting medium when polarization is maximum.

Brewster then worked on the polarization of light reflected by metals, and established the new field of optical mineralogy. By 1819 he had classified most crystals and minerals on the basis of their optical properties. During the 1820s he studied colour in the optical spectrum, finding that it could be divided into red, yellow and blue regions, and worked on absorption spectroscopy of natural substances, greatly extending the number of dark lines identified in spectra.

Brewster was an advocate of the particular theory of light on the basis of his experimental results. Although forced to admit that the wave theory did provide excellent explanations for certain observed phenomena, he clung to his views as he was philosophically unable to accept the wave theory because it assumed the existence of a hypothetical 'ether'.

Bridgman Percy Williams 1882–1961 was a US physicist famous for his work on the behaviour of materials at high temperature and pressure, for which he won the 1946 Nobel Prize for Physics.

Bridgman was born on 21 April 1882 in Cambridge, Massachusetts. His father was a journalist and a social and political writer. He went to Harvard in 1900 and was awarded a PhD in 1908. He then began research work in the Jefferson Research Laboratory and spent his entire research life at Harvard University, starting as assistant professor in 1913. He became professor in 1919, Hollis Professor of Mathematics and Philosophy in 1927 and Higgins Professor in 1950. He retired to become Professor Emeritus in 1954. Suffering from an incurable disease, Bridgman killed himself at his home in Randolph, New Hampshire, on 20 August 1961.

At Harvard, it became clear that Bridgman was an excellent experimentalist and was skilled at handling machine tools and at manipulating glass. His experimental work on static high pressure began in 1908, and was at first limited to pressures of around 6,500 atm/6.5×10^8 Pa. He gradually extended it to more than 100,000 atm/10^{10} Pa and eventually to about 400,000 atm/4×10^{10} Pa. Because this field of research had not been explored before, Bridgman had to invent much of his own equipment. His most important invention was a special type of seal in which the pressure in the gasket always exceeds that in the pressurized fluid. The result is that the closure is self-sealing – without this, his work at high pressure would not have been possible. He was later able to use the new steels and alloys of metals with heat-resistant compounds. His work involved meas-urement of the compressibilities of pressurized liquids and solids, and measurements of physical properties of solids such as electrical resistance. His discoveries included that of new high-pressure forms of ice. With the increase in range of possible pressures, new and unexpected phenom-ena appeared. He discovered that the electrons in caesium rearrange at a certain transition pressure. He pioneered the work to synthesize diamonds, which was eventually achieved in 1955. His technique was used to synthesize many more minerals and a new school of geology devel-oped, based on experimental work at high pressure and temperature. Because the pressures and temperatures that Bridgman achieved simulated those deep below the ground, his discoveries gave an insight on the geophysical processes that take place within the Earth. His book *Phys-ics of High Pressure* (1931) still remains a basic work.

Bridgman was an individualist. He published 260 papers, only two of which were with another author. He devoted himself to his scientific research, refusing to attend faculty meetings or to serve on committees. He disliked lecturing and did it badly. In 1914, however, during a course of lectures on advanced electrodynamics, Bridgman realized that many ambiguities and obscurities exist in the definition of scientific ideas and moved into the philosophy of science, publishing *The Logic of Modern Physics* in 1927.

Briggs Henry 1561–1630 was an English mathematician, one of the founders of calculation by logarithms.

Briggs was born at Warley Wood, in Halifax, Yorkshire, in February 1561. He attended a nearby grammar school, and in 1577 went to St John's College, Cambridge. He was made a scholar in 1579 and received his BA in 1581. He was elected a Fellow of the college in 1588 and was appointed lecturer and examiner in 1592. It was a tribute to his rare abilities that, in an age when the universities paid scant attention to mathematics, he was appointed Professor of Geometry at the newly established Gresham College, London, in 1596. He held the chair until 1620, resigning it because a year earlier he had accepted the invitation of Henry Savile (1549–1622) to succeed him as Professor of Astronomy at Oxford. He was elected a Fellow of Merton College, where he died on 26 January 1630.

In 1616 Briggs wrote a letter to James Ussher, later Archbishop of Armagh, informing him that he was wholly absorbed in 'the noble invention of logarithms, then lately discovered'. Two years earlier John Napier had published his discovery of logarithms and as soon as Briggs learned of it he formed an earnest desire to meet the great man, for which purpose he travelled to Edinburgh in 1616. On this and subsequent visits the two men worked together to improve Napier's original logarithms which, having in modern notation $\log N = 10^7 \log_e (10^7/N')$, was in need of simplification. It seems most probable that the idea of hav-ing a table of logarithms with 10 for their base was originally conceived by Briggs. And although both men published separate descriptions of the advantages of allowing the logarithm of unity to be zero and of using the base 10, the first such logarithmic tables were published by Briggs in 1617. They were published under the title *Loga-rithmorum Chilias Prima*, and were followed in 1624 by the *Arithmetica Logarithmica*, in which the tables were given to 14 significant figures. In fact, the logarithms of Briggs (and of Napier) were logarithms of sines, a reflec-tion of both men's interest in astronomy and navigation, fields in which accurate and lengthy calculations using that trigonometrical function were everyday matters. It is owing to the way that sines were then considered that the large factor 10^9 was important and that Briggs's 1624 tables took the form $10^9 \log_{10} N$.

For that reason Briggs's logarithms were 10^9 times 'larger' than those in modern tables. Despite that, however, and despite the fact that many mathematicians (including Kepler) subsequently calculated their own tables, Briggs's tables remain the basis of those used to this day.

Bright Richard 1789–1858 was a British physician who was the first to describe the kidney disease known as Bright's disease, which is actually a rather vague term – now largely obsolete – sometimes used to denote any of several different kidney disorders that share a number of the same symptoms.

Bright was born on 28 September 1789 in Bristol and was privately educated in Exeter and Edinburgh. In 1809 he began studying medicine at Edinburgh University but interrupted his studies to travel to Iceland. On returning to Britain he resumed his medical training at Guy's Hospital and St Thomas's Hospital in London, receiving his medical degree from Edinburgh University in 1813. After spending several years touring Europe – during which time he worked in several European hospitals – Bright was appointed Assistant Physician at Guy's Hospital in 1820. Four years later he became a full physician at Guy's, a post he held for the rest of his life. In 1837 he was also appointed Physician Extraordinary to Queen Victoria. Bright died in London on 16 December 1858.

Bright's principal interest was disorders of the kidneys, in which area he initiated the use of biochemical studies by working with chemists to demonstrate that urea is retained in the body in kidney failure. He also correlated symptoms in patients with the pathological changes he later found in postmortem examinations of these same people. Using these methods he found that albuminuria (the presence of the protein albumin in the urine) and oedema (accumula-tion of fluid in the body) are associated with pathological changes in the kidneys – a condition that came to be called Bright's disease. Later, however, it was discovered that several different kidney disorders produce these symptoms (although the most common cause is glomerulonephritis – inflammation of the glomeruli) and the term Bright's dis-ease is little used today. Bright first published his findings in *Reports of Medical Cases* (1827) and subsequently in the second volume of the *Reports* (1831) and in the first vol-ume of *Guy's Hospital Reports* (1836), which he helped to establish.

In addition to his studies of kidney disorders, Bright investigated jaundice, nervous diseases and abdominal tumours. He also collaborated with Thomas Addison, a contemporary at Guy's Hospital, in writing *Elements of the Practice of Medicine* (1839); in fact Addison wrote most of this volume and the second volume – to which Bright was to have been the principal contributor – was never published.

Brindley James 1716–1772 was a British engineer who, in spite of the most formidable handicaps, became a pioneer of canal building.

Brindley was born near Buxton, Derbyshire, in 1716. Apprenticed to a millwright at the age of 17, he appears to have been completely devoid of promise even at this humble pursuit until the erection of a paper mill with certain novel features brought out in him a remarkable mechanical sense. As a result he was put in charge of his master's shop.

On his master's death Brindley set up his own business at Leek, Staffordshire, where he was soon running a thriving business repairing old machinery and installing new machines. Wedgwoods, then only a small pottery firm, employed him to construct flint mills. He completed the machinery for a silk mill at Congleton in Cheshire and had some limited success in improving the machinery then used to draw water from mines, although this problem was not effectively solved until the advent of the steam engine.

In 1759 Brindley was engaged by the Duke of Bridgwater to construct a canal to transport coal from the Duke's mines at Worsley to the textile manufacturing centre of Manchester. He pursuaded the Duke to change his plan of using locks and rivers and approve instead a revolutionary scheme, including a subterranean channel extending from the barge basin at the head of the canal into the river, and the Barton Aqueduct – 12 m/40 ft high – carried the canal over the river Irwell. Brindley's mechanical skill enabled him to construct impervious banks by 'puddling' clay, and the canal simultaneously acted as a mine drain, a feeder for the main canal at the summit level, and a barge-carrying canal.

Brindley's achievement, remarkable enough for any man, is made still more noteworthy by his almost complete lack of any formal schooling. He was virtually illiterate, barely able to write his name. He made no calculations or drawings of the tasks he set himself but worked out everything in his head. He was, in a most extreme form, a natural engineer.

The success of the Worsley scheme established Brindley as the leading canal builder in England. His next commission was to construct the Bridgwater canal linking Manchester and Liverpool, after which came others, the most important being the Grand Union Canal which connected Manchester with the Potteries in the Midlands. Derby and Birmingham (completely transforming) the life of the Midlands – the population of the Potteries trebled between 1760 and 1785 the Oxford Canal, the old Birmingham and the Chesterfield Canals, the Staffordshire and Worcestershire Canals, and the Coventry Canal. All these were designed and with only one exception executed by Brindley. In all he constructed 584 km/360 mi of canals.

It is said of Brindley that when faced with an apparently intractable problem he would go home and think it over in his bed. This must not lead one to assume, however, that he took a phlegmatic attitude to his work. He died from his excessive and arduous labours on 27 September 1772, at Turnhurst, Staffordshire.

Brinell Johann August 1849–1925 was a Swedish metallurgist who developed what became known as the Brinell hardness test, a rapid nondestructive method of estimating metal hardness.

Brinell was born at Bringetofta and attended technical school at Boras. On leaving school he worked as a mechanical designer. In 1875 he was appointed chief engineer at the iron works at Lejofors, and it was there that he became interested in metallurgy. In 1882 he became chief engineer at the Fagersta iron works. While at Gagersta he studied the internal composition of steel during heating and cooling, and devised the hardness test which was put on trial at the Paris Exhibition of 1890. The test is based on the impression left by a small hardened steel ball after it is pushed into a metal with a given force. With minor innovations the test remains in use today. Brinell also carried out investigations into the abrasion resistance of selected materials. He died in Stockholm in 1925.

Based on the idea that a material's response to a load placed on one small point is related to its ability to deform permanently, the hardness test is performed by pressing a hardened steel ball (Brinell test) or a steel or diamond cone (Rockwell test) into the surface of the test piece. The hardness is inversely proportional to the depth of penetration of the ball or cone.

Bronowski Jacob 1908–1974 was a Polish-born British scientist, journalist and writer, originally trained as a mathematician, who won international recognition as one of the finest popularizers of scientific knowledge in the twentieth century.

Bronowski was born in Poland on 18 January 1908, fled with his family to Germany when, in World War I, Russia occupied Poland, and moved in 1920 to England and became a naturalized British citizen. He studied mathematics at Jesus College, Cambridge, where he also edited a literary magazine and published some unremarkable verse, graduating as Senior Wrangler (with the highest marks in the final examination.). He was awarded his PhD in 1933 and a year later was appointed Senior Lecturer at University College, Hull. He remained there until after the outbreak of World War II, when in 1942 he joined Reginald Stradling's Military Research Unit at the Home Security Office.

His principal job was to forecast the economic effects of bombing. After the war he conducted statistical research at the Ministry of Works until 1950, when he was appointed Director of the National Coal Board's research establishment. From 1959 to 1963 he served as Director-General of

Process Development for the board. During these years as a government official, he was continually extending the range of his intellectual pursuits. In particular he devoted himself to studying the development of Western science and thought. In 1953 he was visiting Professor of History at the Massachusetts Institute of Technology. His last appointment was a senior Fellow at the Salk Institute for Biological Studies in California, a post he took up in 1964. He died in San Diego on 22 August 1974.

Of his appointment to the Salk Institute Bronowski said that he was a 'mathematician trained in physics, who was taken into the life sciences in middle age by a series of lucky chances'. In fact he will probably be least remembered as a biologist. His true métier was for explaining to a large public the broad canvas of European intellectual history. His first published work of note was *The Poet's Defence* (1939) and at one time it might have been thought that he would come to specialize in literary subjects. After World War II he wrote several plays for radio, most memorably two in 1948, *Journey to Japan* and *The Face of Violence*, the latter of which won the Italia Prize in 1951 as the best radio play in Europe. He never lost his interest in literature (*William Blake and the Age of Revolution* appeared as late as 1965), but from the early 1950s his main interest turned towards broader intellectual and scientific themes.

The Common Sense of Science (1951) represented Bronowski's first attempt to bring the mysteries of science within the ken of nonscientific readers and was notable for the manner in which it displayed the history and workings of science around three central notions: cause, chance, and order. In the early days of the Cold War he used the pages of the *New York Times* to discuss, in accessible language, both the technology of nuclear science and the moral question raised by the development of nuclear weapons. An extension of his newspaper articles was the book *Science and Human Values*, which was published in 1958.

Bronowski was deeply concerned about the general effects on society of the widening division between the arts and the sciences, a phenomenon given great publicity by the famous Leavis–Snow controversy, and he was at pains to do what he could to narrow the divide, while at the same time bringing the specialist conclusions of scholars in both the sciences and the humanities to a wide public. The result was two of his finest popularizing works, *The Western Intellectual Tradition* (1960), an illuminating survey of the growth of political, philosophical and scientific knowledge from the Renaissance to the nineteenth century written with Bruce Mazlish, and the brilliant 13-part BBC television documentary, *The Ascent of Man*, issued as a book in 1973.

When he was president of the British Library Association in 1957–58, Bronowski said, in his inaugural address, that for public libraries to serve in the general expansion of a society's culture, writers must make the language of science comprehensible to the nonspecialist. It is his great distinction that he practised, with great wit and erudition, what he preached.

Brönsted Johannes Nicolaus 1879–1947 was a Danish physical chemist whose work in solution chemistry, particularly electrolytes, resulted in a new theory of acids and bases.

Brönsted was born on 22 February 1879 in Varde, Jutland, the son of a civil engineer. He was educated at local schools before going to study chemical engineering at the Technical Institute of the University of Copenhagen in 1897. He graduated two years later and then turned to chemistry, in which he qualified in 1902. After a short time in industry, he was appointed an assistant in the university's chemical laboratory in 1905, becoming Professor of Physical and Inorganic Chemistry in 1908. In his later years he turned to politics, being elected to the Danish parliament in 1947. He died on 17 December in that year, before he could take his seat.

Brönsted's early work was wide ranging, particularly in the fields of electrochemistry, the measurement of hydrogen ion concentrations, amphoteric electrolytes, and the behaviour of indicators. He discovered a method of eliminating potentials in the measurement of hydrogen ion concentrations, and devised a simple equation that connects the activity and osmotic coefficients of an electrolyte, and another that relates activity coefficients to reaction velocities. From the absorption spectra of chromic – chromium (III) – salts he concluded that strong electrolytes are completely dissociated, and that the changes of molecular conductivity and freezing point that accompany changes in concentration are caused by the electrical forces between ions in solution.

Brönsted related the stages of ionization of polybasic acids to their molecular structure, and the specific heat capacities of steam and carbon dioxide to their band spectra. In 1912 he published work with Herman Nernst on the specific heat capacities of steam and carbon dioxide at high temperatures. Two years later he laid the foundations of the theory of the infrared spectra of polyatomic molecules by introducing the so-called valency force field. Brönsted also applied the newly developed quantum theory of specific heat capacities to gases, and published papers about the factors that determine the pH and fertility of soils.

In 1887 Svante Arrhenius had proposed a theory of acidity that explained its nature on an atomic level. He defined an acid as a compound that could generate hydrogen ions in aqueous solution, and an alkali as a compound that could generate hydroxyl ions. A strong acid is completely ionized (dissociated) and produces many hydrogen ions, whereas a weak acid is only partly dissociated and produces few hydrogen ions. Conductivity measurements confirm the theory, as long as the solutions are not too concentrated.

In 1923 Brönsted published (simultaneously with Thomas Lowry in Britain) a new theory of acidity which has certain important advantages over that of Arrhenius. Brönsted defined an acid as a proton donor and a base as a proton acceptor. The definition applies to all solvents, not just water. It also explains the different behaviour of pure acids and acids in solution. Pure dry liquid sulphuric acid or acetic (ethanoic) acid does not change the colour of indicators nor react with carbonates or metals. But as soon as water is added, all of these reactions occur.

In Brönsted's scheme, every acid is related to a conjugate base, and every base to a conjugate acid. When hydrogen chloride dissolves in water, for example, a reaction takes place and an equilibrium is established:

$$HCl + H_2O \leftrightarrow H_3O^+ + Cl^-$$
Acid 1   Base 2   Acid 2   Base 1

HCl is an acid for the forward reaction, but the hydroxonium ion (H_3O^+) is an acid in the reverse reaction; it is the conjugate acid (acid 2) of water (base 2). Similarly, the chloride ion (Cl^-, base 1) accepts protons in the reverse reaction to form its conjugate acid (HCl, acid 1). In this theory acids are not confined to neutral species or positive ions. For example, the negatively charged hydrogen sulphate ion can behave as an acid:

$$HSO_4^- (aq) + H_2O(l) \leftrightarrow H_3O^+ + SO_4^{2-}(aq)$$

It donates a proton to form the hydroxonium ion.

Brouwer Luitzen Egbertus Jan 1881–1966 was a Dutch mathematician who founded the school of mathematical thought known as intuitionism.

Brouwer was born in Overschie on 27 February 1881. He studied mathematics at the University of Amsterdam, and on receiving his BA was appointed an external lecturer to the university in 1902. He remained at that post until 1912, when he was appointed Professor of Mathematics, a chair which he held until his retirement in 1951. His contribution to mathematics earned him numerous honours, most notably the Knighthood of the Order of the Dutch Lion in 1932. He died in Blaricum on 2 December 1966.

Brouwer's first important paper, a discussion of continuous motion in four-dimensional space, was published by the Dutch Royal Academy of Science in 1904, but the greatest early influence on him was Gerritt Mannoury's work on topology and the foundations of mathematics. This led him to consider the quarrel between Jules Poincaré and Bertrand Russell on the logical foundations of mathematics, and his doctoral dissertation of 1907 came down on the side of Poincaré against Russell and David Hilbert. He took the position that, although formal logic was helpful to describe regularities in systems, it was incapable of providing the foundation of mathematics.

For the rest of his career Brouwer's chief concern remained the debate over the logical, or other, foundations of mathematics. His inaugural address as Professor of Mathematics at Amsterdam in 1912 opened new ground in this debate, which had begun with the work of Georg Cantor (1843–1918) in the early 1880s. In particular Brouwer addressed himself to problems associated with the law of the excluded middle, one of the cardinal laws of logic. He consistently took issue with mathematical proofs (so-called proofs, as he saw them) that were based on the law. In 1918 he published his set theory which was independent of the law, explaining the notion of a set by the introduction of the idea of a free-choice sequence.

Having rejected the principle of the excluded middle as a useful mathematical concept, Brouwer went on to establish the school of intuitional mathematics. Put simply, it is based on the premise that the only legitimate mathematical structures are those that can be introduced by a coherent system of construction, not those which depend upon the mere postulating of their existence. So, for example, the intuitionist principle denies that it makes sense to talk of an actual infinite totality of natural numbers; that infinite totality is something which requires to be constructed.

Brouwer's work did not create an overnight sensation. But when, in the late 1920s, Kurt Gödel (1906–1978) broke down Hilbert's foundation theory, it gained great pertinence. The result of Gödel's endeavours was the theory of recursive functions, and in that field of mathematics Brouwer's work came to be of such fundamental significance that his intuitional theories and analysis have continued to be at the very centre of research into the foundations of mathematics.

Brown Ernest William 1866–1938 was a British mathematician with a particular interest in celestial mechanics and lunar theory.

Brown was born in Hull on 29 November 1866. He was awarded a scholarship to Christ's College, Cambridge, where he was introduced to problems in lunar theory by George Darwin (1845–1912). Brown gained a BA in 1887, and from 1889 until 1895 he held a fellowship at Christ's, although in 1891 he went to the United States to teach mathematics at Haverford College in Pennsylvania. He was Professor of Mathematics at Haverford (1893–1907) and then at Yale University until his retirement as Professor Emeritus in 1932.

Brown's work on the motions of the Moon earned him an international reputation. He received numerous honorary degrees and awards from scientific organizations such as the National Academy of Sciences and the Royal Astronomical Society. He was an active participant within the professional societies of which he was a member. Brown died in New Haven, Connecticut, on 22 September 1938.

The effect of gravity on the motions of the planets and smaller members of the Solar System was the major research interest of Brown's career. His work first focused on lunar motion, and he produced extremely accurate tables of the Moon's movements. Unable to account for the variation in the Moon's mean longitude, he proposed that the observed fluctuations arose as a consequence of a variable rate in the rotation of the Earth.

Brown was also interested in the asteroid belt. It had been proposed that asteroids might at one time have been part of one planet. Some astronomers attempted to compute the possible orbit of such a parent planet on the basis of the distribution of the asteroids, but Brown was highly critical of this approach.

One of his concerns during the later years of his career was the calculation of the gravitational effect exerted by the planet Pluto on the orbits of its nearest neighbours, Uranus and Neptune.

Brown's work on gravity did much to increase our understanding of the relationship of members of the Solar System.

Brown Robert 1773–1858 was a distinguished Scottish botanist whose discovery of the movement of suspended particles has proved fundamental in the study of physics.

Brown was born at Montrose, Scotland, on 21 December 1773, the son of an Episcopalian priest. He studied medicine at Edinburgh University but did not obtain his degree. He subsequently held the position of assistant surgeon in a Scottish infantry regiment, but soon revealed that his true interest lay in botany. In the late 1790s he was introduced to the well-known English botanist Joseph Banks, who allowed him the free use of his library and collections. Shortly afterwards Brown resigned from the army in order to accept the post of naturalist on an expedition under Captain Matthew Flinders, on the *Investigator*, to survey the coast of the lately discovered Australian continent. He voyaged from 1801 to 1805 and on his return to England published, in 1810, the first part of his studies on the flora he had discovered on his Antipodean journey. The poor sales of the book discouraged him and he left the rest unpublished. In the same year, he was appointed librarian to Joseph Banks, a post which he held until Banks's death in 1820. Banks bequeathed to Brown the full use of the library and its collections for life. In 1827, in compliance with the stipulations of Banks's will, he agreed to the transfer of the books and specimens to the British Museum and was appointed curator of the botanical collections there. He died in London on 10 June 1858.

In 1791 Brown submitted his first paper to the Natural History Society. It was a highly detailed classification of the plants he had collected in Scotland, with accompanying notes and observations. This list was to win him many introductions in the scientific world of his day. It was not until 1828, however, that he made one of his greatest contributions to science, published in the *Edinburgh New Philosophical Journal*. The paper was entitled 'A brief account of microscopical observations made in the months of June, July and August 1827 on the particles contained in the pollen of plants, and on the general existence of active molecules in organic and inorganic bodies' and it was in this paper that Brown set out his observations on 'Brownian movement', or 'motion', which perpetuates his name. The concept arose from his observation that very fine pollen grains of the plant *Clarkia pulchella* when suspended in water move about in a continuously agitated manner. This phenomenon is true for any small solid particles suspended in a liquid or gas and can be viewed in a bright light through a microscope. Brown was able to establish that the constant movement was not purely biological in origin because inorganic materials such as carbon and various metals are equally subject to it, although he could not find the cause of the movement. During his lifetime there was no shortage of theories to explain his discovery, but it was not until the twentieth century that the question was answered.

Brown also published papers on Asclepiadaceae (1809) and on Proteaceae (1810), wrote on the propagatory process of the gulf-weed and on the anatomy of fossilized plants. He also described the organs, and mode of reproduction in orchids. In 1831, while investigating the fertilization of both Orchidaceae and Asclepiadaceae, he discovered that a small body which is fundamental in the creation of plant tissues, occurs regularly in plant cells – he called it a 'nucleus', a name which is still used. Another significant revelation Brown made was the identification of the difference between gymnosperms and angiosperms.

Brown's various papers on his findings and opinions in every division of botanical science made him the outstanding authority on plant physiology of his day, and he did much to improve the system of plant classification by describing new genera and families. His observation of Brownian movement was important in showing how molecular motion forms the basis of kinetic theory.

Brunel Isambard Kingdom 1806–1859 was the only son of Marc Isambard Brunel, and pursued a similar career, marked by hugely ambitious projects unparalleled in engineering history, becoming ever more ambitious as the years progressed.

Born in Portsmouth in 1806, at the age of 14 he was sent to France to the College of Caen in Normandy and later to the Henri Quatre school in Paris. Brunel was appointed resident engineer on his father's Thames Tunnel enterprise when only 19. This promising start was abruptly ended when he was seriously injured by a sudden flood of water into the tunnel. While recuperating from this accident at Bristol, he entered a design competition, submitting four designs for a suspension bridge over the river Avon. The judge, Thomas Telford, rejected all of them in favour of his own. After many battles and a second contest, one of Brunel's designs was accepted. Work on the bridge began in 1833, but owing to lack of funds it was not completed until after its designer's death.

In 1833 Brunel was appointed to carry out improvements on the Bristol docks to enable heavily loaded merchantmen to berth more easily. It was while working on this project that Brunel's interest in the potential of railways was fired. The famous Rainhill trials, at which Stephenson's *Rocket* had triumphed, had been held four years previously.

Brunel completed a survey for constructing a railway from London to Bristol which was to be known (with a grandiloquence typical of Brunel) as the Great Western Railway. This characteristic love of the outsize was again evident in Brunel's decision to adopt a broad gauge of 2.1 m/7 ft for his locomotives, a choice which had the advantage of offering greater stability at high speeds. This size was in contrast to Stephenson's 'standard' of 1.44 m/ 4 ft 8 in.

In all, Brunel was responsible for building more than 2,600 km/1,600 mi of the permanent railway of the west of England, the Midlands, and South Wales. He also constructed two railway lines in Italy, acted as advisor on the construction of the Victoria line in Australia and on the East Bengal railway in India. The bridges which Brunel designed for his railways are also worthy of note. Maidenhead railway bridge had the flattest archheads of any bridge in the world when it was opened, and Brunel's use of a compressed-air caisson to install the pier foundations for the bridge helped considerably to win acceptance of the compressed-air technique in underwater and underground constructions.

However, many of Brunel's viaducts were work-a-day timber structures which used cheap, readily available materials and were designed so that renewal of members was quick and simple. They were only replaced with more solid structures when timber for repair rose to an uneconomic price. Of all the railway bridges Brunel produced, the last and the greatest was to be the Royal Albert, crossing the river Tamar at Saltash near Plymouth. This has two spans of 139 m/455 ft and a central pier built on the rock, 24 m/ 80 ft above the high water mark. The bridge was opened in 1859, the year of Brunel's death.

No sooner, it was said, had Brunel provided a new land link between Bristol and London through his Great Western Railway than he decided to extend the link to New York. The means of achieving his aim came in the shape of ships: the *Great Western*, launched in 1837, the *Great Britain* of 1843 and the *Great Eastern* of 1858. Each was the largest steamer in the world at the time of its launch.

Brunel's predilection for large vessels was not simply the outcome of his love for the outsize: there was sound engineering reasoning behind his designs. At the time, there was a strong body of scientific opinion which held that ships could never cross the Atlantic under steam alone because they could not carry enough coal for the journey. And building a larger ship was no remedy, it was argued, because doubling the size of the vessel doubled the drag forces it had to overcome, thus doubling the power needed and therefore the amount of coal required. Brunel was probably the only man of his time who could see the fallacy of this argument. This lay in the fact that the coal-carrying capacity of a ship is roughly proportional to the cube of the vessel's leading dimension, while the water resistance increases only as the square of this dimension. The voyage of the *Great Western* proved Brunel right: when it docked after its first transatlantic voyage it had 200 tonnes of coal left in its bunkers.

The *Great Western*, with 2,340 tonnes displacement, was a timber vessel driven by paddles. Its crossing of the Atlantic in the unprecedented time of 15 days brought, after initial wariness, the most enthusiastic acclaim and established a regular steamship between Britain and America.

Brunel's next ship, the *Great Britain* of 3,676 tonnes displacement, represented a great advance in the design of the steamship. It had an iron hull and was the first ship to cross the Atlantic powered by a propellor. The value of its revolutionary hull was made clear on its first voyage, when it was beached in Dundrum Bay on the Southern Coast of Ireland. It remained there for the best part of a year without suffering serious structural damage. As a passenger ship,

the *Great Britain* was an unprecedented success, remaining in service for 30 years, sailing to San Francisco, journeying regularly to Australia and even serving as a troopship.

Then, in what was thought to be the final chapter of its life, the ship was badly damaged off Cape Horn in 1866, managing to struggle to the Falkland Islands only to be condemned. It lay, a hulk that refused to rot, in Sparrow cove until it was salvaged by the *Great Britain* project, set up in 1968. Through the efforts of the enthusiasts, who towed it to Montevideo and from there to Bristol, the *Great Britain* entered the dock where it was made on 19 July 1970, exactly 127 years from the day it was floated out.

On 31 January, 1858 Brunel witnessed the spectacular sideways launching of his last ship, the *Great Eastern*, which was to remain the largest ship in service until the end of the nineteenth century. Well over ten times the tonnage of his first ship, it was 211 m/692 ft in length, had a displacement of 32,513 tonnes/32,000 tons, and was the first ship to be built with a double iron hull. It was driven by both paddles and a screw propellor.

Initially the *Great Eastern* was beset with problems. There was constant engine trouble and the day after the ship set out on a new commissioning trial Brunel, who was too ill to be on board, was struck down with paralysis. Unable to delegate responsibility, the work and worry of his many other enterprises had finally broken his health. He died at his home in London on 15 September 1859, having heard that an explosion aboard *Great Eastern* had apparently brought this most ambitious enterprise to nothing. Despite the damage this was not the end of his great ship. It was used successfully as a troop ship, and its greatest moment came in 1866 when, under the supervision of the great physicist Lord Kelvin, it was used to lay down the first successful transatlantic cable.

Brunel was elected to the Royal Society in 1830, at the age of 24, and was made a member of most of the leading scientific societies in Britain and abroad. These honours, however, seem scant reward for such a giant in an age which bred such engineering giants. He was the last, and the greatest, of them all.

Brunel Marc Isambard 1769–1849 was a French-born English inventor and engineer who is best remembered for his success in overcoming the age old problem of tunnelling in strata beneath water. His most notable achievement was the construction of the Thames Tunnel.

Born in Hacqueville in Normandy, Brunel served six years in the French navy, but left France in 1793, then at the peak of Revolutionary fervour, because of his royalist sympathies. His new home was the United States where he practised as an architect and civil engineer in New York, finally becoming the city's chief engineer. Here, besides surveying, canal engineering and constructing buildings, he advised on the improvement of the defences of the channel between Staten Island and Long Island, built an arsenal, and designed a cannon foundry.

While he was in the United States, Brunel developed an improved method for manufacturing ships' pulley blocks:

at the time a 74-gun ship used 1,400 blocks, all of which had to be made by hand. Acting on the advice of a friend, Brunel sailed for England in 1799 and took the drawings of his invention to Henry Maudslay in London. Maudslay, a pioneer of machine tools and a former pupil of Joseph Bramah, was impressed by Brunel's designs with the result that the 43 machines operated by ten men produced blocks superior in quality and consistency to those which previously 100 men had made by hand. Historically it was a portent of what was soon to become a universal trend: specialist machine tools, each performing one of a series of operations and taking over almost all the many labours formerly performed by hand.

Brunel was a tireless and prolific inventor. Among the gadgets he devised were machines for sawing and bending timber, bootmaking, knitting stockings, printing, copying drawings, and manufacturing nails.

In 1814 the first of a series of disasters overtook Brunel when fire badly damaged his sawmill at Battersea. The heavy loss this occasioned him brought to the surface the financial incompetence with which his partners were conducting his enterprises. Brunel soon found himself heavily in debt, a condition which worsened when in the following year the final defeat of Napoleon at Waterloo brought peace to Europe, the government accordingly ceasing to pay for a workshop producing army boots. In 1821 Brunel was imprisoned for several months for debt, being released only when friends eventually obtained from the government a grant of £5,000 for his discharge on the condition that he should remain in England.

In 1825 Brunel started work on his last and longest commission: the construction of a tunnel under the Thames in London to link Rotherhithe with Wapping. Brunel took out a patent on a revolutionary tunnelling shield in 1818, a move that hastened his subsequent appointment as engineer of the Thames Tunnel Company. Although the project ended in success it underwent many reverses, took 18 years to complete and shattered Brunel's health long before its completion.

At the time of the tunnel's construction, nothing so ambitious had been attempted before. The shield Brunel had designed covered the area to be excavated and consisted of 12 separate frames comprising altogether 36 cells in each of which a workman could be engaged at the workface independently of the others. The whole device obtained its propulsion from screw power which drove it forwards in 114 mm/4.5 in steps (the width of a brick) as the work progressed.

All too often, however, excavation had to stop. Five times water burst through the thin layer of earth beneath which the diggers worked, forcing a halt. Fortunately the shield always held, preventing total catastrophe. The long delays caused by these reverses put the scheme's finances under a severe strain. At one point, the whole operation came to a halt through lack of funds, the tunnel was bricked up and for seven years no more work was done. When the operation was resumed, a much larger shield was introduced to cover the 120 m/400 ft of the tunnel already

constructed. The excavations took place only 4 m/14 ft under the riverbed at its lowest point.

The Thames Tunnel was opened in 1843, and was the first public subaqueous tunnel ever built. Of a horseshoe cross-section, its total length was 406 m/1,506 ft and its width 11 m/37 ft by height 7 m/23 ft. Brunel, elected to the Royal Society in 1814, was knighted in reward for his labours in 1814. The public flocked to go through the long-awaited tunnel and in three and a half months more than a million people had passed through it. The first trains used the tunnel in 1865. It is now part of the London Underground system.

Buchner Eduard 1860–1917 was a German organic chemist who discovered noncellular alcoholic fermentation of sugar – that is, that the active agent in the reaction is an enzyme contained in yeast, and not the yeast cells themselves. For this achievement he was awarded the 1907 Nobel Prize for Chemistry.

Buchner was born in Munich on 20 May 1860, of an old Bavarian family of scholars. His father was a professor of forensic medicine and obstetrics, as well as being editor of a medical journal. When Buchner graduated from the Realgymnasium he served in the field artillery before going to study chemistry at the Munich Technische Hochschule. His studies were again interrupted – for financial reasons this time – and he spent four years working in the canneries of Munich and Mombach. In 1884, with the assistance of his elder brother Hans, a bacteriologist, he resumed his academic training in the organic section of the chemical laboratory of the Bavarian Academy of Sciences in Munich, where he worked under Johann von Baeyer.

While he was studying chemistry, Buchner also worked in the Institute for Plant Physiology under the Swiss botanist Karl von Nägeli (1817–1891). He obtained his doctorate in 1888 and was appointed teaching assistant to von Baeyer at the Privatdozent. In 1893 he succeeded Theodor Curtius (1857–1928) as head of the section for analytical chemistry at the University of Kiel and became Associate Professor there in 1895. Later professorships included appointments at Tübingen (1896), Berlin (1898), Breslau (1909) and Würzburg (1911). In 1914 he served in the German army as a captain in the ammunition supply unit and was promoted to major in 1916. He was recalled to Würzburg to teach for a short time but returned to the front in Romania on 11 August 1917. He was killed by a grenade four days later at Focşani.

It was while Buchner was working at the Institute for Plant Physiology that he first became interested in the problems of alcoholic fermentation, and in his first paper (1886) he came to the conclusion that Louis Pasteur was wrong in his contention that the absence of oxygen was a necessary prerequisite for fermentation. In 1858 Moritz Traube had proposed that all fermentations were caused by what he termed 'ferments' – definite chemical substances which he thought were related to proteins and produced by living cells; in 1878 Willy Kühne (1837–1900) called these substances enzymes. Many researchers, including Pasteur, had tried to liberate the fermentation enzyme from yeast.

In 1893 Buchner and his elder brother found that the cells of microorganisms were disrupted when they were ground with sand. After yeast had been treated in this way, it was possible to use a hydraulic press to squeeze out a yellow viscous liquid, free from cells. The Buchners were using the liquid for pharmaceutical studies (not for experiments on fermentation) and wished to add a preservative to it. As the juice was being used in experiments on animals, antiseptics could not be used and so Buchner added a thick sugar syrup to stop any bacterial action. He fully expected the sugar to act as a preservative, as it usually does, but to his surprise it had the opposite effect and carbon dioxide was produced. Thus the sugar had fermented, producing carbon dioxide and alcohol, in the same way as if whole yeast cells had been present. He named the enzyme concerned zymase.

Invertase, another enzyme of yeast, has been known since 1860 but zymase is different in that it is less stable to heat and catalyses a more complex reaction. Buchner was fortunate that he chose the correct type of yeast. It was soon realized that the conversion of sugar into alcohol by means of yeast juice is a series of stepwise reactions, and that zymase is really a mixture of several enzymes. It was to be 40 years before the process was fully understood, through the work of Arthur Harden (1865–1940), Otto Meyerhof (1884–1951), and others.

Buchner's other main research concerned aliphatic diazo compounds. Between 1885 and 1905 he published 48 papers that dealt with the synthesis of nitrogenous compounds, especially pyrazole. He also synthesized cycloheptane compounds.

Buckland William 1784–1856 was a principal pioneer of British geology.

Born at Axminster, England, the son of a clergyman, Buckland attended Oxford University and took holy orders. However, from an early age he had shown unbounded enthusiasm for the youthful science of geology. Appointed Reader in Mineralogy in 1813, he became Reader in Geology in 1818, while his ecclesiastical career culminated in his elevation in 1845 to the deanery of Westminster. Always an enthusiastic field-worker and a renowned lecturer, his geological investigations blossomed in three distinct, though related, areas. First, he made major contributions to the descriptive and historical stratigraphy of the British Isles, inferring from the vertical succession of the strata a stage-by-stage temporal development of the globe's crust. In this, he built on the pioneering stratigraphical work of William Smith and the palaeontology of Cuvier. Second, he became a celebrated palaeontologist, using the techniques of Cuvierian comparative anatomy. In this respect, his greatest achievement lay in his reconstruction of *Megalosaurus*, and, in his book *Relics of the Deluge* 1823, in exploring the geological history of Kirkdale Cavern, a hyena cave den in Yorkshire. Third, he long sought to discover evidence for catastrophic transformations of the Earth's surface in the geologically recent past, as indicated by features of relief, fossil bones, erratic boulders and gravel displacement. In his *Geology Vindicated* 1819, he confidently attributed such items to the Biblical Flood – an assertion he felt forced, however, to withdraw some 15 years later in the light of fresh evidence, adduced by Lyell and others, of the power of gradual and regular geomorphological processes. Buckland's interest in such evidence of violent change nevertheless helped him to become a leading and early British exponent of Agassiz's glacial theory.

Buffon Georges-Louis Leclerc, Comte de 1707–1788 was a French naturalist who compiled the vast encyclopedic work *Histoire naturelle générale et particulière*.

Buffon was born in Montbard in France on 7 September 1707 and was educated at the Jesuits' College in Dijon. He graduated in law in 1726 and took the opportunity to study mathematics and astronomy. He travelled a great deal, and spent some time in England where science was undergoing a renaissance. Buffon set himself the task of translating the works of Newton and Hales into French. In 1732 Buffon's mother died and left him a handsome legacy, so the young man was financially stable enough to devote himself entirely to his scientific interests. Buffon was elected Associate of the French Academy of Sciences in 1739 and took up the appointment of Keeper of the Botanical Gardens (Jardin du Roi), a post that stimulated his interest in natural history. He was a prolific writer, and he turned his skills towards compiling what would eventually be a 44-volume work on natural history encompassing both the plant and the animal kingdoms. He became a member of the French Academy in 1753, was made a Count in 1771 and a Fellow of the Royal Society in 1739. He died in Paris on 16 April 1788 after a long and painful illness.

Buffon's encyclopedia was the first work to cover the whole of natural history and it was extremely popular. He wrote in a clear and interesting style – which he regarded as more important than originality – and did not personally originate a great deal of the material. He was aided by several eminent naturalists of the time, and organized the sometimes confusing wealth of material into a coherent form.

Although Buffon's work was inclined to generalizations, he proposed some innovatory and stimulating theories. He suggested that a cosmic catastrophe initiated the Earth's beginnings, and that its existence was far older than the 6,000 years suggested by the Book of Genesis. He observed that some animals retain parts that are vestigial and no longer useful, suggesting that they have evolved rather than having been spontaneously generated. Theories such as these could have caused a furore at a time when it was strongly believed that the creation of the world and humankind occurred as defined in the Bible, and even though Buffon wrote with political care he did upset the authorities and had to recant. It was not until the theories of Charles Lyell and Charles Darwin that people began to take such ideas seriously.

Buffon's encyclopedia did, however, arouse a great

interest in natural science which carried through to the early part of the nineteenth century.

Bullard Edward Crisp 1907–1980 was a British geophysicist who, with the US Maurice Ewing, is generally considered to have founded the discipline of marine geophysics. He received many honours for his work, including a knighthood in 1953 and the Vetlesen Medal and Prize (the earth sciences' equivalent of a Nobel prize) in 1968.

Bullard was born in Norwich on 21 September 1907 into a family of brewers. He was educated at Repton School and Clare College, Cambridge, from which he graduated in physics in 1929. In 1931, after two years research at the Cavendish Laboratory, Cambridge, he became a demonstrator in Cambridge University's Department of Geodesy and Geophysics. During World War II he researched into methods of demagnetizing ships and of sweeping acoustic and magnetic mines. He then became the British Admiralty's Assistant Director of Naval Operational Research under P. M. S. Blackett, and later served on several committees concerned with combating German V-weapons. He continued to advise the Ministry of Defence for several years after the war. At the end of his wartime military service he returned to Cambridge as a Reader in Geophysics and soon became Head of Geophysics but, disillusioned by the university's tardiness in promoting growth in this subject area, went to Canada in 1948 to become Professor of Geophysics at the University of Toronto. He returned to England two years later, however, to take up the directorship of the National Physical Laboratory at Teddington. In 1957 he returned to Cambridge as Head of Geodesy and Geophysics, a position he held until his official retirement in 1974. On retiring he moved to La Jolla, California, where he continued to teach at the University of California (of which he had been a professor since 1963). He also advised the United States government on nuclear waste disposal. In addition to his academic and advisory posts, he played an active part in his family's brewing business and was a director of IBM (UK) for several years. Bullard died on 3 April 1980 in La Jolla.

Bullard's earliest work was to devise a technique (involving timing the swings of an invariant pendulum) to measure minute gravitational variations in the East African Rift Valley. Also before World War II he investigated the rate of efflux of the Earth's interior heat through the land surface and – influenced by the work of Ewing, a professor at Columbia University – he pioneered the application of the seismic method to study the sea floor. After the war, while at Toronto University, Bullard developed his 'dynamo' theory of geomagnetism, according to which the Earth's magnetic field results from convective movements of molten material within the Earth's core. Also in this period – working at the Scripps Institute of Oceanography, California, in his summer vacations – he devised apparatus for measuring the flow of heat through the deep sea floor.

After returning to Cambridge he played a large part in developing the potassium–argon method of rock-dating. He also studied continental drift – before the theory became generally accepted. Using a computer to analyse the shapes of the continents, he found that they fitted together reasonably well, especially if other factors such as sedimentation and deformation were taken into account. Later, when independent evidence for these factors had been found, Bullard's findings lent considerable support to the continental drift theory.

Bunsen Robert Wilhelm 1811–1899 was a German chemist who pioneered the use of the spectroscope to analyse chemical compounds. Using the technique, he discovered two new elements, rubidium and caesium. He also devised several pieces of laboratory apparatus, although he probably played only a minor part (if any) in the invention of the Bunsen burner.

Bunsen was born on 31 March 1811 at Göttingen, son of a librarian and linguistics professor at the local university. He studied chemistry there and at Paris, Berlin, and Vienna, gaining his PhD in 1830. He was appointed professor at the Polytechnic Institute of Kassel in 1836, and subsequently held chairs at Marburg (1838) and Breslau (1851) before becoming Professor of Experimental Chemistry at Heidelberg in 1852. He remained there until he retired. Bunsen never married, and ten years after retiring he died, on 16 August 1899.

Bunsen's first significant work, begun in 1837, was on cacodyl compounds, unpleasant and dangerous organic compounds of arsenic; a laboratory explosion cost Bunsen the sight of one eye and he nearly died of arsenic poisoning. He did, however, stimulate later researches into organometallic compounds by his student, the British chemist Edward Frankland. In 1841 he devised the Bunsen cell, 1.9-volt carbon–zinc primary cell which he used to produce an extremely bright electric arc light. He then (1844) invented a grease-spot photometer to measure brightness (by comparing a light source of known brightness with that being investigated). His contribution to the improvement of laboratory instruments and techniques gave rise also to the Bunsen ice calorimeter, which he developed in 1870 to measure the heat capacities of substances that were available in only small quantities.

Bunsen's first work in inorganic chemistry made use of his primary cell. Using electrolysis, he was the first to isolate metallic magnesium, and demonstrate the intense light produced when the metal is burned in air. But his major contribution was the analysis of the spectra produced when metal salts (particularly chlorides) are heated to incandescence in a flame, a technique first advocated by the American physicist David Alter (1807–1891). Working with Gustav Kirchhoff (1824–1887) in about 1860, Bunsen observed 'new' lines in the spectra of minerals which represented the elements rubidium (which has a prominent red line) and caesium (blue line). Other workers using the same technique soon discovered several other new elements.

The Bunsen burner, probably used to heat the materials for spectroscopic analysis, seems to have been designed by Peter Desdega, Bunsen's technician. Gas (originally coal gas, but any inflammable gas can be used) is released from

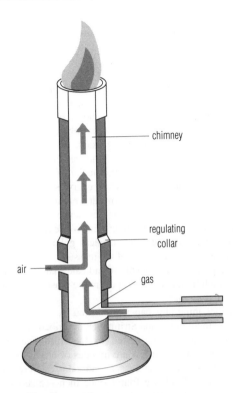

Bunsen *The Bunsen burner.*

a jet at the base of a chimney. A hole or holes at the base of the chimney are encircled by a movable collar, which also has holes. Rotation of the collar controls the amount of air admitted at the base of the chimney; the air–gas mixture burns at the top. With the air holes closed, the gas burns with a luminous, sooty flame. With the air holes open, the air–gas mixture burns with a hot, nonluminous flame (and makes a characteristic roaring sound).

Burali-Forte Cesare 1861–1931 was an Italian mathematician who is famous for the paradox named after him and for his work on the linear transformations of vectors.

Burali-Forte was born at Arezzo on 13 August 1861. He received his BA in mathematics from the University of Pisa in 1884 and then taught for three years at the Technical School in Sicily before being appointed extraordinary professor at the Academia Militare di Artiglieria e Genio in Turin. At Turin he lectured on analytical projective geometry. In the years 1894 to 1896 he served as assistant to Giuseppe Peano (1858–1932) at the University of Turin, and he later did much to make known Peano's work on mathematical logic, especially by his expanded edition (with his own interpolations) of Peano's *Logica Mathematica*. He remained at the Academia Militare until his death, at Turin, on 21 January 1931.

Burali-Forte published his famous paradox in 1897: 'To every class of ordinal numbers there corresponds an ordinal number which is greater than any element of the class'. This discovery dealt a sudden, and severe, blow to the developing science of mathematical logic – that is, to the notion that mathematics (or at least its foundations) could be adequately expressed in purely logical terms. What Burali-Forte had done was to expose a contradiction in Georg Cantor's theory of infinite ordinal numbers, and in 1902 Bertrand Russell demonstrated that this contradiction was of a fundamental logical character and could not be overcome by minor changes in the theory. It was thus Burali-Forte who brought to the fore the threat which such paradoxes posed to the foundations of mathematical logic.

Burali-Forte's chief interest, and major accomplishments, however, lay in the field of vector analysis. Much of this work was done in collaboration with Roberto Marcolongo. In 1904 they published a series of papers on the unification of vector notation, including in this work a comprehensive analysis of all the notations that had been proposed for a minimal system. Five years later they produced their own proposals for a unified system of vector notation. Having thus laid the groundwork, they began in 1909 to study the linear transformation of vectors. Of great importance was Burali-Forte's simplification of the foundations of vector analysis by the introduction of the notion of the derivative of a vector with respect to a point, which led to new applications of the theory of vector analysis and, in particular, to more efficient treatment of such operators as the Lorenz transformations.

In 1912–13 Burali-Forte published more volumes on linear transformations and demonstrated their application to such things as the theory of mechanics of continuous bodies, hydrodynamics, optics, and some problems of mechanics. His great ambition was to produce an encyclopedia of vector analysis and its applications, but he did not live to complete this work. His last contribution, a paper on differential projective geometry, was finished in 1930, shortly before his death.

Burbidge Geoffrey 1925–   and Eleanor Margaret 1922–   are a British husband-and-wife team of astrophysicists distinguished for their work, chiefly in the United States, on nucleosynthesis – the creation of elements in space – and on quasars and galaxies.

Geoffrey Burbidge studied physics at Bristol University (graduating in 1946), then combined lecturing and research for his PhD at University College, London, before going to the United States as Agassiz Fellow at Harvard University. From 1952 to 1953 he was a Research Fellow at the University of Chicago; he then returned to England as a Research Fellow at the Cavendish Laboratories, Cambridge. In 1955 he went back to the United States as Carnegie Fellow at the Mount Wilson and Palomar Universities, Caltech, and in 1957 he joined the Department of Astronomy, University of Chicago, as Assistant Professor. Appointments followed at the University of California, San Diego, as Associate Professor (1962–63) and Professor of Physics (1963–78). He served as Director of the Kitt Peak National Observatory, Arizona (1978–84).

Margaret Burbidge studied at University College, London, and gained her PhD for research at the University of

London Observatory, where in 1948 she became Assistant Director and then Acting Director (1950–51). Travelling to the United States, she held a Fellowship from the International Astronomical Union at the Yerkes Observatory, University of Chicago. From 1955 to 1957, she was a Research Fellow at the California Institute of Technology. Her next appointment was at the Yerkes Observatory again, firstly as Shirley Farr Fellow and later, in 1959, as Associate Professor. Three years later she moved to the University of California as Research Astronomer before coming Professor (in 1964), being granted leave of absence to be Director of the Royal Greenwich Observatory (1972–73).

Together the Burbidges published *Quasi-Stellar Objects* 1967. Their work in nucleosynthesis followed the discovery of the spectral lines of the unstable element technetium in red giant stars by Paul Merrill (1889–1961) in 1952. Because technetium is too unstable to have existed for as long as the stars themselves, the discovery provided the first evidence for the actual creation of elements. The paper published by the Burbidges, William Fowler (1911–), and Fred Hoyle (1915–) in 1957 began with the premise that at first stars consisted mainly of hydrogen and that most of the stars now visible are in the process of producing helium from hydrogen and releasing energy as starlight. They then suggested that as stars age some of their helium is 'burned' to form other elements, such as carbon and oxygen. The carbon and oxygen may trap hydrogen nuclei (protons) to form more complex nuclei, or may trap helium nuclei (alpha particles) to produce magnesium, silicon, sulphur, argon, and calcium. The Burbidges and their colleagues distinguished five additional processes; one is the e-process, in which elements such as iron, nickel, chromium, and cobalt are formed at a high temperature. Up to this point, the 'iron peak', the build-up results in energy being released. Beyond the iron peak more energy is required to create heavier elements. In a supernova, a massive star exploding, this energy is available. Before the supernova, the star has medium-weight elements.

It becomes unstable and nuclei trap neutrons so rapidly (the r-process), that newly formed nuclei do not have time to shed electrons. There is subsequent explosion and heavier elements such as selenium, bromine, krypton, tellurium, iodine, xenon, osmium, iridium, platinum, gold, uranium, and several unfamiliar elements are formed. From theoretical considerations, the collaborators calculated the proportions of the different heavy elements which would be most likely to be formed in a supernova. Observations indicate that the distribution of heavy elements could be explained by their production in supernovae. They believe a slow process (s-process) in red giants also builds up heavy elements.

The Burbidges also researched into quasars, objects originally detected by their strong radio emissions and believed to be travelling away from the Earth at immense speed.

Quasars give off ultraviolet radiation, and the Doppler effect, caused by their receding at great speed, results in the spectrum's being shifted towards the red (red shift), so that only faint points of blue light from the quasar reach Earth.

Determining the spectra of suspected quasars is a laborious task. Margaret Burbidge, together with Kinman, measured the red shifts of several objects found by means of the screening process devised by Martin Ryle (1918–1984) and Allan Sandage (1926–), and in the process she detected objects without radio radiation, but with large red shifts; they were called quasi-stellar objects and are now placed in the general group of quasars.

In 1963, the Burbidges and Sandage reviewed the evidence for intense activity in the nuclei of radio galaxies, quasi-stellar objects and Seyfert galaxies (those with a small bright nucleus, fainter arms and broad spectral emission lines). In 1970, using evidence gained from observations, Geoffrey Burbidge and Wolfe calculated that the stars emitting light in elliptical galaxies could not account for more than 25% of the mass.

They produced arguments to indicate that black holes, from which light cannot escape, are the most likely source of the missing mass.

Another important paper, published by the Burbidges jointly with Solomon and Stritmatter, described the discovery that four quasars (listed in the third Cambridge catalogue) lie within a few arc minutes of bright galaxies. They suggested that the quasars and galaxies were linked in some way. Evidence found since then – examples of quasars located on opposite sides of a galaxy – tends to support this belief, which is accepted by several notable astronomers.

Burkitt Denis Parsons 1911– is a British surgeon who is best known for his description of the childhood tumour named after him, Burkitt's lymphoma. He is also known for stressing the importance of roughage in the diet.

Burkitt was born on 28 February 1911 at Enniskillen, near Lough Erne in Northern Ireland. He was educated at the local school and later at schools in Anglesey and Cheltenham. When he was 18 years old he entered Dublin University to study engineering, but later turned his interests to medicine. He worked as a surgeon in the Armed Forces and became FRCS in 1938 in Edinburgh. In 1946 he was accepted into the Colonial Service and eventually became Senior Consultant to the Ministry of Health in Kampala, in 1961.

In 1957 Burkitt examined a child in Kampala. This was his first case of the lymphoma that typically affects the face and jaw, presenting several swellings. Subsequent observations convinced him that these diversely distributed lumps were of a single tumour type. This has now been histologically confirmed by the presence of 'starry sky' cells within the tumours. Burkitt undertook a 15,000-km/9,300-mi safari with two other doctors, Edward Williams and Clifford Nelson, to discover if there is a geographical correlation with the incidence of Burkitt's disease. Their research eventually showed that the lymphoma is commonest in areas of certain temperature and rainfall, in fact, where malaria is endemic. It is also associated with the presence of antibodies to the Epstein–Barr virus. Burkitt's work in the comparatively new field of geographical pathology was acknowledged in 1972 when he was elected a Fellow of the Royal Society.

He also became well known for his theories that a high-fibre diet prevents many of the common ailments of the Western world, such as appendicitis, diverticular disease, and carcinoma of the bowel, all of which are rarely encountered among the African peoples, for example. Burkitt pioneered the popular trend of high-roughage diets.

Burnet Frank Macfarlane 1899–1985 was an Australian immunologist whose research into viruses inspired his theory that antibodies could be produced artificially in the body in order to develop a specific type of immunity, which led to the concept of acquired immunological tolerance, particularly important in tissue transplant surgery. For this work he was awarded the 1960 Nobel Prize for Physiology or Medicine, which he shared with Peter Medawar.

Burnet was born in Australia on 3 February 1899 at Traralgon, Victoria. After graduating in biology from Geelong College, Victoria, he obtained his medical degree at Melbourne University in 1923 and for the following year was Resident Pathologist at Melbourne Hospital. He studied at the Lister Institute, London, from 1926 to 1927 and in that year gained his PhD from the University of London. He then returned to Australia as Assistant Director of the Walter and Eliza Hall Institute for Medical Research, Melbourne, and held this position until 1944, when he became the Institute's Director. Burnet was knighted in 1951. He was appointed Emeritus Professor of Melbourne University in 1965 and made an Honorary Fellow of the Royal College of Surgeons in 1969.

Early in his career Burnet did extensive research on viruses. He was the first to investigate the multiplication mechanism of bacteriophages (viruses that attack bacteria) and devised a method for identifying bacteria by the bacteriophages that attack them. This work was of immense importance, particularly 20 years later, when bacteriophages were first used as research tools in genetics and molecular biology.

In 1932 Burnet developed a technique for growing and isolating viruses in chick embryos, a technique that was to be used as a standard laboratory procedure for more than 20 years. Burnet's work on the chick embryo increased interest in the specific character of an embryo by which it seemed to be unable to resist virus infection or to produce any antibodies against viruses. Early attempts were made in Burnet's laboratory to use the chick embryo to show that tolerance could be produced artificially.

As a result of his virus research, Burnet became interested in immunology and in 1949 he predicted that an individual's ability to produce a particular antibody to a particular antigen was not innate, but was something that developed during the individual's life. In 1951 Medawar carried out the experiments that confirmed this theory.

Burnet's second major contribution to immunology was made in 1957 – his highly controversial 'clonal selection' theory of antibody formation, which explains why a particular antigen stimulates the production of its own specific antibody. According to Burnet, there is a region in the genes of the cells that produce antibodies that is continually mutating, such that each mutation leads to a new variant of antibody being produced. Normally the cells that produce a particular variant are few, but if the antibody they produce suddenly finds a target, then they multiply rapidly to meet this demand, and the other, useless, variants die out.

In recent years there has been a large amount of research on the immune response and on the many ways of imitating nature's way of reacting without inoculating the embryo. In particular, work has been carried out on the production of tolerance by drugs such as 6-mercaptopurine, in the hope that it will prove effective in surgical organ transplants. Burnet's investigations have stimulated research into the way viruses cause infection. His own research helped to eradicate diseases such as myxomatosis and isolate organisms such as *Rickettsia burneti*, which causes Q fever.

Burnet's publications include *Viruses and Man*, published in 1953, and *The Clonal Selection Theory of Acquired Immunity* (1959).

Burnside William 1852–1927 was a British applied mathematician and mathematical physicist whose interest turned in his later years to a profound absorption in pure mathematics. In this field he was particularly prominent for his research in group theory and the theory of probability.

Burnside was born in London on 2 July 1852, and was orphaned when only a young child. He proved so gifted in mathematics at Christ's Hospital that in 1871 he won a scholarship to St John's College, Cambridge. He transferred to Pembroke College after two years, and graduated with high honours in 1875, winning the first Smith's Prize and being given the post of Fellow and Lecturer at the College. Apart from his normal lecturing duties, Burnside began to give advanced courses on hydrodynamics to other groups outside the College.

In 1885 he left Cambridge to take up the Chair of Mathematics at the Royal Naval College at Greenwich where, in addition to his teaching routine, he often accepted the responsibility of being an examiner for the universities and the civil service. It was at this stage that his interests began to move towards group theory. Even as he formalized the institution of instruction at three different levels at the Royal Naval College, he was putting together his thoughts and his papers for a book that finally appeared in 1897. A revised form of that work is now regarded as a classic. A Fellow of the Royal Society from 1893, Burnside served on the Society's Council from 1901 to 1903 and was awarded its Royal Medal. After World War I he began to reduce his scholarly output, and retired from the Royal Naval College in 1919. He continued to write on mathematical subjects, and a nearly completed manuscript on probability theory was found after his death and published posthumously. He died in West Wickham, Kent, on 21 August 1927.

Although it was well outside the scope of Burnside's undergraduate courses, his first publication (produced while he was at Cambridge) was in the field of elliptic functions. His study of elliptic functions led him, over the years, to study the functions of real variables and the theory of functions in general. One of his most influential papers, written in 1892, was a development of some work by Jules Poincaré on automorphic functions.

Burnside's lectures at the Royal Naval College were on a surprisingly wide range of subjects. He varied the content and the standard of his lectures according to whether he was addressing the Junior (ballistics), the Senior (dynamics and mechanics) or the Advanced (hydrodynamics and kinetics) Section of trainees.

His research interests in the meantime included differential geometry and the kinetic theory of gases. During the early 1890s, references to group theory began to enter his work, and by the middle of the decade his study of automorphic functions had brought him fully into the field. He became particularly concerned with the theory of the discontinuous group of finite order. In 1897 he published the first book on group theory to appear in English. The papers that he and other mathematicians wrote over the ensuing years produced such major advances in the subject that a revised edition was soon necessary. It was issued in 1911, and is today considered to be a standard work.

Bush Vannevar 1890–1974 was a US electrical engineer and scientist who developed several mechanical and mechanical–electrical analogue computers which were highly effective in the solution of differential equations. During World War II he was scientific advisor to President Roosevelt and was instrumental in the initiation of the atomic bomb project. Later he greatly influenced the development of postwar science and engineering in the United States, being instrumental in the setting up and running of the Office of Scientific Research and Development and its successor, the Research and Development Board.

Bush was born in Everett, Massachusetts. His father had first been a sailor but later became a clergyman. During his childhood Vannevar developed an interest in practical things and constructed a radio receiver when they were almost unknown. He attended a local high school and then went on to Tufts College. He was forced to supplement the little money he had to support himself by washing dishes and giving private coaching in mathematics. In 1913 Bush received BSc and MSc degrees, the latter being for a thesis on a machine for plotting the profile of the land. He patented this device, which combined a gear mechanism mounted between two bicycle wheels which recorded the vertical distance the machine had risen while a paper was moved forward proportional to the distance travelled; a pen recorder drew the profile. After graduation he obtained a job as an engineer with the General Electric Company, but was made redundant shortly afterwards. This misfortune inspired him to read for a doctorate in engineering at the Massachusetts Institute of Technology (MIT). He proved himself so able at his task that he gained his degree in 1916 after only one year of study. From 1932 he held senior

positions at MIT, including that of Dean of the Engineering School. He died of a stroke on 28 June 1974.

During World War I he worked on a magnetic device for detecting submarines. In 1919 he returned to MIT and was appointed associate professor of power transmission. In his research on the distribution of electricity, many problems arose which contained differential equations. Such equations, containing differential coefficients, can be difficult and time-consuming to solve in an explicit way, even if a solution is possible – an algebraic solution is of little use to a practising engineer who needs numbers with which to work. In about 1925 Bush began to construct what he called the product integraph. It contained a linkage to form the product of two algebraic functions and represent it mechanically. He also devised a watt-hour meter, a direct ancestor of the electricity meter now found in nearly all homes, which integrated the product of electric current and voltage. By similarly using current and voltage to represent equations, his 'calculus' machine was able to evaluate integrals involved in the solution of differential equations.

The product integraph, while suitable for solving the problems Bush had encountered, was limited to the solution of first-order differential equations. A device capable of the solution of second-order equations would be much more useful to electrical engineers, and to scientists generally. There was, however, a complication in coupling two watt-hour meters together. Bush found the solution by coupling one of the meters to a mechanical device known as a Kelvin integrator, after the physicist Lord Kelvin.

In 1931 Bush began work on an almost totally mechanical, and very much more ambitious machine known as the differential analyser. This machine had six integrators, three input tables and an output table, which showed the graphical solution of an equation. The Bush analyser was the model for developments in the mechanical world, and many similar machines were built. One, a large differential analyser at Manchester University was completed in 1935 and is now an exhibit in the Science Museum, London.

Following the success of the differential analyser, Bush built several specific purpose analysers for solving problems related to electrical networks and for the evaluation of particular types of integrals. Then he built another large analyser, this time with many of the operations electrified, and a tape input that reduced the time taken for setting up a problem from days to minutes.

Bush's machines, and others that had been built in the United States, were used during World War II for military purposes, particularly the calculation of artillery range tables. The same military applications spawned the new machine, the digital electronic computer, which was to be the downfall of mechanical analogue devices.

One of Bush's other inventions was a cipher-breaking machine which played a large part in breaking Japanese codes. One of his less practical devices was a bird perch which dropped the birds the garden owner did not want to encourage. There were also many more patented and unpatented inventions to Bush's name.

C

Cailletet Louis Paul 1832–1913 was a French physicist and inventor who is remembered chiefly for his work on the liquefaction of the 'permanent' gases: he was the first to liquefy oxygen, hydrogen, nitrogen and air, for example.

Cailletet was born in Chatillon-sur-Seine on 21 September 1832, the son of an ironworks owner. He was educated at the college in Chatillon, the Lycée Henry IV in Paris and the Ecole des Mines (also in Paris), after which he returned to Chatillon to manage his father's ironworks. Shortly afterwards he began his metallurgical studies, and later extended his research to the problems of liquefying gases. He died in Paris on 5 January 1913.

Investigating the causes of accidents that occurred during the tempering of incompletely forged iron, Cailletet found that many were due to the highly unstable state of the iron while it was hot and had gases dissolved in it. Also in the field of metallurgy he analysed the gases from blast furnaces. Other scientists had drawn off the gases under conditions that resulted in gradual cooling, which enabled the dissociated components in the gases to recombine. Cailletet, using a new technique by which the gases were cooled suddenly as soon as they had been collected, showed that the gases comprised of a large proportion of finely divided carbon particles, carbon monoxide, oxygen, hydrogen and a small proportion of carbon dioxide – a composition different from that obtained by the old sampling method. As a result of these and other metallurgical studies, Cailletet developed a unified concept of the role of heat in changes of state of metals, and confirmed the views with which Antoine Lavoisier had introduced his *Traité élémentaire de chimie* in 1789. In 1883 Cailletet was awarded the Prix Lacaze of the French Academy of Sciences for his work in metallurgy.

Cailletet's best-known work, however, was on the liquefaction of gases. Until his studies (which were completed by 1878) several gases – including oxygen, hydrogen, nitrogen, and air – were considered to be 'permanent' because nobody had succeeded in liquefying them, despite numerous attempts involving the use of what were then considered to be extremely high pressures. Cailletet realized that the failure of these attempts was due to the fact that the gases had not been cooled below their critical temperatures (the concept of critical temperature – the temperature above which liquefaction of a gas is impossible – was put forward by the Irish chemist Thomas Andrews by 1869). In order to obtain the necessary amount of cooling, Cailletet used the Joule–Thomson effect (the decrease in temperature that results when a gas expands freely), compressing a gas, cooling it, then allowing it to expand to cool it still further. Using this method he liquefied oxygen in 1877 and several other 'permanent' gases in 1878, including hydrogen, nitrogen and air. (Raoul Pictet, working independently, also liquefied oxygen in 1877.)

Cailletet's other achievements included the installation of a 300 m/985 ft-high manometer on the Eiffel Tower; an investigation of air resistance on falling bodies; a study of a liquid oxygen breathing apparatus for high altitude ascents; and the construction of numerous devices, including automatic cameras, an altimeter, and air sample collectors for sounding-balloon studies of the upper atmosphere. He was elected President of the Aéro Club de France for these accomplishments.

Cairns Hugh John Forster 1922–   is a British virologist known for his research into cancer.

Cairns was born on 21 November 1922, the son of Professor Sir Hugh Cairns, a physician and Fellow of Balliol College, Oxford. He attended Edinburgh Academy from 1933 to 1940, when he went to Balliol and gained his medical degree in 1943. His first appointment was in 1945 as Surgical Resident at the Radcliffe Infirmary, Oxford, and the next five years were spent in various appointments in London, Newcastle and Oxford. From 1950 to 1951 Cairns was a virologist at the Hall Institute in Melbourne, Australia, and he then went to the Viruses Research Institute, Entebbe, Uganda. In 1963 he became Director of the Cold Spring Harbor Laboratory of Quantitative Biology in New York, a position he held until 1968. He then took professorships at the State University of New York and with the American Cancer Society. From 1973 to 1981 he was in charge of the Mill Hill laboratories of the Imperial Cancer Research Fund, London. In 1982 he moved to the United

States at the Department of Microbiology of the Harvard School of Public Health, Boston.

One of Cairns's first pieces of research in the early 1950s was into penicillin-resistant staphylococci, and their incidence in relation to the length of a patient's stay in hospital. He found that the rapid rise in their incidence was caused by continuous cross-infection with a few strains of the bacteria rather than repeated instances of fresh mutations. These findings are now generally accepted, although they were not at the time.

Cairns has also done much research on the influenza virus and in 1952 discovered that the virus is not released from the infected cell in a burst – as is a bacteriophage – but in a slow trickle. This evidence has since been found also to be true for the polio virus. In the following year he showed that the influenza virus particle is completed as it is released through the cell surface (also unlike a phage). This discovery has since been confirmed by electron microscopy and isotope incorporation techniques. In 1959 Cairns succeeded in carrying out genetic mapping of an animal virus for the first time. In 1960 he showed that the DNA of the vaccinia virus is replicated in the cytoplasm or protoplasm of the cell (excluding the nucleus) and that each infecting virus particle creates a separate DNA 'factory'.

Cairns's investigations into DNA have also led him to look at the way that DNA replicates itself and to compare the rates of replication of DNA in mammals with those in *Eschericia coli* (a bacterium). He has found that mammalian DNA is replicated more slowly than that of *E. coli*, but is replicated simultaneously at many points of replication.

His later work studied the link between DNA and cancer, some forms of which may be caused by the alkylation of bases in the DNA. He showed that bacteria are able to inhibit the alkylation mechanism in their own cells, and later demonstrated this ability in mammalian cells. A similar mechanism probably prevents a high incidence of DNA mutations in human beings despite the presence of alkylating agents in the environment. Cairns made many important advances in the study of cancer and its relation to society. He has also spent much time in fund-raising for cancer research.

Cajori Florian 1859–1930 was a Swiss-born US historian of mathematics. His books dealt with the history of both elementary and advanced mathematics, as well as the teaching of mathematics (including its importance in education). Cajori was also the author of biographies of eminent mathematicians.

Cajori was born on 28 February 1859 in St Aignan, near Thusius. At the age of 16 he emigrated to the United States, where he took up studies at the University of Wisconsin. He was awarded his bachelor's degree in 1883, and in 1885 was offered the position of Assistant Professor of Mathematics at Tulane University in New Orleans. There he continued his own studies in parallel with his teaching activities, earning his master's degree in 1886, and his PhD in 1894. In 1889 he moved to Colorado College at Colorado Springs in order to take up the position of Professor of

Physics, but returned to Tulane University in 1898 to become Professor of Mathematics, a position he held until 1918. He served also as Dean of the Department of Engineering from 1903 until 1918. In that year Cajori moved to the University of California at Berkeley to take up the Chair of History of Mathematics, a post he retained until his death. He died at Berkeley on 14 August 1930.

Cajori's influence on the modern perception of the development of mathematics was profound, and his works are frequently quoted to this day. His reputation is founded mainly on his many books on the history of mathematics, although a number of his works – notably his edited version of Newton's *Principia Mathematica* (published posthumously) – have been subject to some criticism for their interpretation of historical material. His two-volume *History of Mathematical Notations* (1928–29) is, however, still very much a standard reference text. He also compiled *A History of Physics* (1899).

Callendar Hugh Longbourne 1863–1930 was the British physicist and engineer who carried out fundamental investigations into the behaviour of steam. One of the results was the compilation of reliable steam tables that enabled engineers to design advanced steam machinery.

Callendar was born on 18 April 1863, at Hatherop, Gloucestershire. Callendar was educated at Marlborough where he was joint editor of the school magazine. At Marlborough he spent his spare time reading science although he was a classics scholar. He obtained first-class honours degrees in classics (1884) and mathematics (1885) at Cambridge, then studied physics under J. J Thomson. After becoming a Fellow of Trinity College, Cambridge, in 1886, with a substantial research grant, he started to study medicine, devised his own shorthand system, then entered Lincoln's Inn to study law in 1889.

From 1888 to 1893 he was Professor of Physics at the Royal Holloway College, Egham. Thomson persuaded him to take up science in earnest, and in 1893 he accepted a professorship in physics at McGill College, Montreal. After six months there, he returned to marry Victoria Mary Stewart, whom he had met at Cambridge. They returned to Montreal together and in 1898 the family, which now included three children, left Montreal when Callendar was offered the Chair of Physics at University College, London.

In 1901 much larger chemistry and physics departments were built for the Royal College of Science in South Kensington and Callendar was appointed the new Professor of Physics, a chair which he held for 29 years. In 1905 he went to Spain to observe a total eclipse of the Sun for the Royal Society, using his own design of shielded coronal thermopile.

He moved his family to Ealing in 1906 and was elected President of the Royal Physical Society for the 1910–11 term, having been treasurer for ten years. In 1912 he was President of Section A of the British Association.

During World War I Callender was a consultant to the Board of Inventions which received more than 100,000 'war-winning' ideas. In 1902 he published his great treatise, *The Properties of Steam and Thermodynamic Theory of Turbines*, and was invited by Sir Charles Parsons to become a consultant.

In addition to his main research on steam, he served as Director of Engine Research for the Air Ministry from 1924. Callendar died of pneumonia on 21 January 1930 in London; his wife survived him by 28 years.

Unlike most gases, steam used to power boilers, engines, turbines and other equipment does not follow easily interpreted laws. Its behaviour must therefore be predicted by the use of tables or graphs. Steam tables had been produced in different countries from experimental results but the researches did not take into account the laws of thermodynamics, so the tables were unreliable. Callendar was the first to compile accurate tables, and these accelerated the development of turbines and other equipment dramatically.

While he was at Cambridge Callendar's main research was on the platinum resistance thermometer with which he obtained an accuracy of $0.1°C$ $(14°F)$ in $1,000°C$ $(1,832°F)$ – about 100 times better than previous results. It was not until 40 years later, in 1928, that the method was adopted as an international standard.

This work led to recording temperatures on a moving chart, a principle now fundamental to any branch of science or industry which requires a continuous record of temperature.

Callendar's research topics were varied, most of them connected with thermodynamics. He carried out experiments on the flow of steam through nozzles, producing much information of great value to steam turbine designers. With the collaboration of Howard Turner Barnes in Montreal, he developed his method of continuous electric calorimetry used to measure the specific heat of liquids. He also worked on antiknock additives for fuels.

Callendar was not only an experimental physicist but also a talented engineer and mechanic (he converted the Stanley steam car to run on compressed air). He gave engineers fundamental tools with which they advanced the state of engineering, leading to the highly efficient plant in operation in the early 1980s.

Calne Roy Yorke 1930–   is a British surgeon who has developed the technique of organ transplants in human patients, and pioneered kidney transplant surgery in the United Kingdom.

Calne was born on 30 December 1930 and was educated at Lancing College and later at Guy's Hospital Medical School, London, where he qualified with distinction in 1953. He held a junior post at Guy's Hospital for one year before serving with the Royal Army Medical Corps from 1954 to 1956. After military service Calne spent two years at the Nuffield Orthopaedic Centre, Oxford, as a Senior House Surgeon, and then became Surgical Registrar at the Royal Free Hospital, London, until 1960. From 1960 to 1961 he went to the Peter Bent Brigham Hospital, Harvard Medical School, and on his return to Britain was appointed Lecturer in Surgery at St Mary's Hospital, London. In 1962

he became Senior Lecturer at the Westminster Hospital, where he remained until accepting the appointment of Professor of Surgery at Cambridge University in 1965.

The idea of removing a diseased organ and grafting on a healthy one is ancient. This concept was not finally realized, however, until the middle of the present century; Peter Medawar demonstrated in 1957 how the rejection of tissue grafts could be prevented, and Joseph Murray and his team working in Boston, USA, successfully transplanted a kidney from one identical twin to the other, whose kidneys were afflicted with an incurable disease. Calne further developed the technique of kidney transplants, decreasing the possibility of rejection.

He has also carried out liver transplants. The liver is a complicated organ with many vital functions, and liver transplants presented serious technical problems. These have now been overcome, although liver transplants are still not carried out to the same extent as kidney transplants. One reason is that the patient cannot be kept in reasonable health while waiting for a suitable transplant – unlike a kidney patient who can be treated by dialysis (using a kidney machine).

Once the surgical techniques for tissue transplants were perfected, research centred around the development of specific immunosuppressive drugs to prevent the donor organs being rejected by the antigens produced by the recipient. Calne has persevered with his operations, despite the ethical arguments surrounding this type of surgery, and has encouraged many developments in transplant techniques.

Calvin Melvin 1911–   is a US chemist who worked out the biosynthetic pathways involved in photosynthesis, the process by which green plants use the energy of sunlight to convert water and carbon dioxide into carbohydrates and oxygen. For this achievement he was awarded the 1961 Nobel Prize for Chemistry.

Calvin was born of Russian immigrant parents on 8 April 1911 in St Paul, Minnesota. He graduated from Michigan College of Mining and Technology in 1931 and was awarded his PhD by the University of Minnesota in 1935. For the next two years he did research at the University of Manchester, England, and then returned to the United States as an Instructor at the University of California. He remained there, becoming Assistant Professor (1941), Associate Professor (1945), Professor (1947) and finally University Professor of Chemistry (1971).

Calvin began work on photosynthesis in 1949, using radioactive carbon-14 as a tracer to investigate the conversion of carbon dioxide into starch. It was already known that there were two interdependent processes: the light reaction, in which a plant 'captured' energy from sunlight, and the dark reaction (which proceeds in the absence of light), during which carbon dioxide and water combine to form carbohydrates such as sugar and starch. Calvin studied the latter reaction in a single-celled green alga called *Chlorella*. He showed that there is in fact a cycle of reactions (now called the Calvin cycle) in which the key step is the enzyme-catalysed carboxylation of the phosphate ester of a pentose (5-carbon) sugar, ribulose diphosphate (RuDP), to form the 3-carbon phosphoglyceric acid (PGA). This acid is then reduced to the 3-carbon glyceraldehyde phosphate (GALP), with the formation also of triose phosphate (TP) and hexose and its phosphate. The reduction and phosphorylation of PGA involves a reducing agent, NADPH, and the energy-rich compound adenosine triphosphate, ATP, which are derived from the photochemical light reaction. Finally, another enzyme catalyses the generation of ribulose monophosphate (RuMP) which a second ATP-induced phosphorylation reconverts to ribulose diphosphate (RuDP). The sequence of reactions is also called the reductive pentose phosphate cycle.

Cameron Alastair Graham Walter 1925– is a Canadian-born US astrophysicist responsible for theories regarding the formation of the unstable element technetium within the core of red giant stars and of the disappearance of Earth's original atmosphere.

Born in Winnipeg, Cameron gained a BSc from the University of Manitoba in 1947 and a PhD from the University of Saskatchewan in 1952. While Research Officer for Atomic Energy Canada Ltd, he emigrated to the United States in 1959 (becoming naturalized in 1963). He then successively became Senior Research Fellow at the California Institute of Technology, Pasadena; Senior Scientist of the Goddard Institute for Space Studies in New York; and Professor of Space Physics at Yeshiva University, New York City. In 1973 he became Professor of Astronomy at Harvard University.

Following Paul Merrill's discovery in 1952 of the spectral lines denoting the presence in red giants of technetium – an element too unstable to have existed for as long as the giants themselves (thus indicating the actual creation and flow of technetium in the stellar core) – it was Cameron's suggestion that Tc^{97} (mean lifetime 2.6×10^6 years) might result from the decay of a nucleus of molybdenum, Mo^{97}, a usually stable nuclide that becomes unstable when it absorbs an X-ray photon at high temperatures.

Cameron also suggested that the Earth's original atmosphere was blown off into space by the early solar 'gale' – as opposed to the present weak solar 'breeze' – with its associated magnetic fields.

Campbell William Wallace 1862–1938 was a US astronomer and mathematician, now particularly remembered for his research into the radial velocities of stars.

Born into a farming family in Hancock, Ohio, Campbell taught for a short while after completing his schooling; he then decided to continue his education at the University of Michigan in 1882. Although he enrolled to study engineering, he became keenly interested in astronomy and studied avidly under J. M. Schaeberle, who was responsible for the Michigan University Observatory. Campbell received his degree in 1886 and became Professor of Mathematics at the University of Colorado. He returned to the University of Michigan in 1888 to take up the post of Instructor in Astronomy, then moved again in 1891, this time to the newly established Lick Observatory, California. He served first as a Staff Astronomer (1891–1901) and then as Director of the Observatory (1901–1930).

During Campbell's tenure at the Lick Observatory he was responsible for much of the spectroscopic work undertaken and was an active participant in and organizer of seven eclipse expeditions to many parts of the world. His administrative talents were also exercised during the period from 1923 to 1930, when he served as President of the University of California. He retired from both posts in 1930, and in the following year was elected to a four-term as President of the National Academy of Sciences. His most significant achievement in this office was the establishment of the influential Scientific Advisory Committee which serves to improve links between the National Academy and the US government.

Failing health and the fear of complete loss of his faculties led him to commit suicide on 14 June 1938, in Berkeley, California.

Campbell's talent for observation was apparent from early in his career: one of his earliest interests in astronomy was the computation of the orbits of comets. His spectroscopic observations of Nova Auriga in 1892 enabled him to describe the changes in its spectral pattern with time. He also made spectroscopic studies of other celestial bodies and was active in the design of the Mills spectrograph, which was available for use from 1896.

It was in 1896 that Campbell initiated his lengthiest project, the compilation of a vast amount of data on radial velocities. He was aware that this would be of interest not merely for its own sake, but also for the determination of the motion of the Sun relative to other stars. He did not, however, anticipate that the programme would also lead to the discovery of many binary systems, nor that the data would later be used in the study of galactic rotation, nor that the programme itself would encourage the improvement of several techniques. Campbell published a catalogue of nearly 3,000 radial velocities in 1928.

The project has also led to the establishment of an observatory in Chile, which contributed data for the radial velocities programme from 1910 until 1929.

Campbell was an uncompromising scientist, even when it meant attracting criticism. He went against the popular opinion of the time in reporting his observations on the absence of sufficient oxygen or water vapour in the Martian atmosphere to support life as found on Earth. Other astronomers, who had been less careful in the design of their observations and interpretation of their results disagreed but his findings were supported by later work – most spectacularly by *Viking*'s Mars landing.

Another important result obtained by Campbell was a confirmation of the work done in 1919 by Sir Arthur Eddington (1882–1944) on the deflection of light during an eclipse, which supported the general theory of relativity. The positive result which Campbell obtained in 1922 was arrived at only after two previous attempts (in 1914 and in 1918) which had been frustrated by poor weather conditions and by the use of inadequate equipment.

Campbell's contributions to astronomy spanned several fields, but were perhaps most notable in spectroscopy.

Cannizzaro Stanislao 1826–1910 was an Italian chemist who, through his revival of Avogadro's hypothesis, laid the foundations of modern atomic theory. He is also remembered for an organic reaction named after him, the decomposition of aromatic aldehydes into a mixture of the corresponding acid and alcohol.

Cannizzaro was born on 13 July 1826 in Palermo. He studied chemistry at the universities of Palermo, Naples and Pisa, where in 1845 he became assistant to Raffaele Piria (1815–1865), who worked on salicin (preparing salicylic acid) and glucosides. In 1848 Cannizzaro joined the artillery to fight in the Sicilian Revolution, was condemned to death, but in 1849 escaped to Marseilles and went on to Paris. There he worked with Michel Chevreul (1786–1889, who was aged 103 when he died) and F. Cloëz (1817–1883). In 1851 he synthesized cyanamide by treating an ether solution of cyanogen chloride with ammonia, and in the same year became Professor of Physics and Chemistry at the Technical Institute of Alessandria, Piedmont. It was there that he discovered the Cannizzaro reaction. He was appointed Professor of Chemistry at Genoa University in 1855, followed by professorships at Palermo (1861–71) and Rome. He became a Senator in 1871 and eventually Vice-President, pursuing his interest in scientific education. He died in Rome on 10 May 1910.

Cannizzaro's reaction involves the treatment of an aromatic aldehyde with an alcoholic solution of potassium hydroxide. The aldehyde undergoes simultaneous oxidation and reduction to form an alcohol and a carboxylic acid. It is an example of a dismutation or disproportionation reaction, and finds many uses in synthetic organic chemistry. Cannizzaro also investigated the natural plant product santonin, used as a vermifuge, which he showed was related to naphthalene.

His greatest contribution to chemistry was made in 1858 when he revived Avogadro's hypothesis and insisted on a proper distinction between atomic and molecular weights (relative atomic and molecular masses). The pamphlet he published was distributed at the Chemical Congress at Karlsrühe in 1860. Cannizzaro pointed out that once the molecular weight of a (volatile) compound had been determined from a measurement of its vapour density, it was necessary only to estimate, within limits, the atomic weight of one of its elemental components. Then by investigating a sufficient number of compounds of that element, the chances were that at least one of them would contain only one atom of the element concerned, so that its equivalent weight (atomic weight divided by valency) would correlate with its atomic weight. Despite objections by a group of French chemists led by Sainte-Claire Deville (who studied abnormal vapour densities of substances such as ammonium chloride and phosphorus pentachloride, and were reluctant to account for these in terms of thermal dissociation), Cannizzaro's proposal was soon widely accepted.

Cannizzaro's contribution to atomic theory paved the way for later work on the periodic law and on an understanding of valency. The Royal Society recognized its significance with the award in 1891 of its Copley Medal.

Cannon Annie Jump 1863–1941 was the most honoured US woman astronomer of her day, and is justly famous for her meticulous work in stellar spectral classification, with particular reference to variable stars.

She was born in Dover, Delaware, and attended local schools, showing aptitude for scientific study; she gained her bachelor's degree at Wellesley College in 1884. After a protracted period spent at her home in Dover, Cannon returned to Wellesley College at about the age of 30 to take postgraduate courses. A protégée of Edward Pickering (1846–1919), Director of the Harvard Observatory, Cannon became a special student in astronomy at Radcliffe College in 1895 and was made an assistant at the Harvard College Observatory in 1896 – a post she held until 1911. From then until 1938 Cannon was curator of astronomical photographs at the Harvard Observatory. In 1938 she was appointed William Cranch Bond Astronomer and Curator. She retired in 1940, but continued in active research.

Cannon was the first woman to receive an honorary DSc from Oxford University, and she received several other honorary degrees from other universities in the United States and in Europe.

Cannon's return to academic life in 1894 was to research in physics, rather than astronomy, and into the uses of X-rays, recently discovered by Wilhelm Röntgen (1845–1923). A year later, at Harvard, her interests had inclined towards stellar spectroscopy in the field of astronomy.

One of Cannon's particular interests was the phenomenon of variable stars. Hipparchus had established the concept of a continuous sequence of stellar magnitudes based on the assumption that a star's brightness was constant with time. The observation of variable stars, whose brightness sometimes changed quite dramatically, upset this scheme. (A star's brightness could change for any of several reasons; for example, a bright star might be orbited by a dim luminous star, which obscures it at regular intervals: an eclipsing variable.) Cannon studied photographs to record details of variable stars, and discovered 300 new variable stars. She also kept a detailed index-card record of all her information, which has served as an invaluable tool for many succeeding research astronomers.

Edward Pickering and Williamina Fleming (1857–1911) had in 1890 established a system for classifying stellar spectra. It allowed each spectrum to be allocated to one of a series of categories labelled alphabetically 'A' to 'Q'; the groups are related to the stars' temperatures and their compositions. In 1901 Cannon reformed this system: she subdivided the letter categories into ten subclasses, based on details in the spectra. With time the system became further modified, some letters being dropped, others rearranged. The sequence which Cannon eventually settled for ran O, B, A, F, G, K, M, R, N, and S. Stars in the O, B, A group are white or bluish, those in the F, G group are yellow, those in the K group are orange, and those in the M, R,

N, S group are red. Our Sun, for instance is yellow and its spectrum places it in the G group.

In 1901 Cannon published a catalogue of the spectra of more than 1,000 stars, using her new classification system. She went on to classify the spectra of over 300,000 stars. Most of this work was published in a ten-volume set which was completed in 1924. It described almost all stars with magnitudes greater than nine. Her later work included classification of the spectra of fainter stars.

The ten-volume catalogue of stellar spectra stands as her greatest contribution to astronomy. It enabled Cannon to demonstrate that the spectra of virtually all stars can be classified easily into few categories which follow a continuous sequence. Cannon's work was characterized by great thoroughness and accuracy. Her interest lay primarily in the description of the stars as they were observed; her legacy to astronomy was a vast body of accurate and carefully compiled information.

Cantor Georg Ferdinand Ludwig Philip 1843–1918 was a Danish-born German mathematician and philosopher who is now chiefly remembered for his development of the theory of sets, for which he was obliged to devise a system of mathematics in which it was possible to consider infinite numbers or even transfinite ones.

Cantor was born on 3 March 1843 to Danish parents living in St Petersburg. The family moved to Germany when Cantor was 11, and he was educated at schools in Wiesbaden and Darmstadt, where he showed exceptional talent in mathematics. He then attended the Universities of Zurich and Berlin – obtaining his doctorate in 1867 – before moving to Halle University to take up a position as member of staff in 1869. He remained at Halle for the rest of his life, as Extraordinary Professor from 1872, and as Professor of Mathematics from 1879. He founded the Association of German Mathematicians, was its first President from 1890 to 1893, and was also responsible for the first International Mathematical Congress in Zurich in 1897. Although he received a few honorary degrees and other awards, he did not gain great recognition during his lifetime. Indeed, controversy over some of his work may have contributed to the deep depression and mental illness he suffered towards the end of his life, particularly after 1884. He died in the psychiatric clinic of Halle University on 6 January 1918.

Cantor's early work was on series and real numbers, a popular field in Germany at the time. In a study on the Fourier series – a well-known series that enables functions to be represented by trigonometric series – he extended the results he obtained and developed a theory of irrational numbers. It was in this connection that he exchanged correspondence with Richard Dedekind (1831–1916), who later became famous for his definition of irrational numbers as classes of fractions. With Dedekind's support, Cantor investigated sets of the points of convergence of the Fourier series, and derived the theory of sets that is the basis of modern mathematical analysis (now more commonly called set theory). His work, fundamental to subsequent mathematics and mathematical logic, contains many definitions and theorems that are now referred to in textbooks on topology. For the theory of sets, however, Cantor had had to arrive at a definition of infinity, and also had therefore to consider the transfinite; for this consideration he used the ancient term 'continuum'. He showed that within the infinite there are countable sets and there are sets having the power of a continuum, and proved that for every set there is another set of a higher power – a realization that was of great importance to the continued development of general set theory. Cantor's definitions were necessarily crude: he was breaking new ground. He left refinements to his successors.

Some of Cantor's other ideas and studies were distinctly odd, particularly in the realm of physics. He considered metaphysics and astrology to be a science, for example – a science into which mathematics, and especially set theory, was capable of being integrated. As probably the last Platonist among serious mathematicians, he also insisted that the atoms of the universe were countable.

Cantor's was in its way a unique contribution to the science of mathematics; he opened up a complete new area of research that at the same time was fundamental to basic mathematics.

Carathéodory Constantin 1873–1950 was a German mathematician who made significant advances to the calculus of variations and to function theory.

Carathéodory was born in Berlin on 13 September 1873; his parents were of Greek extraction. He showed an aptitude for mathematics from an early age and attended the Belgian Military Academy from 1891 to 1895. He then worked in Egypt for the British Engineering Corps on the building of the Asyut Dam. Carathéodory returned to Germany in 1900, first attending the University of Berlin and then in 1902 moving to Göttingen University, where Felix Klein, David Hilbert, and Hermann Minkowski had built up an excellent mathematics department.

Carathéodory was awarded his PhD in 1904 and qualified as a lecturer a year later. He taught at the University of Bonn for four years and then in 1909 was appointed to a professorship at the University of Hanover. In 1910 he transferred to a similar position at the University of Breslau (now Wroclaw, Poland). In 1913 he moved back to Göttingen, and five years later he returned to the University of Berlin.

The Greek government invited Carathéodory to supervise the establishment of a new university and he went to Smyrna in 1920, but his efforts were destroyed by fire two years later. He took a post at the University of Athens, where he taught until 1924 before becoming Professor of Mathematics at the University of Munich. He remained there for the rest of his life, apart from one year (1936–37) in the United States as Visiting Professor at the University of Wisconsin. He died in Munich on 2 February 1950.

Carathéodory's work covered several areas of mathematics, including the calculus of variations, function theory, theory of measure and applied mathematics. His first major contribution to the calculus of variations was his proposal of a theory of discontinuous curves. From his work on field theory he established links with partial differential calculus, and in 1937 he published a book on the application to geometrical optics of the results of his investigations into the calculus of variations.

One of Carathéodory's most significant achievements – also the subject of a book (1932) – was a simplification of the proof of one of the central theorems of conformal representation. It formed part of his work on function theory, which extended earlier findings of Picard and Schwarz. In measure theory he developed research begun in the 1890s by Emile Borel (1871–1956) and his student Henri Lebesgue (1875–1941), work which he summarized in 1918 in a text on real functions.

Carathéodory's interest also extended beyond pure mathematics into the applications of the subject, particularly to mechanics, thermodynamics and relativity theory. A mathematician of diverse talents, he can thus be seen to have enlarged the understanding of several disciplines.

Cardozo William Warrick 1905–1962 was a US physician and paediatrician who is remembered for his pioneering investigations into sickle cell anaemia.

William Cardozo was born on 6 April 1905. After attending the public schools in Washington, DC, and Hampton Institute he went to Ohio State University where he received his AB in 1929 and MD in 1933. He was an intern at City Hospital, Cleveland and a resident in paediatrics at Provident Hospital, Chicago, spending a year in each, and then between 1935 and 1937 he had a General Education Board fellowship in paediatrics at Children's Memorial Hospital and Provident Hospital. In 1937 he started private practice in Washington, DC, and was appointed part-time instructor in paediatrics at Howard University College of Medicine and Freedmen's Hospital, later being promoted to clinical assistant professor and then clinical associate professor. In 1942 he was certified by the American Board of Paediatrics and in 1948 became a Fellow of the American Academy of Paediatrics. Cardozo was also a school medical inspector for the District of Columbia Board of Health for 24 years. He died from a heart attack on 11 August 1962; he was married with one daughter.

Cardozo is known for his pioneering research into sickle cell anaemia. His investigations were published in *The Archives of Internal Medicine*, 'Immunologic studies in sickle cell anaemia', in 1937, in which he concluded that the disease was inherited following Mendelian law and almost always occurred in black people or people of African descent; not all persons with sickle cells were necessarily anaemic and not all patients died of the disease. He also contributed articles to the *Journal of Paediatrics on Hodgkin's Disease* (1938) and *The Growth and Development of Black Infants* (1950).

Carnegie Andrew 1835–1919 was a Scottish-born US industrialist whose willingness to adopt new methods of steel-making was instrumental in advancing both the techniques and commercial potential of the iron and steel industry.

Carnegie was born in Dunfermline into a poor weavers's family, which, when he was 13 years old, emigrated to the United States. He was largely self-educated and began work as a telegraph messenger. After having held a variety of jobs and having saved some capital, he bought shares in a railroad company and land containing oil in Pennysylvania. These investments laid the foundation for his eventual huge fortune. After the American Civil War (1861–65) he became an iron manufacturer and built great iron and steel works in Pittsburgh. Apart from his fame as a steel-maker he is best known for his philanthropic activities, giving much of his fortune to the provision of libraries, a craft school, music centre, research trusts, and many other public amenities in the United States, Britain, and Europe. Carnegie scorned the word 'philanthropist', calling himself, in his later years, 'a distributor of wealth for the improvement of mankind.'

The American iron industry received a great impetus from the Civil War. Until this time the country had no great steel industry, but the sudden demand for war materials, railway supplies and the like brought fortunes to the previously struggling ironmasters of Pittsburgh. Carnegie was 30 years old when the war ended and he had not yet begun his work in this field. It was not until 1873 that he concentrated on steel, having made a small fortune in oil and taken several trips to Europe selling railroad securities. His operations in bond selling, oil dealing and bridge-building were so successful that conservative Pittsburgh businessmen regarded him with a mixture of doubt and jealousy. Carnegie's European tours, however, had results of great consequence. He came into close touch with British steel makers – then the world's leaders – and he became closely acquainted with the Bessemer process and formed a friendship with Henry Bessemer (1813–1898), which was maintained until the latter's death.

Bessemer patented his process for the manufacture of steel in 1856 (based on the idea of blowing air through the molten steel to oxidize impurities) which the Carnegie Company adopted with great success. Then in 1867, William Siemens (1823–1883) invented the open-hearth process, after the French engineer Pierre Emile Martin had made the first open-hearth furnace. Always adventurous, and with tremendous foresight, Carnegie scrapped most of the equipment used in the old processes and invested heavily in the new one. Pittsburgh is situated conveniently near to abundant supplies of coal, iron ore, and limestone and has become the leading iron and steel producing centre in the world.

Carnegie's success was the result of optimism, enthusiasm and courage. He was not a gambler; he detested the speculative side of Wall Street. He did make one gamble of titanic proportions, however – and won. He wagered everything he possessed on the industrial future of the United States and its economic potential. He was probably the most daring man in American industry; his insistence on having the most up-to-date machines, his readiness to discard costly equipment as soon as something better appeared has become a tradition in the steel trade.

Carnot Nicolas Léonard Sadi 1796–1832 was a French physicist who founded the science of thermodynamics. He was the first to show the quantitative relationship between work and heat.

Sadi Carnot was born in Paris on 1 June 1796 into a distinguished family (his father Lazare Carnot and other relatives held important government positions). He was educated at the Ecole Polytechnique in Paris from 1812 to 1814 and then at the Ecole Genie in Metz until 1816. He became an army engineer, at first inspecting and reporting on fortifications and in 1819 transferring to the office of the general staff in Paris. Carnot had many interests, carrying out a wide range of study and research in industrial development, tax reform, mathematics and the fine arts. He was particularly interested in the problems of the steam-engine and, in 1824, he published his classic work, *Reflections on the Motive Power of Heat*, which was well received. He was then forced to return to active service in the army in 1827 but was able to resign a year later in order to concentrate on the problems of engine design and to study the nature of heat. In 1831, he began to study the physical properties of gases, in particular the relationship between temperature and pressure. Sadi Carnot died suddenly of cholera on 24 August 1832. In accordance with the custom of the time, his personal effects, including his notes, were burned. Only a single manuscript and a few notes survived, and Carnot's work was virtually forgotten. However, it was rediscovered by Lord Kelvin (1824–1907), who confirmed Carnot's conclusions in his *Account of Carnot's Theorem* in 1849. He and Rudolph Clausius (1822–1888) then derived the second law of thermodynamics from Carnot's work.

In *Reflections*, Carnot reviewed the industrial, political and economic importance of the steam-engine. The engine invented by James Watt (1736–1819), although the best available, had an efficiency of only 6%, the remaining 94% of the heat energy being wasted. Carnot set out to answer two questions:

1. Is there a definite limit to the work a steam-engine can produce and hence a limit to the degree of improvement of the steam-engine?
2. Is there something better than steam for producing the work?

Engineers had worked on problems like these before, but Carnot's approach was new for he sought a theory, based on known principles, that could be applied to all types of heat engines. Carnot's theorem showed that the maximum amount of work that an engine can produce depends only on the temperature difference that occurs in the engine. (In a steam-engine, the hottest part is the steam and the coldest part the cooling water.) It is independent of whether the temperature drops rapidly or slowly or in a number of stages, and it is also independent of the nature of the gas used in the engine. The maximum possible fraction of the heat energy that is capable of being converted into work is represented by Carnot's equation $(T_1 - T_2)/T_2$, where T_1 is

the absolute temperature of the hottest part and T_2 is the temperature of the coldest part. Carnot's equation put the design of the steam-engine on a scientific basis. Using experimental data and his own conclusions, he recommended that steam should be used over a large temperature interval and without losses due to conduction or friction.

In formulating his theorem, Carnot considered the case of an ideal heat engine following a reversible sequence known as the Carnot cycle. This cycle consists of the isothermal expansion and adiabatic expansion of a quantity of gas, producing work and consuming heat, followed by isothermal compression and adiabatic compression, consuming work and producing heat to restore the gas to its original state of pressure, volume, and temperature. Carnot's law states that no engine is more efficient than a reversible engine working between the same temperatures. The Carnot cycle differs from that of any practical engine in that heat is consumed at a constant temperature and produced at another constant temperature, and that no work is done in overcoming friction at any stage and no heat is lost to the surroundings – so that the cycle is completely reversible.

At the time he wrote *Reflections*, Carnot was a believer in the caloric theory of heat, which held that heat is a form of fluid. But this misconception of the nature of heat does not invalidate his conclusions. Some notes which escaped destruction after Carnot's death indicate that he later arrived at the idea that heat is essentially work, or rather work which has changed its form. He had calculated a conversion constant for heat and work and showed he believed that the total quantity of work in the universe is constant. It indicated that he had thought out the foundations of the first law of thermodynamics, which states that energy can never disappear but can only be altered into other forms of energy. Carnot's notes, however, remained undiscovered until 1878.

Carnot's work led Lord Kelvin, in 1850, to confirm and extend it. Rudolph Clausius made some modifications to it and Carnot's theorem became the basis of the second law of thermodynamics, which states that heat cannot flow of its own accord from a colder to a hotter substance.

The application of the science of thermodynamics founded by Sadi Carnot has been of great value to the production of power and also to industrial processes. To give one example, it has been useful in forecasting the conditions under which chemical reactions will or will not take place and the amount of heat absorbed or given out.

Carothers Wallace Hume 1896–1937 was a US organic chemist who did pioneering work on the development of commercial polymers, producing nylon and neoprene (one of the first synthetic rubbers).

Carothers was born on 27 April 1896 in Burlington, Iowa, the son of a teacher. He attended schools in Des Moines, graduating from the North High School in 1914. His further studies were in accountancy and clerical practice at his father's college in Des Moines, until he entered Tarkio College, Missouri, in 1915 and specialized in chemistry. He later gained higher degrees in organic chemistry from the University of Illinois and Harvard, where he was appointed in 1926. In 1928 he accepted the post as head of organic chemistry research at the Du Pont research laboratory in Wilmington, Delaware. For his fundamental work on polymers he was elected to the US National Academy of Sciences in 1931. Carothers suffered from periods of depression which became more prolonged and severe as he grew older. He was deeply affected by the death of his sister in 1936, the same year in which he married. A few months later, on 29 April 1937, he committed suicide in Philadelphia, Pennsylvania.

Carothers began his work on polymerization and the structures of high molecular mass substances while at Harvard. Then at Du Pont he carried out studies on linear condensation polymers, which culminated in 1931 with the development of nylon and neoprene. Much of his research effort was directed at producing a polymer that could be drawn out into a fibre. His first successful experiments involved polyesters formed from trimethylene glycol (propan-1,3-diol) and octadecane dicarboxylic acid (octadecan-1,18-dioic acid). But for finer fibres with enough strength (emulating silk) he turned to polyamides. Early attempts, made by heating amino-caproic acid (hexan-6-amino-1-oic acid), resulted in an unstable product containing ring compounds. The first polymer to be called nylon (strictly the trade-name Nylon 6,6) was made by heating hexamethylene diamine (hexan-1,6-diamine) and adipic acid (hexan-1,6-dioic acid). The product is a linear chain polymer which can be cold-drawn after extrusion through spinnerets to orientate the molecules parallel to each other so that lateral hydrogen bonding takes place. The resultant nylon fibres are strong and have a characteristic lustre.

Carothers also worked on synthetic rubbers. His monomer was chlorobutadiene (but-2-chloro-1,3-diene), which he first had to make by treating vinylacetylene (but-1-en-3-yne) with hydrogen chloride. Using a peroxide catalyst, the chloro compound polymerizes readily by a free radical mechanism to form neoprene. This polymer, first produced commercially in 1932, is resistant to heat, light and most solvents. In the years that followed, a whole range of useful polymers of the nylon and neoprene types were produced.

Carr Emma Perry 1880–1972 was a US chemist, teacher and researcher, internationally renowned for her work in the field of spectroscopy.

Emma Carr was born in Holmesville, Ohio, on 23 July 1880. After attending Coshocton High School, Ohio, she spent a year (1898–99) at Ohio State University and then transferred to Mount Holyoke for a further two years' study; she continued as a chemistry assistant for three years and in 1905 finished her BS at the University of Chicago. Between 1905 and 1908 Carr worked as an instructor at Mount Holyoke, after which she returned to Chicago and completed her PhD in physical chemistry in 1910. From 1910 to 1913 she was associate professor of chemistry at Mount Holyoke after which she became professor of chemistry and head of the department, posts which she held until her retirement in 1946. Carr lived in South Hadley for 18 years until she moved in 1964 to the Presbyterian Home in Evanston, Illinois. She died there in 1972.

In 1913 Carr introduced a departmental research programme to train students through collaborative research projects combining both physical and organic chemistry. She achieved her renown in the research into spectroscopy; under her leadership, she, her colleague Dorothy Hahn and a team of students were among the first Americans to synthesize and to analyse the structure of complex organic molecules using absorption spectroscopy. Her research into unsaturated hydrocarbons and far ultraviolet vacuum spectroscopy led to grants from the National Research Council and the Rockefeller Foundation in the 1930s. She served as a consultant on the spectra for the *International Critical Tables*, and during the 1920s and 1930s she was three times a delegate to the International Union of Pure and Applied Chemistry. She was awarded four honorary degrees and was the first to receive the Garvan Medal, annually awarded to an American woman for achievement in chemistry.

Carrington Richard Christopher 1826–1875 was a British astronomer who was the first to record the observation of a solar flare, and is now most remembered for his work on sunspots.

Carothers *The condensation polymerization of hexamethylene diamene and adipic acid to form nylon.*

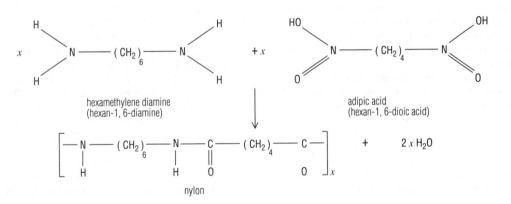

Carrington's family was in brewing, but neither the family business nor the Church (for which he had been intended) attracted him. He very early realized that his interests lay in astronomy and scientific activities. He left Cambridge in 1849 and his first post was as Observer at Durham, from where he made several reports to the *Monthly Notices of the Royal Astronomical Society* and to the *Astronomische Nachrichten* of Altona (mainly dealing with minor planets and comets), work which eventually led to his election as a Fellow of the Royal Astronomical Society in 1851.

However, by 1852 he was impatient with the limited resources at Durham – he had in mind an ambitious programme of observation leading to a catalogue of circumpolar stars. In 1853 he set up his own house and observatory at Redhill, Surrey, with instruments made by W. Simms. One of these was based on a larger Greenwich instrument; its telescope had a 12.7-cm/5-in aperture and a focal length of 1.68 m/5.5 ft.

By 1857 he had completed his *Catalogue of 3,735 Circumpolar Stars*, which was so highly regarded that it was printed by the Admiralty at public expense. The *Catalogue* won him the Gold Medal of the Royal Astronomical Society (1859), and his election to a Fellowship of the Royal Society shortly afterwards was fitting recognition of his qualities as an astronomer.

The death of his father in 1858 meant that Carrington had to take over the management of the Brentford Brewery, a substantial undertaking which entailed a reduction in his research activity. Nevertheless, in 1859 he recorded the first solar flare.

It is felt by some that he was disappointed by his failure to succeed James Challis (his mentor) as Director of the Cambridge Observatory. At any rate his output declined in the 1860s and ill health overtook him in 1865. He sold his business and his Redhill establishment and moved to Churt, near Farnham, Surrey, where he built another observatory containing some large telescopes.

Carrington is best known for his work on sunspots. He pursued the daylight project at Redhill for more than seven years (the original aim was for an eleven-year period), in tandem with the work on the Redhill catalogue. The sunspot cycle – an eleven-year period between maxima of activity – had recently been discovered by Schwabe and the connection with magnetic disturbances had been noted. A study of sunspot activity was therefore highly topical and Carrington was keen to tidy up the mass of observations on the subject which had accumulated in the contemporary literature.

He required a simple yet accurate means of plotting sunspot positions and movements, and with much trial and error he arrived at a simple, elegant method. His system projected an image of the Sun of about 28 cm/11 in diameter using his 11.4-cm/4^1/$_2$-in equatorial telescope. Crosswires at right angles were placed in the focus of the telescope inclined at 45° to the meridian; the exact angles were not important. The telescope was fixed and the Sun's image allowed to pass across the field; the times of contact of the Sun's limbs and spots were recorded. The method allowed the heliographic latitude and longitude of a sunspot to be determined without recourse to micrometers or clockwork mechanisms.

The principal results of this extended work were, first, to determine the position of the Sun's axis and, second, dramatically to show that the Sun's rotation is differential, that is, that it does not rotate as a solid body, but turns faster at the equator than at the poles. This conclusion was the result of observing the great systematic drift of the photosphere as seen in the drift of individual sunspots during the cycle. Carrington also derived a useful expression for the rotation of a spot in terms of heliographical latitude. An extensive account of all the observations was published, with the help of the Royal Society, in 1863. The complete cycle of work was, however, never accomplished by him.

An immensely practical and meticulous man, interested in international cooperation and the mutual contribution of ideas, Carrington in his work and his publications represents the true Victorian ideal of the investigative scientist.

Carson Rachel Louise 1907–1964 was a US biologist, conservationist and campaigner. Her writings on conservation and the dangers and hazards that many modern practices imposed on the environment inspired the creation of the modern environmental movememnt.

Carson was born in Springdale, Philadelphia, on 27 May 1907, and educated at the Pennsylvania College for Women, studying English to achieve her ambition for a literary career. A stimulating biology teacher diverted her towards the study of science, and she went to Johns Hopkins University, graduating in zoology in 1929. She received her Master's degree in zoology in 1932 and was subsequently appointed to the Department of Zoology at the University of Maryland, spending her summers teaching and researching at the Woods Hole Marine Biological Laboratory in Massachusetts. Family commitments to her widowed mother and orphaned nieces forced her to abandon her academic career and she worked for the United States Bureau of Fisheries, writing in her spare time articles on marine life and fish, and producing her first book on the sea just before the Japanese attack on Pearl Harbor. During World War II she wrote fisheries information bulletins for the US Government and reorganized the publications department of what became known after the war as the United States Fish and Wildlife Service. In 1949 she was appointed chief biologist and editor of the Service. She also became occupied with fieldwork and wrote regular freelance articles on the natural world. During this period she was also working on *The Sea Around Us*, which finally appeared in 1951 and was an immediate bestseller, being translated into several languages and winning several literary awards. Given a measure of financial independence by this success she resigned from her job in 1952 to become a professional writer. Her second book *The Edge of the Sea* (1955), an ecological exploration of the seashore, further established her reputation as a writer on biological subjects. Her most famous book *The Silent Spring* (1962) was a powerful denunciation of the effects of the chemical poisons, especially DDT, with which humans were destroying the earth, sea, and sky. Despite denunciations from the influential agrochemical lobby, one immediate effect of Carson's book was a Presidential Advisory Committee on the use of pesticides. By this time Carson was already seriously incapacitated by ill health and she died in Silver Spring, Maryland, on 14 April 1964.

On a larger canvas, *The Silent Spring* alerted and inspired a new worldwide movement of environmental concern. While writing about broad scientific issues of pollution and ecological exploitation, she also raised important issues about the reckless squandering of natural resources by an industrial world.

Cartwright Edmund 1743–1828 was a British clergyman who invented various kinds of textile machinery, the most significant of which was the power loom which helped initiate the Industrial Revolution.

Cartwright was born in Marnham, Nottinghamshire, and received his early education at Wakefield Grammar School. At the age of only 14 he went to University College, Oxford, and the regulations were changed to enable him to be awarded his BA earlier than usual (in 1764). In that same year he was elected a Fellow of Magdalen College, gaining his MA two years later. He received the perpetual curacy of Brampton near Wakefield, and became rector of Goadby Marwood, Leicestershire, in 1779. He was prebendary of Lincoln from 1786 until his death in 1828.

Cartwright was an innovator, always curious and on the look-out for new ways of doing things. At Goadby Marwood he made agricultural experiments on his glebe land, and while on holiday at Matlock he visited the spinning mills of Richard Arkwright (1732–1792) at nearby Cromford. Arkwright had watched cotton weavers working in their homes. They used cotton thread from the side of the loom (the weft), but he noticed that they wove it in and out of Irish linen threads stretching lengthwise. When he asked the reason for this, he was told that they could not spin cotton thread which was fine or strong enough to use for the warp. This motivated Arkwright to invent the spinning frame and, watching it working, Cartwright remarked that Arkwright would have to set his wits to work to invent a weaving mill. Soon after returning home Cartwright himself set about this task, devoting all his spare time and money to experiment.

Cartwright had never seen the working of the hand loom and the first machine he made was an inadequate substitute for it. However, he patented it in 1785 and moved to Doncaster in the same year where his wife had inherited some property. There he continued to improve the simple water-driven machine, and visited Manchester to have it criticized by the local workmen; he also tried to enlist the help of the local manufacturers. Disappointed in this hope and having taken out two more patents for further improvements in his loom, he set up a factory at Doncaster for weaving and spinning. His power loom now worked well and became the parent of all those in use today. It contained

an ingenious mechanism which substituted for the hands and feet of the ordinary hand-loom weaver. There was a beam on which the required number of warp ends was wound side by side, in perfect order. A device called a let-off motion held the warp ends in place and let them go forwards only as required. The ends were threaded through eyes (loops) in sets of cords or wires called healds, and there was an apparatus which raised some sets of healds and lowered others, thus making a tunnel, called a shed, between the lower and upper warp ends. (The healds could be reversed so that the upper and lower layers of warp ends changed places.) The weft was carried to and fro through the shed by the shuttle. There was a device for pressing the weft up tightly against the already woven cloth, and another for keeping the cloth taut and rolling it up as fast as it was woven.

For centuries Yorkshire had been a principal seat of woollen manufacture, and in 1789 at Doncaster Cartwright invented a wool-combing machine which contributed greatly to cutting the cost of manufacture. Even in the earlier stages of its development, one machine did the work of 20 hand-combers. Petitions against its use poured into the House of Commons from the wool-combers – some 50,000 in number – and a committee was appointed to inquire into the matter; nothing came of the wool-combers' agitation.

Cartwright's Doncaster factory was enlarged when a steam engine was erected to power it, and in 1799 a Manchester firm contracted with Cartwright for the use of 400 of his power looms and built a mill where some of these were powered by steam. The Manchester mill was burned to the ground, probably by workmen who feared to lose their jobs, and this catastrophe prevented other manufacturers from repeating the experiment. Cartwright's success at Doncaster was obstructed by opposition and by the costly character of his processes; in 1793, deeply in debt, he relinquished his works at Doncaster and gave up his property to his creditors. In 1807, however, 50 prominent Manchester firms petitioned the government to bestow a substantial recognition of the services rendered to the country by Cartwright's invention of the power loom. Cartwright too petitioned the House of Commons, which in 1809 voted him £10,000.

Carver George Washington *c.*1860–1943 was a US agricultural chemist who revolutionized agriculture in the south of the country. He advocated the diversification of crops, crop rotation, and the cultivation of peanuts, from which he made over 300 products.

Carver was born about 1860 to slave parents near Diamond Grove in Missouri. He was educated at Minneapolis High School in Kansas and, having achieved an outstanding record, he received a scholarship to the Highland University, Kansas. He was, however, later rejected on account of his race. After working on the land and saving money he was accepted in 1887 by Simpson College, Iowa, and in 1891 entered Iowa Agricultural College from where he graduated in 1894 with a BS degree. After graduation, Carver was given an appointment in the faculty teaching agriculture and bacterial botany, and while pursuing research and conducting experiments into plant pathology, he obtained the MS in agriculture in 1896. In 1897 he transferred to the Tuskegee Institute, Alabama. He was made first director of agriculture and was also director of a research and experiment station. Carver remained at Tuskegee until his death from anaemia on 5 January 1943. During his time in Iowa, Carver made important discoveries in the field of plant pathology; in 1897 he reported on new species of fungi which have since been named after him: *Taphrina carveri*, *Collectotrichum carveri* and *Metasphaeria carveri*. At Tuskegee, Carver demonstrated the need for crop rotation and the use of leguminous plants, especially the peanut. Following his advice, farmers planted peanuts, which soon became the principal crop in the farming belt running from Montgomery to the Florida border. They were soon making more money from the peanut and its 325 by-products (including milk, cheese, face powder, printer's ink, shampoo, and dyes) which were developed by Carver, than from tobacco and cotton. In 1921, following Carver's presentation to the Ways and Means Committee, the peanut was included in the Hawley–Smoot Tariff Bill to protect it from foreign competition. Carver also discovered 118 products which could be made from the sweet potato and 75 products from the pecan nut. Carver's other work included developing a plastic material from soya beans which Ford later used in part of his automobile, and extracting dyes and paints from the clays of Alabama. He received three patents, for a cosmetic (1925), a paint and stain (1925), and a process for producing paints and stains (1927).

Carver was also an accomplished artist, and he received many awards and honours for his outstanding work: he was elected a Fellow of the Royal Society of Arts, Manufactures and Commerce of Great Britain (1916); he was awarded the Spingarn Medal (1923), the Theodore Roosevelt Medal 'for distinguished research in agricultural chemistry' (1939); and he was chosen 'man of the year' by the International Federation of Architects, Engineers, Chemists and Technicians (1941). He received honorary ScD degrees from Simpson College (1928) and from the University of Rochester (1941). Three months after his death he was posthumously awarded the New York City Teachers Union Medal.

Cassegrain *c.*1650–1700 was the inventor of the system of mirrors within many modern reflecting telescopes – a system by transference also sometimes used in large refraction telescopes.

Nothing is known for certain about the details of Cassegrain's life – not even his first name. Believed to have been a professor at the College of Chartres, in France, he is variously credited with having been an astronomer, a physician and a sculptor at the court of Louis XIV.

In the same year as he submitted a scientific paper concerning the megaphone to the Academy of Sciences in Paris, Cassegrain presented another paper in which he claimed to have improved on Newton's telescope design.

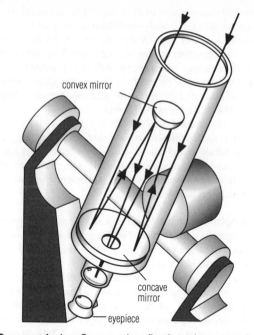

Cassegrain *In a Cassegrain reflecting telescope, a hole in the centre of the concave main mirror allows light reflected by the convex secondary mirror to reach the eyepiece (or a camera).*

Newton himself, however, suggested that the 'improvement' had been strongly influenced by the work of James Gregory (whose telescope had been described in *Optica Promota* in 1663).

Newton's own design used a second, plane mirror to bring the reflected, magnified image out to the eyepiece through the side of the telescope. Cassegrain's telescope used an auxiliary convex mirror to reflect the image through a hole in the objective – that is, through the end of the telescope itself. One intention behind this innovation was further to increase the angular magnification.

An even more advantageous facet of this design was not realized until a century later, when Jesse Ramsden noted that it also partly cancelled out the spherical aberration, the blurring of the image caused by the use of two mirrors. The first practical reflecting telescope based on Cassegrain's design was 'Short's "dumpy"' (focal length 609 mm/ 23.75 in, aperture 152 mm/5.93 in), built by J. Short in the eighteenth century.

Cassini Giovanni Domenico Jean Dominique 1625–1712 was an Italian-born French astronomer with a keen interest in geodesy.

Born in Perinaldo, Cassini studied in Vallebone and Genoa, displaying great talent in astronomy and mathematics. In 1644 he was invited to assist the Marquis Mavasia in his observatory at Panzano, near Bologna, and there he was introduced to the two prominent local astronomers, Giovanni Battistae Riccioli (1598–1671) and Francesco Maria Grimaldi (1618–1663). Six years later, aged only 25,

Cassini was made Professor of Astronomy at the University of Bologna; he remained there for 19 years.

In addition to his teaching duties at the university, Cassini was also called upon to serve a variety of civic and diplomatic duties. These included contributing to hydraulic projects, supervising construction work and mediating in a dispute between Ferrara and Bologna over navigation rights on the River Reno.

In 1669 Cassini departed for Paris at the invitation of King Louis XIV, who had nominated him as a member of the new French Academy of Sciences. Despite Pope Clement IX's insistence that his stay in France be only temporary, Cassini never returned to Italy.

The first task to confront Cassini upon his arrival in Paris was the construction of the Paris Observatory. He had been unable to convince the King or the architect (Claude Perrault) that certain aspects of the design were impractical, but he nevertheless took up the directorship of the observatory and assumed French citizenship. Cassini was extremely active in exploiting the work of astronomers in research outposts around the world, seeking to equip the observatory with the latest instruments and making use of the skills of students of Galileo.

At the end of the century Cassini's health began to fail and his son Jacques took over an increasing share of his work. The elder Cassini lost his eyesight in 1710 and Jacques assumed the directorship of the observatory.

Cassini was renowned for his skills as an observational astronomer, which led him to many important discoveries. He was also extremely conservative in his approach to the more theoretical aspects of astronomy, and this conservatism led him frequently to propound the incorrect view. He refused to accept the Copernican cosmological model and rejected the concept of a finite speed of light (although its proof was demonstrated by Ole Römer (1644–1710) using Cassini's own data; it is likely that Cassini himself considered the possibility even before Römer's work). He also opposed a theory of universal gravitation and insisted (despite critical disagreement by Christiaan Huygens and Isaac Newton) that the Earth was flattened at the Equator rather than at the poles. Despite these errors in judgement Cassini earned a well-deserved reputation as one of the finest astronomers of his day.

The best-known examples of Cassini's early work are a treatise on his observations of a comet made in 1652 and his design work for a meridian constructed at San Petrino in 1653. The meridian was used to make accurate observations of the Sun and enabled Cassini to publish improved tables in 1662.

During the period from 1664 to 1667 Cassini concentrated his efforts on determining the rotation periods of Mars, Jupiter, and Venus. In 1664 he found the rotation period of Jupiter to be nine hours from a study of Saturn. In 1675 he distinguished two zones within what was thought to be the single ring around Saturn. The dark central 'border' has since been named Cassini's division. Cassini correctly suggested that the rings were composed of myriads of tiny satellites, although it was not until the work of

James Clerk Maxwell (1831–1879) in the 1850s that he was proved correct.

From 1671 until 1679 Cassini made many observations of details on the lunar surface which culminated in the production of a beautiful engraving of the Moon, presented to the French Academy in 1679. In 1672 Cassini took advantage of a good opposition of Mars to determine the distance between the Earth and that planet. He arranged for Jean Richer (1630–1696) to make measurements from his base in Cayenne, on the northeastern coast of South America, while Cassini made simultaneous measurements in Paris which permitted them to make a triangulation of Mars with a baseline of nearly 10,000 km/6,000 mi. This derived a good approximation for the distance between the Earth and Mars, from which Cassini was able to deduce many other astronomical distances. These included the Astronomical Unit (AU), which Cassini found to be 138 million km/86 million mi, only 11 million km/7 million mi short.

Cassini's later work included a study (with N. Fatio) on zodiacal light (1683), and a triangulation of the arcs of meridian aimed at resolving a controversy concerning the shape of the Earth.

Cassini's contributions to astronomy were original and plentiful, but his best work was of an observational rather than of a theoretical nature.

Cauchy Augustin-Louis 1789–1857 was a French mathematician who did important work in astronomy and mechanics, but who is chiefly famous as the founder, with Gauss, of the modern subject of complex analysis.

He was born in Paris on 21 August 1789 and received his early education from his father, an accomplished classical scholar and a barrister in the *parlement* of Normandy. When he was little more than an infant he was taken by his family to live in the village of Arceuil, where his father went to escape the terror of the French Revolution in 1793–94. There he grew up with illustrious neighbours, the mathematician Pierre Laplace, and the chemist Claude-Louis Berthollet (1749–1822), who together had established the famous Société d'Arceuil. The story runs that Lagrange, who also met the young Cauchy, quickly recognized the boy's scientific flair, but advised his father to give him a firm literary education before showing him any mathematical texts. True or not, Cauchy's first real introduction to mathematics was delayed until he entered the Ecole Polytechnique, in Paris, in 1805. Two years later he entered the Ecole de Ponts et Chausses to study engineering, leaving in about 1809 to take employment first at the Ourcq Canal works, then at the Saint-Clou bridge, and finally (in 1810) at the Cherbourg harbour naval installations. In 1813 he returned to Paris, apparently for health reasons, and two years later he was appointed to the Ecole Polytechnique, where he was made a full professor in 1816.

In the same year the restoration of the Bourbons to the French throne after the Napoleonic interlude brought a great change in his fortunes. Gaspard Monge (1746–1818) and Lazare Carnot (1753–1823), both of them republicans and Bonapartists, were expelled on political grounds from the Academy of Sciences and Cauchy was elected to fill one of the vacancies. In that year his paper on wave modulation won the Grand Prix of the Academy (renamed the Institut de France). That paper marked the real beginning of his fruitful years as a mathematician, years which gained the reward, some time before 1830, of his appointment to the chair of the faculty of science at the Collège de France.

In 1830 Charles X was overthrown by the July Revolution, and when Cauchy refused to take the new oath of allegiance he was forced to resign his chair. He went into exile at Fribourg, Switzerland, where he lived among a Jesuit community; they recommended him to the king of Sardinia and he was appointed to the Chair of Mathematical Physics at the University of Turin. From 1833 to 1838 he was tutor to Charles X's son at Prague. At last, in 1838, he returned to Paris to resume his professorship at the Ecole Polytechnique. From 1848 to 1852 he was a professor at the Sorbonne. He died at Sceaux, outside Paris, on 23 May 1857.

In 1805 Cauchy provided a simple solution to the problem of Apollonius, namely to describe a circle touching three given circles; in 1811 he generalized Euler's theorem on polyhedra; and in 1816 he published his award-winning paper on wave modulation. His best work, however, was all done in the 1820s and was published in his three great treatises: *Cours d'analyse de l'Ecole Polytechnique* (1821), *Résumé des leçons sur le calcul infinitésimal* (1823), and *Leçons sur les applications de calcul infinitésimal à la géométrie* (1826–28). Although he did other valuable research – in mechanics he substituted the concept of the continuity of geometrical displacements for the principle of the continuity of matter and in astronomy he described the motion of the asteroid, Pallas – his vital contributions were contained in these three treatises.

Cauchy made the principles of calculus clearer by developing them with the aid of limits and continuity, and he was the first mathematician to provide a rigorous proof for the famous theorem of Brook Taylor (1685–1731). Taylor's theorem, enunciated in 1712, expands a function of x as an infinite series in powers of x. Cauchy's proof was of great usefulness, because the theorem is extremely helpful in finding the difference columns in books of tables. More generally, Cauchy's work in the 1820s provided a satisfactory basis for the calculus. Perhaps even more important, for future pure and applied mathematicians alike, was his monumental research into the fundamental theorems of complex functions. He provided the first comprehensive theory of complex numbers (still, at the beginning of the nineteenth century, not accepted by all mathematicians) in his *Cours d'analyse*, and in doing so made a vital contribution to the development of mathematical physics and, in particular, to aeronautics.

During his lifetime Cauchy published seven books and about 800 papers. He has the credit for 16 fundamental concepts and theorems in mathematics and mathematical physics, more than for any other mathematician. For both his creative genius and his prolific output he is remembered as one of the greatest mathematicians in history.

Cavendish Henry 1731–1810 was a British natural philosopher whose main interests lay in the fields of chemistry and physics. His chief experimental work concerned gases, although he also carried out fundamental experiments concerning electricity and gravitation. He made the first determination of the gravitational constant and thereby obtained the first values for the mass and density of the Earth. He is also usually credited with the discovery of hydrogen. Cavendish was one of the few scientists to approach Newton's standard in both mathematical and experimental skills and was a major figure in eighteenth-century science. He devoted his entire life to the acquisition of knowledge, but published only those results that satisfied him completely. Most of his work, especially his experiments with electricity, were unknown for 100 years or more, so the immediate impact of his work was far less than it might have been.

Cavendish was born in Nice, France, on 10 October 1731. He was of aristocratic descent, his paternal grandfather being the Duke of Devonshire and his maternal grandfather the Duke of Kent. Cavendish attended Dr Newcome's Academy in Hackney, London, and then went on to Peterhouse, Cambridge, in 1749. He left in 1753 without a degree, which was not unusual at that time. He spent the rest of his life in London. His father encouraged his scientific interests and introduced him to the Royal Society, of which he became a Fellow in 1760. Despite his active participation in the scientific community, Cavendish was a recluse and shunned most social contact, making no attempt to use a fortune of the order of a million pounds bequeathed to him.

Cavendish published his first paper, which demonstrated the existence of hydrogen as a substance, in 1776. He received the Copley Medal of the Royal Society for this achievement. His subsequent papers were few and far between, and included most notably a theoretical study of electricity in 1771, the synthesis of water in 1784 and the determination of the gravitational constant in 1798. He died alone in London on 24 February 1810.

Little is known of Cavendish's work until the late 1760s, when he began experimenting with 'facticious airs' (gases that can be produced by the chemical treatment of solids or liquids). He studied 'fixed air' (carbon dioxide) produced by mixing acids and bases; 'inflammable air' (hydrogen) generated by the action of acids on metals; and the 'airs' produced during decay and fermentation. He measured the specific gravities of hydrogen and carbon dioxide, comparing them with that of 'common' (i.e. atmospheric) air.

In 1783 Cavendish found that the composition of the atmosphere is the same in different locations and at different times. He also found that a small fraction of 'common air' seems to be inert; 100 years later William Ramsay was to show that this inert gas is mainly argon. A year later Cavendish demonstrated that water is produced when hydrogen burns in air, thus proving that water is a compound and not an element as had been suggested by early Greek scientists. By sending electric sparks through 'common air' he caused the nitrogen in it to combine with oxygen. When the gas produced was dissolved in water it produced nitric acid. He also showed that 'calcareous earth' dissolves in water containing carbon dioxide, to form what is now known as calcium bicarbonate. He distinguished between the two oxides of arsenic, demonstrating that one contains more oxygen than does the other.

Cavendish's most important work in physics was on electricity and gravitation. His 1771 paper on the nature of electricity shows that he believed it to be an elastic fluid. He then worked on electricity for ten years, aiming to produce a sequel to Newton's *Principia* that would explain all electrical phenomena. But although this was his most concentrated research effort, Cavendish published nothing more about it. His fastidious attention to the details of his results and his thorough efforts to understand and unify all his observations frustrated this plan and he was not able to gain the overview that he sought. He tried unsuccessfully to uncover the relationship between force, velocity of current, and resistance, although he found that electric fields obey the inverse square law and was able to produce some valuable work on conductivity.

Much of the work done by Michael Faraday (1791–1867) and Charles Coulomb (1736–1806) and others during the next 50 years is foreseen in this early work by Cavendish; in spite of this, none of his experiments were known until James Clerk Maxwell edited and published them in 1879.

During the latter part of the 1780s, Cavendish worked on the production of heat and determined the freezing points for many materials, including mercury. He relied on some of the early work he had done on latent heats. One of the practical outcomes from these experiments was the explanation for some anomalous readings obtained when using mercury thermometers at low temperatures.

The five papers which Cavendish published during the last 25 years of his life all had an astronomical theme. By far the most important of these appeared in 1798, when he announced his determination of Newton's gravitational constant, thereby deriving the density and mass of the Earth. Newton's law of gravitation contained two unknowns: the gravitational constant and the mass of the Earth. Determining one would give the other. In what has become known as the Cavendish experiment, the gravitational constant was found.

Cavendish used an apparatus that had been devised by John Michell (1724–1793). It consisted of a delicate suspended rod with two small spheres made of lead attached to each end. Two large stationary spheres were placed in a line at an angle to the rod. The gravitational attraction of the large spheres caused the small spheres to twist the rod towards them. The period of oscillation set up in the rod enabled Cavendish to determine the force of attraction between the large and small spheres, which led him to determine the gravitational constant for Newton's equation, and thus the density of the Earth (about 5.5 times that of water) and its mass (6×10^{24} kg). The sensitivity of this apparatus was extraordinary, for the gravitational force involved was 500 million times less than the weight of the

spheres, and Cavendish's results were not bettered for more than a century.

Cavendish was a great scientist and was honoured by the naming of the Cavendish Laboratory at the University of Cambridge in his memory. His contributions to science are notable for their quality and diversity. Had he permitted all his results to be published, the rate of advancement of physical science would undoubtedly have been greatly accelerated. He stands today as one of the giants of modern science.

Cayley Arthur 1821–1895 was a British mathematician who was responsible for the formulation of the theory of algebraic invariants. A prolific writer of scholarly papers, he also developed the study of n-dimensional geometry, introducing the concept of the 'absolute', and devised the theory of matrices.

Cayley was born in Richmond, Surrey, on 16 August 1821, the son of a merchant and his wife who were visiting England from their home in St Petersburg, Russia. Cayley spent the first eight years of his life in Russia, and then attended a small private school in London, before moving to King's College School there. He entered Trinity College, Cambridge, as a 'pensioner' to study mathematics and became a scholar in 1840. He graduated with distinction in 1842. Awarded a Fellowship at the College, he took up law at Lincoln's Inn in 1846 instead, prevented from remaining at Cambridge through his reluctance to take up religious orders – at that time a compulsory qualification. Cayley was called to the Bar in 1849 and worked as a barrister for many years before, in 1863, he was elected to the newly established Sadlerian Chair of Pure Mathematics at Cambridge. He occupied the post until he died in Cambridge on 26 January 1895.

Cayley published about 900 mathematical notes and papers on nearly every pure mathematical subject, as well as on theoretical dynamics and astronomy. Some 300 of these papers were published during his 14 years at the Bar, and for part of that time he worked in collaboration with James Joseph Sylvester, another lawyer dividing his time between law and mathematics. Together they founded the algebraic theory of invariants (in their later lives they drifted apart, until Cayley lectured at Johns Hopkins University, Baltimore, in 1881–82 at Sylvester's invitation).

The beginnings of a theory of algebraic invariants may be traced first in the work of Joseph Lagrange (1736–1813), who investigated binary quadratic form in 1773. Later, in 1801, Karl Gauss (1777–1855) studied binary ternary forms. A final impetus was provided by George Boole (1815–1864), who, in a paper published in 1841, showed that all discriminants – special functions of the roots of an equation, expressible in terms of the coefficients – displayed the property of invariance. Two years later, Cayley himself published two papers on invariants; the first was on the theory of linear transformations. In the second paper he examined the idea of covariance, setting out to find 'all the derivatives of any number of functions which have the property of preserving their form unaltered after any linear transformations of the variables'. He was the first mathematician to state the problem of algebraic invariance in general terms, and his work immediately attracted a lot of interest from other mathematicians.

Over the next 35 years he wrote ten papers on what he called 'quantics' (which later mathematicians refer to as 'form') in which he gave a lively account of the theory as it was being developed. He used the term 'irreducible invariant' and defined it as an invariant that cannot be expressed rationally and integrally in terms of invariants of the same quantic(s) but of a degree lower in the coefficients than its own. At the same time he acknowledged that there are many circumstances in which irreducible invariants and covariants are limited. (His system was eventually simplified and generalized by David Hilbert (1862–1943).)

Cayley developed a theory of metrical geometry that could be identified with the non-Euclidean geometry of such mathematicians as Nikolai Lobachevski, János Bolyai and Bernhard Riemann. His geometry was the geometry of *n* dimensions. He introduced the concept of 'absolute' into geometry, which links projective geometry with non-Euclidean geometry, and together with Felix Klein (1849–1925) distinguished between 'hyperbolic' and 'elliptic' geometry – a distinction that was of great historical significance. When Cayley's 'absolute' was real, his distance function was that of hyperbolic geometry, and when 'absolute' was imaginary, the formulae reduced to Riemann's elliptic geometry.

Cayley also created a theory of matrices which did not need repeated reference to the equations from which their elements were taken, and established the principles for forming general algebraic functions of matrices. He went on to derive many important theorems of matrix theory. He claimed to have arrived at the theory of matrices via determinants, but he always made great use of geometrical analogies in his algebraic and analytical work.

He also laid down in general terms the elements of a study of 'hyperspace', and in 1860 devised a system of six homogeneous coordinates of a line. These are now more often known as Plücker's line coordinates because the same ideas were independently published – five years later – by Julius Plücker (whose assistant was Cayley's former collaborator, Felix Klein).

Cayley wrote on almost every contemporary subject in mathematics, but completed only one full-length book. He clarified many of the theorems of algebraic geometry that had previously been only hinted at, and he was one of the first to realize how many different areas of mathematics were drawn together by the theory of groups. Awarded both the Royal Medal (1859) and the Copley Medal (1881) of the Royal Society, and generally in demand for both his legal and his administrative skills, Cayley played a great part in bringing mathematics in England back into the mainstream and in founding the modern British school of pure mathematics.

Cayley George 1773–1857 was an English baronet who spent much of his life experimenting with flying machines,

particularly kites and gliders. He eventually constructed a human-carrying glider, but never ventured into the realms of powered flight.

Cayley was born at Brompton, in Yorkshire, the son of wealthy parents. He received a good education and from an early age showed a keen observation and an enquiring mind. Throughout his life he could turn his attention to almost any problem with a degree of success. He is particularly associated with aeronautics and the teaching of engineering, and in later life he helped to found the Regent Street Polytechnic in London.

Cayley first began experimenting with flight after patiently observing how birds use their wings. He realized that they have two functions: the first is a sort of sculling action by the wing tips which provides thrust; the second is the actual lift, achieved by the shape of the wing, which we now refer to as an aerofoil. Air rushing faster over the curved surface of the upper wing creates low pressure and a sucking effect. As a result, the higher pressure on the undersurface of the wing gives lift.

His first attempt at a flying invention was a kite fitted with a long stick, a movable tail for some control, and a small weight at the front for balance. His idea was to create a design which would glide safely but with enough speed to give lift. Spurred on by the success of his first design, he wrote in his diary of how nice it was to see it in flight and 'it gave the idea that a larger instrument would be a better and safer conveyance down the Alps than even a sure-footed mule'.

In 1808, Cayley constructed a glider with a wing area of nearly 28 sq m/300 sq ft, and was probably the first person to achieve flight with a machine heavier than air. During the next 45 years he worked on many aspects of flight, including helicopters, streamlining, parachutes and the idea of biplanes and triplanes. Eventually, in 1853, he built a triplane glider which carried his reluctant coachman 275 m/900 ft across a small valley – the first recorded flight by a person in an aircraft. Although delighted with the results he had attained, he realized that control of flight could not be mastered until a lightweight engine was developed to give the thrust and lift required.

The developments from Cayley's experiments are plain for everyone to see in the modern world, with the use of the aeroplane as a common means of transport. The first successful sustained flight was made by du Temple's clockwork model in 1857 (the year Cayley died) and the first actual crew-carrying powered flight was in 1874, but the plane did take off down a slope. It was another 16 years before a piloted plane managed a level-ground take-off, and this was Ader's *Eole*; it hopped about 50 m/160 ft. True success came with the Wright brothers and the key to their success was, as Cayley had predicted, a lightweight engine.

Celsius Anders 1701–1744 was a Swedish astronomer, mathematician and physicist, now mostly remembered for the Celsius scale of temperature.

Celsius was born on 27 November 1701 in Uppsala,

where his father was Professor of Astronomy. In 1723 he became secretary of the Uppsala Scientific Society; by the age of 30 he was himself Professor of Astronomy there. It was at this time that he began to travel extensively in Europe, visiting astronomers and observatories in particular.

On his travels he observed the Aurora Borealis; he published some of the first scientific documents on the phenomenon in 1733. While in Paris he visited Pierre-Louis Maupertuis (1698–1759), who invited him to join an expedition which centred on Torneå in Lapland (now on the Finnish–Swedish border). It confirmed the theory propounded by Newton that the Earth is flattened at the poles. With knowledge and expertise gained in this way from the leading astronomers and scientists throughout Europe, Celsius returned to the University of Uppsala, where he built a new observatory – the first installation of its kind in Sweden.

In 1742 Celsius presented a paper to the Swedish Academy of Sciences containing a proposal that all scientific measurements of temperature should be made on a fixed scale based on two invariable (generally speaking) and naturally occurring points. His scale defined 0° as the temperature at which water boils, and 100° as that at which water freezes. This scale, in an inverted form devised eight years later by his pupil, Martin Strömer, has since been used in almost all scientific work. Generally known in most of Europe under the name of Celsius, in Britain the scale has also commonly been known as centigrade.

Celsius left several other important scientifc works, including a paper on accurately determining the shape and size of the Earth, some of the first attempts to gauge the magnitude of the stars in the constellation Aries, and a study of the falling water level of the Baltic Sea.

Cesaro Ernesto 1859–1906 was an Italian mathematician whose interests were wide-ranging, but who is chiefly remembered for his important contributions to intrinsic geometry. His name is perpetuated in his description of 'Cesaro's curves', first defined in 1896.

Cesaro was born on 12 March 1859 in Naples, where he grew up and completed the first part of his education. At the age of 14 he joined his brother in Liège, Belgium, and entered the Ecole des Mines on a scholarship. After matriculation, he continued studying mathematics and published his first mathematical paper. On the death of his father in 1879, Cesaro returned to his family in Torre Annunziata for three years before going back to Liège on another scholarship. In 1883 he published a major mathematical paper, 'Sur diverses questions d'arithmétique', in the *Mémoires de l'Academie de Liège*. After some sort of disagreement with the educational authorities in Liège, however, he entered the University of Rome in 1884. There he wrote prolifically on a wide range of subjects. Two years later he became Professor of Mathematics at the Lycée Terenzio Mamiani, but left after one month to fill the vacant Chair of Higher Algebra at the University of Palermo, where he remained until 1891. Finally, he became Professor of

Mathematical Analysis at Naples, and held this post until his untimely death on 12 September 1906 as a result of injuries he received in attempting to rescue his son from rough seas near Torre Annunziata.

Cesaro's most important contribution to mathematics was his work on intrinsic geometry. He began his study of the subject while in Paris, and continued to develop it for the rest of his life. His earlier work is summed up in his monograph of 1896, the *Lezione di Geometrica Intrinsica*, in which, commencing with Gaston Darboux's method of a mobile coordinate trihedral (formed by the tangent, the principal normal, and the binormal at a variable point of a curve), Cesaro simplified the analytical expression and made it independent of extrinsic coordinate systems. He stressed the intrinsic qualities of the objects. In elaborating this method later, he pointed out further applications. In the *Lezione* Cesaro described the curves which now bear his name. He later included the curves devised by Koch (which are continuous but have no tangent at any point). The *Lezione* also deals with the theory of surfaces and multidimensional spaces in general. Much later on, Cesaro was able to emphasize the independence of his geometry from the axioms of parallels, and also established other foundations on which to base non-Euclidean geometry.

Cesaro's other work, particularly during his time at the University of Rome, covered topics ranging from elementary geometrical principles to the application of mathematical analysis; from the theory of numbers to symbolic algebra; and from the theory of probability to differential geometry. He also made notable interpretations of James Clerk Maxwell's work in theoretical physics.

Chadwick James 1891–1974 was a British physicist who discovered the neutron in 1932. For this achievement he was awarded the 1935 Nobel Prize for Physics.

Chadwick was born at Bollington, Cheshire, on 20 October 1891. He began his scientific career at Manchester University, graduating in physics in 1911. Chadwick then continued at Manchester and, under Ernest Rutherford (1871–1937), investigated the emission of gamma rays from radioactive materials. To gain further research experience, he went in 1913 to Berlin to work with Hans Geiger (1882–1945), the inventor of the geiger counter, where he discovered the continuous nature of the energy spectrum and investigated beta particles emitted by radioactive substances. Chadwick was then interned as an enemy alien on the outbreak of World War I, living and working in a stable for the duration of the war. He still managed to do original research, however, and investigated the ionization present during the oxidation of phosphorus and the photochemical reaction between chlorine and carbon monoxide.

At the end of the war, Rutherford invited Chadwick to Cambridge. During this period, he determined the atomic numbers of certain elements by the way in which alpha particles were scattered. He also established the equivalence of atomic number and atomic charge. With Rutherford, he produced artificial disintegration of some of the lighter elements by alpha-particle bombardment.

His most famous achievement, the discovery of the neutron, came in 1932 after its existence had been suspected by Rutherford as early as 1920. In experiments in which beryllium was bombarded by alpha particles, an unusually energetic gamma radiation appeared to be emitted. It was more penetrating than gamma radiation from radioactive elements. Measurements of the energies involved and the conservation of energy and momentum suggested to Chadwick that a new kind of particle was being produced rather than radiation. The results pointed towards a neutral particle made up of a proton and an electron. Its mass should thus be slightly greater than that of the proton. Because the mass of the beryllium nucleus had not then been measured, Chadwick designed and carried out an experiment in which boron was bombarded with alpha particles. This produced neutrons, and from the mass of the boron nucleus and other elements and the energies involved, Chadwick determined the mass of the neutron to be 1.0067 atomic mass units, slightly greater than that of the proton.

In the same year, Chadwick became Professor of Physics at the University of Liverpool. He ordered the building of a cyclotron and, from 1939 onwards, used it to investigate the nuclear disintegration of the light elements. During World War II, he was closely involved with the atomic bomb, and much of the research and calculation for the British contribution to the Manhattan Project was carried out at Liverpool under his direction. From 1943, he led the British team with the project in the United States.

In 1945 Chadwick was knighted, and in the same year he returned to Liverpool to continue his own research and to develop a research school in nuclear physics. He returned to Cambridge as Master of Gonville and Caius College in 1948, and stayed in this position until his retirement ten years later. He died on 24 July 1974.

The discovery of the neutron made by Chadwick led to a much deeper understanding of the nature of matter, explaining for example why isotopes of elements exist. It also inspired Enrico Fermi (1901–1954) and other physicists to investigate nuclear reactions produced by neutrons, leading to the discovery of nuclear fission.

Chain Ernst Boris 1906–1979 was a German-born British biochemist who, in collaboration with Howard Florey, first isolated and purified penicillin and demonstrated its therapeutic properties. Chain, Florey, and Alexander Fleming shared the 1945 Nobel Prize for Physiology or Medicine, Chain and Florey for their joint work in isolating penicillin and demonstrating its clinical use against infection, and Fleming for his initial discovery of the *Penicillium notatum* mould. Chain also received many other honours for his work, including a knighthood in 1969.

Chain was born on 19 June 1906 in Berlin, the son of a chemist. He was educated at the Luisen-gymnasium then at the Friedrich Wilhelm University in Berlin, from which he graduated in chemistry and physiology in 1930. After graduation he did research in the Chemistry Department of the Pathological Institute at the Charité Hospital in Berlin, but with the rise to power of Adolf Hitler in 1933, Chain

emigrated to Britain. Initially he worked for a short time at University College, London, and then, on the recommendation of J. B. S. Haldane, he worked under Frederick Gowland Hopkins at the Sir William Dunn School of Biochemistry at Cambridge University from 1933 to 1935. In that year Florey invited Chain to work with him at the Sir William Dunn School of Pathology at Oxford University as University Demonstrator and Lecturer in Chemical Pathology. In 1949 Chain was invited to be Guest Professor of Biochemistry at the Istituto Superiore di Sanità in Rome; in the following year he accepted a permanent position as professor there and was also appointed Scientific Director of the International Research Centre for Chemical Microbiology. In 1961 he returned to Britain as Professor of Biochemistry at Imperial College, London, where he did much to ensure that the laboratories were equipped with modern facilities. On his retirement in 1973, Chain was made Emeritus Professor and Senior Research Fellow of Imperial College. He died on 12 August 1979 in Ireland.

At Oxford University, Chain initially investigated the observation first made by Fleming in 1924 that tears, nasal secretion, and egg white destroyed bacteria. Chain showed that these substances contain an enzyme, lysozyme, which digests the outer cell wall of bacteria. In 1937, while preparing this discovery for publication, Chain found another observation of Fleming's, that the mould *Penicillium notatum* inhibits bacterial growth. In the following year, Chain, in collaboration with Florey, started research to try to isolate and identify the antibacterial factor in the mould. Chain first developed a method for determining the relative strength of a penicillin-containing broth by comparing its antibacterial effect (as shown on culture plates) with that of a standard penicillin solution, 1 cu cm/0.061 cu in of which is defined as containing one Oxford unit of penicillin. Then he developed a method of purifying penicillin without destroying its antibacterial effect. He found that the optimum time for extraction of the penicillin is when the mould is one week old; he also found that free penicillin is acidic and is therefore more soluble in certain organic solvents than it is in water. He then agitated the penicillin broth with acidified ether or amyl acetate, reduced the acidity of the solution until it was almost neutral, removed impurities, and evaporated the purified solution at a low temperature to give a stable form of the active substance. Chain and his coworkers found that 1 mg/0.015 grain of the active substance they had obtained contained between 40 and 50 Oxford units of penicillin and that, in a concentration of only one part per million, it was still able to destroy staphylococcus bacteria. Furthermore, they also showed that their purified penicillin was only minimally toxic and that its antibacterial effect was not diminished by the presence of blood or pus. With E. P. Abraham, Chain then elucidated the chemical structure of crystalline penicillin, finding that there are four different types, each differing in their relative elemental constituents.

Chain also studied snake venoms and found that the neurotoxic effect of these venoms is caused by their destroying an essential intracellular respiratory coenzyme.

Challis James 1803–1882 was a British astronomer renowned in his time for unconventional views concerning the fundamental laws of the universe, but now remembered more for an almost unbelievable lapse in scientific professionalism.

Challis was born in Braintree, Essex, on 12 December 1803. He attended a local school where he showed such promise that he won a place at a London school and later at Trinity College, Cambridge, which he entered in 1821. He graduated with top honours in 1825, and was a Fellow of the College from 1826 to 1831 (and later from 1870 to 1882). Challis was ordained in 1830 and served as Rector at Papworth Everard, Cambridgeshire, from 1830 until 1852. He also succeeded George Airy (1801–1892) to the Plumian Professorship of Astronomy at Cambridge University in 1836 – a post he held until his death – and he served as Director to the Cambridge Observatory from 1836 until 1861. A member of the Royal Astronomical Society and a Fellow of the Royal Society, the author of several scientific publications, he died in Cambridge on 3 December 1882.

In 1844, John Couch Adams (1819–1892) – a young and enthusiastic astronomer and mathematician, a recent graduate of Cambridge University – approached Challis to enlist his aid in obtaining data from Airy at the Greenwich Observatory regarding the known deviations in the orbit of the planet Uranus. These were suspected of indicating the gravitational influence of a planet even farther out. With Challis's mediation, Adams received from Airy all the data the observatory possessed on Uranus for the period 1754 to 1830.

In September 1845 Adams supplied Challis and Airy with an estimated orbital path for the unknown planet and a prediction for its likely position on 1 October 1845. But Challis did not take the calculations seriously (saying later that he could not believe so youthful and inexperienced an astronomer as Adams would arrive at anything like a correct prediction), and Airy, through a series of mishaps, did not even see them until the following year.

By that time, in France, Urbain Leverrier (1811–1877) had performed calculations similar to those of Adams, and he was more successful than Adams in obtaining the co-operation of senior astronomers. Almost immediately after he sent his predictions to Berlin Observatory in September 1846, the new planet was discovered – by Johann Galle (1812–1910) and Heinrich d'Arrest (1822–1875) – later to be called Neptune.

All Challis could do then was lamely to report that if he had indeed conducted a search at Adams's predicted position for 1 October 1845 he would have been within 2° of the planet's actual position and would almost certainly have spotted it.

Chamberlin Thomas Chrowder 1843–1928 was a boldly speculative US geophysicist. A farmer's son born in Illinois and brought up in Wisconsin, Chamberlin always claimed that his native terrain had shaped his geological thinking. Partly self-taught in science, Chamberlin joined the Wisconsin Geological Survey in 1873, and rose to become its chief geologist, publishing the *Geology of Wisconsin* 1877–83. He went on to work for the US Geological Survey before becoming Professor of Geology at Chicago in 1892, working there until his retirement in 1918.

Chamberlin's most important contribution to geological thinking lay in his bold attack on Lord Kelvin. Kelvin had postulated that the Earth was rather young (less than 100 million years), basing his views on the assumption, derived from the nebular hypothesis, that the Earth had steadily cooled from a molten mass. Chamberlin rebuked Kelvin for his dogmatic confidence in extrapolations from a single hypothesis, and stressed that geological reasoning must follow from a plurality of working hypotheses. He also believed geological evidence in any case suggested the Earth to be older than Kelvin had estimated. Chamberlin backed his refutation of the nebular hypothesis by developing (with the aid of the celestial physicist, F. R. Moulton), the planetesimal hypothesis. This postulated a gradual origin, by accretion of particles, for the Earth and other planetary bodies – an origin for these bodies that was therefore cool and solid.

Chandrasekhar Subrahmanyan 1910– is an Indian-born US astrophysicist who is particularly concerned with the structure and evolution of stars. He is well-known for his studies of white dwarfs and the radiation of stellar energy.

Chandrasekhar was born on 19 October 1910 in Lahore, India (now in Pakistan). He grew up in India and went to Presidency College, University of Madras, from which he graduated with a BA in 1930. He continued his studies at Trinity College, Cambridge, gaining a PhD in 1933. There he studied under the physicist Paul Dirac. He left Trinity College in 1936 to take up a position on the staff of the University of Chicago, working in the Yerkes Laboratory. In 1938 he became Assistant Professor of Astrophysics there and in 1952 was promoted to Distinguished Service Professor. The following year he became a United States citizen.

Chandrasekhar's greatest contribution to astronomy was his explanation of the evolution of white dwarf stars, as laid out in his *Introduction to the Study of Stellar Structure* 1939. These stellar objects, which were first discovered in 1915 by Walter Sydney Adams (1876–1956), are similar in size to the Earth. They have a very high density and are therefore very much more massive than the Earth. This enormous density is explained in terms of degeneracy – a consequence of the Pauli exclusion principle in which electrons become so tightly packed that their normal behaviour is suppressed; as stars evolve, they 'burn' their hydrogen which is converted to helium and, eventually, heavier elements. During his work at Cambridge, Chandrasekhar suggested that when a star had burned nearly all its hydrogen, it would not be able to produce the pressure against its own gravitational field to sustain its size and would then contract. As its density increased during the contraction the star would build up sufficient internal energy to collapse its atomic structure into the degenerate state.

Not all stars, however, become white dwarfs. Chandrasekhar believed that – up to a certain point – the greater the mass of a star, the smaller would be the radius of the eventual white dwarf. But he also stated that beyond this point a large stellar mass would not be able to equalize the pressure involved and would explode. He calculated that stellar masses below 1.44 times that of the Sun would form stable white dwarfs, but those above this limit would not evolve into white dwarfs. This limit – known as the Chandrasekhar limit – was based on calculations involving the complete degeneracy of the stellar matter; the limit is now believed to be about 1.2 solar masses.

Stars with masses above the Chandrasekhar limit are likely to become supernovae and rid themselves of their excess matter in a spectacular explosion. The remaining mass may form a white dwarf if the conditions of mass and pressure are suitable, but it is more likely to form a neutron star. Neutron stars were first identified by J. Robert Oppenheimer and his co-workers in 1938. These stars are even more dense than white dwarfs, with an average radius of approximately 15 km/9 mi.

With the Polish astrophysicist Erich Schönberg, Chandrasekhar determined the Chandrasekhar–Schönberg limit of the mass of a star's helium core; if it is more than 10–15% of that of the entire star, the core rapidly contracts, often collapsing. Chandrasekhar has also investigated the transfer of energy in stellar atmospheres by radiation and convection and the polarization of light emitted from particular stars.

Charcot Jean-Martin 1825–1893 was a French neurologist whose studies of hysteria still excite controversy.

Born in Paris, the son of a wheelwright, Charcot studied at the Paris Faculty of Medicine, graduating MD in 1853 with a doctoral thesis on chronic rheumatism and gout. In 1862 he became resident doctor at the Salpétrière, where he built up a leading neurological department. In 1872 he was appointed Professor of Anatomical Pathology at the Faculty of Medicine, ten years later moving to the chair for the study of nervous disorders at the Salpétrière, where the distinguished Joseph Babinski served as his director.

Charcot was an ardent champion of the clinical anatomical method that systematically correlated the symptoms presented by the patient with the lesions discovered at autopsy. He was also committed to the view that all diseases (even apparently strange psychiatric conditions) were regular natural phenomena, whose laws could be discovered by medical science. Widespread observation of multiple cases (simple at a huge institution like the Salpétrière), would thus crack the secrets of diseases. Over the course of a generation, Charcot published a series of memoirs that made him one of the world's most eminent neurologists. As well as portraying the neuropathy that became known as Charcot's disease, he produced classic descriptions of multiple or disseminated sclerosis; of amyotrophic lateral sclerosis; of cerebral haemorrhage; and of tabes dorsalis, a form of neurosyphilis. He studied Parkinson's disease and contributed to the investigation of

poliomyelitis. His *Leçons sur les maladies du système nerveux faits à la Salpétrière* 1872–73 laid his teachings on such subjects before a larger audience.

In his approach to brain function, Charcot vigorously supported the theory of cerebral localization, as developed by Hughlings Jackson. He applied this theory to cases of Jacksonian epilepsy, aphasia and Beard's neurasthenia. During the 1870s, he developed highly publicized work on hysteria. Far from being a psychogenic disorder or just a disease of women, Charcot regarded it as a general malady of neurological origin. Such views proved influential upon his pupil, Sigmund Freud, not least because Charcot was also fascinated by the relations between hysteria and hypnotic phenomena. Critics widely accused Charcot of inadvertently 'training' the young women who were his main hysterical subjects. One of the founders of modern neurology, Charcot thus left the relations between neurology and psychiatry extremely obscure.

Chardonnet Louis-Marie-Hilaire Bernigaud, Comte de 1839–1924 was a French industrial chemist who invented rayon, the first type of artificial silk. He also worked on nitrocellulose (gun cotton).

Chardonnet was born into an aristocratic family on 1 May 1839 at Besançon, Doubs. He trained first as a civil engineer at the Ecole Polytechnique, Paris, and then went to work under Louis Pasteur, who was studying diseases in silkworms. This inspired Chardonnet to seek an artificial replacement for silk which he first patented in 1884. Five years later, at the Paris Exposition, he was awarded the Grand Prix for his invention. He opened his first factory, the Société de la Soie de Chardonnet, at Besançon in 1889 and in 1904 he built a second factory at Sarvar in Hungary. He died in Paris on 12 March 1924 at the age of 85.

Chardonnet began his experiments in 1878 but it was six years before he produced a satisfactory fibre. He prepared nitrocellulose (mainly cellulose tetranitrate) by treating a pulp made from mulberry leaves – the food plant of silkworms – with mixed nitric and sulphuric acids. The cellulose compound was dissolved in a mixture of ether and alcohol and the hot viscous solution forced through fine capillary tubes into cold water. The warm threads were stretched and dried in heated air.

The original nitrocellulose fibre was highly inflammable, and Chardonnet continued working to produce a fireproof version. By 1889 he had developed rayon, so-called because the brightness of the material was thought to resemble the emission of the Sun's rays. He later was able to make 35–40 denier threads (denier is the mass in grams of 9,000 m of yarn; 9,000 m/29,500 ft of 40 denier nylon has a mass of 40 g/1.4 oz) of tensile strength equivalent to that of natural silk.

Rayon was the first artificial fibre to come into common use. It was, admittedly, only modified cellulose but it pointed the way to the totally synthetic fibres developed about 50 years later by Wallace Carothers and others. Today the term rayon is generally used for all types of fibres made from cellulose, although is most often applied to viscose yarns. The cellulose is usually derived from cotton or wood pulp.

As well as his development of artificial fibres, Chardonnet also spent some time working for the French government on the production of gun cotton, the original smokeless powder for cartridges and shells which exploits the material's high inflammability – the very feature that Chardonnet had to eliminate from his nitrocellulose fibre. He also made minor contributions to studies of the absorption of ultraviolet light, telephony, and the behaviour of the eyes of birds.

Charles Jacques Alexandre César 1746–1823 was a French physicist and mathematician who is remembered for his work on the expansion of gases and his pioneering contribution to early ballooning.

Charles was born in Beaugency, Loiret, on 12 November 1746. He became interested in science while working as a clerk in the Ministry of Finance in Paris. Stimulated by Benjamin Franklin's experiments with lightning and electricity, he constructed a range of apparatus which he demonstrated at popular public lectures. He also experimented with gases. He was elected to the French Academy of Sciences in 1795 and later became Professor of Physics at the Paris Conservatoire des Arts et Métiers. He died in Paris on 7 April 1823.

The Montgolfier brothers made their first experiments with uncrewed hot-air balloons at Viadalon-les-Annonay in June 1783. On hearing about them, Charles tried filling a balloon with hydrogen, and with the brothers Nicolas and Anne-Jean Robert made the first successful (uncrewed) experiment in August 1783. In November of that year the Montgolfiers demonstrated their hot-air balloons in Paris, and on 1 December Charles and Nicolas Robert made the first human ascent in a hydrogen balloon. In later flights Charles ascended to an altitude of 3,000 m/9,846 ft. On a tide of public acclaim he was invited by King Louis XVI to move his laboratory to the Louvre – patronage that Charles was to regret ten years later during the French Revolution.

In about 1787 Charles experimented with hydrogen, oxygen and nitrogen and demonstrated the constant expansion of these gases, that is at constant pressure the volume of a gas is inversely proportional to its temperature. He found that a gas expands by 1/273 of its volume at 0°C for each centigrade (Celsius) degree rise in temperature (implying that at −273°C/−459.4°F, now known as absolute zero, a gas has no volume). He did not publish his results, but communicated them to the French physical chemist Joseph Gay-Lussac, who repeated the experiments and made more accurate measurements. Unknown to the two French scientists, John Dalton in England was also about to embark on similar research. Dalton deduced the same gas law in 1802, but the first to publish (six months later) was Gay-Lussac. For this reason, the law became known in France as Gay-Lussac's law but elsewhere it was, and still is, generally known as Charles's law. Incidentally, Gay-Lussac continued to emulate Charles by becoming a pioneer balloonist.

Charles devised or improved many scientific instruments. He invented a hydrometer and a reflecting goniometer, and improved the aerostat of Gabriel Fahrenheit (1686–1736) and the heliostat of W. J. van s'Gravesande (1688–1742).

Charnley John 1911–1982 was a British orthopaedic surgeon who appreciated the importance of applying engineering principles to the practice of orthopaedics. He is best known for his work on degenerative hip disease and his new technique, the total hip replacement, or arthroplasty. He also successfully pioneered arthrodeses for the knee and hip. He was knighted in 1977 and awarded the Gold Medal of the British Medical Association in 1978.

Charnley was born on 29 August 1911 in Bury, Lancashire. He went to the local grammar school and then to Manchester University. His academic achievements in medicine were impressive – he was the only student to pass primary FRCS before graduating MB (in 1935), and he obtained his FRCS in 1936, only a year after qualifying. At the outbreak of World War II Charnley became a major in the Royal Army Medical Corps and spent some time in the Middle East. At Heliopolis he ran the army splint factory, turning out the Thomas splint for treating leg fractures among the soldiers. When the war ended he went back to Manchester as a lecturer, and in 1947 he became Consultant Orthopaedic Surgeon at Manchester Royal Infirmary. He married in 1957 and had two children. In the mid-1960s Charnley retired from the Infirmary in order to devote his time to hip arthroplasty at the Centre of Hip Surgery at Wrightington Hospital, Lancashire, where he became the Director. He built the centre up to become the primary unit for hip replacement in the world, and surgeons from many countries visited Wrightington to observe the latest techniques. The Royal Society acknowledged his contributions to surgery in 1975 when he was made a Fellow. He died suddenly on 5 August 1982.

The replacement of the femoral head and acetabulum (socket) in the hip had been researched and tried by McKee and others to treat the painful condition of degenerative hip disease, but Charnley realized that the fundamental problem was one of lubrication of the artificial joint. He carried out research on the joints of animals and tried using the low-friction substance polytetrafluoroethylene (PTFE or Teflon), with great success at first. Teflon was eventually abandoned, but Charnley had learnt much – including the use of methyl methacrylate cement for holding the metal prosthesis or implant to the shaft of the femur. In 1962 the right high-density polythene was developed, and his results became increasingly successful.

For the treatment of rheumatoid arthritis Charnley devised a system for surgically fusing joint surfaces (arthrodesis) to immobilize the knee joint using an external compression device, which bears his name. A metal pin is passed through the lower femur and another through the upper tibia and these are clamped together externally to hold the bared joint surfaces together until the joint fuses, leaving it immobile but pain-free.

Throughout his career Charnley developed a series of highly practical and successful surgical instruments. In his fight against postoperative infection he used air 'tents' which allowed the surgeon and the wound to be kept in a sterile atmosphere throughout the operation.

Chase Mary Agnes Meara 1869–1963 was a US botanist and suffragist who made outstanding contributions to the study of grasses, despite a lack of higher education and any formal qualifications. During the course of several research expeditions she collected many plants previously unknown to science, and her work provided much important information about naturally occuring cereals and other food crops. This knowledge could then be used by nutrition and agricultural scientists in developing disease-resistant and nutritionally enhanced strains.

Born Mary Agnes Meara on 20 April 1869 in Iroquois County, Illinois, she was the fifth of six children. After her father's death two years later the family moved to Chicago, where she attended public school and worked at various jobs to help with household expenses. Aged 18 she married a newspaper editor named William Chase but was widowed the following year. She returned to employment as a proof reader and, through encouraging a nephew's botanical pursuits, developed an interest in the flora of her local area. In this she was assisted by the Reverend Ellsworth Hill, also a botanist, who guided her collecting and recording and employed her to draw specimens from his own collections. He also helped her to apply for more suitable positions, first as a meat inspector in the Chicago stockyards and in 1903 in Washington working for the United States Department of Agriculture Bureau of Plant Industry and Exploration. In that position she worked with Albert Spear Hitchcock, the principal scientist in the division of agrostology (study of grasses), illustrating Bureau publications and becoming first an assistant botanist and then a botanist in systematic agrostology. In 1936 she succeeded Hitchcock, with whom she had collaborated closely, and became the principal scientist for agrostology. She died on 24 September 1963 in Bethesda, Maryland, having been officially retired since 1939.

Chase was particularly responsible for work in modernizing and extending the national collection of the grass herbarium that had been part of the United States National Herbarium, although it was incorporated into the Smithsonian Institution in 1912. She travelled widely, collecting plants from several regions of North and South America,

and also visiting European research institutes and herbaria during the 1920s. Several of her expeditions were self-financed, and it has been estimated that by the conclusion of her final collecting trip in 1940 she had collected more than 12,000 plants for the Herbarium. She was also consulted by foreign officials and scientists for assistance in identifying grasses, and used these opportunities to acquire duplicates of 'type specimens' (from which the first descriptions of a new plant are made) from foreign collections. She donated her own extensive library to the Smithsonian Institution, wrote important monographs on grasses in the western hemisphere, and was responsible for the authoritative *Manual of the Grasses of the United States* that was published in 1950. She also wrote popular accounts of her work, including the *First Book of Grasses*, which appeared in 1922. She was politically active in various reform movements, especially those for female suffrage, and on this account was jailed and forcibly fed during World War I.

Chevreul Michel Eugène 1786–1889 was a French organic chemist who in a long lifetime devoted to scientific research studied a wide range of natural substances, including fats, sugars and dyes.

Chevreul was born in Angers on 31 August 1786, the son of a surgeon. He went to Paris in 1803 when he was 17 years old to study chemistry at the Collège de France under Louis Vauquelin. He became an assistant to Antoine Fourcroy (1783–1791) in 1809 and a year later took up an appointment as an assistant at the Musée d'Histoire Naturelle. He was Professor of Physics at the Lycée Charlemagne from 1813 until 1830, when he returned to the Musée as Professor of Chemistry, succeeding his old tutor Vauquelin. In 1824 he was made a director of the dyeworks associated with the Gobelins Tapestry Factory, and in 1864 he became Director for life of the Musée d'Histoire Naturelle. He died, aged 103, in Paris on 8 April 1889.

Chevreul's earliest research under Vauquelin was on indigo, a subject he was to return to later. He began his studies of fats in 1809 by first decomposing soaps (which at that time were made exclusively by the action of alkali on animal fats). By treating soaps with hydrochloric acid he obtained and identified various fatty acids, including stearic, palmitic, oleic, caproic and valeric acids. He thus realized that the soap-making process is the treatment of a glyceryl ester of fatty acids (i.e. a fat) with an alkali to form fatty acid salts (i.e. soap) and glycerol.

Soap-making (saponification):

glyceryl stearate	+	sodium hydroxide	→	sodium stearate	+	glycerol
(fat)		(alkali)		(soap)		

Chevreul's acid hydrolysis:

sodium stearate	+	hydrochloric acid	→	stearic acid	+	sodium chloride
(soap)		(acid)		(fatty acid)		(salt)

One of the most useful of the newly discovered acids was stearic acid, and in 1825 Chevreul and Joseph Gay-Lussac patented a process for making candles from stearin (crude stearic acid), providing a cleaner and less odorous alternative to tallow candles. Chevreul determined the purity of fatty acids by measuring their melting points, and constancy of melting point soon became a criterion of purity throughout preparative and analytical organic chemistry. He also investigated natural waxy substances, such as spermaceti, lanolin and cholesterol (which did not yield fatty acids on treatment with hydrochloric acid).

During the many years he was working with fats Chevreul also studied other natural compounds. In 1815 he isolated grape sugar (glucose) from the urine of a patient suffering from diabetes mellitus. At the Gobelin dyeworks he discovered haematoxylin in the reddish-brown dye logwood and quercitrin in yellow oak; he also prepared the colourless reduced form of indigo. His interest in the creation of the illusion of continuous colour gradation by using massed small monochromatic dots (as in an embroidery or tapestry) later influenced the Pointillistes and Impressionist painters.

Cheyne and Stokes John Cheyne 1777–1836 and William Stokes, 1804–1878 were two physicians who practised in Dublin and gave their name to Cheyne–Stokes breathing, or periodic respiration.

John Cheyne was born in Leith on 3 February 1777. He was educated at Edinburgh High School, and was formally apprenticed to his physician father at the age of 13. He qualified in 1795 and joined the British Army as a surgeon. He returned four years later to take charge of an ordnance hospital at Leith and he began to take medicine seriously as a result of working for Charles Bell. He visited Dublin in 1809, and decided to settle there. In 1811 he became physician at Meath Hospital, where his practice flourished. He took the first professorial chair in medicine at the Royal College of Surgeons of Ireland in 1813, and was succeeded by Whitley Stokes, the father of William Stokes. Cheyne died in Newport Pagnell, Buckinghamshire, on 31 January 1836.

William Stokes studied clinical medicine at the Meath Hospital, then became a student in Edinburgh where he graduated in 1825. He went back to Dublin as physician to the Dublin General Dispensary, and later succeeded his father at the Meath Hospital. He died on 10 January 1878.

In 1818 Cheyne described the sign of periodic respiration which occurs in patients with intracranial disease or cardiac disease. His paper described the breathing that would cease entirely for a quarter of a minute or more, then would become perceptible and increase by degrees to quick, heaving breaths that gradually subside again. Stokes referred to Cheyne's paper in his famous book *The Diseases of the Heart and Aorta*, and thus their names became eponymous with the sign.

Stokes's name was also applied to Stokes–Adams attacks after his paper *Observations on Some Cases of Permanently Slow Pulse*, which was published in 1846.

Child Charles Manning 1869–1954 was a US zoologist who tried to elucidate one of the central problems of biology – that of organization within living organisms.

Child was born on 2 February 1869 in Ypsilanti, Michigan, where his grandfather was a physician, then three weeks later was taken home to Higganum, a small village in Connecticut where his father was a farmer. The last-born and only surviving of five sons, Child was taught by his mother until he was nine years old, when he went to Higganum District School. He then attended high school in Middleton, Connecticut, from 1882 to 1886, after which he studied zoology at the Wesleyan University in Middletown, from which he graduated in 1890 and obtained his MSc in 1892. While studying at university he continued to live with his parents so that he could run the farm for his father, who had previously been incapacitated by a cerebral haemorrhage. His parents died in 1892, and two years later Child went to Leipzig University to research for his doctorate, which he gained in the same year.

After returning to the United States, he went to the newly established University of Chicago in 1895 and remained there for almost all of his academic career – as Assistant (1895–96), Associate (1896–98), Instructor (1898–1905), Assistant Professor (1909–16) and Professor (1916–34). After his retirement in 1934, Child was appointed Professor Emeritus. The only interruptions to his association with the University of Chicago were two sabbaticals – to Duke University as Visiting Professor in 1930, and to Tohoku University in Japan as Visiting Professor of the Rockefeller Foundation from 1930 to 1931. On retiring, Child moved to Palo Alto in California and became a guest at Stanford University. He remained there until his death on 19 December 1954.

Child's early work concerned the functioning of the nervous system in various invertebrates, but his interest soon turned to embryology, in which field he did some important research into cell lineage – tracing the fate of each cell in the early embryo. In 1900, however, he began a long series of experiments on regeneration in coelenterates and flatworms, a topic that occupied him for most of his career. Child believed that the regeneration of a piece of an organism into a normal whole resulted from the piece functioning like the missing parts. In 1910 he perceived that there is a gradation in the rate of physiological processes along the longitudinal axis of organisms, and in the following year he developed his gradient theory. According to this theory, each part of an organism dominates the region behind and is dominated by that in front. In general, the region of the highest rate of activity in eggs, embryos and other reproductive regions becomes the apical end of the head of the larval form; in plants it becomes the growing tip of the shoot or of the primary root. Child also pointed out that regeneration is fundamentally the same as embryonic development, in that the dominant apical region is formed first then the other parts of the organism develop in relation to it. In 1915 Child demonstrated that the parts of an organism that have the highest metabolic rates are most susceptible to poisonous substances, but that these parts also have the greatest powers of recovery after damage.

Child's explanation of how the various cells and tissues in organisms are organized – by a gradation in the rate of physiological processes leading to relationships of dominance and subordination – may not be thought to be correct, but it was an important early contribution to the problem of functional organization within living organisms.

Chladni Ernst Florens Friedrich 1756–1827 was a German physicist who studied sound and invented musical instruments, helping to establish the science of acoustics.

Chladni was born in Wittenberg, Saxony, on 30 November 1756. His father insisted that he study law at the University of Leipzig, from which he graduated in 1782. After his father's death in about 1785 Chladni changed to the study of science, concentrating on experiments in acoustics. He died in Breslau, Silesia (now Wroclaw, Poland) on 3 April 1827.

Chladni's interest in sound stemmed from his love of music. In 1786 he began studying sound waves and worked out mathematical formulae that describe their transmission. His best-known experiment made use of thin metal or glass plates covered with fine sand. When a plate was made to vibrate and produce sound (for example by striking it or stroking the edge of the plate with a violin blow), the sand collected along the nodal lines of vibration, creating patterns called Chladni's figures. In 1809 he demonstrated the technique to a group of scientists in Paris.

He also measured the velocity of sound in various gases by measuring the change in pitch of an organ pipe filled with a gas other than air (the pitch, or sound frequency, varies depending on the molecular composition of the gas). He invented various musical instruments, including ones he called the clavicylinder and the euphonium. The latter consisted of rods of glass and metal that were made to vibrate by being rubbed with a moistened finger; he demonstrated it at lectures throughout Europe.

In 1794 Chladni published a book about meteorites (which he collected) and postulated that they come from beyond the Earth as debris of an exploded planet. Nobody accepted his theory until 1803, when the French physicist Jean Biot (1774–1862) confirmed that meteorites do, in fact, fall from the sky.

Christoffel Elwin Bruno 1829–1900 was a German mathematician who made a fundamental contribution to the differential geometry of surfaces, carried out some of the first investigations that later resulted in the theory of shock waves, and introduced what are now known as the Christoffel symbols into the theory of invariants.

Christoffel was born on 10 November 1829 in Montjoie (now Monschau), near Aachen. He studied at the University of Berlin, where he received his doctorate at the age of 27. Three years later he became a lecturer at the University before, in 1862, becoming a professor at the Polytechnicum in Zurich. After seven years there, he returned to Berlin to take the Chair of Mathematics at the Gewerbsakademie. In 1872 he became Professor of Mathematics at the newly founded University of Strasbourg, where he remained until his retirement in 1892. He died on 15 March 1900.

Christoffel's best-known paper investigated the theory of invariants. Called 'Über die Transformation der homogen Differentialausdrücke zweiten Grades' and published in 1869, the paper introduced the symbols that later became known as Christoffel symbols of the first and second order. The series of other symbols of more than three indices, including the four-index symbols already introduced by Bernhard Riemann, are now known as the Riemann–Christoffel symbols. (The symbols of an order higher than four are obtained from those of a lower order by a process called covariant differentiation.)

Christoffel is additionally remembered as the formulator of the theorem that also bears his name, and concerns the reduction of a quadrilateral form; this was later incorporated by Gregorio Ricci-Curbastro (1853–1925) and Tullio Levi-Civita (1873–1941) in their tensor calculus.

Christoffel's contribution to the differential geometry of surfaces is contained in his *Allgemeine Theorie der geodätischen Dreiecke* (1868), in which he presented a trigonometry of triangles formed by geodesics on an arbitrary surface. He used the concept of reduced length of a geodesic arc, stating that when the linear element of the surface can be represented by $ds^2 = dr^2 + m^2 dx^2$, m is the reduced length of the arc r.

Inspired by Riemann – who was of a similar age – Christoffel's papers in 1867 and 1870 described the conformal tracing of a simply connected area bounded by polygons on the area of a circle. In 1880 he showed algebraically that the number of linearly independent integrals of the first order on a Riemann surface is equal to the genus p. Later, in *Vollständige Theorie der Riemannschen θ-Function* (published posthumously), Christoffel gave an independent interpretation of Riemann's work on the subject.

In 1877, Christoffel published a paper on the propagation of plane waves in media with a surface discontinuity, and thus made an early contribution to shock wave theory.

Church Alonso 1903– is a US mathematician who in 1936 published the first precise definition of a calculable function, and so contributed enormously to the systematic development of the theory of algorithms.

Church was born on 4 June 1903. Completing his education at Princeton University, and obtaining his PhD there in 1927, he joined the university staff and remained at Princeton for 40 years, finally occupying the Chair of Mathematics and Philosophy. In 1967 he obtained a similar post at the University of California in Los Angeles. The author of many books on mathematical subjects, Church is a member of a number of academies and learned societies.

The concept of the algorithm, in the development of which Church played such a part, did not properly appear until the twentieth century. Then, as the subject of independent study, the algorithm became one of the basic concepts in mathematics. The term denotes an exact procedure specifying a process of calculation that begins with an arbitrary initial datum and is directed towards a result that is fully determined by the initial datum. The algorithm process is one of sequential transformation of constructive entities: it proceeds in discrete steps, each of which consists of the replacement of a given constructive entity with another. (Familiar examples of algorithms are the rules for addition, subtraction, multiplication, and division in elementary mathematics.)

Luitzen Brouwer (1887–1966) and Hermann Weyl (1885–1955) did some tentative studies in the 1920s, and Alan Turing (1912–1954) later offered the first application of the algorithm concept in terms of a hypothetically perfect calculating machine. The solving of algorithmic problems involves the construction of an algorithm capable of solving a given set with respect to some other set, and if such an algorithm cannot be constructed, it signifies that the problem is unsolvable. Theorems establishing the unsolvability of such problems are among the most important in the theory of algorithms, and Church's theorem was the first of this kind. From Turing's thesis, Church proved that there were no algorithms for a class of quite elementary arithmetical questions. He also established the unsolvability of the solution problem for the set of all true propositions of the logic of prediction.

Since Church's pioneering work, much further progress has been made: Alfred Tarski (1902–   ), for example, has obtained some important results. Today, the theory of algorithms is closely associated with cybernetics, and the concept is fundamental to programmed instruction in electronic computers.

Cierva Juan de la 1895–1936 was a Spanish aeronautical engineer who invented the rotating-wing aircraft known as the autogyro.

Cierva was born at Murcia in southeastern Spain on 21 September 1895, the son of the Conservative politician, Juan de la Cierva y Penafiel (1864–1938). He was educated in Madrid at the engineering school called the Escuela Especial de Caminos, Canales y Puertos. During his six years there he also studied theoretical aerodynamics on his own, especially the work of Frederick Lanchester (1868–1946). Soon after leaving school he followed his father into politics and was elected to the Cortes, the Spanish parliament, in 1919 and 1922. He showed little enthusiasm for politics, however, for his real interest was the designing of flying machines.

In 1919, he entered a competition to design a military aircraft for the Spanish government. His plan was for a three-engined biplane bomber with an aerofoil section of his own design. When it was tested in May 1919 engine failure caused it to stall in mid-air, and the plane crashed. This accident led Cierva to turn away from fixed-wing flying machines and to search for a machine with a rotating-wing mechanism that would be less vulnerable to engine failure. His first three designs, which all had blades fixed to the motor shaft, were unsuccessful. But his fourth design

introduced freely rotating wings. On 19 January 1923 the new gyroplane, to which Cierva gave the name Autogiro, was tested at Getafe, Spain, and it flew for 182 m/600 ft.

The autogyro consisted of one nose-mounted engine driving a conventional propeller, a fuselage, and a large, freely rotating rotor mounted horizontally above the fuselage. Allowing the blades of the motor to pivot on hinges, instead of being rigidly fixed to the shaft, largely solved the problem of uneven lift being generated by the advancing and retreating blades. In order to gain sufficient lift for the aurogyro to take off, rope was wound many times around the rotor shaft and then pulled by a gang of men to turn the rotor quickly.

After the initial success of 19 January three more test flights were quickly arranged and just two days later, on 21 January, the autogyro completed a 4 km/2.5 mi circuit in three-and-a-half minutes. The usefulness of the new machine to the police and maritime rescue services was immediately recognized and, after minor adjustments were made to eliminate teething troubles, full-scale production began in 1925 with the founding of the Cierva Autogyro Company in England. Cierva became technical director and on 18 September 1928 he flew one of the company's aircraft across the English Channel. He then flew one all the way to Spain; and in 1929 he demonstrated his new invention at the National Air Races held at Cleveland, Ohio. Cierva continued to exeriment and to test his own aircraft until he was killed in a crash at Croydon Aerodrome, just south of London, on 9 December 1936.

There is an essential difference between Cierva's machine and the modern helicopter. A helicopter has a powered rotor that rotates horizontally overhead, enabling the machine to rise vertically. As a result, it needs only a small space in which to take off and land. During flight, motion backwards and forwards is achieved by altering the inclination of the rotor blades. To stop the plane from rotating with the rotor, a small secondary rotor is is fitted to the tail; a helicopter can also hover. Cierva's autogyro, on the other hand, was designed in some respects like an ordinary aeroplane, with a propeller to pull it through the air. But instead of fixed wings on each side, it had a revolving rotor – a rotating wing – overhead to provide lift at slow forward speeds and allow it almost vertical descent, although it is not capable of vertical ascent nor the 360° manoeuvrability of the helicopter. It was this difference that constituted the basis of his invention and which made the autogyro the prototype, and provided some of the performance features of the modern helicopter.

Clark Wilfrid Edward Le Gros 1895–1971 was a British anatomist and surgeon who carried out important research that made a major contribution to the understanding of the structural anatomy of the brain.

Clark was born on 5 June 1895 in Hemel Hempstead, Hertfordshire. He went to Blundell's School in Tiverton and entered St Thomas's Hospital, London, in 1912 on an entrance scholarship. He qualified in 1917 with a Conjoint Diploma, and joined the Royal Army Medical Corps without working his house appointments. He served in France until the end of World War I, after which he went back to St Thomas's as house surgeon to Sir Cuthbert Wallace. Clark became a Fellow of the Royal College of Surgeons in 1919. He took the post of Principal Medical Officer in Sarawak, Borneo, to gain experience in practical surgery, and began research into the evolution of primitive primates. After successfully treating several local people for yaws, he became highly venerated and was tattoed on the shoulders with the insignia of the Sea Dyaks as a mark of their esteem. Clark returned to England in 1923 as an anatomy demonstrator at St Thomas's until he moved to St Bartholomew's Hospital in 1924 as a reader, then professor, of anatomy. He returned to St Thomas's as Professor of Anatomy in 1930, accepting the professorship of anatomy at Oxford in 1934, which he held until he retired in 1962. His work on primate evolution resulted in election to the Royal Society in 1935. During World War II his research was connected with the war effort despite his pacifist principles. After the war he created a new Department of Anatomy at Oxford which was finally opened in 1959. He was knighted in 1955, and was Arris and Gale Lecturer (1932), Hunterian Professor (1934 and 1945), and editor of the *Journal of Anatomy*. He died suddenly in Burton Bradstock, Dorset, on 28 June 1971 on a visit to a friend from his student days.

Clark had a profound influence on the teaching of anatomy. He moved away from the popular topographical approach which encouraged students to learn repetitiously, and towards the importance of relating structure to function. His anatomy research was directed mainly towards the brain, and the relationship of the thalamus to the cerebral cortex. He also carried out further studies of the hypothalamus. His work on the sensory (largely visual) projections of the brain remains the basis of contemporary knowledge of this aspect of neuroanatomy.

His chief publications include: *Morphological Aspects of the Hypothalamus* (1938); *The Tissues of the Body* (1939); *History of the Primates* (1949); *Fossil Evidence of Human Evolution* (1955); and his autobiography *Chant of Pleasant Exploration* (1968).

Clausius Rudolf Julius Emmanuel 1822–1888 was a German theoretical physicist who is credited with being one of the founders of thermodynamics, and with originating its second law. His great skill lay not in experimental technique but in the interpretation and mathematical analysis of other scientists' results.

Clausius was born in Köslin in Pomerania (now Koszalin in Poland) on 2 January 1822. He obtained his schooling first at a small local school run by his father, and then at the Gymnasium in Stettin. He entered the University of Berlin in 1840, and obtained his PhD from the University of Halle in 1848. Clausius then taught at the Royal Artillery and Engineering School in Berlin, and in 1855 became Professor of Physics at the Zurich Polytechnic. He returned to Germany in 1867 to become Professor of Physics at the University of Würzburg, and then moved to Bonn in 1869 where he held the chair of physics until his death.

In 1870 the Franco–Prussian war stimulated Clausius to organize a volunteer ambulance service run by his students. He was wounded during the course of these activities, and the injury caused him perpetual pain. This, combined with the death of his wife in 1875, probably served to reduce his productivity during his later years. However, his scientific achievements were rewarded with many honours, including the award of the Royal Society's Copley Medal in 1879. Clausius died in Bonn on 24 August 1888.

Sadi Carnot (1796–1832), Benoit Clapeyron (1799–1864) and Lord Kelvin (1824–1907) had made contributions to the theory of heat and to changes of state. Clausius examined the caloric theory, eventually rejecting it in favour of the equivalence of heat and work. Drawing particularly on Carnot's and Kelvin's concept of the continuous degradation or dissipation of energy, Clausius formulated (in a paper published in 1850) the second law of thermodynamics and introduced the concept of entropy.

The word entropy derives from the Greek word for transformation, of which there are two types according to Clausius. These are the conversion of heat into work, and the transfer of heat from high to low temperature. Flow of heat from low to high temperature produces a negative transformation value, and is contrary to the normal behaviour of heat. Clausius deduced that transformation values can only be zero, which occurs only in a reversible process, or positive, which occurs in an irreversible process. Clausius therefore concluded that entropy must inevitably increase in the universe, a formulation of the second law of thermodynamics. This law can also be expressed by the statement that heat can never pass of its own accord from a colder to a hotter body.

Entropy is considered to be a measure of disorder, and of the extent to which energy can be converted into work. The greater the entropy, the less energy is available for work. Clausius was opposed in his views by a number of scientists, but James Clerk Maxwell (1831–1879) gave Clausius considerable support in scientific argument on the subject.

Clausius did other work on thermodynamics, for example by improving the mathematical treatment of Helmholtz's law on the conservation of energy, which is the first law of thermodynamics; and by contributing to the formulation of the Clausius–Clapeyron equations, which describe the relationship between pressure and temperature in working changes of state.

The second area to which Clausius made important contributions was the development of the kinetic theory of gases, which Maxwell and Ludwig Boltzmann (1844–1906) had done so much to establish. From 1857 onwards, Clausius examined the inner energy of a gas, determined the formula for the mean velocity and mean path length of a gas molecule, provided support for Avogadro's work on the number of molecules in a particular volume of gas, demonstrated the diatomic nature of oxygen, and ascribed rotational and vibrational motion (in addition to translational motion) to gas molecules. Clausius also studied the relationship between thermodynamics and kinetic theory.

Clausius's third major research topic was the theory of

electrolysis. In 1857, he became the first to propose that an electric current could induce the dissociation of materials, a concept which was eventually established by Svante Arrhenius (1859–1927). Clausius also proposed that Ohm's law applies to electrolytes, describing the relationship between current density and the electric field.

Clausius was clearly a brilliant theoretician, but he showed a curious lack of interest in the developments which arose from his work. The results of Boltzmann, Maxwell, Willard Gibbs (1839–1903) and other scientists and the advancements in the field of thermodynamics, statistical mechanics and kinetic theory seem to have gone completely unnoticed by Clausius in his later years. One interesting consequence of the second law of thermodynamics is that as entropy increases in the universe, less and less energy will be available to do work. Eventually a state of maximum entropy will prevail and no more work can be done: the universe will be in a static state of constant temperature. This idea is called 'the heat death of the universe' and although it seems to follow logically from the second law of thermodynamics, this view of the future is by no means accepted by cosmologists.

Clifford William Kingdon 1845–1879 was a British mathematician and scientific philosopher who developed the theory of biquaternions and proved a Riemann surface to be topologically equivalent to a box with holes in it. His name is perpetuated in 'Clifford parallels' and 'Clifford surfaces'. In his philosophical studies he was much preoccupied with theories of evolution.

Clifford was born on 4 May 1845 in Exeter, Devon. Educated locally, he went to King's College, London, at the age of 15. Three years later he won a small scholarship and entered Trinity College, Cambridge, where his academic progress was phenomenal. In 1868 he was made a Fellow of the College, and continued to live there until 1871, when he was appointed Professor of Applied Mathematics at University College, London. He was elected a Fellow of the Royal Society in 1874, and became a prominent member of the Metaphysical Society. In 1876, however, he developed pulmonary tuberculosis and was obliged to live first in Algiers, and then in Spain, for the sake of his health. But his condition continued to deteriorate and, although he was able to make several trips to England for short periods, he went finally to Madeira in 1879, and died there on 3 March of that year.

Despite his connection with the city, Clifford was one of the first mathematicians to protest against the analytical methods of the 'Cambridge school'. Primarily a geometrician, regarding geometry as to all intents and purposes a branch of physics, Clifford had as his fundamental aim in his teaching the compelling of his students to think for themselves. He did much to revolutionize the teaching of elementary mathematics, and was responsible for introducing into England the geometrical and graphical methods of August Möbius, Karl Culmann and others. It was through a generalization of the quaternions (themselves a generalization of complex numbers) formulated by William Hamilton that Church derived his theory of biquaternions, associating them specifically with linear algebra. In this way representing motions in three-dimensional non-Euclidean space, and together with his suggestion in 1870 that matter itself was a kind of curvature of space, Church may be seen to have foreshadowed in some respects Einstein's general theory of relativity.

Clifford continued his studies in non-Euclidean geometry, with reference particularly to Riemann surfaces, for which he established some significant topological equivalences. He further investigated the consequences of adjusting the definitions of parallelism, and found that parallels not in the same place can exist only in a Riemann space – and he proved that they do exist. He showed how three parallels define a ruled second-order surface that has a number of interesting properties (which were subsequently examined by Bianchi and Felix Klein).

Clifford also achieved a measure of renown as an agnostic philosopher.

Cockcroft John Douglas 1897–1967 was a British physicist who, with Ernest Walton (1903–), built the first particle accelerator and achieved the first artificial nuclear transformation in 1932. For this achievement, Cockcroft and Walton shared the award of the 1951 Nobel Prize for Physics.

Cockcroft was born in Todmorden, Yorkshire, on 27 May 1897. He was admitted to Manchester University in 1914 to study mathematics but left a year later to volunteer for war service. He returned to Manchester in 1918 to attend the College of Technology and study electrical engineering, having gained an interest in this subject during his army service as a signaller. He obtained an MSc Tech in 1922 and then went to Cambridge University, taking a BA in mathematics in 1924. Cockcroft then remained at Cambridge, working at the Cavendish Laboratory under Ernest Rutherford (1871–1937). There he collaborated with Peter Kapitza (1894–1984) on the design of powerful electromagnets and then with Walton on the construction of a voltage multiplier to accelerate protons. With this instrument, the first artificial transformation – of lithium into helium – was achieved in 1932. Cockcroft and Walton then worked on the artificial disintegration of other elements such as boron.

During World War II, Cockcroft was closely involved with the development of radar and with the production of nuclear power, directing the construction of the first nuclear reactor in Canada. He returned to Britain in 1946 and was appointed first director of the Atomic Energy Research Establishment at Harwell. Cockcroft was knighted in 1948, and in 1959 became Master of Churchill College, Cambridge. He received the Atoms for Peace Award in 1961 and was made President of the Pugwash Conference, an international gathering of eminent scientists concerned with nuclear developments, shortly before he died at Cambridge on 18 September 1967.

Cockcroft's background in electrical engineering was unusual for a nuclear physicist, but he was able to put it to good use when the need to increase the energy of particles used to bring about nuclear transformations became apparent in the late 1920s. Until then, only alpha particles emitted by radioactive substances were available. Cockcroft and Walton built a voltage multiplier to build up a charge of 710,000 volts and accelerate protons in a beam through a tube containing a high vacuum. They bombarded lithium in this way in 1932, and produced alpha particles (helium nuclei), thus artificially transforming lithium into helium. The production of the helium nuclei was confirmed by observing their tracks in a cloud chamber.

Cockcroft and Walton's voltage multiplier was the first of many particle accelerators. It was soon superseded by the cyclotron, the ring-shaped acclerator developed by E. O. Lawrence, and led to accelerators of a wide range of energies. The development of the particle accelerator was essential to high-energy physics because it provided a means of studying subatomic particles that could not be produced in any other way, and it has also proved to be of great value in the production of radioactive isotopes.

Cockerell Christopher Sydney 1910– is the British engineer who invented the hovercraft. This has led to the establishment of a new industry and has stimulated worldwide interest. He also made a major contribution to aircraft radio navigation and communications.

Educated at Gresham's school, Holt, and Peterhouse, Cambridge, where he read engineering, Cockerell graduated in 1931 and spent two years in Bedford with the engineering firm of W. H. Allen & Sons. He then returned to Cambridge for two years of research into wireless, which was his hobby.

He joined the Marconi Wireless Telegraph Company in 1935, working on VHF transmitters and direction finders, and during World War II, on navigational and communication equipment for bombers, and later on radar. During this period he filed 36 patents, the most interesting of which are frequency division (1935), the linearization of a transmitter by feedback (1937), pulse differentiation (1938) and various navigational systems.

In 1950 Cockerell left Marconi's and started up a boat-hire business on the Norfolk Broads, building boats and caravans. Trained as a development engineer, he set himself the task of trying to make a boat go faster. First experiments were on the air lubrication of a hull and later on a 6.1-m/20-ft launch.

He was appointed consultant (hovercraft) to the Ministry of Supply and has been director of and consultant to a number of firms working in the field. The winner of many awards and distinctions, Cockerell is a Fellow of the Royal Society. In the 1970s and 1980s he has interested himself in the generation of energy by wavepower and has been chairman of Wavepower Ltd since 1974, the year he was made an Honorary Fellow of Peterhouse.

A number of engineers had earlier suggested the use of air lubrication for the reduction of drag. Indeed in the 1870s the British engineer, John Thorneycroft, built test models to check the drag on a ship's hull with a concave bottom in which air could be contained between the hull and the water.

Cockerell concluded after his first experiments with air lubrication that a major reduction in drag could be obtained only if the hull could be supported over the water by a really thick air cushion. This was because the fine structure of the upward pressure of the water, with the craft in motion over waves, varied so much that the pressure peaks broke through the layer of air. With a thin cushion, the best that could be attained was a 50% reduction in drag.

The first concept of a hovercraft was a sidewalled craft with hinged doors to contain the air, capable of lifting over a wave. The next idea was a water curtain discharged inwards and downwards across the front and back of the craft. The third idea was to replace the water curtain by thin air jets across the bow and stern pointing towards the athwartships centreline of the craft.

The next step for a faster amphibious craft, was a peripheral downward and inward-facing thin jet all round the periphery of the craft. Calculations showed that the power could be provided if the cushion pressure was kept down to reasonable limits. The power requirement was confirmed by some primitive experiments using tin cans as a simple way of constructing an annular jet. This concept was tried out in the experimental SR N1 in 1959.

The first true hovercraft was a 762 mm/2 ft 6 in balsawood model weighing 127.6 g/4.5 oz made in 1955 and powered by a model aircraft petrol engine. It could travel at 20.8 kph/13 mph over land or water. Cockerell's first hovercraft patent was filed on 12 December 1955 and in the following year he formed Hovercraft Ltd. He then began the thankless task of trying to interest manufacturers – shipbuilders said it was an aircraft and aircraft manufacturers said it was a ship.

Realizing that the hovercraft would have considerable military potential. Cockerell approached the Ministry of Supply, the Government's procurement agency for defence equipment. The air cushion vehicle was promptly classified 'secret' in 1956. This meant considerable delays but finally the secret leaked out the project was declassified and the National Research Development corporation decided to back the idea in 1958. Saunders Roe Ltd, a manufacturer with flying-boat history (and therefore both aircraft and marine experience) built the SR N1 which crossed the English Channel with the inventor on deck.

In the waiting period (1957) Cockerell had come to the conclusion that the hovercraft could not go fast in a seaway without diving into it. He came up with the idea of flexible skirts which gave rise to much derision because nobody could believe that a piece of fabric could be made to support 100 tonnes/97 tons with a mere 2.873Nm^{-2}/60 lb ft^{-2} compared with say, 206.843Nm^{-2}/30 lb in^{-2} for a motor-car tyre.

Manufacturers and operators in many parts of the world became interested and craft were made in the United States, Japan, Sweden and France while in Britain other companies also began to manufacture. The SR N4 maintained commercial cross-Channel car ferry services in the

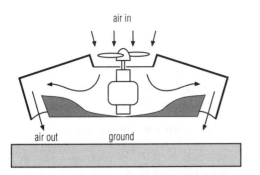

air in

air out   ground

Cockerell *Cockerell's original hovercraft design made use of a peripheral jet of compressed air to achieve lift.*

severest weather conditions and the AP 1–88, the world's first diesel-powered hovercraft, went into service in early 1983.

Apart from passenger- and car-ferry applications, craft have been used for seismic surveys over shallow water or desert and in search and rescue operations. Military uses include troop carrying during amphibious assaults and as logistic support craft. Future military uses include mine countermeasure work, antisubmarine work, aircraft carrying and missile launching.

Air cushion trains have a potential speeds of 480 kph/ 300 mph and the cost of track for these would be much less than that for conventional trains. However, because the air cushion requires considerable energy to maintain it, opinion has swung towards magnetic levitation for advanced tracked transport.

The technology has been applied to load lifters for heavy industrial plant when crossing bridges that were not stressed for the axle loading involved if normal wheeled transport was to be used. In hospitals, airbeds for supporting burned patients use the hovercraft principle: the hovermower is one of the most common applications; hovercraft transport heavy loads in Arctic regions; there are few fields where the hovercraft principle is not showing ever widening applications.

Cockerill Wilham 1759–1832 was a British engineer who is generally regarded as the founder of the European textile machinery industry.

Cockerill was born in Lancashire, and throughout his childhood showed a remarkable talent for anything mechanical. Being brought up in a district greatly involved in the production of cloth, he naturally found an outlet for his particular genius constructing machinery for that use.

His working career began with the building of 'roving billies' and 'flying shuttles', but in 1794 he decided to seek his fortune elsewhere and chose Russia as the most likely place. He found employment in St Petersburg and, under the patronage of Catherine II, enjoyed a fair measure of success for as long as she ruled. On her death, however, her successor, the madman Paul, imprisoned Cockerill for failing to complete a contract within the given time.

Eventually he escaped to Sweden and there tried to arouse an interest in a textile industry. The Swedes rejected his ideas, and so in 1799 he migrated to Belgium, where he established a flourishing business at Verviers as a manufacturer of textile machinery. In 1802 he was joined in partnership by a James Holden, but their association was short-lived and in 1807 Cockerill transferred his interests to Liège. There, together with his three sons William, Charles, and John – who also shared his enthusiasm for constructing machines – he built up a highly successful business making carding machines, spinning frames and looms for the French woollen industry.

After retiring from the firm in 1814 he went to live at the home of his son Charles, at Aix-la-Chapelle (Aachen), and remained there until his death in 1832.

The invention of John Kay's flying shuttle in 1733 had helped to speed up the process of weaving in the English textile industry, but not until the latter half of the eighteenth century were other significant mechanical designs introduced to streamline the production further. Cockerill's childhood saw the coming of James Hargreaves's 'spinning jenny' in 1770, which enabled spinners to spin several threads at once; Richard Arkwright's water frame, which brought water power to the spinning machines; and the invention of a rotary carding machine to make the possibility of mass production a reality. By the time Cockerill left for Russia in 1794 he had mastered all the techniques of manufacturing the new machines.

Basing his designs on the English machines. Cockerill was able to build up a reputation for first-class workmanship and attention to detail. When, after his previous disappointments, he eventually founded his business in Liège he found a ready market in France. In fact, it was during the era of Napoleon that Cockerill achieved his greatest esteem and his efforts were to a large extent reponsible for breaking the monopoly England had previously held over the continental market.

Colles Abraham 1773–1843 was an Irish surgeon who observed and described the fracture of the wrist which bears his name.

Colles was born on 23 July 1773 in County Kilkenny, Ireland; he was educated at Kilkenny Preparatory School and Kilkenny College. In 1790 he entered Trinity College, Dublin, as a student of arts, but swiftly began his clinical training as a resident surgeon at Steeven's Hospital. He was granted his diploma of the Royal College of Surgeons in Ireland in 1795. He then went to Edinburgh and graduated at Edinburgh University in 1797. He returned to Dublin that year to set up in practice and began to teach anatomy and surgery. In 1799 he was appointed Resident Surgeon at Steeven's Hospital (where he eventually became governor) and was elected a member of the Royal College of Surgeons in Ireland. At the age of 29 he became President of the College. In 1804 the Surgeons' School at the College made him Professor of Anatomy and Surgery (a position he held until 1836), and he gained his MA from the University of Dublin in 1832. He died at his home in Dublin on 16 November 1843.

In 1814 the paper on Colles's fracture was published describing the fracture of the distal (carpel) end of the radius bone in the forearm. This common fracture causes deformity and swelling of the wrist, but can be easily and successfully treated once diagnosed. It must be remembered that the diagnosis of fractures in those days was made on purely clinical grounds because X-rays had not then been discovered. Thus his accurate description of the fracture was that much more impressive. In his original paper he advocated the use of tin splints to stabilize the wrist after closed reduction of the fracture. Nowadays the reduction is followed by the use of plaster of Paris casts but exactly the same principles apply.

Although Colles is best remembered for the Colles fracture, he was one of the greatest professors the Royal College of Surgeons in Ireland ever had. He was a brilliant anatomist and an excellent teacher who did much to make Dublin the leading medical centre it had become by the beginning of the nineteenth century.

Colles's other eponyms include Colles's fascia, Colles's space, the Colles ligament (of inguinal hernia), and Colles's law of the communication of (congenital) syphilis.

Colt Samuel 1814–1862 was the US inventor of what was probably the most successful family of revolving pistols of his time, which revolutionized military tactics.

Born on 19 July 1814, at Hartford, Connecticut, Colt had made up his mind as a boy to become an inventor but his early inventions (which included an abortive four-barrel rifle) were somewhat unreliable. One of his discoveries was that it was possible to fire gunpowder using an electric current. He applied this principle to an explosive mine, but after a disastrous public demonstration (which covered all the spectators with mud), he was sent off to Amherst Academy. Unfortunately, as a result of a fire caused by another of his experiments, he was asked to leave there as well. He then became apprenticed as a seaman.

During a journey to India in 1830, aboard the brig *Corlo*, Colt made some observations that were to change his life. In watching the helmsman and the wheel he operated, Colt noticed that whichever way the wheel turned, each of its spokes always lined up with a clutch that locked it into position. Colt conceived the idea that the mechanism of the helmsman's wheel could be applied to a firearm.

Grasping the opportunity presented by this, he left the sea at the age of 18 to raise money for a professionally built prototype. Ever resourceful, he travelled America as 'Dr Coult', giving demonstration of nitrous oxide (laughing gas) to willing audiences in a series of adventures worthy of Huckleberry Finn.

By 1835, Colt had perfected his revolver, which emulated the helmsman's wheel with its rotating breech which turned, locked and unlocked by cocking the hammer. That year, Colt took out patents in Britain and France and, in 1836, in his native United States. On 5 March 1836 Colt set up a company in Paterson, New Jersey, the Patent Arms Manufacturing Company, where he produced the Colt Paterson, a five-shot revolver with folding trigger and a

number of different rifles and shotguns based on the same revolving principle. These models were revolutionary in a market still dominated by the cumbersome 'pepperbox' type of revolver, with their revolving barrels. In the resulting conflict of ideas, and amid apathy among firearms buyers stemming from the price and unreliability of the early models, Colt's company failed in 1842.

Colt temporarily turned his attention to inventing electrically discharged submarine mines and running a telegraph business. Events, however, soon restored him to his original course when, in 1846, the Mexican–American War broke out. Colt Paterson revolvers had come into the possession of the Texas Rangers, who persuaded the government to order Colt revolvers for use in the war. Colt had to be tracked down first, but was of course delighted with an order to make 1,000 pistols for the army. Having no factory of his own, Colt contracted the work out to Eli Whitney at Whitneyville, Connecticut, and before long the US Army was receiving supplies of the new revolver. Captain Walker, who had succeeded in tracing Colt, contributed his own ideas to the .44 calibre revolver, which became known as the Colt–Walker 'Dragoon'. The pioneers then starting to open up the American West adopted the Colt with alacrity, since it was superior to the one- or double-shot pistol against native Americans and wildlife.

Colt used the profits from his early sales to buy land at Hartford, where in 1847 he rented premises and in 1854 built a large factory. Colt's Hartford company is still in existence; the factory burnt down in 1864, but was later rebuilt. Colt could now increase the types of firearms he produced, and by 1855 he had the largest private armoury in the world. One of Colt's new weapons was the .31 calibre Pocket Model first produced in 1848, and in 1851 he produced his famous percussion revolver, the .36 calibre Colt Navy. The Rootes .28 and the Army followed in 1860, and the Police Model in 1862.

By this time, the American Civil War had begun. Colt, whose pre-war sympathies were with the secessionist Confederate South, changed allegiance to the Unionist North once fighting started in 1860 and supplied thousands of guns to the United States government.

While his patents (which he defended vigorously) lasted, Colt had a virtual monopoly of the firearms market. In time, however, his rivals found ways around the restrictions of the patent laws, and eventually the patents simply expired.

In the teeth of bitter opposition from established British arms manufacturers, Colt set up a factory in Pimlico, London, in 1852. He also decided to show his revolver at the Great Exhibition of 1851. Orders to equip the British armed forces, including those fighting in the Crimea, followed, with customers being supplied by a variety of models produced in both Hartford and London. The Pimlico factory closed down in 1856, but Colt retained an agency in London until 1913.

By the time of his death at Hartford, on 10 January 1862, Colt could leave behind a vast and efficient company, and a considerable personal fortune. He had succeeded in marketing a well-conceived product whose success stemmed from the mass production of interchangeable parts which needed little hand finishing in assembly. As a result, he made a major contribution to both the development of repeating firearms and to the process of mass production by machine tools.

Compton Arthur Holly 1892–1962 was a US physicist who is remembered for discovering the Compton effect, a phenomenon in which electromagnetic waves such as X-rays undergo an increase in wavelength after having been scattered by electrons. For this achievement he shared the 1927 Nobel Prize for Physics with the British physicist C. T. R. Wilson. Compton was also a principal contributor to the development of the atomic bomb.

Compton was born at Wooster, Ohio, on 10 September 1892, the son of a philosophy professor at Wooster College who was also a Presbyterian minister. He was educated at his father's college, graduating in 1913 and going on to Princeton University, from which he gained his PhD three years later. He became a physics lecturer at the University of Minnesota, but left after a year to take up an appointment as an engineer in Pittsburgh for the Westinghouse Corporation. He travelled to Britain in 1919 and stayed for a year at Cambridge University, where he worked under Ernest Rutherford. In 1920 he returned to the United States and went to Washington University, St Louis, as Head of the Physics Department, moving to the University of Chicago in 1923 as Professor of Physics. Compton remained at Chicago for 22 years until, after the end of World War II in 1945, he returned to Washington University. He was Chancellor of the University from 1945 until 1953 and then became Professor of Natural Philosophy there, a position he held until 1961, when he became Professor at Large. He died in Berkeley, California, on 15 March 1962.

Compton began studying the scattering of X-rays by various elements in the early 1920s (using blocks of paraffin wax in which the carbon atoms' electrons deflected X-rays) and by 1922 he had noted the unexpected effects that the scattered X-rays show an increase in wavelength. He explained the Compton effect in 1923 by postulating that X-rays behave like particles and lose some of their energy in collisions with electrons, so increasing their wavelength and decreasing their frequency. He calculated that the change in wavelength, $\Delta\lambda$, is given by the equation: $\Delta\lambda = (h/mc)(1 - \cos\theta)$, where h is Planck's constant, m is the rest mass, c is the velocity of light and θ is the angle through which the incident radiation (such as X-rays) is scattered.

The chief significance of Compton's discovery is its confirmation of the dual wave–particle nature of radiation (later extended by Louis de Broglie in his hypothesis that matter can also have wave–particle duality). The behaviour of the X-ray, previously considered only as a wave, is explained best by considering that it acts as a corpuscle or particle – as a photon (Compton's term) of electromagnetic radiation. Quantum mechanics benefited greatly from this convincing interpretation. Further confirmation came from

experiments using C. T. R. Wilson's cloud chamber in which collisions between X-rays and electrons were photographed and analysed. They could be interpreted as elastic collisions between two particles and measurements of the particle tracks in the cloud chamber proved the correctness of the mathematical formulation of the Compton effect.

In the early 1930s several scientists were debating the nature of cosmic rays. Robert Millikan had proposed that the rays are a form of electromagnetic radiation which, if this were true, would be unaffected by their passage from outer space through the Earth's magnetic field and should strike the Earth in undeflected straight lines. Other scientists, such as the German physicists Walther Bothe (1891–1957), postulated that the 'rays' are made up of streams of charged particles which, therefore, should be deflected into curved paths as they traverse the Earth's magnetic field. Compton reasoned that the argument could be resolved by making cosmic-ray measurements at various latitudes, which he carried out during a long series of trips to various parts of the world to measure – or to organize dozens of other scientists to measure – comparative cosmic-ray intensities using ionization chambers. By 1938 he had collated the multitude of results and demonstrated a significant latitude effect by the Earth's field, proving that at least some component of cosmic rays consists of charged particles. He went on to confirm these findings by showing variations in cosmic-ray intensity with the rotation period of the Sun and with the time of day and time of year. One interpretation of these results suggests an extragalactic source for cosmic rays.

During World War II Chicago University was the prime location of the Manhattan Project, the effort to produce the first atomic bomb, and in 1942 Compton became one of its leaders (as Director of the code-named Metallurgical Laboratory). He organized research into methods of isolating fissionable plutonium and worked with Enrico Fermi in producing a self-sustaining nuclear chain reaction, which led ultimately to the construction of the bombs that, used against Japan, contributed to the ending of World War II. Compton's book *Atomic Quest* (1956) summarized this part of his career.

Cook James 1728–1779 was an English sea captain who made notable contributions to hydrography.

An agricultural labourer's son born at Marton, Yorkshire, Cook obtained his early experience of the sea by sailing in a Whitby collier before joining the Royal Navy in 1755 as an ordinary seaman. Ambitious and clever, he quickly rose in the service. After much first-rate hydrographical surveying work, undertaken particularly around Newfoundland and the mouth of the St Lawrence, Cook received in 1768 a commission, via the Royal Society, to take the *Endeavour* to the newly discovered Tahiti with various scientists, including Joseph Banks, to observe the transit of Venus. Observations of transits of the inner planets across the face of the Sun were highly valued among astronomers as a means of calculating the distance of the Sun from the Earth. On that voyage, Cook charted the east coast of Australia (exploring Botany Bay) and the entire coastline of New Zealand, clearly demonstrating that it consisted of two main islands. He brought back fascinating descriptions of the exotic civilization of the Tahitians. Cook formed the belief that the long-suspected Unknown Continent of the southern hemisphere could not exist, or at least, could not be of immense dimensions. (He was under Admiralty orders to secure such a continent for the Crown, should it exist – Cook's voyages thus served both political and scientific purposes.) As a result of sailing at higher latitudes south than any previous captain, Cook's second expedition of 1772–75 further shrank the possible extent of a southern continent. In 1776, shortly after being made a Fellow of the Royal Society, Cook embarked on his third expedition, which aimed to explore the possibility of a northern route between the Atlantic and the Pacific. The expedition reached a sad climax in his death at the hands of natives in Hawaii – ironically, because Cook was generally scrupulous in his treatment of indigenous peoples.

Cook's own energies and talents, aided by improved sextants and Harrison's chronometer, ensured that a greater quantity of high-quality survey work and scientific research was accomplished on Cook's three expeditions than on any comparable expeditions. He set new standards of cartography and hydrography, and was the source of modern maps of the Pacific and its coasts. Cook is also remembered as a sagacious captain, who took great care of his crews. He proved the value of fresh fruit and vegetables, cleanliness and morale, in preventing and treating scurvy, a major problem on long sea voyages at the time.

Coolidge Julian Lowell 1873–1954 was a US geometrician and a prolific author of mathematical textbooks in which he not only reported his results but also described the historical background, together with contemporary developments.

Coolidge was born into a prominent family in Brookline, Massachusetts, on 28 September 1873. Completing his education, he studied at Harvard University where in 1895 he was awarded his bachelor's degree with top honours. Two years later he travelled to Britain and took a degree in natural sciences at Balliol College, Oxford. He then returned to the United States to teach mathematics at the Groton School, where one of his pupils was Franklin D. Roosevelt. In 1900 he became an instructor in mathematics at Harvard, and was made a member of the faculty in 1902. In that same year he was given leave of absence in which to continue his own studies, and returned to Europe, studying in Paris, Griefswald, Turin and Bonn. Having earned a PhD in 1904 at Bonn University, Coolidge went back to teaching at Harvard. Four years later he was made an assistant professor. His work was then disrupted by World War I, in which he served as a major in the army. After the war he was a liaison officer in France, where he organized and taught courses at the Sorbonne for the benefit of US soldiers still stationed in the country.

In 1918 Coolidge had been made a full professor of mathematics at Harvard, and on his return from France he remained there until his retirement in 1940. Even then, however, as his own war effort, he again took to giving courses to military personnel. Many honours were awarded to Coolidge in recognition of his contributions to mathematics. He died in Cambridge, Massachusetts, on 5 March 1954.

Coolidge's work lay in the field of geometry and the history of mathematics. His early research for his PhD thesis and his first book (1909) were on non-Euclidean geometry, and were strongly influenced by the work of Segre and Study; in particular, Coolidge used Study's new method of approach to line geometry. He was especially interested in the use of geometry in the investigation of complex numbers.

Other areas of interest soon began to develop. Coolidge wrote his first paper on probability theory in 1909, in which he also examined certain problems in game theory. He produced several later studies on statistics, and all of this work was included in his 1925 book on probability – one of the first on the subject to be published in English, and as such it received considerable acclaim.

His first work on the algebraic theory of curves appeared in 1915. Stimulated by the topic, Coolidge elaborated on his initial investigation and eventually (1931) published a full-size book detailing his results. Work on two classical geometrical figures – the circle and the sphere – also led to the writing of a book published in 1916. Lectures given at the Sorbonne in 1919 on the geometry of the complex domain were expanded into book form and published in 1924. There were three further books: one in 1940 on geometrical methods, one in 1943 on conic sections, and the last a historical text in 1949. In addition to his books, Coolidge was also the author of many papers, the last of which was written in 1953.

Cooper Leon Niels 1930–   is a US physicist who shared the 1972 Nobel prize with John Bardeen and J. Robert Schrieffer for developing the 'BCS' theory of superconductivity. One of Cooper's key contributions to the theory, which explains why certain metals lose all electrical resistance at very low temperatures, was the suggestion that electrons in the metal form pairs. These are now known as Cooper pairs.

Cooper was born on 28 February 1930 in New York City where he attended the Bronx High School of Physics and Columbia University, obtaining a BA (1951), MA (1953) and PhD (1954). At Columbia Cooper's speciality was quantum field theory – the interaction of particles and fields in subatomic systems. After a year at the Institute for Advanced Study in Princeton, Cooper moved to the University of Illinois in 1955 to work with John Bardeen, who was to share the Nobel prize in 1956 for his work on the transistor. Bardeen was now working on superconductivity. The electrical resistance of all metals decreases as they are cooled. This is because the thermal vibrations of the nuclei – which scatter the electrons carrying the current, and hence give rise to the resistance – decrease with temperature. (Resistance and conductivity are inversely related:

that is, zero resistance is equivalent to infinite conductivity – superconductivity). But whereas the decrease in resistance is gradual in most metals, the resistance of superconductors suddenly disappears below a certain temperature. Experiments had shown that this temperature, the critical temperature, was inversely related to the mass of the nuclei, so Bardeen proposed that superconductivity depended on the interaction of electrons with these variations. Cooper showed that an electron moving through the lattice attracts positive ions, slightly deforming the lattice. This leads to a momentary concentration of positive charge that attracts a second electron. This is the Cooper pair. Although the electrons in the pair are only weakly bound to each other, BCS were able to show that they all formed a single quantum state with a single momentum. The scattering of individual electrons did not effect this momentum, and this lead to zero resistance. Cooper pairs could not be formed above a certain temperature, the critical temperature, and superconductivity broke down. After a year at Ohio State University, in 1958 Cooper moved to Brown University, Rhode Island, and in 1968 published *The Meaning and Structure of Physics*. In his Nobel lecture in 1972, Cooper said that a theory is more than its applications, being 'an ordering of experience that both makes experience meaningful and is a pleasure to regard in its own right'. In 1978 Cooper also became director of the Centre for Neural Science at Brown and worked on developing a theory of the central nervous system. He has also worked on distributed memory and character recognition.

Copernicus Nicolaus (Mikolaj Kopernigk) 1473–1543 was a Polish doctor and astronomer who, against the religion-reinforced tradition of many centuries, finally declared once and for all that the planet Earth is the centre neither of the universe, nor even of the Solar System. He was never a skilled observer, and probably made fewer than 100 observations in all, preferring to rely almost entirely on data accumulated by others. In his own time he was renowned probably more as a medical man and priest than as an astronomer.

Copernicus was born in Toruń in Ermland (under the Polish crown) on 19 February 1473, and at an early age attended St John's School there. After the death of his father in 1483, however, it was arranged, probably by his uncle (and patron) L. Watzenrode, for him to study at the Cathedral School in Wloclawek. From 1491 until 1494 Copernicus studied mathematics and classics at the University of Krakow, under Brudzewski. Encouraged by Watzenrode to continue his studies, he travelled in 1496 to the University of Bologna, Italy, where he studied law and astronomy. In the latter subject he was taught by D. M. di Novara.

Despite his nephew's absence from Poland, Watzenrode (now himself a bishop) was able to arrange for Copernicus to be made Canon of Frombork (Frauenburg), a post he retained for life. This job brought in sufficient income, combined with only light duties, to enable Copernicus to devote a great deal of time to astronomy.

When in 1500 he had completed his studies in Bologna, Copernicus moved on to Padua, where he continued his studies in law and Greek. He gained a doctorate in canon law in 1503 in Ferrara and then returned to Padua to study medicine. In 1506 he went home to Poland (where he remained for the rest of his life). From 1506 until 1512, when Watzenrode died, he not only worked for his uncle as his personal doctor and private secretary, but served in the Cathedral chapter of Frombork.

After the death of Bishop Watzenrode, Copernicus was still not able to devote himself entirely to his ecclesiastical and astronomical interests. He also served on a number of diplomatic missions and as a financial advisor and administrator. Nevertheless, he wrote a brief outline of his new ideas in astronomy in about 1513; he privately circulated a more comprehensive version in 1530. Copernicus had the good sense to realize that he risked being branded a lunatic, or a heretic, and he was therefore reluctant to publish his theory more widely. G. Joachim, known as Rhaeticus, encouraged him to present his work in book form, which he finally agreed to do. The book, *De Revolutionibus Orbium Coelestrium*, prudently dedicated to Pope Paul III, was delayed in publication, so that Copernicus did not receive a copy of it until he was on his deathbed. Copernicus died at Frombork on 24 May 1543.

Copernicus began to make astronomical observations in March 1497, although it was not until about 1513 that he wrote the brief, anonymous text, entitled *Commentariolus*, in which he outlined the material he later discussed more fully. His main points were that the Ptolemaic system of a geocentric planetary model was complex, unwieldy and inaccurate. Copernicus proposed to replace Ptolemy's ideas with a model in which the planets (including the Earth) orbited a centrally situated Sun. The Earth would describe one full orbit of the Sun in a year, while the Moon orbited the Earth. The Earth rotated daily about its axis (which was inclined at 23.5° to the plane of orbit), thus accounting for the apparent daily rotation of the sphere of the fixed stars.

This model was a distinct improvement on the Ptolemaic system for several reasons. It explained why the planets Mercury and Venus displayed only 'limited motion': their orbits were inside that of the Earth's. Similarly, it explained why the planets Mars, Jupiter, and Saturn displayed such curious patterns in their movements ('retrograde motion', loops and kinks). These were all a consequence of their travelling in outer orbits at a slower pace than the Earth. The precession of the equinoxes, as discovered by Hipparchus, could be accounted for by the movement of the Earth on its axis.

Copernicus's theory was not, however, by any means perfect. He was unable to free himself entirely from the constraints of classical thinking and was able to imagine only circular planetary orbits. This forced him to retain the cumbersome system of epicycles, with the Earth revolving around a centre which revolved around another centre which in turn orbited the Sun. It was the work of Kepler, who introduced the concept of elliptical orbits, which rescued the Copernican model. Copernicus also held to the notion of spheres, in which the planets were supposed to travel. It was the Dane, Tycho Brahe (1546–1601), who rid astronomy of that archaic concept.

In his greatest work, the *De Revolutionibus*, Copernicus proposed that the atmosphere (at least part of it) rotated with the Earth about the planetary axis, so that the skies did not constantly flow westwards. He obtained fairly accurate estimates for the distances of the planets from the Sun in terms of the astronomical unit (the distance from Earth to the Sun). He echoed Aristarchos in explaining the inability to observe stellar parallax from the extremes of the Earth's orbit around the Sun as being a consequence of the fact that the diameter of the orbit was insignificant in comparison with the distance from the Earth to the stars. (In fact, stellar parallax is detectable, but only with superior instruments; it was not observed until Friedrich Bessel (1784–1846) finally succeeded in 1838.)

The Lutheran minister Andreas Osiander oversaw the publication of *De Revolutionibus* and inserted a preface (without Copernicus's permission), stating that the theory was intended merely as an aid to the calculation of planetary positions, not as a statement of reality. This served to compromise the value of the text in the eyes of many astronomers, but it also saved the book from instant condemnation by the Roman Catholic Church. *De Revolutionibus* was not placed upon the index of forbidden books until 1616 (it was removed from the list in 1835).

Copernicus relegated the Earth from being the centre of the universe to being merely a planet (the centre only of its own gravity and the orbit of its solitary Moon). This in itself forced a fundamental revision of man's anthropocentric view of the universe and came as an enormous psychological shock to the whole of European culture. Copernicus's model could not be 'proved' right, because it contained several fundamental flaws, but it was the important first step to the more accurate picture built up by Brahe, Kepler, Galileo and later astronomers. It is no exaggeration to view Copernicus's work as the cornerstone of the scientific revolution.

Cori Carl Ferdinand 1896–1984 and Gerty Theresa 1896–1957, husband and wife, were Czech-born US biochemists who worked out the biosynthesis and degradation of glycogen, the carbohydrate stored in the liver and muscles. For this achievement they were jointly awarded the 1947 Nobel Prize for Physiology or Medicine, which they shared with the Argentinian biochemist Bernardo Houssay (1887–1971).

Carl Cori was born on 5 December 1896 in Prague, then in Austrio-Hungary. He was educated in Austria at the Trieste Gymnasium, and graduated in 1920 with a medical degree from the University of Prague. It was while he was a medical student that he met and married his classmate Gerty Radnitz. She was also born in Prague, on 15 August 1896, entered the medical school in 1914 and graduated in the same year as her husband. After having spent several years in war-torn Austria – Carl Cori was in the Austrian army during World War I – they emigrated to the United States in 1922 and became US citizens six years later.

From 1922 to 1931 Carl Cori was a biochemist at the State Institute for Study of Malignant Diseases at Buffalo, New York. In 1931 he was appointed Professor of Biochemistry at Washington University School of Medicine in St Louis, Missouri. In the same year his wife became Fellow and Research Associate in Pharmacology and Biochemistry there, a position she held until 1947. Gerty Cori died in St Louis on 26 October 1957. Carl Cori remained at St Louis until 1967, when he took up the appointment of Biochemist at Massachusetts General Hospital, Harvard Medical School, Boston.

It was during the 1930s that the Coris began their researches on glycogen. Its basic structure was known; it is a polysaccharide, a highly branched sugar molecule composed of several hundred glucose molecules linked by glycosidic bonds. Any excess food in an animal's diet is stored as glycogen or fat, and in times of shortage the animal makes use of these reserves. Glycogen is broken down in the muscles into lactic acid (2-hydroxypropanoic acid), as worked out by Otto Meyerh (1884–1951) about 20 years earlier, which when the muscles rest is reconverted to glycogen. The Coris set out to determine exactly how these changes take place.

It was tempting to assume that glycogen broke down into separate glucose molecules. But this hydrolysis would involve a loss of energy, which would have to be resupplied for the conversion back to glycogen. Gerty Cori found a new substance in muscle tissue, glucose-1-phosphate, now known as Cori ester. Its formation from glycogen involves only a small amount of energy change, so that the balance between the two substances can easily be shifted in either direction. The second step in the reaction chain involves the conversion of glucose-1-phosphate into glucose-6-phosphate. Finally this second phosphate is changed to fructose-1,6-diphosphate, which is eventually converted to lactic acid. The first set of reactions from glycogen to glucose-6-phosphate is now termed glycogenolysis; the second set, from glucose-6-phosphate to lactic acid, is referred to as glycolysis.

Coriolis Gaspard Gustave de 1792–1843 was a French physicist who discovered the Coriolis force that governs the movements of winds in the atmosphere and currents in the ocean. Coriolis was also the first to derive formulae expressing kinetic energy and mechanical work.

Coriolis was born in Paris on 21 May 1792. From 1808, he studied there at the Ecole Polytechnique and the Ecole des Ponts et Chaussées, graduating in highway engineering. He returned to the Ecole Polytechnique in 1816 as a tutor, and then became Assistant Professor of Analysis and Mechanics. In 1829 Coriolis accepted a chair in mechanics at the Ecole Centrale des Arts et Manufactures and then, in 1836, he obtained the same position at the Ecole des Ponts et Chaussées. In 1838 he became Director of Studies at the Ecole Polytechnique. Often in poor health, he died in Paris on 19 September 1843.

From 1829, Coriolis was concerned that proper terms and definitions should be introduced into mechanics so that the principles governing the operation of machines could be clearly expressed and similar advances be made in technology as were being made in pure science. His teaching experience aided him greatly in this endeavour, and he succeeded in establishing the use of the word 'work' as a technical term in mechanics, defining it in terms of the displacement of force through a certain distance. He proposed a unit of work called the dynamode, which was equal to 1,000 kg m/7,200 lb ft, but it did not enter general use.

Coriolis then proceeded to give the name 'kinetic energy' to the quantity he defined as $\frac{1}{2}mv^2$, making the mathematical expression of dynamics much easier.

Coriolis went on to investigate the movements of moving parts in machines relative to the fixed parts and in 1835 made his most famous contribution to physics in a paper on the nature of relative motion in moving systems. Applying his ideas to the general case of rotating systems, Coriolis showed that an inertial force must be present to account for the relative motion of a body within a rotating frame of reference. This force is called the Coriolis force. It explains how the rotation of the Earth causes objects moving freely over the surface to follow a curved path relative to the surface, the Coriolis force turning them clockwise in the northern hemisphere and anticlockwise in the southern hemisphere. The Coriolis force therefore appears prominently in studies of the dynamics of the atmosphere and the oceans, and it is also an important factor in ballistics.

Coriolis was a scientist of acute perception who did much to clarify thought in the field of mechanics and dynamics. But for poor health, he might have realized his full potential and made an even greater contribution to science.

Corliss George Henry 1817–1888 was a US engineer and inventor of many improvements to steam engines, particularly the Corliss valve for controlling the flow of steam to and through the cylinder(s).

Little is known about the early life of Corliss except that he was born on 2 June 1817 in Easton, New York. In 1844 he had moved to Providence, Rhode Island, where he became interested in steam engines, concerning himself particularly with attempts to improve their performance.

He took out the first of his many patents in 1849, for the Corliss valve. In 1856 he founded the Corliss Engine Company which was to become one of the largest steam-engine manufacturers in the United States and, incidentally, helped to make Providence a major tool and machinery centre. His company designed and built the largest steam engine then in existence to power all the exhibits in the Machinery Hall at the 1876 Philadelphia Centennial Exposition. With a mass of 630 tonnes/609 tons and a 9-m/29.5-ft flywheel, the engine was not surprisingly billed as the 'Eighth Wonder of the World'.

In the 1840s the most advanced valves then in use suffered from several disadvantages inherent to their design. Most regulated steam flow by sliding a plate over the port

into the piston chamber in such a way that while steam entered one end of the chamber it was exhausted from the other. Unequal wear on the sliding parts caused them to lose their steam-tight fit; they were also heavy to operate and, with the steam entering and exhausting through the same passage, they suffered considerable heat loss.

An essential feature of the Corliss valve was the separate inlet and exhaust port at each end of the cylinder (giving a total of four valve units per engine in twin cylinder engines). This saved heat loss and also cut down the 'dead' space at the ends of the chamber using older valves where the ports were located along the side of the chamber. Of equal significance to this was Corliss's realization that for maximum efficiency the valves must open and close as quickly as possible. This he achieved by a spring-loaded action.

Like any good invention the basic principle of the valve was very simple. On the first stroke, one inlet valve opened, allowing steam under pressure to enter that end of the cylinder and force the piston to move to where an exhaust valve was open. On the next stroke these two valves closed, and the other inlet and exhaust valves opened again allowing steam to enter the cylinder and force the piston to move back in the other direction. Because of its compact size, the reduced wear on moving parts, and heat saved by having separate steam inlet and outlet paths, this valve greatly improved the efficiency of large steam engines. The success of Corliss's design is best shown by the fact that it continued in use for as long as large steam engines were manufactured, long after its inventor's death on 2 February 1888.

Cornforth John Warcup 1917– is an Australian organic chemist who shared the 1975 Nobel Prize for Chemistry with the Swiss biochemist Vladimir Prelog for his work on the stereochemistry of biochemical compounds.

Cornforth was born on 7 September 1917 in Sydney, Australia. He began his academic training at the University of Sydney and then went to Oxford University, where he obtained his doctorate in 1941. For the next five years he worked with Robert Robinson, who 30 years earlier had worked in Sydney and who was to receive the 1947 Nobel Prize for Chemistry for his work on plant alkaloids. At about this time Cornforth's hearing began to deteriorate and he was soon totally deaf. He worked for the British Medical Research Council from 1946 until 1962, when he became Director of the Milstead Laboratory of Chemical Enzymology, Shell Research Ltd. He remained there until 1975, when he accepted a professorship at the University of Sussex; he had previously been an associate professor at Warwick University (1965–71) and Visiting Professor at Sussex (1971–75).

In his researches, Cornforth studied enzymes, trying to determine specifically which group of hydrogen atoms in a biologically active compound is replaced by an enzyme to bring about a given effect. He painstakingly developed techniques to pinpoint a specific hydrogen component by using the element's three isotopes: normal hydrogen (^{1}H), deuterium (^{2}H), and tritium (^{3}H). Each isotope has a different speed of reaction and, by careful observation and many experiments, Cornforth was able to identify precisely which hydrogen atom was affected by enzyme action. He was able, for example, to establish the orientation of all the hydrogen atoms in the cholesterol molecule.

Cort Henry 1740–1800 was a British inventor who devised a method of producing high-quality iron using a reverberating furnace.

Cort was born in Lancaster, and at the age of 25 he left his home town to seek his fortune in London. He became an agent for the Navy and was responsible for purchasing their guns. At this time (1765) all the best metal suitable for arms manufacture was imported from abroad, mainly from Russia, Sweden, and North America. Cort, realizing the potential, experimented to find a method of making this high-grade metal in England, and in 1775 he set up his own forge at Fontley, near Farnham.

By 1784 he was in a position to apply for a patent on his process of 'puddling and rolling', which allowed bar-iron to be produced on a large scale and of a high quality. The process came at a particularly fortunate time for England's iron production, for, with the advent of the Napoleonic Wars, pig-iron requirements rose from 40,000 tonnes/38,700 tons in 1780 to 400,000 tonnes/387,000 tons by 1820. But he was somewhat unwise in his choice of financial backer. He chose Samuel Jellicoe, a naval paymaster, who without Cort's knowledge obtained the money from public funds. When Jellicoe was found out he committed suicide and Cort, having handed over the rights to his patent as security for the capital, was left bankrupt. His patent was confiscated by the Admiralty and he was forced to watch the iron-masters of England grow wealthy on his hard work. He and his large family were reduced to living off a pension of £200 a year, which the state eventually granted him.

The skill of iron smelting has been known for centuries. Using a blast furnace fired by charcoal, the process remained unchanged until the early eighteenth century, when several people attempted to better the traditional ways. Three main factors limited production. The first was the choice of fuel. Using charcoal meant that the iron works, of necessity, were sited near forests. The second was the source of power to work the bellows and the forge hammer. A water wheel was a natural solution, but its efficiency depended upon the amount of water available and as a result production was often seasonal. The fluctuation of water supply also influenced the third factor, transport, which relied heavily on the rivers.

Coal answered the first problem, the steam-engine was the solution for the power, and canals took over from natural waterways. Coal, however, although cheaper, still produced a poor quality iron and it was not until Abraham Darby used coal, converted to coke, at his Coalbrookdale works (and combined it with the right type of ore) that a pig-iron was made that could be forged into bar-iron. Even so, the quality tended to be unreliable.

Cort experimented and found a solution in his 'puddling and rolling' process, which removed the impurities by stirring the pig-iron on the bed of a reverberatory furnace. (This is a furnace with a low roof, so that the flames in passing to the chimney are reflected down on to the hearth, where the ore to be smelted can be heated without its coming into direct contact with the fuel.) The 'puddler' turned and stirred the mass until it was converted into a malleable iron by the decarburizing action of air circulating through the furnace. The iron was then run off, cooled and rolled into bars with the aid of grooved rollers.

The significance of Cort's work was such that at last England did not have to rely on imported iron and could become self-sufficient. His method of manufacture combined previously separate actions into one process, producing high-class metal relatively cheaply and quickly, allowing iron production to increase to meet the growing needs of the industrial age.

Cotton William 1786–1866 was a British inventor, financier and philanthropist who greatly publicized the work of an earlier hydrographer and inventor, Joseph Huddart. He is probably best known, however, for inventing a knitting machine which paved the way for significant advances in the production of hosiery.

Cotton was born in Leyton, Essex. On leaving school at the age of 15, he entered a counting-house as a clerk and all subsequent progress was as a result of self-education. In 1807 he was admitted as a partner in the firm of Huddart & Co. in Limehouse, London. This business had been founded a few years earlier to promote the large-scale manufacture of an ingenious cordage-making machine designed by Joseph Huddart. The attention of Huddart was drawn to this field when, in the course of travelling for the East India Company, he saw a ship's cable snap. He thereupon worked on a method 'for the equal distribution of the strains upon the yarn'. When put on the market, his machine was to bring him an appreciable fortune.

Cotton did well in Huddart's firm and was eventually entrusted with its general management. He disposed of Huddart's machinery to the government in 1838. In the same year he wrote a memoir of Huddart with an account of his inventions, which was privately printed in 1855. In recognition of this publication he received a silver medal from the Institution of Civil Engineers.

In 1821 Cotton was elected a director of the Bank of England, a position he continued to hold until a few months before his death. He held the governership of the Bank of England over the years 1843 to 1846. A lasting memorial of his tenancy of this high office came about through his invention of the 'governor' and automatic weighing machine for sovereigns. It weighed these coins at a rate of 23 per minute, discharging the full and underweight specimens into separate compartments.

Although Cotton prospered in business, it is as a philanthropist that he is chiefly remembered, notably for the building of schools, churches and lodging houses in the East End of London. After his death he was commemorated by a painted window in St Paul's Cathedral, raised by public subscription.

It was not until 1864 that Cotton secured a patent for the knitting machine to be used in hosiery, the principal invention for which he is remembered. The machine he produced was remarkable for its adaptability: it had a straight-bar frame which automatically made fully fashioned stockings knitted flat and sewn up the back.

Subsequent improvements to Cotton's original machine have largely been directed at either increasing the capacity of the machine or producing fabric of even finer gauge. Increased capacity can be attained only by the use of larger and faster models, and finer gauge work requires finer needles and sinkers. A modern machine making fully fashioned silk or nylon hose may have as many as 40 needles to 2.5 cm/1 in and may produce 32 stockings at once. The reliability of such machines obviously depends on the accuracy with which they are made, and in this respect the hosiery industry owes much to improvements in engineering techniques. Towards the end of the nineteenth century the making of hosiery machines became an important industry in Germany and the United States, and in time the types used in these countries became also widely used in Britain.

Coulomb Charles 1736–1806 was a French physicist who established the laws governing electric charge and magnetism. The unit of electric charge is named the coulomb in his honour.

Coulomb was born in Angoulême on 14 June 1736 and educated at the Ecole du Génie in Mézières, graduating in 1761 as a military engineer with the rank of First Lieutenant. After three years in France, he was posted to Martinique to undertake construction work and remained there from 1764 to 1772. Postings followed within France to Bouchain, Cherbourg, Rochefort and eventually to Paris in 1781.

In 1774, Coulomb had become a correspondent to the Paris Academy of Science. Three years later he shared the first prize in the Academy's competition with a paper on magnetic compasses and, in 1781, gained a double first prize with his classic work on friction. In the same year, he was elected to the Academy. During his years in Paris, his duties were those of an engineering consultant and he had time in hand for his physics research. The year 1789 marked the beginning of the French Revolution and Coulomb found it prudent to resign from the army in 1791. Between 1781 and 1806, he presented 25 papers (covering electricity and magnetism, torsion, and applications of the torsion balance) to the Academy, which became the Insitut de France after the Revolution. He was also a contributor to several hundred committee reports to the Academy on engineering and civil projects, machinery and instruments. In 1801 he became President of the Institut de France; he died in Paris on 23 August 1806.

Coulomb took full advantage of his various postings to pursue a variety of studies, including structural mechanics, friction in machinery and the elasticity of metal and silk fibres. In structural mechanics, he drew greatly on his experience as an engineer to investigate the strengths of materials and to determine the forces that affect materials in beams. In his study of friction, he extended knowledge of the effects of friction caused by factors such as lubrication and differences in materials and loads, producing a classic work that was not surpassed for 150 years. In both these fields, Coulomb greatly influenced and helped to develop engineering in the nineteenth century. He also carried out fundamental research in ergonomics, which led to an understanding not only of the ways in which people and animals can best do work but also, in influencing the subsequent research of Gaspard Coriolis (1792–1843), to a proper understanding of the fundamental nature of work.

Coulomb's major contribution to science, however, was in the field of electrostatics and magnetism, in which he made use of a torsion balance he invented and which is described in a paper of 1777. He was able to show that torsion suspension can be used to measure extremely small forces, demonstrating that the force of torsion is proportional to the angle of twist produced in a thin stiff fibre. This paper also contained a design for a compass using the principle of torsion suspension, which was later adopted by the Paris Observatory.

Coulomb was very interested in the work of Joseph Priestley (1733–1804) on electrical repulsion, and in his paper of 1785 he discussed the adaptation of his torsion balance for electrical studies. He demonstrated that the force between two bodies of opposite charge is inversely proportional to the square of the distance between them. He also stated but did not demonstrate that the force of attraction or repulsion between two charged objects is proportional to the product of the charges on each. Coulomb's next paper, in 1787, produced proof of the inverse square law for both electricity and magnetism, and for both attractive and repulsive forces. A magnetic needle or charged pith ball was suspended from the torsion balance at a measured distance from a second similar needle or pith ball fixed independently, the torsion arm was deflected and the period of the resulting oscillations was timed. The experiment was repeated for varying distances between the fixed and oscillating bodies under test. Coulomb showed that if certain assumptions hold, the forces are proportional to the inverse square of the period and the period will vary directly as the distance between the bodies. The assumptions were (1) that the electrical or magnetic forces behave as if concentrated at a point, and (2) the dimensions of the bodies must be small compared with the distance between them. The results are embodied in Coulomb's law which states that the force between two electric charges is proportional to the product of the charges and inversely proportional to the square of the distance between them.

Coulomb went on to investigate the distribution of electric charge over a body and found that it is located only on the surface of a charged body and not in its interior.

Many of Coulomb's discoveries – including the torsion balance and the law that bears his name – had been made by John Michell (1724–1793) and Henry Cavendish

(a)

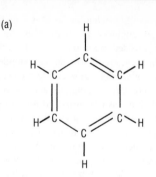

(b)

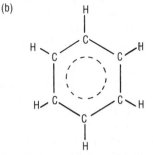

Coulson (a) 'Classical' (Kekulé) structure of benzene with alternate double and single carbon–carbon bonds that would not be equivalent. (b) Molecular orbital structure of benzene with six equivalent bonds; each carbon atom has three complete bonds and two partial bonds to complete its valency of four.

(1731–1810) in Britain. Michell died before completing his work and Cavendish neglected to publish his results, so Coulomb can truly be regarded as the principal scientist of his time. With his researches on electricity and magnetism, Coulomb brought this area of physics out of traditional natural philosophy and made it an exact science.

Coulson Charles Alfred 1910–1974 was a British theoretical chemist whose major contribution to the science was his molecular orbital theory and the concept of partial valency. He also worked – and held professorships – in physics and mathematics and published hundreds of papers on topics as diverse as pure mathematics and the effects of radiation on bacteria.

Coulson was born in Dudley, Yorkshire, on 13 December 1910, one of twin sons. His family moved to Bristol in 1920 and he went to Clifton College, from which he won an open scholarship to Trinity College, Cambridge, in 1928. He graduated in mathematics three years later. He was awarded a research scholarship, and later an open scholarship, and worked with Lennard Jones, the first holder of the chair in theoretical chemistry. It was then that he was introduced to wave mechanics and quantum theory, to which he was to devote himself so successfully for much of his career. He remained at Cambridge until 1938, when

he married and took up an appointment as Senior Lecturer in Mathematics at the University of Dundee. In 1930 he had adopted deep Christian beliefs that were to lead him to become a pacifist; for many years he was chairman of the charity Oxfam (1965–71).

In Dundee, Coulson taught and carried out research. His first book, *Waves* (1941), ran to seven editions and is still in worldwide use. In 1945 he moved to Oxford University to join the Physical Chemistry Laboratory as a 'theoretician', and soon also became a mathematics lecturer. Two years later he accepted the chair of theoretical physics at King's College, London. His second book, *Electricity*, was published in 1948 and was another immediate success.

In 1952 came the appearance of his third bestseller, *Valence*. In the same year he returned to Oxford as a professor of mathematics. But he continued to teach chemistry and physics, beginning his famous summer schools in 1955. In 1963 he became Curator of the Mathematics Institute (where he started the first university computing department), and was appointed its president in 1971. He further demonstrated his versatility by becoming Oxford's first professor of theoretical chemistry. He died suddenly in 1974 from an illness that had begun some years earlier.

Coulson's great contribution to chemistry was the way in which he influenced thinking about forces in molecules. Since J. J. Thomson's discovery of the electron in 1897 and Ernest Rutherford's of the atomic nucleus in 1911, it had been obvious that chemical bonds must involve electrons, but nobody fully understood what form this involvement takes. The quantum theory of the time showed how electrons occupy energy levels in atoms, but it gave no real help with molecular forces. In the 1920s wave mechanics began to revolutionize the subject, but Erwin Schrödinger's famous wave equation becomes hopelessly complicated in all but the simplest cases. Coulson used powerful methods of approximation and computation to obtain some solutions of the wave equation for molecular systems.

The molecular orbital theory that Coulson developed is an extension of atomic quantum theory and deals with 'allowed' states of electrons in association with two or more atomic nuclei, treating a molecule as a whole. He was thus able to explain properly phenomena such as the structure of benzene and other conjugated systems, and invoked what he called partial valency to account for the bonding in such compounds as diborane – neither of which could be accounted for by the alternative valence bond system proposed by Linus Pauling and others.

Practical applications of Coulson's molecular orbital theory have included the prediction of new aromatic systems and accurate forecasting of bond lengths and angles.

Coulson also contributed significantly to the understanding of the solid state (particularly metals), such as the structure of graphite and its 'compounds'. He developed many mathematical techniques for solving chemical and physical problems while retaining the capacity to produce simple, intuitive models of the systems he was studying, and the ability to formulate these models in manipulatable mathematical terms.

Courant Richard 1888–1972 was a German-born US mathematician who taught mathematics from a very early age, wrote several textbooks that are now standard reference works, and founded no fewer than three highly influential mathematical institutes.

Courant was born the son of a Jewish businessman on 8 January 1888 in Lublintz, in Upper Silesia (now in Poland). He attended schools in Glatz, and then Breslau (Wroclaw), where he showed exceptional talent and was soon teaching privately people who were several years above him at school. At the age of 16 Courant was earning so much money as a teacher that he remained in Breslau when his parents moved elsewhere; at school, however, he was told to give up his tutoring or be expelled. Instead, he left of his own accord, and began attending lectures unofficially at the university there. The following year, he entered the university on an official basis, but in 1907, on the advice of one of his friends, he moved on to the University of Göttingen, where he quickly distinguished himself. He became acquainted with David Hilbert (1862–1943), and tutored Hilbert's son, becoming also Hilbert's assistant – a position that allowed him to be at the centre of one of the most thriving contemporary mathematical communities. In 1910 he was awarded his doctorate for an investigation into Dirichlet's principle. The following year he returned to Göttingen, married, and prepared the thesis to qualify him as a university teacher. During World War I, Courant served as an infantryman (being promoted to lieutenant) until he was wounded in 1915. At about this time he interested the military authorities in a device for sending electromagnetic radiation through the earth to carry messages; this was developed, and Courant spent the rest of the war involved with communications. He then returned to Göttingen for a few months before being appointed Professor of Mathematics at the University of Münster. However, he was soon recalled to the vacant chair at Göttingen (which only a few years earlier had been occupied by Felix Klein).

Courant spent more than ten years at Göttingen, gradually building up the mathematics department into an autonomous unit, and raising funds for new buildings. He found his position untenable, however, as a Jew in Adolf Hitler's Germany, and after a visit to the United States in 1931–32, settled there in 1934, joining the teaching staff at New York University. Neither the university nor his position were similar to that which he had been used to in Germany, but from these small beginnings he again built the mathematical department into a renowned centre of research; the success he achieved was even greater than that in Göttingen, in that he was Director of the Institute of Mathematical Sciences of New York University from 1953 to 1958, and in 1962 work was begun on a new institute which was opened in 1965 as the Courant Institute of Mathematical Sciences. Courant retired in 1958 but still retained an interest in mathematics and mathematical education. He died in New Rochelle on 27 January 1972.

Much of what Courant achieved during his life was as a result of his great skill in administration and organization.

At Göttingen, Felix Klein (1849–1925), although he had retired some years before, still enjoyed the privileged status of Emeritus Professor and wielded enormous influence in the structure of its mathematical teaching and organization. It was he who saw in Courant a worthy successor to himself as administrator. In his own characteristic way, Courant then set about developing the few rooms and the autonomous mathematical professors and other staff into a mathematical institute worthy of the reputation it already commanded throughout the world. By subtle changes (including that of the title on the stationery) he separated the mathematics department from the philosophical faculty. Then he persuaded the International Education Board to donate the money for a new building. It was unfortunate that after this was built and Courant appointed its first director in 1929 that Hitler came to power, because four years later Courant, being of Jewish descent, was suspended from duty. Nevertheless, once in the United States, Courant managed to do it all over again, this time more by paying close attention to the nurturing of graduates, and by publicizing the fact that his courses dealt with applied, as well as pure, mathematics.

Courant was also a very able writer on mathematical topics. By treating many of the subjects developed by his mentor David Hilbert, he prepared a book, *Methods of Mathematical Physics* (now universally known as the 'Courant–Hilbert'), which turned out to be just what was needed by physicists in their research on the quantum theory. The first volume was published while he was in Germany, the second after he had settled in the United States. Its influence on the scientific community has been so great that it is still in use today. Another book, *Differential and Integral Calculus*, also in two volumes, has been used for 40 years as a university textbook, and is still in use in an updated form. Like the Courant–Hilbert it was soon translated into English. Courant was also sole or part author of a number of books on more specialized areas of mathematics. One general book still of great interest, however, is his *What is Mathematics?*, written in conjunction with Herbert Robbins and published in 1941.

Cousteau Jacques-Yves 1910–   is the French diver who invented the aqualung and pioneered many of the diving techniques used today.

Cousteau was born on 11 June 1910 at Saint André in the Gironde. He became a gunner in the French navy shortly before the outbreak of World War II, and worked in Marseilles for Naval Intelligence during the Nazi occupation. His interest in diving came in 1936 when he borrowed a pair of goggles and peered beneath the surface of the Mediterranean. He was instantly captivated by what he saw. Throughout the occupation he and his colleagues Frederick Dumas, Phillipe Taillez and Emile Gagnan experimented with diving techniques in comparative peace.

Cousteau initially experimented with available naval equipment. Le Prieur had invented the compressed air cylinder in 1933 that released a continuous flow of air through the face mask. This system restricted the diver to very short periods of time beneath the surface. Cousteau designed an oxygen rebreathing apparatus that would give him longer dives, but after two near-fatal accidents he abandoned his ideas. He also tested the Fernez equipment, which fed compressed air to the diver from the surface, and again was nearly killed in the process.

The turning point came in 1942, when Cousteau met Emile Gagnan, an expert on industrial gas equipment. Gagnan had designed an experimental demand valve for feeding gas to car engines. Together, Gagnan and Cousteau developed a self-contained compressed air 'lung', the aqualung. In June 1943 Cousteau made his first dive with it, achieving a depth of 18 m/ 60 ft.

In 1945, Cousteau founded the French Navy's Undersea Research Group. Much of its early diving work involved locating and defusing the mines left behind by the German navy. He fitted out and commanded the group's research ship, the *Elie Monnier*, on many oceanographic expeditions.

In 1951 he set out in the research ship *Calypso* on a four-year voyage of exploration beneath the oceans the world. He went on to make a series of films about his many voyages aboard the ship, becoming a popular television personality as a result.

The aqualung has changed very little since Cousteau's original design. A free-swimming, air-breathing diver can now descend to depths of over 60 m/200 ft and carry out work. Greater depths require extreme care and sophisticated gas mixtures to avoid danger.

Without Cousteau's contribution to the science of diving, the massive underwater tasks performed offshore for the oil industry and others would have been vastly more difficult. The foundations he laid for modern divers are being built upon constantly, allowing more difficult tasks to be carried out in ever more adverse conditions.

Cowling Thomas George 1906–1990 was a British applied mathematician and physicist who contributed significantly to modern research into stellar energy, with special reference to the Sun.

Cowling was born on 17 June 1906 and educated at the Sir George Monoux School in Walthamstow, Essex. He attended Oxford University, where he studied mathematics, and spent many years teaching in various university posts in London, Swansea, Dundee and Manchester before becoming Professor of Mathematics at University College, Bangor, in 1945. In 1948 he was appointed Professor of Mathematics at Leeds University, where he stayed until his retirement in 1970.

The author of several books on mathematics, Cowling became a Fellow of the Royal Society in 1947, and was made President of the Royal Astronomical Society for a two-year period in 1965.

Cowling's work was of important assistance in the discovery of the carbon–nitrogen cycle by Hans Bethe (1906–) in 1939. Bethe showed that the most significant source of energy in stars was the process by which four hydrogen atoms are converted into one helium atom, with carbon and nitrogen as intermediate products. This process was found to account satisfactorily for the rate of generation of the Sun's energy if its central temperature was 18.5 million kelvin (18.5 million °C/33 million °F) – an estimate that corresponded well with the temperature of 19 million kelvin (19 million °C/34 million °F) calculated following the theory of stellar structure previously proposed by Arthur Eddington (1882–1944).

Less indirectly, Cowling was responsible for demonstrating the existence of a convective core in stars, suggesting thus that the Sun may behave like a giant dynamo whose rotation, internal circulation and convection produce the immensely powerful electric currents and magnetic fields associated with sunspots. (The electric current required to produce the field of a large sunspot may be of the order of 10^{13} amperes; magnetic fields similarly may reach several thousand gauss.) With the Swedish astronomer, Hannes Alfvén (1908–), Cowling showed that such currents and fields would be difficult to initiate in the solar atmosphere and are therefore likely to have existed since the Sun was first formed.

Cramer Gabriel 1704–1752 was a Swiss mathematician who is now chiefly remembered for Cramer's rule, Cramer's paradox, and for the concept of utility in mathematics. He was, however, an influential teacher, personally acquainted with some of the great mathematicians of his age, and a prolific editor of other people's writings.

Cramer was born on 31 July 1704 in Geneva. It was there that he was educated, and there too – at the age of 20 – that he shared the Chair of Mathematics at the Académie de la Rive with his friend Calandrini (he taught geometry and mechanics, and Calandrini taught algebra and astronomy). In 1727 Cramer travelled to Basle, where he met Leonhard Euler (1707–1783) and his friend Daniel Bernoulli (1700–1782), and Daniel's uncle Jean Bernoulli (1667–1748) – all famous mathematicians. During the next 18 months he visited London, Leiden and Paris. (The next time he visited Paris, 20 years later, he took with him his pupil, the young Prince of Saxe-Gotha.) Cramer returned to Geneva, and in 1734 was appointed to the full Chair of Mathematics at de la Rive. In 1750 he was made Professor of Philosophy. In that year also, he published his major work, the *Introduction à l'analyse des lignes courbes algébriques,* in which 'Cramer's rule' provided a method for the solution of linear equations. The following year, however, he had an accident, after which he was advised to go to the south of France and rest. The journey there was itself too much for him; he died on the way, on 4 January 1752.

Cramer's work was acclaimed by his contemporaries and he received many honours. He was made a Fellow of the Royal Society in London, and of the Academies of Berlin, Lyons and Montpellier.

Although Cramer made several original contributions to mathematics, the two he remains famous for are Cramer's rule and Cramer's paradox. His rule, published in 1750, was responsible for a revival in interest over the use of determinants. Determinants exist as part of a method to solve linear equations, and although the German mathematician Gottfried Leibniz is most often credited with their discovery in 1693, there is some evidence that they were known to the Japanese mathematician Seki Kowa some years before. But despite the fact that they were also referred to in the major work of the Scotsman Colin Maclaurin in 1720, determinants had never received general attention until Cramer rediscovered them while working on the analysis of curves. The next year, Alexandre-Théophile Vandermonde (1735–1796) developed Cramer's work; his results were in turn extended by Pierre Laplace and Joseph Lagrange; later, the foundations of modern determinant theory were laid by Augustin Cauchy, Karl Jacobi and many others. Today, determinants are part of matrix theory, and are the means of classifying different systems of linear equations.

Cramer's paradox revolves around a theorem formulated by Colin Maclaurin. He stated that two different cubic curves intersect at nine points. Cramer pointed out that the definition of a cubic curve – a single curve – is that it is determined itself by nine points. But although he attempted an explanation, his was inadequate, and it was Leonhard Euler and others who derived a proper elucidation later.

Cramer's concept of utility now provides a connection between the theory of probability and mathematical economics. (That this was not the only interest Cramer showed in probability is revealed in his correspondence with Abraham de Moivre.)

As an editor of historically mathematical works, Cramer was indefatigable. In the last ten years of his life, he edited and published the collected works of Jean Bernoulli (whom he had met) and his brother Jacques Bernoulli (whom he had not), two volumes of correspondence between Jean Bernoulli and Gottfried Leibniz, and five volumes of the *Elementa* of Christian Wolf.

Crick, Watson, and Wilkins Francis Harry Compton Crick 1916– , James Dewey Watson 1928– , and Maurice Hugh Frederick Wilkins 1916– shared the 1962 Nobel Prize for Physiology or Medicine for their work on determining the structure of DNA, generally considered to be the most important discovery in biology this century.

Crick was born on 8 June 1916 in Northampton and was educated at Hill School. After graduating in physics from University College, London, he worked from 1940 to 1947 for the British Admiralty on the development of radar and magnetic mines. He then chose to study biology and worked at the Strangeways Research Laboratory at Cambridge University from 1947 to 1949, after which he moved to the Medical Research Council's Laboratory of Molecular Biology (also at Cambridge University). It was while working there in 1951 that he met James Watson, a student from the United States, and they performed their Nobel prize-winning research. Crick gained his PhD from Caius College, Cambridge, in 1953 and remained at the university – except for visiting lectureships to several US universities in 1959 and 1960 – until 1977, when he was appointed a Professor at the Salk Institute for Biological

Studies in San Diego, California.

Watson was born on 6 April 1928 in Chicago and, when only 15 years old, entered the University of Chicago to study zoology. He graduated when he was 19, in 1947, and then did postgraduate research on viruses at the University of Indiana, from which he gained his PhD in 1950. In the same year he went to the University of Copenhagen to continue his work on viruses but his interest turned to molecular biology and in 1951 he went to the Cavendish Laboratory at Cambridge University, where he performed the famous work on DNA with Crick. In 1953 Watson returned to the United States, where he was Senior Research Fellow in Biology at the California Institute of Technology until 1955. He then held several positions in the Department of Biology at Harvard University: Assistant Professor from 1955 to 1958, Associate Professor from 1958 to 1961 and Professor from 1961. In 1968 he became Director of the Cold Spring Harbor Laboratory of Quantitative Biology.

Wilkins was born on 15 December 1916 in Pongaroa, New Zealand, the son of a physician. Taken to England at the age of six, he was educated at King Edward VI School, Birmingham, and St John's College, Cambridge, from which he graduated in physics in 1938. He gained his doctorate from the University of Birmingham in 1940 and then joined the Ministry of Home Security and Aircraft Production to work on radar, and later went to the University of California as part of the British team assigned to the Manhattan Project to develop the atomic bomb. Disillusioned with nuclear physics, he became interested in biophysics and in 1945 took up a position on a biophysics project at St Andrews University, Scotland. In the following year he joined the Medical Research Council's Biophysics Research Unit at King's College, London, becoming its Assistant Director in 1950, Deputy Director in 1955 and Director in 1970 (until 1972). He was appointed Professor of Biophysics and Head of Department at King's College in 1970.

Following the discovery in 1946 that genes consist of DNA, Wilkins – working at King's College – began to investigate the structure of the DNA molecule. Studying the X-ray diffraction pattern of DNA, he discovered that the molecule has a double helical structure (one of his colleagues, Rosalind Franklin, also showed that DNA's phosphate groups are situated on the outside of the helix). Wilkins passed on his findings to Crick and Watson at Cambridge, who were trying to elucidate the detailed structure of DNA. Using Wilkins's results, Erwin Chargaff's discovery that nucleic acids contain only four different organic bases (with equal numbers of guanine and cytosine and equal numbers of adenine and thymine), and Alexander Todd's demonstration that nucleic acids contain sugar and phosphate groups, Crick and Watson postulated that DNA consists of a double helix consisting of two parallel chains of alternate sugar and phosphate groups linked by pairs of organic bases. They then built a series of accurate molecular models, eventually making one that incorporated all the known features of DNA and which gave the same diffraction pattern as that found by Wilkins. They envisaged replication occurring by a parting of the two strands of the double helix, each organic base thus exposed linking with a nucleotide (from the free nucleotides within a cell) bearing the complementary base. Thus two complete DNA molecules would eventually be formed by this step-by-step linking of nucleotides, with each of the new DNA molecules comprising one strand from the original DNA and one new strand (so-called semiconservative replication – as opposed to conservative replication, in which one DNA molecule would consist of both the original strands and the replicated molecule would consist of two new strands). Their model also explained how genetic information could be coded – in the sequence of organic bases. Crick and Watson published their work on the proposed structure of DNA in 1953, since when many other researchers have confirmed their hypothetical model and it is now generally accepted as correct.

Later Crick, this time working with Sydney Brenner, demonstrated that each group of three adjacent bases (he called a set of three bases a codon) on a single DNA strand codes for one specific amino acid. He also helped to determine codons that code for each of the 20 main amino acids. Furthermore, he formulated the adaptor hypothesis, according to which adaptor molecules mediate between messenger RNA and amino acids. These adaptor molecules, now known as transfer RNAs, were later independently identified by Paul Berg and Robert Holley.

After working on the structure of DNA, Watson researched into the genetic code and, later, on cancer; and Wilkins applied his X-ray diffraction technique to RNA.

Crompton Rookes Evelyn Bell 1845–1940 was a British engineer who became famous as a pioneer of the dynamo, electric lighting and road transport. During his long lifetime he also contributed to the development of various standards, both electrical and mechanical, including the British Association standards for small screw threads. In his later years, through experiment and design, he contributed to the design of the military tank.

Crompton was born near Thirsk in Yorkshire on 31 May 1845. His early days were spent in nearby Ripon. When he was 11 he accompanied his parents to Gibraltar on HMS *Dragon*, which was commanded by his mother's cousin. He then sailed on to the Crimea as a cadet on the same ship, and visited his brother during the siege of Sebastopol. When he returned to school in the autumn of 1856 – still only 12 years old – he was one of the youngest decorated members of the Royal Navy.

In 1858, after two years at Elstree, he entered Harrow School. During his holidays he built first a model steam-driven road locomotive, and then a full-sized one called *BlueBell*. On leaving school in 1864 he went to India as an army officer. During his service there he continued to develop his road vehicles and transferred out of his regiment so that he could concentrate his efforts on the development of steam locomotives for road haulage. The results were successful both in India and at home, but they

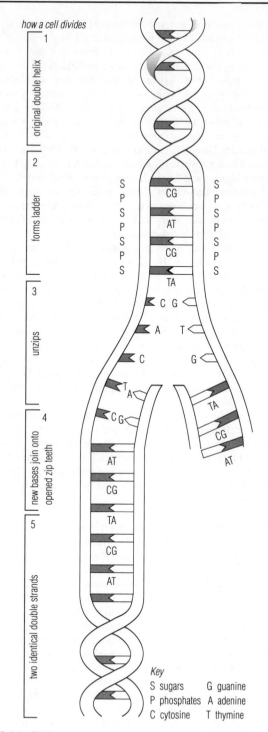

how a cell divides

1   original double helix
2   forms ladder
3   unzips
4   new bases join onto opened zip teeth
5   two identical double strands

Key
S sugars   G guanine
P phosphates   A adenine
C cytosine   T thymine

Crick *DNA replicates by a process that involves the 'unzipping' of the parent double helix.*

were in many ways ahead of their time. The poor quality of roads, the cheapness of other forms of transport and the developing railway system were to eclipse his pioneering work in this area.

On his return from service he became involved with the Stanton Iron Works in Derbyshire, which was owned by a branch of his family. He redesigned the works and brought some dynamos from France to power his electric lighting system. This innovation spread to other firms. At first he acted as a supplier of the lighting sets, and then as a manufacturer through his firm of Crompton and Co. Electrical Engineers. These were the early days of electricity generation and Crompton found an area ready for exploitation. He developed and manufactured generating systems for lighting town halls, railway stations and small residential areas. Direct current electricity of about 400 volts was generated and used with large storage batteries (accumulators) to allow the systems to operate smoothly within a range of demands. Later the supply of electricity became a highly profitable and fast-growing area, particularly with the coming of Swan's incandescent filament lamp, and Crompton's Kensington Court Electric Light Co. Ltd was formed. He believed strongly in his direct-current system with batteries, and there was much competition between him and his younger contemporary Ferranti, with his rival alternating-current system. As time went on Ferranti's method proved the better for large distribution networks, but Crompton's machines fared very well within their limitations.

During the Boer War, Crompton served in South Africa as commandant of the Electrical Engineers' Royal Engineers Volunteer Corps, which was later to become the Royal Corps of Electrical and Mechanical Engineers (REME). At the turn of the century he returned to road transport, and contributed considerably to the principles of automobile engineering and the maintenance and design of roads which would withstand the new demands put upon them. He was a founder member of what is now the Royal Automobile Club.

His firm, which had manufactured all kinds of electrical machinery and instruments, became, after a merger in 1927, Crompton Parkinson Ltd. He continued as a director. In his later years he was internationally respected as an expert electrical engineer and a foremost authority on storage batteries. He used his influence to achieve standardization in industry generally, and was involved in the founding of the National Physical Laboratory and what is now the British Standards Institution. Heavy industry had the Whitworth Standard for screw threads, but there was no such standard for the small screws used in instruments. Crompton was a prime mover in the formation of the British Association committee which eventually developed the well-known BA Standards.

During World War I he was an adviser on the design and production of military tanks, having been appointed by Winston Churchill (who was then First Lord of the Admiralty). Crompton carried out much research of his own but the final vehicle embodied the work of several designers.

Many honours were awarded to him, including honorary membership and medals of the engineering institutions. He was elected a Fellow of the Royal Society in 1933. Crompton died at Azerley Chase, Yorkshire, on 15 February 1940, aged 94 years.

Crompton Samuel 1753–1827 was a British inventor whose machine for spinning fine cotton yarn – the spinning mule – revolutionized the industrial production of high-quality cotton textiles.

Crompton was born 3 December 1753 on a small farm at Firwood near Bolton, Lancashire. His parents were poor and from an early age he was expected to work alongside them on their piece of land. When he was six years old his father died, and his mother was forced to rely heavily upon the cottage industry of spinning and weaving to support the family. Samuel had to help with this too, and he grew up familiar with all the associated problems. These experiences were to influence him for the rest of his life.

At the age of 21, with years of home spinning and weaving behind him, he resolved to try to design a better method for spinning the yarn than James Hargreaves's spinning jenny (patented in 1770). The jenny's yarn tended to break frequently and produced a coarse cloth which was commercially suitable only for the working-class market.

It took Crompton five years of hard work and all his money to develop a machine that span a yarn so fine and continuously that it revolutionized the cotton industry. The machine became known as a local wonder, and people flocked to Crompton's house to catch a glimpse of his spinning mule. Unfortunately, because he had used all his savings, he had no means of patenting the design and he lived in fear of someone stealing his idea. Eventually, he was persuaded to reveal his invention to a group of Bolton's manufacturers, who promised to pay him a generous subscription in return. This never materialized other than the initial payment of £7 6s 6d (£67.32).

Aggrieved at his treatment by the Bolton manufacturers, he tried to set up his own company, but having 'sold' his new machine, he was soon outpaced by the other manufacturers (with better resources) and forced to sell up. By 1812 the boost given to the cotton industry by the introduction of Crompton's mule was at last recognized, and a sum of approximately £500 was raised by subscription. A national award was also sought, and this time £5,000 was offered, but Crompton was far from satisfied and returned home to Bolton from London, a broken man.

Once again he tried to enter business, this time as a partner in a cotton firm. Ultimately this too failed, and he was forced to turn to the generosity of friends for an annuity. He died on 26 June 1827 at Bolton, a saddened man.

Crompton's new invention, which contributed so much to the wealth of the textile industry, was called the mule because, just like the animal of that name, it was a hybrid. His machine used the best from the spinning jenny and from Richard Arkwright's water frame of 1768. The strong, even yarn it produced was so fine that it could be used to weave delicate fabrics such as muslin which became particularly fashionable among the middle and upper classes, creating a new market for the British cotton trade. As a direct result of the better and faster method for spinning, which took the job out of the home and into the factories, coupled with the increase of raw cotton available from America, the cotton trade entered a golden age. The introduction of the power loom in the early 1800s further mechanized production and made the system of using home weavers practically obsolete. The Industrial Revolution had begun in the textile industry.

Cronin James Watson 1931–   . US physicist; *see* Fitch and Cronin.

Crookes William 1832–1919 was a British physicist and chemist who, in a scientific career lasting more than 50 years, made many fundamental contributions to both sciences. He is best known in physics for his experiments with high-voltage discharge tubes and for inventing the Crookes radioscope or radiometer. A major achievement in chemistry was the discovery of the element thallium. Crookes never held a senior academic post, and carried out most of his researches in his own laboratory.

Crookes was born in London on 17 June 1832, the eldest of the 16 children of a tailor and businessman. Little is known of his early education but in 1848 he began a chemistry course at the new Royal College of Chemistry under August von Hofmann, who later made him a junior assistant, promoting him to senior assistant in 1851. Crookes's only academic posts were as Superintendent of the Meteorological Department at the Radcliffe Observatory, Oxford (1854–55), and Lecturer in Chemistry at Chester Training College (1855–56). He left Chester after only one year, dissatisfied at being unable to carry out original research and after having inherited enough money from his father to make him financially independent for life.

He returned to London and became secretary of the London Photographic Society and editor of its *Journal* in 1858. The following year he set up a chemical laboratory at his London home and founded as sole manager and editor the weekly *Chemical News*, which dealt with all aspects of theoretical and industrial chemistry and which he edited until 1906. Among his greatest achievements in chemistry was the discovery in 1861 of the element thallium using the newly developed spectroscope. He also applied this instrument to the study of physics. He was knighted in 1897 and made a Member of the Order of Merit in 1910. He died in London on 4 April 1919.

One of Crookes's earliest pieces of chemical research concerned the preparation and properties of potassium selenocyanate (potassium cyanoselenate (VI)), which he made using selenium isolated from 5 kg/11 lb of waste from a German sulphuric acid works. Then in 1861, while examining selenium samples by means of a spectroscope – the method newly developed by Robert Bunsen (1811–1899) and Gustav Kirchhoff (1824–1887) – Crookes observed a transitory green line in the spectrum and attributed it to a new element which he called thallium (from the Greek *thallos*, meaning a budding shoot). Over the next few years he determined the properties of thallium and its compounds; using thallium nitrate and a specially constructed sensitive balance he measured thallium's atomic weight (relative atomic mass) as 203.715 ± l0.0365 (modern value: 204.39).

This work involved weighing various samples of material in a vacuum, and he observed that when the delicate balance was counterpoised it occasionally made unexpected swings. He began to study the effects of light radiation on objects in a vacuum and in 1875 devised the radioscope (or radiometer). The instrument consists of a four-bladed paddle-wheel mounted horizontally on a pinpoint bearing inside an evacuated glass globe. Each vane of the wheel is black on one side (making it a good absorber of heat) and silvered on the other side (making it a good reflector). When the radioscope is put in strong sunlight, the paddle-wheel spins round. Although little more than a scientific toy, it defied attempts to explain how it works until James Clerk Maxwell correctly showed that it is a demonstration of the kinetic theory of gases. The few air molecules in the imperfect vacuum in the radioscope bounce more strongly (with more momentum) off the heated, black sides of the vanes (than off the cooler, silvered sides), creating a greater reaction which 'pushes' the paddle-wheel around. Crookes's own observations were described in his paper 'Attraction and repulsion resulting from radiation' (1874).

During the 1870s Crookes's studies concerned the passage of an electric current through glass 'vacuum' tubes containing rarified gases; such discharge tubes became known as Crookes tubes. The ionized gas in a Crookes tube gives out light – as in a neon sign – and Crookes observed near the cathode a light-free gap in the discharge, now called the Crookes dark space. He named the ion stream 'molecular rays' and demonstrated how they are deflected in a magnetic field and how they can cast shadows, proving that they travel in straight lines. He made similar observations about cathode rays, but it was left to J. J. Thomson (1856–1940) to understand the true significance of such experiments and to discover the electron (in cathode rays) in 1897. Crookes also noted that wrapped and unexposed photographic plates left near his discharge tubes became fogged, but he did not follow up the observation, which was later the basis of Wilhelm Röntgen's discovery of X-rays in 1895.

In the 1880s Crookes studied the phosphorescent spectra of rare earth minerals, principally substances containing yttrium and samarium. He made the first references to what Frederick Soddy was to call isotopes. While experimenting with radium, he devised the spinthariscope. The instrument consists of a screen coated with zinc sulphide at the end of a tube fitted with a low-powered lens. When alpha particles emitted from a radioactive source (such as radium) hit the screen they produce a small flash of light.

Crookes's interests were very wide. Topics covered by his publications included chemical analysis; the manufacture of sugar from sugarbeet; dyeing and printing of textiles; oxidation of platinum, iridium and rhodium; use of carbolic acid (phenol) as an antiseptic in the treatment of diseases in cattle; the origin and formation of diamonds in South Africa; and the use of artificial fertilizers and their manufacture from atmospheric nitrogen.

Crookes was never afraid of pursuing an idea counter to

the trend of contemporary opinion. For several years, for example, he was very interested in spiritualism and published several papers that described experiments undertaken by a medium. To many of his scientific colleagues this was akin to heresy!

Cugnot Nicolas Joseph 1725–1804 was a French soldier and engineer who was the first to make a self-propelled road vehicle (in 1769). His three-wheeled machine was designed for towing guns: the front wheel was driven by a two-cylinder steam engine, and it could carry four people at a walking pace. As such it was the first automobile.

Cugnot was born at Void, Meuse. As a young man he joined the French army and served for a time in Germany and Belgium, inventing a new kind of rifle used by French troops under Marshal du Saxe, who encouraged him to work on a steam-propelled gun-carriage. After serving in the Seven Years' War, Cugnot returned to Paris in 1763 as a military instructor. He also devoted his time to writing military treatises and exploring a number of inventions he had conceived during his campaigning, obtaining official help from the Duc de Choiseul, then Minister of War. His major invention was a steam-propelled tractor for hauling artillery. He built two models, the first appearing in 1769 and the second in 1770. By that time he had also constructed a truck, now preserved in the Conservatoire des Arts et Métiers, Paris. The truck ran at a speed of 3–5 kph/ 2–3 mph before a large crowd of official spectators, and Cugnot was commissioned to make a larger one. The fall of the Duc de Choiseul led to the project being shelved and the truck was not tested further. Granted a pension in 1779 from the Ministry of War, Cugnot migrated to Brussels. The pension stopped at the outbreak of the French Revolution. In 1798 Napoleon asked the Institut de France to enquire into Cugnot's machine, but nothing came of this.

In 1698 the English inventor Thomas Savery (c.1650– 1715) had invented a steam engine designed for pumping water out of mines. This was the first practical use of the steam engine, but the engine itself was inefficient and wasted a lot of steam. In 1690 the French physicist Denis Papin (1647–c.1712) designed a superior engine, using high-pressure steam expansively without condensation. Cugnot improved on this, making an engine driven by the movements of two piston rods working from two cylinders (which, like the boiler, were made of copper); it carried

no reserves of water or fuel. Although it proved the viability of steam-powered traction, the problems of water supply and pressure maintenance severely handicapped the vehicle.

Cugnot's vehicle was a huge, heavy tricycle and his model of 1769 was said to have run for 20 minutes while carrying four people, and to have recuperated sufficient steam pressure to move again after standing for 20 minutes. Cugnot was an artillery officer and the more-or-less steam-tight pistons of his engine were made possible by the invention of a drill that accurately machined cannon bores. The first post-Cugnot steam carriage appears to have been that built in Amiens in 1790, although followers of Cugnot were soon on the road in other countries, notably in Britain. Steam buses were running in Paris about 1800. Oliver Evans (1755–1819) of Philadelphia ran an amphibious steam dredger through the streets of that city in 1805. English exponents of the new form of propulsion became active, and by the 1830s the manufacture and use of steam road carriages had approached the status of an industry. James Watt's foreman William Murdock (1754–1839) ran a model steam carriage on the roads of Cornwall in 1784, and Robert Fourness showed a working three-cylinder tractor in 1788. Richard Trevithick (1771–1833) developed Murdock's ideas and at least one of his carriages with driving wheels 4.8 m/10 ft in diameter ran in London.

Between 27 February and 22 June 1831, steam coaches ran about 6,500 km/4,000 mi on the regular Gloucester– Cheltenham service, carrying some 3,000 passengers. Thus many passengers had been carried by steam carriage before the railways that were to cause their demise had accepted their first paying passenger. The decline did not mean that all effort in this field would be abandoned, and much attention was given to the steam tractor as a prime mover. Beginning in about 1868, Britain was the scene of a vogue for light, steam-powered personal carriages; if the popularity of these vehicles had not been hindered by legislation, it would certainly have resulted in the appearance of widespread enthusiasm for motoring in the 1860s rather than in the 1890s. Steam tractors, or traction engines, were also used in agriculture and on the roads. It is thus possible to argue that the line from Cugnot's first lumbering vehicle runs unbroken to the twentieth-century steam automobiles which were made as late as 1926.

Curie Marie 1867–1934 was a Polish-born French scientist who, with her husband Pierre Curie 1859–1906 was an early investigator of radioactivity. The Curies discovered the radioactive elements polonium and radium, for which achievement they shared the 1903 Nobel Prize for Physics with Henri Becquerel. Madame Curie went on to study the chemistry and medical applications of radium, and was awarded the 1911 Nobel Prize for Chemistry in recognition of her work in isolating the pure metal.

Madame Curie's Polish maiden name was Manya Sklodowska. She was born in Warsaw on 7 November 1867, at a time when Poland was under Russian domination after the unsuccessful revolt of 1863. Her parents were teachers

Cugnot *Cugnot's steam tractor.*

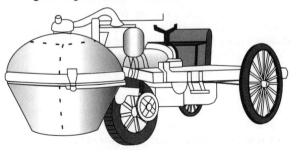

and soon after Manya was born – their fifth child – they lost their teaching posts and had to take in boarders. Their young daughter worked long hours helping with the meals, but nevertheless won a medal for excellence at the local high school, where the examinations were held in Russian. No higher education was available so Manya took a job as a governess, sending part of her savings to Paris to help to pay for her elder sister's medical studies. Her sister qualified and married a fellow doctor in 1891 and Manya went to join them in Paris. She entered the Sorbonne and studied physics and mathematics, graduating top of her class. In 1894 she met the French chemist Pierre Curie and they were married the following year.

Pierre Curie was born in Paris on 15 May 1859, the son of a doctor. He was educated privately and at the Sorbonne, becoming an assistant there in 1878. He discovered the piezoelectric effect and, after being appointed head of the laboratory of the Ecole de Physique et Chimie, went on to study magnetism and formulate Curie's law (which states that magnetic susceptibility is inversely proportional to absolute temperature). In 1895 he discovered the Curie point, the critical temperature at which a paramagnetic substance become ferromagnetic. In the same year he married Manya Sklodowska.

From 1896 the Curies worked together on radioactivity, building on the results of Wilhelm Röntgen (who had discovered X-rays) and Henri Becquerel (who had discovered that similar rays are emitted by uranium salts). Madame Curie discovered that thorium also emits radiation and found that the mineral pitchblende was even more radioactive than could be accounted for by any uranium and thorium content. The Curies then carried out an exhaustive search and in July 1898 announced the discovery of polonium, followed in December of that year with the discovery of radium. They eventually prepared 1 g/0.04 oz of pure radium chloride – from 8 tonnes of waste pitchblende from Austria. They also established that beta rays (now known to consist of electrons) are negatively charged particles.

In 1906 Pierre Curie was run down and killed by a horse-drawn carriage. Marie took over his post at the Sorbonne, becoming the first woman to teach there, and concentrated all her energies into research and caring for her daughters (one of whom, Irène, was to later marry Frédéric Joliot and become a famous scientist and Nobel prizewinner). In 1910 with André Debierne (1874–1949), who in 1899 had discovered actinium in pitchblende, she isolated pure radium metal.

At the outbreak of World War I in 1914 Madame Curie helped to equip ambulances with X-ray equipment, which she drove to the front lines. The International Red Cross made her head of its Radiological Service. Assisted by Irène Curie and Martha Klein at the Radium Institute she held courses for medical orderlies and doctors, teaching them how to use the new technique. By the late 1920s her health began to deteriorate: continued exposure to high-energy radiation had given her leukaemia. She entered a sanatorium at Haute Savoie and died there on 4 July 1934, a few months after her daughter and son-in-law, the Joliot-Curies, had announced the discovery of artificial radio-activity.

Throughout much of her life Marie Curie was poor and the painstaking radium extractions were carried out in primitive conditions. The Curies refused to patent any of their discoveries, wanting them freely to benefit everyone. The Nobel prize money and other financial rewards were used to finance further research. One of the outstanding applications of their work has been the use of radiation to treat cancer, one form of which cost Marie Curie her life.

Curtis Heber Doust 1872–1942 was a US astronomer who became interested in astronomy only after he had begun a career in classics, but who went on to carry out important research into the nature of spiral nebulae.

Curtis was born in Muskegan, Michigan, on 27 June 1872. He attended Detroit High School, where he displayed a flair for languages and an interest in the classics. It was primarily in these subjects that he concentrated his efforts while a student at the University of Michigan. Curtis gained his BA in 1892 and his MA a year later. His first job was teaching Latin at his old high school, and then, aged only 22, he became Professor of Latin at Napa College, California, where he shortly became aware of the availability of a refracting telescope and small observatory.

In 1897 Curtis abruptly changed the entire direction of his career and became Professor of Mathematics and Astronomy at the University of the Pacific. He worked at the Lick Observatory, California, in 1898; in 1900 he became Vanderbilt Fellow at the Leander McCormick Observatory at the University of Virginia; and in 1902 he returned to become an assistant at the Lick Observatory. There he was promoted to Assistant Astronomer in 1904. In 1906, under the auspices of the Observatory, he took charge of work being done at an observatory in Chile. Three years later, he returned to the Lick, where he worked as an astronomer until he retired from research work in 1920. His next appointment was as Director of the Allegheny Observatory, a post which he held until 1930, when he became Director of the Observatory at the University of Michigan. He died in Ann Arbor, Michigan, on 8 January 1942.

Curtis's first astronomical studies were of total solar eclipses in Thomaston, Georgia, in 1900, and in Solok, Sumatra, in 1901. But the main value of Curtis's early work at the Lick Observatory lay in his contributions to the programme for the measurement of stellar radial velocities, undertaken under the direction of William Campbell (1862–1938). He worked on this programme at Mount Hamilton from 1902 until 1906, and then in Chile from 1906 until 1909. For the following 11 years Curtis concentrated his efforts on the photography of spiral nebulae and on research into their nature.

Ever since Charles Messier (1730–1817) had included 'nebulosities' in his catalogue of 1771, their precise composition had been the subject of dispute. There were two main schools of thought: that they were either giant star clusters far beyond our own Galaxy – as proposed by Richard Proctor (1837–1888) – or that they were merely clouds of debris. The scale of the universe itself was central to both points of view. Through his photography of spiral nebulae, Curtis began to appreciate the actual vastnesses of space and to incline towards Proctor's view of 'islands in the universe'.

He also noticed that on photographs of spiral nebulae viewed edge-on there was a dark line along the rim of each nebula. This suggested to Curtis a combination of the two theories: that spiral nebulae might indeed be complex galaxies like our own, and that such galaxies produced a cloud of debris which accumulated in the plane of the galaxy. If such a cloud of debris had also gathered outside our own Galaxy, this would explain the reported 'zone of avoidance' – spiral nebulae never appeared in the Milky Way (i.e. in the plane of our own Galaxy). Spiral nebulae in that position, it now was evident, would simply be obscured by dust.

Following his appointment as Director of the Allegheny Observatory in 1920, Curtis's research output declined. His most important contributions to astronomy remain therefore improving the modern understanding of the nature and position of spiral nebulae.

Cushing Harvey Williams 1869–1939 was a US surgeon who pioneered several important neurosurgical techniques, made famous studies of the pituitary gland, and first described the chronic wasting disease now known as Cushing's syndrome (or disease).

Cushing was born on 8 April 1869 in Cleveland, Ohio, the fourth child in a family of physicians. He studied medicine at Yale College and at the Harvard Medical School, graduating from the latter in 1895. He then spent about four years in practical training at the Massachusetts General Hospital, Boston, and the Johns Hopkins Hospital, Baltimore, where he worked under William Halsted (1852–1922), a great innovator of surgical techniques. At about the turn of the century Cushing studied in Europe – under Emil Kocher (1841–1917) at Berne University and, briefly, under the famous neurophysiologist Charles Sherrington (1857–1952) in England – after which he returned to the Department of Surgery at the Johns Hopkins University. From 1912 to 1932 Cushing was Professor of Surgery at the Harvard Medical School and Surgeon-in-Chief at the Peter Bent Brigham Hospital in Boston. During this period he served in the Army Medical Corps in World War I and in 1918 was appointed Senior Consultant in Neurological Surgery to the American Expeditionary Force. In 1933 he became Sterling Professor of Neurology at Yale University, a post he held until his retirement in 1937. Cushing died on 7 October 1939 in New Haven, Connecticut. He bequeathed his large collection of books on the history of medicine and science to the Yale Medical Library.

Although Cushing is probably best known for his work on Cushing's syndrome, his major contribution was in the field of neurosurgery, which, until he introduced his pioneering techniques, was seldom successful. As a result of

experimenting on the effect of artificially increasing intercranial pressure in animals, Cushing developed new methods of controlling blood pressure and bleeding during surgery on human beings. Moreover, his whole approach to medicine was characterized by painstaking carefulness: before operating he gave each of his patients an extremely thorough physical examination and took a detailed medical history. The operations themselves, which usually lasted for many hours, were performed with meticulous care and, over the years, were increasingly successful.

In addition to developing neurosurgical techniques, Cushing wrote a description – still valid today – of the stages in the development of different types of intercranial tumours, classified such tumours, and published (in 1917) a definitive account of acoustic nerve tumours.

In 1908 Cushing began studying the pituitary gland and, after experimenting on animals, discovered a way of gaining access to this gland which, being situated at the base of the brain and behind the nasal sinuses, is extremely difficult to approach surgically. As a result of this discovery it became possible to treat cases of blindness caused by tumours pressing on the optic nerve in the region of the pituitary gland. Cushing also investigated the effects of abnormal activity of the pituitary gland, establishing that hypopituitarism (undersecretion of pituitary hormones) in a growing person can cause a type of dwarfism and that hyperpituitarism (oversecretion of pituitary hormones) in fully grown adults can cause acromegaly (a form of giantism characterized by excessive growth of the bones of the hands, feet and face). As a result of his extensive studies of the pituitary gland, Cushing discovered the condition now called Cushing's syndrome, a rare chronic wasting disease with symptoms that include obesity of the face and trunk, combined with thin arms and legs; wasting of the muscles; atrophy of the skin, with the appearance of red lines on the skin; weakness; and accumulation of body fluids. Cushing attributed this disorder to a tumour of the basophilic cells of the anterior pituitary gland, but although this is one of the causes, the disorder is now known to be caused by any of several conditions that increase the secretion of glucocorticoids (particularly cortisol) by the adrenal glands, such as a tumour of the adrenal cortex itself.

Cushing was also interested in the history of medicine and in 1925 wrote a biography of William Osler (1849–1920) – one of the leading physicians of the time – which won him a Pulitzer prize.

Cuvier Georges 1769–1832 was a French zoologist eminent for his role in the founding of modern palaeontology.

Born at Montebéliard in the principality of Württemburg, Cuvier, the son of a Swiss soldier, received his training in natural history at Stuttgart, before spending six years as a private tutor in Normandy. He came to Paris in 1795 as assistant to the Professor of Comparative Anatomy at the Natural History Museum. In 1799, he was appointed Professor of Natural History at the Collège de France, and in 1802 Professor at the Jardin des Plantes. Cuvier proved a key figure in trailblazing new classificationary approaches to natural history, including a total rejection of the old idea of the great chain of being). But his supreme contribution lay in systematizing the laws of comparative anatomy and applying them to fossil vertebrates. Thanks to his work in this field, Cuvier was perhaps the most influential palaeontologist of the nineteenth century.

Cuvier's pursuits ranged widely through the animal kingdom. He divided invertebrates into three phyla, and conducted notable investigations into fish and molluscs. In two major texts, *Researches on the Fossil Bones of Quadrupeds* 1812 and the *Animal Kingdom* 1817, he reconstructed such extinct fossil quadrupeds as the mastodon and the palaeotherium, applying the principles and practices of comparative anatomy. Undertaken in collaboration with Brongniart, his stratigraphical explorations of the tertiary rocks of the Paris Basin demonstrated that fossil flora and fauna were specific to particular strata (a parallel discovery to that of William Smith in England). On the basis of these major conclusions in historical geology and vertebrate palaeontology, Cuvier judged that the history of the Earth had involved a chain of revolutions ('catastrophes') that had recurrently swept away entire living populations, their place being taken either by migration or by the creation of new species (in a manner that Cuvier tactfully chose never to specify). This theory, set out in his *Preliminary Discourse* 1812, expressly countered the evolutionary views of Lamarck and the palaeontologist Geoffrey Saint-Hilaire (1772–1844). In later life, Cuvier was much concerned with scientific organization and education. He was Councillor of State under Napoleon and later under Louis Philippe. In 1831 he was raised to the peerage of France, a rare honour for a Protestant.

D

Daimler Gottlieb Willhelm 1834–1900 was the German engineer who designed internal combustion engines of relatively advanced performance for automobiles, developing the motor car possibly more than Karl Benz (who had run a car at an earlier date).

Born at Schorndorf near Stuttgart on 17 March 1834, Daimler's technical education began in 1848 when he became a gunsmith's apprentice. Following a period at technical school in Stuttgart and factory experience in a Strasbourg engineering works, he completed his formal training as a mechanical engineer at the Stuttgart Polytechnic in 1859. He returned to Grafenstadt to do practical work for a while and then, sponsored by a leading Stuttgart benefactor, travelled to England where he worked for Joseph Whitworth. He then moved to France, where he many have seen Lenoir's newly developed gas engine.

Daimler spent the next ten years in heavy engineering. He joined Bruderhaus Maschinen-Fabrik in Reutlingen as manager in 1863, and there met Wilhelm Maybach, with whom he was to be closely involved for the rest of his life.

Daimler's work on the internal combustion engine began in earnest in 1872 when he teamed up with Dr Nikolaus August Otto (later to become famous for the Otto cycle) and Peter Langen, at the Gasmotoren-Fabrik Deutz, where Daimler was technical director. One of Daimler's first moves was to sign up Maybach as chief designer. Daimler was to work with Otto and Langen for the next ten years, studying gas engines (which resulted in Otto's historic patent of 1876) and perhaps also petrol engines.

Differences of opinion led to Daimler leaving the firm in 1881. He bought a house in Cannstatt, a suburb of Stuttgart, and it was in the summer house of this building that Daimler's first engines were built.

When he started work with Maybach, gas engines were being operated at 150–250 rpm. Daimler's first working petrol-fuelled unit, built in 1883, was an air-cooled, single-cylinder engine with a large cast-iron flywheel running at 900 rpm. With four times the number of power strokes per unit time, his engine had a very much greater output for a given size and weight. In itself, the use of petrol was not new: in 1870 Julius Hock in Vienna had built an engine

working on Lenoir's principle. A piston drew in half-a-cylinder-full of mixture, which was fired when the piston was halfway down the cylinder. Without compression, the power produced was very low and the fuel consumption massive.

The genius of Daimler and Maybach lay in the combining of four of the elements essential to the modern car engine: the four-stroke Otto cycle, the vaporization of the fuel with a device similar to a carburettor, low weight and high speeds. Lenoir had used electric ignition, but this proved unreliable; Daimler and Maybach used an igniter tube that was light, worked well and operated independently of engine speed.

Daimler's second engine, which ran later the same year, was a 0.4 kW/0.5 hp vertical unit. It was fitted to a cycle in November 1885 (possibly even earlier), creating the world's first motorcycle. Daimler was apparently not impressed with the possibilities of motorized two-wheelers and went on to try his engine as the power source for a boat.

In 1889, Daimler produced two cars, and obtained a licence to Panhard and Levassor in Paris to sell them. The first was a light four-wheeler with a tubular frame and a vertical, single-cylinder, water-cooled engine in the rear. It also featured a novel four-speed gear transmission to the rear wheels, and engine cooling water circulated through the frame, which acted as a radiator. The second car of 1889 had a belt drive and a vee-twin engine.

The two cars are important in that they show that Daimler had revised his earlier opinion that motorcars would be straight conversions of horsedrawn carriages. (Benz, on the other hand, had conceived his vehicle for motor-drive from the outset; nevertheless, Daimler's models were in many ways more advanced than the contemporary Benz models.)

In 1886, Daimler approached Sarazin, a representative of Otto and Langen at the Deutz works, eager to increase sales of his engines overseas. Sarazin persuaded Panhard and Levassor to manufacture Daimler engines under licence, but died before they went into production. The firm succeeded in entering the motor industry with the Daimler licence, following the marriage of Levassor to Sarazin's widow in 1890.

The Daimler Motoren Gesellschaft was also founded in 1890, but Daimler and Maybach both retired the following year to concentrate on technical and commercial development work, only to rejoin in 1895.

A Daimler-powered car won the 1894 Paris to Rouen race, the first international motor contest, organized to promote the concept of motoring. Six years after this great success, on 6 March 1900, Daimler died from heart disease, and was buried in Cannstatt.

Dale Henry Hallett 1875–1968 was a British physiologist, best known for his work on the chemical transmission of nerve impulses (particularly for isolating acetylcholine), for which he was awarded – jointly with Otto Loewi, a German pharmacologist – the 1936 Nobel Prize for Physiology or Medicine. Dale also received numerous British honours, including the Copley Medal of the Royal Society in 1937, a knighthood in 1943, the Order of Merit in 1944 and, at various times during his career, the presidencies of the Royal Society, the British Association, the Royal Society of Medicine, and the British Council. Furthermore, in 1959 the Society of Endocrinology struck the Dale Medal, an annual award, and in 1961 the Royal Society established the Henry Dale Professorship, bestowed by the Wellcome Trust, of which Dale had been the Chairman from 1938 to 1960.

Dale was born on 9 June 1875 in London, the son of a businessman, and was educated at Tollington Park College, London, then at Leys School, Cambridge. He read natural sciences at Trinity College, Cambridge, graduating in 1898, then succeeded Ernest Rutherford in the Coutts-Trotter Studentship at Trinity College. In 1900 Dale began his clinical training at St Bartholomew's Hospital, London, gaining a bachelor of surgery degree in 1903 and a medical degree in 1907.

While undergoing his clinical training, Dale continued his physiological studies in London between 1902 and 1904 under Ernest Starling and William Bayliss, first as a George Henry Lewes Student and later as a Sharpey Student in the Department of Physiology at University College, London. He also studied under Paul Ehrlich in Frankfurt for several months. In 1904 Dale accepted a post at the Wellcome Physiological Research Laboratories, becoming Director there two years later. In 1914 he was appointed Head of the Department of Biochemistry and Pharmacology of the Medical Research Council, and from 1928 until his retirement in 1942 he was Director of the National Institute for Medical Research. He died in Cambridge on 23 July 1968.

Dale's earliest research, performed while he worked at the Wellcome Physiological Research Laboratories, concerned the chemical composition and effects of ergot (a fungus that infects cereals and other grasses). In 1910, working with G. Barger, he identified a substance in ergot extracts that produced dramatic effects, such as dilation of the arteries. Histamine, as this substance is now called, is found in all plant and animal cells and is one of the irritants in wasp venom, bee stings and stinging nettles.

In 1914 Dale isolated acetylcholine from biological material. Between 1921 and 1926 Otto Loewi and his co-workers showed that stimulation of the parasympathetic nerves in a perfused frog's heart (a heart which has an artificial passage of fluids through its blood vessels) resulted in the appearance of a substance that inhibited the action of a second heart that was receiving the perfused fluid from the first heart. This substance was later shown by Dale and Loewi to be acetylcholine, which is produced at the nerve endings of parasympathetic nerves. This finding provided the first definite proof that chemical substances are involved in the transmission of nerve impulses.

In addition to his research, Dale became concerned in his later years with the social effects of scientific developments. With Thorvald Madson of Copenhagen he was largely responsible for the adoption of an international scheme to standardize drugs and antitoxins. He was also concerned with preserving the apolitical nature of science and with the peaceful use of nuclear energy.

d'Alembert Jean le Rond 1717–1783 was a French mathematician and theoretical physicist who was a great innovator in the field of applied mathematics, discovering and inventing several theorems and principles – notably d'Alembert's principle – in dynamics and celestial mechanics. He devised the theory of partial differential equations and contributed many of the scientific articles that went into the first editions of Denis Diderot's (1713–1784) *Encyclopédie*.

D'Alembert was a foundling, discovered on the doorstep of a Paris church on 16 November 1717. Evidently the illegitimate son of a courtesan, d'Alembert nevertheless grew up well provided for, his accommodation and education financed by the chevalier Destouches (who is therefore generally supposed to have been his father). Following schooling under the Jansenists at Mazarin College, he studied law and was called to the Bar in 1738. However, he then spent a year engrossing himself in medical studies before deciding to devote the rest of his life to mathematics. This he did very successfully, distinguishing himself greatly and becoming personally acquainted with many famous scientists and literary men of the time; in 1741 he was admitted as a member of the Academy of Sciences. He died in Paris on 29 October 1783.

D'Alembert's first published mathematical work was a paper on integral calculus (1739). The subject continued to fascinate him – in the next nine years he wrote two further papers published in the *Mémoires* of the Academy of Berlin, which were fundamental to the development of calculus – and eventually led him to the discovery of the calculus of partial differences. Thereafter he applied his calculus to as many mathematical problems as he encountered.

Nevertheless, it is in the field of dynamics that d'Alembert remains best remembered. The principle that now bears his name was first published in 1743 in his *Traité de dynamique*, and was an extension of the third law of motion formulated by Isaac Newton more than 50 years earlier:

that for every force exerted on a static body there is an equal, opposite force from that body. D'Alembert maintained that the law was valid not merely for a static body, but also for mobile bodies. Within a year he had found a means of applying the principle to the theory of equilibrium and the motion of fluids; previously, such problems had always been solved by means of geometrical calculations. Within a further three years, and by then using also the theory of partial differential equations, he carried out important studies on the properties of sound, and air compression, and had also managed to relate his principle to an investigation of the motion of any body in a given figure.

It was natural for him, therefore, to turn his attention to astronomy. From the early 1750s, together with other mathematicians such as Leonhard Euler (1707–1783), Alexis-Claude Clairaut (1713–1765), Joseph Lagrange (1736–1813) and Pierre Laplace (1749–1827), he applied calculus to celestial mechanics. The problem they set themselves was to determine the motion of three mutually gravitating celestial bodies; in solving it they brought Newton's celestial mechanics to a high degree of sophistication, capable of explaining in detail all the peculiarities of celestial movements shown by contemporarily increasing accuracy of measurements. In particular, d'Alembert worked out in 1754 the theory needed to set Newton's discovery of the precession of the equinoxes on a sound mathematical basis. He determined the value of the precession and explained the phenomenon of the oscillation of the Earth's axis. At about the same time he also wrote an influential paper in which he gave accurate calculations of the perturbations in the orbits of the known planets.

It was also at that time that d'Alembert was persuaded by his friend Denis Diderot to contribute to his *Encyclopédie* – a work that was to contain a synthesis of all knowledge, particularly of new ideas and scientific discoveries. D'Alembert duly wrote on scientific topics, linking, especially, various branches of science. After a few years, however (when at least one volume had already appeared), the Church in France denounced the project, and d'Alembert resigned his editorship. It may have been this, all the same, that spurred him into publishing no fewer than eight volumes of his mathematical investigations over the next 20 years.

Towards the end of his life, d'Alembert's friend Johann Lambert (1728–1777) announced that he had discovered a moon circling the planet Venus, and proposed that it should be named d'Alembert; he declined the honour very diplomatically – but whether because of (entirely justified) suspicions about the existence of such a satellite, or for other reasons, there is no means of knowing.

Dalton John 1766–1844 was an English chemist, one of the founders of atomic theory. Some of his proposals have since proved to be incorrect, but his chief contribution was that he channelled the thinking of contemporary scientists along the correct lines, particularly in his method of using established facts to explain a new phenomenon.

Dalton was born in the village of Eaglesfield near Cockermouth in Cumbria on or about 6 September 1766. He was the third of six children of a weaver, who was a devout Quaker and did not register the date of his son's birth. Dalton attended the village Quaker school and by the age of 12 was running it. He later became headmaster of a school in Kendal, before taking up a post in 1793 to teach mathematics and natural philosophy in Manchester. Dalton was largely self-taught, his Quaker beliefs excluding him from attending Oxford or Cambridge universities (at that time open only to members of the Church of England).

Even before he moved to Manchester, a wealthy Quaker friend, the blind philosopher John Gough, had stimulated in Dalton an interest in meteorology and for 57 years (beginning in 1787) he kept a diary of observations about the weather. He gave lectures on this subject to the Manchester Literary and Philosophical Society, of which he became honorary Secretary and later President. He determined that the density of water varies with temperature, reaching a maximum at 6.1°C/42.5°F (the modern value of this temperature is 4°C). He also lectured about colour blindness, a condition he shared with his brother and which for a time was known as Daltonism. He resigned his lectureship in Manchester in 1799 in order to pursue his own researches, working as a private tutor to make a living. He did, however, remain as the Society's Secretary and was given accommodation in a house they bought for him. This house, still containing many of Dalton's records, was destroyed in a bombing raid in 1940, during World War II. He was awarded a government pension of £150 in 1833, which was doubled three years later. He died in Manchester on 27 July 1844.

From his interest in the weather, atmosphere and gases in general, Dalton in 1803 proposed his law of partial pressures (which states that, in a mixture of gases, the total pressure is the sum of the pressures that each component would exert if it alone occupied the same volume). He also studied the variation of a gas's volume with temperature, concluding (independently of Joseph Gay-Lussac) that all gases have the same coefficient of thermal expansion. Gaseous diffusion and the solubility of gases in water were also the subjects of his experiments.

The work on the absorption of gases led Dalton to formulate his atomic theory – he considered that gases must be made up of particles that can somehow occupy spaces between the particles that make up water, and that in a mixture of gases the different particles must intermingle rather than separate into layers depending on their density. When presented in his book *New System of Chemical Philosophy* (1808), the idea that atoms of different elements have different weights was supported by a list of atomic weights (relative atomic masses) and his newly devised system of chemical symbols. Combinations of element symbols could be made to represent compounds.

Many of the atomic weights (confused with equivalent weights) were incorrect; for example oxygen's was 8 and carbon's 6, but a pattern had been established, introducing order to a science that was hitherto little more than a collection of facts.

Taking into account later work, Dalton's atomic theory may be summarized as follows:

(a) Matter cannot be subdivided indefinitely, because each element consists of indivisible particles called atoms.

(b) The atoms of the same element are alike in every respect, having the same weight (mass), volume and chemical properties; atoms of different elements have different properties.

(c) In chemical combinations of different elements, atoms join together in simple definite numbers to form compound atoms (now called molecules).

His formula for water – one hydrogen atom combined with one oxygen atom – was wrong, although he was more fortunate with carbon monoxide ('carbonic oxide') and carbon dioxide ('carbonic acid'). But nevertheless, he did bring a sort of order to the existing chaos, and provided a foundation for several generations of scientists. Several years later Jöns Berzelius was to supersede Dalton's system with the chemical symbols and formulae still used today.

Throughout his life Dalton retained his Quaker habits and dress, and new acquaintances were often taken aback by his appearance. He continued to keep his diary, which eventually ran to 200,000 entries. He distrusted the results of other workers, preferring to rely on his own experiences. As he grew older he became almost a recluse, with few friends, and deeply involved in his pursuit of knowledge. And although he shunned fame and glory, he became famous even outside the realms of science. When his coffin stood on public display in Manchester Town Hall, more than 40,000 people filed past to pay their respects.

Dancer John Benjamin 1812–1887 was a British optician and instrument maker who applied his knowledge of physics to various inventions, particularly the development of microphotography.

Dancer was born in London on 8 October 1812. Both his father and grandfather were manufacturers of 'optical, philosophical and nautical instruments'. In 1818 the family moved to Liverpool where his father, Joseph Dancer, was one of the founders of and a lecturer at the Liverpool Mechanical Institution.

Dancer often assisted his father, and his interest in science grew. One instrument constructed by Josiah was a large solar microscope, which had a 30-cm/12-in condensing lens. Dancer used this equipment to view aquatic animals and he soon became an expert. On his father's death he took over control of the family business and the public lectures.

Dancer was a popular lecturer and he soon improved many of the standard laboratory practices of the period. He introduced unglazed porous jars in voltaic cells to separate the electrodes (previously the division was made of membranes from animal bladders and ox gullets). During the nineteenth century they were adopted as standard in

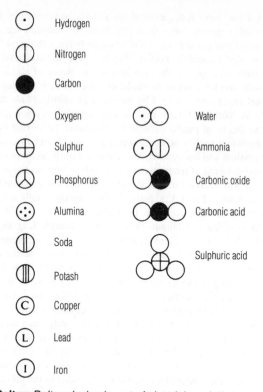

Dalton *Dalton devised a set of pictorial symbols to represent individual elements; these could be combined to represent compounds of these elements.*

Daniell and Leclanché cells, but unfortunately Dancer had not patented his invention.

He also devoted a lot of time to electrolysis during the 1830s; he became particularly expert in the electrodeposition of copper. He perfected a method of preparing a sheet of electrodeposited copper which retained features of a 'master' on which it was plated. Unfortunately for Dancer the invention was developed by Thomas Spencer into an early form of electrotype. Dancer also improved on the Daniell cell by crimping or corrugating its copper plates. The power of the cell was considerably increased because of the greater electrode surface area.

Dancer was particularly interested in electrical circuits – in fact, in anything spectacular for inclusion in his lectures. Dancer used a Faraday voltameter with large platinum electrodes to prepare gases (hydrogen and oxygen) from water by electrodecomposition. By slight modification of the conditions he was able to prepare a colourless gas with a strong odour which caused coughing. The gas was not named initially, but was later (1839) identified as ozone by Christian Schonbein (professor of chemistry at Basle). The induction coil was a popular piece of equipment which was often claimed to have spurious medical benefits. Dancer incorporated the 'magnetic vibrator', similar to the spring make-and-break contact used in electric bells. He again did not patent his invention.

He began a series of experiments in 1839 based on

Daguerre's and Fox Talbot's photography methods. By July 1840 Dancer had developed a method of taking photographs of microscopic objects, using silver plates. The photographic image was capable of magnification up to 20 times before clarity was lost. Dancer also gave 'magic lantern' shows, and these new plates were a considerable improvement over the painted slides then available. In the 1850s Scott Archer introduced a new process, the collodian method, and Dancer was not slow to realize its benefits in improving on his earlier techniques.

By 1856 Dancer had prepared many microphotographs and his work was exhibited throughout Europe. The clarity was excellent and by 1859 he was showing slides which carried whole pages of books 'in one-sixteenth hundredth part of a superficial inch'. Microdot photography was born. In all, the Dancer business produced for sale 512 different microphotographs, mounted on standard microscope slides. They included photographs of distinguished scientists of the era and portraits of the British royal family.

Dancer was an active member of the Manchester Literary and Philosophical Society, at whose meetings he presented many papers on a variety of scientific topics. He was a Fellow of the Royal Astronomical Society and optician in Manchester to the Prince of Wales. He also constructed the apparatus with which James Joule determined the mechanical equivalent of heat.

In his later years he suffered from diabetes, which led to glaucoma of his eyes. Dancer died in November 1887, leaving behind a technique which has led to a method of storing huge quantities of data in a small space – microfilm and microfiche.

Daniell John Frederic 1790–1845 was a British meteorologist, inventor and chemist who is famous for devising the Daniell cell, a primary cell that was the first reliable source of direct-current electricity. He received many honours for his work, including the Royal Society's Rumford and Copley Medals, awarded in 1832 and 1837, respectively.

Daniell was born in London on 12 March 1790, the son

of a lawyer. He received a private education, principally in the classics, after which he began working in a relative's sugar-refining factory. Early in his career he made the acquaintance of William Brande, Professor of Chemistry at the Royal Institution; the two men became lifelong friends, travelling together on several scientific expeditions and through their joint efforts reviving the journal of the Royal Institution. In 1831 Daniell was appointed the first Professor of Chemistry at the newly founded King's College, London, a post he held until his death (at a meeting of the Royal Society) on 13 March 1845.

Daniell made his best-known contribution to science – the cell named after him – in 1836. The Italian physicist Alessandro Volta had invented (in 1797) a copper–zinc battery, which the English physicist William Sturgeon had improved by using zinc and mercury amalgam; but both these simple or voltaic cells suffered the severe disadvantage of producing a current that diminished rapidly. Daniell found that this effect (polarization) was caused by hydrogen bubbles collecting on the copper electrode, thereby increasing the cell's internal resistance. In 1836 he proposed a new type of cell. The Daniell cell consists of a copper plate (anode) immersed in a saturated solution of copper sulphate and separated by a porous barrier (Daniell used a natural membrane), from a zinc rod (cathode) immersed in dilute sulphuric acid. The barrier prevents mechanical mixing of the solutions but allows the passage of charged ions. When the current flows zinc combines with the sulphuric acid freeing positively charged hydrogen ions which pass through the barrier to the copper plate. There they lose their charge and combine with copper sulphate, forming sulphuric acid and depositing copper on the plate. Negatively charged sulphate ions pass in the opposite direction and react with the zinc to form zinc sulphate, therefore no gases are evolved at the electrodes and a constant current is produced. Daniell's cell stimulated research into electricity – including his own investigations into electrolysis – and was later commercially used in electroplating and glyphography (a process giving a raised relief copy of an engraved plate for use in letterpress printing).

Daniell's other work included the development of improved processes for sugar manufacturing; investigations into gas generation by the distillation of resin dissolved in turpentine; and inventing a new type of dewpoint hygrometer for measuring humidity (1820) and a pyrometer for measuring the temperatures of furnaces (1830). He also studied the behaviour of the Earth's atmosphere; gave an explanation of trade winds; researched into the meteorological effects of solar radiation and of the cooling of the Earth; suggested improvements for several meteorological instruments; and pointed out the importance of humidity in the management of greenhouses.

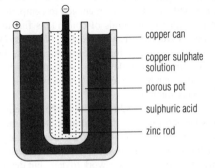

copper can

copper sulphate solution

porous pot

sulphuric acid

zinc rod

Daniell *A Daniell cell has a zinc cathode immersed in zinc sulphate solution contained in a porous pot, surrounded by copper sulphate solution. The whole is contained in a copper can, which also forms the anode.*

Dantzig George Bernard 1914– is a US mathematician who is an expert on (computer) linear programming and operations research. His work is now regarded as fundamental to many university courses in business studies, industrial engineering and managerial sciences.

Born the son of a well-known mathematician, Tobias Dantzig, on 8 November 1914 in Portland, Oregon, Dantzig completed his education at the University of Maryland, gaining his BA in 1936. He then attended the University of Michigan as a Horace Rackham Scholar for a year, earning his MA. During the latter part of World War II he was with the Statistical Control Headquarters of the US Air Force as Chief of the Combat Analysis Branch where, from 1946 to 1952, he then became Mathematical Adviser. For the next eight years he was a research mathematician with the Rand Corporation at Santa Monica, California, transferring in 1960 to the University of California at Berkeley as Professor and Chairman of the Operations Research Center. Since 1966, among other consultative posts, his major position has been as C. A. Criley Professor of Operations Research and Computer Science at Stanford University, Palo Alto, California. The author of two influential books and many technical papers, Dantzig has received several awards and honorary degrees, including the National Medal of Science (1975), the National Academy of Sciences Award in Applied Mathematics and Numerical Analysis (in 1977) and the Harvey Prize (1985).

A fundamental problem in economics involves the optimum allocation of scarce resources among competing activities – a problem that can be expressed in mathematical form. In 1947 Dantzig discovered that many such planning problems could be formulated as linear computer-programs. He compounded this discovery in that at the same time he also devised an algorithm – known as the simplex method – that turned out to be remarkably efficient for the purpose. (His method is still the best way to resolve nearly all linear programs of this type.) Moreover, Dantzig's discovery coincided with the development of the first successful computers, which meant that managers in industry were provided with a powerful and practical method for comparing a large number of interdependent alternative courses of action. Dantzig led the way in developing applications for the new linear programming approach. By 1972, a survey showed that a considerable proportion of all industrial organizations was using the simplex method of linear programming. The system has also had an impact on economics and statistics.

Subsequently, Dantzig has been involved in all the main areas of mathematical programming and other parts of operations research. He has worked on the development of techniques for dealing with large systems, and originated the 'decomposition principle' for solving large systems with block-diagonal structure. In 1971 it was possible by this method, and using an IBM 370/165 computer, to solve a linear program with 282,468 variables and 50,215 equations in $2^1/_2$ hours.

The development of linear programming in the 1950s enabled the mathematical science of decision-making to be developed into the discipline now known as operations research or management science. Most universities at present offer courses on operations research, nearly all emphasizing the importance of linear programming. Operations research is also important in the academic studies of many university departments of business sciences and industrial engineering.

Darboux Jean Gaston 1842–1917 was a French geometrician who contributed immensely to the differential geometry of his time, and to the theory of surfaces. An innovator in much of his research, he was also an able teacher, a capable administrator, and an influential author in writing of his own studies. His name is perpetuated in the Darboux sums and the Darboux integrals.

Darboux was born on 13 August 1842 in Nîmes. Educated locally, at the age of 19 he sat the entrance examinations for both the Ecole Polytechnique and the Ecole Normale Supérieure in Paris. In both he came top; he chose to enter the Ecole Normale. During his studies there, he wrote his first paper (on orthogonal surfaces, in 1864). Two years later he extended this work, for which he received his doctorate from the Sorbonne. Thereafter, Darboux taught at the Lycée Louis le Grand (1866 to 1872) before transferring to a similar post at the Ecole Normale. From 1873 to 1878 he also held the Chair of Rational Mechanics at the Sorbonne, as an assistant to Joseph Liouville (1809–1882, from whom he may have gained his interest in differential geometry). He then became assistant to Michel Chasles, the author of a standard reference work on geometry, and in 1880 succeeded him as Professor of Higher Geometry. He held this post until he died, in Paris, on 25 February 1917.

Elected a member of the Paris Academy of Sciences in 1884 (and for many years later its secretary), Darboux was made a Fellow of the Royal Society in 1902. He also won several other awards and honours.

Darboux's research concentrated on geometry – the major interest throughout his life. Nevertheless, only five years or so after his doctoral thesis on orthogonal surfaces (a subject he returned to time and time again), he published a paper on partial differential equations of the second order. What was particularly novel in his approach was that in order to further his examination he had devised a new method of integration. Within another five years – during which he also formulated the theory of a specific class of surface called a cyclide – he managed to complete a proof of the existence of integrals of continuous functions that had defeated Augustin-Louis Cauchy (1789–1857) a generation before. Continuing his investigations, Darboux succeeded (in 1879) in defining the Riemann integral, in order to do which he derived the 'Darboux sums' and used the 'Darboux integrals'.

Between 1887 and 1896 he published a collection of the lectures he had given at the Sorbonne, under the overall title *Leçons sur la théorie générale des surfaces et les applications géométriques du calcul infinitésimal*. The four volumes described all of his work to date, but dealt mainly with the application of analysis to curves and surfaces, and the study of minimal surfaces and geodesics. In a later work, *Leçons sur les systèmes orthogonaux et les coordonnées curvilignes*, 1898, Darboux applied the theorem on algebraic integrals formulated by the Norwegian Niels

Abel (1802–1829) to orthogonal systems in *n* dimensions. Important among Darboux's other work were his papers on the theory of integrations, the theory of analytical functions, and his research into the problems involving the Jacobi polynomials.

Darby Abraham 1677–1717 was an English ironmaster and engineer who devised a new way of smelting iron using coke rather than the much more expensive charcoal.

Darby was born at Wren's Nest near Dudley, Worcestershire, the son of a farmer. He served an engineering apprenticeship with a maltmill maker in Birmingham, and on completing this in 1698 he set up in business on his own. In about 1704 he visited Holland and brought back with him some Dutch brass founders, establishing them at Bristol (at the Baptist Mill Brass Works) using capital from four associates who left to him the management of the business.

Believing that cast iron might be substituted for brass in some products, he tried with his Dutch workers to make iron castings in moulds of sand. At first the experiment failed but proved eventually successful when he adopted a suggestion made by a boy in his employment, John Thomas, who consequently rose in his service and whose descendents were trusted agents of the Darby family for about 100 years. In April 1708 he took out a patent for a new way of casting iron pots and other ironware in sand only, without loam or clay. This process cheapened utensils much used by poorer people and at that time largely imported from abroad. The decision led to protracted arguments with his associates, who finally refused to risk more money on the new venture. Darby dissolved his connection with them and, drawing out his share of the capital, he took a lease on an old furnace in Coalbrookdale, Shropshire, moving to Madely Court in 1709. There he prospered until his death on 8 March 1717.

In the seventeenth century the development of the iron industry in Britain was limited by two technical difficulties. Firstly, the growing demand for charcoal – then the only satisfactory fuel for blast furnaces – had forced up the price very considerably. The shortage of wood for conversion into charcoal had been caused by the heavy demands of shipbuilders and a rapid growth in the demand for iron-made objects (resulting in the denuding of the large areas of forest which had been a feature of the English landscape for thousands of years). The second difficulty was due to the attempts to improve efficiency by using bigger furnaces; these were frustrated by the fact that charcoal is too soft to support more than a relatively short column of heavy ore. Attempts were made to use coal in place of charcoal, but the presence in most coals of sulphur, which spoils the quality of the iron, resulted in only limited success; the claim put forward by Dud Dudley (1599–1684) that in 1619 he had successfully smelted iron by using coal is not now generally accepted.

Attention was therefore directed to coke – already used in the smelting of copper and lead – since the coking process eliminates the sulphur. Darby had considerable experience of smelting copper with coke at Bristol and had also used coke in malting during his apprenticeship days in Birmingham. (In malting, for different reasons, the presence of sulphur prevents the use of raw coal.) The old furnace which Darby converted to house the Bristol Iron Company was ideally situated on the River Severn, close to good local supplies of iron ore and good coking coal. Initially much of the iron was used for making pots and other hollow-ware. The quality of the molten iron made it possible for him to make thin castings which competed satisfactorily with the heavy brassware then in common use.

The advent of the Newcomen steam-engine gave an important new market; some of the cylinders required for mine-pumping engines weighed as much as 6 tonnes with a length of 3 m/10 ft and a bore of 1.8 m/6 ft. By 1758 more than 100 such cylinders had been cast. The steam-engine, in turn, improved the manufacture of iron by giving a more powerful and reliable blast for the furnaces than water power could supply.

The Coalbrookdale Works were much enlarged, their processes improved and increased and their operations extended under the second Abraham Darby. His son and successor, the third Abraham Darby, took over the management of the Coalbrookdale Works when he was about 18 and is memorable as having constructed the first iron bridge ever erected, the semicircular cast-iron arch across the river Severn near the village of Brosely at Coalbrookdale, opened for traffic in 1779 (and still standing today).

Dart Raymond Arthur 1893–1988 was an Australian-born anatomist and palaeoanthropologist who made a signal contribution to the tracing of the early history of the human species.

Born in Brisbane, Dart studied medicine at the University of Sydney, where he qualified in 1917. He served in France in World War I before being appointed in 1922 to the Chair of Anatomy at the newly formed University of Witwatersrand, Johannesburg, South Africa. Two years later he had a lucky find: one of his students brought him a fossil baboon skull, found in a lime quarry at Taung, Botswana. Excited, Dart took steps to ensure that similar finds would be preserved and despatched to him. Soon he received the skull of a previously unknown hominid, which Dart named *Australopithecus africanus* (southern ape of Africa). Dart saw *Australopithecus* as the 'missing link' between humans and the apes. This view received little support until Dr Robert Broom found further hominid remains in the Transvaal in the 1930s.

Physical anthropologists today believe that *Australopithecus* lived about 1.2–2.5 million years ago. Debate still rages, however, whether modern humans are directly descended from this ancestor, or whether *Australopithecus* only represents a failed evolutionary offshoot from a much earlier shared predecessor.

Darwin Charles Robert 1809–1882 was a British naturalist famous for his theory of evolution and natural selection as put forward in 1859 in *On the Origin of Species*.

Darwin was born in Shrewsbury on 12 February 1809. His father was a wealthy doctor, and his paternal grandfather was Erasmus Darwin, a well-known poet and physician; his mother's father was Josiah Wedgwood. Darwin was educated locally from 1818, and when he left school in 1825 he attended Edinburgh University to study medicine. But he abhorred medicine and the science taught to him there disgusted him, and two years later his father sent him to Christ's College, Cambridge, to study theology – which he did not enjoy either. Natural history was his main interest, which was very much increased by his acquaintance with John Stevens Henslow, who was Professor of Botany at Cambridge. Henslow recommended to the Admiralty that Darwin should accompany HMS *Beagle* as a naturalist on its survey voyage of the coasts of Patagonia, Tierra del Fuego, Chile, Peru and some Pacific islands. His father opposed this idea but, with the support of Wedgwood, Darwin sailed in the *Beagle* from Devonport on 27 December 1831 for a voyage of five years. On his return to England he found that some of his papers had been privately published during his absence and that he was regarded as one of the leading men of science. He published his findings on this epic voyage in *Journal of Researches into the Geology and Natural History of the various countries visited by HMS* Beagle *(1832–1836)* in 1839. In 1838 he was appointed Secretary to the Geological Society, a position he retained until 1844. He married his cousin, Emma Wedgwood, in 1839, and the marriage produced ten children. He spent the rest of his life collating the findings made during the voyage and developing his theory for publication. He died on 19 April 1882 at Down, in Kent.

Before the voyage of the *Beagle* Darwin, like everyone else at that time, did not believe in the mutability of species. But in South America he saw fossil remains of giant sloths and other animals now extinct, and on the Galapágos Islands found a colony of finches that he could divide into at least 14 similar species, none of which existed on the mainland. It was obvious to him that one type must have evolved into many others, but how they did so eluded him. Two years after his return he read Malthus's *An Essay on the Principle of Population* 1798, which proposed that the human population is growing too fast for it to be adequately fed, and that something would have to happen to reduce it, such as war or natural disaster. This work inspired Darwin to see that the same principle could be applied to animal populations and he theorized that variations of a species which survive (while other members of the species do not) pass on the changed characteristic to their offspring. A new species is thereby developed which is fitter to survive in its environment than was the original species from which it evolved. Darwin did not make his ideas public at first, but put them into an essay in 1844 to which only his friend Joseph Hooker and a few others were privy.

In 1856 Darwin began writing fully about evolution and natural selection. Two years later he received a paper from a fellow naturalist, Alfred Wallace, explaining exactly the same theory of evolution and natural selection. Unsure what to do, Darwin consulted his friends Charles Lyell and Hooker, who persuaded him to have the joint papers read in the absence of the authors before the Linnaean Society. The papers caused no stir, but Darwin was forced to speed up the completion of his work.

The abstract of Darwin's findings was published in 1859, and was called *On the Origin of Species by means of Natural Selection or the Preservation of Favoured Races in the Struggle for Life*. It was very widely read, although many fellow scientists criticized it violently. Some considered that the book lacked a foundation of experimental evidence and was based purely on hypothesis; others were simply jealous. Many Christians were shocked by Darwin's work because it implied that the Biblical account of creation – if taken literally – is wrong, and that if evolution works automatically by natural selection then divine intervention plays no part in the lives of plants, animals, or humans.

When Darwin wrote the book, he avoided the issue of human evolution and merely remarked at the end that 'much light will be thrown on the origin of man and his history'. He did not seek the controversy he caused but his ideas soon caught the public imagination. After the publication in 1871 of *The Descent of Man and Selection in Relation to Sex*, in which he argued that people evolved just like other organisms, the popular press soon published articles about the 'missing link' between humans and apes. In fact what Darwin believed was that our ancestors, if alive today, would have been classified among the primates.

Darwin's name remains inseparably linked with the theory of evolution to this day. He never understood what actually caused newly formed advantageous characteristics to appear in animals and plants because he had no knowledge of heredity and mutations. The irony is that the key work on heredity, by the Austrian monk Gregor Mendel, was carried out during Darwin's own lifetime and published in 1865, but was neglected until 1900. Darwin's revolutionary publication, which is still widely read, marked a turning point in many of the sciences, including physical anthropology and palaeontology, and remains a source of strong controversy.

da Vinci Leonardo 1452–1519. Italian artist, inventor and scientist; *see* Leonardo da Vinci.

Davis William Morris 1850–1934 was the leading US physical geographer of the early twentieth century.

The son of a Philadelphia Quaker businessman, Davis studied science at Harvard University. After a spell as a meteorologist in Argentina, Davis served with the US North Pacific Survey before securing an appointment as a lecturer at Harvard in 1877. Becoming a professor, he taught at Harvard until 1912. During those 30 years, Davis became the most prominent US investigator of the physical environment. In three fields of science – meteorology, geology and notably geomorphology – he left enduring legacies. Above all, he proved an influential analyst of landforms. Building on experience gained in a classic study made in 1889 of the drainage system of the Pennsylvania and New Jersey region, he developed the organizing concept of the regular cycle of erosion, a theory that was to dominate geomorphology and physical geography for half a century. Davis proposed a standard stage-by-stage life-cycle for a river valley, marked by youth (steep-sided V-shaped valleys), maturity (flood-plain floors), and old age, as the river valley was imperceptibly worn down into the rolling landscape he termed a 'peneplain'. On occasion these developments, which Davis believed followed from the principles of Lyellian geology, could be punctuated by upthrust, which would rejuvenate the river and initiate new cycles. The Davisian cycle presupposed an explicitly uniformitarian view of Earth history, in which the present was key to the past and piecemeal, and natural causes were paramount.

Davisson Clinton Joseph 1881–1958 was a US physicist who made the first experimental observation of the wave nature of electrons. For this achievement, he was awarded the 1937 Nobel Prize for Physics with George Thomson, who made the same discovery independently of Davisson.

Davisson was born in Bloomington, Illinois, on 22 October 1881. He attended local schools before gaining a scholarship for his proficiency in mathematics and physics in 1902, to the University of Chicago, where he studied under Robert Millikan (1868–1953). In 1905 Davisson was appointed part-time instructor in physics at Princeton University while continuing his studies at Chicago, gaining a BS degree from Chicago in 1908, followed by a PhD in 1911 for a thesis on 'The thermal emission of positive ions from alkaline earth salts'. Davisson then spent six years as an instructor in the Department of Physics at the Carnegie Institute of Technology in Pittsburgh.. He was able to spend the summer of 1913 in the Cavendish Laboratory at Cambridge, England, working under J. J. Thomson (1856–1940). Refused enlistment in the army in 1917, he accepted wartime employment in the engineering department of the Western Electric Company (later Bell Telephone) in New York. At the end of the war, although offered an assistant professorship at the Carnegie Institute, Davisson remained at Bell Telephone where he was able to work on research full-time. He stayed there until his retirement in 1946, when he became a Visiting Professor of Physics at the University of Virginia, Charlottesville. He retired from this position in 1949 and died on 1 February 1958.

Davisson's work at Western Electric was concerned with the reflection of electrons from metal surfaces under electron bombardment. In April 1925, an accidental explosion caused a nickel target under investigation to become heavily oxidized. Davisson removed the coating of oxide by heating the nickel and resumed his work. He now found that the angle of reflection of electrons from the nickel surface had changed. Davisson and his assistant Lester Germer (1896–1971) suspected that the change was due to recrystallization of the nickel, the heating having converted many small crystals in the target surface into several large crystals.

A year later, Davisson attended a meeting of the British Association for the Advancement of Science where he received details of the theory proposed by Louis de Broglie (1892–1987) that electrons may behave as a wave motion. Davisson immediately believed that the effects he had observed were caused by the diffraction of electron waves in the planes of atoms in the nickel crystals, just as X-ray diffraction had earlier been observed in crystals by Max von Laue (1879–1960) and William Henry Bragg (1862–1942) and William Lawrence Bragg (1890–1971).

To resolve this question, Davisson and Germer used a single nickel crystal in their experiments. The atoms were in a cubic lattice with atoms at the apex of cubes, and the electrons were directed at the plane of atoms at 45° to the regular end plane. Electrons of a known velocity were directed at this plane and those emitted were detected by a Faraday chamber. In January 1927, results showed that at a certain velocity of incident electrons, diffraction occurred, producing outgoing beams that could be related to the interplanar distance. The wavelength of the beams was determined, and this was then used with the known velocity of the electrons to verify de Broglie's hypothesis. The first work gave results with an error of 1–2% but later systematic work produced results in complete agreement. Similar experiments at higher voltages using metal foil were carried out later the same year at Aberdeen University, Scotland, by George Thomson (1892–1975), the son of J. J. Thomson. The particle–wave duality of subatomic particles was established beyond doubt, and both men were awarded the 1937 Nobel Prize for Physics.

Davisson's experimental findings that electrons have a wave nature confirmed and established theoretical explanations of the structure of the atom in terms of electron waves. This understanding proved vital to investigations into the nature of chemical bonds, and to the production of high magnifications in the electron microscope.

Davy Humphry 1778–1829 was an English chemist who is best known for his discovery of the elements sodium and potassium and for inventing a safety lamp for use in mines.

Davy was born on 17 December 1778 at Penzance, Cornwall, the son of well-to-do parents. He was educated in Penzance and, from 1793, in Truro, where he studied classics. But his father died a year later and, to help to support the family, the young Davy became apprenticed to a Penzance surgeon-apothecary, J. Bingham Borlase. His interest in chemistry began in 1797 through reading Antoine Lavoisier's *Traité elémentaire*, and by 1799 he was working on the therapeutic uses of gases as an assistant at the Pneumatic Institute in Bristol.

Following Alessandro Volta's announcement in 1800 of the voltaic cell, Davy began his researches in electrochemistry. He moved to the Royal Institution in London in 1801, where he was influenced by Count Rumford (1753–1814) and Henry Cavendish. (1731–1810). He was knighted by the Prince Regent in 1812 and three days later married a wealthy widow, Jane Apreece. In 1813 he took on Michael Faraday as a laboratory assistant, who accompanied him on a tour of Europe. When he returned in 1815 Davy designed his miner's safety lamp, which would burn safely even in an explosive mixture of air and fire damp (methane). He did not patent the lamp, a fact that was to lead to an acrimonious claim to priority by the steam locomotive engineer George Stephenson (1781–1848). Davy was created a baronet in 1818 and two years later he succeeded the botanist Joseph Banks (1743–1820) as President of the Royal Society. He became seriously ill in 1827 and went abroad in 1828 to try to improve his health. He settled in Rome in early 1829 but suffered a heart attack and died in Geneva, Switzerland, on 29 May of that year.

While Davy was working in Bristol (1799) he prepared nitrous oxide (dinitrogen monoxide) by heating ammonium nitrate. He investigated the effects of breathing the gas, showing that it causes intoxication (although it was to be another 45 years before the gas was used as a dental anaesthetic). His early experiments on electrolysis of aqueous solutions (from 1800) led Davy to suggest its large-scale use in the alkali industry. He theorized that the mechanism of electrolysis could be explained in terms of species that have opposite electric charges, which could be arranged on a scale of relative affinities – the foundation of the modern electrochemical series. The climax of this work came in 1807 with the isolation of sodium and potassium metal by the electrolysis of their fused salts. Later, after consultation with Jöns Berzelius (1779–1848), he also isolated calcium, strontium, barium and magnesium. His intensive study of the alkali metals provided proof of Lavoisier's idea that all alkalis contain oxygen. In 1808 he first isolated boron, by heating borax with potassium.

Davy also initially supported Lavoisier's contention that oxygen is present in all acids. But in 1810, after doing quantitative analytical work with muriatic acid (hydrochloric acid) he disproved this hypothesis. He went on to show that its oxidation product, oxymuriatic acid (discovered in 1774 by Karl Scheele), is an element, which he named chlorine. He explained its bleaching action and later prepared two of its oxides and chlorides of sulphur and phosphorus. He also suggested that the element common to all acids is hydrogen, not oxygen.

Davy was reluctant to accept the atomic theory of his contemporary John Dalton, but in the face of mounting evidence finally concurred and attempted to apply the laws of definite and multiple proportions to various compounds. He determined the 'proportional weights' (relative atomic masses), of various elements, including chlorine at 33.9 (actual value 35.5), oxygen 15 (16), potassium 40.5 (39.1) and sulphur 30 (32).

The safety lamp of 1815 was designed after a series of laboratory experiments on explosive mixtures. Davy showed that a flame continues to burn safely in such a mixture if it is surrounded by a fine metal mesh to dissipate heat, if only a narrow air inlet is used, and if the air inside the lamp is diluted with an unreactive gas such as carbon dioxide.

Numerous other achievements can be attributed to this great scientist. Davy introduced a chemical approach to agriculture, the tanning industry and mineralogy; he designed an arc lamp for illumination, an electrolytic process for the desalination of sea water, and a method of cathodic protection for the copper-clad ships of the day by connecting them to zinc plates. But his genius has been described as erratic. At his best he was a scientist of great perception, a prolific laboratory worker and a brilliant lecturer. At other times he was unsystematic, readily distracted and prone to hasty decisions. He was never trained as a chemist, and consequently his excellence in qualitative work was not always matched by quantitative skills. He sought and won many scientific honours, which he then jealously guarded, even going so far in 1824 as trying to oppose the election of his protégé Michael Faraday to the Royal Society.

Dawkins (Clinton) Richard 1941– is an English zoologist and Darwinian.

Dawkins was born on 26 March 1941, and educated at Oundle School and Balliol College, Oxford, where he gained his doctorate under the ethologist Nikolaas Tinbergen (1907–1988), who won the Nobel Prize for Physiology or Medicine in 1973. After two years as Assistant Professor of Zoology at the University of California at Berkeley, Dawkins returned to Oxford as Lecturer in Animal Behaviour, becoming Reader in Zoology in 1989 at the University, and a Fellow of New College. His research has continued in ethology, the study of animal behaviour, and he has written and broadcast his ideas about the evolution of behavioural mechanisms widely. In 1976 he published *The Selfish Gene* in which he argued that genes, not individuals, populations or species, are the driving force of evolution and that animals such as humans are merely machines through which genes survive. He also suggested an analogous system of cultural transmission in human societies, and proposed the term 'mimeme', abbreviated to 'meme' as the unit of such a scheme. The book became an international popular success, and was translated into several languages. Dawkins's comments about religion attracted particular attention as he considered the idea of God to be a meme with a high survival value. His contentions were further developed in *The Extended Phenotype* (1982), primarily an academic work, and in *The Blind Watchmaker* (1986), which achieved wide acclaim from literary as well as scientific critics. Following the success of his books Dawkins has presented scientific programmes on television, given the prestigious Royal Institution Christmas lectures for young people, and become a prominent public spokesman for science in general and evolutionary ethology in particular.

De Beer Gavin Rylands 1899–1972 was a British zoologist known for his important contributions to embryology and evolution, notably disproving the germ layer theory and developing the concept of paedomorphism (the retention of juvenile characteristics of ancestors in mature adults). He received numerous honours for his work, including a knighthood in 1954.

De Beer was born on 1 November 1899 in London and, after military service in World War I, graduated from Oxford University. From 1926 to 1938 he was Jenkinson Memorial Lecturer in Embryology at Oxford University, after which he served in World War II. In 1945 he became Professor of Embryology at University College, London, then in 1950 was appointed Director of the British Museum (Natural History), a post he held until his retirement in 1960. De Beer died on 21 June 1972 in Alfriston, East Sussex.

De Beer's first major work was his *Introduction to Experimental Embryology* (1926), in which he observed that some vertebrate structures, such as certain cartilage and bone cells, are derived from the outer ectodermal layer of the embryo. This finally disproved the germ layer theory, according to which cartilage and bone cells are formed from the mesoderm. Continuing his embryological investigations, De Beer described in *Embryos and Ancestors* (1940) his work, which showed that certain adult animals retain some of the juvenile characteristics of their ancestors, a phenomenon called paedomorphism. This finding refuted Ernst Haeckel's theory of phylogenetic recapitulation, according to which the embryonic development of an organism repeats the adult stages of the organism's evolutionary ancestors.

Turning his attention to evolution, De Beer then suggested that gaps in the fossil records of early ancestral forms are due to the impermanence of the soft tissues in these early ancestors. Also in the field of evolution, his studies of the fossil *Archaeopteryx*, the earliest known bird, led him to propose mosaic evolution – whereby evolutionary changes occur piecemeal – to explain the presence of both reptilian and avian features in *Archaeopteryx*.

De Beer also researched into the functions of the pituitary gland, and applied scientific methods to various historical problems, such as the origin of the Etruscans (which he traced using blood group data) and establishing the route taken by Hannibal in his march across the Alps (for which De Beer used pollen analysis, glaciology and various other techniques).

de Broglie Louis Victor 1892–1987 was a French physicist who first developed the principle that an electron or any other particle can be considered to behave as a wave as well as a particle. This wave–particle duality is a fundamental principle governing the structure of the atom, and for its discovery, de Broglie was awarded the 1929 Nobel Prize for Physics.

De Broglie was born in Dieppe, France, on 15 August 1892, the second son of a noble French family and as such expected to have a distinguished military or diplomatic career. It was perhaps fortunate that his elder brother Maurice (1875–1960), who pioneered the study of X-ray spectra, had pursued his scientific interests to considerable success against the wishes of his family, his only compromise being a relatively short naval career. Nevertheless, in 1909 Louis entered the Sorbonne in Paris to read history. A year later, when he was 18, he began to study physics. It

had been his intention to enter the diplomatic service but he became so interested in scientific subjects, partly through the influence of Maurice, whom he helped in his extensive private laboratory at the family home, that he took a physics topic for his doctoral dissertation rather than that on French history which the Sorbonne first offered. This dissertation was submitted in 1924, but before that de Broglie began exploring the ideas of wave–particle duality. Following the award of his doctorate, de Broglie stayed on at the Sorbonne until 1928. He moved to the Henri Poincaré Institute in 1932 as Professor of Theoretical Physics, retaining this position until 1962. From 1946, he was a senior adviser on the development of atomic energy in France.

In 1922, de Broglie was able to derive Planck's formula $E = hv$, where E is energy, h is Planck's constant and v is the frequency of the radiation, using the particle theory of light. This probably suggested the idea of wave–particle duality to him because it prompted the question of how a particle could have a frequency. Using this idea and Einstein's mass–energy equation $E = mc^2$, he derived $E = mc^2 = hv$. Now, mc is the momentum of the particle and C/v is the wavelength of the associated wave, λ. Hence the momentum is h/λ. This relation between the momentum of the particle and the wavelength of the associated wave is fundamental to de Broglie's theory.

The extension of this idea from light particles (photons) to electrons and other particles was the next step. Niels Bohr (1885–1962) in his model of the atom, found that the angular momentum of an electron in an atom must be $nh/2\pi$, where n is a whole number. De Broglie showed that this expression for the angular momentum of the electron could be derived from his momentum–wavelength equation if an electron wave exactly makes up the circular orbit of the electron with a whole number of wavelengths and produces a standing wave, that is, $n\lambda = 2\pi r$, where r is the radius of the orbit. Otherwise interference would take place and no standing wave would form. De Broglie's idea gave a further explanation of Bohr's model of the atom. Much of this work was described in his doctoral dissertation, although some of it was published in 1923.

If particles could be described as waves then they must satisfy a partial differential equation known as a wave equation. De Broglie developed such an equation in 1926, but found it in a form which did not offer useful information when it was solved. A more useful wave equation was developed by Erwin Schrödinger (1887–1961), later in 1926.

The experimental evidence for de Broglie's theory was obtained by Clinton Davisson (1881–1958) and George Thomson (1892–1975) in 1927. These two men independently produced electron diffraction patterns, showing that particles can produce an effect which had until then been exclusive to electromagnetic waves such as light and X-rays. Such waves are known as matter waves.

De Broglie's discovery of wave–particle duality enabled physicists to view Einstein's conviction that matter and energy are interconvertible as being fundamental to the structure of matter. The study of matter waves led not only to a much deeper understanding of the nature of the atom but also to explanations of chemical bonds and the practical application of electron waves in electron microscopes.

Throughout his life, de Broglie was concerned with the philosophical issues of physics and he was the author of a number of books on this subject. He pondered whether the statistical results of physics are all that there is to be known, or whether there is a completely determined reality which our experimental techniques are as yet inadequate to discern. During many of his years as a professional scientist, he inclined to the former view but his later writing suggests his belief in the latter.

Debye Peter Joseph Willem 1884–1966 was a Dutch-born US physical chemist who in a long career made many important contributions to the science. He was awarded the 1936 Nobel Prize for Chemistry for his work on dipole moments and molecular structure.

Debye was born on 24 March 1884 at Maastricht in the Netherlands. From 1900 to 1905 he went to school at the Technische Hochschule over the border in Aachen, Germany, where he qualified as an electrical engineer. He then became assistant to Arnold Sommerfeld (1868–1951) at the University of Munich, gaining his PhD in 1910. The next few years saw a remarkable progress from one distinguished post to another, starting in Zurich (1910) where he succeeded Albert Einstein as Professor of Theoretical Physics and culminating, via Utrecht (1912) and Göttingen (1914), in a return to Zurich in 1920 as Professor of Experimental Physics and Director of the Physics Institute. By 1927 Debye moved to Leipzig, where he took over from Friedrich Ostwald, and in 1934 he went to Berlin to supervise the building of the Kaiser Wilhelm Institute of Physics, which he renamed the Max Planck Institute.

In the late 1930s Debye found himself in difficulties with the Nazi authorities in Germany – mainly because of his Dutch nationality. He was lecturing at Cornell University in the United States in 1940 when Germany invaded the Netherlands, whereupon he accepted the post of Professor and Head of the Chemistry Department at Cornell. He became a US citizen in 1946 and formally retired in 1952, although he remained active until his death at the age of 82 on 2 November 1966 at Ithaca, New York.

Debye's first major contribution was a modification of Einstein's theory of specific heats to include compressibility and expansivity, leading to the expressions for specific heat capacities, $C_v = aT^3$ etc., as T approaches absolute zero. The 'Debye extrapolation' incorporating these terms acknowledges the action of intermolecular forces. His studies of dielectric constants led to the explanation of their temperature dependence and of their importance in the interpretation of dipole moments as indicators of molecular structure. The unit of dielectric constant is now called the debye.

While he was at Göttingen in 1916, Debye followed on the work of Max von Laue (1879–1960) and the Braggs on X-ray crystallography. He showed that the thermal motion of the atoms in a solid affects the X-ray interfaces and explained (using his specific heat theories) the temperature dependence of X-ray intensities. This work provided the basis for his observation with Paul Scherrer (1890–1969) that randomly orientated particles can produce X-ray diffraction patterns of a characteristic kind. Thus the need for comparatively large single crystals was avoided and powder X-ray diffraction–analysis – now called the Debye–Scherrer method – became a new and versatile analytical tool.

At Zurich, Debye's work was dominated by his interest in electrolysis and the extension in 1923 of Svante Arrhenius's theory of ionization developed by himself and Erich Hückel. The Debye–Hückel theory compares the ordering of ions in solution to the situation in the crystalline state and postulates (a) that ionization is complete in a strong electrolyte and (b) that each ion is surrounded by a cluster of ions of opposite charge. The extent of this ordering is determined by the equilibrium between thermal motion and interionic forces. Also at Zurich, Debye used the quantum theory to derive a quantitative interpretation of the Compton effect (the small change in wavelength that occurs when an X-ray is scattered by collision with an electron). This laid the foundation for other researchers' work on electron diffraction.

In the 1930s Debye moved on to a study of the scattering of light by solutions. He showed that sound waves in a liquid can behave like a diffraction grating and developed techniques in turbidimetry which led to useful molecular weight determinations for polymers. In the last phase of his researches Debye pursued his interest in polymers, investigating their behaviour in terms of viscosity, diffusion and sedimentation, and he was involved in studies of synthetic rubber.

Debye was an excellent teacher as well as a brilliant experimentalist. But perhaps the outstanding feature of his long career was the very clear thinking that enabled him to persist with incomplete or inadequate theories until he had derived important generalizations.

Dedekind Julius Wilhelm Richard 1831–1916 was a German mathematician, a great theoretician, in some respects way ahead of his time, whose work on irrational numbers – in which he devised a system known as Dedekind's cuts – led to important and fundamental studies on the theory of numbers. A pupil and friend of outstanding mathematicians, he taught for many years and published some highly influential books.

Dedekind was born on 6 October 1831, the youngest son of a professional civil servant who worked at the Collegium Carolinum in Braunschweig (Brunswick). Educated locally, Dedekind showed aptitude in the sciences, particularly in physics. Nevertheless, at the age of 17, it was mathematics he went to study at the Collegium Carolinum, for two years before entering the University of Göttingen, where he was taught by the ageing Karl Gauss (1777–1855). Having received his PhD in 1852, he remained as an unpaid lecturer at the university for a few years before taking up the post of Professor of Mathematics at the Zurich

Polytechnicum. In 1862 he returned to Brunswick and became Professor at the Technische Hochschule – a position he held for the remainder of his life. He was Director of the Hochschule from 1872 to 1875, and retired (as Professor Emeritus) in 1894. Dedekind became a member of several academies and received a large number of honorary degrees. He died in Brunswick on 12 February 1916.

At Göttingen, Dedekind developed a friendship with Bernhard Riemann (1826–1866), who later became Professor of Mathematics there. He also met Lejeune Dirichlet (1805–1859) and, with Karl Gauss, the four of them formed a formidable mathematical quartet which profoundly influenced each other's ways of thinking. From Gauss Dedekind learned about the method of least squares (as he was able to recall vividly in lectures after 50 years); probably through Riemann's influence. Dedekind's thesis concerned the theory of integrals devised by Leonhard Euler (1707–1783); and from Dirichlet – himself to succeed Gauss and precede Riemann as Professor of Mathematics at Göttingen – Dedekind learned about the theory of numbers, potential theory, definite integrals and partial differential equations. Both Riemann and Dedekind also gave lectures at the university, Riemann on Abelian and elliptic functions, Dedekind on the new beginnings of group theory (as advanced by Evariste Galois just before his death in 1832). Dedekind was no lecturer, however, and his seminars were very ill-attended.

He was nevertheless outstanding for his original contributions to mathematics. In 1858 he had succeeded in producing a purely arithmetic definition of continuity and an exact formulation of the concept of the irrational number. From this, and from his editing of Dirichlet's lectures in 1871 (to which he added a supplement establishing the theory of algebraic number fields), he derived the subject of the first of his three great publications, *Stetigkeit und irrationale Zahlen*, published in 1872. In it he defined and explained the use of what are now called Dedekind's cuts – a device by which irrational numbers can be categorized as fractions – a completely original idea that has since passed into general use in the real number system.

Number theory continued to fascinate Dedekind. In his second great work, *Was sind und was sollen die Zahlen?*, published in 1888, he elaborated on his attempt to derive a purely logical foundation for arithmetic, and devised a number of axioms that formally and exactly represented the logical concept of whole numbers. (The axioms were later wrongly attributed to the Italian mathematician of a generation ahead, Giuseppe Peano (1858–1932), by whose name they are still known.)

In his third great work, Dedekind returned again to one of his former interests in order to extend his previous research; he described the factorization of algebraic numbers using his new theory of the 'ideal' – the modern algebraic concept. Published in two sections, in 1879 and 1894, the work was fundamental in that Dedekind later further developed his theory of the ideal (determining the number of ideal classes in a field) and the subject was taken up by others.

Dedekind was also responsible for the publication of papers on a variety of other mathematically oriented subjects such as time-relationships and hydrodynamics; in 1897 and 1900 he introduced the concept of dual groups, which was eventually developed (well after his death) into the modern lattice theory.

de Duve Christian René 1917– is a British-born Belgian biochemist who discovered two organelles, the lysosome and the peroxisome. For this important contribution to cell biology he was awarded (jointly with Albert Claude and George Palade) the 1974 Nobel Prize for Physiology or Medicine.

De Duve was born on 2 October 1917 in Thames Ditton, Surrey, and was educated at the University of Louvain, Belgium, from which he graduated in medicine in 1941. He then held positions at the Nobel Institute, Stockholm, and Washington University before returning to Belgium in 1947. In 1951 he was appointed Professor of Biochemistry at the University of Louvain Medical School, a position he retained in 1962 when he became Professor of Biochemistry at the Rockefeller Institute, New York City.

De Duve discovered lysosomes in the cytoplasm of animal cells in 1955, since when a similar organelle has been found in plant and fungal cells. As seen under the electron microscope, lysosomes in animal cells are usually spherical and are surrounded by a unit membrane. They are internally structureless but contain characteristic degradative enzymes, which can digest most known biopolymers, such as proteins, fats and carbohydrates.

After his discovery of the lysosome, de Duve found that the main role of lysosomes in the normal functioning of cells is intracellular digestion and he went on to describe the way in which this occurs. All cells require a supply of essential extracellular raw materials, some of which pass into the cell by diffusion. But with substances that are too large to diffuse into the cell, the cell membrane invaginates and surrounds the extracellular matter, forming a food vacuole; this process is called endocytosis. A lysosome then fuses with the food vacuole and releases its acid digestive enzymes into the vacuole (which at this stage is called a digestive vacuole or a secondary lysosome). The enzymes break down the material in the vacuole into molecules small enough to diffuse through the vacuole's wall into the cytoplasm. The undigested remnants contained within the residual body (as the vacuole is called at this stage) eventually pass out of the cell by exocytosis. In addition to intracellular digestion, lysosomes also play a part in the digestion and removal of dysfunctional organelles, a process known as autophagy.

Since the discovery of lysosomes and the elucidation of their role in intracellular processes, Hers – a collaborator of de Duve – found in 1964 that malfunctioning of the lysosomes (which often involves an absence or insufficiency of one or more of the lysosomal enzymes) is associated with several diseases, some of which are hereditary.

As a result of his research on lysosomes, de Duve suspected the existence of another organelle, and in the 1960s he discovered the peroxisome. Almost identical to the lysosome in structure, the peroxisome is characterized by the enzymes it contains. As yet, however, its function is unknown.

De Forest Lee 1873–1961 was a US inventor and pioneer of wireless telegraphy, and is sometimes known as the father of radio.

Born in Council Bluffs, Iowa, on 26 August 1873, De Forest was raised in Alabama, where his family moved in 1879. Even as a child, De Forest was keenly interested in machinery, building model trains and a blast furnace while still in his early teens. Despite pressure from his father (a Congregational minister who made great efforts to bring education to the local black community), he decided to follow his scientific inclinations, enrolling at the Sheffield Scientific School of Yale University in 1893. He was forced to supplement his scholarship and meagre allowance from his parents by taking menial jobs, but succeeded in gaining his doctorate in 1899. Called 'Reflection of Hertzian waves from the ends of parallel wires', this thesis was probably the first US dissertation to deal with radio.

De Forest's first appointment was with the Western Electric Company in Chicago. In the company's experimental laboratories, he devised ways of rapidly transmitting wireless signals, his system being used in 1904 in the first wireless news report (of the Russo–Japanese War).

Following this and other successes De Forest set up his own wireless telegraph company, but by 1906, following a number of serious misjudgements (he was twice defrauded by business partners), the firm went bankrupt. Then came his biggest breakthrough: the 'audion' detector, which he patented in 1907. This thermionic grid-triode vacuum tube was based on the two-element valve device patented by John Ambrose Fleming in 1904. By adding a grid for a third electrode, De Forest had turned Fleming's valve into an amplifier as well as a rectifier. It made possible radios, radar, television and even the earliest computers.

Even with such a major development behind it, De Forest's second company started to fold in 1909, again through internecine wrangling. He was indicted for attempting to use the US mail to defraud, by seeking to promote the 'worthless' audion tube, a charge of which he was later acquitted. He achieved much more favourable attention in 1910, however, by using his invention to broadcast the singing of Enrico Caruso, at the Metropolitan Opera. In 1912, De Forest realized that by cascading the effect of a series of audion tubes, it would be possible to amplify high-frequency radio signals to a far higher degree than that achieved using single tubes. The use of this effect made possible long-range (consequently weak-signal) radio and telephonic communication. The same year saw his discovery of a way of using feedback to produce a means of transmission capable of sending both speech and music, and in 1916 he set up a radio station and was broadcasting news.

De Forest eventually sold his audion triode to American Telephone and Telegraph for $290,000, which used it to amplify long-distance communication. This company was to purchase many of De Forest's best inventions at very low prices, yet further proof of his lack of business acumen.

In 1923, De Forest demonstrated an early system of motion pictures carrying a soundtrack, called phonofilm. Unfortunately its poor quality, and lack of interest from film makers, led to its demise, despite the fact that the system which later succeeded commercially, was based on the same principles.

Even in the scientific world, De Forest was unlucky, being strongly recommended for the Nobel prize but failing to be awarded it. He ended his days quietly, with his third wife Marie, and died in Hollywood, California, on 30 June 1961.

De Havilland Geoffrey 1882–1965 was an English pioneer of the early aircraft industry. He built his own aircraft six years after the Wright brothers' first flight, and then produced a line of mainly successful machines which culminated in the Comet, the first pure jet passenger airliner and the Comet 4, the first jetliner to fly the Atlantic.

De Havilland was born on 27 July 1882, near High Wycombe, where his father was a curate. As a schoolboy he was an enthusiastic builder of model engines. He graduated to the design and construction of racing steam cars, designed and built motorcycles, one of which became the basis of successful production machines, and in 1905 joined the Wolseley company in Birmingham.

In 1908, De Havilland was a draughtsman at the Motor Omnibus Company of Walthamstow, fascinated by news from abroad of flying. Now that flight had become a reality in the hands of the Wright brothers and Farman, De Havilland became convinced that he could design his own engine and aeroplane and teach himself to fly. He asked his

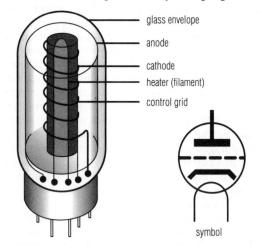

glass envelope
anode
cathode
heater (filament)
control grid

symbol

De Forest *De Forest developed the triode valve, a three-electrode vacuum tube in which a control grid allows the current flowing through the valve to be controlled by the voltage at the grid. The triode was used in amplifying circuits for radio and early computers, until largely superseded by transistors and other solid-state devices.*

friend, former marine engineering apprentice Frank Hearle, to give up his job as a mechanic at another bus company and to join him in this project. Having obtained his legacy of £1,000 in advance from his grandfather, De Havilland rented a room and set about designing his first aero engine.

It was a flat-four water-cooled design giving 34 kW/ 45 hp at 1,500 rpm, for a dry mass of 113 kg/250 lb. The Iris Car Company agreed to make a prototype for £220 and De Havilland rented the attic of a builder's workshop in Fulham, barely big enough for the span of the biplane wings he contemplated. The two men began construction of the aircraft which was based on a tapering whitewood space-frame fuselage projecting almost equally in front and behind the 11-m/36-ft wings. It was to carry a 4.3-m/14-ft pair of elevators each side of the nose and a 3-m/10-ft plane at the tail, above which a rudder was stayed by a thin outrigger from the flat radiator forming the top wing centre-section. The front spar of the main planes differed noticeably in location from all previous biplanes, being more than 30 cm/12 in back from the leading edge. Earlier biplanes, such as those of Wright, Farman and Cody, had used the leading edge as a spar but this gave a high drag entry and a weak spar for the weight. De Havilland's design, using a location at a point of greater wing thickness in a better aerodynamic shape was to prove the ultimate standard.

By November 1909 work was nearly finished. De Havilland hired a lorry to transport the four wings, engine, fuselage and other components to the Hampshire Downs where sheds had been bought at Seven Barrows from Moore-Brabazon (later Lord Brabazon).

De Havilland had never flown before (indeed, he had only ever seen one aircraft flying in the distance), and his first flight rather inevitably ended in the aircraft being wrecked. However, the salvaged aircraft Number One was the basis of the successful Number Two, even though only the engine had been left intact. In 1910 the Downs south of Newbury, Berkshire, saw De Havilland flying consistently for as long as 40 minutes around the countryside. He had done away with the twin propellers and mounted a single wooded screw direct on the crankshaft. The landing gear had been lightened and the main structure made lighter, simpler and more robust.

To raise more money for further trials, De Havilland sold Number Two to the Army Balloon Factory (later called the Army Aircraft Factory) for £4,000, its two builders being taken on the staff. Although officialdom favoured balloons, De Havilland's aircraft was awarded an airworthiness certificate and officially designated FE1. (This stood for 'Faman Experimental No. 1' because, like Faman's machine, it had a pusher propeller.) De Havilland produced FE2 with a Gnome rotary engine, then BE1 ('Blériot Experimental' because it had a tractor propeller like Blériot's machine) and several other designs.

As the war clouds were gathering in 1914, De Havilland joined the Aircraft Manufacturing Company, or Airco, as chief designer and produced DH1, a two-seater pusher, and DH2, a single-seat fighter with a single Lewis gun in the nose which went into quantity production. Then came the DH4, a fast two-seat bomber that could hold its own against most fighters.

After World War I, De Havilland established his own company at the famous Stag Lane Works at Edgeware and during the inter-war period produced a series of extremely successful light transport aircraft. The DH50, a civil version of the DH4, won a competition at Gothenburg, with Alan Cobham at the controls, but it was the Moth series, starting with the Cirrus Moth, which opened up aviation as no other aircraft had done before. To power these new machines De Havilland designed the 75 kW/100 hp Gipsy 1 engine, followed by the 97 kW/130 hp Gipsy Major , and established his own Engine Division. The Tiger Moth became the standard RAF trainer and De Havilland won the 1933 King's Cup air race in the three-seat Leopard Moth at 224 kph/140 mph.

Many records were established in DH aircraft, with the all-wood DH88 Comet racer monoplane, built in 1934 for the MacRobertson England–Australia Air Race, covering the 19,680 km/11,300 mi to Melbourne in 79 hours 59 minutes.

As a private venture, the De Havilland Company designed the famous all-wood Mosquito, which was at first rejected by the Air Ministry, losing six months of precious time. The highly versatile aircraft went into squadron service in September 1941 and a total of 7,781 were built. It was about 30 kph/20 mph faster than the Spitfire and so, like the DH4, could out-fly virtually anything in the air.

After World War II the De Havilland company put a range of jet-powered aircraft into production, many of which used the company's own engines. The world's first jet trainer was the De Havilland Vampire fighter and fighter-bomber, built under licence in Europe. The final development of this twin-boom jet fighter was the Sea Vixen.

The dangers facing pioneers in aeronautics were made tragically apparent to De Havilland on several occasions. On 27 September 1946, the experimental tailless DH108 Swallow broke up over the Thames Estuary, killing his eldest son, Geoffrey. His second son John, had been killed three years earlier in a mid-air collision in a Mosquito. A re-built Swallow subsequently captured the world air speed record, averaging 968.37 kph/605.23 mph on a 100 km/ 62 mi closed circuit with John Derry at the controls. Derry died when a wing of his DH110 folded during a display at the Farnborough Air Show: 28 spectators were killed in the wreckage.

The world's first production jet airliner, the Comet, first flew on 27 July 1949, but after a triumphant entry into service in May 1952, a Comet crashed after take off from Rome in January 1954. A second crashed under the same circumstances in April, and caused the Comets to be withdrawn from service. After the most exhaustive investigation ever carried out on an aircraft, it was established that the pressurized cabin had ruptured due to the then unsuspected problems of low-cycle metal fatigue. The Comet eventually surmounted its problems, Comet 2s entering service with the Royal Air Force while the Comet 4 became the first jet airliner to operate transatlantic scheduled services, beating the Boeing 707 by a narrow margin.

Other civil projects, however, enjoyed notable success. De Havilland died in 1965 having seen the company he had founded absorbed into the Hawker Siddeley conglomerate, which had also swallowed Armstrong Whitworth, Avro, Blackburn, Folland, Gloster and Hawker. He was knighted in 1944.

Dehn Max 1878–1952 was a German-born US mathematician who in 1907 provided one of the first systematic studies of what is now known as topology – that branch of mathematics dealing with geometric figures whose overall properties do not change despite a continuous process of deformation, by which a square is (topologically) equivalent to a circle, and a cube is (topologically) equivalent to a sphere.

Dehn was born on 13 November 1878 in Hamburg, the son of a Jewish family. He studied at Göttingen University under David Hilbert (1862–1943) and received his doctorate in 1900. He then became a teacher. At the outbreak of World War I he joined the army. After the war Dehn became Professor of Pure and Applied Mathematics at Frankfurt University, and remained there until 1935, when he fell victim to Adolf Hitler's anti-Semitism laws and he lost his position. Accordingly, he emigrated in 1940 to the United States, and occupied posts at the University of Idaho, the Illinois Institute of Technology, and St John's College, Annapolis. From 1945, Dehn worked at the Black Mountain College in North Carolina. He died there on 27 June 1952.

Influenced strongly by David Hilbert, Dehn's work was mainly concerned with a study of the geometric properties of polyhedra. His first major contribution was to demonstrate that whereas the postulate of Archimedes – that the sum of the angles of a triangle is not greater than two right-angles – is not provable, a generalization of the related theorem proposed by Adrien Legendre (1752–1833) – that the sum of the angles of any two triangles is identical – *is* provable.

In a famous address in 1900, David Hilbert presented 23 unsolved mathematical problems to the International Congress of Mathematicians. Dehn found a solution to one of them (concerning the existence of tetrahedra with equal bases and heights, but not equal in the sense of division and completeness).

In 1910, Dehn proved an important theorem on topological manifolds. The theorem came to be known as Dehn's lemma, but was later found not to apply in all circumstances. It nevertheless provided stimulation for considerable scientific discussion. Dehn continued to work on topological problems of transformation and isomorphism.

Dehn's later research concerned statistics and the algebraic structures derived from differently axiomatized projective planes. He also made a notable contribution with his published work on the history of mathematics.

De la Beche Henry Thomas 1796–1855 was the British geologist who secured the founding of the Geological Survey.

Born in London, the son of a military officer (Thomas Beach) whose wealth derived from Caribbean slave plantations, De la Beche originally trained for the army at the military school at Great Marlow, and served in the Napoleonic wars. Quitting the army and gentrifying his name, he turned himself into a gentleman amateur geologist, joining the Geological Society of London in 1817 and travelling extensively during the 1820s, through Great Britain and Europe. He was soon publishing widely in descriptive stratigraphy, above all on the Jurassic and Cretaceous rocks of the Devon and Dorset area. He also conducted important fieldwork on the Pembrokeshire coast and in Jamaica. He prided himself upon being a scrupulous fieldworker and a meticulous artist. Works such as *Sections and Views Illustrative of Geological Phenomena* 1830 and *How to Observe* 1835 insisted upon the primacy of facts and sowed distrust of theories – views still being expounded in his final masterpiece, *The Geological Observer* 1851. Volumes such as the *Manual of Geology* 1831 prove he was also an effective textbook author.

In the 1830s he evolved the plan of a government-sponsored geological study of Britain, region by region, on national and economic grounds. He was paid £500 out of state funds for a survey of Devon, which he personally undertook; and he quickly proved successful in persuading the government to put these provisions on a more formal basis. In 1835, on the analogy of the Ordnance Survey, the Geological Survey was founded, with De la Beche as its first director. The Survey flourished and expanded, and De la Beche's own career culminated in the establishment of a Mines Record Office and the opening, in 1851, under the aegis of the Geological Survey, of the Museum of Practical Geology and the School of Mines in Jermyn Street, London. He was knighted in 1842.

de la Rue Warren 1815–1889 was a pioneer of celestial photography. Besides inventing the first photoheliographic telescope, he took the first photograph of a solar eclipse and used it to prove that the prominences observed during an eclipse are of solar rather than lunar origin.

De la Rue was born in Guernsey on 15 January 1815, the eldest son of Thomas de la Rue, a printer. Warren attended the Collège Sainte-Barbe in Paris before joining his father in the printing business. It was then that he first came into contact with science and technology. He was one of the first printers to adopt electrotyping and in 1851 he invented the first envelope-making machine. Initially he saw himself as an amateur chemist and, before making any achievements in the field of astronomy, he invented the silver chloride battery. He became a Fellow of the Royal Society, the Chemical Society and the Royal Astronomical Society. In later life, from 1868 to 1883, he conducted a series of experiments on electric discharges through gases, but his results were inconclusive. He died in London on 19 April 1889.

De la Rue was introduced to astronomy by a friend, James Nasmyth, who, like de la Rue himself, was a successful businessman besides being an inventor and telescope maker. De la Rue began his research career in astronomy with the intention of producing more accurate and detailed pictures of the Moon and neighbouring heavenly bodies. His early observations of the Moon, the Sun and Saturn were superbly drawn and their details were enhanced by de la Rue's innovative techniques in polishing and figuring the mirror of his own 33-cm/13-in reflecting telescope.

De la Rue's interest in new technologies led him to apply the art of photography that had been pioneered by Louis Daguerre (1769–1851) to astronomy. He modified his 33-cm/13-in telescope to incorporate a wet collodion plate. His first photographs were of the Moon, taken with exposures of 10 to 30 seconds. They were remarkably successful, considering that there was no drive fitted to de la Rue's telescope and that he had to guide the instrument by hand to hold the image steady on the photographic plate. However, the lack of a good drive made longer exposures impossible to achieve and so de la Rue postponed further research in celestial photography until he had built and equipped a new observatory.

At the new observatory de la Rue began a regular programme of astronomical photography, including a daily sequence of photographs of the Sun. He designed a photoheliographic telescope in connection with this project and took it to Spain in July 1860 to photograph a total eclipse of the Sun. This expedition was made in collaboration with the italian astronomer Father Angelo Secchi (1818–1878) of the Collegio Romano, in order that photographs of prominences, only seen during the total phase of a solar eclipse, could be taken from two separate stations, 400 km/249 mi apart. The resulting photographic plates showed conclusively that the prominences were attached to the Sun and were not, as had been suggested, either effects of the Earth's atmosphere or the result of some unknown lunar phenomenon.

De la Rue's photoheliograph was subsequently set up at the Kew Observatory, sited about 8 km/5 mi from de la Rue's private observatory and home at Cranford, Middlesex. He used it to map the surface of the Sun and study the sunspot cycle. This work led to his being able to show that sunspots are in fact depressions in the Sun's atmosphere. De la Rue continued to use his reflecting telescope to take photographs of the Moon's surface and, over a period of eight years, his series of plates brought certain details to light that had never been noted before. This sequence turned out to be particularly relevant to a controversy initiated by Julius Schmidt, Director of Athens Observatory, who announced in 1866 that one of the lunar craters had disappeared.

De la Rue's particular talents were his understanding of technology and his innovative flair in designing and crafting instruments. As a result, his observations were so accurate that they contributed to major advances in theoretical astronomy.

de Laval Carl Gustaf Patrik 1845–1913 was a Swedish engineer who made a pioneering contribution to the development of high-speed steam turbines.

De Laval was born at Orsa, into a family that had emigrated to Sweden from France in the early seventeenth century. He was educated at the Stockholm Technical Institute and at Uppsala University. In the vigour and variety of his interests – and of his inventive talent – he has been likened to Thomas Edison. In 1878 he invented a high-speed centrifugal cream separator that incorporated a turbine, and the machine was successfully marketed and used in large dairies throughout the world. He also invented various other devices for the dairy industry, including a vacuum milking machine, perfected in 1913. De Laval's greatest achievement, however, lay in his further contribution to the development of the steam turbine, which he completed in 1890 after several years of experiments. In the absence of reliable data on the properties of steam, de Laval solved the problem of the high velocity by special features in the design of the wheel carrying the vanes of the turbine, and that of the direction of the stream of particles by the form given to the nozzle through which the steam jet was produced. The turbine disc had a hyperbolic profile and was mounted on a flexible shaft, which ran well above the critical whirling speed. The machine had a convergent–divergent ('condi') exit nozzle.

De Laval's other interests ranged from electric lighting to electrometallurgy in aerodynamics. In the 1890s he employed more than 100 engineers in developing his devices and inventions, which are exactly described in the 1,000 or more diaries he kept.

The history of turbines goes back to ancient times, when people first began to use water and wind to perform useful tasks. The first device that could be classified as a steam turbine is generally attributed to Hero of Alexandria, in about the first century AD. Operating on the principle of reaction, this device achieved rotation through the action of steam issuing from curved tubes, or nozzles, in a manner similar to that of water in a rotary lawn sprinkler. Another steam-driven machine, described in about 1629, used a jet of steam impinging on blades projecting from a wheel, causing it to rotate. This 'motor', in contrast to Hero's reaction machine, operated on the impulse principle.

The first steam turbines having any commercial significance appear to have been those built in the United States by William Avery, in 1831. His turbines consisted of two hollow arms attached at right angles to a hollow shaft, through which the steam could issue. The steam vents were openings at the trailing edge of the arms, so that rotation was achieved by the reactive force of the steam. Although about 50 of these turbines were made and used in sawmills, woodworking shops and even on a locomotive, they were finally abandoned because of their difficult speed regulation and frequent need of repair. Among the prominent later inventors working in the steam-turbine field was Charles Parsons (1854–1931). He soon recognized the advantages of using a large number of stages in series, so that the release of energy from the expanding steam could take place in small steps. This principle opened the way for the development of the modern steam turbine. Parsons also developed the reaction-stage principle, in which pressure drop and energy release are equal, through both the stationary and moving blades.

In 1887, de Laval developed his small high-speed turbines with a single row of blades and a speed of 42,000 revolutions per minute. Although several of these were later used for driving cream separators, he did not consider them practical for commercial application. De Laval turned to the development of reliable, single-stage, simple-impulse turbines. He is credited with being the first to use a convergent–divergent type of nozzle in a steam turbine in order to realize the full potential energy of the expanding steam in a single-stage machine. During the period from 1889 to 1897 he built a large number of turbines (ranging in size from about five to several hundred horse-power), and it was de Laval who invented the special reduction gearing which allows a turbine rotating at high speed to drive a propellor or machine at comparatively slow speed, a principle having universal application in marine engineering.

de Lesseps Ferdinand 1805–1894. French civil engineer; *see* Lesseps, Ferdinand de.

Democritus *c.*460–*c.*370 BC was a Greek philosopher who is best known for formulating an atomic theory of matter and applying it to cosmology.

Little is known of Democritus's life. He was probably born in Abdera, Thrace, in about 460 BC. He became a rich man, and travelled far from his native Greece, particularly to Egypt and the East. In his long lifetime, Democritus is reputed to have written more than 70 works, although only fragments of them have survived. He died, aged 90, in about 370 BC.

Democritus believed that all matter – throughout the universe, solid or liquid, living or nonliving – consists of an infinite number of tiny indivisible particles which he called atoms. According to the theory, atoms cannot be destroyed (an idea similar to the modern theory of the conservation of matter) and exist in a vacuum or 'void', which corresponds to the space between atoms. Atoms of a liquid, such as water, are smooth and round and so they easily 'roll' past each other and the liquid is formless and flows. A solid, on the other hand, has angular, jagged atoms that catch onto each other and hold them together as a solid of definite form. Atoms differ only in shape, position and arrangement. Thus the differences in the physical properties of atoms account for the properties of various substances. When atoms separate and rejoin, the properties of matter change, dissolving and crystallizing, dying and being reborn. Even new worlds can be created by the coming together in space of a sufficient number of similar atoms.

It is tempting, with the benefit of modern scientific knowledge, to overemphasize the apparent foresight of Democritus's theories. Remarkable although they undoubtedly were, they were a product of his mind and had no basis in experiment or scientific observation. Eventually they were superseded by other philosophies, and atomism was discarded as a theory until revived 1800 years after Democritus, by such scientists as Galileo Galilei (1564–1643).

de Moivre Abraham 1667–1754 was a French mathematician who, despite being persecuted for his religious faith and subsequently leading a somewhat unstable life, pioneered the development of analytical trigonometry – for which he formulated his theorem regarding complex numbers – devised a means of research into the theory of probability, and was a friend of some of the greatest scientists of his age.

De Moivre was born on 26 May 1667 in Vitry-le-François, Champagne – a Huguenot (Protestant) in an increasingly intolerant Roman Catholic country. Although he first attended a local Catholic school, his next school was closed for being too evidently Protestant, and he then studied in Saumur, and finally (in 1684, at the age of 17) in Paris. With the revocation of the Edict of Nantes in the following year, however, he was imprisoned as a Protestant for 12 months; on his release he went immediately to England. In London he became a close friend of Isaac Newton (1642–1727) and the Secretary to the Royal Society, Edmund Halley (1656–1742). It was Halley who read de Moivre's first paper – on Newton's 'fluxions' (calculus) – to the Royal Society in 1695, and saw to his election to the Royal Society in 1697. (Forty years later he was elected a Fellow of the Berlin Academy of Sciences, and no less than 50 years later – in the year of his death – to the Paris Academy of Sciences.) In 1710 de Moivre was appointed to the Grand Commission through which the Royal Society tried to settle the dispute over the priority for the systematization of calculus between Gottfried Leibniz (1646–1716) and Newton. Although de Moivre was a distinguished mathematician, he spent his whole life in comparative poverty, eking out a precarious living by tutoring and acting as a consultant for gambling syndicates and insurance companies. In spite of the fact that he had powerful friends, he never obtained a permanent position; no matter how he begged his influential associates to help him secure a chair in mathematics, particularly at Cambridge, it was without success. Finally, at the age of 87, he gave in to lethargy and spent 20 hours of each day in bed. He died, nearly blind, on 27 November 1754 in London.

While de Moivre was studying in Saumur, he read mathematics almost secretly, and studied Christiaan Huygens's work on the mathematics of games of chance. It was not until he went to Paris that he received any thorough mathematical instruction and studied the later books of Euclid under the supervision of Jacques Ozanam (1813–1853). But his first view of Newton's *Principia Mathematica* came even later, in London. Fascinated by it, it is said that he cut out the pages and read them as he walked along the street between tutoring one pupil and another.

He dedicated his own first book, *The Doctrine of Chances*, to Newton. (Subsequently, Newton – as he felt himself becoming more infirm with age – took to sending students to de Moivre.) A masterpiece, the work was published first in Latin in the *Philosophical Transactions* of the Royal Society, and then in expanded English versions in 1718, 1738 and 1758. Until 1711, the only texts on probability were the one by Huygens and another by Pierre de Montmort, published in Paris in 1708. When de Moivre first published, Montmort contested his priority and originality. Both men had made an approximation to the binomial probability distribution; now known as the normal or Gaussian distribution, it was the most important single discovery in the formulation of probability theory and was incorporated into statistical studies for the next 200 years. De Moivre was the first to derive an exact formulation of how 'chances' and stable frequency are related. He obtained from the binomial expansion of $(1 + 1)^n$ what is now recognized as $n!$ (the approximation of Stirling's formula). With $n!$, de Moivre could sum the terms of the binomial from any point up to the central term. He seems to have been aware of the standard deviation parameter (σ) although he did not specify it. He also hinted at another approximation to the binomial distribution – which is now attributed to Siméon-Denis Poisson (1781–1840), a century later – but in this case he seems not to have realized its potential in probability theory.

Perhaps with regard to his own constant state of penury, de Moivre also took a great interest in the analysis of mortality statistics, and laid the mathematical foundations of the theory of annuities, for which he devised formulae based on a postulated law of mortality and constant rates of interest on money. He worked out a treatment for joint annuities on several lives, and one for when both age and interest on capital have equal relevance. First published in 1725, his work became standard in textbooks of all subsequent commercial application. Again, however, he had to fight for copyright with Thomas Simpson (1710–1761), who published a work on annuities in 1742, the year before de Moivre republished.

All his life, de Moivre published papers in other branches of mathematics; one of the subjects that particularly interested him was analytical trigonometry. In this field he discovered a trigonometric equation that is now named after him:

$$(\cos z + i \sin z)^n = \cos nz + i \sin nz$$

It was first stated in 1722, although it had been anticipated by related forms in 1707. It entails or suggests a great many valuable identities, and became one of the most useful steps in the early development of complex number theory.

Sadly, de Moivre died a disillusioned man; much of his work was valued only long after his death.

De Morgan Augustus 1806–1871 was a British mathematician whose main field was the study of logic, an interest that led him into a bitter controversy with his contemporary William Hamilton (1805–1865).

De Morgan was born on 27 June 1806 in Madura (now Madurai), India, the son of a colonel in the Indian Army.

He entered Trinity College, Cambridge, in 1823 and graduated with a BA four years later. He disliked competitive scholarship and did not become a candidate for a fellowship; nor would he comply with his parents' wishes that he should enter the Church. De Morgan considered a career in medicine but decided instead to become a barrister, and entered Lincoln's Inn to study for the bar. He changed his mind yet again and in 1828 applied for and obtained the position of the first professor of mathematics at the new University College, London, where he remained for 30 years. He died in London on 18 March 1871. After De Morgan's death, Lord Overstone bought his library of more than 3,000 books and presented them to the University of London.

De Morgan expended most of his energy on writing voluminous articles on mathematical, philosophical and antiquarian matters. A major controversy arose from his tract on 'The structure of the syllogism', read to the Cambridge Philosophical Society in 1846 and subsequently incorporated into his book *Formal Logic*, published a year later. De Morgan had consulted William Hamilton on the history of Aristotelian theory, and Hamilton accused him of appropriating his doctrine of 'quantification of the predicate' (and returned a copy of *Formal Logic* which De Morgan presented to him).

In logic the expression 'every … is' (or 'all … is') is treated as a single syntactically unanalysable term, which in concatenation (linked) with two noun expressions forms a proposition. Traditionally, however, it has been held that 'every' (or 'all') modifies the way in which the subject should be construed. Logicians therefore suggested a similar modification of the predicate. This idea was not new – it had been suggested by Aristotle, only to be abandoned later. Some of his early commentators worked out that such modification could generate 16 different propositions, but that is as far as they went. De Morgan recognized these restrictions and succeeded in expanding syllogistic, when he developed a logic of noun expressions in *Formal Logic* and in his *Syllabus of a Proposed System of Logic*. He also extended his syllogistic vocabulary using definitions, so giving rise to new kinds of inferences, both direct (involving one premise) and indirect (involving two premises). He was thus able to work out purely structural rules for transforming a premise or pair of premises into a valid conclusion.

Inferences that appeared to illustrate principles belonging to the logic of noun expressions – but which could not be accommodated in syllogistic – had been known before. But De Morgan initiated and developed a theory of relations. He devised a symbolism that could express such notions as the contradictory, the converse and the transitivity of a relation, as well as the union of two relations. Together with George Boole (1815–1864), De Morgan can be credited with stimulating the upsurge of interest in logic that took place in the mid-nineteenth century.

Descartes René 1596–1650 was a celebrated and influential French philosopher–mathematician whose work in

Descartes *Folium*

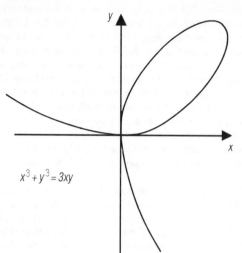

$$x^3 + y^3 = 3xy$$

attempting to reduce the physical sciences to purely mathematical principles – and particularly geometry – led to a fundamental revision of the whole of mathematical thought. So influential was his work that his astronomically erroneous description of the Solar System (in an endeavour not to offend the powerful Roman Catholic Church) held back astronomical research in continental Europe for decades.

Descartes was born at La Haye, Touraine, on 31 March 1596, the third son of Joachim Descartes, a Councillor of the Parliament of Rennes in Brittany. When René was eight years old, he was sent to the Jesuit College at La Flèche, where he spent five years studying grammar and literature and then three years studying science, elementary philosophy and theology; his favourite subject was mathematics. In 1612 he went to the University of Poitiers to study law and graduated four years later. Wanting to see the world, he joined the army of Prince Maurice of Nassau and used his mathematical ability in military engineering. A dream in November 1619 made him think that physics could be reduced to geometry and that all the sciences should be interconnected by mathematical links: he spent the next ten years applying this tenet to algebra.

Returning to France in 1622, Descartes sold his estate in Poitou in order to resume his travels, visiting scientists throughout France and western Europe. He finally settled in the Netherlands in 1629. Twenty years later he was invited to go to Sweden to instruct Queen Christina. On his arrival in Stockholm, he found that the somewhat whimsical queen intended to receive her instruction at 5 o'clock each morning. Unused to the cold of a Swedish winter, Descartes very shortly afterwards caught a severe chill and died on 11 February 1650. His remains were taken back to France and buried in the church of St Geneviève du Mont in Paris.

It was at the Jesuit College that mathematics became Descartes's favourite subject 'because of the certainty of its proofs and the logic of its reasoning'; he was surprised at

how little had been built on such firm and logical foundations. Later, with the rector of Breda – the philosopher and mathematician Isaac Beeckman – he devised a way of approaching physics generally following mathematical principles.

Descartes's great work in mathematics – most of his publications concerned either philosophy or astronomy – was his *Geometry*, published in 1637. Much of the book was revolutionary for its time but has now been long absorbed into standard textbooks of coordinate geometry. In it he provided a basis for analytical geometry – the geometry in which everything is reduced to numbers, so that a point is a set of numbers which are called its (Cartesian) coordinates, and a figure may be considered as an aggregate of points and described by formulae, equations or inequalities. Today, analytical geometry has many practical applications, such as cartography and the construction of graphs.

In establishing analytical geometry, Descartes introduced constants and variables into conventional geometry in order to enable the properties of curves to be expressed as algebraic equations. He used algebra to resolve complicated problems in geometry, and he expressed algebraic results geometrically (in graphs). Although not the first to apply algebra to geometry, he was the first to apply geometry to algebra. He was also the first to classify curves systematically, separating 'geometric curves' (which can be precisely expressed as an equation) from 'mechanical curves' (which cannot). The geometric curves he then further subdivided into three groups of increasing complexity according to the degree of the equation – the simplest group contained the circle, the parabola, the hyperbola and the ellipse. One of his more complex curves was the folium: $x^2 + y^2 = 3axy$, where x and y are variables and a is a constant; the result is a near loop.

Descartes's new geometry led to a concept of continuity, which in turn led to the theory of function and thence to the theory of limits.

In algebra he systematized the use of exponents (where the variable is itself a power; for example, a^x), interpreted the idea of negative quantities, and enlarged on his 'rule of signs' for determining the number of negative and positive roots in (solutions to) an equation. He also resolved the long-standing problem of doubling the cube.

Descartes's work in mathematics was his greatest service to future science; his attempts to 'geometrize' nature and his contributions to pure mathematics were far more permanent in their significance than all his other scientific work.

Descartes set up a general theory of the universe, which was accepted in France for more than 100 years. This was much longer than the theory deserved, but it was accepted because of Descartes's fame as a mathematician and philosopher. It had been Descartes's intention to write a work entitled *On the World*, founded on the Copernican system. But when he heard of the Roman Catholic Church's attack on Galileo, he gave up the idea. Some years later, he resolved his dilemma by proposing that the

Earth did not move freely through space, but that it was carried round the Sun in a vortex of matter without changing its place in respect of surrounding particles. In this way, it could be said (through a slight stretch of the imagination) to be 'stationary'. In 1644 he published *Principia Philosophiae*, in which he assumed space to be full of matter that in the beginning had been set in motion by God, resulting in an immense number of vortices of particles of different sizes and shapes, which by friction had had their corners rubbed off. In this way, two kinds of matter were produced in each vortex: small spheres which continued to move round the centre of motion, with a tendency to recede from it, and fine dust which gradually settled at the centre and formed a star. Those particles which become channelled or twisted, in passing through the vortex, formed sunspots; these, declared Descartes, might eventually dissolve, or might form a comet, or might settle permanently in a part of the vortex that has a velocity equal to its own, and form a planet. In this way he was able to account for the origin of the Moon and other satellites. His theory was of course pure speculation, unsupported by any facts, and was an attempt to explain how the planets move round the Sun, able neither to move away nor to move closer. It does not explain any of the deviations of the planetary orbits, and only with difficulty accounts for the elliptical, not circular, form of the orbits.

Although his theory now seems rather illogical, it was significant in that the considerable following it had partly explains why Newton's theories and models were not generally accepted on the Continent until the middle of the eighteenth century.

de Sitter Willem 1872–1934 was a Dutch mathematician, physicist and astronomer whose wide knowledge and energetic application made him one of the most respected theoreticians of his time. He was particularly influential in English-speaking countries in bringing the relevance of the general theory of relativity to the attention of astronomers.

De Sitter was born in Sneek, in the Netherlands, on 6 May 1872. He attended the high school in Arnhem before going on to the University of Groningen. Although his primary interests were in mathematics and physics, he soon became interested in astronomy on learning of studies being conducted by Haga and Jacobus Kapteyn (1851–1922). De Sitter was invited to take his postgraduate courses at the Royal Observatory, Cape Town, South Africa, by David Gill (1843–1914), and, since conditions for astronomical observations were excellent in South Africa, he sailed for the Cape in 1897.

Within two years de Sitter had collected sufficient data to enable him to return to the Netherlands to write his doctoral dissertation: he was awarded his PhD in 1901. From 1899 to 1908 he served as an assistant to Kapteyn, becoming Professor of Theoretical Astronomy at the University of Leiden in 1908. In 1919 he took on the additional post of Director of the Observatory at Leiden, which was undergoing a programme of redevelopment and expansion. He died of pneumonia in Leiden on 19 November 1934.

De Sitter's early work in Cape Town consisted primarily of photometry and heliometry. Gill suggested that he study the moons of Jupiter, and his subsequent observations of the Jovian satellites were the first of many such investigations, the results of which he published during the course of his career. Taking advantage of data on Jupiter's moons dating back to 1688, in 1925 he produced a new mathematical theory on them. Four years later he published accurate tables describing the satellites. He also obtained an estimate for the mass of the parent planet itself.

Einstein's special theory of relativity appeared in 1905, but few astronomers recognized its importance to their field. In 1911 de Sitter wrote a brief paper in which he outlined how the motion of the constituent bodies of our Solar System might be expected to deviate from predictions based on Newtonian dynamics if relativity theory were valid. After the publication of Einstein's general theory of relativity in 1915 he expanded his ideas and presented a series of three papers to the Royal Astronomical Society on the matter. In the third paper, he introduced the 'de Sitter universe' (as distinct from the 'Einstein universe'). His model later formed an element in the theoretical basis for the steady-state hypothesis regarding the creation of the universe. De Sitter also noted that the solution Einstein presented for the Einstein field equation was not the only possible one. He therefore presented further models of a nonstatic universe: he described both an expanding universe and an oscillating universe (i.e. one which alternately increases and decreases in diameter). One of several theoretical astronomers to develop this field, de Sitter thus contributed to the birth of modern cosmology.

There were two other areas of astronomical research to which he made important contributions. The first of these was the bringing up to date of many of the astronomical constants. He published his work on this subject in two papers, one in 1915 and the other in 1927. A third, incomplete, paper on the subject was published posthumously in

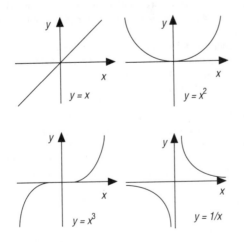

Descartes *The use of graphs*

1938. Using geodetic and astronomical data in his analysis, de Sitter demonstrated that there is some variation in the rotation of the Earth, and presented suggestions for its mechanism. He suggested that tidal friction might affect the rotation of not only the Earth but also the Moon, but that some factors might affect the Earth alone.

Deslandres Henri Alexandre 1853–1948 was a French physicist and astronomer, now remembered mostly for his work in spectroscopy and for his solar studies.

Deslandres was born in Paris on 24 July 1853. He studied at the Ecole Polytechnique in Paris, graduating in 1874 and then entering the army. He retired from the army in 1881 because of his strong interest in science, having attained the rank of Captain in the Engineers. Deslandres worked at the Ecole Polytechnique and the Sorbonne from 1881 until 1889, when he began his astronomical career at the Paris Observatory. In 1897 he moved to the observatory at Meudon, where he became assistant director in 1906 and director in 1907. The Paris and Meudon Observatories were combined in 1926 and in 1927 Deslandres became Director of the Paris Observatory. He retired officially in 1929, but he continued to publish his research until 1947. He died in Paris on 15 January 1948.

Deslandres's early scientific work in spectroscopy led to the formulation of two simple empirical laws describing the banding patterns in molecular spectra; the laws were later found to be easily explained using quantum mechanics. When Deslandres joined the Paris Observatory his task was to organize spectroscopic research. He studied both planetary and stellar spectra, but he soon began to direct his attention to the study of the Sun.

In 1891 George Ellery Hale (1868–1938) and Deslandres independently made the same discovery; both turned to the development of a photographic device for studying the solar spectrum in more detail. Hale's spectrograph was ready a full year before Deslandres's version, but the latter's spectroheliograph was a more flexible device and Deslandres used it to full advantage over the ensuing years. Deslandres found his spectroheliograph particularly well adapted to studying the solar chromosphere.

Deslandres was a member of several expeditions to observe total solar eclipses. In 1902 he predicted that the Sun would be found to be a source of radio waves, and 40 years later this was confirmed. From 1908 onwards much of his work on the Sun was done in collaboration with L. d'Azabuja. After his official retirement Deslandres returned to spectroscopy, with a special interest in the Raman effect. Much of his later work was, however, outside the mainstream of contemporary research..

Desmarest Nicolas 1725–1815 was a French naturalist who became a champion of volcanist geology.

A poor boy educated at the Oratorian school at Troyes, he moved to Paris, worked as a private tutor, and later occupied various minor government offices, spending some time as an industry inspector for the department of commerce and displaying expertise in technology. Gradually he staked out for himself a scientific career, publishing charts of the English Channel and developing skills in experimental philosophy. In the 1760s, he made extensive geological tours in Burgundy, Lorraine, Alsace, Franche-Comté and Gascony, and became interested in the large basalt deposits of central France, discovered by Jean-Etienne Guettard (1715–1786) a decade before. He succeeded in tracing their origins to ancient volcanic activity in the Auvergne region. In 1768 he produced a detailed study of the geology and eruptive history of the volcanoes responsible. His work was important because it asserted categorically that prismatic basalts were igneous in origin – countering the widely held belief, advanced by his contemporary, Werner, that all rocks were sedimentary. Desmarest never, however, became a fully fledged plutonist, in his later career specifically refuting the views of Hutton. He emphasized the critical role of water in the shaping of the Earth's history. He did not believe that all rocks had an igneous origin, and was always cautious about making inferences from the former existence of volcanic activity to the role of central heat in the Earth's history. Indeed, Desmarest liked to see himself as a sound empiricist who eschewed speculative theories of the Earth.

Having written extensively for the *Encyclopédie*, in 1771 he was appointed a member of the Academy of Sciences.

Désormes Charles Bernard 1777–1862 was a French physicist and chemist whose principal contribution was to determine the ratio of the specific heats of gases. He did this and almost all of his scientific work in collaboration with his son-in-law Nicolas Clément (1779–1841).

Désormes was born in Dijon, Côte d'Or, on 3 June 1777. He was a student at the Ecole Polytechnique in Paris from 1794, when it opened, and subsequently worked there as a demonstrator. Désormes met Clément at the Ecole Polytechnique, beginning a scientific collaboration that lasted until 1824. He left the Ecole in 1804 to establish an alum refinery at Berberie, Oise, with Clément and Joseph Montgolfier (1740–1810), who had earlier pioneered balloon flight.

Clément and Désormes published their most famous work on specific heat in 1819. This memoir led to the election of Désormes (but not of Clément) to the Academy of Sciences in Paris. Désormes's scientific productivity subsequently declined and in 1830 he turned his efforts entirely to politics. He was elected counsellor for Oise in 1830 and, after three consecutive attempts to enter Parliament in 1834, 1837 and 1842, finally was elected to the Constituent Assembly in 1848, in which he sat with the Republicans. He died in Verberie on 30 August 1862.

The period of Désormes's life in which he was actively involved in scientific study was comparatively short. It began in 1801, when he commenced his joint work with Clément. In this and the following year, they correctly determined the composition of carbon disulphide (CS_2) and carbon monoxide (CO). There was a certain amount of acrimonious debate with Claude Berthollet (1748–1822) over the work on carbon disulphide, as he contended that it was in fact identical to hydrogen sulphide (H_2S). In 1806, Clément and Désormes made an important contribution in the field of industrial chemistry by the elucidation of all the chemical reactions that take place during the production of sulphuric acid by the lead chamber method, in particular clarifying the catalytic role of nitric oxide in the process. This improved the production of sulphuric acid, which was of great benefit to the chemical industry. In 1813 they made a study of iodine and its compounds.

From 1812 onwards, Clément and Désormes worked together on heat. Their first achievement in this field was to make an accurate estimate of the value for absolute zero. They also studied steam engines, showing that an engine would develop more power if the steam could undergo maximum expansion.

In 1819 Désormes and Clément published a classic paper describing the first determination of γ, the ratio of the specific heat capacity of a gas at constant pressure (C_p) to its specific heat capacity at constant volume (C_v). They did this by allowing a quantity of gas to expand adiabatically, thus lowering its temperature and pressure, and then to heat up to the original temperature without expanding, thus increasing its pressure. The ratio is found by a simple calculation involving the pressures that are produced. This method was not significantly improved until 1929.

Sadi Carnot (1796–1832), who attended Clément's lectures, was probably inspired by the work of Clément and Désormes to develop his fundamental theorem on the motive power of heat.

De Vaucouleurs Gerard Henri 1918– is a French-born US astronomer who has carried out important research into extragalactic nebulae.

Born in France on 25 April 1918, De Vaucouleurs was educated at the University of Paris and obtained his BSc in 1936. From 1945 to 1950, he was a Research Fellow at the Institute of Astrophysics, National Centre for Scientific Research, before he went to Australia, to the Australian National University from 1951 to 1954, and, as Observer, to the Yale–Columbia Southern Station in Australia from 1954 to 1957. He was awarded a DSc from the Australian National University in 1957, the year he went to the United States to become Astronomer at the Lowell Observatory in Arizona. A year later, he was appointed Research Associate at the Harvard College Observatory, and in 1960 he became Associate Professor. From 1965, he was Professor of Astronomy at the University of Texas, at Austin.

The main object of De Vaucouleurs's research has been to find a pattern in the location of nebulae, clusters of stars formerly thought to be randomly scattered. Yet as telescopes become more powerful, there is increasing evidence that in fact nebulae themselves tend to cluster together. Extragalactic nebulae, small and faint as they appear, are more numerous and seem distinctly grouped. The nebulae that are closer to us, of up to the 12th or 13th magnitude, form what are now called superclusters, and many occur in a definite band across the heavens.

In 1952, De Vaucouleurs, then at the Australian Commonwealth Observatory, Mount Stromlo, Canberra, began a thorough re-investigation of a local supercluster, using newer and more accurate data. He redetermined the magnitudes of many of the brighter southern nebulae and revised the magnitudes for most of the brighter catalogued nebulae, following a modern photometric system. His aim was to obtain photometric consistency and completeness over the whole sky to approximately magnitude 12.5.

In 1956, he published and discussed his material in great detail and it seemed to indicate that a 'local supergalaxy' exists, which includes our own Milky Way stellar system. He suggested a model in which the great Virgo cluster might be 'a dominant congregation not too far from its central region'. As evidence for its existence, De Vaucouleurs pointed out the similarity in position and extent of a broad maximum of cosmic radio noise, reported by other researchers both in Britain and the United States. (It may be that a local supercluster of bright nebulae including a galaxy may not be unique, since Harlow Shapley in 1934 reported among the fainter ones a distant double supergalaxy in Hercules.)

De Vaucouleurs's work in the southern hemisphere led him to suspect that there was another supergalaxy, from a great, elongated swarm of nebulae extending through Cetus, Dorado, Fornax, Eridanus and Horologium. He estimated that its distance was only slightly greater than the Virgo cluster and he noted that the relative sizes and separations of this southern supergalaxy and the local supergalaxy appear to be comparable with Shapley's double supergalaxy in Hercules. Thus, it seems that superclustering is a phenomenon that has some essential relevance to the structure of the universe.

de Vries Hugo Marie 1848–1935 was a Dutch botanist and geneticist who is best known for his rediscovery (simultaneously with Karl Correns and Erich Tschermak von Seysenegg) of Gregor Mendel's laws of heredity, and for his studies of mutation.

De Vries was born on 16 February 1848 in Haarlem, the Netherlands. He studied medicine at the universities of Heidelberg and Leyden, graduating from the latter in 1870. He then taught at the University of Amsterdam from 1871 until 1875, when he went to work for the Prussian Ministry of Agriculture in Würzburg. In 1877, he taught at the Universities of Halle and Amsterdam; he was appointed Assistant Professor of Botany at Amsterdam in the following year and became a full professor there in 1881. De Vries remained at the University of Amsterdam – except for a visiting lectureship to the University of California in 1904 – until he retired in 1918. He died near Amsterdam on 21 May 1935.

De Vries began his work on genetics in 1886, when he noticed that some specimens of the evening primrose *Oenothera lamarckiana* differed markedly from others. He cultivated seeds of the various types in his experimental garden, and also undertook detailed research to discover the origin of the plants and the history of the species introduced into Europe. At the time it was believed that *Oenothera lamarckiana* had been introduced from Texas and was known by Jean Lamarck under the name *Oenothera grandiflora*. De Vries, however, discovered that the plant was unknown in the United States, from which he concluded that *Oenothera lamarckiana* was a pure species. Continuing his investigations, he began a programme of plant breeding experiments in 1892. Eight years later, he formulated the same laws of heredity that – unknown to de Vries – Mendel had discovered 34 years previously; but while searching the scientific literature on the subject, de Vries came across Mendel's paper of 1866. De Vries's work went further than Mendel's had, however: he found that occasionally an entirely new variety of *Oenothera* appeared and that this variety reappeared in subsequent generations. De Vries first called these new varieties that appeared suddenly single variations but later called them mutations. He postulated that, in the course of evolution, a species produces mutants only during discrete, comparatively short periods (which he called mutation periods) in which the latent characters are formed. He also distinguished between mutants with useful characteristics (which he called progressive mutants) and those with useless or harmful traits (which he called retrogressive mutants), and proposed that only the progressive mutants contribute to the evolution of the species. Furthermore, he suggested that mutation was the means by which new species originated and that those mutations that were favourable for the survival of the individual persisted unchanged until other, more favourable mutations occurred. He considered the slight variations caused by environmental factors to be insignificant in the evolutionary process, mutation being the most important factor involved. De Vries's work on heredity and mutations – which he summarized in *Die Mutationstheorie* (1901–03; translated into English as *The Mutation Theory*, 1910–11) was soon generally accepted and played an important role in helping to establish Darwin's theory of evolution.

Thomas Hunt Morgan (1866–1945), in his work on the fruit fly *Drosophila*, also encountered unexpected new variations that were capable of breeding true, and he assumed that this was the result of a change in their genes; using de Vries's terminology, Morgan applied the term mutation to this spontaneous change in a gene. It has since been discovered that the mutations found by de Vries resulted from changes in the number of chromosomes in the new species, not from changes to the genes themselves. Today, the term mutation refers to any change in the genetic material and includes many additional types of changes not known to de Vries.

De Vries's other major contribution concerned the physiology of plant cells. Experimenting on the effects of salt solutions of various concentrations on plant cells, he demonstrated in 1877 how plasmolysis of the cells can be used to establish a series of isotonic solutions.

Dewar James 1842–1923 was a British physicist and chemist with great experimental skills that enabled him to carry out pioneering work on cryogenics, the properties of matter at extreme low temperatures. Dewar is also remembered for his invention of the Dewar vacuum flask, which has been adapted for everyday use as the thermos flask.

Dewar was born at Kincardine-on-Forth, Scotland, on 20 September 1842. He went to local schools until he contracted rheumatic fever in 1852, which required a long convalescence at home. He was nevertheless able to enter the University of Edinburgh in 1859, where he studied physical science and later worked as a demonstrator. In 1867, Dewar presented models of various chemical structures at an exhibition of the Edinburgh Royal Society. This work interested Friedrich Kekulé (1829–1896), who invited Dewar to Ghent in Belgium, for the summer. Dewar then returned to Edinburgh, and continued as an assistant until 1873, holding a concurrent post as a lecturer at the Royal Veterinary College from 1869. The Jacksonian Professorship of Natural Experimental Philosophy and Chemistry at the University of Cambridge was offered to Dewar in 1875, followed two years later by the Fullerian Professorship of Chemistry at the Royal Institute in London. Dewar occupied both these chairs until his death. In 1877, he was also elected a Fellow of the Royal Society, which later awarded him the Rumford Medal for his research on the cryogenic properties of matter. Dewar was appointed to a government committee on explosives from 1888 to 1891, and in 1889 invented the important explosive cordite in collaboration with Frederick Abel (1827–1902). Dewar was knighted in 1904 and died in London on 27 March 1923.

Dewar's early work in Edinburgh concerned several branches of physics and chemistry. In 1867, he worked on the structures of various organic compounds and during his visit to Ghent may have suggested the structural formula for benzene that is credited to Kekulé. He also proposed formulas for pyridine and quinoline. In 1872, Dewar invented the vacuum flask as an insulating container in the course of an investigation of hydrogen-absorbed palladium. In collaboration with Peter Tait (1831–1901), he found that charcoal could be used as an absorbent to improve the strength of the vacuum. The vacuum prevented heat conduction or convection, so heat loss could occur only by radiation. The reflective silver layer on the walls of a vacuum flask are designed to minimize this.

When Dewar went to Cambridge in 1875, he began a long-term collaborative project on spectroscopy with George Liveing (1827–1924). They examined the correlation between spectral lines and bands and molecular states, and were particularly concerned with the absorption spectra of metals.

Dewar's move to the laboratories of the Royal Institution in London in 1877 marks the beginning of his main work on the liquefaction of gases and an examination of cryogenic properties. In that year, Louis Cailletet (1832–1913) and Raoul Pictet (1846–1929) in France, had succeeded in liquefying oxygen and nitrogen, two gases that had resisted Faraday's attempts at liquefaction. Dewar first turned to the production of large quantities of liquid oxygen, examining its properties. He found, in 1891, that both liquid oxygen and ozone are magnetic. His vacuum flasks were invaluable in this work because they enabled him to preserve the very cold liquids for longer than would otherwise have been possible.

Dewar developed the cooling technique used in liquefaction by applying the Joule–Thomson effect, whereby expansion of a compressed gas results in a lowering of temperature due to the energy used in overcoming attractive forces between the gas molecules. By subjecting an already chilled gas to this process, he became the first to produce liquid hydrogen in 1895, though only in small quantities; then by using bigger apparatus he made larger amounts in 1898. He proceeded to examine the refractive index of liquid hydrogen, and a year later succeeded in solidifying hydrogen at a temperature of −259°C/−434°F.

Every gas had now been liquefied and solidified except one – helium. Dewar tried to liquefy helium, but was unsuccessful because his sample was contaminated with neon, which froze at the low temperature and blocked the apparatus. Heike Kamerlingh Onnes (1853–1926) managed to liquefy helium using Dewar's techniques in 1908, and Dewar was then able to achieve temperatures within a degree of absolute zero (−273°C/−459°F) by boiling helium at low pressure.

Dewar investigated properties such as chemical reactivity, strength and phosphorescence at low temperature. Substances such as feathers, for example, were found to be phosphorescent at these temperatures. From 1892 to 1895, he studied with John Fleming (1849–1945; the inventor of the thermionic valve) the electric and magnetic properties of metals at low temperatures. They predicted that electrical resistance becomes negligible at extremely low temperatures, this phenomenon of superconductivity subsequently being detected by Onnes in 1911.

During the World War I cryogenic research, being very expensive, had to be suspended. Dewar turned to the study of thin films, and later to the measurement of infrared radiation. He was a scientist with tremendous experimental flair and patient application.

Dicke Robert Henry 1916– is a US physicist who has carried out considerable research into the rates of stellar and galactic evolution. Much of his work has been innovatory and some remains controversial.

Born in St Louis, Missouri on 6 May 1916, Dicke completed his education at Princeton University, graduating in 1938, before obtaining a PhD from the University of Rochester in 1941. That same year, he joined the staff of the Radiation Laboratory, Massachusetts Institute of Technology, where he remained until 1946. He then returned to Princeton, where he rose from being Assistant Professor to Professor of Physics, then to Cyrus Fogg Brachett Professor (1957–75), to Chairman of the Department of Physics and finally (from 1975) to Albert Einstein Professor of Science. Among the many prizes and honours he has received, is NASA's Medal for Exceptional Science Achievement, awarded in 1973.

In 1964, Dicke turned his attention to a version of the Big Bang theory known as the 'hot Big Bang'; he suggested that the present expansion of the universe had been preceded by a collapse in which high temperatures had been generated.

Realizing that he should be able to test this hypothesis by detecting residual radiation in space at a wavelength of a few centimetres, Dicke and his colleagues started to build the equipment they needed. But before they were in a position to begin their measurements, Arno Penzias (1933–   ) and Robert Wilson (1936–   ) of Bell Telephone Laboratories announced they had detected an unexpected and relatively high level of radiation at a wavelength of 7 cm/ 2.8 in, with a temperature of about 3.5K (−270°C/−453°F. Dicke immediately proposed that this was cosmic black-body radiation from the hot Big Bang.

Dicke carried out experiments to verify the supposition of the general theory of relativity, that a gravitational mass is equal to its inertial mass. He was able to establish the equality to an accuracy of one part in 10^{11}. In 1961, he put forward a theory (the Brans–Dicke theory) that the gravitational constant varies with time (by about 10^{-11} per year). Experiment has not supported this idea.

Dicksee Cedric Bernard 1888–1981 was the British engineer who was a pioneer in developing the compression–ignition (diesel) engine into a suitable unit for road transport.

Dicksee received his technical education at the Northampton Engineering College (1906–10). He was chief designer at the Aster Engineering Company Ltd, Wembley, and in 1918 he joined the Austin Motor Company Ltd. The following year he went to Westinghouse Electric and Manufacturing Company in charge of design and development of engines for generating plants.

After seven years in the USA, Dicksee returned to Britain and joined the Association Equipment Company (AEC) Ltd – the manufacturing subsidiary of the London General Omnibus Company – in 1928 as engine designer, later becoming research engineer. When, in 1929, the development of a compression ignition engine suitable for road transport was started, this became his major activity and was the work for which he is best known.

During World War II Dicksee worked on combustion chamber design for the De Havilland series of jet engines. He returned to AEC at about the time the war ended and after leaving became a consultant to Waukesha for several years. He retired to Seaford, East Sussex, where he died.

When Dicksee began work, the London General Omnibus Company (LGOC) had two lorries at their depot fitted with massive three-cylinder Junkers oil engines. These were unsuitable for automotive use. Within a year, Dicksee had his first engine running. It was a six-cylinder 8.1 litre/ 40 cu in unit with an aluminium crankcase and it entered service with the LGOC just before Christmas 1930.

After development with Ricardo and Company, an improved version took to the road in September 1931. A smaller (7.7 litre/35 cu in) engine followed in 1933 and soon became the standard for AEC chassis. A further development of this engine used combustion chambers of a toroidal shape and was also highly successful. This feature was subsequently adopted for larger engines in the company's range.

Dicksee's engines ran at speeds ranging from 1,800 to 2,400 (governed) rpm which were higher than comparable engines of the day. With this performance, the way was opened for the adoption of compression–ignition engines instead of petrol engines for road transport. As a result, in the early 1980s there was scarcely a commercial vehicle of any size which did not have a high-speed compression–ignition engine as its power unit.

Dickson Leonard Eugene 1874–1957 was a US mathematician who gave the first extensive exposition of the theory of fields. A prolific writer, he was also the author of a massive three-volume *History of the Theory of Numbers* (published between 1919 and 1923), now a standard work, in which he investigated abundant numbers, Diophantine equations, perfect numbers and Fermat's last theorem.

Dickson was born in Independence, Iowa, on 22 January 1874. He received his BA at the age of 19 from the University of Texas, where he then taught for some months. Receiving his MA in 1894, he entered the University of Chicago where in 1896 he gained his doctorate in mathematics. In 1897, he became a postgraduate student at Leipzig and in Paris. The following year, Dickson returned to the United States and – for a year at a time – was appointed Instructor in Mathematics at the Universities of California and Texas. In 1900, he was appointed Assistant Professor at the University of Chicago; promoted in 1907 to Associate Professor, he became Professor in 1910. Apart from short periods as visiting Professor at the University of California in 1914, 1918 and 1922, Dickson remained at the University of Chicago until his retirement in 1939. He died on 17 January 1957 in Harlingen, Texas.

For his prolific work in mathematics, Dickson received many awards and honours, and became a member of many influential societies. In 1913, he was elected to the National Academy of Sciences, and in 1916 became President of the American Mathematical Society, from whom he received the Cole Prize for his book *Algebren und ihre Zahlentheorie*.

Dickson's work in mathematics spanned many topics, including the theory of finite and infinite groups, the theory of numbers, algebras and their arithmetics (for his work on which he won a $1,000 prize), and the history of mathematics.

In his work on finite linear groups he generalized the results of Evariste Galois's studies, and those of Ernst Jordan and Jean-Pierre Serre, for groups over the field of n elements to apply to groups over an arbitrary finite field. He proved his modified version of the Chevalley theorem and published the first extensive exposition of the theory of finite fields.

During an investigation of the relationships between the theory of invariants and number theory, Dickson examined

divisional algebra, particularly in the form systematized by Arthur Cayley, and he expanded the theorems of linear associative algebras formulated by Elie Cartan and Wedderburn.

Investigating the history of the theory of numbers, Dickson also studied the work of Diophantus, who lived in Alexandria in the third century. Diophantus assumed that every positive integer is the sum of four squares, a theory for which (in 1770) Waring attempted to derive an extension in the direction of higher powers. Using the results of Ivan Vinogradov, Dickson (in the 1930s) succeeded in proving Waring's theorem and made his contribution to Diophantine analysis, from which the complete criteria were obtained for the solution of :

$$ax^2 + by^2 + cz^2 + du^2 = 0$$

for any nonzero integer a, b, c or d. In addition, Dickson gave the first complete proof that:

$$ax^2 + by^2 + cz^2 + du^2 + ev^2 = 0$$

is always soluble in integers, if the nonzero integers a, b, c, d and e are not all of a like sign.

Diels Otto 1876–1954 was a German organic chemist who made many fundamental discoveries, including (with Kurt Alder) the diene synthesis or Diels–Alder reaction. He shared with Alder the 1950 Nobel Prize for Chemistry.

Diels was born in Hamburg on 23 January 1876 into a talented academic family. His father Hermann Diels (1848–1922) was Professor of Classical Philology at Berlin University and his brothers Paul and Ludwig became professors of, respectively, philology and botany. During his school years Otto Diels became interested in chemistry and carried out a series of experiments with his brother Ludwig. In 1895, he went to his father's university to study chemistry under the great organic chemist Emil Fischer (1852–1919), obtaining his doctorate in 1899. He continued research as Fischer's assistant and became a lecturer in 1904. He moved to the Chemical Institute of the then Royal Friedrich Wilhelm University in 1914 and two years later was invited by the Christian Albrecht University in Kiel to serve as Director of their Chemical Institute, where he remained for 32 years until he retired.

Diels married in 1909 and had three sons and two daughters; two of his sons were killed on the eastern front near the end of World War II. After the destruction of his Institute by enemy action during the war, Diels planned to retire in 1945 but was persuaded to stay on to help with its rebuilding. He finally retired in 1948, and died in Kiel on 7 March 1954.

In 1906, while working in Berlin, Diels discovered carbon suboxide (tricarbon dioxide), which he prepared by dehydrating malonic acid (propandioic acid) with phosphorus pentoxide (phosphorus(V) oxide).

A year later he published the first edition of his textbook of organic chemistry, *Einführung in die organische Chemie*, which, because of its scope and clarity, became one of the most popular books of its kind.

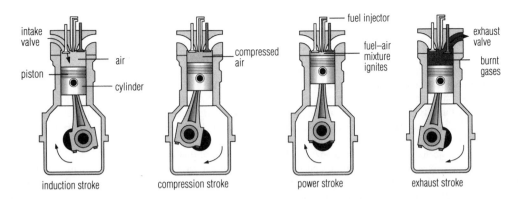

intake valve
air
piston
cylinder

compressed air

fuel injector
fuel–air mixture ignites

exhaust valve
burnt gases

induction stroke   compression stroke   power stroke   exhaust stroke

Diesel *In a diesel engine, fuel is injected on the power stroke into hot compressed air at the top of the cylinder, where it ignites spontaneously. The four stages are exactly the same as those of the four-stroke Otto cycle.*

Diels's investigation of the nature of the biologically important compound cholesterol began as early as 1906. He isolated pure cholesterol from gallstones and converted it to 'Diels acid'. But it was not until 1927 that he successfully dehydrogenated cholesterol (by the drastic process of heating it with selenium at 300°C/572°F) to produce 'Diels hydrocarbon'. This substance was later demonstrated to be 3-methyl-1,2–cyclopentanophenanthrene ($C_{18}H_{16}$), an aromatic hydrocarbon closely related to the skeletal structure of all steroids, of which cholesterol is one. In 1935 he synthesized the $C_{18}H_{16}$ compound and showed it to be identical with Diels hydrocarbon. This work proved to be a turning point in the understanding of the chemistry of cholesterol and other steroids, although it was not until about 1955 that its structure was completely known.

Working with his assistant Kurt Alder (1902–1958), Diels spent much of the rest of his life developing the diene synthesis, which first achieved success in 1928, when they combined cyclopentadiene with maleic anhydride (*cis*-butenedioic anhydride) to form a complex derivative of phthalic anhydride. Generally, conjugated dienes (compounds with two double bonds separated by a single bond) react with dienophiles (compounds with one double bond activated by a neighbouring substituent such as a carbonyl or carboxyl group) to form a six-membered ring. For example, butadiene (but-1,3-diene) reacts with acrolein (prop-2-en-1-al) to give tetrahydrobenzaldehyde (cyclohex-4-en-1-carbaldehyde).

Applications of the Diels–Alder synthesis are numerous throughout organic chemistry and it is of great importance because many reactions of this type occur easily at low temperatures and give good yields.

Diels was considered always to be a reserved man, but one with a good sense of humour, and he was liked and respected by his students who enjoyed his well-planned lectures and experimental demonstrations.

Diesel Rudolph Christian Karl 1853–1913 was a German engineer whose name will always be associated with the compression-ignition internal combustion engines he invented.

Diesel was born in Paris on 18 March 1853, of parents who came from the Bavarian town of Augsburg. His father was a bookbinder. While receiving his early education in Paris Diesel spent much of his spare time in the city's museums. He was particularly attracted to the Museum of Arts and Crafts, which had on permanent display Joseph Cugnot's steam-propelled gun carriage of 1769.

At the outbreak of the Franco–Prussian War in 1870, the Diesel family travelled to England and took up residence in London. Rudolph, however, was sent back to Augsburg to continue his education with a relative who was a teacher at the local trade school. There, he progressed so well that he qualified for the Munich Polytechnic, where he proved to be a brilliant scholar. He shared part of his time there with Carl von Linde (1842–1934), the founder of modern refrigeration engineering, from whom he learnt the basic theory of heat engines.

Diesel became fascinated by engines and came to the conclusion that an engine four times as effective as a steam engine could be made by designing it to carry out the combustion within the cylinder, while utilizing as large a temperature range as possible. Because the temperature obtained depends to a great extent on the pressure, a very high pressure must be used to compress the air before fuel injection, to avoid premature explosion.

His ideas were published in a paper of 1893, one year after he had taken out his first patent in Berlin. In the early stages of his experiments he was lucky to escape being killed when a cylinder head blew off one of his prototype engines, but this incident did not deter him and he carried on to perfect a model which was capable of commercial exploitation.

In 1899, he founded his own manufacturing company at Augsburg, which flourished despite Diesel having little or no business sense. He was, however, talented at putting over his ideas, and he gave a series of lectures in the United States in 1912, which were well attended. In 1913, at the height of his success, he vanished from the decks of a steamer crossing from Antwerp to England while on his way for consultations with the British Admiralty. His body was never found.

Like the petrol engine, the diesel engine is a form of internal-combustion engine but it differs in not having a carburettor to pre-mix air and fuel; nor does it require a spark for ignition. In the diesel engine, as the piston moves down, pure air is drawn into the cylinder and on the up stroke this air is compressed by a ratio of between 12:1 and 25:1 (much higher than that of the petrol engine, which is usually between 6:1 and 10:1). Compression, by increasing pressure, raises the temperature of the air to in excess of 540°C/1,004°F. When the piston nears the top of the compression stroke, an injector admits through a nozzle a fine spray of fuel which mixes with the heated air and ignites spontaneously, the explosion moving the piston downwards. As the rate of inflow of air remains constant in a diesel engine, its power output is governed by the amount of fuel injected.

In his first engine, Diesel is thought to have used coal dust as a fuel, but he later discarded this along with several other types in favour of a form of refined mineral oil. Diesel engines will in fact run on many fuels but the most widely used today is distilled from petroleum closely related to kerosene.

Until the 1920s further development of the diesel engine took place mainly in Germany (principally by Karl Bosch), where it proved particularly useful during World War I for powering submarines. Designers such as Cummins of the United States and Gardner in Britain, eventually adapted the invention for small boats, and later work by Cedric Dicksee (1888–1981) and Harry Ricardo (1885–1979) put it into a practical form for road use.

Diesel's death remains a mystery but his work lives on. A century after his birth more than half the world's tonnage of ships and a large proportion of the railways were diesel driven. The engine has been adapted for buses, tractors and trucks as well as serving as an alternative to coal-burning boilers and fast-flowing water for driving alternators in the generation of electricity.

Diophantus lived *c.* AD 270–280, was an ancient Greek mathematician who, in solving linear mathematical problems, developed an early form of algebra. Particularly innovative was his use of a symbol for an unknown quantity.

Very little is known about him or about his life. It is probable, however, that he was born in Alexandria (now in Egypt), a great Greek cultural centre, where he certainly lived. If the evidence of what is called 'Diophantus' riddle' is to be taken seriously, he lived at least to the age of 84 (which is the answer to the riddle).

To the Greeks, mathematics comprised two branches of study: arithmetic, the science of numbers, and geometry, the science of shapes. Towards the end of the Greek era – indeed, already under a declining Roman domination – Diophantus formulated his theories in a neglected field, the study of unknown quantities. His *Arithmetica*, according to its introduction, was compiled in 13 books (although the six that have come down to us are probably all that were completed); the work was translated by the Arabs, and it is

through their word for his equations – 'the reuniting of separate parts' – that we now call the system 'algebra'. Diophantus would merely have thought of it as abstract arithmetic.

In the solution of equations Diophantus was the first to devise a system of abbreviating the expression of his calculations by means of a symbol representing the unknown quantity. Because he invented only one symbol for the unknown, however, in equations requiring two or more variables his work can become extremely confusing in its repetition of that single symbol.

His main mathematical study was in the solution of what are now known as 'indeterminate' or 'Diophantine' equations – equations that do not contain enough facts to give a specific answer but enough to reduce the answer to a definite type. These equations have led to the formulation of a system for numbers, commonly called the theory of numbers, that is regarded as the purest branch of present-day mathematics. Using his method, the possibility of determining a type of answer rather than a specific one to a given problem has allowed modern mathematicians to approach the properties of various kinds of whole numbers (such as odds, evens, primes and squares) with new insight. By then applying the use of infinite trains of numbers correlated through the Diophantine equation system, mathematicians have come to a new understanding of some of the basic rules which numbers follow.

Dirac Paul Adrien Maurice 1902–1984 was a British theoretical physicist of great international standing. He played a pivotal role in the development of quantum electrodynamics, being responsible for the introduction of concepts such as electron spin and the magnetic monopole. Dirac's most important achievement was to predict the existence of antiparticles. For his contributions to theoretical physics, he shared the 1933 Nobel Prize for Physics with Erwin Schrödinger (1887–1961).

Dirac was born in Bristol on 8 August 1902, the son of a Swiss father. He attended Bristol University from 1918 to 1921, taking a degree in engineering but also attending lectures on philosophy. He then spent two further years at Bristol studying mathematics, and in 1923 transferred to the University of Cambridge, where he was introduced to the work of Niels Bohr (1885–1962) and Ernest Rutherford (1871–1937) on the structure of the atom. Dirac continued to study projective geometry, however, a tool he found most useful in his later research.

Dirac completed his PhD thesis in 1926. He then began to travel, visiting major centres for theoretical physics including the Institute of Theoretical Physics at Copenhagen and the University of Göttingen. There Dirac met and worked with the leading figures in the developing field of quantum mechanics, such as Niels Bohr, Wolfgang Pauli (1900–1958), Max Born (1882–1970) and Werner Heisenberg (1900–1976), with whom Dirac had already built up a lengthy correspondence.

St John's College at Cambridge University elected Dirac a Fellow in 1927, and in 1928 he formulated the relativistic theory of the electron from which he predicted the existence of the positron. The impact of the theory was profound and brought Dirac wide recognition. In 1932, he became Lucasian Professor of Mathematics at Cambridge, a post he held until 1969. In 1933, the positron was discovered and Dirac won the Nobel Prize for Physics. He was awarded the Royal Society's Royal Medal in 1939 and the Copley Medal in 1946. In 1969, he was made the first recipient of the Oppenheimer Prize. From 1971, Dirac was Professor of Physics at Florida State University.

During the late 1920s, there was an intense burst of activity in the field of theoretical physics. In 1923, Louis de Broglie (1892–1987) proposed a model of the electron which ascribed to it wave-like properties. This was a major break from the Bohr model of the atom. Schrödinger extended de Broglie's work by considering the movement of a particle in an electromagnetic field, thereby establishing the subject now known as wave mechanics. Heisenberg also worked on this subject and developed a nonrelativistic model of the electron.

Dirac was perturbed by several features of Heisenberg's theory. He was dissatisfied with its nonrelativistic form and wished to integrate it with special relativity. Consideration of these problems enabled Dirac in 1928 to formulate the relativistic theory of the electron. Apart from its mathematical elegance, an attribute Dirac constantly strove for in his work, the theory was also astonishingly fertile. The model was able to describe many quantitative aspects of the electron, including properties that were entirely new. These included the half-quantum spin and magnetic moment of the electron. Dirac's model was also able to account perfectly for certain anomalies in the hydrogen spectrum. Perhaps the most significant prediction to arise from Dirac's relativistic wave equation of the electron was the proposal of the positive electron, the positron.

Dirac found that the matrices describing the electron contained twice as many states as were expected. He proposed that positive energy states in the matrices described the electron, and that the negative energy states described a particle with a mass equal to that of an electron but with an opposite (positive) charge of equal strength, that is an antiparticle of the electron. He predicted that an electron and its antiparticle, later called the positron, could be produced from a photon. This was soon confirmed by both Carl Anderson (1905–1991) and Patrick Blackett (1897–1974), who independently discovered the positron in 1932 and 1933. This led to the prediction of the existence of other antiparticles, such as the antiproton. This was discovered by Emilio Segrè (1905–1989) and co-workers in 1955. Dirac's relativistic wave equation was also able to give a quantitative explanation for the Compton effect, and furthermore, enabled Dirac to predict the existence of the magnetic monopole. This has proved to be more elusive than the positron and has not been irrefutably detected even to this day.

Dirac was eager to learn new mathematical methods, and he studied the mathematics of eigenvalues and eigenvectors which Schrödinger had found so useful. In examining the uses of these, he noticed that those particles with half-integral spins (for example the electron), which obeyed Pauli's exclusion principle, also obeyed statistical rules different from the other particles. The latter class of particles obeyed statistical rules developed by Satyendranath Bose (1894–1974) and Albert Einstein (1879–1955). Dirac worked out the statistics for the other particles, only to discover that Enrico Fermi (1901–1954) had done very similar work already. Since the two scientists developed the theory independently, it is usually called Fermi–Dirac statistics to honour both. Fermi–Dirac statistics are of great value in nuclear and solid-state physics and are used, for example, to determine the distribution of electrons at different energy levels.

Dirac's later work was related principally to the field of mathematics. He worked on the large number hypothesis. This hypothesis deals with pure, dimensionless numbers, such as the ratio of the electrical and gravitational forces between an electron and a proton, which is 10^{39}. This ratio, perhaps coincidentally or perhaps not, also happens to be the age of the universe when it is expressed in terms of atomic units.

If there is some meaningful connection between these two values, there must be a connection between the age of the universe and either the electric force or the gravitational force. It is possible, therefore, that the gravitational force is not constant but is decreasing at a rate proportional to the rate of ageing of the universe.

Dirac was responsible for the formulation of equations that were essential to the development of quantum theory, and for the reconciliation of relativity with quantum theory resulting in the formulation of the relativistic wave theory for the electron. He was one of the giants of twentieth-century physics.

Dirichlet Peter Gustav Lejeune 1805–1859 was an influential German mathematician, whose work in applying analytical techniques to mathematical theory resulted in the fundamental development of the theory of numbers. Also a physicist interested in dynamics, he knew many of the great scientists of his age and published a seminal book of some historic importance.

Dirichlet was born in Düren on 13 February 1805, the son of the town postmaster. Precociously interested in mathematics (it is said that at the age of 12 he used his pocket money to buy mathematical books), in 1819 he was sent to the University of Cologne – where his teachers included the physicist Georg Ohm – and completed his final examination at the very early age of 16. He was then sent to France, the country of all the contemporarily great mathematicians. Arriving at the Collège de France, in Paris, he attended lectures for little more than a year before being appointed tutor to the children of the renowned General Fay; as such, he was treated as one of the family and met many of the most prominent figures in French intellectual life, particularly Joseph Fourier (1768–1830). In 1825, Dirichlet presented his first paper to the French Academy of Sciences.

General Fay's death, however, led Dirichlet to return to Germany, where he took a research post at the University of Breslau (now Wroclaw). Later he moved to Berlin, teaching initially at the Military Academy, but was soon appointed Professor also at the University of Berlin (still at the age of only 23). Dirichlet spent 27 years as a professor in Berlin, exerting a strong influence on German mathematics. He was an excellent teacher, despite being a modest and retiring man who shunned public appearances – unlike his lifelong friend Karl Jacobi (1804–1851). These two mathematicians stimulated and influenced each other, and when Jacobi's health forced him to move to Italy, Dirichlet spent 18 months there with him, from 1843. Their presence caused a circle of leading German mathematicians to gather round them. In 1855, on the death of the great Karl Gauss (1771–1855), the University of Göttingen offered Dirichlet the prestigious vacant professorship. He accepted, but enjoyed the post for a mere three years, suffering a severe heart attack in the summer of 1858. He died the following spring, shortly after his wife had died.

Dirichlet's first interest was number theory, and much of his work was on this topic. His first paper, written in France, concerned Diophantine equations of the form $x^5 + y^5 = kz^5$. Using the methods of this paper, Adrien Legendre (1752–1833) succeeded only a few weeks later in appending a proof that Pierre de Fermat's famous equation ($x^n + y^n = z^n$) has no integral solution when $n = 5$.

Dirichlet was considerably influenced by Gauss, some of his early work being improvements on Gauss's proofs, but as his abilities developed, Dirichlet's intensive search for a general algebraic number theory substantially advanced this branch of mathematics with a number of very important papers. These included studies on quadratic forms, the number theory of irrational fields (including the integral complex numbers) and the theory of units. In 1837, Dirichlet presented his first paper on analytic number theory, giving a proof to the fundamental theorem that bears his name: any arithmetical series of integers $a \times n + b$, where a and b are relatively prime and $n = 0, 1, 2, 3, \ldots$, must include an infinite series of primes. Later papers included the analytical consideration of quadratic forms, studies of the theory of ideals and the convergence of Dirichlet series, and introduced the deceptively simple *Schubfachprinzip* (the 'box principle') – a principle much used in the logic of modern number theory, which states: if in n boxes one distributes more than n objects, at least one box must contain more than one object.

In 1863 Dirichlet's *Vorlesungen über Zahlentheorie* was published posthumously by his friend and pupil Richard Dedekind (1831–1916). This summary of Dirichlet's work, together with supplements by Dedekind, is now considered one of the foundations on which the theory of ideals – the core of algebraic number theory – is based.

Alongside his theoretical work, Dirichlet also carried out a series of studies on analysis and applied mathematics. Important among these was an analysis of vibrating strings, in which he developed techniques now considered classic for the discernment of convergence. He also began to rewrite the vocabulary of mathematics. Whereas the mathematical concept of a function had previously been as an expression formulated in terms of mathematical symbols, Dirichlet introduced the modern concept of $y = f(x)$ as a correspondence that links each real x value with some unique y value denoted by $f(x)$.

Other papers included applications of Fourier series, a critique of Pierre Laplace's analysis of the stability of the Solar System, boundary values, and the first exact integration of the hydrodynamic equations.

Dirichlet's contributions to mathematics were both numerous and of different kinds; he made many important individual discoveries, but more important still was his method of approach – an essentially modern way of formulating or analysing mathematical problems, especially in number.

Dobzhansky Theodosius 1900–1975 was a Russian-born US geneticist whose synthesis of Darwinian evolution and Mendelian genetics established evolutionary genetics as an independent discipline. He also wrote about human evolution and the philosophical aspects of evolution.

Dobzhansky was born on 25 January 1900 in Nemirov, Russia, the son of a mathematics teacher. His family moved to Kiev in 1910 and Dobzhansky first attended school there. In 1917, he went to Kiev University to study zoology and, after graduating in 1921, remained there to teach zoology until 1924, when he moved to Leningrad University as a teacher of genetics. Also in 1924 he married Natalia Sivertzev, whom he met while teaching at Kiev. In 1927, Dobzhansky went as a Rockefeller Fellow to Columbia University, New York City, where he worked with Thomas Hunt Morgan (1866–1945), one of the pioneers of modern genetics. Morgan moved to the California Institute of Technology in 1928 and, impressed with Dobzhansky's ability, offered him a post teaching genetics there when his fellowship ended in 1929. Dobzhansky, who had become a US citizen in 1937, remained at the California Institute of Technology until 1940, when he returned to Columbia University as Professor of Zoology. He then worked at the Rockefeller Institute (later the Rockefeller University) from 1962 until his official retirement in 1971, after which he moved to the University of California at Davis, where he remained until his death on 18 December 1975.

Dobzhansky's most important contribution to genetics was probably his *Genetics and the Origin of Species* (1937), which was the first significant synthesis of Darwinian evolutionary theory and Mendelian genetics – areas in which there had been much progress since about 1920. This book was highly influential and established the discipline of evolutionary genetics.

Dobzhansky's other major contribution was his demonstration in the 1930s that genetic variability within populations is greater than was then generally thought. Until this work the consensus of opinion was that, in the wild state, most members of a species had the same 'wild-type' genotype and that each of the wild-type genes was homozygous in most individuals. Variant genes were usually deleterious mutants that rapidly vanished from the gene pool. Furthermore, when an advantageous mutation appeared, it gradually – over several generations – increased in frequency until it became the new, normal wild type. Working with wild populations of the vinegar fly *Drosophila pseudoobscura*, Dobzhansky found that, in fact, there is a large amount of genetic variation within a population and that some genes regularly changed in frequency with the different seasons. Continuing this line of research, he also showed that many of the variant genes are recessives and so are not commonly expressed in the phenotype – a finding that disproved the original assumption of a high level of homozygosis in wild populations. Furthermore, he found that heterozygotes are more fertile and better able to survive than are homozygotes and, therefore, tend to be maintained at a high level in the population.

Dobzhansky also investigated speciation and, using the fly *Drosophila paulistorum*, proved that there is a period when speciation is only partly complete and during which several races coexist. In addition, he wrote on human evolution – in which area his *Mankind Evolving* (1962) had great influence among anthropologists – and on the philosophical aspects of evolution in *The Biological Basis of Human Freedom* (1956) and *The Biology of Ultimate Concern* (1967).

Dodgson Charles Lutwidge 1832–1898 as Lewis Carroll is famous as the English author of *Alice's Adventures in Wonderland* and *Through the Looking Glass*, but was in fact also responsible in his publication of mathematical games and problems requiring the use of intelligent mental arithmetic, for a general upsurge of interest in such pastimes. Several of his books of such puzzles suggest an awareness of the theory of sets – the basis on which most modern mathematical teaching is founded – that was being formulated by Dodgson's contemporary, Georg Cantor (1843–1918), but that did not become established until more than 20 years after Dodgson's death.

Dodgson was born in Daresbury, near Warrington, Cheshire, on 27 January 1832, the eldest son in a parish priest's family of 11 children. An acutely shy child with a pronounced stammer, he was educated at home until he reached the age of 12. He was then sent to Rugby School where, under the watchful eye of the Anglican prelate Archibald Tait (1811–1882), he displayed a natural talent for mathematics and an aptitude for divinity. He was awarded a place at Christ Church, Oxford, in 1850 and after taking courses in mathematics and classics, received his BA in 1854. The following year he was appointed Lecturer in Mathematics, and six years later he was ordained deacon in the Church of England (although he never in fact became a priest). Despite a great love of children – particularly girls – possibly also as a result of his shyness, Dodgson never married. Instead, he poured all his enthusiasm into writing and telling stories to the children of his friends. Under the pseudonym Lewis Carroll, he was eventually persuaded to publish two stories he had composed to amuse one little girl he especially favoured, Alice Liddell.

Alice's Adventures in Wonderland (1865) and *Through the Looking Glass* (1872) became immensely popular, hailed as classics in the world of children's fiction. A life Fellow of Christ Church, Dodgson gave up his lectureship in 1881 and concentrated on writing, both of mathematics and children's fantasy. After the *Alice* books, however, he never again achieved such popularity. He died at Guildford, Surrey, on 14 January 1898.

Dodgson's enjoyment of mathematics and his affection for children are both reflected in the papers he wrote on the teaching of mathematics for the young. He was particularly interested in the use of number games, and made a compilation of a wide range of puzzles and brain teasers covering all aspects of the subject (including geometry, algebra and graph work) that call for general intelligence to solve the problems, rather than specialized knowledge. Number games were not a new idea (they probably originated with the ancient Greeks), but some time around the fifteenth century they had re-emerged; the Victorian society of the nineteenth century was ripe for them to regain considerable popularity. Dodgson saw their potential as teaching aids, and wrote about them as such, publishing several books including *Pillow Problems, The Game of Logic* and *A Tangled Tale*. The chessboard featured in some of these games. With the publisher Edouard Lucas (1868–1938), Dodgson was responsible for the continued revival of such puzzles during the latter half of the nineteenth century.

Nevertheless, Dodgson also wrote a considerable number of serious and advanced papers on mathematical subjects (all of which Queen Victoria was apparently dumbfounded to receive, having ordered the author's complete works after reading *Alice in Wonderland*). He produced lengthy general syllabus textbooks, quite a few books on historical mathematics (particularly on Euclid and his geometry) and a number of specialized papers (such as his 'Condensation of determinants').

He also showed a keen interest in photography, and has been described as having an exceptional flair for it.

Dollfus Audouin Charles 1924–   is a French physicist and astronomer whose preferred method of research is to use polarization of light, for which method he is prepared to put up with some discomfort.

Dollfus was born on 12 November 1924; he studied at the Lycée Janson-de-Sailly and at the Faculty of Sciences in Paris, where he gained his doctorate in mathematical sciences. Since 1946, he has been Astronomer of the Astrophysical Section of Meudon Observatory, in Paris.

Before *Viking* landed on Mars, the mineral composition of the Martian deserts was a subject of considerable dispute. Dollfus checked the polarization of light by several hundreds of different terrestrial minerals to try to find one for which the light matched that polarized by the bright Martian desert areas. He found only one, and that was pulverized limonite (Fe_2O_3), which could be oxidized cosmic iron. (Another astronomer, Gerard Kuiper (1905–1973) of the University of Chicago, however, did not agree with Dollfus's findings. In his work, iron oxides gave poor results and he obtained his closest match with brownish fine-grained igneous rocks.)

In pursuit of his detailed investigations into Mars, Dollfus made the first ascent in a stratospheric balloon, in France.

By means of the polarization of light it is possible to detect an atmosphere round a planet or satellite. In 1950, at which time it was thought the planet Mercury, because of its small size, had probably lost its atmosphere through the escape of the molecules into space, Dollfus announced that he had detected a very faint atmosphere from polarization measurements carried out at the Pic-du-Midi Observatory in the French Pyrenees. This was also in contrast to theoretical expectations based on the kinetic theory of gases. Dollfus estimated that the atmospheric pressure at ground level was about 1 mm/0.039 in of mercury. (The nature of the gas making up this atmosphere is unknown, but it must be a dense, heavy gas. It is certain that the atmosphere on Mercury is not more than 1/300 that on Earth.)

Mercury shows faint shady markings, set against a dull whitish background, that were first observed by Giovanni Schiaparelli in 1889. Using the 60-cm/24-in refractor at the Pic-du-Midi Observatory, Dollfus, again in 1950, was able clearly to resolve spots about 300 km/186 mi apart.

Dollfus has also looked at the possibility of an atmosphere around the Moon. The rate of thermal dissipation into space of all but the heavier gases (which are cosmically very scarce) from the Moon is so high that an atmosphere cannot be expected. The most telling evidence is the complete absence of the twilight phenomena on the Moon. Any elongation of the points (or the cusps) of the Moon beyond 90°, caused by scattered sunlight, should be detectable by polarization. But Bernard Lyot (1897–1952), and later Dollfus, proved that there was no detectable polarization.

In 1966 Dollfus discovered Janus, the innermost moon of Saturn, at a time when the rings – to which it is very close – were seen from Earth edgeways on (and practically invisible).

A practical astrophysicist, Dollfus has achieved remarkable results through patient and persistent research.

Domagk Gerhard 1895–1964 was a German bacteriologist and a pioneer of chemotherapy who discovered the antibacterial effect of Prontosil, the first of the sulphonamide drugs. This important discovery led, in turn, to the development of a range of sulphonamide drugs effective against various bacterial diseases, such as pneumonia and puerperal fever, that previously had high mortality rates. For this outstanding achievement Domagk received many honours, including the 1939 Nobel Prize for Physiology or Medicine. At the time, however, Germans were forbidden by Adolf Hitler to accept such awards and Domagk did not receive his Nobel medal until 1947, by which time the prize money had reverted to the funds of the Nobel Foundation.

Domagk was born on 30 October 1895 in Lagow, Brandenburg (now in Poland), and studied medicine at Kiel University, graduating – after a period of military service

during World War I – in 1921. In 1924 he became Reader in Pathology at the University of Griefswald then, in the following year, was appointed to a similar position at the University of Münster. In 1927, he accepted an invitation to direct research at the Laboratories for Experimental Pathology and Bacteriology of I. G. Farbenindustrie, Düsseldorf, a prominent German dye-making company. But he also remained on the staff of Münster University, which appointed him Extraordinary Professor of General Pathology and Pathological Anatomy in 1928 and a Professor in 1958. Domagk died on 24 April 1964 in Burgberg, West Germany.

Following Paul Ehrlich's discovery of antiprotozoon chemotherapeutic agents, considerable advances had been made in combating protozoon infections but bacterial infections still remained a major cause of death. While working for I. G. Farbenindustrie, Domagk began systematically to test the new azo dyes in an attempt to find an effective antibacterial agent. In 1932, his industrial colleagues synthesized a new azo dye called Prontosil red, which Domagk found was effective against streptococcal infections in mice. In 1935, he published his discovery, but it received little favourable response. In the following year, however, the British Medical Research Council confirmed his findings, and shortly afterwards the Pasteur Institute in Paris found that the sulphonilamide portion of the Prontosil molecule is responsible for its antibacterial action. (This latter was an important finding because sulphonilamide is much cheaper to produce than is Prontosil.) Meanwhile Domagk had demonstrated the effectiveness of Prontosil in combating bacterial infections in humans. His daughter had accidentally infected herself while working on the clinical trials of Prontosil and, after the failure of conventional treatments, Domagk had cured her with Prontosil.

From about 1938 other sulphonamide drugs were produced that were effective against a number of hitherto serious bacterial diseases, but antibiotics were discovered shortly afterwards and they came to replace sulphonamides as the normal drugs used to treat bacterial infections. Nevertheless, sulphonamides and chemotherapy were – and still are – of great value, particularly in the treatment of antibiotic-resistant infections. In 1946, Domagk and his co-workers found two compounds (eventually produced under the names of Conteben and Tibione) which, although rather toxic, proved useful in treating tuberculosis caused by antibiotic-resistant bacteria. Subsequently Domagk attempted to find chemotherapeutic agents for treating cancer, but was unsuccessful.

Donati Giovanni Battista 1826–1873 was an Italian astronomer whose principal astronomic interests were the study of comets and cosmic meteorology. He made important contributions to the early development of stellar spectroscopy and to the application of spectroscopic methods to the understanding of the nature of comets.

Donati was born in Pisa on 16 December 1826. He received his university training at the University of Pisa and began his career in astronomy at the observatory in

Florence in 1852. He was first employed as an assistant, but in 1864 he succeeded Giovan Battista Amici (1786–1868) to the directorship of the Observatory.

One of Donati's major responsibilities after assuming the directorship of the Observatory was the supervision of the work at Arcetri, not far from Florence, where a new observatory was being set up. It was formally established a year before Donati's death. He was struck down by the plague and died in Florence on 20 September 1873.

Donati's active research career spanned little more than 20 years, but it was very productive. During the 1850s he was an enthusiastic comet-seeker, with six discoveries to his credit – the most dramatic of these was named after him. Donati's comet, which was first sighted on 2 June 1858, was notable for its great beauty. It had, in addition to its major 'tail', two narrow extra tails. It even featured in William Dyce's painting 'Pegwell Bay'.

Donati then applied his talents and his time to the developing subject of stellar spectroscopy. He compared and contrasted the spectrum of the Sun with those of other stars, and then sought to use this technique to examine the properties and composition of comets. Donati found that when a comet was still distant from the Sun, its spectrum was identical to that of the Sun. When the comet approached the Sun it increased in magnitude (brightness) and its spectrum became completely different. Donati concluded that when the comet was still distant from the Sun, the light it emanated was simply a reflection of sunlight. As the comet approached the Sun the material in it became so heated that it emitted a light of its own, which reflected the comet's composition.

Shortly thereafter, William Muggins (1824–1910) reported that the tail of a comet contained carbon compounds. More definitive analyses of cometary make-up were written over the ensuing years, culminating with Whipple's report published in 1950.

Other areas of interest which engaged Donati's attention were atmospheric phenomena and events in higher zones, such as the aurora borealis. His most important research was, however, his pioneering efforts in the use of spectroscopy to elucidate the nature of comets.

Donkin Bryan 1768–1855 was a British engineer who made several innovations in paper-making, printing and food preservation.

Little is known of Donkin's childhood. He was born in Northumberland and later apprenticed to John Hall, a paper-maker of Dartford, Kent. He went on to work for Hall and perfected a new type of paper-making machine, which had been devised in 1798 by the Frenchman Nicolas Robert and later patented in England by Henry and Sealy Fourdrinier. This success led Donkin to investigate another recent invention from France, the preservation of food by bottling. He established his own company, even gaining royal approval by presenting samples to the Prince Regent in 1813. With this security behind him, he returned to the printing and paper trade, and invented the forerunner of the rotary press. By 1815 he had turned to civil engineering,

and became a founder member of the Instition of Civil Engineers. When he died in London on 27 February 1855, he was remembered as much for his role in founding such official groups as for his engineering skills.

Donkin made his first practical Fourdrinier paper-milling machine at Frogmore Hill, Hertfordshire, in 1803. He set up a factory in Bermondsey, South London, where he manufactured nearly 200 machines in all. Donkin's contribution to food preservation was to take Appert's bottling process and modify it to use metal cans instead of glass bottles. In printing, he tackled the problem of increasing the speed of presses. The original flat-bed press, with its back-and-forth movement, was too slow. Donkin arranged four (flat) formes of type around a spindle – a rudimentary rotary press. He introduced a composition of glue and treacle for the inking rollers, an innovation which was still widely used long after his press had been superseded.

Doppler Johann Christian 1803–1853 was an Austrian physicist who discovered the Doppler effect, which relates the observed frequency of a wave to the relative motion of the source and the observer. The Doppler effect is readily observed in moving sound sources, producing a fall in pitch as the source passes the observer, but it is of most use in astronomy, where it is used to estimate the velocities and distances of distant bodies.

Doppler was born in Salzburg, Austria, on 29 November 1803, the son of a stonemason. He showed early promise in mathematics, and attended the Polytechnic Institute in Vienna from 1822 to 1825. He then returned to Salzburg and continued his studies privately while tutoring in physics and mathematics. From 1829 to 1833, Doppler went back to Vienna to work as a mathematical assistant and produced his first papers on mathematics and electricity. Despairing of ever obtaining an academic post, he decided in 1835 to emigrate to the United States. Then, on the point of departure, he was offered a professorship of mathematics at the State Secondary School in Prague and changed his mind. He subsequently obtained professorships in mathematics at the State Technical Academy in Prague in 1841, and at the Mining Academy in Schemnitz in 1847. Doppler returned to Vienna the following year and, in 1850, became Director of the new Physical Institute and Professor of Experimental Physics at the Royal Imperial University of Vienna. He died from a lung disease in Venice on 17 March 1853.

Doppler explained the effect that bears his name by pointing out that sound waves from a source moving towards an observer will reach the observer at a greater frequency than if the source is stationary, thus increasing the observed frequency and raising the pitch of the sound. Similarly, sound waves from a source moving away from the observer reach the observer more slowly, resulting in a decreased frequency and a lowering of pitch. In 1842, Doppler put forward this explanation and derived the observed frequency mathematically in Doppler's principle.

The first experimental test of Doppler's principle was made in 1845 at Utrecht in the Netherlands. A locomotive was used to carry a group of trumpeters in an open carriage to and fro past some musicians able to sense the pitch of the notes being played. The variation of pitch produced by the motion of the trumpeters verified Doppler's equations.

Doppler correctly suggested that his principle would apply to any wave motion and cited light as an example as well as sound. He believed that all stars emit white light and that differences in colour are observed on Earth because the motion of stars affects the observed frequency of the light and hence its colour. This idea is not universally true as stars vary in their basic colour. However, Armand Fizeau (1819–1896) pointed out in 1848 that shifts in the spectral lines of stars could be observed and ascribed to the Doppler effect and hence enable their motion to be determined. This idea was first applied in 1868 by William Huggins (1824–1910), who found that Sirius is moving away from the Solar System by detecting a small red shift in its spectrum. With the linking of the velocity of a galaxy to its distance by Edwin Hubble (1889–1953) in 1929, it became possible to use the red shift to determine the distances of galaxies. Thus the principle that Doppler discovered to explain an everyday and inconsequential effect in sound turned out to be of truly cosmological importance.

Draper Henry 1837–1882 was one of the USA's outstanding 'amateur' astronomers, noted for his work on stellar spectroscopy and commemorated by the Henry Draper Catalogue of stellar spectral types.

Draper was born in Virginia on 7 March 1837; his father, John William Draper, was a distinguished physician and chemist. He was educated at the University of the City of New York and entered the Medical School at the age of 17. By the time he was 20, he had completed the medical course, but since he had not reached the age required for graduation, he spent the following year travelling in Europe. During this period he visited, and was greatly influenced by, William Parsons and his Observatory in Parsonstown (now Birr) in Ireland. In 1860, Draper was appointed Professor of Natural Science at the University of the City of New York, but the interests that he developed during his travels (telescope-making and photography) were to be woven into his professional career. He died unexpectedly, of double pleurisy, at his home in New York on 20 November 1882.

On returning from his travels in Europe, Draper began preparing his own glass mirror and by 1861 he had installed it in his new observatory on his father's estate at Hastings-on-Hudson, New York. Draper began his research career by making preliminary studies of the spectra of the more common elements and photographing the solar spectrum. By 1873, he had devised a spectrograph that was similar to Huggins's visual spectroscope; it clarified the spectral lines by means of a slit and incorporated a reference spectra so that celestial elements could be identified more easily.

In 1874, Draper was asked to act as director of the photographic department of the US commission to observe the transit of Venus of that year. Draper's work was stimulated by the spectroscopic studies of Huggins and Lockyer in Europe and during the last years of his life he worked towards obtaining high-quality spectra of celestial objects. He studied the Moon, Mars, Jupiter, the comet 1881 III and the Orion Nebula. He also succeeded in obtaining photographs of stars that were too faint to be seen with the same telescope by using exposure times of more than 140 minutes – exemplifying the advantages of photography in astronomy.

After Draper's sudden death, his widow established a fund to support further spectral studies. It was used by a team at Harvard College Observatory as part of a programme, begun in 1886, to establish a useful classification scheme for stars and a catalogue of spectra. The Harvard project was not completed until 1897 but the result was a comprehensive classification of stars according to their spectra, named the Henry Draper Catalogue.

Drew Charles Richard 1904–1950 was a US surgeon chiefly remembered for his research into blood transfusion.

Born on 3 June 1904 in Washington, DC, Charles Drew was educated in the public schools of the city graduating with honours from Dunbar High School in 1922. After receiving a BA from Amherst College in 1926 and as an accomplished athlete, he worked for two years as director of athletics and teacher of biology at Morgan State College. In 1928, he went on to McGill University Medical School and was awarded his MD and CM in 1933. Having completed his internship in Montreal General Hospital, in 1935 he went to Howard University Medical School as an instructor of pathology. In 1938 he was granted a research fellowship by the Rockefeller Foundation and spent two years at Columbia University, New York, and as a resident in surgery in the Presbyterian Hospital connected with Medical School. He received a MedDSc degree from the University in 1940 – the first black person to receive this degree in the country. After working for the American Red Cross, Drew returned to Howard Medical School where in 1942 he was made Professor and Head of the Department of Surgery and Chief of Surgery at the Freedmen's Hospital. In 1944, he was appointed chief of staff of the hospital. He remained there until his death following a motor accident, on 1 April 1950.

While at McGill University, Drew became interested in the problems of blood transfusion and it is his work in this field for which he is remembered. At the Presbyterian Hospital his research demonstrated that plasma had a longer life than whole blood and therefore could be better used for transfusion; he wrote a doctoral thesis, *Banked Blood: A Study in Preservation*, and was supervisor of the blood plasma division of the Blood Transfusion Association of New York City. In 1939, he established a blood bank and was in charge of collecting blood for the British Army at the beginning of World War II. In 1941, he became Director of the American Red Cross Blood Bank in New York City, which collected blood for the American armed forces. Drew resigned, however, when the Red Cross decided to segregate blood according to the race of the donor. Drew is also known for his teaching and training of surgeons and his publication of many papers in medical and scientific journals. For his work, Drew was awarded the Spingarn Medal (1944), honorary DSc degrees from Virginia State College (1945) and Amherst College (1947), and posthumously, the Distinguished Service Medal of the National Medical Association (1950). Several schools and medical centres have been named after him and a stamp was issued in his honour in 1981.

Dreyer John Louis Emil 1852–1926 was a Danish-born astronomer and author, whose working life was spent almost entirely in Ireland. He is best known for a biographical study of the work of the great Danish scientist, Tycho Brahe, and for the meticulous compilation of catalogues of nebulae and star clusters.

Dreyer was born in Copenhagen on 13 February 1852. Educated there, he displayed unusual talents in mathematics, physics and history, although it was not until he was aged 14 that he read a book about Tycho Brahe and became keenly interested in astronomy – an interest that was encouraged by his friendship with Schjellerup, an astronomer at the Copenhagen University. Dreyer began his studies at the university in 1869, and by 1870 he had been given a key that allowed him free access to the instruments in the University Observatory. In 1874, he was appointed assistant at Lord Rosse's Observatory at Birr Castle in Parsonstown in Ireland. Four years later he took up a similar post at Dunsink Observatory at the University of Dublin, and four years later again, in 1882, he became Director of the Armagh Observatory, where he remained until he retired in 1916. On his retirement he went to live in Oxford and, continuing his writing, made use of the facilities of the Bodleian Library. He died in Oxford on 14 September 1926.

Dreyer's earliest formal astronomical publication, published in 1872, was a description of the orbit of the first comet of 1870. After his move to Ireland in 1874, he became increasingly interested in making observations of nebulae and star clusters, a subject which occupied most of his time for the next 14 years. He was acutely aware of the element of error involved in astronomical observations of objects such as nebulae, and he published an important paper on the subject in 1876.

In 1877, Dreyer presented to the Royal Astronomical Society data on more than 1,000 new nebulae and corrections to the original catalogue on nebulae and star clusters compiled by John Herschel (1792– 1871). He extended this work at Armagh, which led to the publication in 1886 of the Second Armagh Catalogue, with information on more than 3,000 stars. The Royal Astronomical Society then invited him to compile a comprehensive new catalogue of nebulae and star clusters to incorporate all the modern data and to supersede Herschel's old catalogue. This enormous task was completed in only two years, but the rapid accumulation of more information necessitated the publication of two supplementary indexes in 1895 and 1908. Together these three catalogues described more than 13,000 nebulae and star clusters and achieved international recognition as standard reference material.

The catalogue completed, Dreyer decided to write a biography of his hero, Tycho Brahe. It was published in 1890 and preceded a 15-volume series (1913–19) detailing all of Brahe's work. Dreyer's other writings included a history of astronomy (at that time the only authoritative and complete historical analysis), and an edition of the complete works of William Herschel (1738–1822).

On his retirement in 1916, Dreyer, a Fellow of the Royal Astronomical Society since 1875, was awarded the Society's highest honour, the Gold Medal. A patient and skilled observational astronomer, Dreyer was also an excellent mathematician, a talented scholar and an accomplished writer. He put all of these attainments to good use during his career, combining them to produce work of enduring quality.

Driesch Hans Adolf Eduard 1867–1941 was a German embryologist and philosopher who is best known as one of the last advocates of vitalism, the theory that life is directed by a vital principle and cannot be explained solely in terms of chemical and physical processes. Nevertheless, he also made several important discoveries in embryology, although these have tended to be overlooked because of his mistaken belief in vitalism.

Driesch was born on 28 October 1867 in Bad Kreuznach, Germany, the son of a prosperous gold merchant. He studied zoology, chemistry and physics at the Universities of Hamburg, Freiburg, Munich and Jena, obtaining his doctorate from Jena – where he studied under Ernst Haeckel (1834–1919) – in 1887. Coming from an affluent family, Driesch had no need of paid employment and he spent the next 22 years privately pursuing his embryological studies. After obtaining his doctorate, he travelled extensively in Europe and the Far East, spending nine years (1891 to 1900) working at the International Zoological Station in Naples. In 1899 he married Margarete Reifferscheidt; they later had two children, both of whom became musicians. Eventually Driesch settled in Heidelberg, and in 1909 he was appointed Privatdozent (unpaid lecturer) in Philosophy at the university there, becoming Professor of Philosophy in 1911. Subsequently he was Professor of Philosophy at Cologne University from 1920 to 1921 and at Leipzig University from 1921 to 1935, when he was forced to retire by the Nazi regime. Driesch died in Leipzig on 16 April 1941.

In 1891 Driesch, experimenting with sea-urchin eggs, discovered that when the two blastomeres of the two-cell stage of development are separated, each half is able to develop into a pluteus (a later larval stage) which is completely whole and normal, although of smaller than average size. Similarly he found that small, whole individuals can be obtained by separating the four cells of the four-cell stage of development. From these findings he concluded that the fate of a cell is not determined in the early developmental stages. Later, other workers discovered the same phenomenon in the early developmental stages of hydroids, most vertebrates and certain insects. (It should be noted, however, that not all animal eggs behave in this way; for example, separation of the early embryonic cells of annelids, molluscs and ascidians results in incomplete embryos.) Subsequently Driesch produced an oversized larva by fusing two normal embryos, and in 1896 he was the first to demonstrate embryonic induction when he displaced the skeleton-forming cells of sea-urchin larvae and observed that they returned to their original positions. These findings provided a great impetus to embryological research but Driesch himself – unable to explain his results in mechanistic terms (principally because at that time very little was known about biochemistry) – came to believe that living activities, especially development, were controlled by an indefinable vital principle, which he called entelechy. After his appointment as Privatdozent in 1909, Driesch abandoned scientific research and devoted the rest of his life to philosophy.

Dubois Marie Eugène François Thomas 1858–1940 was the Dutch palaeontologist who, in 1891, discovered the remains of *Pithecanthropus erectus*, known as Java Man.

Dubois was born in Eijsden, the Netherlands, on 28 January 1858, and studied medicine and natural history at the University of Amsterdam. In 1886, he took the appointment of lecturer in anatomy there. A year later, he joined the Dutch Army Medical Service and was posted to Java – then a Dutch possession – where he was commissioned by the Dutch government to search for fossils. The discoveries he made there brought him worldwide fame and he returned to Europe in 1895. Dubois took up a professorship in palaeontology, geology and mineralogy at the University of Amsterdam in 1899. He retired in 1928 and died at Halen in Belgium on 16 December 1940.

The excavations of Pompeii and Herculaneum in 1748, and later of Troy (1870) were the beginnings of archaeology and encouraged attempts to piece together the history of human development. In 1857 Neanderthal Man was found and later Cro-Magnon Man was discovered in southwestern France. After Darwin published his *On the Origin of Species* in 1859, many people thought that the evolutionary principles he had outlined could equally apply to the origin of the human species.

The skeletons already found were undoubtedly those of an early species of human, but there was still a large gap in evolutionary terms between modern humans and the apes. Having first been involved in the comparative anatomy of vertebrates, Dubois became fascinated with the problem of the 'missing link' in the evolutionary chain and was convinced that somewhere there existed its remains. He reasoned that such a missing link could have lived in an area where apes were still numerous, such as Africa or southeastern Asia. His posting to Java gave Dubois the opportunity to investigate this theory.

The remains of extinct animals had been found in the deposits of volcanic ash on the banks of the Solo River in East Java and it was in the bone beds near the village of Trinil that he concentrated his search. In 1891, Dubois found teeth, a skullcap and a femur. The skullcap was much larger than that of any living ape, and more primitive and apelike than Neanderthal Man; the bones were heavily ridged and the vault of the braincase extremely low, indicating that the size of the brain was far smaller than that of a modern human. The teeth were also intermediate between ape and human. The femur was definitely human, the ends of the bone and the straightness of the shaft suggesting that its owner had walked erect.

Dubois published a scientific description of the fragments he had found in 1894, and named them *Pithecanthropus erectus* ('erect ape-man') after the name given to the intermediate human by the German zoologist Ernst Haeckel (1834–1919).

The femur Dubois had found was discovered some distance away from the skullcap and could therefore have been from some other form of human. Many people were not convinced that Dubois had discovered the missing link, especially when further excavations at Trinil produced no more traces of *Pithecanthropus erectus*. In response to the controversy aroused by his findings, Dubois withdrew his discovery from the public until 1923. But between 1936 and 1939 the German archaeologist, von Koenigswald, was working in the Solo River valley farther upstream from Dubois's original discoveries, and found more skullcaps, a lower and an upper jaw. Later, a child's skull was discovered at Modjokerto on Java and is now believed to be from a young *Pithecanthropus erectus*.

Detailed measurements of casts of the skull of *Pithecanthropus erectus* indicated that the brain cavity had a volume of about 940 cu cm/57 cu in, whereas few apes have more than 600 cu cm/37 cu in and modern humans have about 1,500 cu cm/91.5 cu in. From these figures it was thought that *Pithecanthropus erectus* was almost exactly halfway between ape and human on the evolutionary scale, although it is now considered to be definitely more human and is called *Homo erectus*. In contrast to this belief, in Dubois's later years he changed his ideas and stressed the apelike similarities of his discovery rather than its hominid likenesses.

The history of human descent has been further pieced together as archaeological techniques have been refined and more discoveries have been made. But it cannot be doubted that one of the most important of these contributions has been that of Dubois.

Du Bois-Reymond Emil Heinrich 1818–1896 was one of the leading German physiologists of the nineteenth century.

Du Bois-Reymond's father was a Swiss teacher who had settled in Berlin – the family was French-speaking. Talent ran in the family, his brother Paul (1831–1889) shining as a mathematician, and making contributions to function theory. Du Bois-Reymond studied a wide range of subjects in Berlin for two years before finally choosing a medical career. Working under Johannes Müller, he graduated in 1843, soon plunging into research on animal electricity and especially on electric fishes. All through his career he was closely associated with the leading German investigators of human physiology: Schwann, Schleiden, Ludwig and also the physicist, Helmholtz.

In 1858, he succeeded Müller as Professor of Physiology, and was appointed head of the new Physiological Institute, which opened in Berlin in 1877. Du Bois-Reymond's importance lay in his investigations of the physiology of muscles and nerves, and in his demonstrations of electricity in animals. He was adept in introducing improved techniques for measuring such neuroelectrical effects, first investigated by Galvani. By 1849 he had evolved a delicate multiplier for measuring nerve currents. Using his highly sensitive apparatus, he was able to detect an electric current in ordinary localized muscle tissues, notably contracting muscles. He observantly traced it to individual fibres, finding their interior was negative with regard to the surface. He showed the existence of electrical currents in nerves, correctly arguing that it would be possible to transmit nerve impulses chemically.

Du Bois-Reymond's experimental methods proved the basis for almost all future work in electrophysiology. He held trenchant views about scientific metaphysics. He denounced the vitalistic doctrines that were in vogue among German scientists and denied that nature contained mystical life-forces independent of matter.

Dulong Pierre Louis 1785–1838 was a French chemist, best known for his work with Alexis Petit that resulted in Dulong and Petit's law, which states that, for any element, the product of its specific heat and atomic weight is a constant, a quantity they termed the atomic heat. In modern terms, the product of the specific heat capacity of an element (expressed in joules per gram per Kelvin) and the relative atomic mass is about 25.

Dulong was born on or about 12 February 1785 in Rouen. He studied at the Ecole Polytechnique in Paris, training initially as a doctor. He married young and had to take a teaching post to keep his family and finance his research; one such (1811) was at the Ecole Normale and another (1813) was at the Ecole Vétérinaire at Alfort. He returned to Paris in 1820 to become Professor of Chemistry at the Faculté des Sciences, moving back to the Ecole Polytechnique, becoming its Director of Studies in 1830. He was elected to the physics section of the French Academy of Sciences in 1823. He died in Paris on or about 18 July 1838.

During Dulong's early work in chemistry he was an assistant to Claude Berthollet, and he studied the oxalates of calcium, strontium and barium. In 1811, he discovered the explosive compound nitrogen trichloride, an accident which cost him a finger and the sight in one eye. He resolved the contemporary dispute among chemists about the composition of phosphorus and phosphoric acids, identifying two new acids in the process. In 1815, he began working with Alexis Petit, and at first they applied their researches to the problem of measuring heat. They determined the absolute coefficient expansion of mercury, so

improving the accuracy of mercury thermometers. They then explored the laws of cooling in a vacuum, work later extended by Josef Stefan (1835–1893).

Then in 1818, Dulong and Petit began studying the specific heat capacity of elements, measuring this quantity for sulphur and 12 metals. When they multiplied each result by the element's atomic weight (relative atomic mass), they obtained values that were in close agreement with each other. They showed that an element's specific heat capacity is inversely proportional to its relative atomic mass (Dulong and Petit's law). Its chief application was in the estimation of the atomic masses of new elements.

After Petit died in 1820, Dulong continued to work on specific heat capacities, publishing his findings in 1829. He concluded that, under the same conditions of temperature and pressure, equal volumes of all gases evolve or absorb the same quantity of heat when they are suddenly expanded or compressed to the same fraction of their original volumes. He also deduced that the accompanying temperature changes are inversely proportional to the specific heat capacities of the gases at constant volume. Dulong also collaborated with the French physicist Dominique Arago (1786–1853) on a study of the pressure of steam at high temperatures. In this rather hazardous research, working at pressures up to 27 atmospheres, their results were in agreement with Boyle's law. Dulong's last paper on the heats of chemical reaction was published in 1838, after his death.

Dumas Jean Baptiste André 1800–1884 was a French chemist who made contributions to organic analysis and synthesis, and to the determination of atomic weights (relative atomic masses) through the measurement of vapour densities.

Dumas was born on or about 19 July 1800 in Alais (now Alès), Gard. He was educated at the local college and intended to enter the navy, but changed his mind after the overthrow of Napoleon I, and became instead apprenticed to an apothecary, also in Alais. But he soon moved to Geneva, Switzerland, and in 1816 was working in the laboratory of a pharmacist there (who was investigating plant extracts). He studied also under the Swiss physicist Pierre Prévost (1751–1839) and the botanist Augustin Candolle (1778–1841), and in 1822 accepted an invitation from Alexander von Humboldt to go to Paris, where he took up an appointment at the Ecole Polytechnique and held the Chair in Chemistry at the Lyceum (later Athenaeum), succeeding André Ampère. The following year he became a lecturing assistant to the French chemist Louis Thénard (1777–1857) at the Ecole Polytechnique, whom he succeeded as Professor of Chemistry in 1835. Following the political upheavals of 1848 Dumas abandoned much of his scientific work for politics and public office. He served under Napoleon III as Minister of Agriculture and Commerce, Minister of Education, and Master of the Mint. After the deposition of Napoleon in 1871, Dumas left politics. He died in Cannes on 11 April 1884.

While he was an 18-year-old in Geneva, Dumas was involved in the study of the use of iodine to treat goitre (endemic in Switzerland at that time). Then under Prévost he unsuccessfully investigated the physiological effects of digitalis. They also studied blood and showed that urea is present in the blood of animals from which the kidneys have been removed, proving that one of the functions of the kidneys is to remove urea from the blood, not to produce it.

In 1826, Dumas began working on atomic theory. He determined the molecular weights of many substances by measuring their vapour densities and concluded that 'in all elastic fluids observed under the same conditions, the molecules are placed at equal distances' – that is, they are present in equal numbers.

His important work on the theory of substitution in organic compounds was inspired at a soirée at the Tuileries when the candles gave off irritating fumes. The candle wax had been bleached with chlorine, some of which had been retained and during combustion was converted to hydrogen chloride. Dumas soon proved by experiments that organic substances treated with chlorine retain it in combination, and proposed that the chlorine had displaced hydrogen, atom for atom. He studied the action of chlorine on alcohol (ethanol) to produce chloral (trichloroethanal), which he decomposed with alkali to give chloroform (trichloromethane) and formic acid (methanoic acid). He also chlorinated acetic acid (ethanoic acid), to give trichloracetic acid (trichlorethanoic acid), thus proving his theory of substitution. He had shown that atoms of apparently opposite electrical charge had replaced each other, in opposition to the dualistic theory of chemistry proposed by Jöns Berzelius (1779–1848), who in 1830 was at the height of his fame and influence.

Together with the Belgian chemist Jean Stas (1813–1891), who was at that time his student, Dumas investigated the action of alkalis on alcohols and ethers, which led to a study of the acids produced by the oxidation of alcohols.

In 1833, Dumas worked out an absolute method for the estimation of the amount of nitrogen in an organic compound – which still forms the basis of modern methods of analysis. The nitrogen in a sample of known weight is eliminated in gaseous form and estimated by direct measurement. The sample is heated with cupric oxide (copper(II) oxide) and oxidized completely in a stream of carbon dioxide; the gaseous products of combustion are passed over a heated copper spiral and the nitrogen collected in a gas burette over concentrated potassium hydroxide solution.

Also with Stas, in 1849, he revised the atomic mass of carbon to 12 (from Berzelius's value of 12.24). He went on to correct the atomic masses of 30 elements – half the total number known at that time – referring to the hydrogen value as 1. With Milne Edwards he investigated the way in which bees convert sugar to fat (wax). His last papers were on alcoholic fermentation (1872) and on the occlusion of oxygen in silver (1878).

Dunlop John Boyd 1840–1921 was a Scottish veterinary surgeon who is usually credited with the invention of the pneumatic tyre (originally for bicycles).

Dunlop was born at Dreghorn, Ayrshire, on 5 February 1840. He studied veterinary medicine at Edinburgh University, before setting up practice in Ireland near Belfast, in 1867. He devised a pneumatic tyre twenty years later in an attempt to improve the comfort of his son's tricycle. It was so successful that in the following year (1888) he applied for the British patent and within three years he had founded his own company, with the encouragement of Harvey du Cros (whose sons were keen racing cyclists).

When Dunlop's business was already established, it was discovered that the tyre had previously been patented by another Scotsman, R. W. Thomson of Stonehaven, Kincardineshire. He had made a set of tyres as early as 1846 and fitted them to a horse-drawn carriage. They had been tested over 1,600 km/995 mi before they had needed replacing. Thomson's invention had gone practically unnoticed, whereas Dunlop's arrived at a crucial time in the development of transport. The bicycle was fast becoming recognized as more than a hobby, and the motorcar was about to make its spectacular appearance on the roads. In 1896, after trading for only about five years, Dunlop sold both his patent and his business for £3 million, and Arthur Philip du Cros took over as managing director.

The key to Dunlop's invention was rubber. The rubber industry had become established in Europe in around 1830 with the development of vulcanization (the blending of india rubber with sulphur to produce a workable substance). In 1876, the Englishman Henry Wickham travelled to the Amazon and collected seeds of wild rubber plants, which he brought back to Kew Gardens for propagation. The young plants were taken to Ceylon (Sri Lanka) and Malaya, where they formed the nucleus of the new rubber plantations.

Dunlop realized that rubber was the most suitable material for making tyres because it could stand up to wear and tear while retaining its resilience. His first simple design consisted of a rubber inner tube, covered by a jacket of linen tape with an outer tread also of rubber. The inner tube was inflated using a football pump and the tyre was attached by flaps in the jacket which were rubber-cemented to the wheel. Later, he incorporated a wire through the edge of the tyre which secured it to the rim of the wheel. Whether or not Dunlop's invention was the first, he certainly pioneered the mass production of tyres and Fort Dunlop, the Dunlop Rubber Company's factory in Birmingham, remains today.

Du Toit Alexander Logie 1878–1948 was a South African geologist who helped pave the way for theories of continental drift.

Born near Cape Town of a wealthy family of Huguenot descent, Du Toit studied at Cape Town, Glasgow and the Royal College of Science, London. He spent 17 highly fruitful years of his career from 1903 mapping for the Geological Commission of the Cape of Good Hope. At the height of his powers, and following a visit to South America, he developed a profound interest in Wegener's theory of continental drift. In *A Geological Comparison of South*

America and South Africa 1927, he systematically adumbrated the abundant similarities in the geologies of the two continents, suggesting that they had probably once been joined. Concepts of this kind were most thoroughly and famously stated in his significant *Our Wandering Continents* 1937, in which he maintained that the southern continents had, in earlier times, formed the supercontinent of Gondwanaland, which was distinct from the northern supercontinent of Laurasia. This notion, though initially deprecated, steadily grew in acceptance, and was to form one of the foundations for the synthesis of continental drift theory and plate tectonics that created the geological revolution of the 1960s. The most eminent geologist South Africa has produced, Du Toit was widely hailed as the world's finest field geologist.

Dyson Frank Watson 1868–1939 was an English astronomer especially interested in stellar motion and time determination.

Dyson was born in Ashby-de-la-Zouch, Leicestershire, on 8 January 1868. He attended Bradford Grammar School and Trinity College, Cambridge, from which he graduated in 1889. He became a Fellow of Trinity in 1891 and in 1894 he was made Chief Assistant at the Greenwich Observatory. He left in 1906 to become Astronomer Royal for Scotland, but returned to Greenwich in 1910 to serve as Astronomer Royal for England. He retired in 1933, but remained active in research and writing. In addition to his many research publications, Dyson was the author of several general books on astronomy. He died off the coast of South Africa while on a sea voyage from Australia, on 25 May 1939.

Dyson's early research was concentrated on problems in gravity theory, but as soon as he started his work at the Greenwich Observatory he began a lengthy study of stellar proper motion in collaboration with William Thackeray. He was an active member of several expeditions to study total eclipses of the Sun, and in 1906 he published a book in which he discussed data he had obtained on these occasions on the spectrum of the solar chromosphere.

Dyson was one of a number of astronomers who confirmed the observations of Jacobus Kapteyn (1851–1922) on the proper motions of stars, which indicated that the stars in our Galaxy seemed to be moving in two great streams. These results were later realized to be the first evidence for the rotation of our Galaxy.

The measurement of time has always been an important function of the Greenwich Observatory, and Dyson was passionately interested in this aspect of his work. It was he who initiated the public broadcasting of time-signals by the British Broadcasting Corporation over the radio, in the form of the familiar six-pip signal. (This was first broadcast in 1924 from Rugby.)

Another important research area for the Greenwich Observatory is the study of solar eclipses, and Dyson was active in the organization of expeditions to observe these. The most significant of these expeditions were the two he coordinated for the 1919 eclipse. They served as the occasion for Arthur Eddington's famous confirmation of the gravitational deflection of light by the Sun, as predicted by Einstein's general theory of relativity.

Other areas to which Dyson made important contributions include the study of the Sun's corona and of stellar parallaxes.

E

Eastman George 1854–1932 was the US inventor, businessman and benefactor who founded the Kodak company. He started the 'press button' trend of photography and brought it within range of virtually everybody's skill and pocket.

Born on 12 July 1854, in Waterville, New York, Eastman left school at 14 to earn a living and ease his family's financial hardships. He started as a messenger boy, studying accounting in the evenings. He saved $3,000, in 1879 patented a photographic emulsion coating machine, and the following year began mass production of dry plates.

Eastman had started experimenting with photographic emulsion in 1878. At that time, there were large numbers of photographic processes in use, the most popular of which was collodion-coated glass plates. These had to be prepared, sensitized, exposed and developed in rapid succession, which meant that landscape photographers had to carry a complete darkroom about with them. Enlargement was not practicable so negatives had to be the size of the finished print – thus some cameras were the size of soap boxes.

Various methods to preserve the collodion were used but most were complicated and reduced the sensitivity of the plate. The first practical dry process introduced in 1855 was the collodion albumen plate, which needed six times the exposure of the wet plate. Gelatine emulsion, usually attributed to Maddox, appeared in 1871 and Burgess's gelatino-bromide prepared dry plates appeared in 1873. This ushered in a new era of photography in which the mobile darkroom was no longer required. Eastman began production of his own dry plates in 1880 in a rented loft of a building in Rochester, New York.

Stripping film, in which paper was used only to support the emulsion and was not destined to be part of the negative, was patented in 1855 by Scott Archer, but the technique was not used to any great extent until Eastman introduced it for the Eastman–Walker roller slide in 1886. Eastman's system consisted of a paper base, a layer of soluble gelatine, a layer of collodion and a layer of sensitized gelatine emulsion.

After exposure, the roll was cut up into individual negatives, developed and fixed. The emulsion side was then attached to glass plates and coated with glycerine. Hot water dissolved the gelatine so the paper could be stripped off. The image on the glass had to be transferred to a gelatine sheet for printing.

Eastman's roller slide fitted into existing cameras and he confidently expected that users of glass plates would soon change over to the new film. But this was not to be. He therefore decided to reach the general public with the Kodak camera, launched in 1888. This camera was loaded with a roll of the stripping film large enough for 100 exposures and sold, with shoulder strap and case, for $25. After use, the camera was sent to Rochester, where the film was developed by the complicated process and a new film loaded for $10.

Eastman followed this up in the next year with the first commercially available transparent nitrocellulose (celluloid) roll films. These had been produced since the mid 1850s, but until the advent of celluloid, none was successful. Unfortunately, on 2 May 1887 the Rev. Hannibal Goodwin applied for a patent for a transparent roll film made of celluloid but the patent was not granted until September 1898. Meanwhile Eastman's company had captured the market with its own roll film. A long and complicated law suit between the company which had acquired Goodwin's interests (he had died in 1900), and Eastman's company was finally settled in March 1914 with the Eastman company paying $5 million to the owners of Goodwin's patent. Virtually the only change there has been since then is the replacement of the flammable nitrocellulose by nonflammable cellulose acetate. Eastman's roll film replaced his stripping film in the Kodak camera and ushered in the era of press-button photography.

Eastman continued the popularization of photography by spooling his film so that the camera could be loaded in daylight, and did not have to be returned to the Rochester factory. A pocket-sized box camera was marketed for $1 in 1900. The 8-mm movie camera (which appeared in 1932), colour film, panchromatic film, automatic exposure control, sound film, instant cameras – all these and many more innovations followed.

Eastman was the first person to recognize the importance of the amateur market in photography, which until 1880 had been restricted by the need for darkroom work. Other manufacturers tried to produce rival cameras but due to Kodak's virtual roll-film monopoly, for two or three decades a hand camera, of whatever make, was simply known as a 'Kodak'.

Despite all these successes, Eastman had severe private problems that led to his suicide on 14 March 1932.

Eastwood Alice 1859–1953 was a US botanist who provided critical specimens for professional botanists as well as advising travellers on methods of plant collecting and arousing popular support for saving native species.

Eastwood was born on 19 January 1859 in Toronto, daughter of Colin Skinner, steward of the Toronto Asylum for the Insane. Following the death of her mother, she was taken into care at the age of six by her uncle, William Eastwood. Alice Eastwood's passion for botany started at an early age; she learned the Latin names for plants from her physician uncle and gained gardening knowledge from a French priest at the convent where she lived for six years. After graduating from East Denver High School as valedictorian in 1879, she taught a variety of subjects at the school for the next ten years and studied plants in the Colorado Mountains. In 1890, she resigned from teaching and toured California and remote places in Colorado. In 1892, she received an assistantship at the California Academy of Sciences, San Francisco, and founded and ran the California Botanical Club there. In 1894, she became the curator of the herbarium at the Academy and for a while acted as editor of the journal *Zoe*; she also carried out extensive field work in California. She remained at the Academy until her retirement in 1949. She died from cancer in 1953.

Eastwood's early fieldwork led to the publication of *A Handbook of the Trees of California* (1905), and her time before the San Francisco earthquake of 1906, was spent enlarging the Academy's botanical collection and segregating critical type specimens from the herbarium. Much, however, was destroyed in the earthquake and between 1906 and 1912 she devoted her time to rebuilding the collections, involving field trips to the coastal ranges and the Sierra Nevada and visits to botanical gardens around the world. Between 1912 and her retirement, over 340,000 specimens were added to the herbarium. In 1932, she and her assistant John Thomas Howell founded and edited the journal *Leaflets of Western Botany*; she also assisted in editing the journal *Erythea*. Eastwood herself published over 300 items. She received awards from garden clubs and, in 1950, was elected honorary president of the Seventh International Botanical Congress in Stockholm.

Eastwood Eric 1910–1981 was a British electronics engineer who made major contributions to the development of radar for both military and civilian purposes.

Eastwood was born on 12 March 1910 and educated at Oldham High School, Lancashire. He went to Manchester University and studied physics under Lawrence Bragg (1890–1971), and then moved to Cambridge University to take his PhD. He taught physics for a while at the Collegiate School, Liverpool, before entering the Signals branch of the Royal Air Force during World War II, and becoming involved with the solution of technical problems concerning radar.

In 1945, Eastwood joined the Nelson Research Laboratory of the English Electric Company. In 1948 he transferred to the Marconi Research Laboratory and was its director for many years.

In 1962, he accepted the appointment of Director of Research for the English Electric Group of Companies and, with the merger in 1968 of GEC and AEI, became Director of Research for the new company, General Electric, a position he held until his retirement in 1974.

Radar is a system using pulsed radio waves transmitted to and reflected back from a 'target', to measure its distance and direction. It was developed as a result of research carried out in the late nineteenth and early twentieth centuries. Heinrich Hertz (1857–1894) first discovered radio waves in 1888 and found they could be focused into a beam. Over the next 20 years several other scientists and inventors experimented with their potential, the most successful being Guglielmo Marconi (1874–1937), who in 1901 used these 'Hertzian waves' to send a message in Morse code across the Atlantic.

Further research on radio waves led to the first 'wireless' (radio) programmes of the early 1920s, but it took the probability of a war to rekindle interest in radio waves as a possible means of early detection of hostile aircraft. In 1935, the British Air Ministry asked Robert Watson-Watt (1892–1973) to investigate the concept of pulse radar. By the outbreak of World War II in 1939, there were radar stations along Britain's eastern coast, and these played a major part in the subsequent Battle of Britain.

Eastwood's part in the development of this system, and in the solving of its technical problems, left him ideally qualified to continue in this field of research. At the Marconi Research Laboratory he concentrated on the extension of the laboratory's interest in telecommunications, radar and applied physics. He soon realized that the new radar systems, which the laboratory had been commissioned to develop, could also be used as powerful tools in research. With the aid of the Marconi experimental station at Bushy Hill, Essex, he applied radar methods to the study of various meteorological phenomena (such as the auroras) and carried out extensive investigations into the flight behaviour of birds and migration; his book *Radar Ornithology* was published in 1967.

Eckert John Presper 1919–   is a US electronics engineer who is best known for his pioneering work on the design and construction of digital computers. He also holds patents on numerous other electronic devices.

Eckert was born in Philadelphia, Pennsylvania on 9 April 1919. He attended the William Pema Charter School and went on to graduate in 1941 from the Moore School of Electrical Engineering at the University of Pennsylvania.

He remained at the Moore School for five years as a research associate. During this time he worked on the design of radar ranging systems and then, to help with the complex calculations that these involve, turned to the design of electronic calculating devices. From 1942 to 1946, with John William Mauchly (1907–1980), he devised the Electronic Numerical Integrator And Calculator (ENIAC), one of the first modern computers.

ENIAC was first used in 1947, and proved to be successful although open to various improvements. Eckert had left Pennsylvania University the year before to become a partner in the Electric Control Company, and in 1947 began a three-year term as a Vice-President of the Eckert–Mauchly Computer Corporation. During this period, design improvements were incorporated into new computer models.

In 1950, the company was incorporated in Remington Rand. Eckert became Director of Engineering within the Eckert–Mauchly Division, becoming Vice-President in 1954. A year later the company came under the control of the Sperry Rand Corporation, and Eckert stayed on as Vice-President of the UNIVAC (Universal Automatic Computer Division).

Eckert's important contributions to computer design have been recognized with awards and honorary degrees from various organizations and universities.

The need for improved accuracy and, particularly speed, in routine calculations became most apparent during World War II, when ballistic firing tables had to be rapidly recalculated to suit new weaponry and battle conditions. Eckert realized that the available calculators were ineffective and inefficient. With the assistance of Mauchly he used electronics to construct an integrator, and produced a flexible digital computer which could be used for calculating firing tables and much more.

The ENIAC, although only a prototype of present-day computers, incorporated many modern design features and could perform mathematical functions. It lacked a memory, but could store a limited amount of information. Its major drawbacks were its size (it weighed many tonnes and included thousands of resistors and valves) and its high running cost (it consumed 100 kW of electric power). Nevertheless it formed applications to various military, meteorological and research problems.

ENIAC was superseded by BINAC, also designed in part by Eckert, which (by virtue of even more sophisticated design) was smaller and faster. In the early 1950s, Eckert's group began to produce computers for the commercial market with the construction of the UNIVAC I. Its chief advance was the capacity to store programs.

Eddington Arthur Stanley 1882–1944 was a British astronomer and writer who discovered the fundamental role of radiation pressure in the maintenance of stellar equilibrium, explained the method by which the energy of a star moves from its interior to its exterior, and finally showed that the luminosity of a star depends almost exclusively on its mass – a discovery that caused a complete revision of contemporary ideas on stellar evolution. He also showed

that a ray of light is deflected by gravity, thus confirming one application of Albert Einstein's general theory of relativity.

Eddington was born on 28 December 1882 in Kendal, Cumbria, although he spent his childhood in Weston-super-Mare, Somerset, and was educated there and at Owen's College, Manchester. In 1902, he won an entrance scholarship to Trinity College, Cambridge. Graduating three years later, he taught for a short time before being appointed Chief Assistant at the Royal Observatory, Greenwich. His seven-year stay there saw the beginning of his theoretical work.

In 1909, he was sent to Malta to determine the longitude of the geodetic station there and in 1912, he went to Brazil as the leader of an eclipse expedition. In 1913, Eddington returned to Cambridge to become Plumian Professor of Astronomy; shortly afterwards, he became Director of the University Observatory. Eddington remained at Cambridge for the next 31 years.

During his lifetime, Eddington was considered to be one of the greatest astronomers of the age. In 1906, he was elected Fellow of the Royal Astronomical Society and he was its president from 1921 to 1923. In 1914, he was elected Fellow of the Royal Society. He was knighted in 1938.

In the autumn of 1944, Eddington underwent major surgery, from which he never recovered. He died in Cambridge on 22 November of that year. After his death the Eddington Memorial Scholarship was established and the Eddington Medal was struck as an annual award.

Eddington published a large number of works. His first book, *Stellar Movements and the Structure of the universe* (1914), is considered to be a model of scientific exposition. The final chapter of the book alone, entitled 'Dynamics of the stellar system', marked the founding of an important branch of astronomical research. His book, *The Internal Construction of the Stars* (1926), became one of the classics of astronomy. His report to the Physical Society in 1918, expanded into his *Mathematical Theory of Relativity*, was the work that first gave English-speaking people the chance to learn the mathematical details of Einstein's famous theory of gravitation. Eddington's ability as a writer not only served to introduce a whole generation to the science of astronomy; it also had a stimulating effect on other astronomers.

Eddington involved himself in a great deal of practical work. He was the leader of the expedition to West Africa in 1919, where on 29 May on the island of Principe he observed the total eclipse of the Sun. The data he obtained served to verify one of the predictions contained in Einstein's general theory of relativity, that rays of light are affected by gravitation.

Eddington's first theoretical investigations were concerned with the systematic motion of the stars, but his great pioneering work in astrophysics began in 1916 when he started to study their composition. He established that in a star energy is transported by radiation, not by convection as had hitherto been thought. He also established that the mechanical pressure of the radiation was an important element in the maintenance of the star's mechanical equilibrium. Eddington showed that the equation of equilibrium must take into account three forces: gravitation, gas pressure and radiative pressure. One of the major questions at that time was how the gas of which stars and the Sun are composed was prevented from contracting under the tremendous force of stellar gravity. Eddington decided that the expansive force of heat and radiation pressure countered the contractive force of gravity. He also concluded that, since the pressure of the stellar matter increased rapidly with depth, the radiative pressure must also increase, and the only way in which that could happen was through a rise in temperature. Eddington showed that the more massive a star, the greater the pressure in its interior, and so the greater the countering temperature and radiation pressure, and consequently the greater its luminosity. He had found that the luminosity of a great star depends almost exclusively on its mass and, in 1924, he announced his mass-luminosity law. This work was of outstanding importance and necessitated a complete revision of contemporary notions regarding stellar evolution.

From 1930, Eddington worked on relating the theory of relativity and quantum theory. He believed that he could calculate mathematically, without recourse to observation, all the values of those constants of nature which are pure numbers, for example, the ratio of the mass of the proton to that of the electron. In his posthumously published *Fundamental Theory*, he presented his calculations of many of the constants of nature, including the recession velocity constant of the external galaxies, the number of particles in the universe, the ratio of the gravitational force to the electrical force between a proton and an electron, the fine structure constant, and the velocity of light.

In his later years, Eddington dealt with the philosophy of science and discussed the question of what sort of knowledge it was that science conveys to people. He had himself contributed considerably to that knowledge.

Edinger Tilly (Johanna Gabrielle Ottilie) 1897–1967 was a leading palaeontologist internationally known for her research in the field of palaeoneurology.

Tilly Edinger was born in Frankfurt on November 13 1897, daughter of Ludwig Edinger, a leading neuroanatomist. Between 1916 and 1918 she studied psychology, zoology and geology at Heidelberg and Munich and then returned to Frankfurt to study for her doctorate on *Nothosaurus*, which she completed in 1921. Her interest in the biological interpretation of fossils was influenced by Friedrich Drevermann and after her doctorate she became his research assistant. In 1927, she became the curator of the vertebrate collection at the Natural History Museum of Senckenberg. With Hitler's rise to power, Edinger, of Jewish extraction, was forced to leave Germany. In 1939, Edinger went to London where she worked as a translator; she then went to Cambridge, Massachusetts, in 1940, having been offered work by Alfred S. Romer, Director of the Museum of Comparative Zoology at Harvard. She received

fellowships from the Guggenheim Foundation (1943–44) and the American Association of University Women (1950–51); She also taught at Wellesley college (1944–45). In 1945 she became a US citizen. She died following a traffic accident on 26 May 1967.

Edinger was a leading figure in the field of twentieth-century vertebrate palaeontology and laid the foundations for the study of palaeoneurology. In her two great works *Die fossilen Gehirne* (Fossil Brains) (1929) and *The Evolution of the Horse Brain* (1948) she demonstrated that the evolution of the brain could be studied directly from fossil cranial casts. Her research shed new light on the evolution of the brain and showed that the progression of brain structure does not proceed at a constant rate in a given family but varies over time; also that the enlarged forebrain evolved several times independently among advanced groups of mammals and there was no single evolutionary scale. In 1962, she and three other authors published the two-volume *Bibliography of Fossil Vertebrates, Exclusive of North America, 1509–1927* and until her death she worked on *Paleoneurology 1804–1966*. An annotated bibliography was published posthumously in 1975. For her work, Edinger received honorary doctorates from Wellesley College (1950), the University of Giessen (1957) and a medical doctorate from the University of Frankfurt (1964); after her death the Tilly Edinger Fund at the Museum of Comparative Zoology at Harvard University was established to support the writing of books on vertebrate palaeontology.

Edison Thomas Alva 1847–1931 was a US electrical engineer and inventor. He took out more than a thousand patents, the best known of which were for the phonograph, the precursor of the gramophone, and the incandescent filament lamp.

Edison was born in the small town of Milan, Ohio, on 11 February 1847 and brought up in Michigan. Most of his tuition was provided by his mother – he received only three months' formal public elementary education. His lifelong interest in things technical was soon apparent – by the age of ten he had set up a laboratory in the basement of his father's house. By the age of 12, he was selling newspapers and candy on trains between Port Huron and Detroit, and three years later (1862) he had progressed to telegraph operator, a job he maintained throughout the Civil War (1861–1865) and for a couple of years thereafter. During this period, in 1866 (at the age of 19), he took out a patent on an electric vote recorder, the first of a total of 1,069 patents.

Perceiving the need for rapid communications, made apparent by the recent war, he turned his inventive mind to problems in that field. His first success came with a tape machine called a 'ticker', which communicated stock exchange prices across the country. He sold the rights in this and other telegraph improvements to the Gold and Stock Telegraph Company for $30,000, using the money to equip an industrial research laboratory in Newark, New Jersey, which he opened in 1869.

From telegraphy, the transmission of coded signals across long distances, he then turned his attention to telephony, the transmission of the human voice over long distances. In 1876, he patented an electric transmitter system, which proved to be less commercially successful than the telephone of Bell and Gray, patented a few months later. Typically, he was undeterred and not for the first time he applied his keenly inventive mind to improving someone else's idea. His improvements to their systems culminated in the invention of the carbon granule microphone, which so increased the volume of the signal that despite his deafness he could hear it.

With the money made from this invention he moved to Menlo Park, where he bought a house and equipped the laboratory that was to remain the centre for his research. In the following year, 1877, he invented the phonograph, a device in which the vibrations of the human voice were engraved by a needle on a revolving cylinder coated with tin foil. Thus began the era of recorded sound.

In the 1870s gas was the most advanced form of artificial lighting, the only successful rivals being various clumsy and expensive types of electrically powered arc lamp. While experimenting with the carbon microphone, Edison had toyed briefly with the idea of using a thin carbon filament as a light source in an incandescent electric lamp, an idea he returned to in 1879. His first major success came on 19 October of that year when, using carbonized sewing cotton mounted on an electrode in a vacuum (one millionth of an atmosphere), he obtained a source which remained aglow for 45 hours without overheating, a major problem with all other materials used. Even this success was not enough for him, and he and his assistants tried 6,000 other organic materials before finding a bamboo fibre which gave a bulb life of 1,000 hours. In 1883, he joined forces with Joseph Wilson Swan, a chemist from Sunderland to form the Edison and Swan United Electrical Company Ltd.

To produce a serious rival to gas illumination, a power source was required as well as a cheap and reliable lamp. The alternatives were generators or heavy and expensive batteries. At that time the best generators rarely converted more than 40% of the mechanical energy supplied into electrical energy. Edison made his first generator for the ill-fated Jeannette Arctic Expedition of 1879. It consisted of a drum armature of soft iron wire and a simple bipolar magnet, and was designed to operate one arc lamp and some incandescent lamps in series.

A few months later he built a much more ambitious generator, the largest built to date; weighing 500 kg/1,103 lb, it had an efficiency of 82%. Edison's team were at the forefront of development in generator technology over the next decade, during which efficiency was raised above 90%. To complete his electrical system he designed cables to carry power into the home from small (by modern standards) generating stations, and also invented an electricity meter to record its use.

Edison became involved with the early development of the film industry in 1888. After persuading George Eastman (1854–1932) to make a suitable celluloid film, he developed the high-speed camera and kinetograph, viewing the picture through a peep-hole. Although he had referred to the possibility of projecting the image, he omitted it from his patent – a rare error. He had dropped his interest in kinematography by 1893, but three years later resumed it when Thomas Armat developed a projector. They joined forces but Armat was commercially naive and the machine was advertised as Edison's latest triumph; the resulting split caused considerable patent litigation.

Edison's later years were spent in an unsuccessful attempt to develop a battery-powered car to rival the horseless carriages of Henry Ford. During World War I he produced many memoranda on military and naval matters for the Department of Operational Research.

When he died, aged 84, on 18 October 1931, Edison had come a long way from the ten-year-old boy with a laboratory in his father's basement to being probably the most prolific and practical inventive genius of his age. He was a man whose work has greatly influenced the world in which we live, particularly in the fields of communication and electrical power. On the day following his death, his obituary in the *New York Times* occupied four-and-a-half pages, an indication of the importance of Edison to the twentieth-century world.

Edlen Bengt 1906–   is a Swedish astrophysicist whose main achievement has been to resolve the identification of certain lines in spectra of the solar corona that had misled scientists for the previous 70 years.

Edlen was born in Gusum in Ostergotland, southeastern Sweden, on 2 November 1906. He was educated at Uppsala University and, in 1928, became an assistant in the physics department there. In 1936, he was appointed Assistant Professor of Physics. In 1944, he moved to southern Sweden to become Professor of Physics at Lund University. He held this post until 1973, when he became Emeritus Professor.

During the eclipse of 1869, astronomers recorded the presence of a hitherto unknown series of spectral lines in the Sun's corona. Because they failed to identify the origin of these lines, they ascribed them to the presence of a new element which they called 'coronium'. The origin of the lines was originally recorded as being located high above the Sun's surface, but similar lines were then discovered to originate nearer the Earth; these were accordingly attributed to another new element, which they called 'geocoronium'. Both the new elements were predicted to be much lighter than hydrogen, the lightest element known on Earth.

For 70 years all attempts to associate the coronal lines with known elements on Earth failed. Then, in the early 1940s, Edlen carried out a series of experiments and showed that, if iron atoms are deprived of some of their electrons, they can produce spectral lines similar to those produced by 'coronium'. He established that if half the normal number of 26 electrons of iron are removed, the effect produced is that of the green lines observed on the coronal line. Other lines were identified as iron atoms with different numbers of their electrons removed. Furthermore, it was found that similarly ionized atoms of nickel, calcium and argon produced even more lines.

It was determined that such high stages of ionization would require temperatures of about 1,000,000°C/1,800,000°F and when, in the 1950s, it was verified that such high temperatures did exist in the solar corona, it became accepted that 'coronium' as a separate element did not exist and that the 'coronium' lines owed their existence to ordinary elements being subjected to extreme temperatures – temperatures so high that they caused the corona to expand continuously. It was also established that the lines formerly thought to be caused by the presence of 'geocoronium' are produced by atomic nitrogen emitting radiation in the Earth's upper atmosphere.

Edwards and Steptoe Robert Edwards 1925–　is a British physiologist who, after many years of successful experimental work with embryos, became interested in the problem of human infertility. He and Patrick Steptoe 1913–1988, a skilled and experienced British surgeon, devised a technique for fertilizing a human egg outside the body and transferring the fertilized embryo to the uterus of a woman. A child born following the use of this technique is popularly known as a 'test-tube' baby.

Edwards was educated at the universities of Wales and Edinburgh and served in the British Army from 1944 to 1948. For the next three years he was at the University College of North Wales, Bangor, and from 1951 to 1957 at the University of Edinburgh. That year Edwards went to the California Institute of Technology but returned to England the following year to the National Institute of Medical Research at Mill Hill, remaining there until 1962 when he took up an appointment at Glasgow University. A year later he moved again to the Department of Physiology at Cambridge.

During his research in Edinburgh Edwards successfully replanted mouse embryos into the uterus of a mouse and he wondered if the same process could be used to replant a human embryo into the uterus of a woman.

One common cause of infertility in women is disease or damage to the Fallopian tubes, which prevents eggs from being fertilized. Normally these tubes allow the mature egg, when released from the ovary, to travel to the uterus and if spermatozoa are present the egg may become fertilized on the way. Gregory Pincus (1903–1967), the American biologist who developed the contraceptive pill, showed that human eggs could mature outside the body, when they would be ready for fertilization. Edwards, by then working in Cambridge, was able to obtain human eggs from pieces of ovarian tissue removed during surgery. He found that the ripening process was very slow, the first division beginning only after 24 hours.

During the following year he studied the maturation of eggs of different species of mammals, and in 1965 attempted the fertilization of human eggs. He left mature eggs with spermatozoa overnight and found just one where a sperm had passed through the outer membrane, but it had failed to fertilize the egg. In 1967, Edwards read a paper by Steptoe describing the use of a new instrument, known as the laparoscope, to view the internal organs, which he saw had a possible application to his own research. At about this time, Bavister, a research student at Cambridge, who had been trying to fertilize hamster eggs, devised a successful culture solution. Edwards used some of this solution with the one he used for the culture of human eggs and achieved fertilization.

Patrick Steptoe was educated at King's College and St George's Hospital Medical School, London, qualifying in 1939. During World War II he served in the Royal Naval Volunteer Reserve and was a prisoner of war in Italy from 1941 to 1943. He was appointed Chief Assistant Obstetrician and Gynaecologist at St George's Hospital, London, in 1947, and Senior Registrar at Whittington Hospital in 1949. From 1951 to 1978 he was Senior Obstetrician and Gynaecologist at Oldham General Hospital, and from 1969 Director of the Centre for Human Reproduction.

The paper which interested Edwards described laparoscopy, Steptoe's method of exploring the interior of the abdomen without a major operation. Steptoe inserted the laparoscope through a small incision near the navel and by means of this telescopelike instrument, with its object lens inside the body and its eyepiece outside, he was able to examine the ovaries and other internal organs.

Early in 1968 Edwards and Steptoe met and arranged to collaborate. During the next few months they repeated experiments on the fertilization of human eggs. Steptoe treated volunteer patients with a fertility drug to stimulate maturation of the eggs in the ovary, while Edwards devised a simple piece of apparatus to be used with the laparoscope for collecting mature eggs from human ovaries. The mature eggs were removed and Edwards then prepared them for fertilization using spermatozoa provided by the patient's husband. For a year they continued experiments of this kind until they were sure that the fertilized eggs were developing normally. The next step was to see if an eight-celled embryo would develop to the blastocyst state (the last stage of growth before it implants itself into the wall of the uterus); success was achieved.

In 1971, Edwards and Steptoe were ready to introduce an eight-celled embryo into the uterus of a volunteer patient who hoped to become pregnant, but this and similar attempts over a period of three years were unsuccessful. In 1975 an embryo *did* implant but in the stump of a Fallopian tube where it could not develop properly and was a danger to the mother. It was removed, but it did demonstrate the basic technique to be sound. In 1977 it was decided to abandon the use of the fertility drug and remove the egg at precisely the right natural stage of maturity; an egg was fertilized and then reimplanted (a process called *in vitro* fertilization) in the mother two days later. The patient became pregnant and 12 weeks later the position of the baby was found to be satisfactory and its heartbeat could be heard. The last eight weeks of the pregnancy were kept under close medical supervision and a healthy girl – Louise Brown – was delivered by Caesarean section on 25 July 1978.

Edwards and Steptoe showed how one common cause of infertility may be overcome. In Britain, infertility due to nonfunctional Fallopian tubes affects several thousand women every year, of which only a half can be helped by conventional methods. *In vitro* fertilization has also been used to overcome the infertility in men that is due to a low sperm count. Edwards's research has further added to knowledge of the development of the human egg and young embryo, and Steptoe's laparoscope is a valuable instrument capable of wider application.

Eggen Olin Jenck 1919– is a US astronomer who has spent much of his working life in senior appointments all round the world. His work has included studies of high-velocity stars, red giants (using narrow- and broadband photometry) and subluminous stars, and he has published some research on historical aspects of astronomy.

Born in Orfordville, Wisconsin, on 9 July 1919, Eggen graduated from Wisconsin University before becoming Astronomer at the University of California in 1945. In 1956, he became the Chief Assistant at the Royal Observatory, Greenwich. Maintaining his links with the United States, he was Professor of Astronomy at the California Institute of Technology from 1960 to 1963. He returned to Greenwich for a short time before serving as Astronomer at the Mount Palomar Observatory from 1965 to 1966. In 1966 he went to Australia to take up the post of Director of Mount Stromlo and Siding Spring Observatories, combining with this Professor of Astronomy in the Institute of Advanced Study, the Australian National University. Eggen remained at Mount Stromlo until 1977, when he moved to a position at the Observatory Interamericano de Cerro Tololo, Chile.

During the mid-1970s, Eggen completed a study – based on UBV photometry and every available apparent motion – of all red giants brighter than $V = 5^m0$. As a result he was able to classify these stars, categorizing them as very young discs, young discs and old discs. A few remained unclassifiable (haloes).

He also systematically investigated the efficiency of the method of stellar parallax using visual binaries originally suggested by William Herschel in 1781, and reviewed the original correspondence of John Flamsteed (1646–1719) and Edmond Halley (1656–1742).

Ehrlich Paul 1854–1915 was a German bacteriologist who founded chemotherapy – the use of a chemical substance to destroy disease organisms in the body. He was also one of the earliest workers on immunology, and through his studies on blood samples the discipline of haematology was recognized. In 1908, together with the Russian–French bacteriologist Ilya Mechnikov (1845–1916), he was awarded the Nobel Prize for Physiology or Medicine, for his work on serum therapy and immunity.

Ehrlich was born on 14 March 1854 in Silesia, which was then part of the Austro-Hungarian Empire, in a town called Strehlin (now Strzelin, in Poland). He studied in Breslau and Strasbourg, graduating in 1878 with a medical

degree from the University of Leipzig. For the next six years he was a clinical assistant at the University of Berlin and then became Head Physician at the medical clinic in the Charité Hospital in Berlin; in 1884 he was promoted to Professor there. Ehrlich spent two years in Egypt, from 1886 to 1888, to cure himself of tuberculosis. Successful, he returned to Berlin in 1889 where he set up a small private laboratory. The following year, he took up a professorial appointment at the University of Berlin. In 1891, he joined the Institute of Infectious Diseases, Berlin, as a researcher and five years later became Director of the newly established Institute for the Investigation and Control of Sera, opened in Berlin by the German government. Ehrlich continued working in his laboratories until just before his death on 20 August 1915.

As a student, Ehrlich had shown an unusual fascination with chemistry. Encouraged by his teachers he worked on the use of aniline dyes in microscopic techniques and discovered a few dyes for selectively staining, and therefore simplifying the study of, bacteria. He made histological preparations and stained them with various combinations of dyes to observe the different effects of basic and acidic stains. While at the Charité Hospital Ehrlich was able to distinguish between a number of blood disorders by examining blood cells in his stained preparations. In this way he discovered 'mast cells' (connective tissue cells), and it was also while studying the staining of tubercle cells that he contracted a mild case of tuberculosis.

On his return from his curative stay in Egypt, Ehrlich teamed up with the German bacteriologist Emil von Behring (1854–1917) and the Japanese Shibasaburo Kitasato (1856–1931) to try to find a cure for diphtheria. Ehrlich had studied antigen–antibody reactions using toxic plant proteins on mice, gradually increasing the dose, and found that the mice developed specific antibodies in their blood. Litters bred from these immunized mice possessed a short-lived immunity, sustained by suckling from the immunized mothers. Behring and Ehrlich were able to produce antitoxins obtained from much larger mammals which had been immunized against the diphtheria organism; these antitoxins were concentrated and purified for use in clinical trials, and once Ehrlich had developed the correct dosage, in 1892, the antitoxin was ready for use. In 1894 it was tried on 220 children with diphtheria and achieved great success. Ehrlich then decided that antitoxins should be standardized, their potency described in terms of international units of antitoxin, and the distribution made in dried form in vacuum phials.

At the Institute for the Investigation and Control of Sera, the number of Ehrlich's staff allowed him to investigate his theory that chemical compounds could cure a disease and not merely alleviate the symptoms. This stage was the beginning of chemotherapy (Ehrlich's word). The search progressed for dyes that would stain only bacteria and not other cells, and from this research the team continued synthesizing and testing chemical substances that could seek out and destroy the bacteria without harming the human body. Ehrlich termed these compounds 'magic bullets'.

Ehrlich's first success developed from the use of trypan red to kill trypanosomes (parasitic protozoans which cause sleeping sickness) in infected mice. The results of the tests proved inconclusive, but Ehrlich decided that the active agents in trypan red were nitrogen compounds. Atoxyl – an arsenical organic compound, and therefore similar in chemical properties to its nitrogen analogues – had shown greater success in the treatment of sleeping sickness and Ehrlich believed that it should be possible to make more effective derivatives of the substance. The accepted formula for atoxyl was a benzene ring with one side chain; Ehrlich, however, believed it had two side chains. If the accepted formula was correct, any derivatives would be unstable; but if Ehrlich was correct, they would be stable. Ehrlich proved to be right. He and his staff prepared nearly 1,000 derivatives of arsenic-containing compounds, testing each on animals. In 1907, they reached compound no. 606 (dihydroxydiamino-arsenobenzene hydrochloride), which proved ineffective against trypanosomes and so was forgotten. But it was investigated again in 1909 and discovered to be effective, instead, against spirochaetes, the bacteria that cause syphilis. Ehrlich tried it on himself, without any harm, and in 1910 announced the discovery of the synthetic chemical, now called Salvarsan (arsphenamine), for treating syphilis.

Ehrlich devised scientific techniques for developing chemical cures that opened up new fields of research in twentieth-century medicine, particularly in chemotherapy, haematology and immunology. Innumerable lives have been saved and the economic and social effects of his work continue to be far-reaching.

Eiffel Alexandre Gustav 1832–1923 was the French engineer now known chiefly for the 300-m/98-ft high edifice he erected for the 1889 exhibition commemorating the 100th anniversary of the French Revolution.

Eiffel was born at Dijon in 1832. After attending the Ecole des Arts et Manufactures in Paris he won early fame by specializing in the design of large metal structures, notably the iron railway bridge over the Garonne at Bordeaux. Here he was one of the first to use compresssed air for underwater foundations. In 1867, he set up his own firm which constructed bridges, viaducts, harbour works and other large projects. His great arch bridges include the 159-m/530-ft span over the Douro river in Portugal, and the 165-m/550-ft span of the Garabit Viaduct in France. Other projects included the immense roof that covers the central station at Budapest, the protective ironwork for the Statue of Liberty in New York Harbour, designed by Bartholdi, and the 84-m/277-ft dome for the observatory at Nice.

Eiffel began work on the famous wrought-iron tower for the Champs de Mars in 1886, using his experience of building high-level railway bridges. From a detailed set of plans, the 12,000 metal parts of the tower were all prefabricated and numbered for assembly. The majority of the 2.5 million rivets used were put in place before the structure was erected on the site. Work proceeded so smoothly that not one worker's life was lost through accidents on the scaffolding, and was completed (except for the lifts) in two and a quarter years.

The tower's cross-braced, latticed girder structure offers minimum wind resistance: the estimated movement of the structure with hurricane-force winds is only 22 cm/9 in. It is constructed from over 7,000 tonnes/6,900 tons of wrought iron, resting upon four masonry piers. The piers are set in 2 m/7 ft of concrete on foundations carried down by the aid of caissons and compressed air to about 15 m/44 ft on the side next to the Seine and about 9 m/30 ft on the other side.

The Tower has three well marked stages. Below the first platform, which is placed at a height of 57 m/183 ft the four quadrilateral legs are linked by arches. At the second platform, 115 m/380 ft high, the legs join and the third platform is at a height of 276 m/911 ft. Above this platform are the lantern and the final terrace.

Originally, the Tower was intended to be dismantled at the conclusion of the Exhibition. Many writers and artists deplored its construction, describing it as 'a hideous hollow candlestick'. However, the newly discovered possibilities of the tower as a radio transmitting station finally won the day and it was left standing. For some time it was by far the highest artificial structure in the world. It also showed what could be achieved by correct engineering design and paved the way for yet higher structures in the future.

With subsequent disaster for himself and his country. Eiffel participated in the Panama Canal enterprise, in the course of which he designed and partly constructed some huge locks. When the entire project collapsed in 1893, Eiffel was implicated in the scandal; he went to prison for two years and received a fine of 200,000 francs. In 1900, he took up meteorology and later, using wind tunnels, carried out extensive research in aerodynamics at the Eiffel Tower and afterwards at Auteuil, where he constructed the first laboratory for the new science. He died in Paris in December 1923.

Eigen Manfred 1927–   is a German physical chemist who shared the 1967 Nobel Prize for Chemistry with Ronald Norrish and George Porter for his work on the study of fast reactions in liquids.

Eigen was born in Bochum, Ruhr, on 9 May 1927, the son of a musician. He was educated at Göttingen University and on his 18th birthday, one day after the formal ending of World War II, he was drafted to do military service with an antiaircraft artillery unit. He later returned to Göttingen, gained his doctorate in 1951, and worked as a research assistant there for the next two years. In 1953, he moved to the Max Planck Institute for Physical Chemistry, becoming a research fellow (1958), head of the Department of Biochemical Kinetics (1962), and eventually Director of the Institute.

Eigen studied fast-reaction kinetics by disturbing the equilibria in liquid systems using short changes of temperature, pressure or electric field (whereas Norrish and Porter had used flashes of light to disturb equilibria in gaseous systems). He investigated particularly very fast biochemical reactions that take place in the body, trying to discover how rapidly reactions proceed among the working molecules of life and how a particular sequence of chemical units could come about by chance in the time available. With his colleague Ruthild Winkler he tried to relate chance and chemistry in processes that could have led to the origin of life on Earth. They questioned how molecules with the right kind of properties might form and what would be the simplest combination of molecules that could survive and evolve into the first primitive organisms. Eigen theorized that in the 'primeval soup' of the early Earth, cycles of chemical reactions would have occurred, one reproducing nucleic acids (which possess information but have a very limited chemical function) and one reproducing proteins (which ensured chemical function and reproduction of the information contained in the nucleic acids). He postulated that eventually a number of the nucleic acid cycles and proteins would have come to co-exist and form a 'hypercycle'. By natural selection the best hypercycle would have eventually caused the first organism to evolve – a chance set of molecules coming together in a single drop – providing a possible theory of the chemical transition from nonlife to life.

Eilenberg Samuel 1913–   is a Polish-born US mathematician whose research in the field of algebraic topology led to considerable development in the theory of cohomology. He is also well known for his work in computer mathematics.

Eilenberg was born on 30 September 1913 in Warsaw, where he grew up and completed his education, gaining his master's degree in 1934 and his PhD in mathematics two years later at the University of Warsaw. He then emigrated to the United States where, in 1940, he joined the staff of the University of Michigan as an instructor; by 1946 he was Associate Professor of Mathematics. In that year he was appointed Professor of Mathematics at the University of Indiana, where he remained for three years. After a series of visiting professorships – some in the United States, some in Europe and the Indian subcontinent – Eilenberg became Professor of Mathematics at Columbia University, New York, where he remained for the rest of his academic life.

Eilenberg's main field of work has been that of algebraic topology – a subject on which, with N. Steenrod, he wrote a successful advanced textbook. Topology is the study of figures and shapes that retain their essential proportions even when twisted or stretched; in topology, therefore, a square is (topologically) equivalent to any closed plane figure – such as a circle – and a cube is even (topologically) equivalent to a sphere. Since Henri Poincaré (1854–1912) first developed the subject systematically in a series of papers written between 1895 and 1905, topology theory has been elaborated at a rapid rate, and has considerably influenced other branches of mathematics. Eilenberg has carried out valuable work in an area of topology that, although generally known as algebraic topology, is sometimes called 'combinatorial' topology and is distinctive for the extensive use of algebraic techniques to solve topological problems. The basis on which algebraic topology is founded is homology theory – the study of closed curves, closed surfaces and similar geometric arrangements in a given topological space. Much of Eilenberg's work has been concerned with a modification of homology theory called cohomology theory; cohomology groups have properties similar to homology groups but have several important advantages. It is possible to define a 'product' of cohomology classes by means of which, together with the addition of cohomology classes, the direct sum of the cohomology classes of all dimensions becomes a ring (the cohomology ring). This is a richer structure than is available for homology groups, and allows finer results. Various other very complicated algebraic operations using coho-mology classes can lead to results not provable in any other way – the Poincaré duality theorem, for example, is considerably easier to state precisely if cohomology groups are used.

Einstein Albert 1879–1955 was a German-born US theoretical physicist who revolutionized our understanding of matter, space and time with his two theories of relativity. Einstein also established that light may have a particle nature and deduced the photoelectric law that governs the production of electricity from light-sensitive metals. For this achievement, he was awarded the 1921 Nobel Prize for Physics. Einstein also investigated Brownian motion and was able to explain it so that it not only confirmed the existence of atoms but could be used to determine their dimensions. He also proposed the equivalence of mass and energy, which enabled physicists to deepen their understanding of the nature of the atom, and explained radioactivity and other nuclear processes. Einstein, with his extraordinary insight into the workings of nature, may be compared with Isaac Newton (whose achievements he extended greatly) as one of the greatest scientists to have lived.

Einstein was born in Ulm, Germany, on 14 March 1879. His father's business enterprises were not successful in that town and soon the family moved to Munich, where Einstein attended school. He was not regarded as a genius by his teachers; indeed there was some delay because of his poor mathematics before he could enter the Eidgenössosche Technische Hochschule in Zurich, Switzerland, when he was 17. As a student he was not outstanding and Hermann Minkowski (1864–1909), who was one of his mathematics professors, found it difficult in later years to believe that the famous scientist was the same person he had taught as a student.

Einstein graduated in 1900 and, after spending some time as a teacher, he was appointed a year later to a technical post in the Swiss Patent Office in Berne. Also in 1901 he became a Swiss citizen and then in 1903 he married his first wife, Mileva Marié. This marriage ended in divorce in 1919. During his years with the Patent Office, Einstein worked on theoretical physics in his spare time and evolved the ideas that were to revolutionize physics. In 1905, he published three classic papers on Brownian motion, the photoelectric effect and special relativity.

Einstein did not, however, find immediate recognition. When he applied to the University of Berne for an academic position, his work was returned with a rude remark. But by 1909 his discoveries were known and understood by a few people, and he was offered a junior professorship at the University of Zurich. As his reputation spread, Einstein became full Professor or Ordinariat, first in Prague in 1911 and then in Zurich in 1912, and he was then appointed Director of the Institute of Physics at the Kaiser Wilhelm Institute in Berlin in 1914, where he was free from teaching duties.

The year 1915 saw the publication of Einstein's general theory of relativity, as a result of which Einstein predicted that light rays are bent by gravity. Confirmation of this prediction by the solar eclipse of 1919 made Einstein world famous. In the same year he married his second wife, his cousin Elsa, and then travelled widely to lecture on his discoveries. One of his many trips was to the California Institute of Technology during the winter of 1932. In 1933 Adolf Hitler came to power and Einstein, who was a Jew, did not return to Germany but accepted a position at the Princeton Institute for Advanced Study, where he spent the rest of his life. During these later years, he attempted to explain gravitational, electromagnetic and nuclear forces by one unified field theory. Although he expended much time and effort in this pursuit, success was to elude him.

In 1939, Einstein used his reputation to draw the attention of the President of the United States to the possibility that Germany might be developing the atomic bomb. This prompted US efforts to produce the bomb, though Einstein did not take part in them. In 1940, Einstein became a citizen of the United States. In 1952, the state of Israel paid him the highest honour it could by offering him the presidency, which he did not accept because he felt that he did not have the personality for such an office. Einstein was a devoted scientist who disliked publicity and preferred to live quietly, but after World War II he was actively involved in the movement to abolish nuclear weapons. He died at Princeton on 18 April 1955.

Einstein's first major achievement concerned Brownian motion, the random movement of fine particles which can be seen through a microscope and was first observed in 1827 by Robert Brown (1773–1858) when studying a suspension of pollen grains in water. The motion of the particles increases when the temperature increases but decreases if larger particles are used. Einstein explained this phenomenon as being the effect of large numbers of molecules bombarding the particles. He was able to make predictions of the movement and size of the particles, which were later verified experimentally by the French physicist Jean Perrin (1870–1942). Experiments based on this work were used to obtain an accurate value of the Avogadro number, which is the number of atoms in one mole of a substance, and the first accurate values of atomic size. Einstein's explanation of Brownian motion and its

subsequent experimental confirmation was one of the most important pieces of evidence for the hypothesis that matter is composed of atoms.

Einstein's work on photoelectricity began with an explanation of the radiation law proposed in 1901 by Max Planck (1858–1947). This is $E = hv$, where E is the energy, h is a number known as Planck's constant and v is the frequency of radiation. Planck had confined himself to black-body radiation, and Einstein suggested that packets of light energy are capable of behaving as particles called 'light quanta' (later called photons). Einstein used this hypothesis to explain the photoelectric effect, proposing that light particles striking the surface of certain metals cause electrons to be emitted. It had been found experimentally that electrons are not emitted by light of less than a certain frequency v^0; that when electrons are emitted, their energy increases with an increase in the frequency of the light; and that an increase in light intensity produces more electrons but does not increase their energy. Einstein suggested that the kinetic energy of each electron, $\frac{1}{2}mv^2$, is equal to the difference in the incident light energy hv and the light energy needed to overcome the threshold of emission hv^0. This can be written mathematically as:

$$\tfrac{1}{2}mv^2 = hv - hv^0$$

and this equation has become known as Einstein's photoelectric law. Its discovery earned Einstein the 1921 Nobel Prize for Physics.

Einstein's most revolutionary paper of 1905 contained the idea which was to make him famous, relativity. Up to this time, there had been a steady accumulation of knowledge which suggested that light and other electromagnetic radiation do not behave as predicted by classical physics. For example, no method had been found to determine the velocity of light in a single direction. All the known methods involved a reflection of light rays back along their original path. It had also proved impossible to measure the expected changes in the speed of light relative to the motion of the Earth. The Michelson–Morley experiment had demonstrated conclusively in 1881 and again in 1887 that the velocity of light is constant and does not vary with the motion of either the source or the observer. To account for this, Hendrik Lorentz (1853–1928) and George Fitz-Gerald (1851–1901) independently suggested that all lengths contract in the direction of motion by a factor of $(1 - v^2/c^2)^{1/2}$, where v is the velocity of the moving body and c is the speed of light.

The results of the Michelson–Morley experiment confirmed that no 'ether' can exist in the universe as a medium to carry light waves, as was required by classical physics. This did not worry Einstein, who viewed light as behaving like particles, and it enabled him to suggest that the lack of an ether removes any frame of reference against which absolute motion can be measured. All motion can only be measured as motion relative to the observer. This idea of relative motion is central to relativity, and is one of the two postulates of the special theory, which considers uniform relative motion. The other is that the velocity of light is constant and does not depend on the motion of the observer. From these two notions and little more than school algebra, Einstein derived that in a system in motion relative to an observer, length would be observed to decrease by the amount postulated by Lorentz and FitzGerald. Furthermore, he found that time would slow by this amount and that mass would increase. The magnitude of these effects is negligible at ordinary velocities and Newton's laws still hold good. But at velocities approaching that of light, they become substantial. If a system were to move at the velocity of light, to an observer its length would be zero, time would be at a stop and its mass would be infinite. Einstein therefore concluded that no system can move at a velocity equal to or greater than the velocity of light.

Einstein's conclusions regarding time dilation and mass increase were later verified with observations of fast-moving subatomic particles and cosmic rays. Length contraction follows from these observations, and no velocity greater than light has ever been detected. Einstein went on to show in 1907 that mass is related to energy by the famous equation $E = mc^2$. This indicates the enormous amount of energy that is stored as mass, some of which is released in radioactivity and nuclear reactions, for example in the Sun. One of its many implications concerns the atomic masses of the elements, which are not quite what would be expected by the proportions of their isotopes. These are slightly decreased by the mass equivalent of the binding energy that holds their molecules together. This decrease can be explained by, and calculated from, the famous Einstein formula.

Minkowski, Einstein's former teacher, saw that relativity totally revised accepted ideas of space and time and expressed Einstein's conclusions in a geometric form in 1908. He considered that everything exists in a four-dimensional space–time continuum made up of three dimensions of space and one of time. This interpretation of relativity expressed its conclusions very clearly and helped to make relativity acceptable to most physicists.

Einstein now sought to make the theory of relativity generally applicable by considering systems that are not in uniform motion but under acceleration. He introduced the notion that it is not possible to distinguish being in a uniform gravitational field from moving under constant acceleration without gravitation. Therefore, in a general view of relativity, gravitation must be taken into account. To extend the special theory, he investigated the effect of gravitation on light and in 1911 concluded that light rays would be bent in a gravitational field. He developed these ideas into his general theory of relativity, which was published in 1915. According to this theory, masses distort the structure of space–time.

Einstein was able to show that Newton's theory of gravitation is a close approximation of his more exact general theory of relativity. He was immediately successful in using the general theory to account for an anomaly in the orbit of the planet Mercury that could not be explained by Newtonian mechanics. Furthermore, the general theory

made two predictions concerning light and gravitation. The first was that a red shift is produced if light passes through an intense gravitational field, and this was subsequently detected in astronomical observations in 1925. The second was a prediction that the apparent positions of stars would shift when they are seen near the Sun because the Sun's intense gravity would bend the light rays from the stars as they pass the Sun. Einstein was triumphantly vindicated when observations of a solar eclipse in 1919 showed apparent shifts of exactly the amount he had predicted.

Einstein later returned to the quantum theory. In 1909, he had expressed the need for a theory to reconcile both the particle and wave nature of light. In 1923, Louis de Broglie (1892–1987) used Einstein's mass–energy equation and Planck's quantum theory to achieve an expression describing the wave nature of a particle. Einstein's support for de Broglie inspired Erwin Schrödinger (1887–1961) to establish wave mechanics. The development of this system into one involving indeterminacy did not meet with Einstein's approval, however, because it was expressed in terms of probabilities and not definite values. Einstein could not accept that the fundamental structure of matter could rest on chance events, making the famous remark 'God does not play dice'. Nevertheless, the theory remains valid.

Albert Einstein towers above all other scientists of the twentieth century. In changing our view of the nature of the universe, he has extended existing laws and discovered new ones, all of which have stood up to the test of experimental verification with ever increasing precision. The development of science in the future is likely to continue to produce discoveries that accord with Einstein's ideas. In particular, it is possible that relativity will enable us to make fundamental advances in our understanding of the origin, structure and future of the universe.

Eisenhart Luther Pfahler 1876–1965 was a US theoretical geometrist whose early work was concerned with the properties of surfaces and their deformation; later he became interested in Riemann geometry from which he attempted to develop his own geometry theory. The author of several books detailing his results, he also wrote two books on historical topics.

Eisenhart was born on 13 January 1876 in York, Pennsylvania, the second son of the dentist who was also the founder of the Edison Electric Light and York Telephone Company. Educated locally, and to a high standard, he attended Gettysburg College (in southern Pennsylvania) from 1892 to 1896 where, for the last two years, he studied mathematics independently of his other work, through guided reading. After a year's teaching at the College he went to Johns Hopkins University, Baltimore, in 1897, to carry out graduate studies, obtaining his PhD there three years later. He then began his life's work in mathematical research at Princeton University, retiring from there in 1945 after 45 years' successful study and teaching. He died there, 20 years later, on 28 October 1965.

One of Eisenhart's major achievements was to relate his theories regarding differential geometry to studies bordering on the topological. At the age of 25 he wrote one of the first characterizations of a sphere as defined in terms of differential geometry (the paper had the somewhat daunting title 'Surfaces whose first and second forms are respectively the second and first forms of another surface'). For the next 20 years he continued to develop his research, concentrating particularly on the subject of surface deformation. The theory of the deformation of surfaces was a part of the study of the properties of surfaces and systems of surfaces that was an especially popular area for geometrical research in continental Europe at that time, but Eisenhart was (apparently) the only person in the United States to devote his attention to it. It was he, nevertheless, who managed to formulate a unifying principle to the theory. The deformation of a surface involves the congruence of lines connecting a point and its image. Eisenhart's contribution was to realize that in all known cases, the intersections of these surfaces with the given surface and its image, form a set of curves which have special properties. He wrote his account of the theory in 1923, in *Transformations of Surface*.

His work on surfaces led him then to study Riemann geometry – in which the properties of geometric space are considered locally rather than in one overall framework for the whole space. Eisenhart developed a geometry analogous to Riemann geometry, which he called non-Riemann geometry (although the term has since been used for several other forms of geometry), and wrote *Fields of Parallel Vectors in the Geometry of Paths* in 1922; it was followed by *Fields of Parallel Vectors in a Riemannian Geometry* (1925) and *Riemannian Geometry* (1926).

Interested in history, Eisenhart also wrote several other papers, including ' Lives of Princeton mathematicians' (1931), 'Plan for a university of discoverers' (1947) and 'The preface to historic Philadelphia' (1953).

Elkington George Richards 1801–1865 was a British inventor who pioneered the use of electroplating for finishing metal objects.

Elkington was born in Birmingham and in 1818 he became an apprentice in the local small-arms factory; in due course he became its proprietor. With his cousin Henry Elkington he explored the alternatives to silver-plating from about 1832. The fire-gilding process of plating base metals, or more often silver, with a thick film of gold had been practised from early times. The article was first cleaned and then placed in a solution of mercurous nitrate and nitric acid, so that it acquired a thin coating of mercury. The surface was next rubbed with an amalgam of gold and mercury, in the form of a stiff paste held in a porous fabric bag, until a smooth coating of the pasty mixture had been applied. Finally, the article was heated on a charcoal fire to drive off the mercury, and the residual gold was burnished.

The process of plating base metals with silver and gold by electrodeposition was announced in a patent taken out by the Elkington cousins in 1840. This proposed the use of electrolytes prepared by dissolving silver and gold, or their oxides, in potassium or sodium cyanide solution. The articles to be coated were cleaned of grease and scale and immersed as cathodes in the solution, current being supplied through a bar of metallic zinc or other electropositive metal (anode); later silver or gold anodes were used, brass and German silver were suggested as the most suitable metals for plating.

A second patent, granted in 1842 to Henry Beaumont (an employee of the Elkingtons), covered some 430 additional salts of silver which it was thought might have application to electroplating. In the same year, John Stephen Woolrich obtained a patent for the use of a magnetoelectric machine which depended on Michael Faraday's discovery of electromagnetism in 1830, the plating solution being the soluble double sulphate of silver and potassium. Licences were first issued to the Elkingtons in 1843, but Thomas Prime of Birmingham appears to have been the first to use such plate commercially, by using Woolrich's patented machine; Dr Percy, a famous metallurgist, claims to have conducted Faraday himself round Prime's works in 1845. Subsequently the Elkingtons took over Woolrich's patent, for which they paid him a royalty, and they were thereafter able to command a minimum royalty themselves of £150 from all who practised electroplating. In spite of this, however, electroplate rapidly supplanted Old Sheffield plate; the Sheffield directory of 1852 contained the last entry under this heading, with only a single representative remaining, from the many who had practised the art five years previously.

Taking into partnership Joseph Mason, the founder of Mason College (subsequently Birmingham University), the Elkingtons established a large workshop in Newhall Street, Birmingham, which after a seven-year battle against the older methods of silver plating, at last won acceptance. The Elkingtons also successfully patented their ideas in France. George Elkington established large copper-smelting works in Pembrey, South Wales, additionally providing houses for his workers and schools for their children, but the chief centre for his activities remained in Birmingham. He died in Pool Park in North Wales in 1865.

Ellet Charles 1810–1862 was a US civil engineer who designed the first wire-cable suspension bridge in the United States and became known as the 'American Brunel'.

Ellet's career began when he was appointed as a surveyor and assistant engineer on the Chesapeake and Ohio Canal in 1828, where he remained for three years. He then went to Europe and enrolled as a student at the Ecole Polytechnique in Paris, and continued to gather experience by studying the various engineering works taking place in France, Germany and Britain.

He returned to the United States in 1832 and submitted to Congress a proposal for a 305-m/1,000-ft suspension bridge over the Potomac River at Washington, DC, but the plan was too advanced for its time and failed to receive government support. In 1842, over the Schuylkill River at Fairmount, Pennsylvania, he built his first wire-cable suspension bridge. Ellet introduced there a technique that was common in France, that of binding small wires together to make the cables; five of these cables supported the bridge at each side, the span being 109 m/358 ft.

Between 1846 and 1849 he designed and built for the Baltimore and Ohio Railway the world's first long-span wire-cable suspension bridge, crossing the Ohio River at Wheeling, West Virginia. The central span of 308 m/ 1,010 ft was then the longest ever built. However, the bridge failed under wind forces in 1854 because of its over-all aerodynamic instability. Ellet's towers remained standing, and the rest of the bridge was rebuilt by John Roebling (1806–1869), who later achieved fame by his own record-breaking activities in building long-span suspension bridges of wire cable of his own manufacture. (In 1956 the Wheeling Bridge was again under repair; Ellet's towers and anchorages and Roebling's cables and suspenders were retained, but the deck was entirely renewed.)

In 1847, Ellet received a contract to build a bridge over the Niagara River, only 3.2 km/2 mi below the falls. The result was a light suspension structure, and Ellet subsequently claimed to be the first person to cross the Niagara Gorge on the back of a horse (thanks to the bridge). This was, however, to prove to be another enterprise which turned sour for its promoter. A dispute over money led Ellet to resign in 1848, leaving the project uncompleted.

Following the outbreak of the American Civil War in 1861, Ellet produced a steam-powered ram which was used by the Union (Northern) forces with decisive effect against the Confederate army on the Mississippi River. In June 1862, Ellet personally led a fleet of nine of these rams in the Battle of Memphis. The Union side was victorious, but in the course of the fighting Ellet was fatally wounded.

Elsasser Walter Maurice 1904–1991 was a leading modern geophysicist.

Born in Germany and educated at Göttingen, Elsasser left in 1933 following Hitler's rise to power, and spent three years in Paris where he worked on the theory of atomic nuclei. After settling in 1936 in the USA and joining the staff of the California Institute of Technology, he specialized in geophysics. His magnetical researches in the 1940s yielded the dynamo model of the Earth's magnetic field. In this the field is explained in terms of the activity of electric currents flowing in the Earth's fluid metallic outer core. The theory premises that these currents are magnified through mechanical motions, rather as currents are sustained in power-station generators. It was Elsasser who pioneered analysis of the Earth's former magnetic fields, frozen in rocks. Taken up into the work of Drummond Hoyle Matthews, Tuzo Wilson and others, Elsasser's insights have subsequently proved crucial to the development of modern ideas of oceanic expansion and continental movement.

Elsasser served as Professor of Physics at the University of Pennsylvania from 1947; in 1962 he was made Professor of Geophysics at Princeton, and between 1968 and 1974 he held a research chair at the University of Maryland. A wide-ranging and speculative thinker, Elsasser also produced *The Physical Foundation of Biology* 1958 and *Atom and Organization* 1966.

Eméleus Harry Julius 1903–   is a British chemist who made wide-ranging investigations in inorganic chemistry, studying particularly nonmetallic elements and their compounds.

Eméleus was born in London on 22 June 1903. He attended Hastings Grammar School and Imperial College, London. After graduation he went to Karlsruhe University where he met several of the German exponents of preparative inorganic chemistry. From 1929 to 1931 he worked at Princeton University in the United States, and it was there that he met and married Mary Catherine Horton. He returned to Imperial College to continue his researches and in 1945 became Professor of Inorganic Chemistry at Cambridge University, where he remained until he retired in 1970.

Eméleus began his researches during his first period at Imperial College with a study of the phosphorescence of white phosphorus, showing that the glow was caused by the slow oxidation of phosphorus(III) oxide formed in a preliminary nonluminous oxidation. He also studied the inhibition of the glow by organic vapours. His work continued with spectrographic investigations of the phosphorescent flames of carbon disulphide, ether (ethoxyethane), arsenic, sulphur and the phosphorescence of phosphorus(V) oxide illuminated with ultraviolet light. The results provided new information about the mechanisms of combustion reactions.

While at Princeton Eméleus worked on the photosensitization by ammonia of the polymerization of ethene (ethylene), the photochemical interaction of amines and ethene, and the photochemistry of the decomposition of amines and their reaction with carbon monoxide. This phase of his work, on chemical kinetics, helped to prepare him for the great career in chemistry that lay ahead.

On his return to Imperial College he began investigating the hydrides of silicon, especially the kinetics of the oxidation of mono-, di- and trisilane. He also studied the isotopic composition of water from different sources. He developed a very accurate method of measuring densities and showed that naturally occurring water exhibits a small variation in deuterium content, and that distillation, freezing and adsorption methods can all effect some degree of separation of the two isotopic forms. Continuing his work on silicon hydrides, he prepared tetrasilane, Si_4H_{10} (the silicon analogue of butane), by treating magnesium silicide with dilute hydrochloric acid, and went on to produce alkyl and aryl derivatives of the silanes. In 1938, Eméleus and John Anderson published *Modern Aspects of Inorganic Chemistry*.

When Eméleus moved to Cambridge in 1945 he started studying the halogen fluorides. He showed that the much sought after trifluoroiodomethane, CIF_3 – the key to many synthetic processes – can be made by reacting iodine(V) fluoride with carbon(IV) iodide. He prepared polyhalides of potassium and demonstrated that bromine(III) fluoride could be used as a nonaqueous solvent in the study of acid/base reactions. In 1949, he prepared organometallic fluorides of mercury and went on to make various derivatives containing the methylsilyl group, CH_3SiH_2-. By 1959 he was working with the fluorides of vanadium, niobium, tantalum and tungsten and much of his research in the 1960s concerned the fluoralkyl derivatives of metals. He summarized much of his work in *The Chemistry of Fluorine and its Compounds* (1969).

Eméleus received many honours but remains an extremely modest man, never claiming that the results of his work were of outstanding significance. However, his influence on inorganic chemistry since 1945 has been enormous.

Empedocles *c.*490–430 BC was a Greek philosopher, who was rather a man of many parts, being prominent in poetry, philosophy, politics, mysticism and medicine. He lived in Acragas (Agrigentum) in Sicily, and seems to have been one of the earliest philosophers to embrace the view that terrestrial objects are made up of four elements or basic principles, that is, fire, air, water and earth. He viewed these as united or divided by two forces, attraction and repulsion (or, more poetically, love and strife). Such views of the elemental construction of matter, later more fully developed by Aristotle, proved influential until the mechanical philosophy associated with the scientific revolution of the seventeenth century imposed a doctrine which viewed matter as corpuscular and governed by mathematically calculable mechanical forces.

Empedocles seems to have been influenced by Pythagoreanism and perhaps Orphism. Legend has it that Empedocles ended his life by jumping into the crater of Mount Etna, possibly in the belief that he would demonstrate his immortality. Perhaps for those reasons, he was a figure greatly admired by nineteenth-century Romantic poets.

Encke Johann Franz 1791–1865 was an influential German astronomer whose work on star charts during the 1840s contributed to the discovery of the planet Neptune in 1846. He also worked out the path of the comet which bears his name.

Born in Hamburg on 23 September 1791, Encke was the eighth child of a Lutheran preacher. As a child he was exceedingly proficient at mathematics and at the age of 20 he became a student at the University of Göttingen. His degree studies were interrupted by military service in the Wars of Liberation.

As a student Encke impressed the physicist Karl Gauss (1777–1855), who was instrumental in securing a post for him at a small astronomical observatory at Seeberg near Gotha. There the quality of his work was soon recognized and he rose in seniority from Assistant to Director. Encke then accepted the offer of a professorship at the Academy of Sciences in Berlin and the directorship of the Berlin Observatory in 1825. After 40 years in Berlin, Encke died there on 26 August 1865.

A fine mathematician who carried out continuous research on comets and the perturbations of the asteroids, Encke spent much of his time putting together the information with which to prepare new star charts. The compilation was from both old and new observations and many alternative sources. The charts, taking nearly 20 years to draw up, were completed in 1859 – but were soon improved upon by those of Friedrich Argelander (1799–1875). Nevertheless Encke's charts were of some value in that they pointed out the existence of Neptune and several asteroids.

Encke's most successful piece of work was on what subsequently became known as Encke's Comet. This comet had been reported by Jean Pons (1761–1831), but little was known of its behaviour. Encke showed that the comet had an elliptical orbit with a period of just less than four years.

One of his tasks at Berlin was to oversee the Berlin *Astronomisches Jahrbuch* (Yearbook), of which he was the editor for the period 1830–66. The books included large sections on minor planets, which increased their cost considerably. Encke was prepared to take the risk that high costs would diminish the market for the annual, since he considered that the data were worth publishing. He also used the books to publish many of his own mathematical determinations of orbits and perturbations, although he preferred to publish papers on planets of the Solar System in *Astronomische Nachrichten*.

During the later 1820s and the early 1830s Encke was responsible for the re-equipping and resiting of the Berlin Observatory. After raising the necessary financial support he installed a meridian circle, a large Fraunhofer refractor and a heliometer. The observatory specialized in the observation of moveable stars.

Enders John Franklin 1897–1985 was a US microbiologist who succeeded in culturing viruses in quantity outside the human body. Before this time, progress in research on viruses had been greatly hindered by the fact that viruses need living cells in which to grow. For his work on virus culture he was awarded the 1954 Nobel Prize for Physiology or Medicine, which he shared with Frederick Robbins (1916–) and Thomas Weller (1915–).

Enders was born on 10 February 1897 in West Hartford, Connecticut, the son of a banker. He was educated at Yale University but interrupted his studies to become a flying instructor during World War I and did not graduate until 1920. He then took up a business career but left it to study English at Harvard. Enders changed to study medicine, and finally obtained a doctorate in bacteriology in 1930. He remained at Harvard Medical School, progressing from Instructor in 1930 to Professor in 1962 and Professor Emeritus in 1968. From 1947 to 1972 he was Chief of Research in the Division of Infectious Diseases at the Children's Hospital Medical Center, Boston. From 1972, he was Chief of the Virus Research Unit at the hospital.

Viruses cannot be grown, as bacteria can, in nutrient substances, and so a method had been developed for growing them in a living chick embryo. Enders believed that he could improve on this method and reasoned that it was unnecessary to use a whole organism, but that living cells might be sufficient. In 1948, Enders and his colleagues, Robbins and Weller, prepared a medium of homogenized chick embryo and blood and attempted to grow a mumps virus in it. This experiment had been tried before, unsuccessfully, but at that time penicillin had not been available and it was penicillin that they added to suppress the growth of bacteria in the mixture. The experiment worked and they turned to the growth of other viruses.

The disease poliomyelitis attacked and debilitated many children at that time and they decided to investigate the virus responsible. Previously the polio virus could be grown only in living nerve tissue from primates. But using their method, Enders managed to grow the virus successfully on tissue scraps obtained first from stillborn human embryos, and then on other tissue.

In the 1950s, the threesome produced a vaccine against the measles virus, which was improved and then produced commercially in 1963.

The virus culture technique developed by Enders and his co-workers enabled virus material to be produced in sufficient quantity for experimental work. The use of this technique meant that viruses could be more readily isolated and identified.

Erasistratus lived *c.*250 BC, was a Greek physician and anatomist regarded as the father of physiology. He came close to discovering the true function of several important systems of the body which were not fully understood until nearly 1,000 years later when physiologists had access to far more advanced methods of experimentation and dissection.

Erasistratus was born on the island of Ceos (now the Aegean island of Khios). He learnt his skills in Athens and became court physician to Seleucus I, who governed western Asia. He then moved on to Alexandria where he taught and advanced some of the work of the Greek anatomist Herophilus. But the Egyptians, among whom he worked, were morally against the use of cadavers for dissection, so that after Erasistratus, this type of anatomical research ceased until well into the thirteenth century.

Erasistratus dissected and examined the human brain, noting the convolutions of the outer surface, and observed that the organ is divided into larger and smaller portions (the cerebrum and cerebellum). He compared the human brain with those of other animals and made the correct hypothesis that the surface area/volume complexity is directly related to the intelligence of the animal. He traced the network of veins, arteries and nerves and realized the topographical associations, but his conclusions ran too closely along the lines of popular opinion. He postulated that the nerves carry the 'animal' spirit, the arteries the 'vital' spirit, and the veins blood. He did, however, grasp a rudimentary principle of oxygen exchange, noting that air was taken from the lungs to the heart where it became vital spirit for distribution via the arteries (as vital spirit) to the brain and then via the nerves to the body as animal spirit. (If one reads vital spirit as haemoglobin, he was not far

wrong.) He also put forward the idea of capillaries, explaining that the reason an artery bleeds when cut is because, as the vital spirit flows out, the blood rushes in from the veins through the capillaries to replace the vacuum created. He described the valves in the heart and condemned bloodletting as a form of treatment.

Erasistratus came near to discovering the principle of blood circulation (although he had it circulating in the wrong direction), but this mystery was not to be finally unravelled until Harvey's discoveries of the seventeenth century.

Eratosthenes *c.*276–*c.*194 BC was a Greek scholar and polymath, many of whose writings have been lost, although it is known that they included papers on geography, mathematics, philosophy, chronology and literature.

The son of Aglaos, Eratosthenes was born in Cyrene (now known as Shahhat, part of Libya). He underwent the equivalent of a university education in Athens before being invited, at the age of 30, by Ptolemy III Euergetes to become tutor to his son and to work in the library of the famous museum at Alexandria. On the death of Zenodotus in 240 BC he became the museum's Chief Librarian.

No single complete work of Eratosthenes, a premier scholar of his time, survives. The most important that remains is on geography – a word that he virtually coined as the title of his three-volume study of the Earth (as much as he knew of it) and its measurement. The work was concerned with the whole of the known world and divided the Earth into zones and surface features; parallels and meridian lines were used as a basis for establishing distances between places. It was accepted as the definitive work of its time (although it was criticized by Hipparchus for not making sufficient use of astronomical data). Eratosthenes greatly improved upon the inaccurate Ionian map.

The base line was a parallel running from Gibraltar through the middle of the Mediterranean and Rhodes to the Taurus Mountains in modern Turkey, onward to the Elburz range, the Hindu Kush and to the Himalayas. At right-angles to this line was a meridian passing through Heroe, Syene (now Aswan), Alexandria, Rhodes and the mouth of the River Borysthenes (now Dniepr). The data available were mostly the notes and records of travellers and their estimates of days in transit, although some data about the height and angle of the Sun at Meroe, Alexandria and Marseilles had been collected.

Eratosthenes' measurements of the height of the Sun at Alexandria used the fact that Syene was on the Tropic of Cancer (where at midday on the summer solstice a vertical post casts no shadow, for the Sun is directly overhead). Alexandria and Syene were on the same meridian. Eratosthenes' measurements were made at midday, but with a thin pillar in the centre of a hemispherical bowl. He estimated that the shadow was 1/25 of the hemisphere and so was 1/50 of the whole circle. Since the rays of the Sun can be considered to be striking any point on the Earth's surface in parallel lines, by using alternate angles and incorporating the known distance between Alexandria and

Syene of 5,000 stades, Eratosthenes was able to calculate that the total circumference of the Earth was just over 250,000 stades. There are several errors in his assumptions, but as the first attempt it was given much credit. The conversion of stades into modern units creates additional error, although a value of 46,500 km/28,900 mi is usually quoted. (The modern figure is usually accepted as 40,075 km/24,903 mi at the Equator.)

Eratosthenes also divided the Earth into five zones; two frigid zones around each pole with a radius of 25,200 stades on the meridian circle; two temperate zones between the polar zones and the tropics, with a radius of 21,000 stades; and a torrid zone comprising the two areas from the equator to each tropic, having a radius of 16,800 stades. The frigid zones he described as the 'Arctic' and 'Antarctic' circles of an observer on the major parallel of latitude (approximately 36°N); these circles mark the limits of the circumpolar stars which never rise or set. Eratosthenes' model of the known world had a north–south length of 38,000 stades from the Cinnamon country to Thale, and an east-west length of 77,800 stades from Eastern India to the Straits of Gibraltar.

Eratosthenes was familiar with the Earth and the apparent movement of the Sun, but he did little serious astronomical study. However, he also estimated the obliquity of the ecliptic as 11/83 of a circle, equivalent to 23°51′.

In the mathematical area he was most successful in offering a solution to the famous Delian problem of doubling the cube. For his proof, Eratosthenes proposed an apparatus consisting of a framework of two parallel rulers with grooves along which could be slid three rectangular plates capable of moving independently of each other and able to overlap. In arithmetic he devised a technique called the Sieve for finding prime numbers.

Eratosthenes also spent a considerable period of his life establishing the dates of historical events. In his two major works *Chronography* and *Olympic Victors*, many of the dates he set for events have been accepted by later historians and have never been changed. (For example, the Fall of Troy was in 1184–1183 BC, and the First Olympiad took place in 777–776 BC.) He also wrote many books on literary criticism in a series entitled *On the Old Comedy*. Like much of his output it is referred to by contemporary and later scholars, but it did not survive the passage of time.

Ericsson John 1803–1889 was a Swedish-born US engineer and inventor who is best known for his work on naval vessels.

Ericsson was born on 31 July 1803 at Langban Shyttan in Varmland, Sweden, and between the ages of 13 and 17 served as a draughtsman in the Gotha Canal Works. He was then commissioned into the Swedish Army, where he did map surveys. In 1826, he moved to London to seek sponsorship for a new type of heat engine he had invented (which used the expansion of superheated air as the driving force). This forerunner of the gas turbine was not successful, and Ericsson turned his attention to steam engines.

In 1829, he built the *Novelty*, a steam locomotive which competed unsuccessfully against Stephenson's *Rocket* at the Rainhill Trials for adoption on the Liverpool and Manchester Railway. In 1839, Captain Stockton of the United States Navy placed an order for Ericsson to supply a small iron vessel fitted with steam engines and a screw propellor. The vessel was built and sailed to New York, Ericsson himself sailed out a few months later. He became a US citizen in 1848.

In 1851, he resumed his interest in the heat or 'caloric' engine. It was found to be too heavy for the ship he built for it (immodestly called the *Ericsson*), making the vessel too slow. Only towards the end of his life did Ericsson construct small, efficient engines of this type.

A more successful line of development resulted from his work on the helical screw propellor, an interest he shared with his contemporary, Isambard Kingdom Brunel. Realizing that the paddle steamer was incapable of further development, Ericsson had built two small screw-driven ships in Britain in 1837 and 1839. In 1849, he built the *Princeton*, the first metal-hulled, screw-propelled warship and the first to have its engines below the waterline for added protection.

It was the outbreak of the American Civil War in 1861, which finally gave Ericsson the opportunity to demonstrate his skill as a naval engineer. His turreted iron clad ship, the *Monitor*, was first offered to Napoleon III and only after he refused it did it go to Ericsson's adopted country. Equipped with a low freeboard and heavy guns, it was the first warship to have revolving gun turrets – a practice soon to be adopted universally. In the Battle of Hampton Roads in 1862, the *Monitor* defeated the Confederate ship *Morrimack*.

After the Civil War, Ericsson continued to design warships, torpedoes and a 36-cm/14-in naval gun, but he also devoted time to more peaceful pursuits. Among his inventions were an apparatus for extracting salt from sea water, fans for forced draught and ventilation, a shipboard depthfinder, a steam fire engine and surface condensers for marine engines. Between the years 1877 and 1885, he also explored the possiblity of using solar energy and gravitation and tidal forces as sources of power.

Not always successful in the realization of his ideas, Ericsson was a wilful and impetuous man, often far ahead of other engineers of the day. He died on 8 March 1889 in New York City, and in 1889 the VSS *Baltimore* took his body back to Sweden, in accordance with his last wishes.

Erlang Agner Krarup 1878–1929 was a Danish mathematician who, although he was extremely knowledgeable in many fields, might never have become famous if he had not become scientific adviser – and leader – of the Copenhagen Telephone Company's research laboratory. His application of the theory of probabilities to problems connected with telephone traffic made his name known all over the world, and the 'erlang' is now the unit of traffic flow. A meticulous mathematician, he published many influential papers and was also responsible for constructing a device to measure alternating electric current.

Erlang was born on 1 January 1878 in the village of Lonborg, near Tarm, in Jylland (Jutland), the son of a schoolmaster. Completing his education he studied mathematics and natural sciences at the University of Copenhagen, obtaining his MA in mathematics in 1901 (with astronomy, physics and chemistry as secondary subjects). On leaving the University he worked as a teacher in various schools. During this time he won the award for solving the mathematical prize problem set by the University of Copenhagen which, in that year, was concerned with Huygens's solution of infinitesimal problems. In 1908, at the age of 30, Erlang was appointed scientific collaborator and leader of the Laboratory of the Copenhagen Telephone Company, where he remained for the rest of his life. He died suddenly on 3 February 1929, at the age of only 51. Single all his life, he devoted almost all his time to scientific study. He had a large library and collected mathematical, physical and astronomical works in particular; his knowledge in these subjects was extensive, but he was also well versed in philosophy, history and poetry. A modest man with an original mind, he was of an extremely kind and friendly disposition.

At the Laboratory of the Copenhagen Telephone Company, Erlang came under the influence of Franz Johanssen, the Managing Director of the company and another mathematician who had himself, in the year before Erlang joined, published two short essays in which he dealt with problems of telephone traffic flow – such as congestion and waiting time – and in which he introduced probability calculations. Erlang took to the work immediately. Within a year he had published his first paper on the subject, in which he was able to arrive at an exact solution to another problem posed by Johanssen previously. And over the next few years, Erlang published a number of other papers on the theory of telephone traffic which, because of their meticulous precision, became pioneer works.

It is rare for a telephone caller to get the engaged tone for any reason other than that the receiver at the call's destination is already in use. Yet it is possible through the links of the connection – the many coding selectors and digit selectors – at any one time for the call to find a selector already in use, which would automatically trip in the engaged tone. The fact that this is so rare is a result of the provision of switches in numbers based on the calculation that not more than one call in 500 at each selecting stage will fail because the equipment is engaged. To fulfil this standard, rules determining the amount of traffic to be carried per selector have been established by every telephone authority and company, following Erlang's formulae. Especially important are his formula for the probability of barred access in busy-signal systems – the so-called B-formula – and his formulae for the probability of delay and for the mean waiting time in waiting time systems. These formulae may be considered the most important within the theory of telephone traffic.

The erlang is the unit of telephone traffic flow; it is defined as 'the number of calls originated during a period, multiplied by the average holding time of a call, expressed in terms of the period: one erlang is therefore equivalent to the traffic flow in one circuit continuously occupied'.

Erlang also published studies of other mathematical problems. His work on logarithms and other numerical tables, in which he attempted to reduce the mean error to the lowest figure possible, resulted in the compilation of four- and five-figure tables that are now considered among the best available.

As the leader of the Laboratory of the Telephone Company, Erlang also investigated several assorted physiotechnical problems that cropped up. In particular, he constructed a measuring bridge to meter alternating current (the so-called Erlang complex compensator) which was a considerable improvement on earlier apparatus of similar function. Of equal significance were his investigations into telephone transformers and telephone cable theory.

Esaki Leo 1925– is a Japanese physicist who shared the 1973 Nobel prize for his discovery of tunnelling in semiconductor diodes. These devices are now called Esaki diodes. Esaki spent most of his working life in the United States but recently returned to Japan as President of the University of Tsukuba.

Esaki was born on 12 March 1925 in Osaka. When he graduated from the University of Tokyo in 1947, Esaki wanted to become a nuclear physicist but Japan lacked the particle accelerators needed to compete in this field and he moved to solid-state physics. Working for Sony in 1957, Esaki noticed whilst studying very small heavily doped germanium diodes that sometimes, and unexpectedly, the resistance decreased as current increased. This was caused by 'tunnelling' – a quantum mechanical effect whereby electrons can travel (tunnel) through electrostatic potentials that they would be unable to overcome classically. These barriers have to be very thin for tunnelling to occur and Esaki was able to use this effect for switching and to build ultra-small and ultra-fast tunnel diodes. In 1959, Esaki received a PhD from Tokyo for this discovery. It also earned Esaki the 1973 Nobel Prize for Physics which he shared with Brian Josephson, who predicted supercurrents in superconducting diodes, and Ivar Giaever who discovered electron tunnelling in superconductors. In 1960, Esaki joined IBM's Thomas J Watson Research Center in Yorktown Heights, New York, where he became an IBM fellow, the company's highest research honour, in 1967. He continued to research the nonlinear transport and optical properties of semiconductors, in particular multilayer superlattice structures grown by molecular beam epitaxy techniques. In 1992, Esaki returned to Japan to become president of Tsukuba University. The return of the nation's only living physics Nobel laureate was front-page news in Japan.

Eskola Pentti Eelis 1883–1964 was a Finnish geologist important in the field of petrology.

Born in Lellainen, Finland, the son of a farmer, Eskola was educated as a chemist at the University of Helsinki, before specializing in petrology. Throughout his life he was

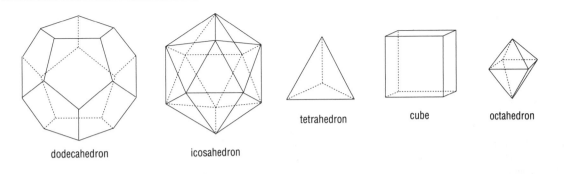

dodecahedron icosahedron tetrahedron cube octahedron

Euclid *The Platonic solids are the only regular convex polyhedra. A tetrahedron is made up of four equilateral triangular faces, a cube is six squares, an octahedron is eight equilateral triangles, a dodecahedron is twelve regular pentagons and an icosahedron is twenty equilateral triangles. They are the basis for Euclidean solid geometry.*

fascinated by the study of metamorphic rocks, taking early interest in the Precambrian rocks of England.

Fascinated by the mineral facies of rocks, Eskola was one of the first to apply physicochemical postulates on a far-reaching basis to the study of metamorphism, thereby laying the foundations of most subsequent studies in metamorphic petrology. Building largely on Scandinavian studies, Eskola was concerned to define the changing pressure and temperature conditions under which metamorphic rocks were formed. His approached enabled comparison of rocks of widely differing compositions in respect of the pressure and temperature under which they had originated.

Euclid lived *c.*300 BC was an ancient Greek mathematician whose works, and the style in which they were presented, formed the basis for all mathematical thought and expression for the following 2,000 years (although they were not entirely without fault). He also wrote books on other scientific topics, but these have survived the passage of the centuries only fragmentarily or not at all.

Very little indeed is known about Euclid. No record is preserved of his date or place of birth, his education, or even his date or place of death. The influence of Plato (*c.*427–*c.*347 BC) is certainly detectable in his work – so Euclid must either have been contemporary or later. Some commentators have suggested that he attended Plato's Academy in Athens but, if so, it is likely to have been after Plato's death. In any case, it has been established that Euclid went to the recently founded city of Alexandria (now in Egypt) in around 300 BC and set up his own school of mathematics there. Fifty years later, however, Euclid's disciple Apollonius of Perga was said to have been leading the school for some considerable time; it seems very possible, therefore, that Euclid died in around 270 BC.

Euclid's mathematical works survived in almost complete form because they were translated first into Arabic, then into Latin; from both of these they were then translated into other European languages. He used two main styles of presentation: the synthetic (in which one proceeds from the known to the unknown via logical steps) and the analytical (in which one posits the unknown and works towards it from the known, again via logical steps). In his

major work, *The Elements*, Euclid used the synthetic approach, which suited the subject matter so perfectly that the method became the standard procedure for scientific investigation and exposition for millennia afterwards. The strictly logical arrangement demanding the absolute minimum of assumption, and the omission of all superfluous material, is one of the great strengths of *The Elements*, in which Euclid incorporated and developed the work of previous mathematicians as well as including his own many innovations. The presentation was one of extreme clarity and he was rigorous, too, about the actual detail of the mathematical work, attempting to provide proofs for every one of the theorems.

The Elements is divided into 13 books. The first six deal with plane geometry (points, lines, triangles, squares, parallelograms, circles, and so on), and includes hypotheses such as 'Pythagoras' theorem' which Euclid generalized. Books 7 to 9 are concerned with arithmetic and number theory. In Book 10 Euclid treats irrational numbers. And Books 11 to 13 discuss solid geometry, ending with the five 'Platonic solids' (the tetrahedron, cube, octahedron, icosahedron and dodecahedron).

Euclid favoured the analytical mode of presentation in writing his other important mathematical work, the *Treasury of Analysis*. This comprised three parts, now known as *The Data*, *On Divisions of Figures* and *Porisms*.

Euclid's geometry formed the basis for mathematical study during the next 2,000 years. It was not until the nineteenth century that a different form of geometry was even considered: 'accidentally' discovered by Saccheri in 1733, non-Euclidean geometry was not in any way defined until Nikolai Lobachevsky (in the 1820s), János Bolyai (in the 1830s) and Bernhard Riemann (in the 1850s) examined the subject. It is difficult to see, therefore, how Euclid's contribution to the science of mathematics could have been more fundamental than it was.

Eudoxus of Cnidus *c.*406–*c.*355 BC was a Greek mathematician and astronomer who is said to have studied under Plato. Himself a great influence on contemporary scientific thought, Eudoxus was the author of several important works. Many of his theories have survived the test of centuries; work attributed to Eudoxus includes methods to calculate the area of a circle and to derive the volume of a pyramid or a cone. He also devised a system to demonstrate the motion of the known planets when viewed from the Earth.

Very little is known about Eudoxus' life, although it is recorded that he spent more than a year in Egypt, some of the time as the guest of the priests of Heliopolis. In a series of geographical books with the overall title of *A Tour of the Earth* he later described the political, historical and religious customs of the countries of the eastern Mediterranean area.

Primarily a mathematician, Eudoxus used his mathematical knowledge to construct a model of homocentric rotating spheres to explain the motion of planets as viewed from the Earth, which was at the centre of the system. The model was later extended by Aristotle (384–322 BC) and Callippus (c.370–c.300 BC), and although superseded by the theory of epicycles, it was still widely accepted during the Middle Ages. Eudoxus was able to give close approximations for the synodic periods of the planets Saturn, Jupiter, Mars, Venus and Mercury. The geometry of the model was impressive, but there are several weaknesses. Eudoxus assumed, for example, that each planet remains at a constant distance from the centre of its orbital circle, and in addition that each retrograde loop as seen from the Earth is identical with the last. Neither assumption complies with observation.

The model of planetary motion was published in a book called *On Rates*. Further astronomical observations were included in two other works, *The Mirror* and *Phenomena*. Subject later to considerable criticism by Hipparchus (lived 2nd century BC), these books nevertheless established patterns where before none existed.

In mathematics Eudoxus' early success was in the removal of many of the limitations imposed by Pythagoras on the theory of proportion. Eudoxus showed that the theory was applicable in many more circumstances. Subsequently he established a test for the equality of two ratios, and noted that it was possible to find a good approximate value for the area of a circle by the 'method of exhaustion': a polygon is drawn within the circle, the number of its sides is repeatedly doubled, and the area of each new polygon is found by using a simple formula. It was to Eudoxus that Archimedes attributed the discovery that the volume of a pyramid or a cone is equal to one-third of the area of the base times the perpendicular height.

Although none of Eudoxus' works has survived the passage of time in complete form, the mathematical skills he practised were important and influential both in his own age and for centuries afterwards.

Euler Leonhard (or Leonard) 1707–1783 was a Swiss mathematician whose power of mental calculation was prodigious; so great was his capacity for concentrating on mathematical computation that even when he became totally blind towards the end of his life, he was able to continue his work without pause. With such ability and with

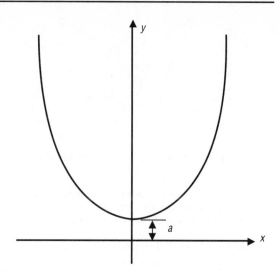

Euler *A catenary is a transcendental curve that may be represented by the equation $y = (a/2)(e^{x/a} + e^{-x/a})$, where a is a constant and e is Euler's number (2.718 ...).*

true scientific curiosity, Euler – a brilliant teacher – expanded the scope of virtually all the known branches of mathematics, devising and formulating a considerable number of theorems and rules now named after him. He also enlarged mathematical notation.

Euler was born on 15 April 1707 in Basle, where he grew up and was educated. At the University of Basle he studied under Jean Bernoulli, obtaining his master's degree at the age of 16, in 1723. He then found it impossible – perhaps because he was so young – to gain a faculty position. Four years later, however, he was invited by his great friend and contemporary Daniel Bernoulli (nephew of Jean) to join him in Russia, at St Petersburg. Euler duly arrived there, spent three years at the Naval College, and then (in 1730) was appointed Professor of Physics at the Academy of Sciences. When Bernoulli returned home in 1733, Euler succeeded him as Professor of Mathematics. Shortly afterwards, through looking at the Sun during his astronomical studies, he lost the sight of his right eye. In 1741, he travelled to Berlin at the request of the Emperor, Frederick the Great, and in 1744 became Director of the Berlin Academy of Sciences. He remained there until 1766 when the Empress Catherine the Great recalled him to St Petersburg to become Director of the Academy of Sciences there. Soon after his return he lost the sight of his other eye, through cataracts, but retained his office, competently carrying out all his duties and responsibilities for another 15 years or more, until he died on 18 September 1783.

Euler contributed to all the classic areas of mathematics, although his most innovatory work was in the field of analysis, in which he considerably improved mathematical methodology and the rigour of presentation. He advanced the study of trigonometry, in particular developing spherical trigonometry. Following the analytical work of his friend Daniel Bernoulli, Euler demonstrated the significance of the coefficients of trigonometric expansions;

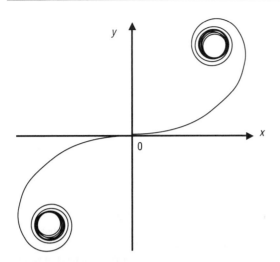

Euler *Euler's spiral is sometimes alternatively called a clothoid.*

Euler's number (e, as it is now called) has various useful theoretical properties and is also used in the summation of particular series.

Euler also studied algebraic series and demonstrated the importance of convergence. He applied algebraic methods instead of geometric ones (as used by Newton, Galileo and Kepler previously), and improved differential and integral calculus, bringing them virtually to their modern forms. He used constant coefficients in the integration of linear differential equations and originated the calculus of partial differentials. And he became very interested in applying mathematical – and particularly analytical – principles to mechanics, especially celestial mechanics.

After his return to Russia in 1766 Euler carried out further research into the motion and positions of the Moon, and the gravitational relationships between the Moon, the Sun and the Earth. His resulting work on tidal fluctuations took him into the realm of fluid mechanics, in which he successfully analysed motion in a perfectly compressible fluid. Further astronomical work brought him an award of £300 from the British government for his development of theorems useful in navigation (he had been a Fellow of the Royal Society since 1746).

A prolific author of influential papers detailing his methods and his results, and a teacher of outstanding authority, Euler made a number of innovations both in mathematical concept – such as Euler's constant, Euler's equations, Euler's line and Euler's variables – and in mathematical notation – he was responsible among other things, for the use of π, e in natural (Naperian) logarithms, i for imaginary numbers, and Σ for summation. Other interests of his included acoustics and optics.

Evans Alice Catherine 1881–1975 was a US microbiologist whose research into the bacterial contamination of milk led to the recognition of the danger of unpasteurized milk. As a result of her research the incidence of brucellosis was greatly reduced when the dairy industry accepted that all milk should be pasteurized.

Alice Evans was born in Neath, Pennsylvania on 29 January 1881. Leaving Susquehanna Collegiate Institute in 1901, she taught in a school for four years, before going to Ithaca, New York on a two-year nature-study course for rural teachers organized by Cornell University. During this time she became interested in science and continued her course, graduating with a BS degree in 1909; she was awarded an MS degree in 1910 by the University of Wisconsin, after which she took a research post at the Dairy Division, Bureau of Animal Industry of the United States Department of Agriculture studying the bacteriology of milk and cheese. In 1913, she transferred to the new laboratories in Washington, DC, where she researched the bacterial contamination of milk products. In 1918, Evans moved to the Hygienic Laboratories of the United States Public Health Service as an assistant biologist to research into epidemic meningitis and influenza; she returned to the study of milk flora at the end of World War I. Evans retired in 1945 and moved to Arlington, Virginia, in 1969 where she lived until her death in 1975 following a stroke.

Evans's renown as a microbiologist resulted from her discovery of a common origin for the disease brucellosis in humans and cattle, previously thought to be two separate diseases. Her research into the Bruce bacillus (affecting humans) and the Bang Bacillus (affecting cattle) showed that for practical purposes, these could be regarded as identical and either may cause the disease 'undulant fever'. Since humans contract the disease by handling infected animals or through drinking their milk, Evans warned of the dangers of drinking unpasteurized milk. Her results were published in the *Journal of Infectious Diseases* in 1918 and were at first ignored in US medical circles. They were, however, later confirmed by other scientists and by the late 1920s numerous cases of human brucellosis had been reported and the disease was recognized as a major threat, not only to those in close contact with animals but also through the food industry. By the 1930s the dairy industry accepted that all milk should be pasteurized.

Evans Oliver 1755–1819 was a US engineer who developed high-pressure steam engines and various machines powered by them. He also pioneered production-line techniques in manufacturing.

Evans was born on 13 September in 1755 in Newport, Delaware, and at the age of 16 he was apprenticed to a wagon-maker. He read books on mathematics and mechanics, and became interested in steam engines. His first invention was a machine for cutting and mounting wire teeth in a leather backing to make devices for carding textile fibres before spinning; his machine produced 1,500 cards a minute. In 1780, he joined his two brothers at a flour mill at Wilmington, where he helped to build the machinery that used water power to drive conveyors and elevators. As a result, he attained a degree of automation that allowed one man to operate the whole mill as a single production line. Other millers remained unimpressed by his invention.

Evans then moved to Philadelphia, where he spent more than ten years trying to develop a steam carriage. For some time he tried to couple one of James Watt's steam engines to a wagon and as early as 1792 he was working on internally fired boilers for steam engines. But he had to abandon the enterprise because of lack of financial support, and eventually he turned his attention to the manufacture of stationary steam engines.

In 1786 and 1787 he successfully petitioned the legislatures of Pennsylvania and Maryland for exclusive patent rights to profit from his inventions, and by 1802 he had developed a high-pressure steam engine with a 15-cm/6-in cylinder, 45-cm/18-in stroke and a 2.3-m/7.5-ft flywheel. It used a 'grasshopper' beam mechanism and had an internally fired boiler and provision for exhausting spent steam into the air. It worked at a pressure of 0.008 Pa/50 psi and ran at 30 revolutions per minute.

Two years later, Evans built a steam dredger for use on the Schuylkill river. It had power-driven rollers as well as a paddle so that it could be moved on land under its own power. This amphibious machine, the *Orukter Amphibole*, was the forerunner of some 50 or so steam engines which he built in the next 15 years.

In 1806, Evans began to develop the Mars works for the manufacture of steam engines. In 1817, he completed his last work, a 17 kW/24 hp engine for a waterworks. On 11 April 1819, his principal workshop was destroyed in a fire started by a grudge-bearing apprentice. The consequent shock to Evans probably hastened his death in New York City a few days later on 15 April 1819.

Like many engineers Evans died with many bold dreams unrealized. In March 1815 he claimed, in the *National Intelligencer* (although many seriously question the claim), that he could have introduced steam carriages as early as 1773 and steam paddle boats by 1778. Less controversially, Evans published the *Young Millwright* and *The Miller's Guide* in 1792.

Ewing William Maurice 1906–1974 was a major innovator in modern geology. Born in Texas, the son of a merchant, Ewing studied at the Rice Institute in Houston from 1923 and developed his geological interests by working for various oil companies. He moved to work at Lehigh University, Pennsylvania in 1929 (first in physics, then in geology), and in 1944, joined the Lamont–Doherty Geological Observatory, New York, the world's foremost geophysical research institution. Using marine 'sound-fixing' seismic techniques (SOFAR – 'sound fixing and ranging'), and pioneering deep-ocean photography and sampling, he ascertained that the crust of the Earth under the ocean is much thinner (5–8 km/3–5 mi thick) than the continental shell (about 40 km/25 mi thick). Ewing also demonstrated that mid-ocean ridges were common to all oceans, and in 1957 further discovered the presence within them of deep central canyons. His studies of ocean sediment showed that its depth increases with distance from the mid-ocean ridge, which gave clear support for the sea-floor spreading hypothesis proposed by Harry Hess (1906–

1969) in 1962. Mainly through his notions of oceanic rift valleys, Ewing provided several of the crucial jigsaw pieces for the plate tectonics revolution in geology in the 1960s.

From 1947 till his retirement, Ewing was Professor of Geology at Columbia University, while also holding a position at the Woods Hole Oceanographic Institute.

Eyde Samuel 1866–1940 was a Norwegian industrial chemist who helped to develop a commercial process for the manufacture of nitric acid, which made use of comparatively cheap hydroelectricity. He was a member of the Norwegian Parliament, and in 1920 Norwegian Minister in Poland.

Eyde was born at Arendal, Norway and received his higher education in Germany at the Charlottenburg High School in Berlin, where he gained a diploma in constructural engineering. He then worked as an engineer in various German cities, principally Hamburg. In partnership with the German engineer C. O. Gleim, he returned to Scandinavia to work on the construction of various railway and harbour installations. But increasingly his interest turned to the possibiltity of developing the electrochemical industry in his native Norway, where hydroelectric schemes were beginning to make cheap electrical energy available.

In 1901, while studying the problem of the fixation of nitrogen (the conversion of atmospheric nitrogen into chemically useful compounds), he met his compatriot Christian Birkeland (1867–1917). The two men set up a small laboratory, where they combined their efforts to discover the conditions necessary for the economic combination of nitrogen and oxygen (from air) in an electric arc to produce nitrogen oxides and, eventually, nitric oxide. Their method, known as the Birkeland-Eyde process, can be summarized by the following chemical equations:

$$N_2 + O_2 \rightarrow 2NO$$

(nitrogen and oxygen combine to give nitric oxide)

$$2NO + O_2 \rightarrow 2NO_2$$

(nitric oxide combines with oxygen to form nitrogen dioxide)

$$4NO_2 + O_2 + 2H_2O \rightarrow 4HNO_3$$

(oxidation of nitriogen dioxide in the presence of water produces nitric acid).

Their experiments resulted in the first commercial success in this area of research. The key to the process was an oscillating disc-shaped electric arc, produced by applying a powerful magnetic field to an arc formed between two metal electrodes by an alternating current. The electrodes were copper tubes cooled by water circulating inside them. In 1903, the Norwegische Elektrishe Aktiengeswllschaft constructed a small-scale plant at Ankerlökken, near Oslo. Two years later full-scale operation began at Notodden. Then interest in the work was shown by the company of Badische und Anilin Soda Fabrik, which, following on from German research on the thermodynamic equilibrium of nitrogen oxides, had employed O. Schonherr to study the nitrogen–oxygen reaction in 1897. In 1904, he patented a method of producing a steadier arc than had been attained by Birkeland and Eyde.

By this time Eyde had obtained the hydroelectric rights on some waterfalls, and in 1904 he became administrative director of an electrochemical company, financed partly from a Swedish source which supported the Birkeland–Eyde process. In the following year, he obtained extensive support from French financiers to found a hydroelectric company, which he directed with great skill until he retired from active participation in 1917.

Eyde died at Aasgaardstrand in Norway in 1940, shortly after writing his autobiography.

Eysenck Hans Jurgen 1916–   is probably Britain's best-known psychologist, renowned for his controversial theories about a wide range of subjects, particularly human intelligence.

Eysenck was born in Germany on 4 March 1916 and educated at various schools in Germany, France and Britain. With the rise to power of Adolf Hitler in the 1930s he left Germany and went to Britain. He studied psychology at London University, graduating in 1938 and gaining his doctorate in 1940. During the rest of World War II he was Senior Research Fellow Psychologist at the Mill Hill Emergency Hospital, London, and in 1946 he was appointed Director of the Psychology Department at the Maudsley Hospital, Surrey. He went to the United States in 1949 as Visiting Professor at the University of Pennsylvania, and on his return to Britain in 1950 became Reader in Psychology at London University's Institute of Psychiatry. In 1954, he again went to the United States, as Visiting Professor at the University of California. In the following year he was appointed Professor of Psychology at London University's Institute of Psychiatry.

Eysenck has investigated many areas of psychology, often producing highly controversial theories as a result of his studies. But it is his theory that intelligence is almost entirely inherited and can be only slightly modified by education that has aroused the greatest opposition. The concept of intelligence is difficult to define and even more difficult to measure – the commonly used intelligence tests, for example, are often criticized for being culturally biased, favouring well-educated white people and penalizing poorly educated black people. Eysenck has attempted to devise a fairer, culture-free method for assessing intelligence. It involves neither problem-solving nor even conscious thought and therefore, Eysenck argues, it cannot be criticized for being culturally biased. Basically his method involves subjecting a person to stroboscopic light flashes and buzzing sounds while simultaneously recording the electrical activity of the person's brain by means of an electroencephalograph. It has been known since the early 1970s that the pattern of brain waves is related to intelligence (as measured by conventional intelligence tests) but the correlation has been too approximate to be of practical use. Eysenck, however, claims to be able to measure the brain waves accurately enough to give a very high correlation with conventional intelligence tests which, if true, would mean that his method of measuring intelligence is as valid and useful as conventional tests. Using his method, Eysenck then did a cross-cultural comparison of intelligence and found that, on average, black people obtained significantly lower intelligence quotients than whites. Combining this finding with his theory that intelligence is predominantly inherited, he claimed that black people are inherently less intelligent than are whites. This claim met with great – sometimes violent – opposition and was widely criticized by educationalists and other psychologists, who opined that, because Eysenck had validated the results obtained by electroencephalography with those from conventional intelligence tests the two methods must contain the same biases and therefore Eysenck's method could not be considered culturally unbiased.

Eysenck has also studied personality traits, anxiety and neurosis, the influence of violence shown on television on behaviour, and the psychology of smoking. In addition, he has written many popular psychology books, in which he presents his contentious ideas in relatively simple, non-technical language.

F

Fabre Jean Henri 1823–1915 was a French entomologist whose studies of insects, particularly their anatomy and behaviour, have become classics.

Fabre was born on 22 December 1823 in Saint-Léons in southern France, the son of a farmer who had left the land and set up a small business. Fabre's family was poor and for much of his early childhood he lived with his grandmother in the country, which fostered his interest in natural history. When he was seven years old he returned to Saint-Léons to attend the village school. Later, after passing through senior school, he won a scholarship to Avignon, from which he gained his certificate of education in 1842. While at school Fabre had to take part-time jobs to help to pay for his education and contribute to his family's income and so, on obtaining his certificate he immediately took a teaching post at the lycée in Carpentras, a small town in northeastern Avignon. After further studies at Montpellier, he gained his teaching licence (which enabled him to teach in higher schools) in mathematics and physics and in 1851 was appointed as physics teacher at a lycée in Ajaccio, Corsica. But soon afterwards he contracted a fever, which forced him to resign and return to the mainland. After he had recovered, he went to Paris to gain a degree, then he returned to Avignon where, in 1852, he became Professor of Physics and Chemistry at the Lycée. He held this post for 20 years, eventually resigning because the authorities would not allow girls to attend his science classes. He then decided to abandon his teaching career and moved to the village of Orange, where he embarked on a serious study of entomology. He was very poor and had virtually no equipment to help him in his studies, but by writing articles for scientific journals he eventually managed, in 1878, to buy a small plot of waste land in Serignan, Provence. He built a wall around the plot, treating it as an open-air laboratory, and remained there for the rest of his life, pursuing his entomological studies and writing about his findings. Towards the end of his life he became world famous as an authority on entomology, and in 1910 many leading scientific figures visited him in Serignan for a celebration given in his honour. Fabre died in Serignan five years later, on 11 October 1915. After his death, the French National Museum of Natural History purchased Fabre's plot of land as a memorial to the man and his work.

Although he had written several previous articles, Fabre's first important paper was published in 1855. It was a detailed account of the behaviour of a type of wasp that paralyses its prey (mainly beetles and weevils), which it then carries to its nest to feed to its young. In 1857, he wrote another important paper describing the life cycle of the Meloidae (oil beetles), hypermetamorphic beetles that begin life as larvae, then hatch into a second larval stage (called a triungulin) and climb onto particular types of flowers frequented by solitary bees of the genus *Anthophora*. When a bee visits the flower, the triungulin attaches itself to the bee and is carried to the bee's nest, where it passes through several other larval stages – feeding on honey – before finally developing into an adult beetle. The significance of this and other early work by Fabre was recognized by Charles Darwin, who quoted Fabre in *On the Origin of Species*, but Fabre himself did not accept the idea of evolution.

Fabre began his most famous work after he settled at Serignan where, in addition to writing numerous entomological papers, he embarked on his 10-volume *Souvenirs Entomologiques*, which took him 30 years to complete. Based almost entirely on observations Fabre made in his small plot, this work is a model of meticulous attention to detail. It became a classic entomological work and also did much to revitalize interest in entomology.

Fabricius ab Aquapendente Hieronymus (Latinized name of Girolamo Fabrizio) 1537–1619 was an Italian anatomist and embryologist who gave the first accurate description of the semilunar valves in the veins and whose pioneering studies of embryonic development helped to establish embryology as an independent discipline.

Fabricius was born on 20 May 1537 in Aquapendente, near Orvieto in Italy, and studied first the humanities then medicine at the University of Padua, where he was taught anatomy and surgery by the eminent Italian anatomist Gabriele Falloppio (1523–1562). After graduating in 1559, Fabricius worked privately as an anatomy teacher and surgeon until 1565, when he succeeded Falloppio as Professor of Surgery and Anatomy at Padua University (the chair had been vacant since Falloppio's death in 1562). Fabricius remained at Padua University for the rest of his career, eventually retiring because of ill health in 1613. During his time at Padua, Fabricius built up an international reputation that attracted students from many countries, including William Harvey (who studied under him from 1597 to 1602). He also helped to establish a permanent anatomical theatre at the university, a structure that still exists today. About 1596, Fabricius started to acquire an estate at Bugazzi where, after retiring, he remained until his death on 21 May 1619.

Fabricius's principal anatomical work was his accurate and detailed description of the valves in the veins. Although they had previously been observed and crudely drawn by other scientists, Fabricius publicly demonstrated them, in 1579, in the veins of the limbs and, in 1603 published the first accurate description – with detailed illustrations – of these valves in *De Venarum Ostiolis/On the Valves of the Veins*. He mistakenly believed, however, that the valves' function was to retard the flow of blood to enable the tissues to absorb nutriment.

Fabricius's most important completely original work was in embryology. In his treatise *De Formato Foetu/On the Formation of the Foetus* 1600 – the first work of its kind – he compared the late foetal stages of different animals and gave the first detailed description of the placenta. Continuing his embryological studies, Fabricius published in 1612, *De Formatione Ovi et Pulli/On the Development of the Egg and the Chick*, in which he gave a detailed, excellently illustrated account of the developmental stages of chick embryos. Again, however, Fabricius made some erroneous assumptions. He believed that the sperm did not enter the ovum, but stimulated the generative process from a distance. He also believed that both the yolk and the albumen nourished the embryo and that the embryo itself was produced from the spiral threads (chalaza) that maintain the position of the yolk. Nevertheless, his embryological studies were extremely influential and helped to establish embryology as an independent science.

Fabricius also investigated the mechanics of respiration, the action of muscles, the anatomy of the larynx (about which he was the first to give a full description) and the eye (he was the first to correctly describe the location of the lens and the first to demonstrate that the pupil changes size).

Fabry Charles 1867–1945 was a French physicist who specialized in optics. He is best remembered for his part in the invention, and then in the investigations of, the applications of the Fabry–Pérot interferometer and etalon.

Fabry was born in Marseilles on 11 June 1867. He and his two brothers all became prominent mathematicians or physicists. Fabry attended the Ecole Polytechnique in Paris from 1885 to 1889, and despite becoming interested in astronomy along with his brothers, his studies under Macé de Lépinay took him into the field of optics. After he had completed his degree at the Ecole Polytechnique, Fabry went on to take a doctorate in physics at the University of Paris in 1892. He then taught at lycées in Pau, Bordeaux, Marseilles, Nevers and Paris, before joining the Faculty of Science at the University of Marseilles in 1894.

The next 25 years saw Fabry concentrating on research and teaching in Marseilles, where he was elected to the Chair of Industrial Physics in 1904. The Ministry of Inventions recalled him to Paris in 1914 in order to investigate interference phenomena in light and sound waves, and in 1921 Fabry returned to Paris on a more permanent basis as Professor of Physics at the Sorbonne. He was later also made Professor of Physics at the Ecole Polytechnique, and the first Director of the Institute of Optics. In 1935, Fabry became a member of the International Committee on Weights and Measures at the Bureau de Longitudes, but he retired in 1937 and died in Paris on 11 December 1945.

Fabry's research was dedicated to devising methods for the accurate measurement of interference effects, and to the application of this technique to a broad range of scientific subjects. His interest in interference was apparent very early in his career, his doctoral thesis having been concerned with various theoretical aspects of interference fringes.

Alfred Pérot (1863–1925) and Fabry invented the interferometer which was named after them in 1896. It is based on multiple beam interference and consists of two flat and perfectly parallel plates of half-silvered glass or quartz. A light source produces rays which undergo a different number of reflections before being focused. When the rays are reunited on the focal plane of the instrument, they either interfere or cohere, producing dark and light bands respectively. When the light source is monochromatic, as for example with a laser or a mercury vapour lamp, the Fabry–Pérot interference fringes appear as sharp concentric rings. A nonmonochromatic source produces a more complex pattern of concentric rings. If the two reflecting plates are fixed in position relative to one another, the device is called a Fabry–Pérot etalon. If the distance between the plates can be varied, the apparatus is a Fabry–Pérot interferometer.

These devices can be used to distinguish different wavelengths of light from a single source, and are accurate to a resolving power of one million. The Fabry–Pérot interferometer has a resolving power 10 to 100 times greater than a prism or a small diffraction grating spectroscope, and is more accurate than the device used by Michelson and Morley in their classic experiment intended to detect the motion of the Earth through the hypothetical ether.

Fabry and Pérot worked together for a decade on the design and uses of their invention. One of their achievements was the setting of a series of standard wavelengths. Then in 1906, Fabry began to collaborate with Henri Buisson. By 1912, study of the spectra of neon, krypton and helium enabled them to confirm a broadening of lines predicted by the kinetic theory of gases. In 1914, they used the interferometer to confirm the Doppler effect for light, not with stellar sources as had until then been necessary, but in the laboratory.

Applications to astronomy were also explored, including the measurement of solar and stellar spectra, and photometry. An issue of practical importance in biology, medicine, astronomy and of course physics was the identification of the material responsible for the filtering out of ultraviolet radiation from the Sun. In 1913, Fabry was able to demonstrate that ozone is plentiful in the upper atmosphere and is responsible for this.

Fahrenheit Daniel Gabriel 1686–1736 was a Polish-born Dutch physicist who invented the first accurate thermometers and devised the Fahrenheit scale of temperature.

Fahrenheit was born at Danzig (now Gdansk), Poland, on 14 May 1686. He settled in Amsterdam in 1701 in order to learn a business, and became interested in the manufacture of scientific instruments. From about 1707, he spent his time wandering about Europe, meeting scientists and other instrument makers and gaining knowledge of his trade. In 1717, he set himself up as an instrument maker in Amsterdam, and remained in the Netherlands for the rest of his life.

Fahrenheit published his methods of making thermometers in the *Philosophical Transactions* in 1742 and was admitted to the Royal Society in the same year. He died at The Hague on 16 September 1736.

The first thermometers were constructed by Galileo Galilei (1564–1642) at the end of the sixteenth century, but no standard of thermometry had been decided upon even a century later. Galileo's was a gas thermometer, in which the expansion and contraction of a bulb of air raised or lowered a column of water. Because it took no account of the effect of atmospheric pressure on the water level, it was very inaccurate. Guillaume Amontons (1663–1705) improved the gas thermometer in 1695, by using mercury instead of water and having the mercury rise and fall in a closed column, thus avoiding effects due to atmospheric pressure.

Fahrenheit's first thermometers contained a column of alcohol which expanded and contracted directly, in the same way as modern thermometers. Fahrenheit came across this instrument in 1708 when he visited Claus Roemer (1644–1710), who had devised the thermometer in 1701. Roemer had developed a scale of temperature in which the upper fixed point, the boiling point of water, was 60° and the lower, the temperature of an ice–salt mixture, was 0°. Body temperature on this scale was $22\frac{1}{2}$° and the freezing point of water came to $7\frac{1}{2}$°.

Fahrenheit took up Roemer's ideas and combined them with those of Amontons in 1714, when he substituted mercury for alcohol and constructed the first mercury-in-glass thermometers. Fahrenheit found that mercury was more accurate because its rate of expansion, although less than that of alcohol, is more constant. Furthermore, mercury could be used over a much wider temperature range than alcohol.

In order to reflect the greater sensitivity of his thermometer, Fahrenheit expanded Roemer's scale so that blood heat was 90° and the ice–salt mixture was 0°; on this scale freezing point was 30°. Fahrenheit later adjusted the scale

to ignore body temperature as a fixed point so that the boiling point of water came to 212° and freezing point was 32°. This is the Fahrenheit scale that is still in use today.

Using his thermometer, Fahrenheit was able to determine the boiling points of liquids and found that they vary with atmospheric pressure. In producing a standard scale for the measurement of temperature as well as an accurate measuring instrument, Fahrenheit made a very substantial contribution to the advance of science.

Fairbairn William 1789–1874 was a Scottish engineer who designed a riveting machine which revolutionized the making of boilers for steam engines. He also worked on many bridges, including the wrought iron box-girder construction used first on the bridge across the Menai Straits in North Wales.

Fairbairn was born into a poor family on 19 February 1789, at Kelso. He received little early education, although he did learn to read at the local parish school. He started work when he was 14 years old when his family moved to a farm owned by the Percy Main colliery near Newcastle-upon-Tyne; Fairbairn became apprenticed to a millwright. He learned mathematics in his spare time and displayed his engineering ingenuity by constructing an orrery (a working model of the Solar System).

He finished his apprenticeship in 1811 having, in the meantime, become a friend of the engineers George and Robert Stephenson. He worked as a millwright at Bedlington, then took a series of jobs in London, Bath, Dublin and Manchester. During this time he invented a sausage-making machine and a machine for making nails. In Manchester, he worked on the construction of the Blackfriar's Bridge and then set up as a manufacturer of cotton-mill machinery. In 1824, Fairbairn erected two watermills in Zurich, and later turned his attention to ship-building and, finally, bridge-building.

In 1862, he invented a self-acting planing machine for dealing with work up to 6 m/20 ft by 1.8 m/6 ft. He became an authority on mechanical and engineering problems and received many honours and awards, including a baronetcy in 1869. Fairbairn died on 18 August 1874, at Moor Park in Surrey and was buried at Prestwick, Northumberland.

From very humble beginnings, Fairbairn used his inventive skills and engineering ability to earn a fortune by the time he was 40 years old. He had acquired a sound reputation for producing machinery for the cotton mills and by that time employed about 300 workmen. Furthermore, his reputation abroad had been enhanced when he solved the problem of an irregular water supply with his watermills in Switzerland.

In 1830, he was commissioned by the Forth of Clyde Company to build a light iron boat to run between Glasgow and Edinburgh. He then concentrated on ship-building, first in Manchester (where he built ships in sections) then, from 1835 in Millwall on the river Thames, where his Millwall Iron Works employed some 2,000 people.

Fairbairn returned to Manchester, and in 1844 designed and built the first Lancashire shellboiler. It was constructed of rolled wrought-iron plates rivetted together by a machine of his own design. It was this expertise that led Robert Stephenson to consult Fairbairn over the building of the Menai railway bridge, which was constructed of wrought iron plates. Built between 1846 and 1850, it was the longest railway bridge at the time with a continuous box girder (in which the trains ran) 461 m/1,511 ft long. Fairbairn's participation ended in 1849 after a misunderstanding about his position (he and Stephenson were both termed Superintendent), and he published his own account of the construction.

Fajans Kasimir 1887–1975 was a Polish-born US chemist, best known for his work on radioactivity and isotopes and for formulating rules that help to explain valence and chemical bonding.

Fajans was born in Warsaw on 27 May 1887. He was educated at Leipzig, Heidelberg (where he gained his PhD in 1909), Zurich and Manchester. From 1911 to 1917 he worked at the Technische Hochschule at Karlsruhe, and between 1917 and 1935 he held appointments at the Munich Institute of Physical Chemistry, where he rose from Assistant Professor to be the Director. In 1936, he emigrated to the United States and served as a professor at the University of Michigan, Ann Arbor. He became a US citizen in 1942. He died, aged 88, on 18 May 1975.

In 1913, Fajans formulated simultaneously with, but independently of, Frederick Soddy (1877–1956) the theory of isotopes – that is, atoms that have the same atomic number but different atomic weights (relative atomic masses). He showed that uranium-X_1 (itself a decay product by alpha-ray emission of uranium-238) disintegrates by beta-ray emission into uranium-X_2, which he called 'brevium' on account of its short half-life; this latter isotope then undergoes further beta-decay to form

uranium-234. An alpha particle is a helium nucleus ($^4_2$He), consisting of two protons and two neutrons; its loss from an element's nucleus results in a different element that is four mass units lighter and two less in atomic number. A beta particle is an electron so that its emission, on the other hand, results in an element of the same mass number but with an atomic number larger by 1. Thus the decay of uranium-238 by emitting first an alpha particle and then two beta particles should produce a new uranium isotope four mass units lighter: uranium-234. Fajans and Soddy were the first to explain this and other radioactive processes in terms of transitions between various isotopes.

Fajans's work in inorganic chemistry was equally important. He formulated two rules to account for the well-known diagonal similarities between elements in the periodic table in terms of the ease of formation of covalencies and electrovalencies (ionic valencies). The first rule states that covalencies are more likely to be formed as the number of electrons to be removed or donated increases, so that highly charged ions are rare or impossible. The removal (or donation) of a second electron must overcome the effect of charge due to the removal (or donation) of the first, and so on for each successive electron. As the number of electrons increases, the work required soon becomes impossibly great for chemical forces, and covalencies result instead.

Fajans's second rule states that electrovalencies are favoured by large cations and small anions. In a large atom, the outer electrons are farther from the attractive force of the positive nucleus and hence are more easily removed to form cations. In a small atom, on the other hand, an electron added to form an anion can approach more closely to the positive field of the nucleus and is therefore more strongly held than in a large anion.

The operation of the first rule in passing from Group I to Group II of the periodic table is approximately balanced by the effect of the second rule in passing from the first period to the second period. The most obvious examples of diagonal similarity are the following:

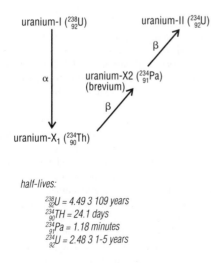

half-lives:
$^{238}_{92}U$ = 4.49 3 109 years
$^{234}_{90}TH$ = 24.1 days
$^{234}_{91}Pa$ = 1.18 minutes
$^{234}_{92}U$ = 2.48 3 1-5 years

Fajans *Fajans' scheme for the decay of uranium-I with modern symbols added. Uranium-X_1 and uranium-X_2 are actually isotopes of thorium and protactinium.*

Lithium, for example, shows its similarity to magnesium in all those points that emphasize its difference from sodium.

In an extension of his work on radioactivity, Fajans estimated the ages of minerals from Norway, by measuring the percentages of lead, the end-product of radioactive decay. In 1919, he did research on the energies of hydration of ions. Thermochemistry, theory of chemical forces, light absorption and photochemistry are other areas in which he became an authority.

Falloppio Gabriele (Gabriel Falloppius) 1523–1562 was a major Italian anatomist in the manner of Andreas Vesalius (1514–1564).

Born in Modena, Falloppio first intended to become a priest, but shifted to medicine, and was taught anatomy by Vesalius in Padua, becoming professor of anatomy at Pisa 1548 and Padua 1551.

He extended Vesalius's work and corrected its details. His discoveries included structures in the human ear and skull, explorations in the field of urology, and researches into the female genitalia. He coined the term 'vagina' and described the clitoris. He was the first to describe the tubes leading from the ovary to the uterus that were subsequently named after him. He failed, however, to grasp the function of the Fallopian tubes; it took a further two centuries before it was recognized that ova are formed in the ovary, passing down these tubes to the uterus. Falloppio also carried out investigations on the larynx, the eye, muscular action, and respiration. He was the teacher of Fabricius ab Aquapendente (1537–1619).

Faraday Michael 1791–1867 was a British physicist and chemist who is often regarded as the greatest experimental scientist of the 1800s. He made pioneering contributions to electricity, inventing the electric motor, electric generator and the transformer, and discovering electromagnetic induction and the laws of electrolysis. He also discovered benzene and was the first to observe that the plane of polarization of light is rotated in a magnetic field.

Faraday was born in Newington, Surrey, on 22 September 1791. His father was a poor blacksmith, who went to London to seek work in the year that Faraday was born. Faraday received only a rudimentary education as a child and although he was literate, he gained little knowledge of mathematics. At the age of 14, he became an apprentice to a bookbinder in London and began to read voraciously. The article on electricity in the *Encyclopedia Britannica* fascinated him in particular, for it presented the view that electricity is a kind of vibration, an idea that was to remain with Faraday. He also read the works of Lavoisier and became interested in chemistry, and carried out what scientific experiments he could put together with his limited resources. He was aided by a manual dexterity gained from his trade, which also stood him in great stead in his later experimental work.

In 1810, Faraday was introduced to the City Philosophical Society and there received a basic grounding in science, attending lectures on most aspects of physics and chemistry and carrying out some experimental work. He also attended the Royal Institution, where he was enthralled by the lectures and demonstrations given by Humphry Davy (1778–1829). He made notes eagerly, assembling them at work into finely bound books. In 1812, Faraday came to the end of his apprenticeship and prepared to devote himself to his trade, not expecting to make a career in science. Almost immediately, however, there came an extraordinary stroke of luck for Faraday. Davy was temporarily blinded by an explosion in a chemistry experiment and asked Faraday to help him until he regained his sight. When he recovered, Faraday sent Davy the bound notes of his lectures. Impressed by the young man, Davy marked him out as his next permanent assistant at the Royal Institution and Faraday took up this post in 1813.

This was remarkably good fortune for Faraday, because Davy was a man of wide-ranging interests and great scientific insight as well as a brilliant exponent of ideas. Furthermore, Davy undertook a tour of France and Italy from 1813 to 1815 to visit the leading scientists of the day, including the pioneer of current electricity Alessandro Volta (1745–1827). Faraday accompanied Davy, gaining an immense amount of knowledge, and on his return to London threw himself wholeheartedly into scientific research.

Faraday remained at the Royal Institution and made most of his pioneering discoveries in chemistry and electricity there over the next 20 years. He became a great popularizer of science with his lectures at the Royal Institution, which he began in 1825 and continued until 1862. His fame grew rapidly, soon eclipsing even that of Davy, who became embittered as a result. But the strain of his restless pursuit of knowledge told and in 1839, Faraday suffered a breakdown. He never totally recovered but at the instigation of Lord Kelvin (1824–1907), returned to research in 1845 and made his important discoveries of the effect of magnetism on light and developed his field theory. In the 1850s, Faraday's mind began to lose its sharp grip, possibly as a result of low-grade poisoning caused by his chemical researches, and he abandoned research and then finally lecturing. He resigned from the Royal Institution in 1862 and retired to an apartment provided for him at Hampton Court, Middlesex, by Queen Victoria. He died there on 25 August 1867.

Faraday was mainly interested in chemistry during his early years at the Royal Institution. He investigated the effects of including precious metals in steel in 1818, producing high-quality alloys that later stimulated the production of special high-grade steels. Faraday's first serious chemical discoveries were made in 1820, when he prepared the chlorides of carbon – C_2Cl_6 from ethane and C_2Cl_4 from ethylene (ethene) – substitution reactions that anticipated the work a few years later by Jean Baptiste Dumas.

In 1823, Faraday produced liquid chlorine by heating crystals of chlorine hydrate ($Cl_2.8H_2O$) in an inverted U-tube, one limb of which was heated and the other placed in a freezing mixture (liquefaction resulted because of the high pressure of the gas cooled below its relatively high critical temperature). He then liquefied other gases, including sulphur dioxide, hydrogen sulphide, nitrous oxide (dinitrogen monoxide), chlorine dioxide, cyanogen and hydrogen bromide. After the production of liquid carbon dioxide in 1835, Faraday used this coolant to liquefy such gases as ethylene (ethene), phosphine, silicon tetrafluoride and boron trifluoride.

In the same year (1835) he made his greatest contribution to organic chemistry, the isolation of benzene from gas oils. He also worked out the empirical formula of naphthalene and prepared various sulphonic acids – later to have great importance in the industries devoted to dyestuffs and detergents. It was also at about this time that Faraday demonstrated the use of platinum as a catalyst and showed the importance in chemical reactions of surfaces and inhibitors – again foreshadowing a huge area of the modern chemical industry.

But Faraday's interest in science had been initiated by a fascination for electricity and he eventually combined the knowledge that he gained of this subject with chemistry to produce the basic laws of electrolysis in 1834. His researches, summed up in Faraday's laws of electrolysis, established the link between electricity and chemical affinity, one of the most fundamental concepts in science. It was Faraday who coined the terms anode, cathode, cation, anion, electrode and electrolyte. He postulated that during the electrolysis of an aqueous electrolyte, positively charged cations move towards the negatively charged cathode and negatively charged anions migrate to the positively charged anode. At each electrode the ions are discharged according to the following rules:

(a) the quantity of a substance produced is proportional to the amount of electricity passed;

(b) the relative quantities of different substances produced by the same amount of electricity are proportional to their equivalent weights (that is, the relative atomic mass divided by the oxidation state or valency).

But his first major electrical discovery was made much earlier, in 1821, only a year after Hans Oersted (1777–1851) had discovered with a compass needle that a current of electricity flowing through a wire produces a magnetic field. Faraday was asked to investigate the phenomenon of electromagnetism by the editor of the *Philosophical Magazine*, who hoped that Faraday would elucidate the facts of the situation following the wild theories and opinions that Oersted's sensational discovery had aroused. Faraday conceived that circular lines of magnetic force are produced around the wire to explain the orientation of Oersted's compass needle, and therefore set about devising an apparatus that would demonstrate this by causing a magnet to revolve around an electric current. He succeeded in October 1821 with an elaborate device consisting of two vessels of mercury connected to a battery. Above the vessels and connected to each other were suspended a magnet and a wire; these were free to move and dipped just below the surface of the mercury. In the mercury were fixed a wire and a magnet respectively. When the current was switched on, it flowed through both the fixed and free wires, generating a magnetic field in them. This caused the free magnet to revolve around the fixed wire, and the free wire to revolve around the fixed magnet.

This was a brilliant demonstration of the conversion of electrical energy into motive force, for it showed that either the conductor or the magnet could be made to move. In this

experiment, Faraday demonstrated the basic principles governing the electric motor and although practical motors subsequently developed had a very different form to Faraday's apparatus, he is nevertheless usually credited with the invention of the electric motor.

Faraday's conviction that an electric current gave rise to lines of magnetic force arose from his idea that electricity was a form of vibration and not a moving fluid. He believed that electricity was a state of varying strain in the molecules of the conductor, and this gave rise to a similar strain in the medium surrounding the conductor. It was reasonable to consider therefore that the transmitted strain might set up a similar strain in the molecules of another nearby conductor – that a magnetic field might bring about an electric current in the reverse of the electromagnetic effect discovered by Oersted.

Faraday hunted for this effect from 1824 onwards, expecting to find that a magnetic field would induce a steady electric current in a conductor. In 1824, François Arago (1786–1853) found that a rotating nonmagnetic disc, specifically of copper, caused the deflection of a magnetic needle placed above it. This was in fact a demonstration of electromagnetic induction, but nobody at that time could explain Arago's wheel (as it was called). Faraday eventually succeeded in producing induction in 1831. In August of that year, he wound two coils around an iron bar and connected one to a battery and the other to a galvanometer. Nothing happened when the current flowed through the first coil, but Faraday noticed that the galvanometer gave a kick whenever the current was switched on or off. Faraday found an immediate explanation with his lines of force. If the lines of force were cut – that is, if the magnetic field changed – then an electric current would be induced in a conductor placed within the magnetic field. The iron core in fact helped to concentrate the magnetic field, as Faraday later came to understand, and a current was induced in the second coil by the magnetic field momentarily set up as current entered or left the first coil. With this device, Faraday had discovered the transformer, a modern transformer being no different in essence, even though the alternating current required had not then been discovered.

Faraday is thus also credited with the simultaneous discovery of electromagnetic induction, though the same discovery had been made in the same way by Joseph Henry (1797–1878) in 1830. However, busy teaching, Henry had not been able to publish his findings before Faraday did, although both men are now credited with the independent discovery of induction.

Faraday's insight enabled him to make another great discovery soon afterwards. He realized that the motion of the copper wheel relative to the magnet in Arago's experiment caused an electric current to flow in the disc, which in turn set up a magnetic field and deflected the magnet. He set about constructing a similar device in which the current produced could be led off, and in October 1831 built the first electric generator. This consisted of a copper disc that was rotated between the poles of a magnet; Faraday touched wires to the edge and centre of the disc and connected them to a galvanometer, which registered a steady current. This was the first electric generator, and generators employing coils and magnets in the same way as modern generators were developed by others over the next two years.

Faraday's next discoveries in electricity, apart from his major contribution to electrochemistry, were to show in 1832 that an electrostatic charge gives rise to the same effects as current electricity, thus proving that there is no basic difference between them. Then in 1837 he investigated electrostatic force and demonstrated that it consists of a field of curved lines of force, and that different substances take up different amounts of electric charge when subjected to an electric field. This led Faraday to conceive of specific inductive capacity. In 1838, he proposed a theory of electricity based on his discoveries that elaborated his idea of varying strain in molecules. In a good conductor, a rapid build-up and breakdown of strain took place, transferring energy quickly from one molecule to the next. This also accounted for the decomposition of compounds in electrolysis. At the same time, Faraday rejected the notion that electricity involved the movement of any kind of electrical fluid. In this, he was wrong (because the motion of electrons is involved) but in that this motion causes a rapid transfer of electrical energy through a conductor, Faraday's ideas were valid.

Faraday's theory was not taken seriously by many scientists, but his concept of the line of force was developed mathematically by Kelvin. In 1845, he suggested that Faraday investigate the action of electricity on polarized light. Faraday had in fact already carried out such experiments with no success, but this could have been because electrical forces were not strong. Faraday now used an electromagnet to give a strong magnetic field instead and found that it causes the plane of polarization to rotate, the angle of rotation being proportional to the strength of the magnetic field.

Several further discoveries resulted from this experiment. Faraday realized that the glass block used to transmit the beam of light must also transmit the magnetic field, and he noticed that the glass tended to set itself at right-angles to the poles of the magnet rather than lining up with it as an iron bar would. Faraday showed that the differing responses of substances to a magnetic field depended on the distribution of the lines of force through them, and not on the induction of different poles. He called materials that are attracted to a magnetic field paramagnetic, and those that are repulsed diamagnetic. Faraday then went on to point out that the energy of a magnet is in the field around it and not in the magnet itself, and he extended this basic conception of field theory to electrical and gravitational systems.

Finally Faraday considered the nature of light and in 1846 arrived at a form of the electromagnetic theory of light that was later developed by James Clerk Maxwell (1831–1879). In a brilliant demonstration of both his intuition and foresight, Faraday said 'The view which I am so bold to put forth considers radiation as a high species of vibration in the lines of force which are known to connect particles, and also masses of matter, together. It endeavours to dismiss the ether but not the vibrations.' It was a bold view, for no scientist until Albert Einstein (1879–1955) was to take such a daring step.

Michael Faraday was a scientific genius of a most extraordinary kind. Without any mathematical ability at all, he succeeded in making the basic discoveries on which virtually all our uses of electricity depend and also in conceiving the fundamental nature of magnetism and, to a degree, of electricity and light. He owed this extraordinary degree of insight to an amazing talent for producing valid pictorial interpretations of the workings of nature. Faraday himself was a modest man, content to serve science as best he could without undue reward, and he declined both a knighthood and the Presidency of the Royal Society. Characteristically, he also refused to take part in the preparation of poison gas for use in the Crimean War. His many achievements are honoured in the use of his name in science, the farad being the SI unit of capacitance and the Faraday constant being the quantity of electricity required to liberate a standard amount of substance in electrolysis.

Feller William 1906–1970 was a Yugoslavian-born US mathematician largely responsible for making the theory of probability accessible to students of subjects outside the field of mathematics through his two-volume textbook on the subject. He was also interested in the theory of limits, but it was his work on probability theory that brought him widespread recognition.

Feller was born on 7 July 1906 in Zagreb, where he grew up, was educated, and attended the university, earning his BA in 1925. He then continued his studies at the University of Göttingen, where he was awarded a PhD in 1926. He moved to Kiel University, and in 1929 was put in charge of the laboratory of applied mathematics there. Four years later, Feller went to the University of Stockholm, where he served as a Research Associate until 1939, working as a consultant for economists, biologists and others interested in probability theory.

At the outbreak of World War II in 1939, Feller emigrated to the United States. His first job was as executive editor of *Mathematical Reviews*, but he was soon appointed to the simultaneous post of Associate Professor at Brown University. In 1945 – by then a naturalized US citizen – he was made Professor of Mathematics at Cornell University, New York, and in 1950 Feller became Eugene Higgins Professor of Mathematics at Princeton University, a post he held until his death in 1970.

Feller had always been fascinated by the study of chance fluctuations. He came to the conclusion, early on, that the traditional emphasis placed on averages meant that insufficient attention was paid to random fluctuations which could bear significant impact on processes under investigation. His serious study of probability theory began soon after his arrival at the University of Stockholm. Much of his effort focused on the nature of Markov processes (a mathematical description of random changes in a system which, for instance, can occur in either of two states). Feller used semigroups to develop a general theory of Markov processes, and was able to demonstrate the applicability of this tool to subjects in which probability theory had not usually previously been employed, for example in the study of genetics. A strong advocate of the renewal method, Feller's preoccupation with problems of methodology greatly influenced the style of his book. The work – on the introduction to and the applications of probability theory – was published in two volumes, the first in 1950 and the second in 1966, and received considerable acclaim. It may fairly be said to have been significant to the further development of study on the subject.

Feller was also keenly interested in limit theory, in which he first formulated the law of the iterated logarithm; he made a real contribution to the study of the central limit theorem.

Ferguson Henry George 1884–1960 was the British inventor and engineer who developed automatic draught control for farm tractors. This improved performance so dramatically that farmers worldwide could buy the small, inexpensive machines, thus precipitating a revolution in farming methods. Virtually every modern tractor incorporates some form of automatic draught control.

Ferguson was born near Growell, near Belfast. He left school at 14 to work on his father's farm. In 1902 he joined his brother in a car and cycle repair business in Belfast. He started to build his own aeroplane in 1908 and flew it on 31 December 1909, becoming one of the first Britons to do so.

He started his own motor business and imported tractors from the United States. In 1917, he and William Sands designed their first plough. Another went into production in the United States, and on 12 February 1925 Ferguson patented the principle of draught contol. The Brown–Ferguson tractor, manufactured by David Brown, was launched in May 1936 and a redesigned version, built by Henry Ford in the United States, was launched in June 1939.

In 1946, with British government backing, the famous TE20 Ferguson tractor, made by the Standard Motor Company in Coventry, was launched. In the United States, Ferguson and Ford fought a massive anti-trust suit, largely over a similar machine produced by Ford. Ferguson set up his own US plant, the first machine coming off the line on 11 October 1948.

After selling out to Massey–Harris in 1953, Ferguson entered into various negotiations and tried to interest motor manufacturers in a revolutionary design of car produced by Harry Ferguson Research. These attempts came to little. Ferguson died at his home at Abbotswood, Stow-on-the-Wold, on 25 October 1960.

Tractive effort on smaller farms around the turn of the century was provided almost exclusively by draught animals. In addition to ploughs and harrows, there were horse-drawn binders and mowers and some barn machinery was driven by oil engines. Mechanized ploughing was carried out by traction engines, such as those produced by Fowler, one at each end of a field pulling a balance plough between them. This and other mechanized equipment, was generally affordable only by large landowners.

In the 1880s, two mobile steam engines for the farm appeared. They were small by the standards of the time but weighed 2 tonnes. Both were expensive, however, and farmers that could afford them preferred the steam-driven tackle. In Canada 10-tonne traction engines were being used. These heavy machines compacted the soil and produced a 'pan' which prevented proper drainage and inhibited root formation.

Lighter, petrol-engined tractors were being made by 1910, but they were simply mechanical horses used to pull trailed equipment. Driving them was dangerous because, if a plough hit an obstruction, they tended to rear up and could crush the driver. For the first Ferguson tractor, Ferguson designed a plough that coupled to the back of the tractor with the 'Duplex' hitch. This overcame the rearing problem, but there was little transfer of plough draught to the rear wheels. It was the principle of draught control which was the complete answer to the problem. The Ferguson System plough was built with two hitching points at about the normal level and a third hitch about 1 m/3 ft above the ground, connecting via a top link to a point about the same height on the tractor. When the plough hit an obstruction, the top link went into compression and tended to push the front of the tractor on to the ground instead of the reverse.

Ferguson incorporated a hydraulic system in the tractor for raising or lowering the plough out of or into work. This system was not new but using it to keep the plough at the correct working depth without the use of a depth wheel was a revolutionary step. The compression in the top link operated a sleeve valve in the hydraulic system, admitting varying amounts of oil to the pump. When the plough bit deep, it tended to tilt forwards, putting the top link in compression and opening the valve. This allowed more oil to be pumped into the hydraulic cylinder.

The cylinder raised the two bottom links, keeping the plough at an even depth. Thus the weight of the furrow slice was transferred to the tractor's back wheels, enabling it to exert greater effective traction. If the plough had a depth wheel, the draught was transferred to this wheel and not to the tractor's wheels. Ferguson's system meant that expensive, heavy machines were no longer necessary.

Ferguson Margaret Clay 1863–1951 was a US botanist who made important contributions to the study of plant genetics, and as a teacher and administrator inspired generations of undergraduate students at Wellesley College.

She was born on 20 August 1863 in Orleans, New York State, the fourth of six children. She attended local schools and began teaching children herself when she was 14, whilst also studying at the local Wesleyan Seminary. She attended Wellesley College from 1888 to 1891 as a special student studying chemistry and botany, and then returned to schoolteaching in Ohio. In 1893, the head of the

Department of Botany at Wellesley, Professor Susan Hallowell, asked her to return as an instuctor in botany. In 1897, she decided to complete her formal education at Cornell University and gained her Bachelor's degree in 1899 and her doctorate in 1901. She then returned to Wellesley, becoming head of the department of botany in 1902 and full professor in 1906, and remained there until 1938. She spent her retirement in New York State and Florida, and moved to California in 1946. She died in San Diego on 28 August 1951.

Ferguson proved to be a tireless teacher and administrator, emphasizing in particular the importance of practical experimental work. She supervised the building of new laboratory accommodation, and extensively developed the college herbarium and library facilities. Significantly, she also designed and organized the erection of college greenhouses and associated laboratories in which students performed experiments in plant genetics, horticulture and plant physiology; after her retirement these buildings were named after her. Ferguson's department became a major centre for botanical education and she has been credited with training more professional botanists and their spouses at Wellesley than any other botanist. One of the most distinguished pupils to emerge from Wellesley was the Nobel prizewinner Barbara McClintock (1902–1992). Ferguson's doctoral research at Cornell had focused on the life history and reproductive physiology of a species of North American pine, which was published by the Washington Academy of Sciences. It was an innovative study of the functional morphology and cytology of a native pine and served as a model for such work for many years. Her research interests shifted, as did her teaching interests, towards plant genetics during the 1920s. She made a particular study of the genetics of *Petunia* emphasising its use as a tool for studying the transfer of heredity from generation to generation. She extensively analysed the inheritance of features such as petal colour, flower pattern, and pollen colour, and built up a major database of genetic information. She continued to publish on the genetics of *Petunia* as research professor at Wellesley from 1930 until her formal retirement in 1932, after which her work was supported by the National Research Council for a further six years. She received many important honours and awards, including election as the first woman president of the Botanical Society of America (1929).

Fermat Pierre de 1601–1675 was a French lawyer and magistrate for whom mathematics was an absorbing hobby. He contributed greatly to the development of number theory, analytical geometry and calculus, carried out important research in probability theory and in optics, and was at the same time a competent classical scholar. Yet it is thanks only to his letters to various scientists and theoreticians that many of his accomplishments did not vanish into obscurity.

Born on 20 August 1601 in Beaumont de Lomagne, Fermat obtained a classical education locally. Between the ages of 20 and 30 he was in Bordeaux, possibly at the University of Toulouse. It was not, however, until he was 30 that he gained his bachelor's degree in civil law from the University of Orléans, set up a legal practice in Toulouse, and became Commissioner of Requests for the local parliament. In that parliament he was gradually promoted, gaining the high rank of King's Counsellor in 1648, an office he retained until 1665. In 1652, however, he suffered a severe attack of the plague after which he devoted much of his time to mathematics, being particularly concerned with reconstructing some of the missing texts of the ancient Greeks such as Euclid and Apollonius. Curiously, he refused to publish any of his achievements, which were considerable despite the occasionally eccentric style in which they were presented. In increasing isolation, therefore, from the rest of the European mathematical community, Fermat lived to an old age. He died in Castres on 12 January 1675.

While Fermat was in Bordeaux, he became fascinated by the work of the mathematician François Viète (1540–1603); it was from then that most of his mathematical achievements were attained. It was through Viète's influence that Fermat came to regard number theory as a 'lingua franca' between geometry and arithmetic, and went on to make many significant discoveries in the field. (His work on the theory was later revived by Leonhard Euler (1707–1783) and continued to stimulate further research well into the nineteenth century.) In 1657, Fermat published a series of problems as challenges to other mathematicians, in the form of theorems to be proved. All of them have since been proved – including 'Fermat's last theorem', which was reportedly proved in 1993 and which states that there is no solution in whole integers to the equation:

$$x^n + y^n = z^n$$

where n is greater than 2.

Fermat's technique in much of his work was 'reduction analysis', a reversible process in which a particular problem is 'reduced' until it can be seen to be part of a group of problems for which solutions are already known. Using this procedure, Fermat turned his attention to geometry. Unfortunately, analytical geometry was developed simultaneously both by Fermat (in letters written before 1636) and by the great René Descartes (who published his *Géométrie* in 1637). There followed a protracted and bitter dispute over priority. The discipline permitted the use of equations to describe geometric figures, and Fermat demonstrated that a second degree equation could be used to describe seven 'irreducible forms', each of which gave complete descriptions for different curves (such as parabolas and ellipses). He tried to extend this system into three dimensions to describe solids (1643), but was unsuccessful in the attempt beyond the establishment (in 1650) of the algebraic foundation for solid analytical geometry.

In 1636, he turned to the concept of 'infinitesimals' and applied it to equations of quadrature, the determination of the maxima and minima of curves, and the method of finding the tangent to a curve. All his work in these fields was superseded within 50 years through the development of calculus by Isaac Newton and Gottfried Leibniz; Newton did, however, acknowledge the importance of Fermat's work in the evolution of his own ideas.

Correspondence between Fermat and Blaise Pascal (1623–1662) resulted in the foundation of probability theory. Their joint conclusion was that if the probability of two independent events is respectively p and q, the probability of both occurring is pq.

In the field of optics, yet another disagreement with Descartes – this time on his law of refraction – led Fermat to investigate it mathematically. Ultimately obliged to confirm the law – but incidentally discovering the fact that light travels more slowly through denser mediums – Fermat also derived what is now known as 'Fermat's principle', which states that light travels by the path of least duration, after making a study of the transmission of light through materials with different refractive indices.

Fermi Enrico 1901–1954 was an Italian-born US physicist best known for bringing about the first controlled chain reaction (in a nuclear reactor) and for his part in the development of the atomic bomb. He also carried out early research using slow neutrons to produce new radioactive elements, for which work he was awarded the 1938 Nobel Prize for Physics.

Fermi was born in Rome on 19 September 1901, the son of a government official. He was educated at the select Reale Scuola Normale Superior in Pisa (which he attended from 1918) and went on to the University of Pisa, receiving his PhD in 1929 for a thesis on X-rays. He then travelled to Göttingen University, where he worked under Max Born, and to Leiden University, where he studied with P. Ehrenfest. He became a mathematics lecturer at the University of Florence in 1924, and two years later he was appointed Professor of Theoretical Physics at Rome University. Fermi married a Jew (in 1928) and during the 1930s became alarmed by increasing antisemitism in Fascist Italy under Benito Mussolini. After the Nobel prize ceremony in Stockholm in 1938, Fermi did not return to Italy but went with his wife and two children to the United States, where he took up an appointment in New York at Columbia University. In 1941, he and his team moved to Chicago University where he began building a nuclear reactor, which first went 'critical' at the end of 1942. He became involved in the Manhattan Project to construct an atomic bomb, working mainly at Los Alamos, New Mexico. At the end of World War II in 1945 Fermi became a US citizen and returned to Chicago to continue his researches as Professor of Physics. He died there, of cancer, on 28 November 1954.

Fermi first gained fame soon after his Rome appointment with his publication *Introduzione alla Fisica Atomica* (1928), the first textbook on modern physics to be published in Italy. His experimental work on beta-decay in radioactive materials provided further evidence for the existence of the neutrino (as predicted by Wolfgang Pauli) and earned him an international reputation. The decay, which takes place in the unstable nuclei of radioactive elements, results from the conversion of a neutron into a proton, an electron (beta particle) and an antineutrino.

Following the work of the Joliot-Curies, who discovered artificial radioactivity in 1934 using alpha-particle bombardment, Fermi began producing new radioactive isotopes by neutron bombardment. He found that a block of paraffin wax or a jacket of water round the neutron source produced slow, or 'thermal', neutrons which are more effective at producing such elements. This was the work that earned him the Nobel Prize. He did, however, misinterpret the results of experiments involving neutron bombardment of uranium, and it was left to Lise Meitner and Otto Frisch to explain nuclear fission in 1938.

In the United States, Fermi continued the work on the fission of uranium (initiated by neutrons) by building the first nuclear reactor, then called an atomic pile because it had a moderator consisting of a pile of purified graphite blocks (to slow the neutrons) with holes drilled in them to take rods of enriched uranium. Other neutron-absorbing rods of cadmium, called control rods, could be lowered into or withdrawn from the pile to limit the number of slow neutrons available to initiate the fission of uranium. The reactor was built on the squash court of Chicago University, and on the afternoon of 2 December 1942 the control rods were withdrawn for the first time and the reactor began to work, using a self-sustaining nuclear chain reaction. Two years later the Americans, through a team led by Arthur Compton (1892–1962) and Fermi, had constructed an atomic bomb, which used the same reaction but without control, resulting in a nuclear explosion.

Element number 100, discovered in 1955 a year after Fermi died, was named fermium, and his name is also honoured in the fermi, a unit of length equal to 10^{-15} m.

Ferranti Sebastian Ziani de 1864–1930 was a British electrical engineer who pioneered the high-voltage AC electricity generating and distribution system still used by most power networks. He also designed, constructed and experimented with many other electrical and mechanical devices, including high-tension cables, circuit breakers, transformers, turbines and spinning machines.

Ferranti was born in Liverpool, Lancashire, on 9 April 1864. From his youth he was fascinated by machines and the principles by which they operate. After moving south he attended St Augustine's College in Ramsgate, Kent, and so impressed his teachers with his mechanical ideas that they set aside a room in which he could experiment. During this time he constructed an electrical generator. He left school in 1881 and took a job at the Siemens works in Charlton near London. He discovered that he could rotate and therefore mix the molten steel in a Siemens furnace by applying an electric current, and within a year he was supervising the installation of electric lighting systems for the company. A year after that – still only 18 years old – he was engineer to his own company which designed and manufactured the Thompson–Ferranti alternator and installed lighting systems. The company was formed in partnership with Lord Kelvin (William Thomson) and a solicitor named Ince.

In 1886, Ferranti became engineer to the Grosvenor Gallery Company in London. The gallery had its own electricity generating system for lighting, and was also selling electricity to outside customers. Ferranti modified the system considerably to meet extra demand and, realizing the business potential, led the Grosvenor Company into the formation of a separate enterprise, the London Electric Supply Corporation Ltd, and suggested building a large generating station at Deptford. Extending an electricity supply to such a large area would, he argued, eventually become more economical and practicable than hundreds of small electrical enterprises serving limited areas. Most of the small systems used direct current of 200 to 400 volts together with storage batteries (accumulators), and were suitable only over short distances and when the demand for electricity fell within suitable limits. To achieve large-scale distribution, Ferranti proposed using alternating current at 10,000 volts which was fed by mains to London, where step-down transformers reduced it to a voltage suitable for its purpose. The idea was revolutionary, because electricity at more than 2,000 volts was considered to be extremely dangerous. The cable for the mains was made to Ferranti's design, and produced in 6 m/20 ft lengths which were spliced together without the use of solder. The Deptford power station and its associated distribution network became the basic model for the future of electricity generation and supply.

In 1888, Ferranti married Gertrude Ince, the daughter of his solicitor partner. Three years later he left the London Electric Supply Corporation and concentrated on work as a consultant and on the development of his own company. This firm went on to design and build all kinds of electrical equipment, most of which was designed by Ferranti himself. He was also involved with heat engines of various kinds, turbines, cotton-spinning machines and, during World War I, the design and manufacture of steel casings.

He became President of the Institution of Electrical Engineers in 1911 and a year later was awarded a DSc degree by the University of Manchester. In 1929, he was elected a Fellow of the Royal Society. He enjoyed motoring and was very proud of his fast journey times. He died after an illness while on holiday in Zurich, Switzerland, on 13 January 1930.

Fessenden Reginald Aubrey 1866–1932 was a Canadian–US physicist whose invention of radio-wave modulation paved the way for modern radio communication.

Born in East Bolton, Québec, Canada on 6 October 1866, Fessenden studied at Trinity College School, Port Hope, Ontario and Bishop's University at Lennoxville, Québec. It was only after he had moved to Bermuda, to take up the position of Principal at the Whitney Institute there, that his interest in science really came alive. With little opportunity to follow up such interests in Bermuda, he left to go to New York, where he met Thomas Edison (1847–1931). By his early twenties, Fessenden had become the chief chemist at Edison's laboratories at Orange, New Jersey.

In 1890, he left to join Edison's great rival, the Westinghouse Electric & Manufacturing Company, where he stayed for two years before returning to academic life, first at Purdue University, Lafayette, as Professor of Electrical Engineering, and then at the Western University of Pennsylvania (now the University of Pittsburgh). It was there that Fessenden began major work on the problems of radio communication.

By 1900, he had overcome some major technical difficulties and succeeded in transmitting speech by radio, taking out the first patent for voice transmission the following year. Two key inventions of Fessenden's turned his 1900 experiment into the forerunner of modern radio communication. The first was that of modulation. In the early days of radio, experimenters such as Guglielmo Marconi (1874–1937) had sent messages using short bursts of signals to mimic Morse code. Fessenden realized that the amplitude of a continuous radio wave could be varied to mimic the variations of more complex wave patterns. By converting sound waves into variations of amplitude, it would therefore be possible to transmit voices and music. This is the principle of amplitude modulation.

Fessenden's other major invention was that of the heterodyne effect. In this, the received radio wave is combined with a wave of frequency slightly different to that of the carrier wave. The resulting intermediate frequency wave is easier to amplify before being demodulated to generate the original sound wave.

Two years after his initial broadcast, Fessenden organized the building of a 50-kHz alternator for radiotelephony by the General Electric Company. This was followed by his building a transmitting station at Brant Rock, Massachussetts. On Christmas Eve, 1906, the first amplitude-modulated radio message was broadcast; both words and music were transmitted. In the same year he established two-way radio communication between Brant Rock and Scotland.

By his death, in Bermuda on 22 July 1932, Fessenden held 500 patents, a figure surpassed only by his former employer Edison. Amongst his patents are those for the sonic depth finder, the loop-antenna radio compass and submarine signalling devices.

Feynman Richard Phillips 1918–1988 was a US physicist who shared the 1965 Nobel prize for his role in the development of the theory of quantum electrodynamics. He also made important contributions to the theory of quarks and superfluidity, was a noted teacher and a self-styled 'curious character'. Feynman diagrams have become a standard way of representing particle interactions.

Feynman was born on 11 May 1918 in New York City. As a child he was fascinated by mathematics and electronics and became known as 'the boy who fixes radios by thinking'. He graduated from the Massachusetts Institute of Technology in 1939 and obtained a PhD from Princeton University in 1942. His supervisor was John Wheeler and his thesis, 'A principle of least action in quantum mechanics', was typical of his first-principles approach to fundamental problems. During World War II, Feynman worked at Los Alamos and was in charge of a group responsible for 'diffusion problems'. This involved large-scale computations (in the days before computers) to predict the behaviour of neutrons in atomic explosions. Working on the bomb, however, did not dampen his enthusiasm for practical jokes. Sadly, Feynman's first wife, Arlene, died of tuberculosis whilst he was working in Los Alamos.

After the war Feynman moved to Cornell University, where Hans Bethe, who had also been at Los Alamos, was building up an impressive school of theoretical physicists. He continued developing his own approach to quantum electrodynamics (QED) before moving to California Institute of Technology in 1950.

Feynman shared the Nobel prize with Julian Schwinger and Sin-Itiro Tomonaga for his work on QED. All three had independently developed methods for calculating the interaction between electrons, positrons (antielectrons) and photons (light). The three approaches were fundamentally the same and QED remains the most accurate physical theory known. In Feynman's 'space-time' approach, different physical processes were represented as collections of diagrams showing how the particles moved from one space-time point to another. Feynman had rules for calculating the probability associated with each diagram, which he added to give the probability of the physical process itself.

Although Feynman only wrote 37 research papers in his career, a remarkably small number for such a prolific researcher, many consider that two discoveries he made at Caltech were also worthy of the Nobel prize. The first is the theory of superfluidity (frictionless flow) in liquid helium developed in the early 1950s; the second is Feynman's work on the weak interaction (with Murray Gell-Mann) and the strong force, and his prediction that the proton and neutron are not elementary particles. Both particles are now known to be composed of quarks.

A series of undergraduate lectures given by Feynman at Caltech, *The Feynman Lectures on Physics* (three volumes with R. Leighton and R. Sands; 1963), quickly became standard reference in physics. At the front of the lectures Feynman is shown indulging in one of his favourite pastimes, playing the bongo drum. Painting was another hobby.

Feynman's fame increased further in 1986 when he was appointed to the commission to investigate the explosion on board the *Challenger* space shuttle in which seven astronauts died. In front of television cameras, he demonstrated how the failure of a rubber 'O-ring' seal, caused by the cold, was responsible for the disaster. His experiences on the commission feature heavily in his *'further adventures of a curious character'* – *What Do You Care What Other People Think?* (1988). Feynman died in Los Angeles on 15 February 1988.

Fibonacci Leonardo (Leonardo of Pisa) *c.* 1180–*c.* 1250 was a mathematician of medieval Pisa whose writings were influential in introducing and popularizing the Indo–Arabic numeral system, and whose work in algebra, geometry and theoretical mathematics was far in advance of the contemporary European standards.

Fibonacci was born in Pisa in about 1180, the son of a member of the government of the Republic of Pisa. When Fibonacci was 12 years old, his father was made administrator of Pisa's trading colony in Algeria, and it was there – in a town now called Bougie – that he was taught the art of calculating, using the commercial North African medium of Indo–Arabic numerals. His teacher, who remains completely unknown, seems to have imparted to him not only an excellently practical and well-rounded fundamental grounding in mathematics, but also a true scientific curiosity.

Having achieved maturity, Fibonacci travelled extensively, both for business and for pleasure, spending time in Italy, Syria, Egypt, Greece and elsewhere. Wherever he went he observed and analysed the arithmetical systems used in local commerce, studying through discussion and argument with native scholars of the countries he visited. He returned to Pisa in about the year 1200 and began his mathematical writings. Little more is known of him, although in 1225 he won a mathematical tournament in the presence of the Holy Roman Emperor Frederick II at the court of Pisa. A marble tablet dated 1240 appears to refer to him as being awarded an annual pension following his valuable accountancy services to the state. He is assumed to have died in Pisa in about 1250.

Two years after finally settling in Pisa, Fibonacci produced his most famous book, *Liber Abaci/The Book of the Calculator*. In four parts, and revised by him a quarter of a century later (1228), it was a thorough treatise on algebraic methods and problems in which he strongly advocated the introduction of the Indo–Arabic numeral system, comprising the figures 1 to 9, and the innovation of the 'zephirum' – the figure 0 (zero). Dealing with operations in whole numbers systematically, he also proposed the idea of a bar (solidus) for fractions, and went on to develop rules for converting fraction factors into the sum of unit factors. (However, his expression of fractions followed the Arabic practice – on the *left* of the relevant integral.) At the end of the first part of the book, he presented tables for multiplication, prime numbers and factoring numbers. In the second part, he demonstrated mathematical applications to commercial transactions. In part three he gave many examples of recreational mathematical problems of the type enjoyed today, leading up to a thesis on series from which, in turn, he derived what is now called the Fibonacci series. This is a sequence in which each term after the first two is the sum of the two terms immediately preceding it – 1, 1, 2, 3, 5, 8, 13, 21, ... for example – and which has been found to have many significant and interesting properties. And in the final part of the book Fibonacci, a student of Euclid, applied the algebraic method. The *Liber Abaci* remained a standard text for the next two centuries.

In 1220, he published *Practica Geometriae*, a book on geometry that was of fundamental significance to future studies of the subject, and that (to some commentators, at least) seems to be based on a work of Euclid now lost. In it, Fibonacci used algebraic methods to solve many arithmetical and geometrical problems. In *Flos/Flower*, published four years later, he considered indeterminate problems in a way that had not properly been carried out since the work of Diophantus, and again demonstrated Euclidean methodology combined with techniques of Chinese and Arabic origin (learned during his travels many years before) in solving determinate problems. In both *Liber Quadratorum/ The Book of Squares* and in a separate letter to the philosopher Theodorus, Fibonacci dealt with some problems set by John of Palermo (one of which was the one he solved in front of the Emperor); his treatments show unusual mathematical skill and originality.

The complete works of Fibonacci were edited in the nineteenth century by B. Boncompagni, and published in two volumes under the title *Scritti di Leonardo Pisano*.

Field George Brooks 1929– is a US theoretical astrophysicist whose main research has been into the nature and composition of intergalactic matter and the properties of residual radiation in space.

Field was born in Providence, Rhode Island, on 25 October 1929. Educated at the Massachusetts Institute of Technology, he graduated in 1951 and four years later gained a PhD at Princeton University. In 1957 he was appointed Assistant Professor in astronomy at Princeton and he progressed to become Associate Professor. He was made Professor of the University of California at Berkeley in 1965 and was Chairman of the Department from 1970 to 1971. In 1972, he became Professor of Astronomy at Harvard University: from 1973 he was also Director of the Center of Astrophysics at the Harvard College Observatory and the Smithsonian Astrophysical Observatory.

One of Field's major areas of research has been to investigate why a cluster of galaxies remains a cluster. It seems evident that such clusters ought to be rapidly dispersing unless they are stabilized in some way, presumably gravitationally by intergalactic matter that contributes from ten to thirty times more material than the galaxies themselves. Such matter has never been detected, although considerable research has been undertaken and is still in progress. From a consideration of the composition of galaxies in clusters, it would seem probable that the most likely substance of such intergalactic matter would be in the form of hydrogen, that around 27% by mass would be helium, and that a negligible fraction would be in the form of heavier elements. There is at present no means of detecting helium in intergalactic space – but intergalactic hydrogen does produce effects that are potentially detectable.

Atomic hydrogen distributed intergalactically (in contrast to ionized hydrogen) would act both as an absorber and an emitter of radiation at a wavelength of 21 cm/8 in. Field first tried to find evidence of this absorption in 1958. He studied the spectrum of the radio source Cygnus A – the brightest extragalactic radio source in the sky – in the region of 21 cm/8 in, taking into account the known red

shift associated with the expansion of the universe. The narrow range of wavelengths over which the intergalactic hydrogen would absorb is called the 'absorption trough'. A wavelength of 21 cm/8 in is remarkably long for an atom to absorb or emit, and from the point of view of the two energy levels involved, the hydrogen is immersed in a heat bath at an extremely high temperature – so that there are nearly as many atoms in the upper state as in the lower, and the absorption trough cannot be very exactly observed. Field's later results have given greater precision.

Field has also carried out research into the spectral lines in the spectra of stars. In the 1930s, A. McKellar found absorption lines corresponding to interstellar cyanogen (CN) in the spectra of several stars. For the star Zeta Ophiuci, McKellar was able to obtain a measure of the relative number of molecules in the ground state and the first rotational state, which he defined in terms of an excitation temperature (the temperature at which the molecules would possess the observed degree of excitation if they were in equilibrium in a heat bath). At that time, however, excitation was assumed to occur only by collisions with other particles or by radiation with no thermal spectrum. But in 1966, microwave measurements of the cosmic background were made that suggested that this background has a blackbody spectrum at all wavelengths. Field then wondered if the CN 'molecules' might after all be in a heat bath. He re-observed the spectrum of Zeta Ophiuci, and obtained an excitation temperature of 3.22 ± 0.15K, which corresponded well with the value of 2.3K ($-270.8°$C/$-454.9°$F) determined by McKellar. A number of other stars have since been analysed and all of them yield a temperature of about 3K ($-270°$C/$-454°$F).

Fischer Emil 1852–1919 was a German organic chemist who analysed and synthesized many biologically important compounds. He was awarded the 1902 Nobel Prize for Chemistry for his work on the synthesis of sugars and purine compounds.

Fischer was born in Euskirchen, near Bonn, on 9 October 1852, the son of a merchant. After leaving school he joined the family business, but in 1871 entered the University of Bonn to study chemistry under Friedrich Kekulé. The following year he went to Strasbourg and graduated from the university there in 1874 with a doctoral thesis which was supervised by Johann von Baeyer. He continued his studies at Munich where he became an unpaid lecturer in 1878 and (paid) Assistant Professor in 1879. He then held professorships at Erlangen (1882), Würzburg (1885) and finally Berlin (1892). Before his last move he married Agnes Gerlach. His wife died young but they had three sons, the eldest of whom, Hermann, also became a distinguished organic chemist; two sons were killed in World War I. Fischer suffered a serious bout of mercury poisoning and, equally seriously, phenylhydrazine poisoning. He contracted cancer and this fact, coupled with the death of his sons, led him to commit suicide on 15 July 1919.

Fischer's early research, carried out with his cousin Otto Fischer, concerned the dye rosaniline and similar com-

Fischer Fischer's indole synthesis.

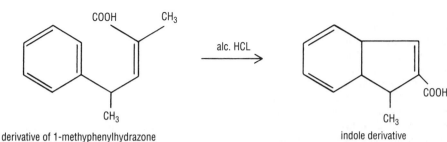

derivative of 1-methyphenylhydrazone   indole derivative

pounds, which they showed have a structure related to that of triphenylmethane. In 1875, Fischer discovered phenylhydrazine, but it was not until 1884 that he found it formed bright yellow crystalline derivatives with carbohydrates, a key reaction in the study of sugars. The derivatives are known as osazones, and Fischer obtained the same osazone from three different sugars – glucose, fructose and mannose – demonstrating that all three have the same structure in the part of their molecules unaffected by phenylhydrazine. He went on to determine the structures of the 12 possible stereoisomers of glucose, the important group of sugars known collectively as hexoses.

From about 1882, he began working on a group of compounds that included uric acid and caffeine. Fischer realized that they were all related to a hitherto unknown substance which he called purine. Over the next few years he synthesized about 130 related compounds, one of which was the first synthetic nucleotide, a biologically important phosphoric ester of a compound made from a purine-type molecule and a carbohydrate. These studies led to the synthesis of powerful hypnotic drugs derived from barbituric acids, including in 1903 5,5-diethyl barbituric acid which became widely used as a sedative.

In 1885, experiments with phenylhydrazine led to what is known as Fischer's indole synthesis, in which he heated a phenylhydrazone with an acid catalyst to produce a derivative of indole. Indole itself cannot, however, be obtained by this method.

Fischer's investigations into the chemistry of proteins began in 1899. He synthesized the amino acids ornithine (1,4-diaminopentanoic acid) in 1901, serine (1-hydroxy-2-aminobutanoic acid) in 1902 and the sulphur-containing cystine in 1908. He then combined amino acids to form polypeptides, the largest of which – composed of 18 amino acid residues – had a molecular weight of 1,213. Later work included a study of tannins, which he carried out with the assistance of his son Hermann.

Fischer was involved with many aspects of organic chemistry. He was a man of considerable insight, as exemplified by his description of the action of enzymes as a lock-and-key mechanism in which the enzyme model fits exactly onto the molecule with which it reacts. But he did not consider himself to be a theoretician; he believed in, and used, the synthetic methods of the practical organic chemist.

Fischer and Wilkinson Ernst Otto Fischer 1918– and Geoffrey Wilkinson 1921– are inorganic chemists who shared the 1973 Nobel Prize for Chemistry for their pioneering work, which they carried out independently, on the organometallic compounds of the transition metals.

Fischer was born in Munich on 10 November 1918, the son of Professor Karl T. Fischer. He was educated at the Munich Technical University, from which he gained a Diploma in Chemistry in 1949 and a doctorate in 1952. He remained at Munich, becoming successively Associate Professor (1957), Professor (1959) and finally Professor and Director of Inorganic Chemistry (1964).

Wilkinson was born on 14 July 1921 and educated at Todmorden Grammar School. In 1939 he obtained a scholarship to Imperial College, London, and from 1943 to 1946 worked as a junior scientific officer for the Atomic Energy Division of the National Research Council in Canada. He then took up various appointments in the United States: at the Radiation Laboratory at Berkeley, California (1946–1950); in the Chemistry Department of the Massachusetts Institute of Technology (1950–1951); and Assistant Professor (1951–1956) and Professor (1956–1978) at Harvard. In 1978 he became the Sir Edward Frankland Professor of Inorganic Chemistry at the University of London, working at Imperial College. Wilkinson was knighted in 1976.

In about 1830 a Danish pharmacist described the compound $PtCl_2.C_2H_4$, which is now known to exist as a dimer with chloride bridges. This and the ion $[C_2H_4PtCl_3]^-$ were the first known organometallic derivatives of the transition metals. In 1951, both Fischer and Wilkinson read an article in the journal *Nature* about a puzzling synthetic compound called ferrocene. Working independently, they came to the conclusion that each molecule of ferrocene consists of a single iron atom sandwiched between two five-sided carbon rings – an organometallic compound. A combination of chemical and physical studies, finally confirmed by X-ray analysis, showed that the compound's structure is as shown in the diagram. In the ferrocene molecule, the two symmetrical five-membered rings are staggered with respect to each other, but in the corresponding ruthenium compound (called ruthenocene) they are eclipsed.

With this work came the general realization that transition metals can bond chemically to carbon, and other ring systems were then studied. The hydrocarbon cyclopentadiene behaves as a weak acid and with various bases forms salts containing the symmetrical cyclopentadienide ion $C_5H_5^-$. It also forms 'sandwich' compounds and, like other ring systems that behave in this way has the 'aromatic' sextet of (six) electrons. All of the elements of the first transition series have now been incorporated into molecules of this kind and all except that of manganese have the ferrocene-type structure. Only ferrocene, however, is stable in air, the others being sensitive to oxidation in the order (of decreasing stability) nickel, cobalt, vanadium, chromium, titanium. The cationic species behaves like a large monopositive ion; $(\pi\text{-}C_5H_5)Co^+$ is particularly stable.

The recognition of the 'sandwich' concept initiated a vast amount of research, not only on cyclopentadienyl derivatives but on similar systems with four-, six-, seven- and even eight-membered carbon rings. All of this work was stimulated by the revolutionary explanation by Fischer and Wilkinson of the previously unknown way in which metals and organic compounds can combine.

Fischer Hans 1881–1945 was a German organic chemist who is best known for his determinations of the molecular structures of three important biological pigments: haemoglobin, chlorophyll and bilirubin. For his work on haemoglobin he was awarded the 1930 Nobel Prize for Chemistry.

Fischer was born in Höchst-am-Main, near Frankfurt, on 27 July 1881, the son of a chemical manufacturer. He studied chemistry at the University of Marburg, gaining his doctorate in 1904. He then went to the University of Munich to study medicine, qualifying as a doctor in 1908. He became a research assistant to his namesake Emil Fischer at the University of Berlin, before taking up an appointment as Professor of Medical Chemistry at the University of Innsbruck in 1915 as successor to Adolf Windaus (1876–1959). Three years later he held a similar post at Vienna, before taking over from Heinrich Wieland at the Munich Technische Hochschule in 1921 as Professor of Organic Chemistry. In 1945 Fischer's laboratories were destroyed in an Allied bombing raid and in a fit of despair, like Emil Fischer before him, he committed suicide. He died in Munich on 31 March 1945.

In 1921 Fischer began investigating haemoglobin, the

oxygen-carrying, red colouring matter in blood. He concentrated on haem, the iron-containing non-protein part of the molecule, and showed that it consists of four pyrrole rings (five-membered heterocyclic rings containing four carbon atoms and one nitrogen atom) surrounding a single iron atom. By 1929 he had elucidated the complete structure and synthesized haem.

He then turned his attention to chlorophyll, the green colouring matter in plants that Richard Willstätter had isolated in 1910. He found that its structure is similar to that of haem, with a group of substituted porphins surrounding an atom of magnesium. This work occupied Fischer for much of the 1930s, after which he began to study the bile pigments, particularly bilirubin (the pigment responsible for the colour of the skin of patients suffering from jaundice). He showed that the bile acids are degraded porphins and by 1944 had achieved a complete synthesis of bilirubin.

Fischer Hermann Otto Laurenz 1888–1960 was a German organic chemist whose chief contribution to the science concerned the synthetic and structural chemistry of carbohydrates, glycerides and inositols.

Fischer was born on 8 December 1888 in Würzburg, Bavaria, the eldest of the three sons of Emil Fischer, who at that time was Professor of Organic Chemistry at the local university. The early death of his mother brought him and his two brothers into closer contact with their illustrious father. The two brothers went on to study medicine, while Hermann followed his father into organic chemistry.

Fischer began his undergraduate career at Cambridge University in 1907, but had to return to Germany the following year for military training. After a brief period in Berlin, he started his doctorial research work at Jena University under Ludwig Knorr. In 1912, he returned to the Chemical Institute of Berlin University to continue research with his father. Two years later the outbreak of World War I interrupted his research. Both his brothers were killed in the war and his father committed suicide in 1919, soon after Hermann returned to Berlin.

In 1922, Fischer married Ruth Seckels, and they had a daughter and two sons. With the rise of Adolf Hitler the Fischers left Berlin in 1932 and went to Basle in Switzerland. Then in 1937, he moved to the Banting Institute in Toronto, Canada, where he stayed until his final move to the United States in 1948, to the biochemistry department of the University of California at Berkeley. He became Chairman of the department and Emeritus Professor before retiring in 1956. Fischer died on 9 March 1960.

While working with Knorr at Jena, Fischer separated the keto and enol tautomers of acetyl-acetone (pentan-2,4-dione) by low-temperature crystallization. Then in Berlin his father (Emil Fischer), who was investigating tannins, assigned him the task of synthesizing some of the naturally occurring depsides (a depside is a condensation product formed from two hydroxy-aromatic acid molecules). He succeeded in producing various didepsides and diorsellinic acids.

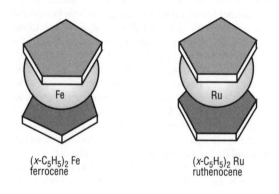

$(x\text{-}C_5H_5)_2$ Fe
ferrocene

$(x\text{-}C_5H_5)_2$ Ru
ruthenocene

Fischer *Structures of ferrocene and ruthenocene.*

Between 1920 and 1932 Fischer pursued two main lines of research. One was the study of quinic acid (tetrahydroxycyclohexanecarboxylic acid). By 1921 he had made various derivatives, but it was not until 1932 that he finally worked out its exact structure. The other research during this period dealt with the difficult chemistry of the trioses glyceraldehyde and dihydroxyacetone (2,3-dihydroxypropanal and 1,3-dyhydroxypropanone) and the related two-, three- and four-carbon compounds. The crowning achievement of this work was the preparation by his assistant Erich Baer of DL-glyceraldehyde-3-phosphate.

Baer accompanied Fischer to Basle in 1932 and there they developed a practical method for the preparation of the enantiomorphous acetonated glyceraldehydes. They were able to make D-fructose and D-sorbose almost entirely free from their isomers D-psicose and D-tagatose, using the aldol reaction between unsubstituted D-glyceraldehyde and its ketonic isomer, dihydroxyacetone.

While at Basle Fischer continued to collaborate with Gerda Dangschat (who had remained at Berlin) on structural and configurational studies of shikimic acid (first isolated in 1885). By a series of degradation reactions they converted shikimic acid into 2-deoxy-D-arabino-hexonic acid, which finally located the position of the double bond in the former (proving it to be 3,4,5-trihydroxycyclohexene-1-carboxylic acid).

From 1937, at the Canadian Banting Institute, Fischer extended his work on glyceraldehydes to glycerides (esters of glycerol, i.e. propan-1,2,3-triol). He prepared the first optically pure α-monoglycerides and α-glycerophosphoric acids (glycerol-1-phosphates) and demonstrated the action of lipase enzymes on these biologically important substances.

Despite the intervention of World War II and the distance between Berlin and Toronto, Fischer and Dangschat continued their work on the inositols (hexahydroxycyclohexanes). They succeeded in establishing the configuration of myoinositol and showed its relationship to D-glucose.

At Berkeley, Fischer carried on research into the inositols and other carbohydrates. He described the 12 years in California as the most pleasant in his life. His warm and friendly personality made him many friends there and among distinguished scientists throughout the world.

Fisher Ronald Aylmer 1890–1962 was a British mathematical biologist whose work in the field of statistics resulted in the formulation of a methodology in which the analysis of results obtained using small samples produced interpretations that were objective and valid overall. His work revolutionized research methods in many areas, and found immediate and widespread use, particularly in genetics and agriculture.

Fisher was born in London on 17 February 1890. He attended Stanmore Park and Harrow schools before going to Gonville and Caius College, Cambridge, in 1909. He graduated in 1912, having specialized in mathematics and theoretical physics, and then spent an additional year at Cambridge studying statistical mechanics and researching into the theory of errors. The next six years were spent in various occupations – his poor eyesight made him unacceptable for military service during World War I – and he worked in an investment brokerage, as a teacher and even as a farm labourer. At the end of the war, however, Fisher obtained a post at the Rothamstead Experimental Station, where he single-handedly ran a statistics department whose main job was to analyse a huge backlog of experimental data that had built up over more than 60 years. It was while at Rothamstead that he evolved many improvements to traditional statistical methods. His textbook on the subject, which appeared in 1925, was a landmark.

At Rothamstead, Fisher was able also to indulge in his second scientific passion, genetics: he bred poultry, mice, snails and other creatures, and in his papers on the subject contributed to the contemporary understanding of genetic dominance. As a result, in 1933 he was appointed to the Galton Chair of Eugenics at University College, London. During World War II, however, his department was (ironically) evacuated to Rothamstead, and eventually disbanded. Fisher then became Balfour Professor of Genetics at the University of Cambridge in 1943. He was knighted nine years later.

He officially retired from Cambridge in 1957, but stayed on until 1959, when a successor was found. Following a visit to the Mathematical Statistics Division of the Commonwealth Scientific and Industrial Research Organisation in Adelaide, Fisher emigrated to Australia. He died in Adelaide on 29 July 1962.

Elected a Fellow of the Royal Society in 1929, Fisher was awarded their Royal Medal in 1938, their Darwin Medal in 1948, and their Copley Medal in 1955.

Fisher's early work concerned the development of methods for the determination of the exact distributions of several statistical functions, such as the regression coefficient and the discriminant function. He improved the Helmut–Pearson χ^2 and Gosset's Z functions, modifying the latter to the now familiar t-test for significance. He evolved the rules for 'decision-making' that are now used almost automatically, and are based on the percentage deviation of the results of an experiment from the 'null hypothesis' (which assumes that events occur on an exclusively random basis). A deviation of between 95% and 99% represented only a suggestive likelihood that the null hypothesis was incorrect; a deviation in excess of 99% indicated strongly that this was so.

Other statistical methods that Fisher originated include the analysis of variance, the analysis of covariance, multivariate analysis, contingency tables, and more. All his mathematical methods were developed for further application in fields such as genetics, evolution and natural selection.

One of Fisher's first studies concerned the importance of dominant genes. A confirmed eugenicist, he looked on the study of human blood as an essential factor in his research. The department he established at the Galton Laboratories to investigate blood types made significant contributions to the final elucidation of the inheritance of rhesus blood groups. Fisher's methods have since been extended to virtually every academic field in which statistical analysis can be applied.

Fitch and Cronin Val Lodgson Fitch 1923– and James Watson Cronin 1931– are US physicists who shared the 1980 Nobel Prize for Physics for their work in particle physics.

Fitch was born in Merriman, Nebraska, on 10 March 1923. He was educated at McGill University, where he obtained his BEng in 1948. He gained a PhD from Columbia University in physics in 1954. From 1954 to 1960 Fitch rose from Instructor to Professor of Physics at Princeton University, and in 1976 he became Cyrus Fogg Bracket Professor of Physics at Princeton. He was a member of the President's Scientific Advisory Committee from 1970 to 1973.

Cronin was born in Chicago, Illinois, on 29 September 1931. He was educated at the Southern Methodist University, obtaining a BS in 1951. He then gained an MS from the University of Chicago in 1953 and a PhD in physics in 1955. From 1955 to 1958, he was an assistant physicist at the Brookhaven National Laboratory. Cronin then went to Princeton University, becoming Professor of Physics in 1965. In 1971, he became Professor of Physics at the University of Chicago.

The discovery for which Fitch and Cronin received the 1980 Nobel Prize for Physics was first published in 1964, and at that time was regarded as a bombshell in the field of particle physics. It is surprising that their work was not rewarded earlier.

Until 1964, it was not possible to distinguish unambiguously between matter and antimatter outside our own galaxy and a few nearby galaxies. In 1964, Cronin, Fitch and their colleagues had set up an experiment with the proton accelerator at the Brookhaven Laboratory in New York to study the properties of K^0 mesons. These are neutral, unstable particles with a mass equal to approximately half that of the proton, and had been discovered earlier in the interactions of particles from outer space with the Earth's atmosphere. K^0 is a mixture of two 'basic states' which have a long and a short lifetime and are therefore called K^0_L and K^0_S respectively. These two basic states can also mix together to form not K^0 but an antimatter particle (anti-K^0), and K^0 can oscillate from particle to antiparticle through either of its basic states. This is a unique phenomenon in the world of particle physics.

A rule called CP-conservation states how K^0_L and K^0_S should decay (C stands for conjugation and P for parity). Charge conjugation changes all particles to antiparticles. Parity concerns handedness and is like a mirror reflection – all positive points in a three-dimensional coordinate system (x, y, z) are changed to $(-x, -y, -z)$ by a parity operation. The conservation rule means that these two operations when applied together do not alter an interaction, such as the decay of an unstable particle. What Cronin and Fitch found to their surprise was that a K^0_L meson can decay in

such a way as to violate CP-conservation about 0.2% of the time. Their results were verified at other accelerators and many strange explanations were provided. The conclusion that had to be drawn was that decays of K^0_L mesons do violate CP-conservation and so are different from all other known particle interactions.

While it was relatively simple to confirm these results, it has proved much more difficult to explain them. The latest theories seem to view particles as being built up from six basic different entities called quarks. Earlier theories based on four quarks did not explain CP-violation satisfactorily. There is firm evidence for the existence of five kinds of quark, but the search for the sixth has so far been unsuccessful. CP-violation could explain why we exist at all. Scientists are puzzled why matter seems to dominate over antimatter, when all the theories suggest that matter and antimatter should have been formed in equal amounts. CP-violation could help to solve this problem, since if antiparticles decay faster than particles, they would totally disappear.

The work of Cronin and Fitch is recorded as a classic piece of research and their findings have had impact on an outstanding controversy about the symmetry of nature. They have shown for the first time that left–right assymmetry is not always preserved when some particles are changed in state from matter to antimatter.

FitzGerald George Francis 1851–1901 was an Irish theoretical physicist who worked on the electromagnetic theory of light and radio waves. He is best known, however, for the theoretical phenomenon called the Lorentz–FitzGerald contraction which he and, independently, Hendrik Lorentz (1853–1928) proposed to account for the negative result of the famous Michelson–Morley experiment on the velocity of light.

FitzGerald was born in Dublin on 3 August 1851. He was a nephew of the physicist George Stoney, and received his initial education at home, tutored by the sister of the mathematician George Boole. He entered Trinity College, Dublin, in 1867 (at the age of only 16) and graduated four years later. FitzGerald went on to study mathematical physics, and in 1877 received a Fellowship of Trinity College. He became Erasmus Smith Professor of Natural and Experimental Philosophy there in 1881, an appointment he retained until he died, in Dublin, on 22 February 1901.

FitzGerald predicted that a rapidly oscillating (that is, alternating) electric current should result in the radiation of electromagnetic waves – a prediction proved correct in the late 1880s by Heinrich Hertz's early experiments with radio, which FitzGerald brought to the attention of the scientific community in Britain.

At that time, one school of thought among physicists postulated that electromagnetic radiation – such as light – had to have a medium in which to travel. This hypothetical medium, termed the 'ether', was supposed to permeate all space and in 1887 the US physicists Albert Michelson and Edward Morley conducted an experiment intended to detect the motion of the Earth through the ether. They measured the velocity of light simultaneously in two directions at right-angles to each other, but found no difference in the values – the velocity was unaffected by the Earth's motion through the ether. FitzGerald proposed in 1892 that the Michelson–Morley result – or lack of result – could be accounted for by assuming that a fast-moving object diminishes in length, and that light emitted by it does indeed have a different velocity but travels over a shorter path, and so seems to have a constant velocity no matter what the direction of motion. (Only an observer outside the moving system would be aware of the reduction in light velocity; within the system the contraction would also affect the measuring instruments and result in no change in the perceived velocity.) He worked out a simple mathematical relationship to show how velocity affects physical dimensions. The idea was independently arrived at and developed by the Dutch physicist Hendrik Lorentz in 1895, and it became known as the Lorentz–FitzGerald contraction. In 1905, four years after FitzGerald's death, the contraction hypothesis was incorporated and given a different interpretation in Albert Einstein's general theory of relativity.

Fizeau Armand Hippolyte Louis 1819–1896 was a French physicist who was the first to measure the speed of light on the Earth's surface. He also found that light travels faster in air than in water, which confirmed the wave theory of light, and that the motion of a star affects the position of the lines in its spectrum.

Fizeau was born in Paris on 23 September 1819 into a wealthy family. He began to study medicine, his father being Professor of Pathology at the Paris Faculty of Medicine, but was forced to abandon his medical ambitions through ill health. Fizeau then turned to physics, studying optics at the College de France and taking a course with François Arago (1786–1853) at the Paris Observatory.

Fizeau's principal contributions to physics and astronomy cover a fairly short period of time and include the first detailed photographs of the Sun in 1845, the proposal that a moving light source such as a star undergoes a change in observed frequency that can be detected by a shift in its spectral lines in 1848, the first reasonably accurate determination of the speed of light in 1849, and the discovery that the velocity of light is greater in air than in water in 1850. Many of his discoveries were made in collaboration with Léon Foucault (1819–1868).

In recognition of his work, Fizeau gained the first Triennial Prize of the Institut de France in 1856. He became a member of the Academy of Sciences in 1860, rising to become its president in 1878, and he was awarded the Rumford Medal of the Royal Society in 1866. Fizeau died at Venteuil on 18 September 1896.

Inspired by Arago, Fizeau began to research into the new science of photography in 1839, improving the daguerrotype process. The fruitful collaboration with Foucault commenced at this time, leading them to develop photography for astronomical observations by taking the first detailed pictures of the Sun's surface in 1845. The two men went on to study the interference of light rays and showed how it is related to the wavelength of the light. They also found, in 1847, that heat rays from the Sun undergo interference and that radiant heat therefore behaves as a wave motion.

These latter results greatly strengthened the view that light is a wave motion, an issue that was being hotly debated at the time following the experiments of Thomas Young (1773–1829) and Augustin Fresnel (1788–1827) that had established the wave theory earlier in the century. In 1838, Arago had proposed an experiment to settle the question by measuring the velocity of light in air and in water. The wave theory demanded that light travels faster in air and the particle theory required it to travel faster in water. Arago's proposal was to reflect light from a rotating mirror, the amount of deflection produced being related to the velocity of light.

In 1847, Fizeau and Foucault split up their partnership. Foucault persevered with Arago's suggestion but Fizeau decided to try a simpler method. In 1849, he sent a beam of light through the gaps in the teeth of a rapidly rotating cog wheel to a mirror 8 km/5 mi away. On returning, the beam was brought to the edge of the wheel, the speed being adjusted so that the light was obscured. This meant that light rays which had passed through the gaps were being blocked on their return by the adjacent teeth as they moved into the position of the gaps. The time taken for the teeth to move this distance was equal to the time taken for light travel 16 km/10 mi to the mirror and back. Fizeau's wheel had 720 teeth and rotated at a rate of 12.6 revolutions per second. By a simple calculation, he obtained the value of 315,000 km/195,741 mi a second for the speed of light. Although 5% too high, this was the first reasonably accurate estimate, previous values having been obtained by inaccurate astronomical methods.

In collaboration with Louis Bréguet (1804–1883), Fizeau now applied himself to Arago's proposal and set about measuring the speed of light in air and water. Using a rotating-mirror apparatus and dividing the light beam so that half passed through air and half through a tube of water, Fizeau showed in 1850 that light travels faster in air. Foucault reached the same conclusion in the same way at the same time, and the wave theory of light was finally confirmed.

Another important experiment conducted by Fizeau in 1851 was one in which he determined the amount of drift of light waves in a transparent medium which is in motion. According to a theory given by Fresnel, the velocity of drift of waves in a medium moving with velocity u is

$$(1 - 1/\eta^2)u,$$

where η is the refractive index of the medium. Fizeau arranged two tubes side by side and water was forced at a considerable speed (as much as 7 m/23 ft per second) along one tube and back by the other, while a beam of light was split into two parts which were sent through the tubes, one with the stream and the other against it. They were then brought together again and tested for interference produced by any difference in time traversed arising from the motion of the water. The result gave exactly the formula quoted above.

Earlier, in 1848, Fizeau made a fundamental contribution to astrophysics by suggesting that the Doppler effect would apply to the light received from stars, motion away from the Earth causing a red shift in the spectral lines and motion towards the Earth producing a blue shift. Fizeau may have been unaware of Doppler's discovery of the effect in sound in 1842 and of his erroneous theory relating the colour of stars to their motion. Fizeau's discovery is now the basis of the principal method of determining the distances of galaxies and other distant bodies.

Fizeau was able to finance his experiments from his personal wealth, and this enabled him to develop experimental methods that were later refined by others, particularly Foucault and Albert Michelson (1852–1931), leading to accurate values for the velocity of light and the consequent development of relativity.

Flamsteed John 1646–1719 was an English astronomer and writer who became the first Astronomer Royal based at Greenwich. His work on the stars, which formed the basis of modern star catalogues, was much admired by contemporary scientists, among whom were Isaac Newton (1642–1727) and Edmond Halley (1656–1742). Like many professional scientists of his time, he was also a clergyman.

Flamsteed was born on 19 August 1646 at Denby, near Derby, the son of a prosperous businessman. At the age of 16, ill health forced him to leave Derby Free School and abandon, at least temporarily, his university ambitions. The next seven years he spent at home, educating himself in astronomy against the wishes of his father (who evidently saw him as his own successor in the family business). In 1670, however, he entered his name at Jesus College, Cambridge, and took an MA degree by letters-patent. He was ordained in the following year. Through the influence of Jonas Moore, a courtier of King Charles II, Flamsteed was made 'Astronomical Observator' at the newly established Greenwich Observatory in 1675. In 1684 he was presented with the living of Burstow, Surrey, by Lord North. He died at Greenwich on 31 December 1719 and was succeeded in his post there by Edmond Halley.

Flamsteed began his astronomical studies at home by observing a solar eclipse on 12 September 1662, about which he corresponded with several other astronomers. When he decided to take his university degree in 1670, he sent his early studies to the Royal Society and they were published in *Philosophical Transactions*. This was enough to gain him general scientific recognition, but it was Jonas Moore who launched him in his career when he gave him a micrometer and promised him a good telescopic lens, thus enabling him to start serious practical work.

At the time Flamsteed's research began, 60 years had passed since Galileo (1564–1642) had made his discoveries, the star catalogue prepared by Tycho Brahe (1546–1601) was still the standard work, and the laws of Johannes Kepler (1571–1630) were only gradually being accepted.

Flamsteed resolved to end the apparent stagnation in astronomical science. In 1672, he determined the solar parallax from observations of Mars when the planet was at its closest to the Sun, using the rotation of the Earth to establish a base-line. Four years later, and only two months after his appointment to Greenwich, he began observations that were to result in a 3,000-star British catalogue. His results improved on Brahe's work by a factor of 15, but at first they were only relative measurements, with no anchor in the celestial sphere. To assist him in his work, Jonas Moore further donated two chronometers and a 2.1-m/7-ft sextant, with which he made 20,000 observations between 1676 and 1689.

Settled in at Greenwich, Flamsteed became interested in the work on lunar theory published by Jeremiah Horrocks (or Horrox; 1619–1641) some decades earlier. He brought Horrocks's constants up to date and revised his calculations no fewer than three times. The models in Flamsteed's third set of calculations were of fundamental importance in some of Newton's theoretical work. Following his revisions of lunar theory, Flamsteed also produced three different sets of tables describing the motion of the Sun. The first was issued before Flamsteed had any original observations on which he could base his parameters. His second gave a new determination of solar eccentricity, at almost the true value of 0.01675, and was published in his book *Doctrine of the Sphere* in 1680. The third was printed in 1707 in Whiston's *Praelectiones Astronomicae*. It included more detailed observations made possible by new equipment purchased after he inherited his father's estate in 1688.

With his improved facilities, including a mural arc, Flamsteed was able to make some very precise measurements: he determined the latitude of Greenwich, the slant of the ecliptic and the position of the equinox. He also worked out an ingenious method of observing the absolute right ascension – a coordinate of the position of a heavenly body. His method, a great improvement on previous systems, removed all errors of parallax, refraction and latitude. Having obtained the positions of 40 reference stars, he then went back and computed positions for the rest of the 3,000 stars in his catalogue.

Flamsteed also produced tables of atmospheric refraction, tidal tables, and supervised the compilation of the first table describing the inequality of the lunar elliptic following Kepler's second law.

A serious-minded man (possibly as a result of his constantly frail health), Flamsteed was never good at dealing with other people. Much of the last 20 years of his life was spent in controversy over the publication of his work. His results were urgently needed by Isaac Newton and Edmond Halley to test their theories, but Flamsteed was determined to withhold them until he was quite certain they were correct. A row with both Halley and Newton in 1704 eventually led to Flamsteed's work being unlawfully printed in 1712, but Flamsteed managed to secure and burn 300 copies of the printed production. Accordingly, the preparation of his great work, *Historia Coelestis Britannica*, was completed by his assistants six years after his death, in 1725, and his *Atlas Coelestis* was published even later, in 1729.

Fleming Alexander 1881–1955 was a British bacteriologist who discovered penicillin, a substance produced by the mould *Penicillium notatum* and found to be effective in killing various pathogenic bacteria without harming the cells of the human body. Penicillin was the first antibiotic to be used in medicine. For this discovery, he shared the 1945 Nobel Prize for Physiology or Medicine with Howard Florey and Ernst Chain, who developed a method of producing penicillin in quantity.

Fleming was born on 6 August 1881 in Lochfield, Ayrshire, Scotland, the son of a farmer. He was educated at Kilmarnock Academy and, after his father died in 1894, his poverty-stricken family sent him to London, where he first studied at the London Polytechnic Institute and then got a job as a clerk in a shipping office. While working there, and encouraged by his brother who was a doctor, he won a scholarship to study medicine at St Mary's Hospital Medical School, London, in 1902. He graduated four years later and remained at St Mary's in the bacteriology department for the rest of his career.

In his early years Fleming assisted the bacteriologist Almroth Wright, an association that continued when the two men were in the Royal Army Medical Corps and worked together in military hospitals during World War I. After the war, in 1918, Fleming returned to St Mary's as a lecturer, becoming Director of the Department of Systematic Bacteriology and Assistant Director of the Inoculation Department in 1920. He was appointed professor there and lecturer at the Royal College of Surgeons in 1928. He was knighted in 1944 and in 1946 became Director of the Wright–Fleming Institute, where he continued to work until he retired in 1954. He died in London on 11 March 1955.

In 1928, Fleming made his major discovery quite by accident. He was working on the bacterium *Staphylococcus aureus* and had put aside some Petri dishes that contained the cultures. He later noticed that specks of green mould had appeared on the nutrient agar and that the bacterial colonies around the specks had disappeared. The effect on the bacteria was 'antibiosis' (against life). Fleming cultured the mould in nutrient broth and it formed a felt-like layer on the surface, which he filtered off. He tested the filtrate on a range of bacteria and found it killed some disease bacteria, but not all of them. He identified the mould as *Penicillium notatum*, a species related to that which grows on stale bread, and named the active substance it produced – the antibiotic element – 'penicillin'. Craddock, one of Fleming's assistants, grew some *Penicillium* in milk and ate the cheeselike product without any ill effects; no harm resulted, either, when mice and rabbits were injected with the material.

The purification and concentration of penicillin was, however, a chemical problem and Fleming was not a chemist. Two of his assistants made some progress but they left the matter unresolved until 1939, when Florey and Chain, in Oxford, isolated the substance and purified it. They published their results in 1940, and work began on the large-scale production of penicillin.

It was generally assumed that the original phenomenon that Fleming had observed was a common event, but Fleming was never able to produce the effect again. It has since been shown that a similar result is achieved only under very precise conditions, which are unlikely to be met during the routine inoculation and incubation of a bacterial plate.

Fleming also developed methods, which are still in use, of staining spores and flagella of bacteria. He identified organisms that cause wound infections and showed how cross-infection by streptococci can occur among patients in hospital wards. He also studied the effects of different antiseptics on various kinds of bacteria and on living cells. His interest in chemotherapy led him to introduce Paul Ehrlich's Salvarsan into British medical practice.

In 1922, Fleming discovered the presence of the enzyme lysozyme in nasal mucus, tears and saliva, where it catalyses the breakdown of carbohydrates surrounding bacteria and kills them. Fleming later showed it to be present in most body fluids and tissues. The enzyme thus helps to prevent infections, and has become a useful research tool for dissolving bacteria for chemical examination.

Penicillin, the first of the antibiotics, has been used with outstanding success in the treatment of many bacterial diseases, including pneumonia, scarlet fever, gonorrhea, diphtheria and meningitis, and for infected wounds. Its discovery led to a scramble for further antibiotics in which streptomycin, chloromycetin and the tetracyclines were discovered. Most antibiotics can now be made synthetically, and penicillin can be modified by chemical means for specific purposes.

Fleming John Ambrose 1849–1945 was a British electrical engineer whose many contributions to science included the invention of the thermionic diode valve – a key electronic component in the early development of radio.

Fleming was born in Lancaster, Lancashire, on 29 November 1849, the son of a parson. When he was four years old his family moved to London, where his father continued his ministry in Kentish Town. Fleming was educated at University College School, and later at University College, London where he was awarded his BSc in 1870. His studies at the University were only part-time because for two years he worked as a clerk with a firm of stockbrokers. From 1872 to 1874 he studied at South Kensington, first under Edward Frankland and later under Frederick Guthrie.

For the next three years he taught science at Cheltenham College, then in 1877 he entered St John's College, Cambridge, having won an entrance exhibition (and saved £400 to pay the fees). At Cambridge he worked in the Cavendish Laboratory and studied electricity and advanced mathematics under James Clerk Maxwell (1831–1879), author of the famous treatise on electricity and magnetism. In 1879,

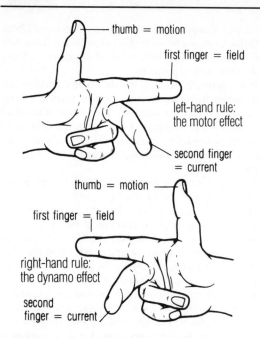

thumb = motion

first finger = field

left-hand rule:
the motor effect

second finger
= current

thumb = motion

first finger = field

right-hand rule:
the dynamo effect

second
finger = current

Fleming *Fleming's rules give the direction of the magnetic field, motion, and current in electrical machines. The left hand is used for motors, and the right hand for generators and dynamos.*

Fleming obtained his DSc (London) and researched into electrical resistances. Three years later he was appointed to the newly created Chair of Mathematics and Physics at University College, Nottingham, but resigned his post the following year to take up consulting work with the Edison Electric Light Company.

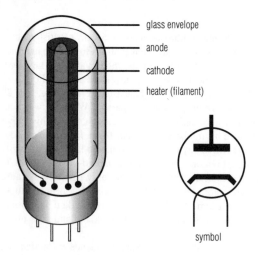

glass envelope

anode

cathode

heater (filament)

symbol

Fleming *Fleming's work on electric lamps led him to invent the diode valve, a two-electrode vacuum tube in which a heated filament causes thermionic emission of electrons from the cathode. The valve was used as a rectifier and a detector in radio receivers until largely superseded by the semiconductor diode.*

In 1883, Fleming was elected Fellow of St John's and in 1885 he was appointed professor at University College, London. Between 1889 and 1898 he published several important papers on the practical problems of the electrical and magnetic properties of materials at very low temperatures. During part of this time he made a careful study of the 'Edison effect' in carbon-filament lamps. In 1904 he produced experimental proof that the known rectifying property of a thermionic valve was still operative at radio frequencies, and this discovery led to the invention and production of what was first known as the 'Fleming valve'.

In 1905, Fleming described his electric wave measurer or cymometer, and demonstrated it to the Royal Society. In 1874, he read the very first paper to the Physical Society on its foundation in that year; 65 years later in 1939, at the age of 90, he read his last paper to the same Society. Fleming died on 18 April 1945 at his home in Sidmouth, Devon.

The value of Fleming's work was widely recognized. In 1892, he was elected a Fellow of the Royal Society and received the Hughes Medal from the Royal Society in 1910. In 1921, he was awarded the Albert Medal of the Royal Society of Arts and, in 1928, the Faraday Medal from the Institution of Electrical Engineers. He was knighted in 1929.

Although an avid experimenter, Fleming did not concern himself merely with the theoretical aspects of electrical science but took an active part in its practical application. As engineer and adviser to the Edison, Swan and Ferranti electric lighting companies between 1882 and 1889, he was responsible for improvements in incandescent lamps, meters and generators. For 26 years he was scientific consultant to the Marconi Wireless Telegraph Company and he designed many parts of their early radio apparatus – particularly those used by Marconi in his pioneering transatlantic transmission in 1901.

In the early 1880s, Fleming investigated the phenomenon known as the Edison effect – the escape of electrons or ions from a heated solid or liquid – but had abandoned the project as being of no practical value. In 1904, this early work on the 'one-way' conductance of electricity in an incandescent lamp led to Fleming's most important practical achievement – the invention of the two electrode thermionic rectifier which became known as the Fleming valve or diode.

Fleming was searching for a more reliable detector of weak electric currents. He recalled his earlier experiments and made a new 'lamp' that had a metal cylinder surrounding the filament in a high vacuum. He found that it was very useful in detecting the very weak currents in radio receiving apparatus because it responded to currents alternating at very high frequencies. Previous instruments he had employed had been unable to do this because the electric 'waves' produced forces that tended to produce additional alternating currents.

He called his invention a valve because it was the electric equivalent of a check valve in a water supply system which allows water to pass in one direction only. In a like manner, the Fleming valve allowed electrical currents to pass in only one direction. It worked by allowing one of the electrodes – the cathode – to be kept hot so that electrons could evaporate from it into the vacuum. The other electrode – the anode – was left cool enough to prevent any appreciable evaporation of electrons from it. It was thus a device that permitted currents to flow essentially in one direction only when an alternating current was applied to it, and it revolutionized the early science of radio.

The Fleming valve has now been superseded by the transistor diode, an electronic device which utilizes the properties of single-crystal semiconductors. However, because he made possible a very significant advance in radio and television, Fleming's work will always remain an important milestone in electronic engineering. And despite the comparative relegation of the importance of his most famous practical contribution, his work in the theoretical and teaching aspects of electrical science remains undiminished.

Fleming Williamina Paton Stevens 1857–1911 was the British-born co-author (with Edward Pickering) of the first general catalogue classifying stellar spectra.

Fleming was born on 15 May 1857 in Dundee, Scotland, where she was educated. She taught for a few years there before she married and emigrated to the United States in 1878. Shortly after her arrival in Boston her marriage broke up. She was then employed, in 1879, as an assistant to Edward Pickering (1846–1919), Director of the Harvard College Observatory. Her work for him was as a 'computer' and copy editor, at which she was so successful that she was soon put in charge of twelve other 'computers'. In 1898, she was appointed curator of astronomical photographs. She died in Boston on 21 May 1911.

The project initiated by Pickering was simple in concept, but required meticulous dedication and patience. Photographs were taken of the spectra obtained using prisms placed in front of the objectives of telescopes. Although the use of the technique was restricted to stars about a certain magnitude, it yielded a wealth of information. In the course of her analysis of these spectra, Fleming discovered 59 nebulae, more than 300 variable stars, and 10 novae (which is even more impressive when it is recalled that at the time of her death in 1911 only 28 novae had been found).

The spectra of the stars observed in this manner could be classified into categories. Fleming designed the system adopted in the 1890 *Draper Catalogues*, in which 10,351 stellar spectra were listed in 17 categories ('A' to 'Q'). The majority of the spectra fell into one of six common categories; only 72 spectra accounted for the other eleven classes. This classification system represented a considerable advance in the study of stellar spectra, although it later was superseded by the work of Annie J. Cannon (1863–1941) at the same observatory.

Fleming's special interest was in the detailed classification of the spectra of variable stars; she proposed a system in which their spectra were subdivided into eleven subclasses on the basis of further detailed spectral characteristics.

Florey Howard Walter 1898–1968 was an Australian-born British bacteriologist who developed penicillin and made possible its commercial production. For this work he received the 1945 Nobel Prize for Physiology or Medicine, which he shared with Alexander Fleming (who discovered penicillin) and Ernst Chain.

Florey was born on 24 September 1898 in Adelaide. He was educated locally and read medicine at the University of Adelaide, qualifying in 1921 and winning a Rhodes scholarship to Oxford University to study physiology and pharmacology, where he worked under Charles Sherrington. He spent a brief period at Cambridge in 1924 before going to the United States to study. In 1926 he returned to Britain as a researcher at the London Hospital, moving back to Cambridge in the following year as a lecturer in Special Pathology and later Director of Medical Studies. Florey was appointed to the Chair of Pathology at Sheffield University in 1932 and became Professor of Pathology at Oxford in 1935 and Head of the Sir William Dunn School of Pathology. He was knighted in 1944 and was President of the Royal Society from 1960 to 1965, the same year that he was made a life peer and member of the Order of Merit. He was elected Provost of Queen's College, Oxford, in 1962 and died in Oxford on 21 February 1968.

At Oxford, Florey conducted investigations on antibacterial substances and during these he successfully purified lysozyme, the bacteriolytic enzyme discovered by Fleming. In 1939, continuing his research, he decided to concentrate on Fleming's unresolved problem of the purification of penicillin. Florey's co-worker, Ernst Chain, an accomplished biochemist, set about growing Fleming's strain of *Penicillium* and extracting the active material from the liquid culture medium. Chain and Florey extracted a yellow powder from the medium, but the process proved to be an extraordinarily difficult task and 18 months later they had collected only 100 mg/0.0035 oz of this substance.

Florey and his team began a series of carefully controlled experiments on mice infected with standard doses of streptococci. The team found that a dilution of one in a million inhibited the growth of streptococci but was harmless to mice, growing tissue cells and leucocytes. It was also discovered that the penicillin did not behave like an antiseptic or an enzyme, but blocked the normal process of cell division. The tests showed conclusively that penicillin could protect against infection but that the concentration of penicillin in the human body and the length of time of treatment were vital factors in the rate of success.

In 1940, during the early part of World War II, the German invasion of Britain seemed imminent; Florey and his colleagues smeared spores of the *Penicillium* culture on their coat linings so that, if necessary, any one of them could continue their research elsewhere. Further research was hindered by the great difficulty in producing enough penicillin for tests, and commercial firms were at that time too committed to vaccine production to participate. Florey's team persevered and improved their techniques, which resulted in a purer product suitable for preliminary trials on human beings. Only desperately ill patients with little hope of recovery were selected. The first patient was a police constable with a rampant infection of the face, head and lungs. Within five days his improvement was miraculous, but he died one month later because it had been impossible to continue treatment long enough – the stock of penicillin was exhausted. The next five patients treated made complete recoveries.

The problem remained of producing penicillin in large quantities. Small-scale production continued in Florey's department, supplemented by minimal contributions from two commercial firms. In 1943, he went to Tunisia and Sicily and used penicillin successfully on war casualties. By 1945, studies had progressed far enough to show that antibacterial activity could take place using a dilution of 1 part in 50 million and, with the war over, large-scale commercial production of penicillin could begin.

Florey and his co-workers resumed their researches on other antibiotics. They discovered cephalosporin C, which later became the basis of some derivatives, such as cephalothin, which can be used as an alternative antibiotic to penicillin.

Florey was a great scientist with abundant energy, experimental skill and a flair for choosing fruitful lines of research. He and his collaborators made penicillin available for therapeutic purposes, and were responsible for ushering in the era of antibiotic therapy.

Flory Paul John 1910–1985 was a US polymer chemist who was awarded the 1974 Nobel Prize for Chemistry for his investigations of synthetic and natural macromolecules. With Wallace Carothers (1896–1937) he developed nylon, the first synthetic polyamide, and the synthetic rubber neoprene.

Flory was born on 19 June 1910 at Sterling, Illinois. He graduated from Manchester College, Indiana, in 1931 and gained his PhD from Ohio State University three years later. He then embarked on a career as an industrial research chemist, working at the Du Pont Experimental Station in Wilmington, Delaware (with Carothers) from 1934 to 1938 and at the Esso Laboratory Standard Oil Development Company in Elizabeth, New Jersey (1940–1943), before becoming Director of Fundamental Research at the Goodyear Tire and Rubber Company (1943–1948). He had held a research associateship at Cincinnati for three years between 1940 and 1943, and in 1948 he again took up an academic post as Professor of Chemistry at Cornell University. He remained there until 1956 when he became Executive Director of Research at the Mellon Institute, Pittsburgh.. In 1961, he was made Professor of Chemistry at Stamford University where he remained, eventually becoming Emeritus Professor.

Flory pioneered research into the constitution and properties of substances made up of giant molecules, such as rubbers, plastics, fibres, films and proteins. He showed the importance of understanding the sizes and shapes of these flexible molecules in order to be able to relate their chemi-

cal structures to their physical properties. In addition to developing polymerization techniques, he discovered ways of analysing polymers. Many of these substances are able to increase the lengths of their component molecular chains and Flory found that one extending molecule can stop growing and pass on its growing ability to another molecule.

Working with Carothers, he prepared the polyamide Nylon 66 by heating a mixture of adipic acid (hexan-1,6-dioic acid) and hexamethylene diamine (hexan-1,6-diamine). They made Nylon 6 from caprolactam, showing it to be a polymer of the type $-[-CO(CH_2)_5NH-]_n-$. Neoprene was made by polymerizing chloroprene (but-2-chloro-1,3-diene), and was soon followed by other synthetic rubbers made by polymerization and co-polymerization of various butenes.

Flory's later researches looked for and found similarities between the elasticity of natural organic tissues – such as ligaments, muscles and blood vessels – and synthetic and natural plastic materials.

Fontana Niccolò c.1499–1557. Italian mathematician and physicist; *see* Tartaglia.

Forbes Edward 1815–1854 was a versatile Scottish naturalist who made significant contributions to oceanography.

A Manxman, Forbes was a banker's son, who showed an early talent for natural history and for drawing. He studied medicine at Edinburgh, but soon grew absorbed in natural history, under the influence of Professor Robert Jameson. Taking a large part in the British Association in its early years, he became curator and later, in 1844, palaeontologist to the Geological Society of London and subsequently Professor of Natural History at Edinburgh; the last post in his tragically brief career was at the Royal School of Mines in London.

Travelling widely in Europe and in the Near East, and in 1841 serving as naturalist on a naval expedition in the eastern Mediterranean, Forbes proved himself a tireless collector of fauna and flora, showing particular interest in the natural distribution of species. Molluscs interested him particularly; he was concerned with their taxonomy, and he also studied their migration habits and their different environments. A passionate palaeobotanist, Forbes divided British plants into five groups, and proposed that Britain had once been joined to the continent by a land-bridge. Plants had crossed over, he believed, in three distinct periods. Forbes also blazed trails in scientific oceanography. He discounted the contemporary conviction that marine life subsisted only close to the sea surface, spectacularly dredging a starfish from a depth of 400 m/1,300 ft in the Mediterranean. His *The Natural History of European Seas* (posthumously published in 1859) was a pioneering oceanographical text. It developed his favourite idea of 'centres of creation', that is, the notion that species had come into being at one particularly favoured location. Though not an evolutionist, Forbes's ideas could be commandeered for evolutionary purposes.

Ford Henry 1863–1947 was a US automotive engineer and industrialist who, in the early twentieth century, revolutionized the motorcar industry and manufacturing methods generally. His production of the Model-T popularized the car as a means of transport and made a considerable social and economic impact on society. His introduction of the assembly line (bringing components to the workers, rather than vice versa) gave impetus to the lagging Industrial Revolution.

Ford was born at Springwells, Michigan, on 30 July 1863. He attended rural schools and soon displayed a mechanical and inventive skill. Moving to Detroit at the age of 16. he obtained a job as a machinist's apprentice. During the next few years he worked for several different companies, and repaired watches and clocks in his spare time. After his apprenticeship Ford worked on the maintenance and repair of Westinghouse steam engines. In 1891. he was appointed chief engineer to the Edison Illuminating Company.

In about 1893, he constructed a one-cylinder petrol (gasoline) engine and went on to build his first car in 1896. Three years later he resigned from Edison's and joined the Detroit Automobile Company. He left there in March 1902 and, with some financial backing, formed the Ford Motor Company on 16 June 1903.

The first Ford car sold almost as soon as it was produced; further orders came in, and production rose rapidly. In 1906, because of a disagreement with his business associates (the Dodge brothers), Ford became the majority shareholder and president of the company, and by 1919 he and his immediate family held complete control of it.

Despite his success with the motorcar, Ford's non-industrial activities met with little success. An expedition to Europe he organized in December 1915 aimed at ending World War I, proved to be a fiasco. His attempt to run as a democrat for a Senate seat in Michigan and subsequent defeat left him bitter about alleged irregularities in his opponent's campaign.

However, other of his activities have brought him a great deal of credit as a benefactor. He created Greenfield Village as a monument to the simple rural world – a world that his automobiles had done so much to destroy. In it, he reconstructed the physical surroundings and the crafts of an earlier era. Near the village is the Henry Ford Museum containing his fine collection of antiques. Ford also restored the Wayside Inn of Longfellow's poem, and his important collection of early motion pictures were donated to the National Archives. He endowed the Ford Foundation, established in 1936, as a private, non-profitmaking corporation 'to receive and adminster funds for scientific, educational and charitable purposes'. Ford finally gave up the Presidency of his company in 1945. He died in Dearborn on 7 April 1947.

Motorcars were in their infancy when Henry Ford produced his first automobile in 1896 and decided to make his reputation in the field of racing cars. His determination to do this led him to leave the Detroit Automobile Company and to work on his own. In a memorable race at Grosse Point, Michigan, in October 1901, his victory brought him the publicity he sought. Barney Oldfield, also driving a Ford racer, added to Ford's reputation and in 1904 Ford himself drove his '999' to a world record of 39.4 sec for 1 mi/1.6 km over the ice on Lake St Clair in January 1904.

The success of Ford's family cars was immediate. From the low-priced Model-N he went on to produce the Model-T, which first appeared in 1908. Over 19 years, 15 million were sold, and the car is regarded as having changed the pattern of life in the United States. It was one of the first cars to be made using assembly-line methods, and Henry Ford's name became a household word the world over.

The car itself was a sturdy black vehicle with a four-cylinder 20 hp engine with magnetic ignition. A planetary transmission eliminated the gear-shift (and the danger of stripping the gears). 'Splash' lubrication was used and vanadium steel, of high tensile strength but easy to machine, was employed in many of the car's parts. Ford himself was responsible for the overall concept, and many of the basic ideas embodied in the construction of the Model-T were his own. In 1914 Ford became the first employer of mass labour to pay $5 a day minimum wage to all his employees who met certain basic requirements.

Dictatorial in his attitude, he later dismissed many key individuals who had helped to build the company's early success. He relinquished presidency of the company in 1909 to his son Edsel but strongly resisted changes in production despite an increasing loss of the market to up-and-coming competitors like the General Motors Corporation and Chrysler. Eventually, Ford acknowledged the inroads the newcomers were making on his Model-T. Characteristically, he set out to beat them with a new design, and in January 1928 he produced the Model-A.

The new car was the first to have safety glass in its windscreen as standard equipment. It was available in four colours and 17 body styles. Four-wheel brakes and hydraulic shock absorbers were incorporated in the car and it became a worthy successor to the Model-T. But Ford's previously undisputed leadership in the industry was not restored. Even the introduction of the V-8 engine – an engineering innovation at the time – did not halt the steady deterioration of Ford's share of the market.

When Edsel Ford died in 1943. Henry Ford resumed presidency of the company but in 1945 he surrendered it, for the last time, to his grandson and namesake, Henry Ford II.

Forsyth Andrew Russell 1858–1942 was a British mathematician whose facility with languages enabled him first of all to keep pace with, and even surpass, mathematical developments elsewhere in contemporary Europe, and then to translate such developments into English for the benefit of British mathematicians. Having done so – in an extremely important book – he was apparently unable then to maintain his precedence. Nevertheless, it was through his influence that in Britain the subject of the theory of functions dominated mathematical research for many years.

Forsyth was born in Glasgow on 18 June 1858, and

obtained his initial education at Liverpool College. From there he won a scholarship to Trinity College, Cambridge, which he entered in 1877; lectures given there by Arthur Cayley (1821–1895) had a profound influence upon him and upon his general approach to mathematics. A dissertation by Forsyth published in the *Proceedings* of the Royal Society then led to the offer of a 'prize' Fellowship at Trinity College, but no subsequent offer of a faculty position was forthcoming, so in 1882 he took up the Chair of Mathematics at Liverpool College instead. Two years later, however, he returned to Trinity College as Lecturer, and remained there for the next 26 years. During that time he wrote several books and translated the works of others – his crucial *Theory of Functions* appeared in 1893 – and he became considerably involved with the day-to-day administration of Cambridge University. In 1895, he was appointed to the Sadlerian Chair of Pure Mathematics.

In 1910, however, at the age of 52, Forsyth left Cambridge. Much of 1912 he spent lecturing in India, returning to England the following year to become Chief Professor of Mathematics at the Imperial College, London. Determined to renew his study of languages he retired early (in 1923), but within two years had reverted to his mathematical interests. After the publication of his last mathematical work in 1935, however, he again returned to linguistic studies. He died in London on 2 June 1942.

During his lifetime Forsyth received many honours and awards. A Fellow of the Royal Society from 1886, he was presented with its Royal Medal in 1897.

Forsyth's early work was to systematize and develop the theory of double theta functions. He succeeded in demonstrating that such functions are related to the square roots of quintic and sextic polynomials in the same way as single theta functions are related to the square roots of cubic and quadratic polynomials. He also formulated a theorem that generalized a large number of identities between double theta functions; because this work was also carried out independently yet simultaneously by Henry Smith (1826–1883), the theorem is now called the Smith–Forsyth theorem.

Forsyth's *Theory of Functions* was intended as an advanced text that would introduce the main strands of continental mathematical study to British mathematicians who were then tending to lag behind in terms of development and innovative creativity. In fact, more importantly, the book not only served to introduce the work of the European schools, but also brought together the work of all the various schools in a single volume – and as such was of considerable importance not merely in Britain but also in continental Europe, where it also achieved success in translation. In Britain, the book led to the introduction of concepts such as symbolic variant theory, Weierstrassian elliptic functions, and many more, and completely changed the nature of mathematical thinking. The developments that the book stimulated were rapid and, sometimes, fundamental, and sadly left Forsyth – who only five years previously had been publicly acknowledged as the most brilliant pure mathematician in the country – far behind.

His skills belonged to older methods. During his later years he wrote a number of books (some on ordinary, linear and partial differential equations, one or two on Einstein's general theory of relativity) but he never again achieved the spectacular acclaim he had once enjoyed.

Fortin Jean Nicholas 1750–1831 was a French instrument maker who made precision equipment for many of the most eminent French scientists of his time, although he is remembered today for the portable mercury barometer, which is named after him.

Little is known of Fortin's life. He was born in Mouchy-la-Ville, Ile de France, on 8 August 1750. He worked in Paris as a member of the Bureau de Longitudes, having been helped in his early career by Antoine Lavoisier, for whom he made several scientific instruments. During his later years Fortin worked for the Paris Observatory, constructing instruments for astronomical studies and surveying. His only known publication is an abridgment of John Flamsteed's *Atlas Celeste*, which was published in 1776. Fortin died in Paris in 1831.

One of Fortin's major early achievements was his construction of a precision balance for Lavoisier; it consisted of a beam, 1 m/3 ft long, mounted on steel knife-edges, and was able to measure masses as little as 70 mg/0.0025 oz. He made the first version of this balance in 1778, and another in 1799 for the Convention Committee on Weights and Measures; this latter version incorporated a comparator for standardizing weights. In the same year he adjusted the weight standard, the platinum kilogram, which was stored in the French National Archives.

Fortin is best known, however, for his barometers, although he did not make many of them. In 1800, he designed a portable mercury barometer that incorporated a mercury-filled leather bag, a glass cylinder in the cistern, and an ivory pointer for marking the mercury level. The mercury level could also be adjusted to the zero mark, and any barometer that possesses this feature is now known as a Fortin barometer. Fortin did not inevent these features but he was the first to use them together in a sensitive portable barometer.

Fortin also made apparatus used by Joseph Gay-Lussac in experiments on gas expansion; for Pierre Dulong and François Arago's investigation of the validity of the Boyle–Mariotte law; for Jean Biot and Arago's expedition to Spain in 1806 to measure the arc of the terrestrial meridian; and numerous other instruments, including various clocks.

Foucault Jean Bernard Léon 1819–1868 was a French physicist who invented the gyroscope, demonstrated the rotation of the Earth, and obtained the first accurate value for the velocity of light.

Foucault was born in Paris on 19 September 1819. He was educated at home because his health was poor, and went on to study medicine, hoping that the manual skills he developed in his youth would stand him in good stead as a surgeon. But Foucault soon abandoned medicine for science and supported himself from 1844 onwards, at first by

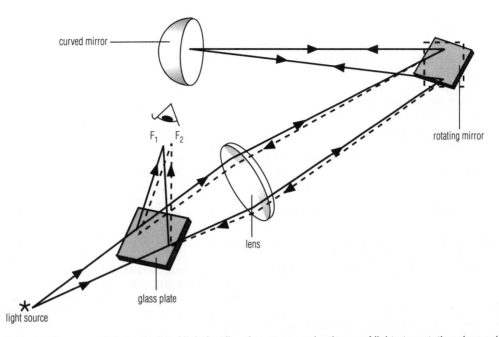

Foucault *Foucault measured the velocity of light by directing a converging beam of light at a rotating plane mirror, situated at the radius of curvature of a spherical mirror which reflected the light back along the same path. The change in angle of the rotating mirror displaced the focus of the returning beam from F_1 to F_2, and from this displacement and the mirror's rotational speed the velocity of light was calculated.*

writing scientific textbooks and then popular articles on science for a newspaper. He carried out research into physics at his home until 1855, when he became a physicist at the Paris Observatory. He received the Copley Medal of the Royal Society in the same year, and was made a member of the Bureau des Longitudes in 1862 and the Académie des Sciences in 1865. He died of a brain disease in Paris on 11 February 1868.

Foucault's first scientific work was carried out in collaboration with Armand Fizeau (1819–1896). Inspired by François Arago (1786–1853), Foucault and Fizeau researched into the scientific uses of photography, taking the first detailed pictures of the Sun's surface in 1845. In 1847, they found that the radiant heat from the Sun undergoes interference and that it therefore behaves as a wave motion. Foucault parted from Fizeau in 1847, and his early work then propelled him in two directions.

Making the long exposures required in those early days of photography necessitated a clockwork device to turn the camera slowly so that it would follow the Sun. Foucault noticed that the pendulum in the mechanism behaved rather oddly and realized that it was attempting to maintain the same plane of vibration when rotated. Foucault developed this observation into a convincing demonstration of the Earth's rotation by showing that a pendulum maintains the same movement relative to the Earth's axis and the plane of vibration appears to rotate slowly as the Earth turns beneath it. Foucault first carried out this experiment at home in 1851, and then made a spectacular demonstration by suspending a pendulum from the dome of the

Panthéon in Paris. From this, Foucault realized that a rotating body would behave in the same way as a pendulum and in 1852, he invented the gyroscope. Demonstrations of the motion of both the pendulum and gyroscope proved important to an understanding of the action of forces, particularly those involved in motion over the Earth's surface.

Foucault's other main research effort was to investigate the velocity of light. Both he and Fizeau took up Arago's suggestion that the comparative velocity of light in air and water should be found. If it travelled faster in water, then the particulate theory of light would be vindicated; if the velocity were greater in air, then the wave theory would be shown to be true. Arago had suggested a rotating mirror method first developed by Charles Wheatstone (1802–1875) for measuring the speed of electricity. It involved reflecting a beam of light from a rotating mirror to a stationary mirror and back again to the rotating mirror, the time taken by the light to travel this path causing a deflection of the image. The deflection would be greater if the light travelled through a medium that slowed its velocity. Fizeau abandoned this method after parting from Foucault and developed a similar method involving a rotating toothed wheel. With this, he first obtained in 1849 a fairly accurate numerical value for the speed of light.

Foucault persevered with the rotating mirror method and in 1850 succeeded in showing that light travels faster in air than in water, just beating Fizeau to the same conclusion. He then refined the method and in 1862 used it to make the first accurate determination of the velocity of light. His value of 298,005 km/185,177 mi per second was within

1% of the correct value, Fizeau's previous estimate having been about 5% too high.

Foucault also interested himself in astronomy when he went to work at the Paris Observatory. He made several important contributions to practical astronomy, developing methods for silvering glass to improve telescope mirrors in 1857 and for accurate testing of mirrors and lenses in 1858. In 1860, he invented high-quality regulators for driving machinery at a constant speed and these were used in telescope motors and also in factory engines.

Foucault's outstanding ability as an experimental physicist brought great benefits to practical astronomy and also, in the invention of the gyroscope, led to an invaluable method of navigation. It is ironic that he missed the significance of an unusual observation of great importance. In 1848, Foucault noticed that a carbon arc absorbed light from sunlight, intensifying dark lines in the solar spectrum. This observation, repeated by Gustav Kirchhoff (1824–1887) in 1859, led immediately to the development of spectroscopy.

Fourier Jean Baptiste Joseph 1768–1830 was a French mathematical physicist whose particular interest was to try to describe the transfer of heat in purely mathematical terms. The formulation of equations in order to achieve this was a complex task that necessitated the development of new mathematical tools, and he was responsible in this way for the discovery of the Fourier series and the Fourier integral theorem, which have together led to the evolution of the modern process now known as harmonic analysis.

Fourier was born in Auxerre on 21 March 1768, the son of a tailor. Orphaned when very young, he obtained his education at the local military academy, and it was there that his interest in mathematics was first aroused. He then went on to a Benedictine school in St Bênoit-sur-Loire, but returned to Auxerre at the outbreak of the French Revolution and taught at his old school. He was arrested in 1794, only to be released a few months later after the execution of Robespierre. He next studied in Paris at the Ecole Normale for a short period, and in 1795 was made an assistant lecturer at the Ecole Polytechnique under Joseph Lagrange (1736–1818) and Gaspard Monge (1746–1818).

In 1798 Fourier was selected to accompany Napoleon on his Egyptian campaign, and there conducted a variety of diplomatic affairs. Returning to France in 1801 he was appointed Prefect of Isère in the south of the country. During this period he continued his mathematical studies on a part-time basis. Napoleon conferred the title of Baron on Fourier in 1808, and later made him a Count. Fourier was then made Prefect of the *département* of Rhône, but resigned the post during Napoleon's Hundred Days in protest against the activities of the regime. Soon afterwards he obtained a post at the Bureau of Statistics and was able to devote all his energies to mathematics. He was elected to the French Academy of Sciences and made joint Secrétaire Perpétuel with Georges Cuvier in 1822; he was also elected to the Académie Française and made a Foreign Member of the Royal Society. He died on 4 May 1830, as an indirect result of a disease he had contracted while serving in Egypt.

One of Fourier's most important contributions to both mathematics and physics was the use of linear partial differential equations in the study of physical phenomena as boundary-value problems. In order to comprehend and explain the conduction of heat under conditions of different temperature gradients, and in materials with different shapes and conductivities, Fourier developed what is now called Fourier's theorem. This enables the equation for the description of heat diffusion to be broken up into a series of simpler (trigonometric) equations, the sum of which equals the original. The Fourier series can be used to describe complex periodic (that is, repeating) functions, and so can be applied to many branches of mathematical physics. Light, sound, and other wave-like forms of energy can be studied using Fourier's theorem, and a developed version of this method is now called harmonic analysis. At the time, such was the creative brilliance of Fourier in using linear partial differential equations to this end, however, that for the following century or more nonlinear differential equations were hardly used at all in mathematical physics.

Fourier contributed to other areas of mathematics as well; for example, he laid the groundwork for the later development of dimensional analysis and linear programming. Fascinated since the age of 16 by the theory of equations, Fourier's work at the Bureau of Statistics also stimulated him to investigate probability theory and the theory of errors.

Fourneyron Benoit 1802–1867 was a French engineer who invented the first practical water turbine.

Fourneyron was born on 31 October 1802 in Saint-Etienne, Loir. As the son of a geometrician, he was well prepared in mathematical sciences before entering the New School of Mines at Saint-Etienne, at 15 years of age. He graduated at the top of his class. His early activities were devoted to developing the mines at Le Creusot, prospecting for oil, laying out a railway and, finally, initiating the fabrication of tin-plate – until then an English monopoly – at Pont-sur-l'Ognon, Haute Saône.

Because the process involved the use of a water wheel, whose efficiency was very low, Fourneyron became obsessed with the idea of producing a high-efficiency waterwheel and, in 1827, he succeeded. In 1855, he produced an improved version and went on to build more than a thousand hydraulic turbines of various forms and for use in different parts of the world. He died in Paris on 8 July 1867.

The idea of using a stream of water to drive a wheel is very old and it is thought that the waterwheel was invented in the first century BC. The first device that operated on the principle of reaction was the steam 'turbine' of Hero of Alexandria in the first century AD.

Improvement in the design and efficiency of waterwheels came slowly. By the early part of the nineteenth century, with the application of mathematics and a growing knowledge of hydraulics, the first reaction wheels of Leonhard Euler and those of Claude Burdin were produced – but it was Fourneyron, one of Burdin's pupils, who first achieved success. His 1827 reaction turbine was 80% efficient and could develop about $4^1/_2$ kW/6 hp.

Fourneyron's machine is generally recognized as opening the modern era of practical water turbines. It was essentially an outward-flow turbine. Water passed through fixed guide passages and hence into guide passages in the moveable outer wheel. When the water impinged on these wheel vanes, its direction was changed and it escaped round the periphery of the wheel. But the outward-flow turbine was essentially unstable because, as water flowed through the fixed and moveable vanes it entered a region of successively increasing volume. The speed regulation of the turbine also presented difficulties.

Fourneyron patented an improved design which incorporated a three-turbine installation in 1832. However, his machine lost favour, being superseded in 1843 by the Jonval axial-flow machine.

Fourneyron's machines were still used in large commercial undertakings. All his earlier designs were of the free-flow efflux type, but he foresaw the advantages of allowing the efflux to flow into a diffuser and, in 1855, he patented an outflow diffuser in the form of the present-day inflow scroll case. Two turbines, each consisting of Fourneyron wheels keyed to one shaft, were used by the Niagara Falls Power Company in 1895. They were built into 49-m/160-ft wheel-pits dug into the supply channel at the top of the Falls.

Fowler William Alfred 1911– is a US physicist and astronomer who has published many papers on the measurement of nuclear reaction rates in the laboratory for application to the study of energy generation and the creation of elements heavier than hydrogen in the Sun and other stars.

Fowler was born in Pittsburgh, Pennsylvania, on 9 August 1911. Obtaining his bachelor's degree in physics at Ohio State University in 1933, he went to the California Institute of Technology, gained a PhD, and became a Research Fellow there in 1936. He has remained there, rising from Assistant Professor to Professor and, in 1970, Instructor Professor of Physics.

Fowler's work has, in the main, concentrated in research into the abundance of helium in the universe. The helium abundance was first defined as the result of the 'hot big bang' theory proposed by Ralph Alpher (1921–), Hans Bethe (1906–) and George Gamow (1904–1968) in 1948, and corrected through the brilliant theoretical work of Chushiro Hayashi (1920–) in 1950. In addition to altering the time-scale proposed in the α-β-γ theory, Hayashi also showed that the abundance of neutrons at the heart of the Big Bang did not depend on the material density but on the temperature and the properties of the weak interreactions. Provided the density is great enough for the reaction between neutrons and protons to combine at a rate faster than the expansion rate, a fixed concentration of neutrons will be incorporated into helium nuclei, however great the material density is – producing a 'plateau' in the relationship between helium abundance and material density.

In 1967, Fowler – together with Fred Hoyle (1915–) and R. Wagoner – made elaborate calculations of the percentage plateau abundance. His calculations took into account all the reactions that can occur between the light elements, and also considered the build-up of heavier elements; 144 different reactions were observed and the results analysed by computer. He and his collaborators claim an accuracy of helium abundance to 1% and found that the percentage abundance of helium in this plateau is between 25% and 29%. Their calculations for the build-up of other elements such as deuterium and lithium agree well with observations.

Fraenkel Abraham Adolf 1891–1965 was a German-born Israeli mathematician who is chiefly remembered for his research and perception in set theory, and for his many textbooks.

Fraenkel was born on 11 February 1891 in Munich, where he grew up, was educated, and first attended University. He also studied at the Universities of Marburg, Berlin and Breslau, Germany (now Wrocław, Poland). In 1916, he became a lecturer at the University of Marburg, and in 1922 was appointed to the position of Professor. Six years later, he taught for a year at the University of Kiel before going to Israel where, from 1929 to 1959, he taught at the Hebrew University of Jerusalem. He was a fervent Zionist, and throughout his life showed a deep interest in Jewish culture, becoming involved in many social and educational activities. Fraenkel died in Jerusalem on 15 October 1965.

From early on, Fraenkel was interested in the axiomatic foundations of mathematical theories, and some of his first work comprised an investigation into the axiomatics (universally accepted facts) of Hensel's p-adic numbers and into the theory of rings. He then became interested in the theory of sets (on which, in 1919, he wrote *Einleitung in die Mengenlehre*, a book that was well received and was reprinted several times).

Fraenkel became very involved with set theory as it had been formulated in 1908, in the axiomatic system put forward by Ernst Zermelo (1871–1953). The axioms, however, included the hitherto unexplained notion of a 'definite property', and Fraenkel determined he should be the first of the several mathematicians to succeed in the attempt to overcome this difficulty.

He put forward his own proposed solution in 1922. Instead of Zermelo's notion of definite property, Fraenkel suggested the use of a notion of function introduced by definition. He also omitted entirely Zermelo's axiom of subsets, which stated that if a property E is definite in a set m, there is a subset consisting of those elements x of m for which $E(x)$ is true. To replace this axiom, Fraenkel said instead that if m is a set and ϕ and ψ are functions, there are subsets mE and $m°E$ consisting of those elements x of m for which $\phi(x)$ is an element of $\psi(x)$ and $\phi(x)$ is not an element of $\psi(x)$ respectively. Using this axiom, Fraenkel showed that the axiom of choice – also first devised in axiomatic form by Ernst Zermelo, in 1904 – can be treated independently by referring to an infinite set of objects that are not sets themselves. It turned out to be extremely complicated to prove this without referring to an external assumption. (It was not, in fact, successfully accomplished until 1963, when P. Cohen proved it for a revised system combining the work of Zermelo, Fraenkel, and Thoralf Skolem – calling it, therefore, the ZFS system.)

It was, however, Skolem's proposal for the explanation of Zermelo's definite property, published in 1923, that was ultimately accepted. His suggestion had the advantage over Fraenkel's in that it led more directly to a logical formulation of Zermelo's axioms (which, till then, existed only as intuitive statement).

Fraenkel nevertheless actively continued his development of the theory of sets, in which he showed considerable perception, evident in his papers and books. In 1953, he published *Abstract Set Theory*, and in 1958 *Foundations of Set Theory*. His research led him to posit an eighth axiom (to follow Zermelo's seventh), an axiom of replacement, which stated that if the domain of a single-valued function is a set, its counter-domain is also a set.

Later, John Von Neumann (1903–1957) – the pioneer in computer mathematics – was to propose a ninth axiom, the axiom of foundation. It states that every non-empty set *a* contains a member *b* such that *a* and *b* have no members in common.

Fraenkel-Conrat Heinz 1910– is a German-born US biochemist who showed that the infectivity of bacteriophages (viruses that infect bacteria) is a property of their inner nucleic acid component, not the outer protein case.

Fraenkel-Conrat was born on 29 July 1910 in Breslau, Germany (now Wroclaw, Poland). He studied medicine at the University of Breslau, graduated in 1933 and then, with the rise to power of Adolf Hitler, left Germany and went to Britain. He did postgraduate work at the University of Edinburgh and obtained his PhD in 1936 for a thesis on ergot alkaloids, after which he went to the United States. He settled there and became a naturalized citizen in 1941. He went to the University of California in 1951, and became a professor there in 1958.

In 1955, Fraenkel-Conrat developed a technique for separating the outer protein coat from the inner nucleic acid core of bacteriophages without seriously damaging either portion. He also succeeded in reassembling the components and showed that these reformed bacteriophages are still capable of infecting bacteria. This work raised fundamental questions about the molecular basis of life. He then showed that the protein component of bacteriophages is inert and that the nucleic acid component alone has the capacity to infect bacteria. Thus, it seemed the fundamental properties of life resulted from the activity of nucleic acids.

Francis James Bicheno 1815–1892 was a British-born hydraulics engineer who spent most of his working life in the USA and played a crucial role in the industrial development of part of New England. He made significant contributions to the understanding of fluid flow and to the development of the Francis-type water turbine for which he is remembered.

Francis was born on 18 May 1815 at Southleigh, Oxfordshire, the son of a railway superintendent and builder. After a short education at Radleigh Hall and Wantage Academy, he became assistant to his father on canal and harbour works. Two years later, he was employed by the Great Western Canal Company.

He travelled to the United States in search of greater opportunities, arriving in New York City in the Spring of 1833. There he was employed by Major George Washington Whistler (1800–1849) on building the Stonington Railroad, Connecticut. A year later when Whistler became chief engineer to 'The Proprietors of the Locks and Canals on the Merrimack River', a corporation known simply as 'The Proprietors', Francis went with him to Lowell, Massachusetts.

In 1837, Whistler resigned and Francis succeeded him. On 12 July the same year Francis married Sarah Wilbur Brownell of Lowell. When The Proprietors decided (in 1845) to develop the river's water-powered facilities, Francis was made chief engineer and general manager. He travelled to England (1849) briefly to study timber preservation methods and on his return turned his attention to developing water turbines. In 1855, his famous work, *The Lowell Hydraulic Experiments*, was published.

Francis wrote more than 200 papers for learned societies and was president of the American Society of Civil Engineers in 1880. He advised on a number of important dam projects, was a member of the Massachusetts state legislature, and was president of the Stonybrook Railroad for 20 years and for 43 years a director of the Lowell Gas Light Company.

He retired from active business in 1885, and was succeeded by one of his sons. Francis died on 18 September, survived by his wife and six children.

The industrialization of New England resulted initially from water power rather than steam. The leading part Francis played in the exploitation of the Merrimack River was thus at the time more important than his work on turbines.

The Proprietors corporation had been formed in 1792, originally to improve navigation. Realizing the potential, a Boston group purchased 400 acres near the Pawtucket Falls, a site which soon developed into the town of Lowell. The company built a 290-m/950-ft dam on the river which produced an 11-m/35-ft head and 29 km/18 mi of backwater, the pondage feeding 11 independent mills.

One of Francis's responsibilities was the measurement of the flows used by each of the manufacturing companies along the river to assess costs. He made numerous tests on sharp-crested weirs, and determined the numerical values in the Francis weir formula, the form of which was suggested by his colleague, Uriah Atherton Boyden (1804–1879). The second (1868) edition of Francis's work included his studies of measurements with weighted floats.

Francis's work on turbines started when The Proprietors acquired, in the late 1840s, an interest in the patent turbine designed by Samuel B. Howd. This was a radial inflow (or 'centre-vent') machine which was effective but inefficient. Significantly, however, Francis had built (in 1847) a model wheel similar to Howd's, and it, too, was somewhat inefficient. Two years later, several inward-flow wheels of 170 kW/230 hp each were built from Francis's design. Tests showed peak efficiencies of nearly 80%.

The Francis wheels of the development days were an improvement on those of Howd, but only to a small degree do the so-called Francis turbines of today resemble Francis's original designs. At the outset they utilized purely radial flow runners and they had neither the familiar scroll case nor the draught tube of modern units. Later engineers developed the design into the forerunner of the modern mixed flow unit.

The reason Francis's name continues to be associated with the design presumably stemmed initially from the wide-spread attention attracted by his book and then from the adoption of the design by the German and Swiss firms which led in its scientific development later in the century.

Francis also devised a complete system of water supply for fire protection and had it working in the Lowell district for many years before anything similar was in operation anywhere else. He designed and built hydraulic lifts for the guard gates of the Pawtucket Canal and between 1875 and 1876 he reconstructed the Pawtucket Dam.

Francis was largely responsible for Lowell's rise to industrial importance. In retrospect, however, this is less notable than the experimental work he did in connection with the flow of fluids over weirs, and the establishment of the Francis formula. His work on the inward-flow turbine was significant and after his death the Canadian Niagara Power Company installed Francis turbines developing 7,650 kW/ 10,250 hp at the famous falls.

Franck James 1882–1964 was a German-born US physicist who provided the experimental evidence for the quantum theory of Max Planck (1858–1947) and the quantum model of the atom developed by Niels Bohr (1885–1962). For this achievement, Franck and his co-worker Gustav Hertz (1887–1975) were awarded the 1925 Nobel Prize for Physics.

Franck was born in Hamburg on 26 August 1882. When he left school, his father (a prosperous banker) sent him to Heidelberg University in 1901 to read law and economics as a preparation for his entry into the family firm, considering the status of scientists to be very lowly indeed. Fortunately, at Heidelberg Franck met Max Born (1882–1970) and a lifelong friendship began. Born, also from a wealthy Jewish family, had full parental approval for his career and this eventually convinced Franck's father to allow his son to follow a scientific career. At first it was to be geology, but this quickly turned to chemistry and then to physics when he went to Berlin University in 1902. It was there that Franck obtained his doctorate in 1906 for research into ionic mobility in gases.

Franck was awarded the Iron Cross during World War I, and from 1916 he worked at the Kaiser Wilhelm Institute of Physical Chemistry under Fritz Haber (1868–1934) on the study of gases, becoming head of the physics division there in 1918. Two years later, Franck became Professor of Experimental Physics at the University of Göttingen, where Born had just taken the Chair of Theoretical Physics. At Göttingen, Franck and Hertz undertook the work on the quantum theory that gained them the 1925 Nobel Prize for Physics.

Franck remained in Göttingen until 1933, when Adolf Hitler came to power. Although allowed to retain his position because of his distinguished war record, Franck was told to dismiss other Jewish members of his Institute in the University. Franck refused to do this and left Germany, going first to Copenhagen and then to the United States, where he became Professor of Physics at Johns Hopkins University in Baltimore in 1935. This was followed by a move to Chicago in 1938, where Franck was appointed Professor of Physical Chemistry.

During World War II, Franck became a US citizen and carried out metallurgical work related to the production of the atomic bomb. He became aware of the devastating power of this weapon and, in a document that became known as the Franck Report, he and other scientists suggested that it should first be demonstrated to the Japanese on unpopulated territory. Franck retired from the University of Chicago in 1949. Numerous honours were accorded him in these late years by academics and universities in both the United States and Europe. The city of Göttingen, as part of its 1,000th anniversary in 1953, made Franck and Born honorary citizens and Franck died there on 21 May 1964, while visiting friends.

In his major contribution to physics, Franck investigated the collisions of electrons with rare-gas atoms and found that they are almost completely elastic and that no kinetic energy is lost. With Hertz, he extended this work to other atoms. This led to the discovery that there are inelastic collisions in which energy is transferred in definite amounts. For the mercury atom, electrons accept energy only in quanta of 4.9 electronvolts. For such collisions to be inelastic, the electrons need kinetic energy in excess of this figure. As the energy is accepted by the mercury atoms, they emit light at a spectral line of $2,537 Å/2.5 \times 10^{-7}$ m. This was the first experimental proof of Planck's quantum hypothesis that $E = h\nu$ where E is the change in energy, h is Planck's constant and ν the frequency of light emitted. These experiments also tended to confirm the existence of the energy levels postulated by Bohr in his model of the atom. For this work, Franck and Hertz shared the 1925 Nobel prize.

Franck also studied the formation, dissociation, vibration and rotation of molecules. With Born he developed the potential energy diagrams that are now common in textbooks of physical chemistry. From the extrapolation of data regarding the vibration of molecules obtained from spectra, he was able to calculate the dissociation energies of molecules. Edward Condon (1902–1974) interpreted this method in terms of wave mechanics, and it has become known as the Franck–Condon principle.

During his later years at Göttingen and at Baltimore, Franck carried out experiments on the photodissociation of diatomic molecules in liquids and solids and this led to an interest in photosynthesis. Research in this field was dominated by organic chemists and biochemists and Franck found himself involved in much controversy. His research led him to believe in a two-stage mechanism within the same molecule for the photosynthetic process, when the established view was that two different molecules are involved.

Franklin Benjamin 1706–1790 was the first great American scientist. He made an important contribution to physics by arriving at an understanding of the nature of electric charge as a presence or absence of electricity, introducing the terms 'positive' and 'negative' to describe charges. He also proved in a classic experiment that lightning is electrical in nature, and went on to invent the lightning conductor. Franklin also mapped the Gulf Stream, and made several useful inventions, including bifocal spectacles. In addition to being a scientist and inventor, Franklin is widely remembered as a statesman. He played a leading role in the drafting of the Declaration of Independence and the Constitution of the United States.

Franklin was born in Boston, Massachusetts, of British settlers on 17 January 1706. He started life with little formal instruction and by the age of ten he was helping his father in the tallow and soap business. Soon, apprenticed to his brother, a printer, he was launched into that trade, leaving home shortly afterwards to try for himself in Philadelphia. There he set himself up as a printer and in 1724 was sent to London to prospect for presses and types. However, this turned out to be a ruse of the city governors to get rid of him – the reason is obscure. Nevertheless, Franklin, without funds or introductions, soon found himself work and put the next two years to good use in becoming a skilled printer.

Back in Philadelphia, Franklin's fortunes progressed. His own business prospered and he was soon active in journalism and publishing. He started the *Pennsylvania Gazette*, but is better remembered for *Poor Richard's Almanack*, a great collection of articles and advice on a huge range of topics, 'conveying instruction among the common people'. Published in 1732, it was a great success and brought Franklin a considerable income. Public affairs also proved to be his metier and gradually Franklin became enmeshed in all sorts of progressive undertakings. He was Clerk of the Pennsylvania Assembly as well as Postmaster of Philadelphia, and founded the American Philosophical Society in 1743 and in 1749 a college that later became the University of Pennsylvania. He was elected to the Pennsylvania Assembly in 1751 and as a politician became concerned with the government of the colony from Britain. These activities by no means prevented his scientific investigations, however, and his major work on electricity was done in this period.

He was awarded the Copley Medal of the Royal Society in 1753 and elected to the Society in 1756, the subscription of 25 guineas being waived in honour of his achievements. In 1757, Franklin travelled again to Britain, this time with proper credentials as the Agent of the Pennsylvania Assembly, and stayed on and off until 1775, attending meetings of the Royal Society as well as campaigning for the independence of the American colonies as their leading spokesman in Britain.

Back in America, Franklin helped to draft the Declaration of Independence in 1776 and was one of its signatories. He then travelled to France to enlist help for the American cause in the Revolutionary War that followed, successfully organizing nearly all outside aid. He played a central part in the negotiation of the peace with Britain, signing a treaty that guaranteed independence in 1783. Franklin, though now well over 70, continued to play an active part in the affairs of the new nation. He became president of Pennsylvania, worked hard to abolish slavery, and in 1787 guided the Constitutional Convention to formulate and ratify the Constitution. He died soon after, in Philadelphia on 17 April 1790.

In 1746, his business booming, Franklin turned his thoughts to electricity and spent the next seven years in the execution of a remarkable series of experiments. Although he had little formal education, his voracious reading gave him the necessary background and his practical skills, together with an analytical yet intuitive approach, enabled Franklin to put the whole topic on a very sound basis. It was said that he found electricity a curiosity and left it a science.

By the time of Franklin's entry to the field, the notions of charged bodies, insulators and conductors were established, though what was being 'charged' or 'conducted' was a matter of speculation. One of Franklin's earliest observations was of the ability of a pointed metal object to discharge an electrified conductor. A bodkin was used to discharge metal shot on a dry glass base. An earthed bodkin discharged the shot either touching it or as much as 20 cm/ 8 in distant, but an insulated bodkin had no effect. A man on an insulated base could electrify a glass tube by rubbing, and 'communicate' the charge to another man similarly insulated. These experiments led Franklin to the fundamental conclusion that electricity is a single fluid that flows into or out of objects to produce electric charges. This naturally led to the introduction of the terms positive and negative, a positive charge being an excess of electricity and a negative charge a corresponding deficiency of electricity. But Franklin had no way of knowing in which direction the electric fluid moved, and he made an arbitrary choice of which bodies became positive and which negative. In fact, electric charge moves to negatively charged bodies, which is why the electron is given a negative charge. However, Franklin made a fundamental discovery when he realized that the gain and loss of electricity must be balanced – the concept of conservation of charge. And his notion that the so-called electrical fire is fundamental to all matter, his 'one-fluid' theory, brings us right up to the twentieth century.

The Leyden jar, which was invented in 1745, was an ideal proving ground for the clarification of these ideas. Franklin was able to show that the two coatings were oppositely charged, and that one had to be earthed while the other was being electrified. And finally he showed that the 'power is in the glass' – an appreciation of the importance of the dielectric later to be fully discussed by Michael Faraday (1791–1867). This work led to the first plate condenser or capacitor, which contained glass sheets between lead plates – the Franklin panes. This device simplified electrical experiments by allowing the glass and lead components to be separated and increased in number, and gave rise to a 'battery' of condensers.

These fundamental ideas led to Franklin's most famous work of all – that on lightning – and his invention of the lightning conductor. Although not the first to wonder about thunder and lightning in connection with electricity, Franklin showed that the thunderclouds are indeed seats of electric charge whose behaviour is like charged bodies and that pointed conductors can dissipate the charge or carry it to earth safely.

The sentry-box or Philadelphia experiment, which was devised by Franklin but first performed in France in 1752, caused a sensation. A man on an insulating base stood inside a sentry box, which had a pointed rod some 10 m/ 33 ft long fixed to its roof and insulated from it. The box itself was sited on a high building. Franklin suggested that a man could draw charge from a thundercloud by presenting an earthed wire to the rod, the man being protected by a wax handle. Sparks a few centimetres long were obtained. A later worker who did not fully observe the safety precautions laid down by Franklin was electrocuted. Later that year (1752), Franklin himself carried out his famous experiments with kites. By flying a kite in a thunderstorm, he was able to charge up a Leyden jar and produce sparks from the end of the wet string, which he held with a piece of insulating silk. The lightning conductor used everywhere today owes its origin to these experiments. Furthermore, some of Franklin's last work in this area demonstrated that while most thunderclouds have negative charges, a few are positive – something confirmed in modern times.

Of the rest of Franklin's work, staggering in its diversity, interest and humanity, we can give only a summary here. His interest in atmospheric electricity led him to recognize the aurora borealis as an electrical phenomenon, postulating good conditions in the rarefied upper atmosphere for electrical discharges, and speculating on the existence of what we now call the ionosphere. An interest in meteorology followed naturally from this. Franklin explained whirlwinds and water-spouts as being due to very rapid air circulation leading to low pressure in axial regions. He also pondered (long before the eruption of Krakatoa) on the effect of volcanic activity on weather.

Franklin's interest in heat and insulation led him to study clothing, air circulation and the cooling effect of perspiration. He was very enthusiastic about health-improving activities, especially swimming, wrote on lead poisoning,

gout, sleep, and was a late convert to inoculation. There was a fashion to apply electric-shock treatment in cases of paralysis but Franklin remained a sceptic and Mesmer and his followers were discredited. In a very modern comment, Franklin wondered about the possible psychological value of such treatment.

The Pennsylvanian fireplace – a stove with underfloor draught – was a great success in 1742. Franklin refused a patent and showed his magnanimity of nature when a London manufacturer made a lot of money out of a 'very similar' model. 'Not the first instance …' he said quietly in his autobiography. It was recognition enough to be imitated.

Franklin was also influential in areas of physics other than electricity. Unfashionably for the time, he rejected the particle theory of light, being unable to account for the vast momentum that the particles should possess if they existed. This view inspired Thomas Young (1773–1829) to his fundamental work on the wave theory early in the following century, and the concept of light particles having momentum was used by Louis de Broglie (1892–1987) to establish the wave nature of the electron more than a century later still. But in considering heat, Franklin's views were less helpful to the course of science. Franklin suggested a famous experiment to demonstrate conductivity in which rods of different metals are heated at one end and wax rings placed at the same distance along the rods fall off at different times as heat spreads through the rods at different rates. This led Franklin to speculate that heat is a fluid like electricity, and aided development of the erroneous caloric theory of heat.

Always interested in the sea, Franklin produced the first chart of the Gulf Stream following observations made in 1770 that merchantmen crossed the Atlantic Ocean from east to west in two weeks less than the mail ships – the former keeping to the side of the current, not fighting it. In 1775, he used a thermometer to aid navigation in the warm waters of the Gulf Stream and, as late as 1785, was devising ways of measuring the temperature at a depth of 30 m/ 98 ft. Finally, Franklin also busied himself with such diverse topics as the first public library, bifocal lenses, population control, the rocking chair and daylight-saving time.

Benjamin Franklin is arguably the most attractive and interesting figure in the history of science, and not only because of his extraordinary range of interests, his central role in the establishment of the United States, and his amazing willingness to risk his life to perform a crucial experiment. By conceiving of the fundamental nature of electricity, he began the process by which a most detailed understanding of the structure of matter has been achieved.

Fraunhofer Joseph von 1787–1826 was a German physicist and optician who was the first person to investigate the dark lines in the spectra of the Sun and stars, which are named Fraunhofer lines in his honour. In so doing, he developed the spectroscope, and he later became the first to use a diffraction grating to produce a spectrum from white light. These achievements were made and used by

Fraunhofer mainly to improve his optical instruments, and his work laid the basis for subsequent German supremacy in the making of high-grade scientific and optical instruments.

Fraunhofer was born in Straubing on 6 March 1787. His education was limited and he started work in his father's glazing workshop at the age of ten. After his father's death in 1798, he was apprenticed to a Munich mirror-maker and glass-cutter, and in 1806 he entered the optical shop of the Munich Philosophical Instrument Company, which produced scientific instruments. There Fraunhofer developed an expertise in both the practice and theory of optics. Under Pierre Guinand, a master glassmaker, Fraunhofer acquired practical knowledge of glassmaking and combined this with his grasp of optical science to improve the fortunes of the company. This success brought him promotion within the firm and, by 1811 he had become a director.

Still highly active in business, Fraunhofer made his discovery of the dark lines in the Sun's spectrum in 1814, and invented the diffraction grating in 1821. In 1823, Fraunhofer accepted the post of Director of the Physics Museum of the Bavarian Academy of Sciences in Munich. But he contracted tuberculosis two years later and died in Munich on 7 June 1826, at the early age of 39.

Although he had little formal education, Fraunhofer sought to understand optical theory and apply it to the practical work of constructing lens combinations of minimum aberration. At that time there was little high-quality crown and flint glass, and methods of determining the optical constants of glass were crude, limiting the size and quality of lenses that could be produced and also confining instrument makers to trial-and-error methods of construction. Fraunhofer approached lens-making according to optical theory and set out to determine the dispersion powers and refractive indices of different kinds of optical glass with greater accuracy. In collaboration with Guinand, he also sought to improve the quality of the glass used to make lenses.

In 1814, Fraunhofer began to use two bright yellow lines in flame spectra as a source of monochromatic light to obtain more accurate optical values. Comparing the effect of the light from the flame with the light from the Sun, he found that the solar spectrum is crossed with many fine, dark lines. A few of these had been seen by William Wollaston (1766–1828) in 1802, but Fraunhofer observed 574 lines between the red and violet ends of the spectrum.

In this way, Fraunhofer was able to make very accurate measurements of the dispersion and refractive properties of various kinds of glass, and in so doing he developed the spectroscope into an instrument for the scientific study of spectra.

In 1821, Fraunhofer examined the patterns that result from light diffracted through a single slit, and related the width of the slit to the angles of dispersion of the light. Extending from a large number of slits, he constructed a grating of 260 parallel wires and made the first study of spectra produced by diffraction gratings. The presence of the solar dark lines enabled him to note that the dispersion

of the spectra is greater with a diffraction grating than with a prism. Fraunhofer examined the relationship between the dispersion and the separation of the wires in the grating, and concluded that the dispersion is inversely related to the distance between successive slits in the grating. By measuring the dispersion, he was able to determine the wavelengths of light of specific colours and the dark lines.

Fraunhofer also constructed reflection gratings, enabling him to extend his studies to the effect of diffraction on oblique rays. By using the wave theory of light, he was able to derive a general form of the grating equation that is still in use today.

Fraunhofer never published his researches into glassmaking nor the methods that he developed for calculating and testing lenses, viewing them as trade secrets. This work enabled him to develop telescopes of unsurpassed quality, leading to important discoveries in astronomy. But Fraunhofer did publish his work on diffraction gratings, which was important in the development of spectroscopy, a vital tool in the elucidation of atomic structure much later. He also published his observations of the dark lines in the spectrum, although he could not provide an explanation of them. This was achieved by Gustav Kirchhoff (1824–1887) in 1859.

The dark lines crossing the Sun's spectrum indicate the presence in the Sun's atmosphere of certain elements in the vapour state. The vapours in the chromosphere are cooler than the lower photosphere of the Sun and they absorb their own characteristic wavelengths from the Sun's continuous spectrum according to Kirchhoff's law – a substance which emits light of a certain wavelength at a given temperature will also absorb light of the same wavelength at that temperature. Some of the wavelengths emitted are absorbed by the vapours in the chromosphere as the Sun's light passes through them, producing the dark lines. It therefore became possible to identify the elements in the Sun's atmosphere from a study of the Fraunhofer lines, and hydrogen and helium were both shown to be present – in fact, this was how helium was first discovered. The same method is also applied to other stars.

Fraunhofer's insistence on high-quality craftmanship thus led to discoveries and developments in science on both a cosmological as well as an atomic scale.

Fredholm Erik Ivar 1866–1927 was a Swedish mathematician and mathematical physicist who founded the modern theory of integral equations, and in his work provided the foundations upon which much of the extremely important research later carried out by David Hilbert (1862–1943) was based. Fredholm's name is perpetuated in several concepts and theorems.

Fredholm was born on 7 April 1866 in Stockholm, where he grew up and was educated. At the age of 19 he entered the Polytechnic Institute there, studying applied mathematics; his particular interest was the solution of problems of practical mechanics. After one year, however, he transferred to the University of Uppsala where in 1888 he received his bachelor's degree. He then returned to

Stockholm. After ten years of further research and work, Fredholm finally obtained his PhD from Uppsala University – for a thesis on partial differential equations – and became a lecturer in mathematical physics at Stockholm University. Within the next five years he wrote his most important paper, on integral equations; for it, in 1903, he received the Wallmark Prize of the Swedish Academy of Sciences, and the Poncelet Prize of the Académie de France. In 1906 he was promoted to Professor of Rational Mechanics and Mathematical Physics, a post he held until his death. He died in Stockholm on 17 August 1927.

Fredholm's success in deriving a theory of integral equations was in some respects a success in combining parts of the work of others with his own creative flair and a novel approach. He founded much of his theory on work carried out by US astronomer George Hill (1838–1914) in 1877, who was investigating lunar motion. In his examination, Hill used linear equations involving determinants of an infinite number of rows and columns. In Fredholm's paper 'Sur une nouvelle méthode pour la résolution du problème de Dirichlet', published in 1900, he first developed the essential part of the theory of what is now known as Fredholm's integral equation; further, he went on to define and solve the Fredholm equation of the second type, involving a definite integral.

Such equations had been under scientific consideration for some years. Niels Abel, Franz Neumann and Vito Volterra had all put forward tentative or incomplete results: Henri Poincaré had even arrived at Fredholm's solution but been unable to derive a proof of it, although in order to carry out his own work on partial differential equations (between 1895 and 1896) he had been obliged to assume that it was correct.

Fredholm's novel approach in continuing his research led to his discovery, also in 1900, of the algebraic analogue of his own theory of integral equations. It was not until 1903, however, that he completed the solution, recognizing the analogous identity between the Fredholm integral equation and the linear-matrix vector equation $(I+F)U = V$, and showing that the analogy was complete.

Shortly afterwards, Fredholm's results were used by David Hilbert, who extended them in deriving his own theories – such as the theory of eigenvalues, and the theory of spaces involving an infinite number of dimensions – that finally contributed fundamentally towards quantum theory.

Frege Friedrich Ludwig Gottlob 1848–1925 was a German logical philosopher and mathematician whose main purpose was to define once and for all an evolutionary connection between the fundamental rigour and mathematics in logic and the fundamental rigour and logic in mathematics. For this purpose he revised certain parts of mathematical notation in order to introduce total precision in logic, and he further revised philosophical vocabulary with the same intent.

His first books describing this work were successful – and to some extent gratifyingly revolutionary in effect. His later development of this in a two-volume work, however,

was over-ambitious and was accounted a failure, unfortunately discrediting much of his earlier and his later work. Nevertheless, his final system of logic is now accepted as one of the greatest contributions in the field put forward in the century surrounding it.

Frege was born on 8 November 1848 in Wismar, Germany, and grew up and received his early education there. He then spent two years (1869–1871) as a student in Jena before transferring to Göttingen University, where he studied physics, chemistry, mathematics and philosophy, earning his PhD in mathematics in 1873. The next year he entered the faculty of philosophy there.

Frege's studies over the following years were crystallized in a book he published in 1879 on a new symbolic mathematical language he had devised, which he called *Begriffschrift*. In that year he returned to Jena to take up a teaching post, and remained there for the rest of his working life. In 1884, Frege published another important book on the foundations of mathematics, and again relied upon his *Begriffschrift*. He attempted to develop his ideas still further in another, ill-fated, two-volume text, the first volume of which appeared in 1893, and the second in 1903. These volumes, entitled *Grundgesetze der Arithmetik*, received a severe blow when, shortly before the second volume was due to be published, Frege was sent a letter which demonstrated to him that the entire mathematical system described in the books was in fact of no value; nobly, he included a postscript to that effect in the second volume. After this personal disaster, Frege continued to study mathematics but never with the same scope or depth. He retired in 1917, still writing further material, extending some of his previous studies in the period between 1918 and 1923. He died in Bad Keinen on 26 July 1925.

At the beginning of what was to become his life's work, Frege was correctly convinced that in terms of absolute precision ordinary language is not sufficiently strict for the expression of mathematical concepts such as the definition of 'number', 'object' and 'function'. Furthermore, he saw that the symbols already available to mathematicians were themselves not adequate for this purpose either, and so it would be necessary to create new ones – a vital step that mathematicians before him had resisted taking.

The resultant *Begriffschrift* (which translates literally as 'idea-script') was intended as a method for the analysis and representation of mathematical proofs. It has since been developed into modern mathematical symbolic logic, and Frege is generally – and only reasonably – credited as its originator. He introduced the symbols for assertions, implications, and their converse notions; he also introduced propositional logic and quantification theory, inventing symbols for 'and', 'or' and 'if ..., then ...,' and so on. Using his new 'language' he was able succinctly and unambiguously to express complex logical relations, and even – when Frege applied it to the theory of sequences – to define the ancestral relation. This represented a major development in mathematical induction, and was later further explored by mathematicians such as Bertrand Russell (1872–1970) and Alfred Whitehead (1861–1947).

Frege incorporated improvements to the *Begriffschrift* into *Grundgesetze*, but was devastated to receive a letter from Bertrand Russell in 1902 – nine years after the appearance of the first volume – in which Russell asked Frege how his logical system coped with a particular logical paradox. To his chagrin, Frege's system was not able to resolve it – and since the system had been intended to be complete and contradiction-free, he was forced to acknowledge his system to be useless.

Although at the time Frege was largely discredited, his work today is seen as of considerable importance. His innovations have been useful in the development of symbolic logic, and even the problem posed by Russell was resolved by later logico-mathematicians.

Frege, nevertheless, in many ways simply stopped at that point. Despite the fact that he carried on working, and for quite a number of years, developments in early twentieth-century mathematics – such as Hilbert's axiomatics – were apparently beyond Frege's scope. He was unable to accept these new ideas, even when David Hilbert (1862–1943) himself tried to clarify the issue for him. Frege was, therefore, a mathematician with the most ambitious plans for the development of a rigorous foundation for mathematics in which, in his own eyes, he did not succeed in his own lifetime.

Fresnel Augustin Jean 1788–1827 was a French physicist who established the transverse-wave theory of light in 1821.

Fresnel was born on 10 May 1788 in Broglie, Normandy. His parents provided his early education themselves and, at the age of 12, he entered the Ecole Centrale in Caen where he was introduced to science. In 1804, Fresnel entered the Ecole Polytechnique in Paris intending to become an engineer. Two years later, he went to the Ecole des Ponts et Chaussées for a three-year technical course which included experience in practical engineering. Completing the course successfully, he became a civil engineer for the government.

Fresnel then worked on road projects in various parts of France, but became interested in optics. When Napoleon returned to France from Elba in 1815, Fresnel deserted his government post in protest. He was arrested and confined to his home in Normandy, taking advantage of this enforced leisure to develop his ideas on the wave nature of light into a comprehensive mathematical theory.

Napoleon's return proved to be shortlived and Fresnel was soon reinstated into government service. His scientific investigations were curtailed from then on and all his work was done in his spare time. But his achievements were great, rewarding Fresnel with unanimous election to the Académie des Sciences in 1823 and, just before he died, the Rumford Medal of the Royal Society. Fresnel had been continually plagued with ill-health, and eventually died of tuberculosis at Ville D'Avray, near Paris, on 14 July 1827 at the early age of 39.

Fresnel began his investigations into the nature of light convinced that the wave theory was correct because of its essential simplicity. He was not aware of earlier work by Christiaan Huygens (1629–1695) and Thomas Young (1723–1829) advocating a wave nature for light, and sought to establish his own ideas in a study of diffraction. Fresnel was soon able to demonstrate mathematically that the dimensions of light and dark bands produced by diffraction could be related to the wavelength of the light producing them if light were to consist of waves.

Huygens had believed that light consists of longitudinal waves like sound waves, and Fresnel endeavoured to refine his mathematical explanation on this basis. He ran into formidable difficulties but then became aware of new discoveries indicating that light may be polarized by reflection. Fresnel plunged into a mathematical explanation for this phenomenon and concluded in 1821, that polarization could occur only if light consists of transverse waves. Few agreed with him because of difficulties in conceiving of the nature of a medium to carry transverse waves. Fresnel was able to show, however, that his theory convincingly explained the phenomenon of double refraction and it rapidly gained acceptance.

Neither the particle theory nor the longitudinal wave theory could explain double refraction. Fresnel showed that light could be refracted through two different angles because one ray would consist of waves oscillating in one plane while the other ray consisted of waves oscillating in a plane perpendicular to the first one. The understanding of polarized light came to have important applications in organic chemistry (in optical isomerism) and was later carried beyond the visible spectrum to lay the groundwork for the theoretical uncovering of a wide range of radiation by James Clerk Maxwell (1831–1879).

Fresnel is also remembered for the application of his new ideas on light to lenses for lighthouses. He produced a revolutionary design consisting of concentric rings of triangular cross-section, varying the overall curvature to produce lenses that required no reflectors to produce a bright parallel beam.

Freud Sigmund 1856–1939 was an Austrian psychiatrist and the father of modern psychoanalysis. He is best known for his use of the free-association method in analysis, and for his ideas on the interpretation of dreams. His theories on child and adult sexuality shocked Europe and greatly influenced later psychology.

Freud was born on 6 May 1856 at Freiberg, in Moravia (now Príbor, Czech Republic), the son of an unsuccessful Jewish wool merchant. The family moved to Vienna when he was four, and at the age of 17 he entered Vienna University to read medicine. He graduated in 1881 and continued neurological research in the university laboratories under Ernest Brücke (1819–92). In 1885 he went to Paris where he studied with Jean-Martin Charcot (1825–1893). The following year he returned to Vienna and set up private practice as a neurologist. He also married that year, and later had six children, one of whom was Anna Freud, a distinguished child psychoanalyst. Although derided by much of the medical profession, he gave psychoanalysis enough of an impetus to warrant the holding of the first psychoanalytic congress, in Salzburg in 1908. He was elected a member of the Royal Society in 1936. Two years later Nazi Germany invaded and occupied Austria and Freud had to leave. He moved to London and a year later, on 23 September 1939, he died of cancer of the jaw, from which he had suffered for 16 years.

While assisting the French neurologist, Charcot, Freud became influenced by his use of hypnosis in trying to find an organic basis for hysteria. Charcot had been investigating areas of the brain responsible for certain nervous functions, and encouraged Freud's interest in the psychological aspects of neurology, in particular hysteria. Freud set up his practice in Vienna to study the psychological basis of nervous disorders. The Viennese physician Josef Breuer (1842–1925) had told Freud of an occasion when he had cured symptoms of hysteria by encouraging a patient to recollect, under hypnosis, the circumstances of the hysteria and to express the emotions that accompanied them. Following the methods used by his colleague, Freud treated his patients with hypnosis and formulated his ideas about the conscious and the unconscious mind. He believed that repressed thoughts were stored in the unconscious mind and affected a person without the source of the effect being known.

Freud used this method until about 1895 when he replaced hypnosis by the technique of free-association (the 'talking cure'), perhaps because he could not master the art of hypnosis. The free-association method allowed the patient to talk randomly and with little guidance. The patient relaxed to such an extent that thoughts came through that were previously hidden from the patient's conscious.

The introduction of free-association marked the beginning of psychoanalysis in the sense used today. That is, that through a succession of periods of analysis, barriers that one puts up against knowledge of oneself are slowly broken down, with the help of the analyst.

Free-association led to the interpretation of dreams. Freud reasoned that dreams represented thoughts in the unconscious mind. He had noticed how often the train of thought in free-association included the recollection of a dream. By using free-association on the subjects of some of his own dreams he explained them as attempts to fulfil in fantasy some desire that he was repressing. The use of dream interpretation thus lay in revealing the contents of the unconscious which are repressed when one is awake.

Freud drew a comparison between the symbolism of dreams and of mythology and religion, stating that religion was infantile (God as the father image) and neurotic (projection of repressed wishes); he claimed that it was unnecessary and retarded social maturity by perpetuating the projection of these desires.

Around 1905, Freud was collecting about him a following of young men such as Alfred Adler (1870–1937) and Carl Jung (1875–1961), whom he influenced deeply. It was at this time also that the most controversial part of his work was publicized, connected with his ideas on sex as a cause of some neurotic disorders. Freud maintained that sexual gratification in childhood could carry through to adulthood resulting in psychological problems. An example he gave was that the first sexual impulses are felt during the sucking phase of an infant's life, and if these impulses become fixated, the child may mature into an adult dependant on the mother. In Freud's terms, the sexual nature of a child's relationship with its parents could result in an Oedipus complex involving dormant sexual feelings towards the mother and jealousy towards the father, feelings which could last through adulthood.

Freud saw the mind as operating on two levels: the primary level, characterized by symbolic thinking, which he called the Id, and which was the source of all basic drives, and the secondary level, marked by logical thinking and characterized by a sense of reality, and critical faculties. The Id suppressed by social mores becomes the Superego, and eventually emerges as the Ego.

Freud produced several publications detailing his theories. The best-known of these books are: *The Physical Mechanism of Hysterical Phenomena* (1893), *Studien über Hysterie/Studies in Hysteria* (1895), *Die Traumdeutung/ The Interpretation of Dreams* (1900), and *Totem und Tabu/ Totem and Taboo* (1913).

In his early years Freud also carried out pioneer work on cocaine and advocated its use as a local anaesthetic for mild pain. Only later, however, was it accepted and introduced into the medical practice.

Freundlich Herbert Max Finlay 1880–1941 was a German physical chemist, best known for his extensive work on the nature of colloids, particularly sols and gels.

Freundlich was born in Berlin-Charlottenburg on 28 January 1880, the son of a German father and a Scottish mother. Shortly after his birth the family moved to Biebrich, where his father had been made director of an iron foundry. At school Freundlich studied classics and showed a great aptitude for music, but by the time he left the Gymnasium in 1898 he had opted for a career in science. In that same year he went to the University of Munich for a preliminary science course, then specialized in chemistry when he moved to the University of Leipzig. His thesis on the precipitation of colloidal solutions by electrolysis gained him a doctorate in 1903.

From 1903 to 1911 he assisted Friedrich Ostwald at his Leipzig Institute for Analytical and Physical Chemistry. He then became Professor of Physical Chemistry at the Technische Hochschule in Brunswick. At the outbreak of World War I in 1914 he was declared unfit for military service, so went to the Kaiser Wilhelm Institut in Berlin-Dahlem where he worked (under Fritz Haber) on the use of charcoal in gas masks. In 1919, he became the head of the institute's section devoted to colloid chemistry. He had married his first wife Marie Mann in 1908, but she died in 1917 during childbirth. Six years later he married Hella Gilbert.

Freundlich resigned his position when Adolf Hitler came to power in 1933 and emigrated to Britain where, sponsored by Imperial Chemical Industries (ICI), he worked at

University College, London. Later, in 1938, he went to the USA and took up the appointment of Distinguished Service Professor of Colloid Chemistry at the University of Minnesota. He died in Minnesota on 30 March 1941.

While he was at Leipzig, Freundlich formulated what has become known as Freundlich's adsorption isotherm, which concerns the accumulation of molecules of a solution at a surface. It can be expressed as:

$$x/m = (kc)^{1/n}$$

where x and m are the masses of material adsorbed and adsorbent, respectively, c is the equilibrium concentration of the solution, and k and n are constants.

Freundlich's other researches were devoted to all aspects of colloid science. He investigated colloid optics, the scattering of light by dispersed particles of various shapes. He studied the electrical properties of colloids, since electrostatic charges are largely responsible for holding colloidal dispersions in place.

He also carried out a major series of investigations on mechanical properties such as viscosity and elasticity, and studied the behaviour of certain systems under other types of mechanical forces. For example, he introduced the term thixotropy to describe the behaviour of gels, the jelly-like colloids that show many strange properties intermediate between those of liquids and solids. One modern industrial application of this work has been the development of non-drip paints.

Freundlich was very much a chemist and was not particularly interested in formulating long mathematical equations to explain or justify his practical results. In his later years he was particularly interested in the commercial use of colloids in the rubber and paint industries.

Friedel

Friedel Charles 1832–1899 was a French organic chemist and mineralogist, best remembered for his part in the discovery of the Friedel–Crafts reaction. Throughout his career, he successfully combined his interests in chemistry and minerals.

Friedel was born on 12 March 1832 in Strasbourg. He was educated locally (where he studied under Louis Pasteur) and at the Sorbonne in Paris. He qualified in both chemistry and mineralogy and in 1856 was made curator of the collection of minerals at the Ecole des Mines. In 1871 he became an instructor at the Ecole Normale and from 1876 was Professor of Mineralogy at the Sorbonne. Following the death of Charles Adolphe Wurtz in 1884 he became also Professor of Organic Chemistry and Director of Research, a post he held until he died at Montaubin on 20 April 1899.

Friedel's dual interests gave him a wide-ranging field for research, which he covered with great success. He made extensive inroads into the mysteries of the various alcohols, a subject that had received little attention until then. In 1871, working with R. D. da Silva, he synthesized glycerol (propantriol) from propylene (propene). Friedel met and became friends with US chemist James Mason Crafts (1839–1917), who arrived in Paris having spent a year doing postgraduate studies in Germany. Crafts returned to the United States to take up a professorship at Cornell University, but went back to Paris in 1874 and began carrying out research with Friedel.

In 1877 they discovered the Friedel–Crafts reaction, which uses aluminium chloride as a catalyst to facilitate the addition of an alkyl halide (halogenoalkane) to an aromatic compound. The reaction has proved to be extremely useful in organic synthesis and is now employed extensively in the industrial preparation of triphenylamine dyes.

From 1879 to 1887 Friedel worked on the attempted synthesis of minerals, including diamonds, using heat and pressure. He established the similarity in properties between carbon and silicon and therefore the similarity in structure of their long-chain polymeric compounds. This work paved the way for a better understanding of silica minerals and the use of silicates in industry.

Friedman

Friedman Aleksandr Aleksandrovich 1888–1925 was a skilled Russian mathematician with a keen interest in applied mathematics and physics. He made fundamental contributions to the development of theories regarding the expansion of the universe.

Friedman was born in St Petersburg, on 29 June 1888. An excellent scholar, both at school and at the University of St Petersburg, where he studied mathematics from 1906 to 1910, Friedman later served as a member of the mathematics faculty staff. In 1914, he was awarded his master's degree in pure and applied mathematics, and served with the Russian air force as a technical expert during World War I. Friedman also worked in a factory (which he later managed) that produced instruments for the aviation industry. From 1918 to 1920 he was Professor of Theoretical Mechanics at Perm University, but he returned to St Petersburg in 1920 to conduct research at the Academy of Sciences. He died prematurely, after a life dogged by ill-health, on 16 September 1925 in Leningrad.

Friedman's early research was in the fields of geomagnetism, hydromechanics and, above all, theoretical meteorology. His work of the greatest relevance to astronomy was his independent and original approach to the solution of Einstein's field equation in the general theory of relativity. Einstein had produced a static solution, which indicated a closed universe. Friedman derived several solutions, all of which suggested that space and time were isotropic (uniform at all points and in every direction), but that the mean density and radius of the universe varied with time – indicating either an expanding or contracting universe. Einstein himself applauded this significant result.

Friese-Greene

Friese-Greene William 1855–1921 was a British inventor and one of the early pioneers of cinema photography.

Friese-Greene was born on 7 September 1855 in Bristol where he was educated at the Blue Coat School. He became interested in photography and in about 1875 opened a portrait studio in Bath. In the early 1880s he met a mechanic J.A.R. Rudge, who asked him to produce slides for a magic lantern (forerunner of the modern slide projector). This work awakened Friese-Greene's interest in moving pictures. In 1885, he opened a studio in London and met Mortimer Evans, an engineer. They decided to collaborate, and in 1889 Friese-Greene patented a camera that could take ten photographs per second on a roll of sensitized paper. Using his own apparatus, he was able to project a jerky picture of people and horse-drawn vehicles moving past Hyde Park Corner – probably the first time a film of an actual event had been projected on a screen. In 1890, he substituted celluloid film for the paper in the camera, and in the next few years he patented improved cameras and projectors. Friese-Green died in London on 5 May 1921.

The early story of moving pictures is obscure, complicated by claims and counter-claims. Certainly several inventors were working on similar lines at about the same time. In 1824, Peter Roget lectured to the Royal Society on the subject of persistance of vision, and projected onto a screen a series of still pictures at the rate of 24 per second which gave the illusion of smooth and continuous movement. In the 1860s and 1870s there were various inventions for similarly projecting a series of stills, such as that of Heyl (in which transparencies were mounted on a glass disc and rotated). Faster camera shutter speeds and improved photographic emulsions enabled people to take sharper pictures of moving objects. Eduard Muybridge took a series of photographs of a racehorse by placing 24 cameras along a track, then projecting the pictures using an apparatus similar to Heyl's. Thomas Edison designed a motion-picture machine that recorded pictures in a spiral on a cylinder, but it was unsatisfactory. In 1889, George Eastman (1854–1932), founder of the Kodak company, produced roll film which solved part of Edison's problem, but the pictures could still be viewed only through a lens and not projected. Edison's invention was improved by others, and in France the Lumière Brothers developed a machine that functioned both as a camera and a projector. They arranged a show in London in 1896, the first time the public had been able to see moving pictures.

Although Friese-Greene's films were only short fragments (consisting of only ten pictures per second) – inadequate to produce a convincing effect of movement – he had taken and projected 'moving' pictures before Edison, and his patent was judged by a United States court to be the master patent. This brought him neither success nor financial gain, however. The same seems to have been true of his other inventions: a three-colour camera, moving pictures using a two-colour process, and machinery for rapid photographic processing and printing.

Frisch Otto Robert 1904–1979 was an Austrian-born British physicist who first described the fission of uranium nuclei under neutron bombardment, coining the term 'fission' to describe the splitting of a nucleus.

Frisch was born in Vienna on 1 October 1904. He displayed surprising skills in mathematics at a very early age, but when he entered the University of Vienna in 1922 it was to study physics, being awarded his PhD in 1926. The next year he began work in the optics division of the

Physikalische Technische Reichsanstalt in Berlin on a new light standard to replace the candle power.

In 1930, Frisch moved to Hamburg to become an assistant to Otto Stern (1888–1969), conducting research into molecular beams. Then in 1933, Frisch, being Jewish, was fired because of the racial laws the Nazis introduced following Adolf Hitler's rise to power. He moved to the department of physics at Birkbeck College, London, and a year later went to the Institute of Theoretical Physics in Copenhagen.

The German occupation of Denmark at the beginning of World War II forced Frisch to move once again and he was offered posts in Britain, first at Birmingham University and then in 1940 at Liverpool, where facilities such as a cyclotron were available for nuclear reseach. In 1943, he became a British citizen, which enabled him to become part of the delegation of British scientists sent to the Los Alamos nuclear research base in the United States. There he supervised the Dragon experiment, which led to the first atomic bomb. At the end of World War II, Frisch returned to Britain to become Deputy Chief Scientific Officer at the newly established Atomic Energy Research Establishment at Harwell, near Oxford. He remained there until 1947, when he became Jacksonian Professor of Natural Philosophy at Cambridge University. He retired in 1971 and died after an accident in Cambridge on 22 September 1979.

In 1938, Frisch was spending a holiday in Sweden with his aunt Lise Meitner (1878–1968), herself a well-known physicist. She received a letter from two of her colleagues, Otto Hahn (1879–1968) and Fritz Strassmann (1902–1980), which stated that the bombardment of uranium nuclei with neutrons had resulted in the production of three isotopes of barium, a much lighter element. Frisch and Meitner realized that this meant that the uranium nucleus had been cleaved. They calculated that the collision of a neutron with a uranium nucleus might cause it to vibrate so violently that it would become elongated. If the nucleus stretched too much, the mutual repulsion of the positive charges would overcome the surface tension of the nucleus and it would then break into two smaller pieces, liberating 200 MeV (2×10^8 electronvolts) of energy and secondary neutrons. Frisch coined the term of nuclear fission for this process, after the use of that term to describe cell division in biology.

During his wartime researches in Britain, Frisch worked on methods of separating the rare uranium-235 isotope that Niels Bohr (1885–1962) had calculated would undergo fission. He also calculated details such as the critical mass needed to produce a chain reaction and make an atomic bomb, and was instrumental in alerting the British government to the need to undertake nuclear research. Then at Los Alamos, Frisch designed the Dragon experiment, which involved the dropping of a slug of fissile material through a hole in a near-critical mass of uranium. This allowed the scientists to study many details of a near-critical mass which they would not otherwise have been able to investigate. At the first test explosion on 16 July 1945, Frisch conducted experiments from a distance of 40 km/25 mi.

When Frisch moved to Harwell after the war he did some mathematical work on chain reactions and supervised the building of the laboratories. At Cambridge he was mainly concerned with the development of automatic devices to evaluate information produced in bubble chambers, which are used to study particle collisions.

Frisch was a talented experimental physicist, who partly by virtue of his fortunate position – both geographical and temporal – was able to contribute to and observe some of the most dramatic developments in the history of science.

Frobenius Georg Ferdinand 1849–1917 was a German mathematician who is now chiefly remembered for his formulation of the concept of the abstract group – a theory that proposed what is now generally considered to be the first abstract structure of 'new' mathematics. His research into the theory of groups and complex number systems was of fundamental significance, and he also made important contributions to the theory of elliptic functions, to the solution of differential equations, and to quaternions.

Frobenius was born on 26 October 1849 in Berlin, where he received his early education. His study of mathematics began when he attended Göttingen University from 1867; in only three years he gained a doctorate. In 1870, he returned to his former school in Berlin as a teacher, and stayed there for a year before moving to a school of higher standard and status in the same city. By this time he had already presented many papers on mathematics – including the publication of his method of finding an infinite series solution of a differential equation at a regular single point, and other papers on Abel's problem in the convergence of series and Pfaff's problem in differential equations – and had earned a fair reputation. The result was that he was appointed Assistant Professor at Berlin University in 1874, and in the following year became full Professor at the Eidgenossische Polytechnikum in Zurich. Seventeen years later, in 1892, Frobenius returned to the University of Berlin in order to take up the post of Professor of Mathematics, where he remained for the rest of his working life. He died on 3 August 1917, in Charlottenberg.

It was in Berlin the first time, in a study of the work of Ernst Kummer (1810–1893) and Leopold Kronecker (1823–1891), that Frobenius became interested in abstract algebra; his major contributions to the subject were published in 1879 (*Über Gruppen von vertauschbaren Elementen*, written in collaboration with Stickelberger) and in 1895 (*Über endliche Gruppen*). Later publications contained further development of group theory, the last of which (in 1906) – *Über die reellen Darstellungen der endlichen Gruppen* – was written with Schur, together with whom Frobenius completed the theory of finite groups of linear substitutions of *n* variables.

By studying the different representations of groups and their elements, Frobenius provided a firm basis for the solving of general problems in the theory of finite groups. His methods were later continued by William Burnside (1852–1927), and his results also proved useful to the development of quantum mechanics.

Froude William 1810–1879 was the English engineer and hydrodynamiscist who first formulated reliable laws for the resistance that water offers to ships and for predicting their stability. He also invented the hydraulic dynameter for measuring the output of high-power engines. These achievements were fundamental to marine development.

Froude was educated at Buckfastleigh, Westminster School and Oriel College, Oxford, where he obtained (1832) a first in mathematics and a third in classics. He remained at Oxford working on water resistance and the propulsion of ships and in 1838 became an assistant to Isambard Kingdom Brunel on the building of the Bristol and Exeter Railway.

In 1839, he married Katherine Holdsworth of Widdicombe. He probably worked on the South Devon Railway and was intimately connected with the ill-fated atmospheric railway. In 1846, he went to live with his father at Dartington Parsonage and began work in earnest on marine hydrodynamics. Brunel consulted him on the behaviour of the *Great Eastern* at sea and, on his recommendation, the ship was fitted with bilge keels.

When his father died in 1859, Froude moved to Paignton where he began his tank-testing experiments. In 1863, he started to build his own house, known as Chelston Cross, at Cockington, Torquay, and helped the local water authority with its supply problems. In 1867, he began his experiments with towed models. After grudging financial assistance from the Admiralty, he built another experimental tank near his home in 1871.

He described his hydraulic dynameter in a paper to the Institution of Mechanical Engineers in 1877, but did not live to see his machine work. It was built in 1878, the year in which Froude's wife died and in which he became seriously ill. He went on a voyage of recuperation to the Cape but caught dysentry and died on 6 May 1879 in Simon's Town, where he was buried.

Froude was elected a Fellow of the Royal Society in 1870 and his work was continued by his sons Richard and Robert. Robert built the towing tank for the Admiralty at Haslar, near Portsmouth, and Richard joined Hammersley Heenan to manufacture the dynamometer commercially.

Generally there are two modes in which vessels can travel: by displacement, in which they force their way through the water, and by planing, or skating on top of the water. In the displacement mode, the propulsive power is absorbed in making waves and in overcoming the friction of the hull against the water.

Froude's first successful experiments started in 1867 when he towed models in pairs, balancing one hull shape against the other. Initially, he incorporated his findings into a single law, now known as Froude's law of comparison. This stated that the entire resistance of similar-shaped models varies as the cube of their dimensions if their speeds are as the square root of their dimensions.

As Froude himself realized, this law becomes increasingly unreliable as the difference in size between the models increases. This is because the frictional resistance and the wave-making resistance follow different laws. His law of comparisons is now only applied to the wave-making component.

To estimate frictional resistance, Froude carried out tests in his tank at Torquay, where he towed submerged (to eliminate wave-making resistance) planks with different surface roughness. He was able to establish a formula which would predict the frictional resistance of a hull with accuracy.

With these two analytical results, using only models and mathematics, Froude had found a reliable means of estimating the power required to drive a hull at a given speed. Model testing had been tried before but was considered unreliable because previous workers had failed to appreciate that the two major components of ship resistance varied differently. Opponents of model testing maintained that the only way to gather the required information was to work on full-sized hulls. It was this anti-model lobby which was largely responsible for Froude's difficulties in persuading the Admiralty to part with £2,000 for building the Torquay tank.

Froude had done a large amount of theoretical work on the rolling stability of ships and when the Torquay tank had been built, he was able to carry out model experiments, relating them to observations made on actual ships. His general deductions were challenged at the time but they were found to be correct and to this day are the standard exposition of the rolling and oscillation of ships.

Engine builders usually need to test their machinery before installation. The friction brake is useful only for lower-power applications, so Froude used hydrodynamic principles to absorb 1,500 kW/2,000 hp at 90 rpm. His brake consisted of a rotor and stator, both of which were shaped in a series of semicircular cups angled at 45° but of opposite pitch. The change is momentum when water passes from rotor to stator and back again, creating a braking reaction. By measuring this braking reaction and the shaft speed, the power of the engine could be calculated.

Froude was a tireless experimenter who put one of the most powerful analytical tools into the hands of marine architects. He has been credited as the founder of ship hydrodynamics. If any one person can be said to have founded anything in engineering, this is probably accurate.

Fuchs Immanuel Lazarus 1833–1902 was a German mathematician whose work on Riemann's method for the solution of differential equations led to a study of the theory of functions that was later crucial to Henri Poincaré in his own important investigation of function theory. Fuchs's main scientific importance may be seen, therefore, as providing a sort of link between the nineteenth- and twentieth-century ideas of mathematical development.

Fuchs was born on 5 May 1833 in Moschin (now in Poland). During his elementary education his mathematical talents were already apparent, and he went on to study at the University of Berlin, under Ernst Kummer (1810–1893) and Karl Weierstrass (1815–1897). He gained his PhD in Berlin in 1858, and began teaching at local schools. In 1865, he became a lecturer at Berlin University, and only

a year later was promoted to Professor. In 1869, he became the Professor of Mathematics at the Artillery and Engineering School at Griefswald. Later he transferred first to the University of Göttingen (in 1874) and then (in 1875) to that of Heidelberg, to take up their Chairs of Mathematics. Returning to Berlin in 1882, he succeeded Weierstrass two years later as Professor of Mathematics there, and over the next 20 years held several administrative and academic posts of responsibility at the University. He died, in Berlin, on 26 April 1902.

Fuchs's interest in the theory of functions was first aroused during his student days in Berlin, but it was not until he took up his first professional appointment that he began to work seriously on his research. He produced a number of papers that were intended to develop Riemann's work on a method for solving differential equations. Fuchs's proposals were in contrast to those put forward earlier by Augustin Cauchy (1789–1857), who used power series.

The first proof for solutions of linear differential equations of order n was developed from this study, as were the Fuchsian differential equations and the Fuchsian theory on solutions for singular points. His work in this field was of great importance to Poincaré's work on automorphic functions. Fuchs also carried out some research into number theory and geometry.

Fuller Solomon Carter 1872–1953 was a US physician, neurologist, psychiatrist and pathologist who is acknowledged as being the first black psychiatrist.

The son of state officials, Solomon Carter Fuller was born in Monrovia, Liberia on 11 August 1872. He went to the United States in 1889 and was educated at Livingstone College, Salisbury, North Carolina, receiving his BA in 1893. After studying medicine at Long Island College Hospital, Brooklyn, New York and later at Boston University School of Medicine, he was awarded his MD in 1897. Following a two-year internship at Westborough State Hospital, Massachusetts, he was appointed as a pathologist at the hospital and as a faculty member of Boston University School of Medicine. He practised medicine in Boston and at his home in Framingham and taught pathology, neurology and psychology at the University until he retired as Professor Emeritus in 1937. Fuller died in 1953; he was married to the sculptor Meta Warrick Fuller and had three sons.

Fuller is best known for his work in neuropathology and psychiatry. His postgraduate studies included training at the Carnegie Laboratory, New York, and, between 1904–1905, at the University of Munich under Dr Alzheimer. He was well known for his work on degenerative brain diseases including Alzheimer's disease, which he attributed to causes other than arteriosclerosis; this was further supported by medical researchers in 1953. Fuller published widely in books and medical journals and in 1913 became the editor of *Westborough State Hospital Papers*, which specialized in mental diseases. He was a member of many medical and psychiatric societies. Fuller's work was acknowledged in the 1970s; in 1971 the Black Psychiatrists of America presented a portrait of Fuller to the American Psychiatric Association in Washington, DC, and in 1973 Boston University memorialized him in an all-day conference; the Solomon Carter Fuller Mental Health Center in Boston is named after him.

Fulton Robert 1765–1815 was a US artist and engineer and inventor who built one of the first successful steamboats, propelled by two paddlewheels.

Fulton was born on 14 November 1765 in Lancaster County, Pennsylvania. His artistic talent was evident at an early age, and he was employed by local gunsmiths to draw designs for their work. At the age of 17, he moved to Philadelphia, where he became a successful portrait painter and miniaturist. Four years later, in 1786, he decided to go to London to study under the American artist Benjamin West. England changed Fulton's life. The country was involved in the Industrial Revolution; roads, canals and bridges were being built, factories were springing up and mining enterprises were getting under way. He was fascinated by all he saw and eventually, in 1793, he gave up art as a vocation, in favour of engineering projects.

When Fulton was 14, he designed a small paddleboat and now he considered designing one with a steam engine to power it. In 1786, John Fitch (1743–1798) had demonstrated a steamboat in the United States, but it had not proved to be a success. The British government had placed a ban on the export of steam engines, but Fulton nevertheless contacted a British company about the possibility of purchasing an engine suitable for boat propulsion. Meanwhile he designed and patented a device for hauling canal boats over difficult country. He also patented machines for sawing marble and twisting hemp (for rope), and he built a mechanical dredger for canal construction.

In 1796, Fulton went to France, where he experimented with fitting steam engines to ships, and by 1801 he had also carried out tests with the *Nautilus*, a submarine he had invented. But he failed to interest the French in his inventions, so by 1804 he tried the British government, again without success. In 1802, Fulton had met Robert Livingstone, a former partner in another steamboat invention and then US Minister to the French government. They joined forces and in 1803 a steam engine was obtained from the British firm of Boulton and Watt; but it took three years to get permission to export it to the United States.

In New York, Fulton worked to install the new engine in a locally-built vessel. Livingstone favoured designing a propulsion system using a jet of water forced out at the stern under pressure, but Fulton settled on a paddlewheel on each side. In 1807, the paddlesteamer *Clermont*, with a 18-kW/24-hp engine fitted into its 30-m/100-ft hull, made its first successful voyage up the Hudson River at an average speed of 8 kph/5 mph. A large boatworks was built in New Jersey, and steamboats came into use along the Atlantic Coast and later in the West. In 1815, Fulton began to build a steam-powered warship for the United States navy.

Fulton died on 24 February 1815 in New York City.

Funk Casimir 1884–1967 was a Polish-born US biochemist, best known for his researches on dietary requirements, particularly of vitamins (a term derived from vitamines – Funk's original name for them). He was the first to isolate nicotinic acid (also called niacin, one of the vitamins of the B complex).

Funk was born in Warsaw on 23 February 1884, the son of a prominent dermatologist. An able student, he was educated at the University of Bern, Switzerland, from which he gained his doctorate in organic chemistry in 1904, when he was still only 20 years old. He then worked at the Pasteur Institute in Paris, the Wiesbaden Municipal Hospital, the University of Berlin, and the Lister Institute in London, where he was assigned to work on beriberi. In 1915, he emigrated to the United States, where he held several industrial and university positions in New York; he became a naturalized US citizen in 1920.

In 1923, the Rockefeller Foundation supported his return to Warsaw as Director of the Biochemistry Department of the State Institute of Hygiene but, because of the country's uncertain political situation, Funk left this post in 1927 and went to Paris. In the following year, he became a consultant to a pharmaceutical company and founded a privately financed research institution, the Casa Biochemica. With the German invasion of France at the outbreak of World War II in 1939, Funk abandoned this venture and returned to the United States, where he worked as a consultant to the United States Vitamin Corporation. In 1940, he became President of the Funk Foundation for Medical Research, a position he held until his death. He died in Albany, New York, on 20 November 1967.

Following the research of Frederick Gowland Hopkins into what we now know to be deficiency diseases, such as beriberi and pellagra, Funk succeeded in 1911 at the Lister Institute in obtaining a concentrate that cured a pigeon disease similar to beriberi. He showed that the anti-beriberi concentrate was an amine, and in 1912 he discussed the causes of beriberi, scurvy and rickets as deficiency diseases. He also suggested that the accessory food factors needed to prevent these disorders are all amines and so he named them vitamines (vital amines). It was later discovered that not all accessory food factors are amines, and Funk's term was shortened to vitamins.

Funk continued to try to find the anti-beriberi factor for human beings and finally isolated nicotinic acid from rice polishings. But it did not cure beriberi, so he discarded it. (Later Otto Warburg discovered that nicotinic acid prevents pellagra.) In 1934 Robert Williams (1886–1965) isolated about 10 g/0.3 oz of the anti-beriberi factor (now known as thiamin or vitamin B_1) from 1,000 kg/2,200 lb of rice polishings, and in 1936 Funk determined its molecular structure and developed a method of synthesizing it.

Funk also did extensive research into animal hormones, particularly male sex hormones, and into the biochemistry of cancer, diabetes and ulcers. He improved the methods used for the commercial manufacture of many drugs, as well as developing several new commercial products in his own laboratories.

G

Gabor Dennis 1900–1979 was a Hungarian-born British physicist and electrical engineer, famous for his invention of holography – three-dimensional photography using lasers – for which he received the 1971 Nobel Prize for Physics.

Gabor was born on 5 January 1900 in Budapest. He was educated at the Budapest Technical University and then at the Technishe Hochschule in Charlottenburg, Berlin. From 1924 to 1926 he was an assistant there, and for the next three years he held the position of Research Associate with the German Research Association for High-Voltage Plants. He was a research engineer for the firm of Siemens and Halske in Berlin from 1927, until he fled Nazi Germany in 1933 to Britain. He then worked as a research engineer with the Thomson–Houston Company of Rugby from 1934 to 1938, and later became a British subject. In 1949, he joined the Faculty of the Imperial College of Science and Technology, London, as a Reader in Electronics. He was Professor of Applied Electron Physics from 1958 to 1967, when he became a Senior Research Fellow. From 1976 until his death, he was Professor Emeritus of Applied Electron Physics of the University of London. He died in London on 8 February 1979.

Gabor first conceived the idea of holography in 1947 and developed the basic technique by using conventional filtered light sources. Because conventional light sources provided too little light or light that was too diffuse, his idea did not become commercially feasible until the laser was demonstrated in 1960 and was shown to be capable of amplifying the intensity of light waves.

Holography is a means of creating a unique photographic image without the use of a lens. The photographic recording of the image is called a hologram. The hologram appears to be an unrecognizable pattern of shapes and whorls, but when it is illuminated by coherent light (as by a laser beam), the light is organized into a three-dimensional representation of the original object. Gabor coined the name from the Greek *holos* ('whole') and *gram* ('message'), because his image-forming mechanism recorded all the optical information in a wavefront of light. In ordinary photography, the photographic image records the variations in light intensity reflected from an object, so that dark areas are produced where less light is reflected and light areas where more light is reflected. Holography records not only the intensity of light, but also its phase, or the degree to which the wavefronts making up the reflected light are in step with each other. The wavefronts of ordinary light waves are not in step – ordinary light is incoherent.

When Gabor began work on the holograph, he considered the possibility of improving the resolving power of the electron microscope, first by using the electron beam to make a hologram of the object and then by examining this hologram with a beam of coherent light. It is possible to obtain a degree of coherence by focusing light through a very small pinhole, but the resulting light intensity is then too low for it to be useful in holography. In 1960, the laser beam was developed. This has a high degree of coherence and also has high intensity. There are many kinds of laser beam, but two have special interest in holography, the continuous-wave (CW) and the pulsed laser. The CW laser emits a bright continuous beam of light of a single, nearly pure colour. The pulsed laser emits an extremely intense, short flash of light that lasts only about 10^{-8} sec. Two US scientists, Emmett Leith and Juris Upatnieks have applied the CW laser to holography with great success, opening the way to many research applications. To achieve a three-dimensional image, the light streaming from the source must itself be photographed. If the waves of this light, with its many rapidly moving crests and troughs, are frozen for an instant and photographed, the wave pattern can then be reconstructed and will show the same three-dimensional character as the object from which the light is reflected.

Pulse laser holography is used in the study of chemical reactions, where optical properties of solutions often change. It is also used in wind-tunnel experiments, where it can be used to record refractive index changes in the air flow, created by pressure changes as the gas deflects around the aerodynamic object. This recording is done interferometrically (by observing interference fringes).

Apart from the invention of holography, Gabor's other work included research on high-speed oscilloscopes,

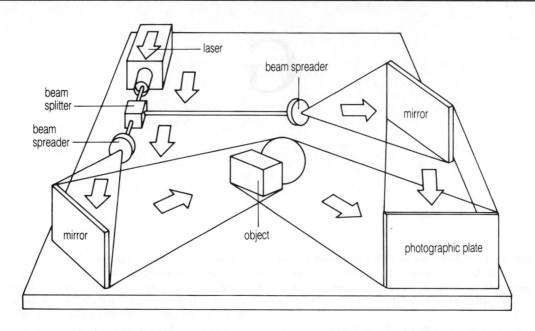

Gabor *Recording a transmission hologram*

communication theory, physical optics and television. He took out more than 100 patents for his inventions and became renowned as an outstanding engineer and physicist of the twentieth century.

Gajdusek Daniel Carleton 1923– is a US paediatrician and virologist who won the Nobel Prize for Physiology or Medicine for his work on identifying and describing slow virus infections in humans, based on his extensive studies of kuru, a disease of neural degeneration found in people in New Guinea.

He was born on 9 September 1923 in Yonkers, New York, into a family of Central European extraction that put a high premium on education. He was educated at the Universities of Rochester and Harvard and graduated in medicine in 1941, when he decided to specialize in paediatrics and completed internship and residencies in several children's hospitals. He also studied physical chemistry at the California Institute of Technology with the Nobel Laureate Linus Pauling (1901–), and won a research fellowship to work in the virology department at Harvard. Drafted in 1952, he served at the Walter Reed Army Institute for Research and continued virological research, also becoming interested in epidemiology, which he continued to study on his return to civilian life. He worked at the University of Tehran in Persia (now Iran) and with the future Nobel prizewinning immunologist Macfarlane Burnet (1899–1985) in Melbourne, Australia. From there he travelled to New Guinea and learned of the disease of kuru, a fatal degeneration of the brain, restricted to a group of people living in a remote area. Gajdusek lived with these people, studying their language and culture and collecting

post mortem samples from kuru victims for laboratory analysis. In 1958, he returned to the United States to the National Institute of Neurology and Communicative Disorders and Stroke in Bethesda, and in 1970 became Head of the Central Nervous Systems Studies Laboratory. His research on kuru continued for many years and he made frequent trips back to New Guinea. The affected people practised a form of ritual cannibalism, in which the women and children consumed the brains of the dead, including those who had died from the disease. Analyses of such brain tissue failed to reveal any signs of infective organisms, and Gajdusek and his collaborators injected extracts from such brains into the brains of chimpanzees. After about a year, the animals began to display signs that characterized the human disease – shaking, trembling and loss of motor function. The experiment was repeated, material from the brains of the ill animals being injected into further animals. In this way the illness was transmitted from animal to animal in a way that was similar to the course of the human disease. Education programmes in New Guinea have eradicated the practices that caused kuru, although there are still a few affected individuals alive. The transmissability of the disease led Gajdusek to propose that kuru was caused by a 'slow virus' that exerts its effects after a very long incubation period and further work by his group showed that Creutzfeldt–Jakob disease was similarly caused by slow viruses. Since then such a mechanism has been proposed for many illnesses, including AIDS and multiple sclerosis. Gajdusek has contributed significantly, not only to further studies of slow virus infections, but also to child psychology and ethnology. Unmarried, he has adopted many children from the Pacific Islands and has received numerous honours and awards, including membership of the National Academy of Sciences, and the Nobel Prize in 1976.

Galen *c*.AD 130–*c*.200 was a Greek physician, anatomist and physiologist whose ideas – as conveyed in his many books, more than 130 of which have survived – had a profound influence on medicine for some 1,400 years.

Galen was born in about AD 129 in Pergamum (now Bergama in Turkey), the son of an architect. The city of his birth contained an important shrine to Asclepius, the god of medicine, and Galen received his early medical training at the medical school attached to the shrine; he also studied philosophy at Pergamum. From about 148 to 157 he continued his studies at Smyrna (now Izmir) on the west coast of present-day Turkey, Corinth in Greece and Alexandria in Egypt, after which he returned home to become chief physician to the gladiators at Pergamum. In 161 he went to Rome where he cured the eminent philosopher Eudemus, who introduced Galen to many influential people. Successful as a physician – and boastful of his successes – he soon became physician to Marcus Aurelius, then co-emperor with Lucius Verus. In about 166, however, Galen returned to Pergamum – according to Galen himself, because of the intolerable envy of his colleagues, although it is also probable that he left Rome to escape the plague brought back by Verus' army after a foreign war. About two years later, Marcus Aurelius recalled Galen to Rome, and when Lucius Verus died of the plague in 169, Galen was appointed physician to Commodus, Marcus Aurelius' son. Little is known of Galen's life after this, although it is thought he remained in Rome until his death in about AD 200.

Galen was an expert dissector and his best work was in the area of descriptive anatomy. The dissection of human beings was then regarded as taboo and Galen made inferences about human anatomy from his many dissections of pigs, dogs, goats and especially Barbary apes – although there is some evidence that he did dissect humans as well. He was a meticulous observer and his detailed descriptions of bones and muscles, many of which he was the first to identify, are particularly good; he also noted that many muscles are arranged in antagonistic pairs. He described the heart valves, distinguished seven pairs of cranial nerves and noted the structural differences between arteries and veins. He also described the network of blood vessels under the brain, but although this network is present in many animals, it is not found in humans, and Galen erroneously assigned it an important role in his tripartite scheme of circulation in the human body. In addition he performed several vivisection experiments: he tied off the recurrent laryngeal nerve to show that the brain controls the larynx; he demonstrated the effects of severing the spinal cord at different points; and he tied off the ureters to show that urine passes from the kidneys down the ureters to the bladder. More important, he demonstrated that arteries carry blood, not air, thus disproving Erasistratus' view, which had been taught for some 500 years.

In physiology, however, Galen was less successful. He thought that blood (the natural spirit) was formed in the liver and was then carried through the veins to nourish every part of the body, forming flesh in the body's periphery. Some of the blood, however, passed through the vena cava to the right-hand side of the heart, from where it then passed to the left-hand side through minute pores (the existence of which was pure conjecture on Galen's part) in the septum. In the left side the blood mixed with air brought from the lungs, and the resulting vital spirit was then carried around the body by the arteries. Some of this fluid went to the head where, in the hypothetical network of blood vessels beneath the brain, the fluid was infused with animal spirit (which was thought to produce consciousness) and distributed to the muscles and senses by the nerves (which Galen thought were hollow). Thus, Galen postulated a tripartite circulation system in which the liver produced the natural spirit, the heart the vital spirit, and the brain the animal spirit. Although this system is now known to be a complete misconception, it did provide a rational explanation of the facts known at the time and, because of the high esteem in which Galen was held until the Renaissance, persisted until William Harvey (1578–1657) proposed his theory of blood circulation in 1628. Another of Galen's mistaken beliefs that inhibited the progress of medical knowledge was his subscription to Hippocrates' theory that human health relied upon a balance between the four humours (phlegm, black bile, yellow bile and blood).

Galen was a prolific author who, in addition to scientific works, also wrote about philosophy and literature. He believed that Nature expressed a divine purpose, a belief that was reflected in many of his books, including his scientific texts. Belief in divine purpose became increasingly popular with the rise of Christianity (Galen himself was not a Christian), which fact, combined with the eclipse of science until the Renaissance, probably ensured the survival of many of Galen's works and accounted for the enormous influence of his ideas.

Galileo properly Galileo Galilei 1564–1642 was an Italian physicist and astronomer whose work founded the modern scientific method of deducing laws to explain the results of observation and experiment. In physics, Galileo discovered the properties of the pendulum, invented the thermometer, and formulated the laws that govern the motion of falling bodies. In astronomy, Galileo was the first to use the telescope to make observations of the Moon, Sun, planets and stars.

Galileo was born in Pisa on 15 February 1564. His full name was Galileo Galilei, his father being Vincenzio Galilei (*c*.1520–1591), a musician and mathematician. Galileo received his early education from a private tutor at Pisa until 1575, when his family moved to Florence. He then studied at a monastery until 1581, when he returned to Pisa to study medicine at the university. Galileo was attracted to mathematics rather than medicine, however, and also began to take an interest in physics. In about 1583 he discovered that a pendulum always swings to and fro in the same period of time regardless of the amplitude (length) of the swing. He is said to have made this discovery by using his pulse to time the swing of a lamp in Pisa Cathedral.

Galileo remained at Pisa until 1585, when he left without taking a degree and returned home to Florence. There he studied the works of Euclid (lived 300 BC) and Archimedes (287–212 BC), and in 1586 extended Archimedes' work in hydrostatics by inventing an improved version of the hydrostatic balance for measuring specific gravity. At this time, Galileo's father was investigating the ratios of the tensions and lengths of vibrating strings that produce consonant intervals in music, and this work may well have demonstrated to Galileo that the validity of a mathematical formula could be tested by practical experiment.

In 1589, Galileo became Professor of Mathematics at the University of Pisa. He attacked the theories of Aristotle (384–322 BC) that then prevailed in physics, allegedly demonstrating that unequal weights fall at the same speed by dropping two cannon balls of different weights from the Leaning Tower of Pisa to show that they hit the ground together. Galileo also published his first ideas on motion in *De Motu/On Motion* in 1590.

Galileo remained at Pisa until 1592, when he became Professor of Mathematics at Padua. He flourished at Padua, and refined his ideas on motion over the next 17 years. He deduced the law of falling bodies in 1604 and came to an understanding of the nature of acceleration. In 1609 he was diverted from his work in physics by reports of the telescope, which had been invented in Holland the year before. Galileo immediately constructed his own telescopes and set about observing the heavens, publishing his findings in *Sidereus Nuncius/The Starry Messenger* in 1610. The book was a sensation throughout Europe and brought Galileo immediate fame. It resulted in a lifetime appointment to the University of Padua, but Galileo rejected the post and later, in 1610, became mathematician and philosopher to the Grand Duke of Tuscany, under whose patronage he continued his scientific work at Florence.

In 1612, Galileo returned to hydrostatics and published a study of the behaviour of bodies in water, in which he championed Archimedes against Aristotle. In the following year he performed the same service for Nicolaus Copernicus (1473–1543), publicly espousing the heliocentric system. This viewpoint aroused the opposition of the Church, which declared it to be heretical and in 1616 Galileo was instructed to abandon the Copernican theory. Galileo continued his studies in astronomy and mechanics without publicly supporting Copernicus, although he was personally convinced. In 1624, Urban VIII became pope and Galileo obtained permission to present the arguments for the rival heliocentric and geocentric systems in an impartial way. The result, *Dialogue Concerning the Two Chief World Systems*, was published in 1632, but it was scarcely impartial. Galileo used the evidence of his telescope observations and experiments on motion to favour Copernicus, and immediately fell foul of the Church again. The book was banned and Galileo was taken to Rome to face trial on a charge of heresy in April 1633. Forced to abjure his belief that the Earth moves around the Sun, Galileo is reputed to have muttered 'Eppur si muove' ('Yet it does move').

Galileo was sentenced to life imprisonment, but the sentence was commuted to house arrest and he was confined to his villa at Arcetri near Florence for the rest of his life. He continued to work on physics, and summed up his life's work in *Discourses and Mathematical Discoveries Concerning Two New Sciences*. The manuscript of this book was smuggled out of Italy and published in Holland in 1638. By this time Galileo was blind, but he still continued his scientific studies. In these last years, he designed a pendulum clock. Galileo died at Arcetri on 8 January 1642.

Galileo made several fundamental contributions to mechanics. He rejected the impetus theory that a force or push is required to sustain motion, and identified motion in a horizontal plane and rotation as being 'neutral' motions that are neither natural nor forced. Galileo realized that gravity not only causes a body to fall, but that it also determines the motion of rising bodies and furthermore that gravity extends to the centre of the Earth. He found that the distance travelled by a falling body is proportional to the square of the time of descent. Galileo then showed that the motion of a projectile is made up of two components: one component consists of uniform motion in a horizontal direction and the other component is vertical motion under acceleration or deceleration due to gravity. This explanation was used by Galileo to refute objections to Copernicus based on the argument that birds and clouds would not be carried along with a turning or moving Earth. He explained that a horizontal component of motion provided by the Earth always exists to keep such objects in position even though they are not attached to the ground.

Galileo came to an understanding of uniform velocity and uniform acceleration by measuring the time it takes for bodies to move various distances. In an era without clocks, it was difficult to measure time accurately and Galileo may have used his pulse or a water clock – but not ironically, the pendulum – to do so. He had the brilliant idea of slowing vertical motion by measuring the movement of balls rolling down inclined planes, realizing that the vertical component of this motion is a uniform acceleration due to gravity while the horizontal component is a uniform velocity. Even so, this work was arduous and it took Galileo many years to arrive at the correct expression of the law of falling bodies. This was presented in the *Two New Sciences*, together with his derivation that the square of the period of a pendulum varies with its length (and is independent of the mass of the pendulum bob). Galileo also deduced by combining horizontal and vertical motion that the trajectory of a projectile is a parabola. He furthermore realized that the law of falling bodies is perfectly obeyed only in a vacuum, and that air resistance always causes a uniform terminal velocity to be reached.

The other new science of Galileo's masterwork was engineering, particularly the science of structures. His main contribution was to point out that the dimensions of a structure are important to its stability: a small structure will

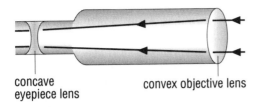

concave
eyepiece lens   convex objective lens

Galileo *In a Galilean refracting telescope, a convex (converging) objective lens and a concave (diverging) lens in the eyepiece together produce a magnified upright image.*

stand whereas a larger structure of the same relative dimensions may collapse. Using the laws of levers, Galileo went on to examine the strengths of the materials necessary to support structures.

Galileo's other achievements include the invention of the thermometer in 1593. This device consisted of a bulb of air that expanded or contracted as the temperature changed, causing the level of a column of water to rise or fall. Galileo's thermometer was very inaccurate because it neglected the effect of atmospheric pressure, but it is historically important as one of the first measuring instruments in science.

Galileo's astronomical discoveries were a tribute to both his scientific curiosity and his ability to devise new techniques with instruments. Within two years of first building a telescope he had compiled fairly accurate tables of the orbits of four of Jupiter's satellites and proposed that their frequent eclipses could serve as a means of determining longitude on land and at sea. His observations on sunspots are noteworthy for their accuracy and for the deductions he drew from them regarding the rotation of the Sun and the orbit of the Earth. He believed, however – following both Greek and medieval tradition – that orbits must be circular, not elliptical, in order to maintain the fabric of the cosmos in a state of perfection. This preconception prevented him from deriving a full formulation of the law of inertia, which he himself discovered although it is usually attributed to the contemporary French mathematician René Descartes. Lacking the theory of Newtonian gravity, he hoped to explain the paths of the planets in terms of circular inertial orbits around the Sun.

Galileo was a pugnacious and sarcastic man, especially when confronted with those who could not or would not admit the validity of his observations and arguments. This characteristic symbolizes his position in the history of science, for with Galileo the idea that experiment and observation can be used to prove the validity of a proposed mathematical description of a phenomenon really became a working method in science. Although he built on the views and work of Archimedes and Copernicus, in his achievements Galileo can be considered to have founded not only classical mechanics and observational astronomy, but the method of modern science overall. In mechanics Isaac Newton (1642–1727) developed Galileo's work into a full understanding of inertia, which Galileo had not quite

grasped, and arrived at the basic laws of motion underlying Galileo's discoveries.

Galle Johann Gottfried 1812–1910 was a German astronomer who was one of the first to observe Neptune and recognize it as a new planet.

Galle was born at Pabsthaus in Prussian Saxony (now in Germany) on 9 June 1812, the son of a turpentine-maker. He was educated at the University of Berlin, which he entered in 1830, and where he first met Johann Encke (1791–1865). Graduating, he joined the staff of the Berlin Observatory, and in 1835 became Assistant Director under Encke. In 1851, he was appointed Professor of Astronomy and Director of the Breslau Observatory (now in Wroclaw, Poland), where he spent the rest of his working life until his retirement at the age of 83. Galle died at the age of 98, in Potsdam, Prussia, on 10 July 1910.

To gain his doctorate in 1845, Galle sent his dissertation to the Director of the Paris Observatory, Urbain Leverrier (1811–1877). By way of reply, Leverrier sent back his calculation of the position of a new planet, predicted mathematically from its apparent gravitational effect on Uranus, then the outermost known planet; Galle received Leverrier's final prediction in September 1846. Within one hour of beginning their search, Galle and his colleague Heinrich d'Arrest had located Neptune, less than 1° away from the predicted position. In fact, they located it more from its absence as a star plotted on one of Encke's new star-charts, but the discovery was acclaimed nevertheless as a triumph for scientific theory as opposed to observation.

Galle was also the first to distinguish the Crêpe Ring, a somewhat obscure inner ring around Saturn. First announced by Galle in 1838, its existence was forgotten until it was rediscovered independently by George Bond (1825–1865) at Harvard and by William Dawes (1799–1868) in England.

He also suggested a method of measuring the scale of the Solar System by observing the parallax of asteroids, first applying his method to the asteroid Flora in 1873. The method was eventually carried out with great success, but not until Galle had been dead for nearly two decades.

Galois Evariste 1811–1832 was a French mathematician who greatly extended the understanding of the conditions in which an algebraic equation is solvable and, by his method of doing so, laid the foundations of modern group theory.

Galois was born in the village of Bourg-la-Reine, on the outskirts of Paris, on 25 October 1811. His family was of the highly respectable bourgeois class that came into its own with the restoration of the monarchy after the Napoleonic Empire. His father was the headmaster of the local boarding school and mayor of the town; his mother came from a family of jurists, and it was she who gave Galois his early (chiefly classical) education. He first attended school at the age of 12, when he was sent to the Collège Louis-le-Grand in Paris as a fourth-form boarder in 1823. Another four years passed before Galois, who gained notoriety for resisting the strict discipline imposed at the school, had his mathematical imagination fired by the lectures of H. J. Vernier. He soon became familiar with the works of Legrange and Legendre and by 1828 he was busy mastering the most recent work on the theory of equations, number theory, and elliptic functions. He quickly came to believe that he had solved the general fifth-degree equation. Like Abel before him, he discovered his error, and that discovery launched him on his search – ultimately successful and ultimately of momentous consequence for the future of mathematics – for a solution to the problem of the solubility of algebraic equations generally. By 1829 he had progressed far enough to interest Augustin Cauchy (1789–1857), who presented Galois's early results to the Academy of Science.

In a remarkably short period of time Galois, at the age of 17, had arrived very near the apex of existing mathematical thought. Then the first of the emotional disruptions which were to darken the few remaining years of his life occurred. His father, the victim of a political plot which unjustly discredited him, committed suicide in July 1829. A month later Galois, partly from his fiery impatience with the examiners' instructions, failed to gain entrance to the Ecole Polytechnique. He had therefore to be content with entering, in the autumn of 1829, the Ecole Normale Supérieure.

It was then that Galois learned that the ideas which were contained in the paper presented to the Academy by Cauchy were not original. Shortly before, very similar ideas had been published by Abel in his last paper. Encouraged by Cauchy, Galois began to revise his paper in the light of Abel's findings. He presented the revised version to the Academy in February 1830, with high hopes that it would gain him the Grand Prix. To his dismay the Academy not only rejected his paper, but the examiner lost the manuscript. Galois was indignant at this ill-treatment, but four months later he succeeded in having a paper on number theory published in the prestigious *Bulletin des sciences mathématiques*. It was this paper that contained the highly original and diverting theory of 'Galois imaginaires'.

Thus, when the July Revolution that drove Charles X off the throne shocked French society, Galois had reason to be proud of his barely acknowledged mathematical genius and cause to feel estranged, both on his own and his father's account, from the stuffy officialdom which had cast its pall over France since 1815. He joined the revolutionary movement. In the next year he was twice arrested, in May 1831 for proposing a regicide toast at a republican banquet (Louis-Philippe having taken Charles's place as king) and again in July for taking part in a republican demonstration. For the second offence Galois was imprisoned for nine months. In prison he continued his mathematical research. But shortly after his release he became involved in a duel, perhaps from political reasons, perhaps from complications arising from a love affair. The event remains mysterious. Galois, it is evident, expected to be killed. On 29 May 1832, in a letter written in feverish haste, he outlined the principal results of his mathematical inquiries. He sent the letter to his friend Auguste Chevalier, almost certainly in the expectation that it would find its way to Karl Gauss and Karl Jacobi. The duel took place on the following morning. Galois was severely wounded in the stomach and died in hospital on 31 May 1832.

So brief was Galois's life – he was only 20 when he died – and so dismissively were his ideas treated by the Academy, that he was known to his contemporaries principally as a rather headstrong republican agitator. Cauchy, who alone sensed the importance of what Galois was doing, was out of France after 1830 and did not see any of the revisions or developments of Galois's first paper. Moreover, in his letter to Chevalier, Galois had time only to put down in concise form – without demonstrations – his most important conclusions. When he died, therefore, Galois's work amounted to fewer than 100 pages, much of it fragmentary and nearly all of it unpublished.

The honour of rescuing Galois from obscurity belongs, in the first instance, to Joseph Liouville (1809–1882), who in 1843 began to prepare his papers for publication and informed the Academy that Galois had provided a convincing answer to the question whether first-degree equations were solvable by radicals. Finally, in 1846, both Galois' 1831 paper and a short notice on the solution of primitive equations by radicals were (thanks to Liouville) published in the *Bulletin*.

Galois's achievement, put tersely, was to arrive at a definitive solution to the problem of the solvability of algebraic equations, and in doing so to produce such a breakthrough in the understanding of fields of algebraic numbers and also of groups that he is deservedly considered as the chief founder of modern group theory. What has come to be known as the Galois theorem made immediately demonstrable the insolubility of higher-than-fourth-degree equations by radicals. The theorem also showed that if the highest power of x is a prime, and if all other values of x can be found by taking only two values of x and combining them using only addition, subtraction, multiplication and division, then the equation can be solved by using formulae similar in principle to the formula used in solving quadratic equations.

Galois's work involved groups formed from the arrangements of the roots of equations and their subgroups, groups which he fitted into each other rather on the analogy of the Chinese box arrangement. This, his most far-reaching achievement for the subsequent development of group theory, is known as 'Galois theory'. Along with other such terms – Galois groups, Galois fields and the Galois theorem – it bears testimony to the lasting influence of his rejected genius.

Galton Francis 1822–1911 was a British scientist, inventor and explorer who made contributions to several disciplines, including anthropology, meteorology, geography and statistics, but who is best known as the initiator of the study of eugenics (a term he coined).

Galton was born on 16 February 1822 near Sparkbrook, Birmingham, the youngest of nine children of a rich banker and a first cousin of Charles Darwin. Galton's exceptional intelligence was apparent at an early age – he could read before he was three years old and was studying Latin when he was four – but his achievements in higher education were unremarkable. Acceding to his father's wish, Galton studied medicine at Birmingham General Hospital and then King's College, London, but interrupted his medical studies to read mathematics at Trinity College, Cambridge. He returned to medicine, however, studying at St George's Hospital in London, but abandoned his studies on inheriting his father's fortune and spent the rest of his life pursuing his own interests.

Shortly after the death of his father, Galton set out in 1850 to explore various uncharted areas of Africa. He collected much valuable information, and on his return to England wrote two books describing his explorations. These writings won him the Royal Geographical Society's annual gold medal in 1853, and three years later he was elected a Fellow of the Royal Society.

Galton then turned his attention to meteorology. He designed several instruments to plot meteorological data, discovered and named anticyclones and made the first serious attempt to chart the weather over large areas – described in his book *Meteorographica* (1863). He also helped to establish the Meteorological Office and the National Physical Laboratory.

Stimulated by Darwin's *On the Origin of Species* (1859), Galton then began his best known work, that concerning heredity and eugenics, which occupied him for much of the rest of his life. Studying mental abilities, he analysed the histories of famous families and found that the probability of eminent men having eminent relatives was high, from which he concluded that intellectual ability is inherited. He tended, however, to underestimate the role of the environment in mental development, since intelligent parents tend to provide their children with a mentally stimulating environment as well as with genes for high intelligence. Galton described his study of mental abilities in eminent families in *Hereditary Genius* (1869), in which he used the term genius to define creative ability of an exceptionally high order, as shown by actual achievement. He formulated the theory that genius is an extreme degree of three traits, intellect, zeal and power of working. He also formulated the regression law, which states that parents who deviate from the average type of the race in a positive or negative direction have children who, on average, also deviate in the same direction but to a lesser extent.

Continuing his research, Galton devised instruments to measure mental abilities and used them to obtain data from some 9,000 subjects. The results were summarized in *Inquiries into Human Faculty and its Development* (1883), which established Galton as a pioneer of British scientific psychology. He was the first to use twins to try to assess the influence of environment on development, but more important was his quantitative analytical approach to his investigations. In order to interpret his results, Galton devised new statistical methods of analysis, including correlational calculus, which has since become an invaluable tool in many disciplines, especially the social sciences.

As a result of his research into human abilities, Galton became convinced that the human species could be improved, both physically and mentally, by selective breeding, and in 1885 he gave the name 'eugenics' to the study of methods by which such improvements could be attained. He defined eugenics as 'the study of the agencies under social control which may improve or impair the racial qualities of future generations, physically or mentally.'

Galton also made a number of other contributions. He demonstrated the permanence and uniqueness of fingerprints and began to work out a system of fingerprint identification, now extensively used in police work. In 1879, he devised a word-association test which was later developed by Sigmund Freud and became a useful aid in psychoanalysis. Also, he invented a teletype printer and the ultrasonic dog whistle.

Galton was knighted in 1909, two years before his death on 17 January 1911 in Haslemere, Surrey. In his will he left a large bequest to endow a Chair in Eugenics at the University of London.

Galvani Luigi 1737–1798 was an Italian anatomist whose discovery of 'animal electricity' stimulated the work of Alessandro Volta and others to discover and develop current electricity.

Galvani was born in Bologna on 9 September 1737. He studied medicine at the University of Bologna, graduating in 1759 and gaining his doctorate three years later with a thesis on the development of bones. In that same year (1762) he married Lucia Galeazzi, a professor's daughter, and was appointed Lecturer in Anatomy. In 1772, he became President of the Bologna Academy of Science and in 1775 he took up the appointment of Professor of Anatomy and Gynaecology at the university. The death of his wife in 1790 left him griefstricken. In 1797, he was required to swear allegiance to Napoleon as head of the new Cisalpine Republic, but Galvani refused and thereby lost his appointment at Bologna University. Saddened, he retired to the family home and died there on 4 December 1798.

Comparative anatomy formed the subject of Galvani's early research, during which he investigated such structures as the semicircular canals in the ear and the sinuses of the nose. He began studying electrophysiology in the late 1770s, using static electricity stored in a Leyden jar to stimulate muscular contractions in living and dead animals. The frog was already established as a convenient laboratory animal and in 1786 Galvani noticed that touching a frog with a metal instrument during a thunderstorm made the frog's muscles twitch. He later hung some dissected frogs from brass hooks on an iron railing to dry them, and noticed that their muscles contracted when the legs came into contact with the iron – even if there was no electrical storm. Galvani concluded that electricity was causing the contraction and postulated (incorrectly) that it came from the animal's muscle and nerve tissues. He summarized his findings in 1791 in a paper 'De viribus electricitatis in motu musculari commentarius' which gained general acceptance except by Alessandro Volta (1745–1827), who by 1800 had constructed electric batteries consisting of plates (electrodes) of dissimilar metals in a salty electrolyte, thus proving that Galvani had been wrong and that the source of the electricity in his experiment had been the two different metals and the animal's body fluids. But for many years current electricity was called Galvanic electricity and Galvani's name is preserved in the word galvanometer, a current-measuring device developed in 1820 by André Ampère (1775–1836), and in galvanization, the process of covering steel with a layer of zinc (originally by electroplating).

Galvani's original paper was translated into English and published in 1953 as 'Commentary on the Effect of Electricity on Muscular Motion'.

Gamow George 1904–1968 was a Russian-born US physicist who provided the first evidence for the Big Bang theory of the origin of the universe. He predicted that it would have produced a background of microwave radiation, which was later found to exist. Gamow was also closely involved with the early theoretical development of nuclear physics, and he made an important contribution to the understanding of protein synthesis.

Gamow was born in Odessa on 4 March 1904. He attended local schools and first became interested in astronomy when he received a telescope from his father as a gift for his thirteenth birthday. In 1922, he entered Novorossysky University, but he soon transferred to the University of Leningrad where he studied optics and later cosmology, gaining his PhD in 1928. Gamow then went to the University of Göttingen, where his career as a nuclear physicist really began. His work impressed Niels Bohr (1885–1962) who invited Gamow to the Institute of Theoretical Physics in Copenhagen. There he continued his work on nuclear physics and he also studied nuclear reactions in stars. In 1929, Gamow worked with Ernest Rutherford (1871–1937) at the Cavendish Laboratory, Cambridge. His work there laid the theoretical foundations for the artificial transmutation of elements, which was achieved in 1932.

Gamow returned to Copenhagen in 1930. From 1931 to 1933 he served as Master of Research at the Academy of Science in Leningrad and then, after being denied permission by the Soviet government to attend a conference on nuclear physics in Rome in 1931, he used the Solvay Conference held in Brussels in 1933 as an opportunity to defect. He settled in the United States where he held the chair of Physics at George Washington University in Washington from 1934 until 1956. He then held the post of Professor of Physics at the University of Colorado until his death. In 1948, Gamow was given top security clearance and worked on the hydrogen bomb project at Los Alamos, New Mexico. Later in life, he gained considerable fame as an author of popular scientific books, being particularly well-known for the 'Mr Tompkins' series. He was awarded the Kalinga Prize by UNESCO in 1956 in recognition of the value of his popular scientific texts. Gamow died in Boulder, Colorado, on the 20 August 1968.

Gamow's work can be broadly divided into two main areas: his theoretical contributions to nuclear physics, and astronomy. Some of his studies fell outside both of these disciplines.

His first major scientific work was his theory of alpha decay, which he produced in 1928 while at the University of Göttingen. He was able to explain why it was that uranium nuclei could not be penetrated by alpha particles that have as much as double the energy of the alpha particles emitted by the nuclei. This, he proposed, was due to a 'potential barrier' which arose due to repulsive Coulomb and other forces. The model represented the first application of quantum mechanics to the study of nuclear structure, and was simultaneously and independently developed by Edward Condon (1902–1974) at Princeton University.

During his visits to Copenhagen in 1928 and 1930, Gamow continued his work on nuclear physics and in particular on the liquid drop model of nuclear structure. This theory held that the nucleus could be regarded as a collection of alpha particles which interacted via strong nuclear and Coulomb forces. His work under Rutherford in 1929 was concerned with the calculation of the energy which would be required to split a nucleus using artificially accelerated protons. This work led directly to the construction of a linear accelerator in 1932 by John Cockcroft (1897–1967) and Ernest Watson (1903–   ), in which protons were used to disintegrate boron and lithium, resulting in the production of helium. This was the first experimentally produced transmutation that did not employ radioactive materials.

In collaboration with Edward Teller (1908–   ), Gamow produced in 1936 the Gamow–Teller selection rule for beta decay. This has since been elaborated, but was one of the first formulations to describe beta decay.

During World War II, Gamow was intimately concerned with the Manhattan Project on the development of the atomic bomb, and later contributed significantly to research at Los Alamos that led to the production of the hydrogen bomb.

His astronomical studies were concerned mainly with the origin of the universe and the evolution of stars. He followed the model devised by Hans Bethe (1906–   ) on the mechanism by which heat and radiation are generated in the cores of stars (thermonuclear reactions), and postulated that a star heats up – rather than cools down – as its 'fuel' is consumed. In 1938, he related this theory to the Hertzsprung–Russell diagram. The following year, in collaboration with M. Schoenberg, he investigated the role of neutrino emission in novae and supernovae. He then turned to the problem of energy production in red giant stars and in 1942 produced his 'shell' model to describe such stars.

Gamow's most famous contribution to astronomy began with his support in 1946 for the Big Bang theory of the origin of the universe proposed by Georges Lemaître (1894–1966). With Ralph Alpher (1921–   ) he investigated the possibility that heavy elements could have been produced by a sequence of neutron-capture thermonuclear reactions. They published a famous paper in 1948, which became known as the Alpher–Bethe–Gamow (or alpha–beta–gamma) hypothesis, describing the 'hot big bang'. They suggested that the primordial state of matter – which they called *ylem*, the term Aristotle had given to the ultimate state of matter – consisted of neutrons and their decay products. This mixture of neutrons, protons, electrons and radiation was almost unimaginably hot. As this matter expanded after the hot big bang, it cooled sufficiently for hydrogen nuclei to fuse to form helium nuclei (alpha particles) – which explained the abundance of helium in the universe (one atom in twelve is helium).

The model's inability to account for the presence – and evident creation – of elements heavier than helium was later vindicated by the work of Geoffrey and Margaret Burbidge and their collaborators, who found some evidence of what they called nucleosynthesis, or nucleogenesis.

The hot big bang model indicated that there ought to be a universal radiation field as a remnant of the intense temperatures of the primordial big bang. In 1964 Arno Penzias (1933–) and Robert Wilson (1936–) detected isotropic microwave radiation. Research at Princeton University confirmed that this was 3K black-body radiation, as predicted by Gamow's model. Gamow had in fact postulated the temperature of this radiation as 25K, but errors were found in his method that explained why his estimate was too high. The detection of this microwave radiation led most cosmologists to support the Big Bang model, in preference to the steady-state hypothesis proposed originally by Fred Hoyle (1915–), Hermann Bondi (1919–) and Thomas Gold (1920–).

In the entirely different field of molecular biochemistry, Gamow contributed to the solution of the genetic code in which virtually all hereditable biological information is stored. The double helix model for the structure of DNA had been published in 1953 by James Watson (1928–) and Francis Crick (1916–). The double helix consists of a twisted double chain of four types of nucleotides, sugar residues and phosphates. Gamow realized that if three nucleotides were used at a time, 64 different triplets could be constructed. These could easily code for the different amino acids of which all proteins are constructed. Gamow's theory was found to be correct, and in 1961 the genetic code was cracked and the meanings of the 64 triplets identified.

Gamow was a talented and wide-ranging scientist who, with his fundamental contribution to cosmology, was responsible for the first clear indication of the origin of the universe.

Gardner Julia Anna 1882–1960 was a US geologist and palaeontologist internationally known for her work on stratigraphic palaeontology.

Julia Gardner was born on 26 January 1882 in Chamberlain, South Dakota. She was educated at Drury Academy, North Adams, Massachusetts, and later at Bryn Mawr College, where she studied geology and palaeontology. She graduated with an AB in 1905 and, following a year's schoolteaching in Chamberlain, she received an AM in 1907. She was awarded a PhD from Johns Hopkins University 1911 with her dissertation 'On certain families of gastropoda from the Miocene and Pliocene of Virginia and North Carolina'. Between 1911 and 1917, Gardner worked as a research assistant in palaeontology in Hopkins and also for the United States Geological Survey (USGS). After joining the war effort in 1917, often serving near the front, she returned to the USA in 1920 and was employed by the USGS Coastal Plain Section, transferring in 1936 to the Palaeontology and Stratigraphy Section. Gardner remained with the USGS until her retirement in 1952, after which she was hired on a yearly contract basis until 1954. She died on 15 November 1960, following a stroke.

Gardner advanced quickly in the USGS; she became associate geologist in 1924 and geologist in 1928. Her work on the Cenozoic stratigraphic palaeontology of the Coastal Plain, Texas and the Rio Grande Embayment in northeast Mexico led to the publication of *Correlation of the Cenozoic Formations of the Atlantic and Gulf Coastal Plain and the Caribbean Region* (1943) co-authored with C. Wythe Cooke and Wendell Woodring; her work was important for petroleum geologists establishing standard stratigraphic sections for Tertiary rocks in the southern Caribbean. During World War II, Gardner joined the Military Geologic Unit where amongst other things, she helped to locate Japanese beaches from which incendiary bombs were being launched, by identifying shells in the sand ballast of the balloons. On her retirement, Gardner was awarded the Interior Department's Distinguished Award. She was president of the Paleontological Society in 1952 and a vice president of the Geological Society of America in 1953.

Gatling Richard Jordan 1818–1903 was a prolific US inventor who is best remembered for his invention of the Gatling gun, one of the earliest successful rapid-fire weapons.

Gatling was born in Hertford County, North Carolina, on 12 September 1818. Son of a well-to-do planter, he showed mechanical aptitude early in life and collaborated with his father on the invention of a machine for sowing cotton and for thinning out the young plants. He followed a varied career, his first job being in the county clerk's office. He then taught for a brief period before becoming a merchant.

In 1844, Gatling moved to St Louis, adapted his cotton-sowing machine for sowing rice and other grains and established his manufacturing centre there. Other factories were later set up in Ohio and Indiana.

Gatling's career took a new turn after an attack of smallpox. He entered Ohio Medical College, in Cincinnati to study medicine, qualifying in 1850 but never practising. Instead, for a time he was concerned with railway enterprises and property, but he was still inventing – a hemp-breaking machine (1850), a steam plough (1857) and a marine steam ram (1862).

Forms of rapid-fire weapons were in use as early as the sixteenth century, but not until about 1860, when breech loading became firmly established, did effective machine guns become possible. One of the first was the Reffye, which had 25 barrels round a common axis.

Gatling's gun, which he patented in 1862, consisted of ten parallel barrels arranged round a central shaft, each barrel firing in turn as it reached the firing position. An operating cam, driven through a hand-cranked worm gear, forced the locks forward and backward. On a single barrel, loading took place in the first position. The second, third and fourth positions were used to force the cartridge into the breech, and the gun was fired into the fifth position. The sixth, seventh, eighth and ninth positions were used for extraction of the empty case, which was rejected at the tenth.

The gun was fed from a drum on top which dropped the cartridges into the locks. Each barrel therefore fired as it approached the bottom of its circular journey. In 1862, the year of Gatling's patent specification, the gun was capable of firing 350 rounds per minute. There was no official interest in his invention, however, so he made further improvements to it, extending its range and increasing its firing rate. This still failed to attract official interest, but a few commanders during the Civil War purchased some with private funds. It was not until 1866, after the end of the war, that the Army Ordnance Department ordered 100. Meanwhile, they were being manufactured by the Colt Company and sold overseas to England, Russia, Austria and South America, and were used in the war of 1870.

In that year, Gatling moved to Hartford, Connecticut, to supervise manufacture and by 1882 his gun could fire 1,200 rounds per minute. Gatling continued to develop his invention, producing greater rates of fire and driving it by different methods.

The Gatling was one of the earliest successful rapid-fire weapons but it cannot be regarded as the forerunner of the modern machine gun as it needed an external power source to drive the loading, firing, extraction and ejection mechanism. The laurels for the first workable, truly automatic machine gun must go to Maxim. The main contribution Gatling made to history was to alert the military mind to the possibilities of such weapons. Indeed, special tactics were developed in the Franco–German war of 1870 to counter Gatling-type weapons by massive artillery barrages. It thus failed to make any serious impact.

Gatling died, aged 84, in New York City on 26 February 1903.

Gauss Karl (or Carl) Friedrich 1777–1855 was a brilliant German mathematician, physicist and astronomer, whose innovations in mathematics proved him to be the equal of Archimedes or Newton.

Gauss was born in Braunschweig (Brunswick), Germany, on 30 April 1777, to a very poor, uneducated family. His father was a gardener, an assistant to a merchant and the treasurer of an insurance fund. Gauss taught himself to count and read – he is said to have spotted a mistake in his father's arithmetic at the age of three. At elementary school, at the age of eight, he added the first 100 digits in his first lesson. Recognizing his precocious talent, his teacher persuaded his father that Gauss should be encouraged to train towards following a profession rather than learn a trade. At the age of eleven, he went to high school and proved to be just as proficient at classics as mathematics. When he was 14 he was present at the court of the Duke of Brunswick in order to demonstrate his skill at computing; the Duke was so impressed that he supported Gauss generously with a grant from then until his own death in 1806. In 1792, with the Duke's aid, Gauss began to study at the Collegium Carolinum in Brunswick, and then from 1795 to 1798 he was taught at the University of Göttingen. He was awarded a doctorate in 1799 from the University of Helmstedt, by which time he had already made nearly all his fundamental mathematical discoveries. In 1801 he decided to develop his interest in astronomy; by 1807 he pursued it so enthusiastically that not only was he Professor of Mathematics, he was also Director of the Göttingen Observatory.

At about this time, he began to gain recognition from other parts of the world: he was offered a job in St Petersburg, was made a Foreign Member of the Royal Society in London, and was invited to join the Russian and French Academies of Science. Nevertheless, Gauss remained at Göttingen for the rest of his life, and died there on 23 February 1855.

Between the ages of 14 and 17, Gauss devised many of the theories and mathematical proofs that, because of his lack of experience in publication and his diffidence, had to be rediscovered in the following decades. The extent to which this was true was revealed only after Gauss's death. There are, nevertheless, various innovations that are ascribed directly to him during his three years at the Collegium Carolinum, including the principle of least squares (by which the equation curve best fitting a set of observations can be drawn). At this time he was particularly intrigued by number theory, especially on the frequency of primes. This subject became a life's work and he is known as its modern founder. In 1795, having completed some significant work on quadratic residues, Gauss began to study at the University of Göttingen, where he had access to the works of Fermat, Euler, Lagrange and Legendre. He immediately seized the opportunity to write a book on the theory of numbers, which appeared in 1801 as *Disquisitiones Arithmeticae*, generally regarded to be his greatest accomplishment. In it, he summarized all the work which had been carried out up to that time and formulated concepts and questions that are still relevant today.

He was still at university when he discovered, in 1796, that a regular 17-sided polygon could be inscribed in a circle, using ruler and compasses only. This represented the first discovery in Euclidean geometry that had been made in 2,000 years.

In 1799, Gauss proved a fundamental theorem of algebra, that every algebraic equation has a root of the form $a + bi$, where a and b are real numbers and i is the square root of −1. In his doctoral thesis, Gauss showed that numbers of the form $a + bi$ (called complex numbers) can be regarded as analogous to points on a plane.

The years between 1800 and 1810 were for Gauss the years in which he concentrated on astronomy. In mathematics he had had no collaborators, although he had inspired men such as Dirichlet and Riemann. In astronomy, however, he corresponded with many, and his friendship with Alexander von Humboldt (1769–1859) played an important part in the development of science in Germany. The discovery of the first asteroid, Ceres – and its subsequent 'loss' – by Guiseppe Piazzi (1746–1826) at the beginning of 1801 gave Gauss the chance to use his mathematical brilliance in another cause. He developed a quick method for calculating the asteroid's orbit from only three observations and published this work – a classic in astronomy – in 1809. The 1,001st planetoid to be discovered was named Gaussia in his honour. He also worked out the theories of perturbations which were eventually used by Leverrier and Adams in their independent calculations towards the discovery of Neptune. After 1817, he did no further work in theoretical astronomy, although he continued to work in positional astronomy for the rest of his life.

Gauss was also a pioneer in topology, and he worked besides on crystallography, optics, mechanics and capillarity. At Göttingen, he devised the heliotrope, a type of heliostat that has applications in surveying. After 1831, he collaborated with Wilhelm Weber (1804–1891) on research into electricity and magnetism, and in 1833 they invented an electromagnetic telegraph. Gauss devised logical sets of units for magnetic phenomena and a unit of magnetic flux density is therefore called after him.

There is scarcely any physical, mathematical or astronomical field in which Gauss did not work. He retained an active mind well into old age and at the age of 62, already an accomplished linguist, he taught himself Russian. The full value of his work has been realized only in the twentieth century.

Gay-Lussac Joseph Louis 1778–1850 was a French chemist who pioneered the quantitative study of gases and established the link between gaseous behaviour and chemical reactions. He also made important discoveries in inorganic chemistry, often parallelling (independently) the work of Humphry Davy in England.

Gay-Lussac was born on 6 December 1778 at Saint-Léonard, Haute Vienne, the son of a judge (who was arrested in 1793 for his aristocratic sympathies). He entered the Ecole Polytechnique in Paris in 1797 and graduated in 1800. His first interest was in engineering but in 1801 the great French chemist Claude Berthollet invited Gay-Lussac to join him as an assistant at his house in Arcueil, the meeting place of many contemporary scientists, including Pierre Laplace. From 1805 to 1806 he accompanied Alexander von Humboldt on an expedition to measure terrestrial magnetism and in 1809, a year after his marriage to Geneviève Rojet, he became Professor of Chemistry at the Ecole and Professor of Physics at the

Sorbonne. He held various government appointments, including that of superintendent of a gunpowder factory (1818) and Chief Assayer to the Mint (1829). In 1832, he was made Professor of Chemistry at the Musée National d'Histoire Naturelle. He was a member of the Chamber of Deputies for a short time in the 1830s, and was made a Peer of France in 1839. He died in Paris on 9 May 1850.

Gay-Lussac began studying gases in collaboration with Louis Thénard. In 1802, he formulated the law of expansion of gases, which states that for a given rise in temperature all gases expand by the same fraction of their volume (Jacques Charles had recognized equal expansion in 1787 but had not published his findings; nevertheless the relationship is now usually known as Charles's law). During balloon ascents in 1804, in which he established an altitude record that was to stand for 50 years, Gay-Lussac showed that the composition of air is constant up to heights in excess of 7,000 m/23,000 ft above sea level. With Humboldt he accurately determined the proportions of hydrogen and oxygen in water, showing the volume ratio to be 2:1; they also established the existence of explosive limits in mixtures of the two gases.

His greatest achievement concerning gases came in 1808 with the formulation of Gay-Lussac's law of combining volumes, which states that gases combine in simple proportions by volume and that the volumes of the products are related to the original volumes. His examples include:

$$HCl \quad + \quad NH_3 \quad \rightarrow \quad NH_4Cl$$
$$1 \qquad\qquad 1 \qquad\qquad 1$$

and

$$2CO \quad + \quad O_2 \quad \rightarrow \quad 2CO_2$$
$$2 \qquad\qquad 1 \qquad\qquad 2$$

Another fruitful area of Gay-Lussac's work with Thénard concerned the alkali metals, where they followed up the discovery of sodium and potassium by Davy by devising chemical means for producing the metals in quantity (heating fused alkalis with red-hot iron). They anticipated (by nine days) Davy's isolation of boron and independently investigated the properties of iodine, which had been discovered in 1811 by Bernard Courtois (1777–1838) but was named by Gay-Lussac in 1813. He first prepared iodine(I) chloride, ICl, and iodine(III) chloride, ICl_3. Studies of the hydrogen halides – hydrogen fluoride, hydrogen chloride and hydrogen iodide – led to the extremely important conclusion that acids need not contain oxygen (as proposed by Antoine Lavoisier) and to the naming of hydrochloric acid (previously known as muriatic acid). From his researches with Prussian blue Gay-Lussac prepared hydrogen cyanide and in 1815 discovered cyanogen, $(CN)_2$; he went on to develop the chemistry of the cyanides.

Another significant area of work was the development of volumetric analysis – which provided more evidence, if any were needed, of Gay-Lussac's quantitative approach to chemistry. In 1832, he introduced the method of estimating silver by titrating silver nitrate against sodium chloride. Later work included the analysis of vegetable and animal substances and a contribution to the manufacture of sulphuric acid, the Gay-Lussac tower, which recovers spent nitrogen oxides from the chamber process.

Geiger Hans Wilhelm 1882–1945 was a German physicist who invented the Geiger counter, a device for detecting radioactivity.

Geiger was born at Neustadt, Rheinland-Pfalz, on 30 September 1882. He studied physics at the Universities of Munich and Erlangen, obtaining his PhD from the latter institution in 1906 with a dissertation on gaseous ionization. In 1906, he became assistant to Arthur Schuster (1851–1934), Professor of Physics at the University of Manchester. Ernest Rutherford (1871–1937) succeeded Schuster in 1907 and Geiger stayed on as his assistant. They were later joined by Ernest Marsden, with whom Geiger collaborated.

In 1912, Geiger became head of the Radioactivity Laboratories at the Physikalische Technische Reichsanstalt in Berlin, after declining the offer of a post at the University of Tübingen. He established a successful research group, which was joined by such eminent scientists as Walther Bothe (1891–1957) and James Chadwick (1891–1974). Geiger served in the German artillery during World War I, and then returned to the Reichsanstalt in Berlin.

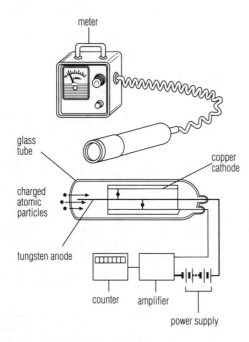

Geiger *A Geiger–Müller counter has a central wire anode in a metal tube, which acts as the (negative) cathode, containing argon at low pressure. Ionizing radiation enters a 'window' made of thin metal foil or mica and ionizes the gas in the tube, giving rise to an electric current which is amplified and made to work a counting device.*

In 1925 Geiger accepted a post as Professor of Physics at the University of Kiel. He moved in 1929 to the University of Tübingen, and in 1936 he became head of Physics at the Technical University of Charlottenberg-Berlin as well as the editor of the *Zeitschrift für Physik*. Illness reduced the pace of his scientific activities during most of World War II and then he lost his home and belongings during the Allied occupation of Germany. Geiger died a few months later on 24 September 1945 at Potsdam.

Geiger's early postdoctoral work in Manchester involved the application of his expertise in gaseous ionization to the field of radioactive decay. One direction which his work took was the development, in 1908, of a device that counted alpha particles. The counter consisted of a metal tube containing a gas at low pressure and thin wire in the centre of the tube. The wire and tube were under a high voltage and each alpha particle arriving through a window at one end of the tube caused the gas to ionize, producing a momentary flow of current. The resulting electric signals counted the number of alpha particles entering the tube. With this counter, Geiger was able to show that approximately 3.4×10^{10} alpha particles are emitted per second by a gram of radium, and that each alpha particle has a double charge. Rutherford later demonstrated that the doubly charged alpha particle was in fact a helium nucleus.

In 1909, Marsden and Geiger studied the interactions of alpha particles from radium with metal reflectors. They found that most alpha particles passed straight through the metal, but that a few were deflected at wide angles to the beam. They showed that the amount of deflection depended on the metal that the reflector was made of and that it decreased with a lowering of atomic weight. They used metal reflectors of different thicknesses to demonstrate that the deflection was not caused merely by some surface effect. This work led directly to Rutherford's proposal in 1911 that the atom consists of a central, positively charged nucleus surrounded by electron shells. The observed deflection of the alpha particles was due to the positively charged particle interacting with the positively charged nuclei in the metal and being deflected by repulsive forces. Geiger and Marsden later published the mathematical relationship between the amount of alpha scattering and atomic weight.

Other subjects which Geiger studied under Rutherford included the relationship between the range of an alpha particle and its velocity in 1910; the various disintegration products of uranium in 1910 and 1911; and the relationship between the range of an alpha particle and the radioactive constant, which was determined with John Nuttall (1890–1958) in 1911.

In 1912, upon his return to Germany, Geiger began the first of many refinements to his radiation counter, enabling it to detect beta particles and other kinds of radiation as well as alpha particles. The modern form of the instrument was finally developed with Walther Müller in 1928, and it is now usually known as the Geiger–Müller counter. It can be operated so that radioactivity causes it to emit audible clicks, which can be recorded automatically. Geiger used this instrument to confirm the Compton effect in 1925 and to study cosmic radiation from 1931 onwards. Cosmic rays became a prolonged source of interest which occupied him for most of the rest of his career.

Geiger's contribution to science in discovering a simple and reliable way of detecting radiation not only enabled discoveries to be made in nuclear physics, but also afforded an instant method of checking radiation levels and finding radioactive minerals.

Gell-Mann Murray 1929– is a US theoretical physicist who was awarded the 1969 Nobel Prize for Physics for his work on the classification and interactions of subatomic particles. He is the originator of the quark hypothesis.

Gell-Mann was born in New York City on 15 September 1929. He showed strong scholastic aptitude at an early age and entered Yale University when he was only 15 years old. He received his bachelor's degree in 1948 and proceeded to the Massachusetts Institute of Technology, where he earned his PhD three years later. He spent the next year as a member of the Institute for Advanced Studies at Princeton University. In 1952, he joined the faculty of the University of Chicago as an instructor to work at the Institute for Nuclear Studies under Enrico Fermi (1901–1954), becoming an assistant professor in 1953. In 1955, Gell-Mann moved to the California Institute of Technology at Pasadena as Associate Professor of Theoretical Physics, becoming Professor a year later and in 1966 Robert Andrews Millikan Professor of Theoretical Physics.

From the early 1930s, the relatively simple concept of the atom as being composed of only electrons and protons began to give way to more complex models involving neutrons and then other particles. By the 1950s, the field was in a state of complete chaos following the discovery of mesons, which have masses intermediate between a proton and an electron, and then hyperons, which are even more massive than the proton. One of the most puzzling things was that the hyperons and some mesons had much longer lifetimes than was predicted by accepted theory at that time, although still only of the order of 10^{-9} sec.

In an endeavour to bring order to the subject, Gell-Mann proposed in 1953 that the longlived particles, and indeed other particles such as the neutron and the proton, should be given a new quantum number called the strangeness number. The strangeness number differed from particle to particle. It could be 0, –1, +1, –2, etc. He also proposed the law of conservation of strangeness, which states that the total strangeness must be conserved on both sides of an equation describing a strong or an electromagnetic interaction but *not* a weak interaction. This work was also done independently by Nishijima in Japan.

The law of conservation of strangeness formed an important theoretical basis for the subsequent theory of associated production, which was proposed by Gell-Mann in 1955. This model held that the strong forces which were responsible for the creation of strange particles could create them only in groups of two at a time (that is, in pairs). As the partners in a pair moved apart, the energy which would be required for them to decay through strong interactions would exceed the energy which had originally gone into their creation. The strange particles therefore survived long enough to decay through weak interactions instead. This model explained their unusually extended lifespan.

Gell-Mann used these rules to group mesons, nucleons (neutrons and protons) and hyperons, and was thereby able to form predictions in the same way that Dmitri Mendeleyev (1834–1907) had been able to make predictions about the chemical elements once he had constructed the periodic table. Gell-Mann's prediction of a particle which he named xi-zero to complete a doublet with xi-minus was soon rewarded with experimental verification.

The law of conservation of strangeness was also important in the formulation of SU(3) symmetry, a scheme for classifying strongly interacting particles which Gell-Mann proposed in 1961. This classification system itself formed part of the basis for yet another classification scheme entitled the eightfold way (named after the eight virtues of Buddhism). This model was devised by Y. Ne'emann and Gell-Mann and published in 1962.

The eightfold way was intended to incorporate all the new particles and the new quantum numbers. It postulated the existence of supermultiplets, or groups of eight particles which have the same spin value but different values for charge, isotopic spin, mass and strangeness. The model also predicts the existence of supermultiplets of different sizes. The strongest support for the theory arose from the discovery in 1964 of a particle named omega-minus, which had been predicted by the eightfold way.

In 1964, Gell-Mann proposed a model for yet a further level of complexity within the atom. He proposed that particles such as the proton are not themselves fundamental but are composed of quarks. Quarks differ from all previously proposed subatomic particles because they have fractional charges of, for instance, $+^2/_3$ or $-^1/_3$. Quarks always occur in pairs or in trios and can never be detected singly, exchanging gluons that bind them together. Gluons are the quanta for interactions between quarks, a process which goes by the exotic name of quantum chromodynamics, and their behaviour is in some ways analogous to the exchange of protons between electrons in electromagnetic interactions (quantum electrodynamics). Six quarks have been predicted, and five have so far been indirectly detected. The names of the quarks are up, down, strange, charm, bottom and top, and they have been ascribed properties of electric charge, strangeness, charm, bottomness, topness, baryon number, and so on.

Gell-Mann's work has been characterized by originality and bold synthesis. His models have been useful not only for their predictive value, but also for the work they have spurred others to do.

Giacconi Riccardo 1931– is an Italian-born US physicist, the head of a team whose work has been fundamental in the development of X-ray astronomy.

Giacconi was born in Genoa on 6 October 1931. Educated in Italy, he obtained a doctorate in 1954 in the

University of Milan, where he then took up the post of Associate Professor. Two years later he emigrated to the USA and became Research Associate at the University of Indiana at Bloomington, before taking up a similar position at Princeton in 1958. He then joined American Scientific and Engineering Inc. as Senior Scientist for a year until his naturalization papers came through in 1960. In 1963, he became Chief of the Space Physics Division.

In June 1962, a rocket was sent up by Giacconi and his group to observe secondary spectral emission from the Moon. But to the surprise of all concerned, it detected extremely strong X-rays from a source evidently located outside the Solar System. X-ray research has since become an important branch of astronomy, leading to the discovery of the existence of many previously unsuspected types of stellar and interstellar material, whose emissions lie in the X-ray band of the electromagnetic spectrum. Because much of the high-energy radiation is filtered out by the Earth's upper atmosphere, research has had to make considerable use of equipment carried aboard balloons or rockets; but the launching of *Uhuru*, in 1970, marked the beginning of a new era in that it was the first satellite devoted entirely to X-ray astronomy.

Under the guidance of Giacconi, the team maintained a lead in the field. They devised and developed a telescope capable of producing X-ray images. Their years of experimental work in the laboratory and the more recent facility to use rocket-borne equipment using solar power have resulted in instruments of increasing efficiency and angular resolution.

Giacconi has also worked a great deal with a Cherenkov detector, by means of which it is possible to observe the existence and velocity of high-speed particles, important in experimental nuclear physics and in the study of cosmic radiation.

Gibbs Josiah Willard 1839–1903 was a US scientist who laid the foundation of modern chemical thermodynamics. He devised the phase rule and formulated the Gibbs adsorption isotherm.

Gibbs was born on 11 February 1839 in New Haven, Connecticut, into an academic family. His father was Professor of Sacred Literature at the Divinity School of Yale University, and Gibbs excelled at classics at school. He attended Yale in 1854, winning prizes for Latin and mathematics before graduating in 1858 at the age of only 19. During the next five years he continued his studies by specializing in engineering and in 1863 gained the first Yale PhD in this subject for a thesis on the design of gears. He then accepted teaching posts at Yale, first in Latin and then in natural philosophy. In 1866, he patented a railway braking system.

Also in 1866 Gibbs went abroad for three years to attend lectures (mainly in physics) in Paris, Berlin and Heidelberg. In 1871, two years after his return to the United States, he was appointed Professor of Mathematical Physics at Yale, a post he retained until his death despite offers from other academic institutions. He never married but

lived with his sister and her family. He died in New Haven on 28 April 1903.

Gibbs did not publish his first papers until 1873, which were preliminaries to his 300-page series *On the Equilibrium of Heterogeneous Substances* (1876–78). In it he formulated the phase rule, which may be stated as:

$$f = n + 2 - r$$

where f is the number of degrees of freedom, n the number of chemical components, and r the number of phases – solid, liquid or gas; degrees of freedom are quantities such as temperature and pressure which may be altered without changing the number of phases. Gibbs did not explore the chemical applications of the phase rule, later done by others who came to realize its importance. In the same work he also described his concept of free energy, which can be used as a measure of the feasibility of a given chemical reaction. It is defined in terms of the enthalpy, or heat content, and entropy, a measure of the disorder of a chemical system. From this Gibbs developed the notion of chemical potential, which is a measure of how the free energy of a particular phase depends on changes in composition (expressed mathematically as the differential coefficient of the free energy with respect to the number of moles of the chemical). The fourth fundamental contribution in this extensive work was a thermodynamic analysis which showed that changes in the concentration of a component of a solution in contact with a surface occur if there is an alteration in the surface tension – the Gibbs adsorption isotherm.

All of these very technical discoveries now form part of the armoury of the physical chemist and thermodynamicist, together with their extension to electrochemistry and the subsequent developments of other scientists. But for many years Gibbs's work was unknown outside the United States, until it was translated into German by Friedrich Ostwald in 1891 and into French by Henri Le Châtelier in 1899.

During his teaching studies Gibbs adapted the work of the mathematicians William Hamilton (1805–1865) and Hermann Grassman (1809–1877) into a vector analysis which was both simple to use and easily applicable to physics, particularly electricity and magnetism. It was left to one of Gibbs's students, E. B. Wilson, to write a textbook on the subject, which was largely responsible for the popularization of vector analysis. Also during the 1880s Gibbs worked on the electromagnetic theory of light. From an entirely theoretical viewpoint and making very few assumptions he accounted correctly for most of the properties of light using only an electrical theory.

In his last major work, *Elementary Principles of Statistical Mechanics*, Gibbs turned his attention to heat and showed how many thermodynamic laws could be interpreted in terms of the results of the movements of enormous numbers of bodies such as molecules. His ensemble method equated the behaviour of a large number of systems at once to that of a single system over a period of time.

Giffard Henri 1825–1882 was a French aeronautical engineer, famous for building the first steerable powered airship and making a successful flight in it.

Giffard, who studied engineering, was particularly interested in balloon flight, which had been pioneered in his native France. In the early 1850s he began to experiment with methods for steering balloons, which hitherto had been entirely at the mercy of the wind. In 1852, Giffard built his airship. It was a sausage-shaped gas bag 44 m/144 ft long and 12 m/52 ft in diameter, with a hydrogen capacity of 2,500 cu m/88,300 cu ft. The gondola was strung from a long pole or keel attached to the gasbag by ropes. It was powered by a small 2-kW/3-hp steam engine driving a three-bladed propellor – itself an innovation at the time. The airship was steered using a rudder, a canvas sail stretched over a bamboo frame and hinged to the keel. On 24 September 1852 Giffard took off from the Hippodrome in Paris and flew to Elancourt, near Trappes. His average speed was only about 5 kph/3 mph and he had problems with the steering.

Giffard gave up his experiments with airships but went on to other inventions, such as an injector to feed water into a steam engine boiler to prevent it running out of steam when not in motion. But his historic flight marked the real beginning of the human conquest of the air.

Gilbert Walter 1932– is a US molecular biologist who shared (with Paul Berg and Frederick Sanger) the 1980 Nobel Prize for Chemistry for his work in devising techniques for determining the sequence of bases in DNA. He also isolated the repressor substance postulated to exist by François Jacob, André Lwoff and Jacques Monod in their theory concerning the regulation of gene action.

Gilbert was born on 21 March 1932 in Boston, Massachusetts. He was educated at Harvard University, from which he graduated in physics in 1954, and at Cambridge University, England, from which he obtained his doctorate in mathematics in 1957. He then returned to Harvard, where he was a National Science Foundation Fellow in Physics from 1957 to 1958, and a lecturer and Research Fellow from 1958 to 1959. In 1960, however, influenced by James Watson, Gilbert changed to biology, becoming Professor of Biophysics (1964 to 1968) then Professor of Molecular Biology (1969 to 1972) at Harvard. In 1972, he became American Cancer Society Professor of Molecular Biology at Harvard.

Gilbert began his first major biological project in 1965, when he attempted to isolate and identify repressor substances involved in the regulation of gene activity. Working with Benno Muller-Hill, he devised a special experimental technique called equilibrium dialysis that enabled him to produce relatively large quantities of the repressor substance, which he then isolated and purified, and by late 1966 he had identified it as a large protein molecule.

After his work on the repressor substance, Gilbert then developed a method of determining the sequence of bases in DNA, which involved using a restriction enzyme that breaks the DNA molecule at specific known points, thereby effectively excising predetermined fragments. It was for developing this important investigative technique that he shared the 1980 Nobel Prize for Chemistry.

Gilbert William 1544–1603 was an English physician and physicist who performed fundamental pioneering research into magnetism and also helped to establish the modern scientific method.

Gilbert was born in Colchester, Essex, on 24 May 1544. He was educated at St John's College, Cambridge, where he graduated in medicine in 1569 and later became a Fellow. In about 1573, he settled in London, where he established a successful medical practice, and in 1599 he was elected President of the Royal College of Physicians. In the following year, he was appointed personal physician to Queen Elizabeth I, and was later knighted for his services to the queen. On Elizabeth's death in March 1603 he was appointed physician to her successor, James I, but Gilbert held this position for only a short time because he died on 10 December of that year.

Although successful as a physician, Gilbert's most important contribution to science was his research into magnetism and electrical attraction, which he described in his book *De Magnete, Magneticisque Corporibus, et de Magno Magnete Tellure/Concerning Magnetism, Magnetic Bodies, and the Great Magnet Earth* (1600). This work gave a detailed account of years of investigations (on which he is thought to have spent some £5,000 of his own money), which were rigorously performed. The importance of this work was recognized by Galileo, who also considered Gilbert to be the principal founder of the experimental method – before the work of Francis Bacon and Isaac Newton. Not only did Gilbert disprove many popular but erroneous beliefs about magnetism – that lodestone (naturally occurring magnetic iron oxide, the only magnetic material available until Gilbert discovered a method for making metal magnets) cured headaches and rejuvenated the body, for example – but he also discovered many important facts about magnetism, such as the laws of attraction and repulsion and magnetic dip. From his studies of the behaviour of small magnets, Gilbert concluded that the Earth itself acts as a giant bar magnet, with a magnetic pole (a term he introduced) near each geographical pole. He also showed that the strength of a lodestone can be increased by combining it with soft iron, that iron and steel can be magnetized by stroking them with lodestones, and that when an iron magnet is heated to red-heat, it loses its magnetism and also cannot be remagnetized while it remains hot. So extensive were his investigations of magnetism that not until William Sturgeon made the first electromagnet in 1825 and Michael Faraday began his studies was substantial new knowledge added to the subject.

Gilbert also investigated static electricity. The ancient Greeks had discovered that amber, when rubbed with silk, can attract light objects. Gilbert demonstrated that other substances exhibit the same effect and that the magnitude of the effect is approximately proportional to the area being rubbed. He called these substances 'electrics' and clearly differentiated between magnetic attraction and electric attraction (as he called the ability of an electrostatically charged body to attract light objects). This work eventually led to the idea of electrical charge, but it was not until 1745 that it was discovered (by the French physicist Charles Du Fay) that there are two types of electricity, positive and negative.

Gilbert also held remarkably modern views on astronomy. He was the first English scientist to accept Copernicus' idea that the Earth rotates on its axis and revolves around the Sun. He also believed that the stars are at different distances from the Earth and might be orbited by habitable planets, and that the planets were kept in their orbits by magnetic attraction.

Gilchrist Percy Carlyle 1851–1935 was a British metallurgist who devised an inexpensive steel-making process.

Gilchrist was born in Lyme Regis, Dorset, on 27 September 1851, the son of a local barrister. He was educated at Felsted School, Essex, and later at the Royal School of Mines, where he acquired a sound knowledge of metallurgy. Between 1875 and 1877, together with his cousin Sidney Gilchrist Thomas, he developed a method of producing low-phosphorus steel from high-phosphorus British ores (reducing the need – and cost – of using special imported ores). Thomas died soon afterwards but Gilchrist, who lived on for more than 50 years, became famous. He was Vice President of the Iron and Steel Institute and in 1891 was elected a Fellow of the Royal Society. He died on 16 December 1935.

Steel was a comparatively rare commodity until the 1850s, because its production was difficult and costly. Then in 1855, Henry Bessemer invented his convector (in which air blown through molten pig iron oxidized impurities to produce a brittle, low-carbon steel). The addition of ferromanganese gave a tougher product, but even so only low-phosphorus iron could be used, and most British ores were too high in phosphorus.

In 1870, Sidney Gilchrist Thomas was a junior clerk working in London for the Metropolitan Police but, like his cousin, he had an intense interest in natural science and metallurgy. After attending a series of lectures at the Birkbeck Institute, he joined Percy Gilchrist in trying to manfacture cheap steel from high-phosphorus ores. They used an old cupola to make a Bessemer-type convector and lined it with a paste made from crushed brick, sodium silicate and water. The pig iron was added and melted before being subjected to prolonged 'blowing'. The oxygen in the blast of air oxidized carbon and other impurities, and the addition of lime at this stage caused the oxides to separate out as a slag on the surface of the molten metal. Continued blowing then brought about oxidation of the phosphorus, raising the temperature of the metal still further. When oxidation was complete (as judged by the colour of the flames coming from the convector), the cupola was tilted and the slag run off, leaving the molten steel to be poured into ingot moulds. The product became known at first as 'Thomas steel', and the age of cheap steel had arrived.

Gill David 1843–1914 was a Scottish astronomer whose precision and patience using old instruments brought him renown before he achieved even greater fame for his pioneer work in the use of photography to catalogue stars.

Gill was born in Aberdeen on 12 June 1843, the eldest surviving son of a well-established clock and watchmaker who intended him eventually to take over the family business and educated him accordingly. After two years at Marischal College – where he attended classes given by the physicist and astronomer James Clerk Maxwell (1831–1879) – he went to Switzerland to study clock-making. There he became expert in fine mechanisms, experienced in business methods and fluent in French, all of which later stood him in good stead.

He returned to Coventry, and later Clerkenwell (London), to continue his studies and then went back to Aberdeen, where he ran his father's business for ten years, gradually developing his interests in astronomy and astronomical techniques. In 1872, he became Director of Lord Lindsay's private observatory at Dun Echt, 12 miles from Aberdeen. Seven years later he was appointed Astronomer at the observatory at the Cape of Good Hope, South Africa. He was made Knight Commander of the Order of the Bath on 24 May 1900. He retired in 1906, for health reasons, and lived in London until he died of pneumonia on 24 January 1914.

On Gill's return to Aberdeen in 1862, he was given the use of a small telescope at King's College Observatory. Through business contacts he acquired a 30-cm/12-in reflector for the College with which he began to try to determine stellar parallaxes. This work was interrupted by Lord Lindsay's invitation to become Director of his new Observatory. Gill's job was to equip and supervise the building of this new Observatory, which Lindsay was determined to make the best possible. In the course of this work, Gill met many important European astronomers and skilled instrument-makers, contacts that were to be very useful in his later position at the Cape. He also mastered the use of a heliometer, a rather old-fashioned instrument that he was to use a great deal and one which required great precision of hand and eye. The heliometer was first designed for measuring the variation of the Sun's diameter at different seasons, but it was also known to be one of the most accurate instruments for measuring angular distances between stars. Gill was perhaps the last great master of the heliometer and his measurements of the parallax of the southern stars were unsurpassed for many years.

In 1872, Gill went on a six-year expedition to Mauritius, with Lord Lindsay and others, in order to measure the distance of the Sun and other related constants particularly during the 1874 transit of Venus. The method Gill used was first proposed by Edmond Halley and involved combining the times of the transit of Venus across the face of the Sun as observed from a number of places as widely spaced around the world as possible. While on the island of Mauritius, Gill used his heliometer to observe the near approach of the minor planet Juno, and he was able to deduce an accurate value of the solar parallax. He used the same method of determining solar parallax on a private expedition, sponsored by the Royal Astronomical Society, on Ascension Island in 1877, when he measured solar parallax by considering the near approach of Mars. On both of these expeditions he used a 10-cm/4-in heliometer.

In 1879, he was appointed to the Observatory at the Cape of Good Hope. This Observatory had been built in 1820 in order to observe the southern hemisphere, but by 1879, when Gill took over, it was in poor condition. Most of the instruments were in need of repair and many of the results which had been obtained had not been published. Gill set to work at once. He was particularly unhappy about the Airy transit circle; although it was identical to the one at Greenwich, Gill designed an improved version which, when built in 1900, became the pattern for most transit circles built afterwards.

But it was the heliometer that Gill used most – at first a 10-cm/4-in one, then a 18-cm/7-in instrument which was installed in 1887. With the co-operation of many other astronomers, he made intensive investigations of several minor planets – Iris, Victoria and Sappho – and in 1901 he made the first accurate determination of the solar parallax: 8.80″. This figure was used in all almanacs until 1968, when it was replaced by a value of 8.794″, derived by radar echo methods and by observations made with a *Mariner* space probe. With the heliometer he also measured the distances of 20 of the brighter and nearer southern stars.

The bright comet of 1882 was of great interest to astronomers. But it was on seeing a photograph of it that Gill realized it should be possible to chart and measure star positions by photography. At once he initiated a vast project, with the help of other observatories, to produce the *Cape Durchmusterung*, which gives the positions and brightness of more than 450,000 southern stars. This was the first important astronomical work to be carried out photographically. Gill also served on the original council for the *International Astrographic Chart and Catalogue*, which was to give precise positions for all stars to the 11th magnitude. It was not completed until 1961, although all the photographs had been taken by 1900.

Gilman Henry 1893–   is a US organic chemist best known for his work on organometallic compounds, particularly Grignard reagents.

Gilman was born on 9 May 1893 in Boston, Massachusetts. He grew up there and attended university in nearby Harvard, from which he graduated with top honours in 1915. After a year abroad, during which he visited academic centres in Zurich, Paris and Oxford, he returned to Harvard to continue his postgraduate studies, and gained his PhD in chemistry in 1918.

He began his career as an instructor at Harvard, but in 1919 was offered a similar post in the chemistry department at the University of Illinois. Then an assistant professorship at the newly established faculty at Iowa State University prompted another move in the same year. He became Professor of Organic Chemistry in 1923 and continued to teach at Iowa until 1947, when his vision

deteriorated to such an extent that he became virtually blind.

Gilman's first paper on organometallic compounds, on Grignard reagents, was published in 1920. He went on to investigate the organic chemistry of 26 different metals, from aluminium, arsenic and barium through to thallium, uranium and zinc, and discovered several new types of compounds. He was the first to study organocuprates, now known as Gilman reagents, and his early work with organomagnesium compounds (Grignard reagents) led to experiments with organolithium compounds, later to play an important part in the preparation of polythene. Organogold compounds found applications in medicine.

The comprehensive study of methods of high-yield synthesis, quantitative and qualitative analysis, and uses of organometallic compounds represent only part – albeit the major one – of Gilman's research achievements. He also developed an international reputation for his work on organosilicon compounds, which again found many industrial applications. Other research interests include heterocyclic compounds, catalysis, resins, plastics and insecticides.

Ginzburg Vitalii Lazarevich 1916–   is a Soviet scientist who has become one of the leading astrophysicists of the twentieth century.

Born in Moscow on 21 September 1916, Ginzburg completed his education at Moscow University, graduating in physics in 1938. He then studied as a postgraduate there and for a further two years at the Physics Institute of the Academy of Sciences. Since that time he has been a member of staff at the Physics Institute, from 1942 as Head of the Sub-Department of Theoretical Physics. A member of the Communist Party from 1944, Ginzburg won several Soviet prizes and honours, notably the Order of Lenin in 1966. He is also the author of books detailing his work.

Ginzburg's first major success was his use of quantum theory in a study of the Cherenkov radiation effect (in which charged particles such as electrons travel through a medium at a speed greater than that of light in that medium) in 1940. This work was important for the development of nuclear physics and the study of cosmic radiation, and Ginzburg continued to theorize in this field. Seven years later, with Igor Tamm (1895–1971), one of the original interpreters of the Cherenkov effect), he formulated the theory of a molecule containing particles with varying degrees of motion, for which he devised the first relativistically invariant wave equation. With Lev Landau (1908–1968), his former teacher, Ginzburg posited a phenomenological theory of conductivity. After 1950, he concentrated on problems in thermonuclear reactions. His work on cosmic rays – he was one of the first to believe that background radio-emission came from farther than within our own Galaxy – led him to a hypothesis about their origin and to the conclusion that they can be accelerated in a supernova.

Glaser Donald Arthur 1926–   is a US physicist who invented the bubble chamber, an instrument much used in nuclear physics to study shortlived subatomic particles. For this development he was awarded the 1960 Nobel Prize for Physics.

Glaser was born on 21 September 1926 in Cleveland, Ohio, and educated there at the Case Institute of Technology. He graduated in 1946 and then went to the California Institute of Technology, where he gained his PhD three years later. He joined the physics department of the University of Michigan in 1949, and was Professor of Physics there from 1957. Two years later he took up a similar appointment at the University of California, Berkeley, and in 1964 became Professor of Physics and Biology.

In the 1920s, the British physicist C.T.R. Wilson developed the cloud chamber for detecting and recording the presence of elementary particles. It contained a saturated vapour which condensed as a series of liquid droplets along the ionized path left by a particle crossing the chamber. With the advances in particle accelerators in the early 1950s, nuclear physicists began to deal with fast high-energy particles which could be missed by the cloud chamber. In 1952, while he was at the University of Michigan, Glaser built his first bubble chamber. It consisted of a vessel only a few centimetres across, containing superheated liquid ether under pressure. When the pressure was released suddenly, particles traversing the chamber left tracks consisting of streams of small bubbles formed when the ether boiled locally; the tracks were photographed using a high-speed camera. In later bubble chambers, liquid hydrogen was substituted for ether and chambers 2 m/ 6.6 ft across are now in use.

In the early 1960s, Glaser turned his attention from physics to molecular biology, as reflected in his change of appointment at Berkeley.

Glauber Johann Rudolf 1604–1670 was a German chemist who lived at a time when chemistry was evolving as a science from the mysticism of alchemy. He prepared and sold what today would be called patent medicines and panaceas, one of which (Glauber's salt, sodium sulphate) still bears his name. He did make some genuine chemical experiments and preparations, although many of the results were misinterpreted (through ignorance) or simply ignored by his contemporaries.

Glauber was born in Karlstadt, Franconia, some time during 1604, the son of a barber. His parents died when he was young and he had little or no formal education, teaching himself and acquiring knowledge on his travels through Germany. He earned a living chiefly from selling medicines and chemicals. He lived for a while in Vienna, then in various places along the Rhine valley, before going to Amsterdam in 1648 and finally settling there in 1655 after another six-year sojourn in Germany. He designed and built a chemical laboratory in an Amsterdam house formerly occupied by an alchemist. He died in Amsterdam on or about 10 March 1670, possibly from the cumulative effects of poisonous substances with which he had worked and, no doubt, tried on himself as medicaments.

Some time about 1625 Glauber prepared hydrochloric acid by the action of concentrated sulphuric acid on common salt (sodium chloride). He found that the other product of the reaction, sodium sulphate, was a comparatively safe but efficient laxative, and could be sold as a cure-all for practically any complaint. He called it *sal mirabile* (miraculous salt), although it soon became known as Glauber's salt – and it is still the chief active ingredient of various patent medicines. A Stassfurt mineral, a double sulphate of sodium and calcium, is known as glauberite. He also prepared nitric acid by substituting saltpetre (potassium nitrate) for salt in the reaction with sulphuric acid. He made many metal chlorides and nitrates from the mineral acids. He discovered tartar emetic (antimony potassium tartrate), so-called because of its medicinal use in the days before antimony's toxicity was fully realized.

In his Amsterdam laboratory, Glauber investigated and developed processes that could have industrial application, pursuing the principles he had outlined in his book *Furni Novi Philosophici* (c. 1646). Many of his experiments were carried out in secret, so that the methods or their products could be sold exclusively. He produced organic liquids containing such solvents as acetone (dimethylketone) and benzene – although he did not identify them – by reacting and distilling natural substances such as wood, wine, vinegar, vegetable oils and coal. One aim of his experiments was to improve the chemical techniques involved, and this aspect of Glauber's work was summarized in his book *Opera Omnia Chymica*, which was published in various versions between 1651 and 1661. He was conscious of the value to any country of its chemical and mineral raw materials, which he thought should be husbanded and exploited only for the good of the whole community. He outlined his views on a possible utopian future for Germany in *Teutschlands Wolfarth/Germany's Prosperity*.

Goddard Robert Hutchings 1882–1945 was the US physicist who pioneered modern research into rocketry. He developed the principle of combining liquid fuels in a rocket motor and the technique has been used subsequently in every practical space vehicle. He was the first to prove by actual test that a rocket will work in a vacuum and he was the first to fire a rocket faster than the speed of sound.

Goddard was born on 5 October 1882 in Worcester, Massachusetts. He was 17 when he began to speculate about conditions in the upper atmosphere and in interplanetary space. He considered the possibility of using rockets as a means of carrying research instruments.

In 1901, while studying for his BSc at Worcester Polytechnic Institute, he wrote a paper suggesting that the heat from radioactive materials could be used to expel substances at high velocities through a rocket motor to provide power for space travel. Two years later, after making his first practical experiments to determine the efficiency of rocket power, he proposed the use of high-energy propellants, such as liquid oxygen and hydrogen.

At Clark University in his home town, and at Mount Wilson, California, Goddard carried out experiments with naval signal rockets. He went on to design and build his own rocket motors. As an instructor in physics at Worcester Polytechnic, Goddard directed his experiments towards the development of rockets to explore the upper atmosphere. It was not until 1917 that he received financial aid from the Smithsonian Institution.

On the USA's entry into World War I, he turned his energies to investigating the military application of rockets, whereupon the US Army provided funds. At Clark University, teaching physics in the early 1920s, Goddard switched his practical research from solid fuel to liquid propellants. On 1 November 1923 he fired the first liquid rocket in a test stand. Two years of development were rewarded with the historic first launching of a liquid-propelled rocket on 16 March 1926 at Auburn, Massachusetts.

During the next three years he improved engine performance and reduced the weight of his design. For the first time, on 17 July 1929, instruments, and a camera to record them, were carried aloft. This significant flight attracted funds for a fuel-scale rocket-testing programme.

The greater altitudes and speeds his rocket attained dictated the need for a precise method of flight control and on 28 March 1935 Goddard successfully fired a rocket controlled by gyroscopes linked to vanes in the exhaust stream.

It was in 1937 that one of his rockets reached 3 km/ 1.8 mi, then the greatest altitude for a projectile. His work in New Mexico continued until World War II. In 1942, with the USA in the War, he joined the Naval Engineering Experimental Station at Annapolis, Maryland, continuing his research until his death on 10 August 1945.

Unlike his contemporaries at the dawn of the space age, Goddard pioneered the science of rocketry by practical experiment. He proved that liquid oxygen and a hydrocarbon fuel, such as petrol or kerosene, was the mixture which would provide the considerable thrust necessary. He built a motor in which to fire the fuels and used it in the first rocket of its kind, nearly five years before the first German experimental flights.

His rocket designs which flew in 1940, although smaller, were more advanced than the menacing V2 missile which was to fly two years later and was to be launched against European cities. Goddard's development of gyroscopically controlled steering vanes also predated German work by a number of years.

Goddard's other practical work led to the development of the centrifugal propellant pump. While in his early years he foresaw the use of the rocket for high-altitude research and wrote about space exploration. He was also aware of the military applications of his work. A rocket he developed for the US Army was fired from a launching tube, but the war ended a few days after its successful demonstration and the need disappeared.

Through much of his career Goddard worked as a 'loner'. Many of his achievements being made single-handedly. As one of the few practitioners of space research, in a world which was still trying to grasp the significance of air travel, his work and writings were not taken seriously until 1929 and the flight of his first instrumented rocket. The finance which this achievement attracted (initially from the industrialist, Daniel Guggenheim) enabled Goddard and his small team to undertake research during a period of 12 years. This was surpassed only (because it had practical military application) by the achievement of the army of German scientists under Dornberger, in their five years' work at Peenemunde to put the V2 strategic missile into service.

The full significance of the German work was not appreciated until the war's end, when Goddard was amazed to find such similarity with his designs. He died only months later and the military and space rocket programmes of both the USA and USSR were to be founded upon this German work.

It was not until 1960, fifteen years after his death, that the US government recognized the value of Goddard's work. It admitted to frequent infringement of many of his 214 patents during the evolution of missile and satellite programmes. In that year his widow received $1,000,000 from Federal funds.

In view of the USA's failure to recognize the significance of his work, at a time when its enemy was almost in a position to influence the course of the war with the rocket, it is ironic to record that Robert Goddard received more financial support for one project than any other single scientist up to World War II.

Gödel Kurt 1906–1978 was an Austrian-born US philosopher and mathematician who, in his philosophical endeavour to establish the science of mathematics as totally consistent and totally complete, proved that it could never be. This realization has been seen as fundamental to both mathematical and philosophical studies and concepts, and at the time to other scientists working in the same endeavour was devastatingly revolutionary.

Gödel was born on 28 April 1906 at Brunn, Moravia (now Brno in the Czech Republic, then in Austria), where his father was a businessman. Several childhood illnesses left him with a lifelong preoccupation with his health, but at school he worked hard and successfully: he is reputed never to have made a mistake in Latin grammar. At the age of 17 he went to the University of Vienna where he was at first unsure whether to read mathematics or physics, but settled for mathematics. At that time Vienna – and particularly its university – was the centre of activity in positivist philosophy and Gödel could not help but be influenced, and although he was apparently unimpressed at the meetings of the prestigious Viennese Circle of philosophers that he attended, his studies drifted from pure mathematics towards mathematical logic and the foundations of mathematics. In 1930, he was awarded his doctorate at the university, and in 1931 he published his most important paper, which was accepted by the university authorities as the thesis on which they licensed him to become an un-salaried lecturer there. Throughout the 1930s Gödel continued to work at the University of Vienna (except for a visit to Princeton in the United States in 1933–1934). Then, in 1938, when Austria became part of Germany, he was not appointed as a paid lecturer with his colleagues because it was thought, erroneously, that he was Jewish. Later the same year he married and travelled to the United States where he spent a short time at the Institute for Advanced Study, Princeton, before returning to Austria. Finally, in 1939, he went back to the United States to settle at Princeton where, in 1953, at the age of 47, he was appointed Professor.

A quiet and unassuming man, Gödel was awarded many honours – although he also refused quite a number, particularly from Germany. He died at Princeton on 14 January 1978.

In 1930, Gödel was granted his doctorate for a dissertation in which he showed that a particular logical system (predicate calculus of the first order) was such that every valid formula could be proved within the system; in other words, the system was what mathematicians call complete. This research represented the way in which his studies were subsequently to take him. He then investigated a much larger logical system – that constructed by Bertrand Russell (1872–1970) and Alfred Whitehead (1861–1947) as the logical basis of mathematics, published as *Principia Mathematica*. Accordingly the title of his licensiate paper, published in 1931, was 'On Formally Undecidable Propositions of *Principia Mathematica* and Related Systems'. In it Gödel dashed the hopes of philosophers and mathematicians alike, and showed that there were systems upon which mathematics was based in which it was impossible to decide whether or not a valid statement or formula was true or false within the system. In more practical terms, it was not possible to show that such areas as arithmetic worked entirely within arithmetic; there was always the possibility of coming across something that could be either true or false.

Mathematics itself was unaffected by this bombshell. But the hopes for an absolutely perfect subject that could justify itself completely to even the most penetrating philosopher, were utterly set back. Gödel himself felt that David Hilbert's Formalist School would save the day, and did not fully anticipate the effect his proof would have. Nevertheless, mathematical logic has subsequently made some progress – albeit mainly by attacking the subject from outside the logical system it embraces.

During his career Gödel made a number of other inspired contributions to the subject of mathematical logic, but none of the same importance as his earlier work. Because of his way of numbering statements in his famous proof, certain numbers were given the name 'Gödel numbers' by fellow mathematicians. The other area of mathematics to which he contributed significantly was general relativity theory; this was probably due in part at least to his close friendship with Albert Einstein (1879–1955) at the Princeton Institute. Gödel solved some of Einstein's equations and constructed mathematical models of the universe. In one of these it was theoretically possible for a person to travel into his own past, but at the cost of such a large consumption of fuel as not to make the journey feasible. (According to the model, however, it might be possible for a person to send a message into his own past.)

Gödel kept detailed diaries that record numerous interests, including the laws of nature, the evolution of life, time travel, and ghosts and demonology, as well as his preoccupation with mathematics and philosophy.

Goeppert-Mayer Maria 1906–1972 was a German-US physicist who shared the 1963 Nobel prize for discovering the shell model of nuclear structure.

Maria Goeppert was born on 28 June 1906 in Kattowitz, Upper Silesia (now Katowice in Poland). When Maria was four years old her father was appointed Professor of Paediatrics at the University of Göttingen and the family moved there. High inflation rates in Germany in the 1920s meant that the only school capable of preparing girls for the *abitur* exam was closed so Maria's education had to be continued in private. She passed the exam in Hanover in 1924. Goeppert entered Göttingen University as a mathematics student but was excited by quantum mechanics – Max Born was on the staff in Göttingen – and quickly changed to physics. She obtained her doctorate in theoretical physics in 1930 and married Joseph Edward Mayer, an American working in Göttingen with James Franck. The couple moved to Johns Hopkins University in Baltimore, USA, where Goeppert-Mayer worked on chemical physics and the colour of organic molecules, and to Columbia University in 1939, where she worked on isotope separation. In 1946, Goeppert-Mayer was appointed a professor in the physics department at Chicago and given a post at the nearby Argonne National Laboratory. Such an appointment came as a relief as she had not been paid at Johns Hopkins and felt she had been kept on the sidelines at Columbia.

In 1945, Goeppert-Mayer had developed a 'little bang' theory of cosmic origin with Edward Teller to explain element and isotope abundances in the universe and so became interested in the stability of nuclei. The 'liquid drop' model of the nucleus explained some properties but nuclei with certain numbers of neutrons (or protons) were known to be exceptionally stable. The stability of some of the light elements – helium (two neutrons and two protons), oxygen (8 each), calcium (20 each) – could be explained, but what about nuclei with 28, 50, 82 and 126 neutrons? The stability of nuclei with these so-called magic numbers of neutrons or protons had been demonstrated in experiments but could not be explained. Goeppert-Mayer and Hans Jensen (1907–1973) independently proposed a shell model in which the nucleons (protons and neutrons) moved in orbits (or shells) around the centre of the nucleus. This gave the neutrons orbital angular momentum which then coupled to their spin (inbuilt angular momentum). A combination of the shell model and this spin-orbit coupling correctly predicted which nuclei were the most stable. Goeppert-Mayer and Jensen shared half of the 1963 Nobel prize (the other half went to Eugene Wigner) and in 1955 they wrote a book *Elementary Theory of Nuclear Shell Structure*. In 1960, Goeppert-Mayer and her husband moved again, to take up professorships in physics and chemistry respectively at the University of California, La Jolla.

Gold Thomas 1920–   is an Austrian-born US astronomer and physicist who has carried out research in several fields but remains most famous for his share in formulating, with Fred Hoyle (1915–) and Hermann Bondi (1919–), the steady-state theory regarding the creation of the universe.

Gold was born in Vienna on 22 May 1920. He received his university training at Cambridge University, where he earned his bachelor's degree in 1942. Elected a Fellow of Trinity College, Cambridge, in 1947, he lectured in physics at Cambridge until 1952. From then until 1956 he was Chief Assistant to Martin Ryle (1918–1984), the discoverer of quasars, later to become Astronomer Royal. In 1956, Gold emigrated to the United States where, two years later, he became Professor of Astronomy at Harvard. Moving to Cornell University in 1959, he took up the posts of Professor of Astronomy, Chairman of the Department, and Director of the Centre for Radiophysics and Space Research. Gold has served as an adviser to NASA and is a member of the Royal Astronomical Society, the Royal Society and the National Academy of Sciences.

The question of the conditions surrounding the beginning of the universe has fascinated astronomers for many centuries. The 'hot Big Bang' theory, developed by George Gamow (1904–1968) and others, was paralleled in 1948 by the steady-state hypothesis put forward by Gold, Bondi and Hoyle.

The steady-state theory assumes an expanding universe in which the density of matter remains constant. It postulates that as galaxies recede from one another, new matter is continually created (at an undetectably slow rate). The implications that follow are that galaxies are not all of the same age, and that the rate of recession is uniform.

Evidence began to accumulate in the 1950s, however, that the density of matter in the universe had been greater during an earlier epoch. In the 1960s, microwave background radiation (at 3K) was detected, which was interpreted by most astronomers as being residual radiation from the primordial 'Big Bang'. Accordingly, the Steady State hypothesis was abandoned by most cosmologists in favour of the Big Bang model.

Gold has also carried out research on a variety of processes within the Solar System, including studies on the rotation of Mercury and of the Earth, and on the Moon. In addition he has published some work on relativity theory.

Goldberg Leo 1913–1971 was a US astrophysicist who carried out research, generally as one of a team, into the composition of stellar atmospheres and the dynamics of the loss of mass from cool stars.

Goldberg was born in Brooklyn, New York, on 26 January 1913. Completing his education, he gained his bachelor's degree in 1934 at Harvard University, where he immediately became Assistant Astronomer. Three years later he received his master's degree and moved to the University of Michigan, where he was Special Research Fellow from 1938 to 1944. At the end of that time he was made a Research Associate both at Michigan University and at the McMath and Hulbert Observatory. Between

1945 and 1960 he rose from Assistant Professor to Professor of Astronomy, and for almost all of that time he was also Chairman of the Department of Astronomy in the university and Director of the Observatory. In 1960, he was appointed Higgins Professor of Astronomy at Harvard, a position he held until 1973. He was Chairman of the Department of Astronomy and Director of Harvard College Observatory from 1966 to 1971. From 1971 to 1977, he was Director – and in 1977 he became Emeritus Research Scientist – of the Kitt Peak National Observatory in Arizona.

Goldberg's main subject of research was the Sun. As one of a team at the McMath and Hulbert Observatory, he contributed towards some spectacular films of solar flares, prominences and other features of the Sun's surface. At Harvard, he and his colleagues designed an instrument that could function either as a spectrograph or as a spectroheliograph (a device to photograph the Sun using monochromatic light), that formed part of the equipment of Orbital Solar Observatory IV, launched in October 1967. He has also carried out research on the temperature variations and chemical composition of the Sun and of its atmosphere, in which he succeeded in detecting carbon monoxide (as predicted by Henry Russell).

Goldring Winifred 1888–1971 was a US palaeontologist who is known for her research on Devonian fossils and popularization of geology.

Goldring was born on 1 February 1888 and was raised near Albany, New York State. She was educated at the New York State Normal School from which she graduated in 1905 as valedictorian, and Wellesley College receiving her AB in 1909 and AM in 1912. She remained at Wellesley until 1914 as an instructor in petrology and geology and assisting in geography and field geology, as well as teaching geography at the Teachers' School of Science. Returning to Albany in 1914, Goldring took a temporary appointment in the Hall of Invertebrate Palaeontology at the New York State Museum. She remained at the museum and in 1920 was appointed associate palaeontologist. In 1939, Goldring was made state palaeontologist, and was the first woman to hold the position which she kept until her retirement in 1954; she died in Albany on 30 January 1971 following a gastrointestinal illness.

Goldring became internationally known in the discipline of palaeobotany with her work on Devonian crinoids started in 1916 and published in 1923 (*The Devonian Crinoids of the State of New York*). During the late 1920s and 1930s, as well as geologically mapping the Coxsackie and Berne quadrangles of New York, she developed and maintained the State Museum's public programme in palaeontology by popularizing geology through her stimulating museum exhibits and the publication of handbooks, notably the two-volume *Handbook of Paleontology for Beginners and Amateurs* (1929, 1931) and her *Guide to the Geology of John Boyd Thacher Park* (1933). Goldring's work led to recognition and honours. She was the first woman president of the Palaeontological Society (1949) and was vice president of the Geological Society of America (1950); she was also awarded honorary degrees from Russell Sage College (1937) and Smith College (1957).

Goldschmidt Victor Moritz 1888–1947 was a Swiss-born Norwegian chemist who has been called the founder of modern geochemistry.

Goldschmidt was born on 27 January 1888 in Zurich but moved to Norway with his family in 1900, where his father became Professor of Physical Chemistry, at the University of Christiania (now Oslo). The family became Norwegian citizens in 1905. Goldschmidt completed his schooling in Christiania and continued at the university there, where he studied under the geologist Waldemar Brøgger and graduated in 1911. He was appointed Professor and Director of the Mineralogical Institute in 1914, a post he held until 1929, when he moved to Göttingen. The rise of anti-Semitism forced Goldschmidt to return to Norway in 1935, but soon World War II engulfed Europe and he had to flee again, first to Sweden and then to Britain, where he worked at Aberdeen and Rothamsted (on soil science). He returned to Norway after the end of the war in 1945, and died in Oslo on 20 March 1947.

Goldschmidt's doctoral thesis on contact metamorphism on rocks is recognized as a fundamental work in geochemistry. It set the scene for a huge programme of research on the elements, their origins and their relationships, which was to occupy him for the next 30 years. He broke new ground when he applied the concepts of Josiah Gibbs's phase rule to the colossal chemical processes of geological time, which he considered to be interpretable in terms of the laws of chemical equilibrium. The evidence of geological change over millions of years represents a series of chemical processes on a scarcely imaginable scale, and even an imperceptibly slow reaction can yield megatonnes of product over the time scale involved.

Shortage of materials during World War I led Goldschmidt to speculate further on the distribution of elements in the Earth's crust. In the next few years he and his co-workers studied 200 compounds of 75 elements and produced the first tables of ionic and atomic radii. The new science of X-ray crystallography, developed by the Braggs after Max von Laue's original discovery in 1912, could hardly have been more opportune. Goldschmidt was able to show that, given an electrical balance between positive and negative ions, the most important factor in crystal structure is ionic size. He suggested, furthermore, that complex natural minerals, such as hornblende $(OH)_2Ca_2Mg_5Si_8O_{22}$, can be explained by the balancing of charge by means of substitution based primarily on size. This led to the relationships between close-packing of identical spheres and the various interstitial sites available for the formation of crystal lattices. He also established the relation of hardness to interionic distances.

At Göttingen (1929) Goldschmidt pursued his general researches and extended them to include meteorites, pioneering spectrographic methods for the rapid determination of small amounts of elements. Exhaustive analysis of results from geochemistry, astrophysics and nuclear physics led to his work on the cosmic abundance of the elements and the important links between isotopic stability and abundance. Studies of terrestrial abundance reveal about eight predominant elements. Recalculation of atom and volume percentages lead to the remarkable notion that the Earth's crust is composed largely of oxygen anions (90% of the volume), with silicon and the common metals filling up the rest of the space.

Goldschmidt was a brilliant scientist, with the rare ability to arrive at broad generalizations which draw together many apparently unconnected pieces of information. He also had a steely sense of humour. During his exile in Britain he carried a cyanide suicide capsule for the ultimate escape should the Germans have successfully invaded Britain. When a colleague asked for one he was told 'Cyanide is for chemists; you, being a professor of mechanical engineering, will have to use the rope.'

Goldstein Eugen 1850–1930 was a German physicist who investigated electrical discharges through gases at low pressures. He discovered canal rays and gave cathode rays their name.

Goldstein was born at Gleiwitz in Upper Silesia (now Gliwice in Poland) on 5 September 1850. He went to the University of Breslau in 1869 but after only one year moved to the University of Berlin to work with Hermann Helmholtz (1821–1894), first as his student and then as his assistant. Before obtaining his doctorate he became employed as a physicist at the Berlin Observatory in 1878. Following the award of his doctorate in 1881 for some of his important work on cathode rays, Goldstein continued at the Observatory until 1890, spending the next six years at the Physikalische Technische Reichsanstalt. There then followed some time when he had a laboratory and assistants in his own home, until he was appointed head of the Astrophysical Section of Potsdam Observatory. Goldstein died in Berlin on 25 December 1930.

As Helmholtz's assistant, Goldstein worked on electrical discharges in gases. Goldstein coined the term cathode rays to explain the glow first observed in an evacuated tube by Julius Plücker (1801–1868) in 1858. His first publication, in 1876, demonstrated that cathode rays could cast shadows and that the rays are emitted perpendicular to the cathode surface, enabling a concave cathode to bring them to a focus. There followed investigations by Goldstein and others which showed how cathode rays could be deflected by magnetic fields. This was the period when such phenomena were under investigation throughout Europe, and Goldstein made many contributions which aided the conclusion that cathode rays were fast-moving, negative particles or electrons. He himself believed that they were waves not unlike light.

In the course of his investigations, Goldstein studied many different arrangements of anode and cathode discharge circuits. In an experiment in 1886, he perforated the anode and observed glowing yellow streamers behind the anode emanating from the perforations. Since they were connected to the holes or canals, he termed them *Kanalst-rahlen* or canal rays. These always moved away from the anode. Wilhem Wien (1864–1928), Goldstein's student, showed in 1898 that they could be deflected by electric and magnetic fields in such a way that they appeared to be positively charged. Earlier, Goldstein had been able to observe only slight deflections. Wien also calculated the charge-to-mass ratio of the rays and found this to be some 10,000 times greater than the corresponding ratio for cathode rays. Later, the rays were called positive rays by J. J. Thomson (1856–1940), who also investigated their nature and led the way to their identification as ionized molecules or atoms of the gaseous contents of the discharge tube. Tubes with different contents formed different positive ions, and the different charge-to-mass ratios were detected by Thomson. Francis Aston (1877–1945) began investigations in this field and developed the mass spectrograph in 1919 to separate positive rays. From these beginnings has developed mass spectrometry, a method extremely useful in chemical analysis.

In 1905, Johannes Stark (1874–1957), another student of Goldstein, used light from canal rays to demonstrate an optical Doppler shift in a terrestrial, rather than a stellar, source of light, for the first time.

Goldstein later investigated the wavelengths of light emitted by metals and oxides when canal rays impinge on them, and observed that alkali metals show their characteristic bright spectral lines. He continued his investigation of the complex phenomena and gaseous discharge at the anode for many years, but his subsequent discoveries were interesting and curious rather than of great importance for physics. He compared his work with the natural phenomenon of the aurora borealis, which he believed was caused by cathode rays originating in the Sun. His last paper, published in 1928, gave a new discovery, the importance of which was almost completely missed. In a discharge tube containing nitrogen and hydrogen and possibly other gases, he observed that a trace of ammonia was present after the discharge. It was only much later that investigations along these lines were begun in earnest to see if biologically important molecules, and hence life, could have originated in this way.

Golgi Camillo 1843–1926 was an Italian cytologist and histologist who pioneered the study of the detailed structure of the nervous system, made possible by his development of a new staining technique. For his outstanding work in this field, Golgi shared the 1906 Nobel Prize for Physiology or Medicine with Santiago Ramón y Cajal.

Golgi was born on 7 July 1843 in Corteno, near Brescia, the son of a physician. He studied medicine at the University of Pavia, graduating in 1865, then worked in a psychiatric clinic, before researching in histology. In 1872, he became principal physician at a hospital in Abbiategrasso near Milan, but managed to continue his histological research. In 1875, he was appointed Lecturer in Histology at the University of Pavia and Professor the following year, a post he held until 1879, when he became Professor of

Anatomy at the University of Siena. In the following year, however, he returned to Pavia University as Professor of Histology, subsequently becoming Professor of General Pathology, and remained there until his retirement in 1918. During his time at the university he took an active part in its administration, as Dean of the Faculty of Medicine and later as President of the university. He also played an important role outside the university, becoming a Senator in the Italian government in 1900. Golgi died on 21 January 1926 in Pavia.

At the time Golgi was studying medicine there were no techniques suitable for the study of nerve cells, with the result that very little was known about their structure and function. In 1873, however, Golgi invented a new technique of staining cells with silver salts, a method he found to be particularly suitable for studying nerve cells because it allowed controlled staining of certain features in the cells, thereby enabling them to be studied in great detail. From 1873 onwards he published many articles on the results of his systematic studies – using his new technique – of the fine anatomy of the nervous system. He verified Wilhelm von Waldeyer's view that nerve cells are separated by synaptic gaps, and he also discovered a type of nerve cell (later called the Golgi cell) that connects many other cells by means of dendrites. In 1896, while studying the brain cells of a barn owl, he detected stacks of flattened cavities in the cytoplasm, near the nuclear membrane. The function of these structures – called Golgi bodies, the Golgi complex or the Golgi apparatus – is still largely unknown, although they are thought to be involved in intracellular secretion and transport.

From his examinations of different parts of the brain, Golgi put forward the theory that there are two types of nerve cells, sensory and motor cells, and that axons are concerned with the transmission of nerve impulses. He showed that there is a fine nerve network in the grey matter but could not say positively whether the filaments were joined or merely interwoven. Golgi made other contributions to the study of the nervous system, including the discovery of tension receptors in the tendons – now called the organs of Golgi.

Between 1885 and 1893 Golgi investigated malaria, the causative agent of which – the protozoon *Plasmodium* – had been discovered by the French physician Charles Laveran in 1880. (And by 1885 two Italian scientists had discovered the stages in the parasite's life cycle.)

Golgi's research verified a number of important facts about the disease and he discovered that different species of *Plasmodium* are responsible for the two main types of intermittent fever – tertian fever, with attacks every three days, and quartan fever, with attacks every four days – and that the fever attacks coincide with the release into the bloodstream of a new generation of the parasites. He also put the results of his work to practical use by establishing a method of treatment that involved determining the type of malaria a person had contracted, and then giving the patient quinine (the only effective antimalarial drug then available) a few hours before the predicted onset of the fever.

Goodyear Charles 1800–1860 was a US inventor who is generally credited with inventing the process for vulcanizing rubber.

Goodyear was born in New Haven, Connecticut, on 18 December 1800. When he came of age he entered his father's hardware business at Naugatuck, Connecticut, where he worked with enthusiasm, inventing various implements, including a steel pitchfork to replace the heavy iron type. The firm became financially unstable, and by 1830 Goodyear realized that he would have to turn to something else. He chose to investigate indiarubber and the problems associated with it, particularly how to make the rubber remain strong and pliable over a range of temperatures. He thought he had found the solution by mixing nitric acid with the rubber, and in 1836 secured a government order for a consignment of mail bags. But they would not stand up to high temperatures and were therefore useless. Goodyear was forced to start again.

In 1837, he bought out the rights of Nathaniel Hayward, who had had some success by mixing sulphur with raw rubber. After much patient experiment – and a deal of luck – Goodyear finally perfected the process he called vulcanization.

He obtained United States patents in 1844, but both Britain and France refused his applications because of legal technicalities. His attempts to set up companies in both countries failed, and for a while he was imprisoned for debt in Paris. Eventually he returned to the United States, where many of his patents had been pirated by associates. Even his son, who had been working for him, decided to leave the ailing firm and Goodyear was forced to face his heavy debts alone. He died lonely and poverty-striken in New York City on 1 July 1860.

Various people throughout the western world tried to make rubber a more commercially viable material. Goodyear discovered the vulcanization process by accident. One day he was mixing rubber with sulphur and various other ingredients when he dropped some on top of a hot stove. The next morning the stove had cooled, the rubber had vulcanized, and he thought that all his problems had been solved.

The invention was highly significant at a time of industrial advancement. The new process made rubber a suitable material for such applications as belting and hoses, for which strength at high temperatures was the governing factor. It was also particularly valuable once the idea of rubber tyres was conceived, at first for bicycles and then for motorcars.

Goodyear's process was eventually superseded by more refined methods and by the development of synthetic rubbers. And although he failed to find wealth in his lifetime, his name still lives on and can be seen on motor tyres throughout the world.

Gosset William Sealey 1876–1937 was a British industrial research scientist, famous for his work on statistics.

Gosset was born at Canterbury on 13 June 1876 and educated at Winchester College, before going to New College, Oxford, to study mathematics and chemistry. He received a first class in his mathematical moderations in 1897 and a first class on receiving his BA in natural sciences (chemistry) in 1899. On leaving Oxford he immediately joined the Guinness brewery firm in Dublin, the firm recently having adopted the policy of hiring a number of university-trained scientists to conduct research into the manufacture of its ale. He remained with the company in Dublin until 1935, when he was posted to London to direct the operations of the new Guinness brewery there. He was appointed the company's head brewer a few months before his death, at Beaconsfield, on 16 October 1937.

When Gosset arrived in Dublin he found that there was a mass of data concerning brewing – on the relationships between the raw materials, hops and barley, and the quality of the finished product and on the methods of production – which had been left almost entirely unanalysed. After a few years with the company he was able to persuade the owners that they would profit from more sophisticated mathematical analysis of a variety of processes from the production of barley to the fermentation of yeast. Accordingly, the company sent him to University College, London, in September 1906 to study for a year under Karl Pearson (1857–1936). It was the experience of that year that turned Gosset into an outstanding statistical theorist, even though all the questions he asked were inspired directly by problems in the brewery trade.

Gosset's statistical techniques were simple: he relied on the mean, the standard deviation and the correlation coefficient as his basic tools. Through all his work ran one theme, expressed in two formulae:

$$\sigma_x^2 - y = \sigma_x^2 + \sigma_y^2 - 2p\sigma_x\sigma_y$$

$$\sigma_x^2 + y = \sigma_x^2 + \sigma y + 2p\sigma_x\sigma_y$$

Gosset's amplification of these formulae opened the door to modern developments in the analysis of variance.

When Gosset began to examine the data at the Guinness brewery, it quickly became apparent to him that what was most needed was improved knowledge of the theory of errors. In 1904, he wrote a report for the company, 'Application of the Law of Error' to the brewing industry. His most famous work, published in 1908, was on the probable error of a mean. More than a century earlier Gauss had worked out a satisfactory method of estimating the mean value of a characteristic in a population on the basis of large samples; Gosset's problem was to do the same on the basis of very small samples, for use by industry when large sampling was too expensive or impracticable. For any large probability, that is one of 95% or more, Gosset was able to compute the error e, such that it is 95% probable that:

$$(x - \mu) \leq e$$

where x is the value of the sample, and μ is the mean. From this t-test was derived what came to be known as Student's t-test of statistical hypotheses. The name of the test comes from the fact that Gosset published all his papers under the pseudonym 'Student'. The test consists of rejecting a hypothesis if, and only if, the probability (derived from t) of erroneous rejection is small.

Gosset hit on the statistic that would become fundamental for the statistical analysis of the normal distribution, and he went further, to make the shrewd observation that the sampling distribution of such statistics is of basic importance in the drawing of inferences. In particular, it opened the way to the analysis of variance, that branch of the subject so important to statisticians who came after him.

Gould Stephen Jay 1941– is a US palaeonotologist and writer who teaches and researches in geology, evolutionary biology and the history of science.

He was born in New York City on 10 September 1941 and graduated in 1963 from Antioch College in Ohio and then took a PhD at Columbia University in 1967. In that year he joined the staff at Harvard University and became Professor of Geology there in 1973 and Alexander Agassiz Professor of Zoology in 1981. He holds a concurrent post in the department of the history of science.

Gould's early research interests were focused on the evolutionary development and speciation of the land snail, and initiated his wider studies of animal form and function and the relationship between ontogeny and phylogeny. In 1972, he proposed the theory of punctuated equilibrium, suggesting that the evolution of species does not occur at a steady rate but can suddenly accelerate, with rapid change occurring over a few hundred thousand years.

Gould has written extensively on several aspects of evolutionary science, in both professional and popular books. His *Ontogeny and Phylogeny* (1977) provided a detailed scholarly analysis of his work on the developmental process of recapitulation. In *Wonderful Life* (1990) he drew attention to the diversity of the fossil finds in the Burgess Shale Site in Yoho National Park, Canada, which he interprets as evidence of parallel early evolutionary trends extinguished by chance rather than natural selection.

Collections of essays published in, for example, *Ever Since Darwin* (1977), *The Panda's Thumb* (1980) and *Hen's Teeth and Horses' Toes* (1983) have all achieved critical acclaim and large audiences.

Graham Thomas 1805–1869 was a British physical chemist who pioneered the chemistry of colloids, but who is best known for his studies of the diffusion of gases, the principal law concerning which is named after him.

Graham was born on 21 December 1805 in Glasgow, the son of a successful local manufacturer. His father had hoped that his son would, after leaving school, enter the Presbyterian ministry but in 1819, when he was only 14 years old, Graham enrolled at Glasgow University to study science. He later transferred to Edinburgh University and graduated in 1824. He returned to Glasgow to teach at the Mechanics Institute, which had been founded a year or two earlier by George Birkbeck for teaching craftsmen the scientific principles of their trades. In 1830, Graham became Professor of Chemistry at Anderson's College, Glasgow. He left Scotland seven years later to take up a similar position at University College, London, where he remained until 1854. In 1841, he became the first President of the Chemical Society of London, itself the first national society devoted solely to the science of chemistry. In 1855, he was appointed Master of the Royal Mint, a position once held by Isaac Newton. He died in London on 16 September 1869.

Graham's early interest was the dissolution and diffusion of gases. In 1826, he discovered that very soluble gases do not obey Henry's law (which states that solubility is proportional to the pressure of the gas). He measured the rates at which gases diffused through a porous plug of plaster-of-Paris, through narrow glass tubes, and through small holes in a metal plate. By 1831, he had formulated Graham's law of diffusion, which states that the rate of diffusion of a gas is inversely proportional to the square root of its density.

In 1829, Graham turned his attention briefly to inorganic chemistry. He studied the glow of phosphorus and observed that it was extinguished by organic vapours and various gases. He went on to examine phosphorus compounds in general, particularly phosphine and salts of the various oxyacids. He distinguished ortho-, meta- and pyro-phosphates, which he prepared by fusing sodium carbonate with orthophosphoric acid. Graham had made the first detailed study of a polybasic acid.

In the 1850s, following his work on gases, Graham investigated the movement of molecules in solutions. He added crystals of a coloured chemical, such as cupric sulphate, to water and noted how long it took for the colour to spread throughout the solution. He observed that different chemicals took different times to disperse and that the dispersion rate increased with increasing temperature.

Then in 1861, he tried a technique similar to that which he had used for gases. He inserted a parchment barrier across a tank of water and added a coloured salt to the water on one side of it. He discovered that some of the coloured substance passed through the barrier. Repeating the experiment using glue or gelatine, he found that these substances did not pass through parchment. All the substances tested that could pass through also formed crystals, and Graham called this category crystalloids. Those that failed to cross the barrier did not form crystals and he called these colloids (from the Greek *kolla*, meaning glue). He distinguished between sols and gels (although he did not use these terms to describe them).

Using the same discovery, Graham developed a method of purifying colloids. The impure colloid was placed in a porous tube suspended in running water. The crystalloids (impurities) were washed away, leaving the purified colloid in the tube. He called the process dialysis, and it has since found a multitude of applications, from desalination equipment to artificial kidney machines.

Grandi Guido 1671–1742 was an Italian mathematician famous for his work on the definition of curves – particularly curves that are symmetrically pleasing to the eye. It was he who devised the curves now known as the 'versiera', the 'rose' and the 'cliela' (after Cliela, Countess Borromeo), and his theory of curves also comprehended the means of finding the equations of curves of known form. He was mainly responsible, in addition, for introducing calculus into Italy.

At his birth, on 1 October 1671 in Cremona, his given names were Francesco Lodovico Grandi. At the age of 16 he entered the religious order of the Camaldolese and changed his first names to Guido. In 1694, he was appointed teacher of mathematics in the order's monastery in Florence, and it was there that he first became acquainted with the *Principia Mathematica* of Isaac Newton (published in England in 1687), which inspired him to devote much of the remainder of his life to the study of geometry. Nevertheless, he became Professor of Philosophy at Pisa in 1700, and seven years later was given the post of honorary mathematician to the Grand Duke. In 1714, he became Professor of Mathematics at Pisa and established a considerable reputation as a teacher. The recipient of several awards and honours – a Fellow of the Royal Society from 1709 – Grandi died in Pisa on 4 July 1742.

In his fascination for the study of curves, Grandi was influenced first by Newton; early in his career he determined the points of inflection in the conchoid curve. He examined also the studies published by Pierre de Fermat (1601–1675), whose treatment he found somewhat limited, and by Christiaan Huygens (1629–1695), who had revealed the most important properties of logistic curves in research based on the work of Evangelista Torricelli (1608–1647). In 1701, Grandi devised a proof for Huygens's theorem.

However, Grandi's name will always be associated with the 'rose', the 'cliela' and the 'versiera' curves; his work on the first two was presented in a paper to the Royal Society in 1723 – ten years after he had corresponded with the great Gottfried Leibniz on the subject. In 1728 he published his complete theory in *Fleores geometrica*, an attempt (among other things) to define geometrically the curves that have the shapes of flowers, particularly multi-petalled roses.

What is today known as the 'rose' curve Grandi called by its Greek name *rhodonea*. The polar equation for such a curve is:

$$r = a \sin k\theta$$

where k is an integer. Depending on whether k is an odd or an even number, there are k or $2k$ 'petals' on the rose; and depending on whether k is a rational or an irrational number, the number of petals is finite or infinite. The Cartesian coordinates for the curve are:

$$(x^2 + y^2)^3 = 4a^2x^2y^2$$

where a is a constant.

The 'cliela' curve Grandi described as the locus of P where P represents a point on a sphere of radius a where ϕ and θ are the longitude and co-latitude of P, with P moving such that $\theta = m\phi$, where m is a constant.

Grandi defined the curve now called the 'versiera' (from the Latin *sinus versus*) by stating: given a circle with diameter AC, let BDM be a straight line perpendicular to AC at B and intersecting the circumference at D; let M be a point determined by the length of BM so that $AC:BM = AB:BD$; the locus of all such points, M, is the versiera. The curve is in some places more commonly known as the 'witch of Agnesi' as a result of a mistranslation and the erroneous attribution of its discovery to Maria Gaetana Agnesi in a treatise of 1748.

Although the study of curves was Grandi's joy and passion, he did make other contributions to mathematics, notably in 1703 when his treatise on quadrature – using Leibniz's methods in preference to those of Francesco Cavalieri and Vincenzo Viviani – was responsible for the introduction of calculus into Italy.

Grandi also did some work in practical mechanics and his observations regarding hydraulics were utilized by the Italian government in such public works as the drainage of the Chiana valley and the Pontine Marshes in central Italy.

Grassmann Hermann Günther 1809–1877 was an extremely gifted German mathematician whose methods of presentation were so unclear that his innovations were never really appreciated during his lifetime despite their importance. He had greater success with his work in a completely different field: comparative linguistics.

Grassmann was born in Stettin, German Pomerania (now Szczecin, Poland), on 15 April 1809. He studied first at home and then in local schools, before going to the University of Berlin, where he studied theology from 1827 to 1830. Instead of becoming a minister, however, he returned to Stettin and began to study mathematics and physics for an examination to enable him to teach at secondary schools. Poor marks in the examination meant that he could teach only younger students, from 1832. It was not until 1840 that he obtained the qualifications necessary to teach more advanced students. In 1842, he joined the staff of the Friedrich Wilhelm School in Stettin; five years later he became Senior Teacher there. Finally the Stettin high school appointed Grassmann to an important teaching post (which carried the title of Professor) in 1852. Grassmann's numerous schoolbooks and his texts on more advanced mathematics and linguistics were not uniformly successful. The international mathematical community was slow to recognize the significance of Grassmann's work in mathematics, although he received much more immediate recognition for his work in linguistics, and in particular on Sanskrit. He was elected to the Göttingen Academy of Sciences in 1871, but the honorary doctorate he received in 1876 from the University of Tübingen was for his work in linguistics. He died in Stettin on 26 September 1877. After his death his work was popularized, partly through the efforts of his children, several of whom became prominent academics.

The examination paper that Grassmann sat in 1840 in order to become eligible to teach advanced students included a paper on tides. In his research at home while framing his answer to the paper, Grassmann discovered a new calculus which enabled him to apply parts of Pierre Laplace's book on celestial mechanics to the paper on tides. The technique clearly had wider applications, so Grassmann decided to devote himself to exploring it further, and during the years from 1840 to 1844, concentrated on developing his method, which he called the theory of extension. It was one of the earliest mathematical attempts to investigate n-dimensional space, where n is greater than 3. He published the method in a book in 1844, but his vocabulary was so obscure that the book had virtually no impact at all in the short term.

Grassmann in the following year applied the method to reformulate Ampère's law, but again through poor exposition the work was ignored. (Thirty years later, Rudolf Clausius independently found a similar improvement to Ampère's law, but gave credit to Grassmann for its earlier discovery.)

The next ten years saw Grassmann investigating the subject of algebraic curves, using his calculus of extension, but the papers he published on the subject once more made little impact. He was awarded a prize for a paper in topology by the Leipzig Academy of Sciences in 1847 – but the published version needed a note of clarification because of the essay's complex format.

The poor response of the mathematical community to Grassmann's work led him also to devote considerable energy to a completely different field of study. He learned and examined many ancient languages, such as Persian, Sanskrit and Lithuanian. His books on this subject were considerably more successful, especially the glossary he published to the Hindu scriptures, the Rig-Veda, in 1873–1875. From his investigations he derived a theory of speech.

Another area of his interest was the mixing of colours. His work in this field attracted some favourable comments from Hermann Helmholtz (1821–1894) – the physiologist and expert on colour vision and colour blindness – whose work Grassmann had in fact criticized.

The book on extension theory that Grassmann published in 1844 was reprinted in a revised form in 1862, and again posthumously in 1878. Gradually it began to receive some recognition as a work of considerable value by German, French and US mathematicians. A scientist on whom it had a particularly strong effect was Josiah Gibbs (1839–1903), who made use of it in his development of vector analysis.

Gray Asa 1810–1888 was a US botanist who was the leading authority on botanical taxonomy in the United States in the nineteenth century and a pioneer of plant geography. He was also the chief US proponent of Charles Darwin's theory of natural selection.

Gray was born on 18 November 1810 in Sauquoit, New York, and studied medicine at Fairfield Medical School, Connecticut, teaching himself botany in his spare time. After graduating in 1831 he taught science at Bartlett's High School in Utica, New York, until 1834 when he became an assistant to John Torrey, a chemistry professor at the College of Physicians and Surgeons in New York.

Also interested in botany, Torrey became a lifelong friend of Gray and the two men collaborated on several botanical projects. In 1835, Gray accepted the post of Curator and Librarian of the New York Lyceum of Natural History, which gave him more time to devote to botany and provided greater financial security than did his previous position. Gray was made botanist to the United States Exploring Expedition in 1836, but he resigned in the following year because of delays in sailing. In 1838, he was appointed a professor at the newly established University of Michigan, but did not commence his duties because, in the same year, he travelled to Europe to acquire books for his library and to study specimens of American plants in European herbaria. On his return to the United States in 1839 he was occupied with writing and organizing the results of his studies. In 1842, Gray accepted the professorship of natural history at Harvard University, on the understanding that he could devote himself to botany. He held this position for 31 years, until he retired in 1873, during which time he made several journeys to Europe – meeting Charles Darwin in 1851 – and donated his priceless collection of plants and books to Harvard University on condition that the university housed his collection in a special building; this led to the establishment of the botany department at Harvard University, and the botanical garden and herbarium were later named after him. Gray died on 30 January 1888 in Cambridge, Massachusetts.

Gray was a prolific author, writing more than 360 books, monographs and papers, but is probably best known for his *Manual of the Botany of the Northern United States, from New England to Wisconsin and South to Ohio and Pennsylvania Inclusive* (1848), often called *Gray's Manual*, which was an extremely comprehensive study of North American flora and which also helped to establish systematic botany in the United States. This work went through several editions and is still a standard text in its subject area. Before the publication of *Gray's Manual*, however, Gray had written several other important works, including the two-volume *Flora of North America* (1838–1843), which he co-authored with Torrey. This work was later expanded, under Gray's direction, and published in 1878 as the first volume of the *Synoptical Flora of North America*.

In addition to these contributions to botanical taxonomy, Gray also investigated the geographical distribution of plants, notably a comparison of the flora found in Japan, Europe, northern America and eastern America. This knowledge of plant distribution proved useful to Darwin, who asked Gray to analyse his plant-distribution data in order to provide him with evidence for his theory of natural selection. Darwin also told Gray about his ideas on natural selection as early as 1857, and after publication of *On the Origin of Species* (1859) Gray became a leading advocate of Darwin's theory. In his own work, Gray reached conclusions about variations that foreshadowed the work of Gregor Mendel and Hugo de Vries.

Gray Henry 1827–1861 was a British anatomist who compiled a book on his subject which, through its various editions and revisions, has remained the definitive work on anatomy for more than 100 years.

Little is known of Gray's early life and education. His father was a private messenger to King George IV and William IV. Gray became a 'perpetual student' at St George's Hospital, London, on 6 May 1845. In 1848, he was awarded the Triennial Prize of the Royal College of Surgeons. When Gray was 25 years old, the Royal Society acknowledged his contribution to anatomy by electing him a Fellow, and the following year, in 1853, he was awarded the Astley Cooper Prize of 300 guineas for a talk on the structure and function of the spleen. He was offered and accepted the post of Demonstrator of Anatomy at St George's Hospital. He became a Fellow of the Royal College of Surgeons, and was curator of the St George's Museum. He was also Lecturer in Anatomy and in line for the post of Assistant Surgeon. At this point his brilliant and promising career was cut short. He tended a nephew suffering from smallpox, and contracted a particularly virulent strain of the disease. He died in London in 1861 at the age of only 34.

The first edition of what is now known as *Gray's Anatomy* was published in 1858. He had painstakingly and methodically learnt all his anatomy on the merits of his own dissections, and had the good fortune to persuade his friend H. Vandyke Carter to do the drawings for him. Vandyke Carter was a demonstrator of anatomy as well as a draughtsman, and undoubtedly his beautiful illustrations were partly responsible for the success of the book. Gray prepared the second edition in 1860.

Gray's Anatomy was fundamentally different from other contemporary works of a similar nature because it was organized in terms of systems, rather than areas of the body. His layout and general philosophy remain in the current edition, although the book has been revised and updated nearly 50 times. Areas such as neuroanatomy have been greatly enlarged but the section that deals with, for example, the skeletal system is almost identical to Gray's original work. It remains a standard text for students and surgeons alike.

Green George 1793–1841 was a British businessman and self-taught mathematician; through hard work and considerable creative and perceptive ability he made significant advances in both the physics and the mathematics of his time, although he was really achieving his proper status and recognition only at the time of his death. He is best remembered for his paper, published before he entered formal education, in which he introduced the term 'potential', now a central concept in electricity.

Green was born in Nottingham on 14 July 1793. He had to leave school at an early age in order to assist at the family mill, but his mathematical interests and abilities spurred him to continue his studies on his own, through reading. The death of his father and the ensuing sale of the family firm eventually made Green financially independent. He moved to Cambridge and in 1833 (at the age of 40) became a student at Caius College in the University. Conducting his own studies in mathematics above and beyond those required by the curriculum, he graduated with honours in 1837. Further private researches led to the award to Green of a Fellowship at Caius College – although he was not to be able to continue his work. Poor health led to his untimely death, in Sneinton, Nottinghamshire, on 31 March 1841.

Green's most famous paper was published in 1828. Only a few copies were printed, and these were circulated only privately and locally in Nottingham. It could easily have simply sunk out of sight without ever attracting the attention it deserved and eventually got. The essay dealt with a mathematical approach to electricity and magnetism. Its two outstanding features were the coining of the term 'potential', and the introduction of Green's theorem – which is still applied in the solution of partial differential equations, for instance in the study of relativity. Green demonstrated the importance of 'potential function' (also known as the Green function) in both magnetism and electricity, and he showed how the Green theorem enabled volume integrals to be reduced to surface integrals.

Green went on to produce other important papers on fluids (1832, 1833), attraction (1833), waves in fluids (1837), sound (1837), and light (1837); his development of earlier work by Augustin Cauchy (1789–1857) on light reflection and refraction (1837) and on light propagation (1839) represented the best of his later work.

Green's work on electricity and magnetism might have been totally lost had it not been for William Thomson (later Lord Kelvin; 1824–1907), an undergraduate at the time of Green's death. He, after completing his studies, showed Green's 1828 essay to a number of prominent physicists whom he knew personally. It stimulated great interest, and went on to influence scientists such as James Clerk Maxwell and Kelvin himself, and was thus significant in the development of nineteenth-century theories of electromagnetism.

Greenstein Jesse Leonard 1909– has made important astronomical discoveries by combining his observational skills with current theoretical ideas and techniques. His early work involved the spectroscopic investigation of stellar atmospheres; later work included a study of the structure and composition of degenerate stars. He took part in the discovery of the interstellar magnetic field and in the discovery and interpretation of quasi-stellar radio sources – quasars.

Greenstein was born in New York City on 15 October 1909. He developed an interest in astronomy from an early age, although when he was young there were few popular astronomy textbooks and almost no professional astronomers in the United States. His grandfather, however, had an excellent library and from the age of eight the young Jesse began to use it; he was also encouraged by the gift of a small telescope. Greenstein admits that its location, overlooking New York City, was not ideal for observational purposes, but he nevertheless retained his childhood enthusiasm for astronomy throughout his life. He studied at Harvard and wrote a thesis on interstellar dust for his PhD, which he received in 1937. For the following two years he was a fellow of the National Research Council. In 1939, he accepted a post at the University of Chicago where, besides his teaching, he continued his research career at the Yerkes Observatory; he held this post for nine years, with a brief interlude during World War II, during which time he was involved in designing specialized optical instruments. In 1948, he joined the California Institute of Technology and also became a staff member of the Mount Wilson and Palomar Observatories. Soon after joining the California Institute of Technology, he initiated its Graduate School of Astrophysics; he became Chairman of the Faculty in 1965 and Lee A. Dubridge Professor of Astrophysics in 1971. He has been awarded several medals, including the Gold Medal of the Royal Astronomical Society, the Gold Medal of the Astronomical Society of the Pacific and NASA's Distinguished Public Service Medal. After his retirement, Greenstein continued to write prolifically and carried on with his observational research work.

Greenstein began his career by spectroscopically studying stellar atmospheres to try to explain anomalies in the spectra of some stars. Using the spectrographs at the McDonald Observatory and later the Mount Wilson and Palomar Observatories (now Hale), Greenstein developed a method of analysing the spectra of 'peculiar' stars by comparing them with the spectra of other average stars such as the Sun. His spectroscopic programme was closely linked with a parallel growth of new ideas in nuclear astrophysics that eventually led to the currently accepted theories of stellar evolution. In connection with the development of these theories, Greenstein independently suggested that the neutron-producing reaction in red giant stars was required for the production of heavy elements.

Greenstein rekindled his early postgraduate interest in the properties of interstellar dust by studying the nature of the interstellar medium with Leverett Davis. He developed the idea that space was pervaded by dominantly regular magnetic fields which aligned non-spherical, rapidly spinning, paramagnetic dust grains and so produced the interstellar polarization of light.

Greenstein was one of the leading figures in the explosive post-World War II development of radioastronomy. The rapid growth in this field resulted in the discovery of quasars in 1964, and Greenstein played a key role in the story of their discovery. He confirmed Maarten Schmidt's hypothesis that the emission lines of these peculiar bluish stellar objects could be explained by a shift in wavelength and he found this to be true of the spectrum of 3 C 48. He found that the lines were due to hydrogen provided that the velocity with which 3 C 48 was receding was just over 110,000 km/70,000 mi per second. In collaboration with Schmidt, Greenstein proposed a detailed physical model of the size, mass, temperature, luminosity, magnetic field and high-energy particle content of quasars. The nature of quasars, the most luminous and enigmatic objects in the universe, still remains debatable, however.

During the 1970s, Greenstein spent most of his time studying white dwarf stars. He collected quantitive information on their size, temperature, motion and composition and by 1978 he had discovered some 500 of these degenerate stars. His research enabled him to pinpoint the problems of explaining the evolutionary sequence that links red giant stars with white dwarfs, thus initiating spectroscopic studies of such stars from space.

Greenstein is a leading figure in modern astronomy. During the 1970s he guided both NASA and the National Academy of Sciences in their future policies and by chairing the board of directors of the Association of Universities for Research in Astronomy from 1974 to 1977 he has been influential in the research programmes of the Kitt Peak and Cerro Tololo Observatories. He is the author of more than 380 technical papers and has edited several books, including *Stellar Atmospheres* (1960). During his career, Greenstein has changed fields of specialization many times, with the personal goal of applying new branches of physics to the study of the universe.

Grignard François Auguste Victor 1871–1935 was a French organic chemist, best known for his work on organomagnesium compounds or Grignard reagents. For this work he shared the 1912 Nobel Prize for Chemistry with Paul Sabatier.

Grignard was born on 6 May 1871 in Cherbourg, the son of a sailmaker, and educated locally and at Cluny. His studies were interrupted for a year in 1892 while he did his military service (rising to the rank of corporal), after which he went to the University of Lyon and took a degree in mathematics. A former classmate then persuaded him to work in an organic chemistry laboratory under P. A. Barbier (1848–1922), where he quickly took to the new subject and began studying organomagnesium compounds, the subject of his PhD thesis in 1901. In 1905, he began lecturing in chemistry at Besançon and a year later he moved to Lyon. He was placed in charge of the organic chemistry courses at the University of Nancy in 1909, becoming a professor there in 1910. During World War I Grignard headed a department at the Sorbonne in Paris, concerned with the development of chemical warfare, going on a scientific trip to the United States in 1917. He succeeded Barbier as Professor of Chemistry at the University of Lyon in 1919, where he remained for the rest of his life. He died in Lyon on 13 December 1935.

Grignard's scientific career can be divided into three main periods. Up to 1914 he was concerned with the discovery, modifications and applications of organomagnesium compounds (Grignard reagents). During the years 1914 to 1918 he was engaged in research geared to the war effort, and afterwards he carried out a variety of researches in organic chemistry and concerned himself with chemistry education.

The general formula of a Grignard reagent is RMgX, where R is an alkyl radical, Mg is magnesium, and X is a halogen – usually chlorine, bromine or iodine. It is used to add an alkyl group (–R) to various organic molecules. Grignard made his first reagents in about 1900 by treating magnesium shavings with an alkyl halide (halogenoalkane) in anhydrous ether (ethoxyethane). The discovery made a dramatic impact on synthetic organic chemistry; by 1905, 200 publications had appeared on the topic of Grignard reagents and within another three years the total had risen to 500.

Grignard reagents added to formaldehyde (methanal) produce a primary alcohol; with any other aldehyde they form secondary alcohols, and added to ketones give rise to tertiary alcohols. They will also add to a carboxylic acid to produce first a ketone and ultimately a tertiary alcohol.

The reagents have many industrial applications, such as the production of tetraethyl lead (from ethyl magnesium bromide and lead chloride) for use as an antiknock compound in petrol.

During World War I Grignard worked on the production of the poisonous gas phosgene (carbonyl chloride) and on methods for the rapid detection of the presence of mustard gas. After the war he started compiling his great *Traité de chimie organique*, a multi-volume work that began publication in 1935. He continued research, investigating the organic compounds of aluminium and mercury; he also studied the terpenes.

Grimaldi Francesco Maria 1618–1663 was an Italian physicist who discovered the diffraction of light and made various observations in physiology.

Grimaldi was born in Bologna on 2 April 1618 and at the age of 14 entered the order of the Society of Jesus, being educated at the Society's houses at Parma, Ferrara and finally, in 1637, back at Bologna. As a Jesuit, his primary studies were in theology, which he finished in 1645. Three years later he became Professor of Mathematics at the Jesuit College in Bologna. There he acted also as assistant to the Professor of Astronomy, Father Giovanni Riccioli (1598–1671), whom he helped with astronomical observations. Grimaldi remained at the Jesuit College in Bologna, and died there on 28 December 1663.

Grimaldi observed muscle action and was the first to note that minute sounds are produced by muscles during contraction. But he is best known for his experiments with light. He let a beam of sunlight enter a darkened room through a small circular aperture and observed that, when the beam passed through a second aperture and onto a screen, the spot of light was slightly larger than the second aperture and had coloured fringes. Grimaldi concluded that the light rays had diverged slightly, becoming bent outwards or diffracted. On placing a narrow obstruction in the light beam, he noticed bright bands at each side of its shadow on the screen. He therefore added diffraction to reflection and refraction, the other known phenomena that 'bend' light rays. In modern terms, Grimaldi had observed a phenomenon that can be explained readily only if light is regarded as travelling in waves – contrary to the then accepted corpuscular theory of light. In 1665, two years after Grimaldi's death, a description of the experiments was published in the large work *Physicomathesis de Lumine, Coloribus, et Iride/Physicomathematical Thesis of Light, Colours and the Rainbow.*

Guettard Jean-Etienne 1715–1786 was a French naturalist who pioneered geological mapping.

Born at Etampes, Guettard rose to become keeper of the natural history collection of the Duc d'Orléans. He displayed a deep interest in many departments of botany and medicine, but he is remembered for his geological work. Extensive research in the field suggested that the rocks of the Auvergne district of central France were volcanic in nature, and Guettard boldly identified several peaks in the area as extinct volcanoes. Though, as a good Baconian empiricist, he later had doubts about this hypothesis, Guettard's studies proved a notable stimulus for later investigations, notably those of Desmarest.

In further work, Guettard played an important part in stimulating the debate on the origin of basalt. Originally he had taken the view that columnar basalt was not volcanic in origin. Visits to Italy in the 1770s, however, led him to modify his earlier views.

Guettard was a humble man who consistently avoided wider speculations regarding the implications of his views for such issues as the age of the Earth. He was also an innovator in geological cartography. In 1746, he presented his first mineralogical map of France to the Académie des Sciences; and in 1766, he and Lavoisier were commissioned to prepare a geological survey of France, though only a fraction was ever completed.

Guillemin Roger Charles Louis 1924– is a French-born US endocrinologist. His research has included the isolation and identification of various hormones, for which he received the 1977 Nobel Prize for Physiology or Medicine, together with his co-worker Andrew Schally (1926–) and the US physicist Rosalyn Yalow (1921–). Yalow developed radioimmunoassays of insulin, a technique that allows the analysis of minute amounts of substances such as hormones.

Guillemin was born in Dijon on 11 January 1924. In 1941, he obtained a BA from the University of Dijon, and the following year a BSc. He read medicine at the University of Lyons and gained his medical degree in 1949. During the following two years he was a resident intern at the University Hospital in Dijon, and then took up a professorial appointment at the Institute of Experimental Medicine and Surgery, Montreal, from which he gained his PhD in 1953. That year, Guillemin went to Baylor College of Medicine, Houston, Texas, and also became a naturalized US citizen. From 1960 to 1963 he was an Associate Director in the Department of Experimental Endocrinology at the Collège de France, Paris. He returned to Baylor College and worked there until 1970 when he joined the Salk Institute in La Jolla, California.

While at the Baylor College, Guillemin attempted to prove the theory put forward by the British anatomist Geoffrey Harris that hormones secreted from the hypothalamus regulate those produced by the pituitary gland. Guillemin found that the brain controls the pituitary gland by means of hormones produced by central neurons – the neurosecretory cells of the hypothalamus.

Guillemin was joined in his studies by the Lithuanian refugee Andrew Schally, who went to Baylor College in 1957. They tried to isolate the substance that regulates the secretion of the adrenocorticotrophic hormone (ACTH), which is produced by the pituitary gland, but they were defeated.

Schally left Baylor College in 1962 and went to New Orleans but both he and Guillemin turned their attention to the isolation of other hormones. In parallel investigations between 1968 and 1973 they isolated and synthesized three hypothalamic hormones which regulate the secretion of the anterior pituitary gland. The first of these was the tripeptide TRH (thyrotropin-releasing hormone), isolated in 1968. This hormone in turn induces the pituitary gland to secrete TSH (thyroid-stimulating hormone). After months of hard work they finally extracted 1 mg/0.000035 oz of the hormone, which they found to be a simple compound and easily synthesized. In 1971, Schally announced his discovery of LHRH (luteinizing-hormone-releasing hormone), which regulates the release in the pituitary gland of the luteinizing and follicle-stimulating hormones that control ovulation in women. Two years later, working at the Salk Institute, Guillemin discovered somatostatin, which inhibits the secretion of the growth hormone somatotropin, and of insulin and gastrin.

Guillemin continued his work at the Salk Institute, and in 1979 investigated the possible presence in the pituitary gland of opiate-like peptides. The peptide structure discovered is called β-endorphin and represents the opiate-like effects of all pituitary extracts.

The discoveries of Guillemin and Schally are highly relevant to endocrinologists. Synthetic TRH is now used to treat conditions connected with deficiencies in the secretion of pituitary hormones; and somatostatin is under investigation for its possible use in treating diabetics, and for curing peptic ulcers.

Gurdon John Bertrand 1933– is a British molecular biologist who is probably best known for his work on nuclear transplantation.

Gurdon was born on 2 October 1933 in Dippenhall, Hampshire, and was educated at Eton College and then Christ Church College, Oxford, from which he graduated in zoology in 1956. From 1956 to 1960 he studied for his doctorate (in embryology) in the Zoology Department at Oxford University, then from 1960 to 1962 he worked at the California Institute of Technology as a Gosney Research Fellow. He then returned to Britain, and from 1963 to 1964 was a departmental demonstrator in Oxford University's Zoology Department. Also in 1963, he was appointed a Research Fellow of Christ Church College. He became a lecturer in Oxford's Zoology Department in 1965, remaining there until 1972, when he was made a staff member of the Laboratory of Molecular Biology at Cambridge; seven years later he was appointed Head of the Cell Biology Division.

Gurdon has worked mostly on the effects of transplanting nuclei and nuclear constituents, such as DNA, into enucleated eggs, mainly frogs' eggs. Using nuclei from somatic cells, he showed how the genetic activity changes in the recipient eggs; for example, after transplanting nuclei from differentiated somatic cells, Gurdon discovered that the somatic nuclei come to resemble the zygote nuclei of fertilized eggs in both structure and metabolism. From this and other research, he concluded that the changes in gene activity induced by nuclear transplantation are indistinguishable from those that occur in normal early development. He also demonstrated how nuclear transplantation and micro-injection techniques can be used to elucidate the intracellular movements of proteins, and has investigated the effects of known protein fractions on gene activity.

Gutenberg Johann c.1398–1468 was a German printer who invented moveable type, often regarded as one of the most significant technical developments of all time.

Gutenberg was born in Mainz, a small town on the river Rhine. His exact birthdate is not known, although it is believed to have been between 1394 and 1399. He served an apprenticeship as a goldsmith, but his interest soon turned to printing. In partnership with his friend Andreas Dritzehn he set up a printing firm in Strasbourg some time in the late 1430s. Dritzehn died soon afterwards, and his brother sued for admission to the partnership in his place. Gutenberg was exposed to legal proceedings and from evidence given during the hearing it is thought that the two men may have already invented moveable type by 1438, although no printed work has survived to substantiate this assumption.

Gutenberg returned to Mainz and persuaded the goldsmith and financier Johann Fust to lend him the money to set up a new press. With the security of financial backing he produced the so-called Gutenberg Bible, now regarded as the first major work to come from a printing press. His security was shortlived, however, when Fust brought a lawsuit to recover his loan. In November 1455, Gutenberg's printing offices were taken over in lieu of payment, and Fust installed his son-in-law Peter Schoeffer to operate the press.

Gutenberg may or may not have set up another press. Certainly an edition of Johann Balbus' *Catholicon* printed in Mainz in 1460 is often attributed to him, along with other lesser works. In 1462, Mainz was involved in a local feud, and in the upheaval Gutenberg was expelled from the city for five years before being reinstated, offered a pension and given tax exemption. He died there on 3 February 1468.

The art of printing using moveable type is thought to have been originally invented nearly four centuries earlier in China, then a much more advanced civilization, but was unknown in the western world until the work of Gutenberg. Previously books had to be printed by the laborious method of carving out each page individually on a wood block, and this meant that only the very wealthy were able to afford them. Gutenberg made his type individually, so that each letter was interchangeable. After a page had been printed, the type could be disassembled and used again. He punched and engraved a steel character (letter shape) into a piece of copper to form a mould which he then filled with molten type metal. The letters were in the Gothic style, nearest to the handwriting of the day, and of equal height.

Using paper made from cloth rags and vellum sheets he printed the famous '42-line' (42 lines to a column) Gutenberg Bible, thought to be the first major work to come from a press of this kind. There are 47 surviving copies of the book, of which 12 are printed on vellum. Many bear no printer's name or date, but from evidence found in two copies in the Paris Library it is concluded that they were on sale before August 1456. By 1500, more than 180 European towns had working presses of this kind, including William Caxton's in Westminster, London, set up in 1476.

H

Haber Fritz 1868–1934 was a German chemist who made contributions to physical chemistry and electrochemistry, but who is best remembered for the Haber process, a method of synthesizing ammonia by the direct catalytic combination of nitrogen and hydrogen. For this outstanding achievement he was awarded the 1918 Nobel Prize for Chemistry (presented in 1919).

Haber was born on 9 December 1868 in Bresslau, Silesia (now Wroclaw, Poland), the son of a dye manufacturer. He was educated at the local Gymnasium and at the universities of Berlin (under August von Hofmann) and Heidelberg (where he was a student of Robert Bunsen). He completed his undergraduate studies at the Technische Hochschule at Charlottenburg, where he carried out his first research in organic chemistry. After spending some time working for three chemical companies, he briefly entered his father's business, but after only a few months Haber returned to organic chemistry research at Jena.

He then obtained a junior post at Karlsruhe Technische Hochschule, becoming an unpaid lecturer there in 1896. He held the professorship of Physical Chemistry and Electrochemistry at Karlsruhe from 1906 until 1911, by which time his reputation had grown to such an extent that he was made Director of the newly established Kaiser Wilhelm Institute for Physical Chemistry at Berlin-Dahlem. During World War I he placed the Institute's resources in the hands of the government and, being a staunch patriot, was bitterly disappointed by Germany's defeat in 1918. Then when Adolf Hitler rose to power in 1933 even Haber's genius and patriotism could not compensate for his Jewish ancestry and he was forced to seek exile in Britain, where he worked for a few months at the Cavendish Laboratory, Cambridge. Shortly afterwards, while on holiday in Switzerland, he died of a heart attack in Basle on 29 January 1934.

During the late 1890s Haber carried out important researches in thermodynamics and electrochemistry, studying particularly electrode potentials and the processes that occur at electrodes. He was an early, but largely unsuccessful, pioneer of fuel cells in which the electric current was produced by the atmospheric oxidation of carbon or carbon monoxide. He also worked on the electrodeposition of iron, which had applications in the making of plates for printing banknotes. His work during this period culminated with the publication of two influential books: *The Theoretical Bases of Technical Electrochemistry* (1898) and *The Thermodynamics of Technical Gas Reactions* (1905).

Haber used his extensive knowledge and experience of thermodynamics in his investigations of the synthesis of ammonia by the catalytic hydrogenation of atmospheric nitrogen. Although from 1904 he made extensive free energy calculations and performed many experiments, it was not until 1909 that the process was ready for industrial development under Karl Bosch (1874–1940). Haber's early successful experiments used osmium or uranium catalysts at a temperature of 550°C/1,022°F (823K) and 150–200 atmospheres pressure ($2.2–2.9 \times 10^3$ psi/$1.5–2.0 \times 10^7$ Pa). Later industrial processes used higher pressures and slightly lower temperatures, with finely divided iron as the catalyst. The nitrogen from air was compressed with producer gas (the source of hydrogen) and the ammonia formed was removed as a solution in water. In this version it is also known as the Haber–Bosch process.

At Karlsruhe Haber continued his work in electrochemistry and in 1909 constructed the first glass electrode (which measures hydrogen ion concentration – acidity, or pH – by monitoring the electrode potential across a piece of thin glass). One of his first tasks after the outbreak of World War I in 1914 was to devise a method of producing nitric acid for making high explosives, using ammonia from the Haber process (to make blockaded Germany independent of supplies of Chile saltpetre – sodium nitrate – from South America). Later he became involved in the gas offensive and superintended the release of chlorine into the Allied trenches at Ypres. When the French later retaliated by using phosgene (carbonyl chloride, whose production was in the hands of François Grignard), Haber was given control of the German Chemical Warfare Service, with the rank of captain, and organized the manufacture of lethal gases. He was also responsible for the training of personnel and developed an effective gas mask.

After the war Haber set himself the task of extracting gold from sea water to help to pay off the debt demanded by the Allies in reparation for the damage caused during the hostilities. Svante Arrhenius had calculated that the sea contains 8,000 million tonnes of gold. The project got as far as the fitting out of a ship and the commencement of the extraction process, but the yields were too low and the project was abandoned in 1928.

Hadamard Jacques 1865–1963 was a French mathematician whose contributions to the subject ranged over so wide a field that he gained recognition as one of the outstanding mathematicians of modern times.

Hadamard was born at Versailles on 8 December 1865 into a family well able to perceive and to encourage his rare intellectual gifts. His father taught Latin at a lycée in Paris and his mother was an accomplished pianist and music teacher. Hadamard spent his undergraduate years at the Ecole Normale Supérieure in Paris, and after four years there gained his BA in mathematics in 1888. In 1890, while he was working on his doctoral thesis on function theory, he began to teach at the Lycée Buffon. He received his doctorate in 1892 and in the following year took up a teaching appointment at Bordeaux. In the next few years he was also an occasional lecturer at the Sorbonne.

In 1909, Hadamard was appointed Professor of Mathematics at the Collège de France in Paris, a Chair which he held, together with subsequent appointments at the Ecole Polytechnique and the Ecole Centrale, until 1937. In order to escape from the disruptions of the German occupation of France during World War II, he went to the United States in 1941, then to London, where he took part in operation research for the Royal Air Force. In 1945, he returned to France, where he lived in retirement – cultivating his tastes as an amateur musician (Einstein once played in an orchestra he assembled) and as a collector of ferns and fungi. He died in Paris on 17 October 1963.

Rare are the mathematicians who have left a heritage as rich as Hadamard did; few are the fields of mathematics which have not been touched by his influence. His first major papers, those written in the mid-1890s and growing out of his doctoral thesis, dealt with the nature of analytic functions – that is, functions which can be developed as power series that converge. Hadamard created the theory of the detection and nature of singularities in the analytic continuation of a Taylor series. He began by giving a proof of the so-called Cauchy test for convergence of a power series ($\Sigma a_n x^n$), and included the definition of an upper limit from first principles. By applying this method to Taylor series he arrived at results that are now standard textbook illustrations for the study of these series.

At about the same time he began to study the Riemann zeta function and in 1896 solved the old and famous problem relating to prime numbers of determining the number of primes less than a given number x. Hadamard was able to demonstrate that this number was asymptotically equal to $x/\log x$, which was the most important single result ever obtained in number theory.

Hadamard also became interested in the work of his friend, Vito Volterra (1860–1940), on the 'functions of lines' – numerical functions that depend upon a curve or an ordinary function as their variable. In an ordinary function such as $y = f(x)$, the variable, x, is a simple number. In the kinds of functions on which Volterra was working – Hadamard named them 'functionals' – one might have, to take a simple example, $y = A(c)$, where A is the area of a closed curve, c. By asking a bold question of these functions, or functionals, Hadamard created a new branch of mathematics. He asked whether it might be possible to extend the theory of ordinary functions to the case where the variable, or variables, would no longer be a number, or numbers. This required a redefinition, or at least a new generalization, of many concepts: continuity, derivative and differential among them. In seeking to provide this new approach Hadamard gave birth to functional analysis, one of the most fertile branches of modern mathematics.

By extension from this work Hadamard came to investigate functions of a complex variable and, in so doing, to define a singularity as a point at which a function ceases to be regular. He was able to show, however, that the existence of a set of singular points may be compatible with the continuity of a function. He named the region formed by such a set a 'lacunary space' and the study of such spaces has occupied mathematicians ever since.

Finally, there is Hadamard's famous concept of 'the problem correctly posed'. Since it is often helpful, or necessary, to find an approximate solution (in physics, for example), a correctly posed problem, according to Hadamard, is one for which a solution exists that is unique for given data, but which also depends continuously on the data. This is the case when the solution can be expressed as a set of convergent power series. The idea has proved to be far more fruitful than Hadamard was able to envisage and has led mathematical analysts to consider different types of neighbourhood and continuity and so been fundamental to the development of the theory of function spaces, functional analysis.

Hadamard was a prolific mathematician. In his lifetime he published more than 300 papers on a great variety of topics. And few mathematicians influenced the growth of mathematics in so many different directions as he did.

Hadfield Robert Abbott 1858–1940 was a British industrial chemist and metallurgist who invented stainless steel and developed various other ferrous alloys.

Hadfield was born in Sheffield, Yorkshire, on 29 November 1858; his father owned a small steel foundry. He received a good education and was trained locally as a chemist before taking up an appointment as an analyst. After a while he joined his father's firm (founded in 1872) and by the age of 24 was its manager. His father died in 1888 and Hadfield became chairman and managing director. He also continued his research into various steel alloys, which he had begun in the early 1880s, eventually publishing more than 150 scientific papers. His findings proved productive and commercially successful, and in recognition of his contribution to the British steel industry he was knighted in 1908: elected a Fellow of the Royal Society a year later, and created a Baronet in 1917. He died in London on 30 September 1940, leaving the well-known company of Hadfields in Sheffield as reminder of the family name.

The basic ingredient of steel is pig iron, the product of iron smelting in a blast furnace, which contains 2.5–4.5% carbon. In making ordinary mild steel, carbon (and other impurities) are oxidized, to lower the carbon content. Hadfield carried out many experiments in which he mixed other metals to the steel. He found, for example, that a small amount of manganese gave a tough, wear-resistant steel suitable for such applications as railway track and grinding machinery. By adding nickel and chromium he produced corrosion-resistant stainless steels. Within 20 years silicon steels were available. Again the basic metal with the addition of 3–4% silicon produced an ideal metal for making transformer cones, for which high permeability is required. Today steels and other alloys can be tailor-made to suit almost any requirements.

Haeckel Ernst Heinrich 1834–1919 was a German zoologist well-known for his genealogical trees of living organisms and for his early support of Darwin's ideas on evolution.

Haeckel was born in Potsdam, Prussia (now East Germany), on 16 February 1834. His father was a lawyer in Merseburg, where Ernst Haeckel was educated. He studied medicine (although his main interest was botany) at the University of Würzburg and obtained his degree from the University of Berlin in 1857. There he was taught by Johannes Müller (1801–1858), who interested him greatly in zoology; he also studied under Rudolph Virchow (1821–1902). He travelled through Italy and then, after practising medicine for a year, became a lecturer at the University of Jena in 1861. The following year he was appointed Extraordinary Professor of Comparative Anatomy at the Zoological Institute in Jena. He founded the Phyletic Museum in Jena and the Ernst Haeckel Haus, which contains his archives as well as many personal mementos. In 1865, he took up the full professorship at Jena, a position he retained until he retired in 1909. He died in Jena on 8 August 1919.

In 1866, Haeckel met Charles Darwin and was completely convinced by his theory of evolution. He went further and, using Darwin's research, developed Haeckel's law of recapitulation, which concerns resemblances between the ontogeny (the development from fertilized egg to adult) of different animals. According to the law of recapitulation, during its ontogeny an individual goes through a series of stages similar to the evolutionary stages of its adult ancestors, which show characteristics of less highly evolved animals. In this way ontogeny recapitulates phylogeny (the developmental history of a species). An example of the evidence on which Haeckel founded this view is the series of gill pouches found both in birds and mammals at the embryo stage. These gill pouches are not present in adult mammals, although the slits are present in full-grown birds and fish from which the embryo forms were descended. Haeckel's theory was thought to make ontogeny relevant to evolution simply because he believed that it should be possible to find out, by studying the development of an individual, what its adult ancestors were like. The concept that Haeckel revived had already been refuted by Karl von Baer, who had shown that embryos resemble the embryos only and not the adults of other species, but Haeckel's claims remained popular to the end of the nineteenth century.

In the same year that he met Darwin, Haeckel introduced a method of representing evolutionary history, or phylogeny, by means of tree-like diagrams. His method is still used today by animal systematists to show degrees of presumed relationship in the various groups and can be traced in present modified zoological classifications.

Haeckel also tried to apply Darwin's doctrine of evolution to philosophy and religion. He believed that just as the higher animals have evolved from the simpler forms of life, so the highest human faculties have evolved from the soul of animals. He denied the immortality of the soul, the freedom of the will and the existence of a personal God. He now occupies no serious place in the history of philosophy, although he was widely read in his own day.

Haeckel held some ideas that are still accepted, one of them being his view that the origin of life lies in the chemical and physical factors of the environment, such as sunlight, oxygen, water and methane. This theory has recently, as a result of laboratory experiments, been shown to be likely. He also believed that the simplest forms of life were developed by a form of crystallization. A further influence Haeckel had on science was the coining of the word 'ecology', to mean the study of living organisms in relation to one another and to the inanimate environment. As a field naturalist, Haeckel was a man of extraordinary energy, and he gave much of interest to biology, even if the theory was tenuous and was typical of the extreme evolutionists of his era.

Hahn Otto 1879–1968 was a German radiochemist who discovered nuclear fission. For this achievement he was awarded the 1944 Nobel Prize for Chemistry.

Hahn was born on 8 March 1879 in Frankfurt-am-Main and studied at the University of Marburg, gaining his doctorate at Marburg in 1901. From 1904 to 1905 he worked in London under William Ramsay (1852–1916), who introduced Hahn to radiochemistry. In 1905, he moved to McGill University at Montreal, Canada, to further these studies with Ernest Rutherford (1871–1937). The next year Hahn returned to Germany to work under Emil Fischer (1852–1919) at the University of Berlin. There he was joined in 1907 by Lise Meitner (1878–1968), beginning a collaboration that was to last more than 30 years. In 1912 Hahn was made a member of the Kaiser-Wilhelm Institute for Chemistry in Berlin-Dahlem and in 1928 was appointed its Director. He held this position until 1944, when the Max Planck Institute in Göttingen took over the functions of the Kaiser-Wilhelm Institute and Hahn became President.

Perturbed by the military uses to which nuclear fission had been put, Hahn was concerned in his later years to warn the world of the great dangers of nuclear weapons. In 1966, he was co-winner of the Enrico Fermi Award with Meitner and Fritz Strassmann (1902–   ). Hahn died in Göttingen on 28 July 1968.

Hahn's first important piece of research was his discovery of some of the intermediate radioactive isotopes formed in the radioactive breakdown of thorium. These were radiothorium (discovered in 1904) and mesothorium (1907). In 1918, Hahn and Meitner discovered the longest-lived isotope of a new element which they called protactinium, and in 1921 they discovered nuclear isomers – radioisotopes with nuclei containing the same subatomic particles but differing in energy content and half-life.

In the 1930s, Enrico Fermi (1901–1954) and his collaborators showed that nearly every element in the periodic table may undergo a nuclear transformation when bombarded by neutrons. In many cases, radioactive isotopes of the elements are formed and slow neutrons were found to be particularly effective in producing many of these transformations. In 1934, Fermi found that uranium is among the elements in which neutron bombardment induces transformations. The identity of the products was uncertain, but it was suspected that artificial elements higher than uranium had been found.

Hahn, Meitner and Strassmann investigated Fermi's work, believing the formation of radium to have occurred, and treated the bombarded uranium with barium. Radium is chemically very similar to barium and would accompany barium in any chemical reaction, but no radium could be found in the barium-treated fractions. In 1938 Meitner, being Jewish, fled from Germany to escape the Nazis and Hahn and Strassmann carried on the search. Hahn began to wonder if radioactive barium was formed from the uranium when it was bombarded with neutrons, and that this was being carried down by the barium he had added. The atomic number of barium is so much lower than that of uranium, however, that the only way it could possibly have been formed was if the uranium atoms broke in half. The idea of this happening seemed so ridiculous that Hahn hesitated to publish it.

Meitner was now in exile in Copenhagen and with Otto Frisch (1904–1979) quickly confirmed Hahn's explanation, showing the reaction to be an entirely new type of nuclear process and suggesting the name nuclear fission for it. Meitner and Frisch communicated the significance of this discovery to Niels Bohr (1885–1962), who was preparing to visit the United States. Arriving in January 1939, Bohr discussed these results with Albert Einstein (1879–1955) and others. The presence of barium meant that uranium had been split into two nearly equal fragments, a tremendous jump in transmutation over all previous reactions. Calculations showed that such a reaction should yield 10 to 100 times the energy of less violent nuclear reactions. This was quickly confirmed by experiment, and later the same year it was discovered that neutrons emitted in fission could cause a chain reaction to occur.

Fearing that Nazi Germany would put Hahn's discovery to use and build an atomic bomb, Einstein communicated

with the President of the United States and urged that a programme be set up to achieve a nuclear chain reaction. Fermi subsequently built the first atomic reactor in 1942 and the first atomic bomb was exploded in 1945, bringing World War II to a close. Hahn had in fact not worked on the production of nuclear energy for Germany, and the German effort to build a reactor and a bomb had not got anywhere.

Otto Hahn made a discovery of enormous importance for the human race for as well as providing a new and immense source of power, it has given us enormous destructive ability, as he was well aware.

Haldane J(ohn) B(urdon) S(anderson) 1892–1964 was a British-born physiologist famous for his work in physiology and genetics.

Haldane was born in Oxford on 5 November 1892, the son of John Scott Haldane, himself a well-known physiologist. From the early age of eight he was introduced to medicine and assisted his father. He went to Eton and was later educated at New College, Oxford. After gaining a degree in mathematics he did equally well in classics and philosophy. During World War I he served on the Western Front and in Mesopotamia, where he was wounded twice; he returned to study physiology at New College in 1919. Two years later he moved to Cambridge to work under the English biochemist Frederick Gowland Hopkins. In 1933, he took up the genetics chair at University College, London, and later the Chair in Biometry. Between 1927 and 1936 he also held a part-time appointment at the John Innes Horticultural Institution at Merton, where he carried on the work of the previous director William Bateson (1861–1926). He was an outspoken Marxist during the 1930s and served for a time as chairman of the editorial board of the London *Daily Worker*. He worked for the Admiralty in World War II and left the Communist Party disappointed by the fame awarded to the Soviet biologist Trofim Lysenko. He emigrated to India in 1957 in protest at the Anglo-French invasion of Suez and was appointed Director of the Genetics and Biometry Laboratory in Orissa, which had excellent facilities. He became a naturalized Indian citizen in 1961, and died of cancer at Bhubaneshwar on 1 December 1964.

Haldane's interest in genetics was first aroused by a lecture in 1901 on the recently discovered work of Gregor Mendel, and in 1910 he began to study the laws of inheritance as shown by his sister's 300 guinea pigs. Some years later he published a paper on gene linkage in vertebrates. In 1922, he formulated Haldane's law and three years later investigated the variation of gene linkage with age, and published a paper on the mathematics of natural selection. He was convinced that natural selection and not mutation is the driving force behind evolution. In 1932, he estimated for the first time the rate of mutation of the human gene and worked out the effect of recurrent harmful mutations on a population. While he was at University College he continued his work on human genetics and in 1936 showed the genetic link between haemophilia and colour blindness.

While Haldane was still at school he helped his father

with research on the physiology of breathing and aspects of respiration concerned with deep-sea diving and safety in mines. Haldane's interest in respiration led him, during World War I, to work with his father yet again, on the improvization of gas masks. After the war, at Oxford, he investigated how carbon dioxide in the bloodstream of human beings enables the muscles to regulate breathing under different conditions. He and his colleague, Peter Davies, consumed quantities of sodium bicarbonate and introduced hydrochloric acid into their blood by drinking ammonium chloride. They also experimented with changes in sugar and phosphate concentration in the blood and urine. During World War II, in 1942, Haldane and a friend spent two days in a submarine to test an air-purifying system. He also simulated conditions inside submarines and subjected himself to extremes of temperature and a concentration of carbon dioxide in the air.

In 1924, having been introduced to enzyme reactions by Hopkins, Haldane produced the first proof that they obey the laws of thermodynamics. In 1930, Haldane published *Enzymes*, which provided an overall picture of how enzymes work.

Haldane is most remembered as a geneticist and a proponent of the unity of the sciences. His papers, lectures and broadcasts made him one of the world's best-known scientists.

Haldane John Scott 1860–1936 was a British physiologist well known for his investigations into the safety measures necessary in stressful working conditions such as coal-mining and deep-sea diving, and for his studies on the exchange of gases during respiration.

Haldane was born in Edinburgh on 3 May 1860, the son of a lawyer and a member of the Cloan branch of the ancient Haldane family of Gleneagles. He was educated at the Edinburgh Academy, and graduated in medicine from Edinburgh University in 1884. He then spent short periods of time at the universities of Jena and Berlin, and was a demonstrator in physiology at the University of Dundee. In 1887, he became a demonstrator in physiology at Oxford University, and from 1907 until he resigned in 1913 was a reader there. He then became Director of the Mining Research Laboratory, Doncaster (which moved to Birmingham in 1921), a position he held until 1928. Haldane lectured at Yale in 1916, at Glasgow University from 1927 to 1929, and at Dublin University in the following year. He was elected to the Royal Society in 1897 and received its Royal Medal in 1916. He died in Oxford on 15 March 1936.

Haldane was concerned about the artificial differences that then existed between theoretical and applied science and strove to combine the two. His first piece of research was on the composition of the air in houses and schools in Dundee. Using the information he gained doing this research about the carbon dioxide content of inspired air and respiratory volume, he began investigating the health hazards to which coal miners were subjected, particularly suffocation. In a report in 1896 he laid stress on the lethal effects of carbon monoxide, which is usually present in mines after a mine explosion. Haldane then researched into the reasons for the toxicity of carbon monoxide and produced a significant paper in which he showed that this gas binds to the haemoglobin in the blood in preference to oxygen. The full clinical implications of this discovery were not appreciated, however, for another 50 years.

Between 1892 and 1900 Haldane devised methods for studying respiration and the blood. In 1898, he developed in principle the Haldane gas analyser and, a few years later, he invented an apparatus for determining the blood gas content from relatively small amounts of blood – the haemoglobinometer. Both methods are still used, although the blood gas apparatus has largely been superseded.

In 1905, Haldane published his idea that breathing is controlled by the effect of the concentration of carbon dioxide in arterial blood on the respiratory centre of the brain. He showed that, except under extreme conditions, the regulation of breathing depends more on the carbon dioxide content of the inspired air than on its oxygen content. Although this was his most influential work, it received little clinical attention until the late 1940s.

Haldane then attempted to unravel the basic problems of caisson disease (the bends), suffered by divers and other workers who have to breathe compressed air at high pressures. In 1907, he announced a technique of decompression by stages which allows a deep-sea diver to rise to the surface safely; it is still used today. In 1911, Haldane undertook an expedition to Pikes Peak, Colorado, with several US physiologists, to study the physiological effects of low barometric pressures.

Haldane's studies of haemoglobin dissociation demonstrated the degree to which oxygenation of haemoglobin affects the uptake of carbon dioxide in the body tissues and its release in the lungs. He also researched the reaction of the kidneys to the water content of the blood, and the physiology of sweating.

The application of many of Haldane's findings contributed much to the safety and health of miners and divers. He also had a great interest in philosophical topics and wrote extensively on the relationship between science and philosophy. In the 25 years he was at Oxford he did more than anyone else to bring the Oxford school of physiology into international prominence. His concern was to show that there is a dynamic and constructive equilibrium between theoretical and applied science.

Hale George Ellery 1868–1938 was one of the finest US astronomers. Much of his life, however, was spent organizing and arranging funds to be made available for the construction of large telescopes, including the 1.5-m/60-in and 2.5-m/100-in reflectors at Mount Wilson, and the 5-m/200-in Mount Palomar telescope. Principally he was an astrophysicist, and he distinguished himself in the study of solar spectra and sunspots. He developed a number of important instruments for the study of solar and stellar spectra, including the spectroheliograph and the spectrohelioscope.

Hale was born in Chicago on 29 June 1868, the son of a man of considerable means. He began his education at the Oakland Public School and continued at Adam Academy, Chicago, before going on to the Massachusetts Institute of Technology to study physics, chemistry and mathematics.

In 1892, Hale was appointed Professor of Astronomy at the University of Chicago, and at once set about the establishment of a new observatory. He persuaded C. T. Yerkes, a Chicago industrialist, to donate a large sum of money which was to be used to build the 1-m/40-in Yerkes refractor telescope. When this was completed in 1897, Hale moved in his astrophysical instruments from his own laboratory at Kenwood.

For a while in 1893, he worked with the physicists Helmholtz, Planck and Kundt at the University of Berlin. However, he eventually abandoned his plans to take a doctorate and never found the time afterwards to work towards this goal. He was, even at this time, a very distinguished man of science and later (1899) he was instrumental in the founding of the American Astronomical Society.

From 1904, Hale was the Director of the Mount Wilson Observatory. Eventually, however, overwork forced him to resign in 1923. Nevertheless, he persisted in pursuit of his goal of a 5-m/200-in telescope. Through his inspiration in the conception of the telescope, and his efforts in securing from the Rockefeller Foundation the vast amount of money necessary for its construction, the 5-m/200-in reflecting telescope on Mount Palomar came into being. He did not see the completion of his greatest dream, for he died in Pasadena on 21 February 1938; but ten years later, when it was completed, the telescope was dedicated to him.

Even when Hale was quite young his interest in astronomy was enormous. Through a friendship with the amateur astronomer Sherburne Burnham (1838–1921) he bought a secondhand 10-cm/4-in refracting telescope. Another friendship with Hough (1836–1909) allowed him to view the heavens, from time to time, through the old 47-cm/18$^1/_2$-in refractor at the University of Chicago. Long before he became a student he constructed spectroscopes. When his father bought him a spectrometer he accurately measured the principal Fraunhofer lines in the solar spectrum. While he was a student he continued his observations and developed the idea of the spectroheliograph – a means of surveying the occurrence of a particular line in the Sun's spectrum. The instrument he constructed did not fit well onto his 10-cm/4-in telescope, so his father bought him a 30-cm/12-in refractor and also equipped a solar laboratory for him. From 1891 to 1895 he used this observatory extensively and made improvements to his instruments.

Hale's work on solar spectra was the stimulus for his construction of a number of specially designed telescopes, the most important of which was the Snow telescope (named after the benefactor, Miss Helen Snow) eventually installed on Mount Wilson. Later Hale was also to seek benefactors to build a 150-cm/60-in reflector (completed in 1908) using the mirror blank his father bought, and later still he managed to persuade J. D. Hooker, a Los Angeles businessman, to finance the construction of the 250-cm/100-in reflector (completed in 1918).

Meanwhile, he surveyed the solar chromosphere as well as the rest of the Sun. In 1905, he obtained photographs of the spectra of sunspots which suggested that sunspots were colder – rather than hotter, as had been suspected – than the surrounding solar surface. Three years later he detected a fine structure in the hydrogen lines of the sunspot spectra. By comparing the split spectral lines of sunspots with those lines produced in the laboratory by intense magnetic fields (exhibiting the Zeeman effect) he showed the presence of very strong magnetic fields associated with sunspots. This was the first discovery of a magnetic field outside the Earth. In 1919, he made another important discovery relating to these magnetic fields by showing that they reverse polarity twice every 22–23 years.

Hale had to resign as Director of the Mount Wilson Observatory because of overwork, but he was not a man to rest, and shortly afterwards he adapted his spectroheliograph to allow an observer to view the spectra with his eye. This adaptation, which created the spectrohelioscope, involved much more than merely the replacement of photographic film by an eyepiece.

As is evident from his creation of numerous telescopes and observatories, Hale was a highly successful organizer. At one time he was elected to the governing body of Throop Polytechnic Institute in Pasadena, and through his influence this initially little known institute with a few hundred students developed into the California Institute of Technology, now famous throughout the world for its research and scholarship.

Hales Stephen 1677–1761 was a British clergyman who devoted his life to the careful investigation of scientific matters. He researched the physiology and growth of plants, in particular transpiration.

Hales was born on 17 September 1677 at Bekesbourne in Kent. He spent 13 years at Cambridge, from 1696 to 1709, studying divinity at what is now Corpus Christi College. His long stay there allowed him to obtain a thorough grounding in all scientific disciplines, as well as to prepare for the priesthood. He was ordained in 1703 and in 1709 took up the post of curate at Teddington, where he remained for the rest of his life. He died there on 4 January 1761.

Hales's experiments on plants took place mainly between 1719 and 1725. In one experiment he cut off a vine at ground level in the spring, before the buds had burst. He attached a glass tube to the cut surface and showed that the sap rose up the tube to a height of 7.6 m/24.9 ft, indicating that the sap was under considerable pressure .

It is now known that this high 'root pressure' establishes a column of water between the roots and the buds in the spring. Once the leaves open, the column extends to the spongy tissues in the leaves, from which water evaporates. The transpirational stream is then set in motion, drawing water from the soil, through the plant and to the atmosphere via the leaves.

Hales also worked on vines in full leaf and was able to

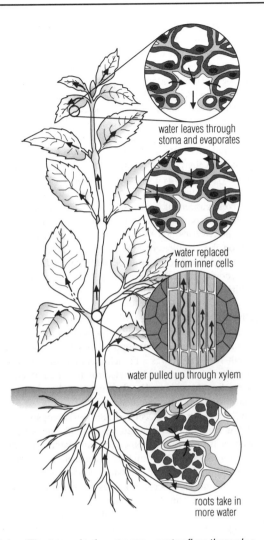

water leaves through
stoma and evaporates

water replaced
from inner cells

water pulled up through xylem

roots take in
more water

Hales *The transpiration stream – water flow through a living plant.*

show that the pressure exerted by water in the transpiration stream is greatest when the plant is illuminated. He demonstrated this phenomenon by attaching his pressure-measuring apparatus to cut side-shoots (or roots) and taking measurements at various times of day. He further showed that water flows in a plant in one direction only, from root to stem to leaf. He used either a straight glass tube to measure the pressure of plant sap or a simple manometer filled with mercury. He calculated in this way the actual velocity of the sap and found that the rate of flow varies in different plants. In investigating the development of plants, Hales was able to show that most of the growth in plant size is made by the younger shoots. On any one section of shoot, he found that it is usually the part of the shoot between the nodes that increases in size the most.

Hales was particularly interested in processes that are common to both plants and animals, such as the nature of growth, and the role of water and air in the maintenance of life. His animal studies involved inserting probes into the veins and arteries of domestic animals and recording the height to which their blood rose up a glass tube connected to the probe, which gave a measure of blood pressure.

He went on to make wax casts of the left ventricles of the hearts of several animals to measure the blood capacities. He also estimated the rate of blood flow through the capillaries by injecting chemicals into various body organs, and watching their movement and the dispersion of the chemicals. He showed that the presence of foreign chemicals can change the rate of flow of blood through an organ and thereby demonstrated (although indirectly) that capillaries are capable of dilation or contraction. As a result of these experiments Hales became convinced of the fallacy of the popular belief that the pressure of the blood itself was powerful enough to effect muscular movement in animals and human beings.

Hales also worked on gases, from both a biological and physical chemical point of view. He made several important discoveries about air but was unable to understand fully the significance of his results because, like his contemporaries, he believed that all gases were a single chemical substance. In his experiments, Hales heated up several materials and noted the change in the volume of air involved in the process, thereby confirming the findings of previous workers that some solids liberate 'air' on heating whereas others absorb it. He devised an apparatus which allowed him to breath solely his own expelled air and found that he could continue to do this for about one minute. If, however, he introduced salt or tartar treated with an alkali, he could use the apparatus for eight-and-a-half minutes. (The alkali mixture absorbed carbon dioxide, thereby delaying for several minutes the response of his lungs to the high carbon dioxide levels.)

Hales's work on air revealed to him the dangers of breathing 'spent' air in enclosed places, and he invented a ventilator for such situations which he had introduced on naval, merchant and slave ships, in hospitals, and in prisons. There was an immediate acknowledgement of the improvement in human health and survival following their introduction, although these favourable reactions foreshadowed an overemphasis on air quality as a factor in disease control in the eighteenth and nineteenth centuries.

Hales carried out many other investigations, such as examining stones taken from the bladder and kidney and suggesting possible chemical solvents for their non-surgical treatment. In the course of his investigations he also invented the surgical forceps.

Hales's research on plants together with his work on the nature and properties of air were published in his famous book *Vegetable Staticks* (1727). This book was subsequently republished in an enlarged form, containing an additional section on his findings on blood and circulation in mammals (in 1733). The enlarged work was entitled *Statical Essays, Containing Haemastaticks, etc.* Hales's emphasis on the need for careful measurement of scientific phenomena may well have had as great an influence on his contemporaries as the results he obtained.

Hall Asaph 1829–1907 was a US astronomer who discovered the two Martian satellites, Deimos and Phobos. He is also noted for his work on satellites of other planets, the rotation of Saturn, the mass of Mars and double stars.

Hall was born in Goshen, Connecticut on 15 October 1829, into a well-established and once prosperous New England family. His father, a clock manufacturer, also called Asaph Hall, died in 1842 and to alleviate the family's financial difficulties the young Asaph became a carpenter's apprentice at the age of 16. However, by 1854 he wanted to continue his education and so he enrolled at the Central College in McGrawville, New York. There he met his future wife, a determined suffragist, Chloe Angeline Stickney, and soon after their marriage in 1856, Hall, who by this time was determined to become an astronomer, took an extremely low-paid job at the Harvard College Observatory in Cambridge, Massachusetts. His status soon rose to that of an observer and being an expert computer of orbits he managed to supplement his meagre income by compiling almanacs and observing culminations of the Moon. In 1862, Hall accepted a post as assistant astronomer at the United States Naval Observatory in Washington and within a year of his arrival he was given a professorship

In 1877, Hall discovered that Mars had two satellites and this achievement won him the Lalande Prize, the Gold Medal of the Royal Astronomical Society in 1879 and the Arago Medal in 1893. He was elected to the National Academy of Sciences in 1875 and served as its Home Secretary for 12 years and its Vice-President for six years. In 1902, he served as President of the American Association for the Advancement of Science and from 1897 to 1907 he was associate editor of the *Astronomical Journal*. Following his retirement from the United States Naval Observatory at the age of 62, he continued to work as an observer in a voluntary capacity until 1898, when he went to Harvard to take up a teaching post in celestial mechanics. After five years of teaching he retired to his rural home in Connecticut where he lived until his death on 22 November 1907.

Hall's first years at the US Naval Observatory were troubled by the unwholesome climate of the American Civil War. He spent this inital period, from 1862 to 1875, as an assistant observer of comets and asteroids. However, in 1875 Hall was given responsibility for the 66-cm/26-in telescope that had been built by Alvan Clark specifically for the United States Naval Observatory. At the time, it was the largest refractor in the world. Hall's first discovery using this telescope was a white spot on the planet Saturn which he used as a marker to ascertain Saturn's rotational period.

In the year 1877, Hall decided to use the superb image-forming qualities of the 66-cm/26-in refractor in a search for possible moons of Mars. It was a particularly good year for such a search because of the unusually close approach of Mars and Hall was guided by the parameters set by his theoretical calculations that indicated the possibility of a Martian satellite being close to the planet itself. Hall's first glimpse of a Martian moon, later named Deimos, occurred on 11 August 1877. By 17 August, Hall had convinced himself that the object was definitely a satellite and on that same day he found the second satellite, Phobos. He disclosed his observations to Simon Newcomb, the Scientific Head of the Observatory, who erroneously believed that Hall failed to recognize that the 'Martian stars' were satellites. Hence Newcomb took the undeserved credit for their discovery in the wide press coverage that followed.

During the following years Hall continued to work on the orbital elements of planetary satellites. In 1884, he showed that the position of the elliptical orbit of Saturn's moon, Hyperion, was retrograding by about 20° per year. In addition to his planetary satellite studies, Hall made numerous investigations of binary star orbits, stellar parallaxes, and the position of the stars in the Pleiades cluster. Owing to his discoveries and to the clarity and precision of his work on the satellites of Mars, Saturn, Uranus and Neptune, which has been compared with that of Bessel, Hall has become generally known as 'the caretaker of the satellites'.

Hall and Héroult Charles Martin Hall 1863–1914 was a US chemist who developed an economical way of making aluminium. His exact contemporary, Paul Louis Toussaint Héroult 1863–1914 was a French metallurgist who simultaneously developed the electrolytic manufacturing process for aluminium. He also invented the electric arc furnace for the production of steels.

Hall was born on 6 December 1863 in Thompson, Ohio, and educated at Oberlin, where he was influenced by F. F. Jewett, a former pupil of the German chemist Friedrich Wöhler (1800–1882) who had succeeded in producing impure aluminium under laboratory conditions.

As a result, Hall became interested in the commercial manufacture of aluminium while still a student, perhaps attracted by the prospect of becoming wealthy and famous by bringing this versatile metal to the industrialized world. Just eight months after graduating in 1855, Hall had found that electrolysis was the best route to achieving his goal, had built his own batteries and, in his home laboratory, had isolated the best compound for commercial production of aluminium: cryolite (sodium aluminium fluoride). By adding aluminium oxides to this compound in its molten state, and using small carbon anodes, a direct current gave a deposit of free molten aluminium on the carbon lining of the electrolytic cell. On 23 February 1886, the 22-year-old Hall was able to present his professor with buttons of aluminium he had himself prepared; these are still preserved by the Aluminum Company of America.

Despite this astonishing achievement, Hall had both financial and technical difficulties in achieving commercial success. A firm which took out a one-year option on the process found that the electrolytic cells failed after just a few days. Financial backing eventually came from the wealthy Mellon family, and the Pittsburgh Reduction Company (later to become the Aluminum Company of America) was formed to build commercially viable units, requiring no external heating during operation.

Meanwhile, in Europe, the French metallurgist Héroult had arrived at the same process and was achieving similar success, the first British plant being based at Patricroft in 1890. Héroult was born at Thury-Harcourt, Normandy, in November 1863. He was still a student when he read Henri DeVille's account of his efforts to isolate aluminium by chemical reactions. Following a year spent at the Paris School of Mines, Héroult began to study the use of electrolysis to extract aluminium from compounds. Like Hall in the United States, Héroult used direct-current electrolysis to find the best combination of material for the production of aluminium, dissolving aluminium oxide in a variety of molten fluorides, and finding cryolite (sodium aluminium flouride) the most promising. Unlike Hall's apparatus, however, Héroult used one large, central graphite electrode in the graphite cell holding the molten material. As with Hall, Héroult succeeded in producing aluminium using his electrolytic appartus at the age of 23. He patented the system in 1886. He also met with the same subsequent difficulty in commercializing the process. The system was taken up, however, by a joint German–Swiss venture at Neuhausen, and large-scale production got underway. Héroult also patented a method for the production of aluminium alloys in 1888.

By 1914, the Hall–Héroult process had brought the price of aluminium down from tens of thousands of dollars a kilogram to 40 cents, in the space of 70 years.

As Hall had been told as a student, the process which he had succeeded in making commercial resulted in his becoming a multimillionaire. On his death at Daytona Beach, Florida, on 27 December 1914, Hall recognized his debt to Oberlin by leaving several million dollars as a gift. Eight months earlier, on 10 April 1914, Héroult had died near Antibes, southern France.

Hall James 1761–1832 was one of the founders of experimental geology.

The son and heir of Sir John Hall of Dunglass, Berwickshire, Hall succeeded to his father's fortune (and baronetcy) in 1776. This enabled him to spend much of the 1780s travelling in Europe. He undertook extensive geological observations in the Alps and in Italy and Sicily, studying Mount Etna. He was also won over to the new chemistry of Lavoisier. Developing a warm friendship with James Hutton, Hall set out to defend Hutton's 'Plutonist' geological theories (the view that heat rather than water was the chief rock-building agent and shaper of the Earth's crust) by providing them with experimental proofs. By means of an innovative series of furnace investigations, he showed with fair success that Hutton had been correct to maintain that igneous rocks would generate crystalline structures if cooled very slowly (a possibility denied by supporters of the Wernerian or Neptunian theory, which attributed stratification to water). Hall also demonstrated that there was a degree of interconvertibility between basaltine and granitic rocks; and that, even though subjected to immense heat, limestone would not decompose if sustained under suitable pressure. All these conclusions helped bolster Hutton's

Plutonian geology and, through being experimental findings, gave it a more modern appearance.

Hall Philip 1904–1982 was a British mathematician who specialized in the study of group theory.

Born on 11 April 1904, in London, Hall attended Christ's Hospital until the age of 18, when he went to King's College, Cambridge. After receiving his degree, he became a Fellow of the College in 1927. Six years later he became a University Lecturer in Mathematics, a post he held until 1951, when he became Reader in Algebra at the University. Hall then became Sadlerian Professor of Pure Mathematics in 1953 and remained as such until his retirement in 1967. As Emeritus Professor, his connections with the University did not cease upon his retirement, and in 1976 he was elected Honorary Fellow of Jesus College.

In addition to his academic duties, Hall was an active participant in professional societies. The London Mathematical Society elected him to its Presidency in 1955, and awarded him the De Morgan Medal and the Lamor Prize in 1965. He was elected a Fellow of the Royal Society in 1942, and awarded its Sylvester Medal in 1961.

Hall's major contributions to mathematics came through his work on group theory. In 1928, he extended some work on group theory done by Sylow in 1872, which led him to a study of prime power groups. From this work he developed his 1933 theory of regular groups. In that paper he also presented material which forms part of the connection between group theory and the study of Lie rings.

An investigation of the conditions under which finite groups are soluble led him in 1937 to postulate a general structure theory for finite soluble groups. In 1954, he published an examination of finitely generated soluble groups in which he demonstrated that they could be divided into two classes of unequal size.

Hall collaborated with G. Higman in 1956 in a study of p-soluble groups; their work generated results that were important to later work on the theory of finite groups, and in particular to the work of J. Thompson and W. Feit.

At the end of the 1950s Hall turned to the subject of simple groups, and later also examined non-strictly simple groups. His researches into the subject of group theory produced many valuable results that have contributed to the further development of this area.

Haller Albrecht von 1708–1777 was a Swiss physiologist, the founder of neurology. By means of skilful investigations into the neuromuscular system he freed it from the remaining myths and superstitions of his day.

Haller was born in Bern on 16 October 1708. A sickly child, he spent most of his time indoors studying. By the age of ten he had compiled a Greek dictionary and written on a number of scholarly topics. He was a medical student of the Dutch physician Hermann Boerhaave (1668–1738) at Leiden, graduating in 1727 and starting his own practice two years later. In 1736, he was appointed Professor of Medicine, Anatomy, Surgery and Botany at the new University of Göttingen, where he remained until he retired in

1753 to write. He died in Bern on 12 December 1777.

During the eighteenth century the idea prevailed that a mysterious force was associated with the nerves, which were believed to be liquid-filled hollow tubes in the body. Haller rejected these notions, especially as they could not be observed experimentally. His own experiments concerning the relationship between nerves and muscles demonstrated the irritability of muscle. He found that if he applied a stimulus to a muscle, the muscle contracted, and that if he stimulated the nerve attached to the muscle there was a stronger contraction. He therefore deduced that it is the stimulation of a nerve that brings about muscular movement.

Haller carried out further experiments to show that tissues are incapable of feeling, but that nerves collect and conduct away impulses produced by a stimulus. Tracing the pathways of nerves, he was able to demonstrate that they always lead to the spinal cord or the brain, suggesting that these regions might be where awareness of sensation and the initiation of answering responses are located. Haller also experimented on the brains of animals and observed the reactions resulting from the damage or stimulation of various parts of the brain.

While carrying out his experiments, Haller discovered several processes of the human body, such as the role of bile in digesting fats. He also wrote a report on his study of embryonic development.

In 1747, Haller published *De Respiratione Experimenta Anatomica/Experiments in the Anatomy of Respiration* and between 1757 and 1766 his eight-volume encyclopedia *Elementa Physiologiae Corporis Humani/The Physiological Elements of the Human Body* was published.

Medicine was just one of Haller's many interests. Botany was another, and he was an avid collector of plants. He devised a botanical classification that equals that of Linnaeus, and wrote a work on the Swiss flora.

Halley Edmond (or Edmund) 1656–1742 was an English mathematician, physicist and astronomer who not only identified the comet later to be known by his name, but also compiled a star catalogue, detected stellar motion using historical records, and began a line of research that – after his death – resulted in a reasonably accurate calculation of the astronomical unit.

Halley was born at Haggerton, near London, on 8 November 1656. The son of a wealthy businessman, he attended St Paul's School, London, and then Oxford University, where he wrote and published a book on the laws of Johannes Kepler (1571–1630) that drew him to the attention of the Astronomer Royal, John Flamsteed (1646–1719). Flamsteed's interest secured for him, despite his leaving Oxford without a degree, the opportunity to begin his scientific career by spending two years on the island of St Helena, charting (none too successfully) the hitherto unmapped stars of the southern hemisphere. The result was the first catalogue of star positions compiled with the use of a telescope. On his return in 1678, Halley was elected to the Royal Society: he was 22 years old. For some years he then

travelled widely in Europe, meeting scientists – particularly astronomers – of international renown, including Johannes Hevelius (1611–1687) and Giovanni Cassini (1625–1712), before finally returning to England to settle down to research. He also became a firm friend of Isaac Newton (1642–1727). It may have been through Newton's influence that Halley, in 1696, took up the post of Deputy Controller of the Mint at Chester. Two years later he accepted command of a Royal Navy warship, and spent considerable time at sea. In 1702 and 1703 he made a couple of diplomatic missions to Vienna before, in the latter year, being appointed Professor of Geometry at Oxford. His study of comets followed immediately. He succeeded Flamsteed as Astronomer Royal in 1720 and held the post until his death, at Greenwich, on 14 January 1742.

In St Helena, Halley first observed and timed a transit of Mercury, realizing as he did so that if a sufficient number of astronomers in different locations round the world also timed their observations and then compared notes, it would be possible to derive the distance both of Mercury and of the Sun. Many years later he prepared extensive notes on procedures to be followed by astronomers observing the expected transit of Venus in 1761. No fewer than 62 observing stations noted the 1761 transit, and from their findings the distance of the Sun from the Earth was calculated to be 153 million km/95 million mi – remarkably accurate for its time (the modern value is 149.6 million km/92.9 million mi).

Astronomy was always Halley's major interest. In the 1680s and 1690s he prepared papers on the nature of trade winds, magnetism, monsoons, the tides, the relationship between height and pressure, evaporation and the salinity of inland waters, the rainbow, and a diving bell; for some of the time he was also helping Newton both practically and financially to formulate his great work, the *Principia*; but these activities were all incidental to the pleasure he took in observing the heavens.

One of Halley's first labours as Professor at Oxford was to make a close study of the nature of comets. Twenty years earlier, the appearance of a comet visible to the naked eye had aroused great popular excitement; yet somehow Newton's *Principia* had ignored the subject. Now Halley, with Newton's assistance, compiled a record of as many comets as possible and charted their progress through the heavens. A major difficulty in determining the paths of comets arose from their being visible for only short periods, leaving by far the greater part of their journey explicable by any number of hypotheses. Some authorities of the time believed that comets travelled in a straight line; some in a parabola (as Newton) or a hyperbola; others suggested an ellipse. In the course of his investigations Halley became convinced that the cometary sightings reported in 1456, 1531, 1607 and 1682, all represented reappearances of the same comet. Halley therefore assumed that such a traveller through space must follow a very elongated orbit around the Sun, taking it at times farther away than the remotest of the planets. On the parabolic path that he calculated it should follow – and making due allowance for deviations from its 'proper' path through the attraction of Jupiter – Halley declared that this comet would appear again in December 1758. When it did, public acclaim for the astronomer (who by that time had been dead 16 years) was such that his name was irrevocably attached to it.

In 1710, Halley began to examine the writings of Ptolemy, the Alexandrian astronomer of the second century AD. Throughout his life Halley was always keenly interested in classical astronomy (he made outstanding translations of the *Conics* of Apollonius and the *Sphaerica* of Menelaus of Alexandria), and having catalogued stars himself, he paid special attention to Ptolemy's stellar catalogue. This was not Ptolemy's original; it was in fact 300 years older than it appeared to be, being a direct borrowing of the list compiled by Hipparchus in the second century BC. For all Ptolemy's shortcomings, however, Halley could not believe that he had been so negligent as to credit the stars with positions wildly at variance with the bearings they now occupied, 15 centuries later. The conclusion that Halley was forced to come to was that the stars had moved. Later Halley was able to detect such movements in the instances of three bright stars: Sirius, Arcturus and Procyon. He was correct, too, in assuming that other stars farther away and consequently dimmer underwent changes of position too small for the naked eye to detect. More than a century had to pass before optical instruments achieved a sophistication sufficient to be able to detect such movement.

Hamilton Alice 1869–1970 was a US physician and social reformer who pioneered the study of industrial toxicology.

She was born in New York State on 27 February 1869 into an affluent family and grew up on the family's estate in Indiana. She was educated at home in an intellectually stimulating family with a strong sense of social responsibility. The failure of her father's provisioning firm meant that she had to earn her living and she decided to study medicine and entered the University of Michigan, graduating in 1893. She then followed a course open to few men, let alone women; she studied bacteriology in Ann Arbor before travelling with her sister for further training in pathology and bacteriology at the Universities of Leipzig and Munich, before returning to the Johns Hopkins Medical School for a final postgraduate year of study. She was then appointed Professor of Pathology at the Woman's Medical College of North Western University. While working in Chicago she tried to combine medical practice and research with social commitment, the latter fostered by her involvement with Hull House – a radical settlement in which she first observed links between environment and disease. She was alerted to the dangers inherent in many industrial practices by an investigative journalist, and she decided to shift her work in this direction. In 1908, she was appointed to the Illinois Commission on Occupational Diseases, and two years later supervised a state-wide survey of industrial poisons, her own researches concentrating on lead, which was extensively used in many industrial processes. By laboriously examining hospital records, factories

and workers themselves, she and her staff identified many hazardous procedures and consequent state legislature introduced safety measures in the workplace and medical examinations for workers at risk. The following year Hamilton was appointed as a special investigator for the United States Bureau of Labor and rapidly became the leading authority on lead poisoning in particular and industrial diseases in general. Her reputation was such that she became the first woman professor at Harvard when she was appointed Assistant Professor of Industrial Medicine, a position she accepted in 1919, almost 30 years before Harvard accepted women as medical students. She continued with her fieldwork for the Department of Labor (as the Bureau had become) and her pioneer book *Industrial Poisons in the United States* (1925) established her reputation worldwide. She retired from Harvard in 1935 to live in Connecticut, but continued to be active in many fields. Her classic textbook *Industrial Toxicology* (1934) was extensively revised in 1949, and she published an autobiography *Exploring the Dangerous Trades* in 1943. She died on 22 September 1970 at the age of 101, much honoured and admired. Her medical work was strongly infused with social concerns and her political beliefs and activities were in turn influenced by many of the conditions she witnessed in the course of her researches. During and after World War I she attended International Congresses of Women and declared herself a pacifist. She was profoundly affected by the conditions of famine she witnessed in postwar Germany and a further visit to Nazi Germany in 1933 disturbed her deeply, but she continued to believe in pacifism until 1940, when she urged US participation in World War II. During the 1940s and 1950s she spoke out on controversial subjects such as contraception, civil liberties and workers rights. Even in the 1960s she was still considered worthy of attention by the Federal Bureau of Investigation when she protested against American involvement in Vietnam.

Hamilton William Rowan 1805–1865 was an Irish mathematician, widely regarded in his time as a 'new Newton', who created a new system of algebra based on quaternions.

Hamilton was born in Dublin, of Scottish parents, on the stroke of midnight, 3/4 August 1805. At about the age of three he was sent by his father, a solicitor, to Trim, where he was raised by his uncle (the local curate) and his aunt. He soon showed a remarkably precocious facility for languages, and made himself expert in Latin, Greek, Hebrew, French, Italian and a number of oriental languages (in which he was encouraged by his father, who wished him to become a clerk with the East India Company). He never lost his interest in languages; throughout his life he wrote poetry and corresponded with Wordsworth, Coleridge and Southey. But at the age of ten Hamilton came upon and read Euclid, and from that moment mathematics became his chief love. By the age of twelve he was reading Newton and when he went to Trinity College, Dublin, to study classics and mathematics, he had mastered the work of the great French mathematicians, including Laplace, and had begun to make experiments in physics and astronomy. In particular, he had begun to investigate the patterns produced by rays of light on reflection and refraction, known as caustics.

At Trinity College Hamilton was far and away the outstanding scholar of his generation. He gained the highest grade of *optime*, not once, which would have been rare enough, but twice (in Greek and in mathematical physics), which was unprecedented. He also twice won the vice-chancellor's gold medal for English verse. In only his second year as an undergraduate he presented a paper on caustics to the Royal Irish Academy in which he showed that light travels by the path of the least action and that its path can be treated as a function of the points through which it travels and expressed mathematically in one 'characteristic function' involving advanced calculus. The paper, which predicted the existence of conical refraction (later proved experimentally by H. Lloyd) was published as 'The Theory of Systems of Rays' by the Academy in 1828.

This early work on caustics so impressed his contemporaries that in 1827, while he was still an undergraduate, Hamilton was appointed Andrews Professor of Astronomy at the university. With the Chair went the title of Royal Astronomer of Ireland and the charge of the Dunsink Observatory. In fact, Hamilton never gave his heart to astronomy; for the rest of his life he devoted himself to mathematical research.

Throughout much of his life Hamilton was closely involved in the work of the British Association for the Advancement of Science, and at a meeting of the association in Dublin in 1835 he was knighted by the Lord-Lieutenant of Ireland. With the knighthood was awarded an annual pension of £200. Many international honours followed, and Hamilton had the distinction of being the first non-American to be elected a fellow of the National Academy of Sciences of the United States. In his latter years Hamilton became something of a recluse, working tirelessly at his home (unfinished meals were found among the piles of paper in his study when he died) and drinking too heavily. He died, of gout, at Dublin on 2 September 1865.

In addition to his work on caustics, Hamilton's most important research was in the field of complex numbers and in the branch of study which he created, the algebra of quaternions. In 1833, he began to seek an improved way of handling complex numbers of the form:

$$a + ib$$

where $i = \sqrt{-1}$ and a and b are real numbers. He thought that, since a and ib were different types of quantity, it was wrong to connect them by a + sign; he also considered $\sqrt{-1}$ to be meaningless. He therefore devised new rules of addition, subtraction, multiplication and division on the basis of considering $a + ib$ as a couple of real numbers represented by (a, b).

He also showed that the sum of two complex numbers could be represented by a parallelogram and that complex numbers could be used, in general, as a useful tool in plane geometry.

From couples Hamilton went on to investigate triples

and this led him to his great work on quaternions. Hamilton first formally announced his definition of quaternions to the Irish Royal Academy in 1844. But it was not until he began to lecture on them in 1848 that his new algebra really began to take shape. The lectures were published in 1853. Hamilton's investigation of triples, with its obvious relevance to three-dimensional geometry, was a much more difficult task than his work on doubles. He found that many mathematical notions had to be sacrificed. Of these by far the most important was the commutative principle. Hamilton found out, during his research, that his quaternions had four components (hence the name), not three, as he had expected. They took the form

$$a + bi + cj + dk$$

where j and k were similar to i. The fundamental formula was:

$$i^2 = j^2 = k^2 = ijk = -1$$

So delighted was Hamilton at this discovery that he scratched the formula on the stonework of a bridge under which he was passing when it came to him. The result of this formula was to reveal that for quaternions the ordinary commutative principle of multiplication (that is $3 \times 4 = 4 \times 3$) did not work, for

$$ij = -ji$$

This was perhaps Hamilton's most important contribution to mathematics, since it forced mathematicians to abandon their belief in the commutative principle as an axiom.

Hamilton hoped that quaternions would find applications in the solution of problems in physics in the way that vectors have proved to do. Indeed, it was Hamilton who coined the word 'vector' and it was he who made it possible to deal with lines in all possible positions and directions and freed them from dependence on Cartesian axes of reference. But although he wrote two large books on the subjects his hopes for quaternions were not fulfilled, and today their usefulness has been superseded, for most problems, by the development of vector and tensor analysis.

Hammick Dalziel Llewellyn 1887–1966 was a British chemist whose major contributions were in the fields of theoretical and synthetic organic chemistry.

Hammick was born in West Norwood, London, the son of a businessman; when he was 12 years old his family moved to the country, which gave him a life-long interest in rural pursuits. He was educated at Whitgift School and then won a scholarship to Magdalen College, Oxford; he graduated in 1908. From Oxford he went to work in the laboratory of Otto Dimroth in Munich, where in 1910 he gained a PhD for his researches on solubility. He returned to Britain and held teaching posts at Gresham's School in Holt, Norfolk, and at Winchester College. In 1921, he returned to Oxford as a Fellow of Oriel College and a lecturer at Corpus Christi College. He remained at Oxford until his retirement; he died there on 17 October 1966.

Hammick's research activities began at Gresham's School, where he investigated the action of sulphur dioxide on various metal oxides. At Oxford he initially continued researching inorganic substances, particularly with regard to their solubilities. He investigated the dimorphism of potassium ethyl sulphate and the ternary system comprising water, sodium sulphite and sodium hydroxide. He also studied sulphur and its compounds (such as carbon disulphide), and suggested structures for liquid sulphur and plastic sulphur (which later workers interpreted in terms of linear polymers).

By the mid-1920s, Hammick's attentions had begun to swing towards organic chemistry. In 1922, he showed that the sublimation of α-trioxymethylene results in the polymer polyoxymethylene; 40 years later this substance was to be used as a commercial polymer. In 1930, Hammick devised a rule to predict the order of substitution in benzene derivatives. Aromaticity and the role of electron attraction and donation of substituents already present in the benzene ring were not yet understood, but Hammick proposed the following useful rule for further substitution of the compound Ph-XY: 'If Y is in a higher group of the periodic table than X or is in the same group (Y is of a lower mass than X), then further substitution will take place meta to the group XY (giving 1,3 di-substitution). In all other cases, substitution goes ortho or para (giving 1,2- or 1,4-di-substitution); the nitroso group is an exception.'

Hammick spent much time investigating organic reaction rates. He studied, for example, the decarboxylation of quinaldinic acid (the alpha-, ortho- or 2-carboxylic acid of quinoline) by aldehydes and ketones. The reaction, now known as the Hammick reaction, is used in the synthesis of larger molecules.

During World War II, when supplies of toluene for making explosives were short, Hammick directed his efforts at syntheses using benzene (obtained from the distillation of coal tar). Using alumina–silica catalysts, he achieved a 30% conversion of benzene to toluene.

Hanbury-Brown Robert 1916–   is a British radio-astronomer who was involved with the early development of radioastronomy techniques and who has since participated in designing a radio interferometer which permits considerably greater resolution in the results provided by radio telescopes.

Hanbury-Brown was born on 31 August 1916 in Aruvankadu, India. After studying engineering at Brighton Polytechnic College (Sussex), he was awarded an external degree by the University of London. He went on to do some postgraduate research at the City and Guilds College before joining a radar research team under the auspices of the Air Ministry in 1936. During World War II Hanbury-Brown took an active part in the radar research programme; a member of the British Air Commission, he also worked at the Naval Research Laboratory, Washington, DC. At the conclusion of the war, he briefly became a private radar consultant, but in 1949 he joined the staff at the Jodrell Bank Observatory in Cheshire, and began to carry out

research into radioastronomy. In 1960 he was made a Fellow of the Royal Society and appointed Professor of Radio-astronomy at the Victoria University, Manchester, where he remained until 1962. He then took up the Chair of Astronomy at the University of Sydney.

Radio waves of cosmic origin were first detected accidentally by Karl Jansky (1905–1950) in 1931 while he was investigating a problem in communications for the Bell Telephone Company. Eighteen years later at Jodrell Bank Observatory, then under the direction of Bernard Lovell (1913–   ), Hanbury-Brown joined a team actively engaged in using radio methods for the investigation of the origin of meteor showers. He became one of the first astronomers to construct a radio map of the sky. Such a map could be compiled using data collected at night or during the day (unlike optical astronomy, which requires clear night-time conditions for observation purposes), and revealed features quite different from those found using optical telescopes.

In addition to the examination of radio emission from structures within the Solar System and our own Galaxy, Hanbury-Brown investigated possible emissions from extragalactic sources. In 1949, with C. Hazard, he detected radio waves emanating from M31, the Andromeda nebula, at a distance of 2.2 million light years. But radio telescopes of the time lacked sufficient resolution to pinpoint a radio source accurately enough to identify that source through an optical telescope. It took three more years for Hanbury-Brown and his colleagues to devise the radio interferometer, which greatly improved resolution. Using the device, Hanbury-Brown measured the size of Cassiopeia A and Cygnus A – both very strong radio sources. Walter Baade (1893–1960) and Rudolph Minkowski (1895–1976) were

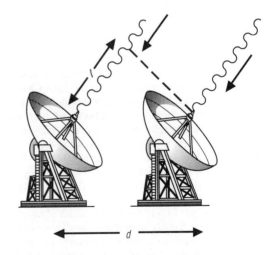

Hanbury-Brown *Hanbury-Brown developed interferometry in radioastronomy, which uses two (or more) radio telescopes receiving waves from the same radio source whose path length differ by l, and which are out of phase, resulting in interference when they are combined. Usually one of the telescopes is mounted on rails so that it can be moved to vary the baseline d.*

then able to relate the more accurate radio locations given to their own optical observations, and as a result Cygnus A became the first radio source traced to a definite optical identification – even though it had a magnitude (brightness) of only 17.9.

In 1956, Hanbury-Brown devised a further refinement to radioastronomy, in the form of the technique of intensity interferometry. Since then he has used the stellar interferometer at Narrabi Observatory (in Australia) to study the sizes of hotter stars.

The early work carried out by Hanbury-Brown at Jodrell Bank contributed to the development of the 76-m/249-ft radio telescope, for a long time the largest steerable radio telescope in the world. Other types of radio telescopes have been developed since then, such as the 300-m/1,000-ft dish at Arecibo, in Puerto Rico, the 5-km/3-mi radio telescope at Cambridge (consisting of eight-in-line 13-m/43-ft dishes), and the VLA (Very Large Array) in the United States.

Hancock Thomas 1786–1865 was a British inventor who was influential in the development of the rubber industry.

Hancock, the son of a timber merchant and brother of Walter Hancock, one of the pioneers of steam road-carriages was born in Marlborough, Wiltshire. Both brothers were enterprising in their own particular fields. Thomas travelled to London in 1820 and took out a patent for applying caoutchouc, (now known as india rubber) to various articles of dress. Having secured the patent, he opened a factory in Goswell Road, London, and began manufacturing on a large scale, assisted at a later date by his brother.

It was a successful venture, and from 1820 to 1847 he took out a further 16 patents connected with working rubber. He also collaborated with Charles Macintosh (1766–1843) of Glasgow, famous for his waterproof cloth, over some aspects of manufacturing.

The Goswell Road factory was transferred to his nephew James Lyne Hancock in 1842, and Hancock himself applied his talents to research in his own private laboratory, which he had set up in Stoke Newington. Like Charles Goodyear (1800–1860) in the United States, he was interested in solving the problems of rubber's tackiness and inconsistency at different temperatures. After seeing some samples of Goodyear's products and reading an article about his work, Hancock adopted the heat process of vulcanization. In 1857 he published *Personal Narrative of the Origin and Progress of the Caoutchouc or India Rubber Manufacture in England.*

Joseph Priestley (1733–1804), the discoverer of oxygen, introduced the term 'rubber' after using a small piece of the substance to erase a pencil mark on his notes. The prefix 'India' came from the place where it was first found, the West Indies. The first description of rubber in use came to Europe via Charles de la Condamine, who in 1735 was commissioned by the Academy of Sciences, Paris, to carry out some research in South America. He noted that the local people made items from a watertight material which they had previously prepared from a milky liquid spread

out in the sun to cure.

One of Hancock's earliest and most important inventions was a 'masticator', a machine which kneaded the raw rubber to produce a solid block. He nicknamed this his 'pickle machine'.

He worked continuously on improvements to rubber, seeking the help and advice of Charles Macintosh. Like Goodyear in the United States, he found that the real problems lay in maintaining the strength and pliable nature of the substance under changing temperatures. Hancock finally mastered the technique after experimenting with sulphur additives and reading an article on Goodyear's work.

Handley Page Frederick 1885–1962 was a British aeronautical engineer who founded a company that achieved a worldwide reputation for constructing successful military aircraft.

Handley Page was born at Cheltenham and as a boy determined to become an electrical engineer: by the age of 21 he was chief designer of an electrical company. In 1907, he presented a paper on the design of electrical equipment to the Institute of Electrical Engineers, and such was the interest created that he was offered a tempting post by the Westinghouse Company in the United States. But Handley Page was fascinated with the potential of the aeroplane and he became an early member of the (later Royal) Aeronautical Society. In 1908, he set up as an aeronautical engineer and a year later, with a capital of £10,000, he established the first private British company of this kind, using some sheds at Barking, Essex as workshops and nearby waste ground as an aerodrome.

The outbreak of World War I in 1914 determined the line he was to follow. The need of the British Admiralty for an aircraft capable of carrying a heavy load of bombs led to his design of the first two-engined bombers. Igor Sikorsky (1889–1972) in Russia and Giovanni Caproni (1886–1957) in Italy had been thinking along similar lines, but Handley Page produced the first large aircraft. It first flew in December 1915 and before the war had ended had been enlarged into a four-engined bomber with a fully laden weight of 13 tonnes. More than 250 of these aircraft were produced although the war ended before many saw active service.

After the War (1918), Handley Page turned his attention to civil aircraft. For four years another company bearing his name used his machines to operate a service on routes from London to Paris and Brussels, later extended to Zurich. All that this experience proved commercially, however, was that civil aviation was not viable without government subsidies, and these were only to be had when Handley Page lost its identity by merging with Imperial Airways, the forerunner of BOAC. The manufacturing side of Handley Page's business continued, due largely to his development of the automatic slot safety device.

During the interwar years, in 1930, Handley Page produced the first 40-seat airliner. This was the Hercules, a four-engined plane which arose at about 145 kph/90 mph. It was extremely comfortable – often being compared with

Pullman trains but the aircraft failed to attract overseas orders.

During World War II Handley Page produced another bomber, the Halifax, of which 7,000 were constructed. Work on the bomber continued after the end of the war in 1945, resulting in a four-engined jet of unusual design, the Victor, which made its first flight in 1952. For his services to the aircraft industry in peace and war, Handley Page was knighted in 1942. In 1960, he was awarded the Gold Medal of the Royal Aeronautical Society.

Hankel Hermann 1839–1873 was a German mathematician and mathematical historian who, in a somewhat brief life, nevertheless made significant contributions to the study of complex and hypercomplex numbers and the theory of functions. A great innovator himself – his name is perpetuated in the Hankel functions – much of his work was also in developing that of others. In turn, his work later received considerable and useful development.

Hankel was born in Halle on 14 February 1839, but it was in Leipzig – where his father was Professor of Physics at the University – that he grew up and was educated. At the age of 21, Hankel went to Göttingen University, but transferred a year later to the University of Berlin, where he studied under Karl Weierstrass (1815–1897) and Leopold Kronecker (1823–1891), receiving his doctorate in only twelve months for a thesis on a special class of symmetrical determinates. In 1867 he was appointed Associate Professor of Mathematics at Leipzig. Taking up a similar post at Tübingen in 1869, he spent only four years there before he died, on 29 August 1873.

Hankel published many important works, the first of which was his *Theorie der complexen Zahlensysteme*, issued in 1867. The work dealt with the real, complex and hypercomplex number systems, and demonstrated that no hypercomplex number system can satisfy all the laws of ordinary arithmetic. Hankel also revised Peacock's principle of the permanence of formal laws and developed the theory of complex numbers as well as the higher algebraic systems of August Möbius, Hermann Grassmann and William Hamilton. Presenting algebra as a deductive science, Hankel was one of the original few to recognize the importance of Grassmann's work. His researches into the foundations of arithmetic promoted the development of the theory of quaternions.

In *Untersuchungen über die unendlich oft oscillerenden und unstetigen Functionen* – another of Hankel's major works, concerning in particular the theory of functions – he presented a method for constructing functions with singularities at every rational point. The method was based on a principle of his own devising, regarding the condensation of singularities. He also explicitly stated that functions do not possess general properties; this work was an important advance towards modern integration theory.

In another work on the same topic Hankel provided an example of a continuous function that was non-differentiable at an infinite number of points.

The Hankel functions provide a solution to the Bessel differential equation, which had originally occurred in connection with the theory of planetary motions. Today the equation holds more relevance to the study of wave propagation, to problems in optics, to electromagnetic theory, to elasticity and fluid motions, and to potential and diffusion problems.

Hankel was also the first to suggest a method for assessing the magnitude, or 'measure', of absolutely discontinuous point sets (such as the set of only irrational numbers lying between 0 and 1). Subsequently developed by Georg Cantor, Emile Borel, Henri Lebesgue and (finally) Andrei Kolmogorov, the 'measure' theory of point sets has now been extensively applied to probability, cybernetics and electronics.

Apart from his purely scientific work, Hankel was also a noteworthy historian of mathematics, concentrating on the mathematics of the classical and medieval periods. One of his assertions was that it was the Brahmins – the learned men of the highest caste – of Hindustan who were the real inventors of algebra, in that they were the first to recognize the existence of negative numbers.

Harden Arthur 1865–1940 was a British biochemist who investigated the mechanism of sugar fermentation and the role of enzymes in this process. For this work he shared the 1929 Nobel Prize for Chemistry with the German biochemist Hans von Euler-Chelpin (1873–1964).

Harden was born in Manchester on 12 October 1865, the third child of a local businessman. With his eight sisters, he was brought up in an austere Nonconformist family environment. At the age of seven he attended a private school in Manchester and then from 1876 to 1881 he was at Tettenhall College, Staffordshire. From there he moved to Owen's College (now the University of Manchester) to study chemistry, graduating in 1885. In the following year he won a research scholarship to Erlangen University and was awarded his PhD in 1888 for his work on purification and properties of β-nitrosonaphthylamine.

On his return to Owen's College Harden became a lecturer and demonstrator, teaching chemistry and writing textbooks. In 1897, he became head of the chemistry section of the British Institute of Preventative Medicine, later called the Jenner Institute. In 1905, the chemistry and biology departments merged under his direction. In 1912, he was appointed Professor of Biochemistry at the University of London. His main researches were interrupted by World War I, when he turned his attention to the production of vitamins B and C as a contribution to the war effort. Harden was knighted in 1936 and died in Bourne End, Buckinghamshire, on 17 June 1940.

Harden began work on sugar fermentation in 1898, soon after joining the Jenner Institute. He used the bacterium *Bacillus coli* (now *Escherichia coli*), hoping to detect metabolic differences that would help to distinguish between the various strains. He started studying the metabolism of yeasts in 1900, three years after Eduard Buchner had produced evidence for the existence of enzymes with his experiments on cell-free alcoholic fermentation of sugar.

Harden showed that the occurrence of fermentation in the cell-free yeast extract even in the absence of sugar was caused by the action of the enzyme zymase on glycogen from the disrupted cells. The extract's loss of enzymic activity on standing, he demonstrated to be caused by a second enzyme, a proteolytic (protein-splitting) substance which breaks down zymase.

Harden and his co-workers went on to show that zymase consists of at least two different substances: one heat-sensitive (probably a protein, the enzyme) and one heat-stable, which he called a coferment (now known as a coenzyme). He separated the two using Thomas Graham's dialysis method, proving that the co-enzyme is not colloidal in nature. Hans von Euler-Chelpin later showed that it is diphosphopyridine nucleotide (DPN).

So far Harden had demonstrated that the fermentation process requires three ingredients: enzyme, coenzyme and substrate. But with only these three the reaction soon slowed down. It was accelerated, however, by the unlikely addition of phosphates. Harden discovered the hexose sugar compound hexosediphosphate in the normal reaction mixture (he later found hexosemonophosphate as well), thus proving that phosphorylation is an intermediate step in the fermentation process. This important finding stimulated great interest in intermediate metabolism, and later researchers were able to show the widespread importance of phosphate groups in many metabolic reactions.

After World War I, Harden continued to investigate the details of fermentation. His approach to work always remained accurate, objective and dispassionate. In fact his unwillingness to speculate combined with a sometimes narrow approach to a problem may have hampered his researches. In later life he devoted much of his energy to launching and promoting the *Biochemical Journal* and was its editor for many years.

Hardy Alistair Clavering 1896–1985 was the British marine biologist who designed the Hardy plankton continuous recorder. His development of methods for ascertaining the numbers and types of minute sea organisms helped to unravel the intricate web of life that exists in the sea.

Hardy was born in Nottingham on 10 February 1896 and was educated at Oundle School and then at Exeter College, Oxford. He served in World War I as a lieutenant and later as a captain, from 1915 to 1919. The following year he studied at the Stazione Zoologica in Naples and returned to Britain in 1921, to become Assistant Naturalist in the Fisheries Department of the Ministry of Agriculture and Fisheries. In 1924, he joined the *Discovery* expedition to the Antarctic as Chief Zoologist and on his return in 1928, he was appointed Professor of Zoology and Oceanography at Hull University, where he founded the Department of Oceanography. In 1942, he was made Professor of Natural History at the University of Aberdeen and, two years later, became Professor of Zoology at Oxford. He held this post for 15 years, from 1946 to 1961, when he served as Professor of Field Studies at Oxford. In recognition of his achievements in marine biology, he was knighted in 1957.

He returned to Aberdeen in 1963 to take up a lectureship at the University.

Before Hardy's investigations, the German zoologist Johannes Müller had towed a conical net of fine-meshed cloth behind a ship and collected enough specimens from the sea to reveal an entirely new sphere for biological research. The really serious study of the sea began in the late nineteenth century with the voyage of HMS *Challenger*, the purpose of which was to investigate all kinds of sea life, and it returned to Britain with an enormous wealth of material. A German plankton expedition was led by Victor Hensen in 1899, who coined the word plankton to describe the minute sea creatures which, it has since been established, form the first link in the vital sea food chain.

Hardy made his special study of plankton on the 1924 *Discovery* expedition. The aim of quantitative plankton studies is to estimate the numbers or weights of organisms beneath a unit area of sea surface or in a unit volume of water. Müller's original conical net of fine mesh is still the basic requirement in any instrument designed to collect specimens from the sea, but the drawback is that it can be used only from stationary vessels for collections at various depths, or from moving vessels whose speed must not exceed two knots. Faster speeds displace the small organisms by the turbulence created. Hardy developed a net that can be used behind faster-moving vessels and which increases enormously the area in which accurate recordings can be made.

The first of these nets was the Hardy plankton recorder which, reduced to its simplest terms, is a high-speed net. It consists of a metal tube with a constricted opening, a fixed diving plane instead of a weight, and a stabilizing fin. The net itself is a disc of 60-mesh silk attached to a ring placed inside near the tail. It was designed for easy handling aboard herring-fishing vessels, to be towed when the skipper was near the grounds chosen for the night's fishing. If the disc showed plenty of herring food, the chances of a good catch were high. The indicator is no longer used for this purpose, having been superseded by more modern echo-sounding equipment, but it served as the basis for Hardy's second invention which is still used and through which it has been possible to map the sea life in the oceans of the world.

This improved instrument is known as the Hardy continuous plankton recorder, and was first developed at the Oceanographic Department which Hardy himself founded at the University of Hull in 1931. The present Edinburgh Oceanographic Laboratory, which is the thriving continuation of Hardy's original Hull laboratory, still employs the improved version. The instrument can be used by unskilled personnel aboard ship after being suitably prepared by scientists ashore and is later returned to the laboratory where scientific observations can be made under more ideal conditions.

While the ship tows it along at normal cruising speed the instrument continuously samples the plankton. As the instrument is towed, the water flowing through it drives a small propeller which, acting through a gearbox, slowly winds a long length of plankton silk mesh and draws it across the path of the incoming water. As the mesh, which is graduated in numbered divisions, collects the plankton, it is slowly wound round a spool. This roll of silk mesh is then met by another roll and both are wound together in sandwich form, with the plankton collected by the first roll safely trapped between the two. The whole is stored in a small tank filled with a solution which preserves the plankton for later detailed study in the laboratory.

The recorder can be used at a depth of 10 m/33 ft and at speeds ranging from eight to sixteen knots. Regular surveys, initiated at Hull and developed in Edinburgh, now annually cover many thousands of kilometres in the Atlantic, North Sea and Icelandic waters. The knowledge of plankton distribution, which is continuously being updated, is of major importance to the fishing industry because there is a vitally close relationship between the occurrence of plankton and the movements of plankton-feeding fish used for human consumption.

Hardy has suggested that if just 25% of the pests that exist in the sea can be eliminated and fish can be allowed to have some 20% of the potential food supply instead of the 2% they have now, then any given area could support ten times the amount of fish it supports at present. Hardy's methods have therefore not only added to our overall knowledge of sea life but have also indicated ways in which it could be put to better use.

Hardy Godfrey Harold 1877–1947 was one of the foremost British mathematicians of the twentieth century; almost all his research was at a very advanced level in the fields of pure mathematics known as analysis and number theory. However, although there are few – if any – well-known results that can be directly linked with his name, his influence on early twentieth-century mathematics was enormous. His book, *A Course in Pure Mathematics*, revolutionized the teaching of mathematics at senior school and university levels, and another book, *An Introduction to the Theory of Numbers*, written in conjunction with E. Wright, was once described as one of the most often quoted in mathematical literature.

Hardy was born on 7 February 1877 at Cranleigh, Surrey, where his father was art master at Cranleigh public school. Hardy attended the school and showed no little talent in mathematics and languages. Later he obtained a scholarship to Winchester School on the basis of his mathematical ability. There, considered too able in the subject to be taught as part of the usual class, he was tutored on his own. Completing his education, Hardy then went to Trinity College, Cambridge, where in 1896 he was awarded the Smith's Prize for mathematics and elected Fellow of the College. His first results were published in 1900 and by 1906, when he became a permanent lecturer in mathematics, he had shown himself to be an accomplished research mathematician. Two years later he published his mighty work, *A Course in Pure Mathematics*, in which his approach was based on emulation of Camille Jordan's spiritedly modern *Cours d'analyse*. Hardy's book has ever

since been a mathematical equivalent of a bestseller. By 1910, when he was elected a Fellow of the Royal Society, his reputation was worldwide. He was elected to the Savilian Chair of Geometry at Oxford in 1919, but returned in 1931 to Cambridge as Sadlerian Professor of Pure Mathematics. His fame as a mathematician was universal and recognized by many honours and awards. He spent two separate years as visiting Professor at Princeton University and at the California Institute of Technology. At the age of 65 he retired from his position at Cambridge, and died on 1 December 1947.

Hardy's researches included such topics as the evaluation of difficult integrals and the treatment of awkward series of algebraic terms. Among his successes in number theory was his new proof of the prime number theorem. According to the theorem – orginally proved by the French mathematicians Hadamard and Poussin – $\pi(x)$, the number of prime numbers not exceeding x, is said to approach $x/\log_e x$ when x approaches infinity. His investigations in this area led to the discovery of further important results in the theory of numbers. Other problems on which he worked, often with John Littlewood (1885–1977), were the ways in which numbers could be partitioned into simpler numbers. (For example, $6 = 5 + 1$, $4 + 2$, and $3 + 3$, and can thus be partitioned into two numbers in three ways. It can also be partitioned into three numbers in three ways, and four numbers in two ways.) Akin to these problems were the ways in which numbers could be 'decomposed' into squares and cubes, and so on. Hardy also worked on the conjecture put forward by Goldbach as to whether every even number is the sum of two prime numbers. (This is still an unsolved problem.) A more geometrical problem, yet one deep in number theory, was to examine the number of points with whole number coordinates (or points at the vertices of an equal square lattice) inside or on the circumference of a given circle.

Apart from the books already cited, Hardy was the sole or joint author of several others, including *Inequalities* and *Divergent Series*, the latter of which was published after his death.

The purest of pure mathematicians, Hardy shunned all connections of mathematics with physics, technology, engineering – and especially with warfare. Even those branches of the subject which had peripheral applications were avoided. The only 'tarnishes' on his work were a note on a problem in genetics, which described what has become known as Hardy's law, and an interest in relativity theory.

Many leading mathematicians were influenced by him and considered him a great teacher. Besides his prolific collaboration with John Littlewood, he also had a fascinating mathematical association with the largely untutored Indian mathematical genius Srinavasa Ramanujan (1887–1920).

Hargreaves James 1720–1778 was a British weaver who invented the spinning jenny, a machine that enabled several threads to be spun at once.

Virtually nothing is known of Hargreaves's early life. He

was probably born in about 1720 in Blackburn, Lancashire, and from 1740 to 1750 seems to have been a carpenter and handloom weaver at nearby Standhill. In about 1760, his skill led Robert Peel (grandfather of the statesman) to employ him to devise an improved carding machine. He is supposed to have invented the spinning jenny in about 1764 when he observed that a spinning-wheel, accidentally overturned by his daughter, continued to revolve, together with the spindle, in a vertical position. Seeing the machine in this unfamiliar upended state apparently made him think that if a number of spindles were placed upright and side by side, several threads might be spun at the same time.

At first the resulting jenny was used only by Hargreaves and his children to make weft for the family loom. But in order to supply the needs of a large family he sold some of his new machines. Spinners with the old-fashioned wheel became alarmed by the possibility of cheaper competition and in the spring of 1768 a mob from Blackburn and its neighbourhood gutted Hargreaves's house and destroyed his jenny and his loom. Hargreaves moved to Nottingham, where he formed a partnership with a Mr James, who built a small cotton mill in which the jenny was used.

Throughout his life Hargreaves was handicapped by having had neither formal education nor what has often proved to be an adequate compensation for it, a sound business sense. He was, however, at least aware of his deficiencies in the latter and with the aid of his partner took out a patent for the spinning jenny in 1770. And when he found out that many Lancashire manufacturers were using the machine, he brought actions for infringement of patent rights. They offered him £3,000 for permission to use it, but he stood out for £4,000. The case continued until his lawyer learned that Hargreaves had sold a number of jennies in Blackburn, and the lawyer withdrew his services.

Hargreaves continued his business partnership until James died in April 1778. Six years later there were 20,000 hand-jennies in use in England compared with 550 of the mechanized (spinning) mules. Hargreaves is said to have left property worth £7,000 and his widow received £4,000 for her share in the business. But after her death some of the children were extremely poor and Joseph Brotherton sought to raise a fund for them and found great difficulty in getting from the wealthy Lancashire manufactures sufficient money to save them from destitution.

The spinning jenny came, like so many successful inventions, at a time when the need for it was at a maximum. The flying shuttle, invented by John Ray and first used in cotton weaving in about 1760, had doubled the productive output of the weaver, whereas that of the worker at the spinning-wheel had remained much the same. The spinning jenny at once multiplied eightfold the output of the spinner and, because of its simplicity, could be worked easily by children. It did not, however, entirely supersede the spinning-wheel in cotton manufacturing (and was itself overtaken by Crompton's mule). But for woollen textiles the jenny could be used to make both the warp and the weft.

There is as about as much uncertainty surrounding the origins of the spinning jenny and its name as there is about its inventor. Some maintain that it was Thomas Highs, and not Hargreaves, who invented the jenny and that Richard Arkwright (another contender in this confused history) stole part of Highs' idea in order to claim the invention was his own. But there is little conclusive evidence that it was Highs or Arkwright, rather than Hargreaves, who invented the jenny. As to the name of the machine, it is said that Highs had a daughter named Jane whereas Hargreaves definitely did not, and that Jane Highs gave her pet name – Jenny – to the machine. Were this so, it is difficult to believe that Hargreaves would have accepted without comment or protest the widespread use of a name that virtually branded his claim to be 'father' of the jenny as nothing less than imposture.

Harrison John 1693–1776 was a British horologist and instrument maker who made the first chronometers that were accurate enough to allow the precise determination of longitude at sea, and so permit reliable (and safe) navigation over long distances.

Harrison was born in Foulky, Yorkshire, in March 1693, the son of a carpenter and mechanic. He learned his father's trades and became increasingly interested in and adept at making mechanisms and instruments. In 1726, he made a compensated clock pendulum, which remained the same length at any temperature by making use of the different coefficients of expansion of two different metals.

In 1714, the British Government's Board of Longitude announced that it would award a series of prizes up to the princely sum of £20,000 to anyone who could make an instrument to determine longitude at sea to an accuracy of 30 minutes (half a degree). By 1728, Harrison had constructed his first marine chronometer. Encouraged by the horologist George Graham (1673–1751) he made various improvements, and in 1735 he submitted one of his instruments for the award. Two other chronometers, smaller and lighter than the first, were finished in 1739 and 1757 and also submitted before the Board had completed trials on the first. Then in about 1760, Harrison perfected his No. 4 Marine Chronometer which was not much larger than a watch of the time and was finally subjected to the ultimate test at sea. On two voyages to the West Indies it kept accurate time to within 5 seconds over the duration of the voyage, equivalent to just over one minute of longitude; after five months on a sea voyage, it was less than one minute (of time) out of true.

A unique feature that contributed to the chronometer's accuracy – and subsequently incorporated into other chronometers – was a device that enabled it to be rewound without temporarily stopping the mechanism. In 1763 the Government awarded Harrison the sum of £5,000, but he did not receive the remainder of the original 'prize' of £20,000 until ten years later, by which time the Admiralty had built its own chronometer to Harrison's design. Three years later Harrison died in London on 24 March 1776.

Harvey Ethel Browne 1885–1965 was a US embryologist and cell biologist, renowned for her discoveries of the mechanisms of cell division, using sea urchin eggs as her experimental model.

Ethel Browne was born on 14 December 1885 in Baltimore, the youngest of five children in a family that was strongly supportive of women's education. Her father was a distinguished gynaecologist and two of her sisters became physicians. Browne was educated at Bryn Mawr School and the Women's College of Baltimore, before attending Columbia University from which she graduated in 1907 and achieved a PhD in 1913 for her work on germ cells in an aquatic insect. This marked the beginning of a lifetime's fascination with cellular mechanisms of inheritance and development. During World War I she worked as a laboratory assistant in biology and histology and in 1916 married Edmund Harvey, a professor of biology at Princeton. She subsequently worked part-time, and visited several marine laboratories, until 1928 when she became an instructor at New York University for three years. Most of her subsequent career was as an independent research worker attached to the biology department of Princeton, largely self-financed. She continued to work in marine laboartories, and was a frequent visitor to the Stazione Zoologica in Naples. Her work in cytology, the study of cells, was internationally recognized and she was awarded, amongst other honours, fellowships of the American Association for the Advancement of Science and the New York Academy of Sciences. She died in Falmouth, Massachusetts on 2 September 1965.

At a time when most research on cell development was directed towards understanding the function of the nucleus, which contained the chromosomes, Harvey's work concentrated on the role in cell fertilization and development of non-nuclear cell components in the cytoplasm. She undertook morphological studies and physiological experiments to examine the factors that affect the process of cell division and was able to stimulate division in fragments of sea urchin eggs that contained no nucleus. This was publicized by the popular press as creation without parental influences, but was an important scientific contribution to unravelling the connections between different cellular structures in controlling cell division and development.

Harvey William 1578–1657 was an eminent English physician who discovered that blood is circulated around the body by pulsations of the heart, a landmark in medical investigations. His work did much to pave the way for modern physiology.

Harvey was born in Folkestone, Kent, on 1 April 1578. He went to the King's School, Canterbury, and then attended Gonville and Caius College at Cambridge in 1593. He graduated with a BA from Caius College in 1597 and extended his studies under Fabricius ab Aquapendente at the university medical school in Padua, Italy, gaining his medical degree in 1602. He returned to London, built up a successful practice, and in 1609 he was appointed Physician to St Bartholomew's Hospital, London, and served as a professor there from 1615 to 1643. In 1618, he became Physician Extraordinary to James I, and then Royal Physician, a position he retained until the death of Charles I in 1649. He was elected President of the College of Physicians in 1654 but was too old to accept, and he died three years later in Roehampton on 3 June 1657.

Harvey was deeply involved in medical research and his spare time was devoted to his consuming interest, the investigation of the movement of blood in the body. He had developed this interest while studying in Padua under Fabricius, who had discovered the valves in the veins but had not appreciated their significance. The old idea about blood movement, established by Galen, was that food turned to blood in the liver, ebbed and flowed in vessels and, on reaching the heart, flowed through pores in the dividing wall (septum) from the right to the left side and was sent on its way by heart spasms. Andreas Vesalius (1514–1564), who secretly dissected corpses, failed to find the pores in the heart's dividing wall, and concluded that Galen could never have dissected a human body.

Harvey was not at all convinced by Galen's explanation either. Examining the heart and blood vessels of about 128 mammals, Harvey found that the valve separating the auricle from the ventricle, on each side of the heart, is a one-way structure, as are the valves in the veins discovered by his tutor, Fabricius. For this reason he decided that the blood in the veins must flow only towards the heart. Harvey tied off an artery and found that it bulged with blood on the heart side; he then tied a vein and discovered that it swelled on the side away from the heart. He also calculated the amount of blood that left the heart at each beat. He worked out that in human beings it was about 60 cu cm/4 cu in per beat, which meant that the heart pumped out 259 litres/57 gal (68 US gal) of blood an hour. This amount would weigh more than 200 kg/441 lb – more than three times the weight of an average man. Clearly that was absurd, and therefore a much smaller quantity of the same blood must be circulating continuously around the body. Harvey demonstrated that no blood seeps through the septum and reasoned that it passes from the right side of the heart to the left through the lungs (pulmonary circulation).

The publication of these findings aroused the hostility Harvey had predicted, because to refute Galen was almost unthinkable in his time. His practice declined but he continued with his studies and, unlike many early scientists who made an outstanding discovery, he lived to see his work accepted.

The great classic Harvey published in 1628, *De Motu Cordis et Sanguinis in Animalibus*/*On the Motion of the Heart and Blood in Animals*, pointed the way for physicians who followed him. He also published *Exercitationes de Generatione Animalium* (1651), translated as *Anatomical Exercitations concerning the Generation of Living Creatures* (1653).

Harvey was one of the first to study the development of a chick in the egg. He also carried out many dissections to find out how mammalian embryos are formed, and many of the animals he dissected were the royal deer put at his disposal by Charles I. Harvey suspected that semen might be involved in the making of an embryo, but did not have the microscopic apparatus, later developed by Leeuwenhoek, needed to study the tiny spermatozoa.

Harvey's discovery of the circulation of the blood marked the beginning of the end of medicine as taught by Galen, which had been accepted for 1,400 years. From then on, experimental physiology was to sweep away many erroneous ideas and replace them with personal observations made by experiment and careful measurement.

Hausdorff Felix 1868–1942 was a German mathematician and philosopher who is chiefly remembered for his development of the branch of mathematics known as topology, in which he formulated the theory of point sets. The author of an influential book on the subject, he wrote philosophical works as well, and contributed extensively to several other fields of mathematics.

Hausdorff was born of Jewish parents in Breslau, Germany (now Wroclaw, Poland), on 8 November 1868. He completed his secondary education in Leipzig and, after studying mathematics and astronomy in Berlin and Freiburg, returned to Leipzig to graduate there in 1891. For the next seven years he wrote and published a number of papers dealing with optics, mathematics and astronomy; he also produced works on philosophical and literary themes. In 1902, he was appointed Professor at Leipzig University, after which the subject of mathematics – and particularly set theory – dominated his output. Then appointed Associate Professor at Bonn University, he published his major work, *Grundzuge der Mengenlehre/Basic Features of Set Theory*. In 1913, Hausdorff took up the position of Professor at Griefswald, but eight years later returned to Bonn for the remainder of his academic life; he retired at the statutory age of 67, in 1935. Even after his retirement he continued to work on set theory and topology; all his work, however, was published outside Germany. Because of his Jewish faith, Hausdorff in 1942 was scheduled to be sent to an internment camp. Rather than let that happen he committed suicide – with his wife and her sister – on 26 January of that year.

Topology is the study of figures and shapes that retain their essential proportions despite being 'squeezed' or 'stretched'. In this way, a square may be considered (topologically) equivalent to any closed plane figure and a cube may even be regarded as (topologically) equivalent to a sphere. In his *Grundzuge der Mengenlehre* Hausdorff formulated a theory of topological and metric spaces into which concepts previously advanced by other mathematicians fitted well. He proposed that such spaces be regarded as sets of points and sets of relations among the points, and introduced the principle of duality. This principle states that an equation between sets remains valid if the sets are replaced by their complements and the symbol for union is exchanged for that of intersection; and that an inequality (inequation) remains valid if each side is replaced by its complement and the symbol for union is exchanged for that of intersection, provided also the inclusion symbol is reversed.

Developing his point set theory, Hausdorff also created

what are now called Hausdorff's neighbourhood axioms; there are four axioms:

1) to each point x there corresponds at least one neighbourhood U_x, each neighbourhood U_x containing the point x;

2) if U_x and V_x are two neighbourhoods of the same point x, their intersection contains a neighbourhood of x;

3) if U_x contains a point y, there is a neighbourhood U_y such that $U_y \subset U_x$;

4) if x is not equal to y, there are two neighbourhoods U_x and U_y such that the intersection of U_x and $U_y = 0$ (that is, such that U_x and U_y have no points in common).

From these axioms Hausdorff derived what are now called Hausdorff's topological spaces. A topological space is understood to be a set E of elements x and certain subsets S_x of E which are known as neighbourhoods of x. These neighbourhoods satisfy the Hausdorff axioms. Later in the book – after dealing with the properties of general space – Hausdorff introduced further axioms involving metric and particular Euclidean spaces.

Hausdorff contributed extensively to several fields of mathematics including mathematical analysis, in which he proved some very important theorems on summation methods and properties of moments. He also investigated the symbolic exponential formula which he himself derived. His main contribution to mathematics, however, was in the field of point set theory. Hausdorff's work in this area led to many results of primary importance, such as the investigation of general closure spaces, and what is now termed Hausdorff's maximal principle in general set theory.

Haüy René-Just 1743–1822 was the founder of modern crystallography.

Born at St Jest-en-Chaussee, Oise, France, the son of an impoverished weaver, Haüy entered the priesthood before developing a passionate love of mineralogy and crystallography. He became Professor of Mineralogy in Paris in 1802. His two major works – the *Treatise of Mineralogy* (1801) and the *Treatise of Crystallography* (1822) – are widely acknowledged as foundational texts of modern crystallography. Haüy remained a priest, receiving protection from anticlerical attacks during the Revolutionary era thanks to Napoleon's patronage.

The seventeenth and eighteenth centuries had generated many explanations as to why crystals had regular forms. Various sorts of shaping forces, sometimes chemical, sometimes physical, had been touted (though this might be seen as explaining the unknown in terms of the unknown). In 1670, Steno had demonstrated the constancy of the angle between corresponding faces in the crystals of one substance, irrespective of crystal size or nature. In his vast crystal studies, Haüy was to build on the theories of Steno and his followers. His approach was principally morphological; he did not seek to explain the causes of the regularly varied forms of crystals, but rather sought to develop a geometrical classification of those forms – ideally, through the geometry of simple relationships between integers. Thus he demonstrated in 1784 that the faces of a calcite crystal could be regarded in terms of the standard stacking of cleavage rhombs, provided that the rhombs were assumed to be so small that the face looked smooth. Similar principles would lead to other crystal shapes being built up from suitable simple structural units. In broad terms, this remains the foundation of the modern view.

Haüy thus interpreted crystals as structured assemblages of secondary bodies (integrant molecules), that grouped themselves according to regular geometric laws. He proposed six primary forms: parallelepiped, rhombic dodecahedron, hexagonal dipyramid, right hexagonal prism, octahedron, and tetrahedron. Most of his research was devoted to establishing and explicating this typology of forms.

Hawking Stephen William 1942–   is a British theoretical physicist and mathematician whose main field of research has been the nature of space–time and those anomalies in space–time known as singularities. He has succeeded in communicating his ideas to wide audiences.

Hawking was born in Oxford on 8 January 1942, and showed exceptional talent in mathematics and physics from an early age. At Oxford University he became especially interested in thermodynamics, relativity theory and quantum mechanics – an interest that was encouraged by his attending a summer course at the Royal Observatory in 1961. When he completed his undergraduate course in 1962 (receiving a First Class Honours degree in physics), he enrolled as a research student in general relativity at the Department of Applied Mathematics and Theoretical Physics at the University of Cambridge.

During his postgraduate programme Hawking was diagnosed as having ALS (amyotrophic lateral sclerosis), a rare and progressive neuromotor disease which handicaps motor and vocal functions, and which has made it necessary for him to carry out mentally long and complex mathematical calculations. He was nevertheless able to continue his studies and to embark upon a distinguished and productive scientific career. He was elected Fellow of the Royal Society in 1974, and became Lucasian Professor of Mathematics at Cambridge University in 1980.

From its earliest stages, Hawking's research has been concerned with the concept of singularities – breakdowns in the space–time continuum where the classic laws of physics no longer apply. The prime example of a singularity is a black hole, the final form of a collapsed star. During the later 1960s, Hawking – relying on a few assumptions about the properties of matter, and incidentally developing a mathematical theory of causality in curved space–time – proved that if Einstein's general theory of relativity is correct, then a singularity must also have occurred at the Big Bang, the beginning of the universe and the birth of space–time itself.

In 1970, Hawking's research turned to the examination of the properties of black holes. A black hole is a chasm in the fabric of space–time, and its boundary is called the event horizon. Hawking realized that the surface area of the event horizon around a black hole could only increase or remain constant with time – it could never decrease. This meant, for example, that when two black holes merged, the surface area of the new black hole would be larger than the sum of the surface areas of the two original black holes. He also noticed that there were certain parallels between the laws of thermodynamics and the properties of black holes. For instance, the second law of thermodynamics states that entropy must increase with time; the surface area of the event horizon can thus be seen as the entropy ('randomness') of the black hole.

Over the next four years, Hawking – with Carter, Israel and Robinson – provided mathematical proof for the hypothesis formulated by John Wheeler (1911–   ), known as the 'no hair theorem'. This stated that the only properties of matter that were conserved once it entered a black hole were its mass, its angular momentum and its electric charge; it thus lost its shape, its 'experience', its baryon number and its existence as matter or antimatter.

Since 1974, Hawking has studied the behaviour of matter in the immediate vicinity of a black hole, from a theoretical basis in quantum mechanics. He found, to his initial surprise, that black holes – from which nothing was supposed to be able to escape – could emit thermal radiation. Several explanations for this phenomenon were proposed, including one involving the creation of 'virtual particles'. A virtual particle differs from a 'real' particle in that it cannot be seen by means of a particle detector, but can be observed through its indirect effects. 'Empty' space is thus full of virtual particles being fleetingly 'created' out of 'nothing', forming a particle and antiparticle pair which immediately destroy each other. (This is a violation of the principle of conservation of mass and energy, but is permitted – and predicted – by the 'uncertainty principle' of Werner Heisenberg.) Hawking proposed that when a particle pair is created near a black hole, one half of the pair might disappear into the black hole, leaving the other half which might radiate away from the black hole (rather than be drawn into it). This would be seen by a distant observer as thermal radiation.

Hawking's present objective is to produce an overall synthesis of quantum mechanics and relativity theory, to yield a full quantum theory of gravity. Such a unified physical theory would incorporate all four basic types of interaction: strong nuclear, weak nuclear, electromagnetic and gravitational.

The properties of space–time, the beginning of the universe, and a unified theory of physics are all fundamental research areas of science. Hawking has made, and continues to make, major contributions to the modern understanding of them all. His book *A Brief History of Time* (1988), explaining his ideas to a popular audience, was an international bestseller, remaining on the UK bestseller list for more than three years, longer than any other title.

Haworth Walter Norman 1883–1950 was a British organic chemist whose researches concentrated on carbohydrates, particularly sugars. He made significant advances in determining the structures of many of these compounds, for which he shared the 1937 Nobel Prize for Chemistry with the German organic chemist Paul Karrer, becoming the first British organic chemist to be so honoured.

Haworth was born in Chorley, Lancashire, on 19 March 1883, the son of a factory manager. He left school at the age of 14 and joined his father's linoleum works, where he became interested in the dyestuffs used in floor coverings. In spite of parental disapproval, he continued his education with a private tutor and in 1903 realized his ambition to enter the chemistry department of the University of Manchester, where he studied under William Perkin Jr. He gained his degree in 1906 and won a scholarship to Göttingen, where he did research with Otto Wallach (1847–1931). He was awarded his doctorate after only one year and returned to Manchester, gaining a DSc there in 1911 in the remarkably short time of only five years since his first degree.

Over the next few years Haworth held various university appointments – at Imperial College, London (1911–1912), St Andrews (1912–1920) and Durham (1920–1925) – before accepting a professorship at Birmingham University, where he remained until he retired in 1948, a year after he received a knighthood. He died in Birmingham in 1950 on his birthday.

Haworth's lifelong interest in carbohydrates began when he was at St Andrews. There he teamed up with T. Purdie and J. C. Irvine, trying to characterize sugars by methylation of their hydroxy groups. The work was interrupted by World War I, when the laboratory was turned over to the production of fine chemicals and drugs, which Haworth supervised. After the war he continued research, concentrating his team's efforts on the disaccharides. Again he used the technique of methylation followed by acid hydrolysis. Lactose, for example, forms an octamethyl derivative which on hydrolysis yields tetramethyl galactose and 2,3,6-trimethyl glucose, indicating the structural formula for lactose.

Later investigations suggested that the carbon atoms in hexoses usually have a ring structure and a great deal of Haworth's work at Birmingham was devoted to establishing the linkages in these rings (known as Haworth formulae). In 1926, he proposed that methyl glucose normally exists as what is now called a pyranose ring. The complex analytical procedures involved oxidation and methylation to give products that were further reacted to form simpler known substances that could be crystallized and identified. He also established that methyl glucose can have an alternative furanose structure. His book *The Constitution of the Sugars* (1929) became a standard work.

Using their knowledge of simple sugars, Haworth's research team was able to investigate the chain structures of polysaccharides, establishing the structures of cellulose, starch and glycogen. Work on sugars led naturally to studies of vitamin C and by 1932, the whole research effort was directed to the determination of the structure and synthesis of this substance, which Haworth named ascorbic acid. The research resulted in an industrial process for synthesizing vitamin C for medical uses.

In the late 1930s, Haworth studied the reactions of polysaccharides with enzymes and certain aspects of the chemistry of the hormone insulin. During World War II (1939–1945) Haworth's laboratory became a primary producer of purified uranium for the war effort, which led to work on the preparation and properties of organic fluorine compounds. Many of the workers went to the newly opened experimental stations at Oak Ridge in the United States and Chalk River in Canada.

Hayashi Chushiro 1920– is a Japanese physicist whose research in 1950 exposed a fallacy in the 'hot big bang' theory proposed two years earlier by Ralph Alpher (1921–), Hans Bethe (1906–) and George Gamow (1904–1968). Since that time, Hayashi has published many papers on the origin of the chemical elements in stellar evolution and on the composition of primordial matter in an expanding universe.

Hayashi was born in Kyoto, Japan, on 25 July 1920. Completing his education at the University of Kyoto, from which he received a BSc in 1942, he became a Research Associate at the University of Tokyo for four years, before returning to Kyoto as Research Associate from 1946 to 1949. He then spent five years as Assistant Professor at Naniwa University, Osaka, until he once again returned to Kyoto University as a member of the Physics Faculty, becoming Professor of Physics in 1957.

It was George Gamow who, in 1946, proposed that the early dense stages of the universe were hot enough to enable thermonuclear reactions to occur. The first detailed calculation of the formation of helium in this way, in the 'hot big bang', was published two years later. The three collaborators hoped that their work might account not only for the observed abundance of helium in the universe but also for the distribution of other, heavier elements. This theory assumed that matter was originally composed of neutrons which, at very high temperatures and if the matter was of great enough density, combine with protons.

But Hayashi pointed out that at times in the Big Bang earlier than the first two seconds, the temperature would have been greater than 10^{10}K, which is above the threshold for the making of electron–positron pairs. The creation of one pair requires about 1 MeV (10^6 electronvolts) of energy, and at 10^{10}K many protons have this much energy. The effect of this is radically to change the timescale of the α-β-γ theory. Neutrons can react with the thermally excited positrons through the so-called weak interactions to produce protons and antineutrinos, and the reaction time for this to happen (at temperatures above 10^{10}K) is less than about one second. Hayashi proposed that at such a temperature there was complete thermal equilibrium between all forms of matter and radiation. Below that temperature, the weak interactions cannot maintain the neutrons in statistical balance with the protons because the concentration of electron pairs is falling abruptly. This means that the ratio of neutrons to protons is 'frozen in' until a few hundred seconds have gone by and neutron decay becomes appreciable. The frozen-in ratio at a temperature below 10^{10}K is about 15% – and such a change in the ratio affects the helium abundances. The frozen-in abundance of neutrons does not depend on the material density but on the temperature and the properties of the weak interactions. So, provided the density is great enough for the combining reaction between neutrons and protons to occur faster than the expansion rate, a fixed concentration of neutrons will be incorporated into helium nuclei, however great the material density is, producing a 'plateau' in the relationship between helium abundance and material density.

Hayashi derived a percentage value for the abundance plateau, but more recent research by physicists such as Fred Hoyle (1915–   ), William Fowler (1911–   ) and R. Wagoner suggests that his value is too high.

Heath Thomas Little 1861–1940 was a British civil servant and mathematical historian of the ancient Greek mathematicians. For his considerable services to the Treasury Office he was knighted; he also received no little acclaim for his historical works – he became a Fellow of the Royal Society in 1912, and of the Royal Academy in 1932 – and for a year he was President of the Mathematical Association.

Heath was born in 1861 in Lincolnshire. As a boy he attended the Caistor Grammar School and later studied at Clifton and Trinity Colleges, Cambridge, where he obtained a First Class Classics Tripos – the first part in 1881 and the second in 1883. It was at this time that he won first place in an open competition for the Home Civil Service, and was appointed a clerk in the Treasury in 1884. In 1885, he was made Fellow of Trinity College, although it was not until 11 years later that he obtained his doctorate. In 1887, he became Private Secretary to the Permanent Secretary at the Treasury, and from 1891 to 1894 held a similar post in relation to successive Financial Secretaries. He became Principal Clerk of the Treasury from 1901 to 1907, and in that year was himself appointed Assistant Secretary. In 1908 he published his first full-scale book, *Euclid's Elements*, a monumental work in three volumes. Five years later he was appointed Joint Permanent Secretary to the Treasury and Auditor of the Civil List, a post he held until 1919, when he was made Comptroller General and Secretary to the Commissioners for the Reduction of the National Debt. He remained as such till 1926, when he retired from the civil service. Meanwhile, in 1921 he produced his two-volume *History of Greek Mathematics*, which has come to be regarded as the standard work on the subject in the English language. After leaving the civil service, Heath was appointed one of the University of Cambridge Commissioners, and was also a member of the Royal Commission on National Museums and Galleries from 1927 to 1929. He died on 16 March 1940 in Ashtead, Surrey.

No doubt Heath's interest in Greek mathematics, which made him one of the leading authorities on the subject, was the result of his university training in mathematics and the classics. He took a special interest in Diophantus, whose *Arithmetica* he edited for an English-language edition. The work, *Diophantus of Alexandria: A Study in the History of Greek Algebra*, was published in 1885, and was more in the style of an essay.

Apollonius of Perga: A Treatise on Conic Sections was published in 1896. It is particularly interesting that in this work Heath used modern mathematical notation. It was followed, a year later, by an edition of the works of Archimedes, in which Heath used the same treatment. (This was translated into German in 1914.) And Heath's version of Archimedes' *Method* appeared in 1912. In 1913, Heath produced *Aristarchus of Samos, the Ancient Copernicus*, a comprehensive account of Aristarchos and his work on astronomy.

In the great *History of Greek Mathematics*, Heath dealt with his subjects mainly according to topics and not in chronological order as others had done before him. In the preface to the book, Heath justified such an approach by reference to the famous problem of doubling the cube, saying, 'If all recorded solutions are collected together, it is much easier to see the relations … between them and to get a comprehensive view of the history of the problem.' Heath rewrote the *History* in 1931 and published it as a *Manual of Greek Mathematics*; in fact, it is a mere condensed version of the original text.

Heathcoat John 1783–1861 was a British inventor of lace-making machinery. His contribution was acknowledged by Marc Isambard Brunel who said that Heathcoat (then aged 24) had devised 'the most complicated machine ever invented'.

Heathcoat was born at Duffield, near Derby. He received an average education and, after completing his apprenticeship as a journeyman in the hosiery trade, he became a master mechanic at Hathorn in about 1803. He then set himself the task of constructing a machine that would do the work of the pillow, the multitude of pins, thread and bobbins, and the fingers of the hand lacemaker and supersede them in the production of lace – just as the stocking loom had replaced knitting needles in stocking-making.

Analysing the component threads of pillow lace, he classified them into longitudinal and diagonal. The former he placed on a beam as a warp. The remainder he reserved as weft with each thread worked separately, twisted around the warp thread to close the upper and lower sides of the mesh. He then devised the necessary mechanical features: bobbins to distribute the thread, the carriage and groove in which they must run, and their mode of twisting round the warp and travelling from side to side of the machine. The first square yard (0.83 sq m) of plain net from the machine was sold for £5. By the end of the nineteenth century its price had fallen to one shilling.

In 1805, Heathcoat settled in Loughborough – as a consequence his improved machine became known as the 'Old Loughborough'. Four years later he went into partnership with the aptly named Charles Lacy, a former point-net maker at Nottingham. Following this amalgamation the machine's capacity was so increased that by 1816 there were as many as 35 frames at work in the Loughborough factory. They also made much money from royalties paid by other companies with permission to use the machines.

On the night of 28 June 1816 an angry crowd of Luddites, fearful that the new machines would deprive them of their jobs, attacked Heathcoat's Loughborough factory and destroyed 35 frames, burning the lace that was on them. The company sued the county for damages and received £10,000 in compensation, on condition that the money was spent locally. Heathcoat refused to accept the condition; he had already received threats to his life and wanted to leave the district for good. He dissolved his partnership with Lacy and left for Tiverton in Devon, forfeiting his right to compensation.

At Tiverton events took a more favourable turn. With a former partner, John Boden, he bought a large water-powered mill on the river Exe. Heathcoat devised new frames which were wider and faster, and by using rotary power he lowered the cost of production. He patented a rotary, self-narrowing stocking frame and put gimp and other ornamental threads into the bobbin by mechanical adjustment.

In 1821, the partnership with Boden was ended. Year by year Heathcoat took out patents for further inventions, continuing to make improvements in the textile trade until he retired in 1843. Also in 1832, with Henry Handley, he patented a steam plough to assist with agricultural improvements in Ireland. On 12 December of that year he was elected to represent Tiverton in the new reformed Parliament and remained MP for the borough until 1859. He died at Tiverton three years later, in his 78th year.

Heaviside Oliver 1850–1925 was a British physicist and electrical engineer who predicted the existence of the ionosphere, which was known for a time as the Kennelly–Heaviside layer. Heaviside made significant discoveries concerning the passage of electrical waves through the atmosphere and along cables, and added the concepts of inductance, capacitance and impedance to electrical science.

Heaviside was born in Camden Town, London, on 18 May 1850. His uncle was Charles Wheatstone (1802–1875), who was a pioneer of the telegraph and may have stimulated in him an interest in electricity. Heaviside was disadvantaged at school because of severe hearing difficulties, and so was mainly self-taught. He took up employment with the Great Northern Telegraph Company at Newcastle-upon-Tyne in 1870, when he was 20. This employment lasted only four years, when he was forced to retire because of his deafness. Shortly after he is believed to have visited Denmark to study telegraph apparatus. On his return he was supported first by his parents and later in life he lived with his brother. Heaviside never obtained an academic position but received several honours, including Fellowship of the Royal Society in 1891 and the first award

of the Faraday Medal by the Institution of Electrical Engineers shortly before he died in Paignton, Devon, on 3 February 1925.

During his early years, Heaviside was absorbed by mathematics. He was familiar with the attempt by Peter Tait (1831–1901) to popularize the quaternions of William Rowan Hamilton (1805–1865), but he rejected the scalar part of the quaternions in his notion of a vector. Heaviside's vectors were of the form:

$$v = ai + bj + ck$$

where i, j and k are unit vectors along the Cartesian x-, y- and z-axes respectively. When Heaviside became involved with work relating to Maxwell's famous theory of electricity, he wrote Maxwell's equations in vector form incorporating some discoveries of his own. Heaviside also used operator techniques in his expression of calculus and made much use of divergent series (those with terms whose sum does not approach a fixed amount). He made great use of these in his electrical calculations when many mathematicians were afraid to venture away from convergent series.

In his twenties and thirties, Heaviside became very interested in electrical science, first carrying out experiments and then developing advanced theoretical discussions. These were, in fact, so advanced that they were eventually dismissed by some mathematicians and scientists, and were classed as too abstract to be published in the electrical journals. He expressed his ideas in the three-volume *Electromagnetic Theory* (1893–1912). Much of this work extended Maxwell's discoveries but he made many valuable discoveries of his own. Heaviside visualized electrical ideas often in mechanical terms and thought of the inertia of machines as being similar to electrical inductance.

When Heaviside became involved with the passage of electricity along conductors, he modified Ohm's law to include inductance and this, together with other electrical properties, resulted in his derivation of the equation of telegraphy. On considering the problem of signal distortion in a telegraph cable, he came to the conclusion that this could be substantially reduced by the addition of small inductance coils throughout its length, and this method has since been used to great effect.

In the third volume of his *Electromagnetic Theory*, Heaviside considered wireless telegraphy and, in drawing attention to the enormous power required to send useful signals long distances, he suggested that they may be guided by hugging the land and sea. He also suggested that part of the atmosphere may act as a good reflector of electrical waves. Arthur Kennelly (1861–1939) in the United States also made a similar suggestion and this layer, which was for some time known as the Kennelly–Heaviside layer, was subsequently shown to exist by Edward Appleton (1892–1965). This part of the atmosphere, about 100 km/ 62 mi above the ground, is now known as the ionosphere.

Although from time to time Heaviside had his detractors, perhaps because he was unorthodox in his approach, he was later known and highly valued by the leading physicists and electrical engineers of his day.

Heine Heinrich Eduard 1812–1881 was a German mathematician and a prolific author of mathematical papers on advanced topics. He completed the formulation of the notion of uniform continuity and subsequently provided a proof of the classic theorem on uniform continuity of continuous functions. This theorem has since become known as Heine's theorem.

Heine was born on 16 March 1812 in Berlin, where his father was a banker. His first education was by tutors at home, and he then attended local high schools. Eventually he went to Göttingen University and studied under Karl Gauss (1777–1855). To complete his education he returned to Berlin and became a pupil of Lejeune Dirichlet (1805–1859), Jakob Steiner (1796–1863) and the more astronomically minded Johann Encke (1791–1865), receiving his PhD in 1842. He then spent a further couple of years at Königsberg, learning from Karl Jacobi (1804–1851) and Franz Neumann (1798–1895). Finally, at the age of 32, Heine became an unpaid lecturer at Bonn University. Four years later, in 1848, he was appointed Professor there, but in the same year he took up a similar post at Halle University. And it was at Halle he remained for the rest of his life; he died there on 21 October 1881.

During his lifetime he received few honours or awards. On the other hand, when offered the Chair in Mathematics at Göttingen University – perhaps the most prestigious appointment possible at the time – in 1875, he turned it down.

In all, Heine published more than 50 papers on mathematics. His specialty was spherical functions, Lamé functions and Bessel functions. He published his most important work – *Handbuch der Kügelfunctionen* – in 1861; it was reissued in a second edition in 1881 and became the standard work on spherical functions for the next 50 years.

Although the name of his theory is sometimes linked with the name of Emile Borel (as the Heine–Borel theorem), it can be confidently argued that Borel's contribution was of lesser value. Borel formulated the covering property of uniform conformity, and proved it; Heine formulated the notion of uniform continuity – a notion that had escaped the attention of Augustin Cauchy (1789–1857), regarded as the innovator of continuity theory – and went on to prove the classic theorem of uniform continuity of continuous functions.

Heisenberg Werner Karl 1901–1976 was a German physicist who founded quantum mechanics and the uncertainty principle. In recognition of these achievements, he was awarded the 1932 Nobel Prize for Physics.

Heisenberg was born on 5 December 1901 in Duisberg. He showed an early aptitude for mathematics and when, after leaving school, he came across a book on relativity by Hermann Weyl (1885–1955) entitled *Space, Time and Matter* (1918), he became interested in the mathematical arguments underlying physical concepts. This led Heisenberg to study theoretical physics under Arnold Sommerfeld (1868–1951) at Munich University. There he formed a lasting friendship with Wolfgang Pauli (1900–1958). He obtained his doctor's degree in 1923 and immediately became assistant to Max Born (1882–1970) at Göttingen. From 1924 to 1926, Heisenberg worked with Niels Bohr (1885–1962) in Copenhagen and then in 1927 he was offered the Chair of Theoretical Physics at Leipzig at the age of only 26.

Heisenberg stayed at Leipzig until 1941 and then from 1942 until 1945 was Director of the Max Planck Institute for Physics in Berlin. During World War II he was in charge of atomic research in Germany and it is possible that he may have been able to direct efforts away from military uses of nuclear power. In 1946, Heisenberg was made Director of the Max Planck Institute for Physics in Göttingen, which in 1958 was moved to his home city of Munich. He held the post of Director until 1970, and died in Munich on 1 February 1976.

In the early 1920s, there were burning questions relating to the quantum theory. Bohr had used the quantum concept of Max Planck (1858–1947) to explain the spectral lines of atoms in terms of the energy differences of electron orbits at various quantum states, but there were weaknesses in the results. In 1922, Pauli began to throw doubt on them and Sommerfeld, although an optimist, could see some flaws in Bohr's work.

In 1923, Heisenberg went to Göttingen to work with Born and used a modification of the quantum rules to explain the anomalous Zeeman effect, in which single spectral lines split into groups of closely spaced lines in a strong magnetic field. Information about atomic structure can be deduced from the separation of these lines.

Heisenberg was concerned not to try to picture what happens inside the atom but to find a mathematical system that explained the properties of the atom – in this case, the position of the spectral lines given by hydrogen, the simplest atom. Born helped Heisenberg to develop his ideas, which he presented in 1925 as a system called matrix mechanics. By mathematical treatment of values within matrices or arrays, the frequencies of the lines in the hydrogen spectrum were obtained. This was the first precise mathematical description of the workings of the atom and with it Heisenberg is regarded as founding quantum mechanics, which seeks to explain atomic structure in mathematical terms.

The following year, however, Erwin Schrödinger (1887–1961) produced a system of wave mechanics that accounted mathematically for the discovery made in 1923 by Louis de Broglie (1892–1987) that electrons do not occupy orbits but exist as standing waves around the nucleus. Wave mechanics was a much more convenient system than the matrix mechanics of Heisenberg and Born and rapidly replaced it, though John van Neumann (1903–1957) showed the two systems to be equivalent in 1944.

Nevertheless, Heisenberg was able to predict from studies of the hydrogen spectrum that hydrogen exists in two allotropes – ortho-hydrogen and para-hydrogen – in which the two nuclei of the atoms in a hydrogen molecule spin in the same or opposite directions respectively. The allotropes were discovered in 1929.

In 1927, Heisenberg made the discovery for which he is best known – that of the uncertainty principle. This states that it is impossible to specify precisely both the position and the simultaneous momentum (mass multiplied by velocity) of a particle. There is always a degree of uncertainty in either, and as one is determined with greater precision, the other can only be found less exactly. Multiplying the degrees of uncertainty of the position and momentum yield a value approximately equal to Planck's constant. (This is a consequence of the wave–particle duality discovered by de Broglie.) Heisenberg's uncertainty principle maintains that the result of an action can be expressed only in terms of the probability that a certain effect will occur. The idea was revolutionary and discomforted even Albert Einstein (1879–1955), but it has remained valid.

Another great discovery was made in 1927, when Heisenberg solved the mystery of ferromagnetism. In it he used the Pauli exclusion principle, which states that no two electrons can have all four quantum numbers the same. Heisenberg used the principle to show that ferromagnetism is caused by electrostatic intereaction between the electrons. This was also a major insight which has stood the test of time.

Heisenberg was a brilliant scientist with a very incisive mind. He will be remembered with gratitude and respect for solving some of the great and complex problems in quantum mechanics which were so crucial at the beginning of this century for our understanding of nuclear physics.

Helmholtz Hermann Ludwig Ferdinand von 1821–1894 was a German physicist and physiologist who made a major contribution to physics with the first precise formulation of the principle of conservation of energy. He also made important advances in physiology, in which he first measured the speed of nerve impulses, invented the ophthalmoscope and revealed the mechanism by which the ear senses tone and pitch. Helmholtz also made important discoveries in physical chemistry.

Helmholtz was born in Potsdam on 31 August 1821. His father was a teacher of philosophy and literature at the Potsdam Gymnasium and his mother was a descendant of William Penn, the founder of the state of Pennsylvania in the United States. Although a delicate child, Helmholtz thrived on scholarship; his father taught him Latin, Greek, Hebrew, French, Italian and Arabic. He attended the local Gymnasium and showed a talent for physics, but as his father could not afford a university education for him, Helmholtz embarked on the study of medicine, for which financial aid was available in return for a commitment to serve as an army doctor for eight years. In 1838, Helmholtz entered the Friedrich Wilhelm Institute in Berlin, where his time was spent not only on medical and physiological studies but also on chemistry and higher mathematics. He also became an expert pianist.

In 1842, Helmholtz received his MD and became a surgeon with his regiment at Potsdam, where he carried out experiments in a laboratory he set up in the barracks. His ability in science as opposed to medicine was soon recognized, and in 1848 he was released from his military duties. The following year, he took up the post of Associate Professor of Physiology at Königsberg. In 1855, Helmholtz moved to Bonn to become Professor of Anatomy and Physiology, and then in 1858 was appointed Professor of Physiology at the University of Heidelberg. In 1871, he took up the Chair of Physics at the University of Berlin, and in 1887 became Director of the new Physico-Technical Institute of Berlin. His health then began to fail, and Helmholtz died at Berlin on 8 September 1894.

Throughout his life, Helmholtz's phenomenal intellectual capacity and grounding in philosophy were at the basis of his work. While training as a doctor, his supervisor was Johannes Müller (1801–1858), a distinguished physiologist who like Helmholtz was interested in the ideas promoted by philosophers. The abstract nature of the subject was obviously attractive to his superb brain. During his student days, Helmholtz rejected many of the ideas offered by the German philosopher Immanuel Kant (1724–1804). Helmholtz's supervisor subscribed to Kant's Nature philosophy, which assumed that the organism as a whole is greater than the sum of its parts. Helmholtz's scientific work in many fields was guided by an effort to disprove these fundamentals. He was convinced, for example, that living things possess no innate vital force, and that their life processes are driven by the same forces and obey the same principles as nonliving systems.

Helmholtz's doctoral submission in 1842 was concerned with the relationship between the nerve fibres and nerve cells of invertebrates, which led to an investigation of the nature and origin of animal heat. In 1850, he measured the velocity of nerve impulses, stimulated by a statement by Müller that vitalism caused the impulses to be instantaneous. Helmholtz measured the impulse velocity by experiments with a frog and found it to be about a tenth of the speed of sound. He went on to investigate muscle action and found in 1848 that animal heat and muscle action are generated by chemical changes in the muscles.

This work led Helmholtz to the idea of the principle of conservation of energy, observing that the energy of life processes is derived entirely from oxidation of food. His deduction was remarkably similar to that made by Julius Mayer (1814–1878) a few years earlier, but Helmholtz was not aware of Mayer's work and when he did arrive at a formulation of the principle in 1847, he expressed it in a more effective way. Helmholtz also had the benefit of the precise experimental determination of the mechanical equivalent of heat published by James Joule (1818–1889) in 1847. He was able to derive a general equation that expressed the kinetic energy of a moving body as being equal to the product of the force and distance through which the force moves to bring about the energy change. This equation could be applied in many fields to show that energy is always conserved and it led to the first law of thermodynamics, which states that the total energy of a system and its surroundings remains constant even if it may be changed from one form of energy to another.

Helmholtz went on to develop thermodynamics in physical chemistry, and in 1882, he derived an expression that relates the total energy of a system to its free energy (which is the proportion that can be converted to forms other than heat) and to its temperature and entropy. It enables chemists to determine the direction of a chemical reaction. In this work, Helmholtz was anticipated independently by Willard Gibbs (1839–1903) and both scientists are usually given credit for it.

Helmholtz's work on nerve impulses led him also to make important discoveries in the physiology of vision and hearing. He invented the ophthalmoscope, which is used to examine the retina, in 1851, and the ophthalmometer, which measures the curvature of the eye, in 1855. He also revived the three-colour theory of vision first proposed in 1801 by Thomas Young (1773–1829), who like Helmholtz was also a physician and physicist. He developed Young's ideas by showing that a single primary colour (red, green or violet) must also affect retinal structures sensitive to the other primary colours. In this way Helmholtz successfully explained the colour of after-images and the effects of colour blindness. He also extended Young's work on accommodation. In acoustics, Helmholtz's contribution was to produce a comprehensive explanation of how the upper partials in sounds combine to give them a particular tone or timbre, and how resonance may cause this to happen – for example in the mouth cavity to produce vowel sounds. He also formulated a theory of hearing, correctly suggesting that structures within the inner ear resonate at particular frequencies to enable both pitch and tone to be perceived.

Helmholtz also interested himself in hydrodynamics and in 1858 produced an important paper in which he established the mathematical principles that define motion in a vortex. Helmholtz will also be remembered for his electrical double layer theories. First proposed in 1879, they were initially applied to the situation of a solid immersed in a solution. He imagined that a layer of ions is held at a solid surface by an oppositely charged layer of ions in the solution. These ideas were later applied to solids suspended in liquids.

Helmholtz dominated German science during the mid-1800s, his wide-ranging and exact work bringing it to the forefront of world attention, a position Germany was to enjoy well into the following century. He served as a great inspiration to others, not least his many students. Foremost of these was Heinrich Hertz (1857–1894), who as a direct result of Helmholtz's encouragement discovered radio waves. Helmholtz took classical mechanics to its limits in physics, paving the way for the radical departure from tradition that was soon to follow with the quantum theory and relativity. When this revolution did occur, however, it was ushered in mainly by German scientists applying the mathematical and experimental expertise upon which Helmholtz had insisted.

Helmont Jan Baptista van 1579–1644 was a Flemish chemist and physician.

He was born into Flemish landed gentry in Brussels on 12 January 1579. He studied chemistry, medicine and mysticism at the University of Louvain and received his MD in 1599. From 1600 to 1605 he travelled to Spain, Italy, France and England to gain more medical knowledge and then carried out private research at his home at Vilvorde. Helmont took great interest in the controversy over the 'weapon salve' and the magnetic cure of wounds (applying ointment to the weapon rather than the wound). In 1621 he published his treatise *De Magnetica Vulnerum ... Curatione* in which, although he did not reject the belief, he insisted that it was a natural phenomenon with no supernatural elements. He was arrested for his views and the church proceedings against him did not formally end until 1642. Helmont died in Vilvorde on 30 December 1644. He had married in 1605.

Although Helmont believed in alchemy, his work represents a transition to chemistry proper. He first emphasized the use of the balance in chemistry, giving him an insight into the indestructibility of matter and the fact that metals dissolved in acid were recoverable, not destroyed or transmuted. His experiments led to his belief that all matter was composed of water and air, which he demonstrated by growing a willow tree over five years in a measured quantity of earth; after this time, after adding only water, the tree had increased its weight by 74 kg/164 lb, while the earth it was in had lost little weight. It is claimed that Helmont was the first to use the term 'gas'; he applied chemical analysis to smoke which he found to be different from air and water vapour and specific to the substance burned. He identified four gases: carbon dioxide, carbon monoxide, nitrous oxide and methane. He was the first to take the melting point of ice and the boiling point of water as standards for temperature and was the first to use the term 'saturation' to signify the combination of an acid and a base. In medicine, Helmont was a follower of Paracelsus and improved on his chemical medicines, notably mercury preparation. He discussed the effects of ether, demonstrated acid as the digestive agent in the stomach, made contributions to the understanding of asthma and was among the founders of the modern ontological concept of disease. He rejected the humoral theory and its traditional therapy and used remedies which specifically considered the type of disease, the organ affected and the causative agent. During his lifetime, Helmont published *Supplementum de Spadanis Fontibus* (1624), *Febrium Doctrina Inaudita* (1642) and *Opuscula Medica Inaudita* (1644). His works were collectively published posthumously by his son as *Ortus Medicinae* (1648).

Henry Joseph 1797–1878 was a US physicist who carried out early experiments in electromagnetic induction.

Henry was born in Albany, New York, on 17 December 1797, the son of a labourer. He had little schooling, working his way through Albany Academy to study medicine and then engineering (from 1825). He worked at the academy as a teacher, and in 1826 was made Professor of Mathematics and Physics. He moved to New Jersey College (later Princeton) in 1832 as Professor of Natural

Philosophy, in which post he lectured in most of the sciences. He was the Smithsonian Institution's first Director (1846) and first President of the National Academy of Sciences (1868), a position he held until his death, in Washington, on 13 May 1878.

Many of Henry's early experiments were with electromagnetism. By 1830, he had made powerful electromagnets by using many turns of fine insulated wire wound on iron cores. In that year he anticipated Michael Faraday's discovery of electromagnetic induction (although Faraday published first), and two years later he discovered self-induction. He also built a practical electric motor and in 1835 developed the relay (later to be much used in electric telegraphy). In astronomy, Henry studied sunspots and solar radiation, and his meteorological studies at the Smithsonian led to the founding of the US Weather Bureau. The unit of inductance was named the henry in 1893.

Henry William 1774–1836 was a chemist and physician, best known for his study of the solubility of gases in liquids, which led to what is now known as Henry's law. He was also an early experimenter in electrochemistry, and he established that firedamp – the cause of many mining disasters – is methane.

Henry was born in Manchester on 12 December 1774. His father was a physician and industrial chemist who introduced the use of chlorine as a bleaching agent for textiles. A childhood accident affected Henry's growth and left him in poor health for the rest of his life. He went to a school run by a local clergyman and then to Manchester Academy, which was an offshoot of the famous Dissenting Academy at Warrington. (The Dissenting Institutions offered a wide curriculum containing scientific, practical and mathematical subjects, unlike the grammar schools of the time which provided an education based almost entirely on Latin.)

On leaving the Academy in 1790, Henry became secretary companion to Dr Thomas Percival, the founder of the influential Literary and Philosophical Society of Manchester. Five years later he went to Edinburgh University to study medicine, but left after a year to help his father. During this time he began his own chemical experiments, and published his first paper in 1797. He returned to Edinburgh in 1805 and graduated in medicine two years later, with a thesis on uric acid.

Henry suffered from a disorder that affected his hands and after unsuccessful surgery in 1824 he was unable to carry out chemical experiments; he returned to a study of medicine. But his continuing poor health led to severe depression and in Manchester on 2 September 1836 he committed suicide.

Henry's early chemical experiments concerned the composition of hydrogen chloride and hydrochloric acid. A series of chemistry lectures which he gave in Manchester in the winter of 1798–1799 were later published as a successful textbook *Elements of Experimental Chemistry*, which he expanded over the following 30 years and which by 1829 had run to its 11th edition. In 1803, he published the results of his research on the solubility of gases, outlining the basis of Henry's law which in its modern form states that the mass of gas dissolved by a given volume of liquid is directly proportional to the pressure of the gas with which it is in equilibrium, provided the temperature is constant. In mathematical terms, $m = kp$, where m is the mass of gas, p the pressure exerted by the undissolved gas, and k a constant. There are many deviations from the law and it is closely obeyed only by very dilute solutions.

In 1805, Henry worked at determining the composition of ammonia. He confirmed that it contains hydrogen and nitrogen in the proportions suggested by Humphry Davy (1778–1829) and Claude Berthollet (1748–1822), but showed that Davy was incorrect in his belief that it contains also oxygen, which for Davy was a necessary characteristic of all alkalis.

At the beginning of the nineteenth century coal gas was being introduced as a possible fuel for illumination. Henry worked for about 20 years on the analysis of inflammable mixtures of gases and attempted to find correlations between chemical composition and illuminative properties. During these investigations he confirmed John Dalton's results of the composition of methane and ethane, and like Dalton showed that hydrogen and carbon combine in definite proportions to form a limited number of compounds. In this work Henry made use of the catalytic properties of platinum, newly discovered by Johann Döbereiner in 1821.

After 1824, Henry returned to medical investigations and studied contagious diseases. He believed that these were spread by chemicals which could be rendered harmless by heating; he used heat to disinfect clothing during an outbreak of Asiatic cholera in 1831. By the time of Henry's death, his son W. C. Henry was already an active chemist, publishing a work in 1836 on the poisoning of platinum catalysts.

Heraklides of Pontus 388–315 BC was an ancient Greek philosopher and astronomer who is remembered particularly for his teaching that the Earth turns on its axis, from west to east, once every 24 hours.

Born at Heraklea, near the Black Sea, Heraklides migrated to Athens and became a pupil of Speusippus, under the direction of Plato, in the Academy. Nearly elected to succeed Speusippus as Head of the Academy, Heraklides was at one stage left temporarily in charge of the whole establishment by Plato. He is said also to have attended the schools of the Pythagorean philosophers, and would thus have come into contact with his contemporary, Aristotle (384–322 BC). Although all his writings are lost, it is clear from those of his contemporaries that his astronomical theories were more advanced than those of other scientists of his age.

In proposing the doctrine of a rotating Earth, Heraklides contradicted the accepted model of the universe put forward by Aristotle. Aristotle had accepted that the Earth was fixed, and said that the stars and the planets in their respective spheres might be at rest. Heraklides thought it highly impossible that the immense spheres of the stars and planets rotated once every 24 hours. The idea of a rotating Earth was not to be accepted for another 1,800 years. He also thought that the observed motions of Mercury and Venus suggested that they orbited the Sun rather than the Earth. He did not completely adopt the heliocentric view of the universe stated later by Aristarchos (in around 280 BC). He proposed instead that the Sun moved in a circular orbit (in its sphere) and that Mercury and Venus moved on epicycles around the Sun as centre.

Herbig George Howard 1920– is a US astronomer who specializes in spectroscopic research into irregular variable stars, notably those of the T-Tauri group.

Born on 20 January 1920 in Wheeling, West Virginia, Herbig was educated at the University of California in Los Angeles, graduating in 1943. In 1944, he became a member of the staff at the Lick Observatory, California, beginning as an assistant. From junior astronomer he progressed to assistant astronomer, astronomer and, from 1960 to 1963, Assistant Director; in 1966 he became Professor of Astronomy.

Herbig's main area of research has been the nebular variables of which the prototype is T-Tauri. It is believed that the members of this group are in an early stage of stellar evolution. Most of them are red and fluctuate in light intensity; their associated nebulosities are also variable in brightness and structure, although the reason for this is not known.

With Bidelman and Preston, Herbig has worked on the spectra of these and other variable stars. In 1960, he drew attention to the fact that many of them have a predominance of lithium lines, similar to the abundance of lithium on Earth and in meteorites (although considerably more abundant than in the Sun). He concluded that both the planetary and T-Tauri abundance of lithium might represent the original level of this element in the Milky Way, but that the lithium in the Sun and other stars may have largely been lost through nuclear transformation. Herbig also showed that there seems to be a conservation of angular momemtum in such young, cool variable stars, and that T-Tauri variables move together in parallel paths within the obscuring cloud in which they were formed.

Herbig also worked on binary stars, which are in relative orbital motion because of their proximity and mutual gravitational attraction. He investigated the binary of shortest known orbital period, VV Puppis, which is an eclipsing binary of period 100 minutes.

Herbig has also investigated the spectra of atoms and molecules that originate in interstellar space. In 1904, Johannes Hartmann (1865–1936) found that the absolute lines of Ca(II) in the spectrum of Delta Orionis do not take part in the periodic oscillation of other lines. Since then other atoms and molecular combinations have been discovered to have originated in interstellar space, such as Na(I), Ca(I), K(I), Ti(II), CN and CH. There are also a number of diffuse interstellar absorption lines which are as yet unidentified but which Herbig has succeeded in resolving into band lines.

Herbrand Jacques 1908–1931 was a French mathematical prodigy who, in a life cut tragically short, still originated some innovatory concepts in the field of mathematical logic. Interested also in class-field theory, he remains best remembered for the Herbrand theorem.

Herbrand was born on 12 February 1908 in Paris, where he grew up and was educated, entering the Ecole Normale Supérieure at the age of 17 already precociously talented in mathematics. Three years later he published his first paper on mathematical logic for the Paris Academy of Sciences, and in the following year – 1929, when he was 21 – for his doctorate, he produced the paper in which he formulated the Herbrand theorem. His military service in the French army then occupied him for 12 months, at the end of which a Rockefeller scholarship enabled him to go to Berlin to study there, and then for a couple of months in Hamburg and Göttingen. He was killed in an Alpine accident on 27 July 1931.

For his doctor's degree, Herbrand introduced what is now considered to be his main contribution to mathematical logic, the theorem to which his name is attached. A fundamental development in quantification theory, the Herbrand theorem established a link between that theory and sentential logic, and has since found many applications in such fields as decision and reduction problems.

Herbrand was also fascinated by modern algebra and wrote a number of papers on class field theory, which deals with Abelian extensions of a given algebraic number field from properties of the field. He contributed greatly to the development of the theory which, until that time, had received scant attention since first being devised by Leopold Kronecker (1823–1891) in the middle of the nineteenth century. After Herbrand's death, however, the theory was the subject of further development by later mathematicians.

Hermite Charles 1822–1901 was a French mathematician who was a principal contributor to the development of the theory of algebraic forms, the arithmetical theory of quadratic forms, and the theories of elliptic and Abelian functions. Much of his work was highly innovative, especially his solution of the quintic equation through elliptic modular functions, and his proof of the transcendence of e. A man of generous disposition and a good teacher, in his work Hermite showed that the brilliance of the previous generation of mathematicians – Karl Gauss, Augustin Cauchy, Karl Jacobi and Lejeune Dirichlet – was not to fade in the next.

Hermite was the sixth of the seven children of a minor businessman in Dieuze, Lorraine, on 24 December 1822. He grew up and was educated in Nancy, however. Completing his education, he went to Paris and first studied physics at the Henry IV Collège before he went to the Lycée Louis le Grand and became fascinated by mathematics. He was always hopeless at passing examinations – but his private reading at night of the works of Leonhard Euler, Karl Gauss and Joseph Lagrange might well have distracted him from other preparation. From 1842 he studied at the Ecole Polytechnique for a year, but because he did not pass the examinations there either (and apparently also because he was congenitally lame), he was asked not to return. So it was not until 1848, at the age of 25, that he finally graduated, although by that time he already had some reputation as an innovative mathematician. He was appointed to a minor teaching post at the Collège de France. In 1856, two things happened: he was elected to the Paris Academy of Sciences, and he caught smallpox. Recovering from the latter, he had to wait until he was 47 years old before he finally attained the status of Professor, first in 1869 at the Ecole Normale, then a year later at the Sorbonne. An honorary member of many societies, he received numerous awards and honours during his lifetime. He died in Paris on 14 January 1901.

Throughout his life Hermite exerted great scientific influence by corresponding with other prominent mathematicians. One of his first pieces of work was to generalize Abel's theorem on elliptic functions to the case of hyperelliptic ones. Communicating his discovery to Karl Jacobi (1804–1851) in August 1844, he also discussed four other papers on number theory.

Between 1847 and 1851 he worked on the arithmetical theory of quadratic forms and the use of continuous variables. Then for ten years between 1854 and 1864 he worked on the theory of invariants.

Today he is remembered chiefly in connection with Hermitean forms (a complex generalization of quadratic forms) and with Hermitean polynomials – work carried out in 1873. In the same year, he showed that e, the base of natural logarithms, is transcendental. (Transcendental numbers are real or complex numbers which are not algebraic; we now know that most real numbers are transcendental.) It was by a slight adaptation of Hermite's proof for this that Ferdinand von Lindemann (1852–1939) in 1882 demonstrated the transcendence of π.

In 1872 and 1877 Hermite solved the Lamé differential equation, and in 1878 he solved the fifth-degree (quintic) equation of elliptic functions.

Hermite's scientific work was collected and edited by the later French mathematician Charles Picard (1856–1941).

Hero of Alexandria lived c.AD 60, variously described as an Egyptian scientist and a Greek engineer, was the greatest experimentalist of antiquity. His numerous writings describe many of his inventions, formulae and theories, some of them centuries ahead of their time. His famous aeolipile which demonstrated the force of steam generated in a closed chamber and escaping through a small aperture (the first rudimentary steam turbine) was not developed further for another 18 centuries.

A gifted mathematician, he adopted the division of the circle into 360°, orginally brought to the western world by Hipparchus of Bithynia. He was also a famous teacher and at Alexandria founded a technical school with one section devoted entirely to research.

He regarded air as a very elastic substance which could

be compressed and expanded, and successfully explained the phenomenon of suction and associated apparatus, such as the pipette and cupping glass; he also used a suction machine for pumping water. His writing on air compressibility and density and his assumption that air is composed of minute particles able to move relative to one another preceded Robert Boyle by 1,500 years.

In mechanics, he devised a system of gear wheels which could lift a mass of 1,000 kg/2,200 lb by means of a mere 5 kg/11 lb. His work entitled *Mechanics*, the only existing copy of which is in Arabic, contains the parallelogram of velocity, the laws of levers, the mysteries of motion on an inclined plane and the effects of friction, and much data on gears and the positions of the centre of gravity of various objects. His construction of a variable ratio via a friction disc has been used to build a motor vehicle with a semi-automatic transmission.

Civil and construction engineers owe him a debt for his tables of dimensions used in building arches and in drilling tunnels and wells. Another of his famous books, *On the Dioptra*, describes a diversity of instruments, including a type of theodolite that uses a refined screw-cutting technique for use in boring tunnels; the book also describes a hodometer for measuring the distance travelled by a wheel.

Hero's book *Metrica* explains the measurement of plain and solid geometrical figures and discusses conic sections, the frustum of a cone, and the five regular or Platonic solids. It includes a formula for calculating the area of a triangle from the lengths of its sides, and a method of determining the square root of a non-square number.

Another of his books, *Pneumatics*, describes numerous mechanical devices operated by gas, water, steam, or atmospheric pressure, and siphons, pumps, fountains and working automata in the likeness of animals or birds devised, it seems, just for amusement.

Hero was a genius in the techniques of measurement of all kinds and in founding the science of mechanics. After his death there was no further progress in physics for centuries. But for the Arabs, it is extremely likely that all record of his works would have vanished.

Héroult Paul Louis Toussaint 1863–1914. French metallurgist; *see* Hall and Héroult.

Herschel Frederick William 1738–1822 was a German-born English astronomer who contributed immensely to contemporary scientific knowledge. Through determined efforts to improve the quality of his telescopes, he was able to use the finest equipment of his time, which in turn permitted him to make many significant discoveries about the nature and distribution of stars and other bodies, both within the Solar System and beyond it. He was the most influential astronomer of his day.

Herschel was born in Hanover, Germany, on 15 November 1738. At the age of 14 he joined the regimental band of the Hanoverian Guards as an oboist (as his father had before him), and four years later he visited England with the band. In 1757, he emigrated to Britain, going first to

Leeds – where he earned his living copying music and teaching – and then, three years later, by commission from the Earl of Darlington, to Durham to become conductor of a military band there. From 1761 to 1765, Herschel worked as a teacher, organist, composer and conductor in Doncaster. He then did similar work in Halifax for a year. In 1766, he was hired as organist at the Octagon Chapel in Bath, where he remained for the next 16 years; during this time he ran a tutorial service that helped to finance his growing interest in astronomy. In 1772, he went to Hanover to bring his sister Caroline back to England. She too became fascinated by astronomy and helped Herschel enormously, both in the delicate task of preparing instruments and in making observations. The serious astronomical work began in 1773, with the building of telescopes and the grinding of mirrors. Herschel's first large reflector was set up behind his house in 1775.

Herschel's discovery of the planet Uranus in 1781 – the first planet to have been discovered in modern times – created a sensation: it signalled that Newton's work had not covered everything there was to know about the universe. Herschel originally named the new planet 'Georgium Sidum' (George's Star) in honour of King George III, but the name Uranus – proposed by Johann Bode (1747–1826) of the Berlin Observatory – was ultimately accepted. Nevertheless, the King was flattered, and Herschel received a royal summons to bring his equipment to court for inspection. In the same year the Royal Society elected Herschel a Fellow, and awarded him the Copley Medal. In 1782, he was appointed court astronomer, a post that carried with it a pension of 300 guineas per year. This enabled him to give up teaching and to move from Bath to Windsor, and then to Slough, although he continued to make telescopes for sale in order to supplement his income until 1788 (when he married a wealthy widow). In addition, the King provided grants for the construction of larger instruments; for instance, he provided £4,000 for a telescope with a focal length of more than 12 m/40 ft and a reflector that was 1 m/3 ft in diameter. (This telescope proved rather cumbersome and was not used after 1811, although for many years it remained the largest in the world.)

Herschel visited Paris in 1801, meeting Pierre Laplace (1749–1827) and Napoleon Bonaparte. Many honours were conferred upon him (he was knighted in 1816), and he was a member of several important scientific organizations. He became ill in 1808, but continued to make observations until 1819. In that year his only son, John (1792–1871), finished his studies at Cambridge and came to take over his father's work. Herschel died in Slough on 25 August 1822.

Herschel's first large telescope was a 1.8-m/6-ft Gregorian reflector that he built himself in 1774. In many ways it was better than other existing telescopes, and Herschel decided that its primary use would be to make a systematic survey of the whole sky. His first review was completed in 1779, when he immediately began a second survey with a 2.1-m/7-ft reflector. In this project he concentrated on noting the positions of double stars, that is, stars that appear to be very close together, perhaps only as a consequence of chance alignment with the observer. His first catalogue of double stars was published in 1782 (with 269 examples); a second catalogue (with 434 examples) appeared in 1785; and a third was issued in 1821, bringing the total number of double stars recorded to 848. Galileo (1564–1642) and others had suggested that the motion of the nearer (and it was presumed therefore brighter) star relative to the more distant (thus fainter) star could be used to measure annual movements of the stars. Herschel's observations in 1793 demonstrated that a correlation between dimness and distance did not apply in all cases, and that in fact some double stars were so close together as to rotate round each other, held together by an attractive force. This was the first indication of gravity acting on bodies outside the Solar System.

Herschel's work on bodies within the Solar System included an accurate determination of the rotation period of Mars (in 1781): 24 h, 39 min and 21.67 sec – only two minutes longer than the period now accepted. By inventing an improved viewing apparatus for his telescope, particularly valuable when little light was available, in 1787 he discovered Titania and Oberon, two satellites of Uranus. Incorporating the 'Herschelian arrangement' into a massive telescope 12 m/39 ft long, Herschel then discovered two further satellites of Saturn (Enceladus and Mimas) on its first night of use. Saturn continued to be an object of great interest to him.

During the 1780s Herschel published a number of papers on the evolution of the universe from a hypothetical uniform initial state to one in which stars were clumped into galaxies (seen as nebulae). Herschel had become interested in nebulae in 1781, when he was given a list of a hundred of these indistinct celestial bodies compiled by Charles Messier (1730–1817). Herschel began looking for more nebulae in 1783, and the improved resolving power of his telescopes enabled him to see them as clusters of stars. His first catalogue of nebulae (citing no fewer than 2,500 examples) was published in 1802; an even longer catalogue (with 5,000 nebulae) appeared in 1820.

In 1800, Herschel examined the solar spectrum using prisms and temperature-measuring equipment. He found that there were temperature differences between the various regions of the spectrum, but that the hottest radiation was not within the visible range, but in the region now known as the infrared. This was the beginning of the science of stellar photometry. Using data published by Nevil Maskelyne (1732–1811) on seven particularly bright stars, Herschel demonstrated that if the Sun's motion towards Argelander (a star in the constellation Hercules) was accepted, then the 'proper motion' of the seven stars was a reflection of the motion of the observer. This relegation of the Solar System from the centre of the universe was, in its way, analogous to the dethronement of the Earth from the centre of the Solar System by Copernicus. Herschel also measured the velocity of the Sun's motion.

An industrious and dedicated astronomer, and a very practical man, Herschel contributed enormously to the advance of scientific progress.

Herschel John Frederick William 1792–1871 was the first astronomer to carry out a systematic survey of the stars in the southern hemisphere and to attempt to meaure, rather than estimate, the brightness of stars. Besides this, he continued his father's studies of double stars, nebulae and the Milky Way.

John Herschel was born in Slough, England, on 7 March 1792, the only child of the astronomer William Herschel. Unsurprisingly, the young Herschel's career was strongly influenced by the fact that he was brought up by his father and his aunt Caroline, who were both devoted to astronomy. From the age of eight, after a short spell at Eton, Herschel was educated at a local private school. He then went to St John's College, Cambridge, to read mathematics. From 1816 to 1850 Herschel had no permanent paid post, but his scientific life was closely bound up with two Royal Societies. He became a Fellow of the Royal Society in 1813, served as its Secretary from 1824 to 1827, and won its Copley Medal in 1821 and 1847 and its Royal Medal in 1833, 1836 and 1840. He received the Lalande Prize of the French Academy in 1825 and the Gold Medal of the Astronomical Society in 1826 for the catalogue of double stars which he compiled with James South. When Herschel died on 11 May 1871 he was mourned by the whole nation, not only as a great scientist, but also as a remarkable public figure.

After completing his studies and taking his MA in 1816, Herschel embarked on his scientific career. As one of his obituarists noted, he may well have taken up astronomy out of a sense of 'filial devotion'. His first paper, on the computation of lunar occultation, appeared in 1822, by which time he had moved on to systematically observing double stars. This work, a continuation of one of his father's projects, was carried out in collaboration with James South, who possessed two excellent refracting telescopes in London. William Herschel had demonstrated that the orbital motion of binary stars was due to their mutual attraction; John Herschel studied new and known double star systems, especially Gamma Virginis, in order to establish a method for determining their orbital elements.

Soon after his marriage to Margaret Brodie Stewart, in 1829, Herschel planned an expedition to the southern hemisphere. He chose to go to the Cape of Good Hope, partly because of its excellent astronomical tradition, built up as a result of Lacaille's work in the 1850s, partly because, being located on the same meridian of longitude as eastern Europe, it made cooperative observations easier. Herschel arrived in Cape Town with his wife and the first three of their twelve children on 16 January 1834 and set himself the mammoth task of mapping the southern skies. By 1838 he had catalogued 1,707 nebulae and clusters, listed 2,102 pairs of binary stars and carried out star counts in 3,000 sky areas. Besides simply mapping the stars, he made accurate micrometer measurements of the separation and position angle of many stellar pairs and produced detailed sketches of the Orion region, the Magellanic Clouds, and extragalactic and planetary nebulae. He also recorded the behaviour of Eta Carinae (whose nature is still not entirely understood) when it underwent a period of dramatic brightening in December 1837.

To ascertain the brightness of the stars he catalogued, Herschel invented a device called an astrometer. It enabled him to compare the light output of a star with an image of the full moon, whose brightness could be varied to match the star under observation with the main telescope. This attempt to ascertain absolute magnitude was a major step forward in stellar photometry. Besides his personal research work, Herschel also collaborated with Thomas Maclear, Director of the Cape Observatory, in the Observatory's routine geodetic and tidal work and in observations of Encke's and Halley's comets.

On his return to London on 15 May 1838, Herschel began to prepare the results of his African trip for publication. At the same time he pursued his interest in the art of photography. His interest in the subject led him to invent much of the vocabulary – positive, negative and snapshot – that is associated with the craft today. His massive *Results of Astronomical Observations Made During the Years 1834–38 at the Cape of Good Hope* was published in 1847; two years later he produced *Outlines of Astronomy*, which became a standard textbook for the following decades. His *General Catalogue of Nebulae and Clusters*, now known as the NGC, still remains the standard reference catalogue for these objects. The last of his ambitious projects, *General Catalogue of 10,300 Multiple and Double Stars*, was published posthumously.

John Herschel was a celebrated scientist throughout his life and his name epitomized science to the public in much the same way as Einstein's did in the following century. After his death his reputation suffered a decline, and it rose again only when astronomers began to realize that he occupied the same commanding and innovative position for astronomers in the southern hemisphere as his father William did for astronomers in the north.

Hertz Heinrich Rudolf 1857–1894 was a German physicist who discovered radio waves. His name is commemorated in the use of the hertz as a unit of frequency, one hertz being equal to one complete vibration or cycle per second.

Hertz was born in Hamburg on 22 February 1857. As a child he was interested in practical things and equipped his own workshop. At the age of 15 he entered the Johanneum Gymnasium and, on leaving school three years later, went to Frankfurt to gain practical experience as the beginning of a career in engineering. Engineering qualifications were governed by a state examination, and he went to Dresden Polytechnic to work for this in 1876. During a year of compulsory military service from 1876 to 1877, Hertz decided to be a scientist rather than an engineer and on his return he entered Munich University. He began studies in mathematics, but soon switched to practical physics, which greatly interested him.

Hertz moved to Berlin in 1878 and there came into contact with Hermann Helmholtz (1821–1894), who immediately recognized his talents and encouraged him greatly. He gained his PhD in 1880 and remained at Berlin for a further three years to work with Helmholtz as his assistant. Then in 1883, Hertz moved to Kiel to lecture in physics, but the lack of a proper laboratory there caused him to take up the position of Professor of Physics at Karlsruhe in 1885. He stayed in Karlsruhe for four years, and it was there that he carried out his most important work, discovering radio waves in 1888. In the same year, he began to suffer from toothache, the prelude to a long period of increasingly poor health due to a bone disease. Hertz moved to Bonn in 1889, succeeding Rudolf Clausius (1822–1888) as Professor of Physics. His physical condition steadily deteriorated and he died of blood poisoning on 1 January 1894, at the early age of 36.

In 1879, a prize was offered by the Berlin Academy for the solution of a problem concerned with Maxwell's theory of electricity, which was set by Helmholtz with Hertz particularly in mind. Hertz declined the task believing that it would take up more time and energy than he could then afford. Instead, for his doctorate he carried out a theoretical study of electromagnetic induction in rotating conductors. It was not until five years later on his move to Karlsruhe in 1885 that Hertz found the facilities and the time to work on the problem set by the Berlin Academy. He used large improved induction coils to show that dielectric polarization leads to the same electromagnetic effects as do conduction currents; and he generally confirmed the validity of the famous equations of James Clerk Maxwell concerning the behaviour of electricity.

In 1888, he realized that Maxwell's equations implied that electric waves could be produced and would travel through air. Hertz went on to confirm his prediction by constructing an open circuit powered by an induction coil, and an open loop of wire as a receiving circuit. As a spark was produced by the induction coil, so one occurred in the open receiving loop. Electric waves travelled from the coil to the loop, creating a current in the loop and causing a spark to form across the gap. Hertz went on to determine the velocity of these waves (which were later called radio waves), and on showing that it was the same as that of light, devised experiments to show that the waves could be reflected, refracted and diffracted. One particular experiment used prisms made of pitch to demonstrate that radio waves undergo refraction in the same way that light is refracted by a glass prism.

From about 1890, Hertz gained an interest in mechanics. He developed a system with only one law of motion – that the path of a mechanical system through space is as straight as possible and is travelled with uniform motion. When this law is developed subject to the nature of space and the constraints on matter, it can be shown that the mechanics of Isaac Newton (1642–1727), Joseph Lagrange (1736–1813) and William Rowan Hamilton (1805–1865), who extended the methods of dealing with mechanical systems considerably, arise as special cases.

Heinrich Hertz made a discovery vital to the progress of technology by demonstrating the generation of radio waves. It is tragic that his early death robbed him of the opportunity to develop his achievements and to see Guglielmo Marconi (1874–1937) and others transform his discovery into a worldwide method of communication.

Hertzsprung Ejnar 1873–1967 was a Danish astronomer and physicist who, having proposed the concept of the absolute magnitude of a star, went on to describe for the first time the relationship between the absolute magnitude and the temperature of a star, formulating his results in the form of a graphic diagram that has since become a standard reference.

Hertzsprung was born in Frederiksberg, Denmark, on 8 October 1873. His father was interested in astronomy and stimulated a similar fascination for the subject in his son, but the poor financial prospects of an aspiring astronomer led Hertzsprung initially to choose chemical engineering as his career. He graduated from the Frederiksberg Polytechnic in 1898 and then went to St Petersburg, where he worked as a chemical engineer until 1901. Returning to Copenhagen via Leipzig, he studied photochemistry under Wilhelm Ostwald (1853–1932) and began to work as a private astronomer at the observatory of the University of Copenhagen and the Urania Observatory in Frederiksberg. Under the generous tutelage of H. Lau, Hertzsprung rapidly acquired the skills of contemporary astronomy.

Following a correspondence with Karl Schwarzschild (1873–1916) at the University of Göttingen, Hertzsprung was invited to take up the post of Assistant Professor of Astronomy at the Göttingen Observatory in 1909. When Schwarzschild moved on to the Potsdam Astrophysical Observatory later that year, Hertzsprung went with him. Willem de Sitter (1872–1934) was appointed Director of the Leiden Observatory (in the Netherlands) in 1919, and he appointed Hertzsprung Head of the Department of Astrophysics. Within a year, Leiden University appointed Hertzsprung a Professor, and on the death of de Sitter he also became Director of the Leiden Observatory. He retired in 1945 and returned to Denmark, but he did not cease his astronomical research until well into the 1960s.

Hertzsprung's outstanding contributions to astrophysics were recognized by his election to many prestigious scientific academies and societies and with the award of a number of honours. He died in Roskilde, Denmark, on 21 October 1967.

It was quite early in his work at the Observatory in Copenhagen that Hertzsprung realized the importance of photographic techniques in astronomy. He was extremely well qualified to apply these methods and did so with great precision and energy. In 1905, he published the first of two papers (the second appearing in 1907) in a German photographic journal on the subject of stellar radiation. He proposed a standard of stellar magnitude (brightness) for scientific measurement, and defined this 'absolute magnitude' as the brightness of a star at the distance of 10 parsecs (32.6 light years). As a further innovation, he described the relationship between the absolute magnitude and the colour – i.e. the spectral class or temperature – of a star. During the following year (1906), Hertzsprung plotted a graph of this relationship in respect of the stars of the Pleiades. Later, he noticed that there were some stars of the same spectral class that were much brighter, and some that were much dimmer, than the Sun. He named these the red giants and the red dwarfs respectively.

Publication of his papers in a photographic journal, and refusal altogether to publish his diagrammatic material (because of diffidence in the quality of his own observations), meant that his discoveries were simply not known by Hertzsprung's fellow astronomers. And in 1913 Henry Russell (1877–1957), an American astronomer, presented to the Royal Astronomical Society a diagram depicting the relationship that Hertzsprung had previously and independently discovered, between the temperature and absolute magnitude of stars. Credit was eventually accorded to both astronomers equally and the diagram named after both of them.

The Hertzsprung–Russell diagram, one of the most important tools of modern astronomy, consists of a log–log plot of temperature versus absolute magnitude. As plotted, the stars range themselves largely along a curve running from the upper left (the blue giant stars) to the lower right (the red dwarf stars) of the graph. This apparent arrangement is simply a reflection of the mass of each star, which is responsible for its temperature and luminosity. Approximately 90% of stars belong to this 'main sequence'; most of the rest are red giants, blue dwarfs, Cepheid variables or novae. The blue giant stars are giant hot stars, the red dwarfs are compact cooler stars. Our Sun lies near the middle of the main sequence, and is classed as a yellow dwarf.

One of the earliest uses of the Hertzsprung–Russell diagram was devised in 1913, when Hertzsprung developed the method of 'spectroscopic parallax' (as distinct from 'trigonometric parallax') for the determination of the distances of stars from the Earth. His method relied on data for the proper motions of the nearest (galactic) Cepheids and on Henrietta Leavitt's data for the periods of the Cepheids (which are variable stars) in the Small Magellanic Clouds (which are, it turned out, extragalactic). He deduced the distances of the nearest Cepheids from their proper motions and correlated them with their absolute magnitude. He then used Henrietta Leavitt's data on the length of their periods to determine their absolute magnitude and hence their distance. He found the Small Magellanic Cloud Cepheids to be at an incredible distance of 10,000 parsecs. His method was excellent, but there was a serious source of error, which led to an overestimation of the distance: he had not accounted for the effect of galactic absorption of stellar light. Nevertheless, this work earned Hertzsprung the Gold Medal of the Royal Astronomical Society.

The Hertzsprung–Russell diagram has also been essential to the development of modern theories of stellar evolution. As stars age and deplete their store of nuclear fuel, they are believed to leave the main sequence and become red giants. Eventually they radiate so much energy that they then cross the main sequence and collapse into blue dwarfs. Larger stars may follow a different pattern and explode into novae, or collapse to form black holes, at the end of their lifespans.

In 1922, Hertzsprung published a catalogue on the mean colour equivalents of nearly 750 stars of magnitude greater than 5.5. This catalogue was notable for the particularly elegant manner in which Hertzsprung managed to analyse the data to uncover a linear relationship.

Most of Hertzsprung's later work was devoted to the study of variable and of double stars. He worked on variable stars (especially Polaris) at Potsdam in 1909, and later in Johannesburg (from 1924 to 1925) and at the Harvard College Observatory (1926) and at Leiden.

Herzberg Gerhard 1904– is a German-born Canadian physicist who is best known for his work in determining – using spectrocopy – the electronic structure and geometry of molecules, especially free radicals (atoms or groups of atoms that possess a free, unbonded electron). He has received many honours for his work, including the 1971 Nobel Prize for Chemistry.

Herzberg was born on 25 December 1904 in Hamburg, where he received his early education. He then studied at the Technische Universität in Darmstadt, from which he gained his doctorate in 1928, and carried out postdoctoral work at the universities of Göttingen (1928 to 1929) and Bristol (1929 to 1930). On returning to Germany in 1930 he became an unsalaried lecturer at the Technische Universität but in 1935, with the rise to power of Adolf Hitler, he fled to Canada, where he became Research Professor of Physics at the University of Saskatchewan, Saskatoon, from 1935 to 1945. He spent the period from 1945 to 1948 in the United States, as Professor of Spectroscopy at the Yerkes Observatory (part of the University of Chicago) in Wisconsin, then returned to Canada. From 1939 until his retirement in 1969, he was Director of the Division of Pure Physics for the National Research Council in Ottawa – a laboratory generally acknowledged as being one of the world's leading centres for molecular spectroscopy.

Herzberg's most important work concerned the application of spectroscopy to elucidate the properties and structure of molecules. Depending on the conditions, molecules absorb or emit electromagnetic radiation (much of it in the visible part of the spectrum) of discrete wavelengths. Moreover, the radiation spectrum is directly dependent on the electronic and geometric structure of an atom or molecule and therefore provides detailed information about molecular energies, rotations, vibrations and electronic configurations. Herzberg, studying common molecules such as hydrogen, oxygen, nitrogen and carbon monoxide, discovered new lines in the spectrum of molecular oxygen; called Herzberg bands, these spectral lines have been useful in analysing the upper atmosphere. He also elucidated the geometric structure of molecular oxygen, carbon monoxide, hydrogen cyanide and acetylene (ethyne); discovered the new molecules phosphorus nitride and phosphorus carbide; proved the existence of the methyl and methylene free radicals; and demonstrated that both neutrons and protons are part of the nucleus. His research in the field of molecular spectroscopy not only provided experimental results of fundamental importance to physical chemistry and quantum mechanics but also helped to stimulate further research into the chemical reactions of gases.

In addition, Herzberg provided much valuable information about certain aspects of astronomy. He interpreted the spectral lines of stars and comets, finding that a rare form of carbon exists in comets. He also showed that hydrogen exists in the atmospheres of some planets, and identified the spectra of certain free radicals in interstellar gas.

Herzog Bertram 1929– is a German-born computer scientist who became one of the major pioneers in the use of computer graphics in engineering design.

Herzog was born in Offenburg, Germany, on 28 February 1929. He went to the United States and became a US citizen, and began his university education at the Case Institute of Technology, from which he graduated with a bachelor's degree in engineering in 1949. In the early 1950s he worked as a structural engineer before taking up appointments as an associate professor, first at Doanbrook and then at the University of Michigan, where he obtained a doctorate in engineering mechanics in 1961.

In 1963, Herzog joined the Ford Motor Company as engineering methods manager, where he extensively applied computers to tasks involved in planning and design. During this time he was engaged in bringing the developing field of computer graphics to the requirements of design problems in the motor industry. Herzog remained as a consultant to Ford, while returning to academic life in 1965 as Professor of Industrial Engineering at the University of Michigan. Two years later he became Director of the computer centre and Professor of Electrical Engineering and Computer Science at the University of Colorado.

Hess Germain Henri 1802–1850 was a Swiss-born Russian chemist, best known for his pioneering work in thermochemistry and the law of constant heat summation named after him.

Hess was born in Geneva on 7 August 1802, the son of a Swiss artist. When he was only three years old his family moved to St Petersburg to enable his father to be a tutor in a rich Moscow family and they adopted a Russian way of life. Hess became known as German Ivanovich Hess. He qualified in medicine in 1825 at the University of Dorpat, where he also received a thorough grounding in chemistry and geology. He went to Stockholm for a short time to study chemistry under Jöns Berzelius, then returned to Russia for a geological expedition to the Urals before settling in Irkutsk in a medical practice. He was elected an adjunct member of the St Petersburg Academy of Sciences in 1828 and two years later was chosen as extraordinary Academician. He then settled in St Petersburg, abandoned his medical practice, and devoted the rest of his life to chemistry.

Hess then took various academic appointments: Commissioner to plan the course in practical and theoretical chemistry at the St Petersburg Technological Institute (leading to a professorship there); at the Mining Institute and Chief Pedagogical Institute; and in 1838, at the Artillery School. One of his students at the Chief Pedagogical Institute was A. A. Voskressenskii, who was later to become the Professor of Chemistry and teach Dmitri Mendeleyev (1834–1907).

In 1834, Hess was made an Academician of the Russian Academy of Sciences. Between 1838 and 1843 he did the research in thermochemistry for which he is most famous, but then became less active in the field of chemical research, concentrating more on education and the seeking of due recognition for other workers. He was responsible, for example, for the granting of the prestigious Demidov Award to Karl Klaus for his discovery of ruthenium. In 1848, Hess's health failed, and he died in St Petersburg on 30 November 1850.

Hess pioneered in the field of thermochemistry which, in a climate concerned mainly with analysis and synthesis in organic and inorganic chemistry, had largely been left to physicists. He had previously worked in various other areas, including the prevention of endemic eye disorders, the analysis of water and minerals from various parts of Russia, the analysis of natural gas from the region of Baku, the oxidation of sugars, and the chemical properties of waxes and resins. His textbook *Fundamentals of Pure Chemistry* (1831) remained the standard work in the Russian language until Mendeleyev's books of the 1860s.

Hess's first paper on thermochemistry, on 'The evolution of heat in multiple proportions', was published in 1838. Two years later he published the full text of Hess's law in both French and German, which states that the heat change in a given chemical reaction depends only on the initial and final states of the system and is independent of the path followed, provided that heat is the only form of energy to enter or leave the system. Every chemical change either absorbs or evolves heat (even if the amount is not enough to cause a measurable temperature change); reactions that absorb heat are called endothermic, those that evolve heat are termed exothermic. According to modern convention, evolved heat is negative in sign and absorbed heat is positive. The symbol ΔH denotes a change in heat content at constant pressure, the unit of heat and energy being the joule.

Suppose a substance A can be converted to substance D either directly, with a heat of reaction of w joules, or indirectly by way of substances B and C, with heats of reaction of the three stages of x, y and z joules respectively. Then for

$$A \rightarrow D, \ \Delta H = -w$$
$$A \rightarrow B, \ \Delta H = -x$$
$$B \rightarrow C, \ \Delta H = -y$$
$$C \rightarrow D, \ \Delta H = -z$$

According to Hess's law the heat of reaction in going from A to D is the same whether the change is achieved directly or through a series of changes, and $w = x + y + z$. Heat changes can thus be added algebraically, allowing the calculation of heats of reaction that it would be impossible to measure directly by experiment. It is in modern terms merely an application of the law of conservation of energy, which was not formulated until 1842. In that year Hess proposed his second law, the law of thermoneutrality, which states that in exchange reactions of neutral salts in aqueous solution, no heat effect is observed. No explanation of this law was forthcoming until the announcement of Svante Arrhenius's theory of electrolytic dissociation in 1887.

After Hess died no other researchers carried on his work, and thermochemistry was neglected for the next decade. Investigations of heats of reaction had to be carried out all over again (by scientists such as Pierre Berthelot in France and Hans Thomsen in Denmark). Final recognition came in 1887, when Friedrich Wilhelm Ostwald began the section on thermochemistry in his *General Chemistry* textbook with a full account of Hess's work.

Hess Harry Hammond 1906–1969 was the US geologist who played the key part in the plate tectonics revolution of the 1960s.

Born in New York, Hess first studied electrical engineering and then trained in geology at Yale and later Princeton. From 1931, he began geophysical researches into the oceans, accompanying F. A. Vening Meinesz on a Caribbean submarine expedition to measure gravity and take soundings. During World War II, Hess continued his oceanographic investigations, undertaking extended echosoundings while captain of the assault transport USS *Cape Johnson*. During the war years he made studies of the flat-topped sea mounts which he called guyots (after Arnold Guyot, an earlier Princeton geologist). In the postwar years, he was one of the main advocates of the Mohole project, whose aim was to drill down through the Earth's crust to gain access to the upper mantle.

The vast increase in sea-bed knowledge (deriving in part from wartime naval activities) led to the recognition that certain parts of the ocean floor were anomalously young. Building on Ewing's discovery of the global distribution of mid-ocean ridges and their central rift valleys, Hess enunciated in 1962 the notion of deep-sea spread. This contended that convection within the Earth was continually creating new ocean floor at mid-ocean ridges. Material, Hess claimed, was incessantly rising from the Earth's mantle to create the mid-ocean ridges, which then flowed horizontally to constitute new oceanic crust. It would follow that the further from the mid-ocean ridge, the older would be the crust – an expectation confirmed by research in 1963 by D. H. Matthews and his student, F. J. Vine, into the magnetic anomalies of the sea floor. Hess envisaged that the process of sea-floor spreading would continue as far as the continental margins, where the oceanic crust would slide down beneath the lighter continental crust into a subduction zone, the entire operation thus constituting a kind of terrestrial conveyor belt. Within a few years, the plate tectonics revolution Hess had spearheaded had proved entirely successful. Hess's role in this 'revolution in the earth sciences' was largely due to his remarkable breadth as geophysicist, geologist and oceanographer.

Hess Victor Francis 1883–1964 was an Austrian-born US physicist who discovered cosmic rays, for which he was jointly awarded the 1936 Nobel Prize for Physics with Carl Anderson.

Hess was born in Waldstein, Austria, on 24 June 1883, the son of a forester. He was educated in Graz, at the Gymnasium then at the University, obtaining his doctorate from the latter in 1906. From 1906 to 1910 he worked at the Vienna Physical Institute as a member of F. Exner's research group studying radioactivity and atmospheric ionization, then from 1910 to 1920 he was an assistant to S. Meyer at the Institute of Radium Research of the Viennese Academy of Sciences. In 1920, Hess was appointed Extraordinary Professor of Experimental Physics at Graz University. From 1921 to 1923, however, he was on a two-year sabbatical in the United States, as Director of the Research Laboratory of the US Radium Corporation in Orange, New Jersey, and also as a consultant to the Department of the Interior (Bureau of Mines). In 1925, two years after his return to Graz University, he became its Professor of Experimental Physics, a post he held until 1931, when he was appointed Professor of Physics at Innsbruck University and Director of its newly established Institute of Radiology; while at Innsbruck he founded a cosmic ray observatory on the Hafelekar mountain, near Innsbruck. After the Nazi occupation of Austria in 1938 Hess – a Roman Catholic himself but with a Jewish wife – emigrated to the United States, where he was Professor of Physics at Fordham University, New York City, until his retirement in 1956, having become a naturalized US citizen in 1944. Hess died in Mount Vernon, New York, on 17 December 1964.

In the early 1900s, it was found that gases in the atmosphere are always slightly ionized, even samples that had been enclosed in shielded containers. The theories to explain this phenomenon included radioactive contamination by the walls of the containers and the influence of gamma-rays in the soil and air; these theories were later proved incorrect by Hess's findings. In 1910, Theodor Wulf, investigating atmospheric ionization using an electroscope on top of the Eiffel Tower, found that ionization at 300 m/1,000 ft above the ground was greater than at ground level, from which he concluded that the ionization was caused by extraterrestrial rays.

Continuing this line of research, Hess – with the help of the Austrian Academy of Sciences and the Austrian Aeroclub – made ten balloon ascents in 1911–12 to collect data about atmospheric ionization. Ascending to altitudes of more than 5,000 m/16,000 ft, he established that the intensity of ionization decreased to a minimum at about 1,000 m/3,000 ft then increased steadily, being about four times more intense at 5,000 m/16,000 ft than at ground level. Moreover, by making ascents at night – and one, on 12 April 1912, during a nearly total solar eclipse – he proved that the ionization was not caused by the Sun. From his findings Hess concluded that radiation of great penetrating power enters the atmosphere from outer space. This discovery of cosmic rays (as this type of radiation was called by Robert Millikan in 1925) led to the study of elementary particles and paved the way for Carl Anderson's discovery of the positron in 1932.

Later in his career Hess investigated the biological effects of exposure to radiation (in 1934 he had a thumb amputated following an accident with radioactive material); the gamma radiation emitted by rocks; dust pollution of the atmosphere; and the refractive indices of liquid mixtures.

Hevelius (or Hewel or Hewelcke) Johannes 1611–1687 was a German astronomer, most famous for his careful charting of the surface of the Moon.

Hevelius was born at Danzig (now Gdansk) in northern Poland on 28 January 1611. A wealthy brewing merchant, he had a well-equipped observatory installed on the roof of his house in 1641, and was one of the most active observers of the seventeenth century. During the daytime he worked in his business and some evenings he took his seat on the City Council, but most of the rest of his free time he was up on his roof, observing, noting and cataloguing. His wife Elizabeth shared his interests and assisted him greatly in the study of the Moon, his catalogue of the stars and his work on comets. After his death, she edited and published his most famous work, *Prodromus Astronomiae* (1690).

Between 1642 and 1645, Hevelius deduced a fairly accurate value for the period of the solar rotation and gave the first description of the bright ideas in the neighbourhood of sunspots. The name he gave to them, *faculae*, is still used. He also made observations of the planets, particularly of Jupiter and Saturn. On 22 November 1644, he observed the phases of Mercury, which had been predicted by Copernicus.

In 1647, Hevelius published the first comparatively detailed map of the Moon, based on ten years' observations. It contained diagrams of the different phases for each day of lunation. He realized that the large, uniform grey regions on the lunar disc consisted of low plains, and that the bright contrasting regions represented higher, mountainous relief. He obtained better values for the heights of these lunar mountains than had Galileo (1564–1642) a generation before. His *Selenographia* also has an appendix that contains his observations of the Sun from 1642 to 1645.

Hevelius was interested in positional astronomy and planned a new star catalogue of the northern hemisphere, which was to be much more complete than that of Tycho Brahe (1546–1601). He began in 1657, but his observatory, with some of his notes, was destroyed by fire in 1679. Nevertheless, his observations enabled him to catalogue more than 1,500 stellar positions. The resulting *Uranographia* contains an excellent celestial atlas with 54 plates, but Hevelius's practice of using only the naked eye to observe positions (despite representations by no less a man than Edmond Halley) considerably reduces the value of his work. Hevelius used telescopes for details on the Moon and planets, but refused to apply them to his measuring apparatus.

Hevelius discovered four comets – he called them 'pseudo-planetae' – and suggested that these bodies orbited in parabolic paths about the Sun. Many later writers have declared that this suggestion indicates that he knew the nature of comets earlier than did either Halley or Newton.

A few of the names he gave to features of the Moon's surface are still in use today, particularly those that reflect geographical names on Earth. For his charting of the lunar formations, Hevelius has come to be known as the founder of lunar topography.

Hevesy Georg von 1885–1966 was a Hungarian-born chemist whose main achievements were the introduction of isotopic tracers (to follow chemical reactions) and the discovery of the element hafnium. For his work on isotopes he was awarded the 1943 Nobel Prize for Chemistry. In 1959, he received the Atoms for Peace Prize.

Hevesy was born in Budapest on 1 August 1885, into a family of industrialists. He was educated there in a Roman Catholic School and then attended the Technische Hochschule in Berlin with the intention of training as a chemical engineer. But he contracted pneumonia and moved to the more agreeable climate of Freiburg. After obtaining his doctorate in 1908 he moved again to study chemistry at the Zurich Technische Hochschule, where he worked under Richard Lorenz on the chemistry of molten salts. Following yet another move, to the Karlsruhe laboratories of Fritz Haber (where he investigated the emission of electrons during the oxidation of sodium–potassium alloys), Hevesy took Haber's advice and went to Manchester in 1911 to learn some of the new research techniques being developed by Ernest Rutherford, particularly those involving radioactive elements.

In 1913, Hevesy went to join Friedrich Paneth in Vienna for a short time to continue these studies and but for the outbreak of World War I in 1914 would have carried on this work with Henry Moseley at Oxford. Instead he continued his researches in Budapest and then in 1920 went to Copenhagen to work in the Institute of Physics under the guidance of Niels Bohr and Johannes Brönsted. He returned to Freiburg in 1926, only to go back to Copenhagen eight years later. In 1943, during World War II, he escaped to Sweden from the German occupation of Denmark and became a professor at the University of Stockholm, where he remained until shortly before his death. During an academic career extending over 58 years Hevesy worked in nine major research centres in seven European countries. He died at a clinic in Freiburg on 5 July 1966.

At Manchester in 1911, Rutherford set Hevesy the task of separating radioactive radium-D from a 100-kg/220-lb sample of lead. After a year's work Hevesy had been unable to achieve any separation – neither could he detect any chemical differences between radium-D and lead (we now know that this was an impossible task using conventional chemical techniques because radium-D is an isotope of lead, lead-210). But he turned this similarity to advantage by mixing some pure radium-D with ordinary lead and following chemical reactions of the lead by detecting its 'acquired' radioactivity. He perfected this technique while working with Paneth in Vienna (1913), using added radium-D to study the chemistry of lead and bismuth salts. This was the beginning of the use of radioactive tracers.

During his first period at Copenhagen (1920–1926), Hevesy successfully separated isotopes of mercury using fractional distillation at low pressure. During an investigation of zirconium minerals using X-rays, he and Dirk Coster in 1922 discovered hafnium (so-called after Hafnia, the Latin name for Copenhagen). Continuing his interest in isotopes, he effected a method of isotopic enrichment of chlorine and potassium.

On his return to Freiburg (1926) he studied the relative abundances of elements on Earth and in the universe, basing his calculations on chemical analyses by means of X-ray fluorescence. During the early 1930s, he commenced experiments with his radioactive tracer technique on biological specimens, noting, for example, the take-up of radioactive lead by plants. The production of an unstable isotope of phosphorus in 1934 enabled the first tracer studies to be made on animals. Hevesy used the isotope to trace the movement of phosphorus in the tissues of the human body. During his second stay at Copenhagen (1934–1943) he took with him a sample of heavy water (donated by Harold Urey in the United States, who in 1923 had spent a year at Bohr's laboratory). Hevesy used the sample to study the mechanism of water exchange between goldfish and their surroundings and also within the human body. He then extended his experiments using the isotopes potassium-42, sodium-34 and chlorine-38.

During his later years in Stockholm he continued to work on the transfer of radioactive isotopes within living material. Using radioactive calcium to label families of mice he showed that of calcium atoms present at birth about 1 in 300 are passed on to the next generation.

Hewish Antony 1924–   is a British astronomer and physicist whose research into radio scintillation resulted in the discovery of pulsars. For this achievement, and for his continued work in the field, he was awarded the 1974 Nobel Prize for Physics (jointly with Martin Ryle).

Hewish was born on 11 May 1924 in Fowey, Cornwall, and attended King's College, Taunton, before going to Gonville and Caius College at Cambridge. Graduating in 1948, Hewish worked at the Telecommunications Research Establishment in Malvern, where he met Martin Ryle (1918–1984), who was later to become Astronomer Royal. With Ryle, Hewish then became part of a team undertaking solar and interstellar research by radio at the Cavendish Laboratory, in Cambridge, and carrying out a series of intensive surveys known respectively, when published, as the *First, Second, Third,* and *Fourth Cambridge Catalogues.* In the later 1960s, he initiated research into radio scintillation at the Mullard Observatory; the discovery of the first pulsar – and then another three – followed. Made Reader at Cambridge University in 1969, he became Professor of Radioastronomy there in 1972.

Before 1950, Hewish's experimental work using radio telescopes was directed in the main to the study of solar atmosphere. He used simple corner reflectors to examine the Sun's outer corona in order to discover the electron density in its atmosphere, and to study the irregular hot gaseous clouds of plasma surrounding the Sun. After 1950, when – mainly through the efforts of Ryle – new instruments became available, radio observations were extended to sources other than the Sun. In particular, Hewish examined the fluctuation in such sources of the intensity of the radiation (the scintillation) resulting from disturbances in ionized gas in the Earth's atmosphere, within the Solar System, and in interstellar space. Engaged in this research at Mullard Observatory one day in 1967, Hewish's attention was drawn by a research student, Jocelyn Bell, to some curiously fluctuating signals being received at regular intervals during the sidereal day. Installation of a more sensitive high-speed recorder revealed, on 28 November 1967, the first indication of a pulsed emission whose fixed celestial direction ruled out the possibility of an artificial source. The November results were confirmed in early December, and verified that Hewish and Bell had discovered pulsating radio stars, or pulsars.

The results of a more detailed investigation by Hewish were published in February 1968, by which time three more pulsars had been identified. This investigation showed that pulsars are sources of radio emission in our galaxy which give out radiation in brief pulses. Although each pulsar emits radiation with nearly constant pulsation periods, the rate of emission differs between pulsars, as does the 'shape' of the pulse. Pulses can be single, double, or even triple peaked. Shapes can differ within each pulsar, but the mean pulse shape changes only very slightly – generally to decrease the interval – over many months. Comparative studies of successive shapes have shown that each pulse itself is made up of two pulsatory constituents: a regular (class 1) one, and another (class 2) pulsating at irregular frequency within the first one.

It is now generally accepted – as originally proposed by Thomas Gold (1920–   ) and others – that pulsars are rotating neutron stars (stars that are nearing the end of their stellar life, having practically exhausted their nuclear energy). It remains less clear how rotational energy is converted to emission of such shapes and such intensity as have been discovered, although many hypotheses have been put forward.

Hewish's initial discovery of four pulsars began a period of intensive research, in which more than 170 pulsars have been found since 1967. In the meanwhile, Hewish has patented a system of space navigation using three pulsars as reference points, that would provide 'fixes' in outer space accurate up to a few hundred kilometres.

Hey James Stanley 1909–   is a British physicist whose work in radar led to pioneering research in radioastronomy.

Hey was born on 3 May 1909 in the Lake District. Reading physics at Manchester University, he gained his master's degree in X-ray crystallography in 1931. From 1940 to 1952 he was on the staff of an Army Operational Research Group, for the last three years as the head of the establishment. He then became a research scientist at the

Royal Radar Establishment, being promoted in 1966 to Chief Scientific Officer. He retired in 1969, although he continued to write about astronomy from his home in Sussex. Hey was made a Fellow of the Royal Society in 1978.

Between 26 and 28 February 1942, during World War II, the British early-warning coastal defence radar became severely jammed. At first the jamming was attributed to enemy counter-measures but Hey, noting that the interference began as the Sun rose and ceased as it set, concluded that the spurious radio radiation emanated from the Sun and that it was related to solar activity. And he further proposed that the radiation – in strength about 10^5 of the calculated black-body radiation – was associated with a large solar flare that had just been reported.

At the end of World War II, research began in earnest at the Royal Radar Establishment in Malvern. For some years, Grote Reber (1911–) had been working along the same lines in the United States, and he had published his discovery of intense radio sources located in the Milky Way, notably in the constellations Cygnus, Taurus and Cassiopeia. The announcement, in 1946, by Hey and his colleagues that they had narrowed down the location of the radiation source in Cygnus to 'a small number of discrete sources' (in fact, to Cygnus A) stimulated a search for other discrete sources around the world. Attempts were also made to devise methods to achieve better resolution in the locating of radio sources so that, eventually, it would be possible to identify such sources as optically observed objects.

Hey and his team also returned to a study of the Sun as a radio source. They discovered that large sunspots were powerful ultra-shortwave radio transmitters and that, although the Sun was constantly emitting radio waves, they were of unexpected strength.

Using radio, the team noted that they could detect and follow meteors more accurately than ever before.

Heyrovský Jaroslav 1890–1967 was a Czech chemist who was awarded the 1959 Nobel Prize for Chemistry for his invention and development of polarography, an electrochemical technique of chemical analysis.

Heyrovský was born in Prague on or about 20 December 1890, the son of a Professor of Law at Charles University, Prague. He studied chemistry, physics and mathematics at his father's university and graduated from there in 1910. He then went to University College, London, to pursue postgraduate research under William Ramsey and Frederick Donnan (1870–1956), who aroused his interest in electrochemistry. He became a demonstrator in the chemistry department in 1913. He served in a military hospital during World War I and in 1920 was appointed an assistant in the Institute of Analytical Chemistry at Prague. He subsequently became a lecturer (1922), assistant professor (1924) and Professor of Physical Chemistry (1926). He remained at the Institute until 1950, when he became Director of the newly founded Polarographic Institute of the Czechoslovak Academy of Sciences. He revisited London in November 1955 to deliver his presidential address

to the Polarographic Society. He died in Prague on 27 March 1967.

Heyrovský began the work that was to lead to the invention of polarography during his student days in London, while he was investigating the electrode potential of aluminium. The technique was perfected in 1922, soon after his return to Prague. It depends on detecting the discharge of ions during electrolysis of aqueous solutions. The solution to be analysed is placed in a glass cell containing two electrodes, one above the other. The lower electrode is simply a pool of mercury. The upper electrode, called a dropping mercury electrode, consists of a fine capillary tube through which mercury flows and falls away as a series of droplets. The growing mercury droplet at the tip of the capillary tube constitutes the actual electrode.

With the dropping mercury electrode made, say, the cathode (to analyse for cations in solution), the voltage between the electrodes is slowly increased and the associated current observed on a galvanometer. When a cation is discharged (reduced) at the cathode, the current increases rapidly and then remains at a constant value. At that time, the voltage is characteristic of the cation concerned and the magnitude of the current is a measure of its concentration.

In 1925, together with M. Shikita, Heyrovský developed the polarograph, an instrument that automatically applies the steadily increasing voltage and traces the resulting voltage-current curve on a chart recorder. Such curves are called polarograms.

Polarography can be used to analyse for several substances at once, because the polarogram records a separate limiting current for each. It is also extremely sensitive, being capable of detecting concentrations as little as 1 part per million. Most chemical elements can be determined by the method (as long as they form ionic species) in compounds, mixtures or alloys. The technique has been extended to organic analysis and to the study of chemical equilibria and the rates of reactions in solutions. It can also be used for endpoint detection in titrations, a type of volumetric analysis sometimes called voltammetry.

Hilbert David 1862–1943 was a German mathematician, philosopher and physicist whose work in all three disciplines was brilliantly innovative and fundamental to further development. An excellent teacher, unsurpassed in lucidity of exposition, his influence on twentieth-century mathematics has been enormous. He is particularly remembered for his research on the theory of algebraic invariants, on the theory of algebraic numbers, on the formulation of abstract axiomatic principles in geometry, on analysis and topology (in which he derived what is now called Hilbert's theory of spaces), on theoretical physics, and finally on the philosophical foundations of mathematics.

Hilbert was born on 23 January 1862 in Königsberg, then in German Prussia, now Kaliningrad in Russia. He grew up and was educated there, attending Königsberg University – an ancient and venerable seat of learning – from 1880 to 1885, when he received his PhD. He then studied further in Leipzig and in Paris before returning to

Königsberg University to become an unsalaried lecturer. Six years later he was appointed professor, and three years later still (in 1895) he was offered the highly prestigious post as Professor of Mathematics at Göttingen University. He accepted, and held the position until he retired in 1930. Although he developed pernicious anaemia in 1925, he recovered, and died in Göttingen on 14 February 1943.

Hilbert's first period of research (between 1885 and 1892) was on algebraic invariants; he tried to find a connection between invariants and fields of algebraic functions and algebraic varieties. Representing the rational function in terms of a square, he eventually arrived at what is known as Hilbert's irreducibility theorem, which states that, in general, irreducibility is preserved if, in a polynomial of several variables with integral coefficients, some of the variables are replaced by integers. He later investigated ninth-degree equations, solving them by using algebraic functions of only four variables. In this way, Hilbert had by the end of the period not only solved all the known central problems of this branch of mathematics, he had in his methodology introduced sweeping developments and new areas for research (particularly in algebraic topology, which he himself returned to later).

In 1897, with some help from his colleague and friend Hermann Minkowski (1864–1909), Hilbert produced *Der Zahlbericht*, in which he gathered together all the relevant knowledge of algebraic number theory, reorganized it, and laid the basis for the developing class-field theory. He abandoned this work, however, when there was still much to be done.

Two years later, having moved to another area of study, Hilbert published his classic work, *Grundlagen der Geometrie*. (It is still available in its latest edition.) In it, he gave a full account of the development of geometry in the nineteenth century, and although on this occasion his innovations were (for him) relatively few, his use of geometry, and of algebra within geometry, to devise systems incorporating abstract yet rigorously axiomatic principles, was important both to the further development of the subject and to Hilbert's own later work in logic and consistency proofs. In the related field of topology, he referred back to his previous work on invariants in order to derive his theory of spaces in an infinite number of dimensions.

In 1900, attending the International Congress of Mathematicians in Paris, Hilbert set the Congress a total of 23 hitherto unsolved problems. Many have since been solved – but solved or not, the problems stimulated considerable scientific debate, research and fruitful development.

In a study of mathematical analysis some years afterwards, Hilbert used a new approach in tackling Dirichlet's Problem, and made other contributions to the calculus of variations. In 1909, he provided proof of Waring's hypothesis (of a century earlier) of the representation of integers as the sums of powers. From that time forward, Hilbert worked on problems of physics, such as the kinetic theory of gases, and the theory of relativity – problems, he said, too difficult to be confined to physics and physicists. His deep research led finally to his critical work on the foundations of mathematical logic, in which his contribution to proof theory was extremely important by itself.

Hill Archibald Vivian 1886–1977 was a British physiologist who studied muscle action in great detail. For this work he received the 1922 Nobel Prize for Physiology or Medicine, which he shared with Otto Meyerhof (1884–1951).

Hill was born on 26 September 1886 in Bristol. He was educated at Blundell's School, Tiverton, and then went to Trinity College, Cambridge. There he excelled at mathematics and was greatly influenced by his tutor Morley Fletcher (1873–1933), who had collaborated with Frederick Hopkins in the discovery of the role of lactic acid in muscle contraction. Graduating with a medical degree in 1907 he remained at Trinity until 1914 when World War I broke out, and he served with distinction. He became Professor of Physiology at Manchester University in 1920 after obtaining his doctorate from Trinity earlier that year. He joined the staff of University College, London, in 1923 and three years later took up a professorship at the Royal Society, a position he held until 1951. During that period he was Secretary to the Royal Society, from 1935 to 1946. He also served as Scientific Advisor to India from 1943 to 1944 and was active in various scientific organizations, including the Marine Biological Association, of which he was President from 1955. He was a member of the War Cabinet Scientific Advisory Committee during World War II. He died in Cambridge on 4 June 1977.

Influenced by his mentor Fletcher, at Cambridge, Hill researched into the workings of muscles as early as 1911. He was not concerned with the chemical details of muscle action, but wanted rather to ascertain the amount of heat produced during muscle activity. To do this he used delicate thermocouples, and discovered that contracting muscle fibres produce heat in two phases. Heat is first produced quickly as the muscle contracts. Then after the initial contraction, further heat is evolved more slowly but often in greater amounts. The thermocouples which Hill used recorded heat changes quickly and minutely in the form of tiny electric currents. He had to modify this apparatus for his particular purpose and was able to measure a rise in temperature of as little as 0.003 degrees over a few hundredths of a second.

By 1913, Hill was aware that heat is produced after the muscles have contracted and he showed that molecular oxygen is consumed after the work of the muscles is over and not during muscular contraction. This discovery was made by proving that if muscle fibre is made to contract in an atmosphere of pure nitrogen, the first phase of heat production is not affected but the second phase does not take place at all. He realized that oxygen is not necessary in the first phase – that is, the chemical reactions immediately involved in the contraction of muscles do not require oxygen. But in phase two, when muscle contraction has taken place, oxygen is needed to produce further energy for subsequent muscle contraction.

Mammals and birds maintain a constant body temperature which in human beings is 37°C/98.6°F. The maintenance of such a constant temperature is under involuntary control, but can be partly attributed to muscular activity. When a warm-blooded animal is cold, it is necessary for the body to produce heat. One of the ways in which it does so is by shivering, an involuntary contraction of muscles resulting in heat generation by the process elucidated by Hill.

Hill Robert 1899–1991 is a British biochemist who greatly advanced our understanding of photosynthesis.

Hill was born on 2 April 1899. After serving in World War I in the Royal Engineers Pioneer Antigas Department, from 1917 to 1918, he studied at Emmanuel College, Cambridge. He remained researching there until 1938, by which time he had become a Senior Research Fellow. From 1943 to 1966 he was a member of the Scientific Staff of the Agricultural Research Council.

The process of photosynthesis has been shown to occur in two separate sets of reactions, those that require sunlight (the light reactions) and those that do not (the dark reactions). Both sets of reactions are dependent on one another. In the light reactions some of the energy of sunlight is trapped within the plant and in the dark reactions this energy is used to produce potentially energy-generating chemicals, such as sugar.

In 1894, Engelman first showed that the light reactions of photosynthesis occur within the chloroplasts of leaves. Hill's experiments in 1937 confirmed the localization of the light reactions within the chloroplast, as well as elucidating in part the mechanism of the light reactions. He isolated chloroplasts from leaves and then illuminated them in the presence of an artificial electron-acceptor. The electron-acceptor he used was a ferric salt (Fe^{3+}) which was reduced to the ferrous form (Fe^{2+}) during the reaction. He showed that during the reaction, oxygen is produced and that this derived oxygen comes from water. (He also demonstrated the evolution of oxygen in human blood cells by the conversion of haemoglobin to oxyhaemoglobin.) This process, known as the Hill reaction, can be summed up by the following equation:

$$4Fe^{3+} + 2H_2O \xrightarrow[\text{chloroplasts}]{\text{light}} 4Fe^{2+} + 4H^+ + O_2\uparrow$$

Much more is now known about the mechanism of the Hill reaction in plants, and the equation has become more complicated:

$$2NADP + ADP + P_i + 2H_2O \xrightarrow[\text{chloroplasts}]{\text{light}} 2NADPH_2$$
$$+ ATP + O_2\uparrow$$

The electron-acceptor in the plant is now known to be NADP (nicotinamide adenine dinucleotide phosphate), and is reduced in the reaction to $NADPH_2$. ADP (adenosine diphosphate) is phosphorylated to ATP (adenosine triphosphate) in the reaction, a process known as photophos-

phorylation. In the plant the energy from sunlight is first stored as ATP and later as sugar, the final product of the dark reaction of photosynthesis which uses energy from ATP. Hill's findings were of great significance in further investigations into photosynthesis.

Hinshelwood Cyril Norman 1897–1967 was a British physical chemist who made fundamental studies of the kinetics and mechanisms of chemical reactions. He also investigated bacterial growth. He shared the 1956 Nobel Prize for Chemistry with the Soviet scientist Nikolay Semenov for his work on chain reaction mechanisms.

Hinshelwood was born in London on 19 June 1897, the son of an accountant. His family emigrated to Canada when he was a child, but returned to England after his father's death. He was educated at the Westminster City School, London, from where he won a scholarship to Balliol College, Oxford. The outbreak of World War I in 1914 interrupted his studies and from 1916 to 1918 he worked in the Department of Explosives at the Royal Ordnance Factory at Queensferry, Scotland. He returned to Balliol in 1919 and took the shortened postwar chemistry course, graduating in 1920 and being immediately elected a Fellow of the college. Subsequent academic appointments were as Fellow of Trinity College, Oxford (1921–1937), Dr Lee's Professor of Chemistry in the University of Oxford (1937–1964) and, on his retirement, Senior Research Fellow at Imperial College, London. Hinshelwood was knighted in 1948; he died in London on 9 October 1967.

While he was at the explosive works, Hinshelwood tried to measure the slow rate of decomposition of solid explosives by monitoring the gases they evolved. His first researches at Balliol pursued this line, and he studied the decomposition of solid substances in the presence and absence of catalysts. He then investigated homogeneous gas reactions. He found initially that the thermal decomposition of the vapours of substances such as acetone (propanone) and aliphatic aldehydes occur by means of first or second order processes. If A and B are the reactants, then the rate of a chemical reaction can be expressed as:

$$R = k[A]^x[B]^y$$

where R is the rate, k is the velocity constant, [A] and [B] are the concentrations of the reactants, and x and y are powers. If $x = 1$ the reaction is said to be first order, if $x = 2$ it is second order, and so on. At that time the activation of chemical reactions was considered only in terms of collision mechanisms – easily applied to bi-molecular reactions but not capable of explaining unimolecular reactions. Hinshelwood showed that even apparently simple decomposition reactions usually occur in stages. Thus for acetone (propanone):

a)
$$(CH_3)_2CO \rightarrow CH_2=CO + CH_4$$
acetone
(propanone)

b)
$$CH_2=CO \rightarrow {}^1\!/_2C_2H_4 + CO$$

This early work centred on the relationship between temperature, concentration and the influence of second and third components.

Hinshelwood went on to demonstrate that many reactions can be explained in terms of a series – a chain – of interdependent stages. At low temperatures the reaction between hydrogen and oxygen, or hydrogen and chlorine, for example, is comparatively slow because the chain reactions involved terminate at the walls of the vessel. But at high temperatures the chain reactions accelerate the process to explosion point. He provided experimental evidence for the role of activated molecules in initiating the chain reaction.

He also investigated reaction kinetics in aqueous and nonaqueous solutions, together with hydrolytic processes, esterification, and acylation of amines. He published his classic work on reaction kinetics, *Kinetics of Chemical Change*, in 1926.

In 1938, Hinshelwood published with S. Daglay the first of more than 100 papers on bacterial growth. Initially he investigated the effects of various nutrients such as carbohydrates and amino acids on simple nonpathogenic organisms. Later he studied the effects of trace elements and toxic substances such as sulphonamides, proflavine and streptomycin. He considered that all the various chemical reactions that occurred in his bacterial growth experiments were interconnected and mutually dependent, the product of one reaction becoming the reactant for the next – a process he termed 'total integration'.

Hinshelwood's contributions to both these fields won universal acknowledgement during his lifetime, and he was the recipient of many honours. He was President of the Chemical Society during its centenary year (1947) and President of the Royal Society at its tercentenary in 1960. He was a capable linguist and artist, and a year after his death his paintings went on exhibition in Goldsmiths Hall, London.

Hinton William Augustus 1883–1959 was a US bacteriologist and pathologist who achieved renown through his work on syphilis, in particular the development of the Hinton test.

William Hinton was born in Chicago on 15 December 1883. Between 1900 and 1902 he studied at the University of Kansas and went on to Harvard College, receiving a BS in 1905. Over the next four years he taught at Walden University, Nashville, Tennessee and in Langston, Oklahoma, and continued his studies in bacteriology and physiology during the summers at the University of Chicago. In 1909, he started at Harvard Medical School and graduated in 1912, having been awarded the Wigglesworth and Hayden Scholarships. For three years he worked at the Wasserman Laboratory and as a volunteer assistant in the Department of Pathology at the Massachusetts General Hospital, and in 1915 he was made chief of the Wasserman Laboratory. In 1918, he was appointed instructor in preventive medicine and hygiene at the Harvard Medical School and between 1921 and 1946 he was instructor in bacteriology and immu-

nology and a lecturer until 1949. In 1949, he was promoted to clinical professor – the first black professor in the university's history – a post which he held until he retired as professor emeritus in 1950. Hinton died in 1959. He had married in 1919 and had two daughters.

Hinton became internationally known in the field of syphilology. He developed a serological test for syphilis based on flocculation, reducing the number of false positive diagnoses of the disease. In 1934, the US Public Health Service showed the Hinton test for syphilis to be the best. He wrote many scientific papers and his book *Syphilis and its Treatment* (1926) was the first medical textbook by a black American to be published.

Hinton was also the discoverer of the Davies–Hinton tests of blood and spinal fluid. He was a member of many societies including the American Medical Association and the American Association of Bacteriologists. In 1938, he turned down the Spingarn medal as he was afraid that his work would not be accepted on its merit if it were known that he was black. In his will, Hinton left $75,000 to establish the Eisenhower Scholarship Fund for graduate students at Harvard. Fifteen years after his death the Serology Laboratory of the State Laboratory Institute Building of the Massachusetts Department of Public Health was named in his honour.

Hipparchus lived 2nd century BC was a Greek astronomer and mathematician whose careful research and brilliant deductions led him to many discoveries that were to be of importance and relevance even 2,000 years later.

Hipparchus was born in Nicaea, in Bithynia (now in Turkey), in about 146 BC. What little is known about him and his life is contained in the writings of Strabo of Amasya and Ptolemy of Alexandria, both of whom were writing well over 100 years after Hipparchus' death. But it is recorded that Hipparchus carried out his astronomical observations in Bithynia, on the island of Rhodes, and in Alexandria.

In 134 BC Hipparchus noticed a new star in the constellation Scorpio, a discovery which inspired him to put together a star catalogue – the first of its kind ever completed. He entered his observations of stellar positions using a system of celestial latitude and longitude, making his measurements with greater accuracy than any observer before him, and taking the precaution wherever possible to state the alignments of other stars as a check on present position.

His finished work, completed in 129 BC, listed about 850 stars classified not by location, but by magnitude (brightness), for which classification he devised a system very close to the modern one. The resulting catalogue was thus so excellent that it was not only plagiarized *en bloc* by Ptolemy, but was used by Edmund Halley 1,800 years later.

Hipparchus was troubled by the fact that the Babylonians had had a different number of days in their year. In trying to resolve this problem he studied the motions not only of the Sun and the Moon, but also of the Earth. In his star catalogue he noted that the star Spica was measured at

6° from the autumn equinox – whereas in the records of Timocharis of Alexandria 150 years earlier it had plainly been at 8°. He came to the conclusion that it was not Spica that was moving, but an east-to-west movement (or precession) of the Earth, and he managed to calculate a value for the annual precession of 45″ or 46″, which, considering the now accepted rate of 50.26″ per year, was an amazing feat of accuracy with the simple instruments available to him at the time. (Ptolemy credits Hipparchus with the invention of an improved kind of theodolite.) The definition of the precession of the equinoxes is usually said to be Hipparchus' most significant work. Using his knowledge of the equinoxes, Hipparchus calculated the terrestrial year to be 365$\frac{1}{4}$ days, diminishing annually by a three-hundredth of a day (in which he was again astonishingly accurate), and the lunar period to be 29 days 12 hours 44 minutes and 2$\frac{1}{2}$ sec (which was only 1 sec too short). From Hipparchus' time, eclipses of the Moon could be predicted to within one hour, and those of the Sun less accurately.

With such knowledge, it is surprising that Hipparchus accepted the notion of the geocentric universe, declaring that the Sun orbited the Earth in a circle of which Earth was not quite at the centre. The Moon was said to do the same, except that the centre of its orbit was itself moving (a 'moving eccentric').

Hipparchus' astronomical work led him to various areas of mathematics. He was one of the earliest formulators of trigonometry, for which he devised a table of chords. The theorem that provides a basis for the field of plane geometry was also proposed by him, although later attributed to Ptolemy (whose name is still generally attached to it).

Until Copernicus (1473–1543), there was no greater astronomer than Hipparchus.

Hippocrates *c*.460–*c*.377 BC is known as the founder of medicine, and in ancient times was regarded as the greatest physician who had ever lived. In contrast to the general views of his time, which considered sickness to be brought about by the displeasure of the gods or by possession by demons, Hippocrates looked upon it as a purely physical phenomenon capable of rational explanation.

Hippocrates was born on the Greek island of Cos, the son of a physician (although the legends surrounding his feats ascribed him to being a member of a family of magicians). Not much is known for certain about his life as he is referred to little by his contemporaries. He travelled throughout Greece and Asia Minor where the cures he achieved, his great skill and humanity, together with his exemplary conduct, soon made him famous. Eventually, on his return to Greece. he founded a medical school, one of the first of its kind, on Cos, where he taught.

Hippocrates concerned himself with the whole patient, regarding the body as a whole organism and not just a series of parts. He used few medicines but considered rather that it was the duty of the doctor to find out by careful observation what Nature was trying to do and to meddle as little as possible with the process of healing, relying on good diet, fresh air, rest and cleanliness in both patient and doctor. Hippocrates did believe, however, that some diseases, such as those resulting from a poor diet, were caused by residues of undigested food which gave off vapours that seeped into the body.

The several generations of doctors who studied under Hippocrates are thought to have contributed to the 72 books known as the *Corpus Hippocraticus/Hippocratic Collection*, the library of the medical school at Cos which was later assembled in Alexandria, in the third century BC. These works comprise textbooks, research reports, lectures, essays and clinical notebooks. They contain few correct anatomical observations, although one book contains accurately observed symptoms with the likely outcome for the patient. Another book, *About the Nature of Man*, describes a theory of 'humours', or body fluids. The theory was based on the belief in the existence of four important fluids – blood, phlegm, and yellow and black bile. If the normal levels of these fluids were unbalanced, illness ensued. This theory persisted among physicians in Europe right through the Middle Ages.

Hitchings George Herbert 1905– is a US pharmacologist who shared the Nobel Prize for Physiology or Medicine in 1988, for his work on therapeutics, especially anticancer agents, and immunosuppressive drugs and antibiotics that were of vital importance in the growing surgical field of transplantation.

Hitchings was born in Hoquiam, Washington on 18 April 1905 and was educated at the University of Washington in Seattle from which he graduated in 1927 and was awarded a master's degree in 1928. He then moved to Harvard University, gaining his doctorate in biochemistry in 1933. He remained at Harvard as an instructor until 1939 when he moved to Case Western Reserve University. In 1942 he joined the biochemical research laboratory of Burroughs Wellcome, where he was joined two years later by Gertrude Elion (1918–), with whom he was to work closely for the rest of his career, and with whom he shared the Nobel prize. In 1955, Hitchings was appointed associate research director at Burroughs Wellcome, he became the research director of chemotherapy in 1963 and vice president of the company in 1967 from which position he retired in 1975. He continues to act as a consultant to the Research Laboratories, and is president of the Burroughs Wellcome Fund, a charitable endowment supported by the company.

Hitchings worked extensively on drug development, initially approaching problems from a basic research perspective, trying to understand fundamental biological mechanisms from which therapeutic strategies could be devised. Experiments in the 1940s, on cellular metabolism, revealed that some bacteria could not produce DNA and could not therefore divide and grow, in the absence of certain chemicals, particularly purines. Using this information, Hitchings and Elion developed an 'antimetabolite' philosophy, and started to synthesize compounds that inhibited DNA synthesis, in the hope that these could be used to prevent the rapid growth of cancer cells.

Their many investigations of the chemistry of purines and pyrimidines, two groups of chemicals that are involved with DNA synthesis, resulted in them jointly holding at one stage 18 pharmaceutical patents related to these two compounds. Clinical trials in patients with leukemia revealed that these powerful drugs that rapidly inhibited the growth of cancerous cells also caused severe side effects, although further refinements of their work lead to the production of clinically significant anticancer drugs. The basic chemical research of Hitchings and Elion provided important new information about DNA metabolism, from which several clinically important compounds were developed, against malaria, for the treatment of gout and kidney stones, and also drugs that suppressed the normal immune reactions of the body, vital tools in transplant surgery and in treating autoimmune diseases such as rheumatoid arthritis. In the 1970s, Hitchings and Elion's research produced an antiviral compound, acyclovir, active against the herpes virus, which preceded the successful development by Burroughs Wellcome of AZT, the anti-AIDS compound.

Hoagland Mahlon Bush 1921– is a US biochemist who was the first to isolate transfer RNA (tRNA), which plays an essential part in intracellular protein synthesis.

Hoagland was born on 5 October 1921 in Boston, Massachusetts. He studied medicine at Harvard University Medical School and obtained his degree in 1948. Following his graduation he worked as a Research Fellow in Medicine in the Huntingdon Laboratory of Massachusetts General Hospital until 1951. He then spent a year at the Carlsberg Laboratory in Copenhagen as a Fellow of the American Cancer Society. From 1953 to 1967 he held several positions in the Huntingdon Laboratory at Harvard Medical School; he joined the laboratory as an Assistant in Medicine, progressed to Assistant Professor of Medicine, then in 1960 became Associate Professor of Bacteriology and Immunology. In 1967, he was appointed Professor of Biochemistry and Chairman of the Biochemistry Department at the Dartmouth Medical School. In 1967, he also became President and Scientific Director of the Worcester Foundation for Experimental Biology in Shrewsbury, Massachusetts.

In the late 1950s Hoagland isolated various types of RNA molecules (now known as tRNA) from cytoplasm and demonstrated that each type of tRNA can combine with only one specific amino acid. Within the cytoplasm a tRNA molecule and its associated amino acid combine to form a complex – amino acyl tRNA. This complex then passes to the ribosome, where it combines with a messenger RNA (mRNA) molecule in a specific way: each tRNA molecule has as part of its structure a characteristic triplet of nitrogenous bases that links to a complementary triplet on the mRNA. A number of these reactions occur on the ribosome, building up a protein one amino acid at a time.

In addition to his research on tRNA and the biosynthesis of proteins, Hoagland has also investigated the carcinogenic effects of beryllium and the biosynthesis of coenzyme A.

Hodgkin and Huxley Alan Lloyd Hodgkin 1914– and Andrew Fielding Huxley 1917– are British physiologists who have contributed much to the understanding of how nerve impulses are transmitted. For their work on the role of ions in the excitation of nerve membranes they shared the 1963 Nobel Prize for Physiology or Medicine with the Australian physiologist John Eccles (1903–).

Hodgkin was born on 15 February 1914, in Banbury, Oxfordshire. He was educated at Gresham School, Holt, and then at Trinity College, Cambridge, from which he graduated in 1936. In 1937 and 1938 he worked at the Rockefeller Institute and at Woods Hole Marine Biological laboratories in Massachusetts, where he began his research on the squid. On his return to Cambridge he began his collaboration with Huxley but their work was interrupted by World War II, during which Hodgkin researched airborne radar for the Air Ministry. After the war he went back to Cambridge and served as a lecturer and Assistant Research Director in the Department of Physiology from 1945 to 1952, when he became Foulerton Research Professor. In 1970, he accepted the biophysics professorship at Cambridge, and the following year took up the appointment of Chancellor of the University of Leicester. Hodgkin was awarded the Royal Medal of the Royal Society in 1958 and was President of the society from 1970 to 1975.

Andrew Huxley was born in London on 22 November 1917, the grandson of the distinguished nineteenth-century scientist Thomas Huxley. He was educated at University College and Westminster Schools and then at Trinity College, Cambridge, from which he graduated in 1938 and gained his MA in 1941. His researches were interrupted by World War II; he was involved in operational research for Antiaircraft Command from 1940 to 1942, and then for the Admiralty until 1945. It was on his return to Cambridge after the war that Huxley collaborated with Hodgkin on their award-winning work. He became a demonstrator and then Assistant Director of Research at Cambridge from 1946 to 1960. He was appointed Director of Studies at Cambridge from 1952 to 1960, and was elected a Fellow of the Royal Society in 1955. In 1960 he moved to University College, London, and in 1969 gained a professorship there in the Department of Physiology as Jodrell Professor of Physiology. He succeeded Alexander Todd as President of the Royal Society in 1980.

The theory concerning the nervous system developed at the end of the nineteenth century stated that nerve cells have as their primary function the transmission of information in the form of changes in electric potential across the cell membrane. Hodgkin first became interested in the mechanics of nerve impulses in the late 1930s and devoted most of his research to the unknown quantities and qualities of the associated electric potentials.

In 1902, Julius Bernstein, an experimental neurophysiologist, had suggested that resting potential was due to the selective permeability of nerve membrane to potassium ions. He had also suggested that the action potential was brought about by a breakdown in this selectivity so that membrane potential fell to zero. In 1945 at Cambridge,

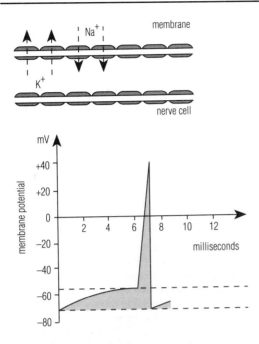

Hodgkin *The potential inside an axon changes rapidly (and reverses) with the passage of a nerve impulse, accompanied by movement of sodium and potassium ions through the cell membrane.*

Hodgkin and Huxley attempted to measure the electrochemical behaviour of nerve membranes. They experimented on axons of the giant squid (*Loligo forbesi*) – each axon is about 0.7 mm/0.03 in in diameter. They inserted a glass capillary tube filled with sea water into the axon to test the composition of the ions in and surrounding the cell, which also had a microelectrode inserted into it. They succeeded in measuring the potential differences between the tip of the microelectrode and the sea water. Stimulating the axon with a pair of outside electrodes, it was shown that the inside of the cell was at first negative (the resting potential) and the outside positive, and that during the conduction of the nerve impulse the membrane potential reversed so the inside became positive and the outside negative. This was the first time that electrical changes across the cell membrane had been recorded and the discovery that the membrane potential exceeds the zero level during the action potential implied that some other process than that proposed by Bernstein must be involved. Hodgkin suggested that this process was a rapid and specific increase in the permeability of the membrane to sodium ions.

Working with Bernhard Katz, another cell physiologist, Hodgkin showed that when there was no current flowing through the membrane, the membrane potential could be given by a formula which acknowledges that sodium ions do play an important part in determining the membrane potential. The theory proposed that an excited membrane becomes permeable to sodium ions (which are positively charged), which on entry to a cell cause its contents also to become positively charged; it is known as the 'sodium theory' and is now accepted to be of fairly general application.

In 1947, Hodgkin concluded that during the resting phase a nerve membrane allows only potassium ions to diffuse into the cell, but when the cell is excited it allows sodium ions to enter and potassium ions to move out. By 1952, Hodgkin had demonstrated that the inside of a cell has a great concentration of potassium ions, and the surrounding solution is rich in sodium ions. In that year, Hodgkin and Katz used a special technique now known as the 'voltage clamp method' to measure the currents flowing across a nerve membrane. They found that during an action potential, the inside of the cell becomes electrically positive by 30–60 millivolts, and that the membrane recovers from an impulse within milliseconds. In 1953, Hodgkin investigated the role played by potassium ions in nerve cells and showed that the internal potassium ions are free to move in an electric field. From this he concluded that almost all the potassium in the axoplasm is effectively in a free solution and that it contributes in some way to the production of the resting potential. Two years later Hodgkin devised an apparatus to measure the extrusion of radioactive sodium from the giant axons of a squid. It was shown that there is a relationship between the efflux of sodium and the time taken for it to diffuse out and, further, that when the axon was surrounded with DNP the amount of sodium efflux fell markedly but recovered when the DNP was washed away. This experiment implied that the extrusion of sodium is probably dependent on the metabolic energy supplied either directly or indirectly in the form of ATP (adenosine triphosphate). It was also discovered that the amount of sodium flowing in equals that of the potassium flowing out.

Hodgkin, with his associates, produced more conclusive evidence of this extrusion dependence in 1960 and also showed that the sodium efflux is dependent upon the external potassium ion concentration. The action potential would not be able to revert to the resting potential inside the cell if the influx of sodium ions were not balanced by the extrusion of another positively charged ion. The decrease from the peak is followed by the exit of potassium ions, so restoring the cell interior to a negative phase. In 1959 Hodgkin had found that the inward and outward flow of sodium in frog muscle was approximately equal, implying that the greater part of the sodium efflux must be dependent upon some active transport process. Keynes has since shown that there seems to be a 'chloride pump' involved, but its function is still obscure.

Many scientists have used Hodgkin and Huxley's methods to study resting and action potentials in various excitable membranes. The 'voltage clamp' method is used to obtain information on the elements affecting nerve conduction. Investigations are also being directed to discover the mechanism that possibly involves an enzyme, which is present in the peripheral cell membrane and breaks down ATP, thus releasing energy, but only if sodium and potassium ions are present.

Hodgkin Dorothy Mary Crowfoot 1910– is a British chemist who has used X-ray crystallographic analysis to determine the structures of numerous complex organic molecules, including penicillin and vitamin B_{12} (cobalamin). She was awarded the 1964 Nobel Prize for Chemistry.

Hodgkin was born on 12 May 1910 in Cairo, where her father, John Crowfoot, was serving with the Egyptian education service. After the family returned to Britain she attended the Sir John Leman School, Beccles, before going to Somerville College, Oxford. In 1928, she indulged her interest in archaeology by accompanying her father on an expedition to Transjordan. After graduation she went to Cambridge University where between 1932 and 1934 she worked on determining the structure of sterols, at a time when X-ray analysis was limited to confirming the correct formula as predicted by organic chemical methods. She developed the technique of X-ray investigation to the point at which it became a very useful analytical method. In 1934 she returned to Oxford to take up a lecturing appointment, and three years later married Thomas Hodgkin, a noted authority on African affairs. She remained at Oxford for a further 33 years until 1970, when she became Chancellor of Bristol University.

While at Cambridge, Hodgkin studied the structures of calciferol (vitamin D_2) and lumisterol and, with C. H. Carlisle, she correctly analysed cholesterol iodide, the first complex organic molecule to be determined completely by X-ray crystallography. On her return to Oxford (1934) she investigated various compounds of physiological importance, especially penicillin, the structure of which she and her co-workers determined before the organic chemists. This work was of national importance at the time, and was to have a lasting effect on the development of antibiotics. She later elucidated the structure of cephalosporin C, an antibiotic closely related to penicillin.

In about 1948 Hodgkin began her work on vitamin B_{12}, a compound essential to the life of red blood cells in the body; the inability to absorb sufficient vitamin B_{12} from the diet leads to pernicious anaemia. Chemical analysis had suggested that this complex compound has an approximate empirical formula of $C_{61-64}H_{86-92}O_{14}N_{14}PCo$, with a single cobalt atom, a cyanide group and a nucleotide-like group. Hodgkin and her collaborators collected photographs and data of X-ray diffraction patterns of both wet and dry B_{12}, a hexacarboxylic acid derivative, and a derivative in which a selenium atom had been introduced. Using Fourier series the data were analysed mathematically by one of the first electronic computers, which were just becoming available. After a lengthy and painstaking step-by-step process the structure was finally worked out and announced in 1957 (the empirical formula was found to be $C_{63}H_{88}O_{14}N_{14}PCo$). Work on the biologically active aspect of the molecule continued into the 1960s.

In addition to the 1964 Nobel prize award, Hodgkin was in 1965 admitted to the Order of Merit, becoming only the second woman to receive this honour (the first was Florence Nightingale).

Hodgkin Thomas 1798–1866 was a British physician who described six cases of malignant reticulosis in his paper 'On some morbid appearances of the absorbent glands and spleen', which was published in 1832. The disease is named after him.

Hodgkin was born in Tottenham, London, on 6 January 1798, and was tutored at home. Some of his education took place on the Continent, where he completed his medical training after a few years at Guy's Hospital, London. He gained his MD from Edinburgh University in 1821. In 1825 he was selected to become one of the first Fellows of the Royal College of Physicians, but he declined the honour. He became curator of the new museum at Guy's and lectured in morbid anatomy; he was the first to give regular tuition in the subject. He was lauded at home and abroad by a number of societies for his work, but in 1837 he was passed over for the post of Assistant Physician at Guy's and was deeply disappointed. He resigned from the hospital and gradually devoted more and more of his time to philanthropic work. He was an excellent linguist, and an active crusader in the Aborigines Protection Society. He contracted dysentery and died at the age of 68 while on a mercy mission to the Jewish people in Jaffa. He is buried in Israel.

Hodgkin was the first to describe a particular type of lymphoma that usually affects young adults and causes malignant inflammation of the lymph glands. The spleen and liver may also become involved. Hodgkin's disease can now be definitely diagnosed by the histological presence of Reed Sternburg cells.

Hodgkin received little recognition of his observations until 1865, when Samuel Wilks of Guy's referred to Hodgkin's account in his paper 'Cases of enlargement of the lymphatic glands and spleen (or Hodgkin's disease)'.

Hodgkin pioneered the use of the stethoscope in Britain after being favourably impressed in France by René Laennec (1781–1826), who invented the instrument in 1816. He was also the first person to stress the importance of postmortem examinations.

Hodgkinson Easton 1789–1861 was a British civil engineer who worked to introduce scientific methods of measuring the strength of materials.

Hodgkinson was born on his father's farm at Anderton, Cheshire, in February 1789. After showing distinct ability in mathematics during his schooldays because of the early death of his father and his mother's consequent penury he was obliged to give up hopes of a professional education and instead to assist his mother in running the family farm. He had little aptitude for this work and pursuaded his mother to invest her limited capital in a pawnbroking business in Salford, Manchester.

He also found time to develop his interest in natural science and became acquainted with the chemist John Dalton (1766–1844) and other gifted men then living in Manchester. In March 1822 he read a paper on 'The Transverse Strain and Strength of Materials' before the Literary and Philosophical Society. In this contribution he recorded a factor which became important in all his subsequent experiments, namely 'set' or the original position of a strained body and the position it assumes when the strain is removed. He fixed the exact position of the 'neutral line' in the section of rupture or fracture and made it the basis for the computation of the strength of a beam of given dimensions. His conception of the true mechanical principle by which the position of the line could be determined has long been generally accepted.

In 1828 he read before the same society an important paper on the forms of the catenary links in suspension bridges, and in 1830 one on his research into the strength of iron beams, one of the most valuable contributions ever made to the study of the strength of materials.

From a theoretical analysis of the neutral line, he devised experiments to determine the strongest beam which resulted in the discovery of what is known as 'Hodgkinson's beam'.

Hodgkinson rendered important services to Robert Stephenson (1803–1859) in the construction of the Britannia (Menai) and Conway tubular bridges by fixing the best forms and dimensions of tubes. He edited the fourth edition of Tredgold's work on the strength of cast iron (1842) and published a volume of his own: *Experimental Researches on the Strength and other Properties of Cast Iron* (1846).

He worked from 1847 to 1849 as one of the Royal Commissioners to inquire into the application of iron to railway structures. Also in 1847 he was appointed Professor of the Mechanical Principles of Engineering at University College, London, where his lectures were somewhat impaired by his hesitancy of speech. He did not live to see an authoritative publication of all his collected papers, dying in Manchester in 1861.

Hoe Richard March 1812–1886 was a US inventor and manufacturer, famous for inventing the rotary printing press.

Hoe was born in New York City on 12 September 1812. He was the eldest son of Robert Hoe (1784–1833), a British-born American who, with his brothers-in-law Peter and Matthew Smith, established in New York a firm manufacturing printing presses. Richard Hoe was educated in public schools before entering his father's firm at the age of 15. When his father retired in 1830, Richard and his cousin Matthew took over the business. Richard proved to have the same mechanical genius as his father, and the application of his ideas revolutionized printing processes. He discarded the old flatbed printing press and placed the type on a revolving cylinder. This was later developed into the Hoe rotary or 'lightning' press, patented in 1846 and first used by the *Philadephia Public Ledger* in 1847.

Under Hoe's management the company grew at a rapid rate. In 1859 he built Isaac Adams's Press Works in Boston. After the Civil War, new premises were built in Grand Street, New York, and the old buildings in Gold Street were abandoned. Between 1865 and 1870, a large manufacturing branch was built up in London, employing 600 people.

In 1871, with Stephen D. Tucker as a partner, Hoe began experimenting and designed and built the Hoe web perfecting press. This press enabled publishers to satisfy the increasing circulation demands of the rapidly growing US population.

While Richard Hoe was the leading influence in the company, he spent much time and money on the welfare of his employees. Quite early in his career he started evening classes for apprentices, at which free instruction was given in those aspects of their work most likely to be of practical use to them. He was addressed by them as 'the Colonel', which dated from his early service in the National Guard.

He died suddenly while on a combined health and pleasure trip to Florence with his wife and a daughter, on 7 June 1886. He was succeeded in the business by his nephew, Robert Hoe.

At the time when Richard Hoe was made responsible for the company, it was making a single small cylinder press. Its capacity was 2,000 impressions per hour and there was demand for a greater speed of output. This prompted Hoe to concentrate on improvements to meet the demand and, in 1837, a double small cylinder press was perfected and introduced. In the next ten years, he designed and put into production a single large cylinder press. This was the first flatbed and cylinder press ever used in the United States. Hundreds of these machines were made in subsequent years and were used for book, job, and woodcut printing. During 1845 and 1846 Hoe was busily engaged in designing and inventing presses to meet the increased requirements of the newspaper publishers. The result was the construction of the revolving machine based on Hoe's patents. The basis of these inventions was a device for securely fastening the forms of type on a central cylinder placed in a horizontal position. The first of these machines, installed in the *Public Ledger* office, had four impression cylinders grouped around the central cylinder. With one boy to each cylinder to feed in blank paper, 8,000 papers could be produced per hour.

Almost immediately newspaper printing was revolutionized, and Hoe's rotary press became famous throughout the world. In 1853, he introduced the cylinder press which had been patented in France by Dutartre and improved on it in the following years for use in lithographic and letterpress work. In 1861, the curved stereotype plate was perfected, and in 1865 William Bullock succeeded in producing the first printing machine that would print on a continuous web or roll of paper. Spurred on by this latest development, Hoe and his partner began experimenting and designed and built a web press. The first of these to be used in the United States was installed in the office of the *New York Tribune*. At maximum speed, this press printed on both sides of the sheet and produced 18,000 papers per hour. Four years later, Tucker patented a rotating, folding cylinder which folded the papers as fast as they came off the press.

In 1881, the Hoe Company devised the triangular former-folder which, when incorporated into the press, together with approximately twenty additional improvements, gave rise to the modern newspaper press. It was with the introduction of this that the 1847-type revolving press was superseded.

Hofmann August Wilhelm von 1818–1892 was a German organic chemist, one of the greatest of the nineteenth century, and had enormous influence on the development of the subject in both Britain and Germany. Much of his work was connected with coal tar and its constituents, particularly aniline (phenylamine) and phenol. He was the first to explore the chemistry of the aliphatic amines (although not the first to synthesize them) and was the discoverer of the quaternary ammonium salts. His students included those (such as William Perkin) who originated and developed the British synthetic dye industry, and as a chemist he discovered and patented a number of dyes of his own.

Hofmann was born on 8 April 1818 in Giessen, the son of an architect. He entered the local university in 1836 to study law and philosophy, but his interests changed when he attended some of Justus von Liebig's lectures and he continued his studies by specializing in chemistry. In 1841, he obtained his doctorate for a thesis based on investigations into the constitution of coal tar. He then neglected his studies and academic advancement to look after his father, who died in 1843.

In that same year Hofmann became Liebig's assistant and in the spring of 1845 he was appointed briefly to a position in the University of Bonn. Later that year he was requested by Prince Albert, Queen Victoria's consort, to become a professor at the new Royal College of Chemistry in London. He held this position for the next 20 years, during which time the college became amalgamated with the Royal School of Mines. In 1863, he was offered chairs of chemistry in both Bonn and Berlin. He accepted the position vacated by Eilhard Mitscherlich in Berlin, after designing new laboratories for both institutions, taking up the appointment in 1865 and remaining there until his death. On his seventieth birthday in 1888 he was made a baron, becoming von Hofmann. He died in Berlin on 2 May 1892.

Hofmann's first paper confirmed the presence of aniline in coal tar. He believed that phenol and aniline (phenylamine) were related, and converted the former into the latter by heating phenol with ammonia in a sealed tube for three weeks. Later he prepared nitrobenzene from the light oil distillate of coal tar and made aniline from it by reduction with nascent hydrogen.

From about 1850, Hofmann explored the behaviour and properties of amines, showing that alkyl halides (halogenoalkanes) react with aniline to give secondary and tertiary amines (although Hofmann used different terms for them). He also prepared alkyl amines by the reaction between ammonia and alkyl halides, reporting that he was replacing the hydrogen atoms of ammonia with alkyl groups and thus producing compounds of the 'ammonia type'. The theory of types held vogue in organic chemistry for a number of years and other organic compounds were classified as the 'water', 'hydrogen' and 'hydrochloric acid' types. Hofmann also set about investigating a 'phosphorus' type based on phosphine, PH_3. In 1865, he published *An Introduction to Modern Chemistry*, which was a textbook based on the theory of types.

Hofmann produced more complex amines from diamines and alkyl halides. He discovered what is now known as the Hofmann degradation, in which an amide is treated with bromine and alkali (hypobromite) to produce an amine with one fewer carbon atoms:

$$CH_3CONH_2 + Br_2 + 4KOH \rightarrow CH_3NH_2 +$$

acetamide   methylamine
(ethanamide)

$$2KBr + K_2CO_3 + 2H_2O$$

2 carbon atoms   1 carbon atom

By the reaction of ethyl iodide (iodoethane) with triethylamine he produced tetraethyl-ammonium iodide, $(C_2H_5)_4N^+ I^-$, the first quaternary ammonium salt. He found that the ion $(C_2H_5)_4 N^+$ behaved like a sodium or potassium ion, as a 'true organic metal'. With silver oxide he was able to form from it the strong base $(C_2H_5)_4NOH$, which exists as a solid.

Hofmann was also the first to investigate the structure and properties of formaldehyde (methanal), which he called 'methyl aldehyd' and prepared by passing methanol vapour and air over heated platinum. Working with A.A.T. Cahours he also discovered the first unsaturated alcohol, prop-2-en-1-ol, $CH_2{=}CH.CH_2OH$.

In 1858, Hofmann obtained the dye known as fuchsine or magenta by the reaction of carbon tetrachloride (tetrachloromethane) with aniline (phenylamine). Later he isolated from it a compound which he called rosaniline and used this as a starting point for other aniline dyes, including aniline blue (triphenyl rosaniline). With alkyl iodides (iodoalkanes) he obtained a series of violet dyes which he patented in 1863. These became known as 'Hofmann's violets' and were a considerable commercial and financial success.

On his return to Germany Hofmann founded the Deutsche Chemische Gesellschaft (German Chemical Society). The many reactions and rules that still bear his name are a testimony to his stature as an organic chemist. The term valence is a contraction of his notion of 'quantivalence' and he devised much of the terminology of the paraffins (alkanes) and their derivatives which was accepted at the 1892 Geneva Conference on nomenclature. Apart from his ammonia and phosphine types he was not a great theorist but became renowned as an experimental chemist. Yet he is said to have been very clumsy with apparatus, the handling of which he usually left to his more dextrous assistants. He was married four times and had eleven children, eight of whom survived him when he died at the age of 74.

Hofstadter Robert 1915–1990 was a US physicist who shared the 1961 Nobel prize for discovering that protons and neutrons contained smaller particles (now known to be quarks). Hofstadter later became active in medical physics and astrophysics.

Hofstadter was born on 5 February 1915 in New York City. He graduated from the City College of New York in

1935, and obtained a PhD from Princeton University in 1938 for research into the infrared spectroscopy of small organic molecules, particularly studies of the hydrogen bond. He remained at Princeton as a postdoctoral student, studying solid-state luminescence and photoconductivity before moving to the University of Pennsylvania where he helped build a large van de Graaff generator. It was at Pennsylvania that his interest in detectors for nuclear physics began. But World War II interrupted and Hofstadter worked on proximity fuses at the National Bureau of Standards before joining Norden Laboratory Corporation, an aerospace company, for three years. In 1946, Hofstadter returned to Princeton to pick up his studies in particle detectors and nuclear physics. In 1948, he made what he considered his most important discovery – that thallium-activated sodium iodide (NaI(Tl)) was an excellent scintillation counter and, with John McIntyre in 1950, that it could be used to measure the energies of gamma rays. Despite being expensive and difficult to work with, NaI(Tl) has been used in gamma-ray spectrometers ever since – in medicine, chemistry, biology and geology in addition to nuclear, particle and astrophysics. In 1950, Hofstadter moved to Stanford University where he began studying the elastic and inelastic scattering of electrons by nuclei. Over the next 20 years these experiments showed how electric charge (and the associated magnetic field) is distributed in the nucleus. Hofstadter also showed that although the neutron is neutral, it has a distribution of electric charge and magnetism within it. This was the first evidence that neutrons and protons were not point-like particles, rather they were made up of other smaller particles (now known to be quarks). Hofstadter was awarded the 1961 Nobel prize for these discoveries (shared with Rudolf Mössbauer). In 1968, Hofstadter began working with Barrie Hughes on the Crystal Ball detector at the Stanford Linear Accelerator Center. This device, which comprised 900 NaI detectors pointing at the point where beams of electrons and positrons collided, made a number of important discoveries in the spectroscopy of mesons containing charmed and bottom quarks. Hofstadter also used the synchrotron radiation given off by the electron storage rings at SLAC for coronary angiography studies in medicine. In 1970, Hofstadter also suggested putting a large gamma-ray spectrometer on a satellite to carry out gamma-ray astronomy. This device was launched aboard NASA's Gamma Ray Observatory (GRO) five months after Hofstadter's death, on 17 November 1990.

Hogben Lancelot Thomas 1895–1975 was a British zoologist and geneticist who, somewhat surprisingly, wrote a very successful book entitled *Mathematics for the Millions.* Although he thereafter sought to use mathematical techniques in his various posts, there were in fact few opportunities to do so until he was put in charge of the medical statistics records for the British army during World War II.

Hogben was born in Southsea, Hampshire, on 9 December 1895. Growing up, he developed a strong interest in natural history. From the Middlesex County Secondary School he won a scholarship to Trinity College, Cambridge, in 1913. Simultaneously he took an external degree course in zoology at London University. Imprisoned, however, as a conscientious objector in 1916 during World War I, he was released when his health deteriorated seriously and briefly worked as a journalist before being appointed Lecturer in Zoology at Birkbeck College, London, in 1917. Two years later, he transferred to the Imperial College and shortly afterwards received his doctorate. He then went to teach in Edinburgh for a couple of years, to Montreal, Canada, for a similar period, and to Cape Town, South Africa, for again the same duration of time. He returned to London University in 1930 to take up the post of Professor of Social Biology. In hospital in 1933, he wrote his famous *Mathematics for the Millions* partly as therapy, partly for self-education; he also became interested in linguistics. Following this latter interest, he was in Norway at the outbreak of World War II, and it was not until 1941 – after travelling through Sweden, the Soviet Union and the United States – that he got back to Britain and became a colonel in the War Office. After the war he became Professor of Medical Statistics at the University of Birmingham, where he remained until he retired in 1961. The author of many scientific papers, and the recipient of many awards and honours – a Fellow of the Royal Society from 1936 – Hogben died in Glyn Ceiriog, Wales, on 22 August 1975.

It was in London in the 1930s – around the time that his popular *Mathematics for the Millions* was published – that Hogben first began to try to apply mathematical principles to the study of genetics, with particular reference to his investigation of generations of the fruitfly *Drosophila* in relation to research on heredity in humans. He was especially concerned to evaluate the validity of statistical methods as applied in the biological and behavioural sciences, and was surprised at the apparent level of ignorance in such methods among the many scientists of those disciplines.

His desire to use mathematical principles was given considerably more scope on his appointment as colonel in charge of the War Office medical statistical records. Asked to reorganize the entire system, Hogben successfully did just that – and it was, after all, in the days when data processing was still a matter of mechanical sorting. He was subsequently promoted to Director of Army Medical Statistics. (One of his investigations into the army's clinical trials of the sulphonamide drugs proved that their indiscriminate use had favoured the selection of resistant strains of the bacteria responsible for causing gonorrhoea.)

Working at Birmingham University after the war, his attempts to reorganize civilian medical records were less successful – partly because of resistance from some senior medical practitioners reluctant to change their ways.

During his last years, Hogben spent much time revising previously published work and writing on other aspects of philosophy and mathematics.

Hollerith Herman 1860–1929 was a US mathematician

and mechanical engineer who invented electrical tabulating machines. He was probably the first to automate the large-scale processing of information, and as such was a pioneer of the electronic calculator, particularly its application to data handling in business and commerce.

Hollerith was born on 29 February 1860 in Buffalo, New York. After his schooldays he attended the School of Mines at Columbia University, from which he graduated in 1879. The following October he became an assistant to W. P. Trowbridge, one of his university lecturers, in the work for the United States census of 1880.

Hollerith worked with physician and librarian John Billings (1838–1913) on the statistics of manufacturing industries, especially those concerned with steam and water power in the iron and steel industries. In the course of processing the many census returns, they became aware that an automated recording process would have considerable labour-saving advantages. Whether it was Hollerith who had the idea of punching holes into cards to indicate quantities or whether Billings made the initial suggestion is not known. What is important is that later Hollerith developed the idea of punching first a continuous roll of paper, and later individual cards the same size as a dollar bill, with holes to represent information. The quantities indicated by the holes were counted when the tape or cards passed through a device in which electrical contact was made through the holes. The passage of an electric current caused electromechanical counters to advance one place for each hole. The realization of these ideas in practical terms did not, however, come about in time for the processing of the 1880 returns.

In 1882, Hollerith obtained a position as instructor in mechanical engineering at the Massachusetts Institute of Technology (MIT). He preferred experimental work to teaching, and after a year he left MIT and went to St Louis until 1884, where he worked on the development of electromechanically operated air-brakes for railways. He was then employed by the Patent Office in Washington, DC, and until the next census in 1890 he developed his recording and tabulating machines.

By 1889, he had developed not only his electrical machines for recording the information on punched cards, but also machines for punching the cards and for sorting them. But by this time he was not the only inventor of data-processing equipment, and a trial was held to decide which system was to be adopted for the 1890 census. The Hollerith system proved to be twice as fast as the best of the other two and was used to handle the returns of 63 million people. In 1891, the Hollerith system was used in the censuses in Austria, Canada and Norway (and in Britain in 1911). The machines were later successfully adapted to the needs of government departments and businesses which handled large quantities of data, and particularly to the tabulating of railway statistics.

Hollerith soon realized that the age of large-scale data handling had begun, and in 1896 he formed the Tabulating Machine Company, to manufacture the machines and the cards they used. With the growth of business in this area,

Hollerith's company was soon merged with two others into the Computing-Tabulating-Recording Company, which later became the International Business Machines Corporation (IBM). IBM has remained one of the foremost companies in the development and manufacture of data-processing systems, and has been an important producer of electronic computers for many years.

Hollerith stayed with IBM as a consultant engineer until his retirement in 1921. He died from heart disease in Washington on 17 November 1929, aged 69.

Holley and Nirenberg Robert William Holley 1922– and Marshall Warren Nirenberg 1927– are two biochemists who shared the 1968 Nobel Prize for Physiology or Medicine (with Har Gobind Khorana) for their work in deciphering the chemistry of the genetic code.

Holley was born on 28 January 1922 in Urbana, Illinois, in the United States. He studied chemistry at the University of Illinois and graduated in 1942. He gained his PhD from Cornell University in 1947, having spent the latter part of World War II engaged on research into penicillin. From 1948 to 1957 he was Assistant Professor and then Associate Professor of Organic Chemistry at the New York State Agricultural Experimental Station at Cornell, when he moved to the Plant Soil and Nutrition Laboratory. In 1962 he became Professor of Biology at Cornell, and in 1966 took up a senior appointment at the Salk Institute for Biological Studies in San Diego, California.

Nirenberg was born in New York City on 10 April 1927. He studied biology at the University of Florida, where he graduated in 1948, gaining his master's degree four years later. He then went to the University of Michigan's Department of Biological Chemistry and gained a PhD in 1957. From 1957 to 1962 he was at the National Institute of Health (Arthritic and Metabolic Diseases), becoming Head of the Laboratory of Biochemical Genetics there in 1962. He later moved to the Laboratory of Biochemical Genetics at the National Heart, Lung and Blood Institute in Bethesda, near Washington, DC.

Holley's early work in the New York State Agricultural Experimental Station at Cornell concerned plant hormones, the volatile constituents of fruits, the nitrogen metabolism of plants, and peptide synthesis. At the Salk Institute he began to study the factors that influence growth in cultured mammalian cells.

At Cornell he obtained evidence for the existence of transfer RNAs (tRNAs) and for their role as acceptors of activated amino acids. In 1958, he succeeded in isolating the alanine-, tyrosine- and valene-specific tRNAs from baker's yeast and approximately one gram of highly purified alanine-specific tRNA was prepared and used in structural studies over the next two-and-a-half years. The technique for elucidating its structure was to break up the molecule into large 'pieces', identify the fragments and then reconstruct the original. Eventually Holley and his colleagues succeeded in solving the entire nucleotide sequence of this RNA.

Nirenberg was interested in the way in which the

nitrogen bases – adenine (A), cytosine (C), guanine (G) and thymine (T) – specify a particular amino acid. To simplify the task of identifying the RNA triplet responsible for each amino acid, he used a simple synthetic RNA polymer. Using an RNA polymer containing only uracil, for example, he obtained phenylalanine, so he concluded that its code must be UUU. Similarly a cytosine RNA polymer produced only the amino acid proline.

In this way Nirenberg built up a tentative dictionary of the RNA code. He found that certain amino acids could be specified by more than one triplet, and that some triplets did not specify an amino acid at all. These 'nonsense' triplets signified the beginning or the end. In this way he assigned values to 50 triplets. He then worked on finding the orders of the letters in the triplets. By labelling one amino acid at a time with carbon-14, and passing the experimental material through a filter that retained only the ribosomes with the tRNA and amino acids attached, he obtained unambiguous results for 60 of the possible codons. Work continued in his laboratory on the role of 'synonym' codons, codon recognition by tRNA, and the mechanics of the rate of protein synthesis during viral infection and embryonic differentiation.

Holmes Arthur 1890–1965 was a British geologist who helped develop interest in the theory of continental drift.

Born at Hebburn, Newcastle-upon-Tyne, Holmes received his scientific education at Imperial College, London, first in physics and mathematics, then in geology. After spells teaching at Imperial College and working for oil companies, Holmes was appointed in 1924 Head of the Geology Department at Durham University, moving in 1943 to Edinburgh University. He distinguished himself in many branches of geology, not least petrology, where his *Petrographic Methods and Calculations* 1921 became a classic, as did his more general survey, *Principles of Physical Geology* 1944.

Holmes is best remembered for his geochronological explorations, giving the first reliable modern estimates of the age of the Earth. Radioactivity had already been recognized as holding out the possibility of a technique for determining the actual ages of minerals; and the work of R. J. Strutt (later Lord Rayleigh), in demonstrating that the profusion of radioactive minerals in the Earth's crust constituted a major heat source had scotched the thermodynamic arguments deployed by Lord Kelvin against what he saw as extravagant estimates of the age of the Earth. Holmes pioneered the use of radioactive decay methods for rock-dating. Painstaking analysis of the proportions of elements formed by radioactive decay, combined with a knowledge of the rates of decay of their parent elements, would yield an absolute age. From 1913 he used the uranium–lead technique systematically to date fossils whose relative (stratigraphical) ages were established but not the absolute age. His *The Age of the Earth* (1913) summarized existing data and developed a time-scale of its own. Publishing extensively, he continued work on the Phanerozoic time-scale until 1959, though his original estimates did not require fundamental modification.

Almost alone in Britain, Holmes was also an early advocate of continental drift. In 1928, he proposed that convection currents within the Earth's mantle, driven by radioactive heat, might furnish the mechanism for the continental drift theory broached a few years earlier by Wegener. In Holmes's view, new oceanic rocks were forming throughout the ocean ridges. Little attention was given to Holmes's ideas on this subject until the 1950s, when palaeomagnetic studies put continental drift theories on a sure footing.

Hooke Robert 1635–1703 was a British physicist who was also active in many other branches of science. He is remembered mainly for the derivation of Hooke's law of elasticity, for coining the term 'cell' as used in biology, and for the invention of the hairspring regulator in timepieces and the air pump.

Hooke was born in Freshwater, Isle of Wight, on 18 July 1635. He was sickly as a child, which prevented him from studying for the Church as his father intended. Left on his own, Hooke constructed all kinds of ingenious toys, developing the great mechanical skill that he later applied to instrument making. Upon the death of his father in 1648, Hooke went to London and was educated at Westminster School. There he was introduced to mathematics, mastering Euclid in only a week. In 1653, he went on to Oxford University as a chorister and there became one of a group of brilliant young scientists, among them Robert Boyle (1627–1691), to whom Hooke became an assistant.

Hooke eventually obtained his MA from Oxford in 1663, but the group broke up in 1659 and he and most of his colleagues moved to London, where they established the Society in 1660 (which became the Royal Society in 1662). Hooke was appointed Curator of the Society, a post that entailed the demonstration of several new experiments at every weekly meeting. In 1664, he also became a lecturer in mechanics at the Royal Society, and in the following year he took up the additional post of Professor of Geometry at Gresham College, London. Hooke retained these positions for the rest of his life, and was also Secretary of the Royal Society from 1677 to 1683. He died in London on 3 March 1703.

Hooke's post of curator at the Royal Society required him to provide a continual stream of new ideas to demonstrate before the members, and he consequently examined all fields of experimental science but made no deep or thorough investigations in any of them. His main contributions were in four main areas – mechanics, optics, geology and instrument making.

In mechanics, Hooke was the first to realize that the stress placed upon an elastic body is proportional to the strain produced. This relationship is known as Hooke's law; it was discovered in 1678. But Hooke was active in mechanics long before this, and in 1658 he found that a spiral spring vibrates with a regular period in the same way as a pendulum. He began to develop this discovery to produce a watch with a spring-controlled balance wheel, but it is uncertain whether he made a working model before Christiaan Huygens (1629–1695) did so in 1674. Hooke can thus be credited with the discovery of the principle of the watch if not the invention of the device itself. Hooke himself claimed priority, however, one of many such disputes that were a feature of his life. His main antagonist in this respect was Isaac Newton (1642–1727), with whom Hooke argued over credit for the discovery of gravitation. The idea that gravity exists between bodies was prevalent at the time, and in 1664, Hooke suggested that a body is continually pulled into an orbit around a larger body by a force of gravity directed towards the centre of the larger body. This was an important step towards an understanding of gravity, and Hooke also suggested in 1679 that the force of gravity obeys an inverse square law. Both these ideas helped Newton, with his immense analytical powers, to build on Hooke's insight and arrive at his law of universal gravitation in 1687.

Another area of contention between Hooke and Newton was in optics. In 1665, Hooke published a book called *Micrographia*. It contained superb accounts of observations that Hooke had made with the microscope and was the first important work on microscopy. From his description of the empty spaces in the structure of cork as cells comes our use of the word 'cell' to mean a living unit of protoplasm. The *Micrographia* also contained Hooke's work on optics. This included the idea that light might consist of waves, which Hooke developed from his observations of spectral colours and patterns in thin films. Newton was again able to build on Hooke's work here, examining the optical effects of thin films in detail.

In geology, Hooke made an important contribution by insisting that fossils are the remains of plants and animals that existed long ago, a daring view in an age dominated by the biblical account of creation. Hooke furthermore held that the history of the Earth would be revealed by a close study of fossils.

As an instrument maker, Hooke made several important inventions and advances. In 1658, while working for Robert Boyle at Oxford, Hooke perfected the air pump, a development that led directly to the derivation of Boyle's law in 1662. He also made considerable advances to the microscope, and invented the wheel barometer, which registered air pressure with a moving pointer; a weather clock, which recorded such factors as air pressure and temperature on a revolving drum; and the universal joint.

Hooke is an unusual figure in the history of science. Although he made no major discoveries himself, his wide-ranging intuition and great experimental prowess sparked off important contributions in others, notably Newton and Huygens.

Hooker Joseph Dalton 1817–1911 was a British botanist who made many important contributions to botanical taxonomy but who is probably best known for introducing into Britain a range of previously unknown species of rhododendron and for his improvements to the Royal Botanical Gardens at Kew.

Hooker was born on 30 June 1817 in Halesworth, Suffolk. He studied medicine at Glasgow University, where his father was Professor of Botany, graduating in 1839. In the same year he obtained the post of assistant surgeon and naturalist on an expedition to the southern hemisphere. The expedition, which was led by Captain James Clark Ross, set out in 1839 and returned in 1843; its main aims were to locate the magnetic South Pole and to explore the Great Ice Barrier, but other places were visited, including the Falkland Islands, Tasmania and New Zealand. On his return to England, Hooker applied for the botany chair at Edinburgh University but was not accepted and so took a job identifying fossils for a geological survey. From 1847 to 1850 he took time off to undertake a botanical exploration of northeastern India, mainly of the Himalayan state of Sikkim and eastern Nepal. In 1855 he became Assistant Director of Kew Gardens, where his father was by this time Director. On the death of his father in 1865 Hooker became the Director, a post he held until 1885, when he retired. He died on 10 December 1911 in Sunningdale, Berkshire.

While Hooker was on the expedition to the southern hemisphere he made extensive notes and sketches of the plants he saw and collected many specimens, which he pressed and mounted. On his return to Britain he produced a six-volume work (published between 1844 and 1860) of his observations and findings, with two volumes each on the flora of Antarctica, New Zealand and Tasmania. This work combined accurate and detailed descriptions of plants with perceptive essays on plant distribution and established Hooker's reputation as a botanist of the highest calibre. The importance of this work was quickly recognized by the Royal Society, which elected Hooker a Fellow in 1847.

In 1854, Hooker published a general account of his travels in the Indian subcontinent, entitled *Himalayan Journals*. He also wrote many scientific works based on his research on the Indian flora; the first of these was about rhododendrons and was published by Hooker's father while Hooker himself was still in India. In addition, Hooker sent back to England many previously unknown species of rhododendrons. His first general botanical work on Indian plants was the single-volume *Flora Indica* (1855), written in conjunction with Thomas Thomson. This was superseded by Hooker's monumental seven-volume *Flora of British India* (1872 to 1897), written jointly with several other scientists. While in India, Hooker became interested in the genus *Impatiens* (a group that includes the Himalayan Balsam, which has since become naturalized in Britain), and gave descriptions of about 300 species of this genus. He supplemented his Indian work by writing volumes four and five of *A Handbook to the Flora of Ceylon* between 1898 and 1900, a work that had remained unfinished since the death in 1896 of its original author, H. Trimen.

As Director of Kew Gardens, Hooker introduced many improvements, with the introduction of the rock garden, the addition of new avenues and an extension of the arboretum. Several other important developments occured during Hooker's directorship. In 1876, T. J. Phillips

Jodrell, a friend of the Hooker family, died and left a bequest for the foundation of a botanical laboratory at Kew. The Jodrell Laboratory is now world famous for the scientific work performed there on the structure and physiology of plants. Kew Gardens also became increasingly important as a repository for collections of pressed plants and as a centre for the propagation and distribution of many crop plants, including rubber, coffee and the oil palm. Furthermore, in 1883 the *Index Kewensis* was founded; this is a list of all scientific plant names, accompanied by descriptions which, since the publication of the first volume in 1892, has become an invaluable aid in preventing duplication and error in the naming of plants.

As well as establishing Kew Gardens as an international centre for botanical research, Hooker also continued his own botanical work. With the botanist George Bentham he published *Genera Plantarum* (1862 to 1883), a complete catalogue of all the known genera and families of flowering plants from all parts of the world. Nevertheless, Hooker did not neglect the British flora. In 1870, he published *Student's Flora of the British Isles*, and from 1887 to 1908 he edited various editions of Bentham's *Handbook of the British Flora*. He was also interested in aspects of botany other than taxonomy, such as the dispersal of plants over large areas and the evolution of new species. After much consideration he became an evolutionist, but his belief in the theory was founded on his own rather specialized knowledge of plants and regional floras and so he contributed little to the popular debate on the subject.

Hooker Stanley 1907–1984 was the British engineer responsible for major aircraft engine development projects which culminated in the successful Proteus turboprop, Orpheus turbojet, Pegasus vectored thrust turbofan, Olympus turbojet and RB211 turbofan.

Hooker was born in Sheerness, Kent, on 30 September 1907 and educated at Imperial College, London, and Brasenose College, Oxford. He joined the Admiralty Scientific and Research Department in 1935, working on anti-aircraft rocket development. In 1938 he moved to Rolls-Royce in charge of the Performance and Supercharger Department.

In 1940 he met Sir Frank Whittle and the following year, as chief engineer of the Rolls-Royce Barnoldswich Division, he was responsible for the development of the Whittle W2B turbojet. In the period 1944–45 the Nene and Derwent engines emerged from Barnoldswick and in 1946 the Meteor, powered by two Derwent V engines, established a world speed record of 970 kph/603 mph.

Hooker joined the Aero Engine Division of the Bristol Aeroplane Company, becoming chief engineer in 1951 and a director the following year. When Bristol Siddeley Engines was formed in 1959, he became technical director of the Aero Division and, following the merger with Rolls-Royce in 1966, technical director of the Bristol Engine Division.

In January 1971, Hooker was appointed group technical director and seconded to Derby in charge of RB211

development. The same year he took his seat on the main board as group technical director.

In his distinguished career Hooker received many honorary degrees, decorations, medals and awards, including the OBE (1946), CBE (1964) and a knighthood (1974). He was elected a Fellow of the Royal Society in 1962 and a Fellow of the Royal Aeronautical Society in 1947.

The Merlin engine, which was to power so many Allied aircraft in World War II, started life in 1935 developing under 670 kW/900 hp. As the need for greater engine power increased, particularly at high altitudes, supercharging was one of the most effective ways of meeting the demand. The Merlin supercharger, developed under Hooker, was so successful that special versions of the engine were delivering up to 1970 kW/2,640 hp by the end of the War.

A turboprop engine uses the exhaust efflux from a turbo jet (usually called a gas generator, in this application) to drive a conventional propeller. The Proteus turboprop in its Mark 705 version entered service in the 100 Series Britannia airliner in January 1957 at 2,900 kW/3,900 tehp (total equivalent hp, being the shaft power delivered to the propeller plus the equivalent of the residual jet).

The turbojet engine passes all the air it draws in through the combustion chambers and expels it in a hot high-velocity jet; it uses the energy of the jet alone to drive the aircraft. It is inefficient at low speeds, but as the aircraft passes through about 1,500 kph/900 mph its rating improves and it is the preferred power unit for military and supersonic applications. The Orpheus turbojet, for which Hooker had overall responsibility, was chosen to power the Fiat G91, selected under the NATO mutual weapons development programme, entering service in May 1958 as the Mark 801 at 18 kN/4,050 lb thrust.

The turbofan engine passes only a proportion of the air through the combustion chambers. This air emerges as a hot high-velocity jet and is mixed with the air which by-passed the core of the engine. The amount of by-pass air varies in different engines.

The Pegasus vectored thrust engine was conceived after discussions between Hooker and Sir Sydney Camm. It is the power unit for the Harrier V/STOL (vertical/short take-off and landing) combat aircraft and is a unique design of turbofan. Its by-pass air is ducted through two forward nozzles and the gas coming from the core of the engine is ducted through two rear nozzles. All four nozzles can be rotated downwards to provide vertical thrust and rearwards to provide forwards thrust. Intermediate settings give the aircraft its short take-off capability. Rotating the nozzles in flight, even forwards to give reverse thrust, makes the Harrier a most manoeuvrable combat aircraft.

The Olympus turbojet, eventually to power the supersonic Concorde airliner, started life in 1946 when Bristol submitted designs to the Ministry of Supply in answer to a tender for an engine of about 36 kN/8,000 lb thrust. The design was then unique, because it was the first time the two-spool concept had been proposed. This layout uses two independent compressors driven by two independent turbines, giving the engine high compression (important for fuel economy) and great adaptability.

The engine underwent extensive development under Hooker and then flew in the prototype and preproduction aircraft, designated 593B, with 156 kN/35,080 lb thrust. In the series aircraft designated 600, it flies with 170 kN/ 38,000 lb thrust.

The RB211 turbofan was designed with one of the highest by-pass ratios of any engine in the early 1980s. This and other design features makes it the quietest and most economical in airline service. Economy encompasses every aspect of operating an engine, not just the fuel consumption. One of the most important aspects of the RB211 operating economics is its modular design. In older concept engines, changing a turbine at the end of its service life needed a complete engine strip-down; the RB211 has a turbine module, a compressor module, and so on, so that replacement is comparatively simple and swift.

From the Merlin supercharger to the Olympus and RB211, Hooker led design teams which introduced some of the most advanced aeroengine concepts in the world, operating to the rigid requirements of aviation.

Hopkins Frederick Gowland 1861–1947 was a British biochemist who was jointly awarded (with Christian Eijkman) the 1929 Nobel Prize for Physiology or Medicine for his work that showed the necessity of certain dietary components – now known as vitamins – for the maintenance of health. He received several other honours for his work, including a knighthood in 1925.

Hopkins was born on 20 June 1861 in Eastbourne, Sussex. He showed no remarkable distinction at school, except in chemistry, and after leaving school he was articled for three years to a consulting analyst in London. He then became an analytical assistant at Guy's Hospital in London, simultaneously studying for an external degree in chemistry at the University of London. In 1888, he became a medical student at Guy's Hospital Medical School, from which he graduated in 1894. He remained at Guy's Hospital until 1898, when he was invited by Michael Foster, Professor of Physiology at Cambridge University, to become a lecturer in chemical physiology at Cambridge. This was an extremely taxing job, the strain of which adversely affected Hopkins's health, and it left little time for original research. In 1914, however, Hopkins was appointed Professor of Biochemistry at Cambridge and was able to devote more time to his own investigations; he held this position until he retired in 1943. He died in Cambridge on 16 May 1947.

Hopkins began his Nobel prizewinning work in 1906, when he realized that animals cannot survive on a diet containing only proteins, fats and carbohydrates. Experimenting on the growth rates of rats fed on diets of artificial milk, he noticed that they failed to grow unless a small quantity of cow's milk was added to the artificial milk. From this he concluded that the cow's milk contained accessory food factors that are required in only trace amounts but which are essential for normal growth. But he failed to isolate

these substances (now called vitamins) and his hypothesis remained controversial for many years, although it had been proved correct by the time he was awarded a Nobel prize in 1929.

Hopkins also made several other important contributions to biological knowledge. He discovered the amino acid tryptophan when one of his students – John Mellanby, who later became Professor of Physiology at Oxford University – failed to obtain the Adamkiewicz colour reaction for proteins (this involves adding acetic acid then strong sulphuric acid to the test solution). This led Hopkins, with the assistance of S. W. Cole, to investigate the reaction, which they found is the result of a reaction between glyoxylic acid – a common contaminant of acetic acid – and tryptophan. This discovery meant that tryptophan is an important constituent of proteins, which then led to the concept of essential amino acids. Hopkins also showed that tryptophan and certain other amino acids cannot be manufactured by the body and must therefore be supplied in the diet. In addition, he helped to lay the foundation for the modern understanding of muscle contraction with his demonstration (in collaboration with Morley Fletcher) that contracting muscle accumulates lactic acid. Hopkins also discovered the tripeptide glutathione, which is important as a hydrogen carrier in the intracellular utilization of oxygen.

Hopper Grace 1906–1992 was a US computer pioneer whose most significant contributions were in the field of software. She created the first compiler and helped invent the computer language COBOL. She was also the first person to isolate a computer 'bug' and successfully debug a computer.

After doing postgraduate work at Yale, Hopper returned to her original university, Vassar, as a member of the mathematics faculty. She volunteered for duty in World War II with the Naval Ordinance Computation Project. This was the beginning of a long association with the Navy, resulting in her being appointed to the rank of rear admiral in 1983. She was then the oldest officer on active duty in the US armed forces. In 1945, Hopper was ordered to Harvard University to assist Howard Aiken (1900–1973) in building a computer. In those days, computers had to be rewired for each new task and Hopper frequently found herself entwined in the wiring of the computer. One day a breakdown of the machine was found to be due to a moth that had flown into the computer. Aiken came into the laboratory as Hopper was dealing with the insect. 'Why aren't you making numbers, Hopper?' he asked. Hopper replied: 'I am debugging the machine!' This first computer 'bug' was removed with tweezers and is preserved at the Naval Museum in Virginia in the laboratory logbook, glued beside the entry for 15.45 on 9 Sept 1945. After the war, Hopper joined John Mauchly and Presper Eckert who had set up a firm, eventually to become the Univac division of Sperry–Rand, to manufacture a commercial computer. Her main contribution was to create the first computer language, together with the compiler needed to translate the instructions into a form that the computer could work with.

In 1959, she was invited to join a Pentagon team attempting to create and standardize a single computer language for commercial use. This led to the development of the COBOL language, still one of the most widely used languages.

Hounsfield Godfrey Newbold 1919–   is a British research scientist whose part in the invention and development of computerized axial tomography (the CAT scanner) was recognized by the award of the 1979 Nobel Prize for Physiology or Medicine.

Hounsfield was born on 28 August 1919 and educated at the Magnus Grammar School, Newark, and later at the City and Guilds College, London and the Faraday House Electrical Engineering College.

After serving with the Royal Air Force during World War II, he joined the Medical Systems Section of EMI in 1951 and has been actively engaged in research with the company ever since. He was made Professorial Fellow of Manchester University (for Imaging Sciences) in 1978 and has been awarded some of the most coveted prizes in the scientific world including the Nobel prize, the Price Phillip Medical Award (1975), the Lasker Award (1975) and the Churchill Gold Medal (1976).

Hounsfield's long career in medical research and engineering has culminated in the invention of the EMI scanner, formerly a computerized transverse axial tomography system for X-ray examination. Its development, using all the most advanced technology available, has led to major improvements in the field of X-ray, enabling the whole body to be screened at one time. It is proving particularly valuable in the detection of cancer.

Howe Elias 1819–1867 was a US engineer who invented a sewing machine, one of the first products of the Industrial Revolution that ultimately eased the burden of domestic work.

Howe was born on 9 July 1819 in Spencer, Massachusetts. As a boy, he worked on the farm his father ran, and in his gristmills and sawmills where he took a particular interest in the machinery. He trained to be a machinist, and having conceived the idea of a sewing machine, spent much of his life developing and patenting one. He died in Brooklyn, New York, on 3 October 1867.

Howe began work on the design of a sewing machine in about 1843. Within a year he had a rough working model, and in September 1846 he was granted a US patent for a practical machine. He was the first to patent a lock-stitch mechanism, and his machine had two other important features: a curved needle with the eye (for the thread) at the point, and an under-thread shuttle (invented by Walter Hunt in 1834).

Howe immediately went to Britain and sold the invention for £250 to a corset manufacturer named William Thomas of Cheapside, London. In December 1846, Thomas secured the English patent in his own name, and engaged Howe (on weekly wages) to adapt the machine for his needs. Howe worked with Thomas from (1847 to 1849), but his career in London was unsuccessful. He pawned his US patent rights, and returned in poverty to the United States, where he found his wife dying.

While Howe had been away, the sewing machine was beginning to arouse public curiosity, and he found that various people were making machines which infringed his patent. The most prominent of these – an inventor in his own right – was Isaac Singer (1811–1875), who in 1851 secured a US patent for his own machine. Howe now became aware of his rights, redeemed his pawned patent, and took out law suits against the infringers. After much litigation, the courts found in his favour and from that time (about 1852) until his patent expired in 1867, Howe received royalties on all sewing machines made in the United States. When he died, he left an estate worth $2 million, an enormous sum in 1867.

The basic invention in machine sewing was the double-pointed needle, with the eye in the centre, patented by C. F. Weisenthal in 1755 which enabled the sewing or embroidery stitch to be made with the needle being inverted. Many of the features of the sewing machine are distinctly specified in a patent secured in England in 1790 by Thomas Saint. The machine he described was for stitching, quilting or sewing but seems to have been chiefly intended for leather-work. If Saint had hit upon the eye-pointed needle, his machine would have completely anticipated the modern chain-stitch machine. A real working machine was invented in France by a poor tailor, Barthélemy Thimmonier, and a patent was obtained in 1830. By 1841 about 40 of these machines were being used in Paris to make army clothing, although the machines were clumsy, being largely made of wood. An ignorant crowd wrecked his establishment and his machines, but undeterred he patented vast improvements on it. The troubles of 1848, however, blasted his prospects and his patent rights for Britain were sold. A machine of his shown at the Great Exhibition of 1851 did not attract any interest and Thimmonier died in 1857, unrewarded.

In about 1832, Walter Hunt of New York constructed a machine with a vibrating arm, at the extremity of which he fixed a curved needle with an eye near its point. The needle formed a loop of thread under the cloth to be sewn, and an oscillating shuttle passed a thread through the loop, thus making the lock-stitch of all ordinary two-thread machines. Howe was apparently unaware of Hunt's invention.

Since that time, thousands of patents have been issued in the United States and in Europe, covering improvements in the sewing machine. Although these numerous attachments and accessories have improved the machine's efficiency and usefulness, the main principles are still the same as they were in the basic machine. There are well over 2,000 types of modern sewing machine, designed for making up garments, boots and shoes, hats, for working embroidery, for edging lace curtains and for sewing buckles on shoes. Most machines are now powered by electricity, although treadle-operated machines are still seen in specialized fields such as shoe-mending. Microelectronics are now providing push-button controls on home sewing machines, but the basic mechanism remains the same as that devised by Howe.

Hoyle Fred 1915– is a British cosmologist and astrophysicist, distinguished for his work on the evolution of stars, the development of the steady-state theory of the universe, and a new theory on gravitation.

Hoyle was born in Bingley, Yorkshire, on 24 June 1915. He attended the local grammar school and then went to Emmanuel College, Cambridge. In 1939, he was elected a Fellow of St John's College, Cambridge, and in 1945 he became a lecturer in mathematics at the University. Three years later he developed the steady-state theory as a cosmological model to explain the structure and properties of the universe. In 1956 he left Britain for the United States to join the staff of the Hale Observatory. He returned to Britain ten years later to become Director of the Institute of Theoretical Astronomy at Cambridge. Having been a Fellow of the Royal Society for some years, he was elected President of the Royal Astronomical Society.

According to Hoyle's new theory on gravitation, matter is not evenly distributed throughout space, but forms self-gravitating systems. These systems may range in diameter from a few kilometres to a million light years and they vary greatly in density. They include galaxy clusters, single galaxies, star clusters, stars, planets and planetary satellites. Hoyle argues that this variety need not imply that the self-gravitating systems were formed in diverse ways, but rather that there is no significant intrinsic difference between one place in the universe and another, or between one time and another.

To explain this structure of a universe composed of clusters of matter of different size and to explain its formation, Hoyle calculated the theoretical thermal conditions under which a large cloud of hydrogen gas would contract under the influence of its own gravitation. He found that a contracting cloud whose temperature is less than that required for it to exist in equilibrium will break up into self-gravitating fragments small enough to be at equilibrium in the new, increased density of the cloud fragments. Moreover, such fragmentation will continue until the formation of opaque fragments dense enough for gravitational contraction to offset radiation loss and to maintain the temperature at the necessary equilibrium.

To account for the origin of the elements in the universe, Hoyle, in collaboration with his colleague, William Fowler (1911–), proposed that all the elements may be synthesized from hydrogen in eight separate processes that occur in different stages of the continual process by which hydrogen is converted to helium by successive fusions of hydrogen with hydrogen. The second stage occurs when the supply of hydrogen is exhausted and the cloud of gas heats up to allow the helium-burning stage, in which helium nuclei interact, to proceed. The third stage is the alpha process, whereby the cloud contracts further to reach temperatures of 1 billion K (10 billion °C/18 billion °F) until neon (^{20}Ne) nuclei built up by the helium-burning stage can interact, releasing particles that in turn are used

to build up nuclei of new elements, and so on, until only the element iron is left. Although alternative approaches continue to be explored, there is considerable evidence confirming such an account for the abundant distribution of the elements.

The steady-state theory was expounded, in collaboration with Thomas Gold and Hermann Bondi, as a model to explain the structure and properties of the universe. It postulated that the universe is expanding, but that its density remains constant at all times and places, because matter is being created at a rate fast enough to keep it so. According to this model the creation of matter and the expansion of the universe are interdependent. It calls for no new theory of space–time. Hoyle was able to propose a mathematical basis for his theory which could be reconciled with the theory of relativity.

New observations of distant galaxies have, however, led Hoyle to alter some of his initial conclusions. According to his theory, because no intrinsic feature of the universe depends on its distance from the observer, the distant parts of the universe are the same in nature as those parts that are near. But the detection of some radio galaxies indicates that in fact the universe is very different at different distances, evidence which directly contradicts the fundamental hypotheses of the steady-state theory.

Hoyle is a prolific writer of science fiction as well as popular books on science; these latter include *Of Man and the Galaxies* (1966), *Astronomy Today* (1975) and *Energy and Extinction: the Case for Nuclear Energy* (1977).

Hubble Edwin Powell 1889–1953 was a US astronomer who studied extragalactic nebulae and demonstrated them to be galaxies like our own. He found the first evidence for the expansion of the universe, in accordance with the cosmological theories of George Lemaître and Willem de Sitter, and his work led to an enormous expansion of our perception of the size of the universe.

Hubble was born in Marshfield, Missouri, on 20 November 1889. He went to high school in Chicago and then attended the University of Chicago where his interest in mathematics and astronomy was influenced by George Hale and Robert Millikan (1868-1953). After receiving his bachelor's degree in 1910, he became a Rhodes Scholar at Queen's College, Oxford, where he took a degree in jurisprudence in 1912. When he returned to the United States in 1913, he was admitted to the Kentucky Bar, and he practised law for a brief period before returning to Chicago to take a research post at the Yerkes Observatory from 1914 to 1917.

In 1917 Hubble volunteered to serve in the United States Infantry and was sent to France at the end of World War I. He remained on active service in Germany until 1919, when he was able to return to the United States and take up the earlier offer made to him by Hale of a post as astronomer at the Mount Wilson Observatory near Pasadena, where the 2.5-m/100-in reflecting telescope had only recently been made operational. Hubble worked at Mount Wilson for the rest of his career, and it was there that he carried out his most important work. His research was interrupted by the outbreak of World War II, when he served as a ballistics expert for the US War Department. He was awarded the Gold Medal of the Royal Astronomical Society in 1940, and received the Presidential Medal for Merit in 1946. He was active in research until his last days, despite a heart condition, and died in San Marino, California, on 28 September 1953.

While Hubble was working at the Yerkes Observatory, he made a careful study of nebulae, and attempted to classify them into intra- and extragalactic varieties. At that time there was great interest in discovering what other structures, if any, lay beyond our Galaxy. The mysterious gas clouds, known as the smaller and larger Magellanic Clouds, which had first been systematically catalogued by Messier and called 'nebulae', were good extragalactic candidates and were of great interest to Hubble. He had been particularly inspired by Henrietta Leavitt's work on the Cepheid variable stars in the Magellanic Clouds; and the later work by Harlow Shapley, Henry Russell and Ejnar Hertzsprung that the distances of these stars from the Earth had demonstrated that the universe did not begin and end within the confines of our Galaxy. Hubble's doctoral thesis was based on his studies of nebulae, but he found it frustrating because he knew that more definite information depended upon the availability of telescopes of greater light-gathering power and with better resolution.

After World War I, with the 2.5-m/100-in reflector at Mount Wilson at his disposal, Hubble was able to make significant advances in his studies of nebulae. He found that the source of the light radiating from nebulae was either stars embedded in the nebular gas or stars that were closely associated with the system. In 1923, he discovered a Cepheid variable star in the Andromeda nebula. Within a

year he had detected no fewer than 36 stars within that nebula alone, and found that 12 of these were Cepheids. These 12 stars could be used, following the method applied to the Cepheids that Leavitt had observed in the Magellanic Clouds, to determine the distance of the Andromeda nebula. It was approximately 900,000 light years away, much more distant than the outer boundary of our own Galaxy – then known to be about 100,000 light years in diameter.

Hubble discovered many gaseous nebulae and many other nebulae with stars. He found that they contained globular clusters, novae and other stellar configurations that could also be found within our own Galaxy. In 1924 he finally proposed that these nebulae were in fact other galaxies like our own, a theory which became known as the 'island universe'. From 1925 onwards he studied the structures of the galaxies and classified them according to their morphology into regular and irregular forms. The regular nebulae comprised 97% of them and appeared either as ellipses or as spirals, and the spirals were further divided into normal and barred types. All the various shapes made up a continuous series, which Hubble saw as an integrated 'family'. The irregular forms comprised only 3% of the nebulae he studied. By the end of 1935, Hubble's work had extended the horizons of the universe to 500 million light years.

Having classified the various kinds of galaxies that he observed, Hubble began to assess their distances from us and the speeds at which they were receding. The radial velocity of galaxies had been studied by several other astronomers, in particular by Vesto Slipher (1875–1969). Hubble analysed his data, and added some new observations. In 1929 he found, on the basis of information for 46 galaxies, that the speed at which the galaxies were receding (as determined from their spectroscopic red shifts) was

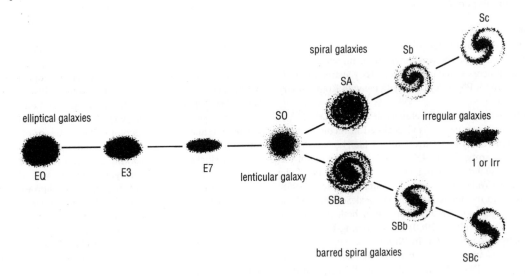

Hubble *Hubble's classification, still used today, is based on a galaxy's shape. Elliptical galaxies range from the almost spherical E0 type to the markedly elliptical E7. Type S0 – a spiral disc without arms – marks the transition from elliptical to spiral types. Types Sa, Sb and Sc are normal spirals with progressively looser arms. Barred spirals, types SBa, SBb and SBc, are also differentiated by the degree of openness of their arms.*

directly correlated with their distance from us. He found that the more distant a galaxy was, the greater was its speed of recession – now known as Hubble's law. This astonishing relationship inevitably led to the conclusion that the universe is expanding, as Lemaître had also deduced from Einstein's general theory of relativity.

These data were used to determine the portion of the universe which we can ever come to know, the radius of which is called the Hubble radius. Beyond this limit, any matter will be travelling at the speed of light, and so communication with it will never be possible. The data on galactic recession were also used to determine the age and the diameter of the universe, although at the time both of these calculations were marred by erroneous assumptions, which were later corrected by Walter Baade (1893–1960). The ratio of the velocity of galactic recession to distance has been named the Hubble constant, and the modern value for the speed of galactic recession is 530 km/330 mi per second – very close to Hubble's original value of 500 km/310 mi per second.

During the 1930s, Hubble studied the distribution of galaxies and his results supported the idea that their distribution was isotropic. They also clarified the reason for the 'zone of avoidance' in the galactic plane. This effect was caused by the quantities of dust and diffuse interstellar matter in that plane.

Among his later studies was a report made in 1941 that the spiral arms of the galaxies probably did 'trail' as a result of galactic rotation, rather than open out. After World War II Hubble became very much an elder statesman of US astronomy. He was involved in the completion of the 5-m/200-in Hale telescope at Mount Palomar, which was opened in 1948. One of the original intentions for this telescope was the study of faint stellar objects, and Hubble used it for this purpose during his few remaining years.

Hubel David Hunter 1926– is a US neurophysiologist who worked with Torsten Wiesel (1924–) on the physiology of vision and the way in which the higher centres of the brain process visual information. They shared the 1981 Nobel Prize for Physiology or Medicine with Roger Sperry (1913–).

Hubel was born on 27 February 1926 in Ontario, Canada, to American parents and qualified in medicine from McGill University in Montreal, after which he was drafted into the US army. From 1955 to 1958 he worked in physiological research, studying the electrical activity of the brain, at the Walter Reed Army Research Institute. In 1958, he moved to the Johns Hopkins Medical School, and the following year to Harvard, working at both places with Stephen Kuffler (1913–1980) and becoming Professor of Neurophysiology in 1965. In 1968 he was appointed Professor of Neurobiology, where he remained until his retirement in 1982. At Harvard he met Wiesel, and they began experiments implanting electrodes into the region of the brain of anaesthetized cats known to be involved in the processing of visual information, and then analysing the responses of individual brain cells to different types of visual stimulation. With painstaking work they correlated the anatomical structure of the visual cortex of the brain with their physiological responses. They built up a complex picture of how the brain analysed visual information by an increasingly sophisticated system of detection by the nerve cells. Later study of the development of the visual system in young animals suggested that eye defects should be treated and corrected immediately, and the then routine ophthalmological practice of leaving a defect to correct itself was abandoned.

Hückel Erich Armand Arthur Joseph 1896–1980 was a German physical chemist who, with Peter Debye (1884–1966), developed the modern theory that accounts for the electrochemical behaviour of strong electrolytes in solution. In his own right he was known for his discoveries relating to the structures of benzene and similar compounds that exhibit aromaticity.

Hückel was born in Berlin-Charlottenberg on 9 August 1896, the son of a doctor of medicine. When he was three years old the family moved to Göttingen and, after leaving school in 1914, he went to the local university to study physics. He interrupted his studies after two years to take a job concerned with aerodynamics in Göttingen University's Applied Mechanics Institute, under the direction of L. Prandtl. At the end of World War I in 1918, Hückel resumed his studies in mathematics and physics and in 1921 was awarded a doctorate based on a thesis (prepared under the direction of Debye) on the diffraction of X-rays by liquids.

He stayed at Göttingen for the next two years, first as a physics assistant to the mathematician David Hilbert and then in a similar role to Max Born, the Nobel prizewinning physicist. In 1922, he left to join Debye again, this time at the Zurich Technische Hochschule. He became an unpaid lecturer there, and in 1925 married Annemarie Zsigmondy, the daughter of the colloid chemist Richard Zsigmondy. In 1928 Hückel was awarded a Rockefeller Foundation Fellowship and worked briefly with Frederick Donnan. He also spent some time with Niels Bohr in Copenhagen, before becoming a Fellow of the Notgemeinschaft der Deutschen Wissenschaft at the University of Leipzig. In 1930 he went to the Technische Hochschule at Stuttgart, where he remained until becoming the Professor of Theoretical Physics at the University of Marburg in 1937.

When Hückel started working with Debye in 1922, the prevailing theory of electrolyte solutions was that of Svante Arrhenius, who held that an equilibrium exists in the solution between undissociated solute molecules and the ions from its dissociated molecules. In strong electrolytes, particularly those formed from strong acids and bases, dissociation was considered to be nearly complete. There were, however, many instances which this theory did not adequately explain.

Debye and Hückel suggested that strong electrolytes dissociate completely into ions, explaining the deviations from expected behaviour in terms of attraction and repulsion between the ions. In a series of highly mathematical investigations they found formulae for calculating the electrical and thermodynamic properties of electrolytic solutions, principally dilute solutions of strong electrolytes. But even taking into account the sizes of the ions concerned, they did not find a complete theory for concentrated solutions.

In 1930, Hückel began his work on aromaticity, the basis of the chemical behaviour of benzene, pyridine and similar compounds. The benzene molecule is held together by electrons that are delocalized and contribute to one large hexagonal bond above and below the plane of the molecule. This accounts for its planar shape and its ability to preserve its structure through chemical reactions. Hückel developed a mathematical approximation for the evaluation of certain integrals in the calculations concerned with the exact nature of the bonding in benzene.

Later in the 1930s, he extended his research to other chemical systems that appear to possess the same kind of aromatic nature as benzene (although usually to a lower degree). From this study emerged the Hückel rule for monocyclic systems, which states that for aromaticity to occur the number of electrons contributing to the correct type of bonding (π-bonding) must be $4n + 2$, where n is a whole number. When $n = 1$, for example, there are six π-electrons, as exemplified by benzene. When $n = 2$, the ten π-electrons occur in derivatives of [10]annulene.

There seem to be exceptions to the rule in large ring systems, where the predicted aromaticity does not occur. In general the degree to which a molecule resembles benzene in aromaticity decreases with increasing ring size.

Hückel also carried out research into unsaturated compounds (those with double or triple bonds) and into the chemistry of free radical compounds (those with a free, non-bonding electron).

Huggins William 1824–1910 was a British astronomer and pioneer of astrophysics. He revolutionized astronomy by using spectroscopy to determine the chemical make-up of stars and by using photography in stellar spectroscopy. With these techniques he investigated the visible spectra of the Sun, stars, planets, comets and meteors and discovered the true nature of the so-called 'unresolved' nebulae.

As a young boy he had been given a microscope, which helped to develop his first interest in physiology. Then when he was about 18 he bought himself a telescope and became increasingly interested in astronomy, although London, even at that time, was a poor place for making observations. It was intended that he should go to Cambridge University, but when the time came his family persuaded him to take charge of their drapery business in the City of London. He diligently followed this trade from 1842 to 1854, when he sold the business and moved with his parents to a new home south of London in Tulse Hill, where he built his own private observatory. Although he had continued to make observations whilst in business, it was only after he moved that he devoted his time entirely to science.

Huggins was elected a Fellow of the Royal Society in 1865 and was President from 1900 to 1905. He also served terms as President of the British Association for the Advancement of Science and the Royal Astronomical Society. His country honoured him by making him a Knight Commander of the Order of Bath and he was also one of the original members of the Order of Merit in 1902. From 1890 he received a Civil List pension (of £150). Huggins died at his home in Tulse Hill on 12 May 1910.

Huggins spent his first two years of research from 1858 to 1860, observing the planets with an 20-cm/8-in refracting telescope. He was looking for a new line of action when, in 1859, Gustav Kirchoff (1824–1887) and Robert Bunsen (1811–1899) published their findings on the use of spectroscopy to determine the chemical composition of the Sun. Huggins, together with his friend W. A. Miller (who was Professor of Chemistry at King's College, London), designed a spectroscope and attached it to the telescope. They began to observe the spectra of the Sun, Moon, planets and brighter stars. They compared the spectral lines which they observed with those produced by various substances in the laboratory, and in this way they were able to draw conclusions about the composition of the stars and planets. They published their results in 1863 in a paper presented to the Royal Society: 'Lines of the spectra of some of the fixed stars'.

It showed that the brightest stars had elements in common with those of the Earth and the Sun, although there was a diversity in their proportions. In the same paper they suggested that starlight originated in the central part of a star and then passed through the hot gaseous envelope that surrounds it.

Following this initial success, other great achievements were to come. Some of the tiny indistinct objects known as nebulae had been observed to be faint clusters of stars. Others could not be resolved but it was believed that eventually more powerful telescopes would show up the individual stars and confirm that they were similar to those already resolved. Huggins realized that those nebulae of stellar origin would give a characteristic stellar spectrum. When he observed the unresolved nebulae in the constellation of Draco in 1864, this was not the case. Only a single bright line was observed. Seeing this, he understood the nature of these objects – they were clouds of luminous gas and not clusters of stars. In the same year he observed the great nebula in Orion and saw two unknown lines in its spectrum. Huggins postulated the existence of a previously undiscovered element, 'nebulium', but in 1927 Ira Bowden (1898–1973) showed that the lines were caused by ionized oxygen and nitrogen. Although Huggins had discovered the gaseous nature of nebulae, which suggested that they may be the 'parents' of stars, his cautious nature prevented him from stating it as a fact. Like other scientists of his day, he believed that chemical elements were unchangeable.

On 18 May 1866 Huggins made his first spectroscopic observation of a nova and found that, superimposed on a solar-type spectrum, there were a number of bright hydrogen lines which suggested that the outburst occurred with an emission of gas with a higher temperature than that of the star's surface. He then showed that the bands of this gas were coincident with those obtained from a candle flame in the laboratory and therefore arose from carbon vapours.

In 1868, Huggins spectroscopically measured the velocity of the star Sirius by observing the Doppler shift of the F line of hydrogen towards the red end of the spectrum. Since the measurements that he was making were tiny, his result of 47.1 km/29.4 mi per sec) was high compared with the modern figure, but it was as accurate as his purely visual observations would allow. In the same year he examined the spectrum of a comet and found that its light was largely due to luminous hydrocarbon vapour. He went on to more observational successes with a pair of large telescopes lent to him by the Royal Society in 1870 and he made the first ultraviolet spectrograph. In 1899, he and his wife jointly published an *Atlas of Representative Stellar Spectra.*

Besides his interest in astronomy, Huggins was also a keen violinist and fisherman. He worked briefly with the chemist and physicist William Crookes (1832–1919) on the investigation of spiritualism, but became disillusioned as his suspicion of trickery increased.

Huggins was a pioneer, and as such he was often hampered by the inadequacy of his instruments and perhaps frustrated because he knew that more accurate observations and measurements could be made when better and more powerful equipment became available.

Part of Huggins's talent lay in his ability to foresee the possibilities in new ideas that were being initiated by his contemporaries. The three fundamental discoveries that made him famous within various fields of astronomy were made with his modest 20-cm/8-in refracting telescope. He established that elements such as hydrogen, calcium, sodium and iron were to be found in the stars and that the universe was made up of well-known elements. He resolved the long debate as to whether all nebulae were clusters of stars by proving that some, like that in Orion, were gaseous and he used the spectroscope to detect the motion of stars and to measure their compositions and velocities.

Humason Milton Lasell 1891–1972 was a US astronomer famous for his investigations, at Mount Wilson Observatory, into distant galaxies. Humason was born at Dodge Centre, Minnesota on 19 August 1891. There is no great list of educational achievements for Humason because he entered astronomy by an unusual route. When he was 14, his parents sent him to a summer camp on Mount Wilson and he so enjoyed it that when he had been back at high school for only a few days, he obtained permission from his parents to take a year off school and return there. He extended his 'year' well beyond 12 months and became one of astronomy's most notable educational dropouts. Sometime between 1908 and 1910, after his voluntary withdrawal from higher education, he became a mule driver for the packtrains that travelled the trail between the Sierra Madre and Mount Wilson, during construction work on the Observatory. He brought up much of the timber and other building materials for the telescope's supporting

structure, the local cottages and the scientists' quarters.

He became engaged to the daughter of the Observatory's engineer and married her in 1911. In the same year he gave up his job as driver of the packtrains and went to be foreman on a relative's range in La Verne. But he still loved Mount Wilson and in 1917, when a janitor was leaving, his father-in-law suggested that this might be an opening for better opportunities. So Humason initially joined the staff of Mount Wilson Observatory as a janitor. His position was soon elevated to night assistant. George Hale (1868–1938), the Director of the Observatory at the time, recognized Humason's unusual ability as an observer and in 1919 he was appointed to the scientific staff. There was considerable opposition to this appointment, partly because Humason had had no formal education after the age of 14 and partly because he was a relative of the Observatory's engineer.

Humason was Assistant Astronomer from 1919 to 1954 and then Astronomer at both the Mount Wilson and Palomar Observatories. In 1947 he was appointed Secretary of the Observatories, a position which involved him in handling public relations and administrative duties. In 1950, he was awarded the honorary degree of Doctor of Philosophy by the University of Lund in Sweden. He was a quiet, friendly man who was often consulted on administrative and personal problems. He died suddenly at his home near Mendocino, California, on 18 June 1972.

At Mount Wilson Observatory Humason took part in an extensive study of the properties of galaxies that had been initiated by Edwin Hubble. The research programme began in 1928 and consisted of making a series of systematic spectroscopic observations to test and extend the relationship that Hubble had found between the red shifts and the apparent magnitudes of galaxies. But because of the low surface brightness of galaxies there were severe technical difficulties. Special spectroscopic equipment, including the Rayton lens and the solid-block Schmidt camera, was designed. With these instruments it became possible to obtain a spectrum of a galaxy too faint to be picked up visually with the telescope used. The method was to photograph the field containing the galaxy and accurately measure the position of its image with respect to two or more bright stars. Guide microscopes were then set up at the telescope with exactly the same offsets from the slit of the spectrograph. It was a tedious task and often took several nights of exposure to produce a spectrum on a 13-mm/ 0.5-in plate.

Humason undertook this exacting programme. He personally developed the technique and made most of the exposures and plate measurements. During the period from 1930 to his retirement in 1957, the velocities of 620 galaxies were measured. He used the 2.5-m/100-in telescope until the 5-m/200-in Hale telescope was completed and the programme transferred to that instrument. Humason's results were published jointly with those of N. U. Mayall and A. R. Sandage in 'Redshifts and magnitudes of extragalactic nebulae', which appeared in the *Astronomical Journal* of 1956. These data still represent the majority of known values of radial velocities for normal galaxies, including most of the large values.

Humason applied the techniques he had developed for recording spectra of faint objects to the study of supernovae, old novae that were well past peak brightness, and faint blue stars (including white dwarfs). One by-product of his studies on galaxies was his discovery of Comet 1961e, which is notable for its large perihelion distance and its four-year period of visibility with remarkable changes in form.

Humason is remembered for his ability to handle instruments with meticulous care and great skill. He provided criteria for studying various models of the universe. He became internationally famous for his work on galaxies and, in spite of his lack of formal training, won a leading role in US astronomy.

Humboldt Friedrich Heinrich Alexander, Baron von 1769–1859 was one of the most eminent geologists and geophysicists of the nineteenth century.

Born in Berlin, Humboldt was the son of a Prussian soldier who wanted him to pursue a political career. Humboldt had other ideas. Preferring science, he studied at Göttingen University, proceeding to Werner's Freiberg mining school and spending two years as a mines engineer. In the 1790s he travelled widely in Europe, before setting out, in 1799, with the French botanist A. Bonpland, on a pioneering and immensely productive expedition across Latin America. Studying physical geography above all, but also collecting vast quantities of geological, botanical and zoological material, Humboldt covered some 9,600 km/6,000 mi. His expedition included travelling up the Orinoco and the Magdalena, passing a year in Mexico and some time in Cuba, visiting the sources of the Amazon and cutting across the continent to the Cordilleras and on to Quito and Lima.

He returned to Europe in 1804, laden with scientific specimens, and spent the next 20 years writing up his results, not least in his *Narrative of Travels* 1818–19. Humboldt aimed to erect a new science, a 'physics of the globe', analysing the deep physical interconnectedness of all terrestrial phenomena. He believed physical relief should be grasped in terms of Earth history, and likewise that geological phenomena were to be understood in terms of more basic physical causes (for example, terrestrial magnetism or rotation). Showing a phenomenal ability to keep abreast of developments in geophysics, **meteorology** and geography, he patiently amassed **evidence of** comparable geological phenomena from every continent.

Humboldt arrived at numerous important findings. On the basis of studies of Pacific coastal currents, he was one of the first to propose a Panama canal. In meteorology, he introduced isobars and isotherms on weathermaps, made a general study of global temperature and pressure, and finally instituted a worldwide programme for compiling magnetic and weather observations. His studies of American volcanoes demonstrated they corresponded to underlying geological faults; on that basis he deduced that vol-

canic action had been pivotal in geological history and that many rocks were igneous in origin. In 1804, he discovered that the Earth's magnetic field decreased from the poles to the equator.

His most popular work, *Cosmos*, begun in 1845 at the age of 76, is a profound and moving statement of our relationship with the Earth, and of the relations between physical environment and flora and fauna. A gracious polymath held in universal respect, Humboldt proved enormously influential in study of the globe, both as an explorer and as a theorist. As one who set out to chart the history of human interrelations with planet Earth, he may be called the founder of ecology.

Hume-Rothery William 1899–1968 was a British chemist who spent his entire scientific life working on the structures of metals and their alloys. When he began his studies, metallurgy was considered to be a branch of physical chemistry; when he retired, it had come to be regarded as a distinct discipline in its own right.

Hume-Rothery was born on 15 May 1899 at Worcester Park, Surrey. He was educated at Cheltenham College before going to the Royal Military Academy, in anticipation of a career in the army. But his training was brought to a sudden stop by a serious illness that left him totally deaf. He then went to Oxford to study natural sciences, specializing in chemistry and graduating in 1926. He did his postgraduate research and gained his PhD at the Royal School of Mines, where he became interested in metallurgy. He soon returned to Oxford and continued his research into metal alloys, first in the Dyson Perrins Laboratory and later in the old Chemistry Department. For most of this period his research was financed by outside organizations such as the Armourers and Braziers Company, and he did not gain a formal university post until 1938. During World War II he supervised many contracts from the Ministry of Supply and Aircraft Production for work on complex aluminium and magnesium alloys.

In 1955, with the offer to Hume-Rothery of the George Kelley Readership, metallurgy finally became recognized as a discipline within Oxford University. Three years later he received the first professorship in the new Department of Metallurgy. He died in Oxford on 27 September 1968.

While he was working in the Dyson Perrins laboratory Hume-Rothery complained bitterly that vapours from the experiments of neighbouring organic chemists spoiled the finish of the specimens he had prepared for study with a microscope. The accepted method of preparing metals in order to study their structure was to smooth the surface by a series of grinding and polishing operations until it had a mirror finish. The surface was then carefully etched in an acid that attacked certain areas preferentially. The specimen was then viewed under a microscope using indirect reflected light. In the laboratory atmosphere the carefully prepared specimens became corroded, interfering with the study. Hume-Rothery was allowed to move.

Many of Hume-Rothery's early ideas about alloys came from interpreting their microstructures. If metals mix to form an alloy, the melting point, physical strength and microstructure vary with the composition of the alloy. During the later part of the 1920s and the 1930s he and his continuous stream of research students established that the microstructure of an alloy depends on the different sizes of the component atoms, the valency electron concentration, and electrochemical differences. They showed that in a binary (two-metal) alloy, a common crystal lattice is possible if the two types of atoms present have a similar size. And if a common lattice forms, the alloy has a uniform structure. With atoms of widely different sizes, at least two types of lattices may form, one rich in one metal and one rich in the other. The presence of two types of structures can increase the strength of an alloy.

Hume-Rothery and his researchers discovered that the solid solubility of one element in another is extremely restricted if the atomic diameters of the two elements concerned differ by more than 15%. Within the 15% limit, solutions are formed provided that other factors are favourable. In such a solution, the solute (alloying) element replaces some of the atoms of the solvent (primary) metal. Nickel and copper have very similar atomic radii and so form a continuous range of solid solutions – no doubt assisted by the fact that they both have face-centred cubic crystal structures. Zinc crystallizes with a hexagonal structure, so that although zinc and copper alloy over the whole range of compositions, most of the alloys involve two crystal structures – making some brasses much stronger than their component metals. If the two elements differ considerably in electronegativity, a definite chemical compound is formed. Thus steel, an 'alloy' of iron and carbon, contains various iron carbides.

All this theory was backed up by experimental work, and Hume-Rothery and his team constructed the equilibrium diagrams for a great number of alloy systems. The experimental procedures became standard practice and work that was noted for its accuracy and a stern appraisal of the source of errors. He went on to make fundamental studies of solid-state transformations and deformation, while continuing his original research, particularly his interest in equilibrium phase diagrams. Today metallurgists can produce 'tailor-made' alloys to suit particular and exacting requirements. Without Hume-Rothery's painstaking, systematic studies this would not have become possible.

Hunter John 1728–1793 was a celebrated Scottish surgeon known for his occasional use of unorthodox methods. He built up a collection of 14,000 anatomical specimens and his memorial brass in Westminster Abbey records his 'services to mankind as the Founder of Scientific Surgery'.

Hunter was born in Long Calderwood, Lanarkshire, on 13 February 1728. He worked on the family farm after the death of his father and at the age of 20 went to London to join his brother, William, who was a distinguished surgeon and obstetrician. John Hunter had no formal university qualification; he assisted in the preparation of anatomical specimens for his brother's lectures and engaged in investigations of his own. He also attended surgical classes at various London hospitals. He was appointed a Master of Anatomy of the Surgeons' Corporation in 1753 and three years later served as house surgeon at St George's Hospital, London (during which time he taught Edward Jenner). He joined the British army in 1759 and in 1760 worked in France and Portugal as an army surgeon. He returned to London in 1763 and set up a private practice, and continued with his research. During the late 1760s he took up a senior surgical post at St George's Hospital and was appointed physician extraordinary to George III. Ten years later he became Deputy Surgeon to the Army and in 1790 became Inspector General of Hospitals. Hunter collapsed and died at a meeting of the board of governors at St George's Hospital on 16 October 1793. He was buried at St Martin's-in-the-fields, but his remains were later taken to Westminster Abbey on 28 March 1859.

While Hunter was working with his brother, he made a detailed study of the structure and function of the lymphatic vessels, and the growth and structure of bone. He made these investigations by collecting a great deal of material from postmortem examinations. Many of the corpses that he dissected he obtained from 'resurrectionists', who raided graveyards at night to sell newly buried corpses to surgeons for dissection.

Hunter also kept a number of animal specimens in his garden for dissection, that at one time included a bull and the carcass of a whale. The knowledge he gained performing these dissections allowed him to improve further the embalming technique by arterial injection, as developed by William Harvey. One of the most interesting of these experiments was the case of the late Mrs Martin van Butchell, whose husband took her body to Hunter for embalming in 1775. Mrs van Butchell had stated in her will that her wealth remained her husband's, as long as her body remained above the ground. Hunter embalmed the body, and Mr van Butchell clothed it and placed it in a glass case for visitors to view – and to meet the conditions of the will.

Hunter's keenness to collect specimens was well known. One of his most prized specimens was the skeleton of Charles Byrne, an Irishman who was 8 feet tall and who, determined not to fall into Hunter's clutches, had arranged to be buried at sea. But Hunter, on hearing of his death, arranged for his body to be seized by the resurrectionists, for a fee of £500.

Surgical training was not easily available in Hunter's day, so when he moved nearer to central London he gave lectures from his own house. Later he had a lecture room and museum built in his garden where he held meetings of the Lyceum Medicum Londinense (London Medical Academy). He had helped to found this student society and encouraged each student to prepare a paper on a medical topic and read it to his fellow students.

Hunter's experiments were wide-ranging, including studies of lymph and blood circulation, the sense of smell, the structure of teeth and bone, tissue grafting and various diseases. He often carried out experiments on himself, such as the occasion when, trying to prove that syphilis and gonorrhoea are types of the same disease, he inoculated himself and later developed syphilis. When Hunter's health began to suffer, he took an army appointment on the surgical staff concerned with the Seven Years' War. Whilst dealing with casualties, he gained the knowledge for his treatise on gunshot wounds.

During his lifetime Hunter published an impressive number of papers on a wide variety of medical and biological subjects such as his *Treatise on the Natural History of the Human Teeth* (1771) in which he describes his experiments on the transplantation of tissues – the best known of these being that of a human tooth fixed into a cock's comb. In 1786, he published his *Treatise on the Venereal Disease*, and his *Treatise on the Blood, Inflammation and Gun-shot Wounds* was published posthumously in 1794. Hunter's collection was handed to the Royal College of Surgeons in 1795 and later formed the basis of the Hunterian Museum.

Hussey Obed 1792–1860 was a US inventor who developed one of the first successful reaping machines and various other agricultural machinery.

Hussey was born into a Quaker family in Maine. The family moved to Nantucket, Massachusetts, and like most boys living there at that time, the young Hussey wanted to go to sea when he grew up. He was a quiet and studious boy, always thoughtful and modest. He liked studying intricate mechanisms and became a skilled draughtsman and inventor. In addition to the reaper, he invented a steam plough, a machine for making hooks and eyes, a grinding mill for maize and a horse-powered husking machine, a sugar-cane crusher and an ice-making machine. On 4 August 1860, while travelling on a train from Boston to Portland, Maine, he got off at a station to fetch a drink of water for a child. The train started as he was getting back on, and he fell beneath the wheels and was killed.

Hussey began work on a reaping machine early in 1833 in a room at the factory of Richard Chenoweth, a manufacturer of agricultural implements in Baltimore, Maryland. The finished prototype was tested later that year near Cincinnati, Ohio, where it so impressed a local businessman, Jarvis Reynolds, that he provided the finance and a factory for manufacturing the reaper. It was patented in December 1833 and pictured in the *Mechanics Magazine* of April 1834 as 'Hussey's Grain Cutter'.

At first Hussey used a reel to gather the grain up to the cutter and throw it on to the platform, but after trials he decided that the reel was an encumbrance and discarded it. The main frame containing the gearing was suspended on two wheels about 1 m/3 ft in diameter; the platform was attached to the rear of the frame and extended nearly 2 m/7 ft on each side. A team of horses attached to the front of the frame walked along the side of the standing grain. The cutting knife consisted of 7.5-cm/3-in wide steel plates, tapered towards the front, rivetted to a flat iron bar to form a sort of saw with very coarse teeth sharpened on both edges. The knife was supported on what Hussey called guards attached to the front of the platform every 7.5 cm/3 in across the whole width of the machine. They projected forwards and had long horizontal slits in which the cutter

oscillated from side to side. They thus supported the grain while it was being cut and protected the blades from damage by large stones and other obstructions. The cutter was attached by means of a Pitman rod to a crank activated by gearing, connected to one or both of the ground wheels. This arrangement gave a quick vibrating motion to the cutter as the machine was pulled along.

The point of each blade oscillated from the centre of one guard, through the space between, to the centre of the next, thus cutting equally both ways.

In his next modification to the machine, Hussey used one large ground wheel instead of two and moved the platform to a position alongside the frame, providing a seat for an operator who faced forward and raked the cut grain off the back of the platform.

During the harvest of 1834, Hussey successfully demonstrated his machine to hundreds of farmers and began to sell the reapers for $150 each. The fame of the machine spread, slowly at first, but it was eventually used in the far West. In September 1851 he went to Britain and demonstrated the reaper at Hull and Barnardscastle. He was invited to show it to Prince Albert who bought two (at £21 each), one for the estate at Windsor and one for Osborne House on the Isle of Wight.

An earlier rival design of reaper had been developed by Cyrus McCormick in 1831, although Hussey's was patented first. Both machines used the principle of a reciprocating knife cutting against stationary guards or figures, although McCormick's design employed a division between the cut and uncut grain and a reel to topple the cut grain on to the platform. Hussey's contribution was summed up succinctly in the title of a book which relates the story: *Obed Hussey: Who of all inventors, made bread cheap.*

Hutton James 1726–1797 was a Scottish natural philosopher who pioneered uniformitarian geology.

Son of an Edinburgh merchant, Hutton studied at Edinburgh University, Paris and Leiden, training first for the law but taking his doctorate in medicine in 1749 (though he never practised). He spent the next two decades travelling and farming in the southeast of Scotland. During this time he cultivated a love of science and philosophy, developing a special taste for geology. About 1768 he returned to his native Edinburgh. A friend of Joseph Black, William Cullen and James Watt, Hutton shone as a leading member of the scientific and literary establishment, playing a large role in early history of the Royal Society of Edinburgh and in the Scottish Enlightenment.

Hutton wrote widely on many areas of natural science, including chemistry (where he opposed Lavoisier), but he is best known for his geology, set out in his *Theory of the Earth*, of which a short version appeared in 1788, followed by the definitive statement in 1795. In that work, Hutton attempted (on the basis both of theoretical considerations and of personal fieldwork) to demonstrate that the Earth formed a steady-state system, in which terrestrial causes had always been of the same kind as at present, acting with comparable intensity (the principle later known as uniformitarianism). In the Earth's economy, in the imperceptible creation and devastation of landforms, there was no vestige of a beginning, nor prospect of an end. Continents were continually being gradually eroded by rivers and weather. Denuded debris accumulated on the sea bed, to be consolidated into strata and subsequently thrust upwards to form new continents thanks to the action of the Earth's central heat. Non-stratified rocks such as granite were of igneous origin. All the Earth's processes were exceptionally leisurely, and hence the Earth must be incalculably old. Though supported by the experimental findings of Sir James Hall, Hutton's theory was vehemently attacked in its day, partly because it appeared to point to an eternal Earth and hence to atheism. It found more favour when popularized by Hutton's friend, John Playfair, and later by Charles Lyell. The notion of uniformitarianism still forms the groundwork for much geological reasoning.

Huxley Andrew Fielding 1917– . English physiologist; *see* Hodgkin and Huxley.

Huxley Hugh Esmor 1924– is a British physiologist whose contribution to science has been concerned with the study of muscle cells.

Huxley was born in Birkenhead, Cheshire, on 25 February 1924 (he is not a member of the Huxley family descended from the nineteenth-century scientist Thomas Huxley). He was educated at Park High School, Birkenhead, and from 1914 at Christ's College, Cambridge. He read natural science there and graduated in 1943. He worked on radar research until the end of World War II and then returned to Cambridge in 1948, where he gained his

PhD in 1952. He spent five years (1956–61) in the University of London biophysics department and was a Fellow of King's College, Cambridge, from 1961 to 1967; when he became a Fellow of Churchill College, Cambridge.

Until Huxley's research, the physiology of muscle structure had been under scientific scrutiny for a long time and it had been thought that muscular contraction involved a process of coiling and contraction, rather like the shortening of a helical spring. But investigations had been held up because adequate equipment, such as the electron microscope, was not available. Once this instrument was built in 1933, studies into the chemical reactions that produce muscle contraction progressed, but the real breakthrough came in the 1950s when Huxley, using the electron microscope and thin slicing techniques, established the detailed structural basis of muscle contraction. It had been ascertained that muscle fibres contain a large number of longitudinally arranged myofibrils, and that two main bands alternately cross the fibrils: the dark anisotropic (A) and the lighter isotropic (I) bands. Each A-band is divided by a lighter region, the H-zone (which in turn is bisected by an M-line), and each I-band by a dark line, the Z-line.

Huxley demonstrated that the myofibrils are composed of interdigitating rows of thick and thin myosin and actin filaments. He found that the thin filaments are attached to the Z-lines and extend through the I-bands into the A-bands, and that the M-line is caused by a further thickening in the middle of the thick filaments. In addition, he proved that the helical spring assumption was wrong. The A-band does not change its length when the muscle is stretched or when it is shortened. Huxley suggested that contraction is brought about by sliding movements of the I-filaments between the A-filaments, and that the sliding is caused by

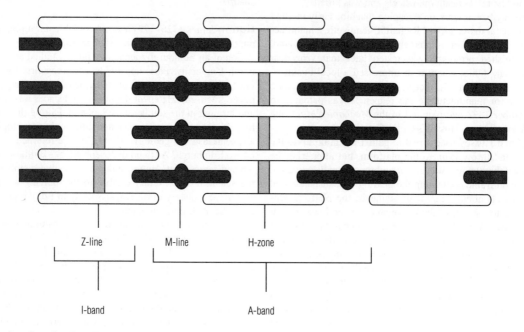

Huxley *Banding in a fibril of striated muscle.*

a series of cyclic reactions between the myosin filaments and active sites on the actin filaments. This proposal is known as the sliding filament theory.

From previous research, Huxley knew that the contractile machinery of muscle cells consists of a small number of different proteins, and that actin and myosin are the two major ones involved. He proceeded to investigate the exact location of the actin and myosin. Using a solution of potassium chloride, pyrophosphate and magnesium chloride, he found that the A-filaments are composed of myosin. He then discovered, using a solution of potassium iodide, that the I-filaments are composed of actin. The Z-lines were unaffected by the solutions which indicated that they are composed of some other substance. Huxley also noticed that the myosin-extracted fibres still retained their elasticity and he suggested that there is yet another set of filaments crossing the H-zone. He called these hypothetical structures S-filaments.

In 1953, Huxley had observed a series of projections which formed a repetitive pattern along the line of the A-filament. The projections appeared to emerge in pairs from opposite sides of the myosin filaments at regular intervals, and he discovered that they are able to aggregate under suitable conditions to form 'artificial' filaments of varying lengths. He proposed that the 'tails' of the myosin molecules become attached to each other to form a filament with the heads projecting from the body of the filament, and that these projecting filaments play an important part in the sliding effect described in the sliding filament theory.

By coincidence, Andrew Huxley, working separately, came to the same conclusions at about the same time. Both Huxleys subscribe to the sliding filament theory although their interpretation of it is slightly different. H. E. Huxley suggests that the movement can be likened to that of a ratchet device or a cog-type operation, whereas A. Huxley proposes that each myosin filament has side pieces which can slide along the main backbone of the filament and that the slides can combine temporarily with sites on adjacent filaments. Many problems remain surrounding this discovery because although the sliding filament theory has been accepted by many scientists when applied to striated muscles, it has not yet been shown to apply to smooth muscles.

Huxley T(homas) H(enry) 1825–1895 was a distinguished British biologist who helped to break down the great barrier of traditional resistance to scientific advance during the mid-nineteenth century, and did a great deal to popularize science.

Huxley was born in Ealing, Middlesex, on 4 May 1825. He went to Ealing School in 1833, where his father taught mathematics, but his schooldays were limited to two years because his father was dismissed from his teaching post in 1835. Despite the lack of a formal education, he received some tuition in medicine from a brother-in-law, and obtained an apprenticeship in 1840 to a medical practitioner in London's East End. Huxley attended botany lectures in his spare time and in 1842, on entering a public competition, won a scholarship to study medicine at Char-

ing Cross Hospital; he graduated in 1845. From 1846 to 1850 he was the assistant ship's surgeon on HMS *Rattlesnake* on its voyage around the South Seas. A year after his return, in 1851, he was elected to the Royal Society. Three years later he took up the appointment of Professor of Natural History at the Royal School of Mines (which in time became the Royal College of Science). From 1881 to 1885 he was President of the Royal Society. He then retired to continue his research and died in Eastbourne, of influenza and bronchitis, on 29 June 1895. His grandchildren include the Nobel prizewinning physiologist Andrew Huxley, the biologist Julian Huxley, and the author Aldous Huxley.

Huxley's original intention was to be a mechanical engineer but his success in the field of medicine dissuaded him. For instance, when he was 19 years old he discovered the structure at the base of the human hair, known today as Huxley's layer.

Huxley established his reputation in the scientific world during the four-year voyage to the South Seas. Each day he dissected, drew and observed, with his microscope tied down against the lurching of the ship. Most naturalists of the time were interested only in collecting specimens which could be dried, stuffed and mounted, but Huxley concentrated on delicate creatures, such as the heteropods, which were difficult to preserve and liable to disintegrate. He sent his detailed recordings back to England from each port the ship docked at, and these observations were published in various influential scientific journals. On his return to England in October 1850, Huxley was acclaimed by men such as Joseph Hooker and Charles Lyell. From being an unknown assistant surgeon, he found himself at the forefront of British science.

European zoology had long been under the shadow of the French anatomist Georges Cuvier (1769–1832). In 1817 Cuvier had formulated a system of classification which placed quadrupeds, birds, amphibians and fish into the class Vertebrata, and insects among the Articulata. Although he recognized the molluscs as a distinct group, he had lumped most other animals together as Radiata. While carrying out his examinations on the ship Huxley realized that Cuvier's classification was not good enough and that there was a vast range of distinctions in minute anatomy of which the great man had not been aware. Huxley therefore reclassified the animal kingdom into Annuloida, Annulosa, Infusoria, Coelenterata, Mollusca, Molluscoida, Protozoa and Vertebrata.

One of the illuminating suggestions Huxley was able to make from his detailed observations was that the inner and outer layers of the Medusae correspond with the two embryonic layers of higher animals. This suggestion provided the base to all later embryological thinking. Huxley also started a fundamental revision of the Mollusca. At that time a wide variety of non-segmented and non-radiate soft-bellied animals (sea squirts, sea mats and lamp shells) were grouped together indiscriminately with the true molluscs. Although his intended project to produce a regular monograph of Mollusca never really materialized, he was able to 'construct' an archetypal cephalous mollusc which was remarkably similar to the evolutionary ancestor of the molluscs deduced by zoologists more than 80 years later.

Huxley became very much embroiled in the great controversy that raged when Darwin published his *On the Origin of Species*, and he was one of Darwin's most outspoken champions. A famous example of his involvement in Darwinism is this extract from a public debate held in 1860 at Oxford University with Bishop Samuel Wilberforce. The bishop had asked whether Huxley traced his descent from the apes through his mother or his father, to which Huxley replied: 'If ... the question is put to me, would I rather have a miserable ape for a grandfather or a man highly endowed by nature and possessed of great means of influence, and yet who employs these faculties and that influence for the mere purpose of introducing ridicule into a grave scientific discussion – I unhesitatingly affirm my preference for the ape.'

Huxley is also credited with the founding of craniology. In his investigations into the true origins of the newly discovered Neanderthal skull, he devised a series of quantitive indices and the first real rationale of the measurement of skulls. He also produced a new system of classification of birds based mainly on the palate and other bony structures. Previously they had been classified according to their feeding habits, foot-webbing and beaks. Huxley raised bird classification to a science and his own classification of birds is the foundation of the modern system.

Although he was never a great experimenter, Huxley's scientific work was distinguished by its critical assessment of both pre-existing and newly acquired knowledge. He produced more than 150 research papers on subjects as varied as the morphology of the heteropod molluscs, the hybridization of gentians, the taxonomy of crayfish and the physical anthropology of the Patagonians. His dissections and observations filled numerous gaps in the knowledge of the animal kingdom, and his attitudes did much to establish the idea that science and the scientific method are the only means by which the ultimate truths of the animal world can be found.

Huygens Christiaan 1629–1695 was a Dutch physicist and astronomer. He is best known in physics for his explanation of the pendulum and invention of the pendulum clock, and for the first exposition of a wave theory of light. In astronomy, Huygens was the first to recognize the rings of Saturn and the discoverer of its satellite Titan.

Huygens was born on 14 April 1629 at The Hague into a prominent Dutch family with a tradition of diplomatic service to the ruling house of Orange, and a strong inclination towards education and culture. René Descartes (1596–1650) was a frequent guest at the house of Huygens's father, Constantijn, who was a versatile and multitalented man, a diplomat, a prominent Dutch and Latin poet, and a composer. So it was natural for the young Christiaan to be educated at home to the highest standard. In 1645 he was sent to the University of Leiden to study mathematics and law, followed in 1647 by two years studying law at Breda.

But, eschewing his expected career in diplomacy, Huygens returned to his home to live for 16 years on an allowance from his father, which enabled him to devote himself to his chosen task, the scientific study of nature. This long period of near seclusion was to be the most fruitful period of his career.

In 1666, on the foundation of the Académie Royale des Sciences, Huygens was invited to Paris and lived and worked at the Bibliothèque Royale for 15 years, until his delicate health and the changing political climate took him back to his home. Here he continued to experiment, only occasionally venturing abroad to meet the other great scientists of the time as in 1689, when he visited London and met Isaac Newton (1642–1727). During his stay in Paris, Huygens twice had to return home for several months because of his health and in 1694 he again fell ill. This time he did not recover, and he died at The Hague on 8 July 1695.

Huygens worked in different areas of science in a way almost impossible in our modern age of vast knowledge and increasing specialization. He made important advances in pure mathematics, applied mathematics and mechanics, which he virtually founded, optics and astronomy, both practical and theoretical, and mechanistic philosophy. He is also credited with the invention of the pendulum clock, enormously important for its use in navigation, since accurate timekeeping is necessary to find longitude at sea. In a seafaring nation such as Holland, this was of particular importance. Typically of Huygens, he developed this work into a thorough study of pendulum systems and harmonic oscillation in general.

At first Huygens concentrated on mathematics. In the age of such revolutionary figures as Descartes, Newton and Gottfried Leibniz (1646–1716), his career may be termed conservative for it contained nothing completely new except the theory of evolutes and Huygens's study of probability, including game theory, which originated our modern concept of the expectation of a variable. His suspicion of the new methods may have been partly due to the secrecy of scientists of his time, and partly to Huygens's fastidiousness, which often led him to delay publishing his observations and theories. The importance of his mathematical work is in its improving on available techniques, and Huygens's application of them to find solutions to problems in science.

The first of his many studies in applied mathematics was a paper on hydrostatics, including much mathematical analysis, published in 1650. Fascinating work on impact and collision followed, motivated by Huygens's disbelief in Descartes's laws of impact. Huygens used the idea of relative frames of reference, considering the motion of one body relative to the other. He discovered the law of conservation of momentum, but as yet the vectorial quantity had little intuitive meaning for him and he did not proceed beyond stating the law as the conservation of the centre of gravity of a system of bodies under impact. In *De Motu Corporum* (1656) he was also able to show that the quantity $^1/_2mv^2$ is conserved in an elastic collision, though again this concept had little intuitive sense for him.

Huygens also studied centrifugal force and showed, in 1659, its similarity to gravitational force, though he lacked the Newtonian concept of acceleration. Both in the early and the later part of his career he considered projectiles and gravity, developing the mathematically primitive ideas of Galileo (1564–1642), and finding in 1659 a remarkably accurate experimental value for the distance covered by a falling body in one second. In fact his gravitational theories successfully deal with several difficult points which Newton carefully avoided. Later, in the 1670s, Huygens studied motion in resisting media, becoming convinced by experiment that the resistance in such media as air is proportional to the square of the velocity. Without calculus, however, he could not find the velocity time curve.

Early in 1657, Huygens developed a clock regulated by a pendulum. The idea was patented and published in the same year, and was a great success; by 1658 major towns in Holland had pendulum tower clocks. Huygens worked at the theory first of the simple pendulum and then of harmonically oscillating systems throughout the rest of his life, publishing the *Horologium Oscillatorium* in 1673. He made many technical and theoretical advances, including the derivation of the relationship of the period T of a simple pendulum to its length l as $T = 2\pi\sqrt{(l/g)}$. Huygens made use of his previously worked out theory of evolutes in this work.

Huygens is perhaps best known for his wave theory of light. This was the result of much optical work, begun in 1655 when Huygens and his brother began to make telescopes of high technical quality.

Through working with his brother, Constantijn, Huygens became skilful in grinding and polishing lenses, and the telescopes that the two brothers constructed were the best of their time. Huygens's comprehensive study of geometric optics led to the invention of a telescope eyepiece that reduced chromatic aberration. It consisted of two thin plano-convex lenses, rather than one fat lens, with the field lens having a focal length three times greater than that of the eyepiece lens. Its main disadvantage was that crosswires could not be fitted to measure the size of an image. To overcome this problem Huygens developed a micrometer, which he used to measure the angular diameter of celestial objects.

Having built (with his brother) his first telescope, which had a focal length of 3.5 m/11.5 ft, Huygens discovered Titan, one of Saturn's moons, in 1655. Later that year he observed that its period of revolution was about 16 days and that it moved in the same plane as the so-called 'arms' of Saturn. This phenomenon had been somewhat of an enigma to many earlier astronomers, but because of Huygens's superior 7-m/23-ft telescope, and a piece of sound if not brilliant deduction, he partially unravelled the detail of Saturn's rings. In 1659, he published a Latin anagram which, when interpreted, read 'It (Saturn) is surrounded by a thin flat ring, nowhere touching and inclined to the ecliptic'. The theory behind Huygens's hypothesis followed later in *Systema Saturnium*, which included observations on the planets, their satellites, the Orion Nebula and the

determination of the period of Mars. The content of this work amounted to an impressive defence of the Copernican view of the Solar System.

The *Traité de la Lumière*, containing Huygens's famous wave or pulse theory of light, was published in 1678. Huygens had been able two years earlier to use his principle of secondary wave fronts to explain reflection and refraction, showing that refraction is related to differing velocities of light in media, and his publication was partly a counter to Newton's particle theory of light. The essence of his theory was that light is transmitted as a pulse with a 'tendency to move' through the ether by setting up a whole train of vibrations in the ether in a sort of serial displacement. The thoroughness of Huygens's analysis of this model is impressive, but although he observed the effects due to polarization, he could not yet use his ideas to explain this phenomenon.

The impact of Huygens's work in his own time, and in the eighteenth century, was much less than his genius deserved. Essentially a solitary man, he did not attract students or disciples and he was also slow to publish his findings. Nevertheless, after Galileo and until Newton, he was supreme in mechanics, and his other scientific work has had a significant effect on the development of physics.

Hyatt John Wesley 1837–1920 was a US inventor who became famous for his invention of Celluloid, the first artificial plastic.

Hyatt was born in Starkey, New York, on 28 November 1837. As a young man he worked in Illinois as a printer, printing boards and playpieces for draughts and dominoes at his plant in Albany. Probably with a view to making such playpieces, he became interested in the material pyroxylin, a partly nitrated cellulose developed in Britain by Alexander Parkes (1813–1890) and D. Spill. Then, in the early 1860s, the New York company of Phelan and Collender, which manufactured billiard balls, offered a prize of $10,000 for a satisfactory substitute for ivory for making the balls. Hyatt remembered pyroxylin and from it he and his brother Isaac developed and, in 1869, patented Celluoid (the US trade name: it was called Xylonite in Britain).

Celluloid consisted of a mouldable mixture of nitrated cellulose and camphor. Its chief disadvantage was its inflammability, but nevertheless for several years it was the favoured material for making a wide range of products from shirt collars and combs to toy dolls and babies' rattles. It was used as a flexible substrate for photographic film and, as a substitute for ivory, for making piano keys – and billiard balls. Celluloid was also used as the central 'filling' in sandwich-type safety glass for car windscreens. It has gradually been superseded in nearly all uses by other synthetic materials, but in one application it continues to be used: the manufacture of table-tennis balls.

Hyatt was never awarded the Phelan and Collender prize money. He continued to patent his inventions – more than 200 of them, including roller bearings and a multiple-stitch sewing machine. He died in Short Hills, New Jersey, on 10 May 1920.

Hyman Libbie Henrietta 1888–1969 was a US zoologist renowned for her studies on the taxonomy (classification) and anatomy of invertebrates.

She was born on 6 December 1888 in Des Moines, Iowa, the third in a family of four children. Unhappy home circumstances encouraged her to indulge in long walks and rambles and she began to collect and classify flowers and butterflies. After graduation with honours from high school in 1905, she passed state examinations as a teacher, but was refused a position because she was considered too young. She therefore worked in a factory, until her former English and German teacher helped her gain a scholarship to the University of Chicago in 1906. She graduated in zoology in 1910 and gained her PhD in 1915, working under the supervision of Charles Manning Child, for whom she then worked as a research assistant until 1930, also teaching undergraduate classes in comparative anatomy. As a young instructor in zoology Hyman had produced a *Laboratory Manual for Elementary Zoology* (1919) and *Laboratory Manual for Comparative Vertebrate Anatomy* (1922). However her main research interests were with invertebrate animals, especially their taxonomy. Initially she worked on flatworms, but soon extended her investigations to a wide spread of invertebrates, and was frequently asked to help colleagues to identify specimens. She recognized the need for a comprehensive reference book on invertebrates, and decided to remedy the lack. A small income generated by the royalties of her books enabled her to resign her university position in 1931 to embark on an independent career. She travelled to several European laboratories, working for a period at the Stazione Zoologica, Naples, before returning to New York City to begin to write a major volume on the invertebrates, for which she was given office and laboratory space, but no salary, by the American Museum of Natural History. She read extensively in the English and foreign-language literature, and combined comparative anatomy, histology and physiology in her assessments. Her major work, *The Invertebrates* was published in six volumes between 1940 and 1968, and provided an encyclopedic account of most phyla of invertebrates. She received several honours and accolades for this achievement, including honorary degrees, the Gold Medal of the Linnean Society of London and membership of the National Academy of Sciences. Severely handicapped by Parkinson's disease towards the end of her life, she was unable to produce volumes on the molluscs or arthropods, and she died in New York City on 3 August 1969.

I

Ingenhousz Jan 1730–1799 was a Dutch biologist and physiologist who discovered photosynthesis and plant respiration.

Ingenhousz was born in Breda, in The Netherlands, on 8 December 1730, the son of a pharmacist. He was educated locally and received his training in medicine and chemistry at the University of Louvain, graduating in 1753; he then studied at the University of Leyden during the following year. He went to universities in Paris and Edinburgh for short periods, after which he set up a private medical practice in Breda. On his father's death in 1765, he left for England, encouraged by a physician from the British army who had befriended his family during the war of the Austrian Succession. In 1766, he worked at the Foundling Hospital, London, where he was responsible for inoculating patients against smallpox (using the hazardous live virus). His methods were reasonably successful and in 1768 he was sent to the Austrian court in Vienna, by George III, to inoculate the royal family. He took up the appointment of court physician there from 1772 to 1779. In that year he returned to England, where he continued his research until he died on 7 September 1799, in Bowood, Wiltshire.

In 1771, Joseph Priestley had found that the flame of a candle burning in a closed space eventually goes out and that a small animal confined under similar conditions soon dies. He also found that plants can restore the capacity of the air to support life, or the burning of a candle. Later Priestley discovered 'dephlogisticated air' (oxygen). It seems likely that Priestley's work inspired Ingenhousz to carry out similar experiments of his own.

Ingenhousz discovered in 1779 that only the green parts of plants are able to 'revitalize' the air, and that they are capable of doing so only in sunlight. His investigations also showed that the active part of the Sun's radiation is not in the heat generated, but rather in the visible light. He found that plants, like animals, respire all the time and that respiration occurs in all the parts of plants.

In the following years Ingenhousz demonstrated that the amount of oxygen released by a plant during photosynthesis is greater than that absorbed in respiration. He suggested that green plants take in carbon dioxide and produce oxygen, whereas animals do the reverse, and therefore that animals and plants are totally dependent on each other.

Ingenhousz believed that this discovery would help to distinguish between animals and plants among the lower orders of life. At that time a controversy existed over the origin of carbon in plants, some scientists believing that it is absorbed in some form by the roots – this belief was termed the humus theory. Ingenhousz, however, was of the opinion that carbon comes from the carbon dioxide absorbed by a plant. This idea explained the disappearance of the gas, and the presence of carbon in plants. Whereas he was right about the source of carbon, he was mistaken about that of oxygen, and it is now known that oxygen given off by plants comes from the water they take in.

Apart from the life of plants, Ingenhousz had various other interests, which led him in 1776 to develop an improved apparatus for generating large amounts of electricity; he also invented a hydrogen-fuelled lighter to replace the tinderbox, and investigated the use of an air and ether vapour mixture as a propellant for an electrically fired pistol.

In 1779, Ingenhousz published his work *Experiments On Vegetables, Discovering their Great Power of Purifying the Common Air in Sunshine, and of Injuring it in the Shade or at Night*. His discovery laid the foundations for the study of photosynthesis, the process upon which most animals ultimately depend for their food.

Ingold Christopher 1893–1970 was a British organic chemist who made a fundamental contribution to the theoretical aspects of the subject, with his explanation for the mechanisms of organic reactions in terms of the behaviour of electrons in the molecules concerned.

Ingold was born in London on 28 October 1893 but moved to Shanklin, Isle of Wight, when he was a few years old because of his father's ill health. He was educated at Sandown Grammar School and then went to study chemistry at Hartley University College, Southampton (later the University of Southampton), where he graduated in 1913.

He began research under J. F. Thorpe at Imperial College, London, investigating spiro-compounds of cyclohexane, but left in 1918 to spend two years as a research chemist with the Cassel Cyanide Company, Glasgow. He returned to Imperial College as a lecturer in 1920, where he met and married Edith Usherwood, a promising young chemist. From 1924 to 1930 he was Professor of Organic Chemistry at the University of Leeds, when he succeeded Robert Robinson as Professor of Chemistry at University College, London. He remained there for the rest of his career, retiring officially in 1961 (although continuing as an active contributor to the work of the department). Ingold was knighted in 1958. He died in London on 8 December 1970.

Much of Ingold's work at University College was carried out in collaboration with E. D. Hughes. For 30 years he specialized in the concepts, classification and terminology of theoretical organic chemistry. In 1926, for example, he put forward the concept of mesomerism, which allows a molecule to exist as a hybrid of a pair of equally possible structures. This work culminated in his classic reference book *Structure and Mechanisms in Organic Chemistry* (1953), whose second edition (1969) ran to 1,266 pages. His ideas, first published in 1932, are still fundamental to reaction mechanisms taught today. They concerned the role of electrons in elimination and nucleophilic aliphatic substitution reactions, which he interpreted in terms of ionic organic species.

Ingold's work removed much of the 'art' from organic chemistry and replaced it with scientific methodology. As he is reputed to have said: 'One could no longer just mix things; sophistication in physical chemistry was the base from which all chemists – including the organic – must start.'

Ipatieff Vladimir Nikolayevich 1867–1952 was a Russian-born US organic chemist who is best known for his development of catalysis in organic chemistry, particularly in reactions involving hydrocarbons.

Ipatieff was born in Moscow on 21 November 1867, the son of an architect. It was intended that he should have a career in the army so he attended a military school, became an officer in the Imperial Russian Army in 1887, and in 1889 won a scholarship to continue his higher education at the Mikhail Artillery Academy in St Petersburg. From 1892, the year in which he married, he gave lectures in chemistry at the Academy, and in 1897 he was given permission to go to the University of Munich for a year to study under Johann von Baeyer. One of his fellow students was Richard Willstätter (1872–1942). After a brief period in France studying explosives, he returned to Russia in 1899 as Professor of Chemistry and Explosives at the Mikhail Artillery Academy. In 1908 he gained his PhD from the University of St Petersburg.

Ipatieff's research work was interrupted by World War I and the Russian Revolution, during which he held various administrative and advisory appointments. He was head of the Chemical Committee during World War I, and increased the monthly output of explosives from 60 tonnes to 3,300 tonnes. But he was not a Communist and when, at the age of 64, he went to Berlin in 1930 to attend a chemical conference he accepted the offer of a post in the United States and did not return to the Soviet Union. He was immediately condemned as a traitor by the Soviet authorities and expelled from the Soviet Academy of Sciences. From 1931 to 1935 he was Professor of Chemistry at Northwestern University, Illinois, and acted as a consultant to the Universal Oil Products Company, Chicago. In 1938, this company funded the building of the Ipatieff High Pressure Laboratory at Northwestern. He died in Chicago on 29 November 1952, just after his birthday and ten days before the death of his wife. In 1965 he was posthumously reinstated to the Soviet Academy of Sciences.

While working as a student in Munich in 1897 Ipatieff synthesized the hydrocarbon isoprene, the basis of the rubber molecule. Back in Russia in 1900 he discovered the specific nature of catalysis in high-temperature organic gas reactions, and how using high pressures the method could be extended to liquids. He developed an autoclave called the Ipatieff bomb – for heating liquid compounds to above their boiling points under high pressure. He synthesized methane and iso-octane (2-methylheptane), and produced polyethylene by polymerizing ethylene (ethene).

In Chicago after 1931, Ipatieff began to apply his high-temperature catalysis reactions to petrol with low octane ratings (which produce 'knock' or pre-ignition in car engines). The result of the catalytic cracking (or 'cat cracking') is petrol with a higher octane rating, and the method became particularly important for the production of aviation fuel during World War II; it is still widely used.

Isaacs Alick 1921–1967 was a British virologist who discovered interferon, an antibody produced by cells when infected by viruses.

Isaacs was born on 17 July 1921 in Glasgow, the first of four sons in a Jewish family of Russian origin (his grandparents were Russian Jews who had emigrated to Scotland in about 1880). He had a conventional Jewish upbringing and was educated at Pollockshields Secondary School in Glasgow, attending classes in Judaism every day after school. His family moved to Kilmarnock in 1939 but Isaacs stayed in Glasgow and enrolled at the university there to study medicine. He was an able student, graduating in 1944 and winning several prizes, but clinical medicine did not greatly interest him and in 1945 he became a McCann research scholar in the Department of Bacteriology at Glasgow University, where he came under the influence of Carl Browning, the professor there. In 1947, Isaacs was awarded a Medical Research Council studentship to research into influenza viruses under Stuart Harris at Sheffield University, and in the following year he went to Australia, having won a Rockefeller Travelling Fellowship to work under Frank Macfarlane Burnet at the Walter and Eliza Hall Institute for Medical Research in Melbourne. Isaacs returned to Britain in 1951 and went to work in the laboratory of the World Influenza Centre at Mill Hill, London, where he remained for the rest of his life. His work on influenza gained him his medical degree from Glasgow University and a Bellahousten Gold Medal. In 1958 to 1959 he suffered a three-month depression but seemed to recover, and in 1961 he took over the directorship of the World Influenza Centre. In 1964, however, he suffered a subarachnoid haemorrhage and died two years later, on 26 January, only 45 years old.

Although Isaacs began investigating influenza in 1947, it was not until 1956 that he discovered interferon. Working with a Swiss colleague, Jean Lindenmann, Isaacs found that chick embryos injected with influenza virus produce minute amounts of a protein that destroys the invading virus and also makes the embryos resistant to other viral infections. Isaacs and Lindenmann named this protein interferon. Further research demonstrated that most living creatures can make interferon, and that even plants react to viral infection in a similar way. When a virus invades a cell, the cell produces interferon, which then induces uninfected cells to make a protein that prevents the virus from multiplying. Almost any cell in the body can make interferon, which seems to act as the first line of defence against viral pathogens, because it is produced very quickly (interferon production starts within hours of infection whereas antibody production takes several days) and is thought to trigger other defence mechanisms.

In the 1950s there was no treatment or cure for viral infections (antibodies are effective only against bacteria and at that time antiviral vaccines were in their early stages of development) and so the discovery of interferon was thought to be a major breakthrough. As a result, the Medical Research Council took out a patent on interferon and established a scientific committee to undertake further research into it. This initial enthusiasm waned, however, when it was found that interferon was species specific, and that it was very difficult and costly to produce. By the time Isaacs died, research into the substance had come to a virtual standstill.

Then in the late 1960s, after Isaacs' death, interest revived as a result of a chance discovery made by Ion Gresser, a US scientist then working in Paris. He found that interferon inhibits the growth of virus-induced tumours in mice and also that it stimulates the production of special cells that attack tumours. In addition, he later showed that interferon can be made in relatively large amounts from human blood cells. These findings stimulated further research, particularly into the use of interferon against cancer, leukaemia and certain viral diseases, such as hepatitis, rabies, measles and shingles. In early 1980 Charles Weissmann, a Swiss scientist, produced by genetic manipulation a strain of bacteria that can make human interferon, the effectiveness of which is still being investigated.

Issigonis Alec 1906– is a naturalized British automotive engineer and the first person to exploit scientific component packaging in the design of small volume-produced motorcars. His designs gave much greater space for the occupants together with greatly increased dynamic handling stability and improved the small-car ride.

Issigonis was born on 18 November 1906 in what was then Smyrna, and came to Britain with his widowed mother after the 1922 war between Turkey and Greece. He studied engineering at Battersea Polytechnic, and began his career working for a small engineering firm which was developing an automatic gearchange.

This work introduced Issigonis to the Humber division of the Rootes Group in 1934, and in 1936 he joined Morris Motors to work on suspension design. During this prewar period he built the ingenious Lightweight Special hill climb and sprint single seater, which demonstrated the potential of all independent suspension with rubber springs. His first complete production motorcar was the Morris Minor, launched in 1948, which brought new standards of steering and stability to small motorcars – and went on to become the first British motorcar to pass the million mark (in 1961). After a spell (1952–1956) work-ing at Avis on an experimental 3.5-litre car with hydrolastic suspension, he returned to what had now become BMC to face, within a short while, his greatest challenge.

Sir Leonard Lord (later to become Lord Lambury), chairman of BMC, asked him to design and produce a small and economical car to counteract the flood of 'bubble cars' which followed the Suez crisis. A period of intensive design and development led to the launch of the Mini in 1959. His other major designs were the 1100 (1962), the 1800 (1964) and the Maxi (1969). He was made CBE in 1964, became a Fellow of the Royal Society in 1967, and received a knighthood in 1969.

The main significance of his work was in taking on car design as a 'vehicle architect', overseeing the separate approaches of styling, interior packaging, body engineering and chassis layout. In this way he conceived the overall package from his knowledge and experience of the major factors affecting the product; specialists in his team then designed and engineered the subsystems of the vehicle. The approach was to make the human factor paramount in selecting design criteria.

This approach was particularly recognizable in the 'wheel at each corner' layout of the Morris Minor and its effect on handling. The vehicle's polar moment of inertia (a measure of directional stability), was made small compared with the magnitude of the tyre's cornering forces. This improved the speed of response to change in direction, and designed-in nose-heaviness allowed quick correction of the car after a side-gust disturbance.

With the layout of the Mini, a degree of interior spaciousness was achieved beyond that available in previous cars of similar exterior size. The technique included repositioning the dash panel to follow the projected line of the curved lower edge of the windscreen, using single skinned doors and rear quarter panels, with large open lockers on the inside. By adopting a transverse engine-over-gearbox/final-drive layout with front-wheel drive, and independent suspension of all wheels, considerable gains were made in front knee-room and space at the rear of the seat base.

The use of compact wheel-location lever-arms acting on rubber suspension springs also gave a substantial packaging advantage, besides giving a well-damped ride, free of static friction. In arranging for fluid correction of the springs, the suspension could be tuned to separate motions of pitch and bounce of the vehicle; tuning virtually eliminated pitch as a prime factor in ride discomfort.

The Mini, for which Issigonis is best known, reigned supreme among small cars until the late 1970s. At the beginning of its life the car attracted a considerable cult following and all through its life it was the basis for a great diversity of modification by specialist firms. In the early 1980s, despite the competition, this 25-year-old design was actually increasing its market share, following an intial dip when the Mini Metro was introduced.

J

Jacob François 1920– is a French cellular geneticist who was jointly awarded the 1965 Nobel Prize for Physiology or Medicine with André Lwoff and Jacques Monod for their collaborative work on the control of gene action in bacteria.

Jacob was born on 17 June 1920 in Nancy. He was educated at the Lycée Carnot and at the University of Paris, from which he gained his medical degree in 1947 – his studies having been interrupted by military service during World War II – and his doctor of science degree in 1954. In 1950 he joined the Pasteur Institute in Paris as a research assistant, becoming Head of Laboratory there in 1956. In 1964, he became Head of the Department of Cellular Genetics at the Pasteur Institute and also Professor of Cellular Genetics at the Collège de France.

Jacob began his Nobel prizewinning work on the control of gene action in 1958. Previous work by Francis Crick, James Watson and Maurice Wilkins had shown that the types of proteins produced in an organism are controlled by DNA, but it was Jacob – working with Lwoff and Monod – who demonstrated how an organism controls the amount of protein produced. Jacob performed a series of experiments in which he cultured the bacterium *Escherichia coli* in various mediums to discover the effect of the medium on enzyme production. He found that *E. coli* grown on a medium containing only glucose produced very little of the enzyme β-galactosidase, whereas *E. coli* grown on a lactose-only medium produced much greater amounts of this enzyme. This phenomenon, which is called enzyme induction, occurs because *E. coli* can metabolize glucose easily but requires β-galactosidase to metabolize lactose. From these findings, Jacob, Lwoff and Monod concluded that increased production of an enzyme (β-galactosidase in this case) occurs when an organism needs that particular enzyme, and they also proposed a theory to explain the mechanisms involved in this increased enzyme production. According to their theory there are three types of genes concerned with the production of each specific protein: a structural or Z-gene, which controls the type of protein produced, and a regulator or R-gene, which produces a repressor substance that binds to the third type of gene, the operator or O-gene. The binding of the repressor substance to the O-gene prevents the production of messenger RNA (mRNA) by the appropriate Z-gene and therefore also

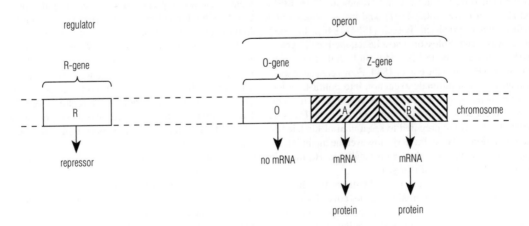

Jacob *A regulator gene (R-gene) produces a repressor substance that prevents an operator (O-gene) from providing messenger RNA, blocking the production of protein.*

prevents the production of the specific protein coded for by that Z-gene. In Jacob's experiments, there was minimal production of β-galactinosidase in *E. coli* grown on glucose because the repressor substance produced by the R-gene was specifically bound to the O-gene, thereby preventing β-galactosidase production by the Z-gene. *E. coli* grown on lactose produced large amounts of β-galactosidase because lactose binds to the repressor substance produced by the R-gene, which alters the molecular conformation of the repressor substance so that it can no longer bind to the O-gene; as a result, the Z-gene can produce mRNA for β-galactosidase production. This theory has since been shown to be correct in the control of protein synthesis in *E. coli* and other bacteria.

Jacobi Karl Gustav Jacob 1804–1851 was a German mathematician and mathematical physicist, much of whose work was on the theory of elliptical functions, mathematical analysis, number theory, geometry and mechanics. An influential teacher, he corresponded with many of the other great mathematicians of his time.

Jacobi was born in Potsdam on 10 December 1804, the second son of a wealthy and well-educated Jewish banker. A child prodigy, he rose to the top class of his local school within months of first entering, and was ready to go to university at the age of 12 – but that the authorities would not permit. He went to Berlin University in 1821 and graduated in the same year, with excellent results in Greek, Latin, history and mathematics. Continuing at the University he studied philosophy and the Classics, reading mathematics privately because, apart from Karl Gauss at Göttingen – with whom he corresponded – there was simply no one who could teach him. At the age of 19 he qualified as a teacher. The following year he received his doctorate, underwent conversion to Christianity (as may have been professionally advisable) and became an unsalaried lecturer at Berlin University. After only one year there, however, in 1826 he joined the staff at Königsberg (now Kaliningrad) University, coming into contact there with people such as the physicists Franz Neumann (1798–1895) and Heinrich Dove (1803–1897), and the astronomer and mathematician Friedrich Bessel (1784–1846). In 1832 Jacobi was made Professor, a post he retained for 18 years. In 1843, however, he fell ill and went to Italy to recover, returning to Berlin at the end of 1844. Four years later, after an unwise and unfortunate excursion into politics, he lost the royal pension on which he had been living and his family was reduced to near-destitution. Luckily, the Prussian government was not prepared to see a man of his talent – and reputation – starve. In 1851, however, he again fell ill, first with influenza and then with smallpox. He died, in Berlin, on 18 February of that year.

There are very few areas in mathematics that Jacobi's researches at one time or another did not cover fully. He is particularly remembered for his work on elliptic functions, in which to some extent, he was competing against a rival in the contemporary Norwegian mathematician Niels Abel (1802–1829). His principal published work on elliptic function theory was *Fundamenta nova theoriae functionum ellipticarum*, produced in 1829, incorporating some of Abel's ideas and introducing his own concept of hyper-elliptic functions. In analysis, Jacobi studied differential equations (the results of which he applied to problems in dynamics) and the theory of determinants. In the latter field he invented a functional determinant – now called the Jacobian determinant – which has since been of considerable use in later analytical investigations, and was even supportive in the development of quantum mechanics. He advanced the theory of the configurations of rotating liquid masses by showing that the ellipsoids now known as Jacobi's ellipsoids are figures of equilibrium.

He had immense knowledge in every branch of mathematics – Jacobi's PhD thesis was an analytical discussion on the theory of partial fractions; he was at the same time writing to Gauss on cubic residues in number theory; his first lecture, nevertheless, was on the theory of curves and surfaces in three-dimensional space. He was also always trying to link together different mathematical disciplines. For instance, he introduced elliptic functions into number theory and into the theory of integration, which in turn connected with the theory of differential equations and his own principle of the last multiplier.

Jacobi's work, and particularly his researches in collaboration with Franz Neumann and Friedrich Bessel, revived an interest in mathematics generally in Germany. He therefore became a highly respected and honoured man. After his death, his great friend Lejeune Dirichlet (1805–1859) delivered a memorial lecture at the Berlin Academy of Sciences on 1 July 1852, in which he described Jacobi as 'the greatest mathematician among members of the Academy since Joseph Lagrange'.

Jacquard Joseph Marie 1752–1834 was a French engineer who developed a loom (originally for carpets), whose complicated patterns were 'programmed' on to punched cards.

Jacquard was born in Lyons on 7 July 1752, the son of a weaver. He was apprenticed when he was quite young, first to a bookbinder and later to a cutler. Throughout his time in these trades he considered ways of improving the crafts and reducing the labour needed. When his parents died, they left him their small weaving business and Jacquard took up that trade. He attempted to weave patterned fabrics, which brought a relatively high profit, but he was not successful at this intricate craft, which required long hours of hard, patient toil to finish even the most modest piece of material. He returned to the cutlery trade after becoming bankrupt as a weaver.

He continued his efforts to devise an improved weaving machine, but his first loom was not made until 1801. It combined a number of innovations which had been used before on various other machines with some ideas of his own. In 1804, he went to Paris to demonstrate a new machine for weaving net. This was welcomed and, as well as receiving a patent, he was allocated a small pension, which enabled him to work on improving looms and weaving at the Conservatoire des Arts et Métiers.

Jacquard's attachment for pattern weaving, which was developed during this time and later improved by others, allowed patterns to be woven without the intervention of the weaver. Weavers had always had to plan the pattern they wished to weave before they began their task. This planning now became the essential feature of the weaver's job. Part of Jacquard's invention was a series of cards with holes punched in them through which tools could come up and pull the warp threads down, so that the shuttle could pass over them. If there was no hole the shuttle went underneath the warp. In this way the correct threads were brought into conjunction to weave an intricate pattern.

During the first decade of the nineteenth century many Jacquard looms were installed in weaving mills. As Jacquard intended, his loom saved a great deal of physical labour, and because of this it was not kindly received by the workers in the weaving trade. In Lyons and elsewhere his machines were smashed, and Jacquard himself suffered violence at the hands of the angry weavers. Even angry workers, however, could not halt mechanical progress and by 1812 it is recorded that there were 11,000 Jacquard looms working in France. During the next few years they were introduced into many other countries, including England.

One pattern, when punched onto cards, could be used over and over again, which gave the new machine a tremendous advantage over the traditional one, but the preparation of the cards was a difficult and tedious task and it was only with the coming of the electronic computer that this labour was reduced.

The idea of the Jacquard cards was to feed a machine with instructions, and it was applicable in principle, at least, to many kinds of machine. One direct application of the cards was in the mechanical computing machine called the 'analytical engine' designed and partly constructed by Charles Babbage in the years following 1833. In this machine the cards directed a sequence of arithmetical operations and the transfer of numbers to and from the store. Other pioneers of automatic computing have also used punched cards and punched tape as a means of storing, sorting and transferring information. Herman Hollerith, for example, used a vast mechanical tabulatory system which relied on punched cards with holes in 288 positions for his 1890 census of the United States.

Only now with the rapid modern development of electronic computers and magnetic storage systems are the descendants of the Jacquard cards declining in use as representations and stores of information. The principle of the Jacquard loom still survives in weaving, even though the machinery and programming are somewhat different.

Jacquard was highly honoured, and eventually even the weavers appreciated his invention. He survived the violence of his times and died at Oullines, France, on 7 July 1834.

Jansky Karl Guthe 1905–1950 was a US radio engineer whose discovery of radio waves of extraterrestrial origin led to the development of radioastronomy.

Jansky was born at Norman, Oklahoma, on 22 October 1905. He was educated at the University of Wisconsin, where his father was a member of the Faculty. He spent a year as an instructor before beginning his career as an engineer with the Bell Telephone Laboratories in 1928. His fundamental discovery of the existence of extraterrestrial radio waves was made and published by 1932. Thereafter Jansky did not continue this kind of scientific work. He became more involved with engineering and left the research in astronomy to others. Jansky died at an early age, of a heart complaint, at Red Bank, New Jersey, on 14 February 1950.

When Jansky joined the Bell Telephone Company as a research engineer, he was assigned the task of tracking down and identifying the various types of interference from which radio telephony and radio reception were suffering. The company was particularly concerned with the interference that occurred at short wavelengths of around 15 m/49 ft (then being used for ship-to-shore radio communications). It was well known that some of the static interference was caused by thunderstorms, nearby electrical equipment or aircraft, but, by building a high quality receiver and aerial system that was set on wheels and could be rotated in various directions, Jansky was able to detect a new and unidentifiable kind of static. After months of study Jansky associated the source of the unidentified radio interference with the stars. He had noticed that the background hiss on a loudspeaker attached to the receiver and antenna system reached a maximum intensity every 24 hours. From overhead, it seemed to move steadily with the Sun but gained on the Sun by four minutes per day. This amount of time correlates with the difference of apparent motion, as seen on Earth, between the Sun and the stars, and so Jansky surmised that the source must lie beyond the Solar System. By the spring of 1932, he had concluded that the source lay in the direction of Sagittarius. This was the direction which the US astronomer, Harlow Shapley (1885–1972), and the Dutch astronomer, Jan Oort (1900–1992), had confirmed as being the direction of the centre of our Galaxy. Jansky published his results in December 1932. In the same month Bell Telephone Laboratories issued a press release on Jansky's discovery to the New York Times and it made front-page news. This was the birth of radioastronomy. There was now a second region of the electromagnetic spectrum in which it was possible to study the stars – a radio window on the universe. There was no immediate follow-up to Jansky's discovery, however, and it was left to an amateur US astronomer, Grote Reber, to pursue the matter. It was not until the end of World War II that the importance of radioastronomy began to be recognized.

Although Jansky died at an early age, he did live long enough to see radioastronomy develop into an important tool of modern astronomical research. By their ability to penetrate the Earth's atmosphere and dust clouds in space, radio telescopes enable the study of celestial objects that are impossible to detect by optical means. In honour of Karl Jansky the International Astronomical Union named the unit of strength of a radio-wave emission a jansky (symbolized Jy), in 1973.

Janssen Pierre Jules César 1824–1907 was a French astronomer, famous for his work in physical astronomy and spectroscopy.

Janssen was born in Paris on 22 February 1824. His father was a musician of Belgian descent and his maternal grandfather was a well-known architect. In his early life he had an accident which left him permanently lame, and as a result he never went to school. Because of financial difficulties he went out to work at an early age. He worked in a bank from 1840 to 1848, but educated himself at the same time and took his Baccalaureat at the age of 25. He studied mathematics and physics at the Faculty of Sciences in Paris and gained his Licence des Sciences in 1852. He worked for a while as a substitute teacher in a lycée, before being sent on the first of his scientific missions – to Peru – in 1857. He returned with a bad attack of dysentery and became a tutor to the wealthy Schneider family who owned iron and steel mills in Le Creusot. He was awarded his doctorate in 1860, and in 1862 began work with E. Follin of the Faculty of Medicine. In 1865 he was made Professor of Physics at the Ecole Spéciale d'Architecture. He was elected to the Academy of Sciences in 1873 and to the Bureau of Longitudes in 1875. He was appointed Director of the new astronomical observatory, established by the French government at Meudon in 1875, a position he held until his death in Paris on 23 December 1907.

Although Janssen travelled to Peru to measure the magnetic equator, his first real research work was in the field of ophthalmology. He went on to work on the construction of an ophthalmoscope at the Faculty of Medicine in Paris. But after hearing that Gustav Kirchoff (1824–1887) had demonstrated the presence of terrestrial elements in the make-up of the Sun in 1859, Janssen had decided that his real passion was for physical astronomy. He built himself an observatory on the flat roof of his wife's house in Montmartre and began to work on the nature of the dark bands in the solar spectrum. He found that these bands were most noticeable at sunrise and sunset. For his research he constructed a special spectroscope with a high dispersive power, to which he attached a device for regulating luminous intensity. By 1862, he was able to show that these bands can be resolved into rays and that they are always present. In 1864 he travelled to Italy and there he was able to show that the intensity of the rays varied during the course of a day in relation to the terrestrial atmosphere. He showed that the origin of the phenomenon was terrestrial and called the rays 'telluric rays'. To prove his point that the intensity of telluric rays would be less in thinner air, he travelled to the Bernese Alps to measure the rays at a height of 2,700 m/8,900 ft. They were even weaker than he expected, and so he attributed the extra effect to the dryness of the air, and went to the shores of Lake Geneva to study the effect of humidity on the strength of the rays. In 1867, he spent time in the Azores, carrying out optical and magnetic experiments which concluded that water vapour was present in the atmosphere of Mars.

In 1868, Janssen went to India to observe the total eclipse of the Sun which occurred on 18 August. Together with other scientists, he noted that there were a number of bright lines in the spectrum of the solar chromosphere. He demonstrated the gaseous nature of the red prominences, but he was unable to correlate the exact positions of the lines in the solar spectrum with wavelengths of any known elements. Janssen continued to observe the unobscured Sun for 17 days after the eclipse and he reported his findings to the French Academy of Sciences by telegram. In October of the same year, Norman Lockyer (1836–1920), an English astronomer and physicist, observed the chromosphere through a special telescope designed by him in 1866. He noted, as Janssen had, that there was a third yellow line in the spectrum, which did not correspond to either of the two known lines of sodium. He immediately reported his findings to the French Academy of Sciences and a letter from Janssen, also reporting that a new element must be responsible for the yellow line, arrived at the same time as the letter from Lockyer. The two scientists became firm friends, because of the coincidence of their discovery, and a medallion was struck by the French Academy of Sciences bearing their profiles and names. Lockyer went on to discover that the new line represented a substance that was later named helium.

In the same year, 1868, Janssen developed a spectrohelioscope so that he could observe the Sun and solar prominences spectroscopically in daylight conditions. With it he attempted to ascertain whether or not the Sun contains oxygen. He realized it would help his research if he could eliminate some of the obscuring effects of the Earth's atmosphere, and for this reason he established an observatory on Mont Blanc. By this time his health was not good, and to overcome his disability he invented a device which enabled him to be carried up Mont Blanc; even so, the journey in this conveyance took 13 hours. Janssen went to any lengths to continue his observations. In order to observe the eclipse on 22 December 1870, he travelled to Algeria by balloon, despite the Franco–Prussian War. He later devised an aeronautical compass which was capable of instantly indicating the direction and speed of flight.

From 1876 to 1903 Janssen summarized the history of the surface of the Sun by means of solar photographs. These were made at the astrophysical observatory in Meudon and were collected together in his *Atlas de photographies solaires* (1904). In order to observe the transit of Venus in 1882, he took a series of photographs in rapid succession which enabled him to measure successive positions of the planet. He obtained a series of separate images, laid out in a circle on the photographic plate. Here Janssen had anticipated one of the operations necessary for cinematography, which was to be invented 20 years later. He also invented another device of historical interest, the photographic revolver, in 1873.

Jeans James Hopwood 1877–1946 was a British mathematician and astrophysicist who made important contributions to cosmogony – particularly his theory of continuous creation of matter – and became known to a wide public through his popular books and broadcasts on astronomy.

Jeans was born at Ormskirk, Lancashire, on 11 September 1877, but moved at the age of three to Tulse Hill, London, where he spent his childhood. His interests in science and his literary bent disclosed themselves at an early age. When he was only nine years old he wrote a handbook on clocks which included an explanation of the escapement principle and a description of how to make a clock from pieces of tin. He was educated at Merchant Taylors' School, London, from 1890 to 1896 and then went to Cambridge to study mathematics at Trinity College. He graduated in 1900 and was awarded a Smith's Prize. In 1901, Jeans was appointed a Fellow of Trinity and from 1905 to 1909 he was Professor of Applied Mathematics at Princeton University in the United States. From 1910 to 1912 he was Stokes Lecturer in Applied Mathematics at Cambridge. Thereafter he held no university post, but devoted himself to private research and writing, although he was a research associate at the Mount Wilson Observatory, California, from 1923 to 1944. In 1928 Jeans stated his belief that matter was continuously being created in the universe. He never developed this idea (a forerunner of the steady-state theory), however, for in that year his career as a research scientist came to an end. After that time he concentrated on broadcasting and writing popular books on science. He was awarded the Royal Medal of the Royal Society in 1919 and was knighted in 1928. He died on 1 September 1946 at Dorking, Surrey.

The first years of Jeans's research career were spent on problems in molecular physics, in particular on the foundations of the kinetic molecular theory. His *Dynamical Theory of Gases* (1904), which contained his treatment of the persistence of molecular velocities after collisions (that is, the tendency for molecules to retain some motion in the direction in which they were travelling before collision), became a standard text.

Jeans next worked on the problems of equipartition of energy in its application to specific heat capacities and black-body radiation. In 1905 he corrected a numerical error in the derivation of the classical distribution of blackbody radiation made by John Rayleigh (1842–1919) and formulated the relationship known as the Rayleigh–Jeans law, which describes the spectral distribution of blackbody radiation in terms of wavelength and temperature. For some time thereafter Jeans investigated various problems in quantum theory, but in about 1912 he turned his attention to astrophysics.

Jeans had been interested in astrophysics since his student days. In an undergraduate prizewinning essay he had treated compressible and incompressible fluids at a deeper level than had Henri Poincaré (1854–1912), and this work led directly to a consideration of the origin of the universe, or cosmogony. Jeans distinguished two extremes: an incompressible mass of fluid and a gas of negligible mass surrounding a mass concentrated at its centre. He argued that if an incomprehensible mass were to contract or be subjected to an increase in angular momentum, it would evolve into an unstable pear-shaped configuration and then split in two. Double stars could be formed in this way. At the other extreme, gas of a negligible mass could evolve through ellipsoidal figures to a lenticular shape and then eject matter from its edge; spiral nebulae could be formed in this way. Jeans concluded that the rotation of a contracting mass could not give rise to the formation of a planetary system. He himself favoured a tidal theory of the Solar System's origins, in which planetary systems were created during the close passage of two stars. This theory is no longer held in much favour, but Jeans did point out the errors in the nebular theory by which Laplace (1749–1827) attempted to explain the origin of the Solar System.

Jenner Edward 1749–1823 was a British biologist who was the first to prove by scientific experiment that cowpox gives immunity against smallpox. He was the founder of virology and was one of the pioneers of vaccination.

Jenner was born in Berkeley, Gloucestershire, on 17 May 1749, the son of a vicar. He was educated locally and in 1761 apprenticed to a surgeon in Sodbury. In 1770, he went to London to study anatomy and surgery under John Hunter, who took Jenner as his first boarding pupil at St George's Hospital in London. Returning to Gloucestershire in 1773, he set up in private practice and remained at Berkeley until he died there of a stroke on or about 26 January 1823.

In 1788, an epidemic of smallpox swept Gloucestershire and inoculations with live vaccine were used in spite of the tragedies that had previously accompanied this practice. The method used was well known in eastern countries and had been brought to England in 1721 by Lady Mary Wortley Montagu, the wife of the British Ambassador to Turkey. It consisted of scratching a vein in the arm of a healthy person and working into it a small amount of matter from a smallpox pustule taken from a person with a mild attack of the disease. This treatment often resulted in the patient fatally contracting smallpox, despite the successes of Jan Ingenhousz.

In the course of his inoculations Jenner noticed that people who had suffered from cowpox, a disease affecting the teats of cows and later the hands of their milkers, remained quite unaffected by the smallpox inoculation, and did not even produce the symptoms of a mild attack of smallpox (as did other patients). Over the course of 25 years he observed that he was unable to infect previous cowpox victims with smallpox, and that where whole families succumbed to smallpox, a previous cowpox victim remained healthy. He also noticed that whereas inoculation with cowpox appeared to protect the patient from smallpox, it did not give immunity against cowpox itself. In his study of cowpox, Jenner was the first to coin the word 'virus'.

In 1796 Jenner carried out an experiment on one of his patients, James Phipps, a healthy eight-year-old boy. Jenner made two small cuts in the boy's arm and worked a speck of cowpox into them. A week later, the patient had a slight fever (the usual reaction) but quickly recovered. Some weeks later, Jenner repeated the inoculation, this time using smallpox matter. The boy remained healthy – vaccination was born (named after *vaccinia*, the medical name for cowpox).

Jenner continued with his experiments and reported his findings to the Royal Society but the Fellows considered that he should not risk his reputation by presenting anything 'so much at variance with established knowledge', so in 1798 Jenner published his work privately. Within a few years vaccination was a widespread practice and Jenner not only improved his technique but also found a way of preserving active vaccine.

Jenner was also interested in natural history, one of his favourite hobbies being birdwatching. He studied the habits of the cuckoo, which often lays its eggs in the nest of the hedge sparrow. It had been thought that the hen hedge sparrow threw out her young from the nest to make room for the developing cuckoo, but Jenner's patient observations revealed that it was the young cuckoo itself that heaved its competitors out of the nest. In 1788, he reported these findings to the Royal Society, who published them.

Jenner's work led to an immediate reduction in mortality from smallpox and, nearly 200 years later, the worldwide eradication of the disease. He may be considered to be one of the pioneers of immunology.

Jensen Johannes Hans Daniel 1907–1973 is a German physicist who shared the 1963 Nobel Prize for Physics with Maria Goeppert-Mayer (1906–1972) and Eugene Wigner (1902–) for work on the detailed characteristics of atomic nuclei.

Jensen was born in Hamburg on 25 June 1907. He studied at the Universities of Hamburg and Freiburg, gaining a PhD in 1932. From 1932 to 1936 he was an assistant in science at the University of Hamburg, and from 1937 to 1941 in charge of courses at the University. In 1941, he joined the Institute of Technology in Hanover as Ordinary Professor, staying until 1948. In 1949 he became Professor at the University of Heidelberg.

With Goeppert-Mayer and Wigner, Jensen proposed the shell theory of nuclear structure in 1949, and explained it in *Elementary Theory Of Nuclear Shell Structure*, written with Goeppert-Mayer in 1955. According to this theory, a nucleus could not be thought of as a random motion of neutrons and protons about a point, but as a structure of shells or spherical layers, each with a different radius and each filled with neutrons and protons.

Careful studies of nuclear energy levels had shown that not only do even numbers of neutrons or protons lead to more stable nuclei than odd numbers, but that nuclei containing certain definite numbers of neutrons or protons, or both, are especially stable (like the electron structures of inert gas atoms). In 1948, Goeppert-Mayer published evidence of the special stability of the following numbers of protons and neutrons: 2, 8, 20, 50, 82 and 126. These are commonly called magic numbers. Similar conclusions

were reached by Jensen and Wigner. It is not only the energy levels that show particularly tight binding of nuclei with a magic number of protons or neutrons, or particularly both; the angular momentum also shows properties similar to those observed in closed shells of electrons in atoms. The value of angular momentum is zero for closed shells, but has a value where there are small numbers of neutrons or protons outside the closed shells. It seems likely that what is being dealt with here is something like the periodic table of the elements.

The three Nobel prizewinners tried to establish the significance of the magic numbers and why they have the values that they do. They concluded that the orbital and spin moments of the particles within the nucleus could be treated to a first approximation by methods used for atomic structure. One had to assume that the central field was nothing like that met in an atom, but consisted of a square well. In addition, the potential was constant throughout the interior of the sphere but had high barriers at the surface, and the energy levels were in a different order from the order of electronic levels in an atom. For example, the two 1s levels came lowest and then the six 2p levels, explaining the magic number 8. If the ten levels of the 3d orbitals were next followed by two 2s levels, we could add 2 and 10 to 8 and get the next magic number, 20. Jensen and his colleagues then supplemented this theory with a second hypothesis – that there is a very strong spin-orbit interaction between the spin and orbit of a particle and that the lower of two states is always the one with angular momentum parallel rather than antiparallel. In this way, magic numbers seem to fall into place, although research is still going on.

Jessop William 1745–1814 was a British civil engineer, a builder of canals and early railways.

Jessop was born in January 1745 in Devonport, Devon, the son of a foreman and shipwright at the local naval dockyard. His father had been associated with John Smeaton in the building of the Eddystone lighthouse. When his father died in 1761, William Jessop, aged 16, became a pupil of Smeaton, who was appointed his guardian.

Jessop worked with Smeaton on the Calder and Hebble, and the Aire and Calder navigations in Yorkshire. Jessop later became England's greatest builder of large waterways; he was responsible for the Barnsley, Rochdale and Trent navigation, the Nottingham and the Grand Junction (Grand Union) canals. The only narrow one he worked on was the Ellesmere canal. In 1773 Smeaton went to Ireland, taking his pupil with him, and the Grand Canal of 50 km/ 80 mi from Dublin to the Shannon (which had been started in 1753) was completed in 1805, just two months after the Grand Junction Canal in England.

It was his work on the building of the Cromford Canal to link Arkwright's Mill at Cromford with the Derbyshire coalfield and Nottingham which led Jessop, together with Benjamin Outram, John Wright and Francis Beresford, to found the Butterley Iron Works Company in 1790. This company later became responsible for many iron bridges.

This was the main reason why the canal engineers Jessop and Outram became much in demand for the development of the iron railways in the Midlands, especially in Nottinghamshire and Derbyshire. The first all-iron rails was laid in the 1790s. In 1792, fishbellied cast-iron rails were designed by Jessop. He was also involved in some spectacular iron bridges, particularly the Pontcysyllte aqueduct, completed in 1805. His son Josias was trained by him and became an important engineer in his own right. William Jessop was held in very high repute and for 16 years from 1774 served as Secretary of the Society of Civil Engineers. He died at Butterley Hall, Derbyshire, on 18 November 1814.

Jessop was the chief engineer, appointed in 1793, on the construction of the Grand Union Canal (which was then called the Grand Junction Canal). This canal was 149.5 km/93.5 mi long from the Oxford Canal at Braunston, via Wolverton, Leighton Buzzard, Tring, King's Langley, to the Thames at Brentford. The canal was completed in 1800, apart from the 2.8-km/1-mi long tunnel at Blisworth, which was not finished until 25 March 1805. While this tunnel was being completed, Jessop built a railway over the hill – the first in Northamptonshire – to enable traffic to operate on the canal in 1799. The canal provided a vital link between London and the Midlands.

Jessop's first tunnel was the 2.8-km/1.7-mi-long Butterley Tunnel on the Cromford Canal, and led to the forming of the Butterley Iron Works. The Cromford and High Peak Railway, engineered by Jessop's son Josias in 1825, was built to connect the Cromford Canal with the Peak Forest Canal at Whaley Bridge.

The Surrey Iron Railway grew out of a proposal in 1799 to open a part railway/part canal route from London to Portsmouth. The Thames at Wandsworth to Croydon was to have been canal, but Jessop and Rennie, retained as consultants, decided that a canal would be harmful to the industries of the Wandle Valley.

The Surrey Iron Railway was incorporated by an Act of Parliament on 21 May 1801, and Jessop was appointed Chief Engineer. The railway was opened in 1802.

The first all-iron rails were laid in the 1790s and spread over the country during the first years of the nineteenth century, coinciding with the Revolutionary and Napoleonic Wars. These wars caused the cost of wood to soar and the price of iron to fall with the intensification of industry.

The fishbellied iron rails which Jessop designed had a broad head on a thin web, deepest in the centre. One end had a flat foot, nailed with a peg onto a stone block. The other end had a round lug which was fitted into the the foot of the next rail. It was the true ancestor of the modern rail.

Jessop is not given enough credit for his involvement with Telford in the construction of the Pontcysyllte aqueduct. This crossed the river Dee and was completed in November 1805. In 1795, Jessop and Telford decided on a 300-m/1,000-ft iron trough, standing on a series of slender stone piers, each solid at the base but hollow from about 21 m/70 ft upwards. The work involved building 19 arches, each with a span of 14 m/45 ft. The work must have been extremely dangerous, carried out high above the swirling river Dee, with huge blocks of stone. A mixture of ox blood, water and lime was used as mortar. The flanged iron plates to make the trough were cast at Plaskynaston Foundry (within view of the aqueduct) and were bolted together, the joints being made watertight with Welsh flannel and lead dipped in boiling sugar! In November 1805, 8,000 people watched the opening. The bridge cost £47,000 with labourers paid 8–12 shillings (40–60 pence) per week. When Telford wrote later about the achievement, he gave himself all the credit, with no mention of Jessop's contribution, Pontcysyllte is the finest aqueduct in Britain and is certainly the most spectacular piece of aerial navigation in the world. Jessop also built the Derwent aqueduct with 24.3-m/80-ft span, and a less successful three-arch masonry aqueduct near Cosgrove.

He worked on the construction of a large wetdock area on the Avon at Bristol, on the West India Docks and the Isle of Dogs Canal in London, on the harbours at Shoreham and Littlehampton, and on many other projects.

By 1799 or 1800, he had abandoned the fishbellied rails in favour of cast-iron sockets fixed to the sleepers in which the rails were supported in the upright position. He produced what is equivalent to the flanged wheel of today, with flanges inside the rails.

The work of Jessop forms the link between that of Smeaton and Rennie. He achieved an incredible amount in a relatively short lifetime and was always ready to help his fellow professionals, for many of whom he acted as a consulting engineer on their projects, and his opinion was frequently sought because of his vast experience.

Joliot-Curie Irène 1897–1956 and Frédéric 1900–1958 received the Nobel Prize for Chemistry in 1935 for their discovery of artificial, or induced, radioactivity, in light elements. Irène was the daughter of Pierre and Marie Curie, who shared the 1903 physics prize with Becquerel. (Marie also won the 1911 chemistry prize outright). Both mother and daughter died of leukaemia caused by over-exposure to radioactivity during their research. Irène Curie was born on 12 September 1897 in Paris and educated privately by tutors including Paul Langevin and Jean Perrin. She obtained her undergraduate degree at Collège Sévigné. During World War I, she worked as a nurse, helping her mother operate radiography equipment, and then studied physics and mathematics at the Sorbonne, gaining a doctorate for studying the range of alpha particles. She then went to work for her mother at the Radium Institute. There she met Frédéric Joliot whom she married in 1926. Frédéric Joliot was born on 19 March 1900 in Paris. He was the son of a tradesman and graduated from the Ecole Supérieure de Physique et de Chimie Industrielle in engineering. He joined the Radium Institute in 1925 and obtained his Phd in 1930. Together the Joliot-Curies worked on radioactivity and the transmutation of elements. Twice they just missed major discoveries: in 1932 when Chadwick beat them to the neutron, and in 1933 when Anderson discovered the positron. However in 1934, whilst bombarding light elements with alpha particles, the Joliot-Curies noticed that although proton production stopped when the alpha particle bombardment stopped, another form of radiation continued. The alpha particles had produced an isotope of phosphorus not found in nature. This isotope was radioactive and was decaying through beta-decay. The Joliot-Curies received the Nobel Prize for Chemistry in 1935 – one year after Marie died. The Joliot-Curies continued to work on fission in heavy elements like uranium. In 1937, Irène was elected a professor at the Sorbonne and Frédéric became professor of nuclear physics at the Collège de France. In 1946, Irène became director of the Radium Institute. Irène died in Paris on March 17 1956 and Frédéric succeeded her at the Radium Institute. The Joliot-Curies were also noted for their left-wing politics. Irène was refused entry to the American Chemical Society because of her political beliefs. In 1950 Frédéric was replaced as director of the French Atomic Energy programme, despite having been appointed by de Gaulle and having built a nuclear reactor without any outside help. Frédéric died in Paris on 14 August 1958.

Jones Harold Spencer 1890–1960 was a British astronomer, the tenth Astronomer Royal. He is noted for his study of the speed of rotation of the Earth, the motions of the Sun, Moon and planets and his determination of solar parallax.

Jones was born in London on 29 March 1890, the third child of an accountant. He attended Hammersmith Grammar School and Latymer Upper School, where he excelled at mathematics. He won a scholarship to Jesus College, Cambridge. Following his graduation in 1913, he was awarded a research fellowship to Jesus College. In the same year he was appointed Chief Assistant at the Royal Observatory, Greenwich, to succeed Arthur Eddington (1882–1944), who had gone to Cambridge as Plumian Professor. Ten years later, in 1923, Jones was appointed His Majesty's Astronomer on the Cape of Good Hope. In 1933, he returned to England to become Astronomer Royal. He held this position until his retirement in 1955. He was President of the International Astronomical Union from 1944 to 1948 and Secretary General of the Scientific Union from 1955 to 1958. He had been President of the Royal Astronomical Society from 1938 to 1939 and was knighted in 1943. Jones died in London on 3 November 1960.

Jones's first period of research began when he went to Russia in 1914 to observe an eclipse. He wrote several papers on the variation of latitude, as observed using an instrument known as the Cookson floating telescope. During this time he also made determinations of the photographic magnitude scale of the North Polar Sequence (the stars located near the Celestial North Pole). While at the Cape of Good Hope, he published an important catalogue containing the radial velocities of the southern stars, calculated the orbits of a number of spectroscopic binary stars, and made a spectroscopic determination of the constant of aberration. In 1924, he made extensive observations of Mars, using a 18-cm/7-in heliometer – a refracting telescope for measuring the angular diameter of celestial objects – and these observations were later used to obtain the value for solar parallax. In 1925, he obtained and described a long series of spectra of a nova which had appeared in the constellation of Pictor.

While Jones was working at the Cape of Good Hope, he had collected more than 1,200 photographic observations towards the solar parallax programme and as a result, by international agreement in 1931, he was entrusted with dealing with all results of the observations of Eros. Eros was discovered in 1898, a minor planet whose orbit brings it to within 24 million km/15 million mi of Earth. From photographic observations in both hemispheres, Jones derived a figure for the solar parallax that corresponded to a distance of 149,670,000 km/93,005,000 mi and published his results in 1941. By 1967, the distance of the Earth from the Sun had been obtained using direct measurements by radar. The result, a distance of 149,597,890 km/92,960,128 mi, is ten times as accurate (though the difference is only 0.05%) as that which Jones had originally derived using parallax. Any improvements on this value hinges on the limitations of our present knowledge of the velocity of light. It was unfortunate that Jones's painstaking and time-consuming work on determining the mean distance of the Earth from the Sun came just before the use of automatic computing equipment made his methods and instruments redundant.

Besides his work on solar parallax, Jones's principal contributions to astronomy were his work on the motions and secular acceleration of the Sun, Moon and planets and the rotation of the Earth. He proved that fluctuations in the observed longitudes of these celestial bodies are due not to any peculiarities in their motion, but to fluctuations in the angular velocity of rotation of the Earth. He also successfully investigated geophysical phenomena such as the rotation of the Earth, its magnetism and oblateness and he estimated the mass of the Moon. As Astronomer Royal, Jones campaigned for the Royal Observatory to be moved from Greenwich, because the increasing smoke and lights of London hindered observation. It was not until after World War II, however, that a new site was procured at Herstmonceux Castle, Sussex, and the new Observatory was not completed until 1958, three years after Jones's retirement.

Jordan Marie Ennemond Camille 1838–1922 was a French mathematician originally trained as an engineer. An influential teacher, he was himself strongly influenced by the work (one generation earlier) of Evariste Galois (1811–1832) and, accordingly, concentrated on research in topology, analysis and (particularly) group theory. He nevertheless also made further contributions to a wide range of mathematical topics.

Jordan was born in Lyons on 5 January 1838. Completing his education, he entered the Ecole Polytechnique in Paris and studied engineering. He devoted most of his spare time to mathematics, however, and although he qualified as an engineer – and even began to work professionally in that capacity – it was as a mathematician that he joined the staff of the Ecole Polytechnique at the age of 35, in 1873. He also gave lectures at the Collège de France. Eight years later he was elected to the Paris Academy of Sciences. In 1912 he retired. In 1919 – at the age of 81 – he became a Foreign Member of the Royal Society. He died in Paris on 20 January 1922.

Before he had even begun to teach mathematics, Jordan was already acknowledged as the greatest exponent of algebra in his day. Pursuing his interest in the work of the ill-fated Evariste Galois, Jordan systematically developed the theory of finite groups and arrived at the concept of infinite groups. An early result of this was the related concept of composition series, and what is now known as the Jordan–Holder theorem (which deals with the invariance of the system of indices of consecutive groups). He investigated Galois's study of permutation groups, in which Galois had considered the consequences of permutating the roots (solutions) of equations, and linked this with the problem of the solution of polynomial equations.

In 1870 Jordan published his famous *Traité des Substitutions et des equations algébriques* which, for the next three decades, was to be the standard work in group theory. Following this major achievement, Jordan concentrated on his theorems of finiteness. In all he developed three, the first of which dealt with symmetrical groups. The second had its origin in the theory of linear differential equations; Immanuel Fuchs (1833–1902) had completed work on such equations of the second order, Jordan reduced the similar problem for equations of the order n to a problem in group theory. Jordan's last finiteness theorem generalized Charles Hermite's results on the theory of quadratic forms with integral coefficients.

In topology, Jordan developed an entirely new approach to what is now known as homological or combinatorial topology by investigating symmetries in polyhedra from an exclusively combinatorial viewpoint. His other great contribution to this branch of mathematics was his formulation of the proof for the 'decomposition' of a plane into two regions by a simple closed curve.

Jordan's work in analysis was published in 1882 in the *Cours d'analyse de l'école polytechnique*, a spiritedly modern book that had a widespread influence, particularly because of Jordan's insistence on what a really rigorous proof should comprise.

Much of Jordan's later work was concerned with the theory of functions, and he applied the theory of functions of bounded variation to the particular curve that bears his name.

Josephson Brian David 1940– is a British physicist who discovered the Josephson tunnelling effect in superconductivity. He shared the 1973 Nobel Prize for Physics with Leo Esaki (1925–) and Ivar Giaever (1929–) for their work on tunnelling in semiconductors and superconductors.

Josephson was born in Cardiff on 4 January 1940. He was educated at Cardiff High School and at Cambridge University where, as an undergraduate, he published an important paper showing that an application of the Mossbauer effect to verify gravitational changes in the energy of photons had failed to take account of Doppler shifts associated with temperature changes.

Josephson became a Fellow of Trinity College, Cambridge, in 1962 and has remained at Cambridge ever since, apart from the years 1965 and 1966, when he had an assistant professorship at the University of Illinois. In 1974, he became Professor of Physics at Cambridge.

In 1962, Josephson saw some novel connections between solid-state theory and his own experimental problems on superconductivity. He then set out to calculate the current due to quantum mechanical tunnelling across a thin strip of insulator between two superconductors, and the current-voltage characteristics of such junctions are now known as the Josephson effect.

Superconductivity, the property of zero resistivity of some metals below a critical temperature, was discovered by Heike Kamerlingh Onnes (1853–1926) in 1911, and in 1933 it was recognized that superconductors are perfect diamagnets, that is, they totally repel magnetic field lines. Although the effect could be described by phenomenological macroscopic equations, no clear understanding of the microscopic mechanism emerged until the BCS theory developed by John Bardeen (1908–1991) and co-workers in 1957. This explained the absence of electron scattering by electrons of opposite spin pairing up so that, via the response of the atoms in the metal, the electrons had an attractive rather than repulsive influence on one another.

Josephson's contribution was to solve the problem of how such superconductivity pairs would behave when confronted by an insulating barrier, typically of thickness 1–2 nanometres (1–2 billionths of a metre). He recognized that the electron pairs could tunnel through barriers in a manner analogous to the behaviour of single particles in alpha decay. One observable prediction is that an alternating current (AC) occurs in the barrier when a steady external voltage is applied to a system comprising a superconductor, for example, lead, with a thin film of oxide at its surface and a superconducting film evaporated on the oxide. The effect is known as the AC Josephson effect. Conversely radiation, particularly in the far infrared or microwave region, will excite an extra potential across the junction when absorbed.

In addition, when a steady magnetic field is applied across an insulating barrier, a steady current flows. This is the DC Josephson effect. A circuit containing two such barriers in parallel shows quantum interference, analogous to the two-slit experiment in optics, when a magnetic flux is applied to the system.

Josephson's predictions were soon verified by J. M. Rowell at the Bell Telephone Laboratories (the direct current–magnetic field characteristic) and by S. Shapiro, who applied an AC voltage to a superconductor–insulator–superconductor junction and observed resonance coupling to the internal AC Josephson supercurrent.

The Josephson effect has the following applications. The frequency of the AC current is very precisely related to fundamental constants of physics, and this has led to the most accurate method of determining h/e, which in turn establishes a voltage standard. The effect may be used as a generator of radiation, particularly in the microwave and far infrared region. Quantum interference effects are used in squids (superconducting quantum interference devices), which may act as ultra-sensitive magnetometers capable of detecting tiny geophysical anomalies in the Earth's magnetic field or even the anomaly caused by the presence of a submarine. In addition, the fast switching properties of such devices can be exploited in logic elements or binary memories, and this has great potential for future generations of computers. Josephson's discovery may thus have important consequences in the development of artificial intelligence.

Joule James Prescott 1818–1889 was a British physicist who verified the principle of conservation of energy by making the first accurate determination of the mechanical equivalent of heat. He also discovered Joule's law, which defines the relation between heat and electricity, and with Lord Kelvin (1824–1907) the Joule–Thomson effect. In recognition of Joule's pioneering work on energy, the SI unit of energy is named the joule.

Joule was born at Salford on 24 December 1818 into a wealthy brewing family. He and his brother were educated at home between 1833 and 1837 in elementary mathematics, natural philosophy and chemistry, partly by John Dalton (1766–1844). Joule was a delicate child and very shy, and apart from his early education he was entirely self-taught in science. He does not seem to have played any part in the family brewing business, although some of his first experiments were done in a laboratory at the brewery.

Joule had great dexterity as an experimenter, and was able to measure temperatures very exactly indeed. At first, other scientists could not credit such accuracy and were disinclined to believe the theories that Joule developed to explain his results. The encouragement of Lord Kelvin from 1847 changed these attitudes, however, and Kelvin subsequently used Joule's practical ability to great advantage. By 1850, Joule was highly thought of among scientists and became a Fellow of the Royal Society. He was awarded the Society's Copley Medal in 1866 and was President of the British Association for the Advancement of Science in 1872 and again in 1887. Joule's own wealth was able to fund his scientific career, and he never took an academic post. His funds eventually ran out, however. He was awarded a pension in 1878 by Queen Victoria, but by that time his mental powers were going. He suffered a long illness and died in Sale, Cheshire, on 11 October 1889.

Joule realized the importance of accurate measurement very early on and exact quantitative data became his hallmark. His most active research period was between 1837 and 1847 and led to the establishment of the principle of conservation of energy and the equivalence of heat and other forms of energy. In a long series of experiments, he studied the quantitative relationship between electrical, mechanical and chemical effects and heat, and in 1843 he was able to announce his determination of the amount of work required to produce a unit of heat. This is called the mechanical equivalent of heat (currently accepted value 4.1868 joules per calorie).

Joule's first experiments related the chemical and electrical energy expended to the heat produced in metallic conductors and voltaic and electrolytic cells. These results were published between 1840 and 1843. He proved the relationship, known as Joule's law, that the heat produced in a conductor of resistance R by a current I is proportional to I^2R per second. He went on to discuss the relationship between heat and mechanical power in 1843. Joule first measured the rise in temperature and the current and the mechanical work involved when a small electromagnet is rotated in water between the poles of another magnet, his training for these experiments having been provided by early research with William Sturgeon (1783–1850), a pioneer of electromagnetism. Joule then checked the rise in temperature by a more accurate experiment, forcing water through capillary tubes. The third method depended on the compression of air and the fourth produced heat from friction in water using paddles which rotated under the action of a falling weight. This has become the best known method for the determination of the mechanical equivalent. Joule showed that the results obtained using different liquids (water, mercury and sperm oil) were the same. In the case of water, 772 ft lb of work produced a rise of 1°F in 472 cu cm/29 cu in of water. This value was universally accepted as the mechanical equivalent of heat. It now has no validity, however, because as both heat and work are considered to be forms of energy, they are measured in the same units – in joules. A joule is basically defined as the energy expended when a force of 1 newton moves 1 metre.

The great value of Joule's work in the establishment of the conservation of energy lay in the variety and completeness of his experimental evidence. He showed that the same relationship held in all cases which could be examined experimentally and that the ratio of equivalence of the different forms of energy did not depend on how one form was converted into another or on the materials involved. The principle that Joule had established is in fact the first law of thermodynamics – that energy cannot be created nor destroyed but only transformed.

Because he had not received any formal mathematical training, Joule was unable to keep up with the new science of thermodynamics to which he had made such a fundamental and important contribution. However, the presentation of his final work on the mechanical equivalent of heat in 1847 attracted great interest and support from William Thomson, then only 22 and later to become Lord Kelvin. Much of Joule's later work was carried out with him, for Kelvin had need of Joule's experimental prowess to put his ideas on thermodynamics into practice. This led in 1852 to the discovery of the Joule–Thomson effect, which produces cooling in a gas when the gas expands freely. The effect is caused by the conversion of heat into work done by the molecules in overcoming attractive forces between them as they move apart. It was to prove vital to techniques in the liquefaction of gases and low-temperature physics.

Joule lives on in the use of his name to measure energy, supplanting earlier units such as the erg and calorie. It is an appropriate reflection of his great experimental ability and his tenacity in establishing a basic law of science.

Joy James Harrison 1882–1973 was a US astronomer, most famous for his work on stellar distances, the radial motions of stars and variable stars.

Joy was born on 23 September 1882 in Greenville, Illinois, the son of a merchant of New England ancestry. He was educated locally and then attended Greenville College, obtaining a PhD in 1903. From 1904 to 1914 he worked at the American University of Beirut, Lebanon, first as a teacher and then as Professor of Astronomy and Director of the Observatory. He returned to the United States in 1914, worked at the Yerkes Observatory as an instructor for a year, and then joined the staff of Mount Wilson Observatory, where he remained until 1952. He was Vice President of the American Astronomical Society in 1946 and President in 1949.

When Joy joined the staff of the Yerkes Observatory in 1914, he took part in a programme of measuring stellar distances, using the 1-m/40-in Yerkes refractor, by direct photography and parallax measurements, an extremely tedious and out-of-date method. After 1916, he began to make spectroscopic observations at Mount Wilson Observatory and he was subsequently invited to take part in their research programme to obtain stellar distances.

The new method of finding the distance to stars was to compare the apparent magnitude of the star with its absolute magnitude. Joy continued with this programme, in collaboration with his colleagues Walter Adams (1876–1956) and Milton Humason (1891–1972), for more than 20 years and their results enabled them to ascertain the spectral type, absolute magnitude, and stellar distance of more than 5,000 stars.

Joy and his colleagues also studied the Doppler displacement of the spectral lines of some stars to determine their radial velocities. By noting the variations in radial velocity, Joy and his team were able to show that many stars are spectroscopic binary stars and that their period and orbit could therefore be elucidated. From their observations of eclipsing binary stars they deduced the absolute dimensions, masses and orbital elements of some specific stars within eclipsing binary systems. Using radial velocity data of 130 Cepheid variable stars, Joy determined the distance and direction of the centre of the Galaxy and calculated the rotation period for bodies moving in circular orbits at the distance of the Sun, with a view to calculating the rotation period of the Galaxy.

Joy also spent many years of his life observing variable stars and classifying them according to their characteristic spectra. While studying the long-period variable star, Mira Ceti, Joy observed the spectrum of a small hot companion object. It was later named Mira B and shown to be a white dwarf; it can be observed visually today. Joy later became interested in the parts of the Galaxy where dark, absorbing clouds of gas and dust exist, and by carefully observing these areas he found examples of a particular kind of variable star, called a T-Tauri star, which is strongly associated with these areas. T-Tauri stars have a wide range of spectral types combined with characteristically low magnitudes. Joy showed that these characteristics indicated that they were very young stars in an early stage of their evolutionary history.

Jung Carl Gustav 1875–1961 was a Swiss psychologist who founded analytical psychology as a deliberate alternative to the psychoanalysis of Sigmund Freud.

Jung was born in Kesswil, near Basel, on 26 July 1875, the son of a Protestant clergyman. Despite an early interest in archaeology – and a strong family background in religion and theology – he went to Basel University in 1895 to study medicine, graduating in 1900. He then attended Zurich University and obtained his MD in 1902, at the same time turning to psychiatry. For the next seven years he worked at the Burghölzi Psychiatric Clinic in Zurich under Eugen Bleuler, an expert on schizophrenia; also from 1905 to 1913 he lectured in psychiatry at the university. In 1907 he met Freud and for five years became his chief disciple, accepting the appointment as the first President of the International Psycho-Analytical Association on its foundation in 1911. But in 1913 following publication of his *Wandlungen und Symbole de Libido* (1912), translated as *The Psychology of the Unconscious* (1916), he broke with Freud, resigned from the Association, and set up his own practice in Zurich. In 1933, he became Professor of Psychology at the Zurich Federal Institute of Technology, a post he held for eight years. In 1943, he resigned almost immediately after being appointed Professor of Medical Psychology at Basel University (when he was 68 years old) because his health began to fail. But he continued to practise until he was over 80, and he died in Küsnacht, near Zurich, on 6 June 1961.

While Jung was at the Psychiatric Clinic in the early 1900s he devised the word-association test as a psychoanalytical technique for penetrating a subject's unconscious mind. He also developed his theory concerning emotional, partly repressed ideas which he termed 'complexes'. The chief reason for his split with Freud – like Alfred Adler's before him – was Freud's emphasis on infantile sexuality. Jung introduced the alternative idea of 'collective unconscious' which is made up of many archetypes or 'congenital conditions of intuition'.

Each person is born with access to these archetypes, which Jung tried to identify by studying cultures such as those of the North American Pueblo Indians and primitive peoples of Africa and India. Mythology and folklore, alchemical writings, religious texts and even dreams were analysed for archetypes.

Jung also studied personality and its importance in human behaviour and in 1921 introduced the concept of 'introverts' and 'extroverts' in his book *Psychologische Typen*. This work also contained his theory that the mind has four basic functions: thinking, feeling, sensations and intuition. Any particular person's personality can be ascribed to the predominance of one of these functions.

Just Ernest Everett 1833–1941 was a US biologist who became internationally known for his pioneering research into fertilization, cell physiology, and experimental embryology.

Just was born on 14 August 1883 in Charlston, South Carolina. He was educated at Kimball Union Academy, Vermont, graduating in 1903 with honours and went on to Dartmouth College, New Hampshire, to study English and Classics, later switching to biology; he received his AB degree in 1907 as the only *magna cum laude* in his class. Just then went as a teacher to Howard University where he remained until 1929. In 1912, he was made Professor and Head of the Department of Zoology, and in 1916 he took leave of absence to complete his PhD in zoology at the University of Chicago. From 1909 until 1929 Just spent his summers conducting research at the Marine Biological Laboratories at Woods Hole, Massachusetts. In 1929, frustrated with the limitations imposed on him by his race, Just went to Europe to conduct research in German laboratories and at French and Italian marine stations. With the German occupation of France, Just returned to his former post at

Howard in 1940, where he died of pancreatic cancer on 27 October 1941. He was married twice and had two daughters and a son.

Just is chiefly known for his research into fertilization and experimental parthenogenesis in marine eggs, on which he published over 60 papers. Through his research at Woods Hole he became the leading authority on the embryological resources of the marine group of animals. His focus of attention was the cell and in particular the ectoplasm which, contrary to popular belief, he stated was just as important as the nucleus, and was primarily responsible for the individuality and development of the cell. He was a co-author of *General Cytology* (1924) and whilst in Europe he published *Biology of the Cell Surface* (1939) and *Basic Methods for Experiments in Eggs of Marine Animals* (1939). He also worked on the philosophical implications of his research, believing that there was a need for the integration of biological science and social philosophy. Just was an associate editor of the *Biological Bulletin*, *Physiological Zoology*, *Protoplasma* and the *Journal of Morphology*. For his work he was awarded the first Spingarn medal (1915), he was vice president of the American Society of Zoologists (1930) and was elected a member of the Washington Academy of Scientists in 1936.

K

Kant Immanuel 1724–1804 is more generally regarded as a philosopher than an astronomer, but his theoretical work in astronomy inspired cosmological theories and many of his conjectures have been confirmed by observational evidence.

Kant was born in Königsberg (now Kaliningrad) on 27 April 1724 and he lived there or nearby for all of his life. His grandfather was an immigrant from Scotland, called Cant, whose name was Germanized. Kant's father was a saddlemaker of modest means and conventional religious beliefs and his mother was a member of a pietist Protestant sect.

At the age of ten Kant entered the Collegium Fridericianum, where he was expected to study theology. Instead, he concentrated on the classics and then went to the University of Königsberg to study mathematics and physics. Six years after he entered university, his father died and he was forced to work as a private tutor to three of Königsberg's leading families in order to support his brothers and sisters. He did not return to the University until 1755, when he gained his PhD and became an unpaid lecturer. Fifteen years later, Kant was offered the Chair in Logic and Metaphysics, a post he held for 27 years. He died in Königsberg on 12 February 1804.

Kant's most important work was a philosophical text, the *Critique of Pure Reason*, published in 1781 when he was in his late fifties. In his early pre-*Critique* work, however, he dealt with various scientific subjects, including a consideration of the origin and natural history of the universe. Kant was not an experimental scientist in any sense and he has sometimes been accused of being merely an 'armchair scientist'. Yet he was not ignorant of the scientific advances of his day and his theoretical works on astronomy were deeply affected by his studies of Newtonian and Leibnizian physics. The hypotheses Kant put forward inspired other theories, notably those propounded by Carl von Weizsäcker (1921–) in the Gifford Lectures of 1959–60 and by Gerard Kuiper (1905–1973) of the Yerkes Observatory in the United States. They also coincided with another more empirical work on the nature of the universe that later became known as the Kant–Laplacian theory.

Besides being indebted to Newton, Kant's thoughts on astronomy and cosmology were also influenced by a work entitled *Original Theory or a New Hypothesis of the Universe* published by Thomas Wright in 1758. Kant's pre-*Critique Universal Natural History and Theory of the Heavens* was based on the theories of Newton and Wright. It also anticipated facts of astronomy that were later confirmed only by advanced observational techniques. Kant proposed that the Solar System was part of a system of stars constituting a galaxy and that there were many such galaxies making up the whole universe. His conjecture that nebulous stars were also separate galaxies similar to our own was confirmed only in the twentieth century.

Although Kant accepted the religious claim that God created the world, in the *Natural History* he concerned himself with explaining how the universe evolved once it had been created. In contrast with the idea that planets were formed as a result of tidal forces set off in the Sun when some large celestial body passed close by, he argued, in the theory that came to be associated with one put forward by Laplace, that planetary bodies in the Solar System were formed by the condensation of nebulous, diffuse primordial matter that was previously widely distributed in space. He used the Newtonian theory of the attraction and repulsion of materials to account for the fact that the Solar System, like other celestial systems, is flattened out like a disc. He also supposed that action at a distance was possible, explaining the gravitational attraction that held the moons and planets in their orbits. Moreover, Kant argued that the universe continues to develop: its parts come together and disperse with the condensation and diffusion of matter in the infinity of time and space.

Kant's most famous work, *Critique of Pure Reason*, inaugurated the 'Copernican Revolution' in philosophy, turning attention to the mind's role in constructing our knowledge of the objective world. He examined the legitimacy and objectivity of cognition and science, an investigation which led him to doubt a number of the presuppositions necessary to the theories put forward in *Natural History*. For example, one of the self-contradictory conclusions considered in the *Critique* showed that we cannot, with certainty, make assertions concerning the finite or infinite nature of space and time. But even with this qualification it is amazing that, based only on Newtonian principles and information gleaned from reading an abstract of the book by Wright, Kant's hypotheses coincide so well with later cosmological theories.

Kapitza Pyotr Leonidovich 1894–1984 was a Russian physicist best known for his work on the superfluidity of liquid helium. He also achieved the first high-intensity magnetic fields. For his achievements in low-temperature physics, Kapitza was awarded the 1978 Nobel Prize for Physics. He shared the prize with two Americans, Arno Penzias (1933–   ) and Robert Wilson (1936–   ), who discovered the cosmic microwave background radiation, predicted by George Gamow (1904–1968).

Kapitza was born in Kronstadt on 8 July 1894. He graduated in 1919 from Petrograd Polytechnical Institute and then travelled to Britain and worked with Ernest Rutherford (1871–1937) in Cambridge, becoming Deputy Director of Magnetic Research at the Cavendish Laboratory in 1924. In 1930, Kapitza became Director of the Mond Laboratory at Cambridge, which had been built for him. Four years later he went to the USSR for a professional meeting as he had done before, but this time he did not return. His passport was seized and he was held on Stalin's orders. In 1936, he was made Director of the S. I. Vavilov Institute of Physical Problems of the Soviet Academy of Sciences in Moscow. The Mond Laboratory was sold to the Soviet government at cost and transported to the Institute for Kapitza's use. In 1946, he refused to work on the development of nuclear weapons and was put under house arrest until after Stalin's death in 1953. He was restored as Director of the Institute in 1955.

For his graduate work, Kapitza went to Rutherford's laboratory and pioneered the production of strong (though temporary) magnetic fields. In 1924, he designed apparatus that achieved magnetic fields of 500,000 gauss, an intensity that was not surpassed for more than 30 years. In his earliest experiments, Kapitza, using especially constructed accumulator batteries and switch gear, passed currents of up to 8,000 amperes through a coil for short intervals of time. In a coil of 1 mm/0.039 in internal diameter, fields of the order of 500,000 gauss/50 tesla for 0.003 sec were obtained in this way. By 1927, he had produced stronger currents by short-circuiting an alternating current generator of special construction, and a field of 320,000 gauss/32 tesla in a volume of 2 cu cm/0.12 cu in was obtained for 0.01 sec.

Kapitza's most renowned work is in connection with liquid helium, which was first produced by Heike Kamerlingh Onnes (1853–1926) in 1908. Kapitza was one of the first to study the unusual properties of helium II – the form of liquid helium that exists below 2.2K (–270.8°C/–455.4°F). Helium II conducts heat far more rapidly than copper, which is the best conductor at ordinary temperatures, and Kapitza showed that this is because it flows so easily. This property of helium is known as superfluidity, and it has far

less viscosity than any other liquid or gas. In 1937, Kapitza persuaded the Soviet physicist Lev Landau (1908–1968) to move to Moscow to head the theory division of his institute. Kapitza published his findings on the superfluidity of helium II in 1941. Landau then continued the work on superfluidity, resulting in the award of the 1962 Nobel Prize for Physics to him for his theory of superfluidity.

In 1939, Kapitza built apparatus for producing large quantities of liquid oxygen. This work and its application was of great importance, particularly to Soviet steel production. He invented a turbine for producing liquid air cheaply in large quantities. His technique for liquefying helium by adiabatic expansion eliminated the presence of liquid hydrogen characteristic of previous methods.

In the 1950s, Kapitza turned his attention partly to ball lightning, a puzzling phenomenon in which high-energy plasma maintains itself for a much longer period than seems likely. His analysis involves the formation of standing waves, and this research turned his interests to controlled thermonuclear fusion, upon which he published a paper in 1969.

Kaplan Viktor 1876–1934 was an Austrian engineer, famous for inventing the turbine that bears his name.

Kaplan was born on 27 November 1876 at Murz, Austria. He was educated at the Realschule in Vienna and then at the Technische Hochschule, where he studied machine construction and gained his engineer's diploma in 1900. After a year's voluntary service in the navy, he became a constructor of diesel engines for the firm of Ganz and Company of Leobersdorf, in Austria. In 1903, he was appointed to the Chair of Kinematics, Theoretical Machine Studies and Machine Construction at the Technische Hochschule in Brunn. In 1918, he became Professor of Water Turbine Construction. He became debilitated and retired early in 1931. He died on 23 August 1934 at Unterach on the Attersee, Austria.

Kaplan published his first paper on turbines in 1908, writing about the Francis turbine and basing it on work that he had done for his doctorate at the Technische Hochschule in Vienna. In connection with this paper, he set up a propeller turbine for the lowest possible fall of water. In 1913 the first prototype of the turbine was completed. The Kaplan turbine was patented in 1920 as a water turbine with adjustable rotor-blades the runner blades can be varied to match the correct angle for a given flow of water. (The Francis-type turbine corresponds to a centrifugal pump, whereas the Kaplan-type turbine resembles a propeller pump.)

The Kaplan turbine in its traditional form has a vertical shaft, although one power-generator design uses a horizontal shaft and an alternator mounted in a 'nacelle' (metal shell) in the water flow. This is particularly suitable for lower-output machines operating on very low heads of water, in which an almost straight water passage is possible. This type of turbine was designed for a French project on the river Rance estuary on the Gulf of St Malo, Brittany, France. Construction was begun in January 1961, on the world's first large-scale tidal plant, and completed late in

1967. It was designed to be made up of 24 hydroelectric power units. Each consisted of an ogive-shaped shell of metal containing an alternator and a Kaplan turbine. These units were installed in apertures in a dam through which the water flows. Each turbine acts both as a turbine and a pump, in both directions of tidal flow and in both directions of rotation, with each hydroelectric power unit generating 20,000 kW.

The physical size and the operating head for which a Kaplan turbine can be used have increased considerably since 1920 when it was first patented. Runner diameters of 9.1 m/30 ft have been used. British manufacturers have developed designs operating at heads of 58 m/190 ft, but it is likely that further progress in increased heads will be limited due to the advent of the Deriaz turbine.

Kapteyn Jacobus Cornelius 1851–1922 was a Dutch astronomer who analysed the structure of the universe by studying the distribution of stars using photographic techniques. To achieve more accurate star counts in selected sample areas of the sky he introduced the technique of statistical astronomy. He also encouraged fruitful international collaboration among astronomers.

Kapteyn was born in Barneveld, the Netherlands, on 19 January 1851. He was born into a large and talented family and displayed great academic abilities early in his life. He began his studies at the University of Utrecht in 1868 when he was 17, although he had already satisfied the University's entrance requirements a year earlier. Kapteyn concentrated his studies on mathematics and physics and earned his doctorate for a thesis on vibration.

Kapteyn's career in astronomy began when he was employed by the astronomical observatory at Leiden in 1875. Three years later he was appointed as the first Professor of Astronomy and Theoretical Mechanics at the University of Groningen. He was a member of the French Academy of Science and a Fellow of the Royal Society. He was active in the organizational work which eventually led to the establishment of the International Union of Astronomers. He retired from his post at Groningen in 1921, but continued to work and publish his results. He died in Amsterdam on 18 June 1922.

Initially there was a lack of appropriate facilities at the University of Groningen, but this did not deter Kapteyn. He proposed a cooperative arrangement with Sir David Gill at Cape Town Observatory in South Africa, whereby photographs of the stars in the southern hemisphere were analysed at Groningen. This early collaborative work with Gill resulted in the publication during the years 1896 to 1900 of the *Cape Photographic Durchmusterung*. It was welcomed as an essential complement to Argelander's monumental *Bonner Durchmusterung* (1859–62), because it presented accurate data on the brightness and positions of nearly 455,000 stars in the less-studied southern hemisphere. The catalogue included all stars of magnitude greater than 10 which lay within 19° of the South Pole, and it encouraged Edward Pickering's team of astronomers from the Harvard Observatory, who set up a research station in Chile, to expand their observations of stars in the southern skies.

Kapteyn's next project was to improve the technique of trigonometric parallax so that he could obtain data of a higher quality on the proper motions of stars. Although proper motion had been observed for many years, its explanation was unknown. It had been assumed by most astronomers that there was no pattern in the proper motion of stars, that they resembled the random Brownian motions of gas molecules. In 1904, Kapteyn reported that there was indeed a pattern. He found that stars could be divided into two streams, moving in nearly opposite directions. The notion that there was this element of order in the universe was to have a considerable impact, although its significance was not realized at the time.

Arthur Eddington and Karl Schwarzschild extended Kapteyn's work and confirmed his results, but they missed the underlying importance of their observations. It was not until 1928 that Jan Oort and Bertil Lindblad, greatly aided by recent discoveries on the characteristics of extragalactic nebulae, proposed that Kapteyn's results could be readily understood if they were considered in the contest of a rotating spiral galaxy. Kapteyn's data had been the first evidence of the rotation of our Galaxy (which at that time was not understood to be in any way distinct from the rest of the universe), although it was not recognized as such at the time.

F. W. Herschel had investigated the structure of stellar systems by counting the number of stars in different directions. Kapteyn decided to repeat this analysis with the aid of modern telescopes. He selected 206 specific stellar zones, and in 1906 he began a vast international programme to study them. His aim was to ascertain the magnitudes of all the stars within these zones, as well as to collect data on the spectral type, radial velocity, proper motion of the stars and other astronomical parameters. This enormous project was the first coordinated statistical analysis in astronomy and it involved the cooperation of over 40 different observatories. The project was extended by Edward Pickering to include 46 additional stellar zones of particular interest.

The analysis of data collected for the programme was a colossal task and took many years to complete. In 1922, only weeks before his death, Kapteyn published his conclusions based on the analysis he had completed. His overall result was strikingly similar to that obtained by Herschel. He envisaged an island universe, a lens-shaped aggregation of stars that were densely packed at the centre and thinned away as empty space outside the Galaxy was reached. The Sun lay only about 2,000 light years from the centre of a structure which spanned some 40,000 light years along its main axis.

A fundamental error in the construction of 'Kapteyn's universe' was his neglect of the possibility that interstellar material would cause stellar light from a distance to be dimmed. This placed a limit on the maximum distance from which data could be obtained. Kapteyn had, in fact, considered this possibility, and he tried unsuccessfully to detect its effect. The effect Kapteyn had sought was observed by Robert Trumpler nearly ten years after Kapteyn's death. Harlow Shapley's model for the structure of our Galaxy soon replaced that proposed by Kapteyn. Today the Galaxy is thought to span some 100,000 light years and to be shaped like a pancake which thickens considerably towards the centre. Our Solar System lies quite a long way out along one of the spiral arms, 30,000 light years from the centre of the Galaxy – which lies in the direction of the constellation Sagittarius.

Karrer Paul 1889–1971 was a Swiss organic chemist, famous for his work on vitamins and vegetable dyestuffs. He determined the structural formulae and carried out syntheses of various vitamins, in recognition of which achievement he shared the 1937 Nobel Prize for Chemistry with Walter Haworth.

Karrer was born in Moscow, of Swiss parents, on 21 April 1889, and when he was three years old his family moved to Switzerland. He studied at the University of Zurich under Alfred Werner, gaining his doctorate in 1911. After a year at the Zurich Chemical Institute he went to Frankfurt to work with Paul Ehrlich at the Georg Speyer Haus. In 1918, he returned to Zurich to be Professor of Chemistry and a year later succeeded Werner as Director of the Institute. Karrer remained there until he retired in 1959. He died in Zurich on 18 June 1971.

Karrer's early work concerned vitamin A and its chief precursor, carotene. Wackenroder had first isolated carotene (from carrots) in 1831 and in 1907 Richard Willstätter determined its molecular formula to be $C_{40}H_{56}$. Karrer worked out its correct constitutional formula in 1930 (although he was not to achieve a total synthesis until 1950). He showed in 1931 that vitamin A (molecular formula $C_{20}H_{30}O$) is related to the carotenoids, the substances that give the yellow, orange or red colour to foodstuffs such as sweet potatoes, egg yolk, carrots and tomatoes and to inedible substances such as lobster shells and human skin. There are in fact two A vitamins, compounds known chemically as diterpenoids; vitamin A_1 influences growth in animals and its deficiency leads to night blindness and a hardening of the cornea. Karrer proved that there are several isomers of carotene, and that vitamin A_1 is equivalent to half a molecule of its precursor β-carotene.

Karrer later confirmed Albert Szent-Györgyi's constitution of vitamin C (ascorbic acid, so-called by Haworth who synthesized it and shared Karrer's Nobel prize). In 1935, he solved the structure of vitamin B_2 (riboflavin), the watersoluble thermostable vitamin in the B complex that is necessary for growth and health. It occurs in green vegetables, yeast, meat and milk (from which it was first isolated and hence is also known as lactoflavin). Karrer made extensive studies of the chemistry of the flavins.

He also investigated vitamin E (tocopherol), the group of closely related compounds which act as anti-sterility factors. In 1938, he solved the structure of α-tocopherol, the most biologically active component, obtained from wheatgerm oil.

For their time, these syntheses were remarkable, being the most advanced so far undertaken. They led to a better understanding of vitamins in metabolism, and acted as a spur to the work of others. In 1930, Karrer published an organic chemistry textbook, *Lehrbuch der Organischen Chemie*, which became the standard work for many years, being reprinted in many editions in several languages during the 1940s and 1950s.

Katz Bernhard 1911–   is a German-born British physiologist who is renowned for his research into the physiology of the nervous system. He has received many honours for his work, including a knighthood in 1969 and the 1970 Nobel Prize for Physiology or Medicine (jointly with Ulf von Euler and Julius Axelrod).

Katz was born on 26 March 1911 in Leipzig and studied medicine at the university there. After graduation in 1934 he did postgraduate work at University College, London, from which he obtained his PhD in 1938 and his doctor of science degree in 1943. He was a Beit Memorial Research Fellow at the Sydney Hospital from 1939 to 1942. He then served in the Royal Australian Air Force until the end of World War II, after which he returned to England. He spent the rest of his academic career at University College, London – as Assistant Director of Research at the Biophysics Research Unit and Henry Head Research Fellow from 1946 to 1950, Reader in Physiology from 1950 to 1951, and Professor and Head of Biophysics from 1952 until his retirement in 1978.

During the 1950s, Katz found that minute amounts of acetylcholine (previously demonstrated to be a neurotransmitter by Henry Dale and Otto Loewi) were randomly released by nerve endings at the neuromuscular junction, giving rise to very small electrical potentials at the end plate; he also found that the size of the potential was always a multiple of a certain minimum value. These findings led him to suggest that acetylcholine was released in discrete 'packets' (analogous to quanta) of a few thousand molecules each, and that these packets were released relatively infrequently while a nerve was at rest but very rapidly when an impulse arrived at the neuromuscular junction. Electron microscopy later revealed small vesicles in the nerve endings and these are thought to be the containers of the packets of acetylcholine suggested by Katz.

Kay John 1704–*c*.1780 was an English inventor of improved textile machinery, the most important of which was the flying shuttle of 1733.

Kay was born in July 1704 at Walmersley, near Bury in Lancashire. Little is known of his early life, although he is thought to have been educated in France. It is thought that on finishing his schooling he was put in charge of a wool factory owned by his father in Colchester, Essex. By 1730 he was established in his home town of Bury as a reedmaker (for looms). In that year he was granted a patent for an 'engine' for twisting and carding mohair, and for twining and dressing thread. At about the same time he improved the reeds for looms by manufacturing the 'darts' of thin polished metal, instead of cane, which were more durable and better suited to weaving finer fabrics.

In 1733, Kay patented his flying shuttle, which was probably the most important improvement that had been made to the loom. Up to that time the shuttle had been passed by hand from side to side through alternate warp threads. In weaving broadcloth two workers had to be employed to throw the shuttle from one end to the other.

The weft was closed up after each 'pick' or throw of the shuttle by a 'layer' extending across the piece being woven. Kay added to the layer a grooved guide, called a 'race-board', in which the shuttle was rapidly thrown from side to side by means of a 'picker' or shuttle driver. One hand only was required, the other being employed in beating or closing up the weft. Kay's improvement allowed the shuttle to work with such speed that it became known as the flying shuttle. The amount of work a weaver could do was more than doubled, and the quality of the cloth was also improved. A powerful stimulus was thus given to improve spinning techniques, notably the inventions of James Hargreaves.

The patent of 1733 also included a batting machine for removing dust from wool by beating it with sticks. Kay's next patent (granted in 1738) was for a windmill for working pumps and for an improved chain-pump, but neither of these proved to be of any practical importance. In this last patent Kay described himself as an engineer.

Kay's biographer, Woodcroft, states that he moved to Leeds in 1738. The new shuttle was widely adopted by the woollen manufacturers of Yorkshire but since they were unwilling to pay royalties an association called The Shuttle Club was formed to defray the costs of legal proceedings for infringement of the patent. Kay found himself involved in numerous law suits and, although he was successful in the courts, the expenses of prosecuting these claims nearly ruined him.

In 1745, Kay was again in Bury and in that year he obtained a patent (in conjuction with Joseph Stell of Keighley) for a small loom worked by mechanical power instead of manual labour. This attempt at a power loom does not seem to have been successful, probably because of his financial difficulties and the hostility of the workers.

In 1753, a mob broke into Kay's house at Bury and destroyed everything they could find; Kay himself barely escaped with his life. Not surprisingly he lost his enthusiasm for working in England under such trying conditions, and left for France where he remained for the next ten years, introducing the flying shuttle there with fair success.

In 1765, Kay invented a machine for making the cardcloth used in carding wool and cotton. The last years of his life are poorly documented. He is said to have been ruined by suits to protect his patents and is believed to have returned to France in 1774 and to have received a pension from the French court in 1778. After this nothing more is known of him.

Keeler James Edward 1857–1900 was a US astrophysicist noted for his work on the rings of Saturn and on the abun-

dance and structure of nebulae.

Keeler was born in La Salle, Illinois, on 10 September 1857. He did not attend school between the ages of 12 and 20, but during these years he developed a keen interest in astronomy. He made a variety of astronomical instruments and spent long hours studying the Solar System. A benefactor enabled him to study at Johns Hopkins University, where he earned his bachelor's degree in 1881. While a student he participated in an expedition to Colorado to study a solar eclipse.

In 1881, he began his career as assistant to Samuel Langley (1834–1906), Director of the Allegheny Observatory, and he took part in that year's expedition to the Rocky Mountains to measure solar infrared radiation.

In 1883, Keeler went to Germany to study at the Universities of Heidelberg and Berlin. A year later he returned to the Allegheny Observatory and in 1886 he went to Mount Witney, the future site of the Lick Observatory. He was appointed Astronomer at the Lick Observatory upon its completion in 1888. Keeler became Professor of Astrophysics and Director of the Allegheny Observatory in 1891, but returned to the Lick Observatory as Director in 1898. He was elected Fellow of the Royal Astronomical Society of London in 1898 and made a member of the National Academy of Sciences in 1900. He died suddenly in San Francisco on 12 August 1900.

Keeler's earliest work was a spectroscopic demonstration of the similarity between the Orion nebula and stars. In 1888, using the 91-cm/35-in refracting telescope at Lick Observatory and an improved spectroscopic grating, he demonstrated that nebulae resembled stars in their pattern of movement. He also studied the planet Mars, but was unable to confirm Schiaparelli's observation that the Martian surface was etched with a pattern of 'canals'. In 1895, Keeler made a spectroscopic study of Saturn and its rings, in order to examine the planet's period of rotation. He found that the rings did not rotate at a uniform rate, thus proving for the first time that they could not be solid and confirming James Clerk Maxwell's theory that the rings consist of meteoritic particles.

After 1898, Keeler devoted himself to a study of all the nebulae that William Herschel had catalogued a hundred years earlier. He succeeded in photographing half of them, and in the course of his work he discovered many thousands of new nebulae and showed their close relationship to stars. Keeler was not only a keen and successful observer of astronomical phenomena; he was also skilled at designing and constructing instruments. These included modifications to the Crossley reflecting telescope and a spectrograph in which spectral lines were recorded with the aid of a camera.

Kekulé Friedrich August von Stradonitz 1829–1896 was a German organic chemist who founded structural organic chemistry and is best known for his 'ring' formula for benzene.

Kekulé was born in Darmstadt on 7 September 1829, a descendant of a Bohemian noble family from Stradonic, near Prague. At first he studied architecture at Giessen, but came under the influence of Justus von Liebig and changed to chemistry. He left Giessen in 1851 and went to Paris, where he gained his doctorate a year later. After working in Switzerland for a while, he went to London in 1854 where he assisted John Stenhouse (1809–1890) and met many leading chemists of the day. When he returned to Germany in the following year he opened a small private laboratory at Heidelberg and became an unpaid lecturer at the university there under Johann von Baeyer. In 1858, he became a professor at Ghent; in 1865 took over the Chair at Bonn vacated by August Hofmann, and remained there until he died. While at Ghent he got married, but his wife died in childbirth. He had three more children by his second marriage of 1876, the same year in which he suffered an attack of measles which left him in poor health for the rest of his life. He was raised to the nobility by William II of Prussia in 1895 and took the name Kekulé von Stradonitz. He died in Bonn on 13 July 1896.

In 1858, Kekulé published a paper in which, after giving reasons why carbon should be regarded as a four-valent element, he set out the essential features of his theory of the linking of atoms. (A similar paper was presented three months later to the French Academy by the Scottish chemist Archibald Couper, but went largely unnoticed.) He postulated that carbon atoms can combine with each other to form chains, and that radicals are not necessarily indestructible, being capable of persisting through one set of reactions only to disintegrate in others. He explained that compounds behave differently under different sets of conditions, although he continued to subscribe to the idea of 'types', which he thought were not inflexible but served to emphasize one or more of their properties. He pictured atoms grouping themselves in space and linking with each other to form compounds.

In 1865, Kekulé announced his theory of the structure of benzene, which he envisaged as a hexagonal ring of six interconnected carbon atoms. To make the structure compatible with carbon's valency of four, he postulated alternate single and double bonds in the same ring:.

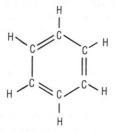

In a dissertation of 1867 outlining his views on molecular diagrams and models, he proposed the tetrahedral carbon atom (that the four valence bonds from a saturated carbon atom are directed towards the corners of a regular tetrahedron), which was to become the cornerstone of modern structural organic chemistry. Kekulé cautiously made it clear that he did not intend all the possible physical implications of his tetrahedral model to be accepted literally. It

later became clear, however, that this model fulfilled strictly chemical necessities that he had not foreseen. In 1890 von Baeyer said that Kekulé's molecular models were 'even cleverer than their inventor' and modern analytical methods have shown that they are true indicators of how the atoms are arranged in space.

There were two fundamental objections to Kekulé's original ring structure for benzene: it does not behave chemically like a double-bonded substance, and there are not as many di-substituted isomers as the formula would predict. Kekulé acknowledged these shortcomings and in 1872 introduced the idea of oscillation or resonance between two isomeric structures:

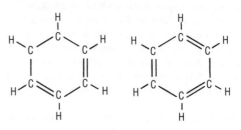

His assistant W. Körner (1839–1925) developed in great detail the implications of these structures in the chemistry of benzene and its compounds. The modern view, worked out by Linus Pauling in 1933 by applying quantum theory to Kekulé's resonant structures, postulates a hybrid (mesomeric) state for the benzene molecule with a carbon–carbon bond length in the ring intermediate between the normal single- and double-bond values.

Kelly William 1811–1888 was a US metallurgist, arguably the original inventor of the 'air-boiling process' for making steel, known universally as the Bessemer process. Kelly remained in relative obscurity in spite of his invention, whereas his rival (Bessemer) received $10 million in the USA and was knighted.

Kelly was born in Pittsburgh, Pennsylvania, on 21 August 1811, the son of a wealthy landowner said to have built the first two brick houses in Pittsburgh. Kelly was educated at local schools. He was inventive and fond of metallurgy, but at the age of 35 found himself working for a dry-goods business in Philadelphia as a junior member of the firm McShane and Kelly. He was sent out collecting debts in Nashville, Tennessee, and it was there that he met his wife and her father, who was a wealthy tobacco merchant. Kelly settled there and with his brother bought nearby iron-ore lands and a furnace known as the Cobb furnace, in Eddyville.

Kelly patented his steel-making process in 1857, and in 1871 had his patent renewed for seven years. Steel under the Kelly patent was first blown commercially in autumn 1864 at the Wyandotte Iron Works near Detroit.

Kelly then moved to Johnstown, Pennsylvania, where he was acclaimed as a genius, and after five years he moved to Louisville, where he founded an axe-making business. He retired from active business at the age of 70 and died in

Louisville on 11 February 1888. His axe-manufacturing business was carried on by his son in Charlestown, Western Virginia. On 5 October the American Society for Steel Treating erected a bronze tablet to Kelly's memory at the site of the Wyandotte Iron Works.

In Eddyville, Kelly developed the Suwanee Iron Works and the Union Forge. He manufactured sugar kettles, which were much in demand by farmers in the area. To do this he used wrought iron made from pig iron, which involved the burning out of the excess carbon. This needed a lot of charcoal, and the local supply soon ran low.

One day he noticed that although the air-blast in his 'finery fire' furnace was blowing on molten iron with no charcoal covering, the iron was still becoming white hot. When he experimented further, he found that contrary to all iron-makers' beliefs, molten iron containing sufficient carbon became much hotter when air was blown on to it. The air-blast can burn out 3–5% of carbon contained in molten cast iron. Here the carbon itself is acting as a fuel, and makes the molten mass much hotter.

Kelly became so obsessed with his discovery that his wife called the doctor, thinking he was ill. Unfortunately, his customers could not be convinced that the steel made by this new, cheaper process was as good and so he had to revert to using charcoal. Meanwhile, with two other iron-makers, he started building a converter secretly. His first attempt was a 1.2-m/4-ft-high brick kettle. Air was blown through holes in the bottom into and through molten pig iron. The method was only partly successful. Between 1851 and 1856 he secretly built seven of these converters. In 1856, he heard that Henry Bessemer of England had been granted a US patent on the same process. Kelly immediately applied for a patent and managed to convince the Patent Office of his priority. On 23 June 1857 he was granted a patent and declared to be the original inventor. Four years later, his patent was renewed for a further seven years, whereas Bessemer's patent renewal was refused.

The financial panic of 1857 had bankrupted him, however, and he sold his patent to his father for $1,000. His father, thinking his son to be an incompetent businessman, bequeathed the patent to his daughters and would not return it to Kelly. Kelly therefore moved away to Johnstown, where a Daniel J. Morrell of the Cambria Iron Works listened to his plight. Kelly built his eighth converter there (the first of a tilting type), and it is still preserved in the office of the Cambria Works, now part of the Bethlehem Steel company. His second attempt at Johnsville was a success and it produced soft steel cheaply for the first time and in large quantities. It was used for rails, bars and structural shapes and marked a milestone in the Steel Age which was just beginning.

Kelly's patent eventually came under the control of Kelly Pneumatic Process Company, with which Kelly had nothing to do. This company later merged with the Company of Alex L. Holley, which was Bessemer's sole US licensee. Bessemer's name came into exclusive use for the Kelly–Bessemer steel-making process, and Kelly remained in relative obscurity.

Kelvin Lord Thomson, William 1824–1907 was a British physicist who first proposed the use of the absolute scale of temperature, in which the degree of temperature is now called the Kelvin. Thomson also made other substantial contributions to thermodynamics and the theory of electricity and magnetism, and he was largely responsible for the first successful transatlantic telegraph cable.

Kelvin was born William Thomson in Belfast on 26 June 1824. His father was Professor of Mathematics at Belfast University and both he and his older brother James Thomson (1822–1892), who also became a prominent physicist, were educated at home by their father. In 1832, Thomson's father took up the post of Professor of Mathematics at Glasgow University, and Thomson himself entered the university two years later at the age of ten to study natural philosophy (science). In 1841, Thomson went on to Cambridge University, graduating in 1845. He then travelled to Paris to work with Henri Regnault (1810–1878) and in 1846 took up the position of Professor of Natural Philosophy at Glasgow, where he created the first physics laboratory in a British university. Among his many honours were a knighthood in 1866, the Royal Society's Copley Medal in 1883, the Presidency of the Royal Society from 1890 to 1894 and a peerage in 1892, when he took the title Baron Kelvin of Largs. Kelvin retired from his chair at Glasgow in 1899 and died at Largs, Ayrshire, on 17 December 1907. He was buried next to Isaac Newton in Westminster Abbey.

Thomson's early work, begun in 1842 while he was still at Cambridge, was a comparison of the distribution of electrostatic force in a region with the distribution of heat through a solid. He found that they are mathematically equivalent, leading him in 1847 to conclude that electrical and magnetic fields are distributed in a manner analogous to the transfer of energy through an elastic solid. James Clerk Maxwell (1831–1879) later developed this idea into a comprehensive explanation of the electromagnetic field. From 1849 to 1859, Kelvin also developed the discoveries and theories of paramagnetism and diamagnetism made by Michael Faraday (1791–1867) into a full theory of magnetism, developing the terms magnetic permeability and susceptibility, and arriving at an expression for the total energy of a system of magnets. In electricity, Kelvin obtained an expression for the energy possessed by a circuit carrying a current and in 1853 developed a theory of oscillating circuits that was experimentally verified in 1857 and was later used in the production of radio waves.

In Paris in 1845, Kelvin was introduced to the classic work of Sadi Carnot (1796–1832) on the motive power of heat. From a consideration of this theory, which explains that the amount of work produced by an ideal engine is governed only by the temperature at which it operates, Kelvin developed the idea of an absolute temperature scale in which the temperature represents the total energy in a body. He proposed such a scale in 1848 and set absolute zero at −273°C/−459°F, showing that Carnot's theory followed if absolute temperatures were used. However, Kelvin could not accept the idea then still prevalent and accepted by

Carnot that heat is a fluid, preferring to see heat as a form of motion. This followed Kelvin's championing of the work of James Joule (1818–1889), whom Kelvin met in 1847, on the determination of the mechanical equivalent of heat. In 1851, Kelvin announced that Carnot's theory and the mechanical theory of heat were compatible provided that it was accepted that heat cannot pass spontaneously from a colder to a hotter body. This is now known as the second law of thermodynamics, which had been advanced independently in 1850 by Rudolf Clausius (1822–1888). In 1852, Kelvin also produced the idea that mechanical energy tends to dissipate as heat, which Clausius later developed into the concept of entropy.

Kelvin and Joule collaborated for several years following their meeting, Joule's experimental prowess matching Kelvin's theoretical ability. In 1852 they discovered the Joule–Thomson effect, which causes gases to undergo a fall in temperature as they expand through a nozzle. The effect is caused by the work done as the gas molecules move apart, and it proved to be of great importance in the liquefaction of gases.

Kelvin was also interested in the debate then taking place about the age of the Earth. Hermann Helmholtz (1821–1894) in 1854 gave a value of 25 million years for the Earth's lifetime, assuming that the Sun gained its energy by gravitational contraction. Kelvin came to a similar conclusion in 1862, basing his estimate on the rate of cooling that would have occurred from the time the Earth formed, and reaching an age of 20 million to 400 million years with 100 million years as the most likely figure. Furthermore, the cooling would have produced volcanic upheavals that would have limited the time available for the evolution of life. Although both estimates were far too low, neither scientists knowing of the nuclear processes that fuel the Sun and the radioactivity that warms the Earth, their figures were taken seriously and helped to bring about theories of mutation to explain evolution.

Kelvin's knowledge of electrical theory was applied with great practical value to the laying of the first transatlantic telegraph cable. Kelvin pointed out that a fast rate of signalling could only be achieved by using low voltages, and that these would require very sensitive detection equipment such as the mirror galvanometer that he had invented. The first cable laid in 1857 broke and high voltages were used in the second cable laid a year later as Kelvin's predictions were not believed. The cable did not work, but a third cable laid in 1866 using Kelvin's ideas was successful, and it was for this achievement that he received a knighthood.

Kelvin was also very concerned with the accurate measurement of electricity, and developed an absolute electrometer in 1870. He was instrumental in achieving the international adoption of many of our present-day electrical units in 1881.

Kelvin was one of the greatest physicists of the nineteenth century. His pioneering work on heat consolidated thermodynamics and his understanding of electricity and magnetism paved the way for the explanation of the electromagnetic field later achieved by Maxwell.

Kendall Edward Calvin 1886–1972 was a US endocrinologist who shared the Nobel Prize for Phsyiology or Medicine in 1950. His work on understanding the physiological chemistry of the secretions of the thyroid and adrenal glands led him to isolate some of their active constitutents. Of these cortisone from the adrenal gland has been found to offer symptomatic relief for sufferers of rheumatoid arthritis.

Kendall was born on 8 March 1886 in South Norwalk, Connecticut and received his university education at Columbia, gaining bachelor's (1908), master's (1909) and doctoral (1910) degrees in chemistry. For a few months he worked for the pharmaceutical firm, Parke Davis and Company, but disliked the lack of research opportunity and regimented atmosphere there, and returned to New York to work in a hospital laboratory, before moving to the Mayo Clinic in Rochester, Minnesota in 1914. In 1921, he became Professor of Physiological Chemistry at the Mayo, and retired in 1951. He was then appointed to a visiting professorship at Princeton University and continued to be associated with laboratory work until his death on 4 May 1972 at Princeton.

Kendall's earliest research in endocrinology was undertaken at St Luke's Hospital in New York where he began to investigate the properties of thyroid gland secretions. In 1913, he isolated a very potent extract and after his move to Rochester he succeeded in crystallizing the substance, which he called thyroxine. For several years he continued to study the chemistry of thyroxine, although it was C. R. Harington (1897–1972) from London who was succesful in first determining its chemical structure and its synthetic pathway. Kendall also started investigating the physiological chemistry of the adrenal glands, directing an interdisciplinary research team that involved physicians, surgeons and scientists. By the beginning of the 1930s, it was becoming accepted that the adrenal gland produced more than one hormone, and research teams around the world, especially in Germany, turned their attentions towards adrenal chemistry. It was not until the late 1940s that Kendall and his team succeeded in isolating and synthesizing cortisone. Unfortunately it seemed to have only limited therapeutic use, and did not attract much attention at the time. However, Kendall had started to work with Phillip Hench (1896–1965) a physician interested in arthritis who was familiar with the experience that in some situations, such as during pregnancy, patients with arthritis improved. The two men discussed whether cortisone, which Kendall's work had shown to have important metabolic effects, was involved in these temporary improvements. They discovered by giving a severely incapacitated patient cortisone that it was an effective treatment for rheumatoid arthritis, for which they shared the Nobel prize.

Kendrew John Cowdery 1917– is a British biochemist who shared the 1962 Nobel Prize for Chemistry with Max Perutz for his determination of the structure of the protein myoglobin.

Kendrew was born in Oxford on 24 March 1917. He won

a scholarship to Trinity College, Cambridge, in 1936 and graduated from there in 1939, just before the outbreak of World War II. During the war he worked for the Ministry of Aircraft Production, returning to Cambridge in 1945. A year later he was appointed Departmental Chairman of the Medical Council's Laboratory for Molecular Biology, Cambridge, where he remained until 1975. In that year he became Director General of the European Molecular Biology Laboratory at Heidelburg. He was knighted in 1974.

In the late 1940s at Cambridge Kendrew began working with Perutz, whose other molecular biologists at that time included Francis Crick. Their research centred on an investigation of the fine structure of various protein molecules; Kendrew was assigned the task of studying myoglobin, the globular protein resembling haemoglobin which occurs in muscle fibres (where it stores oxygen). He used X-ray diffraction techniques to elucidate the amino acid sequence in the peptide chains that form the myoglobin molecule (similar to the work of Maurice Wilkins on DNA).

Hundreds of X-ray diffraction photographs of the crystallized protein were analysed, using electronic computers that were becoming available in the 1950s. By 1960, Kendrew had determined the spatial arrangement of all 1,200 atoms in the molecule, showing it to be a folded helical chain of amino acids with an amino ($-NH_2$) group at one end and a carboxylic ($-COOH$) group at the other. It involves an iron-containing haem group, which allows the molecule to absorb oxygen. In the same year, Perutz determined the structure of haemoglobin.

Kenyon Joseph 1885–1961 was a British organic chemist, best known for his studies of optical activity, particularly of secondary alcohols.

Kenyon was born in Blackburn, Lancashire, the eldest of seven children. He was educated locally and from 1900–1903 worked as a laboratory assistant at Blackburn Technical College. In 1903 he won a scholarship that enabled him to study for a degree, and he graduated two years later. In 1904, he became personal assistant to R. H. Pickard, an industrial chemistry consultant and technical college instructor; in 1906 assistant lecturer and demonstrator at Blackburn Technical College; and in 1907 full lecturer, commencing research under Pickard. He submitted his doctoral thesis to London University in 1914 and in the following year was appointed research chemist to the Medical Research Council. From 1915 to 1916, during World War I, he worked at Leeds University and then until 1920 with William Perkin Jr (1860–1929) at the British Dyestuffs Corporation, Oxford. Finally, in 1920 Kenyon became Head of the Chemistry Department at Battersea Polytechnic, London, a year after Pickard had become its Principal; he remained there until he retired in 1950. In 1951 and 1952 he was Visiting Professor at the Universities of Alexandria and Kansas. He died in Petersham on 11 November 1961.

Kenyon published his research on secondary alcohols in 1911 while still working as an undergraduate with Pickard. The problem was to resolve the optically active

stereoisomers of secondary octyl alcohol (octan-2-ol). Pickard and Kenyon converted it to secondary octyl hydrogen phthalate by heating it with phthalic anhydide, and showed that the phthalate could easily be resolved using the alkaloids brucine and cinchonidine. They went on to obtain an optically pure series of secondary alcohols and were able to relate rotatory power to chemical constitution. The technique was later used to distinguish between inter- and intramolecular rearrangements.

While at Leeds (1915–1916) Kenyon worked on antidotes for gas gangrene, research that led to the development of chloramine-T for this and other purposes. With Perkin (1916–1920) he studied photographic developers and dyes, although most of their results were published only in confidential reports.

At Battersea Polytechnic Kenyon put forward the 'obstacle' theory for the cause of optical activity in certain substituted diphenic acids. He synthesized and attempted to resolve some selenoxides, confirming differences between these and sulphoxides, and investigated the geometric and optical isomerism of the methylcyclohexanols. He pointed out that when the toluene-para-sulphonic acids (methylbenzene-4-sulphonic acids) are prepared by the action of the sulphonic chloride on the corresponding alcohol, the four bonds of the asymmetric carbon atom remain undisturbed – there is no change in configuration. But when the toluene-4-sulphonic esters are converted to the carboxylic esters (by heating them with the alcoholic solutions of the alkali salts of the carboxylic acids), there is an almost 100% inversion of configuration.

He also discovered that the hydroxyl compounds can be converted to the sulphinic esters without inversion (by treatment with toluene-4-sulphinyl chloride) but that treatment of the esters with hypochlorous acid reconverts them into the enantiomers (mirror-images) of the original hydroxyl compounds.

Perhaps Kenyon's most important work in stereochemistry was to prove that in the Beckman, Curtius, Hofmann, Lossen and Schmidt reactions – in which a group migrates from one part of the molecule to another – in no case is the migrating fragment ever kinetically free.

Kepler Johannes 1571–1630 was a German astronomer who combined great mathematical skills with patience and an almost mystical sense of universal harmony. He is particularly remembered for what are now known as Kepler's laws of motion. These had a profound influence on Newton and hence on all modern science. Kepler was also absorbed with the forces that govern the whole universe and he was one of the first and most powerful advocates of Copernican heliocentric (Sun-centred) cosmology.

Kepler was born on 27 December 1571 in Weil der Statt near Stuttgart, Germany. He was not a healthy child and since it was apparently thought that he was capable only of a career in the ministry, he was sent for religious training in Leonberg, Adelberg and Maulbronn. One event that impressed him deeply during his early years was the viewing of the 'great' comet of 1577 and his interest in

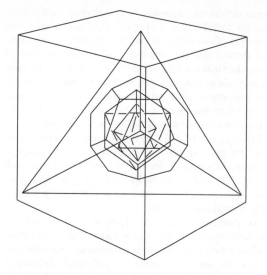

Kepler *Kepler adopted Plato's idea of five nesting regular solids (from the centre: octahedron, icosahedron, dodecahedron, tetrahedron and cube) and suggested that the planetary orbits fitted between them.*

astronomy probably dates from that time. He passed his baccalaureate at the University of Tübingen in 1588 and then returned to Maulbronn for a year. From 1589 to 1591 he studied philosophy, mathematics and astronomy under Michael Mästlin at the University of Tübingen; yet although he showed great aptitude and promise and obtained his MA in 1591, he then embarked on a three-year programme of theological training. This was interrupted in the last year when he was nominated for a teaching post in mathematics and astronomy in Graz. It was during this teaching period that he abandoned his plans for a career in the ministry and concentrated on astronomy. He wrote his first major paper while at Graz and attracted the attention of other notable astronomers of the time, particularly Tycho Brahe and Galileo.

Kepler was a Lutheran and so was frequently caught up in the religious troubles of his age. In 1598 a purge forced him to leave Graz. He travelled to Prague and spent a year there before he returned to Graz; he was expelled again and arrived back in Prague, where he became Tycho Brahe's assistant in 1600. Brahe died a few months later, in 1601, and Kepler succeeded him as Imperial Mathematician to Emperor Rudolph II. On his deathbed Brahe requested that Kepler complete the Rudolphine Astronomical Tables, a task which Kepler finished in 1627.

Kepler lived in Prague until 1612 and produced what was perhaps his best work during those years. He was given a telescope in 1610 by Elector Ernest of Cologne and studied optics, telescope design and astronomy. In 1611, Emperor Rudolph was deposed, but Kepler was retained as Imperial Mathematician. In 1612, upon Rudolph's death, Kepler became District Mathematician for the States of Upper Austria and moved to Linz. But personal problems plagued him for the next ten years – the arrest and trial of

his mother, who was accused of witchcraft in 1615 but exonerated in 1621, being particularly distressing.

It was in Linz that Kepler published three of his major works. In 1628, he became the private mathematician to Wallenstein, Duke of Friedland and Imperial General, partly because of the Duke's promise to pay the debt owed Kepler by the deposed Emperor Rudolph. In 1630 religious persecution forced Kepler to move once again. He fell sick with an acute fever, and died en route to Regensburg, Bavaria, on 15 November 1630.

Kepler's work in astronomy falls into three main periods of activity, at Graz, Prague and Linz. In Graz Kepler did some work with Mästlin on optics and planetary orbits, but he devoted most of his energy to teaching. He also produced a calendar of predictions for the year 1595, which proved so uncanny in its accuracy that he gained a degree of local fame. Kepler found the production of astrological calendars a useful way of supplementing his income in later years, but he had little respect for the art. More important, in 1596 he published his *Mysterium Cosmographicum*, in which he demonstrated that the five Platonic solids, the only five regular polyhedrons, could be fitted alternately inside a series of spheres to form a 'nest' which described quite accurately (within 5%) the distances of the planets from the Sun. Kepler regarded this discovery as a divine inspiration which revealed the secret of the universe. It was written in accordance with Copernican theories and it brought Kepler to the attention of all European astronomers.

Before Kepler arrived in Prague and was bequeathed all Tycho's data on planetary motion, he had already made a bet that, given Tycho's unfinished tables, he could find an accurate planetary orbit within a week. It took rather longer, however. It was five years before Kepler obtained his first planetary orbit, that of Mars. In 1604, his attention was diverted from the planets by his observation of the appearance of a new star, 'Kepler's nova', to which he attached great astrological significance.

In 1609, Kepler's first two laws of planetary motion were published in *Astronomia Nova*, which is a long text and as unreadable as it is important. The first law states that planets travel in elliptical rather than circular or epicyclic orbits and that the Sun occupies one of the two foci of the ellipses. What is now known as the second law, but was in fact discovered first, states that the line joining the Sun and a planet traverses equal areas of space in equal periods of time, so that the planets move more quickly when they are nearer the Sun. This established the Sun as the main force governing the orbits of the planets. Kepler also showed that the orbital velocity of a planet is inversely proportional to the distance between the planet and the Sun. He suggested that the Sun itself rotates, a theory which was confirmed by using Galileo's observations of sunspots, and he postulated that this established some sort of 'magnetic' interaction between the planets and the Sun, driving them in orbit. This idea, although incorrect, was an important precursor of Newton's gravitational theory. The *Astronomia Nova* had virtually no impact at all at the time, and so Kepler turned his attention to optics and telescope design. He published his second book on optics, the *Dioptrice*, in 1611. That year was a difficult one, because Kepler's wife and sons died. Then, in 1612, the Lutherans were thrown out of Prague so Kepler had to move on again to Linz.

In Linz, Kepler produced two more major works. The first of these was *De Harmonices Mundi*, which was almost a mystical text. The book was divided into five chapters and buried in the last was Kepler's third law. In this law he describes in precise mathematical language the link between the distances of the planets from the Sun and their velocities – a feat which afforded him extraordinary pleasure and confirmed his belief in the harmony of the universe. The second major work to be published during his stay in Linz, the *Epitome*, intended as an introduction to Copernican astronomy, was in fact a very effective summary of Kepler's life's work in theoretical astronomy. It was a long treatise of seven books, published over a period of four years, and it had more impact than any other astronomical text of the mid-seventeenth century.

Soon after its publication, Kepler's Lutheran background caused yet another expulsion and this time he went to Ulm, where he finally completed the Rudolphine Tables. They appeared in 1627 and brought Kepler much popular acclaim. These were the first modern astronomical tables, a vast improvement on previous attempts of this kind, and

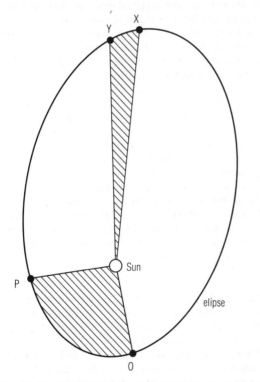

Kepler *Kepler's second law states that the two shaded areas are equal to each other if the planet moves from P to O in the same time that it moves from X to Y. The law says, in effect, that a planet moves fastest when it is closest to the Sun.*

they enabled astronomers to calculate the positions of the planets at any time in the past, present or future. The publication also included other vital information, such as a map of the world, a catalogue of stars, and the latest aid to computation, logarithms.

Kepler wrote the first science-fiction story, *Solemnium*, which described a man who travelled to the Moon. It was published in 1631, a year after his death, although it had been written 20 years earlier. Kepler was a remarkable man and a brilliant scientist. He kept a steady eye on what he saw as his true vocation as a 'speculative physicist and cosmologist' and, despite living in times of political unrest and religious turmoil, was never swayed by religious bigotry or political pressures. His new astronomy provided the basis on which Newton and others were to build, and to this day his three laws of motion are considered to be the basis of our understanding of the Solar System. His work and his strong support of Copernican cosmology mark a fundamental divide between two eras – that of the Ptolemaic Earth-centred view of the universe that had been accepted for the previous 15 centuries, and the new age given birth by the Copernican heliocentric, or Sun-centred, view of the Solar System.

Kerr John 1824–1907 was a British physicist who discovered the Kerr effect, which produces double refraction in certain media on the application of an electric field.

Kerr was born in Ardrossan, Ayrshire, on 17 December 1824 and received his early education in Skye. He entered Glasgow University in 1941, gaining an MA in physical science in 1849. At Glasgow, Kerr became one of the first research students of Lord Kelvin (1824–1907), and received special prizes for work in magnetism and electricity. He next became a divinity student but, although he completed his course, he did not subsequently pursue a career in the Church. Kerr instead took up the post of lecturer in mathematics at the Free Church Training College for Teachers, in Glasgow, in 1857. He stayed in this post for 44 years, even though research facilities were virtually nonexistent. Nevertheless, Kerr set up a modest laboratory and although his publications were small in number, his work was of the highest quality. Kelvin must have had Kerr in mind when in a speech many years later he remarked that theology students made some of the best researchers and became all the better clergymen for their scientific experiences. Kerr's election to the Royal Society in 1890 and the award of its Royal Medal in 1898 were fitting recognition from a wider community of a life of modest excellence. He died at Glasgow on 18 August 1907.

Kerr's first important discovery was in 1875 when he demonstrated that birefringence or double refraction occurs in glass and other insulators when subjected to an intense electric field. A 5-cm/2-in block of glass had collinear holes drilled from each end until they were separated by 6 mm/0.25 in. A beam of polarized light was passed through this narrow band and at the same time an electric field was applied via electrodes inserted into the holes. Nicol prisms were arranged on either side to give extinction of the beam when the glass was free from strain. When the coil producing the field was activated, it was observed that the polarization became elliptical. The effect was strongest when the plane of polarization was 45° to the field, and zero when perpendicular or parallel. Kerr extended the work to other materials including amber and constructed, with great delicacy and manipulative skill, cells in which he could study liquids such as carbon disulphide and paraffin oil. He showed that the extent of the effect, which is more precisely called the electro-optical Kerr effect, is proportional to the square of the field strength.

In 1876, Kerr caused a considerable stir at the Glasgow meeting of the British Association when he demonstrated another remarkable phenomenon of polarized light that we know as the magneto-optical Kerr effect. In this a beam of plane-polarized light was reflected from the polished pole of an electromagnet. When the magnet was switched on the beam became elliptically polarized, the major axis being rotated from the direction of the original plane. The effect depended on the position of the reflecting surface with respect to the direction of magnetization and to the plane of incidence of the light. When both of these were perpendicular, the plane of polarization was turned through a small angle. The mathematical analysis of the effect and its relation to the electromagnetic theory of light was worked out some years later by George FitzGerald (1851–1901).

The electro-optical Kerr effect is now applied in the Kerr cell, which is an electrically activated optical shutter having no moving parts. The cell contains a tube of liquid and crossed polarizers so that it normally transmits no light. Switching on the cell causes an electric field to produce double refraction in the liquid, allowing light to pass through the polarizers. The effect occurs because the field orientates the molecules of liquid to give double refraction, in which incident light is split into two beams that are polarized at right angles. Very fast shutter speeds are obtainable and Kerr cells have been used to make precise determinations of the speed of light.

Kettlewell Henry Bernard David 1907–1979 was a British geneticist and lepidopterist who carried out important research into the influence of industrial melanism on natural selection in moths.

Kettlewell was born in Howden, Yorkshire, on 24 February 1907 and was educated at Charterhouse School, at Godalming in Surrey, and in Paris. He studied medicine at Gonville and Caius College, Cambridge, and at St Bartholomew's Hospital, London, graduating in 1933. After graduation he held several appointments in various London hospitals, including St Bartholomew's, and was an anaesthetist at St Luke's Hospital in Guildford, Surrey. He served as an anaesthetist during World War II. From 1949 to 1952 he investigated methods of locust control at Cape Town University, South Africa, also going on expeditions to the Kalahari Desert, the Belgian Congo, Mozambique and the Knysna Forest. After his return to England, he was awarded in 1952 a Nuffield Research Fellowship in the

Genetics Unit of the Zoology Department at Oxford University. In the following year he was appointed Senior Research Officer in Genetics at Oxford, a post he held until he retired in 1974, when he became an Emeritus Fellow of Wolfson College, Oxford. He died in Oxford on 11 May 1979.

Kettlewell's best known research involved the influence of industrial melanism on the survival of the peppered moth (*Biston betularia*). Until 1845, all known specimens of this moth were light-coloured, but in that year one dark specimen was found in Manchester, then an expanding industrial centre. The proportion of dark-coloured peppered moths increased rapidly, until by 1895 they comprised about 99% of Manchester's entire peppered moth population. This change from light to dark moths – a phenomenon called industrial melanism – corresponds with the increase in industry (and therefore pollution, especially of the atmosphere) since the Industrial Revolution, and today only a few populations of the original light-coloured variant exist in England, being found in unindustrialized rural areas. Kettlewell performed several experiments under natural conditions to demonstrate the significance of colour in protecting the peppered moths from birds, their only predators. He released a known number of light- and dark-coloured moths – specially marked so that they could be identified – in an industrial area of Birmingham (where 90% of the indigenous peppered moths are dark-coloured) and in an unpolluted rural area of Dorset (where there are no dark-coloured moths). After an interval he collected the surviving moths and found that, in Birmingham, birds had eaten a high proportion of the light-coloured moths but a much lower proportion of the dark-coloured ones, whereas in Dorset the reverse had occurred. Thus Kettlewell demonstrated the efficiency of natural selection as an evolutionary force: the light-coloured moths are more conspicuous than the dark-coloured ones in industrial areas – where the vegetation is darkened by pollution – and are therefore easier prey for birds, but are less conspicuous in unpolluted rural areas – where the vegetation is lighter in colour – and therefore survive predation better.

In addition to his research into coloration and natural selection, Kettlewell cofounded the Rothschild–Cockayne–Kettlewell (RCK) Collection of British Lepidoptera, which is now called the National Collection (RCK) and is housed in the British Museum (Natural History) in London.

Khorana Har Gobind 1922–   is an Indian-born US chemist who has worked extensively on the chemistry of nucleic acids, for which he shared the Nobel Prize for Physiology or Medicine in 1968.

Khorana was born on 9 January 1922 in Raipur in the Punjab region of India, now in West Pakistan. His family was very poor, but believed in educating their children, and Khorana was encouraged to study, graduating from Punjab University with bachelor's (1943) and master's (1945) degrees in chemistry. He was awarded a government scholarship to work for his doctorate at Liverpool University, and then moved to Switzerland for postdoctoral work in Zurich, before returning to Britain in 1950 to work in Cambridge where he came under the influence of the future Nobel Laureate Alexander Todd (1907– , later Lord Todd) and became interested in the chemistry of nucleic acids. In 1952, he accepted a position in the division of organic chemistry at the University of British Columbia in Vancouver, Canada, and in 1960 moved to the University of Wisconsin, and was Professor of Biochemistry there from 1962 to 1970. He then became Alfred P. Sloan Professor at the Massachusetts Institute of Technology and was also associated with Cornell University from 1974 to 1980. He is a member of many national academies, including the National Academy of Sciences and the Royal Society of London, and is the receipient of numerous honorary degrees and international awards. Khorana's work has been in unravelling the chemistry of the genetic code and in establishing the steps by which proteins are synthesized. It was known that four nucleotides were involved in determining the genetic code, and Khorana systematically synthesized each possible combination of the genetic signals from these four nucleotides. He showed that a pattern of three nucleotides, called a triplet, specify a particular amino acid, which are the building blocks of proteins. He further discovered that some of the triplets provided 'punctuation' marks in the code, marking the beginning and end points of protein synthesis. For this major discovery Khorana was awarded the Nobel prize. In 1970, he achieved the synthesis of the first artificial gene, and in more recent years has synthesized the gene for bovine rhodopsin, the pigment in the retina which is responsible for converting light energy into electrical energy. His work therefore provides much of the basic knowledge that is important for gene therapy and biotechnology.

Kimura Motoo 1924– is a Japanese biologist who, as a result of his work on population genetics and molecular evolution, has developed a theory of neutral evolution which opposes the conventional neo-Darwinistic theory of evolution by natural selection. He has received many honours for his work, including Japan's highest cultural award, the Order of Culture.

Kimura was born on 13 November 1924 in Okazaki. He studied botany at Kyoto University, from which he gained his master of science degree in 1947, then worked as an assistant there from 1947 to 1949. He spent most of his subsequent career at the National Institute of Genetics in Mishima – as a Research Member from 1949 to 1957, as Laboratory Head of the Department of Population Genetics from 1957 to 1964, and as overall head of the same department from 1964. In 1953, however, he went to the United States as a graduate student; he spent nine months studying under Dr Lush at Iowa State College, then moved to the University of Wisconsin (where he worked under Dr Crow in the Genetics Department), from which Kimura gained his doctorate in 1956. Although he returned to Japan in that year, Kimura continued to collaborate with Crow, jointly writing a book on population genetics, for example.

While a student, Kimura became interested in genetics, particularly the mathematical aspects of genetics and evolution. Stimulated by the work of J. B. S. Haldane and Sewall Wright on population genetics, Kimura began original work in this field in 1949, teaching himself the necessary mathematics. He then began to investigate the fate of mutant genes, how the genetic constitution of living organisms adapts to environmental changes, and the role of sexual reproduction in evolution.

Extending this early research, Kimura then began the work that was to lead to his postulating the theory of neutral evolution in 1968. According to the neo-Darwinistic view, evolution results from the interaction between variation and natural selection; species evolve by accumulating adaptive mutant genes, but these mutant genes increase in the population only if they confer advantageous traits on its individual members. With the advent of molecular genetics, it became possible to compare individual RNA molecules and proteins in related organisms and to assess the rate at which allelic genes (those that occupy the same relative positions on homologous chromosomes) are substituted in evolution. It also became possible to study the variability of genes within a species. Using these techniques, Kimura found that, for a given protein, the rate at which amino acids are substituted for each other is approximately the same in many diverse lineages, and that the substitutions seem to be random. Comparing the amino acid compositions of the alpha and beta chains of the haemoglobin molecules in humans with those in carp, he found that the alpha chains have evolved in two distinct lineages, accumulating mutations independently and at about the same rate over a period of some 400 million years. Moreover, the rate of amino acid substitution observed in the carp–human comparison is very similar to the rates observed in comparisons of the alpha chains in various other animals. These findings led Kimura to his theory of neutral evolution. According to this theory, evolutionary rates are determined by the structure and function of molecules and, at the molecular level, most intra-specific variability and evolutionary change is caused by the random drift of mutant genes that are all selectively equivalent and selectively neutral. Thus Kimura's theory directly opposes the neo-Darwinistic theory of evolution by denying that the environment influences evolution and, as a concomitant of this, also denying that mutant genes confer either advantageous or disadvantageous traits.

Kipping Frederick Stanley 1863–1949 was a British chemist who pioneered the study of the organic compounds of silicon; he invented the term 'silicone', which is now applied to the entire class of oxygen-containing polymers.

Kipping was born at Higher Broughton, near Manchester, on 16 August 1863, the son of a bank employee. He was educated at Manchester Grammar School and in 1882 graduated in chemistry from Owens College, Manchester (later Manchester University), with an external degree from the University of London. After four years as a chemist with the Manchester Gas Department, he went to Johann

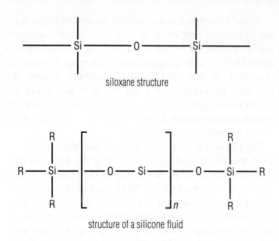

siloxane structure

structure of a silicone fluid

von Baeyer's laboratory in Munich to study under William Perkin Jr, who was to become his close friend and collaborator. Kipping received his doctorate in 1887 and was awarded a DSc degree by the University of London in the same year – the University's first award of this qualification solely for research work.

He then followed Perkin to the Heriot-Watt College, Edinburgh, where he worked as a demonstrator. It was at Edinburgh that the two chemists began work on their classic textbook *Organic Chemistry* (1894), the first to be devoted entirely to this subject and a standard work for the next 50 years. In 1890, Kipping became chief demonstrator at the City and Guilds Institute, London, and seven years later he took up the appointment of Professor of Chemistry at University College, Nottingham (later Nottingham University), where he remained until he retired in 1936. He died in Criccieth, Wales, on 1 May 1949.

In his early research Kipping investigated the preparation and properties of optically active camphor derivatives and nitrogen compounds. This interest in stereoisomerism led him in 1899 to look for such isomerism among the organic compounds of silicon, preparing them using the newly available Grignard reagents. He prepared condensation products – the first organosilicon polymers – which he called silicones. He also tried to make silicon analogues of simple carbon compounds, particularly those containing double bonds, although in this he was not successful.

To Kipping the silicon compounds were mere chemical curiosities and as late as 1937 he could not see any practical applications for them. Yet within a very few years, spurred on by the outbreak of World War II, silicones were being used as substitutes for oils and greases. Their chemical inertness and unusual stability at high temperatures make them useful as lubricants, hydraulic fluids, waterlogging compounds, varnishes, greases, synthetic rubbers and various other hydrocarbon substitutes.

Kirchhoff Gustav Robert 1824–1887 was a German physicist who founded the science of spectroscopy. He also discovered laws that govern the flow of electricity in electrical networks and the absorption and emission of radiation in material bodies.

Kirchhoff was born at Königsberg, Germany (now Kaliningrad) on 12 March 1824. He studied at the University of Königsberg, graduating in 1847. In the following year, he became a lecturer at Berlin and in 1850 was appointed Extraordinary Professor of Physics at Breslau. Robert Bunsen (1811–1899) went to Breslau the following year and began a fruitful collaboration with Kirchhoff. In 1852 Bunsen moved to Heidelberg and Kirchhoff followed him in 1854, becoming Ordinary Professor of Physics there. Kirchhoff stayed at Heidelberg until 1875, when he moved to Berlin as Professor of Mathematical Physics. Illness forced him to retire in 1886 and he died at Berlin on 17 October 1887.

Kirchhoff made his first important contribution to physics while still a student. In 1845 and 1846 he extended Ohm's law to networks of conductors and derived the laws known as Kirchhoff's laws that determine the value of the current and potential at any point in a network. He went on to consider electrostatic charge and in 1849 showed that electrostatic potential is identical to tension, thus unifying static and current electricity. Kirchhoff made another fundamental discovery in electricity in 1857 by showing theoretically that an oscillating current is propagated in a conductor of zero resistance at the velocity of light. This was important in the development in the 1860s of the electromagnetic theory of light by James Clerk Maxwell (1831–1879) and Ludwig Lorenz (1829–1891).

In the 1850s, Bunsen developed his famous gas burner, which gave a colourless flame, and used it to investigate the distinctive colours that metals and their salts produce in a flame. Bunsen used coloured solutions and glass filters to distinguish the colours in a partly successful attempt to identify the substances by the colours they produced. Kirchhoff pointed out to Bunsen that sure identification could be achieved by using a prism to produce spectra of the coloured flames. They developed the spectroscope, and in 1860 discovered that the elements present in the substances each give a characteristic set of spectral lines and set about classifying elements by spectral analysis. In this way, Bunsen discovered two new elements – caesium in 1860 and rubidium in 1861.

Kirchhoff also made another important discovery in 1859 while investigating spectroscopy as an analytical tool. He noticed that certain dark lines in the Sun's spectrum, which had been discovered by Joseph Fraunhofer (1787–1826), were intensified if the sunlight passed through a sodium flame. This observation had in fact been made by Jean Foucault (1819–1868) ten years earlier, but he had not followed it up. Kirchhoff immediately came to the correct conclusion that the sodium flame was absorbing light from the sunlight of the same colour that it emitted, and explained that the Fraunhofer lines are due to the absorption of light by sodium and other elements present in the Sun's atmosphere.

Kirchhoff went on to identify other elements in the Sun's spectrum in this manner, and also developed the theoretical aspects of this work. In 1859, he announced another major law which states that the ratio of the emission and absorption powers of all material bodies is the same at a given temperature and a given wavelength of radiation produced. From this, Kirchhoff went on in 1862 to derive the concept of a perfect black body – one that would absorb and emit radiation at all wavelengths. Balfour Stewart (1828–1887) had reached similar conclusions in 1858 by a consideration of the theory of heat exchanges discovered by Pierre Prevost (1751–1839), but Kirchhoff presented the discovery much more cogently.

Kirchhoff's contributions to physics had far-reaching practical and theoretical consequences. The discovery of spectroscopy led to several new elements and to methods of determining the composition of stars and the structure of the atom. The study of black-body radiation led directly to the quantum theory and a radical new view of the nature of matter.

Kirkwood Daniel 1814–1895 was a US astronomer who is known for his work on asteroids, meteors and the evolution of the Solar System.

Kirkwood was born in Hartford County, Maryland, on 27 September 1814. He became a teacher in 1833 and rose to become Principal of Lancaster High School from 1843 to 1849 and then of the Pottsville Academy from 1849 to 1851. He went on to become Professor of Mathematics at the University of Indiana. During this period of tenure, he had a two-year break which he spent as Professor of Mathematics and Astronomy at Jefferson College in Pennsylvania. In retirement Kirkwood moved to California and in 1891 he became a lecturer in astronomy at the University of Stanford. He died in Riverside, California, on 11 June 1895.

Kirkwood's first astronomical paper was published in 1849. It consisted of a dubious mathematical description of the rotational periods of the planets. He was more interested, however, in the distribution of the asteroids in the Solar System. As early as 1857 he had noticed that three regions of the minor planet zone, sited at 2.5, 2.95 and 3.3 astronomical units from the Sun, lacked asteroids completely. In 1866, he published his analysis of this observation and proposed that the gaps, subsequently named 'Kirkwood gaps', arose as a consequence of perturbations caused by the planet Jupiter. The effect of Jupiter's mass would be to force any asteroid that appeared in one of the asteroid-free zones into another orbit, with the result that it would immediately leave the zone. It has since been proposed that non-gravitational effects, such as collisions, may also play a role in the maintenance of these gaps. Since Kirkwood's time, several more such gaps have been recognized.

Kirkwood used the same theory to explain the non-uniform distribution of particles in the ring system of Saturn. He suggested that Cassini's division was maintained by the perturbing effect which Saturn's satellites – Mimas, Enceladus, Thetys, Dione and, to a lesser extent, Rhea and Titan – had on the orbits of the particulate material making up the rings. The gravitational forces of these satellites would prevent the ring material from entering the Cassini Division, the Encke's Division, and the gap between the Crêpe Ring and Ring B in a manner similar to the way in which Jupiter affected the asteroids.

Kirkwood is best known for his work on the 'Kirkwood gaps', but he also carried out research into comets and meteors, made a fundamental critique of Laplace's work on the evolution of the Solar System, and carried out preliminary studies on families of asteroids.

Kitasato Shibasaburo 1852–1931 was a Japanese bacteriologist who is generally credited with the discovery of the bacillus that causes bubonic plague. He also did much important work on other diseases, such as tetanus and anthrax. His work gained him many international honours, and in 1923 Japan made him a baron.

Kitasato was born on 20 December 1852 in a small mountain village on the island of Kyushu in Japan. Keenly interested in science, he studied medicine at the Kumamoto Medical School and later at Tokyo University, from which he graduated in 1883. After graduating he joined the Central Bureau of the Public Health Department, but in 1885 he was sent by the government to study new developments in bacteriology in Germany and went to work in the laboratory of Robert Koch in Berlin. Under Koch's guidance, Kitasato quickly mastered the new techniques and began his own research, which was so successful that he was made an honorary professor of Berlin University before he returned to Japan in 1891. On his return, Kitasato set up a small private institute of bacteriology, the first of its kind in Japan. Later the institute received financial assistance from the government, but in 1915 it was incorporated into Tokyo University against Kitasato's wishes and he resigned as its director. In the same year he founded another establishment, the Kitasato Institute, which he headed for the rest of his life. Kitasato died on 13 June 1931 in Nakanocho.

Kitasato did his first important work in Koch's laboratory in Germany where, in 1889, he became the first to obtain a pure culture of *Clostridium tetani*, the causative bacillus of tetanus. In the following year, working with Emil von Behring, Kitasato discovered that animals can be protected against tetanus by inoculating them with serum containing inactive tetanus toxin. This was the very important discovery of antitoxic immunity, and Kitasato and von Behring rapidly developed a serum for treating anthrax. In the same year they published a paper on their combined work, giving details of their success with tetanus and similar results with diphtheria, on which von Behring, helped by Paul Ehrlich, had concentrated.

After returning to Japan, Kitasato was sent by his government to Hong Kong in 1894 to investigate an epidemic of bubonic plague. France also sent a small research team led by Alexandre Yersin, a former pupil of Louis Pasteur. The two teams did not collaborate because of language difficulties and there is some doubt as to whether it was Kitasato or Yersin who first isolated *Pasteurella pestis*, the bacillus that causes bubonic plague. But Kitasato published his discovery of the bacillus several weeks before Yersin announced his findings and is therefore generally credited with the discovery.

Kitasato also isolated the causative organism of dysentery in 1898 and studied the method of infection in tuberculosis.

Klaproth Martin Heinrich 1743–1817 was a German chemist famous for his discovery of several new elements and for pioneering analytical chemistry.

Klaproth was born in Wernigerode, Saxony, Germany, on 1 December 1743. His home was destroyed by fire when he was eight years old, leaving the family in poverty. He was apprenticed to an apothecary when he was 16, and after moving from employer to employer, finally in 1771 he became manager of a pharmacy in Berlin. Nine years later he set up on his own, doing chemical research. He took an appointment as a chemistry lecturer at the Berlin School of Artillery in 1792, and when the University of Berlin was founded in 1810 it was Klaproth to whom the authorities offered the first Chair in Chemistry, which he held until his death. He died in Berlin in 1 January 1817.

Klaproth's contributions to the discovery and isolation of new elements began in 1789. From the semi-precious gemstone zircon he prepared an oxide ('earth') containing the new metallic element zirconium. In the same year he investigated pitchblende and from this black ore obtained a yellow oxide containing another new metallic element, which he named uranium after the planet Uranus, discovered by William Herschel in 1781. He distinguished strontia (strontium oxide) from baryta (barium oxide) and in 1795 rediscovered and named titanium, acknowledging the prior isolation of the element four years earlier by William Gregor (1761–1817). He isolated chromium in 1797 independently of Louis Vauquelin, but credited Franz Müller (1740–1825) with the priority for the discovery of tellurium, which Klaproth extracted in 1798 and named after *tellus*, the Latin for earth; he refuted, however, the claim to this discovery by the Hungarian chemist P. Kitaibel (1757–1817).

In 1803, Klaproth identified cerium oxide and confirmed the existence of cerium, discovered by Jöns Berzelius in the same year and named after the newly found asteroid Ceres. He also studied the rare earth minerals researched by Johan Gadolin (1760–1852), confirming that they are complex mixtures of very similar substances.

All of Klaproth's work on minerals and new elements hinged on his outstanding ability as a quantitative analytical inorganic chemist, a branch of the science which he can be credited with helping to found. He was a champion of Antoine Lavoisier's antiphlogiston theory of combustion. But he was not reluctant (unlike some of his contemporaries) to publish anomalous results; he did not 'modify' his findings to make them suit some preconceived theory. He also applied his analytical skills to archaeological finds and antiquities, such as metal artefacts, glassware and coins.

Klein Christian Felix 1849–1925 was an extremely influential German mathematician and mathematical physicist whose unification of the various Euclidean and non-Euclidean geometries was crucial to the future development of that branch of mathematics. Equally skilled, however, in almost every branch of mathematics, Klein had as perhaps his greatest talent the ability to discover relationships between different areas of research, rather than in carrying out detailed calculations. He also possessed great organizational skills, and initiated and supervised the writing of an encyclopedia on mathematics, its teaching and its applications. He became widely known for his lectures and his books on the historical development of mathematics in the nineteenth century. It was to a great extent due to him that Göttingen became the main centre for all the exact sciences – not just mathematics – in all Germany.

Klein was born on 25 April 1849 in Düsseldorf, where he grew up and was educated. In 1865, at the age of 16, he went to the University of Bonn to read mathematics and physics; only three years later he was awarded his doctorate there. He decided to further his education by spending a few months in different European universities, and went to Göttingen, to Berlin, and then to Paris in 1869. At the outbreak of the Franco–Prussian War he was obliged to leave Paris so he entered the military service as a medical orderly.

In 1871, Klein qualified as a lecturer at Göttingen University; the following year he became full Professor of Mathematics at Erlangen University. In his inaugural address there he introduced what he called his 'Erlangen *Programm*' on geometries. From 1875 to 1880 he was Professor at the Technische Hochschule in Munich, and then took up a similar post from 1880 to 1886 at Leipzig. He spent the rest of his active research life at Göttingen, retiring as Professor in 1913 because of ill health. During and after World War I, however, he gave lectures at his home. Klein died in Göttingen on 22 June 1925.

Klein *Klein bottle is a 'bottle' in name only because, although it would be possible to store liquids in it, it is in fact a one-sided surface which is closed and has no boundary.*

Klein's Erlangen programme of 1872 comprised a proposal for the unification of Euclidean geometry with the geometries that had been devised during the nineteenth century by mathematicians such as Karl Gauss, Nicolai Lobachevsky, János Bolyai and Bernhard Riemann. He showed that the different geometries are each associated with a separate 'collection' or 'group' of tranformations. Seen in this way, the geometries could all be treated as individual members of one overall family, and from this very connection conclusions and inferences could be drawn. In the next two years, Klein developed the 'programme' and published papers which demonstrated that every individual geometry could be constructed purely projectively; he produced projective models for Euclidean, elliptic and hyperbolic geometries. (Such projective models are now called Klein models.) Much later in his life, Klein returned to the Erlangen programme to apply it to problems in theoretical physics, with special reference to the theory of relativity.

Klein's early research was on topics in line geometry. In 1870, with Sophus Lie (1842–1899), who was later to succeed him at Göttingen, he discovered fundamental properties of the asymptotic lines of the 'Kummer' surface, which became famous in algebraic line geometry. In his work on number theory, group theory and the theory of differential equations, Klein was greatly influenced by Bernhard Riemann. It was Klein who redefined a Riemann surface so that it came to be regarded as an essential part of function theory, not just a valuable way of representing multi-valued functions. Borrowing concepts from physics (and especially fluid dynamics), Klein revised and developed the theory in such a way that to him it seemed the most important work he ever accomplished.

He was also interested in fifth-degree (quintic) equations, and succeeded in solving the general algebraic equation of the fifth degree by considering the icosahedron. This work led him on to elliptic modular functions.

In the 1890s, he worked on mathematical physics and engineering, and wrote a textbook with Arnold Sommerfeld (1868–1951) on the theory of the gyroscope which is still an important work in this field of mechanics.

In 1900, he became interested in school mathematics, and recommended the introduction of differential and integral calculus and the function concept into school syllabuses. Eight years later, at the International Congress of Mathematicians in Rome, he was elected Chairman of the International Commission on Mathematical Instruction.

Knopf Eleanora Frances Bliss 1883–1974 was a US geologist who is known for her laboratory work on metamorphic rocks and for bringing the technique of petrofabrics to the United States.

Eleanora Frances Bliss was born in Rosemont, Pennsylvania, on 15 July 1883. She was educated at the Florence Baldwin School and from 1900 at Bryn Mawr College, a private women's college, from which she graduated in 1904 with an AB in chemistry and an AM in geology. Between 1904 and 1909 she worked as an assistant curator

in the geological museum and as a demonstrator in the geological laboratory at Bryn Mawr. In 1912, having spent a year at Berkeley (1910–11), she received a PhD in geology from Bryn Mawr and after passing the civil service examinations she went to Washington, DC, as a geologic aide to the United States Geological Survey (USGS); she was promoted to assistant geologist in 1917 and for three years worked with the federal survey and the Maryland State Geological Survey. In 1920 she married the geologist Adolf Knopf (1882–1966) and moved to New Haven where he became professor at Yale. Knopf continued work with the USGS, in the early 1920s studying the Pennsylvania and Maryland piedmont and from 1925 onwards the mountainous region along the New York–Connecticut border. During the 1930s she was also a visiting lecturer at Yale and at Harvard. In 1951, Knopf and her husband moved to Stanford University where she was appointed research associate in the geology department. She retired in 1955 and died of arteriosclerosis on 21 January 1974 in Menlo Park, California.

Knopf's early work on the geology of the area around Bryn Mawr led to several publications including a report in the *American Museum of Natural History Bulletin* in 1913 on her discovery of the mineral glaucophane in Pennsylvania, previously unsighted in America east of the Pacific. It was from her work on the geologically complex Stissing Mountain area on the New York–Connecticut border and her laboratory work on metamorphic rocks, however, that gained her reputation. She brought to the USA the technique of petrofabrics, developed by Bruno Sander of Innsbruck University, and applied it over the next 40 years to the study of metamorphic rocks. Her major work on the subject, *Structural Petrology*, was published in 1938.

Koch Robert 1843–1910 was a German bacteriologist who, with Louis Pasteur, is generally considered to be one of the two founders of modern bacteriology. He developed techniques for culturing, staining and observing microorganisms and discovered the causative pathogens of several diseases – including tuberculosis, for which discovery he was awarded the 1905 Nobel Prize for Physiology or Medicine.

Koch was born on 11 December 1843 in Klausthal, Germany, one of 13 children of a mining official. He studied natural sciences and then medicine at Göttingen University, where he was taught by Friedrich Wöhler and Friedrich Henle, obtaining his medical degree in 1866. After serving as an army surgeon (on the Prussian side) in the Franco–Prussian War, in 1872 Koch became District Medical Officer in Wollstein where, despite having few research facilities, he began important investigations into anthrax. For a brief period in 1879 he was Town Medical Officer in Breslau, before being appointed to the Imperial Health Office in Berlin to advise on hygiene and public health. By 1881 his work was becoming well known and he was invited to speak at the Seventh International Medical Congress in London. In 1882, he announced to the Berlin Physiological Society his discovery of the bacillus that causes tuberculosis, and in the following year, while investigating an outbreak of cholera in the Nile delta, he identified the cholera bacillus. In 1885, he was appointed Professor of Hygiene at Berlin University and Director of the Institute of Hygiene. The Tenth International Medical Congress was held in Berlin in 1890 and Koch was persuaded to announce the discovery of an antituberculosis vaccine; this proved to be premature, however, as the vaccine was ineffective. In 1891, he was appointed Director of the newly established Institute for Infectious Diseases, but he resigned his directorship in 1904 and spent much of the rest of his life advising foreign countries on ways to combat various diseases. Koch died on 27 May 1910 in Baden-Baden, Germany.

Koch started his bacteriological research in the 1870s with the gift of a microscope from his wife, and built up a primitive laboratory in part of his consulting room. Out of necessity Koch devised simple and original methods for growing and examining bacteria. For three years he worked on anthrax in his spare time, developing techniques for culturing the bacteria in cattle blood and in aqueous humour from the eye. He trapped a small smear of blood from an anthrax victim with a drop of aqueous humour between two microscope slides and observed the bacteria grow and divide under the microscope, finding that the bacteria were shortlived but that they formed spores that were resistant to desiccation. He then inoculated animals with the spores and found that they developed anthrax, thus proving that the spores remained infective; this was the first time a bacterium cultured outside a living organism had been shown to cause disease. Koch published his findings, but only after Pasteur's demonstration of an anthrax vaccine in 1882 were Koch's findings accepted.

Koch experimented with various dyes and found some that stain bacteria and make them more visible under the microscope. He also devised an ingenious method of separating a mixture of bacteria, which involved inoculating an animal with the bacteria and passing the resulting infection from one animal to another until, at the end of the experimental chain, only one type of bacterium remained. Using this method he identified the bacteria responsible for several disorders, including septicaemia.

On joining the Imperial Health Office in 1879, Koch was provided with two assistants and for the first time had adequate laboratory facilities. Here he developed the technique of culturing bacteria on gelatin. Using this technique Koch and his assistants isolated several microorganisms and showed that they cause disease. They also investigated the effects of various disinfectants on different bacteria and showed that steam is more effective than dry heat in killing bacteria, a discovery that revolutionized hospital operating theatre practice.

Koch then set out to discover the causative agent of tuberculosis, a common and frequently fatal disease at that time. Initially he was unable to find any microorganisms that might cause the disease, but after developing a special staining technique he identified the bacterium responsible and, despite the difficulties caused by the bacterium's small size and slow rate of growth, managed to culture it in 1882.

In 1883, Koch went to the Nile delta to investigate a cholera epidemic. Finding bacteria in the intestinal walls of dead cholera victims and the same bacteria in the excreta of cholera patients, he succeeded in isolating the causative organism. On a later visit to Calcutta, where cholera was rife, he found similar bacteria in excreta and in supplies of drinking water. On returning to Berlin he advised regular checks on the water supply, made recommendations regarding sewage disposal, and organized courses in the recognition of cholera. And when the disease occurred in Hamburg in 1892, he recommended that the victims should be isolated, all excretory matter should be disinfected and that a special check should be made on the water supply.

Koch made several other important contributions. As a result of his investigations into a bubonic plague epidemic in Calcutta in 1897, he showed that rats are vectors of the disease (although there is no evidence that he knew that the rat flea was the actual vector). He also demonstrated that sleeping sickness is transmitted by the tsetse fly. His isolation of many disease-causing organisms eventually led to the development of vaccines and to the realization of the importance of the public health measures he recommended. Furthermore, many bacteriologists received their training as his assistants, including Georg Gaffky, Friedrich Löffler, Shibasaburo Kitasato, and the Nobel prizewinners Emil von Behring and Paul Ehrlich. Perhaps most important, however, Koch formulated a systematic method for bacteriological research, including various rules – still observed today – for identifying pathogens. According to these rules (called Koch's postulates), the suspected pathogen must be identified in all of the cases examined; the pathogen must then be cultured through several generations; these later generations must be capable of causing the disease in a healthy animal; and the newly infected animal must yield the same pathogen as found in the original victim.

Kolbe Adolf Wilhelm Hermann 1818–1884 was a German organic chemist, generally credited as the founder of modern organic chemistry with his synthesis of acetic acid (ethanoic acid) – an organic compound – from inorganic starting materials. (Previously organic chemistry had been considered as the branch of the science devoted to compounds that occur only in living organisms.)

Kolbe was born in Elliehausen, near Göttingen, on 27 September 1818, the eldest of 15 children of a Lutheran pastor; his mother was the daughter of A. F. Hempel, Professor of Anatomy at Göttingen University. He was educated at the Gymnasium at Göttingen, where he was introduced to chemistry by a student who had studied under Robert Bunsen. In 1838 he entered Göttingen University, where he attended lectures by Friedrich Wöhler (who ten years earlier had synthesized urea from ammonium cyanate, arguably the first inorganic–organic transition) and became a great admirer of Jöns Berzelius. He became an assistant to Bunsen at Marburg University in 1842 and three years later accepted an invitation from Lyon Playfair to work with him at the London School of Mines.

During his two-year stay in London Kolbe met many leading chemists of the day, including Edward Frankland who was at that time developing his theory of valency. In 1847, Kolbe moved to Brunswick to join the editorial team on the *Handwortenbuch der Chemie* (founded by von Liebig, Wöhler and Poggendorf), and in 1851 he was appointed Bunsen's successor as Professor of Chemistry at Marburg – a rapid promotion that did not meet with the full approval of the establishment. By 1865 he had moved to Leipzig and had begun to set up the largest and best equipped laboratory of the time, which had its full complement of students within three years. From 1869 he was editor of the *Journal für practische Chemie* and became notorious for his very personal and often violent criticism of the work of his contemporaries. He continued his theoretical and literary work until he died in Leipzig on 25 November 1884.

Kolbe correctly realized that organic compounds can be derived from inorganic materials by simple substitution. He introduced a modified idea of structural radicals, which contributed to the development of the structure theory, and he predicted the existence of secondary and tertiary alcohols. He is best known for his work on the electrolysis of the fatty (alkanoic) acids, for his important preparation of salicylic acid (2-hydroxybenzenecarboxylic acid) from phenol – called the Kolbe reaction, which was to lead to an easy synthesis of the drug aspirin – and for his discovery of nitromethane.

His early work, with Frankland, was on the conversion of nitriles into fatty acids. He then investigated the action of electric currents on organic compounds. But he became an extremely conservative influence on organic chemistry largely because of his adherence to the ideas of Berzelius. His method of representing molecular structures eventually gave way to the much simpler structural theory based on the work of Friedrich Kekulé, although his unorthodox formulae actually embodied many of the ideas that were developed by Kekulé.

One of the greatest drawbacks of Kolbe's formulae resulted from his refusal to abandon equivalent weights in favour of atomic weights (relative atomic masses). Until 1869, for example, he still followed Berzelius's contention that the equivalent weight of carbon was 6, and that of oxygen 8, so that he had to double the number of atoms of these elements in his formulae. He therefore wrote the methyl group as C_2H_3– and assumed that in acetic (ethanoic) acid the methyl radical was joined to oxalic (ethanedioic) acid and water, which he expressed as $C_2H_3 + C_2O_3 + HO$. He became convinced that methyl groups existed in compounds and could be isolated from them. By electrolysing potassium acetate (ethanoate) he obtained a gas which he thought was 'methyl' (really ethane). Frankland had obtained 'free ethyl' (really butane) by the action of zinc on ethyl iodide (iodoethane) and both chemists were now certain that they had proved the existence of radicals in organic compounds. Their formulae continued to represent these misapprehensions.

Kolbe explained the relationship between aldehydes and ketones and identified a new group – the carbonyl group. He was able to predict their behaviour on oxidation, and in so doing anticipated the existence of secondary and tertiary alcohols. His later work involved the nitroparaffins (nitroalkanes) and the Kolbe synthesis for salicylic acid. But he never could see the similarity of his formula system and that of Kekulé. In 1858, he bitterly opposed the whole idea of structural formulae and ridiculed the theory of structural isomerism put forward by Jacobus van't Hoff and Joseph Le Bel, which are now regarded as fundamental to structural organic chemistry.

Kornberg Arthur 1918–   is a US biochemist who in 1957 made the first synthetic molecules of DNA. For this achievement he shared the 1959 Nobel Prize for Physiology or Medicine with Severo Ochoa. By 1967 he had synthesized a biologically active artificial viral DNA.

Kornberg was born in Brooklyn, New York City, on 3 March 1918. He was educated at local schools and in 1933 graduated from the Abraham Lincoln High School. On a state scholarship he took a pre-medical course at the College of the City of New York, obtaining a BS degree in 1937. A further scholarship enabled him to go on to the University of Rochester School of Medicine from which he gained his medical degree in 1941. Following a year as an intern at the Strong Memorial Hospital in Rochester, he joined the US Coast Guard for a short time. From 1942 to 1945 he worked in the nutritional section of the physiology department of the National Institute of Health in Bethesda, Maryland, becoming Chief of the Enzyme and Metabolic Section from 1947 to 1952. He held senior appointments at the Washington University School of Medicine (1953) and the Stamford University School of Medicine, Palo Alto (1959), before becoming executive head of the Biochemistry Department at Stamford University.

From the beginning of his career, Kornberg was interested in enzymes – not merely what they are but what they do. For many years the fundamental genetic mystery had concerned the ability of a cell to produce one particular enzyme and no other. In 1941, George Beadle and Edward Tatum demonstrated that genes control the processes of life by chemical means and in 1944 Oswald Avery isolated the chemical responsible – the nucleic acid DNA, whose structure was elucidated by Francis Crick and James Watson in 1953. It was known that DNA consists of sugar, phosphate and nucleotides, the 'letters' of a genetic alphabet that spell out the 'recipe' of a particular genetic trait by controlling the production of the appropriate protein. Another nucleic acid, RNA, translated the DNA code in this complex chemical process.

At Washington University Kornberg set himself the task of producing a giant molecule of artificial DNA. To do this he needed a pre-existing DNA molecule as a template to be copied, the four nucleotides adenine, thymine, guanine and cytosine – known as A, T, G and C – and an enzyme to select and arrange the nucleotides according to the directions from the template and link them together to form the

DNA chain. In 1956, he isolated the enzyme DNA polymerase and a year later made an artificial DNA that had all the physical and chemical properties of its natural counterpart, but lacked its genetic activity. One cause of this partial failure was the impurity of the enzyme (which had resulted in errors in the arrangement of nucleotides).

He then took a new and simpler template, the DNA of the virus known as Phi X174, which is single-stranded and in the form of a ring; its activity (infectivity) is lost if the ring is broken. Another enzyme was needed to close the ring and in 1966 it was discovered, and called ligase. Using the natural DNA of Phi X174 as a template, Kornberg and his co-workers mixed the enzyme DNA polymerase, the enzyme ligase, and the four nucleotides. The DNA polymerase ordered the nucleotides into the arrangement dictated by the template and the ligase closed the ring of the artificial DNA so produced.

When the synthetic DNA was added to a culture of bacteria cells (*Escherichia coli*), it infected them and usurped their genetic machinery. Within minutes the cells had abandoned their normal activity and had started to produce Phi X174 viruses. The synthetic DNA had now become the template for a second generation of synthetic viruses identical to the original: the sequence of 6,000 'building blocks' and the arrangement of the 35 atoms within each was precisely the same in the artificial ones as in the natural ones.

By this achievement, Kornberg has opened the way to future progress in the study of genetics, the possibility of curing hereditary defects and controlling virus infections and cancer.

Kornberg Hans Leo 1928–   is a German-born British biochemist who has made important contributions to the understanding of metabolic pathways and their regulation, especially in microorganisms.

Kornberg was born in Herford, Germany, on 14 January 1928, and went to Britain in 1939 as a refugee from Nazi persecution. He attended schools in southern England and in Wakefield, Yorkshire, before entering Sheffield University in 1946. He gained his BSc in 1949 and his PhD in 1953, studying under Hans Krebs who in 1937 had determined the tricarboxylic acid cycle (Krebs cycle), the sequence of energy-generating biological reactions in which glucose is converted into carbon dioxide and water. From 1953 to 1955 Kornberg travelled in the United States, studying first the pentose phosphate pathway (an alternative to the Krebs cycle) under E. Racker at Yale and then helping Melvin Calvin at the University of California to resolve a controversy concerning one step of the 'dark reaction' of photosynthesis in plants.

On his return to England he joined the Medical Research Council Cell Metabolism Unit headed by Krebs in Oxford. From 1960 to 1975 he was Professor and Head of the Biochemistry Department at the University of Leicester, and then he became the Sir William Dunn Professor of Biochemistry at Cambridge University. He was knighted in 1978.

The tricarboxylic acid (Krebs) cycle has two apparently conflicting roles. One is the complete oxidation (breakdown) of glucose to provide the energy required in all cellular processes – a catabolic role. The other is to provide carbon 'skeletons' (backbones of organic molecules) for the various complex compounds essential to cellular function – and anabolic role. Certain simple organisms can survive with only acetate (ethanoate, a two-carbon molecule) as their source of carbon. Kornberg directed his research to seek an answer to the question of how an organism uses acetate to build up larger molecules while at the same time needing to break it down to provide the very energy required for these anabolic functions.

His study led in 1957 to the discovery of the 'glyoxylate cycle' in plants, microorganisms and some worms (glyoxal is ethan-1, 2-dial). The cycle's basic function is to convert the two-carbon molecules of acetate into a four-carbon molecule of succinate (buten-1, 4-dioate). It is achieved by means of a bypass reaction in which the decarboxylation reactions of the normal tricarboxylic acid cycle (in which enzymes remove carbon that ultimately is converted to carbon dioxide) are skipped over.

The glyoxylate cycle is localized in subcellular organelles called glyoxysomes in the germinating seeds of higher plants, which use it to convert stored fatty acids into carbohydrates for the production of the energy needed for rapid growth.

Two years later, in 1959, Kornberg discovered the glycerate pathway and in 1960 the dicarboxylic acid cycle, which play important roles in enabling microorganisms to grow on certain types of nutrients. He investigated the control mechanisms essential to the regulations of these metabolic processes, studying the way in which the cellular economy is balanced to prevent overproduction or waste. He introduced the concept of 'anaplerotic reactions', whereby metabolic processes are maintained by special enzymes that replenish materials syphoned off for anabolic purposes.

Kornberg studied the regulation of enzyme activity at the genetic level, where the production of enzymes is largely controlled, and at the cytoplasmic level, where the reaction rates of existing enzymes can be modified. His later research concentrated on the very first step in the processing of food materials, the selective uptake of compounds across cellular membranes. He used mutant bacterial strains which can be grown on specific, defined substrates, to rigorously control the nature of the materials entering their cells. This combination of genetic and biochemical approaches provides a unique way of studying both the nature of the transport process and the way in which it is regulated.

Korolev Sergei Pavlovich 1906–1966 was the Russian engineer who designed the first crewed spacecraft, interplanetary probes and soft lunar landings. As such, he was one of the founders of practical space flight.

Korolev was born on 30 December 1906 at Zhitomir in the Ukraine, the son of a Russian schoolmaster. His early technical training was received at a building trades school in Odessa; he graduated from there in 1924. Three years later he entered the aviation industry, but continued with his studies at the Moscow School of Aviation. He graduated from the Baumann Higher Technical School in 1930 and, in June of that year, he became a senior engineer at the Central Aerodyamics Institute. After successfully developing a number of gliders he became interested in rocket-type aircraft. In 1931, with Tsander, he formed a group for the study of jet propulsion. This became an official body, and in 1932 Korolev became its head. The group eventually trained many of the men responsible for the subsequent Soviet triumphs in space. It was this group that built the first Soviet liquid-fuel rocket, which was launched in 1933. Korolev was appointed deputy director of the new Institute for Jet Research, which was formed by combining the original Korolev group and the Gas Dynamics Laboratory in 1933. In 1934, he was appointed head of the rocket vehicle department.

Korolev was responsible for designing a number of rockets and, from 1924 to 1946, he worked as deputy chief engine designer in the Specialized Design Bureau. Later, Korolev was appointed head of the large team of scientists who developed high-power rocket systems.

Korolev received many honours. In 1953, he was made corresponding member of the Academy of Sciences of the USSR and a member of the CPSU. He was twice named Hero of Socialist Labour, the first time in 1956 and the second time in 1961. In 1958, he was made full member of the Academy of Sciences, having been awarded the Lenin Prize in 1958. In addition, his name is commemorated for all time: the largest formation on the hidden side of the Moon is named after him. On his death on 14 January 1966 his body was buried in the Kremlin Wall in Red Square, Moscow – one of the highest honours which could be conferred on a Soviet citizen.

Korolev published his first paper on jet propulsion in 1934. By 1939 he had designed and successfully launched the Soviet 212 guided wing rocket. This was followed in 1940 by the RIP-318–1 rocket glider, which made its first piloted flight in 1940. From then on, Korolev was responsible for a large number of historical first achievements in space exploration.

He was responsible, in an executive and scientific capacity, for directing the early research and design work which later led to vast Soviet achievements in space flight. Korolev initiated many scientific and technological ideas which have since been widely used in rocket and space technology, including ballistic missiles, rockets for geophysical research, launch vehicles and crewed spacecraft. The most famous of these is the Vostok series of Soviet single-seater spaceships designed for close-earth orbit in which Soviet cosmonauts made their first flights.

Vostok 1, crewed by Yuri Gagarin, made the first crewed space flight in history, which took place on 12 April 1961. Launched from the Baikonur space-launch area, Gagarin and *Vostok 1* completed one orbit of the earth in just under two hours. This flight was followed by *Vostoks 2, 3, 4, 5*

and *6* (which was crewed by Valentina Tereshkova, the first female astronaut). The technological improvements made by Korolev and his team were such that Tereshkova and *Vostok 6* were able to make 48 orbits in a total time of just over 70 hours.

These close-earth flights were followed by the launching of Sun satellites and the flights of the uncrewed interplanetary probes into the Solar System – including the Elecktron and Molniia series. Korolev was also responsible for the *Voskhod* spaceship, from which the first space-walks were made.

Krebs Hans Adolf 1900–1981 was a German-born British biochemist famous for his outstanding work in elucidating the cyclical pathway involved in the intracellular metabolism of foodstuffs – a pathway known as the tricarboxylic cycle, the citric acid cycle or the Krebs cycle. He received many honours for this work, including the 1953 Nobel Prize for Physiology or Medicine (which he shared with Fritz Lipmann, a German-born US biochemist), a knighthood in 1958, and several medals and honorary degrees.

Krebs was born on 25 August 1900 in Hildesheim, Germany, the son of an ear, nose and throat specialist. He was educated at the universities of Göttingen, Freiburg, Munich, Berlin and Hamburg, gaining his medical degree from the last in 1925. From 1926 to 1930 he worked under Otto Warburg at the Kaiser Wilhelm Institute in Berlin, then taught at the University of Freiburg until 1933, when he moved to England because of the rise to power of Adolf Hitler and the Nazi movement in Germany. On his arrival in England, Krebs went to Cambridge University, as a Rockefeller Research Student from 1933 to 1934, then as a Demonstrator in Biochemistry until 1935. From 1935 to 1954 he worked at Sheffield University, as Lecturer in Pharmacology from 1935 to 1938, Lecturer in charge of the Department of Biochemistry from 1938 to 1945, and Professor of Biochemistry from 1945 to 1954. He then moved to Oxford University as Whitley Professor of Biochemistry and a Fellow of Trinity College, which positions he held until his retirement in 1967.

While working under Warburg, Krebs became interested in the process by which the body degrades amino acids. He discovered that, in amino-acid degradation, nitrogen atoms are the first to be removed (deamination) and are then excreted as urea in the urine. Continuing this line of research, Krebs then investigated the processes involved in the production of urea from the removed nitrogen atoms, and by 1932 he had worked out the basic steps involved in what is now known as the urea cycle. Later workers discovered the details of the cycle, but Krebs's original basic scheme was correct.

Krebs is best known, however, for his discovery of the processes involved in the tricarboxylic acid cycle, by which carbohydrates are aerobically metabolized to carbon dioxide, water and energy. The energy yielded by this pathway is the main source of intracellular energy and therefore of the entire organism. Previous work by Otto Meyerhof and Carl and Gerty Cori had shown that the carbohydrate

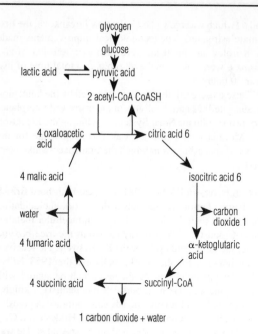

Krebs *The purpose of the tricarboxylic acid (TCA) cycle (Krebs cycle) is to complete the oxidation of glucose begun in glycolysis. In the sequence of reactions from citric acid to oxaloacetic acid, pyruvic acid is oxidized to carbon dioxide and water. During this process considerable amounts of energy are released – 93% of the total energy released in glucose oxidation. (The numbers in the diagram indicate the number of carbon atoms in the principal compounds.)*

glycogen is broken down in the liver to lactic acid by a process that does not require oxygen and that yields very little energy. Krebs continued this work (while at Sheffield University) to show how the lactic acid is then metabolized to carbon dioxide, water and energy. As a result of his investigations – performed on pigeon breast muscle – Krebs proposed the self-regenerating biochemical pathway now known as the Krebs or tricarboxylic acid (TCA) cycle.

In this cycle the two-carbon acetyl coenzyme A (acetyl CoA) – derived from lactic acid – combines with the four-carbon oxaloacetic acid to form the six-carbon citric acid, which then undergoes a series of reactions to be reconverted to oxaloacetic acid. This oxaloacetic acid combines with another molecule of acetyl CoA to form citric acid again, and the cycle is repeated. Carbon dioxide, water and hydrogen are produced by the various reactions in the cycle; the hydrogen combines with oxygen in a complex series of reactions, to produce energy and water. The degradation of glucose to carbon dioxide, water and energy occurs in two stages. The first stage – in which acetyl CoA is formed – occurs in the cytoplasm; the second stage, the tricarboxylic acid cycle, takes place in mitochondria.

In addition to carbohydrate metabolism, the tricarboxylic acid cycle is also involved in the degradation of fats and amino acids, and provides substrates for the biosynthesis of other compounds, such as amino acids. Thus the cycle plays a central role in cell metabolism, and Krebs's elucidation of the steps involved in it was therefore a fundamental contribution to biochemistry.

Kronecker Leopold 1823–1891 was a German mathematician, skilled in many branches of the subject but preeminent in none, who is remembered chiefly for the 'Kronecker delta'.

Kronecker was born at Liegnitz (now Legnica, Poland) on 7 December 1823 and attended secondary school there, where he was taught by the outstanding mathematician Ernst Kummer (1810–1893). Kummer remained a lifelong friend and exercised great influence upon Kronecker, especially in interesting him in number theory. In 1841, Kronecker entered the University of Berlin, where he studied mathematics, chemistry, astronomy and philosophy. In 1843 he spent some time at the universities of Bonn and Breslau (where he had been appointed Professor of Mathematics), before returning to Berlin. In 1845, he received his doctorate from Berlin for a thesis on complex units.

Kronecker was then diverted from mathematical research by pressing family duties – the management of an estate near Liegnitz and the winding-up of an uncle's banking business. Not until 1855, when he took up residence again in Berlin, did he begin a professional career in mathematics. Financially independent, he was able to devote himself to research without taking employment, but his election to the Berlin Academy in 1861 entitled him to give lectures at the university, an occupation which gave him such satisfaction that he turned down an invitation to take the Chair of Mathematics in Göttingen in 1868. By then his reputation had grown to such an extent that he was elected to the French Academy. In 1883, he succeeded Ernst Kummer as Professor of Mathematics at Berlin, a Chair which he held until his death, in Berlin, on 29 December 1891.

Kronecker was a gifted mathematician who never quite made the kind of discovery to place him in the very front rank of mathematical geniuses. His thesis on complex units, for example, came close to hitting upon Kummer's idea of 'ideal numbers', but the actual discovery belongs to his mentor. Much of his life he was engaged in somewhat disputatious and cranky controversy. After 1855 he worked chiefly in the fields of number theory, algebra and elliptical functions, and he achieved a great deal in his endeavour to unify analysis, algebra and elliptical functions. But his progress, and his influence, were somewhat blocked by his obsession with the idea that all branches of mathematics (apart from geometry and mechanics) should be treated as parts of arithmetic. He also believed that whole numbers were sufficient for the study of mathematics – 'The whole numbers has the Dear God made; all else is man's work'. His refusal to grant irrational numbers equal validity with whole numbers, although it placed him in that tradition of nineteenth-century mathematical thought which attempted to define processes such as differentiation and integration in terms of simple arithmetic, led him into what we can now see as time-wasting debate.

One other controversy was more fruitful. His debate with Georg Cantor (1843–1918) and Karl Weierstrass (1815–1897) about the foundations of mathematics, has been carried into the twentieth century in the form of the debate between the formalists and the intuitionists. His system of axioms, published in 1870, has also been useful. His axioms were later shown to govern finite Abelian groups and have proved to be important in the modern development of algebra.

Kronecker remains, however, most famous for his work in linear algebra, especially for the delta named after him. (It is denoted by δ_{rs}, for which $\delta_{rs} = 0$ when $r \neq s$, $\delta_{rs} = 1$ when $r = 1$, where r, s are 1, 2, 3 ...) Kronecker found this a useful delta in the evaluation of determinants, the r and s being concerned with a row and a column of the determinant. The delta is a fairly simple example of a tensor and Kronecker's pupil, Hensel, who edited and published Kronecker's work between 1895 and 1931, was largely responsible for popularizing the notation.

Krupp Alfred 1812–1887 was a German metallurgist who became known as the Cannon King because of his success in manufacturing cast-steel guns. He also invented the first weldless steel tyre for railway vehicles.

Krupp was born in Essen on 26 April 1812. At 14 years of age he was obliged to leave school to help his mother run the small family steel works. The firm remained small until Krupp made his first gun in 1847. In 1851, at the Great Exhibition in London, he exhibited a solid flawless ingot of cast steel weighing 2 tonnes. Soon afterwards, Krupp produced the first weldless steel tyre for railway vehicles. The works expanded its operations in the manufacture of artillery, eventually becoming the largest armaments firm in the world. Krupp was also humanely interested in the social welfare of his workers, and introduced low-cost housing and pension schemes for all his employees. He died in Essen on 14 July 1887.

In 1811 Alfred Krupp's father, Frederick, had founded a firm in Essen for the purpose of making cast steel of a superior quality. His experiments had produced a formula for making a high-quality product and, when he died prematurely in 1826, the secret was left to his son. Four years after taking charge of his father's company, Alfred Krupp extended production to include the manufacture of steel rolls. He also designed and developed new machines. He invented the spoon roll for making spoons and forks, and manufactured rolling mills for use in government mints. In 1847 the firm cast its first steel cannon, a small three-pounder muzzle-loading gun, but it still specialized in making fine steel suitable for dies, rolls and machine-building.

At the Great Exhibition of 1851 in London, Krupp exhibited his solid flawless ingot of cast steel. In 1852, he manufactured the first seamless steel railway tyre and was the first to introduce the Bessemer and open-hearth steel-making processes in Europe. The advent of the railways caused the firm's fortunes to increase. At first, railway axles and springs were the only cast steel products in use, but Krupp's seamless tyre greatly contributed to the speed and safety of railway travel and became extensively used. Then mass production of rails began and the company expanded still further to become a vast organization with collieries, ore-mine, blast furnaces, steel plants and manufacturing shops.

In 1863, Krupp established a physical laboratory for the testing of steels. To prove the quality of his own steel, Krupp turned to making guns and was the first to produce a successful all-steel gun. Drilled out of a single block of cast metal, it was the marvel of its day, and in the Franco–Prussian war of 1870–1871 the performance of Krupp guns won him world renown.

By this time, the firm employed 16,000 workers and had developed into a vast integrated industrialized empire – one of the first in Germany. Although Krupp's reputation is mainly built on the manufacture of weapons of war, the output of goods from his works – even in wartime – was predominantly for peaceful purposes.

Krupp was also a philanthropist. Recognizing early the human problems of industrialization, he created a comprehensive welfare scheme for his workers. As early as 1836 he instituted a sickness fund and, in 1855, he established a pension fund for retired and incapacitated workers. Low-cost housing, medical care and consumer cooperatives all enjoyed company backing. These institutions proved so successful that they acted as models for the social legislation enacted in Germany under Bismark from 1883 to 1889.

Krupp built housing settlements, hospitals, schools and churches for his employers. His social paternalism minimized serious labour troubles and fostered the Krupp employee's pride in being Kruppianer. From employing at one time only seven people, at Krupp's death his enterprise employed 21,000.

Kuhn Richard 1900–1967 was an Austrian-born German organic chemist who worked mainly with carbohydrates and was awarded (but not allowed immediately to accept) the 1938 Nobel Prize for Chemistry for his research on the synthesis of vitamins.

Kuhn was born into a Jewish family in Vienna on 3 December 1900, the son of a hydraulics engineer. He was first educated at home by his mother, who was a schoolteacher, and at the age of only eight began to attend the Döblinger Gymnasium, where one of his fellow pupils was Wolfgang Pauli (1900–1958) who was later to become a Nobel prizewinning physicist. He was introduced to chemistry by Ernst Ludwig, who was Professor of Medical Chemistry at Vienna University. Kuhn was conscripted into the Austrian army in 1917, and went to Vienna University at the end of World War I in 1918. He soon moved to Germany, and completed his university education at the University of Munich where he studied under Richard Willstätter, graduating in 1921 and gaining his PhD a year later with a thesis on enzymes. Kuhn then became Willstätter's assistant, before going in 1926 to teach at the Eidenössische Technische Hochschule in Zurich as Professor of General and Analytical Chemistry. He married one of his students in 1928, and they had six children. In 1929, he became Professor of Organic Chemistry at the University of Heidelberg and Director of the Kaiser Wilhelm (later Max Planck) Institute for Medical Research. He remained there until the late 1930s when he was caught in a Nazi round-up of Jews and imprisoned in a concentration camp. The award to him of a Nobel Prize came after the introduction of Adolf Hitler's policy forbidding any German to accept such an award. But after the end of World War II in 1945 Kuhn received his prize and returned to work in Heidelberg. He became editor of the chemical journal *Annalen der Chemie* in 1948. His health began to fail in 1965 and he died of cancer in Heidelberg on 31 July 1967.

Kuhn's early research concerned the carotenoids, the fat-soluble yellow pigments found in plants which are precursors of vitamin A. In the early 1930s, Kuhn and his co-workers determined the structures of vitamin A and vitamin B_2 (riboflavin) and isolated them from cow's milk, at about the same time as Paul Karrer was doing similar work. Then in 1938, they took about 70,000 l/123,200 pt (148,000 US pt) of skimmed milk and after a painstaking series of extractions isolated from it 1 g/0.035 oz of vitamin B_6 (pyridoxine).

In the 1940s, Kuhn continued to carry out research on carbohydrates, studying alkaloid glycosides such as those that occur in tomatoes, potatoes and other plants of the genus *Solanum*. In 1952, he returned to experiments with milk, extracting carbohydrates from thousands of litres of milk using chromatography. This work led in the 1960s to the investigation of similar sugar-type substances in the human brain; many of these chemicals were synthesized for the first time by Kuhn and his co-workers.

Kuiper Gerard Peter 1905–1973 was a Dutch-born US astronomer best known for his studies of lunar and planetary surface features and his theoretical work on the origin of the planets in the Solar System.

Kuiper was born on 7 December 1905 at Harenkarspel, the Netherlands. After completing his education in his own country, he emigrated to the United States in 1933 and four years later became a naturalized US citizen. He joined the staff of the Yerkes Observatory, which is affiliated to the University of Chicago. Between 1947 and 1949 he was Director of the Observatory, returning for a second term in this office from 1957 to 1960. From 1960 until his death in 1973, he held a similar position at the Lunar and Planetary Laboratory at the University of Arizona, and he was closely linked with the US space programme.

Kuiper's work on the origin of the planets stemmed from the theoretical discrepancies that arose from new twentieth-century hypotheses on galactic evolution. One of the more favoured of these theories is that stars, and presumably planets, are formed from the condensation products of interstellar gas clouds. For this condensation to take place, the gravitational effects pulling the cloud together must exceed the expansive effect of the gas pressure of the cloud. However, calculations of this hypothetical process showed that under given conditions of temperature and density, there was a lower limit to the size of the condensation products. In fact, it was found that this condensation theory was inadequate to account for the temperature or the amount of material that make up the bulk of the planets in the Solar System. To compensate for this, Kuiper and his colleagues proposed that the mass of the cloud from which the planets were formed was much greater than the present mass of the planets, and suggested that the mass of the original interstellar gas cloud was approximately one-tenth the mass of the Sun, or 70 times the total mass of the planets. The results of condensation according to these new conditions would produce 'proto-planets'. But the idea of the formation of proto-planets still appears to be unworkable, partly because it involves the condensation of only 1.5% of the original interstellar gas cloud, thus leaving the rest of the material unaccounted for.

Kuiper's work on planetary features proved to be far more fruitful. In 1948, he predicted that carbon dioxide was one of the chief constituents of the Martian atmosphere – a theory he was able to see confirmed when the era of research into that planet began in 1965 with the *Mariner* space probes to Mars. He also discovered the fifth moon of Uranus in 1948, which he called Miranda; and in 1949 he discovered the second moon of Neptune, Nereid. Compared with Neptune's first moon, Triton, which has a diameter of 3,000 km/1,864 mi, Nereid is a dwarf and has an eccentric orbit; its distance from Neptune varies by several million kilometres. Kuiper's spectroscopic studies of the planets Uranus and Neptune led to the discovery of features subsequently named 'Kuiper bands', found at wavelengths of 7,500Å (7.5×10^{-7} m), which have been identified as being due to the presence of methane.

During his working life, Kuiper instigated many planetary research programmes and he played a vital role in the United States' space-probe programme during the late 1960s and early 1970s. In recognition of his work, the International Astronomical Union has named a ray-crater on the planet Mercury after him.

Kummer Ernst Eduard 1810–1893 was a German mathematician, famous for his work in higher arithmetic and geometry, who introduced 'ideal numbers' in the attempt to prove Fermat's last theorem.

He was born at Sorau (now Zary, Poland) on 29 January 1810 and was educated there before entering the University of Halle in 1828. Although his original intention was to study theology, he soon changed to mathematics and proved himself so brilliant in the subject that in 1831 the university awarded him a doctorate for his prizewinning essay on sines and cosines. From 1832 to 1842 he taught at a school at Liegnitz (now Legnica, Poland), where Leopold Kronecker (1823–1891) was one of his pupils. He was Professor of Mathematics at the University of Breslau from 1842 to 1855, when he was appointed to professorships at the University of Berlin and the Berlin War College. In 1857 he was awarded the Grand Prix of the French Academy of Sciences and in 1868 he was elected Rector for one year of the University of Berlin. In 1882 he astonished his

Berlin colleagues by announcing that failing powers of memory were weakening his ability to carry out logical, coherent thought; he therefore resigned his Chair. He died at Berlin on 14 May 1893.

Kummer made an important contribution to function theory by his investigations – which surpassed those of Gauss – into hypergeometric series. But his two outstanding achievements were his introduction of 'ideal number' to attempt to solve Fermat's last theorem and his research into systems of rays, which led to the discovery of what is now known as the Kummer surface.

Fermat's last theorem is one of the most famous in the history of mathematics. It states that 'there do not exist integers x, y, z, none of which being zero, that satisfy:

$$x^n + y^n = z^n$$

where n is any given integer greater than 2'. Fermat left merely the statement of the theorem, without proof (except for $n = 4$), in his 1637 marginal note to his copy of Diophantus. Euler later proved the theorem for $n = 3$, but no one was able to provide a generalized proof. In general what was needed was a proof that:

$$x^l + y^l + z^l = 0$$

is impossible in non-zero integers x, y and z for any odd prime $l > 3$. In the early nineteenth century Legendre proved this for $l = 5$ and Lebesgue for $l = 7$, but later attempts to find a proof for other values of l came to naught.

Kummer studied previous attempted solutions carefully and came up with one of the most creative and influential ideas in the history of mathematics, the idea of ideal numbers. With their aid he was able to prove, in 1850, that the equation:

$$x^l + y^l + z^l = 0$$

was impossible in non-zero integers for all regular prime numbers (a special type of prime numbers related to Bernoulli numbers). He was then able, by extremely complicated calculations, to determine that the only primes less than 100 which were not regular are 37, 59 and 67. For many years Kummer continued to work on the problem and he eventually was able to prove that the equation is impossible for all primes $l < 100$. It was for his work in this field, especially the introduction of ideal numbers, an invention which some mathematicians rank in importance with the invention of non-Euclidean geometry, that Kummer received the Grand Prix of the French Academy.

Kummer's other great achievement was his discovery of the fourth-order surface known as the Kummer surface. It can be described as the quartic which is the singular surface of the quadratic line complex and involves the very sophisticated and complicated concept of this surface as the wave surface in space of four dimensions.

Kundt August Adolph 1839–1894 was a German physicist who is best known for Kundt's tube, a simple device for measuring the velocity of sound in gases and solids.

Kundt was born at Schwerin, Mecklenburg, on 18 November 1839 and educated at the Universities of Leipzig and Berlin, from which he gained his doctorate in 1864. He began his scientific career as a lecturer in 1867, becoming Professor of Physics at the Polytechnic in Zurich the following year. He subsequently took up the Chair in Physics at the University of Würzburg in 1869, and was one of the founders of the Strasbourg Physical Institute, established in 1872. He ultimately succeeded Hermann Helmholtz (1821–1894) as Professor of Experimental Physics and Director of the Berlin Physical Institute in 1888. Kundt died at Israelsdorf, near Lübeck, on 21 May 1894.

Kundt devised his classic method of determining the velocity of sound in 1866. A Kundt's tube is a glass tube containing some dry powder and closed at one end. Into the open end, a disc attached to a rod is inserted. When the rod is sounded, the vibration of the disc sets up sound waves in the air in the tube and a position is found in which standing waves occur, causing the dust to collect at the nodes of the waves. By measuring the length of the rod and the positions of the nodes in the tube, the velocity of sound in either the rod or air can be found, provided one of these quantities is known. By using rods made of various materials and different gases in the tube, the velocity of sound in a range of solids and gases can be determined. It is also possible using the tube to find the ratio of the molar heat capacities of a gas (and hence the atomicity) and also Young's modulus of elasticity of the material in the rod.

Kundt's later work entailed the demonstration of the dispersion of light in liquids, vapours and metals, and with Wilhelm Röntgen (1845–1923) he experimented with the Faraday effect, showing that the magnetic rotation of the plane of polarizing light occurs in gases. Kundt also published observations on the spectra of lightning and on the electrical properties of crystals.

L

Lacaille Nicolas Louis de 1713–1762 was a French scientist, a positional astronomer who also contributed to advances in geodesy. His best-known work grew out of his four-year expedition to South Africa.

Lacaille was born in Rumigny, France, on 15 March 1713. Being destined for a career in the ministry, he entered the Collège Lisieux, Paris, in 1729, but by chance he became fascinated in books on mathematics and astronomy and decided to change the direction of his career. He contacted the French Academy of Sciences in 1736 and began to make astronomical observations in 1737. He participated in two Academy projects, in 1738 and 1739, and established his reputation as a thorough and talented scientist. This led to his election to the membership of the French Academy of Sciences and his appointment, at the age of only 26, as Professor of Mathematics at the Collège Mazarin, now the Institut de France, Paris. He was subsequently given an observatory for his personal research.

The stars of the southern hemisphere were then relatively unknown compared with those of the northern hemisphere, and so Lacaille proposed to the French Academy that a study be made of the southern skies. His suggestion was accepted, the government provided funding, and Lacaille was put in charge of the project. He left Paris in 1750 and sailed for the Cape of Good Hope. But the conditions under which he worked were poor, and he suffered from overwork and ill-health. He had completed his programme by 1753, and travelled back to France, via Mauritius and Réunion, where he continued his studies.

On his return to Paris in 1754, Lacaille quickly returned to his observatory to analyse the enormous quantity of data which he had collected. He continued his work schedule at a merciless pace and as a result died, prematurely, in Paris on 21 March 1762.

Lacaille's first serious scientific work was the task of obtaining an accurate plot of the French coastline between Nantes and Bayonne. It was assigned him by the French Academy of Sciences and was undertaken with the help of his colleague Giacoma Maraldi (1665–1729), Giovanni Cassini's nephew, in 1738. The next year Lacaille was appointed to the team of scientists who were trying to

resolve a dispute which had been raging for decades. In opposition to Huygens, Newton and their followers, there was a school of thought, headed by Jacques Cassini (1677–1756) that insisted that the Earth was a distorted sphere which was flattened at the equator rather than at the poles.

Lacaille and his team sought to confirm Cassini's measurement of the meridian of France which ran from Perpignan to Dunkirk.

They discovered that Cassini had in fact been in error, and that his conclusions on the shape of the Earth were therefore invalid.

Upon assuming his post as Professor at the Collège Mazarin, Lacaille's primary interest was to collect a huge quantity of positional data for stars in the northern skies. He also devoted considerable energy to writing popular textbooks on mathematics and applied science. The turning point in his career came in October 1750, when he began his expedition to South Africa.

Despite the lack of equipment and poor research conditions Lacaille achieved an extraordinary number of observations. He determined the positions of nearly 10,000 stars, including a large number of seventh-magnitude stars that are invisible to the naked eye. He charted and named 14 new constellations, abandoning the traditional mythological naming system and instead choosing names of contemporary scientific and astronomical instruments. Lacaille measured the lunar parallax, with the aid of Joseph de Lalande (1732–1807) in Berlin. Their simultaneous observations provided them with a base-line longer than the radius of the Earth and gave an accurate measurement of the lunar parallax. It indicated that the average distance from the Earth to the Moon was approximately 60 times the radius of the Earth. Lacaille's determination of the solar parallax was less accurate, being underestimated by about 10%.

Lacaille's other astronomical work during his stay in South Africa included the compilation of the first list of 42 'nebulous stars', estimates for the distances between the Earth and both Mars and Venus, and the observation that Alpha Centauri is a double star. He also performed a number of geodetic investigations, in particular he made the first measurement of the arc of meridian in the southern hemisphere.

Soon after his return from Cape Town, Lacaille published his preliminary star catalogue which included a star map and a description of 2,000 of the stars he had studied in South Africa. Although this catalogue contained nearly five times as many southern stars as its most complete predecessor, it suffered from the inevitable inaccuracy resulting from Lacaille's reliance on small telescopes. The catalogue of all Lacaille's data, *Coelum Australe Stelliferum*, was published in 1763, a year after his death.

Lacaille's later work included a brief catalogue of 400 of the brightest stars, and the first accurate estimate of the distance between the Earth and the Moon in which the fact that the Earth is not a perfect sphere was taken into consideration (1761). His major work during the last years of his life was the analysis of the data he had collected during his visit to South Africa, but he also performed many long computations for other purposes (including the publication of tables of logarithms in 1760).

Laennec René Théophile Hyacinthe 1781–1826 was the French physician who invented the stethoscope in 1816.

Laennec was born at Quimper, Brittany, on 17 February 1781. Because his mother died early, he was sent to live with his uncle who was a physician in Nantes, and there he was introduced to medical work. The French Revolution struck the town with some ferocity and Laennec, only just in his teens, worked in the city hospitals. In 1795, he was commissioned as third surgeon at the Hôpital de la Paix, and shortly afterwards at the Hospice de la Fraternité – he was then just 14 years old. It was there that Laennec became acquainted with clinical work, surgical dressings and the treatment of large numbers of patients.

Because his own health was not good, he abandoned medicine until 1799, when he returned to his medical studies and became a pupil of J. N. Corvisart. He was appointed surgeon at the Hôtel Dieu in Nantes. From there, he entered the Ecole Pratique in Paris and studied dissection in Depuytren's laboratory.

Between June 1802 and July 1804. Laennec published a number of papers on anatomy. His 1804 thesis 'Propositions sur la doctrine d'Hippocrate relativement à la médecin-pratique' resulted in the award of a doctor's degree.

Laennec became particularly interested in pathology. Appointed an associate of the Société de l'Ecole de Médecine in 1808, he founded the Athenée Médical which later merged with the Société Académique de Médecine de Paris. Laennec was appointed personal physician to Cardinal Fesch, uncle of Napoleon I. In 1812–1813, with France at war, Laennec took charge of the wards in the Salpetrière reserved for Breton soldiers. On the restoration of the monarchy, he took the post of physician to the Necker Hospital but, by February 1818, poor health obliged him to retire to his estate in Brittany, where he lived the life of a gentleman farmer. At the end of that year, however, he returned to Paris, but in 1819 he was again forced to retire to his estates

at Kerbourarnec, where he assumed the role of country squire.

In 1822, Laennec was appointed to the chair and a lectureship at the Collège de France. In 1823 he became a full member of the Académie de Médecine and, in 1824, he was made a Chevalier of the Légion d'Honneur. He became internationally famous as a lecturer. However, his health failed him again and he left Paris for the last time in May, 1826. He died on 13 August of the same year.

Laennec's important contribution to medical science was his invention of the stethoscope. In 1816, while walking in a courtyard in the Louvre, he noticed two small children with their ears close to the ends of a long stick. They were amusing themselves by tapping lightly on the stick and listening to the transmitted sound. He immediately realized the diagnostic possibilities and quickly developed an instrument based on the children's activities. It was essentially a wooden tube some 30 cm/12 in long and he used it to listen to sounds in the human body. As it was particularly useful for sounding the lungs and heart, he called it a stethoscope (from the Greek *stethos*, meaning chest).

Laennec was now deeply interested in emphysema, tuberculosis and physical signs of chest diseases. Although auscultation had been known since the days of Hippocrates, it was always done by the 'direct' method – which was often inconvenient. Laennec introduced the 'mediate' method by using his hollow tube for listening to the lungs, and a solid wooden rod to listen to the heart.

By February 1818 he was able to present a paper on the subject to the Académie de Médecine and, by 1819, he was able to classify the physical signs of egophony, rales, rhonchi and crepitations. His book *De l'Auscultation Médiate* was published in August 1819. This was much more than a book on the use of the stethoscope; it was an important treatise on diseases of the heart, lungs and liver, and was acknowledged to represent a great advance in the understanding of chest diseases.

Lagrange Joseph Louis (Giuseppe Lodovico), Comte 1736–1813 was an Italian-born French mathematician who revolutionized the study of mechanics.

Lagrange was born, of a French father and Italian mother, at Turin on 25 January 1736. Almost entirely by teaching himself the subject he made himself a formidable mathematician by the age of 16, and in 1755, when he was just 19, he was appointed Professor of Mathematics at the Royal Artillery School at Turin. In 1759, he organized a mathematical society of his students and colleagues and this ultimately became the Turin Academy of Sciences. He remained at Turin until 1766 when, on the recommendation of Leonhard Euler and Jean d'Alembert, Frederick the Great of Prussia invited him to Berlin to succeed Euler as the Director of the Berlin Academy of Sciences. In 1787, he accepted the invitation of Louis XVI to go to Paris as a member of the French Royal Academy and in 1797 he was appointed Professor of Mathematics at the Ecole Polytechnique.

Like so many brilliant mathematicians, Lagrange had his best ideas as a very young man. By the time that he began to teach at Turin in 1755, he had already conceived the heart of his great work, the *Mécanique Analytique*, and was near to his first important discovery, the solution to the isoperimetrical problem, on which the calculus of variations partly depends. In 1759 he sent his solution to Euler, who had been baffled and held up by the problem for years; Euler wrote back to express his delight at the young man's solution and delayed publishing his own work on the subject until Lagrange – whose proof had greater generality than Euler's – had published his discovery. In that year also Lagrange wrote a paper on minima and maxima and, advancing beyond Newton's theory of sound, settled a fierce dispute on the nature of a vibrating string.

In 1764, Lagrange addressed himself to the famous three-body problem on the mutual gravitational attraction of the Earth, Sun and Moon, and for his work on it was awarded the French Academy's Grand Prix. Two years later he won the award again for his work on the same problem with regard to Jupiter and its four (as were then known) satellites. When he moved to Berlin he diverted himself for a time with problems in number theory. He proved some of Fermat's theorems which had remained unproven for a century; and he solved a problem known since the time of the ancients, namely, to find an integer x so that $(nx^2 + 1)$ is a square where n is a positive integer, but not a square.

All the while, however, his chief concern was mechanics. By 1782 he had completed his *Mécanique Analytique*, although it was not published until 1788, by which time Lagrange had become so uninterested in mathematics that the manuscript had lain untouched and unconsulted for two years. In his epoch-making treatise Lagrange, by considering a four-dimensional space with the aid of calculus, succeeded in reducing the theory of solid and fluid mechanics to an analytical principle. Breaking with conventional methods used since the time of the Greeks. Lagrange published his work without the aid of a single diagram or construction: his discussion of mechanics, that is to say, was entirely algebraic. Of this achievement Lagrange was justly proud, especially since he was able to reflect that the event which had, more than anything else, drawn him into mathematics was his reading, as a young boy, the work of Edmund Halley (1656–1742) in the use of algebra in optics.

Lagrange never again did anything of such significance, although his lectures at the Ecole Polytechnique, published as the *Théorie des Fonctions Analytiques* (1797) and the *Leçons sur le Calcul des Fonctions* (1806), were of much use to Augustin Cauchy and others. But he had still one contribution to make to the world. In 1793, he was appointed president of the commission established to standardize French weights and measures and by persuading the commission to adopt 10, not 12, as the basic unit, was the father of the metric system in use throughout most of the world today. For that, perhaps, but above all for his work on mechanics he was deservedly given a final resting place in the Panthéon.

Laithwaite Eric Robert 1921– is a British electrical engineer, most famous for his development of the linear motor.

Laithwaite was born on 14 June 1921 at Atherton, Yorkshire. He was educated at Kirkham Grammar School and then at Regent Street Polytechnic, London. He read for his BSc at the University of Manchester, but his studies were interrupted by World War II. From 1941 to 1946 he served in the Royal Air Force; for four years he worked at the Royal Aircraft Establishment, Farnborough, researching into automatic pilots for aircraft. He gained his BSc in 1949 and an MSc in the following year. He remained at the University of Manchester until 1964, holding positions of Assistant Lecturer, Lecturer and Senior Lecturer. During that time he was awarded a PhD in 1957, and a DSc in 1964. In that same year he was appointed Professor of Heavy Electrical Engineering at the Imperial College of Science and Technology at the University of London. From 1967 to 1976 he was External Professor of Applied Electricity at the Royal Institution, and in 1970 he became President of the Association for Science Education.

Laithwaite is unusually able to put his ideas across to non-specialists and children. He is one of the most popular contributors to the Christmas Lectures at the Royal Institution. In 1966, the title of his lectures was 'The Engineer in Wonderland' which he followed in 1974 with 'The Engineer Through the Looking Glass'. He has also participated in radio and television broadcasts on general science including: 'Young Scientists of the Year', 'It's Patently Obvious' and 'The Engineer's World'. He has also made a number of films, several of which have won major awards.

The idea of a linear induction motor had been suggested in 1895, but a new principle has been discovered and worked on by Laithwaite: that it is possible to arrange two linear motors, back-to-back, so as to produce continuous oscillation without the use of any switching device. The important feature of the linear motor is that it is a means of propulsion without the need for wheels. This in itself is nothing remarkable. Yachts, rowing boats, surfboards, ice-skates, skis, canoes, gliders – all are examples of propulsion methods without wheels. Even swimming and walking are mechanisms requiring no rotary movement in a pure sense. Our early propulsion systems were designed to use either muscular power or natural power sources such as air currents or the wave energy of a sailing yacht in a steady wind. It is possible to continually extract power from a source, although often continuous motion has to be made up from reciprocating motion – achieved by the use of a ratchet, for example.

Linear motion is very common in industrial processes. Planing machines, looms, saws, knitting machines and pumps all make use of motion in a straight line. In 1947, Laithwaite began research into electric linear induction motors as the shuttle drives in weaving looms. Further investigations were carried out into the use of linear motors for conveyers and as propulsion units for railway vehicles. By 1961, his research had aroused sufficient interest for it to be included in a BBC television programme which added to public interest in the linear induction motor, and development was further stimulated by a lecture which Laithwaite gave to the Royal Institution in March 1962 entitled 'Electrical Machines of the Future'

Probably the project which has aroused the greatest interest in the media is the high-speed transport project, in which hovering vehicles, moving on air cushions or jets, are powered by linear induction motors, used as high-speed propulsion units. These systems are in use in the United States, Germany, Japan and the former Soviet Union for high-speed railways.

Lalande Joseph Jérome le François de 1732–1807 was a French astronomer noted for his planetary tables, his account of the transit of Venus, and numerous writings on astronomy.

Lalande was born at Bourg-en-Bresse on 11 July 1732. He was the only child of relatively wealthy parents, his father being the director of both the local post office and a tobacco warehouse. Lalande was educated at the Jesuit College in Lyons in preparation for his joining the Order. Instead he went to Paris to continue his studies and it was during this time that he first became intrigued by astronomy. He was influenced by Joseph-Nicolas Delisle (1688–1768) and Charles le Monnier, both leading figures in the fields of astronomy and mathematical physics. Lalande was appointed Professor of Astronomy at the Collège de France in 1762 and during his tenure there he published his *Treatise of Astronomy*.

In 1795, he was made Director of the Paris Observatory, where he concentrated on compiling a catalogue of stars in which he noted 47,000 stars in all, including the planet Neptune, without realizing that it was not a star; thus he had observed Neptune 50 years before it was formally discovered. Lalande died in Paris on 4 April 1807.

It was Lalande's involvement in calculating the distance to the Moon that earned him his initial acclaim. In those days the method used to calculate the distance to celestial objects was parallax. This meant that measurements had to be made simultaneously by two people from observatories sited on opposite sides of the Earth. Lalande's mentor, Monnier, generously suggested that Lalande should participate in the parallax programme and so in 1751 he was despatched to Berlin, where he became a guest of the Prussian Observatory. At the same time, in collaboration with Lalande, another French venture led by Nicolas de Lacaille was mounted at the Cape of Good Hope. Its purpose was to note the position of the Moon at the time of its transit from different positions on the same meridian. The combined results of these two ventures greatly improved the accuracy of the value of the Moon's parallax. Having published the results of these observations, Lalande and Monnier disagreed over the allowance to be given to the flattening of the Earth's surface. It was eventually decided, by consensus, that Lalande's results were valid, but Monnier never quite recovered from the slight and chose to remain somewhat distant from his former pupil from that time onwards.

Lalande was at the height of his profession during the 1760s and so was fortunate enough to be able to participate in the observations of the two transits of Venus that occurred in 1761 and 1769. Such transits are a rare phenomenon that occur twice within a period of eight years only every 113 years, and the event offered eighteenth-century astronomers the chance to establish accurately the size of the Solar System. During the transit, which takes approximately five hours, Venus can be seen silhouetted across the face of the Sun; the distance of the Earth from the Sun can be deduced by measuring the different times that the planet takes to cross the face of the Sun when seen from different latitudes on Earth. A host of expeditions was organized with worldwide collaboration, and Lalande was made personally responsible for collecting the results of observations and co-ordinating expeditions to all corners of the world.

Lalande was a controversial person throughout his long life, often falling into arguments with fellow astronomers. Besides his particularly well-documented account of the transit of Venus, Lalande wrote numerous textbooks on astronomy.

Lamarck Jean Baptiste Pierre Antoine de Monet, Chevalier de 1744–1829 was a French naturalist best known for his alternative to Charles Darwin's theory of evolution and the distinction he made between vertebrate and invertebrate animals.

Lamarck was born at Bazantin, Picardy, on 1 August 1744, the 11th child of a large family of poor aristocrats. This noble background restricted his career and he was expected to join either the army or the Church. The first intention was that he should enter the Church, but when his father died in 1760 he joined the army. After serving in the Seven Years' War he left the army in 1766, when his health failed, and eventually went into medicine. He soon became interested, however, in meteorology and botany. He was admitted to the Academy of Sciences in 1778 for his book *Flore Française*, written while he was posted in southern France, and attracted the interest of Georges Buffon (1707–1788), the famous French naturalist. With Buffon's assistance he travelled across Europe, having been appointed botanist to King Louis XVI in 1781. He returned in 1785 and three years later took up a botanical appointment at the Jardin du Roi. In 1793, he was made a Professor of Zoology at the Museum of Natural History, in Paris, where he lectured on the 'Insecta' and 'Vermes' of Linnaeus, which he later called the Invertebrata. He continued his writings, although his sight was failing, with the help of his eldest daughter and Pierre-André Latreille (1762–1833). He died blind and in great poverty, in Paris, on 18 December 1829.

Lamarck's own experiences in both botany and zoology made him realize the need to study living things as a whole, and he coined the word 'biology'. In 1785, he pointed out the necessity of natural orders in botany by an attempt at the classification of plants. He widely propagated the idea, as Buffon had done before him, that species are not unalterable and that the more complex ones have developed from pre-existent simpler forms.

Lamarck's most important contribution was his detailed investigation of living and fossil invertebrates. At the age of 50, while lecturing on Linnaeus' classifications, he began the study of invertebrates. So little was known about them at this time that some scientists grouped snakes and crocodiles with insects. Lamarck was the first to distinguish vertebrate from invertebrate animals by the presence of a bony spinal column. He was also the first to establish the crustaceans, arachnids and annelids among the invertebrates.

In studying and classifying fossils, Lamarck was led to wonder about the effect of environment on development. In 1809 he published one of his most important essays, *Zoological Philosophy*, which tried to show that various parts of the body developed because they were necessary, or disappeared because of disuse when variations in the environment caused a change in habit. He believed that these body changes are inherited by the offspring, and that if this process continued for a long time, new species would eventually be produced.

Lamarck thought that it ought to be possible to arrange all living things in a branching series showing how some species gradually change into others. Unfortunately he chose poor examples to illustrate his ideas and he did not attempt to show how one species might gradually change into another on the basis of the inheritance of acquired characteristics. One of the well-known examples that constitutes the evidence on which Lamarck based his theory is the giraffe which he assumed, as an antelope-type animal, sought to browse higher and higher on the leaves of the trees on which it feeds, and 'stretched' its neck. As a result of this habit continued for a long time in all the individuals of the species, the 'antelope's' front limbs and neck gradually grew longer, and it became a giraffe. Another example is that of birds that rest on the water and which, perhaps to chase and find food, spread out their feet when they wish to swim. The skin on their feet becomes accustomed to being stretched and forms a web between the toes. These examples were regarded as rather naïve, even to his contemporaries, which could explain why Lamarck's ideas were not accepted.

A contemporary of Lamarck, Georges Cuvier (1769–1832), proposed his own theory of evolution which although inferior to Lamarck's achieved considerable acclaim. The work of Lamarck received little recognition in his own lifetime and some of his successors, including Darwin, regarded it with ridicule and contempt. There is no doubt, however, that his zoological classifications have aided further study in that field and that he greatly influenced later evolutionists.

Among the many works he produced, one of Lamarck's greatest is the seven-volume book *Natural History of Invertebrates*, produced between 1815 and 1822.

Lamb Horace 1849–1934 was a British applied mathematician, noted for his many books on hydrodynamics, elasticity, sound and mechanics.

Lamb was born in Stockport, Cheshire, on 27 November 1849. His father was the foreman in a cotton mill; his mother died when he was young and he was brought up by his aunt, who was largely responsible for enrolling him at Stockport Grammar School. There Lamb excelled at Latin and Greek and at the age of 17 he won a classical scholarship to Queen's College, Cambridge. He spent a year at Owens College, Manchester, before going to the university, however, and having gained a scholarship to Trinity College entered there in the autumn of 1868. In 1870 he was elected a Sheepshank (astronomical exhibitioner) at the college and in 1872 he gained his BA in mathematics as second Wrangler and second Smith's prizeman. In the same year he was elected a Fellow and Lecturer at Trinity.

Cambridge still imposed celibacy upon its Fellows and in 1875, having married, Lamb went to Australia to take up the Chair of Mathematics at the University of Adelaide. He remained there until 1885, when he returned to Manchester as Professor of Mathematics at Owens College, a post which he held until his retirement in 1920. His last years were spent in Cambridge. The University established an honorary lectureship for him, called the Rayleigh Lectureship, and Trinity elected him an honorary Fellow. From 1921 to 1927 he was also a member of the Aeronautical Research Committee attached to the Admiralty and until his death he served on an advisory panel which was concerned with the fluid motion set up by aircraft.

Lamb was elected to the Royal Society in 1884 and was awarded the Society's Royal Medal in 1902 and the Copley Medal, its highest honour, in 1923. He was president of the British Association in 1925. He was knighted in 1931, and he died in Cambridge on 4 December 1934.

Lamb was born a generation after men such as George Stokes (1819–1903), James Clerk Maxwell (1831–1879) and John Rayleigh (1842–1919) had begun to revive interest in the mathematical aspect of problems in physics – heat, electricity, magnetism and elasticity. His was an age of brilliant achievement in applied mathematics and he made his own mark in many fields: electricity and magnetism, fluid mechanics, elasticity, acoustics, vibrations and wave motions, seismology, and the theory of tides and terrestrial magnetism. He was particularly adept at applying the solution of a problem in one field to problems in another. His greatest achievement was in the field of fluid mechanics. In 1879 he published his best known work, *A Treatise on the Motion of Fluids*, issued in a new edition as *Hydrodynamics* in 1895, and reissued in five subsequent editions (the last in 1932), each of which incorporated the latest developments in the subject. The book is rightly regarded as one of the most clearly written treatises in the whole of applied mathematics.

Lamb's contributions ranged wide over the field of applied mathematics. A paper of 1882, which analysed the modes of oscillation of an elastic sphere, achieved its true recognition in 1960, when free Earth oscillations during a Chilean earthquake behaved in the way he had described. Another paper of 1904 gave an analytical account of propagation over the surface of an elastic solid of waves generated by given initial disturbances, and the analysis he provided is now regarded as one of the seminal contributions to theoretical seismology. Then, in 1915, in collaboration with Lorna Swain, he gave the first satisfactory account of the marked phase differences of the tides observed in different parts of the Earth's oceans, thereby settling a question which had been a matter of controversy since Newton's time.

Lamb made numerous important discoveries; he published widely; he was, in the words of Rutherford, 'more nearly my ideal of a university professor than anyone I have known'. By many people he is regarded as the finest applied mathematician of the century.

Lanchester Frederick William 1868–1946 was the English engineer who created the first true British motorcar of reliable design and produced the first comprehensive theory of flight.

Lanchester was born at Lewisham, London on 28 October 1868. Two year later his family moved to Hove, in Sussex, where his father set up business as an architect. In 1883, Lanchester went to the Hartley Institution (which later became University College, Southampton). From here he won a scholarship to what is now Imperial College.

After a brief period as a Patent Office draughtsman, he joined the Forward Gas Engine Company of Birmingham as assistant works manager in 1889, becoming works manager within a year.

It was in 1893, following his return from an unsuccessful trip to the United States to sell his gas-engine patents, that Lanchester set up a workshop next to the Forward Company. By 1896 his first motorcar had emerged, with a single cylinder 4-kW/5-hp engine and chain drive. He rebuilt it the following year with a 6-kW/8-hp flat twin engine and a worm gear transmission for which he designed both the worm form and the machine tool to produce it. His second car, built at his Ladywood Road factory, was a 6-kW/8-hp Phaeton with flywheel ignition, roller bearings and cantilever springing.

The car won a gold medal for its performance and design at the Royal Automobile Club trials at Richmond in 1899, and the following year it completed a 1,600-km/1,000-mi tour. The company then went into production with 7.5–9-kW/10–12-hp motorcars, of which some 350 were built between 1900 and 1905. A 15 kW/20 hp four-cylinder and 21 kW/28 hp six-cylinder model were produced in 1904 and 1905 respectively.

As early as 1891 Lanchester had acquired an interest in crewed flight and soon began experimenting with powered and gliding models. He showed that a well-designed aircraft could be inherently stable, a concept that underpins modern airworthiness requirements. On 19 June 1894 he marshalled his theories of flight in a paper of fundamental importance: 'The soaring of birds and the possibilities of mechanical flight', which was given at a meeting of the Birmingham Natural History and Philosophical Society.

Lanchester revised and amplified his paper, and submitted it to the Physical Society, only to have it rejected in September 1897. Nothing daunted, he resurrected his theories and incorporated them in his great work, *Aerial Flight* which was published in two volumes, *Aerodynamics* in 1907 and *Aerodonetics* in 1908. The first volume won Lanchester instant acclaim.

Lanchester and his collaborator Norman A. Thompson designed an experimental aircraft whose wings were covered by an aluminium skin and a sheet steel fuselage. Sadly, the undercarriage collapsed on the first trial flight in 1911, and Lanchester decided to quit the practical side of aviation. A second aircraft, with broadly similar design, did fly successfully in the summer of 1913, and in the following year a seaplane, also incorporating many of Lanchester's ideas, took to the air. This machine was the prototype for a batch of six built for the Admiralty and used during World War I to counter the German submarine menace.

In 1900, Lanchester acquired the Armourer Mills in Montgomery Street, Sparbrook, which were to house the Lanchester Engine Company until 1931. However, the company went into receivership in 1904 and was reconstructed the following year as the Lanchester Motor Company. This was a financial success.

The publication of *Aerial Flight* led to Lanchester's joining Prime Minister Asquith's advisory committee for Aeronautics at its formation on 30 April 1909. The same year he was appointed consulting engineer and technical advisor to the Daimler company.

With the outbreak of World War I, Lanchester's involvement in military work became so great that he moved to London. On 3 September 1919 he married Dorothea Cooper, daughter of the vicar of Field Broughton, near Windermere.

After the war he left the Committee, and his work in the capital tailed off. Having become increasingly busy in the Daimler business in Coventry, Lanchester decided to move back to Birmingham. He founded Lanchester's Laboratories Ltd in 1925, intending to provide research and development services, but the economic depression meant there was little call for his expertise.

Lanchester was also interested in radio and, having patented a loudspeaker and other audio equipment he went into production.

Again he hit difficulties, and these, combined with his failing health, forced him finally to close down his laboratories in January 1934.

His health improved towards the end of the year, and he began to write prolifically both on technical and nontechnical subjects. His works included a collection of poems (under the pseudonym of Paul Netherton-Herries), an explanation of relativity and a treatise on the theory of dimensions.

When World War II broke out, the symptons of Parkinson's disease had become clear. Although his mind was still lively, Lanchester was forced to retire. Finally, plagued by financial worries and losing his sight, Lanchester died on 8 March 1946.

Lanchester's work with motorcars is remarkable, not only because of the number of new ideas which were incorporated into them, but also because he produced them with interchangeable parts. This was the concept later adopted so successfully by mass production manufacturers. In his study of flight he was the first engineer to depart from the empirical approach and to formulate a comprehensive theory of lift and drag. His work on stability was fundamental to aviation.

Lanchester received many honours and awards, including the presidency of the Institute of Automobile Engineers (1910). Fellowship of the Royal Aeronautical Society (1917) and of the Royal Society (1922), and the Gold Medal of the RAS in 1926. Birmingham University made him an honorary doctor of law in 1919.

Landé Alfred 1888–1975 was a German-born US physicist known for the Landé splitting factor in quantum theory.

Landé was born in Elberfeld on 18 December 1888. He studied at the Universities of Marburg, Göttingen and Munich, gaining his DPhil from Munich in 1914. In 1919 he became a lecturer at the University of Frankfurt-am-Main until 1922, when he went to Tübingen as Associate Professor of Theoretical Physics. He visited Columbus, Ohio, USA, in 1929 and again in 1930, and in 1931, because of the rise of the Nazi regime, left Germany forever to settle in Columbus as Professor of Theoretical Physics at Ohio State University.

Landé had just gained his doctorate under Arnold Sommerfeld (1868–1951) when World War I began in 1914. He joined the Red Cross and served for a few years in a hospital until Max Born (1882–1970) placed him as his assistant at the Artillerie-Prüfungs-Kommission in Berlin to work on sound detection methods. He collaborated with Born to conclude that Bohr's model of coplanar electronic orbits must be wrong, and that they must be inclined to each other. When they published this paper in 1918, it caused quite a stir in the scientific world.

In another paper Landé strongly advocated a tetrahedral arrangement for the four valence electrons of carbon. While at Frankfurt, Landé had the opportunity to visit Niels Bohr (1885–1962) in Copenhagen to discuss the Zeeman effect (the splitting of spectral lines in a magnetic field). Then on moving to Tübingen, he found himself at the most important centre of atomic spectroscopy in Germany. In 1923, Landé published a formula expressing a factor known as the Landé splitting factor for all multiplicities as a function of the quantum numbers of the stationary state of the atom. The Landé 'splitting' factor is the ratio of an elementary magnetic moment to its causative angular momentum when the angular momentum is measured in quantized units. It determines fine as well as hyperfine structures of optical and X-ray spectra.

Landé then collaborated with Louis Paschen (1865–1947) and others to analyse in great detail the fine structure of the line spectra and the further splitting of the lines under the action of magnetic fields of increasing strength. He formulated the regularities obeyed by the frequencies and intensities of the lines in terms of the sets of quantum numbers attached to the spectroscopic terms and taking integral or half-integral values.

Landsteiner Karl 1868–1943 was an Austrian-born US immunologist who discovered the ABO blood groups, and the M and N factors and rhesus factor in human blood. For this work he received the 1930 Nobel Prize for Physiology or Medicine.

Landsteiner was born in Vienna on 14 June 1868, the son of a journalist. He attended the University of Vienna from 1885 to 1891, when he graduated with a doctorate in medicine. He spent the next five years studying at various universities in Europe, including a time with the German chemist Emil Fischer (1852–1919). In 1898, he became a research assistant at the Vienna Pathological Institute and remained there until 1908, when he was appointed Professor of Pathology at the Royal Imperial Wilhelminen Hospital in Vienna. Unsatisfactory working conditions there caused him to leave in 1919, and he went to the RK Hospital in the Hague. He was no happier there, and moved to the United States where he joined the Rockefeller Institute in New York in 1922. He became a US citizen in 1929 and died in New York on 26 June 1943.

While at the Vienna Pathology Institute, Landsteiner published a paper in 1900, which included information on the interagglutination of human blood cells by serum from a different human being. He explained this feature as occurring because of antigen differences in the blood between individuals. In the following year Landsteiner described a technique which enables human blood to be categorized into three groups: A, B and O (and later, AB). His sorting method, which is still used, was to mix suspensions of blood cells with each of the sera, anti-A and anti-B. Landsteiner discovered in this way that serum contains only those antibodies that do not cause agglutination with its own blood cells. It was not until 1907, however, that a blood-match was first carried out and blood transfusions did not become a widespread practice until after 1914 when Richard Lewisohn found that the addition of citrates to blood prevents it from coagulating, enabling it to be preserved.

In 1927, Landsteiner and Philip Levine found that, in addition to antigens A and B, human blood cells contain one or other or both of two heritable antigens, M and N. These are of no importance in transfusions, because human serum does not contain the corresponding antibodies, but being inherited they are of value in resolving paternity disputes.

In 1940, Landsteiner and Levine, together with Wiener, injected blood from a rhesus monkey into rabbits and guinea pigs. The resulting antibodies agglutinated the rhesus blood cells because they contain an antigen, named Rh, which is identical or very similar to an antigen present in the rhesus monkeys. It was shown that the Rh factor is inheritable; those people possessing it are termed Rh positive, and those without it, Rh negative. The connection between this factor and a blood disorder of newborn babies (erythroblastosis foetalis) was revealed. An Rh-negative mother pregnant with an Rh-positive foetus forms antibodies against the Rh factor. In a subsequent similar pregnancy the oxygen-carrying capacity of the foetus's red cells is impaired and brain damage may result. Diagnostic tests can now enable the necessary preventive measures to be taken to avoid harm to the baby.

Landsteiner investigated the Wassermann test (a blood test for syphilis involving adding antigen to detect the presence of antibodies in a patient's blood). Together with Ernest Finger he discovered that the antigen, previously extracted from a syphilitic human being, could be replaced with an extract prepared from ox hearts.

Between 1908 and 1922, Landsteiner researched into poliomyelitis. He injected a preparation of brain and spinal cord tissue, obtained from a polio victim, into a rhesus monkey, which then developed paralysis. Landsteiner could find no bacteria in the monkey's nervous system and concluded that a virus must be responsible for the disease. He later developed a procedure for the diagnosis of poliomyelitis.

Langevin Paul 1872–1946 was a French physicist who invented the method of generating ultrasonic waves that is the basis of modern echolocation techniques. He was also the first to explain paramagnetism and diamagnetism, by which substances are either attracted to or repulsed by a magnetic field. Langevin was the leading mathematical physicist of his time in France, and contributed greatly to the dissemination and development of relativity and modern physics in general in his country. The nuclear institute in Grenoble is named after him.

Langevin was born in Paris on 23 January 1872. He attended the Ecole Lavoisier and the Ecole de Physique et de Chimie Industrielles, where he was supervised by Pierre Curie (1859–1906) during his laboratory classes. In 1891 he entered the Sorbonne, but his studies were interrupted for a year in 1893 by his military service. In 1894, Langevin entered the Ecole Normale Supérieure, where he studied under Jean Perrin (1870–1942). Langevin won an academic competition which in 1897 enabled him to go to the Cavendish Laboratories at the University of Cambridge for a year. There he studied under J. J. Thomson (1856–1940), and met scientists such as Ernest Rutherford (1871–1937), Charles Wilson (1869–1959) and John Townsend (1868–1957).

Langevin received his PhD in 1902 for work done partly at Cambridge and partly under Curie on gaseous ionization. He spent a great deal of time in Perrin's laboratory, and was caught up in the excitement accompanying the early years of study on radioactivity and ionizing radiation. Langevin joined the faculty of the Collège de France in 1902, and was made Professor of Physics there in 1904, a post he held until 1909 when he was offered a similar position at the Sorbonne.

During World War I Langevin contributed to war research, improving a technique for the accurate detection and location of submarines, which continued to be developed for other purposes after the war. He became a member of the Solvay International Physics Institute in 1921, and was elected President of that organization in 1928. In 1940, after the start of World War II and the German occupation

of France, Langevin became Director of the Ecole Municipale de Physique et de Chimie Industrielles, where he had been teaching since 1902, but he was soon arrested by the Nazis for his outspoken antifascist views. He was first imprisoned in Fresnes, and later placed under house arrest in Troyes. The execution of his son-in-law and the deportation of his daughter to Auschwitz (which she survived) forced Langevin to escape to Switzerland in 1944. He returned to Paris later that year and was restored to the Directorship of his old school, but died soon after, in Paris on 19 December 1946.

Langevin's early work at the Cavendish Laboratories and at the Sorbonne concerned the analysis of secondary emission of X-rays from metals exposed to radiation, and Langevin discovered the emission of secondary electrons from irradiated metals independently of Georges Sagnac (1869–1928). He also studied the behaviour of ionized gases, being interested in the mobility of positive and negative ions, and in 1903 he published a theory for their recombination at different pressures. He then turned to paramagnetic (weak attractive) and diamagnetic (weak repulsive) phenomena in gases. Curie had demonstrated experimentally in 1895 that the susceptibility of a paramagnetic substance to an external magnetic field varies inversely with temperature. Langevin produced a model based on statistical mechanics to explain this in 1905. He suggested that the alignment of molecular moments in a paramagnetic substance would be random in the absence of an externally applied magnetic field, but would be nonrandom in its presence. The greater the temperature, however, the greater the thermal motion of the molecules and thus the greater the disturbance to their alignment by the magnetic field.

Langevin further postulated that the magnetic properties of a substance are determined by the valence electrons, a suggestion which influenced Niels Bohr (1885–1962) in the construction of his classic model describing the structure of the atom. He was able to extend his description of magnetism in terms of electron theory to account for diamagnetism. He showed how a magnetic field would affect the motion of electrons in the molecules to produce a moment that is opposed to the field. This enabled predictions to be made concerning the temperature-independence of this phenomenon and furthermore to allow estimates to be made of the size of electron orbits.

Langevin became increasingly involved with the study of Einstein's work on Brownian motion and on space and time. He was a firm supporter of the theory of the equivalence of energy and mass. Einstein later wrote that Langevin had all the tools for the development of the special theory of relativity at his disposal before Einstein proposed it himself and that if he had not proposed the theory, Langevin would have done so.

During World War I, Langevin took up the suggestion which had been made by several scientists that the reflection of ultrasonic waves from objects could be used to locate them. He used high-frequency radio circuitry to oscillate piezoelectric crystals and thus obtain ultrasonic waves at high intensity, and within a few years had a practical system for the echolocation of submarines. This method has become the basis of modern sonar and is used for scientific as well as military purposes.

Langmuir Irving 1881–1957 was a US physical chemist who is best remembered for his studies of adsorption at surfaces and for his investigations of thermionic emission. For his work on surface chemistry he was awarded the 1932 Nobel Prize for Chemistry. Unlike most scientists of world renown he did most of his research in a commercial environment, not in an academic institution.

Langmuir was born in Brooklyn, New York City, on 31 January 1881. He attended local elementary schools before his family moved to Paris for three years, where he was a boarder at a school in the suburbs. His interests in the practical aspects of science were fostered by his brother Arthur. In 1895 the family returned to the United States to Philadelphia, and Langmuir went to the Chapel Hill Academy and later to the Pratt Institute in Brooklyn. After High School he entered the School of Mines at Columbia University and graduated in 1903 with a degree in metallurgical engineering. His postgraduate studies were undertaken at the University of Göttingen under the guidance of Hermann Nernst; he gained his PhD for a thesis on the recombination of dissociated gases. After a brief period as a teacher in New Jersey, he joined the General Electric Company in 1909 at their research laboratories at Schenectady and remained there until he retired in 1950; he was its Assistant Director from 1932 until 1950. He died in Falmouth, Massachusetts, on 16 August 1957.

Langmuir's work at Columbia concerned the dissociation of water vapour and carbon dioxide around red-hot platinum wires. His first studies at General Electric involved the thermal conduction and convection of gases around tungsten filaments. Langmuir showed that the blackening on the inside of 'vacuum-filled' electric lamps was caused by evaporation of tungsten from the filament. The introduction of nitrogen into the glass bulb prevented evaporation and blackening, but increased heat losses which were overcome by making the tungsten filament in the form of a coiled coil.

At the same time, Langmuir was carrying out research on electric discharges in gases at very low pressures, which led to the discovery of the Child–Langmuir space–charge effect: the electron current between electrodes of any shape in vacuum is proportional to the 3/2 power of the potential difference between the electrodes. He also studied the mechanical and electrical properties of tungsten lamp filaments to which thorium oxide had been added. He showed that high thermal emissions were caused by diffusion of thorium to form a monolayer on the surface. One consequence was the development of an improved vacuum pump based on the condensation of mercury vapour. This work also initiated his 1934 patent for an atomic welding torch, in which the recombination of hydrogen atoms produced by an electric arc between tungsten electrodes generated heat at temperatures in the order of 6,000°C/11,000°F.

For three years Langmuir considered the problems of atomic structure. Building on Gilbert Lewis's atomic theory and valency proposals, Langmuir suggested that chemical reactions occur as a consequence of a desire by an atom to achieve a full shell of eight outer electrons. He was the first to use the terms electrovalency (for ionic bonds between metals and nonmetals) and covalency (for shared-electron bonds between nonmetals).

During the 1920s, Langmuir became particularly interested in the properties of liquid surfaces. He went on to propose his general adsorption theory for the effect of a solid surface during a chemical reaction. He made the following assumptions:

(a) the surface has a fixed number of adsorption sites; at equilibrium at any temperature and gas pressure, a fraction θ of the sites are occupied by adsorbed particles and a fraction $1-\theta$ are unoccupied;

(b) the heat of adsorption is the same for all sites and is independent of the number of sites occupied;

(c) there is no interaction between molecules on different sites, and each site can hold only one adsorbed molecule.

From this model he formulated Langmuir's adsorption isotherm, which can be expressed as

$$1/\theta = 1 + 1/bP,$$

where θ is the fraction of sites occupied, b is a constant based on rate constants of evaporation and condensation, and P is the pressure.

During World War II Langmuir was responsible for work that led to the generation of improved smokescreens using smoke particles of an optimum size. He later applied this knowledge to particles of solid carbon dioxide and silver iodide which were scattered from aircraft to seed water droplets for cloud formation, in an attempt to make rain.

Laplace Pierre Simon, Marquis de 1749–1827 was a French astronomer and mathematician who contributed to the fields of celestial mechanics, probability, applied mathematics and physics.

Laplace was born on 28 March 1749 at Beaumont-en-Auge, in Normandy, France. The family were comfortably well-off – his father was a magistrate and an administrative official of the local parish and was probably also involved in the cider business. When he was seven Laplace was enrolled at the Collège in Beaumont-en-Auge run by the Benedictines and he attended as a day-pupil until the age of 16. Typically, a young man would go either into the Church or into the army after such an education. His father intended him to go into the Church, but Laplace entered Caen University in 1766 and matriculated in the Faculty of Arts. Rather than take his MA, he went to Paris in 1768 with a letter of recommendation to Jean le Rond d'Alembert (1717–1773). The story is that d'Alembert gave him a problem and told him to return with the solution a week later. It is said that Laplace returned with the solution the next day and that d'Alembert immediately employed him at the Ecole Militaire.

Laplace was elected to the Academy of Sciences in 1773 and was fairly well-known by the end of the 1770s. By the late 1780s, he was recognized as one of the leading people at the Academy. In 1784, the Government appointed him to succeed E. Bezout (1730–1783), the French mathematician, as examiner of cadets for the Royal Artillery. At the same time Gaspard Monge (1746–1818), the French mathematician who invented descriptive geometry, was appointed to examine naval cadets. When Napoleon became first consul in 1799, he appointed Laplace as Minister of the Interior, but dismissed him shortly afterwards and elevated him to the Senate.

In 1814, Laplace voted for the overthrow of Napoleon in favour of a restored Bourbon monarchy, and so after 1815 he became an increasingly isolated figure in the scientific community. He remained loyal to the Bourbons until he died. In 1826 he was very unpopular with the Liberals for refusing to sign the declaration of the French Academy supporting the freedom of the press. After the restoration of the Bourbons, he was made a Marquis. Laplace died on 5 March 1827.

Laplace's first important discovery came soon after he was appointed to the Ecole Militaire. He discovered that 'any determinant is equal to the sum of all the minors that can be formed from any selected set of its rows, each minor being described by its algebraic complement'. This is what is now known as the Laplace theorem.

He then turned his mind to problems in celestial mechanics, beginning in 1773 by examining the unexplained variations in the orbits of Jupiter and Saturn. Jupiter's orbit appeared to be continually shrinking, while Saturn's appeared to be continually expanding. No one had succeeded in explaining the phenomenon within the framework of Newtonian gravitation. In a brilliant three-part paper presented to the Academy between 1784 and 1786, Laplace demonstrated that the phenomenon had a period of 929 years and that arose because the average motions of the two planets are nearly commensurable. The variations, therefore, were not at odds with, but compatible with, Newton's law.

While working on that problem Laplace also published his famous paper, 'Théorie des attractions des sphéroïdes et de la figure des planètes' (1785), in which he introduced the potential function and the Laplace coefficients, both of them useful as a means of applying analysis to problems in physics. Two years later he submitted a paper to the Academy in which he dealt with the average angular velocity of the Moon about the Earth, a problem with which Euler and Lagrange had grappled unsuccessfully. Laplace found that, although the mean motion of the Moon around the Earth depends principally on the mutual gravitational attraction between them, it is slightly reduced by the Sun's pull on the Moon, this action of the Sun depending, in turn, upon the changes in eccentricity of the Earth's orbit due to the perturbations caused by the other planets. The result (although the inequality has a period of some millions of years) is that

the Moon's mean motion is accelerated, while that of the Earth slows down and tends to become more circular.

Between 1799 and 1825, Laplace wrote volumes of *Traité de mécanique céleste*. In the first part he gave the general principles of the equilibrium and motion of bodies. By applying these principles to the motion of the planetary bodies, he obtained the law of universal attraction – the law of gravity as applied to the Earth. By analysis he was able to obtain a general expression for the motions and shapes of the planets and for the oscillations of the fluids with which they are covered. From these expressions, he was able to explain and calculate the ebb and flow of tides, the variations in the length of a degree of latitude and the associated change in gravitational force, the precession of the equinoxes, the way the Moon turns slightly to each side alternately (so that over a period slightly more than half of its surface is visible), and the rotation of Saturn's rings. He tried to show why the rings remain permanently in Saturn's equatorial plane. From the same gravitational theory he deduced equations for the principal motions of the planets, especially Jupiter and Saturn.

In 1796, Laplace published his *Exposition du système du monde*, a popular work which was widely read because it gave a theory designed to explain the origin of the Solar System. It suggested that the Solar System resulted from a primitive nebula which rotated about the Sun and condensed into successive zones or rings from which the planets themselves are supposed to have been formed. This theory was generally accepted through the nineteenth century. Since then, however, it has been shown that it is impossible for planets to form in this way because new eccentricities were later discovered which could not be explained by Laplace's concept. Laplace suggested that, as the nebula rotated and contracted, its speed of rotation would increase until the outer parts were moving too quickly for the nebula's gravitational pull to hold them. The remainder of the nebula would then shrink further, rotate faster and another planet would form. This would be repeated until only enough material to form a stable Sun was left at the centre. This theory of planetary formation does not account for the retrograde spin of Venus and the eccentric orbit of Pluto. Also, the theory fails to explain how the energy from the Sun is being continuously created.

Laplace was one of the most influential of eighteenth-century scientists, whose work confirmed earlier Newtonian theories and finally affirmed the permanence and stability of the Solar System. He also compiled considerably improved astronomical tables of the planetary motions, tables that remained in use until the mid-nineteenth century. Together with Lagrange, he was largely responsible for the development of positional astronomy in the eighteenth and nineteenth centuries. Moreover, the methods by which he achieved his results established celestial mechanics as a branch of analysis and through his work potential theory, the Laplace theorem, the Laplace coefficients, orthogonal functions and the Laplace transform (introduced in his papers on probability) became fundamental tools of analysis.

Lapworth Arthur 1872–1941 was a British organic chemist whose most important work was the enunciation of the electronic theory of organic reactions (independently of Robert Robinson).

Lapworth was born in Galashiels, Scotland, on 10 October 1872, the son of the geologist Charles Lapworth who was teaching there at the time. The family moved to Birmingham on his father's appointment as Professor of Geology at Mason College, which Lapworth also attended to study chemistry. He graduated in 1893 and went to the City and Guilds College, London, to do research, first under H. E. Armstrong on the chemistry of naphthalene and then with Frederick Kipping in studies of camphor. Lapworth, Kipping and their contemporary William Perkin Jr became related by marrying three sisters. From 1895 to 1900, Lapworth was a demonstrator at the School of Pharmacy then Head of the Chemistry Department at Goldsmiths College, London. In 1909 he moved to Manchester, where he spent the rest of his life, first as Senior Lecturer in Organic and Physical Chemistry at the University and finally in 1922, holding the senior professorship in the department. In his latter years he developed a painful illness. He died in Manchester on 5 April 1941.

In Lapworth's early work on camphor he recognized an intramolecular change, related to the pinacol-pinacolone rearrangement, which made possible the acceptance of Bredt's structure for camphor.

A little later he began a study of reaction mechanisms, notably of cyanohydrin formation from carbonyl compounds and the benzoin condensation reaction of benzaldehyde. The results of this work entitle Lapworth to be regarded as one of the founders of modern physical-organic chemistry. He was one of the first to emphasize that organic compounds can ionize, and that different parts of an organic molecule behave as though they bear electrical charges, either permanently or at the moment of reaction.

With the development of theories of valency based on the electronic structure of the atom, Lapworth was able to refine some speculations about 'alternative polarities' in organic compounds into a classification of reaction centres as either anionoid or cationoid, the changes being determined by the influence of a key atom such as oxygen. In the mid-1920s, when he collaborated on these concepts with Robert Robinson, a controversy arose with Christopher Ingold and his school who were developing a similar approach to the problem but using a different terminology (nucleophilic for anionoid and electrophilic for cationoid). Ingold's terminology eventually gained general acceptance and the controversy, although occasionally sharp, was fruitful and Lapworth's last paper in 1931 bore Ingold's name as co-author.

Latimer Louis Howard 1848–1928 was a US inventor who is best known for his work on electric lighting. He worked for a long time with Thomas Edison and was the only black member of the 'Edison Pioneers', an organization of scientists who worked with Edison in his pioneering work in the field of electricity.

The son of an escaped slave, Louis Latimer was born on 4 September 1848 in Chelsea, Massachusetts. Escaping from the farm school to which he was sent, Latimer went to Boston to help support the family by selling copies of Garrison's paper *The Liberator*. During the Civil War he enlisted in the United States Naval Service from which he was honourably discharged in 1865. Latimer went to work as an office boy at a firm of patent solicitors, Crosby and Gould, where he learned mechanical drawing and went on to become chief draughtsman of the company. It was here that he became acquainted with Alexander Graham Bell with whom he worked, executing the drawings and assisting in preparing descriptions for the application for the telephone patent which was issued in 1876. In 1880, Latimer was employed as a draughtsman by Hiram Maxim at the United States Electric Lighting Company, Connecticut. In 1883 he went to serve as an engineer and chief draughtsman to Thomas Edison at the Edison Electric Light Company (later the General Electric Company), New York City. In 1890 he transferred from the engineering department to the legal department defending Edison's patents in court as an expert witness. Between 1896 and 1911 he served as chief draughtsman and expert witness for the Board of Patent Control formed by the General Electric and Westinghouse Companies to protect their patents. Latimer died following a long illness on 11 December 1928; he was married and had two daughters.

Latimer is best known for work on electric lighting. In 1881, he and Joseph V. Nichols were given a patent for their 'electric lamp'; he went on to invent a cheap method for producing long-lasting carbon filaments for the maxim electric incandescent lamp, obtaining a patent for them in 1882; these were used in railway stations both nationally and abroad. Also in 1882 he was awarded a patent for his 'Globe supporter for electric lamps'. His book *Incandescent Electric Lighting. A Practical Description of the Edison System* (1890) was a pioneer on the subject. He also supervised the installation of electric lights in New York, Philadelphia, Canada and London. It was not only for lighting that Latimer obtained patents but also for a 'Water Closet for Railroad Cars' (1874), 'Apparatus for Cooling and Disinfecting' (1886), 'Locking Rack for Hats, Coats and Umbrellas' (1896) and 'Book Supports' (1905). Latimer also wrote poetry and his *Poems of Love and Life* was privately published in 1925 to commemorate his 77th birthday. He was honoured in May 1968 when the Lewis Howard Latimer School in Brooklyn was named after him.

Laue Max Theodor Felix von 1879–1960 was a German physicist who established that X-rays are electromagnetic waves by producing X-ray diffraction in crystals. In recognition of this achievement he was awarded the 1914 Nobel Prize for Physics.

Laue was born at Pfaffendorf, near Koblenz, on 9 October 1879. He entered the University of Strasbourg in 1899, but soon transferred to the University of Göttingen, where he chose to specialize in theoretical physics. He then attended the University of Berlin, where he obtained his

DPhil for a dissertation supervised by Max Planck (1858–1947) on interference theory. In 1903, Laue returned to Göttingen, where he spent two years studying art and obtained a teaching certificate. He then became an assistant to Planck at the Institute of Theoretical Physics in Berlin, cementing a long and fruitful partnership between the two scientists. Laue remained in Berlin for four years, during which time he qualified as a lecturer at the University, and became an early adherent of Einstein's special theory of relativity. He published the first monograph on the subject in 1909, which was expanded in 1919 to include the general theory of relativity and subsequently ran to several editions.

In 1909, Laue moved to the University of Munich, and became a lecturer at the Institute of Theoretical Physics, which was directed by Arnold Sommerfeld (1868–1951). It was here that he began his work on the nature of X-rays and crystal structure for which he was awarded the Nobel prize in 1914, only two years after the research was carried out, a remarkable testimony to the importance of the work.

Laue became a full Professor at the University of Frankfurt in 1914, but spent the years of World War I at the University of Würzburg, working on the development of amplifying equipment to improve communications for the military. After the end of the war he exchanged teaching posts with Max Born (1882–1970) and went to Berlin, where he became Professor of Theoretical Physics and Director of the Institute of Theoretical Physics.

In the 1920s and early 1930s, through the German Research Association, Laue directed the provision of financial resources for physics in Germany, thus helping to bring about the important discoveries in theoretical physics with which Germany led the world at that time. In 1932, he was awarded the Max Planck Medal by the German Physical Society. But the rise to power of Adolf Hitler in 1933 and the consequent persecution of Jewish scientists (such as Einstein) and manipulation of German science by the National Socialist Party was intolerable to Laue. His protests led to a gradual trimming of his influence and he lost his position in the German Research Association. He did, however, continue to act as Professor at the University of Berlin until 1943, when he resigned in protest. Although Laue refused to participate in the German atomic energy project, he was nevertheless interned in Britain by the allies after the war, along with other German atomic physicists.

In 1946, Laue returned to Germany where he helped to rebuild German science. He served as Deputy Director of the Kaiser Wilhelm Institute for Physics, and in 1951 became the Director of the Max Planck Institute for Research in Physical Chemistry. He died on 23 April 1960 as the result of a car accident.

Laue's early work in Berlin under Max Planck was chiefly concerned with interference phenomena. He also worked on the entropy of beams of rays, and developed a proof for Einstein's special theory of relativity based on optics. This used the verification in 1851 by Armand Fizeau (1819–1896) of a theory developed even earlier by Augustin Fresnel (1788–1827) that the velocity of light is affected by the motion of water. This seemed to be at odds with special relativity, but Laue was able to show that Fresnel's theory could be derived from the conclusions of relativity. Fizeau's verification thus also became an experimental confirmation of relativity, contributing greatly to its rapid acceptance.

At the time of Laue's move to Munich in 1909, the subjects of crystal structure and the nature of X-radiation were ill-defined. It had been proposed that crystals consist of orderly arrays of atoms, and that X-rays are a type of electromagnetic radiation like light, but with very short wavelengths. The wavelength of normal monochromatic light was determined by the use of artificial diffraction gratings, but if X-rays did in fact have such short wavelengths, no grating sufficiently fine could possibly be produced with which to diffract them. Laue realized that if crystals are indeed such orderly atomic arrays, then they could be used as superfine gratings.

To test Laue's ideas, Sommerfeld's assistant Walter Friedrich and Paul Knipping, a postgraduate student, began experiments on 21 April 1912 using copper sulphate. They bombarded the crystal with X-rays and produced a photographic plate with a dark central patch representing X-rays that had penetrated straight through the crystal surrounded by a multilayered halo of regularly spaced spots representing diffracted X-rays. The results were announced to the Bavarian Academy of Science less than two weeks later along with Laue's mathematical formulation, called the 'theory of diffraction in a three-dimensional grating'. The experimental demonstration of the nature of crystal structure and of X-rays at one strike was widely acclaimed.

Following World War I, Laue expanded his original theory of X-ray diffraction. He had initially considered only the interaction between the atoms in the crystal and the radiation waves, but he now included a correction for the forces acting between the atoms. This correction accounted for the slight deviations which had already been noticed. Laue did not participate in the development of quantum theory, however, being rather cautious in his approach to it, but he did incorporate the discovery of electron interference into his diffraction theory. He also did some important work on superconductivity, including the effect of magnetic fields on this phenomenon, publishing papers on the subject between 1937 and 1947.

Laue's achievement proved to be one of the most significant discoveries of the twentieth century. Einstein described it as one of the most beautiful discoveries in physics, and it has had a profound influence on physics, chemistry and biology. Two entirely new branches of science arose out of Laue's prizewinning work, although he had no more than a passing interest in them. These were X-ray crystallography, which was developed by William Henry Bragg (1862–1942) and William Lawrence Bragg (1890–1971) and led to the determination of the structure of DNA in 1953, and X-ray spectroscopy, which was developed by Maurice de Broglie (1875–1960) and others for the exact measurement of the wavelength of X-rays and for advances in atomic theory.

Laveran Charles Louis Alphonse 1845–1922 was a French physician who discovered that the cause of malaria is a protozoon, the first time that a protozoon had been shown to be a cause of disease. For this work and later discoveries of protozoon diseases he was awarded the 1907 Nobel Prize for Physiology or Medicine.

Laveran was born in Paris, on 18 June 1845, the son of an army surgeon. The family moved to Algeria from 1850 to 1855 and then returned to Paris, where he continued his education. He studied medicine in Strasbourg and graduated in 1867. When the Franco–Prussian war broke out he became an army surgeon, and was stationed at Metz from 1870 to 1871. In 1874, he was appointed Professor of Military Medicine at the Ecole du Val-de-Grâce. Between 1878 and 1883 he was posted to Algeria, first to Bône and then Constantine. In 1884, he returned to Val-de-Grâce and served as Professor of Military Hygiene, until 1894, when temporary posts took him to Lille and Nantes. In 1896, he left the army to join the Pasteur Institute in Paris and in 1907 he used the money from his Nobel prize to open the Laboratory of Tropical Diseases at the Institute. The following year he founded the Societé de Pathologie Exotique and was its president until 1920. He died in Paris on 18 May 1922.

While in Algeria, Laveran found malaria rife there and he became interested in the disease, although his early research had concentrated on nerve regeneration. Malaria has affected humans for centuries, and at that time was explained by a number of hypotheses, one of the most common being that it was due to bad air, especially over swamps and marshes (*mal aria* are the Italian words for bad air). Scientists were beginning, however, to regard it as a bacterial disease.

It was known that malarial blood contains tiny black granules which are sometimes located in hyaline cysts inside the red blood cells. In 1880, Laveran examined blood samples from malarial patients and while investigating a cyst, saw flagellae pushed out from it, revealing that it was not a granule of pigment (as had been believed), but an amoeba-like organism. Laveran took blood samples from these patients at regular intervals and found that the organisms increased in size until they almost filled the blood cell, when they divided and formed spores. When these spores were liberated from the destroyed blood cell they invaded unaffected blood cells. He noted that the spores were released in each affected red cell at the same time and corresponded with a fresh attack of fever in the patient. The point at which the fever subsided coincided with the invasion of new blood cells. Laveran also found crescent-shaped bodies with black granules lying in the plasma.

Laveran's studies of protozoon diseases included leishmaniasis and trypanosomiasis. His investigations concerning malaria allowed Ronald Ross, after him, to discover that the malarial parasite is transmitted to human beings by the *Anopheles* mosquito and these findings enabled preventive measures to be taken against the disease.

Laveran's publications included *Traité des maladies et épidemies des armées/Treatise on Army Sicknesses and Epidemics* (1875) and *Trypanosomes et trypanosomiasis* (1904).

Lavoisier Antoine Laurent 1743–1794 was a French chemist, universally regarded as the founder of modern chemistry. His contributions to the science were wide ranging, but perhaps his most significant achievement was his discrediting and disproof of the phlogiston theory of combustion, which for so long had been a stumbling block to a true understanding of chemistry.

Lavoisier was born in Paris on 26 August 1743 into a well-off family. His mother died when he was young and he was brought up by an aunt. He received a good education at the Collège Mazarin, where he studied astronomy, botany, chemistry and mathematics. In 1768 he was elected an associate chemist to the Academy of Sciences; he eventually became its Director in 1785 and Treasurer in 1791.

Also in 1768 Lavoisier became an assistant to Baudon, one of the farmers-general of the revenue, and later he became a full member of the *ferme générale*, employed by the government as tax collectors. He married 14-year-old Marie Paulze, daughter of a tax farmer, in 1771 and the following year his father bought him a title. In 1775 he was made *régisseur des poudres* and improved the method of preparing saltpetre (potassium nitrate) for the manufacture of gunpowder. A model farm he set up at Frénchines in 1778 applied scientific principles to agriculture, and he drew up various agricultural schemes as secretary to the committee on agriculture, to which he was appointed in 1785. Two years later he became a member of the provincial assembly of Orléans, in which position he initiated many improvements for the community, such as workhouses, savings banks and canals. He was also a member of various other committees and commissions, including that formed in 1790 to rationalize the system of weights and measures throughout France, which ultimately led to the founding of the metric system.

During the French Revolution, Lavoisier came under suspicion because of his membership of the *ferme générale* (from which he derived a considerable income) and because of his marriage to one of its senior executives, although his wife acted as his scientific assistant, taking notes and even illustrating some of his books. Jean-Paul Marat, an extremist revolutionary whose membership of the Academy of Sciences had been blocked by Lavoisier, accused him of imprisoning Paris and preventing air circulation because of the wall he had built round the city in 1787. He fled from his home and laboratory in August 1792 but was arrested in the following November and sent for trial by the revolutionary tribunal in May 1794. At a cursory trial Lavoisier was one of 28 unfortunates sentenced to death. He was guillotined on 8 May 1794 and buried in a common grave. His widow later (1805) married the US physicist Benjamin Thompson (Count Rumford).

Among Lavoisier's early scientific work were papers of the analysis of the mineral gypsum (hydrated calcium sulphate), on thunder, and a refutation that water changes into 'earth' if it is distilled repeatedly. He helped the geologist Jean Etienne Guettard to compile a mineralogical atlas of France.

But his most significant experiments concerned combustion. He found that sulphur and phosphorus increased in weight when they burned because they absorbed 'air', and reported these results in a note he left with the Academy of Sciences in 1772. He also discovered that when litharge (lead(II) oxide) was reduced to metallic lead by heating with charcoal it lost weight because it had lost 'air'. Then in 1774 Joseph Priestley produced 'dephlogisticated air' and Lavoisier grasped the true explanation of combustion, inventing the name oxygen (acid-maker) for the substance that combined with caloric and formed 'oxygen gas'. He coined the word azote for the 'non-vital air' (nitrogen) that remained after the oxygen in normal air had been used up in combustion. In June 1783, he published his finding that the combustion of hydrogen in oxygen produces water, although unknown to him this fact had already been announced by the British chemist Henry Cavendish. Lavoisier burned various organic compounds in oxygen and determined their composition by weighing the carbon dioxide and water produced – the first experiments in quantitative organic analysis. Lavoisier's findings were universally accepted after the publication of his clear and logical *Traité élémentaire de chimie* in 1789, in which he listed all the chemical elements then known (although some of these were in fact oxides).

After establishing that organic compounds contain carbon, hydrogen and oxygen, Lavoisier showed by weighing, that matter is conserved during fermentation as with more conventional chemical reactions. From quantitative measurements of the changes during breathing, he discovered the composition of respired air and showed that carbon dioxide and water are both normal products of respiration.

Lavoisier also made many studies outside the field of chemistry. With Pierre Laplace he experimented with calorimetry and other aspects of heat. He began to use solar energy for scientific purposes as early as 1772 and observed that 'the fire of ordinary furnaces seems less pure than that of the Sun'. He anticipated later theories about the interdependence of sequential processes in plant and animal life forms, as described in one of his papers discovered only many years after his death. His great contribution to science was summed up by Joseph Lagrange who said, on the day after Lavoisier was guillotined at the Place de la Revolution, 'It required only a moment to sever that head, and perhaps a century will not be sufficient to produce another like it.'

Lawrence Ernest Orlando 1901–1958 was a US physicist who was responsible for the concept and development of the cyclotron. For this achievement he was awarded the 1939 Nobel Prize for Physics. As Director of the Radiation Laboratories at Berkeley, California, he and his co-workers then produced a remarkable sequence of discoveries which included the creation of radioactive isotopes and the synthesis of new transuranic elements.

Lawrence was born in Canton, South Dakota, on 8 August 1901. He went to the University of South Dakota, where he was awarded a BS degree in 1922, and then continued his studies at the University of Minnesota, specializing in physics and gaining an MA the following year. After a further two years spent at the University of Chicago, he went to Yale University where he carried out research on photoelectricity and gained his PhD in 1925. He continued at Yale as a National Research Fellow and later as Assistant Professor of Physics until 1928. During this period he made a precise determination of the ionization potential (the energy required to remove an electron) of the mercury atom, developed methods for spark discharges of minute duration (3×10^{-9} sec), and discovered a new method of measuring the charge-to-mass ratio of the electron (e/m). Lawrence then returned to the University of California as Associate Professor of Physics and in 1930 he became, at 29, the youngest full professor (of physics) at Berkeley. Lawrence retained this position until the end of his life, together with the directorship of the Radiation Laboratories, which he commenced in 1936.

During World War II, Lawrence was involved with the separation of uranium-235 and plutonium for the development of the atomic bomb, and he organized the Los Alamos Scientific Laboratories at which much of the work on this project was carried out. After the war, he continued as a believer in nuclear weapons and advocated the acceleration of their development. He died at Palo Alto, California, on 27 August 1958.

In 1929, Lawrence was pondering methods of attacking the atomic nucleus with particles at high energies. The methods then developed involved the generation of very high voltages of electricity. To Lawrence, these seemed to be very limited in the long run as the electrical engineering necessary seemed likely to be technically very difficult, and their cost would eventually prove to be prohibitive. Early that same year, he came across an article in a German periodical which described a linear device for the acceleration of ions, and Lawrence realized the potential of the design for producing high-energy particles which could be directed at atomic nuclei. On carrying out the calculation to produce a machine that would produce protons with the energy of 1 million elecronvolts (1 MeV), he discovered that the device would be too long for the laboratory space he had available. He therefore considered bending the accelerator so that the particles would travel in a spiral.

This idea led to the cyclotron, which had two electrodes that were hollow and semicircular in shape mounted in a vacuum between the poles of a magnet. In between the two electrodes was the source of the particles. To the electrodes was connected a source of electricity whose polarity could be oscillated, positive and negative, from one electrode to the other at very high frequency. The negative field caused the particles to turn in a circular path in each electrode and each time they crossed between the electrodes they received an impetus which accelerated them further, giving them more and more energy without involving very high voltages. As this happened, the increased velocity carried the particles farther from the centre of the apparatus, producing a spiral path.

The first cyclotrons were made in 1930 and were only a few centimetres in diameter and of very crude design. They were made by Nels Edlefsen, one of Lawrence's research students, and one of them showed some indication of working. Later the same year M. Stanley Livingston, another of Lawrence's research students, constructed a further small, but improved device which accelerated protons to 80,000 eV with 1,000 volts used on the electrodes. Once the principle had been seen to work well, larger and improved devices were made. Each new design produced particles of higher energy than its predecessor and new results were obtained from the use of the accelerated particles in nuclear transformations. One of these was the disintegration of the lithium nucleus to produce helium nuclei, confirming the first artificial transformation made by John Cockcroft (1897–1962) and Ernest Walton (1903–) in 1932, and a 68-cm/27-in model was used to produce artificial radioactivity. With increasing size, the engineering problems associated with power supplies and the production of large magnetic fields increased, and in 1936 the Radiation Laboratory with the necessary engineering and administrative facilities was opened to house the new equipment.

From his laboratory, a multitude of discoveries in chemistry, physics and biology were made under Lawrence's direction. Hundreds of radioactive isotopes were produced, including carbon-14, iodine-131 and uranium-233. Furthermore, over the years, most of the transuranium elements were synthesized. Mesons, which were then little known particles, were produced and studied within a laboratory for the first time at the Radiation Laboratory, and antiparticles were also found. Many of these discoveries owe their origin to other famous scientists who worked within the laboratory, but Lawrence worked tirelessly advising researchers and in making the experiments possible. Fairly early in the development of the cyclotron it became apparent that as particles were accelerated to high speeds their masses increased, as predicted by the special theory of relativity. This increase in mass caused them to lose synchronization with the oscillating electric field. The solution of this problem by Vladimir Veksler (1907–1966) and Edwin McMillan (1907–1991) independently led to the development of the synchrotron and particles of even higher energies.

Lawrence made an important contribution to physics with the invention of cyclotron, for the subsequent development of high-energy physics and the elucidation of the subatomic structure of matter was highly dependent on this machine and its successors. These have also provided radioisotopes of great value in medicine. Appropriately, Lawrence's name is remembered in the naming of element 103 as lawrencium, for it was discovered in his laboratory at Berkeley.

Leakey Louis Seymour Bazett 1903–1972 was a British archaeologist, anthropologist and palaeontologist who became famous for his discoveries of early hominid fossils in East Africa, indicating that humans probably evolved in this part of the world.

Leakey was born in Kabete, Kenya, on 7 August 1903, the son of a British missionary. There were few European settlers in Kenya at that time and he spent his boyhood with the local Kikuyu children, learning their language but receiving little formal schooling. He was sent to Britain when he was 16 years old and entered St John's College, Cambridge, in 1922 to study French and Kikuyu, and later archaeology and anthropology. In 1926 he went to East Africa leading the first of a series of archaeological research expeditions, which continued until 1937. From 1937 until the outbreak of World War II in 1939, he studied and recorded the customs of the Kikuyu people. During the war he was in charge of the African Section at the British Special Branch Headquarters; he was a handwriting expert and remained available as a consultant in this role after the end of the war. During the war he gave freely of his spare time to the Coryndon Memorial Museum in Nairobi, Kenya, and in 1945 he became its curator. He built up a research centre at the museum and at the same time became one of the founder trustees of the Kenya National Parks and Reserves. He resigned from the museum in 1961 and founded the National Museum Centre for Prehistory and Palaeontology. During the 1960s his health began to fail and he spent less time in the field and more time lecturing and fundraising. He died in London on 1 October 1972.

Leakey began excavations at Olduvai Gorge, now in Tanzania, in 1931 and it was to become the site of some of his most important finds. During wartime leave he and his wife Mary, an expert on palaeolithic stone implements, discovered a significant Acheulian site at Olorgesailie in the Rift Valley. The Acheulian culture is characterized by stone hand axes and flourished between 1 million and 100,000 years ago. On another occasion the Leakeys found the remains of 20-million-year-old Miocene apes on the island of Rusinga in Lake Victoria. From 1947 they initiated in Nairobi a Pan-African Congress on Prehistory, which was a great success and proved to be the first of many.

In 1959, the Leakeys returned to their excavations at Olduvai. In that year they found a skull of *Australopithecus boisei* (its massive teeth earned it the name 'Nutcracker Man') and a year later they discovered the remains of *Homo habilis*, established by potassium–argon dating to be 1.7 million years old. A third exciting find was a skull of an Acheulian hand-axe user, *Homo erectus*, which Leakey maintained was an advanced hominid on the direct evolutionary line of *Homo sapiens*, the modern human. In 1961 at Fort Ternan, Kenya, he found jawbone fragments of another early primate, believed to be 14 million years old.

Leakey was a man of many talents with firm convictions that our origins lie in Africa, a view that was totally opposed to contemporary opinion. He discovered sites of major archaeological importance containing vast numbers of bones and artefacts. The status of some of the finds, and Leakey's interpretation of them, is still in dispute, but whatever the eventual outcome he made an unparalleled contribution to knowledge of early humans and their contemporaries. He published a large number of scientific papers as well as some books of wider appeal, such as *Stone Age Africa* (1936) and *White African* (1937).

After Louis Leakey's death his work in Africa was continued by his wife and his sons John and Richard. Richard Leakey (1944–   ) later became a well-known author and broadcaster in his own right.

Leakey Mary Douglas 1913–   is a British anthropologist and archaeologist.

Born Mary Douglas Nicol in London on 6 February 1913, Leakey was the only child of a landscape painter from whom she inherited a talent for drawing that was of considerable importance to her later career. Much travelling during her childhood disrupted her formal education but a visit to prehistoric caves in southwest France, where her father went to paint, kindled an interest in archaeology. A chance meeting with the archaeologist Dorothy Liddell in the late 1920s, convinced her that a career in the subject was possible and she became Liddell's assistant, chiefly as the illustrator at a major dig at a Neolithic site in Devon. She met the anthropologist Louis Leakey at a dinner party and he invited her to illustrate a book he was then working on. She agreed, and the two began to work closely together. Mary Nicol travelled to join him in Kenya at the Olduvai Gorge, and after Louis Leakey's subsequent divorce, they married in 1936. Mary Leakey immediately began excavating a Late Stone Age site north of Nairobi, and with her husband discovered the reamins of an important Neolithic settlement and many important artefacts. During World War II her husband was involved in British intelligence, but Mary continued to develop some of their research projects, and after the war they devoted much of their time to organizing a major Pan-African Congress of Prehistory and Palaeontology in 1947. The success of that conference brought the work of the Leakeys to a wide audience, attention that was continued when Mary discovered an ape-like skull of the human ancestor *Proconsul*, in 1948. Over the following decade and a half the Leakeys together and separately continued major excavations in Kenya, and accumulated evidence that East Africa was the possible cradle of the human race. By the middle of the 1960s Mary was living almost continuously at the permanent camp they had established at the Olduvai Gorge, whilst Louis was based in Nairobi and increasingly travelling around the world on lecture tours, and although they never officially separated, they lived apart until Louis's death in 1972. That event precipitated Mary onto the international stage and she became an accomplished lecturer and writer, whilst continuing to organize excavation work. She has been recognized by several international awards and honorary doctorates, and has published an account of her work *Olduvai Gorge: My Search for Early Man* (1979) and an autobiography *Disclosing the Past* (1984).

Leavitt Henrietta Swan 1868–1921 was a US astronomer, an expert in the photographic analysis of the magnitudes of variable stars. Her greatest achievement was the discovery of the period–luminosity relationship for variable stars, which enabled Ejnar Hertzsprung and Harlow Shapley to devise a method of determining stellar distance.

Leavitt was born in Lancaster, Massachusetts, on 5 July 1868. She studied at the Society for the Collegiate Instruction of Women, which later became Radcliffe College and associated with Harvard. She earned her bachelor's degree in 1892, but continued her studies there for an additional year. Her first work in astronomy was in a voluntary capacity at the Harvard College Observatory, where she was an assistant in the programme which Edward Pickering had initiated on the measurement of stellar magnitudes. In 1902, Leavitt was given a permanent appointment to the staff of the Observatory, and was ultimately appointed Head of the department of photographic stellar photometry. She died prematurely of cancer on 12 December 1921, in Cambridge, Massachusetts.

Leavitt's work was a direct outgrowth of Pickering's overall research programme at the Observatory. She was extensively involved with the establishment of a standard photographic sequence of stellar magnitudes, and besides discovering a total of 2,400 new variable stars, she also discovered four novae. Her most far-reaching discovery, however, was that the periods of some variable stars is directly related to their magnitudes.

Leavitt spent many years studying a kind of variable star called the Cepheid variable, named after the first of the type to be discovered, Delta Cephei in the Magellanic clouds. These star collections were photographed by the astronomers at the Harvard Observatory in Arequipa, Peru. As early as 1908, when she published a preliminary report of her findings, Leavitt suspected that there was a relationship between the length of the period of variation in brightness and the average apparent magnitude of the Cepheids. In 1912 she published a table of the length of the periods, which varied from 1.253 to 127 days, with an average of around five days, and the apparent magnitudes of 25 Cepheids. There was a direct linear relationship between the average apparent magnitude and the logarithm of the period of these variable stars.

The reason Leavitt was able to notice this relationship for the variable stars in the Magellanic clouds, while it had remained unnoticed for other variable stars, lies in the close grouping of these Cepheids in terms of their distances from each other in contrast to their enormous distance from the Earth. A nearby, albeit dim, variable star might seem to be brighter than a more distant star of greater magnitude. It has since become apparent that the period-luminosity relationship discovered by Leavitt applies only for Cepheids which lie in 'dust-free' space.

At the time this undiscovered factor did not prevent Hertzsprung and Shapley from independently applying the period–luminosity relationship to the determination of stellar distances. All of the Cepheids were too far from the Earth for their distances to be determined using the standard parallax method, pioneered by Friedrich Bessel. But Hertzsprung and Shapley were able to convert Leavitt's period–luminosity curve so that it could be read in terms of absolute rather than apparent magnitude. Then, once they had determined the period of any Cepheid, they could return to the curve and read off its absolute magnitude. By comparing a Cepheid's apparent magnitude with its absolute magnitude, the distance of the star from the Earth could be deduced.

Their results were nothing short of astonishing for contemporary astronomers. They found that the Magellanic Clouds were approximately 100,000 light years away, a distance which was almost beyond the comprehension of their colleagues. Shapley revised these figures in 1918, finding the smaller Magellanic Cloud to be 94,000 light years away.

When Leavitt first began studying the stars in the two hazy areas of the sky, the large and small Magellanic Clouds, she did not realize that they were extragalactic. But, due to her work, it is now known that the Magellanic Clouds are in fact two irregular galaxies, companions of our own Galaxy. Thus Leavitt's vital contribution to astronomy was to provide the critical impetus for the discovery of the first technique capable of measuring large stellar distances.

Lebedev Pyotr Nikolayevich 1866–1912 was a Russian physicist whose most important contribution to science was the detection and measurement of the pressure that light exerts on bodies, an effect that had been predicted by James Clerk Maxwell in his electromagnetic theory of light.

Lebedev was born in Moscow on 8 March 1866. He initially received training in business and engineering in Moscow but decided that he would rather be a physicist and so went to the University of Strasbourg to study physics under August Kundt. His teacher moved to Berlin in 1888 and Lebedev tried to follow him but found that he did not possess the qualifications necessary to enrol in Berlin University, and so he returned to Russia in 1891. On his return he joined the physics department of Moscow Univeristy – at the invitation of the head of the department, A. G. Stoletov – and became Professor of Physics there in the following year, although he did not obtain his PhD until 1900. In 1911 government interference in university affairs caused a storm of protest, which led to the resignation of many of the university's staff, including Lebedev. Although he was subsequently offered numerous prestigious positions at other universities, both in Russia and elsewhere, the shock of the upheaval precipitated a collapse of his health and he died shortly afterwards in Moscow, on 14 March 1912.

Lebedev began studying light pressure (now called radiation pressure) in the late 1890s but did not complete his investigations until 1910. Working first with solid bodies and later with gases, he not only observed the minute physical effects caused by the infinitesimal pressure exerted by light on matter but also measured this pressure using extremely lightweight apparatus in an evacuated chamber. His findings provided substantial supportive evidence for Maxwell's theory of electromagnetic radiation, which had predicted the phenomenon of light pressure. Also, Lebedev

suggested that the force of the light pressure emanating from the Sun could balance the gravitational force attracting cosmic dust particles towards it, and that this was the reason why comets' tails point away from the Sun. Later, however, it was discovered that comets' tails point away from the Sun because of the much greater effect of the solar wind. The Swedish chemist Svante Arrhenius (1859–1927) took up the idea of light pressure to explain how primordial spores might have travelled through space in his theory of the extraterrestrial origin of life. This model has not become widely accepted because it does not explain how the spores themselves originated nor how the travelling spores could endure the intense cosmic radiation in outer space.

Lebedev also investigated (while working for his doctorate) the effects of electromagnetic, acoustic and hydrodynamic waves on resonators; demonstrated the behavioural similarities between light and (as they are now known to be) other electromagnetic radiations; detected electromagnetic waves of higher frequency than the radio waves that had been studied by Heinrich Hertz and Augusto Righi; and researched into the Earth's magnetic field.

Lebesgue Henri Léon 1875–1941 was a French mathematician who is known chiefly for the development of a new theory of integration named after him.

Lebesgue was born at Beauvais on 28 June 1875 and educated at the Ecole Normale Supérieure in Paris between 1894 and 1897. From 1899 to 1902 he worked on his doctoral thesis while teaching mathematical science at the lycée in Nancy. He received his doctorate from the Sorbonne in 1902 and in the same year was appointed a lecturer in the faculty of sciences at the University of Rennes. He left there to become a Professor at the University of Poitiers in 1906, remaining at Poitiers until 1910, when he was appointed Lecturer in Mathematics at the Sorbonne. In 1920, he was promoted to the Chair of the application of geometry to analysis, but he left the Sorbonne in the following year to take up his final academic post as Professor of Mathematics at the Collège de France. He died in Paris on 26 July 1941.

Lebesgue was awarded many honours, including the Prix Houllevique (1912), the Prix Poncelet (1914) and the Prix Saintour (1917). He was elected to the French Academy of Sciences in 1922 and to the Royal Society in 1934.

Lebesgue made contributions to several branches of mathematics, including set theory, the calculus of variation and function theory. With Emile Borel (1871–1956) he laid the foundations of the modern theory of the functions of a real variable. His chief work, however, was his creation of a new approach to the theory of integration.

From an early stage in his mathematicial career, Lebesgue was intrigued by problems associated with Riemannian integration and he began to get results in this field in 1902. His introduction of the Lebesgue integral was not only an impressive piece of mathematical creativity in itself, but quickly proved itself to be of great importance in the development of several branches of mathematics, especially calculus, curve rectification and the theory of trigonometric series. Later the integral was also discovered to be of fundamental significance for the development of measure theory.

Leblanc Nicolas 1742–1806 was a French industrial chemist who devised the first commercial process for the manufacture of soda (sodium carbonate), which became the general method of making the chemical for a hundred years.

Leblanc was probably born in Ivoy-le-Pré, Indre, on 6 December 1742, although there is some doubt as to the exact place and date. His father was an ironmaster and paid for his son's education as an apothecary's apprentice; he went on to study medicine, qualifying as a doctor. In 1780, he became physician and assistant to the future duc d'Orléans (Philippe Egalité). Leblanc invented his famous process in the 1780s and in 1791, using capital supplied by the duke, he built a factory for making soda at St Denis, near Paris. But the duc d'Orléans was guillotined in 1793 during the French Revolution and Leblanc was forced to run the factory at no profit, giving all the output to the state. He had no money left to re-establish the process when the factory was handed back to him by Napoleon in 1802. He became a pauper and committed suicide at St Denis on 16 January or February 1806.

Soda was an important industrial chemical in the second half of the eighteenth century, for making glass, soap and paper. It was made by calcining wood, seaweed and other vegetable matter, hence its common name soda ash. Common salt (sodium chloride) was, however, plentiful and in 1775 the French Academy of Sciences offered a cash prize for the first person to devise a commercially practical way of making soda from salt. Leblanc invented his process in 1783 (although he never received the prize money).

In the Leblanc process, salt (sodium chloride) was dissolved in sulphuric acid to form sodium sulphate:

$$2NaCl + H_2SO_4 \rightarrow Na_2SO_4 + 2HCl$$

The large amounts of hydrogen chloride generated in this reaction were released into the atmosphere. The sodium sulphate (called salt cake) was then roasted with powdered coal and crushed chalk or limestone (calcium carbonate) to yield a dark residue ('black ash') which was made up mainly of sodium carbonate and calcium sulphide:

$$2Na_2SO_4 + 2CaCO_3 + 4C \rightarrow 2Na_2CO_3 + 2CaS + 4CO_2$$

The sodium carbonate was dissolved out of the residue with water and recrystallized by heating the solution. The waste calcium sulphide that remained was known as 'galligu'.

Leblanc patented the process in 1791, and at his first factory produced 350 tonnes of soda a year. But the patent was rendered useless by the activities of the Revolutionary government, which confiscated the factory. The process was adopted and used throughout Europe, particularly in England, and earned large amounts of money for the soda manufacturers (who by the 1860s were making 180,000 tonnes of soda a year using the Leblanc process) and for the makers of sulphuric acid, one of the starting materials.

After the development of the ammonia–soda process by Ernest Solvay in the 1860s, the Leblanc process gradually fell into disuse, although from time to time it was given a new lease of life by modifications. In the 1850s, the British industrial chemists Henry Deacon (1822–1876) and Ferdinand Hurter (1844–1898) introduced the improvement of catalytically oxidizing the waste hydrogen chloride to chlorine, which was absorbed by lime to make bleaching powder, then much in demand by the Lancashire textile industry. In the late 1880s Alexander Chance (1844–1917), of the Birmingham glassmaking family, turned his attention to calcium sulphide, the other main waste product of the Leblanc process. He used carbon doxide to react with an aqueous slurry of the obnoxious waste to produce hydrogen sulphide gas (and leave calcium carbonate), which he then oxidized to sulphur in a Claus kiln. By the 1890s in Britain alone the Chance process, as it came to be known, was producing 35,000 tonnes of sulphur each year. Eighty years after Leblanc's death, his process and its adaptations were still providing the chemical industry with some of its most important basic raw materials. The last Leblanc plant (in Bolton, Lancashire) did not close until 1938.

Lebon Phillipe 1767–1804 was a French engineer who was the first person to successfully use 'artificial' gas as a means of illumination on a large scale.

Lebon was born in the charcoal-burning town of Bruchay, near Jonville. He received a sound scientific education, first at Chalon-sur-Sôane and later at the Ecole des Ponts et Chaussées, a famous school for engineers. He graduated in 1792 and received the rank of major, serving as a highway engineer. After a short while in Angouleme, near Bordeaux, he was recalled to Paris to teach mechanics at the Ecole des Ponts et Chaussées.

In about 1797, Lebon became interested in using 'artificial' gas, produced from wood, for heating and lighting purposes. He also made some attempts at perfecting the steam engine and received a national prize of 2,000 livres for the improvements he accomplished. In 1799, he read a paper on his experiments with gas to the Institut de France and was granted a patent on his invention later that year. Further patents were granted in 1801. Lebon's work was, however, cut short by his sudden death on 2 December 1804. He was attacked and stabbed on the Champs Elyssée in Paris, and died of his wounds.

Lighting on a large scale was demonstrated when the streets of Paris were first lit in 1667 using large lanterns containing candles and metal reflectors. An oil lantern was developed in 1744 and, in 1786, Argand completed his invention of an oil lamp with a glass chimney. Carcel introduced an oil pump and such oil lamps continued to be popular for lighting streets and houses. Lebon was the first person to consider using gas as a lighting medium.

In 1797, at Bruchay, he became interested in the extraction of gas from wood for this purpose and began experimenting with sawdust. He placed some sawdust in a glass tube and held it over a flame. The gas given off caught alight as it emerged from the tube – but it smoked badly and emitted a strong resinous smell. On his return to Paris, Lebon discussed his work with several scientists, including Fourcroy. They encouraged him and, in 1799, he read a paper on the subject before the Institut de France. He patented his invention in that year and called his new lamp the Thermolampe (heat lamp) because he intended to use it for heating as well as lighting purposes.

Approaches to the government to interest them in using his invention for public heating and lighting proved of no avail. To publicize his Thermolampe, Lebon leased the Hotel Seignelay in Paris in 1801 and, for several months, he exhibited a large version of the lamp which attracted huge crowds. But, although they admired Lebon's successful attempt to illuminate a fountain in the hotel, because he had been unable to eliminate the repulsive odour given off by the gas, the public decided that his invention was not a practical one.

Further work by Lebon was curtailed by his early death and it was left to William Murdoch (working independently at about the same time in Scotland) to succeed where Lebon had failed, and it was Murdoch who has received the credit for the invention of gas lighting.

Le Châtelier Henri Louis 1850–1936 was a French physical chemist, best known for the principle named after him which states that if any constraint is applied to a system in chemical equilibrium, the system tends to adjust itself to counteract or oppose the constraint.

Le Châtelier was born in Paris on 8 October 1850, the son of France's Inspector-General of Mines. He was educated at the Collège Rollin in Paris and went to study science and engineering at the Ecole Polytechnique, although his studies were interrupted by the Franco–Prussian War (1870–1871). He graduated in 1875 then, after working for two years as a mining engineer, he took up an appointment as Professor of Chemistry at the Ecole des Mines in 1877. In 1898, he moved to the Collège de France as Professor of Mineral Chemistry, before finally settling at the Sorbonne in 1908 as Professor of Chemistry in succession to Henri Moissan (1852–1907). He worked for the Ministry of Armaments during World War I and retired in 1919. He died at Miribel-les-Echelles, Isère, on 17 September 1936.

Le Châtelier's first major contribution was to temperature measurement, a subject that followed naturally from his high-temperature studies of metals, alloys, glass, cement and ceramics. In 1887, he devised a platinum–rhodium thermocouple for measuring high temperatures by making use of the Seebeck effect (the generation of a current in a circuit made up of two dissimilar metals with the junctions at different temperatures; the magnitude of the current is proportional to the difference in temperature). Le Châtelier also made an optical pyrometer which measures temperature by comparing the light emitted by a high-temperature object with a standard light source.

This work involving flames and thermometry led him to thermodynamics, and in 1884 Le Châtelier put forward the first version of his principle, in which he stated that a change in pressure on an equilibrium system results in a movement of the equilibrium in the direction that opposes the pressure change. By 1888 he had generalized the principle as the *Loi de stabilité de l'équilibre chimique* and applied it to any change that affects chemical equilibrium. In its general form Le Châtelier's principle is all-embracing, and includes the law of mass action, as formulated by Cato Guldberg (1836–1902) and Peter Waage (1833–1900) in 1864. It is particularly relevant in predicting the effects of changes in temperature and pressure on chemical reactions: for example, it predicts that a rise in temperature or an increase in pressure should facilitate or accelerate a reaction that is reluctant to take place at normal temperatures and pressures. Industrial chemists, such as Fritz Haber with his process for synthesizing ammonia, were soon to make good use of the principle. It also agreed with the new thermodynamics being worked out in the United States by Josiah Willard Gibbs. Le Châtelier was largely responsible for making Gibbs' researches known in Europe, translating his papers into French and performing experiments to test the phase rule. He also wrote extensively about labour relations and efficiency in industry. In 1895 he put forward the idea of the oxyacetylene torch for cutting and welding steel.

Leclanché Georges 1839–1882 was a French engineer who invented the Leclanché battery or dry cell in 1866. This is the kind of battery used today in torches, calculators, portable radios, and so on.

Leclanché was born in Paris in 1839. Educated at the Ecole Centrale des Arts et Manufactures, he joined the Compagnie du Chemin de l'Est as an engineer in 1860. Six years later, he developed his electric cell, which consisted of a zinc anode and a carbon cathode separated by a solution of ammonium chloride. The following year, in 1867, he gave up his job as a railway engineer to devote all his time to the improvement of the cell's design. He was successful in having it adopted by the Belgian Telegraphic Service in 1868.

The Leclanché cell rapidly came into general use whenever an intermittent supply of electricity was needed and it was later developed into the familiar dry battery in use today. Towards the end of his life, Leclanché diversified his researches and worked for a while on electrical methods of time measurement. He died in Paris on 14 September 1882.

Almost any electric cell with electrodes of different materials can act as a battery. The first cells to be invented were wet cells made by Alessandro Volta (1745–1827) in 1800. They were made of tin or copper and zinc or silver separated by pasteboard or hide soaked in water, vinegar or salt solution. These first electric batteries, although cumbersome, caused a revolution by making comparatively large amounts of electric current available for the first time.

The revolution was carried even further when Leclanché produced the first dry cell, which facilitated both handling and mobility. Leclanché's investigations into new kinds of electrodes and electrolyte enabled him to make the advance that he did. Volta's cell prompted several other types of cell, among them a design by Robert Bunsen (1811–1899) for a battery containing carbon and zinc electrodes and an acid such as chromic acid as the electrolyte. Leclanché produced the most useful design of all by having zinc for the negative pole, carbon for the positive pole and a solution of ammonium chloride as the electrolyte. This first cell was the forerunner of all subsequent dry cells in that it was sealed so that the outside stayed dry.

Although there are other types of dry cells in existence today, the Leclanché cell is still the only one produced on a large scale. In the modern dry Leclanché cell, zinc serves both as the container and anode, the cathode is formed from graphite, and the electrolyte is a paste of ammonium chloride, zinc chloride, manganese dioxide and carbon particles. The cell produces 1.5 volts.

Lederberg Joshua 1925– . US geneticist; *see* Beadle, Tatum, and Lederberg.

Leeuwenhoek Anton van 1632–1723 was a Dutch microscopist, famous for the numerous detailed observations he made using his single-lens microscopes. He was not the first of the many well-known early microscopists, being preceded by Marcello Malpighi for example, nor did he make notable innovations to the microscope itself (the forerunner of the modern compound microscope was developed in Leeuwenhoek's lifetime by Robert Hooke, an English physicist and microscopist). But such was the dramatic nature of Leeuwenhoek's discoveries that he became – and remains – world renowned.

Leeuwenhoek was born on 24 October 1632 in Delft, Holland. Relatively little is known of his early life but it seems that he received scant schooling. His stepfather died when Leeuwenhoek was 16 years old and he was apprenticed to a cloth merchant in Amsterdam. He returned to Delft four years later in 1652 and opened a drapery shop. In 1660 he obtained the sinecure of chamberlain to the sheriffs of Delft. Having guaranteed his financial security, Leeuwenhoek devoted much of his time to his hobbies of lens grinding and microscopy. From 1672 to 1723 he described and illustrated his observations in a total of more than 350 letters to the Royal Society of London, which elected him a Fellow in 1680. Leeuwenhoek continued his work almost until he died, aged 90 years, on 26 August 1723. After his death several of his microscopes were sent to the Royal Society, in accordance with his will.

Leeuwenhoek ground more than 400 lenses, which he mounted in various ways. Most of them were very small (some were about the size of a pinhead) and had magnifying powers of between 50 and about 300 times; but each was meticulously made and the optical excellence of these lenses – combined with Leeuwenhoek's careful observations – undoubtedly helped him to make so many important

discoveries. In 1674 he discovered protozoa, which he called 'animalicules', and calculated their sizes. He was also probably the first to observe bacteria, when he saw tiny structures in tooth scrapings; the first known drawing of bacteria was made by Leeuwenhoek and appeared in the Royal Society's *Philosophical Transactions* in 1683. He made many other important observations: he was the first to describe spermatozoa (1677) and also studied the structure of the lens in the eye, muscle striations, insects' mouthparts, the fine structure of plants, and discovered parthenogenesis in aphids. In 1684 he gave the first accurate description of red blood cells, also noticing that they can have different shapes in different animal species.

Leeuwenhoek became world famous during his lifetime and was visited by several reigning monarchs, including Frederick I of Prussia and Tsar Peter the Great. Many of Leeuwenhoek's observations remained unsurpassed for more than a century – partly because his microscopes were of very high optical quality and partly because he kept secret the details of the techniques he used.

Legendre Adrien-Marie 1752–1833 was a French mathematician who was particularly interested in number theory, celestial mechanics and elliptic functions.

Legendre was born in Paris on 18 September 1752, studied mathematics and natural science at the Collège Mazarin in Paris and, despite having a private income to sustain him in independent research, took employment as a Lecturer in Mathematics at the Ecole Militaire in 1775. He taught there until 1780 without making any impression on the mathematical world. His fortunes began to rise in 1882, when he won the prize awarded by the Berlin Academy for an essay on the path of projectiles travelling through resistant media. A year later he was elected to the French Academy of Sciences and from the year 1783 began to publish important papers. He was appointed Professor of Mathematics at the Institut de Marat in Paris in 1794, the same year in which he became head of the government department established to standardize French weights and measures. From 1799 to 1815 he served as an examiner of students of artillery and in 1813 he succeeded Pierre Laplace (1749–1827) as chief of the Bureau de Longitudes. He remained at that post until his death, in Paris, on 10 January 1833.

Legendre's first published work, a paper on mechanics, appeared in 1774, but a decade passed before his real talent showed itself. In several papers of 1783–1784 he introduced to celestial mechanics what are now known as Legendre polynomials. These are solutions to the second-order differential equation (still important in applied mathematics):

$$(1 - x^2) \frac{d^2 y}{dx^2} - 2x \frac{dy}{dx} + n(n + 1)y = 0$$

where *n* is a non-negative integer. The functions which satisfy this equation are called Legendre functions.

During the 1780s Legendre worked on a number of other

topics, including indeterminate analysis and the calculus of variations, but at the end of the decade his research was interrupted for a time by the outbreak of the French Revolution and the suppression, in 1793, of the French Academy of Sciences. He placed himself at the service of the revolutionary government, however, to direct the project which altered the system for the measurement of angles and other decimalization projects.

His two most important contributions to mathematics were made in the 1790s, although he had begun work on both – number theory and elliptical functions – in the mid-1780s. In number theory his most significant result was the law of reciprocity of quadratic residues, although the credit for establishing the law rigorously belongs to Karl Gauss in 1801. It was Legendre alone, however, who in 1798 gave the law of the distribution of prime numbers; and very late in his career, in 1823, he proved that there was no solution in integers for the equation:

$$x^5 + y^5 = z^5$$

Of even more use to fellow mathematicians was Legendre's long and painstaking work on elliptical functions. In 1786 he made a tentative start on the subject with a paper on the integration of elliptical curves and in 1792 he touched on the theory of elliptical transcendentals in a paper to the Academy. His great achievement, however, was the two-volume textbook on elliptical functions which he published in 1825 and 1826, in which he gave the tables of elliptical functions which he had laboriously compiled.

One other accomplishment of Legendre, of a more mundane character, ought not to be forgotten. In 1794 he published his *Eléments de géometrie*, a reworking and a clarification of Euclid's *Elements*. Among its delights was the single proof of the irrationality of π and the first proof of the irrationality of π^2. The text was translated into several languages and in many parts of the world stood as the basic school text in geometry for the next 100 years.

Leibniz Gottfried Wilhelm 1646–1716 was a German philosopher and mathematician who was one of the founders of the differential calculus and symbolic logic.

Leibniz was born on 1 July 1646 in Leipzig, where his father was Professor of Moral Philosophy at the university. Although he attended the Nicolai School at Leipzig, most of his early education came from his own reading, especially in the classics and the early Christian writers, in his father's library. At the age of 15 he entered the University of Leipzig, where his formal training was chiefly in jurisprudence and philosophy. Privately, he read all the important scientific texts – of Bacon, Galileo, Kepler, Descartes and others. In 1663 he went to the University of Jena, where he was taught Euclidean geometry by Erhard Weigel (1625–1699). He then returned to Leipzig and after three years more study of law applied for the degree of Doctor of Law in 1666. It was refused on the ground that he was too young. He therefore went to Altdorf, where his thesis *De Casibus per plexis in jure* was accepted and the doctorate awarded.

Leibniz turned down the offer of a professorship at Altdorf and decided to travel about Europe. At the end of 1666 he entered the service of the Elector and Archbishop of Mainz; he was employed chiefly in foreign affairs, his special task being to devise plans to preserve the peace of Europe, just then emerging from the Thirty Years' War. He was invited to France by Louis XIV, to present to him his plan for a French invasion of Egypt (and so transfer war from European to African soil) and although, in the event, he never met the king, he remained in Paris for about three years. It was in Paris that his serious work in mathematics began. He met leading scientists, including Christiaan Huygens (1629–1695), made a thorough study of Cartesianism, and began work on his calculating machine. The machine was completed about 1672 and was a marked improvement on Pascal's machine, in that it was able to multiply, divide and extract roots.

The death of the Elector of Mainz in 1673 left Leibniz without an official position. He was offered the post of librarian to the Duke of Brunswick at Hanover, but went instead to London. The visit marked a turning-point in his mathematical life, for it was in London in 1673 that he became acquainted with the work of Isaac Newton and Isaac Barrow and began to work on problems which led him to his independent discovery of differential and integral calculus.

In 1676, Leibniz at last took up the appointment as librarian to the House of Brunswick. He remained in that service for the rest of his life and much of his time was spent in conducting research into the genealogy and history of the Brunswick line.

He also continued to be charged with diplomatic missions and on one of his visits to Berlin he succeeded in persuading the local elector to establish an academy of science. It was founded in 1700 and Leibniz was appointed President for life. From 1712 to 1714 he was an imperial privy councillor at Vienna. In 1714 the Elector of Hanover, Georg Ludwig, Duke of Brunswick, acceded to the English throne as George I. Leibniz asked to be allowed to accompany him to London, but the request was denied. He therefore spent the last two years of his life engaged in genealogical work, embittered by the dispute with Newton over the invention of the calculus and suffering from gout. He died a neglected man – neither the Royal Society nor the Berlin Academy took any notice of the event – on 14 November 1716 in Hanover.

Just as much of his service to princes consisted in the search for a balance of power and international cooperation in Europe, and as he sought to reconcile in much of his philosophical writing Protestantism and Roman Catholicism, so did Leibniz dream of an international community of scholars, served by academies like that of Berlin, freely sharing their discoveries and continually exchanging their ideas. To this end he worked intermittently throughout his life at devising what he called a Universal Characteristic, a universal language that would be accessible to everyone. It is, therefore, a matter of some sorrow that he became embroiled in a long and acrimonious dispute about the authorship of the calculus, a dispute which darkened the last 15 years of his life. In 1699, the Swiss mathematician and Fellow of the Royal Society Fatio de Duillier accused Leibniz of stealing the idea from Newton, a charge which the Royal Society formally upheld in 1711. Leibniz himself never sought to conceal that it was after his 1673 visit to London, by which time Newton had worked out his calculus of fluxions, that he began his investigations into tangents and quadratures, the research that eventually led to his discovery of calculus. But Newton's discovery, probably made in 1665, was not published for many years and there is no doubt that Leibniz arrived at his calculus independently. As he put it, he, Newton and Barrow were 'contemporaries in these discoveries'. Leibniz always communicated his findings to his fellows; most mathematicians of the time were working on the same problems and they all knew the work that had been done on infinitesimal quantities. At any rate, to Leibniz is due the credit for first using the infinitesimals as differences. To him also is due the credit for working out, like Newton, a complete algorithm and for devising a notation so much more convenient than Newton's, that it remains in standard use today.

The idea of the calculus was in the mathematical air. It was Leibniz who expressed its fundamental notions in the most effective manner. That should not be surprising, for Leibniz will always be remembered chiefly as the founder of symbolic logic. Centuries later, it has become clear that his logic, free from all concepts of space and number and hence in his lifetime not recognized as mathematical at all, was the prototype of future abstract mathematics.

Leishman William Boog 1865–1926 was a British army physician who discovered the protozoon parasite that causes kala-azar, a relatively common and potentially fatal infectious disease endemic to the tropics and subtropics that affects the reticulo-endothelial system (particularly the liver, spleen and bone marrow). The genus of protozoons to which the causative microorganism belongs is called *Leishmania*, after Leishman. He received many honours for his work, including a knighthood in 1909.

Leishman was born on 6 November 1865 in Glasgow, a son of the Regius Professor of Midwifery at Glasgow University. He was educated at Westminster School, London, then studied medicine at Glasgow University, from which he graduated in 1886. After graduation he obtained a commission in the Royal Army Medical Corps, in which he remained for the rest of his life. He was posted to India from 1890 to 1897, after which he returned to England to the Army Medical School at Netley, Hampshire, where he was soon appointed Assistant Professor of Pathology under Almroth Wright (1861–1947). After Wright resigned, Leishman succeeded him as Professor in 1903; in the same year the Medical School was transferred to London and Leishman moved with it. In 1914, he became a member of the Army Medical Advisory Board, advising the War Office on tropical diseases then, with the outbreak of World War I, he joined the British Expeditionary Force as Advisor in Pathology. After the war ended he became the first

Director of Pathology at the War Office in 1919. In 1923, he was appointed Director-General of the Army Medical Service, a position he held until his death on 2 June 1926 in London.

Leishman discovered the protozoon parasite that causes kala-azar in 1900, using his modified form of the Romanowsky stain for protozoa and blood cells (this modified stain is now called Leishman's stain) to examine cells from the spleen of a soldier who had died of kala-azar at Netley. He published his findings in 1903 but in the same year Charles Donovan of the Indian Medical Service independently made the same discovery, as a result of which the causative protozoon was called the Leishman–Donovan body. Other workers later discovered that related species of the kala-azar-causing protozoon were responsible for various other diseases; all such similar protozoons were therefore classified as members of the same genus, named *Leishmania* – the protozoon causing kala-azar being called *Leishmania donovani* – and the diseases they cause were grouped under the term leishmaniasis.

Leishman also assisted Wright in developing an effective antityphoid inoculation, and helped to elucidate the life cycle of the spirochaete (*Spirochaeta duttoni*) which causes African tick fever.

Lemaître Georges Edouard 1894–1966 was a Belgian cosmologist who – perhaps because he was also a priest – was fascinated by the Creation, the beginning of the universe, for which he devised what later became known as the 'Big Bang' theory.

Lemaître was born in Charleroi on 17 July 1894. Trained as a civil engineer, he served as an artillery officer with the Belgian army during World War I. After the war he entered a seminary, where he was ordained a priest in 1923. He nevertheless maintained an unwavering interest in science, and from 1923 to 1924 he visited the University of Cambridge, where he studied solar physics and met Arthur Eddington (1882–1944). Afterwards, he spent two years at the Massachusetts Institute of Technology in the United States, and it was while he was there that he became influenced by the theories of Edwin Hubble (1889–1953) and the Harvard astronomer Harlow Shapley (1885–1972) concerning the likelihood of an expanding universe.

Having returned to his native country with better insight into the thinking of his contemporaries, Lemaître was made Professor of Astrophysics at the University of Louvain from 1927. In 1933, he published his *Discussion on the Evolution of the Universe*, which stated the theory of the Big Bang, and he followed this in 1946 with his *Hypothesis of the Primal Atom*. He died at Louvain on 20 June 1966.

The main feature of Lemaître's theory of the beginning of the universe stemmed from his belief in the 'primal atom', formulated first in 1931. He visualized this atom as a single unit, an incredibly dense 'egg' containing all the material for the universe within a sphere about 30 times larger than the Sun. Somewhere between 20,000 and 60,000 million years ago, in his view, this atom exploded, sending out its matter in all directions. There then took place a balancing act between expansion and contraction. Ultimately expansion won, since when (around 9,000 million years ago) the galaxies have been drifting away from each other. The significance of this theory is not so much its affirmation of the expansion of the universe as its positing of an event to begin the expansion.

In 1946 George Gamow (1904–1968) improved on Lemaître's basic theory, considering the Big Bang from just before the event to just after, thus giving the Big Bang itself a definite beginning and a definite end (and a scientific existence in between). However, Lemaître's and Gamow's solutions were for a time somewhat overshadowed by the invention of the radio telescope, which began to reveal aspects of the universe previously unknown. Hermann Bondi (1919–   ), Thomas Gold (1920–   ) and Fred Hoyle (1915–   ), working at Cambridge, put forward their 'steady-state' theory, in which they saw the universe as having no beginning and no end: stars and galaxies were created, went through a life cycle, and died, to be replaced by new matter being created out of 'nothingness' (possibly hydrogen atoms).

For a time during the late 1940s and early 1950s the steady-state theory was a serious rival to Lemaître's and Gamow's Big Bang, but more recently the steady-state theory has been virtually abandoned. Research by Martin Ryle (1918–1984) and others has shown that the universe may simply undergo periods of total expansion and total contraction that will go on indefinitely.

Lenard Philipp Eduard Anton 1862–1947 was a Hungarian-born German physicist who devised a way of producing beams of cathode rays (electrons) in air, enabling electrons to be studied. This led him to the first conclusion that an atom is mostly empty space, and Lenard was awarded the 1905 Nobel Prize for Physics as a result. He also did pioneering research on the photoelectric effect.

Lenard was born on 7 June 1862 at Pozsony, Hungary (now Bratislava, Slovak Republic), but spent most of his life in Germany. He studied physics at Heidelberg under Robert Bunsen (1811–1899) and at Berlin under Hermann Helmholtz (1821–1894), receiving a doctorate in 1886 from Heidelberg. In 1891, he became assistant to Heinrich Hertz (1857–1894) at Bonn, assuming command of Hertz's laboratory on his early death in 1894. He then took an associate professorship at Breslau the same year, then moved in turn to Aachen in 1895, Heidelberg in 1896 and Kiel, where he was appointed Professor of Experimental Physics in 1898. He then held the same post at Heidelberg from 1907 to 1931. Lenard became a follower of Hitler, the only eminent scientist to do so, in 1924 and spent his later years reviling Jewish scientists such as Albert Einstein (1879–1955). He was antagonistic by nature and in his book *Great Men of Science* (1934), he omitted Wilhelm Röntgen (1845–1923) and several other eminent modern scientists who had had priority disputes with him, who had criticized Lenard's work or whose work Lenard considered inferior to his own. Lenard died in Messelhausen, Germany, on 20 May 1947.

Lenard's initial studies were in mechanics investigating the oscillation of precipitated water drops, work no doubt stimulated by his regard for the studies of Hertz. In his youth Lenard had been fascinated by the phenomenon of luminescence and later with his knowledge of science he returned to the topic. He studied the luminosity of pyrogallic acid (1,2,3-trihydroxybenzene) in the presence of alkali and sodium hydrogen sulphite, establishing that the crucial factor was the oxidation of the acid. He continued to look at luminescent compounds for 20 years along with his main field of study.

Lenard's principal contribution to physics can be traced back to work begun in 1892. Inspired by William Crookes (1832–1919), who in 1879 had published a paper on the movement of cathode rays in discharge tubes, Lenard conceived an experiment to examine these rays outside the tube. He was then assistant to Hertz, who had shown that a piece of uranium glass covered with aluminium foil and put inside the discharge tube became luminescent when struck by cathode rays. Lenard developed a technique using aluminium foil as a window in a discharge tube, and showed that the cathode rays could move about 8 cm/3 in through the air after passing through the thin aluminium window. Since the cathode rays were able to pass through the foil, he suggested that the atoms in the metal must consist of a large proportion of empty space. A further conclusion of greater significance to later work was that the part of the atom where the mass was concentrated consisted of neutral doublets or 'dynamids' of negative and positive electricity. This proposal preceded by ten years the classic model of the atom proposed by Ernest Rutherford (1871–1937) in 1911.

Lenard had at his fingertips a great knowledge of electricity. He devised the grid in the thermionic valve that controls electron flow. He also showed that an electron must have a certain minimum energy before it can produce ionization in a gas. From 1902 onwards, Lenard studied photoelectricity and discovered several fundamental effects. He showed that negative electricity can be released from metals by exposure to ultraviolet light and later found that this electricity was identical in properties to cathode rays (electrons). He also discovered that the energy of the electrons produced depends on the wavelength of the light incident on the metal. He explained that the release of electrons was caused by a resonance effect intensified by vibrations induced by the ultraviolet light. Einstein in 1905 provided a correct explanation of photoelectricity by using the quantum theory, which Lenard rejected.

Philipp Lenard was a gifted experimental physicist who was able to build on the earlier work of other scientists, enabling theoretical physicists to use his results to delve deeper into the structure of matter. Many of his discoveries are fundamental to our understanding of atomic physics.

Lenoir Jean Joseph Etienne 1822–1900 was a Belgian-born French engineer and inventor who in the early 1860s produced the first practical internal-combustion engine and a car powered by it. He also developed a white enamel

(1847), an electric brake (1853) and an automatic telegraph (1865). Lenoir was born at Mussy-la-Ville, Belgium, on 12 January 1822; he died in Varenne-St-Hilaire, France, on 4 August 1900.

The first self-propelled road vehicles were steam cars – using an external-combustion steam engine. The French army captain Nicolas Cugnot built a three-wheeled steam tractor for hauling cannon as early as 1769, and a passenger-carrying vehicle was constructed in England in about 1801. By 1830, steam carriages were in regular use but they were noisy and dirty, and the smoke and hot coals often caused fires along the route, making them unpopular with farmers who feared for their crops.

Several people had claimed to have invented an internal-combustion engine before Lenoir. The Reverend W. Cecil read a paper to Cambridge University in 1829 about his experiments, and William Barnett in 1838 also laid claims, but not until Lenoir in 1859 did a practical model become a reality. His engine consisted of a single cylinder with a storage battery (accumulator) for the electric ignition system. Its two-stroke cycle was provided by slide valves, and it was fuelled by coal gas, as used then for domestic purposes and street lighting. Lenoir built a small car around one of his prototypes in 1863, but it had an efficiency of less than 4% and although he claimed it was silent, this was only true when the vehicle was not under load.

The real value of his engine was for powering small items of machinery, and by 1865 more than 400 were in use in the Paris district driving printing presses, lathes and water pumps. Its use for vehicles was restricted by its size and not until some 20 years later, when Gottlieb Daimler and Karl Benz independently devised the four-cycle engine, that internal-combustion engines were successful in vehicles.

Lenz Heinrich Friedrich Emil 1804–1865 was a Russian physicist, known for Lenz's law, which is a fundamental law of electromagnetism.

Lenz was born in Dorpat (now Tartu, Estonia) on 12 February 1804. He graduated from secondary education in 1820 and entered the University of Dorpat, where he studied chemistry and physics. His physics teacher recommended Lenz for the post of geophysical scientist to accompany Otto von Kotzebue (1787–1846) on his third expedition around the world, which lasted from 1823 to 1826. Upon his return, Lenz was appointed scientific assistant at the St Petersburg Academy of Science, later becoming Associate Academician in 1830 and Full Academician in 1834. From 1840 to 1863, Lenz was Dean of Mathematics and Physics at the University of St Petersburg, where he was also elected Rector. He was also the author of a number of very successful books which ran to many editions. Lenz died after suffering a stroke while on holiday in Rome on 10 February 1865.

Lenz's major fields of study were geophysics and electromagnetism. While on his voyage with Kotzebue, he studied climatic conditions such as barometric pressure, finding the areas of maximum and minimum pressure that exist in the tropics and determine the overall climatic pattern. Lenz also made a careful investigation of water salinity, discovering areas of maximum salinity either side of the equator in the Atlantic and Pacific Oceans and establishing the differences in salinity between these oceans and the Indian Ocean. Lenz also made extremely accurate measurements of the temperature and specific gravity of sea water. His geophysical observations of the oceans were not bettered until the following century.

In 1829, Lenz went on a mountaineering expedition to the Caucasus in southern Russia, where he made a study of some of the natural resources of the area as well as making measurements of mountain heights and of the level of the Caspian Sea.

Lenz's studies of electromagnetism date from 1831, following the discovery in that year of electromagnetic induction by Michael Faraday (1791–1867) and Joseph Henry (1797–1878). Lenz's first major discovery was that the direction of a current which is induced by an electromagnetic force always opposes the direction of the electromagnetic force that produces it. This law was published in 1833 and it is known as Lenz's law. When a moving magnet or coil produces a current by induction in a conductor, the induced current flows in such a direction that it in turn produces a magnetic field that opposes the motion of the magnet or coil inducing the current. This means that work has to be done by the magnet or coil to produce current, and Lenz's law is in fact a special case of the law of conservation of energy. If the induced current were to flow in the opposite direction, it would assist the motion of the magnet or coil and a perpetual-motion machine would result because energy would increase without any work being done, which is impossible. Lenz's law later helped Hermann Helmholtz (1821–1894) to formulate the law of conservation of energy, and it is applied today in electrical machines such as generators and electric motors.

Around the same time, Lenz also began to study the relationship between heat and current, and discovered, independently of James Joule (1818–1889), the law now known as Joule's law which describes the proportional relationship between the production of heat and the square of the current. Lenz also found that the strength of a magnetic field is proportional to the strength of the magnetic induction. He also worked on the application of certain theoretical principles to engineering design, on formulating programs for geographical expeditions, and on establishing the unit for the measurement of electrical resistance. A final major area of study was electrochemistry where in the 1840s he made a series of investigations into additivity laws.

Leonardo da Vinci 1452–1519 was a famous Italian artist, inventor and scientist, and is regarded as one of the greatest figures of the Italian Renaissance for the universality of his genius.

Leonardo was born on his father's estate, in Vinci, Tuscany. The illegitimate son of the Tuscan landowner Ser Piero and a peasant girl, he was taken into his father's household in Florence, where he was the only child. He received an elementary education and in about 1467 was apprenticed to the artist Andrea del Verrocchio. There he trained in artistic as well as technical and mechanical subjects. He left the workshop in about 1477 and worked on his own until 1481. He went to Milan the following year and was employed by Ludovico Sforza, the Duke of Milan, as 'painter and engineer of the duke'. In this capacity he advised the duke on the architecture of proposed cathedrals in Milan and nearby Pavia, and was involved in hydraulic and mechanical engineering. After Milan fell to the French in 1499, he fled and began a long period of wandering. In 1500 he visited Mantua and then Venice, where he was consulted on the reconstruction and fortification of the church San Francesco al Monte. Two years later he went into the service of Cesare Borgia as a military architect involved in the designing and development of fortifications. In 1503 he returned to Florence to investigate, on Cesare Borgia's behalf, the possibility of re-routing the river Arno so that the besieged city of Pisa would lose its access to the sea. It was at about this time that he painted his internationally renowned portrait the *Mona Lisa*. He was invited that year by the Governor of France in Milan, Charles d'Amboise, to work for the French in Milan, and in 1506 he took up the offer. There he devised plans for a castle for the governor and for the Adda Canal to connect Milan to Lake Como. In 1513 the French were defeated and forced to leave Milan; da Vinci left with them and went to Rome to look for work. He stayed with Cardinal Giulano de' Medici, the brother of Pope Leo X, but there was little for him to do (although both Michelangelo and Raphael were working there at that time) other than to advise on the proposed reclamation of the Pontine Marshes. Three years later he left Italy for France on the invitation of King François I, and he lived in the castle of Cloux, near the king's summer residence at Amboise. Leonardo spent the rest of his life there, sorting and editing his notes. He died on 2 May 1519.

Leonardo's training in Verrocchio's workshop developed his practical perception, which served him well as a technical scientist, a creative engineer and as an artist. In the years of his first visit to Milan, principally between 1490 and 1495, he produced his notebooks, in mirror-writing. The illustrated treatises deal with painting, architecture, anatomy and the elementary theory of mechanics. The last was produced in the late 1490s and is now in the Biblioteca Nacional, Madrid. In it Leonardo proposes his theory of mechanics, illustrated with sketches of machines and tools such as gear, hydraulic jacks and screw-cutting machines, with explanations of their functions and mechanical principles and of the concepts of friction and resistance.

Leonardo's interest in mechanics developed as he realized how the laws of mechanics, motion and force operate everywhere in the natural world. He studied the flight of birds in connection with these laws and, as a result, designed the prototypes of a parachute and of a flying machine.

During this time, he also developed his ideas about the Renaissance Church Plan, which later were considered favourably by the architect Bramante in connection with the building of the new St Peter's in Rome.

In about 1503, when Cesare Borgia's plan for the diversion of the river Arno failed, Leonardo also devised a project to construct a canal, wide and deep enough to carry ships, which would bypass the narrow portion of the Arno so that Florence would be linked to the sea. His hydrological studies on the properties of water were carried out at this time.

The variety of Leonardo's inventions reflect his passionate absorption in biological and mechanical details and ranged from complex cranes to pulley systems, lathes, drilling machines, a paddlewheel boat, an underwater breathing apparatus and a clock which registered minutes as well as hours. As a military engineer he was responsible for the construction of assault machines, pontoons, a steam cannon and a tortoise-shaped tank. For a castle in Milan he created a forced-air central heating system and also a water-pumping mechanism. His notes and diagrams established him as, arguably, the greatest descriptive engineer and scientist of his age. Despite these achievements, he remains most famous as an artist, unique in the history of the world's greatest painters.

Lesseps Ferdinand, Vicomte de 1805–1894 was a French civil engineer who is remembered as the designer and builder of the Suez Canal, the strategic importance of which, as a trade route between Europe and the East remains until the present day.

Lesseps had the distinction of being born at the Palace of Versailles, being a cousin of the Empress Eugénie. As befitted his noble birth (he was a Vicomte), he was brought up to regard diplomatic service as the natural choice for a career and from 1825 he held posts in various capitals including Lisbon, Tunis and Cairo.

He also had other interests, and these centred around engineering and construction, especially where canal-building was concerned. When it was suggested in 1854 that a passage should be cut to link the Mediterranean with the Red Sea, Lesseps was the ideal man to take charge of the work.

In 1856 permission was granted by the Viceroy of Egypt, Muhammad Said, and the canal was begun in 1860, financed mainly with money put up by the French government and Ottoman Empire. Ten years later the canal opened for traffic, shortening the route between Britain and India by 9,700 km/6,000 mi.

Lesseps received many honours for his achievement, an English knighthood, the Grand Cross of the Legion of Honour and election to the French Academy of Science being chief among them. Unfortunately, all these were rather overshadowed by the disaster of the Panama Canal, the construction of which he reluctantly undertook in 1881. The project met with failure and bankruptcy. Lesseps was sentenced to five years' imprisonment for breach of trust, but was too ill to leave his house. A broken and sick old man, he was allowed to remain there till his death on 7 December 1894.

Since Roman times, thoughts of cutting across the Isthmus of Suez had been discussed at various intervals. Napoleon I, on his visit to Egypt in 1799, saw the possibilities and advantages of a canal, but no practical steps were taken until 1854 when the Suez Canal scheme was conceived. A technical commission met in 1855 and mapped out a suitable route. It then remained for permission to be granted by the Viceroy and for a company to be formed to finance the operation. All this was achieved by 1860 when work began in earnest.

As there was no great difference in the levels of the two seas at either end of the isthmus, locks were unnecessary and the construction, although long – more than 160 km/100 mi – was relatively simple. It linked the cities of Port Said and Suez and was initially 8 m/26 ft deep, 22 m/72 ft wide at the bottom and 70 m/230 ft across at the surface. When it was finished on 17 November 1869 it reduced the journey from the Mediterranean to the Indian Ocean by thousands of miles and removed the necessity for weeks at sea traversing the Cape of Good Hope.

Originally Britain owned none of the shares in the Suez Company, but in 1875 the Khedive of Egypt, Ismail Pasha (who succeeded Muhammad Said in 1863) sold his stock to the British government. From then on the company was mainly controlled by the British and the French, and in 1888 it was agreed that the canal should be opened to all nations, at all times. Constructed as a one-lane canal, it has been widened and deepened in recent years to accommodate the increase of traffic, particularly from the oil fields of the Arab countries. At one time, tankers made up 70% of shipping recorded as passing through the canal.

The Suez Canal was a brilliant piece of engineering carried out efficiently under the watchful eye of Lesseps. The advantages to trade were obvious to our Victorian forebears, and in the modern world that narrow strip of artificial waterway has proved vital to the Western economy.

Leverrier Urbain Jean Joseph 1811–1877 was a French astronomer who, as well as being a trained chemist, became an authority in France on many aspects of astronomy and is chiefly remembered for his contribution to the discovery of the planet Neptune. Leverrier was also instrumental in the establishment of the meteorological network across continental Europe.

He was born in St Lô, Normandy, on 11 March 1811, and went first to local schools before attending the Collège de Caen (1828–1830) and then the Collège de St Louis in Paris. There, in 1831, he won a prize in mathematics and entered the Ecole Polytechnique. After a short time doing research in chemistry under the direction of Joseph-Louis Gay-Lussac (1778–1850) at Administration Tobaccos, he began teaching, both privately and at the Collège Stanislas in Paris. He applied for the post of Demonstrator in Chemistry at the Ecole Polytechnique in 1837, but was instead offered a post in astronomy which, having already published some work in this subject, he accepted. Leverrier

was soon recognized as an astronomer of distinction. After the discovery of Neptune, he was elected to the Academy of Sciences in Paris (in 1846) and made a Fellow of the Royal Society in London (1847), which also awarded him the Copley Medal. A new Chair of Astronomy was created for him at the Faculty of Science in 1847, and in 1849 the Chair of Celestial Mechanics was established for him at the Sorbonne. He was politically active in the revolution of 1848, serving as a member of the legislative assembly in 1849, and later in 1852 as a senator. In 1854, Leverrier took over the Directorship of the Paris Observatory after the death of his friend and colleague Dominique Arago (1786–1853). He was not a popular administrator (evidently he kept a tight rein on both the direction of research and its funding) and he was eventually dismissed from the post in 1870. The untimely death of his successor (Charles Delaunay), however, brought him back to the Observatory, albeit with some restrictions imposed. Thereafter, during the last few years of his life, Leverrier was plagued with a progressive deterioration in his health. He died in Paris on 23 September 1877.

Leverrier's first paper in astronomy, which dealt with shooting stars, was published in 1832. At the Ecole Polytechnique his first major investigation (1838–1840) was into the stability of the Solar System. He made calculations based on minor variations in the planetary orbits and extended them to cover a period of more than 200,000 years, intending to demonstrate how little variation over time does actually occur. Through this exercise, however, he became fascinated by the notion of tracing the cause of the perturbations he had recorded, and he immediately began a study designed to identify the periodic comets and other bodies within the Solar System whose gravitational pulls might affect the planets in their orbits.

It was already known that the point of the planet Mercury's orbit closest to the Sun was progressively 38 seconds per century greater than would be predicted on the basis of Newtonian mechanics. In 1845, Leverrier attempted to resolve this by proposing the existence of a planet – which he named Vulcan – lying 30 million km/19 million mi from the Sun, *inside* the orbit of Mercury. (Leverrier was by no means the only astronomer of the time, or since, to be seeking an 'intramercurial' planet. Heinrich Schwabe (1789–1875), for instance, also had the idea. However, all attempts optically to detect a planet in this location have failed – and in any case, the anomaly in Mercury's orbit was later used by Karl Schwarzschild in 1916 to support Einstein's general theory of relativity, which predicted such an advance in Mercury's perihelion.)

In the same year Dominique Arago pointed out to Leverrier that there was another discrepancy between the predicted and the observed behaviour of a planet, in the Solar System. Alexis Bouvard (1767–1843) had produced tables on the planet Uranus in 1821, and fewer than 25 years later they were already grossly inaccurate. This suggested to Leverrier that some planet outside Uranus's orbit was having a profound influence (although the possibility that another planet might exist beyond Uranus had in fact already been suggested by William Herschel and by Friedrich Bessel). Accordingly, Leverrier published three papers on the subject during 1846, in the last of which he gave a prediction for the position and apparent diameter of the hypothetical planet.

Unknown to Leverrier, a young English astronomer, John Couch Adams (1819–1892), had carried out virtually identical calculations a year earlier at Cambridge and had sent them to the Astronomer Royal, George Airy. For various reasons, Airy had left Adams's communication unread – until he perused Leverrier's second publication on the matter. By that time Leverrier had written to a number of observatories, asking them to test the prediction contained in his third paper. It happened that Johann Galle and Louis d'Arrest at the Berlin Observatory had just received a new and accurate star map of the relevant sector, and in order to test it, were glad to oblige. On the very first night of observation the new planet was found within 1° of Leverrier's coordinates.

The argument that then ensued over who should receive the credit for the planet's discovery was aggravated by somewhat chauvinistic debate in the popular presses of both France and Britain. It extended to the question of naming the new planet.

Dominique Arago wanted it to be named after Leverrier, but (perhaps because of its optically greenish hue) he finally proposed that it should be called Neptune.

Thereafter, at the Paris Observatory, Leverrier saw it as his life's work to compile a comprehensive analysis of the masses and the orbits of the planets of the Solar System, taking special note of their mutual influences. The work on Mercury, Venus, the Earth and Mars was carried out during the 1850s; that on Jupiter, Saturn, Uranus and Neptune during the 1870s. The whole was published only after Leverrier's death, in the *Annals* of the Paris Observatory.

Levi-Civita Tullio 1873–1941 was an Italian mathematician skilled in both pure and applied mathematics whose greatest achievement was his development, in collaboration with Gregorio Ricci-Curbastro (1853–1925), of the absolute differential calculus.

Levi-Civita was born at Padua on 29 March 1873 and received his secondary education there before entering the University of Padua to study mathematics in 1890. There he came strongly under the influence of Ricci-Curbastro, one of his teachers. He was awarded his BA in 1894 and took up employment as a lecturer at the teacher-training college at Pavia. In 1897, he was appointed Professor of Mechanics at the Engineering School at Padua. In 1918, he left Padua to become Professor of Higher Analysis at the University of Rome; in 1920 he was made Professor of Rational Mechanics. He remained there until 1938, when the anti-Jewish laws promulgated by the Fascist government forced him to leave the university; he was also expelled from all Italian scientific societies. His health began to deteriorate rapidly and he died of a stroke, at Rome, on 29 December 1941.

Although Levi-Civita began to publish while he was still an undergraduate, his first important results – the fruit of several years' labour – were first published, with Ricci-Curbastro, in 1900. Together they presented to the mathematical world a completely new calculus, which became known as absolute differential calculus. One of the most important features of this new system was its remarkable flexibility – applicable, as it was, to both Euclidean and non-Euclidean spaces. Most significantly, it could be applied to Riemannian curved spaces, and the tensor system which the paper outlined was fundamental to Einstein's development of the general theory of relativity. Levi-Civita's own most important contribution to the absolute differential calculus was the publication in 1917 of a paper in which he postulated a law of parallel translation of a vector in a Riemannian curved space. This introduction of the concept of parallelism in curved space was Levi-Civita's most brilliant contribution to the history of mathematics. The discussions to which it gave rise eventually allowed absolute differential calculus to develop into tensor calculus, a tool of immense usefulness to mathematicians attempting to derive a unified theory of gravitation and electromagnetism. The idea of parallel displacement has also been of great importance in the field of the geometry of paths.

In addition to this central work, Levi-Civita also published interesting papers on celestial mechanics and hydrodynamics, and in general it may be said that his achievements in both pure and applied mathematics established him as one of the foremost mathematicians of his age.

Levi-Montalcini Rita 1909–   is an Italian neuroscientist whose work has principally been on chemical factors that control the growth and development of cells. In particular she isolated a substance called nerve growth factor that promoted the development of nerve cells. For this she shared the 1986 Nobel Prize for Physiology or Medicine with Stanley Cohen (1922–   ).

Levi-Montalcini was born and educated in Turin, graduating in medicine from the city's University in 1936. She immediately began studying the mechanisms of nerve growth. From 1939 onwards she was unable to hold an academic position in an Italian university because she was Jewish. She constructed a home laboratory for herslf, making microsurgical instruments out of cutlery, and also served as a volunteer physician towards the end of World War II. In 1947 she moved to the USA, to the Washington University in St Louis, where she remained until 1981 having become Professor in 1958. In 1981 she went to Rome. It was whilst in St Louis that she had discovered nerve-growth factor in the salivary glands of developing mouse embryos. She continued to try to find further sources of this factor and established that it was chemically a protein. She also analysed the mechanism of its action in isolated tissues and then in whole neonatal and adult animals. She showed that nerve-growth factor could be identified in many tissues, including mouse cancer cells and snake venom glands; and that it was most effective on cells when they were in the early stages of differentiation. Her work has stimulated important new work into understanding processes of some neurological diseases and possible repair mechanisms; into tissue regeneration; and into cancer mechanisms.

Lewis Gilbert Newton 1875–1946 was a US theoretical chemist who made important contributions to thermodynamics and the electronic theory of valency. He is best known for his explanation of the behaviour of acids and bases.

Lewis was born in Weymouth, Massachusetts, on 23 October 1875. He was educated at the preparatory school of the University of Nebraska and Harvard, from which he graduated in 1896, gaining his MA in 1898 and his PhD a year later for a thesis on the electrochemical and thermochemical relations of zinc and cadmium amalgams. He remained at Harvard for a year as an instructor, before going to Europe on a travelling scholarship to study under Friedrich Ostwald at Leipzig and Hermann Nernst at Göttingen. He then went to the Philippines for a year as Superintendent of Weights and Measures and Chemist at the Bureau of Science in Manila. He returned to the United States in 1905 to join the research team of A. A. Noyes at the Massachusetts Institute of Technology (MIT). In 1912, he became Chairman of the Chemistry Department at the University of California, where he remained until his death. He died in his laboratory at Berkeley on 23 March 1946.

During his seven years at MIT Lewis published more than 30 papers, including fundamental work on chemical thermodynamics and free energies. At Berkeley he set about reorganizing and rejuvenating the department, appointing staff with a broad chemical knowledge rather than specialists.

In 1916, Lewis began his pioneering work on valency. He postulated that the atoms of elements whose atomic mass is higher then helium's have inner shells of electrons with the structure of the preceding rare gas. The valency electrons lie outside these shells and may be lost or added to comparatively easily to form ionic bonds. He went on to state that bonding electrons prefer to pair up – the idea of the covalent bond. Much of this work involved the building of bridges between inorganic and organic chemists, who had often considered that polar (predominantly ionic) and non-polar (predominantly covalent) substances bore little relation to each other. Lewis also drew attention to the unusual properties of molecules that have an odd number of electrons, such as nitric oxide (nitrogen monoxide, NO).

In 1923 Lewis and M. Randall published *Thermodynamics and the Free Energy of Chemical Substances*, which was the culmination of 20 years' research in compiling data on free energies (ΔG). Until the early years of this century it was considered that the heat of reaction (ΔH) could be taken as a measure of chemical affinity and that changes in enthalpy could be used to predict the direction of the reaction. It was later realized that free energy is the correct basis for such predictions. Lewis' treatise listed the

Lewis *A molecule of chlorine with a covalent bond involving the sharing of a pair of electrons by the two atoms, and two hybrid forms of a molecule of nitric oxide (nitrogen monoxide), each with an odd (unpaired) electron.*

free energies of 143 important substances, which could be used to evaluate the outcome of several hundred reactions. This work was linked to the determination of the electrode potentials of more than a dozen elements.

Also in 1923 Lewis published his highly influential book *Valence and the Structure of Atoms and Molecules*. In it he put forward a new definition of a base as a substance that has a lone pair of electrons which may be used to complete the stable group of another atom, and defined an acid as a substance that can use a lone pair from another molecule in completing the stable group of one of its own atoms. In other words, a base supplies a pair of electrons for a chemical bond, and an acid accepts such a pair. This definition, which has stood the test of time, was remarkable because it was the first to suggest that bases include substances which do not produce hydroxyl ions.

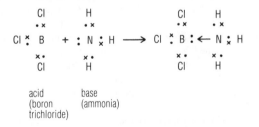

acid
(boron
trichloride)

base
(ammonia)

Lewis *The neutralization of an acid and a base, during which the base supplies a pair of electrons to form a chemical bond.*

Previously, according to the Brönsted–Lowry theory, it was thought that an acid is a substance capable of donating a proton (hydrogen ion) to an acceptor substance, a base. Lewis proved his theory by carrying out acid–base reactions, detected by colour changes to indicators, in non-aqueous, hydrogen-free solvents such as tetrachloromethane (carbon tetrachloride) in which proton transfer was not possible.

For several years during the mid-1930s Lewis' research team carried out investigations on heavy water and deuterium (which had been discovered in 1932 by Harold Urey, one of Lewis' former students). In his later years he carried out studies on the excited electron states of organic molecules, contributing to the understanding of the colour of organic substances and the complex phenomena of phosphorescence and fluorescence. Lewis died in his laboratory while performing experiments on fluorescence.

Libby Willard Frank 1908–1980 was a US chemist best known for developing the technique of radiocarbon dating, for which he was awarded the 1960 Nobel Prize for Chemistry.

Libby was born in Grand Valley, Colorado, on 17 December 1908, the son of a farmer. He received his university education at the University of California, Berkeley, from which he graduated in 1931 and gained his PhD in 1933. He then took a teaching appointment at Berkeley and in 1941, soon after the outbreak of World War II, moved to Columbia University, New York, to work on the development of the atomic bomb (the Manhattan Project). After the war, in 1945, he became Professor of Chemistry at the University of Chicago's Institute for Nuclear Studies. From 1954 to 1959 he was a member of the United States Atomic Energy Commission, then in 1959 he returned to the University of California to become Director of the Institute of Geophysics till his retirement. He died in Los Angeles on 8 September 1980.

During the early 1940s at Columbia, Libby worked on the separation of uranium isotopes for producing fissionable uranium-238 for the atomic bomb. Back in Chicago after the war he turned his attention to carbon-14, a radioactive isotope of carbon that had been discovered in 1940 by Serge Korff. It occurs as a small constant percentage of the carbon in the carbon dioxide in the atmosphere – resulting from cosmic-ray bombardment – and in the carbon in the tissues of all living plants and animals. Carbon-14 has an extremely long half-life (5,730 years) but when the plant or animal dies, it accumulates no more of the radioactive isotope, which steadily decays and changes into nitrogen. Libby reasoned that a determination of the carbon-14 content of anything derived from plant or animal tissue – such as wood, bones, cotton or woollen cloth, hair or leather – gives a measure of its age (or the time that has elapsed since the plant or animal died).

He and his co-workers accurately dated ancient Egyptian relics by measuring the amount of radiocarbon they contained using a sensitive Geiger counter. By 1947 they had further developed the technique so that it could date objects up to 50,000 years old. It has proved to be extremely useful in geology, anthropology and archaeology. In 1946, Libby showed that tritium (a radioactive isotope of hydrogen of mass 3) is formed by the action of cosmic rays and devised a method of dating based on the amount of tritium in the water in an archaeological specimen. In 1952 he published *Radiocarbon Dating*. Later workers have extended the method using other isotopes, such as potassium-40.

Li Cho Hao 1913–   is a Chinese-born US biochemist who is best known for his work on the hormones secreted by the pituitary gland.

Li was born in Canton (Guangzhu) in southern China on 21 April 1913 and educated at the University of Nanking (Nanjing), from which he graduated in 1933. He was an Instructor in Chemistry at Nanking University from 1933 to 1935 and then in 1935 he emigrated to the United States, where he took up postgraduate studies at the University of

California at Berkeley. After obtaining his PhD in 1938, Li joined the staff of the university, becoming in 1950 Professor of Biochemistry and Professor of Experimental Endocrinology and Director of the Hormone Research Laboratory. In 1955 he became a citizen of the United States.

Li spent his entire academic career studying the pituitary gland hormones. In collaboration with various co-workers, he isolated several protein hormones from the pituitary gland, including adreno-corticotrophic hormone (ACTH), which stimulates the adrenal cortex to increase its secretion of corticoids. In 1956, Li and his group showed that ACTH consists of 39 amino acids arranged in a specific order, and that the whole chain of the natural hormone is not necessary for its action. He isolated another pituitary hormone called melanocyte-stimulating hormone (MSH) and found that not only does this hormone produce some effects similar to those produced by ACTH, but also that part of the amino acid chain of MSH is the same as that of ACTH. Li has also studied pituitary growth hormones, finding that they are effective only in the species that produces them – that is, growth hormone from cattle, for example, is ineffective in humans. Continuing this line of research, he discovered in 1966 that human pituitary growth hormone (somatotropin) consists of a chain of 256 amino acids, and in 1970 he succeeded in synthesizing this hormone, thereby setting a record for the largest protein molecule synthesized up to that time.

Lie Marius Sophus 1842–1899 was a Norwegian mathematician who made valuable contributions to the theory of algebraic invariants and who is remembered for the Lie theorem and the Lie groups.

Lie was born at Nordfjordeid, near Bergen, on 17 December 1842. He received his primary and secondary schooling in Moss and then in 1859, at the age of 17, entered the University at Christiania (now Oslo) to study mathematics and science. He graduated in 1865 without, it appears, having formed a determination to become a mathematician. But in the next two or three years, while he earned money by giving private lessons, he read the works of Jean Poncelet (1788–1867) and Julius Plücker (1801–1868) and his imagination was fired by the latter's idea for creating new geometries by using figures, not points, as elements of space. This idea stayed with Lie to influence him throughout his career.

In 1869, Lie was awarded a scholarship to study abroad and he went to Berlin, where he worked under Felix Klein, and then to Paris. In 1870 he was arrested for spying – a false charge – but was released within a month and made his way to Italy just before the German blockade of Paris in the Franco–German war of that year. In 1871, the University of Christiania awarded him a scholarship to do doctoral research and he returned there in 1872. By the end of the year he had gained his doctorate and had a Chair of Mathematics created for him. He remained there until 1886, when he travelled to Leipzig to succeed Klein in the Chair of Mathematics at the university. His years at Leipzig, which lasted until 1898, were broken by a year in a mental hospital, caused by a kind of nervous breakdown (which was then called neurasthenia) in 1889. His last year was spent at Christiania, where another Chair of Mathematics was specially created for him. He died there of pernicious anaemia on 18 February 1899.

Lie shares with Klein the distinction of being the first mathematician to emphasize the importance of the notion of groups in geometry. By using group theory they were able to show that it was possible to decide to which kind of geometry a particular notion belonged. It was also possible to establish the relationships between different kinds of geometry, for instance of non-Euclidean geometry to projective geometry.

Lie's first great discovery, made while he was in Paris in 1870, was that of his contact transformation, which mapped straight lines with spheres and which mapped principal tangent curves into curvature lines. In his theory of tangential transformations occurs the particular transformation which makes a sphere correspond to a straight line. By 1873 Lie had turned away from contact transformation to investigate transformation groups. In this work on group theory he chose a new space element, the contact element, which is an incidence pair of point and line or of point and hyperplane. This led him to his greatest achievement, the discovery of transformation groups known as Lie groups, one of the basic notions of which is that of infinitesimal transformation.

The Lie groups provided the means to deduce from the structure the type of auxiliary equations needed for their integration, and the integration theorem which he developed (and which goes by his name) made it possible to classify partial differential equations in such a way as to make most of the classical methods of solving such equations reducible to a single principle. Moreover, the theorem led to a geometric interpretation of Cauchy's solution to partial differential equations.

The general effect of Lie's discoveries was to reduce the amount of work required in integration, although Lie's own papers on integration in the last 20 years of his life were clumsily presented and repetitive. His chief contribution to mathematics was to provide, in the Lie groups, the foundations of the modern science of topology.

Liebig Justus von 1803–1873 was a German organic chemist, one of the greatest influences on nineteenth-century chemistry. Through his researches and those of his ex-students, he had a profound influence on the science for nearly 100 years. To the schoolchild of today he is best known for the piece of chemical apparatus that he made popular and which still bears his name (the Liebig condenser). A better measure of his status is the fact that his students, assistants and co-workers included such famous chemists as Edward Frankland, Joseph Gay-Lussac, August von Hofmann, Friedrich Kekulé, Friedrich Wöhler and Charles Wurtz.

Liebig was born in Darmstadt, Hesse, on 12 May 1803. His father sold drugs, dyes, pigments and other chemicals and carried out his own chemical experiments, to which Liebig was introduced as a boy. When he was 15 years old he was apprenticed to an apothecary and first went to university to study under Karl Kastner at Bonn (where Liebig was arrested for his liberalist political activity) and then he accompanied Kastner to Erlangen University, where he gained his PhD in 1822 when he was still only 19 years old. Financed by the Grand Duke of Hesse, Liebig went to Paris for two years where Alexander von Humboldt (1769–1859) obtained a position for him in Joseph Gay-Lussac's laboratory at the Arsenal. He also made the acquaintance of Louis Thénard (1777–1857), who with von Humboldt recommended the 21-year-old Liebig for the Chair of Chemistry at the small University of Giessen. He stayed there for 27 years (1825–1852), building up a prestigious teaching laboratory. In 1840 he founded the journal *Annalen der Chemie* and was made a baron in 1845. Then in 1852 he moved to the University of Munich but, because of failing health, did less active research himself and concentrated on lecturing and writing. He remained there for the rest of his life and died in Munich on 18 April 1873.

In the early 1820s Liebig investigated fulminates, at the same time that Wöhler was independently working with cyanates. In 1826 Liebig prepared silver fulminate (modern formula $AgCNO$) and Wöhler made silver cyanate ($AgNCO$). When they reported their results they assigned the same formula to the two different compounds, which stimulated Jöns Berzelius' work that led to the concept of isomers.

Liebig and Wöhler became friends and continued their researches together. In 1832, from a study of oil of bitter almonds (benzaldehyde; phenylmethanal), they discovered the benzoyl radical (C_6H_5CO-). They showed that benzaldehyde can be converted to benzoic acid and made a number of other related compounds, such as benzyl alcohol and benzoyl chloride. The benzene ring had, in fact, conferred unusual stability to the benzoyl grouping, allowing it to persist in the various reactions. Liebig and Wöhler introduced the idea of compound radicals in organic chemistry, although they found no other radicals that as convincingly supported their theory and found themselves in an acrimonius dispute over the matter with Berzelius and Jean Baptiste Dumas. They had, however, tried to introduce a degree of systematization into the confused field of organic chemistry. To facilitate this work, many new methods of organic analysis were introduced by Liebig, and he devised ways of determining hydrogen, carbon and halogens in organic compounds.

From 1838, Liebig's work centred on what we would now call biochemistry. He studied fermentation (but would not acknowledge that yeast is a living substance, a view to which Berzelius also subscribed, and which brought them both into contention with Louis Pasteur) and analysed various body fluids and urine. He calculated the calorific values of foods, emphasizing the role of fats as a source of dietary energy and even developed a beef extract – long marketed as Liebig extract. Liebig also applied his chemical knowledge to agriculture. He demonstrated that plants absorb minerals (and water) from the soil and postulated that the carbon used by plants comes from carbon dioxide in the air rather than from humus in the soil. He also thought, incorrectly, that ammonia in rainwater passed into the soil and provided plants with their sole source of nitrogen. He thus advocated the use of artificial fertilizers in agriculture instead of animal manure, although his original formulation omitted essential nitrogen compounds.

In later life his rather rigid views made Liebig even more dogmatic in his statements – often labelled as arrogance by both his friends and his antagonists – but by then he was an established authority and his opinions were seldom questioned.

It is said that he could be grossly unfair, stimulating controversy and admitting an error only when it no longer mattered; only his lifelong friend Wöhler seems to have continued to have survived Liebig's irascibility.

Lilienthal Otto 1848–1896 was a German engineer whose experiments with gliders helped to found the science of aeronautics. But for his premature death in a flying accident, he might well have beaten the Wright brothers to the achievement of powered flight.

Lilienthal was born in Pomerania (then part of Prussia), and trained as an engineer just before the outbreak of the Franco–Prussian War in 1870. He made exhaustive studies of the flight of birds, especially of the stock, but was aware that it was 'not enough to acquire the art of the bird' it was also necessary to put the whole problem of flight on a scientific basis. From these studies he learnt that curved wings allow horizontal flight without an angle of incidence to the wind, and that soaring is related to air thermals. In 1889, he published his famous book *Der Vogelflug als Grundlage der Fliegekunst /Bird Flight as a Basis for Aviation*, which was to have a great influence on the work of the pioneers of the next vitally important years. He was one of the few who showed conclusively that birds produce thrust by the action of their outer primary feathers. It was perhaps this that kept Lilienthal on the path of the ornithopter – a machine which simulated the winged flight of a bird as the ultimate means of powered flight.

Lilienthal flew the first of his famous series of gliders in 1891, and continued, until his final fatal glide in 1896, to hold to his fundamental conviction that the key to eventual powered flight was in glider-flying, in which pilots could master the elements of control and design. Its relative safety – sadly not safe enough for Lilienthal – allowed the pioneers of flying to come to terms with problems of movement and airflow. The step to powered flight was a straightforward progression from advanced gliding. The Wrights' breakthrough was a simple extension of their own work with gliders, which in turn was greatly helped by Lilienthal's many glides and careful observations. Among other things, he had demonstrated the superiority of cambered wings over flat wings – the principle of the aerofoil.

By 1893, he was flying a cambered-wing monoplane from a springboard near his home. It had a wingspan of 7 m/23 ft and a surface area of about 13 sq m/140 sq ft; its wings folded for transport. Other gliders included two biplanes, with tailplanes in front of the vertical tail fin (the first gliders were tailless). His 1894 model (Number 11 in the offical biographies) became his standard machine. With a wing surface of 14 sq m/150 sq ft this glider had the tailplane integral with the fin and achieved glides of more than 300 m/1,000 ft. By this time his activities had given a boost to gliding as a sport, and enthusiasts began 'sailing' in many countries.

1894 also saw the introduction of a shock-absorbing hoop which, on 9 November, certainly helped to save Lilienthal's life in a crash. A detailed description of this incident caused other workers, including the Wrights, to build elevators in front of their machines, the idea being to prevent damage in the event of a nose-dive. By 1889 Lilienthal had introduced leading-edge flaps to counteract the tendency to nose-dive. Also in that year he introduced the biplanes referred to above, which were found to possess increased stability with reduced wingspan, but he did not develop these machines. (The Wrights' successful flights were with biplanes.) Lilienthal's later machines were flown from an artificial earthwork which he had constructed outside Berlin. These gave a launching height of about 15 m/50 ft and, being conical, allowed independence of wind direction.

In 1895 Lilienthal resumed some earlier work on the idea of a powered machine with moving wing-tips, using a carbon dioxide motor, despite the fact that petrol engines had been made by that time. The machine was not tested. Only in that year did he consider means of control other than body swinging. Up to then the 'pilot' was suspended by his arms – like a modern hang-glider – the rest of him dangling free after the running take-off. Movement was produced by swinging the body, thus altering the centre of gravity. The idea of the body harness was introduced, an echo of something Leonardo da Vinci had sketched some 500 years previously.

The influence of Lilienthal, both before and after his death, was enormous, helped greatly by the developments in photography and printing of that period – the dry-plate negative and the half-tone printing process. These advances allowed a magnificent series of pictures to be given worldwide circulation. They had a tremendous and important impact at a time when even a scientist like Lord Kelvin was saying that he had 'not the smallest molecule of faith' in flying, other than in balloons!

On a fine summer's day in 1896, after several trouble-free glides, an unexpected gust caused Lilienthal's Number 11 to stall and fall to the ground. His spine was broken and he died the following day. His gravestone carries a favourite saying of his: 'Sacrifices have to be made'.

Lindblad Bertil 1895–1965 was a Swedish expert on stellar dynamics whose chief contribution to astronomy lay in his use of the work of Jacobus Kapteyn (1851–1922) and Harlow Shapley (1885–1972) to demonstrate the rotation of our Galaxy.

Lindblad was born in Örebro in southern central Sweden

on 26 November 1895. He completed his education at Uppsala University, where he studied mathematics, astronomy and physics. He graduated in 1920, having earned his PhD for research on radiative transfer in the solar atmosphere. He spent a two-year postdoctoral research period in the United States, visiting the Lick, Harvard and Mount Wilson Observatories. Lindblad returned to Sweden in 1922, and continued his work first in Uppsala and, from 1927, in Stockholm. He was appointed Director of the new Stockholm Observatory in 1927, and made Professor of Astronomy at the Royal Swedish Academy of Sciences.

Lindblad's original research earned him international recognition, and he was accorded many professional honours, including the Gold Medal of the Royal Astronomical Society. He died in Stockholm on 26 June 1965.

Lindblad's early research was concentrated in the field of spectroscopy, but he soon became interested in stellar dynamics. At that time there was a vigorous astronomical debate on the subject of the structure of the galaxy. Jacobus Kapteyn had proposed a model based on his observations on stellar motion, which suggested to him that the Solar System lay near the centre of the Galaxy. Harlow Shapley alternatively proposed that the centre of the Galaxy was some 50,000 light years away in the direction of the constellation Sagittarius.

Lindblad analysed Kapteyn's results and suggested that the two streams of stars which Kapteyn had observed could in fact represent the rotation of all the stars in our Galaxy in the same direction, around a distant centre; he thus confirmed Shapley's hypothesis that the centre lay in the direction of Sagittarius. But he also went on to stipulate that the speed of rotation of the stars in the Galaxy was a function of their distance from the centre (the 'differential rotation theory').

Such an interpretation of Kapteyn's work was supported by an analysis put forward by Jan Oort (1900–1992), published shortly thereafter.

Stellar motion continued to be the dominant theme in Lindblad's research. He inspired many other Swedish astronomers, including his son, Per Olaf Lindblad, who succeeded him as Director of the Stockholm Observatory.

Lindemann Carl Louis Ferdinand 1852–1939 was a German mathematician who is famous for one result, his discussion of the nature of π, which laid to rest the old question of 'squaring the circle'.

Lindemann was born at Hanover on 12 April 1852, did his undergraduate work at Göttingen and Munich, and received his doctorate from the university at Erlangen (where he studied under Felix Klein) in 1873. He then spent a couple of years travelling in England and France, meeting leading mathematicians and carrying on private research. In 1877 he was appointed a lecturer at Würzburg and in 1879 he was promoted to Professor. From 1883 to 1893 he was a professor at Königsberg; from 1893 until his death, he taught at Munich. He died on 6 March 1939.

Lindemann published papers on a number of subjects, including spectrum theory, invariant theory and theoretical

mechanics (his doctoral dissertation was on the infinitely small movements of rigid bodies). He was also a highly acclaimed teacher, supervising more than 60 doctoral candidates during his career and being more than anyone else responsible for introducing the seminar method of teaching into German universities. He also translated and edited the works of Henri Poincaré (1854–1912) and made his mathematics known in Germany.

His fame, nevertheless, rests almost entirely on the paper which he published in 1882 on the nature of π as a transcendental number. Since the time of the Greeks, mathematicians had known of irrational numbers and wondered if they were algebraic: or, in other words, was it possible to define an algebraic equation with rational coefficients in which irrational numbers were the roots? In 1844 Liouville had shown, by the use of continued fractions, that a host of numbers existed which are non-algebraic. These are known as transcendental numbers. Then in 1872 Charles Hermite proved a rigorous demonstration that e, the base of 'natural' logarithms, was a transcendental number. The highly interesting question, whether π was also a transcendental number, had still, however, not received a satisfactory answer.

It was Lindemann who provided, in his 1882 paper, the proof. He demonstrated that, except in trivial cases, every expression of the form:

$$\sum_{i=1}^{n} A_i e^a i$$

where A and a are algebraic numbers, must be non-zero. Therefore, since i is a root of $x^2 + 1 = 0$, and since it was known that:

$$1^{i^\pi} + 1^0 = 0$$

$$\text{(that is } 1^{i^\pi} = -1\text{)}$$

then ip and therefore p (since i is algebraic) must be transcendental.

If p cannot be the root of an equation, it cannot be constructed. Therefore the 'squaring of a circle' is impossible. Lindemann had brought an end to one of the oldest puzzles in mathematics.

Lindemann Frederick Alexander 1886–1957 was a British physicist who was involved with Hermann Nernst (1864–1941) in advancing the quantum theory. He later became Lord Cherwell and played an important role in the administration of British science during and after World War II.

Lindemann was born in Baden-Baden, Germany, on 5 April 1886. He went to school in Scotland and Germany, and attended the University of Berlin, where he gained his PhD in 1910. Lindemann remained at Berlin until 1914, when he returned to Britain and during World War I became Director of the RAF Physical Laboratory, where he was concerned with aircraft stability. In 1919, Lindemann became Professor of Experimental Philosophy at Oxford, building up the Clarendon Laboratory into an important scientific institution. In 1941, he was granted a peerage and took the title of Lord Cherwell. Lindemann was Paymaster-General from 1942 to 1945 and again from 1951 to 1953. In this position he gave valuable advice to the British government, helping to direct scientific research during World War II and to create the Atomic Energy Authority afterwards. Lindemann returned to Oxford from 1945 to 1951 and from 1953 to 1956, when he retired. He was then instrumental in founding Churchill College, Cambridge, to promote the study of technology. Lindemann was awarded the Hughes Medal of the Royal Society in 1956, and died at Oxford on 3 July 1957.

Lindemann's main contributions to physics were made in his early years at Berlin, where he studied with and later assisted Nernst. Nernst and Lindemann together made an important advance in quantum theory in 1911 by constructing a special calorimeter and measuring specific heats at very low temperatures. They confirmed the specific heat equation proposed by Albert Einstein (1879–1955) in 1907, in which he used the quantum theory to predict that the specific heats of solids would become zero at absolute zero. Lindemann also derived a formula which relates the melting point of a crystalline solid to the amplitude of vibration of its atoms.

In World War I, Lindemann developed a theory to explain why the aircraft of that time were liable to spin out of control. He then worked out a way of recovering from a spin and successfully tested it himself.

Linnaeus Carolus (Carl von Linné) 1707–1778 was a Swedish botanist who became famous for introducing the binomial system of biological nomenclature (which is named after him and is universally used today), and for formulating basic principles for classification.

Linnaeus was born on 27 May 1707 in South Råshult, Sweden, the son of a clergyman. He was interested in plants even as a child, but his father sent him to study medicine, first at the University of Lund in 1727 and then at Uppsala University.

In 1730, Linnaeus was appointed Lecturer in Botany at Uppsala and two years later explored Lapland for the Uppsala Academy of Sciences. In 1735, Linnaeus left Sweden for Holland to obtain his MD at the University of Harderwijk. On his return to Sweden in 1738 Linnaeus practised as a physician, with considerable success, and in the following year he married Sara Moraea, a physician's daughter. In 1741, he was appointed Professor of Medicine at Uppsala University but changed this position in 1742 for the Chair of Botany, which he retained for the rest of his life. In 1761 Linnaeus was granted a patent of nobility – antedated to 1757 – by which he was entitled to call himself Carl von Linné. He suffered a stroke in 1774, which impaired his health, and he died on 10 January 1778 in Uppsala Cathedral, where he was buried.

Linnaeus's best known work is probably *Systema*

Naturae, published in 1735. In this book he introduced a simple, yet methodical, system of classifying plants according to the number of stamens and pistils in their flowers. This system overshadowed the earlier work of John Ray and was so convenient that it was a long time before it was replaced by a more natural system – despite the fact that Linnaeus himself recognized its artificiality.

Classification of modern humans

kingdom	Animalia
phylum	Chordata
sub-phylum	Vertebrata
class	Mammalia
order	Primates
family	Hominidae
genus	Homo
species	sapiens

Linnaeus *Linnaeus devised a system of classification that has remained largely unchanged to the present day.*

Linnaeus made his most important contribution – the introduction of the binomial system of nomenclature, by which every species is identified by a generic name and a specific name – in 1753, with the publication of Species Plantarum. Even today the starting point in the nomenclature of all flowering plants and ferns is internationally agreed to be the first edition of Species Plantarum, together with the fifth edition of Genera Plantarum (1754; first edition 1737). In these works he became the first person to formulate the principles for defining genera and species and to adhere to a uniform use of specific names. In 1758, he applied his binomial system to animal classification. With the rapid discovery of previously unknown plants and animals that was occurring in the eighteenth century, the value of Linnaeus's system was soon recognized and it had become almost universally adopted by the end of his life. The survival of the system to the present day is probably due to its great flexibility; Linnaeus himself believed that species were immutable and that he was classifying Creation (although he later modified this viewpoint slightly), but so adaptable was his system that it was able to accommodate modifications that later resulted from the introduction of evolutionary principles to taxonomy.

In addition to his books on classification, Linnaeus wrote many other works, including Flora Laponica (1737), the results of his journey to Lapland; Hortus Cliffortianus (1738), a description of the plants in the garden of George Clifford, a merchant with whom Linnaeus stayed during much of his time in Holland; and Flora Suecica (1745) and Fauna Suecica (1746), accounts of his biological observations during his travels in Sweden. After Linnaeus's death, his widow sold his manuscripts and natural history collection to James Edward Smith (1759–1828), the first President of the Linnaean Society (founded in 1788), who took them to England. When Smith died the Society purchased Linnaeus's manuscripts and specimens and they are now preserved by the Society in Burlington House, London.

Linnett John Wilfred 1913–1975 was a British chemist of wide-ranging interests, from spectroscopy to reaction kinetics and molecular structure.

Linnett was born in Coventry on 3 August 1913. He was educated at the local King Henry VIII School and in 1931 won a scholarship to read chemistry at St John's College, Oxford, from which he graduated in 1934. He was awarded his doctorate three years later for a thesis on the spectroscopy and photochemistry of the metal alkyls; he continued this work in 1938 during a visit to Harvard. From 1938 to 1965 Linnett was at Oxford, becoming a Reader in 1962. In 1965 he was appointed Professor of Physical Chemistry at Emmanuel College, Cambridge, where he remained for the rest of his research life. He became Master of Sidney Sussex College, Cambridge, in 1970, Deputy Vice-Chancellor of Cambridge University in 1971 and Vice-Chancellor from 1973 to 1975. He died on 7 November 1975.

The beginning of Linnett's career coincided with the outbreak of World War II in 1939. During that period he participated in a broad project aimed at providing methods of protection from gas attacks, such as developing catalysts that would oxidize carbon monoxide. After the war he studied molecular force fields, the measurement of burning velocities in gases, the recombination of atoms at surfaces, and theories of chemical bonding.

His work on explosion limits concentrated on the reaction between carbon monoxide, hydrogen and oxygen, which led to the study of atomic reactions on surfaces, such as the efficiency of the surface recombination of hydrogen atoms on palladium–gold, palladium–silver and copper–nickel alloys. In the early 1960s he was using mass spectroscopy for the direct sampling of reacting systems. In 1967, using this technique, he discovered that the HCO radical attracts the hydrogen from the methyl group of acetaldehyde (ethanal):

$$\cdot HCO + CH_3CHO \rightarrow CO + H_2 + \cdot CH_2CHO$$

The following year he extended the method to study the pyrolysis of mixtures of methyl iodide (iodomethane, CH_3I) and nitric oxide (nitrogen monoxide, NO), and attempted to account for the formation of N–N bonds in the reaction of methyl radicals with excess nitric oxide.

In 1960, while visiting Berkeley on a Cherwell Memorial Fellowship, Linnett originated his important modification to the Lewis–Langmuir octet rule concerning valency electrons. He proposed that the octet should be considered as a double quartet of electrons rather than as

four pairs, and in this way he was able to explain the stability of 'odd electron' molecules such as nitric oxide.

Nitrogen has five valence electrons and oxygen has six; there are $2\frac{1}{2}$ bonds (or electron pairs) linking the two atoms. In the dimer N_2O_2, however, there are five bonds with two sets of coincident electrons, whereas in the five bonds of 2NO the quartets of each spin do not have the same orientation. The 2NO version should therefore have the lower electron–electron repulsion energy. Linnett and his co-worker Hirst described this analysis as the non-pairing method to distinguish it from the valence-bond and molecular-orbital methods.

Linnett published more than 250 scientific papers and two textbooks, in one of which (*Wave Mechanics and Valency*) he explains to the experimental chemist the processes and techniques involved in the application of wave mechanics to the electronic structures of atoms and molecules.

Liouville Joseph 1809–1882 was a French mathematician who wrote prolifically on problems of analysis, but who is famous chiefly as the founder and first editor of the learned journal popularly known as the *Journal de Liouville*.

Liouville was born at St Omer, Pas-de-Calais, on 24 March 1809 and studied at Commery and Toul before entering the Ecole Polytechnique in Paris in 1825. In 1827, he transferred to the Ecole des Ponts et Chaussées, where he received his baccalaureate in 1830. For the next 50 years, beginning with his appointment to the Ecole Polytechnique in 1831, Liouville taught mathematics at all the leading institutions of higher learning in Paris. While he was lecturing at the Ecole Centrale des Arts et Manufactures (1833–1838), he received his doctorate in 1836 for a thesis on Fourier series.

In 1838, he was elected to the Chair of Analysis at the Ecole Polytechnique, where he remained until 1851, when he became a professor at the Collège de France. He stayed at the Collège until 1879, although in the years from 1857 to 1874 he held concurrently the post of Professor of Rational Mechanics at the Sorbonne. He was also for a time the Director of the Bureau de Longitudes. Quite unexpectedly, having shown no political ambition previously, he became infected by the revolution of 1848 and was elected as a moderate republican to the constituent assembly in April 1848. A year later he was defeated in the elections to the new Legislative Assembly and his political career closed as suddenly as it had opened. He died in Paris on 8 September 1882.

Although Liouville's early interest lay in problems associated with the study of electricity and heat, and although he was elected to the Academy of Sciences in 1839 as a member of the astronomy section, the chief mathematical interest of his career was in analysis. In that field he published more than 100 papers between 1832 and 1857. In collaboration with Charles-François Sturm (1803–1855) he published papers in 1836 on vibration, which were of considerable importance in laying the foundations of the theory of linear differential equations. He also provided the

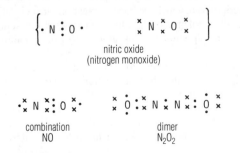

nitric oxide
(nitrogen monoxide)

combination
NO

dimer
N_2O_2

first proof of the existence of transcendental functions and in a paper of 1844 laid down that the irrational numbers, e and e^2 were transcendental, since they could not be used to solve any second-degree polynomial equation. The proof of this, however, had to await Charles Hermite's demonstration in 1873.

More important than his research was Liouville's founding of the *Journal des Mathématiques Pures et Appliqués* in 1836. Since the demise of Gergonne's *Annales de Mathématiques* in 1831 French mathematicians had been deprived of a receptacle for their research papers. Liouville's journal filled the gap. He edited it from the issue of the 1st number in 1836 down to the 39th number in 1874. When he retired from the editorship, the mathematical spark died in him and he produced nothing more of importance.

Lipschitz Rudolf Otto Sigismund 1832–1903 was a German mathematician of wide-ranging interests, who is remembered for the so-called Lipschitz algebra and the Lipschitz condition.

Lipschitz was born at Königsberg on 14 May 1832. He does not appear to have become seriously interested in mathematics until, in 1847, he entered the University of Königsberg, where one of his teachers was Franz Neumann. After graduating from Königsberg he continued his studies at the University of Berlin, chiefly under Lejeune Dirichlet. He was awarded his doctorate in 1853. For the next few years he taught at schools in Königsberg and Elbinc, before becoming a lecturer at the University of Berlin in 1857. He moved to Breslau in 1862, but was there for only two years. In 1864, he was appointed Professor of Mathematics at the University of Bonn. He remained there for the rest of his career, so contented with his work and life there that he turned down an invitation to become a professor at the more prestigious University of Göttingen in 1873. He died in Bonn on 7 October 1903.

Lipschitz did extensive work in number theory, Fourier series, the theory of Bessel functions, differential equations, the calculus of variations, geometry and mechanics. He was also much interested in the fundamental questions concerned with the nature of mathematics and of mathematical research, and German higher education was much indebted to him for his two-volume *Grundlagen der Analysis* (1877–1880), a synthetic presentation of the foundations of mathematics and their applications. The work provided a comprehensive survey of what was then known of the theory of rational integers, differential equations and function theory.

Among his more specific contributions to mathematical knowledge, several stand out. His work in basic analysis provided a condition now known as the Lipschitz condition, subsequently of great importance in proofs of existence and uniqueness, as well as in approximation theory and constructive function theory. He has a place in the history of number theory, too, as the developer of a hyper-complex system which became known as Lipschitz algebra. In investigating the sums of arbitrarily many squares, Lipschitz derived computational rules for certain symbolic expressions from real transformations. Even more important were the investigations which he began in 1869 into forms of n differentials, for these led to his most valuable contribution to mathematics – the Cauchy–Lipschitz method of approximation of differentials. Finally, there was his work on co-gradient differentiation, which he conducted parallel to, but independently of, similar research by Elwin Christoffel (1829–1900). Lipschitz showed that the vanishing of a certain expression is a necessary and sufficient condition for a Riemannian manifold to be Euclidean, and further research into Riemann's mathematics enabled him to produce what is now the chief theorem concerning mean curvature vectors. Lipschitz's two papers on this subject, taken together with one written by Christoffel, formed a vital ingredient of what became the tensor calculus of Ricci-Curbastro and Tullio Levi-Civita.

Partly because he spread himself so wide, Lipschitz's star does not shine as brightly as some others in the mathematical firmament; but he was one of the most industrious and most technically proficient of nineteenth-century mathematicians.

Lipscomb William Nunn 1919–   is a US chemist whose main interest is in the relationships between the geometric and electronic structures of molecules and their chemical and physical behaviour. Much of his work has been carried out with the boron hydrides.

Lipscomb was born in Cleveland, Ohio, on 9 December 1919. He graduated from the University of Kentucky in 1941 and gained his PhD from the California Institute of Technology in 1946. During World War II he was associated with the Office of Scientific Research and Development Projects. He taught at the University of Minnesota from 1946 to 1959, and at the University of Harvard after 1959.

Lipscomb studied the boron hydrides and their derivatives to elucidate problems about electron deficient compounds in general. He developed low-temperature X-ray diffraction methods to study simple crystals and established the structures of these compounds, which are not readily described using the usual electron pair-bonding method. He and other research workers related them to the polyhedral structures of borides. Using the simpler members of the series, they developed bonding theories that account for filled electron shells in terms of three-centre two-electron bonds. They also proposed molecular orbital descriptions in which the bonding electrons are delocalized over the whole molecule. Much of this work was summarized in Lipscomb's book *Boron Hydrides*, published in 1963.

Lipscomb went on to investigate the carboranes, $C_2B_{10}H_{12}$, and the sites of electrophilic attack on these compounds using nuclear magnetic resonance spectroscopy (NMR). This work led to the theory of chemical shifts. The calculations provided the first accurate values for the constants that describe the behaviour of several types of molecules in magnetic or electric fields. They also gave a theoretical basis for applying quantum mechanics to complex molecules, with wide potential for both inorganic and organic chemical problems. Lipscomb and his co-workers developed the idea of transferability of atomic properties, by which approximate theories for complex molecules are developed from more exact calculations for simpler but chemically related molecules, using high-speed computers. With Pitzer, Lipscomb made the first accurate calculation of the barrier to internal rotation about the carbon–carbon bond in ethane.

Lipscomb's team developed X-ray diffraction techniques for studying simple crystals of nitrogen, oxygen, fluorine and other substances that are solid only below liquid nitrogen temperatures. They also determined the molecular structure of cyclo-octatetraene iron and the tricarbonyl complexes of natural products. One of these, leurocristine, is used in leukaemia therapy. Lipscomb also elucidated the three-dimensional structure of carboxypeptidase A, one of the largest globular proteins with a molecular mass of 34,400.

Lissajous Jules Antoine 1822–1880 was a French physicist who developed Lissajous figures for demonstrating wave motion.

Lissajous was born at Versailles on 4 March 1822. He was educated at the Ecole Normale Supérieure from 1841 to 1847, when he became Professor of Physics at the Lycée Saint-Louis. He became Rector of the Academy of Chambéry in 1874, and then took up the same position at Besançon in 1875. In 1879 Lissajous was elected a member of the Paris Academy, and he died at Plombières on 24 June 1880.

Lissajous was interested in acoustics and from 1855 developed Lissajous figures as a means of visually demonstrating the vibrations that produce sound waves. He first reflected a light beam from a mirror attached to a vibrating object such as a tuning fork to another mirror that rotated. The light was then reflected onto a screen, where the spot traced out a curve whose shape depended on the amplitude and frequency of the vibration. Lissajous then refined this method by using two mirrors mounted on vibrating tuning forks at right angles, and produced a wider variety of figures. By making one of the forks a standard, the acoustic characteristics of the other fork could be determined by the shape of the Lissajous figure produced.

Lissajous figures can now be demonstrated on the screen of an oscilloscope by applying alternating currents of different frequencies to the deflection plates. The curves produced depend on the ratio of the frequencies, enabling signals to be compared with each other.

Lister Joseph 1827–1912 was a skilful British surgeon with a great interest in histology and bacteriology. He introduced the concept of antiseptic surgery and was a pioneer, in Britain, of preventive medicine.

Lister was born on 5 April 1827, in Upton, Essex, the son of the British physicist Joseph Jackson Lister (1786–1869). He was educated at various Quaker schools and at University College, London, the only university then open to dissenters. He first studied arts and after graduation took up medicine at University College, where he was taught by the eminent physiologist William Sharpey, and graduated in 1852. In 1856 he was a surgeon at the Edinburgh Royal Infirmary as assistant to James Syme, and three years later was appointed Regius Professor of Surgery at the University of Glasgow. In 1861, he took charge of the surgical wards at the Royal Infirmary, Glasgow, and in 1869 became Professor of Clinical Surgery at Edinburgh. Eight years later he took up the Chair in Clinical Surgery at King's College, London. He was knighted in 1883 and was raised to the peerage in 1897. Nearing retirement in 1891, he became Chairman of the newly formed British Institute of Preventive Medicine (later the Lister Institute) and served as President of the Royal Society from 1895 to 1900. He died on 10 February 1912 in Walmer, Kent.

Nearly half the patients who underwent major surgery at that time died as the result of post-operative septic infection. Sepsis was thought to be a kind of combustion caused by exposing moist body tissues to oxygen – an assertion put forward by the German chemist Justus von Liebig in 1839. Great care was therefore taken to keep air from wounds, by means of plasters, collodion or resins. Lister doubted the explanation and these methods; he regarded wound sepsis as a form of decomposition.

In 1865, Louis Pasteur suggested that decay is caused by living organisms in the air, which enter matter and cause it to ferment. Lister immediately saw the connection with wound sepsis. In addition, the previous year he had heard that carbolic acid (phenol) was being used to treat sewage in Carlisle, and that fields irrigated with the final effluent were freed of a parasite that was causing disease in cattle. Lister began to use a solution of carbolic acid for wound cleansing and dressings, and also experimented with operating under a spray of carbolic acid solution. In 1867 he announced to a British Medical Association meeting that his wards in the Glasgow Royal Infirmary had remained clear of sepsis for nine months. At first his new methods met with hostility or indifference, but gradually doctors began to support his antiseptic techniques.

Continuing his studies in histology and bacteriology Lister became interested in Robert Koch's work, carried out between 1876 and 1878, on wound infections. In Germany, Koch was demonstrating that steam was a useful sterilizer for surgical instruments and dressings, and German surgeons were beginning to practise aseptic surgery, keeping wounds free from microorganisms by using only sterilized instruments and materials. Lister realized that both methods relied on destroying pathogenic microorganisms, and believed that, in the future, more emphasis would be placed on preventive medicine. He strove for the establishment of an institute of preventive medicine, which he saw opened in 1891.

Lobachevsky Nikolai Ivanovich 1792–1856 was a Russian mathematician, one of the founders of non-Euclidean geometry, whose system is sometimes called Lobachevskian geometry.

Lobachevsky was born at Nizhni-Novgorod (now Gorki) on 2 November 1792. About eight years later, when his father died, he moved with his family to Kazan, where he was educated at the local school. In 1807, he entered the University of Kazan to study mathematics and in 1814 he was appointed to the teaching staff there. In 1822 he was made a full professor and in 1827 he was elected Rector of the university. He also took on administrative work for the government, serving as assistant trustee for the Kazan educational district from 1846 to 1855. For reasons which remain obscure (perhaps to compel him to devote himself to his government work), the government relieved him of his posts as professor and rector in 1847. Earlier it had recognized his talent by raising him to the hereditary nobility in 1837. In his later days Lobachevsky suffered from cataracts in both his eyes, and he was nearly blind when he died at Kazan on 24 February 1856.

Lobachevsky's whole importance rests on the system of non-Euclidean geometry which he developed between 1826 and 1856. Karl Gauss and János Bolyai were working on Euclid's fifth postulate and formulating their own non-Euclidean geometries at the same time; but in 1826, when Lobachevsky first gave the outline of his system to a meeting of colleagues at Kazan, neither Gauss nor Bolyai had uttered a public word, and Lobachevsky's first published paper on the subject appeared in 1829, three years before Bolyai's appendix to his father's *Tentatem*. The clearest statement of his geometry was made in the book *Geometrische Untersuchungen zue Theorie der Paralellinien*, which he published in Berlin in 1840. His last work on the subject, the *Pangéométrie*, was published just before his death.

Ever since the time of Euclid it had been believed that no geometry could be constructed without his fifth postulate, or in other words that any set of axioms other than Euclid's must, in the course of the geometry's development, produce contradictory consequences which would invalidate the geometry. Like Gauss and Bolyai, Lobachevsky abandoned the fruitless search for a proof to the fifth postulate. He came to see – this was the starting-point of his invention – that it was not contradictory to speak of a geometry in which all of Euclid's postulates *except* the fifth held true. His new geometry, by analogy with imaginary numbers, he called 'imaginary geometry'. By including imaginary numbers geometry became more general, and Euclid's geometry took on the appearance of a special case of a more general system.

The chief difference between the geometry of Euclid and that of Lobachevsky may be pointed up by the fact that, in Euclid's system, two parallel lines will (as a consequence of the fifth postulate) remain equidistant from each other, whereas in Lobachevskian geometry, the two lines will approach zero in one direction and infinity in the other. Another example of the difference is that in Euclidean geometry the sum of the angles of a triangle is always equal to the sum of two right angles; in Lobachevskian geometry,

the sum of the angles is always less than the sum of two right angles. Lobachevskian space is such a different concept from that of Euclid's, that in the former triangles can be defined as functions of their angles, which determine the length of the sides. In Lobachevskian space, also, two geometric figures cannot have the same shape but different sizes.

The work of Lobachevsky, as of Gauss and Bolyai, demonstrated that it was useless to attempt to prove Euclid's fifth postulate by showing all other alternatives to be impossible. It demonstrated that different geometries, self-contained and self-consistent, were logically possible. The notion, prevalent since the time of Euclid, that geometry offered *a priori* knowledge of the physical world was destroyed (somewhat ironically it is the Lobachevskian model, not the Euclidean, which today seems closer to the actual world of space). Non-Euclidean geometry destroyed, once and for all, the notion of empirical mathematics. It represented, William Clifford has said, a revolution in the history of human thought as radical as the revolution begun by Copernicus. Largely unrecognized in his lifetime, Lobachevsky and his fellow revolutionaries are still, outside the highest mathematical circles, too little recognized today.

Locke Joseph 1805–1860 was a British railway engineer, a contemporary and associate of Isambard Brunel and the Stephensons.

Lock was born on 9 August 1805, the youngest of four sons of a colliery manager at Attacliffe Common, Sheffield, Yorkshire. His father was a man of strong views and for various reasons changed employers several times while Joseph was still a child. He did, however, receive a 'sort of an education' at the grammar school in Barnsley and left at the age of 13 knowing a little but, as he said, 'knowing that little well'.

Locke found employment carrying letters at Pelaw, County Durham, under the watchful eye of William Stobart, colliery viewer for the Duke of Norfolk, but after two years he returned to Barnsley. He was then sent to Rochdale to work for a land surveyor named Hampson. The situation looked promising until he was asked to look after the Hampson baby. He gave up the job and once again returned home – this time after only a fortnight.

His family resigned themselves to the fact that he was rather idle and left him to find work for himself. He might well have remained unsettled had not luck interceded. His father's old friend from the Attacliffe colliery called to see them – George Stephenson, the civil engineer. When he saw the plight of his friend's young son he suggested Joseph should be sent to work for him, and this began what was to prove a fortunate career as a railwayman.

From 1823 to 1826 Locke learnt much about surveying, railway engineering and construction. He loved the life and learnt with enthusiasm, something which had been lacking in his earlier attempts at a career. After three years he was to emerge as one of the foremost engineers of the railway era.

Locke's first task undertaken alone was the construction of a railway line from the Black Fell colliery to the River Tyne. This he managed so successfully that Stephenson confessed to have 'complete confidence' in Locke's ability. He was immediately asked to begin surveys for lines running between Leeds and Hull, Manchester and Bolton, and Canterbury and Whitstable.

The hectic era of railway expansion was under way when George Stephenson (now joined by his son Robert), Brunel and Locke were in great demand. By 1842 more than 2,988 km/1,857 mi of track had been laid and Locke had made a reputation for himself as a man who built as straight as possible, used the terrain, and avoided the expense of tunnels whenever he could. He was asked to tackle the London to Southampton line and although others favoured tunnelling through the chalk Downs, he chose to cut through them, leaving steep embankments on either side of the line. His method proved successful and gave few problems after the line was opened on 11 May 1840.

Locke took over part of the construction of the Grand Junction railway from Stephenson and the Sheffield-to-Manchester route from Vignoles. The latter was a complicated operation which cut a 4,850-m/15,900-ft bore through the millstone grit of the Pennines. In 1841 he began work as Chief Engineer on the Paris to Rouen line and completed it on schedule in May 1843. This was to be the first of several contracts in France and it added even more prestige to his name.

Turning once more to work in Britain, he carried out construction of lines on behalf of the Lancaster and Carlisle Railway and the Caledonian Railway, but during 1846 he made some important decisions regarding his own life. He bought the manor of Honiton in Devon and fulfilled a boyhood ambition to enter Parliament. Many accused him of buying his way in, although he did in fact stand for election in 1847 and remained Liberal member for Honiton until his death 13 years later.

All three of the railway 'giants' spent their remaining years in well-earned comfort, while yet being fully employed. The construction of new railways continued relentlessly, and their skills were always in great demand. Contemporaries to the point of all being born within the same three-year period, strangely they all died within the same space of time: Brunel on 14 September 1859, Robert Stephenson, his friend of long-standing, on 12 October 1859, and Locke himself on 18 September 1860.

Lockyer Joseph Norman 1836–1920 was a British scientist whose interests and studies were wide-ranging, but who is remembered mainly for his pioneering work in spectroscopy, through which he discovered the existence of helium, although it was not to be isolated in the laboratory until nearly 30 years later.

Lockyer was born at Rugby on 17 May 1836. After his schooling in the Midlands he worked briefly as a civil servant in the War Office. The high reputation he was meanwhile gaining as an amateur astronomer led to his becoming (temporarily) secretary to the Duke of Devon-

shire's commission on scientific instruction. He was then appointed to a permanent post in the Science and Art department and in 1890 he became Director of the Solar Physics Observatory in South Kensington. He remained in this post until 1911, when he resigned rather than move with the Observatory to Cambridge.

Elected to the Royal Society in 1869 – the year in which he founded the scientific journal *Nature*, which he was to edit for 50 years – he was awarded its Rumford Medal in 1874. He was knighted in 1897, after the element he had named helium so many years before, had finally been isolated in the laboratory by William Ramsay (1852–1919). Lockyer died in Salcombe Regis, Devon, on 16 August 1920.

A primary influence on Lockyer's researches was the newly discovered science of spectroscopy initiated in 1859 when Gustav Kirchhoff (1824–1887), together with his colleague Robert Bunsen (1811–1899), showed how the lines in the spectrum of a substance could indicate the actual composition of that substance. Throughout his life Lockyer was especially interested in solar phenomena. In 1868 he attached a spectroscope to a 15-cm/6-in telescope and made a major breakthrough by observing solar prominences at times other than during a total solar eclipse. The success of Lockyer's experiment was ensured by his use of an instrument that could breach the spectrum of the diffused sunlight in the atmosphere, and thereby make visible the bright lines of the prominence spectrum. Although Lockyer had been the first to think of it, the same idea had occurred to Pierre Janssen (1824–1907) – then working in India – and both men, in mutual ignorance, decided to put their theory to the test during the same eclipse. Accordingly, the French Academy of Sciences experienced the surprising coincidence of receiving a message from each man confirming the success of their experiments within minutes of each other. This remarkable event was duly commemorated with the issuing of a medal by the French government which bore the likenesses of both astronomers.

Almost simultaneously with their recording of prominence spectra, Lockyer and Janssen (this time working together) announced a more momentous discovery. While studying the spectrum of the Sun during the eclipse, Janssen had noticed a line he had not seen before. He forwarded his observations to Lockyer who, after comparing the reported position of the line with that of the known elements, concluded that it originated in some previously unknown element that possibly did not exist on Earth. This idea did not receive widespread support among the chemists of the day. Spectroscopy was a new science which in the opinion of many had still to prove the bold claims that were being made for it. Lockyer's claim, however, was to prove an exception to the general record of contemporary illusory 'discoveries'. He named the unknown element 'helium', after the Greek word for the Sun.

In 1881, Lockyer declared that certain lines produced in a laboratory became broader when an element was strongly heated. It was his belief that at very high temperatures atoms disintegrated into yet more elementary forms. The truth was not so simple, but in the next 20 years it was discovered that the atom has a complex internal structure and that it can acquire an electrical charge through the systematic removal of electrons. Lockyer was also the first astronomer to study the spectra of sunspots.

Further subjects of Lockyer's interest and investigation were the mysterious megalithic monuments that occur in Brittany and England, the most celebrated being those at Carnac and Stonehenge. It had long been believed that these erections were primarily of religious significance, but Lockyer noticed that the geometrical axis of Stonehenge is oriented towards the northeast, the direction in which the Sun rises at the time of the summer solstice. In the case of Stonehenge the central 'altar stone' seemed out of alignment by some 1° 12″. Lockyer believed, however, that the original builders had not been guilty of any inaccuracy but that the apparent error could be explained by a gradual change in position of the solstitial sunrise. And because the only possible source of change was the minute but regular variation in the progress of the Sun's ecliptic, it would be possible to calculate how many years were needed to have achieved a difference of 1° 12″. By this means Lockyer dated Stonehenge from the year 1840 BC (plus or minus 200 years) – a reckoning that was virtually confirmed in 1952 when, by radiocarbon dating of charred wood found in post-holes, a date of 1848 BC (plus or minus 275 years) was indicated.

Lodge Oliver Joseph 1851–1940 was a British physicist who was among the pioneers of radio. He also proved that the ether does not exist, a discovery that proved fundamental to the development of relativity.

Lodge was born at Penkhull, Staffordshire, on 12 June 1851, where his father supplied clay and other materials for the pottery industry. As the eldest son, Lodge entered his father's firm at the age of 14 and worked there for six years. In 1871 he began to study science at London University, at first while still working part-time for his father. He obtained his BSc in 1873 followed by a DSc on electrical topics in 1877.

Lodge became the first Professor of Physics at the University of Liverpool on its founding in 1881, and there carried out his most important experimental work. In 1900, he moved to the University of Birmingham to become its first Principal. Among many subsequent honours were a knighthood in 1902 and the presidency of the British Association for the Advancement of Science in 1913. Lodge retired in 1919. Following the death of his son in World War I, he became interested in psychic phenomena and devoted much of his later life to psychical research. Lodge died at Lake near Salisbury on 22 August 1940.

Lodge's first significant paper concerned the shape of lines of force and equipotential lines between two electrodes applied to conducting surfaces. He also proposed how a Daniell cell could be modified in order that it might be used as a standard for e.m.f. measurements. During the 1880s, Lodge devoted a great deal of time to experiments involving the discharge of electricity from Leyden jars.

Two of his sons extended this work, which was then used as the basis of the high-tension electric ignition system used in early internal-combustion engines.

Lodge also began work on the production of electromagnetic waves, following the prediction by James Clerk Maxwell (1831–1879) that such waves must exist. He came close to achieving Maxwell's prediction, but was just anticipated by Heinrich Hertz (1857–1894) in 1888. Lodge then turned to methods of detecting the waves and invented a coherer, a device consisting of a container packed with metal granules whose electrical resistance varies with the passage of electromagnetic radiation. This was developed into a detector of radio waves in the early investigations of radio communication, with which Lodge was closely involved.

Lodge's other major contribution to science was made in 1893 following the classic Michelson–Morley experiments of 1881 and 1887. These had failed to detect the ether that was postulated as a medium for the propagation of light waves. This result could be explained by the ether moving with the Earth, but Lodge disproved this unlikely cause in a clever experiment in which light rays were passed between two rotating discs. The resulting interference effects showed the ether does not exist, providing one of postulates – that all motion is relative – on which Albert Einstein (1879–1955) built the special theory of relativity.

Longuet-Higgins Hugh Christopher 1923– is a versatile British theoretical chemist whose main contributions have involved the application of precise mathematical analyses, particularly statistical mechanics, to chemical problems.

Longuet-Higgins was born at Lenham, Kent, on 11 April 1923. He was educated at Winchester College and won a scholarship to Balliol College, Oxford, where he worked under Charles Coulson and obtained his doctorate in 1947. He continued at Oxford as a Research Fellow, before spending a year studying molecular spectroscopy with Robert Mulliken (1896–) in Chicago. On his return to Britain in 1949, Longuet-Higgins was appointed a Lecturer and Reader in Theoretical Chemistry at the University of Manchester, and it was there that he turned his attention to statistical mechanics, work which he pursued further while he was Professor of Theoretical Physics at King's College, London, from 1952 to 1954. In that year he became Professor of Theoretical Chemistry at Cambridge, where he stayed for 13 years. In 1967, he took a Royal Society Research Fellowship at Edinburgh University to study artificial intelligence and information-processing systems, which he thought had a closer bearing on true biology than purely physio-chemical studies. Then in 1974 he moved to Sussex University, where he expanded this field into studies of the mechanisms of language and the perception of music.

Longuet-Higgins made his first contribution to theoretical chemistry when he was only 20 years old and still an undergraduate. He overthrew the previously held views about the structures of the boron hydrides, the simplest of which, diborane (B_2H_6), was thought structurally to resemble ethane. Longuet-Higgins pointed out that spectroscopic evidence suggested a bridged structure:

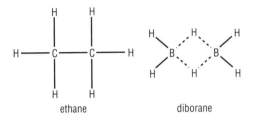

ethane   diborane

This hypothesis later proved to be correct, but not before he had also predicted the structures of other borohydrides and the then unknown beryllium hydride. He returned to this work after 23 years and predicted the existence of the ion $(B_{12}H_{12})^{2-}$, whose stability was strikingly verified several years later.

For his doctorate in 1947 Longuet-Higgins developed, with Coulson, the orbital theory of conjugated organic molecules, deriving theoretically results that had been known experimentally for decades. He continued this work in Chicago, showing how the properties of conjugated systems can be derived by combining into molecular orbital theory a study of 'non-bonding' orbitals. This work led directly to Michael Dewar's famous linear combination of molecular orbital theory. A later collaboration between Dewar and Longuet-Higgins resulted in the discovery of a system (biphenylene) in which the molecular orbital theory and the then more fashionable resonance theory gave contrary predictions. They published their findings in 1957 and several years later experimental results confirmed the molecular orbital predictions.

The work in statistical mechanics which Longuet-Higgins began at Manchester in 1949 made important contributions to many fields. He formulated a theory to describe the thermodynamic properties of mixtures, which he later extended to polymer solutions. He also investigated the optical properties of helical molecules and continued his work on electronic spectra.

At Cambridge, from 1954, he used mathematical techniques to make theoretical chemical predictions. He predicted, for example, that cyclobutadiene (which had defeated all attempts to prepare it) should exist as a ligand attached to an atom of a transition metal; such a compound was successfully prepared three years later.

In a larger piece of work, he applied group theory to define the elements of symmetry of non-rigid molecules, such as hydrazine (N_2H_4), and thus was able to classify the individual quantum levels of the molecule. This, in turn, allowed analysis of the spectra of such molecules and the evaluation of their molecular characteristics.

Other work of this nature involved the study of a group of important organic reactions known as electrocyclic rearrangements, of which the best known is the Diels–Alder reaction. With W. Abrahamson, Longuet-Higgins discovered symmetry principles in combination with molecular

orbital theory, which permit clear predictions to be made about the outcomes of such reactions, some of which seem to be quite contrary to others. In this and his other contributions Longuet-Higgins demonstrated the large part he has played in advancing chemistry from a science of largely practical experiment to one of predictive theory.

Lonsdale Kathleen 1903–1971 was an X-ray crystallographer who rose from the most humble background to become one of the best known workers in her field, being among the first to determine the structures of organic molecules. She also paved the way in a male-dominated world for the many women who followed in her footsteps.

Lonsdale was born Kathleen Yardley, in Newbridge, Ireland, on 28 January 1903, the youngest of the ten children of the local postmaster. The family was desperately poor – her father drank heavily – and four of her brothers died in infancy. An elder brother, unable to take up a scholarship to secondary school because he had to work to help to support the family, later became one of the first wireless operators and founded a school of wireless telegraphy in the north of England.

The family moved to England in 1908 and settled in Seven Kings, Essex. Kathleen went to the local elementary school until 1914, when she won a scholarship to the County High School for Girls in nearby Ilford. For her last two years there she had to attend classes at the boys' school as the only girl to study chemistry, physics and higher mathematics. At the age of 16 she went to Bedford College for Women in London, switching from mathematics to physics after one year. She graduated as top student in the University in 1922 and W. H. Bragg immediately offered her a place in his research team at University College, London, and then later at the Royal Institution.

In 1927 she married Thomas Lonsdale and moved to Leeds. There she had three children and worked in the Physics Department of Leeds University. She moved back to London in 1931 and carried out research for 15 years at the Royal Institution, first under Bragg and then with Henry Dale. In May 1945 she and Marjory Stephenson became the first women to be elected to the Royal Society. In 1946, after World War II, Lonsdale became Professor of Chemistry and Head of the Department of Crystallography at University College, London, and only then (at the age of 43) did she start university teaching and developing her own research school. She was made a Dame in 1956 and in 1968 became the first woman President of the British Association for the Advancement of Science. She died in Bexhill-on-Sea, Sussex, on 1 April 1971.

In her first post under Bragg she worked with W. T. Astbury, trying to relate space group theory to the phenomenon of X-ray diffraction by crystals. She assembled her own apparatus and the first organic crystal she measured was succinic acid (butanedioic acid). At Leeds she used a grant from the Royal Society to buy an ionization spectrometer and electroscope and correctly solved the structure of crystals of hexamethylbenzene provided by Christopher Ingold, who was then Professor of Chemistry at Leeds. Her solution for hexachlorobenzene was less complete, but was important as the first investigation using Fourier analysis. Lonsdale was a competent mathematician and did all her own calculations, aided only by logarithm tables. When she returned to work with Bragg in London she derived the structure factor formulae for all space groups.

At the Royal Institution she researched many subjects. She was interested in X-ray work at various temperatures and thermal motion in crystals. She also used divergent beam X-ray photography to investigate the textures of crystals. Lonsdale continued this work at University College, while also studying solid-state reactions, the pharmacological properties and crystal structures of methonium compounds, and the composition of bladder and kidney stones.

Lonsdale's attitudes were influenced by the Society of Friends (Quakers) and at the outbreak of World War II in 1939 she did not register for employment, regarding all war as evil. On being fined £2 she refused to pay and was sent to prison at Holloway for one month. One result of this experience was a commitment to pacifism and a lifelong interest in prison visiting.

Lorentz Hendrik Antoon 1853–1928 was a Dutch physicist who helped to develop the theory of electromagnetism, which was recognized by the award (jointly with his pupil Pieter Zeeman) of the 1902 Nobel Prize for Physics.

Lorentz was born in Arnhem, Holland, on 18 July 1853. He was educated at local schools and at the University of Leyden, which he left at the age of 19 to return to Arnhem as a teacher while writing his PhD thesis on the theory of light reflection and refraction. By the time he was 24 he was Professor of Theoretical Physics at Leyden. He remained there for 39 years, before taking up the Directorship of the Teyler Institute in Haarlem where he was able to use the museum's laboratory facilities. He died in Haarlem on 4 February 1928.

Much of Lorentz's work was concerned with James Clerk Maxwell's theory of electromagnetism, and his development of it became fundamental to Albert Einstein's special theory of relativity. Lorentz attributed the generation of light by atoms to oscillations of charged particles (electrons) within them. This theory was confirmed in 1896 by the discovery of the Zeeman effect, in which a magnetic field splits spectral lines.

In 1904, Lorentz extended the work of George FitzGerald to account for the negative result of the Michelson–Morley experiment and produced the so-called Lorentz transformations, which mathematically predict the changes to mass, length and time for an object travelling at near the speed of light.

Lorenz Konrad Zacharias 1903–1989 was an Austrian zoologist who is generally considered to be the founder of modern ethology. He is best known for his studies of the relationships between instinct and behaviour, particularly in birds, although he also applied his ideas to aspects of human behaviour, notably aggression. He received many honours for his work, including the 1973 Nobel Prize for Physiology or Medicine, which he was awarded jointly with Karl von Frisch and Nikolaas Tinbergen.

Lorenz was born in Vienna on 7 November 1903, the son of an orthopaedic surgeon. From an early age he collected and cared for various animals, and kept a detailed record of his bird observations. He was educated at the High School in Vienna, then in 1922, following his father's wishes, went to the United States to Columbia University and studied medicine. After two years he returned to Austria and continued his medical studies at the University of Vienna, from which he graduated in 1928. In the previous year he had married Margarethe Gebhardt, and they later had a son and two daughters. After graduation he studied comparative anatomy as an assistant in the Anatomy Department of Vienna University, where he remained until 1935 – having gained his doctorate in 1933. In 1936 the German Society for Animal Psychology was founded and in the following year Lorenz was appointed Coeditor-in-Chief of the society's new journal *Zeitschrift für Tierpsychologie*, which became one of the world's leading ethology journals; he held the post for many years. Also in 1937 he became lecturer in Comparative Anatomy and Animal Psychology at Vienna University, remaining there until 1940, when he was appointed Professor and Head of the Department of General Psychology at the Albertus University in Königsberg. From 1942 to 1944 he was a physician in the German army, but was captured in the Soviet Union and spent four years as a prisoner of war there. He returned to Austria in 1948 and in the following year was appointed Head of the Institute of Comparative Ethology at Altenberg. In 1951, he established the Comparative Ethology Department in the Max Planck Institute at Buldern, becoming its Codirector in 1954. He then worked at the Max Planck Institute of Behavioural Physiology in Seewiesen from 1958 to 1973 (as its Director after 1961), when he was appointed Director of the Department for Animal Sociology at the Austrian Academy of Sciences' Institute for Comparative Ethology.

Lorenz made most of his observations and basic discoveries during the late 1930s and early 1940s. From 1935 to 1938 he carried out intensive studies on bird colonies he had established, including jackdaws and greylag geese, and published a series of papers on his observations, which gained him worldwide recognition. In 1935, he described the phenomenon for which he is perhaps best known: imprinting. He discovered that many birds do not instinctively recognize members of their own species but that they do possess an innate ability to acquire this capacity. He observed that during a brief period after hatching a young bird treats the first reasonably large object it sees as representative of its species – the object becomes imprinted. Normally this object is the bird's parent but Lorenz found that it is possible to substitute almost any other reasonably sized object, such as a balloon or a human being, in which case the bird does not respond in the usual manner to other members of its species. There has since been evidence that imprinting may also occur in human children, although this is still a matter of controversy because it is extremely difficult to differentiate between innate and learned responses, especially in humans and other higher animals.

After this research, Lorenz collaborated with Nikolaas Tinbergen on further studies of bird behaviour. They showed that the reactions of many birds to birds of prey depend on attitudes or gestures made by the predators and on a particular feature of their shapes – the shortness of their necks, which is common to all birds of prey. Lorenz and Tinbergen found that the sight of any bird with a short neck – or even a dummy bird with this feature – causes other birds to fly away.

On the subject of instinct and behaviour, Lorenz hypothesized that every instinct builds up a specific type of 'desire' in the central nervous system. If there is no appropriate environment that helps to release the behaviour pattern corresponding to the desire, then tension gradually increases, eventually reaching such a level that instincts take control, even when the correct stimulus is lacking. For example, a pregnant ewe acts in a maternal manner towards a newborn lamb, although the ewe herself has not yet given birth.

In his later work Lorenz supplied his ideas to human behaviour, most notably in his book *On Aggression* (1966), in which he argued that aggressive behaviour in human beings has an innate basis but, with a proper understanding of instinctual human needs, society can be changed to accommodate these needs and so aggression may be diverted into socially useful behaviour.

Lorenz Ludwig Valentin 1829–1891 was a Danish mathematician and physicist who made important contributions to our knowledge of heat, electricity and optics. He is, however, relatively little known, partly because he published most of his work in Danish and partly because of his idiosyncratic mathematical style.

Lorenz was born in Elsinore on 18 January 1829. He graduated in civil engineering from the Technical University in Copenhagen. He then taught at a teacher-training college and at the Military Academy in Copenhagen, becoming Professor of Physics at the latter in 1876. After 1887, however, he received sufficient financial support from the Carlsberg Foundation to be able to devote himself entirely to research. He died in Copenhagen on 9 June 1891.

Lorenz is perhaps best known for his work on optics and electromagnetic theory. Early in his investigations he found that the theoretical models describing the nature of light were contradictory and incompatible, and he therefore concentrated his efforts on studying the transmission of light rather than its nature. He investigated the mathematical description for light propagation through a single homogeneous medium and also described the passage of light between different mediums. His most famous discovery was the mathematical relationship between the refractive index and the density of a medium. This formula was published by Lorenz in 1869 and by Hendrick Lorentz

(who discovered it independently) in 1870 and is therefore called the Lorentz–Lorenz formula. Today it is usually given as:

$$R = \frac{(n^2 - 1)}{(n^2 + 2)} \times \frac{M}{\rho}$$

where R is the molecular refractivity, n is the refractive index, M is the molecular mass and ρ is the density. Also in 1869 Lorenz published an experimental verification (for water) of the formula. In addition, he studied birefringence (double refraction) and wrote a paper on the electromagnetic theory of light in 1867 – after James Clerk Maxwell's theory had been published in Britain but at a time when it was almost unknown elsewhere in Europe. Lorenz's approach was different from that of Maxwell and he was able to derive a correct value for the velocity of light using his theories.

Lorenz's other main major contribution to science was his discovery of what is now called Lorenz's law. Relating a metal's thermal conductivity (λ) and its electrical conductivity (σ) to its absolute temperature (T), the law is usually stated mathematically as:

$$(\lambda/\sigma) \propto T$$

A believer in combining theory and experiment, he performed a series of experiments to test his law and thereby confirmed its validity.

Lorenz also did basic work on the development of the continuous loading method for reducing current losses along cables, and helped to establish the ohm as the internationally accepted unit of electrical resistance. His work in pure mathematics included studies of prime numbers and Bessel functions, which are widely used in mathematical physics.

Lovell Alfred Charles Bernard 1913–   is a British astrophysicist and author whose experience with radar during World War II led to his applying radar to the detection of meteors and to his energetic instigation of the construction of the radio telescope at Jodrell Bank in Cheshire, where he has been Director for more than 30 years.

Lovell was born at Oldland Common, Gloucestershire, on 31 August 1913, the son of a lay preacher. Educated at the Kingswood Grammar School, Bristol, he then attended Bristol University where he read physics, graduating in 1933. Three years later he became Assistant Lecturer in Physics at Manchester University. During World War II he was in the Air Ministry Research Establishment in Malvern where, under his guidance, centimetric airborne radar was developed for use on 'blind bombing' air-raids and submarine defence. At the end of the war, Lovell returned to Manchester as Lecturer in Physics and immediately began pressing the authorities to set up a radioastronomy station at Jodrell Bank (about 32 km/ 20 mi south of Manchester). He was appointed Senior Lecturer in 1947 and Reader in 1949, all the while agitating for his dream of a radio telescope at Jodrell Bank to be made a reality. Finally, in 1951,

Manchester University created a special Chair of Radio-astronomy for him and, with the Government guaranteeing part of the financing of his radio telescope, made the Directorship of Jodrell Bank an official post. He was elected a Fellow of the Royal Society in 1955 and received the Royal Medal of the Society in 1960. He was knighted in 1961. The author of several books, a number of which popularized radioastronomy, he has also published many articles in scientific journals.

Lovell's first postwar research used radar to show that echoes could be obtained from daylight meteor showers invisible to the naked eye. Significantly, he proved the worth of such radio techniques by observing the meteor shower as the Earth passed through the tail of a comet in 1946 – a meteor shower that was visible to the naked eye. Having established the value of radio in this way, he showed by further studies that it was possible to make determinations of the orbits and radiants of meteors and thus prove that all meteors originate within the Solar System. With the same equipment, Lovell investigated the loud solar radio outburst in 1946, and in 1947 began to examine the aurora borealis.

In 1950, Lovell discovered that galactic radio sources emitted at a constant wavelength and that the fluctuations ('scintillation') recorded on the Earth's surface (the subject of considerable scientific speculation) were introduced only as the radio waves met and crossed the ionosphere.

The year 1951 saw the beginning of the construction of the Jodrell Bank radio telescope. Taking six years to build, under Lovell's close personal supervision, the gigantic dish has an alt-azimuth mounting with a parabolic surface of sheet steel; it remains the largest completely steerable radio telescope in existence. It was completed just in time to track the Soviet *Sputnik 1* (the first artificial satellite), thus confounding the criticism that too much money had been spent on the project. The Jodrell Bank radio telescope (now part of the Nuffield Radio Astronomy Laboratory) is still probably the most useful instrument in the world for tracking satellites.

From 1958 Lovell became interested in radio emission from flare stars. After two years at work, when his results were still inconclusive, he began a collaboration with Fred Whipple (1906–   ) of the Smithsonian Astrophysical Observatory in the United States. A joint programme was arranged for simultaneous radio and optical observations of flare stars using Baker Nunn cameras from the Smithsonian satellite tracking network. The first results were published in *Nature* in 1963; they opened up new avenues for the study of large-scale processes occurring in a stellar atmosphere. It was also shown at that time that the integrated radio emission from the flare stars may account for a few per cent of the overall emission from the Milky Way. These combined optical and radio observations have also led to the establishing of a new value for the constancy of the relative velocity of light and radio waves in space.

Lowell Percival 1855–1916 was a US astronomer and mathematician, the founder of an important observatory in the United States, whose main field of research was the planets of the Solar System. Responsible for the popularization in his time of the theory of intelligent life on Mars, he also predicted the existence of a planet beyond Neptune which was later discovered and named Pluto.

Lowell was born in Boston, Massachusetts, on 13 March 1855. His interest in astronomy began to develop during his early school years. In 1876, he graduated from Harvard University, where he had concentrated on mathematics, and then travelled for a year before entering his father's cotton business. Six years later, Lowell left the business and went to Japan. He spent most of the next ten years travelling around the Far East, partly for pleasure, partly to serve business interests, but also holding a number of minor diplomatic posts.

Lowell returned to the United States in 1893 and soon afterwards decided to concentrate on astronomy. He set up an observatory at Flagstaff, Arizona, at an altitude more than 2,000 m/6,600 ft above sea level, on a site chosen for the clarity of its air and its favourable atmospheric conditions. He first used borrowed telescopes with diameters of 30 cm/12 in and 45 cm/18 in to study Mars, which at that time was in a particularly suitable position. In 1896, he acquired a larger telescope and studied Mars by night and Mercury and Venus during the day. Overwork led to a deterioration in Lowell's health, and from 1897 to 1901 he could do little research, although he was able to participate in an expedition to Tripoli in 1900 to study a solar eclipse.

He was made non-resident Professor of Astronomy at the Massachusetts Institute of Technology in 1902, and gave several lecture series in that capacity. He led an expedition to the Chilean Andes in 1907 which produced the first high-quality photographs of Mars. The author of many books, and the holder of several honorary degrees, Lowell died in Flagstaff on 12 November 1916.

The planet Mars was a source of fascination for Lowell. Influenced strongly by the work of Giovanni Schiaparelli (1835–1910) – and possibly misled by the current English translation of 'canals' for the Italian *canali* ('channels') – Lowell set up his observatory at Flagstaff originally with the sole intention of confirming the presence of advanced life forms on the planet. Thirteen years later the expedition to South America was devoted to the study and photography of Mars. Lowell 'observed' a complex and regular network of canals and believed that he detected regular seasonal variations which strongly indicated agricultural activity. He found darker waves that seemed to flow from the poles to the equator and suggested that the polar caps were made of frozen water. (The waves were later attributed to dust storms and the polar caps are now known to consist not of ice, but mainly of frozen carbon dioxide. Lowell's canal system also seems to have arisen mostly out of wishful thinking; though part of the system does indeed exist, it is natural not artificial and it is apparent only because of the chance apposition of dark patches on the Martian surface.)

Lowell also made observations at Flagstaff of all the other planets of the Solar System. He studied Saturn's rings, Jupiter's atmosphere and Uranus's rotation period. Finding that the perturbations in the orbit of Uranus were not fully accounted for by the presence of Neptune, Lowell predicted the position and brightness of a planet that he called Planet X, but was unable to discover. (Nearly 14 years after Lowell's death Clyde Tombaugh found the planet – Pluto – on 12 March 1930; the discovery was made at Lowell's observatory and announced on the 75th anniversary of Lowell's birth.)

Lowell is remembered as a scientist of great patience and originality. He contributed to the advancement of astronomy through his observations and his establishment of a fine research centre and he did much to bring the excitement of the subject to the general public.

Ludwig Karl Friedrich Wilhelm 1816–1895 was a German physiologist – one of the great teachers and experimenters in the history of physiology.

Ludwig was born at Witzenhausen, Hesse, on 29 December 1816. He completed his schooling at Hanau Gymnasium and then went to Marburg University in 1834. He was compelled to leave the university as a result of his political activities, but after studying at Erlangen and at the surgical school in Bamberg, was allowed to return to Marburg to complete his studies. He received his medical degree in 1840, and in 1841 became Professor of Anatomy. He was appointed Associate Professor Extraordinarius at Marburg in 1846 and continued with teaching and research there until 1849 when he moved to Zurich and became Professor of Physiology and Anatomy. Six years later he went to Vienna and served as Professor of Anatomy and Physiology at the Josephinum, the Austrian military medical academy which had been founded the previous year, in 1854. In 1865, he accepted the newly created Chair of Physiology at Leipzig and set out to develop it into an important teaching centre for physiology. He held this post until his death in Leipzig on 23 April 1895.

While studying at Marburg, Ludwig investigated the mechanism of secretion. In 1844, he was studying renal secretion and, on examining the structure of the kidney tubules (glomeruli), he recognized that during the first stage of secretion the surface membrane of the glomeruli acts as a filter. Liquid diffuses through it as a result of the pressure difference on either side of the membrane. Ludwig also devised a system of measuring the level of nitrogen in urine to quantify the rate of protein metabolism in the human body. Continuing his research on secretion, he showed that secretion from the salivary glands is dependent on secretory nerve stimulation and not on blood supply.

Following William Harvey's findings on the circulation of blood it was believed that blood was moved by an unseen vital force. Ludwig was against this idea and in 1847 developed the kymograph (an instrument that continuously records changes in blood pressure and respiration). Ludwig could thus prove that blood is moved by a mechanical force.

In 1856, when studying the effects of certain drugs on the heart, Ludwig discovered that a frog's heart, removed after the death of the animal, could be revived and that organs could be kept alive *in vitro*. This was the first time this operation had been performed successfully and was done by perfusing the coronary arteries under pressure with blood or a salt solution which resembles the composition of the saline medium of the blood.

Ludwig was also the first to discover the depressor and accelerator nerves of the heart, and in 1871, working with Henry Bowditch, a US physiologist, he formulated the 'all-or-none' law of cardiac muscle action. This law postulates that when a stimulus is applied to a few fibres of heart muscle, the whole heart muscle contracts to the extent that with increased stimulation there is no further increase in the contraction. The 'all-or-none' law is most evident in cardiac muscle although it can occur elsewhere.

In 1859, in a paper published by his student Sechenov, Ludwig described his invention of the mercurial blood-gas pump, which enabled him to separate gases from a given quantity of blood taken directly *in vivo*. This invention led to later understanding of the part played by oxygen in the purification of the blood. Ludwig also invented the *stromuhr*, a flowmeter which measures the rate of the flow of blood in the veins.

Ludwig's ingenuity and inventiveness, combined with a good knowledge of physics and chemistry, made him important in the development of modern physiology. The kymograph with its subsequent modifications has become the standard tool for the recording of experimental results, and much of what is now known about the mechanism of cardiac activity is based on his work. Ludwig published a textbook for his students, *Das Lehrbuch der Physiologie/A Physiology Textbook*, the first volume in 1852 and the second in 1856, which was the first modern text on physiology.

Lumière Auguste Marie Louis Nicolas 1862–1954 and Lumière, Louis Jean 1864–1948 were two brothers who became famous as inventors of the cinematograph.

Louis Lumière, the younger of the two and ultimately the one who spent his whole career in photography, was born on 5 October 1864 at Besançon in eastern France. The family owned and operated a photographic firm in Lyons. When the brothers were old enough to join the business they did so with enthusiasm and contributed several minor improvements to the developing process, including in 1880 the invention of a better type of dry plate.

In 1894, their father purchased an Edison Kinetoscope (a cine viewer) which impressed them greatly, although more for the potential it represented than for itself. They borrowed some of the ideas and developed a new all-in-one machine – camera and projector – which they patented in 1895, calling it a cinematograph. To advertise their success they filmed delegates arriving at a French photographic congress and 48 hours later projected the developed film to a large audience.

When the brothers ultimately separated, Auguste went on to do medical research leaving Louis to continue in the photographic industry. Louis was associated with several other improvements, among them a photorama for pano-ramic shots and in 1907 a colour printing process using dyed starch grains. During World War I he worked on air-craft equipment and afterwards carried on with photography, branching out into stereoscopy and three-dimensional films. He died at Bandol, France, on 6 June 1948.

Edison had patented his design for the kinetograph in 1888, after persuading George Eastman to make suitable celluloid film. Unfortunately the pictures could be viewed only through a peep-hole and, although Edison had seen the possibility of projecting them so they could be viewed by a large audience, he had omitted this from his patent, which was for him a rare error.

The Lumière brothers took the best of Edison's ideas and developed them further, inventing a combined camera and projector which weighed much less than the Edison model. The film passed through the camera at the rate of 16 frames per second, slower than the modern equivalent, while a semicircular shutter cut off the light between the lens and the film. By the end of 1895 the brothers had opened their first cinema in Paris, attracting large crowds to see the real-life films. Other establishments soon followed and a whole new industry was founded, bringing film entertainment within the reach of the average person.

In the United States, Edison had seen how the error in his patent had been exploited and had resumed his interest in the moving-picture phenomenon, joining forces with Tho-mas Armat who had devised a projector. To the Lumières in France and Edison in the United States can be attributed the honours for the origin of the motion picture industry of the twentieth century.

Lummer Otto Richard 1860–1925 was a German physi-cist who specialized in optics and is particularly remembered for his work on thermal radiation. His investi-gations led directly to the radiation formula of Max Planck (1858–1947), which marked the beginning of quantum theory.

Lummer was born in Jena, Saxony, on 17 July 1860. He attended a number of different German universities, which was a common practice at the time, and in 1884 wrote his doctoral dissertation at Berlin on work conducted in the department of Hermann Helmholtz (1821–1894). Lummer then became an assistant to Helmholtz and moved with him to the newly established Physikalische Technische Reich-sanstalt in Berlin in 1887. Lummer became a member of this research institute in his own right in 1889, and in 1894 was made professor there. In 1904, he was appointed Pro-fessor of Physics at Breslau (now Wroclaw, Poland), and he died there on 5 July 1925.

Lummer's early research was on interference fringes produced by internal reflection in mica plates. These fringes had in fact been noted before, but were nevertheless named Lummer fringes. His next area of study concerned the establishment of an international standard of luminos-ity. This work was done in collaboration with Eugen Brodhun, and they designed a photometer named the Lummer–Brodhun cube. This represented a considerable improvement on the grease-spot photometer which Robert Bunsen (1811–1899) had originated.

The study of radiant heat then attracted Lummer's atten-tion. As a black body is a perfect heat absorber, a black-body radiator was believed to be the most efficient trans-mitter of thermal radiation. However, until that time, it was considered to be merely a theoretical object. Lummer and Wilhelm Wien (1864–1928) made a practical black-body radiator by making a small aperture in a hollow sphere. When heated to a particular temperature, it behaved like an ideal black body.

In 1898, Lummer and Ernst Pringsheim (1859–1917) began a quantitative study on emission from black bodies. They were able to confirm Wien's displacement law, which is used to determine the temperature of a black body spec-troscopically and has been applied to stellar temperature determinations, but found an anomaly in Wien's radiation law. Their result was confirmed by another research group headed by Heinrich Rubens (1865–1922) and Ferdinand Kurlbaum (1857–1927), who extended the work. These studies were of critical significance in the development of Planck's radiation formula, which was presented to the Berlin Physical Society on 19 October 1900 and marks the beginning of quantum theory.

Lummer's early work on interference fringes led him in 1902 to the design of a high-resolution spectroscope. Ernst Gehrcke improved it by adding a prism, and the Lummer–Gehrcke interference spectroscope was a great advance on its predecessor, the interferometer designed by Charles Fabry (1867–1945) and Alfred Perot (1863–1925).

Other areas which interested Lummer were the subject of solar radiation and the problem of obtaining a source of monochromatic illumination. In the former area, he man-aged to obtain an estimate for the temperature of the Sun, and in the latter he designed a mercury vapour lamp, which is still used when monochromatic light is required, for instance in fluorescence microscopy.

Lwoff André Michael 1902–   is a French microbiologist who was awarded the 1965 Nobel Prize for Physiology or Medicine for his research into the genetic control of enzyme activity. He shared the prize with his fellow researcher Jacques Lucien Monod and François Jacob.

Lwoff, of Russian-Polish descent, was born in Ainy-le-Château, Allier, on 8 May 1902. He studied natural sci-ences, graduating in 1921 and taking a post at the Pasteur Institute; in 1927 he received doctorates in medicine and science. During World War II he was an active member of the French Resistance movement, for which his country awarded him the Legion of Honour. From 1959 to 1968 he was Professor of Microbiology at the University of Paris and from 1968 to 1972 he was Head of the Cancer Research Institute at Villejuif.

In his early research carried out in the 1920s, Lwoff demonstrated the coenzyme nature of vitamins. He also discovered the extranuclear genetic control of some char-acteristics of protozoa. In the early 1940s the US geneticist

George Beadle had done important work that showed that genes are responsible for the production of the enzymes that moderate biochemical processes. Towards the end of the decade Lwoff and his co-workers proved that enzymes produced by some genes regulate the functions of other genes. He worked out the mechanism of lysogeny in bacteria, in which the DNA of a virus becomes attached to the chromosome (DNA) of a bacterium, behaving almost like a bacterial gene. It is therefore replicated as part of the host's DNA and so multiplies at the same time. But certain agents (such as ultraviolet radiation) can turn the 'latent' viral DNA, called the prophage, into a vegetative form which multiplies, destroys its host, and is released to infect other bacteria.

Lyell Charles 1797–1875 was a British geologist who succeeded in turning the opinion of his time away from the theory that the Earth was produced literally along the lines expounded in the Book of Genesis towards the principle of an unlimited, gradual effect of natural forces. His beliefs became known in geology as 'uniformitarianism'.

Lyell was born in Kinnordy, Forfarshire, Scotland, on 14 November 1797, the son of a lawyer and amateur botanist. When he was still a child the family moved to Hampshire. Lyell always had an interest in natural history; he was a keen lepidopterist, and his interest in geology was stimulated by Bakewell's book on the subject. Lyell went to Oxford University to study classics, but also attended lectures given by William Buckland, the Professor of Geology. Buckland was of the opinion that the different strata in rocks result from silt being laid down under water over a long period of time: he was a 'Neptunist'. Lyell made his first tentative geological observations during family holidays in Britain and from 1818 on the Continent, and he began to believe more fully in the principles of uniformitarianism. (In fact the geologist James Hutton (1726–1797) had postulated similar theories 50 years earlier, but Lyell formed his conclusions independently; it was only when he later read Hutton's work that he realized that their views were similar.) Lyell continued his education by studying law, was called to the Bar in 1822, and started to practise in 1825. In 1823, he became involved in the running of the Geological Society as its Secretary and later as Foreign Secretary; he was twice its President some 15 years later. He also set up the finance for the Lyell Medal and the Lyell Fund. He made a trip to Paris in 1823 and met Georges Cuvier (1769–1832), the eminent French anatomist who had stuck rigidly to the geological theories of 'catastrophism', despite his brilliant understanding in other fields. Lyell also met Alexander von Humboldt (1769–1859), the German naturalist; both men influenced his eventual ideas. In 1831, he became Professor of Geology at King's College, London, and a year later he married the daughter of the geologist Leonard Horner. Lyell was knighted in 1848 and created a baronet in 1864. He died in London on 22 February 1875.

Lyell did not originate much material, but he expounded the theories of Hutton and organized them into a popular and coherent form. His masterpiece *The Principles of Geology* was published in three volumes from 1830 to 1833 and was revised regularly until 1875. It laid out evidence to support the theory that the Earth's geological structure evolved slowly through the continuous action of forces still at work today, including the erosive action of the wind and weather. Lyell conceded very little to catastrophism, although modern geologists accept that some 'catastrophies' must have occurred – for instance, at the time of the disappearance of the dinosaurs. Lyell classified some geological eras – subdividing the Tertiary into the Eocene, Miocene, Pliocene and Pleistocene – and suggested that some of the oldest rocks may be as much as 240 million years old. People were astonished by such a time scale, even though present-day geologists think that ten times that number may be nearer the probable truth.

The conservative scientists were alarmed by Lyell's theories, but his book was popular and stimulated other geologists to investigate along similar lines. Charles Darwin, a colleague and friend of Lyell's, was deeply impressed, and Lyell in turn eventually embraced the theory of evolution as outlined in Darwin's *On the Origin of Species* (1859) – it was Lyell and Joseph Hooker who in 1858 presented to the Linnaean Society the original papers on natural selection by Darwin and Alfred Wallace. Lyell then went further than Darwin had been prepared to do in an attempt to trace human origins, and used archaeological findings as the key to his book *The Geological Evidence of the Antiquity of Man with Remarks on Theories of the Origin of Species by Variation* (1863).

Lyman Theodore 1874–1954 was a US physicist famous for his spectroscopic work in the ultraviolet region.

Lyman was born in Boston on 23 November 1874. His father was a marine biologist devoted to the interests of Harvard College, where Lyman also spent his scientific career. Lyman came from a wealthy family and after 1885 lived at the ancestral mansion built by his grandfather at Brookline, Massachusetts. Lyman entered Harvard in 1893 with tastes leaning towards physical sciences. In his first year, he took and passed a course in physics. In his second and third years, he worked hard at electrical engineering and might well have gone on to make a career of it had it not been for the influence of Wallace Sabine (1868–1919), who lectured Lyman in optics. Lyman was offered an assistantship in physics and gained a BA in 1897 followed by a PhD in 1900. During the winter of 1901–1902 he studied under J. J. Thomson (1856–1940) at Cambridge and then spent the summer of 1902 in Göttingen. On his return to Harvard he became an instructor, progressing to assistant professor in 1907. In 1921, he became Hollis Professor of Mathematics and Natural Philosophy. He held this position for five years and then retired Emeritus in 1925, fifteen years before the official retiring age. It was generally thought that he retired to have more freedom. From 1910 to 1947, he also held the position of Director of the Jefferson Physical Laboratory. Lyman received many honours, among them the presidencies of the American Physical

Society from 1921 to 1922 and the American Academy of Arts and Sciences from 1924 to 1927. He died after a long illness at Brookline on 11 October 1954.

Lyman's scientific work was confined to the spectroscopy of the extreme ultraviolet region. When he was commencing his research, the observable spectrum had been extended into the ultraviolet as far as 1,260 Å (1Å = 10^{-10}m) by Victor Schumann (1841–1913), who enclosed the entire spectroscope in an evacuated chamber to eliminate absorption by air and by using fluorite for windows, lenses and prisms. Although Schumann had obtained a wealth of new spectroscopic lines beyond 2,000 Å, the wavelengths could not be exactly determined without knowledge of the index of refraction of the fluorite of his prism.

Sabine suggested that Lyman should try a concave ruled grating instead of a fluorite prism. Success took seven years. False lines of a new kind were found in the Schumann range due to light in the visible region. These came to be called Lyman ghosts and gave Lyman the subject for his PhD thesis. By 1906, he had published reports of more than 300 lines due to hydrogen in the 1,675–1,228 Å region and another 50 probably due to hydrogen in the 1,228–1,030 Å region. He established the limit of transparency of fluorite to be 1,260 Å and looked at the absorbency of other suitable solids but found none better. He also examined the absorption of various gases.

At the beginning of the century, spectroscopic analysis had received an enormous boost because of the discovery of the mathematical regularities shown by many spectra. This was very important in substantiating the quantum theory of atomic structure proposed by Niels Bohr (1885–1962) in 1913. Two series of hydrogen spectra had been found, the Paschen series and the Balmer series. The wave number ν of each line in these series is given by the expression:

$$\nu = R \left(\frac{l}{m^2} - \frac{l}{n^2} \right)$$

where R is the Rydberg constant and n is an integer greater than m, which has the value 3 for the Paschen series and 2 for the Balmer series. Johann Balmer (1825–1898), who discovered this relationship, had predicted the existence of a series of lines in the ultraviolet in which m would be equal to 1. Lyman announced the discovery of the first three members of this series in 1914, and it was named the Lyman series. He predicted that the first line, the Lyman alpha, would be present in the Sun's spectrum but absorption by the Earth's atmosphere prevented Lyman from observing it. The line was eventually photographed by a rocket five years after Lyman's death, and ultraviolet observation of the Sun by satellite has since become very important in solar research.

By 1917, Lyman had extended the spectrum to 500 Å. War service and then college administration subsequently reduced his scientific output, and others continued to extend the spectrum further. Lyman turned to spectroscopic

examination of elements other than hydrogen, the first being helium. In 1924, seven lines were tabulated in the principal series of helium and Lyman made the rare observation of the continuous spectra beyond the limit of this series due to the recapture of free electrons of different kinetic energies. Papers followed on a series of spectra in the ultraviolet region of aluminium, magnesium and neon. Lyman's last paper was published in 1935 on the transparency of air between 1,100 and 1,300 Å. Towards the end of his life, Lyman suffered a very sore hand from excess exposure to ultraviolet radiation, and impaired eyesight.

Lyman was also a keen traveller and collector. His most important trip was to the Altai mountains of China and Mongolia. He brought back the first specimen of a gazelle which became named *Procapra altaica* and also 13 previously unknown smaller mammalian species. A stoat became known as Lyman's stoat, *Mustela lymani*.

Lynden-Bell Donald 1935– is a British astrophysicist particularly interested in the structure and dynamics of galaxies.

Lynden-Bell was born the son of an army officer on 5 April 1935. Educated at Marlborough College, he went to the University of Cambridge, where he graduated from Clare College. In 1960, having completed his PhD, he was elected Harkness Fellow of the Commonwealth Fund and joined the California Institute of Technology (CIT) to work with Allan Sandage (1926–), at the Hale Observatory, on the dynamics of galaxies. Two years later he returned to Cambridge to take up an appointment as Assistant Lecturer in Applied Mathematics. Afterwards he became Fellow and Director of Studies in Mathematics at Clare College. In 1972 he became Professor of Astrophysics and Director of the Institute of Astronomy. The author of a number of significant papers, he is a Fellow of the Royal Astronomical Society.

During his second period at CIT, Lynden-Bell published a paper on 'Galactic nuclei as collapsed old quasars' (1969), which proposed that quasars were powered by massive black holes. Later, continuing this line of thought, he postulated the existence of black holes of various masses in the nuclei of individual galaxies. The presence of these black holes – objects optically invisible because their light is 'imprisoned' by the object's own tremendously strong gravitational attraction – as power centres of galaxies would account for the large amounts of infrared energy that emanate from a galactic centre. Lynden-Bell further argued that in the dynamic evolution of star clusters the core of globular star clusters evolves independently of outer parts, and that it is necessary to postulate a dissipative collapse of gas to account for that evolution. This is certainly compatible with the presence of a central black hole within a stellar system.

Lynen Feodor 1911–1979 was a German biochemist known for his research into the synthesis of cholesterol in the human body and into the metabolism of fatty acids. For this work he shared the 1964 Nobel Prize for Physiology or

Medicine with the US biochemist Konrad Bloch.

Lynen was born in Munich, Bavaria, on 6 April 1911. He studied at the University of Munich, gaining his doctorate in 1937. In the same year he married the daughter of his professor, Heinrich Wieland. He remained on the academic staff at Munich as a lecturer (1942–1946), Associate Professor (1947–1953) and Professor of Biochemistry (from 1953). Between 1954 and 1972 he was Director of the Max Planck Institute for Cell Chemistry in Munich, and from 1972, Director of the Max Planck Institute for Biochemistry.

Cholesterol is a key substance in the body, the starting material for adrenal cortical hormones, sex hormones and other steroids. Lynen in Munich and Bloch in the United States studied the complicated mechanism by which cholesterol is formed. Bloch found that the basic unit for cholesterol synthesis is the simple acetate (ethanoate) ion, a chemical fragment containing only two carbon atoms. Fritz Lipmann postulated that the substance known as co-enzyme A, which he isolated in 1947, might be the carrier of the fragment. In 1951, Lynen isolated 'active acetate' from yeast and found it to be identical to acetyl coenzyme A, a combination of coenzyme A and a two-carbon fragment, thus confirming Lipmann's hypothesis. Bloch then found an intermediate compound, squalene – a long hydrocarbon containing 30 carbon atoms. Lynen and Bloch corresponded and worked out the 36 steps involved in the synthesis of cholesterol. The final stage was found to be the transformation of the carbon chain of squalene ($C_{30}H_{50}$) into the four-ring molecule of cholesterol.

Lynen also worked on the biosynthesis of fatty acids, isolating from yeast an enzyme complex which acts as a catalyst in the synthesis of long-chain fatty acids from acetyl coenzyme A and malonyl coenzyme A. His study of fatty acids also elucidated a series of energy-generating reactions that occur when fatty acids from food are respired to form carbon dioxide and water. From this research has resulted the more general conclusion that repeated condensation of two carbon fragments originating from acetate is the basis of the synthesis of many natural substances.

Lyot Bernard Ferdinand 1897–1952 was a French astronomer and an exceptionally talented designer and constructor of optical instruments. He concentrated on the study of the solar corona, for which he devised the coronagraph and the photoelectric polarimeter, and he proved that some of the Fraunhofer lines in the solar spectrum represent ionized forms of known metals rather than undiscovered elements.

Lyot was born in Paris on 27 February 1897. He graduated from the Ecole Supérieure d'Electricité in 1917 and in 1918 he was awarded a diploma in engineering. He worked under Alfred Pérot at the Ecole Polytechnique in Paris as a demonstrator in physics from 1918 to 1929, and from 1920 he held the post of Assistant Astronomer at the Meudon Observatory. In 1930, Lyot was made joint Astronomer at the Observatory. Lyot's advances in the study of polarized and monochromatic light soon earned him an international

reputation. He was elected to the French Academy of Sciences in 1939 and in the same year was awarded the Gold Medal of the Royal Astronomical Society in London. He published several books that outlined his discoveries and innovations. Having become Chief Astronomer at the Meudon Observatory in 1943, Lyot travelled to the Sudan in 1952 to observe a total eclipse of the Sun. He suffered a heart attack on the train journey home and died near Cairo, Egypt, on 2 April 1952.

Most of Lyot's research during the 1920s was devoted to the study of polarized light, reflected to the Earth from the Moon and from other planets. In addition to designing a polariscope of greatly improved sensitivity, Lyot made a number of observations about the surfaces and atmospheric conditions on other planets. In 1924, he reported that the Moon was probably covered by a layer of volcanic ash and that duststorms were a common feature of the Martian surface. (He also claimed to have detected water vapour on the surface of Venus, but it was later demonstrated that the tiny amount of water vapour actually there, could not have been seen using Lyot's instruments.)

For centuries, astronomers wishing to study the solar corona had been restricted to the rare and brief occasions of total eclipses of the Sun. The main problem involved in using optical instruments at other times had always been the 'scattering' of light – by even the slightest particle of dust or the minutest fault in the object lens – so that the corona, which has only one-millionth of the brilliance of the solar disc, was totally obscured. In 1930, Lyot designed a 'coronagraph', which included three lenses. The object lens was as perfect as he could make it, with a diameter of 8 cm/3 in and a focal length of 2 m/6.5 ft. He took the instrument to the clear air of the Pic du Midi Observatory, in the Pyrenees, at an altitude of 2,870 m/9,420 ft, and for the first time in the history of astronomy was able to observe the corona in broad daylight.

During the 1930s Lyot improved upon his coronagraph: he increased the size and focal length of the object lens and fitted the device with a sophisticated monochromatic filter designed to enable him to concentrate on the most important wavelengths in the coronal light. By increasing the length of time during which the solar corona could be observed, the coronagraph also permitted the observation of continuous changes in the corona. This meant that the corona could be filmed, as Lyot demonstrated for the first time in 1935. Lyot also reported the rotation of the corona in synchrony with the Sun.

The coronagraph was also essential to Lyot's realization that some of the lines in the solar spectrum, believed to represent the unknown elements 'coronium', 'geocoronium' and others, did in fact represent the highly ionized forms of elements well-known on Earth.

Lyot's later work included the construction of the photoelectric polarimeter, which facilitated further research on the solar corona.

Lysenko Trofim Denisovich 1898–1976 was a botanist who dominated biology in the Soviet Union from about the mid-1930s to 1965. During this period he was virtual dictator of biology in the Soviet Union and his theories, although largely rejected outside his own country, were officially adopted within the Soviet Union. He actually contributed very little to scientific knowledge and his importance has been attributed to his friendship with the Soviet political leaders Josef Stalin and Nikita Khrushchev, who awarded Lysenko many honours: he was made a Hero of Socialist Labour, and received the Order of Lenin eight times and the Stalin Prize three times.

Lysenko was born on 29 September 1898 in Karlovka in the Russian Ukraine, the son of a peasant, and was educated at the Uman School of Horticulture. After graduating in 1921, he went to the Belaya Tserkov Selection Station then to the Kiev Agricultural Institute to study for his doctorate, which he gained in 1925. He was stationed at the Gandzha (now Kirovabad) Experimental Station from 1925 to 1929, when he became the senior specialist in the Department of Physiology at the Ukrainian All-Union Institute of Selection and Genetics in Odessa. In 1935, he became Scientific Director of this institute and in the following year he was promoted to Director, a post he held until 1938. With the increase of his political influence, Lysenko rose rapidly to the top of the scientific hierarchy, becoming Director of the Institute of Genetics of the USSR Academy of Sciences in 1940. As a result of Khrushchev's fall from power in 1964, Lysenko's influence diminished considerably and in 1965 he was removed from his post and stripped of all authority. He died on 20 November 1976, his ideas discredited both within the Soviet Union and by the Western world.

Lysenko rose to prominence as a result of his advocating vernalization to increase crop yields. In vernalization – a practice well known since the nineteenth century – seeds are moistened just sufficiently to allow germination to begin then, when the radicles start to emerge, the seeds are cooled to slightly above 0°C/32°F, thereby halting further growth. When the seeds are planted in spring, they mature quickly; this is particularly useful in the Soviet Union because large areas of the country have only a very short growing season. Using this method Lysenko achieved considerable increases in crop yields, which gained him substantial political support. A succession of important appointments followed, and by 1935 Lysenko had become a powerful influence in Soviet science.

As Lysenko's influence increased, so he enlarged the scope of his theories, using his authority to remove any opposition. He innovated the doctrine of the phasic development of plants, claiming that all plants develop in recessive phases, each with different requirements. He stated that by altering any stage of development, changes could be caused in successive stages. This doctrine was opposed by Nikolai Vavilov (1887–1942/3), an internationally respected Soviet geneticist, but Lysenko used his political influence to have Vavilov arrested and banished to Siberia, where he died in exile in 1942 or 1943.

Expanding his theories still further, Lysenko defined heredity as the capacity of an organism to require specific conditions for its life and development, and to respond in different ways to various conditions. Moreover, he believed that when an organism is subjected to abnormal environmental conditions, it develops in such a way as to take advantage of these conditions and that the offspring of this organism also tend to develop in the same way as the parent. This idea was, in fact, a restatement of the Lamarckian doctrine of the inheritance of acquired characteristics.

As leader of the Soviet scientific world, Lysenko encouraged the defence of mechanistic views about the nature of heredity and speciation. These views – termed Michurin biology after the prominent Soviet scientist I. V. Michurin – became an integral part of Soviet scientific thought and created an environment conducive to the spread of unverified facts and theories, such as the doctrine of the noncellular 'living' substance and the transformation of viruses into bacteria. To many people, this period represented the dark ages of Soviet science and research in several areas of biology came to a halt.

Although Lysenko's views were imposed on Soviet scientists, his ideas were widely criticized by many European and US scientists and, encouraged by this external support, the struggle to counteract Lysenkoism gained strength.

Lyttleton Raymond Arthur 1911–   is a British astronomer and theoretical physicist whose main interest is stellar evolution and composition, although he has extended this in order to investigate the nature of the Solar System.

Lyttleton was born in Warley Woods, near Birmingham. Educated at King Edward School, Birmingham, he then went to Clare College, Cambridge. As a Visiting Fellow at Princeton University in the United States from 1935–1937, Lyttleton worked with Henry Russell (one of the originators of the Hertzsprung–Russell diagram) and was inspired while there to propose a theory of planetary formation that at the time received some critical acclaim. Upon his return to Cambridge in 1937, Lyttleton was awarded his PhD and appointed Research Fellow of St John's College; together with Fred Hoyle (1915–   ) he established an active research school there in theoretical astronomy. During World War II he served as an Experimental Officer with the Ministry of Supply (1940–1942) and as a Technical Assistant to Sydney Chapman as Scientific Adviser to the Army Council War Office (1942–1945). Lyttleton was then appointed Lecturer in Mathematics at Cambridge University, becoming Stokes Lecturer in Mathematics in 1954, and Reader in Theoretical Astronomy in 1959. During the 1960s he held a number of scientific posts including a position as Research Associate at the Jet Propulsion Laboratory in California (1960), and various visiting professorships. In 1967 he was an original member of the Institute of Astronomy at Cambridge. Lyttleton is the author of many papers and several books on astronomical subjects and a member of leading scientific societies. He was elected Fellow of the Royal Society in 1955, from whom he received a Royal Medal in 1965.

Lyttleton's research has spanned most areas of theoretical astronomy. His earliest work in the subject, his theory of planetary formation, was formulated at Princeton. It involved the possibility of a binary companion star to the Sun, dealt with the rotation of the planets and their satellites, and showed that Pluto may be an escaped satellite of Neptune.

Upon his return to Cambridge, Lyttleton began his long and fruitful association with Fred Hoyle, with whom he contributed to the growing knowledge of stellar evolution. In the early 1940s they applied the new advances in nuclear physics, as developed by Hans Bethe (1906–) and others, to the problem of energy generation in stars. They also published, in 1939, a paper which demonstrated the presence of interstellar hydrogen on a large scale, at a time when most astronomers believed space to be devoid of interstellar gas.

In 1953, Lyttleton published a book on cometary forma-tion and evolution, based upon the accretion theory. In the same year he published an important monograph on the stability of rotating liquid masses.

Lyttleton also made important contributions to geophysics. He postulated that the Earth's liquid core was produced by a phase-change resulting from the combined effects of intense pressure and temperature. This is of great significance in the determination of the rate of change of Earth's volume, and would be of considerable relevance to the mechanics of mountain-formation. Lyttleton also stressed the hydrodynamic significance of the liquid core in the processes of precession and nutation.

Lyttleton's other interests include celestial mechanics, and the electrostatic theory of the expanding universe, which he proposed in 1959 with Hermann Bondi (1919–).

M

McAdam John Loudon 1756–1836 was a British civil engineer whose system of road-building – and particularly, surfacing – made a major contribution to the improvement of road transport during the nineteenth century. Indeed, his methods continued to be used after the development of motorized road vehicles, well into the twentieth century.

McAdam was born in Ayr, Scotland, on 21 September 1756. At the age of 14 he emigrated to the United States, where he settled in New York City. He went to work for his uncle, and eventually became a successful businessman in his own right. McAdam returned to Ayrshire in 1783, and 15 years later moved to Falmouth in Devon. He was appointed paving commissioner in Bristol in 1806 and ten years later he became Surveyor-General of the roads in that region, and of all the roads in Britain in 1827. He died on 26 November 1836.

The success of McAdam's technique came not so much from making wide, straight roads but from the excellent road surfaces, the method that was to become known as 'macadamizing'. He also drew up basic rules for highway management. Before McAdam's time there was some organization of roads repairs in Britain, but it seldom worked well. Each parish was responsible for its own repairs, but the parishes were small (and sometimes poor) and often not particularly interested in maintaining roads largely for the use of travellers from outside the parish. Also there was little skill and experience among either the labourers or the supervisors.

Towards the end of the eighteenth century the usual method of making a road was to plough the area, smooth the surface, and put down loose sand, gravel or pebbles. McAdam raised the road above the surrounding terrain, compounding a surface of small stones bound with gravel on a firm base of large stones. A camber, making the road slightly convex in section, ensured that rainwater rapidly drained off the road and did not penetrate the foundation. By the end of the nineteenth century, most of the main roads in Europe were using this method.

McAdam was also responsible for reforms in road administration which were as important as his innovatory road surfacings. He encouraged people to enter the profession of road-making by offering them suitable salaries, thus ensuring that more able and intelligent workers would be attracted to the task. He was in a better position to wield such influence after he was appointed Surveyor-General of metropolitan roads in 1827.

It was, in fact, many years before this that McAdam began to take an interest in road construction – indeed it became almost an obsession with him. In the ten years from 1810, he had spent his fortune on his 'hobby' and he petitioned Parliament for reward. The roads he had made in Bristol and other areas, and the three books he had written on road-making, had ensured that his name was very well known. At the time of his appointment as Surveyor-General he was also granted £10,000 by a grateful government.

Prior to the ultimate job in highways maintenance, McAdam became (in 1816), Surveyor to the Bristol Turnpike Trust; he remade the roads there and his advice was widely sought. Further to his belief in the importance of good road administration, he recommended the strengthening of turnpike trusts – but under the control of Parliament – and was advisor to many of these trusts. Turnpike roads were introduced in the late seventeenth century to provide better roads, but the costs were transferred to the road-users. McAdam ensured that public roads became the responsibility of the Government, financed (out of taxes) for the benefit of everyone.

MacArthur Robert Helmer 1930–1972 was a Canadian-born US ecologist who did much to change ecology from a descriptive discipline to a quantitative, predictive science.

MacArthur was born on 7 April 1930 in Toronto, Canada. He studied mathematics at Marlboro College, Vermont, graduating in 1951, and at Brown University, Providence, from which he gained his master's degree in 1953. He then went to Yale University to do his doctorate in mathematics but changed to zoology at the end of his second year and began to study ecology under G. Evelyn Hutchinson, one of the leading US ecologists of the time. MacArthur's studies were interrupted by a two-year period of military service, after which he returned to Yale

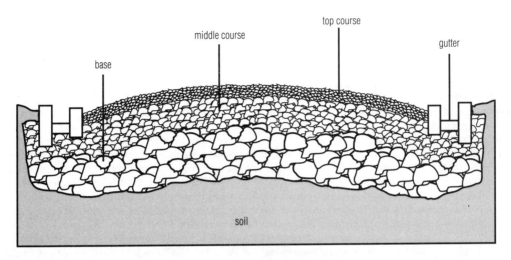

McAdam *In McAdam's method of building roads, a base course of large stones was laid on a compacted, cambered soil footing. A middle course of smaller stones was capped by a top course of small ones or gravel which traffic made into a fairly smooth, (but still cambered) surface. Gutters at each side carried away rainwater.*

to complete his doctoral dissertation, which won him the Mercer Award for the best ecology paper of 1957–58. He then joined the staff of the University of Pennsylvania, first as Assistant Professor of Zoology (from 1958 to 1961), then as Associate Professor of Zoology (1961 to 1964). In 1965, he was appointed Professor of Biology at Princeton University, where he remained until he died of cancer on 1 November 1972, only 42 years old.

In his doctoral thesis, MacArthur studied the relationship between five species of warbler that coexist in the New England forests; these species (now known as MacArthur's warblers) are ecologically very similar and it was suspected that they violated the competitive exclusion principle. MacArthur discovered, however, that there are subtle differences in the foraging strategies used by each species.

During the remainder of his life, MacArthur devoted himself to investigating population biology. He examined how the diversity and relative abundance of species fluctuate over time and how species evolve – particularly the evolution of communities – and the strategies that coexisting species evolve under the pressures of competition and natural selection. In these studies he attempted to interrelate several important factors: the structure of the environment, the morphology of the species, the economics of species' behaviour and the dynamics of population change. But perhaps his most important contribution was to quantify some of the many factors involved in the ecological relationships between species. For example, complex habitats such as forests support more species of birds than do grasslands, but it was only after MacArthur had devised his index of vegetational complexity (called foliage height diversity) in 1961 that it became possible to translate the observation about bird species' diversity into a definite equation whereby habitat structures can be compared and their bird species' diversity predicted.

McBain James William 1882–1953 was a Canadian physical chemist whose main researches were concerned with colloidal solutions, particularly soap solutions.

McBain was born in Chatham, New Brunswick, and entered the University of Toronto at the age of 17, graduating in chemistry and mineralogy. He spent the winter of 1904–1905 at the University of Leipzig, which at that time was at the height of its academic activity in physical chemistry. His first academic appointment was as a lecturer at the University of Bristol in 1906, a post he held until he became the first Leverhulme Professor of Chemistry there in 1919. In 1926, he went to the United States to become Professor of Chemistry at the University of Stanford, California; he became Emeritus Professor in 1947, six years before he died.

McBain's first research concerned the rate of oxidation of ferrous (iron (II)) salts on exposure to air. But he was soon attracted to colloid chemistry and a study of simple soaps. As early as 1910 he showed that aqueous solutions of soaps such as sodium palmitate are good electrolytic conductors. He discovered the interesting anomaly that there are maxima and minima in the conductivity versus concentration curves. In his next investigation he examined the degrees of hydrolysis for various concentrations of soap. He postulated the existence of highly mobile carriers of negative electricity with mobilities similar to that of the citrate ion. This led to the concept of the 'association ion' or 'ionic micelle', which has since proved invaluable in elucidating the properties of a large class of colloidal systems, including soaps, detergents and dyes.

McBain and his co-workers also developed new apparatus for their studies. They improved on the method of Northrop and Anson for determining the diffusion constants of substances in solution by passing them through discs of sintered glass. They developed a simple and elegant transparent air-driven ultra-centrifuge. McBain pointed out that in sedimentation of an equilibrium system such as that which exists in soap solutions above the micelle point, the rapidly moving micelles dissociate as they leave their normal environment of the equilibrium concentration of monomer, but as the material collects at the periphery of the centrifuge rotor, micelles naturally reform. To determine the thermodynamic properties of soap solutions he examined the possibility of using various methods: osmotic pressures, lowering of vapour pressure, and lowering of freezing point. Ultimately he developed his own method based on the lowering of the dew point.

In aqueous solution the micelle (association particle) encloses the non-polar or chain portions of the molecules, with the carboxyl or hydrophilic portions on the outside of the sphere. McBain pointed out that invert micelles should exist, with the polar heads clustered together at the centre, and these should be capable of dissolving substances that are usually insoluble. Such micelles in aqueous solutions should be able, for example, to dissolve hydrocarbons. He gave the name 'solubilization' to this phenomenon and spent much time and energy on experiments in this field. He also introduced the term cosolvency to describe the process of effecting solution by means of a mixture of liquid solvents.

McBain also questioned the idea that the surface phase in simple solutions was only one monolayer thick (as implied by the experimental work of Frederick Donnan and the theoretical proposals of Irving Langmuir). Above the micelle point there are several distinct species of potentially active materials (anions, undissociated molecules and micelles), all of which could accumulate at the surface. McBain proved that orientated underlayers do exist beneath the monolayers of soap, and he devised an ingenious apparatus for determining their actual composition. William Hardy came to a similar conclusion: that molecular orientation is not confined to a monolayer of liquid in contact with a solid surface, but that long-range forces give rise to cybotactic layers of considerable thickness (cybotactics are regions of 'order' of molecules within the structure of a liquid).

Another phenomenon studied by McBain was the adsorption of gases and vapours by solids. Various processes can take place simultaneously: physical sorption, chemisorption, and – by 'activating' the solid or forming it into a skeletonized structure – permeation of the solid by the gas or vapour. McBain introduced the generalized term sorption to include all such cases. He devised the McBain–Bakr spring balance, which provides a continuous record of the quantities and rate of sorption by direct weighing.

McClintock Barbara 1902–1992 was a US geneticist who was awarded a Nobel Prize for Physiology or Medicine in 1983. Very unusually the award was not shared with another scientist.

She was born on 16 June 1902 in Hartford, Connecticut but grew up in Brooklyn, New York. She developed an interest in science whilst still at school and went to Cornell University in 1919. She achieved her PhD in 1927, for

work on the genetics of maize. She moved from Cornell in 1931 to the California Institute of Technology and then to the University of Missouri but failed to achieve tenure at any of these institutes, and in 1941 moved to the Cold Spring Harbour Laboratory where she remained until formal retirement in 1967, although she continued to work in the laboratory almost until her death. Her early research was on the chromosomes of maize which she carefully analysed and described. Her work on the cytology of the chromosomes provided an important tool for her later analysis, using breeding experiments, of the genetics of the plant. She utilized X-rays to induce chromosomal abberations and rearrangements and examined the ways in which chromosomes repair such damage. This information from maize plants helped other scientists understand the problems of radiation sickness after the explosion of the atom bomb at Hiroshima, Japan. Further genetic and cytological analysis led McClintock to discover regions of chromosomes that could be transposed from one chromosome to another. These so called 'jumping genes' acted as regulators and were later discovered in bacteria and in fruit flies, which redirected attention to McClintock's work, which was then little known outside the world of plant geneticists.

McCollum Elmer Verner 1879–1967 was a US biochemist and nutritionist best known for his work on vitamins, and for originating the letter system of naming them.

McCollum was born on 3 March 1879 in Fort Scott, Kansas, the fourth child of an initially prosperous farming family. But his father became ill when McCollum was still very young and he had to take a succession of jobs while at school and university to pay for his education and to help to support his parents. McCollum studied at the University of Kansas, graduating in 1903, then gained a postgraduate scholarship to Yale University, from which he obtained his doctorate in 1906. In the following year he joined the faculty of the University of Wisconsin, where he remained for ten years. In 1917, he was appointed Professor of Biochemistry in the newly formed School of Hygiene and Public Health at Johns Hopkins University, a post he held until his retirement in 1944. McCollum died in Baltimore on 15 November 1967.

In his first years at Wisconsin University, McCollum worked on analysing the food and excreta of cattle. He soon decided, however, that it would be much easier to perform this work using a more convenient laboratory animal, and he chose the albino rat. This led to the establishment of the first rat colony for nutrition experiments, and it is largely through McCollum's efforts that albino rats have today become one of the most used animals for research. Investigating the nutritional requirements of his albino rats, McCollum discovered in the early 1910s that growth retardation results from a diet deficient in certain fats and that such deficiencies can be compensated for by providing a specific extract from either butter or eggs. He called this essential component 'fat-soluble A', because it dissolves in lipids. This was the start of the alphabetical naming system for vitamins. McCollum then showed that there is another essential dietary component that is not found in lipids but which is water soluble; he called this component 'water-soluble B'. Initially he thought that the two essential components were single compounds, but he later showed that they are in fact complexes.

McCollum continued his nutritional research at Johns Hopkins University, where he collaborated in the discovery of vitamin D. He also investigated the way in which sunlight prevents rickets, but it was not shown until later that the anti-rickets effect of sunlight is caused by the conversion of fats in the skin to vitamin D by the ultraviolet component of sunlight. In addition to his work on vitamins, McCollum also researched into the role of minerals in the diet.

McCormick Cyrus Hall 1809–1884 was a US inventor best known for developing the first successful mechanical reaper.

Son of the inventor Robert McCormick, Cyrus McCormick was born in Virginia on 15 February 1809. Under the guiding influence of his father, he was encouraged to use the talents of his own inventive mind, and in 1831 produced a hillside plough, the first of several agricultural implements with which he was associated. In the same year he invented the prototype for his reaping machine, and although it was another nine years before it was perfected, when the machine was put into production a ready market existed.

With the Midwest opening up there was a need for mechanization to cope with the huge acreages involved. McCormick was invited to Chicago to demonstrate his machine and, as a result, began manufacturing there in 1847. But in 1848, when his patent expired, he had to face strong competition and only his good business sense kept him from being overwhelmed by other manufacturers who had been waiting to encroach on his markets. He survived and prospered, introduced his reaping machine into Europe and opened up an entirely new market, winning several prizes (including one at the Great Exhibition of 1851).

An inventor first and foremost, he also had the intelligence to spread his interests in other directions, including mining, railways and newspaper publishing. He died in Chicago on 13 May 1884.

McCormick's reaping machine consisted of seven basic mechanisms which were principally the same as in its modern equivalent: the divider, reel, straight reciprocating knife, guards, platform, main wheel and gearing, and side-draft propulsion. The working was simple. A pulley at the side of the one road-driving wheel was connected by a band to another pulley above, turning the circular wooden frame. The blades of this were slightly twisted so that they gradually bent the corn down. Bevelled gears turned a small cranked shaft, which gave the movement to the cutting knife, working it backwards and forwards. The machine was designed purely as a reaper, and it left the corn lying flat, ready to be raked up and tied into sheaves. In the original machine the raker had to walk backwards while raking to avoid standing on the corn; in later models

a seat was incorporated into the machine so that the raker could sit.

It was estimated that a McCormick reaper operated by a two-person crew and drawn by a single horse could cut as much corn as could 12 to 16 people with reap hooks. In a nation like the United States, with a rapidly expanding population (from 13 to 60 million in 40 years) needing to be fed, it was essential that agriculture should be lifted from the manual to the mechanical wherever possible. McCormick's invention was the first of many which were to increase the production of cereals and guarantee enough food for all.

McCrea William Hunter 1904– is an Irish theoretical astrophysicist and mathematician whose main interest has been the evolution of galaxies and planetary systems. He has proposed several theories that have aroused considerable scientific speculation.

McCrea was born in Dublin on 13 December 1904. He was educated in England, first at Chesterfield Grammar School, Derbyshire, and then at Trinity College, Cambridge. Graduating in 1926, McCrea then travelled to the University of Göttingen as part of his postgraduate research programme, before returning to receive his PhD at Cambridge in 1929. He lectured in mathematics at Edinburgh University 1930–32, then moved to London to take up the position of Reader and Assistant Professor in the Department of Mathematics at Imperial College.

In 1936, McCrea became Professor of Mathematics at Queen's University, Belfast. Although he formally held this post until 1944, McCrea was a temporary Principal Experimental Officer with Operational Research of the Admiralty from 1943 to 1945. After World War II McCrea became Professor of Mathematics at the Royal Holloway College of the University of London, a post he held – with some visiting Professorships – until 1966, when he was appointed Research Professor of Theoretical Astronomy at the University of Sussex. He became Emeritus Professor there in 1972, but remained active and travelled widely. A Fellow of the Royal Society, and a recipient of the Gold Medal of the Royal Astronomical Society, McCrea wrote several influential books on physics.

In pursuit of his investigations into the evolution of galaxies, McCrea had studied the factors that would influence the earliest stages of this evolution, when 'protostars' condense out of the primordial gas cloud (formed predominantly of hydrogen, but with some helium) and then disintegrate to form globular clusters that in turn disperse as older (or 'population II') stars, located mainly in dust-free zones of space. He has particularly focused his attention on what might happen to this process if it encountered interstellar matter that was itself in a state of turbulence. Such an encounter would be critical in regulating the instability of the condensing material. McCrea has also analysed the effect of angular momentum, since the spinning of the interstellar matter would tend to counter the gravitational forces which promote condensation. Magnetic forces may also be of significance in this process. As a corollary of this research, McCrea has proposed a theory for the mechanism by which planets and other satellites may form.

Together with Edward Milne (1896–1950), McCrea was the founder of modern 'Newtonian' cosmology. Milne had devoted much effort to investigating alternatives to relativistic cosmology, and with McCrea he found that Newtonian dynamics could be advantageously applied to the analysis of the primordial gas cloud. The model relied on the assumption that the gas cloud would be 'very large' rather than of infinite size, although for the purposes of observation it would be 'infinite'.

McCrea has also contributed to discussion of Paul Dirac's 'large number hypothesis', which deals – among other things – with the ratio of the electrical and gravitational forces between an electron and a proton. This number comes to 10^{39}, which is – strikingly – the age of the universe in terms of atomic units. This may suggest some meaningful connection between the age of the universe (which is always increasing) and either the (hitherto presumed to be constant) electrical or gravitational forces – or it may be coincidental.

Other areas of interest to McCrea include the formation of molecules in interstellar matter (the formation of clouds of hydrogen in a gas cloud originally made up predominantly of monoatomic hydrogen), and the composition of stellar atmospheres. He has also investigated the fact that the emission of neutrinos from the Sun appears to be less than would be expected on the basis of predictions about the nuclear reactions taking place inside. This may simply mean, however, that thermal conditions in the Sun's interior are different from what has been so far deduced.

In physics the most notable of McCrea's contributions has been his work on forbidden (low-probability) transitions of electrons between energy states, analyses of penetration of potential barriers (for instance by 'tunnelling'), and his writings on relativity theory.

McCrea's work has covered many scientific fields, including mathematics, quantum mechanics, stellar astronomy and cosmology. He has made fundamental contributions to all of these fields and is an important figure in contemporary science.

Mach Ernst 1838–1916 was an Austrian physicist whose name was given to the Mach number, the velocity of a body in a medium relative to the speed of sound in that medium. Mach also made an important contribution to science in a fundamental reappraisal of scientific thought. He sought to understand knowledge in the context of the physiological, sensory and psychological processes that govern and limit its acquisition. This led him to question mechanical explanations of matter and the universe that could not be adequately observed, and to favour more conceptual or mathematical explanations. This approach to science had a profound influence on Albert Einstein (1879–1955) in the formulation of the general theory of relativity.

Mach was born in Turas, then in Austria–Hungary but now in the Czech Republic, on 18 February 1838. His family moved to Unter Siebenbrunn near Vienna in 1840. He was almost entirely educated at home by his parents until the age of 15, when he entered the local Gymnasium. He was most impressed by the study of natural science and, in particular, enjoyed the lessons of his teacher of natural history, a man named F. X. Wessely. Mach began his studies at the University of Vienna in 1855 and was awarded his PhD in 1860 for a thesis on electricity. While in Vienna he was influenced by Gustav Fechner (1801–1887), who worked on 'psychophysics' or the physiology of perception.

Mach lectured at the University, and also earned extra money by giving popular lectures on a variety of scientific topics. He published two books in 1863, one on physics for medical students and the other on psychophysics. He moved to Graz in 1864 where he became Professor of Mathematics and, in 1866, Professor of Physics. He then went to the University of Prague in 1867, where he was appointed Professor of Experimental Physics.

Mach remained in Prague for 28 years, conducting many research projects and publishing many books and lectures. He served as Rector to the University from 1882 to 1884. In 1895, he became Professor of History and Theory of Inductive Sciences at the University of Vienna. He suffered a stroke in 1897, which paralysed his right side and from which he made only a slow recovery. He retired from the University in 1901, and was appointed to the upper chamber of the Austrian parliament, a post he held for 12 years. In 1913, Mach moved to his son's home in Vaterstetten, near Munich, Germany. He continued to write books and died there on 19 February 1916.

Mach's early work in Vienna was in the department of Andreas von Ettingshausen, who had succeeded Johann Doppler (1803–1853) as Professor of Experimental Physics. Accordingly, Mach's research was aimed at investigating Doppler's then controversial law which described the relationship between the perceived frequency of sound and light and the motion of the observer relative to that of the source. Mach also investigated vibration and resonance. His interest in subjects which his colleagues saw as being only of peripheral interest, such as perception, was active even in these early days. His book on psychophysics, published in 1863, examined the complex physiological problems associated with vision and Mach concluded that the reductionist mechanistic approach had not given satisfactory explanations of phenomena.

At Graz, Mach was forced to fund his research out of his own pocket and, lacking equipment for investigations into physics, he continued to study subjects like vision, hearing and our sense of time. He investigated stimulation of the retinal field with spatial patterns and discovered a strange visual effect called Mach bands. This was subsequently forgotten, and was rediscovered in the 1950s.

In Prague, Mach was able to turn to issues of a more physical nature. He made a long theoretical study of mechanics and thermodynamics. He did not, however, abandon his other interests. Among his projects were a study of the kinesthetic sense and how it responds to body movements, more work on Mach bands, investigations of hearing and vision, various aspects of optics, and wave phenomena of mechanical, electrical and optical kinds. One of Mach's most important contributions arose from his work, published in 1887, on the photography of projectiles in flight, which showed the shock wave produced by the gas around the tip of the projectile. The Mach angle describes the angle between the direction of motion and the shock wave, and Mach found that it varies with the speed of the projectile, the flow of gas changing its character when the projectile reaches the speed of sound. This came to be very important in aerodynamics and particularly in supersonic flight, and in 1929 the term Mach number came into use to describe the ratio of the velocity of an object to the speed of sound in the medium in which the object is moving. An aircraft flying at Mach 2 is therefore flying at twice the speed of sound in air at that particular height.

Mach's philosophy of science – although he would never have described it in those terms, feeling himself to be a scientist and not a philosopher – included an investigation of modern science based on an analysis of the sequence of previous developments. He felt that the very order in which discoveries had been made altered their content and that this was an important factor to bear in mind when assessing their value and meaning. Mach was very suspicious of any model which could not be tested in at least an indirect fashion. Because they could not be observed, he was forced to reject the theory of atoms (which was in his day quite inaccurate, but became less crude towards the end of his life) as being merely a hypothesis which had got out of hand. He was considered quite eccentric by many of his colleagues for whom atomism was an exciting development. However, in that it subsequently became increasingly difficult to explain atoms as concrete objects, and theoretical physicists turned to mathematical and statistical treatments of the energies and positions of atomic particles to obtain more valid descriptions of the atom, Mach can be said to be vindicated in his approach and his views did in fact aid the development of quantum mechanics.

One of Mach's most important books was *Die Mechanik* which appeared in 1863. This gave rise to an enduring debate on Mach's principle, which states that a body could have no inertia in a universe devoid of all other mass as inertia depends on the reciprocal interaction of bodies, however distant. This principle influenced Einstein who tried to find a mathematical formulation to describe it, playing a role in Einstein's thinking which culminated in his explanation of gravity in the general theory of relativity.

Mach was greatly displeased to find that he was being hailed as a predecessor of relativity, a model which he rejected. He intended to write a book criticizing Einstein's theory, but died before this was possible.

Mach postulated that all knowledge is mediated by perception, and believed that the greatest scientific advances would only arise through a deeper understanding of this process. His stubborn scepticism forced him into a somewhat isolated position, but he was subsequently respected by scientists as great as Einstein. Mach is remembered as a scientist skilled not only at experimental design, execution and interpretation but also at thinking about the wider implications of his work – a rare quality.

Maclaurin Colin 1698–1746 was a Scottish mathematician who first presented the correct theory for distinguishing between the maximum and minimum values of a function and who played a leading part in establishing the hegemony of the Newtonian calculus in eighteenth-century Britain.

He was born at Kilmoden in February 1698. His father died when he was six weeks old and his mother before he was ten, and he was raised by his uncle, the incumbent in the parish of Kilfinnan, in Argyllshire. In 1709, he entered the University of Glasgow to study divinity, but within a year – influenced largely by Robert Simson, the Professor of Mathematics – he abandoned divinity for mathematics. In 1713, he presented a paper entitled 'On the Power of Gravity' and received his MA. Four years later, aged 19, he was appointed Professor of Mathematics at the Marischal College of Aberdeen.

On a visit to London during the vacation in 1719 Maclaurin met Isaac Newton and was elected to the Royal Academy. He visited Newton again in 1721 and a year later left Aberdeen, without official leave, to become travelling tutor to the eldest son of the English diplomat, Lord Polwarth. The next three years were spent chiefly in France and Maclaurin's paper on 'The Percussion of Bodies' (later included in substance in his treatise on fluxions) won him a prize from the French Academy of Sciences. When his pupil died in the autumn of 1724 Maclaurin returned to Scotland, but having forfeited the goodwill of his colleagues at Aberdeen, moved to Edinburgh. Thanks to the kindness of Newton, who wrote a letter on his behalf and promised to provide £20 annually towards the stipend, he was appointed Deputy Professor of Mathematics at the university there. A few months later he succeeded to the Chair.

Although Maclaurin's chief work for the next 20 years was on fluxions, he was a popular lecturer on many subjects and won the admiration of Edinburgh society for his public lectures and demonstrations in experimental physics and astronomy. He also wrote a paper on the gravitational theory of tides which won a prize in 1740 from the French Academy. In 1745, during the Jacobite rebellion, he organized the defence of Edinburgh, and the exertions and exposure to cold which the effort entailed ruined his health. He died of oedema at Edinburgh on 14 June 1746.

Maclaurin's reputation rests on his great *Treatise of Fluxions*, published in 1742. It was an attempt to prove Newton's doctrine of prime and ultimate ratios and to provide a geometrical framework to support Newton's fluxional calculus. Indeed, so highly praised and so influential was the treatise, that it may be accounted, one of the major contributions to the ascendancy of Newtonian mathematics which cut off Britain from developments on the continent and left it to the continental mathematicians, for the next three generations, to make the running in the establishment of those methods of analysis which are the foundation of modern mathematical analysis.

The treatise did also include, however, important solutions in geometry and in the theory of attractions. It contained Maclaurin's development of his paper on the gravitational influence on tides, and proved that an ellipsoid of evolution is formed whenever a homogeneous fluid mass under the action of gravity revolves uniformly about an axis. The treatise also included the method of defining the maximum and minimum points on a curve.

The specific theorem for which Maclaurin is most famous, and which is named after him, is a special case of the Taylor theorem. It can be used to find the series for functions such as $\log_e (1 + x)$, e^x, $\sin x$, $\cos x$, $\tan x$, etc., where e is the base of natural logarithms. The theorem, however, is of minor significance in the history of mathematics. Maclaurin's real distinction lies in being the first person to publish a logical and systematic exposition of the methods and principles of Newtonian mathematics.

McNaught William 1813–1881 was a Scottish mechanical engineer who invented the compound steam engine. This type of engine extracts the maximum energy from the hot steam by effectively using it twice – once in a high-pressure cylinder (or cylinders) and then, when exhausted from this, in a second low-pressure cylinder.

The evolution of the steam engine had been taking place for nearly 100 years when McNaught developed his method of compounding at his Glasgow works in 1845. The technique became known as 'McNaughting', and was to prove to be a valuable energy-saving system for many years to come. The firm of J. and W. McNaught was eventually established in 1860, and in 1862 acquired premises at St George's Foundry, Rochdale, Lancashire, where they set about manufacturing small steam engines. Later the business branched out into the construction of larger types and moved to a bigger building, where it remained in operation until 1914.

In all the McNaughts were responsible for the making of some 95 engines, many of which were for used in the northern textile industry. At least two of the engines have been preserved as fine examples of their type, one at the Glasgow Kelvingrove Museum and the other at the Bradford Industrial Museum.

The forerunner of the steam engine was invented by Thomas Savery, an English military engineer, in 1698. The idea was taken up in 1712 by Thomas Newcomen, a blacksmith, and became known as an atmospheric engine. But the first real 'steam engine' is credited to James Watt, a Scottish instrument maker, in 1765. Watt unfortunately met with various technical problems after his initial success and, faced by lack of resources, was about to give up when he met Matthew Boulton. Together they formed the partnership of Boulton and Watt and were concerned with the making of more than 500 engines which supplied power to all manner of equipment during the crucial years of the Industrial Revolution.

Many steam engines were used to power the textile mills of the cotton and wool trade in the north of England. By 1845, the majority of these mills found that even by the introduction of higher steam pressures, their engines (most of which were the Boulton and Watt design) could no longer keep up with the increasing demand for more power. This meant that the mill owners faced the costly business of replacing the engines, but naturally they were reluctant to do so. It was at this point that McNaught supplied a very acceptable answer, by offering a conversion to compound action.

His conversion consisted of a high-pressure cylinder exhausting into the original low-pressure cylinder. The new cylinder was connected to the opposite end of the 'beam', halfway between its pivot and the connecting-rod end, making it necessary for the stroke to be shorter than on the original cylinder. Three types of new engine were to emerge using McNaught's principle. The first used cylinders mounted side by side, the second had cylinders in a line (a tandem compound), and the third and rarer type had the high-pressure cylinder enclosed by the low-pressure one. As a measure of the success of the McNaught's design, the last beam engine made for use in a mill in 1904 was partly based on his principles. It can be said, therefore, that he was responsible for a vast and enduring improvement in the performance of the factory 'steam giants'.

Magendie François 1783–1855 was a French doctor who pioneered modern experimental physiology.

Born in Bordeaux, Magendie graduated in medicine in Paris in 1808, subsequently practising and teaching medicine in Paris, and becoming physician to the Hôtel Dieu. Elected a member of the Académie des Sciences in 1821, he became its president in 1837. In 1831, he was appointed Professor of Anatomy at the Collège de France.

He was a distinguished pioneer of scientific pharmacology. Using extensive vivisection and a certain amount of self-experimentation, he conducted trials on plant poisons, deploying animals to track precise physiological effects. He demonstrated that the stomach's role in vomiting was essentially passive, and analysed emetics. Through such researches, he helped to introduce into medicine the range of plant-derived compounds now known as alkaloids, many of which possess outstanding pharmacological properties. Magendie demonstrated the medicinal uses of strychnine (derived from the Indian vomit nut), morphine and codeine (derived from opium), and quinine (obtained from cinchona bark).

Magendie's studies were remarkably comprehensive. He investigated the role of proteins in human diet; he was interested in olfaction; and he studied the white blood cells. He worked protractedly on the nerves of the skull – a canal leading from the fourth ventricle is now known as the 'foramen of Magendie'. His demonstration of the separate pathways of the spinal nerves extended the findings of Pierre Flourens and Charles Bell. His numerous works included the *Elements of Physiology* 1816–17. Magendie's style of investigation avoided speculation and stuck to data. He has a good claim to be called the founder of experimental physiology.

Maiman Theodore Harold 1927–   is a US physicist who is best known for constructing (in 1960) the first working laser.

Maiman was born in Los Angeles on 11 July 1927, the son of an electrical engineer. After a period of military service (1945 to 1946) in the United States Navy, he studied engineering and physics at Columbia University, obtaining his BS degree in 1949. He then moved to Stanford University to do postgraduate work, gaining his MS in electrical engineering in 1951 and his PhD in physics in 1955. From 1955 to 1961 – the period in which he built the first laser – he worked at the Hughes Research Laboratories. In 1962 he founded his first company, the Korad Corporation (of which he was the President until 1968), which became a leading developer and manufacturer of lasers. In 1968, he founded Maiman Associates, a laser and optics consultancy, and four years later he cofounded the Laser Video Corporation, of which he was Vice President until 1975. In that year he joined the TRW Electronics Company, Los Angeles, as Director of Advanced Technology.

On moving to the Hughes Research Laboratories in 1955, Maiman began working on the maser, the first working model of which had been built in 1953 by Charles Townes. Maiman made a number of design improvements that increased the practicability of the solid-state maser, then set out to develop an optical maser, or laser (an acronym for *l*ight *a*mplification by *s*timulated *e*mission of *r*adiation). Although Townes and A. L. Schawlow published a paper in which they demonstrated the theoretical possibility of constructing a laser and started to build one themselves, it was Maiman who actually constructed the first working laser in 1960. His laser consisted of a cylindrical, synthetic ruby crystal with parallel, mirror-coated ends, the mirror coating at one end being semitransparent to allow the emission of the laser beam. A 'flash lamp' provided a burst of intense white light, part of which was absorbed by the chromium atoms in the ruby which, as a result, were stimulated to emit noncoherent red light. This red light was then reflected back and forth by the mirrored ends of the ruby until eventually some of the light emerged through the semitransparent end as an intense beam of coherent, monochromatic red light – laser light. Maiman's apparatus produced pulses of laser light; the first continuous-beam laser was made in 1961 by Ali Javan at the Bell Telephone Laboratories.

Since its original development, numerous improvements have been made to the laser and, using materials other than synthetic ruby (gases, for example), it is now possible to generate laser light of almost any frequency; even tunable lasers have been constructed. Moreover, the laser has found many practical applications – as a 'light scalpel' in microsurgery, in astronomy, spectroscopy, chemistry and physics, in holography and in communications, to give but a few examples.

Malpighi Marcello 1628–1694 was an Italian physician who discovered, among other things, blood capillaries, and pioneered the use of the microscope in the study of tissues.

Malpighi was born in Crevalcore, Italy, on 10 March 1628. He attended the University of Bologna from 1646 to 1653 and graduated as doctor of medicine and philosophy. He first lectured in Logic at Bologna, and then accepted the Chair in Theoretical Medicine at the University of Pisa in 1656. There he met and befriended the mathematician Giovanni Borelli. Malpighi found that the climate in Pisa did not suit his health and he returned to Bologna after three years, to lecture in theoretical and practical medicine. In 1662, he took up the offer of the Chair in Medicine at the University of Messina, but four years later was back in Bologna. In 1667, the Royal Society invited him to submit his research findings to them and made him an honorary member – the first Italian to be thus elected – and also supervised the printing of his later works. In 1691, Malpighi moved to Rome and retired there as chief physician to Pope Innocent XII. He died in Rome on 30 November 1694.

In Malpighi's time the microscope was a new invention and he became absorbed in using it to study animal and insect tissue, as did Anton van Leeuwenhoek. One of Malpighi's early investigations, in 1661, concerned the lungs of a frog. These organs were previously thought to have been fleshy structures, but Malpighi found them to consist of thin membranes containing fine blood vessels covering vast numbers of small air sacs. This discovery made it easier to explain how air (oxygen) seeps from the lungs to the blood vessels and is carried around the body. Malpighi traced the network of capillaries and found that they provide the means of blood travelling from the small arteries to the small veins. These findings filled the gap in the theory of blood circulation proposed by William Harvey.

Malpighi also investigated the anatomy of insects and found the tracheae, the branching tubes that open to the outside in the abdomen and supply the insect with oxygen for respiration. Turning to the dissection of plants, Malpighi found what he took to be tracheae in the stem – long tubes with rings of thickening. In fact he was looking at young vessels in the xylem. He also discovered the stomata in leaves but had no idea of their function.

Malpighi included the various structures and organs of the human body in his examinations. He indentified the sensory receptors (papillae) of the tongue, which he thought could be nerve endings. He also investigated the spinal cord and nerves and found them to be composed of the same fibres, but did not put forward a correct theory of their function. He proved that bile was uniform in colour, not yellow and black (as had been believed), and also indentified the urinary tubules in the kidney.

Chick embryos also fascinated Malpighi and in his microscope studies he recorded their neural folds and neural tube, the aortic arches, the optic vesicles and feather follicles.

Malpighi was a pioneer in the field of microscopy, and studied such a wide range of material that the curiosity of many scientists was aroused. Their combined efforts laid the foundations for further studies in a number of directions including histology, embryology, and the anatomy of organisms until then too small to observe.

Malthus Thomas Robert 1766–1834 was a British economist who made the first serious study of human population trends, although his views of the future of the human race enraged many thinkers of his day.

Malthus was born near Dorking, Surrey, on or about 14 February 1766. He went to Cambridge University, was ordained in 1788 and in 1796 became a curate in Albury, Surrey. In 1805 he was invited to accept a professorial post at Haileybury College, where he became Professor of History and Political Economy. During this time Malthus produced his major books on economics, *An Inquiry into the Nature and Progress of Rent* (1815) and *Principles of Political Economy* (1820). His work was acknowledged by many foreign academies, and he became a Fellow of the Royal Society in 1819. He died near Bath, Somerset (now Avon), on 23 December 1834.

Malthus's most controversial work was *An Essay on the Principle of Population*, which was published anonymously in 1798. In it he set out his reasons for believing that the human population of the world will increase at such a rate that it will eventually outstrip the Earth's resources. He postulated, that the situation will ultimately become resolved as the numbers are whittled away by starvation and disease, or war. In a later revision of the work Malthus conceded that moral constraints on sexual intercourse and marriage could stabilize population growth.

Malthus's theories brought a storm of protest at the time, which only slightly abated with publication of the revised work. Now it is accepted that Malthus was probably correct up to a point. Populations throughout the plant and animal kingdoms also tend to increase faster than the resources to support them until some check builds up sufficiently. After reading Malthus's work, Charles Darwin postulated that the transformation or extinction of species depends on their response (a function of variability) to changing environmental factors.

Malus Etienne Louis 1775–1812 was a French physicist who discovered polarized light. He found that light is not only polarized on passing through a double-refracting crystal but also on reflection from a surface.

Malus was born in Paris on 23 June 1775. He received private instruction in the classics and mathematics until the age of 18. After a year in military service, he became one of the first students to enter the Ecole Polytechnique in Paris. He studied there for only two years without receiving any diplomas, and then became a sublieutenant in the engineers' corps in 1796. Malus was sent on Napoleon's campaign in Egypt and Syria in 1798, but contracted a serious infection and was sent home to Marseilles in 1801. He then undertook postings to Lille in 1802, Anvers in 1804, and Strasbourg in 1806. During this period, Malus was able to resume his scientific activities. He became an examiner for the Ecole Polytechnique in geometry, analysis and physics, which caused him to make frequent visits to Paris. This allowed him to renew his contacts with many other scientists.

Malus was recalled to Paris permanently in 1808, and in 1810 was elevated to the rank of major. His scientific career reached its peak in the same year with the award of a prize from the French Institute and his election to its first class. In addition Malus was awarded the Rumford Medal by the Royal Society in 1811. However, his health had been seriously undermined by the illness he suffered abroad and he died of tuberculosis in Paris on 23 February 1812 at the early age of 37.

Malus's research in the field of optics may be traced as far back as his period in Egypt under Napoleon, when he found the time to work on the nature of light. His first important publication appeared in 1807. It was entitled *Traité d'Optique*, and was notable for its mathematical style. In it he put forward the equation which was later known as the Malus theorem, which described some of the differences in the properties of light following a second reflection or refraction as distinct from its properties following the first reflection or refraction. This theorem was given a full mathematical proof by William Rowan Hamilton (1805–1865) in 1824, who also extended the work.

Malus began doing experiments on double refraction in 1807. This phenomenon, which was first found in 1669 by Erasmus Bartholin (1625–1698), causes a light beam to split in two on passing through Iceland spar and certain other crystals. The French Institute offered a prize in 1808 for an experimental and theoretical explanation of double refraction, and this encouraged Malus to continue his work in this area. Christiaan Huygens (1629–1695) had found empirical laws to describe double refraction, but they were based on the assumption that light is wave-like in character, a theory that was not popular with those scientists who held with Newton's belief in its particulate nature. However, Malus's results confirmed Huygens's laws, and in 1810 he was awarded the French Institute's prize for his experiments and also for his theoretical deduction of the laws. The method he used in his analysis was the same as that which Pierre Laplace (1749–1827) had applied to the same problem a year previously.

A more important result of Malus's investigations was announced soon after its discovery in 1808. Malus held a piece of Iceland spar up to some light reflecting off the windows of the Luxembourg Palace in Paris. To his surprise, the light beam emanating from the crystal was single, not double. He further noted that of the two beams that normally emerge from the crystal, only one was reflected from a water surface if the crystal was held at a certain angle. The other passed into the water and was refracted. If the crystal was turned perpendicular, the second beam was reflected and the first refracted. The effect was most intense at an angle of reflection that varied from one medium to another. Malus concluded that light reflected from a surface behaves in the same way as the beams of light emerging after double refraction in a crystal. Although he had confirmed Huygens's laws of double refraction, Malus still believed that light consisted of particles which have sides or poles, and he thought that in the two refracted beams, the particles were lined up in planes perpendicular to each other. He described the light as being 'polarized' in perpendicular directions and thus concluded that light is also polarized on reflection from surfaces. He found too that the reflected ray is polarized in one plane while the refracted ray that passes into the surface is polarized in the perpendicular plane. Malus also discovered the law of polarization that relates the intensity of the polarized beam to the angle of reflection.

Malus's work on polarization was continued by David Brewster (1781–1868), Jean Biot (1774–1862) and François Arago (1786–1853) after his untimely death. His explanation of polarization was rationalized by Augustin Fresnel (1788–1827) who, in 1821, concluded that light exists as transverse waves.

Malus was a talented scientist who made a major discovery but whose research was curbed by his devotion to his military duties and by his poor health and early death.

Mandelbrot Benoit B. 1924– is a Polish-born French mathematician who coined the term 'fractal' to describe a curve or surface generated by the repeated subdivision of a mathematical pattern.

He was born in Warsaw, Poland, on 20 November 1924, and was educated at Ecole Polytechnique, Paris, graduating in 1947. He went on to the California Institute of technology, receiving an MS in 1948 and returned to Paris where he was awarded his PhD from the Sorbonne in 1952. Between 1949 and 1957 he was a staff member at the Centre National de la Recherche Scientifique, Paris, during which time he spent a year at the Institute of Advanced Study, New Jersey (1953–54) and two years at the University of Geneva as Assistant Professor of Mathematics (1955–57). Mandelbrot then went as junior Professor of Applied Mathematics to Lille University and of Mathematical Analysis at Ecole Polytechnique, Paris. In 1958 he was appointed a research staff member at IBM Thomas J. Watson Research Centre in New York and was made an IBM Fellow in 1974. In 1987, he became the Abraham Robinson Professor of Mathematical Science at Yale University, New Haven. He held positions on long-term leave as visiting professor at Harvard in 1962–64, 1979–80 and 1984–87. Mandelbrot's research has provided mathematical theories for erratic chance phenomena and self-similarity methods in probability. He has also carried out research on sporadic processes, thermodynamics, natural languages, astronomy, geomorphology, computer art and graphics, fractals and the fractal geometry of nature. His books include *Logique, Langage et Théorie de l'Information*, with L. Apostel and A. Morf (1957), *Fractals: Form, Chance and Dimension* (1977) and *The Fractal Geometry of Nature* (1982); he has also published numerous scientific papers, and is on the editorial boards of several journals. For his research, Mandelbrot has received honorary degrees from several universities and numerous awards and medals, including Chevalier, L'ordre de la Légion d'Honneur (1989).

Mannesman Reinhard 1856–1922 was a German iron-founder who invented a method of making seamless steel tubes.

Mannesman was born on 13 May 1856. Both he and his brother Max were ironmasters and, like other inventors in the second half of the nineteenth century, they increasingly turned their attentions from machine tools to heavy equipment for the production of metals. They perfected a way of making a seamless tube, which was much more accurate than the welded tube. The idea behind the invention had been conceived by their father in 1860, while he was working at Remscheid.

The Mannesman process, as it was later to be called, involved the passing of a furnace-heated bar between two rotating rolls. Because of their geometrical configuration, the rolls drew the bar forward and at the same time produced tensions in the hot metal that caused it to tear apart at the centre. A stationary, pointed mandrel caused the ingot of metal to open out and form a tube. The tubes were later forged to size – again on mandrels in rolling mills with grooved rolls that varied in diameter like cams. Between the compression phases, in which the tube is advanced by the rolls, the activating mechanism rotates and then partially returns the tubes. Thus the process is intermittent, taking place in a series of short steps. This later technique involved the tube stock being fed to pilger mills from the Mannesman machines; thus the process was called pilgering.

The first Mannesman plant and pilger mills were installed in Swansea, S Wales, in 1887 and operated by the Landore Siemens Steel Company. The plant consisted of six Mannesman machines, the largest being capable of piercing solid billets of up to 25 cm/10 in in diameter. In 1891 the process was granted a patent. It was this invention that most impressed the US inventor Thomas Edison during his visit to the World Exhibition in Chicago in 1893.

The Landore Siemens Steel Company remained the major manufacturer of seamless steel tubes until near the end of the nineteenth century. Then a plant designed and created at Youngstown, Ohio, took over as the most important manufacturer in the world.

Manson Patrick 1844–1922 has a good claim to be called the founder of modern tropical medicine.

Born at Old Meldrum, Aberdeenshire, the son of a bank manager, Manson qualified in medicine at Aberdeen in 1865. He travelled East, becoming medical officer in Formosa, and later moving to Hong Kong. He worked in the Far East for 23 years, amongst other things being active in the introduction of vaccination. Manson addressed himself to the problems of tropical diseases; after 1876 particularly studying filarial infection in humans. Having gained a clear idea of the life history of the invading parasite, he correctly conjectured that the disease was transmitted by an insect, a common brown mosquito, setting out his views of the crucial importance of the insect vector in *The Filaria Sanguinis Hominis and Certain New Forms of Parasitic Diseases* 1883. Manson went on to study other parasitic infections, for instance, the fluke parasite, ringworms, and guinea worm. But he is best remembered for his part in the discovery of the life-history of the malarial parasite – work begun in the Far East and continued after his return to Britain. By analogy with his work on filaria, he developed the thesis that malaria was also spread by a mosquito (1894). Manson and Ronald Ross between them developed the work that proved this (later they wrangled fiercely). It became axiomatic for Manson that the key to understanding and controlling many tropical diseases involved an arthropod vector that was indispensable as a host to the life cycle of the parasite.

After his return to Britain in 1892, Manson delivered at the Seaman's Hospital Society in London's dockland the first courses of lectures in tropical medicine given in England. It was largely through his initiatives, supported by Joseph Chamberlain at the Colonial Office, that the London School of Tropical Medicine was founded in 1899; he taught there till his retirement in 1914. He was knighted in 1903.

Marconi Guglielmo 1874–1937 was a physicist who saw the possiblity of using. radio waves – long wavelength electromagnetic radiation – for the transmission of information. He was the first to put such a service into operation a service on a commercial scale, and was responsible for many of the developments which have made radio and telegraph services into major industries.

Marconi was born in Bologna, Italy, into a wealthy family on 25 April 1874. His education consisted largely of private tutoring, although he was sent for a brief period to the Technical Institute in Livorno where he received instruction in physics. He studied under a number of prominent Italian professors but never enrolled for a university course.

His studies of radio transmission began on his father's estate in the 1890s. By 1897 he had established a commercial enterprise in London, based on developments from his early work. He became famous in 1901 when he succeeded in sending a transatlantic coded message.

In 1909, he was awarded, jointly with K. F. Braun, the Nobel Prize for Physics, and was honoured by the receipt of the Albert Medal from the Royal Society of Arts in 1929. He was also being made a Knight Grand Cross of the Royal Victorian Order, and was given the title of Marchese by the Italian Government. From 1921 he lived aboard his yacht *Elettra*, which served as a home, laboratory and receiving station. Marconi was given a state funeral after his death in Rome on 20 July 1937.

Marconi's researches began in 1894, the year of Heinrich Hertz's death, when he read a paper on the possible technical applications of the electromagnetic waves discovered by Hertz in 1886. Marconi realized that the waves could be used in signalling and began experiments with Professor Righi of the University of Bologna to determine how far the waves would travel.

Marconi based his apparatus on that used by Hertz, but used a coherer to detect the waves. (The coherer was designed to convert the radio waves into electric current.) Marconi improved Hertz's design by earthing the transmitter and receiver, and found that an insulated aerial enabled him to increase the distance of transmission. During 1895 he slowly increased the distance over which he was able to transmit a signal, first from the house into the garden and eventually to about 2.5 km/1.5 mi – the length of the family estate.

The Italian government was not interested in the device, so Marconi travelled to London, where he enlisted the help of relatives to enable him to obtain a patent and to introduce his discovery to the British government. He obtained his patent in June 1896 for the use of waves similar to those discovered by Hertz but of longer wavelength, for the purpose of wireless telegraphy. Marconi enabled Queen Victoria to send a message to the Prince of Wales aboard the Royal yacht, and increased his transmission distance to about 15 km/9 mi, then to 30 km/18 mi. The first commercial 'Marconigram' was sent from Lord Kelvin to Stokes.

The Wireless Telegraph Company was founded in London in 1897, and later became the Marconi Wireless Telegraph Co. Ltd in Chelmsford in 1900. In 1899, Marconi went to the United States and sent reports about the presidential election taking place there. On 12 December 1901, after many hold-ups, Marconi succeeded in sending a radio signal in Morse across the Atlantic Ocean from Pondhu in Cornwall to St Johns in Newfoundland, Canada.

Marconi became increasingly involved with the management of his companies from 1902, but he attracted many distinguished scientists to work with him. Some of the important developments were: the magnetic detector (1902), horizontal direction telegraphy (1905) and the continuous wave system (1912).

During World War I Marconi worked on the development of very short wavelength beams, which could be used for many purposes including enabling a pilot to fly an aircraft 'blind'. After the war these short wavelength beams contributed to communication over long distances. In 1932, Marconi discovered that he could detect microwave radiation, that is waves with very high frequencies. These wavelengths were soon to form the basis of radar.

Markov Andrei Andreevich 1856–1922 was a Russian mathematician, famous for his work on the probability calculus and for the 'Markov chains'.

Markov was born into the minor gentry class at Ryazan on 14 June 1856, a somewhat sickly child who was dependent upon crutches until the age of ten. He studied mathematics at the University of St Petersburg, where he was fortunate to have Pafnuty Tchebychev as a teacher, from 1874 to 1878. His BA dissertation of 1878, on the integration of differential equations by means of continued fractions (a special interest of Tchebychev), was awarded a gold medal. He began to tutor at the university in 1880 and four years later was awarded a doctorate for his thesis on continued fractions and the problem of moments.

For the next 25 years Markov continued his research and teaching at St Petersburg, where he became an extraordinary professor at the age of 30 and a full professor in 1893. At the same time he became involved in liberal political movements and perhaps only his academic eminence saved him from punitive government measures when he protested against the Tsar's refusal to accept Maxim Gorky's election to the St Petersburg Academy in 1902. He also refused to accept Tsarist decorations and in 1907 renounced his membership of the electorate when the government dissolved the fledgling representative *duma*, or parliament. In the harsh winter famine months of 1917 he worked, at his own request and without pay, teaching mathematics in a secondary school in Zaraisk, deep in the Russian interior. Shortly afterwards his health began to fail, and he died at Petrograd (St Petersburg) on 20 May 1922.

Markov's early work was devoted primarily to number theory – continued fractions, approximation theory, differential equations, integration in elementary equations – and to the problem of moments and probability theory. Throughout he used the method of continued fractions, most notably applying it to evaluate as precisely as possible the upper and lower boundaries of quantities such as quadratic forms.

In the latter 1890s he began to concentrate upon the probability calculus. During the middle years of the century there had been considerable development of the law of large numbers and of the central limit theorem invented by Pierre Laplace and Abraham de Moivre. But there were still no proofs with satisfactorily wide assumptions; nor had the limits of their applicability been discovered. Together, Tchebychev, Alexander Lyapunov (1857–1918) and Markov explored these problems far enough to bring about what amounted to the modernization of probability theory. Tchebychev's argument was extended by Markov in a paper of 1898 entitled 'The law of large numbers and the method of least squares' (written, in fact, as a series of letters to Alexander Vassilyev). Two years later he published his book, *Probability Calculus*, which was based on the method of moments. Markov's method was less flexible than Lyapunov's, who a year later published his proof of limit theorems using a method of characteristic functions; and although Markov attempted over the next eight years to establish the superiority of his method, and did succeed in proving Lyapunov's conclusions by means of it, the method of moments remains cumbersome, more complex and less general than the method of characteristic functions.

The chief fruit of Markov's endeavours to justify his method was the discovery of the important sequence of random variables named after him, the Markov chains. Put in informal language, a Markov chain may be described as a chance process which possesses a special property, so that its future may be predicted from the present state of affairs just as accurately as if the whole of its past history were known. Markov seems to have believed that the only real examples of his chains were to be found in literary texts, and he illustrated his discovery by calculating the alteration of vowels and consonants in Pushkin's *Eugeny Onegin*. Markov chains are now used, however, in the social sciences, in atomic physics, in quantum theory and in genetics. They have proved to be his most valuable contribution to twentieth-century thought.

Martin Archer John Porter 1910–   is a British biochemist who shared the 1952 Nobel Prize for Chemistry with his co-worker Richard Synge for their development of paper chromatography.

Martin was born in London on 1 March 1910, the son of a doctor of medicine. He was educated at Bedford School and then went to Cambridge University, where he graduated in 1932 and gained his PhD three years later, having been influenced by J. B. S. Haldane. He worked at the Dunn Nutritional Laboratories for two years, leaving in 1938 to join the Wool Industries Research Association at Leeds. In 1946 he became Head of the Biochemistry Division in the research department of the Boots Pure Drug Company in Nottingham, and he held this position until 1948 when he was appointed to the staff of the Medical Research Council. He worked first at the Lister Institute of Preventive Medicine and then at the National Institute for Medical Research, where he was Head of the Division of Physical Chemistry from 1952 to 1956 and Chemical Consultant from 1956 to 1959. He became Director of the Abbotsbury Laboratory from 1959 to 1970, Consultant to Wellcome Research Laboratories (1970–1973), at the University of Sussex (1973–1978) and in 1980 Invited Professor of Chemistry at the Ecole Polytechnique at Lausanne, Switzerland.

In his first researches, at the Dunn Nutritional Laboratory, Martin investigated problems relating to vitamin E. After 1938, at Leeds, he was involved in a study of the felting of wool. The work for which he was to become famous began in 1941, when he and Synge began the development of partition chromatography for separating the components of complex mixtures (of amino acids). In their method a drop of the solution to be analysed is placed at one end of a strip of filter paper and allowed to dry. That end of the strip is then immersed in a solvent which as it moves along the strip carries with it, at different rates, the various components of the mixture, which thus become separated and spread out along the strip of paper. Their positions are revealed by spraying the dried strip with a reagent that produces a colour change with the components; Martin and Synge used ninhydrin to record the positions of amino acids. The 'developed' strip is called a chromatogram.

The technique combines two different principles, adsorption chromatography (devised by the Russian botanist Mikhail Tswett in 1903 and later revived by Richard Willstätter) and counter-current solvent extraction for partitioning components between solvents – hence the name partition chromatography. It also has the advantage of requiring only a small sample of material and has proved to be a powerful tool in analytical chemistry, particularly for complex biochemicals.

In 1953, Martin and A. T. James began working on gas chromatography, which separates chemical vapours by differential adsorption on a porous solid. The versatility of both techniques has been extended by using radioactive tracers in the mixture to be analysed, when the positions of components in the resulting chromatogram can be found by using a counter.

Martin James 1893–1981 was the Ulster-born British aeronautical engineer whose pioneering work in the design and manufacture of ejection seats has saved the lives of thousands of military jet aircrew since 1949.

Martin was born on 11 September 1893 at Crossgar, County Down. The son of a farmer, he had designed, made and sold various machines while still a teenager. He scorned conventional education, but by practical work and study he had become an accomplished engineer before the age of 20.

In his early 20s he designed a three-wheeled enclosed car as a cheap runabout. Small oil engines and specialized vehicles were among his first products when he set up a one-man business in Acton, London. In 1929, Martin established the works at Denham, Middlesex that have served as a permanent home for his company until the present day. His first aircraft, a two-seater monoplane, which had the engine mounted amidships, driving a propeller through an extension shaft, did not reach completion.

In the early 1930s he designed and built a small, cheap, two-seater monoplane which used an ingenious construction – round-section thin-gauge steel tubing throughout. The design of this machine, Martin–Baker MB1, marked the start of the partnership between Martin and Captain Valentine Henry Baker, the company's chief test pilot. During the next ten years to 1944, Martin was to design three significant fighter aircraft, each of which could have been developed into an outstanding machine for the RAF but no orders were forthcoming.

Martin is remembered best for his devices that help to save the lives of aircrew. The first of these devices was a barrage-balloon cable cutter, designed before World War II, which was produced in large quantities and employed with great success by most RAF Bomber Command aircraft.

Martin's quick response to an urgent need during the Battle of Britain heralded his pioneering work on ejection seats. This was to make it possible to jettison the cockpit canopy of the Spitfire to improve the pilot's chances of escape by parachute. In 1943, increasing aircraft speeds, and the prospect of even greater speeds with the jet engine, led to considerable interest in means for improving the chances of escape for aircrew.

Martin's first response was demonstrated in model form in 1944. Using a powerful spring and a swinging arm to lift the pilot from his aircraft, this scheme generated sufficient official interest to enable Martin to pursue the much more elegant solution of the seat and its occupant being forceably ejected from the aircraft by means of an explosive charge.

The first successful dummy ejections took place in May 1945 and Martin developed his seat until it reached the level of sophistication, reliability and universal application that it has today. Most notable landmarks were the first live ejection on 24 July 1946, the first live ground-level ejection on 3 September 1955, ejection from an aircraft with rear-facing seats on 1 July 1960, and a zero speed, zero altitude ejection on 1 April 1961.

Martin was appointed a Commander of the British Empire in 1957 and made a Knight Bachelor in 1965. He continued to develop the ejection seat for use at higher speeds, greater altitudes, vertical take-off, multiple crew escape and underwater ejection.

When he died in January 1981 he was Managing Director and chief designer of the company he had founded. About 50,000 seats had been delivered and in February 1983, the 5,000th life was saved.

Martin's MB2 fighter of 1938 had a performance as good as that of contemporary fighters. It could be produced quickly and cheaply, because of the simplicity of its structure and easy assembly – features not shared by the Spitfire. The MB5, developed at the war's end, attained the highest standards in piston-engined fighter aircraft. Despite the praise it attracted from fighter pilots, it is thought to have been viewed officially as 'decadent', and was never put into production.

Martin's development of the ejection seat was pioneering because there was no previous knowledge of the effects on the human body of violent acceleration. He experimented first with sandbags and then human volunteers, firing them up test rigs. So successful and significant was this work that the decision was taken to install Martin–Baker seats in new British military jet aircraft as early as June 1947. The experience from their service pointed the way for the numerous improvements, such as automatic separation of pilot from seat and automatic parachute deployment.

In the mid 1950s, when most US carrier-borne aircraft were fitted with US ejection seats, ejection at take-off was usually fatal. So convinced was the US Navy that the Martin–Baker seat would increase the chances of pilot survival that, against great opposition from politicians and industrialists, it took Martin and a team to the USA in August 1957 to demonstrate conclusively the efficiency of the ground-level ejection system.

Martin–Baker seats were then fitted retrospectively to the whole inventory of US carrier-borne aircraft. Similarly, Martin–Baker was called in to fit its seat to existing NATO aircraft, including the F-86 Sabre, RF Thunderstreak and F-100 Super Sabre.

Like Rolls-Royce, Martin–Baker's is one of the few British companies to leave its mark on the US aerospace industry. At the time of his death it was reckoned that about 35,000 ejection seats were in service with the air forces and navies of 50 countries.

Maskelyne Nevil 1732–1811 was an influential British astronomer. He was the founder of the *Nautical Almanac* and he became Astronomer Royal at the age of 32.

Maskelyne was born in London on 6 October 1732. Educated at Westminster School, he studied divinity at Trinity

College, Cambridge, where he received his bachelor's degree in 1754. He was ordained a year later, but instead of taking up a living, he went to the Greenwich Observatory as an Assistant to James Bradley (1693–1762). He was awarded his master's degree by Trinity College in 1757 and elected a Fellow of the college. In the following year he was made a Fellow of the Royal Society, which sent him, with R. Waddington, to the island of St Helena to observe the 1761 transit of Venus. Four years later he was appointed Director of the Greenwich Observatory and (the fifth) Astronomer Royal, although he continued to carry out his clerical duties in such parishes as Shrawardine, Shropshire, and North Runcton, Norfolk. Awarded the Royal Society's Copley Medal in 1775, he received a doctorate in divinity from Cambridge University in 1777 and was elected to the French Academy of Sciences in 1802. He died in Greenwich on 9 February 1811.

It was probably his early interest in solar eclipses that led Maskelyne into his career as an astronomer. His first major project in observational astronomy was the excursion to St Helena, under the auspices of the Royal Society, in order to study the solar parallax during the 1761 transit of Venus and thereby determine accurately the distance of the Earth from the Sun. At the appropriate moment, however, the weather turned bad and, in any case, he had lost confidence in the instruments he had brought with him. (It was not until 1772 that Maskelyne perfected his technique for observing transits – by which time another transit of Venus had occurred, in 1769.) Nevertheless, on the sea journey to and from St Helena, he developed an interest in marine navigation by astronomical methods (an interest that was to colour most of what he later achieved), and spent a considerable amount of effort trying to devise a better means of determining longitude at sea.

Maskelyne's interests were perfectly represented by his foundation, in 1767, of the *Nautical Almanac*. This comprised a compendium of astronomical tables and navigational aids and included many of the results of Maskelyne's studies of the Sun, the Moon, the planets and the stars. His observations of the proper motions of several of the brighter stars were used by William Herschel (1738–1822) to demonstrate the movement of the Sun, which until 1783 had been presumed stationary.

Maskelyne's experiment on plumb-line deflection in 1774 aroused great interest among fellow geodeticists, although it was one of many attempts to determine the gravitational constant, solve Newton's gravitational equation, and thus deduce the density of the Earth. To measure the gravitational effect an isolated mountain might exert, he travelled to Schiehallion, a mountain in Perthshire, and determined the latitude both north and south of the mountain both by using a plumb-line and by direct survey. He found that the mass of the mountain between the two points of measurement caused the plumb-line to be deflected, so that the separation of the points was 27% greater than was found by direct geographical measurement. Making certain assumptions about the mass and volume of the mountain, Maskelyne deduced from the magnitude of the plumb-

line's deflection a gravitational constant and came to the conclusion that the Earth had a density of between 4.56 and 4.87 times that of water. This was reasonably close to the value now accepted (approximately 5.52 times that of water).

Maskelyne's work in astronomy contributed to a number of fields of study, but perhaps his most enduring contribution was the establishment of the *Nautical Almanac*.

Mästlin Michael 1550–1631 was a German scholar, author and teacher who was one of the first influential people to accept the theories of Copernicus (1473–1543), and to transmit them. One of Mästlin's pupils was Johannes Kepler (1571–1630).

Mästlin was born on 30 September 1550 in Goppengin, Germany. He attended the University of Tübingen and gained both a bachelor's and a master's degree there before joining a theological course. While still a student Mästlin compiled astronomical tables and wrote learned essays that were read by influential scientists of the age; he also began to put together the information for a popular textbook. His education completed, Mästlin became Assistant to Apian, the Professor of Mathematics at Tübingen. When Apian went on extended leave in 1575, Mästlin took over his duties. But the arrangement proved unsatisfactory and after only a year he became a local pastor instead. In 1580, however, he was appointed Professor of Mathematics at Heidelberg and when, four years later, Apian was dismissed from Tübingen for refusing to sign the oath of Protestant allegiance, Mästlin was reinstated there. The religious oath did not present Mästlin with a problem: never a fervent believer, his religious views were compatible enough with those required for professorship. He had already advised Protestant governments against accepting the Gregorian calendar because it seemed to him a papal scheme to regain power over lost territories. Consequently, all of Mästlin's books and writings appeared on the Index of Pope Sixtus V in 1590.

Mästlin taught at Tübingen for 47 years, during which period he was elected Rector of the College of Arts and Sciences no fewer than eight times. He remained at Tübingen until his death in 1631.

In 1573, Mästlin published an essay concerning the nova that had appeared the previous year. He had taken some care to establish the position of the nova, and its location in relation to known stars convinced him that the nova was a new star – which implied, contrary to traditional belief, that things could come into being in the spheres beyond the Moon. Mästlin's essay made ingenious use of relatively simple observations and it impressed his contemporary Tycho Brahe (1546–1601) enough for him to incorporate it into his own *Progymnasmata*.

Mästlin's great popular work, the *Epitome of Astronomy*, represented the fruits of his researches as a student. A general introduction to the subject for the layperson, it ran quickly through seven editions after its publication in 1582. Yet it propounded a severely traditional cosmology based on the ancient system of Aristotle (because this was easier to teach) despite the fact that Mästlin's own research had convinced him that Aristotle was wrong.

Later, Mästlin was explicitly to argue against the ideas of Aristotle, on the basis of his own observations, not only of the 1572 nova, but also of the 1577 and 1580 comets. Traditionally it was supposed that comets were merely meteorological phenomena, existing between the Moon and the Earth. Observation of the comet of 1577, however, showed no perceptible parallax: changes in observation position should have resulted in a particular and apparent displacement of the comet. Together with other observations, this led Mästlin to the conclusion that the comet was located beyond the Moon, probably in the sphere of Venus. These conclusions, in turn, seemed to Mästlin to be better explained on the basis of Copernicus's cosmology than on the basis of the one propounded by Ptolemy. Mästlin's subsequent expositions of the superiority of Copernican cosmology were delivered as lectures at Tübingen. Kepler attended these lectures as a young man, was deeply influenced by them, and, while Mästlin remained cautious in his acceptance, went on to embrace and develop the new cosmology quite fully. As a result the teacher–pupil relation between Mästlin and Kepler matured into a lifelong, affectionate friendship.

Masursky Harold 1922–   is a US geologist who has conducted research into the surface of the Moon and the other planets of the Solar System. From the early years of the US space programme, Masursky has been a senior member of the team at National Aeronautics and Space Administration (NASA) responsible for the surveying of lunar and planetary surfaces, particularly in regard to the choice of landing sites. He and his colleagues – it is a field in which teamwork is especially important – have participated from the very first in the Ranger, Apollo, Viking, Pioneer and Voyager programmes.

Masursky was born on 23 December 1922 at Fort Wayne, Indiana. He graduated from Yale University in 1943 and gained his master's degree in geology there in 1951. At the end of that year, he joined the US Geological Survey, working in its fuels branch in the search for petroleum. Eleven years later, still with the Survey, he transferred to the branch for 'astrogeological' studies and began work at NASA. Since then Masursky has received four medals from NASA for Exceptional Scientific Achievement, and in 1979 was made a member of the Space Science Advisory Committee of NASA's Advisory Council. An associate editor of a popular astronomical journal, he has also published more than 100 technical papers and edited books for NASA on the Apollo missions.

As early as 1964, Masursky was a member of the *Ranger 9* site selection programme and coordinated the *Ranger 8* and *9* Science Reports published a year later. His interest in the lunar surface led him, two years afterwards, to become a member of the Lunar Orbiter site selection working group and of the Apollo Group for Lunar Exploration Planning that monitored and guided the Moon landing. The results gained from *Apollo 8* and *10* about the chemical and

'geological' composition of the Moon were studied by the Lunar Science Working Group under Masursky's chairmanship. His work on *Apollo 14, 15, 16* and *17* was soon followed by involvement with teams exploring Mars and Venus.

Masursky participated in the Mariner Orbital (1971) and Viking Lander (1975) explorations of Mars. He led the team that selected and monitored observations of Mars made by the Mariner Orbital Craft, then selected landing sites on Mars for the Viking landing. Contrary to expectation, observations by Mariner and Viking showed the existence of craters and very high mountains (Nix Olympia rising 27 km/17 mi above the mean surface level of Mars, compared with Mount Everest's 9 km/6 mi). A more surprising discovery concerned thousands of small channels on the planet's surface. These were not the 'canals' observed by Giovanni Schiaparelli, but something new. They were mostly a few kilometres wide; themselves sinuous and twisting, they also had tributary systems and, in some photographs, were almost indistinguishable from orbital observations of rivers on Earth. According to Masursky, only the assumption that rainfall occurred on Mars can adequately account for the observed nature of these channels. Some of the tributaries are reckoned to be as young as a few hundred million years, suggesting further to Masursky that the process of their creation occurred repeatedly.

In 1978, Masursky joined the Venus Orbiter Imaging Radar Science Working Group. The surface of Venus, hidden from visual or televisual observation by its thick layer of cloud, was mapped on the basis of radar readings taken from Pioneer.

The pictures gained from Voyager's passage past Saturn provided data for which 'geological' interpretation has continued for a number of years.

Mauchly John William 1907–1980 was a US electronics engineer who, with John Eckert, became co-inventor of one of the first electronic computers (the ENIAC). He played an active role in the development of more advanced machines (EDVC, BINAC and UNIVAC) and in encouraging the appreciation of their enormous potential for government, military, scientific and business purposes.

Mauchly was born in Cincinnati, Ohio, on 30 August 1907. He attended Johns Hopkins University, entering as an engineering student but transferring into the physics department. He was awarded his PhD in physics in 1932 and went into teaching. He became Professor of Physics at Ursinus College in Collegeville, Pennsylvania. He attended a course in electronics at the Moore School of Electrical Engineering of the University of Pennsylvania in the summer of 1941. The faculty there requested him to stay on as an instructor. He learnt of the work being done at the Moore School under contract from the Ballistics Research Laboratory as part of the war effort. He submitted a memorandum in 1942 on the design of a computing machine, and the Moore School was awarded a contract for its construction. The project lasted from 1943 until 1946, with Mauchly as principal consultant and Eckert as chief engineer. A. W. Burks, T. K. Sharpless and R. Shaw were three of the many other scientists involved in the project.

The ENIAC (Electrical Numerical Integrator and Computer) was announced in 1946 and put into operational service a year later. A dispute over patent policy with the Moore School caused Mauchly and Eckert to leave the institute, although they had been in the middle of the development of the EDVAC (Electronic Discrete Variable Automatic Computer). Their departure greatly hindered the development of this machine, which did not make its debut until 1951. Mauchly and Eckert set up a partnership, which became incorporated in 1948 (the Eckart–Mauchly Computer Coroporation). An unfortunate accident in 1949 meant that they lost their financial backing, so in 1950 they sold the company to Remington Rand Inc, which merged with the Sperry Corporation in 1955. From 1950 unti 1959 Mauchly served as director of UNIVAC (Universal Automatic Computer) applications research. He left the Sperry Rand Corporation in 1959 in order to set up a consulting company called Mauchly Associates, and set up a second consulting organization (Dynatrend) in 1967. Mauchly returned to the Sperry Rand Corporation as a consultant in 1973.

The importance of Mauchly's work is so great that it is all the more surprising that he is not better known outside the ranks of his own profession. He received many awards and honours, but was not famous with the general public. He died during surgery for a heart ailment in Abington, Pennsylvania, on 8 January 1980.

Mauchly's early research, at Collegeville, was on meteorology. It involved many laborious calculations, so he was deeply interested in finding methods of speeding up the process. He realized the possibility of using an electronic apparatus, constructed with vacuum tubes (valves), and this was why he attended the course in 1941 at the Moore School; to improve his understanding of electronics.

The United States Army found that new battle conditions and new types of weapons required that its artillery range tables (which enabled the gunners to aim and fire effectively) needed to be recalculated. This was a mammoth task, and so Mauchly and the Army had a community of interest: both needed methods for rapid calculation. Mauchly's 1942 memorandum on computer design initiated the project which culminated in 1946 with the ENIAC.

The ENIAC was one of the first electronic computers. It could perform addition in 200 microseconds and multiplication in 300 microseconds. Although it was primarily designed for trajectory calculation, it could also be used for other purposes, including the solution of partial differential equations. Its drawbacks were that it was huge, and had vast power requirements and running costs. Its input and output consisted of punched cards; it had limited storage, and no memory.

Work on the EDVAC began in 1944. It had improved storage, John von Neumann helping with the design of the storage system.

The ENIAC was first tested in 1947 at the Aberdeen Proving Ground in Maryland, for ballistics purposes.

In 1947 Mauchly and Eckert obtained a contract from the Northrop Aircraft Company to design a small-scale binary computer. The BINAC (Binary Automatic Computer) was completed in 1949, and was more economical to use as well as being faster and more compact. A new feature was the replacement of punch cards with magnetic tape, and the computer's capacity to use internally stored programs.

The UNIVAC was designed during the early 1950s, and was first tried out in 1951 by the US Bureau of Census; it was designed to serve the business community. Mauchly's role in its development was primarily in the design of the software.

Maudslay Henry 1771–1831 was a British engineer and toolmaker who, in an age when mechanical engineering lagged behind other crafts, improved the metal-working lathe to the point that it could be employed for precise screw cutting. He also desiged a bench micrometer, which became the forerunner of the modern instrument.

Maudslay was born on 22 August 1771 at Woolwich, London, where his father was a joiner at the Royal Arsenal. He had little formal education and at the age of 12 went to work at the Arsenal filling cartridges, but after two years went first to the joiner's shop and then to the metalworking shop as an apprentice. By the time he was 18 his skill as a craftsman was renowned. He joined the firm of Joseph Bramah, the pioneer of hydraulics and the inventor of the Bramah press and lock. Maudslay eventually became manager of the workshop but after a disagreement over pay he left to start his own business.

In his works, just off Oxford Street, London, he developed a method of cutting screw threads on a lathe. Previously large threads had been forged and filed, and small threads had been cut by hand by the most skilled of craftsmen. Maudslay's new screw-cutting lathe gave such precision as to allow previously unknown interchangeability of nuts and bolts and standardization of screw threads. He was also able to produce sets of taps and dies.

Using his new device he cut a long screw with 50 threads per inch (about 20 per cm) and made this the basis of a micrometer, which came into daily use as an instrument to check the standard of the work he produced. In the period 1801–08, in conjunction with Marc Brunel, he constructed a series of machines for making wooden pulley blocks at Portsmouth dockyard. The A3 special-purpose power-driven machines, operated by ten unskilled workers, did the work formerly done by 110 skilled workers using hand methods.

Maudslay's firm, with him as its chief working craftsman, went on to produce marine steam engines. The first was a 17 hp/13 kW model: later, engines of 56 hp/42 kW were built. In 1838, after Maudslay's death, the company built the engines for the first successful transatlantic steamship, Isambard Brunel's *Great Western*, which developed 750 hp/560 kW. Early in 1831 Maudslay caught a chill on his return to Britain after a trip to France and he died on 15

February that year. He was buried in a cast-iron tomb at Woolwich.

Maury Antonia Caetana de Paiva Pereira 1866–1952 was a US expert in stellar spectroscopy who specialized in the detection of binary stars. She also formulated a classification system to categorize the appearance of spectral lines, a system that was later seen to relate to the appearance of the stars themselves.

Maury was born on 21 March 1866, into a family that had already produced several prominent scientists. Educated at Vassar, she became an Assistant at the Harvard College Observatory even before she graduated in 1887. Working under the direction of Edward Pickering (1846–1919) she rapidly mastered spectroscopy, to the extent that within four years she had devised her new classification scheme for spectral lines, and in 1896 published the results of her work, based on the examination of nearly 5,000 photographs and covering nearly 700 bright stars in the northern sky. For many years following the publication of her scheme, Maury lectured in astronomy in various US cities. She accepted private pupils and occasionally also took on teaching jobs. At 42 she returned to Harvard as a Research Associate to study the complex spectrum of Beta Lyrae. Following her retirement in 1935, Maury became Curator of the Draper Park Museum at Hasting on Hudson while continuing her study of Beta Lyrae. She died at Dobbs Ferry, New York, on 8 January 1952.

Maury's first spectroscopic work was to assist Pickering in establishing that the star Mizar (Zeta Ursae Majoris) was in fact a binary star, with two distinct spectra. That successfully accomplished, Maury was the first to calculate the 104-day period of this star. In 1889, she discovered a second such star, Beta Aurigae, and established that it had a period of only four days. During the next year she was engaged in a project studying the spectra of bright stars. Previously the great variety in types of star, as judged by their spectra, had been classified according to the mere absence or presence of the Fraunhofer lines. Maury now found this system inadequate for representing all the characteristics she observed: it was also possible to make classifications according to the appearance of the lines – the intensity, the distinctions and the line width, for example. Maury assumed the existence of three major divisions among spectra, depending upon the width and distinctness of the spectral lines. She defined class (a) as having normal lines; class (b) as having hazy or blurred lines; and class (c) as having exceptionally distinct lines. These divisions and their combinations represented in many ways a better classification of stellar spectra than both the system previously used, and the 'improved' version proposed – also at Harvard – by Annie Cannon (1863–1941) almost simultaneously.

Maury's system enabled Ejnar Hertzsprung (1873–1967), co-originator of the Hertzsprung–Russell diagram, to verify his discovery of two distinct varieties of star: dwarfs, which in Maury's scheme fall under (a) and (b), and giants (c).

Maxim Hiram Stevens, 1840–1916 was a US-born British inventor, chiefly remembered for the Maxim gun, the first fully automatic, rapid-firing machine gun.

Maxim was born on 5 February 1840 in Sangerville, Maine, the son of a farmer and wood-turner. He spent his early life in various apprenticeships before exhibiting his talent for invention at the age of 26 with a patent for a curling iron in 1866. More significant ideas followed, and in 1878 he was appointed chief engineer to the United States Electric Lighting Company. While working for this company he came up with a way of manufacturing carbon-coated filaments for the early light bulbs that ensured that each filament was evenly coated.

It was during a trip to the Paris Exhibition of 1881 that Maxim made the statement that was to change his life. He declared that work on armaments would prove the most profitable sector of invention. Taking his own advice, he left the United States to settle in Britain, where he thought the most opportunity for his work existed. He set up a small laboratory at Hatton Garden, London, and set to work on improving the design of current guns.

In 1884, he had produced the first fully automatic machine gun, which used the recoil from the shots to extract, eject, load and fire cartridges. With a water-cooled barrel, the Maxim gun used a 250-round ammunition belt to produce a rate of fire of ten rounds per second. Its efficiency was further improved by Maxim's development of a cord-like propellant explosive, cordite. His own company, set up on the unveiling of the Maxim gun, became absorbed into Vickers Limited. By 1889, the British Army had adopted the gun for use. It gave a decided advantage to the British in conflicts in Africa and Asia. By World War I both sides were armed with machine guns. Maxim lived just long enough to see the devastation caused. It was not until the appearance of tanks on the battlefront that the power of the Maxim weapon eclipsed significantly. Maxim became a naturalized British subject in 1900, was knighted by Queen Victoria in 1901, and later also became a Chevalier of the Legion of Honour of France.

Although chiefly remembered for the gun, Maxim is responsible for many other, lesser, inventions, having taken out well over 100 patents in both the United States and Britain for devices ranging from mousetraps to gas-powered engines. He was particularly interested in powered flight, his experiments being described in his book *Artificial and Natural Flight.* In 1889, he started investigations into the relative efficiencies of aerofoils and airscrews driven by steam-powered engines. Five years later he had produced a steam-driven machine whose engine produced 6 hp/4.5 kW for every mass of 1 lb/0.4 kg weight. After three trials the aircraft succeeded in leaving the tracks along which it ran. Although impressive, it was clear that the massive amount of feed-water needed for longer flights would add an impossible weight burden to the craft.

Maxim wrote an autobiography in 1915, a year before his death in Streatham, South London, on 24 November 1916. He was survived by his inventor son, Hiram Percy Maxim (1869–1936), who developed a 'silencer' for rifles.

Maxwell James Clerk 1831–1879 was a British physicist who discovered that light consists of electromagnetic waves and established the kinetic theory of gases. He also proved the nature of Saturn's rings and demonstrated the principles governing colour vision.

Maxwell was born at Edinburgh on 13 November 1831. He was educated at Edinburgh Academy from 1841 to 1847, when he entered the University of Edinburgh. He then went on to study at Cambridge in 1850, graduating in 1854. He became Professor of Natural Philosophy at Marischal College, Aberdeen, in 1856 and moved to London in 1860 to take up the post of Professor of Natural Philosophy and Astronomy at King's College. On the death of his father in 1865, Maxwell returned to his family home in Scotland and devoted himself to research. However, in 1871 he was persuaded to move to Cambridge, where he became the first Professor of Experimental Physics and set up the Cavendish Laboratory, which opened in 1874. Maxwell continued in this position until 1879, when he contracted cancer. He died at Cambridge on 5 November 1879, at the early age of 48.

Maxwell demonstrated his great analytical ability at the age of 15, when he discovered an original method for drawing a perfect oval. His first important contribution to science was made from 1849 onwards, when Maxwell applied himself to colour vision. He revived the three-colour theory of Thomas Young (1773–1829) and extended the work of Hermann Helmholtz (1821–1894) on colour vision. Maxwell showed how colours could be built up from mixtures of the primary colours red, green and blue, by spinning discs containing sectors of these colours in various sizes. In the 1850s, he refined this approach by inventing a colour box in which the three primary colours could be selected from the Sun's spectrum and combined together. Maxwell confirmed Young's theory that the eye has three kinds of receptors sensitive to the primary colours and showed that colour blindness is due to defects in the receptors. He also explained fully how the addition and subtraction of primary colours produces all other colours, and crowned this achievement in 1861 by producing the first colour photograph to use a three-colour process. This picture, the ancestor of all colour photography, printing and television, was taken of a tartan ribbon by using red, green and blue filters to photograph the tartan and to project a coloured image.

Maxwell worked on several areas of enquiry at the same time, and from 1855 to 1859 took up the problem of Saturn's rings. No one could give a satisfactory explanation for the rings that would result in a stable structure. Maxwell proved that a solid ring would collapse and a fluid ring would break up, but found that a ring composed of concentric circles of satellites could achieve stability, arriving at the correct conclusion that the rings are composed of many small bodies in orbit around Saturn.

Maxwell's development of the electromagnetic theory of light took many years. It began with the paper 'On Faraday's Lines Of Force' (1855–56), in which Maxwell built on the views of Michael Faraday (1791–1867) that electric and magnetic effects result from fields of lines of force that surround conductors and magnets. Maxwell drew an analogy between the behaviour of the lines of force and the flow of an incompressible liquid, thereby deriving equations that represented known electric and magnetic effects. The next step towards the electromagnetic theory took place with the publication of the paper 'On Physical Lines Of Force' (1861–62). Here Maxwell developed a model for the medium in which electric and magnetic effects could occur (which might throw some light on the nature of lines of force). He devised a hypothetical medium consisting of an incompressible fluid containing rotating vortices responding to magnetic intensity separated by cells responding to electric current.

By considering how the motion of the vortices and cells could produce magnetic and electric effects, Maxwell was successful in explaining all known effects of electromagnetism, showing that the lines of force must behave in a similar way. However, Maxwell went further, and considered what effects would be caused if the medium were elastic. It turned out that the movement of a charge would set up a disturbance in the medium, forming transverse waves that would be propagated through the medium. The velocity of these waves would be equal to the ratio of the value for a current when measured in electrostatic units and electromagnetic units. This had been determined by Friedrich Kohlrausch (1840–1910) and Wilhelm Weber (1804–1891), and it was equal to the velocity of light. Maxwell thus inferred that light consists of transverse waves in the same medium that causes electric and magnetic phenomena.

Maxwell was reinforced in this opinion by work undertaken to make basic definitions of electric and magnetic quantities in terms of mass, length and time. In *On The Elementary Regulations Of Electric Quantities* (1863), he found that the ratio of the two definitions of any quantity based on electric and magnetic forces is always equal to the velocity of light. He considered that light must consist of electromagnetic waves, but first needed to prove this by abandoning the vortex analogy and arriving at an explanation based purely on dynamic principles. This he achieved in *A Dynamical Theory Of The Electromagnetic Field* (1864), in which he developed the fundamental equations that describe the electromagnetic field. These showed that light is propagated in two waves, one magnetic and the other electric, which vibrate perpendicular to each other and to the direction of propagation. This was confirmed in Maxwell's *Note On The Electromagnetic Theory of Light* (1868), which used an electrical derivation of the theory instead of the dynamical formulation, and Maxwell's whole work on the subject was summed up in *Treatise On Electricity And Magnetism* in 1873.

The treatise also established that light has a radiation pressure, and suggested that a whole family of electromagnetic radiations must exist, of which light was only one. This was confirmed in 1888 with the sensational discovery of radio waves by Heinrich Hertz (1857–1894). Sadly, Maxwell did not live long enough to see this triumphant vindication of his work. He also did not live to see the ether (the medium in which light waves were said to be propagated) disproved with the classic experiments of Albert Michelson (1852–1931) and Edward Morley (1838–1923) in 1881 and 1887, which Maxwell himself suggested in the last year of his life. However, this did not discredit Maxwell as his equations and description of electromagnetic waves remain valid even though the waves require no medium.

Maxwell's other major contribution to physics was to provide a mathematical basis for the kinetic theory of gases. Here he built on the achievements of Rudolf Clausius (1822–1888), who in 1857–58 had shown that a gas must consist of molecules in constant motion colliding with each other and the walls of the container. Clausius developed the idea of the mean free path, which is the average distance that a molecule travels between collisions. As the molecules have a high velocity, the mean free path must be very small, otherwise gases would diffuse much faster than they do, and would have greater thermal conductivities.

Maxwell's development of the kinetic theory was stimulated by his success in the similar problem of Saturn's rings. It dates from 1860, when he used a statistical treatment to express the wide range of velocities that the molecules in a quantity of gas must inevitably possess. He arrived at a formula to express the distribution of velocity in gas molecules, relating it to temperature and thus finally showing that heat resides in the motion of molecules – a view that had been suspected for some time. Maxwell then applied it with some success to viscosity, diffusion and other properties of gases that depend on the nature of the motion of their molecules.

However, in 1865, Maxwell and his wife carried out exacting experiments to measure the viscosity of gases over a wide range of pressure and temperature. They found that the viscosity is independent of the pressure and that it is very nearly proportional to the absolute temperature. This later finding conflicted with the previous distribution law and Maxwell modified his conception of the kinetic theory by assuming that molecules do not undergo elastic collisions as had been thought but are subject to a repulsive force that varies inversely with the fifth power of the distance between them. This led to new equations that satisfied the viscosity–temperature relationship, as well as the laws of partial pressures and diffusion.

However, Maxwell's kinetic theory did not fully explain heat conduction, and it was modified by Ludwig Boltzmann (1844–1906) in 1868, resulting in the Maxwell–Boltzmann distribution law. Both men thereafter contributed to successive refinements of the kinetic theory and it proved fully applicable to all properties of gases. It also led Maxwell to an accurate estimate of the size of molecules and to a method of separating gases in a centrifuge. The kinetic theory, being a statistical derivation, also revised opinions on the validity of the second law of thermodynamics, which states that heat cannot of its own accord flow from a colder to a hotter body. In the case of two connected containers of gases at the same temperature, it is statistically possible for the molecules to diffuse so that the faster-moving molecules all concentrate in one container while the slower molecules gather in the other, making the first container hotter and the second colder. Maxwell conceived this hypothesis, which is known as Maxwell's demon. Even though this is very unlikely, it is not impossible and the second law can therefore be considered to be not absolute but only highly probable.

Maxwell is generally considered to be the greatest theoretical physicist of the eighteenth century, as his forebear Faraday was the greatest experimental physicist. His rigorous mathematical ability was combined with great insight to enable him to achieve brilliant syntheses of knowledge in the two most important areas of physics at that time. In building on Faraday's work to discover the electromagnetic nature of light, Maxwell not only explained electromagnetism, but paved the way for the discovery and application of the whole spectrum of electromagnetic radiation that has characterized modern physics. In developing the kinetic theory of gases, Maxwell gave the final proof that the nature of heat resides in the motion of molecules.

Maybach Wilhelm 1847–1929 was a German engineer and inventor who worked with Gottlieb Daimler on the development of early motorcars. He is particularly remembered for his invention of the float-feed carburettor which allowed petrol (gasoline) to be used as a fuel for internal combustion engines – most of which up to that time had been fuelled by gas.

From 1862, when Maybach was still a teenager, he was a great friend and associate of Daimler. In 1882, they went into partnership at Cannstatt, where they produced one of the first petrol engines. In 1895, Maybach became technical director of the Daimler Motor Company. While working with Daimler, in 1901, Maybach designed the first Mercedes car – named after Mercedes Jellinec, the daughter of their influencial associate, the Austro–Hungarian consul in Nice.

Maybach invented the spray-nozzle or float-feed carburettor in 1893. An adjusting screw controlled the rate of maximum fuel flow and a second screw varied the area of the choke. When the inlet valve was opened, air sent along the choke passage caused a drop in pressure, making the fuel enter through a jet as a fine spray. The vaporized fuel mixed with air to produce a combustible mixture for the engine's cylinders.

Maybach's other inventions included the honeycomb radiator – still used in some Mercedes cars – an internal expanding brake (1901) and an axle-locating system for use with independent suspensions. He left Daimler's in 1907 to set up his own factory for making engines for Zeppelin's airships.

Mayer Christian 1719–1783 was a Moravian Jesuit priest, an astronomer, mathematician and physicist. His work was seriously interrupted by the Pope's dissolution of the Jesuit order in 1773, although he managed to continue his astronomical studies, researching particularly into double stars.

Mayer was born on 20 August 1719 in a district now known as Mederizenlin, in modern Czechoslovakia. Various sources suggest that he was educated in many centres of learning around Europe and that he excelled in Greek, Latin, philosophy, theology and mathematics, but all that is definitely known is that he left home in his early twenties because his father disapproved strongly of his determination to become a Jesuit. Mayer entered the novitiate at Mannheim, and by the age of 33 he had had such success in his chosen career that he was appointed Professor of Mathematics and Physics at Heidelberg, although his main interest remained astronomy. Consequently, when the Elector Palatine, Karl Theodor, built an observatory first at Schwetzingen, then a larger one at Mannheim, Mayer was appointed Court Astronomer and given responsibility for equipping both with the best available instruments. The effects of Pope Clement XIV's dissolution of his order, however, rendered Mayer's court position untenable and he was relieved of his duties before he had completed the furnishing of the observatories. He managed to continue his own astronomical work and became well known in Europe – and even in the United States – for the careful presentation of his discoveries and observations in international journals. He died in Heidelberg on 12 April 1783.

Mayer carried out important astronomical research before, during, and after Karl Theodor's patronage. His studies included measurement of the degree of the meridian, based on work conducted in Paris and in the Rhenish Palatinate, and observations of the transits of Venus in 1761 and 1769. (The latter observation was conducted in Russia at the invitation of Catherine II.)

Mayer also studied double stars. His equipment was unable to distinguish true binary stars (in orbit round each other) from separate stars seen together only by the coincidence of Earth's viewpoint, but Mayer was the first to investigate and catalogue stars according to their apparent 'binary' nature. Later work was more critical and therefore more successful, but Mayer's pioneering contribution is important to the history of astronomy.

In the late 1770s Mayer turned his attention to observing the companions of fixed stars, mistakenly thinking that he had discovered more than a hundred planets of other stars. The controversy arising from this claim marks the inauguration of a period of methodologically more sound observation in the study of stars.

Mayer Johann Tobias 1723–1762 was a German cartographer, astronomer and physicist who in his short life did much to improve contemporary standards of observation and navigation, although a considerable amount of his research was superseded shortly after his death.

Mayer was born in Marbach, near Stuttgart, on 17 February 1723, the son of a cartwright and the youngest of six children. Shortly after his birth, his family moved to Esslingen, where he grew up. When he was six both his parents died and Mayer went to live in an orphanage. He developed some skill in architectural drawing and surveying and,

under the direction of a local artillery officer, in 1739 produced plans and drawings of military installations. Mayer's map of Esslingen and its surroundings – the oldest such map still extant – was made in the same year. Having taught himself mathematics (a subject not studied at his Latin school), he published his first book two years later, on the application of analytical methods to the solution of geometrical problems; he was 18 years old. Within the next few years he also acquired some knowledge of French, Italian and English. In 1746, he began work for the Homann Cartographic Bureau, in Nuremberg, devoting much of his time to collating geographical and astronomical facts contained in Homann's archives. He became so interested in the astronomical side of his work that in 1750 he published a compilation called *Kosmographische Nachrichten und Sammlungen auf das Jahr 1748*. His reputation both as cartographer and astronomer earned him an invitation to take up the post of Professor at the Georg August Academy in Göttingen, which he accepted. Eleven years later he contracted gangrene, and died on 20 February 1762.

At the Homann Cartographic Bureau, Mayer's most important work was the construction of some 30 maps of Germany. These established exacting new standards for using geographical data in conjunction with accurate astronomical details to determine latitudes and longitudes on Earth. To obtain some of the astronomical details, he observed lunar oscillations and eclipses using a telescope of his own design. But it was in Göttingen that he decided to produce a map of the Moon's surface, which entailed both theoretical and practical work never undertaken previously. By observing the Moon he concluded that it had no atmosphere, and continuous observations with his repeating (or reflecting) circle produced Mayer's Lunar Tables in 1753, the accuracy of which (correct to one minute of arc) gained him international fame. After 1755 he used a superior instrument, a 1.8-m/6-ft radius mural quadrant made by John Bird, with which he made improvements on his earlier stellar observations, enabling him to introduce correction formulae for meridian transits of stars.

Mayer also invented a simple and accurate method for calculating solar eclipses, compiled a catalogue of zodiacal stars and studied stellar proper motion. In the process of devising a method for finding geographical coordinates without using astronomical observations, he arrived at a new theory of the magnet, which provided a convincing demonstration of the validity of the inverse-square law of magnetic attraction and repulsion.

Acting upon one of his last requests, shortly after Mayer died, his widow submitted to the British Admiralty a method for computing longitude at sea. Although the tables resulting from this method were superseded not long after by more accurate data being compiled by James Bradley at Greenwich, Mayer's widow was awarded £3,000 by the British Government for her husband's claim to a prize offered for such a venture.

Mayer Julius Robert 1814–1878 was a German physicist who was the first to formulate the principle of conservation of energy and determine the mechanical equivalent of heat.

Mayer was born in Heilbronn on 25 November 1814 and received a classical and theological education there. In 1832 he went to the University of Tübingen to study and in spite of being expelled in 1837 for membership of a student secret society, he managed to qualify as a doctor with distinction in the following year. In 1840, he took a position as a ship's physician and sailed to the East Indies for a year. During this time he made his first observations on body heat. He then settled in his native city and built up a prosperous medical practice. In 1842, Justus von Liebig (1803–1873) published Mayer's classic paper on the conservation of energy in his *Annalen der Chemie*. This was followed by two papers that Mayer himself published. In 1845, he extended his ideas into the world of living things and in 1848 to the Sun, Moon and Earth. Disappointed at the lack of recognition and despairing because others were making the same discoveries independently and gaining priority, Mayer tried to kill himself in 1850. He was then confined to mental institutions during the 1850s, but from 1858 he at last began to gain the credit he deserved, principally by the efforts of Hermann Helmholtz (1821–1894), Rudolf Clausius (1822–1888) and John Tyndall (1820–1893). As a result, Mayer was recognized by the scientific world, receiving the Royal Society's Copley Medal in 1871. Mayer died in Heilbronn on 20 March 1878.

Mayer's conviction that the various forms of energy are interconvertible arose from an observation made during his voyage to the East. He found while blood-letting that the venous blood of the European sailors was much redder in the tropics than at home. Mayer put this down to a greater concentration of oxygen in the blood caused by the body using less oxygen, as less heat was required in the tropics to keep the body warm. From this, Mayer made a conceptual leap to the idea that work such as muscular force, heat such as body heat and other forms of energy such as chemical energy produced by the oxidation of food in the body are all interconvertible. The amount of work or heat produced by the body must be balanced by the oxidation of a certain amount of food, and therefore work or energy is not created but only transformed from one form to another.

This in essence is the principle of conservation of energy, which Mayer conceived as an idea but without any evidence to prove it in 1840. On returning to his home, he set about learning physics in order to prove his conviction. In fact, other scientists had been working towards the same conclusions but no one had yet formulated them so widely as Mayer. In 1839, Marc Seguin (1786–1875) had made a rough estimate of the mechanical equivalent of heat, assuming that the loss of heat represented by the fall in temperature of steam on expanding was equivalent to the mechanical work produced by the expansion. He also remarked that it was absurd to suppose that 'a finite quantity of heat could produce an indefinite quantity of mechanical action and that it was more natural to assume that a certain quantity of heat disappeared in the very act of producing motive power'.

In 1842, Mayer stated the equivalence of heat and work more definitely. He published an attempt to determine the mechanical equivalent of heat from the heat produced when air is compressed. He made the assumption that the whole of the work done in compressing the air was converted into heat, and by using the specific heats of air at constant volume and constant pressure, he was able to reach a value for the mechanical equivalent solely on theoretical grounds. He concluded that the heat required to raise the temperature of 1 kg/2.2 lb of water by 1°C/1.8°F would be equivalent to the work done by a mass of 1 kg/2.2 lb in falling 365 m/1,200 ft. This result is about 15% too low, but this was mainly due to the inaccuracy of the values for specific heats that Mayer had to use. James Joule (1818–1889) made the first accurate derivation experimentally in 1847, and credit for the determination of the mechanical equivalent of heat goes to him. But Mayer had clearly demonstrated the validity of the principle of conservation of energy and is generally regarded as its founder, though Helmholtz expressed it in a much clearer and more specific way in 1847.

In 1845, Mayer extended the principle of the interconvertibility and conservation of energy to magnetism, electricity, chemical energy and to the living world. He described the energy conversions that take place in living organisms, realizing that plants convert the Sun's energy into food that is consumed by animals to provide a source of energy to power their muscles and provide body heat. By insisting that living things are powered solely by physical processes utilizing solar energy and not by any kind of innate vital force, Mayer took a daring step forward and laid the foundations of modern physiology.

Mayer's paper of 1848 was concerned with astronomy. He realized correctly that chemical reactions could not provide enough energy to power the Sun, and proposed incorrectly that the Sun produces heat by meteoric bombardment. But he also calculated that tidal friction caused by the Moon's gravitation gradually slows the Earth's rotation, an effect now known to exist. However, like most of Mayer's ideas, this was later proposed independently by others.

Mayer was a very bold thinker and it is unfortunate that his lack of experimental ability and his position outside the scientific community prevented him from securing the proofs and immediate recognition that his ideas deserved. However, in conceiving the principle of conservation of energy and applying it to all processes in the universe both living and nonliving, he made a discovery of fundamental importance to science.

Mechnikov Ilya (Elie) Ilich 1845–1916 was a Russian-born French zoologist who discovered phagocytes, amoeba-like blood cells that engulf foreign bodies. For this discovery he was awarded (jointly with Paul Ehrlich) the 1908 Nobel Prize for Physiology or Medicine.

Mechnikov was born on 15 May 1845 in Kharkov, Russia, the son of an officer of the Imperial Guard. He was educated at the University of Kharkov, from which he graduated in 1864. He then travelled to Germany to pursue his studies but returned to Russia in 1867, becoming Professor of Zoology and Comparative Anatomy at the University of Odessa. In 1882, he inherited sufficient money to make him financially independent and moved to Messina in Italy to continue his research. In 1886, he accepted the post of Director of the Bacteriological Institute in Odessa but remained there only a short time, being invited to join the Pasteur Institute, where he remained for the rest of his life, becoming Director on Pasteur's death in 1895. Mechnikov died in Paris on 16 July 1916.

Mechnikov first noticed phagocytes while he was in Messina studying the transparent larvae of starfish; he observed that certain cells surrounded and engulfed foreign particles that had entered the bodies of the larvae. He then deliberately introduced bacteria into starfish larvae and fungal spores into water fleas (*Daphnia*) and again observed that special amoeba-like cells moved to where these foreign bodies were and engulfed them. Mechnikov continued this line of research at Odessa and later at the Pasteur Institute, where he demonstrated that phagocytes exist in higher animals. In humans about threequarters of the white blood cells or leucocytes are phagocytic, and they form the first line of defence against acute infections, moving to the site of infection and engulfing the invading bacteria. This work opposed the theory of the time, which postulated that leucocytes actually helped the growth of bacteria, but it is now accepted that leucocytes are one of the body's basic defence mechanisms against disease.

In his later years Mechnikov became interested in longevity, and he spent the last decade of his life trying to demonstrate that lactic acid-producing bacteria in the intestine increase a person's lifespan.

Medawar Peter Brian 1915–1987 was a British zoologist best known for his important contributions to immunology, for which he shared the 1960 Nobel Prize for Physiology or Medicine with Frank Macfarlane Burnet. He was also renowned for his elegant essays – collected in, for example, *The Hope of Progress* (1972), *The Art of the Soluble* (1967), and *The Limits of Science* (1985) – which described the nature of science to the general reader.

Medawar was born on 28 February 1915 in Rio de Janeiro, Brazil. He was educated at Magdalen College, Oxford, from which he graduated in 1939. From 1938 to 1944 and from 1946 to 1947 he was a Fellow of Magdalen College; and from 1944 to 1946 he also held a fellowship at St John's College, Oxford. In 1947, he was appointed Mason Professor of Zoology at Birmingham University and he held this position until 1951, when he became Jodrell Professor of Zoology and Comparative Anatomy at University College, London. Medawar remained at University College until 1962, when he was appointed Director of the National Institute for Medical Research at Mill Hill in London, becoming Director Emeritus in 1975. In 1977 he was appointed Professor of Experimental Medicine at the Royal Institution.

Medawar began his Nobel prizewinning research in the early 1950s. Acting on Burnet's hypothesis that an animal's

ability to produce a specific antibody is not inherited, Medawar inoculated mouse embryos of one strain (strain A) with cells from mice of another strain (strain B). He found that the strain A embryos did not produce antibodies against the strain B cells. When the strain A embryos had developed sufficiently to be capable of independent existence, he grafted onto them skin from strain B mice. Again he found that the strain A embryos did not produce antibodies against the strain B tissue. Thus Medawar confirmed Burnet's hypothesis that the ability of an animal to produce a specific antibody develops during the animal's lifetime and is not inherited. This finding suggests that an animal's immune system can be influenced by external factors, which may have significant implications in the field of transplant techniques.

Meitner Lise 1878–1968 was an Austrian-born Swedish physicist who was one of the first scientists to study radioactive decay and the radiations emitted during this process. Her most famous work was done in 1938 in collaboration with Otto Frisch (1904–1979), her nephew, on describing for the first time the splitting or fission of the uranium nucleus under neutron bombardment. This publication provoked a flurry of research activity and was of pivotal importance in the development of nuclear physics.

Meitner was born in Vienna on 7 November 1878. She became interested in science at an early age, but before studying physics, her parents insisted that she should first qualify as a French teacher in order to be certain of supporting herself. She passed the examination in French and then entered the University of Vienna in 1901. She obtained her PhD from that University, becoming only the second woman to do so, in 1905, her thesis work on thermal conduction having been supervised by Ludwig Boltzmann (1844–1906).

In 1907, Meitner entered the University of Berlin to continue her studies under Max Planck (1858–1947). She met Otto Hahn (1875–1966) soon after arriving in Berlin. He was seeking a physicist to work with him on radioactivity at the Kaiser Wilhelm Institute for Chemistry. Meitner joined Hahn, but their supervisor Emil Fischer (1852–1919) would not allow Meitner to work in his laboratory because she was a woman, and they had to set up a small laboratory in a carpenter's workroom. Despite this inauspicious start, Meitner became a member of the Kaiser Wilhelm Institute in 1912, and in 1917 was made joint Director of the Institute with Hahn and was also appointed head of the Physics Department.

In 1912, Meitner had also become an assistant to Planck at the Berlin Institute of Theoretical Physics. She was appointed Docent at the University of Berlin in 1922, and then made extraordinary Professor of Physics in 1926. Meitner, who was Jewish, remained in Berlin when the Nazis came to power in 1933 because she was protected by her Austrian citizenship. However, the German annexation of Austria in 1938 deprived her of this citizenship and placed her life in danger. With the aid of Dutch scientists, she escaped to Holland and soon moved to Denmark as the guest of Niels Bohr (1885–1962). She was then offered a post at the Nobel Physical Institute in Stockholm, where a cyclotron was being built, and Meitner accepted. It was shortly after her arrival in Sweden that Meitner and Frisch, who was working at Bohr's Institute in Copenhagen, made the discovery of nuclear fission.

During World War II, Meitner was invited to participate in the development programme for the construction of the nuclear bomb, but she refused in the hope that such a weapon would not be feasible. In 1947, a laboratory was established for her by the Swedish Atomic Energy Commission at the Royal Institute of Technology, and she later moved to the Royal Swedish Academy of Engineering Science to work on an experimental nuclear reactor. In 1949, she became a Swedish citizen. In 1960 she retired from her post in Sweden and settled in Cambridge, England. The Fermi Award was given jointly to Meitner, Hahn and Fritz Strassmann (1902–1980) in 1966, Meitner being the first woman to be so honoured. She died in Cambridge on 27 October 1968.

Meitner's early work in Berlin with Hahn concerned the analysis of physical properties of radioactive substances. Hahn's primary interest lay in the discovery of new elements, but Meitner's work was concerned with examining radiation emissions. They were able to determine the beta line spectra of numerous isotopes, leading to the discovery of protactinium in 1918. During the 1920s, Meitner studied the relationship between beta and gamma irradiation. She examined the basis for the continuous beta spectrum, her results leading Wolfgang Pauli (1900–1958) to postulate the existence of the neutrino. She was the first to describe the emission of Auger electrons, which occurs when an electron rather than a photon is emitted after one electron drops from a higher to a lower electron shell in the atom.

The rapid developments in physics during the 1930s, such as the discovery of the neutron, artificial radioactivity and the positron, did not leave Meitner behind. In 1933, she used a Wilson cloud chamber to photograph positron production by gamma radiation and in the following year, she began to study the effects of neutron bombardment on uranium with Hahn. They were interested in confirming the results of Enrico Fermi (1901–1954) that suggested the production of transuranic elements, that is, elements with atomic numbers higher than that of uranium (92). In 1935, Meitner and Hahn used a hydrogen sulphide precipitation method to remove elements with atomic numbers between 84 and 92 from their neutron-irradiated sample of uranium. They thought that they had found evidence for elements with atomic numbers of 93, 94, 95 and 96. Then in 1938, after Meitner was forced to flee from Germany, Hahn and Strassmann found that the radioactive elements produced by neutron bombardment of uranium had properties like radium. From Sweden, Meitner requested firm chemical evidence for the identities of the products. Hahn and Strassmann were surprised to find that the neutron bombardment had produced not transuranic elements but three isotopes of barium, which has an atomic number of 56.

Meitner and Frisch realized that these results indicated that the uranium nucleus had been split into smaller fragments. They predicted correctly that krypton would also be found among the products of this splitting process, which they named fission. A paper describing their analysis appeared in January 1939, and immediately set in motion a series of discoveries leading to the first nuclear reactor in 1942 and the first atomic bomb in 1945.

Meitner had also found evidence for the production of uranium-239 by the capture of a neutron by uranium-238. Beta decay of this new, heavy isotope would yield a transuranic element with the atomic number 93. This element was found by Edwin McMillan (1907–1991) and Philip Abelson (1913–   ) in 1940 and named neptunium.

Meitner continued to study the nature of fission products and contributed to the design of an experiment whereby fission products of uranium could be collected. Her later research concerned the production of new radioactive species using the cyclotron, and also the development of the shell model of the nucleus.

Meitner was a distinguished scientist who made important contributions to nuclear physics despite having to overcome both sexual and racial discrimination.

Mellanby Kenneth 1908–   is a British entomologist and ecologist who is best known for his work on the environmental effects of pollution.

Mellanby was born on 26 March 1908 and was educated at the Barnard Castle School. He gained a scholarship to read natural sciences at London University and graduated in 1929. In the following year he became a research worker at the London School of Hygiene and Tropical Medicine, from which he gained his doctorate in 1933, and remained a member of staff there until 1945. During this period he went to East Africa to study the tsetse fly and, while doing his World War II military service, investigated scrub typhus in Burma and New Guinea. In 1945, he was appointed Reader in Medical Entomology at London University, and in 1947 became the Principal of University College Ibadan – Nigeria's first university, which Mellanby played a substantial part in creating. He was then appointed head of the Entomology Department at Rothamsted Experimental Station in 1955, a position he held until 1961, when he founded and became the Director of the Monks Wood Research Station (now called the Institute of Terrestrial Ecology) at Huntingdon in Cambridgeshire. He remained in this post until he officially retired. In 1978, he visited Australia to advise the newly established Association for the Protection of Rural Australia, and in 1979 he went to Peru to advise on the ecological effects of the Montaro Transfer Scheme, a plan to divert water from the Montaro River to Lima.

Mellanby's early career was spent in entomological research but his most important contributions are probably his pioneering investigations into the effects of pollution, particularly by pesticides. Shortly after he had established the Monks Wood Research Station, he drew attention to the deleterious effects of pesticides – before Rachel Carson had published her famous book on this theme, *Silent*

Spring. Continuing his research, Mellanby undertook a comprehensive study of pesticides, concluding that although these chemicals damage the environment, pests destroy huge amounts of food and other essential materials and must therefore be controlled. Instead of pesticides, however, Mellanby advocated the use of biological control methods, such as introducing animals that feed on pests. He has written several books about entomology, ecology and pollution, and edited *Environmental Pollution*, one of the main research journals in this subject area.

Mendel Gregor Johann 1822–1884 was an Austrian monk who discovered the basic laws of heredity, thereby laying the foundation of modern genetics – although the importance of his work was not recognized until after his death.

Mendel was born Johann Mendel on 22 July 1822 in Heinzendorf, Austria (now Hyncice in the Czech Republic), the son of a peasant farmer. He studied for two years at the Philosophical Institute in Olmütz (now Olomouc), after which, in 1843, he entered the Augustinian monastery in Brünn, Moravia (now Brno in Czechoslovakia), taking the name Gregor. In 1847 he was ordained a priest. During his religious training, Mendel taught himself a certain amount of science and for a short time he was a teacher of Greek and mathematics at the secondary school in Znaim (now Znojmo) near Brünn. In 1850 he tried to pass an examination to obtain a teaching licence but failed, and in 1851 he was sent by his Abbot to the University of Vienna to study physics, chemistry, mathematics, zoology and botany. Mendel left the university in 1853 and returned to the monastery in Brünn in 1854. He then taught natural science in the local Technical High School until 1868, during which period he again tried, and failed, to gain a teaching certificate that would have enabled him to teach in more advanced institutions. It was also in the period 1854 to 1868 that Mendel performed most of his scientific work on heredity. He was elected Abbot of his monastery at Brünn in 1868, and the administrative duties involved left him little time for further scientific investigations. Mendel remained Abbot at Brünn until his death on 6 January 1884.

Mendel began the experiments that led to his discovery of the basic laws of heredity in 1856. Much of his work was performed on the edible pea (*Pisum* sp.), which he grew in the monastery garden. He carefully self-pollinated and wrapped (to prevent accidental pollination by insects) each individual plant, collected the seeds produced by the plants, and studied the offspring of these seeds. He found that dwarf plants produced only dwarf offspring and that the seeds produced by this second generation also produced only dwarf offspring. With tall plants, however, he found that both tall and dwarf offspring were produced and that only about one-third of the tall plants bred true, from which he concluded that there were two types of tall plants, those that bred true and those that did not. Next he cross-bred dwarf plants with true-breeding tall plants, planted the resulting seeds and then self-pollinated each plant from this second generation. He found that all the offspring in the first generation were tall but that the offspring from the

R = dominant red　　　　　　w = recessive white

Mendel *Mendel's law of segregation applied to an initial cross between pea plants with red flowers and pea plants with white flowers. The second generation has red-flowered plants to white-flowered plants in the ratio of 3:1.*

self-pollination of this first generation were a mixture of about one-quarter true-breeding dwarf plants, one-quarter true-breeding tall plants and one-half non-true-breeding tall plants. Mendel also studied other characteristics in pea plants, such as flower colour, seed shape and flower position, finding that, as with height, simple laws governed the inheritance of these traits. From his findings Mendel concluded that each parent plant contributes a factor that determines a particular trait and that the pairs of factors in the offspring do not give rise to an amalgamation of traits. These conclusions, in turn, led him to formulate his famous law of segregation and law of independent assortment of characters, which are now recognized as two of the fundamental laws of heredity.

Mendel reported his findings to the Brünn Society for the Study of Natural Science in 1865 and in the following year he published *Experiments with Plant Hybrids*, a paper that summarized his results. But the importance of his work was not recognized at the time, even by the eminent botanist Karl Wilhelm von Nägeli, to whom Mendel sent a copy of his paper. It was not until 1900, when his work was rediscovered by Hugo De Vries, Carl Erich Correns and Erich Tschermak von Seysenegg, that Mendel achieved fame – 16 years after his death.

Mendeleyev Dmitri Ivanovich 1834–1907 was a Russian chemist whose name will always be linked with his outstanding achievement, the development of the periodic table. He was the first chemist to understand that all elements are related members of a single ordered system. He

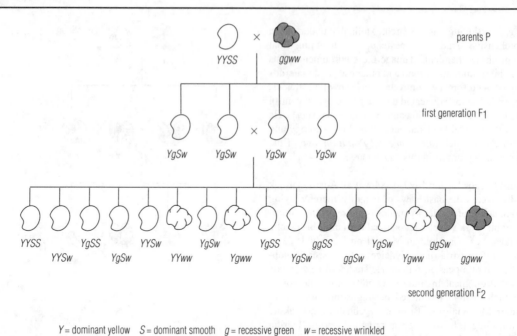

Y = dominant yellow   S = dominant smooth   g = recessive green   w = recessive wrinkled

Mendel *Mendel's law of independent assortment applied to an initial cross between pea plants with smooth yellow seeds and pea plants with green wrinkled seeds. The second generation includes four visibly different types of seeds – yellow smooth, yellow wrinkled, green smooth and green wrinkled in the ratio 9:3:3:1.*

converted what had hitherto been a highly fragmented and speculative branch of chemistry into a true, logical science. The spelling of his name has been a source of confusion to students and frustration to editors for more than a century, and the forms Mendeléeff, Mendeléev and even Mendelejeff can all also be found in print.

Mendeleyev was born in Tobol'sk, Siberia, on 7 February 1834, the youngest of the 17 children of the head of the local high school. His father went blind when Mendeleyev was still a child, and the family had to rely increasingly on their mother for support. He was educated locally but could not gain admission to any Russian university (despite his mother's attempts on his behalf with the authorities at Moscow) because of the supposedly backward attainments of those educated in the provinces. In 1855, he finally qualified as a teacher at the Pedagogical Institute in St Petersburg. He took an advanced degree course in chemistry, and in 1857 obtained his first university appointment.

In 1859, he was sent by the government for further study at the University of Heidelberg where he made valuable contact with the Italian chemist Stanislao Cannizzaro, whose insistence on a proper distinction between atomic and molecular weights influenced Mendeleyev greatly. In 1861, he returned to St Petersburg and became Professor of General Chemistry at the Technical Institute there in 1864. He could find no textbook adequate for his students' needs and so he decided to produce his own. The resulting *Principles of Chemistry* (1868–70) won him international renown; it was translated into English in 1891 and 1897.

Mendeleyev began work on his periodic law in the late 1860s, and he went on to conduct research in various other

fields. Then in 1890 he chose to be a spokesman for students who were protesting against unjust conditions. For these allegedly improper activities he was retired from the university and became controller of the Bureau for Weights and Measures, although from 1893 he received no other professorial appointment. He died in St Petersburg on 2 February 1907, five days before his seventy-third birthday. His nomination for the 1906 Nobel Prize for Chemistry failed by one vote (the award went to Henri Moissan) but his name became recorded in perpetuity 50 years later when element number 101 was called mendelevium.

Before Mendeleyev produced his periodic law, understanding of the chemical elements had long been an elusive and frustrating task. The attempts by various chemists to put the whole field into some intelligible reference system had acted rather like the progressively stronger lenses of a microscope in bringing a sensed but unseen object into clear vision. According to Mendeleyev the properties of the elements, as well as those of their compounds, are periodic functions of their atomic weights (relative atomic masses). In 1869, he stated that 'the elements arranged according to the magnitude of atomic weights show a periodic change of properties'. Other chemists, notably Lothar Meyer in Germany, had meanwhile come to similar conclusions, Meyer publishing his findings independently.

Mendeleyev compiled the first true periodic table, listing all the 63 elements then known. Not all elements would 'fit' properly using the atomic weights of the time, so he altered indium from 76 to 114 (modern value 114.8) and beryllium from 13.8 to 9.2 (modern value 9.013). In 1871, he produced a revisionary paper showing the correct

			Ti = 50	Zr = 90	? = 180
			V = 51	Nb = 94	Ta = 182
			Cr = 52	Mo = 96	W = 186
			Mn = 55	Rh = 104.4	Pt = 197.4
			Fe = 56	Ru = 104.4	Ir = 198
			Ni = Co	Pd = 106.6	Os = 198
			= 50		
H = 1			Cu = 63.4	Ag = 108	Hg = 200
	Be = 9.4	Mg = 24	Zn = 65.2	Cd = 112	
	B = 11	Al = 27.4	? = 68	Ur = 116	Au = 197?
	C = 12	Si = 28	? = 70	Sn = 118	
	N = 14	P = 31	As = 75	Sb = 122	Bi = 210?
	O = 16	S = 32	Se = 79.4	Te = 128?	
	F = 19	Cl = 35.5	Br = 80	I = 127	
Li = 7	Na = 23	K = 39	Rb = 85.4	Cs = 133	Te = 204
		Ca = 40	Sr = 87.6	Ba = 137	Pb = 207
		? = 45	Ce = 92		
		?Er = 56	La = 94?		
		?Yt = 60	Di = 95		
		?In = 75.6	Th = 118?		

Mendeleyev *Mendeleyev's original (1869) periodic table, with similar elements on the same horizontal line.*

repositioning of 17 elements.

Also in order to make the table work Mendeleyev had to leave gaps, and he predicted that further elements would eventually be discovered to fill them. These predictions provided the strongest endorsement of the periodic law. Three were discovered in Mendeleyev's lifetime: gallium (1871), scandium (1879) and germanium (1886), all with properties that tallied closely with those he had assigned to them.

Far-sighted though Mendeleyev was, he had no notion that the periodic recurrences of similar properties in the list of elements reflect anything in the structures of their atoms. It was not until the 1920s that it was realized that the key parameter in the periodic system is not the atomic weight but the atomic number of the elements – a measure of the number of nuclear protons or electrons in the stable atom. Since then great progress has been made in explaining the periodic law in terms of the electronic structures of atoms and molecules.

Among Mendeleyev's other investigations were the specific volumes of gases and the conditions that are necessary for their liquefaction. Following visits to the oilfields of the Caucasus and in the United States he examined the origins of petroleum. He was convinced that the future held great possibilities for human flight, and in 1887 he made an ascent in a balloon to observe an eclipse of the Sun. He farmed a small estate and applied his scientific knowledge to improve the yield and quality of crops, an endeavour invaluable for Russia's predominantly agricultural economy.

Menzel Donald Howard 1901–1976 was a US physicist and astronomer whose work on the spectrum of the solar chromosphere revolutionized much of solar astronomy. He was one of the first scientists to combine astronomy with atomic physics and, as a teacher and writer he had a considerable influence on the development of astrophysics during the twentieth century.

Menzel was born on 11 April 1901 at Florence, Colorado, where his father was a railroad agent. When Menzel was four his father moved to Leadville, a remote mining centre where Menzel then lived until the age of 16, when he enrolled at the University of Denver. After graduation, he joined the staff of Princeton University as a Graduate Assistant and came under the benign influence of Henry Russell (1877–1957), the co-originator of the Hertzsprung–Russell diagram, who taught a course on basic astrophysics there. Menzel soon became fascinated by the combination of atomic physics, mathematics and relativity theory. Nevertheless, to gain more practical astronomical experience, he became Assistant Astronomer at the Lick Observatory in California in 1924. There priority was given to measuring visual binary stars and stellar radial velocity, but of more interest to Menzel was William Campbell's collection of solar chromospheric spectra and he took the opportunity to develop a quantitative spectroscopy using recently gained knowledge of the spectra of complex atoms and wave mechanics.

In 1932, Menzel joined Harvard University Observatory (where he was to become Director some 30 years later). After four years he began experimental work to study the solar corona outside eclipses. The coronagraph he constructed for this purpose was the beginning of High Altitude Observatory, which has since been developed into one of the leading institutions participating in solar physics research. At Harvard, Menzel established a course that included study of radiative transfer, the formation of spectra in stellar atmospheres and gaseous nebulae, and atomic physics and statistical mechanics, together with the study of dynamics, classical electromagnetic theory and relativity. (This comprehensive syllabus was later published as *Mathematical Physics*.)

During World War II Menzel served with the US Navy, advising on the effect of solar activity on radio communication and radar propagation. After the war, he became involved in raising funds for a number of observatories. The war had shown the value of radio communication, and the need for information about solar activity led the US Air Force to construct a solar observatory in New Mexico. Menzel supervised the design of several instruments for the laboratory, using the results obtained to further study solar prominences, low temperatures in sunspots and the origin of solar flares. Menzel was also Chairman of the National Radio Astronomy Observatory Advisory Committee; he participated in setting up the Kitt Peak National Observatory in Arizona; he was instrumental in bringing the Smithsonian Astrophysics Observatory from Washington to Cambridge; and he was a key figure in finding independent funding for the Solar Observatory in South Africa.

Menzel retired from Harvard in 1971 to become scientific director of a company manufacturing antennae for communications and radioastronomy. He died on 14 December 1976 after a prolonged illness.

At Princeton, Menzel and Henry Russell held a virtual monopoly on theoretrical astrophysics. But that was by no means Menzel's sole interest. He devised a technique for computing the temperature of planets from measurements of water cell transmissions and he made important contributions to atmospheric geophysics, radio propagation and even lunar nomenclature. He also held patents on the use of gallium in liquid ball bearings and on heat transfer in atomic plants. Further work included the development of a fluid clutch and the investigation of solar energy conversion into electricity.

In his early days at Harvard, Menzel returned to his favourite subject, the Sun. In observing eclipses he developed a means of taking photographs with very good height resolution. Examining more than 800 spectra, he then established a wholly new theoretical approach to the structure of the gaseous envelope surrounding the Sun and other stars. With Perkins, Menzel also calculated the theoretical intensities of atomic hydrogen's spectral lines – a work that became a standard reference on the subject. It was followed by a series of papers on *Physical Processes in Gaseous Nebulae*, which provided a framework for the quantitative analysis of nebular spectra.

Mergenthaler Ottomar 1854–1899 was a German-born US inventor who devised the Linotype machine, which greatly speeded typesetting and revolutionized the world of printing and publishing.

Mergenthaler was born in Germany on 11 May 1854 and was apprenticed to a watchmaker before emigrating to the United States in 1872. He settled in Washington, DC, and found employment in a factory manufacturing scientific instruments. In 1876, he moved to Baltimore and began working for James O. Clephane, remedying faults in a recently invented prototype of a writing machine. Although a full-sized machine was constructed, it was not a success and the two men abandoned the idea. They combined their talents and in 1884 produced what was to become known as the Linotype machine.

By 1886, the first machines were in use and a company was formed to spearhead their production with Mergenthaler as one of the directors. He resigned from the position in 1888 but continued to show an interest in the firm, contributing over the years as many as 50 new modifications to the original design. He died in Baltimore on 28 October 1899.

Before Linotype, printing was carried out by hand-setting, a long and laborious process. Mergenthaler's invention speeded this operation and made printed matter, from books to penny news sheets, cheaper to produce. The design of the machine enabled a line of type (hence the name) to be composed at one time and cast as a single piece of metal. The machine was rather like a large typewriter, about 2 m/6^1/$_2$ ft high, with a store of matrices (moulds) at the top. The operator selected the letters by means of rods controlled by the 'typewriter' keys, and these letters fell through tiny trapdoors to drop into position in a line setting. As each line was completed it was passed on to the 'metal pot' area where a cast was made to form a 'slug' with the letters in relief on one side. This then fitted into a page of type ready for printing, while the matrices were returned to the store at the top of the machine for reuse.

A person operating one of Mergenthaler's keyboards could set type up to three or four times faster than by hand-setting, cutting the labour cost of production to a fraction of before. The machine heralded a new age for printing in which books became affordable, and newspapers could really claim to carry up-to-the-minute information, for the Linotype made it possible to change and reset copy to within minutes of going to press.

Meselson and Stahl Matthew Stanley Meselson 1930– and Franklin William Stahl 1929– are US molecular biologists who are best known for their collaborative work that confirmed Francis Crick and James Watson's theory that DNA replication is semi-conservative (that is, that the daughter cells each receive one strand of DNA from the original parent cell and one newly replicated strand, in contrast to conservative replication in which one daughter cell would receive both of the parental DNA strands and the other daughter would receive both the newly replicated strands).

Meselson was born on 24 May 1930 in Denver, Colorado. He studied liberal arts at the University of Chicago, gaining his doctorate in 1951, then physical chemistry at the California Institute of Technology, obtaining a second doctorate in 1957. He remained at the California Institute of Technology until 1976 – as Assistant Professor of Chemistry from 1958 to 1960, as Associate Professor of Chemistry from 1958 to 1960, then as Associate Professor and later Professor of Biology. In 1976, he became Cabot Professor of

Natural Sciences at Harvard University. Since 1963 he has also been a consultant to the United States Arms Control and Disarmament Agency.

Stahl was born on 8 October 1929 in Boston, Massachusetts. He was educated at Harvard University, from which he graduated in 1951, and at the University of Rochester, from which he gained his doctorate in biology in 1956. He was a Research Fellow in Biology at the California Institute of Technology from 1955 to 1958, when he moved to the University of Missouri, where he was Associate Professor of Zoology from 1958 to 1959 then Associate Professor of the University until 1970. Since 1970 he has been Professor of Biology at the University of Oregon and a Research Associate of the Institute of Molecular Biology.

Meselson and Stahl began their research that demonstrated the semi-conservative nature of DNA replication in 1957. After unsuccessfully experimenting with viruses, they turned their attention to the bacterium *Escherichia coli* (*E. coli*). They grew the bacteria on a culture medium containing nitrogen-15 (a heavy isotope of nitrogen) as the only nitrogen source, so that the nitrogen-15 would become incorporated into the nitrogenous bases of the bacterial DNA. They then transferred the bacteria to a medium containing nitrogen-14 (the normal nitrogen isotope) as the only nitrogen source. After the bacteria had reproduced several times, they were centrifuged to extract their DNA, and the density of the DNA was determined by equilibrium density gradient centrifugation, in which samples of different densities separate into discrete bands; the concentration of DNA in each band can then be determined by ultraviolet absorption spectrography. From these processes Meselson and Stahl obtained three different types of DNA; one containing only nitrogen-14, one containing only nitrogen-15 and a hybrid containing both nitrogen isotopes. On heating, the hybrid separated into two halves, one from the parental DNA and one that had been newly synthesised. These findings demonstrated that the double helix of DNA splits into two strands when the DNA replicates, with each of the single strands acting as a template for the synthesis of a complementary strand; the final result is two DNA molecules, each comprising one strand from the parent molecule and one newly synthesised strand. Thus Meselson and Stahl confirmed the hypothesis of semi-conservative DNA replication, one of the most important concepts in modern molecular biology.

In 1961, working with Sidney Brenner and François Jacob, Meselson and Stahl demonstrated that ribosomes require instructions in order to be able to manufacture proteins, and that ribosomes can make proteins which are different from those normally produced by a particular cell. They also showed that messenger RNA supplies the instructions to the ribosomes.

In addition to their collaborative work, Stahl has researched into the genetics of bacteriophages and has written a book on the mechanism of inheritance; and Meselson has investigated the molecular biology of nucleic acids, the mechanisms of DNA recombination and repair, and the processes of gene control and evolution.

Messerschmitt Willy (Wilhelm Emil) 1898–1978 was the German aircraft designer and industrialist, the foremost of only a handful of people who shaped the development of the aircraft industry in his country from the early 1920s to the early 1980s.

Messerschmitt was born in Frankfurt am Main on 26 June 1898. In Bamberg, in 1909, he attended the Realschule, a secondary school with a scientific bias. He was fascinated by the fast-developing world of aviation, and in 1912 met the architect and gliding pioneer Friedrich Harth. It was their building and experimentation with primitive canard (tail first) gliders that decided Messerschmitt on his future career.

As a student at the Oberealschule in Nuremburg during 1915, Messerschmitt undertook the detailed design and construction of his first glider. Given a military discharge on medical grounds in 1917, he continued his education at the Technische Hochschule in Munich.

A glider he designed with Harth achieved an unofficial world duration record in 1921. The following year they set up a flying school and in 1923, while still a student, Messerschmitt formed his own company in Bamberg. Its first product, the S14 cantilever monoplane glider, formed the subject of his Dipl Ingenieur thesis.

After experiments with gliders powered by auxiliary engines, he produced his first powered aircraft in 1925, the ultra-light sports two-seater M17. In 1926 came the flight of the M18 small transport, the production version of which was adapted to largely metal construction.

Working in a country in which the manufacture of military aircraft was forbidden under the Treaty of Versailles, Messerschmitt pursued designs for light sporting aircraft, none of which attracted many orders. When the Nazis came to power in 1933, however, the German air ministry gave great stimulus to aviation, and an order to Messerschmitt to develop a new touring aircraft.

The resulting design, the M37 (or Bf108 in production) became the archetypal low-wing four-seater cabin monoplane, with retractable landing gear and flaps. It was the success of this aircraft which kept Messerschmitt in the industry, for in 1934 he had given serious thought to taking up a professorship in Danzig Technical College.

In taking part in the German air ministry's 1934 design competition for a fighter aircraft to equip the 'secret' Luftwaffe, Messerschmitt risked all by building a machine of advanced concept and employing the most modern techniques. The resulting Bf109 was to become the standard fighter of the Luftwaffe and several other air forces. It was to be equated with the legendary Spitfire.

By 1938, firmly established as a designer of advanced fighter aircraft, Messerschmitt was appointed chairman and general director of the company manufacturing his designs – Bayerische Flugzeugwerke – which was then renamed Messerschmitt AG. He and his company went on to produce numerous designs for fighter, bomber and transport aircraft, many of which served the Luftwaffe in large numbers throughout World War II.

At the war's end Messerschmitt was taken prisoner by

the British who with the French and Americans, asked him to work for them as an adviser, but he declined all proposals. On return to Germany he was arrested by US troops and held in custody for two years. Banned from manufacturing aircraft in Germany, Messerschmitt went on to design and produce components for prefabricated houses, sewing machines, and a popular tandem, two-seat, three-wheel 'bubble' car.

He took up aircraft design again in 1952 under an advisory contract with the Spanish manufacturer Hispano, which was already building his '109' fighter under licence. His HA200 Saeta, first flown in 1955, served in the early 1980s as the standard advanced trainer of the Spanish Air Force.

Between 1956 and 1964 Messerschmitt worked in association with the German Bolkow and Heinkel companies and developed the VJ101 supersonic V/STOL (vertical/short take-off and landing) combat aircraft. His company merged with Bolkow in 1963 and they were later joined by Hamburger Flugzengbau. In 1969, Messerschmitt became chairman of the supervisory board of the resulting Messerschmitt Bolkow Blohm (MBB) group.

MBB is identified today with the Tornado multi-role combat aircraft being built for NATO countries, and the A300 Airbus airliner, both designs being initiated before Messerschmitt's death on 15 September 1978, when he was honorary chairman.

Many of the great names in aircraft manufacture were from the generation before Messerschmitt's. They built their companies on the sales of vast numbers of military biplanes for World War I. Messerschmitt, however, never willingly designed biplanes. To him they were a retrograde step. The efficient cantilever monoplanes he designed from the early 1920s failed to secure large sales mainly because they were launched on to a market convinced of the need for visible struts and bracing wires.

The M18 small transport and the M19 sporting aircraft, both of the mid-1920s, embodied characteristics to be found in any aircraft designed or supervised by him. These were: simple concept, minimum weight and aerodynamic drag, and the possibility of continued development.

The last-mentioned characteristic was taken to the extreme in his '109' fighter which, in 20 years and more than 35,000 examples (making the second largest production of any aircraft), was developed in more than 50 variants. Messerschmitt's devotion to innovation in construction was exemplified by the hundreds of personal patents that were used throughout his designs.

Many of Messerschmitt's aircraft, when first flown, were the most advanced of their kind, having considerable performance advantage over competitors. Performance, coupled with elegance, characterizing his earlier designs, was to be seen also during World War II in the Me 262 twinjet fighter. In operation at the same time as the Gloster Meteor jet, Messerschmitt's machine featured a triangular minimal cross-section fuselage, high aspect ratio swept wings, heavier armament and a 210-kph/130-mph speed advantage over the staid British design.

Messerschmitt's '109' fighter, which gave employment to (or forced it upon) as many as 81,000 people in 25 plants, was the backbone of the Luftwaffe fighter force for the whole of the war. Production topped an incredible 2,000 per month by the end of 1944. Messerschmitt's name and work live on in what is today Germany's largest aerospace enterprise, Messerschmitt Bolkow Blohm. Employing more than 20,000 people, MBB makes satellites, helicopters, missiles and transport systems, as well as the Tornado and Airbus aircraft.

Messier Charles 1730–1817 was a French astronomer whose work on the discovery of comets led to a compilation of the locations of nebulae and star clusters – the Messier Catalogue – that is still of some relevance 200 years later.

Messier was born on 26 June 1730 in Badonviller, in the province of Lorraine. Little is known about his life until he joined the Paris Observatory as a draughtsman and astronomical recorder under the duration of one of the most famous men of the time, Joseph-Nicolas Delisle (1688–1768). Later he searched the night sky for comets, but he was continually hampered by encountering other rather obscure forms which he came to recognize as nebulae. During the period 1760 to 1784, therefore, Messier set about compiling a list of these nebulae and star clusters in order that he and other astronomers could more easily pinpoint (and thus ignore) these celestial features, in this way not only saving time but reducing the risk of any confusion with possible new comets.

At an early stage in Messier's career, his work was already being acknowledged for its importance and thoroughness, and in 1764 he was elected to the Royal Society of London. Six years later, in 1770, he was duly honoured by his own country and made a member of the prestigious Academy of Sciences in Paris. He died in Paris on 12 April 1817.

Initially, Messier's interest in comets stemmed from the predicted return of Halley's Comet; before he died (in 1740), Halley calculated that the comet's reappearance would take place around 1758 to 1759. Messier duly sighted its return on 21 January 1759, an experience which was to inspire him with the desire to go on discovering new comets for the rest of his life. (Although he is attributed with being the first person to resight the Halley comet from French soil, a German amateur astronomer is believed to have been the first to actually see it, on Christmas Day, 1758.)

Messier certainly earned his nickname, the 'Comet Ferret', from Louis XV. He spent long hours of painstaking search, over many years, to discover ultimately between 15 and 21 new comets. (The actual numbers vary according to the source of information, but it is thought not in any case to be less than 15.) In the beginning, he was frustrated by nebulae and star clusters which were, on occasions, readily confused with comets and had to be investigated continuously, a process that was also time-consuming. Messier decided that the most sensible idea was to compile a list

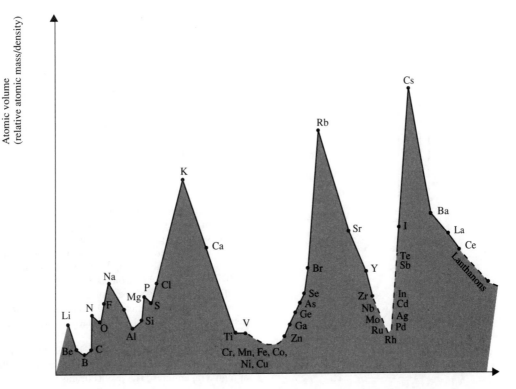

Meyer *Part of the Lothar Meyer curve (with modern additions).*

identifying each of these permanent objects, numbering them and noting their position.

The task he undertook was an extremely difficult one, given the equipment available at that time. Although a vast improvement had then recently been achieved with the development (during the first half of the 1700s) of the compound lens, the range and capability of the telescope was still in its infancy. Nevertheless, Messier began this work in earnest in 1760, and by 1771 he had completed a preliminary list of 45 nebulae and galaxies, giving each one an identifying M number. Within ten years he had compiled the majority of his catalogue, and by 1784 the list consisted of 103 numbers.

Basically, Messier's original catalogue is still relevant today with some additions – Pierre Méchain (1744–1804) added six more during Messier's lifetime – although doubts inevitably exist as to the reality of some he registered. But none can doubt the presence of such famous astronomical features as the Crab Nebula M1, Andromeda M31, and the Pleiades or Seven Sisters M45.

Meyer Julius Lothar 1830–1895 was a German chemist who, independently of Dmitri Mendeleyev, produced a periodic law describing the properties of the chemical elements.

Meyer was born in Varel, Oldenburg, on 19 August 1830, the son of a doctor of medicine. He began his university career by studying medicine at Zurich. He then took courses in chemistry at Heidelberg (under Robert Bunsen), in physics at Königsberg and in pathology at Würzburg, where he qualified as a physician in 1854. Four years later he gained his PhD from the University of Breslau. On the basis of this wide combination of interests he began his career as a science educator in 1859, holding various appointments until he became Professor of Chemistry at Karlsruhe Polytechnic in 1868. In 1876, he was appointed the first Professor of Chemistry at Tübingen University, where he remained for the rest of his life. He died in Tübingen on 11 April 1895.

In his book *Modern Chemical Theory* (1864), a lucid exposition of the contemporary principles of the science of chemistry, Meyer drew up a table which presented all the elements according to their atomic weights (relative atomic masses), relating the weights to chemical properties. In 1870 he published the results of his further researches in the form of a graph of atomic volume (atomic weight divided by density) against atomic weight, which demonstrated the periodicity in the variation of the elements' properties. He showed that each element will not combine with the same numbers of hydrogen or chlorine atoms, establishing the concept of valency. He coined the terms univalent, bivalent, trivalent and so on, according to the

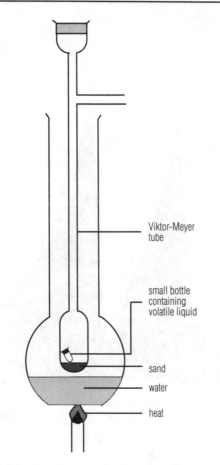

Viktor-Meyer tube

small bottle containing volatile liquid

sand

water

heat

Meyer *Viktor Meyer's apparatus for measuring vapour density.*

number involved. Consequently it became customary to divide the elements into groups defined by the number of hydrogen atoms with which a given element can combine (or displace).

Despite the fact that Meyer had reached his conclusions quite independently of Mendeleyev, he never claimed priority for his findings. And unlike Mendeleyev, he made no predictions about the composition and properties of any elements still to be discovered.

Meyer Viktor 1848–1897 was a German organic chemist best known for the method of determining vapour densities (and hence molecular weights) named after him. He was also the discoverer of the heterocyclic compound thiophene.

Meyer was born in Berlin on 8 September 1848. His undergraduate studies were carried out partly at Heidelberg (under Robert Bunsen), at Berlin (under Johann von Baeyer) and at Würtemberg. He gained his PhD at Heidelberg in 1867 while only 19 years old and three years later was appointed Professor of Chemistry at Stuttgart Polytechnic. In 1872 he progressed to the Technische Hochschule at Zurich and in 1885 he became a professor at

Göttingen. When Bunsen died in 1889, Meyer succeeded to his professorship at Heidelberg. A long period of working with iodine and bromine at high temperatures undermined his health and, in a fit of depression, he committed suicide by taking cyanide on 8 August 1897.

Meyer's discovery about the nature of vapour densities was based largely on previous work in this field, notably that of the Italian physical chemist Amadeo Avogadro who as early as 1811 had indicated the difference between atoms and molecules. Later, in his celebrated law, Avogadro maintained that equal volumes of two gases contain the same numbers of molecules (when the temperatures and pressures are also equal). When the two volumes of gas combine chemically, the individual atoms join to form molecules, whose volume depends on the proportion in which the atoms combine.

This welcomingly straightforward definition of one of the key processes in experimental chemistry met with little acceptance from the leading chemists of the day until, in 1871, Meyer gave an incontestable demonstration of its validity. He determined the molecular weights of volatile substances formed in this way by measuring their vapour densities (molecular weight, or relative molecular mass, is twice the vapour density).

He went on to a series of pyrotechnical studies in which he determined the vapour densities of inorganic substances at high temperatures. The results of this work, undertaken with his brother Karl, were published in 1885 in their book *Pyrotechnical Research*. At about the same time, Meyer described his organic vapour density studies in his *Textbook of Organic Chemistry* (2 vols, 1883–96).

In 1883, in the course of a lecture demonstration on the nature of benzene, Meyer was surprised to discover that the substance did not react in the way he had predicted. Benzene obtained from petroleum reacted as expected, whereas a purer sample synthesized from benzoic acid did not. From the impure benzene Meyer isolated thiophene, a heterocyclic compound containing sulphur, which much later was to become an important component of various synthetic drugs.

In the Viktor Meyer method for determination of vapour density (and hence relative molecular mass or molecular weight, which is twice the vapour density), the apparatus shown is used. A known mass of a volatile liquid is rapidly vaporized by raising its temperature well above the boiling point and the volume occupied by the resulting vapour is determined by measuring the volume of air it displaces. The inner tube (Viktor-Meyer tube) contains a little sand to break the fall of the small bottle of liquid introduced at the top and the heating jacket causes the liquid to boil, pushing off the ground-glass stopper of the small bottle. The heavy vapour remains at the bottom of the V-M tube and an equal volume of air is displaced and collected 'over water'. This volume is converted to standard temperature and pressure and, by using the fact that the relative molecular mass in grams occupies 22.4 l/39.4 pt at standard temperature and pressure (or the vapour density in grams occupies 11.2 l/ 19.7 pt), these values can be found.

The Viktor Meyer method is simple to carry out and adaptable, only small amounts of the volatile liquid being needed, and the method can be used over a wide temperature range.

Meyerhof Otto Fritz 1884–1951 was a German-born US biochemist who shared (with Archibald Hill) the 1922 Nobel Prize for Physiology or Medicine for his research into the metabolic processes involved in the action of muscles.

Meyerhof was born on 12 April 1884 in Hanover, the son of a merchant. He attended school in Berlin but was often absent because of a kidney disorder, as a result of which he received much of his early education at home. Later he became a medical student at the universities of Freiburg, Berlin, Strasbourg and Heidelberg. In 1909, he graduated from Heidelberg and worked in a medical laboratory there until 1912, when he moved to the University of Kiel as an assistant in the department of physiology, becoming a professor at the university in 1918. Shortly after receiving his Nobel prize, he was offered a Chair in Biochemistry in the United States. To keep him in Germany, Meyerhof was appointed the head of a new department specially created for him at the Kaiser Wilhelm Institute for Biology in Berlin-Dahlem; he held this position from 1924 to 1929, when he became Head of the Department of Physiology in a new institute for medical research at the University of Heidelberg. As a result of Adolf Hitler's rise to power in the 1930s, Meyerhof left Germany in 1938 and went to Paris, where he became Director of Research at the Institut de Biologie Physiochimique. In 1940, when France fell to Germany in the early part of World War II, he fled to the United States, where he was appointed Research Professor of Physiological Chemistry at the University of Pennsylvania, Philadelphia, a position he held for the rest of his life. Meyerhof became a US citizen in 1948, and died three years later on 6 October 1951.

Meyerhof's early work concerned energy exchanges in nitrifying bacteria, about which he published three papers between 1916 and 1917. He then became interested in the mechanism by which energy from food is released and utilized by living cells. In 1920, he showed that, in anaerobic conditions, the amounts of glycogen metabolized and of lactic acid produced in a contracting muscle are proportional to the tension in the muscle. He also demonstrated that 20–25% of the lactic acid is oxidized during the muscle's recovery period and that energy produced by this oxidation is used to convert the remainder of the lactic acid back to glycogen. Meyerhof introduced the term glycolysis to describe the anaerobic degradation of glycogen to lactic acid, and showed the cyclic nature of energy transformations in living cells. The complete metabolic pathway of glycolysis – known as the Embden–Meyerhof pathway (after Meyerhof and a co-worker) – was later worked out by Carl and Gerty Cori. Despite these and later revisions, Meyerhof's work remains the basic contribution to our knowledge of the very complex processes involved in muscular action.

Continuing his research on intracellular energy metabolism, between 1926 and 1927 Meyerhof demonstrated that glycolysis is not the result of bacterial activity, and in 1928, working with Lohmann, discovered that 50,000 joules per gram molecule of phosphate are liberated during the hydrolysis of creatine phosphate. In the following year Lohmann discovered adenosine triphosphate (ATP) in muscle and, with Meyerhof, began to study the new concept of oxidative phosphorylation. In the 1940s, Meyerhof found in the microsomes in muscle cells a new ATPase enzyme that is magnesium activated.

Michelson Albert Abraham 1852–1931 was a German-born US physicist who, in association with Edward Morley (1838–1923), performed the classic Michelson–Morley experiment for light waves. Michelson also made very precise determinations of the velocity of light. This work entailed the development of extremely sensitive optical equipment, and in recognition of his achievements in optics and measurement, Michelson was awarded the 1907 Nobel Prize for Physics, becoming the first American to gain a Nobel prize.

Michelson was born in Strelno in Germany, now Strzelno in Poland, on 19 December 1852. His family emigrated to the United States while he was a young child, and eventually settled in Nevada. Michelson went to school in San Francisco and then tried unsuccessfully to enter the US Naval Academy at Annapolis. However, he was allowed to plead his case with the President of the United States, Ulysses S. Grant, who directed that Michelson be admitted in 1869. After graduating in 1873, Michelson spent two years at sea and was then appointed as an instructor in physics and chemistry at the Academy. From 1880 to 1882, Michelson undertook postgraduate study at Berlin under Hermann Helmholtz (1821–1894) and at Paris. On returning to the United States, he left the navy and was appointed Professor of Physics at the Case School of Applied Science in Cleveland, Ohio. In 1889, Michelson moved to Clark University at Worcester, Massachusetts, and then in 1892 became Professor of Physics at the University of Chicago, a position he held until 1929. Among his honours were the presidency of the American Physical Society from 1901 to 1903 and the award of the Royal Society's Copley Medal in 1907. Michelson died on 9 May 1931.

Michelson began his scientific career in 1878, when he improved the rotating-mirror method of Jean Foucault (1819–1868) to determine the velocity of light. This led to the construction of an interferometer to detect any difference in the velocity of light in two directions at right angles. In this instrument, a beam was split into two beams at right angles. Each was sent to a mirror and the reflected beams were made to interfere with each other. The interference pattern produced was recorded and the interferometer turned through a right angle. If there was any difference in the velocity of light, it would cause the light to take different times to traverse the same paths and a change would occur in the interference pattern produced. Michelson built the interferometer to make an experiment suggested by

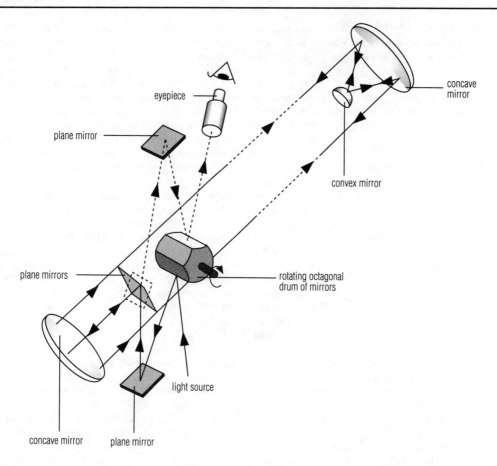

Michelson *Michelson determined the velocity of light by using a rotating drum faced with eight or more mirrors to time a light beam over a distance of more than 70 km/43 mi.*

James Clerk Maxwell (1831–1879) to detect the motion of the Earth through the ether, which was believed to exist as the medium in which light waves were propagated. If the Earth were moving through the ether, light would travel slower in the direction of the Earth's motion than at right angles to it. The amount of change in the interference pattern produced in the interferometer would indicate how fast the Earth was moving through the ether.

Michelson first undertook this experiment at Berlin in 1881. He could detect no change in the interference pattern and so later set about constructing a much more sensitive interferometer in collaboration with Morley, with whom Michelson began to work in 1885. The experiment was repeated in 1887 and still no effect due to the ether was found. This was a disappointment to Michelson, and he set about finding other uses for the interferometer. However, the experiment, which is now known as the Michelson–Morley experiment, marked a turning point in theoretical physics, for its negative result demonstrated that the velocity of light is constant whatever the motion of the observer. Two explanations could account for this. One was that the ether does exist but moves with the Earth; this was disproved by Oliver Lodge (1851–1940) in 1893. The other

was that the ether does not exist but that moving objects contract slightly in the direction of their motion. This was put forward by George FitzGerald (1851–1901) and became part of the special theory of relativity proposed by Albert Einstein (1879–1955) in 1905. Thus the Michelson–Morley experiment was of crucial importance in the formulation of relativity, though Michelson himself remained sceptical.

Michelson's other achievements were concerned with extremely precise measurement. In 1892 and 1893, he and Morley redefined the length of the standard metre kept at Paris in terms of a certain number of wavelengths of monochromatic light, using a red line in the cadmium spectrum. This method of defining the standard unit of length was finally adopted in 1960, and a krypton line is now used.

Michelson also developed his interferometer into a precision instrument for measuring the diameters of heavenly bodies and in 1920 announced the size of the giant star Betelgeuse, the first star to be measured. Michelson then undertook extremely precise determinations of the velocity of light, measuring it over a 35-km/22-mi path between two mountain peaks in California. In 1926, he obtained the value 299,796 ± 4 km per second, the range including the present value of 299,792.5 km/186,291 mi per second.

Michelson later returned to the Michelson–Morley experiment and repeated it yet again to try and produce a positive result. He did not succeed a third time, and this helped to verify Einstein's theories. No one else has succeeded since.

Michelson's gift for extremely precise experimental work enabled him to make one of the most important discoveries in physics. It is ironic that only such ability could yield the negative result that was required for a new understanding of the nature of the universe.

Midgley Thomas 1889–1944 was a US industrial chemist and engineer who discovered that tetraethyl lead is an efficient antiknock additive to petrol (preventing pre-ignition in car engines) and introduced Freons as the working gases in domestic refrigerators. Today the most commonly used is Freon 12, which is difluorodichloromethane, CF_2Cl_2.

Midgley was born in Beaver Falls, Pennsylvania, on 18 May 1889, the son of an engineer and inventor. He studied mechanical engineering at Cornell University, graduating in 1911. Early in World War I he worked on torpedo control systems. In 1916, he went to Ohio to work in the research department of an engineering firm, the Dayton Engineering Laboratories Company. He became a vice president of the Ethyl Corporation in 1923 and ten years later was a director of the Ethyl-Dow Chemical Company. In 1940, he contracted poliomyelitis and became paralysed. He died in Worthington, Ohio, on 2 November 1944.

While working for the Dayton Engineering Company, Midgley discovered empirically that ethyl iodide (iodoethane) prevents pre-ignition in car engines using low-octane fuel. He spent several years teaching himself the relevant chemistry and looking for a less expensive additive. In 1921, he experimented with tetraethyl lead, showing it to be an efficient antiknock agent. It became a standard additive, but its use was questioned from the 1970s because of the hazardous effects of airborne lead compounds emitted in exhaust fumes, and from the 1980s unleaded petrol became increasingly common.

Midgley used ethyl bromide (bromoethane) in his research, which led him to investigate the bromine in sea water and demonstrate a method for its extraction. Dayton Engineering was taken over by General Motors and he continued to work in their research laboratories.

In 1930 Midgley introduced Freon (CF_2Cl_2) as a non-inflammable, non-toxic refrigerant, which rapidly replaced ammonia, methyl chloride (chloromethane) and sulphur dioxide as the volatile but liquefiable gas in the mechanisms of domestic refrigerators. This gas and closely related compounds are still used universally for this purpose in freezers, fridges and air-conditioning units. For many years after World War II, Freons were also extensively used as propellants in aerosol containers. A hypothesis that continued release of these compounds into the atmosphere could damage the Earth's ozone layer (which acts as a filter to prevent excessive ultraviolet radiation reaching the surface of the Earth) led to restrictions or bans on the use of Freon propellants in the 1970s.

Miller Stanley Lloyd 1930– is a US chemist who carried out a key experiment that demonstrated how amino acids, the building blocks of life, might have arisen in the primeval oceans of the primitive Earth.

Miller was born in Oakland, California, on 7 March 1930. He graduated from the University of California in 1951 and three years later gained his PhD from the University of Chicago. From 1954 to 1955 he was a postdoctoral Jewett Fellow at the California Institute of Technology; he then worked for five years in the Department of Biochemistry at the Columbia College of Physicians and Surgeons, first as an Instructor in Biochemistry and then as Assistant Professor. From 1960 he held appointments at the University of California in San Diego as Assistant Professor, Associate Professor and finally Professor of Chemistry.

While working for his PhD under Harold Urey, Miller set himself the task of trying to account for the origin of life on Earth. He chose as the experimental substances the components that had been proposed for the Earth's primitive atmosphere by Urey and Alexandr Oparin. He used sterilized and purified water under an 'atmosphere' of methane, ammonia and hydrogen, which he circulated for a week past an electric discharge (to simulate the likely type of energy source). He then analysed the mixture and surprisingly, after such a relatively short experimental period, found organic compounds such as simple amino acids. These were of great significance because of their ability to combine together to form proteins.

It is important when assessing the significance of the results (which have been duplicated by various other workers since) to bear in mind the time scale of the experiment in comparison with that for the origination of life on Earth. If the conditions of Miller's experiment had been continued for millions of years (instead of just seven days) and other plausible prebiotic syntheses added in, the oceans of the primordial Earth – analogous to the water used in the experiment – would have become rich with a whole range of different types of organic molecules: the so-called prebiotic soup.

There are many other steps that are needed to develop a system that is capable of self-replication. Some of these have also been worked out in the laboratory, such as the prebiotic syntheses of purines, pyrimidines and sugars needed to make up RNA or DNA. And we still do not know how the nucleic acids first began to self-replicate.

Millikan Robert Andrews 1868–1953 was a US physicist who made the first determination of the charge of the electron and of Planck's constant. For these achievements, he was awarded the 1923 Nobel Prize for Physics.

Millikan was born at Morrison, Illinois, on 22 March 1868. He showed no interest in science as a child and became interested in physics only after entering Oberlin College in 1886, where he obtained his BA in 1891 and MA in 1893. Millikan then went to Columbia University to continue his studies in physics, gaining his PhD in 1895.

He then went to Germany to study with Max Planck (1858–1947) at Berlin and Hermann Nernst (1864–1941) at Göttingen before taking up an assistantship in physics at the University of Chicago in 1896.

Millikan remained at Chicago until 1921. At first he concentrated on teaching but from 1907, when he became Associate Professor of Physics, he took more interest in research and began his experiments to find the electronic charge, completing this work in 1913. In 1910, Millikan became Professor of Physics at Chicago and during World War I was director of research for the National Research Council, which was concerned with defence research. In 1921, Millikan moved to Pasadena to become Director of the Norman Bridge Laboratory at the California Institute of Technology. He retained this position until 1945, when he retired. In addition to the 1923 Nobel Prize for Physics, Millikan received many honours, including the award of the Royal Society's Hughes Medal in 1923 and the Presidency of the American Physical Society from 1916 to 1918. Millikan died at Pasadena on 19 December 1953.

Millikan employed a method to determine the electronic charge that was simple in concept but difficult in practice, and a satisfactory result took him the five years from 1908 to 1913 to achieve. He began by studying the rate of fall of water droplets under the influence of an electric field. Millikan conjectured correctly that the droplets would take up integral multiples of the electronic charge, which he would be able to compute from the strength of field required to counteract the gravitational force on the droplets. By 1909, Millikan had arrived at an approximate value for the electronic charge. However, the droplets evaporated too quickly to make precise determination possible and Millikan switched to oil droplets. These were far less volatile and furthermore Millikan was able to irradicate the suspended droplets to vary their charge. In 1913, Millikan finally announced a highly accurate value for the electronic charge that was not bettered for many years.

Millikan also worked on a study of the photoelectric effect during this period, investigating the interpretation of Albert Einstein (1879–1955) that the kinetic energy of an electron emitted by incident radiation is proportional to the frequency of the radiation multiplied by Planck's constant. Millikan took great pains to improve the sensitivity of his apparatus and announced in 1916 that Einstein's equation was valid, thereby obtaining an accurate value for Planck's constant.

After World War I, Millikan moved into two new areas of research. In the 1920s, he investigated the ultraviolet spectra of many elements, extending the frequency range and identifying many new lines. Millikan also undertook a thorough programme of research into cosmic rays, a term that he coined in 1925, when he proved that the rays do come from space. Millikan did this by comparing the intensity of ionization in two lakes at different altitudes. He found that the intensity was the same at different depths, the absorptive power of the difference in the depth of water being equal to the absorptive power of the depth of atmosphere between the two altitudes. This proved that the rays

producing the ionization must have passed through the atmosphere from above and could not have a terrestrial origin.

Millikan went on to assert that cosmic rays were electromagnetic waves, a theory disproved by Arthur Compton (1892–1962) in 1934, when he demonstrated that they consist of charged particles. However, in the course of his research, Millikan directed Carl Anderson (1905–1991) to study cosmic rays in a cloud chamber and, as a result, Anderson discovered the positron.

Millikan's achievements are of fundamental value in the history of science. His determination of the charge on the electron was very important because it proved experimentally that electrons are particles of electricity, while the determination of Planck's constant was vital to the development of quantum theory.

Mills William Hobson 1873–1959 was a British organic chemist famous for his work on stereochemistry and on the synthesis of cyanine dyes.

Mills was born in London on 6 July 1873, although his family came from Lincolnshire and he was educated at Spalding Grammar School. He then went to Uppingham School and Jesus College, Cambridge, where he began research under Thomas Easterfield and became a Fellow of the college in 1899. Also in 1899 he went for two years to Tübingen in Germany to work under Hans von Pechmann; while there he met the British inorganic chemist Nevil Sidgewick, who was to become his lifelong friend. He returned to England in 1902 as Head of Chemistry at the Northern Polytechnic Institute, London, and ten years later he was appointed Professor of Natural Philosophy at Jesus College. In 1919, he became a university lecturer and in 1931 the university created for him a Readership in Stereochemistry, a post he held until he retired in 1938. He died in Cambridge on 22 February 1959.

Mills' early research concerned stereochemistry, particularly optical isomerism – the phenomenon in which pairs of (usually organic) compounds differ only in the arrangements of their atoms in space. Certain oximes (derivatives of aldehydes and ketones) were known to exist in two or more isomeric forms. In 1931, with B. C. Saunders, Mills prepared the ortho-carboxyphenylhydrazone of β-methyltrimethylenedithiolcarbonate and resolved it into two optically active forms. This work confirmed the theory of Arthur Hantzsch and Alfred Werner that optical isomerism of oximes is caused by the non-planar orientation of the nitrogen atom valencies.

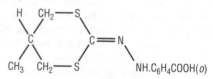

It had also been recognized that a spirocyclic compound consisting of two carbon rings linked together by a common carbon atom should show optical activity if the rings possessed appropriate substituents to ensure molecular dissymmetry. In 1921, Mills and C. R. Nodder synthesized and resolved the first compound of this kind, the ketodilactone of benzophenone-2,4,2′,4′-tetracarboxylic acid.

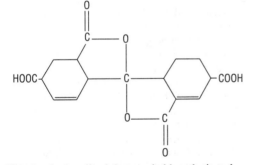

Chemists had realized that a suitably substituted compound of allene (propadiene) would also be dissymmetric, but had been unable to synthesize any such compound with acidic or basic groups for resolution. After six years of patient research Mills and Maitland synthesized an allene derivative with two phenyl and two naphthyl substitutents.

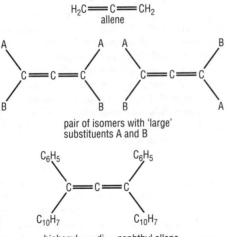

pair of isomers with 'large' substituents A and B

αγ-biphenyl-αγ-di-α-naphthyl allene

By extending these general methods, they went on to show that the nitrogen atom of a quaternary ammonium salt has a tetrahedral configuration (like a carbon atom), and is not situated at the centre of a square-based pyramid. On the other hand, in 1935 Mills and his co-workers produced stereochemical evidence that a four coordinated platinum atom has a planar as opposed to a tetrahedral configuration.

Other workers were puzzled by the fact that a biphenyl molecule shows optical activity if there are suitable substituents in the 2, 2′, 6 and 6′ positions. In 1926, Mills was the first to point out that the size of the substituents can prevent the free rotation of the two phenyl groups about their common axis. He then became interested in the general problem of restricted rotation and investigated substituted derivatives of naphthalene, quinolene and benzene.

Mills's other field of research was into the synthesis of cyanine dyes. In 1905, German chemists had synthesized pinacyanol, which when added to photographic emulsions extended their sensitivity into the red region of the spectrum, as well as into the blue, violet and ultraviolet regions. Mills and his co-workers investigated similar dyestuffs for preparing photographic emulsions, mainly for use by the military in World War I.

After his retirement Mills had more time to devote to his interest in natural history, especially to the British bramble plant, *Rubus fructiosus*. His collection of 2,200 specimens of 320 of the 389 'microspecies' of this plant is arranged in systematic order and now housed in the Botany Department of Cambridge University.

Milne Edward Arthur 1896–1950 was a brilliant British astrophysicist, mathematician and theoreticist, most famous for his formulation of a theory of relativity parallel to Einstein's general theory which he called kinematics relativity.

Milne was born in Hull, Yorkshire, on 14 February 1896. After attending local schools, he won scholarships to Hymer's College, Hull (where he showed exceptional talent in mathematics), and then to Trinity College, Cambridge, which he entered in 1914. Poor eyesight prevented him from taking up active service on the outbreak of World War I, but in 1916 he began work at the Anti-Aircraft Experimental Section of the Munitions Inventions Department, and carried out important research there, surrounded by many of the country's leading mathematicians and physicists. It was during these years that Milne became interested in atmospheric theory. His research contributions during the war years were later recognized in the award of an MBE.

Milne returned to Cambridge in 1919 and wrote three papers that derived, in part, from the work he had been engaged in during the war. On the strength of these papers he was elected Prize Fellow of Trinity College in 1919 (a post he held until 1925). He was offered the post of Assistant Director of the Solar Physics Observatory at Cambridge in the same year, but deferred acceptance for a year while he made additional studies of the subject. From 1921 to 1924 he served as Lecturer in Astrophysics and from 1924 to 1925 as Lecturer in Mathematics at Cambridge. Milne then moved to Manchester University to take up the Chair of Applied Mathematics. He was elected Fellow of the Royal Society in 1926, at the age of 30. Three years later, he accepted the Rouse Ball Chair of Mathematics at Oxford University, where he was elected a Fellow of Wadham College. He held both of these posts until his death.

From 1939 to 1944, during World War II, Milne served as a member of the Ordnance Board at Chislehurst, part of the Ministry of Supply. There he carried out research very similar to the sort he had done during World War I. His later years were affected by deteriorating health and he died suddenly while at a conference in Dublin on 21 September 1950.

Milne received many honours in recognition for the contributions he made to astrophysics and to cosmology, including the Royal Medal from the Royal Society in 1941. A member of many prominent international scientific organizations, he was also the author of several books.

There were three main phases in the development of Milne's work: his research on stellar spectra and radiative equilibrium at Cambridge and Manchester; his study of stellar structure during his early years at Oxford; and his formulation and development of kinematic relativity during the later years at Oxford.

At Cambridge Milne collaborated with Ralph Fowler to extend earlier theoretical work by Meghnad Saha (1894–1956) on thermal ionization. Together they developed a temperature scale for stellar spectra that has been used to improve the understanding of stellar surface conditions. It was then recommended to Milne that he investigate stellar atmospheres, and he turned to the primary work done by Karl Schwarzschild (1873–1916) and Arthur Schuster (1851–1935) on radiative equilibrium. Interested in the balance between radiation pressure and gravity in the Sun's chromosphere, Milne found that under certain conditions there would be great instability, and that such instability could lead to the emission of atoms from the Sun at speeds as high as 1,000 km/600 mi per second. Milne derived a mathematical method to describe the net amount of radiation passing through the atmosphere.

He continued to investigate the structure of stellar atmospheres and he was able to relate the optical depth (or opacity) to observations on spectral frequency. At the time of his move to Oxford, Milne began to re-examine some of the work done by Arthur Eddington (1882–1944) on stellar structure. His proposals for a number of changes to the theory were not generally accepted, but his critical approach to the subject prompted serious scientific reappraisal. It was at about this time that Milne suggested that a decrease in luminosity might cause the collapse of a star, and that this would be associated with nova formation.

In 1932, Milne began the development of his theory of kinematic relativity. He felt that kinematics could be used to explain the properties of the universe, thus providing at least an equivalent, a parallel rendering, of Einstein's general theory of relativity. Basing his theory on Euclidean space and on Einstein's special theory of relativity alone, Milne was able to formulate a system of theoretical cosmology (in which he was also able to derive a gravitational theory) and systems of dynamics and electrodynamics. Furthermore, he provided a more acceptable estimate for the overall age of the universe (10,000 million years) than that provided by the general theory of relativity.

Kinematic relativity has not in fact replaced the general theory, largely because it cannot be used to resolve detailed issues. Nor does it provide any genuinely new insights into the physical nature of the universe. Its approach to the problem of time and space–time did, however, stimulate considerable and creative interest.

Milstein César 1927– is an Argentinian molecular biologist who has performed important research on the genetics, biosynthesis and chemistry of immunoglobulins (antibodies), developing a technique for preparing chemically pure monoclonal antibodies.

Milstein was born on 8 October 1927 in Bahia Blanca, Argentina. From 1939 to 1944 he was educated at the Collegio Nacional, Bahia Blanca, from which he gained his Bachiller; in 1945 he went to the University of Buenos Aires, from which he graduated in 1952. He remained at Buenos Aires University until 1963, initially to study for his doctorate (which he obtained in 1957), then as a member of the staff of the Institute of Microbiology. During a period of leave of absence, Milstein worked in the Department of Biochemistry at Cambridge University, from which he gained his second doctorate in 1960. He returned to Cambridge in 1963 as a member of the staff of the Medical Research Council Laboratory of Molecular Biology and later became head (with F. Sanger) of its Protein Chemistry Division.

Milstein and his colleagues were among the first to determine the complete sequence of the short, low-molecular-weight part of the immunoglobulin molecule (known as the light chain). He then determined the nucleotide sequence of a large portion of the messenger RNA for the light chain and found that there is only one type of messenger RNA for both domains within that chain. The separate domains within the heavy (high molecular weight) and light chains are called constant and variable, and Milstein deduced that although the genes for the constant and variable domains may be separate in the germ line, these genes must have come together in the antibody-producing cells. This finding led Milstein to develop his technique for preparing chemically pure monoclonal antibodies. Working with Köhler in 1975, Milstein succeeded in fusing myeloma cells (which are easily cultured but produce only their own predetermined immunoglobulin) with spleen cells (which cannot be cultured but can produce immunoglobulin against any antigen with which an animal has been injected) to produce hybrid cells that can be cultured and that can produce antibodies against a wide range of antigens. Moreover, by selecting clones Milstein and Köhler obtained cell lines that secreted only one chemically homogenous antibody. This was a revolutionary piece of research, because the permanent cultures derived from one clone can be propagated indefinitely and can therefore provide an unlimited supply of a specific immunoglobulin. The technique was later extended to human cells and can also be used to prepare purified antibodies against impure antigens.

Minkowski Hermann 1864–1909 was a Lithuanian mathematician whose introduction of the concept of space–time was essential to the genesis of the general theory of relativity.

Minkowski was born at Alexotas, near Kaunas, on 22 June 1864, but moved to Germany with his family in 1872 and settled in Königsberg (now Kaliningrad in Russia). There he was educated, receiving his doctorate from the University of Königsberg in 1885. He lectured at the University of Bonn from 1885 to 1894, when he returned to

Königsberg as Associate Professor of Mathematics. After only two years there, he accepted the post of Professor of Mathematics at the Federal Institute of Technology at Zurich. His last years were spent at Göttingen University, where he held the Chair created for him by David Hilbert from 1902 until his death on 12 January 1909.

Minkowski's genius first declared itself when, at the age of 19, he shared the French Academy's Grand Prix for a paper on the theory of quadratic forms with integral coefficients. Quadratic forms remained his chief interest, although he failed to find a method, like many before and after him, of generalizing Karl Gauss's work on binary quadratic forms so as to describe 'n-ary' forms. His passion for pure mathematics led him also into problems in number theory, to which his concept of the geometry of numbers constituted an important addition. This was achieved by his use of the lattice in algebraic theory, and it was his research into that topic which led him to consider certain geometric properties in a space of *n* dimensions and so to hit upon his notion of the space–time continuum. The principle of relativity, already put forward by Jules Poincaré and Albert Einstein, led Minkowski to the view that space and time were interlinked. He proposed a four-dimensional manifold in which space and time became inseparable, a model which has not received much acceptance; but the central idea – contained in his *Raum und Zeit* (1909) – was, as Einstein allowed, necessary for the working out of the general theory of relativity.

So, in a manner not uncommon in the history of science, Minkowski, who was moved chiefly by a love of pure mathematics, made his most brilliant discovery in the field of mathematical physics.

Minkowski (or Minkowsky), Rudolph Leo 1895–1976 was a German-born US astrophysicist, responsible for the compilation of the incomparably valuable set of photographs found in every astronomical library, the National Geographic Society Palomar Observatory Sky Survey. A lead-ing authority on novae and so-called planetary novae, he was a pioneer in the science of radioastronomy.

Minkowski was born in Strasbourg (then in Germany, now in France) on 28 May 1895. His family was of Polish extraction, and after he had been educated in local schools, he attended the University of Breslau (now Wroclaw) and gained a PhD in optics in 1921. After a year on the physics teaching staff at Göttingen, he became Professor of Physics at Hamburg University. Increasing oppression of those of his racial background in Germany in 1935, however, caused him to emigrate to the United States, where he joined his friend and former colleague, Walter Baade (1893–1960), at Pasadena. Shortly afterwards, he found a position as Research Assistant at the Mount Wilson Observatory, California; comfortably settled into a regular staff position, he remained there for more than 20 years before transferring to Palomar Observatory. Minkowski retired twice, first from the Mount Wilson and Palomar Observatories in 1959, then, after spending a year at the University of Wisconsin and another five at the University

of California at Berkeley, again in 1965. He continued his observations and investigations, publishing scientific papers until well into the 1970s. He died on 4 January 1976.

Minkowski became one of the world's leading investigators of the universe's more violent phenomena. One of his central interests during a long career was the examination of supernovae. He quickly distinguished between the two principal types of supernovae and studied the spectra of many individual types in other galaxies. In collaboration with Walter Baade he studied the remnants of the few supernovae known to have appeared in our own Galaxy. The Crab nebula was the subject of a particularly stringent examination. Its importance in astrophysics has increased as it has been discovered successively to be a radio source, an X-ray source and a pulsar.

It was the collaboration between Minkowski and Baade that first identified a 'discrete radio source', Cygnus A, in 1951 – the first time that an extragalactic radio source was optically identified, albeit as an extremely distant object. But because of the distance, the radio emission was seen to be of immense power. It was Baade's view (in which Minkowski, at least at the beginning, concurred) that Cygnus A represented the collision of two galaxies; the fact that two apparent nuclei were identifiable by their radio signals seemed evidence enough. (Later investigation, however, has failed to find the expected corroborative evidence and the theory of colliding galaxies is at the moment discredited.)

The nature of planetary nebulae (dense stars in the process of becoming white dwarfs and shedding mass to do so) was another of Minkowski's long-term interests. In addition to his analysis of these objects he set up a survey, using a 25-cm/10-in telescope, that more than doubled the number of planetary nebulae then known. It was at this time that he took over supervision of the National Geographic Society Palomar Observatory Sky Survey.

During Minkowski's study of supernovae and disturbed galaxies, his investigation into the internal motions within elliptical galaxies was important; but it was superseded only 12 years later when new and improved observing equipment became available. Minkowski determined the optical red-shift of the radio source 3C 295 (which remained the farthest point on the velocity–distance diagram of cosmology for 15 years) in his last observing run at the Palomar 500-cm/200-in telescope.

Mitchell Peter Dennis 1920–1992 was a British biochemist who performed important research into the processes involved in the transfer of biological energy. He received many honours for his work, including the award of the 1978 Nobel Prize for Chemistry.

Mitchell was born on 29 September 1920 in Mitcham, Surrey. He was educated at Queen's College, Taunton, and then at Jesus College, Cambridge, from which he graduated in 1943 and gained his doctorate in 1950. He worked in the Biochemistry Department of Cambridge University from 1943 to 1955, becoming a demonstrator in 1950. In

1955 he was appointed Director of the Chemical Biology Unit in the Department of Zoology at Edinburgh; he was later promoted to Senior Lecturer then, in 1962, to Reader. In the following year he established and became Director of the privately run Glynn Research Institute at Bodmin, Cornwall.

In the early 1960s the way in which the synthesis of ATP (adenosine triphosphate) is linked with the transfer of electrons was still unknown. At the intracellular level, metabolic energy is stored in the form of ATP, produced by phosphorylating the diphosphate ADP (in effect the metabolic energy is stored in ATP's extra chemical bond). The energy needed for this reaction is produced by the transfer of electrons along a chain of proteins attached to the double membrane of mitochondria. In 1961, Mitchell postulated that, simultaneously with this electron transfer, protons are expelled from the outer surface of the inner mitochondrial membrane and are transmitted through the outer membrane by osmosis, thus setting up an ion-concentration gradient (and therefore a potential difference). He further suggested that this potential difference is the form in which the energy needed to convert ADP to ATP is stored. It has since been shown that other cellular processes that require energy involved the mechanism proposed by Mitchell and that the chemical equilibrium in the phosphorylation of ADP to ATP is highly sensitive to the concentration of ions, which is controlled by the mitochondrial membrane. Although greeted with great scepticism at first, Mitchell's theory is now considered to be correct and it has been established as the basic principle in the science of bioenergetics.

Mitchell Reginald Joseph 1895–1937 was a British aeronautical engineer who, in Britain at least, was largely responsible for the development of the high-speed monoplane. His early work on the design of seaplanes for the Schneider Trophy competition led to ideas for the construction of a single-engined fighter; this became the Spitfire.

Mitchell was born in Talke, near Stoke-on-Trent, on 20 May 1895. His father was a schoolmaster and later a printer. His early childhood was spent at Longton, a few miles from his birthplace, and he went first to the village school; and later to Hanley High School. On leaving school he became apprenticed to a firm of locomotive builders in Stoke. Mitchell realized the importance of sound scientific principles in engineering, and continued his education in the evenings at technical colleges. It was not an easy way to get the scientific and practical expertise needed to become a design engineer, but he succeeded. He soon passed from the workbench to the drawing office, and in 1916 he moved to the Supermarine Aviation Company at Southampton. He had always had an interest in aeroplanes and had built models to test his own theories. Now he was able to test his ideas on a larger scale.

Aviation was still in its infancy, but was rapidly developing. Jacques Schneider, the son of a wealthy armaments manufacturer, was impressed by the flights of Wilbur Wright in France in 1908, and decided at the Aero Club of France in 1912 to offer a trophy (Le Coupe d'Aviation Maritime Jacques Schneider) to the national aero club of the country which produced the fastest seaplane. If a country won the trophy three times within five years, then it was to retain it.

Mitchell first became involved with producing aeroplanes for the trophy for the 1919 race (the third contest). Partly under his direction, the Supermarine Sea Lion (G-EALP) was adapted from a standard design. No aeroplane completely finished the course, Mitchell's Sea Lion having retired during the first lap.

A year later Mitchell became chief engineer for Supermarine. The firm at that time specialized in flying boats, and later models of the Sea Lion flew with distinction in the 1922 race, when Sea Lion II came first with a speed of 234.4 kph/145.7 mph and in 1923 when Sea Lion III came third at 252.9 kph/157.17 mph. For the next contest, in 1924, Mitchell was allowed to design his own aeroplane. This was the Supermarine S4. It was a monoplane and the whole wing section was made in one piece. The radiator for the engine was on the underside of the wing and the fuel was carried in the floats. The S4 looked fast and elegant and, was probably too fast for its design for it crashed during the trial before the race, previously having captured the world seaplane speed record at 365.055 kph/226.752 mph. In the 1927, 1929 and 1931 contests, Mitchell's S5, S6 and S6B – the direct decendents of the ill-fated S4 – took the trophy with speeds finally increased to 547.30 kph/340.08 mph. The Schneider Trophy was won forever for Britain and it now has a place of honour in the Science Museum, London. This was not quite the end of the story: the S6B S1585 set a speed record of 668.19 kph/415.2 mph on 29 September 1931 with George Stainforth at the controls.

To fulfil an Air Ministry contract for a single-engined fighter, Mitchell used and extended the ideas that had been embodied in the Schneider Trophy winners. This was not a simple task because he had to develop a reliable aeroplane that could fly time and again carrying a heavy load of fuel and ammunition at high speeds. His first aeroplane built directly to the specifications is said to have been mediocre, but when he was allowed full rein for his own ideas he produced the prototype of the Spitfire (K 5054). This was first flown by Captain J. Summers on 5 March 1936.

The Spitfire came into being when the storm clouds of war were gathering and it proved to be Britain's most important, but by no means only, fighter aeroplane. The early Spitfires had a top speed of 557 kph/346 mph, but the later ones were capable of about 740 kph/460 mph. More than 19,000 Spitfires were built, adapted to every role and circumstance appropriate to a high-speed single-engined aeroplane. Although faster single-engined fighters were eventually built, the manoeuverability and all-round adaptability of the Spitfire made it one of the most successful aeroplanes ever.

Mitchell saw very little of his greatest creation after its first flight. When it took to the air he was already seriously ill with tuberculosis. He died on 11 June 1937, aged 42 years.

Mitscherlich Eilhard 1794–1863 was a German chemist, famous for his discovery of isomorphism (the phenomenon in which substances of analogous chemical composition crystallize in the same crystal form). He also synthesized many organic compounds for the first time.

Mitscherlich was born in Neuende, Jever (now part of Wilhelmshaven) on 17 January 1794, the son of a minister. He was educated at Jever and in 1811 entered Heidelberg University to study Oriental languages, later continuing this study at Paris, with the intention of becoming a diplomat. The fall of Napoleon ended that prospect and he returned to Germany, enrolling at Göttingen to read science and medicine. He thought that he might reach the Orient as a ship's doctor (since he could not do so as a diplomat) but became increasingly interested in chemistry and in 1818 went to Berlin to work in the laboratory of the botanist Heinrich Link. There he began his study of crystallography.

In 1819 Mitscherlich met Jöns Berzelius while he was visiting Berlin, and when the Prussian Minister of Education offered Berzelius the professorship of chemistry at Berlin University following the death of Martin Klaproth, he declined but recommended Mitscherlich for the appointment. But at only 25 years old Mitscherlich was thought to be too young and as a compromise he was sent to Stockholm to work with Berzelius for two years, to widen his knowledge of chemistry. He returned to Berlin in 1822 as Assistant Professor of Chemistry, becoming Professor three years later (finally succeeding Klaproth) and retaining the appointment until his death. He died in Berlin on 28 August 1863.

Mitscherlich began studying crystals in Link's laboratory in 1818. He observed that crystals of potassium phosphate and potassium arsenate appear to be nearly identical in form. He asked the mineralogist Gustav Rose (1798–1873) to instruct him in exact crystallographic methods so that he could make precise measurements, and then applied spherical trigonometry to the data he obtained. In his first publication he announced that the sulphates of various metals, as well as the double sulphates of potassium and ammonium, crystallize in like forms provided that they have the same amounts of water of crystallization. He further stated his hope that 'through crystallographic examination the composition of bodies will be determined with the same certainty and exactness as through chemical analysis'.

Berzelius immediately recognized the importance of Mitscherlich's work and applied it to his determinations of atomic weights; he was later able to correct the atomic weights (relative atomic masses) of 27 elements. While he was in Stockholm, Mitscherlich extended his researches to phosphates, arsenates and carbonates, publishing the results in 1822 and introducing the term isomorphism. He continued the work after his return to Berlin, refining his isomorphism law and establishing more and more classes of compounds to which it applies. In 1827, during the course of this work, he discovered selenic acid.

In 1834, Mitscherlich synthesized benzene (introducing the name, which he termed *Benzin*) by heating calcium benzoate (the calcium salt of benzene carboxylic acid). Two years earlier he had synthesized nitrobenzene, and he went on to prepare azobenzene, benzophenone and benzene sulphonic acid. He recognized that the part played by the oxides of nitrogen in the chamber process for making sulphuric acid is that of a catalyst, and showed that yeast (which in 1842 he identified as a microorganism) can invert sugar in solution. He maintained his early interest in geology and mineralogy, and was particularly concerned with the production of artificial minerals, achieving valuable experimental results in this area.

In 1829, Mitscherlich published his influential *Lehrbuch der Chemie*, which in less than 20 years had run to four further new editions in German as well as two in French and one in English. This work continued Mitscherlich's lectures on all aspects of pure and applied chemistry (and included a considerable amount of physics), characterized by their exemplary clarity and ingenious experiments.

Mitscherlich's youngest son also became a chemist and, with help from his father, developed the Mitscherlich process for extracting cellulose from wood pulp by boiling it with calcium bisulphite (calcium hydrogen sulphite), which became the basis of the German cellulose industry.

Möbius August Ferdinand 1790–1868 was a German mathematician and theoretical astronomer, whose name is attached to several discoveries, most notably the 'barycentric calculus'.

Möbius was born at Schulpforta, near Naumburg, on 17 November 1790. His father was a dancing master and his mother a descendant of Martin Luther. Until the age of 13 he was educated at home by his father and, after his father died in 1798, by his uncle. Although he showed a talent for mathematics at school (1803–1809), he went to the University of Leipzig in 1809 to study law. He soon abandoned law for mathematics and astronomy and in 1813 he went to Göttingen University to receive two semester's instruction from Karl Gauss. He received his doctorate from Leipzig in 1814 and in the following year was appointed an instructor in astronomy there. In 1816, he was promoted to Extraordinary Professor of Astronomy and was appointed as an observer at the Leipzig Observatory. He spent time in the next few years visiting observatories throughout Germany and his recommendations for the refurbishing and reconstruction of the Leipzig Observatory were carried out in 1821. In 1844, he was made a full Professor of Astronomy and Higher Mechanics and in 1848 Director of the observatory. He died at Leipzig on 26 September 1868.

Möbius made important contributions to both astronomy and mathematics. His chief astronomical work was *Die Elemente der Mechanik des Himmels* (1843), a somewhat novel book, in that it provided a thorough mathematical discussion of celestial mechanics without resort to higher mathematics. His most important mathematical work, now regarded as something of a classic, was *Der Barycentrische Calkul* (1827). Möbius had formulated his idea of the barycentric calculus in 1818. The word 'barycentric', formed

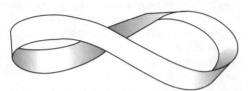

Möbius *The Möbius strip has only one side and one edge. It is made by attaching one end of a strip to the other – but with a half-twist first.*

from the Greek *barys* for 'heavy', means 'pertaining to the centre of gravity'. Möbius began with the well-known law of mechanics that several weights positioned along a beam can be replaced by a single weight, equal to the sum of the other weights, at the centre of the beam's gravity. From that law he constructed a mathematical system in which numerical coefficients were assigned to points. The position of any point in the system could be expressed by varying the numerical coefficients of any four or more non-coplanar points. This calculus proved helpful in a number of geometrical problems.

The treatise also introduced the principle of duality and gave a thorough treatment to the cross ratio; in addition, it included a discussion of the 'Möbius net', a concept later of great value in the development of projective geometry. A number of other discoveries bear his name: the Möbius tetrahedrons, two tetrahedrons which mutually circumscribe and inscribe each other, which he described in 1828; the Möbius function in number theory, published in 1832; and the Möbius strip.

The Möbius strip, presented in a paper discovered only after his death, was devised by him to illustrate the properties of one-sided surfaces, and consists of a length of paper connected at its ends with a half-twist in the middle. Credit for its discovery is shared with Johann Benedict Listing, a nineteenth-century German mathematician who is now known to have discovered it independently at the same time. The strip is one of Möbius's two contributions to the pre-natal stage of the history of topology. The other is his five-colour problem, given as a poser during a lecture of 1840. The problem he set was to find the least number of colours required on a plane map to distinguish political regions, given that each boundary line should separate two differently coloured regions. He drew maps requiring at least two, three or four colours. Neither Möbius nor anyone else has ever found a five-colour solution (though on a torus five-colour maps are easily drawn). Recently, it has been proved by computer analysis that four colours will always suffice.

Mohs Friedrich 1773–1839 was a major nineteenth-century German mineralogist.

Born at Gernrode in the Saxon Hartz Mountains, Mohs studied at Halle and at the Freiberg Mining Academy under the great teacher, A. G. Werner. After almost obtaining a post at a proposed mining school in Dublin (the undertaking fell through), he was appointed Professor of Mineralogy at Graz in 1812, and later at Freiberg – though by then he had abandoned Werner's doctrines. He finally occupied a chair in Vienna from 1826.

During his lifetime, Mohs achieved eminence for his system of the classification of minerals, dividing these into genera and species in the manner of Linnaeus. Though recognized as useful, the strategy was widely criticized on account of its failing to take sufficient account of chemical composition.

Mohs is chiefly remembered for having given his name to a scale of relative hardness. The Mohs scale extends from talc (1) to diamond (10). Higher numbered minerals will scratch anything beneath them or equal to them in hardness. Despite first appearances, the scale is not linear (the simple unit differences on Mohs's scale do not correspond straightforwardly to differences in hardness) and it possesses only limited applicability to crystals (the hardness of crystals tends to differ in separate crystal directions). The main use of the scale has been in field mineralogy.

Mond Ludwig 1839–1909 was a German-born chemist who established many industrial chemical processes in Britain. He gave his name to a method of extracting nickel from nickel carbonyl, one of its volatile organic compounds.

Mond was born at Kassel on 7 March 1839, the son of a well-to-do Jewish merchant. He studied chemistry at Marburg (under Adolf Wilhelm Hermann Kolbe) and Heidelberg (under Robert Bunsen) but did not proceed to a doctorate, instead embarking in 1858 on a career of short-term employment in the chemical industry. In 1859, he was working in a small soda works at Ringkuhl, near Kassel, where he initiated a new process for the recovery of sulphur. This attitude, that of making good commercial use of hitherto wasted by-products, was to help to shape his career as an industrial chemist. In 1862, he accepted an invitation from a Lancashire industrial chemist to apply his idea for the recovery of sulphur from the alkali waste from the Leblanc process for making soda. For several years, Mond travelled regularly between John Hutchinson's works in Widnes and Utrecht in the Netherlands, where another Leblanc soda works was being constructed. He died in London of heart disease on 11 December 1909.

From 1867 Mond made his home in Britain and increased his involvement in the alkali industry, to meet the ever-growing needs of textile manufacturers. Then in 1872, he met Ernest Solvay, the Belgian industrial chemist who had devised the ammonia–soda (Solvay) process for making sodium carbonate from ammonia and salt. Using the royalties from his sulphur-recovery process, Mond and his partner John Brunner became the sole British licensees of the Solvay process. His works at Winnington near Northwich in Cheshire eventually superseded those still using the now outdated Leblanc process, although it was several years before it came into full production. The Brunner, Mond Company became the most successful in its field, employing more than 4,000 people by the end of the century, with the world's largest output of alkali.

In 1879, Mond became interested in the production of ammonia, an intermediate in the Solvay process which was increasingly being used as an artificial fertilizer. One outcome was the development of the Mond producer gas process, in which carbon monoxide and hydrogen are produced by alternately passing air and steam over heated coal or coke (and the hydrogen used to convert nitrogen into ammonia). By the early 1900s, Mond's Dudley Port Plant in Staffordshire was using 3 million tonnes of coal each year to make producer gas (Mond gas). An interesting extension of this work was the attempt, with the assistance of K. Langer, to turn the energy of the fuel directly into electrical energy. This early fuel cell, which used porous plates moistened with sulphuric acid, was not developed further at that time. Mond also hoped to recover the chlorine from the waste calcium chloride from the Solvay process, but this endeavour was also largely unsuccessful.

One unexpected result of the producer gas process came from Mond's observation that nickel is corroded by gases containing carbon monoxide. This led, in 1889, to the discovery of nickel carbonyl, $Ni(CO)_4$, one of the first organometallic compounds. After two years' work, with Langer and Quincke, he had developed a new extraction process for nickel. The Mond Nickel Company was founded at Clydach, Swansea, to produce nickel from Canadian ores by the thermal decomposition of nickel carbonyl; it later became the British manufacturing plant of Imperial Nickel Limited.

Mond became a rich man. He was generous with his wealth during his lifetime, and among various trusts he founded was one for the building of the Davy Faraday Laboratory at the Royal Institution, London. He lived in Italy towards the end of his life, and bequeathed a valuable collection of paintings to the British National Gallery. His son Robert (1867–1938) became a scientist and Egyptologist and was knighted, and his other son Alfred (later Lord Melchett) (1868–1930) was a successful politician and the first Chairman of Imperial Chemical Industries when it was founded in the 1920s. Mond's great-grandson Julian Mond, the third Lord Melchett, became the first Chairman of the British Steel Corporation in 1967.

Monge Gaspard 1746–1818 was a French mathematician and chemist, who was famous for his work in descriptive and analytical geometry and is generally regarded as the founder of descriptive geometry.

Monge was born at Beaune, into a merchant family, on 9 May 1746, and received his first education at the local Oratory. From 1762 to 1764 he studied at the Collège de la Trinité at Lyons, where he displayed such advanced scientific knowledge and such instinctive skill that he was placed in charge of the physics course. He returned to Beaune in 1764 and never again received formal scientific education. In the summer of 1764 he made a graphic sketch of his native town, the excellence of which won the praise of an officer at the Ecole Royale du Génie at Mézières and gained Monge his first appointment, as a draughtsman and

technician, at the school. His official duties were to prepare plans of fortifications and other architectural models, but in his spare time he devoted himself to geometrical research. It was in these years, 1766 to 1771, that he made his most fruitful insights into the nature of descriptive geometry.

In 1769, Monge succeeded the Professor of Mathematics at the Ecole Royale, although he was not given the title of Professor. A year later he was also appointed an instructor in experimental physics. In 1771, he met Condorcet and d'Alembert, by whom he was drawn into the scientific circle attached to the Academy of Sciences in Paris. In that year he submitted four papers to the Academy, of which the most significant was the 'Mémoire sur les développées des rayons de courbure et différents genres d'inflexions des courbes à double courbure', which was read to the Academy in 1771, but was not published until 1785. That paper was Monge's first important original work; it was followed in 1776 by another paper in infinitesimal geometry in which Monge introduced lines of curvature and the congruences of straight lines.

In 1780, Monge was elected to the Academy of Sciences and was given official duties as assistant geometer to the Academy. By then he was showing as much interest in the physical sciences as in mathematics. In 1783, independently of Lavoisier, he synthesized water. He also collaborated with J. F. Clouet to liquefy sulphur dioxide. In 1785, he was appointed examiner of naval cadets by the French government, an appointment which marked the beginning of his long participation in French public life. The burdens of the job, added to his duties at the Academy, forced him to resign his professorship at Mézières in 1784. For the next eight years Monge divided his time between inspecting naval schools throughout the country and participating in the activities of the Academy.

By the time the French Revolution broke out in 1789, Monge was one of the most celebrated of French scientists. He was an earnest supporter of the radicals and joined several revolutionary clubs and societies. In 1792, he was appointed Minister of the Navy, but as the revolution took its speedy course towards the Terror, he was discovered (despite his association with the left-wing Jacobins) to be a moderate and he resigned his post in April 1793. Thereafter he held no overt political position, although he was a member of the Committee on Arms in 1793–1794 and did important work in supervising the Paris armaments workshops and in helping to develop military balloons. He also served on the commission established to standardize French weights and measures. In March 1794 he was appointed to the commission set up to establish the Ecole Centrale des Travaux Publics and was its instructor in descriptive geometry when the new school opened in 1795. It soon changed its name to the Ecole Polytechnique.

In 1796, Monge's friendship with Napoleon began. Having conquered Italy, the revolutionary French government decided to plunder the country of its artistic and scientific treasures and Monge was sent, as a member of the Commission des Sciences et des Arts en Italie, to assist in the selection of objects to be removed to France. He met

Napoleon briefly, but was then recalled to France in 1797 to take up a new appointment as director of the Ecole Polytechnique. He then went back to Italy in 1798, this time as a member of a mission to inquire into the country's political organization. While he was there he was invited by Napoleon to assist in the preparation for the Egyptian campaign; he then accompanied Napoleon on the expedition to Egypt and was appointed president of the Institut d'Egypte established at Cairo in 1798. Monge also went with Napoleon on the expeditions to the Suez region and Syria in 1799, before returning to Paris at the end of the year.

He had scarcely begun to resume his duties as director of the Ecole Polytechnique when the *coup d'état* of 18 Brumaire placed Napoleon in control of the French government. Two months later Napoleon appointed him a senator for life and he resigned the directorship of the Ecole Polytechnique. For the rest of Napoleon's ascendancy Monge assumed the role of the foremost scientific supporter of the imperial regime. He was rewarded by being made a Grand Officer of the Legion of Honour in 1804, President of the senate in 1806, and the Count of Péluse in 1808. His creative scientific life was now a thing of the past, but in the leisure which freedom from onerous official appointments allowed, Monge brought together his life's work in a number of publications: the *Géométrie Descriptive* (1799), the *Feuilles d'Analyse Appliquées à la Géométrie* (1801), its expanded version, *Application de l'Analyse à la Géométrie* (1807), and several smaller works on infinitesimal and analytical geometry.

In the last decade of his life Monge was painfully afflicted by arthritis, which forced him to abandon his teaching at the Ecole Polytechnique in 1809. Thereafter, he lived in semi-retirement, although he went to the district of Liège in 1813 to organize the defence of the region against the allied armies then making their final victorious penetration into France. When Napoleon was finally overthrown in 1815, Monge was discredited. In 1816 he was expelled from the Institut (the renamed Academy of Sciences) and on 28 July 1818 he died in Paris.

Monge was one of the most wide-ranging scientists and mathematicians of his age. In the years between 1785 and 1789, for example, he submitted to the Academy of Sciences papers or notes on an astonishing variety of subjects: the composition of nitrous acid, the generation of curved surfaces, finite difference equations and partial differential equations (1785); double refraction, the composition of iron and steel and the action of electric sparks on carbon dioxide (1786); capillary phenomena (1787); and the physiological aspects of optics (1789). He holds an honoured place in the history of chemistry, not simply for his independent synthesis of water, but also for working with Lavoisier in 1785 in the epoch-making experiments on the synthesis and analysis of water. Although his own research had not led him to break entirely with the phlogiston theory, he was readily converted to the new chemistry by Lavoisier and played an energetic part in getting it accepted. Such was his standing as a chemist that he was one of the founders of the *Annales de chimie*.

It is, nevertheless, as a geometer that Monge gains his place in the scientific pantheon. Probably because he began his career as a draughtsman, he was always able to combine the practical, analytical and geometrical aspects of a problem. It was from his work on fortifications and architecture that he developed the basic principles of his descriptive geometry. The *Géométrie Descriptive* was based on lectures given at the Ecole Polytechnique in 1795. Its achievement was to translate the practical graphic procedures used by draughtsmen into a generalized technique, elegantly ordered and based upon rigorous geometrical reasoning. So popular was the book that it was quickly being used throughout Europe. It established Monge as the founder of descriptive geometry and, more, made a vital contribution to the nineteenth-century renaissance in geometry.

Throughout his life Monge's principal interest was infinitesimal geometry, especially in so far as it provided him with the opportunity to improve the rigour and enhance the status of analytical geometry. Monge broke with Cartesian tradition and asserted the autonomy of analytical geometry as a separate branch of mathematics. In particular, he devoted himself to two subjects: the families of surfaces as defined by their mode of generation (which he studied in relation to their corresponding partial differential equations); and the properties of surfaces and space curves. Much of this work was a development of the theory of developable surfaces outlined by Euler in 1772, although as early as 1769, Monge had defined the evolutes of a space curve and demonstrated that such curves are the geodisics of the developable envelope of the family of planes normal to a given curve. Monge established the distinction between ruled surfaces and developable surfaces and he provided a simple method of determining, from its equation, whether a surface *is* developable. His work in analysis was less original and less rigorous, but he made valuable contributions to the theory of both partial differential equations and ordinary differential equations. He introduced the notions, fundamental to later work, of the characteristic curve, the integral curve, the characteristic developable and the characteristic cone. He also created the theory of equations of the type $Ar + Bs + Ct + D = 0$ which are now known as 'Monge equations'.

He also introduced contact transformations, which were to be generalized by Lie a century later.

By his development of descriptive geometry and his application of analysis to infinitesimal geometry Monge not only paved the way for much of the geometrical flowering of the nineteenth century; he also proved himself to be one of the most original mathematicians of his time.

Monod Jacques Lucien 1910–1976 was a French biochemist who is best known for his research into the way in which genes regulate intracellular activities. For this research Monod and his co-workers André Lwoff and François Jacob were jointly awarded the 1965 Nobel Prize for Physiology or Medicine.

Monod was born in Paris on 9 February 1910 and was

educated at the university there. He graduated in 1931 and carried out research on the origins of life on Earth. He became Assistant Professor of Zoology at the University of Paris in 1934, and gained his doctorate in 1941. He worked for the French Resistance during World War II. From 1945 to 1953 he was laboratory chief at the Pasteur Institute, and it was during this period that he collaborated with Lwoff and Jacob on the work that was to gain them their Nobel prize. In 1953, Monod became Director of the Department of Cellular Biochemistry at the Pasteur Institute and also a professor in the faculty of sciences at the University of Paris. In 1971, he was appointed Director of the entire Pasteur Institute. Monod died in Cannes on 31 May 1976.

Working on the way in which genes control intracellular metabolism in microorganisms, Monod and his colleagues postulated the existence of a class of genes (which they called operons) that regulate the activities of the genes that actually control the synthesis of enzymes within the cell. They further hypothesized that the operons suppress the activities of the enzyme-synthesizing genes by affecting the synthesis of messenger RNA (mRNA). This theory has since been proved generally correct for many types of microorganisms but there is some doubt as to whether or not it applies in more complex plants and animals.

In 1971, Monod published his well-known book *Chance and Necessity*, a wide-ranging biological and philosophical work in which he summoned contemporary biochemical discoveries to support the idea that all forms of life result from random mutation (chance) and Darwinian selection (necessity). In the conclusion to this book, Monod stated his belief that there is no overall plan to human existence and that the human race must choose its own values in a vast and indifferent universe.

Montgolfier Joseph Michel 1740–1810 and Jacques Etiènne 1745–1799 pioneered manned flight by building the first hot-air balloons.

The Montgolfier brothers were sons of a paper manufacturer from Annonay, near Lyons, France. The eldest, Joseph, was born on 26 August 1740 and although he was given to playing truant from school, he developed a strong interest in chemistry, mathematics and natural science, even to the point of setting up his own small laboratory. Jacques, the younger by nearly five years, was born at Vidalon on 7 January 1745. He became a successful architect before joining his father's company, but once involved with it invented the first vellum paper.

Both brothers were fascinated by the views and theories on the possibility of flight by early scholars like the fourteenth-century Augustinian monk Albert of Saxony and the seventeenth-century Jesuit priest Father Francesco de Lana de Terzi, who designed a 'ship of the air' supported by evacuated copper spheres. But neither made any attempt at invention until 1782.

It was then, purely by chance, that they noticed the effects of rising smoke on particles of unburnt paper. Using this as a basis for their work, they progressed within two years from inflating paper bags to the first crewed free balloon flight in history in 1783.

Their success story really began in June 1783, when the brothers demonstrated their invention to an admiring crowd in the marketplace of Annonay. A large fire of straw and wood was built and the brothers placed over it a sphere of their own design made of paper and linen, allowing the heat to lift it gently into the air. Encouraged by their achievement, they took their invention to Versailles and in the grounds of the Palace before a large audience (which included Louis XVI and Marie Antionette) their balloon ascended, carrying a sheep, a cock and a duck, and made an eight-minute flight of approximately 3 km/2 mi.

A month later, on 15 October 1783, the first crewed ascent was made by François Pilâtre de Rozier (1757–1785) in a tethered balloon, paving the way for the first crewed free balloon flight on 21 November of that year. The crew of two, Pilâtre de Rozier and the Marquis d'Arlandes, ascended from the gardens of the Château la Muette, near Paris, and travelled over the city, completing a journey of 9 km/6 mi).

Originally the brothers thought they may have discovered a new gas, but this was obviously not so and they discarded the idea. The success of their invention could not be denied, however, and other people were quick to set up in competition, the main rivals being the Robert brothers, who used hydrogen to inflate their balloon under the instructions of the French scientist Jacques Charles (1746–1823). The first crewed flight in this type of balloon took place on 1 December 1783.

Jacques Montgolfier died at Servières on 2 August 1799 and, after his death, his brother Joseph devoted himself entirely to scientific research with varying degrees of success. He developed a type of parachute, a calorimeter and hydraulic ram and press. He was elected to the French Academy of Science and honoured by Napoleon when he was created Chevalier of the Legion of Honour. Outliving his brother by 11 years, he died at Balaruc-les-Bains on 26 June 1810.

Moore Patrick 1923– is a British broadcaster, writer and popularizer of astronomy.

An extrovert character, Moore has presented the BBC Television series *The Sky At Night* since 1968, as a result of which he has become a national celebrity, in constant demand as a speaker and raconteur.

Moore was born on 4 March 1923. Privately educated because of his poor health, he nevertheless served with the RAF during World War II, and for seven years after the war he assisted at a training school for pilots. After 1952 his interest in astronomy and his talent for communication inspired him to become a freelance author, although for three years, beginning in 1965, he was Director of the Armagh Planetarium, Northern Ireland. He was awarded the OBE in 1968 and the Royal Astronomical Society's Jackson-Gwilt Medal in 1977. He has also received many honours from abroad, such as honorary membership of the Astronomic-Geodetic Society of the Soviet Union (1971).

To Moore, astronomy was once merely an exciting

hobby for the evenings and other times off work. Part of his immense appeal today is the fact that he has managed to retain the air of the enthusiastic amateur, while at the same time boldly entering areas of study and research that to most lay people are fraught with complexities. He has never himself been employed as an astronomer, but it is very evident from his books and his broadcasts that he is consistently up to date with all modern facets of astronomy and that he understands his subject so thoroughly as to be able to make even the most difficult concepts clear to ordinary people.

His many books include *Moon Flight Atlas* (1969), *Space* (1970), *The Amateur Astronomer* (1970), *Atlas of the Universe* (1970), *Guide to the Planets* (1976), *Guide to the Moon* (1976), *Can You Speak Venusian?* (1977), *Guide to the Stars* (1977) and *Guide to Mars* (1977).

Morgagni Giovanni Battista 1682–1771 is generally seen as the founder of pathology.

Born in Forlì and graduating in Bologna, Morgagni taught anatomy there and later in Padua. Active throughout his life in anatomical research, his great work *De Sedibus et Causis Morborum per Anatomen Indagatis* (1761) was not published until he was 80. It was grounded on over 600 post mortems and was written in the form of 70 letters to an anonymous medical confrère. Case by case, Morgagni described the clinical aspects of the illness during the patient's lifetime, before proceeding to detail the postmortem findings. His goal was to relate the illness to the lesions established at autopsy. Morgagni did not use a microscope and he regarded each organ of the body as a composite of minute mechanisms. His book may be seen as a crucial stimulus to the rise of morbid anatomy, especially if physicians also availed themselves of the techniques of percussion, developed by Auenbrugger, and auscultation, pioneered by Laennec.

Morgagni himself made significant discoveries. He was the first to delineate syphilitic tumours of the brain and tuberculosis of the kidney. He grasped that where only one side of the body is stricken with paralysis, the lesion lies on the opposite side of the brain. His explorations of the female genitals, of the glands of the trachea, and of the male urethra also broke new ground. Judged the founder of pathological anatomy, his work was later developed by Bichat, Baillie and others.

Morgan Ann Haven 1882–1966 was a US zoologist and ecologist.

Morgan was born on 6 May 1882 in Waterford, Connecticut and entered Wellesley College in 1902 to study biology. She later transferred to Cornell University and graduated in 1906. After a brief period as an instructor in zoology at Mount Holyoke College she returned to Cornell and received her PhD in zoology in 1912 for a dissertation on mayflies. She returned to Mount Holyoke in that year as an instructor in zoology, became an associate professor in 1914 and full professor in 1918, and served as Chairman of the department of zoology from 1914 until her formal

retirement in 1947. She died on 5 June 1966 at her home in South Hadley, Massachusetts.

Her research interests covered a broad range of biological problems, including the zoology of aquatic insects, the comparative physiology of hibernation, and conservation and ecology, and she spent her summers at a variety of research laboratories, including the Marine Biological Laboratory at Woods Hole in Massachusetts, and also worked at the Tropical Research Laboratory of British Guiana. She was a keen communicator of her subject, and her *Field Book of Ponds and Streams: An Introduction to the Life of Fresh Water* (1930) was an important book in attracting and encouraging amateur naturalists, as well as providing an authoritative taxonomic guide for professional collectors and zoologists. Further books on hibernation and animal behaviour, and an educational film brought her work and interests to a wide audience, and after her retirement she devoted a considerable amount of energy to reform of the scientific curriculum, especially in promoting the study of ecology and conservation. She was a member of the National Committee on Policies in Conservation Education and produced annual summer schools for teachers. She remained active in science education and local conservation movements until the end of her life.

Morgan Garrett A. 1875–1963 was a US inventor who patented the safety hood, later known as the gas mask, and the automatic three-way traffic signal.

Garrett Morgan was born in Paris, Kentucky, on 4 March 1875. He received only elementary school education and at the age of 14 he left home to work as a handyman in Cincinnati, Ohio. In 1895 he moved to Cleveland as a sewing machine adjuster. He remained in Cleveland, where he patented his inventions until his death on 27 July 1963. Morgan married in 1908 and had three sons. After opening his own shop selling and repairing sewing machines in 1907 and a tailoring shop in 1909, Morgan went on to establish the G. A. Morgan Hair Refining Company in 1913 as a result of discovering a human hair-straightening process. He is best known however, for the invention of his safety hood, for which he obtained a patent in 1914. After rescuing men trapped in a tunnel following an explosion at Cleveland Water Works in 1916, requests came from fire and police departments and mining companies for demonstrations of the hood; Morgan then set up a company to manufacture and sell them. During World War I the design was improved and it became part of the standard field equipment of United States soldiers. Morgan is not only known for his invention of the gas mask, in 1923 he obtained the patent for his invention of the three-way automatic traffic signal; he was also awarded the British and Canadian patents. He sold the right to the traffic signal to the General Electric Corporation for $40,000.

Morgan also invented a woman's hat fastener, a round belt fastener and a friction drive clutch. Morgan was very much concerned with black welfare, and in 1920 he started his own newspaper, the *Cleveland Call*, later the *Call and Post*. He served as treasurer of the Cleveland Association of Colored Men from 1914 until it merged with the National Association for the Advancement of Colored People of which he remained an active member all his life. In 1931, he ran as an independent candidate for the City Council in Cleveland but was unsuccessful. Morgan received several awards for his inventions; he was awarded the First Grand Prize Golden Medal by the National Safety Device Company at the Second International Exposition of safety and sanitation in 1914, and following the rescue in 1916 he was awarded the Carnegie Medal and a medal for bravery by the city of Cleveland and honorary membership of the International Association of Fire Engineers. In 1976 a public school in Harlem was named after him.

Morgan Thomas Hunt 1866–1945 was a US geneticist and embryologist famous for his pioneering work on the genetics of the fruit fly *Drosophila melanogaster* – now extensively used in genetic research – and for establishing the chromosome theory of heredity. He received many honours for his work, including the 1933 Nobel Prize for Physiology or Medicine.

Morgan was born on 25 September 1866 in Lexington, Kentucky, the son of a US diplomat. He was educated at the State College of Kentucky, graduating in 1886, and then at Johns Hopkins University, from which he gained his PhD in 1890. In the following year he joined the staff of Bryn Mawr College, near Philadelphia, as Associate Professor of Zoology and remained there until 1904, when he became Professor of Experimental Zoology at Columbia University. In 1928, he was appointed Director of the Laboratory of Biological Sciences at the California Institute of Technology, a post he held until his death in Pasadena on 4 December 1945.

Morgan's early work was in the field of embryology, investigating such phenomena as fertilization in nucleated and unnucleated egg fragments, the development of embryos from separated blastomeres, and the effect of salt concentration on the development of unfertilized and fertilized eggs. In about 1907, however, his interest turned to the mechanisms involved in heredity (following the rediscovery of Gregor Mendel's work), and in 1908 he began his famous research on the genetics of *Drosophila* – initially to test Mendel's laws, about which Morgan was sceptical. After breeding several generations of *Drosophila*, Morgan noticed many small phenotypic variations, some of which could not be accounted for by Mendel's law of independent assortment – he discovered, for example, that the *Drosophila* variant now known as white eye is confined almost entirely to males. From his findings he postulated that certain characteristics are sex-linked, that the X-chromosome carries several discrete hereditary units (genes) and that the genes are linearly arranged on chromosomes. Morgan also demonstrated that sex-linked characters are not invariably inherited together, from which he developed the concept of crossing-over and the associated idea that the extent of crossing-over is a measure of the spatial separation of genes on chromosomes. (From these ideas A. H. Sturtevant – one of Morgan's student collaborators – drew up in 1911

the first chromosome map, which showed the positions of five sex-linked genes.) Morgan realized that his findings proved that Mendel's 'factors' have a physical basis in chromosomes and revised his earlier scepticism of Mendelian genetics. In 1915, in collaboration with Sturtevant and his other student co-worker A. B. Bridges and Hermann Muller, Morgan published a summary of his work in *The Mechanism of Mendelian Heredity*, which had a profound influence in genetic research and evolutionary theory.

In the following years Morgan and various co-workers continued to elaborate the chromosome theory of heredity. Towards the end of his life, however, he returned to embryological investigations, trying to support with experimental evidence the theoretical links between embryological development and genetic theory. But it is his early work that is the most important, providing one of the cornerstones of modern genetic theory. Moreover, largely as a result of Morgan's experimentation, *Drosophila* became one of the principal experimental animals used for genetic investigations.

Morley Edward Williams 1838–1923 was a US physicist and chemist who is best known for his collaboration with Albert Michelson (1852–1931) in the classic Michelson–Morley experiment that disproved the existence of the ether as a medium that propagates light waves.

Morley was born in Newark, New Jersey, on 29 January 1838. He was taught at home by his parents until 1857, when he entered Williams College. He obtained a bachelor's degree from there in 1860, and then began theological training at the Andover Seminary, intending to become a Congregational minister like his father. At the same time he continued his studies at Williams College, which awarded him a master's degree in 1863.

Morley completed his theological training in 1864, and worked for a year with the Sanitary Commission in Fort Monroe in Virginia, before returning to the Andover Academy for another year of study. In 1866, he got a teaching job at the South Berkshire Academy in Marlboro, Massachusetts.

In 1868, Morley was offered a ministry in Twinsburg, Ohio, and a teaching post in the chemistry department of Western Reserve College in Hudson, Ohio. He accepted the latter, but only on the condition that he be permitted to preach at the University chapel. Morley remained at Western Reserve College, later called Adelbert College when the college moved to Cleveland, Ohio, for the rest of his career. He became Professor of Chemistry there in 1869 and retired in 1906. From 1873 to 1888, he simultaneously held the post of Professor of Chemistry and Toxicology at Cleveland Medical College. Morley's contributions to the fields of physics and chemistry were recognized with the award of several distinguished honours, including the Davy Medal of the Royal Society in 1907, and the presidency of the American Association for the Advancement of Sciences from 1895 to 1896. Morley died in West Hartford, Connecticut, on 24 February 1923.

All of Morley's research involved the use of precision

instruments. During the early part of his scientific career, he was involved with the design of such equipment. One of his instruments was a eudiometer, which he applied to test a theory concerning the origin of cold waves of air. Using the eudiometer, which was accurate to within 0.0025%, Morley was able to confirm on the basis of the oxygen content of the air and meteorological data that, under certain conditions, cold air derives from the downward movement of air from high altitudes rather than from the southward movement of northerly cold air.

Morley's most important contribution to science came with his collaboration with Michelson, which began in 1885. Michelson too was concerned with precision measurement, and the two men improved the interferometer that Michelson had first used in 1881 in an unsuccessful attempt to detect the motion of the Earth through the ether. The confirmation of this negative result in 1887 by Michelson and Morley led to the conclusion that the velocity of light is constant irrespective of the motion of the observer, which is one of the postulates of the special theory of relativity.

Morley also achieved wide recognition with the publication in 1895 of his accurate determinations of the densities of hydrogen and oxygen, and the ratios for their combination. Morley also investigated a controversial and long-standing theory known as Prout's hypothesis (proposed by William Prout (1785–1850) in 1815), which suggested that all the elements of the periodic table are multiples of the hydrogen atom. Morley decided to test this theory by taking the atomic weight of hydrogen to equal one, and then determining the relative atomic weight (relative atomic mass) of oxygen by chemical means. He found that the ratio of their atomic weights was 1:15.879, not 16 as expected if oxygen was an integral multiple of hydrogen. Morley felt that he had finally disproved Prout's hypothesis. However, the discovery of isotopes later led scientists to realize that chemical methods of determining atomic weight provide only an average result for the weights of all the different isotopes present in the sample. Prout's idea was essentially correct and he was honoured in 1920 when Ernest Rutherford (1871–1937) suggested that the subatomic particle which forms the hydrogen nucleus be named the *pro*ton.

Morley continued doing analytical research after his retirement. His constant preoccupation with precision and accuracy produced a rich yield of results during the course of his career, and their importance was quickly recognized.

Morris Desmond John 1928– is a British zoologist, well known for his publications and films on animal and human behaviour.

Morris was born on 21 January 1928 and was educated at Birmingham University where he gained as BSc in zoology. He then went to Magdalen College, Oxford, to read for a PhD, working on animal behaviour under Nikolaas Tinbergen (1907–1988). From 1954 he worked in the Department of Zoology at Oxford before becoming Head of the Granada Television and Film Unit at the Zoological Society in London in 1956. Three years later he was

appointed Curator of Mammals at the Zoological Society of London and from 1967 to 1968 served as Director of the Institute of Contemporary Arts in London.

Morris has been a prolific writer, as well as having made several films and presented television programes on social behaviour; many of his works have been produced jointly with his wife Ramona Morris. Probably the best known of his books are *The Naked Ape* (1967), in which he examines the human animal in a brutally objective way, and *The Human Zoo* (1969), which follows on in that he scrutinizes the society that the naked ape has created for itself. Morris compares civilized humans with their captive animal counterparts and shows how confined animals seem to demonstrate the same neurotic behaviour patterns as human beings often do in crowded cities. He believes the urban environment of the cities to be the human zoo.

He has done much to popularize sociology and zoology, and is an entertaining presenter of programmes with some social influence.

Morse Samuel Finley Breese 1791–1872 was a US artist who invented an electric telegraph and gave his name to Morse code.

Morse was born on 27 April 1791, in Charlestown, Massachusetts, the eldest son of a clergyman. Brought up to appreciate all forms of culture, he chose to devote himself to art, and after studying at Yale University, from which he graduated in 1810, he studied at the Royal Academy in London.

On returning to the United States he achieved a fair amount of success, particularly in sculpture and in the painting of miniatures on ivory. Unfortunately the financial rewards brought him no more than a bare living, and in 1829 he returned once more to Europe in the hope of establishing a firm artistic reputation. From 1829 to 1832 he travelled widely, but after three years he once more set sail for home aboard the ship *Sully*. It was on this voyage that fate took a turn in the shape of fellow passenger Charles Jackson, who had recently attended lectures on electricity in Paris. He had with him an elctromagnet and Morse, having made his acquaintance, became fascinated with the talk of electricity and the possiblities of this new idea.

For the remainder of the voyage he spent hours making notes and became fired with the concept of communication by electric current 'telegraph'. As soon as he landed he set about turning his theories into working models and became so obsessed with the invention that he more or less abandoned his art career, relying for means of support on his appointment as Professor of the Art of Design, at the University of New York.

By 1836 he had devised a simple relay system and having enlisted the help of Professor Gale, a chemist, and Alfred Vail, he improved his apparatus to the point where it became a commercial proposition.

After a further four years of political wrangling, his invention was finally given the backing of Congress and with a salary to give him freedom from money problems, Morse undertook the superintending of the first line between Baltimore and Washington. A trial run was telegraphed on 23 May 1843 and the line was eventually opened for public use on 1 April 1845.

Morse's idea hinged on the idea that an electric current could be made to convey messages. The signal current would be sent in an intermittent coded pattern and would cause an elctromagnet to attract intermittently to the same pattern on a piece of soft iron to which a pencil or pen would be attached and which in turn would make marks on a moving strip of paper. Once a suitable code had been worked out and batteries with sufficient power located, the technical problems mainly concerned the conductors, which on the first line were carried above ground on poles. Although underground conductors were envisaged, telegraph poles became a familiar feature of the countryside for the next 100 years.

The original receiver recorded signals by indentation. The Morse inker was a later innovation, invented mainly because Morse disliked the operators 'reading' the messages by ear (they became so adept they could identify letters from the sound of the clicks made by the machine), and also because he felt a written proof of receipt by the machine was a necessity.

Morton William Thomas Green 1819–1868 was a US dentist who in 1846 gave the first public demonstration of ether anaesthesia during surgery. Although he was neither the discoverer of the painkilling effects of ether nor the first to use it during a surgical operation, it was largely as a result of his efforts that anaesthesia became quickly and widely adopted by surgeons and dentists.

Little is known about Morton's early years. He was born in Charlton City, Massachusetts, on 9 August 1819, the son of a smallholder and shopkeeper, and is reputed to have graduated from the Baltimore College of Dentistry, although this is uncertain. After a brief period in partnership with Horace Wells, another pioneer in the use of anaesthesia, Morton set up his own dental practice in Boston in 1844 and began investigating ways to deaden pain during dental surgery. After numerous unsuccessful attempts, he consulted the chemist and former physician Charles Jackson (1805–1880), who advised him to try ether as an anaesthetic. (This was not a new idea: Crawford Long (1815–1878) had in 1842 successfully used ether anaesthesia during an operation to remove a tumour from a patient's neck, although he did not publish this work until 1849 – by which time he had been pre-empted by Morton. In addition, Wells used nitrous oxide anaesthesia in a public demonstration – attended and partly arranged by Morton – in 1845, but had failed to convince his spectators of the gas's effectiveness as an anaesthetic.) In 1846, Morton successfully extracted a tooth from a patient under ether and later in the same year staged a public demonstration of ether anaesthesia in an operation (also successful) to remove a facial tumour; this operation was performed at the Massachusetts General Hospital – where Wells had earlier failed in his demonstration. So successful was Morton's demonstration that ether anaesthesia was rapidly adopted by surgeons and

dentists in the United States and Europe.

Morton derived no financial benefits from his work. Morton and Jackson had jointly patented the process of ether anaesthesia but Morton attempted to claim sole credit as its discoverer. Jackson strongly contested this claim and Morton spent the rest of his life in costly litigation with him. A large fund was raised for Morton in Britain as an award for his discovery of anaesthesia but the offer was withdrawn in the face of strong opposition from Jackson. The French Academy of Medicine offered another monetary award to both men jointly, but Morton refused to accept it. Likewise, a bill to award Morton $100,000 in recognition of his work failed to pass in the United States Congress in 1852, 1853 and again in 1854. Meanwhile, official recognition of priority had been accorded to Wells (for discovering nitrous oxide anaesthesia) and Long (for discovering ether anaesthesia). Finally Morton died, bitter and poor, on 15 July 1868 in New York City. Eventually, however, his role in the development of anaesthesia was recognized when he was elected in 1920 to the American Hall of Fame.

Moseley Henry Gwyn Jeffreys 1887–1915 was a British physicist who first established the atomic numbers of the elements by studying their X-ray spectra. This led to a complete classification of the elements, and also provided an experimental basis for an understanding of the structure of the atom.

Moseley was born at Weymouth on 23 November 1887. He came from a family of prestigious scientists, his father being Professor of Anatomy at Oxford and a member of the great *Challenger* expedition that surveyed the world's oceans from 1872 to 1876. Moseley was educated at Eton and in 1906 entered Trinity College, Oxford, where he obtained a degree in physics in 1910. Immediately after graduation, he was appointed lecturer in physics in the laboratory of Ernest Rutherford (1871–1937) at Manchester, becoming a research fellow in 1912. Moseley remained with Rutherford until 1913. He then worked privately at Oxford with a view to obtaining a professorship there, and in 1914 he visited Australia. On the outbreak of World War I, he returned home and immediately enlisted. Fearing for his safety, Rutherford endeavoured unsuccessfully to secure him scientific duties. Moseley was sent to the Dardanelles and was killed at Gallipolli on 10 August 1915.

Moseley was a researcher only from 1910 until 1914, and yet he made discoveries which were of fundamental importance to the development of both physics and chemistry. When he joined Rutherford's group in 1910, Rutherford was researching the phenomena associated with natural radioactivity. Moseley at first helped Rutherford in this work, but when reports of the diffraction of X-rays by Max von Laue (1879–1960) reached him in 1912, Moseley persuaded Rutherford to allow him to study X-ray spectra. He received instruction in X-ray diffraction from Lawrence Bragg (1890–1971) and in 1913 Moseley introduced X-ray spectroscopy to determine the X-ray spectra of the elements.

In a series of brilliant investigations, Moseley allowed the X-rays produced from various substances used as a target in an X-ray tube to be diffracted by a crystal of potassium ferrocyanide. The glancing angles were measured accurately and the position of the diffracted beams determined to obtain the wavelengths and frequencies of the X-rays emitted. Moseley examined metals from aluminium to gold and he found that their X-ray spectra were similar but with a deviation that changed regularly through the series. He found that a graph of the square root of the frequency of each radiation against the number representing the element's position in the periodic table gave a straight line. He called this number the atomic number of the element, which has since been shown to be the positive charge on the nucleus and thus the number of protons in the nucleus. Since the atom is electrically neutral, the atomic number is also the number of electrons surrounding the nucleus.

It was as a direct result of this work that atomic numbers were placed on a sound experimental foundation. Moseley found that when the elements are arranged in the periodic table according to their atomic numbers, all irregularity caused in the older system of grouping elements by their atomic weight (relative atomic mass) disappeared. Now that the elements were numbered, the rare earth elements could be sorted out, a process that Moseley began at Oxford towards the end of his life. The numbering system also enabled Moseley to predict that several more elements would be discovered, namely those with atomic numbers of 43, 61, 72, 75, 87 and 91. These were all found in due course.

Although the number of elements which Moseley was able to examine was limited, the equation relating the square root of the frequency to the atomic number has been found to hold in all cases. The equation is known as Moseley's law. It has enabled scientists to identify a total of 105 elements in a continuous series of atomic numbers. Any further elements that might be produced by nuclear reactions can only have greater atomic numbers.

In 1913 and 1914, the young physicist published his findings in two remarkable papers in the *Philosophical Magazine* and entitled them 'The High-Frequency Spectra of the Elements'. Moseley's discovery told how many electrons were present in any element, and tied in nicely with the quantum theory of the hydrogen atom which was published in 1913 by Niels Bohr (1885–1962).

Moseley's fundamental discovery was a milestone in our knowledge of the constitution of the atom, and we are left to ponder on what this great brain might have discovered had he not been so tragically killed at so young an age.

Mössbauer Rudolf Ludwig 1929– is a German physicist who was awarded (with Robert Hofstadter) the 1961 Nobel Prize for Physics for his discovery of the Mössbauer effect. The effect, which involves the recoil-free absorption of gamma radiation by an atomic nucleus, provided experimental verification for Albert Einstein's theory of relativity.

Mössbauer was born in Munich on 31 January 1929, and was educated there. He graduated from the Munich Institute of Technology in 1952 and was awarded his PhD in 1958, the same year that he announced the Mössbauer effect, which he discovered during his postgraduate research in Heidelberg at the Max Planck Institute for Medical Research. In 1960 he went to the United States and a year later became Professor of Physics at the California Institute of Technology, Pasadena. He remained in this position while simultaneously holding a professorship at Munich.

Mössbauer began research into the effects of gamma rays on matter in 1953. The absorption of a gamma ray by an atomic nucleus usually causes it to recoil, so affecting the wavelength of the re-emitted ray. Mössbauer found that at low temperatures crystals will absorb gamma rays of a specific wavelength and resonate, so that the crystal as a whole recoils while the nuclei do not. This recoilless nuclear resonance absorption became known as the Mössbauer effect. A lengthening of the gamma-ray wavelength in a gravitational field, as predicted by the general theory of relativity, was observed experimentally in 1960, first in England then in the United States.

Mott Nevill Francis 1905– is a British physicist whose work on semiconductors won him the 1977 Nobel Prize for Physics. He shared the prize with Philip Anderson (1923–), who has also worked on crystalline materials, and John Van Vleck (1899–1980), who played a central role in the development of the laser.

Mott was born on 30 September 1905. He was educated at Clifton College and then at St John's College, Cambridge. He became a lecturer at Manchester University in 1929 and at Gonville and Caius College, Cambridge, in 1930. In 1933, he moved to Bristol to become Melville Wills Professor of Theoretical Physics and held this position until 1948, when he became Director of the Henry Herbert Wills Physical Laboratories at Bristol. From 1954 to 1971, he was Cavendish Professor of Physics at Cambridge University and then Senior Research Fellow at Imperial College, London, from 1971 to 1973. He is now Emeritus Professor at Cambridge and has retired to Milton Keynes. His honours include a knighthood in 1972 and the award of the Royal Society's Copley Medal also in 1972.

Mott is a leading authority on solid state physics, particularly on the theory of electrons in metals and on dislocations and other defects in the crystalline structure. He was the first, with R. W. Gurney, to put forward a comprehensive theory of the process involved when a photographic film is exposed to light. They postulated that the incident light produces free electrons and holes which wander about the crystal. The electrons become trapped at imperfections, which might be dislocations or possibly foreign atoms, and attract interstitial silver ions to form silver atoms and make the latent image. When a developer is present, the entire grain may be catalysed into free silver by the initially formed specks of silver.

Mott's work on semiconductors, which won him the

Nobel prize, has shown how a cheap and reliable material can be used to improve the performance of electronic circuits, increasing the memory capacity of computers by several times. More efficient photovoltaic cells can now be produced which are capable of converting solar energy into electricity. The use of carefully prepared crystalline materials as semiconductors has led to a revolution in the transistor industry. Mott and his colleagues have discovered special electrical characteristics in glassy semiconductors and have laid down fundamental laws of behaviour for their materials. Their subject has been described as the last frontier of solid state physics and their and Anderson's research opens the way for a wide range of new developments in electronics, including cheaper methods of solar heating.

Mottelson Ben Roy 1926– . US physicist; *see* Bohr, Mottelson, and Rainwater.

Muir Thomas 1844–1934 was a British mathematician, who is famous for his monumental and pioneering work in unravelling the history of determinants.

Muir was born at Stonebrye, Lanarkshire, on 25 August 1844 and grew up in the nearby town of Biggar, where his father was a shoemaker. He was educated at Wishaw Public School and Glasgow University, where he excelled at Greek and mathematics. He then spent some time in Berlin, studying and book-collecting, before returning to Scotland in 1868 to take up a post as mathematics tutor at College Hall, St Andrews. From 1871 to 1874 he was Assistant Professor of Mathematics at Glasgow and from 1874 to 1891 Head of the Mathematics and Science Department at Glasgow High School. That he chose to leave university life for secondary-school teaching revealed the deep interest he took in the general educational standards of the community. In 1891 his wife, who suffered poor health, was advised to move to a warmer climate and together they emigrated to South Africa where, in 1892, Muir accepted the post of Superintendent-General of Education in the colony. He held the post until his retirement in 1915. From 1892 to 1901 he was also Vice Chancellor of the University of Cape Town. In addition to his mathematical work, Muir took a keen interest in geography and he was elected a Fellow of the Royal Geographical Society in 1892 and of the Royal Scottish Geographical Society in 1899. He was knighted in 1915. In 1916 he received the Gunning–Victoria Prize for his contribution to science. He died in Cape Town on 21 March 1934.

Muir earned the gratitude of the South Africans for the part that he played – by far the most important part – in raising educational standards in the country and securing a proper place for science in the curriculum. In the history of mathematics he is remembered for his work on determinants, first discovered by Gottfried Leibniz in 1693. In all he published 307 papers, most of them on determinants and allied subjects, in addition to his books, *A Treatise on the Theory of Determinants* (1882), *The Theory of Determinants in its Historical Order of Development* (1890) and,

above all, the magisterial five-volume treatise on the history of determinants. The first volume appeared in 1906, the last in 1930, and when it was completed it was widely acclaimed as one of the most thorough treatments of the history of any branch of theoretical knowledge. Muir's work made the results of Laplace, Cauchy, Schweins and a host of other mathematicians accessible to scholars. Muir himself was not a creative mathematician but his book lay behind many an algebraic discovery.

Muller Hermann Joseph 1890–1967 was a US geneticist famous for his discovery that genetic mutations can be artificially induced by X-rays, for which he was awarded the 1946 Nobel Prize for Physiology or Medicine.

Muller was born on 21 December 1890 in New York City and was educated at Morris High School in the Bronx district of the city. In 1907, he won a scholarship to Columbia University, from which he graduated in 1910. He remained at Columbia to do postgraduate research on genetics – under Thomas Hunt Morgan – and gained his PhD in 1916. Muller then spent three years at the Rice Institute in Houston, Texas, at the invitation of Julian Huxley, followed by a brief period as an Instructor at Columbia University. Then in 1920 he joined the University of Texas, Austin, initially as Associate Professor of Zoology and later as Professor of Zoology. The next 12 years at the University of Texas were the most scientifically productive in Muller's life but eventually the pressure of work, the ending of his marriage to the mathematician Jessie Marie Jacob (whom he married in 1923), and the constraints on his freedom to express his socialist political views all combined to produce a nervous breakdown, and in 1932 Muller left the United States to work at the Kaiser Wilhelm Institute in Berlin. In 1933, he moved to Leningrad to become – at the invitation of Nikolai Vavilov – Senior Geneticist at the Institute of Genetics; the institute was transferred to Moscow in the following year and Muller moved with it. But in the mid-1930s the false ideas of Trofim Lysenko began to dominate Soviet biological research; Muller openly criticized Lysenkoism but so great was Lysenko's political influence that Muller was forced to leave the Soviet Union in 1937. After serving in the Spanish Civil War he worked at the Institute of Animal Genetics in Edinburgh, where in 1939 he met and married Dorothea Kantorowitz, a German refugee. Muller returned to the United States in 1940. He held various posts at Amherst College, Massachusetts, from 1941 to 1945, when he was appointed Professor of Zoology at Indiana University. He remained there, and died on 5 April 1967 in Bloomington, Indiana.

Muller began his research on genetics while working for his doctorate under Morgan, in the course of which he made several important contributions to the understanding of the arrangements and recombinations of genes. During this period he became particularly interested in mutations, and when he began independent research he attempted to find techniques for accelerating mutation rates. In 1919 he found that the mutation rate was increased by heat, and that heat did not always affect both of the chromosomes in a chromosome pair. From this he concluded that mutations involved changes at the molecular or submolecular level. Next Muller experimented with X-rays as a means of inducing mutations, and by 1926 he had proved the method successful. This was an important finding because it meant that geneticists could induce mutations when required, rather than having to wait for the considerably slower process of natural mutation, and it also showed that mutations are nothing more than chemical changes.

Muller's research had convinced him that almost all mutations are deleterious. He realized that in the normal course of evolution deleterious mutants die out and the few advantageous ones survive but he also believed that if the mutation rate is too high, the number of imperfect individuals may become too large for the species as a whole to survive. Consequently, he began to concern himself with the social effects of genetic mutations. He campaigned against the needless use of X-rays in diagnosis and treatment, and pressed for safety regulations to ensure that people who were regularly exposed to X-rays were adequately protected. He also opposed nuclear bomb tests, arguing that the radioactive fallout could burden future generations with an excessive number of deleterious mutations. Furthermore, he advocated the establishment of sperm banks, in which the sperm of gifted men could be preserved for use by later generations so that the human gene pool would be improved.

Müller Johannes 1436–1476 known as 'Regiomontanus' was a German astronomer who compiled astronomical tables, translated Ptolemy's *Almagest* from Greek into Latin, and assisted in the reform of the Julian Calendar.

Müller was born in Königsberg on 6 June 1436, the son of a miller. Nothing is known of him until he enrolled at the University of Vienna on 14 April 1450, under the name of 'Johannes Molitoris de Künigsberg'. The name Regiomontanus is derived from a latinization of his birth place: Regio Monte, meaning King's Mountain. At the age of 15 he was awarded his bachelor's degree and, in 1457, was appointed to the Faculty of Astronomy at the University of Vienna. He died, probably of the plague, in Rome in 1476.

At Vienna Regiomontanus (as he was called by then) became a close friend and colleague of Georg von Purbach (Peurbach), under whom he had studied astronomy. The course of their lives was deeply affected in 1460 by the arrival in Vienna of Cardinal Bessarion, who, as part of his campaign to bring ancient Greek authors to the attention of intellectuals in the Latin west, persuaded Purbach to translate Ptolemy's *Almagest* from Greek into Latin. But Purbach failed to finish this mammoth task and on his deathbed, in April 1461, he pledged Regiomontanus to complete the project. Regiomontanus complied with this last wish and, in addition to translating the work, he added more recent observations, revised some computations and added his own criticisms. Regiomontanus named the complete translation *Epitome*, but it was not printed until 20 years after his death. Copernicus had been aware of many

of the inaccuracies in Ptolemy's system (which had prevailed for more than 1,300 years) and it was the critical reflections in the *Epitome* that led him to overthrow the Ptolemaic system and so lay the foundations of modern astronomy.

In 1467, Regiomontanus started compiling trigonometric and astronomical tables. He began computing his *Tables of Directions*, which gave the longitudes of celestial objects in relation to the apparent daily rotation of the heavens and was relevant to observers as far north as 60 degrees (although the finished work was not published until 1490). In 1468, he completed his sine tables, which facilitated the making of astronomical observations before the advent of logarithms, but these too were not published until more than 50 years after his death.

In 1471, Regiomontanus moved to Nuremberg, where he installed a printing press in his own house and so became one of the first publishers of astronomical and scientific literature. Among his first publications were Purbach's *New Theory of the Planets* and his own *Ephemerides*, which was issued in 1474. This was the first publication of its kind to be printed; it gave the positions of the heavenly bodies for every day from the year 1475 to 1506. It acquired some historical interest when it was used by Christopher Columbus in Jamaica to predict a lunar eclipse that frightened the hostile Indians into submission.

According to the *Nuremberg Chronicle*, Regiomontanus went to Rome in 1475 in response to a papal invitation to assist in amending the notoriously incorrect ecclesiastical calendar. Unfortunately, he died within a year of leaving Nuremberg, probably falling victim to the plague that swept through Rome after the Tiber overflowed in 1476.

After Regiomontanus's death, the statement 'the motion of the stars must vary a tiny bit on account of the motion of the Earth' was found to be written in his handwriting. This fact has led some people to believe that Regiomontanus was a Copernican before Copernicus. It has also been suggested that Regiomontanus sent the letter containing this statement to Novara (who was Copernicus's teacher), who in turn communicated it to Copernicus. Thus some people infer that the revolutionary geocentric doctrine was first conceived by Regiomontanus.

Müller Paul Hermann 1899–1965 was a Swiss chemist, known for his development of DDT as an insecticide, for which he was awarded the 1948 Nobel Prize for Physiology or Medicine.

Müller was born in Olten, Solothurn, on 12 January 1899. He received his early education in Basle and worked in the electrical and chemical laboratories of several industrial firms before continuing his academic studies. He gained his doctorate in chemistry at Basle in 1925 and then went to work for J. R. Geigy, researching principally into dyestuffs and tanning agents; he subsequently joined the staff of Basle University. He died in Basle on 12 October 1965.

In 1935, Müller started work on a research project designed to discover a substance that would kill insects quickly but have little or no poisonous effect on plants and animals, unlike the arsenical compounds then in use. He concentrated his search on chlorine compounds and in 1939 synthesized dichlorodiphenyl trichloroethane (DDT) – which had first been prepared 65 years earlier by the German chemist Othmar Zeidler, who had not been aware of its insecticidal properties.

The Swiss government successfully tested DDT against the Colorado potato beetle in 1939 and by 1942 it was in commercial production. Its first important use was in Naples, where a typhus epidemic broke out soon after the city had been captured by American forces in 1943; in January of the following year the population of Naples was sprayed with DDT to kill the body lice that are the carriers of typhus. A similar potential epidemic was arrested in Japan in late 1945 after the American occupation of the country.

For the following 20 years the use of DDT was to have a profound effect on the health of the world, both by killing insect vectors such as the mosquitoes that spread malaria and yellow fever and by combating insect pests that feed on food crops. Gradually the uses of DDT in public hygiene and in agriculture became limited by increasing DDT-resistance in insect species, and it has been supplanted by new synthetic insecticides. Also DDT is a very stable chemical compound; it does not break down and tends to accumulate in the environment, disrupting food chains and presenting a hazard to animal life. By the 1970s its use had been banned in several countries.

Murchison Roderick Impey 1792–1871 was one of the most prolific British geologists of the nineteenth century.

Descended from an ancient Scottish Highland family, Murchison entered the army in 1815 and briefly fought in the Peninsular War. He then married and settled near Durham, where he pursued gentlemanly interests, notably fox-hunting. His love of geology was whetted through friendship with Humphry Davy. Murchison soon embarked upon a series of arduous geological field explorations, often accompanied by Adam Segwick or Charles Lyell, exploring in Scotland, France and the Alps. In 1839, he produced his major book, *The Silurian System*, on the basis of his studies of the 'greywackes' (old slatey rocks) of South Wales. (A follow-up work, *Siluria* came 15 years later.) The Silurian system (that is, those strata of the lower palaeozoic beneath the Old Red Sandstone) contained, in Murchison's view, remains of the earliest life-forms, though no fossils of vertebrates or land plants were to be expected. Following controversy with De la Beche, Murchison, with Sedgwick's cooperation, also established the Devonian system in southwest England. An expedition to Russia in 1841 led him to define yet another worldwide system, the Permian, named after the strata of the Perm region.

Murchison had a distinct philosophy of geology. He believed in a universal order of the deposition of strata, indicated by fossils rather than solely by lithological features. Fossils showed a clear progression in complexity

from Azoic (pre-life) times to invertebrates, and thence up to vertebrate forms, humans being created last of all: this progression was seen in respect of the Earth's cooling.

Knighted in 1846, Murchinson became a 'professional' in 1855, succeeding De la Beche as Director of the Geological Survey. He was one of the founders of the British Association. An ardent imperialist, for many years he was also president of the Royal Geographical Society, encouraging African exploration and annexation. A haughty, dogmatic and increasingly tetchy man, Murchison quarrelled sooner or later with most leading geologists. His noisy campaign against Lyell's uniformitarianism turned into a rigid denial of Darwinian evolution.

Murdock William 1754–1839 was a British engineer who introduced gas lighting and was a pioneer in the development of steam engines and their application. Murdock was born in Auchinleck, Ayrshire. In 1777, he entered the engineering firm of Matthew Boulton and James Watt at their Soho works in Birmingham, and two years later was sent to Cornwall to supervise the fitting of Watt's engines in mines. He lived in Redruth and proved an invaluable help to Watt, and references to him are numerous in the Soho correspondence. According to documents at Soho he signed an agreement on 30 March 1800 to act as an engineer and superintendant of the Soho foundry for a period of five years. He was, however, constantly despatched to various parts of the country, and he frequently visited Cornwall after he had ceased to reside there permanently. His connection with Boulton and Watt's firm continued until 1830, when he virtually retired. He died in 1839 at his house at Sycamore Hill, within sight of the Soho foundry.

An industrious but modest man, Murdock's subsequent fame has been somewhat overshadowed by that of his employers, Matthew Boulton (1728–1809) and James Watt (1736–1819). In about 1782, while residing in Cornwall, he began experiments on the illuminating properties of gases produced by distilling coal, wood, peat and so on. In 1792, he lighted his house at Redruth by this means, and in 1892 the centenary of gas lighting was duly celebrated, but on the evidence now available it appears that the decisive breakthrough came several years later than 1792. Murdock succeeded in producing coal gas in large iron retorts and conveying it 21 m/70 ft through metal pipes. After returning to Birmingham in about 1799 he perfected further methods for making, storing and purifying gas.

In 1802, in celebration of the Peace of Amiens with France, part of the exterior of the Soho factory was illuminated by gas, and a year later the factory interior was similiarly lit. Thereafter apparatus was erected ensuring that a part of the Soho foundry was regularly lighted this way, and the manufacture of gas-making plant seems to have been commenced about this period, probably in connection with apparatus for producing oxygen and hydrogen for medical purposes. In 1804, George Lee of the firm of Phillips and Lee, cotton-spinners of Manchester, ordered an apparatus for lighting his house with gas. Subsequently Phillips and Lee decided to light their mills. On 1 January

1806, Murdock wrote informing Boulton and Watt that 'fifty lamps of the different kinds' were lighted that night with satisfactory results. There was, Murdock added, 'no Soho Stink' – an expression which seems to show that the method of purification used at the foundry was somewhat primitive.

In February 1808, Murdock read a paper before the Royal Society in which he gave a full account of his investigations, and also by the saving effected by the adoption of gas lighting at Phillips and Lee's mill. The paper is the earliest practical essay on the subject. The Rumford Gold Medal, bearing the inscription *Ex fume clare lucem*, was awarded to Murdock for his paper, which concludes with these words: 'I believe I may, without presuming too much, claim both the first idea of applying and the first actual application of the gas to economical purposes.'

Murdock also made important improvements to the steam engine. He was the first to devise an oscillating engine, of which he made a model in about 1784. In 1786, he was bringing to the attention of Boulton and Watt, who both remained highly sceptical of such possibilities, the idea of a steam carriage or road locomotive, an enterprise which was not successful. In 1799, he invented the long-D slide-valve. He is generally credited with inventing the so-called sun and planet motion, a means of making a steam engine give continuous revolving motion to a shaft provided with a fly wheel. Watt, however, patented this motion in 1781. Murdock also experimented with compressed air, and in 1803 constructed a steam gun.

N

Nagell Trygve 1895–   is a Norwegian mathematician whose most important work was in the fields of abstract algebra and number theory.

Nagell was born in Oslo on 13 July 1895 and was educated at the University of Oslo, where he received his MA in mathematics in 1920. He was appointed to the mathematics faculty of the University in that year and was awarded his PhD in 1926. He was promoted to the rank of associate professor in 1930, but a year later left Oslo to become Professor of Mathematics at the University of Uppsala, in Sweden, where he remained until his retirement in 1962. Since 1962 he has continued to live in Uppsala. For his work in mathematics he was awarded the Norwegian Order of St Olav and made Knight Commander in the Swedish Order of the North Star.

Nagell first made his name with a series of papers in the early 1920s on indeterminate equations, investigations which led to the publication of a treatise on indeterminate analysis in 1929. He also published papers and books, in Swedish and English, on number theory. From the late 1920s onwards his chief interest was the study of algebraic numbers, and his 1931 study of algebraic rings was perhaps his most important contribution to abstract algebra.

Napier John, 8th Laird of Merchiston 1550–1617 was a Scottish mathematician who invented logarithmic tables.

Napier was born at Merchiston Castle, near Edinburgh, in 1550, into a family of influential landed nobility and statesmen who were staunchly attached to the Protestant cause. As a young boy he was educated chiefly at home, although he may have spent some time at the Edinburgh High School and, less probably, studying in France. At the age of 13 he was sent to St Salvator's College, in the University of St Andrews. There he studied mainly theology and philosophy, gained a reputation for his quick temper, and left without taking his degree. He may then have passed a few years studying on the continent. He was, at any rate, in Scotland in 1571. He built a castle at Gartnes, on the banks of the Endrick, and lived there with his wife, whom he married in 1572, until the death of his father in 1608 brought him the inheritance of Merchiston.

Napier was an aristocratic scientific and literary amateur. He never occupied any professional post. But he became known as the 'Marvellous Merchiston' for his varied accomplishments. He made advances in scientific farming, especially by the use of salt as a fertilizer; he invented a hydraulic screw and revolving axle by means of which water could be removed from flooded coalpits and obtained the patent for its sole manufacture and use in 1597; he also published, in 1593, a violent denunciation of the Roman Church entitled *A Plaine Discovery of the Whole Revelation of St John*, a popular work which ran through several editions and was translated into French, Dutch and German. He had scarcely a moment of idleness, and overwork – combined with the gout from which he suffered in his later years – brought him to his death, at Merchiston, on 4 April 1617.

Napier's favourite intellectual pursuit was astronomy, and it was via astronomy that he was led to make his great invention. He performed many calculations in the course of his observations and research. He found the lengthy calculations, involving the use of trigonometric functions (especially sines) a tiresome burden, and over the course of about 20 years the idea of logarithmic tables slowly gestated in his mind. In 1614, he explained his new invention and printed the first logarithmic table in the *Mirifici Logarithmorum Canonis Descriptio*. The word 'logarithm' he formed from the Greek *logos* for 'expression' and *arithmos* for 'number'. The best statement of his invention, however, was given in the posthumously published *Constructio* (1619).

Napier's publication was immediately recognized by mathematicians for the great advance that it was. In particular, it excited the English mathematician Henry Briggs (c. 1556–1631), who went to Edinburgh in 1616 (and a couple of times thereafter) to discuss the new tables with Napier. Together they worked out improvements – such as the idea of using the base ten – and the result, Briggs's tables of 1617, was the production of the standard form of logarithmic tables in use until the present day (although they have largely been replaced by electronic computers and calculators).

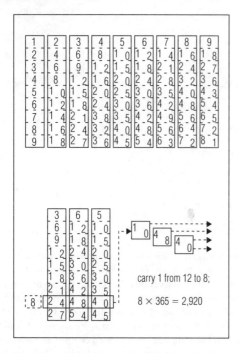

Napier *Napier's bones consist of a set of (originally ivory) rods with numbers marked off on them. Using them, multiplication became merely a process of reading off the appropriate figures and minor additions. (For multipliers greater than 9, however, the user had to resort to factorization.)*

Napier himself has a claim as the inventor of the first mechanical calculator, albeit one of a wholly primitive kind. His last work, *Rabdologiae* ('numeration by little rods'), he published in 1617. In it he explained his system of multiplying and dividing by the use of rods – usually made of bones or ivory, and hence known as Napier's Bones – and showed also how square roots could be extracted by the manipulation of counters on a chessboard.

As a footnote, and a testament to the splendid practical inventiveness of the man, it should be remembered that it was Napier, too, who first used and then popularized the decimal point to separate the whole number part from the fractional part of a number.

Nasmyth James 1808–1890 was a British engineer who contributed greatly to the design and production of tools. He is particularly known for his invention of the steam hammer and powered milling, shaping, slotting and planing machines. One of his lesser-known inventions, which he did not choose to patent, is the flexible drive shaft for drilling and other machines.

Nasmyth was born in Edinburgh on the 19 August 1808; his father was the well-known Scottish portrait painter and amateur engineer Alexander Nasmyth (1758–1840). James Nasmyth attended Edinburgh High School, and on leaving there at the age of 12 he devoted his time to building

engines and other mechanical devices. His success was such that he attempted to build a steam road carriage.

When he was 21 he travelled to London with his father, and there met Henry Maudslay (1771–1831), himself famous for work in tool and engine construction. Nasmyth showed Maudslay a model steam engine he had built, and so impressed him that he was taken on as his assistant. Over the next two years he learned Maudslay's techniques and developed his own accurate and rapid means of producing hexagonal-heading nuts. Also during this period he devised a flexible shaft of coiled spring steel for drilling holes in awkward places. (Later in life, during a visit to his dentist, he was told that this was the latest American invention.) Shortly after the death of Maudslay he returned to Edinburgh and built his own small workshop, which included a Maudslay lathe and a number of other machine tools.

On seeing better prospects for an engineer in Manchester, he moved there in 1834 to a small workshop in which power was available. With the success of his machine tool business he moved to a site on the Bridgwater Canal at Patricroft. In his Bridgwater foundry Nasmyth continued to manufacture machine tools and began to build railway locomotives and other machinery. His famous steam hammer was invented there in 1839, initially to forge the driving shaft of the steamship *Great Britain*. This device speeded up the production of large forgings without the loss of accuracy. It was a very successful tool in the great age of machine building.

Nasmyth devised many other tools, including a vertical cylinder-boring machine which speeded up the production of steam engines, and all sorts of milling, shaping, slotting and planing machines. Apart from the obvious effect these had on the accurate repetition of old techniques at an increased rate, they handled metal in new ways: all manner of lateral, transverse and rotating cutting machines were devised. His shaping machines in particular were a financial success. Generally these devices speeded up the rate of production in the engineering industry in an unprecedented manner. Even with his hand-held tools, such as taps, he devised means of producing more accurate work with increased ease.

As the building of locomotives, and other heavy engineering projects, overshadowed the machine-tool aspect of his business, his personal interest declined and in 1856, at the age of 48, he retired from engineering. This was not the end of his scientific curiosity, for he devoted more time to astronomy, a hobby which he pursued in a serious way. At his foundry he had built a number of telescopes, the largest being a reflector with a 50-cm/20-in mirror. During his retirement he lived at Penshurst in Kent, and used this instrument to make an extensive study of the Sun and Moon. In the course of his solar observations he was particularly concerned with sunspots. He chartered the lunar surface and developed his own volcanic theory of the origin of the craters. In 1851, his maps of the Moon received a prize at the Great Exhibition in London. Later he wrote a book about his lunar discoveries and theories.

Nasmyth died in London on 7 May 1890.

Nathans and Smith Daniel Nathans 1928– and Hamilton Othanel Smith 1931– are US microbiologists who shared (with Werner Arber, a Swiss microbiologist) the 1978 Nobel Prize for Physiology or Medicine for their work on restriction enzymes, which are special enzymes that can cleave genes into fragments.

Nathans was born on 30 October 1928 in Wilmington, Delaware. He was educated at the University of Delaware, from which he graduated in 1950, and at Washington University, St Louis, from which he gained his medical degree in 1954. He then worked as a Clinical Associate at the National Cancer Institute until 1957, when he became a resident physician at the Columbia-Presby Medical Center. From 1959 to 1962 he was a guest investigator at the Rockefeller University, New York City, after which he held several positions at the Johns Hopkins University, Baltimore: Assistant Professor then Professor of Microbiology at the School of Medicine there between 1962 and 1976, Director of the Department of Microbiology since 1972 and Boury Professor of Microbiology since 1976.

Smith was born on 23 August 1931 in New York City. He was educated at the University of California, Berkeley, from which he graduated in mathematics in 1952, and at the Johns Hopkins University, from which he obtained his medical degree in 1956. From 1956 to 1957 he was a junior resident physician at Barnes Hospital, then carried out research at the Henry Ford Hospital, Detroit, from 1959 until 1962, when he became a research fellow in microbial genetics at the University of Michigan. Since 1964 he has been at the Johns Hopkins University, as a Research Assistant, Assistant Professor, Associate Professor and, from 1973, Professor of Microbiology. He was also Guggenheim Professor there between 1975 and 1976.

Arber discovered restriction enzymes in *Escherichia coli* in the 1960s. These enzymes cleave genes at specific sites on the DNA molecules and thus enable the order of genes on the chromosomes to be determined. The gene fragments can also be used to analyse the chemical structure of genes as well as to create new gene combinations. Smith, working at the Johns Hopkins University independently of Arber, verified Arber's findings and was also able to identify the gene fragments. Smith collaborated with Nathans on some of his work, but Nathans also performed much original research of his own in this field. Using the carcinogenic SV40 virus, he showed in 1971 that it could be cleaved into 11 specific fragments, and in the following year he determined the order of these fragments.

As a result of the work of Nathans, Smith and Arber, it is now possible to determine the chemical formulae of the genes in animal viruses, to map these genes and to study the organization and expression of genes in higher animals.

Natta Giulio 1903– is an Italian chemist who shared the 1963 Nobel Prize for Chemistry with Karl Ziegler for his work on the production of polymers.

Natta was born in Imperia, near Genoa, on 26 February 1903, the son of a judge. He obtained his doctorate in chemical engineering from the Polytechnic Institute in Milan in 1924 and then held professorships in general chemistry at the University of Pavia, in physical chemistry at Rome, and in industrial chemistry at Turin. In 1938, he returned to Milan Polytechnic as Professor of Chemistry and Director of the Industrial Chemistry Research Institute. There he was charged by the government to investigate the problems of producing artificial rubber, because the supply of natural rubber had ceased with the approaching imminence of World War II.

In 1953, Natta began intensive studies of macromolecular chemistry. These investigations were initiated by knowledge obtained through a licence arrangement between Ziegler and the Italian company of Montecatini, of which Natta was a consultant. Ziegler had discovered how to synthesize linear polythene of high molecular weight at low pressures using as a catalyst a resin containing ions of titanium or aluminium. Because the polymer was made up of unbranched chains it was tougher and had a higher melting point than previous types of polythene.

Natta used these catalysts to polymerize propylene (propene, $CH_3CH=CH_2$). Early in 1954, he found that part of the polymer is highly crystalline and realized that it must have an ordered structure. He confirmed this using X-ray crystallography and coined the term 'isotactic' to describe the polymer's symmetrical structure. He also postulated that the surface of the catalyst must be highly regular to give rise to isotactic polymers.

After 1954, he continued to study the mechanism of the reaction and its stereo-specific aspects. He made other similar catalysts and produced new polymers, such as those *cis*-buta-1,4-diene and copolymers of ethylene (ethene) and propylene (propene), both potentially important synthetic rubbers.

Natta's early work on heterogeneous catalysts formed the basis for modern industrial syntheses of methyl alcohol (methanol), of formaldehyde (methanal) from methyl alcohol, of propionaldehyde (butanal) from propylene (propene) and carbon monoxide, and of succinic acid (butandioic acid) from acetylene (ethyne), carbon monoxide and synthetic gas.

The isotactic polymers he discovered after 1954 showed remarkable and unexpected properties of commercial importance, such as high melting point, high strength, and an ability to form films and fibres. It was realized that a new type of polymerization, called coordination polymerization, was involved. The growth of the polymer chain occurs by insertion of monomer between the existing chain and the solid surface of the catalyst, which controls the geometry of the reaction. (In other types of polymerization the catalyst remains remote from the growing end of the chain and therefore has no effect on the reaction geometry.)

Needham Joseph 1900– is a British biochemist, science historian and Orientalist whose most important scientific contribution was in the field of biochemical embryology. In the later part of his career his interests turned to the history of science, particularly the development of science in China.

Needham was born in London in December 1900, the son of one of the first Harley Street specialists in anaesthesia. He attended Oundle School and during the school holidays assisted in military hospitals, which were greatly understaffed due to the influx of casualties from World War 1. From Oundle he went on to study natural sciences at Gonville and Caius College, Cambridge. On graduation in 1921 he was offered a place in the research laboratory of Frederick Gowland Hopkins, and subsequently gained his doctorate and was elected a Fellow of Gonville and Caius College. In the same year he married Dorothy Moyle, a fellow student at Hopkins's laboratory; she was a talented scientist and her work on the biochemistry of muscles led to her election as a Fellow of the Royal Society in 1948 (Needham himself had become a member in 1941).

In 1928, Needham was appointed University Demonstrator in Biochemistry at Cambridge, then in 1933 he became Dunn Reader in Biochemistry (also at Cambridge), a post he held until 1966. It was during this latter period that he progressively reduced his scientific work and became increasingly devoted to studying Chinese science and culture. He learned Chinese, and in 1942 accepted an invitation to head the British Scientific Mission to China; he spent the next four years travelling through the country. From 1946 to 1948 he was head of the Division of Natural Sciences at the United Nations, after which he returned to Cambridge. He was elected Master of Gonville and Caius College in 1966 and held this post until 1976, when he retired in order to pursue his Oriental studies at the East Asian History of Science Library in Cambridge. Needham travelled extensively throughout his career, visiting and lecturing in numerous universities in the United States, Europe and Asia.

Needham's principal scientific contributions were made in the first half of his academic career. Initially he worked on the biochemistry of embryonic development, trying to discover the processes underlying the development of a fertilized egg from a mass of undifferentiated cells into a highly differentiated complex organism. In his three-volume *Chemical Embryology* (1931) Needham surveyed the morphogenetic changes and the various attempts to explain them, concluding that embryonic development is controlled chemically – in contrast to the traditional vitalistic view of Otto Driesch and others, which held that some indefinite principle (called entelechy) caused embryonic changes. The discovery of morphogenetic hormones that control embryonic development confirmed Needham's mechanistic view, and he proceeded – in collaboration with Conrad Waddington (1905–   ) to hypothesize (before the discovery of DNA) that only structural chemistry could fully explain the complex changes that occur during an organism's development. In *Order and Life* (1935) Needham foresaw the importance of organelles, anticipating some of the discoveries about the microstructure of living cells that later resulted from electron microscopy.

From about the mid-1930s Needham became increasingly interested in the history of science, particularly of Chinese science, and he progressively reduced his scientific investigations in order to devote himself to a comprehensive study of the development of Chinese science and culture. The first volume of *Science and Civilization in China* was published in 1954 and has since been followed by several more volumes in this huge synthesis of history, science and culture in China.

Nernst Hermann Walther 1864–1941 was a German physical chemist who made basic contributions to electro-chemistry and is probably best known as the discoverer of the third law of thermodynamics. He was awarded the 1920 Nobel Prize for Chemistry.

Nernst was born in Briessen, East Prussia (now Wabreźno, Poland), on 25 June 1864, the son of a civil servant and judge. He was educated at Grandenz Gymnasium, and went on to read natural sciences at university. He continued his studies with Albert von Ettinghausen at Graz, took a PhD degree under Friedrich Kohlrausch at Würzburg in 1886, and became an assistant to Friedrich Wilhelm Ostwald at Leipzig in the following year. Under their influence his interests narrowed into aspects of physical chemistry. In 1891, he became a Reader in Physics at Göttingen University and three years later, after a new laboratory had been built, the first Professor of Physical Chemistry. In 1905 he moved to a similar position in Berlin as successor to Hans Landolt (1831–1910), and remained there until 1922. Nernst then became President of the Physikalisch-Technische Reichsanstalt for two years, but relinquished this post to return to Berlin University as Professor of Physics and Director of the Physical Laboratory, where he stayed until he retired in 1934. No subscriber to the politics of Nazi Germany (two of his daughters married Jews), he spent his latter years in farming and agriculture on his country estate at Zibelle near the Polish border. He had a heart attack on 18 November 1941 and died at Muskan, near Berlin; his body was later reinterred in Göttingen, where his academic career began.

Nernst's first publication (1886) described his work with von Ettinghausen at Graz. What became known as the Ettinghausen–Nernst effect concerns the establishment of a potential difference across a metal plate along which there is a temperature gradient and a magnetic field. These experiments were significant in the development of the electronic theory of metals, according to which both thermal and electrical conduction are caused by the motion of electrons.

Working with Kohlrausch, an expert in electrochemistry, Nernst studied solution chemistry. Theories presented in the late 1880s are still in use today; every pH measurement depends on them, as does the use and theory of indicators and buffer solutions.

Nernst's theory concerning solids in contact with a liquid developed from the supposition that metals go into solution only as positive ions, the driving force being a 'solution pressure' which is opposed by the osmotic pressure of its ions in solution. The solution acquires a positive charge and the metal a negative one, if the metal is high in the electrochemical series. (A metal low in the electrochemical series has a low solution pressure and collects metal ions to become positively charged.) The Nernst equation relates ionic concentration, c, electrode potential, E:

$$E = E_0 \pm \frac{RT}{zF} \log_e c$$

where R is the gas constant, T the temperature, z the valency of the ion, F Faraday's constant, and E_0 the standard electrode potential. Nernst was also the first to advocate that the electrochemical standard be based on the hydrogen electrode. This work led to the theory of solubility product (1890), which had been initiated by Ostwald who had shown that it could be used as the basis for a system of qualitative and quantitative analysis.

The work on thermodynamics was carried out in Berlin in 1905–06. Developing the theories of Hermann von Helmholtz, Nernst formulated the third law of thermodynamics. It may be stated in various ways, one expression being:

If the entropy of each element in a crystalline state be taken as zero at the absolute zero of temperature, then every substance has a finite positive entropy but at absolute zero the entropy may become zero and does so in the case of a perfect crystalline substance.

Nernst and his students collected accurate thermodynamic data to substantiate the law. In 1911, with Frederick Lindemann (later Lord Cherwell), Nernst constructed a special calorimeter for measuring specific heats at low temperatures.

Two other significant contributions to physical chemistry concerned chemical equilibria and photochemistry. With Fritz Haber he studied equilibria in commercially important gas reactions, such as the reversible reaction between hydrogen and carbon dioxide to form water and carbon monoxide. He also examined the hydrogen–nitrogen reaction at high pressures (the basis of the Haber process). In 1918, Nernst investigated reactions that are initiated by light. He proposed that the fast reaction between chlorine and hydrogen begins when light causes chlorine molecules to dissociate into atoms:

$$Cl_2 + h\nu \rightarrow 2Cl$$

A chlorine atom then reacts with a hydrogen molecule to form hydrogen chloride and a hydrogen atom:

$$Cl + H_2 \rightarrow HCl + H$$

The hydrogen atom reacts with a chlorine molecule to produce another chlorine atom, and the process continues as a chain reaction:

$$H + Cl_2 \rightarrow HCl + Cl$$

Like many great scientists of the time, Nernst did not restrict himself to one narrow field. In 1897, he invented an electric lamp which, instead of a carbon filament, had a 'glower' made from zirconium oxide and some rare earth oxides. It was a good source of infrared and highly successful until superseded ten years later by the tungsten filament lamp. Even so, Nernst sold his patent for a million marks, and used the money to become a pioneer motorist. Many early automobiles had difficulty climbing hills, but Nernst devised a method of injecting nitrous oxide (dinitrogen monoxide) into the cylinders when the engine got into difficulties. In the 1920s he invented a 'Neo-Bechstein' piano which amplified sounds produced at low amplitudes. Although acoustically correct, it did not find favour with musicians or concert audiences.

During an academic career of about 50 years, Nernst published 157 papers and 14 books, one of which, *Theoretische Chemie* (1895), became the recommended text for a generation of physical chemists throughout the world.

Neugebauer Gerald 1932– is an American astronomer whose work has been crucial in establishing infrared astronomy.

Neugebauer was born on 3 September 1932 and was educated at Cornell University. He received his PhD from the California Institute of Technology in 1960.

Having completed his education, he served in the US army, from 1960 to 1962 and was stationed at the Jet Propulsion Laboratory. In 1962, he accepted a post as Assistant Professor of Physics at the California Institute of Technology, being promoted to Associate Professor in 1965. In 1965, he was also appointed a Staff Associate of Mount Wilson and Palomar Observatories. Since 1970 he has been Professor of Physics at the California Institute of Technology and since 1981 he has been Director of the Palomar Observatory.

During his professional career. Neugebauer has been closely involved with NASA's interplanetary missions and the design of new infrared telescopes. From April 1969 to July 1970 he was a member of the NASA Astronomy Missions Board and from 1970 to 1973 he was appointed as the Principal Investigator of the infrared radiometers carried aboard the Mariner missions to Mars and the Infrared Explorer Satellite. He was also the team leader of the Infrared radiometer for the Large Space Telescope Definition Study. Since 1976 he has been the US Principal Scientist on the Infrared Astronomical Satellite.

During the mid-1960s Neugebauer and his colleagues began to establish the first infrared map of the sky. As their telescope was designed to pick up radiation of in the region of 2.2 μm (2.2×10^{-6} m), this project became known as the 'Two Micron Survey'. The results of this survey were astounding; from the part of the sky that can be mapped from the top of Mount Wilson, some 20,000 new infrared sources were detected and most of these did not coincide with known optical sources. The survey also highlighted a large number of curious objects that demanded further study, among the most interesting of which are cool objects that are immersed in thick warm dust. These are thought to be stars that are in the process of formation. Among the brightest and strangest of these sources is an object known as the Becklin–Neugebauer object, named after its discoverers. It is located in the Orion Nebula, but it cannot be seen in photographs taken in visible light. Carbon monoxide, detected as being associated with this object, is blowing outward from it at a high velocity. This phenomenon is being interpreted as a strong stellar wind blowing from a young star that only began the process of nuclear fusion as recently as 10,000 or 20,000 years ago.

Newcomb Simon 1835–1909 was a Canadian-born US mathematician and astronomer who compiled charts and tables of astronomical data with phenomenal accuracy. His calculations of the motions of the bodies in the Solar System were in use as daily reference all over the world for more than 50 years, and the system of astronomical constants for which he was most responsible is still the standard.

Newcomb was born in Wallace, Nova Scotia, on 12 March 1835. He had little or no formal education, although his father later claimed he was a mathematical prodigy. At the age of 16 he was apprenticed to a quack doctor in Salisbury, New Brunswick, but after two or three years he ran away and settled in Maryland, in the United States, as a country schoolmaster. Deciding that his talents lay in mathematics, he became a 'computer' at Cambridge, Massachusetts, in 1857. He enrolled in the Lawrence scientific school of Harvard University and received a degree in 1858. In 1861 he applied for, and received, a commission in the corps of Professors of Mathematics in the United States Navy, where he was assigned to the United States Naval Observatory at Washington, DC. Sixteen years later he was put in charge of the American Nautical Almanac office, then also in Washington. In 1884, he obtained the additional appointment of Professor of Mathematics and Astronomy at Johns Hopkins University, Baltimore, but continued to live in Washington. When he reached the compulsory retiring age for captains in 1897, he received the unusual distinction of retirement with the rank of rear admiral.

For many years the editor of the *American Journal of Mathematics*, Newcomb was also one of the founders of the American Astronomical Society and its first President (1899–1905). He received honorary degrees from 17 universities and was a member of 45 foreign societies; he was awarded the Gold Medal of the Royal Astronomical Society in 1874, the Copley Medal of the Royal Society in 1890, and the Schubert Prize of the Imperial Academy of Sciences, St Petersburg, in 1897. He wrote several popular books on astronomy, one or two on finance and economics, and even published some fiction. Altogether his books and papers totalled an amazing 541 titles. Newcomb died on 11 July 1909 and was buried in Arlington National Cemetery.

Assigned to the US Naval Observatory in Washington in 1861, Newcomb worked for more than ten years determining the positions of celestial bodies with the meridian instruments, and for two years with the new 66-cm/26-in refractor. When he was put in charge of the American Nautical Almanac office, he started the great work that was to occupy most of his time for the rest of his life: the calcula-

tion of the motions of the bodies in the Solar System. His most important work appeared in *Astronomical Papers Prepared for the Use of the American Ephemeris and Nautical Almanac*, a series of memoirs which he founded in 1879. Newcomb was the principal author of 25 out of 37 articles in the first nine volumes. Among them were his tables of data concerning the Sun, Mercury, Venus, Mars, Uranus and Neptune, together with Hill's tables concerning Jupiter and Saturn. This series of papers is of virtually unsurpassable standard; hardly a figure or statement in them has been found to be incorrect, and they are still widely used to calculate daily positions of celestial objects.

Newcomb's most far-reaching contribution, however, was his establishment, jointly with Arthur Matthew Weld Downing (1850–1917), Superintendent of the British Nautical Almanac office, of a universal standard system of astronomical constants. Until then there had been a considerable diversity in the fundamental data used by astronomers of different countries and institutions. In May 1896 a conference was held in Paris for the directors of the astronomical almanacs of the United States, Great Britain, France and Germany. They came to the resolution that after 1901 a single set of constants, mainly Newcomb's, should be used by each country. Although some of Newcomb's work was not complete then, time has proved the decision to have been a wise one. A similar conference held in 1950 decided that the system of constants that had been adopted in 1896 was still preferable to any other for practical use.

Newcomen Thomas 1663–1729 was an English blacksmith who developed the first really practical steam engine, which was principally to power pumps in the tin mines of Cornwall and the coal mines of northern England.

Newcomen was born in Dartmouth in 1663, and christened there on 24 February of that year. His family were probably merchants. His education was probably obtained from a Noncomformist called John Favell. It is thought that he was eventually apprenticed to an ironmonger in Exeter, and on completion of his training went to Lower Street, Dartmouth, where he established his own business as a blacksmith and ironmonger. Newcomen is known to have been an ardent Baptist and a lay preacher.

In his trade Newcomen was assisted by a plumber called John Calley. It is not known exactly when they built the first steam engine. The first authenticated Newcomen engine was erected in 1712 near Dudley Castle, Wolverhampton. This engine was, however, much more than a prototype and it is believed that a number of earlier machines must have been operated to develop the engine to this point. The whole situation is confused by a patent granted to Thomas Savery to 'raise water by the force of fire' and most steam machines of the time were referred to as 'Saverys'. Also Newcomen may have infringed Savery's patent and deliberately not advertised the fact. In later years Newcomen paid royalties to Savery and, after his death, to a syndicate which bought up the patent.

The Newcomen engine was used to draw water from mines and so operated a pump. The engine consisted of a

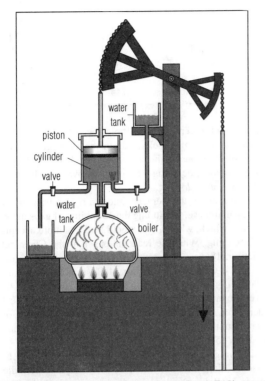

Newcomen *In Newcomen's steam engine (1712), steam from the boiler entered the cylinder while the weight of the pump-rod pulled the piston to the top of the cylinder. Water from a header tank was sprayed into the cylinder, condensing the steam and creating a vacuum so that air pressure forced down the piston and activated the pump. Condensed water flowed into a second tank. The various inlet and outlet stages were controlled by valves.*

boiler and cylinder, with the cylinder mounted above the boiler. In the cylinder was a piston, sealed as closely as possible to the wall with a leather flap. The piston was, in turn, attached to one end of a large wooden beam, the other end being attached to the water pump. Steam from the boiler entered the cylinder at little more than atmospheric pressure when the piston had been pulled to the top of the cylinder by the weight of the beam. The steam valve was shut and water injected into the cylinder to condense the steam. As the steam condensed a vacuum was created, and the pressure of the atmosphere pushed the piston to the bottom of the cylinder, thus operating the pump. Gravity operating on the beam again raised the piston ready for the next stroke.

Newcomen's engine consumed an enormous amount of coal because fresh hot steam had to be raised for each stroke. This made the machine more popular in the coal mines. The early engines were very expensive because the cylinder was made of brass which could be founded and machined more accurately than iron. Later, iron cylinders were produced, but they were thick-walled and consequently much less efficient in terms of coal consumed. This

was, of course, of little importance as long as they were used in coal mines. As mining operations changed, the pumps were moved to new locations and sometimes modified. The practical success and usefulness of the Newcomen engine was such that after later refinements, several of which were introduced by Watt in the 1800s, they became popular for adoption anywhere water was to be raised. Some are known to have remained in operation until the early part of the twentieth century when they were superceded by the electric pump.

It was with the Newcomen engine that the art and science of steam was begun. Tables were drawn up in 1717 by Henry Beighton to show the size of cylinder required to raise a particular quantity of water through a particular height. For water drawn from great depths several lifts could be used in series, one raising through the first level and then another taking it from there upwards, and so on.

Thomas Newcomen did not live to see the widespread adoption of his long-serving engine or its effect on developing British industry. He died of a fever in Southwark, London on 5 August 1729.

Newlands John Alexander Reina 1837–1898 was a British chemist who preceded Dmitri Mendeleyev in formulating the concept of periodicity in the properties of chemical elements, although his ideas were not accepted at the time.

Newlands was born in Southwark, London, on 26 November 1837, the second son of a Presbyterian minister; his mother, born Maria Reina, was of Italian descent. He was educated by his father and in 1856 entered the Royal College of Chemistry, London, where he studied for a year under August Hofmann. He then became assistant to J. T. Way, the Royal Agricultural Society's chemist. He stayed with Way until 1864, except for a short time in 1860 when he served as a volunteer with Giuseppe Garibaldi in Italy. In 1864, Newlands set up in practice as an analytical chemist, supplementing his income by teaching chemistry. He seems to have made a special study of sugar chemistry, and in 1868 became chief chemist in a sugar refinery belonging to James Duncan, with whom he developed a new system for cleaning sugar and introduced a number of improvements in processing. The business declined as a result of foreign competition, so he left the refinery and again set up as an analyst, this time in partnership with his brother B.E.R. Newlands. The brothers revised an established treatise on sugar growing and refining, in collaboration with C.G.W. Lock, one of the original authors. Newlands died in London on 29 July 1898.

Newlands's early papers on organic compounds (the first suggesting a new nomenclature and the second proposing the compilation of tables to show the relationships between compounds) were hampered by the absence of any clear ideas about structure and valency. But they did show his inclination towards systematization. His first communication to *Chemical News* in February 1863, on the numerical relationships between atomic weights (relative atomic masses) of similar elements was a summary, with some of

Group		No.		No.		No.		No.		No.
a	N	6	P	13	As	26	Sb	40	Bi	54
b	O	7	S	14	Se	27	Te	42	Os	50
c	F	8	Cl	15	Br	28	I	41	–	–
d	Na	9	K	16	Rb	29	Cs	43	Tl	52
e	Mg	10	Ca	17	Sr	30	Ba	44	Pb	53

his own observations, of what had been pointed out by others, of whom he credited only Jean Baptiste Dumas. Two main phenomena had been observed: that there existed groups of three elements (the 'triads' of Johann Döbereiner) of similar properties, the atomic weight of the middle one being the mean of those of the other two; and that the difference between the atomic weights of analogous elements seemed often to be multiples of eight.

Like many of his contemporaries, Newlands first used the terms equivalent weight and atomic weight without any distinction in meaning, and in this first paper he used the values accepted by his predecessors. Then in 1864 he employed Alexander Williamson's values (based on the system of Stanislao Cannizzaro) in a table of the 61 then known elements in order of their 'new' atomic weights. In a second table he grouped 37 elements into ten classes, most of which contained one or more triads. He attributed the incompleteness of the table to uncertainty regarding the properties of some of the more recently discovered elements, and to the possible existence of additional, undiscovered elements. For example he considered that silicon (atomic weight 28) and tin (atomic weight 118) were the extremities of a triad, the middle member of which was unknown. Thus his later claim to have predicted the existence of germanium (atomic weight 73, the mean of 28 and 118) before Mendeleyev is valid.

He went on to number the elements in the order of their atomic weights, giving the same number to any two having the same weight, and observed that elements with consecutive numbers frequently either belonged to the same group or occupied similar positions in other groups. He set out the list as a table (shown above).

The difference in number between the first and second members of a group was seven. In Newlands's words: 'The eighth element starting from a given one is a kind of repetition of the first, like the eighth note in an octave of music.' One or two transpositions were made to give acceptable groupings; the omitted element (number 51) would have been mercury, which he clearly could not group with the halogens.

A year later, in 1865, he again drew attention to the

difference of seven (or a multiple thereof) between the ordinal numbers of the elements in the same group, and termed the relationship the law of octaves. This time he put all 62 elements (including the newly discovered indium) in a table (set out below).

This forcing of the elements into too rigid a framework weakened his case, precluding the possibility of gaps in the sequence which, when filled, would lead to a more acceptable grouping. When he read his paper to the Chemical Society in 1866 he was severely criticized and even ridiculed – G. C. Foster, Professor of Physics at University College, London, is reputed to have asked Newlands if he had ever examined the elements when listed in alphabetical order. More seriously, he pointed out the unacceptability of any system that separated chromium from manganese, and iron from cobalt and nickel.

Discouraged, Newlands did no more work on his theories until after the publication of Mendeleyev's periodic table in 1869. Newlands claimed priority, particularly after the award of the Davy Medal of the Royal Society to Mendeleyev and Lothar Meyer in 1882. His persistence was eventually rewarded in 1887, when the medal was awarded to him nearly 25 years after he had first published his work.

Newton Isaac 1642–1727 was a British physicist and mathematician who is regarded as one of the greatest scientists ever to have lived. In physics, he discovered the three laws of motion that bear his name and was the first to explain gravitation, clearly defining the nature of mass, weight, force, inertia and acceleration. In his honour, the SI unit of force is called the newton. Newton also made fundamental discoveries in optics, finding that white light is composed of a spectrum of colours and inventing the reflecting telescope. In mathematics, Newton's principal contribution was to formulate the calculus and the binomial theorem.

Newton was born at Woolsthorpe, Lincolnshire, on 25 December 1642 by the old Julian calendar, but on 4 January 1643 by modern reckoning. His birthplace, Woolsthorpe Manor, is now preserved. Newton's was an inauspicious beginning for he was a premature, sickly baby born after

H	1	F	8	Co, Ni	22	Br	29	Pd	36	I	42	Pt, Ir	50	Cl	15
Li	2	Na	9	Cu	23	Rb	30	Ad	37	Cs	44	Tl	53	K	16
Be	3	Mg	10	Zn	25	Sr	31	Cd	38	Ba, V	45	Pb	54	Ca	17
B	4	Al	11	Y	24	Ce, La	33	U	40	Ta	46	Th	56	Cr	19
C	5	Si	12	In	26	Zr	32	Sn	39	W	47	Hg	52	Ti	18
N	6	P	13	As	27	Di, Mo	34	Sb	41	Nb	48	Bi	55	Mn	20
O	7	S	14	Se	28	Rh, Ru	35	Te	43	Au	49	Os	51	Fe	21

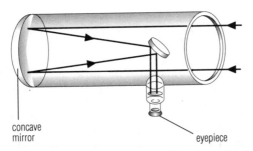

concave
mirror   eyepiece

Newton *In 1673 Newton constructed the first reflecting telescope, which uses a concave mirror to collect and focus incoming light.*

his father's death, and he was not expected to survive. When he was three, his mother remarried and the young Newton was left in his grandmother's care. He soon began to take refuge in things mechanical, reputedly making water-clocks, kites bearing fiery lanterns aloft, and a model mill powered by a mouse, as well as innumerable drawings and diagrams. When Newton was 12, he began to attend the King's School, Grantham, but his schooling was not to last. His mother, widowed again, returned to Woolsthorpe in 1658 and withdrew him from school with the intention of making him into a farmer. Fortunately, his uncle recognized Newton's ability and managed to get him back to school to prepare for university entrance. This Newton achieved in 1661, when he went to Trinity College, Cambridge, and began to delve widely and deeply into the scholarship of the day.

In 1665, the year that he became a Bachelor of Arts, the university was closed because of the plague and Newton spent eighteen months at Woolsthorpe, with only the occasional visit to Cambridge. Such seclusion was a prominent feature of Newton's creative life and, during this period, he laid the foundations of his work in mathematics, optics, dynamics and celestial mechanics, performing his first prism experiments and reflecting on motion and gravitation.

Newton returned to Cambridge in 1666 and became a minor Fellow of Trinity in 1667 and a major Fellow the following year. He also received his Master of Arts degree in 1668 and became Lucasian Professor of Mathematics – at the age of only 26. It is said that the previous incumbent, Isaac Barrow (1630–1677), resigned the post to make way for Newton. Newton remained at Cambridge almost 30 years, studying alone for the most part, though in frequent contact with other leading scientists by letter and through the Royal Society in London, which elected him a Fellow in 1672. These were Newton's most fertile years. He laboured day and night in his chemical laboratory, at his calculations, or immersed in theological and mystical speculations. In Cambridge, he completed what may be described as his greatest single work, the *Philosophae Naturalis Principia Mathematica /Mathematical Principles of Natural Philosophy*. This was presented to the Royal Society in 1686, who subsequently withdrew from publishing it through shortage of funds. The astronomer Edmund Halley

(1656–1742), a wealthy man and friend of Newton, paid for the publication of the *Principia* in 1687. In it, Newton revealed his laws of motion and the law of universal gravitation.

After the *Principia* appeared, Newton appeared to become bored with Cambridge and his scientific professorship. In 1689, he was elected a member of Parliament for the university and in London he encountered many other eminent minds, notably Christiaan Huygens (1629–1695). The excessive strain of Newton's studies and the attendant disputes caused him to suffer severe depression in 1692, when he was described as having 'lost his reason'. Four years later he accepted the appointment of Warden of the London Mint, becoming Master in 1699. He took these new, well-paid duties very seriously, revising the coinage and taking severe measures against forgers. Although his scientific work continued, it was greatly diminished.

Newton was elected President of the Royal Society in 1703, an office he held until his death, and in 1704, he summed up his life's work on light in *Opticks*. The following year, Newton was knighted by Queen Anne. Although he had turned grey at 30, Newton's constitution remained strong and it is said he had sharp sight and hearing, as well as all his teeth, at the age of 80. His later years were given to revisions of the *Principia*, and he died on 20 March 1727. Newton was accorded a state funeral and buried in Westminster Abbey, an occasion that prompted Voltaire to remark that England honoured a mathematician as other nations honoured a king.

Any consideration of Newton must take account of the imperfections of his character, for the size of his genius was matched by his ambition. A hypersensitivity to criticism and possessiveness about his work made conflicts with other scientists a prominent feature of his later life. This negative side of Newton's nature may be well illustrated by his dispute with Gottfried Leibniz (1646–1716). These two great mathematicians worked independently on the development of a differential calculus, both making significant advances. No one today would seriously question Leibniz's originality and true mathematical genius, but Newton branded him a plagiarist and claimed sole invention of the calculus. When Leibniz appealed to the Royal Society for a fair hearing, Newton appointed a committee of his own supporters and even wrote their, supposedly impartial, report himself. He then further proceeded to review this report anonymously, later remarking that 'he had broken Leibniz's heart with his reply to him'. The partisan and 'patriotic' views that resulted from this controversy served to isolate English mathematics and to set it back many years, for it was Leibniz's terminology that came to be used.

A similar dispute arose between Newton and Robert Hooke (1635–1703), one of the more brilliant and versatile members of the Royal Society, who supported Huygens's wave theory of light. Although, in the past, he had collaborated with Hooke, Newton published results without giving credit to their originator. Hooke, however, was notably disputatious and better able to stand up for himself than

Leibniz. On the other hand, Newton remained faithful to those he regarded as friends, appointing several to positions in the Mint after he took charge, and part of his quarrel with the Astronomer Royal, John Flamsteed (1646–1719), was that Flamsteed had fallen out with Newton's friend Halley.

Newton's work itself must be considered in many parts: he was a brilliant mathematician and an equally exceptional optical physicist, he revolutionized our understanding of gravity and, throughout his life, studied chemistry and alchemy, and wrote millions of words on theological speculation and mysticism.

As a mathematician, Newton developed unusually late, being well through his university career when he studied Pierre de Fermat (1601–1675), René Descartes (1596–1650) and others, before returning to Euclid (lived 300 BC), whom he had previously dismissed. However, in those two plague years of 1665 and 1666, Newton more than made up for this delay, and much of his later work can be seen as a revision and extension of the creativity of that period. To quote one of his own notebooks: 'In the beginning of the year 1665 I found the method for approximating series and the binomial theorem. The same year I found the method for tangents of Gregory and in November had the direct method of fluxions [differential calculus] and in January [1666] had the theory of colours [of light] and in May following I had entrance into the inverse method of fluxions [integral calculus] and in the same year I began to think of gravity extending to the orb of the moon ...'

The zenith of his mathematics was the *Principia*, and after this Newton did little mathematics, though his genius remained sharp and when Bernouilli and Leibniz composed problems with the specific intention of defeating him, Newton solved each one the first day he saw it. Both in his own day and afterwards, Newton influenced mathematics 'following his own wish' by 'his creation of the fluxional calculus and the theory of infinite series', which together made up his analytic technique. But he was also active in algebra and number theory, classical and analytical geometry, computation and approximation and even probability. For three centuries, most of his papers lay buried in the Portsmouth Collection of his manuscripts and only now are scholars examining his complete mathematics for the first time.

Newton's work in dynamics also began in those two years of enforced isolation at Woolsthorpe. He had already considered the motion of colliding bodies and circular motion, and had arrived at ideas of how force and inertia affect motion and of centrifugal force. Newton was now inspired to consider the problem of gravity by seeing an apple fall from a tree – a story that, according to Newton himself, is true. He wondered if the force that pulled the apple to the ground could also extend into space and pull the Moon into an orbit around the Earth. Newton assumed that the rate of fall is proportional to the force of gravity and that this force is inversely proportional to the square of the distance from the centre of the Earth. He then worked out what the motion of the Moon should be if these

assumptions were correct, but obtained a figure that was too low. Disappointed, Newton set aside his considerations on gravity and did not return to them until 1679.

Newton was then able to satisfy himself that his assumptions were indeed true and he also had a better radius of the Earth than was available in the plague years. He then set to recalculating the Moon's motion on the basis of his theory of gravity and obtained a correct result. Newton also found that his theory explained the laws of planetary motion that had been derived earlier that century by Johannes Kepler (1571–1630) on the basis of observations of the planets.

Newton presented his conclusions on dynamics in the *Principia*. Although he had already developed calculus, he did not use it in the *Principia*, preferring to prove all his results geometrically. In this great work, Newton's plan was first to develop the subject of general dynamics from a mathematical point of view and then to apply the results in the solution of important astronomical and physical problems. It included a synthesis of Kepler's laws of planetary motion and Galileo's laws of falling bodies, developing the system of mechanics we know today, including the three famous laws of motion. The first law states that every body remains at rest or in constant motion in a straight line unless it is acted upon by a force. This defines inertia, finally disproving the idea which had been prevalent since Aristotle (384–322 BC) had mooted it, that force is required to keep anything moving. The second law states that a force accelerates a body by an amount proportional to its mass. This was the first clear definition of force and it also distinguished mass from weight. The third law states that action and reaction are equal and opposite, which showed how things could be made to move.

Newton also developed his general theory of gravitation as a universal law of attraction between any two objects, stating that the force of gravity is proportional to the masses of the objects and decreases in proportion to the square of the distance between the two bodies. Though, in the years before, there had been considerable correspondence between Newton, Hooke, Halley and Kepler on the mathematical formulation of these laws, Newton did not complete the work until the writing of the *Principia*.

'I was in the prime of my age for invention' said Newton of those two years 1665 and 1666, and it was in that period that he performed his fundamental work in optics. Again it should be pointed out that the study of Newton's optics has been limited to his published letters and the *Opticks* of 1704, its publication delayed until after Hooke's death to avoid yet another controversy over originality. No adequate edition or full translation of the voluminous *Lectiones Opticae* exists. Newton began those first, crucial experiments by passing sunlight through a prism, finding that it dispersed the white light into a spectrum of colours. He then took a second prism and showed that it could combine the colours in the spectrum and form white light again. In this way, Newton proved that the colours are a property of light and not of the prism. An interesting byproduct of these early speculations was the development of the reflecting telescope. Newton held the erroneous opinion that optical dispersion was independent of the medium through which the light was refracted and, therefore, that nothing could be done to correct the chromatic aberration caused by lenses. He therefore set about building a telescope in which the objective lens is replaced by a curved mirror, in which aberration could not occur. In 1668, Newton succeeded in making the first reflecting telescope, a tiny instrument only 15 cm/6 in long, but the direct ancestor of today's huge astronomical reflecting telescopes. In this invention, Newton was anticipated to some degree by James Gregory (1638–1675) who had produced a design for a reflecting telescope five years earlier but had not succeeded in constructing one.

Other scientists, Hooke especially, were critical of Newton's early reports, seeing too little connection between experimental result and theory, so that, in the course of a debate lasting several years, Newton was forced to refine his theories with considerable subtlety. He performed further experiments in which he investigated many other optical phenomena, including thin film interference effects, one of which, 'Newton's rings', is named after him.

The *Opticks* presented a highly systematized and organized account of Newton's work and his theory of the nature of light and the effects that light produces. In fact, although he held that light rays were corpuscular in nature, he integrated into his ideas the concept of periodicity, holding that 'ether waves' were associated with light corpuscles, a remarkable conceptual leap, for Hooke and Huygens, the founder of the wave theory, both denied periodicity to light waves. The corpuscle concept lent itself to an analysis by forces and established an analogy between the action of gross bodies and that of light, reinforcing the universalizing tendency of the *Principia*. However, Newton's prestige was such that the corpuscular theory held sway for much longer than it deserved, not being finally overthrown until early in the 1800s. Ironically, it was the investigation of interference effects by Thomas Young (1773–1829) that led to the establishment of the wave theory of light.

Although comparatively little is known of the bulk of Newton's complete writings in chemistry and physics, we know even less about his chemistry and alchemy, chronology, prophecy and theology. The vast number of documents he wrote on these matters have never yet been properly analysed, but what is certain is that he took great interest in alchemy (including attempts to transmute other metals into gold), performing many chemical experiments in his own laboratory and being in contact with Robert Boyle (1627–1691). He also wrote much on ancient chronology and the authenticity of certain biblical texts.

Newton's greatest achievement was to demonstrate that scientific principles are of universal application. In the *Principia Mathematica*, he built logically and analytically from mathematical premises and the evidence of experiment and observation to develop a model of the universe that is still of general validity. 'If I have seen further than other men,' he once said with perhaps assumed modesty, 'it is because I have stood on the shoulders of giants.' Newton was certainly able to bring together the knowledge of his forebears in a brilliant synthesis. Newton's life marked the first great flowering of the scientific method, which had been evolving in fits and starts since the time of the ancient Greeks. But Newton really established it, completing a scientific revolution in Europe that had begun with Nicolaus Copernicus (1473–1543) and ushering in the Age of Reason, in which the scientific method was expected to yield complete knowledge by the elucidation of the basic laws that govern the universe. No knowledge can ever be total, but Newton's example brought about an explosion of investigation and discovery that has never really abated. He perhaps foresaw this when he remarked: 'To myself, I seem to have been only like a boy playing on the seashore, and diverting myself in now and then finding a smoother pebble or a prettier shell than ordinary, whilst the great ocean of truth lay all undiscovered before me.'

With his extraordinary insight into the workings of nature and rare tenacity in wresting its secrets and revealing them in as fundamental and concise a way as possible, Newton stands as a colossus of science. In physics, only Archimedes (287–212 BC) and Albert Einstein (1879–1955), who also possessed these qualities, may be compared to him.

Nice Margaret Morse 1883–1974 was a US biologist and ornithologist.

Margaret Morse was born on 6 December 1883 in Amherst, Massachusetts, the fourth of seven children of an intellectual family. In 1901, she entered Mount Holyoke College to study languages, but then switched to natural sciences. She graduated in 1906 and moved to Clark University in Worcester, Massachusetts to study biology. In 1909, she married a fellow graduate student at Clark, Leonard Blaine Nice, who studied medicine at Harvard University before moving to the University of Oklahoma where he was head of the Department of Physiology. While bringing up five daughters, born between 1910 and 1923, Margaret Nice undertook ornithological research relying on close observation, a technique she later applied to studying child phsychology in which she graduated from Clark in 1915, writing several papers on the subject from observations on her own children. However her interests in bird behaviour continued and she devoted herself to a detailed study of the birds of Oklahoma. In 1927, her husband moved to Ohio State University, and Margaret Nice began an extensive study of the life history of the sparrow, recording the behaviour of individual birds over a long period of time, which established her as one of the leading ornithologists in the world. A further family move to Chicago provided fewer opportunities for Margaret Nice to study living birds, and she spent more of her time writing, and became particularly involved in conservation issues, such as campaigning against the indiscriminate use of pesticides. Despite never having a faculty appointment she achieved a unique position in American ornithology, and received several honours and an honorary doctorate from Mount Holyoke. She died in Chicago on 26 June 1974, shortly after her husband.

Nicol William 1768–1851 was a British physicist and geologist who is best known for inventing the first device for obtaining plane-polarized light – the Nicol prism.

Very little is known of Nicol's life. He was born in Scotland in 1768 and seems to have spent most of his career lecturing at the University of Edinburgh. Nicol did not publish any of his research findings until 1826, when he was 58 years old, with the result that his work made relatively little impact during his lifetime. He died in Edinburgh on 2 September 1851.

Nicol invented the Nicol prism, as it is now called, in 1828. Consisting of Iceland spar (a naturally occurring, transparent crystalline form of calcium carbonate), it utilized the phenomenon of double refraction discovered by Erasmus Bartholin in 1699. Nicol made his prism by bisecting a parallelepiped of Iceland spar along its shortest diagonal then cementing the two halves back in their original position with Canada balsam, which has a refractive index between the two indices of the double-refracting Iceland spar. Light entering the prism is refracted into two rays, one of which is reflected by the Canada balsam (an example of total internal reflection) out of the side of the prism, the other ray being transmitted, with very little deviation, through the prism and emerging as plane-polarized light. This second, plane-polarized ray can then be passed through another Nicol prism aligned parallel to the first. Rotating the second prism causes the amount of light transmitted through it to decrease, reaching a minimum (with no light transmitted) when the second prism has been rotated through 90°, then increasing again to a maximum when the second prism has been rotated through 180°. Furthermore, if a solution of an organic substance is placed between the two prisms, the second prism must be turned through a specific angle to allow maximum light transmission; this angle represents the degree of refraction of the polarized light. Nicol prisms greatly facilitated the study of refraction and polarization, and later played an essential part in the development of polarimetry, especially in the use of this technique to investigate molecular structures and optical activity of organic compounds.

In 1815, Nicol, who was primarily a geologist, developed a method of preparing extremely thin sections of crystals and rocks for microscopical study. His technique (which involved cementing the specimen to a glass slide and then carefully grinding until it was extremely thin) made it possible to view mineral samples by transmitted rather than reflected light and therefore enabled the minerals' internal structures to be seen. Nicol also used this technique to examine the cell structure of fossil woods, and the information he obtained from these studies was later used as a basis for identification and classification. But because of his reluctance to publish his work, Nicol's slide-preparation technique did not become widely used until after 1853, when Henry Sorby demonstrated its usefulness for studying mineral structures.

Nirenberg Marshall Warren 1927–   . US biochemist; *see* Holley and Nirenberg.

Nobel Alfred Bernhard 1833–1896 was a Swedish industrial chemist and philanthropist who invented dynamite and endowed the Nobel Foundation, which after 1901 awarded the annual Nobel prizes.

Nobel was born in Stockholm, Sweden, on 21 October 1833, the son of a builder and industrialist. His father, Immanuel Nobel, was also something of an inventor, and his grandfather had been one of the most important Swedish scientists of the seventeenth century. Alfred Nobel attended St Jakob's Higher Apologist School in Stockholm before the family moved to St Petersburg, Russia, where he and his brothers were taught privately by Russian and Swedish tutors, always being encouraged to be inventive by their father. From 1850 to 1852 Nobel made a study trip to Germany, France, Italy and North America, improving his knowledge of chemistry and mastering all the necessary languages.

During the Crimean War (1853–1856), Nobel worked in St Petersburg in his father's company, which produced large quantities of munitions. After the war his father went bankrupt, and in 1859 the family returned to Sweden. During the next few years Nobel developed several new explosives and factories for making them, and became a rich man. He spent the latter years of his life in San Remo, and died there on 10 December 1896.

Gun cotton, a more powerful explosive than gunpowder, had been discovered in 1846 by the German chemist Christian Schönbein. It was made by nitrating cotton fibre with a mixture of concentrated nitric and sulphuric acids. A year later the Italian Ascanio Sobrero discovered nitroglycerin, made by nitrating glycerin (glycerol). This extremely powerful explosive gives off 1,200 times its own volume of gas when it explodes, but for many years it was too dangerous to use because it can be set off much too easily by rough handling or shaking. Alfred and his father worked independently on both explosives when they returned to Sweden, and in 1862 Immanuel Nobel devised a comparatively simple way of manufacturing nitroglycerin on a factory scale. In 1863, Alfred Nobel invented a mercury fulminate detonator for use with nitroglycerin in blasting.

In 1864, the nitroglycerin factory blew up, killing Nobel's younger brother and four other people. Nobel turned his attention to devising a safer method of handling the sensitive liquid nitroglycerin. After many experiments he patented dynamite (in Sweden, Britain and the United States) in 1867. It is an easily handled, solid, ductile explosive consisting of nitroglycerin absorbed by keiselguhr, a porous diatomite mineral.

Guhr dynamite, as it was known, had certain technical weaknesses. Continuing his research, Nobel in 1875 created blasting gelatin or gelignite, a colloidal solution of nitrocellulose (gun cotton) in nitroglycerin, which in many ways proved to be an ideal explosive. Its power was somewhat greater than that of pure nitroglycerin, and it was easier to work with because less sensitive to shock, and it was strongly resistant to moisture.

The Nobels had long been trying to improve blasting powder. In 1887, the younger Nobel produced a nearly smokeless blasting powder called ballistite, a mixture of nitroglycerin and nitrocellulose with camphor and other additives. Upon ignition it burned with almost mathematical precision in concentric layers. Nobel's last development was progressive smokeless powder, a further product of ballistite devised in his San Remo laboratory.

Nobel's interests as an inventor were not confined to explosives. He worked in electrochemistry, optics, biology and physiology and helped to solve many problems in the manufacture of artificial silk, leather and rubber, and of artificial semi-precious stones from fused alumina. In his will, made in 1895, he left almost all his fortune to a foundation that would award annual prizes to 'those who, during the preceding year, shall have conferred the greatest benefit on mankind'. In 1958 the new element number 102 was named nobelium in his honour.

Noether Emmy (Amalie) 1882–1935 was a German mathematician who became one of the leading figures in modern abstract algebra.

Noether was born in Erlangen on 23 March 1882, the eldest child of the famous mathematician Max Noether, Professor of Mathematics at the University of Erlangen. After completing her secondary education, she came up against the rule then prevailing in Germany which barred women from becoming fully fledged students at a university. She was, however, allowed to attend lectures in languages and mathematics without student status for two years from 1900 to 1902; eventually she was accepted as a student and awarded a doctorate in 1907 for a thesis on algebraic invariants.

Once more the rules blocked her way, this time barring her from a post in the university faculty. She nevertheless persisted with her research independently and at the request of David Hilbert (1862–1943) was invited to give lectures at Göttingen University in 1915–16. There she worked with Hilbert and Felix Klein (1849–1925) on problems arising from Einstein's theory of relativity, and thanks to Hilbert's constant nagging the university eventually, in 1919, gave her the status of 'unofficial associate professor'. In 1922, her position was made official, and she remained at Göttingen until the Nazi purge of Jewish university staff in 1933. The rest of her life was spent as Professor of Mathematics at Bryn Mawr College in Pennsylvania, where she died from a post-surgical infection on 14 April 1935.

Noether first made her mark as a mathematician with a paper on noncommutative fields which she published in collaboration with Schmeidler in 1920. For the next few years she worked on the establishment, and systematization, of a general theory of ideals. It was in this field that she produced her most important result, a generalization of Dedekind's prime ideals and the introduction of the concept of primary ideals. Modern work in this field dates from her papers of the early 1920s. After 1927 she returned to the subject of noncommutative algebras (in which the order in which numbers are multiplied affects the result), her chief investigations being conducted into linear transformations of noncommutative algebras and their structure.

No other woman has achieved the mathematical eminence of Emmy Noether and for that reason, as also for her breaking down the male grasp on German universities, her life had an importance stretching beyond the history of mathematics.

Noether Max 1844–1921 was a German mathematician who contributed to the development of nineteenth-century algebraic geometry and the theory of algebraic functions.

Noether was born at Mannheim on 24 September 1844 and educated locally until the age of 14, when he contracted polio, which left him permanently handicapped and for two years deprived him of the use of his legs. He was tutored at home until he entered the University of Heidelberg in 1865 to study mathematics. He was awarded his doctorate in 1868 and was appointed to the Heidelberg faculty. In 1874, he became an associate professor, but a year later he moved to the University of Erlangen, where he was promoted to a Professorship in 1888. He retired, with an emeritus professorship, in 1919 and died at Erlangen on 13 December 1921. His daughter Amalie (Emmy) also became a famous mathematician.

Noether published books on algebraic curves (1882) and algebraic functions (1894), as well as several biographies of mathematicians, but his reputation rests principally on his work in the early 1870s. Much of his initial inspiration came from the work of Antonio Cremona (1830–1903). In 1871 he published a proof (independently found at about the same time by William Clifford and J. Rosanes) that a Cremona transformation can be constructed from quadratic transformations. Then, two years later in 1873, he published his one outstanding result, the theorem concerning algebraic curves which contains the 'Noether conditions'. The theorem runs as follows. Given two algebraic curves, $\phi(x,y) = 0$ and $\psi(x,y) = 0$, which intersect at a finite number of isolated points, the equation of an algebraic curve that passes through all the points of intersection may be expressed as:

$$A\Phi + B\Psi = 0$$

where A and B are polynomials in x and y if, and only if, certain conditions (the 'Noether conditions') are satisfied.

Noether said that his result could be extended to surfaces and hypersurfaces, but he never succeeded in demonstrating this. It was left to Julius König to generalize the theorem to n dimensions in 1903.

Norrish Ronald George Wreyford 1897–1978 was a British physical chemist who studied fast chemical reactions, particularly those initiated by light. For his achievements in this area, he shared the 1967 Nobel Prize for Chemistry with his co-worker George Porter and the German chemist Manfred Eigen.

Norrish was born in Cambridge on 9 November 1897. He was educated at Perse School, Cambridge, and won a scholarship to the University. His studies where interrupted by World War I, during which he was an officer in the artillery. He graduated from Cambridge two years after returning in 1919, gaining his PhD in 1924. He was made a Fellow of Emmanuel College, Cambridge, in 1925 and became Professor of Physical Chemistry in 1937. He retired in 1965 and died, in Cambridge, on 7 June 1978.

Norrish began working in photochemistry in 1923 with E. K. Rideal, studying the reactions of potassium permanganate solution. For the next few years he investigated the photochemistry of nitrogen dioxide. Then in 1928 his paper on the photochemistry of glyoxal (ethan-1,2-dial, $(CHO)_2$) announced his studies of various aldehydes and ketones. This led to the recognition of what became known as the Norrish type I and type II reactions, which may be generalized as:

Type I $\left\{ \begin{array}{l} R^1COR^2 + h\nu \rightarrow \cdot R^1 + \cdot COR^2 \\ \rightarrow \cdot R^1 + \cdot R^2 + CO \end{array} \right.$

Type II $\quad R^1COCH_2CH_2CH_2R^2 + h\nu \rightarrow R^1COCH_3 +$
$$CH_2{=}CHR^2$$

in which the alkyl radicals R^1 and R^2 contribute one or two carbon atoms.

Up to the mid-1930s, Norrish studied the correlation between photodecomposition and physical phenomena such as spectral character and phosphorescence. Then his studies were again interrupted, this time by World War II. During this period his department contributed to the war effort, investigating methods of suppressing the flash from guns and developing incendiary materials.

Norrish's interest in using intense flashes of light to initiate photochemical reactions seems to have been stimulated by his work during the war with his student George Porter. Flash photolysis makes use of a powerful 'photoflash' to bring about the rapid dissociation of a compound into radicals or ions. A second spectroscopic flash triggered at a precise time interval after the first allows the transient species to be observed. By varying the time delay between the two flashes, Norrish was able to study the kinetics of the formation and decay of such shortlived entities, even if they existed only for microseconds.

Norrish went on to apply these techniques to the study of chain reactions. He established, for example, that the retarding effect of hydrogen chloride in a pure system and one containing oxygen can be attributed to the process:

$$\cdot H + HCl \rightarrow H_2 + \cdot Cl$$

In his investigation of the combustion of hydrocarbons, he studied the transition between slow reactions and ignition, demonstrating the existence of degenerate or delayed branching. He also made pioneering studies of the kinetics of polymerization. He and his co-workers discovered the gel-effect, which occurs in the later stages of free-radical polymerization and results from the steadily decreasing rate of chain termination that sets in when the viscosity becomes high. He also correlated the photolysis of certain polymers with his type I and II reactions.

Norrish was largely responsible for the advance of reaction kinetics to a distinct discipline within physical chemistry. He was one of the first to realize the power of absorption spectroscopy for identifying intermediates and products of thermal and photochemical gas reactions and to introduce high vacuum techniques for handling gases.

Nyholm Ronald Sydney 1917–1971 was an Australian inorganic chemist famous for his work on the coordination compounds (complexes) of the transition metals. He was also interested in science education, and was responsible for many of the changes in chemistry-teaching methods in British schools.

Nyholm was born in Broken Hill, New South Wales, on 29 January 1917, the fourth of the six children of a railway employee whose father had emigrated to Australia from Finland. He was educated at local schools and in 1934 won a scholarship to study natural sciences at Sydney University. He worked for a short time as a research chemist with the Ever Ready Battery Company near Sydney and in 1940 became a member of the staff of Sydney Technical College, where he worked on the coordination compound of rhodium. In 1947, after World War II, he was awarded an ICI Fellowship to University College, London, where he studied under Christopher Ingold and became a lecturer. He returned to Austrialia in 1951 as Associate Professor of Inorganic Chemistry at the University of New South Wales, Sydney. Four years later he again went to Britain to take up a professorship at University College, and in 1963 he was made Head of the Chemistry Department. He was granted a knighthood in 1967 for his services to science. He was killed in a car accident on the outskirts of Cambridge on 4 December 1971.

Nyholm was introduced to coordination chemistry during his final undergraduate year at Sydney by George Burrows, with whom he worked on the reactions between ferric (iron(III)) chloride and the simple arsines. Then Nyholm and F.P.J. Dwyer studied the coordination compounds of rhodium, again using arsines as ligands.

Diarsine had been synthesized in 1937 by Chatt and Mann, and Nyholm met Chatt in London in 1948 and realized that the arsenic compound might help with his own work. He prepared the complex formed between diarsine and palladium chloride, and went on to use the same ligand to prepare stable compounds of transition metals in valence states that previously had been thought to be unstable. For example, he prepared an octahedral complex of nickel(III), in which the nickel has a coordination number of six.

He also made the diarsine complexes of the tetrachloride and tetrabromide of titanium(IV), the first example of an 8-coordination compound of a first-row transition metal.

Nyholm also systematically exploited physical methods to study the structures and properties of coordination compounds. He used potentiometric titrations to determine oxidation states, and electrical conductivity measurements to discover the nature of the charged species in which the coordinated transition metal was contained. He also employed X-ray crystallography and nuclear magnetic resonance spectroscopy (NMR), and found that magnetic moment seemed to give the closest connection between electronic structure, chemical structure and stereochemistry. He always insisted that the three main branches of chemistry – physical, inorganic and organic – are closely interwoven and that methods from one could often be used to solve problems in another.

Nyholm maintained his interest in the teaching side of chemistry, and was a member of the Science Research Council from 1967 to 1971. It was partly as a result of his influence that the Nuffield Foundation set up the Science Teaching Project. As Chairman of the Chemistry Consultative Committee he was largely responsible for the Nuffield O-Level chemistry course in British schools and for changes to the O- and A-level syllabuses. He was a strong advocate of an integrated approach to the teaching of chemistry, particularly at the introductory levels.

O

Ochoa Severo 1905–1993 was a Spanish-born US biochemist who reproduced in the laboratory the way in which cells synthesize nucleic acids by their use of enzymes. For this achievement, he shared the 1959 Nobel Prize for Physiology or Medicine with Arthur Kornberg, who synthesized DNA.

Ochoa was born in Luarca on 24 September 1905, the youngest son of a lawyer. He graduated from the University of Málaga in 1921 and obtained a degree in medicine from the University of Madrid eight years later. He lectured at Madrid from 1931 to 1936. He spent a year in Germany at the University of Heidelberg in 1936 before going to Britain for three years at Oxford University. He then went to the USA in 1940 and was an Instructor and Research Associate at Washington University from 1941 to 1942. Ochao moved to New York University, first as a Research Associate in the College of Medicine and then from 1954 to 1975 as a professor in the Department of Biochemistry. He joined the Roche Institute of Molecular Biology in 1975. Ochoa became a US citizen in 1956. He died in Madrid on 1 November 1993.

Ochoa's early work concerned biochemical pathways in the human body, especially those involving carbon dioxide. But his main research has been into nucleic acids and the way in which their nucleotide units are linked together, either singly (as in RNA) or to form two helically wound strands (as in DNA).

In 1955, Ochoa obtained an enzyme from bacteria that was capable of joining together similar nucleotide units to form a nucleic acid, a type of artificial RNA. (Nucleic acids containing exactly similar nucleotide units do not occur naturally, but the method of synthesis used by Ochoa was the same as that employed by a living cell.) He also found that strands of similar nucleotides form random small fibres, but when mixed with a similar preparation made from a different nucleotide, two-stranded helixes form, one strand from each preparation.

Ochoa's synthesis of an RNA was the result of outstanding experimental work. Research by other workers soon yielded further important results. For example Arthur Kornberg, working independently, isolated an enzyme that will link different nucleotides to form nucleic acids that closely resemble natural ones.

Oersted Hans Christian 1777–1851 was a Danish physicist who discovered that an electric current produces a magnetic field.

Oersted was born at Rudkøbing, Langeland, on 14 August 1777. He had little formal education as a child but on moving to Copenhagen in 1794, Oersted entered the university and gained a degree in pharmacy in 1797, proceeding to a doctorate in 1799. He then worked as a pharmacist before making a tour of Europe from 1801 to 1803 to complete his studies in science. On his return, Oersted gave public lectures with great success, for he was a very good teacher, and then in 1806 became Professor of Physics at Copenhagen. He retained this position until 1829, when he became director of the Polytechnic Institute in Copenhagen. As a teacher and writer, Oersted was instrumental in raising science in Denmark to an international standard, founding the Danish Society for the Promotion of Natural Science in 1824. Oersted remained at the Polytechnic Institute until his death, which occurred at Copenhagen on 9 March 1851.

Oersted made his historic discovery of electromagnetism in 1820, but he had been seeking the effect since 1813 when he predicted that an electric current would produce magnetism when it flowed through a wire just as it produced heat and light. Oersted made this prediction on philosophical grounds, believing that all forces must be interconvertible. Others were also seeking the electromagnetic effect, but considered that the magnetic field would lie in the direction of the current and had not been able to detect it. Oersted reasoned that the effect must be a lateral one and early in 1820, he set up an experiment with a compass needle placed beneath a wire connected to a battery. A lecture intervened before he could perform the experiment and, unable to wait, Oersted decided to try it out before his students. The needle moved feebly, making no great impression on the audience but thrilling Oersted. Because the effect was so small, Oersted delayed publication and investigated it more fully, finding that a circular magnetic field is produced around a wire carrying a current. He communicated this momentous discovery to the major scientific journals of Europe in July 1820.

In 1822, Oersted turned to the compressibility of gases and liquids, devising a useful apparatus to determine compressibility. He also investigated thermoelectricity, in 1823. Oersted may have wanted to continue work on electromagnetism, but his sensational discovery resulted in an explosion of activity by other scientists and Oersted possibly felt unable to compete. Major theoretical and practical advances were made by André Ampère and Michael Faraday soon afterwards, Oersted thereby providing the basis for the main thrust of physics in the 1800s.

Ohm Georg Simon 1789–1854 was a German physicist who is remembered for Ohm's law, which relates the current flowing through a conductor to the potential difference and the resistance.

Ohm was born at Erlangen, Bavaria, on 16 March 1789. He received a basic education in science from his father, who was a master locksmith, and in 1805 he entered the University of Erlangen. However, he left after a year and until 1811 was a school teacher and private tutor in Switzerland. He then returned to Erlangen and gained his PhD in the same year. Ohm then became a privatdozent (unpaid lecturer) in mathematics at Erlangen, but after a year was forced to take up a post as a schoolteacher in Bamberg. In 1817, Ohm moved to Cologne to teach at the Jesuit Gymnasium and in 1825 he decided to pursue original research in physics. He obtained leave in 1826 and went to Berlin, where he produced his great work on electricity, *Die Galvanische Kette*, in 1827.

Ohm hoped to obtain an academic post on the strength of his achievements, but found that his work was little appreciated, mainly because of its mathematical rigour which few German physicists understood at that time. He stayed in Berlin as a schoolteacher, and then in 1833 moved to Nuremberg where he became Professor of Physics at the Polytechnic Institute, which was not the prestigious university appointment that Ohm both desired and deserved. However, Ohm's work began to achieve recognition, especially in Britain where he was awarded the Royal Society's Copley Medal in 1841. His ability eventually also made its mark in Germany and in 1849 he became Extraordinary Professor of Physics at Munich, acceding to the chair of physics proper in 1852. Ohm thus achieved his ambition but had little time left to savour his success. He died at Munich on 6 July 1854.

Ohm began the work that led him to his law of electricity in 1825. He investigated the amount of electromagnetic force produced in a wire carrying a current, expecting it to decrease with the length of the wire in the circuit. He used a voltaic pile to produce a current and connected varying lengths of wire to it, measuring the electromagnetic force with the magnetic needle of a galvanometer. Ohm found that a longer wire produced a greater loss in electromagnetic force.

Ohm continued these investigations in 1826 using a thermocouple as the source of current because it produced a constant electric current unlike the voltaic pile, which fluctuated. He found that the electromagnetic force, which is in fact a measure of the current, was equal to the electromotive force produced by the thermocouple divided by the length of the conductor being tested plus a quantity that Ohm called the resistance of the remainder of the circuit, including the thermocouple itself. From this, Ohm reached the more general statement that the current is equal to the tension (emf or potential difference) divided by the overall resistance of the circuit, thus expressing the law in the form known as Ohm's law.

Ohm went on to use an electroscope to measure how the tension varied at different points along a conductor to verify his law, and presented his arguments in mathematical form in his great work of 1827. He made a useful analogy with the flow of heat through a conductor, pointing out that an electric current flows through a conductor of varying resistance from one tension or potential to another to produce a potential difference, just as heat flows through a conductor of varying conductivity from one temperature to another to produce a temperature difference. Ohm used the analytical theory of heat published by Jean Fourier (1768–1830) in 1822 to justify this approach, believing partly from his use of thermocouples that there was an intimate link between heat and electricity.

Ohm's derivation of a basic law of nature from experiment was a classic piece of scientific deduction. Together with the laws of electrodynamics discovered by André Ampère (1775–1836) at about the same time, Ohm's law marks the first theoretical investigation of electricity. It is fitting that his name is remembered in both the unit of resistance, the ohm, and the unit of conductivity, the mho (ohm spelt backwards).

Olbers Heinrich Wilhelm Matthäus 1758–1840 was a German doctor, mathematician and astronomer who is now chiefly remembered for his work on the discovery of asteroids and the formulation of Olbers's method for calculating the orbits of comets. He also caused considerable scientific controversy by asking the basic question, why is the night dark?

Olbers was born near Bremen on 11 October 1758. He attended the local school where, at the age of 16, his mathematical and astronomical interests were so advanced that he computed the time of a solar eclipse. In 1777, he went to Göttingen to study medicine, but attended lectures also in physics and mathematics, coming under the influence of Kästner, the Director of the small observatory there. Olbers's lifelong concern with comets began two years later, when he used his observations of Bode's comet to calculate its orbit according to a method devised by Euclid. In 1781, he received his degree in medicine, settled in Bremen, and soon acquired an extensive medical practice from which he retired only at the age of 64, in 1823. Astronomy had in the meantime become a consuming hobby, to satisfy which he had early on installed all the equipment for a full observatory on the second floor of his house – refractors, a reflector, a heliometer and three comet-seekers. He collected the finest private library of literature on comets (now part of the Pulkovo collection). Olbers died in Bremen on 2 March 1840.

In 1796, Olbers discovered a comet and calculated its parabolic orbit with a new method, simpler than that used earlier by Pierre Laplace (1749–1827). Laplace had given formulae for the computation of a parabola through successive approximations, but the procedure was cumbersome and unsatisfactory. It had been assumed that when three observations of a comet had been obtained within a short period of time, the radius vector of the middle observation would divide the chord of the orbit of the comet from the first to the last observation in relation to the traversed time. Olbers's contribution was to establish that this assumption could be applied with equal advantage to the three positions of the Earth in its orbit. After reading his treatise on this, Baron von Zach used it to compute the orbit of the comet of 1779. In publishing his work the Baron thus established Olbers among the foremost astronomers of his time; his method was used throughout the nineteenth century.

For some years astronomers had searched the apparent 'gap' in the Solar System between the planets Mars and Jupiter – a gap emphasized by the formulation of Bode's law. Then the first asteroid was discovered in 1801 by Giuseppe Piazzi (1749–1826), who noticed a starlike object that moved during successive days. He communicated this news to other astronomers, but although it was soon realized that this must be a new planet, which Piazzi named Ceres, it disappeared before further observations could be made. (At that time it was still impossible to compute an orbit from such a small arc without having to make assumptions about the eccentricity.) However, the young astronomer Karl Gauss (1777–1855) determined the orbit and Olbers, in January 1802, refound the new planet, very near where Gauss had calculated it would be. This was the beginning of a lifelong friendship and collaboration between the two men.

While following Ceres, Olbers discovered a second asteroid, Pallas, in 1802; a third, Juno, was discovered by Karl Harding (1765–1834) at Lilienthal in 1804. The orbits of these small planets suggested to Olbers that they had a common point of origin and might have originated from one large planet. For years Olbers searched the sky, where the orbits of Ceres, Pallas and Juno approached each other; this resulted in his discovery of Vesta in March 1807.

Olbers's main interest remained the search for comets, however, and his efforts were rewarded with the discovery of four more. Of particular interest is the comet which he discovered in March 1815, which has an orbit of 72 years, similar to Halley's. Olbers calculated the orbits of 18 other comets. Noticing that comets consist of a starlike nucleus and a parabolic cloud of matter, he suggested that this matter was expelled by the nucleus and repelled by the Sun.

In a publication of 1823, Olbers discussed the paradox that now bears his name, that if we accept an infinite, uniform Universe, the whole sky would be covered by stars shining as brightly as our Sun. Olbers explained the darkness of the night sky by assuming that space is not absolutely transparent and that some interstellar matter absorbs a very minute percentage of starlight. This effect is sufficient to dim the light of the stars, so that they are seen as points against the dark sky. (In fact darkness is now generally accepted as a by-product of the red shift caused by stellar recession.)

Although interested in the study of comets, Olbers was also interested in the influence of the Moon on the weather, the origin of meteorite showers and the history of astronomy. A very modest man, he encouraged many young astronomers and claimed that his greatest contribution to astronomy had been to lead Friedrich Bessel (1784–1846) to become a professional astronomer, after Bessel had approached him in 1804 with his calculation of the orbit of Halley's comet.

Onnes Heike Kamerlingh 1853–1926 was a Dutch physicist who is particularly remembered for the contributions he made to the study of the properties of matter at low temperatures. He first liquefied helium and later discovered superconductivity, gaining the 1913 Nobel Prize for Physics in recognition of his work.

Onnes was born in Groningen on 21 September 1853 and attended local schools. In 1870 he entered the University of Groningen to study physics and mathematics. He won two prizes at the university during his early years there for his studies on the nature of the chemical bond. In 1871, Onnes travelled to Heidelberg to study under Gustav Kirchhoff (1824–1887) and Robert Bunsen (1811–1899). He then returned to Groningen and in 1876 completed the research for his PhD, on new proofs for the Earth's rotation, but the degree was not awarded until 1879. In the meantime, Onnes served as an assistant at the Polytechnic in Delft and later lectured there. In 1882, he was appointed Professor of Experimental Physics at the University of Leiden, a post he held for 42 years. In 1894, he founded the famous cryogenic laboratories at Leiden, which became a world centre of low-temperature physics. Onnes died in Leiden on 21 February 1926.

During his years at Delft, Onnes met Johannes van der Waals (1837–1923), who was then Professor of Physics in Amsterdam. Their discussions led Onnes to become interested in the properties of matter over a wide temperature range, and in particular at the lower end of the scale. Onnes felt that he might thereby be able to obtain experimental evidence to support van der Waals's theories concerning the equations of state for gases.

Onnes applied the cascade method for cooling gases that had been developed by James Dewar (1842–1923), and in 1908 succeeded in liquefying helium, a feat which had eluded Dewar. He was not content merely to report the technical achievements involved in these processes, he also wanted to investigate how matter behaves at very low temperatures. In 1910, Onnes succeeded in lowering the temperature of liquid helium to 0.8K (−272.4°C/−458.2°F). Lord Kelvin (1824–1907) had postulated in 1902 that as the temperature approached absolute zero, electrical resistance would increase. In 1911, Onnes found the reverse to be the case, showing that the electrical resistance of a conductor tends to decrease and vanish as temperature approaches absolute zero. He called this phenomenon supraconductivity, later renamed superconductivity.

Onnes made a particular study of the effects of low temperature on the conductivity of mercury, lead, nickel and manganese–iron alloys. He found that the imposition of a magnetic field eliminated superconductivity even at low temperatures. Special alloys have now been produced which do not behave in this manner.

Onnes was also interested in the magneto-optical properties of metals at low temperature, and in the technical application of low-temperature physics to the industrial and commercial uses of refrigeration.

Onsager Lars 1903–1976 was a Norwegian-born US theoretical chemist who was awarded the 1968 Nobel Prize for Chemistry for his work on reversible processes.

Onsager was born in Oslo on 27 November 1903. He was educated at the local High School and then at the Norges Tekniske Høgskole in Trondheim, where he studied chemical engineering. During the five years at Trondheim he acquired the mathematical skills he was to use later on, and developed an interest in electrolytes. After qualifying in 1925 he went to Zurich to work as a research assistant to Peter Debye. He emigrated to the United States in 1928 and was appointed an Associate in Chemistry at Johns Hopkins University, Baltimore. He was not, however, a success as a lecturer (because his course did not attract a sufficient number of students) and he soon moved to Brown University, where he was a research instructor from 1928 to 1933.

In 1933, he went to Europe to visit the Austrian electrochemist Falkenhagen, and while there met and married Falkenhagen's sister. His lectures at Brown University were no more comprehensible to the students, who named his course 'Sadistical Mechanics'. Also in 1933 he took up an appointment at Yale, becoming Assistant Professor in Chemistry a year later. The university authorities were disconcerted to have a plain 'Mr' as a professor, and urged Onsager to submit one of his published papers as a PhD thesis. Onsager chose 'Solutions to the Mathieu equation of period 4π and certain related functions'. The chemistry department were unable to make anything of it so they passed it to the physics department, which in turn passed it to the mathematics department. The final outcome was an award of a PhD by the bemused chemists. It is little wonder that the students at Yale were no kinder than those at Brown and described Onsager's lectures as 'Advanced Norwegian I and II'. His almost total failure as a lecturer probably came from the fact that he could not appreciate that others were unable to understand the topics that interested him. He did, however, remain at Yale for the rest of his career.

Onsager was not required to do military service during World War II because he had not yet become an American citizen and his wife was Austrian; he adopted US nationality in 1945. After retirement he bought a farm at Tilton,

New Hampshire, and grew his own crops. He died in Coral Gables, Florida, on 5 October 1976.

In 1925, Peter Debye and Erich Hückel had put forward a new theory of electrolytes based on the idea that the electrostatic field of a dissolved ion is effectively screened by surrounding ions of opposite charge. They were able to calculate the activity coefficient for any ion in dilute solution although the calculated values for conductivity differed considerably from experimental values, particularly for strong electrolytes. When Onsager went to Zurich in 1925 he told Debye that he thought the electrolyte theory was incorrect – and was offered a research assistantship. Debye had assumed that one particular ion should be thought of as moving uniformly in a straight line, while all other ions undergo Brownian motion. Onsager showed that this constraint should be lifted and the result, known as the Onsager limiting law, gave better agreement between calculated and actual conductivities. He went on to investigate dielectric constants of polar liquids and solutions of polar molecules.

At Brown University Onsager submitted a PhD thesis on what is now a classic work on reversible processes, but the authorities turned it down. It was published in 1931 but ignored until the late 1940s; in 1968 it earned Onsager a Nobel Prize. He then turned his attention to the equilibrium states in the muta-rotation (change in optical rotation) of sugars. Riiber had shown in 1922 that galactose exists in at least three tautomeric forms (interconvertible stereoisomers). Onsager proposed that the equilibrium states between these forms must conform to the principle of 'detailed balancing', as conceived by Gilbert Lewis, and showed that this idea is thermodynamically equivalent to the principle of 'least dissipation' used by Hermann von Helmholtz in his theory of galvanic diffusion cells and by Lord Kelvin in his theories about thermoelectric phenomena.

Onsager also looked at the connection between microscopic reversibility and transport processes. He found that the key to the problem is the distribution of molecules and energy caused by random thermal motion. Ludwig Boltzmann had shown that the nature of thermal equilibrium is statistical and that the statistics of the spontaneous deviation is determined by the entropy. Using this principle Onsager derived a set of equations known as Onsager's law of reciprocal relations, sometimes called the fourth law of thermodynamics. It has many applications to cross-coefficients for the diffusion of pairs of solutes in the same solution, and for the various interactions that can occur between thermal conduction, diffusion and electrical conduction. He announced these ideas in the late 1930s, but not until 1960 did the theory receive experimental confirmation.

During and after World War II Onsager made calculations concerning the two-dimensional Ising lattice (an assembly of particles or 'spin' located at the vertices of an infinite space lattice; the simplest case is a two-dimensional planar square lattice). It can be used as a model to describe ferromagnetism and anti-ferromagnetism of gaseous condensations, phase separations in fluid mixtures and metallic alloys. Onsager's treatment showed that the specific heat approaches infinity at the transition point. In 1949, he published a paper which established a firm statistical basis for the theory of liquid crystals, and at Cambridge in 1951–1952 he put forward a theory concerned with diamagnetism in metals.

Oort Jan Hendrik 1900–1992 was a Dutch astrophysicist whose main area of research has been the composition of galaxies. In his investigation of our own galaxy, he used data provided by other scientists to demonstrate the position of the Sun within the rotating Galaxy, and through the use of the radio telescope, he traced our Galaxy's spiral arms by the detection of interstellar residual hydrogen.

Oort was born in Franeker, the Netherlands, on 28 April 1900, the son of a doctor. Completing his education at the University of Groningen, he studied under Jacobus Kapteyn (1851–1922), from whom, perhaps, he derived his great interest in galactic structure and movement. In 1926 he received his PhD at Groningen and went immediately (for a short time) to work at Yale University in the United States. He then returned to join the staff of Leiden University, in the Netherlands, where he became Professor of Astronomy in 1935. Ten years later he was made Director of the Leiden Observatory, retiring finally in 1970.

Oort's teacher, Jacobus Kapteyn, published in 1904 the results of an investigation to find the centre of our Galaxy. He preferred to believe that the Sun itself was at or very near the middle, but he was puzzled by two definite 'streams' of galactic stars apparently moving in a linear sequence in two opposing directions; moreover, a line connecting the streams would follow the Milky Way. The Swedish astronomer Bertil Lindblad (1895–1965), using these data in the year that Oort was receiving his PhD, suggested that if the Sun were not at the centre, the two streams could represent stars going to the centre and stars returning, and on this basis he worked out that the centre of the Galaxy was somewhere in the direction of the constellation Sagittarius – incidentally agreeing with independent calculations by the American Harlow Shapley (1885–1972). Oort's major success was to provide confirmation for Lindblad and Shapley, although he located the centre of the Galaxy at a distance of 30,000 light years rather than at Shapley's 50,000.

Oort went on to show that the streams of stars were not in fact linear, but very much like the planets revolving round a sun, in that the stars nearer the centre of the Galaxy revolved faster round the centre than those farther out. Noting the Sun's position in the Galaxy, and calculating its period of revolution as slightly more than 200 million years, Oort derived a calculation also of the mass of the Galaxy: about 100,000 million (10^{11}) times that of the Sun.

The beginning of radioastronomy during and just after World War II was of great assistance to Oort's investigations. After considerable theoretical work on the structure of the hydrogen atom, over the passage of time and under different circumstances, he and Hendrik van de Hulst (1918–) discovered in 1951 the radio emission at a wavelength of 21 cm/8 in of interstellar neutral hydrogen. The fact that hydrogen occurs between galactic stars, not in open space (so to speak), meant that the shape of the Galaxy could now be traced by the shape of the hydrogen between the stars. For the first time it was possible accurately to chart the spiral arms of our Galaxy. It was also possible to study the centre of the Galaxy in detail, by monitoring its radio waves.

At about the same time, Oort put forward an ingenious theory concerning the origin of comets. He suggested that at a great distance from the Sun – a light year, say – there was an enormous 'reservoir' of comets in the form of a cloud of particles. Gravitational perturbations by passing stars could, he suggested, every now and then cause one of them to be hurled into the Solar System and become a comet.

In 1956, Oort and Theodore Walraven studied the radiation emitted from the Crab Nebula and found it to be polarized, indicating synchrotron radiation produced by high-speed electrons in a magnetic field.

Oparin Alexandr Ivanovich 1894–1980 was a Soviet biochemist who made important contributions to evolutionary biochemistry, developing one of the first of the modern theories about the origin of life on Earth. He received many honours for his work, particularly from the Soviet Union.

Oparin was born on 3 March 1894 in the small village of Uglich, north of Moscow, the youngest of three children. When he was nine years old, his family moved to Moscow because Uglich had no secondary school. He studied plant physiology at Moscow State University, where he was influenced by K. A. Timiryazev, a plant physiologist who had known Charles Darwin. After graduating in 1917, Oparin researched in biochemistry under A. N. Bakh, a botanist, then in 1929 became Professor of Plant Biochemistry at Moscow State University. In 1935 he helped to found, and began working at, the Bakh Institute of Biochemistry in Moscow, which was established in honour of his former teacher. Oparin became Director of the Bakh Institute in 1946 and held this post until his death in April 1980.

Oparin first put forward his ideas about the origin of life in 1922 at a meeting of the Russian Botanical Society. His theory contained three basic premises: that the first organisms arose in the ancient seas, which contained many already formed organic compounds that the organisms used as nutriment (thus his hypothetical first organisms did not synthesize their own organic nutrients but took them in ready-made from the surrounding water); that there was a constant, virtually limitless supply of external energy in the form of sunlight (thus conditions in which the first forms of life arose did not constitute a closed system and were not limited by the second law of thermodynamics); and that true life was characterized by a high degree of structural and functional organization, an idea that was contrary to the prevailing view that life was basically molecular.

Oparin's theory did not explain how complex molecules could have arisen in the primordial seas, nor how his primitive organisms could reproduce. Later research, however, suggested that a degree of order in the structure of proteins might have occurred as a result of the restrictions imposed on the coupling of amino acids due to their different shapes and distributions of electric charge. Regarding reproduction, later experiments with microscopic coacervate drop-lets of gelatin and gum arabic demonstrated that these droplets repeatedly grow and reproduce by budding. Oparin then showed that enzymes function more efficiently inside such synthetic cells than they do in ordinary aqueous solution.

Oparin's theory (which was first published in 1924, although it reached its widest audience after 1936, when he published *The Origin of Life on Earth*) stimulated much research into the origin of life, perhaps the most famous of which is Stanley Miller's attempt in 1953 to reproduce primordial conditions in the laboratory. In his experiment, Miller put sterile water, methane, ammonia and hydrogen (simulating the primordial atmosphere) in a sealed container and subjected the mixture to electrical discharges (simulating lightning). After one week he found that the solution contained simple organic compounds, including amino acids. In addition, C. Ponnamperuma, using slightly different experimental conditions, demonstrated that nucleotides, dinucleotides and ATP can be formed from simple ingredients.

Although best known for his pioneering work on the origins of life, Oparin also researched into enzymology and did much to provide a technical basis for industrial biochemistry in the Soviet Union.

Öpik Ernst Julius 1893–1985 was an Estonian astronomer whose work on the nature of meteors and comets was instrumental in the development of heat-deflective surfaces for spacecraft on their re-entry into the Earth's atmosphere.

Öpik was born in Kunda, a coastal village near Rakvere, Estonia, on 23 October 1893. He completed his education at the Tartu State University, and in 1916 began working at the Tashkent Observatory in Uzbekistan; he then moved in 1918 to the Observatory at the University of Moscow, where he worked as an Assistant and Instructor. From 1920 to 1921 he served as a Lecturer at Turkistan University, before returning to Tartu University as a Lecturer in Astronomy. Apart from four years as a Research Associate and visiting Lecturer at the Harvard College Observatory, in the United States, Öpik remained at Tartu University until 1944. He was then appointed Research Associate at the University of Hamburg; he became Professor and Estonian Rector at the German Baltic University in 1945. Three years later, Öpik moved to Northern Ireland where, initially appointed as a Research Associate, he eventually became Director of the Armagh Observatory. From 1956 onwards he held a concurrent post as a visiting Professor (and later Associate Professor) at the University of Maryland, in the United States, where in 1968 he was appointed to the Chair of Physics and Astronomy. Öpik received several awards, including the Gold Medal of the Royal Astronomical Society (1975).

Öpik's early research was devoted to the study of meteors; he was the originator of the 'double-count' method for counting meteors, a method which requires two astronomers to scan simultaneously. His theories on surface events in meteors upon entering the Earth's atmosphere at high speed (the ablation, or progressive erosion, of the outer layers) proved to be extremely important in the development of heat shields and other protective devices to enable a spacecraft to withstand the friction and the resulting intense heat upon re-entry.

Much of Öpik's other work was directed at the analysis of comets that orbit our Sun. He postulated that the orbit of some of these comets may take them as far away as 1 light year. He also made studies of double stars, cosmic radiation and stellar photometry. His interests and contributions thus covered a broad range of astronomical disciplines.

Oppenheimer J(ulius) Robert 1904–1967 was a US physicist who contributed significantly to the growth of quantum mechanics and played a critical role in the rapid development of the first atomic bombs. In his later years, he was a prominent advocate of international control of nuclear technology and of a cautious approach to the escalation of the military applications of that knowledge.

Oppenheimer was born in New York on 22 April 1904, into a wealthy family. He attended the Ethical Culture School and was a very serious, studious child. His interest in minerals and geology led to his admission to the Minerological Club of New York at the mere age of 11. In 1922, he went to Harvard University, graduating after only three of the usual four years in 1925. He spent the next year in Cambridge with Ernest Rutherford (1871–1937), Werner Heisenberg (1901–1976) and Paul Dirac (1902–1984), and in 1926 at the invitation of Max Born (1882–1970) went to Göttingen where he worked on quantum theory of molecules and *bremsstrahlung* (continuous radiation emission). He received his PhD from Göttingen in 1927. Oppenheimer then returned to the United States and became a National Research Fellow, and in 1929 he was appointed Assistant Professor at both the California Institute of Technology (CalTech) and the University of California at Berkeley. He spent the next 13 years commuting between these two campuses, becoming Associate Professor in 1931 and Professor in 1936.

From 1943 to 1945, Oppenheimer was Director of the Los Alamos Scientific Laboratories in New Mexico, where he headed the research team that produced the first atomic bombs. After World War II he returned briefly to California and then in 1947 was made Director of the Institute of Advanced Study at Princeton University. Oppenheimer also served as Chairman of the General Advisory Committee to the Atomic Energy Commission from 1946 to 1952. He retired as Director of the Institute of Advanced Study in 1966, but continued as Professor there until his death at Princeton on 18 February 1967.

Oppenheimer's most important research was to investigate the equations describing the energy states of the atom that Dirac had formulated in 1928. In 1930, Oppenheimer was instrumental in showing that the equations indicated that a positively charged particle with the mass of an electron could exist. This particle was detected in 1932 and called the positron.

During the 1930s at California, Oppenheimer built up a formidable research group around him. Theoretical physics had never before been studied with such intensity in the United States. When World War II broke out in 1939 in Europe, it was immediately apparent to Oppenheimer that the newly discovered phenomenon of nuclear fission could have great military significance. Work on the various aspects of bomb construction began in many places, one of them being Berkeley. The Manhattan Project, the code name for the programme aimed at the production of the atomic bomb, was formally initiated in 1942. Oppenheimer suggested that the work on the various related projects be brought together to one site and proposed a remote location in the Pecos Valley, New Mexico. The Los Alamos research centre was set up there in 1943 with Oppenheimer as Director.

Oppenheimer did not take on personal responsibility for the development of any single aspect of the bomb programme, but concentrated first on gathering together the finest scientists he could find and secondly on instilling in the whole team a sense of urgency and intense excitement. The success of the Los Alamos project in so quickly overcoming the enormous number of new problems is in no small measure attributable to his efforts. Oppenheimer also served as one of the four scientists consulted on the decision of how to deploy the bomb, the scientists recommending that a populated 'military target' be selected.

Oppenheimer was also involved in the decision to develop the hydrogen bomb after the war. The majority of the AEC Advisory Committee, of which Oppenheimer was chairman, was opposed to this but the unexpected explosion of a nuclear device by the Soviet Union in the summer of 1949 led to President Truman overriding this advice. Oppenheimer's offer of resignation from the committee was not accepted, but in 1953, at the height of the McCarthy era, Oppenheimer was informed that President Eisenhower had ordered that his security clearance (permission for access to secret information) was to be withdrawn on the basis of suspicions concerning his loyalty. Oppenheimer opposed this and underwent a gruelling quasi-judicial procedure which did not lead to his clearance being restored. There is little doubt that those who had disagreed with Oppenheimer's position regarding the H-bomb were at least partly responsible for this.

In 1963 President Johnson publicly awarded Oppenheimer the Enrico Fermi Prize, the highest award the AEC can confer, as an attempt to make amends for the unjust treatment he had received.

Oppolzer Theodor Egon Ritter von 1841–1886 was an Austrian mathematician and astronomer whose interest in asteroids and comets and eclipses led to his compiling meticulous lists of such bodies and events for the use of other astronomers.

Oppolzer was born in Prague (then in the Austrian Empire now in the Czech Republic) on 26 October 1841. He displayed keen mathematical abilities from an early age and was a top student at the Piaristen Gymnasium in Vienna (1851–1859). Although he followed his father's wishes and studied to qualify as a doctor – he was awarded his medical degree in 1865 – Oppolzer devoted most of his spare time to carrying out astronomical observations. In 1866, he became Lecturer in Astronomy at the University of Vienna. He was promoted to Associate Professor in 1870, made Director of the Austrian Geodetic Survey in 1873, and in 1875 became Professor of Geodesy and Astronomy at Vienna. In addition to his teaching and research activities, Oppolzer was active in many European scientific societies. He was elected to the Presidency of the International Geodetic Association in 1886, shortly before his death in Vienna on 26 December 1886.

Oppolzer was fortunate to possess a private observatory, which permitted him to make accurate and thorough investigations of the behaviour of comets and asteroids. (Asteroids were in fact only discovered by Giuseppe Piazzi in 1801, and the details of their orbits were of considerable interest to astronomers.) Oppolzer methodically sought, by observation and calculation, to confirm and amend, where necessary, the putative orbits of these bodies. He was the originator of a novel technique for correcting orbits he found to be inaccurate. His two-volume text on the subject (1870–1880) provides a clear description of this work.

In 1868, Oppolzer participated in an expedition to study a total eclipse of the Sun and his interest in eclipses dates from that time. He decided to calculate the time and path of every eclipse of the Sun and every eclipse of the Moon for as long a period as possible. The resulting *Canon der Finsternisse* was published posthumously in 1887. It covered the period 1207 BC–AD 2163, an astonishing total of 3,370 years.

Oppolzer's contributions to astronomy were characterized by their great thoroughness and by the accuracy of his mathematical procedures.

Ore Oystein 1899–1968 was a Norwegian mathematician whose studies, researches and publications concentrated on the fields of abstract algebra, number theory and the theory of graphs.

Ore was born on 7 October 1899 in Oslo, where he grew up and was educated, entering Oslo University in 1918; he received his BA in 1922. He then paid a fleeting visit to the University of Göttingen – that great centre of mathematics – where he met the influential mathematician Emmy Noether (1882–1935) who was at that time building up an active research group in abstract algebra. The work being done at Göttingen undoubtedly exercised a strong influence over Ore.

Returning to Scandinavia, he worked from 1923 until 1924 at the Mittag-Leffler Mathematical Institute in Djursholm, Sweden, at the end of which time he was awarded his PhD in mathematics by the University of Oslo. Two years later (and after another short visit to Göttingen), Ore was appointed Professor of Mathematics at Oslo University. Only 12 months later, however, he moved to take up an equivalent position at Yale University in the United States; there, he was rapidly promoted to Associate (1928) and then to full Professor (1929), serving as Chairman of his department from 1936 to 1945. He then returned to Norway. A Knight of the Order of St Olav, Ore wrote many books on mathematical subjects.

During the 1920s and 1930s Ore's primary research interest was abstract algebra. Among other topics, he investigated linear equations in noncommutative fields; most of his work on this subject was summarized in a book he published in 1936 on abstract algebra. He then turned to an examination of number theory, and in particular of algebraic numbers. His investigations in this subject were contained in a text on number theory and its origins which he published in 1948.

Much of Ore's later research dealt with the theory of graphs. He took a special interest in the four-colour problem, the theory that maps require no more than four colours for each region of the map to be coloured but with no zone sharing a common border with another zone of the same colour. Ore wrote a book detailing his investigations into the problem; it appeared in 1967.

Another of Ore's interests was the history of mathematics. He wrote biographies of the sixteenth-century Italian Gerolamo Cardano and of his fellow Norwegian, the ill-starred Niels Abel; they were published in 1953 and 1957 respectively.

Osborn Henry Fairfield 1857–1935 was a palaeontologist who did much to promote the acceptance of evolutionism in the United States.

The son of a wealthy businessman, Osborn studied at Princeton, developed an interest in natural history, and began a long career of palaeontological exploration in 1877 with an expedition to Colorado and Wyoming. From 1881 he occupied a post at Princeton, moving in 1891 to Columbia University.

Osborn's main work was in vertebrate palaeontology. He eagerly accepted evolutionary theory and was concerned to fill out its main trends and details. He continued Edward Drinker Cope's work on the evolution of mammalian molar teeth and wrote an influential textbook, *The Age of Mammals* 1910. Osborn's evolutionary studies focused on the problem of the adaptive diversification of life. He was particularly concerned with the parallel but independent evolution of related lines of descent, and with the explanation of the gradual appearance of new structural units of adaptive value. He placed great emphasis on the fact that evolution was the resultant of pressures from four major directions: external environment, internal environment, heredity and selection. A convinced Christian, the interpretation of evolution in religious and moral terms concerned him greatly.

Ostwald Friedrich Wilhelm 1853–1932 was a Latvian-born German physical chemist famous for his contributions to solution chemistry and to colour science. He was awarded the 1909 Nobel Prize for Chemistry for his work on catalysis, chemical equilibria and reaction velocities.

Ostwald was born of German parents in Riga, Latvia, on 2 September 1853; his father was a master cooper. He was good at handicrafts when he was a boy, a skill that was later to stand him in good stead when he had to make his own chemical apparatus. He was educated at the Realgymnasium in Riga and in 1872 went to the University of Dorpat (Tartu) in Estonia to study chemistry under Carl Schmidt and Johann Lemberg. He also studied physics under Arthur von Oettingen, and was awarded his PhD in 1878. While working for his doctorate, he lectured at Dorpat on the theory of chemical affinity and was Oettingen's assistant. He married in 1880; his son Wolfgang (1883–1943) was also to become a notable chemist, and his daughter Grete wrote her father's biography in 1953.

In 1881, Ostwald was appointed Professor of Chemistry at Riga. Six years later he accepted what was then the only Chair in Physical Chemistry in Germany, at the University of Leipzig, and in 1898 celebrated the official dedication of the new Physico-chemical Institute there, of which he was made Director. He retired in 1906, having been appointed the first German exchange professor to Harvard (1905–1906). In 1901, he had moved his family and his huge library to 'Landhaus Energie', a house in Grossbothen, and after 1909 devoted an increasing amount of his time to philosophy. He died in Leipzig on 4 April 1932.

As a student Ostwald worked out that the magnitude of a chemical change could be calculated from any measurable change in a physical property that accompanies it. In 1887, he determined the volume changes that take place during the neutralization of acids by bases in dilute solutions, and in 1879 proposed that the rate at which compounds such as zinc sulphide and calcium oxalate are dissolved by various acids be used as a measure of the acids' relative affinities. He read the memoir by Svante Arrhenius on the 'galvanic conductibility of electrolytes' in 1884 and became an enthusiastic supporter of the new theory of ionic dissociation. He was then able to redetermine the affinities of the acids using Arrhenius's electrolytic conductivity method.

In 1885 and 1887 Ostwald published the two volumes of his ambitious textbook *Lehrbuch der allgemeinen Chemie*. Also in 1887, together with Jacobus van't Hoff and Arrhenius, he founded the journal *Zeitschrift für physikalische Chemie*.

In 1888, he proposed the Ostwald dilution law, which relates the degree of dissociation of an electrolyte, α, to its total concentration c expressed in moles per litre (dm^3). It states that:

$$k = \alpha^2 c/(1 - \alpha)$$

The constant, k, neglects the activity coefficient and is therefore not a true thermodynamic constant K. The equation is important historically because it was the form in which the law of mass action was first applied to solutions of weak organic acids and bases. Ostwald then worked on the theory of acid–base indicators.

Ostwald turned his attention to catalysis in 1900. He discovered a method of oxidizing ammonia to convert it to oxides of nitrogen (for making nitric acid) by passing a mixture of air and ammonia over a platinum catalyst. By means of this technique (using ammonia from the Haber process), and by later developments connected with it, Germany became independent of supplies of Chilean nitrates and was able to continue the manufacture of explosives during World War I after the Allies had blockaded its ports. Ostwald patented the Ostwald–Bauer process for the manufacture of nitric acid from ammonia.

From 1909 Ostwald became interested in the methodology and organizational aspects of science, in a world language, in internationalism and in pacifism. He enlarged the premises at 'Landhaus Energie' and built a laboratory for colour research. His studies of colour theory and the techniques of painting are noteworthy, and his book *Grosse Männer* on the lives of famous scientists shows great insight into the factors that make for great people.

Otis Elisha Graves 1811–1861 was a US engineer who pioneered the development of the lift. Although mechanical lifts and hoists had been known since early in the nineteenth century, it was the invention by Otis of a safety device which caused them to be generally adopted in commerce and slightly later as passenger-carrying machines. He also invented and patented a number of other important machines.

Otis was born on 3 August 1811 on his father's farm in Halifax, Vermont. His father was a Justice of the Peace and served four terms in the Vermont State Legislature. Elisha's education was received in Halifax. When he was 19 he went to Troy, New York, to work as a builder. Illness and enterprise led him to take up a haulage business. As he accumulated capital he engaged in other businesses with varying degrees of success: first as a miller, and when this failed he converted his mill into a factory to make carriages and wagons. Again ill health caused him to abandon his livelihood. This time he became employed in Albany, New York, as a master mechanic in a factory making beds. Again he acquired capital and opened his own machine shop, which was successful for a time. Misfortune struck again when the Albany authorities diverted the stream which drove the water turbine he had invented to supply the workshop with power. He returned to his old job as a master mechanic, in another factory making beds, this time in New Jersey, where after a time he was put in charge of the preparation of a new factory at Yonkers, New York. During the construction of the building, in 1852, he had to make a hoist and, as many serious accidents had been caused by out-of-control lifting platforms, he sought a way of making this one absolutely safe.

Otis built the frame of his hoist with a ratchet into which could be slotted a horizontal wagon spring attached to the 'cage' of the lift. The rope of the hoist was fixed to the centre of the spring and kept it in tension, so preventing it from engaging with the ratchet. If, however, the rope broke,

the spring was released and jammed into the ratchet, immediately immobilizing the lift.

Interest was shown in his lifts, and when the bed factory went out of business he took over its plant to fulfill the initial three orders he received for his invention. Further orders came slowly until Otis exhibited his patented lift at the second season of the Crystal Palace Exposition in New York in 1854. As part of his demonstration he climbed onto his elevator, was hoisted into the air, and then a mechanic cut the hoisting rope. This was a grand advertisement and the orders started to come in. In 1857, the first public passenger lift was opened by the New York china firm of E. V. Haughwout and Company. Generally the lifts were powered by steam engines and in 1860 Otis patented and improved the double oscillatory machine specially designed for his lifts. Also from the workshops of his company, Otis invented and patented railway trucks and brakes, a steam plough and a baking oven.

Otto Nikolaus August 1832–1891 was a German engineer who developed the first commercially successful four-stroke internal-combustion engine, even today, the four-stroke cycle is still sometimes referred to as the Otto cycle.

Otto was born at Holzhausen, Nassau, the son of a farmer. He left school at the age of 16 to work in a merchant's office, and later moved to Cologne where he became greatly interested in the gas engines developed by Jean Lenoir (1822–1900). In 1861 he built a small experimental gas engine, and three years later joined forces with Eugen Langen, an industrialist trained at the Karlsruhe Polytechnic, to form a company to market such engines. He received valuable help also from a former fellow student, Franz Reuleaux. At the Paris Exhibition of 1867 the firm's product won a gold medal in competition with 14 other gas engines. Further capital was raised, and a new factory, the Gasmotorenfabrik, was built at Deutz near Cologne in 1869. Otto concentrated on the administrative side of the business, leaving Langen, with his new recruits Gottlieb Daimler (1834–1900) and Wilhelm Maybach (1847–1929), to develop the engineering side.

In 1876, Otto described the four-stroke engine for which his name is famous. Unfortunately his patent was invalidated in 1886 when his competitors discovered that Alphonse Beau de Rochas had described the principle of the four-stroke cycle in an obscure pamphlet. In the period 1860 to 1865 Lenoir sold several hundred of his small double-acting gas engines, but technical weaknesses – especially low compression – limited their potential. Otto's much more efficient and relatively quiet engine, the so-called 'silent Otto' was well received and sold extensively in the first ten years of manufacture. Otto died in Cologne in 1891.

Otto first designed a successful vertical atmospheric gas engine in 1867. Some ten years later he introduced a horizontal engine, the operation of which was closely similar to the cycle of Beau de Rochas. Almost certainly, however, Otto reached his results independently of Rochas and the system has ever since been referred to as the Otto cycle.

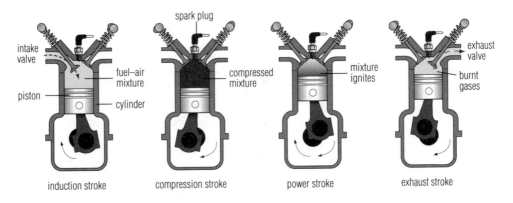

intake valve · spark plug · exhaust valve · piston · fuel–air mixture · cylinder · compressed mixture · mixture ignites · burnt gases

induction stroke · compression stroke · power stroke · exhaust stroke

Otto *The Otto cycle, devised originally for a gas engine, is the basis of all four-stroke internal-combustion engines.*

In the Otto four-stroke cylinder cycle the explosive mixture, in the first stroke of the piston towards the crankshaft, is drawn into the cylinder (the induction stroke). It is compressed on the return (compression) stroke. Ignition is then effected at or about the top dead centre position and the burning mixture drives the piston during the third stroke of the cycle. Finally, on the fourth (exhaust) stroke, the burnt gases are driven out of the cylinder. The cycle is then repeated.

The superiority of Otto's new engine over other types was soon apparent. Thousands of them, manufactured by Otto and Langen, were installed throughout the world in a very few years. Other types of engine were still made, such as the low-power vertical Bischop engine, in which expansion of the exploded gases raised the piston, which was then driven downwards by atmospheric pressure during its downward stroke.

Owen Richard 1804–1892 was one of the leading British naturalists of the Victorian era.

Born at Lancaster, Owen was apprenticed to a local surgeon before studying medicine at Edinburgh University and completing his medical studies at St Bartholomew's Hospital (Bart's), London. Showing great proficiency in working with specimens, he was soon appointed Curator at Bart's, and later became the first Hunterian Professor at the Royal College of Surgeons. Owen was elected Fellow of the Royal Society in 1834 and served as Fullerian Professor of Physiology and Comparative Anatomy at the Royal

Institution from 1858 to 1862. In 1856, he was made the first Superintendent of the Natural History Departments of the British Museum, and was later promoted to Director when the collections were moved to South Kensington. On retirement in 1884 this very eminent and highly public scientist was knighted.

Owen's early career was marked by his phenomenal output of high-quality zoological identification, description and classification in the manner of Cuvier. Owen published more than 360 detailed monographs on recent and fossil invertebrates and vertebrates, the most important being concerned with the pearly nautilus, the moa and other birds of New Zealand, the dodo from Mauritius, and the *Archaeopteryx* – his reconstruction of that extinct bird on comparative anatomical principles being regarded as a classic. His *History of British Fossil Reptiles* 1849–84 was a staggering work of patient erudition; he also produced a popular textbook, *Palaeontology* (1860).

Owen never accepted Darwinian evolution. Amongst other objections, he contended on taxonomic grounds that the doctrine could not explain human beings; he turned them into the single example of a special subclass of Mammalia. (In response, T. H. Huxley demonstrated that the anatomical grounds for Owen's taxonomy were illusory.) Owen also fiercely attacked Darwin's mechanism for evolution, denying natural selection. Insofar as Owen was prepared to countenance evolution, it was only in terms of the unfolding of a grand plan of Nature, as suggested by the framework of German *Naturphilosophie*.

P

Paget James 1814–1899 was a British surgeon, one of the founders of pathology. He is best remembered for describing two conditions named after him: Paget's disease of the nipple, a precancerous disorder, and Paget's disease of the bone, or osteitis deformans. From a fairly humble beginning he rose to be one of the greatest and most respected surgeons of his time.

Paget was born in Great Yarmouth on 11 January 1814. He went to school in Yarmouth, but had to leave at the age of 13 when his father's business ran into hard times. At 16 he became apprenticed to the local surgeon apothecary, Charles Costerton, and in 1834 his elder brother George paid for him to go to St Bartholomew's Hospital in London, where despite the poor standard of teaching at the time, he gained his MRCS in 1836. In his struggle to make up his income, Paget took pupils, worked as a subeditor on the *Medical Gazette*, reported lectures, reviewed books and translated works. It was not until he became warden of the students' residential college at St Bartholomew's in 1843 that he was able to give up his journalism. He was one of the original 300 Fellows of the Royal College of Surgeons of England in 1843, where he later became Professor of Anatomy and Surgery (1847–52). At the age of 33 he became Assistant Surgeon at St Bartholomew's Hospital, and four years later in 1851 had his own practice in Cavendish Square. In 1878, he tended the Princess of Wales and his fame spread. He became a rich man, and was appointed Surgeon Extraordinary to Queen Victoria and became a close friend of the Royal Family. During a postmortem in 1871, Paget contracted a severe infection through a cut, which left him unable to continue his hospital work, although he did maintain his consulting rooms. He received a baronetcy in 1871 and was a member of the General Medical Council, the Senate of the University of London, and President of the Royal College of Surgeons (1874). He died on 30 December 1899, and his funeral was held at Westminster Abbey.

Paget's original clinical descriptions of the two conditions named after him were so accurate that virtually nothing has needed to be added to them since. Paget's disease of the nipple was described in 1874 and is an eczematous skin eruption that indicates an underlying carcinoma of the breast, although the eruption is not simply an extension of the cancer cells inside the breast. Histologically Paget cells can be identified and are pathognomonic of the condition.

When Paget described the disease of the bone in 1877, he referred to it as osteitis deformans. This implies an inflammation of the bone, which is not accurate and it is now called osteodystrophia deformans. Paget did, however, accurately describe this idiopathic condition which can affect the bones of the elderly, particularly the femora and tibiae in the legs and the bones of the skull. The bones soften, giving rise to deformity of the limbs, which may also fracture easily. If the skull is affected bony changes cause enlargement of the head, and pressure on the VIIIth cranial nerve can cause deafness.

Paneth Friedrich Adolf 1887–1958 was an Austrian chemist known for his contribution to the development of radiotracer techniques and to organic chemistry.

Paneth was born in Vienna on 31 August 1887, the second of the three sons of the well known physiologist Joseph Paneth. He was educated in Vienna and attended the universities of Munich and Glasgow before obtaining his PhD from Vienna in 1910. From 1912 to 1918 he worked as assistant to Stefan Meyer at the Vienna Institute for Radium Research – in 1913 he spent a short time with Frederick Soddy in Glasgow and visited Ernest Rutherford's laboratory in Manchester. He spent two years (1917–19) at the Prague Institute of Technology and three (1919–22) at the University of Hamburg, before going to the University of Berlin. From 1929 until 1933 he was Professor and Director of the Chemical Laboratories at the University of Königsberg.

Because of the growth of the Nazi movement, Paneth left Germany in 1933 and went to Imperial College, London, as a reader and guest lecturer. Six years later he moved to Durham University as Professor of Chemistry, where he stayed until 1953. During World War II he was head of the chemical division of the Joint British and Canadian Atomic Energy Team in Montreal, and from 1949 to 1955 served as

President of the Joint Commission on Radioactivity, an organization of the International Council of Scientific Unions. In 1953, Paneth returned to Germany to become Director of the Max Planck Institute for Chemistry in Mainz. He died in Vienna on 17 September 1958.

One of Paneth's first chemical papers was concerned with the acid-catalysed rearrangement of the two organic compounds quinidine and cinchonidine. But he soon became much more involved with radioactive substances. He unsuccessfully tried to chemically separate radium D and thorium B from lead, and eventually realized that they must both be isotopes of lead. In collaboration with the Hungarian chemist Georg von Hevesy he extended this work into research on using radium D and thorium B as indicators – radioactive tracers – to determine the solubility of the slightly soluble compounds lead sulphide and lead chromate.

A similar attempt to separate the radioactive products of thorium decay led to the preparation and isolation of bismuth hydride, BiH_3, and to the realization that radium E and thorium C are isotopes of bismuth. It was only by the use of radioactive isotopes that the minute quantities of bismuth hydride formed could be detected. In order to decompose the hydride and concentrate the metal from the unstable hydrides he studied, such as those of bismuth, lead, tin and polonium, he used a method known as the mirror deposition technique. The metal formed a metallic mirror on the inside of a heated tube through which the hydride was passed. He then developed a better method involving the electrolysis of the metal sulphate. He prepared several grams of a new tin hydride, SnH_4, and made intensive investigations of its properties.

The work on metal hydrides led, in turn, to that on free radicals. In 1929, while at the University of Berlin, Paneth and Wilhelm Hofeditz announced the preparation and identification of the free methyl radical from lead tetramethyl.

Also in the period up to 1929, he developed sensitive methods for determining trace amounts of helium. Using spectroscopy and, later, mass spectroscopy, he determined the helium content of natural gas from various sources, measured its rate of diffusion through glass, measured the amount of helium in rocks and meteorites, and unsuccessfully tried to measure the helium produced by attempted transmutations from light elements into helium. From 1929 to the end of his life, meteorites dominated his interests. He estimated the ages of iron meteorites to be in the range 10^8 to 10^9 years, and speculated that they were formed within the Solar System.

In the late 1930s Paneth succeeded in obtaining measurable amounts of helium by the neutron bombardment of boron; he had induced an artificial transmutation. He then began to investigate the trace elements in the stratosphere. He determined the helium, ozone and nitrogen dioxide content of the atmosphere and investigated the extent of gravitational separation of the components of the atmosphere. He found none below 60 km/37 mi, but discovered a measurable change in relative concentration above this altitude. He then went back to studying free radicals and

explored the use of radioactive isotopes to combine with them in a mirror removal technique.

Papin Denis 1647–c.1712 was a French physicist and technologist who invented a vessel which was the forerunner of the pressure cooker or autoclave.

Papin was born in Blois, France, on 22 August 1647 and studied medicine at Angers University, where he obtained his MD in 1669. His first job was as an assistant to Christiaan Huygens in Paris, and then in 1675 he went to London to assist Robert Hooke (then responsible for the running of the Royal Society), to write letters for a payment of two shillings (10 p) each. Papin was not appointed a Fellow of the Royal Society until late in the 1680s. In 1680, he returned to Paris to work with Huygens and in 1681 he went to Venice for three years as Director of Experiments at Ambrose Sarotti's academy. In 1684, he tried to secure the position of Secretary of the Royal Society, but Edmund Halley got it and Papin was appointed 'temporary curator of experiments' with a salary of £30 a year. In November 1687, he was appointed to the Chair of Mathematics at Marburg University and stayed there until 1696 when he moved to Cassel to take up a place in the Court of the Landgrave of Hesse. He returned to London in 1707, but by then his friends had gone and there was no position for him at the Royal Society. He drifted into obscurity and died in about 1712 (certainly not later than 1714).

Papin's first scientific work as Huygens's assistant was to construct an air pump and to carry out a number of experiments. Later, with Robert Boyle, he introduced a number of improvements to the air pump and invented the condensing pump. In 1680, they published *A Continuation of New Experiments*. It was while he was with Boyle that he invented the steam digester – a closed vessel with a tightly fitting lid in which water was heated. This was the prototype of the modern pressure cooker, but it is used more extensively in the chemical industry as the autoclave. In Papin's prototype, the steam was prevented from escaping so that a high pressure was generated, causing the boiling point of the water to rise considerably. Papin invented a safety valve to guard against excessive rises in pressure. This safety valve was of technical importance in the development of steam power. He showed his invention to the Royal Society in May 1679.

In the early 1680s, when he was employed by the Royal Society, he carried out numerous experiments in hydraulics and pneumatics, which were published in *Philosophical Transactions*.

In 1690, Papin suggested that the condensation of steam should be used to make a vacuum under a piston previously raised by the expansion of the steam. This was the earliest cylinder-and-piston steam engine. His idea of using steam later took shape in the atmospheric engine of Thomas Newcomen. His scheme was unworkable, however, because he proposed to use one vessel as both boiler and cylinder. He proposed the first steam-driven boat in 1690 and in 1707 he built a paddle-boat, but the paddles were turned by human power and not by steam.

Also during this period, he considered the idea of a piston ballistic pump with gunpowder and discussed the project with Huygens. In a letter to Leibniz on 6 March 1704, he claimed the idea as his own. In 1705, Leibniz sent him a sketch of Thomas Savery's high-pressure steam pump for raising water. In 1707, Papin devised a modification of this, which was workable but not as productive as the original piston model.

Pappus of Alexandria lived *c*. AD 300–350 was a Greek mathematician, astronomer and geographer whose chief importance lies in his commentaries on the mathematical work of his predecessors.

Nothing is known of his life and many of his writings survive only in translations from the original Greek. According to the *Suda Lexicon*, he lived in the time of the emperor Theodosius I (reigned AD 379–395), but the compiler of the *Suda Lexicon* was notoriously unreliable, and from other sources, it appears that Pappus lived rather earlier. The most important piece of evidence is found in his commentary on Ptolemy, in which he writes of an eclipse of the Sun that took place in AD 320 in language which strongly suggests that he himself witnessed the event as a grown man. It is usual, therefore, to place Pappus in the first half of the fourth century.

Pappus' chief works are the *Synagogue* (commonly referred to as the *Collection*), the commentary on Ptolemy's *Syntaxis* or *Almagest*, and the commentary on Euclid's *Elements*. He also wrote a commentary on Euclid's *Data* and one on the *Anelemma* of Diodorus; neither of these works, however, survives. Pappus may also have written the section from the fifth book onwards of the commentary on Ptolemy's *Harmonica* which was chiefly the work of Porphyry. More convincingly established is Pappus' authorship of the *Description of the World*, a geographical treatise which has come down to us only in the form of a book written in Armenian and bearing the name of Moses of Khoren as its author. That the treatise does not exist in Greek is no hindrance to attributing it to Pappus because, of all his works, only the *Collection* and the commentary on the *Almagest* survive in their original form. Some scholars believe that the *Description* should be attributed to Anania Shirakatsi, but the consensus is that the work is either a direct translation or a very close paraphrase of Pappus. It should also be mentioned, in a list of Pappus' work, that a twelfth-century Arabic manuscript attributes to Pappus the invention of an instrument to measure the volume of liquids. If Pappus did invent such an instrument – and there seems to be no good reason to doubt it – he may have written a treatise on hydrostatics that has been lost to posterity.

By far the most important of Pappus' works is the *Collection*. Without it, much of the geometrical achievement of his predecessors would have been lost for ever. The *Collection* is written in eight parts, of which the first two have never been found; Pappus may have intended to extend it to twelve parts and may indeed have done so. The *Collection* deals with nearly the whole body of Greek geometry, mostly in the form of commentaries on texts which it is assumed the reader has to hand. It reproduces known solutions to problems in geometry; but it also frequently gives Pappus' own solutions, or improvements and extensions to existing solutions. Thus Pappus handles the problem of inscribing five regular solids in a sphere in a way quite different from Euclid; gives a broader generalization than Euclid to the famous Pythagorean theorem; and provides a demonstration of squaring the circle which is quite different from the method of Archimedes (who used a spiral) or that of Nicomedes (who used the conchoid).

Perhaps the most interesting part of the *Collection*, measured by its influence on modern mathematics, is Book VII, which is concerned with the problems of determining the locus with respect to three, four, five, six or more than six lines. Pappus' work in this field was called 'Pappus' problem' by René Descartes, who demonstrated that the difficulties which Pappus was unable to overcome could be got round by the use of his new algebraic symbols. Pappus thus came to play an important, if minor, role in the founding of Cartesian analytical geometry. And it is another mark of his originality and skill that he spent much time working on the problem of drawing a circle in such a way that it will touch three given circles, a problem sophisticated enough to engage the interest, centuries later, of both François Viète and Isaac Newton.

For his own originality, even if his chief importance is as the preserver of Greek scientific knowledge, Pappus stands (with Diophantus) as the last of the long and distinguished line of Alexandrian mathematicians.

Paracelsus Philippus Aureolus Theophrastus Bombast von Hohenheim 1493–1541 was a Swiss physician and chemist whose works did much to overthrow the accepted scientific authorities of his day (such as Galen) and to establish the importance of chemistry in medicine. He adopted the name Paracelsus as a claim to superiority over the Roman physician Celsus, whose works had recently been translated.

Paracelsus was born on or about 10 November 1493 in Einsiedeln, Switzerland, the son of a doctor. On his mother's death, he and his father moved to Villach, Austria, where he attended the Bergschule. This local school specialized in teaching mineralogy to students who would later work in the mines nearby. In 1507 he became, like many of his contemporaries, a wandering scholar. He is said to have obtained a baccalaureate in medicine from the University of Vienna, to have been to the University of Basel and to have studied at several universities in Italy, and may have received a doctorate from Ferrara. He was a military surgeon in Venice in 1521 and then continued his travels as far as Constantinople. He returned to Villach in 1524 and by 1525 had set himself up in medical practice. He was successful enough to be elected Professor of Medicine at Basel from 1527 to 1528. Here he scandalized other academics by lecturing in the vernacular and by his savage attacks on the accepted medical texts. In 1527, he burned the works of Galen and Avicenna in public and the next year was forced to leave Basel. He spent several years travelling once more and then returned to Villach, when he was appointed physician to Duke Ernst of Bavaria, in 1541. He died in Salzburg on 24 September of that year.

Paracelsus substituted the traditional medical theories with an animistic view of nature, believing that all matter possesses its characteristic spirits or life substances, which he called *entia*. He emphasized the importance of the observation of the properties of all things and it was this principle that led him to discover new remedies and means of treatment for many illnesses, some of which he characterized accurately for the first time.

Chemical therapy had been used in the ancient world, although chiefly externally. Paracelsus, however, realized the therapeutic power of chemicals taken internally, although he imposed strict control on their use, dosage and purity. Paracelsus was extremely successful as a doctor. His descriptions of miners' diseases first identified silicosis and tuberculosis as occupational hazards. He also recognized goitre as endemic and related to minerals in drinking water, and originated a medical account of chorea, rather than believing this nervous disease to be caused by possession by spirits. Paracelsus was the first to recognize the congenital form of syphilis, and to distinguish it from the infectious form. He showed that it could be successfully treated with carefully controlled doses of a mercury compound.

Paracelsus' study of alchemy helped to develop it into chemistry. His investigations produced new, nontoxic compounds for medicinal use; he discovered new substances arising from the reaction of metals and described various organic compounds, including ether. He was the first to devise such advanced laboratory techniques as the concentration of alcohol by freezing. Paracelsus also devised a specific nomenclature for substances already known, but not precisely defined. Paracelsian chemicals were introduced into the *London Pharmacopoeia* of 1618, and his attempt to construct a chemical system, grouping chemicals according to their susceptibility to similar processes, was the first of its kind.

Paracelsus' concept of man as a 'microcosm' of the natural world led to his theory of an external agency being the source of disease, overturning contemporary views which regarded illness as an imbalance of the four humours (blood, phlegm, choler and spleen) within the body. His ideas encouraged new modes of treatment supplanting, for example, bloodletting, and opened the way for new ideas on the source of infection.

Despite Paracelsus' mystical preoccupations such as astrology and the use of magic seals and amulets, his importance to the development of science is substantial. He can be regarded as a founder of modern medicine, as he was the first to demand that a doctor should master all those arts then divided between barbers, field-surgeons, apothecaries, alchemists and local 'wise women'. His revolutionary views and vitriolic nature continually involved him in clashes with authority, and his investigations into 'forbidden fields' led to frequent accusations against him of sorcery and heresy. Nevertheless, within his works lie the stepping stones between ancient and modern science.

Parsons Charles Algernon 1854–1931 was a British mechanical engineer who designed and built the first practical steam turbine and developed its use as the motive power for the generation of electricity in power stations, in centrifugal pumps and, particularly, as a source of power for steamships. In most areas of its application it replaced the much less efficient reciprocating steam engine. To accompany this invention Parsons developed more efficient screw propellers for ships and suitable gearing to widen the turbine's usefulness, both on land and sea. In his later life he designed searchlights and optical instruments, and developed methods for the production of optical glass. The firm of Grubb Parsons is still of world renown in the design and servicing of the optical systems for large telescopes.

Parsons, the sixth and youngest son of the Earl of Rose, was born in London on 13 June 1854. He came from a talented family. His grandfather had been Vice President of the Royal Society and his father was its President. During his youth in Ireland, Parsons was educated by private tutors. When he was 17 he went to Trinity College, Dublin, where he spent two years. From there he went to St John's College, Cambridge, where he gained high honours in mathematics. He began his training in engineering on leaving Cambridge by becoming an apprentice at William Armstrong and Company near Newcastle-upon-Tyne. After further experience as an engineer he went to the firm of Clarke, Chapman and Company at Gateshead as a junior partner, in charge of their electrical section. It was there that he succeeded in making a practical steam turbine and applied it to the generation of electricity. Parsons died while on a cruise to the West Indies on 11 February 1931, aged 76 years.

In a steam turbine a cylindrical bladed rotor is enclosed within a casing with static blades, and the steam passes between the two, contacting first one set of turbine blades then being directed on to another set designed to work with the same steam at a slightly lower pressure. At the same time the work done by the expansion of the steam aids the rotation produced. Parsons's first machine was used as a turbogenerator for electricity and can now be seen in the Science Museum, London.

In 1889, he formed his own company, which still exists, near Newcastle-upon-Tyne. He developed turbogenerators of various kinds and, as time went on, increasing capacities, which formed the basic machinery for national (and much of international) electricity production.

From about 1894 Parsons, with the formation of a new company at Wallsend-upon-Tyne, applied his turbine to various uses. His first venture, the 48-m/160-ft steamship *Turbinia* (displacement 44 tonnes) achieved initially a speed of 20 knots. Even at this high speed he calculated that the output from the turbine (about 2,000 hp/1,500 kW) was not being used efficiently. He designed a propulsion system which incorporated three shafts, each with three propellers. With this new machinery the *Turbinia*, with Parsons at the controls, reached a record-breaking speed of $34^1/_2$ knots when it sailed up and down the assembled rows of British and foreign warships at the Naval Review of 1897, to celebrate the Diamond Jubilee of Queen Victoria. *Turbinia* can still seen in the Museum of Science and Industry at Newcastle-upon-Tyne.

Prompted by the difficulties of realizing the full power output from the turbine at sea, Parsons investigated the loss of efficiency of screw propellers and the phenomenon of cavitation caused by the water not adhering to the propeller blades at high speeds – and the accompanying damage to the blades. Using steam turbines instead of reciprocating steam engines, high speeds – with less vibration – were to be had by ocean liners, and greater efficiency – with increased fuel economy – was obtained by the slower trading vessels. Parsons's turbines fitted to the liners *Lusitania* and *Mauritania* developed some 70,000 hp/52,000 kW.

Parsons's work on searchlights and specialist marine and other optical instruments was taken up by the Royal Navy and various maritime trading companies. He also revitalized the British optical glass industry and safeguarded its production for possible military applications at a time when it was about to be eclipsed by German companies. He spent much of his life attempting, without success, to make artificial diamonds by the crystallization of carbon at high pressures and temperatures.

Parsons was a man of considerable courage in so far as he often undertook to produce machines far beyond the current limits of design and expertise. He was, without a doubt, one of the greatest engineers of the late nineteenth and early twentieth centuries and was recognized as such by his election to the Royal Society, the award of a knighthood and various honorary degrees and in being the first engineer to be admitted to the Order of Merit.

His ingenuity has lasted and benefitted humanity in that now there are turbines in use in a large variety of roles, providing the most convenient, useful and efficient means of converting power into motion.

Parsons William, Third Earl of Rosse 1800–1867 was an Irish politician, engineer and astronomer, whose main interest was in rediscovering the techniques used by William Herschel (1738–1822) to build bigger and better telescopes. After considerable expense and dedicated effort he succeeded, and with his new instruments made some important observations, particularly of nebulae.

Parsons was born in York on 17 June 1800, but it was at his ancestral home, Birr Castle in County Offaly, that he received his early education. He then attended Trinity College, Dublin, for a year before going up to Magdalen College, Oxford, in 1819, and graduating with a First in mathematics at the age of 22. A political career commensurate with his family's land ownership and title was virtually obligatory for Parsons, as the eldest son, and even while still an undergraduate he was elected to the House of Commons to represent King's County, a seat he then held

at Westminster for 13 years. In 1831 he became Lord Lieutenant of County Offaly, and although he retired from parliamentary life in 1834, seven years later he was back, at the House of Lords, an elected Irish representative peer, having succeeded to the title of Rosse on his father's death. During and after the potato famine of 1846, Parsons worked to alleviate the living conditions of his tenants. It was work that Irish landowners were more or less forced to undertake when the Government in London delayed aid; but his tenants were grateful to Parsons, and when he died in Monkstown on 31 October 1867, thousands of them attended his funeral.

The work of William Herschel was a source of fascination to Parsons. Early on he decided that he too would construct enormous telescopes and make great astronomical discoveries. Accordingly, he learned to grind mirrors, made a few small ones, and then began to seek a material capable of being cast as a large mirror. An alloy using copper and tin was considered but found difficult to cast directly. An attempt at another solution, using sectional mirrors surrounding a central disc soldered on to a brass disc, proved unsatisfactory for instruments with an aperture larger than 46 cm/18 in. Subsequently, Parsons developed a way of casting solid discs, designing a mould ventilator that permitted the even cooling of the metal forming the mirror, in an annealing oven. Thirteen years after his experiments began, Parsons was able to construct a 92-cm/36-in mirror in sections; a solid mirror of the same size was completed a year later. And in 1842 Parsons cast the first 1.8-m/72-in disc, the 'Leviathan of Parsonstown', which weighed nearly 4 tonnes and was incorporated into a telescope with a focal length of 16.2 m/54 ft. It took three years to put together, including setting it up on two masonry piers.

At last Parsons was ready for the observational side of his work. And during the next 13 years, when the Irish weather allowed, he made a number of important observations. His telescope was, after all, the largest in contemporary use, and with it he researched, particularly into nebulae. He was the first to remark that some were shaped in a spiral – in fact he went on to find 15 spiral nebulae – and resolved others into clusters of stars. It was he who named the famous Crab Nebula.

In constructing his 'Leviathan', Parsons designed a mechanism (since copied by many others) for grinding and polishing metal mirrors. He also invented a clockwork drive for the large equatorial mounting of an observatory. He was even among the first to take photographs of the Moon.

His other interests included a study of problems in constructing iron-armoured ships.

Pascal Blaise 1623–1662 was a French mathematician, physicist and religious recluse who was not only a scientist anxious to solve some of the problems of the day but also a gifted writer and a moralist. He is remembered not so much for his original creative work – as are his contemporaries René Descartes (1596–1650) and Pierre de Fermat

(1601–1675) – but for his contributions to projective geometry, the calculus of probability, infinitesimal calculus, fluid statics, and his methodology in science generally. Much of his work has become appreciated only during the last 150 years.

Pascal was born on 19 June 1623 at Clermont-Ferrand, the son of a civil servant in the local administration. His mother died when he was only three, and his father – also a respected mathematician – looked after the family and saw to the education of the children. In 1631 they all moved to Paris, where Pascal's sister Jacqueline showed literary talent and Pascal himself displayed mathematical ability. By 1639, when Pascal was 16, he was already participating in the scientific and philosophical meetings run at the Convent of Place Royale by its Director, Father Marin Mersenne (1588–1648); some of these meetings were attended also by Descartes, Fermat and other celebrated figures (such as Thomas Hobbes). The illness and eventual death of his father led Pascal to commit himself to a more spiritual mode of life, one from which he was at times terrified of lapsing. Converted to the rigorous form of Roman Catholicism known as Jansenism in 1646, he finally experienced a fervently spiritual 'night of fire' on 23 November 1654, and from then on wrote only at the direct request of his spiritual advisers, the order of monks at Port Royal. Five years later his health had become poor enough to prevent him from working at all. After 1661, when his sister died, Pascal became even more solitary and his health deteriorated further. His last project was to design a public transport system for Paris. The system was actually inaugurated in 1662, the year Pascal died. He died from a malignant ulcer of the stomach on 19 August 1662 in Paris, aged only 39.

Pascal's first serious work was actually on someone else's behalf. In 1639, Gérard Desargues (1593–1662) published a work entitled *Brouillon project d'une atteinte aux événements des rencontres du cone avec un plan/Experimental project aiming to describe what happens when the cone comes in contact with a plane*, but its content baffled most of the mathematicians of that time because of its style and vocabulary, and the refusal of Desargues to use Cartesian algebraic symbols. Pascal became Desargues's main disciple, and in the following year published his *Essai pour les coniques* in explanation of the subject. The paper was an immediate success in the mathematical world; that in itself, coupled with the fact that his own algebraic notational system now had strong competition, left René Descartes smarting rather, and he thenceforward regarded Pascal as something of an opponent.

Grasping the significance of Desargues's work, Pascal used its basic ideas – the introduction of elements at infinity, the definition of a conic as any plane section of a circular cone, the study of a conic as a perspective of a circle, and the involution determined on any straight line by a conic and the opposite sides of an inscribed quadrilateral – and went on to make his first great discovery, now known as Pascal's mystic hexagram. He stated that the three points of intersection of the pairs of opposite sides of a hexagon

inscribed in a conic are collinear. By December 1640 he had deduced from this theorem most of the propositions now known to have been contained in the *Conics* of the ancient Greek mathematician Apollonius. It was not until 1648, however, that Pascal found a geometric solution to the problem of Pappus (which Descartes had used in connection with demonstrating the strength of his new analytical geometry in 1637). Pascal's solution was important because it showed that projective geometry might prove as effective in this field as the Cartesian analytical methods.

The full treatise that Pascal wrote covering the whole subject was never published; the manuscript was seen later only by Gottfried Leibniz (1646–1716). And in fact, because the work of Desargues was so complicated, it was not until the nineteenth century, with the researches of Jean-Victor Poncelet (1788–1867), that attention was drawn to the work of Pascal.

In 1642, to help his father in his work, Pascal decided to construct an arithmetical machine that would mechanize the processes of addition and subtraction. He devised a model in 1645, and then organized the manufacture and sale of these first calculating machines. (At least seven of these 'computers' still exist. One was presented to Queen Christina of Sweden in 1652.)

Pascal kept up a long correspondence with Fermat on the subject of the calculus of probabilities. Their main interest was in the study of two specific problems: the first concerned the probability that a player will obtain a certain face of a dice in a given number of throws; and the second was to determine the (portion of the) stakes returnable to each player of several if a game is interrupted. Pascal was the first to make a comprehensive study of the arithmetical triangle (now called the Pascal triangle) that he then used to derive combinational analysis. Together with Fermat, he provided the foundations for the calculus of probability in 1657. In 1658 and the next year, Pascal perfected what he called 'the theory of indivisibles' (which he had first referred to in 1654). This was in fact the forerunner of integral calculus, and enabled him to study problems involving infinitesimals, such as the calculations of areas and volumes.

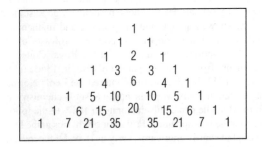

Pascal *Pascal's triangle is an array of numbers in which (a) each is the sum of the two immediately above, left and right; (b) each is the sum of all those in each diagonal, starting from either immediately above, left or right; and (c) constitutes an arithmetical series.*

Pascal's work in hydrostatics was inspired by the experiment of Evangelista Torricelli (1608–1647) in 1643, which demonstrated that air pressure supports a column of mercury only about 76 cm/30 in high. In 1647, Pascal succeeded in repeating Torricelli's experiment, but this time using wine and water in tubes 12 m/39 ft high fixed to the masts of ships. He confirmed that a vacuum must exist in the space at the top of the tube, and set out to prove that the column of mercury, wine or water is held up by the weight of air exerted on the container of liquid at the base of the tube. Pascal suggested that at high altitudes there would be less air above the tube and that the column would be lower. Unable through poor health to undertake the experiment himself, he entrusted it to his brother-in-law who obtained the expected results using a mercury column in the mountains of the Puy de Dôme in 1648.

Pascal's proof that the height of a column of mercury does depend on air pressure led rapidly to investigations of the use of the mercury barometer in weather forecasting. Pascal however turned to a study of pressure in liquids and gases, and found that it is transmitted equally in all directions throughout a fluid and is always exerted perpendicular to any surface in or containing the fluid. This is known as Pascal's principle and it was propounded in the treatise on hydrostatics that Pascal completed in 1654. This principle is fundamental to applications of hydrostatics and governs the operation of hydraulic machines.

Pascal's pioneering work on fluid pressure laid the foundations for both hydraulics and meteorology. In his honour, the SI unit for pressure is called the pascal. It is equal to one newton per square metre.

Pasteur Louis 1822–1895 was a French chemist and microbiologist who became world famous for originating the process of pasteurization and for establishing the validity of the germ theory of disease, although he also made many other scientific contributions. Regarded as one of the greatest scientists in history, he received many honours during his lifetime, including the Legion of Honour, France's highest award.

Pasteur was born on 27 December 1822 in Dôle in eastern France, the son of a tanner. While he was still young, his family moved to Arbois, where he attended primary and secondary schools. He was not a particularly good student, but he showed an aptitude for painting and mathematics and his initial ambition was to become a professor of fine arts. He continued his education at the Royal College in Besançon, from which he gained his BA in 1840 and his BSc in 1842. In 1843, Pasteur entered the Ecole Normale Supérieure in Paris, where he began to study chemistry and from which he gained his doctorate in 1847. In the following year he was appointed Professor of Physics at the Dijon Lycée but shortly afterwards, early in 1849, he accepted the post of Professor of Chemistry at the University of Strasbourg. In the same year he married Marie Laurent, the daughter of the university's rector; they were to have five children, only two of whom survived beyond childhood. In 1862, Pasteur was elected to the French Academy of

Sciences, and in 1863 to a chair at the Ecole Normale Supérieure, a position that was created for him so that he could institute an original teaching programme that related chemistry, physics and geology to the fine arts. Also in 1863 he became Dean of the new science faculty at Lille University, where he initiated the novel concept of evening classes for workers. Meanwhile, in 1857 he had been appointed Director of Scientific Studies at the Ecole Normale Supérieure. Because of the pressure of his research work, Pasteur resigned from the directorship in 1867 but, with financial assistance from Emperor Napoleon III, a laboratory of physiological chemistry was established for him at the Ecole. Pasteur suffered a stroke in 1868 but, although partly paralysed, continued his work. In 1873 he was made a member of the French Academy of Medicine, and in the following year the French parliament granted him a special monetary award to guarantee his financial security while he pursued his research. In 1882, he was elected to the Academic Française. In 1888, the Pasteur Institute was created in Paris for the purpose of continuing Pasteur's pioneering research into rabies, and he headed this establishment until his death, in Paris, on 28 September 1895.

Pasteur first gained recognition through his early work on the optical activity of stereoisomers. In 1848 he presented a paper to the Paris Academy of Sciences in which he reported that there are two molecular forms of tartaric acid, one that rotates plane polarized light to the right and another (a mirror image of the first) that rotates it to the left. In addition, he showed that one form can be assimilated by living microorganisms whereas its optical antipode cannot.

Pasteur began his biological investigations – for which he is best known – while at Lille University. After receiving a query from an industrialist about wine- and beer-making, Pasteur started researching into fermentation. Using a microscope he found that properly aged wine contains small spherical globules of yeast cells whereas sour wine contains elongated yeast cells. He also proved that fermentation does not require oxygen, but that it nevertheless involves living microorganisms and that to produce the correct type of fermentation (alcohol-producing rather than lactic acid-producing) it is necessary to use the correct type of yeast. Pasteur also realized that after wine has formed, it must be gently heated to about 50°C/122°F to kill the yeast and thereby prevent souring during the ageing process. Pasteurization – as this heating process is called today – is now widely used in the food-processing industry.

Pasteur then turned his attention to spontaneous generation, a problem that had once again become a matter of controversy, despite Lazzaro Spallanzani's disproof of the theory about a century previously. Pasteur showed that dust in the air contains spores of living organisms which reproduce when introduced into a nutrient broth. Then he boiled the broth in a container with a U-shaped tube that allowed air but not dust to reach the broth. He found that the broth remained free of living organisms, thereby again disproving the theory of spontaneous generation.

In the mid-1860s the French silk industry was seriously threatened by a disease that killed silkworms and Pasteur was commissioned by the government to investigate the disease. In 1868, he announced that he had found a minute parasite that infects the silkworms, and recommended that all infected silkworms be destroyed. His advice was followed and the disease eliminated. This stimulated his interest in infectious diseases and, from the results of his previous work on fermentation and spontaneous generation, Pasteur developed the germ theory of disease. This theory was probably the most important single medical discovery of all time, because it provided both a practical method of combating disease by disinfection and a theoretical foundation for further research.

Continuing his research into disease, in 1881 Pasteur developed a method for reducing the virulence of certain pathogenic microorganisms. By heating a preparation of anthrax bacilli he attenuated their virulence but found that they still brought about the full immune response when injected into sheep. Using a similar method, Pasteur then inoculated fowl against chicken cholera. He was thus following the work of Edward Jenner, who first vaccinated against cowpox in 1796. In 1882, Pasteur began what proved to be his most spectacular research: the prevention of rabies. He demonstrated that the causative microorganism (actually a virus, although their existence was not known at that time) infects the nervous system and then, using the dried tissues of infected animals, he eventually succeeded in obtaining an attenuated form of the virus suitable for the inoculation of human beings. The culmination of this work came on 6 July 1885, when Pasteur used his vaccine to save the life of a young boy who had been bitten by a rabid dog. The success of this experiment brought Pasteur even greater acclaim and led to the establishment of the Pasteur Institute in 1888.

Pauli Wolfgang 1900–1958 was an Austrian-born Swiss physicist who made a substantial contribution to quantum theory with the Pauli exclusion principle. For this achievement, Pauli was awarded the 1945 Nobel Prize for Physics. He also postulated the existence of the neutrino.

Pauli was born in Vienna on 25 April 1900. He made rapid progress in science while at school and by the time he entered the University of Munich to study under Arnold Sommerfeld (1868–1951), he had already mastered both the special and general theories of relativity proposed by Albert Einstein (1879–1955). Recognizing his ability, Sommerfeld gave Pauli the task of preparing a comprehensive exposition of relativity, as Einstein had not done so. Pauli produced a monograph of extraordinary clarity, an amazing achievement for a student of 19.

Pauli obtained his doctorate in 1922 and then went to Göttingen as an assistant to Max Born (1882–1970). He soon moved to Copenhagen to study with Niels Bohr (1885–1962) and then in 1923 went to Hamburg as a privatdozent (unpaid lecturer). In 1928, Pauli was appointed Professor of Experimental Physics at the Eidgenössische Technical University, Zurich. He retained this position until the end of his life, but spent World War II in the United States at the Institute for Advanced Study, Prince-

ton. Pauli became a Swiss citizen on his return, and died at Zurich on 15 December 1958.

In the early 1920s, the quantum model of the atom proposed by Bohr in 1913 and elaborated by Sommerfeld in 1916 was in some disarray. Observations of the magnetic anomaly in alkali metals and of the fine spectra of alkaline earth metals did not accord with the Bohr–Sommerfeld model and it was suggested that the quantum numbers used to describe the energy levels of electrons in atoms in this model would have to be abandoned. Pauli realized that the situation could be explained if a fourth quantum number were added to the three already used (n, l and m). This number, s, would represent the spin of the electron and would have two possible values. He further proposed that no two electrons in the same atom can have the same values for their four quantum numbers. This is the Pauli exclusion principle, which was announced in 1925.

The exclusion principle means that the energy state of each electron in an atom can be defined by giving a unique set of values to the four quantum numbers. It not only accounted for the unusual properties of the elements that had been observed, but also gave a means of determining the arrangement of electrons into shells around the nucleus that explained the classification of elements into related groups. The successive shells could contain a maximum of 2, 8, 18, 32 and 50 electrons, with most elements containing outer valence electrons and inner complete shells. This discovery was of great importance, for it gave an explanation of the similarities in the properties of elements and revealed the significance of ordering elements by their atomic number.

Pauli's other main contribution to physics was made in 1930. The production of beta radiation in a continuous spectrum had puzzled physicists as theory demanded that the spectrum should be discontinuous. It appeared to be caused by a loss of energy when beta particles (electrons) were emitted from atoms, but no explanation could be found for such a loss. It was suggested that the theory of conservation of energy would have to be abandoned, but Pauli proposed that the emission of an electron in beta decay is accompanied by the production of an unknown particle. This particle would have unusual properties, having no charge and zero mass at rest – hence the fact that it had not been observed. Pauli persisted in this opinion, and in 1934 Enrico Fermi (1901–1954) confirmed Pauli's view and called the particle the neutrino. It was eventually detected in 1956.

Pauli made substantial advances in our understanding of the nature of the atom by adhering to accepted theories that he felt he could not abandon. He demonstrated exceptional insight, for such an approach usually blinds scientists to progress. Pauli summed up his life in a characteristically succinct fashion when he declared 'In my youth I believed myself to be a revolutionary; now I see that I was a classicist.'

Pauling Linus Carl 1901–   is a US theoretical chemist and biologist whose achievements rank among the most important of any in twentieth-century science. His main contribution has been to molecular structure and chemical bonding. He is one of the very few people to have been awarded two Nobel prizes: he received the 1954 Nobel Prize for Chemistry (for his work on intermolecular forces) and the 1962 Peace prize. Throughout his career his work has been noted for the application of intuition and inspiration; he has often carried over principles from one field of science and applied them to another.

Pauling was born in Portland, Oregon, on 28 February 1901, the son of a pharmacist. He began his scientific studies at Oregon State Agricultural College, from which he graduated in chemical engineering in 1922. He then began his research at the California Institute of Technology, Pasadena, gaining his PhD in 1925. From 1925 to 1927 he was a postdoctoral fellow in Europe, where he met the chief scientists of the day who were working on atomic and molecular structure: Arnold Sommerfeld in Munich, Niels Bohr in Copenhagen, Erwin Schrödinger in Zurich, and William Henry Bragg in London. He became a full professor at Pasadena in 1931 and left there in 1936 to take up the post of Director of the Gates and Crellin Laboratories, which he held for the next 22 years. He also held university appointments at the University of California, San Diego, and Stanford University, and during the 1960s spent several years on a study of the problems of war and peace at the Center for the Study of Democratic Institutions at Santa Barbara, California. His last appointment was as Director of the Linus Pauling Institute of Science and Medicine at Menlo Park, California.

Pauling's early work reflects his European experiences. In 1931, he published a classic paper, 'The nature of the chemical bond', in which he used quantum mechanics to explain that an electron-pair bond is formed by the interaction of two unpaired electrons, one from each of two atoms, and that once paired these electrons cannot take part in the formation of other bonds. It was followed by the book *Introduction to Quantum Mechanics* (1935), of which he was co-author. He was a pioneer in the application of quantum mechanical principles to the structures of molecules, relating them to interatomic distances and bond angles by X-ray and electron diffraction, magnetic effects and thermochemical techniques.

It was Pauling who introduced the concept of hybrid orbitals in molecules to explain the symmetry exhibited by carbon atoms in most of its compounds. The electrons in the ground state and in the excited state of the carbon atom can be represented as follows:

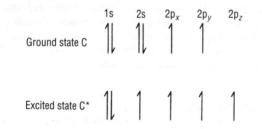

One of the 2p electrons can then form sp hybrid orbitals with the 2s electron; two linear 2p atomic orbitals remain:

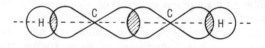

In acetylene (ethyne), for example, overlap of two sp hybrid orbitals between two carbon atoms results in a linear molecule. A hydrogen atom is bonded to each end by overlap between the carbons' sp orbitals and the s orbitals of the hydrogens. (The remaining carbon p orbitals also overlap to form two π bonds which, together with the bond just described, account for the traditional triple bond in this molecule.)

The structures of many other organic molecules can be explained in a similar way.

Pauling also investigated electronegativity of atoms and polarization (movement of electrons) in chemical bonds. He assigned electronegativities on a scale up to 4.0. A pair of electrons in a bond are pulled preferentially towards an atom with a high electronegativity.

In hydrogen chloride, HCl, for example, hydrogen has an electronegativity of 2.1 and chlorine of 3.5. The bonding electrons are pulled towards the chlorine atom, giving it a small excess negative charge (and leaving the hydrogen atom with a small excess positive charge), polarizing the hydrogen–chlorine bond:

$$H-Cl \qquad H \overset{..}{\underset{..}{\times}} Cl : \qquad H \rightarrow Cl \qquad \overset{\delta+}{H} \; \overset{\delta-}{-Cl}$$

electron pair pulled towards chlorine atom   *polarized molecule*

Electronegativity values can be used to show why certain substances, such as hydrochloric acid, are acid, whereas others, such as sodium hydroxide, are alkaline.

electronegativities of some elements

H			
2.1			
C	N	O	F
2.5	3.0	3.5	4.0
Si	P	S	Cl
1.8	2.1	2.5	3.0
Ge	As	Se	Br
1.8	2.0	2.4	2.8

For compounds whose molecules cannot be represented unambiguously by a single structure, Pauling introduced the idea of resonance hybridization. An example is carbon dioxide, CO_2:

$$O = C = O \qquad O \leftarrow C \equiv O \qquad O \equiv C \rightarrow O$$

Resonance hybrids (canonical forms) of carbon dioxide.

The true structure is regarded as an intermediate between two or more theoretically possible structures, which are termed canonical forms.

These, and Pauling's other ideas on chemical bonding, are fundamental to modern theories of molecular structure. Much of this work was consolidated in his book *The Nature of the Chemical Bond* (1939).

In the 1940s, Pauling turned his attention to the chemistry of living tissues and systems. He applied his knowledge of molecular structure to the complexity of life, principally to proteins in blood. With Robert Corey, he worked on the structures of amino acids and polypeptides. They proposed that many proteins have structures held together with hydrogen bonds, giving them helical shapes. This concept assisted Francis Crick and James Watson in their search for the structure of DNA, which they eventually resolved as a double helix.

In his researches on blood, Pauling investigated immunology and sickle-cell anaemia. Later work confirmed his hunch that the disease is genetic and that normal haemoglobin and the haemoglobin in abnormal 'sickle' cells differ in electrical charge. Throughout the 1940s he studied living materials; he also carried out research on anaesthesia. At the end of this period he published two textbooks, *General Chemistry* (1948) and *College Chemistry* (1950), which are still bestsellers.

Like many of his contemporaries, Pauling became concerned about the proliferation of nuclear weapons and their atmospheric testing during the 1950s. He presented to the United Nations a petition signed by 11,021 scientists from throughout the world urging an end to nuclear weapons testing, a view reinforced in his book *No More War* (1958). For these efforts he was awarded the Nobel Peace Prize in 1962.

Pavlov Ivan Petrovitch 1849–1936 was a Russian physiologist, best known for his systematic studies of the conditioning of dogs and other animals. For his observations on the gastrointestinal secretion in animals he received the 1904 Nobel Prize for Physiology or Medicine.

Pavlov was born in Ryazan on 24 September 1849. He decided to follow in the footsteps of his father, the local priest, and entered a theological college. In 1870, however, he left the seminary to study chemistry and physiology at the University of St Petersburg. There he was taught by the Russian chemists Dmitri Mendeleyev (1834–1907) and Alexander Butlerov (1828–1886). He received his medical degree in 1879 from the Imperial Medical Academy, St Petersburg, and his PhD from the Military Academy there in 1883. From 1884 to 1886 he studied cardiovascular and gastrointestinal physiology under Karl Ludwig in Leipzig

and Rudolf Heidenhain in Breslau. He researched at the Botkin laboratory in St Petersburg from 1888 to 1890 and in that year was appointed Professor of Physiology at the Imperial Medical Academy, where he remained until 1924. He died in Leningrad on 27 February 1936.

Pavlov's first unaided research was on the physiology of the circulatory system, studying cardiac physiology and the regulation of blood pressure. Using experimental animals, he became a surgeon of some distinction, a typical example of his experiments being the dissection of the cardiac nerves of a living dog to show how the nerves that leave the cardiac plexus control heartbeat strength.

During the years from 1890 to 1900, Pavlov investigated the secretory mechanisms of digestion. He developed an operation to prepare an ancillary miniature stomach or pouch, isolated from salivary and pancreatic secretions but with its vagal nerve supply intact. In this way he was able to observe the gastrointestinal secretion of a living animal.

Pavlov then went on to develop the idea of the conditional reflex – the discovery for which he is most famous. Pavlov confined a dog in a sound-proof room, in order to ensure that there were no distracting influences such as extraneous sounds and smells. The dog was held in a loose harness so that it could not move about too much. Food was delivered to it by an automatic apparatus operated from outside the room, so that the dog was fed at an appropriate moment without direct interference from the person directing the experiment. The flow of saliva from the dog's parotid gland was collected in a small measuring tube attached to the animal's cheek. The experiment was continued until the dog became used to the artificial situation. Pavlov discovered that if a neutral stimulus, such as a bell, was presented simultaneously with a natural stimulus to salivate (such as the sight of food) and the combination repeated often enough, the sound of the bell alone caused salivation.

Pavlov termed salivation the 'unconditioned reflex' and food the 'unconditioned stimulus'. The sound of the bell is the 'conditioned stimulus' and the salivation caused by the bell alone, the 'conditioned reflex'. Many inborn reflexes may be conditioned by Pavlov's method, including responses of the skeletal muscles (knee-jerking and blinking) and of the smooth muscles and glands.

A similar approach was developed in Pavlov's work relating to human behaviour and the nervous system, all the time emphasizing the importance of conditioning. He deduced that the inhibitive behaviour of a psychotic person is a means of self-protection. The person shuts out the world and, with it, all damaging stimuli. Following this theory, the treatment of psychiatric patients in Russia involved placing a sick person in completely calm and quiet surroundings.

Pavlov's study of the normal animal in natural conditions enabled him to add greatly to scientific knowledge. He also demonstrated the necessity of providing the right situation for completely objective study and measurement of behaviour, and greatly improved operative and postoperative conditions for animals.

In 1897, Pavlov summarized his findings in his Nobel prizewinning work, *Lectures on the Work of the Principal Digestive Gland.*

Payne-Gaposchkin Cecilia Helena 1900–1979 was a British-born US astronomer and author whose interest in stellar evolution and galactic structure led to important research in the study of variable, binary and eclipsing stars.

Payne-Gaposchkin was born Cecilia Helena Payne in Wendover, Buckinghamshire, on 10 May 1900. She attended schools in Wendover and London before entering Newnham College at Cambridge University in 1919. Although her scientific interests were not at first clearly defined, contact with such eminent astronomers as Arthur Eddington (1882–1944) quickly induced her to choose astronomy as her main interest and career. After graduation in 1923, Payne-Gaposchkin went to the Harvard College Observatory in Cambridge, Massachusetts, to continue her studies under Harlow Shapley (1885–1972). For her research into stellar atmospheres she was awarded her PhD in 1925, and in 1927 was appointed an Astronomer at the Observatory. In 1938, she was made Phillips Astronomer at Harvard, before being awarded the Chair in Astronomy there in 1956 (the first woman to receive such an appointment at Harvard).

The author of many scientific papers, Payne-Gaposchkin also wrote a number of successful books on astronomical topics, ranging from introductory texts for the layperson to erudite academic monographs on specialized subjects. Her introductory textbook, first published in 1953, was revised with the help of her daughter in 1970. She died on 7 December 1979.

Stellar astronomy held a prominent place in Payne-Gaposchkin's research from an early stage in her career. One of her earliest significant findings, published in 1925, was the discovery of the relationship between the temperature and spectral class of a star. She continued to employ a variety of spectroscopic techniques in the investigation of stellar properties and composition, and her further investigation of stellar atmospheres during the 1920s led her to encounter some of the first indications of the overwhelming abundance of the lightest elements (hydrogen and helium) in the Galaxy.

During the 1930s Payne-Gaposchkin concentrated increasingly on the study of variable stars; she was particularly concerned to use the information she obtained towards an improvement in the understanding of galactic structure. Much of this work, especially the studies of the Large and Small Magellanic Clouds, was carried out in collaboration with her husband, S. Gaposchkin.

Other major areas of her interest included the devising of methods to determine stellar magnitudes, the position of variable stars on the Hertzsprung–Russell diagram, and novae.

Peano Giuseppe 1858–1932 was an Italian mathematician who applied the rigorous and axiomatic methods used in mathematics to his study of logic. He is chiefly remem-

bered for his concise logical definitions of natural numbers, devised – not entirely by himself – in order to derive a complete system of notation for logic, and for his discovery in analysis of a curve that fills topological space.

Peano was born on 27 August 1858 in Spinetta, near Cuneo. Living on a farm, he was educated locally until the age of 12 or 13, when he was sent to Turin to receive private lessons. In 1876, he won a scholarship to the Collegio delle Provincie, Turin University. On graduating, he joined the staff of the University and remained there for the rest of his life, first becoming a professor there in 1890. By that time he had already held a professorial appointment for four years teaching at Turin Military Academy; he retired from this concurrent post in 1901. During his lifetime he received several honours and awards, and participated actively within the Turin Academy of Sciences. He died of a heart attack on 20 April 1932 and, by his request, was buried in Turin General Cemetery; his remains, however, were removed to the family tomb in Spinetta in 1963.

Peano was a pioneer in symbolic logic and a fervent promoter of the axiomatic method – but he himself considered that his work in analysis was more important. It was carried out mainly in the 1880s while Peano was investigating the integrability of functions. In 1886, he was the first to show that the first-order differential equation $dy/dx = f(x,y)$ was solvable using only the one assumption that f is continuous. In 1890 he generalized this result, and in his published study gave the first explicit statement of the axiom of choice.

His first work in logic was published in 1888. It contained his rigorously axiomatically derived postulates for natural numbers which received considerable acclaim, although he studiously acknowledged his debt for some of the work to Richard Dedekind (1831–1916) who had also published during the same year. The postulates are nevertheless now known as the Peano axioms. They formed the basis on which Peano went on to found his system of mathematical notation for logic; *Formulario Mathematico*, comprising his work and that of collaborators, was published between 1895 and 1908, and contained 4,200 theorems. Later, Bertrand Russell was to say that he reached a turning point in his own life on meeting Peano; part of Peano's work was used in Russell's *Principia Mathematica.*

Peano also applied the axiomatic method to other fields, such as geometry, first in 1889 and again in 1894. A treatise on this work contained the beginnings of geometrical calculus.

His name is in addition particularly associated with the discovery of the 'Peano curves' that fill a space (such as a square), and that are used in topology. He provided new definitions of the length of an arc of a curve and of the area of a surface. He determined an error term in Simpson's formula and became interested in errors in numerical calculations; and he developed a theory of gradual operations which led to new methods for resolving numerical equations.

After 1900, he changed his interests slightly, and created

an international language that never really caught on. Between 1914 and 1919 he organized conferences for secondary mathematics teachers in Turin. Finally he became a mathematical historian and recorded many exact origins of mathematical terms and first applications of symbols and theorems.

Pearson Karl 1857–1936 was a British mathematician and biometrician who is chiefly remembered for his crucial role in the development of statistics as applied to a wide variety of scientific and social topics.

Pearson was born in London on 27 March 1857. He was tutored at home, except for a period between 1866 and 1873 when he attended the University College School. He began his university studies at King's College, Cambridge, in 1875, where an indication of Pearson's somewhat uncompromising and unconventional spirit is found in his successful pressuring of the authorities to abolish the mandatory classes in divinity for undergraduates. Pearson graduated with high honours in 1879, and was awarded a Fellowship at the College from 1880 to 1886 that gave him financial independence without obligation and enabled him to travel and study as he pleased. He visited universities in Germany, took a degree in law in 1881 (although he never practised), and was awarded his master's degree in 1882. In 1884 – still officially a Fellow at King's College – Pearson was named Goldsmid Professor of Applied Mathematics and Mechanics at University College, London; he was to hold this post until 1911 although his most productive work during the period was carried out elsewhere. He was also appointed Lecturer in Geometry at Gresham College, London, in 1891, which required him to give a short series of lectures each year. And it was from them on that Pearson became interested in the development of statistical methods for the investigation of evolution and heredity. His efforts in this aim were most fruitful, and were recognized by his election as Fellow of the Royal Society in 1896 (which awarded him its Darwin Medal in 1898). He then founded and became editor (until his death) of *Biometrika*, a journal established to publish work on statistics as applied to biological subjects. His work on eugenics led to his appointment as Head of the Laboratory of Eugenics at London University upon Francis Galton's retirement in 1906. In 1911, he became the first Galton Professor of Eugenics, a post he retained until 1933, when he retired to become Emeritus Professor. (His department was then split into two sections, one of which was headed by Pearson's son.) Pearson continued to work in his department until his death in Coldharbour, Surrey, on 27 April 1936.

During the early years of Pearson's career he did little work in mathematics, concentrating instead on law and political issues. His appointment to the Goldsmid Chair required him to focus his attentions on academic duties and on writing. A further marked change overtook his life in the early 1890s with the publication of Francis Galton's book *Natural Inheritance*, and with Pearson's exposure to the ideas of Walter Weldon, the newly appointed Professor of Zoology at University College.

Weldon was interested in the application of Galton's methods for correlation and regression to the investigation of the validity of Darwin's model of natural selection. Pearson threw himself into this project with great vigour, examining graphical methods for data presentation, studying probability theory and concepts such as standard deviation (a term he himself introduced in 1893, although the idea was by then nearly a century old), and more complex distribution patterns. He submitted many papers on statistical methods to the Royal Society of London, but encountered some stiff opposition to his mathematical approach to biological material. This prompted him to launch his own journal, *Biometrika*.

The major achievement in the early part of his investigation was the elucidation of a method for finding the values for the parameters essential for the description of a particular distribution, and also the classification of the different types of curves produced in the plotting of data into general types. This contributed to putting the Gaussian (or 'normal') distribution into more realistic perspective.

Pearson's discoveries included the Pearson coefficient of correlation (1892), the theory of multiple and partial correlation (1896), the coefficient of variation (1898), work on errors of judgement (1902), and the theory of random walk (1905). The last theory has since been applied to the study of random processes in many fields. In addition, Pearson's *Biometrika* for 1901 is a book of tables of the ordinates, integrals and other properties of Pearson's curves, and was of great practical use in rendering accessible statistical methods to a large number of scientists.

Perhaps the most familiar of Pearson's achievements was his discovery in 1900 of the χ^2 (chi-squared) test applied to determine whether a set of observed data deviates significantly from what would have been predicted by a 'null hypothesis' (that is, totally at random). Pearson also demonstrated that it could be applied to examine whether two hereditary characteristics (such as height and hair colour) were inherited independently.

Weldon's death in 1906 dealt a severe blow to Pearson's work in the field of mathematics as applied to biology; Pearson himself lacked the biological background to keep up with the increasingly sophisticated developments in the field of genetics. A great controversy had grown up around the approach of Weldon – who believed in gradual but continuous evolution – as against that of the followers of Gregor Mendel, such as William Bateson – who believed in intermittent variation. Pearson felt that Mendel's results were not incompatible with a statistical approach, although many Mendelians were convinced that they were. But it was the equally celebrated statistician Ronald Fisher (1890–1962) who was ultimately able to bring about the beginnings of a reconciliation between the two approaches.

During the rest of Pearson's career he concentrated on the establishment of a thriving department dedicated to the training of postgraduate students so that statistical techniques might be applied to subjects in many areas of academic study. He also worked on eugenics, examining the relative importance of environment and heredity in disorders such as tuberculosis and alcoholism, and in the incidence of infant mortality.

Peierls Rudolf Ernst 1907– is a German-born British physicist who made contributions to quantum theory and to nuclear physics.

Peierls was born in Berlin on 5 June 1907. He was educated at the Humboldt School, Oberschöneweide, Berlin and then at the University of Berlin. He studied with Arnold Sommerfeld (1868–1951) in Munich, with Werner Heisenberg (1901–1976) in Leipzig, and with Wolfgang Pauli (1900–1958) in Zurich. From 1929 to 1932, he worked as Pauli's research assistant at the Federal Institute of Technology in Zurich. He held a Rockefeller Fellowship in Rome and Cambridge between 1932 and 1933, and was an Honorary Research Fellow at Manchester University from 1933 to 1935. He then became an Assistant in Research for the Royal Society at the Mond Laboratory in Cambridge in 1935 and in 1937, he was appointed Professor of Mathematical Physics (formerly called Applied Mathematics) at the University of Birmingham. Peierls worked on the Atomic Project in Birmingham from 1940 to 1943 and in the United States from 1943 to 1946. He then returned to Birmingham until 1963, when he became Wykeham Professor of Physics at the University of Oxford. From 1974 to 1977, he was Professor of Physics at the University of Washington, Seattle. Peierls is now an Honorary Fellow of New College, Oxford. His many honours include a knighthood in 1968 and the award of the Royal Medal from the Royal Society in 1959, the Max Planck Medal of the Association of German Physical Societies in 1963, and the Enrico Fermi Award of the US Department of Energy in 1980.

Peierls began his research in 1928, a very exciting time in the development of quantum theory. The laws of quantum mechanics had been formulated consistently and now was the time to begin to apply them to the many unexplained phenomena in the fields of atomic, molecular and solid-state physics. His first work was on the theory of solids and concerned a phenomenon called the Hall effect. Then, in 1929, he developed the theory of heat conduction in nonmetallic crystals, concluding that the thermal conductivity of a large perfect crystal should grow exponentially at low temperatures. This was not verified experimentally until 1951. Peierls also proposed a general theory of the diamagnetism of metals.

In nuclear physics, Peierls contributed to the early theory of the neutron–proton system and, in 1938, gave the first complete treatment of resonances in nuclear collisions. During World War II, he was concerned with atomic energy. Uranium fission and the emission of secondary neutrons had been discovered and the possibility of releasing nuclear energy was of wide interest. In 1940, Otto Frisch (1904–1979) and Peierls showed that a simple estimate of the energy released in a chain reaction indicated that a fission bomb would make a weapon of fantastic power. They drew the attention of the British government to this in 1940, and Peierls was placed in charge of a small theoretical group concerned with isotope separation by more precise methods for evaluating the chain reaction and its efficiency. In 1943, when Britain decided not to continue its work on nuclear energy, Peierls moved to the United States to help in the work of the Manhattan Project, first in New York in consultation with designers of the isotope separation plans and then at Los Alamos.

Pelletier Pierre-Joseph 1788–1842 was a French chemist whose extractions of a whole range of biologically active compounds from plants founded the chemistry of the alkaloids, the most important of which he discovered was quinine.

Pelletier was born in Paris on 22 March 1788, the son of a distinguished chemist and pharmacist. He studied at the Ecole de Pharmacie, qualifying in 1810. He was awarded his doctorate in 1812 and three years later was appointed Assistant Professor of Natural History and Drugs at the Ecole. He was promoted to full Professor in 1825 and in 1832 made Assistant Director of the school itself. He was elected to the French Academy in 1840, the same year in which illness forced him to retire from the Ecole. He died in Paris on 19 July 1842.

In his early career, Pelletier was concerned with the analysis of gum resins and the colouring matter in plants. In 1813, working with the physiologist François Magendie (1783–1855), he produced reports on opopanax, sagapenum, asafoetida, myrrh, galbanum and caranna gum. His first major success came in 1817 when he discovered the emetic substance in ipecacuanha root; he called it emetine.

The following four years were particularly productive. Together with Joseph Caventou he investigated the action of nitric acid on the nacreous material of human gallstones and the green pigment in leaves, which they named chlorophyll. In 1818 Pelletier obtained crotonic acid from croton oil and analysed carmine from cochineal. Pelletier and Caventou then isolated ambrein from ambergris, but it was their discovery of plant alkaloids that brought international fame: strychnine in 1818, brucine and veratrine in 1819, and – most important of all – quinine in 1820. Quinine is the chief alkaloid in cinchona bark and for the next hundred years was the only effective treatment for malaria, representing the first successful use of a chemical compound in combating an infectious disease.

During the following 20 years Pelletier continued his alkaloid and phytochemical research. In 1823, with Jean Baptiste Dumas, he obtained firm evidence for the presence of nitrogen in alkaloids, something he had failed to confirm in earlier work with Caventou. He later carried out researches on strychnine and developed procedures for its extraction from nux vomica. He also made chemical examinations of upas, improved the method of manufacturing quinine sulphate, and isolated cahinca acid, the bitter crystalline substance from cahinca root.

In 1832, Pelletier reported his discovery of a new opium alkaloid, narceine; he also claimed to have been the first to isolate thebaine (which he called paramorphine). In partnership with Walter, he went on to study the oily hydrocarbons obtained by the destructive distillation of amber and bitumen. In a similar study (1837–1838) of an oily by-product of pine resin – used in the manufacture of illuminating gas – he discovered a substance he called retinaphte, later known as toluene (now methylbenzene).

Few people have discovered as many pharmaceutically important natural products as Pelletier did. Their powerful effects and their use in medical practice introduced specific chemical compounds into pharmacology instead of the imprecise plant extracts and mixtures used previously.

Pelton Lester Allen 1829–1918 was a US engineer who developed a highly efficient water turbine used to drive both mechanical devices and hydroelectric power turbines using large heads of water.

After spending his youth in Vermillion, Ohio, the 20-year-old Pelton, then a carpenter, went to California in search of gold. He failed (like thousands of others) in his quest, but it was while involved in gold-mining that he made a discovery which led him to his invention.

Waterwheels were used at the mines to provide power for machinery. The energy to drive these wheels was supplied by powerful jets of water which struck the base of the wheel on flat-faced vanes. In time, these vanes were replaced by hemispherical cups, with the jet striking at the centre of the cup. Pelton noticed that one of the water wheels appeared to be rotating faster than usual. It turned out that this was because the wheel had become loose on its axle, and the water jet was striking the inside edge of the cups, rather than the centre.

Pelton went away to reconstruct what he had seen, finding again that the wheel rotated more rapidly, and hence developed more power. Working on the construction of stamp mills at Camptonville in California, Pelton found that by using split cups the effect could be enhanced, and by 1879 he had tested a prototype at the University of California. This was so successful that he was awarded a prize. A patent was granted in 1889, and he later sold the rights to the Pelton Water Wheel Company of San Francisco.

By 1890, Pelton wheels developing hundreds of horsepower at efficiencies of more than 80% were in operation. Efficiencies of more than 90% were being achieved using the wheel in hydroelectric schemes of thousands of horsepower by the time of Pelton's death in 1910. The Pelton wheel remains the only hydraulic turbine of the impulse type in common use today.

Pennington Mary Engle 1872–1952 was a US chemist known chiefly for her pioneering work into food refrigeration.

Mary Pennington was born in Nashville, Tennesse on 8 October 1872. She gained an interest in chemistry at an early age and after graduating from high school in 1890 she went to the Towne Scientific School of the University of Pennsylvania where she studied chemistry and biology. She completed the course for a BS in 1892 but was denied a degree on account of her sex and was awarded a Certificate of Proficiency. She completed her PhD at the University of Pennsylvania in 1895, majoring in chemistry and then worked for two more years at the University as a fellow in botany. After a year at Yale as a fellow in physiological chemistry, Pennington returned to Philadelphia and in 1898 opened her own Philadelphia Clinical Laboratory. She was given a post as a lecturer at the Women's Medical College at Pennsylvania which she held until 1906. She passed civil service examinations and in 1907 was appointed as a bacterial chemist in the Bureau of Chemistry of the United States Department of Agriculture, being made head the new Food Research Laboratory in 1908. In 1919, she moved to New York City to the American Balsa Company, manufacturers of insulating materials, leaving after three years to be an independent consultant on the storage, handling and transportation of perishable goods. Pennington never retired and died in New York on 27 December 1952.

Pennington achieved renown for her pioneering work on food preservation. In Philadelphia, from research into the preservation of dairy products, she developed standards of milk inspection which were later used by health boards across the United States. During World War I she conducted experiments into railroad refrigeration cars and recommended the standards which remained in use into the 1940s. As a consultant, she turned her interest to frozen food, not only carrying out research into food processing but also designing both industrial and household refrigerators. She published widely in technical journals and government reports and co-authored the book *Eggs* (1933). Through her research, Pennington was awarded the Garvan medal, awarded annually to an American woman chemist, and was the first female member of the American Society of Refrigerating Engineers. She was vice president of the American Institute of Refrigeration.

Pennycuick Colin James 1933–   is a British biologist who is best known for his extremely detailed studies of flight.

Pennycuick was born in Virginia Water, Surrey, on 11 June 1933, the son of an army officer. He graduated from Merton College, Oxford, in 1956 and obtained his doctorate from Peterhouse College, Cambridge, in 1962. In 1964 he became a lecturer in zoology at Bristol University, with a break between 1971 and 1973 while he was researching at Nairobi University in Kenya.

Pennycuick's research is unusual in that it interrelates an extremely large number of factors and therefore gives a very detailed account of the various processes involved in flight. In flying vertebrates, for example, he has investigated the mechanics of flapping; the aerodynamic effects of the feet and tail; the physiology of gaseous exchange; heat disposal; the relationship between the size and anatomy of a flying creature and the power it develops; and the frequency of wing beats. In applying the results of his research to bird migration he discovered that many migratory birds have minimal energy reserves and must stop to feed at regular intervals. Therefore the destruction of the intermediate feeding places of these birds could lead to their extinction, even if their summer and winter quarters are conserved. Pennycuick has also hypothesized that migratory birds navigate using the Sun's altitude and its changing position.

Penrose Roger 1931–   is a British mathematician who, through his theoretical work, has made important contributions to the understanding of astrophysical phenomena. He has examined especially those anomalies in space–time, the singularities known as black holes, which occur when a sufficiently large mass is contained within a sufficiently small volume so that its gravitational pull prevents the escape of any radiation.

Penrose was born the son of an eminent British human geneticist in Colchester, Essex, on 8 August 1931. He grew up amid a family tradition of scholarship and creativity, and completed his education at University College, London. Even as he worked for his doctorate at Cambridge in 1957, Penrose and his father were devising geometrical figures, the construction of which is three-dimensionally impossible. (Published the following year in the *British Journal of Psychology*, they became well known when incorporated by the Dutch artist H. C. Escher into a couple of his disturbing lithographs.) A series of lecturing and research posts followed, both in Britain (London and Cambridge) and the United States (Princeton, Syracuse (New York) and Texas). In 1966, Penrose was made Professor of Applied Mathematics at Birkbeck College, London. In 1973 he became Rouse Ball Professor of Mathematics at Oxford University.

Penrose's early work in mathematics included the formulation of some of the fundamental theorems that describe black holes. The explanation of the occurrence of black holes in terms of gravitational collapse is now usually given in a form which owes a great deal to Penrose's work in stressing the importance of space–time geometry. A model of the behaviour of stars that collapse upon themselves had first been proposed by Oppenheimer and Snyder in 1939 and their results have been proved valid to a remarkable degree by later work. Their model of spherical collapse, together with an interest in gravitational collapse stemming from study of black holes, led to vigorous research on the dynamics and the inevitability of collapse to a singularity. The most important result of such research was a set of theorems formulated by Penrose and Stephen Hawking (1942–   ) in 1964, which extend the dynamics of simple spherical collapse to the much more complex situation of gravitational collapse. Singularities in any physical theory might naturally be taken to indicate the breakdown of the theory, but using techniques developed jointly with Hawking and Geroch, Penrose has established that once gravitational collapse has proceeded to a certain degree, assuming the truth of general relativity theory that gravitation is always attractive, singularities are inevitable. These techniques are now famous as the singularity theorems.

The existence of a trapped surface within an 'event horizon' (the interface between the black hole and space–time), from which little or no radiation or information can escape, implies that some events remain hidden to observers outside the black hole. But it remains unknown whether all singularities must be hidden in this way. Penrose has put forward the hypothesis of 'cosmic censorship' – that they are all so hidden – which is now widely accepted.

On moving to Oxford University, Penrose began developing an intuition that first occurred to him in Texas in 1964. This is a model of the universe whose basic building blocks are what he calls 'twistors'. The model arises in response to a dichotomy in physics, in that calculations in the macroscopic world of ordinary objects (including Einstein's theory of gravity and the general theory of relativity) use real numbers, whereas the microscopic world of atoms and quantum theory often requires a system using complex numbers, containing imaginary components that are multiples of the square root of -1. Penrose holds that, as everything is made up of atoms, and as energy exists as discrete quanta bundles, all calculations about both the macroscopic and microscopic worlds should use complex numbers. Logically to maintain such a hypothesis would require reformulation of the major laws of physics and of space–time.

Penzias Arno Allan 1933–   is a German-born US radio engineer who, with Robert Wilson (1936–   ), was the first to detect isotropic cosmic microwave background radiation. This radiation had been predicted on the basis of the 'hot big bang' model of the origin of the universe, and it represents some of the strongest evidence in favour of this model.

Penzias was born in Munich, Germany, on 26 April 1933. For political reasons his parents emigrated to the United States, taking their young son with them; all were later naturalized. Studying at the City College of New York (CCNY), Penzias earned his bachelor's degree in physics in 1954. He then continued his studies at Columbia University in New York, where he was awarded his master's degree in 1958 and his doctorate in 1962. Since 1961, Penzias has been associated with the Radio Research Laboratories of the Bell Telephone Company. From then to 1972 he was a staff member of the radio research department, and from 1972 to 1974 he was head of the technical research department. He then became head of the radio-physics research department before, in 1976, becoming Director of the Radio Research Laboratory. In 1979, he became Executive Director of Research and Communication Sciences at Bell Telephone Laboratories.

In addition to his posts in the telecommunications industry, Penzias has also concurrently held a series of academic positions. The first of these was as Lecturer in the Department of Astrophysical Science of Princeton University (1967–1982). He was appointed an Associate of the Harvard College Observatory in 1968, visiting Professor at Princeton University in 1972, adjunct Professor at the State University of New York (SUNY) at Stony Brook in 1975, and made Trustee of Trenton State College in New Jersey in 1976.

Penzias's many important contributions to the field of

radioastronomy have brought him widespread acclaim. He received the Henry Draper Medal of the National Academy of Sciences in 1977, the Herschel Medal of the Royal Astronomical Society in 1977, and in 1978, with Robert Wilson and Pyotr Kapitza (1894–1984), the Nobel Prize for Physics.

In 1963, Penzias and Wilson were assigned by the Bell Telephone Company to the tracing of radio 'noise' that was interfering with the development of a communications programme involving satellites. By May 1964 the two had detected a surprisingly high level of radiation at a wavelength of 7.3 cm/2.9 in, which had no apparent source (that is, it was uniform in all directions, or isotropic). They excluded all known terrestrial sources of such radiation and still found that the noise they were detecting was one hundred times more powerful than could be accounted for. They also found that the temperature of this background radiation was 3.5K (–269.7°C/–453.4°F), later revised to 3.1K (–270°C/–454.1°F).

They took this enigmatic result to Robert Dicke, Professor of Physics at Princeton University. Dicke was interested in microwave radiation and had predicted that this sort of radiation should be present in the universe as a residual relic of the intense heat associated with the birth of the universe following the 'hot big bang'. His department was then in the process of constructing a radio telescope designed to detect precisely this radiation. at a wavelength of 3.2 cm/1.3 in, when Penzias and Wilson presented their data.

Since then, background radiation has been subjected to intense study. Its spectrum conforms closely to a blackbody pattern and its temperature is now known to be just under 3K (– 270°C/–454°F). For cosmologists this constitutes the most convincing evidence in favour of the 'hot big bang' model for the origin of the universe, although it also raises some fundamental questions, such as the possible 'oscillation' of the universe between total contractions and total expansions.

Penzias's later work has been concerned with developments in radioastronomy, instrumentation, satellite communications, atmospheric physics and related matters. It was, however, for his work on the black-body background radiation that he received the Nobel prize.

Peregrinus Petrus born c.1220, was a medieval French scientist and scholar about whom little is known except for his seminal work – based largely on experiment – on magnetism. His real name was Peregrinus de Maricourt and he may have been given the name 'Peter the Pilgrim' because he was a crusader. He was an engineer in the French army under Louis IX and was active in Paris in the middle of the thirteenth century, where he advised Roger Bacon. He took part in the siege of Lucera in Italy in 1269 and in that year published his findings about magnets in *Epistola de Magnete*. In it he described a simple compass (a piece of magnetized iron on a wooden disc floating in water) and outlined the laws of magnetic attraction and repulsion. His ideas were taken up 250 years later by William Gilbert.

Perkin William Henry 1838–1907 was a British chemist who achieved international fame for his accidental discovery of mauve, the first aniline dye and the first commercially significant synthetic dyestuff.

Perkin was born in Shadwell, South London, on 12 March 1838, the son of a builder. His father wanted him to become an architect and he was educated at the City of London School, where he became interested in chemistry; encouraged by Thomas Hall, one of his teachers, he carried out experiments at home. At the early age of 15 he persuaded his father to let him enter the Royal College of Chemistry, London, and two years later became an assistant to August Hofmann, who was Professor of Chemistry there. Perkin's discovery of mauve occurred during the Easter vacation of 1856 when he was still only 18. With the help of his father he set up a factory to manufacture the dye, and so founded the British synthetic dyestuffs industry. He was knighted in 1906 on the fiftieth anniversary of his famous discovery. He died in Sudbury, Middlesex, on 14 July 1907.

In one of his early home experiments Perkin looked at the reduction products of dinitrobenzene and dinitronaphthalene and obtained a coloured substance initially named nitrosonaphthyline. It was the first example of the group of azo dyes produced from naphthalene.

In 1856, Perkin set himself the ambitious task of trying to synthesize quinine. He used chromic acid to oxidize toluidine (4-methylphenylamine) and obtained only a dirty dark precipitate. He then repeated the experiment using aniline (phenyl amine) and again produced a dark precipitate. But extracted with alcohol it gave an intensely purple solution, which contained the new dye which Perkin called aniline purple – later named mauve by French textile manufacturers. He sent a sample to the dyestuff company of Pullars in Perth, Scotland, who reported favourably once they had found a satisfactory mordant.

Despite advice to the contrary by Hofmann, Perkin decided to develop the dye himself commercially. His father put up the money and his brother, T. D. Perkin (1831–1891), helped with the laboratory work. Without any connections in the textile industry and with only a small experimental quantity of the dye, they built a new factory at Greenford Green (now Perivale), to the west of London in Middlesex, which was opened in 1859. Initially they had difficulty getting supplies of benzene (to make aniline, which was at that time a rare substance, found only in a few research laboratories) and nitric acid (to prepare nitrobenzene). Perkin patented the process, after establishing that someone under the age of 21 could do so; it was soon copied but involving more expensive oxidizing agents than the simple acidified potassium dichromate he used.

In the following years Hofmann also patented several commercial dyes based on the methylation and ethylation of magenta (discovered by Verguin in 1859), producing violets and rosanilines. At the company of Roberts and Dale in Manchester, H. Caro (1834–1911) discovered a new way of making mauve and with Martins introduced

Manchester brown and Martins yellow. Other new dyes included crysaniline, rosolic acid, aniline green, aniline black and diphenylamine blue. Perkin's chance discovery had resulted in a new dyestuffs industry based on coal tar (the source of benzene and aniline). Commercial demand for mauve died within ten years and it was superseded. Perkin's factory introduced new dyes based on the alkylation of magenta, and in 1868 he established a new route for the synthesis of alizarin (independently of Caro and Karl Graebe, who in the same year patented a process for making alizarin). The natural dye was derived from the madder plant, and within a few years the growing of the crop was no longer a commercial proposition in Europe.

The starting point for alizarin was anthracene, another coal-tar derivative. Eventually one tonne of alizarin was prepared during 1869, and by 1871 Perkin's company was producing one tonne every day. Perkin's business acumen was considerable and he became a wealthy man. He sold the factory and retired from industry in 1874 at the age of 36 to continue his academic research.

Even before this time Perkin had investigated various other organic compounds. In 1860, with B. F. Duppa (1828–1873), he established that glycine (aminoethanoic acid) can be obtained by heating bromoacetic acid (bromoethanoic acid) with ammonia. They also showed that tartaric acid (2,3-dihydroxybutanoic acid), fumaric acid and maleic acid (*trans-* and *cis-*ethene-1,2-dicarboxylic acids) are related and they synthesized racemic acid from dibromosuccinic acid (2,3-dibromobutan-1,4-dioic acid).

In the late 1860s, Perkin prepared unsaturated acids by the action of acetic anhydride on aromatic aldehydes, a method known as the Perkin synthesis. In 1868 he synthesized coumarin, the first preparation of a synthetic perfume. He also investigated the effects of magnetic fields on the chemical structures of substances.

In 1906, on the fiftieth anniversary of the discovery of mauve a Jubilee celebration was held at the Royal Institution in London, attended by major chemists from throughout Europe. Pride of place at the dinner which followed was a specimen of benzene first isolated by Michael Faraday in 1825, the parent substance upon which the dyestuffs industry was founded.

Perkin married twice and had three sons and four daughters. The eldest son, also named William Henry (1860–1929), became a Professor of Chemistry at the University of Manchester, where he established a research team devoted to organic chemistry; he later moved to Oxford University. Another son, Arthur George (1861–1937), was also a skilled organic chemist; between 1916 and 1937 he was Professor of Colour Chemistry and Dyeing at the University of Leeds.

Perrin Jean Baptiste 1870–1942 was a French physicist who made the first demonstration of the existence of atoms (by quantitative observation of Brownian motion). In recognition of this achievement, Perrin was awarded the 1926 Nobel Prize for Physics. He also contributed to the discovery that cathode rays are electrons.

Perrin was born in Lille on 30 September 1870. He was educated in Lyons and Paris and in 1891 entered the Ecole Normale Supérieure, gaining his doctorate six years later. He was then appointed to a readership in physical chemistry at the Sorbonne in 1897 and in 1910 he became Professor of Physical Chemistry there. Perrin retained this position until 1940, when his outspoken antifascism caused him to flee the German occupation. He went to New York, where he died on 17 April 1942.

Perrin's first work of significance was done in 1895 on the nature of cathode rays. The cathode rays generated in discharge tubes were allowed to penetrate thin sheets of glass or aluminium and collected in a hollow cylinder. In this way the rays were shown to carry a negative charge, an indication that they consisted of negatively charged particles. The rays could thus be retarded by an electric field, and Perrin carried out further experiments which included imposing a negative charge on a fluorescent screen onto which various rays were focused. As the negative charge was increased, the intensity of fluorescence fell. He was able in this way to establish crude values for e/m, the charge-to-mass ratio of an electron, which were improved upon by the classic studies of J. J. Thomson (1856–1940) in 1897, which established the existence of the electron.

Perrin's more important contribution to scientific knowledge was in papers published in 1909 on atomic theory. Perrin extended the work of Robert Brown (1773–1858), who in 1827 had reported that pollen grains suspended in water and observed by microscope appeared to be in rapid random motion. It was believed that Brownian motion occurs because the tiny pollen grains are jostled by moving water molecules. Perrin suggested that, this being the case, the principles of the kinetic theory of gases were applicable as the grains and water molecules would both behave like gas molecules even though the pollen grains were much greater in size than water molecules.

The experiments that Perrin performed to measure Brownian motion are classics. One system involved gamboge (gum resin) particles obtained from vegetable sap, and required the isolation of 0.1 g/0.035 oz of particles of the necessary size from a sample weighing 1 kg/2.2 lb the isolation took several months. In the experiment, the distribution of suspended particles in a container was analysed by depth. It was found that their number decreases exponentially with height and, using principles proposed by Albert Einstein (1879–1955), Perrin was able to deduce a definite value for Avogadro's number which agreed substantially with experimental values obtained in other ways. This showed that Perrin's assumption was correct and that Brownian motion is due to molecular bombardment. Perrin's work on Brownian motion came as close as was possible then to detecting atoms without actually seeing them and it was accepted as a final proof of the existence of atoms.

Perutz Max Ferdinand 1914– is an Austrian-born British molecular biologist who shared the 1962 Nobel Prize for Chemistry for his solution of the structure of the haemoglobin molecule; his co-worker John Kendrew, who had determined the structure of myoglobin, was the other winner of the prize.

Perutz was born in Vienna on 19 May 1914. Both his parents came from families of textile manufacturers and expected their son to study law before entering the family business. But at school at the Theresianum in Vienna he became interested in chemistry and in 1932 entered the University of Vienna to study the subject. A course in organic biochemistry, given by F. von Wessely, fired his imagination and after graduation he tried (but failed) to get a place at Cambridge University to study under Frederick Gowland Hopkins.

In 1936, Perutz became a research student at the Cavendish Laboratory, Cambridge, where he worked on X-ray crystallography under John Bernal (1901–1971); he has remained at Cambridge for the rest of his academic career. In 1939, he received a grant from the Rockefeller Foundation and was appointed research assistant to William Lawrence Bragg; he gained his PhD a year later. Perutz continued his researches and in 1947 was appointed head of the newly constituted Molecular Biology Unit of the Medical Research Council. In 1957, he formally proposed to the Council the idea of a new laboratory, backed by N. F. Mott, Bragg's successor as Cavendish Professor. Five years later Perutz became Chairman of the new Laboratory of Molecular Biology and held this post until his retirement in 1979.

Perutz first applied the methods of X-ray diffraction to proteins at the Cavendish Laboratory. Following a conversation with F. Haurowitz in Prague in 1937, he began work on determining the structure of haemoglobin. There were enormous difficulties and it was not until 16 years later, in 1953, that he discovered a suitable method. He found that if he added a single atom of a heavy metal such as gold or mercury to each molecule of protein the diffraction pattern was altered slightly. Kendrew, who had joined Perutz in 1945, used a similar technique for the smaller molecule of myoglobin.

Using high-speed computers, which were just becoming available, they analysed hundreds of X-ray pictures and in 1958 Perutz published his first findings on the structure of haemoglobin. By 1960 they had worked out the precise structures of both proteins. Haemoglobin turned out to have 574 amino-acid units in four folded chains, each similar to the single chain of myoglobin. Their basic helical structure confirmed the prediction made ten years earlier by Linus Pauling that protein strands are twisted.

In his later work on haemoglobin Perutz tried to interpret the mechanism by which the molecule transports oxygen in the blood in terms of its molecular structure. He became especially interested in the effect of the protein globulin on the iron-containing haem group, which is related to the effect of protein on the catalytic properties of metals and coenzymes.

The sequences of the 20 different amino-acid residues along the α and β globin chains are determined genetically; occasionally mutations lead to an alteration in one of the sequences. People who carry some of these mutations may suffer from a lack of red blood cells (anaemia) or have too many red cells, because the stability of the oxygen-carrying mechanism of the haemoglobin molecule is impaired. This was the first time that the symptoms of an inherited disorder had been interpreted in terms of the molecular structure of a biochemical. It holds the hope that a treatment may be found for the most common inherited haemoglobin disorder, sickle-cell anaemia.

Petit Alexis-Thérèse 1791–1820 was a French scientist who worked mainly in physics but whose collaboration with Pierre Dulong resulted in a discovery that was to play an important part in chemistry in the determination of atomic weights (relative atomic masses).

Petit was born in Vesoul, Haute-Saône, on 2 October 1791. He went to school at Besançon and at the age of only 16 entered the Ecole Polytechnique in Paris. He graduated in 1809 – after only two years – and a year later was appointed Professor of Physics at the Lycée Bonaparte. He was awarded his doctorate in 1811 for a thesis on capillary action. In 1814, he became an assistant professor at the Ecole, succeeding to a full professorship a year later. The last years of his life were darkened by the death in 1817 of his wife (sister of the physicist Dominique Arago) and by illness. He contracted tuberculosis and died, in Paris, on 21 June 1820 in only his 29th year. He was succeeded at the Ecole Polytechnique by his friend and colleague Dulong.

Petit's early research was conducted in collaboration with his brother-in-law Dominique Arago (1786–1853). They examined the effect of temperature on the refractive index of gases. Their results led Petit to doubt the validity of the then accepted corpuscular theory of light and to become an early supporter of the wave theory.

In 1815, the offer of a prize in a scientific competition on the measurement of temperature and cooling laws stimulated Petit and Dulong to begin their fruitful, albeit short, collaboration. Their results established the importance of the gas thermometer (they won the competition and were awarded the prize in 1818). They continued working in this area, examining the specific heats (specific heat capacities) of various solids and in 1819 announced the famous Dulong–Petit law of atomic heats. This stated that, for most solid elements, the product of the specific heat and atomic weight (termed the atomic heat) is a constant, equal to 5.97. The modern expression of the law is that the product of the specific heat capacity and relative atomic mass is approximately constant and equal to three times the universal gas constant (R), or 25.07 joules per mole per Kelvin ($J\ mol^{-1}\ K^{-1}$).

The law applies at room temperature only, and not to lighter solid elements such as boron and carbon; the constant tends to zero as temperature falls towards absolute zero. Chemists who at that time were having difficulty determining atomic weights (and distinguishing them from equivalent weights) now had a method of estimating the approximate weight merely by measuring the specific heat of a sample of the element concerned.

Pfeffer Wilhelm Friedrich Philipp 1845–1920 was a German physiological botanist who is best known for his contributions to the study of osmotic pressures, which is important in both biology and chemistry.

Pfeffer was born in Grebenstein, near Kassel, on 9 March 1845, the son of a pharmacist. He went to Göttingen University to study botany and chemistry, and gained his doctorate in 1865. He then went to Marburg University, where he spent several years studying botany and pharmacy. From 1867 to 1870 he studied botany as a private assistant to Nathanael Pringsheim (1823–1894), an algae botanist in Berlin, and then from 1870 to 1871 he worked under the plant physiologist Julius von Sachs (1832–1897) at the University of Würzburg. In 1871 Pfeffer returned to Marburg as a privatdozent (an official but unpaid lecturer) and in 1873 he became Extraordinarius at Bonn University, a post he held until 1877, when he was appointed to a professorship at Basel University. In the following year Pfeffer accepted a professorship at Tübingen University, then in 1887 became a professor at Leipzig University. He died on 31 January 1920 in Leipzig.

Pfeffer made the first ever quantitative determinations of osmotic pressure in 1877. The apparatus he used consisted of a semipermeable container of sugar solution immersed in a vessel of water. He connected a mercury-filled manometer to the top of the semipermeable container to measure the osmotic pressure after the solute and water had reached equilibrium. Using these pressure measurements he also showed that osmotic pressure varies according to the temperature and concentration of the solute. Other scientists later made independent determinations of osmotic pressure and confirmed his results. In addition, Jacobus van't Hoff established that the osmotic pressure of a solution is analogous to gaseous pressure: a solute (if not dissociated) exerts the same osmotic pressure as the gaseous pressure it would

exert if it were a gas occupying the same volume at the same temperature. Pfeffer's work on osmosis led to the modern understanding of osmometry and was of fundamental importance in the study of cell membranes, because semi-permeable membranes surround all cells and play a large part in controlling the internal environment of cells.

In addition to his work on osmosis, Pfeffer also studied respiration, photosynthesis, protein metabolism, and transport in plants. He published more than 100 scientific papers and books, and his three-volume *Handbuch der Pflanzenphysiologie* (translated in 1906 as *Physiology of Plants*) was an important text for many years.

Piaget Jean 1896–1980 was a Swiss psychologist famous for his pioneering studies of the development of thought processes, particularly in children. He is generally considered to be one of the most important figures in modern developmental psychology and his work has had a great influence on educational theory and child psychology. He received many international honours for his work, including seven scientific prizes and 25 honorary degrees.

Piaget was born on 9 August 1896 in Neuchâtel, Switzerland. He published his first scientific article (about an albino sparrow) when he was only 10 years old, and by the age of 15 he had gained an international reputation for his work on molluscs. Subsequently he studied at the universities of Neuchâtel, Zurich and Paris, obtaining his doctorate from Neuchâtel in 1918. His interest then turned to psychology and he spent two years at the Sorbonne researching into the reasons why children fail intelligence tests. The results of this research gained him the directorship of the Institut J. J. Rousseau in Geneva in 1921. During his subsequent career Piaget held many academic positions, some of which were concurrent. He was Professor of Philosophy at Neuchâtel from 1926 to 1929; Professor of Child Psychology and History of Scientific Thought at Geneva University from 1929 to 1939; Director of the Institut Universitaire des Sciences de l'Education in Geneva from 1933 to 1971; Professor of Psychology and Sociology at Lausanne University from 1938 to 1951; Professor of Sociology (1938 to 1952) and of Experimental Psychology (1940 to 1971) at Geneva University; Professor of Genetic Psychology at the Sorbonne from 1952 to 1963; and Director of the International Bureau of Education in Geneva from 1929 to 1967. In 1955, with the help of the Rockefeller Foundation and the Swiss National Foundation for Scientific Research, Piaget founded the International Centre of Genetic Epistemology at Geneva University, which he continued to direct after he retired in 1971. He also held several positions with UNESCO at various times during his life. Piaget died in Geneva on 16 September 1980.

Piaget's work on concept formation in children falls into two main phases: an early phase (from 1924 to 1937) in which he established the basic differences between thought processes in children and those in adults, and a late phase (after 1937) in which he carried out detailed investigations of thought development and evolved his theories about

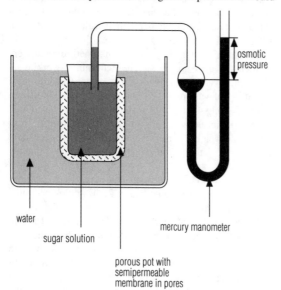

osmotic pressure

water

sugar solution

mercury manometer

porous pot with semipermeable membrane in pores

Pfeffer *Pfeffer's apparatus for measuring osmotic pressure.*

concept formation in children – his best known work.

In his early work Piaget showed how radically different are the mental processes of children from those of adults: according to his theory – which resembles Sigmund Freud's ideas about the development of the id, ego and superego – children's mental processes are dominated by an egocentric attitude, being influenced mainly by the wishes and inner needs of the child, but as the child develops, its thinking becomes increasingly dominated by the influence of the external environment.

After 1937 Piaget carried out much more rigorous investigations into the origin and development of logical and mathematical concepts in children, and attempted to trace the growth of reasoning capacities from birth to maturity. In developing his famous theory of concept development, Piaget invented a new type of logic (called psycho-logic) in an attempt to apply pure logic to experimental psychology, and as a result of this his writings are highly technical. Stated simply, however, Piaget's theory postulates four main stages in the development of mental processes: sensorimeter, preoperational, concrete operational and formal operational.

In the sensorimeter stage, which lasts from birth to the age of about two years, infants obtain a basic knowledge of objects by empirical experimentation. Gradually a child forms concepts of objects, and learns that they continue to exist even when out of sight. The preoperational stage lasts from the age of two years to the age of seven years. In this stage a child learns to imitate and begins to represent concrete objects with words – language starts to develop. From seven to 12 years old a child is in the concrete operational stage; he or she develops the concept of number, begins to classify objects according to their similarities and differences, and can distinguish between past and present. Finally, from 12 onwards, a child is in the formal operational stage, which is characterized by the development of logical thought and mathematical ability; thinking also becomes more flexible – hypotheses are formed and experimented with, for example.

Although it has been criticized for being based on observations of only a small number of subjects, Piaget's work on the development of mental processes is generally considered to be a major achievement and has greatly influenced child psychology and educational theory and practice, particularly the teaching of mathematics.

Piazzi Giuseppe 1749–1826 was an Italian monk originally trained in theology, philosophy and mathematics, who nevertheless was put in charge of an observatory, where he carried out astronomical studies of considerable importance. He is, for example, credited with the discovery of the first asteroid, or 'planetoid', as he more logically termed it.

Piazzi was born in Ponte di Valtellina (then part of Italy, now in Switzerland) on 16 July 1746. He studied in various Italian cities and in 1764 entered the Theatine Order in Milan, where he lived as a monk for several years. He continued his studies of philosophy and mathematics in Milan

and Rome and was awarded a doctorate in both subjects. For ten years after 1769 he worked as a teacher in Genoa and in Malta, before he became Professor of Higher Mathematics at the Palermo Academy in Sicily. During the latter part of the 1780s the Bourbons, then the rulers of the independent Kingdom of Naples, decided to establish observatories in Palermo and Naples. Put in charge of the one at Palermo, Piazzi travelled to observatories in England and France to obtain advice and equipment, and met such great astronomers as William Herschel, Jesse Ramsden and Nevil Maskelyne. He examined their equipment and commissioned from Ramsden a 1.5-m/5-ft vertical circle that he intended to use to determine star positions. The device was installed in Palermo in 1789 and still exists.

The Palermo Observatory opened in 1790 and Piazzi served as its Director until his death. He conducted many astronomical studies and became a Fellow of the Royal Society of London (1804). He also took on additional responsibilities, including the reformation of the Sicilian system of weights and measures (1812), and finally in 1817 he was put in charge of the other observatory at Naples. For a time Piazzi split his time between the two observatories, but eventually he moved to Naples in 1824 because his health was failing. He died in Naples on 22 July 1826.

Piazzi's first astronomical publication appeared in 1789, but his great project on mapping the positions of the fixed stars did not begin until the 1790s. He was fortunate in working at the southernmost observatory in Europe in a favourable climate. These conditions, together with the quality of his equipment, enabled him to produce new and accurate measurements.

Piazzi was examining the apparent 'gap' in the Solar System between Mars and Jupiter, long a source of speculation for astronomers, when, on 1 January 1801, he detected a faint body that had not previously been noted. He followed it for six weeks, until it could no longer be detected because of its position relative to the Sun. Using Piazzi's data, the German Karl Gauss (1777–1855) managed to calculate the orbit of the body, and sent his prediction to Baron von Zach of the Gotha Observatory. The body was rediscovered by Heinrich Olbers (1758–1840), just where Gauss had predicted it would be found; its dimness, considering its distance from the Earth, indicated that it was very small. (Herschel calculated that it was only 320 km/199 mi in diameter and did not therefore warrant being called a planet. He proposed the name 'asteroid', a term that became popular, although it is now no longer quite so accepted a term.) The body that Piazzi discovered was named Ceres, and is now known to have a diameter of 780 km/485 mi. Three more 'asteroids' were discovered before Piazzi's death, and there are probably in all more than 40,000 of them. When the thousandth was discovered it was named Piazzia as a tribute to him.

In 1803, Piazzi published his first catalogue of fixed stars, which located 6,748 stars with unprecedented accuracy. His second catalogue, produced in collaboration with N. Cacciatore, appeared in 1813. It described 7,646 stars. Both publications won prizes.

Picard Charles Emile 1856–1941 was a distinguished French mathematician whose work in analysis – and particularly in analytical geometry – brought him deserved fame. Responsible for the formulation of 'Picard's little theorem' and 'Picard's big theorem', he was also an excellent teacher, interested in applying mathematical principles as much as possible to other branches of science, particularly physics and engineering.

Picard was born on 24 July 1856 in Paris, where his father was the director of a silk factory. He showed talent as a pupil at the Lycée Henry IV, where his excellent memory contributed to his outstanding results. Completing his education, he entered the Ecole Normale Supérieure in 1874. Within three years he had already made some important algebraic discoveries, been consistently placed first among his contemporary fellow students, and had earned his doctorate. He was then retained as an assistant instructor at the Ecole between 1877 and 1878. In 1879 he was appointed professor (at the age of 23) in Toulouse. Two years later, however, he returned to Paris as Lecturer in Physics and Experimental Mechanics at the Sorbonne and, simultaneously, again at the Ecole Normale Supérieure. Also in 1881 his name was put forward for election to the Paris Academy of Sciences (although his election did not actually occur until 1889). In 1885, he took the Chair of Differential and Integral Calculus at the Sorbonne, and served as his own assistant until he reached the prescribed age of 30, at which he was officially able to hold the post. In 1897, at his own request, he exchanged this post for the Chair of Analysis and Higher Algebra because he wanted a position in which he could train students for research. He was made a member of the Académie Française in 1924, and received the Grande Croix de la Légion d'Honneur in 1932. He also won the Mittag-Leffler Gold Medal from the Swedish Academy of Sciences, received honorary degrees from five foreign universities, and was a member of 37 Academies and learned societies. Highly respected for his administrative capability, and an excellent teacher, Picard died on 11 December 1941 in the Palais de l'Institut in Paris, where he was living as permanent secretary.

Picard's work was mainly in the fields of mathematical analysis and algebraic geometry. In 1878 he studied the integrals of differential equations by making successive substitutions with equations having suitable partial derivatives. A year later, he proved the theorem now known by his name, that an integral function of the complex variable takes every finite value, with one possible exception. He expressed it in this way:

Let $f(z)$ be an entire function. If there exist two values of A for which the equation $f(z) = A$ does not have a finite root, then $f(z)$ is a constant. From this it follows that if $f(z)$ is an entire function that is not a constant, there cannot be more than one value of A for which $f(z) = A$ has no solution.

In the following year he stated a second theorem:

Let $f(z)$ be a function, analytic everywhere except at a where it has an essential isolated singularity; the equation $f(z) = A$ has in general an infinity of roots in any neighbourhood of a. Although the equation can fail for certain exceptional values of the constant A, there cannot be more than two such values.

From these results, generalizations were worked out that are now known as 'Picard's little theorem' and 'Picard's big theorem'.

Picard created a theory of linear differential equations analogous to the Galois theory of algebraic equations. (This work was later extended by his pupil, Ernest Vessiot (1865–1952).) His work on the integrals attached to algebraic surfaces, together with the associated topological questions, developed into an area of algebraic geometry that had applications in topology and function theory. Much of Picard's work was recorded in a three-volume book entitled *Traité d'analyse*.

He also applied his method of analysis to the theories of elasticity, heat and electricity, in theoretical physics, and produced a solution to the problem of the propagation of electrical impulses along a cable. When he was over 80 he presented a paper to the Academy of Sciences on questions of homogeneity and similarity encountered by physicists and engineers.

Pickering Edward Charles 1846–1919 was a US astronomer, one of the most famous and hard-working of his time, who was a pioneer in three practical areas of astronomical research: visual photometry, stellar spectroscopy and stellar photography. As Director of the Harvard College Observatory for more than 40 years, he was instrumental in educating and inspiring an entire generation of young astronomers; unusually for his generation, he was also keen to encourage women to take up astronomy as a career.

Pickering was born on 19 July 1846 in Boston, Massachusetts. He began his academic career as an Assistant Instructor of Mathematics at the Lawrence scientific school at Harvard, but after two years he was appointed Assistant Professor of Physics at the newly founded Massachusetts Institute of Technology. During his subsequent ten years in this post he revolutionized the teaching of physics. Then, in 1876, he was appointed Director of the Harvard College Observatory, a post he was to hold for 42 years. In that time he received honorary doctorates from six US and two European universities; in addition he was awarded the Rumford Gold Medal of the American Academy of Arts and Sciences and was twice a recipient of the Gold Medal of the Royal Astronomical Society. He died in Cambridge, Massachusetts, on 3 February 1919.

As a basis for the photometric work he carried out, Pickering made two critical decisions. First, he adopted the magnitude scale suggested by Norman Pogson (1829–1891) in 1854, on which a change of one magnitude represented a change of a factor of 2.512 in brightness. Second, choosing the Pole Star (Polaris), then thought to be of constant brightness, as the standard magnitude and arbitrarily assigning a value of 2.1 to it, he redesigned the photometer to reflect a number of stars round the meridian at the same time so that comparisons were immediately visible. The photometric work that followed continued for nearly a quarter of a century. (Unfortunately Polaris has since been found to vary in brightness to a small degree.)

The first great catalogue of magnitudes, containing 4,260 stars, was published in 1884. It was known as the *Harvard Photometry*. Pickering never wearied of the routine work this procedure involved; he is estimated to have made more than 1.5 million photometric readings. The brightness of every visible star was measured, then taken again to be sure of the greatest possible accuracy. The photometric studies culminated in 1908 with the publication of the *Revised Harvard Photometry*. Printed as Volumes 50 and 54 of the *Annals* of Harvard College Observatory, it tabulates the magnitudes of more than 45,000 stars brighter than the seventh magnitude. It remained the standard reference until photometric methods had largely supplanted visual ones.

A further production of the Harvard College Observatory was the *Henry Draper Catalogue*, a classification of stellar spectra. Pickering's researches into stellar spectroscopy were made possible by a practical invention of his whereby the spectra of a number of stars could be surveyed simultaneously and by the establishment of the Henry Draper fund in 1886. Draper's widow supplied the financial backing for Pickering and his assistants to photograph, classify and measure the spectra of the stars and publish the resulting catalogue in the *Annals* as a memorial volume to her husband. Finally issued in 1918, the complete *Draper Catalogue* contained the spectra of no fewer than 225,000 stars, classified according to the new, improved alphabetical system devised by Pickering's pupil and colleague, Annie Cannon (1863–1941).

Pickering's interest in stellar photography was responsible for the first *Photographic Map of the Entire Sky*, published in 1903. It comprised 55 plates of stars down to the twelfth magnitude, taken both at Harvard and at its sister station in the southern hemisphere, at Arequipa in Peru, where Pickering's younger brother, William (1858–1938), was Director. (The Pickerings had established the Peruvian observatory in 1891.) In addition, Pickering photographed large areas of the sky on clear nights, building up a 300,000-plate Harvard photographic library that has since proved invaluable to astronomers searching for changes in the brightness and position of celestial objects.

One of the most important products of Pickering's researches was the creation of the astronomical colour index: a measure of the apparent colour of a star and thus of its temperature. Cooler stars emit more light at longer wavelengths and so appear redder than hot stars. The colour index is expressed as the difference in a star's brightness when measured on two selected wavelengths. The international colour index, defined by Pickering in about 1890, is the difference between the photographic magnitude (blue light) and the photovisual magnitude (yellow light); it is zero for white stars, positive for red stars, and negative for blue stars. Magnitudes are now seldom measured photographically; instead, colour filters on photoelectric cells measure the colour index between the two wavelengths. The widely used UBV system utilizes the ultraviolet, blue and yellow (visual) images.

Pippard Alfred Brian 1920–   is a British physicist who has carried out important work in superconductivity.

Pippard was born on 7 September 1920 in London. He was educated at Clifton College, Bristol and then at Clare College, Cambridge, obtaining his BA in 1941. He then worked as a Scientific Officer at the Radar Research and Development Establishment at Great Malvern until 1945, when he returned to Cambridge. He gained his PhD in 1949, then became a Lecturer in Physics in 1950 and a Reader in 1959. He was then appointed John Humphrey Plummer Professor of Physics at Cambridge in 1960 and in 1971 he became Cavendish Professor of Physics there. Pippard's honours include a knighthood in 1975, and the award of the Hughes Medal of the Royal Society in 1958.

Pippard is the son of an eminent professor of engineering, and as a boy he was attracted by the fascination of low-temperature physics. He was deterred from this though, thinking his mathematical ability was inadequate. He intended to become a chemist but after graduating in 1941, his work on the design of radar aerials during the war deflected him towards physics. After the war, he became involved in low-temperature work and decided to concentrate his efforts on the task of applying microwaves to the study of superconductors. He worked on the way in which electric currents flow without resistance in a thin layer at the surface of the metal. He measured the thickness (about $1000 \text{ Å}/10^{-7}$ m) of this penetration layer and examined variations with temperature and purity. He found that when he tried to change the properties at one point by applying a disturbance, he influenced the metal over a distance which in pure metals is usually greater than the penetration layer thickness. Because of this, he said that the electrons of superconductors possess a property which he called coherence. From this starting point, he worked out an equation relating current to magnetic field. Similar work was being done at the same time in Russia by Lev Landau (1908–1968) and others. In 1957, a definitive theory of superconductivity was derived by John Bardeen (1908–1991), Leon Cooper (1930–) and John Schrieffer (1931–) and it was found to provide a consistent explanation for both Pippard's and Landau's work. Soon after this, Pippard's guess that impurities in the metal could shorten the coherence length was confirmed, and 'dirty' superconductors were produced with the important technological property of carrying currents and generating extremely strong magnetic fields without their resistance reappearing.

Pippard also discovered that absorption of microwaves at the surface of a given metal at low temperatures is governed by one particular characteristic of the conduction electrons – the shape of the Fermi surface. This initiated experiments and theoretical work in many laboratories, which transformed understanding of the dynamical laws governing the motion of electrons in metals. It also clarified the understanding of the way in which one metal differs from another in the details of this motion.

Pixii Hippolyte 1808–1835 was a French inventor who made the first practical electricity generator.

Following Michael Faraday's announcement to the Royal Society (on 24 November 1831) of his discovery of electromagnetic induction and his suggestions for making a simple dynamo, Pixii (an instrument maker, who had learned the craft from his father) set out to construct a practical electricity generator. Shortly afterwards he made a device that consisted of a permanent horseshoe magnet, rotated by means of a treadle, and a coil of copper wire above each of the magnet's poles. The two coils were linked and the free ends of the wires connected to terminals, from which a small alternating current was obtained when the magnet rotated. This device was first publicly exhibited at the French Academy of Sciences in Paris on 3 September 1832. Later, at the suggestion of André-Marie Ampère, a commutator (a simple switching device for reversing the connections to the terminals as the magnet is rotated) was fitted so that Pixii's generator could produce direct-current electricity. This revised generator was taken to England in November 1833 by Count de Predevalli and exhibited in London. Pixii himself died two years later, only 27 years old, in 1835.

Although Pixii's machine generated only a small current, it was nevertheless the first practical electricity generator and therefore the forerunner of all modern generators.

Planck Max Karl Ernst Ludwig 1858–1947 was a German physicist who discovered that energy consists of fundamental indivisible units that he called quanta. This discovery, made in 1900, marked the foundation of the quantum theory that revolutionized physics in the early 1900s. For this achievement, Planck gained the 1918 Nobel Prize for Physics.

Planck was born at Kiel on 23 April 1858. In 1867, his family moved to Munich, where Planck studied at the Maximilian Gymnasium before entering the University of Munich in 1874. Planck studied mathematics and physics, spending some time in 1877 and 1878 with Gustav Kirchhoff (1824–1887) and Hermann Helmholtz (1821–1894) at the University of Berlin. Planck gained his PhD at Munich with a dissertation on thermodynamics in 1879 and became a lecturer there in the following year. In 1885, he was appointed Extraordinary Professor of Physics at Kiel and then in 1888 moved to Berlin, where he became Assistant Professor of Physics and Director of the newly founded Institute for Theoretical Physics. Planck rose to become Professor of Physics at Berlin in 1892, a position he retained until 1926. In 1930, he was appointed President of the Kaiser Wilhelm Institute but resigned in 1937 in protest at the Nazi's treatment of Jewish scientists. In 1945, the Institute was renamed the Max Planck Institute and moved to Göttingen. Planck was reappointed its President, a position he retained until he died at Göttingen on 4 October 1947.

In 1862, Kirchhoff had introduced the idea of a perfect black body that would absorb and emit radiation at all frequencies, reaching an equilibrium that depended on temperature and not the nature of the surface of the body.

A series of investigations were then undertaken into the nature of the thermal radiation emitted by black bodies following the discovery by Josef Stefan (1835–1893) in 1879 that the total energy emitted is proportional to the fourth power of the absolute temperature. Measurements of the frequency distribution of black-body radiation by Wilhelm Wien (1864–1928) in 1893 produced the result that the distribution is a function of the frequency and the temperature. A plot of the energy of the radiation against the frequency resulted in a series of curves at different temperatures, the peak value of energy occurring at a higher frequency with greater temperature. This may be observed in the varying colour produced by a glowing object. At low temperatures, it glows red but as the temperature rises the peak energy is emitted at a greater frequency, and the colour become yellow and then white.

Wien found an expression to relate peak frequency and temperature in his displacement law, and then attempted to derive a radiation law that would relate the energy to frequency and temperature. He discovered a radiation law in 1896 that was valid at high frequencies only, while Lord Rayleigh (1842–1919) later found a similar equation that held for radiation emitted at low frequencies. Planck was able to combine these two radiation laws to arrive at a formula that represented the observed energy of the radiation at any given frequency and temperature. This entailed making the assumption that the energy consists of the sum of a finite number of discrete units of energy that he called quanta, and that the energy ε of each quantum is given by the equation $\varepsilon = h\nu$, where ν is the frequency of the radiation and h is a constant now recognized to be a fundamental constant of nature and called Planck's constant. By thus directly relating the energy of a radiation to its frequency, an explanation was found for the observation that radiation of greater energy has a higher frequency distribution.

Classical physics had been unable to account for the distribution of radiation, for Planck's idea that energy must consist of indivisible particles was revolutionary, totally contravening the accepted belief that radiation consisted of waves. But it found rapid acceptance because an explanation for photoelectricity was provided by Albert Einstein (1879–1955) in 1905 using Planck's quantum theory, and in 1913, Niels Bohr (1885–1962) applied the quantum theory to the atom and evidence was at last obtained of the behaviour of electrons in the atom. This was later developed into a full system of quantum mechanics in the 1920s, when it also became clear that energy and matter have both a particle and wave nature. Thus the year 1900 marked not only the beginning of a new century but, with the discovery of the quantum theory, the end of the era of classical physics and the founding of modern physics.

Plaskett John Stanley 1865–1941 was a Canadian engineer whose work in instrument design and telescope construction led to his becoming Director of the Observatory in Victoria, British Columbia. There, he used the large reflecting telescope that he designed to carry out important research into binary stars and stellar radial velocities.

Plaskett was born at Hickson, near Woodstock, Ontario, on 17 November 1865. After completing his schooling locally, he was employed as a mechanic in various parts of North America. In 1889, however, he became a mechanic in the Department of Physics at Toronto University, where he decided to take up undergraduate studies; he eventually became a Lecturer there. From 1903 Plaskett was in charge of astrophysical work at the new Dominion Observatory in Ottawa and he initiated comprehensive programmes of research into stellar radial velocities using the observatory's 38-cm/15-in reflector. The spectroscope he produced for the reflector so improved the instrument that it was comparable with the best in North America. Having repeatedly urged the Canadian Parliament to sanction the construction of a 1.8-m/72-in reflector, Plaskett was finally appointed to supervise its creation for the Dominion Astrophysical Observatory in Victoria; he was also appointed the observatory's first Director in 1917, and remained there until he retired in 1935, at which time he was elected a Fellow of the Royal Society and President of the Royal Astronomical Society of Canada. He then supervised the construction of the 205-cm/82-in mirror for the MacDonald Observatory at the University of Texas. He died at Esquimalt, near Victoria, on 17 October 1941.

Using his new telescope at Victoria in conjunction with a spectrograph of high sensitivity, Plaskett discovered many new binary stars, including 'Plaskett's Twins', previously thought to be a single, massive star (B.D.+6° 1309, for many years considered the most massive star known). His work on the radial velocities of galactic stars enabled him to confirm the contemporary discovery of the rotation of the Galaxy and to indicate the most probable location of the gravitational centre of the Galaxy. In turn, this led to a study of the motion and distribution of galactic interstellar matter, particularly involving the detection of spatial calcium.

Plato *c.*427–*c.*347 BC was a Greek mathematician and philosopher who founded an influential school of learning in which the basic precept was not so much one of practical experimentation, as of striving to find mathematical and intellectual harmony. In consequence, most of Plato's astronomical theories involved the most idealistic forms of mathematical wishful thinking, the most fundamental premise being that the Earth, a perfect sphere, was at the centre of a universe in which all other celestial bodies described perfectly circular orbits.

Plato's real name was Aristocles; he was called Plato, 'broad-shouldered', from an early age, however. Born into a patrician Athenian family, he was naturally expected to take up a political career, but as a pupil of Socrates he came to regard politicians with ever-increasing scepticism, and following the trial and death of his mentor in 399 BC, he resolved to become a philosopher and teacher. He travelled for some years, probably to Cyrene, certainly to Sicily, perhaps also to Egypt, before returning at last to Athens in 388 BC where he set up a school that became known as the Academy on part of the premises of a gymnasium. (In one

form or another, the Academy continued to exist for about 900 years.) In about 367 BC, Plato returned to Sicily as tutor to King Dionysus II, but after a few years he became disgusted with the sybaritic lifestyle of the court and (it is thought) went back again to Athens; nothing more is known of his life.

The extant works of Plato, believed on good authority to be genuine, consist of philosophical dialogues among which are *Timaeus*, the *Symposium*, the *Republic* and the *Laws*. In all of them an idealized form of Socrates appears as one of the speakers. Plato divided philosophy into three branches: ethics, physics and dialectics. The basic tenet behind all his arguments is the doctrine of ideas. True science, he reasoned, investigates the nature of those purer and more perfect patterns which were the models after which all created things were formed by the great original intelligence.

Accordingly, it was in particular the science of geometry – with its premise of symmetry and the irrefutable logic of its axioms – that had the most appeal to Plato. In several of his works he therefore presents a picture of the world that is purely conceptual in form. Little thought was given to the idea of observing phenomena before putting forward a theory to explain them. In this legend Plato had been strongly influenced by the ideas of Pythagoras of Samos (*c.*580–*c.*500 BC), the most famous of Greek geometricians. For both Pythagoras and Plato, the ideal of mathematical harmony in the attainment of the perfection of the Creator's original intentions simply meant that the universe had to be spherical because the sphere was the perfect volume; for the same reason the movements of the heavenly bodies had to be circular and uniform. Moreover, the Earth, which lay at the exact centre of the cosmos, was a sphere and was surrounded by a band of crystalline spheres which held in place the Sun, the Moon and the planets.

At one stage, Plato asked one of his pupils, Eudoxus of Cnidus, to make a model showing the circular movements of all celestial bodies. Eudoxus, a skilled mathematician, managed to construct one that demonstrated the movements of Mercury and Venus as epicycles round the Sun, thereby taking the first step towards a heliocentric system.

Despite the high degree of interest shown in such revolutionary concepts, Plato never accepted any concept but that the Earth was at the centre of the universe. This strongly conservative outlook and the unequalled authority which Plato's name gave to it resulted in the acceptance of a geocentric universe until it was invalidated by the findings of Nicolaus Copernicus 19 centuries later.

Pliny Gaius Plinius Secundus AD 23–79 was a Roman military officer of wide interests. Prudently retiring from his commission during the troubled times of the Emperor Nero (54–68), Pliny devoted his energy to a massive compilation of all the known sciences of his day. He is usually now called Pliny the Elder to distinguish him from his nephew and biographer Pliny the Younger.

Born into a wealthy provincial family at Como in the

year 23, Pliny completed his studies in Rome and, in his early twenties, took up a military career in Germany, where he became a cavalry commander and friend of Vespasian. He kept out of harm's way while Nero was on the imperial throne, and (it is assumed) spent much of his time writing. But when in 69 his old comrade Vespasian was made Emperor, Pliny returned to Rome – where his routine included a daily visit to the Emperor to talk of this and that – and took up various public offices. It was in the course of his duties that Pliny's life came to a tragic and untimely end. In the year 79 Pliny was in command of the fleet at Misenum, in the bay of Naples, when the famous eruption of Vesuvius that destroyed the towns of Pompeii and Herculaneum took place. Observing a strange cloud formation, subsequently found to have resulted from the eruption of the volcano, Pliny made for Stabiae where he landed. Here, however, he was fatally overcome by a cloud of poisonous fumes. It is possible that he saw himself as doing his duty, but it is equally possible that he died a martyr to science, his curiosity at this critical moment having been greater than his fears.

Virtually all pursuits, human or scientific, interested Pliny and in his early years he produced a grammar, a history of Rome, a biography of Pomponius Secundus, a report on the Roman military campaign in Germany and a manual on the use of the lance in warfare. All these texts have long since diappeared, but there remains intact what is by far his most ambitious and large-scale work, the *Historia Naturalis/Natural History*, in which he surveys all the known sciences of his day, notably astronomy, meteorology, geography, mineralogy, zoology and botany. At the commencement of the work he states that he has covered 20,000 subjects of importance drawn from 100 selected writers to whose observations he has added many of his own.

All the important assumptions of classical astronomy are described in Book II of the *Natural History*. It is of special interest in that it presents not only the author's opinions, but also those of Hipparchus (lived 2nd century BC) and Eratosthenes (c.276–c.94 BC), major figures in the early history of the science. According to Pliny the Earth lay on the pivot of the heavens and was surrounded by the seven stars: the Sun, the Moon, Mercury, Venus, Mars, Jupiter and Saturn. Nevertheless, he saw the Sun as the ruler of the heavens, providing the Earth with light and with the changing pattern of the seasons. Ascribing to the Sun a zodiacal orbit round the Earth, divided into 12 equal parts then occupied by the zodiacal constellations, Pliny goes on to describe in turn the different orbits of the seven stars, correctly adjudging Saturn to be farthest from the Earth and therefore taking 30 years to complete its circuit. Jupiter, being much nearer, is able to finish its journey in only 12 years, and Mars in about two. After the remaining planets have been described, Pliny again discusses the Sun which, in order that it may concur with a mathematically desirable end, is represented as taking 360 days to complete its circuit, to which a surplus of $5\frac{1}{4}$ days has to be added to ensure that this great star is seen to rise at the identical point each

successive year. Such were the specifications of the Julian calendar, established by Julius Caesar in 46 BC. (It was to result in the error of overestimating the time of the Sun's journey by 11 minutes, 14 seconds, but in Pliny's time this was too small a miscalculation to have had time to become apparent.) The Moon is the last of the seven stars to be described, its puzzling progress of waning being explained as a result of the slant of the zodiacal sky and highly sinuous nature of its course.

Pliny, having no knowledge of distances in the universe, assumed that the Moon is larger than the Earth, for otherwise he could not see how the entire Sun could be obscured from the Earth during an eclipse by the coming of the Moon between them. The Sun, however, he judged to be of far greater size than the Earth for the reasons, among others, that 'the shadow that it throws of rows of trees along the edges of fields are at equal distances apart for very many miles, just as if over the whole of space the Sun lay in the centre' and that 'during the equinoxes it reaches the vertical simultaneously for all the inhabitants of the southern region'. By his knowledge of the night sky at various latitudes, observed during his military journeys, Pliny even reasoned that the Earth must be a globe. In this connection, he cites the experience of Eratosthenes, who reported that in summer the days are considerably longer the farther north the observer travels. He then gives Eratosthenes' famous calculation of the overall circumference of the Earth. As for gravitation, to Pliny the world consisted of four elements – earth, air, fire and water. Of this number the 'light' substances were prevented from rising by the weight of the 'heavy' ones, while the latter were prevented from falling by the countervailing pressures from the more buoyant elements. Such was the earliest hypothesis on the nature of gravity.

In a discussion of comets, Pliny dismisses the popular belief that their arrival portended dramatic events in the fortunes of the Roman world. He alludes to the great impact the appearance of a comet had on Hipparchus and how in order that the appearance of a new star in the heavens could more surely be assessed he made a catalogue of all the stars visible to the naked eye, in the process systematizing a classification that lasted for more than 17 centuries.

The other books of the *Natural History* are just as detailed. Books 3–6 record the geography and ethnography of the then known world, in which frequent references are made to great cities which have since disappeared. Book 7 is concerned with human physiology; Books 8–9 with that of fishes and other marine animals; Book 10 of birds, and Book 11 of insects. Books 12–19 are concerned with botany, agriculture and horticulture, the subjects that appear to have awakened Pliny's keenest interest (he is one of our chief sources of information on early Roman gardens and ancient botanical writings). Books 20–27 cover medicine and drugs, Books 28–32 medical zoology, and Books 33–37 minerals, precious stones and metals, especially those used by Roman jewellers and craftsmen.

The scientific value of this great undertaking varies. The

further the subjects covered are removed from Pliny's own experiences and observations, the more credulous and even silly he becomes, particularly in reporting the existence of strange animals with patently fabulous qualities. His fluency with the Greek language also induced him to translate from it into Latin too freely, often thereby blurring a critical distinction in mathematical or technical passages. The *Natural History* is, however, invaluable in many instances as the only surviving record of people's early reactions to the physical world and the gradual advance of careful observation and systematic classification of natural orders. This great undertaking appears to have been completed in the year 77, only two years before Pliny's death.

Plücker Julius 1801–1868 was a German mathematician and physicist. He made fundamental contributions to the field of analytical geometry and was a pioneer in the investigations of cathode rays that led eventually to the discovery of the electron.

Plücker was born in Elberfeld on 16 June 1801. He attended the Gymnasium in Düsseldorf, and studied at the universities of Bonn, Heidelberg, Berlin and Paris. He was awarded his PhD from the University of Warburg in 1824, and he then took up a lectureship at the University of Bonn, where he became a Professor of Mathematics in 1828. In 1833, he moved to Berlin where he held simultaneous posts as extraordinary Professor at the University of Berlin, and as a teacher at the Friedrich Wilhelm Gymnasium. A personality conflict with one of his colleagues prompted him to move to the University of Halle in 1834, where he held the Chair of Mathematics for two years. In 1836 Plücker became Professor of Mathematics at the University of Bonn, a post he held until 1847, when he became Professor of Physics there. He held this post until his death.

Plücker was the author of a number of books on advanced mathematics, and he also published many papers on his work in experimental physics. He did not gain much standing at home among other German physicists, but his talents were quickly recognized abroad. He was elected Fellow of the Royal Society of London in 1855, and was awarded the Copley Medal by that body in 1868. He was also made a Member of the French Academy of Sciences in 1867. Plücker died in Bonn on 22 May 1868.

The first half of Plücker's career was devoted to the study of mathematics. He found a valuable vehicle for the dissemination of his ideas in a newly founded German mathematical journal. He published books on analytical geometry in 1828, 1831, 1835 and 1839. He resumed the study of geometry shortly before his death, and his latest studies were published in 1868. His mathematics was characterized by elegance and clarity. He was able to correct some errors in work published by Leonhard Euler (1707–1783) in 1748. The work he did towards the end of his life was completed by his assistant Felix Klein (1849–1925). During these years Plücker introduced six equations of higher plane curves which have been named Plücker's coordinates. His work led to the foundation of line geometry.

When Plücker became Professor of Physics at Bonn in 1847, he turned away from his studies on geometry and plunged into theoretical and experimental physics. He studied optics and gas spectroscopy, recognizing early the potential of the latter technique in analysis. He found the first three hydrogen lines well before Robert Bunsen and Gustav Kirchhoff had begun their experiments in this field.

Plücker was strongly influenced by Michael Faraday (1791–1867), who had experimented with electrical discharge in gases at very low pressures. Plücker took up this work and, unlike Faraday, had the advantage of being able to use high-vacuum tubes, which were developed by Heinrich Geissler (1814–1879) in 1855. He found in 1858 that the discharge causes a fluorescent glow to form on the glass walls of the tube, and that the glow could be made to shift by applying an electromagnet to the tube, thus creating a magnetic field. Johann Hittorf (1824–1914), Plücker's student, continued this work and showed in 1869 that the glow was produced by cathode rays formed in the tube, and in 1897 J. J. Thomson (1856–1940) demonstrated that the rays consist of electrons.

Poincaré Jules Henri 1854–1912 was an innovative French mathematician and prolific mathematical writer. His interests and achievements were wide-ranging, although he is probably best known for his introduction of automorphic functions in pure mathematics, of ergodicity in the theory of probability, and of some of the understanding of the dynamics of the electron later attributed to Albert Einstein in the theory of relativity. He was also renowned for his study of celestial mechanics.

Poincaré was born on 29 April 1854 in Nancy, the son of a doctor. A brilliant student, he won first prize in an open competition between lycée students from throughout France. Completing his education, he entered the Ecole Polytechnique in Paris in 1873, where he graduated. He then studied engineering at the Ecole des Mines, but it was in mathematics that in 1879 his doctoral thesis was successfully composed. Immediately afterwards, Poincaré was appointed to a teaching post at the University of Caen, and only two years later he became Professor of Mathematics at Paris University. In 1887, he was elected to the Academy of Sciences, and during the remainder of his lifetime he received many other honours and awards. He died in Paris on 17 July 1912.

Poincaré's first great work was in pure mathematics, where he generalized the idea of functional periodicity in his theory of automorphic functions that are invariant under a denumerably infinite group of linear fractional transformations. He showed how these functions could be used to express the coordinates of any point on an algebraic curve as uniform functions of a single parameter, and could also be used to integrate linear differential equations with rational algebraic coefficients. Developing his investigations, he found that one class of automorphic functions – which he called Fuchsian, after the German mathematician Immanuel Fuchs (1833–1902) – were associated with transformations arising in non-Euclidean geometry. The originator of the study of algebraic topology, Poincaré has sometimes been compared with Karl Gauss in terms of the innovatory nature of his discoveries and the genuine desire for rigorous and precise presentation of data.

Poincaré contributed to the theory of the figures of equilibrium of rotating fluid masses and discovered the pear-shaped figures used in the researches of George Darwin (the great Charles's second son) and others. But perhaps his greatest contribution to mathematical physics was his paper on the dynamics of the electron, published in 1906, in which he obtained many of the results of the theory of relativity later credited to Albert Einstein. Poincaré worked quite independently of Einstein; his treatment was based on the full theory of electromagnetism and limited to electromagnetic phenomena, whereas Einstein developed his theory from elementary considerations involving light signalling. Poincaré's studies of mathematical physics led him inevitably to investigations in the field of celestial mechanics. He made important contributions to the theory of orbits, particularly with the classic three-body problem – the mutual gravitational and other effects of three bodies close together in space – which he generalized to a study of n bodies. In the course of his work he developed powerful new mathematical techniques, including the theories of asymptotic expansions and integral invariants. He made important discoveries about the behaviour of the integral curves of differential equations near singularities, and wrote a massive three-volume treatise on his new mathematical methods in astronomy. From his theory of periodic orbits he developed the entirely new subject of topological dynamics.

Poincaré wrote on the philosophy of science. He believed that some mathematical ideas precede logic, and made an original analysis of the psychology of mathematical discovery and invention, in which he stressed the role played by convention in scientific method.

He was said, very early in his career, to be a 'mathematical giant'. Certainly Poincaré's output of writings was gigantic – he produced, in all, more than 30 books and 500 papers. But one outstanding quality of his authorship was that it appealed not merely to scientists but to educated people in all walks of life. When he was elected to the Académie Française in 1908, it was to fill the position left vacant following the death of the poet René Sully Prudhomme – a writer, not a scientist.

Poiseuille Jean Léonard Marie 1799–1869 was a French physiologist who made a key contribution to our knowledge of the circulation of blood in the arteries. He also studied the flow of liquids in artificial capillaries.

Poiseuille was born on 22 April 1797 in Paris. Little is known of the positions he held, but he is known to have attended the Ecole Polytechnique in Paris 1815–1816 and to have received a doctorate in 1828. In 1842, he was elected to the Académie de Médicine in Paris and the Société Philomathique, both in Paris. He also received the Montyon Medal for his physiological researches in 1829, 1831, 1835 and 1843. Poiseuille is best known for his stud-

ies of the circulation of the blood through arteries. He improved on earlier measurements of blood pressure by using a mercury manometer and filling the connection to the artery with potassium carbonate to prevent coagulation. He used this instrument, known as a hemodynamometer, for his dissertation to show that blood pressure rises during expiration (breathing out) and falls during inspiration. He also discovered that the dilation of an artery fell to less than $1/_{20}$ of its normal value during a heartbeat. Poiseuille was also interested in the flow of liquids through other small pipes and capillaries. Experiments with distilled water led him to state laws for the volume of liquid discharged per unit time from a capillary to its diameter, length, temperature and pressure difference between the ends of the pipe. At first this relationship was named after Poiseuille but later it was realized that Hagen had discovered it independently a year earlier. It is now known as the Hagen–Poiseuille law. Poiseuille died in Paris on 26 December 1869.

Poisson Siméon-Denis 1781–1840 was a French mathematician and physicist. He is mainly remembered for Poisson's ratio in elasticity, which is the ratio of the lateral contraction of a body to its longitudinal extension. The ratio is constant for a given material. Poisson also made contributions to mathematics, especially in probability theory, and to astronomy, in which he investigated planetary and lunar motion.

Poisson was born on 21 June 1781 in Pithiviers, Loirel, where his father was a civil servant in the local administration. Initially training as a surgeon, but discovering he had neither the manual dexterity for nor any interest in the profession, he entered the Ecole Centrale in Fontainebleau in 1796 to study mathematics. Two years later he continued his studies in Paris at the Ecole Polytechnique (coming first in the entry examination), where he studied under Pierre Laplace (1749–1827) and Joseph Lagrange (1736–1813). After only twelve months there, in 1799 he submitted a paper on the theory of equations that enabled him not only to graduate in 1800 but to begin teaching at the Ecole himself. Two years later he was named Deputy Professor and, in 1806, became Professor. In 1808, he was appointed Astronomer at the Bureau des Longitudes, and the following year he was appointed Professor of Mechanics at the Faculty of Sciences. In 1815, he became an examiner at the Ecole Polytechnique. Nominated Conseil Royal de l'Université in 1820, he became an administrator at the highest level in France's educational system and, as such, played a particularly prominent part in the 'defence' of science against the conservative policies of the government of the day. Seven years later, he was appointed Mathematician at the Bureau des Longitudes in succession to Laplace. And in 1837 Poisson became a nobleman on accepting the offer of a baronetcy. He died in Paris on 25 April 1840.

Much of Poisson's work involved applying mathematical principles in theoretical terms to contemporary and prior experiments in physics, particularly with reference to electricity and magnetism but also with special regard to heat and sound. Quite early in his career, Poisson adopted the 'two-fluid' theory of Jean Nollet (1700–1770), according to which the like fluids of electricity repelled and the unlike fluids attracted, and showed that Joseph Lagrange's potential function would be constant over the surface of an insulated conductor. He went on to give an ingenious proof of the formula for the force at the surface of a charged conductor.

Charles Coulomb (1736–1806) had already carried out experimental work involving the surface densities of charge for two spherical magnets placed any distance apart. Poisson produced theoretical results which were in agreement with those obtained experimentally by Coulomb and, in 1824, gave a very complete theory of magnetism using Coulomb's model – again incorporating two 'fluids'. Poisson derived a general expression for the magnetic potential at any point: the sum of two integrals due to volume and surface distribution of magnetism respectively.

In his own experiments on the elasticity of materials, Poisson deduced the ratio between the lateral and longitudinal strain in a wire; this is now known as Poisson's ratio. Poisson summed up his work in physics in several books towards the end of his life. They include *Treatise on Mechanics* (1833) and *Mathematical Theory of Heat* (1835). Poisson also published *Researches on the Movement of Projectiles in Air* (1835), which builds on the work of Gaspard Coriolis (1792–1843) and was the first account of the effects of the Earth's rotation on motion. It inspired the famous pendulum experiment carried out in 1850 by Jean Foucault (1819–1868) that first demonstrated the Earth's rotation.

His significant work in probability theory was considered at first to be a mere popularization of the work of Laplace. Poisson's formula for the great asymmetry between opposite events, such that the prior probability of either event is very small, was not used until the end of the nineteenth century, when its importance was finally recognized. Poisson was also responsible for a formulation of the 'law of large numbers', which he introduced in his important work on probability theory, *Recherches sur la Probabilité des Jugements/Researches on the Probability of Opinions* (1837).

Polanyi Michael 1891–1976 was a Hungarian-born British physical chemist, particularly noted for his contributions to reaction kinetics. In later life he diverted his attention to social philosophy, in which he became equally renowned. Throughout his career he voiced his firm belief in the right of the scientist to seek the truth unhampered by external constraints.

Polanyi was born in Budapest on 12 March 1891. He entered the University of Budapest in 1909 to study medicine, but after graduation went to the Technische Hochschule at Karlsruhe as a student of chemistry under Georg Bredig. After service as a medical officer during World War I, he returned briefly to Karlsruhe before joining the Kaiser Wilhelm Institute of Fibre Chemistry in Berlin. In 1923, at the invitation of Fritz Haber, Polanyi moved to the Institute

for Physical and Electro-Chemistry. But he became increasingly disturbed by the influence of the Nazi Party, especially its dismissal of Jewish scientists, and in 1933 he accepted the Chair of Physical Chemistry at Manchester, England.

During the 1940s, Polanyi made the decision to concentrate on philosophy and in 1948 he transferred to the newly created Chair of Social Studies at Manchester. On retiring from this position in 1958 he moved to Merton College, Oxford, as Senior Research Fellow. He died at Northampton on 22 February 1976.

Polyani's early researches in chemical physics resulted in several papers on the adsorption of gases by solids. He introduced the idea of the existence of an attractive force between a solid surface and the atoms or molecules of a gas; he also suggested that the adsorbed surface is a multilayer and not subject to simple valency interactions. His other work of about that time extended the theory of Hermann Nernst (which stated that the entropy of a system approaches zero as the temperature decreases towards absolute zero). Polanyi showed that an increase in pressure must have the same effect, although in practice the highest attainable laboratory pressure is less effective than a very modest temperature increase.

At Berlin Polanyi's interest turned to X-ray analysis, using the newly developed rotating crystal method. He and his co-workers improved the technique and applied it to the determination of the structure of cellulose fibres. He also investigated the physical and mechanical properties of various materials; he grew crystals of metals and devised a special apparatus to measure their shear and rupture strengths.

Even as early as 1920 Polanyi recognized that the current theories of chemical reaction rates were simplifications of the truth. The collision theory postulated that only molecules with a certain critical energy would react. Working first under Haber and then at Manchester, Polanyi extended this idea and produced theories of rates of association and dissociation based on the angular momenta of the colliding particles. Then quantum mechanics presented the kineticist with a powerful new tool. Reactions were considered in terms of the variation in potential energy of a system, which could be plotted as a function of the distance between reacting nuclei, to produce a diagram somewhat resembling a contour map. The configuration of the components at the 'mountain pass' was defined as the activated complex, and the 'height' of the pass represented the activation energy.

Polanyi and Eyring investigated the reaction between a hydrogen atom and a hydrogen molecule:

$$H + H_2(para) \rightarrow H + H_2(ortho)$$

and made the first reasonable accurate determination of its energy surface (ortho- and parahydrogen are isomers that differ only in the direction of spin of their nuclei).

Polanyi also played a part in solving a problem that had long been puzzling kineticists. It was known that in the hydrogen–iodine equilibrium:

$$H_2 + I_2 \leftrightarrow 2HI$$

the rate of reaction is given by the equation:

$$\frac{d\,[\text{HI}]}{dt} = K\,[\text{H}_2]\,[\text{I}_2]$$

where K is a constant. But in the apparently analagous reaction between hydrogen and bromine, experiment showed the reaction rate to be given by:

$$\frac{d\,[\text{HBr}]}{dt} = \frac{k\,[\text{H}_2]\,[\text{Br}_2]^{1/2}}{m + [\text{HBr}]\,/\,[\text{Br}_2]}$$

where k and m are constants. This expression implies that the velocity of the reaction is inhibited by the presence of the product HBr. Polanyi and others proposed a chain mechanism for the reaction:

initiation:		Br_2	$\to$	2Br
propagation:	$\text{Br} + \text{H}_2$		$\to$	$\text{HBr} + \text{H}$
	$\text{H} + \text{Br}_2$		$\to$	$\text{HBr} + \text{Br}$
inhibition:	$\text{H} + \text{HBr}$		$\to$	$\text{H}_2 + \text{Br}_2$
termination:		2Br	$\to$	Br_2

In his new philosophical role at Manchester Polanyi was active in the Society for Freedom in Science. He advocated that scientific research need not necessarily have a pre-stated function and expressed the belief that a commitment to the discovery of truth is the prime reason for being a scientist. His principal work was an investigation of the processes by which high-level skills such as craftsmanship and connoisseurship are acquired and the means by which such skills are shared and extended. His move to Oxford in 1958 coincided with the publication of his book *Personal Knowledge*, of which he said 'The principal purpose of this book is to achieve a frame of mind in which I may firmly hold what I believe to be true, even though I know it may conceivably be false.'

Polya George 1887–1985 was a Hungarian mathematician, one of the founders of the Hungarian School of Mathematics, best known for his work on function theory, probability and applied mathematics.

Polya was born in Budapest on 13 December 1887. He studied at the Eotvos Lorand University and was awarded his PhD in mathematics by the University of Budapest in 1912. While there he was a member of a thriving community of mathematicians, but he then chose to devote two years to postgraduate study abroad. He attended courses at the University of Göttingen and in Paris before in 1914 accepting the offer of a position as Assistant Professor of Mathematics at the Swiss Federal Institute of Technology in Zurich. He was promoted to Associate Professor and in 1928 to Full Professor of the Institute. He served as Dean and Chairman of the Mathematics Department from 1938, but in 1940 left to go to the United States. Brown University in Providence, Rhode Island, offered him the post of

Visiting Professor which he held for two years before moving to Smith College, Northampton, Massachusetts, as Professor of Mathematics. In 1946, Polya became Professor of Mathematics at Stanford University, Palo Alto, California, where he remained until his retirement as Emeritus Professor in 1953. He frequently made lecture tours to universities throughout North America and Europe after his 'retirement'. From 1963 he also served on the Research Council of Greater Cleveland.

Polya was a member of numerous scientific and mathematical organizations, including the National Academy of Sciences, the American Academy of Arts and Sciences, and the London Mathematical Society. He was awarded several honorary degrees and wrote numerous books.

One of Polya's best known achievements was his discovery in 1920 of the theorem since named after him. Polya's theorem is a solution of a problem in combinatorics theory and method. Much of his other early work was on function theory, and he published studies on analytical functions in 1924 and on algebraic functions in 1927. He also worked on linear homogeneous differential equations (1924) and transcendental equations (1930). One of his studies in mathematical physics was an investigation into heat propagation published in 1931. He extended some of the previous results obtained by Andrei Markov (1856–1922) on the limit of probability, and probability theory became one of Polya's major research areas.

Other subjects he examined included the study of complex variables, polynomials and number theory. His contributions to mathematics can thus be seen to be notable both for their breadth and their depth.

Poncelet Jean-Victor 1788–1867 was a French military engineer who, to pass the time during two years as a prisoner of war, revised all the mathematics he could remember and went on to make fresh discoveries, particularly in projective geometry. He was among the leaders of those who initiated and developed the concept of duality.

Poncelet was born in Metz on 1 July 1788, the illegitimate (although later recognized) son of Claude Poncelet, a rich landowner and an *advocat* at the *Parlement* of Metz. He was sent to live with a family in Saint-Avold, and they were responsible for his earliest education. At the age of 16 he returned to Metz and attended the lycée. In 1807, he went to the Ecole Polytechnique in Paris, and stayed there for three years. He then fell behind with his studies, however, because of ill health, and in 1810 joined the Corps of Military Engineers. He graduated from the Ecole d'Application in Metz in February 1812, and went to work on the fortification of the Dutch island of Walcheren. In June of the same year he became lieutenant of engineers and, attached to the staff of the Engineer-General, took part in the campaign against the Russians. He was captured at the Battle of Krasnoy, and was imprisoned in Saratov, a city on the Volga, until 1814. He then returned to France in September of that year and became Captain of the Engineering Corps in Metz. From then until 1824 he was engaged on projects in military engineering there. At the end of that time he was appointed Professor of Mechanics applied to machines at the Ecole d'Application de l'Artillerie et du Génie in Metz, six years later becoming a member of Metz Municipal Council and Secretary of the Conseil-Général of the Moselle. He was elected to the Mechanics section of the Académie in 1834, and from 1838 to 1848 was Professor to the Faculty of Science at Paris. From 1848 to 1850 he was Commandant of the Ecole Polytechnique with the rank of General. He died in Paris on 22 December 1867.

Poncelet's first great work was done while he was imprisoned at Saratov. With no textbooks at his disposal, he reconstructed the elements of pure and analytical mathematics (specifically geometry) from memory before undertaking some original research on the systems and properties of conics. It was projective geometry that interested him most, and it was his study of this aspect of conics that established the basis for his later important work, the treatise entitled *Traité des Propriétés Projectives des Figures* published in 1822 (with a second edition of two volumes, published in 1865–66). Poncelet had been a pupil of Gaspard Monge (1746–1818), who was the originator of modern synthetic geometry – synthetic geometry's viewing of figures as they exist in space is an alternative mathematical tool to the equation in the analytical method – but was equally conversant with either discipline. In fact, he used both methods and ranks as one of the greatest of those who contributed to the development of the relatively new synthetic (projective) geometry.

Poncelet became the centre of controversy over the principle of continuity. He also discovered the circular points at infinity, although the concept of points at infinity goes back to Gérard Desargues (1591–1661), and many of the individual ideas of projective geometry go back considerably further. But it was Poncelet who first developed them as a distinct branch of the mathematical science. His rather forceful presentation – and his occasionally wild accusations of plagiarism by other geometrists – antagonized the young German, Julius Plücker (1801–1868), to such an extent that he turned from using synthetic methods and became himself one of the greatest of analytical geometrists.

The principle of duality was first recognized and publicized by Poncelet in the *Journal für Mathematik* of 1829, although previously formulated by Joseph Gergonne in 1825–27. (It can be illustrated by considering a statement capable of two meanings, both true, one obtained from the other simply by interchanging two words. In projective geometry, in two dimensions, this is achieved by interchanging the words 'point' and 'line'. In three-dimensional geometry, there is a corresponding duality between points and planes; in this case the line is self-dual, in that it is determined by any two distinct points on it or by any two distinct planes through it.) Much of higher geometry is concerned with duality, and every new application practically doubles the extent of existing knowledge.

His engineering skills were much used between 1814 and 1840, for the first ten years of which he was engaged on projects in topography and the fortification and organization of an engineering arsenal. In 1821, he developed a new model of a variable counterweight drawbridge, which he described and publicized in 1822. His most important technical contributions were concerned with hydraulic engines, such as Poncelet's waterwheel, with regulations and with dynamometers, as well as in devising various improvements to his own previous fortification techniques. In applied mechanics he worked in three interrelated fields: experimental mechanics, the theory of machines, and industrial mechanics.

Pond John 1767–1836 was a British astronomer whose meticulous observations at his private observatory led to his discovering errors in data published by the Royal Observatory in Greenwich. When he himself became Astronomer Royal, he therefore implemented a vigorous programme of renovation and reorganization at Greenwich, that restored the Observatory to its former standards of excellence.

Pond was born in London in 1767 – no more exact date is known. His scientific talents were apparent from an early age and he entered Trinity College, Cambridge, in 1783 to study chemistry. Forced by poor health to leave the university before he could take his degree, he travelled in warmer climates to recover his strength. He went to several Mediterranean and Middle Eastern countries, making astronomical observations wherever possible. When he returned to England in 1798, he established a small private observatory in Westbury, near Bristol. From there he published observations of considerable astronomical interest, and in 1807 he was elected a Fellow of the Royal Society. This prompted him to return to London, and in 1811 he was appointed Astronomer Royal. He immediately set about reorganizing and modernizing the Greenwich Observatory, which until that time had only one assistant and a collection of equipment that was sadly in need of repair. In 1835, however, he was forced to retire from all professional duties because of ill health. He was awarded the Lalande Prize of the French Academy of Sciences in 1817, and the Copley Medal of the Royal Society in 1823. He died in Blackheath on 7 September 1836.

Pond first demonstrated his skills as an astronomer at the age of 15. He noticed errors in the observations being made at the Greenwich Observatory and made a thorough investigation of the declination of a number of fixed stars. By 1806 he had clearly and publicly demonstrated that the quadrant at Greenwich, designed by Bird, had become deformed with age and needed replacing. It was this in particular that prompted his programme to modernize the whole observatory.

One of the first results of the revitalization programme was his 1813 catalogue of the north polar distances of 84 stars. These data were obtained with the new mural circle designed by Edward Troughton and were highly esteemed by Pond's contemporaries. Pond was able to dispute, in 1817, the validity of J. Brinkley's observations which were ostensibly on the parallax of a number of fixed stars. Pond held that Brinkley had not in fact detected stellar parallax, which was being sought by numerous astronomers as a proof of Copernican cosmology, and he was later proved to be right. The interest that the controversy generated contributed to Friedrich Bessel's later successful efforts in this field.

Another controversy, of an unpleasant kind, surrounded Pond's work a few years later – surprisingly, considering how meticulous (even pedantic) he was known to be. A committee of enquiry, set up by the Royal Society, found that two of Pond's assistants were responsible for work that was less than accurate or conscientious, and reprimanded them; Pond was cleared.

Instituting new methods of observation, Pond went on to produce a catalogue of more than 1,000 stars in 1833. His work continued to be admired by many of his fellow astronomers in Britain and in Europe. Nevertheless, he remains remembered most for his modernization of the Greenwich Observatory.

Pons Jean-Louis 1761–1831 was a French astronomer who, in a career that began at a comparatively late age, nevertheless discovered more than 35 comets and became Director of the Florence Observatory.

Pons was born on Christmas Eve in 1761 at Peyre, near Dauphine. The son of a poor family, he was not well educated and held several labouring jobs until the age of 28, when he became a porter and doorkeeper at the Marseilles Observatory. Noting his interest in astronomy, the directors of the Observatory gave him instruction in the subject, paying particular attention to Pons's training in practical observation. Pons learned quickly; knowledge of the sky together with excellent eyesight and considerable patience stood him in good stead. And as a result of his diligence and achievement, Pons was named *astronome adjoint* at the Marseilles Observatory in 1813. Five years later he became its Assistant Director. His achievements were recognized outside the Marseilles Observatory when, in 1819, on the recommendation of Baron Frederich von Zach, Pons became Director of a newly constructed observatory at Lucca, in northern Italy. Three years later, when the observatory was closed, Pons was invited by the Grand Duke Leopold of Tuscany to become Director of the Florence Observatory. Before failing eyesight finally forced him to give up much of his observational work, he received many honours and awards (including no fewer than three Lalande Prizes). He retired from the Observatory a few months before his death on 14 October 1831.

The first Lalande Prize that Pons was awarded by the French Academy of Sciences was for his discovery in 1818 of three small, tailless comets, among which was one that Pons claimed had first been seen in 1805 by Johann Encke of the Berlin Observatory. Alerted to this possibility, Encke carried out further observations and calculations, and finally ascribed to it a period of 1,208 days – which meant that it would return in 1822. Its return was duly observed, in Australia, only the second instance ever of the known return of an identified comet. Encke wanted the comet to be named after Pons, but it continued to be called after its

discoverer. Encke received the Gold Medal of the Royal Astronomical Society of London and Pons the Silver.

At Lucca Pons discovered a number of new comets, for one of which he received his second Lalande Prize. His third Lalande Prize followed his discovery of more comets at the Florence Observatory, raising the total number of his discoveries to 37.

Porsche Ferdinand 1875–1951 was the Bohemian engineer who designed and built the first mass-produced European people's car and later helped to develop small, high-performance luxury cars destined for series production. He also invented synchronizing gearboxes and torsion-bar suspension.

Porsche was born the son of a tinsmith in Bohemia. From 1923 to 1929 he served as technical director of the German Daimler company which became Daimler Benz in 1926. During that period his most notable contribution was the Mercedes SSK racing car. He rose from relative obscurity to form a limited liability company on 25 March 1931, registered in Stuttgart in the name of Dr Ing H. C. F. Porsche Ltd.

From 1931 to 1933, Porsche developed his first small-car prototypes for Zündapp and NSU, under contract. In 1932, he devised the first torsion-bar suspension system and in the same year visited Russia to make an extensive study trip through all centres of the Soviet vehicle industry of that period. He was allowed to visit any factory which interested him and was said to have been shown all their designs for vehicles, aircraft and tractors. Probably no other European has gained the insight into the Russian industry which Porsche possessed at that time. At the end of this journey, the Russians in Moscow offered him the job of chief national designer for the land, to be accompanied by a wealth of authority and privilege. But Porsche declined; the language barrier, in particular, deterred him.

In 1936, he received a contract from the German government to develop the Volkswagen and plan the factory where it would be built. Just before this he conceived a racing car, without contract. The project was taken over by Auto Union and the car subsequently claimed victories on virtually every race track in Europe between 1934 and 1937, as well as many class records.

The first VW prototypes were on the road by the end of 1935; the years to follow, up to the outbreak of World War II, were years of the utmost concentration on the Volkswagen for Porsche. Some other design jobs, however, had appeared alongside the major Volkswagen contract to cause Porsche to expand his company. During June 1938 the design offices moved from Kronenstrasse 24 in Stuttgart to Spitalwaldstrasse 2 in Zuffenhausen, where he led the development of light tractors. Those tractors built to Porsche licence and under the firm's supervision after World War II can be traced to these designs. Concepts were also developed by him for aviation engines as well as plans and designs for wind-driven power plants – large windmills with automatic sail adjustment which delivered electric current via generators.

The war cut short further development of the Volkswagen so Porsche designed the Leopard and Tiger tanks used by German Panzer regiments and helped to develop the V-1 flying bomb. During this period he was awarded an honorary professorship.

After internment following the end of the war, Porsche joined his son in Gmünd, Carinthia, with the firm which had been moved from Zuffenhausen to help develop the first Type 256 Porsche roadster, later to become the 911 model. He was considerably weakened by his 22-month term of imprisonment, however, and his health caused him to withdraw from engineering by the end of the 1940s. He died in 1951.

The Volkswagen was conceived in 1934 as a utility car of low weight to be 'achieved by new basic measures'. The first prototypes were built in 1935 and 1936, before obtaining the contract to build. A lightweight 'flat-four' cylinder air-cooled engine of substantial magnesium alloy construction was rear-mounted in the vehicle. It was united with a similarly constructed gearbox. Independent suspension of all four wheels was by swing axles with torsion-bar springing.

Substantial weight-savings resulted in the absence of axle beams and the combined weight of the rear-mounted engine and gearbox unit was insufficient to cause handling instability. The disposition of major components allowed an aerodynamically efficient saloon car body of lightweight chassis-less construction to be fitted, giving adequate leg room within the small exterior envelope.

Porsche's first production Volkswagen rolled off the assembly lines in 1945 at the Wolfsburg plant. On 17 February 1972 it became the car with the longest and biggest production run in the history of the automobile, outpacing the Model T Ford at 15,007,033 units.

Porsche was a brilliant engineer whose genius reached into many disciplines. It has been said that the torsion-bar suspension alone would have sufficed to establish a monument to his name in the automotive industry. He can be considered one of the pioneers of aircooled engines in the industry. The sports cars bearing his name were developed by his son Ferry and the first, the 356, was based on the incredibly versatile VW Beetle.

Porter George 1920– is a British physical chemist who developed the technique of flash photolysis for the direct study of extremely fast chemical reactions, for which achievement he shared the 1967 Nobel Prize for Chemistry with Ronald Norrish and Manfred Eigen. He has also inspired others – particularly young people – by his television appearances and lectures at the Royal Institution.

Porter was born in Stainforth, Yorkshire, on 6 December 1920. He graduated from Leeds University in 1941, during World War II, and spent the next four years as a Radar Officer in the Royal Navy. He then went to Cambridge University where he carried out research from 1945 to 1949 under Ronald Norrish. From 1952 to 1954 he was Assistant Director of Research at Cambridge and Assistant Director of the British Rayon Research Association for a year in 1955. He then became Professor of Physical Chemistry at Sheffield University and was made Head of the Chemistry Department in 1963. In 1966, Porter became Director and Fullerian Professor of Chemistry at the Royal Institution. He also served as Director of the Davy Faraday Research Laboratory. He was created a life peer, taking the title 'Lord Porter of Luddenham', in 1990.

In 1947, while working with Norrish, Porter began using quick flashes of light to study transient species in chemical reactions, particularly free radicals and excited states of molecules. He studied very fast reactions having short-lived intermediates. In 1950 he could detect entities that exist for less than a microsecond; by 1967 he had reduced the time limit to a nanosecond, and by 1975 he could detect species that lasted for as little as a picosecond (10^{-12} sec). His early work dealt with reactions involving gases (mainly chain reactions and combustion reactions), but he later extended the technique to solutions. He developed a method of stabilizing free radicals by trapping them in the structure of a supercooled liquid (a glass), a technique called matrix isolation. He enabled flash photolysis to be applied to organic chemistry, biochemistry and photobiology. Using laser beams, he extended the technique to study reactions beyond the microsecond range. Today photochemical methods are used to synthesis hydrocarbons for fuels and chemical feedstocks.

One of Porter's main interests since the early 1960s has been the mechanism of photosynthesis in plants, which proceeds via 'light' and 'dark' stages. He studied the light-harvesting mechanisms of chloroplasts and the primary processes that occur in the first nanosecond of photosynthesis.

Few modern scientists of Porter's calibre have devoted so much time to the education of young people and non-specialists about the importance and excitement of scientific studies. Part of this involvement has been spent maintaining the great tradition of the Royal Institution and contributing to many television programmes.

Porter Rodney Robert 1917–1985 was a British immunologist well known for his contribution to the identification of the structure of antibody molecules. For this work he received the 1972 Nobel Prize for Physiology or Medicine, which he shared with the American Gerald Edelman (1929–).

Porter was born in Liverpool on 8 October 1917 and was educated at Ashton-in-Makerfield Grammar School, and then at Liverpool University, where he gained a BSc in 1939. From 1940 to 1946, including much of World War II, he was in military service. He then returned to Cambridge (gaining his PhD in 1948) and continued his research, aided by Frederick Sanger. In 1949, he was appointed to the staff of the National Institute for Medical Research, a position he held until 1960 when he became Pfizer Professor of Immunology at St Mary's Hospital Medical School, London. From 1967 he was Whitley Professor of Biochemistry at Oxford, and Honorary Director of the Medical Research Council's Immunochemistry Unit.

When Porter started his research after the war, he often referred to Karl Landsteiner's book *The Specificity of Serological Reactions*, and it was from this work that he learnt the technique for preparing certain antibodies. Some aspects of the structural studies of immunoglobins, or antibodies, had been completed, such as those for several human myeloma proteins and some rabbit immunoglobulins. Some work had also been done on the structural basis of the combining specificity of antibodies and in the solution of the genetic origins of antibodies. Porter's major scientific interests have been the structural basis of the biological activities of antibodies – in 1962 he proposed a structure for gamma globulin – and worked on the structure, assembly and activation mechanisms of the components of a substance known as complement. This is a protein normally present in the blood, but which disappears from the serum during most antigen–antibody reactions. Porter also investigated the way in which immunoglobulins interact with complement components and with cell surfaces.

Powell Cecil Frank 1903–1969 was a British physicist who developed photographic techniques for studying subatomic particles and who discovered the pi-meson (pion). For this discovery he was awarded the 1950 Nobel Prize for Physics.

Powell was born in Tonbridge, Kent, on 5 December 1903, the son of a gunsmith. He won a scholarship to Cambridge University in 1921 and graduated four years later. He then went to the Cavendish Laboratory to do research under Ernest Rutherford and C.T.R. Wilson, working on methods of taking improved photographs of particle tracks in a cloud chamber. Powell gained his PhD in 1928 and moved to the Wills Physics Laboratory at Bristol University as research assistant to A. M. Tyndall (1881–1961). He remained at Bristol, becoming Professor of Physics in 1948 and Director of the laboratory in 1964.

In the early 1930s Tyndall and Powell studied the mobility of ions in gases and the way water droplets condense on ions in a cloud chamber. Then in 1938, instead of photographing the cloud-chamber tracks, Powell made the ionizing particles trace paths in the emulsions of a stack of photographic plates. The technique received a boost with the development of more sensitive emulsions during World War II and in 1947 Powell used it in his discovery of the pi-meson (predicted by Hideki Yukawa in 1935). Powell collaborated with the Italian physicist Giuseppe Occhislini, and together they published *Nuclear Physics in Photographs* (1947), which became a standard text on the subject.

Powell John Wesley 1834–1902 was the most romantic figure in nineteenth century US geology.

The son of intensely pious Methodist immigrants, Powell was intended by his farmer father for the Methodist ministry, but he developed early a love for natural history. In the 1850s he became secretary of the Illinois Society of Natural History, travelling widely and building up his natural history collections and his geological expertise. Fighting in the Civil War, he had his right arm shot off, but continued in the service, rising to the rank of colonel.

After the end of the war, Powell occupied various chairs in geology in Illinois, while continuing with intrepid fieldwork (he was one of the first to steer a way down the Grand Canyon). In 1870, Congress appointed him to lead an official survey of the natural resources of the Utah, Colorado and Arizona area, the findings of which were published in his *The Exploration of the Colorado River* (1875) and *The Geology of the Eastern Portion of the Uinta Mountains* (1876).

Powell's enormous and original studies produced lasting insights on fluvial erosion, volcanism, isostasy and orogeny. His greatness as a geologist and geomorphologist stemmed from his capacity to grasp the interconnections of geological and climatic causes. In 1881, he was appointed director of the US Geological Survey. He encouraged most of the great US geologists of the next generation, including Grove Karl Gilbert, Clarence E. Dutton and W. H. Holmes.

Powell drew attention to the aridity of the American southwest, and for a couple of decades campaigned for massive funds for irrigation projects and dams, and for the geological surveys necessary to implement adequate water strategies. He also asserted the need in the drylands for changes in land policy and farming techniques. Failing to win political support on such matters, he resigned in 1894 from the Geological Survey.

Poynting John Henry 1852–1914 was a British physicist, mathematician and inventor whose various contributions to science included an equation by which the rate of flow of electromagnetic energy (now called the Poynting vector) can be determined, and the measurement of Newton's gravitational constant. He received many honours for his work, including the Royal Society's Royal Medal in 1905.

Poynting was born in Monton, Lancashire, on 9 September 1852. He attended his father's school until 1867, when he entered Owens College in Manchester (later Manchester University), where he gained an external BSc from London University in 1872. He then studied at Trinity College, Cambridge, from 1872 to 1876. After graduation he served as a physics demonstrator – under Balfour Stewart (1828–1887) – at Owens College until 1878, when he was appointed a Research Fellow at Trinity College where he researched in the Cavendish Laboratories under James Clerk Maxwell. In 1880, he became Professor of Physics at Mason College, Birmingham (which became Birmingham University in 1900), a post he held until his death – caused by a diabetic attack – in Birmingham on 30 March 1914.

Poynting's first publications, of the late 1870s, were in the field of mathematics and included such subjects as statistical studies of alcoholism in England. After moving to the Cavendish Laboratories in 1878 he began a long series of experiments to determine Newton's gravitational constant (from which can be calculated the Earth's mean density) which was the subject of a competition at Cambridge. He published his most accurate results in 1891, having obtained them using an ordinary beam balance; his figures differ only slightly from those now generally accepted. His experimental method was refined in 1895 by Charles Boys, who used a quartz fibre torsion balance. Poynting recognized the value of Boys's apparatus and used it for several later studies of his own, such as the measurement of radiation pressure.

Poynting's best known work concerned the transmission of energy in electromagnetic fields. In *On the Transfer of Energy in the Electromagnetic Field* (1884) he published an equation (which he worked out using Maxwell's electromagnetic field theory) by which the magnitude and direction of the flow of electromagnetic energy – the Poynting vector – can be determined. This equation is usually expressed as:

$$S = (1/\mu)EB \sin \theta$$

where S is the Poynting vector, μ is the permeability of the medium, E is the strength of the electric field, B is the strength of the magnetic field, and θ is the angle between the vectors representing the electric and magnetic fields.

Poynting, in collaboration with W. Barlow, also did important work on radiation. In 1903, he suggested the existence of an effect of the Sun's radiation that causes small particles orbiting the Sun to gradually approach it and eventually plunge in. This idea was later developed by the American physicist Howard Robertson – who related it to the theory of relativity – and it is now known as the Poynting–Robertson effect. Poynting also devised a method for measuring the radiation pressure from a body; his method can be used to determine the absolute temperature of celestial objects.

Poynting's other work included a theoretical analysis of the solid to liquid phase change (1881); a statistical analysis of changes in commodity prices on the Stock Exchange (1884); a study of osmotic pressure; and the construction of a saccharometer and a double-image micrometer.

Prandtl Ludwig 1875–1953 was a German physicist who put fluid mechanics on a sound theoretical basis. In particular, he originated the boundary layer theory and did pioneering work in aerodynamics.

Prandtl was born in Freising on 4 February 1875, and entered the Technische Hochschule at Munich in 1894. He specialized in engineering, obtaining his first degree in 1898 and a doctorate in 1900 with a thesis on the lateral instability of beams in bending. In the same year, Prandtl went to work in the Maschinenfabrik Augsburg-Nürnberg, where he became interested in fluid flow. In 1901, he gained a professorship at the Technische Hochschule in Hanover, and then became Professor of Applied Mechanics at Göttingen in 1904. There he constructed the first German wind tunnel in 1909 and built up an important centre for aerodynamics. Prandtl continued to work at Göttingen until his death. A US investigation team arriving there after World War II found that Prandtl, though still active, had not contributed greatly to the German effort. Prandtl died at Göttingen on 15 August 1953.

Prandtl's interest in fluid flow was triggered by his first task in industry, which was to improve a device for the removal of shavings. His studies revealed many weaknesses in the understanding of fluid mechanics. The current theory was unable to explain the observation that in a pipe a fluid would separate from the wall in a sharply divergent section instead of completely filling the pipe. In 1904, Prandtl published a very important paper which proposed that no matter how small the viscosity of a fluid, it is always stationary at the walls of the pipe. This thin static region or boundary layer has a profound influence on the flow of the fluid, and an understanding of the effects of boundary layers was developed to explain the action of lift and drag on aerofoils during the following half century.

In 1906, Prandtl was joined at Göttingen by Theodore von Kármán (1881–1963), who arrived as a student and subsequently became Prandtl's collaborator. In 1907, Prandtl investigated supersonic flow, extending the pioneering work of Ernst Mach (1838–1916) to slender bodies. From 1909, Prandtl's aerodynamic studies developed apace as the world became interested in flight. He turned to the problem of drag, which could not be fully explained by the skin friction produced by the boundary layer on a wing. In 1911 and 1912, Karman and Prandtl discovered how vortices cause drag. Prandtl's major contribution being an explanation of induced drag, which he showed was caused by lift inducing a trailing vortex. Much of this work was published after World War I and resulted in major changes in wing design and streamlining of aircraft.

The efforts of Prandtl and Karman in the 1920s led to a greater understanding of turbulent flow. In 1926, Prandtl developed the concept of mixing length – the average distance that a swirling fluid element travels before it dissipates its motion – and was able to produce a plausible theory of turbulence. The concept of mixing length can be thought of as being similar to the mean free path in the kinetic theory of gases. Prandtl's work on turbulence is still the basis of present day theory.

Modern-day aircraft with their high degree of streamlining and swept-back wings all owe their shapes to the work of Prandtl and Karman.

Prelog Vladimir 1906–   is a Bosnian-born Swiss organic chemist famous for his studies of alkaloids and antibiotics, and for his work on stereochemistry. He shared the 1975 Nobel Prize for Chemistry with John Cornforth.

Prelog was born in Sarajevo on 23 July 1906. He spent his early years in Zagreb and went to Czechoslovakia to study chemistry at the Institute of Technology in Prague, where he also did postgraduate research. From 1929 to 1934 he worked in Prague as a chemist in a laboratory for the preparation of fine chemicals. Then in 1935 he went back to Yugoslavia to become a lecturer and later Associate Professor of Organic Chemistry in the University of Zagreb's Technical Faculty. But in 1941, after the German occupation at the beginning of World War II, Prelog moved to Zurich to lecture at the Swiss Federal Institute of Technology. He became an Associate Professor in 1947, and ten years later he was made a full Professor, succeeding the 1939 Nobel prizewinner Leopold Ruzicka. Prelog retired in 1976.

Alkaloids were the subject of Prelog's early research, and one of his first achievements was the determination of the structure of the antimalarial quinine alkaloids. In 1945, he showed that Robert Robinson's formulae for strychnine alkaloids were incorrect (they were later rectified), and he derived the structures of steroid alkaloids from plants of the genera *Solanum* and *Veratrum* (the latter with Derek Barton). He also investigated many other alkaloids using classical organic chemistry, confirming the findings by X-ray crystallography.

In the 1940s, after his move to Zurich, Prelog studied many lipoid extracts from animal organs – the work with Ruzicka that resulted in the discovery of various steroids and the elucidation of their structures. Then, with W. Keller-Shlierlein, he investigated metabolic products of microorganisms and with a number of other researchers isolated various new complex natural products that have interesting biological properties. These include antibiotics and bacterial growth factors.

Many of these classes of metabolites have molecules that contain large rings. Prelog became interested in their stereochemistry, and looked at the relationships between the spatial structure and chemistry of many multiple-membered rings. He researched the steric course of asymmetric syntheses and succeeded in determining the then unknown absolute configurations of the steroids and terpenes. He used asymmetric synthesis as a sensitive tool for studying the details of reaction mechanisms, such as the synthesis of cyanhydrin.

Prelog demonstrated experimentally that some microorganisms have the ability to reduce the carbonyl group of certain alicyclic substrates in a highly stereospecific way. Together with Cahn and Christopher Ingold, he developed a widely used system for defining chirality (or handedness) in organic compounds and of stereoisomerism in general. The comprehensive molecular topology that evolved from this work is gradually replacing classical stereochemistry.

Prévost Pierre 1751–1839 was a Swiss physicist who first showed that all bodies radiate heat, no matter how hot or cold they are.

Prévost was born in Geneva on 3 March 1751. His father, a Calvinist minister, was very keen that his son should have the best possible education and Prévost was sent to study classics, science and theology, but ultimately he turned to the study of law and gained his degree in 1773. After graduating he became a teacher and held various positions in Holland, Lyons and Paris. On the death of his father he returned to Geneva and for a time was Professor of Literature there, but after a year he eventually decided to go back to Paris and work on the translation of a Greek drama.

In 1786, Prévost left Paris once more for Geneva, where he became active in politics. It was at this time that his interest in science re-emerged and he devoted much of his research to the problems of magnetism and then to heat. He was appointed Professor of Philosophy and General Physics at Geneva in 1793 and remained in this post until his retirement in 1823. In his later years, Prévost chose to abandon the earlier research in which he had won such a reputation in favour of studying the human ageing process. He used himself for his observations, noting down in detail every sign of advancement which his mind, body and mirror showed. A man of many talents, he used his ability for experimental research to the very end. He died in Geneva on 8 April 1839.

Prévost made his classic analysis of heat radiation in 1791. He conducted experiments to determine the heat properties of different kinds of objects under identical conditions. He found that dark, rough-textured objects give out more radiation than smooth, light-coloured bodies, given that both are at the same temperature. He also found that the reverse is true – that dark, rough objects absorb more heat radiation than light, smooth ones. From these experiments, Prévost conceived of heat as being a fluid composed of particles and that during radiation, the particles streamed out in the form of rays between radiating bodies.

This led to Prévost's theory of heat exchanges. If several objects at different temperatures are placed together, they exchange heat by radiation until all achieve the same temperature. All the objects can then remain at this temperature, however, only if they are receiving as much heat from their surroundings as they radiate away.

Even though Prévost's idea of heat being a fluid, which was the caloric theory current at that time, was erroneous, it did not prevent his basic ideas on heat radiation from being correct. In challenging the notion then prevalent that heat and cold are separate, cold being produced by the entry of cold into an object rather than by an outflow of heat, Prévost made a basic advance in our knowledge of energy. Further consideration of the subject a century later led to the quantum theory of Max Planck (1858–1947) and to the idea that heat and other forms of energy are in fact particulate in nature.

Priestley Joseph 1733–1804 was an English chemist and theologian. He entered chemistry when it was making the transition from alchemy to a theoretical science. An outstanding practical scientist, he combined experimental flair with quantitative accuracy, skills which led him to discover several new gases, including oxygen. He was less dynamic as a theorist; his conservatism made him a lifelong supporter of Georg Stahl's phlogiston theory of combustion despite mounting evidence – much of it provided by Priestley himself – refuting the principle. Outside his scientific work his life was far from harmonious. He was an outspoken man of radical views which brought him notoriety, and eventually drove him to leave his native country.

Priestley was born in Fieldhead, near Leeds, on 13 March 1733, the son of a cloth-dresser. His mother died when he was only seven years old and he was brought up by an aunt, who introduced him to Calvinism. In 1752, he attended the Dissenting Academy at Daventry, and three years later he entered the ministry as Presbyterian Minister at Needham Market, Suffolk. He moved to Nantwich, Cheshire, in 1758 and in 1761 became tutor in languages at Warrington Academy. A year later he married May Wilkinson, sister of the ironmaster John Wilkinson. On a trip to London in 1766 he met the American scientist Benjamin Franklin (1706–1790), who aroused in Priestley an interest in science; thereafter he combined scientific research with his clerical and social duties.

In 1767, Priestley returned to Leeds as minister of a chapel at Mill Hill. He did his most productive work between 1773 and 1780, when he was librarian and literary companion to Lord Shelburne, whom he accompanied on a journey to France in 1774. While in Paris he met Antoine Lavoisier and told him of his experiments with 'dephlogisticated air' (soon to be named oxygen by Lavoisier).

By 1780, Priestley's outspoken criticisms as a Dissenter had become an embarrassment to Lord Shelburne, who retired his companion on a small pension. Priestley moved to Birmingham to become minister of a chapel called the New Meeting. He also joined the Lunar Society, in company with the inventors James Watt and Matthew Boulton, Josiah Wedgwood, Erasmus Darwin (grandfather of Charles Darwin) and a number of less notable inventors and scientists. In Birmingham Priestley continued to voice loudly his opposition to the Established Church and his support of the French Revolutionaries. In 1791, on the second anniversary of the storming of the Bastille, the people of Birmingham rioted and vented some of their wrath on Priestley and other Dissenters, whose homes were ransacked. Priestley escaped to London and settled for a while in Hackney, but his unpopularity mounted, exacerbated by an offer of citizenship from France (by the very people who executed Lavoisier in 1794). In that same year Priestley emigrated to America, to Northumberland in Pennsylvania. He rejected the offer of a professorship at the University of Pennsylvania, preferring to live a life of comparative solitude in Northumberland, where he died on 6 February 1804.

Influenced by Franklin, Priestley's early work of 1767 onwards was in physics, particularly electricity and optics. He established that electrostatic charge is concentrated on the outer surface of a charged body and that there is no internal force. From this observation he proposed an inverse square law for charges, by analogy with gravitation. Priestley's house in Leeds was near a brewery, and it was his interest in the process of fermentation that turned him to chemistry, particularly gases. He experimented with the gas produced during fermentation – the layer of 'fixed air' (carbon dioxide) over a brewing vat – and showed it to be the same as that reported by Joseph Black in 1756. He dissolved the gas under pressure in water, beginning a European craze for soda water.

At Lord Shelburne's estate at Calne, Wiltshire, Priestley continued experimenting with gases. He used a large magnifying glass to focus the Sun's rays to produce high temperatures. He invented the pneumatic trough for collecting gases over water, and overcame the problem of handling water-soluble gases by collecting them over mercury.

An early discovery, in 1772, was 'nitrous air' (nitric oxide, or nitrogen monoxide, NO). Priestley found that a sample of the gas left in contact with iron filings and sulphur decreased in volume and that the new gas produced supported combustion. He had reduced nitric oxide to nitrous oxide (dinitrogen monoxide, N_2O), Humphry Davy's 'laughing gas'. In the same year he became the first person to isolate gaseous ammonia by collecting it over mercury (previously ammonia was known only in aqueous solution).

It had long been known that burning sulphur gives off a choking gas. In 1774, Priestley made the same gas by heating oil of vitriol (concentrated sulphuric acid) with mercury. He also produced it by heating the acid with copper turnings, a method still used today to make sulphur dioxide (SO_2).

Priestley's most famous discovery was that of oxygen. In 1772, he had shown that a gas necessary to animal life is liberated by plants. Two years later he prepared the same gas by heating red calyx of mercury (mercury(II) oxide, HgO) or minium (red lead, Pb_3O_4). His investigation of the properties of the new gas showed it to be superior to common air. A mouse trapped in a container of it stayed conscious twice as long as in ordinary air, and breathing it had no adverse effects (apart from leaving a peculiar light feeling in the chest). When he mixed the new gas with nitrous air (NO) there was a diminution in volume and yet another, red gas (nitrogen dioxide, NO_2) was formed. From all of these observations Priestley concluded that he had prepared dephlogisticated air – that is, air from which the fiery principle of phlogiston had been removed. The Swedish chemist Karl Scheele independently prepared oxygen in 1772, but his tardiness in publication resulted in Priestley being credited with the discovery.

Prigogine Ilya 1917– is a Russian-born Belgian theoretical chemist who was awarded the 1977 Nobel Prize for Chemistry for widening the scope of thermodynamics from the purely physical sciences to ecological and sociological studies.

Prigogine was born in Moscow on 22 January 1917. When he was four years old his parents emigrated to Western Europe, and settled in Belgium in 1929. He studied at the University of Brussels, gaining his doctorate in 1941. In 1951, he became Professor of the Université Libre de Bruxelles and in 1959 was appointed Director of the Instituts Internationaux de Physique et de Chemie. From 1961 to 1966 he was Professor of the Department of Chemistry at the Enrico Fermi Institute for Nuclear Studies and the Institute for the Study of Metals at the University of Chicago. From 1967 he held the position of Director of the Center for Statistical Mechanics and Thermodynamics at the University of Texas in Austin, concurrently with his professorship in Brussels.

Prigogine's work has been concerned with applying

thermodynamic principles to new disciplines. Observation of many physical systems has shown that there is a general tendency to assume the state in which they are most disordered. This occurs by means of processes that dissipate energy and which can in principle produce work. But it was not understood how it is possible for a more orderly system, such as a living creature, to arise spontaneously from a less orderly system and yet maintain itself despite the tendency towards disorder. It is now known that order can be created and preserved only by processes that flow 'uphill' in the thermodynamic sense. They are compensated by 'downhill' events. These interrelated occurrences owe their existence to the absorption of energy from the surroundings and are consistent with thermodynamic laws.

During the late 1940s Prigogine developed mathematical models of what he called dissipative systems of this kind, to show how they might have come about. His models demonstrated how matter and energy can interact creatively, forming organisms that can sustain themselves and grow in opposition to the general drift towards universal chaos. Dissipative systems can exist only in harmony with their surroundings.

Prigogine showed that dissipative systems exhibit two types of behaviour: close to equilibrium, their order tends to be destroyed; far from equilibrium, order can be maintained and new structures formed. The probability of order arising out of disorder, by pure chance, is infinitesimal; but the formation of an ordered dissipative system makes it possible to create order out of chaos. These ideas have been applied to examine how life originated on Earth, to the dynamic equilibria in ecosystems, to the preservation of world resources, and even to the prevention of traffic jams.

Pringsheim Ernst 1859–1917 was a German physicist whose experimental work on the nature of thermal radiation led directly to the quantum theory.

Pringsheim was born in Breslau, Germany (now Wroclaw, Poland) on 11 June 1859. He studied at gymnasia in Breslau and in 1877 entered the University of Heidelberg, moving on to Breslau in 1878 and Berlin in 1879. He gained his PhD at Berlin in 1882 and became a lecturer there in 1886. Pringsheim became Professor of Physics at Berlin in 1896 and in 1905 moved to Breslau to take up the post of Professor of Experimental Physics. He died at Breslau on 28 June 1917.

In 1881, Pringsheim developed an infrared spectrometer that made the first accurate measurements of wavelengths in the infrared region. He put the instrument to its most important use from 1896 onwards when he began to collaborate with Otto Lummer (1860–1925) on a study of blackbody radiation. This led to a verification of the Stefan–Boltzmann law that relates the energy radiated by a body to its absolute temperature. Pringsheim and Lummer then proceeded to make careful measurements of the distribution of energy with frequency at various temperatures. They confirmed Wien's displacement law, which relates the peak frequency of maximum energy to the temperature, but in 1899 found anomalies in radiation laws that had been devised to express the energy of the radiation in terms of its frequency and temperature. The results encouraged Max Planck (1858–1947) to find a new radiation law that would account for the experimental results and in 1900, Planck arrived at such a law by assuming that the energy of the radiation consists of indivisible units that he called quanta. This marked the founding of the quantum theory.

Proust Joseph Louis 1754–1826 was a French chemist who discovered the law of constant composition, sometimes also called Proust's law, which states that every true chemical compound has exactly the same composition no matter how it is prepared.

Proust was born in Angers on 26 September 1754, the son of an apothecary. He was brought up to follow his father's profession, studying chemistry under Guillaume Rouelle at the Jardin du Roi and working as an apothecary-chemist in La Salpêtrière Hospital in Paris and lecturing at the Palais Royal. In the 1780s, before the beginning of the French Revolution, he went to Spain and spent the next 20 years in Madrid. He taught at various academies and carried out his own research in a well-equipped laboratory (The Royal Laboratory) provided by his patron, King Charles IV of Spain. In 1808, Napoleon invaded Spain and French soldiers wrecked Proust's laboratory. He returned to France a poor man, was elected to the Academy of Sciences in 1816, and eked out his retirement on a small pension provided to him as an Academician by Louis XVIII. He died in Angers on 5 July 1826.

Proust's reputation as a chemist rests on his extraordinary ability as an analyst. He identified grape sugar (glucose) and distinguished between it and sugar from other sources. Before his work on chemical compounds in the early 1800s in Madrid, the prevailing view in chemistry was that of Claude Berthollet who had stated (and in 1803 published in his *Statique Chimique*) that the composition of compounds could vary over a wide range, depending on the proportions of reactants used to produce them. In 1799 Proust prepared and analysed copper carbonate produced in various ways and compared the results with those obtained by analysing mineral deposits of the same substance; he found that they all had the same composition. Similar results with other compounds led Proust to propose the law of constant composition (he ascribed the errors in Berthollet's experiments to impurities and inaccurate analyses). Proust's law influenced John Dalton's thinking about atomic theory – in 1808 Dalton proposed the law of definite proportions. The proportions of the elements in a compound result from the linking of definite (usually small) numbers of atoms to form molecules, giving the compound a constant composition. After a long controversy Berthollet finally conceded that Proust was right. Both chemists did agree, however, that the rate of a chemical reaction does depend on the masses of the reactants.

Prout William 1785–1850 was a British chemist who pioneered physiological chemistry, but who is best known for formulating Prout's hypothesis, which states that the atomic weights (relative atomic masses) of all elements are exact multiples of the atomic weight of hydrogen. And since at that time (1815) the atomic weight of hydrogen was taken to be 1.0, the hypothesis implied that all atomic weights are whole numbers. Prout has often been confused with his contemporary the French chemist Joseph Proust (and Prout's hypothesis has been mistaken – in name – for Proust's law of constant composition).

Prout was born in Horton, Gloucestershire, on 15 January 1785 into a prosperous and well-established West Country family. He began by studying medicine at Edinburgh University, qualifying in 1811. He then set up a medical practice in London and established a private chemical laboratory. From 1813 he wrote about and gave lectures in 'animal chemistry' and began his own researches into the chemistry of physiological processes. He published his atomic weight hypothesis anonymously in 1815. He continued to experiment, widening his interests to include some physics. He died in London on 9 April 1850.

In his early researches Prout studied various natural secretions and products, including blood, urine, gastric juices, kidney and bladder stones, and even cuttlefish ink. He became convinced that the products of secretion derive from the chemical breakdown of body tissues. In 1818, he isolated urea and uric acid for the first time, and six years later he found hydrochloric acid in digestive juices from the stomach. In 1827, he became the first scientist to classify the components of food into the three major divisions of carbohydrates, fats and proteins.

His anonymous paper of 1815 was comprehensively entitled 'On the Relation between the Specific Gravities of Bodies in their Gaseous State and the Weight of Their Atoms'. From the determinations of atomic weights that had been made, Prout observed that many were whole-number multiples of that of hydrogen. The hypothesis implied that all other elements were in some way multiples or compounds of hydrogen, which was therefore the basic building block of matter. In 1920 Ernest Rutherford named the proton (hydrogen nucleus) which is a constituent of every atomic nucleus and is, therefore, indeed a 'basic building block of matter', after Prout (*proton*).

Prout's hypothesis gave even more stimulus to the making of accurate determinations of atomic weights – work that inevitably proved the hypothesis to be wrong (for example chlorine has a relative atomic mass of 35.5). The idea was therefore largely abandoned until the work on isotopes by Frederick Soddy and others more than a century later, finally accounted correctly for non-integral atomic weights (resulting from a natural mixture of isotopes).

Prout also studied the gases of the atmosphere and in 1832 made accurate measurements of the density of air. He devised a barometer for making precise atmospheric pressure measurements, and the Royal Society adopted its design for a national standard barometer.

Ptolemy Claudius Ptolemaeus lived AD 2nd century, was an astronomer, astrologer, geographer and philosopher, probably of Egyptian extraction, working in a centre of Greek culture technically under Roman domination. His collected works on astronomical themes – known generally as the *Almagest* (although he called it *Syntaxis*) – influenced astronomical and religious conceptions for at least 13 centuries after his death.

Almost nothing is known of Ptolemy's life. His name suggests to some commentators that he was born in the city of Ptolemais Hermii, on the banks of the Nile. Certainly it was at Alexandria, in the Nile Delta, that he mostly lived and worked, setting up his observatory on the top floor of a temple in order to view the heavens with greater clarity. The exact date of his death is unknown, though there is some evidence that he may have lived to around the age of 78, and died in either AD 141 or 151.

Ptolemy had often been accused of plagiarizing the theories of Hipparchus, the Greek astronomer and philosopher of two centuries earlier, and of using Plato as his authority for adapting Hipparchus where necessary to maintain his own point of view. It is sometimes forgotten, however, how many sources Ptolemy had at his disposal. Living in Alexandria, whose library was the repository of the greatest store of knowledge in the world, Ptolemy had to hand the entire accumulated wealth of information compiled by Greek and other scholars of the civilized world during the previous four centuries. Some of those scholars, particularly Greek ones (such as Aristarchos), had proposed a Sun-centred (heliocentric) Solar System. Even Hipparchus favoured a heliocentric universe. Yet Ptolemy managed to put together a cosmology that it would be difficult to make less accurate – and for various reasons (at least as far as Western Europe was concerned) the science of astronomy stuck fast at that point.

It was Plato (*c.*427–*c.*347 BC) who loved symmetry and sought mathematical perfection in the workings of nature,

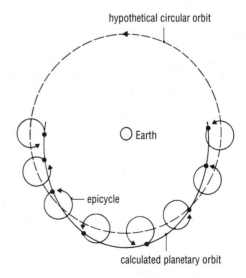

Ptolemy *Ptolemy developed the idea of epicycles, first proposed by Aristarchos, to explain the observed departure of planetary orbits from perfect circles.*

who probably inspired Ptolemy as much as Hipparchus did. Ptolemy began with the premise that the Earth was a perfect sphere. Because gravity brought things demonstrably towards the centre of the Earth, all celestial bodies must likewise conform to it. Thus the Earth was at the centre of the Cosmos, and the Moon, Mercury, Venus, the Sun, Mars, Jupiter, Saturn, and the stars (in that order), in their various spheres, orbited the Earth. All orbits were circular, but Earth was not always at the centre, and indeed of Mercury and Venus, and possibly Mars (Ptolemy was not sure), the orbits were epicyclic (the planets orbited a point which itself was orbiting the Earth). The sphere of the stars comprised a dome with points of light attached or pricked through.

Apart from Ptolemy's interest in astronomy and his desire to find mathematical answers to natural problems, he also studied geography in great detail, producing vivid maps of Asia and large areas of Africa. These, together with his notes on longitude and latitude, were combined into a collection known as his *Geography*, and it was from these maps that Christopher Columbus many centuries later decided that it ought to be possible to reach India by sailing west across the Atlantic.

The work of Ptolemy that is most available today is his thesis on astrology, the *Tetrabiblios*, in which he suggests that some force from the stars may have considerable influence over the lives and events in the human experience.

Ptolemy's legacy to the world was an attempt at a complete understanding of all he could observe. It has not proved accurate, but it was popular and it endured for hundreds of years.

Purkinje Jan Evangelista 1787–1869 was a Czech histologist and physiologist whose pioneering studies were of great importance to our modern knowledge of vision, the functioning of the brain and heart, pharmacology, embryology and histology.

Purkinje was born on 17 December 1787 in Libochovice, Bohemia (now in the Czech Republic), and was educated there by piarist monks (members of a religious congregation established in 1597 to educate the poor). Before being ordained a priest, however, he went to Prague University to study philosophy but changed to medicine, in which he graduated in 1819 with his famous thesis on the visual phenomenon now known as the Purkinje effect. After graduating he worked as an assistant in the Department of Physiology at Prague University until 1823, when he was appointed Professor of Physiology and Pathology at the University of Breslau (now Wroclaw in Poland) – perhaps through the influence of the German poet Goethe, who had previously befriended Purkinje. At Breslau University Purkinje founded the world's first official physiological institute. In 1850, he returned to Prague University as Professor of Physiology, a post he held until his death on 28 July 1869 (in Prague).

In his famous graduation thesis Purkinje described the visual phenomenon in which, as the light intensity decreases, different coloured objects of equal brightness in high light intensities appear to the eye to be unequally bright – blue objects appear brighter than red objects; this phenomenon is now called the Purkinje effect. In 1832, he was the first to describe what are now known as Purkinje's images: a threefold image of a single object seen by one person reflected in the eye of another person. This effect is caused by the object being reflected by the surface of the cornea and by the anterior and posterior surfaces of the eye lens.

Probably Purkinje's best known histological work was his discovery in 1837 of large nerve cells with numerous dendrites found in the cortex of the cerebellum; these cells are called Purkinje cells. Two years later he discovered the Purkinje fibres – atypical muscle fibres lying beneath the endocardium that conduct the pacemaker stimulus along the inside walls of the ventricles to every part of the heart. Also in 1839, in describing the contents of animal embryos, Purkinje was the first to use the term protoplasm in the scientific sense.

Purkinje made numerous other important discoveries and observations. In 1823, he recognized that fingerprints can be used as a means of identification. In 1825, while examining birds' eggs, he discovered the germinal vesicle, or nucleus, of unripe ova; this structure is now sometimes called the Purkinje vesicle. He discovered the sweat glands in skin in 1833 and, in 1835, described in detail the structure of the skin. In that year he also described ciliary motion. In 1836, he observed that pancreatic extracts can digest protein, and in 1837, he outlined the principal features of the cell theory – before Theodor Schwann and Matthias Schleiden enunciated this theory in detail. Purkinje also described the effects on the human body of camphor, opium, belladonna and turpentine. And he did much to improve microscopical techniques, being the first to use the microtome, and using glacial acetic acid, potassium dichromate and Canada balsam in the preparation of tissue samples; moreover, he was one of the first to teach microscopy as part of his university course.

In addition to his scientific work, Purkinje also translated the poetry of Goethe and Schiller.

Pye John David 1932– is a British zoologist who has performed important research in the field of ultrasonic bioacoustics, particularly in bats.

Pye was born on 14 May 1932 in Mansfield, Nottinghamshire. He was educated at Queen Elizabeth's Grammar School, University College of Wales at Aberystwyth (from which he graduated in 1954), and London University (from which he obtained his doctorate in 1961). From 1958 to 1964 he was a zoology research assistant, after which he became a lecturer at King's College, London. In 1970, he was appointed Reader at King's College, then in 1973 he joined the staff of Queen Mary College, London, initially as Professor of Zoology then from 1977 as Head of the Zoology Department.

A surprisingly large number of animals use ultrasound (which has a frequency above about 20 kHz and is inaudible to humans) – bats, the Cetaceae (whales, porpoises and dolphins) and many insects, for example. Because of the lack of sufficiently sophisticated detection devices, the phenomenon was not discovered until 1935 (by Pierce, then Professor of Physics at Harvard University), although the first indication that bats use a system other than sight for navigation came with Lazzaro Spallanzani's discovery in 1794 that blinded bats still managed to find food. Even today, with sensitive electronic instruments widely available, biological ultrasound is relatively little studied. Pye is one of the few investigators in this field and has examined the use of ultrasound in many different animals, although he is best known for his work on echolocation in bats. In 1971, he calculated the resonant frequencies of the drops of water in fog and found that these frequencies coincided with the spectrum of frequencies used by bats for echolocation. For this reason, fog absorbs the ultrasound emitted by bats and renders useless their echolocation systems; this is probably the reason why bats avoid flying in fog. Pye also found that ultrasound seems to be important in the social behaviour of rodents and insects which, since many of these creatures are pests, raises the possibility of developing novel control measures.

Pyman Frank Lee 1882–1944 was a British organic chemist, famous for his contributions to pharmaceutics and chemotherapy.

Pyman was born in Malvern on 9 April 1882. He was educated at Dover College, where his interest in chemistry began, and in 1899 entered Owens College of Victoria University, Manchester, at a time when it was the centre of organic chemistry research in Britain. He graduated in 1902 and went to Zurich Polytechnic for two years, but because Zurich University did not at that time recognize Polytechnic students he submitted his PhD thesis to Basle University, which granted him the degree in 1904.

On his return to Britain, Pyman took a job in the Experimental Department of the Wellcome Chemical Works at Dartford, Kent, in 1906. During World War I he worked on the preparation of drugs needed to treat British troops overseas. In 1919 he was appointed Professor of Technological Chemistry in Manchester University and Head of the Department of Applied Chemistry at the College of Technology. He stayed at Manchester for eight years then in 1927 took up the appointment of Director of Research at the Boots Pure Drug Company's laboratories at Nottingham, where he remained until he died, in Nottingham, on 1 January 1944.

Pyman began research at the Wellcome Chemical Works under Jowett and worked with him for nine years, resulting in his lifelong interest in the glyoxalines (glyoxal is ethanedial), cyclic amidines with therapeutic properties. He was particularly interested in the relationship between their chemical constitution and their physiological action. He later tried to relate chemical constitution with the local anaesthetic action of the substituted amino alkyl esters. After 1915 Jowett moved to the Imperial Institute, but he continued to send Pyman samples for examination. From one such (bark of *Calmatambin glabrifolium*) Pyman

isolated a glycoside, which led him to study the constitution of the anhydro-bases made from it by the Hofmann degradation – which converts an amide to an amine with one carbon atom less. This work placed him in the forefront of British organic chemistry and revealed the existence of a substance whose molecules contained a ten-membered heterocyclic ring. He also examined alkaloids, and became the first person to isolate a natural substance containing an asymmetric nitrogen atom.

Pyman undertook the preliminary processing work in connection with the preparation of the drugs Salvarsan (arsphenamine) and Neosalvarsan, which were needed at the outbreak of World War I to deal with syphilis. He also synthesized an alkaloidal compound used in the treatment of amoebic dysentery among troops during the war. Another alkaloid was used as a uterine haemostat under the name 'Lodal'.

Among compounds synthesized by Pyman were histidine and other simple bases of biological importance, such as guanidine. He continued to work with glyoxalines, synthesizing them and testing their effectiveness as antiseptics, pressor drugs, antimalarials and hypoglycaemic substances. He also investigated arsenicals, acridines (used as powerful antiseptics) and the organic salts of bismuth. He studied the relationship between chemical constitution and the pungency of amides, and examined the preservative properties of hops.

Pythagoras lived $c.580–c.500$ BC, was an ancient Greek religious philosopher, part of whose mystic beliefs entailed an intense study of whole numbers, the effect of which he sought to find in the workings of nature. He founded a famous school which lived as a cultic community governed by what might now be considered eccentric – if not downright primitive – rules, but which during and after his lifetime discovered an astonishing number of facts and theorems, some immortal.

Very little is known about the life of Pythagoras, other than that he was born on the island of Samos and (possibly) obliged to flee the despotism of its ruler, Polycrates, he (probably) travelled extensively. His work seems to show the influence of contemporary ideas in Asia Minor; nevertheless, Pythagoras is next authoritatively recorded in southern Italy, in the Dorian colony of Crotona, in about 529 BC. There he became the leader of a religious community that had political pretensions to being an association for the moral reform of society. The Pythagorean brotherhood flourished; as a mathematical and philosophical community it was extending science rapidly, and as a political movement it was extending its influence over several western Greek colonies. More distant colonies put up some physical resistance, however, and it was probably one act of suppression in particular – led by one Cylon – that saw Pythagoras exiled (yet again) to Metapontum until he died, possibly around 500 BC. The school continued for something like another 50 or 60 years before being finally and totally suppressed.

Pythagoras and his community looked for numerical

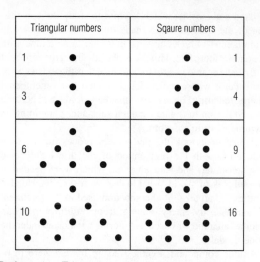

Triangular numbers	Sqaure numbers

Pythagoras *Triangular numbers and square numbers were part of the foundations of number theory investigated by the Ancient Greeks who noted, for example, that the sum of two adjacent triangular numbers always equals a square.*

values in all they saw around them, and strove to create relationships in the values they found. In elementary pure mathematics they studied the properties of the numbers themselves, and their practice of representing numbers as lines, triangles or squares of pebbles has given us our word calculate (from *calculus*, the Greek for pebble). It also led

directly to a firm basis for geometrical considerations. In this way, they established that the addition of each successive odd number after 1 to the preceding ones results in a square ($1 + 3 = 2^2$; $1 + 3 + 5 = 3^2$, and so on) and Pythagoras himself is supposed ultimately to have arrived at the theorem to which his name is attached, regarding right-angled triangles. (In fact he is supposed to have proved it from a more general equation he is said also to have formulated:

$$m^2 + \left\{ \frac{1}{2}(m^2 - 1) \right\}^2 = \left\{ \frac{1}{2}(m^2 + 1) \right\}^2$$

and to have noted that if the triangle in question is isosceles, the ratio of the hypotenuse to either side is the irrational number 2.)

Using geometrical principles, the Pythagoreans were able to prove that the sum of the angles of any regular-sided triangle is equal to that of two right angles (using the theory of parallels), and to solve any algebraic quadratic equations having real roots. They formulated the theory of proportion (ratio), which enhanced their knowledge of fractions, and used it in their study of harmonics upon their stringed instruments: the harmonic of the octave was made by touching the string at $\frac{1}{2}$ its length, of a fifth at $\frac{2}{3}$ its length, and so on. Pythagoras himself is said to have made this the basis of a complete system of musical scales and chords.

He is said also to have taken a keen interest in astronomy, seeking numerical consistency among the celestial movements and objects.

R

Rainwater (Leo) James 1917– . US physicist; *see* Bohr, Mottelson, and Rainwater.

Raman Chandrasekhara Venkata 1888–1970 was an Indian physicist who discovered that light is scattered by the molecules in a gas, liquid or solid so as to cause a change in its wavelength. This effect is known as Raman scattering and the Raman spectra produced are used to obtain information on the structure of molecules. For this, Raman was awarded the 1930 Nobel Prize for Physics.

Raman was born in Trichinopoly, Madras, on 7 November 1888. He studied at the A.V.N. College in Vizagapatam, where his father was Professor of Mathematics and Physics, and at the Presidency College of the University of Madras where he obtained a BA in 1904 and an MA in 1907. Although aged only 16 and 19 respectively, Raman gained first-class degrees of great distinction. However, he was unable to continue his scientific education as this would have meant leaving India, and because there were no opportunities for a scientific career in India, Raman entered the financial division of the civil service in 1907, working as an accountant in Calcutta for ten years.

During this time, Raman pursued his studies privately, using the facilities at the laboratories of the Indian Association for the Cultivation of Science in Calcutta. He concentrated his efforts on investigations into vibration in sound and the theory of musical instruments, an interest that continued throughout his life. His work on these subjects and on diffraction prompted an offer of the professorship of physics at the University of Calcutta, which Raman took up in 1917.

Raman remained at Calcutta until 1933, and it was during this period of his life that he made his most important contributions to physics. In 1926, he established the *Indian Journal of Physics* and in 1928 was made President of the Indian Science Congress. Further honours came with the award of a knighthood by the British government in 1929 and the Hughes Medal of the Royal Society in 1930, in addition to the Nobel Prize for Physics in the same year.

In 1934, Raman became head of the physics department at the Indian Institute of Science, a post he held until 1948,

also serving as President of the Institute from 1933 to 1937. The Indian government then built the Raman Research Institute for him at Bangalore and he became its first Director in 1948. Raman took his duties as an educator very seriously and trained a great number of young scientists who later rose to positions of responsibility in science. Among his colleagues at the Institute was Homi Bhabha (1909–1966). Raman remained Director of his Institute until he died at Bangalore on 21 November 1970.

Raman was inspired to work on the scattering of light in 1921, when returning to India by sea from a conference in Britain. He was struck by the intense blue colour of the Mediterranean Sea, which he could not reconcile with the explanation put forward by Lord Rayleigh (1842–1919) that attributed the blue colour to the scattering of light by particles suspended in the water. Raman began to investigate this phenomenon upon his arrival in Calcutta. He showed that the blue colour of the sea is produced by the scattering of light by water molecules.

In 1923, Arthur Compton (1892–1923) discovered the Compton effect, in which X-rays are scattered on passing through matter and emerge with a longer wavelength. This was explained by assuming that X-ray particles or photons had collided with electrons and lost some energy. In 1925, Werner Heisenberg (1901–1976) predicted that this effect should be observed with visible light. Raman had already come to the same conclusion independently and had in fact made a preliminary observation of this light-scattering effect in 1923. He then refined his experiments and in 1928 was able to report the existence of the scattering effect for monochromatic light in dust-free air and pure liquids. The Raman spectra produced show lines displaced to either side of the normal line in a gas and a continuous band in dense liquids. The effect is caused by the internal motion of the molecules encountered, which may impart energy to the light photons or absorb energy in the resulting collisions. Raman scattering therefore gives precise information on the motion and shape of molecules.

Raman's other research included the effects of sound waves on the scattering of light in 1935 and 1936, the vibration of atoms in crystals in the 1940s, the optics of gemstones, particularly diamonds, and of minerals in the 1950s, and the physiology of human colour vision in the 1960s.

Raman's discovery is important not only because it affords a method for the analysis of molecular structure but also because in demonstrating conclusively that light may behave as particles (photons), it confirms the quantum theory. However, Raman is also to be remembered as a pioneer of Indian science.

Ramanujan Srinavasa Ayengar 1887–1920 was an Indian mathematician who, virtually unaided and untaught, by his own independent endeavour reached an exceptionally high standard in particular aspects of mathematics. His work and his evident desire for further study impressed the Cambridge mathematician Godfrey Hardy (1877–1947) so much that he was offered a scholarship to Trinity College. Despite the brevity of the remainder of his life, Ramanujan established a worldwide reputation.

Ramanujan was born in Erode, near Kumbakonam, in Tanjore, Madras, on 22 December 1887. His family was poor, although of the high Brahmin caste. At school Ramanujan did himself and his teachers credit until, at the age of 15, he read and became fascinated by Carr's textbook *A Synopsis of Elementary Results in Pure and Applied Mathematics* (published in 1880). Obsessed by the section on pure mathematics – which he soon learned by heart – he won a scholarship to the state college at Kumbakonam, but failed at the end of his first year because he had devoted too much of his time and energy to mathematics, to the neglect of other subjects (English in particular). Unemployed, he nevertheless developed his own theorems and hypotheses from the mathematics he knew.

Finances became even worse in 1909 when he got married. In the next three years, however, he was given a small grant by Ramachaudra Rao that enabled him to maintain his mathematical investigations and begin publishing his results, and he found work as a clerk in the offices of the Madras Port Trust. Then began his correspondence with Godfrey Hardy at Cambridge University. Deeply impressed both by the results of Ramanujan's work to date, and by the surprising gaps in his knowledge of certain aspects (later found to be gaps in Carr's textbook), Hardy eventually managed to persuade the somewhat reluctant Indian to take up the scholarship offered. Ramanujan arrived in 1914, and published 25 papers in European journals over the next five years. During that time, however, he contracted tuberculosis, and although the pleasure afforded him by his election as Fellow of the Royal Society in 1918 temporarily alleviated his condition, he was obliged to return to the more favourable climate of India in the following year.

But although he kept on working, his health deteriorated further, and he died in Chetput, near Madras, on 26 April 1920. His collected papers were edited by Godfrey Hardy and published in 1927.

Carr's textbook, and particularly the section on pure mathematics, was the firm foundation on which Ramanu-

jan based all his original work. In fact, he first scrutinized the book in detail from start to finish, checking its theorems and examples. From the knowledge he gained he was able to proceed beyond the material published and develop his own results in many fields, but particularly in function theory and number theory. When he arrived at Cambridge, Hardy was amazed at the standard he had reached in these areas – although Ramanujan knew nothing of subjects that had not been covered in Carr's book. Hardy was thus obliged to ensure that Ramanujan was given a speedy, but thorough, grounding in (for example) doubly periodic functions, the techniques of rigorous mathematical proofs, Cauchy's theorem and other items that were all new ground. Hardy was also astonished to find that Ramanujan did much of his calculation mentally, and had achieved many of his creditable results by working through a large number of examples to check correlation.

It was perhaps in number theory that Ramanujan made his most enduring contribution. With Hardy he published a theory on the methods for partitioning an integer into a sum of smaller integers, called summands. In function theory he found accurate approximations to π, and worked on modular, elliptic and other functions. Because of his background, much of his work was of extraordinary originality – although he had the tendency also, unwittingly, to 'discover' theorems and statements first formulated by other notable mathematicians from Pythagoras onwards.

Ramón y Cajal Santiago 1852–1934 was a Spanish histologist whose research revealed that the nervous system is based on units of nerve cells (neurons). For his discovery he shared the 1906 Nobel Prize for Physiology or Medicine with Camillo Golgi.

Ramón y Cajal was born on 1 May 1852, in Petilla de Aragon, Spain, the son of a country doctor. At the insistence of his father he studied medicine at the University of Zaragoza, but he took little interest in the course other than anatomy. He qualified in 1873 and then joined the army medical service, which sent him to Cuba. There he caught malaria and was discharged; he returned to Zaragoza and obtained a doctorate in anatomy in 1877. In 1884, he was appointed Professor of Descriptive Anatomy at the University of Valencia and in 1887 he took up the Histology professorship at the University of Barcelona. From 1892 to 1921 he served at the University of Madrid as Professor of Histology and Pathological Anatomy becoming Director of the new Instituto Nacional de Higiene in 1900. In 1921 he retired from the university to become Director of the Cajal Institute in Madrid, founded in his honour by King Alfonso XIII, and retained this position until he died, in Madrid, on 17 October 1934.

When Ramón y Cajal commenced his research, the path of a nervous impulse was unknown. In his investigations he used potassium dichromate and silver nitrate to stain sections of embryonic tissue, improving on the procedure developed by Golgi.

By this means he demonstrated that the axons of neurons end in the grey matter of the central nervous system and never join the endings of other axons or the cell bodies of other nerve cells. He considered that these findings indicate that the nervous system consists entirely of independent units and is not a network as was previously thought. In 1897, Ramón y Cajal investigated the human cerebral cortex using methylene blue (also used by Paul Ehrlich) as well as Golgi's silver nitrate stain. He described several types of neurons and demonstrated that there were distinct structural patterns in different parts of the cerebral cortex. His findings indicated that structure might well be related to the localization of a particular function to a specific area. In 1903, he found that silver nitrate stained structures within the cell body, which he identified as neurofibrils, and that the cell body itself was concerned with conduction.

During his years at Madrid University, Ramón y Cajal concerned himself with the generation and degeneration of nerve fibres. He demonstrated that when a nerve fibre regenerates it does so by growing from the stump of the fibre still connected with the cell body. In 1913, he developed a gold sublimate to stain nerve structures, which is now valuable in the study of tumours of the central nervous system.

Modern neurology has its foundations in Ramón y Cajal's meticulous work because his investigations are the basis of modern understanding of the part played by the neuron in the nervous function, and of the nervous impulse. He published numerous scientific papers and books, among them the classic *Structure of the Nervous System of Man and other Vertebrates* (1904) and *The Degeneration and Regeneration of the Nervous System* (1913–1914).

Ramsay William 1852–1919 was a British chemist famous for his discovery of the rare gases, for which achievement he was awarded the 1904 Nobel Prize for Chemistry.

Ramsay was born in Glasgow on 2 October 1852, the son of an engineer (whose father, Ramsay's grandfather, founded the Glasgow Chemical Society). Despite this technical and scientific background, Ramsay received a classical education and entered the University of Glasgow in 1866, when he was only 14 years old, to take an arts course. Two years later he went to work in the laboratory of the Glasgow City Analyst, where he soon made up the

followed by a post at the University. In 1880, he was appointed Professor of Chemistry at the newly created University College of Bristol (later Bristol University) and a year later became Principal of the College. Then in 1887 he moved to become Professor of Chemistry at University College, London, as successor to Alexander Williamson, where he remained until he retired in 1912. He was knighted in 1902. After retirement he moved to a house near High Wycombe, Buckinghamshire, where he continued some research in converted stables. At the outbreak of World War I in 1914 he became busy as a member of various committees. But his health deteriorated and he died at High Wycombe on 23 July 1916.

At University College, London, his first action was to reorganize the out-of-date laboratory. He and his students investigated diketones, the metallic compounds of ethylene (ethene), and the atomic weight (relative atomic mass) of boron. Ramsay became interested in an article in *Nature* (September 1892) by Lord Rayleigh in which he reported finding a difference in the densities of samples of nitrogen extracted from air and from chemical sources. After corresponding with Rayleigh, Ramsay undertook to study the problem. With the help of his assistant Percy Williams, he passed nitrogen from air over heated magnesium (to form magnesium nitride). After this treatment, about 6% of the gas still remained; further treatments reduced the volume of the residual gas even further, until they were left with an unknown gas of density 20 (oxygen's is 16). Despite losing the sample in a laboratory accident, they finally established that it was the new element argon (which had contaminated the nitrogen derived from air).

Early in 1895, Ramsay became interested in helium, a gas known from spectrographic evidence to be present on the Sun but yet to be found on Earth. W. F. Hillebrand had reported that certain uranium minerals produced an inert gas on heating, and Ramsay repeated these experiments and obtained sufficient of the gas to send a sample to William Crookes for spectrographic analysis. Crookes confirmed that it was helium. Ramsay and his co-workers soon made the connection between helium and argon and in his book *The Gases of the Atmosphere* (1896) he repeated his earlier suspicion that there was an eighth group of new elements at the end of the periodic table. He drew up a table with gaps for the unknown elements:

hydrogen	1.01	helium	4.2	lithium	7.0
fluorine	19.0	?		sodium	23.0
chlorine	35.5	argon	39.2	potassium	39.1
bromine	79.0	?		rubidium	85.5
iodine	126.9	?		caesium	132.0
?	169.0	?		?	170.0

deficiencies in his science education, and in 1870 he left for Germany to carry out research in organic chemistry under Rudolf Fittig (1835–1910) at Tübingen, gaining his PhD in 1873.

Ramsay then returned to Glasgow as an assistant at Anderson's College (later the Royal Technical College),

During the next decade Ramsay and Morris Travers sought the remaining rare gases by the fractional distillation of liquid air. Ramsay often used demonstrations and public lectures to show the existence of these gases, and he announced the discovery of neon in 1894 at a meeting in Toronto. Krypton and xenon took until 1898 to isolate. The last member of the series, radon, is a product of radioactive decay. It was identified in 1901 from a minute sample prepared by Ramsay and Robert Whytlaw-Gray (1877–1958).

Rankine William John Macquorn 1820–1872 was a Scottish engineer and physicist who was one of the founders of the science of thermodynamics, especially in reference to the theory of steam engines.

Rankine was born in Edinburgh on 5 July 1820, the son of an engineer. He trained as a civil engineer under John Benjamin MacNeil, and was eventually appointed to the Chair of Civil Engineering and Mechanics at the University of Glasgow in 1855, a position he continued to hold until his death on 24 December 1872. In 1853, he became a Member (and later a Fellow) of the Royal Society, whose official catalogue credits him with 154 papers.

One of Rankine's earliest scientific papers, on metal fatigue in railway axles, led to improved methods of construction in this field. In about 1848, he commenced the series of researches on molecular physics which occupied him at intervals during the rest of his life and which are among his chief claims to distinction in the realm of pure science. In 1849 he delivered two papers on the subject of heat. His first work, 'On an Equation between the Temperature and the Maximum Elasticity of Steam and other Vapours' was published in the *Edinburgh New Philosophical Journal* and at the end of the year he sent to the Royal Society of Edinburgh his paper, 'On a Formula for Calculating the Expansion of Liquids by Heat'.

In 1853, together with James Robert Napier, he projected and patented a new form of air engine, but this patent was afterwards abandoned. It was in 1859 that he produced what is perhaps his most influential work. *A Manual of the Steam Engine and other Prime Movers*, the thirteenth attempt at a treatment of the steam engine theory. In it he described a thermodynamic cycle of events (the so-called Rankine cycle) which was used as a standard for the performance of steam-power installations where a considerable vapour provides the working fluid. Rankine here explained how a liquid in the boiler vaporized by the addition of heat converts part of this energy into mechanical energy when the vapour expands in an engine. As the exhaust vapour is condensed by a cooling medium such as water, heat is lost from the cycle. The condensed liquid is pumped back into the boiler. This concept of a power cycle is useful to engineers in developing equipment and designing power plants.

Besides writing in various newspapers, Rankine contributed many papers to scientific journals, the most significant ones being on the subject of thermodynamics. The application of the doctrine that 'heat and work are convertible' to the discovery of new relationships among the properties of bodies was made about the same time by three scientific men, William Thomson (afterwards Lord Kelvin), Rankine and Clausius. Lord Kelvin cleared the way with his account of Carnot's work on *The Motive Power of Heat* and pointed out the error of Carnot's assumption that heat is a substance and therefore indestructible. Rankine in 1849 and Clausius

in 1850 showed the further modifications which Carnot's theory required. Lord Kelvin in 1851 put the foundations of the theory in the form they have since retained.

Rankine also made an important contribution to soil mechanics and his work on earth pressure and the stability of retaining walls was a notable advance. He collaborated with others to produce *The Imperial Dictionary of Universal Biography* and was the corresponding and general editor of *Shipbuilding, Theoretical and Practical* (1866).

Ray John 1627–1705 was a British naturalist whose plant and animal classifications were the first significant attempts to produce a systematic taxonomy based on a variety of structural characteristics, including internal anatomy. He was also the first to use the term 'species' in the modern sense of the word.

Ray was born on 29 November 1627 in the small Essex village of Black Notley. His father was a blacksmith and his mother was an amateur herbalist and medical practitioner. After attending the grammar school in nearby Braintree, Ray spent two years at Catherine's Hall, Cambridge (now St Catherine's College). He transferred to Trinity College, Cambridge in 1646, and graduated in 1648. He was elected a Fellow of Trinity in the following year and remained at the college for the next 13 years, initially teaching Greek, mathematics and the humanities. In 1650, however, he suffered a serious illness and, while recuperating, spent much time walking through the surrounding countryside; this stimulated his interest in natural history, which thereafter became his main academic pursuit. Ray took holy orders in 1660, but the restoration of Charles II in that year changed the country's religious climate and Ray was obliged to leave Trinity in 1662 because he refused to sign an agreement to the Act of Uniformity, which required from all clergymen a declaration of assent to everything contained in the Prayer Book of Queen Elizabeth I and to conform to the Liturgy of the Church of England. Thus Ray lost his livelihood and for the rest of his life he depended on financial support from his friends, particularly from Francis Willughby, an affluent younger contemporary at Cambridge who shared Ray's interest in natural history. From 1663 to 1666 Ray and Willughby toured Europe to study the flora and fauna and collect specimens. On their return to England, Ray lived at Willughby's home, where they collaborated on publishing the results of their natural history studies. In 1672, Willughby died unexpectedly, but Ray remained at his family home, supported by an annuity left him by Willughby and by his position as tutor to the Willughby children. In 1673, Ray married a governess in the Willughby household. In 1678, however, Willughby's widow forced Ray and his wife to leave, and they returned to Black Notley, where Ray remained for the rest of his life. He died there on 17 January 1705.

Ray's first publication was *Catalogue Plantarum Circa Cantabrigiam Nascentium/Catalogue of Plants around Cambridge* (1660), compiled from his observations during his walks while convalescing from illness. The work listed

558 species and was the best attempt then available at cataloguing plants. His first attempt at a genuine classification, however, was a table of plants that he contributed to John Wilkin's book *Essay towards a Real Character* (1668). In 1670 Ray, with Willughby's help, published *Catalogus Plantarum Angliae/Catalogue of English Plants*. Ray and Willughby then began producing a definitive catalogue and classification of all known plants and animals, with Ray responsible for the botany and Willughby for the zoology. But Willughby died while this work was in its early stages and Ray assumed the task of completing the entire project, since he was a competent zoologist and was familiar with Willughby's material. This task occupied Ray for the rest of his life.

As a tribute to Willughby, Ray published *F. Willughbeii ... Ornithologia* (1676; translated in 1678 as *The Ornithology of F. Willughby*) and *F. Willughbeii ... de Historia Piscium* (1685; *History of Fish*) under Willughby's name, although most of the work was Ray's own. Ray continued his botanical studies and, in *Methodus Plantarum Nova* (1682), developed a clearcut taxonomic system based on plant physiology, morphology and anatomy. In his system he laid great emphasis on the division of plants into cryptogams (flowerless plants), monocotyledons and dicotyledons, a basic categorization that is still used today. Within the dicotyledons Ray defined 36 family groupings, many of which are also still used. He also established the species as the fundamental unit of taxonomy, although he mistakenly believed that species are immutable.

The culmination of Ray's work, however, was *Historia Generalis Plantarum* , a monumental three-volume treatise (published between 1686 and 1704) in which he attempted to produce a complete, natural classification of plants. The book covered about 18,600 species (most of which were European) and, in addition to plant classification, it contained much information on the morphology, distribution, habitats and pharmacological uses of individual plant species as well as general aspects of plant life, such as diseases and seed germination. In it he modified his belief in the immutability of species.

Ray also wrote several books on zoology under his own name, notably *Synopsis Methodica Animalium Quadrupedum et Serpentini Generis/Synopsis of Quadrupeds* (1693), *Historia Insectorum/History of Insects* (published posthumously in 1710) and *Synopsis Methodica Avium et Piscium/Synopsis of Birds and Fish* (published posthumously in 1713). In all of these works Ray followed the same format he had used in his *Historia Generalis Plantarum*, giving details of individual species in addition to classification.

Furthermore, Ray also believed that fossils are the petrified remains of dead animals and plants – a concept that, surprisingly, appeared in his theological writings and did not gain general acceptance until the late eighteenth century.

Although it was not possible to devise a natural classification system until Charles Darwin and Alfred Wallace formulated evolutionary theory, Ray's system approached that ideal far more closely than those of any of his contemporaries and remained the best attempt at classification until superseded by Carolus Linnaeus' taxonomic work in 1735.

Rayet George Antoine Pons 1839–1906 was a French astronomer who, in collaboration with Charles Wolf, detected a new class of peculiar white or yellowish stars whose spectra contain broad hydrogen and helium emission lines. These stars were subsequently named Wolf–Rayet stars, after their discoverers.

Rayet was born near Bordeaux on 12 December 1839. He did not attend school until the age of 14, when his family moved to Paris. In 1859, he was admitted to the Ecole Normale Supérieure and graduated with a degree in physics three years later. After teaching for a year he obtained a post as a physicist in the new weather forecasting service created by Urbain Leverrier (1811–1877) at the Paris Observatory. In 1873, Leverrier entrusted the running of the meteorological service to Rayet, but within a year the two men disagreed over the practical forecasting of storms and this led to Rayet's dismissal. Rayet then became a lecturer in physics at the Faculty of Sciences of Marseilles and in 1876 he was appointed as Professor of Astronomy at Bordeaux.

As a result of a government proposal to construct several new observatories in France, Rayet was asked to organize a collective survey of the history and equipment of the world's observatories. Subsequently, he was offered the appointment of Director of the new observatory to be built at Floirac, near Bordeaux, and from 1879 he held this post along with his appointment at Bordeaux. During his last years Rayet was troubled by a serious lung complaint, and died in Floirac on 14 June 1906.

At the Paris Observatory, Rayet collaborated with Charles Wolf and their first joint success came in 1865 when they photographed the penumbra of the Moon during an eclipse. On 4 May 1866 a nova appeared, and while observing it on 20 May, after its brilliance had significantly diminished, Rayet and Wolf discovered bright bands in its spectrum – a phenomenon that had never been noticed in stellar spectra before. The bands were the result of a phase that can occur in the later stages of evolution of a nova. The two astronomers went on to investigate whether permanently bright stars exhibit this phenomenon and in 1867 they discovered three such stars in the constellation of Cygnus. These stars, with characteristically broad and intense emission lines, are now known to be relatively rare and are called Wolf–Rayet stars. They are about twice the size of the Sun, very hot, with an expanding outer shell, and a disparity between the energy produced in the interior and the radiated energy.

In 1868, Rayet took part in an expedition to the Malay Peninsula to observe a solar eclipse and was given responsibility for the spectroscopic work of the expedition. His observations of solar prominences provided valuable information and in conjunction with other observations made during the same eclipse, notably those of Pierre Janssen, they contributed to the establishment of the first precise data on the Sun.

As Director, Rayet equipped the new observatory in Floirac with the most modern astrometric equipment and began a programme of accurately measuring the coordinates of stars, the positions of comets and nebulae, and the components of double stars. He was one of the first supporters of the *Carte internationale photographique du ciel* which was established in order to map the entire sky using identical telescopes around the world, simultaneously. He also had the satisfaction of being able to publish the first volume of the Bordeaux Observatory's *Catalogue photographique*, a year before his death. Despite his poor state of health, Rayet took part in a 1905 expedition to Spain to study a solar eclipse.

Rayleigh Lord (John William Strutt) 1842–1919 was a British physicist who, with William Ramsay (1852–1919), discovered the element argon. For this achievement Rayleigh was awarded the 1904 Nobel Prize for Physics while Ramsay gained the 1904 Nobel Prize for Chemistry. Rayleigh possessed a remarkable grasp of all the fundamental areas of classical physics (electromagnetic theory, thermodynamics, statistical mechanics), producing important contributions in all of these fields.

Rayleigh was born John William Strutt at Langford Grove, Essex, on 12 November 1842. He received private tutoring until 1861 when he entered Trinity College, Cambridge, to study mathematics, graduating in 1865. He was then elected to a Fellowship of Trinity College, which he held until his marriage in 1871. After graduating from Cambridge he travelled to the United States until 1868, and then set up a laboratory in the family home at Terling Place. On the death of his father in 1873, Strutt inherited his title and became Lord Rayleigh. He continued his scientific research, which was considered rather eccentric for a person in his privileged position.

The same year marked Rayleigh's election to the Fellowship of the Royal Society, which he later served as Secretary from 1885 to 1896 and President from 1905 to 1908, and which awarded him the Rumford and Copley Medals. In 1879, he succeeded James Clerk Maxwell (1831–1879) as Cavendish Professor of Experimental Physics at the University of Cambridge, a post he held until his return from the Montreal meeting of the British Association for the Advancement of Science in 1884, at which Rayleigh had held the Presidency.

After 1884, Rayleigh confined most of his research activities to his personal laboratory and study at home. He did serve as Professor of Natural Philosophy at the Royal Institution in London from 1887 to 1905, but his duties there did not require him often to leave home. In 1900, he contributed to the foundation of the National Physical Laboratories at Teddington, Middlesex. In 1908, Rayleigh was appointed Chancellor of Cambridge University, a post he held until his death at Terling Place on 30 June 1919.

Rayleigh's early researches at home during the late 1860s and early 1870s were largely concerned with the properties of waves in the fields of optics and acoustics. He did work on resonance, extending the studies of Hermann Helmholtz (1821–1894), and on vibration. Rayleigh also studied light and colour. In 1871, he explained that the blue colour of the sky arises from the scattering of light by dust particles in the air, and was able to relate the degree of scattering to the wavelength of the light. Rayleigh was also interested in diffraction gratings, and made the first accurate definition of the resolving power of diffraction gratings. This led to improvements in the spectroscope, which in the 1870s was becoming an important tool in the study of solar and chemical spectra. Rayleigh was also responsible for the invention of the optical zone plate.

At Cambridge, Rayleigh's professorial responsibilities enabled him to improve the teaching of experimental physics, causing an explosion in the popularity of the subject which had farreaching consequences for physics in Britain. In 1884, he completed the project initiated by his predecessor Maxwell that sought the accurate standardization of the three basic electrical units, the ohm, ampere and volt. His insistence on accuracy led to the designing of more precise electrical instruments.

After leaving Cambridge, Rayleigh continued to do research in a broad range of subjects including light and sound radiation, thermodynamics, electromagnetism and mechanics. He was very conscientious at keeping up with developments as they appeared in the scientific literature, and having a keen critical eye, he was often able to devise new experiments to rectify weaknesses he found in the work of others.

The work which most caught the public eye began with Rayleigh's careful study of the measurement of the density of different gases. He originally became interested in this problem because of the old theory known as Prout's hypothesis that all the elements are multiples of hydrogen and would have atomic weights (relative atomic masses) in integers. Rayleigh's results did not support this theory, as others had found, but he made the incidental observation that whenever he measured the density of nitrogen in air it was 0.5% greater than the density of nitrogen from any other source. He immediately eliminated all the possible suggestions for the source of the impurity causing the increase, but could not solve the problem. In desperation he published a short note in *Nature* in 1892, asking for suggestions. Over the next three years, Rayleigh and Ramsay both studied the problem and in the end jointly announced the discovery of a new element: argon. The gas had escaped detection by chemists because of its extreme inertness, which renders it virtually devoid of any chemical properties. Argon might have been discovered a century earlier if Henry Cavendish (1731–1810) had followed up his observation of a residual quantity of gas in air that he simply could not oxidize.

The discovery of argon was followed by one of Rayleigh's most important contributions to physics. This was the Rayleigh–Jeans equation, published in 1900, which described the distribution of wavelengths in blackbody radiation. This equation could only account for the longer wavelengths, and Wilhelm Wien (1864–1928) later produced an equation to describe the shorter wavelength radiation.

This inconsistency led to the formulation shortly after of the quantum theory by Max Planck (1858–1947), which accounted for the distribution of all wavelengths by assuming that energy exists in indivisible units called quanta. Rayleigh was not enthusiastic about such a revolutionary solution, which overturned classical ideas of radiation. He was likewise unable to agree with the explanation of the hydrogen spectrum produced by Niels Bohr (1885–1962) in 1913, which developed from the quantum theory, and he also found the special theory of relativity proposed by Albert Einstein (1879–1955) in 1905 distasteful because it dispensed with the ether as a medium for light waves. He had in fact made an attempt to detect the ether in 1901, and his negative result only added to the evidence for relativity.

However, such conservatism should not blind us to Rayleigh's achievements in advancing and consolidating the branches of classical physics that remained valid. And in bringing about the discovery of argon, he initiated the uncovering of a whole family of elements (the rare gases) that are of great importance in themselves and to an understanding of chemical bonding.

Reber Grote 1911– is a US radioastronomer – indeed, at one time he was probably the world's only radioastronomer – who may truly be said to have pioneered the new aspect of astronomical science from its inception. His major project has been to map all the extraterrestrial sources of radio emission that can be traced.

Reber was born in Wheaton, Illinois, on 22 December 1911. Completing his education at the Illinois Institute of Technology, he became a radio engineer. After Karl Jansky (1905–1950) stumbled on the existence of cosmic radio waves, Reber was among the first to explore this new field. Since then his research has taken him all over the world. From Illinois, Reber moved his telescope to Virginia in 1947, where he was appointed chief of the University's Experimental Microwave Radio Section. Four years later he moved to Hawaii, where a new telescope, sensitive to lower frequencies than he had been able to detect previously, was constructed. In 1954 he moved yet again, this time to Australia, where he joined the Commonwealth Scientific and Industrial Research Organization in Tasmania. Although he then went to the National Radio Astronomy Observatory at Green Back, in West Virginia, in 1957 to work with the 43-m/141-ft radio telescope installed there, he returned to Tasmania in 1961 to help complete the mapping project he had helped to begin.

After a first, unsuccessful, attempt, Reber finally completed the construction of a bowl-shaped reflector 9 m/350 in diameter, with an antenna at its focus, in the backgarden of his Illinois home in 1957. He immediately began to map radio sources in the sky, noting particularly that many seemed to come from the direction of the Milky Way. He took his first results to the Yerkes Observatory for discussion with the astronomers there. Satisfactory explanations of radio emission and radio sources were to come later, with the further development of radioastronomy, but at the time, and for a number of years, Reber's was probably the only radio telescope in existence.

Despite the fact that the resolution of his homemade apparatus was no better than 12°, which meant that he could identify only a general direction from which radio waves were coming, he compiled a map of the sky, noting as he did so how many radio sources seemed to have no optically identifiable presence. The most intense radiation he recorded emanated from the direction of Sagittarius, near the centre of the Galaxy.

The radio telescope in Hawaii represented a great improvement in facilities for him, since it was sensitive to lower frequencies. But it was the equipment in Tasmania that really held Reber's attention. The project there was to complete a map of radio sources emitting waves around 144 m/473 ft in length, and it was to this work that he devoted most of the rest of his professional career.

Redman Roderick Oliver 1905–1975 was a British astronomer who was chiefly interested in stellar spectroscopy and solar physics. A practical man, he also established a thriving solar observatory in Malta and organized the re-equipping of the Cambridge Observatories after World War II.

Redman was born on 17 July 1905 in Rodborough, Gloucestershire. At the age of 18 he won an Exhibition scholarship from Marling School in Stroud to St John's College at Cambridge University, where he studied until 1929. By that time he was working under Sir Arthur Eddington (1882–1944) at the University Observatory for his PhD in astronomy. The doctorate was awarded in 1930, while Redman was serving as Assistant Astronomer at the Dominion Astrophysics Observatory in Victoria, British Columbia. The following year, Redman returned to Cambridge to become Assistant Director under F. Stratton at the Solar Physics Observatory. He became Chief Assistant of the Radcliffe Observatory in 1937 and moved to the new Observatory site in Pretoria in 1939. World War II prevented the Observatory from being fully equipped so, after being elected Fellow of the Royal Society in 1946, Redman returned again to Cambridge in 1947 and was appointed Professor of Astrophysics and Director of the Observatory. He retained these positions until 1972, when he was made Director of the newly amalgamated Observatories and Institute of Theoretical Physics. He retired later that same year, and died not long after, on 6 March 1975.

From the beginning of his career as an astronomer, Redman was interested in spectroscopic analysis. During his years as a research student under Sir Arthur Eddington, Redman contributed to the early analysis of the Hertzsprung–Russell diagram by studying absolute stellar magnitudes. Upon his return to Cambridge in 1931, Redman concentrated his efforts in stellar spectroscopy, and in the development of spectroscopic techniques.

He applied the method of photographic photometry to the study of elliptical galaxies and later, in South Africa, also to the study of bright stars, for which he developed the 'narrow band technique' which was of great value in stellar photometry.

The Sun was a source of fascination for Redman. He devoted considerable time and energy to studies and analyses of the solar spectrum, and went all over the world in order to observe total eclipses, during which he was able to identify thousands of the emission lines in the chromospheric spectrum and to investigate the question of chromospheric temperature.

Redman's final contribution to astronomy was his initiation of a large stellar photometry programme.

Regnault Henri Victor 1810–1878 was a German-born French physical chemist who is best known for his work on the physical properties of gases. In particular he showed that Boyle's law applies only to ideal gases. He also invented an air thermometer and a hygrometer, and he discovered carbon tetrachloride (tetrachloromethane).

Regnault was born in Aachen, Germany (Aix-la-Chapelle) on 21 July 1810. His father, an officer in Napoleon's army, was killed in the Russian campaign of 1812 and his mother died a few months later. His education was supervised by a friend of his father, who found him a job in a draper's shop in Paris. Although he was very poor, Regnault managed to take lessons and in 1830 was admitted to the Ecole Polytechnique in Paris, from which he graduated two years later. He spent two more years at the Ecole des Mines before leaving France to study mining techniques and metallurgical processes in various parts of Europe. After short periods of research under Justus von Liebig at Giessen and Jean-Baptiste Boussingault at Lyons, he returned to the Ecole Polytechnique in 1836 as an assistant to Joseph Gay-Lussac and in 1840 succeeded him to the Chair in Chemistry. In the same year he was elected to the chemical section of the Académie des Sciences but his interests were already turning to physics and he became Professor of Physics at the Collège de France in 1841, where over the next 13 years he performed his most important experimental work.

From 1854 Regnault lived and worked at Sèvres as Director of the famous porcelain factory and was still engaged on research there when, in 1870, all his instruments and books were destroyed by Prussian soldiers. This blow and the death of his son late in the Franco–Prussian War left him a broken man and his last years were clouded by grief and personal disability. He died in Auteuil, near Paris, on 19 January 1878.

In his chemical work, nearly all of which dates from between 1835 and 1839, Regnault followed no unified programme of research. His major contributions were to organic chemistry. He studied the action of chlorine on ethers, leading to the discovery of vinyl chloride (monochloroethene), dichloroethylene (dichloroethene), trichloroethylene (trichloroethene) and carbon tetrachloride (tetrachloromethane).

Regnault was encouraged by Jean Baptiste Dumas, who had long advocated the measurement of specific heats as a means of investigating atomic composition. He began by measuring the specific heats of a wide range of substances, during which work (1839 to 1842) he conclusively demonstrated the approximate nature of Dulong and Petit's law and confirmed the validity of F. E. Neumann's extension of the law from elements to compounds.

In 1842, he was commissioned by the Minister of Public Works to redetermine all the physical constants involved in the design and operation of steam engines. This led Regnault to begin the research on the thermal properties of gases, for which he is now best known. He found that nearly all ordinary gases behave in much the same way and that the nature of this behaviour could generally be described by the perfect gas laws, which can define the volume of a gas in terms of its pressure, temperature and number of molecules (Boyle's law, Charles' or Gay-Lussac's law, and Avogadro's law). He painstakingly measured the coefficients of expansion of various gases and by 1852 had shown how real gases depart from the 'ideal' or 'perfect' behaviour required by Boyle's law – it was left to Johannes van der Waals to formulate a mathematical statement of the variation ten years later. Regnault also calculated that absolute zero is at $-273°C/-459°F$. He redetermined the composition of air, and performed experiments on respiration in animals.

Remington Philo 1816–1889 was a US mechanical engineer largely responsible for perfecting the Remington breech-loading rifle and the Remington typewriter.

Born in Litchfield, New York State, the son of a small-arms manufacturer, Remington entered his father's business, spending 25 years as superintendent of the manufacturing department. It was in this post that he did the work for which he is now remembered.

During the American Civil War, business boomed for the Remingtons but, following the victory of the Union in 1865, the firm looked to diversification to meet the future; in 1873 a perfect answer materialized. After work carried out in the Kleinstuber machine shops in Milwaukee, Wisconsin, the US inventor Christopher Sholes had designed a typewriter capable of being mass-produced. This development attracted the attention of two businessmen, who bought Sholes' patent and then went to Remington with a proposal that the typewriter be manufactured by the company.

The contract was signed in 1873, and what soon became known as the Remington typewriter – based on Sholes's original design – was marketed the following year. This first typewriter had no facility for upper- and lower-case type; only capitals were available. This was soon solved by the simple device of putting both types on each typebar, and providing a mechanism (the shift key) for lifting the key to make contact with the paper. The first typewriter carrying the shift key, the Remington Mark II, appeared in 1878.

To overcome initial market inertia, the Remington machine was loaned to over 100 firms free of charge. Mark Twain bought one, becoming the first author to provide his publisher with a typescript.

Together with his father Eliphalet, Remington made many improvements to guns and their manufacture. This work went back to 1816, when Remington's father had made a flintlock rifle at his own father's forge in Utica, New York, with an accuracy that soon attracted great attention. That led to the setting up of E. Remington & Son at Ilion, beside the Erie Canal. Among the advances made by the father-and-son team was a special lathe for the cutting of gunstocks, a method of producing extremely straight gun barrels, and the making of the first American drilled rifle barrel from cast steel.

Reynolds Osborne 1842–1912 was a British research engineer, one of the first people to approach engineering as an academic rather than a practical subject. He is particularly remembered for his investigation into turbulent flow in fluids.

Reynolds was born in Belfast on 23 August 1842. His father was a mathematician who became a schoolteacher and then rector of the parish of Debach-with-Boulge, Suffolk (as were his grandfather and greatgrandfather before). Most of Reynold's early education was at Dedham Grammar School, where his father was a teacher. He left school at the age of 19 and entered an engineering workshop, where he applied his mathematical knowledge to further the understanding of the action and design of machines. He left these practical pursuits to study mathematics at Cambridge University. Like his father he attended Queens' College, from which he graduated in 1867 and was immediately awarded a Fellowship. Soon afterwards he went to work in London for John Lawson, a civil engineer.

In 1868, Reynolds was elected to the new Chair of Engineering at Owens College (now Manchester University). His course of lectures on civil and mechanical engineering included all the mechanical, structural and thermodynamic principles involved in the subjects and gave rise to a new generation of scientific engineers who were largely to take over from the older, purely practical men.

At the time, academic engineering was not fully divorced from the pure sciences, and much of Reynolds's early research into the behaviour of comets, electrical phenomena in the atmosphere and properties of materials would today be regarded as part of physics. His first real engineering research was concerned with the efficiency of screw propellers for ships, and with their stability. He used scale models, a technique which was to prove of value in his later work.

During the 1880s Reynolds made his famous contribution to hydrodynamics by studying the motion of water at various velocities in parallel channels. Poiseuille and Darcy had already carried out research in this area but obtained conflicting results. Reynolds showed that for the flow of a liquid to be non-turbulent, it needed a high viscosity, low density, and an open surface. If these and other criteria were not met then the flow was turbulent – a phenomenon he often demonstrated using dyes. The two states, Reynolds found, are separated by a critical velocity,

itself dependent on viscous forces (tending to produce stability) and inertial forces (tending to produce instability). The ratio of the two forces, which gives a measure for any given system, is now known as Reynolds's number (equal to $\rho v l / \mu$, where ρ is the density, v the velocity, l the radius of the tube and μ the velocity).

Reynolds applied much of what he had learned about turbulent flow to the behaviour of the water in river channels and estuaries. For one study he made an accurate model of the mouth of the River Mersey, and pioneered the use of such models in marine and civil engineering projects. His later studies led to discoveries about the forces involved in lubrication with oil and the design of an improved bearing that used this knowledge. He also worked on multistage steam turbines, but was disappointed by their low efficiency (not realizing that efficiency increases with size and that large efficient turbines are practicable). Using the experimental 75 kW/100 hp steam engine which he designed for his engineering department, he determined the mechanical equivalent of heat so accurately that it has remained one of the classical determinations of a physical constant.

Reynolds was elected a Fellow of the Royal Society in 1877, and during the following decade was honoured with many medals and honourary degrees. He was married twice, his first wife dying a year after the wedding.

Ricardo Harry Ralph 1885–1974 was the English engineer who played a leading role in development of the internal-combustion engine. His work was particularly significant during World War I and World War II, enabling the British forces to fight with the advantage of technically superior engines. His work on combustion and detonation led to the octane rating system for classifying fuels for petrol engines.

Born on 26 January 1885 in London, Ricardo was the eldest son of Halsey Ricardo, an architect, and Catherine Rendel, daughter of Alexander Rendel, a consultant civil engineer. He was educated at Rugby and designed and built a steam engine, putting it into production when he was 12.

Ricardo went up to Cambridge in 1903 where he designed and built a motorcycle, with its power unit. With it, he won a fuel economy competition and this success led directly to his concentration on internal-combustion engine research.

In the summer of 1905 he designed and built a 2-cylinder, 2-stroke engine, called the Dolphin, for automotive applications but in the event, the first production power unit was a 4-cylinder version to power his uncle's large car, Ricardo qualified for association membership of the Institution of Civil Engineers and was given charge of a small mechanical engineering department to design site equipment for his grandfather's firm.

During World War I he worked on aircraft engines and designed the engine for the Mk V tank. In July 1917, Ricardo set up his research and consultancy company. Between the wars the company worked on engine development and categorization of fuels according to their ease of detonation. From 1932 until the outbreak of World War II, it was almost exclusively engaged in the design of light high-speed diesel engines.

When war broke out, Ricardo was living in Tottington Manor, near Henfield. The house was requisitioned and the company was moved to a less vulnerable site in Oxford. After the war, it moved to Shoreham and Ricardo settled at Woodside, the family house at Graffham. It was here that he fell and broke his leg, resulting in his death six weeks later on 18 May 1974.

Ricardo was elected Fellow of the Royal Society in 1929 and was president of the Institution of Mechanical Engineers for the 1944–45 term. In 1948 he was made Knight Bachelor and during his life received 12 medals from various learned bodies.

Ricardo is probably best known for his work on combustion and detonation in spark ignition engines. In 1912, it was believed widely that the ringing knock in engines with high compression ratios was due to preignition – the firing of the charge in the cylinder before the correct time. Ricardo demonstrated in 1913 that the knock was due to the spontaneous combustion of a part of the charge.

One of the factors influencing the tendency for a fuel to ignite spontaneously was its stability, the paraffinic being the worst and the aromatic the best. An early contract won by Ricardo's embryonic company was one from Shell to investigate the behaviour of various fuels. The work led to the evolution of the Toluene Number System of categorizing fuels. This later became the octane rating.

Ricardo paid great attention to combustion-chamber design and put his ideas into practice during a job to improve the performance of the popular, cheap, side-valve engine. In this he succeeded, so spectacularly, that the combustion chamber configuration was produced under licence, or pirated, throughout the world. This led to a patent infringement case, which Ricardo won in 1932.

One of the reasons for the success of the single sleeve-valve engine was that the cylinder head, having no valve ports, could be made the best shape for correct combustion. The principle was used for aircraft engines, which were so successful that manufacturers adopted the Ricardo design. This type of engine became the last and best of the British piston engines, powering aircraft of World War II.

Zeppelins bombed Britain from heights of 5,000–6,000 m/16,000–20,000 ft, out of reach of both aircraft and anti-aircraft guns. Ricardo's supercharged engine was redesigned in 1915 and 1916 to power an aircraft to tackle this menace. On its type test, the first engine developed 270 kW/360 hp on supercharged air at ground level and would have been able to intercept the Zeppelins with ease.

The first engine ran on the test bed in March 1917, exceeding its rated 110 kW/150 hp, developing 125 kW/168 hp at 1,200 rpm and just over 15 kW/200 hp at 1,600 rpm. The first Mk V tank ran in June 1917. When the weight of armour and armaments on tanks was increased, Ricardo designed 168 kW/225 hp and 224 kW/300 hp units to provide the extra power.

Towards the end of World War II Me 410s (fast twin-engined fighter bombers) were making hit-and-run raids on British cities and were outdistancing British fighters. Ricardo proposed oxygen enrichment of the Merlin engine and after Mosquitos had been equipped, the losses of Me 410s increased dramatically and the raids came to an end.

When Ricardo began his work on internal-combustion engine development, the techniques used were more akin to art than to science. He led the revolution in engine design, and his painstaking work laid a firm foundation for the meticulous programmes which have resulted in smaller engines producing more power on less fuel than ever before.

Ricci-Curbastro Gregorio 1853–1925 was an Italian mathematician who is chiefly remembered for his systematization of absolute differential calculus (also now called the Ricci calculus), which later enabled Albert Einstein (1879–1955) to write his gravitational equations, to express the principle of the conservation of energy, and thereby fully to derive the theory of relativity.

Although his complete surname was Ricci-Curbastro, he is often known simply as Ricci. He was born on 12 January 1853 in Lugo, Romagna, where he grew up and was educated at home by private tutors. At the age of 16 he went for a year to Rome University to study mathematics and philosophy, before returning to Lugo. Two years later he enrolled at the University of Bologna. After a year there, in 1872 he enrolled at the Scuola Normale Superiore in Pisa. In 1875, he received his doctorate in the physical and mathematical sciences. After winning a scholarship to travel and study, he spent a year (1877–1878) at Munich, where he met Felix Klein (1849–1925) and Enrico Betti (1823–1892). During 1879, Ricci stayed in Pisa and worked as an assistant to the mathematician U. Dini.

In December 1880 Ricci was appointed Professor of Mathematical Physics at the University of Padua, and stayed there for the remainder of his life. He published his major work – on the absolute calculus – between 1888 and 1892, but after being asked in 1891 to teach other aspects of mathematics in addition to those he was appointed to, he also published works on higher algebra, infinitesimal analysis and the theory of real numbers. He received many honours and awards, and was elected to membership of a surprising number of Academies. He also made valuable contributions to his local administration: as a magistrate and councillor he was not only concerned with problems of water supply, swamp drainage and public finance, he was also able to encourage the close collaboration of the government with science and its use in the service of the state. He died in Bologna on 6 August 1925.

Ricci's invention of absolute differential calculus was the result of ten years of research from 1884 to 1894. By introducing an invariant element – an element that can also be used in other systems – Ricci was able to modify the existing differential calculus so that the formulae and results retained the same form regardless of the system of variables used. Starting with the idea of covariant derivation formulated by Elwin Christoffel (1829–1900), the work of Bernhard Riemann (1826–1866) on the theory of invariants of algebraic form, and with Rudolf Lipschitz's work on quadratic forms, Ricci realized that the methods used by these mathematicians could be generalized. In 1893 he published his *Di alcune applicazione del calculo differenziale assoluto all teoria delle forme differenzial quadratiche*, in which he gave a specific form to his ideas.

In 1896, he applied the absolute calculus to the congruencies of lines on an arbitrary Riemann variety; later, he used the Riemann symbols to find the contract tensor, now known as the Ricci tensor (which plays a fundamental role in the theory of relativity). Ricci also discovered the invariants that occur in the theory of the curvature of varieties.

Later still, he collaborated with a former pupil of his, Tullio Levi-Civita (1873–1941), to produce *Méthodes de calcul différentiel absolu et leurs applications*. This work contains references to the use of intrinsic geometry as an instrument of computation dealing with normal congruencies, geodetic laws and isothermal families of surfaces. The work also shows the possibilities of the analytical, geometric, mechanical and physical applications of the new calculus. In application to mechanics, Ricci solved the Lagrange equations with respect to the second derivatives of the coordinates. The application to physics includes equations in electrodynamics, the theory of heat and the theory of electricity.

Although for some time, however, the new calculus found few applications, when Albert Einstein came to formulate his general theory of relativity Ricci's method proved to be an apt means. Einstein's use of it has since encouraged the intensive study of differential geometry based on Ricci's tensor calculus.

Ricci continued to take an active part in mathematical studies into his very last years. His final major contribution was in 1924, when he presented a paper to the International Congress of Mathematicians held in Toronto, Canada; his subject then was the theory of Riemann varieties.

Richards Theodore William 1868–1928 was a US chemist who gained worldwide fame for his extremely accurate determinations of atomic weights (relative atomic masses). For this work he was awarded the 1914 Nobel Prize for Chemistry.

Richards was born in Germantown, Pennsylvania, on 31 January 1868, the son of a painter father and author mother. He received his early education at home, and then in 1882 he went to Haverford College, initially to study astronomy but changing to chemistry because of his poor eyesight. He graduated three years later and went to Harvard, where he gained his chemistry degree in 1886. He remained at Harvard to do research under Josiah Cooke (1827–1894), who set him the task of testing Prout's hypothesis (that all atomic weights are whole numbers) by determining the ratio by weight of hydrogen to oxygen in water. He was awarded his PhD for this work in 1888. Richards was then granted a travelling fellowship and visited several European universities, where he came into contact with such influential chemists as Viktor Meyer and Lord Rayleigh.

He became an Instructor in Chemistry at Harvard in 1894; he also met Friedrich Wilhelm Ostwald and Hermann Walther Nernst on a second trip to Europe in 1895. He declined the offer of a professorship at Göttingen University in 1901 in favour of the position of Professor of Chemistry at Harvard, which he retained for the rest of his career. He died in Cambridge, Massachusetts, on 2 April 1928.

Richards's determinations of atomic weights were based on painstakingly precise quantitative measurements, for which he introduced various new analytical techniques. He devised a method of keeping samples sealed and dry so that they could not absorb moisture before or during weighing. For accurately determining the endpoint in silver nitrate titrations (which usually depend on the first appearance or last disappearance of a precipitate) he invented the nephelometer, a means of comparing the turbidity of two solutions. He made accurate atomic weight measurements for 25 elements; his co-workers determined 40 more, improving on the 'standard' values obtained by Jean Stas in the 1860s. In 1913, he detected differences in the atomic weights of ordinary lead and samples extracted from uranium minerals (which had arisen by radioactive decay) – one of the first convincing demonstrations of the uranium decay series and confirming Frederick Soddy's prediction of the existence of isotopes. It also revealed a germ of truth in Prout's defunct hypothesis. Richards also investigated the physical properties of the elements, such as atomic volumes and the compressibilities of nonmetallic solid elements.

Ricketts Howard Taylor 1871–1910 was a US pathologist who discovered the *Rickettsia* (which are named after him), a group of unusual microorganisms that have both viral and bacterial characteristics. The ten known species in the *Rickettsia* genus are all pathogenic in human beings, causing diseases such as Rocky Mountain spotted fever and forms of typhus.

Ricketts was born on 9 February 1871 in Findlay, Ohio. He was educated at the University of Nebraska, from which he graduated in 1894, and Northwestern University, Chicago, from which he obtained his medical degree in 1897. He then became a junior resident doctor at Cook County Hospital, Chicago, before moving to the Rush Medical College (then affiliated with Chicago University) in 1899 as an instructor in cutaneous pathology. In 1901 Ricketts travelled to Europe, performing laboratory work there. On his return in 1902, he became instructor and later Associate Professor in the Pathology Department of Chicago University. In 1909 Ricketts went to Mexico City to investigate typhus; while there he became fatally infected with the disease and died on 3 May 1910.

Ricketts began studying Rocky Mountain spotted fever in 1906 and discovered that the disease is transmitted to human beings by the bite of a particular type of tick that inhabits the skins of animals. In 1908, he found the causative microorganisms in the blood of infected animals and in the bodies and eggs of ticks. This microorganism is now called *Rickettsia ricketsii*, after its discoverer. In his studies of typhus in Mexico Ricketts demonstrated that this disease is also caused by a type of *Rickettsia* and that the micro-organisms are transmitted to humans by the body louse. Before he died from the disease, Ricketts also showed that typhus can be transmitted to monkeys, and that, after recovery, they are immune to further attacks.

Riemann Georg Friedrich Bernhard 1826–1866 was a German mathematician whose work in geometry – both in combining the results of others and in his own crucial and innovative research – developed that branch of mathematics to a large degree. His concepts in non-Euclidean and topological space led to advances in complex algebraic function theory and in physics; later Albert Einstein (1879–1955) was to make use of his work in his own theory of relativity. Riemann's study of analysis was also fundamental to further development. Despite the brevity of his life, he knew most of the other great mathematicians of his time, and was himself a famed and respected teacher.

Riemann was born on 17 September 1826 in Breselenz (in Hanover), one of the six children of a Lutheran pastor. From a very early age he showed considerable talent in mathematics. Nevertheless, on leaving the Lyceum in Hanover in 1846 to enter Göttingen University, it was to study theology at the behest of his father. However, he soon obtained his father's permission to devote himself to mathematics and did so, being taught by the great Karl Gauss (1777–1855). Riemann moved to Berlin in 1847 and there came into contact with the equally prestigious teachers Lejeune Dirichlet (1805–1859) and Karl Jacobi (1804–1851); he was particularly influenced by Dirichlet. Two years later he returned to Göttingen and submitted a thesis on complex function theory, for which he was awarded his doctorate in 1851.

During the next two years Riemann qualified as an unsalaried lecturer, preparing original work on Fourier series, and working as an assistant to the renowned physicist Wilhelm Weber (1804–1891). At Gauss's suggestion, Riemann also wrote a paper on the fundamental postulates of Euclidean geometry, a paper that was to open up the whole field of non-Euclidean geometry and become a classic in the history of mathematics. (Although Riemann first read this paper to the University on 10 June 1854, neither it nor the work on Fourier series was published until 1867 – a year after Riemann's death.) Riemann's first course of lectures concerned partial differential equations as applied to physics. The course was so admired by physicists that it was reprinted as long afterwards as in 1938 – 80 years later. He published a paper on hypergeometric series and in 1855–56 lectured on his (by now famous) theory of Abelian functions, one of his fundamental developments in mathematics, which he published in 1857. When Karl Gauss died in 1855, Dirichlet was appointed in his place, and Riemann was appointed Assistant Professor in 1857. On Dirichlet's untimely death in 1859 Riemann became Professor. Three years later, Riemann married; but in the same year he fell seriously ill with tuberculosis and he

spent much of the following four years trying to recover his health in Italy. But he died there, in Selasca, on 16 June 1866 at the age of 39.

Riemann's first work (his thesis, published when he was already 25) was a milestone in the theory of complex functions. Augustin Cauchy (1789–1857) had struggled with general function theory for 35 years, had discovered most of the fundamental principles, and had made many daring advances – but some points of understanding were still lacking. Unlike Cauchy, Riemann based his theory on theoretical physics (potential theory) and geometry (conformal representation), and could thus develop the so-called 'Riemann surfaces' which were able to represent the branching behaviour of a complex algebraic function.

He developed these ideas further in a paper of 1857, which continued the exploration of Riemann surfaces as investigative tools to study complex function behaviour, these surfaces being given such properties that complex functions could map conformally onto them. In the theory of Abelian functions there is an integer p associated with the number of 'double points' – a double point may be represented on a graph as where a curve intersects itself. Riemann formed, by extension, a multiconnected many-sheeted surface that could be dissected by cross-cuts into a singly connected surface. By means of these surfaces he introduced topological considerations into the theory of functions of a complex variable, and into general analysis. He showed, for example, that all curves of the same class have the same Riemann surface. Extensions of this work become highly abstract – the genus p was in fact discovered by Niels Abel in a purely algebraic context; it is not simply the number of double points. Not until much more work had followed, by Henri Poincaré (1854–1912) in particular, did Riemann's ideas in this field reach general understanding. However, his work considerably advanced the whole field of algebraic geometry.

Often his lectures and papers were highly philosophical, containing few formulae, but dealing in concepts. He took into account the possible interaction between space and the bodies placed in it; hitherto space had been treated as an entity in itself, and this new point of view – seized on by the theoretical physicist Hermann Helmholtz (1821–1894) among others – was to become a central concept of twentieth-century physics.

Riemann's most profound paper on the foundation of geometry (presented in 1854), also had consequences for physics, for in it he developed the mathematical tools that later enabled Albert Einstein to develop his theory of relativity. The three creators of 'hyperbolic geometry' – Karl Gauss, Nikolai Lobachevsky and János Bolyai – all died in the 1850s with their work unacknowledged, and it was Riemann's paper that initiated the revolution in geometry. In 1799 Gauss had claimed to have devised a geometry based on the rejection of Euclid's fifth postulate, which states that parallel lines meet at infinity. Another way of expressing this is to consider a parallelogram in which two opposite corners are right-angled – but in which the other two angles are less than 90°; this is 'hyperbolic' geometry.

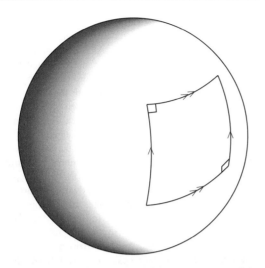

Riemann *In spherical geometry, although a parallelogram may have two opposite angles at 90°, it does not necessarily follow that the other two angles are also 90°, as pointed out by Riemann.*

One possible case for the two angles to be less than 90° would come about according to two-dimensional geometry on the surface of a sphere. Straight lines are in those circumstances sections of great circles, and a 'rectangle' drawn on the surface of the sphere may have angles of less than 90°. In this way, Riemann invented 'spherical' geometry, which had previously been overlooked (and which can be more disconcerting than hyperbolic geometry). In 1868, Eugenio Beltrami developed this, considering 'pseudospheres' – surfaces of constant negative curvature which can realize the conditions for a non-Euclidean geometry on them.

Many other ideas were contained in Riemann's paper. He took up the discussion – and returned to it again later in life – of the properties of topological variabilities (manifolds) with an arbitrary number of (*n*) dimensions, and presented a formula for its metric, the means of measuring length within it. He took an element of length d*s* along a curve, and defined it as:

$$ds^2 = \Sigma g_{ij}{}^2 \, dx^i \, dx^j \, (i, j = 1, 2, ..., n)$$

The structure obtained by this rule is called a Riemann space. It is in fact exactly this sort of concept that Einstein used to deal with time as a 'fourth dimension', and to talk about the 'curvature of space'. Euclidean geometry requires the 'curvature' to be zero. The definition of the curvature of a directed space curve at a point was introduced in this same paper, and implicitly introduces the concept of a tensor. From generalization of this work, and despite opposition at the time, *n*-dimensional geometries began to be used, especially to examine the properties of differential forms with more than three variables.

Riemann's career was short, and not prolific, but he had a profound and almost immediate effect on the development of mathematics. Everything he published was of the highest quality, and he made a breakthrough in conceptual understanding within several areas of mathematics: the theory of functions, vector analysis, projective and differential geometry, non-Euclidean geometry and topology. Twentieth-century mathematics bears witness to the extreme fruitfulness of Riemann's ideas.

Ritter Johann Wilhelm 1776–1810 was a German physicist who carried out early work on electrolytic cells and who discovered ultraviolet radiation.

Ritter was born in Samnitz, Silesia (now in Poland), on 16 December 1776. He worked first as a pharmacist's apprentice and then after four years, in 1795, went to Jena to study medicine. Until 1804 he also taught at the University of Jena and at Gotha, before moving to Munich as a member of the Bavarian Academy of Science. He died in Munich six years later on 23 January 1810.

In 1800, Ritter electrolysed water to produce hydrogen and oxygen and two years later developed a dry battery, both of which phenomena convinced him that electrical forces were involved in chemical bonding. He also compiled an electrochemical series. At about the same time he was studying the effect of light on chemical reactions, and from the darkening of silver chloride in light he discovered ultraviolet radiation.

Roberts Richard 1789–1864 was a Welsh engineer who became one of the greatest mechanical inventors of the nineteenth century. His inventions included a screw-cutting lathe and a planing machine.

Roberts was born on 22 April 1789 at Carreghofa, Montgomeryshire (now part of Powys), the son of a shoemaker. He had very little formal education and at an early age went to work on the newly opened Ellesmere Canal. From there he obtained work as a labourer in a limestone quarry. At 20 years of age he went to seek work in Liverpool and, later, Manchester – where he became a toolmaker.

To evade military service, Roberts moved to London, where he worked as an apprentice to Henry Maudslay, one of the foremost engineers of the day. Roberts returned to Manchester, which was then the centre of the cotton industry, and in 1814 he set up in business for himself. In 1817 he designed a machine for planing metal – one of the first in his long series of inventions. In 1824, at the request of some manufacturers, he built a self-acting spinning mule which was a vast improvement on that devised by Samuel Crompton in 1779. He took only four months to complete the task.

In 1828, Roberts went into partnership with Thomas Sharpe to found the firm of Sharpe, Thomas and Company. Thomas was the director of the machine-making section. The firm manufactured machines to his design which he continued to improve. For example in 1832 he improved his mule by adding a radial arm for winding the yarn. When the Liverpool and Manchester Railway opened in 1830 the firm ventured into the business of building locomotives. Their products were bought by both British and Continental railway companies. In 1834, they produced a steam carriage with a differential drive to the back wheels. Thomas also designed a steam-brake and a system of standard gauges to which all his work was constructed.

The partnership ended when Sharpe died in 1842. The firm split up – the Sharpe family continued to control the manufacture of locomotives, while Roberts retained the remaining part of the company known as the Globe Works. In 1845 he invented an electromagnet – of which examples were placed in the Manchester museum at Peel Park and one was kept by the Scottish Society of Arts. In 1848 Roberts invented a machine for punching holes in steel plates. Incorporating the Jacquard method, he devised a machine for punching holes of any pitch or pattern in bridge plates and boiler plates. He later invented a machine for simultaneously shearing iron and punching both webs of angle iron to any pitch. As with his improvement to the self-acting mule a quarter of a century before, Roberts built his hole-punching machine in response to requests by company owners who this time found themselves faced with labour difficulties as a result of the strike during the building of Thomas Telford's Conway suspension bridge.

At the Great Exhibition of 1851 Roberts received a medal for a turret clock he had made. Recognition of his contribution to engineering took other forms. He was elected member of the Institution of Civil Engineers in 1838 and was a founder member of the Mechanics Institute in 1824. But, despite his genius as an inventor, Roberts's financial position deteriorated almost to poverty and just before he died on 16 March 1864 a public subscription was being raised to support him.

Robertson Robert 1869–1949 was a British chemist whose main work was concerned with improvements to explosives for military use.

Robertson was born in Cupar, Fife, on 17 April 1869. He was educated at the Madras Academy and St Andrews University, where he took extra lessons in chemistry in order to gain a BSc degree as well as his MA. After a short period as assistant to the Glasgow City Analyst he obtained a post at the Royal Gunpowder Factory at Waltham Abbey, Essex (at the time when the military were changing from black powder to smokeless nitro compounds as propellant explosives for cartridges). The results of his work on nitrocellulose were incorporated into his doctoral thesis of 1897. He was put in charge of the main laboratory in 1900 and seven years later was appointed Superintending Chemist in the Research Department at Woolwich Arsenal. The work of the Department increased tremendously during World War I, and in 1918 Robertson was knighted for his services. On the retirement of James Dobbie in 1921 Robertson became the Government Chemist, a position he held until he left government service in 1936 and went to work at the Davy Faraday Laboratory at the Royal Institution. At the outbreak of World War II in 1939 his offer to return to the Explosives Research Department at Woolwich was gladly accepted; he went back to the Royal Institution in 1945. He died in London on 28 April 1949.

From 1900, at Waltham Abbey, Robertson studied the Will test for measuring the rate of decomposition of guncotton (nitrocellulose) and in 1906 introduced an improved method of purifying nitrocellulose. As a result of his work the propellant in British ammunition was changed from Mark I Cordite to the more stable MD Cordite. His appointment to Woolwich in 1907 coincided with the analysis of defects in British ammunition that had been revealed during the South African War. The new explosive tetryl (trinitrophenylmethylnitramine) was developed, as were detonators for Lyddite (picric acid, 2,4,6-trinitrophenol); work began on the use of TNT (trinitrotoluene, 2,4,6-trinitromethylbenzene) as a high explosive for military purposes. Robertson continued his investigations into the stability of Cordite, showing that the presence of impurities can be a critical factor (and leading to improved methods of manufacture and storage).

At the beginning of World War I the British government became concerned about supplies of TNT, which were limited by the availability of toluene (methylbenzene), manufactured from coal tar or extracted from Borneo petroleum, which is rich in aromatic hydrocarbons. Robertson's researchers solved the considerable problems of 'diluting' TNT with ammonium nitrate to produce a new high explosive, Amatol.

After Robertson became the Government Chemist in 1921 he welcomed the opportunity to carry out research unfettered by the secrecy that had inevitably surrounded his previous work. Many of his investigations were carried out to assist the Department of State. These included the carriage of dangerous goods by sea, the determination of sulphur dioxide and nitrous gases in the atmosphere, the elimination of sulphur dioxide from the gaseous products of combustion at power stations, the possible effects on health of tetraethyl lead additives to petrol, the determination of iodine in biological substances, and the preservation of photographic reproductions of valuable documents. He was also concerned with the determination of carbon monoxide and nitrous oxide (dinitrogen monoxide) and in an investigation of the extraction of minerals (such as potassium chloride and bromine) from the waters of the Dead Sea.

Later work on infrared absorption spectroscopy in collaboration with J. J. Fox and E. S. Hiscocks greatly stimulated research in this field. Robertson's improvements in spectrographic equipment also permitted his study of diamonds from various natural sources.

Robinson Robert 1886–1975 was a British organic chemist who, during a long and distinguished career, made many contributions to the science. Among the many and wide-ranging topics he researched were alkaloids, steroids and aromatic compounds in general. For his work on alkaloids and other biologically significant substances derived from plants he was awarded the 1947 Nobel Prize for Chemistry.

Robinson was born in Bufford, near Chesterfield, on 13 September 1886, the son of a local manufacturer. It was intended that he should enter his father's business but after graduating in chemistry from Manchester University in

1905, he embarked on an academic career. He obtained his doctorate in 1912 and then held the Chair in Organic Chemistry successively at Sydney University (1912–15), Liverpool (1915–20), St Andrews (1920–22), Manchester (1922–29) and Oxford (1929–55). He was knighted in 1939. In 1957 he founded the influential journal *Tetrahedron*. He died in Great Missenden on 8 February 1975.

Robinson's lifelong interest in plant materials began with his study of the colourless material brazilin and its red oxidation product brazilein, which occur in brazilwood. This work led on to an investigation of anthocyanins (red and blue plant pigments) and anthoxanthins (yellow and brown pigments). He studied their composition and synthesis, and related their structure to their colour. His first synthesis in this area, that of callistephin chloride, was carried out in 1928.

In his research on alkaloids he worked out the structure of morphine in 1925 and by 1946 he had devised methods of synthesizing strychnine and brucine – using only 'classical' techniques of organic chemistry – and so influenced all structural studies of natural compounds that contain nitrogen. He also suggested biosynthetic pathways for the production of such substances in nature. While not always correct, these proposals confirmed his relentless and convincing assertion that, since nature involves chemical substances, it must obey laws recognized by chemists.

He began research on steroids at Oxford and his studies of the sex hormones, bile acids and sterols were fundamental to the general methods now used to investigate compounds of this type. His discovery that certain synthetic steroids could produce the same biological effects as do the natural oestrogenic sex hormones led to the preparation of stilboestrol, hexoestrol and dienoestrol, paving the way for pharmaceutical applications such as the contraceptive pill and treatments for infertility in women.

In 1942, spurred on by the needs of World War II, Robinson investigated the properties of the antibiotic penicillin and elucidated its structure. His methods were later applied to structural investigations of other antibiotics.

Throughout his career Robinson was also concerned with the theoretical aspects of organic chemistry. He began by studying the polarization (electron displacement) in the carbon–chlorine covalent bond and progressed to investigating conjugate systems, which involve alternate single and double carbon–carbon bonds. He showed that if the original double bonds are sufficiently weak, pairs of electrons can transfer from them to intervening single bonds along the chain. He introduced the method of representing such electron transfer by means of 'curly arrows' on a structural formula. This theory is particularly relevant to aromatic compounds, in which the presence of a substituent in the benzene ring influences further substitution and how fast such substitution takes place. Robinson worked out the theory that governs this important type of reaction and how it is affected by the nature of the reagent concerned: electrophilic reagents attack preferentially positions in which there are an excess of electrons; nucleophilic reagents attack electron-deficient positions.

In his later years Robinson became interested in the composition of petroleum and how it originated on Earth, and he suggested routes by which petroleum products may have originated from amino acids and other chemicals present before life as we know it began.

He was Professor of Organic Chemistry at St Andrews in the early 1920s at the same time as his predecessor, Walter Norman Haworth, was at Durham University. Between them they led contemporary work on an extremely wide range of natural products (Haworth's main area of study was carbohydrates) and many of today's structural and synthetic methods are based on their pioneering work.

Roe Alliot Verdon 1877–1958 was a British aircraft designer. The first Englishman to construct and fly an aeroplane, he also designed the famous Avro series of aircraft.

Roe was born on 26 April 1877 at Patricroft, near Manchester. He was educated at St Paul's School, London, but at the age of 14 he went to British Columbia. Returning from Canada a year later, he served a five-year apprenticeship at the Lancashire and Yorkshire Railway Company's locomotive sheds. He then spent two years at sea, after which he entered the motor industry. He became interested in aircraft design and, in 1908, flew a distance of 23 m/75 ft in a biplane of his own design. This feat was accomplished nearly a year before the first officially recognized flight in England by John Moore-Brabazon (1884–1964).

In 1910, with his brother Humphrey, Roe founded the firm of A. V. Roe and Company. It became one of the world's major aircraft companies and the builder of one of the most famous aircraft of its time – the Avro 504. In 1928, Roe severed all ties with the company and turned his attention to the design of flying-boats. He became associated with the firm of S. E. Saunders and founded Saunders-Roe Company at Cowes, Isle of Wight, and became its president. Roe was knighted in 1929, and became known as Sir (Edwin) Alliot Verdon-Roe. He died at his home near London on 4 January 1958.

The first aircraft from the Manchester works was the Avro 500, of historic importance because it was one of the first machines to be ordered for use by the British army. The order – for three two-seater biplanes fitted with 37 kW/50 hp Gnome engines – was placed by the Government in 1912. Before the Gnome engines appeared, Roe had completed a two-seater biplane powered by a 45 kW/60 hp ENV engine, and it was this aircraft that provided the basis for a succession of very successful planes.

The 500 was an equal-span, two-bay biplane of wooden construction. The undercarriage was of an original design, with wheels mounted on a transverse leaf-spring which attached to a central skid (there was no tail-skid). The machine rested its rear end on the base of a suitably reinforced rudder; a modified rudder, still acting as a tailskid, was fitted later.

The first of the 500s went into service on 12 May 1912, and two were used to form the strength of the Central Flying School of the Royal Flying Corps, which opened at Upavon on 7 August that year. Several of these aircraft were later used by the CFS and also by the naval and military wings of the RFC.

The original order was increased to 12 and this contract enabled the firm of A. V. Roe to become solidly established. In 1913 the company produced its first sea-plane type. It was a large biplane originally known as the Avro 503.

The construction of the Avro 504, which followed the 503, began in April 1913. It made its debut in September of that year. The design, construction and performance were all considerably in advance of its contempories, and it was this aircraft which helped lay the foundations for decades of safe flying instruction. Adapted to a float-plane version in 1914, it is regarded as one of the first production machines. Although not basically a military aircraft, it was used extensively in World War I and took part in some famous actions – the attack on the Zeppelin sheds at Freidrichshaven being one. Further modification led to a long succession of successful aircraft, from the 504A through to the 504H, which was the first aircraft to be successfully launched by catapult.

Romer Alfred Sherwood 1894–1973 was a US palaeontologist and comparative anatomist who is best known for his influential studies of vertebrate evolution and as the author of several books on the anatomy and evolution of the vertebrates.

Romer was born on 28 December 1894 in White Plains, New York, the son of a journalist. His family moved frequently during his early years, but in 1909 Romer returned to White Plains to live with his grandmother and a more settled phase began. He left high school in 1912 and, because his family was poor, spent a year doing odd jobs to earn money for a college education. In 1913, he entered Amherst College to study history and German literature but his interest soon turned to palaeontology, and he took a course in evolution. He graduated in 1917 after the start of World War I and joined the American Field Service in France; later that year he enlisted in the United States Army, and remained in Europe until 1919. On returning to the United States, he did postgraduate work at Columbia University, from which he gained his doctorate in 1921 with a thesis on comparative myology that is still a classic in its field. He then taught anatomy at the Bellevue Hospital Medical College, New York, until 1923, when he was appointed an Associate Professor in the Department of Geology and Palaeontology at the University of Chicago. He held this post for 11 years, during which period he married Ruth Hibbard; they had three children. In 1934, Romer was appointed Professor of Biology at Harvard University; he also became Director of the Biological Laboratories in 1945 and of the Museum of Comparative Zoology in the following year. He held these three posts until 1965, when he officially retired, although he continued to work and lecture for the rest of his life. Romer died on 5 November 1973 in Cambridge, Massachusetts.

Romer spent almost all of his career investigating vertebrate evolution. Using evidence from palaeontology, comparative anatomy and embryology, he traced the basic structural and functional changes that took place during the evolution of fishes to primitive terrestrial vertebrates and from these to modern vertebrates. In these studies he emphasized the evolutionary significance of the relationship between the form and function of animals and the environment.

One of the most important figures in palaeontology since the 1930s, Romer wrote several well-known books on vertebrates, including *Man and the Vertebrates* (1933), *Vertebrate Palaeontology* (1933), which was widely influential in its field for several decades, and *The Vertebrate Body* (1949), a comprehensive study of comparative vertebrate anatomy which is still a standard textbook today. He also collected an extensive range of fossils from his field trips to South Africa, Argentina and Texas.

Römer Ole (or Olaus) Christensen 1644–1710 was a Danish astronomer and civil servant who, through the precision of both his observations and his calculations, first derived a rate for the speed of light. This was all the more remarkable in that most scientists of his time considered light to be instantaneous in propagation. A practical man, Römer was also talented at designing scientific instruments.

He was born at Århus, in Jutland, on 25 September 1644. Educated in Copenhagen, he attended the university there, where he studied under the Bartholin brothers – Thomas (1616–1680, Professor of Mathematics and Anatomy) and Erasmus (1625–1698, physicist and astronomer). First as a student of mathematics and astronomy, then as personal assistant, Römer lived at the house of Erasmus Bartholin, whose daughter he eventually married. In 1671 he collaborated with Jean Picard (1620–1682), who had been sent by the French Academy to verify the exact position of Tycho Brahe's observatory. Evidently impressed by Römer's work, Picard invited him to come back to Paris with him once the investigations were over; Römer gladly accepted. At the Academy in Paris, Römer worked initially as an assistant, but within a year was made a member in his own right. He was also appointed tutor to the Crown Prince. There then followed several years in which he conducted observations, designed and improved instruments, and submitted various papers to the Academy, all culminating, in 1679, in the exposition of his calculation of the speed of light. With his new-found fame, Römer then visited England and met some of the greatest astronomers of his age – Sir Isaac Newton, Sir Edmond Halley and the Astronomer Royal, John Flamsteed. He returned to Denmark in 1681 to take up the dual post of Astronomer Royal to King Christian V and Director of the Royal Observatory in Copenhagen. He also accepted a number of civic duties. He died in Copenhagen on 23 September 1710.

It was while he was in Paris that Römer carried out his famous research which not only demonstrated that light travels at a finite speed but also put a rate to it. His observations of the satellites around Jupiter, especially of the innermost one, Io, led him to notice that the length of time between eclipses of Io by Jupiter was not constant. He found that when the distance between the Earth and Jupiter was least, the interval between eclipses was also smallest. He therefore measured the inter-eclipse period when the two planets were closest and then announced in September 1679 that the eclipse of Io by Jupiter predicted for 9 November would occur 10 minutes later than expected on the basis of all previous calculations.

Römer's prediction was borne out; his interpretation of the delay provoked a sensation. He said that the delay was caused by the time it took for the light to traverse the extra distance across the Earth's orbit when the positions of Jupiter and the Earth were such that they were not as close to each other as they sometimes were. This meant that light did not traverse space instantaneously, but travelled at a finite speed. Römer estimated that speed to be 225,000 km/140,000 mi per second – which is remarkably close (considering that it was the first estimate ever) to the modern value of 299,792 km/186,291 mi per second. Römer's interpretation of his observations was not accepted by all of his contemporary astronomers, particularly not in France, but was confirmed 60 years later – after Römer's death – by James Bradley (who was also able to improve upon Römer's estimate, obtaining a value of 294,995 km/183,310 mi per second).

Römer's later work was on optics, instrument design and the systematization of weights and measures. Unsatisfied by the Copenhagen Observatory, he established his own private observatory, which he named the Tuscalaneum and which possessed the first telescope attached to a transit circle, a device of his own invention.

Röntgen Wilhelm Konrad 1845–1923 was a German physicist who discovered X-rays. For this achievement, he was awarded the first Nobel Prize for Physics in 1901.

Röntgen was born in Lennep, Prussia, on 27 March 1845. He received his early education in Holland and then entered the Polytechnic in Zurich, Switzerland, in 1866 to study mechanical engineering. Röntgen received a diploma in this subject in 1868 and a PhD in 1869. He then became an assistant to August Kundt (1839–1894), who was Professor of Physics at Zurich, and changed direction towards pure science. Under Kundt's guidance, Röntgen made rapid progress in physics. In 1871, he moved to Würzburg and then in 1872 to Strasbourg both times in order to remain with Kundt. He then obtained the position of Professor of Physics and Mathematics at the Agricultural Academy of Hohenheim in 1875, but returned to Strasbourg the following year to teach physics. He was then Professor of Physics at Giessen from 1879 to 1888, when he was appointed to the same position at Würzburg, also becoming Director of the Physical Institute there. Finally in 1900, Röntgen became Professor of Physics and Director of the Physical Institute at Munich. He retired in 1920, and died in poverty on 10 February 1923 at Munich, following runaway inflation in Germany.

During his career, Röntgen's worked on such diverse topics as elasticity, heat conduction in crystals, specific heat capacities of gases and the rotation of plane-polarized light. In 1888, Röntgen made an important contribution to electricity when he confirmed that magnetic effects are produced by the motion of electrostatic charges, following a demonstration of the effect made by Henry Rowland (1848–1901) in 1875, but not since repeated.

It was during his period in Würzburg that Röntgen made his most momentous discovery. In November 1895 he was investigating the properties of cathode rays emitted by a high-vacuum discharge tube, particularly the luminescence produced when such rays impinged on certain chemicals. One of the substances with which he was experimenting was barium platinocyanide. Surprisingly, he found that it glowed even when the tube was encased in black cardboard. He then removed the chemical to an adjacent room; as before it glowed whenever the tube was activated. Such high penetrating power led Röntgen to the conclusion that the radiation was entirely different from cathode rays. Unable to establish the nature of the radiation, which in fact came from the glass walls of the tube when struck by the cathode rays, he coined the term 'X-rays'. Röntgen was not in doubt that his discovery was unique and therefore decided to look more closely at the nature of his X-rays before publicly announcing their existence. He established that they pass unchanged through cardboard and thin plates of metal, travel in straight lines and are not deflected by electric or magnetic fields.

By January 1896, Röntgen was ready to reveal his work to the general public. As X-rays are absorbed by bone, he illustrated one lecture with an X-ray photograph of a man's hand. The impact was tremendous. Soon the use of X-ray equipment and photography in medical work developed both in Europe and America. The discovery also served as an impetus to other scientists. In France it led Henri Becquerel (1852–1908) to the discovery of radioactivity the same year, which in turn caused a revolution in ideas of the atom, while the demonstration of X-ray diffraction in 1912 by Max von Laue (1879–1960) brought about methods of investigating atomic and molecular structure.

It is likely that X-rays had been produced by others before Röntgen, because experiments with cathode rays had been going on ever since Julius Plücker (1801–1868) first produced them in 1858. William Crookes (1832–1919) had noticed that photographic plates kept near cathode-ray tubes became fogged, an effect almost certainly due to X-rays. However, it was Röntgen who first realized the existence of X-rays and first investigated their properties. It was only right that he was acclaimed by the world for a discovery that has brought immense benefits both in medicine and science. It is perhaps fitting that it was made by a man of total integrity who refused to make any financial gain out of his good fortune, believing that the products of scientific research should be made freely available to all.

Ross Ronald 1857–1932 was a British physician who proved that malaria is transmitted to human beings by the bite of the *Anopheles* mosquito. He also devoted much of his time to public health programmes concerned with the prevention of the disease. For his significant contribution to

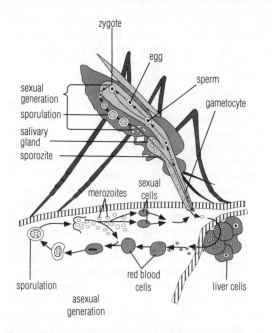

zygote

egg

sperm

gametocyte

sexual generation

sporulation

salivary gland

sporozite

merozoites

sexual cells

sporulation

red blood cells

liver cells

asexual generation

Ross *The reproductive cycle of a malarial parasite is split between* Anopheles *mosquito and human.*

the battle against malaria, which has plagued people for centuries, he received the 1902 Nobel Prize for Physiology or Medicine.

Ross was born at Almora, India, on 13 May 1857, the son of a British army officer serving there. He was educated in Britain and received his medical training at St Bartholomew's Hospital (Bart's), London, graduating in 1879. He was unenthusiastic about medicine, his interests at that time being in the arts and mathematics. When he joined the Indian Medical Service in 1881, however, he gradually became absorbed with medical problems. During his first leave in England, from 1888 to 1889, he obtained a Diploma in Public Health and took a course in bacteriology. On retiring from the Indian Medical Service in 1899, he returned to Britain and lectured at the new School for Tropical Medicine in Liverpool, later holding the Chair in Tropical Medicine there. He was knighted in 1911 and a year later moved to London, where he established a consulting practice at King's College Hospital. During World War I he was consultant on malaria to the War Office and when the Ross Institute of Tropical Diseases was opened in his honour, in 1926, he became its first Director. He died in London on 16 September 1932.

While on leave in England in 1894, Ross became acquainted with Patrick Manson, who demonstrated that the blood of malarial patients contained pigmented bodies and parasites. (The discovery that malaria was caused by a protozoon in the bloodstream had been made by Charles Laveran in 1883.) Manson's suggestion was that malaria was spread by mosquitoes and Ross returned to India determined to investigate this hypothesis and to identify the mosquito responsible. The shortage in India of literature on

the subject delayed the identification of species of mosquitoes and parasites, and Ross was also cut off from the work of others. He missed, for example, Albert King's suggestion that malaria might be transmitted via mosquito bites.

Ross refused to believe the popular idea that malaria was caused by bad air (*mal aria*) or contaminated water, and he continued to collect mosquitoes, identifying the various species and dissecting their internal organs. In the stomachs of some insects he found 'motile filaments' which, although Ross was unaware of it, were the malarial parasite's gametes. He thought that the filaments might develop to further stages, and dissected a few mosquitoes that had fed on malarial patients. In August 1897, he discovered in an *Anopheles* mosquito a cyst containing the parasites that had been found by Laveran in the blood of malarial patients. These sporozoites of the malarial parasite remain in the human blood for only an hour after a bite, before invading liver cells, and their subsequent developmental stages are distinct from those in the mosquito. The biting mosquito may suck up various stages of the parasite with the blood, but all are digested except those that produce gametes. After fertilization the zygote bores through the stomach wall of the mosquito and forms an external cyst. Sporozoites are formed in the cyst, which migrate to the mosquito's salivary glands, ready for injection into a victim. Later, using caged birds with bird malaria, Ross was able to show that the 'motile filaments' do develop to further stages. The life history of the parasite inside a mosquito was thus revealed and the mode of transmission to the victim was identified as taking place through a mosquito bite.

Roux Wilhelm 1850–1924 was a German anatomist and zoologist who is famous for his work on developmental mechanics in embryology. He founded the first-ever journal of experimental embryology and also helped to produce a dictionary of experimental morphology.

Roux was born on 9 June 1850 in Jena. He attended the Oberrealschule in Meiningen then studied at the universities of Jena – where he was a student of Ernst Haeckel – Berlin and Strasbourg. In 1873, he matriculated from the medical faculty, and passed his state medical examinations in 1877. After spending several years as an assistant at Franz Hofmann's Institute of Hygiene in Leipzig, Roux moved to the University of Breslau (now Wroclaw in Poland), where he eventually became Director of his own Institute of Embryology. He was Professor of Anatomy at the University of Innsbruck from 1889 to 1895, when he became Director of the Anatomical Institute at the University of Halle, where he remained until his retirement in 1921. Roux died in Halle on 15 September 1924.

Roux's embryological investigations were performed mainly on frogs' eggs. He performed a series of experiments in which he punctured the eggs at the two-cell stage of development (a technique Roux pioneered) and found that they grew into half-embryos; this finding led him to conclude that the fate of the parts had already been determined at the two-cell stage. In the field of embryology

Roux also researched into the earliest structures in amphibian development. In 1894, he founded the first journal of experimental embryology; it is still published today and is now called *Roux Archiv für Entwicklungsmechanik*.

Roux also investigated the mechanisms of functional adaptations, examining the physical stresses that cause bones, cartilage and tendons to adapt to malformations and diseases. In addition, he collaborated with two botanists and an anatomist to produce a dictionary of experimental morphology, which provided a valuable compendium of definitions and historical notes in this discipline.

Roux was not the only scientist who performed embryological experiments at that time, but he nevertheless made substantial contributions to the knowledge of embryological development. In his own judgement, however, he grossly overestimated the importance of his work, claiming, for example, that he alone was the first to progress from causal manipulation to a causal analysis of development – an absurd contention because, at the very least, it ignored the work of his contemporaries Oskar Hertwig (1849–1922) and Eduard Pfluger (1829–1910).

Rowland Henry Augustus 1848–1901 was a US physicist who is best known for the development of the concave diffraction grating, which heralded a new era in the analysis of spectra.

Rowland was born in Honesdale, Pennsylvania, on 27 November 1848. As a child he displayed a clear interest in science, and entered the Rensselaer Polytechnic Institute at Troy, New York, in 1865, graduating with a degree in civil engineering in 1870. Rowland spent the next year working on surveys as a railway engineer. In 1871, he became a teacher of natural science at Wooster College in Ohio, but returned to the Rensselaer Polytechnic in 1872 to serve first as an Instructor, and from 1874 as Assistant Professor.

The newly established Johns Hopkins University in Baltimore, Maryland, offered Rowland the Chair of Physics in 1876. He accepted, and spent the next year travelling in Europe meeting scientists such as James Clerk Maxwell (1831–1879) and Hermann Helmholtz (1821–1894), conducting some research and purchasing equipment to make the physics laboratories at Johns Hopkins among the best equipped in the world. Rowland remained as Professor of Physics there until his death at Baltimore on 16 April 1901. He was also the founder and first President of the American Physical Society.

Rowland felt that he did not comfortably fit into the category of either experimental or theoretical physicist. He designed his experiments so that their results might shed light on questions of theoretical interest, but he was also extremely talented at the practical design of experimental apparatus.

His earliest research, in 1873 and 1874, concerned the measurement of the magnetic permeability of nickel, iron and steel, and the demonstration of its variation with the strength of the magnetizing force. This led to one of his most noteworthy experiments, which was conducted while on a brief visit to Helmholtz's laboratory in Berlin in 1875.

The question of whether a moving electrically charged body would demonstrate certain properties of an electric current, such as the ability to create a magnetic field, had arisen from the work of Michael Faraday and Maxwell. If it did, then an electric current could be regarded as a sequence of electric charges in motion. Rowland was able to provide the first demonstration that this is in fact so by showing that a rapidly rotating charged body was able to deflect a magnet. Such was the delicacy of this experiment that it was not confirmed until Wilhelm Röntgen (1845–1923) succeeded in repeating it in 1888.

Upon his return to the United States, Rowland began a series of studies aimed at the accurate determination of certain physical constants. He first investigated the value for the ohm, the unit of electrical resistance. The value which he published in 1878 is very close to that accepted today. He then repeated Joule's determination of the mechanical equivalent of heat in 1879, and was able to improve the accuracy of the result.

The development of the concave diffraction grating in 1882 stands as Rowland's major contribution to physics. The science of spectroscopy was well advanced from the days when prisms had been relied upon for the production of spectra, due particularly to the introduction by Joseph Fraunhofer (1787–1826) of ruled gratings in place of prisms. The quality of the results obtainable depended on the accuracy and density of the scratched rulings on the diffraction gratings. Rowland saw that a critical limiting factor in the ability to produce these lines was the availability of a perfect screw for the ruling machine. He tackled the problem using his engineering expertise, and was eventually able to produce lines at a density of more than 16,000 per cm/40,000 per in of grating. This gave a much greater revolving power, but Rowland went on to improve spectroscopy even further by the introduction of a concave metal or glass grating. This had the advantage of being self-focusing and thus eliminated the need for lenses, which absorbed some wavelengths of the spectrum.

Rowland's concave diffraction gratings were a sensational success. He sold over one hundred of the devices, which enabled scientists to accomplish in a few hours what had previously taken days to achieve. Rowland then put his invention to use in the years 1886 to 1895 by remapping the solar spectrum, publishing the wavelengths for 14,000 lines with an accuracy ten times better than his predecessors had managed.

Royce Frederick Henry 1863–1933 was the British engineer who, in cooperation with Charles Rolls, produced the famous Rolls–Royce series of high-performance motorcars.

Royce was born on 27 March 1863 at Alwalton, Huntingdonshire (now Cambridgeshire). On leaving school he became an apprentice engineer with the Great Northern Railway. He later (1882–83) worked on the pioneer scheme to light London's streets with electricity, and as chief electrical engineer on the project to light the streets of Liverpool. In 1884 he founded the firm of F. H. Royce and

Co Ltd, Mechanical Engineers of Manchester, and manufactured electric cranes and dynamos. In 1904, he built his first car and in 1907, with Rolls, he founded the firm of Rolls–Royce Ltd, motorcar and aeroengine builders, of Trafford Park, Manchester (and later of Derby and London). Royce became the company's director and chief engineer.

By 1914, the firm had established a reputation for building 'the best cars in the world'. In 1920 an American factory was opened at Springfield, Massachusetts, and the Rolls-Royce Ghost was made there until 1926. The Ghost was followed by the Phantom, manufactured from 1927 to 1931. In that year the company bought Bentley Motors and two years later the Bentley luxury model car, based on a Rolls-Royce three-litre engine, made its appearance. It maintained its reputation as a first-class luxury limousine until 1950.

Royce was made a baronet in 1930. He was also awarded the OBE and was a member of the Institute of Mechanical Engineers, the Institute of Aeronautical Engineers and the Institute of Electrical Engineers. He died in Sussex on 22 April 1933.

Like all other industrialists at the turn of the century, Rolls at first knew nothing about motorcars and could only copy and improve on the good designs of others. In 1903 France led the world in automobile design. Taking a French car, the $7^1/_2$ kW/10 hp Decauville as his model, Royce used imported parts and set about improving the design. Because he was a craftsman-mechanic and possessed special skills, the cars he built were just that much better than the others being made at that time. They became noted for the extremely fine workmanship, which gave them exceptional running qualities.

The first Royce-built car appeared in 1904. It had mechanically operated overhead inlet valves – a great improvement on the Decauville's automatic inlets. Because it was so well made, the Royce car was also very much quieter and smoother-running. Three of these first models were built and, by chance, one of them was bought by a man called Edmunds who was a director of a car importing firm. Also working for the firm was Charles Rolls. Although he worked as a salesman, Rolls was rich, aristrocratic and well-known as a pioneer motoring and aviation enthusiast.

Rolls drove and tested the car and was so impressed by its performance that he undertook to sell Royce's entire output. The two men became friends, pooled their talents and started the world-famous partnership. The Rolls-Royce range was expanded to include three-litre, three-cylinder 'light' and 'heavy' versions of the four-cylinder Twenty and Six, to sell at £900.

In 1906 a light Twenty, driven by Rolls, won the Tourist Trophy and also broke the Monte Carlo-to-London record. In that year, also, the partnership embarked upon its one-model policy based on 30–37 kW/40–50 hp six-cylinder car which was later to win immortality as the Silver Ghost. The car's reputation was assured after a successful 24,000-km/15,000-mi RAC-observed trial in 1907, and it was in

that year that the Rolls–Royce firm of car-makers officially came into existence. Over 6,000 of these cars were made at Manchester and later at the Derby works.

Rolls was killed in a flying accident in 1912, but the firm continued to prosper. An armoured version of the Ghost saw service both during and after World War I. Later, the firm was to win a reputation for its production of aero-engines as high as that for its motorcars.

Ruffini Paolo 1765–1822 was an Italian mathematician, philosopher and doctor who made valuable contributions in all three disciplines. In his mathematical work he is remembered chiefly for what is now known as the Abel–Ruffini theorem.

Ruffini was born on 22 September 1765 at Valentano, Viterbo. While he was very young his family moved to Modena, and it was there that he grew up and remained for the rest of his life. Studying medicine, philosophy, mathematics and literature at the University, Ruffini was an exceptional student. During his own final year of instruction, he also taught one of the courses in mathematics. In 1788, he obtained degrees in philosophy, medicine and mathematics, and was appointed Professor of the Foundations of Analysis. Three years later he became Professor of the Elements of Mathematics.

Napoleon entered Modena in 1796, and Ruffini found himself obliged to take up an appointment as an official of the Republic, until permitted to return to his teaching two years later. Then, however, he refused to swear the oath of allegiance to the Republic and was immediately barred from teaching or holding any public office. He busied himself with his medical work and other research, and was soon recalled to the University. After the fall of Napoleon, Ruffini held the Chair of Applied Mathematics as well as the Chair of Clinical Medicine, and in 1814 he was also appointed Rector. Three years later he contracted typhus during an epidemic in the city, but with true scientific detachment he observed the progress of the disease and afterwards wrote a paper on the symptoms and treatment of typhus. He died, in Modena, on 9 May 1822.

In the year of his return to the University after being barred, Ruffini published a theorem – later to become known as the Abel–Ruffini theorem – which stated that it was impossible to give a general solution to equations of greater than the fourth degree using only radicals (such as square roots, cube roots, and so on). To try to clarify this, he published a demonstration of the theorem in a paper entitled *The General Theory of Equations*; it was not well received. In 1813 he published it in revised form, again to little avail. His methods of proof were regarded as lacking in scientific rigour. In 1824 the Norwegian mathematician Niels Abel (1802–1829) demonstrated more convincingly the same theorem, having come independently to the identical conclusion.

Ruffini also made a substantial contribution to the theory of equations, developing the so-called theory of substitutions which was the forerunner of modern group theory. His work became incorporated into the general theory of the solubility of algebraic equations developed by the ill-starred Evariste Galois (1811–1832).

In addition, Ruffini brought his mathematical insight to bear on philosophical and biological matters, publishing a number of papers between 1806 and 1821. He considered the possibility that living organisms had come into existence as the result of chance, thus anticipating more modern work on probability.

Rühmkorff Heinrich Daniel 1803–1877 was a German-born French instrument-maker, who invented the Rühmkorff induction coil, a type of transformer for direct current that outputs a high voltage from a low-voltage input.

Rühmkorff was born in Hanover on 15 January 1803, but virtually nothing is known of his life before he went to Paris in 1819 and became a porter in the laboratory of the eminent French physicist Charles Chevalier (1804–1850). There Rühmkorff became interested in electrical equipment and soon began to manufacture scientific instruments. He opened his own workshop in 1840, but although he eventually became famous throughout Europe for his scientific apparatus, his factory remained small (after his death it was auctioned off for just £42).

Rühmkorff's first notable invention was a thermoelectric battery (1844). He then turned his attention to developing an induction coil, the principles of which had been worked out by Michael Faraday in 1831. After a long series of experiments, Rühmkorff eventually produced his induction coil in 1851. It consisted of a central cylinder of soft iron on which were wound two insulated coils – an inner primary coil comprising only a few turns of relatively thick copper wire, and an outer secondary coil with a large number of turns of thinner copper wire; an interrupter automatically makes and breaks the current in the primary coil, thereby inducing an intermittent high voltage in the secondary coil.

Rühmkorff demonstrated his invention at the 1855 Paris Exhibition, where it won him a decoration and a medal, and at the 1858 French Exhibition of Electrical Apparatus, where he was awarded the first prize of 50,000 francs. Later, however, the originality of his invention was contested by C. G. Page (1812–1868) who claimed to have invented a similar device some 20 years earlier in the United States.

The Rühmkorff induction coil played an important part in many later advances, including the development of gas discharge tubes and, indirectly, the discovery of cathode rays and X-rays. Today, coils working on the same principle are still used to provide the ignition spark in internal-combustion engines.

Widely respected and a great philanthropist, Rühmkorff was made an Honorary Member of the French Physical Society – despite his lack of early education. He died in Paris on 19 December 1877.

Rumford Count (Benjamin Thompson) 1753–1814 was an American-born physicist who first demonstrated conclusively that heat is not a fluid but a form of motion.

Thompson was born into a farming family at Woburn, Massachusetts, on 26 March 1753. At the age of 19 he became a schoolmaster as a result of much self-instruction and some help from local clergy. He moved to Rumford (now Concord, New Hampshire) and almost immediately married a wealthy widow many years his senior. Thompson's first activities seem to have been political. When the War of Independence broke out he remained loyal to the Crown and acted as some sort of secret agent. Obliged to flee to London in 1776 (having separated from his wife the year before), he was rewarded with government work and the appointment as Lt Colonel of a regiment in New York. After the war, he retired from the army and lived permanently in exile in Europe. Thompson moved to Bavaria and spent the next few years with the civil administration there, becoming war and police minister as well as grand chamberlain to the Elector.

In 1781, Thompson was made a Fellow of the Royal Society on the basis of a paper on gunpowder and cannon vents. He had studied the relationship between various gunpowders and the apparent force with which the cannon balls were shot forth. This was a topic, part of a basic interest in guns and other weapons, to which he frequently returned, and in 1797 he produced a gunpowder standard.

In 1791, Thompson was made a Count of the Holy Roman Empire in recognition of all his work in Bavaria. He took the title from Rumford in his homeland, and it is by this name that we know him today.

Rumford was greatly concerned with the promotion of science and in 1796 established the Rumford Medals in the Royal Society and in the Academy of Arts and Science, Boston. These were the best endowed prizes of the time, and they still exist today.

In 1799, Rumford returned to England and with Joseph Banks (1743–1820) founded the Royal Institution, choosing Humphry Davy (1778–1829) as lecturer. Its aim was the popularization of science and technology, a tradition that still continues there.

Two years later Rumford resumed his travels. He settled in Paris and married the widow of Antoine Lavoisier (1743–1794), who had produced the caloric theory of heat that Rumford overthrew. However, this was a second unsuccessful match and after separating from his wife, Rumford lived at Auteuil near Paris until his death on 21 August 1814. In his will he endowed the Rumford chair at Harvard, still occupied by a succession of distinguished scientists.

Rumford's early work in Bavaria shows him at his most versatile and innovative. He combined social experiments with his lifelong interests concerning heat in all its aspects. When he employed beggars from the streets to manufacture military uniforms, he was faced with a feeding problem. A study of nutrition led to the recognition of the importance of water and vegetables and Rumford decided that soups would fit the requirements. He devised many recipes and developed cheap food emphasizing the potato. Meanwhile soldiers were being employed in gardening to produce the vegetables, Rumford's interest in gardens and landscape giving Munich its huge Englischer Garten, which he planned and which remains an important feature of the city today. The uniform enterprise led to a study of insulation and to the conclusion that heat was lost mainly through convection; thus clothing should be designed to inhibit this.

No application of heat technology was too humble for Rumford's scrutiny. He devised the domestic range – the 'fire in a box' – and special utensils to go with it. In the interests of fuel efficiency, he devised a calorimeter to compare the heats of combustion of various fuels. Smoky fireplaces also drew his attention, and after a study of the various air movements, he produced designs incorporating all the features now considered essential in open fires and chimneys such as the smoke shelf and damper.

His search for an alternative to alcoholic drinks led to the promotion of coffee and the design of the first percolator.

The work for which Rumford is best remembered took place in 1798. As military commander for the Elector of Bavaria, he was concerned with the manufacture of cannon. These were bored from blocks of iron with drills, and it was believed that the cannons became hot because as the drills cut into the cannon, heat was escaping in the form of a fluid called caloric. However, Rumford noticed that heat production increased as the drills became blunter and cut less into the metal. If a very blunt drill was used, no metal was removed yet the heat output appeared to be limitless. Clearly heat could not be a fluid in the metal, but must be related to the work done in turning the drill. Rumford also studied the expansion of liquids of different densities and different specific heats and showed by careful weighings that the expansion was not due to caloric taking up the extra space.

Although the majority of Rumford's scientific work related to heat and its applications, he had an important subsidiary interest in light. He invented the Rumford shadow photometer and established the standard candle, which was the international unit of luminous intensity right up to the 1900s. Transmission of light and shadows cast by coloured sources interested him, and he also looked into concepts of complementary colours. Photosynthesis drew his passing glance and he was probably one of the first to try to relate heat and light in their effects on chemical reaction – the beginnings of the science of photochemistry.

Rumford's contribution to science in demolishing the caloric theory of heat was very important because it paved the way for the realization of the fact that heat is a form of energy and that all forms of energy are interconvertible. However, it took several decades to establish the view that caloric does not exist, as the caloric theory readily explained the important conclusions on heat radiation made by Pierre Prévost (1751–1839) in 1791 and on the motive power of heat made by Sadi Carnot (1796–1832) in 1824.

Russell Bertrand Arthur William 1872–1970 was a British philosopher and mathematician who, during a long and active life, made many contributions to mathematics and wrote about morals and politics, but is best remembered as one of the founders of modern logic. He was a prolific writer and was awarded the 1950 Nobel Prize for Literature.

Russell was born in Trelleck, Monmouthshire, on 18 May 1872 into a family that had long been prominent in British social and political life. His grandfather was Lord John Russell, who introduced the 1832 Reform Bill in Parliament and went on to serve twice as Prime Minister. Russell's parents died when he was a child and it was his grandfather who, disapproving of arrangements made by his parents, undertook to bring up Russell and his brother Frank. Instead of being educated by the Nonconformist thinkers his parents had nominated, he was educated at home until he was 18 by governesses and tutors.

In 1890 Russell entered Trinity College, Cambridge, to study mathematics and philosophy. Two years later he was elected to the Apostles – a small informal group that regarded itself as being made up of the best minds in the university. Another member, Alfred North Whitehead (1861–1947), was then a Lecturer in Mathematics. Russell was elected to a Prize Fellowship at Trinity in 1895 and in 1910 to a Lectureship in Logic and the Philosophy of Mathematics.

Russell was a pacifist at the outbreak of World War I in 1914, a stand that was bitterly resented by many people of his class. A leaflet published in 1916, in which he protested at the harsh treatment of a conscientious objector, led to his prosecution and dismissal from the Trinity lectureship. Two years later Russell was imprisoned for six months when his article in *The Tribunal*, a pacifist weekly, was judged to be seditious. While serving the sentence in Brixton Prison, he wrote *Introduction to the Philosophy of Mathematics*.

In 1922, after World War I, Russell stood as a Labour Party candidate in the General Election, but was defeated (two other attempts at entering Parliament were also unsuccessful). Soon afterwards he wrote two books that reflected contemporary developments in science: *The ABC of Atoms* (1923) and *The ABC of Relativity* (1925). In 1931 he succeeded his brother and became the 3rd Earl Russell. He was appointed to the staff of City College, New York, in 1940, but was dismissed by a State Supreme Court Order following a wave of protest against the liberal views on sex expressed in his book *Marriage and Morals*. He did, however, lecture in philosophy at other US universities in Chicago and Philadelphia.

During the Cold War that followed World War II Russell even more vehemently expressed his views on pacifism, and in 1958 he was the president of the Campaign for Nuclear Disarmament. He was briefly imprisoned once more in 1961, when he was aged 89.

Russell received many honorary degrees, was elected a Fellow of the Royal Society in 1908 and made a member of the Order of Merit in 1949. He married four times, three of his marriages ending in divorce. He died in Plas Penrhyn, Merionethshire, on 2 February 1970.

Russell published the first volume of *Principles of*

Mathematics in 1903, expounding his belief that the basic principles of mathematics can be founded on fundamental logical concepts so that all the propositions of pure mathematics can be deduced from this basis using only a few of the most important logical principles. This work was indebted to a symbolic calculus developed by George Boole (1815–1864).

But Russell's most significant publication concerning mathematical logic, derived from his collaboration with Whitehead, was the three-volume *Principia Mathematica* (1910, 1912, 1913), whose title reflected that of Isaac Newton's fundamental work. The arguments proposed in the *Principia* were explained, without the use of technical logical symbolism, in *Introduction to the Philosophy of Mathematics*. The book is a model of clarity and of the lucid expression of complex and abstract ideas. It was followed by a second edition of the *Principia* which incorporated an introduction that set out how the system must be modified as a result of the work of other mathematicians such as Leon Chwistek (1884–1944) and Ludwig Wittgenstein (1889–1951).

Russell Frederick Stratten 1897–1984 was a British marine biologist best known for his studies of the life histories and distribution of plankton. He received many honours for his work, including a fellowship of the Royal Society of London in 1938, a Gold Medal from the Linnean Society in 1961, and a knighthood in 1965.

Russell was born on 3 November 1897 in Bridport, Dorset, and was educated at Oundle School and then at Gonville and Caius College, Cambridge. His studies were interrupted by World War I, during which he served with distinction in the Royal Naval Air Service. He graduated in 1922 and then worked for two years for the Egyptian government as Assistant Director of Fisheries Research, after which he returned to Britain and joined the scientific staff of the Marine Biological Association at Plymouth. He remained there for the rest of his working life (although he went on an expedition to the Great Barrier Reef in Australia from 1928 to 1929, and served in Air Staff Intelligence during World War II), becoming its Director in 1945. From 1950 until he retired in 1965 Russell was also a member of the National Oceanographic Council, and from 1962 to 1975 he was Chairman of the Advisory Panel on Biological Research to the Central Electricity Generating Board. He was knighted in 1965.

While working for the Egyptian government, Russell studied the vertical distribution in the sea of the eggs and larvae of marine fish and their migratory movements. Continuing this line of research at the Marine Biological Association, he investigated the different types of behaviour of individual species of fish at various times of the year. He went on to investigate the distribution of *Calanus* (a crustacean copepod) and *Sagitta* (a chaetognath worm). By combining the results of his research, Russell established the value of certain types of plankton as indicators of different types of water in the English Channel and the North Sea. He also offered a partial explanation for the difference in abundance of herring in different areas. In addition, Russell discovered a means of distinguishing between different species of fish shortly after they have hatched, when they are almost identical in appearance. Russell's studies of plankton and of water movements were extremely valuable in providing information on which to base fishing quotas, the accuracy of which is essential to prevent overfishing and the depletion of fish stocks.

Russell also elucidated the life histories of several species of medusa by rearing the hydroids from parent medusae, and he published the two-volume *The Medusae of the British Isles* (1953, 1970).

Russell Henry Norris 1877–1957 was a US astronomer who was chiefly interested in the nature of binary stars, but who is best remembered for his publication in 1913 of a diagram charting the absolute magnitude of stars plotted against their spectral type. Ejnar Hertzsprung (1873–1967) had in fact published similar results previously in a photographic journal that was seen by few astronomers (and noted by none), but Russell was the first to put in graphic form what became known as the Hertzsprung–Russell diagram. Its impact on the scientific world was enormous and the diagram remains of great importance for research into stellar evolution, although some of Russell's initial extrapolations have since been superseded.

Born in Oyster Bay, New York, on 25 October 1877, Russell was five years old when his parents pointed out to him the transit of Venus that inspired him to become an astronomer. He received his early education at home, and then went on to Princeton University, from which he graduated, with distinction, in 1897. He remained at the University, working in the Astronomy Department, until he received his PhD in 1899, for devising a new way of determining the orbits of binary stars. Russell then travelled to England, and joined Arthur Hinks at the Cambridge University Observatories. He worked with Hinks on stellar photography and evolved a technique for the measurement of stellar parallax. Further research into binary stars followed before, in 1905, Russell returned to Princeton, where he was made Professor and Director of the University Observatory. In 1921, he was appointed Research Associate of the Mount Wilson Observatory, and in 1927 he received the appointment of C. A. Young Research Professorship. Russell was awarded many honours from academies and universities. Described as the Dean of American Astronomers by a contemporary, he died at Princeton in 1957.

Like Hertzsprung, Russell concluded from his early works on stellar distances that stars could be grouped in two main classes, one much brighter than the other. By plotting the luminosities against the spectra of hundreds of stars in a diagram, Russell managed to show a definite relationship between true brightness and type of spectrum. (He used Annie Cannon's system of spectral classification, which also indicated surface temperature.) Most of the stars were grouped together in what became known as the 'main sequence' that appeared to run from the top left of

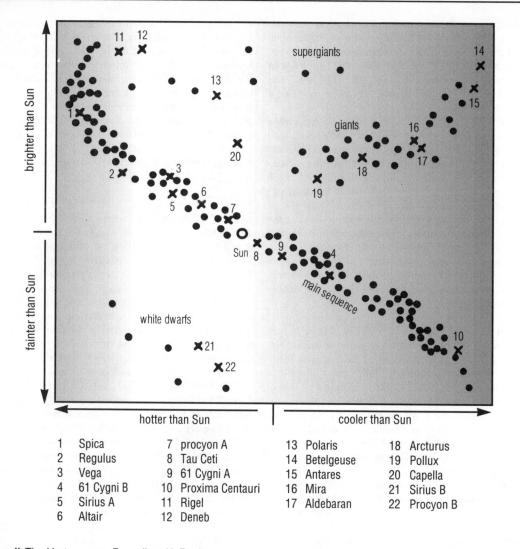

1	Spica	7	procyon A	13	Polaris	18	Arcturus
2	Regulus	8	Tau Ceti	14	Betelgeuse	19	Pollux
3	Vega	9	61 Cygni A	15	Antares	20	Capella
4	61 Cygni B	10	Proxima Centauri	16	Mira	21	Sirius B
5	Sirius A	11	Rigel	17	Aldebaran	22	Procyon B
6	Altair	12	Deneb				

Russell *The Hertzsprung–Russell, or H–R₁ diagram relates the absolute magnitudes of stars to their spectral class or temperature. The Sun is a typical main-sequence star.*

the diagram to the bottom right. But there was a group of very bright stars plotted above the main sequence – indicating that stars of similar spectral type could have very different magnitudes. This was a significant discovery, yet the fact that there was a 'main sequence' at all was profoundly stimulating to Russell, who immediately founded a theory of stellar evolution upon it. He proposed that all stars progress at one time or another either up or down the 'main sequence', depending on whether they are contracting (and therefore becoming hotter) or expanding (thus cooling), and he tried to derive a logical progression. The progression he finally proposed, however, was discredited within a decade – although it provoked considerable scientific research. Stellar evolution is now known to be far more complex than a simple progression.

Russell's lifelong study of binary stars resulted in a method for calculating the mass of each star from a study of its orbital behaviour. He pioneered a system using both orbits and masses in order to compute distance from Earth, and his research into eclipsing binary stars (one of which moves in front of the other, from the viewpoint of Earth) led to the amassing of valuable data on variations in light emission.

Finally, his investigation of the solar spectrum and his analysis of the composition of the solar atmosphere led him to a general speculation on the composition of stars. For the first time he suggested that there was a considerable abundance of hydrogen present, and hesitantly put forward a fairly high percentage figure. The 'Russell mixture' turned out in fact to have *under*estimated the abundance of hydrogen (more than 80% of the Sun's volume).

Rutherford Ernest 1871–1937 was a British physicist who first explained that radioactivity is produced by the disintegration of atoms and discovered that alpha particles consist of helium nuclei. For these achievements, Rutherford was awarded the 1908 Nobel Prize for Chemistry. Rutherford went on to make two more discoveries of fundamental importance to nuclear physics. He was the first to determine the basic structure of the atom and show that it consists of a central nucleus surrounded by electrons, and he also produced the first artificial transformation, thereby changing one element into another.

Rutherford was born near Nelson, New Zealand, on 30 August 1871. His father was a wheelwright and farmer who, like his mother, had emigrated from Britain to New Zealand when a child. Rutherford did not show any great aptitude for science as a child and when he entered Nelson College in 1887, he exhibited an all-round ability. He went on to Canterbury College, Christchurch, in 1889, receiving a BA degree in 1892. He then embarked on a study of mathematics and physics, gaining his MA in 1893 and then a BSc in 1894. Rutherford investigated the magnetic properties of iron by high-frequency electric discharges for his science degree, and constructed a very sensitive detector of radio waves as a result of his research. This was only six years after Heinrich Hertz (1857–1894) had discovered radio waves, and the same year that Guglielmo Marconi (1874–1937) began his radio experiments.

In 1895, Rutherford went to Britain to study at the Cavendish Laboratory, Cambridge. There he became the first research student to work under J. J. Thomson (1856–1940). Armed with his radio detector, Rutherford made a big impact on Cambridge, but under Thomson's guidance, he soon turned to the work in atomic physics that was to become his career. In 1898, helped by Thomson, Rutherford obtained his first academic position with a Professorship in Physics at McGill University, Montreal, Canada, which then boasted the best-equipped laboratory in the world. He was attracted back to Britain in 1907, when he succeeded Arthur Schuster (1851–1934) at Manchester, Schuster declaring that he would resign his chair only for Rutherford. Rutherford built up a renowned laboratory at Manchester, and it was there that he made his momentous discoveries of the nuclear atom and artificial transformation.

During World War I, Rutherford worked for the Admiralty on methods of locating submarines and then in 1919 moved to Cambridge to become Professor of Physics and Director of the Cavendish Laboratory in succession to Thomson. He retained this position for the rest of his life, and was also Professor of Natural Philosophy at the Royal Institution from 1921 onwards. Many honours were accorded to Rutherford in addition to the 1908 Nobel Prize for Chemistry. They included the Royal Society's Copley Medal in 1922, the Presidency of the Royal Society from 1925 to 1930, a knighthood in 1914, the Order of Merit in 1925, and a peerage in 1921, Rutherford taking the title Baron Rutherford of Nelson. In his last years, Rutherford was active in helping refugee scientists who had escaped from Nazi Germany. Rutherford died at Cambridge on 19 October 1937 and is buried in Westminster Abbey.

When Rutherford first came to the Cavendish Laboratory, Thomson put him to work to study the effect that X-rays have on the discharge of electricity in gases. This was early in 1896, only a few weeks after Wilhelm Röntgen (1845–1923) had discovered X-rays. Rutherford found that positive and negative ions are formed, and measured the mobility of the ions produced. In 1897 he went on to make a similar study of the effects of ultraviolet light and the radioactivity produced by uranium minerals, which had been discovered by Henri Becquerel (1852–1908) the year before. Rutherford then became fascinated by radioactivity and began a series of investigations to explore its nature. In 1898, he found that there are two kinds of radioactivity with different penetrating power. The less penetrating he called alpha rays and the more penetrating beta rays. In 1900, Rutherford discovered a third type of radioactivity with great penetrating power, which he called gamma rays. (Alpha and beta rays were later found to consist of streams of particles and so are now known as alpha particles and beta particles. Gamma rays were found to be electromagnetic waves of very high frequency and so are called gamma rays or gamma radiation.)

When Rutherford moved to Montreal in 1898, he began to use thorium as a source of radioactivity instead of uranium. He found that thorium produces an intensely radioactive gas, which he called emanation. This was a decay product of thorium and Rutherford discovered several more, including thorium X. To identify these products, Rutherford enlisted the aid of Frederick Soddy (1877–1956), who was later to discover isotopes. Analysis of the decay products enabled Rutherford and Soddy in 1903 to explain that radioactivity is an atomic phenomenon, caused by the breakdown of the atoms in the radioactive element to produce a new element. Rutherford found that the intensity of the radioactivity produced decreases at a rate governed by the element's half-life. The idea that atoms could change their identity was revolutionary, yet so compelling was Rutherford's explanation of radioactivity that it was accepted immediately with very little opposition.

Rutherford was now concerned to identify alpha rays, which he was sure consisted of positively charged particles and specifically either hydrogen or helium ions. Deflection of the rays in electric and magnetic fields proved in 1903 that they are positive particles, but Rutherford was unable to determine the amount of charge because his apparatus was not sensitive enough.

In 1904, with Bertram Boltwood (1870–1927), Rutherford worked out the series of transformations that radioactive elements undergo and showed that they end as lead. They were able to estimate the rates of change involved and in 1907 Boltwood calculated the ages of mineral samples, arriving at figures of more than a thousand million years. This was the first proof of the age of rocks, and the method of radioactive dating has since been developed into a precise way of finding the age of rocks, fossils and ancient artefacts.

On returning to Britain in 1907, Rutherford continued to explore alpha particles. In conjunction with Hans Geiger

(1882–1945), he developed ionization chambers and scintillation screens to count the particles produced by a source of radioactivity, and by dividing the total charge produced by the number of particles counted, arrived at the conclusion that each particle has two positive charges. The final proof that alpha particles are helium ions came in the same year, when Rutherford and Thomas Royds succeeded in trapping alpha particles in a glass tube and by sparking the gas produced showed from its spectrum that it was helium.

Rutherford's next major discovery came only a year later in 1909. He suggested to Geiger and a gifted student named Ernest Marsden that they investigate the scattering of alpha particles by gold foil. They used a scintillation counter that could be moved around the foil, which was struck by a beam of alpha particles from a radon source. Geiger and Marsden found that a few particles were deflected through angles of more than 90° by the foil. Rutherford was convinced that the explanation lay in the nature of the gold atoms in the foil, believing that each contained a positively charged nucleus surrounded by electrons. Only such nuclei could repulse the positively charged alpha particles that happened to strike them to produce such enormous deflections. But Rutherford needed proof of this theory. He worked out that the nucleus must have a diameter of about 10^{-13} cm – 100,000 times smaller than the atom – and calculated the numbers of particles that would be scattered at different angles. These predictions were confirmed experimentally by Geiger and Marsden, and Rutherford announced the nuclear structure of the atom in 1911.

Few were convinced that the atom could be almost entirely empty space as Rutherford contended. However, among those who agreed with Rutherford was Niels Bohr (1885–1962). He went to Manchester to work in 1912 and in 1913 produced his quantum model of the atom which assumed a central positive nucleus surrounded by electrons orbiting at various energy levels. Also in 1913, another of Rutherford's co-workers, Henry Moseley (1887–1915), announced his discovery of the atomic number, which identifies elements, and showed that it could only be given by the number of positive charges on the nucleus, and thus the number of electrons around it. Rutherford's view of the nuclear atom was thereby vindicated, and universally accepted.

Several more important discoveries were made at Manchester. In 1914, Rutherford found that positive rays consist of hydrogen nuclei. Also in 1914, Rutherford and Edward Andrade showed that gamma rays are electromagnetic waves by diffracting them with a crystal. They measured the wavelengths of the rays and found that they lie beyond X-rays in the electromagnetic spectrum. Becquerel in 1900 had identified beta rays with cathode rays, which were shown to be electrons, and so the nature of radioactivity was now revealed in full.

Rutherford's work was now interrupted by war and he did not return to physics until 1917, when he made his last great discovery. He made it unaided, unlike most of his earlier discoveries, because all his colleagues and students were still engaged in war work. Rutherford followed up earlier work by Marsden in which scintillations were noticed in hydrogen bombarded by alpha particles well beyond their range in the gas. These were due to hydrogen nuclei knocked on by the alpha particles, which was not unexpected. However, Rutherford now carried out the same experiment using nitrogen instead of hydrogen, and he found that hydrogen nuclei were still produced and not nitrogen nuclei. Rutherford announced his interpretation of this result in 1919, stating that the alpha particles had caused the nitrogen nuclei to disintegrate, forming hydrogen and oxygen nuclei. This was the first artificial transformation of one element into another. Rutherford found similar results with other elements and announced that the nucleus of any atom must be composed of hydrogen nuclei. At Rutherford's suggestion, the name proton was given to the hydrogen nucleus in 1920. He also speculated in the same year that uncharged particles, which were later called neutrons, must also exist in the nucleus.

Rutherford continued work on artificial transformation in the 1920s. Under his direction, Patrick Blackett (1897–1974) in 1925 used a Wilson cloud chamber to record the tracks of disintegrated nuclei, showing that the bombarding alpha particles combines with the nucleus before disintegration and does not break the nucleus apart like a bullet. Bombardment with alpha particles had its limits as large nuclei repelled them without disintegrating, and Rutherford directed the construction of an accelerator to produce particles of the required energy. The first one was built by John Cockcroft (1897–1967) and Ernest Walton (1903–   ), and went into operation at the Cavendish Laboratory in 1932. In the same year, another of Rutherford's colleagues, James Chadwick (1891–1974), discovered the neutron at the Cavendish Laboratory.

Rutherford was to make one final discovery of great significance. In 1934, using some of the heavy water recently discovered in the United States, Rutherford, Marcus Oliphant and Paul Harteck bombarded deuterium with deuterons and produced tritium. This may be considered the first nuclear fusion reaction.

Rutherford may be considered the founder of nuclear physics, both for the fundamental discoveries that he made and for the encouragement and direction he gave to so many important physicists involved in the development of this science.

Rutherfurd Lewis Morris 1816–1892 was a US spectroscopist and celestial photographer.

Rutherford was born in Morrisania, New York, on 25 November 1816, into a well-established family of Scottish descent. He showed an early interest in science, and during his student days at Williams College, Massachusetts, he was made an assistant to the Professor of Physics and Chemistry. After graduating, however, he went on to study law. His independent means was augmented by his marriage to Margaret Stuyvesant Chanler, which freed him from the need to practice law and allowed him to travel abroad for seven years. When he returned to the United States in 1856, he had his own observatory built and spent the rest of his life working on astronomical photography and spectroscopy. He died in Tranquility, New Jersey, on 30 May 1892.

From 1858 Rutherfurd produced many photographs that were widely admired of the Moon, Jupiter, Saturn, the Sun and stars down to the fifth magnitude, using a 29-cm/11$\frac{1}{2}$-in achromatic refracting telescope built by Henry Fitz. He went on to map the heavens by photographing star clusters and to enable him to analyse the information of his stellar photographs he devised a new micrometer that could measure the distances between stars more accurately. After 1861 he became more and more interested in the spectroscopic work of Robert Bunsen and Gustav Kirchoff and from 1862 he began to make spectroscopic studies of the Sun, Moon, Jupiter, Mars and 16 fixed stars. From his stellar studies, he independently produced a classification scheme of stars based on their spectra that turned out to be remarkably similar to Angelo Secchi's star classification. Apparently, at the January 1864 meeting of the National Academy of Sciences, he displayed an unpublished photograph of the solar spectrum that had three times the number of lines that had been noted by Kirchoff and Bunsen. To help his spectroscopic work he began devising more sophisticated diffraction gratings and his innovative skill in producing these became generally well known to other contemporary astronomers, providing them with diffraction gratings with up to 6,700 lines per cm/17,000 lines per in.

Rutherfurd made his last observations in 1878. In 1883 he donated his equipment and large collection of photographic plates to the Observatory of the University of Columbia.

Rydberg Johannes Robert 1854–1919 was a Swedish physicist who discovered a mathematical expression that gives the frequencies of spectral lines for elements. It includes a constant named the Rydberg constant after him.

Rydberg was born in Halmstad on 8 November 1854. He was educated at the Gymnasium in Halmstad and in 1873 entered the University of Lund, graduating in mathematics in 1875 and gaining a doctorate in this subject in 1879. He remained at Lund for the rest of his life, beginning work as a lecturer in mathematics in 1880. He became a lecturer in physics in 1882 and was appointed Professor of Physics provisionally in 1897 and permanently in 1901. He retained this position until a month before his death on 28 December 1919.

Rydberg worked in the field of spectroscopy. He did not do a great amount of experimental investigation, but was concerned to organize the mass of observations of spectral lines into some kind of order. Rydberg began by classifying the lines into three types: principal (strong, persistent lines), sharp (weaker but well-defined lines) and diffuse (broader lines). Each spectrum of an element consists of several series of these lines superimposed over each other. He then sought to find a mathematical relationship that would relate the frequencies of the lines in a particular series. In 1890, he achieved this by using a quantity called

the wave number, which is the reciprocal of the wavelength. The formula expresses the wave number in terms of a constant common to all series (the Rydberg constant), two constants that are characteristic of the particular series, and an integer. The lines are then given by changing the value of the integer.

Rydberg then found that Johann Balmer (1825–1898) had already developed a formula for a series of lines of the hydrogen spectrum in 1885. However, it turned out to be a special case of his own more general formula. Rydberg then went on to produce another formula which he believed would express the frequency of every line in every series of an element. This formula is:

$$N = R[1/(n + a)^2 - 1/(m + b)^2]$$

where N is the wave number, R is the Rydberg constant, n and m are integers, m being greater than n, and a and b are constants for a particular series.

Rydberg's intuition proved to be correct and five more series of hydrogen lines were subsequently discovered that fitted the formula. Although it was an empirical discovery and Rydberg was unable to explain why his formula expressed the spectral lines, Niels Bohr (1885–1962) was able to show that it gave the energy states required for the theory of atomic structure that he presented in 1913.

In 1897, Rydberg put forward the idea of atomic number and the importance of this in revising the periodic table. His views were vindicated in the brilliant research conducted into X-ray spectra by Henry Moseley (1887–1915) in 1913. The Rydberg formula also proved to be of use in determining the electronic shell structure of elements, which provided a proper basis for the classification of the elements.

Ryle Martin 1918–1984 was a British astronomer whose main work was the organization, construction, development and use of the radio telescopes at the famous establishment in Cambridge. For this work he received the 1974 Nobel Prize for Physics (jointly with his friend and colleague Antony Hewish, the discoverer of pulsars).

Ryle was born on 27 September 1918. He attended a public school, then studied at Oxford University, graduating in 1939. His interest in radio was already pronounced, and on the outbreak of war he joined the Telecommunications Research Establishment in Malvern, where he was involved in the development of radar and other systems. After the war he was invited by John Ratcliffe (1902–1987), a former colleague at Malvern, to join him at the Cavendish Laboratory in Cambridge, working on the study of solar radio frequency emissions. As a result of his research, Ryle was appointed to a lectureship in physics in 1948 and, the following year, to a Fellowship at Trinity College. Election to a Fellowship at the Royal Society came in 1952, and seven years later Ryle became the first Cambridge University Professor of Radio Astronomy. A knighthood conferred in 1966 was followed by his appointment as Astronomer Royal in 1972.

At the Cavendish Laboratory, Ryle's first task as one of

the first radioastronomers was to study solar emission at 1-m/3-ft wavelengths. This necessitated the building of a suitable instrument. His choice lay between a steerable parabolic dish (like the one in use at Jodrell Bank) and an interferometer-type instrument (which consists, in essence, of two separated aerials, each receiving the same signal from the same source, coupled to a receiver). Ryle chose the interferometer variety that the Cambridge site is now famous for. This type of telescope was first constructed merely to distinguish and measure sources of different angular size, identifying 'compact' emitters and distinguishing them from diffuse sources, but it was then seen also to allow positions to be measured with much greater accuracy than was possible with a single parabolic reflector. The earliest instruments, built at Grange Road on the outskirts of the city, used a spacing of about 0.8 km/0.5 mi between the two parabolic dishes used as aerials.

One of the interferometers was built specifically to study the compact source of radio emission in the constellation Cygnus noted by James Hey (1909–   ) during the war. On the first night of its operation Cassiopeia-A, the most intense compact source in the sky, was also found – a great discovery. Optical astronomers at Mount Palomar were able to identify both these sources only because of the extreme accuracy (to within one minute of arc) of the later radio measurements: Cygnus as a very distant galaxy and Cassiopeia as a supernova within our own Galaxy.

In 1949, the use of this type of aerial led to a breakthrough in radio telescope design. By making observations from a number of different spacings to construct a radio map of the Sun, the forerunner of the 'synthesis' aerials was produced. This work was extended in 1954 to provide the first radio map of the sky. In the earlier Sun-mapping a certain symmetry of the source had been assumed so as to simplify the complex mathematical treatment of the signal data. Now, with an arbitrary distribution of sources, the accurate Fourier inversion required involved a precise locating of the sites of the aerial components and enough computer capacity – which was just becoming available – to interpret the signal information. In 1956, the group (generously assisted by Mullard Ltd) extended their activities to the Lord's Bridge site as more and larger instruments demanded increasing ground space and freedom from electrical interference with the weak signals.

With such equipment it was now possible to begin the First Cambridge Catalogue Survey (1C): a map of the most powerful known radio sources in the northern sky. By increasing yet further the collecting area of the detectors, the resolution and sensitivity of the instrument are also increased – as are also, unfortunately, the problems of alignment and scale of construction. So for the Second Survey, 2C – the first really comprehensive radio mapping of the sky, completed in 1955 – a collecting area of one acre was used. The definitive 3C survey, published in 1959 and. now used as reference by all radioastronomers, was made with a similar arrangement of parabolic reflectors, but working at a shorter wavelength (1.7 m/5.6 ft) which gave better resolution.

As this work went on, the first large-aperture synthesis aerial was being built, and it was that which was used to complete the 4C survey. This survey catalogued no fewer than 5,000 sources.

It was in turn superseded by an extension in concept to 'supersynthesis', in which a fixed aerial maps a band of the sky using solely the rotation of the Earth, and another aerial maps successive rings out from it concentrically. The 'One Mile' telescope was built in 1963 on this principle, and was designed both for the 5C survey – a programme to scan only part of the sky, but in considerable depth and detail – and for the compilation of radio maps of individual sources.

A 5-km/3-mi instrument was completed in 1971, incorporating further advances in design. It can provide a resolution of up to one third of a second of arc – 0.001 of a degree – to give very detailed maps of known sources. The variety of programmes for which it is now in use includes, most importantly, the mapping of extragalactic sources – radio galaxies and quasars – and the study of supernovae and newly born stars. It can provide as sharp a picture as the best groundbased optical telescopes.

Ryle was personally responsible for most of these developments. The site at Cambridge has gained its renown in large measure through his efforts and the results that his team obtained.

S

Sabatier Paul 1854–1941 was a French organic chemist who investigated the actions of catalysts in gaseous reactions. He is particularly remembered for his work, with his assistant Abbé Jean-Baptiste Senderens (1856–1936), on catalytic hydrogenation of gaseous hydrocarbons. For this research he shared the 1912 Nobel Prize for Chemistry with François Grignard.

Sabatier was born in Carcassone, Aude, on 5 November 1854, and was educated locally and at Toulouse. In 1887, he entered the Ecole Normale Supérieure in Paris, graduating three years later. He spent the next year as a teacher at the Lycée in Nîmes and then became assistant to Pierre Berthelot at the Collège de France. He was awarded his doctorate in 1880 for a thesis on metallic sulphides. After a year of further research in Bordeaux he went in 1882 to Toulouse, where he was appointed Assistant Professor of Physics, later transferring to a similar position in chemistry. When in 1884 he reached the age of 30, the minimum for a full professorial appointment, he was duly installed as Professor of Chemistry. He retained the position for the rest of his long life, declining the offers of the Chairs at the Sorbonne and the Collège de France in 1907. He died in Toulouse on 14 August 1941.

Sabatier's early researches were concerned with physical and inorganic chemistry. Among many investigations he correlated the colours of chromates and dichromates with their acidity, and was the first chemist to prepare pure hydrogen sulphide. In 1895, he became interested in the role of metal catalysts through the work on nickel carbonyl by Ludwig Mond. In the following year he produced somewhat similar compounds by the action of the oxides of nitrogen on finely divided metals.

In his next series of experiments, begun in 1897, he studied the reaction of ethylene (ethene) and hydrogen on a heated oxide of nickel. He found that the reduced nickel formed catalysed the hydrogenated acetylene (ethyne) in a similar way, and converted benzene vapour into cyclohexane. With his assistant Sanderens he extended the method to the hydrogenation of other unsaturated and aromatic compounds. He also synthesized methane by the hydrogenation of carbon monoxide using a catalyst of finely divided nickel. He later showed that at higher temperatures the same catalysts can be used for dehydrogenation, enabling him to prepare aldehydes from primary alcohols and ketones from secondary alcohols.

Sabatier later explored the use of oxide catalysts, such as manganese oxide, silica and alumina. Different catalysts often gave different products from the same starting material. Alumina, for example, produced olefins (alkenes) with primary alcohols, which yielded aldehydes with a copper catalyst.

He explained the action of catalysts in terms of 'chemisorption', the formation of unstable compounds on the surface of a catalyst. He cited an improvement in catalytic action with decreasing particle size and the poisoning effect of impurities as evidence for this theory. He also postulated that the suitability of a catalyst for a particular reaction depends on its chemical nature as well as its physical properties (because different catalysts give different products from the same reaction).

Later catalytic hydrogenation was applied to liquid hydrocarbons by Vladimir Ipatieff, leading to applications such as the hardening of natural oils by hydrogenation and the development of the margarine and modified fats industry.

Sabin Albert Bruce 1906–1993 was a Russian-born US virologist who devoted his long and distinguished career to the development of protective vaccines. He is particularly associated with the oral poliomyelitis vaccine.

Sabin was born on 26 August 1906 in Bialystok, Russia (now in Poland). In 1921 he and his family emigrated to the United States where he attended New York University from 1926 to 1931, graduating with a medical degree. He then served as House Physician to the Bellevue Hospital in New York from 1932 to 1933. Between 1935 and 1937 he was an Assistant at the Rockefeller Institute and an Associate from 1937. Two years later he was made Associate Professor of Research in Pediatrics at the University of Cincinnati College of Medicine and, after serving as a medical officer in the army during World War II, he became Research Professor of Pediatrics there from 1946 to 1960.

He held the position of Distinguished Service Professor from 1960 to 1970 and was President of the Weizmann Institute of Science, Israel, from 1970 to 1972. On his return to the United States he joined the staff of the Medical College of South Carolina, and between 1973 and 1974 he was Expert Consultant to the National Cancer Institute.

Sabin became interested in polio research while working at the Rockefeller Institute. In 1936, he and Peter Olitsky were able to make polio viruses from monkeys grow in tissue cultures from the brain cells of a human embryo that had miscarried. At the same time they were unsuccessful in their attempts to cultivate the virus in other human tissues, which gave weight to the existing theory that the virus attacked nerve cells only. (This theory was disproved in 1949 by Thomas Weller, John Enders and Frederick Robbins.)

Jonas Salk had become engaged in producing an inactive polio vaccine, but Sabin was not convinced that the Salk technique of using dead virus was adequate. He concentrated on developing a live-virus oral vaccine because he felt that the inactive vaccine could be nothing more than a temporary measure for protection and would require the patient to be revaccinated at fairly frequent intervals. Sabin believed that only a living virus could be counted on to produce the necessary antibodies over a long period. Also, the living virus could be taken orally because it would multiply and invade the body of its own accord and would not, like the Salk vaccine, have to be injected. Sabin succeeded in finding virus strains of all three types of polio, each producing its own variety of antibody which is too feeble to produce the disease itself. Sabin's vaccine is known as the live-attenuated vaccine and exerts its effect by inducing a harmless infection of the intestinal tract, thus simulating natural infection without causing any disease. It operates by multiplying in the tissues, giving rise to a mild and invisible infection with subsequent antibody formation and immunity.

The vaccine has the advantage of inducing immunity rapidly, a property that is particularly valuable in the face of an epidemic. When given to a significant proportion of a population, the vaccine induces community protection by rendering the alimentary tract of those vaccinated resistant to reinfection by the polio virus.

It took Sabin many years of patient research but by 1957 he had enough confidence after trying the vaccine out on himself, his family and numerous volunteers, to offer it for field trials to the medical community. They tested it on a massive scale, the tests were successful, and by 1961 the vaccine was available for commercial use.

Today, the 'sugar lump' (as it is now known) has become an accepted and easy method of vaccination against polio. Its success can be measured only by the marked decline of the once prevalent disease which killed so many young people or crippled them for life.

Sabine Edward 1788–1883 was a British geophysicist who made important studies of terrestrial magnetism.

Sabine was born in Dublin, Ireland, on 14 October 1788.

He received a military education at the Royal Military Academy at Woolwich, and never attended any university courses in science. In 1803, Sabine was commissioned in the Royal Artillery, with which he served his military career, rising to the rank of Major General in 1859. In 1818, he was elected to the Royal Society, serving in a variety of positions and ultimately becoming President 1861–71.

In 1818, Sabine accompanied the expedition headed by Sir John Ross (1777–1856) to explore the Northwest Passage as official astronomer. The following year he went with William Parry (1790–1855) to the Arctic. While on this voyage he planned a trip in the southern hemisphere which he undertook at the behest of the Royal Society from 1821 to 1822. Sabine then collaborated with Charles Babbage (1792–1871) in 1826 on a survey of magnetism in Britain, a project which was not completed until the 1830s and which was repeated by Sabine himself in the late 1850s. Sabine was also active in the British Association, serving as Secretary to the organization from 1838 to 1859, with the exception of the year 1852, when he was President. From 1849 to 1871, he was deeply involved with the administration of the King's Observatory at Kew. Sabine was awarded several honours, including the Copley Medal of the Royal Society, and was knighted in 1869. He died in Richmond, Surrey, on 26 June 1883.

During the 1820s, one of Sabine's major concerns was the use of the pendulum apparatus designed by Henry Kater (1777–1835) to determine the shape of the Earth. It was for these studies, which involved analysing data from 17 field stations around the globe, that Sabine was awarded the Copley Medal. His conclusions have since been somewhat modified because he failed to consider the fact that the density of the crust is not uniform and so obtained an erroneous result for the shape of the Earth.

Sabine's greatest scientific interest was the study of terrestrial magnetism, and he used his considerable influence to instigate a British 'magnetic crusade'. An expedition to establish observatories in the southern hemisphere was sent out in 1839 and Sabine managed to extend its duration for two three-year periods beyond its original duration, and thereby accumulated an enormous bulk of data. With this, Sabine in 1851 discovered a 10–11-year periodic fluctuation in the number of magnetic storms. He then brilliantly correlated this magnetic cycle with data Samuel Schwabe (1789–1875) had collected on a similar variation in solar activity. Sabine was thereby able to link the incidence of magnetic storms with the sunspot cycle. This work was also done independently by Johann von Lamont (1805–1879) about the same time. Other results also grew out of Sabine's huge mass of data, and in 1851 he was able to demonstrate that the daily variation in magnetic activity comprised two different components.

Sabine's enthusiasm for the subject of terrestrial magnetism yielded important new findings in that field, but it is considered that he caused too great a proportion of the available funding for research to be diverted to fields relevant to that subject and may have compromised other more important fields of research.

Sagan Carl Edward 1934–   is a US astronomer and popularizer of astronomy whose main research has been on planetary atmospheres, including that of the primordial Earth. His most remarkable achievement has been to provide valuable insights into the origin of life on our planet.

Sagan was born on 19 November 1934 in New York City. Completing his education at the University of Chicago, he obtained his bachelor's degree in 1955 and his doctorate in 1960. Then, for two years, he was a Research Fellow at the University of California in Berkeley, before he transferred to the Smithsonian Astrophysical Observatory in Cambridge, Massachusetts, lecturing also at Harvard, where he became Assistant Professor. Finally, in 1968, Sagan moved to Cornell University, in Ithaca (New York), and took up a position as Director of the Laboratory for Planetary Studies; in 1970 he became Professor of Astronomy and Space Science there.

The editor of the astronomical journal *Icarus*, Sagan has published a number of popular books on planetary science and the evolution of life; he has also dabbled in UFOlogy and, in 1980, presented *Cosmos*, a very successful television series of astronomical programmes.

In the early 1960s Sagan's first major research was into the planetary surface and atmosphere of Venus. At the time, although intense emission of radiation had shown that the dark-side temperature of Venus was nearly 600K (327°C/621°F), it was thought that the surface itself remained relatively cool – leaving open the possibility that there was some form of life on the planet. Various hypotheses were put forward to account for the strong emission actually observed: perhaps it was due to interactions between charged particles in Venus' dense upper atmosphere; perhaps it was glow discharge between positive and negative charges in the atmosphere; or perhaps emission was due to a particular radiation from charged particles trapped in the Venusian equivalent of a Van Allen Belt. Sagan showed that each of these hypotheses was incompatible with other observed characteristics or with implications of these characteristics.

The positive part of Sagan's proposal was to show that all the observed characteristics were compatible with the straightforward hypothesis that the surface of Venus was very hot. On the basis of radar and optical observations the distance between surface and clouds was calculated to be between 44 km/27 mi and 65 km/40 mi; given the cloud-top temperature and Sagan's expectation of a 'greenhouse effect' in the atmosphere, surface temperature on Venus was computed to be between 500K (227°C/440°F) and 800K (527°C/980°F) – the range that would also be expected on the basis of emission rate.

Sagan then turned his attention to the early planetary atmosphere of the Earth, with regard to the origins of life. One way of understanding how life began is to try to form the compounds essential to life in conditions analogous to those of the primeval atmosphere. Before Sagan, Stanley Miller (1930–   ) and Harold Urey (1893–1981) had used a mixture of methane, ammonia, water vapour and hydrogen, sparked by a corona discharge which simulated the effect of lightning, to produce amino and hydroxy acids of the sort found in life forms. Later experiments used ultraviolet light or heat as sources of energy, and even these had less energy than would have been available in Earth's primordial state. Sagan followed a similar method and, by irradiating a mixture of methane, ammonia, water and hydrogen sulphide, was able to produce amino acids – and, in addition, glucose, fructose and nucleic acids. Sugars can be made from formaldehyde under alkaline conditions and in the presence of inorganic catalysts. These sugars include five carbon sugars which are essential to the formation of nucleic acids, glucose and fructose – all common metabolites found as constituents of present-day life forms. Sagan's simulated primordial atmosphere not only showed the presence of those metabolites, it also contained traces of adenosine triphosphate (ATP) – the foremost agent used by living cells to store energy.

In 1966, in work done jointly with Pollack and Goldstein, Sagan was able to provide evidence supporting a hypothesis about Mars put forward by Wells, who observed that in regions on Mars where there were both dark and light areas, the clouds formed over the lighter areas aligned with boundaries of adjacent dark areas. Wells suggested that they were lee clouds formed by the Martian wind as it crossed dark areas. The implication, that dark areas mark the presence of ridges, was given support by Sagan's finding that dark areas had a high radar reflectivity that was slightly displaced in longitude. Sagan concluded that these dark areas were elevated areas with ridges of about 10 km/6 mi and low slopes extending over long distances.

During the 1970s Sagan studied the present atmosphere of the Earth; more recently, with Toon and Pollack, he has been examining how volcanic activity affects atmospheric temperature.

Saint-Claire Deville, Henri Etienne 1818–1881 was a French inorganic chemist who worked on high-temperature reactions and is best known for being the first to extract metallic aluminium in any quantity.

Saint-Claire Deville was born on 11 March 1818 in the West Indies on the island of St Thomas, Virgin Islands (then Danish territory), the son of the French Consul there. He was educated in France and studied science and medicine, learning chemistry under Louis Thénard (1777–1857). He gained his medical degree in 1844 and a year later he became Dean and Professor of Chemistry at the newly established University of Besançon. In 1851, he followed Antoine Balard (an ex-assistant of Thénard) as Professor of Chemistry at the Ecole Normale in Paris and took over from Jean Baptiste Dumas at the Sorbonne in 1859. He died in Boulogne-sur-Seine on 1 July 1881.

In 1827, Friedrich Wöhler had isolated small quantities of impure aluminium from its compounds by the drastic method of heating them with metallic potassium. Saint-Claire Deville substituted the safer sodium. He first had to prepare sufficient sodium metal but by 1855 he had obtained enough aluminium to cast a block weighing 7 kg/15 lb. The process was put into commercial production and within four years the price of aluminium had fallen to one-hundredth of its former level. (It was to decrease even further 27 years later when Charles Hall in the United States and Paul Héroult in France independently discovered the method for electrolytically extracting aluminium.)

Saint-Claire Deville also investigated the chemistry and metallurgy of magnesium and platinum, made the first preparation of a monobasic acid 'anhydride' when he made nitrogen pentoxide (dinitrogen pentoxide, N_2O_5), and studied the high-temperature decomposition of gases into atomic species. In organic chemistry he made one of the first extractions of toluene (methylbenzene; phenylmethane), in 1841, while experimenting with tolu balsam and turpentine oil.

Sakharov Andrei Dmitriyevich 1921–1989 was a Russian physicist who was closely involved with the development of Soviet thermonuclear weapons. He later repudiated this work and became an active campaigner for nuclear disarmament and human rights. This resulted in the award of the 1975 Nobel Peace Prize and in his banishment in 1980.

Sakharov was born in Moscow on 21 May 1921 and followed in his father's footsteps to become a physicist. He showed exceptional scientific ability early on in his education and studied at Moscow State University, graduating in 1942. In 1945, Sakharov joined the staff of the P. N. Lebedev Institute of Physics in Moscow as a physicist and spent his research life there. In 1953, at the age of 32, he became the youngest man ever to be made a Full Member of the Soviet Academy of Sciences. Sakharov's other Soviet honours include the Hero of Socialist Labour, the Order of Lenin, and Laureate of the USSR. In addition to the 1975 Nobel Peace Prize, his foreign honours include Membership of the American Academy of Arts and Sciences in 1969 and the National Academy of Sciences in 1972, the Eleanor Roosevelt Peace Award in 1973, and the Cino del Duca Prize and the Reinhold Niebuhr Prize from Chicago University in 1974.

In 1980, Sakharov was stripped of his Soviet honours and banished to Gorky (now Nizhny-Novgorod) for criticizing Soviet action in Afghanistan. He continued to be given honours abroad, receiving the Fritt Ord Prize in 1980 and becoming a Foreign Associate of the French Academy of Science in 1981.

At the P. N. Lebedev Institute of Physics, Sakharov worked with Igor Tamm (1895–1971), who was to win the 1958 Nobel Prize for Physics with Pavel Cherenkov (1904–1990) and Ilia Frank (1908–1990) for their work on Cherenkov radiation. In 1948, Sakharov and Tamm published a paper in which they outlined a principle for the magnetic isolation of high temperature plasma, and in so doing significantly altered the course of Soviet research into thermonuclear reactions. From 1948 to 1956, their work was solely on research connected with nuclear weapons and was top secret. It led directly to the explosion of the first Soviet hydrogen bomb in 1953 but by 1950, Sakharov and Tamm had also formulated the theoretical basis for controlled thermonuclear fusion – the means by which thermonuclear power could be used for the generation of electricity and other peaceful ends.

By the end of the 1950s, Sakharov was becoming interested in wider issues and in 1958, with Yakov B. Zeldovich, he published an article in *Pravda* advocating educational reforms which contradicted official pronouncements. Sakharov pointed out that mathematicians and physicists are most creative early on in their careers and proposed early entry into university. Many of Sakharov's ideas were incorporated. In the early 1960s, he joined the controversy between the genetics of Gregor Mendel (1822–1884) and that of Trofim Lysenko (1898–1976) and was instrumental in breaking Lysenko's hold over Soviet science and in giving science some political immunity. His scientific papers in the 1960s concerned the structure of the universe. In 1965 he wrote on the initial stages of an expanding universe and the appearance of nonuniformity in the distribution of matter, and in the following year on quarks, which are thought by some physicists to be the basic components of protons.

After the mid-1960s, Sakharov's activities showed a shift of interest. In 1968, he published a famous essay entitled *Progress, Peaceful Co-existence and Intellectual Freedom*, which argued for a reduction of nuclear arms by all nuclear powers, an increase in international cooperation and the establishment of civil liberties in Russia. This was followed in 1970 by the founding of a Committee for Human Rights by Sakharov and other Soviet physicists to promote the principles expressed in the Universal Declaration of Human Rights. Sakharov's subsequent publications on this theme included *Sakharov Speaks* (1974), *My Country and the World* (1975) and *Alarm and Hope* (1979) and made him an international figure. But his fame also brought increasing harassment from the Soviet authorities, and Sakharov was not allowed to go to Norway to collect his Nobel prize. In 1980, he was removed from Moscow and banished to Gorky, but was allowed to return to Moscow at the end of 1986 and resume his place in the Soviet Academy of Sciences. He was elected to the Congress of the USSR People's Deputies (CUPD) 1989, where he emerged as leader of its radical reform grouping before his death later the same year.

Sakharov's fundamental work on controlled thermonuclear fusion may one day result in the production of cheap energy. His bravery in resisting oppression and calling for the peaceful use of science demonstrated a rare strength of character.

Salam Abdus 1926–   is a Pakistani theoretical physicist who shared the 1979 Nobel Prize for Physics for jointly developing a unified theory of weak and electromagnetic interactions. He also set up the International Centre for Theoretical Physics in Italy.

Salam was born on 29 January 1926 in Jhang Maghiana, Pakistan, then part of British India. He attended Government College in Lahore before going to Cambridge University in England where he graduated in mathematics and physics. Salam then started experimental research at the Cavendish Laboratory – working on tritium–deuterium scattering – but, according to his Nobel lecture, 'soon I knew the craft of experimental physics was beyond me'. Instead he switched to theory and he started to work on quantum field theory with Nicholas Kemmer in Paul Dirac's department. Salam returned to Government College as a mathematics professor in 1951, received his PhD from Cambridge in 1952 and became a lecturer in mathematics at Cambridge in 1954. Salam was appointed Professor of Theoretical Physics at Imperial College in London in 1957.

Salam shared the Nobel prize with Sheldon Glashow and Steven Weinberg for unifying the theories of electromagnetism and the weak force, the force responsible for a neutron transforming into a proton, electron and neutrino (during radioactive decay). It was known that the electromagnetic force was caused by the exchange of photons (particles of light with no mass or charge) whilst the weak force was transmitted by massive charged particles known as the W^+ and W^- boson. Glashow had shown that a satisfactory 'electroweak' theory required a fourth neutral particle and, independently in 1967, Salam and Weinberg calculated how the mass of the W^+, W^- and this new particle (the Z^0) were related. The theory actually involves two new particles (the W^0 and B^0) which combine in different ways to form either the photon or the Z^0. One prediction of the theory was that the exchange of Z^0 particles in certain reactions, such as electron–neutrino scattering, would cause a weak neutral current (similar to the well-known weak charged current caused by the W particles). This neutral current was successfully measured at CERN, the European particle physics laboratory near Geneva, in 1973. Weinberg and Salam also predicted that the electroweak interaction should violate left–right symmetry and this was confirmed by experiments at Stanford University in California. The W and Z particles themselves were not detected until 1983, four years after he received his prize. Salam was also instrumental in setting up the International Centre for Theoretical Physics in Trieste, Italy. The ICTP, which is administered by the International Atomic Energy Agency (part of the United Nations), aims to stimulate science and technology in developing countries. Salam was also chief scientific advisor to the president of Pakistan between 1961 and 1974.

Salk Jonas Edward 1914– is a US microbiologist who produced the first successful vaccine against the paralytic disease poliomyelitis.

Salk was born in New York, the son of Polish–Jewish immigrants. He graduated in surgery from the College of the City of New York in 1934, and then became a research fellow at the New York University College of Medicine, where he studied the chemistry of proteins. In 1939, he was awarded a doctorate in medicine and during the next three years worked at the Mount Sinai Hospital in New York, before joining the research staff of the Virus Research Unit in the University of Michigan, where he worked on influenza vaccines until 1944. The next two years were spent in consultation regarding the protection of the armed forces from epidemics. In 1946, he became an Assistant Professor in Epidemiology at Michigan and the following year he was invited by the University of Pittsburgh to join a special medical research unit there to carry out a three-year programme on the causes and treatment of virus diseases. The development of the Salk vaccine against poliomyelitis was announced in 1955. He was appointed Director of the Salk Institute for Biological Studies, San Diego, in 1962.

The major obstacle to research on the preparation of vaccines in the 1940s was the difficulty of obtaining sufficient virus. Unlike bacteria, which may be grown in culture, viruses need living cells on which to grow. A breakthrough came when it was found that viruses could be grown in live chick embryos. John Enders improved on this technique with the use of mashed embryonic tissue, supplied with nutrients, and with the addition of penicillin to keep down the growth of bacteria.

Once the method of preparing sufficient quantities of virus was available, Salk set about finding a way of treating the polio virus so that it was unable to cause the disease but was still able to produce an antibody reaction in the human body. He collected samples of spinal cord from many polio victims and grew the virus in the new live-cell culture medium. He studied the reaction of the virus to various chemicals and found that there were three distinct types of virus that cause the disease. Salk experimented with formaldehyde (methanol) to render the virus inactive. By 1952 he had produced a vaccine effective against the three common strains of polio virus in the United States; he tested it on monkeys, which are also susceptible to polio, and found that it worked. Next he tried the vaccine on children who had recovered from polio and were immune to the disease, and he found an increase in the antibody content of their blood. Afterwards, he tried it on his family and children who had not had polio, and again antibodies were formed in the blood.

Salk needed a large scale clinical trial, however, because a large number of people would need to receive the preventive vaccine if any useful results were to be obtained. The vaccine had to be prepared on a commercial scale and licences were issued to five companies who were instructed in the technique of vaccine production and were responsible for their own quality control, because Salk's laboratory could not cope with the volume of work that testing would involve. In 1955, in a big publicity campaign, some vaccine was prepared without adequate precautions and about 200 cases of polio, with eleven deaths, resulted from the clinical trials. Salk recommended that the vaccine should be tested by the public health service in future and more stringent control prevented further disasters.

Salk was the first to make use of Enders's method of growing viruses to prepare a vaccine against poliomyelitis. It saved many people from the crippling and often fatal effects of the disease and prompted Albert Sabin to prepare a polio vaccine that can be administered orally rather than by injection.

Sanger Frederick 1918– is a British biochemist who worked out the sequence of amino acids in various protein molecules. For his work on insulin he was awarded the 1958 Nobel Prize for Chemistry. For determining the sequence in the DNA molecule, he became one of the very few scientists to receive two Nobel prizes when in 1980 he shared the Chemistry Prize with the US molecular biologists Paul Berg (1926–) and Walter Gilbert (1932–).

Sanger was born in Rendcomb, Gloucestershire, on 13 August 1918, the son of a doctor. He was educated at Bryanston School and at St John's College, Cambridge, from which he graduated in 1939. He then began research in biochemistry, gaining his PhD in 1943 and working as a Research Fellow until 1951. In that year he joined the staff of the Medical Research Council and in 1961 became Head of the Protein Chemistry Division of the Council's Molecular Biology Laboratory at Cambridge.

Beginning in 1943 Sanger and his co-workers determined the sequence of 51 amino acids in the insulin molecule, using samples of the hormone obtained from cattle pancreases. By 1945 he had discovered a compound, Sanger's reagent (2,4-dinitrofluorobenzene), which attaches itself to amino acids and so he was able to break the protein chain into smaller pieces and analyse them using paper chromatography. By 1953 he had determined the sequence for insulin, and even shown that there are small but precise differences between the structures of insulins from different animals. He also worked out the structures of other proteins, including various enzymes.

In the late 1950s, Sanger turned his attention to the sequence of the nucleotides that link to form the protein strands in the nucleic acids RNA and DNA. The double-helix structure of DNA had been determined by Francis Crick and James Watson in 1953, and within the next few years other workers had identified enzymes that can join nucleotides to form chains and others that can cut existing chains into shorter pieces. Sanger used the chain-cutting type of enzyme to identify nucleotides and their order along the chain, and in 1977 he and his co-workers announced that they had established the sequence of the more than 5,000 nucleotides along a strand of RNA from the bacterial virus called R17. They later worked out the order for mitochondrial DNA, which has approximately 17,000 nucleotides.

Savery Thomas *c*.1650–1715 was the British inventor and engineer who is generally credited with producing the world's first practical steam-driven water pump. It had only limited success and was not adopted widely, probably due to faulty materials and workmanship.

Much of Savery's life is unrecorded. He came from a Devon family and may have been born at Shilston near Modbury. He might have been a captain in the Military Engineers but it is more likely that he was 'captain' of a mine – a title sometimes used in mining circles. The first definite record comes with the grant of a patent in 1696 for a machine for cutting, grinding and polishing mirror glass. His patent for a fire engine for raising water from mines, for

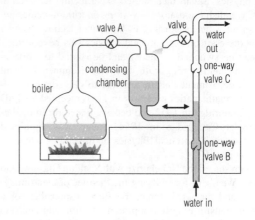

Savery *In Savery's steam pump (1696), steam from the boiler entered a condensing chamber where, after the valve (A) was closed, a spray of water condensed it. The vacuum created in the chamber lifted water into it past the one-way valve (B). Steam again admitted to the condensing chamber forced the water out past the second one-way valve (C).*

which he is best known, was granted in July 1698. In 1705 he was appointed Treasurer for Sick and Wounded Seamen, a post he held until 1714. In 1706 he was elected Fellow of the Royal Society, and the same year he applied for a patent for a bellows for foundry work.

It is probable that he made a trip to Hanover in connection with his engine and on his return he applied, in 1710, for a patent for an improved oven. He also invented a mechanism for measuring the distance sailed by a ship. In 1714, Savery was appointed surveyor of the waterworks at Hampton Court and he designed a pumping system, driven by a water wheel, for supplying the fountains.

Obituaries place Savery's death in May 1715 and his will, dated 15 May, was proved on 19 May by his widow, Martha – his executrix and sole legatee. It is not known where he was buried.

For many years up to the first decades of the eighteenth century, pumping out British mines had been a major problem. Various suggestions had been made for using steam power but the Miners' Friend, as Savery's invention was called, showed most promise of significant advance.

His invention consisted of a boiler heated by an open fire, so it was technically an 'atmospheric engine' and not a 'steam engine' as we know it today. The engine was connected via a regulator and steam pipe to the top of a receiver, the bottom of which was mounted on a hollow box or 'engine tree'. Also connected to the engine tree was the suction pipe, which extended into the mine, and a force (delivery) pipe. There were one-way valves (opening upwards) in the suction and force pipes.

To work the Miners' Friend, the operator opened the regulator, letting steam into the receiver, and driving air up the force pipe through the one-way valve. When the receiver was full of steam, the regulator was closed then doused with cold water to condense the steam it contained. Atmospheric pressure forced water up the suction pipe through the one-way valve into the receiver. The operator then opened the regulator, and the steam pressure forced the water up the force pipe. The steam in the receiver was again condensed, and the cycle repeated.

A contemporary writer described a basic engine built in 1711 or 1712 at Campden House, Kensington, which had a boiler holding 182 l/40 gall (48 US gall), a receiver of 59 l/13 gall (16 US gall), and 76-mm/3-in wooden, suction and delivery pipes, the former 5 m/16 ft long and the latter 13 m/42 ft. It could pump 14,180 l/3,120 gall (3,747 US gall) per hour and cost £50.

A model exhibited to the Royal Society had two receivers working alternately to deliver water continuously and it was on this concept that Savery based his description of The Miners' Friend. He reckoned it could lift water 6–8 m/20–26 ft and then force it to a height of 18–24 m/60–80 ft. In deep mines two or more engines, one above the other, were to be used. Savery claimed that the draught to the furnace would help mine ventilation and that the water delivered above ground could be used to turn a water wheel.

There are no records, however, of any engines being installed in mines. A two-receiver engine built at York Buildings waterworks had continuous problems with blowing steam joints. Savery acquired facilities for production of his engines but it appears that poor workmanship and materials made them unworkable.

There is little doubt that Savery produced the world's first, practical, working steam engine. It was probably never used for the purpose for which it was intended and another Devonian, Thomas Newcomen, solved the problem in a different way. Savery's engine did work, however, and for this he deserves to take his place in history.

Scheele Karl William 1742–1786 was a Swedish chemist, arguably the greatest chemist of the eighteenth century. He anticipated or independently duplicated much of the pioneering research that was taking place at that time in France and England, and isolated many elements and compounds for the first time. Among his many discoveries were the elements oxygen and chlorine, although delays in publishing his findings often resulted in other chemists being credited with priority. He never took a university appointment, doing all his research privately while practising as an apothecary.

Scheele was born on 9 December 1742 in Stralsund, Pomerania, at a time when it belonged to Sweden (it was more often, as now, German territory), the seventh of eleven children of a poor family. He received little education until 1756 when at the age of 14 he became an apothecary's apprentice at Göteborg (Gothenburg) and learned basic chemistry through reading, observation and experiment. He progressed from position to position, moving to Malmö in 1765, Stockholm in 1768 and finally Uppsala in 1770, still practising as an apothecary. In Uppsala he met Johann Gahn (1745–1818) who introduced him to the famous Swedish chemist Torbern Bergman (1735–1784), who in turn recognized Scheele's talents and publicized his work. In 1775, Scheele moved to run a pharmacy in the small town of Köping on Lake Malären in Västmanland, where he remained for the rest of his life despite opportunities to take academic posts in Germany and England, and an offer from the Prussian Frederick II to serve as court chemist. Also in 1775 he was elected to the Swedish Royal Academy of Sciences, a unique honour for an apothecary's assistant. As Scheele approached middle age his health – never good – began to fail and he suffered from rheumatic pain and possibly the toxic effects of some of the chemicals he experimented with; long hours of intense work also took their toll. He married on his deathbed and died, at Köping, aged only 43 on 21 May 1786.

Scheele's research did not appear to follow a particular plan, and he seems to have experimented in an indiscriminate way with the various substances he came across in his work. But even if his approach lacked system, he nevertheless made a huge contribution to chemistry. His original discoveries alone make a formidable list, and the major ones are worth itemizing; ignoring chronology and combining inorganic discoveries with those in organic chemistry, they include the following:

arsenic acid	hydrogen sulphide
arsine	lactic acid
barium oxide (baryta)	malic acid
benzoic acid	manganese and
calcium tungstate	manganates
(scheelite)	molybdic acid
chlorine	nitrogen
citric acid	oxalic acid
copper arsenite	oxygen
(Scheele's green)	permanganates
gallic acid	silicon tetrafluoride
glycerol	tartaric acid
hydrogen cyanide	tungstic acid
and hydrocyanic acid	uric acid
hydrogen fluoride	

Undoubtedly, the most significant of these are chlorine and oxygen. Scheele also discovered that the action of light modifies certain silver salts (50 years before they were first used in photographic emulsions). He isolated phosphorus from calcined bones and obtained uric acid from bladder stones. He studied molybdenum disulphide (molybdenite, MoS_2) and showed how it differs from graphite (molybdena) – both substances have similar physical properties and both are still used as solid lubricants. He demonstrated the different oxidation states of copper, iron and mercury.

Scheele's discovery of oxygen began – as did Antoine Lavoisier's – with a study of air. He first showed that air consists of two main gases, one of which ('fire air') supports combustion and one of which ('vitiated air' or 'foul air') does not. In various experiments on air he consumed the 'fire air' component to leave only the 'vitiated air' – Scheele was a staunch believer in the phlogiston theory of combustion. Then in a series of preparations during 1771 and 1772 he produced 'fire air' (oxygen) in various ways chemically: by the action of heat on saltpetre (potassium nitrate) or manganese dioxide; by heating heavy metal nitrates (and absorbing the nitrogen dioxide also formed in lime water), and by heating mercuric (mercury(II)) oxide (Joseph Priestley's method of 1774). He showed that oxygen is involved in the respiration of plants and fish. Scheele described these experiments in his only major publication, *A Chemical Treatise on Air and Fire*, written in about 1773 in German, the language that Scheele normally used in speech and writing. The introduction to the book was written by Bergman, who took so long to provide his text that publication did not actually take place until 1777. This was three years after Priestley had prepared oxygen and published his findings, and credit is now usually given to him. (The name oxygen – meaning 'acid producer' – was coined by Lavoisier, who mistakenly thought that all acids contain oxygen.)

Following on from this work Scheele isolated chlorine in 1774 by heating manganese dioxide with hydrochloric acid, but he thought that it was a compound of oxygen – 'dephlogisticated muriatic (hydrochloric) acid' – and did not recognize it as an element. This distinction was made a generation later by Humphry Davy, who through his work on hydrochloric acid also discredited Lavoisier's theory that all acids contain oxygen.

Scheiner Christoph 1573–1650 was a German astronomer who carried out one of the earliest and most meticulous studies of sunspots and who made significant improvements to the helioscope and the telescope.

Scheiner was born in Wald, near Mindelheim, on 25 July 1573. He attended the Jesuit Latin School at Augsburg and then the Jesuit College at Landsberg, In 1600, he was sent to Ingolstadt, where he studied philosophy and mathematics, and from 1603 to 1605 he taught humanities in Dillingen. It was during this period that he invented the pantograph – an instrument that can be used for copying plans and drawings to any scale. He returned to Ingolstadt to study theology and after completing his studies, in 1610, he was appointed as Professor of Mathematics and Hebrew at Ingolstadt University. It was there that he began to make his astronomical observations and, as well as carrying out his own research, he also organized public debates on current issues in astronomy. In 1616, Scheiner accepted an invitation to take up residence at the court in Innsbruck and the following year he was ordained to the priesthood. From 1633 to 1639 he lived in Vienna and from then until his death on 18 June 1650 he lived in Neisse (now Nysa in Poland).

Scheiner built his first telescope in 1611 and began making astronomical observations of the Sun. He was among the earliest observers of the Sun and, using one of the first properly mounted telescopes, he was sensible enough to project the image of the Sun onto a white screen so that it would not damage his eyes. Within a matter of weeks of observation, he detected spots on the Sun, but his religious superiors did not wish him to publish his observations under his own name in case he was mistaken and might thus bring discredit on the Society of Jesus. He communicated his discovery to his friend, Marc Wesler, in Augsburg, who had Scheiner's letters printed under a pseudonym and sent copies to Galileo and Kepler. Scheiner believed that the spots were small planets circling the Sun and in a second series of letters, published in the same year under the same false name, Scheiner discussed the individual motion of the spots, their period of revolution and the appearance of brighter patches, or 'faculae', on the surface of the Sun. Having observed the lower conjunction of Venus with the Sun, he concluded that Venus and Mercury revolve around the Sun. But because of his religious beliefs he upheld the traditional view that the Earth is at rest at the centre of the universe.

Although Scheiner had tried to conceal his identity, Galileo identified him and claimed priority for the discovery of sunspots, hinting that Scheiner was guilty of plagiarism. It seems that this criticism was unfair, however, because the sunspots were observed independently, not only by Galileo in Florence and Scheiner in Ingolstadt, but also by Thomas Harriot in Oxford and Johann Fabricius in Wittenberg. In his *Solellipticus*, published in 1615, and *Refractiones Caelestes*, published in 1617, which were both dedicated to Maximillian, the Archduke of Tirol, Scheiner drew attention to his observations of the elliptical form of the Sun near the horizon, which he explained as being due to the effects of refraction. In his major work, *Rosa ursina sive sol*, which was published between 1626 and 1630, Scheiner described the inclination of the axis of rotation of the sunspots to the plane of the ecliptic, which he accurately determined as having a value of $7°30'$, the modern value being $7°15'$.

Unfortunately, Scheiner lived in an age when observational astronomy posed grave confrontations to his theological principles and this not only affected Scheiner's scientific career, but also hindered the progress of science as a whole.

Schiaparelli Giovanni Virginio 1835–1910 was an Italian astronomer whose long experience of rather ancient equipment led to considerable caution in making his discoveries known, but who nevertheless carried out significant research into the nature of comets and the inner planets of the Solar System. He is best known – and most misunderstood – for allegedly discovering 'canals' on Mars.

Schiaparelli was born in Savigliano, a village in Cueno Province, on 14 March 1835. After leaving school, he first trained as a civil engineer. Only after he had graduated from the University of Turin and was teaching mathematics at the University did Schiaparelli begin a study of modern languages and astronomy. Support from the Piedmont government permitted him to engage in advanced studies at the Observatories of Berlin and Pulkovo. On his return in 1860, Schiaparelli was appointed Astronomer at the Brera Observatory in Milan; he remained there until his retirement in 1900. His first observations, using only the primitive instruments then available to him, resulted in the discovery of the asteroid Hesperia. For most of the next 15 years, Schiaparelli's major interest was in comets. In later years, when more sophisticated instruments became available at Milan, Schiaparelli turned his attention to the planets. Towards the end of his working life, he made use of his linguistic skills by assisting in the translation from the Arabic of a historic work on astronomy; it was published after his death, which occurred on 4 July 1910 in Milan.

Schiaparelli's study of comets began with theoretical research into the nature of a comet's tail. Two years later his work on the tail was extended to a consideration of other features of a comet, including the histories of comets recorded in years previously. From these, and from his own notes on the bright comet of 1862, Schiaparelli was led to conjecture that all meteor showers are the result of the disintegration of comets. Such a hypothesis had been put forward before, but Schiaparelli could point to his compilation of extensive observational evidence to show that all meteors moved in elliptical or parabolic orbits around the Sun, orbits which (as would have to be the case for Schiaparelli's hypothesis to be true) were identical with or similar to those of comets. Schiaparelli also argued that meteors became visible as luminous showers falling from a determinable position in the celestial sphere. Pietro Secchi's observations of the 1862 comet were to confirm this hypothesis; and, in the years following, observations by Peters, Galle, Weiss, von Biela and d'Arrest (among others) served to confirm Schiaparelli's theory.

In 1877, using new equipment far superior to the instruments he had used before, Schiaparelli began detailed observation of Mars, preparatory to drawing a map of the fundamental features of its surface. In noting what Pietro Secchi had previously called 'channels' (*canali*), and while introducing further nomenclature involving 'seas' and 'continents', Schiaparelli made it quite clear, in publishing his results, that such terms were for convenience only and did not represent terrestrial actuality. Nevertheless, possibly through mistranslation – and certainly through wishful thinking – fanciful stories of advanced life on Mars proliferated, especially in France, the United States and England, for the next 40 years or more.

Schiaparelli's observations of Mars continued through the next few years, being more frequent during the seven oppositions between 1879 and 1890. All manner of variations, of inclination to axis, of the apparent diameter, of geometric declination, of observational conditions and so on, allowed Schiaparelli to build up a complex picture of Mars. During the 1877 opposition Schiaparelli noted that when Mars moved away from the Earth, its diameter appeared to decrease. He also observed that certain canals seemed to be splitting into two parts and he suggested that this 'gemination' had serious implications for the understanding of the physical constitution of Mars. His observations of 1888 occurred under such good atmospheric conditions that Schiaparelli found it impossible to represent all the features of Mars in adequate detail or

colour. New geminations, absent from the last observation, led him to propose that they were the effect of a periodic phenomenon related to the solar year on Mars. Schiaparelli also observed that split canals were visible for a few days or weeks before becoming single canals or disappearing entirely. He continued his observations until 1890, and their detail and accuracy far surpassed observations made by others using similar instruments.

Schiaparelli also observed Mercury, studying the dark spots that form shadowy bands on the surface of the planet. He concluded that Mercury revolved around the Sun in such a way as always to present the same side to the Sun. He came to exactly the same conclusion about the rotation of Venus. Other observations included a study of binary stars in order to deduce their orbital systems.

Schiaparelli's role in the history of astronomy extends beyond the part he played through his observations. He also made a noteworthy contribution in helping Nallino to translate the only existing Arabic text of al-Battani's *Opus Astronomicum* into Latin and in writing explanatory notes to many chapters. The translation is a major landmark in understanding the development of astronomy in the Arabic world. Schiaparelli intended to compile a major work on the history of ancient astronomy and he published a number of monographs on the subject.

Schleiden Matthias Jakob 1804–1881 was a German botanist who, with Theodor Schwann, is best known for the establishment of the cell theory.

Schleiden was born on 5 April 1804 in Hamburg and studied law at Heidelberg University from 1824 to 1827. After graduating, he practised as a barrister in Hamburg but soon returned to university, taking courses in botany and medicine at the universities of Göttingen, Berlin and Jena. After graduating in 1831 he was appointed Professor of Botany at Jena, where he remained until he became Professor of Botany at the University of Dorpat, Estonia, in 1862. He returned to Germany after a short time, however, and from 1864 began teaching privately in Frankfurt. Schleiden died in Frankfurt on 23 June 1881.

Although the existence of cells had been known since the seventeenth century (Robert Hooke is generally credited with their discovery in 1665), Schleiden was the first to recognize their importance as the fundamental units of living organisms when, in 1838, he announced that the various parts of plants consist of cells or derivatives of cells. In the following year Schwann published a paper in which he confirmed for animals Schleiden's idea of the basic importance of cells in the organization of organisms. Thus Schleiden and Schwann established the cell theory, a concept that is common knowledge today and which is as fundamental to biology as atomic theory is to the physical sciences.

Schleiden also researched into other aspects of cells. He recognized the importance of the nucleus (which he called the cytoblast) in cell division, although he incorrectly believed that new cells budded off from the nuclear surface. In addition, he noted the active movement of intracellular material in plant tissues, calling this movement protoplasmic streaming. The phenomenon is well known today, although the intracellular material is now called cytoplasm.

Schmidt Bernhard Voldemar 1879–1935 was an Estonian lens- and mirror-maker who devised a special sort of lens to work in conjunction with a spherical mirror in a reflecting telescope. The effect of this was to nullify 'coma', the optical distortion of focus away from the centre of the image inherent in such telescopes, and thus to bring the entire image into a single focus, useful for general surveys of the night sky or for photography.

Schmidt was born on the island of Naissaar, Estonia, on 30 March 1879, the child of poor parents. Inclined to scientific pursuits, one of his earliest successes was to make a convex lens by grinding the bottom of a bottle using fine sand. Another experiment had disastrous consequences, however: he made some gunpowder, packed it tightly into a metal tube and ignited it. The explosion caused him to lose most of his right arm. At the age of 21 he began to study engineering at the Institute of Technology at Gothenburg (Göteborg). After one year he went on to the Institute at Mittweida in Germany. After graduating in 1904 he stayed in Mittweida making lenses and mirrors for astronomers. One of his early accomplishments, in 1905, was a 40-cm/27-in mirror for the Potsdam Astrophysical Observatory. Schmidt worked independently, producing optical equipment of very high quality until 1926, when Schorr, the Director of the Hamburg Observatory, asked him to move into the Observatory and work there. Schmidt

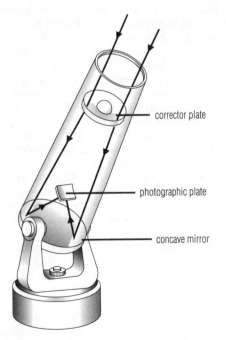

Schmidt *In a Schmidt camera, a corrector plate allows a distortion-free image to be formed on a curved photographic plate or film.*

accepted the invitation. He worked on the mountings and drives of the telescopes, as well as on their optics. It was in Hamburg that he perfected his lens and built it into the Observatory telescope, specifically for use in photography. He died in Hamburg, at an asylum for the insane, on 1 December 1935.

It is usual for reflecting telescopes to have parabolic mirrors, rather than spherical ones; spherical ones are subject to an optical distortion known as spherical aberration. Parabolic mirrors too, however, suffer from their own optically distortive effect, 'coma'; but they provide an image that is at least centrally clear and focused. What Schmidt devised was a means of correcting the image formed by a spherical mirror – a disc-shaped lens thicker at the centre and edges than at half-radius. By replacing the parabolic mirror of a telescope with a spherical one plus his lens – his 'corrector plate', as he called it – he could produce an image that was sharply focused at every point (generally on a curved photographic plate, although on later models Schmidt used a second lens to compensate for the use of a flat photographic plate).

Later astronomers, opticians and engineers improved on Schmidt's basic designs to produce such instruments as the super-Schmidt meteor camera which has been used to great effect at Las Cruces, New Mexico and the large 120-cm/48-in Schmidt on the same site as the 500-cm/200-in Mount Palomar reflecting telescope, and used in the Palomar Sky Survey.

Schoenheimer Rudolf 1898–1941 was a German-born US biochemist who first used isotopes as tracers to study biochemical processes.

Schoenheimer was born in Berlin on 10 May 1898. After graduating in medicine from the University of Berlin in 1923, he spent the next ten years in various teaching posts in Germany. In 1933 he emigrated to the United States, where he became a member of the College of Physicians and Surgeons at the University of Columbia. He committed suicide in New York City on 11 September 1941, while still at the peak of his career.

Schoenheimer introduced the use of isotopic tracers into biochemical research in 1935. Deuterium (heavy hydrogen) had become fairly easily available for the first time, thanks mainly to the work of Harold Urey, who was also at Columbia. Schoenheimer used deuterium to replace some of the hydrogen atoms in molecules of fat which he fed to laboratory animals. It had previously been thought that fat stored in body tissues remained immobile, just lying there until starvation demanded its use. On analysing the body fat of rats four days after feeding them deuterated fat, he found that about half of the labelled fat was being stored – that is, ingested fat was being stored by the animal and stored fat was being used. There was a rapid turnover and the body constituents, far from being static, were changing constantly and dynamically.

Urey prepared the isotope nitrogen-15 at about this time, and Schoenheimer soon used it to label amino acids, the basic building blocks of proteins. In a series of experiments, in which he fed a single labelled amino acid to an animal, he traced the fate of that acid in the animal's proteins. He again found that there is constant action, even though the overall movement may be small, with the protein molecules constantly changing and shifting. He had thus established that many component molecules of the body are continually being broken down and built up. He summarized his findings in his book *The Dynamical State of Bodily Constituents*.

After World War II, researchers such as Melvin Calvin went on to use radioactive isotopes, such as those of carbon and phosphorus, to investigate biochemical pathways in living animals. These techniques were developed from the pioneering work of Rudolf Schoenheimer.

Schrödinger Erwin 1887–1961 was an Austrian physicist who founded wave mechanics with the formulation of the Schrödinger wave equation to describe the behaviour of electrons in atoms. For this achievement, he was awarded the 1933 Nobel Prize for Physics with Paul Dirac (1902–1984) and Werner Heisenberg (1901–1976), who also made important advances in the theory of atomic structure.

Schrödinger was born in Vienna on 12 August 1887. His father was an oilcloth manufacturer who had studied chemistry and his mother was the daughter of a chemistry professor. Apart from a few weeks when he attended an elementary school in Innsbruck, Schrödinger received his early education from a private tutor. In 1898, he entered the Gymnasium in Vienna where he enjoyed mathematics, physics and ancient languages. He then attended the University of Vienna, specializing in physics. Schrödinger obtained his doctorate in 1910 and a year later he became an assistant in the University's Second Physics Institute. His early research ranged over many topics in experimental and theoretical physics.

During World War I, Schrödinger served as an artillery officer and then returned to his previous post at Vienna. Conditions were difficult in Austria after the war and Schrödinger decided to go to Germany in 1920. After a series of shortlived posts at Jena, Stuttgart and Breslau, he became Professor of Physics at Zurich in 1921.

Schrödinger's most productive work was done at Zurich and it resulted in his succeeding Max Planck (1858–1947) as Professor of Theoretical Physics at Berlin in 1927. He remained there until the rise of the Nazis in 1933, when Schrödinger went to Oxford, England, where he became a fellow of Magdalen College. Homesick, he returned to Austria in 1936 to take up a post at Graz, but the Nazi takeover of Austria in 1938 placed Schrödinger in danger. The intervention of the Prime Minister of Ireland, Eamon de Valera (1882–1975), led to his appointment in 1939 to a post at the Institute for Advanced Studies in Dublin. Schrödinger continued work in theoretical physics there until 1956, when he returned to Austria to a chair at the University of Vienna. In the following year Schrödinger suffered a severe illness from which he never fully recovered. He died in Vienna on 4 January 1961.

The origin of Schrödinger's great discovery of wave mechanics began with the work of Louis de Broglie (1892–1987), who, in 1924, using ideas from Einstein's special theory of relativity, showed that an electron or any other particle has a wave associated with it. The fundamental result was that:

$$\lambda = h/p$$

where λ is the wavelength of the associated wave, h is Planck's constant and p is the momentum of the particle. An immediate deduction from this discovery was that if particles, and particularly electrons, have waves then their behaviour should be capable of description by a particular type of partial differential equation known as a wave equation in the same way as sound and other kinds of waves. These ideas were taken up by both de Broglie and Schrödinger and in 1926 each published the same wave equation which, when written in relativistic terms, is:

$$\frac{1}{c^2}\frac{\delta^2\psi}{\delta t^2} = \frac{\delta^2\psi}{\delta x^2} + \frac{\delta^2\psi}{\delta y^2} + \frac{\delta^2\psi}{\delta z^2} - \frac{4\pi^2 m^2 c^2}{h^2}\psi$$

where ψ is the wave function, t is the time, m is the mass of the electron, c is the velocity of light, h is Planck's constant and x, y and z represent the position of the electron in Cartesian coordinates. Unfortunately, while the equation is true, it was of very little help in developing further facts and explanations.

Later the same year, however, Schrödinger used a new approach. After spending some time studying the mathematics of partial differential equations and using the Hamiltonian function, a powerful idea in mechanics due to William Rowan Hamilton (1805–1865), he formulated an equation in terms of the energies of the electron and the field in which it was situated. His new equation was:

$$\frac{\delta^2\psi}{\delta x^2} + \frac{\delta^2\psi}{\delta y^2} + \frac{\delta^2\psi}{\delta z^2} + \frac{8\pi^2 m}{h^2}(E - V)\psi = 0$$

where E is the total energy of the electron and V is the potential of the field in which the electron is moving. This equation neglects the small effects of special relativity. Partial differential equations have many solutions and very stringent conditions had to be fulfilled by the individual solutions of this equation in order for it to be useful in describing the electron. Among other things, they had to be finite and possess only one value. These solutions were associated with special values of E, known as proper values or eigenvalues. Schrödinger solved the equation for the hydrogen atom, where:

$$V = -e^2/r$$

e being the electron's charge and r is its distance from the nucleus, and found that the values of E corresponded with those of the energy levels given in the older theory of Niels Bohr (1885–1962). Also, to each value of E there corresponded a finite number of particular solutions for the wave function ψ, and these could be associated with lines in the spectrum of atomic hydrogen. In the hydrogen atom the wave function describes where we can expect to find the electron, and it turns out that while it is most likely to be where Bohr predicted it to be, it does not follow a circular orbit but is described by the more complicated notion of an orbital, a region in space where the electron can be found with varying degrees of probability.

Atoms other than hydrogen and also molecules and ions can be described by Schrödinger's wave equation but such cases are very difficult to solve. In certain cases approximations have been used, usually with the numerical work being carried out on a computer.

Schrödinger's mathematical description of electron waves found immediate acceptance because these waves could be visualized as standing waves around the nucleus. In 1925, a year before Schrödinger published his results, a mathematical system called matrix mechanics, developed by Max Born (1882–1970) and Werner Heisenberg (1901–1976) had also succeeded in describing the structure of the atom but it was totally theoretical and gave no picture of the atom. Schrödinger's vindication of de Broglie's picture of electron waves immediately overturned matrix mechanics, though it was later shown that wave mechanics is equivalent to matrix mechanics.

During his later years, Schrödinger became increasingly worried by the way quantum mechanics, of which wave mechanics is a part, was interpreted, in particular with the probabilistic nature of the wave function. Schrödinger believed he had given an important description of the atom in the same way that Newton's laws described mechanics and Maxwell's equations described electrodynamics, only to find that the structure of the atom became increasingly more difficult to describe explicitly with each new discovery. Much of his later work was concerned with philosophy, particularly as applied to physics and the atom.

Schrödinger made a fundamental contribution to physics in finally producing a solid mathematical explanation of the quantum theory first advanced by Planck in 1900, and the subsequent structures of the atom formulated by Bohr and de Broglie.

Schwabe Samuel Heinrich 1789–1875 was a German chemist and astronomer who was the first person to measure the periodicity of the sunspot cycle.

Schwabe was born in Dessau on 25 October 1789; his father was a doctor and his mother ran a pharmacy. Educated in Berlin, he entered his mother's business as a pharmacist at the age of 17. Three years later he returned to Berlin and took up pharmaceutical studies at the University there, under Martin Klaproth. It was at the University that he became absorbed in both astronomy and botany. His two-year course over, Schwabe went back to the pharmacy in 1812. Seventeen years later his amateur astronomical research – particularly his study of sunspots – became engrossing. He sold his pharmacy business and became an astronomer. As a result of his work he was presented with the Royal Astronomical Society's Gold Medal in 1857 and was elected to the Royal Society in 1868. During his life-

time he published no fewer than 109 scientific papers, and after his death – which occurred on 11 April 1875 – 31 volumes of his astronomical data were presented to the Royal Astronomical Society. They are stored in its archives.

Wanting to commence his astronomical research while still a pharmacist, Schwabe looked for some branch of astronomy that would occupy him during the daytime. His first thought was that he might find a new planet close to the Sun – inside the orbit of Mercury – spotting it as it passed in front of the Sun's disc. He began to watch the Sun in 1825 with a small 5-cm/2-in telescope and noticed sunspots. After a while he forgot his hopes of discovering an intramercurial planet and concentrated on the sunspots, making daily counts of them for most of the rest of his life. Day after day, year after year he tabulated his results under four headings: the year, the number of sunspots, the number of days free from sunspots, and the number of days when observations were made. Schwabe realized that with his modest apparatus, in a private observatory, numerical determinations were problematical. He acknowledged that on days when there were large numbers of sunspots, it was more than probable that he underestimated the total. Nevertheless, after carefully collating his results, his patience was rewarded when, in 1843, he was able to announce a periodicity. He declared that the sunspots waxed and waned in number according to a ten-year cycle.

His discovery was ignored at the time, but in 1851 it was republished by the explorer and naturalist, Alexander von Humboldt, in his book *Kosmos* and given the recognition it deserved. Immediately afterwards, Rudolf Wolf of Berne collated all existing support data in other astronomical records, recalculated the value of the periodicity more accurately, and fixed it at 11.1 years.

Schwabe's revelation of the periodicity – which may be considered as marking the precise beginning of solar physics – was all the more remarkable because Joseph de Lalande (1732–1807) and Jean-Baptiste Delambre (1749–1822) had previously considered such an investigation and decided it would not be profitable. In 1851, the year the more accurate value for the periodicity was set, the period was first linked with the occurrence of magnetic storms. It was not long before Johann von Lamont showed that the sunspot cycle had further effects upon the Earth in terms of magnetic disturbances, weather conditions, and plant and animal growth rates. Nor was it long before astronomers began using the photo-heliograph in order to keep daily counts of sunspots.

Although Schwabe was preoccupied with his sunspot counts on every sunny day, he nevertheless found the time for other astronomical research. In December 1827 he rediscovered the eccentricity of Saturn's rings. Four years later he drew a picture of the planet Jupiter on which the Great Red Spot was shown for the first time. And he also found time to write scientific papers on the phenomena of frost patterns, haze and rock sources.

Schwann Theodor 1810–1882 was a German physiologist who, with Matthias Schleiden, is credited with formulating the cell theory, one of the most fundamental of all concepts in biology. Schwann also did important work on digestion, fermentation and histology.

Schwann was born on 7 December 1810 in Neuss. He was educated at the Jesuit college in Cologne then studied medicine at the universities of Bonn, Würzburg and Berlin, graduating from the last in 1834. He spent the next four years – the most scientifically productive period in his life – working as an assistant to the German physiologist Johannes Müller at the Museum of Anatomy in Berlin. In 1839, however, Schwann's work on fermentation attracted so much adverse criticism that he left Germany for Belgium, where he was Professor of Anatomy at the Roman Catholic University in Louvain from 1839 to 1848 then held the same post at the University of Liège until his death in Cologne on 11 January 1882.

In 1834, Schwann began to investigate digestive processes and two years later isolated from the lining of the stomach a chemical responsible for protein digestion, which he called pepsin. This was the first enzyme to be isolated from animal tissue, although Anselme Payan, a French chemist, had isolated an enzyme from malt in 1833. Schwann then studied fermentation and between 1836 and 1837 showed that the fermentation of sugar is a result of the life processes of living yeast cells (he later coined the term metabolism to denote the chemical changes that occur in living tissue). This work on fermentation was later criticized heavily, especially by the German chemists Friedrich Wöhler and Justus von Liebig, and this led to Schwann leaving Germany. It was not until Louis Pasteur's work on fermentation in the 1850s that Schwann was proved correct. Meanwhile, however, Schwann investigated putre-faction in an attempt to disprove the theory of spontaneous generation (which had once again become a matter of debate) repeating, with improved techniques, Lazzaro Spallanzani's earlier experiments. Like Spallanzani, Schwann found no evidence to support the theory, despite which it was still believed by some scientists.

In 1839, Schwann published *Mikroskopische Untersuchungen über die Ueberreinstimmung in der Struktur und dem Wachstum der Tiere und Pflanzen* (translated in 1847 as *Microscopical Researches on the Similarity in the Structure and Growth of Animals and Plants*) in which he formulated the cell theory. In the previous year Matthias Schleiden – whom Schwann knew well – had stated the theory in connection with plants, but it was Schwann who extended the theory to animals and enunciated it in its clearest form.

Schwann and Schleiden are therefore generally credited as coformulators of the cell theory. Giving numerous examples from many different types of animal tissues, Schwann in his *Microscopical Researches* concluded that all organisms (both animals and plants) consist entirely of cells or of products of cells and that the life of each individual cell is subordinated to that of the whole organism. The cell theory soon became widely accepted and is today recognized as being one of the most important concepts in biology.

Schwann also discovered the cells (now called Schwann cells) that make up the myelin sheath surrounding peripheral nerve axons, and the striated muscle in the upper region of the oesophagus. In addition, he noted that an egg is a single cell that eventually develops into a complex organism – a basic principle in embryology.

Schwarzschild Karl 1873–1916 was a German astronomer and theoretician who achieved great things despite a short lifespan. In addition to the conceptual work he carried out, he was a practical man who designed and constructed some of his own instruments and devised considerable improvements in the use of photography for astronomical purposes.

Schwarzschild was born in Frankfurt on 9 October 1873, the eldest of a family of five sons and one daughter. His father was a prosperous member of the Jewish business community in Frankfurt, and Schwarzschild spent a happy childhood surrounded by relatives who were talented in art and music. The first in his family to be scientific, he was educated at the local municipal school; in 1891 he went to Strasbourg Univeristy and spent two years there. He then continued his studies at Munich University. After graduating, he became an Assistant at the Kuffner Observatory in Ottakring, Vienna, and in 1901 he was appointed Associate Professor at the University of Göttingen, where the Observatory had been equipped by Karl Gauss (80 years previously). In the following year he was appointed full Professor at the age of only 28, and he was also made Director of the Observatory. He left Göttingen in 1909 to succeed Hermann Vogel (1841–1907) as Director of the Astrophysical Observatory at Potsdam. At the outbreak of World War I he volunteered for service and was sent to Belgium to work at a weather station. He was then transferred to France to calculate trajectories for long-range shells, and to the Eastern Front, in Russia, where he contracted pemphigus, a metabolic disease of the skin that was then incurable. He was invalided out of the army but died in Potsdam, Germany, on 11 May 1916. By his own request he was buried in Göttingen. For his war work he was awarded a posthumous Iron Cross, and in 1960 the Berlin Academy honoured him as the greatest German astronomer of the preceding century.

At secondary school, Schwarzschild bought himself some lenses in order to make a telescope. Seeing his interest, his father introduced him to J. Epstein, a mathematician who owned a private observatory, and it was with Epstein's son – later to become Professor of Mathematics at the University of Strasbourg – that Schwarzschild learned to use the telescope and studied advanced mathematics and celestial mechanics. His first published work was a paper on celestial orbits, written at the age of 16. His thesis for his PhD was on the applications of Poincaré's theory of stable configurations in rotating bodies to some astronomical problems. He investigated the tidal deformation in satellites and the validity of Pierre Laplace's theory on the origin of the Solar System. Even before he graduated, he devised a multi-slit interferometer and used it to measure the separation of close double stars. Between 1896 and 1899 he gave lectures that conveyed an infectious natural enthusiasm to non-astronomers and were to become famous.

In observational astronomy Schwarzschild was the first to apply precise methods using photographic photometry, substituting a photographic plate at the telescope in place of the eye and measuring densities with a photometer. He photographed an aggregate of 367 stars and presented the results to the University of Munich as credentials to entitle him to teach there.

In 1900, he suggested that the geometry of space was possibly not in conformity with Euclidean principles. (This was 16 years before the publication of Einstein's general theory of relativity.)

He introduced the concept of radiative equilibrium in astrophysics and was probably the first to see how radiative processes were important in conveying heat in stellar atmospheres. In 1906, he published work on the transfer of energy at and near the surface of the Sun. He observed the total solar eclipse on 30 August 1905, and obtained spectrograms, using a camera fitted with an objective prism, which gave information on the chemical composition of regions at various heights on the Sun. He also developed methods and techniques later to become standard in the preparation of stellar statistics.

In 1910, he measured photographs of Halley's comet taken by the Potsdam expedition to Tenerife, and suggested that fluorescent radiation occurs in the tails of comets. In spectroscopy, he designed a spectrographic objective that provided a quick, reliable way to determine the radial velocities of stars. He then made further important contributions to geometric optics and to the theory behind the design of optical instruments.

Although primarily an astronomer, he was also a theoretical physicist and was one of the great promoters of Niels Bohr's theory of atomic spectra (1913). As he lay dying, he completed a famous paper, in which he developed the 'rules of quantization'. (Work carried out independently by Arnold Sommerfeld gave the theory of the Stark effect and the quantum theory of molecular structure.) These last papers also dealt with the gravitational field of a point mass in empty space and gave the first exact solution of Einstein's field equations.

Schwarzschild Martin 1912–   is a German-born US astronomer whose most important work has been in the field of stellar structure and evolution.

Schwarzschild was born in Potsdam on 31 May 1912, the son of the astronomer and mathematician, Karl Schwarzschild. Schwarzschild the younger was educated at the University of Göttingen, where he obtained a PhD in astronomy in 1935. He then emigrated to the United States, eventually becoming a naturalized US citizen. He was Nansen Fellow at the University of Oslo (1936–37) and a Littauer Fellow at Harvard University (1937–40). In 1940 he was made a Lecturer in Astronomy at Columbia University, later becoming Assistant Professor (1944) and

Professor (1947). Since 1951 Schwarzschild has held the position of Huggins Professor of Astronomy at Princeton University. He is a member of the National Academy of Sciences and the American Astronomical Society, of which he was President from 1970 to 1972.

Schwarzschild's research has been primarily concerned with the theory of stellar structure and evolution. He has written numerous articles on the internal constitution of stars and has made astronomical observations with telescopes carried by balloons into the stratosphere. In 1959, he obtained structural details of the surface of the Sun and photographs of sunspot penumbrae by using a balloon-supported solar telescope at 24,385 m/80,032 ft.

Arthur Eddington was interested in the fact that whereas stars differ greatly in brightness, density and in some physical properties, they differ relatively little in mass. By 1926 the range of known stellar masses lay from $1/6$ to 100 times that of the Sun. Since then the upper limit has been reduced and much of the work on assessing these limits has been done by Schwarzschild. It is now thought that the upper limit may be only 65 solar masses. The smallest stellar masses known are about $1/100$ that of the Sun or about 10 times the mass of Jupiter.

Schwarzschild has also worked out a quantity (Z_{He}) for the total mass density of the elements heavier than helium, using the density of hydrogen as one unit. The values of Z_{He} are smallest for old stars (0.003) and largest for young stars (0.04), implying that the most recently formed stellar objects were formed out of a medium of interstellar gas and dust that was already enriched with heavy elements. These elements were probably produced in stellar interiors and expelled by the oldest stars.

Schwarzschild has been involved with what is known as pulsation theory. In 1879, even before variations in radial velocities were known, Arthur Ritter had considered the periodic expansions and contractions of a star which are termed radial pulsations. In 1938, Schwarzschild suggested that the star's deepest interior pulsates, but that in the outermost regions the elements of gas do not all vibrate in unison, causing a lag in the light curve by the observed amount.

Throughout his distinguished career, Schwarzschild has made an enormous contribution to our understanding of the dynamics and structure of stellar objects.

Scott Peter Markham 1909–1989 was a British ornithologist and artist best known for his superb bird paintings, book illustrations and wildlife conservation work, including the foundation of the Wildfowl Trust at Slimbridge, Gloucestershire. He received numerous honours from many different countries, including several honorary degrees and in 1973, a knighthood.

Scott was born on 14 September 1909, the son of Captain Robert Falcon Scott (1868–1912), the Antarctic explorer. He was educated at Oundle School, from which he went to Trinity College, Cambridge, then to the Munich State Academy and finally to the Royal Academy School, London. In 1936, he represented Britain in the Olympic

Games, gaining a bronze medal for the single-handed sailing event. During World War II he served with the Royal Navy, and after the war founded the Wildfowl Trust in 1946. In 1949 he led his first expedition, which was to explore the uncharted Perry River area in the Canadian Arctic, and in 1951 and 1953 he led expeditions to Iceland to mark geese. In addition, Scott also led ornithological expeditions to Australasia, the Galápagos Islands, the Seychelles and the Antarctic. From 1961 to 1967 he was the first president of the World Wildlife Fund. In 1963 he became Chairman of the Survival Service Commission of the International Union for the Conservation of Nature and Natural Resources, and in 1969 he was made President of the Wildlife Youth Service. He became Chancellor of Birmingham University in 1974.

Scott did much to promote wildlife conservation, particularly of birds. The Wildfowl Trust contains hundreds of species of birds and attracts thousands of visitors each year. In addition to his conservation work, he made numerous television appearances and wrote several books – including *Key to the Wild Fowl of the World* (1949), *Wild Geese and Eskimos* (1951) and *The Eye of the Wind* (1961) – and illustrated many others, most notably *The Snow Goose*, a novel by Paul Gallico, and *The Swans* (in collaboration with the Wildfowl Trust).

Seaborg Glenn Theodore 1912– is a US physical chemist who is best known for his researches on the synthetic transuranic elements. For this work he shared the 1951 Nobel Prize for Chemistry with his co-worker Edwin McMillan (1907–1991).

Seaborg was born in Ishpeming, Michigan, on 19 April 1912 into a Swedish immigrant family; his father was a machinist. When he was ten years old the family moved to Los Angeles, where he graduated from High School in 1929. He went to study literature at the University of California but changed to science and graduated in 1934. He then went to study at Berkeley under Gilbert Lewis, gaining his PhD in 1937 and spending a further two years as one of Lewis's research associates; he became an instructor in 1939. During part of World War II Seaborg was a section chief at the metallurgical laboratory at Chicago University, where much of the early work on the atomic bomb was carried out. After the war, in 1945, he was appointed Professor of Chemistry and Associate Director of the Radiation Laboratory at Berkeley, becoming Chancellor of the campus from 1958 until 1961. In that year he was made Chairman of the US Atomic Energy Commission and held the appointment for ten years. He returned to the Lawrence Berkeley Laboratory in 1971.

The transuranic elements are all those that lie beyond uranium in the periodic table, that is, all elements of atomic number higher than 92. They constitute the majority of the actinides (elements 89 to 103), so-called by analogy with the lanthanides or rare earths. They are all radioactive and none occurs to any appreciable extent in nature; they are synthesized by transmutation reactions. Of the 13 synthetic transuranic elements known in the early 1970s, Seaborg

was involved in the identification of nine: plutonium (atomic number 94), americium (95), curium (96), berkelium (97), californium (98), einsteinium (99), fermium (100), mendelevium (101) and nobelium (102).

Seaborg and his collaborators discovered plutonium in 1940 by bombarding uranium with deuterons in the Berkeley 152-cm/60-in cyclotron. The first isotope found had a mass of 238, and the more important (because it is fissionable) plutonium-239 was discovered in 1941 (by neutron bombardment of U-238). In 1944 helium bombardment of Pu-239 yielded Cm-242, the first isotope of curium. Americium, as Am-241, was identified by Seaborg and others at the Metallurgical Laboratory in 1944–1945. Helium bombardment of Am-241 at Berkeley produced berkelium (as Bk-249) at the end of 1949, and three months later the minute amount of Cm-242 available was also bombarded with helium to form californium-245. Einsteinium was identified in the debris from the 'Mike' nuclear explosion staged by the Los Alamos Scientific Laboratory in November 1952, where it arose from the radioactive decay of heavy uranium isotopes. Another decay product, fermium-255, was discovered in January 1953. Helium bombardment was again used in early 1955 to create mendelevium-256 out of Es-253. Nobelium, element 102, was discovered in spring 1957 at the Nobel Institute of Physics in Stockholm.

As Chairman of the Atomic Energy Commission, Seaborg encouraged the rapid growth of the US nuclear power industry. Many of the isotopes he discovered have also found other uses in industry and in medicine.

Secchi Angelo Pietro 1818–1878 was an Italian astronomer and physicist famous for his work on solar phenomena, stellar spectroscopy and spectral classification.

Secchi was born on 18 June 1818 in Reggio. At the age of 25 he joined the Society of Jesus and trained to become a Jesuit priest before becoming lecturer in physics and mathematics at the Collegio Romano in 1839. In 1848 he was driven into exile for being a Jesuit and went first to Stonyhurst College, England, then to Georgetown University in Washington, DC, where he continued his mathematical and scientific work. In 1849, he was appointed Director of the Gregonia University Observatory at the Collegio Romano and Professor of Astronomy. He was made a Fellow of the Royal Society in 1856. He died on 26 February 1878 in Rome.

While the Collegio Romano was in papal hands there was no lack of funds for the Observatory. There were plenty of assistants and good equipment. Secchi's memoirs testify to the variety of fields in which he researched, and his position gave him the facilities for gaining the widest publicity for his work. He wrote many papers in astronomy, magnetism and meteorology.

Secchi's interest in solar physics was aroused in America, where he assisted in the first experiments on the heat radiated at different locations on the Sun's disc. His interest in spectroscopy dates from a visit of Pierre Janssen (1824–1907) to Rome.

When Secchi returned to Rome in 1849 he equipped his new observatory with a Merz refractor and with this he carried out research into stellar spectroscopy, terrestrial magnetism and meteorology. With William Huggins (1824–1910), Secchi was the first person to adapt spectroscopy to astronomy in a systematic manner and he made the first spectroscopic survey of the heavens. He pointed out that stellar spectra differ from one another and that stars differ in other respects than brightness, position and colour. He proposed that the differences in stellar spectra reflected differences in chemical composition.

In 1867, Secchi suggested the establishment of spectral classes of stars and he divided the spectra he had studied into four groups. Data accumulated since Secchi's day have necessitated a considerably more complex division, but his classification led to schemes of stellar evolution. His groups were based on stars like Sirus, with strong hydrogen lines; stars similar to the Sun, with numerous fine spectral lines; stars of the Herculis type, with nebulous bands towards the red end of the spectrum; and carbon stars with bands in the violet end of the spectrum. The modern system of spectral classification was based on these four groups.

Secchi's other work included photographing the solar eclipse of 1860 in Spain and observing one in Sicily in 1870. In 1867, he demonstrated his universal meteorograph in the Paris Exhibition and gave lectures, some of which eventually formed the basis of his book on the Sun. He also proved that prominences are appendages of the Sun and he determined many features of their behaviour. He was an active observer of double stars and, with Warren De La Rue and William Cranch Bond, was among the first to use the new technique of photography for astronomical purposes. By 1859 he had a complete set of photographs of the Moon. Towards the end of his life, Secchi founded the Società degli Spettroscopisti Italiani, a society formed for the recording of daily spectroscopic observations of the Sun, mainly from various observatories in Italy.

Secchi achieved a great deal during his lifetime. Before outlining his stellar classification he examined 4,000 stars. He observed comets, meteors and planets, in particular Jupiter, Saturn and Mars. He published his findings in many volumes and he belonged to most scientific societies of the day. Much of our modern knowledge of astronomy has its roots in Secchi's findings.

Sedgwick Adam 1785–1873 was one of the founders of the English school of geology.

Son of the curate of Dent in northwest Yorkshire, Sedgwick attended Trinity College, Cambridge, where he studied mathematics. He became a Fellow of the College in 1810, and was ordained priest in 1818. Though supposedly knowing no geology, he was appointed Woodwardian professor in 1818, holding the chair until his death, 55 years later. He soon made himself, however, one of the most eminent British geologists. An energetic champion of fieldwork, Sedgwick explored such diverse districts as the Isle of Wight, Devon and Cornwall, the Lake District, and northeast England. In the 1830s, he unravelled the strati-

graphic sequence of fossil-bearing rocks in North Wales, naming the oldest of them the Cambrian period (now dated at 500–570 million years ago). In South Wales, his companion Murchison had concurrently developed the Silurian system. The question of where the boundary lay between the older Cambrian and the younger Silurian sparked a celebrated dispute that rumbled on for almost 40 years and was resolved only after their deaths, when in 1879 Charles Lapworth coined the term Ordovician for the middle ground. With Murchison, Sedgwick also identified the Devonian system in southwest England.

Sedgwick commended a highly Baconian approach to geological investigation, seeing facts founded on fieldwork as the bedrock of the science. He combined this with passionate Christian convictions. He deplored Hutton's and Lyell's uniformitarianism; he was suspicious of glacial theory; and completely rejected all theories of evolution, not least those of this former pupil and friend, Charles Darwin. Despite these blindspots, Sedgwick shone in two spheres. He was a supreme student of palaeontology, especially of Palaeozoic fossils; and he contributed greatly to understanding the stratigraphy of the British Isles, using fossils as an index of relative time, and assuming relatively distinct fauna and flora for each period.

Segrè Emilio 1905–1989 was an Italian physicist who shared the 1959 Nobel prize for discovering the antiproton (the antiparticle of the proton). He also discovered many new radioactive elements and pioneered the use of thermalized neutrons in experiments.

Segrè was born on 1 February 1905 in Tivoli near Rome. He went to the University of Rome to study engineering but switched to physics in his fourth year, having met Enrico Fermi, who had just arrived in Rome. Segrè's contemporaries in Fermi's group included Amaldi and Majorana. Segrè obtained his PhD in 1928 for research into the spectroscopy of lithium. At this time the Rome group were building up expertise in atomic and molecular physics – Segrè had been to Pieter Zeeman's laboratory in Amsterdam and Otto Stern's laboratory in Hamburg – but Fermi reckoned the most interesting physics would be found in the nucleus. In 1934, the Joliot-Curies proved him right with their discovery of artificial radioactivity using alpha-particle bombardment. Fermi predicted neutron bombardment would be even better and, as Segrè said, 'we made a discovery practically every week'. Segrè was appointed professor at the University of Palermo in 1936 and quickly he discovered a new element in some old parts from an accelerator at Berkeley that had been heavily radiated by deuterons. The parts were made from molybdenum (atomic number 42) and Segrè called the new element, which had atomic number 43, technetium (from the Greek for artificial). However in 1938, during a visit to Berkeley, Mussolini's anti-Semitic laws forced Segrè, a Jew, to remain in the USA. With the exception of wartime research at Los Alamos, Segrè remained at Berkeley for the rest of his career, and was appointed a professor in the physics department in 1947. In 1944 he became a US citizen.

Working with Dale Corson and Kenneth Mackenzie in 1940, he discovered another new element, now called astatine (atomic number 85). In December he again met up with Fermi, now at Columbia University in New York, to discuss using plutonium-239 instead of uranium-235 in atomic bombs. Segrè began working on the production of plutonium at Berkeley and then moved to Los Alamos to study the spontaneous fission of uranium and plutonium isotopes. In 1947, Segrè started work on proton–proton and proton–neutron interaction at the new 467-cm/184-in cyclotron accelerator at Berkeley, switching to the more energetic Bevatron machine in the early 1950s. The Bevatron accelerated protons to 6 billion electronvolts and when this proton beam struck a metal target inside the accelerator, it produced a negative beam composed mostly of pions, muons and electrons. Using time-of-flight techniques Segrè, Owen Chamberlain, Clyde Wiegand and Tom Ypsilantis, were able to detect the antiproton among these other particles. Antiparticles were a prediction of Paul Dirac's combination of special relativity and quantum mechanics. Although antielectrons (or positrons) had been observed a mere four years after their prediction by Dirac in 1928 – Carl Anderson detected them in cosmic rays in 1932 – physicists had been waiting more than 20 years for the antiproton. Segrè and Chamberlain's discovery, for which they shared the 1959 Nobel prize, confirmed that Dirac's relativistic quantum theory was correct. Segrè was also well-known for his editing and writing, including a biography of Fermi. He died in Berkeley on 22 April 1989.

Seguin Marc 1786–1875 was a French engineer who, in 1825, built the first successful suspension bridge in Europe using cables of iron wire. He also invented the tubular boiler.

Seguin was born in Annonay, France on 20 April 1786. He obtained his early education at a small boarding school in Paris, but was self-taught in engineering science. He had arrived in Paris at the age of 13 where his interest in engineering was stimulated by his close contact with Joseph Montgolfier, his great-uncle. Shortly after Montgolfier's death in 1810, Seguin returned to Annonay.

In 1825, in association with Henri Dufour, using wire cables, Seguin erected the first suspension bridge of its kind in Europe. This bridge was built at Geneva and, over the next 20 years, Seguin and his brothers erected other cable suspension bridges in France – beginning with the one over the river Rhône at Tournon in 1827.

Seguin, again with his brothers, tried to establish a steamboat service on the Rhône, and later turned his attention to railways. He was successful in establishing France's first modern railway between Lyon and St Etienne, completed in 1832. He discovered that the Stephenson steam engine then available was not capable of generating enough power for the high-speed operation he desired, so he invented a new type of boiler – the multitabular or fire-tube boiler.

By the time Seguin retired from active work in engineering in 1838, he had produced engineering projects which provided some of the earliest examples of large scale civil engineering in France. In recognition of his services he was elected corresponding member of the Académie des Sciences in 1845.

Besides practical engineering, Seguin showed a great interest in the problems involved with heat and light. He published his first statements on the subject, made in 1824 and 1825, in a Scottish journal. He argued that matter consisted of small, dense molecules constantly on the move in miniature solar systems and he maintained that magnetic, electrical and thermal phenomena were the result of their particular velocities and particular orbits. He identified heat as molecular velocity and explained the conversion to a mechanical effect by stating that this occurs when the molecules transmit their velocities to external objects. In 1839 he published his *De l'influence des chemins de fer* in which he rejected the calorific theory then dominant in France because it implied perpetual motion due to the supposed existence of heat as a fluid conserved in all processes. Seguin assumed that a certain amount of heat disappeared in the very act of the production of mechanical power and that the converse was equally true. He tried to determine the numerical relationship between heat and mechanical power and, in a table of results, he showed that the heat-loss as measured by a thermometer was not a true indication of the heat lost by the steam producing the power. However, he was unable to specify a relationship between temperature loss and loss of heat content. He could not, therefore, define a unit of heat or state its mechanical equivalent.

When James Joule succeeded in determining the mechanical equivalent of heat in 1847, Seguin supported his conclusions. Later Seguin attempted to claim priority over Joule, but it was decided that only in retrospect could the suggestions that Seguin made in 1839 be interpreted as a mechanical equivalent of heat. In 1853, Seguin published a weekly scientific magazine called *Cosmos*, and this became an important vehicle for the popularization of science in France. It also served as a forum for Seguin's theories including his particle theories for heat, light, electricity and magnetism.

Seki Kowa c. 1642–1708 was a Japanese mathematician who did much to change the role of mathematics in his society; from being an art form indulged in by intellectuals at leisure, he made it a science. To do this he not only created a basic social paradigm embodying scientific curiosity, he even found himself obliged to create a new mathematical notation system. Using his new techniques, Seki discovered many of the theorems and theories that were being – or were shortly to be – discovered in the West. It is likely, for example, that he derived the determinant (part of a method to solve linear equations) before Gottfried Leibniz (1646–1716) did in 1693. In his own country Seki is sometimes referred to as 'the sacred mathematician'.

Seki was born probably in Huzioka in about the year 1642 – the exact date is unknown, and even his birthplace is the subject of some doubt. His father's name was Nagaakira Utiyama, but he was adopted by the Seki family and was known either as Seki Kowa or Seki Takakazu. Nothing is known about his life, other than his efforts to popularize mathematics, except that the date of his death has been definitely established as 24 October 1708; he died in what is now called Tokyo.

Much of Seki's reputation stems from the social reform he introduced in order to develop the study of mathematics in Japan. By the time of his death, anyone could take an interest in the subject, and could teach others – although the books available for instruction were still couched in the formal (and possibly condescending) terms of the *literati* of society. Seki himself was much influenced by the book of Chu Shih-Chieh, which dealt with the solution of problems by transforming them into a one-variable algebraic equation. The challenge of the book was that it contained problems which the author declared to be unsolvable; Seki solved many of these using methods of his own devising. He introduced Chinese ideograms to represent unknowns and variables in equations, and although he was obliged to confine his work to equations up to the fifth degree – his Tenzen Zyutu algebraic alphabet was not suitable for general equations of the nth degree – he was able to create equations with literal coefficients of any degree and with several variables, and to solve simultaneous equations.

In this way he was able to derive the equivalent of $f(x)$, and thereby to arrive at the notion of a discriminant – a special function of the root of an equation expressible in terms of the coefficients.

Another of Seki's important contributions was the mathematically rigorous definition (rectification) of the circumference of a circle; he obtained a value for π that was correct to the 18th decimal place. Further work included the rectification of a circular arc and the curvature of a sphere. He established a theorem relating to the solid resulting from the revolving of a segment of a circle about a straight line that is in the same plane as the segment. (This theorem was substantially the same as the well-known theorem of Pappus.)

Seki is also credited with major discoveries in calculus. He developed a method of finding the approximate value of the root of a numerical equation and also evolved a method of determining the coefficients of an expression in the form

$$y = a_1 x + a_2 x^2 + \ldots + a_n x^n$$

which was similar to the method of finite difference.

Semenov Nikolay Nikolayevich 1896–1986 was a Soviet physical chemist who studied chemical chain reactions, particularly branched-chain reactions which can accelerate with explosive velocity. For his work in this area he shared the 1956 Nobel Prize for Chemistry (the first Soviet Nobel prizewinner) with the British physical chemist Cyril Hinshelwood.

Semenov was born on 3 April 1896 in Saratov, Russia. In 1913 he went to the University of Petrograd (now St Petersburg) and despite the turmoil of World War I and the Russian Revolution he graduated in 1917. During the next 25 years he held appointments at various research establishments in Leningrad (as Petrograd had become). From 1920 to 1931 he worked at the A.F.I or Physical–Technical Institute, becoming a professor in 1928. From 1931 he directed the Institute of Chemical Physics at the Soviet Academy of Sciences before moving to the Moscow State University in 1944, where he became Head of the Department of Chemical Kinetics.

In 1913, Max Bodenstein introduced the idea of a chain reaction to account for various gas reactions. Semenov developed this theory in the 1920s and showed how certain violently explosive reactions – particularly those involving combustion – can be explained in terms of branching chains: each branch in the reaction pathway starts more than one new reaction, thus rapidly accelerating the overall effect. He summarized his results in 1934 in his influential book *Chemical Kinetics and Chain Reactions* (English translation, 1935).

Semenov also played an important part in resisting narrow interpretations of Marxist–Leninism in its application to chemistry. In this way he helped to keep Soviet chemistry progressing and avoiding unprofitable detours such as that caused by Lysenkoism in biology.

Seyfert Carl Keenan 1911–1960 was a US astronomer and astrophysicist whose interests in photometry, the spectra of stars and galaxies, and the structure of the Milky Way resulted in the identification and study of the type of galaxy that now bears his name.

Seyfert was born on 11 February 1911 in Cleveland, Ohio. He graduated from Harvard with a BSc in 1933, gaining an MA two years later. His PhD in astronomy was awarded in 1936 while he was Parker Fellow at Harvard. Before joining the Mount Wilson Observatory as National Research Fellow in 1940, Seyfert worked at the McDonald Observatory in Chicago for four years, carrying out research on the spectra of stars in the Milky Way. He was Director of Barnard Observatory (1946–1951) and then he was appointed Professor of Astronomy at Vanderbilt University, where he became Director of the Arthur S. Dyer Observatory.

Apart from his research work in astronomy, Seyfert held a number of administrative posts, the most important of which was the civilian member of the National Defence Research Committee. He was a member of the International Astronomical Union and the American Astronomical Society and was a Fellow of the Royal Astronomical Society. He died in 1960.

In 1943, Seyfert was studying a series of 12 active spiral galaxies which possess barely perceivable arms and exceptionally bright nuclei. His investigations showed that these galaxies contain small, unusually bright nuclei, often bluish in colour, and have distinctive spectral lines denoting the emission of radio waves and infrared energy. The highly excited spectra of these galaxies showed that they contain hydrogen as well as ionized oxygen, nitrogen and neon. Sulphur, iron and argon were also common to such galaxies. On the basis of their spectra, Seyfert divided the galaxies into two types, I and II.

Radiation from the nuclei of the galaxies is due to the very hot gases that they contain at their centres. The gases are subject to explosions which cause them to move violently, with speeds of many thousands of kilometres per second relative to the centre of the galaxy in the case of type I, and of several hundreds of kilometres per second in the case of type II galaxies. Seyfert galaxies also emit a fairly large quantity of X-rays and differ from other active galaxies in that they exhibit substantial amounts of non-thermal emission.

Only a small percentage of galaxies show these and related phenomena. Seyfert's original list has been extended and these galaxies are still the subject of research. The most intensively studied are NGC 1068, 1275 and 4151. Their spectra are so rich, however, that it is not possible to construct any single model that will satisfactorily account for all the known characteristics.

In 1951, Seyfert began a study of the objects now known as Seyfert's Sextet – a group of diverse extragalactic objects, of which five are spiral nebulae and one an irregular cloud. One member of the group is moving away from the others at a velocity nearly five times that at which the others are receding from each other. Seyfert's original proposal, however, that the six were grouped together because of a chance meeting between objects at different distances is not now the accepted explanation.

Shannon Claude Elwood 1916–   is a US mathematical engineer, whose work on technical and engineering problems within the communications industry led him to fundamental considerations on the nature of information and its meaningful transmission. His mathematical theory to describe this process was sufficiently general for its applications to other areas of communication to be immediately appreciated. He is therefore regarded as one of the founders of information theory.

Shannon was born in Gaylord, Michigan, on 30 April 1916. He earned his bachelor's degree at the University of Michigan in 1936, and then continued his studies at the Massachusetts Institute of Technology (MIT). There he became a Bowles Fellow in 1939, and a year later was awarded both his master's degree and his doctorate in mathematics. Shannon then worked for a year as a National Research Fellow at Princeton University before becoming in 1941 a staff member at the Bell Telephone Laboratories. The work that he carried out during the 1940s led him to postulate a theory for communication, which he published in book form in 1949 (*The Mathematical Theory of Communication*). This theory brought him considerable acclaim, and he was presented with several notable awards and honours. In 1956 Shannon – still technically working for Bell Telephone – became Visiting Professor of Electronic Communications at the Massachusetts Institute of Technology; a year later he became Professor of Communications Science and Mathematics there; and in 1958 he finally and officially left Bell Telephone to become Donner

Professor of Science.

As early as 1938 Shannon was examining the question of a mathematical approach to language. At the laboratories of Bell Telephone he was given the task of determining which of the many methods of transmitting information was the most efficient, in order to enable the development of still more efficient methods. For this Shannon produced a model in which he reduced a communications system to its most simple form, so that it included only the most essential components. He also reduced the notion of information to a binary system of a series of yes/no choices, which could be presented by a 1/0 binary code. Each 1/0 choice, or piece of information, he called a 'bit' (now a technical term in talking of computers). In this way complex information could be organized according to strict mathematical principles.

An important feature of Shannon's theory was the prominence given to the concept of entropy, which he demonstrated to be equivalent to a shortage in the information content (a degree of uncertainty) in a message. One consequence of this work was the demonstration of the redundancy in most messages constructed using ordinary language; it became evident that many sentences could be significantly shortened without losing their meaning. His methods, although devised in the context of engineering and technology, were soon seen to have applications not only to computer design but to virtually every subject in which language was important, such as linguistics, psychology, cryptography and phonetics; further applications were possible in any area where the transmission of information in any form was important.

Shapley Harlow 1885–1972 was a US astronomer who made what Otto von Struve called 'the most significant single contribution toward our understanding of the physical characteristics of the very close double stars'.

Shapley was born on 2 November 1885 in Missouri, the son of a farmer. By the age of 16 he was working as a reporter on a newspaper in Kansas, having received a limited education. He then attended Carthage Presbyterian Collegiate Institute, graduating after two years with the intention of enrolling in the University of Missouri's School of Journalism. The School did not open for another year and Shapley, not wanting to waste time, took up astronomy. After graduating from a three-year course at Laws Observatory, he became a teaching assistant there and gained an MA a year later.

In 1911, Shapley moved to Princeton where he worked with Henry Russell (1877–1957) on eclipsing binary stars. Using a new method of computing and a polarizing photometer with a 58-cm/23-in refractor, Shapley obtained nearly 10,000 measurements of the sizes of stars in order to analyse some 90 eclipsing binaries. He also showed that Cepheid variable stars were pulsating single stars, not double stars.

In 1914, having completed his PhD thesis, Shapley moved to Mount Wilson Observatory. In 1921, he was appointed Director of the Hale Observatory at Harvard.

Under Shapley the Observatory became an important centre for astronomical research. He introduced a graduate programme whose alumni included Carl Seyfert (1911–1960), Jesse Greenstein (1909–) and Leo Goldberg (1913–1971).

Shapley continued as Director until 1952. He was active in retirement, being involved in the grants committee of the American Philosophical Association and undertaking a number of lecture tours. He was subpoenaed by the House of Representatives Committee on Unamerican Activities, having been named by Senator Joseph McCarthy in 1950 as one of five alleged Communists associated with the State Department. Shapley was exonerated, however, by the Senate Foreign Relations Committee.

Shapley received a number of honours during his career, including the Draper Medal, the Rumford Medal of the American Academy of Art and Science, the Gold Medal of the Royal Astronomical Society and the Pope Pius XI Prize. He played a part in the setting up of UNESCO and was one of the American representatives who participated in drafting its charter. He died on 20 October 1972 in Boulder, Colorado.

While he was at the Mount Wilson Observatory, Shapley began observations of light changes from the variable stars in globular clusters. His studies required a great deal of detailed work, gaining and collating information from these very remote stellar systems. The systems were spherical, containing a concentration of tens of thousands of stars. Shapley discovered many previously unknown Cepheid variables and he devised a method, based on the fact that brighter stars have longer cycles of light variation, to measure distances across space. For this relationship between cycle, period and luminosity to be useful, it was first necessary to determine the luminosity of one Cepheid. The great distances involved prevented Cepheids from being measured by direct trigonometric methods. Shapley devised a statistical procedure to establish the distance and luminosity of a Cepheid variable.

Shapley's research served to overthrow previous conceptions about the shape and size of the Milky Way. Jacobus Kapteyn (1851–1922) had argued that the Sun was at the centre of a flat stellar assemblage in which a high proportion of stars were within a boundary some 10,000 light years in diameter. Shapley proposed that Kapteyn's stellar assemblage was only a small part of a much larger galactic system, extending far beyond the visible stars to which Kapteyn had limited it. The centre of Shapley's system was a congregation of globular clusters some 60,000 light years away in the direction of the constellation of Sagittarius, and the whole system was said to have an equatorial diameter of about 300,000 light years.

Next, Shapley turned his attention to the debate about whether spiral nebulae were satellites of our Galaxy or independent stellar systems, similar to the Milky Way, but located well outside our galactic system. The luminosity of novae discovered in spiral nebulae could be used to measure their distances. The distance of 1,000,000 light years that Shapley proposed for the Andromeda Galaxy is close to the figure now accepted, although Shapley withdrew his results soon after publishing them.

In 1919, on Edward Pickering's death, Shapley was offered, but declined, the position of Director of the Hale Observatory at Harvard; he did, however, take up the position two years later. He encouraged completion of the Draper Catalogue, preferring its extension by Annie Cannon to the new, more sophisticated spectral catalogue constructed by Antonia Maury (1866–1952). Shapley's study of the Magellanic Clouds was also begun at Harvard. The Observatory had maintained a southern station in Peru, keeping photographic records that went back many years. Shapley was able to use these to revise his estimate of the distance of the Clouds to 100,000 light years. He also conducted a study of the giant emission nebula 30 Doraches and published the first photographs of the obscured cluster.

Shapley increasingly turned his attention to galaxies, carrying out surveys that recorded the presence of tens of thousands of them in both hemispheres. The surveys showed the irregular distribution of galaxies, a point Shapley used to refute the homogeneity necessary in Edwin Hubble's cosmological model. As a result of these surveys, Shapley also identified two dwarf systems in the constellations of Sculptor and Fornax.

Shapley's contributions to early twentieth-century astronomy are indisputable, especially with regard to galaxies and the structure of the universe. He can be considered one of the founders of modern cosmology.

Sharpey-Schafer Edward Albert 1850–1935 was a British physiologist and endocrinologist who discovered the effects of the hormone epinephrine, also known as adrenaline (although the actual hormone was not isolated until five years after his discovery). He received the Royal Medal of the Royal Society in 1902 and its Copley Medal in 1924.

He was born Edward Albert Schäfer in London on 2 June 1850, the son of a merchant. He went to University College, London, in 1871 and graduated in medicine three years later. His Professor of General Anatomy and Physiology was William Sharpey, who deeply impressed him by his skills. Schäfer became Assistant Professor when Sharpey retired in 1874, and eventually became Jodrell Professor at University College in 1883. In 1876, Schäfer was one of the founder members of the Physiological Society (he wrote a history of the Society in 1927). In 1899, he left University College to take the post of Professor of Physiology at Edinburgh University, which he held until his retirement in 1933. In 1913 he was knighted. He had named one of his sons after his mentor, Sharpey, but after both sons were killed in World War I he affixed Sharpey's name to his own and was thereafter known as Sharpey-Schafer. He died in North Berwick, Scotland, on 29 March 1935.

Sharpey-Schafer's most significant contribution to medical research occurred in 1894. He was working with George Oliver (1841–1915) and they discovered that an extract from the central part of an adrenal gland injected into the bloodstream of an animal caused a rise in blood pressure by vasoconstriction. They also noted that the smooth muscles of the animal's bronchi relaxed. These effects were caused by the action of the hormone adrenaline which is produced by the medulla of the adrenal gland; it was later isolated in 1901 by the Japanese–American chemist Jokichi Takamine (1854–1922).

Sharpey-Schafer also suspected that another hormone was produced by the islets of Langerhans in the pancreas. He adopted for it the name insulin (from the Latin for island), a name which eventually persisted, although the scientists who isolated it in 1922 at first called it 'isletin'.

In 1903, Sharpey-Schafer devised the classic position for artificial respiration, the supine position, which was adopted as standard by the Royal Life Saving Society. He was also an ardent supporter and fighter for equal opportunities for women in the world of medicine.

Shaw William Napier 1854–1945 was a British meteorologist who, in the late 1800s and early 1900s, did much to establish the then young science of meteorology. He is probably best known, however, for introducing in 1909 the millibar as the meteorological unit of atmospheric pressure (not used internationally until 1929) and for inventing the tephigram, a thermodynamic diagram widely used in meteorology, in about 1915. He received numerous honours for his work, including the 1910 Symons Gold Medal (the Royal Meteorological Society's highest award) and a knighthood in 1915.

Shaw was born in Birmingham on 4 March 1854, the third son of a manufacturing goldsmith and jeweller. He was educated at King Edward VI School, Birmingham, then won a scholarship to Emmanuel College, Cambridge, from which he graduated in 1876. In the following year he was elected a Fellow of Emmanuel College and was also appointed a lecturer in experimental physics at the Cavendish Laboratory (part of Cambridge University); he became Assistant Director of the Cavendish Laboratory in 1898. In 1900, he was appointed Secretary of the Meteorological Council which, because of the large workload involved, necessitated his resigning from the Cavendish Laboratory; he retained his fellowship until 1906, however. In 1905 he was made Director of the Meteorological Office, a post he held until his official retirement in 1920; under his directorship the Meteorological Office was transferred to the Air Ministry. From 1907 until 1920 he was also Reader in Meteorology then, on retiring in 1920, was appointed the first Professor of Meteorology at the Royal College of Science of the Imperial College of Science and Technology (part of London University), where he remained until 1924. In 1923 he married Sarah Dugdale, a lecturer at Newnham College, Cambridge. Even after retiring (for the second time), Shaw continued his meteorological writings. He died in London on 23 March 1945.

In addition to his introduction of the millibar and invention of the tephigram – both of which are still used in meteorology – Shaw made several other important contributions to the science. While at the Meteorological Office he pioneered the study of the upper atmosphere by using instruments carried by kites and high-altitude balloons. In 1906, working with R. Lempfert, he measured the rate of descent of air in two anticyclones (arriving at figures of 350 m/1,150 ft and 450 m/1,480 ft per day) and, in the case of one particular depression, calculated that 2 million million tonnes of air must have moved to account for the pressure drop. From these studies – described in *Life History of Surface Air Currents* (1906) – Shaw came near to proposing the polar front theory of cyclones later put forward by Jacob Bjerknes.

Again in collaboration with Lempfert, Shaw wrote *Weather Forecasting* (1911) and *The Air and its Ways* (1923), in which they described their work in determining the paths of air (by means of synoptic charts) in and around the North Atlantic pressure system. This work on pressure fronts formed the basis of a great deal of later work in the field. As a successor to *Forecasting Weather*, Shaw wrote the four-volume *Manual of Meteorology* (1926–1931), his most important book and still a valuable standard reference work.

Shaw also studied hygrometry, evaporation, and ventilation, and (with J. S. Owens), wrote *The Smoke Problem of Great Cities* (1925), an early work on atmospheric pollution.

Sherrington Charles Scott 1857–1952 was a British neurologist who is renowned for his research on the physiology of the nervous system, and his laboratories came to be regarded as the best in the world for teaching and research in neurophysiology. For his innovative work on the function of the neuron he was awarded the 1932 Nobel Prize for Physiology or Medicine, which he shared with Edgar Adrian.

Sherrington was born on 27 November 1857 in Islington, London. His father died when he was young, and his mother remarried. Sherrington's stepfather was Dr Caleb Rose, a classical scholar and archaeologist who influenced the young boy and interested him in medicine. He went to Ipswich Grammar School and then entered St Thomas's Hospital, London, in 1876. He interrupted his studies in 1880 to go to Cambridge as a non-collegiate student, where he became a demonstrator in the physiology department and a member of Gonville and Caius College. There he studied under the British physiologist Michael Foster (1836–1907).

In 1881, the International Medical Congress was held in London, and it was there that Sherrington was introduced to and became interested in experimental neurophysiology. He also met Ramón y Cajal whose interests were similar. The following year Sherrington went to Spain as a member of a research team to study a cholera outbreak. He gained his medical degree at Cambridge in 1885 and the next year qualified as a doctor and published his first paper, on the nervous system. In the same year he travelled to Italy to study cholera and then went to Berlin where he visited the pathologist Rudolf Virchow and Robert Koch. He returned to St Thomas's as a lecturer in physiology and in 1891 was appointed Professor-Superintendent of the Brown Institute, London University's veterinary hospital. He took up the Physiology professorship at Liverpool University in 1895, where he developed many of his original ideas on practical teaching. In 1913 he became Professor of Physiology at Oxford, a position he retained, although with interruptions, until 1935. During World War I he was heavily involved in government committees on the study of industrial fatigue and for three months he worked incognito as a labourer in a munitions factory. The observations he made there did much to improve safety for factory workers. He was elected President of the Royal Society in 1920 and was knighted two years later. Sherrington was also a poet and philosopher, and published his writings. He was made President of the British Association in 1922. He died in Eastbourne on 4 March 1952.

One of Sherrington's important findings, published in 1894, was that the nerve supply to muscles contains 25–50% sensory fibres, as well as motor fibres concerned with stimulating muscle contraction. The sensory fibres carry sensation to the brain so that it can determine, for example, the degree of tension in the muscles. His discovery helped to explain some of the disorders of the nervous system in which there is a deterioration in muscular coordination.

Sherrington then went on to study reflex actions and formulated theories of the way in which antagonistic muscles coordinate behaviour. He showed that reflex actions do not occur independently, as a result of reflex arcs, but in a movement integrated with the movement of other muscles (that is, when one set of muscles is activated, the opposing set is inhibited). This theory of reciprocal innervation is known as Sherrington's law.

Sherrington divided the sense organs into three groups: interoceptive, characterized by taste receptors; exteroceptive, such as receptors that detect sound, smell, light and touch; and proprioceptive, which involve the function of the synapse (Sherrington's word) and which respond to events inside the body. In 1906, he investigated the scratch reflex of a dog using an 'electric flea' and found that the reflex stimulated 19 muscles to beat rhythmically five times a second, and brought into action a further 17 muscles which kept the dog upright. The exteroceptive sensors initiated the order to scratch, and the proprioceptors initiated the muscles to keep the animal upright. Sherrington then removed the cerebrum of the dog and cut the epidermal tactile receptors and found that the proprioceptors still worked, against gravity, and activated the muscles to keep the dog upright.

Sherrington also plotted the motor areas of the cerebral cortex of the brain and identified the regions that govern movement and sensation in particular parts of the body. He experimented on the brain of a live gorilla, which caused an observer to comment that he did not know whether to admire most the skill or the courage of the experimenter.

In 1893, while Sherrington was in charge of the Brown Institute, he investigated diphtheria antitoxins. While experimenting for the first time on a horse (used to produce the antitoxin), an urgent message reached him that a young relative was desperately ill with diphtheria. He bled the horse, prepared the antitoxin, and on reaching the boy found that he had only a few hours to live. He injected the child with the antitoxin, and the boy recovered. It was the first use of diphtheria antitoxin in Britain.

Sherrington carried out significant work in the development of antitoxins, particularly those for cholera and diphtheria. In addition his observations of the nervous function in animals, described in *The Integrative Action of the Nervous System* (1906), greatly influenced modern neurophysiology, particularly brain surgery and the treatment of nervous disorders.

Shrapnel Henry 1761–1842 was a British artillery officer who invented the artillery shell which bears his name. He also invented the brass tangent slide and some types of fuses, compiled range tables, and improved the construction of mortars and howitzers.

Shrapnel was born at Bradford-on-Avon, Wiltshire, on 3 June 1761. He received a commission in the Royal Artillery in 1779, and in the following year he went to Newfoundland. In 1781 he was promoted to First Lieutenant, and on returning to England two years later he commenced his investigations into the problems connected with hollow spherical projectiles filled with bullets and bursting charges and with their discharge from light and heavy ordnance. He was promoted to captain and served in the Duke of York's unsuccessful campaign against the French in 1793, being wounded in the seige of Dunkirk.

Promotions followed fairly regularly and by 1804 he was regimental lieutenant-colonel. In that year also, he was appointed inspector of artillery at the Royal Arsenal at Woolwich and, while he was there, succeeded in perfecting many of his inventions connected with ordnance. Further promotion followed and in 1814 he became regimental colonel. By this time, Shrapnel had spent more than 30 years and several thousand pounds of his own money in perfecting his inventions. The treasury granted him a pension of £1,200 a year for life, but he was disappointed not to receive a baronetcy. In 1819 he was promoted to major-general and, in 1837, lieutenant-general. Shrapnel died at his home in Southampton on 13 March 1842.

When Shrapnel invented his new shell, he introduced a new name into artillery nomenclature. Although shells had been used for more than 400 years, Shrapnel's shell was different. It was fused and filled with musket balls – plus a small charge of black powder which was just sufficient to explode the container after a predetermined period of time. When the fuse acted, the container – or shell-case – was blown open and the musket-balls it had contained scattered in all directions, causing great damage to anyone or anything with which they came into contact. The Duke of Wellington reported the success of Shrapnel's case-shot and wrote telling him of the performance of his invention against the enemy at Vimiera in 1808. Other generals, in other actions, also acclaimed the new weapon.

Although the first shells used as containers for the musket-balls (or shrapnel as they came to be known) were round, later they were of an elongated form with added velocity. Shrapnel's shells continued to be used right up to World War I. They proved to be especially effective against large infantry units on open ground. They were less effective, however, against dispersed or protected personnel.

After World War I there came a period of rapid development and improvement in the design of artillery ammunition. Technological advances in metallurgy, chemistry and electronics led to the Shrapnel shell being replaced by more powerful projectiles containing bursting charges of TNT, Armatol, Explosive D and RDX. As a result it was no longer necessary to pack the shell with steel fragments – the disintegration of the shell case itself provided the same effect. Because the final effect was still achieved in the way Shrapnel had envisaged – by the blast of the explosion and the spraying of steel fragments – the term 'shrapnel' continued to be used, although erroneously, so perpetuating the association of Shrapnel's name.

Sidgewick Nevil Vincent 1873–1952 was a British theoretical chemist best known for his contributions to the theory of valency and chemical bonding.

Sidgewick was born in Oxford on 8 May 1873 into a talented family. His father and two of his uncles were faculty members at the universities of Oxford and Cambridge, and another uncle was Archbishop of Canterbury. He was educated at home until he was 12 years old, when he went to Rugby School to study classics and science. His application for a classical scholarship to Oxford in 1891 was unsuccessful, but the following year he was offered a scholarship in natural sciences to Christ Church College. His tutor, the physical chemist Vernon Harcourt, was one of the first people to study reaction kinetics in physical chemistry. Sidgewick graduated in natural sciences in 1895 and went on to perform the extraordinary feat of graduating also in classics two years later. After a year as a laboratory demonstrator, he went to Germany in 1899 to study under Georg Bredig in Wilhelm Ostwald's department at Leipzig. He returned to Britain for a while to recover from an illness, then went back to Germany for two years to work under von Pechmann in Tübingen. He was awarded his doctorate in 1901 and became a Fellow at Lincoln College, Oxford. He remained at Oxford for the rest of his life, becoming a Reader in 1924 and a supernumerary Professor of Chemistry in 1935, although he did travel abroad frequently in the 1920s and 1930s. He died in Oxford on 15 March 1952.

Sidgewick did little significant work before 1920, spending much of his time teaching and writing his successful and readable book *The Organic Chemistry of Nitrogen* (1910). On a sea voyage to attend a meeting of the British Association in Australia in 1914 he travelled with Ernest Rutherford, and the two scientists forged a lifelong friendship. Sidgewick became absorbed by the study of atomic structure and its importance in chemical bonding, although this work was interrupted by World War I, during which he acted as an unpaid consultant to the Department of Explosive Supplies.

After the war Sidgewick's productivity increased. He extended Gilbert Lewis's ideas on electron sharing to explain the bonding in coordination compounds (complexes) then being studied by Alfred Werner, with a convincing account of the significance of the dative bond. Together with his students he demonstrated the existence and wide-ranging importance of the hydrogen bond. He summarized this stage of his work in *The Electronic Theory of Valency* (1927).

In 1931 Sidgewick made his first visit to the United States as Baker nonresident lecturer in chemistry at Cornell University. He assimilated the new advances in theoretical chemistry such as Erwin Schrödinger's wave mechanics and Werner Heisenberg's uncertainty principle. He also took notice of the new techniques for the determination of physical forces between atoms and the structures of molecules. These advances were surveyed in his 1933 book *Some Physical Properties of the Covalent Link in Chemistry*.

World War II sharply reduced the extent of Sidgewick's overseas travel and the amount of academic research being carried out throughout the world. This permitted him to catch up on the vast amount of literature that had been published in the 1930s and to produce another monumental, definitive two-volume work, *The Chemical Elements and their Compounds*, which was published in 1950. Once more he demonstrated his ability to consider and systematize the diverse work of other people and to provide an insight into a broad subject area for the benefit of scholars and students.

Siemens Ernst Werner von 1816–1892 was the German electrical engineer who discovered the dynamo principle and who organized the construction of the Indo-European telegraph system between London and Calcutta via Berlin, Odessa and Teheran.

Siemens was born on 13 December 1816 at Lenthe, near Hanover. In 1832, he entered the Gymnasium in Lubeck. Three years later he became an officer cadet at the artillery and engineering school in Berlin, and from 1835 to 1838 studied mathematics, physics and chemistry.

As a serving officer he continued his studies and, in his spare time, he made many practical scientific inventions. He invented a process for gold- and silver-plating and a method for providing the wire in a telegraph system with a seamless insulation using gutta-percha. Other inventions included the ozone tube, an alcohol meter, and an electric standard or resistance based on mercury.

In 1847, Siemens founded with Johan Halske, the firm of Siemens–Halske to manufacture and construct telegraph systems. The company was responsible for constructing extensive systems in Germany and Russia. In 1870, the firm laid the London–Calcutta telegraph line and later became involved in underwater cable telegraphy. In Britain, Siemens became scientific consultant to the British government and helped to design the first cable-laying ship, the *Faraday*.

Valuing the contribution science was already making to

technological advancement, Siemens helped to establish scientific standards of measurement and was mainly responsible for establishing the Physickalische-Technische Reichsastalt in Berlin in 1887. He was also cofounder of the Physical Society. His contributions to science were rewarded with a honorary doctorate from the University of Berlin in 1860. He was elected member of the Berlin Academy of Science in 1873 and became a German nobleman in 1888. In 1889, Siemens retired from active involvement with his firm. He died on 6 December 1892 in Charlottenberg.

Siemens's genius for invention was developed on a very wide scale through the firm he established in Germany with Halske and also through the firm of Siemens Brothers which had been established in England. In 1846 he succeeded in improving the Wheatstone telegraph, making it self-acting by using 'make-and-break' contacts. He subsequently developed an entire telegraph system which included the seamless insulation of the wire. The firm obtained government contracts to provide extensive telegraph networks in Germany. But because of disagreements these Prussian contracts were cancelled in 1850, so Siemens went to Russia and established an extensive telegraph network there – one which included the line used during the Crimean War.

His greatest single achievement was the discovery of the dynamo principle. Siemens announced his discovery to the Berlin Academy of Science in 1867. He had already introduced the double-T aramature and had succeeded in connecting the armature, the electromagnetic field and the external load of an electric generator in a single current. This enabled manufacturers to dispense with the very costly permanent magnets previously used. Unlike other workers in the field, including Wheatstone and Varley, Siemens forsaw the use of his dynamo in machines involving heavy currents. This enabled his companies to become pioneers in the development of electric traction in such applications as streetcars and mini-locomotives and also electricity generating stations.

Again unlike some inventor-engineers of his time, Siemens valued the contribution that science could make to practical engineering advancement. He advocated that technique should be based on scientific theory. He often published analyses of his telegraph and cable-laying technology in reports to the Berlin Academy of Science. He also maintained that a nation, in times of harsh international competition, would never maintain its status in the world if it did not base its technology on continuing research work.

Sikorsky Igor 1889–1972 was a Russian-born US aeronautical engineer, one of the great pioneers of aircraft design. He built the first multi-engined aeroplane, was the designer of a famous series of large passenger flying-boats, and built the first practical helicopter.

Sikorsky was born in Kiev, Russia (now in Ukraine) on the 25 May 1889. His father was Professor of Psychology at Kiev University. He was brought up in a cultured family atmosphere and developed an interest in art and a particular fondness for the life and works of Leonardo da Vinci. During his studies of Leonardo he came across the well-known design for a helicopter. Early in his life he developed an interest in model aeroplanes, and when he was 12 he built a small rubber-powered helicopter which could actually fly. In 1903, at the age of 14, he entered the Russian Naval Academy for a career as an officer in the navy. Three years later he resigned because he wanted to devote his time to practical mechanical pursuits. He went first to Paris, where he studied briefly, and then entered the Kiev Polytechnic Institute. After only a year he left these studies because he wished to be a practical engineer, and he found the gap between the theoretical studies of physics and practical engineering too wide. His family was rich and he was not short of money; he equipped his own experimental workshop and attempted to develop his ideas.

In 1908, he spent the summer in France and came into contact with Wilbur Wright (1867–1912) and the new interest in aviation which his French trip had stimulated. He returned to Kiev, believing that he could build a helicopter with a horizontal rotor which would rise straight up into the air, and show that fixed wings were not necessary for flight. The following January he went to Paris to buy a suitable engine for his new machine, and in May 1909 he began to construct his first helicopter. Structural difficulties and the weight of the engine led him to begin work on a second, improved machine. This helicopter did not fly either, and he abandoned his attempts until materials and engines became available which would make his designs practicable.

He decided to concentrate his efforts on fixed-wing aeroplanes and soon built his first biplane, the S1, with a 11-kW/15-hp engine. It was underpowered, but with a larger engine his S2 made a short flight. With these early designs he began the practice of taking the controls on the first flight, which he was to continue throughout his career as a designer and builder of aeroplanes. The same year, 1910, two more designs were built. In 1911 his S5 aeroplane with a 37-kW/50-hp engine flew for more than an hour and achieved altitudes of 450 m/1,480 ft. His cross-country flights enabled him to obtain an International Pilot's Licence (No. 64). By now he had achieved recognition and his next aeroplanes had military applications: the S6 was offered to the army. His famous aeroplanes *Le Grand* and the even larger *Ilia Mourometz* had four engines, upholstered seats, an enclosed cabin for crew and passengers and even a toilet. They became the basis for the four-engined bomber that Russia used during World War I.

After the war and the Russian Revolution, Sikorsky emigrated to the United States but found it difficult to gain a foothold in American aviation. After a number of years as a teacher and lecturer he founded the Sikorsky Aero Engineering Corporation on a farm on Long Island, and when this company showed promise it was taken over by the United Aircraft Corporation. This arrangement allowed Sikorsky a lot of freedom and gave him the money to build new aeroplanes. In 1929 he produced the twin-engined S38

Amphibian, and by 1931 the S40 American Clipper was in production. This large flying-boat allowed Pan American Airways to develop routes in the Carribean and South America. In 1937 the even larger S42 Clipper III was built. But with the coming of World War II and the demise of the flying-boat as the most popular method of passenger transport, Sikorsky again took up the idea of a helicopter.

1939 saw the construction of the VS300 helicopter. The new materials and expertise made his design practical, and on the 14 September 1939 Sikorsky piloted his helicopter a few feet into the air; the machine had a 4-cylinder 56-kW/75-hp air-cooled engine, driving a 8.2-m/28-ft three-bladed rotor. In May 1941 an endurance record of more than 90 minutes was set by a Sikorsky helicopter, and in 1943 the R3, the world's first production helicopter, was flown. There quickly followed a whole series of production designs using one, then two, piston engines. With their usefulness for rescue and transport work in inaccessible or densely populated places helicopters came into their own and were designed for specific purposes. Later Sikorsky models had distinguished military service in Korea. During the late 1950s piston engines were replaced by the newly developed gas turbine engines as the source of power, and helicopters were built for even more rugged duties.

In 1957, Sikorsky retired as engineering manager from his company. During his lifetime he had received many honours. He married in 1924 and had been proud to become a US citizen in 1928. After a lifetime devoted to aviation he died at Easton, Connecticut, on the 26 October 1972.

Simon Franz Eugen 1893–1956 was a German-born British physicist who developed methods of achieving extremely low temperatures and who also established the validity of the third law of thermodynamics.

Simon was born in Berlin on 2 July 1893. In 1903, he entered the Kaiser Friedrich Reform Gymnasium, Berlin, and received a classical education. However, Simon's interest in science enabled him in 1912 to enter the University of Munich to read physics, chemistry and mathematics. He spent a year in Munich under Arnold Sommerfeld (1868–1951) and then a term at Göttingen before he was called up for military service in 1913. Simon was wounded twice and became one of the first poison gas casualties, but was awarded the Iron Cross first class. In 1919, he resumed his studies at Berlin, obtaining his DPhil in 1921 under Hermann Nernst (1864–1941). In 1922, Simon became Assistant to Nernst at the Physical Chemical Institute of the University of Berlin and remained there for ten years, becoming a privatdozent (unpaid lecturer) in 1924 and Associate Professor in 1927.

In 1931, Simon took over the Chair of Physical Chemistry at the Technical University in Breslau. With the rise to power of Hitler in 1933, Simon foresaw trouble because he was Jewish. He resigned his post in that year, and accepted an invitation from Frederick Lindemann (Lord Cherwell; 1886–1957) to work at the Clarendon Laboratory, Oxford. In 1936, Simon became a Reader in Thermodynamics

there. He then decided to stay in Britain and became a British citizen in 1938. During World War II, Simon worked on the atomic bomb project, being concerned with the separation of uranium isotopes by gaseous diffusion.

After the war, Simon became Professor of Thermodynamics at Oxford in 1945 and held this post until 1956, when he was appointed Dr Lee's Professor of Experimental Philosophy at Oxford, succeeding Lindemann. He held this position for only a few weeks and then died at Oxford on 31 October 1956. Among the honours Simon received were the Rumford Medal of the Royal Society in 1948 and a knighthood in 1955.

Simon's first scientific work was his PhD thesis under Nernst on the study of specific heats at low temperatures. While he was at the Physical Chemistry Institute at Berlin University, he built up the low temperature department. He worked on the solidification of helium and other gases by the use of high pressure, and discovered the specific heat anomaly in solid orthohydrogen. In 1930, he installed a new hydrogen liquefier into his laboratory to his own design and this design was adopted by many other laboratories. In 1932, Simon worked out a method for generating liquid helium by single-stroke adiabatic expansion, and this too came to be used widely for its simplicity and cheapness.

From 1933, when Simon went to Britain, his work was concerned with magnetic cooling and investigations below 1K (−272°C/−458°F), the properties of liquid helium and specific heats. He made a major contribution to physics by showing that the third law of thermodynamics is obeyed at very low temperatures. This law was proposed by Nernst in 1905 and it states that the entropy of a substance approaches zero as its temperature approaches absolute zero. There were subsequently doubts that this is a universal law applicable in all cases, but Simon's low temperature work dispelled them and the expression of the third law in this form is largely due to his efforts.

Simon also did a lot of work on helium. Helium is the only substance which, under its vapour pressure, remains liquid down to absolute zero. Liquid helium is unique too in that at low temperatures, its internal energy is smaller than that of the solid in equilibrium with it. Simon succeeded in establishing the helium vapour pressure scale below 1.7K (− 271.5°C/−456.6°F) and heat transport in liquid helium below 1K. He worked on the properties of fluids at high pressure and low temperature, and showed that helium could be solidified at a temperature ten times as high as its liquid/gas critical point. He also worked on heat conductivity in connection with radiation damage.

In the magnetic field, Simon is most noted for his lead in using magnetic cooling to open up a new range of very low temperatures for a wide variety of physical experiments. In Britain in the 1930s, he developed adiabatic demagnetization to investigate properties of substances below 1K. He then went on to investigate nuclear cooling, showing that the cooling effect is limited by interaction energies and that if, instead of a paramagnetic salt, a substance is chosen whose paramagnetism is due to nuclear spins, then lower temperatures can be reached as the interaction energies for nuclear magnetic moments are smaller than for electron moments. In this way, Simon finally achieved temperatures nearly one millionth of a degree above absolute zero.

Simpson George Clark 1878–1965 was a British meteorologist whose numerous contributions to meteorology included studies of atmospheric electricity and of the effect of radiation on the polar ice, and standardization of the Beaufort Scale of wind speed. Among the many honours he received for his work were the Symons Gold Medal from the Royal Meteorological Society in 1930 and a knighthood in 1935.

Simpson was born in Derby on 2 September 1878, the son of a successful tradesman. In about 1887 he was sent to the Diocesan School, Derby, a small school (with a limited curriculum) for educating the sons of tradesmen and artisans. He left school in 1894 and entered his father's business. After reading popular science books, however, he became interested in optics and began to attend evening classes. On his father's suggestion, he then decided to try to further his education at Owens College, Manchester (later Manchester University) and, after private coaching, passed the entrance examination and began studying at the college in 1897. Because of his lack of higher education, he was initially advised to study for an ordinary degree but, at the instigation of Arthur Schuster (1851–1934), changed to an honours degree course. He gained a first class honours degree in 1900 and became an unsalaried tutor at Owens College. In 1902, he won a travelling scholarship and went to Göttingen University to study under Emil Wiechert (1861–1928), after which he visited Lapland to investigate atmospheric electricity before returning to England. In 1905 Schuster set up a small meteorology department at Manchester University (as Owens College had become) and placed Simpson in charge; this position was the first lectureship in meteorology at a British University. In the same year Simpson also accepted an invitation to assist William Shaw, the newly appointed Director of the British Meteorological Office. While holding these posts Simpson travelled widely, spending a period (from 1906) with the Indian Meteorological Office inspecting meteorological stations throughout India and Burma, travelling to the Antarctic in 1910 as a meteorologist on Robert Scott's last expedition, and visiting Mesopotamia (from 1916) as a meteorological adviser to the British Expeditionary Force. In 1920, when in Egypt as a member of the Egyptian government's Nile Project Commission, he was summoned back to England to succeed Shaw as Director of the Meteorological Office in London, a post he held until officially retiring in 1938. In the following year, however, with the outbreak of World War II, he returned from retirement to take charge of Kew Observatory, continuing research into the electrical structure of thunderstorms until 1947. Simpson spent his last years in Westbury-on-Trym, near Bristol, and died in Bristol on 1 January 1965.

Simpson's early work – carried out while at Owens College – concerned magnetism and electricity. At Göttingen University, however, his interest turned to meteorology and he demonstrated that the Earth's permanent negative charge cannot be maintained by the absorption of negative ions from the atmosphere. Continuing this line of research in Lapland, he measured dissipation of atmospheric electricity and ionization and radioactivity in the atmosphere; he described the results of these studies in *Atmospheric Electricity in High Altitudes* (1905). He also investigated ionization, radioactivity and potential gradients while on the Antarctic expedition.

When Simpson became Shaw's assistant at the Meteorological Office, he began working on the Beaufort Scale. This scale, which was devised in 1805 by the British Admiral and hydrographer Francis Beaufort (1774–1857), described and classified wind force at sea but made no reference to actual wind speeds (as originally formulated it was based on the effect of the wind on a fully rigged man o'-war sailing ship). Simpson devised a revised form of the Beaufort Scale in which the numbers of the original scale were assigned wind speeds measured by a freely exposed anemometer at a height of 11 m/36 ft above ground level. In 1921 he was asked by the International Meteorological Committee to reconcile international differences between sets of Beaufort Scale equivalent wind speeds, and in 1926 the Committee accepted the scale Simpson himself had devised earlier. In 1939, however, the Committee adopted a new scale based on wind speed measurements from an anemometer at 6 m/20 ft above ground level, although for some time the United States and British weather services continued to use Simpson's 11-m/36-ft elevation scale.

As Director of the Meteorological Office, Simpson investigated the effects of solar radiation on the polar ice caps. He concluded that excessive solar radiation would increase the amount of cloud and that the resultant increase in precipitation would lead to enlargement of the ice caps. He then investigated the possibility that this conclusion might explain the ice ages. In a series of papers on this subject he showed that, in glaciated regions, the initial effect of increased radiation is greater precipitation (in the form of snow) which, in turn, leads to enlargement of the ice sheet, followed by recession of the ice as the radiation increases further. Then, when the radiation decreases, the resultant drop in temperature causes the precipitation to fall as snow again, which leads to a second advance of the ice.

Simpson George Gaylord 1902–1984 was a US palaeontologist who studied the evolution of mammals and applied population genetics to the subject and to analyse the migrations of animals between continents.

Simpson was born on 16 June 1902 in Chicago, Illinois, the son of a lawyer. He attended the University of Colorado and after graduation went to Yale University, where he gained his PhD in 1926 with a thesis on Mesozoic mammals. A year later he took a post in New York City at the American Museum of Natural History, where he remained for 32 years, continuing his palaeontological research and becoming curator in 1942. He took a professorial appointment at Columbia University in 1945, and from 1959 to

1970 he was Alexander Agassiz Professor of Vertebrate Palaeontology at the Museum of Comparative Zoology, Harvard. He went to Tucson in 1967 to take up an appointment as Professor of Geosciences at the University of Arizona.

Simpson's chief work in the 1930s concerned early mammals of the Mesozoic, Palaeocene and Eocene, which entailed many extensive field trips throughout the Americas and to Asia to study fossil remains. This led him to consider the taxonomy of mammals, and in the 1940s he began applying genetics to mammalian evolution and classification. Much of his work was summarized in a series of textbooks, including *Tempo and Mode in Evolution* (1944), *The Meaning of Evolution* (1949), *The Major Features of Evolution* (1953) and *The Principles of Animal Taxonomy* (1961), which were influential in establishing the neo-Darwinian theory of evolution.

Simpson James Young 1811–1870 was a British obstetrician and one of the founders of gynaecology, who pioneered the use of chloroform as an anaesthetic.

Simpson was born at Bathgate near Linlithgow, Scotland, on 7 June 1811, the son of a village baker. He was a brilliant pupil at school and at the age of only 14 he went to Edinburgh University to study medicine, and graduated in 1832 to become assistant to one of the university professors on the merit of his exceptional thesis. He became Professor of Midwifery in 1840 at the age of 29, and seven years later his skill as an obstetrician was acknowledged when he was requested to attend Queen Victoria during her stays in Scotland. By this time he had a thriving private practice and was making pioneering advances in modern gynaecology; he was eventually appointed physician to Queen Victoria. He was made a baronet in 1866, and died in London on 6 May 1870.

Although Simpson made great advances in gynaecology, his most famous work was in the field of anaesthesia. In 1846, the American dentist William Morton had successfully extracted a tooth painlessly by using ether as an anaesthetic. Simpson was impressed and began experimenting himself, but he was not particularly successful with ether, although he did use it on a patient in childbirth in early 1847. He then heard of the work of the French physiologist Jean Flourens (1794–1867), who was experimenting with chloroform on animals, and of the successful use of chloroform in surgery by Robert Liston at University College Hospital, London. In November 1847 Simpson introduced the use of chloroform in his practice, particularly to relieve the pain of childbirth. He described his cases in *Account of a New Anaesthetic Agent* (1847). This caused a storm of opposition from Calvinists, who regarded labour pains as God-given and that to relieve them was heresy. It was not until royal intervention in 1853 that the controversies died down. Queen Victoria, who never pretended that pregnancy and childbirth were anything but loathsome, accepted the use of chloroform during the birth of Prince Leopold, her seventh child. She described the new drug as 'miraculous' and her praises of

Simpson knew no bounds. Thus criticism of his techniques abated, and they were soon universally adopted.

Simpson Thomas 1710–1761 was a British mathematician and writer who, after a somewhat erratic start in life, contributed greatly to the development of mathematics in the eighteenth century. He is particularly remembered for Simpson's rule, which simplifies the calculation of areas under graphic curves.

Born at Market Bosworth, Leicestershire, on 20 August 1710, Simpson was the son of a weaver. Uninterested in his studies, he left home and lodged at the house of a widow in Nuneaton, whom he married in 1730 (although she was twenty years older than he was). A few years later, following an eclipse of the Sun, Simpson became obsessed by astrology and gained a reputation in the locality for divination. But after he had apparently frightened a girl into having fits by 'raising a devil' from her, he was obliged to flee with his wife to Derby. In 1735 or 1736 he moved to London and worked as a weaver at Spitalfields, teaching mathematics in his spare time. It was there that he published his first mathematical works, which created a better sort of reputation and even won some acclaim. Soon after 1740 he was elected to the Royal Academy of Stockholm. And in 1743 he was appointed Professor of Mathematics at the Royal Academy at Woolwich, largely through the interest of William Jones (1675–1749). Finally, on 5 December 1745 he was elected a Fellow of the Royal Society. From then on he was a constant contributor to *The Ladies' Diary*, acting as its editor from 1754 to 1760. He died in Market Bosworth on 14 May 1761, and was buried at Sutton Cheynell, Leicestershire. (Mrs Simpson survived him and received a pension from the Crown after her husband's death, until she herself died in 1782 – at the age of 102.)

Simpson's first mathematical work, in 1737, was to study Edmund Stone's translation of L'Hôpital's *Analyse des Infiniements Petits*, and from it to write a new treatise on 'fluxions' (calculus). This was an important contribution to the subject – although it also showed up the defects in the mathematical training that Simpson had received. In 1740 he wrote *The Nature and Laws of Chance*, and some essays on several subjects in speculative and general mathematics providing a solution to Kepler's problem. In 1742 *The Doctrine of Annuities and Reversions* appeared; in 1743, *Mathematical Dissertation on a Variety of Physical and Analytical Subjects. A Treatise of Algebra* followed in 1745, and in its appendix contained some extremely ingenious solutions to algebraic problems. Ingenuity also distinguished his *Elements of Geometry*, published in 1747. His next works were *Trigonometry, Plane and Spherical* (1748) and *Select Exercises in Mathematics* (1752). In a paper he produced in 1755, Simpson proved that the arithmetic mean of n repeated measurements (which was already in limited use) was preferable to a single measurement in a precisely specifiable case. His final major publication was *Miscellaneous Tracts on Some Curious Subjects in Mechanics, Physical Astronomy and Special Mathematics* (1757). He also contributed several papers to the *Transactions* of the Faraday Society, most of which have been republished.

Simpson is most famous for having devised a method for determining approximately the area under a curve, known as Simpson's rule. The method is to join the extremities of the ordinates by parabolic segments. In a general parabola whose axis is parallel to the *y*-axis between two coordinates (for example) $x = -h$ and $x = h$ each side of zero, and whose equation is (for example):

$$y = ax^2 + bx + c$$

Simpson found that the approximate area under the curve between those ordinates is:

$$\int_{-h}^{h} (ax^2 + bx + c)\, dx = [\tfrac{1}{3}ax^3 + \tfrac{1}{2}bx^2 + cx]_{h^{-1}}^{h}$$

which he then reduced to:

$$\tfrac{1}{3}h\,[y_3 + 4y_2]$$

This result gives the area under a quadratic curve exactly, and gives an approximation to the area under a curve which is not parabolic. Greater accuracy can be obtained by dividing the range of integration into a larger number of intervals and successively applying the rule for three ordinates.

On another occasion, Simpson worked out a way to calculate the volume of a prismoid, a solid bounded by any number of planes, two of which are parallel and contain all the vertices. The two parallel faces are called the bases. He said that the volume:

$$V = \tfrac{1}{6}h\,(B + B' + 4M)$$

where M is a section made by a plane parallel to the bases B and B', and midway between them, and h is the distance between the bases. This formula can be used to find the volume of any solid bounded by a ruled surface and two parallel planes.

Singer Isaac Merrit 1811–1875 was a US inventor. His name will always be associated with the domestic sewing machine, which became such a feature of 'every good home' in the late 1800s.

Singer was born in Pittstown, New York, on 27 October 1811. He began his career simply enough as an apprentice machinist, and during his early working life patented a rock-drilling machine and, later, a metal- and wood-carving one. It was while he was employed in a machine shop in Boston that his main chance arrived. One day he was asked to carry out some repairs to a Lerow and Blodgett sewing machine. Singer not only did that, but at the same time decided he could add many improvements to the design. Eleven days later he produced a new model which he patented under his own name.

Litigation against his patent followed from Elias Howe, another maker, who claimed Singer had used his patented stitch method. Although Howe eventually won the case,

Singer was already in production with his machine, having formed the I. M. Singer Company in 1851 and, as a measure of his machine's success, his business went from strength to strength. Singer had formed a partnership with an Edward Clark in the June of 1851, and by 1860 the company was the largest sewing machine manufacturers in the world. In 1863, Singer and Clark formed the Singer Manufacturing Company and Singer retired to England where he settled in Torquay, Devon.

Sewing as a means of constructing clothes dates back to prehistoric times when bone needles were used to stitch skins together. From that, people progressed to using an awl with which to first make a hole, followed by a type of crochet needle for the stitch. The use of a steel sewing needle appears to have originated in the sixteenth century and hand sewing continued as the only method of stitching up to the end of the eighteenth century, when inventive minds turned their attention to this domestic chore.

Barthélemy Thimonnier of France is credited with devising the first practical sewing machine as a means of speeding up the production of army uniforms in 1841. By 1845, Elias Howe in the United States had designed his answer, the first lockstitch machine, using a threaded needle and shuttle (bobbin). When Singer patented his improved machine in 1851, he used the best of Howe's design and altered some of the other features. The basic mechanism was the same, however, as the handle turned, the needle paused at a certain point in its stroke so that the shuttle could pass through the loop formed in the cotton. When the needle continued the threads were tightened, forming a secure stitch.

Singer's machines were very popular (and still are) and his marketing, aimed at the ordinary family – he introduced the first instalment plan payment system – was such a success that by 1869 more than 110,000 sewing machines had been produced for the US market alone.

Skinner Burrhus Frederic 1904–1990 was a US psychologist famous for his staunch advocacy of behaviourism, which attempts to explain human behaviour solely in terms of observable responses to external stimuli. He is also well known for his controversial ideas about the relationship between individuals and society, for inventing the Skinner box and the teaching machine, and for developing programmed learning.

Skinner was born on 20 March 1904 in Susquehanna, Pennsylvania, and was educated at Harvard University. After obtaining his doctorate in 1931 he remained at the university as an instructor until 1936, when he moved to the University of Minnesota, Minneapolis – initially as an instructor, then as assistant professor from 1937 to 1939. During World War II he was an associate professor in a research programme for the United States Office of Scientific Research and Development. In 1945, after the war, he was appointed Professor of Psychology at Indiana University, Bloomington, then in 1948 he returned to Harvard as Professor of Psychology, becoming Edgar Pierce Professor of Psychology there in 1958.

Skinner's best known research work concerns operant conditioning which, in general, involves influencing voluntary behaviour patterns by means of rewards or punishments or a combination of both. In his research, Skinner took a firm behaviouristic standpoint, believing that behaviour can be studied properly only by objective experimentation and observation of reactions to definable stimuli, and that all subjective phenomena should be discounted. Although many psychologists consider Skinner's ideas to be rather extreme, he succeeded in bringing a considerable degree of methodological rigidity to psychological experimentation.

In the field of operant conditioning, Skinner conducted many highly original experiments, mainly using pigeons. For example, during World War II he trained pigeons to pilot bombs and torpedoes, although the pigeons were never actually used as missile guides; and later, at Harvard, he taught pigeons to play table tennis. In the course of his work on training animals he developed the Skinner box which, in its basic form, comprises a box with a lever-operated food delivery device inside; when the experimental animal, a rat for example, presses the lever, a pellet of food is delivered. More sophisticated versions of the Skinner box have since been developed and have proved extremely useful in studying the behaviour of a wide variety of animals.

The step-by-step training of experimental animals led Skinner to develop teaching machines and the associated concept of programmed learning. Similar in many respects to the Skinner box, a teaching machine presents information to a student at a pace determined by the student himself, and then tests the student on the material previously presented; correct answers are 'rewarded,' thereby reinforcing learning.

Skinner first gained public attention, however. with his invention in the mid-1940s of the Air-Crib, a large, air-conditioned, soundproof box intended to serve as a mechanical baby minder and designed to provide the optimum environment for child growth during the first two years of life. But Skinner aroused the greatest controversy with *Walden Two* (1948), a fictional description of a modern utopia, and *Beyond Freedom and Dignity* (1971), a non-fiction work in which, using the results of modern psychological research to support his case, Skinner presents his ideas for the improvement of society. The central theme in both of these books is essentially the same: an ideal society can be attained and maintained only if human behaviour is modified – by means of such techniques as conditioning – to fit society instead of society adapting to the needs of individuals.

Skolem Thoralf Albert 1887–1963 was a Norwegian mathematician who did important work on Diophantine equations and who helped to provide the axiomatic foundations for set theory in logic.

Skolem was born at Sandsvaer on 23 May 1887 and was educated at the University of Oslo, where he studied mathematics, physics and life sciences. He graduated with the highest distinction in 1913, having written a dissertation on the algebra of logic. By that time he had for four years been working as an assistant to Otto Birkeland, with whom he published his first papers, on the subject of the 'northern lights'. In 1918 he was made an Assistant Professor of Mathematics at Oslo, eight years before he finally submitted, in 1926, at the age of 40, a doctoral thesis. From 1930 to 1938 Skolem was able to conduct his research, without teaching duties, as a fellow of the Chr Michaelsons Institute in Bergen. In 1938 he was appointed Professor of Mathematics at Oslo, where he lectured on algebra and number theory to the virtual exclusion of mathematical logic, his area of specialization. He retired in 1957, but continued to do research until his death at Oslo, just before a projected lecture tour of the United States, on 23 March 1963.

Skolem was a retiring man, devoted almost entirely to mathematical research. He wrote 182 scientific papers and, unusually for a mathematician, most of them were written after he had reached the age of 40. It was not, indeed, until he was 33 that he published his first papers in mathematical logic, but those papers of 1920 elevated him at once to a leading place in the field. To the chagrin of his foreign colleagues Skolem published chiefly in Norwegian journals (a number of which he edited) and which they found inaccessible; lectures given on his frequent visits to the United States went some way to break down this barrier.

Skolem's early work was in the highly abstruse field of formal mathematical logic. From papers published in the 1920s emerged what is now known as the Löwenheim–Skolem theorem, one consequence of which is Skolem's paradox. It takes the following form. If an axiomatic system (such as Ernst Zermelo's axiomatic set theory, which intends to generate arithmetic, including the natural numbers, as part of set theory) is consistent (that is, satisfiable), then it must be satisfiable within a countable domain; but Georg Cantor had shown the existence of a neverending sequence of transfinite powers in mathematics (that is, uncountability). How to resolve this paradox? Skolem's answer was that there *is* no complete axiomatization of mathematics. Certain concepts must be interpreted only relatively; they can have no 'absolute' meaning.

In this work Skolem was ahead of his time, so much so that in the 1930s he had to take pains to summarize his work of a decade earlier in order to bring it to the attention of mathematicians. As his papers remained largely unread (partly because they were written in Norwegian), he became somewhat dispirited and turned away from mathematical logic to more conventional topics in algebra and number theory. In particular, he began to work on Diophantine equations, publishing what was for many years the definitive text on the subject.

Nevertheless, his main field remained the logical foundations of mathematics. Before such subjects as model theory, recursive function theory and axiomatic set theory had become separate branches of mathematics, he introduced a number of the fundamental notions which gave rise to them.

Slipher Vesto Melvin 1875–1969 was a US astronomer whose important work in spectroscopy increased our knowledge of the universe and paved the way for some of the most important results obtained in more recent astrophysics.

Slipher was born on 11 November 1875 in Mulberry, Indiana. He attended Indiana University and then in 1902, shortly after graduating, joined the Lowell Observatory in Arizona at the request of Percival Lowell (1855–1916). There he began the spectroscopic analyses that led to important conclusions about planetary and nebular rotation, planetary and stellar atmospheres, and diffuse and spiral nebulae.

Slipher's academic and administrative positions show the range of his achievements. He received an MA in 1903 and a PhD in 1909 from the University of Indiana, and he was awarded honorary degrees from a number of universities. He became acting Director of Lowell Observatory in 1916 and was Director from 1926 until his retirement in 1952. He instigated the search that resulted in the discovery of the planet Pluto by Clyde Tombaugh (1906–   ) in 1930. Slipher was active in the International Astronomical Union and the American Association for the Advancement of Science, and he was a member of a number of other astronomical and scientific societies. He died in Flagstaff, Arizona, on 8 November 1969.

Slipher studied Venus, Mars and Jupiter. The lack of surface detail on Venus made calculation of the rotation period difficult. Slipher's method was to measure changes in the inclination of the spectral lines while keeping a spectrograph perpendicular to the terminator. The 26 photographs that he took gave a result that was close to the figure now generally accepted on the basis of more modern methods of computation. In the years following Slipher also published measurements of the period of rotation for Mars, Jupiter, Saturn and Uranus, and in 1933 he was awarded the Royal Astronomical Society's Gold Medal for his work on planetary spectroscopy.

Slipher was responsible for a number of planetary discoveries. His work on Jupiter first showed the existence of bands in the planet's spectrum, and he and his colleagues were able to identify the bands as belonging to metallic elements, including iron and copper. He also showed that the diffuse nebula of the Pleiades had a spectrum similar to that of the stars surrounding it and concluded that the nebula's brightness was the result of light reflected from the stars.

Another of Slipher's discoveries was instrumental to work done by Edwin Hubble (1889–1953), Ejnar Hertzsprung (1873–1967) and others on emission and absorption nebulae. This work depended on Slipher's recognition of the existence of particles of matter in interstellar space. His discovery of a non-oscillating calcium line in the spectra of various celestial objects showed that there was gas between the stars and the Earth.

Slipher's most significant contribution to astronomy concerned spiral nebulae. His investigations paved the way for an understanding of the motion of galaxies and for cosmological theories that explained the expansion of the universe. While Hubble must be given the credit for formulating the relationship between velocity and distance in interstellar space, his work used results gained by Slipher in his research into spiral nebulae. In 1912 Slipher gained a set of spectrographs that showed that the Andromeda spiral nebula was approaching the Sun at a velocity of 300 km/985 mi per second. He continued his observations of this and other nebulae, looking at the Doppler shifts for 14 spirals. By 1925 Slipher's catalogue included measurements of the radial velocities of nearly all the 44 known spirals. His results suggested that spirals were external to our Galaxy as their radial velocities could not be contained within the Milky Way system. His work influenced not only Heber Curtis (1872–1942) and Hubble, who each put forward an account of the nature of the phenomenon, but also other astronomers who were interested in discovering the relationship between velocity and distance in interstellar space. Slipher's contribution to this field of study was outstanding and fundamental.

Smeaton John 1724–1792 was a British civil engineer who rebuilt the Eddystone Lighthouse in 1759, but who was also greatly influential in directing the scientific research which was being carried out in the mid-eighteenth century. It was he who first adopted the term 'civil engineer' in contra-distinction to the fast-growing number of military engineers graduating from the Military Colleges.

Smeaton was born of Scottish ancestry in Austhorpe, near Leeds. He was encouraged to practise law, and after a good elementary education he served in his father's firm of solicitors. Later he went to London for further training in the Courts of Justice, but his natural inclination for mechanical science led him to leave law and become a maker of scientific instruments.

He soon introduced many technical innovations – one of which was a novel instrument with which he was able to measure and study the expansion characteristics of various materials. From 1756 to 1759 he was engaged in the rebuilding of the Eddystone Lighthouse. He was also a consultant in the field of structural engineering, and from 1757 onwards he was responsible for many engineering projects including bridges, power stations operated by water or wind, steam engines, and river and harbour facilities.

He was a charter member of the first professional engineering society founded in 1771, the Society of Civil Engineers which, after his death, became known as the Smeatonian Society. He was a Fellow of the Royal Society and, in 1759, received its Copley Medal. He died on 28 October 1792, at Austhorpe.

Although Smeaton's best known achievement was the rebuilding of Eddystone, his main contribution to engineering was his innovative ability to combine engineering with applied science. His work on waterwheels and windmills served to underline the importance of scientific research to practical engineering problems. It was his own research work that led him to question the relative efficiency of the then firmly established undershot waterwheel (which operates through the action of the flow of water against blades in the wheel) and the overshot wheel (which is operated by water moving the wheel by the force of its weight).

Experimenting with models, Smeaton showed that overshot wheels were twice as efficient as undershot ones. He went further and speculated on the cause of this difference in efficiency. From his experiments, he concluded that the loss of 'mechanic power' in the undershot wheel was caused by turbulence, which he described as the loss of power by water and other non-elastic bodies in changing their 'figure' in consequence of their 'stroke'. Thus, only part of their original power is communicated when acting by impulse or collision.

In 1759, Smeaton presented his important paper to the Royal Society, 'An experimental enquiry concerning the natural power of water and wind to turn mills and other machines depending on a circular motion'. This paper was followed by two others, one on the necessary mechanical power to be employed in giving different degrees of velocity to heavy bodies from a state of rest, and the other on some 'fundamental experiments upon the collision of bodies'.

Smeaton's work, with its emphasis on scientific investigation into practical engineering problems, provided one of the first examples of the interdependence of engineering and applied science. This led to other designers adopting his approach. (One early result of this was that the undershot waterwheel was abandoned as uneconomical.) It also lent a sense of urgency to the recurrent controversy raging at the time over the measure of force, in the discussions of which Smeaton's own research findings played a prominent role.

Later on in life Smeaton performed extensive tests on the experimental steam engine of Thomas Newcomen (1663–1729). These tests led to significant improvements in its design and efficiency.

Smith Francis Graham 1923– is a British astronomer, one of the leaders in radioastronomy since its earliest postwar days.

Smith was born in Roehampton, Surrey, on 25 April 1923. He studied at Epsom College before entering Downing College, Cambridge, in 1941 to read natural sciences. His undergraduate studies were interrupted in 1943, when he was assigned to the Telecommunications Research Establishment at Malvern as a Scientific Officer. In 1946, Smith returned to Cambridge to complete his degree, and he then became a research student in the radio research department at the Cavendish Laboratories. He was awarded the 1851 Exhibition Scholarship in 1951 and received a PhD the following year.

In 1952, Smith went to the Carnegie Institute in Washington, DC, where he spent a year as a research fellow before returning to Cambridge. He was appointed Professor of Astronomy at the University of Manchester in 1964 and then moved to Jodrell Bank, where he worked under Bernard Lovell for the next ten years. In 1974, Smith was made Director-designate of the Royal Greenwich Observa-

tory, and in 1975 he was appointed Visiting Professor of Astronomy at the University of Sussex. He became full Director of the Royal Greenwich Observatory in 1976. In 1981, he moved back to Jodrell Bank to become Director there. In 1982, Smith was appointed Astronomer Royal.

As Director of the Royal Greenwich Observatory, Smith's interests were divided between the running of the Observatory itself and the supervision of the early stages of the Northern Hemisphere Observatory. He was active in the choice of site (Las Palmas in the Canary Islands), the specificiations of the new observatory and in setting up the team to run it. He also organized the equipping of the site with new telescopes. The Observatory is a genuinely international venture and is dedicated to both optical and radio research.

In addition to his many academic papers, Smith wrote successful books on radioastronomy. He is a Fellow of the Royal Society and a Fellow of the Royal Astronomical Society. He served the latter body as Secretary (1964–1970) and as President (1975–1977).

The disruptive effect on Smith's education of his assignment to the Telecommunications Research Establishment in 1943 was more than compensated for by the valuable opportunity it afforded him of learning the sophisticated techniques used in radar research. These methods were to be most useful in his later work in radioastronomy. His earliest experiments at the Cavendish Laboratories were conducted with a small group of scientists that included Martin Ryle (1918–1984). Their initial interest lay in studying radio waves emanating from the Sun, using the radio interferometer that had been designed by Ryle. Methods evolved during this and other projects were soon applied to the study of other sources of radio waves.

Other early radioastronomers – most notably Karl Jansky (1905–1950), Grote Reber (1911–), and James Hey (1909–) – had established that there were powerful localized sources of radio waves in the sky. In 1948, Smith and Ryle set out to investigate the source that had been found in the constellation of Cygnus. They set up a radio interferometer and recorded oscillations in receiver output that indicated that they had found two sources of radio waves. One of these was the Cygnus source. Smith analysed the timing and duration of the second signal in order to work out its position. This source lay in the constellation Cassiopeia.

The interferometer that Smith and Ryle used to obtain this important result had not been aligned with great accuracy; so Smith set himself the task of determining the precise location of both sources. He used interferometers that were more correctly aligned and devised methods of calibrating the interferometers more accurately. It was not until 1951 that he began to seek the assistance of an experienced optical astronomer. He approached D. Dewhirst at the Cambridge Observatory and asked him to attempt to correlate the radio location with an observable optical feature. Dewhirst gave, as a tentative identification, a faint nebulous structure in Cassiopeia. He was not able to provide any details about the nebula.

Smith then wrote to Jan Oort (1900–1992), who put him in touch with Walter Baade (1893–1960) and Rudolph Minkowski (1895–1976) at Mount Palomar. These two astronomers had at their disposal the most powerful optical instrument in the world, the Hale telescope, and they were able to pinpoint optical counterparts. Cassiopeia A, as the source is now known, was shown to derive from a Type II supernova explosion within our Galaxy; and the Cygnus A is a double radio galaxy.

These discoveries provided a powerful impetus to the development of radioastronomy. They presented a valuable new method for observing strong signals from sources that would otherwise have been invisible and inaccessible to study. In essence, this opened up a whole new dimension to the universe.

During the 1950s Smith participated in a systematic search for radio sources, which culminated in 1959 in the publication of the 3C catalogue. This still provides the standard system of nomenclature for the field. Smith and Ryle were the first to publish (in 1957) a paper on the possibility of devising an accurate navigational system that depended on the use of radio signals from an orbiting satellite. This was proposed in the wake of the first Sputnik satellite.

During the early 1960s Smith was active in early scientific experiments that used artificial satellites. In 1962 he installed a radio receiver in *Aeriel II*, one of a series of joint US–UK satellites. It was able to make the first investigation of radio noise above the ionosphere.

At Jodrell Bank from 1964 to 1974 Smith studied radio waves from our Galaxy and from pulsars. His most important discoveries were the strongly polarized nature of radiation from pulsars (1968), an estimate of the strength of the magnetic field in interstellar space, and a theory of the mechanism of radiation in pulsars (1970). This theory is known as the theory of 'relativistic beaming'. Current opinion in the field of pulsar research is divided between supporting this theory and the 'polar cap' theory.

Smith's contributions to radioastronomy have been to the experimental, theoretical and administrative aspects of the field. He has seen the subject grow from a hardly recognized discipline to an integral part of modern astronomy.

Smith Hamilton Othanel 1931–   . US microbiologist; *see* Nathans and Smith.

Smith William 1769–1839 commonly known as 'Strata Smith' is widely viewed as the founder of English geology.

Born at Churchill, Oxfordshire, to a farming family, Smith received little formal education and became a drainage expert, a canal surveyor and a mining prospector – occupations affording him abundant opportunity to examine the varying landforms and outcrops of much of England and Wales. In 1791, he moved to the coal-mining district of Somerset, and became an expert in the construction of canals. In 1794, he undertook a six weeks' tour by postchaise to the north of England which permitted him to see the rocks over wide stretches of terrain. During the 1790s

his geological ideas matured, and by 1799 he was able to set out a detailed list of the secondary strata of England. This led him to the construction of geological maps, beginning with the Bath area. Smith was not the first geologist to recognize the principles of stratigraphy, or the usefulness of type fossils. The real measure of his achievement was less theoretical than practical, for it was he who actually determined in greater particularity than previously the succession of English strata, across the whole country, from the Carboniferous up to the Cretaceous. He also established their fossil specimens. Beyond this, his primary accomplishment lay in mapping. Smith ingeniously viewed the map as the perfect medium for presenting stratigraphical knowledge. In developing a map type showing outcrops in block, he set the essential pattern for geological cartography throughout the nineteenth century.

Always slow to publish, in 1815 he brought out, after much delay, *A Delineation of the Strata of England and Wales*, a geological map using a scale of five miles to the inch. Between 1816 and 1824 he published *Strata Identified by Organized Fossils*, which displayed the fossils characteristic of each formation, and his *Stratigraphical System of Organized Fossils* a descriptive catalogue. He also issued various charts and sections, and geological maps of 21 counties.

Smith's relations with the elite British geological community remained ambiguous and often tense. Ever in financial difficulties, Smith felt scorned and snubbed. In 1831, however, his work was belatedly recognized, when the Geological Society of London made him the award of the first Wollaston Medal.

Snell Willebrord 1580–1626 was a Dutch physicist who discovered the law of refraction. He also founded the method of determining distances by triangulation.

Snell was born in Leiden in 1580. His father was Professor of Mathematics at the University of Leiden, and Snell studied law there. From 1600 to 1604 he travelled throughout Europe, working with and meeting such scientists as Tycho Brahe (1546–1601) and Johannes Kepler (1571–1630). In 1613, Snell succeeded to his father's position at the University of Leiden and there taught mathematics, physics and optics. He died at Leiden on 30 October 1626.

Snell developed the method of triangulation in 1615, starting with his house and the spires of nearby churches as reference points. He used a large quadrant over 2 m/7 ft long to determine angles, and by building up a network of triangles, was able to obtain a value for the distance between two towns on the same meridian. From this, Snell made an accurate determination of the radius of the Earth.

Snell is best known for Snell's law of refraction in optics, which states that the ratio of the sines of the angles of the incident and refracted rays to the normal is a constant. This constant is now called the refractive index of the two media involved. Snell discovered this law after much experimental work in 1621, and he formulated it in terms of the lengths of the paths traversed by the rays rather than the sines of the angles. Snell did not publish his law, and it first appeared in *Dioptrique* (1637) by René Descartes (1596–1650). There Descartes expressed the law using the sines of the angles, which could easily be derived from Snell's original formulation. Whether Descartes knew of Snell's work or discovered the law independently is not known.

Snyder Solomon Halbert 1938– is a US pharmacologist and neuroscientist who has studied the chemistry of the brain, and co-discovered the receptor mechanism for the body's own opiates, the encephalins.

Snyder was born in Washington, DC, on 26 December 1938 and graduated in medicine from Georgetown University in 1962. He joined the National Institutes of Health in Bethesda as a research associate, and in 1965 moved to the Johns Hopkins Medical School, becoming Professor of Psychiatry and Pharmacology in 1970, and Distinguished Service Professor in 1977. His doctoral research under the supervision of the Nobel Laureate Julius Axelrod (1912–) was the study of neurotransmitters, naturally occuring chemicals that are involved in the working of the nervous system, and the effects of drugs on them. In the early 1970s, in collaboration with his research student Candace Pert (1946–), Snyder realized that the very specific effects of synthetic opiates given in small doses suggested that they must bind to highly selective target receptor sites. Using radioactively labelled compounds they located such receptors in specialized areas of the mammalian brain and from this finding arose the suggestion that there might therefore be natural opiate-like substances in the brain that used these sites. These chemicals, the encephalins were discovered by Hans Kosterlitz (1903–) and John Hughes (1942–) shortly afterwards. In 1978, Snyder, Hughes and Kosterlitz were awarded the Lasker award for this work. Snyder continues to examine the relationships of chemicals to neural functioning and has made a particular study of the naturally occuring receptor sites for the benzodiazepine drugs that are widely used in psychiatry.

Soddy Frederick 1877–1956 was a British chemist who was responsible for major advances in the early developments of radiochemistry, being mainly concerned with radioactive decay and the study of isotopes. For this work he was awarded the 1921 Nobel Prize for Chemistry. He was also a controversial character, holding firm views – with which very few people agreed – about the relationship between science and society.

Soddy was born in Eastbourne, Sussex, on 2 September 1877, the youngest of seven children. He attended Eastbourne College and became much influenced by his chemistry teacher R. E. Hughes, with whom he published his first scientific paper in 1894 (at the age of only 17). He went to the University College of Wales at Aberystwyth for a year after leaving school, winning an open scholarship to Merton College, Oxford, in 1895. He graduated with top honours three years later; William Ramsay was his external examiner. He spent two years doing research at Oxford, but achieved little of note.

Then in 1900, at the age of 23, he applied for but was refused the Professorship of Chemistry at the University of Toronto in Canada. He followed this up with a personal visit, which did little to promote his case, and visited Montreal on his way back to Britain. There he was offered a junior demonstrator's post at McGill University, in Ernest Rutherford's department. Soddy accepted and formed a fruitful partnership with Rutherford.

Soddy returned to London in 1902 and worked with Ramsay at University College. In 1904, he went on a brief tour of Australia as an extension lecturer for London University and on his return took up an appointment as a lecturer in physical chemistry at the University of Aberdeen, where he developed the theory of isotopes. In 1914, he was promoted to the Chair in Chemistry, finally achieving the professorship he had striven for since 1900. During World War I he was involved in research aimed at contributing to the war effort.

Then in 1919 he was appointed Dr Lees Professor of Chemistry at Oxford, in the hope that he would build up an active research group in the field of radiochemistry. Soddy was instrumental in modernizing the laboratories and active in teaching, but he did little further original research. His interests turned increasingly to political and economic theory and, although he wrote prolifically on these subjects, he was unable to raise the interest or enthusiasm of others, particularly the university authorities. He retired early, in 1936, soon after the death of his wife, which affected him deeply. He travelled in Asia for a while, visiting thorium mines. During and after World War II he became increasingly concerned with how atomic energy was being put to use (as early as 1906 he had realized the tremendous potential in the energy locked up in uranium), and tried to arouse a more active sense of social responsibility among his fellow scientists to halt what he saw as a dangerous trend in the development of human society. He died in Brighton on 22 September 1956.

Soddy's first major scientific contribution, the disintegration law, was the result of his work with Rutherford in Montreal. They postulated that radioactive decay is an atomic or subatomic process, a theory that was immediately accepted. They proposed that there are two radioactive decay series beginning with uranium and thorium and both ending in lead, in which a parent radioactive element breaks down into a daughter element by emitting either an alpha particle or a beta particle. Soon a third series, beginning with actinium, was also demonstrated; it too ends in lead. (A fourth series beginning with neptunium and ending with bismuth was not discovered until after World War II.)

Soddy and Rutherford also predicted that helium should be a decay product of radium, a fact that Soddy and Ramsay proved spectrographically in 1903. In 1911, Soddy published his alpha-ray rule, which states that the emission of an alpha particle from an element results in a reduction of two in the atomic number (Russel's beta-ray rule holds that the emission of a beta particle causes an increase of one in atomic number). The displacement law, introduced by Soddy in 1913, combines these rules and explains the changes in atomic mass and atomic number for all the radioactive intermediates in the decay processes.

Also in 1913 Soddy and Theodore Richards independently demonstrated the occurrence of different forms of lead in minerals from different sources. These could be added to the plethora of chemically inseparable 'elements' which displayed different radioactive properties – there were far more new elements than there were available places in the periodic table. Then Soddy brought order to chaos by proposing that the inseparable elements are in fact identical substances (in the chemical sense), differing only in atomic weight (relative atomic mass) but having the same atomic number. He named the multiple forms isotopes, meaning *same place* because they occupied the same place in the periodic table. It is now known that all the elements above bismuth (atomic number 83) have at least one radioactive isotope, as do many lighter elements (such as phosphorus). The existence of isotopes also explained anomalies in atomic weight determinations, which were often found to be caused by the existence of isotopes in elements that were neither radioactive nor formed by radioactive decay.

Soddy was a scientist of great foresight; he predicted the use of isotopes in geological dating and the possibility of harnessing the energy of radioactive nuclei. He was capable of thorough experimentation and dramatic interpretation of the results, having the courage to propose unifying hypotheses. The change in interest that overtook him in middle life was a consequence of what he regarded as the disturbing events that were taking place in the world around him.

Solvay Ernest 1838–1922 was a Belgian industrial chemist who invented the ammonia-soda process, also known as the Solvay process, for making the alkali sodium carbonate.

Solvay was born in Rebecq-Rognon, near Brussels, on 16 April 1838, the son of a salt refiner. He was not a healthy child and had little formal education; by his late teens he was working as a bookkeeper for his father. In his spare time he carried out chemical experiments in a small home laboratory. In 1860 he went to Schaarbeek to work for an uncle who directed a gasworks, and there learned about the industrial handling of ammonia both as a gas and as an aqueous solution. Within a year he had discovered and patented the reactions that are the basis of the Solvay process. Trial production at a small plant in Schaarbeek failed; for two years Solvay knew he had the chemistry right, but could not solve the considerable problems of chemical engineering (problems that had nearly bankrupted several other industrial chemists earlier in the century). With the help of his brother Alfred he raised the necessary capital to build a full-scale works at Couillet, which was opened in 1863. By the summer of 1866 the process was well established, and a second factory was opened at Dombasle in 1873.

Solvay was as much a businessman as a chemist and he soon realized that there was more money to be made from granting licences to other manufacturers than there was in making soda. (One of the licensees was the Brunner Mond Company in Cheshire, England, whose alkali division was later to become part of Imperial Chemical Industries.) Throughout the world, the Solvay process replaced the old Leblanc process, which required more energy (heat) and produced obnoxious waste materials, releasing huge quantities of hydrogen chloride into the atmosphere. Towards the end of the nineteenth century the price of soda fell dramatically. Solvay became a very rich man and entered politics, becoming a member of the Belgian Senate and a Minister of State. He endowed many educational institutions throughout Belgium. During World War I he helped to organize food distribution. He lived to be 84 years old and died in Brussels on 26 May 1922.

The Solvay process uses as raw materials sodium chloride (common salt), calcium carbonate (limestone) and heat energy; ammonia is also used as a carrier of carbon dioxide, but is theoretically not consumed by the process. First the limestone is heated to yield calcium oxide (lime) and carbon dioxide:

$$CaCO_3 \rightarrow CaO + CO_2$$

The ammonia is dissolved in a solution of sodium chloride (brine), and the ammoniacal brine allowed to trickle down a tower against an upflow of carbon dioxide. The products of the resulting reaction are ammonium chloride (which stays in solution) and sodium bicarbonate (sodium hydrogen carbonate), which forms a precipitate:

$$NH_3(aq) + CO_2(g) + NaCl(aq) + H_2O \rightarrow NaHCO_3(s) + NH_4Cl(aq)$$

Finally the sodium bicarbonate is filtered off and heated to yield sodium carbonate:

$$2NaHCO_3 \rightarrow Na_2CO_3 + CO_2 + H_2O$$

Ammonia is recovered from the filtrate by reacting it with the calcium oxide from the heated limestone, producing calcium chloride as the only waste product. The key technical development is the use of countercurrent carbonating towers, which are usually used in series to get maximum yield.

Sommerfeld Arnold Johannes Wilhelm 1868–1951 was a German physicist who made an important contribution to the development of the quantum theory of atomic structure. He was also a gifted teacher of science.

Sommerfeld was born in Königsberg, Prussia (now Kaliningrad in Russia) on 5 December 1868. After attending the local Gymnasium, he entered Königsberg University in 1886 and opted for mathematics. He received his PhD in 1891 and then obtained an assistantship at the Mineralogical Institute in Göttingen from 1893 to 1894. In 1895, he became a privatdozent (unpaid lecturer) in mathematics at Göttingen University and in 1897, was appointed Professor of Mathematics at the Mining Academy in Clausthal. In 1900, Sommerfeld became Professor of Technical Mechanics at the Technical Institute in Aachen and in 1906 moved to Munich University as Director of the Institute of Theoretical Physics, which was established there for him. Sommerfeld built his institute into a leading centre of physics and attracted many gifted scientists as students, notably Peter Debye (1884–1966), Wolfgang Pauli (1900–1958), Werner Heisenberg (1901–1976) and Hans Bethe (1906–   ). It was also under Sommerfeld's direction that Max von Laue and his colleagues made the famous discovery of X-ray diffraction in 1912. Sommerfeld also helped to spread and advance physics by publishing several important books that summed up current knowledge in several fields. The most influential of these were *Atombau und Spektrallinien/Atomic Structure and Spectral Lines* (1919) and *Wellenmechanischer Ergänzungsband/Wave Mechanics* (1929).

Sommerfeld remained at Munich until 1940, when he retired. He courageously defended Albert Einstein (1879–1955) and other Jewish scientists in the Weimar and Nazi periods when antisemitism was prevalent in Germany. Sommerfeld returned to his post at the Institute of Theoretical Physics in Munich after World War II. He died in Munich on 26 April 1951 after being struck by a car.

Sommerfeld was active as a theoretician in several fields both in physics and engineering, and he produced a four-volume work on the theory of gyroscopes from 1897 to 1910 in association with Felix Klein (1849–1925). However, his principal contribution to science was in the development of quantum theory. Sommerfeld promoted the theory as a fundamental law of nature, and inspired Niels Bohr (1885–1962) to apply the theory to the structure of the atom. Sommerfeld took the quantum model of the atom that Bohr proposed in 1913 and worked out in 1915 that electrons must move in elliptical and not circular orbits around the nucleus. This led him in 1916 to predict a series of spectral lines based on the relativistic effects that would occur with elliptical orbits. Friedrich Paschen (1865–1945) immediately undertook the spectroscopic work required and confirmed Sommerfeld's predictions. This evidence for Bohr's ideas led to a rapid acceptance of the quantum theory of the atom.

Sommerville Duncan MacLaren Young 1879–1934 was one of the leading geometers of the early twentieth century who made significant contributions to the study of non-Euclidean geometry.

Sommerville was born on 24 November 1879 at Beawar, Rajasthan, India, the son of Scottish parents. He was educated in Scotland, first at Perth Academy and then at the University of St Andrews. From 1902 to 1904 he was a lecturer in the mathematics department there. In 1911 he became President of the Edinburgh Mathematical Society, which he had helped to found.

In 1915, Sommerville went to New Zealand as Professor of Pure and Applied Mathematics at Victoria University College, Wellington. He also became the first executive secretary of the Royal Astronomical Society of New Zealand, of which he was a founder-member. He presided over the mathematical section of the Australasia Association for the Advancement of Science in 1924 and four years later was awarded the Hector Medal of the Royal Society of New Zealand. He retained his post of Professor of Mathematics in Wellington until he died there on 31 January 1924.

Sommerville wrote numerous papers, nearly all of them on geometrical topics. The first appeared in 1905 under the title 'Networks of the Plane in Absolute Geometry', which was followed by 'Semi-regular Networks of the Plane in Absolute Geometry' (1906). He also wrote two other papers in that year which gave a pure mathematical treatment to the questions of a statistical nature that arose from the biometric researches of Karl Pearson (1857–1936).

Sommerville was an accomplished teacher and taught in both Scotland and New Zealand. His contribution to this area was enhanced by the publication of four textbooks on non-Euclidean geometry. Two of these indicated his major research specialties: *Elements of non-Euclidean Geometry* and *An Introduction to the Geometry of* n *Dimensions*, which included concepts that Sommerville himself had originated. He explained how non-Euclidean geometries arise from the use of alternatives to Euclid's postulate of parallels, and showed that both Euclidean and non-Euclidean geometries – such as hyperbolic and elliptic geometries – can be considered as sub-geometries of projective geometry. He stated that projective geometry is the invariant theory associated with the group of linear fractional transformations. He studied the tessellations of Euclidean and non-Euclidean space and showed that, although there are only three regular tesselations in the Euclidean plane, there are five congruent regular polygons of the same kind in the elliptical plane and an infinite number of such patterns in the hyperbolic plane. The variety is even greater if 'semi-regular' networks of regular polygons of different kinds are allowed (because the regular patterns are topologically equivalent to the non-regular designs). In his later work on n-dimensional geometry, Sommerville generalized his earlier analysis to include 'honeycombs' of polyhedra in three-dimensional spaces and of polytopes in spaces of 4, 5, …, n dimensions – including both Euclidean and non-Euclidean geometries.

Sommerville also studied astronomy, anatomy and chemistry. His interest in crystallography played a significant part in motivating him to investigate repetitive space-filling geometric patterns. He was also a skilful artist, as was exhibited in the models he constructed to illustrate his abstract conceptions.

Sorby Henry Clifton 1826–1908 was a British geologist who made huge advances in the field of petrology.

The only son of a wealthy Sheffield tool manufacturer, Sorby was privately educated and soon displayed a single-minded enthusiasm for science. After a brief interest in agricultural chemistry he turned to geology. His father's death allowed the 21-year-old Sorby the leisure to spend his life as an independent scientific researcher (he never married). He became fascinated with the microscopic study of rocks. Sir David Brewster had earlier explored the molecular structure of minerals by investigating the passing of light through them, but had been limited to investigating well-crystallized specimens. Operating from his own laboratory, Sorby overcame this problem by drawing upon the art of thin-slicing of hard minerals, analysing the specimens obtained under the microscope. Microscopic examination of thin sections of rocks had the virtue of enabling the constituent minerals to be scrutinized in transmitted light. Sorby also recognized the value of the Nicol prism for distinguishing the different component minerals in terms of the effect they produced on polarized light. His interest in studying meteorites led him in 1863 to discover the crystalline nature of steel, thus instituting the study of metallography. Though his early papers met with hostility, Sorby's work became credited with opening up a new science – though it proved to be more energetically cultivated in Germany, notably by Zirkel, Rosenbusch and Vogelsand, than in Britain.

Sorby's interests were wide. He used his microscopic techniques to investigate the structures of iron and steel under stress. He attempted to utilize quantitative techniques for geological phenomena, extrapolating from laboratory models and small-scale natural processes in the expectation of explaining vast events in geological history. He was elected a Fellow of the Royal Society in 1857, and he was to play a large role in the foundation of Firth College (later the University of Sheffield), where he endowed a chair of geology and a research fellowship.

South James 1785–1867 was a British astronomer noted for the observatory that he founded and his observations of double stars.

South was born in London in October 1785. He first studied medicine and surgery and became a member of the Royal College of Surgeons before renouncing medicine at the age of 31 in order to devote himself to astronomy. His marriage in 1816 made him wealthy enough to establish observatories in London and in Paris and to equip them with the best telescopes then available.

South became a Fellow of the Royal Society of London in 1821 and he held a variety of positions in the Astronomical Society of London. When the latter body gained a royal charter in 1831, a technicality barred South from serving as its first president and he resigned from the Society. He was knighted in that year and two years later was awarded an honorary LLD by Cambridge University. He was also a member of a number of scientific organizations in Scotland, Ireland, France, Belgium and Italy. He died on 19 October 1867 in London.

South's contribution to astronomy and the development of scientific work in England has been obscured by his argumentative temperament. His public criticism of the Royal Society for participating in the decline of the sciences in Britain offended other Fellows. He published criticisms of other works, including the *Nautical Almanac*, finding it inferior to continental work. None of this endeared him to his peers. South's quarrel with Troughton

about the quality of the latter's workmanship was consistent with South's apprehension about declining standards, but it led to a law suit which South lost. He then publicly destroyed the equipment that Troughton had made for him.

Despite such quarrels, South continued to work until his retirement. He is perhaps best remembered for his work with John Herschel (1792–1871) in observing double stars. They charted and catalogued changes in the positions of some 380 such stars. Their work was presented to the Royal Society in 1824 and rewarded with the Gold Medal of the Astronomical Society and the first prize of the Institut de France. In 1826, South completed another catalogue of double stars and for this he was awarded the Copley Medal of the Royal Society.

Spallanzani Lazzaro 1729–1799 was an Italian physiologist who is famous for disproving Needham's theory of spontaneous generation. In his later years Spallanzani became widely renowned for his biological investigations and received many academic honours, including a fellowship of the Royal Society of London in 1768.

Spallanzani was born on 12 January 1729, the son of a distinguished lawyer. He attended the local school until he was 15, when he went to the Jesuit college at Reggio. He was invited to join the Jesuit order, but declined. He then studied law at the University of Bologna where, under the influence of his cousin Laura Bassi, who was the Professor of Physics and Mathematics, Spallanzani became interested in science and broadened his education to include mathematics chemistry, natural history and French. In 1754, after obtaining his doctorate, he was appointed Professor of Logic, Metaphysics and Greek at Reggio College. Three years later he was ordained a priest, but performed his priestly duties irregularly and devoted himself almost entirely to his scientific studies – which were greatly facilitated by the moral protection and financial assistance provided by the Church. Spallanzani was Professor of Physics at Modena University from 1760 to 1769, when he became Professor of Natural History at the University of Pavia, a position he held for the rest of his life. In his later years Spallanzani travelled widely in order to further his scientific investigations. He died in Pavia on 11 February 1799.

Spallanzani is best known for finally disproving the theory of spontaneous generation. Francesco Redi's experiments on fly maggots in 1668 proved that complex animals do not arise spontaneously, but until Spallanzani's investigations, it was still generally believed that simple forms of life were generated spontaneously. After performing hundreds of experiments in which he boiled infusions of vegetable matter in hermetically sealed flasks, Spallanzani reported in 1765 that microorganisms do not arise spontaneously.

Spallanzani also investigated many other biological problems, such as the physiology of blood circulation. In 1771, while examining the vascular network in a chick embryo, he discovered the existence of vascular connections between arteries and veins – the first time this connection had been observed in a warm-blooded animal. He also studied the effects of growth on the circulation in chick embryos and tadpoles; the influence of gravity and the effects of wounds on various parts of the vascular system; and changes that occur in the circulation of dying animals. In addition, Spallanzani showed that the arterial pulse is caused by sideways pressure on the expansile artery walls from heartbeats transmitted by the bloodstream.

Spallanzani also studied digestion and, after administering food samples in perforated containers to a wide variety of animals then recovering the containers and examining them, concluded that the fundamental factor in digestion is the solvent property of gastric juice – a term first used by him. In his investigations of reproduction, he showed that the clasp reflex in amphibians persists after the male has been severely mutilated or even decapitated. (The clasp reflex is an automatic action on the part of the male in which he tightly holds the female during mating.) And in 1765 he performed an artificial insemination of a dog. Spallanzani's other biological investigations included the resuscitation of rotifers; the regeneration of decapitated snails' heads; the migration of swallows and eels; the flight of bats; and the electric discharge of torpedo-fish. In his later years Spallanzani studied respiration, proving that tissues use oxygen and give off carbon dioxide.

In addition to his biological work, Spallanzani also studied various problems in physics, chemistry, geology and meteorology, as well as pioneering the science of vulcanology.

Spemann Hans 1869–1941 was a German embryologist who discovered the phenomenon now called embryonic induction – the influence exerted by various regions of an embryo that controls the subsequent development of cells into specific organs and tissues. For this outstanding achievement he was awarded the 1935 Nobel Prize for Physiology or Medicine. In carrying out his embryological research, Spemann also pioneered techniques of microsurgery.

Spemann was born on 27 June 1869 in Stuttgart, the eldest of the four children of a bookseller. He attended school at the Eberhard-Ludwigs-Gymnasium, after which he was obliged to do a year of military service with the Kassel Hussars. When his military service ended, Spemann went to Heidelberg University to study medicine, but soon abandoned it in order to study zoology. On his graduation in 1894 he went to Würzburg University to study for his doctorate. and in 1898 was appointed Lecturer in Zoology there. Spemann remained at Würzburg until 1908, when he was appointed to the Chair of Zoology at Rostock University. From 1914 to 1919 he was Director of the Kaiser Wilhelm Institute of Biology in Berlin-Dahlem, after which he became Professor of Zoology at the University of Freiburg-im-Breisgau, a position he held until his retirement in 1935. Spemann died in Freiburg on 12 September 1941.

Spemann's Nobel prizewinning research was carried out on newt embryos. Previous workers had already shown that, as a newt embryo develops, an outgrowth of its brain comes into contact with the ectoderm and that this outgrowth develops into the retina of the eye while the area of ectoderm it has come into contact with develops into the lens. By carefully destroying the outgrowths at an early stage in their development, Spemann found that neither the retina nor the lens subsequently develop. This finding led him to the conclusion that the stimulus causing ectoderm to develop into lens tissue comes from the brain outgrowth. In his next series of experiments, Spemann – using delicate microsurgical techniques that he himself had developed – removed the piece of ectoderm that would normally become the lens and replaced it with a piece of ectoderm from elsewhere in the embryo. He found that, regardless of its site of origin, the transplanted ectoderm develops into a lens if it is in contact with the developing retina.

Spemann continued his line of research by investigating the effect of ligaturing embryos into halves. Embryos at an early stage of development either died or developed into a whole embryo: there were no half embryos formed. Similar results were obtained using embryos in the blastula stage (when the embryo is a hollow ball of cells), but when performed after gastrulation and invagination, ligaturing resulted in half embryos. It seemed, therefore, that as the embryo developed, the fates of different parts became determined. Spemann next began to search for the cause of specific aspects of embryonic development. Working with Otto Mangold, he transplanted various embryonic parts to other areas of the embryo and to different embryos. They discovered that any part of the ectoderm that comes into contact with the mesoderm during gastrulation eventually develops into the central nervous system. By transplanting mesoderm from the dorsal lip region of one embryo into an intact second embryo, Spemann managed to induce the development of a second central nervous system. Thus Spemann and Mangold demonstrated that one area of embryonic tissue influences the development of neighbouring tissues. Spemann named these influential regions organizers.

To investigate whether or not there was any predetermination within embryos, Spemann next conducted a series of experiments in which he exchanged tissue between newt and frog embryos. He found that embryonic tissue from newts always gives rise to newt organs, even when transplanted into a frog embryo, and that frog tissue always develops into frog organs in a newt embryo. Thus Spemann demonstrated that embryonic tissue responds to induction from foreign tissue but has the potential to develop only into the organs of the species from which it originated and is therefore predetermined to some extent.

Sperry Elmer Ambrose 1860–1930 was the US inventor and engineer who exploited the technology of the gyroscope to develop the first commercially successful gyrostabilizer for ships, and the gyrocompass and gyro-controlled autopilot for aircraft and ships. In doing this work he laid the foundations for modern control theory, cybernetics and automation.

Sperry was born on 21 October 1860, the son of Stephen Sperry, a farmer of Cortland county, New York, and Mary Burst from nearby Cincinnatus who died soon after his birth. He attended the Cortland Normal School until January 1880 and then made informal arrangements to sit in on lectures at nearby Cornell and at the same time to develop his first invention (a generator with characteristics suited to arc lighting).

Sperry married Zula Goodman on 28 June 1887 and the following year set up his own research and development enterprise. He formed a mining machinery company and, moving to Cleveland, Ohio in 1893, developed and manufactured streetcars. He produced a superior storage battery (accumulator), teaching himself chemistry in the process. He perfected a process for the production of caustic soda and was closely involved in a complicated process for the de-tinning of scrap from tin-can manufacture.

The family moved to Brooklyn in 1907 and in the same year his interest in the gyroscope began to show itself, leading to the work for which he is best known. Sperry's gyroscope company, from a research and development concern, gradually evolved into a manufacturing organization and World War I brought a dramatic upsurge in foreign sales.

Sperry was also active in internal combustion engine research. He spent $1 million on a compounded diesel but the idea never came to fruition. He developed a track recorder car for detecting substandard railway track and went on to invent a device for revealing defective rails which were undetectable visually.

Sperry resigned as President of the Sperry Gyroscope Company in 1926 to become Chairman of the Board. During his lifetime he and his company filed about 360 patent applications which finally matured as patents. His wife died on 11 March 1930 and Sperry died in Brooklyn on 16 June the same year.

A gyroscope's major active component is a rapidly spinning wheel. This follows the natural laws of motion and inertia so that if its spindle is moved in one direction it will respond by forcing the spindle to move at right angles to the original direction of motion. This behaviour is called precession.

A considerable amount of work had already been done in Europe and the USA on gyrostabilization and in 1908 Sperry filed a massive patent application for a ship's gyroscope. It encapsulated the principle of an active system rather than the passive systems already used by Ernst Otto Schlick, a well-known naval engineer from Hamburg.

Sperry's scheme mounted the gyro with its axis vertical in the hold of the ship. The axis was free to move in a fore and aft direction, but not from side to side. He used an electric motor to precess the gyro (tilt its rotor) artificially just as the ship began to roll. The gyro responded by exerting a force to one side or the other. Since it was fixed rigidly to the ship in the plane, the ship's roll was largely counteracted.

In 1912, a full-sized prototype with two gyro wheels, each weighing 18,00 kg/4,000 lb, was installed in the USS *Worden*, a 433-ton torpedo-boat destroyer. The installation cut a total roll of 30° to about 6°.

A gyrocompass feels the force of gravity and precesses until the axis of rotation of the spinning wheel is parallel to the axis of rotation of the Earth. On a ship, however, the pendulous gyro will also sense the ship's motion so it precesses away from the meridian. Sperry solved this problem by incorporating a servomechanism and a mechanical analogue computer which together compensated for the unwanted inputs.

Hermann Anschutz-Kaempfe (1872–1931) installed a system in the German fleet's flagship, *Deutschland*, in 1908. The prototype Sperry unit was fitted to the Dominion Line ship, *Princess Anne*, early in 1911 and the first production unit went into the USS *Utah* in November the same year.

The Sperry aircraft stabilizer had two components – one providing roll control and the other pitch control. Each unit opened compressed air valves, admitting air to slave cylinders. The pistons acted on the aircraft's controls to correct the error.

The roll component of the equipment was first flown in a Curtiss aircraft, with Curtiss himself at the controls, on Thanksgiving Day 1912. In the winter of 1913–1914 the Sperrys, father and son, designed and built an improved unit in which the four gyros were mounted together on a single stable platform. This arrangement has since become an essential component of guidance systems for missiles, aircraft and submarines.

The problems of blind-flying instruments proved to be extrememly tough and by the end of World War I virtually the only workable gyro instrument was for indicating rate of turn. The all-important gyrocompass and artificial horizon eluded Sperry himself and his research teams due to the rapid accelerations of aircraft in bumpy air.

At the instigation of the US forces, Sperry designed the control equipment for a specially built Curtiss pilotless aircraft which was to deliver a 450 kg/1,000 lb load of explosives to a target 80–160 km/50–100 mi away. This was officially called the 'flying bomb', thus anticipating the V-1s of World War II and the cruise missiles of the early 1980s.

After the War Sperry turned his attention to marine autopilots and a trial installation on an oil tanker performed well. Sales of the equipment increased until 1,000 merchant ships were so equipped by 1932.

Sperry introduced many of the concepts which are common in modern control theory. He did not use present-day jargon to describe his innovations but nevertheless he was one of the first in the fields of cybernetics and automation.

Spitzer Lyman 1914– is a US astrophysicist who has made important contributions to cosmogony.

Spitzer was born on 26 June 1914 in Toledo. He graduated from Yale in 1935 and then went to Cambridge University to work with Arthur Eddington. In 1936 he returned to Yale, where he gained a PhD two years later.

Spitzer stayed at Yale until 1947, when he moved to Princeton as head of the Astronomy Department.

Spitzer's initial interest was in star formation. One hypothesis proposed that stars were formed when gases and dust in interstellar space fused together under the influence of weak magnetic forces. A satisfactory understanding of this hypothesis required an appreciation of the fusion power of gases at temperatures as high as 100 million degrees, by which point hydrogen gas fuses to form helium. Spitzer proposed that only a magnetic field could contain gases at these temperatures, and he devised a figure-of-eight design to describe this field. His model remained important to later attempts to bring about the controlled fusion of hydrogen.

Spitzer's interest in the origin of planetary systems led him to criticize the tidal theory, proposed by James Jeans (1877–1946) amongst others. This was the idea that our planetary system is the result of an encounter between the Sun and a passing star: the star's closeness set up a tidal effect on the Sun, causing it to give off gaseous filaments that subsequently broke off from the main body to become planetary fragments. Spitzer showed that such a theory overlooked the fact that a gas would be dispersed into interstellar space long before it had cooled sufficiently to condense into planets. This objection also applied to Fred Hoyle's hypothesis that the Sun was a binary star whose companion long ago exploded as a supernova, leaving a gas cloud that condensed into planets.

Stahl Franklin William 1929–   . US molecular biologist; *see* Meselson and Stahl.

Stahl Georg Ernst 1660–1734 was an early German chemist and physician who founded the phlogiston theory of combustion. This theory was one of the great dead-ends of chemistry which was to dominate – and mislead – the science for nearly a century. Nevertheless it was instrumental in stimulating much thought and experiment, and helped to bring about the change from alchemy to chemistry.

Stahl was born in Ansbach, Franconia, on 21 October 1660, the son of a Protestant clergyman. He studied medicine under Georg Wedel (1645–1721) at the University of Jena, where a fellow student was Friedrich Hoffmann (1660–1742); Stahl occupied a teaching post at Jena a year before he gained his medical degree in 1684. He became a physician to the Duke of Sachsen Weimar in 1687 and seven years later, on the recommendation of Hoffmann, moved to the new University of Halle as its first Professor of Medicine, where his course included lectures in chemistry. In 1716, he moved again to Berlin to become personal physician to King Frederick I of Prussia, a position he retained until his death. He died in Berlin on 14 May 1734.

The phlogiston theory had its beginnings in 1667 in the ideas of Joachim Becher (1635–1681), who thought that combustible substances contain an active principle which he termed *terra pinguis* (fatty earth). Jan van Helmont called the combustible element phlogiston, but Stahl formulated the theory. The phlogiston theory is simple: when

a substance is burned or heated it loses phlogiston; reduction of the products of combustion (with, say, charcoal) reverses the process and phlogiston is restored. For example, when lead is heated it forms a powdery calx (so the metal must have been a combination of calx and phlogiston). Then when the calx is heated with charcoal, it absorbs phlogiston from the charcoal and becomes lead again. When charcoal is heated on its own, it leaves hardly any ash (calx) and so must be particularly rich in phlogiston.

The theory was the first attempt at a rational explanation for combustion (and what we would term oxidation), and had obvious appeal to the chemists of the time who were familiar with the reduction processes – often using charcoal – associated with the smelting of metals. The first doubts about the phlogiston theory came when chemists began weighing the reactants and products of such reactions. When metals are calcined (oxidized by heating in air) they get heavier – and yet they should *lose* phlogiston and become lighter. Also, a calx demonstrably becomes lighter when it is reduced back to metal, instead of getting heavier as it once more takes up phlogiston. Stahl himself made such quantitative determinations, but accounted for the observations by stating that phlogiston is weightless or can even have negative weight; it might be as insubstantial as 'caloric' (heat) and flow from one substance to another. The falsity of these assumptions and of the whole phlogiston theory was finally proved by Antoine Lavoisier in the 1770s with his experiments on combustion (and by the discovery of oxygen by Karl Scheele and Joseph Priestley).

Stark Johannes 1874–1957 was a German physicist who is known for his discovery of the phenomenon (now called the Stark effect) of the division of spectral lines in an electric field, and for his discovery of the Doppler effect in canal rays (positively charged ions produced by an electric discharge in a rarefied gas; impelled towards the cathode, the particles pass through it – if it is perforated – to form a beam). For this work he received the 1919 Nobel Prize for Physics.

Stark was born in Schickenhof, Bavaria, on 15 April 1874. He studied chemistry, physics, mathematics and crystallography at the University of Munich, from which he graduated in 1898 and where he spent the following two years working at the Physical Institute. In 1900, he became a lecturer at the University of Göttingen, then in 1906 was appointed Extraordinary Professor at the Technische Hochschule in Hanover. In 1909, he moved to the Technische Hochschule in Aachen, where he held a professorship until 1917. From 1917 to 1922 he was Professor of Physics at the University of Greifswald; between 1920 and 1922 he also lectured at the University of Würzburg. He then attempted to set up a porcelain factory in northern Germany but this scheme failed, largely because of the depressed state of the German economy, and so he tried – unsuccessfully – to return to academic life. A vehement antisemite, Stark was attracted to the rising Nazi movement and joined the Nazi party in 1930. Three years later he became President of the Reich Physical-Technical Institute and also President of the German Research Association. But his attempts to become an important influence in German physics brought him into conflict with the authorities and he was forced to resign in 1939. Nor did his decline end there: because of his Nazi background, in 1947 he was sentenced to four years internment in a labour camp by a German denazification court. Stark died on 21 June 1957 in Traunstein, West Germany.

In 1902, Stark predicted that the high-velocity canal rays produced in a cathode-ray tube should exhibit the Doppler effect (the change in the observed frequency of a wave resulting from movement of the wave's source or the observer or both). Three years later he proved his prediction correct by demonstrating the frequency shift in hydrogen canal rays.

His next major work was his discovery (announced in 1913) of the spectral phenomenon now called the Stark effect. Following Pieter Zeeman's demonstration in 1896 of the division of spectral lines caused by the influence of a magnetic field, Stark succeeded in showing (by photographing the spectrum emitted by canal rays – consisting of hydrogen and helium atoms – as they passed through a strong electric field) that an electric field produces a similar splitting effect on the spectral lines. The Stark effect is produced because the electric field causes the radiating atom's electron cloud to alter its position with respect to the nucleus, which distorts the electron orbitals. Light is emitted when an excited electron moves from a high to a lower energy orbital, so distortion of the orbitals also distorts the emitted light, this effect being manifested as splitting of the spectral lines.

Stark is also known for his modification (in 1913) of the photo-equivalence law proposed by Albert Einstein in 1906. Now called the Stark–Einstein law, it states that each molecule involved in a photochemical reaction absorbs only one quantum of the radiation that causes the reaction.

Starling Ernest Henry 1866–1927 was a British physiologist, remembered for his work on the heart, his studies of body functions and on hormones (which he first named).

Starling was born in London on 17 April 1866; his father was a barrister who worked in India, and whom he rarely saw. He was educated at King's College, London, from 1880 to 1882, and then at Guy's Hospital. He gained his medical degree there in 1889, having spent a summer in Heidelberg in 1885 working in the laboratories of the German physiologist Willy Kühne, and was a demonstrator at Guy's in 1887. In 1889, he was appointed a lecturer in physiology at Guy's and retained the position until 1899. During that period he was a part-time researcher at University College, London, in 1890, where he got to know William Bayliss, with whom he was to work a few years later. In 1892 he went to Breslau where he spent some time in the laboratories of Rudolf Heidenhain. From 1899 to 1923 he was Professor of Physiology at University College. World War I interrupted this appointment and in 1914 he became Director of Research at the Royal Army Medical Corps College, where he investigated antidotes to

poisonous gases. From 1917 to 1919 he served as Chairman of the Royal Society's Food Committee and as scientific adviser to the Ministry of Food. He retired from University College in 1922 and became a Research Professor of the Royal Society. He died while on a Caribbean cruise, in Kingston, Jamaica, on 2 May 1927.

Starling spent several years studying the conditions that cause fluids to leave blood vessels and enter the tissues. In 1896, he demonstrated the Starling equilibrium – the balance between hydrostatic pressure, causing fluids to flow out of the capillary membrane, and osmotic pressure, causing the fluids to be absorbed from the tissues into the capillary. The most important plasma protein in this fluid exchange, which helps to generate intravascular pressure, was found to be albumin.

When he started working at University College with Bayliss, Starling researched the nervous mechanisms that control the activities of the organs of the chest and abdomen, and together they discovered the peristaltic wave in the intestine. Their most important discovery, however, was in 1902, when they found the hormone secretin. This substance is found in the epithelial cells of the duodenum and excites the pancreas to secrete its digestive juices when acid chyme passes from the stomach into the duodenum – hence the name 'secretin'. It was the first time that a specific chemical substance had been seen to act as a stimulus for an organ at a distance from its site of origin. Starling and Bayliss coined the word 'hormone' in 1905, to characterize secretin and other similar substances produced internally and carried in the bloodstream to other parts of the body where they affect the function of organs.

Starling is probably best known for his work on the heart and on circulation. In 1918, he devised a heart-lung preparation by which the heart was isolated from all the other organs except the lungs, and attached only by the pulmonary blood vessels. The blood circulation in the heart was recorded by manometers. In this experiment Starling demonstrated the mechanism by which the heart is able to increase automatically the energy of each contraction in proportion to the mechanical demand made upon it and how it can adapt its work to the needs of the body independently of the nervous system. This mechanism, which Starling called the law of the heart, states that the more the heart is filled during diastole (relaxation), the greater is the following systole (contraction), that is, that the one is directly proportional to the other. This mechanism enables the heart to adjust the strength of its beat to variations in blood-flow without changing its rate. If a heart is impaired, it has to dilate more to achieve the same amount of work as it did when undamaged. Constant dilation of the heart is therefore used as a primary indication that it is damaged. This physiological phenomenon is not a feature exclusive to cardiac muscle but occurs in all contractile tissues whether heart, skeletal or plain muscle, although in the heart the function is more immediately vital.

In 1924, Starling succeeded in maintaining the mammalian kidney in isolation from the body, and found that substances lost in the excretory filtrate, such as carbonates, glucose and chlorides, are reabsorbed in the lower parts of the glomeruli.

Stas Jean Servais 1813–1891 was a Belgian analytical chemist who is remembered for making the first accurate determinations of atomic weights (relative atomic masses).

Stas was born in Louvain on 21 August 1813. He initially studied medicine, and although he qualified as a doctor he never practised. After graduation he went to Paris as an assistant to Jean Baptiste Dumas, working mainly in organic chemistry. In 1840, he was appointed Professor of Chemistry at the Ecole Royale Militaire in Brussels, and he advised the Belgian government on military topics related to chemistry. In middle age he developed a disorder of the throat, which made it difficult for him to give lectures. He left the Military School in 1869 and three years later he became Commissioner of the Mint, but his liberalist views did not coincide with the monetary policy of the government and he left the post in 1872. He spent the rest of his life in retirement, although he still voiced his anticlerical opinions and was openly critical of the part played by the Church in education. He died in Brussels in 13 December 1891.

While he was working with Dumas in Paris, Stas helped to redetermine the atomic weights of oxygen and carbon, showing them both to be almost exactly whole numbers (and that of carbon to be 12, not 6 as had previously been assumed). These results gave new support to William Prout's hypothesis of 1815 (that all atomic weights are whole numbers). Beginning in the mid-1850s, Stas spent more than ten years measuring accurately the atomic weights of many elements, using oxygen = 16 as a standard. He gradually found more and more elements with nonintegral atomic weights, and finally he discredited completely Prout's hypothesis. His results provided the foundation for the work of Dmitri Mendeleyev and others on the periodic system, and remained the standards of accuracy until they were superseded 50 years later by the determinations of the American chemist Theodore Richards.

Staudinger Hermann 1881–1965 was a German organic chemist who pioneered polymer chemistry. His contribution was finally recognized when he was 72 years old with the award of the 1953 Nobel Prize for Chemistry.

Staudinger was born in Worms, Hesse, on 23 March 1881, the son of a physician. His university education included studies at Halle (where he obtained his PhD in 1903), Munich and Darmstadt. He taught in Strasbourg, at the Technische Hochschule in Karlsruhe from 1908 to 1912 as Professor of Organic Chemistry in association with Fritz Haber, and as Professor of General Chemistry at Zurich from 1912 to 1926, where he succeeded Richard Willstätter. In 1926, he was appointed Professor of Chemistry at the University of Freiburg-im-Breisgau, where he remained until he retired in 1951. In 1940, he was made Director of the Chemical Laboratory and Research Institute for Macromolecular Chemistry. After 1951 his department

at Freiburg became the State Research Institute for Macromolecular Chemistry and he was made an emeritus professor. He died in Freiburg on 8 September 1965.

Staudinger's first research, under D. Vorländer at Halle, concerned the malonic esters of unsaturated compounds. Then in 1907 under Johannes Thiele (1865–1918) at Strasbourg he made the unexpected discovery of the highly reactive ketenes, the substances that give the aroma to coffee.

It was in Karlsruhe that Staudinger began the work for which he was to become famous, the study of the nature of polymers. He devised a new and simple synthesis of isoprene (the monomer for the production of the synthetic rubber polyisoprene) and with C. L. Lautenschläger prepared polyoxymethylenes. All this work was done at a time when most chemists thought that polymers were disorderly conglomerates of small molecules. From 1926 Staudinger put forward the view – not immediately accepted – that polymers are giant molecules held together with ordinary chemical bonds. To give credence to the theory he made chemical changes to polymers that left their molecular weights almost unchanged; for example he hydrogenated rubber to produce a saturated hydrocarbon polymer.

To measure the high molecular weights of polymers he devised a relationship, now known as Staudinger's law, between the viscosity of polymer solutions and their molecular weight. Viscometry is still widely used for this purpose in the plastics industry and in polymer research. Eventually X-ray crystallography was to confirm some of his predictions about the structures of polymers, particularly the long-chain molecular strands common to many of them.

Although Staudinger had no conception of how information is stored in nucleic acids or how such information is transferred to proteins, in 1936 he made a remarkably accurate prediction: 'Every gene macromolecule possesses a quite different structural plan which determines its function in life.' In his book *Macromolekulare Chemie und Biologie* (1947) he anticipated the molecular biology of the future.

Stebbins Joel 1878–1966 was a US astronomer, the first to develop the technique of electric photometry in the study of stars.

Stebbins was born on 30 July 1878 in Omaha. He was interested in astronomy from an early age, and pursued his studies in the subject at the Universities of Nebraska, where he received a BA in 1899, Wisconsin and California. He worked at the Lick Observatory from 1901 until 1903, when he earned a PhD. He was then appointed instructor in astronomy at the University of Illinois and was made Assistant Professor in 1904. After a sabbatical year at the University of Munich (1912–1913), he became Professor of Astronomy at the University of Illinois and was made Director of the University Observatory.

In 1922, Stebbins became Director of the Washburn Observatory and Professor of Astronomy at the University of Wisconsin. He retired as Professor Emeritus in 1948, but continued active research at the Lick Observatory until

1958. Stebbins was a member of the National Academy of Sciences, the American Association for the Advancement of Science and other prominent scientific organizations. He was the recipient of the Draper Medal of the National Academy of Sciences (1915), the Gold Medal of the Royal Astronomical Society (1950) and other honours. He died in Palo Alto on 16 March 1966.

Stebbins's earliest astronomical research was in spectroscopy and photometry. In 1906 he began attempting to use electronic methods in photometry. The results were encouraging, although at first only the brightest objects in the sky (such as the Moon) could be studied in this manner. From 1909 to 1925 he devoted much of his time to improving the photoelectric cell and using it to study the light curves of eclipsing binary stars. As the sensitivity of the device was increased, it could be used to measure the light of the solar corona during total eclipses. Stebbins discovered that although there was no detectable variations in the light output of the Sun, he could observe variations in the light of cooler stars.

During the 1930s Stebbins applied photoelectric research to the problem of the nature and distribution of interstellar dust and its effects on the transmission of stellar light. He analysed the degree of reddening of the light of hot stars and of globular clusters. His discoveries contributed to an understanding of the structure and size of our Galaxy. He investigated whether interstellar material absorbed light of all wavelengths equally, and found that over a range from the infrared as far as the ultraviolet absorption was constant, but that absorption of ultraviolet light itself was less strong.

Stebbins's other work included studies using photoelectric equipment of the magnitudes and colours of other galaxies. He demonstrated that his method was more accurate than those that relied on the eye or photography, and the advent of the photomultiplier extended the usefulness of his technique even further.

Steele Edward John 1948– is an Australian immunologist whose research into the inheritance of immunity has lent a certain amount of support to the Lamarckian theory of the inheritance of acquired characteristics, thus challenging modern theories of heredity and evolution.

Steele was born in Darwin on 27 October 1948 and educated at the University of Adelaide, South Australia, from which he graduated in molecular biology in 1971 and gained his doctorate in 1975. In 1976, he began his postdoctoral work at the John Curtin School of Medical Research, Canberra, studying naturally occurring autoimmune disease. He moved to Canada in 1977 and continued his research at the Ontario Cancer Institute. In 1978 he became interested in evolutionary theory, and early in 1980 he moved to the Clinical Research Centre of the Medical Research Council at Harrow, Middlesex.

The modern neo-Darwinistic theory of evolution is a synthesis of Darwin's idea of survival of the fittest by means of natural selection, and Mendelian genetics: according to neo-Darwinism chance, in the form of genetic mutations, plays an important role in evolution; and competition for limited resources – natural selection – eventually kills off the 'bad' mutations. These ideas are in direct opposition to Jean Lamarck's theory of the inheritance of acquired characteristics, according to which, various parts of the body develop or disappear as necessitated by changes in habits resulting from environmental changes. Lamarck, however, believed that these acquired changes could be passed on to subsequent generations and that, if continued for a long time, a new species would eventually develop. Steele, working with Reginald Gorczynski, found that mice which have been made immune to certain antigens can pass on this acquired immunity to first and second generations of their offspring. This finding suggests that although Lamarck's original ideas are too unsophisticated to be correct, in a subtler, more refined form they may make a valuable contribution towards a better understanding of evolutionary processes.

Stefan Josef 1835–1893 was an Austrian physicist who first determined the relation between the amount of energy radiated by a body and its temperature. This expression is usually known as the Stefan–Boltzmann radiation law because Ludwig Boltzmann (1844–1906) gave a theoretical explanation of it.

Stefan was born at Klagenfurt on 24 March 1835. He studied at the Gymnasium there and entered the University of Vienna in 1853. After completing his studies, Stefan became a lecturer at the University in 1858, rising to Professor of Higher Mathematics and Physics in 1863. He retained this position for the rest of his life, and was also Director of the Institute for Experimental Physics at Vienna from 1866. From 1885 to his death, Stefan served as Vice President of the Imperial Academy of Sciences. He died at Vienna on 7 January 1893.

Stefan's discovery of his radiation law was made in 1879. For a long time, it had been known that Newton's law of cooling, which held that the rate of cooling of a hot body is proportional to the difference in temperature between the body and its surroundings, is incorrect for large temperature differences. Many people had found that the amount of heat given out by a very hot body is far greater than expected by Newton's law.

Stefan followed up experimental work by John Tyndall (1820–1893), who had measured the amount of radiant heat produced by a platinum wire heated to varying degrees of incandescence by an electric current. He found that over the greatest range, from about 525°C/977°F up to about 1,200°C/2,192°F, the intensity of radiation increased by 11.7 times. Stefan realized that this increase in intensity was equal to the ratio of the absolute temperatures raised to the fourth power, and from this deduced his radiation law. This states that the total energy radiated in a given time by a given area of a body, E, is equal to σT^4, where T is the absolute temperature and σ is a constant known as the Stefan constant. Stefan confirmed the validity of his law over wide temperature ranges.

In 1884, Boltzmann, a former student of Stefan, gave a

theoretical explanation of Stefan's law based on thermo-dynamic principles and the kinetic theory developed by James Clerk Maxwell (1831–1879). Boltzmann pointed out that it held only for perfect black bodies, and Stefan had in fact been able to derive the law because platinum approximates to a black body. However, from his law, Stefan was able to make the first accurate determination of the surface temperature of the Sun, obtaining a value of approximately 6,000°C/11,000°F.

Stefan also made several other contributions to physics. He was an able experimental physicist, and produced accurate measurements of the conductivities of gases that helped to confirm Maxwell's kinetic theory. He also investigated diffusion in gases, making theoretical derivations that confirmed experimental results.

Steiner Jakob 1796–1863 was a Swiss mathematician, the pre-eminent geometer of the nineteenth century and the founder of modern synthetic, or projective, geometry.

Steiner was born at Utzenstorf, near Bern, on 18 March 1796. Because he had to help out in the family business his early education was neglected and colleagues in later years were astonished to discover that he did not learn to read and write until the age of 14. When he was 18 he left home to enrol at the Pestalozzi school in Yverdon; it was run on the monitorial system and in a very short time Steiner was employed as an instructor in mathematics. For a while he maintained himself by teaching and in 1821 received a teacher's certificate in Berlin. He had also attended some mathematical lectures at Heidelberg and acquired sufficient skill and knowledge to be given a place as a student at the University of Berlin in 1822. By 1825 he was teaching at the university and in 1834 a Chair of Geometry was created for him; he held the Chair for the rest of his life. His health began to deteriorate in the 1850s and he died at Bern on 1 April 1863.

When Steiner began to publish his epoch-making geometrical discoveries in the 1820s, mathematics was in that exciting transitional stage which has produced the modern explosion in mathematics. Steiner played an important role in the transition. His first published paper, which appeared in *Crelle's Journal* in 1826, contained his discovery of the geometrical transformation known as inversion geometry. His most important work, the *Systematische Entwicklung der Abhängigkeit geometrischer Gestalten von Einander* appeared in 1832. The work was notable for its full discussion and examination of the principle of duality and for the wealth of fundamental concepts and results in projective geometry which it contained. In it are to be found the two discoveries to which he had lent his name, the Steiner surface (also called the Roman surface), which has a double infinity of conic sections on it, and the Steiner theorem. The theorem states that two pencils (collections of geometric objects) by which a conic is projected from two of its points are projectively related. The book also included a supplement listing problems to be solved, and it remained a rich quarry for geometers for the next 100 years.

Steiner's other principal result was the theorem now known as the Steiner–Poncelet theorem, an extension of work done by Jean Poncelet in 1822. Steiner proved that any Euclidean figure could be generated using only a straight rule if the plane of construction had a circle with its centre marked drawn on it already. (In 1904 it was shown by Francesco Severi (1879–1961) that only the centre of the circle and a small arc of its circumference were necessary.)

Steiner's consistent aim was to discover fundamental principles from which the rest of geometry could be derived in an orderly and coherent manner. His work, in general, was marked by his disdain of using analysis and algebra; he preferred to rely on entirely synthetic methods. It is for that reason that whether he was, as some believe, the finest geometer since Apollonius of Perga, he is universally recognized as the founder of projective geometry.

Steno Nicolaus 1638–1686 was a versatile Danish naturalist widely regarded as one of the founders of stratigraphy.

Born a Protestant in Copenhagen, Steno converted to Catholicism and settled in Florence. Having studied medicine in Leiden, in 1666 he was appointed personal physician to Ferdinand, Grand Duke of Tuscany, before becoming royal anatomist at Copenhagen in 1672. He was ordained a priest in 1675, and gave up science on being appointed vicar-apostolic to North Germany and Scandinavia. He is buried in the Medici crypt in Florence. As a physician he discovered Steno's duct of the parotid gland, and investigated the workings of the ovaries. A passionate anatomist, Steno showed that a pineal gland resembling the human one is found in other creatures, using this finding to challenge Descartes's claim that the gland was the seat of the uniquely human soul. Steno's examination of quartz crystals disclosed that, despite differences in the shapes, the angle formed by corresponding faces is invariable for a particular mineral. This constancy (Steno's law), follows from internal molecular ordering.

Steno is perhaps best remembered for his contributions to geology and palaeontology. Having found fossil teeth (glossopetrae) far inland closely resembling those of a shark he had dissected, in his *Sample of the Elements of Mylogy* 1667 he championed the organic origin of fossils against those who postulated they were 'sports of nature' or similar concretions. On the basis of his palaeontological views, he also set out a view of geological history, contending that sedimentary strata had been deposited in former seas. He assumed six successive periods of Earth history: first, an age of deposition of non-fossiliferous strata from an ocean; then several periods of undermining and collapse; then another epoch of the deposition of strata, this time fossiliferous; followed by further undermining and collapse. This explained why the deepest strata contained no fossils but were overlain by fossiliferous strata, and also why certain strata were found horizontal while others were tilted to the horizon. To accompany his ideas, Steno sketched what are generally regarded as the earliest geological sections. Making significant advances in anatomy, geology, crystallography, palaeontology and mineralogy, Steno was one of the great all-rounders of seventeenth-century science.

Stephan Peter 1943– is a doctor of homeopathic medicine, known for his work in the field of therapeutic immunology.

Stephan was born in Middlesborough, Yorkshire (now Cleveland), on 18 September 1943, the son of Ernest Stephan, a pioneer of cell therapy and the first person to introduce the method into British medical practice (in 1952). Peter Stephan was brought up by his mother and educated at Wallace Tutors, London. He then worked with his father, learning about cell therapy. His father died in 1964, leaving his son to run the Harley Street clinic at the age of only 21. He studied homeopathic medicine and graduated in 1970.

The cells that make up the organs and systems of the human body are continually dying and are replaced by healthy cells. But as the body ages, its cells become less healthy and the replacement cells are also inferior; there is also an increase in poorer cells in someone who is ill. The concept of cell therapy is that by injecting healthy cells into the body, the general state of health of the body as a whole can be improved.

In 1931, Paul Niehans made the first successful transfer of healthy cellular material from an animal to a human being by injection as a method of treatment. He injected whole cells, calling the technique cell replacement therapy. But cells are antigens and when 'foreign' cells are injected the body's immune system produces antibodies, a reaction that may cause discomfort to the patient for a few days. Dyckerhoff discovered that a cell's condition is governed by its constituent RNA and that it is necessary only to inject RNA. Also an RNA injection causes no antigen–antibody reaction because RNA is organ-specific not species-specific; for instance, human kidney RNA is the same as kidney RNA from another animal. Jean Thomas developed a third type of treatment that involves the injection of tissue-specific antisera. An antiserum is an antibody, and its introduction either enlivens sluggish cells or kills them; this treatment is called serocytology.

Stephan's treatment involves injecting an organ-specific RNA to boost the cellular RNA and then injecting tissue-specific antisera. These antisera travel to the 'sick' cells and kill them while at the same time they stimulate the body's immune system, and the healthy cells become active and reproduce.

Stephan's recent work has been concerned with the substance that is formed before the production of the antibody from which he prepares his sera. This substance seems to have beneficial effects which may be used therapeutically. In 1981, he developed Omnigen, a total cellular extract and serum for treating premature cellular degeneration.

Stephenson George 1781–1848 was a British engineer who pioneered the building of the first railways in Britain and of steam locomotives to run on them.

Stephenson was born at Wylam, a village near Newcastle-upon-Tyne, the son of a fireman. He received no formal education and was employed when a boy to look after cattle. When he was 14 years old he became an assistant fireman to his father at Darley Colliery. He was illiterate until the age of 17 when, frustrated by his inability to follow the daily newspaper reports of the Napoleonic Wars, he attended evening classes and learned to read and write.

In 1808, at a time when his prospects seemed so bleak that he seriously thought of emigrating, Stephenson took a contract to work the engine of the Killingworth pit. While there he regularly took the steam pumping engine to pieces in order to understand its construction. The reward came with his success in modifying a Newcomen engine which was performing its pumping function inefficiently, when in 1812, he was appointed engine-wright to the colliery at £100 a year.

Stephenson's first invention was a safety lamp for miners. This device avoided the dangers of combustion by allowing the air to enter along narrow tubes. He demonstrated the success of his discovery by entering gas-infested tunnels at the Killingworth pit with perfect safety. The simultaneous development of a safety lamp by Humphry Davy (1778–1829) produced fierce and sometimes acrimonius controversy as to who was the real inventor, before it was accepted that both men had reached the same goal independently and by different approaches.

In 1811, John Blenkinsop (1783–1831) constructed a steam locomotive for hauling coal wagons in a Yorkshire colliery but the machine, which used toothed wheels on a racked track, was ponderous and unwieldy. At Wylam colliery they were now anxious to introduce steam power on the horse tramways, and Stephenson produced a smooth-wheeled locomotive called *The Blucher* which, in 1814, successfully drew 30 tonnes of coal in eight wagons up an incline of 1 in 150 at 7 kph/4 mph . Not satisfied, Stephenson introduced the 'steam blast' by which exhaust steam was redirected into the chimney through a blast pipe, bringing in air with it and increasing the draught through the fire. This further development made the locomotive truly practical.

Subsequently Stephenson started to make experiments with various gradients. He found that a slope of 1 in 200, common enough on roads, reduced the haulage power of a locomotive by 50% (on a completely even surface, a tractive force of less than 5 kg/11 lb would move a tonne). Furthermore he discovered that friction was virtually independent of speed. The obvious conclusion from such findings was that railway gradients should always be as low as possible. Cuttings, tunnels and embankments were therefore necessary. In 1819 the proprietors of Hetton Colliery, under Stephenson's direction, laid down a railway 11 km/8 mi long. It was opened in 1822, with traction provided partly by stationary engines and partly by locomotives.

In 1821, Stephenson succeeded in persuading Edward Pease, chief promotor of the scheme, that Stockton and Darlington could be better connected by steam locomotion than by the proposed horse traction. He was appointed engineer to the project and advocated the use of malleable iron rails instead of the cast iron which had been used hitherto. The gauge for the new railway was assessed by Stephenson at 1.4 m/4 ft 8 in, a historic decision, since this has remained the 'standard gauge' for railways throughout the world.

Stephenson then induced Pease and Michael Longridge to support him and his son Robert in establishing locomotive works at Newcastle, where the engines were to be made for the Stockton and Darlington Railway. On 27 September 1825 the world's first public railway came into operation with the opening of the line. Stephenson's engine *Locomotion* took a party of passengers from Darlington at a top speed of 24 kph/15 mph.

Before the Stockton and Darlington Railway was opened, Stephenson had been engaged to design a railway from Manchester to Liverpool, the trade between the towns having already grown too great to be accommodated on the existing canals. Surveyors for the projected railway met fierce opposition from the farmers and landowners through whose estates the railway was to run. A Bill to implement the scheme was thrown out by its opponents in Parliament, but a second Bill, introduced in 1826 (and showing an improved overall plan), was accepted. The greatest physical obstacle to Stephenson's plans was a large area of marshy ground known as Chat Moss. By distributing the load over a considerable surface of the bog Stephenson was able to take his line over the treacherous ground.

Having eventually convinced the railway directors that locomotives and not fixed engines should operate on the line, Stephenson took part in an open competition to discover the most efficient locomotive for the railway, the prize being £500. A mean speed of 16 kph/10 mph was to be attained and steam pressure was not to exceed 345,000 pascals/50 psi.

Stephenson saw that if he was to succeed he had to find some means of increasing the heating surface of the boilers of his locomotives. He therefore adopted tubes passing through the cylindrical barrel and connecting the fire-box with the smoke-box. The engine Stephenson produced for the great trial, the *Rocket*, was built at the Newcastle works under the direct supervision of his son Robert, and after many failures the problem of securing the tubes to the tube-plates was overcome. The locomotive had a weight of 4.2 tonnes, half the weight of *Locomotion*.

Three other engines were entered for the competition but basic inadequacy or ill-luck overtook them. On the testing day, the *Rocket*, the only engine ready on time, ran 19 km/12 mi in 53 minutes, and was duly awarded the prize. On 15 September 1830 the line was opened with great ceremony. If the Stockton and Darlington Railway announced the arrival of the steam locomotive, the Manchester and Liverpool Railway, providing the first regular passenger service, showed that it had come to stay. An unwelcome event on the opening day was the first railway accident in which William Huskisson, President of the Board of Trade, was killed. Ironically he had been Stephenson's most influ-

ential supporter in Parliament.

For the remainder of his life Stephenson worked as a consultant engineer to several newly emerging railway companies, all in the north of England or the Midlands. There was hardly a railway scheme in which he was not consulted or an important line constructed without his help and advice. The last great issue with which Stephenson was concerned arose in 1845, in the battle between the supporters of the locomotive and those who advocated the atmospheric railway system developed by Stephenson's great contemporary Isambard Brunel (1806–1859). The dispute arose in connection with the extension of the railway from Newcastle to Berwick. Although Brunel had many influential friends in the Board of Trade, Stephenson's supporters in Parliament were the more numerous and carried the day. This ended the last attempt to challenge the advent of the steam locomotive.

Stephenson Robert 1803–1859 the only son of George Stephenson, was an outstanding nineteenth-century British engineer and builder of many remarkable railway bridges.

Born at Willington Quay, near to Newcastle-upon-Tyne, Stephenson began his working life assisting his father in the survey of the Stockton and Darlington Railway in 1821. The following year he spent six months at Edinburgh University studying mathematics. He then returned to Newcastle to manage the locomotive factory which his father had established there. His health began to decline seriously, however, and seeking a warmer climate he accepted an offer to superintend some gold and silver mines in Columbia in South America. He was away three years, but difficulties in the management of the Newcastle locomotive factory led to a request for his return. He returned to England in 1827, when the controversy over the most suitable form of traction for the Liverpool and Manchester railway was at its height. The successful *Rocket* steam locomotive was eventually built under his personal direction, as were subsequent improvements to it.

In 1833, a scheme to construct a railway from Birmingham to London was introduced, and Stephenson became engineer for the line. The project was very important, being the first railway to be taken into the capital. Stephenson overcame many of the obstacles encountered with outstanding engineering skill, notably with the Blisworth cutting and the Kilsby tunnel. The railway was completed in 1838, and from then on he was engaged on railway work for the rest of his life.

Probably the most important and certainly the most conspicuous of Stephenson's achievements were the various railway bridges which he built in Britain and elsewhere. The High Level Bridge over the river Tyne in Newcastle and the Victoria Bridge over the river Tweed at Berwick are among his earliest and most striking achievements. The former, spanning the river between Gateshead and Newcastle, comprises six iron arches; James Nasmyth's newly invented steam hammer was used to drive in its foundations. In 1844 construction began, under Stephenson's supervision, of a railway line from Chester to Holyhead.

He gave long and detailed thought to the best type of bridges for crossing the river Conway and the Menai Straits. For the Menai Straits he designed a bridge in which the railway tracks were completely enclosed in parallel iron tubes. This proved so successful when put into service that the same plan was adopted for other bridges. One such, the Victoria Bridge over the St Lawrence at Montreal, was constructed by Stephenson between 1854 and 1859, and was for many years the longest bridge in the world.

Steptoe Patrick 1913–1988. English surgeon; *see* Edwards and Steptoe.

Stern Otto 1888–1969 was a German-born US physicist who showed that beams of atoms and molecules have wave properties. He also determined the magnetic moment of the proton. For these achievements, Stern gained the 1943 Nobel Prize for Physics.

Stern was born in Sohrau, Upper Silesia, then in Germany but now Zory in Poland, on 17 February 1888. His family moved to Breslau in 1892. Stern entered the Johannes Gymnasium in Breslau in 1897, and obtained his Arbitur in 1906. Following the custom of the time, Stern then attended a number of German universities. He went to Freiburg im Breisgau, Munich and Breslau, giving him the opportunity to attend lectures by such prominent scientists as Arnold Sommerfeld (1868–1951), Otto Lummer (1860–1925) and Ernst Pringsheim (1859–1917). Stern received his PhD from Breslau in 1912 for a dissertation on the kinetic theory of osmotic pressure in concentrated solutions with special reference to solutions of carbon dioxide.

Although Stern's thesis work had been in the field of physical chemistry, he was fascinated by theoretical physics. He arranged to join Albert Einstein (1879–1955) in Prague as a postdoctoral associate. When Einstein moved to Zurich in 1913, Stern followed him and took up a post as privatdozent (unpaid lecturer) in physical chemistry at the Technical High School in Zurich.

Stern served in the army in World War I and was assigned to meteorological work on the Russian front. He was able to work on theoretical physics in his spare time, and was then transferred to more congenial laboratory work in Berlin towards the end of the war. In 1918, he took up the position of privatdozent in the department of theoretical physics at the University of Frankfurt. Stern obtained his first professional position in 1921, when he was made Associate Professor of Theoretical Physics at the University of Rostock, but he soon moved to the University of Hamburg where, in 1923, he became Professor of Physical Chemistry and Director of the Institute of Physical Chemistry.

The next ten years marked the peak of Stern's career. He made important contributions to the understanding of quantum theory, and built up a thriving research group. Many of his associates were Jewish as indeed was Stern himself, and life became increasingly difficult during the early 1930s as the Nazis rose to power. The sacking of Stern's longtime associate Immanuel Estermann and an order to remove a portrait of Einstein from the laboratory were the last straws. Stern resigned his post in 1933 and went to the Carnegie Technical Institute in Pittsburgh in the United States, where he was given the post of Research Professor and set up a new department for the study of molecular beams.

Stern was granted US citizenship in 1939 and worked as a consultant to the US War Department during World War II. After the war he retired and moved to Berkeley, California, where he maintained contact with the scientific community on a private basis only. He died in Berkeley of a heart attack on 17 August 1969.

The first phase of Stern's scientific career covers the period from 1912 until 1918. During these years he was absorbed in purely theoretical work and in learning how to choose the really central issues for study in an experimental context. His theoretical work was partly in the field of statistical thermodynamics, in which he proposed an elegant derivation of the entropy constant.

After World War I, Stern began to concentrate on the molecular beam method, which had been discovered by Louis Dunoyer (1880–1963) in 1911. This method consisted of opening a tiny hole in a heated container held inside a region of high vacuum. The vapour molecules inside the heated container flowed out to form a straight beam of moving particles suffering virtually no collisions in the vacuum. The beam could be narrowed still further by the use of slits, and a system of rotating slits could be used to select only those particles travelling at particular velocities. This was a powerful tool in the study of the magnetic and other properties of particles, atoms and molecules.

Stern's first experiments with this device were completed in 1919. They dealt with the measurement of molecular velocities and confirmed the Maxwellian distribution of velocities. He then began an experiment with Walther Gerlach (1899–1979) that was intended to measure the magnetic moment of metal atoms and also to investigate the question of spatial quantization.

Developing the quantum model of the atom proposed by Niels Bohr (1885–1962) in 1913, Sommerfeld had derived a formula for the magnetic moment of the silver atom and predicted that silver atoms in a magnetic field could orient in only two directions with respect to that field. This latter idea was not compatible with classical theory. The Stern–Gerlach experiment would determine which of these theories was correct.

The experiment consisted of passing a narrow beam of silver atoms through a strong magnetic field. Classical theory predicted that this field would cause the beam to broaden, but spatial quantization predicted that the beam would split into two distinct separate beams. The result, showing a split beam, was the first clear evidence for space quantization. Stern and Gerlach also obtained a measurement of the magnetic moment of the silver atom.

Stern then went on to improve this molecular beam technique and in 1931 was able to detect the wave nature of particles in the beams. This was an important confirmation of wave–particle duality, which had been proposed in 1924 by Louis de Broglie (1892–1987) for the electron and extended to all particles.

In 1933, Stern measured the magnetic moment of the proton and the deuteron. The magnetic moment of the proton had been predicted by Paul Dirac (1902–1984) to be one nuclear magneton. Stern's group, despite much advice not to attempt the difficult experiment whose result was in any case already 'known', caused much astonishment when they demonstrated that the proton's magnetic moment was 2.5 times greater than expected. The explanation for this discrepancy lies partly in parallel proton movements.

Stern's experiments with molecular beams came at a critical time in the development of quantum theory and nuclear physics as they provided the firm experimental evidence that was needed for theories in quantum mechanics that were hitherto highly controversial. Stern's great talent lay in his experimental foresight, which was based on a solid theoretical grounding, coupled with an almost fanatical obsession with experimental and technical detail.

Stevens Nettie Maria 1861–1912 was a US biologist whose researches concentrated on the role of chromosomes and their relationship to heredity.

Stevens was born in Cavendish, Vermont on 7 July 1861, the daughter of a carpenter. After local schooling she became a librarian at the Free Public Library of Chelmsford, Massachusetts, until 1892. In that year, aged 31, she returned to education by attending classes at a High School, and in 1896 moved to Stanford University as an undergraduate, receiving a bachelor's degree in physiology in 1899 and a master's degree the following year. She was already interested in marine biology and registered at Bryn Mawr College for her PhD which she was awarded in 1903 for a dissertation on ciliate protozoa. She spent research periods at marine laboratories in Europe and also visited the Zoological Institute at Wurzburg where she met Theodor Boveri (1862–1915). Whilst serving as a research fellow at Bryn Mawr, she was awarded a fellowship to work at the Stazione Zoologica, Naples, in 1905. She became an associate professor in the same year, and apart from a further year in Wurzburg in 1908, remained at Bryn Mawr until her early death from carcinoma on 4 May 1912.

Her early work on the morphology of ciliate protozoa developed her interests in cytology in general and she studied regenerative processes in lower invertebrates. From there she moved on to working on the development of the roundworm, examining its regenerative properties after exposure to ultraviolet radiation, and showed that even in very early embryonic life cells were restricted in their regenerative capabilities. Her most significant work was in examining the relationship between chromosomes and the units of heredity postulated by the work of Gregor Mendel. Stevens's experiments with a species of beetle showed that sex was determined by a specific chromosome, a conclusion that the biologist Edmund Wilson arrived at independently. This was the first direct evidence that Mendel's factors were associated with chromosomes.

Stevinus Simon c.1548–1620 was a Flemish scientist who, in physics, developed statics and hydrodynamics and who introduced decimal notation into mathematics.

Stevinus was born in Bruges (now in Belgium) in about 1548. He began work in Antwerp as a clerk and then entered Dutch Government service, using his engineering skills to become Quartermaster-General to the army. (For Prince Maurice of Nassau he designed sluices that could be used to flood parts of Holland to defend it from attack.) He married very late in life and died in The Hague in the early part of 1620.

Stevinus's interests were wide-ranging. In statics he made use of the parallelogram of forces and in dynamics he made a scientific study of pulley systems. In hydrostatics Stevinus noted that the pressure exerted by a liquid depends only on its height and is independent of the shape of the vessel containing it. He is supposed to have anticipated Galileo in an experiment in which he dropped two unequal weights from a tall building to demonstrate that they fell at the same rate.

Stevinus wrote in the vernacular (a principle he advocated for all scientists). Even so his book on mechanics *De Beghinselen der Weeghcoust* (1586), written in Flemish, was translated into Latins 20 years later (as *Hypomnemata Mathematica*) by Willebrord Snell (1591–1626).

Stieltjes Thomas Jan 1856–1894 was a Dutch mathematician who contributed greatly to the theory of series and is often called the founder of analytical theory.

Stieltjes was born at Zwolle on 29 December 1856, the son of a distinguished civil engineer. Almost all his scientific and mathematical training was received at the Ecole Polytechnique in Delft. He graduated from the school in 1877 and was appointed to a post at the Leiden Observatory. He remained there until 1883, and though little is known of his mathematical doings in those years, it seems probable that he continued to do research, since in 1884 he was appointed to the Chair of Mathematics at the University of Groningen. In that year he was also awarded an honorary doctorate by the University of Leiden. Two years later he was elected to the Dutch Academy of Science. In 1886 he went to France, became a naturalized French citizen, and for the rest of his short life taught mathematics at the University of Toulouse. Although he was never elected to the French Academy (despite being nominated in 1892), he won the Academy's Ormoy prize in 1893 for his work on continued fractions. He died in Toulouse on 31 December 1894.

In a short working life Stieltjes studied almost all the problems in analysis then known – the theory of ordinary and partial differential equations, Euler's gamma functions, elliptic functions, interpolation theory and asymptotic series. But his lasting reputation rests on his investigations into continued fractions. The fruit of those researches is contained in his last memoir, *Recherches sur les Fractions Continues*, completed just before he died. The memoir was a milestone in mathematical history. Before its appearance only special cases of continued fractions had been considered. Stieltjes was the first mathematician to give a general treatment of continued fractions as part of complex analytical function theory. He did so, moreover, in a book of exemplary clarity and beauty.

A continued fraction is derived from a sequence of ordinary fractions:

$$\frac{a_1}{b_1}, \frac{a^2}{b_2}, \frac{a^3}{b_3}$$

which are called partial quotients, by adding each fraction to the denominator of the preceding fraction. In such series, a_1, a_2, a_3, etc, are called partial numerators; b_1, b_2, b_3, etc, are called partial denominators. They may be real or complex numbers. To this day one of the most important continued fractions in analytical theory remains the Stieltjes-fraction, or S-fraction.

In this fraction the values of k are constants other than zero and z is a complex variable. Taking the values of k as real and positive, Stieltjes was able to solve what is known as the Stieltjes moment problem – that is, the problem of determining a distribution of mass which has preassigned moments. His solution has been very fruitful, as has the very problem itself, extended as it has been into many fields.

Stieltjes's researches also raised the mathematical status of discontinuous functions and divergent series. He advanced the theory of Riemann's function, especially by the appearance, in his last great paper, of the integral:

$$\int_a^b f(u)\,\delta g(u)$$

which is a generalized form of the Riemann integral and is now known as the Stieltjes integral. Stieltjes came upon it in his search for a way to express the limit of a certain sequence of analytic functions. What he did was to replace lengths of intervals (in the approximating sums for Riemann integration) by masses spread on them. He introduced this distribution of masses by means of a non-decreasing function, g, which gives the increment $g(b) - g(a)$ of the function g to every interval $[a, b]$. From this he was able to obtain his integral.

Stieltjes's analysis of continued fractions has had immense influence in the development of mathematics. His ideas greatly helped David Hilbert in his working out of the theory of quadratic forms in infinitely numerous variables. They were also used by Felix Hausdorff in his work on divergent series. Indeed, so varied are the fields which have profited from Stieltjes's creative imagination – number theory, the theory of equations, the theory of integration, infinite matrices, the theory of functions and, in the physical sciences, dynamics and the construction of electrical networks – that he well deserves to be known as the chief pioneer of modern analysis.

Stirling Robert 1790–1878 was the Scottish minister who is credited with the invention of a working hot-air engine. The principle has a large number of inherent advantages that could make it as important as the internal-combustion engine, so intensive research is being carried out on the Stirling engine.

Stirling was born in Cloag, Scotland on 25 October 1790. He attended Glasgow and Edinburgh universities, studying advanced Latin and Greek, logic and mathematics, metaphysics and rhetoric. He was licensed to preach by the Presbytery of Dunbarton in 1815 and was ordained to the Ministry in 1816. In the same year he took out his patent on the air engine and heat regenerator. The patent was also signed by his younger brother, James, who was in fact a mechanical engineer. It seems probable, therefore, that Robert had the idea for the engine and James developed it.

Robert also designed and made scientific instruments and various other patents relating to air engines in the names of R. and J. Stirling were granted over the years, the last being in 1840.

In 1819 Robert Stirling married Jane, eldest daughter of William Rankine, a wine merchant at Galston, and five years later became minister of the church there. Early in 1840 St Andrew's University awarded him an honorary degree of Doctor of Divinity. He remained at Galston for the rest of his life, retiring in 1876 and dying two years later.

There were several patents for air engines before Stirling's first patent of 1816 but it is doubtful if any of them would have worked, with the probable exception of that of Sir George Cayley.

The Stirling cycle engine differs from the internal-combustion engine in that the working fluid (in Stirling's case, air) remains in the working chambers. The heat is applied from an external source, so virtually any fuel, from wood to nuclear fuel, can be used. It also means that combustion can be made to take place under the best conditions, making the control of emissions (pollution) considerably easier. The burning of the fuel is continuous, not intermittent as in an internal combustion engine, so there is less noise and vibration.

Another advantage is the high theoretical thermal efficiency; in practice thermal efficiencies of Stirling engines of different designs are better than those of conventional diesel engines.

There are many arrangements for Stirling engines but the essential factor is that they use what is effectively two pistons to push the working fluid between two working spaces. One space is kept at a high temperature by the heat source and the other at a low temperature. Between these two spaces is a regenerator which alternately receives and gives up heat to the working fluid.

The pistons are connected to a mechanism which keeps them out of phase (usually by 90°). It is this differential motion which moves the working fluid from one space to the other. On its way to the hot space the fluid passes through the regenerator, gaining heat. In the hot space it gains more heat and expands, giving power. After the power stroke the fluid is pushed back through the regenerator, where it gives up its residual heat into the cold space and is ready to start the cycle again.

Stirling's first engine appeared in 1818. It had a vertical cylinder about 60 cm/2 ft in diameter. It produced about 1.5 kW/2 hp) pumping water from a quarry and ran for two years before the hot sections of the cylinder burnt out. This burning out is a problem which has plagued virtually every engine of the type ever since.

In 1824, the brothers started work on improved engines and in 1843 converted a steam engine at a Dundee factory to operate as a Stirling engine. It is said to have produced 28 kW/37 hp and to have used less coal per unit of power than the steam design it replaced. In any event, the hot parts burned out continually after a few months, and after several replacements it was re-converted to steam.

The type lived on, however, until well into the next century and was used extensively for powering small pumps and similar domestic applications. Improved Otto cycle engines and the standardization and spread of electricity supplies helped to establish the small electric motor and led to the Stirling engine's eventual demise.

Steel was not to become common until after 1860 when Bessemer built his steelworks in Sheffield. Now even steel is not considered suitable for the high temperatures the engine requires for greatest efficiency.

The renaissance of the Stirling engine began in 1938 in the Philips laboratories in the Netherlands. In the early 1980s most effort was directed towards developing the principle for automotive applications. A joint Ford–Philips programme resulted in a Stirling-engined vehicle which was cleaner and had a fuel economy 9–35% better than the conventionally engined car.

The Stirling engine has yet to live up to its promise. Possibly the greatest hurdle is the huge investment already made in conventional internal-combustion engines. For them to be deposed, the Stirling principle will have to show substantial advantages in most respects.

Stock Alfred 1876–1946 was a German inorganic chemist best known for his preparations of the hydrides of boron (called boranes) and for his campaign for better safety measures in the use of mercury in chemistry and industry.

Stock was born in Danzig (now Gdańsk, Poland) on 16 July 1876. He studied chemistry at the University of Berlin under Emil Fischer, and after receiving his doctorate became Fischer's assistant. In 1909 he moved to Breslau, to join the staff of the Inorganic Chemistry Institute. After a period at the Kaiser Wilhelm (later Max Planck) Institute in Berlin, he became Director of the Chemistry Department at the Technische Hochschule in Karlsruhe in 1926, where he remained until he retired ten years later. He died in Karlsruhe on 12 August 1946.

Stock began studying the boron hydrides – general formula B_xH_y – in 1909 at Breslau. By treating magnesium boride (Mg_3B_2) with an acid he produced B_4H_{10}. He went on to prepare several other hydrides and in 1912 devised a high-vacuum method for separating mixtures of them. Many contain more hydrogen atoms in their molecules than ordinary valency rules will allow, at least if normal covalent bonds are involved. Their structures were finally

worked out by Linus Pauling, Hugh Christopher Longuet-Higgins, William Lipscomb and others. In the 1960s boron hydrides found their first practical use as additives to rocket fuels.

In 1921, Stock prepared beryllium (scarcely known before in the metallic state) by electrolysing a fused mixture of sodium and beryllium fluorides. This successful extraction method made beryllium available for industrial use, as in special alloys and glasses and for making windows in X-ray tubes. By 1923 Stock was suffering from chronic mercury poisoning caused by prolonged exposure to the liquid metal and its vapour – a fate previously shared by many other chemists. He introduced sensitive tests for mercury and devised improved laboratory techniques for dealing with the metal to minimize the risk of accidental poisoning.

Stokes George Gabriel 1819–1903 was a British physicist who is mainly remembered for Stokes's law, which relates the force moving a body through a fluid to the velocity and size of the body and the viscosity of the fluid. He also made important contributions in optics, particularly in the field of fluorescence, a term that he coined.

Stokes was born at Skreen, Sligo, in Ireland on 13 August 1819 and went to school in Dublin and to college in Bristol, England, before entering Cambridge University in 1837. He graduated in mathematics in 1841, and became a fellow of Pembroke College. In 1849, Stokes was appointed Lucasian Professor of Mathematics at Cambridge, a position he retained until his death. Among the honours accorded to him were the Presidency of the Royal Society from 1885 to 1890, a knighthood in 1889, and the Royal Society's Copley Medal in 1893. He died at Cambridge on 1 February 1903.

Much of Stokes's reputation was gained through the work he carried out on the theory of viscous fluids from 1845 to 1850. He derived the equation later known as Stokes's law, which determines the movement of a small sphere through viscous fluids of various density. The equation can be stated as $F = 6\pi\eta rv$, where F is the force acting on the sphere, η the coefficient of viscosity, r the radius of the sphere and v its velocity. Stokes's law enables the resistance of fluids to motion to be assessed and the terminal velocity of a body to be calculated. It was to be of enormous importance as the basis of the famous oil-drop experiments to determine the charge on the electron performed by Robert Millikan (1868–1953) from 1909 to 1913.

Stokes's investigation into fluid dynamics led him to consider the problem of the ether, the hypothetical medium that was believed to exist for the propagation of light waves. In 1848, Stokes showed that the laws of optics held if the Earth pulled the ether with it in its motion through space and from this assumed the ether to be an elastic substance that flowed with the Earth. The classic Michelson–Morley experiments of 1881 and 1887 did not totally negate this contention; it was shown to be untrue by Oliver Lodge (1851–1940) in 1893 and the existence of the ether was finally disproved.

Stokes made another important contribution with the first explanation of fluorescence in 1852. He had examined the blue light that is emitted in certain circumstances by a solution of quinine sulphate that is normally colourless. Stokes showed that ultraviolet light was being absorbed by the solution and then re-emitted as visible light. He was then able to use fluorescence as a method to study ultraviolet spectra. Stokes went on in 1854 to realize that the Sun's spectrum is made up of spectra of the elements it contains and concluded that the dark Fraunhofer lines are the spectral lines of elements absorbing light in the Sun's outer layers. He did not, however, develop this important idea. It was left to Robert Bunsen (1811–1899) and Gustav Kirchhoff (1824–1887) to propose the method of spectrum analysis in 1860.

Another field in which Stokes made his mark was geodesy and in 1849 he published an important study of the variation of gravity at the surface of the Earth. Stokes was also a gifted mathematician and helped to develop Fourier series.

Stokes William 1804–1878. Scottish physician; *see* Cheyne and Stokes.

Stopes Marie Charlotte Carmichael 1880–1958 was an early British advocate of birth control who, in 1921, founded the first instructional clinic for contraception in Britain.

Marie Stopes was born on 15 October 1880 in Edinburgh. Her mother was a feminist and one of the first woman members of Edinburgh University; her father was a brewing engineer from Essex. She read botany at University College, London, graduating in 1902, then went to the University of Munich, from which she gained her doctorate in 1904. She was awarded her DSc from London University in 1905, when only 25 years old, and then taught at the University of Manchester – the first woman to be appointed to the science staff there. For several years she continued her palaeobotanical research into fossil plants and primitive cycads and became one of the foremost investigators in her field. In 1911, she married Reginald Ruggles Gates, a Canadian botanist, and left Manchester University. But the marriage was not consummated and was annulled in 1916.

The breakdown of her marriage stimulated Marie Stopes's interest in the subject of sexual intercourse, personal relationships and marriage, and in 1918 she published *Married Love*, the underlying theme of which is that women should be able to enjoy sexual intercourse on the basis of equality with men; in the book she also referred briefly to contraceptive methods. This topic was extremely controversial in Britain at that time and she had great difficulty in finding a publisher. And after it was published, *Married Love* met with considerable opposition: for example, C. P. Blacker (later to help in the creation of the International Planned Parenthood Foundation) said that the book was 'responsible for printing instructions to girls of initially dubious virtues as to how to adopt the profession

of more or less open prostitution'. Nevertheless, Marie Stopes received many requests for more information and advice about contraception from women who had read her book, so later in the same year she wrote and published *Wise Parenthood*, in which she attempted to answer the queries she had received.

Also in 1918 Marie Stopes married for the second time; her husband was Humphrey Verdon Roe, the cofounder of the A. V. Roe aircraft company. Roe supported Marie Stopes's ideas and sponsored her birth control clinic, the first one in Britain, which opened in 1921 in Marlborough Road, Holloway, London. This event re-aroused vehement opposition, especially from the Roman Catholic Church, and Marie Stopes spent the next few years both promoting and defending the idea of contraception. In 1934, she published another book *Birth Control Today*, in which she voiced her disapproval of abortion as a means of population control, describing women who sought abortions as 'a danger to the human race'. She continued to champion the cause of birth control in her later years, travelling to many different countries to do so. She died on 2 October 1958 near Dorking, Surrey, having brought about a considerable change in general attitudes towards a more widespread acceptance of contraception, a trend that continued after her death.

Strömgren Bengt Georg Daniel 1908–1987 was a Swedish astronomer best known for his hypothesis about the so-called 'Strömgren spheres' – zones of ionized hydrogen gas surrounding hot stars embedded in interstellar gas clouds.

Strömgren was born on 21 January 1908 in Göteborg, the son of Elis Strömgren, who was also an astronomer of distinction. In 1927 he received an MA from the University of Copenhagen and he was awarded a PhD from the same university in 1929. He was appointed a lecturer there, but in 1936 he moved to the University of Chicago as an Assistant (and later Associate) Professor. In 1938 he became Professor of Astronomy at the University of Copenhagen and in 1940 he succeeded his father as Director of the Observatory there.

In 1946, Strömgren served as Visiting Professor to the University of Chicago. He was appointed a special lecturer in astronomy at the University of London in 1949 and was Visiting Professor to both the California Institute of Technology and to Princeton University in 1950. From 1951 to 1957 he was Professor at the University of Chicago and Director of both the Yerkes and McDonald observatories. From 1957 to 1967 he was a member of the Institute of Advanced Study at Princeton. He was a member of many scientific associations in Europe and the United States and was awarded the Bruce Medal of the Astronomical Society of the Pacific in 1959. Strömgren retired in 1967.

Some gaseous nebulae that can be observed within our Galaxy are luminous. In 1940, Strömgren proposed that this light was caused by hot stars embedded within obscuring layers of gas in the nebulae. He suggested that these stars ionize hydrogen gas and that the dimensions of the ionized zone (the H II zone) depend on both the density of the surrounding gas and the temperature of the star.

Strömgren's calculations of the sizes of these H II zones or 'Strömgren spheres' have been shown by observations to be largely correct. This concept was fundamental to our understanding of the structure of interstellar material. Strömgren's other work included an analysis of the spectral classification of stars by means of photoelectric photometry, and research into the internal make-up of stars, all these areas being fundamental to the development of modern astronomy.

Strutt John William 1842–1919. English physicist; *see* Rayleigh, Lord.

Struve Friedrich Georg Wilhelm von 1793–1864 was a German astronomer who was an expert on double stars and one of the first astronomers to measure stellar parallax. He was also the founder of a dynasty of famous astronomers that spanned four generations.

Struve was born in Altona, in Schleswig-Holstein, on 15 April 1793. To avoid conscription into the German army, he fled to Dorpat (now Tartu) in Estonia in 1808. He entered the University of Dorpat and graduated in 1810. His interest in astronomy led to his appointment as an observer at the Dorpat Observatory in 1813. In the same year he was awarded his PhD and became Extraordinary Professor of Mathematics and Astronomy at the University of Dorpat.

From 1817 onwards Struve served as Director of the Dorpat Observatory, but after 1834 he was primarily concerned with the construction and equipping of an observatory at Pulkovo near St Petersburg, which was opened in 1839. He retired in 1862 and was succeeded as Director of the Pulkovo Observatory by his son. Struve died in Pulkovo on 23 November 1864. At the time of his death he was a member of virtually every European scientific academy.

Struve's earliest research dealt with questions of geodesy and stellar motion. His primary interest, however, was in the discovery and measurement of double stars. In 1822 he published a catalogue of about 800 known double stars, and he instigated an extensive observational programme. The number of such stars known had increased to more than 3,000 by 1827. Struve published a paper in 1843 in which he described more than 500 multiple stars in addition to his earlier work on double stars.

Struve was one of the first astronomers to detect stellar parallax successfully. His interest in that subject dated from 1822, and in 1830 he measured the parallax of Alpha Lyrae. Other work of particular note included his observations, published in 1846, of the absorption of stellar light in the galactic plane, which he correctly deduced to be caused by the presence of interstellar material. He also investigated the distribution of stars in space. In addition to his work in astronomy, Struve made significant contributions to geodesy with his survey of Livonia (1816–19) and his measurements of the arc of meridian (1822–27).

Struve Gustav Wilhelm Ludwig Ottovich von 1858–1920 the younger brother of Karl Hermann von Struve and son of Otto Wilhelm von Struve, was an expert on the occultation of stars and stellar motion.

Struve was born on 1 November 1858 at Pulkovo, near St Petersburg, Russia. He attended school at Vyborg and followed the family tradition by studying astronomy at the University of Dorpat (now Tartu, Estonia). After his graduation in 1880 he began research at the Pulkovo Observatory near St Petersburg. From 1883 to 1886 Struve travelled through Europe, visiting observatories in many countries.

When he returned to Pulkovo in 1886 Struve continued his work at the Observatory. He wrote a thesis on the constant of precession and was awarded his doctorate in 1887. In 1894 he moved to the University of Kharkov, where he was made Extraordinary Professor of Astronomy. In 1897 he became a full Professor and Director of the University Observatory. In 1919 he left Kharkov for Simferopol, where he was appointed Professor at the Tauris University. He died in Simferopol on 4 November 1920.

Struve's father had done excellent work on determining the constant of precession. Struve was also interested in this subject and he investigated the whole question of motion within the Solar System. This led him on to work on the positions and motions of stars, and to an estimation of the rate of rotation of the Galaxy.

Struve was best known for his expert knowledge about the occultation of stars during a total lunar eclipse. Much of his early work on this subject was done in the 1880s, but his interest in it continued until the end of his career.

Struve Karl Hermann von 1854–1920, third in the line of famous astronomers, was an expert on Saturn. His other work was largely concerned with features of the Solar System, although he also shared the family interest in stellar astronomy.

Struve was born on 30 October 1854 at Pulkovo. He studied at Karlsruhe, Vyborg and Reval (Tallinn), before enrolling at the University of Dorpat (now Tartu) in 1872. On completing his undergraduate studies in 1877, Struve travelled in Europe and visited major centres of astronomical research. He was awarded his PhD from Dorpat in 1882.

A year later Struve was appointed Astronomer at Pulkovo and he became Director there in 1890. In 1895 he left to become Professor of Astronomy at the University of Königsberg. He was later appointed Director of the Observatory of Berlin-Babalsberg (1904), and in 1913 he became the founder-Director of the Neubabalsberg Observatory. He was a member of numerous scientific societies and the recipient of the 1903 Gold Medal of the Royal Astronomical Society. He died in Herrenalb, Germany, on 12 August 1920.

Among the many features of the Solar System studied by Struve were the transit of Venus, the orbits of Mars and Saturn, the satellites (especially Iapetus and Titan) of Saturn, and Jupiter and Neptune. Struve's best work was his 1898 paper on the ring system of Saturn. Data in this publication formed the basis of much of his later research.

Struve Otto von 1897–1963, son of Gustav Wilhelm Ludwig von Struve, was the last of four generations of a family of eminent astronomers. He contributed to many areas of stellar astronomy, but was best known for his work on interstellar matter and stellar and nebular spectroscopy.

Struve was born on 12 August 1897 at Kharkov in Russia. He studied at the Gymnasium at Kharkov before entering a school for artillery training in Petrograd (as St Petersburg was known from 1914 to 1924) in 1915. He served in the Imperial Russian Army on the Turkish front during World War I. After the war he studied at the University of Kharkov, where he was awarded a degree with top honours in 1919. He was conscripted into the counterrevolutionary White Army during the Civil War in 1919, but he fled to Turkey in 1920.

With the aid of E. B. Frost, Director of the Yerkes Observatory, Struve went to the United States in 1921. He became an assistant at the Observatory and studied for his doctorate, which was awarded in 1923. He then rose to the ranks of Instructor (1924), Assistant (1927) and Associate (1930) Professor, and Assistant Director of the Observatory (1931). He became Professor of Astrophysics at the University of Chicago in 1932. When Frost retired in 1932, Struve was made Director of the Yerkes Observatory. He was also the founder-Director of the McDonald Observatory in Texas.

Struve taught at the University of Chicago until 1950, when he became Professor of Astrophysics and Director of the Leuschner Observatory at the University of California at Berkeley. He left in 1959 to become Director of the newly established National Radioastronomy Observatory at Green Bank, West Virginia. He retired because of ill-health in 1962, but was appointed joint Professor of the Institute of Advanced Studies and California Institute of Technology.

Struve was a member of the National Academy of Sciences, the Royal Society and the Royal Astronomical Society. He was the recipient of numerous honours including the Gold Medal of the Royal Astronomical Society (1944) and the Draper Medal of the National Academy of Sciences (1950). He died on 6 April 1963 in Berkeley.

Struve's early work was on stellar spectroscopy and the positions of comets and asteroids. Spectroscopic analysis of interstellar space had fascinated him from early in his career, as had double stars. He did early work on stellar rotation and demonstrated the rotation of blue giant stars and the relationship between stellar temperature (and hence spectral type) and speed of rotation. In 1931 he found, as he had anticipated, that stars that spun at a high rate deposited gaseous material around their equators.

In 1936, together with C. T. Elvey, Struve developed a nebular spectrograph that was used to study interstellar gas clouds. In 1938 they were able to demonstrate for the first time that ionized hydrogen is present in interstellar matter. They also determined that the interstellar hydrogen is concentrated in the galactic plane. These observations had important implications for later work on the structure of our Galaxy and for radioastronomy.

Struve was also interested in theories of the evolution of stars, planetary systems and the universe as a whole. He believed that the establishment of a planetary system should be thought of as the normal course of events in stellar evolution and not a freak occurrence. Struve's contributions to astronomy were of fundamental importance to the fast-growing science of the present century, just as his forefathers' work had been in their time.

Struve Otto Wilhelm von 1819–1905 was an active collaborator with his father, Friedrich von Struve, in many astronomical and geodetic investigations. He is best known for his accurate determination of the constant of precession.

Struve was born in Dorpat (now Tartu), Estonia, on 7 May 1819. He took his degree at the University of Dorpat in 1839, although he had begun work as an astronomer two years earlier. From 1839 to 1848 he worked at the Pulkovo Observatory, becoming its deputy Director in 1848, and its Director in 1862. From 1847 to 1862 he held a concurrent post as a military adviser in St Petersburg. In 1889, he retired from the Pulkovo Observatory and moved to Karlsruhe. He was a member of numerous scientifc academies and received the Gold Medal of the Royal Astronomical Society. He died in Karlsruhe on 16 April 1905.

One of the most ambitious observational programmes initiated at the Pulkovo Observatory in its early days was a systematic survey of the northern skies for the purpose of discovering and observing double stars. Struve was one of the most active participants in this programme and has been credited with the discovery of about 500 double stars. He also made detailed measurements of binary systems.

Struve was interested in the Solar System and he made a careful study of Saturn's rings. He discovered a satellite of Uranus and calculated the mass of Neptune. He also concerned himself with the measurement of stellar parallax, the movement of the Sun through space and the structure of the universe, although he was among those astronomers who erroneously believed our Galaxy to be the extent of the whole universe. His determination of the constant of precession, which served as the best estimate for nearly half a century, served to seal his reputation as an astronomer of distinction.

Sturgeon William 1783–1850 was a British physicist and inventor who made the first electromagnets.

Sturgeon was born on 22 May 1783, at Whittington, Lancashire, the son of a shoemaker. Sturgeon was himself apprenticed to a shoemaker in 1796, but in 1802 went into the army (militia) and two years later enlisted in the Royal Artillery. He studied natural sciences in the evenings, but in 1820 (at the age of 37) he went back to shoemaking and set up a business in Woolwich. He became a member of the Woolwich Literary Society and in 1824 became lecturer in science and philosophy at the East India Royal Military

College of Addiscombe. In 1832 he was appointed to the lecturing staff of the Adelaide Gallery of Practical Science, and in 1840 he moved to Manchester to become superintendent of the Royal Victoria Gallery of Practical Science. At this time, as an itinerant lecturer he was able to support his family from an income. He became a member of the Manchester Literary and Philosophical Society, and through the influence of the president of the society received a grant of £200 from Lord Russell. In 1849 he was granted an annuity of £50 to promote his work on electromagnetism, but died in the following year on 4 December at Prestwick, Manchester.

Sturgeon was the founder of electromagnetism, and the first English-language journal to be devoted wholly to electricity was started by him. His scientific work only began in earnest when he returned to civilian life at the age of 37, although he had carried out occasional electrical experiments in Woolwich where he had developed various mechanical skills which were useful for making scientific apparatus, and was often in demand to lecture to schools and other groups.

In 1828 he put into practice the idea of a solenoid, first proposed by André Ampère (1775–1836), by wrapping about 18 turns of wire round an iron core so that they became magnetic when a current was passed through them. He found that each coil reinforced the next, since they effectively formed a set of parallel wires with the current moving in the same direction through all of them. He also noticed that the magnetic field seemed to be concentrated in the iron core and that it disappeared as soon as the current was switched off. He varnished the core to insulate it and keep it from short-circuiting the wires, and also tried using a core that was bent into a horseshoe shape. His device was capable of lifting 20 times its own weight.

He later invented an important new galvanometer. Sturgeon was one of a small group of lecturers and instrument makers who worked at demonstrating electrical science in new ways. In 1836 he established a monthly periodical *Annals of Electricity*, which ran into ten volumes, ending in 1843. He then founded *Annals of Philosophical Discovery* and *Monthly Report of Progress of Science and Art* which terminated at the end of the same year. They were nevertheless landmarks, being the first electrical journals ever to be published in English.

Suess Eduard 1831–1914 was an Austrian geologist who helped pave the way for modern theories of the continents.

Born in London, though of Bohemian ancestry, Suess was educated in Vienna and at the University of Prague. He moved back to Vienna in 1856 and became Professor of Geology there in 1861. As well as his geological interests, he occupied himself with public affairs, serving as a member of the Reichstag for 25 years. His geological researches took several directions. As a palaeontologist, he investigated graptolites, brachiopods, ammonites and the fossil mammals of the Danube Basin. He wrote an original text on economic geology. He undertook important research on the structure of the Alps, the tectonic geology of Italy and

seismology. The possibility of a former landbridge between North Africa and Europe caught his attention.

The outcome of these interests was *The Face of the Earth* 1885–1909, a massive work devoted to analysing the physical agencies contributing to the Earth's geographical evolution. Suess offered an encyclopedic view of crustal movement, of the structure and grouping of mountain chains, of sunken continents, and of the history of the oceans. He also made significant contributions to rewriting the structural geology of each continent.

In many respects, Suess cleared the path for the new views associated with the theory of continental drift in the twentieth century. In view of geological similarities between parts of the southern continents, Suess suggested that there had once been a great supercontinent, made up of the present southern continents; this he named Gondwanaland, after a region of India. Wegener's work was later to establish the soundness and penetration of such speculations.

Sutherland Earl Wilbur 1915–1974 was a US biochemist who was awarded the 1971 Nobel Prize for Physiology or Medicine for his work with cyclic adenosine monophosphate (cyclic AMP), the chemical substance that moderates the action of hormones.

Sutherland was born in Burlingame, Kansas, on 19 November 1915. He graduated from Washburn College, Topeka, in 1937 and received his MD from Washington University Medical School, St Louis, in 1942. After serving as an army officer during World War II he took an appointment at Washington University to do research on hormones under Carl and Gerty Cori. In 1953 he became Director of the Department of Medicine at Western Reserve (now Case Western Reserve) University in Cleveland. Ten years later he was appointed Professor of Physiology at Vanderbilt University, Nashville, and from 1973 until his death he was a member of the faculty of the University of Miami Medical School. He died in Miami on 9 March 1974.

Sutherland began working with hormones at Washington under the Coris and then spent the 1950s doing research on his own – other workers took little interest in his studies. At that time it was thought that hormones, carried in the bloodstream, activated their target organs directly. Sutherland showed that the key to the process – the activating agent of the organ concerned – is cyclic adenosine 3,5 monophosphate (cyclic AMP). The arrival of a hormone increases the organ's cellular level of cyclic AMP, which in turn triggers or inhibits the cellular activity.

Cyclic AMP is present in every animal cell and therefore affects 'everything from memory to toes', as Sutherland himself said, so the implications of his discovery were enormous. The 1971 Nobel prize committee commented that it is rare for such a discovery to be credited to only one person. By the end of the 1960s Sutherland was no longer alone in his research; hundreds of scientists throughout the world were keen to do research on the newly discovered substance.

Sutherland Gordon 1907–1980 was a British physicist who is best known for his work in infrared spectroscopy, particularly the use of this technique for studying molecular structure. He received many honours for his work, including a knighthood in 1960.

Sutherland was born in Caithness, Scotland, on 8 April 1907. He was educated at the Morgan Academy, Dundee, and graduated from St Andrew's University, then did postdoctoral work at Trinity College, Cambridge, where he joined one of Eric Rideal's research groups working on infrared spectroscopy. From 1931 to 1933 he was in the United States as a Commonwealth Fund Fellow at the University of Michigan, Ann Arbor (one of the leading centres for spectroscopy studies), working with D. Dennison. After returning to Britain he was elected to the Stokes Studentship at Pembroke College, Cambridge, in 1934 and was made a Fellow of the College in the following year. During his fellowship – which he held until 1949 – he spent a year as an assistant to the Director of Production of the Ministry of Supply, was a Proctor of the University from 1943 to 1944, and served on the Council of the University Senate from 1946 to 1949; he was also appointed University Reader in Spectroscopy in 1947. While at Cambridge he established a successful research group, first in the Department of Physical Chemistry and later in the newly founded Department of Colloid Science. In 1949 he left Cambridge to take up the position of Professor of Physics at the University of Michigan, a post he held until 1956, when he returned to Britain to become Director of the National Physical Laboratory at Teddington. In 1962, he visited China with one of the first postwar delegations from the Royal Society. From 1964 until his retirement in 1977 Sutherland was Master of Emmanuel College, Cambridge. He died in Cambridge on 27 June 1980.

Sutherland spent most of his career using infrared spectroscopy to elucidate molecular structure. Working with William Penney (1909–1991), he showed that the four atoms of hydrogen peroxide (H_2O_2) do not lie in the same plane but that the molecule's structure resembles a partly opened book, with the oxygen atoms aligned along the spine and the O–H bonds lying across each cover. Later, during World War II, Sutherland and his research group at Cambridge analysed fuel from crashed German aircraft in order to discover their sources of oil. At Michigan University he was one of the first to use spectroscopy to study biophysical problems; he also continued his investigations into simpler molecules and crystals.

As Director of the National Physical Laboratory, Sutherland reorganized the institution along similar lines to that of the US National Bureau of Standards and established departments of pure physics, applied physics and standards. The scientific reputation of the laboratory was greatly enhanced under Sutherland's management and, while there, he also researched into the structure of a wide range of substances from proteins to the different forms of diamonds. After his appointment as Master of Emmanuel College, he became less active in fundamental research and turned his attention to educational issues.

Sutherland Ivan Edward 1938– is a US electronics engineer who pioneered the development of computer graphics, the method by which computers display pictorial (as opposed to alphanumeric) information on a visual display unit (VDU).

Sutherland was born in Hastings, Nebraska, on 16 May 1938. His university education began at the Carnegie Institute of Technology, where he obtained a bachelor's degree in 1959, and continued at the Massachusetts Institute of Technology (MIT) where he obtained a master's degree in 1960 and a PhD in electrical engineering in 1963. He was then called into the US Army Signals Corps. In 1964 he became director of information processing techniques at the Advanced Research Projects Center of the Department of Defense. He stayed in this position until 1966 when he was appointed associate professor of electrical engineering at Harvard University. Two years later he moved to Salt Lake City, to a position as associate professor at the University of Utah, and there founded the Evans and Sutherland Computer Corporation. He became a full professor at Utah in 1972, but left in 1976 to become Fletcher Jones Professor of Computer Science and head of department at the California Institute of Technology.

From 1960 to 1963 he worked in the Lincoln Laboratory at the MIT on the 'Sketchpad' project. This was the first system of computer graphics which could be altered by the operator in the course of its use for calculation and design. Sketchpad used complex arrangements of the data fed into the computer to produce representations of the objects in space as well as fine geometrical detail. Programs could be altered using light-pens which touch the surface of the VDU.

While at Utah, Sutherland was engaged in the design of a colour graphics system able to represent fine distinctions of colour as well as accurate perspective. The image could be moved, rotated, expanded or made smaller to give a realistic image of the object, rendering the computer suitable for use in engineering and architectural design.

Sutton-Pringle John William 1912–1982 was a British zoologist best known for his substantial contribution to our knowledge of insect flight.

Sutton-Pringle was born on 22 July 1912 and was educated at Winchester College and King's College, Cambridge. After graduation he was appointed Demonstrator in Zoology at Cambridge in 1937, a post he held for two years. From 1938 to 1945 he was a Fellow of King's College Cambridge. In 1945 he was appointed to a lectureship at Cambridge, and became a Fellow of Peterhouse College in the same year. He was a Reader in Experimental Cytology from 1959 to 1961, when he became Emeritus Fellow at Peterhouse. In 1977 he was elected President of the Society for Experimental Biology.

Sutton-Pringle helped to establish much of our present knowledge of the anatomical mechanisms involved in insect flight. Most insects have two hindwings and two forewings, and in many species each hindwing is linked to its anterior forewing, thus enabling each pair of wings to

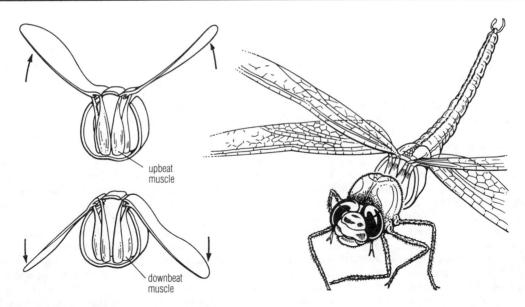

Sutton-Pringle *Sutton-Pringle showed that the rapid wing-beats necessary for flight in insects are achieved by alternate sets of muscles.*

act in unison. Not all species use both pairs of wings for flight, however; in the housefly, for example, the hindwings are reduced in size and serve as balancing organs during flight.

Insect flight is achieved by simple up-and-down movements of the wings. In aphids, for example, these wing movements are brought about by the contractions of two separate sets of muscles: contraction of the longitudinal muscles results in depression of the wings; contraction of the dorso-ventral muscles causes elevation of the wings. When moving through the air, the anterior edge of the wings remains rigid while the posterior edge bends. On the downward stroke, the posterior edge is displaced upwards, and on the upward stroke the posterior edge is displaced downwards. This, in turn, causes the development of a localized region of high pressure air behind the insect, which propels the insect forwards. The faster the rate of wing-beats, the greater the displacement of the posterior wing edges, the greater the pressure exerted on the insect from behind, and therefore the faster the insect flies.

Svedberg Theodor 1884–1971 was a Swedish physical chemist who invented the ultracentrifuge to facilitate his work on colloids. For his contributions to colloid chemistry he was awarded the 1926 Nobel Prize for Chemistry.

Svedberg was born in Fleräng, near Gävle, on 30 August 1884. As a secondary-school student he became interested in natural sciences (particularly botany), and resolved to study chemistry in the belief that many of the unsolved problems in biology could be explained as chemical phenomena. He entered Uppsala University in 1904 and remained associated with it for the rest of his life. He obtained a BSc in 1905 and a PhD two years later, with a thesis on his studies of colloidal solutions. In 1912 he was

appointed to the first Chair in Physical Chemistry in Sweden. When he retired from this post in 1949, he became Head of the new Gustav Werner Institute of Nuclear Chemistry. He resigned in 1967 and died in Örebro on 25 February 1971.

Colloid chemistry was Svedberg's interest for 20 years. By 1903 he was already influenced by the work of Hermann Nernst, Richard Zsigmondy and, particularly, Georg Bredig. Bredig had devised a method of preparing metal sols by passing an electric arc between metal electrodes submerged in a liquid. Svedberg used alternating current with an induction coil having its spark gap in a liquid. In this way he prepared a number of new organosols from more than 30 metals, which were more finely dispersed and much less contaminated than Bredig's. Also the method was reproducible, so that such sols could be used for quantitative analyses in physico-chemical studies. Using an ultramicroscope, he studied the Brownian motion of particles in these sols and correlated the observations with the effects of temperature, viscosity and the nature of the original solvent. These experiments confirmed Albert Einstein's theories about Brownian motion.

Svedberg also had a continuing interest in radioactive processes and, with D. Stronholm, experimented with isomorphic coprecipitation of various radioactive substances. He discovered that thorium-X crystallizes with lead and barium salts (but not with others), anticipating Frederick Soddy's demonstration of the existence of isotopes. By about 1923 Svedberg had also investigated a totally different subject, the chemistry involved in the formation of latent images in photographic emulsions.

In 1924 Svedberg constructed the first ultracentrifuge, a development which made a timely and significant contribution to the study of large molecules. Its ability to sort particles by weight can reveal the presence of contaminants in a sample of a new protein, or distinguish between various long-chain polymers in substances such as cellulose and other natural polymers. His other researches in the 1930s confirmed his view that these substances consist of well defined uniform molecules.

During World War II the Swedish government asked Svedberg to investigate methods of producing synthetic rubber (at that time, polychloroprene). This research led to the establishment of a small manufacturing plant in the north of Sweden. He also studied other synthetic polymers, introducing electron microscopy to study natural and regenerated cellulose, X-ray diffraction techniques to investigate cellulose fibres, and electron diffraction to analyse colloidal micelles and crystallites.

In the late 1930s Svedberg's interest in radiochemistry prompted a need to increase the capacity for making radioactive isotopes. Finance from the Swedish industrialist Gustaf Werner was used to build a large cyclotron, founding the Gustaf Werner Institute of Nuclear Chemistry.

After his official retirement in 1949, Svedberg became the head of the institute and recruited the staff. One group worked on the biological and medical applications of the cyclotron, while another group investigated the effects of radiation on macromolecules, together with problems in radiochemistry and radiation physics.

Swammerdam Jan 1637–1680 was a Dutch naturalist who investigated many aspects of biology but who is probably best known for his outstanding microscope observations, his detailed and accurate anatomical descriptions, and his studies of insects. He is considered by many to be a founder of both comparative anatomy and entomology.

Swammerdam was born in Amsterdam on 12 February 1637. He was the son of an apothecary whose hobby was a museum of curiosities, a hobby that stimulated the younger Swammerdam's interest in natural history, particularly in insects. He graduated with a medical degree from Leiden University in 1667 but never practised as a physician, preferring to pursue his interest in natural history. Subsequently his father, who wanted his son to become a priest, withdrew his financial support – despite which the younger Swammerdam continued his biological studies, although he suffered severe privations and became chronically ill, both physically and mentally. In 1673 Swammerdam came under the influence of the religious zealot Antoinette Bourignon and became increasingly embroiled in religious controversy until he died, only 43 years old, in Amsterdam on 15 February 1680.

Swammerdam made many important contributions to biological knowledge but most of his studies were of insects. He accurately described and illustrated the life cycles and anatomies of many species, including bees, mayflies and dragonflies. In mayflies and dragonflies the change from the last nymph stage to the winged adult is outwardly the most striking, but Swammerdam showed that rudimentary wings occur in the aquatic nymphs some time before the final moult. He also showed that caterpil-

lars develop wings and adult-type legs shortly before pupating. From his observations of their metamorphic development Swammerdam classified insects into four major groups, three of which are still used in a modified form in modern insect classification. In addition, he disproved many false beliefs about insects – for example, that their bodies are structureless, fluid-filled cavities without fully formed internal organs.

Swammerdam also studied vertebrates, about which he provided a substantial body of new knowledge, most of which was correct. He showed that the lungs of newly born mammals sink in water when the lungs are taken from the animals before breathing has started but that lungs taken from young animals whose respiration has been established float. He also erroneously believed, however, that the movements of the chest in mammals are unrelated to inhalation and exhalation but are associated with transferring air from the lungs to the heart. Swammerdam demonstrated that muscles removed from a frog could be stimulated to contract and that when muscle (including heart muscle) contracts it does not increase in volume. Furthermore, he anticipated the discovery of the role of oxygen in respiration by postulating that air contained a volatile element that could pass from the lungs to the heart (contributing to the respiration operation of the heart) and then to the muscles, providing the energy for muscle contraction. Investigating the anatomy of the frog, he observed that the frog's egg passes through a stage when it consists of four joined globules (now known to be the second cleavage of the fertilized egg). He was also probably the first to discover red blood cells when he observed oval particles in frog's blood in 1658.

In his work on human and mammalian anatomy, Swammerdam discovered valves in the lymphatic system; these valves are now called Swammerdam valves. He also investigated the human reproductive system and was one of the first to show that female mammals produce eggs, analogous to birds' eggs. In addition he perfected a technique for injecting dyes into dissected cadavers in order to display anatomical details.

Swammerdam's work – particularly his insect studies – had a profound impact on scientific thinking, although his manuscripts were not published in full until 1737, when Hermann Boerhaave published *Biblia Naturae/Bible of Nature*, a two-volume Latin translation of Swammerdam's Dutch text that included illustrations engraved from Swammerdam's own drawings. The work is one of the finest collections of biological observations ever published and, even today, many of Swammerdam's illustrations remain unsurpassed.

Swan Joseph Wilson 1828–1914 was a British inventor and electrical engineer – trained originally as a chemist who invented an incandescent electric lamp. He also made major contributions to photographic processing, and lesser ones to electroplating methods, electric cells, and the production of artificial fibres.

Swan was born on 31 October 1828 in Sunderland. For

a time during his childhood his parents allowed him to roam about Sunderland, where he became fascinated with the busy industries, towns and ports. Eventually he was sent to a 'dame school' run by three old ladies, and from there he went to a large school under the direction of Dr Wood, a Scottish minister. When he left school he became apprenticed to a Sunderland firm of retail chemists. Before his apprenticeship was finished both partners in the business died. Joseph then went to join his brother-in-law, John Mawson, in his chemical firm at Newcastle-upon-Tyne. During these times he was fascinated by scientific and engineering inventions, and he attended lectures and read books and journals which described them. It was then that he learned of the early interest in the electric lamp of J. W. Starr and W. E. Staite.

Swan quickly proved his worth and soon the firm was producing photographic chemicals. He developed a deep interest in the photographic process, making an experimental study of the various methods. One particular wet process for producing photographic prints, using a gelatine film impregnated with carbon or other pigment granules and photosensitized using potassium dichromate, was patented by Swan in 1864. This was known as the carbon or autotype process. A few years later, in 1879, he invented and patented a bromide printing paper, now a standard photographic medium.

Swan's interest in electric lighting is said to stem from a lecture given by Staite in Sunderland in 1845. From about 1848 he began making filaments by cutting strips of cardboard or paper and baking them at high temperatures to produce a carbon fibre. The recipes he used were often exotic, sometimes entailing cooking with syrup of tar. These filaments were made into coils or circles. In making the first lamps he connected the ends of a filament to wire (itself a difficult task), placed the filament in a glass bottle, and attempted to evacuate the air and seal the bottle with a cork. Usually the filament burned away very quickly in the remaining air, blackening the glass at the same time.

In 1865 the German chemist Hermann Sprengel (1834–1906) invented a mercury vacuum pump, which was used to evacuate the air from radiometer tubes produced by William Crookes (1832–1919). Swan read of Crookes's work and saw the pump as a means of producing an improved vacuum for his lamps. He came across Charles Stearn, who had become familiar with the technique of producing a vacuum using Sprengel's pump, when he read a newspaper advertisement. Swan them produced filaments for Stearn to use to make lamps using the Sprengel pump. Although their first experiments were not very successful they found that, if after first producing the best possible vacuum a strong current was passed to make the filament burn brightly, and if the bulb was further evacuated (thus drawing out the products of the combustion of the carbon with the remnants of the air), then a fairly durable incandescent lamp was produced. Swan demonstrated his electric light and exhibited it throughout the northeast of England in 1878 and 1879, at the same time producing a new type of filament from cotton thread partly dissolved by sulphuric

acid. He patented the process in 1880.

From the summer of 1880 onwards, Swan's electric lamps were manufactured. First they were made in a factory in Benwell, Newcastle-upon-Tyne, but soon a larger London company was formed. Thomas Edison (1847–1931), famous for his electrical inventions (including the phonograph), developed an electric light on a similar principle to Swan's. Edison was quick to take out patents while Swan hesitated. In 1882 he initiated litigation for patent infringement against Swan, but this was dismissed and the joint company Edison and Swan United Electric Light Company came into being in 1883.

In the 1880s electrical supply was in its infancy. There were few electric companies, and those which did exist distributed over very small areas; usually electricity users had their own generators. But the availability of the electric lamp had a great stimulus on the electricity industry, first in public buildings and the private residences of a few notables and then, within a few years, in shops, factories, offices and ships of the Merchant and Royal Navy. Also its potential for advertising was soon recognized.

Swan did not rest to merely reap the financial rewards of his invention. He made a miner's electric safety lamp which, although far too costly to be adopted at the time, was the ancestor of the modern miner's lamp. In the course of this invention he devised a new lead cell (battery) which would not spill acid. He had a lifelong interest in electrical cells, and attempted to make an early type of fuel cell.

In the course of developing a method of producing uniform carbon filaments, Swan devised a process in which nitrocellulose (made by nitrating cotton) was dissolved in acetic acid and extruded through a fine die. This process had obvious advantages in producing lighting filaments, but it was also seen by Swan to be capable of producing an artificial silk. His wife and daughter crocheted some of the material and it was exhibited as 'artifical silk' at an exhibition in 1885, but Swan never considered commercial production.

Throughout his life Swan was an ardent and determined experimenter, with interests ranging from photography to electroplating. He carried out a very extensive study of the best conditions for the electrodeposition of copper and a number of other metals. Many of his discoveries were of practical use, and he took out more than 70 patents. He was elected a Fellow of the Royal Society in 1894, and ten years later he was knighted. He held high offices in a number of professional socities and was in receipt of medals from both the Royal Society and the Royal Photographic Society. He died at home in Warlingham, Surrey, on 27 May 1914, aged 85.

Swinburne James 1858–1958 was one of the leading pioneers in electrical engineering and plastics.

Swinburne was born at Inverness, Scotland, on 28 February 1858. He was the third of six sons of a naval captain and spent most of his childhood on the lonely island of Eileen Shona, in Loch Mordart, where the common spoken language was Gaelic. Eventually he was sent to study at

Clifton College, where the accent was on science subjects, a bias Swinburne's talents were quick to appreciate.

Swinburne began his career in engineering with an apprenticeship to a locomotive works in Manchester, after which he travelled to Tyneside and found employment with an engineering firm. It was during this time that his lifelong interest in electrical engineering developed, and in 1881 Sir Joseph Swan engaged him to take responsibility for setting up a new factory in Paris for the manufacture of Swan's electric lamps. Swinburne carried out the task so successfully that in the following year Swan sent him to the United States to set up a factory there. From 1885 Swinburne was employed as technical assistant to Rookes Crompton, at his Chelmsford factory. He worked on many aspects of electrical engineering and was particularly involved in the development of dynamos and the well-known 'hedgehog' transformer.

In 1894, Swinburne decided to set up his own laboratory and made himself available as a consultant. Some of the research carried out in his laboratory focused on the reaction between phenol and formaldehyde and its commercial potential. Unfortunately when Swinburne came to patent the product in 1907 he was beaten to the idea by the Belgium chemist Leo Baekeland (with his invention of Bakelite). Swinburne was able to obtain a patent on the production of a laquer, however, and set up his own manufacturing concern, Damard Laquer Company, in Birmingham. Baekeland bought him out in the early 1920s and formed Bakelite Limited, Great Britain, of which Swinburne became the first chairman, a position he maintained until 1948. From then until his retirement in 1951 he remained on its board of directors.

Swinburne lived to be 100 years old and was greatly respected by the scientific and industrial world, being affectionately known as 'The Father of British Plastics'

It was in the 1880s that the great march of electrical progress began to gather momentum. Swan and his filament lamp inspired the idea of using electricity for domestic purposes, and other great engineers concentrated on the development of heavier equipment such as dynamos and transformers. When Swinburne joined Crompton's team the impetus was reaching its peak, with many of the problems in electrical engineering being solved and with modifications being made to existing ideas.

Swinburne's own contributions were wide-ranging and included the invention of the watt-hour meter and the 'hedgehog' transformer for stepping up medium voltage alternating current to high voltages for long-distance power transmission. He was also responsible for numerous smaller inventions and for work on the theory of dynamos. The words 'motor' and 'stator', thought to have been coined by him, are now in common use in electrical engineering.

Swings Pol F. 1906– is a Belgian astrophysicist with a particular interest in cometary spectroscopy.

Swings was born in Ransart on 24 September 1906. He studied mathematics and physics at the University of Liège, where he earned his doctorate in mathematics in 1927. He served as an assistant at the University from 1927 to 1932, although he spent much of his time abroad conducting his postdoctoral research. In 1930 he was awarded a special DSc in physics and he was appointed Professor of Astrophysics at the University of Liège in 1932.

During the 1930s and 1940s, Swings spent several years in the United States as Visiting Professor at a number of universities, including Chicago and California. He has been a member of numerous professional bodies including the International Astronomical Union, which he served both as vice president and president, the National Academy of Sciences in the United States, the Royal Astronomical Society in London and the Royal Belgian Academy of Sciences. He received the Francqui Prize in 1947 and the Decennial Prize in 1958, the highest Belgian honours. Swings was co-author of an atlas of cometary spectra, published in 1956, and the recipient of several honorary degrees.

Swings's early astronomical studies concentrated on celestial mechanics, but he soon displayed an interest in the more modern discipline of spectroscopy. At first he approached the subject in an experimental laboratory fashion, but later he applied his expertise in spectral analysis to an investigation of the constitution of a number of types of celestial bodies. His most influential work dealt with the study of cometary atmospheres, and he is credited with the discovery of the Swings bands and the Swings effect. Swings bands are emission lines resulting from the presence of certain atoms of carbon; the Swings effect was discovered with the aid of a slit spectrograph and is attributed to fluorescence resulting partly from solar radiation.

Swings has also made spectroscopic studies of interstellar space and has investigated stellar rotation, as well as nebulae, novae and variable stars.

Sylow Ludwig Mejdell 1832–1918 was a Norwegian mathematician who is remembered for his fundamental theorem on groups and for the special type of subgroups which are named after him.

Sylow was born at Christiania (now Oslo) on 12 May 1832 and attended the cathedral school there until 1850, when he entered the University of Christiania to study mathematics. After graduating, he trained to become a teacher and received his certificate in 1856. For the next 40 years he taught in a school at Halden, although he travelled around Europe on a scholarship in 1861 and lectured at the University of Christiania in 1862. From 1873 to 1877 he was given a leave of absence from his teaching duties to collaborate with Sophus Lie on producing an edition of the works of Niels Abel. A year earlier Sylow had published his theorem, the first major advance in group theory since Cauchy's work of the 1840s, and still regarded as essential for work on finite groups. At the same time he introduced the concept of subgroups, now known by his name. The edition of Abel's work prepared by Sylow and Lie was published in 1881; it was followed by Sylow's edition of Abel's letters in 1901. Through Lie's influence, Sylow was rewarded with a Chair of Mathematics at the University of Christiania in 1898, where he remained until his death on 7 September 1918.

Sylvester James Joseph 1814–1897 was a British mathematician, one of the pre-eminent algebraists of the nineteenth century and the discoverer, with Arthur Cayley, of the theory of algebraic invariants.

Sylvester was born at London on 3 September 1814, the sixth of nine children in the family of Abraham Sylvester, a Jew from Liverpool. He was educated at a school for Jewish boys in north London and in 1828 entered the new University of London, founded especially for Dissenters who were still unable to take degrees at the ancient universities, and dedicated to establishing the sciences on a finer basis in the university curriculum. After only five months there he was expelled for attempting to wound a fellow student with a table knife. In 1829 he enrolled at the Royal Institution school at Liverpool, where he won a prize of £250 for a paper on arrangements before running away to Dublin, apparently to escape from antisemitic persecution. He was sent back to England by a cousin of his mother and in 1831 entered St John's College, Cambridge. In the Tripos of 1837 he emerged as second Wrangler but, unable to subscribe to the Thirty-nine Articles of the Church of England, he was unable to compete for the Smith's Prize or to take his degree. He therefore went to Trinity College, Dublin, where he gained his BA in 1841.

In the meantime he had been appointed to the Chair of Natural Philosophy at University College, London, in 1837 and elected to the Royal Society in 1839. In 1841, he went to the United States to become Professor of Mathematics at the University of Virginia, but after some sort of personal squabble (the truth has never been established) resigned the Chair and returned to England in 1845. For the next ten years he abandoned academic life, although he took in private pupils, including Florence Nightingale. He worked for the Equity and Law Life Assurance Company from 1845 to 1855, entered the Inner Temple in 1846 and was called to the bar in 1850. At about this time he also founded the Law Reversionary Interest Society.

In 1855 he returned to academic life, becoming Professor of Mathematics at the Royal Military Academy at Woolwich and editor of the *Quarterly Journal of Pure and Applied Mathematics*, over which he presided from its first number in 1855 to 1877. He remained at Woolwich until 1877, when he again went to the United States to become Professor of Mathematics at the newly founded Johns Hopkins University at Baltimore. During his tenure there he founded, in 1883, the American *Journal of Mathematics*. In that year he took up his last academic post as Savilian Professor of Geometry at Oxford University, where he was elected a Fellow of New College. He died at London, three years after his retirement, on 15 March 1897.

Sylvester was a prolific writer of mathematical papers, but he was unmethodical, and his brilliant inventiveness was somewhat dulled by his failure to provide rigorous proofs for the ideas which, born of his creative intuition, he

was confident in asserting. He is remembered chiefly for his algebra, especially for laying the foundations with Arthur Cayley (with whom he did not collaborate) of modern invariant algebra. He also wrote two long memoirs (1853 and 1864) on the nature of roots in quintic equations and did brilliant work on the theory of numbers, especially in partitions and Diophantine analysis. He introduced the concept of a denumerant and he coined the term 'matrix' (in 1850) to describe a rectangular array of numbers out of which determinants can be formed. His high achievements were acknowledged by the Royal Society, which awarded him the Royal Medal in 1861 and the Copley Medal in 1880.

Synge Richard Lawrence Millington 1914– is a British biochemist who has carried out research into methods of isolating and analysing proteins and related substances. He is best known for the work on paper chromatography he did with Archer Martin, for which they shared the 1952 Nobel Prize for Chemistry.

Synge was born in Liverpool on 28 October 1914, the son of a stockbroker. He attended Winchester School from 1928 to 1933 and then went to Trinity College, Cambridge, graduating in 1936; he gained his PhD five years later. He went to work as a biochemist at the Wool Industries Research Association in Leeds, and then in 1943 moved to the Lister Institute of Preventive Medicine in London. In 1948 he was put in charge of protein chemistry at the Rowett Research Institute in Aberdeen. He spent the year 1958–59 with the New Zealand Department of Agriculture at its Ruakura Animal Research Station. Since 1967 he has been employed as a biochemist at the Food Research Institute of the Agricultural Research Council in Norwich.

In the early 1940s there were crude chromatographic techniques for separating proteins in a reasonably large sample, but no sufficiently refined method existed for the separation of individual amino acids. Martin and Synge, who worked together both at Cambridge and Leeds, evolved the technique of using porous filter paper in chromatography. A spot of mixed amino-acid solution is placed at the end of a strip of filter paper and allowed to dry. The paper is then dipped in a solvent which either creeps up it by capillary action (ascending chromatography) or down the paper if it hung below the level of the solvent. As the solvent passes the mixture the various amino acids move with it, but at different rates. The filter paper is then dried, and sprayed with a 'developer' such as ninhydrin solution. On heating the paper the positions of the amino acids are revealed as dark spots and can be identified by comparing them with the spots produced by known amino acids. Several mixtures can be analysed at once by applying several spots to a wide piece of paper.

The technique described is one-dimensional paper (or partition) chromatography, because the solvent spreads the amino acids in only one direction. If the chromatogram is dried, but before being treated with ninhydrin is rotated through 90° and dipped in solvent again (either the same or a different one), the amino acids can be resolved even more clearly. This version is known as two-dimensional chromatography.

Martin and Synge announced their method in 1944 and it became an immediate success, being applied widely and adapted to many experimental problems. It was soon demonstrated that not only the type but the concentration of each amino acid can be determined. Synge was able to work out the exact structure of Gramicidin-S, a simple antibiotic peptide, which led in 1953 to Frederick Sanger's elucidation of the complete sequence of insulin. Other chromatographic techniques since developed include gas, thin-layer, ion exchange; gel filtration and, most recently, high-pressure liquid chromatography.

Szent-Györgyi Albert 1893–1986 was a Hungarian-born US biochemist who studied the physiology of muscle contraction and has carried out research into cancer. He is best known, however, for his work on vitamin C, for which he was awarded the 1937 Nobel Prize for Physiology or Medicine.

Szent-Györgyi was born in Budapest on 16 September 1893, into a family of scientists. He completed his early education in Budapest and entered the Medical School at the university there in 1911. During his first year he began research in his uncle's laboratory, and three years later he had published a series of papers on the structure of the vitreous body in the eye. During World War I he served in the Austro–Hungarian army on the Russian and Italian fronts, and was decorated for bravery. But he soon left the army with a (self-inflicted) wound and returned to his studies in Budapest, gaining his medical degree in 1917. During the 1920s he studied at various universities in Germany, the Netherlands, Belgium, the United States and Britain. He obtained his PhD from Cambridge University in 1927 and returned to Hungary in 1937 to the University of Szeged. Szent-Györgyi was active in the anti-Nazi underground movement during World War II; after the war he became Professor of Biochemistry at the University of Budapest. Unhappy with the Soviet regime, in 1947 he emigrated to the United States where he joined the staff of the Marine Biological Laboratories at Woods Hole, Massachusetts. He became a US citizen in 1955. For 32 years he was Director of the National Institute of Muscle Research and was Scientific Director of the National Foundation for Cancer Research from 1975.

Szent-Györgyi published his first significant piece of research in 1928 while he was at Cambridge working under Frederick Hopkins. He isolated a substance from the adrenal glands and called it hexuronic acid, because the molecule appeared to contain six carbon atoms. He isolated the same substance from cabbages and oranges, both rich sources of vitamin C. Back in Hungary in the early 1930s he discovered that paprika – a major crop in the locality around Szeged – is an extremely rich source of the acid, which he prepared as pure white crystals and in 1932 proved to be the same as the substance (first discovered in 1907) that prevents scurvy in human beings; his announcement was anticipated by only two weeks by Charles King. Szent-Györgyi suggested that the acid be called antiscorbutic acid, although it finally became known as ascorbic acid. His work made it possible for Walter Haworth and Paul Karrer to synthesize ascorbic acid (vitamin C), for which they were awarded the 1937 Nobel Prize for Chemistry, the same year in which Szent-Györgyi received the Physiology Prize.

Szent-Györgyi also studied the uptake of oxygen by minced muscle tissue. Left undisturbed, the tissue gradually absorbed less and less oxygen as some substance in it was used up. In 1935, he found that activity was restored by adding any one of four closely related four-carbon compounds: fumaric acid, malic acid, succinic acid, or oxaloacetic acid. This discovery was later used by Hans Krebs in working out the Krebs (tricarboxylic acid) cycle.

In 1940 Szent-Györgyi isolated two kinds of muscle protein from myosin which, until then, had been thought to be the single basic component of muscle tissue. One was composed of rod-shaped particles and the other was in the form of minute globular beads. The former retained the name myosin and the latter was called actin; he renamed the combined compound actomyosin. When adenosine triphosphate (ATP) is added to it, a change takes place in the relationship of the two components which results in the contraction of the muscle itself.

During the late 1940s Szent-Györgyi made further investigations into the chemistry of the citrus fruits and extracted the so-called vitamin P from lemon peel. It is a complex compound of three flavonoids whose function is to reduce the fragility of capillary blood vessels. The breakdown of capillaries is a common result of prolonged radiation therapy in cancer patients, and can be countered by administering 'vitamin P'.

In the 1960s Szent-Györgyi began studying the thymus gland, which had been shown to play a part in the setting up of the body's immunological system. He isolated several compounds from the thymus that seem to be involved in the control of growth. At the National Foundation for Cancer Research he carried out research on the processes of cell division. In 1976, in his 84th year he published a book entitled *Electronic Biology and Cancer*.

T

Tabor David 1913– is a British physicist who has worked mainly in tribology, the study of the effects between solid surfaces.

Tabor was born in London on 23 October 1913. He was educated at Regent Street Polytechnic, London, from 1925 to 1931, and then read physics at the Royal College of Science, London, obtaining his BSc in 1934. From 1934 to 1936, he researched under George Thomson (1892–1975) and then from 1936 to 1939 under Frank Bowden (1903–1968) at Cambridge. He gained his PhD in 1939, and then went to Australia in 1940 to work in tribophysics at a Division of the Commonwealth Scientific Research Organization in Melbourne. In 1946, he returned to Cambridge, and worked in the section of the Cavendish Laboratory concerned with the physics and chemistry of solids. He was Assistant Director of Research until 1961, when he was made a lecturer in physics. He subsequently became a Reader in Physics in 1964 and Professor of Physics in 1973, a position he held until 1981. He is now Emeritus Professor. Among his many awards is the Guthrie Medal of the Institute of Physics in 1974.

Tabor's research interests have been very wide. Early investigations were into the friction and lubrication of metals, low-energy and high-energy electron diffraction techniques and high vacuum methods for studying clean surfaces, the adsorption of vapours and the first stages of conversion of a chemisorbed film to a chemically formed film. Bowden introduced the idea of using mica to study the contact between two surfaces since the cleavage faces are molecularly smooth. Tabor extended this to a study of forces between two curved mica surfaces and has shown that normal Van der Waals forces operate for separations less than 100 Å (10^{-8} m), and retarded Van der Waals forces for separations larger than 500 Å (5×10^{-8} m).

As a by-product of an investigation of the action of windscreen wiper blades, Tabor has been able to study directly the repulsive forces between electrically charged double-layers. These and the attractive Van der Waals forces are of basic importance in colloid stability.

Tabor has also looked at the friction and transfer of polymers. This work includes the effect of temperature and

speed and he has attempted to correlate the behaviour with visco-elastic properties. He has shown that the low friction of teflon is not due to poor adhesion but to molecular structure, and has worked on the self-lubrication of polymers by incorporating surface materials into the polymer itself. A study of the friction of rubber led to the introduction of high-hysteresis rubber into automobile tyres as means of increasing their skid-resistance. Tabor has also examined the effect of hydrostatic pressure on the viscoelastic properties of polymers and has shown that by decreasing the free volume, hydrostatic pressure increases the glass transition temperature.

Tabor has researched into the shear properties of molecular films of long-chain organic molecules as an extension of earlier work on the mechanism of boundary lubrication, and has shown that the shear strength of these materials rises sharply when they are subjected to high pressure. This sheds light on the mechanism of thin film lubrication.

Tabor's work on the hardness of solids includes an explanation of the indentation hardness of metals in terms of their basic yield properties, the first account of plastic indentation and elastic recovery which explains rebound in terms of plastic and elastic properties (rebound hardness), the effect of temperature and loading time on indentation hardness and the first correlation of the hardness behaviour with the creep properties of the material (hot-hardness), a study of scratch hardness and a simple physical explanation of the Mohs' scale used in the testing of minerals.

He has also carried out a broad study of the creep of polycrystalline ice over a wide range of temperatures and strain rates, and has explained the behaviour in terms of various dislocation and grain-boundary properties. These results have a bearing on the flow of glaciers.

Tabor has looked at diffusion in polymers, studying the diffusion of suitably tagged organic molecules through a polymer matrix. By increasing the length of the diffusant, he has provided experimental evidence for the first time for the process of reptation, whereby the diffusant worms its way through the free volume of the polymer. He has also examined the effect of hydrostatic pressure on diffusion since this shows how the polymer matrix restricts the movement of the diffusant.

Tabor has also studied the machining of tools using a transparent tool so that the sliding of the chip over the tool face can be directly observed. Finally, he has studied the adhesion of steel to cement and has demonstrated the important role of shear stresses in compacting the cement.

Talbot William Henry Fox 1800–1877 was the English classical scholar, mathematician and scientist who invented the calotype (or talbotype) photographic process. This was a negative-to-positive process on paper which laid the foundation for modern photography.

Talbot was born on 11 February 1800 at Melbury, Dorset, to William Davenport Talbot, an officer in the Dragoons, and Lady Elizabeth Fox Strangeways, daughter of the second Earl of Ilchester. He was educated at Harrow and Trinity College, Cambridge, and was elected Liberal Member of Parliament for Chippenham, taking his seat in 1833. During a trip to Italy he resolved to try to capture the images obtained in a *camera obscura* and by 1835 had succeeded in fixing outlines of objects laid on sensitized paper. Images of his home, Lacock Abbey, followed and Talbot then appeared to give up his experiments until the announcement of Daguerre's (1789–1851) success on 7 January 1839.

Talbot rushed into publication in case Daguerre's process was the same as his own (there were similarities). He exhibited his work at the Royal Institution and again at the Royal Society on 31 January where he presented a hastily prepared paper. Some of the exhibits were positives and were probably taken between 1835 and 1839.

To publicize his process, Talbot set up a laboratory for printing calotypes in Reading. This produced *The Pencil of Nature*, the first book in the world to be illustrated by photographs. It came out irregularly in six parts, beginning 29 June 1844, each part containing from three to seven photographs. Other books followed.

Talbot patented an enlarger in June 1843 and took the first successful photograph by electric light. This was also the first successful motion-freezing photograph taken by flash and it was demonstrated at the Royal Institution in 1851. He applied the principle of dichromate and gelatine to photoglyphic engraving and tried to lay claim to the collodion process. Squabbling over this claim culminated in a court case in which on 20 December 1854 it was found that collodion photography did not infringe calotype patents.

In parallel with his photographic work, Talbot was giving papers at Royal Society meetings, many of them on mathematical and scientific subjects. He published papers on archaeology and, with Henry Rawlinson and Dr Edward Hincks, was one of the first to decipher the cuneiform inscriptions of Nineveh. He was also the author of *English Etymologies*. He died at Lacock Abbey on 17 September 1877.

Talbot's calotype process was patented on 8 February 1841. Good quality writing paper was coated successively with solutions of silver nitrate and potassium iodide, forming silver iodide. The iodized paper was made more sensitive by brushing with solutions of gallic acid and silver nitrate, and then it was exposed (either moist or dry). The latent image was developed with an application of gallo-silver nitrate solution, and when the image became visible the paper was warmed for one to two minutes. It was fixed with a solution of potassium bromide (later replaced by sodium hyposulphite). Calotypes did not have the sharp definition of daguerreotypes and were generally considered inferior.

In the decade to 1851, Talbot took out four patents, many of which contained previously published claims. He stirred up considerable resentment by his activities, which are considered to have hindered the development of photography in England. However, on 30 July 1852 he announced that he wished only to retain licensing on professional portraiture. This cleared the way for amateurs to use processes developed in other countries.

Tansley Arthur George 1871–1955 was a British botanist who was a pioneer in the science of plant ecology. He helped to promote the subject through his teaching, and by writing and editing textbooks and journals (including contributing to and editing the first major book on the vegetation of the British Isles). He was also instrumental in the formation of organizations devoted to the study of ecology and the protection of wildlife. Tansley's contributions to botany were recognized by the award of several honours, including being elected a Fellow of the Royal Society of London in 1915, the Gold Medal of the Linnean Society in 1941 and a knighthood in 1950.

Tansley was born in London on 15 August 1871. He was educated at Highgate School from 1886 to early 1889 (and later commented on the inadequacy of the science teaching of that time), after which he attended science classes at University College, London, where he received his first instruction in botany from Francis Oliver. In 1890, he went to Trinity College Cambridge, graduating in 1894. He combined his last year of study with a teaching post at University College and, after graduation, returned there as an assistant to Oliver and Demonstrator in Botany, a position he held until 1906. Between 1900 and 1901 Tansley visited Ceylon (now Sri Lanka), the Malay peninsula and Egypt to study their flora. On his return, he found that there was no suitable journal in which to publish his findings, so he founded *The New Phytologist* in 1902, remaining its editor for 30 years. In 1907 he was appointed University Lecturer in Botany at Cambridge University.

After World War I, however, his interest turned temporarily towards psychology, and in 1923 he resigned his lectureship in order to study under Sigmund Freud in Austria. Tansley returned to Britain the following year, and in 1927 was appointed Sherardian Professor of Botany at Oxford University, a position that carried with it a fellowship at Magdalen College. He remained at Oxford until his retirement in 1939. Tansley continued to be active after retiring. He was Chairman of the Nature Conservancy from 1949 to 1953 and President of the Council for the Promotion of Field Studies (now called the Field Studies Council) from 1947 to 1953, having played a large part in the establishment of these organizations.

Most of Tansley's work concerned British plants and plant communities. He coordinated a large project (which lasted from 1903 to 1907) to map the vegetation of the British Isles; the surveys that were completed are still models of vegetation mapping technique. The scientists involved in this project published their findings in *Types of British Vegetation* (1911), of which Tansley was the editor and major contributor. Although this book was a masterly summary of British flora, Tansley and the other scientists felt that a wider approach to plant ecology was needed, so on 12 April 1913 the group founded the British Ecological Society, with Tansley as its president. The society founded the *Journal of Ecology*, which Tansley edited from 1916 to 1938.

While Professor of Botany at Oxford, Tansley enlarged and rewrote *Types of British Vegetation*. The new work, *The British Islands and their Vegetation* (1939), was Tansley's greatest single achievement. In it he showed how vegetation is affected by soil, climate, the presence of wild and domesticated animals, previous land management and contemporary human activities. He also reviewed all known accounts of British flora and then linked the two themes, thereby demonstrating which factors are important in influencing the various types of vegetation. In 1949, he published *Britain's Green Mantle*, a shorter and more popular version.

Tansley also helped to promote the study of plant ecology through his teaching; by writing several practical guides, such as *Practical Plant Ecology* (1923); and by campaigning for the establishment and formation of ecological organizations.

Tartaglia *c.*1499–1557 was a Renaissance Italian mathematician, mathematical physicist and writer who, despite an unpromising youth, worked hard eventually to find fame for his work in mathematics, topography and mechanical physics.

Tartaglia was born Niccolò Fontana in Brescia, Lombardy, in either 1499 or 1500. His father was a postman and his family was very poor. When he was 12, the French marched in and sacked Brescia, seriously injuring the boy in the process. Only the careful nursing by his mother of the savage sword-thrust wound in his mouth saved his life, but ever thereafter he was called Tartaglia – 'stammerer' – because of the speech defects the wound caused. Virtually self-educated, Tartaglia developed a true scientific curiosity and absorbed knowledge from every source he could find, particularly in mathematics and physics. In 1516 he moved to Verona and became a teacher of the abacus. Later he took charge of a school there from 1529 to 1533. After that he went on to Venice, a city in which he was to remain for the rest of his life. Although he gained a position of Professor of Mathematics, and published many papers on his work during the later stages of his life, he never made much money from his skills and died alone and poor in humble dwellings near the Rialto Bridge on 15 December 1557.

With the coming of the sixteenth century, a revival took place in most branches of science, and Tartaglia was perhaps one of the more important contributors to it. Mathematics, in particular, had made little progress since the Greek scholars set down the basic rules; succeeding generations had evidently been content with simple counting and the ability to undertake elementary explanations to problems. Capable of applying his mind to most things, Tartaglia read all he could and then chose – among other enquiries – to explore the complexities of third-degree equations, to investigate the cubic. (Solutions had by then already been found for the linear and quadratic equations.) He was only one among many concerned with the task, however, most of the others also being Italian. There was keen rivalry, and Tartaglia more than once entered into a public contest of skills with another mathematician. In one particular confrontation with a certain Antonio Fior, Tartaglia emerged the victor by applying his methods and solving all the problems set for him by Fior, whereas Fior could not solve those set by Tartaglia.

He next turned his attention to solving the problem of calculating the volume of a tetrahedron from the length of its sides. Successful in that endeavour, he attempted Malfatti's problem – to inscribe within a triangle three circles tangent to one another, and managed to do that too.

Although Tartaglia spent much of his time teaching mathematics, he was also responsible for translating Euclid's *Elements* into Italian (1543) – the first translation of Euclid into a contemporary European language.

His greatest love, however, appears to have been in using his mathematical (and other relevant) skills to solve military problems. He delighted in planning the disposition of artillery, surveying the topography in relation to the best means of defence, and in designing fortifications. He also attempted a study of the motion of projectiles, and formulated what is now generally known as Tartaglia's theorem: the trajectory of a projectile is a curved line everywhere, and the maximum range at any speed of its projection is obtained with a firing elevation of 45°.

Tatum Edward Lawrie 1909–1975. US microbiologist; *see* Beadle, Tatum, and Lederberg.

Telford Thomas 1757–1834 was a Scottish civil engineer, famous for building roads and bridges. He was also the first president the Institute of Civil Engineers.

Telford was born at Westerkirk, Dumfries, on 9 August 1757, the son of a shepherd. He began his career as a stonemason, but despite his humble origins he had strong ambitions and educated himself in architecture in his spare time. In search of work he travelled to London and found employment building the additions to Somerset House in the Strand under the supervision of Sir William Chambers. Recognizing his talents, the rich and famous were soon consulting him about their own buildings and consequently he was launched upon a career which was to make him into one of the outstanding civil engineers of that time.

In 1786, Telford was appointed Official Surveyor to the

county of Shropshire. This proved to be the start of 30 years' intense construction work for different organizations, dealing with anything and everything requiring his particular skills. He was responsible for reconstructing roads, building canals, aqueducts and harbours, erecting suspension bridges and surveying railway lines, but above all he will be remembered for the Menai Bridge, which must remain his most famous achievement.

The state of the roads in Britain during the eighteenth century depended very much on the individual turnpike trust and whether it chose to pay for repairs. Telford and his contemporary, John McAdam, set about rebuilding the existing Roman routes, digging down to the foundations of these roads and levelling them to meet the need for faster travel. It was recorded that after such work had been carried out on a road, a mail coach could attain speeds on it up to 19 kph/12 mph and could average at least 13 kph/8 mph.

In 1793, Telford was appointed engineer to the Ellesmere Canal Company, where he was responsible for the building of aqueducts over the Ceirog and Dee valleys in Wales, using a new method of construction consisting of troughs made from cast-iron plates and fixed in masonry. Ten years later he was asked to take charge of the Caledonian Canal project, an enormous conglomeration of plans which included not only the canal but harbour works at Aberdeen and Dundee and more than 1,450 km/900 mi of link roads with several bridges.

Throughout his entire career, bridges had formed an important part of his construction work. In his early days in Shropshire he had built three over the River Severn (at Montford, Buildwas and Dewdley) and later he tackled a complicated structure over the river Conway, but by far the most impressive was the Menai Bridge. Built as a suspension bridge over the Menai Straits to join Anglesey to the Welsh mainland, it took from 1819 to 1826 to erect and had a finished span of 176 m/580 ft with huge wrought-iron links supporting it.

Although Telford was well aware of the need for faster communications and improvements to the transport system, he chose to almost disregard the railways which were just beginning to be accepted as a means of transport, and concentrated entirely on roads and canals. Nevertheless, his contribution to the advance of civil engineering was considerable during the latter half of the eighteenth century and in the early part of the nineteenth, and it was only just that the Institute of Civil Engineers should reward him with the Presidency.

Teller Edward 1908– is the Hungarian-born US physicist widely known as 'the Father of the hydrogen bomb'. He is also known for his vigorous promotion of nuclear weapons, opposition to communism, and for testifying against J. Robert Oppenheimer at the security hearings of 1954.

Teller was born on 15 January 1908 in Budapest, where he attended the Institute of Technology. He also attended a series of universities in Germany, including Leipzig where he obtained his PhD in 1930. He left Germany in 1933

when Hitler came to power and after periods in Copenhagen and London, was appointed Professor of Physics at George Washington University, Washington, DC, in 1935. There he developed the Gamow–Teller selection rule for beta decay. In 1941 he became a US citizen and joined the staff of Columbia University and then the University of Chicago 1942 to work with Enrico Fermi on atomic fission.

Between 1942 and 1946, Teller was a member of the Manhattan Project, Los Alamos, New Mexico, which developed the first atomic bombs. He worked initially on the fission bomb, but later convinced Los Alamos director, Robert Oppenheimer, to let him work on the fusion, or hydrogen (H-), bomb. By the end of World War II, Teller had designed an H-bomb, but it required a refrigeration plant to make it work and in 1949 the General Advisory Committee (GAC), chaired by Oppenheimer, advised the Atomic Energy Commission not to pursue the project. The situation changed on 29 August 1949, however, when the USSR exploded its first fission bomb. Teller, now a professor at the University of Chicago, and Ernest Lawrence of the Radiation Laboratory in Berkeley, lobbied President Truman and on 31 January 1951, Truman announced a crash programme to build the H-bomb, with Teller appointed Assistant Director of Weapons Development at Los Alamos with responsibility for the bomb. The Super was successfully tested on Eniwetok Atoll in the Pacific Ocean in November 1952.

Teller's exact role in the development of the H-bomb has, however, been surrounded with controversy, for he has been repeatedly accused of downplaying, or ignoring, the contribution of Polish-born mathematician Stanislaw Ulam (1909–1985) to the bomb's design. The original idea of using a fission explosion to ignite a thermonuclear (fusion) explosion in deuterium (heavy hydrogen) came from Enrico Fermi. Ulam and others had proved Teller's original ideas to be unworkable. Ulam then suggested a configuration in which shock waves from the fission explosion would compress and heat the deuterium, causing it to explode. Teller later modified this idea to use X-rays from the first explosion, rather than shock waves. (Teller and Ulam's report on the design remains classified.)

After the war, Teller successfully championed the establishment of a second nuclear weapons research facility, which opened in 1952 as the Lawrence Livermore Laboratory near Berkeley, California. He was Livermore's Associate Director from 1954 until his retirement in 1975, and has been Director Emeritus ever since. He has also been a professor at the University of California since 1953.

Oppenheimer (and the GAC) had opposed setting up a second weapons laboratory just as he had opposed the H-bomb, but in 1954 his security clearance was revoked. It was the height of the McCarthy era and Oppenheimer was suspected of being a communist and Soviet spy. Teller was called to give evidence at his security hearing and, although he testified that he did not believe Oppenheimer to be disloyal, concluded: 'I would feel personally more secure if public matters would rest in other hands.'

Oppenheimer never regained his security clearance and many physicists never forgave Teller.

In the 1980s Teller returned to the public eye when he convinced President Reagan of the feasibility of placing fission-bomb-powered X-ray lasers in space to destroy incoming Soviet nuclear missiles. Billions of dollars were spent on the Strategic Defense Initiative ('Star Wars') before technical problems, and the disintegration of the former Soviet Union, rendered it obsolete. However, Teller remains in favour of nuclear technology and defence, having suggested 'brilliant pebbles' - thousands of intelligent missile-interceptors based in space - and the use of nuclear explosions to prevent asteroids hitting the Earth.

Temin Howard Martin 1934–1994 was a US virologist concerned with cancer research. For his work on the genetic inheritance of viral elements he received the 1975 Nobel Prize for Physiology or Medicine, which he shared with David Baltimore (1938–) and Renato Dulbecco (1914–).

Temin was born in Philadelphia on 10 December 1934 and educated at Swarthmore College, Pennsylvania, where he gained a BA in 1955. He then obtained his PhD at the California Institute of Technology, Pasadena, in 1959 for a thesis on animal virology. During the following year he was a postdoctoral fellow there. Between 1960 and 1964 he became Assistant Professor of Oncology at the University of Wisconsin, and then Associate Professor from 1964 to 1969. He was appointed Professor of Oncology at Wisconsin in 1969 and Professor of Cancer Research in 1971. In 1974 he became Professor of Viral Oncology and Cell Biology at Wisconsin. He died in Madison, Wisconsin, on 9 February 1994.

In the early attempts at organ transplants in the human body, the patient did not live long after the operation because the transplanted organ was rejected by the recipient's body as foreign tissue. Peter Medawar experimented with mice and found that if the intended recipient was injected at the embryo stage with cells from the future donor, its body did not reject the transplant, but accepted the foreign cells as its own. He also found that this acquired immunological tolerance can be handed down to a second generation. Following on Medawar's experiments, Macfarlane Burnet found that the ability of the human body to produce antibodies is not learnt through a process of evolution but he suggested that the genes of the antibody-producing cells contain a region which is continually mutating, with each mutation leading to the production of a new antibody variant.

Temin's prizewinning research was on a virus that has a mechanism which incorporates its material into mammalian genes. He discovered that beneficial mutations outside the germ line are naturally selected and that the mechanism adds genetic information from outside into the germ line.

It has been suggested since that this inheritance might operate in the blood cells, the gut lining and perhaps in the nervous system, from where it enters the germ line. Experimental evidence suggests that acquired characteristics can be passed on through genes, in agreement with Lamarck. For instance, the incidence of chemically induced brain tumours has been found to be very high in the offspring of affected parents.

Tesla Nikola 1856–1943 was a Croatian-born US physicist and electrical engineer who was one of the great pioneers of the use of alternating current electricity. In particular, he invented the alternating current induction motor and the high-frequency coil which bears his name.

Tesla was born at midnight between 9 and 10 July 1856 in Smiljan, Croatia (then part of Austria–Hungary). His father was a priest of the local Serbian Orthodox Church. Tesla was very clever as a child and grew up with a liking for writing poetry and experimentation. It was intended that he should follow his father and become a priest, but Tesla developed an interest in scientific pursuits while he was at the Real Gymnasium in Karlovac. On leaving school he studied engineering at the Technical University at Graz, Austria. In 1880, he went to the University of Prague to continue his studies but the death of his father caused him to leave without graduating.

In 1881, Tesla went to Budapest as an engineer for a telephone company and a year later took up a similar position in Paris. He went to the United States in 1884, and worked for Thomas Edison (1847–1931) for a year before setting up on his own. From 1888, Tesla was associated with George Westinghouse (1846–1914), who bought and successfully exploited Tesla's patents, leading to the introduction of alternating current for power transmission. Tesla became a US citizen in 1889 and after 1892, when his mother died, he became increasingly withdrawn and eccentric. In 1912 both he and Edison were proposed for the Nobel Prize for Physics but Tesla refused to be associated with Edison, who had campaigned for the adoption of direct current. In the event, neither received the prize. Tesla neglected to patent many of his discoveries and made little profit from them. He lived his last years as a recluse and died in New York on 7 January 1943.

In 1878, during his student days at Graz, Tesla saw a direct current electric dynamo and motor demonstrated and felt that the machine could be improved by eliminating the commutator and sparking brushes, which were sources of wear. His idea for the induction motor came to him in Budapest four years later. He had the notion of an iron rotor spinning between stationary coils which were electrified by two out-of-phase alternating currents producing a rotating magnetic field. Like many of his other ideas, Tesla mentally developed his motor for all kinds of practical applications before ever a model was built.

Tesla built his first working induction motor while he was on assignment in Strasbourg in 1883. However, he found that he could raise little interest in his inventions in Europe so he set off for New York, where he eventually set up his own laboratory and workshop in 1887, to develop his motor in a practical way. Only months later he applied for and was granted a complicated set of patents covering the generation, transmission and use of alternating current electricity. At about the same time he lectured to the American Institute of Electrical Engineers on his polyphase alternating current system. After learning about the talk, Westinghouse quickly bought Tesla's patents.

Westinghouse was able to back Tesla's ideas and as a demonstration, employed his system for lighting at the 1893 World Columbian Exposition in Chicago. Months later Westinghouse won the contract to generate electricity at Niagara Falls, using Tesla's system to supply local industries and deliver polyphase alternating current to the town of Buffalo 35 km/22 mi distant.

After 1888 Tesla's interests turned to alternating currents at very high frequencies, which he felt might be useful for lighting and for communication. After first using high-frequency alternators for these purposes, he designed what has come to be known as the Tesla coil. This is an air core transformer with the primary and secondary windings tuned in resonance to produce high-frequency, high-voltage electricity. Using this device, Tesla produced an electric spark 40 m/135 ft long in 1899. He also lit more than 200 lamps over a distance of 40 km/25 mi without the use of intervening wires. Gas-filled tubes are readily energized by high-frequency currents and so lights of this type were easily operated within the field of a large Tesla coil. Characteristic of his way of working, Tesla soon developed all manner of coils which have since found numerous applications in electrical and electronic devices.

Tesla was very interested in the possibility of radio communication and as early as 1897 he demonstrated remote control of two model boats on the lake in Madison Square Gardens in New York. He extended this to guided weapons, in particular a remote control torpedo. In 1900 he began to construct a broadcasting station on Long Island in the hope of developing 'World Wireless'; this eventually proved too expensive for his backers and was abandoned. However, many of his ideas have come to fruition at the hands of others. Tesla also outlined a scheme for detecting ships at sea which was later developed as radar. One of his most ambitious ideas was to transmit alternating current electricity to anywhere in the world without wires by using the Earth itself as an enormous oscillator. There were many other inventions, including electrical clocks and turbines, but often they remained in his head, there being no money to put them into practice.

Tesla gave the world one of the most practical devices of all time, the alternating current induction motor. This, coupled with the distinct advantage that alternating current can be transmitted over much greater distances than direct current, has given the motive power for most of our present-day machines.

Thales of Miletus c.624–c.547 BC was a Greek philosopher who was among the first early Greek philosophers to reject mythopoetic forms of thought for a basically scientific approach to the world.

Thales was born in Miletus, the son of Examyas and Cleobuline, both of whom were members of distinguished Miletan families. Thales' precise date of birth is unknown, but his peak of activity is traditionally dated as 585 BC, the year in which he predicted an eclipse of the Sun. It is his prediction of this eclipse that is the basis of his reputation as a scientist.

Claims are made also for his ingenuity and practicality as an astronomer, mathematician, statesman and businessman. Aristotle records that, when he was reproached for being impractical, Thales, having predicted that weather conditions the next year would be conducive to a large olive harvest, bought up all the olive presses in Miletus and exploited his monopoly to make a large profit. Herodotus gives another example of Thales' belief in the control of physical nature when he recounts that Thales diverted the River Halys to enable Croesus' army to cross. Such engineering required that any divine connotations attaching to the flow of rivers be ignored and objects and events in the world treated much as they are in present-day scientific research.

Thales sought unifying and general hypotheses that relied on the relationships between natural phenomena to explain natural events. The order he believed to be inherent in the world was not to be explained by reference to divine or mystical forces, but was to be discovered and articulated in terms of natural causes.

In astronomy Thales is credited with defining the constellation of Ursa Minor (Little Bear) and with writing a work on navigation in which the Little Bear is commended for its usefulness in this regard. But his reputation as an astronomer rests on a number of doubtful sources which, among others, ascribe to him the introduction of Egyptian mathematics to Greece and the use of a 'Babylonian saros' (a cycle of 223 lunar months) to calculate the solar eclipse. It has been argued that the Babylonian saros was the invention of Edmond Halley (1656–1742) and could not have been the basis of Thales' calculation.

Cosmological views attributed to Thales are known almost entirely from the writings of Aristotle. Thales is said to have proposed that water was the material constituent of all things and that the Earth floats on water. He explained the occurrence of earthquakes by reference to this idea of a floating Earth. Thales' explanation of events is couched in terms of natural phenomena and objects, not in the usual terms of activity by the god Poseidon. His cosmology may be reminiscent of Near Eastern mythology, but the Greek thinkers who followed the path he established by seeking explanations of natural events in terms of natural agencies were the precursors of modern scientists and astronomers.

Thales' greatest achievement (again, it should remembered, a matter for doubt) was his geometry. If Proclus' account is right, Thales was an important innovator in geometry, particularly for introducing the notion of proof by the deductive method, whereas his Babylonian and Egyptian predecessors had not progressed beyond making generalizations from experience on a rough-and-ready inductive principle. In five fundamental propositions Thales laid down the foundations on which classical geometry was raised. (1) A circle is bisected by its diameter.

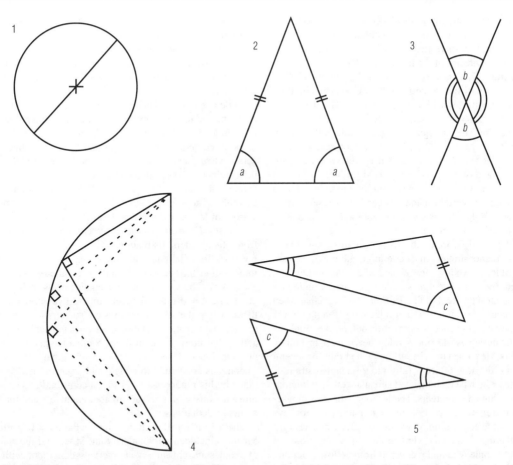

Thales of Miletus *Some of the basic rules of geometry were first laid down by Thales: (1) a circle is bisected by its diameter; (2) in an isosceles triangle, the two angles opposite the equal sides are themselves equal; (3) when straight lines cross, opposite angles are equal; (4) the angle in a semicircle is a right angle; and (5) two triangles are congruent if they have two angles and one side identical.*

(2) In an isosceles triangle the two angles opposite the equal sides are themselves equal to each other. (3) When two straight lines intersect, four angles are produced, the opposite ones being equal. (4) The angle in a semicircle is a right angle. (5) Two triangles are congruent if they have two angles and one side that are respectively equal to each other.

Thales' proofs were not always very rigorous. But they had the nub of logical unassailability in them. Consider the proof of the proposition that the angle in a semicircle is a right angle. It is obvious that the diagonals of a rectangle are of equal length and that they bisect one another. It is obvious, also, that any quadrilateral possessing these properties is a rectangle. If we draw two diameters, AOC and BOD, of a circle with centre O, we therefore get a rectangle. Omit the dotted lines in the figure and we are left with the proposition that, if A is any point on the arc of a semicircle with the diameter BD, then the angle BAD is necessarily a right angle. Without this method of reasoning – so like Euclid's – the original proposition would not appear obvious. It is abstract reasoning of this kind that makes Thales such an important figure in the history of European thought. It is entirely in the proper spirit that Thales, on making his discovery of his fourth proposition, sacrificed a bull to the gods in thanks.

Theiler Max 1899–1972 was a South African-born US microbiologist who developed an effective vaccine against yellow fever, an achievement that gained him the 1951 Nobel Prize for Physiology or Medicine. He also researched into various other diseases, including Weil's disease and poliomyelitis.

Theiler was born on 30 January 1899 in Pretoria, South Africa, the son of a Swiss-born veterinary surgeon. After doing his preliminary medical training at Rhodes University College and the University of Cape Town, Theiler went to Britain and completed his training at St Thomas's Hospital and the London School of Hygiene and Tropical Medicine, from which he received his medical degree in 1922. After graduating, Theiler went to the United States and joined the Department of Tropical Medicine at Harvard University Medical School. In 1930, he moved to the

Rockefeller Institute for Medical Research (now Rockefeller University), New York City, becoming Director of the Virus Laboratory there in 1950. In the same year he was also made a Director of the Division of Medicine and Public Health. Theiler was appointed Professor of Epidemiology and Microbiology at Yale University in 1964 and held this post until he retired in 1967. He died on 11 August 1972 in New Haven, Connecticut.

By the time Theiler began his research into yellow fever, Walter Reed (1851–1902), a US army doctor, had proved in 1900 that the disease is carried by the *Aëdes aegypti* mosquito, and Stokes, Bauer and Hudson had discovered that the causative agent is a virus. Furthermore, it had been established that certain mammals can act as reservoirs of infection and that *Aëdes aegypti* is not the only species of mosquito that can transmit yellow fever.

Theiler's early work on the disease – carried out at Harvard – demonstrated that common albino mice are susceptible to yellow fever, and that when they are infected, the virus undergoes certain changes. Continuing his research at the Rockefeller Institute, he found that when the mice are given intracerebral injections of yellow fever virus they develop encephalomyelitis but, unlike monkeys and human beings, do not develop heart, liver or kidney disorders. He passed the virus through the brains of several mice in order to produce a fully mouse-adapted strain of the virus, which he then injected subcutaneously into monkeys and human volunteers. Theiler found that this mouse-adapted strain produced full, active immunity in monkeys but still affected the kidneys in humans. But when he combined the mouse-adapted viral strain with serum from the blood of people who had recovered from yellow fever and injected the mixture into humans, he found that it produced immunity without affecting the kidneys; he had succeeded in producing a safe vaccine against yellow fever.

This method of making the vaccine is not suitable for large-scale production, however, because human serum containing antibodies against yellow fever is difficult to obtain. Theiler therefore began working on a method of producing yellow fever vaccine that did not require human serum, and in 1937 he developed vaccine 17-D, still the main form of protection against yellow fever.

In 1931 US pathologist Ernest Goodpasture (1886–1960) developed a method of culturing viruses by injecting them into chick embryos. Theiler made vaccine 17-D by combining Goodpasture's technique with his own method of making mouse-adapted viral strain. He passed yellow fever virus successively through 200 mice then cultured the virus in 100 chick embryos; this yielded a mutant form of the virus that caused only mild symptoms when injected into humans yet gave complete immunity against yellow fever.

In addition to his work on yellow fever, Theiler also investigated Weil's disease, amoebic dysentery and poliomyelitis. He hypothesized that the poliomyelitis virus is widespread in the intestines, where it is harmless, and that symptoms are produced only in the rare instances where the virus enters the central nervous system.

Theon of Smyrna lived *c.*AD 130, was a mathematician and astronomer who is remembered chiefly for his *Expositio rerum mathematicarum ad legendum Platoneum utilium*. Most of the text of this work, at present in Venice, is in two manuscripts, one on mathematics and one on astronomy and astrology.

Little is known of Theon's birth or life. The latest thinkers he names in the *Expositio* are Thrasyllus and Adrastus. Assuming that Theon tried to take account of the most modern contemporary writers, and as Thrasyllus was active when Tiberius was Emperor of Rome and Adrastus died in the late 100s, Theon is generally taken to have been active in the early second century AD. He is also referred to by other writers: Ptolemy's work includes Theon's observations of Venus and Mercury between 127 and 132.

Theon's *Expositio* is useful as a source of quotations from other writers. Its main task, in keeping with the suppositions of Platonic philosophy, is to articulate the interrelationships between arithmetic, geometry, music and astronomy. The section on mathematics deals with prime, geometrical and other numbers in the Pythagorean pantheon, while the section on music considers instrumental music, mathematical relations between musical intervals and the harmony of the universe. Neither the mathematical nor the musical sections exhibit any great originality. Theon was concerned to collate and organize discoveries made by his predecessors. His own contribution seems at times to consist of a mysticism about numbers and mathematical calculations.

The astronomical section is by far the most important. Music is included: the planets, Sun, Moon and the sphere of fixed stars are all set at intervals congruent with an octave. The Earth is a sphere standing at the centre of the universe, surrounded by several circles of the heavens. The text sets out the different explanations of the order of heavenly bodies and of deviations in latitude of the Sun, Moon and planets. Theon also shows how different writers, including Aristotle (384–322 BC), Callipus (*c.*370–*c.*330 BC) and Eudoxus (*c.*406–*c.*355 BC), accounted for the workings of the system of rotating spheres, and he puts forward what was then known about conjunctions, eclipses, occultations and transits. Other subjects covered include descriptions of eccentic and epicyclic orbits, and estimates of the greatest arcs of Mercury and Venus from the Sun.

No other works of Theon have survived, although he is known to have written a commentary on Plato's *Republic*.

Theophrastus *c.*372–*c.*287 BC was an ancient Greek thinker who wrote on a wide range of subjects – science, philosophy, law, literature, music and poetry, for example – but who is best known as the founder of botany.

Theophrastus was born in Eresus on the Greek island of Lesbos. He studied under Plato at the Academy in Athens, where Aristotle was also working, and then under Aristotle after he had left the Academy in 348 BC. When Aristotle returned to Athens and founded the Lyceum in 335 BC, Theophrastus became his chief assistant, eventually succeeding him as head of the Lyceum in 323 BC, when

Aristotle again left Athens. Enrolment at the Lyceum reached its peak under Theophrastus, and he employed the pupils, many of whom came from distant parts of Greece, to make botanical observations near their homes. He remained head of the Lyceum until his death in c.287 BC.

Theophrastus was a prolific writer: more than 200 books are attributed to him but most are known only by their titles. The most important of his surviving works are those on botany. In these books he covered most aspects of the subject – descriptions of plants, classification, plant distribution, propagation, germination and cultivation. He described and discussed more than 500 species and varieties of plants from lands bordering the Atlantic and Mediterranean and from India, referring in his descriptions to information from people who had been on Alexander the Great's campaigns and so had first-hand knowledge of foreign flora. On the basis of his collection of information about a wide range of flora, Theophrastus classified plants into trees, shrubs, undershrubs and herbs. In his detailed study of flowers, he noted that some flowers bear petals whereas others do not, and observed the different relative positions of the petals and ovary. He also distinguished between two major groups of flowering plants – dicotyledons and monocotyledons in modern terms – and between flowering plants and cone-bearing trees – angiosperms and gymnosperms. In his work on propagation and germination, Theophrastus described the various ways in which specific plants and trees can grow – from seeds, from roots, from pieces torn off, from a branch or twig, or from a small piece of cleft wood. He also accurately described the germination of seeds.

Theophrastus provided an excellent foundation for the science of botany – his writings are comprehensive and are generally very accurate, even by modern standards, although he was not infallible – for example, like all pre-seventeenth-century scientists Theophrastus believed in spontaneous generation. But he usually substantiated statements with observed facts, thus helping to establish a sound method of scientific investigation.

Thom René 1923– is a French mathematician who is a leading specialist in the fields of differentiable manifolds and topology and is famous for his model popularly known as 'catastrophe theory'.

Thom was born at Montbéliard on 2 September 1923. He studied in Paris at the Ecole Normale Supérieure, where he received a degree in mathematical sciences, from 1943 to 1946 and then spent four years at the Centre National de la Recherche Scientifique. In 1951 he was awarded a doctorate by the University of Paris for his thesis on algebraic topology. From 1954 to 1963 he was a professor in the faculty of science at the University of Strasbourg and in 1963 he became a professor at the Institute of Advanced Scientific Studies at Bures-sur-Yvette. He was awarded the Fields Medal at the Edinburgh Congress of 1958, the Brouwer Medal of the Academy of Sciences of the Netherlands in 1970, and the Grand Prix of Science by the city of Paris in 1974.

Since his student days Thom's major interest has been in problems of topology. Much of his early work was done just at the time that the operations of the US mathematician, Norman Steenrod, had been discovered. In his doctoral thesis he related Steenrod's definition of powers to the action of a cyclic group of permutations. From there he went on, with the help of H. Cartan, to formulate a precise series associated with 'space spherical bundles' and to demonstrate that the fundamental class of open spherical bundles showed topological invariance and formed a differential geometry. This led him to his famous theorem, the theorem of signature, which states that in order that a directed form M^{4K} of dimension 4K may become cuspal, it is necessary that the quadratic curve defined by the cusp produced on $H^{2K}(M^{4K}, Q)$ contains as many positive points as negative points (in order to be of zero index). In his work on the theory of forms Thom has shown that there are complete homological classes that cannot be the representation of any differential form. He was able to improve the understanding of this Steenrodian problem (although it is far from being completely understood) by formulating auxiliary spaces. These are now known as 'Thom spaces'.

In 1956, Thom developed the theory of transversality, and contributed to the examination of singularities of smooth maps. This work laid the ground for his later statement of the 'catastrophe theory', which he first published in 1968. The theory is, in fact, a model (not yet an explanation) for the description of processes which proceed by sudden changes, so that their action is not continuous, but discontinuous. They are thus not able to be described by means of mathematical tools derived from calculus. Catastrophe theory seeks to describe sudden changes from one equilibrium state to another and to do so Thom proposed seven 'elementary catastrophes' which he hoped would be suffcient to describe processes within human experience of space–time dimensions. The seven elementary catastrophes are the fold, cusp, swallow-tail, butterfly, hyperbolic,

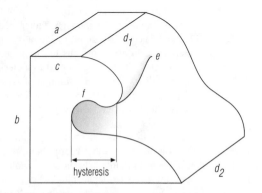

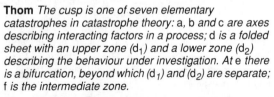

Thom *The cusp is one of seven elementary catastrophes in catastrophe theory:* a, b *and* c *are axes describing interacting factors in a process;* d *is a folded sheet with an upper zone* (d$_1$) *and a lower zone* (d$_2$) *describing the behaviour under investigation. At* e *there is a bifurcation, beyond which* (d$_1$) *and* (d$_2$) *are separate;* f *is the intermediate zone.*

elliptic and parabolic. Of these the cusp is both the simplest and the most likely to provide immediate applied use.

The important features of the cusp are the following:

a, b and c: axes that describe interacting factors in a process;

d: a folded sheet, divided into an upper zone, d_2, and a lower zone, d_2, which describes the behaviour of a substance, person or group under investigation;

e: the bifurcation point, a singularity, beyond which d_1 and d_2 separate;

f: the intermediate zone between d_1 and d_2.

The region of overlap between d_1 and d_2 is called the hysteresis. It is the existence of this overlap which makes possible sudden and dramatic changes from the equilibrium state d_1 and the equilibrium state d_2.

Thom is perhaps best known for his book published in 1972, *Structural Stability and Morphogenesis*. It is there that he introduced the notion of the 'universal unfolding' of a singularity. The book itself is not so much a mathematical treatise as a discussion of how the concept of structural stability may be applied to the appearance and evolution of natural forms in physics and the life sciences. In particular, the book attempts to use the model of catastrophe theory to analyse embryogenesis. The model has been used also to study nerve transmission and the heartbeat, to analyse change of phase in physics (for example, from liquid to gas), and even, by sociologists, to study human events such as the outbreak of riot or war. Whether the catastrophe theory will ultimately prove of great moment in the history of mathematics (it is, after all, fundamentally simply an argument by analogy) may be doubtful, but Thom's work has been of central importance to research in topology since 1945.

Thomas Sidney Gilchrist 1850–1885 was a British metallurgist and inventor who, with his cousin Percy Gilchrist, developed a process for removing phosphorus impurities from the iron melted during steel manufacture.

Thomas was born at Canonbury, London. His father died before he was 17, and the need for Thomas to earn a living became imperative. He spent a few months teaching and then in that same year, 1867, he obtained a post as a clerk at Marlborough Street Police Court. In the summer of 1868 he was transferred to a similar position at the Thames Court, Arbour Square, Stepney where, at a very modest salary, he remained until 1879.

His deep interest was in industrial chemistry, and he made his discoveries in his spare time. From about 1870 onwards one crucial problem in this field monopolized his attention – namely, the need to dephosphorize pig-iron when it was loaded into a Bessemer converter. Both the Bessemer and the Siemens–Martin processes (the most popular methods for converting pig-iron into steel) suffered from the serious drawback that in neither were phosphorus impurities removed. The steel produced from such phosphoric ores was brittle and of little use. Because only non-phosphoric ores could be used, the great mass of

British, French, German and Belgian iron ore was unusable for converting into steel. From 1860 onwards Henry Bessemer (1813–1898) and a great host of experimentalists looked unsuccessfully for the solution.

Thomas devoted the whole of his spare time to this question, experimenting systematically at home and attending the laboratories of various chemistry teachers. Towards the end of 1875 he arrived at a theoretic and provisional solution to the problem. The key was the chemical nature of the lining of the converter or furnace, which varied in composition but always contained silica. The phosphorus in the pig-iron was rapidly oxidized during the process to form phosphoric acid which, because of the silicous character of the slag was reduced back to phosphorus and re-entered the metal. The answer lay in providing a substance which would combine with the phosphoric acid and incorporate it into the slag. A long series of experiments had led Thomas to the conclusion that the material which could best withstand the intense heat of the furnace as well as providing durability which would make it economical to use was lime, or the chemically similar magnesia or magnesian limestone. Thomas foresaw that by using such a lining he was removing phosphorus from the pig iron and also that phosphorus 'deposited' in the basic slag would itself prove to be of immense commercial use (as a fertilizer).

In establishing this theory, Percy Gilchrist (1851–1935), a cousin of the inventor and a chemist at a large ironworks in Blaenavon, proved to be of the greatest assistance to Thomas. In March 1878, at a meeting of the Iron and Steel Institute of Great Britain, Thomas announced that he had successfully dephosphorized iron in a Bessemer converter. This announcement was disregarded, but the complete specification of his patent was filed in May 1878, to be followed by successive patents over the next few years.

Two notable developments followed. One was that the high-phosphorus iron ore of Lorraine and other areas in Europe could now be used for steel-making. The other was that by lining furnaces with a lime or other alkaline material, the basic slag which formed could be an important by-product, with application, in the developing artificial fertilizer industry.

Thomas did not long enjoy the fruits of his discovery. Not until 1879 did he give up his clerkship, but he was already suffering from a lung infection. Following a cruise round the world he died in Paris in February 1885. After providing for his next of kin (he was unmarried), he left all his considerable fortune to charitable institutions.

Thompson D'Arcy Wentworth 1860–1948 was a British biologist and classical scholar who is best known for his book *On Growth and Form*, first published in 1917. He also studied fisheries and oceanography and published several works on classical science.

Thompson was born on 2 May 1860 in Edinburgh; his father was an authority on ancient Greece. He was educated at the Edinburgh Academy, Edinburgh University and Trinity College, Cambridge. In the late 1870s, while studying medicine at Edinburgh University, he came under the influence of Charles Thomson (1830–1882) – the university's Professor of Natural History, who had recently returned from the Challenger Expedition for the exploration of the ocean depths – and became interested in biology and oceanography. In 1884, Thompson was appointed Professor of Biology at University College, Dundee, then became Senior Professor of Natural History at St Andrews University in 1917 – a post he held for the remainder of his career. Concerned about conservation, in 1896 and 1897 Thompson went on expeditions to the Pribilof Islands as a member of the British–American commission on fur-seal hunting in the Bering Sea. He was also one of the British representatives on the International Council for the Exploration of the Sea and a member of the Fishery Board of Scotland. Thompson died on 21 June 1948 in St Andrews.

Thompson's principal contribution to biology was his highly influential *On Growth and Form* (1917), in which he interpreted the structure and growth of organisms in terms of the physical forces to which every individual is subjected throughout its life. He also hypothesized that the evolution of one species into another results mainly from major transformations involving the entire organism – a view contrary to the traditional Darwinistic theory of species arising as a result of numerous minor alterations in the body parts over several generations. In the 1942 revised edition of his book, however, Thompson admitted that his evolutionary theory did not adequately account for the cumulative effect of successive small modifications.

In addition to his theoretical work on growth, form and evolution, Thompson wrote many papers on fisheries and oceanography. He also published works on classical natural history, notably *A Glossary of Greek Birds* (1895) and an edition of Aristotle's *Historia Animalium* (1910).

Thomson Benjamin 1753–1814. American-born physicist; *see* Rumford, Count.

Thomson George Paget 1892–1975 was a British physicist who in 1927 first demonstrated electron diffraction, a confirmation of the wave nature of the electron. Clinton Davisson (1881–1958) also made the same discovery independently using a different method and in recognition of their achievement, Thomson and Davisson shared the 1937 Nobel Prize for Physics.

Thomson was born in Cambridge on 3 May 1892, the son of J.J. Thomson (1856–1940). He was educated at Perse School, Cambridge, and then at Trinity College, Cambridge, where he took a first-class honours degree in mathematics and in physics in 1913. For a year he joined his father's research team but on the outbreak of World War I joined the infantry and saw active service in France. Thomson then transferred to the Royal Air Force and spent the rest of the war investigating the stability of aeroplanes. In 1919, he returned to Cambridge to become a fellow and lecturer at Corpus Christi College. Then in 1922 he became Professor of Natural Philosophy at the University of Aberdeen, holding the post until 1930, when he moved to Imperial College, London, as Professor of Physics.

During World War II, Thomson headed many government committees, including the British Committee on Atomic Energy which investigated the possibility of atomic weapons. For part of 1942 he was Scientific Liaison Officer at Ottawa, where he built up close contact with the United States atomic bomb project, and later in the war he became Scientific Adviser to the Air Ministry. Thomson returned to Imperial College after the war, interesting himself in thermonuclear fusion and undertaking vital research in this field. He retained his professorship there until 1952, when he became master of Corpus Christi College, Cambridge. Among the many honours he received were a knighthood in 1943 and Presidency of the Institute of Physics from 1958 to 1960 and of the British Association for the Advancement of Science in 1960. Thomson retired in 1962 and died at Cambridge on 10 September 1975.

During the mid-1920s, Thomson carried out a series of experiments hoping to verify a hypothesis initially presented in 1924 by Louis de Broglie (1892–1987) that electrons possess a duality, acting both as particles and as waves. The basis of the experiment involved bombarding very thin metal (aluminium, gold and platinum) and celluloid foils with a narrow beam of electrons. The beam was scattered into a series of rings in a similar manner to the rings associated with X-ray diffraction in metals. Applying mathematical formulae to measurements of the rings together with a knowledge of the crystal lattice, Thomson showed in 1927 that all the readings were in complete agreement with de Broglie's theory. Electrons were therefore shown to possess wave–particle duality. Several modifications to the initial experiments proved that the scattering effects were not the result of X-rays produced by the impact of electrons on the metal foil.

George Thomson's achievements parallel those of his father to an extraordinary degree. Both made fundamental discoveries concerning the electron, J. J. Thomson proving the existence of the electron as a particle in 1897 and his son demonstrating its wave nature 30 years later. Both men also became professors by the age of 30, both gained knighthoods and both were awarded the Nobel prize.

Thomson J(oseph) J(ohn) 1856–1940 was a British physicist who is famous for discovering the electron and for his research into the conduction of electricity through gases, for which he was awarded the 1906 Nobel Prize for Physics. He also received several other honours, including a knighthood in 1908 and the Order of Merit in 1912.

Thomson was born at Cheetham Hill, near Manchester, on 18 December 1856, the son of an antiquarian bookseller. At the age of 14 he went to Owens College, Manchester (later Manchester University), to study engineering. When his father died two years later, however, his family could not afford the premium for engineering training and so, with the help of a scholarship, Thomson studied physics, chemistry and mathematics. In 1876 he won a scholarship to Trinity College, Cambridge, where he remained – except for visiting lectureships to Princeton University in 1896 and to Yale University in 1904 – for the rest of his life.

After graduating in mathematics in 1880, he worked at the Cavendish Laboratory under Lord Rayleigh, succeeding him as Cavendish Professor of Experimental Physics in 1884. In 1919 he resigned the Cavendish professorship (being succeeded in turn by his student Ernest Rutherford), having been elected Master of Trinity College the previous year. He held this post until his death (in Cambridge) on 30 August 1940.

Thomson's first important research concerned vortex rings, which won him the Adam's Prize in 1883. It was then thought that atoms may be in the form of vortex rings in the 'ether', an idea which although untrue, led Thomson to begin his investigations into cathode rays and these, in turn, led to his famous discovery of the electron in 1897. At the end of the nineteenth century there was considerable debate as to whether cathode rays were charged particles or whether they were some undefined process in the ether. Hertz had apparently shown that cathode rays were not deflected by an electric field (a finding which indicated that they were not particulate) but Thomson proved this to be incorrect in 1897, and demonstrated that Hertz's failure to obtain a deflection was caused by his use of an insufficiently evacuated cathode-ray tube. Having proved the particulate nature of cathode rays, Thomson went on to determine their charge-to-mass ratio, finding it to be constant – irrespective of the gas in the cathode-ray tube – and with a value nearly 1,000 times smaller than that obtained for hydrogen ions in liquid electrolysis. He also measured the charge of the cathode-ray particles and found it to be the same in the gaseous discharge as in electrolysis. Thus he demonstrated that cathode rays are fundamental, negatively charged particles with a mass much less than the lightest atom known. He announced these findings during a lecture at the Royal Institution on 30 April 1897, calling the cathode-ray particles 'corpuscles'. The Dutch physicist Hendrik Lorentz later named them 'electrons' (a term first used – with a slightly different meaning – by the Irish physicist George Stoney), which became generally accepted.

Thomson then spent several years investigating the nature and properties of electrons, after which he began researching into 'canal rays', streams of positively charged ions, which Thomson named positive rays. Using magnetic and electric fields to deflect these rays, he found (in 1912) that ions of neon gas are deflected by different amounts, indicating that they consist of a mixture of ions with different charge-to-mass ratios. The British chemist Frederick Soddy had earlier proposed the existence of isotopes and Thomson proved this idea correct when he identified – also in 1912 – the isotope neon-22. This work was later continued by Francis Aston (one of Thomson's students and, later, the developer of the mass spectrograph), A. J. Dempster and other scientists, and led to the discovery of many other isotopes.

In addition to his pioneering research, Thomson wrote several notable works that were widely used in British universities, including *Notes on Recent Researches in Electricity and Magnetism* (1893), *Elements of the Mathematical Theory of Electricity and Magnetism* (1895) and,

with John Poynting, the four-volume *Textbook of Physics*. Equally important, Thomson developed the Cavendish Laboratory into the world's leading centre for subatomic physics in the early twentieth century. Furthermore, he trained a generation of scientists who subsequently made fundamental contributions to physics: seven of his research assistants – including his only son, George Thomson – later won Nobel prizes.

Thomson James 1822–1892 was a British physicist and engineer who discovered that the melting point of ice decreases with pressure. He was also an authority on hydrodynamics and invented the vortex waterwheel.

Thomson was born on 16 February 1822 at Belfast into a family of scientists. His father was Professor of Mathematics at Belfast and his younger brother was William Thomson, later Lord Kelvin (1824–1907). Thomson received his early education from his father and then at the age of only ten began to attend Glasgow University, obtaining an MA in 1839. He subsequently held a succession of engineering posts, and then settled in Belfast in 1851 as a civil engineer. In 1857, Thomson became Professor of Civil Engineering at Belfast and in 1873 moved to Glasgow to take up the same position there. He resigned his chair in 1889 and died in Glasgow on 8 May 1892.

As a young man Thomson was most interested in engineering projects, especially those involving paddles and waterwheels, and while still a student he invented a device for feathering paddles. In 1850, Thomson invented the vortex waterwheel, which was a smaller and more efficient turbine than those in use at the time, and it came into wide use. On moving to Belfast, Thomson continued his investigations into whirling fluids, making improvements to pumps, fans and turbines.

Thomson's most important discovery was made in 1849, when he determined the lowering of melting point caused by pressure on ice. This led him to reach an understanding of the way in which glaciers flow. Thomson also carried out painstaking studies of the phase relationships of solids, liquids and gases, and was involved in both geology and meteorology, producing scientific papers on currents and winds.

Tinbergen Niko(laas) 1907–1988 was a Dutch-born British ethologist who studied many animals in their natural environments but is best known for his investigations into the courtship behaviour of sticklebacks and the social behaviour of gulls. He did much to revitalize the science of ethology for which, among many other honours, he shared the 1973 Nobel Prize for Physiology or Medicine with Konrad Lorenz (with whom he worked on several research projects) and Karl von Frisch.

Tinbergen was born on 15 April 1907 in The Hague and was educated at Leiden University, from which he gained his doctorate in 1932. During 1931 and 1932 he went with a scientific expedition to the Arctic and on his return became a lecturer at Leiden University. Except for a period of military service during World War II, Tinbergen remained at Leiden – becoming Professor of Experimental Zoology in 1947 – until 1949, when he was appointed Professor of Zoology at Oxford University. He established a school of animal behaviour studies at Oxford and remained at the university until he retired in 1974.

One of Tinbergen's best known studies was of the three-spined stickleback, described in *The Study of Instinct* (1951). During the spring mating season the male stickleback marks out its individual territory and attacks any other male stickleback that enters its domain. If a female stickleback enters his territory, however, the male does not attack but courts her instead. Tinbergen showed that the aggressive behaviour of the male is stimulated by the red coloration on the underside of other males (this red patch develops on the underbelly of males during the mating season but does not appear on females). Tinbergen also demonstrated that the courtship dance of the male is stimulated by the sight of the swollen belly of a female that is ready to lay eggs. If the female responds to the male's zig-zag courtship dance, she follows him to the nest that he has already prepared. The female enters the nest first and is followed by the male who pokes her under the tail with his snout, causing her to release eggs and to emit a chemical that stimulates the male to ejaculate sperm into the water.

In *The Herring Gull's World* (1953) Tinbergen described the social behaviour of gulls, again emphasizing the importance of stimulus–response processes in territorial behaviour. Defence of territory is particularly important for birds such as herring gulls, which nest in large, densely populated colonies. Although these gulls possess a large vocabulary of warning calls, Tinbergen's findings suggest that gestures are equally, if not more, important in warning off other gulls. For example, he found that when two male herring gulls meet at territorial boundaries, one or both of the gulls tugs at the grass with his beak, rather than immediately fighting. Another similar behaviour pattern – the choking gesture – is performed by male and female gulls to warn off invaders: again, rather than fight, the gulls lower their heads, open their mouths and appear to choke. Tinbergen hypothesized that these gestures are normal behaviour patterns (grass-pulling, for example, occurs during nest-building) that have become adapted to denote aggression – and are recognized as aggressive by other birds – in order to prevent fighting and the possible injury or death that could result.

Tinbergen investigated other aspects of animal behaviour, such as the importance of learning in the feeding behaviour of oystercatchers, and also studied human behaviour, particularly aggression, which he believes is an inherited instinct that developed when humans changed from being predominantly herbivorous to being hunting carnivores.

Tiselius Arne Wilhelm Kaurin 1902–1971 was a Swedish physical biochemist who discovered the complex nature of proteins in blood serum and developed electrophoresis as a technique for studying proteins. For this work he was awarded the 1948 Nobel Prize for Chemistry.

Tiselius was born in Stockholm on 10 August 1902 into an academic family. His father died when he was only four years old, and the remaining family moved to Göteborg. At school he became interested in chemistry and biology and in 1921 he went to Uppsala University to study under Theodor Svedberg, the leading Swedish physical chemist at that time. In 1924 he gained an MA in chemistry, physics and mathematics and later the same year submitted his doctorial thesis on electrophoresis (the migration of charged colloidal particles in an electric field). He became an assistant to Svedberg and remained associated with the university for the rest of his career. He joined the faculty in 1930 and eight years later he was made director of the new Institute of Biochemistry. From 1934 to 1935 he worked in the United States at the Frick Chemical Laboratory at Princeton University. He retired in 1968 and died in Stockholm on 29 October 1971.

Tiselius began his research in Svedberg's laboratory in 1925. At that time Svedberg was developing the ultracentrifuge and Tiselius used the new machine to study the sizes and shapes of protein molecules. He observed that many substances that appeared to be homogeneous in the ultracentrifuge could be separated by electrophoresis. This was particularly true of serum proteins, but he changed the direction of his research and did not return to this problem for a number of years.

He then investigated zeolite minerals, which have a unique capacity to exchange their water of crystallization for other substances, the crystal structure remaining intact even after the water has been removed under vacuum. Tiselius studied the optical changes that occur when the dried crystals are rehydrated. He did this work at Princeton, and while he was there his discussions with such scientists as Karl Landsteiner (1868–1943) and Leonor Michaelis (1875–1949) made him reopen his research into more effective separation methods for biochemistry.

On his return to Uppsala he reconstructed his electrophoresis apparatus and used it to separate the proteins in horse serum. He obtained four protein bands with different mobilities. The fastest-moving band corresponded to the serum albumin boundary and the next three revealed for the first time the existence of three electrophoretically different components which he named α-, β- and γ-globulin. He observed the bands optically by measuring changes in their refractive indices.

He also became interested in adsorption methods of separation and devised a new quantitative optical technique for observing the eluate (previously adsorbed material washed out of the chromatography column in solution). In 1943, he showed that 'tailing' during elution can be prevented by adding to the eluting solution a substance of higher adsorption affinity than any of the components in the mixture. He called this method 'displacement analysis'. In 1954, he used calcium phosphate in the hydroxyl-apatite form as an adsorbent for proteins with phosphate buffers as eluting agents.

Tiselius made a decisive contribution to chromatography. His characteristic method of working was to take well recognized, qualitative experimental phenomena and establish their theoretical basis. As a result he was able to introduce improvements to existing techniques and to devise new ones. During the last ten years of his life he became very concerned about the possible threat to humanity by the advance of science. He believed that the Nobel Foundation was in a unique position to be able to bring pressure to bear on the most pertinent problems facing humanity, and he founded the Nobel Symposia which take place every year in each of the five prize fields to discuss the social, ethical and other implications of the award-winning work.

Todd Alexander Robertus, Lord 1907–   is one of the leading chemists of this century and has made outstanding contributions to the study of natural substances. For his work on nucleotides and coenzymes he was awarded the 1957 Nobel Prize for Chemistry.

Todd was born in Glasgow on 2 October 1907. He was educated at Allan Glen School and at Glasgow University, graduating in 1929. He went to the University of Frankfurt for two years, gaining his doctorate in 1931. From 1931 to 1934 he studied at Oxford University under Robert Robinson. After leaving Oxford he spent two years at the University of Edinburgh as Assistant in Medical Chemistry, and from there he went to the Lister Institute of Preventive Medicine in London. In 1937, he became Reader in Biochemistry at the University of London and a year later was appointed the Sir Samuel Hall Professor of Chemistry and Director of the Chemical Laboratories at Manchester University. He took up the Professorship of Organic Chemistry at Cambridge University in 1944, where he remained until he retired in 1971. He was knighted in 1954 and in 1962 was created Baron Todd of Trumpington; he became Master of Christ's College, Cambridge, in 1963.

Todd began his research in 1931 with Robinson, investigating the synthesis of the water-soluble plant pigments called anthocyanins. In 1936, he began his work on vitamins with the synthesis of the water-soluble vitamin B_1 (aneurin or thiamin), deficiency of which causes the disease beriberi. It is essential for the correct metabolism of carbohydrates; its diphosphate forms the coenzyme of carboxylase. He went on to study pantothenic acid, the so-called 'filtrate factor' of B vitamins which has been of therapeutic value in treating certain anaemias. Todd later worked on the structure of the fat-soluble vitamin E (tocopherol), deficiency of which affects fertility or muscular activity. In 1955, Todd and his co-workers, with Dorothy Hodgkin, established the structure of vitamin B_{12} (cyanocobalamin), deficiency of which causes pernicious anaemia.

In the late 1940s and early 1950s Todd also worked on nucleotides – compounds of a base (such as purine), a pentose sugar and phosphoric(V) acid. The term is applied to certain coenzymes, such as nicotinamide adenine dinucleotide (NAD); to compounds formed by partial hydrolysis of nucleic acids; and to nucleic acids themselves, which can be regarded as polynucleotides. He synthesized adenosine triphosphate (ATP) and adenosine diphosphate (ADP), the key substances in the energy-generating biochemical process in the body. He developed new methods for the synthesis of all the major nucleotides and their related coenzymes, and established in detail the chemical structures of the nucleic acids, such as DNA (deoxyribonucleic acid), the hereditary material of cell nuclei. During the course of this work, which provided the essential basis for further developments in the fields of genetics and of protein synthesis in living cells, Todd also devised an approach to the synthesis of the nucleic acids themselves.

Tolansky Samuel 1907–1973 was a British physicist who made important contributions to the fields of spectroscopy and interferometry.

Tolansky was born in Newcastle-upon-Tyne on 17 November 1907. He attended Snow Street School and Rutherford College and then in 1925 went to Armstrong College, part of Durham University, where he was awarded his BSc in 1928. He undertook scientific research at Armstrong College from 1929 to 1931, when he spent a year at the Physikalische Technische Reichsanstalt in Berlin under Friedrich Paschen (1865–1947).

On his return to England, Tolansky went to London to spend two years doing research at Imperial College and received his PhD in 1934. He then moved to the physics department of Manchester University, where he served as an assistant lecturer under Lawrence Bragg (1890–1971). In 1936, Tolansky was awarded a DSc by Manchester University, and he subsequently became a lecturer in 1937 and senior lecturer in 1945, and in 1946 was given the post of Reader in Physics. In the following year, Tolansky moved to Royal Holloway College, part of London University, to take up the Chair of Physics. He retained this position until he died in London on 4 March 1973.

Tolansky's early research centred on the study of line and band spectra and at Paschen's department in Berlin, Tolansky was exposed to studies in virtually all areas of spectroscopy. His main interest until his move to London in 1947 was the analysis of hyperfine structures in atomic spectra. He made a particular study of the spectrum of mercury, but the occurrence of multiple isotopes in his samples complicated the analysis. He also studied the hyperfine structure of the spectra of halogen gases such as chlorine and bromine, and of arsenic, iron, copper and platinum. He used the analysis of spectra to investigate nuclear spin, and magnetic and quadruple moments.

During World War II, Tolansky was requested by Tube Alloys (a euphemistic name for the body which was coordinating British research into the possibilities of constructing an atomic bomb) to obtain information on the spin of uranium-235, which is the isotope capable of fission in a nuclear chain reaction. Tolansky was unable to obtain any material in which the proportion of uranium-235 was enriched, so he had to use samples in which the proportion was only 0.7%. He was nevertheless able to demonstrate that the spin is in excess of 1/2, and suggested that it was most likely to be 5/2 or 7/2. The latter figure is the value which is currently accepted.

After the war, Tolansky concentrated on the applications of multiple beam interferometry. He used it to investigate the fine details of surface structure as the method is able to resolve structure as small as 15 Å (15×10^{-9} m) in height. He examined the vibration patterns in oscillating quartz crystals and the microtopography of many different crystals, particularly diamonds. He was especially interested in the growth features of crystals such as growth spirals. A quantitative means of assessing the hardness of materials grew out of Tolansky's work on interferometry as the depths of indentations made by the impact of a standard hardness tester could be accurately assessed by using interferometry. Another application for the technique lay in the measurement of the thickness of thin film.

Tombaugh Clyde 1906–   is a US astronomer whose painstaking work led to the discovery of Pluto.

Tombaugh was born on 4 February 1906 in Streator, Illinois. He had a deep fascination for astronomy and constructed a 23-cm/9-in telescope out of parts of old machinery on his father's farm. His family could not afford to send him to college and he joined the Lowell Observatory in 1929 in the hope that he would learn about the subject while working there as an assistant. In 1933, after his discovery of Pluto, he won a scholarship to the University of Kansas, from where he obtained an MA in 1936.

Several astronomers had shown that the orbital motions of Uranus and Neptune exhibited gravitational perturbations suggestive of the existence of a planet beyond them. Percival Lowell (1855–1916) made the first generally accepted calculation of the new planet's likely position and, as Director of the Lowell Observatory, he set up a team to look for the planet. This work continued when Vesto Slipher (1875–1969) became Director in 1926.

On joining the Observatory, Tombaugh worked on Slipher's team. There were various problems. The new planet would be too dim for a telescope to reveal without bringing thousands of dim stars into view also and, because of its distance from the Earth, any visible motion would be very slight. Tombaugh devised a technique by means of which he compared two photographs of the same part of the sky taken on different days. Each photograph could be expected to show anything between 50,000 and 500,000 stars. Tombaugh looked to see if any of the spots of light had moved in a way not expected of stars or the then known planets. Any movement would be noticeable if the different photographic plates were focused at a single point and alternately flashed rapidly on to a screen. A planet moving against the background of stars would appear to move back and forth on the screen.

It was a painstaking process. On 18 February 1930 Tombaugh discovered a moving light in the constellation of Gemini. On 13 March the discovery of the new planet was announced and it was named after the god of the nether darkness, which seemed appropriate for the most distant planet.

There was some doubt that this could be the planet that Lowell had predicted, since it was only one-tenth as bright and too small to account for all the gravitational perturbations, but the doubt was soon dispelled and Tombaugh was recognized for his great contribution to the furthering of our knowledge of the Solar System.

Torricelli Evangelista 1608–1647 was an Italian physicist and mathematician who is best known for his invention of the barometer.

Torricelli was born on 15 October 1608 in Faenza and was educated, mainly in mathematics, at the Sapienza College, Rome. He was impressed by the works of Galileo and the respect became mutual when Galileo read Torricelli's *De motu* (1641), which dealt with movement. In that same year Galileo invited him to Florence where he served as Galileo's secretary for the 3 months till his death. After Galileo's death the following year, he became Professor of Mathematics at Florence, where he remained for the rest of his life. He died there on 25 October 1647.

Galileo had been puzzled why a lift pump could not lift a column of water more than about 9 m/30 ft – current explanations were based on Nature's supposed abhorrence of a vacuum. Torricelli realized that the atmosphere must have weight, and the height of the water column is limited by atmospheric pressure. In 1643 he filled a long glass tube, closed at one end, with mercury and inverted it in a dish of mercury. Atmospheric pressure supported a column of mercury about 76 cm/30 in long; the space above the mercury was a vacuum. Mercury is nearly 14 times as dense as water, and the mercury column was only about one-fourteenth the height of the maximum water column.

Torricelli also noticed that the height of the mercury column varied slightly from day to day and finally came to the conclusion that this was a reflection of variations in atmospheric pressure. Thus by 1644 he had developed the mercury barometer.

Townes Charles Hard 1915–   is a US physicist who is best known for his investigations into the theory, and subsequent invention, of the maser (1953). He has received numerous honours for this work, including the 1964 Nobel Prize for Physics, which he shared with Nikolay Basov and Aleksandr Prokhorov, two Soviet scientists who independently invented the maser in 1955.

Townes, the son of a lawyer, was born in Greenville, South Carolina, on 28 July 1915. He was educated at Furman University, Greenville, from which he graduated in 1935, studied for his master's degree at Duke University and then moved to the California Institute of Technology to work for a PhD, which he gained in 1939. From 1939 to 1947 he worked at the Bell Telephone Laboratories designing radar bomb-aiming systems, after which he joined the physics department of the University of Columbia, becoming Professor of Physics in 1950. Next, he worked at the Massachusetts Institute of Technology from 1961 until 1967, when he was appointed Professor of Physics at the University of California, Berkeley.

In about 1950 Townes became interested in the possibility of constructing a device that could generate high-intensity microwaves, and by 1951 he had concluded that it might be possible to build such a device based on the principle that molecules emit radiation when they move from one energy level to a lower one. Ammonia molecules can occupy only two energy levels and, Townes argued, if a molecule in the high energy level can be made to absorb a photon of a specific frequency, then the molecule should fall to the lower energy level, emitting two photons of the same frequency and producing a coherent beam of single frequency radiation.

Using ammonia molecules, the radiation emitted would have a wavelength of 1.25 cm/0.5 in and would therefore be in the microwave region of the electromagnetic spectrum. But in constructing a device using this principle of the amplification by stimulated emission of radiation Townes had to develop a method (now called population inversion) for separating the relatively scarce high-energy molecules from the more common lower energy ones. He succeeded by using an electric field that focused the high-energy ammonia molecules into a resonator, and by 1953 he had constructed the first working maser (an acronym for *m*icrowave *a*mplification by *s*timulated *e*mission of *r*adiation).

Masers quickly found numerous applications – in atomic clocks (still the most accurate timepieces available) and radiotelescope receivers, for example. In the late 1950s solid-state masers (in which molecules of a solid were used instead of gaseous ammonia) were made and were used to amplify weak signals reflected from the Echo I passive communications satellite and radar reflections from Venus.

In 1958 Townes published a paper that demonstrated the theoretical possibility of producing an optical maser to produce a coherent beam of single-frequency visible light, rather than a microwave beam. But the first working optical maser (or laser, from *l*ight *a*mplification by *s*timulated *e*mission of *r*adiation) was built by Theodore Maiman in 1960.

Townsend John Sealy Edward 1868–1957 was an Irish mathematical physicist who was responsible for the development of the study of the kinetics of electrons and ions in gases. He was the first to obtain a value for the charge on the electron and to explain how electric discharges pass through gases.

Townsend was born in Galway on 7 June 1868. He attended Corrig School and in 1885 entered Trinity College, Dublin. There he studied mathematics, physics and experimental science, graduating in 1890. He spent the next four years teaching and lecturing, especially on mathematics, and in 1895 Townsend went to England, where he remained for the rest of his life. He gained entry to Cambridge University as a research student, and he and Ernest Rutherford (1871–1937) became the first such students to work under J. J. Thomson (1856–1940). Townsend was soon recognized as a scientist of rare quality, and when Oxford University established a new chair of experimental physics in 1900 with the intention of improving the standard of instruction in the fields of electricity and magnetism, Townsend was invited to become the first Wykeham Professor. He accepted, and most of the rest of his professional career was spent in Oxford.

During his early years at Oxford, Townsend was deeply involved with the construction of a permanent Electrical Laboratory. This was completed in 1910, but Townsend's research was then disrupted by the outbreak of World War I. He spent the war years at Woolwich, researching into radio telegraphy for the Royal Naval Air Service. Townsend resumed his work on gaseous ionization at Oxford after the war and continued there until 1941, when he was knighted and retired from his university duties. He did not, however, cease working. He spent a period teaching at Winchester School, and then returned to Oxford where he wrote a number of monographs. Townsend died in Oxford on 16 February 1957.

Townsend's earliest research at Cambridge was in the field of magnetism, but he soon turned to the subject which would occupy him for the rest of his career, namely the properties of ionized gases. He studied the conductivity of gases ionized by the newly discovered X-rays and in 1897 developed a method for producing ionized gases using electrolysis. This method enabled Townsend to obtain the first estimate for the electron charge in 1898. It was later modified by J. J. Thomson and Charles Wilson (1869–1959), and formed part of the basis for the famous oil drop experiments performed by Robert Millikan (1868–1953) from 1909 to 1912.

In 1898, Townsend began the first study of ionic diffusion in gases. He studied diffusion in gases which had been ionized (or electrified) by means of the so-called Townsend discharge. This involved the passage of a weak current through low-pressure gases. He found that the coefficient of diffusion of an ion was only a quarter of that of a neutral molecule, indicating a smaller mean free path for ions.

Around the time of Townsend's move to Oxford, he began to develop his collision theory of ionization. He based this theory on results he had himself obtained and those of Aleksandr Stoletov (1839–1896), which indicated that the ionization potential of molecules was tenfold less than was generally accepted. This meant that collisions by negative ions (electrons) could induce the formation of secondary ions, thus carrying an electric charge through a gas.

The study of the multiplication of charges in a gas was Townsend's major preoccupation for the next few years. He determined the coefficient of multiplication, also called Townsend's first ionization coefficient, which describes the number of pairs of ions produced by the movement of one electron through 1 cm/0.4 in in the direction of an electric field.

Townsend also studied the electrical conditions which lead to the production of a spark in a gas, and on the confusing role played in this by the positive ions that are produced simultaneously with the electrons.

During the 1920s, Townsend was involved with the measurement of the average fraction of energy lost by an electron in a single collision. He discovered in 1924 that this fraction was very small when the electrons were passed through a monatomic gas such as helium or argon. Carl Ramsauer (1879–1955) was independently engaged in a similar study and their results, which are called the Ramsauer–Townsend effect, were later found to be analogous to electron diffraction and were important in understanding the wavelike nature of the electron.

Tradescant John 1570–1638 and John 1608–1662 father and son, were English horticulturalists and pioneers in the collection and cultivation of plants. Linnaeus named the flowering plant genus *Tradescantia* (the spiderworts) after them.

Tradescant senior is generally considered to be the earliest collector of plants and other natural history objects. In 1604, he became gardener to the Earl of Salisbury, who, in 1610, sent him to Belgium to collect plants; the Brussels strawberry was one of the plants that he brought back. The following year he went to France, visiting Rouen and the Apothecary Garden in Paris, and in 1618 went to Russia with an official party sent by James I. Two years later Tradescant accompanied another official expedition – this time one against the North African Barbary pirates – and brought back to England gutta-percha, mazer wood and various fruits and seeds. Between 1625 and 1628 he was employed by the Duke of Buckingham and during this period joined an expedition to La Rochelle, France. Later, when he became gardener to Charles I, Tradescant set up his own garden and museum in Lambeth, London, which he stocked from his private collection of plants and other natural history specimens. In 1624, he published a catalogue of 750 plants grown in his garden; the only known copy of this catalogue is now in the library of Magdalen College, Oxford.

The younger Tradescant, sharing his father's enthusiasm for plants, became a member of the Company of Master Gardeners of London when only 26 years old. Three years later, in 1637, he went to Virginia, America, to collect plants and shells for his father's museum. After his father's death in 1638, the younger Tradescant succeeded him as gardener to Charles I, and went on several plant-collecting expeditions, adding to his father's collection. He published two lists of plants in the collection – in 1634 and 1656, the latter incorporating his father's specimens – most of which were unknown in England before 1600.

The younger Tradescant died on 22 April 1662 in Lambeth. His collection was eventually incorporated with that of Elias Ashmole (1617–1692), who, in turn, gave it to Oxford University in 1683, thereby founding the Ashmolean Museum.

Travers Morris William 1872–1961 was a British chemist famous for his association with William Ramsay on the discovery of the rare gases.

Travers was born in Kensington, London, on 24 January 1872, the second of four sons of a London physician. He was educated at Ramsgate (1879–1882) and Woking

(1882–1885), before going to Blundell's School in Tiverton, Devon, because it had a good chemistry laboratory. He went to University College, London, in 1889 and graduated in chemistry in 1893. He was then advised to study organic chemistry, and in 1894 began research at Nancy University with Alban Haller. But after a few months he returned to London to University College and became a demonstrator under William Ramsay. In 1898 he gained his DSc and became an Assistant Professor, being promoted to Professor of Chemistry in 1903 as successor to S. Young. He went to Bangalore in 1906 as Director of the new Indian Institute of Scientists.

He returned to Britain at the outbreak of World War I in 1914 and directed the manufacture of glass at Duroglass Limited, Walthamstow, becoming President of the Society of Glass Technology. In 1920 he became involved with high-temperature furnaces and fuel technology, including the gasification of coal. He was made Honorary Professor of Chemistry at Bristol University in 1927, and retired in 1937. During World War II he served as an adviser and consultant to the Explosives Section of the Ministry of Supply. He died at his home in Stroud, Gloucestershire, on 25 August 1961.

Travers returned to London to assist Ramsay in 1894, at the time of the discovery of argon, the first of the rare gases to be found. Then after the discovery of helium a year later, he helped Ramsay to determine the properties of both new gases. They also heated minerals and meteorites in the search for further gases, but found none. Then in 1898 they obtained a large quantity of liquid air and subjected it to fractional distillation. Spectral analysis of the least volatile fraction revealed the presence of krypton. They examined the argon fraction for a constituent of lower boiling point, and discovered neon. Finally xenon, occurring as an even less volatile companion to krypton, was found and identified spectroscopically.

The physicist James Dewar (1842–1923) had succeeded in liquefying hydrogen in 1898, but Travers independently constructed the necessary apparatus. Using liquid hydrogen, Ramsay and Travers were able to condense the neon fraction from air while the helium fraction remained gaseous. In this way they obtained enough neon by mid-1900 to complete their study of the rare gases. Travers continued his researches in cryogenics and made the first accurate temperature measurements of liquid gases. He also helped to build several experimental liquid air plants in Europe.

In his later work at Bristol, Travers studied the thermal decomposition of organic vapours and investigated gas and heterogeneous reactions. In 1956, when he was over 80 years old, he published *Life of Sir William Ramsay*, a biography of his partner of 50 years earlier.

Trésaguet Pierre-Marie-Jérôme 1716–1796 was a French civil engineer best known for his improved methods of road-building.

Trésaguet was born into a family that was connected with all aspects of engineering. He grew up absorbing the general principles of the subject, and spent many years

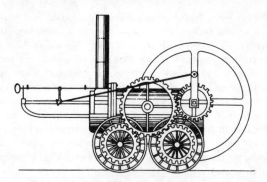

Trevithick *Trevithick's locomotive of 1803, with a small condenserless steam engine and smooth iron wheels that ran on iron rails, embodied most of the features that were to become standard on early steam locomotives.*

working for the Corps des Ponts et Chaussées, where his learning was put to good use in civil engineering projects. At first he was a sub-inspector in Paris, then Chief Engineer in Limoges, before he finally became Inspector General.

In the seventeenth and eighteenth centuries most roads consisted of stretches of old Roman roads and rough tracks, well worn over the years by the passage of horses and farm carts. Pitted, pot-holed and often completely impassable for several months each winter because of mud, they presented an awesome prospect to anyone wishing to travel a long distance. Turnpike roads, with the levying of tolls to pay for maintenance, effected some improvements, although often the tolls were simply pocketed and the repairs left undone. Even for new roads, the emphasis was on providing a suitable surface. Trésaguet realized that the key to the problem was to provide a solid foundation, one which could withstand winter rains and frost, and the effects of traffic. He chose to dig out the road-bed to a depth of about 25 cm/10 in and lay first a course of uniform flat stones, laid on edge to permit drainage. Well hammered in, they provided a solid base on top of which he spread a layer of much smaller stones for a smoother surface. His roads were built 5.4 m/18 ft wide, with a crown that rose 15 cm/ 6 in above the outside edge.

His method was first used for a major road that ran from Paris to the Spanish border, via Toulouse. It proved so successful that many other countries copied the idea, including Britain. The Scottish civil engineer Thomas Telford put the principle into practice when he was surveyor to the county of Shropshire.

Trevithick Richard 1771–1833 was a British engineer who constructed one of the first steam railway locomotives.

Trevithick was born on 13 April 1771, at Illogan, Cornwall, the son of a mine manager. As a boy he was fascinated by the mining machinery and the large stationary steam engines which worked the pumps. In 1797 he made a model of a steam road locomotive which he ran round a table in his home. He built various full-sized engines in the early 1800s and by 1808 he had built *Catch-me-who-can*, and ran it in London as a novelty for those willing to pay a

shilling for a ride. Then in 1816 he left England for Peru. When he returned, after making and losing a fortune, he found that in his absence others had developed steam transport until it was a thriving concern, with the first railway being authorized to carry passengers by an Act of Parliament in 1823. Trevithick was never able to resume his earlier position at the forefront of transport invention, and he died a poor man at Dartford, Kent on 22 April 1833.

The first steam-powered road vehicle was invented by the French army officer Nicolas Cugnot in 1769, for hauling cannon. It probably did not carry passengers, and credit for first doing this is usually given to Trevithick's *Puffing Devil* of 1801. It made its famous debut on Christmas Eve, but unfortunately burnt out while Trevithick and his friends were celebrating their success at a nearby inn. He then made a larger version which he drove from Cornwall to London in the following year, at a top speed of 19 kph/12 mph. The feasibility of a steam road vehicle was proven, but it was still unpopular. Passengers feared it might explode, and the weight of the vehicle damaged the road surface.

By 1804 Trevithick had produced his first railway locomotive, able to haul 10 tonnes and 70 people for 15 km/ 9.5 mi on rails used by horse-trains at Penydarren Mines, near Merthyr Tydfil in Wales. A few years later he hired some land near Euston, London, where he set up his money-making novelty ride. He also applied his inventive genius to many other machines, including steamboats, river dredgers and threshing machines. But he will always be best remembered for his steam locomotives.

Trumpler Robert Julius 1886–1956 was a Swiss astronomer who is known for his studies and classification of star clusters found in our Galaxy. He also carried out observational tests of the general theory of relativity.

Trumpler was born in Zurich on 2 October 1886, the third son of a family of ten children of the Swiss industrialist, Wilhelm Trumpler. He attended the University of Zurich for two years and then transferred to the University of Göttingen, where he gained his PhD in 1910. For the following four years, Trumpler work on the Swiss Geodetic Survey determining longitudes and latitudes, and it was during this time that he developed a personal interest in the annual proper motion of the stars in the Pleiades cluster. The latter interest coincided with that of Frank Schlesinger (1871–1943), Director of the Allegheny Observatory, near Pittsburgh, Pennsylvania. A meeting between the two men led Trumpler in 1915 to accept an invitation by Schlesinger to work at the Allegheny Observatory on comparative studies of galactic star clusters. In 1919, Trumpler was invited to work with William Campbell (1862–1938) at the Lick Observatory, near Chicago, to assist with tests of the general theory of relativity. Trumpler was appointed as Professor of Astronomy at the University of California in 1930. He was elected to the National Academy of Sciences in 1932 and in the same year he became President of the Astronomical Society of the Pacific, being re-elected to the Presidency in 1939. He retired in 1951 and died five years later in Oakland, California, on 10 September 1956.

At the Allegheny Observatory Trumpler showed that galactic star clusters contain different classes of stars, with no observable regularity in the occurrence of blue stars, yellow stars or red giants, and these observations paved the way for later theories about stellar evolution. In 1930 he showed that interstellar material was responsible for obscuring some light from galaxies, which had led to over-estimations of their distances from Earth; this work supported Harlow Shapley's research on the size of our Galaxy.

Three years after joining Campbell at the Lick Observatory, Trumpler took part in a test of the theory of relativity in which stars near the totally eclipsed Sun were photographed from Australia and compared with photographs taken simultaneously from Tahiti. Readings showed that light suffered an outward deflection of 4.45010.02270 cm/ 1.7734 in at the edge of the Sun, compared with Einstein's prediction that the amount of deflection would be 4.4323 cm/1.7463 in.

Trumpler also used the refractor at the Lick Observatory to study the planet Mars, concluding that, while most of Giovanni Schiaparelli's observations were incorrect, there was still a possibility that some of the supposed observations of 'canals' could be volcanic faults. This conclusion exemplifies the accuracy of Trumpler's observational astronomy because his hypothesis was made in 1924 and it only gained real support on the return of the photographs taken by the *Mariner 9* space probe to Mars, more than 50 years later.

Tsiolkovskii Konstantin 1857–1935 was a Russian theoretician and one of the pioneers of space rocketry, hailed by his country as the 'Father of Soviet Cosmonautics'.

Tsiolkovskii was born on 17 September 1857 in the village of Izheskaye, in the Spassk District. His parents were peasants and he had very little formal education. Despite this disadvantage and despite being impeded by deafness, he showed a marked aptitude for scientific invention and from boyhood was particulary intrigued by anything connected with flight.

During the 1880s, Tsiolkovskii earned his living as a schoolteacher, but contined to work in his spare time on theoretical heavier-than-air flying machines and the possibility of flight into outer space. In 1883, he defined the 'principle of rocket motion', which proved that it is feasible for a rocket-propelled craft to travel through the vacuum of space. His brilliant conceptions on rocketry and space flight covered all aspects, including the use of 'high-energy' propellants. He never actually constructed a rocket, but his theories, sketches and designs were fundamental in helping to establish the reality of space flight as we know it today. He died on 19 September 1935, just at the time when Robert Goddard in the United States launched his first liquid-fuelled rocket.

When Tsiolkovskii began his theoretical research, early flight in the form of balloons, airships and fragile uncrewed aircraft had already determined many things about the Earth's atmosphere and gravitation. He calculated that in order to achieve flight into space, speeds of 11.26 km/7 mi per second or 40,232 kph/25,000 mph would be needed – the so-called escape velocity for Earth. Known solid fuels were too heavy for such rocket propulsion, so Tsiolkovskii concentrated on potential liquid-fuels as propellants. In his notes he recorded: 'In a narrow part of the rocket tube the explosives mix, producing condensed and heated gases. At the other, wide end of the tube, the gases – rarefied and, consequently, cooled – escape through the nozzle with a very high relative velocity.' He also emphasized the value of what he called 'rocket trains' as a means of interplanetary travel. He suggested the 'piggyback' or step principle, with one rocket on top of another. When the lower one was expended, it could be jettisoned (reducing the weight) while the next one fired and took over.

Tsiolkovskii was one of three important pioneers working independently on the possibility of space flight. In the 1920s Hermann Oberth in Germany wrote a book about his ideas and stimulated much interest, while in the United States Goddard carried the work further by designing and actually constructing a liquid-fuelled rocket (in 1935). It attained a speed of more than 1,125 kph/700 mph and rose to a height of over 300 m/1,000 ft. How much of today's achievement in space travel can be attributed directly to the self-taught Russian genius is debatable, although Tsiolkovskii's work must at least have had a profound effect on thinking in his own country.

Tswett (or Tsvett) Mikhail 1872–1919 was an Italian-born Russian scientist who made an extensive study of plant pigments and developed the technique of chromatography to separate them.

Tswett was born in Asti, Italy, on 14 May 1872 of a Russian father and an Italian mother. His father was a civil servant in Russia, where his mother had grown up; they had stopped in Asti en route to Switzerland when he was born. His mother died soon afterwards and his father left the baby with a nurse in Lausanne when he returned to Russia. Tswett spent his childhood and youth in Lausanne and Geneva and in 1891 he entered the Department of Mathematics and Physics at Geneva University. He graduated in physical and natural sciences in 1893 and obtained his doctorate three years later. His first scientific publication, of 1894, was on plant anatomy. In 1897, he went to Russia to do research at the Academy of Sciences and the St Petersburg Biological Laboratory. His foreign degrees were not acceptable in Russia so he took an MSc degree at Kazan University in 1901. For the next six years he worked at Warsaw University before being offered a teaching appointment at the Warsaw Veterinary Institute. In 1908 he moved to the Warsaw Technical University and obtained a doctorate in botany in 1910. During World War I he organized the work of the Botany Department of the Warsaw Polytechnic Institute, which was evacuated to Moscow and Gorky. In 1917, he was appointed Professor of Botany and Director of the Botanical Gardens at Yuriev University (Estonia), but under threat of German invasion had to move once again to Voronezh. He died there on 26 June 1919.

Tswett opposed the view that green leaves contain only two plant pigments, chlorophyll and xanthophyll. In 1900 he showed that there are two types of chlorophyll, termed chlorophyll a and chlorophyll b, which differ in colour and absorption spectra. He obtained a pure sample of chlorophyll a, but isolating the b type proved to be troublesome. By 1906 he had devised an adsorption method of separating the pigments. He ground up leaves in petroleum ether and let the liquid trickle down a glass tube filled with powdered chalk or alumina. As the mixture seeped downwards, each pigment showed a different degree of readiness to attach itself to the absorbent, and in this way the pigments became separated as different coloured layers in the tube. To get samples of single pigments, he pushed the adsorbent out of the tube, cut off the coloured pieces of the column with a knife, and extracted the pigment from each piece separately using a solvent. Tswett called the new technique chromatography from the fact that the result of the analysis was 'written in colour' along the length of the adsorbent column. Eventually he found six different pigments. The new method attracted little attention until it was rediscovered by scientists such as Richard Willstätter and Richard Kuhn (1900–1967). By the late 1940s it had become one of the most versatile methods of chemical analysis, particularly in biochemistry, especially after Archer Martin and Richard Synge had developed paper chromatography.

Tull Jethro 1674–1741 was an English lawyer and inventor of agricultural machinery, whose development of the seed-drill revolutionized farming methods and helped to initiate the agricultural revolution.

Tull was born at Basildon, Berkshire. He attended St John's College, Oxford, and later became a law student at Gray's Inn, London, where he was called to the Bar in May 1699. Ill health, which had dogged him throughout much of his life, eventually forced him to return to the country, where he began to look with new eyes at the long-established methods of farming. He realized much could be done to improve the old system and his first invention, the seed drill of 1701, was an attempt to bring some order and economy into the process. Although local farmers were intrigued by his invention, they were reluctant to alter their ways until the advantages had been thoroughly proven. For the next 30 years, Tull was to struggle to bring his new ideas and inventions to agriculture, trying to change the system by simple reasoning and example. All of this he set down in his classic book on farming *The New Horse Houghing Husbandry*, which was published in 1733.

When Tull turned to farming at the beginning of the eighteenth century, the strip system was still generally practised in most parts of Britain. Three common fields in a settlement were cultivated in rotation – two in use, one fallow. The fields were then divided up into approximately half-acre (about 0.2 hectare) strips, each one separated from the next by a grass balk or pathway; each person had the right to a strip in all three common fields.

Tull's first invention was the seed-drill, designed to surmount the problems related to broadcast sowing which was the traditional method. It was a revolutionary piece of equipment, designed to incorporate three previously separate actions into one: drilling, sowing and covering the seeds. The drill consisted of a seed-box capable of delivering the seed in a regulated amount, a hopper mounted above it for holding the seed, and a plough and harrow for cutting the drill (groove in the soil) and turning over the soil to cover the sown seeds.

Tull then experimented with implements for ploughing, paying particular attention to the different requirements for different types of soil. He devised the 'common two-wheeled plough' or 'plow', as he called it, which could cut a furrow to a depth of about 18 cm/7 in on the lighter lands of the Midlands and southern England. It had a single coulter for making the vertical cut and a share for the horizontal one. This type of plough was not suitable for the heavy clay lands of the eastern region, and for these he developed a more sturdy swing-plough with a longer horizontal beam and no wheels.

Although these ploughs proved successful to a certain extent, they did not solve one of the most difficult problems the farmer had to face: removing the top layer of grass and weeds.

Often it was necessary to use a breast plough to cut through these before the main ploughing could take place. Tull solved this eventually by making a four-coulter plough, with blades set in such as way that the grass and roots were pulled up and left on the surface to dry. This was the final stage in the development of the plough, and basically the design is much the same today.

Tull was perhaps the initiator of the farming revolution which during the next 100 years was to raise the productivity of the land beyond the requirements of the immediate village. He and such contempories as Charles Townsend (1674–1738), nicknamed 'Turnip Townsend' because he was responsible for the introduction of turnips as a field crop, for feeding livestock, enabling farmers to keep them alive through the winter, and Robert Bakewell of Leicestershire, who did much to improve animal breeding, the foundation of the new agriculturalism, by which enough food would be provided to feed the masses in the coming Industrial Revolution.

Turing Alan Mathison 1912–1954 was a British mathematician who worked in numerical analysis and played a major part in the early development of British computers.

Turing was born at London on 23 June 1912 into a family distinguished by its diplomats and engineers, three of whom had been elected to the Royal Society. He was educated at Sherborne School from 1926 to 1931, when he went to King's College, Cambridge to study mathematics. After receiving his BA in 1935, he was elected a Fellow of the college on the strength of his paper 'On the Gaussian Error Function', which won a Smith's Prize in mathematics in 1936. The paper was a characteristic example of the headstrong, but brilliant, nature of Turing's mathematical method throughout his life. He 'discovered' the central limit theorem in utter ignorance of the fact that it had already been discovered and proved.

In 1936, Turing went to the United States for two years to work at Princeton University with the mathematical logician, Alonso Church. There he worked on the theory of computation and in 1937 he presented to the London Mathematical Society the paper, 'On Computable Numbers', which was his most famous contribution to mathematics. It constituted a proof that there exist classes of mathematical problems which are not susceptible of solution by fixed and definite processes, that is to say, by automatic machines. He returned to King's in 1938 and after the outbreak of World War II in 1939 was employed by the government Code and Cipher School at Bletchley Park. For his work in designing machines to break the German Enigma Codes he was awarded an OBE in 1946.

After the war Turing joined the mathematics division of the National Physical Laboratory at Teddington, where he began immediately to work on the project to design the general computer known as the Automatic Computing Engine, or ACE. Although he left the project in 1947 to return to Cambridge, Turing played an important part in the theoretical work for the production of the ACE; a pilot version of the machine was in operation by 1950 and the mature version (like most computers of the time quickly rendered obsolete by newer machines) by 1957.

In 1948, Turing was appointed reader in the theory of computation at the University of Manchester and was made assistant director of the Manchester Automatic Digital Machine (MADAM). Two years later he published in *Mind* his trenchant discussion of the arguments against the notion that machines were able to think, 'Computing Machinery and Intelligence'. His conclusion was that, by his definition of 'thinking', it was possible to make intelligent machines.

In his last years at Manchester much of his work was done at home. All his life he had been concerned with mechanistic interpretations of the natural world and he now devoted himself to attempting to erect a mathematical theory of the chemical basis of organic growth. In this he was partly successful, since he was able to formulate and solve complicated differential equations to express certain examples of symmetry in biology and also certain phenomena such as the shapes of brown and black patches on cows. He died from taking poison at his home on 7 June 1954, either deliberately, which was the coroner's verdict, or by accident, as his mother (knowing that he kept poisons in unmarked containers such as tea cups and sugar bowls) believed.

Turing's place in the history of mathematics rests on the theory of computation which he worked out in 1936 and 1937. He suggested a basic machine which was not a mechanical device, but an abstract concept representing the operation of a computer. Quite simply, it was a paper tape, divided into squares, with a head for erasing, reading or writing on each square and a mechanism for moving the tape to either the left or the right. The tape could have instructions already written on it and it was of either limited or unlimited length. So Turing's concept contained, in embryonic form, the now familiar notions of program, input, output and, by implication, the processing of information. Turing machines were therefore of two types: machines designed to carry out a specific function and process information in a specified way and machines of a universal function capable of carrying out any procedure.

Turner Charles Henry 1867–1923 was a US biologist and teacher who was internationally known for his research into insect behaviour patterns.

Charles Turner was born on 3 February 1867 in Cincinnati. On graduating from Cincinnati High School he went to study biology at the University of Cincinnati, receiving a BS degree in 1891 and an MS in 1892. He remained at the university for a year as an assistant instructor in the biology laboratory and then moved to Clark College, Atlanta, as Professor of Biology until 1895. Turner went on to teach in public schools in Evansville, Indiana, and Cincinnati and in 1906 was principal of College Hill High School, Cleveland. From 1907 to 1908 he was Professor of Biology at Haines Normal and Industrial Institute, Augusta, Georgia, after which he moved to St Louis, Missouri, and taught biology at Sumner High School until his death in 1923. Turner was an outstanding teacher but he is chiefly known for his research into insect behaviour. From 1892 until his death he conducted experiments on ants, bees, moths, spiders and cockroaches and published over 50 papers on neurology, animal behaviour and invertebrate ecology. In 1907, with his dissertation 'The homing of ants: an experimental study of ant behaviour', he was awarded a PhD by the University of Chicago. Turner was the first to prove that insects can hear and distinguish pitch and that roaches learn by trial and error; in French literature the turning movement of the ant towards its nest was given the name 'Turner's circling'. As well as scientific research, Turner also wrote nature stories for children and was active in the civil rights movement in St Louis. After his death, a school for the physically handicapped was erected in St Louis and named the Charles H. Turner School, renamed the Turner Middle School in 1954.

Twort Frederick William 1877–1950 was a British bacteriologist, the original discoverer of bacteriophages (often called phages), the relatively large viruses that attack and destroy bacteria. He also researched into Johne's disease, a chronic intestinal infection of cattle. The only major honour that Twort received for his work was his election as a Fellow of the Royal Society in 1929.

Twort was born on 22 October 1877 in Camberley, Surrey, the son of a physician. He studied medicine at St Thomas's Hospital, London, obtaining his degree in 1900. He then spent a year as Assistant Superintendent of the Clinical Laboratory at St Thomas's before becoming Assistant Bacteriologist at the London Hospital in 1902. In 1909, Twort was appointed Superintendent of the Brown Institute, a pathology research centre. He remained there (except for a period of war service during World War I) until he retired 35 years later, having also been made Professor of Bacteriology at the University of London in 1919. He died in Camberley on 20 March 1950.

Twort is probably best known for his discovery in 1915 of what is now called a bacteriophage. While working with cultures of *Staphylococcus aureus* (the bacterium that causes the common boil), he noticed that colonies of these bacteria were being destroyed. Twort isolated the substance that produced this effect and found that it was transmitted indefinitely to subsequent generations of the bacterium. He then suggested that the substance was a virus, a prediction that was later proved correct. He was unable to continue his work, however, and his discovery aroused little interest at the time. Two years later, in 1917, the same discovery was made independently by a Canadian bacteriologist, Félix d'Hérelle, who named the active substance bacteriophage. Again, bacteriophages were virtually ignored and it was not until the 1950s, with the work of Heinz Fraenkel-Conrat and others, that their importance was recognized. Since then bacteriophages have been widely used in microbiology, mainly for studying bacterial genetics and cellular control mechanisms.

Twort was also the first to culture the causative bacillus of Johne's disease, and showed that a specific substance is essential for the growth of this bacillus. In addition, he discovered that vitamin K is needed by growing leprosy bacteria, which opened a new field of research into the nutritional requirements of microorganisms.

Tyndall John 1820–1893 was an Irish physicist who is mainly remembered for the Tyndall effect, which is the scattering of light by very small particles suspended in a medium. The discovery of this effect enabled Tyndall to explain the blue colour of the sky.

Tyndall was born at Leighlinbridge, Carlow, on 2 August 1820. He went to school in Carlow and then held a succession of surveying and engineering jobs in Ireland and England. In 1847, he became a teacher of mathematics at Queenswood College, Hampshire, where he was a colleague of the chemist Edward Frankland (1825–1899). Drawn to the study of science by Frankland, Tyndall left with him in 1848 to enter the University of Marburg, Germany. There Tyndall studied physics and mathematics, and also chemistry under Robert Bunsen (1811–1899). Obtain-

ing his doctorate in 1850, he returned to Queenwood in 1851 and in 1853 became Professor of Natural Philosophy at the Royal Institution. From 1859 to 1868, Tyndall was also Professor of Physics at the Royal School of Mines.

In 1867, Tyndall became Superintendent of the Royal Institution, succeeding Michael Faraday (1791–1867). In his position as lecturer at the Royal Institution and as a journalist and writer, he did much to popularize science in Britain and also in the United States, where he toured from 1872 to 1873. Tyndall also championed those he believed had been wrongly treated, and was responsible for the recognition in Britain of the pioneering work done on the conservation of energy by Julius Mayer (1814–1878). In 1886, Tyndall became seriously ill and he retired from the Royal Institution in the following year. He died from an accidental overdose of drugs at Hindhead, Surrey, on 4 December 1893.

The discovery of the Tyndall effect was made in 1869 when Tyndall was investigating the passage of light through liquids. He found that light passes unimpeded through a pure solvent or an ordinary solution, but that the beam becomes visible in a colloidal solution. Tyndall realized that the colloidal particles although invisible to the eye were big enough to scatter the light, and reasoned that a similar suspension of dust particles in the atmosphere causes the blue light in the sunlight passing through the atmosphere to be scattered more than the red, producing a blue sky. This explanation was confirmed by Lord Rayleigh (1842–1919) in 1871, who showed that the scattering is inversely proportional to the fourth power of the wavelength.

This work led Tyndall to consider the likelihood that the air contains living microorganisms, and he was able to show that pure air devoid of any suspended particles as indicated by the absence of the Tyndall effect did not produce putrefaction in foods. The air must therefore contain bacteria. This result was important in confirming the work of Louis Pasteur (1822–1895) that rejected the spontaneous generation of life, and it also inspired Tyndall to develop methods of sterilizing by heat treatment.

Tyndall also carried out experimental work on the absorption and transmission of heat by gases, especially water vapour and atmospheric gases, which was important in the development of meteorology.

U

Urey Harold Clayton 1893–1981 was a US chemist who in 1931 discovered heavy water and deuterium, the isotope of hydrogen of mass 2. For this extremely significant discovery, which was to have a profound effect on future research in chemistry, physics, biology and medicine, he was awarded the 1934 Nobel Prize for Chemistry.

Urey was born in Walkerton, Indiana, on 29 April 1893 and educated at schools in De Kelb County, Kendallville and Walkerton. He went to Montana State University, Missoula, in 1917, gaining his BS degree three years later. During 1918 and 1919 he worked as a research chemist at the Barrett Chemical Company in Philadelphia, where he helped to produce war materials. From 1919 to 1921 he was an Instructor in Chemistry at Montana State University, and then went to the University of California in Berkeley, where he developed his interest in physical and mathematical chemistry. He received his PhD in 1923 and during the following year was a Fellow of the American Scandinavian Foundation and attended the Institute of Theoretical Physics at the University of Copenhagen. He studied there under Niels Bohr, who was engaged on his pioneering work on the theory of atomic structure.

After Urey returned to the United States he worked for five years as an Associate in Chemistry at Johns Hopkins University in Baltimore, Maryland. Then from 1929 to 1934 he was Associate Professor of Chemistry at Columbia University, New York City, and was Ernest Kempton Adams Fellow there from 1933 to 1936. He was appointed full Professor of Chemistry in 1934. During World War II he was Director of Research of the Substitute Alloy Materials Laboratories at Columbia, which became part of the Manhattan Project for the development of the atomic bomb. In 1945, Urey became Professor of Chemistry at the Institute of Nuclear Studies at the University of Chicago. In 1958 he was named Professor-at-Large of Chemistry at the University of California in La Jolla; he was also a member of the Space Science Board of the Academy of Sciences. He died in La Jolla on 5 January 1981.

Urey discovered deuterium (heavy hydrogen, symbol D) in 1931 with F. G. Brickwedde and G. M. Murphy. He predicted that it would be possible to separate hydrogen (H_2) from HD (whose molecules contain one hydrogen atom and one deuterium atom) by the distillation of liquid hydrogen, taking advantage of the difference in their vapour pressures. In heavy water, D_2O, both hydrogen atoms of normal water (H_2O) are replaced by deuterium atoms. Two years after its discovery by Urey, the US chemist Gilbert Newton obtained nearly pure heavy water by fractional electrolysis of water. Today it is manufactured by a process that involves isotopic chemical exchange between hydrogen sulphide and water. Its chief use is as a moderator to slow down fast neutrons in a nuclear reactor.

Urey went on to isolate heavy isotopes of carbon, nitrogen, oxygen and sulphur. His group provided the basic information for the separation of the fissionable isotope uranium-235 from the much more common uranium-238 by gaseous diffusion of their fluorides. After World War II he worked on tritium (another isotope of hydrogen, of mass 3) for use in the hydrogen bomb.

The evolution of the Earth was another topic that exercised Urey's mind. It was traditionally believed a molten Earth had formed by processes similar to those that occur in oil-smelting furnaces. Today there is evidence to suggest that the Earth and other planets were formed by condensation and accumulation from a dust cloud at low temperatures. Urey theorized that the final accumulation of the Earth had occurred at 0°C/32°F from small planetary particles containing metallic iron, carbon, iron carbide, titanium nitride and some ferrous (iron(II)) sulphide. He considered that most gases had been lost during the previous high-temperature phase, leaving a primitive atmosphere consisting of hydrogen, ammonia, methane, water vapour, nitrogen and hydrogen sulphide. In 1952, he suggested that some of these molecules could have united spontaneously to form the basic 'building blocks' of life. The iron core of the Earth would have accumulated slowly through geological history from a mixture of metallic iron and silicates, meaning that the Earth was not molten at the time when its materials had accumulated. Urey also contributed to theories about the origin of the Moon, subscribing to the view that it had not formed from the Earth but had a separate origin.

V

Van Allen James Alfred 1914– is a US physicist who was closely involved with the early development of the US space program and discovered the magnetosphere, the zone of high levels of radiation around the Earth caused by the presence of trapped charged particles. This region is popularly known as the Van Allen belt.

Van Allen was born in Mount Pleasant, Iowa, on 7 September 1914. He grew up there and then went to the Iowa Wesleyan College, where he received his BS degree in 1935. He then went to the University of Iowa, where he earned his MS in 1936 and his PhD in 1939. From 1939 to 1942, Van Allen worked as a research fellow at the Carnegie Institute in Washington, DC, in the department of terrestrial magnetism. During World War II, from 1942 to 1946, Van Allen served as an Ordnance and Gunnery Officer with the US Navy. He participated in the development of the proximity fuse, which was a device attached to a missile such as an anti-aircraft shell so that it detonated when it neared the target and the radio waves it emitted were reflected back to it with sufficient intensity from the target. This meant that detonation could occur even when the missile had not been aimed sufficiently accurately for a direct hit, and this greatly increased the effectiveness of anti-aircraft weapons.

Van Allen's wartime work gave him experience with the miniaturization of sophisticated equipment, and it was put to good use in his next post from 1946 to 1950 as Supervisor of the High Altitude Research Group and Proximity Fuse Unit in the Applied Physics Department at Johns Hopkins University. During the period from 1949 to 1957, Van Allen organized and led scientific expeditions to Peru 1949, the Gulf of Alaska 1950, Greenland 1952 and 1957, and Antarctica 1957 to study cosmic radiation.

In 1951, Van Allen became Professor of Physics and Head of the Physics Department at the University of Iowa. From 1953 to 1954 he participated in Project Matterhorn, which was concerned with the study of controlled thermonuclear reactions, and he also served as a research associate at Princeton University. Van Allen was then closely involved with the organization of International Geophysical Year, which took place from July 1957 to December

1958. In 1959 he was also made Professor of Astronomy at Iowa, and since 1972, Van Allen has been Carver Professor of Physics there. He lives in Iowa City.

Van Allen's work immediately after the end of World War II involved utilization of the unused German stock of V-2 rockets and of Aerobee rockets for research purposes. Devices for the measurement of the levels of cosmic radiation were sent to the outer regions of the atmosphere in these rockets, and the data were radioed back to Earth. In 1949, Van Allen conceived of rocket-balloons (rockoons) which began to be used in 1952. They consisted of a small rocket which was lifted by means of a balloon into the stratosphere and then fired off, thus being able to reach heights attainable otherwise only by a much larger rocket. Van Allen's experience in the miniaturization of electronic equipment enabled him to include a maximum amount of instrumentation in the limited payload of these rockets, and this was to be crucial to the early stages of the US space program.

The possibility of sending a rocket into orbit around the Earth, thereby creating an artificial satellite, was finally given serious consideration by the US government in the mid-1950s. President Eisenhower announced the Vanguard Program in 1955, which promised to put an artificial satellite into orbit within two years to coincide with the scheduled International Geophysical Year, which was itself timed to coincide with a peak in solar activity. Van Allen was given the responsibility for the instrumentation of the proposed satellites.

Van Allen was on a scientific expedition to the Antarctic when the news of the world's first artificial satellite, the Russian *Sputnik 1*, broke on 4 October 1957. This precipitated great activity in the US camp, and the first US satellite, *Explorer 1*, was launched on 31 January 1958. Part of *Explorer*'s payload was a geiger counter which Van Allen had intended to measure the levels of cosmic radiation, but at a height of about 800 km/500 mi the counters registered a radiation level of zero. This was an absurd reading, and instrument failure was suspected. When the same result was recorded by *Explorer 3*, which was launched on 26 March 1958, Van Allen realized that the

Van Allen *Van Allen belts*

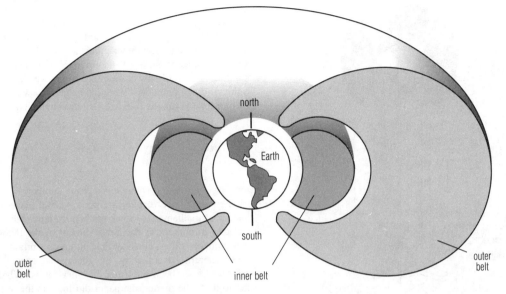

zero reading could have resulted from the counters being swamped with very high counts of radiation. The counter which was sent up with *Explorer 4* on 26 July 1958 was shielded with lead in order to allow less radiation to penetrate, and this showed clearly that parts of space contained much higher levels of radiation than had previously been suspected.

Van Allen studied the size and distribution of the high radiation zones. He found that it consists of two toroidal belts around the Earth, which arise by the trapping of charged particles in the Earth's magnetic field. They were named the Van Allen belts, but are now more usually known as the magnetosphere.

van de Graaff Robert Jemison 1901–1967 was a US physicist who designed and built the electrostatic high-voltage generator named after him.

Van de Graaff was born in Tuscaloosa, Alabama, on 20 December 1901. He studied to be an engineer, obtaining a BS degree from the University of Alabama in 1922 and an MS a year later. He worked for a short while for the Alabama Power Company, but left in 1924 to attend the Sorbonne in Paris for further study. There van de Graaff was deeply impressed by lectures given by Marie Curie (1867–1934), which focused his attentions on the exciting developments in atomic physics. In 1925, he went to study at Oxford under John Townsend (1868–1957), receiving a BSc in 1926 and a PhD in 1928. Van de Graaff conceived the idea of his generator at this time and returned to the United States in 1929 to develop it. He built a working model at Princeton University, where he served as a National Research Fellow from 1929 to 1931. He then undertook further development at the Massachusetts Institute of Technology (MIT), where he was appointed Research Associate in 1931. Van de Graaff then became

Associate Professor of Physics at MIT in 1934, and retained his position until 1960, when he resigned to devote all his time to the High Voltage Engineering Corporation (HVEC), a company he had set up in 1946 with John Trump. Van de Graaff died in Boston on 16 January 1967.

Van de Graaff's main work at Oxford concerned the mobility of ions in the gaseous state, but he saw the need for beams of high-energy subatomic particles in the study of the properties of atoms. He realized that a high potential could be built up by storing electrostatic charge within a hollow sphere and that this could be achieved by depositing charges on to a moving belt that carries the charges into the sphere, where a collector transfers them to the outer surface of the sphere. Van de Graaff's first working model, constructed in 1929, operated at 80,000 volts and he soon upgraded this to 2 million volts, and eventually to 5 million volts. At MIT, the generator was developed so that it could be used to accelerate subatomic particles to very high velocities, and this technology was then immediately applied to nuclear investigations and also to clinical research. Trump and van de Graaff collaborated to modify the generator so that it could be used in the generation of hard X-rays for use in radiotherapy in treating internal tumours and the first machine, a 1-MeV (1 million electronvolts) X-ray generator, was installed in a Boston hospital in 1937.

During World War II, the US Navy commissioned the production of five generators with 2-MeV (2 million electronvolts) capacity for use in the X-ray examination of munitions. The experience gained during these years enabled van de Graaff to set up HVEC for the commercial production of generators. The company developed the van de Graaff generator for the wide variety of scientific, medical and industrial research purposes in which it is used today. The tandem principle of particle acceleration and a

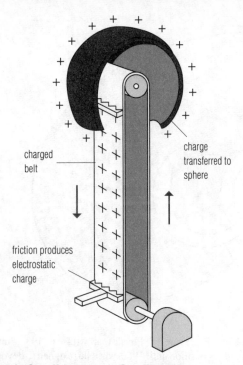

charged belt

charge transferred to sphere

friction produces electrostatic charge

van de Graaff *In a van de Graaff generator, a conveyor belt carries charge from a high-voltage supply to a hollow domed terminal, where it accumulates on the outside.*

new insulating core transformer invented by van de Graaff contributed to these advances.

Vandermonde Alexandre-Théophile 1735–1796 was a French musician and musical theorist who wrote original and influential papers on algebraic equations and determinants.

Vandermonde was born in Paris on 28 February 1735. Being a somewhat sickly child he was educated privately – at first chiefly in music, the sphere in which he was expected to make his career. But the tuition of the famous French geometer, Alexis Fontaine des Bertins (1705–1771), awakened his mathematical interest, and although he later wrote some skilful papers on musical composition, the early years of his maturity were devoted to mathematics. Fontaine introduced him to leading members of the French Academy of Sciences and in 1771, before he had written a single scientific paper or made any very remarkable discovery, he was elected to the Academy. In the next two years he presented to the Academy the only four mathematical papers he ever wrote. He played a part in the founding of the Conservatoire des Arts et des Métiers and served as its Director after 1782. In the 1780s he also collaborated with his close friend Gaspard Monge (1746–1818), and with the chemist Claude Berthollet (1748–1822) in an analysis of the difference between pig iron and steel, published in 1786. He died in Paris on 1 January 1796.

Of Vandermonde's four mathematical papers, the most

celebrated are the first and the fourth. The first, regarded by some mathematicians as his best, considered the solvability of algebraic equations. Vandermonde found formulae for solving general quadratic equations, cubic equations and quartic equations. Lagrange published similar results at about the same time, but Vandermonde's two methods of solution for solving lower order equations – called substitution and combination – were independently worked out by him. In addition he found the solution to the equation:

$$x^{11} - 1 = 0$$

and stated, without giving a proof, that:

$$x^n - 1 = 0$$

must have a solution where n is a prime number.

The fourth paper occupies a controversial place in the history of mathematics. Some mathematicians regard it as a decisive moment in the establishment of the theory of determinants; others consider that, although it might be the first coherent statement of the theory, its content had mostly already been published in other forms. Whatever the truth of the matter, the paper did include the Vandermonde determinant, his best known contribution to mathematics.

The second paper may have influenced Gauss in his work on electrical potentials; the third was a relatively unimportant work on factorials. It is really on the strength of two papers, written within a year of each other, that Vandermonde (who subsequently contributed nothing at all to mathematics) earned a permanent place in the subject's history.

van der Waals Johannes Diderik 1837–1923 was a Dutch scientist whose theoretical work on gases made an important contribution to chemistry and physics, a fact recognized by the award to him of the 1910 Nobel Prize for Physics. His theories about interatomic forces also added to knowledge about molecular structure and chemical bonding.

Van der Waals was born in Leyden (Leiden), the Netherlands, on 23 November 1837, the son of a carpenter. He began his career as primary school teacher, then entered Leyden University in 1862 to study physics, while at the same time working as secondary school physics teacher, becoming headmaster of the school at The Hague in 1866. Following on from the work of Rudolf Clausius and other molecular theorists, van der Waals laid the foundation for most of his future studies in his doctoral thesis of 1873, 'Over de continuiteit van den gasen vloeistoftoestand'/'On the continuity of the gaseous and liquid states'. It received immediate recognition and was soon translated into other European languages. In 1887 he became Professor of Physics at the new University of Amsterdam (formerly the Amsterdam Athenaeum), and remained there until he retired in 1907; he was succeeded by his son. He died in Amsterdam, after a long illness, on 8 March 1923.

Using fairly simple mathematics, van der Waals's 1873 thesis explained in molecular terms various phenomena of

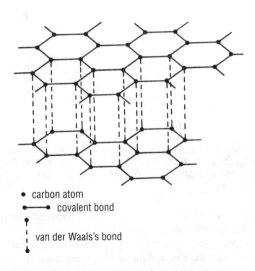

- • carbon atom
- •—• covalent bond
- ┊ van der Waals's bond

van der Waals *Van der Waals' s forces in graphite.*

vapours and liquids that had been observed experimentally by Thomas Andrews and others, especially the existence of critical temperature (above which a gas or vapour cannot be liquefied by pressure alone, no matter how great). The law of corresponding states which van der Waals developed some years later gave a somewhat better 'fit' between the theory and the experimental data and became a useful guide in work on the liquefaction of the so-called permanent gases.

The van der Waals equation of state attempts to explain the behaviour of real gases, as opposed to the 'ideal' or 'perfect' gas laws of Robert Boyle and Jacques Charles (or Joseph Gay-Lussac), which combined to give the equation:

$$PV = RT$$

It still links pressure (P), volume (V) and absolute temperature (T), but introduces two other constants a and b (R remains the universal gas constant):

$$(P + a/V^2)(V - b) = RT$$

The term a/V^2 accounts for intermolecular attraction, determined by integrating over an 'attraction sphere' that extends round each molecule. The constant b accounts for the non-overlapping of molecules and their finite size. Both constants, a and b, are different for different gases. Van der Waals was also able to work out equations for isotherms (how the volume of a gas or liquid changes with pressure at a particular temperature) and calculate the parameters of the critical point.

By extension, van der Waals's results were also applied to other thermodynamic quantities and phenomena, such as the Joule–Thomson effect, saturated vapour pressures, supercooling, and so on.

The cohesive attraction between molecules in a liquid became known as van der Waals's forces. The same forces are postulated for molecular crystals such as those of graphite and naphthalene. In graphite, for example, normal covalent bonds hold together hexagonal arrays of carbon atoms in planes, which are themselves bonded in parallel 'layers' by van der Waals' forces. A shearing force can fairly readily overcome these weak forces, allowing the planes to slip or slide over each other, which accounts for the existence of well defined cleavage planes in solid graphite and its effectiveness as a lubricant.

van't Hoff Jacobus Henricus 1852–1911 was a Dutch theoretical chemist who made major contributions to stereochemistry, reaction kinetics, thermodynamics and the theory of solutions. In 1901 he was awarded the first Nobel Prize for Chemistry.

Van't Hoff was born in Rotterdam on 30 August 1852, the son of a doctor. At school he showed great ability at mathematics, but decided to study chemistry at the Polytechnic at Delft. In 1871 he attended the University of Leiden, and a year later went to Bonn to study under Friedrich Kekulé, who was unimpressed by his Dutch student, so van't Hoff moved to Paris and worked with Charles Adolphe Wurtz at the Ecole de Médecine. He returned to the Netherlands and obtained his doctorate at Utrecht in 1874, taking up a lectureship in physics at the Veterinary College there two years later. In 1878 he became Professor of Chemistry, Mineralogy and Geology at the University of Amsterdam and stayed there until 1896. In that year he moved to Berlin to become a Professor to the Prussian Academy of Sciences, with an honorary professorship at Berlin University as well. He remained there until he died, in Berlin, on 1 March 1911.

Van't Hoff had made his first major contribution to chemistry even before he was awarded his doctorate. In 1874 he announced the results of his research into conformational analysis of organic compounds, which hinged on what we would now call the stereochemistry of the carbon atom. He postulated that the four valencies of a carbon atom are directed towards the corners of a regular tetrahedron. This allows it to be asymmetric (connected to four different atoms or groups) in certain compounds, and it is these compounds that exhibit optical activity. Van't Hoff ascribed the ability to rotate the plane of polarized light to the asymmetric carbon atom in the molecule, and showed that optical isomers are left- and right-handed forms (mirror images) of the same molecule. A similar idea was put forward independently in Paris two months later by Joseph Le Bel, who had studied under Wurtz at the same time as had van't Hoff.

Van't Hoff's first ideas about chemical thermodynamics and affinity were published in 1877, and consolidated in his *Etudes de Dynamique Chimique* (1884), translated into English in 1886. He applied thermodynamics to chemical equilibria, developing the principles of chemical kinetics and describing a new method of determining the order of a reaction. He deduced the connection between the equilibrium constant and temperature in the form of an equation known as the van't Hoff isochore. He generalized it in the form of what he called the principle of mobile equilibrium, which is a special case of Le Châtelier's principle, which

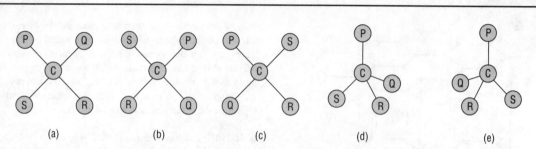

(a)   (b)   (c)   (d)   (e)

van't Hoff *Van't Hoff introduced the idea of a tetrahedral carbon atom. The three square structures (a), (b) and (c) are all equivalent (rotating (b) 90° anticlockwise gives (a); rotating (c) 180° along the axis P–R also gives (a)). The tetrahedral structures (d) and (e) are mirror images of each other, but one cannot be rotated in any way to make it the same as the other; they represent a pair of (optical) stereoisomers.*

he had independently formulated in the same year. It may be stated thus:

If a change occurs in one of the factors (such as temperature or pressure) under which a system is in equilibrium, the system will adjust itself so as to annul, as far as possible, the effect of that change.

In the field of reaction kinetics, van't Hoff announced his findings at the same time as, but independently of, Cato Guldberg and Peter Waage in 1867. They all developed the fundamental law of reaction kinetics which assumes that, at constant temperature, the rate of any simple chemical reaction is proportional to the product of the concentrations of the various reacting substances – a statement of the law of mass action.

Van't Hoff also introduced the modern concept of chemical affinity as the maximum work obtainable as the result of a reaction, and he showed how it can be calculated from measurements of osmotic pressure, gas pressure, and the electromotive force of reversible galvanic cells. In 1886, he published the results of his study of dilute solutions and showed the analogy between them and gases, because they both obey equations of the type $PV = RT$.

In the 1880s he became a friend of Svante Arrhenius. The theories of van't Hoff on osmotic pressure and those of Friedrich Ostwald on the affinity of acids accorded well with Arrhenius's views on electrolytes and the three scientists worked together to get their new theories accepted. In 1887, they started the important journal *Zeitschrift für physikalische Chemie*, whose first volume contained the famous paper by Arrhenius on electrolytic dissociation and the fundamental paper of van't Hoff. He was the first to apply thermodynamics systematically to solutions, although the treatment could have been more generally applicable had he used the thermodynamic system derived by Josiah Gibbs between 1875 and 1879.

After he moved to Berlin in 1896 van't Hoff studied the behaviour of the various salts from the deposits at Stassfurt.

Van Vleck John Hasbrouck 1899–1980 was a US physicist who made important contributions to our knowledge of magnetism, as a result of which he is widely considered to be one of the founders of modern magnetic theory. He received many honours for his work, including the 1977

Nobel Prize for Physics, which he was awarded jointly with Philip Anderson and Nevill Mott.

Van Vleck was born in Middletown, Connecticut, on 13 March 1899, the son of a mathematics professor. He was educated at the University of Wisconsin, from which he graduated in 1920, and at Harvard University, from which he gained his master's degree in 1921 and his doctorate in 1922. He then stayed at Harvard as an instructor in physics until 1923, when he went to the University of Minnesota as an assistant professor. In 1927 he married Abigail Pearson, and was promoted to full Professor of Physics at Minnesota. He was Professor of Physics at Wisconsin University from 1928 to 1934 – during which period he was a Guggenheim Foundation Fellow 1930 – then returned to Harvard. He remained at Harvard until formally retiring in 1969, having been Hollis Professor of Mathematics and Natural Philosophy since 1951; on retiring he was made an Emeritus Professor. While at Harvard he held various other positions in addition to the Hollis professorship: Dean of Engineering and Applied Physics from 1951 to 1957, Lorentz Professor at Leiden University, the Netherlands, in 1960; and Eastman Professor at Oxford University from 1962 to 1963. Van Vleck died in Cambridge, Massachusetts, on 27 October 1980.

Van Vleck devoted most of his career to studying magnetism, particularly its relationship to atomic structure. With the coming of quantum wave mechanics in the early 1930s he began investigating its implications for magnetism and devised a theory – using wave mechanics – that gives an accurate explanation of the magnetic properties of individual atoms in a series of chemical elements. He also introduced the idea of temperature-independent susceptibility in paramagnetic materials – a phenomenon now called Van Vleck paramagnetism – and drew attention to the importance of electron correlation (the interaction between the movements of electrons) in the appearance of localized magnetic moments in metals. His most important work, however, was probably his formulation of the ligand field theory, which is still one of the most useful tools for interpreting the patterns of chemical bonds in complex compounds. This theory explains the magnetic, electrical and optical properties of many elements and compounds by considering the influences exerted on the electrons in particular atoms by other atoms nearby.

Van Vleck also worked on radar during World War II. He found that water molecules in the atmosphere absorb radar waves with a wavelength of about 1.25 cm/0.5 in, and that oxygen molecules have a similar effect on 0.5-cm/0.2-in radar waves. This finding was important for the development of effective radar systems and, later, in microwave communications and radioastronomy.

Vauquelin Louis Nicolas 1763–1829 was a French chemist who worked mainly in the inorganic field analysing minerals and is best known for his discoveries of chromium and beryllium. He rose from humble origins to be one of the most influential scientists of his time.

Vauquelin was born in Saint-André d'Héberôt, Calvados, on 16 May 1763, the son of a Normandy farm labourer. He too worked on the land as a boy until 1777 when he became apprenticed to an apothecary in Rouen. Two years later he moved to Paris and eventually became a laboratory assistant at the Jardin du Roi under Antoine Fourcroy (1755–1809), who recognized Vauquelin's ability, befriended him and began a nine-year collaboration. In 1791 he was made a member of the Academy of Sciences and from that time he helped to edit the journal *Annales de Chimie*, although two years later he left the country for a while during the height of the French Revolution. On his return in 1794 he became Professor of Chemistry at the Ecole des Mines in Paris. He described various analytical techniques in his *Manuel de l'essayeur* (1799), which led in 1802 to his being appointed to the position of Assayer to the Mint. On Fourcroy's death in 1809 he succeeded him as Professor of Chemistry at the University of Paris (and gave a home to Fourcroy's two elderly sisters). He was elected to the Chamber of Deputies in 1828 and died a year later, in his birthplace, on 14 November 1829.

Vauquelin did most of his important work in inorganic chemistry in the late 1790s while he was at the Ecole des Mines. He analysed various minerals, often using specimens supplied by the mineralogist René Haüy (1743–1822). In a Siberian lead mineral called crocolite he discovered chromium in 1797, naming the element from the Greek *chroma* ('colour') because so many of its compounds are brightly coloured. (Martin Klaproth made the same discovery a few months later.) Vauquelin also examined emeralds and the mineral beryl and recognized that they contained another new element which was eventually called beryllium, although at first he called it glucinium because of the sweet taste of some of its salts. The element itself was not isolated until 1828, by Friedrich Wöhler.

In organic chemistry, Vauquelin also made some significant discoveries. In 1806, working with asparagus, he isolated the amino acid aspargine, the first one to be discovered. He also discovered pectin and malic acid in apples, and isolated camphoric acid and quinic acid.

Vening Meinesz Felix Andries 1887–1966 was a Dutch geophysicist who originated the method of making very precise gravity measurements in the stable environment of a submarine. The results he obtained were important in the fields of geophysics and geodesy.

Vening Meinesz was born at The Hague on 30 July 1887, and took a degree in civil engineering at the Technical University of Delft in 1910. He was first employed by the government to take part in a gravimetric survey of the Netherlands. Over the next decade he took measurements at over 50 sites, and was particularly concerned with overcoming the problem of inaccuracies caused by unstable support for his apparatus. His PhD thesis, submitted to the Technical University of Delft in 1915, dealt with this problem. Vening Meinesz developed a device which required the measurement of the mean periods of two pendulums which swung from the same apparatus. The mean of the two periods is not affected by disturbances in the horizontal plane, and so can be used to determine the local gravitational force accurately.

From 1923 to 1939, Vening Meinesz undertook 11 scientific expeditions at sea in submarines. He was fortunate in receiving the cooperation of the Royal Netherlands Navy. The outbreak of World War II put a temporary stop to such peaceful uses of submarines, but expeditions were resumed in the late 1940s. By then it was Vening Meinesz's younger associates who actually put out to sea, but his pendulum apparatus was the only one suited to marine gravimetric measurement until the end of the 1950s. Spring gravimetric devices have since been developed which can be mounted on stable platforms in surface vessels, but Vening Meinesz's method has been applied in thousands of measurements around the world.

Vening Meinesz became Professor of Cartography at the University of Utrecht in 1927, and in 1935 was also made Professor of Geophysics at the same university. He held both these posts until 1957, and from 1938 to 1957 he also held the simultaneous post of Professor of Physical Geodesy at the University of Delft. Vening Meinesz died in Amersfort, the Netherlands, on 10 August 1966.

During the course of his early research, Vening Meinesz realized that measurements of the Earth's gravitational field could yield important indications of the nature of the internal features of the Earth itself. A prerequisite of such study would be the ability to make accurate measurements of the Earth's gravitational field in as many different places as possible. Values could be obtained on land with relative ease, although they might be subject to large distortions by the presence of local features such as mountains. Measurement at sea was made impossible by turbulence caused by waves and wind, but was most desirable because so much of the Earth's surface is covered by ocean and because there are fewer problems with local geographic features as the sea bed is the nearest source of distortion.

It was suggested to Vening Meinesz that gravimetric experiments could be conducted below the sea's surface in submarines because disruption from waves decreases exponentially with the depth of the submarine. On his first voyage, Vening Meinesz found that once he had modified the design of his apparatus to compensate for the craft's rolling, he was indeed able to make very accurate measurements.

One of the earliest important findings to arise from these studies was the discovery of low-gravity belts in the Indonesian Archipelago running from Sumatra to the Mindanao Deep. Vening Meinesz proposed this to have arisen as a result of a downward buckling of the crust causing light sediments to fill the resulting depressions. This is the origin of the concept of the geosyncline. Vening Meinesz later returned to the question of crustal deformation and suggested it to be a consequence of convection currents arising from the cooling of the Earth at its surface. He elaborated this into a shortlived hypothesis that denied the existence of continental drift, but abundant evidence was soon amassed to support the view that continental drift does occur and is driven by convection currents in the mantle.

The results Vening Meinesz obtained on his expeditions at sea were also applied in the field of geodesy, which is concerned with the analysis of the Earth's shape. He showed that the equilibrium shape of a rotating ellipsoid is disrupted by irregularly distributed masses, but that the height of these anomalous masses does not exceed 40 km/ 25 mi. He was able to discount the model of the Earth's shape which proposed a flattening at the equator.

Venn John 1834–1923 was a British logician whose diagram, known as the Venn diagram, is much used in the teaching of elementary mathematics.

Venn was born at Drypool, Hull, on 4 August 1834 into a prominent Anglican family. When he was a young boy his father moved to London to take up an appointment as honorary secretary to the Church Missionary Society and he was educated first at Sir Roger Cholmley's School (now Highgate School). In 1853, he matriculated at Gonville and Caius College, Cambridge. He was elected a mathematical scholar in 1854 and gained his BA in 1857. In the same year he was elected a Fellow of the College; he remained a Fellow all his life.

Given his family background it was no surprise that Venn was ordained as a deacon at Ely in 1858 and as a priest in 1859. He then held curacies at Cheshunt, Hertfordshire, and Mortlake, Surrey, before returning to Cambridge in 1862 as a college lecturer in moral sciences. For the next 30 years his chief interest was in logic, the subject to which he contributed a trilogy of standard texts: *The Logic of Chance* (1866), *Symbolic Logic* (1881) and *The Principles of Empirical Logic* (1889). While he was pursuing his logical research, Venn became infected by the crisis of belief prevalent in high Victorian, post-Darwinian England and, although he never lost his faith, he found himself unable any longer to subscribe to the Thirty-nine Articles of the Church of England and abjured his clerical orders in 1883. In the same year he was elected to the Royal Society and awarded a DSc by the University of Cambridge. After publishing his third volume of logic, he turned increasingly to the history of the university and the three-volume biographical history of Gonville and Caius (1897) was almost entirely his work. He died at Cambridge on 4 April 1923.

Venn is remembered chiefly for his logical diagrams. The use of geometrical representations to illustrate syllogistic logic was not new (they had been used consistently by Gottfried Leibniz), and Venn came to be highly critical of the various diagrammatic methods prevalent in the nineteenth century, in particular those of George Boole and Augustus de Morgan. Neither Boole's algebraic logic nor Morgan's formal logic satisfied him and it was largely in order to interpret and correct Boole's work that he wrote his *Symbolic Logic*. In it appeared the diagrams which have since become universally known as Venn diagrams.

Before publishing the *Symbolic Logic*, Venn had adopted the method of illustrating propositions by means of exclusive and inclusive circles; in the new book he added the new device of shading the segments of the circles to represent the possibilities which were excluded by the

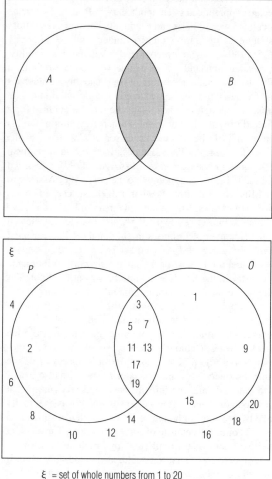

ξ = set of whole numbers from 1 to 20
O = set of odd numbers
P = set of prime numbers

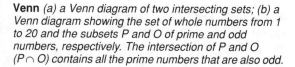

Venn *(a) a Venn diagram of two intersecting sets; (b) a Venn diagram showing the set of whole numbers from 1 to 20 and the subsets P and O of prime and odd numbers, respectively. The intersection of P and O (P ∩ O) contains all the prime numbers that are also odd.*

propositions at issue. It was these diagrams, more than his attempt to clarify what he considered to be inconsistencies and ambiguities in Boole's logic, which constituted the real merit of the book. His diagrams were based on his belief that those which merely represented the relations between two classes or two propositions were not sufficiently general; later, he extended his method by proposing a series of circles dividing the plane into compartments, so that each successive circle should intersect all the compartments already existing. It was this idea, taken up and refined by Charles Dodgson (1832–1898), which led to the use of the closed compartment enclosing the whole diagram to define the universe of discourse – or, what is now known as the universal set.

Vernier Pierre 1584–1638 was a French engineer and instrument-maker who devised the precision measuring scale now named after him.

Vernier inherited an interest in scientific measuring instruments from his father, who was castellan of the château of Ornans, a lawyer by training and probably also an engineer. Vernier worked as a military engineer for the Spanish Habsburgs, then the rulers of Franche-Comté, and he realized the need for a more accurate way of reading angles on the surveying instruments he used in map-making. He was aware of the work of the Portuguese mathematician and astronomer Filde Nunez Salaciense (Nonius) who, working on the same problem in about 1540, hit upon the idea of engraving on the face of an astrolabe a series of fixed scales laid out around concentric circles. On any one circle the scale was determined by dividing its circumference into an equal number of parts, one fewer than those dividing the next circle out but one more than those on the next circle in. Thus a line of sight (to, say, a star) would inevitably fall very close to a whole division on one of the scales. By calculation or by using tables it was possible accurately to determine the angle in degrees, minutes and seconds. But in practice it was extremely difficult to engrave with precision a different scale on each of the circles. As Tycho Brahe remarked, Nunez's method failed to live up to its promise.

Fifty years later Clavius (a former student of Nunez) found a way of engraving the scales, and his associate Jacobus Curtins further simplified the system so that angles could be read off directly (although 60 separate scales were required). The method was described in Clavius's *Geometrica Practica* of 1604, which Vernier probably studied.

Vernier's solution to the problem was to use a single moveable scale, rather than a series of fixed ones, thus avoiding the difficulties of multiple engraving. The mobile scale was graduated with nine divisions equalling the space occupied by ten graduations on the main scale. In this way the accuracy of the reading was increased by a factor of ten. By this time (1630) he had acquired a reputation as an outstanding engineer and was appointed Conseiller et Général des Monnaies to the Count of Burgundy. He made a special journey to Brussels to present his invention to Isabelle-Clair-Eugénie, the Infanta of Spain and ruler of Franche-

Comté, who told him to publish a description of it. After doing so in 1631 Vernier returned to Dôle where he designed and directed the building of fortifications. His other engineering projects included the design of a building for the harquebusiers of Dôle, but in 1636 ill health forced him to give up engineering and he returned to Ornans, where he died a few years later.

Vesalius Andreas 1514–1564 was a Belgian physician who was a founder of modern anatomy.

Vesalius was born in Brussels on 13 December 1514, to a family of physicians from Wesel (the derivation of Vesalius), in Germany. His father was the royal pharmacist to Charles V of Germany, and from an early age Vesalius showed an inclination to follow in the family tradition, by dissecting dead birds and mice. He was educated at the University of Louvain and then studied medicine in Paris. After a period as a military surgeon, he moved to Padua where he gained his medical degree in 1537, and was then appointed lecturer in surgery and anatomy. At Padua he published his famous book *De humani corporis fabrica* (1543), which met with vigorous opposition and led to bitter controversy. In a fit of despondency, he gave up anatomy and resigned his chair at Padua. He became court physician to Charles V, and later to his son Philip II of Spain. On his way back from a pilgrimage to Jerusalem, on 15 October 1564, he died in a shipwreck off the island of Zante (now called Zakinthos) near Greece.

Vesalius was taught anatomy in the Galenist tradition. Galen had never dissected a human body – all his accounts of the human anatomy were based on his research of the Barbary ape – although he was regarded as infallible and was venerated until the Renaissance: Vesalius was therefore taught principles of anatomy that had not been questioned for 1,300 years.

The artists of Vesalius's time encouraged the study of anatomy because they wanted accurate representations of the human body. The greatest of these were Leonardo da Vinci (1452–1519), Albrecht Dürer (1471–1528) and Michelangelo (1475–1564). Da Vinci, who made more than 750 anatomical drawings, paved the way for Vesalius. Vesalius became dissatisfied with the instruction he had received and resolved to make his own observations, which disagreed with Galen's. For instance, he disproved that men had a rib less than women – a belief that had been widely held until then. He also believed, contrary to Aristotle's theory of the heart being the centre of the mind and emotion, that the brain and the nervous system are the centre.

Between 1539 and 1542 Vesalius prepared his masterpiece, a book that employed talented artists to provide the illustrations. The finished work, published in 1543 in Basel, is one of the great books of the sixteenth century. The quality of anatomical depiction introduced a new standard into all illustrated works and especially into medical books. The text, divided into seven sections, is of great importance in expressing the need to introduce scientific method into the study of anatomy. The *De humani corporis fabrica* did for anatomy what Copernicus's book did for astronomy (they were both published in the same year). Vesalius upset the authority of Galen and his book, the first real textbook of anatomy, marks the beginning of biology as a science.

Viète François 1540–1603 was a French mathematician, the first extensively to use letters of the alphabet to represent numerical quantities and the foremost algebraist of the sixteenth century.

Viète was born at Fontenay-le-Comte, in the Poitou region, in 1540. He was educated locally until 1556, when he entered the University of Poitiers to study law. He graduated in 1560 and for the next four years or so practised law at Fontenay, studying cryptography and mathematics in his spare time. From about 1566 to 1570 he was tutor to Catherine of Oarthenay at La Rochelle. In 1570, he moved to Paris and entered the service of Charles IX. He remained in the royal employment until 1584, when persecution of the Huguenots forced him to flee to Beauvoir-sur-Mer. The years at Beauvoir provided him with the opportunity to devote himself to serious mathematical work and it was in these years that his most fruitful algebraic research was carried out. On the accession of Henry IV in 1589, Viète returned to the royal service, and in the succeeding years of the continued war against Spain he gave valuable service to the French crown by deciphering letters of Phillip II of Spain. He was dismissed from the court in 1602 and died at Paris on 13 December 1603.

Although his stock of astronomical knowledge was slight, Viète's mathematical achievements were the result of his interest in cosmology. One of his first major results was a table giving the values of six trigonometrical lines based on a method originally used by Ptolemy. The table, which worked out the values of trigonometric lines from degree to degree and the length of the arc expressed in parts of the radius, appeared in his *Canon mathematicus seu ad triangula* (1579), almost certainly the first book to treat systematically of plane and spherical triangle solution methods by means of all six trigonometric functions. Viète was the first person to use the cosine law for plane triangles and he also published the law of tangents. He was, above all, eager to establish trigonometry as something more than a poor relation of astronomy and his work went a good way towards doing so.

Viète is most celebrated for introducing the first uniformly symbolic algebra. His *In artem analytica isogoge* (1591) used letters for both known and unknown quantities, an innovation which paved the way for the development of seventeenth-century algebra. More specifically, his use of vowels for unknown quantities and consonants for known ones was a pointer to the later development of the concepts of variables and parameters.

In 1593, Viète gave the first explicit explanation of the notion of 'contact'. He asserted that 'the circle may be regarded as a plane figure with an infinite number of sides and angles, but a straight line touching a straight line, however short it may be, will coincide with that straight line and will not form an angle'. His last important work, *De aequationum recognitione et emandatione*, appeared in 1615, more than a decade after his death. In it Viète gave solutions to the lower-order equations and established connections between the positive roots of an equation and the coefficients and the different powers of the unknown quantities. He is credited with introducing the term 'coefficient' into algebra to denote either of the two rational factors of a monomial.

Mathematics to Viète was little more than a hobby, yet by providing the first systematic notation and, as a general tendency, substituting algebraic for geometric proofs in mathematics, he gave algebra a symbolic and analytical framework.

Vogel Hermann Carl 1842–1907 was a German astronomer who became the first Director of the Potsdam Astrophysical Observatory. He is renowned for his work on the spectral analysis of stars and his discovery of spectroscopic binary stars.

Vogel was born in Leipzig on 3 April 1842, the sixth child of Carl Christoph Vogel, the principal of a Leipzig grammar school. Vogel attended his father's school up to the age of 18 and then continued his education at the Polytechnical School in Dresden. In 1863 he returned to Leipzig to study Natural Science at the University and, shortly after beginning his course, he became second assistant at the University Observatory. He showed remarkable dexterity at manipulating instruments and as a result the Director of the Observatory, Karl Brühns, asked him to take part in the Astronomische Gesellschaft's 'zone project'. This was part of a much larger programme that aimed to scan the northern skies and to ascertain the coordinates of all stars down to the ninth magnitude. Vogel's contribution was to observe all nebulae within a specific zone and this work formed the basis of his inaugural dissertation. In 1870, at the joint recommendation of Karl Brühns and Johann Karl Friedrich Zöllner, Professor of Leipzig Observatory, Vogel was appointed Director of an Observatory at Bothkamp, near Kiel, that belonged to the amateur astronomer, F. G. von Bülow. Vogel's work on the spectra of planets, undertaken while he was at the Bothkamp Observatory, won him a prize from the Royal Danish Academy of Sciences. In 1874, he was asked to become an observer at the proposed Astrophysical Observatory at Potsdam, near Berlin. The Potsdam Observatory was officially opened in 1879 and in 1882 he was appointed the first Director of the Observatory, a position he held until his death on 13 August 1907.

During his time at the Bothkamp Observatory, Vogel was given complete scientific freedom, sole discretion in determining the Observatory's research programme and excellent equipment. He worked intensively on the spectroscopic properties of stars, planets, nebulae, the northern lights, comet III 1871, and the Sun. With the aid of a reversible spectroscope, he attempted to ascertain the rotational period of the Sun and, following Huggins's attempts in England (1868), he also attempted spectroscopically to determine the radial velocity of fixed stars.

Having been appointed as an observer to the future Potsdam Observatory in 1874, Vogel became increasingly interested in spectrophotometry. He used this technique to study Nova Cygni in 1876 and his results provided the first evidence that changes occur in the spectrum of a nova during its fading phase. He also began an extensive study of the solar spectrum, but the results of his painstaking measurements were soon superseded by more precise tables compiled by using diffraction gratings. At this point in his career Vogel decided to specialize in spectroscopy, and in response to a proposal made by Angelo Secchi he examined the spectra of some 4,000 stars. His original intention was to classify the stars according to their spectra, a procedure which he believed would reflect their stage of development. However, he became dissatisfied with his findings and abandoned this work to pursue a problem that had intrigued him since his days at Bothkamp Observatory – measuring the Doppler shift in the spectral lines of stars to ascertain their velocity.

Vogel's use of photography to record stellar spectra led to his most sensational discovery – spectroscopic binary stars. His success was based on a study of the periodic displacements of the spectral lines of the stars Algol and Spica, eclipsing binary stars whose components could not, at the time, be detected as separate entities by optical means. From his spectrographs, Vogel derived the dimensions of this double star system, the diameter of both components, the orbital velocity of Algol, the total mass of the system and in 1889, he derived the distance between the two component stars from each other.

Vogel's work ended the controversy over the value of Doppler's theory for investigating motion in the universe. His discovery of spectroscopic binary stars not only led to the realization that such systems are a relatively common feature of the universe; it also played an important role in the discovery of interstellar calcium absorption lines.

Volhard Jacob 1834–1910 was a German chemist who is best remembered for various significant methods of organic synthesis. He also made contributions to inorganic and analytical chemistry.

Volhard was born at Darmstadt on 4 June 1834. He completed his undergraduate studies and worked under Justus von Liebig at the University of Giessen. He then held professorial appointments at three German universities: Munich (1864–1879), Erlangen (1879–1882) and Halle (1882–1910). He remained at Halle until his death on 16 June 1910.

Volhard is best known for his development of a method of quantitatively analysing for an element via silver chloride. A chloride solution of the element to be determined is titrated with an excess of standard silver nitrate solution, and the residual silver nitrate analysed against standard ammonium (or potassium) thiocyanate solution. Bromides can also be determined using this technique. The end-point is usually detected by using a ferric (iron(III)) indicator, such as iron alum.

During the 1860s Volhard developed methods for the syntheses of the amino acids sarcosine (N-methylaminoethanoic acid) and creatine, and the heterocyclic compound thiophen; he also did research on guanidine and cyanimide. His method of preparing halogenated organic acids has become known as the Hell–Volhard–Zelinsky reaction, in which the acid is treated with chlorine or bromine in the presence of phosphorus (iodine is not sufficiently reactive to take part in the reaction). It is specific for α-hydrogen atoms (generating the α-chloro- or α-bromo-acid) and can therefore be used to detect their presence. The α-hydrogens are replaced selectively, and the reaction can be stopped at the mono- or di-halogen stage by using the correct amount of halogen; with excess halogen, all the α-hydrogens are substituted. The reaction is also useful for syntheses because the substituted halogen atom(s) can easily be replaced by a cyanide group (by treatment with potassium cyanide), which in the presence of an aqueous acid and ethyl alcohol (ethanol) yields the corresponding malonic ester (diethylpropandioate), from which barbiturate drugs can be synthesized.

Volta Alessandro 1745–1827 was an Italian physicist who discovered how to produce electric current without using animal tissues, and built the first electric cell. He also invented the electrophorus as a ready means of producing charges of static electricity.

Volta was born at Como on 18 February 1745. He received his early education at various religious institutions but showed a flair for science, particularly the study of electricity which had been brought to the forefront of attention by the experiments and theories of Benjamin Franklin (1706–1790). Volta began to experiment with static electricity in 1765 and soon gained a reputation as a scientist, leading to his appointment as Principal of the Gymnasium at Como in 1774 and, a year later, as Professor of Experimental Physics there. In 1778, Volta took up the same position at Pavia and remained there until 1819. In 1799, the conflict between Austria and France over the region caused him to lose his post but he was reinstated following the French victory in 1800. Volta then travelled to Paris in 1801 to demonstrate his discoveries in electricity to Napoleon, who made him a count and awarded him a pension. Volta retained his academic post when Austria returned to power in 1814 and retired in 1819. He died at Como on 5 March 1827.

Volta's first major contribution to physics was the invention of the electrophorus in 1775. He had researched thoroughly into the nature and quantity of electrostatic charge generated by various materials, and he used this knowledge to develop a simple practical device for the production of charges. His electrophorus consisted of a disc made of turpentine, resin and wax, which was rubbed to give it a negative charge. A plate covered in tinfoil was lowered by an insulated handle on to the disc, which induced a positive charge on the lower side of the foil. The negative charge that was likewise induced on the upper surface was removed by touching it to ground the charge, leaving a positive charge on the foil. This process could then be repeated to build up a greater and greater charge. Volta went on to realize from his electrostatic experiments that the quantity of charge produced is proportional to the product of its tension and the capacity of the conductor. He developed a simple electrometer similar to the gold-leaf electroscope but using straws so that it was much cheaper to make. This instrument was very sensitive and Volta was able to use it to measure tension, proposing a unit that was equivalent to about 13,500 volts.

Volta's next important work did not concern electricity but the air and gases. In 1776, he discovered methane by isolating and examining the properties of marsh gas found in Lake Maggiore. He then made the first accurate estimate of the proportion of oxygen in the air by exploding air with hydrogen to remove the oxygen. Later, in about 1795 to 1796, Volta recognized that the vapour pressure of a liquid is independent of the pressure of the atmosphere and depends only on temperature. This anticipated the law of partial pressures put forward by John Dalton (1766–1844) in 1801.

Volta's greatest contribution to science began with the discovery by Luigi Galvani (1737–1798) in 1791 that the muscles in dead frogs contract when two dissimilar metals (brass and iron) are brought into contact with the muscle and each other. Volta successfully repeated Galvani's experiments using different metals and different animals, and he also found that placing the two metals on his tongue produced an unpleasant sensation. The effects were due to electricity and in 1792, Volta concluded that the source of the electricity was in the junction of the metals and not, as Galvani thought, in the animals. Volta even succeeded in producing a list of metals in order of their electricity production based on the strength of the sensation they made on his tongue, thereby deriving the electromotive series.

Volta's tongue and Galvani's frogs proved to be highly sensitive detectors of electricity – much more so than Volta's electrometer. In 1796, Volta set out to measure the electricity produced by different metals, but to register any deflection in the electrometer he had to increase the tension by multiplying that given by a single junction. He soon hit upon the idea of piling discs of metals on top of each other and found that they had to be separated by a moist conductor to produce a current. The political upheavals of this period prevented Volta from proceeding immediately to construct a cell but he had undoubtedly achieved the 'voltaic pile', as it came to be called, by 1800. In that year, he wrote to the President of the Royal Society, Joseph Banks (1743–1820), and described two arrangements of conductors that produced an electric current. One was a pile of silver and zinc discs separated by cardboard moistened with brine, and the other a series of glasses of salty or alkaline water in which bimetallic curved electrodes were dipped.

Volta's discovery was a sensation, for it enabled high electric currents to be produced for the first time. It was quickly applied to produce electrolysis, resulting in the discovery of several new chemical elements, and then led throughout the 1800s to the great discoveries in electromagnetism and electronics which culminated in the invention of the electrical machines and electronic devices that we enjoy today. Volta's genius lay in an ability to construct simple devices and in his tenacity to follow through his convictions. He was not a great theoretician and did not attempt to explain his discovery. However, he did see the need for establishing proper measurement of electricity, and it is fitting that the unit of electric potential, tension or electromotive force is named the volt in his honour.

Volterra Vito 1860–1940 was an Italian mathematician whose chief work was in the fields of function theory and differential equations.

Volterra was born at Ancona on 3 May 1860, but after his father's death, when he was two, he lived at Florence with his mother and her brother. He attended the Scuola Tecnica Dante Alighieri and the Istituto Tecnico Galileo Galilei. As a young boy he gave signs of a distinct flair for mathematics and physics and by the time he had reached his early teens he had studied a number of sophisticated mathematical texts, including Legendre's Eléments de géométrie. Not only was he able to solve difficult problems, he set himself interesting new ones. So, at the age of 13, after reading Jules Verne's *From the Earth to the Moon* (published eight years before), he became interested in projectile problems and came up with a plausible determination for the trajectory of a spacecraft which had been fired from a gun. His solution was based on the device of breaking time down into small intervals during which it could be assumed that the force was constant. The trajectory could thus be viewed as a series of small parabolic arcs. This was the essence of the argument which he developed in detail 40 years later in a series of lectures at the Sorbonne.

When Volterra was still at school, his mother wished him to abandon his academic studies and become a bank clerk, in order to supplement the small family income. His teacher, Antonio Roiti, averted this by finding him a job as an assistant at the physics laboratory of the University of Florence. He was thus able to continue at the Istituto and in 1878 he entered the University of Florence to study natural sciences. Two years later he entered the Scuola Normale Superiore at Pisa, and there he began to work with Enrico Betti (1823–1892), who was to have a profound influence on his career. It was Betti who turned him in the direction, first of function theory (while still a student, Volterra published papers offering solutions to functions previously believed to be nonderivable), then on mechanics and mathematical physics.

In 1882, Volterra was awarded a doctorate and in the following year he was appointed Professor of Mechanics at the University of Pisa. He remained there until 1892, when he moved to the University of Turin. In 1900, he succeeded Eugenio Beltrami (1835–1899) in the Chair of Mathematical Physics at the University of Rome. During World War I he established the Italian Office of War Inventions, where he played an important part in designing armaments for airships. He also proposed that helium be used in place of hydrogen dirigible airships. After the war he became increasingly involved in politics, speaking in the Senate and openly voicing his opposition to the Fascist regime. For his views he was eventually dismissed from his Chair at Rome in 1931 and banned from taking part in any Italian scientific meeting, although he was elected to the Pontifical Academy of Sciences in 1936 on the nomination of Pope Pius XI. From 1931 onwards, although deprived of an official post, he continued to write papers and to lecture abroad, principally at the Sorbonne. He died in Rome on 11 October 1940.

Volterra's achievements were numerous, but most of them involved function theory and differential equations. He contributed especially to the foundation of the theory of functionals, the solution of integral equations with variable limits, and the integration of hyperbolic partial differential equations. His chief method, hit upon as a young boy, was based on dividing a problem into a small interval of time and assuming one of the variables to be constant during each time-period. Thus his papers on partial differential equations of the early 1890s included the solution of equations for cylindrical waves:

$$\frac{d^2u}{dt^2} = \frac{d^2u}{dx^2} + \frac{d^2u}{dy^2}$$

He also brought his knowledge of mathematics to bear on biological matters. One example of this is his construction of a model for population change, in which the prey, x, and the predator, y, interact in a continuous manner expressed in these differential equations:

$$\frac{dx}{dt} = x(g - ky)$$

and

$$\frac{dy}{dt} = y(-d + kx)$$

Volterra's mathematics was given its broadest statement in his two most important publications, *The Theory of Permutable Functions* (1915) and *The Theory of Functionals and of Integral and Integro-differential Equations* (1930).

von Baer Karl Ernest Ritter 1792–1876 was an Estonian embryologist famous for his discovery of the mammalian ovum, and who made a significant contribution to the systematic study of the development of animals.

Von Baer was born on 29 February 1792, at Piep, in Estonia, on his father's estate. The size of the family – he was one of ten children – forced his parents to send him to live with his paternal uncle and aunt, although his father was a wealthy landholder and district official. On his return home at the age of seven, he was privately tutored until 1807, when he attended a school for members of the nobility for three years.

He then went to the University of Dorpat, the local university, where he was taught by Karl Burdach (1776–1847), the Professor of Physiology there, who had a great influence on his life. He graduated with a medical degree in 1814 and went to Vienna for a year. During the following year he spent some time studying comparative anatomy at the University of Würzburg, where he met Ignaz Döllinger (1770–1841), who was the Professor of Anatomy (and father of the well-known Catholic theologian). Döllinger first introduced him to embryology. In 1817, at the invitation of his old teacher, Burdach, he joined him at the University of Königsberg, where he taught zoology, anatomy and anthropology. Two years later he was appointed Assistant Professor of Zoology. In 1820, he married Auguste Medem, and later had six children. He became restless at Königsberg, and in 1834 moved to St Petersburg, where he took up the appointment of Librarian of the Foreign Division, at the Academy of Sciences. In 1837 he led the first of many expeditions into Novaya Zemlya, in Arctic Russia, where he was the first naturalist to collect plant and animal specimens. He later led expeditions to Lappland, the Caucasian and the Caspian Seas. In 1846 he became the Professor of Comparative Anatomy and Physiology at the Medico-Chirurgical Academy in St Petersburg. He retired from the Academy in 1862 but continued working for them until 1867, as an honorary member. He died on 28 November 1876 at Dorpat, in Estonia.

At Würzburg, Döllinger had suggested that von Baer study the blastoderm of chick embryos removed from the yolk, but the cost of a sufficient number of eggs for observation and someone to look after the incubator was too high and he left the investigation to his more affluent friend, Christian Pander (1794–1865). Von Baer carried on Pander's research and applied it to all vertebrates. In 1817 Pander had described the formation of three layers in the vertebrate embryo – the ectoderm, endoderm and mesoderm. Von Baer developed a theory regarding these germ layers in which he conceived that the goal of early development is the formation of these three layers, out of which all later organs are formed. At the same time he proposed the 'law of corresponding stages', which contradicted the popular belief that vertebrate embryos develop in stages similar to adults of other species. Instead, he suggested that the younger the embryos of various species are, the stronger is the resemblance between them. He demonstrated this fact by deliberately leaving off the labels of embryo species and saying: 'I am quite unable to say to what class they belong. They may be lizards, or small birds, or very young mammalia, so complete is the similarity in the mode of formation of the head and trunk in these animals. The extremities are still absent, but even if they existed, in the earliest stage of the development we should learn nothing, because all arise from the same fundamental form.' From this demonstration he formed his concept of epigenesis, that an embryo develops from simple to complex, from a homogenous to a heterogeneous stage.

In 1827, von Baer published the news of his most significant discovery – the mammalian ovum. William Harvey before him had tried to find it, dissecting a deer, but had searched for it in the uterus. Von Baer found the egg inside the Graafian follicle in the ovary of a bitch belonging to Burdach, which had been offered for the experiment. Von Baer's publication stated that 'every animal which springs from the coition of male and female is developed from an ovum, and none from a simple formative liquid'.

In his observations of the embryo, von Baer discovered the extraembryonic membranes – the chorion, amnion and allantois – and described their functions. He also identified for the first time the notochord, a gelatinous, cylindrical cord which passes along the body of the embryo of vertebrates. In the lower vertebrates it forms the entire back skeleton, whereas in the higher ones the backbone and skull are developed around it. He revealed the neural folds, and suggested that they were the beginnings of the nervous system, and described the five primary brain vesicles.

On his expeditions, von Baer made a significant geological discovery concerning the forces that cause a particular formation on riverbanks in Russia. His study of fishes, made at the same time, in his *Development of Fishes* (1835) stimulated the development of scientific and economic interest in fisheries in Russia.

Von Baer collected skull specimens for his lectures on physical anthropology, the measurements of which he recorded. In 1859, the same year that Darwin published *On the Origin of Species*, he published independently a work that suggested that human skulls might have originated from one type.

Von Baer's publication *De Ovi Mammalium et Hominis Genesi/On the Mammalian Egg and the Origin of Man*, (1827), and his *Über Entwicklunggeschichte der Thiere/On the Development of Animals* (volume one, 1828, volume two, 1837) paved the way for modern embryology and gave a basis for new and scientific interpretation of embryology and biology.

von Baeyer Johann Friedrich Wilhelm Adolf 1835–1917. German organic chemist; *see* Baeyer, Johann von.

von Braun Wernher 1912–1977 was a German-born US rocket engineer, who was instrumental in the design and development of German rocket weapons during World War II and who, after the war, was a prime mover in the early days of space rocketry in the United States.

Von Braun was born on 23 March 1912 in Wirsitz, Germany (now Poland), the son of a German baron. He was educated in Zurich and at Berlin University where he was awarded his PhD in 1934.

In 1930, he had joined a group of scientists who were experimenting with rockets, and within four years they had developed a solid-fuelled rocket that reached an altitude of about 2.5 km/1.5 mi.

In 1938, von Braun became Technical Director of a military rocket establishment at Peenemünde on the Baltic coast. In that year a rocket was produced which had a range of about 18 km/11 mi. In 1940 he joined the Nazi party, and two years later the first true missile – a liquid-fuelled rocket with an explosive warhead – was fired at Peenemünde. The

next developments were the V-1 (a flying bomb powered by a pulse-jet engine) and the V-2 (a supersonic liquid-fuelled rocket), both of which were launched against England from sites on the western coast of the European mainland. Some 4,300 V-2s were fired, 1,230 of which hit London. The rocket weapons came too late, however, to affect the outcome of the war.

In the last days of the war in 1945 von Braun and his staff, not wishing to be captured in the Russian-occupied part of Germany, travelled to the West to surrender to US forces. They went to the United States and soon afterwards von Braun began work at the US Army Ordinance Corps testing grounds at White Sands, New Mexico. In 1952, he moved to Huntsville, Alabama, as Technical Director of the US Army's ballistic missile programme.

In October 1957 the Soviet Union launched its first *Sputnik* satellite and American anxiety was not eased for several months until 31 January of the following year, when von Braun and his team sent up the first artificial satellite *Explorer 1*. Von Braun became a deputy associate administrator of the National Aeronautical and Space Administration (NASA) in March 1970, but two years later left the organization to become an executive in a private business. He died on 16 June 1977.

von Frisch Karl 1886–1982 is an Austrian-born German ethologist who is best known for his studies of bees, particularly for his interpretation of their dances as a means of communicating the location of a food source to other bees. He was awarded many honours for his research into animal behaviour, including the 1973 Nobel Prize for Physiology or Medicine (shared with Konrad Lorenz and Nikolaas Tinbergen).

Von Frisch was born on 20 November 1886 in Vienna, Austria. He was educated at the Schottengymnasium in Vienna then at the universities of Vienna and Munich, gaining his doctorate from Munich in 1910. In the same year he became a research assistant there, and then lecturer in zoology and comparative anatomy in 1912. From 1921 to 1923 he was Professor and Director of the Zoological Institution at the University of Rostock, then held the same post at the University of Breslau.

He returned to the University of Munich in 1925 and established a Zoological Institution there. But the Institution was destroyed during World War II, and so von Frisch moved to the University of Graz in 1946. He returned again to Munich in 1950, however, and remained there until his retirement in 1958.

Von Frisch's most renowned work concerned communication among honey bees. By marking bees and following their movements in special observation hives, he found that foraging bees that had returned from a food source often perform a dance, either a relatively simple round dance when the food is within 100 m/330 ft of the hive, or a more complex waggle dance when the food is farther away. He also found that bees dance only when they have encountered a rich food source, and concluded that these dances are a means of communicating information to other bees

The round dance

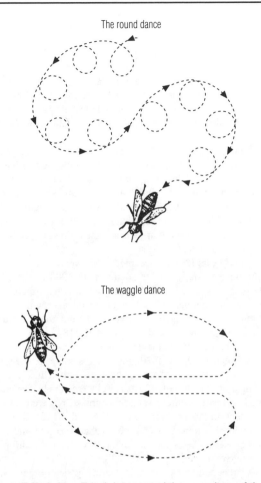

The waggle dance

von Frisch *Von Frisch interpreted the meanings of the 'dances' performed on the vertical honeycomb by returning foraging bees.*

about food sources. In the round dance, the returned forager moves round and round in a tight circle on the honeycomb. Other bees cluster round and touch the dancing bee with their antennae, thereby detecting the scent of the feeding place (bees' scent organs are located on their antennae). According to von Frisch, the round dance tells the bees only that there is food within about 100 m/330 ft of the hive; it does not give the exact location. If the food is more than 100 m/330 ft away, the returned forager performs the waggle dance, in which the bee moves through a figure of eight, its abdomen waggling vigorously during the straight crosspiece of the figure of eight. Von Frisch claims that the waggle dance conveys both the distance of the food from the hive and the direction of the food relative to the hive: the distance is proportional to the time taken to complete one circuit of the dance, and the direction of the food in relation to the Sun is indicated by the angle between the vertical (the dance is performed on a vertical honeycomb) and the crosspiece of the figure of eight.

Von Frisch published his findings in 1943, but doubt has since been thrown on his interpretation of the significance of the dances. Other researchers have found that bees emit various sounds, some of which seem to relate to the location of food. This does not imply that bees do not dance: they do, but their dances may not be the means by which they communicate.

In addition to his studies of bees' dances, von Frisch also researched into the visual and chemical sense of bees and fishes. In 1910, he began an investigation which demonstrated that fishes can distinguish between differences in brightness and colour (the subject of his PhD thesis). And in 1919 he showed that bees can be trained to discriminate between various tastes and odours.

von Guericke Otto 1602–1686 was a German physicist who invented the air pump and carried out the classic experiment of the Magdeburg hemispheres that demonstrated the pressure of the atmosphere. He also constructed the first machine for generating static electricity.

Von Guericke was born at Magdeburg on 20 November 1602 into a wealthy family, which afforded him wide educational opportunities. He attended several universities, going to Leipzig from 1617 to 1620, Helmstedt in 1620, Jena in 1621 and 1622 and Leiden from 1622 to 1625. He studied law, science and engineering, and on his return to Magdeburg was made an alderman of the city.

In 1631, Magdeburg sustained tremendous war damage, most of the population of 40,000 being butchered, and he moved to Brunswick and Erfurt where he worked as engineer to the Swedish government and later for that of Saxony as well. He was able to serve Magdeburg as envoy to various occupying powers, for as the fortunes of war swayed back and forth, French, Habsburg and Swedish forces gained control. Guericke, as he then was, represented Magdeburg at various conferences and diets which marked at least a lessening of hostilities in that part of Europe.

In 1646, Guericke became mayor of Magdeburg and his scientific achievements were made from this time onwards into the 1650s. He was ennobled and became von Guericke in 1666 and remained mayor until 1676. In 1681 he retired and went to live with his son in Hamburg, where he remained for the rest of his life. Von Guericke died at Hamburg on 11 May 1686.

It is remarkable that, amidst all the political turmoil surrounding him, von Guericke found time for deep philosophical speculation about the nature of space. At the time, Aristotle's idea that 'nature abhors a vacuum' – that a vacuum could not exist – was being hotly debated. The phrase had been taken up by René Descartes (1596–1650) and had great currency, but in 1643 Evangelista Torricelli (1608–1647) had demonstrated that a vacuum could exist in the space above an enclosed column of mercury. His experiment could be explained by the atmosphere possessing weight, and von Guericke decided to resolve matters by experiment. He set about constructing an air pump in an attempt to pump the air from a vessel and produce a vacuum. He postulated that if Descartes's ideas (the plenist assumption that space is matter) were correct, then an evacuated vessel should collapse. Von Guericke's first air pump was rather like a large bicycle pump and in 1647, after much trouble with seals, he succeeded in imploding a copper sphere from which he pumped the air via an outlet at the bottom.

Although it seemed that Descartes was vindicated, von Guericke still had his doubts. He built a stronger vessel and succeeded in evacuating it without causing it to collapse. He also demonstrated that the sound of a bell is muffled and that a candle is extinguished as the air is removed, and gradually the theory that a vacuum cannot exist was discarded.

Von Guericke then turned to demonstrating the immense force that the pressure of the atmosphere exerts on an evacuated vessel. One of his most famous experiments was in 1654. A cylinder containing a well fitting piston was connected to a rope passing over a pulley. Fifty strong men were unable to move the piston more than halfway up the cylinder, and when the closed end was evacuated they could not prevent the piston from descending.

Von Guericke went on to make his classic air-pressure experiment, which was probably first performed at Magdeburg in 1657, in which two copper hemispheres with tight-fitting edges were placed together and the air pumped out. Two teams of horses were then harnessed to the hemispheres and proved unable to pull them apart. On admitting air to the hemispheres, they fell apart instantly.

Von Guericke's experiments and other observations on the elasticity of air led him to investigate the decrease of pressure with altitude, which Blaise Pascal (1623–1662) had demonstrated in his famous Puy de Dôme excursion in 1648. A natural consequence was the establishment of the link between atmospheric pressure and the weather, which von Guericke pursued as far as making weather forecasts with a water barometer and proposing a string of meteorological stations contributing data to a forecasting system – all this in 1660.

Von Guericke also interested himself in cosmology and magnetism. While experimenting with a globe of sulphur constructed to simulate the magnetic properties of the Earth, he discovered that it produced static electricity when rubbed and went on to develop a primitive machine for the production of static electricity. Von Guericke also demonstrated the magnetization of iron by hammering in a north-south direction. As part of his astronomical studies he put forward the idea that comets made periodic returns. This was to be substantiated by Halley 20 years later.

Another 'first' with which he is credited is the observation of coloured shadows; these he observed in the early morning when an object illuminated by a candle cast a blue shadow on a white surface.

Von Guericke's remarkable blend of political and scientific activities was to be echoed by such figures as Count Rumford (1753–1814) and Benjamin Franklin (1706–1790) in the next century – although in the latter's case on an altogether grander scale. Von Guericke's discovery that the pressure of the atmosphere can be made to exert considerable force by the creation of a vacuum led directly to the

development of steam engines in the eighteenth century.

von Kármán Theodore 1881–1963 was the Hungarian-born aerodynamicist who worked first in Germany and finally in the USA. He made a fundamental contribution to rocket research, enabling the USA to acquire a lead in this technology. He possessed remarkable teaching and organizational abilities, and coordinated many famous research institutions.

Von Kármán was born on 11 May 1881, in Budapest, Hungary, of intellectual parents. He would have been a mathematical prodigy if his father, Mór (Maurice), a professor at Budapest University and an authority on education, had not guided him towards engineering. He completed his undergraduate studies at the Budapest Royal Polytechnic University in 1902 and the following year was appointed assistant professor there.

Three years later he accepted a two-year fellowship at Göttingen to study for his doctorate. In 1908 he went to Paris and witnessed a flight by Farman, which inspired him to concentrate his efforts on aeronautical engineering. Returning to Göttingen he worked on a wind tunnel for airship research with Ludwig Prandetl (1875–1953) and received his doctorate there in 1909.

Between 1913 and 1930, with the exception of the war years when he directed aeronautical research in the Austro-Hungarian Army, von Kármán built the newly founded Aachen Institute into a world-recognized research establishment. In 1926 he went to the USA and in 1928 decided to divide his time between the Aachen Institute and the Guggenheim Aeronautical Laboratory of the Califonia Institute of Technology (GALCIT).

As the political climate in Germany deteriorated, he became the first director of GALCIT, becoming Professor Emeritus at the California Institute of Technology. At the peak of his career, he became the leading figure in a number of international bodies, such as the International Union of Theoretical and Applied Mechanics, the Advisory Group for Aeronautical Research and Development, International council of Aeronautical Sciences, and International Academy of Astronautics.

Von Kármán never married and his sister, Pipö, first with the help of their mother, Helen (born Konn), managed his households in Aachen and in Pasadena. He received many honorary degrees and medals including the US Medal for merit 1946, the Franklin gold medal 1948 and the first National Medal for Science 1963. He died on 7 May 1963 at the Aachen home of his close friend, Bärbel Talbot, while on holiday. His body was flown by the US Air Force to California to be buried beside those of his mother and sister.

Von Kármán's work concerning the flow of fluids round a cylinder is among his best known. He discovered that the wake separates into two rows of vortices which alternate like street lights. This phenomenon is called the Kármán vortex street (or Kármán vortices) and it can build up destructive vibrations. The spiral flutes on factory chimneys are there to prevent the vortices forming and the

Tacoma Narrows suspension bridge was destroyed in 1940 by vortices which induced huge vibrations when the wind speed was exactly 67.6 kph/42 mph.

In 1938, the US Army Air Corps asked the National Academy of Sciences to devise some means of helping heavily laden aircraft to take off. Von Kármán played a prominent part in the first work on jet assisted take-off (JATO) and in 1943, authored with H. S. Tsien and F. J. Molina, a memorandum which resulted in the first research and development programme on long-range rocket-propelled missiles. The programme called for the design and building of prototype test vehicles, basic research on propellants and construction materials, missile guidance systems and missile aerodynamics.

In aerodynamics and flight mechanics he developed a method of generating aerofoil profiles and produced papers on flight at high subsonic, transonic and supersonic speeds. He also worked on boundary layer and compressibility effects, stability and control at high speeds, propeller design, helicopters and gliders.

Von Kármán was one of the world's great theoretical aerodynamicists. He laid the foundations for the rocket and aircraft programmes which placed the USA in the lead in these fields. Many of the establishments he brought together, such as the famous Jet Propulsion Laboratory at Pasadena, show every sign of carrying on his work.

von Liebig Justus 1803–1873. German organic chemist; *see* Liebig, Justus von.

von Mises Richard 1883–1953 was an Austrian mathematician and aerodynamicist who made valuable contributions to statistics and the theory of probability.

Von Mises was born on 19 April 1883 at Lemberg (now Lvov, Ukraine). He was educated at the University of Vienna, where he received his doctorate in 1907, and was Professor of Applied Mathematics at the University of Strassburg (now Strasbourg, in France) from 1909 to 1918. In 1920 he was appointed Professor of Applied Mathematics and Director of the Institute for Applied Mathematics at the University of Berlin. With the coming to power of Adolf Hitler in 1933, he emigrated to Turkey and taught at the University of Istanbul until 1939. In that year he went to the United States to join the faculty of Harvard University. He was made Gordon McKay Professor of Aerodynamics and Applied Mathematics in 1944, the Chair which he held until his death, at Boston, on 14 July 1953.

Von Mises looked upon himself principally as an applied mathematician and was especially proud of having founded in 1921, and edited until 1933, the journal *Zeitschrift für angewandte Mathematik und Mechanik*. His first interest was fluid mechanics, especially in relation to the new and exciting field of aerodynamics and aeronautics. He learned to fly and in the summer of 1913 gave what is believed to be the first university course in the mechanics of powered flight. In the next year or so he made significant improvements in boundary-layer-flow theory and aerofoil design and, in 1915, built a 454-W/600-hp aeroplane for the Austrian military. During World War I he served as a pilot and in 1916 published a book on flight which formed the basis of his later book, *Theory of Flight*, published with scientists in England towards the end of World War II.

Von Mises' chief contribution to pure mathematics was in the field of probability theory and statistics. He was drawn into this subject by his association (from 1907 until the 1920s) with the Viennese school of logical positivism. The development of the frequency theory of probability had proceeded slowly, in opposition to the classical theory of Laplace, during the second half of the eighteenth and the first half of the nineteenth century. It then took a leap forward with John Venn's imaginative stroke in equating probability with the relative frequency of the event 'in the long run'. Venn introduced the concept of a mathematical limit and the infinite set, but von Mises came to the conclusion that a probability cannot be simply the limiting value of a relative frequency. He added the proviso that any event should be irregularly or randomly distributed in the series of occasions in which its probability is measured. This emphasis on the idea of random distribution, in other words bringing the notion of the Venn limit and that of a random sequence of events together, was von Mises' outstanding contribution to the frequency theory of probability.

Von Mises' ideas were contained in two papers which he published in 1919. Little noticed at the time, they have come to influence all modern statisticians. The consistency of the mathematics of von Mises' theory has been called into question by a number of mathematicians, but it seems doubtful that any of the proposed alternatives will prove to be more satisfactory. After 1919 von Mises achieved very little of a highly creative or original nature, but his *Probability, Statistics and Truth* (1928) is both historically sound and, for the lay reader, stimulating.

Von Neumann John (Johann) 1903–1957 was a Hungarian-born US physicist and mathematician who originated games theory and developed the fundamental concepts involved in programming computers. He also made a valuable contribution to quantum mechanics.

Von Neumann was born at Budapest, Hungary, on 28 December 1903. He received private instruction until 1914, when he entered the Gymnasium, and even then continued to be tutored outside school in mathematics because of the exceptional ability that he displayed in the subject. Von Neumann left Hungary in 1919 in the wake of the disruptions arising as a result of the defeat suffered by Austria-Hungary in World War I. He attended several universities, studying in Berlin from 1921 until 1923 and then in Zurich until 1925. He received a degree in chemical engineering from the Zurich Institute in 1925, and was awarded a PhD in mathematics from the University of Budapest a year later. He then studied in Göttingen, where he worked with Robert Oppenheimer (1904–1967), and from 1927 until 1929 held the post of privatdozent (unpaid lecturer) in mathematics at the University of Berlin.

After a year as lecturer at the University of Hamburg, Von Neumann travelled to the USA in 1930. Holding first the position of Visiting Professor, he became Full Professor of Mathematics at the University of Princeton in 1931. He held this position until 1933, when he was invited to become the youngest member of the newly established Institute of Advanced Studies at Princeton University. Von Neumann was a member of this Institute for the rest of his career, although he also held a number of important advisory posts with the US government from 1940 to 1954.

During World War II, Von Neumann served as a consultant for various committees within the Navy, Army, the Atomic Energy Commission and the Office for Scientific Research and Development. He was associated with research projects at Los Alamos from 1943 until 1955, working under Oppenheimer on the A-bomb project and also under Edward Teller (1908–) on the H-bomb project. The importance of Von Neumann's scientific contributions brought him widespread recognition, including the award of the Medal of Freedom, the Albert Einstein Award and the Enrico Fermi Award – all in 1956. His health had begun to fail in 1955, and he died of cancer in Washington, DC, on 8 February 1957.

Pure mathematics was Von Neumann's primary scientific interest during the first years of his career. He made important contributions to the subjects of mathematical logic, set theory and operator theory and his theory of rings of operators is highly regarded.

In 1928, Von Neumann read a paper on games theory to a scientific meeting in Göttingen. This was an entirely new subject originated by Von Neumann himself. It consisted of proving that a quantitative mathematical model could be constructed for determining the best strategy, that is, the one which, in the long term, would produce the optimal result with minimal losses for any game, even one of chance or one with more than two players. The sort of games for which this theory found immediate use were business, warfare and the social sciences – games in which strategies must be worked out for defeating an adversary. Games theory has also found application in the behavioural sciences. Von Neumann began work on this subject in the late 1920s, but did not devote himself exclusively to it by any means.

Applied mathematics in the field of theoretical physics held a strong fascination for Von Neumann. He began to work on the axiomitization of quantum mechanics in 1927, and in 1932 published a book entitled *The Mathematical Foundations of Quantum Mechanics*. This defended mathematically the uncertainty principle of Werner Heisenberg (1901–1976). In 1944, Von Neumann made another important contribution to quantum mechanics when he showed that the systems of matrix mechanics developed by Heisenberg and Max Born (1882–1970) and of wave mechanics developed by Erwin Schrödinger (1887–1961) were equivalent. Von Neumann also published papers with Subrahmanyan Chandrasekhar (1910–) on gravitational fields. Work on pure mathematics also continued – he collaborated with Francis Murray in an investigation of noncommutative algebras during the latter part of the 1940s.

Work on the atom bomb project brought a variety of new problems, many of them completely different from those Von Neumann had previously encountered. The necessity for quickly producing approximate models for complex physical problems encouraged Von Neumann to examine and develop improvements for the available computing machines. During the war he contributed to work on hydrodynamics and on shock waves, and afterwards he spent a great deal of effort designing and then supervising the construction of the first computer able to use a flexible stored program (named MANIAC-1) at the Institute of Advanced Study in 1952. This work laid the foundations for the design of all subsequent programmable computers.

Von Neumann also developed his games theory, and in 1944 published *Theory of Games and Economic Behavior* with Oskar Morgenstern, his major work on the subject. Games theory was also used during the H-bomb project in the early 1950s, in which Von Neumann played an active role. The possibility of developing automata to the point where they might be self-producing, and the similarities between the nervous system and computers were areas which absorbed much of Von Neumann's interest during his last years.

Von Neumann was a mathematician with an exceptional talent for absorbing the essential features of all the important branches of both mathematics and theoretical physics, and he was also adept at the applications of his theoretical work. He demonstrated great originality and imagination in his pioneering efforts in the areas of computer design and especially in games theory.

von Stradonitz Friedrich August Kekulé 1829–1896. German organic chemist; *see* Kekulé, Friedrich August.

von Welsbach Freiherr (Carl Auer) 1858–1929 was an Austrian chemist and engineer who discovered two rare earth elements and invented the incandescent gas mantle.

Von Welsbach was born Carl Auer on 1 September 1858, in Vienna. He was the son of the Director of the Imperial Printing Press. He went to Heidelberg for his university training and studied under the German chemist Robert Bunsen, where he developed a strong interest in spectroscopy and lighting. He showed that didymium, previously thought to be an element, actually consisted of the two very similar but different elements praseodymium and neodymium. He also found that another rare earth element, cerium, added as its nitrate salt to a cylindrical fabric impregnated with thorium nitrate, produced a fragile mantle that glowed with white incandescence when heated in a gas flame. The 'Welsbach mantle' was patented in 1885.

The major development in artificial lighting in the nineteenth century was the introduction of methods of generating and distributing coal gas in urban communities. The first successful experiments in this field are usually credited to Witham Murdoch, a Scottish engineer working in Cornwall who lit his own home by gas before being commissioned to install gas lighting in London. For several decades the source of light was from the yellow gas flame itself, produced by a simple 'bat's-wing' burner. During the 1820s a new type of burner was introduced, in which a controlled amount of air was admitted to the gas current (as in a laboratory Bunsen burner): producing a high-temperature but non-luminous flame that heated a refractive material. At very high temperatures this material became the light source. But the method proved to be expensive and unreliable until the invention of the Welsbach mantle which, in its final form of a woven net (usually spherical or cylindrical) impregnated with the salts of thorium and cerium, could be made sufficiently durable to be transported. (When a mantle was first used, the cotton burned away to leave a mesh of metal oxides which was fragile but continued to function as long as it was not roughly handled.)

Thomas Edison's electric lamp eventually replaced gas lamps for nearly all applications, although mantles are still used for kerosene lamps and for portable lamps powered by 'bottled' gas. More often than not, such lamps are lit using a cigarette lighter or automatic gas lighter which makes use of a 'flint', another of von Welsbach's discoveries. Most lighter flints consist of what he called Mitschmetal, a pyrophoric mixture containing about 50% cerium, 25% lanthanum, 15% neodymium, and 10% other rare metals and iron. When it is struck or scraped it produces hot metal sparks. Mitschmetal is also used as a deoxidizer in vacuum tubes and as an alloying agent for magnesium.

When the Emperor of Austria conferred upon him the title Freiherr, von Welsbach chose as his baronial motto 'more light'. This has proven to be a prophetic choice, for both von Welsbach's discoveries are still used in the production of light.

Vorontsov-Vel'iaminov Boris Aleksandrovich 1904– is one of the most prominent Russian astronomers and astrophysicists. He was born on 14 February 1904 and in 1934 was appointed Professor at the University of Moscow. Besides being the author of several successful textbooks on astronomy, he compiled astronomical catalogues and published several advanced specialized tests. In 1962 he was honoured by being awarded the Bredikhin Prize.

In 1930, independently of Robert Trumpler (1886–1956), Vorontsov-Vel'iaminov demonstrated the occurrence of the absorption of stellar light by interstellar dust. This fact had not been taken into consideration in 1922 by Jacobus Kapteyn (1851–1922) in his model of the universe, one of the most serious flaws in his calculations. The significance of Vorontsov-Vel'iaminov's discovery was that it became possible to determine astronomical distances and, in turn, the size of the universe more accurately.

Vorontsov-Vel'iaminov devoted considerable energy to the study of gaseous nebulae, the observations of novae and analysis of the Hertzsprung–Russell diagram, with reference to the evolution of stars. He made particularly important contributions to the study of the blue-white star sequence, which was the subject of a book he published in 1947.

In 1959 Vorontsov-Vel'iaminov compiled a list and recorded the positions of 350 interacting galaxies which are clustered so closely that they seem to perturb each other slightly in structure. Besides this catalogue, he compiled a more extensive catalogue of galaxies in 1962, in which he listed and described more than 30,000 examples.

Wald George 1906– is a US biochemist who has investigated the biochemical processes of vision that take place in the retina of the eye. For this work he shared the 1967 Nobel Prize for Physiology or Medicine with the Swedish physiologist Ragnar Granit (1900–1991) and his fellow-American Haldan Hartline (1903–1983).

Wald was born in New York City on 18 November 1906 and educated at the university there, graduating in 1927. He gained his PhD from Columbia University five years later. From 1932 to 1934 he studied in Europe as a National Research Council fellow, first under Otto Warburg in Berlin and then with Paul Karrer in Zurich, who did the pioneering work on vitamin A. When Wald returned to the United States he joined the staff at Harvard University and remained there for the rest of his career; he was appointed Professor of Biology in 1948. In the 1970s Wald rose to fame outside the area of science for his outspoken comments against United States involvement in the Vietnam War.

Wald began his work on the chemistry of vision in the early 1930s. The key to the process is the pigment called visual purple, or rhodopsin, which occurs in the rods (dim-light receptors) of the retina. In 1933, he discovered that this substance consists of the colourless protein opsin in combination with retinal, a yellow carotenoid compound that is the aldehyde of vitamin A. Rhodopsin molecules are split into these two compounds when they are struck by light, and the enzyme alcohol dehydrogenase then further reduces the retinal to form vitamin A. In the dark the process is reversed and the compounds recombine to restore the rhodopsin to the retinal rods. The process does not work with 100% efficiency and over a period of time some of the retinal is lost. This deficiency has to be made up from the body's stores of vitamin A (which is supplied through the diet), but if the stores are inadequate the visual process in dim light is affected and night blindness results.

Wald and his co-workers went on to investigate how these biochemical changes trigger the electrical activity in the retina's nerves and the optic nerve. In the 1950s they found the retinal pigments that detect red and yellow-green light, and a few years later identified the pigment for blue light. All of these – the three primary colour pigments – are related to vitamin A, and in the 1960s Wald demonstrated that the absence of one or more of them results in colour blindness.

Wallace Alfred Russel 1823–1913 was a British naturalist who is best known for proposing a theory of evolution by natural selection independently of Charles Darwin.

Wallace was born in Usk, in Wales, on 8 January 1823. After a rudimentary education (he left school when he was 14 years old), he joined an elder brother in a surveying business. In 1844, however, he became a teacher at the Collegiate School, Leicester, where he met Henry Bates, who interested him in entomology. Together they planned a collecting trip to the Amazon, and arrived in South America in 1848; Bates remained there for 11 years but Wallace returned to England in 1852. Unfortunately, the ship sank on the return voyage and although Wallace survived, all his specimens were lost, the only remaining ones being those previously sent to England. In 1853, he published an account of his experiences in South America in *A Narrative of Travels on the Amazon and Rio Negro*.

From 1842 to 1862 Wallace explored the Malay Peninsula and the East Indies, from which he collected more than 125,000 specimens. During this expedition he observed the marked differences that exist between the Australian and Asian faunas and later, when writing about this phenomenon, drew a hypothetical line that separates the areas in which each of these two distinct faunas exist. This line, now called the Wallace line, follows a deep-water channel that runs between the larger islands of Borneo and Celebes and the smaller ones of Bali and Lombok. In 1855, while in Borneo, he wrote *On the Law Which Has Regulated the Introduction of New Species*, in which he put forward the idea that every species had come into existence coincidentally, both in time and place, with a pre-existing, closely allied species. He also believed that the Australian fauna was less highly developed than the Asian fauna, the survival of the Australian fauna being due to the separation of Australia and its nearby islands from the Asian continent before the more advanced Asian fauna had developed.

These ideas then led Wallace to the same conclusion that Charles Darwin had reached (although had not published at that time) – the idea that species evolve by natural selection. Then, while suffering from malaria, in 1858, Wallace wrote an essay outlining his ideas on evolution and sent it to Darwin, who was surprised to find that Wallace's ideas were the same as his own. The findings of the two men were combined in a paper read before the Linnean Society on 1 July 1858. Wallace's section, entitled *On the Tendency of Varieties to Depart Indefinitely from the Original Type*, described how animals fight to survive, the rate of their reproduction and their dependence on supplies of suitable food. In the conclusion to his section Wallace wrote 'those that prolong their existence can only be the most perfect in health and vigour; … the weakest and least perfectly organized must always succumb'. He described his work more fully in *The Malay Archipelago*, published in 1869; Darwin's *On the Origin of Species* had appeared in 1859.

Wallace continued to gather evidence to support the theory of evolution and in 1870, while on an expedition to Borneo and the Molucca islands, published *Contributions to the Theory of Natural Selection*. In this work he diverged from Darwin's views: both thought that the human race had evolved to its present physical form by natural selection but, in keeping with his spiritualistic beliefs, Wallace was of the opinion that humans' higher mental capabilities had arisen from some 'metabiological' agency. Wallace also differed from Darwin about the origins of the brightly coloured plumage of male birds and the relative drabness of female birds; Wallace believed that it was merely natural selection that had led to the development of dull, protectively coloured plumage in females, rather than subscribing to Darwin's idea that females are attracted by brightly coloured plumage in males.

Wallace also studied mimicry in the swallowtail butterfly and wrote a pioneering work on zoogeography, *Geographical Distribution of Animals* (1876). In addition he spent much time promoting socialism by, for example, campaigning for women's suffrage and land nationalization.

Public recognition of Wallace's important work on evolution came late in his life; he was elected a Fellow of the Royal Society in 1893 and received the Order of Merit in 1910. He died in Broadstone, Dorset, on 7 November 1913.

Wallis Barnes Neville 1887–1979 was a British aircraft designer who was responsible for devising the unique 'geodesic' construction of airframes. He also gained fame during World War II for his association with the development of a 'bouncing' bomb. used by the Allies to attack dams in Germany.

Wallis, later known as Barnes Wallis, was born in Derbyshire on 26 September 1887. He was educated at Christ's Hospital School and trained as a marine engineer at J. S. White and Company, shipbuilders of Cowes, Isle of Wight. In early 1911 he joined the Vickers Company, and from 1913 to 1915 he worked in the design department of Vickers Aviation: During World War I he served briefly as a private in the Artist's Rifles, then returned to Vickers as chief designer in the airship department of their works at Barrow-in-Furness.

From 1916 to 1922 he was Chief Designer for the Airship Guarantee Company, a subsidiary of Vickers. From 1923 onwards he was Chief Designer of Structures at Vickers's Weybridge works where he stayed as a designer until the end of World War II. In 1945 he became Chief of Aeronautical Research and Development at the British Aircraft Corporation Division at Weybridge – a post he held until his retirement in 1971.

Wallis was widely honoured by his fellow engineers and by a grateful government. He received the CBE in 1943 and was made a Fellow of the Royal Society in 1945; he held honorary degrees from London, Bristol, Oxford and Cambridge Universities. In 1965 he was made an honorary fellow of Churchill College, Cambridge, and in 1968 he was knighted. He died on 31 October 1979.

Towards the end of 1911 Wallis left shipbuilding to join H. B. Pratt as an apprentice designer. (Pratt had been commissioned by Vickers to design a new rigid airship after the *Mayfly* debacle of 1911.) A series of successful designs followed, and in 1913 the British Government initiated the famous R series of airships – starting with the R26. Wallis himself designed the R80; and the performance of this machine was a significant advance on that of all the others.

In 1924, Wallis began designing the R100, and it was during this work that he got the initial ideas which he later incorporated in his geodetic structures for aircraft. Despite the success of the R100 and its proven structural strength, the disaster which befell the government-built (and differently designed) sister-ship, the R101, brought an end to rigid airship building in Britain.

Following the cessation of work on rigid airships, Wallis was transferred to the Vickers Weybridge works as Chief Structures Designer, and one of his first tasks was to design a lighter wing structure for the Vickers-built Viastra 2, which was being used commercially in Australia. The experience Wallis gained in this job – together with that gained in other structural research, particularly the design of the A1-30 torpedo-carrier and the 'wandering web' of the Vivid and Valiant – later enabled him to evolve the now-famous geodesic system for fixed-wing aircraft.

The breakaway from established structural design practice in airframes was made when a Wallis-designed fuselage was incorporated in the G4-31 biplane in 1932. In this design, Wallis sought to dispense with the primary and secondary members by substituting a lattice-work system of main members only – an idea he had originally derived from the wire-netting used to contain the gas-bags on the airship R100. An exploratory structure of this type had been used on the ill-fated M1-30 torpedo-carrier. The G4-31 structure represented a halfway stage between that and the full geodetic structure employed in the monoplane which was later to achieve worldwide fame as the Wellesley bomber.

A geodetic line is, by definition, the shortest distance between two points on the Earth – that is, on a sphere.

Applied to airframes, the theory had many advantages, one being that the load transfer from member to member was by the shortest possible route. In the Wallis lattice pattern, if one series of members was in tension, the opposite members were in compression – thus the system was stress-balanced in all directions. The Wellesley was responsible for the great technical advance in design which took place in the mid-1930s and which eventually produced the Wellington bomber of World War II.

Critics of the geodetic form of construction condemned it on the grounds that, for quality production, it was impractical. Wallis and his team at Weybridge rebutted this criticism by devising the necessary tools and methods: a complete Wellington airframe could be assembled in 24 hours. But the real proof of the essential simplicity of the Wellington structure came when nearly 9,000 airframes were produced for the war effort at Blackpool and Chester using only a minimum of skilled personnel – much of the workforce was made up of semi-skilled workers new to aircraft production.

In all, 11,461 Wellington bombers were produced – the largest number of any British bomber. The Wellington served throughout World War II in almost every role possible for a twin-engined aircraft and in conditions as widely different as those in Iceland and the Middle East. Without it, the war might well have taken a different course.

By applying his considerable knowledge of aerodynamic streamlining to ballistic problems, Wallis also devised new means of attacking the enemy by devising novel, but very effective, bombs. Among others, he designed the 544-kg/1,200-lb Tallboy and the 10-tonne Grandslam. But his most notable invention in this field was the famous Wallis 'bouncing bomb'. The bomb itself was used to destroy vital and hitherto practically impregnable targets – the huge German dams. The dams were of paramount military importance to Germany because they fed the waterways essential for the production of war material.

To smash the dams, Wallis designed a cylindrical bomb to be hung crosswise in the bomb-bay of the aircraft. Each one, 1.5 m/5 ft in length and almost the same in diameter, was designed so that if it was dropped at the correct height and speed, it would skim across the water enabling it to reach the target presented by the wall of the dam. To achieve this, ten minutes before being dropped, the bomb was given a back-spin of 500 revolutions per minute by an auxillary motor. Upon impact with the water, the backward spin caused the bomb to skim across the surface, bouncing along, in shorter and shorter leaps, until it hit the wall of the dam. Then instead of rebounding away from the wall, the back-spin caused a downwards 'crawl' to a depth of 9 m/30 ft, where a hydrostatic fuse caused it to explode. The raids on the dams were successfully carried out by Lancaster aircraft of RAF 617 squadron.

Wallis John 1616–1703 was an English mathematician who made important contributions to the development of algebra and analytical geometry and who was one of the founders of the Royal Society.

Wallis was born in Ashford, Kent, on 23 October 1616. He began his education at Ashford, but was sent to boarding school at Ley Green, near Tenterden, on the outbreak of plague at Ashford in 1625. In the years 1630 to 1631 he attended the Martin Holbeach School in Felsted, Essex. There he learned Greek, Latin and Hebrew and was introduced to the elements of logic. It was his brother who, in the Christmas vacation of 1631, instructed him in the fundamentals of arithmetic. In 1632, he went to Emmanuel College, Cambridge, to study physics, medicine and moral philosophy. He received his BA in 1637. In 1640 he was ordained in the Church of England and for the next four years earned his living as a private chaplain in Yorkshire and later in Essex. On his mother's death in 1643 he came into a large inheritance which left him financially independent.

In 1644, Wallis was elected a Fellow of Queens' College, Cambridge, but was compelled to relinquish the fellowship a year later when he married. He moved to London and assisted the parliamentary side by deciphering captured coded letters during the Civil War. In gratitude for that work Oliver Cromwell overlooked Wallis's signature on the 1648 remonstrance against the execution of Charles I and, in 1649, appointed him Savilian Professor of Geometry at Oxford University. It was then that Wallis began to study mathematics in earnest. In 1655 he published his most famous work, the *Arithmetica Infinitorum*, which immediately raised him to international scientific eminence. In 1658 he was appointed Keeper of the University Archives. In 1660 Charles II chose him as his royal chaplain. In the meantime Wallis had, ever since 1649, been meeting regularly with other lovers of science, notably Robert Boyle (1627–1691), at the discussions that led to the founding of the Royal Society in 1660.

In the second half of his life Wallis's chief publications were his *Mechanica* (1669–1671), which was the fullest treatment of the subject then existing, and his *Algebra* (1685). He also became involved in a long and acrimonious dispute with Thomas Hobbes over what he considered to be the anti-Christian tendencies of Hobbes's philosophy – a quarrel which lasted for 25 years and ended only with Hobbes's death in 1679. He also conducted experiments in speech and attempted to teach, with some success, deaf-mutes to speak. His method was described in his *Gramatica Linguae Anglicanae* (1652). After the revolution of 1688–89 which drove James II from the throne, he was employed by William III as a decipherer. He held the Savilian Chair until his death, at Oxford, on 28 October 1703.

Wallis's two great works established him as one of the foremost mathematicians of the seventeenth century. The *Arithmetica* was the most substantial single work on mathematics yet to appear in England. In it he applied Cartesian analysis to Torricelli's method of indivisibles and (in an appendix) applied analysis for the first time to conic sections as curves of the second degree. The treatise also introduced the symbol ∞ to represent infinity, the germ of the differential calculus and, by an impressive use of interpolation (the word was Wallis's invention), the famous value for π and the celebrated formula:

$$\frac{4}{\pi} = \frac{3 \times 3 \times 5 \times 5 \times 7 \times 7}{2 \times 4 \times 4 \times 6 \times 6 \times 8} \cdots$$

The *Algebra* was the first treatise ever to attempt to combine a full exposition of the subject with its history and was important for introducing the principles of analogy and continuity into mathematics.

Walton Ernest Thomas Sinton 1903– is an Irish physicist best known for his work with John Cockcroft (1897–1967) on the development of the first particle accelerator, which produced the first artificial transmutation in 1932. In recognition of this achievement, Walton and Cockcroft were awarded the 1951 Nobel Prize for Physics.

Walton was born in Dungarvan, County Waterford, on 6 October 1903. He attended the Methodist College in Belfast and went on to Trinity College, Dublin, where he earned his bachelor's degree in 1926. He was later awarded an MSc in 1928 and an MA in 1934 from the same institution. In 1927, Walton continued his studies at the Cavendish Laboratories at Cambridge University, remaining there until 1934. He obtained his PhD in 1931, and in 1934 returned to Ireland to become Fellow of Trinity College, Dublin. In 1947, Walton also became Professor of Natural and Experimental Philosophy at Trinity College. After his retirement in 1974, Walton held the post of Emeritus Fellow of Trinity College. He received many honours in recognition of his scientific achievements. In particular, Walton was the joint recipient with Cockcroft of the Hughes Medal of the Royal Society in 1938 as well as the Nobel Prize for Physics in 1951.

Before Walton and Cockcroft began their investigation of artificial transmutation, the only means of changing one element into another was by bombarding it with alpha particles from a natural source, that is, a radioactive substance. This placed a variety of constraints on the possible procedures in such work. Walton investigated several approaches for the production of fast particles. Two of the methods he attempted in 1928 failed because of difficulties in focusing the particles, and these methods were developed later by others to build the first successful betatron and the early linear accelerators.

In 1929, Walton and Cockcroft were joined at the Cavendish Laboratory by George Gamow (1904–1968). Gamow made calculations on the feasibility of producing transmutations by bombardment with proton beams. His results encouraged them to investigate the production of a high voltage multiplying the circuit which would enable them to produce particles of uniform energy. Protons were preferable to alpha particles because their smaller charge makes them less susceptible to repulsive interactions with the target nuclei. Furthermore, hydrogen gas is more readily ionized than helium, so protons are produced more easily than are alpha particles.

The device that Cockcroft and Walton constructed, the first successful particle accelerator, used an arrangement of condensers to build up very high potentials and was completed in 1932. It could produce a beam of protons accelerated at over 500,000 electronvolts. In 1932, Walton and Cockcroft used the proton beam to bombard lithium. They observed the production of large quantities of alpha particles, showing that the lithium nuclei had captured the protons and formed unstable beryllium nuclei which instantaneously decayed into two alpha particles travelling in opposite directions. Walton and Cockcroft detected these alpha particles with a fluorescent screen. They later investigated the transmutation of other light elements using proton beams, and also deuterons (nuclei of deuterium) derived from heavy water.

These experiments were a pioneering venture, representing the first investigations of artificial transmutation. The type of generator built by Cockcroft and Walton suffers electric breakdown at very high voltages, and has a limit of 1.5 MeV (1.5 million electronvolts). Other generators, particularly that designed by Robert van de Graaff (1907–1967) and the cyclotron developed by Ernest Lawrence (1901–1958) soon superseded it. The particle accelerator thereby became a vital research tool at an extremely turbulent time in the development of nuclear physics, and has been developed to a far more advanced state in recent years.

Wankel Felix 1902–1988 was a German engineer known for his research into the development of the rotary engine which bears his name.

Wankel was born in Luhran on 13 August 1902. He attended Vohlschule and Gymnasium before becoming employed by Druckerie-Verlay in Heidelberg in 1921. In

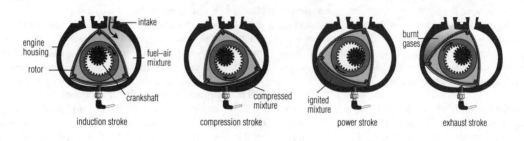

induction stroke   compression stroke   power stroke   exhaust stroke

Wankel *The rotary Wankel engine uses the same four stages as the four-stroke Otto cycle: induction, compression, combustion and exhaust.*

1927 he became a partner in an engineering works before opening his own research establishment, where he carried out work for Bayerisch Motorenwerke (BMW) in 1934. Later he carried out work for the German Air Ministry. At the end of World War II in 1945 he began to work for a number of German motor manufacturers at the Technische Entwicklungstelle in Lindon. In 1960 he was made Director of this institute.

During the 1930s Wankel carried out a systematic investigation of internal-combustion engines. He became particularly interested in rotary engines and gradually began to make mechanical sense of the number of possible arrangements and engine cycles. Although many of his early ideas turned out to be only marginally successful, the German motor firm NSU sponsored the development of his engine with a view to its possible use in motorcycles. Eventually, after the war, and with the help of Froeda, he rearranged his early designs and produced a successful prototype of a practical egine in 1956.

The Wankel engine consists of one chamber of epitrochoidal shape (the path described by a point not on a circle rolling round another circle), which looks like an elongated circle with two dents in it. Inside this chamber is a rotor in the shape of an equilateral triangle with rounded sides. In one revolution the rotor successively isolates various parts of the chamber, allowing for the intake of fuel and air, compression, ignition; expansion and exhaust to take place. While this driving cycle is taking place relative to one face of the rotor, two more are taking place relative to the other two faces; as a result the crankshaft rotates three times for every turn of the rotor. Thus the Wankel engine produces more power for its weight than the more conventional Otto and Diesel engines. There are no separate valves, because the induction and exhaust of vapours is controlled by the movement of the rotor. The great problem with the Wankel engine has been the effective and efficient sealing of the chamber into three parts by the rotor.

Wankel engines are easily connected together in pairs. There are very few moving parts compared with an ordinary motorcar engine; there are no piston rods or camshafts. The saving in engine weight means that slightly less power is required from engines of this type when they are used in cars.

Over the years a number of motor companies have produced cars with this engine, including NSU in Germany and Toyo Kogyo of Japan. Other companies throughout the world have bought the rights to manufacture and use the Wankel engine. As yet it has not displaced the conventional engine, but this is mainly because of the difficulty of efficiently sealing the chambers for a long duration, and because of the now very stringent tests for pollution which any new engine has to undergo in many countries.

Warburg Otto Heinrich 1883–1970 was a German biochemist who made several important discoveries about metabolic processes, particularly intracellular respiration and photosynthesis, and pioneered the use of physicochemical methods for investigating the biochemistry of cells.

Probably his most important contribution was his outstanding work on respiratory enzymes, for which he was awarded the 1931 Nobel Prize for Physiology or Medicine.

Warburg was born on 8 October 1883 in Freiburg-im-Breisgau, the son of a physics professor at the University of Berlin. He studied chemistry under Emil Fischer (1852–1919) at the University of Berlin, obtaining his doctorate in 1906, and then read medicine at the University of Heidelberg, gaining his medical degree in 1911. In 1913 he went to the Kaiser Wilhelm (later Max Planck) Institute for Cell Physiology in Berlin-Dahlem, becoming a professor there in 1918 and its Director in 1931 (having served in the Prussian Horse Guards during World War I). In 1941 Warburg, being part-Jewish, was removed from his post but such was his international prestige and so important was his research that he was soon reinstated and he remained the Director of the institute for the rest of his life. In 1944, he was nominated for a second Nobel Prize but was not allowed to accept the award, because Germans were forbidden to do so under Adolf Hitler's regime. Warburg died in West Berlin on 1 August 1970.

One of Warburg's chief interests was intracellular respiration, and in the early 1920s he devised a method for determining the uptake of oxygen by living tissue using a manometer. Continuing this general line of research, he began to investigate oxidation–reduction reactions involved in intracellular respiration. Warburg and the German chemist Heinrich Wieland (1877–1957) held opposite views about the mechanism of the reaction:

$$AH_2 \ + \ B \ \xrightarrow{\text{catalyst}} \ A \ + \ BH_2$$

(hydrogen donor)   (hydrogen acceptor)   (oxidized)   (reduced)

Warburg believed that the hydrogen acceptor had to be activated and made capable for accepting hydrogen, whereas Wieland thought that the hydrogen donor was activated and made to yield its hydrogen to a hydrogen carrier. Warburg demonstrated that his proposed mechanism was possible and postulated that, in living cells, an iron-containing enzyme activated oxygen (the hydrogen acceptor) and rendered it capable of accepting the hydrogen. He noted that animal charcoal, produced by heating blood, catalyses the oxidation of many organic compounds – oxygen being consumed in the process – whereas vegetable charcoal produced by heating sucrose does not behave in this way. From this finding Warburg concluded that the difference in the behaviour of the two types of charcoal was due to the presence of iron in the blood. He discovered that charcoal systems and living cells behave in similar ways in some respects: in each case, the uptake of oxygen is inhibited by the presence of cyanide or hydrogen sulphide, both of which combine with heavy metals and inhibit respiration. He also showed that, in the dark, carbon monoxide inhibits the respiration of yeast but does not do so in the light. Warburg was aware that heavy metals form complexes with carbon monoxide and that the iron complex is dissociated by light, which provided further evidence for the existence of an iron-containing respiratory enzyme. He then investigated the efficiency of light in overcoming the carbon monoxide inhibition of respiration, finding that the light's efficiency depended on its wavelength. And by plotting the wavelength against the light's efficiency, he determined the photochemical absorption spectrum of the respiratory enzyme, which proved to be a haemoprotein (a protein with an iron-containing group) similar to haemoglobin; he called it iron oxygenase. It was for this work that Warburg was awarded a Nobel prize in 1931.

Meanwhile, the British biologist David Keilin (1887–1963) had discovered cytochromes (the hydrogen carriers in intracellular respiration whose existence had been postulated by Wieland) and cytochrome oxidase, believed to be identical to Warburg's iron oxygenase. Therefore both Wieland and Warburg had been correct in their views about the mechanisms involved in intracellular respiration; they had merely been investigating different stages in this extremely complex pathway.

Warburg also studied coenzymes and he and his collaborators isolated NADP (nicotinamide adenine dinucleotide phosphate) in 1935 and FAD (flavine adenine dinucleotide) in 1938, both of which are important in respiration.

Working on photosynthesis Warburg showed that, given suitable conditions, it can take place with almost total thermodynamic efficiency – that is, virtually 100% of the light energy can be converted to chemical energy. Later he discovered the mechanism of the conversion of light energy to chemical energy that occurs in photosynthesis. In addition, he studied cancer and was the first to discover that malignant cells require less oxygen than do normal cells. This finding was unique at the time, and even today remains one of the relatively few facts that applies to all types of cancer.

Warming Johannes Eugenius Bülow 1841–1924 was a Danish botanist whose pioneering studies of the relationships between plants and their natural environments established plant ecology as a new discipline within botany.

Warming was born on 3 November 1841 on the island of Mandø, Denmark. He studied at the University of Copenhagen but, while still a student, spent the years 1863 to 1866 at Lagoa Santa, Brazil, assisting the Danish zoologist P. W. Lund in a project involving the excavation of fossils. During this expedition, Warming undertook a thorough study of tropical vegetation, the results of which took 25 years to publish fully, although a summary (*Lagoa Santa, a Contribution to Biological Phytogeography*) appeared in 1892. After his return from Brazil, Warming studied for a year at Munich University under Karl von Nägeli (1817–1891) and then spent another year at Bonn University under J. L. von Hanstein. In 1871, Warming gained his doctorate from the University of Copenhagen and he taught botany there from 1873 to 1882, when he became Professor of Botany at the Royal Institute of Technology in Stockholm, Sweden. He went on an expedition to Greenland in 1884 and to Norway in 1885, after which he returned to

Copenhagen to become Professor of Botany at the university and Director of the Botanical Gardens, positions he held until his retirement in 1911. His last major expedition, which lasted from 1890 to 1892, was to the West Indies and Venezuela. Warming died in Copenhagen on 2 April 1924.

Warming's most important contribution to botany was in the area of plant ecology. Ernst Haeckel coined the term 'ecology' in 1866 and it was introduced into botany by Reiter in 1885, but it was Warming who provided the foundation for the study of plant ecology. He investigated the relationships between plants and various environmental conditions, such as light, temperature and rainfall, and attempted to classify types of plant communities (he defined a plant community as a group of several species that is subject to the same environmental conditions, which he called ecological factors). Warming set out the results of his work in *Plantesamfund* (1895), in which he not only provided a theoretical basis for the study of plant ecology but also formulated a programme for future research into the subject, including the investigation of factors responsible for the congregation of plants into communities, and of the evolutionary and environmental pressures that lead to the development of particular habits and habitats in each plant species. *Plantesamfund* was translated into several languages, including English (as *Oecology of Plants* in 1909), and had a tremendous impact in stimulating research into plant ecology.

Warming also investigated a wide range of other areas in botany – including tropical, temperate and arctic flora (about which he provided a vast amount of data), purple bacteria, flower ovules, and the classification of flowering plants.

Waterston John James 1811–1883 was a British physicist who first formulated the essential features of the kinetic theory of gases. Unfortunately, his work was not recognized during his lifetime and Waterston was considered to be an eccentric scientist of no outstanding merit, best known for his work on solar heat. This assessment of his abilities was considerably revised after his death when his contribution to kinetic theory came to light, but by that time the theory had been independently developed by others.

Waterston was born in Edinburgh in 1811. He attended Edinburgh High School and then joined a civil engineering firm. His employers permitted him to attend courses at Edinburgh University, where he studied a broad range of scientific and medical subjects. He showed particular aptitude in mathematics and physics, and began publishing scientific papers as early as 1830.

The expanding British railway system attracted many young engineers to its service, and Waterston moved to London in 1833 to do surveying and draughtsmanship work for it. He wrote papers on matters arising from his work and also continued to do experiments and research into pure science in his own time. He became affiliated with the Institute of Civil Engineers, and took a less time-consuming job in the Hydrographers' Department of the Admiralty in order to spend more time on science.

A complete change in Waterston's life came in 1839 when he went to India to take up a post as teacher of the East India Company's cadets in Bombay. He spent his spare time in the library of the Grant College in Bombay, and continued to write scientific papers which he sent back to Britain – including the ill-fated submission on kinetic theory.

Waterston spent nearly 20 years in Bombay, but when he had saved enough money he retired and in 1857 he returned to Edinburgh to devote all his efforts to research. He wrote papers on a variety of topics covering a broad scientific spectrum, but had difficulty in getting them published. He became increasingly bitter over this and towards the end of his life withdrew completely from all contacts with the rest of the scientific community. On 18 June 1883 he disappeared from his home and was never heard of again.

Waterston's first scientific paper, published when he was only 19 years old, concerned a model which he proposed might explain gravitational force without the necessity for postulating an effect which operated at great distances. This was a topic of lively debate at the time, and although he did not carry his work along these lines any further, the paper did contain some formative ideas which were developed in his kinetic theory of gases 15 years later.

In 1843, Waterston wrote a book on the nervous system in which he attempted to apply molecular theory to physiology. It included several fundamental features of the kinetic theory of gases, among them the idea that temperature and pressure are related to the motion of molecules. However, Waterston's ideas on neurophysiology and psychology were ahead of their time and, perhaps for this very reason, the book aroused little interest.

The work of James Joule (1818–1889) on heat stimulated Waterston to consider more fully its application to the properties of gases. He produced a paper which contained the first formulation of the equipartition theory and was thus an early application of statistical mechanics, and submitted it to the Royal Society in 1845.

The Society, following its standard practice, had the paper read by two scientists who both commented unfavourably on the efforts of this obscure scientist writing from the other side of the globe, and the paper was firmly rejected. An abstract of the paper did appear in the *Transactions* of the Royal Society in 1846, and a short note was published on it by the British Association in 1851.

Waterston was unable to retrieve the paper from the Society, and as he had not kept a copy for himself, he was unable to publish his theory elsewhere. A longer abstract of the paper was the best he could provide, and that purely for private circulation. Hermann Helmholtz (1821–1894) published an abstract of his theory in a German journal, which may conceivably have stimulated interest in kinetic theory in Germany. But it was Rudolf Clausius (1822–1888), James Clerk Maxwell (1831–1879) and Ludwig Boltzmann (1844–1906) who received the credit for the development of the kinetic theory from 1857 onwards. The rejection of Waterston's work had delayed progress to the kinetic theory by about 15 years. Waterston's priority was

finally established by Lord Rayleigh (1842–1919), who discovered his 1845 paper on the kinetic theory in the Royal Society's vaults in 1891 and published it in the following year with an introduction warning young scientists of the resistance to new ideas often displayed by scientific societies.

Waterston did gain some recognition for a paper he presented to the British Association in 1853 on the basis for the generation of solar heat. He proposed that meteoric material falling into the Sun would produce heat. In 1857 he published an estimate of the solar temperature at 13 million degrees, which is not far from the present value accepted for the Sun's interior. Waterston wrote other papers on sound, capillarity, latent heat, and various aspects of astronomy. The rejection of two of his papers by the Royal Astronomical Society in 1878 may well have been related to his resignation from that society shortly thereafter, after having been a member for over a quarter of a century, and his subsequent progressive withdrawal from science.

Waterston must therefore be regarded rather as a tragic figure, a scientist whose originality and talent were frustrated to the detriment not only of himself but also of science as a whole.

Watson James Dewey 1928– . US biologist; *see* Crick, Watson, and Wilkins.

Watson-Watt Robert Alexander 1892–1973 was a British physicist and engineer who was largely responsible for the early development of radar. He patented his first 'radiolocator' in 1919, and perfected his equipment and techniques from 1935 through the years of World War II. His radar was employed in the deployment of British fighter aircraft during the Battle of Britain.

Watson-Watt was born at Brechin, Angus, and educated at the University of St Andrews where he also taught from 1912 to 1921. His interest in the reflection of radio waves was first aroused during his time at the university and this subject became his life's work.

The first patent for a radar-like system had been granted in several countries to a German engineer, Christian Hulsmeyer, in 1904. Evolving from the search for a means of detecting radio waves from ships, his system worked, was demonstrated to the German Navy, but never accepted. The principles used in Hulsmeyer's system had been discovered much earlier through the experimental work of the British physicist Michael Faraday (1791–1867) and the mathematical investigations of the Scottish physicist James Clerk Maxwell (1831–1879), who predicted the existence of radio waves and formulated the electromagnetic theory of light. The German physicist Heinrich Hertz (1957–1894) tested Maxwell's theories experimentally, and in 1886 proved the existence of radio waves.

When Watson-Watt started his work it was known that radio waves can be reflected, for it was their reflection from ionized layers in the upper atmosphere that made long-distance broadcasting possible. The reflection was sharper as wavelength decreased. The device which Watson-Watt patented in 1919 was concerned with radio-location by means of short-wave radio waves. It was based on quite simple principles. Radio waves travel at an accurately known velocity (the velocity of light) which for purposes of approximation may be taken as 300,000 km/186,000 mi per second. When radio waves are radiated from a transmitting antenna and are interrupted by any object – such as a ship, plane or even a mountain – part of the energy is reflected back toward the receiver. The direction from which the echo is obtained is the direction of the obstacle.

Watson-Watt continued his experiments, and by 1935 had patented improvements that made it possible to follow an aeroplane by the radio-wave reflections it sent back. The system was called 'radio detection' and 'ranging' and from this comes the present abbreviation, 'radar'. Research and development was being conducted during the 1930s in Britain, France, Germany and the United States, although as the decade wore on only the British researched with any great energy. Watson-Watt's work was heavily subsidized by the British government, and was carried out in great secrecy. By 1938, when it was apparent that war was inevitable, radar stations were in operation, and during the Battle of Britain in 1940 radar made it possible for the British to detect incoming German aircraft as easily by night as by day, and in all weathers including fog. Early in 1943 microwave aircraft-interceptor radars were operational, ending night-bombing raids on Britain.

The first radar sets specifically designed for airborne surface-vessel detection had been flown early in 1943. Wartime pressures and the enthusiastic hard work of Watson-Watt had given Britain a clear lead in the field of radar, and before the United States joined the war Watson-Watt visited the United States to advise on the setting up of radar systems. In 1942 he was knighted.

Radar as a navigational aid, and for collision avoidance, has become a standard accessory on ships and aircraft. Harbour surveillance sets now guide ships when visibility is poor. Radar on planes detects storms and helps pilots to avoid bad weather and a phenomenon known as clear-air turbulence. A familiar police application is the use of radar to determine the speed of a car by the Doppler shift in frequency of the reflected signal. In weather forecasting the tracking of distant clouds and their development is accomplished by the use of radar, while in the military field it is used in the steering of guided missiles and in warning of their approach. The great radio telescope at Jodrell Bank, Cheshire, which has a bowl-shaped reflector 76.2 m/250 ft across that can be pointed to any part of the sky, can be used as a radar set, the reflector being used to direct the pulses of microwaves in a narrow beam towards a chosen object. In this way echoes have been obtained from the Moon and the paths of satellites and spacecraft have been tracked.

Much research and development of computer-controlled and computer-linked radar continues. Development of higher power sources also continues and it is probable that further improvements in radar performance will be achieved by advances in receiving techniques. Here again,

the computer will play an important role, storing information, separating signals from noise and generally increasing the sensitivity of the receiver. But however sophisticated, all uses of radar owe a debt to Watson-Watt's original invention.

Watt James 1736–1819 was a Scottish mechanical engineer who is popularly credited with inventing the steam engine. In fact he modified the engine of Thomas Newcomen (1663–1729) to the extent that it became a practical, efficient machine capable of application to a variety of industrial tasks. In particular he devised the separate condenser and eventually made a double-acting machine which supplied power with both directions of the piston; this was a great help in developing rotary motion. He also invented devices associated with the steam engine, artistic instruments and a copying process, and devised the horsepower as a description of an engine's rate of working. The modern unit of power, the watt, is named after him.

Watt was born in Greenock on 19 January 1736, the son of a chandler and joiner. Throughout his life he suffered from serious attacks of migraine, and at school both his peers and his teachers took a poor view of this 'weakness'. His great delight was to work in his father's workshop, where a corner had been set aside for him with his own forge and workbench. Soon he developed great skill, and he wished to become an instrument-maker. In his attempt to find an apprenticeship in this trade, he went first to Glasgow, where he worked with an optician and odd-job man for a year. Then, on advice from a friend, he went to London. Eventually, he secured a position with very unfavourable conditions. He did, however, learn the skills of instrument-making before illness forced him to return home to Greenock. After recovering, he set up in business as an instrument-maker in Glasgow and in 1757 obtained work from Glasgow University that allowed him to work in a room within its precincts, and he proudly described himself as 'Instrument Maker to Glasgow University'.

During this period he was asked to repair a small working model of Newcomen's steam engine. The machine proved to be temperamental and difficult to operate without air entering the cylinder and destroying the vacuum. He set about investigating the properties of steam and making measurements of boilers and pistons in the hope of improving Newcomen's machine which was, at best, slow, temperamental, inefficient and extremely costly to run in terms of the coal required to keep a sufficient head of steam in a practical engine. During a short period of inspiration, in the course of a Sunday afternoon walk, he had the idea of a separate condenser (separate from the piston). In Newcomen's engine, the steam in the cylinder was condensed by a jet of water, thus creating a vacuum which, in turn, was filled during the power stroke by the atmosphere pressing the piston to the bottom of the cylinder. On each stroke the cylinder was heated by the steam and cooled by the injected water, thus absorbing a tremendous amount of heat. With his separate cylinder, Watt could keep the cylinder hot, and the condenser fairly cold by lagging, thus

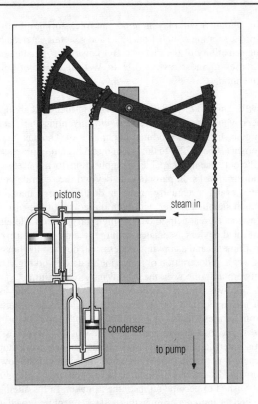

pistons

steam in

condenser

to pump

Watt *Watt's steam engines, dating from 1769, were an improvement on that of Thomas Newcomen in that they had a separate condenser and permitted steam to be admitted alternately on both sides of the piston.*

improving the thermal efficiency of the machine and the economics of its operation. Watt's original engine of 1765 is now in the Science Museum, London. It was only a working model, and reveals the haste in which it was built.

As far as practical engines were concerned, Watt had a great deal of trouble in efficiently lagging the cylinder so that heat was retained and at the same time allowing the piston to move freely. He was helped with facilities and labour by John Roebuck of Kinneil, who eventually employed Watt's engine to pump water from his mines. In 1767, Watt again travelled to England, this time to patent his engine (patent granted in 1769). On his way back to Scotland he visited some friends of Roebuck in Birmingham and met Matthew Boulton. Boulton was a major manufacturer in Birmingham and had the finance to exploit Watt's engine. Because of the patent arrangements between Watt and Roebuck it was not until the latter got into severe financial difficulties that Boulton could buy him out and begin manufacturing the engine. In fact, between 1767 and 1774, Watt made his living as a canal surveyor. Although he was successful at this, his health was not up to an outside job in harsh weather and he suffered accordingly.

From 1775, financial difficulties being solved, Boulton and Watt went into partnership and manufactured Watt's engines at the famous Soho Foundry, near Birmingham. In 1782 Watt improved his machine by making it double-

acting. By means of a mechanical linkage known as 'parallel motion' and an extra set of valves, the engine was made to drive on both the forward and backward strokes of the piston, and a 'sun-and-planet' gear (also devised by Watt) allowed rotatory motion to be produced. This new and highly adaptable engine was quickly adopted by cotton and woollen mills. A universally practical means of producing power for the evolving British industry was therefore at hand, with the consequent rapid rise in the adoption of larger machines.

During this same period, 1775 to 1790, Watt invented an automatic centrifugal governor, which cut off the steam when the engine began to work too quickly and turned it on again when it had slowed sufficiently. He also devised a steam engine indicator, which showed steam pressure and the degree of vacuum within the cylinder. Because of the secretarial duties connected with the business, Watt invented a way of copying letters and drawings; this was a chemical process and was displaced only with the advent of the typewriter and photocopier. Although his steam engines were usually built for specific purposes and individually priced, it was important to have a rational method upon which charges could be made. For this he considered the rate at which horses worked and, after many experiments, concluded that a 'horsepower' was 15,000 kg/ 33,000 lb raised through 0.3 m/1 ft each minute. He rated his engines in horsepower and in the English-speaking world this method of describing the capability of an engine continued, until recent years.

In 1785, Watt was elected a Fellow of the Royal Society. During the last decade of the eighteenth century the active management of the Soho Works was taken over increasingly by Boulton and Watt's sons and in 1800, when the patent rights to the engine expired, Watt retired. He then kept an attic workshop and busied himself designing and constructing copying machines.

Watt died on 25 August 1819, aged 83, leaving the legacy of high, useful machine power for the development and proliferation of industry. His name has become immortalized as the unit of power; a watt is one joule per second, and one horsepower is equivalent to about 746 watts.

Weber Heinrich 1842–1913 was a German mathematician whose chief work was in the fields of algebra and number theory.

Weber was born in Heidelberg on 5 May 1842. He entered the university there to study mathematics and physics and, with a year's interval at the University of Leipzig, received his doctorate in 1863. He then went to Königsberg, where for three years he worked with Franz Neumann (1798–1895) and Friedrich Richelot (1808–1875). He began to teach at the University of Heidelberg in 1866 and was made an Extraordinary Professor in 1869. Thereafter he taught at a number of institutions – the Zurich Polytechnic, the University of Königsberg, the Technical High School at Charlottenburg and the universities of Marburg and Göttingen – before taking up his last post at Strasbourg in 1895. He died in Strasbourg on 17 May 1913.

The three years that Weber spent at Königsberg as a young man had a decisive influence on his mathematical career. The influence of Karl Jacobi (1804–1851) was then very powerful there and much of Weber's work shows his solid grounding in Jacobian methods. His early work was based on the theory of differential equations, and he was encouraged by Neumann to apply his knowledge to problems in physics. Yet it was not until late in his career that Weber's work in such subjects as heat, electricity and electrolytic dissociation was published. Most of it was contained in his *Die partiellen Differentialgleichungen der mathematischen Physik* (1900–01), which was essentially a reworking of, and a commentary upon, a book of the same title, based on lectures given by Bernhard Riemann and written by Karl Hattendorff.

In the decades before that book appeared Weber produced most of his important work, especially his most outstanding contribution to mathematics, his demonstration of Abel's theorem in its most general form. Another brilliant result was his proof of Kronecker's theorem that the absolute Abelian fields are cyclotomic – that is, that they are derived from the rational numbers by the adjunction of roots of unity. This work on Abel's mathematics reflected the influence of Richelot, under whose guidance Weber became an expert in the manipulation of algebraic functions.

In the 1890s Weber produced several important results, chief among them his demonstration of the critical importance of linking analysis and number theory in investigating problems involving complex multiplication. In 1896 there appeared his culminating work in algebra, the two-volume *Lehrbuch der Algebra*, which for a generation was the standard algebra text. That book, as also his editorship of the three-volume *Enzylopädie Elementär-Mathematik* (1903–07), commended Weber not just to higher mathematicians but also to the humbler host of teachers and students who came after him.

Weber Wilhelm Eduard 1804–1891 was a German physicist who made important advances in the measurement of electricity and magnetism by devising sensitive instruments and defining electric and magnetic units. In recognition of his achievements, the SI unit of magnetic flux density is called the weber. Weber was also the first to reach the conclusion that electricity consists of charged particles.

Weber was born at Wittenberg on 24 October 1804. The family was highly gifted, for Weber's father was Professor of Theology at the University of Wittenberg and his older brother Ernst Weber (1795–1878) became a pioneer in the physiology of perception. In 1814, the family moved to Halle, where Weber entered the University in 1822. He obtained his doctorate in 1826, and then became a lecturer at Halle, rising to an assistant professorship in 1828.

In 1831, Weber moved to Göttingen to become Professor of Physics and there began a close collaboration with Karl Gauss (1777–1855). He lost this post in 1837 following a protest at the suspension of the constitution by the new ruler of Hanover. However, Weber managed to continue working at Göttingen and then in 1843 he obtained the position of Professor of Physics at Leipzig. In 1849, Weber returned to his former post at Göttingen. He remained in this position until he retired in the 1870s, and died at Göttingen on 23 June 1891.

Weber's work in magnetism dates from the time when he joined Gauss at Göttingen. They conceived absolute units of magnetism that were defined by expressions involving only length, mass and time, and Weber went on to construct highly sensitive magnetometers. He also built a 3-km/2-mi telegraph to connect the physics laboratory with the astronomical observatory where Gauss worked, and this was the first practical telegraph to operate anywhere in the world. It was not subsequently developed as a commercial invention, and from 1836 to 1841 Gauss and Weber organized a network of observation stations to correlate measurements of terrestrial magnetism made around the world.

In 1840, Weber extended his work on magnetism into the realm of electricity. He defined an electromagnetic unit for electric current that was applied to measurements of current made by the deflection of the magnetic needle of a galvanometer. In 1846, he developed the electrodynamometer, in which a current causes a coil suspended within another coil to turn when a current is passed through both. This instrument could be used to measure alternating currents. Current could also be measured by the Coulomb torsion balance in electrostatic units, and in an experiment carried out with Rudolph Kohlrausch in 1855, Weber found that the ratio of the electromagnetic unit to the electrostatic unit is a constant equal to the velocity of light. They did this by discharging a condenser through a torsion balance and a ballistic galvanometer. Weber and Kohlrausch attached no great importance to their strange result, but it later proved to be vital to James Clerk Maxwell (1831–1879) in his development of the electromagnetic theory of light.

In 1852, Weber defined the absolute unit of electrical resistance and also began to conceive of electricity in terms of moving charged particles of positive and negative electricity, resistance being produced by a combining of the two particles that prevents current flow. In 1846, he had produced a general law of electricity that attempted to express electrical effects mathematically and although it was not successful, it did help Weber to develop his ideas on the nature of electricity. In a remarkable piece of foresight, he put forward in 1871 the view that atoms contain positive charges that are surrounded by rotating negative particles and that the application of an electric potential to a conductor causes the negative particles to migrate from one atom to another. It was not until 1913, with the proposal of the Rutherford–Bohr model of the atom, that Weber's ideas were seen to be essentially correct. Weber also provided similar explanations of thermal conduction and thermoelectricity that were later fully developed by others.

Weber also did important work in acoustics with his brother Ernst Weber. In 1825, they made the first experimental study of interference in sound. Their findings were important in enabling Hermann Helmholtz (1821–1894) to achieve explanations of the perception of sound and mechanism of hearing.

Weber's insistence on precise experimental work to produce correct definitions of electrical and magnetic units was very important to the development of these sciences and to electromagnetism. His farreaching views on the nature of electricity were influential in creating a climate for the acceptance of such ideas when evidence for them was later found.

Wedderburn Joseph Henry 1882–1948 was a Scottish mathematician who opened new lines of thought in the subject of mathematical fields and who had a deep influence on the development of modern algebra.

Wedderburn was born at Forfar, Scotland, on 26 February 1882 and was educated at the University of Edinburgh, which he entered in 1898. He received a degree in mathematics in 1903 and in the following year was awarded a Carnegie fellowship to study at the University of Chicago. In 1905, he returned to Scotland, where he was appointed a lecturer in mathematics at the University of Edinburgh and editor of the *Proceedings of the Edinburgh Mathematical Society*. In 1908, he was awarded a doctorate and in 1909 he went back to the United States to teach at Princeton University. During World War I he saw active duty in France as a soldier in the British army. He then returned to Princeton, where he remained until his retirement in 1945. For the last 20 years of his life he was in poor health and he stopped publishing in 1938. He died at Princeton on 9 October 1948.

The first paper that Wedderburn published, 'Theorem on Finite Algebra' (1905), was a milestone in algebraic history. Before it appeared little was known about hypercomplex numbers and their roles in algebra. The classification of semi-simple algebras had been investigated only for fields composed of real or complex numbers. Wedderburn was able to show, by introducing new methods, that it was possible to arrive at a complete understanding of the structure of these algebras over any field.

From that foundation he went on to derive the two theorems to which his name has become attached. The first was contained in his paper, 'On Hyper-Complex Numbers' (1907), in which he demonstrated that a simple algebra consists of matrices of a given degree with elements taken from a division of algebra. This paper marked the beginning of a new approach to this type of algebra. The first Wedderburn theorem states that 'if the algebra is a finite division algebra (that is, that it has only a finite number of elements and always permits division by a non-zero element), then the multiplication law must be commutative, so that the algebra is actually a finite field'.

Wedderburn's second theorem states that a central-simple algebra is isomorphic to the algebra of all $n \times n$ algebras. He arrived at it by an investigation of skew fields with a finite number of elements. When he started, all commutative fields with a given number of elements had been classified; but it was assumed that no noncommutative field

existed, because none had ever been found. Wedderburn's discovery that every field with a finite number of elements is commutative under multiplication thus led to a complete classification of all semi-simple algebras with a finite number of elements.

The modern study of mathematical fields owes an enormous debt to Wedderburn, who may rightly be regarded as one of the creative geniuses of his age.

Wedgwood Josiah 1730–1795 was a British pottery craftsman, one of the most celebrated and influential of all time.

Wedgwood was born about June 1730 at Burslam, Staffordshire, the youngest son of Thomas Wedgwood, who was also a renowned potter. After his father's death in 1739 the young Josiah worked in the family business at Churchyard Works, Burslam, and in 1744 was apprenticed to his brother Thomas. At about this time he contracted smallpox, and had to have his right leg amputated. During the period of forced inactivity he studied books about pottery and did much experimental work. He was refused a partnership with his brother in 1749, but shared one with John Harrison of Stoke-on-Trent, which lasted until 1753. A year later he joined Thomas Wheildon of Fenton Low, Staffordshire, who was also a leading potter of his day.

The partnership flourished, which enabled Wedgwood to become a master of the art and continue with what he termed his *Experimental Book*, which proved to be an invaluable source of information on the production of Staffordshire pottery. He also invented and produced his improved 'green glaze', which has remained popular until the present day.

Wedgwood then set up in business on his own at the Ivy House Factory in Burslem, and there he perfected cream-colonial earthenware, which became known as Queen's Ware because of the interest and patronage of Queen Charlotte in 1765. In 1768 he went into partnership with the Liverpool businessman Thomas Bentley and they expanded the company into the Brick House Bell Works Factory. They produce unglazed stoneware in various colours, formed and decorated in the popular Neo-Classical style. Wedgwood also continued with his black-basalts which, with added red acaustic painting, allowed him to imitate Greek red-figure vases and fine-grained Jaspar ware. He then built the Etruria Factory, using his engineering skills in the design of machinery and the high-temperature beehive-shaped kilns, which were more than 4 m/12 ft wide. He named the factory after Etruria in northern Italy where coincidentally he died on 3 January 1795.

Wegener Alfred 1880–1930 was the German geologist who perhaps did most to revolutionize scientific study of the Earth in the twentieth century.

Born in Berlin in 1880, Wegener studied at Heidelberg, Innsbruck and Berlin. He obtained his doctorate in astronomy in Berlin in 1905. Before World War I he taught at Marburg, specializing in meteorology. From 1924 he held a specially created chair in meteorology and geophysics at the University of Graz, Austria.

From 1910 Wegener began developing a theory of continental drift. Empirical evidence for such displacement lay, he thought, in the close jigsaw-fit between coastlines on either side of the Atlantic, and notably in palaeontological similarities between Brazil and Africa. Wegener was also convinced that geophysical and geodetical factors would corroborate the conjecture of a flight from the poles and of wandering continents – though he himself was rather confused about the causes of such displacement, believing partly in tidal forces.

Wegener supposed that a united supercontinent, Pangaea, had existed in the Mesozoic. This had developed numerous fractures and had drifted apart, some 200 million years ago. During the Cretaceous, South America and Africa had largely been split, but not until the end of the Quaternary had North America and Europe finally separated; the same was true of the break between South America and Antarctica. Australia had been severed from Antarctica during the Eocene.

Wegener's hypothesis met with widespread hostility. Only with the development of a satisfactory mechanism for displacement, that is, with the rise of plate tectonics since World War II, has the modified hypothesis won support. An intrepid polar explorer, Wegener died while crossing the Greenland ice sheet on his fourth expedition. Wegener may well be remembered as the most influential geologist of the twentieth century. Though not the first theorist of continental drift, his accounts of the theory, especially his *Origin of Continents and Oceans* 1929, gave the premise scientific plausibility.

Weierstrass Karl Theodor Wilhelm 1815–1897 was a German mathematician who is remembered especially for deepening and broadening the understanding of functions.

Weierstrass was born at Ostenfelde, Westphalia, on 31 October 1815. Because of his family's frequent change of residence he attended many schools as a young boy and it was not until he entered the Roman Catholic School at Paderborn in 1829 that he began to reveal his mathematical ability. In deference to his father's wish that he should pursue a 'respectable career', he entered the University of Bonn in 1834 to study law, administration and finance. By 1837 he was certain that mathematics alone interested him and in that year he left Bonn, without taking a degree, to enter the Theological and Philosophical Academy at Münster, with the dual intention of gaining a teacher's certificate and devoting himself to the study of advanced mathematics.

He received his teaching certificate in 1841. From 1842 to 1848 he taught at a secondary school in Deutsch-Krone, in west Prussia, then moved to Braunsberg, where he became a lecturer at the Roman Catholic School. In his spare time he did his mathematical research and in 1854 he

Wegener: *continental drift*

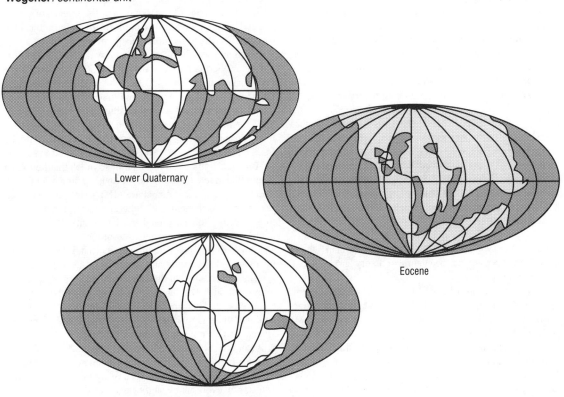

Lower Quaternary

Eocene

Upper Carboniferous period

published in *Crelle's Journal* a paper on Abelian integrals that established his reputation and earned him the award of an honorary doctorate from the University of Königsberg. The strain of combining his research with a heavy teaching load (physics, botany and history in addition to mathematics) told on his health. From 1850 he suffered from debilitating attacks of vertigo and for the rest of his life he was rarely in good health. He was therefore happy to be given leave of absence in 1855 to complete his work on function theory, begun in the 1854 paper.

That paper marked a turning-point in his academic life. Until the age of 40 Weierstrass had worked in isolation. Then, in 1856, he was appointed Professor of Mathematics at the Royal Polytechnic School in Berlin and, jointly with it, Associate Professor at the University of Berlin, although he did not begin actually to lecture at the university until 1864. In 1861, he suffered a complete physical breakdown, and although a year of convalescence brought an end to the attacks of vertigo, he thereafter suffered from bronchitis and phlebitis. By 1894 he was confined to a wheelchair; he died of pneumonia in Berlin on 19 February 1897.

Weierstrass's most important work was in function theory, the subject which he first treated seriously in the examination paper for his teacher's certificate in 1841. This was followed up by the 1854 paper, which solved the inversion of hyperelliptic integrals, and a second paper of 1856. His greatest achievement was the demonstration, published in 1871, that there exist continuous functions in an interval which have derivatives nowhere in the interval. In fact he published very little. Much of his reputation rested on his lectures at Berlin, which ranged over the whole of mathematics and which became famous for their 'Weierstrassian rigour'. In particular, Weierstrass did much (again more in lectures than in publications) to clarify the meaning of basic concepts such as 'function' 'derivative' and 'maximum'. His development of the modern theory of functions was described in his *Abhandlungen aus der Funktionlehre* (1886), a text derived chiefly from his students' lecture notes. In the 1890s Weierstrass planned the publication of his life's work, again to be compiled from lecture notes. Two volumes were published before his death and five more appeared during the next three decades. Three volumes of the projected ten-volume set remain unpublished, but even the incomplete work remains a rich quarry for the present-day mathematician.

Weil André 1906–   is a French mathematician whose main field of activity has been number theory, group theory and algebraic geometry.

Weil was born on 6 May 1906 in Paris, where he grew up, was educated, and attended the University, receiving his doctorate in 1928. In 1930 he went to India for two years as Professor of Mathematics at Aligarh Muslim University. Returning to France, he took up a similar post at Strasbourg University and remained there until 1940. Then, after a year's lecturing in the United States, he moved to Brazil where in 1945 he became Professor of Mathematics at the University of São Paolo. From 1947 to 1958 he was Professor of Mathematics at the University of Chicago, and then transferred to a similar post at the School of Mathematics at the Institute of Advanced Studies of the University of Princeton.

One of Weil's earliest contributions to number theory came in 1929, when he extended some earlier work by Henri Poincaré (1854–1912). This resulted in the postulation of what is now called the Mordell–Weil theorem, a theorem that is closely connected to the theory of Diophantine equations.

Weil worked on quadratic forms with algebraic coefficients and extended Emil Artin's work on the theory of quadratic number fields. He also contributed to the generalization of algebraic geometry.

In addition, he was a founder member of that secretive and esoteric club called the Bourbaki Group, which until recently was still in existence, and which has published brilliant and entertaining mathematical papers under the blanket pseudonym Nicolas Bourbaki. (Although there was no limit to the number of members in the Group, its membership was always changing because retirement was compulsory on reaching the age of 50.)

Weil's own major work, *Foundations of Algebraic Geometry*, was published in 1946.

Weismann August Friedrich Leopold 1834–1914 was a German zoologist who is best known for his germ-plasm theory of heredity and for his opposition to Jean Lamarck's doctrine of the inheritance of acquired characteristics. Weismann was one of the founders of the science of genetics, and many of the ideas he put forward on the subject are essentially correct, including his germ-plasm theory.

Weismann was born on 17 January 1834 in Frankfurt-am-Main, Germany, the son of a classics professor at the Gymnasium there. In 1852 he went to the University of Göttingen to study medicine and graduated in 1856, after which he briefly held several positions, including those of a doctor in the Baden army (Baden was then an autonomous Grand Duchy) and private physician to Archduke Stephen of Austria. In 1860 Weismann visited Freiburg-im-Breisgau, which impressed him so much that he felt he would like to live there. In the following year a brief period studying under Karl Leuckart (1822–1898) in Giessen reawakened Weismann's childhood interest in natural history, and in 1863 he joined the University of Freiburg's medical faculty as a teacher of zoology and comparative anatomy. He persuaded the university to build a zoological institute and museum, and he became its first director. Weismann remained at the University of Freiburg until his retirement in 1912. During his later years he travelled extensively and became famous for his lectures on heredity and evolution. He was extremely patriotic and renounced all the British honours awarded him when World War I broke out. He died in Freiburg on 5 November 1914.

In his early years at Freiburg, Weismann studied insect metamorphosis and the sex cells of hydrozoa. In the mid-1860s, however, his eyesight began to deteriorate and he was unable to perform the microscope work necessary for this research. Although his eyesight improved briefly the improvement was only temporary and by the mid-1880s he was forced to abandon the observational part of his work and to concentrate on theory.

Although an admirer of Charles Darwin, Weismann began by questioning pangenesis (Darwin's theory that every cell of the body contributes minute particles – gemmules – to the germ cells and therefore participates in the transmission of inherited characteristics) and then proceeded to attack the Lamarckian theory of the inheritance of acquired characteristics. Weismann's early work on hydrozoan sex cells led him to postulate that every living organism contains a special hereditary substance, the germ plasm, which controls the development of every part of the organism and is transmitted from one generation to the next in an unbroken line of descent. Furthermore, he realized that repeated mixing of the germ plasm at fertilization would lead to a progressive increase in the amount of hereditary material, and therefore predicted that there must be a type of nuclear division at which each daughter cell receives only half of the original germ plasm. This prediction was proved correct by the cytological work of Oskar Hertwig (1849–1922) and others, which then led Weismann to propose that the germ plasm was situated in what were later called the chromosomes of the egg nucleus.

Weismann's germ-plasm theory is still basically true today, although we now use the terms chromosomes, genes and DNA to refer to the hereditary material Weismann called germ plasm. In one important respect, however, Weismann was not completely correct: he believed that the germ plasm cannot be altered by the action of the environment and that variations among individuals arise from different combinations and permutations of the germ plasm. Although different combinations of the hereditary material do give rise to individual variations, the genetic material can be modified by environmental influences – as later demonstrated by Hugo de Vries and Hermann Muller.

Weizsäcker Carl Friedrich von 1912– is a theoretical physicist who has made fundamental contributions to astronomy by investigating the way in which energy is generated in the cores of stars. He is also known for his theory on the origin of the Solar System.

Weizsäcker was born in Kiel, Germany, on 28 June 1912. He earned his PhD at the University of Leipzig in 1933 and from 1934 to 1936 he was an assistant at the Institute of Theoretical Physics there. He worked at the Kaiser Wilhelm Institute of Physics in Berlin-Dahlem and lectured at the University of Berlin from 1936 to 1942, when he was appointed to a chair at the University of Strasbourg. Weizsäcker returned to the Kaiser Wilhelm Institute in 1944. During World War II, he was a member of the German research team investigating the feasibility of constructing nuclear weapons and harnessing nuclear energy. One of his overriding concerns in this work was that his team should not develop a nuclear weapon which might be placed at the disposal of the Nazi government.

In 1946, Weizsäcker became Director of a department in the Max Planck Institute of Physics in Göttingen, holding an honorary professorship at the University of Göttingen. He became Professor of Philosophy at the University of Hamburg in 1957 and retired in 1969 with an appointment to an honorary chair at the University of Munich. In 1970 he became a Director of the Max Planck Institute.

In 1938, Weizsäcker and Hans Bethe independently proposed the same theory of stellar evolution, one which accounted both for the incredibly high temperatures in stellar cores and for the production of ionizing and particulate radiation by stars. They proposed that hydrogen atoms fused to form helium via a proton–proton chain reaction. Bethe and his collaborators went on to outline the complex of reactions which might follow to produce the energy created by a star. Weizsäcker turned to a study of the atomic reactions which take place in the fission of uranium and the problems of constructing an atomic pile.

In 1944, Weizsäcker revived an old cosmogenic theory, the so-called 'nebular hypothesis' of Kant and Laplace, which had carried much weight during the nineteenth century, but which by the end of the century had given way to the 'collision' theory. The collision theory proposed that planets were produced after another star approached our Sun so closely that a proportion of the solar mass became detached in the form of a wisp of hot gas and dust and then coalesced to form the planets. The chief difficulty with this theory was that, since stellar interactions of that nature must be exceedingly rare, planetary systems must be exceedingly uncommon. Furthermore, the stability of the solar material disturbed in such a manner was unlikely to be sufficient to permit the formation of planets.

Weizsäcker suggested that multiple centres, or vortices, formed in the spinning gaseous discoid mass which preceded our Solar System, and that from them the planets condensed. This model indicated that planetary systems were formed as a natural by-product of stellar evolution, and that they were therefore not likely to be so rare as would have been predicted by the collision theory.

Weizsäcker's other research has dealt with topics in atomic and nuclear physics and philosophy.

Wells Horace 1815–1848 was a US dentist who discovered nitrous oxide anaesthesia and, in 1844, was the first to use the gas in dentistry – although ether anaesthesia had previously been used in other surgical operations.

Wells was born in Hartford, Vermont, on 21 January 1815. He was educated at private institutions in Massachusetts and New Hampshire then, at the age of 19, began to study dentistry in Boston. He subsequently set up a dental practice in Hartford, Connecticut, initially in partnership with William Morton – who later pioneered the use of ether as an anaesthetic – and with John Riggs (1810–1885) as one of his students. In late 1844, while watching an exhibition of the effects of laughing gas (nitrous oxide) staged by a travelling show, Wells observed that the gas induced anaesthesia – an effect also noticed previously by Crawford Long (1815–1878), another pioneer of anaesthesia. Wells then arranged to have one of his wisdom teeth extracted by

Riggs while the showman administered nitrous oxide. Having felt no pain during this operation, he subsequently used nitrous oxide anaesthesia to perform painless extractions on his patients.

In January 1845 Wells went to Boston where, with the help of Morton (then no longer his partner), the chemist Charles Jackson (1805–1880) and the surgeon John Warren (1778–1856), he arranged to demonstrate a painless tooth extraction using nitrous oxide anaesthesia to students at the Massachusetts General Hospital. During the demonstration, however, the patient cried out and, although the patient later claimed to have felt no pain, the audience believed that the demonstration had failed. After this débâcle Wells gave up his dental practice and became a travelling salesman, selling canaries and then showerbaths in Connecticut. In 1846, Morton gave a successful demonstration of ether anaesthesia in the same operating theatre that Wells had used, and in the following year Wells went to Paris to try to establish his priority in using anaesthesia. At about this time he also began experimenting on himself with nitrous oxide, ether and various other intoxicating chemicals; as a result he became addicted to chloroform and mentally unstable. In 1848, having returned to the USA, he was imprisoned in New York City for throwing acid in the face of a prostitute and, while in his prison cell, committed suicide. Ironically, during his imprisonment the Paris Medical Society accepted Wells's claim to priority in the discovery of anaesthesia.

Wenner-Gren Axel Leonard 1881–1961 was a Swedish industrialist who developed a modern monorail system.

Wenner-Gren was born on 5 June 1881 in Uddevalla. He was educated in Germany, and began his working career as a salesman for the Swedish Electric Lamp Company. Over a period of years he gradually gained promotion and eventually became a majority shareholder. In 1921 he founded the Electrolux Company to manufacture vacuum cleaners and, later, refrigerators. With the success of his first company well established, he widened his interests to include many aspects of Swedish industry, including the ownership of one of the country's largest wood pulp mills and of the Bofors munition works. From the profits, he donated a large sum of money for the foundation of an institute for the development of scientific research in Sweden, which became known as the Wenner-Gren Foundation for Nordic Cooperation and Research.

Engineers have been experimenting with monorails for more than 150 years but almost always their construction has proved to be more complicated than expected. The first monorail patent was taken out in 1821 by Henry Robinson Palmer, an engineer to the London Dock Company, who built a line that ran between the Royal Dock Victualling Yard and the river Thames. It consisted of an elevated rail made from wooden planks set on edge and capped by an iron bar to take the wear and tear of the horse-drawn 'car' wheels. The 'cars' were in two parts, which hung down on each side of the rail like saddle-bags. Similar later designs were built by the Frenchman Lartigue and in 1869 in Syria,

by J. L. Hadden. An electric version of the Lartigue line was operated in France in the late 1890s. Later systems used gyroscopes to stabilize the train.

Wenner-Gren's monorail, the Alweg line, consisted of a concrete beam carried on concrete supports. The cars straddled the beam on rubber-tyred wheels, and there were also horizontal wheels in two rows on each side of the beam. The system proved to be commercially successful, and it was used for the 13.3-km/8.3-mi line in Japan from Tokyo to Haneda Airport.

Werner Abraham Gottlob 1749–1817 was the German geologist who developed the first influential models of Earth structure and history.

Born in Silesia into a family involved for generations in mines engineering, Werner studied between 1769 and 1771 at the Mining School at Freiberg, Saxony before proceeding to the University of Leipzig. An early product of his studies was *On the External Characteristics of Fossils* (1774) an examination of minerals in the light of their surface features. In 1775, he was appointed to the Freiberg Akademie, where he continued to teach for the rest of his long life. He proved easily the most influential instructor in the history of geology, most of the leading students of the next generation learning their science under him.

Though now judged largely erroneous, Werner's geology was of cardinal significance in its day for establishing a physically based stratigraphy, grounded on precise mineralogical knowledge. Werner proposed a general succession of the creation of rocks, beginning with Primary Rocks (precipitated from the water of a universal ocean), then passing through Transition, Flötz (sedimentary), and finally Recent and Volcanic. The oldest rocks had been chemically deposited; they were crystalline and without fossils. Later rocks had been mechanically deposited. Formed out of the denuded debris of the first creations, they were fossiliferous and superincumbent.

Werner's approach was particularly important for linking the order of the strata to the history of the Earth, and relating studies of mineralogy and strata. Its 'Neptunism' (that is, belief in water as the chief strata-forming agent) was attacked by the Huttonian 'Plutonists' who saw heat or fire as playing that role. Nineteenth-century geology drew on both Werner and Hutton to develop a balanced view of the significant geomorphological forces.

Werner Alfred 1866–1919 was a French-born Swiss chemist who founded the modern theory of coordination bonding in molecules (formerly inorganic coordination compounds were known by the generic term 'complexes'). For this achievement he was awarded the 1913 Nobel Prize for Chemistry.

Werner was born on 12 December 1866 in Mulhouse, Alsace, when it was part of France (four years later at the end of the Franco–Prussian War it became German territory). He was the son of a foundry worker and, despite his parents' French sympathies, received a German education. He began studying chemistry and experimenting on his own when he was 18 years old, and in 1885, while doing military service with the German army, attended lectures at the Karlsruhe Technische Hochschule. A year later he entered the Zurich Polytechnic, graduating with a diploma in chemistry in 1889. He spent 1890 with Pierre Berthelot in Paris, but returned to Zurich the following year, becoming a full professor in 1895. He had to cease work in 1915 because of severe arteriosclerosis, and he died in Zurich on 15 November 1919.

Although Werner is recognized mainly for his discoveries in inorganic chemistry, his first major success was in 1890 in organic chemistry. With his teacher Arthur Hantzsch (1857–1935) he described the structure and stereochemistry of oximes (organic compounds containing the =N–OH group, prepared by adding hydroxylamine, NH_2OH, to aldehydes or ketones). He showed that these compounds could exhibit geometrical isomerism in the same way as a compound containing a carbon–carbon double bond, and explained their structures by suggesting that the nitrogen bonds are directed in space tetrahedrally (extending the theories of Jacobus van't Hoff and Joseph Le Bel about the carbon atom).

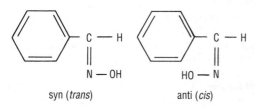

syn (*trans*)   anti (*cis*)

Werner *The two stereoisomers of benzaldoxime.*

Werner developed his theory about bonding in coordination compounds as part of his Habilitation thesis for obtaining a university teaching position. In addition to ionic and covalent bonds, Werner proposed the existence of a set of coordination bonds resulting from an attractive force from the centre of an atom acting uniformly in all directions. The number of groups or 'ligands' that can thus be bonded to the central atom depends on its coordination number and determines the structure (geometry) of the resulting molecules. Common coordination numbers are 4, 6 and 8. Neutral ligands (such as ammonia and water) leave the central atom's ionic charge unchanged; ionic ligands (such as chloride or cyanide ion) alter the central charge accordingly.

A typical Werner-type coordination compound is hexamminocobalt(III) chloride, $[Co(NH_3)_6]Cl_3$, in which the cobalt has a coordination number of 6; there are six coordinate bonds from it to the six ammonia molecules. The cobalt's valence (oxidation state) of + 3 is balanced by the three negatively-charged chloride ions. The six ammonias are located at the corners of a regular octahedron. Werner was able to demonstrate the theory experimentally by preparing geometrical isomers of compounds of the type diamminoplatinum chloride, which is square planar in shape.

More complicated molecules, particularly those having

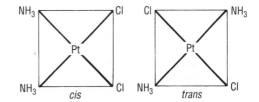

Werner *Stereoisomers of diamminoplatinum chloride, $Pt(NH_4)_2CL_2$, each with a coordination number of 4.*

as a ligand ethylene diamine (1,2-diaminoethene, abbreviated to 'en'), exhibit optical isomerism through a pair of mirror-imaged stereoisomers.

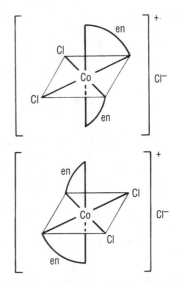

Werner *A pair of optically isomeric coordination compounds: mirror images of cis- $[Co (en)_2Cl_2]^+Cl^-$, each with a coordination number of 6.*

Ethylene diamine is bi-functional and acts as a bridge between a pair of coordinate bonds. Some ligands of this type, such as ethylene diamine tetra-acetic acid (EDTA) and its salts, are used to 'mop up' metal ions – for example, as antidotes for poisoning by heavy metals such as copper and lead.

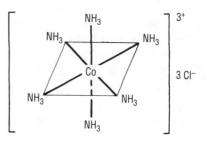

Werner *Hexamminocobalt chloride, $[Co(NH_3)_6] Cl_3$, with a coordination number of 6.*

Westinghouse George 1846–1914 was one of the fore-most engineer–industrialists in the United States at a time of rapid commercial expansion. His early fame came from his invention of a safe and speedy braking system for rail-way trains, but later his ideas and influence spread to gas and electricity distribution systems, and to the electrical industry generally.

Westinghouse was born on 6 October 1846 at Central Bridge, New York. He was the eighth of ten children, and his father was a manufacturer of agricultural implements. When he was 15 he ran away from school to fight for the North in the American Civil War. His parents soon brought him home again, but 18 months later allowed him to join the Union Army. In 1864 he joined the navy, but left the following year, having achieved the rank of acting third-assistant engineer. Westinghouse then spent a period of three months as a student at Union College, Schenectady, New York, after which he decided his place was in his father's workshop. In October of the same year (1865), he took out a patent for a railway steam locomotive – the first of his more than 400 patents. During the next four years he concentrated on inventing railway devices, the most impor-tant of which was his famous air brake system, patented in 1869.

Up to this time railway trains were slowed only by brakes on the locomotive, and then by manually applied brakes on each individual truck, or carriage, if necessary, Westinghouse's invention allowed the driver of the loco-motive to apply the brakes on all the cars simultaneously – a so-called continuous braking system. This allowed the train to brake smoothly and rapidly, and consequently it was safe for them to travel at much higher speeds. In the same year Westinghouse formed the Westinghouse Air Brake Company.

The increased efficiency of the railways was quickly realized. Westinghouse further helped both himself and the nation with his pioneer efforts to standardize railway com-ponents and systems, including the development of a completely new signalling system. This needed electricity and electrical components, and he went on to invent and manufacture devices in this area. As his business grew he formed more home and foreign companies to manufacture his inventions.

Gas was then a source of power to industry and required efficient distribution; Westinghouse developed a system of gas mains whereby the gas was fed into the network at high pressure but was at the required pressure when it was received by the consumer. His initial high-pressure mains were narrow diameter pipes which fed into wider pipes (with a consequent drop in pressure), which in turn fed into even wider pipes until the gas was at the correct pressure in the right place.

During the 1880s electricity as well as gas was impor-tant, and Westinghouse built a single-phase alternating current distribution system in Pittsburgh with French trans-formers and generators. He quickly realized its potential, and got his own engineers to design equipment suitable for a new high-tension (voltage) system. Westinghouse saw

that high tension distribution avoided much of the loss through electrical resistance, which limited moderate volt-age systems to a small distribution area, and enabled the building of large electrical networks. Alternating current (AC) was necessary because step-down transformers were needed to bring the voltage to a suitable level for use.

At about this time he secured the services of Nikola Tesla (1856–1943). He also bought the patents for the AC polyphase induction motor which made his AC system even more useful, because up to that time AC motors had to be started by rotating them at their running speed before the current was switched on. In 1893 the Westinghouse Electric Company lighted the world's first Columbia Expo-sition in Chicago. Two years later the same company harnessed the Niagara Falls to generate electricity for the lights and trams of the town of Buffalo, 35 km/22 mi away.

Many of the Westinghouse industries were based in Pittsburgh and nearby in Turtle Creek Valley, where in 1889 Westinghouse built a model town for his workers. During the period 1907 to 1908 a series of financial crises and takeovers caused him to lose control of the Westing-house Industries. He returned to active experimentation, designing a new steam turbine and reduction gear system and an air spring for use in motorcars. He also spent time reorganizing a large insurance society. From 1913 he suf-fered increasing ill health from heart disease, and he died in New York on the 12 March 1914.

Weyl Hermann 1885–1955 was a German mathematician and mathematical physicist whose range of research and interests was remarkably wide, and found expression also in published works on philosophy, logic and the history of mathematics. A one-time pupil of the renowned David Hilbert (1862–1943) and colleague of Albert Einstein (1879–1955), it was probably inevitable that Weyl is remembered chiefly for his studies on topological space and Riemannian geometry.

Weyl was born in Elmshorn on 9 November 1885. Attending the high school in Altona (now a suburb of Ham-burg), he evinced a strong interest in mathematics and in philosophy. At the age of 18 he entered Göttingen Univer-sity where, four years later, he received his doctorate for a dissertation on singular integral equations. Until 1913 he then became an unsalaried lecturer at the University, but at that time turned down a professorship in order to take up the Chair of Mathematics at the Technische Hochschule in Zurich. He held the post until 1930 – despite temporary conscription into the German army during World War I and a year as Visiting Professor at Princeton University in the United States in 1928. He then succeeded David Hilbert to the Chair of Mathematics at the University of Göttingen, but the unfavourable political climate of Nazi Germany prompted him to move to the Institute of Advanced Studies back at Princeton, where he then held a permanent post as Professor of Mathematics. After his retirement in 1951, Weyl divided his time between Princeton and Zurich. Dur-ing his lifetime Weyl's mathematical talents and con-tributions were recognized by the award of many honorary degrees and his election to the membership of prestigious scientific societies such as the National Academy of Sciences and the Royal Society. In addition to his many articles published in various journals, Weyl also wrote 15 books that were published both in the United States and Europe. He died in Zurich on 8 December 1955.

Weyl's early university days were strongly influenced by two great mathematicians in his department at Göttingen: David Hilbert and Herman Minkowski (1864–1909). Weyl was to develop interests in fields that included and went beyond those of these two important men.

The whole of Weyl's mathematical career is permeated by his abilities in analysis, but it was during his earliest years, from 1908 to around 1915, that these interests were dominant. He first examined the problems of singular eigenvalues for differential equations, and then turned his attention to oscillations in structures such as membranes and elastic bodies.

As a colleague of Albert Einstein during 1913, he became interested in the developing general theory of relativity and differential geometry. He lectured on both topics, and consolidated these lectures into a book he published in 1918 entitled *Raum-Zeit-Materie* ('space–time–matter'), which in five years ran to five editions. Weyl's interest in relativity theory took him to the point where he believed (erroneously) that he had found a way to a grand unification of gravitation and electromagnetism – something towards which Einstein was to strive for the rest of his life (and which is still being sought). Also in theoretical physics, Weyl was able to anticipate the non-conservation of parity, which has now been found to be characteristic of weak interactions between leptons (a class of subatomic particles which obeys Fermi–Dirac statistics).

Weyl's lectures on Riemann surfaces prompted him to write a book on the subject which was first published in 1913 (and then republished in 1955). His most important work in this field was the definition of the complex manifold of the first dimension, which has been important in all later work on the theory of both complex and of differential manifolds. This work demanded skills in topology, geometry and other areas quite different from those he had previously applied in his work in analysis.

The number of papers Weyl produced on the subject of number theory was small – but their impact was great. In 1916 he published one of his best papers, on the definition of the uniform distribution modulo 1, which was to be of great significance in the later work of Godfrey Hardy (1877–1947) and John Littlewood (1885–1977) on number theory. Other productive research areas during Weyl's fruitful tenure at the Technische Hochschule in Zurich included calculus, continuous groups and Lie groups. He applied his work on group theory to the analysis of the atom in a book on quantum mechanics he published in 1928.

During the rest of Weyl's career he returned to these subjects and explored them further. He published books on classic groups in 1939 and on number theory in 1940. His interest in the philosophy of mathematics led him to write

a book on this theme as well, in 1949. A mathematician of profound and diverse talents, Weyl made significant contributions to several fields of both mathematics and mathematical physics.

Wheatstone Charles 1802–1875 was a British physicist who is known principally for the Wheatstone bridge, a method of making accurate determination of electrical resistance. However, Wheatstone did not invent this method but popularized it. He in fact made several important but lesser-known original contributions to science, including the invention of the stereoscope and the rheostat, and the first determination of the velocity of electricity along a wire. He was also a pioneer of the telegraph, and he invented the concertina.

Wheatstone was born in Gloucester on 6 February 1802. His family were involved in the manufacture of musical instruments and he received no formal education in science. Wheatstone entered the music business as an apprentice to an uncle in 1816. He became an instrument maker, inventing the concertina in 1829, and also investigated the acoustic properties of instruments. Wheatstone's work in acoustics led to his appointment as Professor of Experimental Physics at King's College, London, in 1834, a position he retained for the rest of his life. His interests then extended to electricity and optics and in 1837, Wheatstone formed a partnership with William Cooke (1806–1879) to develop a commercial telegraph. This too continued for life and led to Wheatstone being knighted in 1868, Cooke receiving the same honour in the following year. Wheatstone died on a visit to Paris on 19 October 1875.

Wheatstone's early interest in acoustics was stimulated by his family's business, which was principally concerned with the manufacture of flutes. In 1827, he invented an ingenious device called the kaleidophone which visually demonstrated the vibration of sounding surfaces by causing an illuminated spot to vibrate and produce curves by the persistence of vision. Wheatstone went on to investigate the transmission of sound in instruments and subsequently discovered modes of vibration in air columns 1832 and vibrating plates 1833.

Wheatstone's contributions to light and optics are also important. In 1860, he demonstrated how the visual combination of two similar pictures in a stereoscope gives an illusion of three dimensions. However, work of greater significance came much earlier with a paper presented at the British Association meeting in Dublin in 1835 entitled *Prismatic Analysis of Electric Light*. There Wheatstone showed that different spectra are produced by spark discharges from metal electrodes, analysis of the spectra proving that different lines and colours are formed by different electrodes. He showed the sensitivity of the method and predicted correctly that with development, it would become a technique for the analysis of elements. Wheatstone said: 'We have a mode of discriminating metallic bodies more readily than by chemical examination and which may hereafter be employed for useful purposes.' Many years were to pass before the classic development of spectroscopy by Robert Bunsen (1811–1899) and Gustav Kirchhoff (1824–1887) in 1860.

In 1848, Wheatstone described a polar clock, which was an instrument for telling the time based on the change in plane of polarization of sunlight in the direction of the Pole. With this clock, it was possible to tell the hour of the day by the light from the sky though the Sun might be obscured.

In 1834, Wheatstone began a series of important experiments with Cooke on the transmission of electricity along wires. The early work involved the pursuit of pure scientific knowledge, such as the determination of the velocity of electricity related below, but later work was directed towards the public transmission of messages, that is, commercial telegraphy. A series of successes was achieved, the first being a patent in 1837 for the five-needle telegraph. This was an instrument in which five magnetic needles on a panel were deflected by electric signals to various positions that indicated letters of the alphabet, thus transmitting written messages. This device was developed by Wheatstone and Cooke, culminating in a portable two-needle telegraph in 1845 and a single-needle telegraph the following year. These instruments were direct ancestors of the teleprinter.

In the field of electricity, Wheatstone made several important contributions. In 1834, he made the first determination of the velocity of electricity through a wire by using a rotating mirror to determine the delay that occurred between sparks as a current produced several spark gaps in a long loop of wire. The result was very high – 30% greater than the speed of light – possibly because the wire was not straight, but the method was taken up and used by Armand Fizeau (1819–1896) in 1849 to make the first accurate determination of the velocity of light.

Wheatstone also improved on early versions of the dynamo invented by Michael Faraday (1791–1867) by combining several armatures on one shaft so that current was generated continuously; previously only intermittent generation was possible. He also recognized the theoretical and practical importance of Ohm's law, developing the rheostat (variable resistance) and Wheatstone bridge method for the accurate determination of resistance. This important method was published in the *Transactions of the Royal Society* in 1843, but it had been invented ten years earlier by Samuel Christie (1784–1865). It involves placing an unknown resistance in a simple circuit with three known resistances and varying one of them until no current flows through a galvanometer connected like a bridge to all four resistances. The unknown resistance can then be determined from the values of the other three resistances.

It is ironic that Wheatstone is mainly remembered for a device that he himself did not invent, and this must not blind us to the extraordinary range of his talents and great value of his original work in telegraphy, electricity, optics and acoustics.

Whipple Fred Lawrence 1906–   is a US astronomer known for his discoveries of comets and his contribution to the understanding of meteorites.

Whipple was born in Red Oak, Iowa, on 5 November 1906. He was educated in California and received his bachelor's degree from the University of California at Los Angeles in 1927; he went on to become a teaching Fellow at the University of California at Berkeley. He spent some time at the Lick Observatory, and received his PhD in 1931. In the same year he was appointed to the staff of the Harvard College Observatory and a year later he became an Instructor at Harvard. He was gradually promoted to posts of increasing responsibility until, in 1950, he was made Professor, a post he held until 1977. During World War II, Whipple carried out radar research at the Office of Scientific Research and Development, and in 1948 he was awarded the Presidential Certificate of Merit for his work there. He also held a number of advisory posts on scientific bodies such as the Rocket and Satellite Research Panel. In 1955 he became Director of the Smithsonian Institution Astrophysics Observatory, a post he retained until 1973, when he became the Senior Scientist at the Observatory. He was awarded the Donahue Medal six times and in 1971 he received the Kepler Medal of the American Association for the Advancement of Science.

In addition to discovering six new comets, Whipple contributed to the understanding of the constitution and behaviour of comets. In 1949, he produced the 'icy comet' model. He proposed that the nucleus of a comet consisted of a frozen mass of water, ammonia, methane and other hydrogen compounds and that embedded within it was a quantity of silicates, dust and other materials. As the comet's orbit brought it nearer to the Sun, solar radiation would cause the frozen material to evaporate, thus producing a large amount of silicate dust which would form the comet's tail. This explanation has found general acceptance among astronomers.

Whipple has also worked on ascertaining cometary orbits and defining the relationship between comets and meteors. His other interests include the evolution of the Solar System, stellar evolution and planetary nebulae. He actively participated in the organization of the International Geophysical Year in 1957–58 and since the 1950s he has been extremely active in the programme to devise effective means of tracking artificial satellites.

White Gilbert 1720–1793 was a British naturalist and clergyman who is remembered chiefly for his book *The Natural History and Antiquities of Selborne* (1789), a classic work in which White vividly records his acute observations of the flora and fauna in the area of Selborne (now in Hampshire).

White was born on 18 July 1720 in Selborne and, after attending schools in Farnham and Basingstoke, went to Oriel College, Oxford, in 1740. He graduated in 1743, was elected a Fellow of his college in 1744 and obtained a master's degree in 1746. Although many of his associates believed him capable of a successful academic career, White chose to become a clergyman and took deacon's orders in 1747, subsequently becoming a curate in his uncle's parish of Swarraton, near Selborne. White was ordained a priest in 1750 and, after refusing several positions, eventually accepted a post at Moreton Pinkney, Northamptonshire, where his living was paid by Oriel College. But he was strongly attracted to the countryside of his birthplace, so he left the care of the Moreton Pinkney parish to a curate and returned to the family home in Selborne. White subsequently held curacies in several parishes in the neighbourhood but continued to live in Selborne itself until his death on 26 June 1793.

After his return to Selborne, White spent much of his time studying the wildlife in the area. He kept a diary of his observations and also wrote of his findings to his two naturalist friends Thomas Pennant and Daines Barrington. These letters, which cover a period of about 20 years, form the basis of *The Natural History of Selborne*. Elegantly written, it is characterized by acute observations of a wide variety of natural history subjects, as well as descriptions of rural life in eighteenth-century England. The book is more than merely a record, however, it also contains White's theories and speculations and several important discoveries, such as the migration of swallows, the recognition of three distinct species of British leaf warblers, and the identification of the harvest mouse and the noctule bat as British species. On its publication in 1789, *The Natural History of Selborne* was widely praised by leading naturalists; it also has great popular appeal and has been reprinted many times.

White also wrote *Calendar of Flora and the Garden* (1765), an account of observations he made in his garden in 1751, and *Naturalist's Journal* (begun in 1768), a similar but more sophisticated work.

Whitehead Alfred North 1861–1947 was a British mathematician and philosopher whose research in mathematics involved a highly original attempt – incorporating the principles of logic – to create an extension of ordinary algebra to universal algebra, a meticulous re-examination of the relativity theory of Albert Einstein (1879–1955) and, in work carried out together with his friend and collaborator Bertrand Russell (1872–1970), the production of what is probably one of the most famous mathematics books of the century, *Principia Mathematica*. This book endeavoured to show that logic could be regarded as the basis of mathematics.

Whitehead was born in Ramsgate, Kent, on 15 February 1861, son of a school headmaster who, when Whitehead was aged six, became the vicar of a large parish near Ramsgate. From 1875 to 1880 the boy went to Sherborne School; he then went to Trinity College, Cambridge. Receiving his doctorate in 1884, he was elected to a Fellowship of the College, and was appointed lecturer. One of his pupils there was Bertrand Russell; the two struck up a friendship and together attended the 1900 International Congress of Philosophy in Paris. The result of this excursion was the publication of their joint work, *Principia Mathematica*, in 1910. In that year, however, Whitehead left Cambridge and held a number of teaching positions at

University College, London, before being appointed Professor of Applied Mathematics at the Imperial College in 1914. There he produced his book on relativity theory. It gave a thorough, but obscure, analysis and projected further development of the subject. But his interests were already turning more and more to actual philosophy, and the book is a complicated mixture of mathematics, physics and philosophy. In 1924 – at the age of 63 – he was invited to become Professor of Philosophy at Harvard University in New York. During his subsequent years in the United States he produced many fine contributions to philosophy but very little in the way of mathematics. At Harvard he still considered himself an Englishman, and his services to his home country were recognized in the presentation to him of the country's highest award, the Order of Merit, in 1945. A Fellow of the Royal Society since 1903, Whitehead died at Harvard on 30 December 1947.

Even in his mathematical studies, Whitehead preferred to think of the philosophical aspects; he was convinced that to regard the various branches of mathematics in a theoretical way, as manifestations of human intellectuality, was far preferable to considering them as real, concrete structures, as was the prevailing fashion at the turn of the century. It was from this viewpoint that he produced his first major work *A Treatise of Universal Algebra* (published in 1898), an attempt at a total expansion of algebraic principles that also demonstrated his growing interest in philosophy itself.

At the International Congress of Philosophy in 1900, Whitehead and Russell heard Giuseppe Peano (1858–1932) describe the rigorously axiomatic method by which he had arrived at the axioms concerning the natural numbers – for which, despite the acknowledged prior work by Richard Dedekind (1831–1916), Peano is now famous. He was using logical methods to derive mathematical instruments. It occurred, apparently first to Bertrand Russell, that similar methods to those of Peano could be used to deduce mathematics from logic in a general and fundamental way. The collaboration of the two on the project lasted ten years, and culminated in the publication of the very complicated three-volume treatise *Principia Mathematica*. The book did not quite accomplish its objectives – in fact, in some ways, some of it was nullified by events of its own instigation – but it did have a momentous influence on mathematical thought about the foundations of the subject. It had a considerable effect on many individual branches, from the development of Boolean algebra to Kurt Gödel's work in which he demonstrated – devastatingly – that arithmetic, and hence mathematics, can never be proved to be consistent.

There was to have been a fourth volume, by Whitehead alone, devoted entirely to geometry, but it was never completed.

Whitehead John Henry Constantine 1904–1960 was a British mathematician who achieved eminence in the more abstract areas of diffential geometry, and of algebraic and geometrical topology.

Whitehead was born on 11 November 1904 in Madras, India, where his father was the Anglican Bishop; he was the nephew of Alfred North Whitehead. Educated first in Oxford, then at Eton, he entered Balliol College, Oxford, and gained a BA degree in mathematics. After graduating he joined a firm of stockbrokers, but left after 18 months to return to Oxford. For his mathematical researches he was in 1929 awarded a scholarship to the University of Princeton, New Jersey, where he met and worked with Oswald Veblen (1880–1960). Four years later he returned again to Oxford and was elected Fellow of Balliol College; he took up the Wayneflete Chair of Pure Mathematics there in 1947. On a visit to Princeton in 1960, he suffered a heart attack and died in May of that year.

Whitehead's early research was on differential geometry, and the application of differential calculus and differential equations to the study of geometrical figures. It was the results of his years of study in these fields that he took with him to Princeton, to meet a man he had already known by reputation for some considerable time, Oswald Veblen. Yet Veblen's primary interest at the time was in topology – the study of shapes and figures that retain their essential properties despite being 'stretched' or 'squeezed'. Nevertheless, each mathematician influenced the other. Veblen learned so much about differential geometry that he and Whitehead wrote a textbook together on the subject: *Foundations of Differential Geometry* was published in 1932.

Whitehead, in turn, learned a great deal about topology. Combining his interests thereafter, much of his mathematics became very complex and advanced. Some of his most significant work, however, was in the study of knots. In geometry a knot is a two-dimensional representation of a three-dimensional curve that because of its dimensional reduction (by topological distortion) appears to have nodes (to loop onto itself).

Whitehead Robert 1823–1905 was a British engineer, best known for his invention of the self-propelled torpedo.

Whitehead was born on 3 January 1823 in Bolton, Lancashire, one of the eight children of the owner of a cotton-bleaching business. He was educated at the local grammar school and at the age of 14 was apprenticed to a Manchester engineering company: Richard Ormond and Son. His uncle, William Smith, was the manager of the works, and Whitehead received a thorough grounding in practical engineering. He also trained in draughtsmanship by studying at evening classes at the Mechanics Institute in Manchester. His uncle became manager of the works of Philip Taylor and Sons of Marseilles, and in 1844 Whitehead joined him there. Three years later he set up in business on his own in Milan, designing machinery.

Whitehead took out various patents under the Austrian Government, but these were annulled by the revolutionary government of 1848. He moved to Trieste and worked for the Austrian Lloyd Company for two years. From 1850 to 1856 he was manager there of the Studholt works, and then, at the neighbouring naval port of Fiume, he began working at the Stabilimento Tecnico Fiumano. It was there that in 1866 he invented the torpedo.

In 1872, with his son-in-law George Hoyos, Whitehead bought the Fiume establishment, devoting the works to the construction of torpedoes and accessory equipment. His son John became the third partner. In 1890, a branch was founded at Portland Harbour, England, under Captain Payne-Gallway and in 1898 the original works at Fiume were rebuilt on a larger scale.

Whitehead was presented with a diamond ring by the Austrian Empire for having designed and built the engines of the ironclad warship *Ferdinand Max* which rammed the *Re d'Italia* at the Battle of Lissa. In May 1868 he was decorated with the Austrian Order of Francis Joseph in recognition of the excellence of his engineering exhibits at the Paris Exhibition of 1867. He also received honours and awards from many other European countries. In his later years he farmed a large estate at Worth, Sussex. He died on 14 November 1905 near Shriveham, Berkshire, and was buried at Worth.

While in business in Milan (1847) Whitehead designed pumps for draining part of the Lombardy marshes and made improvements to silkweaving looms. From 1856, in Fiume, he built naval marine engines (for Austria) and in 1864 he was invited to cooperate in perfecting a 'fireship' or floating torpedo designed by Captain Lupins of the Australian Navy. But in secret, with his son John and a mechanic, he carried out a series of experiments which led in 1866 to the invention of the Whitehead torpedo. It travelled at 7 knots for up to 630 m/2,070 ft, but had difficulty in maintaining a uniform depth. Within two years he had remedied this defect by an ingenious but simple device called a balance chamber, which for many years was guarded as the torpedo's 'secret'. A typical torpedo of this time was about 4 m/13 ft long, weighed 150 kg/330 lb (including a 9-kg/20-lb dynamite warhead) and was powered by a compressed-air motor driving a simple propellor.

Nonexclusive manufacturing rights were bought by the Austrian government in 1870, by the British in 1871, the French in 1872 and by Germany and Italy in 1873; by 1900 the right to build the torpedoes had been purchased by almost every country in Europe, the United States, China, Japan and some of the South American republics.

In 1876 Whitehead developed a servo-motor, which controlled the steering gear and gave the torpedo a truer path through the water. Speed and range were gradually improved, so that by 1889 the weapons could maintain 29 knots for 900 m/3,000 ft. He devised methods of accurately firing torpedoes either above or below water from the fastest ships no matter what the speed or bearing of the target. The weapon was finally perfected in 1896 by the addition of a gyroscope, invented by an Austrian naval engineer, which was coupled to the servo and steering mechanism. Any doubts about the torpedo's usefulness were dispelled when, on 9 February 1904, outside Port Arthur, a few Japanese destroyers armed with torpedoes reduced the Russian battlefleet to impotence.

Whitney Eli 1765–1825 was a US lawyer who invented the cotton gin, a machine for plucking cotton fibres off the seeds on which they grow.

Whitney was born on 8 December 1765 at Westboro, Massachusetts, the son of a farmer. His exceptional mechanical ability was evident at an early age, and when he was 15 he began manufacturing nails in a small metal-working shop on the family farm. Then in 1789 he went to Yale University to study law. After graduating three years later he moved to Savannah, Georgio, intending to become a tutor, but he continued to make various mechanical contrivances for domestic use. He became acquainted with Mrs Nathaniel Green, the widow of the Revolutionary General, who pointed out to Whitney the local need for a machine for picking cotton. He devised such a machine, the cotton gin, and in May 1793 formed a partnership with Phineas Millar to manufacture the invention. The cotton gin was patented in March the following year. It was widely copied, and although the courts vindicated his patent rights in 1807, five years later Congress refused to extend the patent. Meanwhile, in January 1798, Whitney had secured a government order to make 10,000 guns. It took him eight years to complete the two-year contract, but even so he received a second order in 1812 and his manufacturing methods were later adopted by both the Federal armouries. Whitney died in New Haven, Connecticut, on 8 January 1825.

Even at Yale, Whitney earned pocket money by making and mending mechanical bits and pieces. In Georgia, Mrs Green introduced him to cotton growers, who realized that a mechanical picker could expand the cotton industry and bring wealth to the Southern States. Whitney's machine had metal 'fingers' that separated the cotton from the seeds, and he called it a cotton gin ('gin' being short for engine). It consisted of a wooden cylinder bearing rows of slender spikes set 1.3 cm/0.5 in apart, which extended between the bars of a grid set so closely together that only the cotton lint (and not the seeds) could pass through. A revolving brush cleaned the cotton off the spikes and the seeds fell into another compartment. The machine was hand-cranked, and one gin could produce about 23 kg/50 lb of cleaned cotton per day – a fifty-fold increase in a worker's output.

The introduction of a machine to clean cotton led to the expansion of the plantations and the demand for more labour, which in turn led to the increasing use of slaves – indeed the cotton gin has been cited as a contributory factor to the American Civil War. Courts in South Carolina awarded Whitney and Millar a $50,000 grant, with which they built a factory at New Haven, Connecticut. But the gin was so easy to copy and simple to manufacture that county blacksmiths constructed their own machines. Whitney spent so much of his money in legislation to defend his patent, that he eventually gave up the struggle and in 1798 turned to the manufacture of firearms.

In his arms factory, also in New Haven, Whitney used skilled craftsmen and machine tools to make arms with fully interchangeable parts. It is said that he once took a batch of unassembled rifles and threw then at a government official's feet, inviting him to pick out parts at random and build up a working firearm. This time Whitney made a fortune and kept it.

In 1818, he made a small milling machine, with a power-driven table that moved horizontally beneath and at right angles to a rotating cutter. This device is all that remains today of the machinery with which he launched what became known in Europe as the American system of manufacture. He also introduced division of labour in his musket factory, the beginnings of a production line and mass production.

Whittle Frank 1907– is a British engineer and inventor of the jet engine. The developments from his original designs were used first in the Gloster Meteor fighter at the end of World War II. Direct descendents of these engines are now the sources of power for all kinds of military and civil aircraft.

Whittle was born in Coventry on 1 June 1907. When he was ten the family moved to Leamington Spa where he attended secondary school. His aptitude and interest in engineering and invention was encouraged and often demanded by his father, who was himself a designer and craftsman of sufficient skill to run his own workshop. The workshop stimulated Whittle's interests further than his father's ideas. He became interested in aeronautics from a practical as well as a scientific point of view, and at about the same time he developed a preoccupation with the idea of finding a source of power superior to that of the piston engine. On leaving school he joined the Royal Air Force as an apprentice. His ability was such that he later entered the RAF College, Cranwell as a cadet, and trained as a fighter pilot. During the training, which included all aspects of the theory of flight and motive power for aircraft, he was able to further his ideas for an improved aero-engine and impress his instructors with the high standard of his flying.

In about 1928, the idea of using a jet of hot gas to cause motion according to Newton's third law (every action produces an equal and opposite reaction) seemed promising. He realized that a gas turbine could be incorporated to drive a compressor to compress the air entering an engine and the air would rapidly expand when fuel was ignited in it, causing a thrust of exhaust gases and hot air to push the aeroplane through the sky. The faster the turbine turned, the more the air was compressed and the greater the expansion and combustion, and thus increased thrust was obtained.

This idea was beautiful and simple; perhaps too simple since it took Whittle a long time, and considerable frustration, to convince the Air Ministry that the jet was a source of power with greater potential than the piston engines, which were still in the process of considerable development to new heights of performance. After initially being turned down, Whittle's idea sank into oblivion until about 1935 when, with some support, encouragement and financial backing, he formed the Power Jets Company. The RAF now gave him support and a partial release from his duties. In 1936, he began experiments with his engine at British Thomson-Houston, Rugby. His first engine looked like something between an ancient gramophone and an old vacuum cleaner, but soon his designs took a form that would be recognizable today as a jet engine. His intention was to create a power plant that would propel a small aeroplane at about 800 kph/500 mph. These early assemblies vibrated alarmingly and emitted a terrifying noise.

The prospect of war in 1939 finally prompted the Air Ministry to encourage Whittle's invention by giving Power Jets a contract to power an air frame designed and constructed by Gloster Aircraft. These developments culminated in the maiden flight of the experimental Gloster E28/39 in May 1941. By the end of 1944 the twin-engined Gloster Meteor, the production model developed from the E28/39, was coming into service. Although it was the first Allied jet aeroplane, the Germans had produced the Messerschmitt Me 262 slightly earlier.

In the years following the end of the war, the jet engine made possible the new generation of supersonic fighters and eventually the very high-speed civil air travel that is now commonplace. The power and thrust of jets have increased by very large factors since Whittle's original designs, but his ideas, developments and innovations have influenced the invention's whole development to a degree which is very rare for original and early ideas.

For his inventions, Whittle received considerable financial compensation and a knighthood, but eventually Power Jets was removed from his grasp. In 1918 he retired from the RAF with the relatively senior rank of Air Commodore. He went to the United States, where he took up a university appointment. The famous prototype jet aeroplane, the Gloster E28/39, can still be seen hanging from the ceiling on the top floor of the Science Museum, London.

Whitworth Joseph 1803–1887 was a British engineer who established new standards of accuracy in the production of machine tools and precision measuring instruments. He devised standard gauges and screw threads, and introduced new methods of making gun barrels.

Whitworth was born in Stockport, Cheshire on 20 December 1803, the son of a schoolmaster. He was educated first at his father's school and then, after the age of 12, at Idle, near Leeds. When he was 14 he went to work in his uncle's cotton mill in Derbyshire. He was so fascinated by the mill machinery that, without permission, he left his uncle's employment and took a job as a mechanic with a firm that made machinery. When he was 22 (and newly married) he went to work for Henry Maudslay in his London workshops. In 1833, Whitworth moved to Manchester and set up in business as a toolmaker in a rented room which had access to power from a stationary steam engine. From this small beginning the large Whitworth works developed.

At that time it was the usual practice to build every machine and each machine part separately. Parts were not interchangeable and screw threads (particular-dimensions) were often used by only one firm or workshop. Whitworth, recognizing the advantages, brought standardization to his company and the engineering industry as a whole by

developing means of measuring to tolerances never before possible, so that shafts, bearings and gears could be interchanged. The well-known Whitworth standard for screw threads was part of this process of modernization.

The Whitworth company produced many machines which were needed to work to his new standards and to cope with the new methods of production. There were machines for cutting, shaping and gear-cutting, but of particular importance for the rapid production of high-quality goods was the planning machinery, which substantially reduced the time spent by skilled workmen on routine tasks. Whitworth also designed a knitting machine and a horsedrawn mechanical roadsweeper.

During the later part of the nineteenth century steel became available on a large scale from the Bessemer and Siemens open-hearth processes. At the Whitworth works, steel ingots of hexagonal cross-section were used to make gun barrels, axles, propellor-shafts and the like. Whitworth cast these under pressure applied by a hydraulic press so that imperfections caused by bubbles of air or variable cooling rates were minimized. Guns of all sizes were produced, and Whitworth supervised many experiments to investigate the forces acting on the breech and barrel of a gun and the amount of barrel wear (which affected the accuracy of the weapon). He also made advances in the design of rifling for the barrels of small-calibre weapons.

Whitworth was also concerned with the training, lives and leisure time of his workers. He created 30 Whitworth scholarships for university engineering students. He donated large sums of money to the various Manchester colleges (now Manchester University) and other educational organizations. He received many honours, including a knighthood, Fellowship of the Royal Society, and admission to the Légion d'Honneur. Whitworth died after a long illness on 22 January 1887 while on holiday in Monte Carlo.

Wieland Heinrich Otto 1877–1957 was a German organic chemist, particularly noted for his work on determining the structures of steroids and related compounds. He also studied other natural compounds, such as alkaloids and pterins, and contributed to the investigation of biological oxidation. For his work on steroids he was awarded the 1927 Nobel Prize for Chemistry.

Wieland was born in Pforzheim on 4 June 1877, the son of a chemist, and educated at the local grammar school. Beginning in 1896 he attended several universities: Munich, Berlin, Stuttgart and finally Munich again under Friedrich Thiele. He obtained his PhD in 1901 and spent most of his career in the chemistry faculty of the University of Munich. He became a lecturer there in 1904, and Senior Lecturer in Organic Chemistry in 1913. During World War I, in 1917, he moved to the Technische Hochschule (also in Munich) and almost immediately took a year's leave of absence to work on chemical warfare research under Fritz Haber at the Kaiser Wilhelm Institute of Chemistry at Berlin-Dahlem. After the war Wieland returned to his post as professor at the Technische Hochschule, moving to the

University of Freiburg in 1921. He finally returned to Munich in 1925 as Professor of Chemistry and Director of the Baeyer Laboratory as successor to Richard Willstätter. He remained there until he retired in 1950, when he was made Emeritus Professor. He died in Munich on 5 August 1957.

Most of Wieland's early work was concerned with organic nitrogen compounds. He investigated the addition of nitrogen oxides to double bonds in compounds such as terpenes (proposing the existence of nitrogen–nitrogen bonds in dimeric nitrogen oxides). In 1909, he published a method for, and described the mechanism whereby, fulminic acid could be prepared from ethyl alcohol (ethanol), nitric acid and mercury. Two years later he gave the first demonstration of the existence of nitrogen free radicals by oxidizing diphenylamine to form tetraarylhydrazine, and decomposing it by heating it in toluene.

After World War I Wieland's interest turned to the chemistry of biologically important compounds. He had begun studying the bile acids as early as 1912, showing that the three newly discovered acids have similar structures related to that of cholesterol (at that time being investigated by Adolf Windaus at Göttingen). The importance of steroids became even more apparent with the realization that vitamin D and the gonadotrophic hormones also belong to this class of compound. Using classical chemical methods Wieland later worked out what he thought was the basic skeleton of a steroid molecule (for which he was awarded the Nobel prize), but it was found to be incorrect. In 1932 he collaborated with O. Rosenheim (who used X-ray analysis) and H. King to produce the somewhat modified structure which is still accepted today.

Wieland did other work with the bile acids, demonstrating their role in converting fats into water-soluble cholic acids (a key process in digestion). He was the first to prepare the carcinogen methylcholanthrene, and went on to study the poisons produced in the skins of some species of toads, which are chemically similar to the bile salts. He also determined the structures of, and synthesized many, toadstool poisons, such as phalloidine from the deadly *Amanita* fungus. This led him to an investigation of alkaloids from both plant and animal sources. He isolated and determined the structures of the *Lobelia* alkaloids, and made incomplete studies of the structures of strychnine and curare. At the suggestion of one of his students he began research into the composition and synthesis of pterins, the pigments that give the colour to butterflies' wings (and one of which is the precursor of the essential human dietary factor folic acid).

Also after World War I Wieland began work on biological oxidation – the process within living tissues by which food substances such as glucose are converted to carbon dioxide and water with the liberation of energy for metabolism. He held the view that the oxidation was in fact a catalytic dehydrogenation. Using palladium as a catalyst in the absence of oxygen he was able to prove experimentally that this was the case; he also experimented with anaerobic microbial systems. This proposal was in direct opposition to the findings of Otto Warburg (1873–1970), who had shown that biological oxidation was an addition of oxygen moderated by iron-containing enzymes. The controversy sparked a long and lively debate, and stimulated a great deal of research. In the end both workers were shown to be correct; both catalytic dehydrogenation and oxidation steps do occur in the complex biochemical pathway of energy production in tissues.

Wiener Norbert 1894–1964 was a US statistician whose main interest lay in devising the means to describe continuously changing conditions and phenomena. His work in a number of fields involving such random processes led him to develop, and later to popularize, the theory of cybernetics, and to contribute fundamentally towards an understanding of the concept of decision-making.

Wiener was born in Columbia, Missouri, on 26 November 1894. A child prodigy, he was put under pressure by his father to excel academically. He was reading fluently at the age of three, entered high school when only nine, and completed the four-year high school course in two years. He went on to take his bachelor's degree at Tuft's College at the age of 14. The following year, Wiener began postgraduate studies at Harvard University, intending to concentrate on zoology; he transferred to Cornell University in 1910, but returned to Harvard in 1911, having finally chosen to specialize in the philosophy of mathematics. He earned his master's degree in 1912, and his PhD a year later. A travelling Fellowship then enabled Wiener in 1913 to visit England and go to Cambridge University, where he studied logic under Bertrand Russell (1872–1970), and afterwards to go to the University of Göttingen, Germany, to work with David Hilbert (1862–1943).

On his return to the United States, Wiener took up successive teaching posts at Columbia, Harvard and Maine Universities; worked for a year as a staff writer for the *Encyclopedia Americana* in Albany, New York; spent another year working as a journalist for the *Boston Herald*; and then took up his first appointment at the Massachusetts Institute of Technology (MIT). He worked first as an Instructor in the mathematics department (1919–24), then as an Assistant Professor (1924–28), as Associate Professor (1928–32), and finally as full Professor from 1932 until his retirement in 1960 as Emeritus Professor.

Many professional awards were bestowed on Wiener in recognition of the value of his contribution to mathematics; a member of prestigious international mathematical and scientific organizations, he was also the author of both academic and popular texts. He died in Stockholm, Sweden, on 18 March 1964.

Wiener's commitment to pure mathematics developed somewhat slowly, and it was only in 1918 (at the age of 24) that he began to take a serious interest in integral and differential equations. From the beginning, he worked in areas that had some application to physical processes. His results on the Lebesgue integral are important in the study of wave mechanics. He produced a mathematical theory for Brownian motion (1920), and the enthusiasm of the engineering department at MIT spurred him then to examine problems in harmonic analysis. This led him to make certain deductions about information flow along a wave.

Newtonian analytical methods are not amenable to the investigation of continuously changing processes, so Wiener devoted much of his efforts to methodology, developing mathematical approaches that could usefully be applied to such phenomena. During the 1930s he carried out some work in collaboration with others on Fourier transformations, on Tauberian theorems, on radiation equilibrium, and on the application of mathematics to the study of physiology.

During World War II, Wiener worked on the control of anti-aircraft guns (which required him to consider factors such as the machinery itself, the gunner, and the unpredictable evasive action on the part of the target's pilot), on filtering 'noise' from useful information for radar, and on coding and decoding. His investigations stimulated his interest in information transfer and processes such as information feedback. He related the occurrence of these processes in, for instance, firing of anti-aircraft weapons and in mental processes.

The statistically random (stochastic) components of this process made it familiar material for Wiener. He published a book in 1948 about his ideas on the communication of information and its control. He thereby established and named a new branch of science: cybernetics, the theory of which involves a mathematical description of the flow of information. It is applied to many areas outside mathematics, including neurophysiology, computer design and biochemical regulation. Wiener's book had an immediate impact; the common usage of such terms as 'input', 'feedback' and 'output' in everyday speech is, to a large measure, due to him.

A revised edition of the book, *Cybernetics*, was published in 1961. But soon after it was first produced, Wiener stopped doing research and decided to devote the rest of his life to awakening world leaders to the inevitable prospect of automation in many spheres of ordinary life.

Wigglesworth Vincent Brian 1899–1994 was a British entomologist whose research covered many areas of insect physiology but who is best known for his investigations into the role of hormones in growth and metamorphosis. He received many honours for his contributions towards an understanding of insect physiology, including the Royal Medal of the Royal Society of London in 1955 and a knighthood in 1964.

Wigglesworth was born on 17 April 1899 in Kirkham, Lancashire, the son of a doctor. He was educated at Repton, then won a scholarship to Gonville and Caius College, Cambridge. From 1917 to 1918 he served in France in the Royal Field Artillery before resuming his university education and graduating in physiology and biochemistry. He did two years' research under Frederick Gowland Hopkins, during part of which time he worked with J.B.S. Haldane, and then qualified in medicine at St Thomas's Hospital, London. In 1926, he was appointed Lecturer in Medical

Entomology at the London School of Hygiene and Tropical Medicine, and then Reader in Entomology at London University from 1936 to 1944. Wigglesworth became Director of the Agricultural Research Council Unit of Insect Physiology at Cambridge in 1943 and remained there until he retired in 1967. During his directorship, he also held the post of Quick Professor of Biology at Cambridge from 1952 to 1966. He died on 12 February 1994.

Wigglesworth's work on insect metamorphosis was carried out mainly on the bloodsucking insect *Rhodnius prolixus*, which was brought from South America by E. Brumpt (1877–1951) and proved suitable for experimentation. In 1917 it had been demonstrated that the hormone responsible for growth and moulting is secreted only when the insect's brain is present; decapitated insects live but do not moult. By transplanting various parts of the brain into decapitated *Rhodnius* specimens, Wigglesworth proved that this hormone (which he called moulting hormone) is produced in the region of the brain containing the neurosecretory cells. In addition, he showed that another hormone – one that prevents the development of adult characteristics until the insect larva is fully grown – is also produced in the head. He then demonstrated that this second hormone (which he called juvenile hormone) is secreted by the *corpus allatum*, an endocrine gland near the brain. Wigglesworth investigated the effects of the two hormones and found that insect larvae exposed to juvenile hormone grow but remain in the larval form; that adult insects exposed to moulting hormone moult again and, when also exposed to juvenile hormone, some of their organs partly regress to the larval forms. He also found that juvenile hormone is necessary for normal reproduction in many insects.

Although best known for his work on insect growth and metamorphosis, Wigglesworth also investigated many other aspects of insect anatomy and physiology, including the mechanisms involved in hatching; the mode of action of adhesive organs in walking; the role of the outer waxy layer on insects' bodies in preventing water loss; the respiration of insect eggs; insect sense organs and their use in orientation; and the functions of insect blood cells. His book *The Principles of Insect Physiology* (1939) has been reprinted in several editions as the standard general text on insect physiology.

Wilcox Stephen 1830–1893 was a US inventor who, with Herman Babcock, designed a steam-tube boiler which was developed into one of the most efficient sources of high-pressure steam, which remained so from the latter part of the nineteenth century until the demise of steam engines in the present century.

Wilcox was born on 12 February 1830 in Westerley, Rhode Island, the son of a prosperous banker and anti-slavery campaigner. He went to a local school and developed an interest in mechanisms in his spare time. After leaving school he went to work on improving old machines and inventing new ones – such as a hot-air engine for operating fog signals.

In about 1856, with his first partner D. M. Stillman, Wilcox patented a steam boiler in which slightly bent water tubes were set an angle in the firebox. It was not entirely successful, largely because of difficulties in making joints that were water- and steam-tight at the high pressures involved. Ten years later, with a new partner and boyhood friend Herman Babcock, he designed an improved safety water-tube boiler.

The new boiler had straight tubes, although they were still inclined to the horizontal. Banks of tubes were connected together at their ends, through which the hot water gradually rose by convection. The firebox surrounded the tubes to give rapid heating, and there was a reservoir of hot water above the firebox and tubes, with steam above the water. A patent was granted in 1867 and Babcock and Wilcox formed a company to manufacture the boiler. Their steam engines were used in the first American electricity generating stations and played an important part in the subsequent development of electric lighting.

Wilcox continued research into steam engines and boilers for the rest of his life. During his later years he worked on a marine version of the boiler with his assistant and nephew William Hoxie. Much of this work was carried out using his yacht *Reverie* for sea trials. Wilcox died in Brooklyn, New York, on 27 November 1893.

Throughout his inventing career Wilcox acquired nearly 50 patents and accrued a considerable fortune. He endowed a public library at Westerley and, after his death, his widow used the money to aid the building of parks and schools.

Wilkes Maurice Vincent 1913– is a British mathematician who led the team at Cambridge University which built the EDSAC (Electronic Delay Storage Automatic Calculator), one of the earliest of the British electronic computers.

Wilkes was born on 26 June 1913. He attended King Edward's School, Stourbridge, and then went on to St John's College, Cambridge. He graduated in mathematics and then carried out research in physics at the Cavendish Laboratories. After a short period as a university demonstrator, he became involved with war work in the development of radar and operational research. At the end of World War II he returned to Cambridge, first as a lecturer and Acting Director of the Mathematical Laboratory. The following year he was appointed Director of the Mathematical Laboratory.

In the late 1940s Wilkes and his team began to build the EDSAC. At the time electronic computer developments were in their infancy, with only the American ENIAC having come into operation. There were many rival ideas concerning the principles on which a computer should be designed and how data should be stored. Of the rival serial and parallel systems, Wilkes chose the serial mode in which the information in the computer is processed in sequence (and not several parts at once, as in the parallel type). This choice of design involved the incorporation of mercury delay lines as the elements of the memory.

A means of delaying the passage of information was developed at the Radiation Laboratory, Massachusetts

Institute of Technology, from an original idea of William Shockley (1910–1989) of Bell Telephones, who was later to share a Nobel prize as one of the inventors of the transistor. It was originally intended that the device would be used in radar equipment. The delay lines were made up from tubes about 1.5 m/5 ft long and filled with mercury. At each end of each tube was a suitably cut quartz crystal. By the piezoelectric effect, when alternating electric current met one of the crystals it altered its shape slightly; this sent a small ripple through the mercury, which in turn struck the other crystal and in disturbing it produced the current again. The time taken for the ripple to travel through the mercury was sufficient to store the signal, and the process could be repeated indefinitely provided suitable auxilliary amplification equipment was used.

At the time that EDSAC was being built there was considerable interest in this type of memory device, and a number of American computers were being built which incorporated them. It was by no means certain that either the delay line or the computer would work satisfactorily, but on 6 May 1949 it ran its first program and became the first delay-line computer in the world. From early 1950 it offered a regular computing facility to the members of Cambridge University, the first service of its kind which involved a general-purpose computer. Much time was spent by the research group on the development of programming and in the compilation of a library of programs. The EDSAC was in operation until July 1958, although it had been modified during its period of operation. The expertise which Wilkes and his colleagues had gained was passed on to T. R. Thompson and J.M.M. Pinkerton who, with the Lyons Catering Company, developed the LEO (Lyons Electric Office) as a commercial computer.

In the mid 1950s EDSAC II was built, and it came into service in 1957. This, however, was a parallel machine and the delay line was abandoned in favour of the superior magnetic storage methods.

Maurice Wilkes has played a leading part in subsequent computer developments and has written many books and papers on the subject. In 1950 he was elected a Fellow of St John's College, Cambridge; he was elected a Fellow of the Royal Society in 1956, and in 1957 he became the first President of the British Computer Society.

Wilkins Maurice Hugh Frederick 1916– . New Zealand-born British biophysicist; *see* Crick, Watson, and Wilkins.

Wilkinson Geoffrey 1921– . English chemist; *see* Fischer and Wilkinson.

Wilks Samuel Stanley 1906–1964 was a US statistician whose work in data analysis enabled him to formulate methods of deriving valid information from small samples. He also concentrated on the developments and applications of techniques for the analysis of variance.

Wilks was born on 17 June 1906 in Little Elm, Texas. Educated locally, he studied architecture at the North Texas State College, obtaining his bachelor's degree there in

1926, and his master's two years later at the University of Texas. A two-year scholarship then took him to Iowa University, where he studied statistics and received his doctorate in 1931. Three further scholarships enabled him to continue studying, in New York, London, and Cambridge (England), before he finally joined the staff of Princeton University. Initially an Instructor, he became Assistant Professor in 1936, Associate Professor in 1938, and Professor of Mathematical Statistics in 1944. Active as a government advisory panellist, he received a number of awards and honours; he was also the author of several textbooks on statistics. He died in Princeton on 7 March 1964.

Much of Wilks's early work concerned problems associated with the statistical analysis of data obtained from small samples, such as those derived from experiments in psychology. His investigations of the analysis of variance were devoted especially to multivariate analysis. Two of his most original contributions were the Wilks criterion and his multiple correlation coefficient.

The US College Entrance Examination Board, which carries out extensive educational tests, found his assistance invaluable in analysing their results. Seeking also to apply these methods to industrial problems, Wilks did fundamental work in the establishment of the theory of statistical tolerance.

Williams Frederick Calland 1911–1977 was an electrical and electronics engineer with many accomplishments and inventions to his credit. He is particularly remembered for his pioneering work on electronic computers at the University of Manchester.

Williams was born on 26 June 1911, at Romiley near Stockport, Cheshire. After attending a private primary school near his home he went on to Stockport Grammar School, and in 1929 he was awarded an entrance scholarship to the University of Manchester. He studied engineering and graduated in 1932. The following year he carried out research and was awarded an MSc. After working for a short time with Metropolitan Vickers Electrical Company, he was awarded the Ferranti Scholarship of the Institution of Electical Engineers, which he took up at Oxford University by doing research on 'noise' in electronic circuits and valves. For this work he was awarded a doctorate in 1936. Also in that year he returned to Manchester as an assistant lecturer, where he stayed until the outbreak of World War II in 1939.

During the war Williams was involved in many applications of circuit design. He played a major part in the development of radar and allied devices, and in the design of the feedback systems known as servomechanisms, which had applications in aircraft controls and gunnery. In 1945, at the end of the war, he visited the Radiation Laboratory at the Massachusetts Institute of Technology (MIT) where he worked on circuitry and also learned of attempts to use cathode-ray tubes to store information. On his return to Britain after his second visit to MIT in 1946 he began to develop cathode-ray tube storage devices, in which information was coded and stored as dots on the screen. The

phosphor in the tubes allowed an image to persist for only a fraction of a second, and so the system needed considerable development. At first he transferred information to and from two tubes, but later he designed the appropriate circuitry to repeat the dots in one tube so that they would persist indefinitely.

In December 1946, Williams was appointed to the Chair of Electro-technics at Manchester University and began to work with M.H.A. Newman, who had a grant from the Royal Society for the development of computers. Their first machine began operation on 21 June 1948, the first stored-program computer to do so. The cathode-ray storage tubes allowed for immediate access to data. After modification, the machine went into production with Ferranti Limited, the first of several such computers. Williams's tubes, as they came to be known, were in great demand during the early 1950s because of their simplicity and cheapness. They were adopted in many computers in Britain and the United States; in particular they were a feature of the early 700 series of IBM.

Williams turned from computers when, because of their circuitry, they ceased to be a challenge to him and in the 1950s he began work on electrical machines, principally induction motors and induction-excited alternators. During his later years he worked on an automatic transmission for motor vehicles. He also played an increasingly large part in the administration of the University of Manchester. He received many awards, including election as a Fellow of the Royal Society in 1950 and knighthood in 1976. He died on 11 August 1977.

Williamson Alexander William 1824–1904 was a British organic chemist who made significant discoveries concerning alcohols and ethers, catalysis, and reversible reactions.

Williamson was born in London on 1 May 1824, of Scottish parents; his father was a clerk in the East India Company. A boyhood accident cost him an arm and the sight of one eye; when he was 16 years old his father retired and the family moved to the Continent. Williamson went to Heidelberg University to study medicine, but was persuaded by Leopold Gmelin (1788–1853) to change to chemistry. He then studied under Thomas Graham at University College, London, and Justus von Liebig at Giessen University, from which he gained his PhD in 1846. After a period in Paris studying methematics, he became Professor of Chemistry at University College, London, in 1849, where he remained until he retired in 1887. He died in Haslemere, Surrey, on 6 May 1904.

Beginning in 1850 Williamson studied alcohols and ethers and showed that they are both of the same type – the theory of types was fast gaining ground in organic chemistry. For example, amines are regarded as belonging to the 'ammonia type' (NH_3), with one or more of ammonia's hydrogen atoms replaced by organic (alkyl) radicals. Williamson ascribed alcohols and ethers to the 'water type'. In an alcohol, one of the hydrogen atoms of water (H_2O or HOH) is replaced by an alkyl radical (R) to give a compound $R \cdot OH$, such as ethyl alcohol (ethanol) C_2H_5OH. In an ether, both of water's hydrogens are replaced, either by the same alkyl radical (R) or by two different ones (R and R) to give compounds of the form R·O·R, such as diethyl ether (ethoxyethane), $(C_2H_5)_2O$, or of the form R·O·R , such as methyl ethyl ether (methoxyethane), $CH_3OC_2H_5$. Williamson was the first to make 'mixed' ethers, with two different alkyl groups, and his method is still known as the Williamson synthesis. It involves treating an alkoxide with an alkyl halide (haloalkane). For example:

$$CH_3I + C_2H_5ONa \rightarrow CH_3OC_2H_5 + NaI$$

The original way of making diethyl ether is by treating ethyl alcohol (ethanol) with sulphuric acid. In 1854, Williamson suggested that the reation takes place in two stages. First the substances react to form ethyl sulphate:

$$2C_2H_5OH + H_2SO_4 \rightarrow (C_2H_5)_2SO_4 + 2H_2O$$

Then the ethyl sulphate reacts with further alcohol to form diethyl ether and liberate sulphuric acid:

$$(C_2H_5)_2SO_4 + 2C_2H_5OH \rightarrow 2(C_2H_5)_2O + H_2SO_4$$

The sulphuric acid turns up unchanged at the end of the reactions but has been essential to them: it has acted as a catalyst. This was the first time that anyone had explained the action of a catalyst in terms of the the formation of an intermediate compound.

The theory of types has now outlived its usefulness. But it was important in the mid-nineteenth century because through the work of Williamson and others it established some sort of order among the confusion that then prevailed in organic chemistry.

Some of the reactions of alcohols and ethers are reversible (that is, the products of a reaction may recombine to form the reactants), a phenomenon first noted and described by Williamson in the early 1850s. Using modern notation, if P and Q are the reactants and R and S are the products, the reversible reaction is written:

$$P + Q \leftrightarrow R + S$$

An example is the esterification reaction between ethyl alcohol (ethanol) and acetic acid (ethanoic acid) to form ethyl acetate (ethanoate):

$$C_2H_5OH + CH_3COOH \leftrightarrow CH_3COOC_2H_5 + H_2O$$

If the rate of the forward reaction is the same as that of the reverse reaction, all four compounds (P, Q, R and S) coexist and the system is said to be in dynamic equilibrium (a term also introduced by Williamson).

Willstätter Richard 1872–1942 was a German organic chemist, best known for his investigations of alkaloids and plant pigments, such as chlorophyll, for which he was awarded the 1915 Nobel Prize for Chemistry.

Willstätter was born in Karlsruhe on 13 August 1872, the son of a textile merchant. He studied chemistry at the Munich Technische Hochschule under Johann von Baeyer, and after graduation worked under A. Einhorn (1857–1917) and was awarded his doctorate in 1894 for a thesis

on the structure of cocaine. He worked as Baeyer's assistant for several years, and then in 1905 became a professor at the Technische Hochschule in Zurich. From 1912 to 1916 he was Director of the chemistry section of the Kaiser Wilhelm Institute at Berlin-Dahlem, but his work was interrupted by World War I and at the request of Fritz Haber he turned his attention to the design of an effective gas mask. In 1916 he succeeded Baeyer as a full professor at Munich, but resigned in 1925 because of mounting anti-semitism. He continued working privately – supervising over the telephone some research at the university. At the start of World War II in 1939 he finally left Germany and went to live in exile in Switzerland. He died in Locarno on 3 August 1942.

Willstätter's first research work was on alkaloids. Following his doctoral study of cocaine in 1894, he went on to investigate tropine and atropine, and by 1898 had determined their structures and syntheses. From the pomegranate alkaloid pseudo pelletierine he prepared cyclooctatetraene, an eight-carbon ring compound with alternate single and double bonds, analagous to benzene.

He also worked on quinones, and by following William Perkin's method of oxidizing aniline (phenylamine) with chromic acid determined the structure of the dyestuff aniline black.

Willstätter then began his research into blood pigments and plant pigments. He showed that chlorophyll is not a single homogeneous substance but is made up of four components: two green ones, chlorophyll a ($C_{55}H_{72}O_5 N_4Mg$) and chlorophyll b ($C_{55}H_{70}O_6N_4Mg$), and two yellow ones, carotene ($C_{40}H_{56}$) and xanthophyll ($C_{40}H_{56}O_2$). He found that the blue-green chlorophyll a and the yellow-green chlorophyll b exist in the ratio of 3:1, and that the ratio of xanthophyll to carotene is 2:1. In order to separate the complex substances he redeveloped the technique of chromatography, first used in studies of chlorophyll by Mikhail Tswett in Russia in 1906, at about the same time as Willstätter was doing his work in Switzerland. It came as a surprise that the chlorophylls contain magnesium, later shown to be linked to four pyrrole rings like the iron atom in the haem group of the red blood pigment haemoglobin.

Soon after his return to Germany Willstätter had to abandon his research into plant pigments because during World War I the large quantities of solvents needed for chromatography became unobtainable. After the war he took up the study of enzymes and of catalytic hydrogenation, particularly in the presence of oxygen. He worked on the degradation of cellulose, investigated fermentation and pioneered the use of hydrogels for absorption. He tried to prove, incorrectly, that enzymes are not proteins.

Wilson Charles Thomson Rees 1869–1959 was a British physicist who invented the Wilson cloud chamber, the first instrument to detect the tracks of atomic particles. For this achievement, Wilson shared the 1927 Nobel Prize for Physics with Arthur Compton (1892–1962), who gained his award for the discovery of the Compton effect.

Wilson was born on 14 February 1869 at a farmhouse

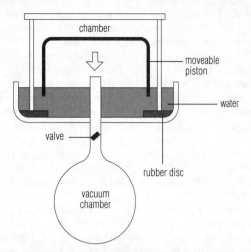

Wilson *The cloud chamber devised by C.T.R. Wilson consisted originally of a cylindrical glass chamber fitted with a hollow piston which was connected, via a valve, to a large evacuated flask. The piston falls rapidly when the valve is opened, and water vapour condenses along the tracks of any particles in the chamber.*

near Glencorse, just outside Edinburgh. His father, a successful farmer, died when Wilson was four and the family moved to Manchester. After studies at Owens College (now the University of Manchester), he gained a BSc in 1887 and then attended Sidney Sussex College, Cambridge, obtaining a BA in 1892 in natural sciences. After four years' teaching at Bradford Grammar School, Wilson returned to Cambridge where he was the Clerk Maxwell Student from 1896 to 1899. In 1900 he became a lecturer at Sidney Sussex College and in 1925 was appointed Jacksonian Professor of Natural Philosophy at Cambridge. Wilson held this post until 1934 and retired from Cambridge two years later. He continued to be active in research, producing his last paper (on thundercloud electricity) at the age of 87, by which time he was the oldest Fellow of the Royal Society. Wilson gained many honours, including the Society's Hughes Medal in 1911, the Royal Medal in 1922 and the Copley Medal in 1935. He died at Carlops, near Edinburgh, on 15 November 1959.

Wilson's great discovery of the cloud chamber was brought about by a thorough study of atmospheric cloud formation, an interest that began with observations of clouds on Ben Nevis (Britain's highest mountain) while on holiday in 1894. The wonderful optical effects fascinated Wilson, and on his return to Cambridge he tried to simulate them in his laboratory. In 1895, he began to construct a device to reproduce cloud formation and was successful in the production of an artificial cloud by causing the adiabatic expansion of moist air. From 1895 to 1899, Wilson carried out many experiments and established that the nucleation of droplets is able to take place in the absence of dust particles. Until this experiment, it was considered essential for each water droplet to form on a nucleus of dust. Wilson demonstrated that once supersaturated with

water vapour, nucleation can occur and that furthermore it is greatly improved by exposure to X-rays. This showed that ions are the nucleation sites on which water droplets form in the absence of dust.

Another unforgettable experience on Ben Nevis in 1895 had inspired a lifelong interest in atmospheric electricity. In Wilson's own words: 'Mist hid the top of Ben Nevis; there was a faint muttering of distant thunder. Suddenly I felt my hair stand up; I did not await any further developments but started to run ... the storm broke overhead with a bright flash and loud thunder just after I had left the summit.' During the years from 1900 to 1910, Wilson's attentions were devoted to the study of electrical conduction in dust-free air. He devised a gold-leaf electroscope in which surface leakage from the leaf to the case was impossible. With this instrument, he was able to show that some electrical leakage always occurs in air, and that the conductivity of the air inside the electroscope is the same in daylight as in the dark, and is independent of the sign of the charge for leaf potential. Wilson hoped to establish some link with an outside source of ions, and he even undertook some experiments underground to collect data. Later work by others was to show cosmic radiation to be the probable explanation.

Wilson returned to the cloud chamber in 1910, having realized that it could possibly show the track of a charged particle moving through it because water droplets could condense along the particle's path. Applying a magnetic field to the chamber would cause the track to curve, giving a measure of the charge and mass of the particle. In 1911, after many attempts, Wilson succeeded in producing a working model of his cloud chamber. The chamber was constructed in the form of a short cylinder in which supersaturation was achieved and controlled by the movement of a piston through a determined distance. The condensation effects were monitored through the other end. The Wilson cloud chamber immediately became vital to the study of radioactivity. It was used to confirm the classic alpha-particle scattering and transmutation experiments first performed by Ernest Rutherford (1871–1937), and Rutherford is reputed to have said it was 'the most original apparatus in the whole history of physics'.

The Wilson cloud chamber and its successors proved to be an indispensable tool in the investigation into the structure of the nucleus. Wilson himself was a lone researcher and did not contribute greatly to the development of his discovery. Modifications by Patrick Blackett (1897–1974) and Carl Anderson (1905–1991) brought about the discovery of the positron in 1932 and meson in 1936. The development of the cloud chamber into the bubble chamber by Donald Glaser (1926–   ) in 1952 produced a highly sensitive detector that has revealed whole families of new particles.

Wilson Robert Woodrow 1936–   is a US radioastronomer who, with Arno Penzias (1933–   ), detected the cosmic microwave background radiation, which is thought to represent a residue of the primordial 'Big Bang' with which the universe is believed to have begun.

Wilson was born on 10 January 1936. He studied at Rice University and earned his bachelor's degree in 1957. He began postgraduate research at the California Institute of Technology and was awarded his PhD in 1962. In 1963 he joined the technical staff of the Bell Telephone Laboratories in Holmdel, New Jersey; he was made Head of the Radiophysics Department in 1976. Wilson detected the 'three-degree background radiation' in 1965, only two years after he began working at the Bell Telephone Laboratories, and the sensational news of this discovery rapidly earned him international acclaim. He was awarded the Henry Draper Award of the National Academy of Sciences in 1977 and the Herschel Award of the Royal Astronomical Society in the same year. He received the 1978 Nobel Prize for Physics, jointly with Penzias and Pyotr Kapitza (1894–1984), for their work on microwave radiation.

Some of the earliest quantitive work on the 'Big Bang' hypothesis, the theory of the origin of the universe, was proposed and carried out by Ralph Alpher, Hans Bethe and George Gamow in 1948. They predicted the existence of isotropic cosmic background radiation, at a temperature of 25K (–248°C/–414°F). But this estimate, which turned out to be too high, could not be confirmed at the time, because the radio wavelengths in the shorter range of the radio spectrum would have been swamped by other sources and radioastronomy had not been developed sufficiently to be applied to this problem. The theory required a number of modifications and so its prediction of microwave radiation was soon forgotten as other issues assumed more importance in astronomy.

In 1964, Wilson and Penzias tested a radio telescope and receiver system for the Bell Telephone Laboratories with the intention of tracking down all possible sources of static that were causing interference in satellite communications. They found a high level of isotropic background radiation at a wavelength of 7.3 cm/2.9 in, with a temperature of 3.1K (276°C/529°F) – their initial result had been a temperature of 3.5K (277°C/500°F). This radiation was a hundred times more powerful than any that could be accounted for on the basis of any known sources.

As they could not explain this residual signal, Wilson and Penzias contacted Robert Dicke (1916–   ), at Princeton University, who was also interested in microwave radiation. Independently of Gamow, Dicke had followed a somewhat different line of reasoning to predict that microwave radiation, which was the remnant of the tremendous heat generated by the primordial fireball, was theoretically detectable. At the time his research team was constructing a radio telescope designed specifically to pick up radiation at a wavelength of 3.2 cm/1.3 in. When he heard of Wilson's and Penzias's observations, he immediately realized that their findings confirmed his predictions. The discovery of microwave radiation by Wilson and Penzias was almost a repeat of Karl Jansky's discovery, also at the Bell Telephone Laboratories, of radioastronomy itself in the 1930s.

Microwave radiation has since become the subject of intense investigation. Its spectrum appears to conform closely with that of black-body radiation and its temperature is only three degrees above absolute zero (–273°C/–459°F). It is thought to be the direct consequence of an explosively expanding universe; the temperature theoretically decreases as the universe continues to expand. Radio-astronomers continue to scrutinize this radiation for any evidence of anisotropy, which may indicate an unsuspected pattern of distribution of matter in the universe.

Winsor Frederick Albert 1763–1830 was a German inventor, one of the pioneers of gas lighting.

Winsor was born in Brunswick and educated in Hamburg, where he learned English. He went to Britain before 1799 and became interested in the technology and economics of fuels. In 1802, he went to Paris to investigate the 'thermo-lamp' which Philippe Lebon (1767–1804) had patented in 1799.

He returned to Britain at the end of 1803 and began a series of lectures at the Lyceum Theatre, London. A retired coach-maker named Kenzie lent him premises near Hyde Park to use as a gasworks. In 1804 Winsor was granted a patent for 'an improved oven, stove or apparatus for extracting inflammable air, oil, pitch, tar and acid and reducing into coke and charcoal all kinds of fuel' (there was no specific mention of coal, although that was the 'fuel' mainly used).

In 1806, he moved to Pall Mall where, during the following year, he lit one side of the street with gas lamps. In that year he was also granted a second patent for a new gas furnace and purifier, followed in 1808 and 1809 by others for refining gas to reduce its smell. Another 1809 patent was for a 'fixed and moveable telegraph lighthouse for signals of intelligence in rain, storm and darkness'.

1809 also saw an application to Parliament for a charter for the Light and Heat Company. The Bill was thrown out but the Westminster Gas-Light and Coke Company was finally incorporated, with the blessing of Parliament, in June 1810, although this time Winsor was not involved. He went to Paris in 1815 and tried to form a similar company there. He made a point of mentioning that he had been one of the first to credit Lebon with the original invention of the gas oven (on his 1802 visit). In 1817, he lit up 'Passage des Panoramas' with gas, but his company made little progress and was liquidated in 1819.

Winsor died in Paris on 11 May 1830 and was buried in the cemetery of Pére Lachaise. A cenotaph was erected to his memory in Kensal Green Cemetery, London, with the inscription 'At evening time it shall be light' (Zach. XIV 7). He had a son, F. A. Winsor Junior (1797–1874), of Shooters Hill, London, born in Vienna and called to the Bar in London in January 1840, who obtained a patent for the production of light as late as 1843.

In 1802, Winsor published *Description of the Thermo-lamp Invented by Lebon of Paris* with remarks by himself. It was published in English, French and German and re-issued in English in 1804 as *An Account of the Most Ingenious and Important Discovery of Some Ages*. There was also *Analogy between Animal and Vegetable Life, Demonstrating the Beneficial Application of the Patent Light Stoves to all Green and Hot Houses* (1807), and others.

Winsor started his career as a company-promoting expert. At the time of his visit to Paris in 1802, William Murdock (1754–1839) had been working in Britain on lines similar to Winsor's and his experiments had first yielded gas as a practical illuminant between 1792 and 1798, when he built gasworks at the Soho factory of Boulton and Watt near Birmingham. There had been similar projects by Archibald Cochrane between 1782 and 1783, but apart from those of Murdock and Lebon experiments in gas-lighting had not progressed further than 'philosphical fireworks'. These were exhibited by a German named Diller (d. 1789) in London and Paris. Similar fireworks were shown by Edmund Cartwright (1743–1828) at the London Lyceum in May 1800, and the population was sceptical when Winsor advertised 'The superiority of the New Patent Coke over the use of coals in Family Concerns, displayed every evening at the Large Theatre, Lyceum, Strand, by the New Imperial Patent Light Stove Company'. He kept secret how he obtained and purified the gas, but he demonstrated how he carried it to the different rooms of a house. He showed a chandelier, in the form of a long flexible tube hanging from the ceiling, connected to a burner in the shape of a cupid grasping a torch with one hand and holding the tube with the other. He explained how the flame could be modified and showed how it did not go out in wind and rain, produced no smoke and did not scatter dangerous sparks. He was a man with plenty of perseverance but with less chemical knowledge and mechanical skill. He was undeterred by the fact that his gas was sneered at as offensive, dangerous, expensive and unmanageable. The distilling retort he used consisted of an iron pot with a fitted lid. The lid had a pipe in the centre leading to the conical condensing vessel, which was compartmented inside with perforated divisions to spread the gas to purify it of hydrogen sulphide and ammonia. The device was not very successful, and the gas being burnt was impure and emitted a pungent smell. He also tried lime as a purifier, with a little success.

When Winsor applied to Parliament for a charter he was opposed by Murdock and Watt, and Walter Scott wrote that he was a madman to propose to light London with smoke. The Corporation of the Westminster Gas-Light and Coke Company were from then on advised by Samuel Clegg, an old disciple of Murdock's, and not by Winsor.

Withering William 1741–1799 was a British physician, botanist and mineralogist, best known for his work on the drug digitalis (from the foxglove plant), which he initially used as a diuretic to treat dropsy (oedema).

Withering was born on 17 March 1741 in Wellington, Shropshire, where his father was an apothecary. He went to the Edinburgh Medical School and graduated in 1766, taking the post of physician at Stafford Infirmary for the next nine years. In 1775, Erasmus Darwin suggested that Withering should take over a practice in Birmingham that had

belonged to William Small (1734–1775), founder of the Lunar Society, who had just died. Withering did so, and also became a member of the Lunar Society where he met Matthew Boulton, Joseph Priestley and other contemporary scientists. His practice in Birmingham did well and he became physician at Birmingham General Hospital. He contracted tuberculosis and retired in 1783; he was elected a Fellow of the Royal Society a year later. He publicly expressed his sympathies with the French Revolution, and in 1791 his house was attacked by a mob (as was Priestley's). He visited Portugal for a year between 1792 and 1793, and died in Birmingham on 6 October 1799.

In 1785, Withering published *Account of the Foxglove*, which detailed the controlled use of the drug digitalis for the treatment of dropsy. The drug was made from the leaves of *Digitalis purpurea*. He had begun studying digitalis in 1775, after noting its use in traditional herbal remedies. He worked out precise dosages of dried foxglove leaves. He also suggested the possible use of the drug in the treatment of heart disease, and he was correct because digitalis increases the heart's output without increasing the heart rate, thus clearing the interstitial fluid that is responsible for dropsy. Digitalis, in the form of digoxin, is still one of the most widely used drugs for treating heart failure.

Withering also made a name for himself in the field of botany after the publication of his *Botanical Arrangement* (1776), which was based on Linnaeus and became a standard work, and his activities in geology are remembered through the mineral ore witherite (barium carbonate), which was named after him.

Wittgenstein Ludwig Josef Johann 1889–1951 was an Austrian mathematician and philosopher who is best known for his philosophical theories of language.

Wittgenstein was born in Vienna on 26 April 1889, the youngest boy of eight children. His father was prominent in the iron and steel industry, and his mother was the daughter of a banker. He was educated at home until he was 14, and in 1903 went to school in Linz to study mathematics and the physical sciences, going on to the Technische Hochschule at Charlottenburg in Berlin to study engineering. He went to Britain in 1908 and registered as a research student in aeronautical engineering at the University of Manchester.

On the advice of Friedrich Frege (1848–1925), Wittgenstein moved to Trinity College, Cambridge, in 1912 to study philosophy under Bertrand Russell (1872–1970). In late 1913 he went to live on a farm in Skjolden, Norway, to have seclusion in which to develop his ideas on logic. On the outbreak of World War I in 1914 Wittgenstein volunteered for the Austrian Army and saw service on the eastern front and in southern Tyrol. He continued his philosophical work, however, and when he was captured in 1918 he had with him the completed manuscript of *Tractatus Logico-Philosophicus*. Through a mutual friend, the economist John Keynes (1883–1946), the manuscript was delivered to Russell in Cambridge and eventually published in 1921.

After the war Wittgenstein abandoned philosophy – saying his mind was 'no longer flexible' – and he became a schoolteacher in various Austrian villages. Then in January 1929 he returned to Cambridge, his interest in philosophy having apparently been rekindled by a lecture given in Vienna by Luitzen Brouwer (1881–1966) on the foundations of mathematics. From 1930 to 1936 he gave lectures in Cambridge, and then spent a year in Norway, returning to Britain in 1937. Two years later he succeeded G. E. Moore as Professor of Philosophy.

During World War II Wittgenstein worked first as a hospital porter in Guy's Hospital, London, and then as a laboratory assistant at the Royal Victoria Infirmary in Newcastle-upon-Tyne. He went back to Cambridge in 1944 and following his retirement in 1947 moved to Ireland to live. Two years later he was found to have cancer and during his last days he returned to Cambridge, to his doctor's house, where he died on 24 April 1951.

Wittgenstein's *Tractatus* is concerned with the presuppositions and conditions of ordinary everyday language. He argued that a sentence literally represents the world in such a way that any significant proposition in language can be 'logically analysed' into simple 'elementary propositions' which are a nexus of names that stand for definite objects. The structuring of the parts of a proposition depicts a possible real combination of elements. Logical and mathematical propositions are exceptions – lacking reference to real or possible states of affairs and yet not being nonsensical.

Wittgenstein began work on his *Philosophical Investigations* in 1936, although it was not published until 1953, after his death. It took a more subtle and complex view than the *Tractatus*, retaining an interest in the problem of meaning but arguing that the meanings of words depends on their role within particular 'language games'. He postulated that no single feature may be supposed to be present in all forms of language, just as no single definitive feature is common to all of the practices we call games. The unity of language games is like a family resemblance, and to understand the workings of a language we must understand the complex of usages present in language games: words are used to do things rather than merely stand for objects, and the meaning of a concept depends on how it is used in a particular language game. So that in mathematics, the force of a proof – the necessity of mathematical proof – is based on convention or agreement about what is to count as following a rule in the language game of mathematics. But we do not agree *because* the proof follows a rule; rather, agreement with a mathematical calculation fixes its meaning as a proof or a rule, and human practice determines what the rules shall be.

Wittig Georg 1897–1987 was a German chemist, best known for his method of synthesizing olefins (alkenes) from carbonyl compounds, a reaction often termed the Wittig synthesis. For this achievement he shared the 1979 Nobel Prize for Chemistry with the British-born US chemist Herbert Brown.

Wittig was born in Berlin on 16 June 1897. He was educated at the Wilhelms-Gymnasium and then at Kassel and Marburg Universities. He was a lecturer at Marburg from 1926 to 1932, and then Head of Department at the Technische Hochschule in Brunswick from 1932 to 1937. He became a Special Professor at the University of Freiburg in 1937, and then from 1944 until 1956 was Professor and Institute Director at Tübingen. He became a professor at Heidelberg in 1956, and remained there until 1967, when he became Emeritus Professor.

In the Wittig reaction, which he first demonstrated in 1954, a carbonyl compound (aldehyde or ketone) reacts with an organic phosphorus compound, an alkylidene-triphenylphosphorane, $(C_6H_5)_3P=CR_2$, where R is a hydrogen atom or an organic radical. The alkylidene group $(=CR_2)$ of the reagent reacts with the oxygen atom of the carbonyl group to form a hydrocarbon with a double bond, an olefin (alkene). In general:

$$(C_6H_5)_3P=CR_2 + R_2'CO \rightarrow (C_6H_5)_3PO + R_2C=CR_2$$

The reaction is widely used in organic synthesis, for example to make squalene (the synthetic precursor of cholesterol) and vitamin D_3.

Wöhler Friedrich 1800–1882 was a German chemist who is generally credited with having carried out the first laboratory synthesis of an organic compound, although his main interest was inorganic chemistry.

Wöhler was born at Eschersheim, near Frankfurt-am-Main, on 31 July 1800, the son of a veterinary surgeon in the service of the Crown Prince of Hesse-Kassel. He entered Marburg University in 1820 to study medicine, and after a year transferred to Heidelberg, where he studied in the laboratory of Leopold Gmelin (1788–1853). He gained his medical degree in 1823, but Gmelin had persuaded Wöhler to study chemistry and so he spent the following year in Stockholm with Jöns Berzelius, beginning a lifelong association between the two chemists. From 1825 to 1831 he occupied a teaching position in a technical school in Berlin, and from 1831 to 1836 he held a similar post at Kassel. In 1836, he became Professor of Chemistry in the Medical Faculty of Göttingen University, as successor to Friedrich Strohmeyer (1776–1835), and remained there for the rest of his career, making it one of the most prestigious teaching laboratories in Europe. He died in Göttingen on 23 September 1882.

In Wöhler's first research in 1827 he isolated metallic aluminium by heating its chloride with potassium; he then prepared many different aluminium salts. In 1828 he used the same procedure to isolate beryllium. Also in 1828 he carried out the reaction for which he is best known. He heated ammonium thiocyanate – a crystalline, inorganic substance – and converted it to urea (carbamide), an organic substance previously obtained only from natural sources. Until that time there had been a basic misconception in scientific thinking that the chemical changes undergone by substances in living organisms were not governed by the same laws as were inanimate substances; it was thought that these 'vital' phenomena could not be described in ordinary chemical or physical terms. This theory gave rise to the original division between inorganic (non-vital) and organic (vital) chemistry, and its supporters were known as vitalists, who maintained that natural products formed by living organisms could never be synthesized by ordinary chemical means. Wöhler's synthesis of urea was a bitter blow to the vitalists and did much to overthrow their doctrine. It involved an isomerization reaction:

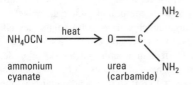

Wöhler worked with Justus von Liebig on a number of important investigations. In 1830, they proved the polymerism of cyanates and fulminates, and two years later announced a series of studies of benzaldehyde (benzenecarbaldehyde) and the benzoyl (benzenecarboxyl) radical. In 1837, they investigated uric acid and its derivatives. Wöhler also discovered quinone (cyclohexadiene-1,4-dione), hydroquinone or quinol (benzene-1,4-diol) and quinhydrone (a molecular complex composed of equimolar amounts of quinone and hydroquinone).

In the inorganic field Wöhler isolated boron and silicon and prepared silicon nitride and hydride. He prepared phosphorus by the modern method, and discovered calcium carbide and showed that it can be reacted with water to produce acetylene (ethyne):

$$CaC_2 + 2H_2O \rightarrow Ca(OH)_2 + C_2H_2$$

He demonstrated the analogy between the compounds of carbon and silicon, and just missed being the first to discover vanadium and niobium. He also obtained pure titanium and showed the similarity between this element and carbon and silicon. He published little work after 1845, but concentrated on teaching.

Wolf Maximilian Franz Joseph 1863–1932 was a German astronomer particularly noted for his application of photographic methods to observational astronomy.

Wolf was born in Heidelberg on 21 June 1863. He was fascinated by astronomy from an early age and his father, who was fairly wealthy, built a small observatory for him in 1885 which he used for most of his research until 1896. He studied at the University of Heidelberg and was awarded his PhD in 1888 for a thesis on celestial mechanics. He then spent a year in Stockholm, before returning to Heidelberg to join the staff of the University there as a lecturer. In 1893, he travelled to the United States and was then appointed Extraordinary Professor of Astrophysics at Heidelberg. With the assistance of wealthy benefactors, Wolf supervised the construction of a new observatory at Königstuhl, near Heidelberg. He was appointed Director of

the Observatory in 1896 and made Professor of Astronomy in 1901. Wolf held both these posts until his death, on 3 October 1932, in Heidelberg.

Wolf made his first important astronomical discovery in 1883, when he observed a comet which has an orbital period of 7.7 years. In Wolf's honour it now bears his name. He devoted much of his time to developing new photographic methods for application to astronomy. One of his most successful innovations was a technique for discovering large numbers of asteroids. Until this time, asteroids had always been discovered visually, the first, Ceres, being noticed by Giuseppe Piazzi (1749–1826) in 1801; more and more of these bodies had gradually been discovered over subsequent years. Wolf arranged for time-lapse photographs to be taken, using a camera mounted on a telescope whose clock mechanism followed as exactly as possible the proper motions of the 'fixed stars'. On the developed plate, the stars would appear as discrete spots – the size being a function of the star's magnitude – whereas any asteroids present would appear as short streaks in the foreground.

The first asteroid Wolf discovered using this technique was number 323, afterwards named Brucia. He subsequently discovered more than 200 other asteroids using the same method. In September 1903, he discovered a special asteroid, number 588, later named Achilles. Its particular significance was that it was the first of the so-called 'Trojan satellites' whose orbits are in precise synchrony with that of Jupiter's; they form a gravitationally stable configuration between Jupiter and the Sun. The possibility of the existence of this kind of triangular three-bodied system was first analysed and predicted theoretically by Lagrange in the 1770s and Wolf's discovery of the 'Trojan satellites' provided the observational evidence to substantiate Lagrange's theory.

Wolf also detected several new nebulae, both within the Milky Way and outside our Galaxy. Independently of Edward Barnard (1857–1923) he discovered that the dark 'voids' in the Milky Way are in fact nebulae which are obscured by vast quantities of dust, and he studied their spectral characteristics and distribution. Among the nebulae that Wolf himself discovered is the North America Nebula (NCG 7000), which resembles that continent in shape and is found in the constellation of Cygnus.

Wolf, with Pulfrich, invented the stereo-comparator, which was used for various kinds of observational astronomy. He also carried out research on the sunspot cycle and was the first to observe Halley's Comet when it approached the Earth in 1909. Besides being a keen observational astronomer and a dedicated and inspiring teacher, Wolf is best remembered for his innovative photographic techniques which led to the discovery of many new asteroids.

Wolff Heinz Siegfried 1928– is a biomedical engineer who works on high technology instruments and the application of technology to medicine.

Wolff was born on 29 April 1928. He worked in the physiological laboratory at Oxford University from 1946 to

1950 and then for a year in the Medical Research Unit in Glamorgan. He then went to University College, London University, to study physiology, and after graduating in 1954 he was employed in the Division of Human Physiology at the Medical Research Council's National Institute for Medical Research, where he specialized in the development of instrumentation suitable for field work. In 1965, he was appointed head of the institute's Division of Biomedical Engineering. He then became Director of the Biomedical Division of the MRCs Clinical Research Centre at Harrow, Middlesex.

Wolff's interests range from the invention of new high technology instruments to the widespread and sensible application of technology to the problems of the elderly and the disabled. He believes that small, specialized pieces of equipment that can be worked by doctors and nurses might be preferable to large centralized units to which patients have to go for tests or treatment. Machines should be simple to use and should show when they are not working properly and be capable of repair on the spot by the operator. The researcher should collaborate more with the manufacturer, and manufacturers should spend some resources on creating a market for the next generation of equipment as well as fulfilling existing market requirements. Technology should also be used to make the chronically ill more comfortable and to allow the partly disabled to remain in their own homes for as long as possible.

Wolff Kaspar Friedrich 1733–1794 was a German surgeon and physiologist who has become regarded as the founder of embryology, although his findings were largely ignored for more than 50 years.

Wolff was born in Berlin on 18 January 1733 and studied at Halle and the Berlin Medical School. He graduated in 1759 and became an army surgeon during the Seven Years' War (1756–1763), and then lectured in pathology in Berlin. Despite the success of his lectures he was not offered a professorship, and so in 1766 he accepted an invitation from Catherine II of Russia to take the post of Academician for Anatomy and Physiology in St Petersburg. He remained there until his death on 22 February 1794.

Wolff produced his revolutionary work *Theoria generationis* in 1759. Until that time it was generally believed that each living organism develops from an exact miniature of the adult within the seed or sperm – the so-called preformation or homunculus theory. Wolff introduced the idea that cells which are initially unspecialized later differentiate to produce the separate organs and systems of the plant or animal body with their distinct types of tissues. In fact Wolff's view that plants and animals are composed of cells was still a subject of controversy. His name is also associated with, among other parts of the anatomy, the Wolffian body, a structure in an animal embryo that eventually develops into the kidney.

Wollaston Hyde William 1766–1828 was a British chemist and physicist who developed the technique of powder metallurgy and discovered rhodium and palladium, two elements similar to platinum.

Wollaston was born in East Dereham, Norfolk, on 6 August 1766, one of 17 children of an academic family. His father Francis Wollaston was a clergyman and amateur astronomer, and his elder brother Francis John Hyde Wollaston became a Professor of Chemistry at Cambridge University. Wollaston was educated at Charterhouse School and at Gonville and Caius College, Cambridge, from which he graduated in medicine in 1793. He practised as a doctor for seven years and then in 1800 moved to London and devoted the rest of his life to scientific research. He died of a brain tumour on 22 December 1828.

Wollaston initiated the technique of powder metallurgy when working with platinum and trying to get it into a workable, malleable form. Using aqua regia (a mixture of concentrated nitric and hydrochloric acids) he dissolved the platinum from crude platina, a mixed platinum–iridium ore. He then prepared ammonium platinichloride, which he decomposed by heating to yield fine grains of platinum metal. The grains were worked using heat, pressure and hammering to form sheets, which he sold to industrial chemists for making corrosion-resistant vessels; manufacturers of sulphuric acid were willing to pay high prices for such a useful metal. Wollaston kept his method secret, and made £30,000 from selling platinum, much of which he donated to various scientific societies to help finance their researches.

While investigating platinum ores Wollaston isolated two new elements. He discovered palladium in 1803. Within a year he also found rhodium, another metal with similar properties to platinum.

He was a great supporter of John Dalton's atomic theory and published several papers based on the law of multiple proportions. In 1808, he suggested that a knowledge of the arrangements of atoms in three dimensions would be a great leap forward (although a century was to pass before this became possible). He advocated the use of 'equivalents' in quantitative chemical calculations, from which the concept of normality (now superseded by molarity) was developed.

Wollaston also worked in various areas of physics. In optics he suggested the total reflection method for measuring refractivity (later developed by Pulfrich and Abbe) and drew attention to the dark lines (later called Fraunhofer lines) in the solar spectrum. In 1807 he developed the *camera lucida*, which was to inspire William Fox Talbot to his discoveries in photography. He also invented a reflecting goniometer for accurately measuring the angles of crystals in minerals.

In 1801, Wollaston established the important physical principle that 'galvanic' and 'frictional' electricity are the same. He also stated that the action in the common voltaic cell is due to the oxidation of the zinc electrode. In early 1821 he reported that there is 'a power ... acting circumferentially round the axis of a wire carrying a current' and tried in Humphry Davy's laboratory at the Royal Institution to make a current-carrying wire revolve on its axis. These experiments were unsuccessful and the final demonstration of the effect is now attributed to Michael Faraday. In 1824 the British Weights and Measures Act incorporated the Imperial gallon, equivalent to 10 pounds weight of water as suggested by Wollaston in 1814. (But as a member of a Royal Commission in 1819 Wollaston was instrumental in the rejection of the decimal system of weights and measures.)

Woodger Joseph Henry 1894–1981 was a British biologist known principally for his theoretical work on the underlying philosophical basis of scientific methodology in biology, especially for his attempt to provide biology with a strict and logical foundation on which observations, theories and methods could be based.

Woodger was born on 2 May 1894 in Great Yarmouth, Norfolk, and was educated at Felsted School then University College, London (where he studied under J. P. Hill, one of the leading British zoologists of the time). Woodger enrolled in the army in 1915 and after World War I spent a brief period in Baghdad. He returned to London in 1919 and began investigating newly discovered cell organelles, including Golgi bodies and mitochondria. In 1924, he was appointed Reader in Biology at the Middlesex Hospital Medical School in London, and became a professor there in 1947. He retired in 1959.

Woodger's transition from practical work to the theoretical and philosophical aspects of biology dated from his appointment to the readership in biology. Confronted for the first time with having to teach, he found no textbooks that gave an adequate grounding in the fundamental scientific principles involved in biology and medicine, so he wrote one – *Elementary Morphology and Physiology* (1924). He became increasingly concerned with the problems of methodology and interpretation in biological experimentation and with the need for 'a critical sorting of fundamental concepts to promote a strictness of thought equal to the strictness of investigation required in the new biology'.

There was at that time no generally accepted (biological) theory of organism, only a host of facts and conflicting ideas ranging from total mechanism to vitalism. The new biology required a causal description for each vital process and a means of understanding how such processes as nutrition, development and behaviour are related to the life of a particular organism or species.

Woodger began by teaching himself philosophy, with particular reference to Whitehead and Russell's *Principia Mathematica*. He developed the idea that one of the characteristics of a living system is the organization of its substance, and that this order is of a hierarchical nature. Thus the components of an organism can be classified on a scale of increasing size and complexity: molecular, macromolecular, cell components, cells, tissues, organs and organisms. Each class exhibits specif-ically new modes of behaviour, which cannot be interpreted as being merely additive phenomena from the previous class.

Woodger's major work, *Biological Principles: a critical study* (1929), examines the fundamental requirements of theories in biology. He was able to show how to resolve many of the apparent antitheses in biology (such as those between structure and function, preformation and development, and mechanism and vitalism). He also made the first attempt to put embryological ideas on a logistic basis by analysing at length, and in strict logical form, the process of cell division. He demonstrated that living matter shows not only spatial hierarchical order but also divisional hierarchies (each cell or group of cells has a parent cell), and that many difficulties in biological theory arose originally through the abstraction from time – that is, viewing an organism as a series of spatially ordered components only. He summarized many of these ideas in his book *The Technique of Theory Construction* (1939).

Woodward Robert Burns 1917–1979 was a US organic chemist famous for his syntheses of complex biochemicals. For his outstanding contribution to this area of science he was awarded the 1965 Nobel Prize for Chemistry.

Woodward was born in Boston on 10 April 1917. He went to the Massachusetts Institute of Technology in 1933, while he was still only 16 years old, gaining his BS degree three years later and his PhD a year after that. He went to Harvard in 1938, at the age of 21, and remained there for the rest of his academic career. He held various appointments, including Assistant Professor (1944–46), Associate Professor (1946–50), full Professor (1950–53), Morris Loeb Professor (1953–60) and Donner Professor of Science (from 1960). From 1973 to 1974 he was also Todd Professor of Chemistry and Fellow of Christ's College, Cambridge.

Woodward's first important research was in collaboration with William Doering when in 1944 they achieved a total synthesis of quinine (from simple starting materials). In 1947 he worked out the structure of penicillin and two years later that of strychnine. In the early 1950s, he began to synthesize steroids, such as cholesterol and cortisone (1951) and lanosterol (1954). In that same year he synthesized the poisonous alkaloid strychnine and lysergic acid, the basis of the hallucinogenic drug LSD. In 1956 he made reserpine, the first of the tranquillizing drugs, and four years later he prepared chlorophyll. Turning his attention again to antibiotics, he and his co-workers produced a tetracycline in 1962 and cephalosporin C in 1965. In 1971 came the culmination of ten years' work, involving collaboration with Swiss chemists – the synthesis of vitamin B_{12} (cyanocobalamin). Many of these syntheses were among the most complicated ever attempted in organic chemistry, involving many stages that had to be carefully selected for stereospecificity and maximum yield. Yet so well worked out were they that many have become the basis of commercial manufacture of drugs and other useful biochemicals.

Woolley Richard van der Riet 1906–1986 was a British astronomer, best known for his work on the dynamics of the Galaxy, observational and theoretical astrophysics, and stellar dynamics.

Woolley was born in Weymouth, Dorset, on 24 April 1906, his father being Rear-Admiral Charles Woolley. He was educated at Allhallows School, Honiton, and at the Universities of Cape Town and Cambridge. He was a Commonwealth Fund Fellow at Mount Wilson Observatory from 1929 to 1931 and an Isaac Newton Student at Cambridge from 1931 to 1933, when he was appointed Chief Assistant at the Royal Observatory, Greenwich. He remained at Greenwich for four years, until he was appointed First Assistant Observer – John Couch Adams Astronomer – at the University Observatory in Cambridge. In 1939, he became the Director of the Commonwealth Observatory at Mount Stromlo, Canberra, Australia. Finally, in 1955, Woolley was appointed as the eleventh Astronomer Royal, succeeding Spencer-Jones at the Royal Observatory; he retained this position until his retirement in 1970. Even after his retirement, he held the post of first Director of the South African Astronomical Observatory until he reached the age of 70. He received a knighthood in 1963 and was President of the Royal Astronomical Society from 1963 to 1965, receiving its Gold Medal in 1971.

While Woolley was Chief Assistant at the Royal Observatory, he became well acquainted with the more traditional aspects of astronomy. His duties included meridian astronomy with the Airy Transit Circle, time service control with small reversible transits and pendulum clocks, double star observations, solar spectroscopy and spectrohelioscope observations. The Observatory at Mount Stromlo was devoted mainly to solar physics and so while he was Director there he devoted much of his time to the study of photospheric convection, emission spectra of the chromosphere, and the solar corona. Besides these investigations, he pioneered the observation of monochromatic magnitudes and constructed colour magnitude arrays for globular clusters. Under Woolley's control, the Commonwealth Observatory grew in stature and its equipment was updated to include a 188-cm/74-in reflecting telescope.

When Woolley was appointed Astronomer Royal, the Royal Observatory had recently been moved from Greenwich to Herstmonceux in Sussex. At the time of his appointment the establishment seemed to lack purpose and direction; observations were being amassed using obsolete equipment and methods that had been devised by George Airy (1801–1892) many years before. Woolley's first priorities were to press for an agreement on the design of the new Isaac Newton telescope and to redistribute resources for astrophysical research projects. During the 15 years that he spent as Astronomer Royal he was noted for the balance that he maintained between theoretical studies and observation. His personal interests during this period were globular clusters, the evolution of galactic orbits, improvements of radial velocities and a re-evaluation of RR Lyrae luminosities. He appointed several physicists to the staff whose studies involved galactic cluster fields, elemental abundance and the evolution of galaxies.

Wright Almroth Edward 1861–1947 was a British bacteriologist who did pioneering work in the field of immunology, notably in the development of a vaccine against typhoid fever. He received numerous honours for his work, including a knighthood in 1906.

Wright was born on 10 August 1861 in Middleton Tyas, a small village near Richmond in Yorkshire. His father was an Irish Presbyterian clergyman who, between 1863 and 1885, held ministries in Dresden, Boulogne and Belfast, and Wright received his early education from his parents and private tutors. While his family was in Belfast, however, he attended the Royal Academic Institution and then in 1878 entered Trinity College Dublin, from which he graduated in modern literature in 1882 and in medicine in the following year. Subsequently he studied – on a travelling scholarship – pathological anatomy at the universities of Leipzig and Marburg and physiological chemistry at Strasbourg. He then went to Australia, where he was a demonstrator of physiology at Sydney University from 1889 to 1891. After his return to England, Wright worked for a short time in the laboratories of the College of Physicians and Surgeons in London until, in 1892, he was appointed Professor of Pathology at the Army Medical School in Netley, Hampshire. It was while he held this post that he developed a vaccine against typhoid, but he disagreed with the army authorities over the use of his vaccine and he resigned his professorship in 1902. In the same year he became Professor of Pathology at St Mary's Hospital in London and held this post until he retired in 1946; in 1908 he also became responsible for the Department of Therapeutic Inoculation (later called the Institute of Pathological Research) at St Mary's. In 1911, Wright went to South Africa, where he introduced prophylactic inoculation against pneumonia for workers in the Rand gold mines. On returning to England, he was appointed Director of the Department of Bacteriology of the newly founded Medical Research Committee (later Council) at the Hampstead Laboratory, London. During World War I he served in France as a temporary Colonel in the Army Medical Service, afterwards returning full-time to his professorship at St Mary's Hospital. Wright died at his home in Farnham Common on 30 April 1947.

Wright first began bacteriological research while he was Professor at the Army Medical School, and by 1896 he had succeeded in developing an effective antityphoid vaccine, which he prepared from killed typhoid bacilli. Preliminary trials of the vaccine on troops of the Indian Army proved its effectiveness and the vaccine was subsequently used successfully among the British soldiers in the Boer War. In addition to this important development, Wright established a new discipline within medicine, that of therapeutic immunization by vaccination, which was aimed at treating microbial diseases rather than preventing them. He proved that the human bloodstream contains bacteriotropins (opsonins) in the serum and that these substances can destroy bacteria by phagocytosis. He researched into wound infections; his work led to the use of salt solution as an osmotic agent to draw lymph into wounds, thereby accelerating their closure. And he also originated vaccines against enteric tuberculosis and pneumonia.

Wright Louis Tompkins 1891–1952 was a US physician, surgeon and civil rights leader, known for his work on fractures – especially head injuries, venereal disease and cancer.

Born on 23 July 1891 in LaGrange, Georgia, Louis Tompkins Wright was educated at Clark University, Atlanta, from which he received a BA, graduating as valedictorian in 1911, and at Harvard Medical School, receiving his MD *cum laude* in 1915. After his internship at Freedmen's Hospital, Washington, DC, and having taken medical examinations for licensing in Maryland, Georgia and New York, in 1917 he joined the US Army as First Lieutenant in the Medical Corps. Returning to New York after World War I, he combined private practice and serving on the staff of the Harlem Hospital – in 1919 as a clinical assistant visiting surgeon rising to director of surgery 23 years later and director of the medical board of Harlem Hospitals in 1948. He was the first black doctor to be appointed to a municipal hospital position in New York City. In 1929, after passing the civil service examinations, Wright was also appointed as the first black police surgeon in the history of the city. Wright remained in New York until his death on 8 October 1952; he had married in 1918 and had two daughters.

Wright is best known for his work as a physician and surgeon. His criticism of the Schick test for diphtheria carried out at the Freedmen's Hospital is said to be the first original work to be published from the hospital. Whilst in the army he originated the intradermal method of vaccination against smallpox, published in the *Journal of the American Medical Association* (1918). Wright specialized in head injuries and fractures; he devised a brace for neck fractures, a blade for the treatment of fractures of the knee joint and a plate out of tantalum for the repair of recurrent hernias. As an authority on the subject he wrote the chapter on head injuries in C. Scudder's *Treatment of Fractures* (11th edition, 1938) and published many articles in medical journals. Wright also became an authority on the venereal disease, lymphogranuloma venereum and was the first physician to experiment with the new antibiotics Aureomycin (the 'wonder drug') and Terramycin, publishing thirty and eight papers respectively on the two drugs between 1948 and 1952. In 1948, Wright moved into the field of cancer research and founded the Harlem Hospital Cancer Research Foundation, where he dealt with the effectiveness of chemotherapeutic drugs. As well as his professional activities, Wright was involved with civil rights activities, promoting better health assurance and expanding equal opportunities for black people. He was active on the local branch of the NAACP and was chairman of its national board of directors for 17 years. His work led to many honours and awards including the Purple Heart in World War I, a DSc from Clarke University (1938) and the Spingarn medal (1940); He was made a Fellow of the New York Surgical Society (1949) and a Fellow of the International College of Surgeons (1950) and was also a fellow of the American Medical Association and the National Medical Association. In 1952 the Louis T. Wright Library of Harlem Hospital was named after him, as was a public school in Harlem in 1976.

Wright Orville 1871–1948 and Wilbur 1867–1912 were two US aeronautical engineers famous for their achievement of the first controlled, powered flight in a heavier-than-air machine (an aeroplane) and for their design of the aircraft's control system.

Wilbur, the elder of the two brothers, was born in Millville, Indiana, on 16 April 1867 and Orville was born in Dayton, Ohio, on 19 August 1871. Their father was a bishop of the United Brethren Church. He brought up his sons to think for themselves, to have initiative and to have the enterprise to use their abilities and express themselves fully. They both went to the local high school in Dayton. During and after their schooldays they developed an interest in mechanical things. They taught themselves mathematics and read as much as they could about current developments in engineering. After some attempts at editing and printing small local newspapers, the two brothers formed the Wright Cycle Company in 1892. For the next ten years they designed, built and sold bicycles.

Like any other serious innovative engineers, the Wright brothers must have considered the possibility of powered flight and how to achieve it, but the exploits of Otto Lilienthal, the German pioneer of gliders who was killed in 1896, brought home to them that much might be achieved in this direction with perseverance and daring.

Lilienthal's death convinced Wilbur that in order to avoid dangerous accidents it was important that they not only build successful aeroplanes but that they also learn to fly them correctly. Control of direction and stability were problems that occupied them for the next few years, in August 1899 they flew a kite with a wingspan of about 1.5 m/5 ft with controls for warping the wings to achieve control of direction and stability. Their wing-warping method was the forerunner of the later idea of ailerons. They discovered that the kite would fly with the horizontal tailplane either forward or aft of the wings.

The following year, 1900, the Wrights built a larger kite of 5-m/17-ft wingspan to carry a pilot. After taking advice as to where to find steady winds together with suitable sandy banks, they decided on Kitty Hawk, North Carolina. The kite flew well and Wilbur achieved a few seconds of piloted flight. They also flew a glider with the tailplane in front of the wings. The following July they returned to Kitty Hawk and built a wooden sled at Kill Devil Hills, where there were large sand dunes. Their new machine was longer and had a different wing camber to the previous model; it also had a hand-operated elevator attached to the tailplane. Again they achieved encouraging results, particularly after further alterations to the wing camber. There were, however, still problems with stability and control.

During the following winter the Wrights built a small wind tunnel and tested various wing designs and cambers. In the course of these tests they compiled the first accurate tables of lift and drag, the important parameters that govern flight and stability. The new glider had a 9.6-m/32-ft wingspan and had, at first, a double vertical fin mounted behind the wings. Turning was still difficult, however, and this was converted to a single moveable rudder operated by the wing-warping controls. This configuration proved so successful that they decided to attempt powered flight the following summer. During the winter of 1902 they searched in vain for a suitable engine for their craft and for knowledge of propeller design. They eventually constructed their own 9-kW/12-hp motor and made their own very efficient propeller. After some initial trouble with the propeller shafts, the Wright biplane took to the air and made a successful flight on 17 December 1903 at Kill Devil Hills near Kitty Hawk. The aeroplane had a wingspan of 12 m/40 ft and weighed 340 kg/750 lb with the pilot. The two brothers took it in turns to fly. Wilbur, in the last of the flights, stayed in the air for 59 seconds and travelled 260 m/852 ft at a little under 16 kph/10 mph relative to the ground.

The following year the Wrights incorporated a 12-kW/16-hp engine and separated the wing-warping and rudder controls. They flew their new model at their home town of Dayton, learning to make longer flights and tighter turns.

In 1905 the Wrights were sufficiently confident of their design to offer it to the United States War Department. The following year they patented their control system of elevator, rudder and wing-warping. Although they spent time patenting and finding markets for their machines during the next few years, they did not feel sufficiently confident to exhibit them publicly until 1908. That year Wilbur flew in France and Orville in the United States. In 1909 Wilbur flew in Rome and Orville in Berlin. Their aeroplanes were now sufficiently well controlled and stable to allow Wilbur to make a flight of 32 km/20 mi in the United States.

During the next few years they and their Wright Company built aeroplanes, but by 1918 their competitors had gained ground and their patents were under pressure. During the ensuing litigation Wilbur caught typhoid fever and died at Dayton, Ohio, on the 30 May 1912.

Orville sold his interest in the Wright Company in 1915 and later pursued aviation research. He eventually became a member of the National Advisory Committee on Aeronautics. He died at Dayton on 30 January 1948 having received during his lifetime many awards and honours as tokens of the momentous achievement of the Wright brothers.

Wright Thomas 1711–1786 was an English philosopher whose interests spread over a wide range of subjects, including astronomy; some of his theoretical work in astronomy anticipated modern discoveries.

Wright was born in Byer's Green, near Durham, on 22 September 1711. He left school at the age of 13 to take up an apprenticeship, which he did not complete, with a clockmaker. Although he was interested in instrument-making, he held a number of jobs outside this field. His most successful work was as a private mathematics teacher and lecturer on popular scientific subjects. In 1742 he was offered the Chair of Mathematics at the Academy of Sciences at St Petersburg. He refused this position, however, on the grounds that the salary offered was inadequate. He spent the rest of his days in Byer's Green as a writer and teacher. He died on 25 February 1786.

The main drive behind most of Wright's scientific work was his desire to reconcile his religious beliefs with his telescopic observations and with the knowledge he obtained through his extensive reading. In his early works Wright described the universe as a series of concentric spheres – hardly an original concept – in which the centre was occupied by some divine power. Then in 1750 he wrote his most influential book, *An Original Theory or New Hypothesis of the Universe*. It did not express his final thoughts on the subject. He wrote another manuscript on the subject, but it was never published. His first essay was, however, a classic work and it created sufficient interest to warrant a new edition to be published in the United States (Philadelphia) in 1937. In it, Wright describes the Milky Way as a flattened disc, in which the Sun does not occupy a central position. Furthermore, he stated that nebulae lay outside the Milky Way. These views were more than 150 years ahead of their time and did not become accepted by the scientific community until they were substantiated by observational evidence in the 1920s.

However, Wright's model does not completely conform to modern views because he persisted in his belief that the centre of the system was occupied by a divine presence. Wright's other work included thoughts on the particulate nature of the rings of Saturn, anticipating the writings of James Clerk Maxwell (1831–1879) by nearly a century, and thoughts on such diverse fields as architecture and reincarnation.

Wurtz Charles Adolphe 1817–1884 was a prominent French organic chemist, best known for his synthetic reactions and for discovering ethylamine and ethylene glycol (1,2-ethanediol). His major contribution, however, was the elevation of the standard of organic chemistry research in mid-nineteenth-century France to a level that challenged the excellence of the German universities.

Wurtz was born in Wolfisheim, near Strasbourg, on 26 November 1817, the son of a clergyman. Given the choice of studying theology or medicine, he chose the latter as being nearer his real interest, chemistry. He attended the University of Strasbourg and graduated with a medical degree in 1843. He spent a year at Giessen, where he met and studied with Justus von Liebig and August Hofmann. In 1844 he moved to Paris and became an assistant to Jean Baptiste Dumas at the Faculty of Medicine. He later succeeded him as Lecturer in Organic Chemistry (1849), Professor of Organic Chemistry (1853) and Dean of the Faculty of Medicine. In 1874, he accepted the Professorship of Organic Chemistry at the Sorbonne and was able to relinquish the heavy administrative duties of Dean and concentrate on teaching and writing. He also held public office as Mayor of the 7th Arrondissement of Paris and as a Senator. He died in Paris on 12 May 1884.

Wurtz initially worked on the oxides and oxyacids of phosphorus; in 1846 he discovered phosphorus oxychloride ($POCl_3$). He later turned to organic chemistry, at first studying aliphatic amines. In 1849 he made his famous discovery of ethylamine, the first organic derivative of ammonia (whose existence had been predicted by Liebig), and went on to prepare various amines and diamines.

In 1855, Wurtz discovered a method of producing paraffin hydrocarbons (alkanes) using alkyl halides (halogenoalkanes) and sodium in ether. The method was named the Wurtz reaction (sometimes also known as the Wurt–Fittig reaction) and has subsequently been used to synthesize hydrocarbons as high up the homologous series as $C_{60}H_{122}$. It was adapted by Edward Frankland, who substituted zinc for sodium.

Wurtz discovered ethylene glycol (1,2-ethanediol) in 1856, which led to methods of preparing glycolic acid (hydroxyethanoic acid), choline and other substances. He discovered aldol (3-hydroxybutanal) while investigating the polymerization of acetaldehyde (ethanal), devised a method of making esters from alkyl halides, and in 1867, with Friedrich Kekulé, synthesized phenol from benzene.

Throughout his work in organic chemistry, Wurtz enthusiastically applied the theory of types as propounded by Charles Gerhardt (1816–1856). He was active at a time when chemistry was undergoing great upheavals, and took clear stands on the many issues under debate.

Y

Yalow Rosalyn 1921– is a US medical physicist who won half of the 1977 Nobel Prize for Physiology or Medicine for the development of the radioimmunoassay (RIA) technique. The Nobel citation called Yalow's work 'a spectacular combination of immunology, isotope research, mathematics and physics.' Although Yalow's research concentrated on the human endocrine system – the glands and the hormones they secrete into the blood to control the organs – the RIA technique has been applied throughout medicine.

Rosalyn Sussman was born on 19 July 1921 in New York City. She graduated in physics from Hunter College, New York, in 1941 and obtained a PhD in experimental nuclear physics from the University of Illinois in 1945. There she met Aaron Yalow whom she married in 1943. Rosalyn then returned to Hunter College to teach; she also started working part-time in the Radioisotope Unit of the Veterans Administration (VA) Hospital in the Bronx. In 1950 Yalow began working full-time at the VA. For the first 22 years, Yalow worked with a medical doctor called Sol Berson. Recalling their partnership, she said: 'He wanted to be a physicist and I wanted to be a doctor.' When Berson died in 1972, their laboratory was renamed the Solomon A. Berson Research Laboratory and Yalow was appointed director. In the Radioisotope Unit at the VA, radioactive isotopes were used in diagnosis, therapy and research. Yalow and Berson started studying the thyroid gland with radioactive iodine and developed methods to measure the amount of blood in circulation and the rate of removal of proteins from the blood. Next they studied the behaviour of insulin (a small protein or peptide hormone) in the blood. They injected diabetic and non-diabetic volunteers with radioactive-labelled insulin and discovered that the insulin disappeared more slowly from the blood of the diabetics. This was surprising because diabetes was thought to result from the absence of insulin. However, the diabetics had a history of taking insulin and this 'foreign' insulin was triggering the production of antibodies in the blood. The insulin was binding to the antibodies. In developing methods to measure the antibody concentration, Yalow and Berson invented RIA. To measure the concentration of a natural hormone, a solution containing a known amount of the radioisotope-labelled form of the hormone and its antibody is prepared. When a solution containing the natural hormone is added to the first solution, some of the labelled hormone is displaced from the hormone–antibody complex. The fraction of labelled hormone displaced is proportional to the amount of the natural hormone (which is unknown). The hormone–antibody complex can then be removed from the solution and the amount of labelled hormone in each sample determined from measurements of radioactivity. This in turn enables the amount of natural hormone to be calculated. This technique is known as radioimmunoassay.

Young Charles Augustus 1834–1908 was a US astronomer who made some of the first spectroscopic investigations of the Sun.

Young was born in Hanover, New Hampshire, on 15 December 1834. From the age of 14 he was educated at Dartmouth College in Hanover, and he graduated in 1852. Initially his ambitions lay elsewhere and it may only have been family tradition that eventually drew Young towards a career in astronomy. His maternal grandfather, Ebenezer Adams, had held the Chair in Mathematics and Philosophy at Dartmouth College and had been succeeded to the Chair by his son-in-law, Ira Young, in 1833, the year before Charles was born. Young's first appointment was at the Phillips Academy, Massachusetts, teaching classics. In 1855 he enrolled, part-time, at the Andover Theological Seminary to train as a missionary. He abandoned this idea a year later, however, when he became Professor of Mathematics and Natural Philosophy at the Western Reserve College in Hudson, Ohio. Apart from a brief interlude during the American Civil War, Young stayed at the Western Reserve College until 1866, when he accepted the Chair at Dartmouth College that had previously been held by his father and grandfather. Eleven years later Young moved from Hanover to the College of New Jersey, which has since become Princeton University. Three years after retiring from his post at Princeton he died, at Hanover, on 3 January 1908.

Most of Young's serious researches in astronomy were carried out during his years at Dartmouth College. The facilities and modern equipment provided by the College inspired Young's interest in the recently developed field of spectroscopy. He was particularly interested in the Sun and many of his investigations were carried out during solar eclipses. He was the first person to observe the spectrum of the solar corona and he also discovered the reversing layer in the solar atmosphere in which the dark hues of the Sun's spectrum are momentarily reversed, but only at the moment of a total solar eclipse. Young published a series of papers relating his spectroscopic observations of the solar chromosphere, solar prominences and sunspots. He also compiled a catalogue of bright spectral lines in the Sun and used these to measure its rotational velocity.

Besides his solar research, Young wrote several excellent textbooks that introduced astronomy to succeeding generations of astronomers. His first book, *General Astronomy*, was published in 1888 and a more basic text for younger students, *Lessons in Astronomy*, was published in 1891. By 1910, 90,000 copies of these two books had been sold, making them bestsellers of their day. Young's most famous work was his *Manual Astronomy* (1902), which was aimed at a more intermediate level. It was so popular that it underwent numerous reprints and in 1926 was republished in an edition revised by Henry Russell (1877–1957).

Young John Zachary 1907– is a British zoologist whose discovery of, and subsequent work on, the giant nerve fibres in squids contributed greatly to knowledge of nerve structure and function. He also did research on the central nervous system of octopuses, demonstrating that memory stores are located in the brain. Young is probably most widely known, however, for his zoological textbooks, *The Life of Vertebrates* (1950) and *The Life of Mammals* (1957). As a result of his work, he received many honorary university degrees, and in 1967 was awarded the Royal Medal of the Royal Society of London.

Young was born on 18 March 1907 in Bristol and was educated at Wells House, Malvern Wells, and at Marlborough College, Wiltshire. He graduated from Magdalen College, Oxford, in 1928 then went to Naples as a Biological Scholar. After his return to England, Young was elected a Fellow of Magdalen College in 1931 and remained there until 1945, when he became Professor of Anatomy at University College, London – the first non-medical scientist in Britain to hold a professorship in anatomy. In the same year, he was also made a Fellow of the Royal Society. He remained at University College until he retired in 1974.

Young began his work on the nerves in squids before World War II. He discovered that certain of their nerve fibres are exceptionally thick – up to 1 mm/0.04 in in diameter (about 100 times the diameter of mammalian neurons) – and are covered with a relatively thin myelin sheath (unlike mammalian nerve fibres, which have thick sheaths), properties that make them easy to experiment on. For example, almost all the intracellular contents can be extracted without destroying the fibre's ability to conduct nerve impulses, and electrodes can easily be inserted into the fibres because of their large diameter. Moreover, extracting the contents of the giant fibres is still the only way of obtaining intracellular nerve material uncontaminated by the myelin sheath or other cells. Young's work on the giant nerve fibres in squids has been invaluable, not only because of his own findings, but also because these fibres are extremely useful for experimentation and have been used by many other researchers in their investigations of nerves.

During the war, Young set up a unit at Oxford to study nerve regeneration in mammals and, with Peter Medawar and others, devised a method of rejoining small severed nerves by using intracellular plasma as a 'glue'. Young also researched into the rates of neuron growth and the factors that determine neuron size.

After the war Young turned his attention to the central nervous system, using octopuses as research animals. Working with Brian Boycott, he showed that octopuses can learn to discriminate between different orientations of the same object – when presented with horizontal and vertical rectangles, for example, the octopuses attacked one but avoided the other. He also demonstrated (this time working with M. J. Wells) that octopuses can learn to recognize objects by touch. In addition, Young proved that the memory stores are located in the brain and proposed a model to explain the processes involved in memory.

Young Thomas 1773–1829 was a British physicist and physician who discovered the principle of interference of light, showing it to be caused by light waves. He also made important discoveries in the physiology of vision and is also remembered for Young's modulus, the ratio of stress to strain in elasticity. In addition, Young was also an Egyptologist and was instrumental in the deciphering of hieroglyphics.

Young was born in Milverton, Somerset, on 13 June 1773. He was an infant prodigy, learning to read by the age of two, whereupon he read the whole Bible twice through. Young was largely self-taught, although he did attend school. He developed great ability at languages, taking particular interest in the ancient languages of the Middle East, and mastered mathematics, physics and chemistry while still a youth. He also showed great mechanical dexterity and for a short time made optical instruments.

In 1792, under the guidance of his great-uncle Richard Brocklesby (1722–1797), a distinguished London physician, Young commenced studying for the medical profession at St. Bartholomew's Hospital. At the early age of 21 he became a member of the Royal Society. He moved to Edinburgh to continue his medical studies in 1794 and then to Göttingen in 1795, where he was awarded his MD in 1796. He travelled throughout Germany for several months, visiting museums before settling in Cambridge in 1797 for two years. He resided as a fellow-commoner at Emmanuel College, where he became known as 'Phenomenon Young'. At Cambridge Young used his time to pursue original scientific studies, and made his early investigations into interference.

Brocklesby left Young his London house and a fortune, and in 1800 Young returned to London and opened a medical practice there. In 1801, he was appointed Professor of Natural Philosophy at the Royal Institution and although he delivered many lectures, he was a disappointing lecturer to popular audiences. He resigned this post in July 1803 to concentrate on his practice as a physician. He was awarded an MB in 1803 from Cambridge University and an MD five years later. In 1811, Young became a physician at St. George's Hospital, London, and held this post until his death. He also held several other appointments, including Secretary of the Commission on Weights and Measures, Secretary of the Board of Longitude and Foreign Secretary of the Royal Society. Young died in London on 10 May 1829.

Young's early work was concerned with the physiology of vision. In 1793, he recognized that the mechanism of accommodation of the eye in focusing the eye on near or distant objects is due to a change of shape in the lens of the eye, the lens being composed of muscle fibres. Young confirmed this view in 1801 after obtaining experimental proof of it. He also showed that astigmatism, from which he himself suffered, is due to irregular curvature of the cornea. In 1801, Young was also the first to recognize that colour sensation is due to the presence in the retina of structures which respond to the three colours red, green and violet, showing that colour blindness is due to the inability of one or more of these structures to respond to light. Young's work in this field was elaborated into a proper theory of vision by Hermann Helmholtz (1821–1894) and James Clerk Maxwell (1831–1879).

Young also made a thorough study of optics at the same time. There was great controversy as to the nature of light – whether it consisted of streams of particles or of waves. In Britain, the particulate theory was strongly favoured because it had been advanced by Isaac Newton (1642–1727) a century before. In 1800, Young reopened the debate by suggesting that the wave theory of Christiaan Huygens (1629–1695) gave a more convincing explanation for the phenomena of reflection, refraction and diffraction. He assumed that light waves are propagated in a similar way to sound waves and are longitudinal vibrations but with a different medium and frequency. He also proposed that different colours consist of different frequencies.

In the following year, Young announced his discovery of the principle of interference. He explained that the bright bands of fringes in effects such as Newton's rings result from light waves interfering so that they reinforce each other. This was a hypothetical deduction and over the next two years, Young obtained experimental proof for the principle of interference by passing light through extremely narrow openings and observing the interference patterns produced.

Young's discovery of interference was convincing proof that light consists of waves, but it did not confirm that the waves are longitudinal. The discovery of the polarization

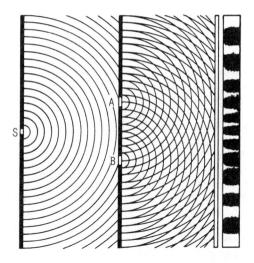

Young *Young demonstrated the interference of light by passing monochromatic light through a narrow slit S (he originally used a small hole) and then letting it pass through a pair of closely spaced slits (A and B). Reinforcement and cancellation of the wave trains as they reached the screen produced characteristic interference fringes of alternate light and dark bands.*

of light by Etienne Malus (1775–1812) in 1808, led Young to suggest in 1817 that light waves may contain a transverse component. In 1821, Augustin Fresnel (1788–1827) proved that light waves are entirely transverse, and the wave nature of light was finally established.

In mechanics, Young established many important concepts in a course of lectures published in 1807. He was the first to use the terms 'energy' for the product of the mass of a body with the square of its velocity and the expression 'labour expended' (that is, work done) for the product of the force exerted on a body 'with the distance through which it moved'. He also stated that these two products are proportional to each other. He introduced absolute measurements in elasticity by defining the modulus as the weight which would double the length of a rod of unit cross-section. Today it is usually referred to as Young's modulus, and equals stress divided by strain in the elastic region of the loading of a material. Young also published in 1805 a theory of capillary action and accounted for the angle of contact in surface-tension effects for liquids.

From 1815 onwards, Young published papers on Egyptology, mostly concerning the reading of tablets involving hieroglyphics. He was one of the first to interpret the writings on the Rosetta Stone, which was found at the mouth of the Nile in 1799. The hieroglyphic vocabulary that he established of approximately 200 signs has stood the test of time remarkably well.

Thomas Young was a scientist who possessed an extraordinary range of talents and a rare degree of insight, and he was able to initiate important paths of investigation that others were to take up and complete. His discoveries in vision are fundamental to an understanding of visual problems and to colour reproduction, and his discovery of interference was a vital step towards a full explanation of the nature of light.

Yukawa Hideki 1907–1981 was a Japanese physicist famous for his important theoretical work on elementary particles and nuclear forces, particularly for predicting the existence of the pi-meson (or pion) and the short-range strong nuclear force associated with this particle. He received numerous honours for his work, including the 1949 Nobel Prize for Physics, being the first Japanese to receive a Nobel prize.

Yukawa was born in Kyoto on 23 January 1907 and was educated at Kyoto University (where his father was Professor of Geology), graduating in 1929. He then moved to Osaka University, where he both taught and studied for his doctorate, which he gained in 1938. In the following year he returned to Kyoto University and, except for various visiting professorships, remained there for the rest of his career – as Professor of Theoretical Physics from 1939 to 1950, Emeritus Professor from 1950 to 1953, and Director of the university's newly created Research Institute for Fundamental Physics from 1953. His visiting professorships included one to the Institute for Advanced Study, Princeton University (at the invitation of Julius Oppenheimer) from 1948 to 1949, and one to Columbia University from 1949 to 1953.

Yukawa proposed his new theory of nuclear forces in 1935 at a time when there was much controversy over what holds the nucleus together. In 1932, James Chadwick had discovered the neutron but, Yukawa pointed out, this meant that because the nucleus contains only positively charged protons and neutral neutrons, the protons should repel each other and disrupt the integrity of the nucleus. He therefore postulated the existence of a nuclear 'exchange force' that counteracted the mutual repulsion of the protons and therefore held the nucleus together. He predicted that this exchange force would involve the transfer of a particle (the existence of which was then unknown), that the force would be of very short range (effective over a distance of only about 10^{-8} m) and that it would be strong enough to overcome the repulsive forces between the protons, yet decrease in intensity with sufficient rapidity so as to have a negligible effect on the innermost electrons. He also calculated – using quantum theory – that his predicted exchange particle would have a mass of about 200 times that of an electron (but would have the same charge as an electron) – about one-ninth the mass of a proton or neutron – and that the particle would be radioactive, with an extremely short half-life. At the time no particle having these characteristics had been found. In the following year (1936) Carl Anderson discovered the muon (or mu-meson) which possesses several of the properties of Yukawa's predicted particle but not all of them. In 1947, however, Cecil Powell discovered the pion (or pi-meson), a particle similar to Anderson's muon but which fulfilled all of the requirements of Yukawa's exchange particle.

Meanwhile, in 1936 Yukawa predicted that a nucleus could absorb one of the innermost orbiting electrons and that this would be equivalent to emitting a positron. These innermost electrons belong to the 1K electron shell, and this process of electron absorption by the nucleus is known as K capture. In 1947, Yukawa began working on a more detailed theory of elementary particles.

Z

Zassenhaus Hans Julius 1912– is a German mathematician whose main area of study has been in the field of group theory and number theory. Perhaps, however, his life has been more concerned with teaching than with publishing his own results.

Born in Coblenz on 28 May 1912, Zassenhaus completed his education by entering the University of Hamburg, from which he received his doctorate in 1934, qualifying as a lecturer there in the same year. For 15 years he worked there and at Rostock University; then he spent ten years as Redpath Professor of Mathematics in Montreal, Canada. Transferring to the United States, he finally became Professor of Mathematics at Ohio State University in 1964.

Zassenhaus's research work has concentrated on group theory – the study of systems in which the product of any two members of a system results in another member of the same system (for example, even numbers) – and on number theory. His most significant results have been obtained in his investigations into finite groups, a class of which has been named Zassenhaus groups and forms part of the basis for the contemporary development of finite group theory. The work of Zassenhaus and Schur on group extensions led to the postulation of what has become known as the Schur–Zassenhaus theory.

Zassenhaus has also made contributions to the study of Lie algebra, the geometry of numbers, and applied mathematics.

Zeeman Erik Christopher 1925– is a British mathematician noted for his work in topology and for his research into models of social behaviour that accord with the relatively recent formulation of catastrophe theory.

Born on 4 February 1925, Zeeman was educated at Christ's Hospital, Sussex. At the age of 18, during World War II, he joined the Royal Air Force and became a Flying Officer. After the war he took up a scholarship to Christ's College, Cambridge, receiving his bachelor's degree in 1948, his master's degree in 1950, and a doctorate in 1954. From then for ten years he was a Fellow of Gonville and Caius College, Cambridge, and a lecturer there. Thereafter,

he became Professor of Mathematics at the University of Warwick, where he founded – and became Director of – the Mathematics Research Centre. He has held visiting Professorships in various countries in Europe, in the United States and in Brazil. Elected a Fellow of the Royal Society in 1975, the following year he was made Senior Fellow of the Science Research Council.

Between 1951 and 1958 Zeeman researched into algebraic topology; between 1958 and 1967 he studied geometric topology and brain modelling; and his interests since then have been concerned with catastrophe theory and its applications to the physical, biological and behavioural sciences, and with dynamical systems.

Probably Zeeman's most significant work has been on the recently formulated catastrophe theory. Mathematical analysis cannot easily be used for things that change suddenly, or that have intermittent fits and starts. Where the change is smooth and continuous, it can be described in terms of differential equations. (Newton's laws of motion and gravitation, and James Clerk Maxwell's theory of electromagnetism, can all be described in terms of differential equations. Even Albert Einstein's general theory of relativity culminates in a set of differential equations.) A mathematical method for dealing with discontinuous and divergent phenomena was only developed in 1968, and derived from topology – the branch of geometry concerned with the properties of shapes and figures that remain unchanged even when the shape or figure is deformed. This method, the catastrophe theory, has great potential; it can describe the evolution of forms in all aspects of nature and so has great generality. It can be applied to situations where gradually changing forces lead to abrupt changes in behaviour. Many events in physics have now been identified as examples of catastrophe theory, but ultimately its most important applications may be in the 'inexact' sciences – biology and social science.

The theory was first devised by René Thom (1923–), who saw it as a development of topology in that the underlying forces in nature can be described as the smooth surfaces of equilibrium, and when equilibrium breaks down, catastrophe occurs. The problem for catastrophe theory then is to describe shapes of all the possible equilibrium surfaces. Thom showed that for processes controlled by not more than four factors, there are two elementary catastrophes. His proofs are very complicated, but the results are easier to understand and can be applied to scientific problems without reference to the proof.

If the two elementary catastrophes (two conflicting behavioural drives, for example) are plotted as axes on the horizontal plane – called the control surface – and the complementary result (the resulting behaviour, to carry the example further) is plotted on a third axis perpendicular to the first two, from most likely result to next most likely, and so on, resultant points can be plotted for the entire control surface, and when connected form a surface of their own. Concerning behavioural results, the application in which Zeeman was particularly interested, this surface is known as the behaviour surface. Catastrophe theory reveals that in the middle of the surface is a pleat, without creases, which becomes narrower towards the back of the surface, and it is this pleat which gives the model its most interesting characteristics. For Zeeman, all the points on the behaviour surface represent the most probable behaviour, with the exception of those on the pleated middle part, which represent the least likely behaviour. At the edge of the pleat, the sheet on which the behaviour points have been travelling folds under and is wiped out. The behaviour state falls to the bottom sheet of the graph and there is a sudden change in behaviour. In an argument, for instance, an aggressive protagonist may waver in his opinion, abandon his position and apologize; the timid opponent may make repeated concessions, then lose his temper and become aggressive.

Zeeman, with J. Hevesi, a psychotherapist, has in particular worked on the problems of anorexia nervosa, an illness especially common in adolescent girls, which involves sudden changes from dieting to obsessive fasting. Catastrophe theory has been able to predict behaviour patterns in this disorder and help towards treating the sufferers.

The theory is still a young offshoot of science. But it has already been applied to the propagation of shock waves, the minimum areas of surfaces, to nonlinear oscillations, to scattering and to elasticity. Zeeman has constructed catastrophe models of heartbeat, the propagation of nerve impulses, and the formation of the gastrula and of somites in an embryo. Considerable research on the subject still remains to be done.

Zeeman Pieter 1865–1943 was a Dutch physicist who discovered the Zeeman effect, which is the splitting of spectral lines in an intense magnetic field. This achievement was important in determining the structure of the atom and Zeeman shared the 1902 Nobel Prize for Physics with Hendrik Lorentz (1853–1928), who had predicted the Zeeman effect.

Zeeman was born at Zonnemaire, Zeeland, on 25 May 1865. He was educated at local schools and at the Gymnasium in Delft before entering the University of Leiden in 1885. There Zeeman studied under Heike Kamerlingh Onnes (1853–1926) and Lorentz, gaining his PhD with a

dissertation on the Kerr effect in 1893. He then remained at Leiden as a tutor and in 1897 moved to the University of Amsterdam to take up a lectureship. Zeeman became Professor of Physics at Amsterdam in 1900. He retained this position until he retired in 1935, and in 1923 became director of a new laboratory that was named the Zeeman Laboratory. In addition to the Nobel prize, among his many honours was the award of the Royal Society's Rumford Medal in 1922. Zeeman died at Amsterdam on 9 October 1943.

While Zeeman was at Leiden, Lorentz proposed that light is caused by the vibration of electrons and suggested that imposing a magnetic field on light would result in a splitting of spectral lines by varying the wavelengths of the lines. Zeeman undertook the first experimental work in search of this in 1896. Using a sodium flare between the poles of a powerful electromagnet and producing spectra with a large concave diffraction grating, Zeeman was able to detect a broadening of the spectral lines when the current was activated. A similar effect was achieved with the sodium absorption spectra, so the changed shape of the flame was not responsible. In 1897, Zeeman refined the experiment and was successful in resolving the broadening of the narrow blue-green spectral line of cadmium produced in a vacuum discharge into a triplet of three component lines. Later work led Zeeman to evaluate the ratio e/m for the oscillating particles involved, which was in agreement with the value obtained for the electron by J. J. Thomson (1856–1940). This confirmed that the magnetic field was affecting the forces which control the electrons within the atom. However, Zeeman's subsequent experimental work suffered as a consequence of his promotion to the University of Amsterdam, where the facilities were much poorer than in Leiden. By the time he was able to acquire a purpose-built laboratory in 1923, many other workers had overtaken Zeeman's team in expertise and experience.

Study of the Zeeman effect led to important theoretical advances in physics. Zeeman's observations confirmed Lorentz's electromagnetic theory and later investigators were able to show that the spectral effects are caused by electron spin. As a result, the quantum theory was expanded to include these findings. Spectral observations of the Sun carried out by George Hale (1868–1938) in 1908 led him to believe that light emitted by sunspots is affected in a similar fashion to Zeeman's laboratory observations. The conclusion reached was that sunspots must be associated with intense magnetic fields within the Sun.

In later years, Zeeman's attention turned to the velocity of light in moving media and his experiments involved glass and quartz. Many difficulties were successfully overcome and Zeeman was able to show that the results were in agreement with the theory of relativity. He also studied isotopes, particularly argon, in which he identified a new isotope with a mass number of 38.

Zel'dovich Yakov Borisovich 1914–1987 was a Soviet astrophysicist who was originally a specialist in nuclear physics, but who became interested in particle physics and cosmology during the 1950s.

Zel'dovich was born in Minsk on 18 March 1914. He studied at the University of Leningrad, and when he graduated in 1931 he began his work at the Soviet Academy of Sciences in the Institute of Chemical Physics. He was made a Corresponding Member of the Academy of Sciences in 1946 and became a full Academician in 1958. During World War II he contributed research towards the war effort and was awarded the Stalin Prize in 1943. He later worked at the Institute of Cosmic Research at the Space Research Institute of the Soviet Academy of Sciences in Moscow.

During the 1930s, Zel'dovich participated in a research programme that was aimed at discovering the mechanism of oxidation of nitrogen during an explosion, and he and his colleagues reported the results of this work in a book published in 1947. During the 1940s, he also maintained an interest in, and wrote about, the chemical reactions of explosions, the subsequent generation of shockwaves and the related subjects of gas dynamics and flame propagation. Zel'dovich, together with Y. B. Khariton, participated in the early work on the mechanism of fission during the radioactive decay of uranium, one of the most significant discoveries of the late 1930s. Their calculations on the chain reaction in uranium fission were published in 1939 and 1940. Besides this, Zel'dovich was also interested in the role played by slow neutrons in the fission process.

It was not until the 1950s that he began to develop an interest in cosmology and Zel'dovich's later writings dealt with such diverse subjects as quark annihilation, neutrino detection, and the applicability of relativistic versus Newtonian theories to the study of the expanding and evolving universe and the earliest stages of the universe – the quantum, hadron and lepton eras. In 1967, together with C. W. Misher, A. G. Doroshkevich and I. D. Novikov, he proposed that in its initial stages the universe was highly isotropic, but that as it has expanded, this isotropy has diminished.

Zel'dovich was a prolific writer and carried out extensive research in the fields of physics and astronomy. His later cosmological theories led to more accurate determinations of the abundance of helium in older stars.

Zeppelin Ferdinand von 1838–1917 was a German soldier and builder of airships.

Zeppelin was born in Constance, Baden, on 8 July 1838. He was educated in Stuttgart and trained to enter the army, which he did as an infantry officer in 1858. In the early 1860s he was appointed to Potamac's forces. This enabled him to join an expedition to explore the sources of the river Mississippi, and in 1870 at Fort Snelling, Minnesota, he made his first ascent in a (military) balloon. After returning home Zeppelin rose to the rank of Brigadier-General. He finally retired from the army in 1891 and turned his energies to the design and construction of an airship.

After many setbacks, and helped by royal patronage and public subscriptions, he launched his first craft in July 1900. Eight years later he made a 12-hour flight to Lucerne, Switzerland. His exploits aroused the enthusiasm of the German people, who raised a national fund of more than 6 million marks with which he founded the Zeppelin Institution. Many airships – now called Zeppelins – were built in the period leading up to World War I, which offered a chance to test their military potential. But enthusiasm for the machine rapidly waned in 1914, the first year of the war, when 13 Zeppelins were destroyed in action and many lives lost. Their chief targets – they were used as bombers – were in Belgium and England. Their vulnerability to anti-aircraft guns and the rapid development of faster and more manouevrable aeroplanes sealed the fate of the airship as an effective weapon of war. Zeppelin died before the end of the war, on 8 March 1917, at Charlottenburg near Berlin.

The early airships – the LZ5, for example – had chain-driven propellors and, in the stern, multiple rudders and elevators. The main principle of Zeppelin's invention was streamlining the all-over envelope, inside of which separate hydrogen-filled gasbags were raised inside a steel skeleton structure. In some craft, as in the design of the LZ1, a balancing rod ran the whole length of the ship below the envelope, and could be used for horizontal trimming. On the LZ18, German naval engineers incorporated several improvements by providing a covered-in passageway and fully enclosed cars directly beneath the hull. The disastrous end to the *Hindenburg*, which burst into flames over Lakehurst on its momentous flight via New York, effectively put an end to any airship being inflated with hydrogen. Since then the non-inflammable helium has almost always been used.

Zermelo Ernst Friedrich Ferdinand 1871–1953 was a German mathematician who made important contributions to the development of set theory, particularly in developing the axiomatic set theory that now bears his name.

Zermelo was born on 27 July 1871 in Berlin, the son of a university professor. Educated locally, he passed his final examinations in 1889 and went to study mathematics, physics and philosophy first at Halle University, then at Freiburg University. Among his teachers were Georg Frobenius (1849–1917), Immanuel Fuchs (1832–1902) and Max Planck (1858–1947). Zermelo received his doctorate at Berlin in 1894 and went to Göttingen where, five years later, he was appointed as an unsalaried lecturer. In 1904, he supplied a proof of the well-ordering theorem, and in the following year he was appointed professor. A few years later he moved to Zurich after accepting a Professorship there; poor health forced him to resign this post in 1916. A gift of 5,000 marks from the Wolfskehl Fund enabled him to live quietly in the Black Forest for the next ten years, restoring himself once more to full health. In 1926 he was appointed Honorary Professor at the University of Freiburg in Breisgau. He stayed there for nine years, but in 1935 resigned his post once more, this time in protest against the Nazi regime. He was reinstated in 1946 at his own request, and remained in Freiburg until he died, on 21 May 1953.

Zermelo's first research was on the applications of the theorem formulated by Henri Poincaré (1854–1912). He detected some apparent anomalies in kinetic theory, following Poincaré's research, notably that the recurrence theorem arising from Poincaré's work in mechanics seems to make any mechanical model such as the kinetic theory incompatible with the second law of thermodynamics.

It was in 1900 that Zermelo turned his attention to set theory. He provided an ingenious proof to the well-ordering theorem, which states that every set can be well ordered (that is, can be arranged in a series in which each subclass – not being null – has a first term).

He said that a relation $a < b$ (a comes before b) can be introduced such that for any two statements a and b, either $a = b$, or $a < b$ or $b < a$. If there are three elements a, b and c, then if $a < b$ and $b < c$, then $a < c$. This gave rise to the Zermelo axiom that every class can be well ordered.

Zermelo subsequently pointed out that for any infinite system of sets, there are always relations under which every set corresponds to one of its elements. Because of a storm of criticism from other mathematicians, however, he felt obliged to produce a second proof.

In 1904, Zermelo defined the axiom of choice, the use of which had previously been unrecognized in mathematical reasoning. The first formulations of axioms for set theory – an axiom system for Georg Cantor's theory of sets – were made by Zermelo in 1908. This system has since proved of great value in the development of mathematics. There are seven axioms (which found written formulation rather later), in which only two terms are used: the set, and the element of the set. Every set except the null set is an object of B for which there is another object b of B such that $a \in b$. The axioms state:

(1) that $m = n$ if, and only if, $a \in m$ is equivalent to $a \in n$;

(2) that there is a null set;

(3) that if a property E is definite for the elements of a set m, then there is a subset m_E of m consisting of exactly those elements of m for which E holds;

(4) that for any set m there is a set $P(m)$ that has the subsets of m for its elements;

(5) that for any set m there is a set $\cup m$ (the union of m) consisting of the elements of m;

(6) that if m is a set of disjoint non-void sets, then $\cup m$ contains a subset n that contains exactly one element from every set of m;

(7) and that there is a set A that has the null set as an element and has the property that, if x is an element of A, then is also an element of A.

However, Zermelo's axioms involved an unexplained notion of a 'definite property'. This difficulty was overcome later by Abraham Fraenkel (1891–1965). Fraenkel had criticized Zermelo's conclusions, particularly axiom 7, because of weakness. Zermelo afterwards improved on them by describing the set of definite properties as the smallest set containing the basic relations of the domain B and satisfying certain close conditions. A logical formula-tion of Zermelo's axioms was later achieved by Thoralf Skolem in 1923.

Ziegler Karl 1898–1973 was a German organic chemist famous for his studies of polymers, for which he shared the 1963 Nobel Prize for Chemistry with Giulio Natta.

Ziegler was born at Helsa, near Kassel, on 26 November 1898, the son of a clergyman. He gained his doctorate from Marburg University in 1923 and then held teaching appointments at Frankfurt-am-Main, Heidelberg and Halle. In 1943, he became Director of the Kaiser Wilhelm (later Max Planck) Institute for Coal Research at Mülheim, where he remained for the rest of his career. He died in Mülheim on 12 August 1973.

In 1933, Ziegler discovered a method of making compounds that contain large rings of carbon atoms, later used to synthesize musks for making perfumes, and in 1942 he developed the use of N-bromosuccinimide for brominating olefins (alkenes). In 1945, after World War II, he began research on the organic compounds of aluminium. He found a method of synthesizing aluminium trialkyls from aluminium metal, hydrogen and olefins and demonstrated that it is possible to add ethylene (ethene) stepwise to the aluminium–carbon bond of aluminium trialkyls to make higher aluminium trialkyls. These higher compounds can be converted into alcohols for use in the manufacture of detergents. He also discovered that nickel will catalyse the exchange of groups attached to aluminium by ethylene and liberate higher olefins. Using electrochemical techniques, he prepared various other metal alkyls from the aluminium ones, the most important of which is tetraethyl lead, which is used as an anti-knock additive to petrol.

His most important discovery came in 1953 when he and a student (E. Holzkamp) were repeating a preparation of higher aluminium trialkyls by heating ethylene with aluminium trialkyl. To their surprise the ethylene monomer ($CH_2=CH_2$) was completely converted to the dimer butylene (but-1,2-ene, $CH_3CH_2CH=CH_2$). They found the explanation to be that the autoclave used for the experiment contained traces of colloidal nickel left from a previous catalytic hydrogenation experiment. This led them to the discovery that organometallic compounds mixed with certain heavy metals polymerize ethylene at atmospheric pressure to produce a linear polymer of high molecular weight (relative molecular mass) and with valuable properties, such as high melting point. All previous processes have the disadvantage of needing high pressures and produced low-melting, partly branched polymers.

Also in 1953, Ziegler and Natta discovered a family of stereo-specific catalysts which are capable of introducing an exact and regular structure to various polymers. They found that they could use the Ziegler-type catalyst triethyl aluminium combined with titanium tetrachloride (titanium(IV) chloride) to polymerize isoprene so that each molecule in the long-chain polymer formed is in a regular position and almost identical to the structure of natural rubber. This discovery formed the basis of nearly all later developments in synthetic plastics, fibres, rubbers and films derived from such olefins as ethylene (ethene) and butadiene (but-1,2:3,4-diene).

Zsigmondy Richard Adolf 1865–1929 was an Austrian-born German colloid chemist who invented the ultramicroscope. For this achievement, and his other work with colloids, he was awarded the 1925 Nobel Prize for Chemistry.

Zsigmondy was born in Vienna on 1 April 1865, the son of a dentist. His early education and first year at University were at Vienna, then he went to Munich and obtained his PhD in organic chemistry in 1889. He became a research assistant in Berlin and then a lecturer in chemical technology at the Technische Hochschule in Graz. In 1897, he joined the Glass Manufacturing Company in Jena, but left in 1900 to carry out his own private research. From 1908 he was Professor of Inorganic Chemistry at Göttingen University. He retired a few months before his death, in Göttingen, on about 23 September 1929.

In Berlin, Zsigmondy worked with the physicist August Kundt (1839–1894) on inorganic inclusions in glass. At Jena he became concerned with coloured and turbid glasses and he invented the famous Jena milk glass. This was the work that aroused his interest in colloids, because it is colloidal inclusions that give glass its colour or opacity. He recognized that the red fluids first prepared by Michael Faraday by the reduction of gold salts are largely colloidal analogues of ruby glass and worked out a technique for preparing them reproducibly. His belief that the suspended particles in such gold sols are kept apart by electric charges was generally accepted, and the sols became model systems for much of his later work on colloids.

In 1903, working with H.F.W. Siedentopf, he constructed the first ultramicroscope, with which it is possible to view individual particles in a colloidal solution. Unlike a conventional microscope, in which the illumination is parallel with the instrument's axis, the ultramicroscope uses perpendicular illumination. With such dark-field illumination, individual particles become visible by scattering light (the Tyndall effect), much as moving dust particles are illuminated by a sunbeam. Furthermore the technique makes it possible to detect particles much smaller than the resolving power of the microscope. Ernst Abba, Director of the Jena Glass Company, put all the company's facilities at Zsigmondy's disposal to develop the apparatus, even though at that time he had no formal links with the firm and no professional attachments at all. It was for this work that Zsigmondy was made a professor at Göttingen.

Using the ultramicroscope Zsigmondy was able to count the number of particles in a given volume and indirectly estimate their sizes; he could detect particles down to a diameter of 3 nm (3×10^{-9} m). Much of his research continued to centre on gold sols ('purple of Cassius'). Several noted chemists had already studied such sols, but it was not known whether they contained a mixture or a compound. In 1898, he showed that it is a mixture of very small gold and stannic acid particles. He also showed that colour changes in sols reflect changes in particle size caused by

coagulation when salts are added, and that the addition of agents such as gelatin stabilizes the colloid by inhibiting coagulation. At Göttingen he investigated ultrafiltration and its use for colloids; he also studied such systems as silica gels and soap gels.

Zsigmondy's work began a study of importance to the understanding of all sols, smokes, fogs, foams and films. His conclusions clarified problems in biochemistry, bacteriology and soil physics. The ultramicroscope has remained of great importance in colloid research, although somewhat superseded by the electron microscope.

Zwicky Fritz 1898–1974 was a Swiss astronomer and astrophysicist who was distinguished for his discoveries of supernovae, dwarf galaxies and clusters of galaxies and also for his theory on the formation of neutron stars.

Zwicky was born in Varna, Bulgaria, but his parents were Swiss and he retained his Swiss nationality throughout his life. He was educated in Switzerland, gaining his BA and his PhD by 1922 from the Federal Institute of Technology at Zurich. He was awarded a Fellowship from the International Education Board in 1925 and left Switzerland for the United States to join the California Institute of Technology (Caltech). He was appointed Assistant Professor at Caltech in 1927 and continued to work there until his retirement in 1968, by which time he had been promoted to the position of Professor of Astronomy. After his retirement, Zwicky continued to live in the United States. He received the Royal Astronomical Society's Gold Medal in 1973. He died on 8 February 1974.

Zwicky began his research by scouring our neighbouring galaxies for the appearance of a supernova explosion, hoping to discover one that was bright enough for its spectrum to be studied. But since the time when Johannes Kepler and Tycho Brahe observed their rare sightings of such events, no other supernovae have been seen to appear in our Galaxy. Zwicky therefore calculated that only one supernova appears every 300–400 years in any galaxy. He was among the first to suggest that there is a relationship between supernovae and neutron stars. He suggested that the outer layers of a star that explodes as a supernova leave a core that collapses upon itself as a result of gravitational forces. He put forward this theoretical model in the early 1930s, when there seemed to be no hope of actually observing such a phenomenon.

In 1936, Zwicky began a study of galaxy clusters. He used the 4-cm/18-in Schmidt telescope at Mount Palomar Observatory to photograph large areas of the sky. This telescope was specially designed to provide a relatively wide field of view, so that a large portion of the sky could be viewed at one glance without sacrificing a high resolution of separate images. Zwicky observed that most galaxies occur in clusters, each of which contains several thousand galaxies. The nearest is the Virgo cluster, which is also the most conspicuous of large clusters. It contains a number of spiral galaxies and Zwicky's spectroscopic studies of the Virgo and the Coma Berenices clusters showed that there is no evidence of any systematic expansion or rotation of clusters. Zwicky also calculated that the distribution of galaxies in the Coma Berenices cluster was similar, at least statistically, to the distribution of molecules in a gas when its temperature is at equilibrium. He compiled a six-volume catalogue of galaxies and galaxy clusters in which he listed 10,000 clusters located north of declination −30°. He completed the catalogue shortly before his death and it is still generally regarded as the classic work in this field.

Zwicky's research interests were not limited to astronomy, but extended to the study of crystal structure, superconductivity, rocket fuels, propulsive systems and the philosophy of science. But his work on galaxies, galaxy clusters, interstellar matter and supernova stars outweighs these other interests in importance and has made a vital contribution to the field of astronomy.

Zworykin Vladimir Kosma 1889–1982 was a Russian-born US electronics engineer and inventor whose major inventions – the iconoscope television camera tube and the electron microscope – have had ramifications far outside the immediate field of electronics. Electronic television has become the major entertainments medium, and the electron microscope has proved to be a key tool in the development of molecular biology and microbiological research.

Zworykin was born in Murom, Russia, and received his higher education at the St Petersburg Institute of Technology, from which he graduated with a degree in electrical engineering in 1912. He then went to Paris to do X-ray research at the College of France, but at the outbreak of World War I in 1914 he returned to Russia, where he remained for the next four years working as a radio officer. When the war ended in 1918 he travelled widely throughout the world, before deciding to settle in the United States.

Having learnt to speak English, Zworykin joined the Westinghouse corporation in Pittsburgh, Pennsylvania, and in 1923 took out a patent for the iconoscope, followed a year later by the kinescope (a television receiver tube). In 1929 he demonstrated an improved electronic television system and was then offered the position of Director of Electronic Research for the Radio Corporation of America (RCA) at Camden, New Jersey. He subsequently moved to the nearby Princeton University to continue the development of television. He obtained his PhD degree from the University of Princeton in 1926.

In 1967, Zworykin was awarded the National Medal of Science by the National Academy of Sciences for his contributions to science, medicine and engineering and for the application of electronic engineering to medicine.

Among the first of Zworykin's developments was an early form of an electric eye. Somewhat later, he invented an electronic image tube sensitive to infrared light which was the basis for World War II inventions for seeing in the dark.

In 1957, Zworykin patented a device which uses ultraviolet light and television, thereby permitting a colour picture of living cells to be thrown upon a screen, which opened up new prospects for biological investigation. It is, however, the electron microscope – to the development of which Zworykin also contributed – that represents the greatest boon that physics has given to biology.

Vastly extending the range of detail covered by an optical microscope, the electron microscope uses a beam of electrons to form a magnified image of a specimen. Useful magnifications of 1 million times can be obtained on the viewing screen of a powerful electron microscope, producing an amplification sufficiently large to disclose a disarranged cluster of atoms in the lattice of a crystal. (Optical microscopes have a useful magnification range of only several thousand times.)

The high degree of magnification of an electron microscope comes from its extremely low resolution. If, for example, a microscope has a resolution of a thousandth of a millimetre, it can reveal objects larger than that size. Smaller objects however appear blurred or distorted. The electron microscope image has a resolution several thousand times better than that achieved by the best optical microscope because the resolving power of the latter is limited by the wavelength of light, whereas the effective wavelength of an electron beam is several thousand times shorter.

Electron microscopes may be classified as either transmission or scanning instruments. With the former the image is usually produced by an electron beam which has passed through the specimen. In scanning instruments a finely focused electron beam sweeps through the specimen, and the image is formed by a process similar to that used in television.

Appendices

Nobel Prize for Chemistry

prizewinners

1901	Jacobus van't Hoff (Netherlands): laws of chemical dynamics and osmotic pressure
1902	Emil Fischer (Germany): sugar and purine syntheses
1903	Svante Arrhenius (Sweden): electrolytic theory of dissociation
1904	William Ramsay (UK): inert gases in air and their locations in the periodic table
1905	Adolf von Baeyer (Germany): organic dyes and hydroaromatic compounds
1906	Henri Moissan (France): isolation of fluorine and adoption of electric furnace
1907	Eduard Buchner (Germany): biochemical researches and discovery of cell-free fermentation
1908	Ernest Rutherford (New Zealand): atomic disintegration and the chemistry of radioactive substances
1909	Wilhelm Ostwald (Germany): catalysis and principles of equilibria and rates of reaction
1910	Otto Wallach (Germany): alicyclic compounds
1911	Marie Curie (Poland): discovery of radium and polonium, and the isolation and study of radium
1912	Victor Grignard (France): discovery of Grignard reagent
	Paul Sabatier (France): catalytic hydrogenation of organic compounds
1913	Alfred Werner (Switzerland): bonding of atoms within molecules
1914	Theodore Richards (USA): accurate determination of the atomic masses of many elements
1915	Richard Willstäter (Germany): research into plant pigments, especially chlorophyll
1916–17	no prizes awarded
1918	Fritz Haber (Germany): synthesis of ammonia from its elements
1919	no prizes awarded
1920	Walther Nernst (Germany): work on thermochemistry
1921	Frederick Soddy (UK): work on radioactive substances, especially isotopes
1922	Francis Aston (UK): mass spectrometry of isotopes of radioactive elements, and enunciation of the whole-number rule
1923	Fritz Pregl (Austria): microanalysis of organic substances
1924	no prizes awarded
1925	Richard Zsigmondy (Austria): heterogeneity of colloids
1926	Theodor Svedberg (Sweden): investigation of dispersed systems
1927	Heinrich Wieland (Germany): constitution of bile acids and related substances
1928	Adolf Windaus (Germany): constitution of sterols and related vitamins
1929	Arthur Harden (UK) and Hans von Euler-Chelpin (Germany): fermentation of sugar and fermentative enzymes
1930	Hans Fischer (Germany): analysis of haem (the iron-bearing group in haemoglobin) and chlorophyll and the synthesis of haemin (a compound of haem)
1931	Carl Bosch (Germany) and Friedrich Bergius (Germany): invention and development of chemical high-pressure methods
1932	Irving Langmuir (USA): surface chemistry
1933	no prizes awarded
1934	Harold Urey (USA): discovery of deuterium (heavy hydrogen)
1935	Irène and Frédéric Joliot-Curie (France): synthesis of new radioactive elements
1936	Peter Debye (Netherlands): work on molecular structures by investigation of dipole moments and the diffraction of X-rays and electrons in gases
1937	Norman Haworth (UK): work on carbohydrates and ascorbic acid (vitamin C)
	Paul Karrer (Switzerland): work on the structure of carotenoids, flavins, retinol (vitamin A) and riboflavin (vitamin B_2)
1938	Richard Kuhn (Austria): carotenoids and vitamins
1939	Adolf Butenandt (Germany): work on sex hormones
	Leopold Ruzicka (Switzerland): polymethylenes and higher terpenes
1940–42	no prizes awarded
1943	Georg von Hevesy (Sweden): use of isotopes as tracers in chemical processes
1944	Otto Hahn (Germany): discovery of nuclear fission
1945	Artturi Virtanen (Finland): agriculture and nutrition, especially fodder preservation
1946	James Sumner (USA): crystallization of enzymes

	John Northrop (USA) and Wendell Stanley (USA): preparation of pure enzymes and virus proteins
1947	Robert Robinson (UK): biologically important plant products, especially alkaloids
1948	Arne Tiselius (Sweden): electrophoresis and adsorption analysis and discoveries concerning serum proteins
1949	William Giauque (USA): chemical thermodynamics, especially at very low temperatures
1950	Otto Diels (Germany) and Kurt Alder (Germany): discovery and development of diene synthesis
1951	Edwin McMillan (USA) and Glenn Seaborg (USA): chemistry of transuranic elements
1952	Archer Martin (UK) and Richard Synge (UK): invention of partition chromatography
1953	Hermann Staudinger (West Germany): discoveries in macromolecular chemistry
1954	Linus Pauling (USA): nature of chemical bonds, especially in complex substances
1955	Vincent Du Vigneaud (USA): investigations into biochemically important sulphur compounds and the first synthesis of a polypeptide hormone
1956	Cyril Hinshelwood (UK) and Nikolay Semenov (USSR): mechanism of chemical reactions
1957	Alexander Todd (UK): nucleotides and nucleotide coenzymes
1958	Frederick Sanger (UK): structure of proteins, especially insulin
1959	Jaroslav Heyrovsky´ (Czechoslovakia): polarographic methods of chemical analysis
1960	Willard Libby (USA): radiocarbon dating in archaeology, geology and geography
1961	Melvin Calvin (USA): assimilation of carbon dioxide by plants
1962	Max Perutz (UK) and John Kendrew (UK): structures of globular proteins
1963	Karl Ziegler (West Germany) and Giulio Natta (Italy): chemistry and technology of high polymers
1964	Dorothy Crowfoot Hodgkin (UK): crystallographic determination of the structures of biochemical compounds, notably penicillin and cyanocobalamin (vitamin B_{12})
1965	Robert Woodward (USA): organic synthesis
1966	Robert Mulliken (USA): molecular orbital theory of chemical bonds and structures
1967	Manfred Eigen (West Germany), Ronald Norrish (UK) and George Porter (UK): investigation of rapid chemical reactions by means of very short pulses of energy
1968	Lars Onsager (USA): discovery of reciprocal relations, fundamental for the thermodynamics of irreversible processes
1969	Derek Barton (UK) and Odd Hassel (Norway): concept and applications of conformation
1970	Luis Federico Leloir (Argentina): discovery of sugar nucleotides and their role in carbohydrate biosynthesis
1971	Gerhard Herzberg (Canada): electronic structure and geometry of molecules, particularly free radicals
1972	Christian Anfinsen (USA), Stanford Moore (USA) and William Stein (USA): amino-acid structure and biological activity of the enzyme ribonuclease
1973	Ernst Fischer (West Germany) and Geoffrey Wilkinson (UK): chemistry of organometallic sandwich compounds
1974	Paul Flory (USA): physical chemistry of macromolecules
1975	John Cornforth (Australia): stereochemistry of enzyme-catalysed reactions
	Vladimir Prelog (Yugoslavia): stereochemistry of organic molecules and their reactions
1976	William N. Lipscomb (USA): structure and chemical bonding of boranes (compounds of boron and hydrogen)
1977	Ilya Prigogine (USSR): thermodynamics of irreversible and dissipative processes
1978	Peter Mitchell (UK): biological energy transfer and chemiosmotic theory
1979	Herbert Brown (USA) and Georg Wittig (West Germany): use of boron and phosphorus compounds, respectively, in organic syntheses
1980	Paul Berg (USA): biochemistry of nucleic acids, especially recombinant-DNA
	Walter Gilbert (USA) and Frederick Sanger (UK): base sequences in nucleic acids
1981	Kenichi Fukui (Japan) and Roald Hoffmann (USA): theories concerning chemical reactions
1982	Aaron Klug (UK): crystallographic electron microscopy: structure of biologically important nucleic-acid–protein complexes
1983	Henry Taube (USA): electron-transfer reactions in inorganic chemical reactions
1984	Bruce Merrifield (USA): chemical syntheses on a solid matrix
1985	Herbert A. Hauptman (USA) and Jerome Karle (USA): methods of determining crystal structures
1986	Dudley Herschbach (USA), Yuan Lee (USA) and John Polanyi (Canada): dynamics of chemical elementary processes
1987	Donald Cram (USA), Jean-Marie Lehn (France) and Charles Pedersen (USA): molecules with highly selective structure-specific interactions

1988   Johann Deisenhofer (West Germany), Robert Huber (West Germany) and Hartmut Michel (West Germany): three-dimensional structure of the reaction centre of photosynthesis
1989   Sydney Altman (USA) and Thomas Cech (USA): discovery of catalytic function of RNA
1990   Elias James Corey (USA): new methods of synthesizing chemical compounds
1991   Richard R. Ernst (Switzerland): improvements in the technology of nuclear magnetic resonance (NMR) imaging
1992   Rudolph A. Marcus (USA): theoretical discoveries relating to reduction and oxidation reactions
1993   Kary Mullis (USA): invention of the polymerase chain reaction technique for amplifying DNA
             Michael Smith (Canada): techniques for reprogramming the genetic code using synthetic DNA

Nobel Prize for Physiology or Medicine

prizewinners

1901   Emil von Behring (Germany): discovery that the body produces antitoxins, and development of serum therapy for diseases such as diphtheria

1902   Ronald Ross (UK): role of the *Anopheles* mosquito in transmitting malaria

1903   Niels Finsen (Denmark): use of ultraviolet light to treat skin diseases

1904   Ivan Pavlov (Russia): physiology of digestion

1905   Robert Koch (Germany): investigations and discoveries in relation to tuberculosis

1906   Camillo Golgi (Italy) and Santiago Ramón y Cajal (Spain): fine structure of nervous system

1907   Charles Laveran (France): discovery that certain protozoa can cause disease

1908   Ilya Mechnikov (Russia) and Paul Ehrlich (Germany): work on immunity

1909   Emil Kocher (Switzerland): physiology, pathology, and surgery of the thyroid gland

1910   Albrecht Kossel (Germany): study of cell proteins and nucleic acids

1911   Allvar Gullstrand (Sweden): refraction of light through the different components of the eye

1912   Alexis Carrel (USA): techniques for connecting severed blood vessels and transplanting organs

1913   Charles Richet (France): allergic responses

1914   Robert Bárány (Austria): physiology and pathology of the equilibrium organs of the inner ear

1915–18   no prizes awarded

1919   Jules Bordet (Belgium): work on immunity

1920   August Krogh (Denmark): discovery of mechanism regulating the dilation and constriction of blood capillaries

1921   no prizes awarded

1922   Archibald Hill (UK): production of heat in contracting muscle

Otto Meyerhof (Germany): relationship between oxygen consumption and metabolism of lactic acid in muscle

1923   Frederick Banting (Canada) and John Macleod (UK): discovery and isolation of the hormone insulin

1924   Willem Einthoven (Netherlands): invention of the electrocardiograph

1925   no prizes awarded

1926   Johannes Fibiger (Denmark): discovery of a parasite *Spiroptera carcinoma* that causes cancer

1927   Julius Wagner-Jauregg (Austria): use of induced malarial fever to treat paralysis caused by mental deterioration

1928   Charles Nicolle (France): role of the body louse in transmitting typhus

1929   Christiaan Eijkman (Netherlands): discovery of a cure for beriberi, a vitamin-deficiency disease

Frederick Hopkins (UK): discovery of trace substances, now known as vitamins, that stimulate growth

1930   Karl Landsteiner (USA): discovery of human blood groups

1931   Otto Warburg (Germany): discovery of respiratory enzymes that enable cells to process oxygen

1932   Charles Sherrington (UK) and Edgar Adrian (UK): function of neurons (nerve cells)

1933   Thomas Morgan (USA): role of chromosomes in heredity

1934   George Whipple (USA), George Minot (USA) and William Murphy (USA): treatment of pernicious anaemia by increasing the amount of liver in the diet

1935   Hans Spemann (Germany): organizer effect in embryonic development

1936   Henry Dale (UK) and Otto Loewi (Germany): chemical transmission of nerve impulses

1937   Albert Szent-Györgyi (Hungary): investigation of biological oxidation processes and of the action of vitamin C (ascorbic acid)

1938   Corneille Heymans (Belgium): mechanisms regulating respiration

1939   Gerhard Domagk (Germany): discovery of the first antibacterial sulphonamide drug

1940–42   no prizes awarded

1943   Carl Dam (Denmark): discovery of vitamin K

Edward Doisy (USA): chemical nature of vitamin K

1944   Joseph Erlanger (USA) and Herbert Gasser (USA): transmission of impulses by nerve fibres

1945 Alexander Fleming (UK): discovery of the bactericidal effect of penicillin

Ernst Chain (UK) and Howard Florey (Australia): isolation of penicillin and its development as an antibiotic drug

1946 Hermann Muller (USA): discovery that X-ray irradiation can cause mutation

1947 Carl Cori (USA) and Gerty Cori (USA): production and breakdown of glycogen (animal starch)

Bernardo Houssay (Argentina): function of the pituitary gland in sugar metabolism

1948 Paul Müller (Switzerland): discovery of the first synthetic contact insecticide, DDT

1949 Walter Hess (Switzerland): mapping areas of the midbrain that control the activities of certain body organs

Antonio Egas Moniz (Portugal): therapeutic value of prefrontal lobotomy in certain psychoses

1950 Edward Kendall (USA), Tadeus Reichstein (Poland) and Philip Hench (USA): structure and biological effects of hormones of the adrenal cortex

1951 Max Theiler (South Africa): discovery of a vaccine against yellow fever

1952 Selman Waksman (USA): discovery of streptomycin, the first antibiotic effective against tuberculosis

1953 Hans Krebs (UK): discovery of the citric acid cycle

Fritz Lipmann (USA): discovery of coenzyme A, a nonprotein compound that acts in conjunction with enzymes to catalyse metabolic reactions leading up to the citric acid cycle

1954 John Enders (USA), Thomas Weller (USA) and Frederick Robbins (USA): cultivation of the polio virus in the laboratory

1955 Hugo Theorell (Sweden): nature and action of oxidation enzymes

1956 André Cournand (USA), Werner Forssmann (Germany) and Dickinson Richards Jr (USA): technique for passing a catheter into the heart for diagnostic purposes

1957 Daniel Bovet (Switzerland): discovery of synthetic drugs used as muscle relaxants in anaesthesia

1958 George Beadle (USA) and Edward Tatum (USA): discovery that genes regulate precise chemical effects

Joshua Lederberg (USA): genetic recombination and the organization of bacterial genetic material

1959 Severo Ochoa (USA) and Arthur Kornberg (USA): discovery of enzymes that catalyse the formation of RNA (ribonucleic acid) and DNA (deoxyribonucleic acid)

1960 Macfarlane Burnet (Australia) and Peter Medawar (UK): acquired immunological tolerance of transplanted tissues

1961 Georg von Békésy (USA): investigations into the mechanism of hearing within the cochlea of the inner ear

1962 Francis Crick (UK), James Watson (USA) and Maurice Wilkins (UK): discovery of the double-helical structure of DNA and of the significance of this structure in the replication and transfer of genetic information

1963 John Eccles (Australia), Alan Hodgkin (UK) and Andrew Huxley (UK): ionic mechanisms involved in the communication or inhibition of impulses across neuron (nerve cell) membranes

1964 Konrad Bloch (USA) and Feodor Lynen (West Germany): cholesterol and fatty-acid metabolism

1965 François Jacob (France), André Lwoff (France) and Jacques Monod (France): genetic control of enzyme and virus synthesis

1966 Peyton Rous (USA): discovery of tumour-inducing viruses

Charles Huggins (USA): hormonal treatment of prostatic cancer

1967 Ragnar Granit (Sweden), Haldan Hartline (USA) and George Wald (USA): physiology and chemistry of vision

1968 Robert Holley (USA), Har Gobind Khorana (USA) and Marshall Nirenberg (USA): interpretation of genetic code and its function in protein synthesis

1969 Max Delbruck (USA), Alfred Hershey (USA) and Salvador Luria (USA): replication mechanism and genetic structure of viruses

1970 Bernhard Katz (UK), Ulf von Euler (Austria) and Julius Axelrod (USA): storage, release and inactivation of neurotransmitters

1971 Earl Sutherland (USA): discovery of cyclic AMP, a chemical messenger that plays a role in the action of many hormones

1972 Gerald Edelman (USA) and Rodney Porter (UK): chemical structure of antibodies

1973 Karl von Frisch (Austria), Konrad Lorenz (Austria) and Nikolaas Tinbergen (Netherlands): animal behaviour patterns

1974 Albert Claude (USA), Christian de Duve (Belgium) and George Palade (USA): structural and functional organization of the cell

1975   David Baltimore (USA), Renato Dulbecco (USA) and Howard Temin (USA): interactions between tumour-inducing viruses and the genetic material of the cell

1976   Baruch Blumberg (USA) and Carleton Gajdusek (USA): new mechanisms for the origin and transmission of infectious diseases

1977   Roger Guillemin (USA) and Andrew Schally (USA): discovery of hormones produced by the hypothalamus region of the brain
Rosalyn Yalow (USA): radioimmunoassay techniques by which minute quantities of hormone may be detected

1978   Werner Arber (Switzerland), Daniel Nathans (USA) and Hamilton Smith (USA): discovery of restriction enzymes and their application to molecular genetics

1979   Allan Cormack (USA) and Godfrey Hounsfield (UK): development of the CAT scan

1980   Baruj Benacerraf (USA), Jean Dausset (France) and George Snell (USA): genetically determined structures on the cell surface that regulate immunological reactions

1981   Roger Sperry (USA): functional specialization of the brain's cerebral hemispheres
David Hubel (USA) and Torsten Wiesel (Sweden): visual perception

1982   Sune Bergström (Sweden), Bengt Samuelson (Sweden) and John Vane (UK): discovery of prostaglandins and related biologically reactive substances

1983   Barbara McClintock (USA): discovery of mobile genetic elements

1984   Niels Jerne (Denmark), Georges Köhler (West Germany) and César Milstein (UK): work on immunity and discovery of a technique for producing highly specific, monoclonal antibodies

1985   Michael Brown (USA) and Joseph L. Goldstein (USA): regulation of cholesterol metabolism

1986   Stanley Cohen (USA) and Rita Levi-Montalcini (Italy): discovery of factors that promote the growth of nerve and epidermal cells

1987   Susumu Tonegawa (Japan): process by which genes alter to produce a range of different antibodies

1988   James Black (UK), Gertrude Elion (USA) and George Hitchings (USA): principles governing the design of new drug treatment

1989   Michael Bishop (USA) and Harold Varmus (USA): discovery of oncogenes, genes carried by viruses that can trigger cancerous growth in normal cells

1990   Joseph Murray (USA) and Donnall Thomas (USA): pioneering work in organ and cell transplants

1991   Erwin Neher (Germany) and Bert Sakmann (Germany): discovery of how gatelike structures (ion channels) regulate the flow of ions into and out of cells

1992   Edmond Fisher (USA) and Edwin Krebs (USA): isolating and describing the action of the enzyme responsible for reversible protein phosphorylation, a major biological control mechanism

1993   Richard Roberts (UK) and Phillip Sharp (USA): discoveries concerning the structure and functioning of genetic material, in particular, that DNA contains units which carry the genetic code with noncoding sequences in between

Nobel Prize for Physics

prizewinners

1901	Wilhelm Röntgen (Germany): discovery of X-rays
1902	Hendrik Lorentz (Netherlands) and Pieter Zeeman (Netherlands): influence of magnetism on radiation phenomena
1903	Antoine Becquerel (France): discovery of spontaneous radioactivity
	Pierre Curie (France) and Marie Curie (Poland): researches on radiation phenomena
1904	John Strutt (Lord Rayleigh, UK): densities of gases and discovery of argon
1905	Philipp von Lenard (Germany): work on cathode rays
1906	J. J. Thomson (UK): theoretical and experimental work on the conduction of electricity by gases
1907	Albert Michelson (USA): measurement of the speed of light through the design and application of precise optical instruments such as the interferometer
1908	Gabriel Lippmann (France): photographic reproduction of colours by interference
1909	Guglielmo Marconi (Italy) and Karl Braun (Germany): development of wireless telegraphy
1910	Johannes van der Waals (Netherlands): equation describing the physical behaviour of gases and liquids
1911	Wilhelm Wien (Germany): laws governing radiation of heat
1912	Nils Dalen (Sweden): invention of light-controlled valves, which allow lighthouses and buoys to operate automatically
1913	Heike Kamerlingh Onnes (Netherlands): studies of properties of matter at low temperatures
1914	Max von Laue (Germany): discovery of diffraction of X-rays by crystals
1915	William Bragg (UK) and Lawrence Bragg (UK): X-ray analysis of crystal structures
1916	no prizes awarded
1917	Charles Barkla (UK): discovery of characteristic X-ray emission of the elements
1918	Max Planck (Germany): formulation of quantum theory
1919	Johannes Stark (Germany): discovery of Doppler effect in rays of positive ions, and splitting of spectral lines in electric fields
1920	Charles Guillaume (Switzerland): precision measurements through anomalies in nickel–steel alloys
1921	Albert Einstein (Switzerland): theoretical physics, especially law of photoelectric effect
1922	Niels Bohr (Denmark): structure of atoms and radiation emanating from them
1923	Robert Millikan (USA): discovery of the electric charge of an electron, and study of the photoelectric effect
1924	Karl Siegbahn (Sweden): X-ray spectroscopy
1925	James Franck (USA) and Gustav Hertz (Germany): laws governing the impact of an electron upon an atom
1926	Jean Perrin (France): confirmation of the discontinuous structure of matter
1927	Arthur Compton (USA): transfer of energy from electromagnetic radiation to a particle
	Charles Wilson (UK): invention of the Wilson cloud chamber, by which the movement of electrically charged particles may be tracked
1928	Owen Richardson (UK): thermionic phenomena and associated law
1929	Louis Victor de Broglie (France): discovery of wavelike nature of electrons
1930	Venkata Raman (India): discovery of the scattering of single-wavelength light when it is passed through a transparent substance
1931	no prizes awarded
1932	Werner Heisenberg (Germany): creation of quantum mechanics
1933	Erwin Schrödinger (Austria) and Paul Dirac (UK): development of quantum mechanics
1934	no prizes awarded
1935	James Chadwick (UK): discovery of the neutron
1936	Victor Hess (Austria): discovery of cosmic radiation
	Carl Anderson (USA): discovery of the positron
1937	Clinton Davisson (USA) and George Thomson (UK): diffraction of electrons by crystals
1938	Enrico Fermi (USA): use of neutron irradiation to produce new elements, and discovery of nuclear reactions induced by slow neutrons
1939	Ernest O. Lawrence (USA): invention and development of cyclotron, and production of artificial

radioactive elements

1940–42   no prizes awarded

1943   Otto Stern (Germany): molecular-ray method of investigating elementary particles, and discovery of magnetic moment of proton

1944   Isidor Isaac Rabi (USA): resonance method of recording the magnetic properties of atomic nuclei

1945   Wolfgang Pauli (Austria): discovery of the exclusion principle

1946   Percy Bridgman (USA): development of high-pressure physics

1947   Edward Appleton (UK): physics of the upper atmosphere

1948   Patrick Blackett (UK): application of the Wilson cloud chamber to nuclear physics and cosmic radiation

1949   Hideki Yukawa (Japan): theoretical work predicting existence of mesons

1950   Cecil Powell (UK): use of photographic emulsion to study nuclear processes, and discovery of pions (pi-mesons)

1951   John Cockcroft (UK) and Ernest Walton (Ireland): transmutation of atomic nuclei by means of accelerated subatomic particles

1952   Felix Bloch (USA) and Edward Purcell (USA): precise nuclear-magnetic measurements

1953   Frits Zernike (Netherlands): invention of phase-contrast microscope

1954   Max Born (Germany): statistical interpretation of wave function in quantum mechanics
Walther Bothe (Germany): coincidence method of detecting the emission of electrons

1955   Willis Lamb (USA): structure of hydrogen spectrum
Polykarp Kusch (USA): determination of magnetic moment of the electron

1956   William Shockley (USA), John Bardeen (USA) and Walter Houser Brattain (USA): study of semiconductors and discovery of transistor effect

1957   Yang Chen Ning (USA) and Lee Tsung-Dao (China): investigations of weak interactions between elementary particles

1958   Pavel Cherenkov (USSR), Ilya Frank (USSR) and Igor Tamm (USA): discovery and interpretation of Cherenkov radiation

1959   Emilio Segrè (Italy) and Owen Chamberlain (USA): discovery of the antiproton

1960   Donald Glaser (USA): invention of the bubble chamber

1961   Robert Hofstadter (USA): scattering of electrons in atomic nuclei and structure of protons and neutrons
Rudolf Mössbauer (Germany): resonance absorption of gamma radiation

1962   Lev Landau (USSR): theories of condensed matter, especially liquid helium

1963   Eugene Wigner (USA): discovery and application of symmetry principles in atomic physics
Maria Goeppert-Mayer (USA) and Hans Jensen (Germany): discovery of the shell-like structure of atomic nuclei

1964   Charles Townes (USA), Nikolai Basov (USSR) and Aleksandr Prokhorov (USSR): quantum electronics leading to construction of oscillators and amplifiers based on maser–laser principle

1965   Sin-Itiro Tomonaga (Japan), Julian Schwinger (USA) and Richard Feynman (USA): quantum electrodynamics

1966   Alfred Kastler (France): development of optical pumping, whereby atoms are raised to higher energy levels by illumination

1967   Hans Bethe (USA): theory of nuclear reactions, and discoveries concerning production of energy in stars

1968   Luis Alvarez (USA): elementary-particle physics and discovery of resonance states, using hydrogen bubble chamber and data analysis

1969   Murray Gell-Mann (USA): classification of elementary particles and study of their interactions

1970   Hannes Alfvén (Sweden): magnetohydrodynamics and its applications in plasma physics
Louis Néel (France): antiferromagnetism and ferromagnetism in solid-state physics

1971   Dennis Gabor (UK): invention and development of holography

1972   John Bardeen (USA), Leon Cooper (USA) and John Robert Schrieffer (USA): theory of superconductivity

1973   Leo Esaki (Japan) and Ivar Giaver (USA): tunnelling phenomena in semiconductors and superconductors
Brian Josephson (UK): theoretical predictions of the properties of a supercurrent through a tunnel barrier

1974   Martin Ryle (UK) and Antony Hewish (UK): development of radioastronomy, particularly aperture-synthesis technique and the discovery of pulsars

1975   Aage Bohr (Denmark), Ben Mottelson (Denmark) and James Rainwater (USA): discovery of connection between collective motion and particle motion in atomic nuclei and development of theory of nuclear structure

1976   Burton Richter (USA) and Samuel Ting (USA): discovery of the psi meson

1977   Philip Anderson (USA), Nevill Mott (UK) and John Van Vleck (USA): electronic structure of magnetic and disordered systems

1978   Pyotr Kapitza (USSR): low-temperature physics
Arno Penzias (Germany) and Robert Wilson (USA): discovery of cosmic background radiation

1979   Sheldon Glashow (USA), Abdus Salam (Pakistan) and Steven Weinberg (USA): unified theory of weak and electromagnetic fundamental forces and prediction of the existence of the weak neutral current

1980   James W. Cronin (USA) and Val Fitch (USA): violations of fundamental symmetry principles in the decay of neutral kaon mesons

1981   Nicolaas Bloemergen (USA) and Arthur Schawlow (USA): development of laser spectroscopy
Kai Siegbahn (Sweden): high-resolution electron spectroscopy

1982   Kenneth Wilson (USA): theory for critical phenomena in connection with phase transitions

1983   Subrahmanyan Chandrasekhar (USA): theoretical studies of physical processes in connection with structure and evolution of stars
William Fowler (USA): nuclear reactions involved in the formation of chemical elements in the universe

1984   Carlo Rubbia (Italy) and Simon van der Meer (Netherlands): contributions to the discovery of the W and Z particles (weakons)

1985   Klaus von Klitzing (Germany): discovery of the quantized Hall effect

1986   Erns Ruska (Germany): electron optics, and design of the first electron microscope
Gerd Binnig (Germany) and Heinrich Rohrer (Switzerland): design of scanning tunnelling microscope

1987   Georg Bednorz (Germany) and Alex Müller (Switzerland): superconductivity in ceramic materials

1988   Leon M. Lederman (USA), Melvin Schwartz (USA) and Jack Steinberger (Germany): neutrino-beam method, and demonstration of the doublet structure of leptons through discovery of muon neutrino

1989   Norman Ramsey (USA): measurement techniques leading to discovery of caesium atomic clock
Hans Dehmelt (USA) and Wolfgang Paul (Germany): ion-trap method for isolating single atoms

1990   Jerome Friedman (USA), Henry Kendall (USA) and Richard Taylor (Canada): experiments demonstrating that protons and neutrons are made up of quarks

1991   Pierre-Gilles de Gennes (France): work on disordered systems including polymers and liquid crystals; development of mathematical methods for studying the behaviour of molecules in a liquid on the verge of solidifying

1992   Georges Charpak (Poland): invention and development of detectors used in high-energy physics

1993   Russell Hulse (USA) and Joseph Taylor (USA): proof of the existence of gravitational waves

Glossary

A

Abelian functions functions of the form $\int f(x,y)\mathrm{d}x$, where y is an algebraic function of x, and $f(x,y)$ is an algebraic function of x and y.

aberration defect in the image formed by a lens, mirror or optical system. Spherical aberration results when different rays of light are brought to more than one focus, producing a blurred image or ⇨coma; chromatic aberration when different wavelengths within a ray of light are brought to more than one focus, producing an image distorted by coloured fringes. Aberration in lenses can be overcome by the use of an ⇨achromatic lens or a combination of lenses made of glasses of different ⇨refractive indices.

aberration of starlight difference in a star's apparent position in the sky from the apparent position it would have if the Earth were stationary. Such displacement caused by the Earth's ⇨sidereal motion results in an optical positioning difference of up to about 20.5 seconds of arc, much greater than any displacement observed by ⇨parallax.

ablation in astronomy, progressive burning away of the outer layers (for example, of a meteor) by friction with the atmosphere.

absolute magnitude the ⇨magnitude of a celestial body as would be apparent from a standard distance of 10 ⇨parsecs. An absolute magnitude of 0.0 represents an actual ⇨luminosity of 95 times that of the Sun.

absolute zero the temperature at which a molecule has zero thermal energy; practically impossible to achieve. Absolute zero occurs at a temperature of about –273˚C/–459˚F, and is used to define the temperature corresponding to zero on the ⇨Kelvin scale of temperature.

absorption trough range of ⇨wavelengths (around 21 cm/ 8 in) at which atomic hydrogen absorbs (or emits) radiation; this is a concept used in the attempt to detect ⇨intergalactic matter.

abstract algebra generalization of ⇨algebra; the word 'abstract' merely draws attention to the level of generality.

abstract group generalization of a group; the word 'abstract' merely indicates that a group is to be considered with reference not to a specific example but to its more general properties (which are shared therefore by other examples).

abundance of helium see ⇨helium abundance.

abundant number a natural number that is less than the sum of its divisors (factors).

acceleration the rate of change of velocity. Typical units: metres per second per second (m s^{-2}) or feet per second per second (ft s^{-2}).

acceleration due to gravity the acceleration experienced by a body falling freely in a vacuum under the influence of ⇨gravity alone. For the Earth, it has a value of approximately 9.81 m s^{-2}/32.17 ft s^{-2}.

accelerator nerve a nerve that conducts impulses to the heart. On stimulation of the cardiac sympathetic nerves, the rate and strength of the heartbeat increases.

accretion collection of material together, generally to form a single body.

accumulator or *storage battery* a device for storing electrical energy. Usually, accumulators are in the form of electric cells linked together. By connecting an accumulator up to an electric ⇨current, it is possible to charge up the cells by ⇨electrolysis. Once this process is complete, the accumulator can be connected to an electric motor to perform some work. A familiar example is the lead–acid car battery.

acetylcholine (ACh) a nerve transmitter chemical. When a nerve impulse arrives at the end of a nerve cell (⇨neuron), it stimulates the release of ACh which chemically transmits the impulse across the ⇨synapse to the next nerve cell.

acetylene the former name of the unsaturated hydrocarbon ethyne (C_2H_2) and generic name of the members of a ⇨homologous series of organic compounds of general formula C_nH_n, where $n = 2$ or more. See ⇨alkyne.

achromatic lens lens (or combination of lenses) that brings different ⇨wavelengths within a ray of light to a single focus, thus overcoming chromatic ⇨aberration.

acid a substance that has a tendency to lose ⇨protons and forms ⇨hydrogen ions in solution. The hydrogen atom(s) of an acid may be replaced by a metal atom (or atoms) or an ammonium ion to form a ⇨salt. An acid with two replaceable hydrogens (such as sulphuric acid, H_2SO_4) is termed dibasic; an acid with more than two replaceable hydrogens (such as orthophosphoric acid, H_3PO_4) is termed polybasic. Solutions of acids have a ⇨pH of less than 7; their presence may be detected by using an ⇨indicator, such as litmus (which turns red in the presence of an acid). An acid may be neutralized by reacting it with a ⇨base.

acoustics the science and study of sound, its generation, properties and reception (detection).

actin a protein that occurs in muscles. See ⇨myosin.

actinide or *actinon* any of the group of elements that follow actinium in the periodic table (that is, that have an ⇨atomic number greater than 89). Many actinides have radioactive ⇨isotopes.

action potential the characteristic momentary electric potential across a nerve cell membrane during the passage of a nerve impulse.

addition reaction in organic chemistry, a reaction in which a new compound is formed by the addition of an atom or group of atoms to an existing compound (as opposed to a ⇨substitution reaction).

adenosine diphosphate (ADP) a nucleotide ⇨coenzyme found in living organisms, essentially involved in the conversion of energy (either light energy in ⇨photosynthesis or energy derived from metabolic processes) into a more available form – that of ATP.

adenosine triphosphate (ATP) a nucleotide ⇨coenzyme that occurs in all organisms. It is derived from ADP and is the principal energy source for biochemical reactions. Energy is provided by a combination of ⇨enzyme action and the subsequent transfer of a phosphate group from ATP to a substance.

adiabatic change a process that takes place without any heat entering or leaving the system.

ADP abbreviation for ⇨adenosine diphosphate.

adrenocorticotrophic hormone (ACTH) a ⇨hormone that stimulates the adrenal cortex to release its hormones. ACTH is released into the bloodstream from the anterior lobe of the pituitary gland at the base of the brain.

aeon or *eon* in astronomical terms, 1,000 million years.

aerial a length of conducting material able to transmit and/or receive ⇨electromagnetic waves, such as ⇨radio waves or ⇨microwaves.

aerobic describing an atmosphere of free (gaseous or dissolved) oxygen. An aerobic organism requires such an atmosphere for respiration and life.

aerodynamics the study of the behaviour of bodies in air streams.

aeronautics or *aeronautical engineering* the discipline concerned with the engineering design, testing, building and maintenance of aircraft systems.

afferent nerve a nerve that carries impulses towards the central nervous system (that is, the spinal cord or brain).

aileron flaps on the trailing edge of an aircraft wing that produce ⇨roll in the aircraft.

air pump a device used to pump air from one vessel to another, or to evacuate a vessel altogether to produce a ⇨vacuum.

albinism a pigment-free condition found in animals and plants caused by a lack of a ⇨melanin-synthesizing ⇨enzyme. Albino animals are characterized by white fur or hair, pink skin and pupils, and a high sensitivity to the Sun's radiation.

alcohol any of a class of organic compounds in which a hydrogen atom (or atoms) of a hydrocarbon has been replaced by a hydroxyl (–OH) group (or groups); their names end in -ol. For example, replacing a hydrogen in methane, CH_4, produces methanol, CH_3OH (former name methyl alcohol). An alcohol with two hydroxyl groups is termed a diol; glycerol (1,2,3-propanetriol, $CH_2(OH)CH(OH)CH_2OH$) is a triol. Alcohols react with acids to form esters. Substitution of a hydroxyl group for a hydrogen atom in benzene or other aromatic compounds results in the formation of a ⇨phenol.

aldehyde any of a class of organic compounds in which a hydrogen atom of a hydrocarbon has been replaced by an aldehyde (–CHO) group; their names end in -al. For example, replacing a hydrogen in methane, CH_4, produces ethanal, CH_3CHO (former name acetaldehyde). Aldehydes therefore have the general formula R·CHO, where R is an ⇨alkyl or ⇨aryl group. An exception to the above rules is formaldehyde, HCHO.

algebra study of structural or manipulative properties in mathematics. At an elementary level, algebra investigates the properties of such operations as addition or multiplication in the general context of real numbers, and recognizes similarities between this context and, for example, the addition of vectors.

algebraic curve geometrical curve that can be precisely described by an (algebraic) equation.

algebraic numbers numbers that satisfy a polynomial equation with rational coefficients: for example, $\sqrt{2}$ solves

$x^2 - 2 = 0$. ⇨Real numbers that are not algebraic are called ⇨transcendental numbers. Although there is an infinity of algebraic numbers, there is in fact a 'larger' infinity of transcendental numbers.

algebraic topology study of ⇨surfaces and similar but more general objects in higher dimensions, using algebraic techniques. It is based upon ⇨homology.

algorithm any basic mathematical operation or method of calculation; for example, in elementary mathematics, the rules for addition, subtraction, multiplication or division. Almost the same as a computer program, an algorithm may be described as a scheme of calculations commonly designed to be applied repeatedly, so that the result of a first calculation is used as the starting point of another calculation that uses identical procedures. A further example is therefore Euclid's algorithm for finding the highest common factor of two natural numbers.

aliphatic describing any organic compound that has a linear or branched chain of carbon atoms, or a ring or rings of completely bonded atoms (termed an alicyclic compound), as opposed to the closed rings of partly unsaturated atoms of ⇨aromatic compounds.

alkali a solution of a metal hydroxide, especially a hydroxide of an alkali metal such as potassium or sodium, or ammonium hydroxide. The presence of an alkali may be detected using an ⇨indicator such as litmus (which turns blue in the presence of an alkali). Alkalis produce hydroxyl (OH^-) ions in solution and the term is often extended to other compounds, such as hydrogen carbonate (bicarbonate) salts, which give an alkaline reaction in solution. Alkalis are a type of ⇨base.

alkaloid any of a number of physiologically active and frequently poisonous substances contained in some plants. They are usually organic ⇨bases and contain nitrogen. They form salts with acids and, when soluble, give alkaline solutions.

alkane any of a class of saturated hydrocarbons, formerly called paraffins. Their names end in -ane and they form a ⇨homologous series of general formula C_nH_{2n+2}, where n is a whole number. The first five members of the alkane series are methane (CH_4), ethane (C_2H_6), propane (C_3H_8), butane (C_4H_{10}) and pentane (C_5H_{12}).

alkene any of a class of unsaturated hydrocarbons, formerly called olefins, characterized by the presence of one or more carbon–carbon double bonds. Their names end in -ene and they form a ⇨homologous series of general formula C_nH_{2n}, where $n = 2$ or more. The first four members of the alkene series are ethene (ethylene) C_2H_4, propene (propylene) C_3H_6, butene (butylene) C_4H_8 and pentene C_5H_{10}. An alkene with two double bonds is called a diene.

alkyl any of a series of hydrocarbon ⇨radicals derived from the ⇨alkanes and having the general formula C_nH_{2n+1}. Examples include the methyl radical (–CH_3) and the ethyl radical (–C_2H_5).

alkylation a chemical process by which an ⇨alkyl group is incorporated into a compound, replacing a hydrogen atom.

alkyne any of a class of unsaturated hydrocarbons, formerly called acetylenes, characterized by the presence of one or more carbon–carbon triple bonds. Their names end in -yne and they form a ⇨homologous series of general formula C_nH_{2n-2}, where $n = 2$ or more. The first four members of the alkyne series are ethyne (acetylene) C_2H_2, propyne C_3H_4, butyne C_4H_6 and pentyne C_5H_8.

allantois a bladder in the embryo of reptiles, birds and mammals, which grows outside the embryo into the wall of the yolk-sac of reptiles and birds and under the ⇨chorion of mammals. In mammals, blood vessels in the allantois carry blood to the ⇨placenta; in reptiles and birds the blood vessels permit respiration. As they develop, the vessels become the umbilical vein and arteries.

allele or *allelomorph* a pair of ⇨genes, located at the same sites (loci) on paired ⇨chromosomes, that determine specific characteristics. Alleles are denoted by a double-letter symbol, a capital letter indicating a ⇨dominant gene and a lower case letter indicating a ⇨recessive gene (for example, Bb).

alloy a material consisting of a mixture of two or more metals, for example bronze is an alloy of mainly copper and tin.

Almagest arabic title for Ptolemy of Alexandria's *Syntaxis*, the writings in which he combined his own astronomical researches with those of others. Although much of the work is inaccurate even in premise, until Nicolaus Copernicus published his results fourteen centuries later the *Almagest* remained the standard reference source in Europe.

alpha-beta-gamma theory (α-β-γ theory) explanation of the ⇨Big Bang model in terms of nuclear physics, proposed by Ralph Alpher, Hans Bethe and George Gamow in 1948; it was later slightly corrected by Chushiro Hayashi.

Alpha Centauri bright binary star in which both components contribute to a magnitude of -0.27; it is also the nearest of the bright stars (at a distance of 4.3 light years).

alpha chain a particular secondary structure of the polypeptide chain (of a protein) brought about by hydrogen bonding between adjacent ⇨peptide units. Alpha chains occur, for example, in the ⇨haemoglobin molecule.

alpha decay or *alpha particle decay* spontaneous emission by a heavier element (such as uranium) of positively charged helium nuclei – alpha particles – comprising two ⇨protons and two ⇨neutrons. The result of this radioactive decay is that the original element is very gradually converted into another element, with a decreased atomic number and mass. Alpha particle emission may be simultaneous with ⇨beta decay.

alpha particle a positively charged particle consisting of a helium nucleus, He^{2+}; that is, a combination of two ⇨protons and two ⇨neutrons. A stream of alpha particles is called an alpha ray.

alt-azimuth comprising a means of measuring or precisely locating in coordinates the position of objects at any ⇨altitude or ⇨azimuth. The term is now used mainly to describe a type of mounting for a telescope.

alternating current (AC) a flow of electricity that, on reaching a maximum value in one direction, reverses, repeating the cycle continuously.

altitude angular distance above the horizon.

amalgam any ⇨alloy of mercury. Some amalgams are used in dentistry to fill cavities in teeth.

amatol a compound used in explosives, comprising four parts ammonium nitrate and one part trinitrotoluene (TNT).

amide any of a class of organic compounds containing the grouping $-CONH_2$. The simplest is acetamide, CH_3CONH_2.

amine any of a class of organic compounds containing one or more amino ($-NH_2$) groups. They can be considered as being formed by replacing one or more hydrogen atoms of ammonia (NH_3) by organic radicals such as ⇨alkyl or ⇨aryl groups.

Substitution of one hydrogen results in a primary amine, general formula RHN_2 (for example, aminomethane (methylamine) CN_3NH_2); disubstitution produces a secondary amine, R_2NH, and trisubstitution results in a tertiary amine, R_3N. A quaternary amine $(R_4N)^+$ is the organic analogue of the ammonium ion (NH_4^+).

amino acid any of a series of organic acids that contain one or more amino groups ($-NH_2$). There are more than 100 different amino acids, of which about 20 (called the alpha amino acids) are found – joined by ⇨peptide linkages – in proteins. The alpha amino acids have the general formula $RCH(NH_2)$-COOH, where R is an ⇨alkyl group. In higher animals, essential amino acids have to be supplied by the diet (because they cannot be synthesized by the liver, as can nonessential amino acids).

ammeter a device that measures ⇨current flowing in a circuit. A perfect ammeter has no electrical ⇨resistance.

ammonia (NH_3) a pungent ⇨gas obtained synthetically by the ⇨Haber process, consisting of three atoms of hydrogen to one of nitrogen. It is used extensively in the production of fertilizers.

ammonite extinct marine cephalopod mollusc of the order Ammonoidea, related to the modern nautilus. The shell was curled in a plane spiral and made up of numerous gas-filled chambers, the outermost containing the body of the animal. Many species flourished between 200 million and 65 million years ago, ranging in size from that of a small coin to 2 m/7 ft across.

amnion the innermost embryonic membrane in reptiles, birds and mammals. In human beings, for example, the amnion suspends the foetus in fluid until about the eighth week, when it becomes incorporated into the ⇨chorion to form the amniochorionic sac.

ampere the ⇨SI unit of electric ⇨current, defined by that current that, if flowing through two infinitely long ⇨conductors of negligible cross-section separated by 1 m/3.3 ft, will produce within the two conductors a ⇨force of 2×10^{-7} ⇨newtons per metre of length.

Ampère's law the strength of the magnetic field B produced by a ⇨current I flowing through a conductor of length l is proportional to the product of the current and the length.

amphoteric describing a chemical compound that reacts as a ⇨base with strong acids and as an ⇨acid with strong bases. For example, aluminium oxide is amphoteric; it forms aluminium salts with strong acids but forms aluminates with strong bases.

amplification the boosting of the strength of a signal.

amplitude for any wave motion, the maximum deviation from equilibrium undergone during the motion; for example, for a pendulum, the amplitude is half the length of the swing.

amplitude modulation (AM) a process in the transmission of ⇨electromagnetic radiation, especially ⇨radio, in which the ⇨amplitude of the ⇨carrier wave is ⇨modulated in relation to the amplitude of the signal being transmitted.

anaerobic describing an atmosphere lacking any free (gaseous or dissolved) oxygen. An anaerobic organism can live in such an atmosphere, obtaining its oxygen from the chemical breakdown of oxygen compounds.

anaesthesia the diminution or absence of sensation or awareness in the whole body (general anaesthesia) or in part of the body (local anaesthesia). It may result from a disorder or can be induced artificially using drugs. General anaesthesia is achieved by blocking nervous impulses in the brain; local anaesthesia affects the sodium–potassium ion interchange across specific nerve cell membranes when an impulse attempts to pass.

analogue in mathematics, part or all of a second statement that can be proved to correspond in operation or effect, though in a different context, with part or all of the original statement, theorem or hypothesis.

analogue computer a computer that performs calculations by taking physical analogues of quantities involved (such as electric current and fluid pressures), the mathematical operations being performed by passing these analogues through suitable devices.

analysis historically came to mean referring to operations that deal with numbers, or that convert other systems so that they can be considered numbers – for example, analytical geometry reduces figures in two or more dimensions to coordinates of points, and thus formulae or equations. By extension, analysis now includes the branch of mathematics in which the central tools are defined by ⇨limit processes such as differentiation and integration, or expansion by infinite series.

analytical geometry see ⇨analysis.

analytical method in geometry, investigation carried out using a coordinate system. (In other mathematical contexts the term merely implies the application of ⇨analysis.)

Andromeda galaxy the largest ⇨galaxy in the ⇨Local Group, also known as the Great Spiral and M31. It is about one and a half times the size of our own Galaxy, and contains at least 300 ⇨globular clusters. Two smaller, elliptical galaxies lie close to it.

anemometer a device for measuring the speed of a moving gas, especially the wind.

angiogram an X-ray photograph in which blood vessels are made visible by injecting them with a radio-opaque substance (such as an iodine compound).

angiosperm a flowering plant, a member of a subdivision of the Spermatophyta. In contrast to a ⇨gymnosperm, an angiosperm has a covered seed carried in an enclosed megasporophyll; the female gametophyte develops inside the megaspore. Also the ⇨xylem in the stem of an angiosperm usually has vessels (whereas that in a gymnosperm usually does not). The angiosperm group includes the ⇨monocotyledon and ⇨dicotyledon classes.

angle of parallelism in non-Euclidean geometry involving the application of classical geometrical principles to nonflat surfaces, the angle at which a line perpendicular to one of two parallel lines meets the other – which may also be 90°, or may be less.

angular momentum the rotational analogue of ⇨momentum in linear motion. It is the product of the moment of inertia of a body and its angular velocity. In quantum mechanics it is quantized in units of ⇨Planck's constant h; for example, the ⇨spin of the electron is $h/2$.

anion a negatively charged ⇨ion, such as chloride (Cl^-), sulphate (SO_4^{2-}) or phosphate (PO_4^{3-}). During ⇨electrolysis, anions migrate towards the (positively charged) ⇨anode. In ionic compounds such as salts, the charges of the anions are usually balanced by those of an appropriate number of ⇨cations.

anisotropic or *aelotropic* describing a substance that has different physical properties in different directions. Some crystals, for example, have different refractive indices in different directions.

anode the positive ⇨electrode in ⇨electrolysis, a cell (battery) or ⇨discharge tube. Current leaves the electrolyte and flows into the external circuit from, and ⇨anions are discharged at, the anode.

antagonistic describing a pair of muscles that act in opposition to each other so that when one contracts, the other relaxes (for example, the biceps and triceps muscles of the upper arm).

antibiotic a chemical substance, often used as a drug in medical treatment, that kills or prevents the growth of microorganisms such as ⇨bacteria. Antibiotics are produced by fungi or bacteria, or they may be chemically synthesized. Antibiotic therapy blocks a chemical reaction that is essential to a bacterial parasite; penicillin, for example, obstructs the development of a protective cell wall in some pathogenic bacteria.

antibody a protein formed by plasma cells in the spleen or lymph nodes whose production is stimulated by the presence of an ⇨antigen (a parasite or other 'foreign' substance). Antibodies circulate in the body's fluids and combine chemically with the specific antigen that stimulated their formation. ⇨Immunity to disease can be induced (for example, by ⇨vaccination) by injecting specific antigens into the body and thus stimulating the formation of the required antibodies.

antiferromagnetism a form of magnetism in which the neighbouring ⇨dipoles in a substance lie in opposite directions.

antigen a substance that gains access to the body and in so doing stimulates the formation of ⇨antibodies. Many antigens are, like antibodies, proteins; others (such as pollen) are not themselves proteins but act by modifying existing proteins which then behave as antigens and stimulate the defensive response; often this mechanism is the cause of an allergy.

antimatter matter made up of ⇨antiparticles, such as anti-⇨protons, and positrons (anti-⇨electrons).

antiparticle a particle of the same mass as one particular particle, but with equal and opposite charge (and other characteristics).

antiseptic any chemical substance that prevents the growth of, or destroys, (pathogenic) microorganisms.

antiserum serum extracted from the blood of human beings or other animals that have been infected by a ⇨pathogen (such as a virus or bacterium) contains ⇨antibodies against a specific ⇨antigen and may be used in treatment. Many antisera have been superseded by ⇨antibiotic therapy, although sera are still used to treat some poisons and bacterial diseases.

antitoxin a naturally produced ⇨antibody that neutralizes specific ⇨toxins produced by ⇨bacteria (such as those released by tetanus and diphtheria bacteria). Antitoxins derived from the serum of infected human beings or other animals may be used as immunizing ⇨vaccines.

aperture the size of the opening admitting light into an optical instrument such as a telescope.

aphelion point representing the greatest distance from the Sun of a body in an elliptical orbit.

Apollonius' problem problem set by Apollonius of Perga, to describe a circle touching three other given circles.

Apollo space programme successful US lunar exploitation programme, in which Apollo spacecraft 1 to 6 were uncrewed and reconnoitred potential landing areas; 7 to 10 were crewed but did not land; and 11, 12 and 14 to 17 landed and returned safely. (*Apollo 13* was an aborted mission.) The first men to land on the Moon were Neil Armstrong and Edwin Aldrin, from *Apollo 11*, on 20 July 1969. The final Apollo flight (17) lasted from 7 to 19 December 1972, and left a considerable quantity of exploratory devices on the lunar surface.

Appleton layer a region in the upper atmosphere where there is a large concentration of free ⇨electrons.

approximation common mathematical device, deriving instead of the exact result one that is 'close enough', that is, differing by less than a specified amount. Iterative techniques proceed by calculating approximations successively closer to the exact answer, and will approach it in the ⇨limit.

aqualung or *scuba* (*s*elf-*c*ontained *u*nderwater *b*reathing *a*pparatus) a device that enables air to be fed to a diver working independently, without restrictions of supply cables, and which can get rid of the carbon dioxide produced by respiration.

aqueduct a structure, usually in the form of a channel raised above the ground, for long-distance supply of water.

aqueous humour a transparent watery solution containing trace salts and albumin that is contained between and nourishes the cornea and lens of the vertebrate eye.

arc the bright ⇨discharge produced when two ⇨electrodes are subjected to a large electric current. The effect, which produces heat of several thousands of degrees as well as light, is used in the smelting of metal, and also welding.

Archaean or *Archaeozoic* the earliest ⇨aeon of ⇨geological time; the first part of the Precambrian, from the formation of Earth up to about 2,500 million years ago. It was a time when no life existed, and with every new discovery of ancient life its upper boundary is being pushed further back.

Archimedes' principle when a body is immersed in a fluid, it experiences an upward ⇨force equal to the weight of fluid displaced.

Arcturus (Alpha Boötis) major star in Boötes; its (apparent) ⇨magnitude is –0.06.

Argand diagram method of representing ⇨complex numbers geometrically, in which the real and imaginary parts of a complex number form respectively the first and second coordinate.

armature the rotating coils of wire in an electric motor or ⇨dynamo, or any electrically powered device where a voltage is induced by a ⇨magnetic field.

armillary sphere ancient Greek, arabic and medieval ⇨alt-azimuth device, comprising a calibrated ring fixed in the ⇨meridian plane, within which a second concentric ring, also calibrated, was mobile around a vertical axis.

aromatic describing an organic compound that has one or more closed rings of atoms which are partly unsaturated and displays aromatic character as typified by benzene and its derivatives (as opposed to ⇨aliphatic compounds, which have open chains of carbon atoms or rings of completely bonded atoms).

artillery large-⇨calibre guns used in warfare.

aryl any organic ⇨radical derived from an ⇨aromatic compound; for example, the phenyl radical ($-C_6H_5$) is derived from benzene (C_6H_6).

association the grouping together of atoms or molecules, often in the vapour state or in solution, to form conglomerations of unexpectedly high relative molecular mass. See also ⇨dissociation.

asteroid or *planetoid* or **minor planet** one of several tiny planets, most of which orbit the Sun between Mars and Jupiter. The largest – and the first discovered – is Ceres, with a diameter of 1,003 km/623 mi. It is estimated that there may be altogether no fewer than 40,000. A few have very elliptical orbits and cross the orbits of several other (major) planets. One or two even have their own satellites (moons).

astigmatism an optical defect, in which rays in mutually perpendicular planes are out of focus relative to one another.

astrolabe ancient Arabic and medieval ⇨alt-azimuth device comprising two or more flat, metal, calibrated discs, attached so both or all could rotate independently. For early navigators and astronomers it acted as star chart, compass, clock and calendar.

astrology divination using the positions of the planets, the Sun and the Moon as seen against the stars in the constellations of the ⇨zodiac. Although at one stage in history astrology and astronomy were almost synonymous, the latter has advanced so far during the last three centuries that the two now bear little relation to each other.

astronomical colour index difference in a star's brightness when measured on two selected ⇨wavelengths, in order to determine the star's temperature. Cooler stars emit more light at longer wavelengths (and so appear redder than hot stars). Modern methods involve ⇨photoelectric filtering and the ⇨UBV photometry system.

astronomical unit (AU) Mean distance between the Earth and the Sun: 149,598,500 km/92,960,507 mi.

astrophysics combination of astronomy and various branches of physics.

asymptote in geometry, the asymptote to a curve may be represented by a straight line which the curve approaches more and more closely. By extension, 'asymptotic' describes curves or calculations approaching a specific value but not achieving it, although finally attaining a limit value in the difference.

atmosphere mantle of gases round a star, planet or moon, sometimes even forming the apparent surface of the body. For a body to retain an atmosphere depends on the body's ⇨gravity, and the temperature and composition of the gases. The atmosphere of the Earth is, by volume, 78% nitrogen and 21% oxygen (with 1% of other gases); mean atmospheric pressure at the surface is 101 kilopascals, and is also referred to as 1 atmosphere.

atmosphere a unit of pressure, defined as the pressure that will support a column of mercury 760 mm in height. It is equivalent to 101 kilopascals/1,013 millibars (14.7 lb per sq in), the ⇨SI unit of pressure.

atom the smallest part of an element capable of taking part in a chemical reaction. Atoms consist of a central ⇨nucleus, comprising ⇨neutrons and ⇨protons, and one or more orbiting ⇨electrons.

atomic bomb a nuclear weapon whose explosive power is the result of ⇨fission occurring within it. Two masses of a fissile material, such as uranium-235, are brought together at detonation, the two masses combined being large enough to support a ⇨chain reaction. This proceeds explosively.

atomic heat the product of atomic weight and specific heat (relative atomic mass and specific heat capacity). It is approximately constant for many solid elements and equal to 6 calories per gram atom per degree (25.2 joules per mole per Kelvin).

atomic number the number (symbol Z) of ⇨protons in the nucleus of an atom. This is equal to the number of orbiting ⇨electrons on a neutral atom.

atomic orbital a region in space occupied by an electron associated with the nucleus of an atom. Atomic orbitals have various shapes, depending on the energy level of the electron and the degree of ⇨hybridization. Atomic orbitals overlap to form molecular orbitals, or chemical bonds between atoms.

atomic radius the effective radius of an atom. Atomic radii vary periodically with atomic number, being largest for the alkali metals and smallest for the rare gases.

atomic volume the volume of one gram-atom of an element.

atomic weight see ⇨relative atomic mass.

ATP abbreviation for ⇨adenosine triphosphate.

attenuation in biology and medicine, the reduction in the virulence of a pathogenic microorganism by culturing it in unfavourable conditions, by drying or heating it, or by subjecting it to chemical treatment. Attenuated viruses, for example, are used in vaccines.

AU symbol for ⇨astronomical unit.

audiometer an instrument for testing hearing at various frequencies and loudness levels.

Auger electron an electron emitted by an ⇨atom as a result of a change from an excited state to a lower energy state, without the emission of either ⇨X-rays or ⇨gamma radiation.

aurora spectacular array of light in the night sky, caused by charged particles from the Sun hitting the Earth's upper atmosphere. The aurora borealis is seen in the north of the northern hemisphere; the aurora australis in the south of the southern.

autoclave a reactor vessel constructed to allow chemical reactions to take place at high temperature and pressure. Such vessels are used, for example, in the manufacture of chemicals and materials for the construction industry.

autogyro ⇨V/STOL aircraft that uses horizontal rotor to achieve take-off and sustain height, and a forward propeller to provide forward motion through the air.

automorphic function function that in relation to a ⇨group of ⇨transformations has a value on the transformed point identical with the value on the original point.

autophagy a process in which a cell synthesizes substances and then metabolizes and absorbs them for its own sustenance.

autumnal equinox see ⇨equinoxes.

Avogadro's hypothesis the number of molecules in two equal volumes of gases are the same, provided the two gases are at the same temperature and pressure.

axiom or *axiomatic principle* in a rigorous treatment of an area of mathematics, a statement that is assumed without proof. Generally, axioms describe the basic properties of a concept in terms of an established 'working definition'.

axiom of choice in set theory, an assumption that amounts to the possibility of an infinite number of choices. Formally, the axiom states that for a family of nonempty sets X_i, there is a set A such that A meets each X_i in precisely one point.

axis theoretical straight line through a celestial body, around which it rotates.

axle the structure connecting the wheels of a vehicle. It can either transmit driving ⇨power provided by a motor to each wheel, or simply connect the wheels together.

axon a long nerve fibre (usually covered by a ⇨myelin sheath) that conducts impulses away from a nerve cell body. Most axons end in ganglia, effector organs or ⇨synapses.

axoplasm the ⇨cytoplasm of an ⇨axon, containing the ⇨neurofibrils of the nerve fibre.

azimuth directional bearing around the horizon, measured in degrees from north (0°).

B

bacillus a rod-shaped bacterium. See ⇨bacteria.

background radiation or background black-body radiation, is the ⇨isotropic residual ⇨microwave radiation in space left from the primordial ⇨Big Bang. At a ⇨wavelength of 7 cm/2.8 in it represents a temperature of about 3K (−270°C/−484°F).

bacteria a group of uni- or multicellular microorganisms that exist in colonies of the same species. They contain no ⇨chlorophyll and therefore cannot photosynthesize, although in other respects they are similar to plants. They are microscopic and classified mainly by their distinctive shapes: a coccus is spherical (irregular clusters are staphylococci, chains of cocci are streptococci), a spirillium is spiral, a spirochaete forms a flexible spiral, and a bacillus has a rodlike shape. Bacteria may be nonmotile (atrichous) or move by means of motile filaments. They reproduce asexually by binary fission (sometimes extremely rapidly). Some process food from inorganic sources, but most feed off organic matter as parasites or saprophytes, and most are ⇨aerobic.

bacteriophage a virus that infects or parasitizes ⇨bacteria. It has a polyhedral head containing DNA and an elongated hollow tail of protein. It attaches its tail to the cell wall of a bacterium and produces ⇨enzymes that break through the wall. The phage then modifies the bacterial cell's metabolism, resulting in the production of more bacteriophages which are released when the host cell disintegrates.

bacteriotropin or *opsonin* a substance in blood serum that helps to make ⇨bacteria more vulnerable to ⇨leucocytes, which can then engulf and destroy them.

Baily's beads bright 'beads' of sunlight showing through the valleys between mountains on the rim of the Moon just before and just after a total eclipse of the Sun, seen from Earth.

Bakelite tradename for a material used for insulating purposes, and in the manufacture of plastics and paints. Produced by condensing cresol or phenol with formaldehyde, the resin is brittle.

ballistics the study of the trajectories and general behaviour of projectiles, such as bullets, artillery shells, and missiles.

Balmer series the visible ⇨spectrum of hydrogen, consisting of a series of distinct ⇨spectral lines with wavelengths in the visible region.

bandwidth the range of ⇨frequencies over which the capability of a receiver or other electric device does not differ from its peak by a given amount.

Bardeen–Cooper–Schrieffer theory or *BCS theory* a theory that accounts for the zero electrical resistance of some ⇨superconducting metals by invoking an attractive interaction between electrons binding them into 'Cooper pairs'.

Barnard's star star that had – until 1968 – the greatest known ⇨proper motion of any. Seen from the Earth it moves just over 10 seconds of arc per year, but even this is deceptive because it is approaching Earth at a rate estimated to be more than 100 km/62 mi per second.

barometer a device for measuring atmospheric ⇨pressure.

barycentric calculus coordinate geometry calculations using a coordinate system devised by August Möbius in which numerical coefficients are assigned to points on a plane, giving the position of a general point by reference to four or more noncoplanar points. Described in this way, the general point thus represents a centre of gravity for a distribution of mass at the four (or more) points proportional to the assigned numbers.

base in chemistry, a substance that has a tendency to gain ⇨protons and forms ⇨hydroxyl ions in solution. A base reacts with (is neutralized by) an acid to form a salt and water. See also ⇨alkali.

base in molecular biology, one of the nitrogen-containing compounds that (together with a phosphate group and deoxyribose or ribose) form ⇨nucleotides, the constituents of DNA and RNA. In DNA the bases are adenine and guanine (purine compounds), and cytosine and thymine (pyrimidine compounds); RNA has the same bases except it contains the pyrimidine uracil instead of thymine.

battery a number of primary ⇨cells connected together to produce a higher voltage (cells in series) or a higher current capacity (cells in parallel). A set of secondary cells is called an accumulator. A single cell is commonly, but mistakenly, called a battery.

behaviourism a theory that regards animal behaviour as a consequence determined by conditioned responses (both nervous and hormonal).

Bernoulli numbers (*B numbers*) sequence of ⇨rational numbers that may be represented by the symbolic form $B_n = (B+1)^n$, corresponding to the sequence:

$B_0 = 1$ $B_1 = -1/2$ $B_2 = 1/6$
$B_3 = 0$ $B_4 = -1/30$ $B_5 = 0$
$B_6 = 1/42$ $B_7 = 0$ $B_8 = -1/30$

and so on.

B numbers of odd order (except B_1) are zero; B numbers of even order alternate in sign.

Bernoulli's theorem the sum of the ⇨pressure, ⇨potential energy, and ⇨kinetic energy of a fluid flowing along a tube is constant, provided the flow is steady, incompressible and nonviscous.

Bessemer process a method of producing steels from cast iron, using air blasted into the molten mixtures to remove carbon impurities, by ⇨oxidation. The molten iron is produced in a ⇨blast furnace, which delivers its charge to the Bessemer converter in which the steel is produced.

beta chain a particular secondary structure of the polypeptide chain (of a protein) which makes it adopt the shape of a pleated sheet. See also ⇨alpha chain.

beta decay or *beta particle decay* spontaneous emission by a heavier element (such as uranium) of negatively charged high-speed ⇨electrons – beta particles. The result of this radioactive decay is that the original element is very gradually converted into another element. Beta particle emission may take place at the same time as ⇨alpha decay.

beta particle an ⇨electron or ⇨positron emitted in the decay of an unstable atomic nucleus. The electron is the result of the conversion within the nucleus of a ⇨neutron into a ⇨proton, whereas the positron is the result of the change of a nuclear proton into a neutron.

Betelgeuse (Alpha Orionis) brightest genuinely ⇨variable star in the sky.

Bethe–Weizsäcker cycle see ⇨proton– proton cycle.

Betti numbers numbers characterizing the connectivity of a ⇨variety.

Bhabha scattering a scattering process involving ⇨electrons and ⇨positrons.

Big Bang model a model in ⇨cosmology in which the universe is taken as beginning in an enormous explosion which initiated the subsequent ⇨Hubble expansion of the galaxies. The theory is based upon Einstein's general theory of ⇨relativity. Since the nuclear physics involved has been explained, and various supporting evidence – notably ⇨helium abundance and the sources of radio emission – has been discovered, the theory is almost universally accepted (although at one time the ⇨steady-state theory rivalled it in popularity).

bile a digestive juice secreted by the liver and stored in the gall bladder which emulsifies fats and facilitates their digestion. It also stimulates the secretion of digestive ⇨enzymes and aids the absorption of fat-soluble ⇨vitamins.

bimetallic strip a length of two metals of different thermal expansivities welded together. When subjected to heat, the strip will deflect in a certain way, for example, to complete an electric ⇨circuit.

binary star system two stars that orbit around a common centre of mass; the shapes of such orbits vary enormously.

binary system simplified form of numeration specifically appropriate for use by computers, involving only two distinct digits; the system is thus to base 2, as opposed to the ordinary standard base 10. Commonly, the two digits are represented by 0 and 1, and the system begins:

decimal 0 is represented by binary 0

1	1
2	10
3	11
4	100
5	101
6	110
7	111
8	1000

biochemistry the branch of chemistry that is concerned with chemical processes that take place within living organisms.

bioluminescence the emission of light by an organism as a result of chemical processes taking place within it.

biosynthesis or *anabolism* the natural formation of more complex biochemical compounds from less complex molecules (as in the synthesis of proteins from ⇨amino acids). This building-up process requires energy, stored in compounds such as ⇨adenosine triphosphate (ATP) which are formed during catabolism, the opposite of biosynthesis.

biplane an aircraft with two parallel wings, set one above the other.

biquaternions type of ⇨quaternions devised by William Clifford to use specifically in association with linear algebra to

represent motions in three-dimensional non-Euclidean space.

birefringence another name for ⇨double refraction.

black body in astronomy, a body with ideal properties of radiation absorption and emission, against which less perfect actual stars and celestial objects can be measured. Black-body radiation has a continuous spectrum governed solely by the body's temperature: for any particular temperature there is a specific wavelength at which radiation emission is greatest.

black-body radiation radiation of a thermal nature emitted by a theoretically perfect emitter of radiation at a certain temperature.

black hole a ⇨singularity in space, surrounded by an ⇨event horizon, caused by the collapse of a small but massively dense star through the effects of its own increasing gravity. By the time the state of singularity is reached, the remnants of the star may be minimal, but the gravitational force is so strong it prevents even light from escaping. Black holes may form the 'power centres' of galaxies, thus explaining ⇨infrared radiation detected in several galactic centres. The properties of matter entering a black hole are the theme of John Wheeler's ⇨no hair theorem.

blast furnace a vessel for the ⇨smelting of iron from iron oxide ore, from which ⇨steel can be made. A mixture of the ore, coke and limestone is heated using a blast of hot air, and the consequent chemical reactions produce molten iron and slag, consisting of nonmetallic impurities. This ⇨cast iron (pig iron) is used to produce steel in the ⇨Bessemer process.

blastocyst a developmental stage in a mammalian embryo following cleavage, when the embryo consists of a thin outer layer of cells (⇨trophoblast) enclosing a central mass of cells.

blastoderm a sheet of cells that grows on the surface of a fertilized ovum. In mammals it forms a disc of cells that eventually develops into the embryo between the amniotic cavity and the yolk sac. The ⇨endoderm, ⇨mesoderm and ⇨ectoderm also develop from the blastoderm.

blastula a stage in the embryonic development of lower animals that corresponds to the ⇨blastocyst in mammals.

blood group a method of categorizing blood types. The most commonly used classification of human blood is the A, B, AB and O group system, which depends on the presence or absence of one or two ⇨antigens (in the blood cells) and one or two agglutins (in the serum).

blue dwarf, blue giant high-temperature stars (as opposed to red stars). Blue giants are generally on or near the ⇨main sequence of the ⇨Hertzsprung–Russell diagram; blue dwarfs represent the very dense, but very small, near-final form of what was once a ⇨red giant.

***B* number** abbreviation for ⇨Bernoulli number.

bobbin in weaving, the spool onto which yarn is wound.

Bode's law or the Titius-Bode rule, was first devised in 1772 and comprised the series $0 + \frac{4}{10}$, $3 + \frac{4}{10}$, $6 + \frac{4}{10}$, $12 + \frac{4}{10}$, $24 + \frac{4}{10}$ and so on, which was found to describe fairly accurately the distance in ⇨astronomical units of the then known planets from the Sun. After the discovery of Neptune (to which the 'law' does not apply at all), the law was somewhat discredited, although later still, the positioning of Pluto (which corresponds approximately) made it seem possibly more than coincidental.

Bohr model a semi-classical model of the atom, in which the electrons orbit the central nucleus, but are allowed to do so only in certain orbits, where their angular momentum is quantized. Emission and absorption of radiation is represented by jumps of electrons between certain orbits, the frequency of the radiation being given by another quantum relation.

boiling point the temperature at which the saturated ⇨vapour pressure of a liquid is equal to atmospheric pressure. Standard boiling points are quoted for an atmospheric pressure of 760 mm/30 in of mercury. For a solution, the dissolved substance (solute) raises the boiling point of the solvent; such elevation of boiling point in dilute solutions may be used to determine the ⇨relative molecular mass of the solute.

Bok's globule small, circular dark spot in a ⇨nebula, with a mass comparable to that of the Sun. Bok's globules are possibly gas clouds in the process of condensing into stars.

bond a chemical bond, or valency bond, is the linkage that holds atoms together to form molecules. Most bonds involve a sharing or exchange of electrons.

Boolean algebra algebraic structure formulated by George Boole initiating algebraic study of the rules of logic. His basic process was to subdivide objects into separate classes, each with a different property.

Bose–Einstein statistics a treatment of particles which have integral ⇨spin in ⇨statistical mechanics, enabling properties of systems made up of such particles to be calculated.

boson particle that obeys ⇨Bose–Einstein statistics, such as a ⇨photon.

boundary value in applied mathematics a natural phenomenon in a given region may be described by functions that satisfy certain differential equations in the interior of the region and take specific values on the boundary of the region. The latter are referred to as boundary values.

Bourdon gauge a device for measuring a wide range of ⇨pressures, which exploits the change in cross section of a freely hanging tube through which fluid at a certain pressure is passed. The tendency for the tube to straighten out is measured by a scale calibrated for pressure.

box girder a girder made from four lengths of metal sheet formed into a box cross section.

Boyle's law the volume of a gas held at constant temperature is inversely proportional to its pressure. This is strictly true only for a gas consisting of perfectly elastic molecules, each of which has zero volume and no attractive or repulsive interaction with its neighbours – that is, an ⇨ideal gas.

brachiopod or *lamp shell* any member of the phylum Brachiopoda, marine invertebrates with two shells, resembling but totally unrelated to bivalves. There are about 300 living species; they were much more numerous in past geological ages. They are suspension feeders, ingesting minute food particles from water. A single internal organ, the lophophore, handles feeding, aspiration, and excretion.

Bragg's law the maximum intensity of X-rays diffracted (see ⇨X-ray diffraction) through a crystal occurs when the sine of the complement of the angle of incidence of the X-rays onto the crystal satisfies the relation:

$$n\lambda = 2d\sin\theta$$

where λ is the wavelength of the radiation, d is the lattice spacing, and n is an integer.

braid theory part of the study of ⇨nodes in three-dimensional space, first devised by Emil Artin.

Bremsstrahlung ⇨electromagnetic radiation produced by the rapid deceleration of charged particles such as electrons, as occurs in the collison between electrons and nuclei.

Brewster's law the ⇨refractive index of a medium is given by the tangent of the angle at which maximum ⇨polarization occurs.

bridge a structure spanning a river, road, and so on, allowing transportation to take place. See ⇨cantilever bridge, ⇨suspension bridge.

British Standard Whitworth thread see ⇨Whitworth standard.

broad gauge the rail track introduced by Isambard Kingdom Brunel, where the inside edges of the steel rails were separated by a distance of 7 ft/2.2 m.

Brownian motion motion affecting smoke and similarly sized microscopic particles as the result of impacts of molecules upon them. It appears as a random and erratic movement in the particles when viewed through a microscope.

bubble chamber a means of detecting and measuring the trajectories of high-energy ionizing particles. By keeping a substance such as liquid hydrogen heated to just above its boiling point, but holding it under ⇨pressure to prevent boiling, particles can be detected by the bubbles they form if the pressure is released just before they enter the chamber.

Bunsen burner a gas-powered burner fitted with a valve to allow different mixtures of air and gas to be burnt, and thus alter the temperature of the flame.

Burali-Forte's paradox paradox stating that to every collection of ordinal numbers there corresponds an ordinal number greater than any element of the collection. In particular it would follow that the collection of all ordinal numbers is itself an ordinal number. This contradiction demonstrated the need for a rigorous exposition of set theory in which not all collections may be accepted as valid subjects of discourse. There is thus no such thing as a 'set of all sets', nor a 'set of all ordinals' (indicating that the ⇨foundations of mathematics cannot be expressed in purely logical terms). One therefore distinguishes between sets, which may be manipulated, and ⇨classes, which may not (except in the simplest of circumstances).

C

CAD acronym for ⇨computer-aided design

calculus general term for a body of useful calculatory procedures. See ⇨differential calculus and ⇨integral calculus; see also ⇨calculus of variations.

calculus of variations method of calculation for solving problems in which one of the unknowns cannot be expressed as a number or a finite set of numbers, but is representable as a curve, a function or a system of functions. (A classic problem in the subject is to show that a circle, among all curves of fixed length, encloses the maximum area.)

calibre the internal diameter of a bore or pipe.

Callisto fifth (known) moon out from ⇨Jupiter, and its second largest.

caloric theory the theory that ⇨heat consists of a fluid called 'caloric' that flows from hotter to colder bodies. It was abandoned by the mid-eighteenth century.

calorimeter an instrument used in the determination of quantities of heat change.

calorimetry the measurement of the heat-related constants of material, for example, thermal capacity, ⇨latent heat of vaporization.

calotype a photographic technique using silver iodide.

camber the curved surface of an ⇨aerofoil, road, and so on.

Cambrian period of ⇨geological time 570–510 million years ago; the first period of the Palaeozoic era. All invertebrate animal life appeared, and marine algae were widespread. The earliest fossils with hard shells, such as trilobites, date from this period.

Cambridge Catalogues the results of five intensive radio-astronomical surveys (1C, 2C, 3C, 4C and 5C) under the direction of Martin Ryle and Antony Hewish, during the 1950s, 1960s and 1970s, at Cambridge.

camera obscura a darkened room in which images of objects outside are projected onto a screen using a long-focus ⇨convex lens.

canal rays streams of positively charged ⇨ions produced from an ⇨anode in a discharge tube, in which gas is subjected to an electric discharge.

cantilever bridge a bridge consisting of a long central span formed by the joining of two shorter self-supporting spans extending from piers.

capacitance the ability of a system of electrical conductors and insulators to store charge. The SI unit of capacitance is the farad (symbol F).

carbohydrate any of a large group of biologically important organic compounds composed only of carbon, hydrogen and oxygen, with the general formula $C_n(H_2O)_n$. There are three main types of carbohydrates. *Monosaccharides*, or simple sugars, contain between three and nine carbon atoms; examples are glucose, fructose and dextrose. *Disaccharides* (also sugars) consist of two monosaccharides linked together (for example, sucrose, or normal table sugar, consists of a molecule of glucose linked to a molecule of fructose). *Polysaccharides* consist of up to about 10,000 monosaccharides linked together; examples are starch and cellulose.

Carboniferous period of ⇨geological time 363–290 million years ago, the fifth period of the ⇨Palaeozoic era. In the USA it is divided into two periods: the Mississippian (lower) and the Pennsylvanian (upper). Typical of the lower-Carboniferous rocks are shallow-water limestones, while upper-Carboniferous rocks have delta deposits with coal (hence the name). Amphibians were abundant, and reptiles evolved during this period.

carbon–nitrogen cycle use of carbon and nitrogen as intermediates in the ⇨nuclear fusion process of the Sun. Cooler stars undergo the ⇨proton–proton cycle.

carboxylic acid any of a class of organic compounds that contain the carboxyl (–COOH) group. The acids' names end in -oic; the simplest are methanoic (formic) acid, HCOOH, and ethanoic (acetic) acid CH_3COOH.

carburettor the device in an ⇨internal-combustion engine that mixes air with a jet of petrol to produce a combustible mixture, which, when ignited, produces driving power via pistons and crankshaft.

carcinogenesis the production of cancer, or a carcinoma.

Carnot's cycle a cycle of operations for a heat engine involving four distinct steps. Firstly, an ⇨isothermal expansion, then an ⇨adiabatic expansion, then an isothermal compression and finally an adiabatic compression.

Carnot's theorem a theorem in ⇨thermodynamics stating that the efficiency of any (reversible) heat engine depends only on the temperature range through which the machine operates.

carrier wave a wave of ⇨electromagnetic radiation of constant ⇨frequency and ⇨amplitude used in ⇨radio communication. ⇨Modulation of the wave allows information to be carried by it.

Cassegrain telescope reflecting telescope devised by Cassegrain in which an auxiliary convex mirror reflects the magnified image, upside down, through a hole in the centre of the main objective mirror – that is, through the end of the telescope itself. It was, however, no improvement on the ⇨Gregorian telescope invented probably slightly earlier.

Cassini Division gap between ⇨Saturn's Rings A and B.

cast iron or *pig iron* a brittle, impure form of iron, containing a few per cent of carbon. Produced in a ⇨blast furnace, such material is turned into steel via a ⇨Bessemer converter.

catalyst a substance that brings about or accelerates a chemical reaction while remaining unchanged at the end of it; the phenomenon is called catalysis. An ⇨enzyme is a type of catalyst in biochemical reactions.

catastrophe theory theoretical model devised in an attempt to investigate processes that are not continuous but subject to sudden – and possibly violent – changes (catastrophes). The model consists of, for example, a ⇨surface that incorporates some kind of discontinuity or fold so that a point travelling on it continuously under gravity may suddenly come off it in a manner much like 'falling off a precipice' and land again on the surface. Plotted by René Thom, the seven elementary catastrophes are: the fold, cusp, 'swallow-tail','butterfly', hyperbolic, elliptic and parabolic.

catastrophism a theory that regards the variations in fossils from different geological strata as having resulted from a series of natural catastrophies that gave rise to new species.

catenary a ⇨transcendental curve that can be represented by an equation of the general form:

$$y = \frac{a}{2} (e^{x/a} + e^{-x/a})$$

where a is a constant and e is ⇨Euler's number. It is the shape assumed by an ideal chain suspended from two points that are not immediately vertical one above the other.

cathode the negatively charged electrode in a cell, discharge tube or ⇨thermionic tube.

cathode rays fast-moving streams of electrons emitted by the ⇨cathode of an evacuated ⇨discharge tube. A tube made specifically for the production of cathode rays is called a cathode-ray tube, and is used for the visual display in computer equipment and in television and radar receivers.

cation a postively charged ⇨ion, such as a sodium ion (Na^+), ammonium (NH_4^+) or aluminium ion (Al^{3+}). During ⇨electrolysis, cations migrate towards the (negatively charged) ⇨cathode. In ionic compounds such as salts, the charges of the cations are usually balanced by those of an appropriate number of ⇨anions.

caustic curve curve formed by the points of intersection of rays of light reflected or refracted from a curved surface.

cavitation the void left behind a body moving through a fluid.

celestial of the heavens; in the sky; in space.

celestial equator projection of the Earth's equator as a line across the sky (so that to an observer actually on the equator, such a line would pass through the ⇨zenith). The directional bearing of a star is given in terms of its ⇨right ascension round the celestial equator.

celestial mechanics study of the movements and physical interactions of objects in space; astrophysical mathematics.

cell in biology, a discrete, membrane-bound portion of living matter, the smallest unit capable of an independent existence. All living organisms consist of one or more cells, with the exception of viruses. Bacteria, protozoa, and many other microorganisms consist of single cells, whereas a human is made up of billions of cells. Essential features of a cell are the membrane, which encloses it and restricts the flow of substances in and out; the jellylike material within, often known as ⇨protoplasm; the ⇨ribosomes, which carry out protein synthesis; and the ⇨DNA, which forms the hereditary material.

cell, electric or *voltaic cell* an apparatus in which chemical energy is converted into electrical energy; the popular name is ⇨battery, but this actually refers to a collection of cells in one unit. An electric cell contains two conducting ⇨electrodes (a cathode and an anode) immersed in an ⇨electrolyte, usually consisting of an acid or a paste or solution of a salt. A potential difference (voltage) is set up between the electrodes.

A *primary cell* uses an irreversible chemical reaction and cannot be recharged; in a *secondary cell*, the reaction can be reversed by applying an external voltage across its electrodes, so recharging the cell. A group of connected cells is called an ⇨accumulator or storage battery. In a ⇨fuel cell, the reaction is the oxidation of a fuel.

cell, electrolytic a device to which electrical energy is applied in order to bring about a chemical reaction; see ⇨electrolysis.

cell theory a theory (proposed by Schleiden and Schwann in 1838–39) which regards all living things as being composed of cells and that their replication and growth result from cell division.

celluloid a ⇨thermoplastic that is strong, elastic and capable of being made into thin sheets. Used in early forms of photographic film.

Cenozoic or *Caenozoic* era of ⇨geological time that began 65 million years ago and is still in process. It is divided into the Tertiary and Quaternary periods. The Cenozoic marks the emergence of mammals as a dominant group, including humans, and the formation of the mountain chains of the Himalayas and the Alps.

centre of gravity the point within an object through which the resultant force of ⇨gravity always passes, for all orientations of the object.

centre of mass the point within an object at which all the object's ⇨mass may be taken as concentrated for dynamical purposes.

centrifugal force a force that must be invoked to satisfy Newton's third law of ⇨motion in rotating systems. It acts in a direction radially outward from the centre of the circular path, and opposite to the ⇨centripetal force.

centrifugal governor a ⇨governor that controls the behaviour of an engine by means of a ⇨feedback process involving ⇨centrifugal forces. As the engine speed increases, the arms of the governor are lifted by the centrifugal force, the power supply to the engine is thereby reduced, cutting the speed.

centrifugation a method of separating substances suspended in a liquid using the force of gravity induced by spinning a sample at high speed.

centripetal force the force that constrains an object moving in a circular path to maintain the given path. It acts towards the centre of the circular path, and opposes the ⇨centrifugal force. Gravity is the centripetal force acting on an object in orbit about the Earth, for example.

centrosome an area of ⇨cytoplasm, near the nucleus of a cell, which contains the centriole.

Cepheid type of regular visible star, of which the power–luminosity curve is particularly valuable in gauging distances in the universe. On the ⇨Hertzsprung–Russell diagram they are found on what has become known as the Cepheid instability strip. They range, however, from massive young (⇨population I) stars to old (⇨population II) stars with comparatively low masses. Delta Cephei was the first of the type to be recognized, although they are now known to be located in many galaxies.

chain reaction in chemistry, a reaction consisting of a sequence of two or more reactions in which the product(s) of one becomes the reactant(s) of another. Such reactions can take place rapidly and, if they involve a branched chain reaction, may accelerate to explosive speeds.

chain reaction in nuclear physics, a chain reaction occurs during nuclear ⇨fission in a reactor or fission bomb. Neutrons produced by the disintegration of fissile material (such as isotopes of uranium or plutonium) initiate the distintegration of others, and the process rapidly accelerates, because the number of neutrons produced is greater than that required to initiate it.

Chandrasekhar limit mass less than which a star eventually evolves into a stable ⇨white dwarf through using all its available energy, and above which it becomes a ⇨supernova and then, probably, a ⇨neutron star or, improbably, a white dwarf. The limit is now reckoned as 1.2 solar masses.

Chandrasekhar–Schönberg limit mass above which the helium core of a star begins to contract (eventually to collapse altogether). The limit is now reckoned as 10–15% of the star's total mass.

charge conservation a feature of ⇨quantum mechanics, in which reactions between ⇨elementary particles occur in such a way that there is no change in the total charge of the system after the event has occurred.

charge electric see ⇨electric charge.

Charles's law the volume of a fixed mass of gas at constant ⇨pressure is directly proportional to the temperature of the gas.

chemotherapy the treatment of an infection or other disorder, particularly cancer, using drugs.

Cherenkov detector apparatus through which it is possible to observe the existence and velocity of high-speed particles, important in experimental nuclear physics and in the study of ⇨cosmic radiation. It was originally built to investigate the Cherenkov radiation effect, in which charged particles travel through a medium at a speed greater than that of light in that medium.

chi-squared function (χ^2 function) function that in ⇨probability theory provides a test for deviation from a ⇨null hypothesis. It is usually represented as being made up of:

$$\frac{(\text{observed frequency of result} - \text{expected frequency of result})^2}{\text{expected frequency of result}}$$

in which the top line indicates the (squared) deviation from the expected.

Chladni figures visual manifestations of sound waves resulting from the behaviour of small particles on plates vibrated in sympathy to a particular sound.

chlorophyll a green pigment that occurs in all green plants, contained in the ⇨chloroplasts (except in blue-green algae). Chlorophyll absorbs light during ⇨photosynthesis and is involved in the transformation of light energy into chemical energy (stored as ATP) which the plant uses to manufacture carbohydrates (and oxygen). Of the four known chlorophylls, a, b, c and d, only chlorophyll a is found in all green plants.

chloroplast a small subcellular pigment-containing body found in photosynthetic plants (except ⇨bacteria and blue-green algae). In addition to ⇨chlorophyll, chloroplasts contain other pigments such as carotene and xanthophyll.

choke a valve in the ⇨carburettor of an ⇨internal-combustion engine that can reduce the amount of air being mixed into the petrol jet.

chordate any animal that belongs to the phylum Chordata, which at some developmental stage has a supporting rod of tissue (⇨notochord) running down its body. Vertebrates constitute a subphylum of the chordates.

chorion the outer layer of the embryonic membrane formed from the ⇨trophoblast and the ⇨mesoderm. In its early stages it forms the outer cell wall of the ⇨blastocyst and later gives rise to the chorionic villi, which are eventually involved in the formation of the ⇨placenta.

chromatic aberration an optical effect commonly found in simple lens instruments in which coloured fringes are seen around an image as a result of the ⇨wavelength-dependence of the ⇨refractive index for glass.

chromatin a nucleoprotein found in ⇨chromosomes and thought to be the molecular substance of heredity. It is readily stained by basic dyes and is therefore easily identified and studied under the microscope.

chromatography a method of chemical analysis in which a solution of the mixture to be analysed is allowed to move through an absorbent material (such as a glass tube packed with powdered chalk or a strip of filter paper). The components of the mixture move at different speeds and become separated into layers or bands along the column or paper.

chromosome a body that exists in the nucleus of a cell, made up mainly of nucleoproteins and bearing the ⇨genes which constitute the hereditary material. Each species has a constant number of chromosomes; human beings, for example, have 23 pairs of chromosomes in each cell (except in gametes), whereas the fruit fly has only four pairs. Gametes (eggs and sperm) have half the normal number of chromosomes, so that union of gametes produces a fertilized ovum with the full complement of chromosomes, half derived from the mother and half from the father. See also ⇨allele.

chromosome map a description of the position of ⇨genes along a ⇨chromosome.

chromosphere the part of the Sun's atmosphere immediately above the surface (the ⇨photosphere) and beneath the ⇨corona. It appears pinkish-red during eclipses of the sun.

circuit the total path travelled by an electric ⇨current.

cis- a prefix used in ⇨stereochemistry to distinguish an ⇨isomer that has two substituents or groupings on the same side of the main axis or plane of the molecule. The isomer with the two on opposite sides is denoted by the prefix **trans-**.

civil engineering the discipline dealing with the design and construction of bridges, roads, tunnels, canals, dams and other large construction projects.

class notion used in set theory to distinguish between collections of objects made up solely with regard to a 'collecting property', and those that may validly be manipulated without risk of inconsistency. See also ⇨Burali-Forte's paradox.

class field theory theory involving the mathematical structure known as a ⇨field, dealing specifically with those that extend a given field in a special kind of way.

Clausius–Clapeyron equation the relationship between the pressure, temperature and latent heat in a change of state, for example, liquid to gas.

clone a cell (which may develop into a whole organism) descended from another cell by asexual reproduction and theoretically having a genetic make-up identical to that of the original cell.

closed curve curve of which the end point coincides with the initial point, for example, a circle or an ellipse.

cloud chamber a detection system that enables the tracks of ionizing particles to be detected. Its operation depends on the fact that water vapour contained within it condenses onto the charged ions more easily than onto the uncharged particles within the chamber.

Clouds of Magellan see ⇨Magellanic Clouds.

cluster group of stars or of ⇨galaxies, usually with some recognizably systematic configuration. It appears that both types of cluster are a structural feature of the universe, which form over the passage of time.

coacervate a collection of particles in an emulsion that can be reversed into droplets of liquid before they flocculate.

coal gas a fuel gas comprising mostly hydrogen and methane and carbon monoxide produced by processing coal.

coccus any spherical bacterium (see ⇨bacteria).

codon the sequence of three ⇨nucleotides that code for a specific ⇨amino acid in (or at the termination of) a polypeptide chain, especially in DNA or messenger RNA.

coefficient of correlation see ⇨correlation coefficient.

coefficient of expansion a quantity that describes the amount of expansion undergone by a material for a degree rise in temperature. It is expressed as the increase in length per unit length, per degree centigrade. For example, the metal with the lowest coefficient of expansion is ⇨Invar, at 2.3×10^{-6} m of expansion for every metre in length, per degree centigrade rise in temperature.

coenzyme an organic compound that activates some ⇨enzymes to catalyse biochemical reactions. It is weakly attached to the enzyme and may be chemically changed in the reaction, during which it may act as a carrier. It is regenerated by further reactions.

coherent radiation a ray of radiation such as light in which all the radiation waves are in ⇨phase with each other.

cohomology or *cohomology theory* algebraic study, using group theory, of geometric objects with specific reference to the operation of finding a boundary. Cohomology theory represents a modification of ⇨homology in which it is possible both to add and to multiply ⇨classes.

coil, electric a length of conducting wire that has been formed into a long series of loops. Used to form transformers, electromagnets, and so on.

coke the solid, porous material produced from the carbonization of coal, all the volatile material having been driven off. It is used in the production of steel.

collinear on a single straight line.

collodion process a photographic technique that uses iodized cellulose tetranitrate to coat a plate after being sensitized to light using ⇨silver nitrate solution.

colloid a solution consisting of particles of a material of sizes in the range of one ten-millionth to one billionth of a metre suspended in a solvent. The behaviour of the substance depends critically on the existence of these size particles.

colour index see ⇨astronomical colour index.

coma inherent optical distortion in ⇨reflecting telescopes, with the effect of decreasing focal clarity away from the centre of an image. The corrective lens or plate invented by Bernhard Schmidt nullifies coma in Schmidt telescopes.

combustion a chemical reaction accompanied by the evolution of heat and light. Usually the reaction is oxidation involving oxygen from air accompanied by flame – that is, burning.

comet small planetoid in highly elliptical orbit around the Sun. At its ⇨aphelion it is a frozen mass of water and other hydrogen compounds covered in space dust, up to one light year away from the Sun; at ⇨perihelion it may be inside the orbit of ⇨Venus and streaming a thawing 'tail' of silicate dust behind it. (The tail – or tails – always points away from the Sun, and is made visible by reflected light.) There is now thought to be a 'reservoir' of potentially cometary material orbiting the Sun at a distance of about 1 light year; gravitational ⇨perturbations cause individual comets to leave the reservoir and assume an orbit.

commutative principle principle obeyed by the operations of ordinary addition and multiplication, to the effect that:

$$a + b = b + a \text{ and } ab = ba$$

This principle need not hold true in certain structures, for example, ⇨quaternions.

compact spaces and *bicompact spaces* special kinds of topological space exhibiting the property that, internally, every family of open sets whose union is the whole space necessarily contains a finite subfamily whose union is already the whole space. An alternative definition, first formulated with regard to a special class of such spaces, requires every sequence of points to have a converging subsequence. Russian mathematicians initially used the word *bicompact* (not now used in the West) to distinguish between the specialized and the general definitions.

companion star either one of a ⇨binary star system (although usually the less massive), sometimes only detectable by ⇨spectroscopy.

competitive exclusion principle a principle of natural selection (in evolution) whereby similar species are forced to specialize ever more minutely so as not to overlap with each other in a particular niche; if they do not specialize adequately, they die and become extinct.

complement a protein substance in blood serum which reacts with almost all antibody–antigen systems, lysing (breaking down) the antigen from the antibody and disappearing in the process.

complement of an angle in geometry, the number of degrees that with it would make a right angle. In set theory, all those elements of a fixed set not belonging to a specified subset.

complex in chemistry, any of a class of substances with a characteristic structure in which a central metal atom (often a ⇨transition element) is surrounded by – and bonded to – several nonmetallic atoms or groups of atoms (⇨ligands). Complexes are also called ⇨coordination compounds.

complex in psychoanalysis, an association of mental factors in the unconscious which relate to an emotional experience involving something unacceptable to the individual, often affecting the individual's behaviour.

complex number the sum of a ⇨real and an ⇨imaginary number (the latter being a multiple of $\sqrt{-1}$). Complex numbers may be represented geometrically by means of an ⇨Argand diagram, and are useful in the study of differential equations.

complex number astrophysics the basis of ⇨twistor theory.

complex variable a ⇨variable representing a ⇨complex number.

compound a substance consisting of two or more elements in chemical combination; it is made up of ⇨molecules. Unlike a mixture, it can be separated into its components only by undergoing one or more chemical reactions.

Compton effect the collison and subsequent behaviour of a ⇨photon of electromagnetic radiation and a charged particle, such as an ⇨electron. In such a collision, the photon's energy is partly passed on to the massive charged particle, so there is a drop in ⇨frequency of the radiation, and a simultaneous rise in energy of the particle.

computer a device that can be programmed to carry out given logical processes on data, and give the results of the processing to an end user. There are two main types of computer: ⇨analogue computers and ⇨digital computers.

computer-aided design (CAD) the use of a computer to design buildings, vehicles, and so on, allowing prototypes to be checked and tested before proceeding further to model-making.

computer, analogue see ⇨analogue computer.

computer, digital see ⇨digital computer.

computerized axial tomography medical technique using a *body scanner* to build up a complete picture of a cross-section through the body using a narrow X-ray beam that scans through the patient.

concave an object, especially a lens, whose thickness is greater at its periphery than at its centre.

concentric having the same centre.

conchoid curve an ⇨algebraic curve represented by an equation of the general form:

$$x^2y^2 = (x - a)^2(c^2 - x^2)$$

conditioned reflex a (behavioural) reflex acquired by repetitive training and conditioning with an originally neutral stimulus.

conditioned stimulus an originally neutral stimulus applied in conditioning experiments which evokes a trained or conditioned response. In Pavlov's classical experiments with dogs, the neutral stimulus (the sound of a bell) originally evoked no salivation reflex; but after being presented for a time with an ⇨unconditioned stimulus (food), it became the conditioned stimulus which evoked the conditioned response.

conduction the transport of thermal energy in a substance as a result of the transfer of higher ⇨kinetic energy of molecules at higher temperatures to those molecules with lower energies and temperatures.

conduction, electrical the ability of ⇨electrons to flow in a medium.

conductivity, electrical a measure of the ease with which a material conducts electricity; the reciprocal of resistivity. The conductivity of an ⇨electrolyte depends on the presence of ⇨ions.

conductor a material that supports the passage of an electric ⇨current.

congruence in geometry two sets are congruent if either can be transformed by translations and rotations into the other.

conics study initiated by Apollonius of Perga of how a cone can be 'cut' so as to produce circles, ellipses, parabolas and hyperbolas; he stated 'a conic section is the locus of a point that moves so that the ratio of its distance *f* from a fixed point, to its distance *d* from a straight line, is constant'. Whether the

constant *c* is greater than, equal to, or less than one determines the type of curve the section represents.

conjugation the presence in an organic compound of two carbon–carbon double bonds separated by a single bond. Such an arrangement favours certain addition reactions in which the double bonds become single, and the single bond becomes double.

conservation of mass and energy important physical principle and one of the basic laws of physics, stating that matter is neither created nor destroyed (although mass may become energy, the energy quantitatively represents the mass). One exception to this principle is a ⇨singularity; another follows from the theory of ⇨virtual particles.

constant of precession see ⇨precession of the equinoxes.

continental drift a theory in geophysics which maintains that the continents, being part of vast tectonic plates, are capable of moving across the face of the Earth as the plates on which they shift move relative to one another.

contingency table table listing information classified as variable according to two or more independent attributes. Such tables are used commonly in commerce, notably by insurance companies.

continued fraction development of any ⇨real number in the form of a sequence of integers from which approximations to the number may be calculated successively; for example:

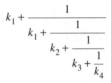

and so on. The sequence can be finite or infinite. The development will be finite in the case of a rational number, and the calculation will then terminate on reaching the rational number. In the case of an irrational number, a termination will be reached only as the ⇨limit of the sequence of values calculated.

continuous function or *continuity* of ⇨transformations, represents continuous motion, uniform variation. More precisely, a function *f* is said to be continuous at an argument value *x* if the function value for arguments close to *x* can be held down to a value as near to *f(x)* as required by keeping the argument close enough to *x*.

convection process in the Sun (and possibly other stars) perhaps caused by ⇨solar rotation, which produces the immensely powerful electrical and magnetic fields associated with ⇨sunspots.

convection the transfer of ⇨heat through a fluid by motion of the fluid itself, as a result of the reduction in ⇨density produced by the higher temperature. See ⇨thermal.

convergence of a sequence or series, describes its approaching to a ⇨limit; it is the opposite of divergence.

convex an object, especially a lens, whose thickness is greatest at its centre than at its periphery.

Cooper pairs see ⇨Bardeen–Cooper–Schrieffer theory.

coordination compound any of a diverse group of complex compounds characterized by a structure in which several ⇨ligands surround – and are covalently bonded to – a central metal atom. Such compounds may be electrically neutral, or positive or negative ions. Similarly the central metal atom may be neutral, anionic or, rarely, cationic, but it is always one that is able to accept an electron pair(s) to form a coordinate bond(s). The total number of bonds between the central atom and the ligands is the coordination number, which, in general, ranges from two to twelve; four and six are the most common.

coordination number see ⇨coordination compound.

Copernican model of the universe ⇨heliocentric model that replaced the geocentric ⇨Ptolemaic model, and was thus a considerable improvement. The model, however, still involved ⇨epicycles and the ⇨spheres.

cordite an explosive consisting of nitroglycerine and cellulose nitrate.

Coriolis force a special force that has to be invoked in dealing with the motion of bodies in certain types of rotating ⇨frames of reference.

corona upper atmosphere of the Sun, above the ⇨chromosphere. Normally observable only during a total eclipse, it has no definite shape or limit but merely rarefies into space in all directions. Its temperature in some regions reaches 1,000,000°C/1,800,000°F; it rotates synchronously with the ⇨photosphere.

coronagraph device for studying the solar ⇨corona at any time of the day. It was first invented by Bernard Lyot.

corpuscle in biology, a free-floating or fixed cell or body such as a blood cell or the enclosed end of a sensory nerve.

correlation coefficient in statistics and probability theory, a measure of how closely two variables, generally within the same numerical framework, can be related or identified. For complete lack of similarity, the correlation coefficient (commonly expressed as r) is 0; for perfect positive linear correlation, $r = 1$; for perfect but negative (opposite) linear correlation, $r = -1$.

cortex the outer layer of an organ or plant. For example, the cerebral cortex constitutes the outer nerve cells of the brain.

corticoid or **corticosteroid** any of various steroid substances (mainly ⇨hormones) secreted by the adrenal cortex, or a synthetic substance closely resembling them.

cosecant (cosec) see ⇨trigonometric functions.

cosine (cos) see ⇨trigonometric functions.

cosmic censorship theory that the hidden interior within all ⇨event horizons is the same and is always, necessarily, hidden.

cosmic radiation, cosmic rays high-speed particles that reach the Earth from outside the Solar System. Heavier cosmic ray particles – such as those sought in ⇨X-ray astronomy – are ordinarily filtered out by the Earth's upper atmosphere.

cosmic year time the Sun takes to 'orbit' in ⇨galactic rotation: about 225 million years.

cosmological principle theory behind most modern interpretations of cosmology, effectively that the universe is essentially the same everywhere.

cosmology the scientific study of the structure and evolution of the universe.

coulomb the ⇨SI unit of electric ⇨charge, defined as the quantity of charge required to flow for one second to produce one ⇨ampere of electric current.

Coulomb field the field of force surrounding an electric charge. Its intensity can be deduced from ⇨Coulomb's law.

Coulomb's law the ⇨force between two charged bodies varies directly as the product of the two charges, and inversely as the square of the distance between them.

covalency a type of chemical bonding in which the bond involves the sharing of two electrons, one supplied by one of the combining atoms and one by the other. A two-electron molecular orbital is formed by the overlap of two one-electron ⇨atomic orbitals.

CP violation the breaking of a fundamental ⇨quantum theory conservation rule by some unstable particles undergoing decay into other particles. C represents charge conjugation that relates particles to ⇨antiparticles and P stands for ⇨parity.

Cramer's paradox that although two different cubic curves intersect at nine points, part of the definition of a single cubic curve is that it is itself determined by nine points.

Cramer's rule method of solving a simultaneous system of linear equation by using ⇨determinants.

crankshaft the main shaft of an engine, by which energy produced in the ⇨piston cylinders is transmitted to the rotary ⇨power unit.

Crêpe Ring rather transparent inner ring (Ring C) of the ⇨Saturn ring system. Its diameter measures about 149,300 km/92,800 mi.

cresol or **hydroxytoluene** an important constituent of explosives, plastics, and dyestuff intermediates.

Cretaceous period of ⇨geological time 146–65 million years ago. It is the last period of the ⇨Mesozoic era, during which angiosperm (seed-bearing) plants evolved, and dinosaurs reached a peak before their almost complete extinction at the end of the period. Chalk is a typical rock type of the second half of the period.

critical temperature the temperature above which a gas cannot be liquefied by ⇨pressure only.

critical velocity the velocity above which the flow of a liquid ceases to be smooth following ⇨streamlines, and becomes ⇨turbulent.

Crooke's radiometer an instrument consisting of an evacuated glass dome in which sits a freely rotating system of vanes, whose opposite sides are white and black respectively. The rotation of the vanes when put near a source of heat is a demonstration of the ⇨kinetic theory of gases.

crossing-over the interchange between chromatids of one or more pairs of allelic ⇨genes on homologous ⇨chromosomes which results in an overall recombination of genetic material. It occurs during the ⇨meiosis that characterizes ⇨gamete formation.

cross ratio in projective geometry, a ratio expressing a relationship between two other ratios determined by four points on a given line, namely the ratio:

$$CA/CB : DA/DB$$

For the purposes of projection, the 'point at infinity' may be any one of A, B, C or D; if the given line is then projected onto another, the cross ratio of the projected points will remain the same – hence the importance of the cross ratio.

crust the outermost part of the structure of Earth, consisting of two distinct parts, the oceanic crust and the continental crust. The *oceanic* crust is on average about 10 km/6 mi thick and consists mostly of basaltic types of rock. By contrast, the *continental* crust is largely made of granite and is more complex in its structure. Because of the movements of ⇨plate tectonics, the oceanic crust is in no place older than about 200 million years. However, parts of the continental crust are over 3 billion years old.

cryogenics the study of phenomena that occur at temperatures close to ⇨absolute zero.

crystal lattice the regular system of points in space (for example, the corners of a cube) about which atoms, molecules or ions in solids vibrate.

crytalloid a substance which, when dissolved in a solvent, can pass through a ⇨semipermeable membrane (as opposed to a ⇨colloid, which cannot).

cubic curve geometrical curve in three-dimensional space. It may be parametrized after a change of variables as $(x, y, z) = (at^3, bt^2, ct)$ and may thus be said to be determined by nine points. See also ⇨Cramer's paradox.

cupola furnace a brick-lined furnace used in the conversion of pig iron (cast iron) into iron castings. Air is driven in underneath a charge of heated material; raising the temperature sufficiently to bring about the transformation.

current an electric current is the result of the motion within a particular substance of the ⇨electrons or ions within it, as the result of some potential difference. Current is measured in ⇨amperes.

current, alternating see ⇨alternating current.

current, direct see ⇨direct current.

cybernetics the study of communication and control mechanisms in machines (including humans).

cyclotron a particle accelerator which operates by imposing a magnetic force on charged particles held within a circular container.

Cygnus-A source in the constellation Cygnus of strong radio emissions, possibly caused by the collisions of interstellar dust within two colliding galaxies. It was nevertheless identified optically in 1952 as a tiny area of magnitude 17.9.

cyst a sac containing fluid or solid matter which forms or causes a swelling. In animals, cysts commonly occur in the skin or organs (such as the ovary) and may result from an obstruction (in a duct) or from a cell disorder.

cytoblast the ⇨nucleus of a cell.

cytochrome any of a group of protein compounds containing incorporated iron, as in ⇨haemoglobin. Cytochromes are found in all aerobic plant and animal tissues, where they have an important role in oxidation–reduction reactions.

cytoplasm the viscous fluid inside a cell, including particles such as the ⇨Golgi apparatus and ⇨mitochondria but excluding the nucleus. The cytoplasm and the nucleus together form the ⇨protoplasm of a cell.

D

daguerreotype an early method of creating photographic images by exposing silver iodide coated plates to the light, and developing the image in mercury vapour.

Daniell cell a primary cell which uses a ⇨cathode of zinc, immersed in sulphuric acid contained in a porous pot which itself stands in a container of copper sulphate, in which the copper ⇨anode stands. An ⇨electromotive force of about 1.1 volts is produced by this cell.

dark reaction a series of reactions in ⇨photosynthesis (which do not require light) in which carbon dioxide is incorporated into three-carbon sugar phosphate molecules; it depends on the ⇨light reaction (which does require light).

deamination the removal of an amino ($-NH_2$) group from an organic compound. In mammals, for example, deaminizing ⇨enzymes in the liver and kidneys oxidize ⇨amino acids (and the ammonia formed is converted into urea by the liver).

de Broglie hypothesis a cornerstone of quantum physics which relates the wave nature of systems to their particlelike characteristics. For a particle with velocity v and mass m, one can associate a de Broglie wave of wavelength λ given by the de Broglie equation:

$$\lambda = h/mv$$

where h is ⇨Planck's constant.

declination angular distance above (positive) or below (negative) the ⇨celestial equator. One of the coordinates, with ⇨right ascension, that defines the position of a heavenly body.

decomposition general term with its meaning set by its context: for example, in number theory, the ⇨factorization of an integer into other integers that are of the second or third degree (squares or cubes); in ⇨plane geometry, the dividing of a plane into two regions by a simple ⇨closed curve.

Dedekind's cuts mathematical device by which ⇨irrational numbers can be referred to by means of sets of fractions (rational numbers).

deflection of light gravitational effect that bends a ray of light. Such an effect was predicted within the general theory of ⇨relativity, although previously considered impossible.

degradation the breaking down of compounds into simpler molecules; for example, the action of enzymes brings about the degradation of proteins to amino acids.

delta (Δ, δ) as a capital, Δ, the term means 'difference'; Δx represents the difference between consecutive x values according to context. Leopold Kronecker's delta is a symbol used in the evaluations of ⇨determinants (in matrix theory), to the effect that $\delta(i, j) = 1$ if $i = j$, otherwise it equals zero. (It thus measures whether i and j are different.) The d used by

Gottfried Leibniz in his notation for differential calculus, as in dy/dx, was based on an intended association with the delta. As a lower-case (small) letter, another delta (d) commonly represents a partial ⇨derivative in ⇨partial differential calculus.

de Moivre's equation statement that for integers n $(\cos z + i \sin z)^n = \cos nz + i \sin z$ where $i = \sqrt{-1}$.

dendrite one of the many branched protoplasmic projections of a nerve cell (⇨neuron). Dendrites carry impulses into the nerve cell body and connect via ⇨synapses to ⇨axons of other nerve cells.

denominator in an ordinary division operation, the number or expression by which the numerator is divided.

density the mean density of a celestial body is generally reckoned as its ⇨mass divided by its volume, expressed either in comparison with the density of water, in kilograms per cubic metre or pounds per cubic foot, or in relation to some other known density. The mean density of the Earth is thus 5.5 times that of water, that is, 5.5×10^3 kg m^{-3}, and is just less than four times that of the Sun. Yet the mean density of rocks at the surface is about half the overall mean value, and that of the Earth's central core is perhaps 2·5 times the overall value.

depressor nerve a nerve that, when stimulated induces reflex vasodilation and thus slows the heartbeat (resulting in a fall in blood pressure).

derivative a ⇨function derived from another by the application of differentiation or partial differentiation. A derivative of a derivative is called a derivative of the second order.

descriptive geometry branch of mathematics in which three-dimensional objects are represented as two-dimensional (plane) figures, using any of many types of projection.

determinant in solving linear equations, a number corresponding in a specified way within a square-shaped array of numbers arranged in rows and columns; for example:

$$\begin{vmatrix} a & b \\ c & d \end{vmatrix} = ad - bc$$

determinates 'known' values, as opposed to indeterminates, 'unknown', values.

deuterium a 'heavy' isotope of hydrogen; its atomic nucleus (deuteron) comprises one ⇨proton plus one ⇨neutron (as opposed to hydrogen's single proton).

Devonian period of ⇨geological time 408–360 million years ago, the fourth period of the ⇨Palaeozoic era. Many desert sandstones from North America and Europe date from this time. The first land plants flourished in the Devonian period, corals were abundant in the seas, amphibians evolved from air-breathing fish, and insects developed on land.

diacaustic a ⇨caustic curve formed by refraction.

dialysis a method of separating dissolved ⇨colloids from noncolloids (⇨crystalloids) using a ⇨semipermeable membrane, which does not allow the passage of colloids. It is the main working principle of a kidney machine for removing waste products from blood.

diamagnetism a form of magnetism induced in one substance by the magnetic field of another. Its basic cause lies in the shift of the orbital motion of the ⇨electrons of a substance resulting from the external magnetic field of the other substance. It occurs in all materials.

diaphragm a thin membrane found in the human ear, and in telephone receivers and microphones onto which ⇨sound waves impinge, being converted into electrical impulses by a device that takes its input from the diaphragm movements.

diastole a period of relaxation in the rhythm of the heartbeat when the fibres of the heart muscle lengthen and the ventricles dilate and fill with blood.

dicotyledon any member of the larger of the two subdivisions of the ⇨angiosperms (the other being the ⇨monocotyledons). Dicotyledon plants are characterized by having two seedleaves (cotyledons), flower parts arranged in fours or fives (or multiples thereof), net-veined leaves, and vascular bundles in the stem arranged in a ring.

dielectric substance (an insulator such as ceramic, rubber, or glass) capable of supporting electric stress.

dielectric constant the capacitance of a capacitor (condenser) with a certain dielectric divided by the capacitance of the same capacitor with a vacuum (or, in practice, air) as a dielectric. It has also been called relative permittivity or specific inductive capacity.

diesel engine an ⇨internal-combustion engine that uses heavy (diesel) oil as the fuel in the ignition and compression power cycle that produces motive power.

difference equation equation that relates the value of a function at time t to its values at a specified number of past times, from among $t - 1, t - 2, t - 3, \ldots$.

differential calculus or *differentiation* branch of mathematics concerned with the behaviour of a function, f, in the vicinity of a point. The central operation is the evaluation of the derivative (written dy/dx or $f(x)$) of the function at the point x. The derivative gives the slope of the tangent to the graph of f at x, and may be calculated as limit, as h tends to 0, of:

$$\frac{f(x + h) - f(x)}{h}$$

The importance of the tangent is that it is the straight line that best approximates to the graph in the vicinity of the point $(x, f(x))$. Higher-order approximations may then be obtained by repeated use of the limit process above. Generalizations to functions of several variables involve similar calculations made separately for each variable (partial differentiation); the overall process of obtaining the derivative is called differentiation; and the reverse of differentiation is called integration, a process that represents the calculation of area under a graph.

differential drive see ⇨differential gear.

differential equations equations involving ⇨derivatives (see ⇨differential calculus). In a linear differential equation, the unknown function and its derivatives never appear in a power other than one. Partial differential equations involve unknown functions of several variables, and partial derivatives do therefore appear.

differential gear a ⇨gear that allows two shafts to rotate at different rates. Such a system is used in cars to allow the wheels on the outside edge of a corner to rotate relative to those on the inside edge.

differential geometry investigation of geometrical surfaces using ⇨differential calculus.

differential rotation of a stellar cluster or galaxy, the 'orbiting' of stars nearer the centre faster than those at the edge. Of a single body (such as the Sun or a gaseous planet), the axial rotation of equatorial latitudes faster than polar latitudes.

differentiation in biology, the transformation of cells (or tissues) of a uniform type into cells (or tissues) that have distinct and specialized structures and functions.

differentiation in mathematics, see ⇨differential calculus.

diffraction a phenomenon, resulting from the wave nature of ⇨light, which takes the form of patterns of light and dark bands on a screen produced when a beam of light from a narrow slit is allowed to fall onto it. A form of ⇨interference.

diffraction grating polished metallic surface (usually a metallic mirror on a block of glass or quartz) on which has been ruled a great number (in thousands) of parallel lines, used to split light to produce a ⇨spectrum.

diffusion the tendency of the molecules of a gas or mixture of gases to mix uniformly and spread to occupy evenly the whole of any vessel containing them. A gas diffuses through a porous substance at a rate that is inversely proportional to the square root of its density. Molecules or ions of a dissolved substance (the solute) also tend to diffuse through the solvent to produce a solution of uniform concentration.

digital computer a computer that performs calculations on digits representing the quantities involved, converting these into a form that can be processed by electronic devices.

dimorphism the property of a chemical substance that allows it to crystallize in two different forms.

diode a ⇨thermionic valve used to rectify and demodulate signals (see ⇨rectification and ⇨modulation). It only allows the passage of current in one direction, and is so-called because it comprises simply an ⇨anode and ⇨cathode.

Diophantine equations algebraic equations involving one or more unknowns (indeterminates) with ⇨integers (whole numbers) as coefficients, to which one or more solutions are sought, also in integers. The classic form is:

$$ax + by = c$$

Part of the significance of this is that even if not enough information is given to derive a single solution, enough is given to reduce the answer to a definite type. Diophantus, who lived in the third century AD, thus began the investigations into ⇨number theory that still continue.

diphosphate a chemical compound containing two phosphate groups, as in ⇨adenosine diphosphate (ADP).

dipole an electric or magnetic system consisting of two opposing charges or poles separated by a small distance.

direct current (DC) a steady flow of ⇨electrons in one direction, such as that produced by an ⇨electric cell.

discharge, electric a release of a stored electric ⇨charge.

discharge tube a device containing (usually two) electrodes and a vacuum or gas at low pressure; a (high) voltage applied to the electrodes causes an electric discharge to take place between them. A gas-filled tube (such as a neon tube) may emit visible light and other forms of radiation.

discriminants special functions of the coefficients of an equation, used to find roots of a polynomial equation.

disjoint sets see set theory, ⇨universal set.

disproportionation the splitting of a molecule into two or more simpler molecules; it is also known as dismutation.

dissociation the reversible splitting of a molecule into two or more simpler portions. An ionic compound (such as a salt) dissociates in solution to form ions; this is termed electrolytic dissociation. If a compound splits into two others under the action of heat, the phenomenon is termed thermal dissociation.

distributive law law expressing the principle operative in the equation:

$$a(b + c) = ab + ac$$

divergent of a sequence or series, describes the fact that there is no ⇨limit for it to approach; it is the opposite of convergent.

DNA (*deoxyribonucleic acid*) a naturally occurring compound found in the ⇨chromosomes of plant and animal cells, which carries genetic information (and therefore functions as the biochemical basis for heredity). Its molecule consists of two long chains of ⇨nucleotides linked at regular intervals like the rungs of a ladder; the whole 'ladder' is twisted into a double helix. Each nucleotide is composed of the sugar deoxyribose, a phosphate group and one of four nitrogenous ⇨bases. DNA stores the genetic code in the sequential arrangement of these bases; each linear sequence of three bases constitutes the code for one ⇨amino acid (see ⇨codon).

domain set of objects within a mathematical structure on which operations are to be performed. In a simpler context it is the set of arguments (inputs) for which a function is defined.

dominant describing a ⇨gene that is expressed physically whether it is present as both or as only one of the pair that make up an ⇨allele. See also ⇨recessive.

Doppler effect or *Doppler shift* effect on the wavelengths of light (or sound) emitted by a source at a distance that is increasing or decreasing in relation to the observer. If the distance is increasing, the wavelengths are 'stretched' (the light received shifts towards the red end of the ⇨spectrum; sound received goes down in pitch). If the distance is decreasing, the wavelengths are 'squeezed' (the light received shifts towards the blue end of the spectrum; sound received goes up in pitch).

double refraction or *birefringence* a crystal is said to permit double refraction if an unpolarized ray of light entering the crystal is split into two ⇨polarized rays, one of which does obey ⇨Snell's law of refraction, and the other does not. Calcite is such a crystal.

double star a 'system' of two stars that appear – because of coincidental alignment when viewed from Earth – to be close together; it is, however, an optical effect only, and therefore not the same as a ⇨binary star system (although until the twentieth century there were few means of distinguishing double and binary stars).

double theta functions ⇨elliptic functions in the form of ⇨theta functions of higher degree.

doubling the cube ancient Greek problem in geometrical construction, to derive a cube of exactly twice the volume of another cube of given measurement, using ruler and compass only.

dry cell a source of ⇨electromotive force which does not contain a liquid. Usually taken to imply the ⇨Leclanché type of metal and paste cell.

duality principle that a law or theorem remains valid if one particular element within that law is exchanged for another equally pertinent element. In projective geometry, a statement of two-dimensional proposition remains valid if the word 'point' is exchanged for 'line' (and vice versa); a three-dimensional proposition likewise if 'point' is exchanged for 'plane'.

ductility the ability of a substance to be drawn out into long wires. Gold is the most ductile metal, with 1 g/0.035 oz having been drawn out to a length of 2.4 km/1.5 mi.

dy/dx see ⇨differential calculus.

dynamics the study of the action of ⇨forces upon bodies, which produce a change in the motion of the body. Classical dynamics is the study of such phenomena when velocities are much less than that of light. Relativistic dynamics, based on both special and general theories of ⇨relativity, is the corresponding study when velocities are close to that of light.

dynamo a device that produces electrical energy from mechanical energy by causing an electric current to flow in a ⇨conductor that is made to move across a ⇨magnetic field. See ⇨induction.

dynamometer a device for measuring the ⇨power developed by a device.

dynamo theory a theory of the origin of the magnetic fields of the Earth and other planets having magnetic fields in which the rotation of the planet as a whole sets up currents within the planet capable of producing a weak magnetic field.

E

e symbol for ⇨Euler's number.

earth pressure the ⇨pressure exerted horizontally by earth behind a retaining wall.

earthquake shaking of the Earth's surface as a result of the sudden release of stresses built up in the Earth's crust. The study of earthquakes is called seismology. Most earthquakes occur along faults (fractures or breaks) in the crust. ⇨Plate tectonic movements generate the major proportion: as two plates move past each other they can become jammed and deformed, and a series of shock waves (seismic waves) occur when they spring free.

eccentricity in astronomy, the extent to which an elliptical orbit departs from a circular one. It is usually expressed as a decimal fraction, regarding a circle as having an eccentricity of 0.

echolocation in zoology, a method by which an animal locates an object by means of echoes reflected from it of high-pitched sounds emitted by the animal. Mammals that use echolocation include various species of bat and dolphin.

eclipse occultation of one celestial body by another which passes between it and the observer. The solar eclipse is caused by the passing of the Moon between the Sun and the Earth in this way; such an eclipse may be complete (total) or incomplete (partial). ⇨Eclipsing binary stars also accord with this pattern. Alternatively – and exceptionally – a lunar eclipse is caused by the passage of the Earth between the Sun and the Moon, so that the Earth's shadow falls across the Moon, again either totally or partially, depending upon the position of the observer.

eclipsing binary a ⇨binary star where, from the viewpoint of Earth, one of the two bodies regularly passes in front of the other. The resulting variation in perceived ⇨luminosity of some eclipsing binaries has led to their classification as ⇨variable stars.

ecliptic apparent linear path through the 12 constellations of the zodiac that the Sun seems to take during one Earth year, also representing therefore the 'edge' of the plane of Earth's orbit. Because the equator of the Earth is at an angle of more than 22° to the plane of its orbit, the ecliptic is at the identical angle to the ⇨celestial equator, intersecting it at two points: the vernal and autumnal ⇨equinoxes.

ectoderm the outer layer of cells in an embryo and all the tissues that it gives rise to.

Edison effect electrical ⇨conduction between a negatively charged filament, and a positively charged ⇨electrode kept together, though separated, in a vacuum chamber.

efferent nerve a nerve that conducts impulses away from the central nervous system (brain and spinal cord). Most efferent nerves are ⇨motor nerves and run to effector organs.

efficiency for a machine, the ratio of the amount of ⇨energy obtained from a given energy input. This ratio is always less than unity because of losses caused by friction.

efflux in ⇨aerodynamics, the combination of combustion products and air forming the propulsive medium of a ⇨jet or ⇨rocket engine.

ego the part of an individual's personality that recognizes reality and maintains the balance between primitive, instinctive wishes (the ⇨id), the restraint exercised by the conscience (the ⇨superego) and reality.

eigenvalue for a matrix A, the number λ is said to be an eigenvalue of the matrix if there is a nonzero ⇨vector x such that $Ax = \lambda x$. Eigenvalues are used to derive a change of base to simplify the matrix to one that has entries only on its diagonal. More generally, the number λ is an eigenvalue of a linear transformation T if there is a nonzero vector x so that $T(x) = \lambda x$.

eightfold-way a ⇨quantum mechanical scheme for the classification of ⇨elementary particles into families, grouped according to common properties as expressed by various ⇨quantum numbers.

elasticity the ability of a body to resume its former shape after having been deformed by an external ⇨force. Such an ability is retained by a body only as far as its elastic limit, beyond which any deformation becomes permanent. See ⇨Hooke's law.

electrical conduction see ⇨conduction, electrical.

electrical energy the ⇨energy associated with moving electrical ⇨charges. A common unit of measurement is the kilowatt-hour (kWh), used in determining the size of domestic electricity bills.

electric charge that property of a substance that permits the substance to be influenced by the ⇨electromagnetic force. The smallest unit of charge is, for practical purposes, that on the ⇨electron, that is, 1.6×10^{-19} coulomb.

electricity the general term applied to the phenomena caused by the effect of electric ⇨charge, whether static or flowing.

electric motor a device that converts electrical energy into mechanical energy, that is, a dynamo in reverse. It works through the ⇨force exerted on a central ⇨conductor around which a ⇨coil of ⇨current-carrying wire has been positioned.

electrochemical series a list of elements in order of their ⇨electrode potentials (with the highest negative electrode potentials at the top of the series). Any element (usually hydrogen is the only nonmetallic element listed) will displace from solution all those elements below it. It is also known as the electromotive series.

electrochemistry the branch of chemistry that is concerned with chemical reactions that involve electricity. Electrical energy may initiate a chemical process (as in ⇨electrolysis) or a chemical reaction may generate electricity (as in a cell or battery).

electrode any terminal by which an electric current passes into or out of a conducting substance; for example, the anode or cathode in an electric or electrolytic cell or the carbons in an arc lamp. The terminals that emit and collect the flow of electrons in thermionic valves (electron tubes) are also called electrodes: for example, cathodes, plates, and grids.

electrode potential the electric potential between an element and its ions in solution.

electrodynamics the study of the effects of magnetic and electric ⇨forces.

electroencephalograph (EEG) an instrument that records the electrical activity in the brain (brain waves). Electrodes attached to the surface of the scalp monitor changing electric potentials in the cerebral cortex and the signals are amplified and displayed on a cathode-ray oscilloscope or recorded on a multitrace chart recorder.

electrolysis the decomposition of chemical substances by the passing of an electric current through them. This is the result of the migration of ⇨ions in the liquid state of these substances towards an electrode with the opposite charge, where the ions give up their charges.

electrolysis, laws of the first of (Faraday's) laws states that the chemical effect of an electric current is directly proportional to the amount of electricity that passes. The second states that the weights of substances deposited, or liberated, by a given quantity of electricity are directly proportional to the chemical equivalent of the substances.

electrolyte a substance that in the molten state or dissolved in a solvent consists of ions and can conduct an electric current. Typical electrolytes include acids, bases (in the chemical sense) and salts; their solutions are also termed electrolytes.

electric cell device in which chemical energy is converted into electrical energy; see ⇨cell, electric.

electrolytic cell device in which an externally applied voltage brings about a chemical reaction; see ⇨electrolysis.

electromagnet a device in which an electric current flows in a ⇨coil about a central core of soft iron, thus turning the soft iron into a temporary ⇨magnet while the current flows.

electromagnetic force one of the four ⇨fundamental forces of nature, the electromagnetic force occurs between charged objects, such as the ⇨electron and the ⇨proton.

electromagnetic induction the process by which an ⇨electromotive force can be generated in a circuit when the ⇨magnetic flux through a circuit changes.

electromagnetic induction, Faraday's law of the induced ⇨electromotive force is equal to the rate of decrease of ⇨magnetic flux.

electromagnetic interaction the interaction between two charged particles (for example an ⇨electron and a ⇨proton) which appears as a ⇨force (attractive if the two charges are different in sign). In ⇨quantum theory, the electromagnetic interaction is carried between particles by ⇨photons.

electromagnetic radiation 'waves' of electrical and magnetic 'disturbance', radiated as visible light, radio waves, or any other manifestation of the ⇨electromagnetic spectrum. The distance between successive crests of each wave is known as the wavelength, and varies considerably over the ⇨electromagnetic spectrum. The velocity of such radiation in a vacuum is the ⇨speed of light. The units of electromagnetic radiation are quanta or photons ('packets' of energy).

electromagnetic spectrum the range of ⇨frequencies (or, equivalently, wavelengths) over which ⇨electromagnetic radiation is propagated. The lowest frequencies are usually taken as those of ⇨radio waves (less than 10^{12} hertz) and the highest as ⇨gamma radiation, with frequencies upwards of 10^{19} hertz.

electromagnetic wave a wave of ⇨electromagnetic energy. Such waves consist of separate waves of electric and magnetic energy at right angles to each other, and are ⇨transverse.

electromotive force (emf) the rate at which electrical ⇨energy is drawn from its source to flow into an external circuit, when unit ⇨current flows. It is essentially the driving ⇨force behind the electrons making up the current.

electron a negatively charged fundamental particle having a mass of about 9×10^{-31} kg, and carrying a negative electric charge of 1.6×10^{-19} C. It is found in the neutral atoms of all elements, where it orbits round the central nucleus. The combined charge of the orbiting electrons is balanced by the charge of an equal number of positively charged ⇨protons in the atomic nucleus. An electron is also the fundamental unit of electricity; an electric current consists of a flow of electrons along a conductor. Beta rays (see ⇨beta particle) and ⇨cathode rays consist of fast-moving streams of electrons.

electron acceptor a chemical compound that can accept an electron and is reduced in doing so; it can therefore take part in oxidation–reduction reactions.

electronegative describing ⇨radicals that behave like negatively charged ions. The electronegativity of a radical is a measure of its ability to attract an electron pair in a chemical bond. On Pauling's electronegativity scale, fluorine is the most electronegative element and caesium is the least.

electronics the applied science concerned with the theory, design and use of electron control devices, such as transistors.

electron microscope a means of achieving very high magnification of objects. It uses high-energy electrons, whose characteristic wavelength is much shorter than that of visible light used in ordinary microscopes. Its ⇨resolving power is thus far higher.

electrophilic describing a reagent that readily accepts electrons during a chemical reaction. Such reagents, therefore, typically react at centres of high electron density.

electrophoresis the movement of electrically charged solute particles in a ⇨colloid towards the oppositely charged electrode when a pair of electrodes is immersed in the colloidal solution and connected to an external source of direct-current electricity. It is used as an analytical technique similar to ⇨chromatography.

electrophorus a means of inducing a charge in one body into another. Usually consists of a metal plate with an insulating handle, the plate being capable of carrying ⇨electrostatic charges.

electroplating the use of ⇨electrolysis to coat a metal with another metal, for protection from corrosion, or decorative purposes. The metal to be coated is made the ⇨cathode of the ⇨electrolytic cell, which contains a solution of the coating metal as the ⇨electrolyte.

electropositive possessing a positive electric charge. In chemistry, the term is used to describe ⇨radicals that behave like positive ions, that is, radicals that tend to give up electrons in forming chemical bonds. There is no electropositivity scale; instead elements are classified according to their electronegativity because, at least in theory, any element (except those at the extremes of the scale) may be electropositive or ⇨electronegative, according to the electronegativity of the entity to which it is bonded.

electrostatics the study of stationary electric charges.

electrovalency the power of an atom to form electrovalent bonds, which are formed by the transfer of electrons from one atom to another in contrast to covalent bonds, which involve the sharing of a pair of electrons. In electrovalent bonding the atom that gains an electron becomes a negative ion (⇨anion), and the atom that loses an electron becomes a positive ion (⇨cation). See also ⇨valency.

element a substance that is made up of atoms of the same ⇨atomic number; it cannot be converted chemically into any simpler substance. An element may, however, consist of a mixture of ⇨isotopes of different masses (see ⇨atomic weight).

element in set theory, a member of a set.

elementary particle one of a group of particles that make up all matter. Some, such as the ⇨electron, are broadly stable, but many more decay very rapidly after being produced in high-energy nuclear reactions. Elementary particles are also involved in the description of ⇨forces in ⇨quantum physics.

Elements title of the major mathematical work by Euclid. Various other mathematical authors have since also used the title.

elevator a moveable part of a horizontal aerodynamic surface that changes the aircraft's ⇨pitch.

ellipse in plane geometry, a ⇨closed curve that has two foci (centres), unlike a circle, which has a single focus. Its typical form is represented by the equation:

$$\frac{x^2}{a^2} + \frac{y^2}{b^2} = 1$$

elliptic function integral of the general form $\int f(x, \sqrt{R})dx$, where f is any rational function of x and R is a quartic polynomial corresponding to:

$$a_0x^4 + a_1x^3 + a_2x^2 + a_3x + a_4$$

with no multiple roots. Adrien Legendre proved that any elliptic integral can be reduced to the sum of an elementary function and of scalar multiples of three special functions. ⇨Abelian functions and ⇨theta functions are both extensions. Elliptic functions are used in the integration of the square root of a cubic or a ⇨quartic, and are thus important to many mathematical operations.

elliptic geometry system of ⇨non-Euclidean geometry developed as the initial form of ⇨Riemann geometry, and regarding all geometrical operations as carried out in 'curved' space, for example, as though on the surface of an ellipsoid or sphere. A 'straight line' is thus defined (then) as the shortest curve (geodesic) on the curved surface joining two points.

ellipticity of the shape of a planet or ⇨galaxy, the amount of distortion by which it departs from a perfect sphere. The overall ellipticity of the Earth is given as $^1/_{299}$. One class of galaxy is defined in terms of ellipticity, subdivided E0 to E7, according to degree.

elliptic modular functions functions defined in the upper half of an Argand plane that are automorphic relative to a group of modular transformations, that is, transformations T such as:

$$T(z) = \frac{az + b}{cz + d}$$

subject to the condition that $ad - bc = 1$.

elution the removal of material absorbed on a surface by treatment with a solvent. It is a technique used in ⇨chromatography.

empirical formula the chemical formula of a substance in which only the relative proportions of each of its constituent elements are given. The empirical formula does not necessarily reflect a substance's ⇨molecular formula nor its structure. The empirical formula of benzene, for example, is CH, whereas its molecular formula is C_6H_6.

emulsion a light-sensitive material, such as a suspension of silver bromide in gelatin, used to record photographic images.

enantiomorph a compound that has two asymmetric structures, each a mirror image of the other. Enantiomorphs (also called antimers, optical antipodes or enantisomers), such as the optically active forms of lactic acid, have identical chemical and physical properties, except in reactions with other enantiomorphs or in interactions with polarized light. See ⇨optical isomerism.

Encke's Division gap within Saturn's Ring A.

endocytosis the ingestion of material by a cell, including ⇨phagocytosis and ⇨pinocytosis. Phagocytosis is the engulfment and ingestion by a white blood cell of bacteria or other foreign particles; pinocytosis involves the absorption and ingestion by a cell of surrounding fluid by the folding-in of the cell membrane to form a vesicle which (eventually) releases some of its contents into the cell's ⇨cytoplasm.

endoderm the innermost of the three germ layers of an embryo.

endoplasmic reticulum a complex network of tubes in the ⇨cytoplasm of a cell, which are covered with membranes that are thought to control the exchange of matter in and out of the tubes; rough-surfaced tubes carry ⇨ribosomes and smooth-surfaced ones do not.

endoscope an instrument for making an internal examination of a body cavity. It consists of a tube with an optical system and a light source and is inserted into a natural opening or a small surgical incision. It may also have room for a catheter through which to introduce or remove various substances.

endothermic describing a process or reaction that involves the absorption of external energy (usually in the form of heat). See also ⇨exothermic.

endplate a mass of motor nerve endings that penetrate a muscle fibre.

end-point the point during a ⇨titration when the two reagents involved are at exact equivalence, that is, when all of each of the reagents has reacted and there is no excess of either.

energy the capacity of a body to do ⇨work. It comes in various forms, which are interconvertible. For example, the ⇨potential energy lost by a falling body is converted into ⇨kinetic energy of motion.

energy, electrical see ⇨electrical energy.

energy, kinetic see ⇨kinetic energy.

energy, law of conservation of a fundamental principle of physics stating that energy can neither be created nor destroyed, but only changed from one form to another in a closed system.

energy levels, electronic the series of specific, discrete energy states that electrons orbiting a nucleus can occupy. In certain processes an electron may absorb external energy and move to a higher energy level (in which case the electron is said to be excited) or it may release energy (usually in the form of light) and move to a lower energy level. Because the energy levels are discrete, these movements of electrons to different energy levels involve specific amounts (quanta) of energy. See ⇨quantum theory.

energy, magnetic see ⇨magnetic energy.

energy, potential see ⇨potential energy.

engine a device for transforming one form of energy (such as ⇨electrical energy) to another (such as ⇨kinetic energy).

engineering the study of the design, construction and operation of mechanical, electrical, and hydraulic devices.

enteric describing something concerning the intestinal tract.

enthalpy that part of the energy change during a chemical reaction which can be measured as the absorption or emission of heat or light.

entropy a mathematical quantity (symbol S) used in ⇨thermodynamics that may be interpreted as a measure of the amount of disorder in a given system. In any process occurring in a closed system, the amount of entropy (that is, of disorder) can never decrease; the total entropy of the universe is always increasing. Together with ⇨enthalpy, entropy makes up the ⇨free energy.

enzyme any of a large number of protein substances produced by living organisms that catalyse chemical reactions involved in various biological processes (see ⇨catalyst). Most enzymes are intracellular, although many occur in the digestive tract (where they act on and break down the components of food into simpler chemical substances). Enzymes act by combining temporarily with a substrate and many are specific to only one reaction; some require activation by a ⇨co-enzyme.

enzyme induction the stimulation of enzyme formation by the presence of its ⇨substrate or a derivative of the substrate.

epicycle circular orbit of a body round a point that is itself in a circular orbit round a parent body. Such a system was formulated to explain some planetary orbits in the Solar System before they were known to be elliptical.

epithelium a close-knit network of cells in tissues that line the surfaces of the body's cavities and tubes. It may be a single layer of cells or consist of several layers of flattened, cuboidal or columnar cells, often with a secretory function.

epitrochoid the ⇨locus of a point on a rolling circle moving round the circumference of another circle, that is not on the circumference of the rolling circle. Also known as the hypotrochoid, the locus is of importance in the ⇨Wankel engine.

equilibrium a state of balance between opposing effects or ⇨forces.

equilibrium constant a numerical value that expresses the position of a chemical equilibrium at a given temperature and pressure. It is given by the product of the concentrations of the reactants divided by the product of the concentrations of the products.

equinox one of two points in the sky that represent where the Sun appears to cross the plane of the Earth's equator. From the Earth's viewpoint, therefore, the Sun reaches one point at a quarter, the other at threequarters, of the way through the ⇨sidereal year: the vernal (spring) equinox is thus on or

around 21 March, the autumnal on or around 22 September. The actual points in the sky change slightly every year through a process called ⇨precession.

equivalent weight the mass of a substance that exactly reacts with, or replaces, an arbitrarily fixed mass of another substance in a particular reaction. The combining proportions (by mass) of substances are in the ratio of their equivalent masses (or a multiple of that ratio) and a common standard has been adopted: for elements, the equivalent weight is the quantity that reacts with, or replaces, 1.00797 g/0.035279 oz of hydrogen or 7.9997 g/0.28215 oz of oxygen, or the mass of an element liberated during ⇨electrolysis by the passage of 1 faraday (96,487 coulombs per mole) of electricity. The equivalent weight of an element is given by its gram ⇨atomic mass divided by its ⇨valency. For oxidizing and reducing agents, the equivalent weight is the gram ⇨molecular mass divided by the number of electrons gained or lost by each molecule. Some substances have several equivalent weights, depending on the specific reaction in which they are involved.

ergodics in dynamics, study of the mathematical principles involved in the kinetic theory of gases.

Erlangen programme expression used by Felix Klein to denote his unification and classification of geometries Euclidean and non-Euclidean as 'members' of one 'family', corresponding to the transformations found in each.

erosion wearing away of the Earth's surface, caused by the breakdown and transportation of particles of rock or soil. Agents of erosion include the sea, rivers, glaciers, and wind. Water, consisting of sea waves and currents, rivers, and rain; ice, in the form of glaciers; and wind, hurling sand fragments against exposed rocks and moving dunes along, are the most potent forces of erosion. People also contribute to erosion by bad farming practices and the cutting down of forests, which can lead to the formation of dust bowls.

error theory or *theory of errors* in statistics and probability theory, method of evaluating the effects and the significance of errors, for example when obtaining a mean value from a small sample.

escape velocity speed a satellite must attain in order to free itself from returning to the parent body under the effects of gravity.

ester any of a class of organic compounds in which a hydrogen atom of an acid has been replaced by an organic ⇨radical or group. Esters are considered as being derived from the reaction between a ⇨carboxylic acid (RCOOH) and an ⇨alcohol (R′OH) and have the general formula RCOOR′. Low molecular weight esters are liquids with pleasant odours and are widely used as flavouring essences. Most high molecular weight esters are solids and they include waxes and fats. ⇨Hydrolysis of an ester with an alkali yields a salt and the free alcohol. With long-chain esters this process is called saponification, and its products are crude ⇨soap and glycerol.

etalon an instrument that uses ⇨interference phenomena to make possible very high resolution observations of ⇨spectral lines.

ether any of a class of organic compounds formed by the condensation of two ⇨alcohol molecules; they have the general formula R-O-R′. The compound commonly called ether is diethyl ether ($C_2H_5OC_2H_5$).

ether the now discarded concept of all-pervasive medium filling the entire universe, allowing ⇨electromagnetic radiation to propagate. Such a medium is not required by the present understanding of ⇨electromagnetism, based on ⇨Maxwell's equations.

ethology the scientific study of animal behaviour and habits in a normal environment.

Euclidean geometry see ⇨geometry.

Euclid's fifth postulate states that parallel lines meet only at infinity.

eugenics the study and manipulation of genetic qualities in the human population.

Euler's number (e) the limit of the sequence:

$$a_n = 1 + \frac{1}{1!} + \frac{1}{2!} + \frac{1}{3!} + \dots + \frac{1}{n!}$$

An irrational number introduced originally by Leonhard Euler, e may be represented to the sixth decimal place as 2.718282; it has useful theoretical properties in differential calculus and serves as a natural base for logarithms (known as 'natural logarithms').

event horizon the 'edge' of a ⇨black hole; the interface between four-dimensional space and a ⇨singularity.

excitation in physiology, the stimulation of a sense receptor or nerve; in chemistry, the injection of energy into an atom (which may be part of a radical or molecule) which raises it or one of its components to a higher energy level.

excited state, electronic the condition of an electron that has absorbed external energy and, as a result, moved from its normal, ⇨ground state energy level to a higher energy level. The excitation energy is the difference in energy between the ground state and the excited state.

excluded middle, law of the law in logic, that a statement is either true or false, leaving no room for any further alternatives. There are nonclassical systems of logic that distinguish between true = proven true and false = proven false, and so allow intermediate values (such as 'possibly true'). Boolean-valued logic systems attach probability values to statements that may therefore also have intermediate values other than merely true or false; the values form a ⇨Boolean algebra.

exocytosis the ejection from a cell of undigested remnants of material.

exothermic describing a process or reaction that involves the release of energy (usually in the form of heat). Combustion, for example, is an exothermic reaction. See also ⇨endothermic.

expansion coefficient see ⇨coefficient of expansion.

Explorer spacecraft a US series of over 50 satellites, many of which remain in orbit round the Earth fulfilling scientific functions. *Explorer 1* was the first US orbital satellite (launched on 31 January 1958) and was instrumental in discovering the inner ⇨Van Allen radiation belts.

exponent power to which a number is raised, commonly represented as the unknown variable x; for example, 4^x.

exteroceptive describing receptors that receive stimuli from outside the body, such as those of the ear and the eye.

F

factorial the product (symbol !) of the all positive whole integers up to and including the one quoted; factorial 5 (or 5!) thus represents $1 \times 2 \times 3 \times 4 \times 5$; $n!$ represents the series extended to n.

factorization reduction into constituent factors (which when multiplied together produce the original number or expression).

faculae bright areas on the face of the Sun, commonly in the vicinity of ⇨sunspots. Named by Johannes Hevelius, they are thought to be caused by luminous hydrogen clouds close to the ⇨photosphere. They last on average about 15 Earth-days.

Faraday effect the rotation of the plane of ⇨polarization of polarized light by passing it through a transparent medium subjected to a transverse magnetic field.

fat any of a class of organic compounds (collectively called ⇨lipids) that consist of ⇨esters of glycerol (glycerides) with various ⇨fatty acids. Fats, oils (excluding mineral oils, which are hydrocarbons) and waxes are chemically similar; the term fat is usually applied to lipids that are solid at and below 20°C/68°F, whereas oils are liquids at this temperature. Waxes are also solid but they are slick to the touch, rather than greasy, and consist of esters of longer-chain fatty acids. Fats and oils are used as energy-storage materials by plants and animals and are an essential part of the human diet.

fatigue failure of metals under cyclic applications of a ⇨stress.

fatty acid any of a class of saturated, unsaturated or polyunsaturated monobasic aliphatic ⇨carboxylic acids having the general formula RCOOH. Saturated fatty acids contain no double bonds, unsaturated fatty acids contain one or two double bonds, and polyunsaturated fatty acids have three or more double bonds. The lower fatty acids are corrosive, pungent liquids, soluble in water; the intermediate members of the series are unpleasantly smelling oily liquids, slightly soluble in water; and the higher members are mainly solids, insoluble in water but soluble in ethanol. Fatty acids occur widely in living organisms, usually in the form of glycerides in fats and oils.

feedback the coupling of the output of a system to the input. *Negative feedback* results in a decrease in the input energy if the output energy increases (for example, a ⇨centrifugal governor). *Positive feedback* boosts the input energy, and is a source of ⇨amplification.

Fermat's last theorem states that the equation:

$$x^n + y^n = z^n$$

is not solvable in integers if n is greater than 2. Fermat's own proof has never been found and the theorem remained unproved until 1993, when Andrew Wiles of Princeton University, claimed to have verified it.

fermentation a slow decomposition process brought about in organic substances by microorganisms (for example, yeast and bacteria) as a result of ⇨enzyme action. A common example is the alcoholic fermentation resulting from the action of zymase (an enzyme produced by yeast) on sugars, producing alcohol and carbon dioxide.

Fermi–Dirac statistics a mathematical treatment of particles with half-integer ⇨spin in ⇨statistical mechanics, enabling properties of systems made up by such particles to be calculated.

fermions particles obeying ⇨Fermi–Dirac statistics, such as electrons.

ferromagnetism a property of certain materials (principally iron, cobalt and nickel, and alloys of these elements) whereby they are capable of being relatively strongly magnetized by weak magnetic fields. Ferromagnetism results from unbalanced electron spin in the inner orbitals of the substances concerned, which gives the atom a magnetic moment. In ferromagnetic crystals, the ionic spacing is such that the individual magnetic moments of groups of atoms are arranged in magnetic domains. In an unmagnetized ferromagnetic material, these domains are oriented randomly; the application of an external magnetic field causes the magnetic axes of the domains to align, making the material strongly magnetic. A given ferromagnet loses its ferromagnetic properties at a specific temperature, called the Curie temperature. See also ⇨paramagnetism.

Fibonacci series sequence in which each term after the first two is the sum of the two terms immediately preceding it; it begins 1, 1, 2, 3, 5, 8, 13, 21, ... and has a variety of important applications (for example, in search algorithms).

field in physics, the region in which an electrically charged, magnetized or massive body can exert an influence.

field strength see ⇨magnetic field strength.

field theory in mathematics, theory involving the mathematical structure known as a field, which displays the operations of addition and multiplication and their inverses (subtraction and division). An elementary example other than the real numbers is constituted by the rational numbers; there are many other types. A ⇨ring is much like a field but does not include the inverse operations. A ⇨group is a more restricted concept still.

filament a thin wire, usually of tungsten, that is capable of acting as a light source when an electric ⇨current is passed through it.

filariasis collective term for several diseases, prevalent in tropical areas, caused by certain roundworm (nematode) parasites.

fine structure a splitting of individual ⇨spectral lines seen when viewed under high resolution. This can be produced by external magnetic fields acting on ⇨atoms, for example.

finite in mathematics generally, may be labelled by the natural numbers; not infinite. Of a set, may be arranged in a (finite) list that can therefore be labelled with the natural numbers less than some fixed number. Of a real number, not plus or minus infinity. Of a line segment, of finite length.

first order in differential equations, involving only the first ⇨derivative.

fission in nuclear physics, the splitting of the atomic nucleus of a heavy element, resulting in the emission of nuclear energy and possibly causing a chain reaction (with similar results) within a mass of the element.

fixation of nitrogen the conversion of free nitrogen in the atmosphere into nitrous compounds suitable for the manufacture of chemicals such as fertilizers, and explosives. See ⇨Haber process.

flare star dim red dwarf star that suddenly lights up with great – but brief – luminosity, corresponding to an equally powerful but shortlived burst of radio emission. The cause is thought to be a sudden and intense outburst of radiation on or above the star's surface.

fluid generic term for ⇨gases and ⇨liquids.

fluorescence an ability of some materials to absorb light of one ⇨frequency and emit it at a different one. The result is often a change in colour (wavelength) of the two types of light.

flux density see ⇨magnetic flux density.

fluxions see ⇨calculus.

flywheel a massive wheel fixed to a shaft, either to store rotational energy for propulsive purposes, or to smooth fluctuations in rotation rates.

food chain a scheme that shows the interdependence of organisms in an ecosystem; each organism in the chain is the source of food of at least the next member of the chain.

force when an object is acted upon in such a way that its state of rest or uniform motion in a straight line is changed, a force is said to act. The SI unit of force is the newton (symbol N). By ⇨Newton's laws of motion, for a mass m to attain an acceleration a, a force F given by:

$$F = ma$$

must have acted on the body.

force, centrifugal see ⇨centrifugal force.

force, centripetal see ⇨centripetal force.

force, electromagnetic see ⇨electromagnetic force.

force, electromotive see ⇨electromotive force.

force, fundamental see ⇨fundamental forces.

formaldehyde a pungent gas, made by the oxidation of methanol, that is used as a disinfectant, and in the manufacture of plastics and dyes.

fossil remains of an animal or plant preserved in rocks. Fossils may be formed by refrigeration (for example, Arctic mammoths in ice); carbonization (leaves in coal); formation of a cast (dinosaur or human footprints in mud); or mineralization of bones, more generally teeth or shells. The study of fossils is called palaeontology.

Foucault pendulum a pendulum consisting of a long wire to which is attached a heavy weight, which is then free to swing in any plane. Once it starts swinging in one particular plane, the plane slowly rotates, this being the result of the rotation of the Earth beneath the pendulum.

foundation the part of a structure that ensures stability by providing mechanical fixing with solid earth.

foundations of mathematics or *foundations of arithmetic* subject of attempts to derive the basic precepts of elementary mathematics from a standpoint of pure logic (thence to derive the more complex principles).

Fourier analysis see ⇨Fourier theorem.

Fourier series series in which the terms comprise multiples of the cosine and/or sine of multiple angles. Represented by the formula:

$$\frac{1}{2}a_0 + \Sigma(a_n \cos nx + b_n \sin nx)$$

it is used to analyse periodic functions (that is, functions whose graph repeats itself periodically).

Fourier theorem adaptation of the process developed by Joseph Fourier as the ⇨Fourier series to the investigation of energy propagated in the form of waves (particularly heat, sound and light). A further developed version of this method is known as harmonic analysis. Use of the theorem in investigating wave forms is known as Fourier analysis.

four-stroke engine an ⇨internal-combustion engine using the ⇨Otto cycle.

fraction ratio describing as one integer above another the number of parts (the top figure, or numerator) of the whole (the bottom figure, or denominator, in the same units); a number over itself thus represents unity, and a top figure higher than a bottom figure represents more than unity.

frame of reference a set of axes fixed in such a way as to define uniquely the position of an object in space.

Fraunhofer lines dark lines crossing the solar ⇨spectrum. They are caused by the absorption of light from hot regions of the Sun's surface by gases in the cooler, outer regions.

free energy the sum of ⇨enthalpy and ⇨entropy, that is, the capacity of a system to perform work. The change in free energy accompanying a chemical reaction is a measure of its completeness.

freezing point the temperature at which there is equilibrium between the solid and liquid phases of a substance at standard pressure (760 mm/30 in of mercury); it is the same as the melting point of the solid. A solution freezes at a lower temperature than does the pure solvent, a phenomenon called depression of the freezing point. The amount of this depression is proportional to the number of ⇨moles of the solute dissolved in unit weight of the solvent.

frequency the time required for one complete cycle of a wave or oscillatory motion. Measured in hertz (Hz), with one hertz being one cycle per second.

frequency modulation a means of ⇨radio transmission in which the ⇨frequency of the ⇨carrier wave is modulated, the ⇨amplitude remaining constant.

friction the ⇨force that resists the relative movement of two surfaces in contact with each other.

front in meteorology, the boundary between two air masses of different temperature or humidity. A *cold front* marks the line of advance of a cold air mass from below, as it displaces a warm air mass; a *warm front* marks the advance of a warm air mass as it rises up over a cold one. Frontal systems define the weather of the mid-latitudes, where warm tropical air is constantly meeting cold air from the poles.

Froude brake a ⇨dynamometer that measures power by determining ⇨torques.

fuel cell a type of cell that produces direct-current electricity by oxidation of a fuel. A simple fuel cell comprises gaseous hydrogen and oxygen brought together over catalytic ⇨electrodes; other fuel cells use ammonia, methanol or hydrazine to provide hydrogen. Fuel cells are recharged by adding more fuel, rather than by passing electricity through them. See also ⇨cell.

fuel injection a system used in high-performance ⇨internal-combustion engines whereby pure fuel is injected into the cylinder during the intake phase of the cycle. This eliminates the need for a ⇨carburettor and gives greater control over the burning mixtures.

fulcrum the point of support o-f a ⇨lever, about which the lever can pivot in lifting and lowering loads applied to it.

function in mathematics, dependence of a quantity or number on one or more other constituent variable quantities or numbers; y is a function of x if a value of y can be calculated for each value of x. The classic form is thus $y = f(x)$, in which x is called the argument and f the function.

functions analysis see ⇨function theory.

functions of complex variables, theory of involves ⇨functions of which the arguments are ⇨complex numbers.

functions of real variables, theory of involves ⇨functions of which the arguments are ⇨real numbers.

function theory or *theory of functions* use of ⇨functions, primarily in order to denote mathematical relationships, but also in application to other sciences. Functional analysis, for example, considers problems in which an unknown function is to be found; at this stage ⇨variables may represent functions as opposed to numerical values.

fundamental forces the basic ⇨forces that underlie all interactions in the universe. There are thought to be four such forces. ⇨Gravity and the ⇨electromagnetic force are of particular importance to the macroscopic world whereas the so-called strong and weak interactions are of great importance in the subatomic world. In strength, the forces can be ranked as strong, electromagnetic, weak and gravity, the latter being the weakest of all. Current physics suggests that the electromagnetic and weak forces are in fact manifestations of one single force.

furnace a device in which very high temperatures are produced to bring about chemical reactions.

fuse a device that prevents the passage of electric ⇨current above a predetermined level in a ⇨circuit, by melting through the consequent temperature rise, and thus breaking the circuit.

fusion in nuclear physics, the combining of the atomic nuclei of lighter elements to form nuclei of a heavier element. Such a process involving the atomic nuclei of elements lighter than iron is accompanied by the emission of energy; for fusion of heavier elements, energy must be supplied. The process is thought to contribute to the condensation of stars from interstellar gas and dust. See also ⇨nuclear fusion.

G

gain the ratio of the output power of an electronic device, such as an amplifier, to the input power.

galactic centres phenomena that are now thought to comprise ⇨black holes – which would explain why the centre of our Galaxy appears strangely obscure, and emits only ⇨infrared radiation.

galactic rotation the revolving of a galaxy round its central nucleus even as it continues its ⇨proper motion. Such rotation, however, is not uniform but ⇨differential. One revolution of the Sun within our own Galaxy takes about 225 million years, or 1 cosmic year.

galaxy vast system of celestial objects, typically consisting of between 10^6 and 10^{12} stars, plus interstellar gas and dust. There are three basic types: spiral (further subdivided into normal spirals and spirals with a 'bar' at the centre, and yet further subdivided according to the 'openness' of the spiral arms), elliptical (subdivided according to ⇨ellipticity) and irregular (subdivided according to whether they are made up of ⇨population I or ⇨population II stars). Another not uncommon type of galaxy is a lenticular form midway between the spiral and the elliptical.

Galaxy or *Milky Way* the system of approximately 100,000 million stars, of which our Sun is one. It is a normal spiral ⇨galaxy of class Sb, with a diameter now reckoned to be probably less than 100,000 light years, and a strong but obscure energy source at the centre (emitting ⇨infrared radiation). It is undergoing ⇨galactic rotation. Possibly one tenth of the Galaxy's total ⇨mass – estimated at 1.8×10^{41} solar masses – comprises interstellar gas and dust.

gallic acid or *trihydroxybenzoic acid* a yellow crystalline substance used in making inks.

galvanometer an instrument that detects small changes in electric ⇨current, by making use of the magnetic effect of the current.

gamete a sexually differentiated mature reproductive cell – in animals, typically a ⇨spermatozoon (male gamete) or an ⇨ovum (female gamete). The fusion of their two nuclei in fertilization results in a ⇨zygote which develops into an embryo.

game theory construction of mathematical models for determining the best strategy to win a game, that is, to achieve optimal results with minimal losses. Devised first by John Von Neumann, the theory has many applications to the behavioural sciences.

gamma globulin one of a group of proteins or ⇨immunoglobulins in the blood that act as ⇨antibodies to specific infections. Gamma globulins extracted from the blood of a patient who has recovered from an infection may be used as ⇨vaccines to stimulate artificial immunity in others.

gamma radiation ⇨electromagnetic radiation with a ⇨wavelength shorter than that of ⇨X-rays, at about a million-millionth of a metre. Gamma rays are emitted in discrete units (photons) by the nuclei of radioactive atoms. See also ⇨radioactivity.

gas a state of matter in which the constituent molecules (or atoms) move randomly relative to one another, but where each molecule (or atom) retains its identity.

gas mantle an illumination device made from impregnating a dome-shaped piece of rayon with compounds of thorium and cerium, which are decomposed by heat.

gastrula a stage in embryonic development following the ⇨blastula stage in which ⇨gastrulation occurs.

gastrulation cell movements during embryonic development (after cleavage) in which cells move to the positions in which they eventually give rise to the organs of the growing embryo.

gas turbine an engine that uses internal combustion to convert the chemical energy locked up in a liquid fuel into mechanical energy. Mostly used in aircraft and trains.

gauge the distance between the inside edges of a railway for trains. Also, the diameter of wires, rods, and so on.

gauss cgs unit (symbol Gs) of magnetic induction or magnetic flux density, replaced by the SI unit, the ⇨tesla, but still commonly used. It is equal to one line of magnetic flux per square centimetre. The Earth's magnetic field is about 0.5 Gs, and changes to it over time are measured in gammas (one gamma equals 10^{-5} gauss).

Gaussian distribution another name for ⇨normal distribution.

gear a system of moving parts that transmits motion of one part of a device to another. Gears typically take the form of notched wheels that interlink with each other.

gegenschein faint oval patch of light, opposite the Sun, visible from Earth only at certain times of the year. Its nature and cause are still not known. It is sometimes known as 'counterglow'.

Geiger counter an instrument capable of detecting alpha, beta and gamma particle emission from materials. The incoming radiation creates ions within the counter, which are accelerated by electrical means and their effects counted.

gel a colloidal solution (see ⇨colloid) that has set to a jelly-like consistency. The viscosity of gels is so high that they behave like solids in many respects, despite the fact that some gels may contain as little as 0.5% of solid matter.

gelatine a soluble protein-based substance with the ability to form a jelly on cooling. It is used in photography and the making of glues.

gemmule in early genetic theory, minute particles thought to consist of miniature copies of all parts of the body carried in the blood to the ⇨gametes and from which their larger forms eventually developed. In modern usage, a gemmule is a bud formed on a sponge that may break free and develop into a new animal.

gene basic unit of heredity, encoded by a strand of ⇨DNA and transcribed by ⇨RNA. In higher organisms, genes are located on the ⇨chromosomes. Genes that occur as ⇨alleles control particular hereditary characteristics, each allelic partner occupying the same site on a pair of chromosomes (and one or both of which may be ⇨dominant or ⇨recessive).

gene pool all the genes found in a single interbreeding population of a plant or animal.

generator an electric power producer operating on the principle of the ⇨dynamo, that is, by converting mechanical energy into electrical energy.

genetic engineering the artificial manipulation of the nucleic acids DNA or RNA to produce new, modified species of plants or, possibly, animals.

genetics the study of heredity and species variation in organisms.

genotype the genetic make-up of an organism as opposed to its physical characteristics (see ⇨phenotype).

geocentric having the Earth at the centre.

geodesic the shortest route between two points on any surface.

geodesy measurement and study of the Earth's size and shape.

geological time time scale embracing the history of the Earth from its physical origin to the present day. Geological time is traditionally divided into aeons (⇨Phanerozoic, ⇨Proterozoic, and ⇨Archaean), which in turn are divided into eras, periods, epochs, ages, and finally chrons.

geometric curve curve that can be precisely expressed by an equation (unlike a mechanical curve); for example, a circle, parabola or hyperbola.

geometry study of the properties and relations of lines, surfaces and solids in space. The basic elementary geometry is Euclidean, involving two or three dimensions only, and in which lines may be straight, curved, parallel, finite or infinite; in which surfaces are two dimensional, corresponding to regular coordinates; and in which solids are three dimensional. Other modern considerations in geometry include ⇨elliptic geometry and ⇨hyperbolic geometry.

germ plasm theory in early genetics, the theory that of the two tissue types in multicellular animals (somatoplasm in body cells and germ plasm in reproductive cells), only the integrity of germ plasm is necessary for the inheritance of characteristics.

gland specialized organ of the body that manufactures and secretes enzymes, hormones, or other chemicals. In animals, glands vary in size from small (for example, tear glands) to large (for example, the pancreas), but in plants they are always small, and may consist of a single cell. Some glands discharge their products internally, endocrine glands, and others, exocrine glands, externally.

globular cluster spherical, densely populated ⇨cluster of older stars. There are a number of such clusters round the edge of our Galaxy.

gluon an ⇨elementary particle which is used to explain the transmission of ⇨forces between ⇨quarks.

glycerine or *glycerol* a thick viscous substance obtained from fats, used in the manufacture of explosives, plastics and as an antifreezing agent.

glycolysis the transformation of glucose, by enzyme action, into ⇨lactic acid or pyruvic acid. This process usually occurs during the preliminary stages of ⇨fermentation and tissue respiration and provides energy for short-term bursts of activity.

Golgi apparatus or *Golgi body* a scattered structure of smooth double-membraned vesicles found in the ⇨cytoplasm of cells which may have secretory and/or transport roles in metabolism.

gondola the personnel-carrying car of an airship or balloon.

Gondwanaland or *Gondwana* southern landmass formed 200 million years ago by the splitting of the single world continent ⇨Pangaea. (The northern landmass was Laurasia.) It later fragmented into the continents of South America, Africa, Australia, and Antarctica, which then drifted slowly to their present positions. The baobab tree found in both Africa and Australia is a relic of this ancient landmass.

governor a device that regulates the ⇨speed of a motor or engine. See ⇨centrifugal governor.

gramophone a device for translating the variations in shape of grooves on a disc or cylinder into audio signals.

graph pictorial plot of a function f, consisting of points on the plane of the form $(x, f(x))$. Such a representation provides the means of solving for x equations of the form $y = f(x)$ for given y value(s). Points of intersection of the graph with the horizontal line corresponding to the ordinate value y give solutions to the equation. It is commonly convenient, however, to rewrite $y = f(x)$ as, for example, $g(x) = h(x)$ so as to have two separate functions of x, one on each side of the equation and so that y is involved in the constitution of g and h; where the graphs of g and h then meet again give the required solution(s).

grating see ⇨diffraction grating.

gravitation one of the four fundamental ⇨forces governing the universe, and the most important on the cosmological scale, despite its being the weakest of the four forces. All bodies with mass experience the gravitational force, and according to the general theory of ⇨relativity, which views gravitation as curvature of space and time, massless objects such as ⇨photons can also be affected. For regions of relatively weak gravitational fields, Newton's law may be used. This states that the force of gravitation between two masses varies directly as the product of the masses, and inversely as the square of the distance between them. The constant of proportionality is g, Newton's ⇨gravitational constant. However, in regions such as exist near black holes for example, Newton's law breaks down, and general relativity must be used.

gravitational constant a fundamental constant (symbol g) of physics that relates the force of ⇨gravity produced by a body to the masses involved and the separation. According to some physicists, its present value of 6.67×10^{-11} m³/(kg s²) is decreasing at a rate proportional to the age of the universe.

gravitational force, gravity as described first by Isaac Newton, gravity is a force that exists between bodies of any ⇨mass whatever (from particles to stars) in proportion to the product of their masses, and in inverse proportion to the square of the distance between them. The weakest of the four natural forces (the other three being the electromagnetic and the two nuclear interactive forces), its real nature is still not fully understood. Einstein's general theory of ⇨relativity presented another viewpoint.

gravity see ⇨gravitation.

gravity, acceleration due to see ⇨acceleration due to gravity.

gravity, centre of see ⇨centre of gravity.

greenhouse effect phenomenon of the Earth's atmosphere

by which solar radiation, trapped by the Earth and re-emitted from the surface, is prevented from escaping by various gases in the air. The result is a rise in the Earth's temperature. The main greenhouse gases are carbon dioxide, methane, and water vapour. Fossil-fuel consumption and forest fires are the main causes of carbon-dioxide build- up; methane is a byproduct of agriculture (rice, cattle, sheep).

Greenwich site now in London of the first ⇨Royal Greenwich Observatory.

Gregorian calendar calendar established with the authority of the Roman Catholic Church by Pope Gregory XIII in 1582. Correcting at a stroke the 10-day accumulated margin of error of the ⇨Julian calendar, the main difference was in fact that century years were discounted as leap years unless they were divisible by 400.

Gregorian telescope reflecting telescope devised – but never constructed – by James Gregory, in which an auxiliary concave mirror reflects the magnified image, the right way up, through a hole in the centre of the main objective mirror, that is, through the end of the telescope itself. The ⇨Cassegrain telescope is similar but produces an inverted image.

grid an ⇨electrode in the form of a mesh that, when placed between the anode and cathode of a ⇨thermionic valve, controls the flow of electrons.

grid, national see ⇨national grid.

ground state (electronic) the state of an electron in its lowest energy level. When all the electrons orbiting a nucleus are in their lowest energy levels, the atom as a whole has its minimum possible energy and is therefore in its most stable state.

group theory investigation and classification of the properties of the mathematical structures known as groups. A group possesses two operations - 'multiplication' and 'inverting of an element' – and a further designated element called 'unity'. An example is provided by the set of nonzero real numbers with ordinary multiplication and reciprocation (that is, being multiplied or becoming reciprocals). Another example is provided by the integers for which the operation of multiplication (n x m), is to be addition ($n + m$), with the inverse of n being – $1/n$, and the 'unity element' is in fact zero. A third example is provided by a collection of ⇨transformations of the plane where $S \times T$ effectively means applying transformation T and then S; such a collection is a group if it includes the identity transformation I (the transformation that leaves all points unchanged) and for each transformation T its inverse, T^{-1} (the transformation of which the effect is to return each point to its original position before it was transformed by T).

gunpowder a combination of sulphur, powdered charcoal and ⇨saltpetre that, when ignited, produces explosions within confined spaces.

gymnosperm any plant of a subdivision of the Spermatophyta distinguished from the other subdivision (the ⇨angiosperms) by having ovules relatively unprotected on the surface of the megasporophylls, which usually take the form of cones. Also (unlike angiosperms) very few gymnosperms have vessels in the ⇨xylem of their stems.

gyroscope a wheel mounted in such a way that it may rotate about any axis. Once set in rotation, the support of the wheel may be moved in any direction without changing the direction of the wheel relative to space. In addition, the application of a ⇨force will make the wheel precess (rotate about an axis at right angles to the axis about which the force is applied).

H

Haber process a method for the ⇨fixation of nitrogen in the atmosphere, to produce industrial amounts of ⇨ammonia-based compounds. A mixture of nitrogen and hydrogen is passed under pressure and high temperature over a ⇨catalyst, the gases combining to form ammonia.

haemoglobin the red respiratory pigment in red blood cells, consisting of a protein containing iron (see ⇨haemoprotein).

haemoprotein a protein containing an iron porphyrin group. The green plant pigment ⇨chlorophyll and the red blood pigment ⇨haemoglobin are both haemoproteins.

halide any compound (including organic compounds) of one of the ⇨halogen elements. Common salt – sodium chloride (NaCl) – is the best known metal halide. Most metal halides are ionic; most nonmetal halides (for example, chloroform, trichloromethane) are covalent.

Hall effect an electric ⇨current-carrying conductor develops a ⇨potential difference across it when placed in a strong transverse magnetic field. This potential difference is at right angles to both the magnetic field and the direction of current flow.

Halley's comet comet that orbits the Sun every 76 years, named after Edmond Halley who calculated its orbit. It is the brightest and most conspicuous of the periodic comets, and recorded sightings go back more than 2,000 years. It will next reappear 2061.

halo nebulous quality round a celestial body (particularly round a red giant); the galactic halo, however, describes the spherical collection of stars forming a surrounding 'shell' for our otherwise compact, discoid Galaxy.

halogen an element belonging to Group VII of the periodic table. There are five halogens – fluorine, chlorine, bromine, iodine and astatine – the last of which is the only halogen without a stable isotope (astatine's most stable isotope, At-210 has a half-life of only about 8.3 hours). All the halogens have one electron vacancy in the outer orbital and their main ⇨oxidation state is −1.

halogenoalkane any of a group of organic compounds (formerly called alkyl halides) formed by the halogenation of (addition of a halogen to) an alkane. In the presence of ultraviolet light, alkanes react with halides by substitution. In the chlorination of methane, for example, one, two, three or all four of the methane's hydrogens may be substituted by chlorine – depending on how far the reaction is allowed to proceed – with the release of hydrogen chloride at each substitution; the resulting halogenoalkanes are chloromethane (methyl chloride, CH_3Cl), dichloromethane (methylene chloride, CH_2Cl_2), trichloromethane (chloroform, $CHCl_{3)}$ and tetrachloromethane (carbon tetrachloride, CCl_4).

hang glider an unpowered crewed glider that achieves great manoeuvrability through a basically triangular-shaped expanse of fabric stretched over cross-struts.

harmonic analysis see ⇨Fourier theorem.

head of water the vertical distance between the top of a water stream and the point at which its energy is to be extracted.

heat the ⇨energy possessed by a substance by virtue of the ⇨kinetic energy of its constituent particles. Heat can be transported by one of three means: ⇨convection, conduction, and ⇨radiation.

heat capacity the amount of heat energy required to raise the temperature of a substance through one degree of temperature. See ⇨specific heat capacity.

heat death of the universe according to the second law of thermodynamics, the ⇨entropy of the universe, considered a closed system, is always increasing. It follows from this that there will eventually come a time in the history of the universe when there is no further energy available, all processes having come into thermal equilibrium. The universe will then no longer consist of hot bodies such as stars surrounded by cold space, all such disequilibrium having been eliminated. The resulting lack of any physical processes is called the heat death of the universe.

heat engine a device that converts heat energy into mechanical energy.

Heaviside layer see ⇨Kennelly–Heaviside layer.

helical screw a screw whose thread is in the form of a helix.

heliocentric having the Sun at the centre.

heliograph device for recording the positions of ⇨sunspots.

heliometer instrument to measure the apparent diameter of the Sun at different seasons, also used to measure angular distances between stars.

helium element which, after hydrogen, is the second lightest and second most abundant in the universe. Its atom comprises two protons, two neutrons and two electrons; its nucleus is sometimes called an alpha particle. Helium is the product of the ⇨nuclear fusion of hydrogen in most stars, but this does not explain the overall ⇨helium abundance.

helium abundance presence – and dominance – of ⇨helium atoms in the universe. The fact that about 8% of *all* atoms are helium can be traced, through the ⇨alpha-beta-gamma theory, to the primordial ⇨Big Bang.

helix a spiral, mathematically defined by a spiral curve that, when wound about a cylinder, crosses the axis of the cylinder at the same angle on each revolution about the cylinder.

hematite naturally occurring ferric oxide; an ore from which iron is extracted industrially.

hermaphroditism a condition in which both male and female reproductive organs are present in the same organism.

hertz the unit (symbol Hz) of ⇨frequency of a wave, defined by one complete oscillation of a wave per second. For example, alternating current in the UK has a frequency of 50 Hz.

Hertzsprung–Russell diagram (H–R diagram) graphic chart plotting the relationship between the ⇨absolute magnitude (implicating the mass) and the ⇨spectral classification (implicating the temperature) of thousands of stars. The majority of stars form a band from top left to bottom right of the diagram, known as the ⇨main sequence. Various theories on the evolution of stars relate to information contained in the diagram.

heterodyne effect the superimposition of two waves of different ⇨frequency in a ⇨radio receiver, one of which is being received, the other transmitted within the device itself, producing an intermediate frequency that can be demodulated.

heterozygous describing a pair of allelic ⇨genes in which one is different from the other. See also ⇨allele, ⇨homozygous, ⇨zygote.

high-energy particles particles of ⇨electromagnetic radiation that contain high energies, measured in terms of electronvolts. The energy in gamma radiation is of the order of 8×10^7 to 8×10^5 electronvolts and in X-rays of 8×10^3 to 8×10^1 electronvolts.

hodometer a device that enables the ⇨acceleration of an object moving with known velocity over a path to be determined.

hoist an engine used to ⇨power a wire-round drum, whose cable is used to lift heavy objects, usually using a ⇨jib and ⇨pulleys.

holography a means of creating permanent three dimensional images of objects using coherent light from ⇨lasers. The beam of ⇨coherent radiation from the laser is split into a reference beam (which falls directly onto the photographic plate), the other being ⇨diffracted by the object being photographed before falling onto the plate. The two beams interfere with each other on the plate, creating ⇨interference patterns that form the holographic, three-dimensional image.

homeopathy an unorthodox form of medical treatment in which very small doses of drugs that cause the symptoms of an illness are administered to treat it.

hominids humans and their humanlike ape predecessors, which together constitute the family Hominidae.

homologous series a series of organic compounds of the same chemical type, each member of which differs from its preceding member by having an additional $-CH_2$ group in its molecule; the ⇨relative molecular masses of members of a series therefore increase in steps of 14, and each series can be represented by a general formula. The members of a series show a gradual, regular change of physical properties with increasing molecular mass and have similar chemical properties. Examples of homologous series include the ⇨alkanes, ⇨alkenes, ⇨alkynes, amd ⇨alcohols.

homology branch of ⇨topology involving the study of ⇨closed curves, closed surfaces and similar geometric arrangements in two- to n-dimensional space, and investigating the ways in which such spatial structures may be dissected. The formulation of the homological theory of dimensionality led to several basic laws of ⇨duality (relating to topological properties of an additional part of space).

homozygous describing a pair of identical allelic ⇨genes. See also ⇨allele, ⇨heterozygous, ⇨zygote.

Hooke's law within the elastic limit of a body possessing ⇨elasticity, the stress acting on the body produces a strain directly proportional to the size of the applied stress. See ⇨elasticity.

hormone any of a group of substances present in plants (phytohormones) and animals which play essential roles in growth, development, function and behaviour. In animals, hormones are released from endocrine (ductless) glands into the blood-stream and take effect on more or less specific target organs or tissues.

horsepower British unit of ⇨power, equivalent to 745.7 ⇨watts.

hot big bang later, but fundamental, concept within the ⇨Big Bang model, that the primordial explosion occurred in terms of almost unimaginable heat. The concept, formulated by George Gamow, led to considerable study of thermonuclear reactions and the search for ⇨background radiation.

Hubble expansion the apparent recession of galaxies as seen from any point within the universe, the velocity of recession being proportional to the distance of the galaxy from the observer.

Hubble's law states that the more distant a ⇨galaxy is, the greater is its speed of ⇨recession. *Hubble's constant* applies to the rate of increase in that speed. It was originally calculated as 530 km sec^{-1} per 10^6 parsecs, but has since been estimated at about a tenth of that value.

hybridization the process by which two or more simple orbitals in the same quantum shell come together then redistribute themselves as an equal number of equivalent hybrid orbitals, this system possessing the optimum energy distribution for the molecule involved. For example, one s orbital and three p orbitals form four sp^3 orbitals; these hybrids can form ⇨sigma bonds with other atoms and are arranged so that the part of each hybrid orbital capable of bond formation is directed towards the corner of a regular tetrahedron. Similarly, one s and two p orbitals hybridize to form three sp^2 orbitals, which lie in a plane, the major axes of the orbitals being at 120° to each other. Thus hybrid orbitals are responsible for the basic geometry of the molecules in which they occur.

hybrid orbital see ⇨hybridization

hydration a special type of solvation in which water molecules are attached – either by electrostatic forces or by coordinate (covalent) bonds – to ions or molecules of a solute. Some salts, called hydrates, retain associated water molecules in the solid state (this water is called the water of crystallization); in solid copper(II) sulphate, for example, the hydrated ion is $[Cu(H_2O)_4]^{2+}$ or $[Cu(H_2O)_6]^{2+}$.

hydraulic press a device that uses an ⇨incompressible fluid, such as water or oil, to transmit a small downward force applied to a piston of small area to a larger area piston, which then produces a proportionately larger upward force. Such a press is a demonstration of ⇨Pascal's law of pressures.

hydraulics the study of the theory and application of fluid flow to engineering problems, such as the design of ⇨hydraulic presses.

hydrocarbon an organic compound consisting only of hydrogen and carbon.

hydrodynamics the study of the effects that forces have on liquids in motion.

hydroelectricity the generation of electric energy by the conversion of energy contained in a stream of water. The conversion is achieved using a ⇨dynamo, the mechanical driving force being provided by a water-driven ⇨turbine.

hydrogen element that is the lightest and the most abundant in the universe. Its atom comprises one proton and one electron. The element occurs both in stars and as interstellar clouds, in regions where it may be neutral (H I regions) or ionized (H II regions).

hydrogen bomb a nuclear weapon in which the heat produced by an initial detonation of a fission, or ⇨atom bomb, is sufficiently high to allow an even greater release of energy from nuclear ⇨fusion to take place.

hydrogen bond a weak electrostatic bond that forms between covalently bonded hydrogen atoms and a strongly ⇨electronegative atom with a lone electron pair (for example, oxygen, nitrogen and fluorine). Hydrogen bonds (denoted by a dashed line —) are of great importance in biochemical processes, particularly the N–H– – –H bond, which enables proteins and nucleic acids to form the three-dimensional structures necessary for their biological activity.

hydrogen carrier a compound that accepts hydrogen ions in biochemical reactions and is therefore important in oxidation–reduction reactions such as the intracellular use of oxygen.

hydrogen ion concentration the number of grams of hydrogen ions per litre of solution; denoted by [H^+]. It is a measure of the acidity of a solution, in which context it is normally expressed in terms of pH values, given by pH = $\log_{10}$ (1/[H^+]).

hydrography the ⇨surveying of oceans, lakes, and rivers.

hydrolysis the chemical reaction of a compound with water, resulting in decomposition into two or more other compounds; the water itself is also decomposed. Hydrolysis occurs with salts of weak acids, weak bases or both. It also occurs with ⇨esters (the reverse of esterification) to produce an alcohol and an acid.

hydrometer an instrument capable of determining the density (relative or absolute) of a liquid. It makes use of ⇨Archimedes' principle of hydrostatics.

hydrostatics the study of the behaviour of ⇨liquids under the action of ⇨forces and ⇨pressures when the liquid is at rest.

hydroxyl ion a hydroxyl group (–OH) with a negative charge, that is, the OH⁻ ion. The presence of hydroxyl ions gives ⇨alkalis their characteristic chemical properties.

hygrometer an instrument capable of determining the humidity of the air as determined by the ratio of the pressure of water vapour in the air at a given time to the pressure expected if the atmosphere was saturated at the same temperature.

hyperbola double-branched plane curve; that is, a curve consisting of two separate but similar and related (infinite) pieces. The equation for the general form is represented by:

$$\frac{x^2}{a^2} - \frac{y^2}{b^2} = 1$$

where a and b are two real numbers.

hyperbolic geometry system of ⇨non-Euclidean geometry developed by Bernhard Riemann, complementary to ⇨elliptic geometry and comprising the geometry of ⇨geodesics in the neighbourhood of a point on the (curved) surface at which a tangential plane intersects the surface in a hyperbolic curve.

hypercomplex numbers numbers that expand on ⇨complex numbers, for example, ⇨quaternions.

hyperfine structure very fine splitting of individual lines in a ⇨spectrum, which can be the result of the presence of different ⇨isotopes of an element in the source.

hyperon one of a group of unstable ⇨elementary particles with masses greater than that of the ⇨neutron.

hypo another name for ⇨sodium thiosulphate.

hysteria according to the work of Sigmund Freud, the conversion of a psychological conflict or anxiety feeling into a physical symptom, such as paralysis, blindness, recurrent cough, vomiting, and general malaise. The term is little used in diagnosis today.

I

ice age any period of glaciation occurring in the Earth's history, but particularly that in the Pleistocene epoch, immediately preceding historic times. On the North American continent, glaciers reached as far south as the Great Lakes, and an ice sheet spread over N Europe, leaving its remains as far south as Switzerland.

iconoscope a type of television camera where a beam of ⇨electrons scans a special mosaic, which can store an optical image electrically, and converts the image stored to electrical signals for transmission.

id one of the three aspects of personality (as used by Freud) and representing the primitive, instinctive aspect that is constantly seeking expression but which is repressed by the ⇨ego (acknowledgement of reality) and the ⇨superego (social conditioning).

ideal in projective geometry and algebra, describes a point (one on every line) at infinity in such a way that the point has a coordinate position. In number theory, the term describes a collection of elements in a ⇨ring that has specific properties within a ⇨universal set, that is, that form a closed system under addition (among themselves) and under scaling by any element from the universal set. For example, the even numbers form an ideal within the universal set of integers in that when added (or multiplied) together, even numbers result.

ideal gas or *perfect gas* a gas obeying the gas laws of Boyle, Charles, and Joule exactly. This would imply that the gas consists of perfectly elastic molecules, each of which has zero volume and no attractive or repulsive interaction with its neighbours.

igneous rock rock formed from cooling magma or lava, and solidifying from a molten state. Igneous rocks are classified according to their crystal size, texture, chemical composition, or method of formation. They are largely composed of silica (SiO_2) and they are classified by their silica content into groups: acid (over 66% silica), intermediate (55–66%), basic (45–55%), and ultrabasic (under 45%). Igneous rocks that crystallize below the Earth's surface are called plutonic or intrusive, depending on the depth of formation. They have large crystals produced by slow cooling; examples include dolerite and granite. Those extruded at the surface are called extrusive or volcanic. Rapid cooling results in small crystals; basalt is an example.

image see ⇨transformation.

imaginary number that part of a ⇨complex number that is a multiple of $\sqrt{-1}$.

immunity an organism's resistance to infection. Immunity may be natural (as a result of the presence of natural ⇨anti-bodies) or acquired (as a result of ⇨antigens introduced by an infection, immunization or vaccination). Active immunity is induced by the introduction of antigens, and passive immunity is conferred by the introduction of antibodies from another organism.

immunization the technique of artificially conferring ⇨immunity to a disease by the introduction into the body of 'live' ⇨antigens, usually given in repeated small doses.

immunoglobulin any globulin (a protein) that acts as an ⇨antibody.

immunosuppressive describing any drug that serves to suppress the body's natural immune response to an ⇨antigen.

imprinting a process of learning that takes place in the highly impressionable period soon after birth in which a pattern is set for the recognition of, and reaction to, particular objects (each with a particular function).

impulse in nerves, the electrical signal that is transmitted along a nerve fibre which has been sufficiently stimulated.

impulse in physics, the quantity used to measure the total change in ⇨momentum produced by a force acting on a body for a very short time. Given by the product of the force and the time interval for which it acts.

incandescence the emission of light caused by high temperatures. The hotter the temperature, the shorter the wavelength of the light; the relation is not, however, linear.

inclination in astronomy, the angle between one plane and another. The (equatorial) inclination of a planet is the angle between the plane of its equator and that of its orbit. The inclination of the orbit of a planet in the Solar System other than Earth is the angle between the plane of that orbit and the ⇨ecliptic.

incoherent light light that is not of a single ⇨phase. Daylight is an example.

incompressible fluid a fluid that resists changes in density – for example, oil.

indeterminate problems problems involving one or more unknown or variable quantities. See also ⇨Diophantine equations.

indicator a substance that indicates – usually by a sharp colour change – the completion of a chemical reaction. Indicators are often used to determine the end-point in titrations. Litmus, which is red in the presence of acids and blue with alkalis, is a commonly used indicator.

induction the production of an ⇨electromotive force by a change in the ⇨magnetic flux of a circuit. The effect is used in ⇨dynamos and other electrical devices.

induction motor a device that produces rotation by induction. An alternating current is fed to a winding of wires which thus induces electrical currents to flow in a second set of windings in a central rotor. Interaction between the two currents and the magnetic flux involved causes rotation.

industrial melanism dark or highly pigmented colouring in a 'variety' of a species that evolves in a region with high atmospheric pollution.

inertia the property of a body measuring its reluctance to be affected dynamically by a force. Defined by ⇨Newton's second law of motion.

infinite in mathematics, having no specified end, no highest or lowest value; not ⇨finite.

infinitesimal number that is not zero but is less than any finite number. Infinitesimal numbers clearly do not exist in the conventional system of real numbers, but it is of interest to know that modern developments in logic allow the use of an extended system of numbers that includes infinitesimal numbers. Calculations with these adhere to certain restrictions and require a good understanding of ⇨limits.

infinitesimal calculus original name for 'calculus', that is, ⇨differential and ⇨integral calculus, and so called because it was thought to rely on 'infinitely small' quantities. (It is now seen to be based upon a precise theory of ⇨limits.)

infinity term with a number of precise uses that, in naive contexts, corresponds to placing some kind of object at the 'end' of a mathematical object that is ⇨infinite. For example, there may be a point on a line at infinity, or a conventional number (written ∞) to follow all real numbers; orders of infinity occur in set theory (see ⇨transfinite).

infrared radiation part of the ⇨electromagnetic spectrum, corresponding to wavelengths between those of microwaves and visible light. Thus the wavelength band for infrared is about 10^{-6} m to 10^{-3} m. Invisible to the human eye, infrared radiation is perceived as heat. Infrared spectroscopy is an important technique in analytical chemistry.

ingot a mould shape into which metal is poured in a molten state for further processing.

ingot iron iron that has been produced in such a way as to reduce the amount of impurities within it, in the form of carbon, manganese and silicon.

inoculation the injection of microorganisms, toxin or infected material (that is, ⇨antigens) to stimulate immunity to a particular infection. See also ⇨vaccination.

insemination the introduction, by natural or artificial means, of sperm into the female reproductive tract (or the transfer of a fertilized ovum from one female to another).

insulator a nonconductor of heat and/or electricity.

integer any whole number. Integers may be positive or negative; 0 is an integer and is often considered positive.

integral calculus or *integration* method of calculation corresponding to a means of determining the area under the graph of a function, a basic $f(x)$, a basic problem in analysis. Suppose $f(x) \neq 0$, the integral

$$\int_a^b f(x)\, dx$$

of f between points a and b (area) is defined by a limiting process through approximate sums representing the sum of rectangles inscribed beneath the curve, all of whose horizontal width grow progressively thinner. It can be shown that if:

$$f(x) = \frac{dF(x)}{dx}$$

for some function F (known as the primitive), then the integral equals the change inbetween a and b. Integration is thus the opposite of ⇨differentiation.

integral equations equations involving integrals of the unknown function.

integral number an ⇨integer.

integration see ⇨integral calculus.

intelligence quotient (IQ) an intelligence rating ascertained through answers to a test, which are expressed as a score and placed on an index of scores. Formerly, IQ was defined as (mental age/calendar age) × 100.

intensity interferometry the use of two telescopes linked by computer to study the intensity of light received from a star. Analysis of the combined results has enabled measurement of the diameters of stars as apparently small as 2×10^{-4} sec of arc.

interference the effect produced from the combining of waves. Constructive interference is said to be produced when the crest of one wave coincides with the crest of another. Such phenomena can be observed for all ⇨electromagnetic radiation, in particular light.

interferometry technique for studying sources of ⇨electromagnetic radiation (light or radio waves) through interference patterns caused when two waves are combined.

intergalactic matter hypothetical material within a ⇨cluster of galaxies, whose gravitational effect is to maintain the equilibrium of the cluster. Theoretically comprising 10–30 times the mass of the galaxies themselves (in order to have the observed effect), it has yet to be detected in any form – although the most likely form is as ⇨hydrogen.

internal combustion the transformation of the chemical energy of a fuel into mechanical energy in controlled combustion in an enclosed cylinder sealed at one end by a piston.

internal-combustion engine an engine that uses ⇨internal combustion to provide motive power, for example, a petrol engine of a motor vehicle.

interoceptive describing a receptor that receives stimuli that originate within the body.

interstellar hydrogen the presence of hydrogen gas between the stars of a galaxy, thus 'filling out' the shape of the galaxy in a way that can be detected by spectral analysis and radio monitoring.

interstellar space space between the stars of a galaxy. It is generally not, however, a void vacuum, and is the subject of considerable spectral research.

intrinsic geometry study of a surface without reference to any point, condition or space outside it. All measurements and operations carried out on the surface are therefore in terms of its own (intrinsic) form.

intuitional mathematics or *intuitionism* an alternative foundational basis for mathematics that adopts a stricter logic in its approach to proofs concerning the infinite. For example, it dismisses the law of the ⇨excluded middle, and so disregards 'proofs' derived by double negatives if the relevant positive statement has not actually been demonstrated to be true. Although this may be considered a sort of philosophical puritanism, the outlook leads to a more refined classification of proof material than merely 'true' or 'false'. The fact that such a critical attitude can itself be formalized in a mathematically sound system is an important achievement. Nevertheless, most practising mathematicians remain unconcerned by this logical analysis.

invagination the formation of an inner pocket within a layer of cells by part of the layer pushing inwards to form a cavity that remains open to the original surface. See also ⇨endocytosis.

Invar an ⇨alloy of approximately 64% iron, 36% nickel, and a small amount of carbon, which possesses a very low ⇨coefficient of expansion. As a result, it is used in devices such as pendulums whose correct operation depend on maintaining constant length despite temperature changes.

invariant as a general term, describes a property that is preserved through specified mathematical operations.

inverse function for a function $f(x)$, a function $g(y)$ such that $g(f(x)) = x$, that is, that inverts the transformation f.

inverse square law the dependence of several of the most important interactions in physics including ➪gravitation and ➪electrostatic Coulomb force, obey the inverse square law, which implies that the strength of the force varies inversely as the square of the distance between the two interacting bodies.

in vitro describing the experimental observation of an organism (or part of an organism) in an artificial environment.

in vivo describing the experimental observation of a biological process in an organism (or part of it) in its natural environment.

ion atom, or group of atoms, that is either positively charged (➪cation) or negatively charged (➪anion), as a result of the loss or gain of electrons during chemical reactions or exposure to certain forms of radiation. Many crystalline substances are composed of ions held in regular lattice arrangements by the mutual attraction of oppositely charged particles. Ions migrate under the influence of electrical fields, and are the conductors of current in electrochemical reactions. See also ➪electrolysis; ➪ionization.

ionic bond or *electrostatic bond* chemical bond based on the electrostatic attraction between oppositely charged ➪ions in a compound. In the ionic compound sodium chloride, for example, the sodium atom loses one of its outer electrons to the chlorine atom (because of the greater ➪electronegativity of chlorine), resulting in the formation of a sodium cation (Na^+) and a chloride anion (Cl^-), which are mutually attracted – and therefore form an ionic bond – because of their opposite charges.

ionic radius the effective radius of an ion. In positively charged ➪cations, the ionic radius is less than the ➪atomic radius (because the electrons are more tightly bound); in ➪anions the ionic radius is more than the atomic radius. Some elements, such as the transition metals, can have several different ionization states and their ionic radii vary according to the state involved.

ionization any process by which an ion is formed. Ionization can occur in several ways: by the reaction of two neutral atoms, as occurs when sodium reacts with chlorine to form a sodium ion (Na^+) and a chloride ion (Cl^-) (ionically bonded as sodium chloride); by the combination of an already existing ion with other particles, for example, the addition of a hydrogen ion to an ammonia molecule to form an ammonium ion (NH_4^+); by the breaking of a covalent bond in such a way that each of the electrons of the bond is associated with one of the entities, for example, the ➪dissociation of a water molecule to form a hydrogen ion (H^+) and a hydroxyl ion (OH^-); and by the passage of energetic charged particles, electricity or radiant energy through gases, liquids, or solids, for example, the passage of X-rays, beta particles, gamma rays, ultraviolet

radiation, or electric discharges through gases.

ionosphere a region in the Earth's upper atmosphere at a height of between about 60 km/40 mi and 500 km/300 mi in which there are a large concentration of free electrons and ions. These are produced by the disruption of molecules at those heights by the ultraviolet and X-ray radiation produced by the Sun. The region is able to reflect radio waves as a result. Two regions of interest within the ionosphere are named the ➪Appleton layer and the ➪Kennelly–Heaviside layer, after the researchers who predicted and investigated their existence.

iron a white, metallic chemical element that, in the form of compounds such as ➪hematite ore is used to make steel and other valuable alloys. See also ➪cast iron and ➪wrought iron.

iron lung a respirator that mechanically assists the breathing of those whose natural mechanism has ceased to function.

irrational number a ➪real number that cannot be represented as a fraction; for example, $\sqrt{2}$ and π. See also ➪algebraic number and ➪transcendental number.

isomer see ➪isomerism.

isomerism the existence of two or more different substances that have the same chemical compositions but different arrangements of their atoms. Butane, for example, has two isomers; each has the same molecular formula (C_4H_{10}) but one form is a straight, four-carbon chain whereas the other isomer consists of a three-carbon chain with a methyl ($-CH_3$) group attached to the middle carbon. There are two main types of isomerism: structural isomerism (of which butane is an example), including ➪tautomerism; and ➪stereoisomerism, including optical isomerism and geometric isomerism. See also ➪*cis-*; ➪*trans-*; ➪enantiomorph.

isomorphism the existence of identical or similar crystalline forms in different – although often chemically similar – compounds.

isoperimetry branch of geometry involving the study and measurement of figures with equal perimeters.

isostasy the theoretical balance in buoyancy of all parts of the Earth's ➪crust, as though they were floating on a denser layer beneath. High mountains, for example, have very deep roots, just as an iceberg floats with most of its mass submerged.

isothermal change a process which takes place without any change in temperature. See ➪adiabatic change.

isotope species of the same chemical element (that is having the same ➪atomic number) that differ in their mass numbers (and therefore in the number of neutrons in the atomic nucleus). An element's isotopes have identical chemical and physical properties, except those determined by the mass of the atom. Most elements have several isotopes, some of which (principally those of the elements with high atomic numbers) are radioactive. See ➪radioisotope.

isotropic having equal and uniform properties at all points and in all directions. In astronomy the term describes microwave ➪background radiation.

J

Jacquard system the use of punched cards to direct the operation of a ⇨loom in weaving patterns.

jet engine a ⇨gas turbine in which air taken in through the front is compressed, and then used to provide oxygen for the combustion of fuel. The consequent backward flow of heated, expanding gas provides propulsion, and also drives a turbine, which powers the compressor bringing the air into the combustion chamber.

jib the boom of a crane, made from a framework of girders in most cases. Half its length approximately defines the range of operation of the crane, the circle in which loads can be lifted and deposited.

Josephson effect a superconducting (see ⇨superconductor) ring interrupted by a thin layer of insulating material gives rise to an ⇨alternating current (AC) in the barrier when a steady external voltage is applied to it. This AC effect has a direct-current analogue, occurring when a steady magnetic field is applied to the insulating material.

joule the ⇨SI unit of ⇨energy and ⇨work, defined as the work done by a force in moving the point of application of the force through a distance of one metre.

Joule's electrical law the heat H in joules produced by the passing of a ⇨current of I amperes through a resistance of R ohms for a time t seconds is given by:

$$H = I^2 Rt$$

Joule's thermal law the internal energy of a gas at constant temperature is independent of its volume, provided the gas is ⇨ideal.

Joule–Thomson effect or *Joule–Kelvin effect* a means of cooling or heating a gas by passing it through a porous plug, under pressure. The temperature difference arises on the expansion of the gas on passing through the plug, and is due to the deviations of the gas involved from being an ⇨ideal gas, that is, obeying ⇨Joule's thermal law and ⇨Boyle's law exactly.

Julian calendar calendar established by Julius Caesar in 46 BC, which overestimated the duration of the ⇨sidereal year by 11 minutes and 14 seconds. It was replaced, from 1582, by the ⇨Gregorian calendar, by which time it was inaccurate by a total of 10 days.

Jupiter fifth and largest major planet out from the Sun.

Jurassic period of ⇨geological time 208–146 million years ago; the middle period of the ⇨Mesozoic era. Climates worldwide were equable, creating forests of conifers and ferns; dinosaurs were abundant, birds evolved, and limestones and iron ores were deposited.

K

Kelvin scale a scale of temperature (symbol K) whose zero point can be taken as ⇨absolute zero, so that 0K corresponds to a temperature of about –273˚C/–459˚F.

Kennelly–Heaviside layer a region in the ⇨ionosphere capable of reflecting radio transmissions.

Kepler's laws of motion 1. A planet's orbit is elliptical round the Sun, with the Sun at one of the foci. 2. Planets accelerate when nearer the Sun, with the result that a radius vector (imaginary line joining planet and Sun) describes equal areas in equal times. 3. The square of the orbital period of a planet is proportional to the cube of its mean distance from the Sun.

kerosene a mixture of hydrocarbons produced by the distillation of petroleum, and used as a fuel in internal-combustion engines.

Kerr cell a device making use of the ⇨Kerr effect. The cell consists of a transparent container of a special liquid in which there are two electrodes, placed between two ⇨polarizing materials in the container. Only if the planes of polarization of all the various layers in the cell are aligned will light pass, and so the Kerr cell can be used as a shutter device.

Kerr effect the elliptical ⇨polarization of light as a result of the beam of light being reflected from a pole of an electromagnet. A similar effect also exists for liquids, if a ⇨potential difference is applied to the liquid itself, the angle of polarization depending on the size of the potential difference.

kinematic relativity theory proposed by Edward Milne as a viable alternative to Einstein's general theory of ⇨relativity, and based generally on kinematics (the science of pure motion, without reference to matter or force), from which Milne successfully derived new systems of dynamics and electrodynamics.

kinematics a branch of ⇨mechanics that relates ⇨accelerations to the velocities and changes in distance they produce, without considering the forces that generate the accelerations involved. See also ⇨dynamics.

kinetic energy the energy possessed by a body by virtue of its motion. For speeds much less than the speed of light, the kinetic energy KE possessed by a body of mass m moving at a speed v is given by:

$$KE = \tfrac{1}{2}mv^2$$

See also ⇨potential energy.

kinetic theory a theory which attempts to understand the properties of ⇨gases by considering them to consist of a vast number of ⇨molecules moving relative to one another. For example, the temperature of a gas is seen to be a measure of the velocity of its constituent particles.

Kirkwood gap one of several (apparent) zones in the asteroid belt that are free of ⇨asteroids, probably caused by the gravitational effects of Jupiter.

knocking premature explosion of the fuel–air mixture within the piston cylinder of an ⇨internal-combustion engine, due to over compression of the mixture.

knot in two- and three-dimensional geometry, a ⇨closed curve that loops over or through itself; representations of such structures are commonly presented as congruency problems.

Kuiper band one of a number of bands in the ⇨spectra of Uranus and Neptune at ⇨wavelengths of 7,500 Å (7.5×10^{-7}m), indicating the presence of ⇨methane.

kymograph a recording produced by an instrument that detects variations such as small muscular contractions or slight changes in arterial blood pressure.

L

lactic acid an organic acid, chemical formula $CH_3CH(OH)COOH$ (hydroxypropanoic acid), which exists as three ⇨stereoisomers. It is formed in the body during ⇨glycolysis (the breakdown of glucose derived from glycogen), and occurs in sour milk (derived from lactose) and other foods where it is produced by the action of microorganisms.

Lamarckism theory of evolution, now discredited, advocated during the early 19th century by French naturalist Jean Lamarck. Lamarckism is the theory that acquired characteristics are inherited. It differs from the Darwinian theory of evolution.

Landé splitting factor a calculational device used in ⇨quantum theory that enables the ⇨fine and ⇨hyperfine structure of a ⇨spectral line to be determined by a knowledge of the various ⇨spin and orbital angular momenta of the electrons in the atom.

lanthanide or *lanthanon* any of the group of rare metallic elements with atomic numbers from 57 (lanthanum) to 71 (lutetium) inclusive. The properties of all these elements are similar and resemble those of aluminium. The lanthanides constitute all but two of the elements that are commonly called the ⇨rare earth elements (the nonlanthanide rare earths are scandium and yttrium).

laparoscope a type of ⇨endoscope used for abdominal investigations.

large-number hypothesis a theory in ⇨cosmology that tries to understand the basic reason for an apparent coincidence of size in ratios of fundamental quantities in atomic and cosmological theory. If the apparent radius of the universe is divided by the radius of the ⇨electron, the resulting large number (about 10^{43}) is remarkably similar to the ratio of the strengths of the ⇨electrostatic and ⇨gravitational force between the ⇨electron and the ⇨proton. The reason for this is not clear at present, but if the relationship is to hold true for all time, it can be shown that on a ⇨Big Bang model of the universe the ⇨gravitational constant g must decrease with time, that is, the strength of gravity must decrease.

large numbers, laws of theorems in probability theory that predict that the observed frequencies of events for a large number of repeated trials are more and more likely to approach their theoretical probability as the number of repetitions increases.

laser (acronym for *l*ight *a*mplification by *s*timulated *e*mission of *r*adiation) a device that consists of an optically transparent cylinder one end of which is reflecting, the other partly reflecting. The atoms of the cylinder (which can be a solid such as ruby, or a gas or liquid) are excited by exposing them to an ⇨incoherent source of ⇨electromagnetic radiation, with the result that the atoms are put into a higher energy state. When they return to the lower level again, they give out a pulse of highly ⇨coherent narrow beam radiation. Continuous emission using rare gases as the central medium is possible. See also ⇨maser.

latent heat the amount of heat needed to change the state of a substance from solid to liquid (latent heat of fusion) or from liquid to vapour (latent heat of vaporization) without changing the substance's temperature. Each substance has characteristic latent heat values for each of its phase changes, corresponding to the amount of energy required to break the intermolecular attractions in the solid or liquid. In reversing the process (that is changing from vapour to liquid or from liquid to solid) heat is liberated, the amount being equal to the latent heat of vaporization or fusion, depending on the phase change involved.

lathe a tool used to produce objects with cylindrical symmetry, such as bars, screws, barrels.

lattice in chemistry, the arrangement of positions in, for example, a crystal in which the atoms, ions or molecules remain virtually stationary.

lattice in mathematics, a structure consisting of objects within an order corresponding to size or value, such that any two possess an element that is a minimum and similarly an element that is a maximum. For example, if there is a point on a plane with integer coordinates (n, m), and if (k, l) is larger than (n, m) in that $k > n$, and $l > m$, joining these points by horizontal and vertical lines creates a 'lattice' pattern, from which the theory gets its name.

law of universal attraction Isaac Newton's formulation of the law of ⇨gravity.

leading edge the edge of an ⇨aerofoil that first encounters the oncoming air stream.

least squares, method of method of deriving as exact an average value as possible from a set of approximate or inaccurate values by introducing the errors as unknown variables and requiring the sum of their squares to be minimized. The method was devised by Karl Gauss as a precise way of best fitting a straight line through a set of plotted data points that are not collinear.

Leclanché cell a primary cell in which the positive electrode (anode) is in the form of a rod of carbon in a mixture of manganese dioxide and carbon particles in a porous pot. The pot itself stands in a solution of ammonium chloride, in which stands the negative zinc electrode (cathode). The resultant electromotive force is about 1.5 volts. Special types of Leclanché cell are possible in which the constituents are all nonliquid.

Legendre functions functions that satisfy the second-order differential equation:

$$(1 - x^2)\frac{d^2y}{dx^2} - 2x\frac{dy}{dx} + n(n+1)y = 0$$

lemniscate curve represented by the equation:
$$(x^2 + y^2)^2 = a^2(x^2 - y^2)$$
where a is constant and x and y are variables.

Lenz's law a law explaining an ⇨induction-related phenomenon. The induced electric current in a circuit relative to which a magnetic field is moving itself, produces a magnetic field that tries to oppose the relative motion.

Leonid meteor shower shower of meteors emanating from an apparent point in Leo every 33 years; the next is due in about 1999.

leucocyte a white blood cell (with no ⇨haemoglobin) which serves chiefly to destroy foreign cells, such as pathogenic microorganisms. There are three main types: granulocytes (neutrophils, basophils and eosinophils), lymphocytes and monocytes. They are produced in the bone marrow, although lymphocytes are also produced in the lymph nodes, thymus and spleen, and monocytes are formed in the cell walls of blood vessels and various organs.

lever a rigid beam provided with a ⇨fulcrum at some point along its length. This enables a force applied at one end to be transmitted to a point on the other side of the fulcrum. A lever in thus a very simple ⇨machine.

Leyden jar an early form of condenser consisting of a glass jar with an interior and exterior coating of metal foil, used to store static electricity.

libration the 'turning' of the Moon so that although the same face is presented to Earth at all times, the overall surface of the Moon visible is 59% of the total. Libration is described as latitudinal, longitudinal and diurnal.

Lie groups or *Lie rings* collections of mathematical objects in groups or rings that have further (topological) structure under which the collective operations are ⇨continuous, for example, vectors in the plane (where open discs define a topological structure).

ligand an atom or molecule attached to the central atom (usually a metallic element) in a complex or ⇨coordination compound. Most ligands are electron-pair donors in the bond formed with central atom, for example, CN^-, Cl^- and OH^-. Occasionally they can be electron pair acceptors, for example, NO^+ and $N_2H_5^+$. Other common ligands include H_2O, NH_3 and CO. Some organic compounds act as ligands with more than one point of attachment to the central atom.

light ⇨electromagnetic radiation visible to the naked eye, lying in the ⇨wavelength range 4×10^{-7} m (violet) to 8×10^{-7} m (red).

light pen a device that, when pointed at a piece of data on a screen or paper can transmit the data to another device such as a computer.

light, pressure of see ⇨radiation pressure.

light reaction the part of the ⇨photosynthesis process in green plants that requires sunlight (as opposed to the ⇨dark reaction, which does not). During the light reaction light energy is used to generate ATP (by the ⇨phosphorylation of ADP), which is necessary for the dark reaction.

light, speed of (symbol c) a fundamental constant of nature, the speed of light is the limiting velocity that any body can travel at. It is equal to 2.997925×10^8 m /186,180 mi per second, and is the same for all observers, no matter how fast they move themselves.

light, theories of the nature of before the advent of modern physics, theories of the basic nature of light fell into two camps: those who viewed light as made up of a stream of particles (corpuscular theory), and those who viewed it as a wave motion (wave theory). Each attempted to explain all phenomena in optics, such as ⇨reflection, ⇨refraction, and so on, on the basis of these two viewpoints. It is now known that light exists in quanta known as photons that exhibit both corpuscular and wavelike behaviour in certain circumstances.

light year distance travelled at the ⇨speed of light for one Earth-year: 9.46 million million km/5.88 million million mi.

limit an important general term, of which the intuitive content may be summed up as 'the end value towards which a process proceeds, which it may or may not ever achieve'. Its importance lies in giving a rigorous basis for any mathematical calculations that involve approximation. Examples are: for the sequence 0.9, 0.99, 0.999, 0.9999 and so on, the terms are closer and closer to unity the further they are taken – and unity is thus the limiting value; regular polygons inscribed in a circle approach the shape of the circle as the number of sides is increased – and the limiting shape of the polygons is thus a circle. A proper mathematical definition of the term 'limit' involves a notion of indexing in an ordered fashion (corresponding to discrete steps, as for instance in calculations of better and better approximations, or to continuous processes, as for example the movement of a point in space with time), an association of some value to each index, and a measurement of closeness.

linear differential equations see ⇨differential equations.

linear function or *linear transformation* in its simplest context, a transformation such as $y = mx$ that may be depicted as a line through the origin. More generally, a transformation T defined on a vector space with the property that:

$$T(\alpha x + \beta y) = \alpha T(x) + \beta T(y)$$

for x, y vectors and α, β scalars.

linear motor a device that uses ⇨induction to produce forward motion along a track.

line spectrum a ⇨spectrum made up of discrete lines of intensity at certain ⇨wavelengths, characterizing an ⇨atom in a particular state.

linkage group all the ⇨genes on a given ⇨chromosome; the genes are more or less linked to each other and, as a result, tend to be inherited together.

lipid any of a group of diverse organic compounds that are ⇨esters of ⇨fatty acids. Typically they are oily or greasy and insoluble in water (but soluble in ether, alcohol and other organic solvents). There are three main types; simple lipids, which are fatty-acid esters of glycerol and include oils, fats and waxes; compound lipids, which are fatty-acid esters of glycerol and phosphoric acid (or one of its derivatives) and include the phospholipids and glycolipids; and derived lipids, a group of complex lipids that includes the ⇨steroids (of which cholesterol is the best known example), carotenoids and lipoproteins. Lipids occur in all plant and animal cells and are essential to life – as an energy source, in biosynthesis and other metabolic processes, and as structural components.

liquid a state of matter intermediate between that of ⇨solid and ⇨gas, where the interatomic and molecular forces are greater than those in the gas state. Liquids are virtually ⇨incompressible fluids.

Lissajous figures the path followed by a point subjected to two or more simultaneous simple wave motions, for example, at right angles to one another.

lithium lightest of all solid elements, third in the periodic table after hydrogen and helium. Its atom comprises one proton and three electrons. One of the electrons is at a higher energy level than the other two.

ln symbol for ⇨natural logarithm.

load generally, the burden inflicted on a system. Thus, in

⇨mechanics, the load is the weight supported by a structure, whereas in electrical engineering it is the output of an electrical device such as a ⇨transformer.

Local Group the ⇨cluster of ⇨galaxies of which our own, the ⇨Milky Way, is one. Its radius is estimated at 10^6 parsecs. Largest of the Group is the ⇨Andromeda spiral galaxy.

lock a section of canal that separates two stretches of water at different heights. By using barriers with sluices at each end of the lock, a vessel can be transported from one level to another as the sluices transfer water from one stretch to another.

locomotive a source of ⇨power for transporting cargo or passengers on a railway.

locus the path described by a point whose position in space changes. Hence, an ⇨orbit is a locus; so is a ⇨helix.

logarithm or *log* the ⇨exponent or index of a number to a specified base – usually 10. For example, the logarithm to the base 10 of 1,000 is 3 because $10^3 = 1,000$; the logarithm of 2 is 0.3010 because $2 = 10^{0.3010}$. Before the advent of cheap electronic calculators, multiplication and division could be simplified by being replaced with the addition and subtraction of logarithms.

logic systematic study of the laws and uses of reasoning. From very early times logic has been considered closely associated in methodology – if not by effectively parallel derivation – with the science of mathematics.

logistic curve curve that represents logarithmic functions, from which logarithms of ordinary numbers can be read off.

longitude the longitude of a particular location is determined by angle between the great circle passing through it and the two poles of the Earth, and the circle passing through the poles and Greenwich, England (the Greenwich meridian), the angle being measured along the equator.

longitudinal wave a wave in which the direction of vibration takes place in the same direction as that in which the wave is travelling. ⇨Sound waves are longitudinal. See also ⇨transverse wave.

loom a machine for weaving textiles.

Lorentz transformation a means of relating measurements of times and lengths made in one ⇨frame of reference to those made in another frame of reference moving at some velocity relative to the first. These formulae, which can be derived directly from Einstein's special theory of ⇨relativity, predict that at velocities approaching that of light, lengths appear to contract (the Lorentz contraction) and time intervals appear to increase (time dilation), as measured in a stationary frame of reference.

lubrication the use of substances to reduce the frictional ⇨forces between two adjacent surfaces.

luminescence the emission of light from an object as a result of any process apart from direct heating.

luminosity brightness of a celestial body, measured in terms of (apparent) ⇨magnitude, absolute magnitude, or using the Sun's brightness as 1 on a solar scale. The luminosity of a star corresponds with its internal ⇨radiation pressure, which in turn depends on its ⇨mass.

lunar of the Moon.

luteinization the development of an ovum in a ruptured Graafian follicle within the ovary, initiated by oestrogen which in turn is activated by luteinizing ⇨hormone.

Lyman series the series of lines in the ⇨spectrum of hydrogen that lie in the ⇨ultraviolet region of the spectrum.

lysogeny the presence of nonvirulent or temperate ⇨bacteriophages in a bacterium, that do not lyse it (damage the outer cell membrane). The phage does not replicate after entering the bacterial cell, although its DNA combines with that of the bacterium and is reproduced with it every time the bacterium multiplies. The basic characteristics of the host bacterium remain unchanged.

lysosome a small double-membraned vesicle that occurs in the ⇨cytoplasm of certain animal (tissue) cells. Lysosomes contain various digestive enzymes or cell nutrients.

M

machine a device that enables ➪force exerted at one point to be applied at another. A ➪lever is a very simple example of such a device.

Mach number the ratio of the ➪speed of a vehicle travelling in a medium to the ➪speed of sound in the medium under the same conditions. Thus supersonic speeds are those with Mach numbers exceeding 1.

macromolecule a very large molecule, typically with a diameter between about 10^{-8} m and 10^{-6} m (most ordinary molecules have diameters of less than about 10^{-9} m). ➪Polymers – natural (for example cotton) and synthetic (for example plastics) – and many biologically important molecules, such as ➪proteins, are macromolecules.

Magellanic Clouds two relatively small, nebulous star ➪clusters visible only in the southern hemisphere; the larger is, however, the brightest 'nebular' object in the sky. Both are members of the ➪Local Group of galaxies, and in fact seem to be associated, though detached, with parts of the ➪Milky Way system.

magic numbers atomic nuclei that contain a magic number of ➪protons or ➪neutrons are very stable against ➪radioactive decay. The numbers are 2, 8, 20, 28, 50, 82 and 126.

magma molten rock material beneath the Earth's surface from which ➪igneous rocks are formed. Lava is magma that has reached the surface and solidified, losing some of its components on the way.

magnetic detector a device used in early ➪radio systems, in which high-frequency currents were detected through their demagnetizing effect on a magnetized ➪iron core surrounded by the wire-carrying currents.

magnetic energy property of a magnet described by the multiplication of the ➪flux density by the ➪field strength on the ➪demagnetization curve of a permanent magnet.

magnetic field area surrounding a magnet in which an object is affected by the ➪magnetism of the central source.

magnetic field strength the property measured to define the strength of a ➪magnetic field. The ➪SI unit is the ➪ampere per metre.

magnetic flux the product of the area of a circuit, for example, and the magnetic field strength at right angles to that area. It is measured by the ➪SI unit, the ➪weber.

magnetic flux density the ➪magnetic flux passing through one square metre of area of a magnetic field in a direction at right angles to the magnetic force. The ➪SI unit is the ➪tesla.

magnetic moment the product of the strength and length of a magnet.

magnetic monopole a prediction of one ➪quantum theory (proposed by Dirac) that involves the existence of individual magnetic poles, analogous to the individual charges found in ➪electrostatics (that is, ➪electrons). Such an entity has not yet been definitely observed to exist, but is expected to be rare in any case.

magnetic permeability the ➪magnetic flux density in a body, divided by the external magnetic field strength producing it. It can be used as a way of classifying materials into different types of magnetism, such as ➪ferromagnetic.

magnetism a property of certain materials to exert a magnetic force upon other materials, particularly iron-based substances. This ability is the result of imbalances of certain properties of electrons in the atoms of the substances concerned.

magnetohydrodynamics the study of the behaviour of ➪plasmas under the influence of magnetic fields.

magnetosphere the extent of a planet's magnetic field. The Earth's magnetosphere is shaped roughly like a teardrop, with the point opposite the Sun; this is due to the effect of the ➪solar wind.

magnet, permanent see ➪permanent magnet.

magnitude the measured brightness of a celestial body. Dim objects have magnitudes of high numbers, bright objects have magnitudes of low or even negative numbers. Seen from Earth, stars of (apparent) magnitude 6 or higher cannot be detected with the naked eye. The full Moon has a magnitude of −11, and the Sun one of −26.8. In order to standardize measurements of the brightness of more distant objects, the system of ➪absolute magnitude is used. A measure of the radiation at *all* wavelengths emitted by a star is known as the bolometric magnitude.

main sequence band that runs from top left to bottom right on the ➪Hertzsprung–Russell diagram representing the majority of stars, including the Sun. Stars off the main sequence are in some way uncharacteristic and include ➪red giants, ➪blue dwarfs, ➪Cepheids, and ➪novae.

make-and-break circuit a circuit that contains a device which, when ➪current flows in one part of the circuit, causes the device to break that part of the circuit, and allow current to flow in another part, and vice versa.

Malfatti's problem medieval European problem, to inscribe within a triangle three circles tangent to one another.

malleability the ability of a substance to be hammered into thin sheets. Gold has an extremely high malleability.

Malus's law a law giving the intensity of light after having been ➪polarized through a certain angle.

mandrel an accurately turned cylinder onto which a bore-tube can be fitted for further turning and milling.

manifold in two-dimensional space, a regular ➪surface that locally looks like a flat plane slightly distorted. It can be represented by differentiable functions. There are analogues in spaces of more dimensions.

manometer a device for measuring liquid or gaseous pressure, consisting classically of a glass U-tube containing a liquid (such as mercury).

mapping in mathematics, another name for ➪transformation.

marine engineering the branch of ➪engineering devoted to the design and production of propulsive devices and other mechanical devices for marine vessels such as ships and submarines.

Mariner spaceprobes series of US spaceprobes launched to explore the planets of the Solar System, particularly Mercury, Venus and Mars.

Markov processes or *Markov chains* random, but mutually dependent, changes within a system that exists in potentially more than one state. As part of ⇨probability theory, they are defined as sequences of mutually dependent random variables for which any prediction about $x_n + 1$, knowing $x_1, \dots x_n$, can be based on just x_n without any loss.

Mars fourth major planet out from the Sun.

mascon (contraction of *mass concentration*) one of a number of apparent regions on the surface of the Moon where gravity is somehow stronger. The effect is presumed to be due to localized areas of denser rock strata.

maser (acronym for *m*icrowave *a*mplification by *s*timulated *e*mission of *r*adiation) a system for boosting microwave signals which operates in a similar way to the ⇨laser, except that the final burst of energy has a single wavelength which lies in the microwave region of the ⇨electromagnetic spectrum.

mass the quantitative property of an object due to the matter it contains. (Weight, in contrast, describes a force with which a body is attracted towards a gravitational focus.) Units of mass are grams and kilograms.

mass, centre of see ⇨centre of mass.

mass number the number of ⇨neutrons and ⇨protons in the nucleus of an ⇨atom.

mass spectrograph an instrument for determining the masses of individual atoms by means of positive-ray analysis, which involves deflecting streams of positive ⇨ions using electric and magnetic fields. Ions with different masses are deflected by different amounts and can be detected (for example, photographically) to produce a mass spectrum. ⇨Isotopes were first discovered in this way.

mass spectrometry a means of analysing a substance in order to determine its basic chemical composition which makes use of the way in which the deflection of an ⇨ion in a magnetic and electric field depends on its mass to charge ratio. A mass spectrometer measures the relative abundances of particles of each mass rather than their individual masses.

mast cell a granular cell common in fatty tissue that secretes heparin and histamine. It plays an important part in stimulating the coagulation of blood when the tissue is injured.

mathematical logic view of mathematics and logic that relates the two disciplines, and to do so uses mathematical or similar notation in the expression of axiomatic statements. ⇨Boolean algebra was the original form, and led to Gottlob Frege's symbolic logic.

mathematical philosophy see ⇨philosophy of mathematics.

mathematical structure collection of objects that display a) one or more relationships, and b) one or more operations of which the properties may be summarized as a list of ⇨axioms; for example, a group, vector, space or ring.

matrix in algebra, a rectangular array of numbers, used for example to represent the coefficients of a system of simultaneous linear equations, each row of the array corresponding to one equation. Matrices may be multiplied following specific rules relating to the change in coefficients of the simultaneous

equations when a change of variables is effected.

matrix mechanics a mathematical description of subatomic phenomena which views certain characteristics of the particles involved as being matrices, and hence obeying the rules of matrix mathematics which differ in significant ways from the rules obeyed by ordinary arithmetic.

matter, continuous creation of a phenomenon invoked in certain cosmological theories, especially the ⇨steady-state model of the universe. In that particular theory, matter is considered to be constantly created, either evenly throughout space, or in localized regions of creation, so as to make up for the diluting effect the ⇨Hubble expansion of the universe has on the average density of matter. By continually creating matter, it is then possible for the universe to maintain a steady-state appearance.

Maxwell–Boltzmann distribution a mathematical function used in the ⇨kinetic theory of gases and derivable on the basis of ⇨statistical mechanics, giving the distribution of particle velocities in a gas at a particular temperature.

Maxwell's equations a set of four ⇨vector equations showing the interdependence of electricty and magnetism. The concepts of charge conservation are built into them, as are the experimental results of Faraday, Gauss and Ampère.

mean value average value of a number of values, obtained by totalling the values and dividing by the number of values. See also the method of ⇨least squares.

measure theory extension of the notion of length, area or volume (as appropriate) to general sets of points on the line, plane or in space. Used commonly in analysis (especially integration theory), functional analysis, probability theory and game theory (in the assessment of the size of coalitions), its definitive form was derived by Henri Lebesgue.

mechanical advantage the ratio of the load lifted by a ⇨machine to the ⇨force required needed to maintain the machine at constant speed.

mechanical curve curve that cannot be precisely expressed as an equation (unlike a geometric curve).

mechanical engineering the branch of ⇨engineering devoted to the study and production of devices, such as tools and vehicles, that are capable of carrying out tasks.

mechanical equivalent of heat the constant by which the number of units of ⇨heat being converted completely into ⇨work must be multiplied to calculate the amount of work obtained. In ⇨SI units, both are measured in ⇨joules, so the constant is 1.

mechanics the study of the behaviour of bodies under the action of ⇨forces.

medium in bacteriology, an environment in which microorganisms can be cultured. Common mediums include agar, broth and gelatine, often with added salts and trace elements.

meiosis two successive special cell divisions of a diploid (paired-chromosome) cell to form four haploid (unpaired-chromosome) daughter cells such as ⇨gametes. See also ⇨mitosis.

melanin a dark brown pigment present as granules in cells, responsible for the yellow, brown or black colour of, for example, skin and hair.

melanocyte a cell that produces ⇨melanin.

membrane potential the potential difference that exists across a membrane or cell wall, such as that across the wall of a nerve cell (⇨neuron).

Mercury first major planet out from the Sun.

meridian theoretical north–south line on the Earth's surface, or an extension of that line onto the night sky, connecting the observer's ⇨zenith with the celestial pole and the horizon. The meridian is used to state directional bearings. Devices and structures – such as meridian arcs – marking the meridian were once common in observatories.

mesoderm the central layer of embryonic cells between the ⇨ectoderm and the ⇨endoderm.

meson an unstable type of ⇨elementary particle, with a mass intermediate between that of the ⇨proton and ⇨electron. A number of such particles have been discovered (with the mu-meson now being considered not to belong to the meson family). The pi-mesons, or pions, make up a family of three mesons, with zero, positive and negative charges, and roughly the same masses.

Mesozoic era of ⇨geological time 245–65 million years ago, consisting of the ⇨Triassic, ⇨Jurassic, and ⇨Cretaceous periods. At the beginning of the era, the continents were joined together as ⇨Pangaea; dinosaurs and other giant reptiles dominated the sea and air; and ferns, horsetails, and cycads thrived in a warm climate worldwide. By the end of the ⇨Mesozoic era, the continents had begun to assume their present positions, flowering plants were dominant, and many of the large reptiles and marine fauna were becoming extinct.

messenger RNA (mRNA) a single-stranded nucleic acid (made up of ⇨nucleotides) found in ⇨ribosomes, ⇨mitochondria and nucleoli of cells that carries coded information for building chains of ⇨amino acids into polypeptides. See ⇨ribonucleic acid.

Messier Catalogue list of the locations in the sky of more than a hundred ⇨galaxies and ⇨nebulae, compiled by Charles Messier between 1760 and 1784. Some designations he originated are still used in identification; M1 is the Crab nebula (in Taurus).

metabolism the chemical processes that take place in an organism and result in the breakdown of large or complex organic molecules into simpler ones (catabolism) with the release of energy, or result in the building-up of larger organic molecules from simpler ones (anabolism) and the storage of energy. These processes are usually moderated by ⇨enzymes.

metallurgy the study of the extraction, purification and properties of metals.

metamorphic rock rock altered in structure and composition by pressure, heat, or chemically active fluids after original formation. (If heat is sufficient to melt the original rock, technically it becomes an ⇨igneous rock upon cooling.)

meteor fragment or particle that enters the Earth's atmosphere and is then destroyed through friction, becoming visible as this occurs as a momentary streak of light. At certain times of the year, meteors apparently emanating from a single area of the sky (a ⇨radiant) form meteor showers. They are thought to originate within the Solar System. See also ⇨meteorite.

meteorite object that enters the Earth's atmosphere and is too large to be totally destroyed by friction before it hits the surface. Meteorites may in some way be connected with ⇨asteroids. See also ⇨meteor.

meteorology the study of the characteristics of the Earth's atmosphere (such as pressure and temperature) in order to understand the weather, and predict conditions at a later date.

methane gaseous hydrocarbon, one of the alkanes, in which every carbon atom is surrounded by four hydrogen atoms.

method of least squares see ⇨least squares, method of.

metric in special mathematical structures, a means of measuring distance or discrepancy. It has the characteristic property of satisfying the triangle inequality: that is, that the distance x to y does not exceed the distance x to z plus the distance z to y for any z.

micelle or *association particle* A loosely bound aggregation of tens or hundreds of atoms, ions or molecules in a continuous medium (usually a liquid), forming a colloidal particle.

Michelson–Morley experiment an ⇨interference based experiment which attempted to find the absolute velocity of the Earth, that is, its velocity relative to the ⇨ether, which would show up as a change in the ⇨velocity of light when measured at different angles to the direction of the Earth's motion. Despite this experiment's great accuracy, no such change was found, indicating that the speed of light was the same for all observers. This finding is the cornerstone of Einstein's special theory of ⇨relativity.

micrometer a device that can measure very small distances, or on a telescope, small angles.

microorganism an (usually unicellular) organism that is too small to be seen with the naked eye. Common microorganisms include ⇨bacteria, ⇨viruses, various fungi and protozoa; some are pathogenic (disease-causing).

microphone a device that converts ⇨sound energy into electrical impulses suitable for transmission by electrical means.

microsome a minute particle occurring in the cytoplasm of a cell composed of vesicles with attached ⇨ribosomes, which are thought to derive from the ⇨endoplasmic reticulum. Microsomes are also thought to give rise to ⇨mitochondria.

microtome a device for cutting extremely thin slices of tissue for microscopic examination. The tissue is embedded in wax or a synthetic resin (or is frozen) for ease of handling.

microwave radiation radiation in the ⇨electromagnetic spectrum between ⇨infrared and radio waves. This range has ⇨wavelengths of between about 20 cm/8 in and about 1 mm/0.04 in. Radiation of this type was detected as ⇨background radiation.

Milky Way our own ⇨Galaxy, the second largest in the ⇨Local Group. Also, the band of luminescence cutting the night sky in two, the faint light being the result of the presence within it of a vast number of stars. These stars occupy one arm of our Galaxy.

mimicry the similarity in behaviour or appearance of one animal to another as a form of protection; it is commonest among insects. In Batesian mimicry, a harmless animal is protected by its similarity in appearance to a toxic or distasteful animal. In Müllerian mimicry, the similarity in appearance between two or more equally dangerous species gives them mutual protection.

mineral naturally formed inorganic substance with a particular chemical composition and a regularly repeating internal structure. Either in their perfect crystalline form or otherwise, minerals are the constituents of rocks. In more general usage, a mineral is any substance economically valuable for mining (including coal and oil, despite their organic origins).

minor in the theory of ⇨determinants, smaller determinant obtained by deleting one of the rows and one of the columns of a ⇨matrix.

minor planet another name for ⇨asteroid.

mitochondrion a microscopic double-membraned body that occurs in the ⇨cytoplasm of nearly every type of cell (except bacteria and blue-green algae). The inner membrane is folded into cristae which carry oxidative enzymes and some DNA. Mitochondria are the sites of much of the ⇨metabolism necessary for the production of ATP (including the citric acid cycle); they are also involved in the metabolism of lipids and ⇨protein synthesis.

mitosis the normal process of cell division in which the (paired) chromosomes duplicate at the beginning of the process and each of the two daughter cells formed has pairs consisting of one original and one new (replicated) chromosome. See also ⇨meiosis.

Mizar a ⇨double star in Ursa Major.

M number designation used in the ⇨Messier Catalogue.

modulation the modification of a property of a wave (such as its ⇨frequency) in accordance with some characteristic of another wave.

modulo in number theory, two numbers are said to be equivalent modulo a fixed number if their difference is divisible by the fixed number.

molarity the concentration of a solution expressed in terms of the number of ⇨moles of solute dissolved in one litre of solution. Thus a 0.5 molar solution of sodium chloride contains 29.22 g of NaCl (molecular mass 58.44) per litre of solution. Molarity is the concentration of a solution expressed as the number of moles of solute dissolved in one kilogram of solvent.

mole (symbol mol) the ⇨relative molecular mass of a substance in grams; for example, the molecular mass of oxygen is 31.9988, therefore one mole of oxygen equals 31.9988 g. The number of molecules in one mole is the same for all substances and is approximately 6.023×10^{23} (Avogadro's constant).

molecular biology the study of the chemical and physical properties of molecules that occur in living organisms.

molecular formula the formula of a chemical compound showing the type and number of atoms of each type present in one molecule of that compound – in contrast to the ⇨empirical formula, which indicates only the relative proportions of each atom present. The molecular formula does not, however, show the structural arrangement of the constituent atoms.

molecular orbital a type of orbital resulting from the overlap of two ⇨atomic orbitals, forming a (usually covalent) chemical bond between the atoms involved.

molecular weight see ⇨relative molecular mass.

molecule the smallest unit of a substance capable of independent existence and retaining the properties of that substance. A molecule may consist of a single ⇨atom, as in a molecule of helium (He), or an aggregation of similar or dissimilar atoms held together by valence forces and acting as a single unit – oxygen (O_2) and water (H_2O), for example.

moment in statistics and probability theory, a generalization to a higher power of the mean. The rth moment of a random variable is the expected value of the rth power of the variable less its mean.

moment of inertia a quantity for a body that represents the sum of the ⇨mass of each particle within it multiplied by the square of the particle's distance from a given axis. Of great importance in rotational ⇨dynamics.

momentum the quantity given by the ⇨mass of a body, multiplied by its velocity. It is used in ⇨dynamics, as the quantity is conserved under certain circumstances, and its rate of change gives the amount of ⇨force acting on the body. See ⇨motion, laws of.

monochromatic literally, one colour, this is the property of some sources of radiation to emit waves of one ⇨frequency (or wavelength). ⇨Lasers and ⇨masers are examples.

monoclonal describing genetically identical cells produced from one ⇨clone.

monocotyledon any member of the smaller of the two subdivisions of the ⇨angiosperms (the other being the ⇨dicotyledons). Monocotyledon plants are characterized by having a single seedleaf (cotyledon), flower parts arranged in threes (or multiples thereof), parallel-veined leaves, and closed vascular bundles arranged randomly in the stem tissue.

monoplane an aircraft (powered or glider) that has only one major aerofoil structure providing lift.

monorail a means of transport whereby the vehicle is constrained to the path described by a single continuous, usually elevated, rail.

monotonic of a sequence, that its terms are increasing or decreasing along the sequence.

Mordell's equation is represented by $y^2 = x^3 + k$.

morphogenesis the development of forms and structures in an organism.

morphology the study of form and structure in an organism.

Morse code a communication system in which the letters of the alphabet are represented by strings of dots and dashes, the most commonly used letters (such as e, t, and a) being simpler strings than those of the letters, j, z, and q. Much used in simple ⇨telegraphy.

Moseley's law the frequency of the characteristic ⇨X-ray line spectra for elements is directly proportional to the square of the ⇨atomic number, Z.

Mössbauer effect the ability for crystals at low temperatures to absorb ⇨gamma rays of a certain ⇨wavelength in such a way that the crystal as a whole may recoil, so that the gamma rays are emitted in an exceedingly narrow range of wavelengths, essentially unchanged from the original.

motion, laws of devised by Isaac Newton, these three laws are the foundation of classical dynamics and statics. The first law states that a body remains in its state of rest or of uniform motion in a straight line, unless acted upon by a force. The second states that the rate of change of momentum of a body is directly proportional to the force applied, and takes place in the direction of the force. For a body of constant mass, this statement implies that:

$$F = ma$$

where F is the size of the force producing an acceleration a in a mass m. The third law, that of action and reaction, states that to every action there is an equal and opposite reaction.

motor nerve a nerve (in the peripheral nervous system) that carries impulses from the central nervous system (brain and spinal cord) to an effector organ.

mu-meson see ⇨meson.

mural arc astronomical apparatus, used from the sixteenth to the nineteenth century, comprising a carefully oriented wall on which a calibrated device was fixed, by which the altitudes of celestial objects could be measured.

mutant a ⇨gene, organism or population that has undergone a change in character because of ⇨mutation.

mutation a spontaneous or artificially induced qualitative or quantitative change in a ⇨gene or ⇨chromosome. A mutation to a gene in a gamete is inherited and expressed in the next generation. Evolution occurs by ⇨natural selection of random mutations (although most mutations are harmful, often leading to the death of the organism concerned).

myelin a fatty substance that forms a sheath around the nerve fibres of vertebrates.

myeloma a malignant or nonmalignant tumour that forms in bone-marrow cells.

myofibril one of many minute fibrils that together make up a fibre of smooth or striped muscle, running along the length of the muscle.

myosin a protein made up of a chain of polypeptides that forms filaments in smooth (or striped) muscle fibrils. During muscle contraction it combines with ⇨actin (another muscle protein, contained in thinner filaments) to form actomyosin; the actin filaments are pulled into the myosin filaments, which shortens the ⇨myofibrils.

N

n undefined positive whole number larger than three, usually representing the unknown variable at the end of a series or sequence.

n! factorial *n*, that is, $1 \times 2 \times 3 \times 4 \times \ldots n$.

nacelle a streamlined housing on the fuselage of an aircraft, containing the engine intakes, radio antenna, and so on.

Napier's bones primitive mechanical calculation device created by John Napier and consisting of a set of little bone rods with which to multiply or divide.

NASA (acronym for *N*ational *A*eronautics and *S*pace *A*dministration) US government body set up in 1958, under which the Space Center at Houston, Texas, and the Space Center at Cape Canaveral, Florida, are responsible for crewed and uncrewed space flights.

national grid the UK's system of high-tension cables distributing electric ⇨power from power stations to end-users.

natural logarithms system of logarithms based on ⇨Euler's number instead of the (more elementary) base 10. In place of the abbreviation 'log', the expression 'ln' is used unless the context itself makes clear what base is intended

natural number a whole, positive ⇨integer.

natural selection the preservation of favourable and rejection of unfavourable variations within a species, such variations resulting from ⇨mutations (of genes or chromosomes). It is the principal process of evolution; variants with favourable mutations survive and therefore tend to leave more offspring, whereas variations with unfavourable or harmful mutations tend to die out (leaving fewer progeny).

n-**dimensional** having an unstated but finite number of dimensions. A typical example is the set of all runs of *n*-numbers (that is, $x_1, \ldots, x_n$), which is the basis of a coordinate geometry generalizing three-dimensional coordinates (x_1, x_2, x_3).

nebula cloud of interstellar gas and/or dust, but containing no actual stars. A spiral nebula, however, is a type of ⇨galaxy, and a ⇨planetary nebula is a type of star. Nebulae may in fact represent the initial form of a star ⇨cluster (that may condense out of it) or the final form of a ⇨supernova.

nebula variable star or *T Tauri variable* variable star of ⇨spectral classification F, G or K (a giant above the ⇨main sequence on the ⇨Hertzsprung–Russell diagram) that loses an appreciable proportion of its mass in its (irregular) more luminous periods, and is thus surrounded by volumes of gas and dust.

negative feedback see ⇨feedback.

neighbourhood in topology, subsets of points in a ⇨topo-logical space defining a 'locality' round a specific point, which includes an open set to which the specific point belongs. Axioms describing neighbourhoods were first formulated by Felix Hausdorff.

Neptune eighth major planet out from the Sun, discovered in 1846 by Johann Galle and Louis d'Arrest, following predictions calculated by Urbain Leverrier. Similar predictions had been made a year earlier by John Couch Adams but were not followed up.

neural fold one of two longitudinal (ectodermal) ridges along the dorsal surface of a vertebrate embryo. The ridges fuse to form the ⇨neural tube.

neural tube a tube in a vertebrate embryo formed by the fusion of the ⇨neural folds. It has a prominent bulge at the anterior end, which eventually develops into the brain; the rest of the tube gives rise to the spinal cord and the nerves of the peripheral nervous system.

neurofibril one of the many fibrils in the cytoplasm of a ⇨neuron, which extend into its ⇨axon and ⇨dendrites.

neuron or *nerve cell* the structural unit of the nervous system of animals. Each neuron is composed of a cell body (perikaryon) containing cytoplasm (which encloses a nucleus) and an ⇨axon and ⇨dendrites. Impulses pass from neuron to neuron across a ⇨synapse between the axon of one nerve cell and the dendrites of an adjacent cell.

neurosecretory cell a nerve cell that secretes a chemical substance, such as a ⇨hormone.

neurosis a collective term for several mental disorders that result from psychological disturbances, rather than from a physiological illness. A neurotic patient is trying to resolve an unconscious conflict, which may manifest itself as, for example, hysteria, hypochondriasis or hyperanxiety.

neurotransmitter chemical that diffuses across a ⇨synapse, and thus transmits impulses between ⇨neurons (nerve cells), or between neurons and effector organs (for example, muscles). Common neurotransmitters are noradrenaline (which also acts as a hormone) and ⇨acetylcholine (ACh), the latter being most frequent at junctions between nerve and muscle. Nearly 50 different neurotransmitters have been identified.

neutrino elementary particle with no mass and no electric charge. The Sun is said to give off great quantities of them, but experiments to trap them as they pass straight through the Earth have failed.

neutron a fundamental particle found in the atomic nuclei of all elements, except normal hydrogen (whose nucleus contains only a single ⇨proton). A neutron has no electrical charge and its mass is 1,838.65 times that of the ⇨electron – slightly greater than the mass of the ⇨proton. The mass number of an ⇨isotope of an element equals the number of neutrons plus the number of protons.

neutron star remnant of a star after it has exploded as a ⇨supernova. Usually optically dim, a neutron star sends out regular or irregular radio emissions and is therefore also called a pulsar. The density of such a star may be unimaginably great although the diameter is generally around only 10 km/6 mi; the gravitational and magnetic forces are correspondingly vast. It is called a neutron star because in such density, protons fuse with electrons to form neutrons, of which the star is almost entirely composed.

newton the ⇨SI unit of ⇨force, defined as the force that, when applied to a mass of one kilogram, produces in that mass an ⇨acceleration of one metre per second, per second.

Newton's law of cooling the rate of loss of ⇨heat from a body is directly proportional to the instantaneous temperature difference between the body and the surroundings. This leads to an exponential law of temperature decline.

Newton's laws of motion see ⇨motion, laws of.

Newton's rings an ⇨interference effect occurring in observations made of thin films separated by an air gap, ⇨reflections between various surfaces being allowed.

Nicol prism a device which uses the optical properties of calcite to produce plane ⇨polarized light by passing rays of light through it.

nitrogen fixation see ⇨fixation of nitrogen.

node in two- and three-dimensional geometry, where a curve intersects itself.

no hair theorem proposed by John Wheeler, it states that the only properties of matter conserved after entering a black hole are its mass, its angular momentum and its electrical charge; it thus becomes neither matter nor ⇨antimatter.

noise spurious voltages occurring in a circuit as a result of random motions of electrons within it, or vibrations in the components of the circuit.

noncommutative in which the principle of commutation (see ⇨commutative principle) does not apply.

non-Euclidean geometry the study of figures and shapes in three-or-more-dimensional (or curved) space, in which Euclid's postulates may not apply fully or at all (see ⇨geometry). There are now many forms of non-Euclidean geometry, probably the best known being those propounded by Bernhard Riemann; the first proponents of such systems, however, were Karl Gauss, Nikolai Lobachevsky and János Bolyai.

normal distribution or *Gaussian distribution* special probability density function, ordinarily represented as a graph that is symmetrically bell-shaped, starting with a near-zero value low on the left, rising to a smooth-domed peak of probability in the centre, and sinking again to low on the right. Devised first by Abraham de Moivre, it is now fundamental to ⇨probability theory. See also ⇨standard deviation.

normality a measure of the concentration of a solution, expressed in terms of the number of gram equivalents (the ⇨equivalent weight in grams) of a reagent per litre of solution. For example, a solution that contains 2 gram equivalents per litre is a twice-normal (or 2N) solution.

North Polar sequence or circumpolar stars, comprises those stars which never set, from the viewpoint of an observer on Earth.

notochord a skeletal rod of connective tissue that occurs in the embryos of ⇨chordates. It lies dorsally along the length of the embryo and is eventually replaced by the vertebral column and skull.

nova originally faint dwarf star which suddenly experiences a tremendous increase in ⇨luminosity, giving off enormous quantities of gas; the luminosity then fades away. Novae are now thought to be the result of collapsing ⇨binary systems. A ⇨supernova is thus not at all the same; ⇨pulsating novae, however, may form a link with ⇨variable stars.

nuclear fusion process by which the Sun (and other stars) radiates energy. The nucleus of an atom fuses with the nuclei of other atoms to form new, heavier atoms, at the same time releasing large amounts of energy. In the Sun, hydrogen atoms

are converted into helium by this process, with carbon and nitrogen as intermediates. Cooler stars undergo the ⇨proton–proton cycle with a similar result.

nuclear isomer atoms of an element of a given mass that differ in their rates of ⇨radioactive decay are known as nuclear isomers.

nucleated describing a cell that has a ⇨nucleus.

nucleic acid one of many large organic molecules made up of long chains of ⇨nucleotides which occur in cells, particularly in their nuclei. Deoxyribonucleic acid (DNA) and ribonucleic acid (RNA) are the most important nucleic acids and are responsible for transmitting hereditary characters. The genetic code that controls ⇨protein synthesis is stored in triplet sequences (⇨codons) of three nitrogenous ⇨bases along the DNA and RNA strands.

nucleophilic describing an atom, molecule or ion that seeks a positive centre (for example, the atomic nucleus) during a chemical reaction. Nucleophiles react at centres of low electron density because they have electron pairs available for bonding. Common nucleophiles include the hydroxide ion (OH^{-1}), ammonia (NH_3), water (H_2O) and halide anions. See also ⇨electrophilic.

nucleosynthesis cosmic production of all the species of chemical elements by large-scale nuclear reactions, such as those in progress in the Sun or other stars. One element is changed into another by reactions that change the number of protons or neutrons involved.

nucleotide any of a group of organic compounds consisting of a nitrogenous base (derived from pyrimidine or purine), a pentose sugar and a phosphate group. Nucleotides are of fundamental importance in living organisms: adenosine triphosphate (ATP), a compound essential for biological energy production, is a nucleotide, and polynucleotide chains make up ⇨nucleic acids, the carriers of genetic information.

nucleus in biology, the central vital body of a plant or animal cell. Each nucleus is surrounded by a membrane and contains mainly nucleoprotein and ⇨DNA in the form of ⇨chromosomes. The nucleus is therefore essential to cell reproduction and the control of the whole cell's ⇨metabolism.

nucleus, atomic the positively charged central core of an atom. Although relatively small in comparison to the volume of the entire atom, the nucleus has nearly all of the atom's mass. It comprises one or more positively-charged ⇨protons and (except in the case of hydrogen) one or more electrically neutral ⇨neutrons, held together by the 'strong force'. The number of protons in the nucleus is equal to the atomic number. The number of neutrons is given by the difference between the atomic number and the mass number.

null hypothesis in probability theory, assumes that events occur on a purely random (chance) basis.

null set conventional ⇨set that has no members.

null vector another name for a zero ⇨vector.

number theory or *theory of numbers* branch of pure mathematics concerned with numbers: algebraic, complex, even, irrational, natural, negative, odd, positive, prime, rational, real and transcendental, in sequence or in series, in any operation and expressed even in other forms.

nutation slight but recurrent oscillation of the axis of the Earth, caused by the Moon's minutely greater gravitational effect on the Earth's equatorial 'bulge'.

O

objective the system of lenses in a telescope or microscope nearest to the object being observed.

occlusion the process by which some solids (mainly metals) absorb certain gases. It may occur in any of three main ways: by the formation of a chemical compound, by the condensation of the gas on the surface of the solid; or by the formation of a solid solution (a homogeneous mixture of two or more substances, the resultant substance being in the solid state).

occultation eclipse of a star by another celestial body.

octane rating a system of establishing the ability of a fuel to avoid ⇨knocking in an internal-combustion engine. Based on the percentage volume of iso-octane in a mix of this with normal heptane that will give the same knocking characteristics as the fuel tested.

ohm the ⇨SI unit of electrical ⇨resistance, defined as the resistance that allows a ⇨potential difference of one ⇨volt produce a ⇨current of one ⇨ampere within a conductor.

Ohm's law law that states that the current flowing in a metallic conductor maintained at constant temperature is directly proportional to the ⇨potential difference in volts, between its ends.

oil see ⇨fat.

Olbers's paradox the fact that the night sky is dark despite the presence of so many 'suns'. The reason now accepted is that it is caused as a by-product of the ⇨red shift due to the stars' ⇨recession.

olefins the former name for the class of unsaturated hydrocarbons now called ⇨alkenes.

ontogeny the history of an individual's development and growth.

open-hearth process or *Siemens– Martin process* a method of producing steel in which cast iron and steel scrap or iron ore are heated together in measured amounts with ⇨producer gas on a hearth in a furnace.

operant conditioning the conditioning of an individual's response (to a stimulus) by means of a reward so that the individual eventually behaves so as to be rewarded (see ⇨conditioned reflex).

opposition of a planet occurs when the Earth comes directly between that planet and the Sun; it can thus only happen in relation to the ⇨superior planets and the ⇨asteroids.

optical activity the property possessed by some substances (and solutions of these substances) of rotating the plane of vibration of plane-polarized light. A substance that rotates the light in a clockwise direction (as viewed facing the light source) is described as dextrorotatory and is prefixed by the symbol d; a substance that rotates the light in an anticlockwise direction is laevorotatory and is prefixed by l.

optical fibre a thin thread of glass so constructed that it permits light to be transmitted down its length, even round corners, by ⇨total internal reflection.

optical isomer or *enantiomorph* one of a pair of compounds whose chemical composition is similar but whose molecular structures are mirror images of each other. The presence of an asymmetric (usually carbon) atom makes each isomer optically active (that is, its crystals or solutions rotate the plane of polarized light); the direction of rotation is different (left or right, denoted by d or l) for each isomer.

optical isomerism a type of ⇨stereoisomerism in which the isomers differ in their ⇨optical activity because of the different spatial arrangements of their atoms. ⇨Enantiomorphs (isomers with asymmetrical structures, each isomer being a mirror image of the other) are optical isomers.

optic vessel one of two bulges on each side of the anterior expansion of the ⇨neural tube in a vertebrate embryo, from which arise the essential nervous structures of the eyes.

orbit the path of an object under the influence of a radially acting ⇨force.

orbital see ⇨atomic orbital; ⇨molecular orbital.

ordinary number see ⇨set theory.

Ordovician period of ⇨geological time 510–439 million years ago; the second period of the ⇨Palaeozoic era. Animal life was confined to the sea: reef-building algae and the first jawless fish are characteristic.

ore a mineral containing a metal that is extracted from it. Hematite is an ore from which iron is obtained using a ⇨blast furnace.

organelle any subcellular structure with a specialized function, such as ⇨mitochondria, ⇨lysosomes or the ⇨Golgi apparatus.

organizer cell in embryonic tissue that induces the morphological development of tissues in other parts of it.

organ of Golgi an elongated structure that occurs at the junction of a muscle and a tendon, which responds to ⇨proprioceptive stimuli.

organometallic compound any of a group of substances in which one or more organic ⇨radicals are chemically bonded to a metallic atom – excluding the ionic salts of metals and organic acids. A typical organometallic compound is tetraethyl lead, the 'antiknock' substance commonly added to petrol.

orogeny or *orogenesis* the formation of mountains. It is brought about by the movements of the rigid plates making up the Earth's crust (described by ⇨plate tectonics). Where two plates collide at a destructive margin rocks become folded and lifted to form chains of fold mountains (such as the young fold mountains of the Himalayas).

orthogonal in geometry, having a right angle, right-angled or perpendicular.

orthohydrogen ⇨molecules of hydrogen in which the ⇨spins of the two atoms are parallel. See ⇨parahydrogen.

oscilloscope or *cathode-ray oscilloscope* (CRO) an instrument that allows one to see how electrical quantities vary with respect to each other, and also to time. The device operates by having the electrical charge pass through onto deflection plates that alter the way ⇨electrons in the cathode tube strike the screen to give a visible image.

osmosis the spontaneous diffusion of a solvent (often water) through a ⇨semipermeable membrane (that is, one that permits the passage of solvents and crystalloids but not of colloids). If two solutions of unequal concentrations are separated by a semipermeable membrane, there will be a net osmotic flow of solvent from the more dilute to the stronger solution, until the two solutions are of equal concentration. Osmosis is of fundamental importance in controlling the concentration of fluids within living cells. See also ⇨osmotic pressure.

osmotic pressure the pressure that must be applied to a solution so that it no longer takes up pure solvent (usually water) across a membrane that is permeable to solvent but not to solute (the dissolved substance). See also ⇨osmosis.

Otto cycle the cycle that powers the ⇨internal-combustion engine intake of fuel–air mixture, compression explosion at constant volume, expansion and exhaust of gases, then back to intake stroke.

Otto engine an ⇨internal-combustion engine that produces power by four strokes: intake, compression, ignition and expansion, followed by exhaust and further intake. This involves two revolutions of the ⇨crankshaft.

ovum an unfertilized egg cell (a female ⇨gamete) which, in sexual reproduction develops into a new individual after fertilization (with a male gamete to form a ⇨zygote).

oxidation any reaction in which oxygen is combined with another substance, hydrogen is removed from it, or in which an atom or group of atoms lose electrons. In the last case – an example of which is the change of a ferrous, or iron(II), ion (Fe^{2+}), to a ferric, or iron(III), ion (Fe^{3+}) – the electrons lost by the oxidized entity are taken up by another substance, which is thereby reduced; such combined oxidation and reduction reactions are called redox reactions. See also ⇨reduction.

oxidation state or *oxidation number* a value given to an element that represents the electrical charge on its atoms in a chemical compound; it equals the difference between the number of electrons associated with the element in the compound and the number associated with it in its pure form. A positive oxidation number indicates a relative electron deficiency; a negative number indicates a relative excess. An element's oxidation number is the same as its ⇨valency (although this latter is not given a positive or negative sign). In monatomic ions the oxidation number equals the electrical charge of the ion. In covalent compounds the shared electron pair is assigned to the atom with the greatest ⇨electronegativity. Some elements have the same oxidation number in different compounds whereas others, such as the ⇨transition elements, have different oxidation numbers depending on the precise compound concerned.

oxygenation the combination of (gaseous or dissolved) oxygen with a substance, such as with the blood in the lungs during respiration.

oxyhaemoglobin haemoglobin carrying oxygen. The ⇨oxygenation of haemoglobin takes place in the lungs (where the oxygen pressure is relatively high) and oxyhaemoglobin carries the oxygen in the bloodstream to the where the oxygen is released.

P

paedomorphosis the persistence of embryonic or juvenile characteristics into an adult form.

pair production the creation of an ⇨electron and antielectron (⇨positron) pair by the interaction between a high-energy particle or ⇨photon and the ⇨electrostatic field around a nucleus. It can also be used to describe the creation out of the vacuum state of particle–antiparticle pairs as allowed by Heisenberg's ⇨uncertainty principle.

Palaeozoic era of ⇨geological time 570–245 million years ago. It comprises the Cambrian, Ordovician, Silurian, Devonian, Carboniferous, and Permian periods. The Cambrian, Ordovician, and Silurian constitute the Lower or Early Palaeozoic; the Devonian, Carboniferous, and Permian make up the Upper or Late Palaeozoic. The era includes the evolution of hard-shelled multicellular life forms in the sea; the invasion of land by plants and animals; and the evolution of fish, amphibians, and early reptiles. The earliest identifiable fossils date from this era.

panchromatic film photographic film that is reasonably sensitive to all ⇨frequencies within the visible ⇨light spectrum.

Pangaea or *Pangea* single landmass, made up of all the present continents, believed to have existed between 250 and 200 million years ago; the rest of the Earth was covered by the Panthalassa ocean. Pangaea split into two landmasses – Laurasia in the north and ⇨Gondwanaland in the south – which subsequently broke up into several continents. These then drifted slowly to their present positions.

pangenesis an erroneous theory that stated that bodies in ⇨gemmules transported by the blood to a parent's reproductive cells represented invisible (but exact) copies of the rest of the organism. After fertilization and combination with the other parent's gemmules these bodies were supposed to develop and grow into the 'adult' forms.

papilla a small growth from the surface of a tissue, such as the papillae on the surface of the tongue.

parabola in plane geometry, an infinite open curve about a single line known as its axis. The equation for the general form is represented by:

$$y^2 = 2px$$

where p is a positive real number.

paraffin former name for the class of saturated hydrocarbons now called ⇨alkanes.

parahydrogen ⇨molecules of hydrogen in which the ⇨spins of the two atoms are opposed. See ⇨orthohydrogen

parallax angle subtended by the apparent difference in a star's position when viewed from the Earth either simultaneously from opposite sides of the planet, or half such an angle, measured after a gap of six months from opposite sides of the planet's orbit; the nearer the celestial body, the greater the parallax.

parallel and *parallelism* in Euclidean geometry, describes two (or more) lines in a plane that neither approach nor diverge from each other, however far extended, but remain at a constant distance; a perpendicular drawn to one crosses the other also at 90°. Such lines, according to Euclid, meet only at infinity. For the non-Euclidean analogue, however, see ⇨angle of parellelism.

parallelogram of forces a means of adding the effects of two nonparallel ⇨forces of given magnitude and direction to give their resultant effect in magnitude and direction.

paramagnetism a property of most elements and some compounds (but excluding the ferromagnetic substances – iron, cobalt, nickel and their alloys) whereby they are weakly magnetized by relatively strong magnetic fields. See also ⇨ferromagnetism.

parity a concept in ⇨quantum theory related to the mirror-symmetry of the functions describing mathematically the behaviour of the particles.

parsec distance at which a celestial body would subtend an angle of one second of arc from a base line of one astronomical unit (that is, the mean distance between the Earth and the Sun). One parsec is 3.26 light years.

parthenogenesis unsexual reproduction by multiplication of an ⇨ovum, in which an ovum develops into a new individual without being fertilized by a male ⇨gamete. The ovum is usually diploid (with paired ⇨chromosomes), and the offspring is usually identical to the parent.

partial differentiation form of ⇨differential calculus in which instead of a function $y = f(x)$ – which has one variable – a function of two or more variables is considered; for example, $z = f(x,y)$. Such functions represent a surface in three-dimensional space.

partitioning of an integer in number theory, breaking the integer down into its constituent parts in as many ways as possible; for example, the number 6 can be partitioned in three ways: $5 + 1$, $4 + 2$, and $3 + 3$. Each of these contributory numbers is known as a summand.

pascal the ⇨SI unit of ⇨pressure, defined as a ⇨force of one newton applied over an area of one square metre.

Pascal's pressure law ⇨pressure applied within a fluid is transmitted equally in all directions, the ⇨force per unit area being everywhere the same.

Pascal's principle a law of ⇨hydrostatics stating that the application of ⇨pressure to a fluid results in that pressure being equally transmitted throughout the fluid, in all directions.

pasteurization a method of treating a liquid, such as milk or wine, in which it is heated (to 62–65°C/144–149°F for 30 minutes or to 72°C/161°F for 15 seconds) and then quickly cooled. The heat kills microorganisms that do not form spores (such as *Salmonella* bacteria).

patent a government-endorsed document conferring on its holder the sole right to proceeds from the manufacture, use or sale of the invention covered by the patent

pathogen any disease-causing microorganism.

Pauli's exclusion principle a law of ⇨quantum theory that states that no two electrons in an uncharged atom can have the same set of ⇨quantum numbers.

Peano axioms axioms that formally introduce the properties of the positive whole numbers (originally devised, despite the term, by Julius Dedekind).

penumbra less than full shadow (umbra).

peptide any organic compound whose molecules are made up of ⇨amino acids joined by a peptide linkage (–NH–CO–) between the carboxyl group (–COOH) of one acid and the amino group (–NH$_2$) of the other. A long chain formed in this way is called a polypeptide. Proteins consist of one or more polypeptide chains cross-linked in various ways.

perfect number number that is equal to the sum of all its factors (except itself); for example, 6 is a perfect number, being equal to $1 + 2 + 3$.

perfusion the ⇨*in-vitro*-induced passage of blood or a nutrient fluid through the blood vessels of an organism to keep it supplied with oxygen and nutrients.

perihelion point representing the nearest to the Sun that a body approaches in an elliptical orbit.

periodic law the generalization that there is a recurring pattern in the properties of elements when they are arranged in order of increasing atomic number. The law is most apparent when the elements are arranged in the ⇨periodic table, in which the elements in each vertical column (group) show similar properties.

periodic table an arrangement of the chemical elements in order of their ⇨atomic numbers such that each vertical column contains a group of elements with similar properties, that is, the elements in each column demonstrate the ⇨periodic law. In the modern periodic table the elements in each group have the same number of electrons in their outer orbitals and therefore share the same ⇨valency, which accounts for the similarity of their chemical properties.

period–luminosity curve graph depicting the variation in ⇨luminosity of a ⇨Cepheid variable star with time. In general, the longer the period, the greater the luminosity. By measuring the period it is possible thus to derive an ⇨absolute magnitude; comparison of this with the star's observed (apparent) ⇨magnitude gives an indication of the distance.

peristalsis a directional involuntary muscular contraction that occurs in body tubes, especially those of the alimentary canal, to move fluid or semi-fluid contents. The rhythmic contraction is produced by smooth muscle fibres and controlled by autonomic nerves.

permanent magnet a ⇨magnet that retains its ⇨magnetism permanently, and not just when subject to some external ⇨energy source (as is the case with an ⇨electromagnet).

permeability the ratio of the ⇨magnetic flux density in a medium to the magnetizing force producing it.

Permian period of ⇨geological time 290–245 million years ago, the last period of the ⇨Palaeozoic era. Its end was marked by a significant change in marine life, including the extinction of many corals and trilobites. Deserts were widespread, and terrestrial amphibians and mammal-like reptiles flourished. Cone-bearing plants (gymnosperms) came to prominence.

permutation group ⇨group consisting of all the transformations (operations) permutating a fixed number of objects among themselves. Such groups were studied first by Evariste

Galois in connection with permutations of roots (solutions) of a polynomial equation.

perpetual motion the notion that a device can operate indefinitely, without using an external ⇨power source. One suggested example is the linking together of an ⇨electric motor and a ⇨dynamo. However, the impossibility of such everlasting motion is shown by the laws of ⇨thermodynamics.

persistence of vision the brief retention of the sensation of light by the brain, once the initial stimulus has been removed. Essential phenomenon for the success of televisual and cinematographic images.

perturbation apparent irregularity in an orbit, or occasionally in a star's ⇨proper motion, caused by the gravitational effects of a nearby celestial body.

pH a measure of the acidity or alkalinity of a solution in terms of its ⇨hydrogen ion concentration. The pH of a solution is given by $pH = \log_{10} (1/[H^+])$, where $[H^+]$ is the hydrogen ion concentration. Pure water is neutral and has a pH of 7. Solutions with a pH of less than 7 are acidic, those with a pH of greater than 7 are alkaline.

phagocyte a white blood cell that ingests foreign substances, such as bacteria and dead cells or tissue. Those that remove debris are called macrophages; bacteria are ingested by microphages.

phagocytosis the process by which ⇨phagocytes surround foreign particles (by an amoeboid movement), engulf and digest them. See also ⇨endocytosis.

Phanerozoic aeon in Earth history, consisting of the most recent 570 million years. It comprises the ⇨Palaeozoic, ⇨Mesozoic, and ⇨Cenozoic eras. The vast majority of fossils come from this aeon, owing to the evolution of hard shells and internal skeletons. The name means 'interval of well-displayed life'.

phase in astronomy, difference in the appearance of the Moon, in particular, but also of Mercury and Venus, caused by the Earth observer's seeing only a part of the body lit by the Sun.

phase in physics, the proportion of a cycle of a periodic motion completed by a given time, from a certain starting time. Often expressed in degrees.

phenol member of a class of ⇨aromatic compounds containing at least one hydroxyl group (–OH) attached directly to a benzene ring. The compound called phenol is the simplest of the phenols and has the formula C_6H_5OH.

phenotype the total characteristics (physical and behavioural) displayed by an organism, excluding its genetic make-up (see ⇨genotype).

philosophy of mathematics has three main aspects: the logical (in which mathematics is simply a branch of logic), the formalist (involving the study of the structure of objects and the property of symbols), and the intuitional (grounded on the basic premise of the possibility of constructing an infinite series of numbers).

phloem connective tissue that occurs in vascular plants which carries processed foods (such as sugars, minerals and proteins). It is characterized by the presence of sieve-tubes, parenchyma and companion cells.

phonograph US name for the ⇨gramophone.

phosphor a chemical substance capable of storing energy received by it in the form of ⇨electromagnetic radiation (such as light), and re-emitting it. Used in fluorescent tubes.

phosphorescence a type of luminescence in which a substance exposed to radiation emits light, this emission continuing after the radiation has been removed.

phosphorylation the chemical addition of a phosphate group ($-PO_3^{2-}$) to a (organic) molecule. One of the most important biochemical phosphorylations is the addition of phosphate to ADP to form the energy-rich ATP. Phosphorylation involving light (as in ⇨photosynthesis) is termed photophosphorylation.

photochemistry the study of chemical reactions that are initiated or accelerated by exposure to visible, ultraviolet or infrared radiation. ⇨Photosynthesis is an example of a biological photochemical process.

photoelectric cell a device that produces an electrical output if exposed to light. ⇨Selenium is often used in such devices.

photoelectric effect the transfer of energy from light rays falling onto a substance to the ⇨electrons within the substance. If the ⇨frequency, and hence the energy, of the radiation is high enough, it is possible to 'boil' electrons, then known as photoelectrons, out of the substance. Being charge-carriers, these electrons can constitute a photoelectric current.

photoelectric filtering means of measuring the ⇨astronomical colour index of a star, involving colour filters on photoelectric cells to define the colour index between two set wavelengths. The filters correspond to the ⇨UBV photometry system.

photoelectricity the phenomenon whereby certain materials, such as ⇨selenium, can produce electrical output if exposed to light.

photography the use of a system of lenses and emulsion to capture light images permanently.

photolysis the decomposition of a compound into smaller units as the result of exposure to light. Flash photolysis is an experimental technique used to study shortlived intermediates formed during many photochemical processes.

photometer an instrument used to measure the relative intensities of light sources. This is often achieved using the ⇨photoelectric effect.

photometry measurement of the ⇨magnitudes of celestial bodies, originally carried out by expertise of eye alone, but now generally making use of photographic or photoelectric apparatus.

photomultiplier device used in ⇨photometry for the amplification of light by the release and acceleration of electrons from a sensitive surface. The result is a measurable electric current that is proportional to the intensity of received radiation.

photon the ⇨quantum of ⇨electromagnetic radiation. A photon can be considered as an ⇨elementary particle with zero mass and charge, which 'carries' the ⇨electromagnetic force between charged particles.

photosphere the 'surface'of the Sun. Granular in appearance, it comprises spicules of gaseous ⇨helium at an average temperature of 6,000°C/11,000°F. Each spicule averages 7,000 km/4,400 mi in height, but lasts for less than 8 minutes. The ⇨sunspots are cooler depressions in the photosphere.

photosynthesis a process in which green plants transform light energy into chemical energy and manufacture ⇨carbohy-

drates and oxygen from a combination of water and carbon dioxide. The reactions take place within ⇨chlorophyll-containing ⇨chloroplasts and the energy is stored as ATP, formed by the ⇨phosphorylation of ADP. Photosynthesis takes place in stages (see ⇨light reaction; ⇨dark reaction).

phylogeny the history of the development of a species or other group of organisms.

piezoelectric effect the property of some crystals to produce an electric ⇨potential difference across their faces when subjected to ⇨pressure.

pigeon-hole principle see ⇨Schubfachprinzip.

pig iron another name for ⇨cast iron.

pinocytosis the ingestion of the contents of a vesicle by a cell (see ⇨endocytosis).

pion see ⇨meson.

Pioneer spaceprobes series of US spaceprobes the first nine of which concentrated predominantly on solar exploration and research. From then on, Pioneer probes have been sent to the outer planets of the Solar System.

piston a device that, when fitted into a chamber in which an explosion takes place, transmits the ⇨force of the explosion via its motion to a ⇨crankshaft, to provide motive ⇨power. Used in ⇨internal-combustion engines.

pitch in aeronautics, the movement of the nose of an aircraft in the vertical plane.

pitch in engineering, the distance between adjacent crests of a thread on a screw, measured parallel to the axis of the thread.

placenta in mammals (excluding egg-laying monotremes) a flattened structure in the uterus that allows interaction between the blood circulation of the mother and the developing foetus to permit respiration and nutrition. It is formed by fusion of the ⇨allantois and ⇨chorion with the wall of the uterus. In plants, the placenta is the (fleshy) part of the wall of the carpel to which ovules become attached.

Planck's quantum constant a constant (symbol h) with a value of about 6.63×10^{-34} joule seconds, which is the fundamental unit of ⇨spin angular momentum.

Planck's radiation law the energy of ⇨electromagnetic radiation of a certain ⇨frequency is given by the product of the frequency and ⇨Planck's constant, h.

plane in mathematics, either two-dimensional (as in plane geometry) or a two-dimensional (flat) surface of the 'thickness' of a point.

plane-polarized light light in which the electric and magnetic vibrations of the waves are restricted to a single plane, the plane of the magnetic vibration being at right angles to that of the electric one.

planetary nebula small, dense, hot star – neither a planet nor a ⇨nebula – that is surrounded by a spherical cloud of gas which it is shedding in the process of becoming a ⇨white dwarf.

planetoid another name for ⇨asteroid.

plankton a colony of plants and animals that live near the surface of the sea or fresh water. It is a vital food source for larger organisms.

plasma in astronomy and physics, matter in which the constituent atoms are in a state of ⇨ionization; that is, it comprises positively charged ⇨protons or atomic nuclei and negatively charged ⇨electrons moving freely.

plasma in biology, the liquid (noncellular) component of blood and lymph in which blood cells and platelets are suspended. It consists of water containing various dissolved substances, such as proteins, sugars and gases.

plasmolysis in a plant cell, the shrinkage of the cell contents away from the cell wall as a result of ⇨osmosis in which water diffuses out of the cell.

plastics substances that are organic, and stable in shape within certain temperature ranges, but can be moulded or extruded under certain conditions of temperature and pressure. Most plastics are ⇨polymers.

plate tectonics theory formulated in the 1960s to explain the phenomena of ⇨continental drift and ⇨seafloor spreading, and the formation of the major physical features of the Earth's surface. The Earth's outermost layer is regarded as a jigsaw of rigid major and minor plates up to 100 km/62 mi thick, which move relative to each other, probably under the influence of convection currents in the mantle beneath. Major landforms occur at the margins of the plates, where plates are colliding or moving apart – for example, volcanoes, fold mountains, ocean trenches, and ocean ridges.

platinum resistance thermometer a thermometer that uses the change in the electrical ⇨resistance of a platinum coil with temperature to allow measurements over a very wide range (over 1,300 degrees).

Platonic solids five regular three-dimensional polyhedra that can be bordered by other, congruent polyhedra (that is, that can pack with identical others into an overall solid shape). They are: the tetrahedron (a pyramid with base and three sides as equilateral triangles), the hexahedron (a cube), octahedron (an 8-faced solid), the dodecahedron (12-faced) and the icosahedron (20-faced). Each can be inscribed and circumscribed by concentric spheres. As two-dimensional figures in profile, they have similar properties.

pluteus in echinoderms, an advanced larval stage characterized by bilateral symmetry.

Pluto ninth major planet out from the Sun, discovered by Clyde Tombaugh in 1933.

pneumatics the study and production of devices that rely on air ⇨pressure for their operation.

point sets in geometry or topology, sets comprising some or all of the points of the space under study.

poisoning (of a catalyst) the reduction in effectiveness of a ⇨catalyst as a result of its being contaminated by a reactant or a product of the reaction it catalyses. Although in theory a catalyst is unaffected by the reaction it catalyses, in practice particles ('poisons') accumulate on the surface of the catalyst and reduce its effectiveness.

Poisson's ratio the ratio of the lateral ⇨strain to the longitudinal strain in a wire held under tension.

polar body in ⇨meiosis, any of three (haploid) egg nuclei that develop from a secondary oocyte (the fourth nucleus is the ⇨ovum); the three polar bodies degenerate.

polarimeter device that measures the ⇨polarization of any form of ⇨electromagnetic radiation, particularly light.

Polaris (Alpha Ursa Minoris) star visible to the naked eye that is at present nearest the North Pole (and therefore also known as the Pole Star). Very slightly variable, its apparent magnitude is 1.99, and its absolute magnitude is –4.6; it is about 680 light years distant.

polariscope an instrument that enables the rotation of the plane of polarization of ⇨polarized light to be determined.

polarization of light reduction of light, considered to travel in three-dimensional transverse waves (vibrating in all directions perpendicular to the direction in which it is travelling), to two dimensions. To achieve this a filter is used. The results may vary from a beam of light in which the waves vibrate in one plane only (plane-polarized light) to one in which the plane rotates but the amplitude is constant (circular polarization). Because light is also polarized by reflection, investigation of polarized light reflected from, for example, the lunar surface enables that surface to be analysed.

polar planimeter device invented by Jakob Amsler-Laffon to measure area on a curved surface; it could be used to determine Fourier coefficients and was thus particularly valuable to ship-builders and railway engineers.

polybasic describing an ⇨acid that has more than two atoms of replaceable acidic hydrogen in each of its molecules.

polyhedron (plural *polyhedra*) many-sided three-dimensional figure. About them, Leonhard Euler stated that the number of vertices v + the number of faces f = the number of edges E + 2, so long as the figures are regular or at least simple (otherwise the constant, 2, needs to be amended).

polymer a large molecule made up of up to many thousands of repeating units derived from a small number of simple molecules, called monomers. There are two main types: *addition polymers*, in which several identical monomer subunits link to form a polymer that has the same ⇨empirical formula as the monomer; and *condensation polymers*, in which the monomers are joined during a condensation reaction (with the elimination of water or other simple compounds) to form a polymer with a different empirical formula to that of the monomer. Copolymers are composed of two or more different types of monomers.

polymerization a reaction in which a number of relatively small molecules (monomers) combine to produce a much larger molecule (a ⇨polymer).

polysaccharide any naturally occurring large molecule consisting of many molecules of simple sugars (monosaccharides) joined together. Polysaccharides are one of the main types of ⇨carbohydrates and they play an important part in many biological processes; for example, starch is used as an energy store and cellulose is the main structural material of plants.

population genetics the study of the genetic make-up of a population, involving an analysis of the stability or fluctuation of various gene frequencies in an interbreeding group.

population I stars younger stars, generally formed towards the edge of a ⇨galaxy, of the dusty material in the spiral arms, including the heavy elements. The brightest of this population are hot, white stars.

population II stars older stars, generally formed towards the centre of a ⇨galaxy, containing few heavier elements. The brightest of this population are red giants.

porism old-fashioned term for a conclusion or hypothesis following directly upon a statement or proposition.

porosity the percentage of empty space existing within a material.

Portland cement a widely used building material comprising chalk or limestone mixed with clay or shale, burnt in a kiln after being finely broken up and mixed together.

positive feedback see ⇨feedback and ⇨amplification.

positron the equivalent of a negatively charged ⇨electron in an ordinary atom of matter, the positron in an atom of ⇨antimatter is positively charged and spins in an opposite direction.

potential difference (pd) the phenomenon that, when present in a ⇨conductor, such as wire, allows an electric ⇨current to flow. Defined as the ⇨work done to pass unit positive electric ⇨charge from one point to another between which a potential difference exists. The unit of potential difference is the ⇨volt.

potential energy the ⇨energy possessed by a body by virtue of its location within a gravitational ⇨field. For a mass m, a height h above the surface of the Earth, it may be taken as mgh, where g is the ⇨acceleration due to gravity.

power ordinarily, an indication of the number of times a number (or equivalent term) is multiplied by itself, expressed as a superior figure after the principal; for example, $2^1 = 2$, $2^2 = 2 \times 2$, $2^3 = 2 \times 2 \times 2$, and $2^n = 2 \times 2 \times \ldots n$ times. The power 0 represents unity; that is, $x^0 = 1$. Minus powers represent reciprocal positive powers; that is $2^{-1} = {}^1/_2$, $2^{-2} = ({}^1/_2)^2 = {}^1/_4$. Fractional powers represent roots; that is, $2^{1/2} = \sqrt{2}$ and $2^{4/3} = {}^3\sqrt{2^4}$.

power the rate of doing ⇨work. If a ⇨force acts over a certain distance over a certain time, then the power developed is the product of the force and the rate of change of distance. The unit is the watt, equal to one joule per second.

power series an infinite series of the general form:

$$a_0 + a_1x^1 + a_2x^2 + a_3x^3 + \ldots + a_nx^n + \ldots$$

This is called a power series in x. Examples of power series are the exponential series:

$$e^x = 1 + x + \frac{x^2}{2!} + \frac{x^3}{3!} + \ldots$$

the logarithmic series, and the series for the trigonometric functions (sine, cosine and tangent).

Poynting–Robertson effect the interaction of dust particles in interplanetary space with solar radiation causing a loss in orbital velocity of the dust around the Sun. This causes the dust particles to spiral into the Sun, if the effect is unopposed. However, under certain circumstances, ⇨radiation pressure is large enough to oppose the effect.

Precambrian in ⇨geological time, the time from the formation of Earth (4.6 billion years ago) up to 570 million years ago. Its boundary with the succeeding ⇨Cambrian period marks the time when animals first developed hard outer parts (exoskeletons) and so left abundant fossil remains. It comprises about 85% of geological time and is divided into two periods: the ⇨Archaean, in which no life existed, and the ⇨Proterozoic, in which there was life in some form.

precession the behaviour of a ⇨gyroscope when subjected to a ⇨force tending to alter the direction of its axis is to turn about an axis at right angles to that about which the force is applied.

precession of the equinoxes apparent movement per year of the two points in the sky representing the ⇨equinoxes: 50.26″ per year, also called the constant of precession. Precession is caused mainly by the gravitational effect of the Moon on the Earth's equatorial 'bulge'.

predicate calculus in logic, part of the theory of devising models involving deduction and the use of variables and negatives in systems of sentences; the term occurs commonly in ⇨mathematical logic.

pre-operational describing a stage of human development between the ages of two and seven years in which a child learns to imitate and acquires language to describe concrete objects.

pressure ⇨force per unit area. Expressed in ⇨newtons per square metre, otherwise known as ⇨pascals.

pressure, saturated vapour see ⇨saturated vapour pressure.

pressure, vapour see ⇨vapour pressure.

primary factors most reduced (lowest) form of numbers, which, when multiplied together, produce the principal numbers or expression.

prime number any whole integer of which the only whole-number factor less than itself is 1.

Principia Mathematica title of two works: the first (full title *Philosophiae Naturalis Principia Mathematica*), by Isaac Newton, published in 1687, was a compilation of mathematics combined with physics, and was a landmark in both disciplines; the second, by Bertrand Russell and Alfred North Whitehead, was an attempt to derive the foundations of mathematics, published 1910–13.

probability theory mathematical tool for assessing the likelihood of an event's occurrence. Its practical significance is its ability to predict outcomes often enough (as opposed to always). In general, the probability that n particular events will happen out of a total of m possible events is n/m. A certainty has a probability of 1; an impossibility has a probability of 0. Empirical probability is defined as the number of successful events divided by the total possible number of events.

One of the most remarkable discoveries in relation to probability theory is the central limit theorem, which defines a ⇨limit (as sample size increases) in terms of the ⇨normal distribution graph. See also the finull hypothesis, the ⇨chi-squared test and ⇨standard deviation.

Procyon major star in Canis Minor; its (apparent) magnitude is 0.37. It has an extremely dim ⇨companion, a ⇨white dwarf, that orbits every 40 years.

producer gas a gas used in ⇨furnaces and the generation of ⇨power formed by the partial combustion of coal, coke or anthracite in a blast of air and steam.

product the result of multiplying one quantity by another.

projective geometry form of two- and three-dimensional geometry concerned with the geometrical properties that remain constant (invariant) under projection, that is, extended on a single plane, or projected from one plane onto another. Perspective – two-dimensional representation of three-dimensional reality – uses the basic theory of projective geometry.

proof rigorously defined and complete demonstration using accepted and specific ⇨axioms of the correctness of a statement, law, or theorem. In mathematics, the expression of the proof may sometimes be even more useful and constructive than the original assertion, in that it may lend itself to further application.

proper motion actual speed and direction in space of a celestial body, measurable only in relation to other celestial bodies.

prophage the DNA of a nonvirulent ⇨bacteriophage that has become linked with the bacterial host's DNA and which is replicated with it. See ⇨lysogeny.

prophylaxis preventative medical treatment.

proprioceptive describing receptors (such as those in muscles) that respond to internal stimuli.

propulsion the source of motive ⇨force in a device.

protein any of a class of naturally occurring ⇨polymers composed of hundreds or thousands of ⇨amino acids (carboxylic acids that contain the amino group –NH₂) joined together by ⇨peptide linkages (–NH–CO–) between the carboxyl group of one acid and the amino group of the adjacent one. Proteins have very high molecular masses – between about 18,000 and 10 million – and a variety of complex molecular shapes – helical in wool, for example, and globular in haemoglobin. About 20 different amino acids occur in proteins and most protein molecules contain all of them; it is the sequence of the amino acids and the type of cross-linkages between their strands that give the individual proteins their specific properties. Proteins are fundamental to life; in addition to forming the main structural components in most animal cells, they are also involved (as ⇨enzymes) in every metabolic pathway.

protein synthesis the building-up of a ⇨protein, one ⇨amino acid at a time, which takes place at the ⇨ribosome of a cell. It involves the production of ⇨transfer RNA (which has complementary bases to those of ⇨messenger RNA) to determine the correct order of amino acids.

Proterozoic aeon of ⇨geological time, possible 3.5 billion to 570 million years ago, the second division of the ⇨Precambrian. It is defined as the time of simple life, since many rocks dating from this aeon show traces of biological activity, and some contain the fossils of bacteria and algae.

proton a stable subatomic particle with unit positive charge and a mass 1,836.12 times greater than that of the electron. The proton is a normal hydrogen nucleus (that is, a hydrogen ion) and is a constituent of every atomic nucleus. In a neutral atom the number of protons equals the number of electrons and is the ⇨atomic number of the element concerned.

proton–proton cycle process of ⇨nuclear fusion by which relatively cooler stars produce and radiate energy; hotter stars commonly achieve the same result by means of the ⇨carbon–nitrogen cycle.

proto-planet early stage in the formation of a planet according to the theory by which planetary systems evolve through the condensation of gas clouds surrounding a young star. The theory is not, however, generally accepted.

protoplasm the contents of a living cell. It consists of the ⇨nucleus and ⇨cytoplasm bounded by a membrane (the cell wall) which mediates the passage of substances into and out of the cell.

proto-star early stage in the formation of a star according to the theory by which ⇨globular clusters of stars evolve through the condensation of a vast (predominantly hydrogen) gas cloud, then to disperse and become ⇨population II stars. The theory, propounded by William McCrea, has gained some acceptance.

prototype an experimental version of a system, where the initial design can be tested and improved upon in the light of tests carried out.

protozoon a member of the subkingdom or phylum Protozoa, which includes the simplest and smallest organisms in the animal kingdom. Most are unicellular, and many reproduce asexually by some sort of fission or budding process; many live as parasites on other animals.

Prout's hypothesis the now defunct proposal that all the chemical ⇨elements were made from the combination of different numbers of hydrogen ⇨atoms.

psychoanalysis a technique used to treat mental disorders (such as ⇨neuroses) by bringing unconscious problems into the patient's conscious mind.

psychotic describing a severe mental disturbance or a person suffering from it, in which the individual loses touch with reality and suffers from delusions, hallucinations or mental confusion.

Ptolemaic model of the universe a ⇨geocentric model in which the Earth remained stationary as the other planets, the Sun, the Moon and the stars orbited it on their ⇨spheres. It was eventually replaced by the ⇨Copernican model.

pulley a simple ⇨machine consisting of a system of one or more grooved wheels that enables a ⇨mechanical advantage to be achieved, for example, to lift heavy objects.

pulmonary circulation the movement of deoxygenated blood from the right ventricle of the heart through the pulmonary arteries to the lungs, where it is oxygenated. The oxygenated blood is then transported by the pulmonary veins to the left atrium for circulation (via the left ventricle) round the body.

pulsar celestial source that emits pulses of energy at regular intervals, now identified with a ⇨neutron star.

pulsating nova or *recurrent nova* a ⇨variable star, probably not a true ⇨nova, in which the change between more and less luminous stages is extreme.

pulsating universe or *oscillating universe* theory that the universe constantly undergoes a ⇨Big Bang, expands, gradually slows and stops, contracts, and gradually accelerates once more to a Big Bang. Alternative theories include an ever expanding universe and the ⇨steady-state universe.

putrefaction the breakdown of organic matter, particularly by ⇨bacteria or fungi. The decomposition of ⇨proteins, especially, gives rise to bad-smelling amines and poisonous substances such as ptomaines and hydrogen sulphide.

pyrometer a device for measuring very high temperatures; for example, a ⇨platinum resistance thermometer.

pyrophoric finely powdered metals, or mixtures of metals and their oxides, which have a tendency to burst into flame, or oxidize when exposed to air.

Pythagoras' theorem states that in a right-angled triangle, the square on the hypotenuse is equal to the sum of the squares on the other two sides. Named after Pythagoras, who provided a proof for it, the effect was nevertheless known to the Babylonians up to a millenium earlier.

Q

quadrant type of early sextant, with which the observer's latitude could be calculated.

quadratic involving second powers, for example, x^2.

quadratic equation equation involving an expression of the second degree; the classic form is represented by:
$$x^2 + px + c = 0$$

quadratic extension extension of a ⇨field to a larger field by adjoining the root of a quadratic equation, for example, the set of all $p + q\sqrt{2}$ with both p and q as rational numbers is a quadratic extension of the field of rationals.

qualitative analysis the identification of the elements or groups of elements present in a sample, without taking into account the relative proportions of each of the constituents.

quantitative analysis the determination of the amount or proportion of one or more constituents of a sample.

quantum the name given to the smallest, indivisible unit of energy; every amount of energy is made up of an integral number of these 'packets'.

quantum chromodynamics (QCD) the branch of high-energy physics that seeks to understand the nature of the ⇨forces between ⇨quarks.

quantum electrodynamics (QED) the theory of the electromagnetic field that includes ⇨quantum concepts, enabling the behaviour of ⇨electrons in atoms to be predicted to very high accuracy.

quantum mechanics the system of mechanics based on the concept of the ⇨quantum, which must be used when dealing with atomic and subatomic systems.

quantum number integral or half-integral numbers defining the various properties of a system such as an ⇨atom in quantum mechanical terms. ⇨Angular momentum, ⇨spin and ⇨strangeness are examples.

quantum physics the branch of physics that takes into account the quantum nature of matter, energy and radiation.

quantum theory a general mathematical theory based on Max Planck's discovery that radiant energy is quantized, that is, emitted in discrete quanta ('packets') of energy. The original theory has been extended to interpret a wide range of physical phenomena; for example, quantum mechanics and wave mechanics are now extensively used to give quantitative accounts of the behaviour of small particles, such as electrons.

quark one of a set of particles that may be the fundamental constituents of all matter. By taking different permutations of quarks, and corresponding antiquarks, it appears possible to explain the various common characteristics of certain ⇨elementary particles (as expressed in the ⇨eightfold way) in terms of the quarks they contain. Despite strenuous efforts, it has not so far been possible to observe a quark in isolation, outside its host particle.

quartan fever a type of fluctuating malarial fever that peaks every fourth day.

quartic to the power of four; it is occasionally replaced by the word 'biquadratic'.

quasar a radio source, travelling at immense speed away from the Earth. It may be a type of ⇨galaxy; it may also be 'powered' by a ⇨black hole at its nucleus. Because of its speed and the ⇨Doppler effect, its ⇨ultraviolet radiation is seen on Earth as faint blue light.

Quaternary period of ⇨geological time that began 1.64 million years ago and is still in process. It is divided into the Pleistocene and Holocene epochs.

quaternary ammonium ion an ion in which the hydrogen atoms of the normal ammonium ion $(NH_4)^+$ have been replaced by organic ⇨alkyl or ⇨aryl radicals; it therefore has the formula $(NR_4)^+$.

quaternion one of an extended system of ⇨complex numbers, representable in the generalized form:
$$a + bi + cj + dk$$
where a, b, c and d are real numbers, and in which i, j and k are additional objects that multiply according to specific rules such that:
$$i^2 = j^2 = k^2 = -1, \ ij = -ji = k, \ jk = -kj = i \text{ and } ki = -ik = j$$

quintic to the power of five.

quotient the result after dividing one quantity by another.

R

racemic mixture a mixture of equal quantities of two ⇨enantiomorphs (isomers with mirror-image molecular structures). Because the ⇨optical activity of each component exactly cancels that of the other, the racemic mixture as a whole is optically inactive.

radar (acronym for *radio detection and ranging* the use of ⇨microwaves to locate such vehicles as ships, aircraft or missiles. Also used in navigation. The objects, on crossing the path of a radar beam, reflect the pulse of microwaves, the reflection being picked up by a suitable ⇨antenna.

radial pulsation periodic expansion and contraction of a star that may be merely an optical effect of ⇨recession.

radial velocity speed of a celestial body away from an observer (written as a minus quantity if the body is actually approaching).

radiant in astronomy, the point in the sky from which a ⇨meteor shower appears to emanate.

radiation emission of energy. See also ⇨electromagnetic radiation.

radiation of heat the transfer of thermal (heat) energy by ⇨electromagnetic waves of wavelengths lying within the ⇨infrared region of the electromagnetic spectrum, that is, between about 10^{-6} m and 10^{-3} m.

radiation pressure the ⇨pressure exerted on an object by the ⇨momentum of ⇨photons of radiation striking its surface, giving rise to a ⇨force in accordance with ⇨Newton's second law of motion.

radiative equilibrium in a star, represents an even process by which energy (heat) is transferred from the core to the outer surface without affecting the overall stability of the star.

radical in chemistry, an atom or group of atoms containing an unpaired electron. Most radicals are incapable of independent existence (for example, the ammonium radical $-NH_4$ and the organic ethyl radical $-C_2H_5$) but maintain their identity during reactions that affect the rest of the molecule of which they are a part.

radical in mathematics, operation principle in the extraction of roots; that is, square roots, cube roots, and so on.

radio the transmission of ⇨electromagnetic radiation of ⇨frequencies from about 10^4 hertz to 10^{11} hertz. Transmission of communication signals frequently involves the use of either ⇨amplitude or ⇨frequency-modulated carrier waves.

radioactive decay the process by which unstable atomic nuclei spontaneously lose some of their excess energy by disintegrating – accompanied by the emission of ⇨alpha particles, ⇨beta particles or ⇨gamma rays – into more stable nuclei. Several of the heavier radioactive elements, notably uranium and thorium, decay through a series of unstable ⇨radioisotopes before finally achieving a stable end-product, which is often an isotope of lead. In certain circumstances it is possible to induce artificial radioactivity, by bombarding the nuclei with particles such as ⇨neutrons.

radioactive fallout airborne radioactive material resulting from a natural phenomenon or from an artificial occurrence, such as the explosion of a nuclear bomb.

radioactivity the spontaneous disintegration undergone by certain unstable types of atomic nuclei; it is accompanied by the emission of ⇨alpha particles, ⇨beta particles or ⇨gamma rays. Alpha particles are emitted only by certain radioactive isotopes of the heavier elements (for example uranium-238); alpha emission results in the daughter nucleus having an atomic number two less than that of the parent nucleus and a mass number of four less. In beta emissions, the atomic number changes by one and the mass number remains the same. Gamma-ray emission almost invariably accompanies alpha or beta emission and is therefore associated with the changes in atomic mass and/or number that these latter produce, although gamma-ray emission by itself affects neither of these quantities because it is merely electromagnetic radiation. Natural radioactivity is the disintegration of naturally occurring radioactive isotopes (such as those of uranium, actinium and thorium). Artificial radioactive isotopes of many elements can be prepared by bombarding them with high-energy particles.

radioimmunoassay a technique that involves the radioactive labelling (using a radioactive isotope) of antigens, whose movements can then be followed using a radiation detector.

radio interferometer type of ⇨radio telescope that relies on the use of two or more aerials at a distance from each other to provide a combination of signals from one source which can be analysed by computer. Such an analysis results in a ⇨resolution that is considerably better than that of a parabolic dish aerial by itself because of the greater effective diameter.

radioisotope (contraction of *radioactive isotope*) any of several species (each with a different atomic mass) of the same chemical element whose atomic nuclei are unstable and radioactive. Every element has at least one radioisotope, although those of many elements can be obtained only by bombarding the elements with high-energy particles or in nuclear reactions.

radio map of the sky celestial chart depicting sources and intensities of radio emission.

radiometer a device for the detection (and also measurement) of radiant ⇨electromagnetic radiation.

radiomicrometer an extremely sensitive detector of ⇨infrared radiation.

radio scintillation the ⇨scintillation in received radio emission; the equivalent of 'twinkling' in visible light from the stars.

radio telescope nonoptical telescope (of various types) which, instead of focusing light received from a distant object, focuses radio signals onto a receiver-amplifier.

radiotherapy the use of ionizing radiation in medical treatment, particularly X-rays or radioactive isotopes.

radio waves waves of ⇨electromagnetic radiation with wavelengths in the range of tens of thousands of metres to a millimetre. Used in communication, such waves can be generated by the ⇨acceleration of electrons in electric circuits.

radius in astronomy, an old instrument for measuring the angular distance between two celestial objects.

radius vector in astronomy, an imaginary line connecting the centre of an orbiting body with the centre of the body (or point) that it is orbiting.

Raman effect in spectroscopy, the change in the ⇨wavelength of light scattered by molecules.

Ranger spaceprobes series of nine US spaceprobes, only the final three of which were successful. All were meant to photograph the surface of the Moon before crashing onto it.

Rankine cycle a cycle used in steam ⇨power plants, where water is introduced under ⇨pressure into a boiler, evaporation taking place followed by expansion without loss of heat to end in condensation and a repeat of the cycle.

rare earth element member of a series of 17 elements whose compounds often occur together naturally and exhibit markedly similar chemical properties. The series comprises the 15 ⇨lanthanide elements (atomic numbers 57 to 71 inclusive) plus scandium (atomic number 21) and yttrium (atomic number 39).

ratchet wheel a wheel with inclined teeth on its rim, used in gearing systems.

rational number fraction; the ratio of integral numbers.

Rayleigh–Jeans law a formula giving the intensity of ⇨black-body radiation at long ⇨wavelengths for a radiator at a certain temperature. It is thus an approximation to Planck's full formula for the black-body intensity based on ⇨quantum concepts.

RDX or *cyclonite* or *hexogen* a very powerful explosive compound.

reaction the equal and oppositely directed ⇨force generated by the application of force to a system or surface.

real number number that can be expressed using a decimal point. Such numbers include ⇨natural (or positive) ⇨integral numbers, ⇨algebraic numbers, ⇨rational numbers, ⇨irrational numbers and ⇨transcendental numbers. The ⇨complex numbers are a ⇨quadratic extension of the real numbers; see also ⇨imaginary numbers.

reaper a device used to collect crops such as wheat, while it grows in the field.

recession in astronomy, motion (increasing distance) away.

recessive describing a ⇨gene or genetic character that is the converse of ⇨dominant. A recessive gene has no phenotypic expression unless it is ⇨homozygous.

reciprocal innervation the reciprocal action of paired sets of nerves that have opposite effects on the same organ.

reciprocator a device that uses a ⇨piston moving cyclically within a system to carry out some task, such as pumping water.

rectification the conversion of ⇨alternating current (AC) into ⇨direct current (DC).

rectifier a device capable of converting ⇨alternating current to ⇨direct current by having a much greater resistance to electric current flowing in one direction than that flowing in the other, for example, a ⇨diode.

recursive in mathematics, a very general description of a function with ⇨natural numbers as arguments, corresponding to the intuitive notion of computability. The basic recursive functions are $x + y$, xy, $f(x, y, z, …) = x$, $g(x, y, z, …) = y$, and a restricted form of the taking of a minimum.

red giant large, highly luminous but relatively cool star that has reached a late stage in its life. It is running out of nuclear fuel and has accordingly expanded greatly and become less dense. Many also become ⇨variable stars of long periodicity. Its next evolutionary stage is to become a ⇨white dwarf, in developing into which the star has to cross the ⇨main sequence on the ⇨Hertzsprung–Russell diagram.

red shift the ⇨Doppler effect on the ⇨spectrum of a receding celestial body. Because of the recession there is a shift towards the red end of the spectrum.

Red Spot of Jupiter huge, generally reddish oval area within the belts and zones of Jupiter's southern hemisphere. Apparently a cyclonic swirl of atmospheric gases, it nevertheless seems to be a permanent feature despite continuous latitudinal drift and changes in size (at its largest it has a surface area greater than the Earth's). The reason for its coloration remains a mystery.

reduction any reaction in which oxygen is removed from a substance, hydrogen is added to it, or in which an atom or group of atoms gain electrons. In many reactions the reduction of one participant is accompanied by the ⇨oxidation of another; such combined oxidation and reduction reactions are called redox reactions.

reflecting telescope or *reflector* telescope that uses mirrors to magnify and focus an image onto an eyepiece.

reflection the reversing of the direction of ⇨electromagnetic radiation by a surface, according to specific laws.

reflection, laws of the incident ray of light, the reflected ray of light, and the normal to the reflecting surface all lie in the same plane. Secondly, the angle between the incident ray and the normal is the same as that between the reflected ray and the normal to the surface.

reflex action an involuntary response to a stimulus.

reflex arc the route of nervous impulses from the point of stimulation, along ⇨sensory nerve fibres to the central nervous system, and back along ⇨motor nerve fibres to the effector organ or muscle.

refracting telescope or *refractor* telescope that uses lenses to magnify and focus an image onto an eyepiece.

refraction deflection (or 'bending') of light – or any ray as it passes from one medium into another of greater or lesser density, representing a change in overall speed of the ray. ⇨Refracting telescopes rely on the refraction of light through lenses. The *refractive index* of a medium (such as glass) is a measure of the medium's 'bending' power.

refraction, laws of the incident ray, the refracted ray and the normal to the surface between the two media at the point of incidence lie in the same plane. In addition, ⇨Snell's law states that the ratio of the sines of the angles of incidence and refraction is equal to a constant. See ⇨refractive index.

refractive index the ratio of the speed of light in a vacuum to the speed of light in a given material.

refractive material materials able to stand very high temperature, such as brick and concrete.

refractometer a device for determining the ⇨refractive index of a substance.

refrigerator a device that uses an external ⇨power source to absorb heat at a low temperature, and reject it at a higher one. This is made possible through the evaporation of special ⇨fluids in the device's coils.

rejection in medicine, the destruction of transplanted tissues or organs by the immune system of the host.

relative atomic mass the mass of an atom relative to one-twelfth the mass of an atom of carbon-12. It depends on the number of protons and neutrons in the atom, the electrons having negligible mass.

relative molecular mass the mass of a molecule calculated relative to one-twelfth the mass of an atom of carbon-12. It is found by adding the relative atomic masses of the atoms which make up the molecule.

relativity, general theory final form of the theory of ⇨relativity formulated by Albert Einstein. In his earlier special theory (see ⇨relativity, special theory), Einstein had outlined a four-dimensional structure of space–time (much helped by his knowledge of ⇨Riemann geometry). The general theory further provided an understanding of gravitation as a curvature within that four-dimensional structure, of defined mathematical properties.

relativity, special theory theory formulated by Albert Einstein comprising two basic yet very original propositions: that a spaceship (or other enclosed vessel) travelling at uniform speed through space contains its own ⇨space–time continuum, and that a ray of light passes an observer at the speed of light no matter how (uniformly) fast nor in what direction the observer is travelling. One consequence of this theory was the equation of mass (m) with energy (E), formulated as:

$$E = mc^2$$

(where c is the speed of light). Ten years later, Einstein produced his general theory of relativity.

replication a process that results in the exact duplication of a biochemical molecule. In the replication of DNA, for example, the two helical strands part and by complementary base-pairing a new strand is built onto each old half, to form two identical double-stranded DNA molecules.

repressor substance a substance produced by a DNA regulator ⇨gene. When the repressor is inactivated by an inducer, the DNA structural genes are freed for the synthesis of ⇨messenger RNA.

resistance the property of a device to restrict the flow of current. Measured in ⇨ohms.

resolution of a telescope, the clarity of the final presentation to the observer (in image, radio picture or X-ray read-out).

resolving power the ability of an optical device to discern two closely spaced light sources as independent entities.

resonance in chemistry, the concept that, in certain molecules, the electrons involved in linking the constituent atoms are not associated with a specific bond (or bonds) but oscillate between atoms. Thus such molecules are not represented by a single valence-bond structure but by two or more alternative structures; the molecule 'resonates' between these alternative structures – that is, its structure is a resonance hybrid of the alternatives. The best known example of resonance is benzene, which, according to Kekulé's original formulation, resonates between two forms in which the double and single bonds are transposed. In Robinson's later modification, the six carbon atoms are linked by single bonds and the extra electrons are distributed equally among the carbon atoms, this being represented diagrammatically by a circle within the hexagonal carbon ring.

resonance hybrid see ⇨resonance.

respiration the metabolic process by which living organisms obtain energy by breaking down foodstuff molecules (for example carbohydrates) to simpler molecules (for example carbon dioxide and water). Most organisms require oxygen to respire (aerobic respiration) but some, chiefly microorganisms such as bacteria, can respire without oxygen (anaerobic respiration). The term respiration is also applied to the way in which oxygen is transported from the atmosphere to the individual cells.

resting potential the potential difference across a nerve or muscle membrane in the absence of a stimulus.

retaining wall a structure, usually of concrete, forming the wall of a structure sunk below the level of the ground, that is, holding back the outer volume of earth surrounding the structure.

reticulo-endothelial system a group of cells that exist in continual contact with the blood and lymph, that is, in the bone marrow, spleen, liver and lymph nodes. They ingest bacteria, other foreign particles and dead tissue, and aid tissue repair.

retort a glass vessel in the form of a glass bulb with an extended, fluted neck. Generally, any vessel in which a chemical reaction takes place.

retrograde in a backwards direction; in astronomy this means in a direction corresponding to east-to-west.

reversing layer lower ⇨chromosphere of the Sun, a comparatively cool region in which radiation at certain wavelengths is absorbed from the continuous spectrum emitted from the Sun's ⇨photosphere.

revolver a pistol with a revolving magazine carrying the bullets and their charges, enabling five or six shots to be taken in succession before reloading.

Reynold's number a dimensionless number formed from certain quantities of a ⇨liquid whose size dictates whether the velocity of flow is fast enough to cause ⇨turbulence within the liquid column.

rhesus factor a substance present in the blood of most people, who are termed rhesus positive (Rh^+); people lacking the factor are rhesus negative (Rh^-). Rh^- people do not possess ⇨antibodies specific to the rhesus ⇨antigen, but can acquire them by blood transfusion.

ribonucleic acid (RNA) a cellular substance found mainly in ⇨microsomes, mitochondria and nucleoli. It has long molecules, usually consisting of a single chain of ⇨nucleotides formed from the sugar ribose, a phosphate group, and one of the four nitrogenous bases that occur in DNA (except for thymine, which is replaced by uracil). RNA has an important function in ⇨protein synthesis; messenger RNA (mRNA) carries the coded information from the ⇨chromosome (DNA) to the ⇨ribosomes, where protein is manufactured. Ribosomes are composed mainly of RNA, as are many ⇨viruses.

ribosome a minute granular particle composed of protein and RNA and present in the ⇨cytoplasm of plant and animal cells. Often associated with ⇨endoplasmic reticulum, ribosomes occur singly or in clusters (called polyribosomes). They play an important part in ⇨protein synthesis.

Riemann geometry system of ⇨non-Euclidean geometry devised by Bernhard Riemann, developed primarily as ⇨elliptic geometry, but then extended to ⇨hyperbolic geometry.

Riemann hypothesis statement that has as yet not been proved (or disproved), that the ⇨zeta function takes the value zero in the right-half plane of the ⇨Argand diagram only for complex numbers of the form $^1/_2 + ia$, where $i = \sqrt{-1}$ and a is real.

Riemann space a non-Euclidean geometry, using n-dimensional coordinates $(x_1, ..., x_n)$ and calculating length according to the formula:

$$ds^2 = \sum g_{ij} dx^i dx^j$$

where ds is the limiting incremental length along a curve, dx^i is 'a limiting increment in the i coordinate', and i, j run through the values 1, 2, 3, ..., n.

Riemann surface in non-Euclidean geometry, a multi-connected many-sheeted ⇨surface that can be dissected by cross-cuts into a singly connected surface. Such a representation of a complex algebraic function is used to study the 'behaviour' of other complex functions as they are mapped conformally (transformed) onto it. A Riemann surface has been described as topologically equivalent to a box with holes in it.

right ascension directional bearing of a celestial body measured eastwards from a point on the ⇨celestial equator representing the ⇨vernal equinox. It is expressed in units of time corresponding to the time taken for the located position to reach the vernal equinox by means of the Earth's rotation.

ring a mathematical structure that constitutes a restricted form of ⇨field in which division might be unavailable.

rivet a fixing device that pins together two sheets of material, usually metal, by insertion through a hole through the sheets, and expansion of the heads by striking with a hammer or other tool.

RNA abbreviation for ⇨ribonucleic acid.

rocket an engine that is powered by fuel and a supply of oxygen that it carries within itself. The hot, expanding, exhaust gases are expelled in one direction to provide motion in the other direction.

roll movement of any aircraft about the axis running down the centre of the aircraft.

root either the solution to an (algebraic) equation in connection with equations involving a real or complex unknown, or, of a number, another number that when multiplied by itself to an indicated extent, provides the first number as product.

rose symmetrical curve represented by the equation:

$$(x^2 + y^2)^3 = 4a^2x^2y^2$$

where a is constant.

rotary engine or *Wankel engine* or *epitrochoidal engine* a form of ⇨internal-combustion engine in which an approximately triangular 'piston' is driven epicyclically in an elliptical combustion chamber containing the air–fuel mixture, ignited by a sparking plug.

rotation of a single body in space: spinning on an ⇨axis. Of a planetary system, rotation is generally planar in relation to the parent star. ⇨Galactic rotation, however, is usually differential.

Royal Greenwich Observatory primary national observatory in Great Britain, first sited at Greenwich in 1675, but in 1958 moved to Herstmonceux, Sussex, and in 1990, relocated to Cambridge. From the first, Directorship of the Observatory has entailed appointment as Astronomer Royal. It also operates the Northern Hemisphere Observatory in Las Palmas, the Canary Islands.

RR Lyrae variable type of short-period ⇨variable star. Spectrally classified as A to F giants, RR Lyrae variables were once called *cluster-Cepheids* (see ⇨Cepheid).

rudder part of the tailplane of an aircraft that moves about a vertical axis perpendicular to the wings, controlling the yaw motion.

Rydberg's constant a constant that relates the ⇨wavelength of spectral lines in hydrogen-like elements to the inverse-squares of integers.

S

saccharide a little-used general term for any ⇨carbohydrate.

salt a compound formed when one or more hydrogen atoms of an acid are replaced by a metal atom(s) or by an ⇨electropositive ion such as ammonium. Most salts are crystalline ionic compounds; soluble salts dissociate into ⇨ions in solution.

saltpetre or *potassium nitrate* or *nitre* a white crystalline substance that acts as a strong oxidizing ⇨element and is used in fertilizers and explosives.

saponification see ⇨ester.

satellite body orbiting a planet. Since 1957 the term has also been applied to artificial satellites; many astronomers make the distinction by calling natural satellites moons (and the Earth's natural satellite the Moon).

saturated vapour pressure the ⇨pressure exerted by a ⇨vapour that exists in equilibrium with its liquid.

Saturn sixth major planet out from the Sun. The most spectacular of the Solar System, it is circled by a series of concentric rings.

scalar a mathematical representation of a quantity requiring just its magnitude for its complete definition. Examples of scalar quantities are temperature, speed, and mass. See also ⇨vector space.

Schlieren photography a means of photographing turbulences in fast-moving ⇨fluids, through the change of ⇨density and ⇨refractive index that such turbulence produces.

Schmidt camera telescopic camera incorporating an internal corrective lens or plate that compensates for optical defects and chromatic faults in the main mirror. The system was invented by Bernhard Schmidt.

Schrödinger's equation an equation that considers the ⇨electron in terms of a wave of probability, and which enables the behaviour of the electron in atoms and electric potentials to be calculated, and also the ⇨spectra of atoms to be predicted. It is the basis of ⇨wave mechanics.

Schubfachprinzip or *pigeon-hole principle* principle that states that if in *n* boxes one distributes more than *n* objects, at least one box must contain more than one object. The principle is used in the logic of number theory.

Schwann cell in vertebrates, the neurilemma cell of myelinated peripheral nerve fibres, important in the manufacture of ⇨myelin. On myelin-coated fibres a Schwann cell occurs between each pair of adjacent nodes.

scintillation in radioastronomy, a rapid oscillation in the detected intensity of radiation emitted by stellar radio sources, caused by disturbances in ionized gas at some point between the source and the Earth's surface (usually in the Earth's own upper atmosphere).

screw a cylinder or cone onto which has been cut a helical thread.

scuba (acronym for *s*elf-*c*ontained *u*nderwater *b*reathing *a*pparatus) another name for an ⇨aqualung.

seafloor spreading growth of the ocean ⇨crust outwards (sideways) from ocean ridges. The concept of seafloor spreading has been combined with that of continental drift and incorporated into ⇨plate tectonics.

secant (sec) see ⇨trigonometric functions.

second order of a differential equation, involving only first and second ⇨derivatives. The term is occasionally used in algebraic contexts to mean 'of the second degree', that is, involving expressions raised to at most the power of two (squares).

secular in astronomy, gradual, taking ⇨aeons to accomplish.

secular acceleration of the Moon, of the Sun apparent acceleration of the Moon and Sun across the sky, caused by extremely gradual reduction in speed of the Earth's rotation (one 50-millionth of a second per day).

sedimentary rock rock formed by the accumulation and cementation of deposits that have been laid down by water, wind, ice, or gravity. Sedimentary rocks cover more than two-thirds of the Earth's surface and comprise three major categories: clastic, chemically precipitated, and organic (or biogenic). Clastic sediments are the largest group and are composed of fragments of pre-existing rocks; they include clays, sands, and gravels.

Chemical precipitates include some limestones and evaporated deposits such as gypsum and halite (rock salt). Coal, oil shale, and limestone made of fossil material are examples of organic sedimentary rocks.

Seebeck effect the ability of some metals, when joined, to give rise to an electric ⇨current when the opposite ends of the wires are kept at different temperatures.

selenium a nonmetallic ⇨element used in light-operated devices because of its ability to conduct ⇨electricity when exposed to ⇨light.

self-induction ⇨current carried in a wire produces a ⇨magnetic field that cuts the wire itself. Any change in the current produces a change in the magnetic field, which produces an electromotive force (emf) by 'self-induction' that resists the change in current.

self-pollination the pollination of a plant by itself, whether intra- or interfloral.

semiconductor an electrical conducting material whose resistance decreases with temperature and the presence of impurities. Germanium and silicon have atomic structures permitting them to behave in this way, as do a variety of other metals. Such materials are used in ⇨electronic devices, such as the ⇨transistor.

semipermeable membrane a membrane that allows certain substances in solution, such as ⇨crystalloids, to pass through it but is impervious to others, such as ⇨colloids. Semi-permeable membranes are used in ⇨dialysis.

sensori-motor phase the first stage in human mental development from birth to about two years of age in which reflex actions lead to an awareness of the permanence of objects.

sensory nerve an afferent nerve of the peripheral nervous system, made up of sensory ⇨neurons, which carries impulses to the central nervous system.

sepsis the invasion of tissue by pathogenic microorganisms and their products. Sepsis involving bacteria throughout the bloodstream is commonly termed blood poisoning.

sequence a list of mathematical objects indexed by the natural numbers, following one another in some defined relationship (but with no mathematical operation implied). Sequences are said to increase (to higher values) or decrease (to lower), and may be finite (if the list terminates), convergent (to a limit) or divergent. If the values increase or decrease along the sequence, the sequence is said to be monotonic or monotone.

series the sum of a list that constitutes a ⇨sequence. Series may be represented in an abbreviated form using the summation sign – for example, the series $a_1 + a_2 + a_3 + \ldots + a_n$ to infinity may be represented as:

$$\sum_{k=1}^{\infty} a_k$$

The word 'series' is for historical reasons occasionally misused to mean 'sequence', as in the Fibonacci series. Strictly speaking, however, a series is the limit of the sequence of partial sums.

servo mechanism a mechanism that uses relatively low ⇨power to control the behaviour of a much larger output device in a proportionate way.

set rigorous manifestation of the notion of collecting together various objects into one new entity if they all possess some property of interest – for instance, *P*. The resultant object is referred to as the set of all objects possessing *P*, commonly expressed as $\{x : x \text{ has property } P\}$. The description $x \in S$ means that *x* belongs to the set *S*. Difficulties may arise if restrictions are not placed on the collecting process, although the process can be continued beyond the finite numbers, giving 'transfinite' numbers characterized by the fact that each consists of all the numbers preceding it. Because these numbers may be used as a standard for indexing purposes, they are called ordinal numbers. Cardinal numbers are those ordinals whose members cannot be indexed by means of a smaller set of ordinals, and the cardinality of a set is the smallest ordinal that can be used to index a set. Its existence is guaranteed by the ⇨axiom of choice. Set theory provides a foundation for mathematics – and is the basis for most modern elementary mathematical education – yet is known to be incomplete (so that, for example, the cardinality of the set of real numbers may be assumed to be either a large or small order of infinity). Two sets are said to be disjoint if they contain no common elements although their elements are part of the same ⇨universal set. A null set has no members.

sextic to the power of six.

Seyfert galaxy galaxy with a small, bright nucleus, comparatively faint spiral arms, and broad spectral emission lines. Some are strong radio sources; all are subject to violent explosions of gas.

shooting star another name for a⇨meteor or ⇨meteorite.

shrapnel strictly, an artillery shell filled with small spheres with an explosive charge gives rise to shrapnel. It now also covers general fragments of metal produced from an exploding device.

shuttle a device for carrying thread on a ⇨loom.

sidereal in astronomy, relating to the period of time based on the apparent rotation of the stars, and therefore equivalent to the rotation of the body from which the observation is made. Thus on Earth a sidereal year is 365.256 times the sidereal day of 23 hours, 56 minutes and 4 seconds.

Siemens–Martin process another name for the ⇨open-hearth process

sigma (Σ) the ⇨summation symbol.

sigma bond a type of chemical bond in which an electron pair (regarded as being shared by the two atoms involved in the bond) occupies a molecular orbital situated between the two atoms; the orbital is located along a hypothetical line linking the atoms' nuclei. See also ⇨hybridization.

silicone any organosilicon polymer containing the SiR_2O group, in which R is an ⇨alkyl or ⇨aryl radical. The silicones are heat stable, chemically inert substances used as lubricants, hydraulic fluids, water repellants and synthetic rubbers.

Silurian period of ⇨geological time 439–409 million years ago, the third period of the ⇨Palaeozoic era. Silurian sediments are mostly marine and consist of shales and limestone. Luxuriant reefs were built by coral-like organisms. The first land plants began to evolve during this period, and there were many ostracoderms (armoured jawless fishes). The first jawed fishes (called acanthodians) also appeared.

silver nitrate a white, crystalline substance used in chemical analysis and inks.

simplex method in linear computer-programming, an algorithm designed to find the optimum solution in a finite number of steps.

sine (sin) see ⇨trigonometric functions.

single phase electrical ⇨power transmission involving a single sinusoidally varying ⇨potential difference.

singularity in astrophysics, the point in ⇨space–time at which the known laws of physics break down. A singularity is predicted to exist at the centre of a black hole, where infinite gravitational forces compress the infalling mass of a collapsing star to infinite density. It is also thought, according to the Big Bang model of the origin of the universe, to be the point from which the expansion of the universe began.

Sirius (Alpha Canis Majoris) the Dog Star, brightest star in the northern sky, with a luminosity 26 times that of the Sun, a ⇨magnitude of –1.42 and an ⇨absolute magnitude of –1.45. It is 8.7 light years away and has a companion star called Sirius B or the Pup.

SI units (French *Système International d'Unités*) standard system of scientific units used by scientists worldwide. Originally proposed in 1960, it is based on seven base units: the metre (m) for length, kilogram (kg) for mass, second (s) for time, ampere (A) for electrical current, kelvin (K) for temperature, mole (mol) for amount of substance, and candela (cd) for luminosity.

smelting the extraction of a metal from its ⇨ore by heat. The process usually involves several steps in order to refine the final product. See, for example, ⇨open-hearth process and ⇨Bessemer process.

smoke a suspension of a solid in a gas; the solid is in the form of extremely small particles and the smoke may be a ⇨colloid.

Snell's law see ⇨refraction, laws of.

soap an alkali metal salt of a high molecular mass ⇨fatty acid, typically the sodium or potassium salt of stearic or palmitic acid. Soaps are made by the ⇨hydrolysis (saponification) of fats (glyceryl ⇨esters of fatty acids) using caustic soda or potash, yielding glycerol as a by-product.

sodium hyposulphite see ⇨sodium thiosulphate.

sodium pump a hypothetical mechanism that maintains the asymmetry of the ionic (concentration) balance across a nerve cell membrane, reflected in the cell's ⇨resting potential.

sodium theory a theory which proposes that the excitation of a nerve results from momentary changes in the selective permeability of a nerve cell membrane, which admits sodium and chloride ions into the cell and allows potassium ions to diffuse out. The ion movements briefly reverse the polarization of the membrane, resulting in an ⇨action potential which constitutes the nerve impulse.

sodium thiosulphate correct chemical name for hypo and sodium hyposulphite, the substance used to fix photographic images after developing.

soft iron a form of iron with a low carbon content that does not retain ⇨magnetism once the current in a ⇨coil surrounding it has been removed.

soil mechanics a branch of civil engineering in which the properties (such as ⇨density, ⇨porosity) of soil at a site is determined.

solar of the Sun.

solar constant mean radiation received from the Sun at the top level of Earth's atmosphere: 1.95 calories per sq cm per minute.

solar energy is produced by ⇨nuclear fusion and comprises almost entirely ⇨electromagnetic radiation (particularly in the form of light and heat); particles are also radiated forming the ⇨solar wind.

solar flare sudden and dramatic release of a huge burst of ⇨solar energy through a break in the Sun's ⇨chromosphere in the region of a ⇨sunspot. Effects on Earth include ⇨auroras, magnetic storms and radio interference.

solar parallax the ⇨parallax of the Sun, now measured as 8.794″.

solar prominence mass of hot hydrogen rising from the Sun's ⇨chromosphere, best observed during a total ⇨eclipse. Eruptive prominences are violent in force and may reach heights of 2,000,000 km/1,243,000 mi; quiescent prominences are relatively pacific but may last for months.

solar rotation is differential, the equatorial rotation taking less time than the polar by up to 9.4 Earth-days.

Solar System system of bodies that orbit our Sun: the planets and their moons, asteroids, and comets.

solar wind stream of charged particles flowing from the Sun at a speed of between 300 km/200 mi per second and 1,000 km/600 mi per second . It is the effects of the solar wind that produce ⇨auroras in the Earth's upper atmosphere, that cause the tails of comets to stream back from the Sun, and that distort the symmetry of planetary ⇨magnetospheres.

solder an ⇨alloy that is heated and used to join two metals together.

solenoid a device consisting of a series of wires wound around a cylinder, which produces a magnetic field within the cylinder when a ⇨current flows through the wire windings. The intensity of the magnetic field depends directly on the number of turns of the wire.

solid the state of matter where the strength of the intermolecular and atomic ⇨forces is such that there is no translational motion within the substance. The ⇨molecules do, however, vibrate about their average positions in the ⇨crystal lattice.

solid-state physics the study of materials in the solid state, investigating the magnetic, thermal and electrical properties, for example.

solstice one of the two points on the ⇨ecliptic at which the Sun appears to be farthest away from the ⇨celestial equator (representing therefore mid-summer or mid-winter).

soluble group notion introduced to extend theorems concerning commutative groups to a wider class of groups that, in an intuitive sense, can be constructed out of commutative 'pieces'. Given a group G, the commutator subgroup C is introduced (generated by elements of the form $xyx^{-1}y^{-1}$); G can then be 'collapsed' to an ⇨Abelian group by a process that reduces C to the identity element; C is then collapsed in an identical fashion, and the process continually repeated. If the process, through a finite number of steps, eventually leads to a last group that is already commutative, G is said to be soluble.

solute the dissolved substance in a ⇨solution, the liquid part of which is the ⇨solvent.

solution a homogeneous mixture of two or more substances, most commonly a gas or solid (the ⇨solute) in a liquid (the ⇨solvent). Liquids dissolved in liquids (that is, any miscible liquids) and solids dissolved in solids (as in most alloys) are strictly also solutions. The components of a solution can be separated by physical means, such as evaporation and crystallization (solid in liquid), heating (gas in liquid) or distillation (liquid in liquid).

solvent the liquid part (or major component) of a ⇨solution, the other component of which is the ⇨solute. In solid solutions (as in most alloys) the solvent is taken to be the major component.

somatic cell any cell in an organism, excluding the reproductive cells.

sound ⇨pressure waves occurring in a medium such as air. Unlike ⇨electromagnetism, the waves are ⇨longitudinal, rather than ⇨transverse.

space collection of mathematical objects (referred to as points) with an associated structure resembling (or analogous to) the properties of the space of everyday experience.

space constant term characterizing the line in ⇨non-Euclidean geometry that relates the ⇨angle of parallelism of two lines – a concept formulated by János Bolyai.

space–time continuum Einsteinian concept of the universe in accordance with his theories of ⇨relativity; four-dimensional actuality, in which any anomaly is known as a ⇨singularity.

spar a beam running the length of a wing or tail-plane.

speciation the formation and development of species.

specific heat capacity the amount of heat energy required to raise the temperature of unit mass (for example a kilogram) of a substance through one degree of temperature.

spectral of a ⇨spectrum.

spectral classification commonly, the system devised by Annie Cannon combining the perceived colour of a star with its spectral characteristics. Very generally, of the overall sequence O B A F G K M R N S, stars in the group O B A are white or blue and display increasing characteristics of the presence of hydrogen; those in F G are yellow and show increasing calcium; those in K are orange and strongly metallic; and those in M R N S are red and indicate titanium oxide through carbon to zirconium oxide bands. The groups are numerically subdivided, according to other characteristics, and there are further small classes for very unusual categories of star. Different methods of classification exist but are not in such common use.

spectral line dark line visible in an absorption ⇨spectrum, or one of the bright lines that make up an emission spectrum. Spectral lines are caused by the transference of an ⇨electron in an atom from one energy level to another; strong lines are produced at levels at which such transference occurs easily, weak where it occurs with difficulty. ⇨Ionization of certain elements can affect such transferences and cause problems in spectral analysis.

spectroheliograph device with which spectra of the various regions of the Sun are obtained and photographed.

spectroscope an instrument that produces a spectrum for study or analysis. An object that produces radiation, such as a heated substance, forms an emission spectrum (see also ⇨absorption spectrum). Elements have characteristic spectra and spectroscopy is used in chemical analysis to identify the elements in a substance or mixture. Molecules or their constituent atoms or components of atoms can be made to absorb various types of energy in a characteristic way and give rise to such analytical techniques as infrared, ultraviolet, X-ray and nuclear magnetic resonance spectroscopy. See also ⇨mass spectroscope.

spectroscopy the use of a ⇨spectroscope to investigate the chemical composition of materials, and their state.

spectrum (plural *spectra*) generally, the range of colours (representing different ⇨wavelengths) that a beam of white light is composed of (a continuous spectrum); particularly, in astronomy, however, the characteristic bright lines (emission spectrum) or dark lines (absorption spectrum) produced by individual elements in the light source. By analysis of such lines, the composition of the light source may be determined.

speed the rate of change of distance with time, expressed in units such as kilometres per hour, metres per second, or miles per hour, feet per second.

speed of light see ⇨light, speed of.

spermatozoon or *sperm* a male germ cell (⇨gamete). It possesses a flat oval head (containing the nucleus), a middle piece (containing ⇨mitochondria), and a long tail of cytoplasm; it moves using a whiplike motion of the tail. At fertilization, the nucleus of the sperm fuses with that of an ⇨ovum to form a ⇨zygote.

sphere concept probably older than the ancient Greeks, in which the Sun, Moon, planets and the stars were thought to orbit the Earth, travelling on their own crystalline but – except for that of the stars – transparent spheres.

spherical aberration an optical error occurring when a lens or curved mirror does not bring all the incident rays of light to a sharp focus. See also ⇨aberration.

spherical collapse initial stage in the collapse of a star, followed by gravitational collapse and finally ⇨singularity.

spherical geometry system of non-Euclidean geometry devised by Bernhard Riemann as an extension of ⇨elliptic geometry and comprising two-dimensional geometry as effected on the outer surface of sphere.

spin a ⇨quantum number that is related to the amount of ⇨angular momentum, which is quantized in ⇨quantum mechanics in units of ⇨Planck's constant divided by 2π, that a particle has intrinsically, as opposed, for example, to orbital angular momentum. It is used to classify particles into various categories: all ⇨fermions have half-integer spin (for example the ⇨electron, with spin $\pm\frac{1}{2}$), all ⇨mesons have integral spin (for example pions, with spin 0).

spin–orbit interaction an interaction between ⇨electrons orbiting a ⇨nucleus in an ⇨atom that arises from the magnetic field produced by the nucleus interacting with the spinning electron is to split the individual ⇨spectral lines in a ⇨spectrum into a number of components.

spiral nebula a spiral ⇨galaxy – not really a ⇨nebula at all (although many do appear nebulous).

spirillium a bacterium with a spiral shape (see ⇨bacteria).

spirochaete a bacterium with a flexible spiral shape (see ⇨bacteria).

spontaneous generation or *abiogenesis* a concept, now discredited, which proposed that living matter can originate from nonliving matter.

spore a (usually) unicellular reproductive cell in ⇨bacteria, plants and some ⇨protozoa. Often spores are released in enormous numbers to ensure that at least some of them survive. Bacterial spores are extremely hardy and can survive prolonged exposure to extreme temperatures.

sporozoite a ⇨protozoon of the class Sporozoa, such as the malarial parasite ⇨Plasmodium.

Sputnik 1 first artificial Earth satellite, launched by the Soviet Union on 4 October 1957.

squaring the circle ancient Greek problem in geometrical construction, to describe a square of exactly the same area as a given circle, using ruler and compass only. Ferdinand von Lindemann established that π was a ⇨transcendental number; unable thus to be the root of an equation, it cannot be constructed by ruler and compass – and the problem is therefore not solvable.

squid (acronym for *s*uperconducting *q*uantum *i*nterference *d*evice) device that makes use of the ⇨Josephson effect to produce a highly sensitive magnetic field detection system, or an ultra-fast switching device for use in computers.

staining the selective pigmentation of microscopic objects and tissues so that they or their structure may be more easily seen.

standard deviation parameter in ⇨probability theory, a numerical assessment (symbol σ) of the amount by which an event is likely to deviate from the expected. It is measured as the square root of the average (either summed or integrated) of the squares of all possible deviations.

standard gauge the distance between the steel rails on the majority of the world's railways; 4 ft 8 in/1.435 m.

staphylococcus a bacterium that takes the form of irregular clusters of spherical cocci (see ⇨bacteria).

Stark effect the splitting of ⇨spectral lines into a number of components by a strong electric field.

starlight energy (seen as light) produced by a star through ⇨nuclear fusion.

statics the study of the ⇨forces that act on a system, when those forces are in ⇨equilibrium.

statistical mechanics the science that deals with the properties and behaviour of atoms and molecules using mathematical analysis applied to the parameters that describe atoms and their component particles. It is an extension of ⇨quantum theory.

steady-state theory proposition, now largely discredited, that the mass and density of the universe is maintained at a constant by continual creation of matter. It was put forward by Fred Hoyle, Thomas Gold and Hermann Bondi as an alternative to the now generally accepted ⇨Big Bang model. Although the later discovery of radio emissions from stars, ⇨background radiation and ⇨helium abundance led to its discrediting, the theory's explanation of the creation of elements within stars is universally accepted.

steam turbine a ⇨turbine that is powered by a jet of high-⇨pressure steam.

steel iron containing 0.1–1.5% carbon impurities. Produced from iron ⇨ore using the ⇨open-hearth and ⇨Bessemer processes.

Stefan–Boltzmann's law the relationship between the total energy E emitted over all ⇨wavelengths by a ⇨black body at temperature T per second per unit area, which is given by:

$$E = \sigma T^4$$

where σ is Stefan's constant.

stellar of a star, of the stars.

stereochemistry the branch of chemistry that is concerned with a study of the shapes of molecules.

stereoisomer one of two (or more) isomers that have the same molecular structure but which differ in the spatial arrangement of their atoms.

stereoisomerism a type of ⇨isomerism in which two or more substances differ only in the way that the atoms of their molecules are oriented in space.

stereoscope a device that can produce the effect of three-dimensional images using only two-dimensional images, using the human brain to carry out the merging necessary.

steroid any of a class of naturally occurring organic compounds that share a common basic structure based on the phenanthrene molecule and typified by the substance cholesterol. Steroids include such important compounds as the sex hormones and various plant poisons.

STOL see ⇨V/STOL.

stone the common name for a calculus, a hard accretion of organic or inorganic salts that precipitate and grow in the kidneys, urinary tract or gall bladder. Calculi are often associated with, or cause, infection in the organs concerned.

strain the dimensional change produced by a ⇨stress, divided by the original dimension. Thus for a wire to which a weight is applied at one end, the strain is the difference between the stretched and unstretched lengths, divided by the original unstretched length.

strangeness a ⇨quantum number assigned to certain unstable ⇨elementary particles that decay much more slowly than was originally expected. Stable particles, such as ⇨protons, have a strangeness quantum number of zero. Others, such as the ⇨hyperons, have nonzero strangeness quantum numbers.

stratigraphy branch of geology that deals with the sequence

of formation of ⇨sedimentary rock layers and the conditions under which they were formed. Stratigraphy in the interpretation of archaeological excavations provides a relative chronology for the levels and the artefacts within them. The basic principle of superimposition establishes that upper layers or deposits have accumulated later in time than the lower ones.

streamline a line in a fluid such that the ⇨tangent to it at every point gives the direction of flow, and its ⇨speed, at any instant.

streptococcus a bacterium that takes the form of a chain of spherical cocci (see ⇨bacteria).

stress a ⇨force applied over a given area that produces a ⇨strain within the material directly proportional to it (within certain ranges of stress). Stresses are usually measured in ⇨newtons per square metre (⇨pascals).

Strömgren sphere or *H II zone* zone of ionized hydrogen gas that surrounds hot stars embedded in interstellar gas clouds.

strong nuclear force the strongest interaction in nature, and the one that binds the ⇨protons and ⇨neutrons together in the ⇨nucleus despite the mutual ⇨electrostatic repulsion positively charged protons feel for each other. The interaction is very short range, being essentially zero outside the nucleus, and is related to the exchange of pi- ⇨mesons between protons and neutrons.

subarachnoid describing the fluid-filled region between the arachnoid membrane and the piamater membrane which surround the brain and spinal cord.

subset in ⇨set theory, set that is completely contained in another set. It is generally written using the symbol $\subset$, for example, $C \subset D$ (the set C is contained in the set D).

substitution reaction in organic chemistry, a general type of reaction in which a substance is formed by substituting a new atom or group of atoms for an atom or group in an existing compound (as opposed to an ⇨addition reaction).

substrate in biochemistry, a substance on which an ⇨enzyme acts.

sugar any mono-, di- or polysaccharide (all of which are ⇨carbohydrates) that forms crystals and dissolves in water; solutions of sugars have a sweet taste. The common sugar extracted from sugar cane or sugar beet is sucrose, a disaccharide.

sum the result of adding one quantity and another.

summation symbol sign (Σ) representing the sum taken over all instances that accompany the sign, indicated above and below it. Thus:

$$\sum_{i=1}^{n} a_i$$

means $a_1 + a_2 + a_3 + ... + a_n$. The symbol is the capital form of the Greek letter sigma.

sunspot comparatively dark spot on the Sun's photosphere, commonly one of a (not always obvious) group of two. The centre of a vast electrostatic field and a magnetic field of a single polarity (up to 4,000 ⇨gauss), a sunspot represents a comparatively cool depression (at a temperature of approximately 4,500°C/8,100°F). Sunspots occur in cycles of about 11 Earth-years in period, although their individual duration – a matter of Earth-days only – is affected by the ⇨differential rotation of the Sun; they tend to form at high latitudes and drift towards the solar equator. They are also sources of strong ultra-shortwave radio emissions.

supercharger in ➪internal-combustion engines, the use of a compressor to supply air or fuel–air mixtures at a high ➪pressure to the ➪piston cylinders; in aero engines, a device to maintain ground-level pressures in the engine inlet pipe when flying at high altitude.

supercharging heating of a liquid above its boiling point.

supercluster unusually large cluster of ➪galaxies; many appear to occur in a definite band across the sky.

superconductor a metal (such as tin) that, when cooled to a temperature just above ➪absolute zero, loses its electrical resistance, becoming a perfect conductor.

superego the part of an individual's personality that includes self-criticism (and therefore conscience), and which enforces the rules for social behaviour. See also ➪id; ➪ego.

superfluid a fluid with an anomalously high ➪thermal conductivity capable of flowing virtually without ➪friction. Cryogenically cooled helium can have such characteristics.

superheterodyne or *supersonic heterodyne* a means of receiving ➪radio transmissions involving the changing of the ➪frequency of the ➪carrier wave to an intermediate frequency above the limit of audible sound by a heterodyne process. In this, the received wave is combined with a slightly different frequency wave produced within the receiver. Once the intermediate frequency has been formed, the combined waves are amplified and the signal taken off by a demodulator.

superior planet planet that is farther from the Sun than the Earth is (that is, Mars to Pluto).

supernova brilliant explosion of a massive star in a cataclysmic burst of energy, leaving only a dissipating gas cloud round a ➪neutron star (pulsar). The Crab nebula (in Taurus) is considered an example.

supersonic velocities greater than the speed of sound in the particular conditions prevailing, that is greater than ➪Mach 1.

supersynthesis a ➪radio interferometer system in which two ➪synthesis aerials are used; one is static and utilizes the rotation of the Earth to provide a field of scan, the other is mobile.

surface in three-dimensional geometry, the equivalent in three dimensions to an area in two dimensions.

surface tension an effect seen on the surface of liquids caused by the attraction between ➪molecules of the liquid. The effect is to produce a 'film' of tension over the surface.

surveying the use of optical devices to determine the angular and spatial relationships of objects on the Earth, enabling them to be accurately depicted on paper or other permanent media.

suspension bridge a bridge in which the main central span is suspended from two towers erected on either side of the space being bridged.

synapse the point of contact between two nerve cells (➪neurons). Impulses are usually transmitted across the synaptic gap by ➪nerve transmitter chemicals.

synchrotron a particle accelerator which uses a system of constant electric field and modulated magnetic field to accelerate charged particles to high energy.

synchrotron radiation polarized form of ➪radiation produced by high-speed ➪electrons in a magnetic field; it is this radiation that is emitted by the Crab nebula (in Taurus).

syndrome a collection of various symptoms that together characterize a particular condition or disorder.

synodic period time between one ➪opposition and the next, of any ➪superior planet or asteroid.

synthesis aerial a ➪radio interferometer system utilizing a number of small aerials to achieve the effect of an impossibly large single one.

systole a phase in the cycle of heart muscle action in which it contracts, emptying the ventricles (into the arterial system) and allowing the atria to fill.

T

tangent (tan) see ⇨trigonometric functions.

tangent a line that touches a curve at one point only, without cutting the curve.

tautomerism a type of ⇨isomerism in which a compound exists as a mixture of two readily interconvertible isomers in equilibrium. When one of the isomers is removed from the mixture, part of the other isomer converts so as to restore the equilibrium. Nevertheless, each of the isomers can form a stable series of derivatives.

Taylor's theorem or *Brook Taylor's theorem* expands a function of x as an infinite power series in powers of x.

telecommunication communication between two places by electric or electromagnetic means.

telegraph a means of communication in which pulses of electric ⇨current are sent down wires connecting the transmitting and receiving stations. ⇨Morse code is often used.

telegraphy a means of long-distance communication by transmitting electric currents along wires.

telephony the conversion of audio signals into electrical impulses for transmission and subsequent reconversion back into sound.

tensor calculus position in ordinary space usually requires specification of three coordinates, singly indexed, for example, x_1, x_2, x_3. In describing mathematical objects more complicated than position, a generalized type of coordinate system may be used: for example, x_{ijk} where i, j, k can each take values 1, 2 or 3. Tensor calculus is a systematized use of such awkward objects.

terminal velocity when a ⇨force is velocity dependent, there is a specific velocity for which the corresponding force equals the force opposing the motion; this is the ⇨terminal velocity, which the body under the action of the two forces will then maintain, undergoing no further acceleration. An example is the air-resistance force, which depends on the square of the velocity. For a falling body, this force opposes that of ⇨gravity, and a terminal velocity is reached.

terpene any of a class of organic compounds, originally derived from plant oils, that contain only carbon and hydrogen and are empirically regarded as derivatives of isoprene (C_5H_8). They are classified according to the number of isoprene units in the molecule – for example, monoterpenes contain two isoprene units and have the formula $C_{10}H_{16}$, sesquiterpenes contain three units ($C_{15}H_{24}$), and diterpenes contain four units ($C_{20}H_{32}$). Turpentine consists of a mixture of several monoterpenes. Rubber is a polyterpene with between 1,000 and 5,000 isoprene units.

terrestrial of the Earth.

tertian fever a type of fluctuating malarial fever that peaks every third day.

Tertiary period of ⇨geological time 65–1.64 million years ago, divided into five epochs: Palaeocene, Eocene, Oligocene, Miocene, and Pliocene. During the Tertiary, mammals took over all the ecological niches left vacant by the extinction of the dinosaurs, and became the prevalent land animals. The continents took on their present positions, and climatic and vegetation zones as we know them became established. Within the geological time column the Tertiary follows the ⇨Cretaceous period and is succeeded by the ⇨Quaternary period.

tesla the ⇨SI unit of ⇨magnetic flux density. It is defined as a magnetic flux density of one ⇨weber of ⇨magnetic flux per square metre.

tesselation covering of a plane surface by regular congruent quadrilaterals in a side-by-side pattern; the first quadrilateral is derived by joining the mid-points of the sides of a given (regular or irregular) quadrilateral.

tetrahedron (plural *tetrahedra*) four-sided figure, each of which is a triangle.

theodolite a device used in ⇨surveying for measuring the relative angles and positions of objects, so that plans may be drawn up.

thermal a column of warm air, which is of a lower ⇨density than its surroundings, and contains rising currents of air.

thermal conductivity the rate of ⇨conduction of heat through a material.

thermionic tube or *thermionic valve* an evacuated metal or glass container enclosing a system of electrodes. The ⇨cathode emits ⇨electrons when heated, and these are attracted to a positively charged ⇨anode. Perforated grid electrodes within the tube can be used to control the electron ⇨current.

thermochemistry the branch of chemistry that deals with the heat changes that accompany chemical reactions.

thermocouple a device comprising two wires of different metals (such as copper and iron) joined, and with their other ends held at different temperatures. A small ⇨current is set up within the wire proportional to the size of the temperature difference between the two ends. This makes thermocouples useful in the determination of temperature.

thermodynamic equilibrium a system is said to be in thermal ⇨equilibrium if no heat flows between its component parts.

thermodynamics the study of the laws that govern heat as a form of energy and its transformation and conservation.

thermodynamics, laws of the first states the ⇨law of conservation of energy for a closed system. The second states that heat cannot be transferred by any self-sustaining process from a colder to a hotter body; equivalently, the ⇨entropy of a closed system can never decrease. The third law states that ⇨absolute zero is unattainable; equivalently, the entropy change for a chemical reaction involving crystalline solids is zero at absolute zero.

thermopile a device consisting of a number of ⇨thermocouples; it is used to detect and measure heat (infrared) radiation.

thermoplastic a plastic that can be repeatedly remelted without losing its overall characteristics.

theta functions four types of ⇨elliptic function devised by Carl Jacobi. Each function is defined as a Fourier series, and written θ_1, θ_2, θ_3, and θ_4; any θ can be converted by translation of the argument into another θ multiplied by a simple factor. (The quotient of any two θ is then periodic twice.)

thixotropy the property of a substance that enables it to form a jelly-like ⇨colloid that reverts to a liquid on mechanical agitation. Nondrip paints are common thixotropic materials.

thread, Whitworth see ⇨Whitworth standard.

three-body problem mathematical problem in astronomy, to describe the gravitational effects of three interacting celestial bodies on each other, and the shape of their orbits round each other.

thrust the propulsive ⇨pressure exerted by, for example, a jet or rocket engine.

tide effect of the Moon's ⇨gravity on the Earth's seas, such that an oceanic 'bulge' each side of the Earth follows the Moon's progress around the planet.

Titan seventh (known) moon out from ⇨Saturn, and its largest. It is possibly also the largest satellite in the Solar System (although Neptune's ⇨Triton may be proved to be larger). It is 20% larger than the planet Mercury and is known to have an ⇨atmosphere.

Titania fourth (known) moon out from Uranus, and probably its largest.

titration a technique in ⇨volumetric analysis in which a liquid is added from a graduated burette to a precisely known volume of a second liquid until a chemical reaction between them is just complete (this ⇨end-point is usually detected using an ⇨indicator). From the volumes involved and the concentration of one of the solutions, the concentration of the second solution can be calculated.

TNT (abbreviation for *trinitrotoluene*) a pale-yellow material, used as a high explosive.

tolerance in biology, the ability (of an organism) to sustain the effects of a drug or poison without harm. Tolerance can be acquired, and built up, by repeated small doses of the drug or poison.

tolerance the range in the physical dimensions of an object within which the true dimensions lie. Often expressed in the form of, for example, ± 3 mm for a length.

tomography see ⇨computerized axial tomography.

topological equivalence see ⇨topology.

topology study of figures and shapes (and other mathematical objects) with particular regard to the properties that are retained even when the figures or shapes are 'stretched' or 'squeezed'. For example, if a solid ball is stretched and squeezed back into the form of an identical ball (so that effectively the ball has been transformed onto itself) there is a point that is transformed into itself. Fixed point theorems like this are important in the study of differential equations. Furthermore, in topology, two objects are regarded as identical (topologically equivalent) if either can be transformed into the other by a transformation that is continuous and that has a continuous inverse – for example, a cube and a solid ball (of which both have at least eight identical coordinate points).

torque a ⇨force that produces rotation.

torus three-dimensional solid resembling a ring doughnut or an inflated inner-tube.

total internal reflection the phenomenon caused when a ray of light, in passing from a dense to a less dense medium (for example, from glass to air) is bent sufficiently by ⇨refrac-
tion so as not to leave the denser medium. Occurs when the angle of incidence exceeds a certain critical angle.

toxicity a measure of the poisonous nature of a substance.

toxin a poison (often a protein), usually produced by ⇨bacteria. Bacterial toxins are classified as endotoxins, which affect only local tissues, or as exotoxins, which are released into the bloodstream and can cause damage far from their site of origin.

tracer a radioactive isotope introduced into an experiment to follow the path of a chemical (often biochemical) reaction.

trans- a prefix used in ⇨stereochemistry to indicate that two groups or substituents lie on opposite sides of the main axis or plane of a molecule (as opposed to ⇨*cis-*, which indicates that they are on the same side).

transcendental curve curve for which there is no representative algebraic equation; examples are logarithmic curves and trigonometric curves.

transcendental number ⇨real number that is not an ⇨algebraic number, and can therefore not be expressed as a root (solution) of an algebraic equation with integral coefficients; an example is ⇨Euler's number (e).

transfer RNA (tRNA) a relatively small molecule of ⇨ribonucleic acid, the function of which is to carry ⇨amino acids to ⇨ribosomes where ⇨protein synthesis occurs. Each amino acid is borne by a different tRNA molecule. tRNA is complementary to ⇨messenger RNA (mRNA).

transfinite see ⇨set theory.

transformation in genetics, the substitution of one section of ⇨DNA by another. It requires at least two crossovers (or breaks) in the DNA and is a source of genetic variation.

transformation or *mapping* in mathematics, operation between two spaces, particularly between two ⇨vector spaces, also defined as an operation allotting to each point P of one space a point (usually written as P') of the other space. P is then known as the argument, and P' the image point.

transformer a device that uses ⇨induction to convert high voltage ⇨alternating current into low voltage, without change in ⇨frequency, and vice versa.

transistor a ⇨semiconductor capable of amplifying electric ⇨currents.

transit crossing of the Sun's face by either ⇨Mercury or ⇨Venus, as seen from Earth. The term is also used to describe the crossing of the ⇨meridian of an observer by a celestial body.

transit circle large instrument for the accurate observation and measurement of a ⇨transit.

transition element any of a group of elements of similar physical and chemical properties that are characterized by having variable ⇨valency (oxidation number). They are grouped together at the centres of the long periods of the periodic table.

transpiration the evaporation of water vapour from plants, mostly through the stomata and sometimes the cuticle. The movement of water within the plant transports mineral salts via the ⇨xylem to the leaves, where transpiration also helps to cool the plant.

transuranic element an ⇨element whose ⇨atomic number exceeds that of uranium – that is, it is greater than 92. Such elements – the first of which is neptunium, with an atomic number of 93 – do not occur naturally in the universe.

transverse wave a wave in which the vibrations take place in a plane at right angles to the direction of propagation. ⇨Electromagnetic waves are transverse waves. See also ⇨longitudinal wave.

Triassic period of ⇨geological time 245–208 million years ago, the first period of the ⇨Mesozoic era. The continents were fused together to form the world continent ⇨Pangaea. Triassic sediments contain remains of early dinosaurs and other reptiles now extinct. By late Triassic times, the first mammals had evolved.

tribology the study of ⇨friction and similar surface effects.

tricarboxylic acid cycle or *Krebs' cycle* the final stages in the biochemical breakdown of ⇨carbohydrates, whose oxidation releases energy, which take the form of a succession of enzyme-catalysed reactions.

trigonometric functions represent the relationships between the sides and angles of a right-angled triangle, such that of an angle a:

sine a	=	opposite side/hypotenuse
cosine a	=	adjacent side/hypotenuse
tangent a	=	opposite side/adjacent side
secant a	=	hypotenuse/adjacent side
cosecant a	=	hypotenuse/opposite side

trigonometric series expression of the sine and cosine ⇨trigonometric functions as convergent ⇨power series.

trim in ⇨aerodynamics, the slight actions on the controls needed to achieve stability in a particular mode of flight.

triode a ⇨thermionic valve consisting of an ⇨anode, a ⇨cathode and a ⇨grid.

tripeptide a sequence of three ⇨amino acids, often occurring in the biochemical synthesis or breakdown of ⇨proteins.

triplane an aircraft that features three wings stacked vertically above each other.

Triton very large and close moon of ⇨Neptune. Probably the largest satellite in the Solar System (although it is not definitely established as larger than Saturn's ⇨Titan), it is likely to be about twice the size of the planet Pluto, and 25% larger than Mercury.

trophoblast epithelial cells that surround all the structures of a placental embryo, forming the outer layer of the ⇨chorion and the embryonic side of the ⇨placenta.

trypanosome a protozoon of the genus *Trypanosoma*, many of which are parasitic on insects which transmit them as ⇨pathogens to human beings. Sleeping sickness (trypanosomiasis) is caused by a trypanosome carried by the tsetse fly.

T Tauri variable another name for ⇨nebular variable star.

turbine a system of a wheel with vanes on its rim, connected via a shaft to a device that either uses the energy developed by the turbine directly (see ⇨jet engine) or indirectly, through a ⇨dynamo for example. The turbine is powered by the ⇨force of water, steam, or wind, striking its vanes.

turbofan an aero engine in which part of the ⇨power produced by the gas turbine engine is used to drive an intake fan inside a duct.

turboprop an aircraft propulsion unit in which the ⇨power produced by burning gases is transmitted to a propeller via a ⇨turbine and ⇨gear system.

turbulent flow the flow in a column of fluid that does not follow ⇨streamlines. Such flow sets in at a velocity dependent on the ⇨Reynold's number.

turnpike road road with a gate or barrier preventing access until a toll (tax) had been paid for its use and maintenance. Such roads were common in the UK from the mid-sixteenth to the nineteenth centuries.

twistor theory model of the universe proposed by Roger Penrose, based on the application of complex numbers (involving the square root of −1) used in calculations in the microscopic world of atoms and quantum theory to the macroscopic ordinary world of physical laws and relativity. The result is an eight-dimensional concept of reality.

Tyndall effect scattering of light to produce a visible beam.

U

UBV photometry measurement of the ⇨astronomical colour index of a star, utilizing the ultraviolet, blue and yellow visual images over two pre-set wavelengths obtained by ⇨photoelectric filtering. Other standardized filter wavebands are also used.

ultrasonics the study of sound waves (that is, ⇨longitudinal pressure waves) with ⇨frequencies above approximately 20 kHz, the audible frequency limit.

ultrasound sound waves at a frequency beyond the range of normal human hearing (more than 20 KHz). Bats, cetaceans and some insects emit ultrasound.

ultraviolet excess screening technique devised by Martin Ryle and Allan Sandage to measure the spectral ⇨red shift of suspected ⇨quasars. It was this process that resulted in the discovery of ⇨quasi-stellar objects.

ultraviolet radiation ⇨electromagnetic radiation with a ⇨wavelength shorter than that of visible ⇨light, lying just beyond the violet part of the visible ⇨spectrum (hence its name). Typical wavelengths are of the order one hundred-millionth of a metre.

umbra full shadow.

uncertainty principle, Heisenberg's a principle of ⇨quantum physics stating the impossibility of simultaneously determining the position and ⇨momentum of a particle with limitless accuracy.

unconditioned reflex a behavioural reflex or response that is natural and not acquired by training or conditioning (see ⇨conditioned reflex).

unconditioned stimulus a natural stimulus unassociated with behavioural training; it evokes a natural or unconditioned reflex or response (see ⇨conditioned stimulus).

unified field theory an attempt to describe the laws of ⇨electromagnetism and ⇨gravitation in a single set of equations.

uniformitarianism in geology, the principle that processes that can be seen to occur on the Earth's surface today are the same as those that have occurred throughout geological time. For example, desert sandstones containing sand-dune structures must have been formed under conditions similar to those present in deserts today.

unit theory or *theory of units* in a ⇨field or ⇨ring, involves an element that possesses an inverse. In a field, every nonzero element is a unit; in a ring, 1 and −1 represent a unit.

universal attraction see ⇨law of universal attraction.

universal set in set theory, with regard to a ⇨mathematical structure, a set of objects in the structure. More generally, it represents the 'universe of discourse' appropriate to the discourse – for example, the set of vowels and the set of consonants are disjoint sets within the universal set comprising the alphabet.

Uranus seventh major planet out from the Sun, the first to be discovered in modern times (by William Herschel in 1781).

urea cycle or *Krebs–Henseleit cycle* a succession of reactions in which nitrogen (in which nitrogen (in the form of ammonia) is metabolized to produce urea.

V

vaccination ⇨inoculation with a killed or attenuated virus or bacterium to confer resistance by stimulating the immune response to it (that is, the production of the appropriate ⇨antibodies).

vaccine a suspension of killed or attenuated pathogenic microorganisms (such as viruses or bacteria) used for the treatment or prevention of an infectious disease.

vacuole a fluid-filled vesicle that occurs in the ⇨cytoplasm of cells.

vacuum a region devoid of atoms of any substance. Hence, the ⇨pressure within a container enclosing a vacuum is zero.

valency or *valence* number that describes the combining power of an atom or group of atoms in terms of the number of bonds it forms with hydrogen or its equivalent. Thus chlorine in hydrogen chloride (HCl) has a valency of one; oxygen in water (H_2O) has a valency of two; nitrogen in ammonia (NH_3) has a valency of three; carbon in methane (CH_4) has a valency of four; phosphorus in phosphorus pentoxide (P_2O_5) has a valency of five; and so on. Increasingly the term valency is being replaced by ⇨oxidation state.

valve, thermionic see ⇨thermionic valve.

vanadium a very hard, white metal used in the making of tough ⇨steel alloy.

Van Allen belts two belts of energetic charged particles about 4,000 km/2,500 mi and 16,000 km/10,000 mi above the Earth. The inner belt is filled with particles produced by collisions between ⇨cosmic rays and atmospheric ⇨molecules that have become trapped in the Earth's magnetic field, whereas the higher belt contains particles captured from the ⇨solar wind.

van de Graaff generator an accelerator of ⇨elementary particles that uses an ⇨electrostatic field to raise the particles to high energies.

Van der Waals's force an interatomic and intermolecular force of attraction resulting from the distortion of the ⇨electron distribution in one atom or molecule by another.

vapour a substance in the gaseous state, which may be liquefied by raising the ⇨pressure, leaving the temperature the same.

vapour density the density of a gas or vapour usually in terms of that of hydrogen (that is, the weight of a given volume of the gas divided by the weight of an equal volume of hydrogen at the same temperature and pressure). The ⇨molecular weight of a gas is approximately equal to twice its vapour density.

vapour pressure the pressure exerted by a gas or vapour in an enclosed space. The ⇨saturated vapour pressure is the pressure of the vapour of a substance in contact with its liquid form; it varies with temperature.

variable in mathematics, an expression representing either a general point taken from a specific set of values, which may therefore be substituted by any selected point of the specified set, or a name for an unknown point to be found within a specified set of possible values. It is often regarded as fixed at least through the duration of a mathematical train of thought.

variable star star whose ⇨luminosity changes over periods of time; there are many reasons and many types. Periods vary widely in length and even regularity. ⇨Novae and ⇨supernovae are classed as variables. The present brightest variable star is Betelgeuse (Alpha Orionis).

variance in probability theory, the square of the ⇨standard deviation.

variety in algebra, the set of solutions of a simultaneous system of equations with a fixed number of variables. In two dimensions, examples are a circle, an ellipse, and a parabola. In three dimensions, a variety is a surface, for example, the surface of a sphere.

vascular describing the systems that conduct fluids in plants (⇨xylem and ⇨phloem) and animals (the blood and lymph systems).

vasoconstriction the constriction of vessels especially blood vessels.

VDU (abbreviation for *visual display unit* a television monitor that gives a visible display of the output of data from a computer.

vector in medicine, a carrier (often an insect) of a ⇨pathogenic microorganism, which may be parasitic on the vector and transmitted to other animals through the vector's bite.

vector in mathematics, a representation of a quantity having both magnitude and direction in its definition, for example ⇨force, velocity and acceleration. More generally, a vector is an element of a ⇨vector space. See also ⇨scalar.

vector space or *linear space* a mathematical structure comprising two types of objects: vectors and scalars. Vectors can be added by themselves; scalars lengthen or shorten them (that is, scale the vector length up or down) and are commonly either real or complex numbers. In general, scalars of a vector space comprise a ⇨field. The prime example of a vector space is the collection of elementary vectors (see ⇨vector) in the two- or three-dimensional space of everyday experience. (The operation of addition follows the parallelogram law: to add two vectors, complete the parallelogram defined by the two lines – the sum is given by the diagonal through the common origin.) Vector spaces provide a framework for the study of ⇨linear transformations – which can in finite-dimensional spaces be represented by ⇨matrices – and are important in mathematical modelling of complicated systems (engineering, biological, and so on) where general transformations are approximated by linear ones, with recourse to the apparatus of ⇨differential calculus.

Vega major star in Lyra; its (apparent) magnitude is 0.04.

Venn diagrams in elementary set theory, use of overlapping (or similar closed curves) to illustrate how some members of one set may also be members of another set.

Venus second major planet out from the Sun; in many ways it is similar to the Earth.

vernal equinox the spring ⇨equinox, on or around 21 March.

vernalization a technique for controlling the flowering times of plants, particularly crop plants (cereals) that are normally sown in winter. It involves moistening the seed so that it just germinates and exposing the seedling's radicles for a few weeks to temperatures just above freezing. The plant then behaves like a spring-sown plant and crops relatively early in the summer of the same year, with obvious economic advantages.

vernier a scale that, when used in conjunction with a scale measuring larger amounts of the property in question (such as length), allows subdivisions of the basic unit on the large scale to be read off accurately.

vesicle a small fluid-filled bladder or blister.

vestigial describing a partial, diminished or redundant structure in an organism.

Viking spaceprobes series of two US spaceprobes that successfully effected landings on Mars and relayed data back to Earth.

virtual particle theory theory devised by Stephen Hawking to account for apparent thermal radiation from a ⇨black hole (from which not even light can escape). It supposes that space is full of 'virtual particles' in a particle–antiparticle relationship, being created out of 'nothing' and instantly destroying each other. At an ⇨event horizon, however, one particle may be gravitationally drawn into the ⇨singularity, and the other appear to radiate as heat.

virus a parasitic microorganism (often ⇨pathogenic that consists of a strand of ⇨nucleic acid covered with a protein or lipoprotein layer. Viruses that parasitize animals contain DNA or RNA; those that live on plants contain RNA only; and most ⇨bacteriophages contain DNA only. Viruses infect cells by injecting their nucleic acid into them, where it activates further synthesis of virus cells. They may be spread by ⇨vectors; in air, soil or water; or by physical contact with an infected organism.

viscosity a property of a ⇨fluid that causes it to resist motion within it. It determines other properties of the fluid, such as the velocity of flow at which ⇨turbulence sets in; see ⇨Reynold's number.

viscous force the drag that occurs on an object placed in a viscous medium. For example, air causes a viscous force on an aircraft travelling through it. For two parallel layers of fluid close to each other, the viscous force is proportional to the difference in velocity between the two layers.

vision, persistence of see ⇨persistence of vision.

visual display unit see ⇨VDU.

vitalism a now discredited belief that the life, development and functions of organisms result from a vital force that only living organisms possess.

vitamin an accessory food factor, one of a group of organic substances (excluding fats, carbohydrates, proteins and minerals) that play an important part in the development, metabolic processes and health of the body. Most vitamins cannot be synthesized in the body and must be obtained from the diet; exceptions include vitamin A (formed from the plant pigment carotene), vitamin D (manufactured by a process involving the action of ultraviolet light on the skin) and vitamin K (which is synthesized by intestinal bacteria).

volcano crack in the Earth's crust through which hot magma (molten rock) and gases well up. The magma becomes known as lava when it reaches the surface. A volcanic mountain, usually cone shaped with a crater on top, is formed around the opening, or vent, by the build-up of solidified lava and ashes (rock fragments). Most volcanoes arise on plate margins (see ⇨plate tectonics), where the movements of plates generate magma or allow it to rise from the mantle beneath. However, a number are found far from plate-margin activity, on 'hot spots' where the Earth's crust is thin.

volt the ⇨SI unit of ⇨potential difference, defined as the potential difference that exists between two points of a ⇨conductor in which a current of one ⇨ampere flows, such that energy flow dissipated between the two points equals one ⇨joule per second, that is, one ⇨watt.

voltaic cell another name for an electric ⇨cell.

volumetric analysis any method of chemical analysis that involves measuring volumes of liquids or gases (as opposed to gravimetric analysis, which involves mainly weighing). ⇨Titration is a common technique of volumetric analysis.

Voyager spaceprobes series of US spaceprobes launched to carry out exploration of the outer planets. In 1982 *Voyager 2* returned some remarkable pictures of Saturn.

V/STOL (abbreviation for *v*ertical/*s*hort *t*ake *o*ff and *l*anding) term applied to aircraft capable of taking off and landing without the need for a runway.

vulcanization the hardening of rubber by heating it with sulphur.

W

Wankel engine another name for a ⇨rotary engine

warp threads in weaving stretched lengthwise across the fabric. Compare ⇨weft.

water gas a mixture of carbon monoxide and hydrogen gas, produced by passing steam over hot coke.

watt the ⇨SI unit of ⇨power, equal to one ⇨joule of energy expended in one second.

wave, carrier see ⇨carrier wave.

wave, electromagnetic see ⇨electromagnetic wave.

wavefront the line of points in a wave motion that are all of equal ⇨phase.

wavelength the distance between each crest of a wave (or each trough). Measured in metres, the wavelength λ is related to frequency f by the velocity of the wave, v, according to the formula:

$$\lambda = v/f$$

wave, longitudinal see ⇨longitudinal wave.

wave mechanics the branch of ⇨quantum theory that derives the various properties of ⇨atoms on the basis of every particle having an associated wave existing in a multidimensional space, representing probabilities of certain properties of the particles involved. ⇨Schrödinger's equation is the basis of wave mechanics, which has been shown to be equivalent to ⇨matrix mechanics.

wave, transverse see ⇨transverse wave.

wax see ⇨fat.

weak interaction one of the four ⇨fundamental forces in nature. It is responsible for the decay of some ⇨elementary particles and of ⇨atomic nuclei.

weber the ⇨SI unit of ⇨magnetic flux, defined as the flux required to flow in a circuit such that, if it changes at a rate of one weber per second, will induce in the circuit an ⇨electromotive force of 1 volt.

weft the threads across the width of the material to be formed by weaving.

welding the use of heat to melt and fuse together two metal surfaces.

Wheatstone bridge a circuit that is divided into sections enabling the relative ⇨resistances or devices placed in the sections to be deduced.

white blood cell another name for a ⇨leucocyte.

white dwarf final stage of a star, at which the nuclear energy is exhausted. Cool, and becoming a dead, black body, it is nevertheless extremely dense.

Whitworth standard a standard of screw thread used before the advent of the metric standard, in which the pitch of the ⇨helix is standardized relative to the diameter of the bar on which the thread is cut.

Wien's law the relationship between the intensity and frequency of ⇨black-body radiation at the high-frequency end of the ⇨spectrum. Wien's displacement law relates the wavelength of maximum intensity to the temperature of the black body.

wind tunnel an enclosure that contains a large ⇨turbine capable of sending air streams over any object (such as a model aircraft) whose ⇨aerodynamic performance is to be assessed.

wireless see ⇨radio.

work the work done by a ⇨force is given by the product of the force and the distance moved by its point of application.

working fluid a ⇨fluid that is used in such a way that its internal ⇨energy is converted into external energy. One example is water in ⇨hydroelectricity generation.

wrought iron iron from which essentially all carbon impurities have been eliminated.

X

X-ray penetrating short-wavelength electromagnetic radiation produced when a stream of ⇨cathode rays (electrons) strike matter (such as the 'target' in an X-ray tube). When X-rays strike a substance, it may emit secondary X-rays which are characteristic of the elements in it and so may be used as a method of analysis. Their ability to penetrate many materials make them very useful in inspecting products for cracks and flaws.

X-ray astronomy detection of stellar and interstellar X-ray emission. Because X-rays are almost entirely filtered out by the Earth's upper atmosphere, the use of balloon- and rocket-borne equipment is essential.

X-ray diffraction a method of determining the structure of molecules, particularly those of crystals, by analysing the way in which their atoms diffract a beam of X-rays.

xylem the vascular tissue in a plant that carries water and dissolved substances (such as minerals) from the roots and throughout the plant. Primary xylem develops by differentiation from the procambium and is divided into protoxylem (produced before its surrounding area becomes elongated) and metaxylem (formed after elongation); secondary xylem develops from the cambium. Xylem tissue is composed of basic cells (tracheids), vessels, fibres and undifferentiated tissue (parenchyma). The woody stems of mature plants consist mainly of xylem.

Y

yarn thread that has been spun.

yeast any of the many species of unicellular fungi of the class Ascomycetes of the division Ascomycota. Yeasts reproduce asexually by alternation of (diploid and haploid) generations, which results in a spore- forming ⇨zygote. Yeasts contain enzymes which are exploited commercially (in brewing and baking, for example) because of their ability to initiate ⇨fermentation (for example, of sugars to produce alcohol). They are also a dietary source of protein and vitamin B.

ylem hypothetical primordial state of matter – neutrons and their decay products (protons and electrons) – that might have existed before the Big Bang. The term was taken from Aristotle and forms part of the ⇨alpha-beta-gamma theory.

Z

Zeeman effect the splitting of ⇨spectral lines of a substance placed in an intense magnetic field.

zenith exact point in the sky vertically above the observer.

zero vector see ⇨vector.

zeta function function that may be represented as the value of the infinite series:

$$\zeta(s) = 1 + \frac{1}{2^s} + \frac{1}{3^s} + \frac{1}{4^s} + \dots$$

where s is a complex number. The function was significant in Leonhard Euler's study of prime numbers. See also the ⇨Reimann hypothesis.

Z_{He} symbol for the total mass density (in a star) of elements heavier than helium; values of Z_{He} are small for ⇨population II stars, large for ⇨population I stars.

zodiac twelve constellations originally only representing a calendar of the Sun's apparent progress in the heavens during one Earth-year. The principal planets are to be found along much the same path (the ⇨ecliptic) and so, probably early in human history, caused each constellation to become a focus for divination according to the pseudoscience of ⇨astrology. Since then, however, the millennia that have passed have taken the Sun out of phase with the original calendar. The 12 constellations of the zodiac are: Aries, Taurus, Gemini, Cancer, Leo, Virgo, Libra, Scorpio, Sagittarius, Capricorn, Aquarius and Pisces.

zodiacal light faint, cone-shaped glow seen briefly in the sky shortly before sunrise or after sunset. Its cause is the reflection of light by particles lying along the plane of the ⇨ecliptic.

zone of avoidance apparent lack of distant ⇨galaxies in the plane of our own Galaxy, now explained as being caused by optical interference of dust and interstellar debris on the rim of the Galaxy.

zygote a fertilized ⇨ovum formed by the union of male and female ⇨gametes. It contains paired ⇨chromosomes, half of each pair being derived from each of the gametes.

Index

Where a name appears in capitals, the scientist has an entry in the dictionary

A

ABBE, ERNST: Zsigmondy
Abbe's sine condition: Abbe
Abbe's substage condenser: Abbe
Abel, Frederick: Dewar
ABEL, NIELS HENRIK: Fredholm,
Galois, Jacobi, Ore, Ruffini, Sylow,
Weber, H.
Abelian functions: Baker, A., Dedekind,
Herbrand, Hermite, Kronecker,
Riemann, Weierstrass
Abel–Ruffini theorem: Ruffini
Abelson, Philip: Meitner
aberration:
chromatic: Abbe, Newton
of light: Bradley, Jones
spherical: Abbe
ABETTI, GIORGIO
Abraham, E. P: Chain
Abrahamson, W: Longuet-Higgins
absolute differential calculus:
Christoffel, Levi-Civita,
Ricci-Curbastro
absolute right ascension: Flamsteed
absolute temperature: Kelvin
absolute zero: Désormes, Kelvin,
Regnault, Simon
absorption spectrography, ultraviolet:
Meselson
absorption spectrum: Brönsted, Norrish,
Robertson
abstract algebra: Artin, Betti, Bourbaki,
Nagell, Noether, A., Ore
abstract groups: Frobenius
abundance plateau: Hayashi
accelerator nerves: Ludwig
accessory food factors: Funk
Account of Carnot's Theorem: Carnot
accumulator: Crompton, R.E.B., Ferranti,
Sperry
acetabulum: Charnley
acetates: Kornberg, H. L., Lynen
acetic acid: Bloch, Kolbe, Volhard
acetylcholine (ACh): Dale, Katz
acetyl coenzyme A: Krebs, Lynen
acetylene: Wöhler
Achilles: Lagrange, Wolf
achromatic lens: Fraunhofer
acids:
discovery of: Scheele
electron acceptor theory: Lewis
electronegativity and: Pauling
hydroxonium ions and: Brönsted
ionic dissociation and: Brönsted
metabolic role: Williamson
oxygen-containing theory of: Berthollet,
Davy, Gay-Lussac, Scheele
polybasic: Brönsted, Graham
preparation of: Dumas, Glauber,
Scheele, Volhard, Wieland
proton donor theory of: Brönsted
reactions of: Brönsted, Dumas, Glauber,
Lewis

acoustics: Chladni, Coulomb, Helmholtz,
Rayleigh, Weber, W. E., Wheatstone
acromegaly: Cushing
ACTH *see* adrenocorticotrophic hormone
actin: Huxley, H. E., Szent-Györgyi
actinides: Seaborg
actinium: Curie, Soddy
action potential: Hodgkin, A. L.
activated molecule: Arrhenius, Polanyi
activation energy: Arrhenius, Polanyi
activity coefficient: Brönsted, Ostwald
actomyosin: Szent-Györgyi
acyclovir: Hitchings
Adamkiewicz colour reaction: Hopkins
ADAMS, JOHN COUCH: Challis,
Gauss, Leverrier
ADAMS, WALTER SYDNEY: Babcock,
H. D., Chandrasekhar, Joy
adaptor hypothesis: Crick
adaptor molecules: Crick
ADDISON, THOMAS: Bright
Addison's disease: Addison
adenine: Crick, Holley, Kornberg, A.
adenosine diphosphate (ADP):
in carbohydrate metabolism: Cori
in photosynthesis: Calvin
synthesis of: Todd
adenosine triphosphate (ATP): Hill, R.,
Hodgkin, A. L., Mitchell, P. D., Oparin
carbohydrate metabolism: Cori
muscles: Calvin
photosynthesis: Meyerhof,
Szent-Györgyi
synthesis of: Todd
adiabatic demagnetization: Simon
adiabatic effects: Carnot, Désormes,
Simon
ADLER, ALFRED: Freud, Jung
ADP *see* adenosine diphosphate
adrenal glands: Addison, Cushing,
Kendall, Sharpey-Schafer
cortex: Addison, Cushing, Li Cho Hao
adrenaline: Sharpey-Schafer
adrenocorticotrophic hormone
(ACTH): Guillemin, Li Cho Hao
ADRIAN, EDGAR DOUGLAS:
Sherrington
adsorption:
isotherms: Freundlich, Gibbs, Langmuir
kinetics of: Gibbs, Langmuir, Polanyi
physical processes in: Langmuir,
McBain
reactions: Langmuir
adsorption chromatography: Martin,
A.J.P., Tiselius, Tswett, Willstätter
Advancement of Learning: Bacon, F.
Aëdes aegypti: Theiler
Aeriel II: Smith, F. G.
aerodynamics: Mach, Prandtl, von
Kármán
aerofoil: Lilienthal
aeroplane: De Havilland, Lanchester,
Maxim, Mitchell, R. J., Whittle

aerosols: Midgley
aerostat: Charles
African tick fever: Leishman
AGASSIZ, JEAN LOUIS RODOLPHE
agglutination: Landsteiner
AGRICOLA, GEORGIUS
AIDS: Gajdusek, Hitchings
AIKEN, HOWARD HATHAWAY:
Babbage
Air-Crib: Skinner
air liquefaction: Cailletet
air pump: Hooke, von Guericke
air resistance induced: Prandtl
law of: Huygens
terminal velocity: Galileo, Stokes, G. G.
air tents: Charnley
air thermometer: Regnault
airliner: De Havilland
airship: Giffard, Wallis, B. N., Zeppelin
AIRY, GEORGE BIDDELL: Adams,
J. C., Challis, Leverrier
AITKEN, ROBERT GRANT
al-Battani: Schiaparelli
albumen: Fabricius
albumin: Bright, Starling
albuminaria: Bright
alchemy: Bacon, R., Helmont, Newton
alcoholic fermentation: Buchner,
Haworth
alcohols:
Cannizzaro's reaction: Cannizzaro
catalytic dehydrogenation of: Sabatier
chlorine and: Dumas
ether synthesis: Williamson
fatty acids and: Berthelot
optical activity of: Kenyon
oxidation of: Dumas
sugars and: Berthelot
synthesis of: Bergius, Berthelot,
Grignard
unsaturated: Hofmann
aldehydes:
Cannizzaro's reaction: Cannizzaro
formaldehyde: Hofmann
Grignard reagents and: Grignard
isomers of: Fischer, H.O.L.
Norrish's reactions and: Norrish
preparation of: Fischer, H.O.L.,
Grignard, Hofmann, Sabatier
Wittig's reaction and: Wollaston
ALDER, KURT: Diels
aldol: Fischer, H.O.L., Wurtz
aldosterone: Barton
ALEKSANDROV, PAVEL
SERGEEVICH
Alexander the Great: Aristotle,
Theophrastus
ALFVEN, HANNES OLOF GOSTA:
Cowling
Alfvén theory: Alfvén
algebra: Artin, Barrow, Betti, Boole,
Descartes, Dickson, Diophantus,
Dodgson, Fibonacci, Hamilton, W. R.,

Hankel, Herbrand, Jordan, Kronecker, Kúrosh, Noether, A., Noether, M., Simpson, T., Viète, Wallis, J., Weber, H., Wedderburn, Whitehead, A. N.

algebraic curves: Coolidge, Cramer, Descartes, Grandi, Grassmann, Noether, M.

algebraic invariance, invariants: Cayley, A., Hilbert, Lie, Ricci-Curbastro, Sylvester

algebraic numbers: Baker, A., Dedekind, Dirichlet, Galois, Hilbert, Lindemann, C.L.F., Nagell, Ore

algebraic topology: Eilenberg, Picard, Poincaré, Thom, Whitehead, J.H.C., Zeeman, E. C.

Algol: Vogel

algorithm: Church, Dantzig

ALHAZEN

alizarin: Baeyer, Perkin

alkali metals: Davy, Gay-Lussac

alkalis:
carbonates and: Black, J.
commercial production of: Mond, Solvay
electrolysis and: Davy
electronegativity and: Pauling
oxygen containing theory: Davy
proton acceptors: Brönsted

alkaloids:
biosynthesis of: Robinson
composition of: Pelletier
isolation of: Pelletier, Wieland
steroid: Prelog
structure of: Prelog, Pyman, Wieland, Willstätter
synthesis of: Pyman, Robinson, Willstätter, Woodward

alkanes: Bergius, Sabatier, Wurtz

alkenes: Wittig, Ziegler

alkoxide: Williamson

alkyl halides: Hofmann, Williamson, Wurtz
Friedel–Craft reaction: Friedel

alkylation: Cairns, Grignard, Perkin

alkylidene: Wittig

All-or-None law: Ludwig

allantois: von Baer

alleles: Kimura

ALLEN, JAMES ALFRED VAN *see* VAN ALLEN

alloys: Hume-Rothery

Almagest: Müller, J., Ptolemy

α–β–γ **(alpha–beta–gamma) theory:** Alpher, Fowler, Gamow, Hayashi

Alpha Centauri: Lacaille

alpha chain: Kimura

Alpha Lyrae: Struve, F.G.W.

alpha particles: Bethe, Blackett, Bragg, Crookes, Fajans, Joliot-Curie, Soddy
alpha decay and: Gamow
as helium nuclei: Rutherford
atomic number and: Chadwick
atomic transmutation and: Cockcroft, Walton
charge of: Geiger
detection of: Crookes

alpha-ray rule: Soddy

ALPHER, RALPH: Fowler, Gamow, Hayashi

Alpher–Bethe–Gamow theory *see* α– β– γ theory

AL-SUFI

alt-azimuth: Airy

ALTER, DAVID

alternating current: Braun, K. F.
and electromagnetism: FitzGerald
production of: Tesla

aluminium: Hall, C. M., Sainte-Claire Deville, Wöhler, Ziegler

aluminium tri-alkyls: Ziegler

ALVAREZ, LUIS WALTER

Alzheimer's disease: Fuller

Amatol: Robertson

Amazon, exploration of: Bates

AMBATZUMIAN, VICTOR AMAZASPOVICH

ambrein: Pelletier

American Astronomical Society: Newcomb

americium: Seaborg

amines: Hofmann, Wurtz
photochemical reactions: Emeléus
vital amines: Funk

amino acid: Berg, Brenner, Crick, Hoagland, Hopkins, Li Cho Hao, Oparin
degradation: Krebs
detection techniques: Martin, A.J.P., Sanger, Schoenheimer, Synge
discovery of: Vauquelin
genetic code and: Holley, Khorana
origin of life and: Miller
polypeptides and: Fischer, E., Pauling
proteins and: Holley, Kendrew, Pauling, Sanger, Schoenheimer
rate of substitution: Kimura
RNA and: Holley
structure of: Pauling
synthesis of: Fischer, E., Miller, Volhard

amino acyl tRNA: Hoagland

ammeter: Ayrton

amnion: von Baer

ammonia: Bosch
catalytic oxidation of: Ostwald
commercial production of: Haber, Mond, Solvay
composition: Henry, W.
isolation: Priestley
nitric acid manufacture and: Ostwald
synthesis of: Haber

ammonia–soda process: Leblanc, Mond, Solvay

amoebic dysentery: Pyman

Amontons, Guillaume: Fahrenheit

AMPERE, ANDRE

Ampère's law: Ampère

Amphioxus: Boveri, Dumas, Galvani, Sturgeon

amplifier: De Forest

amplitude modulation *see* modulation

AMSLER-LAFFON, JAKOB

anabolism: Kornberg, H. L., Schoenheimer

anaerobic respiration: Meyerhof,

Wieland

anaesthetic: Simpson, J. Y.
cocaine: Freud
ether: Morton
nitrous oxide: Wells

analysis, analytical method: Artin, Bolzano, Cesaro, Dirichlet, Euclid, Euler, Hardy, G. H., Hausdorff, Hilbert, Jacobi, Jordan, Kronecker, Liouville, Maclaurin, Monge, Picard, Riemann

analytical geometry: Descartes, Fermat, Monge, Wallis, J.

anaplerotic reactions: Kornberg, H. L.

ANAXIMANDER THE ELDER

ANDERSON, CARL DAVID

Anderson, John: Emeléus

ANDERSON, PHILIP WARREN

Anderson localization: Anderson, P. W.

Andrade, Edward: Rutherford

ANDREWS, THOMAS

Andromeda nebula: Baade, Bond, W. C., Messier
Cepheids in: Baade, Hubble
distance: Hubble, Shapley
populations: Baade
pulsating nebulae: Arp
radial velocity: Slipher
radio source: Hanbury-Brown

angiograms: Barnard, C. N.

angiosperms: Brown, R., Theophrastus

Anglo-Australian telescope: Boksenberg, Bondi

ANGSTROM, ANDERS JONAS

angular momentum: McCrea

aniline dyes: Ehrlich, Fischer, H.O.L., Hofmann, Perkin, Willstätter

Animal Kingdom: Cuvier

animals, experiments on:
amphibians: Spallanzani
conger eel: Adrian
dogs: Banting, Barnard, C. N., Pavlov, Sherrington, Spallanzani
Drosophila (fruit flies): Beadle, Dobzhansky, de Vries, Hogben, Morgan, T. H.
fowl: Pasteur
frogs: Adrian, Galvani, Gurdon, Hodgkin, A. L.
gorilla: Sherrington
guinea pigs: Landsteiner
mice: Edwards, Ehrlich, Fleming, A., Florey, Isaacs, Medawar, Theiler
nematode worms: Boveri
octopuses: Young, J. Z.
pigeons: Skinner
rabbits: Fleming, A., Landsteiner
rats: Hopkins, McCollum
Rhesus monkeys: Landsteiner
roundworms: Brenner
squid: Hodgkin, A. L., Young, J. Z.

anions (negative ions): Faraday, Lapworth

Annalen der Chemie: Liebig

annelids: Lamarck

Anopheles: Driesch, Lamarck, Laveran, Ross

annuities: de Moivre, Simpson, T.

AUDUBON, JOHN JAMES LAFOREST
AUER, CARL *see* VON WELSBACH
Auger electrons: Meitner
aurora borealis: Celsius, Donati, Lovell
aurorae: Alfvén
spectra of: Angström, Goldstein
Australia: Banks, Bok, Bragg, Brunel, I. K., Burnet, Cairns, Cook, Cornforth, De Vaucouleurs, Eggen, Hanbury-Brown, Isaacs, Lamb, Nyholm, Pons, Tabor, Trumpler, Wallace, Wegener, Woolley
Australopithecus spp:
A. africanus: Dart
A. boisei: Leakey, L.S.B.
autoclave: Ipatieff, Papin
autogyro: Cierva
automation: Jacquard
automobile *see* car
automorphic functions: Burnside, Fuchs, Poincaré
autophagy: de Duve
AVERY, OSWALD THEODORE: Kornberg, A.
Avicenna: Paracelsus
AVOGADRO, AMEDEO, CONTE DE QUAREGNA: Meyer, V.
Avogadro's constant: Avogadro
Avogadro's hypothesis: Ampère, Clausius, Einstein, Perrin
Avogadro's law: Avogadro, Berzelius, Cannizzaro, Meyer, V., Regnault
Axelrod, Julius: Snyder
axiom, axiomatic principles: Bernays, Fraenkel, Peano, Skolem, Zermelo
axiom of choice: Fraenkel, Peano, Zermelo
axon: Golgi, Ramón y Cajal, Schwann
giant: Hodgkin, A. L., Young, J. Z.
axoplasm: Hodgkin, A. L.
AYRTON, WILLIAM EDWARD
azo-diesters: Alder
azo dye: Domagk, Perkin
azote: Lavoisier
AZT: Hitchings

B

BAADE, WALTER: Hanbury-Brown, Hubble, Minkowski, R. L., Smith, F. G.
BABBAGE, CHARLES: Sabine
BABCOCK, GEORGE HERMAN: Wilcox
BABCOCK, HAROLD DELOS
Bachman, John: Audubon
bacillus:
anthrax: Pasteur
background radiation: Bondi, Dicke, Gamow, Gold, Penzias, Wilson, R. W.
BACON, FRANCIS
BACON, ROGER
bacteria: Beadle, Burnet, Cairns, Chain, Domagk, Ehrlich, Enders, Fleming, A., Fraenkel-Conrat, Leeuwenhoek, Mechnikov, Twort
anthrax: Koch
gene action in: Jacob
growth of: Hinshelwood

leprosy: Twort
manufacture of interferon: Isaacs
mechanism of lysogeny: Lwoff
bacteriophages (phages): Beadle, Burnet, Cairns, Fraenkel-Conrat, Meselson, Twort
bacteriotropins: Wright, A. E.
BAEKELAND, LEO HENDRIK: Baeyer, Swinburne
Baer, Erich: Fischer, H.O.L.
BAER, KARL ERNST RITTER VON *see* VON BAER
BAEYER, JOHANN FRIEDRICH WILHELM ADOLF VON: Baekeland, Buchner, Fischer, E., Ipatieff, Kekulé, Kipping, Meyer, V., Willstätter
Baeyer's stain theory: Baeyer
BAILY, FRANCIS
Baily's beads: Baily
BAINBRIDGE, KENNETH TOMPKINS
BAIRD, JOHN LOGIE
Bakelite: Baekeland, Baeyer
BAKER, ALAN
BAKER, BENJAMIN
Baker, Valentine Henry: Martin, J.
Bakh, A. N: Oparin
Balanoglossus: Bateson
Balard, Antoine: Berthelot, Sainte-Claire Deville
ball lightning: Kapitza
ballistite: Nobel
balloons:
hot-air: Montgolfier
hydrogen: Charles
observation balloons: Dollfus, Schwarzschild, M.
BALMER, JOHANN JAKOB
Baltimore, David: Temin
Banked Blood: a Study in Preservation: Drew
BANKS, JOSEPH: Brown, R., Davy, Volta
BANNEKER, BENJAMIN
BANTING, FREDERICK GRANT
BARDEEN, JOHN: Cooper
Barbier, P. A: Grignard
barbituric acid: Baeyer, Fischer, E., Volhard
Bardeen–Cooper–Schrieffer theory: Bardeen, Cooper
and impurities: Anderson, P. W.
Barger, G: Dale
barium: Davy
BARKLA, CHARLES GLOVER
Barlow, W: Poynting
BARNARD, CHRISTIAAN NEETHLING
BARNARD, EDWARD EMERSON: Antoniadi
Barnard's star: Barnard, E. E.
Barnell, Miles: Appleton
barometer: Fortin, Pascal, Prout, Torricelli
wheel: Hooke
BARR, MURRAY LLEWELLYN

Barr body: Barr
Barrington, Daines: White
BARROW, ISAAC: Leibniz, Newton
Bartholin, Erasmus: Malus, Römer
BARTLETT, NEIL
BARTON, DEREK HAROLD RICHARD: Prelog
bases (chemical): Brönsted, Lewis
bases, nitrogenous: Cairns; Crick; Gilbert, Walter; Holley; Pyman; Todd
basilar membrane: Békésy
Basov, Nikolay: Townes
Bassi, Laura: Spallanzani
Bateman, Thomas: Addison
BATES, HENRY WALTER: Wallace
BATESON, WILLIAM: Haldane, J.B.S.
bats: Pye
battery, electric *see* accumulator, electric cell
battle fatigue: Adrian
BAYLISS, WILLIAM MADDOCK: Dale, Starling
BCS theory of superconductivity *see* Bardeen–Cooper–Schrieffer theory
BEADLE, GEORGE WELLS: Beadle, Tatum, and Lederberg (joint biography); Kornberg, A.; Lwoff
Beagle, HMS: Darwin
BEAUFORT, FRANCIS: Simpson, G. C.
Beaufort scale: Simpson, G. C.
BEAUMONT, WILLIAM
BECHE, HENRY THOMAS DE LA *see* DE LA BECHE
Becher, Jochim: Stahl, G. E.
Becklin–Neugebauer object: Neugebauer
BECQUEREL, (ANTOINE) HENRI: Andrews, Curie, Röntgen, Rutherford
BEER, GAVIN RYLANDS DE *see* DE BEER
bees: von Frisch
BEG, ULUGH
behaviour: Morris
conditioned: Pavlov
human: Jung
inhibitive: Pavlov
neurotic: Adler
response to external stimuli: Skinner
social, of gulls: Tinbergen
social, of humans: Lorenz, K. Z.
territorial: Tinbergen
BEHRING, EMIL VON: Ehrlich, Kitasato, Koch
Beighton, Henry: Newcomen
BEIJERINCK, MARTINUS WILLEM
BEKESY, GEORG VON
BELL, ALEXANDER GRAHAM: Latimer
BELL, CHARLES: Cheyne, Magendie
BELL, PATRICK
BELL BURNELL, JOCELYN: Hewish
BELTRAMI, EUGENIO
bends, the: Haldane, J. S.
Beneden, Edouard van: Boveri
Bentham, George: Hooker, J. D.
BENZ, KARL FRIEDRICH: Daimler
benzene:

Poisson, Ricci-Curbastro, Riemann, Thom, Whitehead, J.H.C.

differential projective geometry: Burali-Forte

differential rotation theory: Lindblad, Oort

differentiation (biological): Gurdon, Levi-Montalcini, Needham, Wolff, K. F.

differentiation (mathematical) *see* differential calculus

diffraction: Bragg
discovery of: Grimaldi
of electrons: Davisson, Thomson, G. P.
and spectra: Fraunhofer
and transverse wave motion: Fresnel

diffraction analysis: Debye, Lipscomb, Perutz, Polanyi, Svedberg

diffraction gratings: Rutherfurd
invention: Fraunhofer, Rayleigh
concave: Ritter, Rowland

diffuse nebula: Slipher

diffusion: Graham, McBain, Onsager, Urey

digestion: Bernard, Beaumont
gastric juice: Spallanzani
intracellular: de Duve
processes: Schwann
secretory mechanisms: Bernard, Pavlov

Digitalis purpurea: Withering

digoxin: Withering

dilution: Bredig

Dimroth, Otto: Hammick

diode valve: Fleming, J. A.

Dione: Cassini, Kirkwood

Diophantine equations, Diophantine problems: Baker, A., Dickson, Diophantus, Dirichlet, Skolem, Weil

DIOPHANTUS: Heath

Dioptrice: Kepler

Dioptrique: Snell

diphtheria: Ehrlich, Fleming, A., Kitasato, Sherrington

Diplococcus pneumoniae: Avery

dipole moments: Debye

DIRAC, PAUL: McCrea, Salan

direct current: Daniell

directional antenna: Braun, K. F.

DIRICHLET, PETER GUSTAV LEJEUNE: Dedekind, Gauss, Heine, Hermite, Jacobi, Lipschitz, Riemann

Dirichlet's principle: Courant

discharge tube: Aston, Crookes

Disclosing the Past: Leakey, M. D.

discontinuity, theory of: Bateson

Discovery, HMS: Haldane, J.B.S.

discriminants: Cayley, A., Fisher, R. A., Tartaglia, Seki Kowa, Sylvester

Discussion on the Evolution of the Universe: Lemaître

disintegration law: Soddy

dismutation reaction: Cannizzaro

disorder *see* entropy

dispersion: Graham

displacement analysis: Tiselius

displacement law: Soddy

disproportionation reaction: Cannizzaro

dissection: Galen, Harvey, W., Hunter,

Pavlov

dissipative systems: Prigogine

dissociation acidity: Brönsted

dissociation energy: Haber

Distribution of the Stars in Space, The: Bok

divergent series: Borel, Hardy, G. H., Stieltjes

diving: Cousteau

DNA (deoxyribonucleic acid): Avery; Berg; Blakemore; Cairns; Crick; Gilbert, Walter; Gurdon; Lwoff; Meselson; Weismann
composition: Kornberg, A., Sanger

DNA polymerase: Kornberg, A.

Döbreiner, Johann: Henry, W., Newlands

DOBZHANSKY, THEODOSIUS

Doctrine of the Sphere: Flamsteed

DODGSON, CHARLES: Venn

Doering, William: Woodward

DOLLFUS, AUDOUIN CHARLES

Döllinger, Ignaz: von Baer

DOMAGK, GERHARD

DONATI, GIOVANNI BATTISTA

Donati's Comet: Bond, G. P., Donati

DONKIN, BRYAN

Donnan, Frederick: Heyrovsky, Hückel, McBain

Donovan, Charles: Leishman

DOPPLER, JOHANN CHRISTIAN

Doppler effect: Adams, W. S., Burbidge, Doppler, Fizeau, Goldstein, Huggins, Joy, Stark, Vogel

Doraches, 30: Shapley

Dorado: De Vaucouleurs

Doroshkevich, A. G: Zel'dovich

double bond, conjugated: Alder, Diels, Robinson

double refraction: Fresnel, Lorenz, L. V.
Kerr effect and: Kerr
laws of: Malus
use of: Nicol

double stars: Herschel, F. W., Herschel, J.F.W., Mayer, C.
see also binary star systems

double-theta functions: Forsyth

doubling the cube: Descartes, Heath

Downing, A: Newcomb

Draco (constellation): Huggins
Gamma Draconis: Bradley
nebula in: Huggins

dragonflies: Swammerdam

DRAPER, HENRY: Argelander

Draper Catalogue: Draper, Fleming, W.P.S., Pickering, Shapley

dreams: Adrian, Freud, Jung

Drevermann, Friedrich: Edinger

DREW, CHARLES RICHARD

DREYER, JOHN LOUIS EMIL

DRIESCH, HANS ADOLF EDUARD: Nathans

droplet nucleation: Wilson, C.T.R.

dropping mercury electrode: Heyrovsky

dropsy: Withering

Drosophila

(fruit flies): Beadle, Dobzhansky, de Vries, Hogben, Morgan, T. H.

dry cell: Leclanché

D series: Fischer, E.

dualistic theory (chemistry): Avogadro, Berzelius, Dumas

duality: Aleksandrov, Dedekind, Eilenberg, Möbius, Poncelet, Steiner

DUBOIS, MARIE EUGENE FRANCOIS THOMAS: Dulbecco, Temin

DU BOIS-REYMOND, EMIL HEINRICH

du Cros, Arthur Phillip: Dunlop

du Cros, Harvey: Dunlop

Dudley, Dud: Darby

Dufour, Henri: Seguin

DULONG, PIERRE LOUIS: Berzelius, Petit

Dulong and Petit's law: Dulong, Petit, Regnault

DUMAS, JEAN BAPTISTE ANDRE: Andrews, Berthelot, Faraday, Liebig, Newlands, Pelletier, Regnault, Sainte-Claire Deville, Stas, Wurtz

Duncan, James: Newlands

DUNLOP, JOHN BOYD

Dunoyer, Louis: Stern

duodenum: Bayliss, Starling

Dupps, B. F: Perkin

Durchmusterung: Gill, Kapteyn

Dürer, Albrecht: Vesalius

DU TOIT, ALEXANDER LOGIE

Dutton, Clarence E: Powell, J. W.

DUVE, CHRISTIAN RENE DE *see* DE DUVE

Dwyer, F.P.J: Nyholm

Dyckerhoff: Stephan

dyes: Baeyer, Chevreul, Fischer, E., Friedel, Hofmann, Karrer, Mills, Perkin

dynameter: Froude

Dynamical State of Bodily Constituents, The: Schoenheimer

dynamics: d'Alembert, Dirichlet, Galileo, Hertz, Jacobi, Newton, Poincaré, Stevinus, Stieltjes

dynamite: Nobel

dynamo: Crompton, R.E.B., Evelyn Bell, Siemens

dynamo theory for terrestrial magnetism: Bullard

dynamode: Coriolis

dysentery: Kitasato

DYSON, FRANK WATSON

E

e *see* Euler's number

e-process: Burbidge

ear: Békésy
semicircular canals: Galvani

eardrum: Békésy

Earth: Hevesy, Humboldt, Urey
age of: Buffon, Holmes, Hutton, Kelvin, Lyell
atmosphere: Copernicus
chemical composition of crust: Goldschmidt
circumference: Eratosthenes, Pliny
density of: Baily, Boys, Cavendish

optical activity, stereoisomerism, tautomers
isomorphic coprecipitation: Svedberg
isomorphism: Mitscherlich, Svedberg
isoprene: Ipatieff, Staudinger, Ziegler
isotactic: Natta
isothermal processes: Ampère, Carnot
isotope(s): Bainbridge, Thomson, J. J., Zeeman, P.
    artificial production: Cockcroft, Fermi, Lawrence
    atomic mass: Aston, Prout, Richards, Soddy
    biological activity: Cornforth
    dating, use in: Libby
    discovery of: Aston, Richards, Soddy, Svedberg, Urey
    exchange reactions between: Urey
    formation of: Libby
    incorporation techniques: Cairns, Meselson
    isotopic enrichment: Hevesy
    mass spectrography: Aston
    physiological studies, use in: Hevesy
    preparation: Schoenheimer, Seaborg
    Prout's hypothesis and: Morley
    radioactive decay and: Hahn
    separation of: Hevesy, Paneth, Urey
    theory of: Fajans, Soddy
    in water: Emeléus, Urey
    see also radioisotopes, tracers
isotropic cosmic microwave background radiation: Gamow, Gold, Penzias, Wilson, R. W.
Israel, W: Hawking
ISSIGONIS, ALEC
Ivanovski, Dimitri: Beijerinck

J

Jackson, Charles: Morton, Wells
Jackson, Hughlings: Charcot
JACOB, FRANCOIS: Gilbert, Walter; Lwoff; Meselson; Monod
Jacob, Jessie Marie: Muller
JACOBI, KARL GUSTAV JACOB: Cramer, Dirichlet, Galois, Heine, Hermite, Riemann, Weber, H.
Jacobi polynomials: Darboux
JACQUARD, JOSEPH MARIE: Roberts
Jameson, Robert: Forbes
JANSKY, KARL GUTHE: Hanbury-Brown, Wilson, R. W.
JANSSEN, PIERRE JULES CESAR: Lockyer, Rayet
Janus: Dollfus
Java Man: Dubois
Javan, Ali: Maiman
JEANS, JAMES HOPWOOD: Fisher, Spitzer
Jellicoe, Samuel: Cort
JENNER, EDWARD: Hunter, Jenner, Pasteur
JENSEN, JOHANNES HANS DANIEL: Goeppert-Mayer
JESSOP, WILLIAM
jet engine *see* gas turbine

Jodrell Bank: Lovell
Johne's disease: Twort
joints:
    hip: Charnley
JOLIOT-CURIE, IRENE: Curie, Joliot-Curie (joint biography with Frédéric Joliot-Curie)
JOLIOT-CURIE, (JEAN) FREDERIC: Joliot-Curie (joint biography with Irène Joliot-Curie)
JONES, HAROLD SPENCER
Jones, Lennard: Coulson
JORDAN, MARIE ENNEMOND CAMILLE: Hardy, G. H.
JOSEPHSON, BRIAN DAVID: Esaki
Josephson effect: Josephson
JOULE, JAMES PRESCOTT: Seguin
Joule–Kelvin effect *see* Joule–Thomson effect
Joule's law: Joule, Lenz
Joule–Thomson effect: Cailletet, Dewar, Joule, Kelvin
JOY, JAMES HARRISON
JUNG, CARL GUSTAV: Adler, Freud
Juno: Gill, Olbers
Jupiter:
    diameter: Bradley
    eclipse of Io and calculation of speed of light Römer
    mass: de Sitter
    movements: Beg
    orbit: Laplace
    Pliny's observations: Pliny
    Red Spot: Cassini, Schwabe
    rotation: Cassini, Slipher
    satellites of: Barnard, E. E., Bradley, Cassini, de Sitter, Galileo, Lagrange, Römer, Struve, K.H.O.
    spectra: Slipher
    synodic period: Eudoxus
JUST, ERNEST EVERETT
juvenile hormone: Wigglesworth

K

kala-azar: Leishman
kaleidoscope: Brewster
KANT, IMMANUEL: Helmholtz, Weizsacker
Kant–Laplacian theory: Kant
Kantorowitz, Dorothea: Muller
KAPITZA, PYOTR LEONIDOVICH: Penzias, Wilson, R. W.
KAPLAN, VIKTOR
KAPTEYN, JACOBUS CORNELIUS: Babcock, H. D., de Sitter, Dyson, Lindblad, Shapley, Vorontsov-Vel'iaminov
Kapteyn's universe: Kapteyn, Vorontsov-Vel'iaminov
KARMAN, THEODORE VON *see* VON KARMAN
KARRER, PAUL: Haworth, Kuhn, Szent-Györgyi, Wald
Kastner, Karl: Liebig
Kater, Henry: Bond, W. C., Sabine
Kater pendulum: Sabine
KATZ, BERNHARD: Hodgkin, A. L.

KAY, JOHN: Arkwright
K-capture: Kukawa
KEELER, JAMES EDWARD
Keilin, David: Warburg
KEKULE, FRIEDRICH AUGUST VON STRADONITZ: Baeyer, Fischer, E., Kolbe, Liebig, van't Hoff, Wurtz
Keller-Shlierlein, W: Prelog
KELLY, WILLIAM
KELVIN, LORD (WILLIAM THOMSON): Bethe, Ferranti, Lilienthal, Marconi, Onsager, Rankine, Thomson, J. J.
Kelvin scale: Kelvin
Kemmer, Nicholas: Salam
KENDALL, EDWARD CALVIN
KENDREW, JOHN COWDERY: Perutz
Kennelly–Heaviside layer: Appleton, Heaviside
KENYON, JOSEPH
KEPLER, JOHANNES: Brahe, Copernicus, Descartes, Flamsteed, Mästlin, Newton, Scheiner, Snell, Zwicky
Kepler's laws: Flamsteed, Kepler, Newton
Kepler's star: Kepler
KERR, JOHN
Kerr cell: Kerr
Kerr effect: Kerr
ketenes: Staudinger
ketones: Grignard, Norrish, Wittig
KETTLEWELL, HENRY BERNARD DAVID
Keynes, John Maynard: Wittgenstein
KHORANA, HAR GOBIND
Khrushchev, Nikita: Lysenko
kidney: Bright, Galen
    in vitro: Starling
    organ-specific RNA: Stephan
    reaction to blood water content: Haldane, J. S.
    stones: Hales
    transplants: Calne
    tubules: Ludwig
    urinary tubules: Malpighi
Kieselguhr: Nobel
KIMURA, MOTOO
Kinematic Relativity: Milne
kinetic energy: Coriolis
    collisions and: Huygens
kinetic molecular theory: Jeans
Kinetics of Chemical Change: Hinshelwood
kinetic theory of gases: Bernoulli, D., Boltzmann, Clausius, Maxwell
    Brownian motion and: Perrin
King, Albert: Ross
King, Charles: Szent-Györgyi
King, H: Wieland
KIPPING, FREDERICK STANLEY: Lapworth
KIRCHHOFF, GUSTAV ROBERT: Crookes, Fraunhofer, Huggins, Janssen, Lockyer, Rutherfurd
Kirchhoff's circuit laws: Kirchhoff
KIRKWOOD, DANIEL

Lempfert, R: Shaw
LENARD, PHILIPP EDUARD ANTON
LENOIR, JEAN JOSEPH ETIENNE:
　Otto
lenses: Spemann
　aplanatic: Abbe
　apochromatic: Abbe
　magnifying: Bacon, R.
LENZ, HEINRICH FRIEDRICH
　EMIL
Lenz's law: Lenz
LEONARDO DA VINCI: Vesalius
Leonardo Pisano *see* Fibonacci
Leonid meteor shower: Adams, J. C.
Lernberg, Johann: Ostwald
LESSEPS, FERDINAND, VICOMTE
　DE
Lessons in Astronomy: Young, C. A.
Leuckart, Karl: Weismann
leucocytes: Florey, Mechnikov
leukaemia: Isaacs
lever: Archimedes
　and structures: Galileo
LEVERRIER, URBAIN JEAN
　JOSEPH: Adams, J. C. , Challis, Galle,
　Gauss, Rayet
　discovery of Neptune: Adams, J. C.,
　Leverrier
LEVI-CIVITA, TULLIO: Christoffel,
　Lipschitz, Ricci-Curbastro
LEVI-MONTALCINI, RITA
Levine, Philip: Landsteiner
LEWIS, GILBERT NEWTON:
　Langmuir, Onsager, Seaborg, Sidgewick
Lewis–Langmuir octet rule: Linnett
Lewisohn, Richard: Landsteiner
Leyden jar: Franklin, Galvani
LIBBY, WILLARD FRANK
LI CHO HAO
Liddell, Dorothy: Leakey, M. D.
LIE, MARIE SOPHUS: Klein, Sylow
LIEBIG, JUSTUS VON: Arrhenius,
　Baeyer, Hofmann, Kekulé, Kolbe,
　Lister, Mayer, J. R., Regnault, Schwann,
　Williamson, Wöhler, Wurtz
Liebig condenser: Liebig
Lie rings, Lie groups: Hall, P., Lie, Weyl
life, origin of: Eigen, Miller, Prigogine,
　Urey
　Arrhenius's spore theory: Arrhenius
Life on Earth: Attenborough
lift (elevator): Otis
ligand: Longuet-Higgins, Nyholm,
　Werner, A.
ligand field theory: Van Vleck
ligase: Kornberg, A.
light: Adrian
　aberration: Bradley, Jones
　absorption: Hertzsprung, Olbers,
　Stebbins, Struve, F.G.W., Trumpler,
　Vorontsov-Vel'iaminov
　corpuscular theory of: Arago, Compton,
　Newton, Raman
　electromagnetic theory of: Gibbs, Petit
　gravitational deflection: Eddington,
　Trumpler
　nature of: Newton, Römer

　pressure exerted by: Lebedev, Poynting
　scattering of: Raman, Rayleigh, Tyndall
　wave theory of: Arago, Gibbs, Grimaldi,
　Hooke, Huygens, Petit, Young, T.
　see also polarization of light
light, speed of: Arago, Bradley, Cassini,
　Eddington, Einstein, Fizeau, Foucault,
　Jones, Kerr, Lovell, Michelson, Romer
　drift and: Fizeau
　moving media and: Zeeman, P.
　as ratio between units: Weber, W. E.
　theory of: Malus
light curves (of stars): Arp, Baade,
　Schwarzschild, M
lighthouse: Smeaton
lightning:
　ball: Kapitza
　conductor: Banks, Franklin, Galvani
　spectra of: Kundt
　types of: Arago
light reaction (in photosynthesis): Calvin
LILIENTHAL, OTTO: Wright, O.
limit(s), theory of limits: Cauchy,
　Descartes, Feller, Hadamard, Markov,
　Venn, von Mises
Limits of Science, The: Medawar
LINDBLAD, BERTIL: Kapteyn, Oort
Lindblad, Per Olof: Lindblad
Lindemann, Carl Louis Ferdinand von:
　Hermite, Lindemann, C.L.F.
LINDEMANN, FREDERICK
　ALEXANDER (LORD
　CHERWELL): Nernst, Simon
Lindenmann, Jean: Isaacs
linear differential equations: Birkhoff,
　Euler, Forsyth, Fuchs, Jordan, Liouville,
　Picard, Poincaré, Polya, Volterra
　see also differential calculus
linear motor: Laithwaite
linear programming: Dantzig, Fourier
linear transformations: Burali-Forte,
　Cayley, A., Noether, A.
　see also transformation
Lines of Spectra of Some of the Fixed
　Stars: Huggins
Link, Heinrich: Mitscherlich
linkage: Bateson, Haldane, J.B.S.
LINNAEUS, CAROLUS: Lamarck, Ray,
　Tradescant, Withering
LINNETT, JOHN WILFRED
LIOUVILLE, JOSEPH: Baker, A.,
　Darboux, Galois, Lindemann, C.L.F.
LIPSCHITZ, RUDOLF OTTO
　SIGISMUND: Ricci-Curbastro
LIPSCOMB, WILLIAM NUNN: Stock
liquefaction of gases *see* gases
liquid crystals: Onsager
LISSAJOUS, JULES ANTOINE
Lissajous figures: Lissajous
LISTER, JOSEPH: Bridgman
lithium:
　abundance: Herbig
　spectra of: Balmer
litmus test: Boyle
Little, Henry: Bloch
Littlewood, John: Hardy, G. H., Weyl
Liveing, George: Dewar

liver: Bernard, Bloch, Calne, Galen,
　Hodgkin, T., Laennec, Leishman
Living Planet, The: Attenborough
LOBACHEVSKY, NIKOLAI
　IVANOVICH: Cayley, A., Euclid,
　Klein, Riemann
local supergalaxy: De Vaucouleurs
lock: Bramah
lock-and-key hypothesis (enzymes):
　Fischer, E.
LOCKE, JOSEPH
LOCKYER, JOSEPH NORMAN:
　Janssen
locomotive: Ericsson, Roberts,
　Stephenson, G., Stephenson, R.,
　Trevithick, Westinghouse
LODGE, OLIVER JOSEPH
Loewi, Otto: Dale, Katz
Löffler, Friedrich: Koch
logarithms: Babbage, Briggs, Erlang,
　Napier
logic: Bernays, Peano, Russell, B.A.W.,
　Venn, Weyl, Whitehead, A. N.
　mathematical: Boole, Gödel, Herbrand,
　Hilbert, Skolem, Von Neumann,
　Whitehead, A. N.
Logic of Modern Physics, The: Bridgman
Logique, langage et théorie de
　l'information: Mandelbrot
Loligo forbesi: Hodgkin, A. L.
Long, Crawford: Morton, Wells
longitude: Bacon, R.
　determination of: Harrison
LONGUET-HIGGINS, HUGH
　CHRISTOPHER: Stock
LONSDALE, KATHLEEN
loom: Bigelow, Cartwright, Jacquard, Kay
Lord, Leonard: Issigonis
LORENTZ, HENDRIK ANTOON
　Einstein, FitzGerald, Lorenz, L. V.,
　Thomson, J. J., Zeeman, P.
Lorentz contraction:
　ether and: Lorentz, Michelson
　lengths: FitzGerald
　theory of: Einstein
Lorentz–Lorenz formula: Lorenz, L. V.
LORENZ, KONRAD ZACHARIAS:
　Tinbergen, von Frisch
LORENZ, LUDWIG VALENTIN
Lorenz, Richard: Hevesy
Loschmidt, Joseph: Boltzmann
Lothar Meyer curve: Meyer, J. L.
Louis XVI, king of France: Lamarck
louse: Ricketts
LOVELL, ALFRED CHARLES
　BERNARD: Hanbury-Brown
LOWELL, PERCIVAL: Antoniadi,
　Tombaugh
Lowry, Thomas: Brönsted
LSD (lysergic acid diethylamide):
　Woodward
L-series: Barkla
lubricants: Bowden, Coulomb, Tabor
Luddites: Heathcoat
Ludwig, Ernst: Kuhn
LUDWIG, KARL FRIEDRICH
　WILHELM: Du Bois-Reymond,

Pavlov

LUMIERE, AUGUST MARIE LOUIS NICOLAS: Friese-Green, Lumière (joint biography with Louis Lumière)

LUMIERE, LOUIS JEAN: Friese-Green, Lumière (joint biography with August Lumière)

luminescence: Lenard, Röntgen

luminosity: Eddington, Russell, H. N.

LUMMER, OTTO RICHARD

Lummer–Brodhun cube: Lummer

Lummer fringes: Lummer

Lummer–Gehrke interference spectroscopy: Lummer

lunar theory: Adams, J. C., Brown, E. W., Flamsteed

Lush, Dr: Kimura

luteinizing-hormone-releasing hormone (LHRH): Guillemin

LWOFF, ANDRE MICHAEL: Gilbert, Walter; Jacob; Monod

Lyddite: Robertson

LYELL, CHARLES: Agassiz, Buffon, Darwin, Hutton, Huxley, T. H., Murchison

LYMAN, THEODORE

Lyman ghosts: Lorentz

Lyman series: Balmer, Lyman

lymphoma: Hodgkin, T.
Burkitt's: Burkitt

LYNDEN-BELL, DONALD

LYNEN, FEODOR: Bloch

Lyons, Israel: Banks

LYOT, BERNARD FERDINAND: Dollfus

lyphobic sols: Bredig

LYSENKO, TROFIM DENISOVICH: Haldane, J.B.S., Muller, Sakharov

lysergic acid: Woodward

lysogeny: Lwoff

lysosome: de Duve

lysozyme: Chain, Fleming, A., Florey

LYTTLETON, RAYMOND ARTHUR

M

MCADAM, JOHN LOUDON: Telford

MACARTHUR, ROBERT HELMER

MacArthur's warblers: MacArthur

MCBAIN, JAMES WILLIAM

McCarthy, Maclyn: Avery

MCCLINTOCK, BARBARA: Ferguson

MCCOLLUM, ELMER VERNER

MCCORMICK, CYRUS HALL: Hussey

MCCREA, WILLIAM HUNTER

MACH, ERNEST

Mach angle: Mach

Mach bands: Mach

Mach number: Mach

machine gun: Gatling, Maxim

machine tools: Bramah, Brunel, M. I., Maudslay, Nasmyth, Roberts, Whitney, Whitworth

MacIntosh, Charles: Hancock

McIntyre, John: Hofstadter

Mackenzie, Kenneth: Segrè

MACLAURIN, COLIN: Cramer

MacLeod, Colin: Avery

MacLeod, John: Banting

McMillan, Edwin: Lawrence, Meitner, Seaborg

MCNAUGHT, WILLIAM

Macquer, Pierre: Berthollet

Macromolekulare Chemie und Biologie: Staudinger

Madson, Thorvald: Dale

Magellanic Clouds: Arp, Herschel, F. W., Hubble, Leavitt, Payne-Gaposchkin, Shapley
Cepheids: Hubble, Leavitt
distance: Leavitt, Shapley
magnetism: Airy, Gauss, Van Allen
Small: Hubble

MAGENDIE, FRANCOIS: Bernard, Pelletier

magenta: Perkin

'magic bullets': Ehrlich

magic numbers (in chemical nuclei): Goeppert-Mayer, Jensen

magnesium: Bunsen, Davy, Willstätter

magnetic fields: Alfvén, Ampère, Anderson, P. W.
cooling and: Simon
radioactivity and: Becquerel
terrestrial: Appleton; Blackett; Bullard; Gauss; Gilbert, William; Sabine; Weber, W. E.

magnetic induction: Lenz

magnetic monopole: Dirac

magnetic permeability: Kelvin

magnetic solids: Anderson, P. W.

magnetic susceptibility: Curie, Kelvin
temperature dependence of: Langevin

magnetism: Ampère; Arago; Coulomb; Faraday; Gilbert, William; Lipscomb
electric current and: Oersted
temporary: Arago
terrestrial: Adams, J. C.
theory of types: Kelvin
types of: Faraday

magnetohydrodynamics: Alfvén

magnetosphere: Alfvén
discovery of: Van Allen

magnitude (of a star):
absolute: Adams, W. S., Hertzsprung, Joy, Redman, Russell, H. N.
apparent: Argelander, Cannon, Joy, Pickering
photographic determination: Bond, G. P., Jones, Leavitt, Pickering
photovisual: Pickering
UBV system: Pickering

Magnitude and Distance of the Sun and Moon, On the: Aristarchos

MAIMAN, THEODORE HAROLD

main-sequence stars: Hertzsprung, Russell, H. N.

malaria: Burkitt, Golgi, Laveran, Manson, Pelletier, Ross

malic acid: Scheele, Szent-Györgyi, Vauquelin

malignant reticulosis: Hodgkin, T.

malonyl coenzyme A: Lynen

MALPIGHI, MARCELLO:

Leeuwenhoek

MALTHUS, THOMAS ROBERT: Darwin

MALUS, ETIENNE LOUIS

Malus's theorem: Malus

Mammals, The Age of: Osborne

Manchester brown: Perkin

MANDELBROT, BENOIT B

manganese: Scheele

Manhattan Project *see* atomic bomb

manifold (in topology): Dehn, Riemann, Thom

MANNESMAN, REINHARD

manometer: Hales, Pfeffer, Stopes, Warburg

MANSON, PATRICK: Ross

Manual of Astronomy: Young, C. A.

mapping (mathematical) *see* transformation

maps: Airy, Anaximander the Elder, Cook, Du Toit, Eratosthenes, Ericsson, Mayer, J. T., Ptolemy, Tansley
see also astronomical maps, genetic maps, geological maps

MARCONI, GUGLIELMO: Appleton, Armstrong, E. H., Braun, K. F., Fessenden, Hertz, Rutherford

Marcus Aurelius: Galen

marine geophysics: Bullard

MARKOV, ANDREI ANDREEVICH

Markov processes, Markov chains: Feller, Markov

Mars:
atmosphere: Campbell, Janssen, Kuiper, Sagan
'canal' system: Antoniadi, Keeler, Lowell, Masursky, Schiaparelli, Trumpler
channels: Masursky
clouds: Sagan
deserts: Dollfus
diameter: Jones, Schiaparelli
first photographs: Lowell
life on: Lowell, Schiaparelli
life-supporting capability: Campbell
mapping: Schiaparelli
Mariner space probes: Gill, Kuiper, Masursky, Trumpler
movement: Beg
orbit: Kepler
period: Huygens
Pliny's observations: Pliny
rotation: Antoniadi, Cassini, Herschel, F. W., Slipher
sandstorms: Lyot
satellites: Hall, A., Kant
seasons: Cassini, Lowell
surface features: Masursky, Sagan, Schiaparelli, Trumpler
synodic period: Eudoxus
Viking lander: Campbell, Dollfus, Masursky
water vapour: Janssen

Marsden, Ernest: Geiger

marsupials: Banks

MARTIN, ARCHER JOHN PORTER: Synge, Tswett

nematode worms: Brenner
neo-Darwinism: Kimura, Simpson, G. G.,
  Steele
neon: Aston, Ramsay, Travers
neoprene: Carothers, Flory
Neosalvarsan: Pyman
nephelometer: Richards
Neptune:
  discovery: Adams, J. C., Bessel, Bode,
    Challis, Encke, Galle, Gauss
  mass: Struve, O. W.
  methane on: Kuiper
  Pluto's effect on orbit: Brown, E. W.,
    Lowell
  satellites: Kuiper, Lyttleton, Struve,
    K.H.O.
  spectra: Kuiper
Neptunism: Werner, A. G.
neptunium: Meitner, Soddy
Nereid: Kuiper
NERNST, HERMANN WALTHER:
  Arrhenius, Bergius, Brönsted,
  Langmuir, Lewis, Lindemann, F. A.,
  Millikan, Polanyi, Richards, Simon
Nernst equation: Nernst
nerve cells see neurons
nerve impulses: Adrian, Golgi
  chemical transmission of: Dale
  mechanics of: Hodgkin, A. L.
  produced by stimulus: Haller
  role in nervous function: Ramón y Cajal
  transmission of: Hodgkin, A. L.
nerves:
  accelerator: Ludwig
  cranial: Galen
  depressor: Ludwig
  endings: Adrian, Békésy, Dale, Katz,
    Malpighi
  fibres: Adrian, Bell, C., Sherrington
  growth factor: Levi-Montalcini
  parasympathetic: Dale
  relationship with muscles: Haller
  secretory: Ludwig
  squid: Young, J. Z.
nervous system
  effects of rabies: Pasteur
  human: Adrian, Golgi, Hodgkin, A. L.,
    Ramón y Cajal, Sherrington, Theiler,
    Vesalius
  invertebrates: Child
  nematode: Brenner
  octopuses: Young, J. Z.
  physiology of: Katz
NEUGEBAUER, GERALD
Neumann, F. E: Regnault
NEUMANN, JOHN VON see VON
  NEUMANN
neural folds: Malpighi, von Baer
neural tube: Malpighi
neurofibrils: Ramón y Cajal
neurology: Bell, C., Freud, Haller, Ramón
  y Cajal
neurons (nerve cells): Adrian, Barr,
  Guillemin, Hodgkin, A. L., Ramón y
  Cajal, Sherrington, Young, J. Z.
  sensory and motor: Golgi
neuroses: Adler, Eysenck

Neurospora crassa: Beadle
neurosurgery: Cushing
neurotransmitters: Katz
neutral evolution: Kimura
neutralization, heats of: Andrews
neutrino: Fermi, Pauli
  beta decay and: Meitner
  emission: Gamow, McCrea
  supernovae and: Gamow
neutron: Seaborg
  discovery of: Chadwick
  fission of: Frisch, Meitner
  radioactivity and: Fermi
neutron stars: Bell Burnell, Bethe,
  Burbridge, Chandrasekhar, Hewish,
  Zwicky
NEWCOMB, SIMON
NEWCOMEN, THOMAS: McNaught,
  Smeaton, Watt
**NEWLANDS, JOHN ALEXANDER
  REINA**
New System of Chemical Philosophy:
  Dalton
Newton, Gilbert: Urey
NEWTON, ISAAC: Adams, J. C.,
  Barrow, Bernoulli, J., Buffon, Cassini,
  Celsius, de Moivre, Fermat, Flamsteed,
  Galileo, Gauss, Grandi, Halley,
  Hamilton, W. R., Kant, Kepler, Leibniz,
  Maclaurin, Zeeman, P.
  gravitational theory: Adams, J. C.,
  Galileo, Huygens, Kepler, Newton
  laws of motion: Newton
  Principia Mathematica: Halley, Newton
  telescope: Cassegrain, Newton
Newton's cooling law: Stefan
Newton's rings: Newton
Newtonian cosmology: McCrea
Nibel, Hans: Benz
NICE, MARGARET MORSE
nickel: Mitscherlich, Sabatier, Ziegler
NICOL, WILLIAM
Nicol prism: Nicol
**nicotinamide adenine dinucleotide
  phosphate** see NADP
nicotinic acid: Funk
Niehans, Paul: Stephan
niobium: Wöhler
**NIRENBERG, MARSHALL
  WARREN:** Holley and Nirenberg (joint
  biography)
Nishijima, I: Gell-Mann
nitric acid: Eyde, Glauber, Haber,
  Ostwald, Wieland
nitric oxide: Priestley
  as a catalyst: Désormes
nitriles: Kolbe
nitrobenzene: Mitscherlich
nitrocellulose: Chardonnet, Nobel,
  Robertson
nitrogen: Haldane, J.B.S., Ludwig
  ammonia production: Haber
  atmospheric: Lavoisier, Ramsay
  atoms: Krebs
  bonding in: Werner, A., Wieland
  catalytic hydrogenation: Haber
  discovery of: Scheele

  discovery of rare gases: Ramsay
  fertilizers: Liebig
  free radicals of: Wieland
  liquefaction of: Cailletet
  nitrogen-14: Meselson
  nitrogen-15: Meselson, Schoenheimer,
nitroglycerin: Nobel
  organic compounds: Dumas, Wieland
  radioactive decay: Libby
nitrogen bases: Holley
nitrogen cycle: Berthelot, Liebig
nitrogen trichloride: Dulong
nitromethane: Kolbe
nitrous oxide: Davy, Morton, Nernst,
  Priestley, Robertson, Wells
Nix Olympia: Masursky
no hair theorem: Hawking
NOBEL, ALFRED BERNHARD
Nobel, Immanuel: Nobel
nobelium: Nobel, Seaborg
Nodder, C. R: Mills
NOETHER, AMALIE (EMMY): Ore
NOETHER, MAX: Noether, A.
noncommutative (fields, algebras):
  Noether, A., Ore, Von Neumann,
  Wedderburn
non-Euclidean geometry, space: Bolyai,
  Cayley, A., Cesaro, Clifford, Coolidge,
  Eisenhart, Euclid, Gauss, Klein, Lie,
  Lobachevsky, Poincaré, Riemann,
  Sommerville
  see also geometry, Riemann geometry
non-Riemann geometry: Eisenhart
nonstoichiometric compounds:
  Berthollet
normal (Gaussian) distribution: de
  Moivre, Gosset, Pearson
normality: Wollaston
**NORRISH, RONALD GEORGE
  WREYFORD:** Eigen, Porter, G.
Norrish reactions: Norrish
North American Nebula: Wolf
Northern Hemisphere Observatory:
  Smith, F. G.
**North Polar Sequence (circumpolar
  stars):** Jones
Nothosaurus: Edinger
notochord: Bateson, von Baer
Nova Auriga: Campbell
Nova Cygni: Vogel
novae: Fleming, W.P.S.
  formation: Milne
  light curves: Arp
  luminosity: Shapley
  magnitude: Arp
  in nebulae: Hubble
  neutrino emission: Gamow
  pulsating: Arp
  spectra: Huggins, Jones
Novikov, I. D: Zel'dovich
Novum Organum: Bacon, F.
Noyes, A. A: Lewis
nuclear division: Weismann
nuclear energy: Dale
nuclear fission: Becquerel, Bohr, N.H.D.,
  Chadwick, Cockcroft, Fermi,
  Joliot-Curie, Libby, Seaborg

discovery of: Frisch, Hahn, Meitner
nuclear fusion: Alfvén, Bohr, A. N.,
  Thomson, G. P.
  controlled: Kapitza, Sakharov
  first reaction: Rutherford
  stellar: Bethe
nuclear isomers: Hahn
nuclear magnetic resonance (NMR):
  Lipscomb, Nyholm
nuclear physics: Atkinson, Bethe, Fowler,
  Gamow, Giacconi, Ginzburg, Hayashi,
  Lyttleton, Weizsäcker, Zel'dovich
*Nuclear Shell Structure, Elementary
  Theory of:* Goeppert-Mayer
nuclear transplantation: Gurdon
nucleic acids: Crick, Fraenkel-Conrat,
  Khorana
  biosynthesis: Ochoa
  genetics and: Kornberg, A., Staudinger
  linkages: Ochoa
  protein biosynthesis: Kornberg, A.
  structure of: Sanger, Todd
  synthesis of: Kornberg, A., Ochoa
  see also DNA, life, nucleotides, RNA
nucleophiles: Ingold, Lapworth, Robinson
nucleoproteins: Barr
nucleosynthesis: Alpher, Burbidge,
  Gamow
nucleotides: Holley, Kornberg, A., Ochoa,
  Sanger, Todd
  coenzymes and: Harden, Todd
  composition of: Todd
  DNA: Holley, Kornberg, A., Ochoa,
  Sanger
  genetic code: Holley, Khorana,
  Kornberg, A.
  protein biosynthesis: Holley, Khorana,
  Kornberg, A.
  RNA: Holley, Ochoa, Sanger
  synthesis of: Fischer, E., Todd
  see also nucleic acids
nucleus, atomic: Anderson, C. D., Barr,
  Becquerel, Bethe, Boveri, Brown, R.,
  Cairns, Gurdon, Purkinje, Schleiden
  collective and particle motion: Bohr,
  A. N.
  composition of: Herzberg
  discovery of: Rutherford
  disintegration of: Cockcroft, Fermi,
  Lawrence, Rutherford, Walton
  gamma ray absorption: Mössbauer
  liquid-drop model: Bohr, N.H.D.,
  Gamow
  magic number and: Jensen
  magnetic and quadruple moments:
  Tolansky
  shell model of: Bohr, A. N.
  spin of: Heisenberg, Polanyi, Tolansky
  paramagnetism: Simon
nucleus, cell: Brown, Schleiden,
  Weismann
number theory, theory of numbers:
  Cauchy, Cesaro, Dedekind, Dickson,
  Diophantus, Dirichlet, Euclid, Fermat,
  Fuchs, Galois, Gauss, Hadamard,
  Hardy, G. H., Hermite, Jacobi, Klein,
  Kronecker, Lagrange, Legendre,

Lipschitz, Markov, Minkowski, H.,
  Nagell, Ore, Ramanujan, Stieltjes,
  Sylvester, Tartaglia, Von Neumann,
  Weber, H., Weil, Weyl, Zassenhaus
nutation: Bradley, Lyttleton
Nutcracker Man: Leakey, L.S.B.
Nuttall, John: Geiger
NYHOLM, RONALD SYDNEY
nylon: Carothers, Flory

O

Oberon: Herschel, F. W.
Oberth, Hermann: Tsiolkovskii
objective: Abbe
observational astronomy: Barnard, E. E.,
  Cassini, Dreyer
Occhislini, Giuseppe: Powell, C. F.
occlusion: Graham
occultation: Struve, G.W.L.O.
oceanography: Cousteau
OCHOA, SEVERO: Kornberg, A.
octane: Ipatieff
  rating: Ricardo
octaves, law of: Newlands
octopuses: Young, J. Z.
oedema: Bright, Withering
Oedipus complex: Freud
Oenothera grandiflora, lamarckiana: de
  Vries
OERSTED, HANS CHRISTIAN
oestrogens: Robinson
Oettingen, Arthur von: Ostwald
O'Grady, Marcella: Boveri
OHM, GEORG SIMON
Ohm's law: Clausius, Heaviside,
  Kirchhoff, Ohm, Wheatstone
oil beetles: Fabre
oil, extraction from coal: Alter
**OLBERS, HEINRICH WILHELM
  MATHAUS:** Bessel, Piazzi
Olbers's paradox: Olbers
*Olduvai Gorge: My Search for Early
  Man:* Leakey, M. D.
Oliver, Francis: Tansley
Oliver, George: Sharpey-Schafer
ONNES, HEIKE KAMERLINGH
ONSAGER, LARS
Onsager's law of reciprocal relations:
  Onsager
Onsager's limiting law: Onsager
ontogeny: Haeckel
OORT, JAN HENDRIK: Jansky,
  Kapteyn, Lindblad, Smith, F. G.
OPARIN, ALEXANDR IVANOVICH:
  Miller
Opera Omnia Chymica: Glauber
operant conditioning: Skinner
operons: Monod
ophthalmometer: Helmholtz
ophthalmoscope: Helmholtz
opiates: Pelletier
  endogenous: Guillemin, Snyder
OPIK, ERNST JULIUS
OPPENHEIMER, JULIUS ROBERT:
  Chandrasekhar, Penrose
**OPPOLZER, THEODOR EGON
  RITTER VON**

opsin: Wald
opsonins: Wright, A. E.
Optica Promota: Cassegrain
Opticae Thesaurus: Alhazen
optical activity:
  carbohydrate nomenclature and:
  Fischer, E.
  coordination compounds, in: Werner, A.
  isomerism: Berzelius, Kenyon, Mills,
  van der Waals, Werner, A.
  molecular structure: Mills, van der
  Waals
  muta-rotation in sugars: Onsager
  obstacle theory of: Kenyon
  oximes: Mills
  racemates: Berzelius
  in secondary alcohols: Kenyon
  in tartarates: Berzelius
  theories of: Kenyon, Mills
optical dilatometer: Abbe
optical identification of radio source:
  Arp
Opticks: Newton
optic nerve: Cushing
optics: Abbe, Alhazen, Arago, Bacon, R.,
  Bragg, Brewster, Fabry, Fizeau,
  Foucault, Fraunhofer, Fresnel, Gabor,
  Galileo, Grimaldi, Helmholtz, Hooke,
  Huygens, Kepler, Lorenz, L. V., Malus,
  Michelson, Morley, Newton, Nicol,
  Rayleigh, Rowland, Schmidt, Snell,
  Young, T.
optic vesicles: Malpighi
Opus Astronomicum: Schiaparelli
Opus Majus: Bacon, R.
Opus Minus: Bacon, R.
Opus Tertium: Bacon, R.
Orbital Solar Observatory IV: Goldberg
**orbital theory of conjugated organic
  molecules** Longuet-Higgins
ordinal numbers: Burali-Forte, Cesare
Ordovician: Sedgwick
ORE, OYSTEIN
organelles: de Duve, Nathans, Woodger
Organic Chemistry: Kipping
Organic Chemistry, Textbook of:
  Meyer, V.
organometallic compounds: Emeléus,
  Fischer, E. O. and Wilkinson, G.,
  Gilman, Ziegler
organosilicon polymers: Kipping
organosols: Svedberg
organs: Sherrington
  transplants: Burnet, Calne, Temin
Origin of Continent and Ocean: Wegener
Origin of Species, On the: Darwin,
  Dubois, Fabre, Galton, Gray, A.,
  Huxley, T. H., Lyell, von Baer, Wallace
*Original Theory or a New Hypothesis of
  the Universe:* Kant, Wright, T.
Orion Nebula: Bond, W. C., Herschel, F.
  W., Huggins, Huygens, Neugebauer
ortho-hydrogen: Heisenberg
  specific heat anomaly: Simon
orthopaedics: Charnley
Ortus Medicinae: Helmont
osazones: Fischer, E.

Davy, Dumas, Mitscherlich, Petit, Prout, Richards, Stas
Dulong and Petit's law: Dulong, Petit
equivalent weight: Cannizzaro, Kolbe, Newlands, Petit
hypothesis of: Prout, Stas
isotopes: Fajans, Soddy
molecular mass: Cannizzaro
octave law: Newlands
periodicity: Mendeleyev, Meyer, J. L., Newlands
radioactive decay: Fajans, Richards, Soddy
specific heat capacity: Dulong, Petit
universal gas constant: Petit
vapour density: Cannizzaro
relative molecular mass: Cannizzaro, Debye, Dumas, Meyer, V.
relativistic beaming theory: Smith, F. G.
relativity, general theory of: Adams, W. S., Bondi, Campbell, de Sitter, Dicke, Dyson, Eddington, Einstein, Friedmann, Hubble, Leverrier, Mach, Milne, Penrose, Schwarzschild, K., Trumpler
relativity, kinematic: Milne
relativity, mathematical theory of:
Eisenhart, Forsyth, Gödel, Hawking, Hilbert, Hogben, Klein, Levi-Civita, Minkowski, H., Noether, A., Poincaré, Ricci-Curbastro, Riemann, Weyl, Whitehead, A. N., Zeeman, P.
relativity, special theory of: Bhabha, Blackett, Einstein, FitzGerald
electromagnetism and: Lorentz
ether and: Lodge
mass increase and: Chadwick
proposal of: Langevin
verification of: Laue, Michelson, Mössbauer, Zeeman, P.
Relics of the Deluge: Buckland
REMINGTON, PHILO
replication: Avery, Cairns, Meselson
conservative: Crick
of DNA in lysogeny process: Lwoff
semiconservative: Crick
reproduction, sexual *see* sexual reproduction
reserpine: Woodward
resonance (chemical): Kekulé, Longuet-Higgins
resonance hybridization: Pauling
resonator: Lebedev
respiration: Cheyne, Fabricius, Lavoisier, Lynen, Szent-Györgyi, Wieland
deep-sea diving: Haldane, J.B.S., Haldane, J. S.
enclosed spaces: Hales
intracellular: Warburg
lungs: Swammerdam
mines: Haldane, J. S., Hales
resting potential: Hodgkin, A. L.
restriction enzymes: Nathans
reticulo-endothelial system: Leishman
retina: Wald
retrogressive mutants: de Vries
reversible processes (chemical):
Onsager, Williamson

REYNOLDS, OSBORNE
Rhea: Cassini, Kirkwood
rhesus factor: Fisher, Landsteiner
rhodium: Nyholm, Wollaston
Rhodnius prolixus: Wigglesworth
rhododendron: Hooker, J. D.
rhodopsin: Wald
bovine: Khorana
riboflavin (vitamin B$_2$): Karrer, Kuhn
ribonucleic acid *see* RNA
ribosomes: Hoagland, Meselson
RICARDO, HARRY RALPH
Riccati's differential equation:
Bernoulli, D.
RICCI-CURBASTRO, GREGORIO:
Christoffel, Levi-Civita, Lipschitz
Riccioli, Giovanni Battista: Cassini
RICHARDS, THEODORE WILLIAM:
Soddy, Stas
Richer, Jean: Cassini
rickets: Funk, McCollum
RICKETTS, HOWARD TAYLOR
Rickettsia: Ricketts
R. burneti: Burnet
R. rickettsii: Ricketts
Rideal, E. K: Norrish
RIEMANN, GEORG FRIEDRICH BERNHARD: Artin, Betti, Cayley, A., Christoffel, Dedekind, Euclid, Fuchs, Gauss, Klein, Ricci-Curbastro, Weber, H.
Riemann geometry: Eisenhart, Riemann, Weyl
Riemann integral, integration: Darboux, Lebesgue, Stieltjes
Riemann surface, Riemann space(s):
Christoffel, Clifford, Levi-Civita, Lipschitz, Riemann, Weyl
rifle: Remington
Riggs, John: Wells
Righi, Augusto: Lebedev
Rittenberger, David: Bloch
Ritter, A: Schwarzschild, M.
RITTER, JOHANN WILHELM
rivetting machine: Fairbairn
Rizicka, Leopold: Prelog
RNA (ribonucleic acid): Berg, Blakemore, Holley, Kimura, Kornberg, A., Milstein, Ochoa, Sanger, Stephan
messenger: Crick, Hoagland, Jacob, Meselson, Monod
transfer: Berg, Crick, Hoagland
see also DNA; life, origin of
road-building: McAdam, Telford, Trésaguet
Robbins, Frederick: Ehrlich, Sabin
Robert, Nicolas: Charles, Donkin
ROBERTS, RICHARD
Robertson, Howard: Poynting
ROBERTSON, ROBERT
ROBINSON, ROBERT: Cornforth, Ingold, Lapworth, Prelog, Todd
rockets: Goddard, Korolev, Tsiolkovskii, Van Allen, von Braun, von Kármán
fuel: Stock
rockoon: Van Allen
Rocky Mountain spotted fever: Ricketts

rods, retinal: Wald
ROE, ALLIOT VERDON
Roebling, John: Ellet
Roebuck, John: Watt
Roemer, Claus: Fahrenheit
roll film: Eastman, Friese-Green
Rolls, Charles: Royce
Romanowsky stain: Leishman
ROMER, ALFRED SHERWOOD:
Edinger
ROMER, OLE (OLAUS) CHRISTENSEN: Bradley, Cassini
RONTGEN, WILHELM KONRAD:
Cannon, Curie
rosaniline: Fischer, H.O.L., Hofmann, Perkin
Rose, Caleb: Sherrington
Rosenheim, O: Wieland
Rosing, Boris: Baird
Ross, James Clark: Hooker, J. D.
Ross, John: Sabine
ROSS, RONALD: Laveran, Manson
roundworm: Boveri
Roux, Emil: Behring
ROUX, WILHELM
Rowell, J. M: Josephson
ROWLAND, HENRY AUGUSTUS:
Röntgen
Royal Astronomical Society: Baily
Royal Institution, founding of: Rumford
Royal Society of London: Banks, Boyle, Davy, Hooke, Leibniz, Newton, Papin, Wallis, J.
ROYCE, FREDERICK HENRY
Royds, Thomas: Rutherford
Rozier, Francois Pilâtre de *see* de Rozier
rubber: Dunlop, Goodyear, Hancock
composition: Flory, Ipatieff, Ziegler
high-hysteresis: Tabor
synthetic: Carothers, Flory, Staudinger, Svedberg, Ziegler
Rubens, Heinrich: Lummer
rubidium: Bunsen
Rudolphine Astronomical Tables: Kepler
RUE, WARREN DE LA *see* DE LA RUE
RUFFINI, PAOLO
RÜHMKORFF, HEINRICH DANIEL
RUMFORD, COUNT (BENJAMIN THOMPSON): Davy, Lavoisier
RUSSELL, BERTRAND ARTHUR WILLIAM: Bernays, Boole, Brouwer, Burali-Forte, Frege, Gödel, Peano, Whitehead, A. N., Wiener
RUSSELL, FREDERICK STRATTEN
RUSSELL, HENRY NORRIS:
Goldberg, Hertzsprung, Hubble, Menzel, Shapley, Young, C. A.
ruthenium: Hess, G. H.
ruthenocene: Fischer, E. O. and Wilkinson, G.
RUTHERFORD, ERNEST: Atkinson, Dale, Gamow, Hevesy, Paneth, Sidgewick, Soddy
RUTHERFURD, LEWIS MORRIS
RV Tauri: Aristotle
RYDBERG, JOHANNES ROBERT
Rydberg constant: Rydberg

parallax: Flamsteed, Gill, Halley, Jones, Maskelyne
periodicity: Schwabe
photographic analysis: Hale, Schwabe, Schwarzschild, M.
photography: Foucault, Janssen
plasma clouds: Hewish
prominences: Goldberg, Janssen, Lockyer, Secchi
radiation pressure: Milne
radio emission: Deslandres, Hey, Lovell, Ryle, Smith, F. G.
radio map: Ryle
rotation: Adams, W. S., Carrington, Hevelius, Kepler, Lyot, Young
'Russell mixture': Russell, H. N.
secular acceleration: Jones
shape: Dicke
solar flares *see* solar flares
spectra: Adams, W. S., Babcock, H. D., Bowen, I. S., Deslandres, Donati, Edlen, Hale, Herschel, F. W., Huggins, Janssen, Lockyer, Lyot, Menzel, Redman, Russell, H. N., Secchi, Vogel, Young
structure: Donati, Eddington, Schwarzschild, M.
sunspots: de la Rue, Galileo, Henry, J., Scheiner
surface: Foucault, Janssen, Schwarzschild, M.
telluric rays: Janssen
temperature: Hale
superexchange: Anderson, P. W.
supercluster: De Vaucouleurs
superconductors: Anderson, P. W., Dewar, Josephson
coherence and: Pippard
discovery of: Onnes
magnetic fields and: Laue
microwaves and: Pippard
theory of: Bardeen
superego, the: Freud, Piaget
superfluidity: Anderson, P. W., Feynman
of helium: Kapitza
supergalaxies: De Vaucouleurs
superheterodyne receivers: Armstrong, E. H.
supernovae: Burbidge, Chandrasekhar, Minkowski, R. L., Zwicky
in Cassiopeia: Brahe, Ryle
cosmic-ray acceleration: Ginzburg
distance: Baade
neutrino emission: Gamow
radio emission: Minkowski, R. L., Oort, Ryle
remnants: Minkowski, R. L.
spectra: Humason, Minkowski, R. L., Zwicky
supersynthesis (radio interferometer system): Ryle
Supplementum de Spandanis Fantibus: Helmont
supraconductivity: Cooper, Onnes
suprarenal capsules: Addison
Surely You're Joking, Mr Feynman!: Feynman
surface(s): Cesaro, Christoffel, Darboux, Eisenhart, Jacobi

surface tension: Boys
surgery: Colles, Cushing, Hunter
antiseptic: Lister
aseptic: Lister
cardiothoracic: Barnard, C. N.
dental: Morton
ether anaesthesia used in: Morton
kidney transplant: Calne
tissue transplant: Burnet
susceptibility, magnetic: Kelvin
SUTHERLAND, EARL WILBUR
SUTHERLAND, GORDON
SUTHERLAND, IVAN EDWARD
SU(3) theory: Gell-Mann
SUTTON-PRINGLE, JOHN WILLIAM
SVEDBERG, THEODOR: Bredig, Tiselius
SWAMMERDAM, JAN
Swammerdam valves: Swammerdam
SWAN, JOSEPH WILSON: Edison
sweat glands: Boerhaave, Purkinje
SWINBURNE, JAMES
SWINGS, POL F
Swings bands: Swings
Swings effect: Swings
Swinton, Campbell A. A: Baird
SYLOW, LUDWIG MEJDELL
SYLVESTER, JAMES JOSEPH
symbolic logic: Boole, Frege, Leibniz, Peano, Venn
see also logic, mathematical logic
synapses: Sherrington
synchrotron particle accelerators: Lawrence
synchrotron radiation: Oort, Seyfert
SYNGE, RICHARD LAWRENCE MILLINGTON: Martin, A.J.P., Tswett
Syntaxis: Ptolemy
synthesis aerial: Ryle
synthesis reactions: Carothers, Ipatieff, Ziegler
syphilis: Ehrlich, Hinton, Pyman
congenital: Colles, Paracelsus
similarity with gonorrhoea: Hunter
Wassermann test: Landsteiner
Syphilis and Its Treatment: Hinton
Systema Naturae: Linnaeus
Systema Saturnium: Huygens
systole: Starling
SZENT-GYORGYI, ALBERT: Karrer

T

TABOR, DAVID
tabulating machine: Hollerith
Tait, Peter: Dewar
Takamine, Jokichi: Sharpey-Schafer
TALBOT, WILLIAM HENRY FOX: Wollaston
Tamm, Igor: Ginzburg, Sakharov
tanks, military: Crompton, R.E.B., Leonardo da Vinci, Porsche, Ricardo
tannins: Fischer, E.
TANSLEY, ARTHUR GEORGE
TARAGAY, MUHAMMAD *see* BEG
Tarski, Alfred: Church

TARTAGLIA (NICCOLO FONTANA): Fisher, R. A.
tartar emetic: Glauber
tartaric acid: Pasteur, Scheele
TATUM, EDWARD LAWRIE: Beadle, Tatum, and Lederberg (joint biography); Kornberg, A.
tautomers: Fischer, H.O.L., Onsager
taxonomy:
biological: Linnaeus
botanical: Gray, A., Hooker, J. D.
crayfish: Huxley, T. H.
mammals: Simpson, G. G.
systematic based on structural characteristics Ray
Tay Bridge: Baker, B.
Taylor's theorem, series: Cauchy, Hadamard, Maclaurin
tea plant: Banks
technetium: Burbidge, Cameron
Teflon *see* PTFE
telegraphy: Alter, Bell, A. G., Edison, Heaviside, Marconi, Morse, Siemens, Wheatstone
cable: Brunel, I. K.
cable, transatlantic: Kelvin
first practical: Weber, W. E.
relay in: Henry, J.
wireless telegraphy: Branly, Braun, K. F., Heaviside
telephone: Bell, A. G.
switching, traffic (flow): Boole, Erlang
telescopes, optical:
Anglo-Australian: Boksenberg, Bondi
Cassegrain: Cassegrain
Crossley reflector: Keeler
Galileo's: Galileo
Hale: Babcock, Boksenberg, Hubble, Smith, F. G.
Herschelian arrangement: Herschel, F. W.
Herschel's: Herschel, F. W.
Huygens's: Huygens
Newtonian: Cassegrain, Newton
Parsons's Leviathan: Parsons, W.
Schmidt: Schmidt
Short's Dumpy: Cassegrain
telescopes, radio:
Baade, Hanbury-Brown, Hewish, Jansky, Lovell, Reber, Ryle
television: Baird
camera: Zworykin
TELFORD, THOMAS: Brunel, I. K., Jessop, Roberts, Trésaguet
TELLER, EDWARD: Gamow, Goeppert-Mayer, Von Neumann
telluric rays: Janssen
tellurium: Klaproth
TEMIN, HOWARD MARTIN
temperature: Cannon, Cowling, Goldberg, Hertzsprung, Milne, Payne-Gaposchkin, Secchi
measurement: Le Châtelier, Travers
scales: Celsius, Fahrenheit, Kelvin, Milne
see also absolute temperature, critical temperature